2011 The ARRL Handbook

For Radio Communications

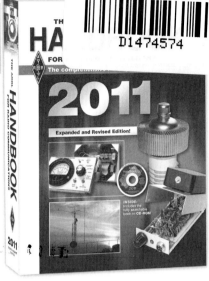

Eighty-Eighth Edition

Published by:
ARRL

the national association for Amateur Radio™
Newington, CT 06111 USA

Cover Info: The cover of the 2011 *ARRL Handbook* highlights the joy of building – everything from test equipment, to transmitters, to antennas and more.

Cover Photography: Clockwise from lower left corner: K6LRG antennas at sunset by Chris Tate, N6WM; a microprocessor-controlled SWR monitor – project and photo by Larry Coyle, K1QW (see Chapter 24); a 4CX1000A tetrode RF power tube (see Chapter 17); the MicroT2 SSB transmitter by Rick Campbell, KK7B (see Chapter 13).

Editor
H. Ward Silver, NØAX

Contributing Editors
Steven R. Ford, WB8IMY
Mark J. Wilson, K1RO

Editorial Assistant
Maty Weinberg, KB1EIB

Technical Consultants
Michael E. Gruber, W1MG
Edward F. Hare, Jr, W1RFI
Zachary H.J. Lau, W1VT

Cover Design
Sue Fagan, KB1OKW
Bob Inderbitzen, NQ1R

Production
Michelle Bloom, WB1ENT
Nancy G. Hallas, W1NCY
Elaine Lengyl
Jodi Morin, KA1JPA
David F. Pingree, N1NAS
Ed Vibert

Additional Contributors to the 2011 Edition
Bob Allison, WB1GCM
Alan Applegate, KØBG
Jim Brown, K9YC
Jeremy Campbell, KC8FEI
Kok Chen, W7AY
Larry Coyle, K1QW
Tim Duffy, K3LR
John Fitzsimmons, W3JN
Doug Grant, K1DG
Dale Grover, KD8KYZ
Joel R. Hallas, W1ZR
Roger Halstead, K8RI
Rick Hilding, K6VVA
Ron Hranac, NØIVN
Carl Leutzelschwab, K9LA
Jim Lux, W6RMK
Mike Martin, K3RFI
Ulrich Rohde, N1UL
Mark Steffka, WW8MW
Rich Strand, KL7RA
Jim Tonne, W4ENE
Paul Wade, W1GHZ
Ed Wetherhold, W3NQN
Gerald Youngblood, K5SDR

Copyright © 2010 by
The American Radio Relay League, Inc.

Copyright secured under the Pan-American Convention

International Copyright secured

All rights reserved. No part of this work may be reproduced in any form except by written permission of the publisher. All rights of translation are reserved.

Printed in the USA

Quedan reservados todos los derechos

ISBN: 978-0-87259-095-3 Softcover

ISBN: 978-0-87259-096-0 Hardcover

Eighty-Eighth Edition

Contents

A detailed Table of Contents is included at the beginning of each chapter.

INTRODUCTION

1 **What is Amateur (Ham) Radio?**
   1.1 Do-It-Yourself Wireless
   1.2 Joining the Ham Radio Community
   1.3 Assembling Your Station
   1.4 *Hello, World!* — Getting on the Air
   1.5 Your Ham Radio "Lifestyle"
   1.6 Public Service
   1.7 Ham Radio in the Classroom
   1.8 Resources
   1.9 Glossary

FUNDAMENTAL THEORY

2 **Electrical Fundamentals**
   2.1 Introduction to Electricity
   2.2 Resistance and Conductance
   2.3 Basic Circuit Principles
   2.4 Power and Energy
   2.5 Circuit Control Components
   2.6 AC Theory and Waveforms
   2.7 Capacitance and Capacitors
   2.8 Inductance and Inductors
   2.9 Working with Reactance
   2.10 Impedance
   2.11 Quality Factor (Q) of Components
   2.12 Practical Inductors
   2.13 Resonant Circuits
   2.14 Transformers
   2.15 Heat Management
   2.16 Radio Mathematics
   2.17 References and Bibliography

3 **Analog Basics**
   3.1 Analog Signal Processing
   3.2 Analog Devices
   3.3 Practical Semiconductors
   3.4 Analog Systems
   3.5 Amplifiers
   3.6 Operational Amplifiers
   3.7 Analog-Digital Conversion
   3.8 Miscellaneous Analog ICs
   3.9 Analog Glossary
   3.10 References and Bibliography

4 **Digital Basics**
   4.1 Digital vs Analog
   4.2 Number Systems
   4.3 Physical Representation of Binary States
   4.4 Combinational Logic
   4.5 Sequential Logic
   4.6 Digital Integrated Circuits
   4.7 Microcontrollers
   4.8 Personal Computer Interfacing
   4.9 Glossary of Digital Electronics Terms
   4.10 References and Bibliography

PRACTICAL DESIGN AND PRINCIPLES

5 **RF Techniques**
   5.1 Introduction
   5.2 Lumped-Element versus Distributed Characteristics
   5.3 Effects of Parasitic Characteristics
   5.4 Ferrite Materials
   5.5 Semiconductor Circuits at RF
   5.6 Impedance Matching Networks
   5.7 RF Transformers
   5.8 Noise
   5.9 Two-Port Networks
   5.10 RF Design Techniques Glossary
   5.11 References and Further Reading

6 **Computer-Aided Circuit Design**
   6.1 Circuit Simulation Overview
   6.2 Computer-Aided Design Examples
   6.3 Limitations of Simulation at RF
   6.4 CAD for PCB Design
   6.5 References and Bibliography

7 Power Supplies
7.1 The Need for Power Processing
7.2 AC-AC Power Conversion
7.3 Power Transformers
7.4 AC-DC Power Conversion
7.5 Voltage Multipliers
7.6 Current Multipliers
7.7 Rectifier Types
7.8 Power Filtering
7.9 Power Supply Regulation
7.10 "Crowbar" Protective Circuits
7.11 DC-DC Switchmode Power Conversion
7.12 High-Voltage Techniques
7.13 Batteries
7.14 Glossary of Power Supply Terms
7.15 Reference and Bibliography
7.16 Power Supply Projects

8 Modulation
8.1 Introduction
8.2 Analog Modulation
8.3 Digital Modulation
8.4 Image Modulation
8.5 Modulation Impairments
8.6 Modulation Glossary
8.7 References and Bibliography

9 Oscillators and Synthesizers
9.1 How Oscillators Work
9.2 Phase Noise
9.3 Oscillator Circuits and Construction
9.4 Designing an Oscillator
9.5 Quartz Crystals in Oscillators
9.6 Oscillators at UHF and Above
9.7 Frequency Synthesizers
9.8 Present and Future Trends in Oscillator Application
9.9 Glossary of Oscillator and Synthesizer Terms
9.10 References and Bibliography

10 Mixers, Modulators and Demodulators
10.1 The Mechanism of Mixers and Mixing
10.2 Mixers and Amplitude Modulation
10.3 Mixers and Angle Modulation
10.4 Putting Mixers, Modulators and Demodulators to Work
10.5 A Survey of Common Mixer Types
10.6 References and Bibliography

11 RF and AF Filters
11.1 Introduction
11.2 Filter Basics
11.3 Lumped-Element Filters
11.4 Filter Design Examples
11.5 Active Audio Filters
11.6 Quartz Crystal Filters
11.7 SAW Filters
11.8 Transmission Line Filters
11.9 Helical Resonators
11.10 Use of Filters at VHF and UHF
11.11 Filter Projects
11.12 Filter Glossary
11.13 References and Bibliography

12 Receivers
12.1 Introduction
12.2 Basics of Heterodyne Receivers
12.3 The Superheterodyne Receiver
12.4 Superhet Receiver Design Details
12.5 Control and Processing Outside the Primary Signal Path
12.6 Pulse Noise Reduction
12.7 VHF and UHF Receivers
12.8 UHF Techniques
12.9 References and Bibliography

13 Transmitters
13.1 Introduction
13.2 Early Transmitter Architectures
13.3 Modulation Types and Methods Applied to Transmitter Design
13.4 Modern Baseband Processing
13.5 Increasing Transmitter Power
13.6 Transmitter Performance and Measurement
13.7 References and Bibliography

14 Transceivers
14.1 The Transceiver Appears
14.2 Early SSB Transceiver Architecture Capabilities
14.4 Selecting an HF Transceiver
14.5 Transceivers With VHF/UHF Coverage
14.6 Transceiver Control and Interconnection
14.7 Transceiver Projects
14.8 References

15 DSP and Software Radio Design
15.1 Introduction
15.2 Typical DSP System Block Diagram
15.3 Digital Signals
15.4 Digital Filters
15.5 Miscellaneous DSP Algorithms
15.6 Analytic Signals and Modulation
15.7 Software-Defined Radios (SDR)
15.8 Glossary
15.9 References and Bibliography

16 Digital Modes
16.1 Digital "Modes"
16.2 Unstructured Digital Modes
16.3 Fuzzy Modes
16.4 Structured Digital Modes
16.5 Networking Modes
16.6 Glossary
16.7 References and Bibliography

17 RF Power Amplifiers
  17.1 High Power, Who Needs It?
  17.2 Types of Power Amplifiers
  17.3 Vacuum Tube Basics
  17.4 Tank Circuits
  17.5 Transmitting Device Ratings
  17.6 Sources of Operating Voltages
  17.7 Tube Amplifier Cooling
  17.8 Amplifier Stabilization
  17.9 Design Example: A High Power Vacuum Tube HF Amplifier
  17.10 Solid-State Amplifiers
  17.11 A New 250-W Broadband Linear Amplifier
  17.12 Tube Amplifier Projects
  17.13 References and Bibliography

18 Repeaters
  18.1 A Brief History
  18.2 Repeater Overview
  18.3 FM Voice Repeaters
  18.4 D-STAR Repeater Systems
  18.5 Glossary of FM and Repeater Terminology
  18.6 References and Bibliography

ANTENNA SYSTEMS AND RADIO PROPAGATION

19 Propagation of Radio Signals
  19.1 Fundamentals of Radio Waves
  19.2 Sky-Wave Propagation and the Sun
  19.3 MUF Predictions
  19.4 Propagation in the Troposphere
  19.5 VHF/UHF Mobile Propagation
  19.6 Propagation for Space Communications
  19.7 Noise and Propagation
  19.8 Glossary of Radio Propagation Terms
  19.9 References and Bibliography

20 Transmission Lines
  20.1 Transmission Line Basics
  20.2 Choosing a Transmission Line
  20.3 The Transmission Line as Impedance Transformer
  20.4 Matching Impedances in the Antenna System
  20.5 Baluns and Transmission-Line Transformers
  20.6 Using Transmission Lines in Digital Circuits
  20.7 Waveguides
  20.8 Glossary of Transmission Line Terms
  20.9 References and Bibliography

21 Antennas
  21.1 Antenna Basics
  21.2 Dipoles and the Half-Wave Antenna
  21.3 Vertical (Ground-Plane) Antennas
  21.4 T and Inverted-L Antennas
  21.5 Slopers and Vertical Dipoles
  21.6 Yagi Antennas
  21.7 Quad and Loop Antennas
  21.8 HF Mobile Antennas
  21.9 VHF/UHF Mobile Antennas
  21.10 VHF/UHF Antennas
  21.11 VHF/UHF Yagis
  21.12 Radio Direction Finding Antennas
  21.13 Glossary
  21.14 References and Bibliography

EQUIPMENT CONSTRUCTION AND MAINTENANCE

22 Component Data and References
  22.1 Component Data
  22.2 Resistors
  22.3 Capacitors
  22.4 Inductors
  22.5 Transformers
  22.6 Semiconductors
  22.7 Tubes, Wire, Materials, Attenuators, Miscellaneous
  22.8 Computer Connectors
  22.9 RF Connectors and Transmission Lines
  22.10 Reference Tables

23 Construction Techniques
  23.1 Electronic Shop Safety
  23.2 Tools and Their Use
  23.3 Soldering Tools and Techniques
  23.4 Surface Mount Technology (SMT)
  23.5 Electronic Circuits
  23.6 Mechanical Fabrication

24 Station Accessories
  24.1 A 100-W Compact Z-Match Antenna Tuner
  24.2 A Microprocessor Controlled SWR Monitor
  24.3 A 160- and 80-M Matching Network for Your 43-Foot Vertical
  24.4 Switching the Matching Network for Your 43-Foot Vertical
  24.5 An External Automatic Antenna Switch for Use With Yaesu or ICOM Radios
  24.6 A Low-Cost Remote Antenna Switch
  24.7 Audible Antenna Bridge
  24.8 A Trio of Transceiver/Computer Interfaces
  24.9 A Simple Serial Interface
  24.10 USB Interfaces For Your Ham Gear
  24.11 The Universal Keying Adapter
  24.12 The TiCK-4 — A Tiny CMOS Keyer
  24.13 The ID-O-Matic Station Identification Timer
  24.14 An Audio Intelligibility Enhancer
  24.15 An Audio Interface Unit for Field Day and Contesting

25 Test Equipment and Measurements
25.1 Test and Measurement Basics
25.2 DC Instruments and Circuits
25.3 AC Instruments and Circuits
25.4 Frequency Measurement
25.5 Other Instruments and Measurements
25.6 Receiver Performance Tests
25.7 Spectrum Analyzer
25.8 Transmitter Performance Tests
25.9 Test Equipment and Measurements Glossary

26 Troubleshooting and Maintenance
26.1 Test Equipment
26.2 Where to Begin
26.3 Testing Within a Stage
26.4 Typical Symptoms and Faults
26.5 Troubleshooting Hints
26.6 Components
26.7 After the Repairs
26.8 Professional Repairs
26.9 Repair and Restoration of Vintage Equipment
26.10 References and Bibliography

27 RF Interference
27.1 Managing Radio Frequency Interference
27.2 FCC Rules and Regulations
27.3 Elements of RFI
27.4 Identifying the Type of RFI Source
27.5 Locating Sources of RFI
27.6 Power-line Noise
27.7 Elements of RFI Control
27.8 Troubleshooting RFI
27.9 Automotive RFI
27.10 RFI Projects
27.11 RFI Glossary
27.12 References and Bibliography

STATION ASSEMBLY AND MANAGEMENT

28 Safety
28.1 Electrical Safety
28.2 Antenna and Tower Safety
28.3 RF Safety

29 Assembling a Station
29.1 Fixed Stations
29.2 Mobile Installations
29.3 Portable Installations
29.4 Remote Stations
29.5 References and Bibliography

30 Space Communications
30.1 Amateur Satellite History
30.2 Satellite Transponders
30.3 Satellite Tracking
30.4 Satellite Ground Station Antennas
30.5 Satellite Ground Station Equipment
30.6 Satellite Antenna Projects
30.7 Satellite References and Bibliography
30.8 Earth-Moon-Earth (EME) Communication
30.9 EME Propagation
30.10 Fundamental Limits
30.11 Building an EME Station
30.12 Getting Started with EME
30.13 EME References
30.14 Glossary of Space Communications Technology

31 Digital Communications
31.1 Sound Card Modes
31.2 Packet Radio
31.3 The Automatic Packet/Position Reporting System (APRS)
31.4 PACTOR
31.5 High Speed Multimedia (HSMM)
31.6 Automatic Link Establishment (ALE)
31.7 D-STAR
31.8 APCO-25
31.9 HF Digital Voice
31.10 EchoLink, IRLP and WIRES-II
31.11 Glossary of Digital Communications Terms
31.12 Bibliography and References

32 Image Communications
32.1 Fast-Scan Amateur Television Overview
32.2 Amateur TV Systems
32.3 ATV Applications
32.4 Video Sources
32.5 Glossary of ATV Terms
32.6 Slow-Scan Television (SSTV) Overview
32.7 SSTV Basics
32.8 Analog SSTV
32.9 Digital SSTV
32.10 Glossary of SSTV Terms
32.11 Bibliography and References

Advertiser's Index

Index

Project Index

Author Index

Foreword

Technology both enables and challenges the amateur. Digital computing is finding its way into nearly every aspect of Amateur Radio, making possible functions and features that are entirely new — with amateurs leading the way — or that were previously exclusive to government and industry. At the same time, increasing miniaturization of electronics requires the practitioner of the "roll your own" tradition to learn new skills and make use of new tools. More than ever, the amateur needs up-to-date references and guidance.

Now in its second year of a cycle of renewal and expansion, the *2011 ARRL Handbook* rises to the challenge, building on last year's milestone edition by remaking an entire chapter, presenting a novel and useful new project and including many useful updates. Several supporting papers and monographs have been added to the extensive collection of material on the CD-ROM included with the book, as well, including a new version of Jim Tonne, W4ENE's, filter design software, *ELSIE*. Ulrich Rohde, N1UL, contributed several advanced and detailed papers on RF circuit simulation, mixers, and oscillators for the advanced reader — these are provided on the CD-ROM. Responding to reader feedback, the Table of Contents now includes an extra level of headings to help you quickly find the material you need.

One of Amateur Radio's largest challenges is addressed in Chapter 27, RF Interference. RFI is the bane of many amateur's existence, perhaps as much as antenna restrictions. No longer just the "Tennessee Valley Indians" (TVI) of years past, RFI now occurs in both directions — from the ham's strong signal to consumer electronics and from stray RF signals and electrical noise radiated by consumer electronics and power-line equipment to the ham's receiver. The old chapter was completely rewritten to address this new reality. In addition, a big new section on Automotive RFI was contributed for the multiplying mobile operators.

Lift the cover of any modern radio and many accessories — what do you see? Microprocessors! The use of these powerful computing devices enables designers to create features that wouldn't have been practical to implement with analog circuitry. The signature new project in this edition — Chapter 24's "A Microprocessor Controlled SWR Monitor" by Larry Coyle, K1QW — exemplifies the power of hybrids of analog and digital. It's easy to build and should appear in many stations.

Another type of hybrid is growing outside the shack as amateurs facing station-building restrictions at home "take to the hills," building remote HF stations linked to the home PC via the Internet. Chapter 29, Assembling A Station, includes a brand-new section written by remote station veteran Rick Hilding, K6VVA. Because every remote station is highly customized, Rick leads you through a list of items that you need to consider as you design your own shack in the country or on a hilltop.

Hand-in-hand with the new construction technologies, free or low-cost software tools are now available for every ham with a computer and Internet connection. This makes it possible to design and order custom circuit boards for personal or club projects at reasonable cost. Dale Grover, KD8KYZ, has contributed a detailed discussion of this process to help the reader take advantage of these new services.

Not completely devoted to the latest technology, there are some how-to's for much-loved tube equipment. Roger Halstad, K8RI, explains how to tune up a triode or tetrode amplifier — a skill any ham with a tube-based amplifier needs to understand. Amplifier care and troubleshooting are addressed, too. John Fitzsimmons, W3JN, describes how to restore and repair vintage equipment in the Troubleshooting and Maintenance chapter.

As you will see, the *ARRL Handbook*'s team of authors, editors, and reviewers are working hard to carry forward the traditional technology of Amateur Radio while at the same time remaining faithful to the FCC's Basis and Purpose for the Amateur Service in advancing the individual's skills and practices. We are confident you'll find this edition to have met that standard.

David Sumner, K1ZZ
Chief Executive Officer
Newington, Connecticut
September 2010

The Amateur's Code

The Radio Amateur is:

CONSIDERATE...never knowingly operates in such a way as to lessen the pleasure of others.

LOYAL...offers loyalty, encouragement and support to other amateurs, local clubs, and the American Radio Relay League, through which Amateur Radio in the United States is represented nationally and internationally.

PROGRESSIVE...with knowledge abreast of science, a well-built and efficient station and operation above reproach.

FRIENDLY...slow and patient operating when requested; friendly advice and counsel to the beginner; kindly assistance, cooperation and consideration for the interests of others. These are the hallmarks of the amateur spirit.

BALANCED...radio is an avocation, never interfering with duties owed to family, job, school or community.

PATRIOTIC...station and skill always ready for service to country and community.

—*The original Amateur's Code was written by Paul M. Segal, W9EEA, in 1928.*

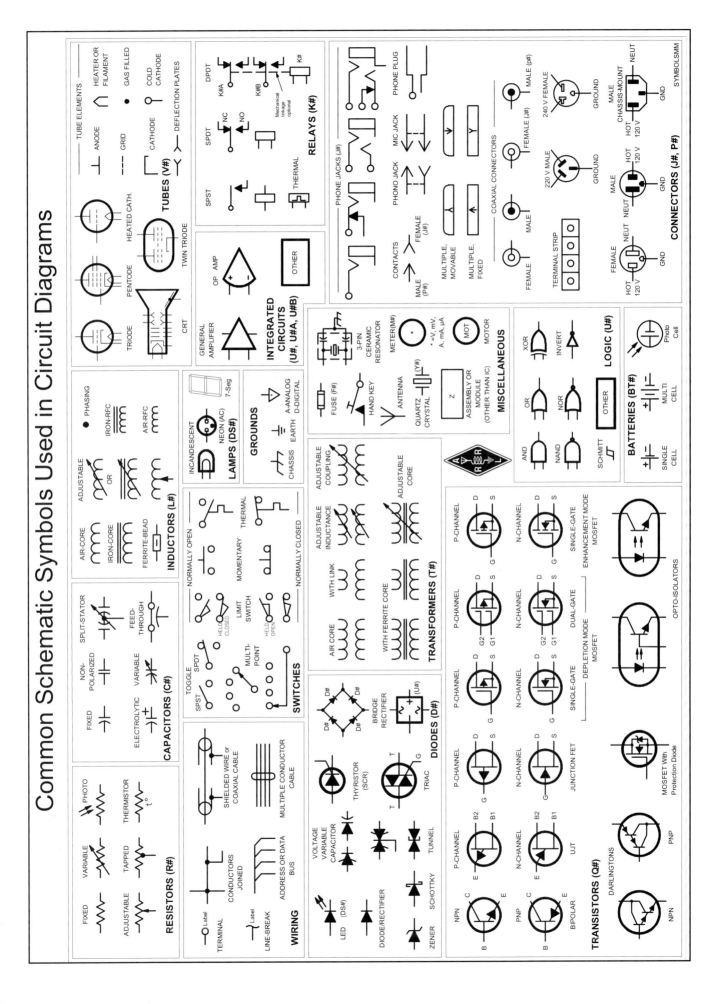

Guide to ARRL Member Services

ARRL, 225 Main Street ◆ Newington, Connecticut 06111-1494, USA

Tel: 860-594-0200, Mon-Fri 8 AM
to 5 PM ET (except holidays)
FAX: 860-594-0303
e-mail: hqinfo@arrl.org
ARRLWeb: www.arrl.org

**VISITING ARRL
HEADQUARTERS AND W1AW**
Tours Mon-Fri at 9, 10, 11 AM; 1, 2, 3 PM
W1AW guest operating 10 AM to noon,
and 1 to 3:45 PM (bring your license).

**JOIN or RENEW
or ORDER Publications**
tel. **Toll Free 1-888-277-5289** (US)
International callers
tel. +1 (860) 594-0355

**INTERESTED IN
BECOMING A HAM?**
www.hello-radio.org
e-mail: newham@arrl.org
tel. 1-800-326-3942

Public Service

Advocacy

News Center
ARRLWeb: www.arrl.org
ARRL Letter:
www.arrl.org/arrlletter

Public Relations/Advocacy
**Government Relations and
Spectrum Protection:**
www.arrl.org/regulatory-advocacy
e-mail: govrelations@arrl.org
Public and Media Relations:
www.arrl.org/media-and-public-relations

Membership Benefits
Membership Benefits (all):
www.arrl.org/membership
Awards: www.arrl.org/awards
Contests: www.arrl.org/contests
FCC License Renewal / Modification:
www.arrl.org/fcc-license-info-and-forms
QSL Service: www.arrl.org/qsl-bureau
Regulatory Information
www.arrl.org/national
Technical Information Service
www.arrl.org/technology
e-mail: tis@arrl.org
tel. 860-594-0214

Contributions, Grants and Scholarships
ARRL Development Office:
www.arrl.org/donate-to-arrl
e-mail: mhobart@arrl.org
tel. 860-594-0397
- ARRL Diamond Club/Diamond Terrace
- Spectrum Defense Fund
- Education & Technology Fund
- Planned Giving/Legacy Circle
- Maxim Society

ARRL Foundation Grants and Scholarships:
www.arrl.org/the-arrl-foundation

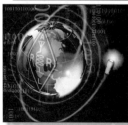

Education

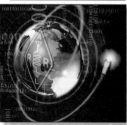

Technology

Public Service
Public Service Programs:
www.arrl.org/public-service
Amateur Radio Emergency Service® (ARES®):
www.arrl.org/ares
ARRL Field Organization:
www.arrl.org/field-organization

Clubs, Exams, Licensing and Teachers
Find an Affiliated Club: www.arrl.org/find-a-club
Mentor Program:
www.arrl.org/mentoring-online-courses
Find a Licensing Class:
www.arrl.org/find-a-class
Support to Instructors:
www.arrl.org/volunteer-instructors-mentors
Find an Exam Session:
www.arrl.org/finding-an-exam-session
Volunteer Examiner Coordinator (VEC):
www.arrl.org/volunteer-examiners

Publications & Education
QST — Official Journal of ARRL:
www.arrl.org/qst
e-mail: qst@arrl.org
QEX — Forum for Communications Experimenters:
www.arrl.org/qex
e-mail: qex@arrl.org
NCJ — National Contest Journal:
www.arrl.org/ncj
e-mail: ncj@arrl.org
Books, Software and Operating Resources:
tel. 1-888-277-5289 (toll-free in the US);
www.arrl.org/shop
Advertising:
www.arrl.org/business-opportunities-sales
e-mail: ads@arrl.org
Continuing Education / Online Courses:
www.arrl.org/online-courses

Membership

The American Radio Relay League, Inc.

The American Radio Relay League, Inc. is a noncommercial association of radio amateurs, organized for the promotion of interest in Amateur Radio communication and experimentation, for the establishment of networks to provide communication in the event of disasters or other emergencies, for the advancement of the radio art and of the public welfare, for the representation of the radio amateur in legislative matters, and for the maintenance of fraternalism and a high standard of conduct.

ARRL is an incorporated association without capital stock chartered under the laws of the State of Connecticut, and is an exempt organization under Section 501(c)(3) of the Internal Revenue Code of 1986. Its affairs are governed by a Board of Directors, whose voting members are elected every three years by the general membership. The officers are elected or appointed by the directors. The League is noncommercial, and no one who could gain financially from the shaping of its affairs is eligible for membership on its Board.

"Of, by, and for the radio amateur," the ARRL numbers within its ranks the vast majority of active amateurs in the nation and has a proud history of achievement as the standard-bearer in amateur affairs.

A *bona fide* interest in Amateur Radio is the only essential qualification of membership; an Amateur Radio license is not a prerequisite, although full voting membership is granted only to licensed amateurs in the US.

Membership inquiries and general correspondence should be addressed to the administrative headquarters: ARRL, 225 Main Street, Newington, Connecticut 06111-1494.

About the ARRL

The seed for Amateur Radio was planted in the 1890s, when Guglielmo Marconi began his experiments in wireless telegraphy. Soon he was joined by dozens, then hundreds, of others who were enthusiastic about sending and receiving messages through the air—some with a commercial interest, but others solely out of a love for this new communications medium. The United States government began licensing Amateur Radio operators in 1912.

By 1914, there were thousands of Amateur Radio operators—hams—in the United States. Hiram Percy Maxim, a leading Hartford, Connecticut inventor and industrialist, saw the need for an organization to band together this fledgling group of radio experimenters. In May 1914 he founded the American Radio Relay League (ARRL) to meet that need.

Today ARRL, with approximately 155,000 members, is the largest organization of radio amateurs in the United States. The ARRL is a not-for-profit organization that:
- promotes interest in Amateur Radio communications and experimentation
- represents US radio amateurs in legislative matters, and
- maintains fraternalism and a high standard of conduct among Amateur Radio operators.

At ARRL headquarters in the Hartford suburb of Newington, the staff helps serve the needs of members. ARRL is also International Secretariat for the International Amateur Radio Union, which is made up of similar societies in 150 countries around the world.

ARRL publishes the monthly journal *QST*, as well as newsletters and many publications covering all aspects of Amateur Radio. Its headquarters station, W1AW, transmits bulletins of interest to radio amateurs and Morse code practice sessions. The ARRL also coordinates an extensive field organization, which includes volunteers who provide technical information and other support services for radio amateurs as well as communications for public-service activities. In addition, ARRL represents US amateurs with the Federal Communications Commission and other government agencies in the US and abroad.

Membership in ARRL means much more than receiving *QST* each month. In addition to the services already described, ARRL offers membership services on a personal level, such as the Technical Information Service—where members can get answers by phone, email or the ARRL website, to all their technical and operating questions.

Full ARRL membership (available only to licensed radio amateurs) gives you a voice in how the affairs of the organization are governed. ARRL policy is set by a Board of Directors (one from each of 15 Divisions). Each year, one-third of the ARRL Board of Directors stands for election by the full members they represent. The day-to-day operation of ARRL HQ is managed by an Executive Vice President and his staff.

No matter what aspect of Amateur Radio attracts you, ARRL membership is relevant and important. There would be no Amateur Radio as we know it today were it not for the ARRL. We would be happy to welcome you as a member! (An Amateur Radio license is not required for Associate Membership.) For more information about ARRL and answers to any questions you may have about Amateur Radio, write or call:

ARRL—The national association for Amateur Radio
225 Main Street
Newington CT 06111-1494
Voice: 860-594-0200
Fax: 860-594-0259
E-mail: **hq@arrl.org**
Internet: **www.arrl.org/**

Prospective new amateurs call (toll-free):
800-32-NEW HAM (800-326-3942)
You can also contact us via e-mail at **newham@arrl.org**
or check out *ARRLWeb* at **www.arrl.org/**

Contents

1.1 Do-It-Yourself Wireless
    1.1.1 Something for Everyone
1.2 Joining the Ham Radio Community
    1.2.1 Getting Started
    1.2.2 Study Guides
    1.2.3 Taking the Test
    1.2.4 Your "Elmer"
    1.2.5 Your Ham Radio Identity
1.3 Assembling Your Station
    1.3.1 How Much Will It Cost?
    1.3.2 Computers and Ham Radio
1.4 Hello, World! — Getting on the Air
    1.4.1 Voice Modes
    1.4.2 Morse Code
    1.4.3 FM Repeaters
    1.4.4 Digital Modes
    1.4.5 Image Communication
1.5 Your Ham Radio "Lifestyle"
    1.5.1 Ham Radio Contesting — Radiosport
    1.5.2 Chasing DX
    1.5.3 Operating Awards
    1.5.4 Satellite Communication
    1.5.5 QRP: Low-Power Operating
    1.5.6 Operating Mobile
    1.5.7 VHF, UHF and Microwave Operating
    1.5.8 Vintage Radio
    1.5.9 Radio Direction Finding (DF)
1.6 Public Service
    1.6.1 Emergency Communication
    1.6.2 Emergency Communication Organizations
    1.6.3 Public Service and Traffic Nets
1.7 Ham Radio in the Classroom
    1.7.1 ARRL Amateur Radio Education & Technology Program
1.8 Resources
1.9 Glossary

Chapter 1

What is Amateur (Ham) Radio?

For more than a century, a growing group of federally licensed radio hobbyists known as Amateur Radio — or "ham radio" — operators has had a front-row seat as radio and electronics have broadened our horizons and touch virtually all of our lives. Hams pioneered personal communication, even in the days before the telephone and household electricity were commonplace and the Internet not yet conceived. The word "radio" — or "wireless" — still evokes awe.

Today we enjoy wireless amenities that range from the ubiquitous cell phone to diminutive "netbook" PCs than can go just about anywhere. Personal communication is the goal. Wireless radiocommunication lets us reach out via voice or text message or tells us — literally — how to navigate in strange surroundings. Printers and scanners now can communicate wirelessly with your PC. Radio signals can even automatically set your clock to the precise time. In this chapter, Rick Lindquist, WW3DE, provides a broad overview of Amateur Radio activities and licensing requirements.

1.1 Do-It-Yourself Wireless

Ever evolving, Amateur Radio has always been what its participants bring *to* it and what they make *of* it. "Make" is a key word in ham radio, since many enthusiasts still enjoy building their own radio communication equipment and electronics. It is in such "hands-on" activities that this *Handbook* often comes into play.

Hams also "make" contact with each other using equipment they've bought or built, or a combination of the two, over a wide range of the radio spectrum, without the need for any external infrastructure — such as the wired or cellular telephone network or the Internet.

The methods hams use to keep in touch range from the venerable Morse code — no longer a licensing requirement, by the way — to modern digital modes and television. The marriage between Amateur Radio and computer technology grows stronger by the day as hams invent ever more creative ways to make computers and the Internet essential station components. The wonder of software defined radio (SDR) techniques has even made it possible to create *virtual* radio communication gear. SDRs require a minimum of physical components; sophisticated computer software does the heavy lifting!

1.1.1 Something for Everyone

Amateur Radio offers such a wide range of activities that everyone can find a favorite niche. As one of the few truly *international hobbies*, ham radio offers the ability to communicate with other similarly licensed aficionados all over the world. The competition of "radiosport" — just to pick one activity many hams enjoy — helps operators to improve their skills and stations. Further, ham radio offers opportunities to serve the public by supporting communication in disasters and emergencies, and it's a platform for scientific experimentation.

Ham radio has extended its horizon into space. The International Space Station boasts a ham radio station, and most ISS crew members are Amateur Radio licensees. Thanks to the Amateur Radio on the International Space Station (ARISS) program, properly equipped hams can talk directly with NASA astronauts in space. Hams contact each other through Earth-orbiting satellites designed and built by other radio amateurs, and they even bounce radio signals off the moon and to other hams back on Earth.

Hams talk with one another from cars, while hiking or biking in the mountains, from remote campsites or while boating. Through a plethora of activities, hams learn a lot, establish lifelong friendships and, perhaps most important, have a *lot* of fun. Along the way, radio amateurs often contribute some of the genius behind the latest technological innovations.

In all likelihood, you're already a ham or at least have experimented with radio and electronics yourself and are thinking about becoming one. This *Handbook* is an invaluable resource that reveals and explains the "mysteries" governing electronics in general and in radio — or wireless — communication in particular, especially as it pertains to Amateur Radio.

HAMS ARE EVERYWHERE!

Amateur Radio is open and accessible to everyone. Hams are mothers, fathers, sons and daughters of all ages, ethnic backgrounds and physical abilities who are part of a unique worldwide community of licensed radio hobbyists. These individuals come from all walks of life. Some are even well-known celebrities. They have one thing in common, however.

Fig 1.1 — Sue Cook, AI6YL enjoys operating RTTY. She is shown here as P4ØYL, operating from the island of Aruba in the 2010 ARRL RTTY Roundup contest. (*AI6YL photo*)

Fig 1.2 — Ham radio contests or "radiosport" cross borders around the world. Here, David Hodge, XE1/N6AN and Ramon Santiago V, XE1KK operate during the IARU HF World Championships, held every year in July. (*XE1KK photo*)

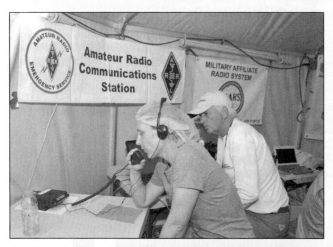

Fig 1.3 — Hams assisted with emergency communications following the Haitian earthquake in January of 2010. Here, Jack Satterfield W4GRJ is helping Dr Elizabeth Greig arrange transportation for an injured survivor to the University of Miami medical station in Port-Au-Prince. (*N4LDG photo*)

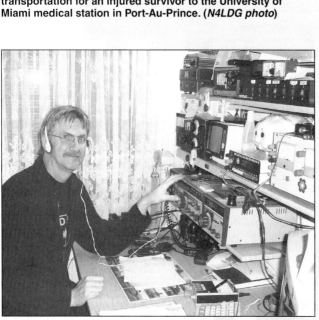

Fig 1.4 — German radio amateur Peter Ernst, DL4SZB, hunts DX from his well-equipped station in Goldberg. (*WW3DE photo*)

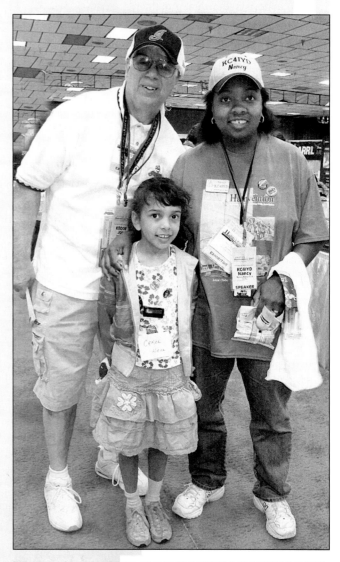

Fig 1.5 — Amateur Radio as a social activity: Nancy Rabel Hall, KC4IYD, and her daughter Carol meet up with Joe Phillips, K8QOE (SK), at a ham radio convention. (*N1BKE photo*)

All find joy and excitement by experiencing radio communication and electronics on a very personal level.

How can you spot a radio amateur? Sometimes it's easy. The driver of that car in front of you sporting an "odd-looking" antenna may be a ham who's equipped his vehicle for mobile operation. Your neighbor on the next block with the wires strung between trees or, perhaps, a tower supporting what looks like a very large television antenna probably is one too.

Modern technology continues to make ham radio more accessible to all, including those on a tight budget or confronting physical challenges. People who are not as mobile as they'd like to be find the world of Amateur Radio a rewarding place to make friends — on the next block or around the globe. It's possible for a ham to control a transmitting and receiving station via the Internet using a laptop computer, even if that station is thousands of miles distant.

For many radio amateurs, a relaxing evening at home is having a two-way radio conversation with a friend in Frankfort, Kentucky — or even Frankfurt, Germany. Unlike any other hobby, Amateur Radio recognizes no international or political boundaries, and it brings the world together as good friends.

Hams at the Forefront

Over the years, the military and the electronics industry have often drawn on the ingenuity of radio amateurs to improve designs or solve problems. Hams provided the keystone for the development of modern military communication equipment, for example. In the 1950s, the Air Force needed to convert its long-range communication from Morse code to voice, and jet bombers had no room for skilled radio operators. At the time, hams already were experimenting with and discovering the advantages of single sideband (SSB) voice equipment. With SSB, hams were greatly extending the distances they could transmit.

Air Force Generals Curtis LeMay and Francis "Butch" Griswold, both radio amateurs, hatched an experiment that used ham radio equipment at the Strategic Air Command headquarters. Using an SSB station in an aircraft flying around the world, LeMay and Griswold were able to stay in touch with Offutt Air Force Base in Nebraska from around the globe. The easy modification of this ham radio equipment to meet military requirements saved the government millions of dollars in research costs.

More recent technological experimentation has focused on such techniques as software defined radio (SDR). This amazing approach enables electronic circuit designers to employ software to replace more costly — and bulkier — hardware components. It's no coincidence or surprise that radio amateurs have been among those investigators doing the ground-level research and experimentation to bring this technology from the laboratory to the marketplace. Transceivers built on the SDR model now are making inroads within the Amateur Radio community and represent the likely wave of the future in equipment design.

Affirming the relationship between Amateur Radio and cutting-edge technology, in 2009, Howard Schmidt, W7HAS, was appointed White House Cybersecurity Coordinator. ARRL member Schmidt is one of the world's leading authorities on computer security, with some 40 years of experience in government, business and law enforcement. Schmidt credits ham radio with helping him launch his career. "Building ... computers to support my ham radio hobby gave me the technical skills that I need to ... start doing computer crime investigations and work on the early stages of computer forensics, in turn enabling me to start working on cybersecurity issues." Hams are often found in industry and the military as technology presses ahead.

1.1.2 What's in it for Me?

As a community *of* communities, Amateur Radio can be whatever *you* want it to be. Whether you are looking for relaxation, excitement, enjoyment or a way to stretch your mental (and physical) horizons, Amateur Radio can provide it — even for those with time and money constraints. However it happens, communication between or among individuals is at the core of nearly all ham radio activities. In its most basic form, ham radio is two people saying "Hello!" to each other over the air, perhaps using inexpensive handheld transceivers or even homemade gear. In "Hamspeak," a two-way, on-the-air communication is known as a "QSO" — an old radiotelegraph, or Morse code, abbreviation often pronounced "CUE-so."

It's nearly as simple for *groups* of hams with common interests to gather on the airwaves to share their thoughts and even pictures. These on-the-air get-togethers are called "nets" or "roundtables," depending on their formality. When hams meet and engage in extended on-the-air conversations, they call it "ragchewing."

Nets often provide an on-the-air venue to find other hams with similar interests both inside and outside of Amateur Radio. Topics may be as diverse as vintage radio, chess, gardening, rock climbing, railroads, computer programming, teaching or an interest in certain types of radio equipment. Religious groups and scattered friends and families may also organize nets. Nets form when like-minded hams gather on the air on a regular schedule. You can find your special interest in *The ARRL Net Directory* on the *ARRLWeb*.

With your ham radio license in hand, you can meet new friends, win awards, exchange "QSL cards" to confirm radio contacts by mail, challenge yourself and others, learn and educate, contribute to your community, travel, generate international goodwill and continue a century-old wireless communication tradition. Let's take a closer look.

1.2 Joining the Ham Radio Community

Although you no longer have to learn the Morse code to become an Amateur Radio licensee, you must have a license granted by the Federal Communications Commission (FCC) to operate an Amateur Radio station in the United States, in any of its territories and possessions or from any vessel or aircraft registered in the US. There is no age requirement, nor do you have to be a citizen to obtain a license a US Amateur Radio license. Children as young as four and five have passed ham radio exams!

In the US, there are three classes — or levels — of Amateur Radio license. From the easiest to the most difficult, they are Technician, General and Amateur Extra class. You must only take and pass a written examination for each license. The higher you climb up the ladder, the more challenging the test and the more generous the privileges the FCC grants. To reach the top — Amateur Extra — you must pass the examinations for all three license classes.

1.2.1 Getting Started

Most people start out in Amateur Radio by getting a Technician license or "ticket," as a ham license is sometimes called. Technicians enjoy a wide, but somewhat limited, range of voice and digital radio operating privileges. These include access to some "high frequency" (HF) or short-wave spectrum (hams have access to HF "bands" or frequency segments in the range from 1.8 MHz to 29.7 MHz), where most direct international communication happens. Most Technician privileges are in the VHF-UHF and microwave spectrum, however. These allow operation on widely available FM voice repeaters. A repeater greatly extends the communication range of low-power, handheld transceivers or mobile

Ham Radio's Rules of the Airwaves

International and national radio regulations govern the operational and technical standards of all radio stations. The International Telecommunication Union (ITU) governs telecommunication on the international level and broadly defines radio services through the international *Radio Regulations*. In the US, the Federal Communication Commission (FCC) is the federal agency that administers and oversees the operation of nongovernmental and nonmilitary stations — including Amateur Radio. Title 47 of the *US Code of Federal Regulations* governs telecommunication. The Amateur Radio Service is governed by Part 97.

Experimentation has always been the backbone of Amateur Radio and the Amateur Service rules provide a framework within which amateurs have wide latitude to experiment in accordance with the "basis and purpose" of the service. The rules should be viewed as vehicles to promote healthy activity and growth, not as constraints that lead to stagnation. The FCC's rules governing Amateur Radio recognize these five aspects, paraphrased below, as the basis and purpose of the Amateur Service.

• Amateur Radio's value to the public, particularly with respect to providing emergency communication support

• Amateur Radio's proven ability to contribute to the advancement of the radio art

• Encouraging and improving the Amateur Service through rules that help advance communication and technical skills

• Maintaining and expanding the Amateur Service as a source of trained operators, technicians and electronics experts

• Continuing and extending the radio amateur's unique ability to enhance international goodwill

The Amateur Radio Service rules, Part 97, are in six sections: General Provisions, Station Operation Standards, Special Operations, Technical Standards, Providing Emergency Communication and Qualifying Examination Systems. Part 97 is available in its entirety on the ARRL and FCC Web sites (see the Resources section at the end of this chapter for further information).

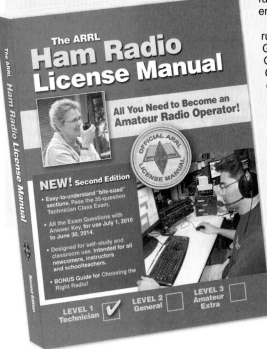

Fig 1.6 — ARRL's *Ham Radio License Manual* contains all the information you need to study for your Technician license.

stations too far apart to transmit to each other directly. The "Tech ticket" is a great introduction to the fun and excitement of ham radio and to the ways of the hobby.

The sole requirement for the Technician license is passing a 35-question written exam. It covers FCC rules and regulations that govern the airways, courteous operating procedures and techniques and some basic electronics. The privileges granted give Technicians plenty of room to explore and activities to try. For some, the Technician is the only ham license they'll ever want or need.

By upgrading to General class, a Technician licensee can earn additional operating privileges, such as access to all of the Amateur Radio HF bands. Upgrading to General entails passing another 35-question written exam. In addition to Technician privileges, General-class hams enjoy worldwide communication using voice, digital, image and television techniques.

Reaching the top rung of the Amateur Radio ladder — Amateur Extra class — means passing a more-demanding 50-question examination. Amateur Extra licensees enjoy privileges on all frequency bands and modes available to hams. The exam may be challenging, but many hams consider it well worth the effort!

1.2.2 Study Guides

You can prepare for the exam on your own, with a group of friends or by taking a class sponsored by a ham radio club in your area. ARRL has materials and lesson plans for hams who wish to teach Amateur Radio licensing classes for newcomers. Anyone can set up license classes. Many Amateur Radio clubs hold periodic classes, usually for the Technician license. The ARRL supports Registered Amateur Radio Instructors, but registration is not necessary to conduct a class. Check the ARRL Web site, **www.arrl.org**, for classes, clubs or volunteer examiners (VEs) in your area (more on VEs below).

Help is available at every step. The ARRL publishes study materials for all license classes. Visit the ARRL Web site or contact ARRL's New Ham Desk for more information on getting started. The Resources section at the end of this chapter includes telephone number and address information. The ARRL can help you find ham radio clubs in your area as well as ARRL-registered instructors and local VE teams. A lot more information is on the ARRL Web site, including frequencies hams can use, popular operating activities and how to order the latest ARRL study guide.

For newcomers seeking to obtain a Technician license, *The ARRL Ham Radio License Manual* includes the complete, up-to-date question pool with the correct answers and clear explanations. The manual assumes no prior electronics background. It delves into the details behind the questions and answers, so you will understand the material, rather than simply memorize the correct answers.

If you already have some electronics background or just want brief explanations of the material, you might decide that *ARRL's Tech Q&A* manual is a more appropriate choice. It also includes the entire Technician question pool to help you prepare.

When you are ready to upgrade to a General class license, *The ARRL General Class License Manual* or *ARRL's General Q&A* can help you prepare. In like fashion, *The ARRL Extra Class License Manual* and *ARRL's Extra Q&A* will guide your study efforts for the Amateur Extra class license. Check the ARRL Web site for detailed information on these and other options to study for your license.

1.2.3 Taking the Test

While the FCC grants US Amateur Radio licenses, volunteer examiners (VEs) have taken over the task of administering Amateur Radio test sessions. Other countries also have adopted volunteer-administered exam systems. Many ham radio clubs sponsor regular exam sessions, so you shouldn't have to wait long or travel far once you're ready. Exam sessions often are available on weekends or evenings. Most volunteer examiner teams charge a small fee to recover the cost of administering the test and handling the FCC paperwork.

The ARRL is a Volunteer Examiner Coordinator (VEC) and supports the largest VE program in the nation. More information about the VE program is available on the ARRL Web site.

The questions for each 35 or 50-question test come from a large "question pool" that's specific to each license class. All three question pools — Technician, General and Amateur Extra — are available to the public in study guides and on the Internet. If you're studying, make sure you're working with the latest version, since question pools are updated on a set schedule. The Resources section at the end of this chapter has more information on where to find the question pools.

1.2.4 Your "Elmer"

Ham radio operators often "learn the ropes" from a mentor. In ham radio parlance, such an experienced ham who is willing to help newcomers is called an "Elmer." This individual teaches newcomers about Amateur Radio, often on a one-to-one basis. Your local ham radio club can pair you up with an Elmer, who

ARRL —
The National Association for Amateur Radio

The American Radio Relay League (ARRL) is the internationally recognized society representing Amateur Radio in the US. Since its founding in 1914, the ARRL — the national association for Amateur Radio (as it now is known) — has grown and evolved along with Amateur Radio. ARRL Headquarters and the Maxim Memorial Station W1AW are in Newington, Connecticut, near Hartford. Through its dedicated volunteers and a professional staff, the ARRL promotes the advancement of the Amateur Service in the US and around the world.

The ARRL is a nonprofit, educational and scientific organization dedicated to the promotion and protection of the many privileges that ham radio operators enjoy. Of, by and for the radio amateur, ARRL numbers some 150,000 members — the vast majority of active amateurs in North America. Licensees can become Full Members, while unlicensed persons are eligible to become Associate Members with all membership privileges except for voting in ARRL elections. Anyone with a solid interest in Amateur Radio belongs in the ARRL.

The ARRL volunteer corps is called the Field Organization. Working at the state and local level, these individuals tackle ARRL's goals to further Amateur Radio. They organize emergency communication in times of disaster and work with agencies such as American Red Cross and Citizen Corps. Other volunteers keep state and local government officials abreast of the good that hams do at the state and local level.

When you join ARRL, you add your voice to those who are most involved with ham radio. The most prominent benefit of ARRL membership is its monthly journal *QST*, the premiere Amateur Radio magazine. *QST* contains stories you'll want to read, articles on projects to build, announcements of upcoming contests and activities, reviews of new equipment, reports on the role hams play in emergencies and much more.

Being an ARRL member is far more than a subscription to *QST*. The ARRL represents your interests before the FCC and Congress, sponsors operating events and offers membership services at a personal level. A few are:
  • Low-cost ham equipment insurance
  • the Volunteer Examiner program
  • the Technical Information Service (which answers your questions about Amateur Radio technical topics)
  • the QSL Bureau (which lets you exchange postcards with hams in foreign countries to confirm your contacts with them)

For answers to any questions about Amateur Radio, e-mail, call or write ARRL Headquarters. See the Resources section at the end of this chapter for contact information.

Fig 1.7 — ARRL Field Day is an opportunity for hams to operate from temporary locations as an emergency communications exercise, engage in some friendly competition, and expose Amateur Radio to the public. Here Tony Brock-Fisher, K1KP, helps a young visitor make his first-ever radio contacts at the W1HP Pentucket Amateur Radio Club operation. (*KB1NYS photo*)

Fig 1.8 — W1AW, the station operated by the ARRL in Newington, Connecticut, is known around the world as the home of ham radio. W1AW memorializes Hiram Percy Maxim, one of the founders of the ARRL. Visitors are welcome and often operate the station.

will be there for you as you study, buy your first radio and set up your station — which many hams call "the radio shack" or "ham shack." Elmers also are pleased and proud to help you with your first on-the-air contacts.

Elmers belonging to the international Courage Handi-Hams organization focus on making study materials and ham radio station operation accessible to those with physical disabilities. Local Handi-Hams assist such prospective radio amateurs in getting licensed, and the Handi-Ham System may lend basic radio gear to get the new ham on the air.

1.2.5 Your Ham Radio Identity

When you earn your Amateur Radio license, the FCC assigns you a unique *call sign* (some hams shorten this to simply "call"). With your license and your unique call sign, you have permission to operate an Amateur Radio station. Your call sign not only identifies your station on the air, it's a personal ham radio identity, and many hams become better known by their call signs than by their names! US call signs, for example, begin with W, K, N or A followed by some combination of letters and one numeral. Each combination is different. One well-known ham radio call sign is W1AW, assigned to the Hiram Percy Maxim Memorial Station at ARRL Headquarters in Newington, Connecticut.

FCC-assigned call signs come in several flavors, with the shortest — and typically most desired — combinations available only to Amateur Extra class licensees. The FCC typically assigns the longest call sign format to new Technician licensees. These call signs start with two letters, a numeral from 0 to 9, and three more letters. The first part of a call sign including the numeral is called a *prefix*. The part following the numeral is called a *suffix*. Each country in the world issues its own distinctive set of call sign prefixes, which can make it easy to tell where a particular station is. For example, typical prefixes in Canada are VE and VA. The common prefix in Mexico is XE.

At one time, the numeral indicated a US station's geographical region — 1 for New England, 6 for California and 9 or Ø (zero) for the Midwest. The FCC has made ham radio call signs portable, however — just like telephone numbers. So a call sign with "1" following the prefix may belong to a ham located in Florida.

You don't have to keep the call sign the FCC assigns. For a modest fee, the FCC's vanity call sign program permits a ham to select a new personalized call sign from among the database of certain unassigned call signs based on the applicant's license class.

1.3 Assembling Your Station

Amateur Radio costs as much or as little as your budget permits and your enthusiasm dictates. Many thrifty hams carry their stations with them, in the form of relatively inexpensive handheld transceivers, some the size of a typical cell phone. Without requiring any antenna beyond the one on the radio itself, such radios can come along when you're away from home or on a trip. On the other end of the scale, radio amateurs who are serious about radio contesting or DXing (contacting distant stations in other countries) often invest in the latest equipment and large antenna systems. The majority of hams fall somewhere in between, however. They have modest stations, simple wire antennas between two trees and maybe a small "beam" (directional antenna) on a backyard tower. Either way, you'll talk to the world.

Fig 1.9 — Riley, K4ZDH, demonstrates that a "ham shack" need not look like one or be relegated to the basement or attic. (*WW3DE photo*)

1.3.1 How Much Will It Cost?

Early radio amateurs generally built their own gear — mainly out of necessity; there were no well-stocked radio emporiums in the early 20th century. Constructing ham radio equipment from kits gained a lot of popularity in the mid-20th century. Many manufacturers still provide parts kits and circuit boards to make building even easier. Some hams continue to enjoy designing and building their own equipment (called "homebrewing") and maintain a low fiscal profile as a result. Many of these amateurs proudly stand at the forefront of technology and keep up with advances that may be applied to the hobby. Indeed, the projects you'll find in this *Handbook* provide a wide variety of equipment and accessories that make ham radio more convenient and enjoyable.

Fig 1.12 — Computers are an important part of most amateur stations these days. Michael McCarty, KE5RJJ, uses his to record contacts in a station log.

By and large, today's radio amateurs start out using off-the-shelf commercially made transceivers purchased new or on the used market, perhaps at a ham radio flea market or hamfest or from an Internet auction or classified ad site. A plethora of ham gear is readily available, and there's something out there to meet your needs while not breaking the bank. Used VHF or VHF-UHF (or "dual-band") handheld transceivers can be purchased for $100 to $200, and sometimes for far less.

Those interested in HF work can get in on the ground floor with a used, but serviceable, transceiver in the $250 to $500 range, and

Fig 1.10 — John Merritt, K4KQZ, operates mobile from his car and uses a temporary clip-on gooseneck lamp base to mount his radio's control head for clear visibility and safe driving. (*K4KQZ photo*)

there's a good selection of new transceivers available in the $500 to $1500 range. Flea-power transceivers covering single bands sell new for less than $100 in kit form. An HF antenna can be a simple backyard dipole. You'll find some great equipment choices advertised in ARRL's monthly journal, *QST*. In addition to advertising new ham gear, *QST* includes comprehensive equipment reviews.

Antennas and accessories can add appreciably to the cost of your station, but less-expensive alternatives are available, including building your own.

1.3.2 Computers and Ham Radio

If computers are your favorite facet of today's technology, you'll soon discover that connecting your PC to your ham station not only can make operating more convenient but can open the door to additional activities on the ham bands. A majority of radio amateurs have computers in their home stations, often one that's dedicated to ham radio tasks. Probably the most common use for a computer in the ham shack is for logging — keeping a record of — contacts. This is especially true for contesters, where speed and accuracy are paramount.

While there's no legal requirement to maintain a logbook of your on-the-air activities, many hams still keep one, even for casual operating, and computer logging can make the task less tedious (see sidebar "Keeping a Log").

Interfacing a PC with your transceiver also can let you control many of your radio's functions, such as frequency or band selection, without having to leave your logging program. It's also possible to control various accessories, such as antenna rotators or selection switches, by computer.

In addition, a soundcard-equipped computer allows you to enjoy digital modes with nothing more than a couple of connections, the right software (often free) and an interface. RTTY and PSK31 are two of the most popular HF digital

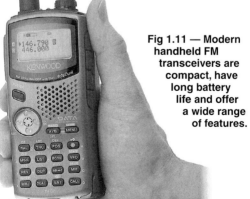

Fig 1.11 — Modern handheld FM transceivers are compact, have long battery life and offer a wide range of features.

Keeping a Log

Keeping a log — on paper or using your computer — of your on-air activity is optional, but there are some important reasons for doing so. These include:

Legal protection — Good record keeping can help you protect yourself if you are ever accused of intentional interference or have a problem with unauthorized use of your call sign.

Awards tracking — A log helps you keep track of contacts required for DXCC, WAS and other awards. Your log lets you quickly see how well you are progressing toward your goal.

An operating diary — A log book is a good place for recording general information about your station. For example, you may want to include comparisons between different antennas or pieces of equipment based on reports from other stations. Your log is also a logical place to make a note of new acquisitions (complete with serial numbers in case your gear is ever stolen). You can track other events as well, including the names and call signs of visiting operators, license upgrades, contests, propagation and so forth.

Paper and Computer Logs

Many hams, even some of those with computers, choose to keep "hard copy" log books. Paper logs do not require power, are flexible and can survive a hard-drive crash! Preprinted log sheets are available, or you can create your own. Computers with word processing and publishing software let you create customized log sheets in no time.

On the other hand, computer logs offer many advantages, especially if you enjoy contesting, DXing or awards chasing. For example, a computerized log can instantly indicate whether you need a particular station for DXCC or WAS. Contesters use computer logs to manage contact data during the contest and to weed out duplicate contacts in advance. Most major contest sponsors prefer to receive computer log files, and some prohibit the submission of paper logs altogether. Computer logs can also tell you at a glance how far along you are toward certain awards and help with printing QSL labels. Computer logs also make it easy to submit your contacts to ARRL's online Logbook of The World (see sidebar).

Computer logging programs are available from commercial vendors. Some are general-purpose programs, while others are optimized for contesting, DXing or other activities. Check the ads in *QST* and take a look at capabilities and requirements before you choose.

Hamfests

Amateur Radio's social world extends beyond on-the-air acquaintances. Regular ham radio gatherings, usually called "hamfests," offer opportunities for those attending to get to know each other in person — called "an eyeball contact" in ham parlance. Hams also enjoy buying, selling and trading ham radio equipment and accessories in the hamfest "flea market." *Every* ham loves a bargain. Others take advantage of classroom sessions or forums to learn more about particular aspects of the hobby.

Hamfests are great places to get good deals on gear — some vendors offer substantial hamfest discounts — and to expand your knowledge. Thousands of radio amateurs from the US and around the world each spring gather at Dayton Hamvention in Ohio. This truly international event epitomizes the goodwill that exists among the world's Amateur Radio enthusiasts.

The annual Dayton Hamvention (www.hamvention.com) attracts upward of 25,000 hams from around the world to meet and greet, learn, buy, sell and trade.

modes. PSK31 lets you talk great distances with a very modest ham station, typically at very low power levels.

Computers also can alert you to DX activity on the bands, help you practice taking Amateur Radio license examinations or improve your Morse code abilities. Many ham radio organizations, interest groups and even individuals maintain Web sites too. There's software for many ham radio applications, from logging and award tracking to antenna and circuit design.

1.4 *Hello, World!* — Getting on the Air

Amateur Radio is a *social* activity as well as a technical pursuit. It's a way to make new friends and acquaintances over the air that you may later meet in person. Some ham radio relationships last a lifetime even though the individuals sometimes never meet face to face. Ham radio can be the common thread that keeps high school and college friends in touch through the years, with or without e-mail.

Amateur Radio also can cement relationships between radio amateurs of different nationalities and cultures, leading to greater international goodwill and understanding. When you become an Amateur Radio operator, you also become a "world citizen." In return you can learn about the lives of the radio amateurs you contact in other countries.

"*What do hams say to each other?*" you might wonder. When they meet for the first time on the air, hams exchange the same sorts of pleasantries that anyone might when meeting for the first time. Ham radio operators give their name, their location (abbreviated "QTH" in Hamspeak) and — specific to hamming — a report indicating how well they're hearing (or "copying") the other station's radio signal. This name-location-signal report pattern typically applies regardless of ham radio mode. With the preliminaries out of the way, hams' conversations often focus on their equipment or other interests.

Although English is widely used and understood on the ham bands, English speakers can make a favorable impression on hams in foreign countries if they can speak a little of the other person's language — even if it's as simple as *danke*, *gracias* or *sayonara*.

1.4.1 Voice Modes

We've mentioned the use of voice (or phone) and Morse code (or CW) on the amateur bands. Although more hams are embracing digital modes every day, phone and CW by far remain the most popular Amateur Radio communication modes. Ham voice modes are amplitude modulation (AM), which includes the narrower-bandwidth single sideband (SSB), and frequency modulation (FM). For the most part, SSB is heard on HF, while FM is the typical voice mode employed on VHF, UHF and microwave bands.

The great majority of ham radio HF phone (short for "radiotelephone") operators use SSB (subdivided further into upper sideband and lower sideband), but a few still enjoy and experiment with the heritage "full-carrier AM." Once the primary ham radio voice mode, this type of AM still is heard on the standard broadcast band (530 to 1710 kHz). Today's AM buffs enjoy its warm, rich audio quality, and the simplicity of circuit design encourages restoring or modifying vintage radios or building from scratch. For more information about AM operation, visit **www.arrl.org/am-phone-operating-and-activities**.

1.4.2 Morse Code

Morse code was the first radio transmission mode, although it wasn't long before early experimenters figured out how to send the human voice and even music over the airwaves. Morse is also the original digital mode; the message is transmitted by turning a radio signal on and off (a "1" and a "0" in digital terms) in a prescribed pattern to generate individual letters, numerals and characters. This pattern is the International Morse Code, sometimes called the "radio code," which varies in many respects from the original Morse-Vail Code (or "American Morse") used by 19th century railroad telegraphers.

Federal regulations once required that prospective radio amateurs become proficient in sending and receiving Morse code. Although this is no longer the case for any class of Amateur Radio license in the US, many hams still embrace CW as a favorite mode and use it on a regular basis to communicate with one another. Hams typically send Morse code signals using a manual telegraph key or a CW "paddle" and an automatic keyer. Most hams receive Morse code "by ear," either writing down the letters, numerals and characters as they come through the receiver's headphones or speaker or simply reading it in their heads. Accessories, some computer based, have become available that will translate CW signals without the need to learn the code.

Hams who enjoy CW cite its narrow bandwidth — a CW signal takes up very little space — simpler equipment and the ability of a CW signal to "get through" noise and interference with a minimum of transmitting power. CW is a common low-power (QRP) mode.

1.4.3 FM Repeaters

Hams often make their first contacts on local voice repeaters. These devices, which can greatly extend the useful range of a typical handheld FM transceiver, carry the vast

Fig 1.13 — Teenager Ashby Bird, KJ4EGJ, enjoys operating on voice modes from her home in Tennessee. (*KU4ME photo*)

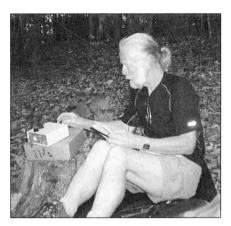

Fig 1.14 — Morse code (CW) operation is especially popular for low power operation. K1YPP carried this compact CW station during a hike along the Appalachian Trail in Virginia.

majority of VHF/UHF traffic, making local mobile communication possible for many hams. Located on hilltops, tall buildings or other high structures, repeaters strengthen signals and retransmit them. This provides communication over distances much farther than would be possible without a repeater.

Typically, hams use repeaters for brief contacts, although socializing and "ragchewing" are routine on some "machines," as repeaters are often called. All repeater users give priority to emergency communications.

Most repeaters are maintained by clubs or groups of hams. If you use a particular repeater frequently, you should join and support the repeater organization. Some hams set up their own repeaters as a service to the community.

The best way to learn the customs of a particular repeater is to listen for a while before transmitting. Most repeaters are *open*, meaning that any amateur may use the repeater, although repeaters typically require users to transmit an access tone (which you can set on any modern FM transceiver). A few repeaters are *closed*, meaning that usage is restricted to members. Some repeaters have *autopatch* capability that allows amateurs to make telephone calls through the repeater. The *ARRL Repeater Directory* shows repeater locations, frequencies, capabilities and whether the repeater is open or closed.

1.4.4 Digital Modes

Radioteletype (RTTY — often pronounced "Ritty") is a venerable data communication mode that remains in wide use today among radio amateurs. While RTTY does not support the features of newer computer-based data modes, it is well suited for keyboard-to-keyboard chats with other stations. It also is the most popular mode for worldwide digital contests and remains in common use among DXers and DXpeditions. RTTY was originally designed for use with mechanical teleprinters, predating personal computers by several decades. Today, Amateur Radio RTTY uses soundcard-equipped computers and dedicated RTTY software.

An HF digital mode for general domestic and DX contacts is PSK31. This mode is particularly effective for low-power (QRP) communication. It's easy to transmit and receive PSK31 using a PC and software — typically free — that operates via a PC soundcard. Quite a few different digital modes use the same basic PC and soundcard setup, and new ones are developed on a regular basis.

Another digital mode, packet radio, is much less popular in today's ham radio world than it was in the 1990s. Packet's most important applications include networking and unattended operation. The most common uses are the worldwide DX Cluster network, the Automatic Packet/Position Reporting System (APRS) and regional or local general-purpose networks. Would you like to see what 20 meter stations in Asia have been worked or heard recently by stations in your area? Log onto your local DX Cluster node and find out. Is your friend out of range of your 2 meter packet radio? Send your message through the packet network.

APRS

Developed by Bob Bruninga, WB4APR, APRS or the Automatic Packet/Position Reporting System (**aprs.org**) is used for tracking stations or objects in motion or in fixed positions and for exchanging data. It uses the unconnected packet radio mode to graphically indicate objects on maps displayed on a computer monitor. Unconnected packets permit all stations to receive each transmitted APRS packet on a one-to-all basis rather than the one-to-one basis required by connected packets.

As with other packet transmissions, APRS data are relayed through stations called

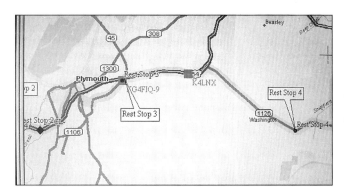

Fig 1.15 — APRS is a popular ham radio digital mode that allows tracking of mobile stations transmitting beacons via packet radio as they travel.

digipeaters (digital repeaters). Unlike standard packet radio, APRS stations use generic digipeater paths, so no prior knowledge of the network is needed. In addition, the Internet is an integral part of the system that is used for collecting and disseminating current APRS data in real time.

Virtually all VHF APRS activity occurs on 2 meters, specifically on 144.39 MHz, the recognized APRS operating channel in the US and Canada. On UHF, you'll find APRS activity on 445.925 MHz.

Many groups and individuals that participate in public service and disaster communications find APRS a useful tool. Others use it to view real-time weather reports.

1.4.5 Image Communication

Several communication modes allow amateurs to exchange still or moving images over the air. Advances in technology have brought the price of image transmission equipment within reach of the average ham's budget. This has caused a surge of interest in image communication.

Amateur TV (ATV) is full-motion video over the air, sometimes called "fast-scan TV." Amateur Radio communication takes on an exciting, new dimension when you can actually *see* the person you're communicating with. In addition, ATV has proved to be very useful in emergency and disaster communication situations. Amateur groups in some areas have set up ATV repeaters, allowing lower-power stations to communicate over a fairly wide area. Since this is a wide-bandwidth mode, operation is limited to the UHF bands (70 cm and higher).

Digital ATV folds nicely into a recent Amateur Radio technological initiative called high-speed multimedia (HSMM) radio. The ham bands above 50 MHz can support computer-to-computer communication at speeds high enough to support multimedia applications — voice, data and image. One approach adapts IEEE 802 technologies, particularly 802.11b, operating on specific Amateur Radio frequencies in the 2400-2450 MHz band.

SSTV or "slow-scan TV" is a narrow-bandwidth image mode that has remained popular for many years in Amateur Radio. Instead of full-motion video, SSTV is for the exchange of photographs and other images. Individual SSTV pictures take anywhere from 8 seconds to about 2 minutes to send depending on the transmission method. These days most SSTV operation is done in color using computers and soundcards. Images are converted into a series of audio tones representing brightness level and colors. Since SSTV is a narrow-band mode, it is popular on HF on the same frequencies used for voice operation.

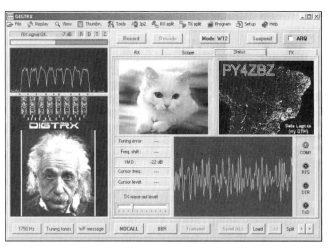

Fig 1.16 — Slow-scan television (SSTV) allows hams to exchange pictures over the air.

1.5 Your Ham Radio "Lifestyle"

After some on-the-air experience, many Amateur Radio enthusiasts focus on a particular mode or operating style and may identify themselves as contesters, DXers, CW ops or VHF-UHF enthusiasts. Others center their operating on such activities as specialized or experimental modes, mobile ham radio, low-power operating (known as "QRP") and radio direction finding.

1.5.1 Ham Radio Contesting — Radiosport

Ham radio contesting, called "radiosport," is growing in worldwide popularity. Hardly a weekend goes by when there isn't a ham radio contest of some sort. These on-the-air competitions range from regional operating events with a few hundred participants to national and worldwide competitions with thousands on the air.

Objectives vary from one event to another, but ham radio contests typically involve trying to contact — or "work" — as many *other* contest participants on the air within a specified period. In each contact, participants exchange certain information, often a signal report and a location, as the contest's rules dictate. A lot of contest scoring schemes place a premium on two-way contacts with stations in certain countries, states or zones. Top scorers in the various entry categories usually get certificates, but a few events bestow sponsored plaques and trophies.

There are contests for nearly every mode and operating preference available to Amateur Radio — voice, Morse code and digital modes. Some members of the contesting community are earnest competitors who constantly tweak their stations and skills to better their scores. Others take a more casual approach. All have lots of fun.

In the ARRL International DX Contest, for example, you'll try to contact as many DX (foreign) stations as possible over a weekend.

During such a weekend, experienced hams with top-notch stations easily contact more than 100 different countries! Even a modest operator can contact dozens of countries in a single weekend.

Other popular contests include state QSO parties, where the goal is to contact stations in as many of the sponsoring state's counties. ARRL November Sweepstakes (SS) is a more high-energy full-weekend contest that attracts thousands of operators each fall. One weekend is dedicated to CW, another to phone. VHF, UHF and microwave contests focus on making contacts using our highest-frequency bands. Digital-mode contests have gained in popularity in recent years, thanks to computer soundcards, radios that offer digital-mode capabilities and often-free software.

You can find information on contests each month in ARRL's monthly membership journal *QST;* the contest calendar on the ARRL Web site also provides up-to-date information on upcoming operating events. The ARRL's bi-monthly publication *National Contest Journal* (*NCJ*) focuses on topics of particular interest to contesting novices and veterans alike. For timely contest news and information, check "The ARRL Contest Update" e-newsletter at **www.arrl.org/contest-update-issues**, available every other week via e-mail and on the ARRL Web site.

Fig 1.17 — Ham radio operators who are serious about radiosport — ham radio competitions — often have extensive antenna systems, such as the "aluminum farm" at K3LR in western Pennsylvania.

ARRL FIELD DAY

An emergency communications training exercise with some elements of a contest, ARRL Field Day, held each year on the fourth full weekend in June, attracts thousands of participants. Portable gear in tow, hams take to the hills, forests, campsites, parking lots and even emergency operations centers or vans to participate in Field Day. Tracing its origins to the 1930s, Field Day started out as a way to publicly demonstrate and exercise the ability of hams to operate "in the field" and "off the grid." The goal is not only to make lots of contacts but to operate successfully under conditions you could face in the aftermath of a disaster or emergency.

Most stations are set up outdoors and use emergency power sources, such as generators, solar panels, wind turbines and occasionally even water wheels. Creativity reigns when it comes to power sources! Over the years, Field Day's contest-like nature has led to plenty of good-natured competition among clubs and groups. Field Day stations range from simple to elaborate. If a real disaster were to strike, these stations could be set up quickly wherever needed, without having to rely on commercial power that could fail.

1.5.2 Chasing DX

"DX" typically refers to stations in distant countries around the world. Hams who concentrate on making contacts with faraway stations are called "DXers." Chasing DX is a time-honored ham radio tradition, and many hams concentrate their efforts toward working stations in far-flung and rare locations. Often they have as their goal getting on the DXCC "Honor Roll" or entering the ARRL's annual DX Challenge.

Working DX does not necessarily require an expensive radio and huge antenna system. It's possible to work a lot of DX — even halfway around the world — with very low power or with modest antennas, including mobile antennas.

Some hams specialize in certain bands to work DX, such as 160 meters, where DXing can be challenging due to its low frequency, frequent noise and infrequent DX propagation. Others prefer "the high bands," such as 20, 15 and 10 meters.

DXPEDITIONS

DXers who have run out of new countries to work sometime *become* the DX! "DXpeditions" travel to countries with few or no hams. Participants often make thousands of contacts in the space of a few days. They not only have a great time but often can spread international goodwill.

Some DXpeditions are huge productions. In 2008, an international team activated Ducie Island, a tiny atoll in the South Pacific with no indigenous population. In the space of two weeks, the team completed nearly 184,000 two-way contacts with other hams around

Fig 1.18 — Tens of thousands of hams participate in ARRL Field Day each June. It's the largest Amateur Radio event on the planet! Here are members of the WØAK Des Moines (Iowa) Radio Amateur Association taking the late shift near Polk City, Iowa. (*KØDSM photo*)

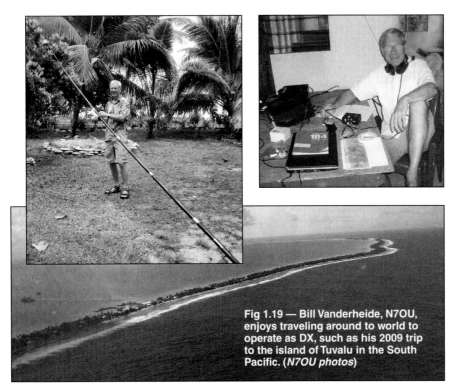

Fig 1.19 — Bill Vanderheide, N7OU, enjoys traveling around to world to operate as DX, such as his 2009 trip to the island of Tuvalu in the South Pacific. (*N7OU photos*)

the globe. Ducie Island counts as a separate country (or "entity") for the DXCC award.

Most DXpeditions are smaller affairs in which one or two operators combine a vacation with some on-air fun. Activity often peaks in conjunction with the major DX contests. If you don't want to drag your radio and antennas along, fully equipped DX stations are available for rental.

DX SPOTS AND NETS

The beginning DXer can get a good jump on DXCC by frequenting the DX spotting sites on the Internet. A DX spotting Web site is essentially an Internet clearing house of reports — or "spots" — posted by other DXers of stations actually heard or worked. The DX Cluster offers much the same information via packet. Users from around the world post spots nearly in real time, and each spot gives the station posting the spot and the DX station's frequency.

DX nets offer another DX gateway. On DX nets, a net control station keeps track of which DX stations have checked into the net. The net control station then allows a small group of operators to check in and work one of the DX stations. This permits weaker stations to be heard instead of being buried in a pileup.

1.5.3 Operating Awards

Earning awards that reflect Amateur Radio operating accomplishments is a time-honored tradition. Literally hundreds of operating awards are available to suit your level of activity and sense of accomplishment.

WORKED ALL STATES

One popular and longstanding award is the Worked All States (WAS) award, sponsored by the ARRL. To earn the "basic" WAS a radio amateur must confirm two-way radio contacts with operators in each of the 50 United States. If you enjoy operating different bands and a greater challenge, try for the ARRL's 5-Band WAS (5BWAS) award by confirming contacts with all 50 US states on each of the 80, 40, 20, 15 and 10 meter bands.

A new twist on the WAS award is the ARRL's Triple Play Award. Introduced in 2009, it became an instant hit with hams. To earn the Triple Play Award, an amateur must contact other amateurs in each of the 50 US states using voice, Morse code and a digital mode such as RTTY or PSK31. All qualifying contacts must be confirmed via the ARRL's Logbook of the World (LoTW — see sidebar, "Logbook of The World). The Triple Play Award is available to hams all over the world.

DX CENTURY CLUB

The most prestigious and popular DX award is the DX Century Club (DXCC), sponsored by the ARRL. Earning DXCC is

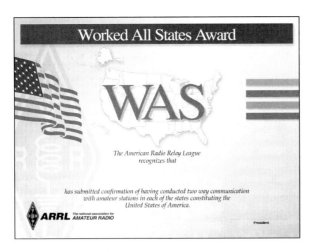

Fig 1.20 — The ARRL Worked All States (WAS) certificate is awarded for confirming contacts with hams in all 50 states. Confirmations may be made by traditional paper QSL cards, or electronically through ARRL's Logbook of The World online database.

Logbook of The World

Rather than exchange and collect paper QSL cards, more and more radio amateurs are taking advantage of the ARRL's Logbook of The World to confirm contacts for award credit. LoTW is a repository of individual radio contact records submitted by users from around the world. When both participants in a QSO submit matching QSO records to LoTW, the result is a virtual QSL that can be applied toward ARRL award credit. Uploading contact data costs nothing; users only pay to "redeem" their QSO credits for an award, such as DXCC or WAS.

Once signed up as a Logbook user, you can submit new contact records whenever you wish. Your contacts will be matched against the logs of other Logbook users. Whenever a match occurs, you receive instant credit for the contact.

To minimize the chance of fraudulent submissions, all LoTW QSO records are digitally "signed" by the licensee, who must hold an LoTW certificate. LoTW began operation in 2003 and has tens of thousands of users, with several hundred million resulting QSL records! Visit the Logbook of The World Web site, **www.arrl.org/logbook-of-the-world**, to learn more.

QSL Cards

Long before the Internet and e-mail, hams began the custom of exchanging postcards that became known as QSL cards or simply QSLs. "QSL" is another radiotelegraph, or Morse code, abbreviation that means "I confirm receipt of your transmission." Exchanging QSL cards can enhance your ham radio enjoyment and even lead to a regular correspondence.

Hams still take great pride in having distinctive QSL cards to exchange following a contact, although today, thanks to the Internet, electronic means exist, such as ARRL's Logbook of The World (see sidebar, "Logbook of The World") to confirm contacts. A QSL card contains information that verifies a two-way contact took place.

DX stations, especially those in very rare places, are often inundated with QSL cards and requests from US hams. To ease the cost and administrative burden, most DX QSLs travel via QSL bureaus, which ship cards in bulk, then sort and distribute them on the receiving end. The Outgoing QSL Service is available to ARRL members at nominal cost. The incoming QSL bureaus are available to all amateurs. Bureau instructions and addresses are on the ARRL Web site.

quite a challenge. You must confirm two-way contact with stations in 100 countries (or "entities," as they're known in the DXCC program). Hams with very simple stations have earned DXCC. Operating in various DX contests when stations all over the world are looking for QSOs is a good way to combine DXing and contesting, and to get a leg up on earning DXCC. There's also a 5-Band DXCC (5BDXCC) for earning DXCC on each of five bands.

Top-rung DX enthusiasts have been challenging themselves and each other through the ARRL DXCC Challenge. This activity, which is ongoing, involves confirming contacts with DXCC entities on all bands from 160 through 6 meters.

The Worked All Continents (WAC) certificate is a good starting point for newcomers. Sponsored by the International Amateur Radio Union (IARU), earning WAC requires working one station on each of six continents (excluding Antarctica).

1.5.4 Satellite Communication

Amateur Radio has maintained a presence in space since 1961 with the launch of OSCAR 1 (OSCAR is an acronym for Orbiting Satellite Carrying Amateur Radio). Since then, amateurs have launched some five dozen satellites, most of the low-earth orbit variety and a small number in the high-earth orbit category. The history of Amateur Radio satellites and information on which ones are in operation is available on the AMSAT Web site, **www.amsat.org**.

Amateurs have pioneered several developments in the satellite industry, including low-earth orbit communication "birds" and PACSATs — orbiting packet bulletin board systems. Operating awards, such as VUCC (VHF/UHF Century Club), WAS and DXCC, are available from ARRL and other organizations specifically for satellite operation.

Satellite operation is neither complex nor difficult; it's possible to work through some satellites with nothing more than a dual-band (VHF/UHF) handheld transceiver and perhaps a small portable antenna. More serious satellite work requires some specialized equipment. You may be able to work several Amateur Radio satellites (OSCARs) with the equipment that's now in your shack!

AMATEUR RADIO IN SPACE

In 1983, the first ham-astronaut made history by communicating from the space shuttle *Columbia* with ground-based hams. NASA subsequently made Amateur Radio a part of many shuttle missions and later as a permanent presence aboard the International Space Station. The space agency promotes ham activity because of its proven educational and public relations value — and because ham radio could be used for backup communication in a pinch.

Today, the Amateur Radio on the International Space Station (ARISS) program allows for digital and analog/voice communication with ISS crew members and with other terrestrial amateurs via an onboard packet mailbox and APRS digipeater (**www.arrl.org/amateur-radio-on-the-international-space-station**). Most ISS crew members are Amateur Radio licensees. Scheduled contacts are frequently made with demonstration stations in school classrooms around the world. Questions and answers are exchanged between astronauts and schoolchildren via FM voice on VHF.

1.5.5 QRP: Low-Power Operating

One very active segment of the Amateur Radio community enjoys operating with a minimum of transmitting power. They call themselves "QRP operators," after the Morse code abbreviation for "I shall decrease transmitter power." While the FCC allows hams with the proper license to run up to 1500 W (watts) PEP and many hams run 100 W, QRPers typically use 5 W or less — some *far* less (one ham achieved WAS while running 2 milliwatts — that's two-thousandths of a watt!). Operating QRP is a challenge; for starters, you won't be heard as easily, so patience becomes a real virtue both for the low-power enthusiast and the station on the other end of the contact! With effective antennas and skillful operating, however, QRPers make contacts around the world. This operating style has become so popular that many contests now include an entry category for stations running 5 W or less output power.

One of the best reasons to operate QRP is that low-power equipment typically is lightweight and less expensive. Many QRP

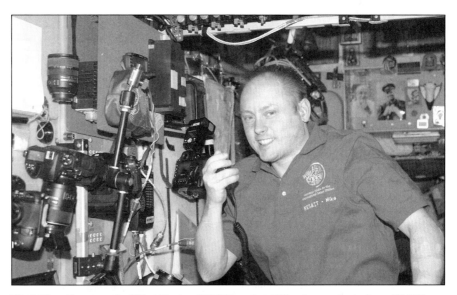

Fig 1.21 — Commander Mike Fincke, KE5AIT, makes FM voice contacts from NA1SS aboard the orbiting International Space Station. (*NASA photo*)

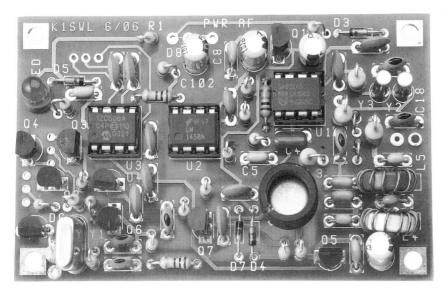

Fig 1.22 — Some hams enjoy building their own equipment to learn more about radio and electronics. This is a complete low power (QRP) CW transceiver assembled from a kit.

operators enjoy designing and building their own "flea-power" transceivers, and various organizations promoting low-power operating offer kits, circuits and advice. A few commercial manufacturers also market QRP equipment and kits.

1.5.6 Operating Mobile

Many hams enjoy operating on the fly — usually from a car or truck but sometimes from a boat, a motorcycle, a bicycle and even while on foot! Operating "mobile" from a car is the most common form, and manufacturers today offer a wide range of ham radio gear, including antennas, just for this type of operating. A mobile station can be something as simple as a basic VHF or dual-band VHF/UHF transceiver and a little antenna attached magnetically to the roof, or as complex as an HF station and a more substantial antenna system. Some mobile stations are very sophisticated, with capabilities that rival those of many fixed stations.

While most hams who operate mobile use FM or SSB, a growing number operate CW while on the road. It takes a bit of practice, in part because the operator must learn to understand Morse code without having to write it down.

Hams on bicycle treks or hikes can easily carry along lightweight radio gear. A lot of cyclists or hikers pack a small ham radio transceiver and wire antenna along with their sleeping bag, food and water.

1.5.7 VHF, UHF and Microwave Operating

Hams use many modes and techniques to extend the range of their line-of-sight signals on the VHF, UHF and microwave bands. Those who explore the potential of VHF/UHF communications are often known as "weak-signal" operators to differentiate them from FM operators — although signals are often not really *weak*. These enthusiasts probe the limits of propagation with the goal of discovering just how far they can communicate.

They use directional antennas (beams or parabolic dishes) and very sensitive receivers. In some instances, they employ considerable transmitter output power, too. As a result of their efforts, distance records are broken almost yearly. On 2 meters, for example, conversations between stations hundreds and even thousands of miles apart are not uncommon. The distances decrease as frequencies increase, but communication regularly can span several hundred miles, even at microwave frequencies.

Weak-signal operators for many years depended on SSB and CW, but computer/sound card based digital modes are now part of their arsenal. These modes use state-of-the-art digital signal processing (DSP) software for transmitting and receiving very weak signals, often below levels that the human ear can detect.

EME

EME (Earth-Moon-Earth) communication, also known as "moonbounce," fascinates many amateurs. The concept is simple: use the moon as a passive reflector of VHF and UHF signals. Considering the total path of some 500,000 miles, EME is the ultimate DX — at least to date. The first two-way amateur EME contacts took place in 1960.

Traditionally EME was a CW mode activity requiring large antennas and high power. Advances in technology, such as low noise receivers and digital signal processing (DSP) tools, are making EME contacts possible for more and more amateurs with modest stations.

METEOR SCATTER

As a meteor enters the Earth's atmosphere, it vaporizes into an ionized trail of matter. Such trails are often strong enough to reflect VHF radio signals for several seconds. During meteor showers, the ionized region becomes large enough — and lasts long enough — to sustain short QSOs. It's exciting to hear

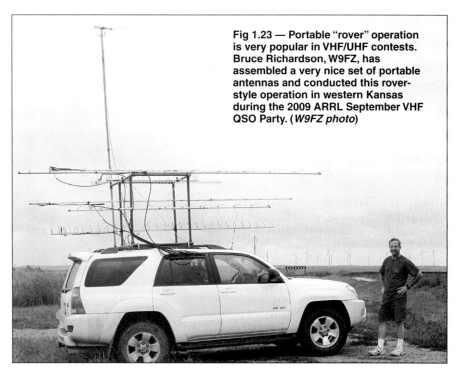

Fig 1.23 — Portable "rover" operation is very popular in VHF/UHF contests. Bruce Richardson, W9FZ, has assembled a very nice set of portable antennas and conducted this rover-style operation in western Kansas during the 2009 ARRL September VHF QSO Party. (*W9FZ photo*)

a signal from hundreds of miles away pop out of the noise for a brief period!

Amateurs experimenting with meteor-scatter propagation use transmitter power of 100 W or more and beam antennas. Most contacts are made using SSB, CW or digital modes. Although most SSB and CW QSOs are made during annual meteor showers, digital mode contacts are possible any day of the year.

1.5.8 Vintage Radio

Present-day commercial Amateur Radio equipment has reached a level of complexity that often requires specialized test and troubleshooting equipment to repair or align. Modern component manufacturing technology such as surface-mount devices (SMDs) has come into such common use that a modular approach to equipment repair has now become customary. Rather than troubleshoot and replace a defective component, many manufacturers now prefer to swap out an entire module.

Many amateurs still prefer to repair and adjust their own equipment and covet the days when it was easier to do so. This is but one reason for the surge in attention to vintage radio operating and collecting. Some others have to do with equipment cost, availability, rarity and, of course, nostalgia. Many of these radios are affectionately called "boat anchors" by their vintage radio aficionados, since early radio gear tends to be relatively large and heavy.

Some enthusiasts enjoy the challenge of collecting and restoring older radios, sometimes striving to bring the equipment back to its original factory condition. Other vintage radio enthusiasts may have a parallel interest in conventional AM voice transmission. These activities take vintage radio fans back to an era when amateurs knew how their equipment worked and repaired it when it didn't.

1.5.9 Radio Direction Finding (DF)

DFing is the art of locating a signal or noise source by tracking it with portable receivers and directional antennas. Direction finding is not only fun, it has a practical side as well. Hams have been instrumental in hunting down signals from illegal jammers and malfunctioning transmitters in addition to locating noise (interference) sources.

"Fox hunting" — also called "T-hunting," "radio-orienteering" or "bunny hunting" — is ham radio's answer to hide-and-seek. One player is designated the fox; he or she hides a transmitter, and the other players attempt to find it. Rules vary, but the fox must generally place the transmitter within certain boundaries and transmit at specific intervals.

Fox hunts differ around the world. American fox hunts often employ teams of fox hunters cruising in vehicles over a wide area. European and other fox hunters restrict their events to smaller areas and conduct fox hunts on foot. *Radiosport* competitions typically follow the European model.

DF techniques can come into play when tracking down sources of interference on the ham bands — intentional or inadvertent — or when there's a suspected "pirate" (unlicensed ham station) in the area.

Fig 1.24 — A lot of hams enjoy tinkering with older Amateur Radio equipment, still widely available at hamfests. (*WW3DE photo*)

1.6 Public Service

The ability to provide communication as a free public service has been a traditional responsibility of Amateur Radio from the start. Today, this most often involves ham radio's volunteer efforts during disasters and emergencies. When floods struck North Dakota and Minnesota in the spring of 2009, the Amateur Radio community rallied to supplement overly stressed public safety systems. Hams volunteered around the clock to provide communication for sandbagging operations and evacuation efforts, as well as to link hospitals, emergency operations centers and non-government relief agencies. Public service also can take less dramatic forms: Hams also step forward to provide communication for walkathons, marathons, bike races, parades and other community events.

Practice makes perfect. On any given weekend, hams associated with emergency communication teams might be found supporting radio communication in the aftermath of a simulated disaster or weather emergency, to sharpen their emergency-preparedness skills.

1.6.1 Emergency Communication

The FCC rules governing Amateur Radio list emergency communication as one of the purposes of the Amateur Radio Service. The ability to provide emergency communication is a big justification for Amateur Radio's existence.

Despite the proliferation of cell phones and other personal communications devices, Amateur Radio continues to prove its value, since it can operate without an existing infrastructure. Ham radio doesn't need the cellular network or the Internet. When all else fails, Amateur Radio works. Battery-powered equipment allows hams to provide essential communication even when power is knocked out. If need be, hams can make and install antennas on the spot from available materials. In the wake of hurricanes, forest fires, earthquakes and other natural disasters that cripple or compromise normal communications, hams may be called upon to handle thousands of messages in and out of the stricken region. The work that hams do during crisis situations cultivates good relations with neighbors and with local governments.

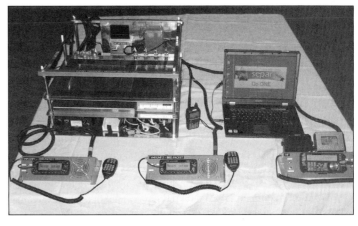

Fig 1.25 — When disaster strikes, amateurs are ready to provide emergency communications by preparing compact "grab and go" kits of radio gear, such as the three-radio, multi-band, multi-mode package built for the Surrey (British Columbia, Canada) Emergency Program Amateur Radio team by VE7IO, VE6XS, VA7DRW, and VA7XB.

Amateur Radio operators have a long tradition of operating from backup-power sources. Through events such as Field Day, hams have cultivated the ability to set up communication posts wherever they are needed. Moreover, Amateur Radio can provide computer networks (with over-the-air links as needed) and other services such as video that no other service can deploy on the fly, even on a wide scale.

1.6.2 Emergency Communication Organizations

Many hams join a local volunteer Amateur Radio emergency communication organization and regularly practice their communication skills. Should an emergency arise, volunteer teams from these organizations cooperate with emergency personnel such as police and fire departments, the Red Cross and medical personnel to provide or supplement communication. Hams sometimes are called upon to fill the communication gap between agencies whose radio systems are incompatible with one another.

ARES AND RACES

The Amateur Radio Emergency Service (ARES) and the Radio Amateur Civil Emergency Service (RACES) are the umbrella organizations of Amateur Radio emergency communication. ARES is sponsored by ARRL and handles many different kinds of public service activities. RACES is administered by the Federal Emergency Management Agency (FEMA) and works with government agencies to maintain civil preparedness and provide communication in times of civil emergency. RACES is activated at the request of a local, state or federal official.

Amateurs serious about emergency communication typically are active in ARES or RACES or may carry dual ARES/RACES membership. FCC rules make it possible for ARES and RACES to use many of the same frequencies, so an ARES group also enrolled in RACES can work within either organization as circumstances dictate.

MILITARY AFFILIATE RADIO SYSTEM (MARS)

MARS is administered by the US armed forces and exists for the purpose of transmitting communications between those serving in the military and their families. This service has existed in one form or another since 1925.

There are three branches of MARS: Army MARS, Navy/Marine Corps MARS and Air Force MARS. Each branch has its own membership requirements, although all three branches require members to hold a valid US Amateur Radio license and to be at least 18 years old (amateurs from 14 to 18 may join with the signature of a parent or legal guardian).

MARS operation takes place on frequencies adjoining the amateur bands and usually consists of nets scheduled to handle traffic or to handle administrative tasks. Various MARS branches may also maintain repeaters or packet systems.

While MARS usually handles routine traffic, the organization is set up to handle official and emergency traffic if needed.

1.6.3 Public Service and Traffic Nets

The ARRL was formed to coordinate and promote the formation of message-handling nets, so public service and traffic nets are part of a tradition that dates back almost to the dawn of Amateur Radio. In those early days, nets were needed to communicate over distances longer than a few miles. From their origination point, messages (also called "traffic") leapfrogged from amateur station to amateur station to their destination — thus the word "relay" in American Radio Relay League. It still works that way today, although individual stations typically have a much greater range.

Some nets and stations are typically only active in emergencies. These include Amateur Radio station WX4NHC at the National Hurricane Center, the Hurricane Watch Net, SKYWARN (weather observers), The Salvation Army Team Emergency Radio Network (SATERN), The Waterway Net and the VoIP (voice over Internet protocol) SKYWARN/Hurricane Net.

THE NATIONAL TRAFFIC SYSTEM (NTS)

The National Traffic System (NTS) exists to pass formal written messages from any point in the US to any other point. Messages, which follow a standard format called a "Radiogram," are relayed from one ham to another, using a variety of modes, including voice, Morse code, RTTY or packet. An NTS operator who lives near the recipient typically will deliver the message by telephone.

During disasters or emergencies, Radiograms are used to communicate information critical to saving lives or property or to inquire about the health or welfare of disaster victims. At these times, the NTS works in concert with the Amateur Radio Emergency Service (ARES) and other emergency and disaster-relief organizations, such as the American Red Cross and The Salvation Army.

The NTS oversees many existing traffic nets, which meet daily. Most nets are local or regional. Handling routine message traffic such as birthday and holiday greetings keeps NTS participants prepared for emergencies.

1.7 Ham Radio in the Classroom

Amateur Radio can complement any school curriculum and give students a chance to make a direct and immediate connection with their studies. For example, the math and science used in Amateur Radio apply equally in the classroom. Even geography takes on a new meaning when students are able to contact other countries around the globe and speak with people who live in them!

Local volunteers are important to establishing an active Amateur Radio presence in schools. An HF or satellite station or even a VHF or UHF handheld transceiver tuned to the local repeater can prove an exciting and educational experience for pupils and volunteers alike.

Thanks to the Amateur Radio on the International Space Station (ARISS) program, amateurs all over the nation have made it possible for students to speak directly with astronauts in space via ham radio. Many individuals began their path towards careers in electronics and wireless communication thanks to their experiences with Amateur Radio as children and teenagers.

1.7.1 ARRL Amateur Radio Education & Technology Program

Through the ARRL Amateur Radio Education & Technology Program (ETP), Amateur Radio can become a significant resource for the classroom teacher. Applying Amateur Radio as part of the class curriculum offers students a new dimension to learning. The ARRL has developed an education project to introduce teachers to this resource and enable them to make the most effective use of it in their classrooms. The program's goal is to improve the quality of education by providing an educationally sound curriculum focused on wireless communication.

Amateur Radio emphasizes self-challenge, the value of lifelong learning and the importance of public service. From a more practical standpoint, future employers will be looking for candidates who are familiar not only with computers but with the sorts of wireless communication concepts used in Amateur Radio.

The ETP offers a range of resources to encourage educators. These include publications related to the use of technology in wireless communications; workshops, tips and ideas for teaching wireless technology in schools, community groups and clubs; and lesson plans and projects to help provide authentic, hands-on technological experiences for students.

The ETP emphasizes the integration of technology, math, science, geography, writing, speaking and social responsibility within a global society. Schools interested in incorporating Amateur Radio into their curricula, using it as an enrichment program or as a club activity may apply to become Project schools. See **www.arrl.org/education-technology-program** for more information on the ARRL Education & Technology Program.

Fig 1.27 — Ham radio on "Mars:" (L-R) Emily Colvin, KI4MSM, Andrea Hartlage, KG4IUM, and Elisha Sanders, KJ4CQM, install a solar-powered ham radio repeater to provide communication at the Mars Desert Research Station in Utah. The research initiative is intended to simulate an expedition to Mars. The Utah desert imitates the Martian wilderness, and the MDRS Habitat is largely self-contained — powered by generators, batteries, and solar panels as an actual Mars habitat would need to be. (KJ4CQI photo)

Fig 1.26 — Granite Bay (California) Montessori School students of the Wireless Technology Class assemble robot kits, a crystal radio, an Elecraft K2 transceiver, and are shown here assembling a GAP Titan DX antenna. Listen for them as K6GBM! (Teri Brown photo)

1.8 Resources

ARRL—the National Association for Amateur Radio
225 Main St
Newington, CT 06111-1494
860-594-0200
Fax: 860-594-0259
e-mail: **hq@arrl.org**
Prospective hams call 1-800-32 NEW HAM (1-800-326-3942)
www.arrl.org
Membership organization of US ham radio operators and those interested in ham radio. Publishes study guides for all Amateur Radio license classes, a monthly journal, *QST*, and many books on Amateur Radio and electronics.

Amateur Radio Service Rules & Regulations — FCC Part 97
Available on the ARRL Web site: **www.arrl.org/part-97-amateur-radio**
Available as a PDF file on the FCC Web site: **www.fcc.gov/Bureaus/Engineering_Technology/Documents/cfr/1998/47cfr97.pdf**

AMSAT NA (The Radio Amateur Satellite Corporation Inc)
850 Sligo Ave — Suite 600
Silver Spring, MD 20910
888-322-6728 or 301-589-6062
www.amsat.org
Membership organization for those interested in Amateur Radio satellites.

Courage Handi-Ham System
3915 Golden Valley Rd
Golden Valley, MN 55422
763-520-0512 or 866-426-3442
www.handiham.org
Provides assistance to persons with disabilities who want to earn a ham radio license or set up a station.

The ARRL Ham Radio License Manual
www.arrl.org/ham-radio-license-manual
Complete introduction to ham radio, including the exam question pool, complete explanations of the subjects on the exams. Tips on buying equipment, setting up a station and more.

The ARRL's Tech Q&A
www.arrl.org
Contains all of the questions in the Technician class question pool, with correct answers highlighted and explained in plain English. Includes many helpful diagrams.

General Information and Other Study Material
The ARRL Web site (**www.arrl.org**) carries a wealth of information for anyone interested in getting started in Amateur Radio. For complete information on all options available for study material, check out the "Welcome to the World of Ham Radio" page, **www.arrl.org/what-is-ham-radio**, and click on the WeDoThat-Radio.org link. You can also use the ARRL Web site to search for clubs, classes and Amateur Radio exam sessions near you.

1.9 Glossary

AM (Amplitude modulation) — The oldest voice operating mode still found on the amateur bands. The most common HF voice mode, SSB, is actually a narrower-bandwidth variation of AM.

Amateur Radio — A radiocommunication service for the purpose of self-training, intercommunication and technical investigation carried out by licensed individuals interested in radio technique solely with a personal aim and without pecuniary interest. (*Pecuniary* means payment of any type, whether money or goods.) Also called "ham radio."

Amateur Radio operator — A person holding an FCC license to operate a radio station in the Amateur Radio Service.

Amateur Radio station — A station licensed by the FCC in the Amateur Radio Service, including necessary equipment.

Amateur (Radio) Service — A radiocommunication service for the purpose of self-training, intercommunication and technical investigations carried out by licensed individuals interested in radio technique solely with a personal aim and without pecuniary interest.

AMSAT (Radio Amateur Satellite Corporation) — An international membership organization that designs, builds and promotes the use of Amateur Radio satellites, which are called "OSCARs."

APRS—Automatic Packet/Position Reporting System, a marriage of an application of the Global Positioning System and Amateur Radio to relay position and tracking information.

ARES (Amateur Radio Emergency Service) — An ARRL program specializing in emergency communication.

ARISS — An acronym for Amateur Radio on the International Space Station. NASA, ARRL and AMSAT cooperate in managing the ARISS Program.

ARRL — The national association for Amateur Radio in the US; the US member-society in the *IARU*.

ATV (Amateur television) — A mode of operation that amateur radio operators use to share video.

Band — A range of frequencies in the radio spectrum, usually designated by approximate wavelength in meters. For example, 7.0 to 7.3 MHz (megahertz) is the 40 meter amateur band. Hams are authorized to transmit on many different bands.

Bandwidth — In general, the width of a transmitted signal. FCC definition: "The width of a frequency band outside of which the mean power of the transmitted signal is attenuated at least 26 dB below the mean power of the transmitted signal within the band."

Beacon — An amateur station transmitting communication for the purposes of observation of propagation and reception or other related experimental activities.

Beam antenna — A type of ham radio antenna with directional characteristics that enhances the transmitted signal in one direction at the expense of others. A "rotary beam" can be pointed in any direction.

Broadcasting — Transmissions intended for reception by the general public, either direct or relayed. Amateur Radio licensees are not permitted to engage in broadcasting.

Call sign — A series of unique letters and numerals that the FCC assigns to an individual who has earned an Amateur Radio license.

Contact — A two-way communication between Amateur Radio operators.

Contest — An Amateur Radio operating activity in which hams and their stations compete to contact the most stations within a designated time period.

Courage Handi-Ham System — Membership organization for ham radio enthusiasts with various physical disabilities and abilities.

CW — Abbreviation for "continuous wave," today a synonym for radiotelegraphy (ie, Morse code by radio.

Digital communication — Computer-based communication modes such as RTTY, PSK31, packet and other radio transmissions that employ an accepted digital code to convey intelligence or data.

Dipole antenna — Typically, a wire antenna with a feed line connected to its center and having two legs, often used on the high-frequency (HF) amateur bands.

DSP (digital signal processing) — Technology that allows software to replace electronic circuitry.

DX — A ham radio abbreviation that refers to distant stations, typically those in other countries.

DXCC — DX Century Club, a popular ARRL award earned for contacting Amateur Radio operators in 100 different countries or "entities."

DX PacketCluster — A method of informing hams, via their computers, about DX activity.

DXpedition — A trip to a location — perhaps an uninhabited island or other geographical or political entity — which has few, if any, Amateur Radio operators, thus making a contact rare.

Elmer — A traditional term for someone who enjoys helping newcomers get started in ham radio; a mentor.

Emergency communication — Amateur Radio communication during a disaster or emergency that support or supplants traditional means of telecommunication.

FCC (Federal Communications Commission) — The government agency that regulates Amateur Radio in the US.

Field Day — A popular, annual Amateur Radio activity sponsored by the ARRL during which hams set up radio stations, often outdoors, using emergency power sources to simulate emergency operation.

Field Organization — A cadre of ARRL volunteers who perform various services for the Amateur Radio community at the state and local level.

FM (Frequency modulation) — A method of transmitting voice and the mode commonly used on ham radio repeaters.

Fox hunt — A competitive radio direction-finding activity in which participants track down the source of a transmitted signal.

Fast-scan television — A mode of operation that Amateur Radio operators can use to exchange live TV images from their stations. Also called ATV (Amateur Television).

Ham band — A range of frequencies in the radio spectrum on which ham radio communication is authorized.

Ham radio — Another name for Amateur Radio.

Ham radio operator — A radio operator holding a license granted by the FCC to operate on Amateur Radio frequencies.

HF (high frequency) — The radio frequencies from 3 to 30 MHz.

HSMM (high-speed multimedia) — A digital radio communication technique using spread spectrum modes primarily on UHF to simultaneously send and receive video, voice, text, and data.

IARU (International Amateur Radio Union) — The international organization made up of national Amateur Radio organizations or societies such as the ARRL.

Image — Facsimile and television signals.

International Morse Code — A digital code in which alphanumeric characters are represented by a defined set of short and long transmission elements — called "dots and dashes" or "dits and dahs" — that many Amateur Radio operators use to communicate.

ITU (International Telecommunication Union) — An agency of the United Nations that allocates the radio spectrum among the various radio services.

Mode — A type of ham radio communication, such as frequency modulation (FM voice), slow-scan television (SSTV), SSB (single sideband voice) or CW (Morse code).

Morse code — A communication mode characterized by on/off keying of a radio signal to convey intelligence. Hams use the International Morse Code.

Net — An on-the-air meeting of hams at a set time, day and radio frequency.

Packet radio — A computer-to-computer radio communication mode in which information is broken into short bursts called packets. These packets contain addressing and error-detection information.

PEP (peak envelope power) — The average power supplied to the antenna transmission line by a transmitter during one RF cycle at the crest of the modulation envelope taken under normal operating conditions.

Phone — Emissions carrying speech or other sound information, such as FM, SSB or AM.

Public service — Activities involving Amateur Radio that hams perform to benefit their communities.

QRP — An abbreviation for low power.

QSL bureau — A system for sending and receiving Amateur Radio verification or "QSL" cards.

QSL cards — Cards that provide written confirmation of a communication between two hams.

QST — The monthly journal of the ARRL. *QST* means "calling all radio amateurs."

RACES (Radio Amateur Civil Emergency Service) — A radio service that uses amateur stations for civil defense communication during periods of local, regional or national civil emergencies.

RF (Radio frequency) — Electromagnetic radiation.

Radio (or Ham) shack — The room where Amateur Radio operators keep their station.

Radiotelegraphy — See **Morse code**.

Receiver — A device that converts radio signals into a form that can be heard.

Repeater — An amateur station, usually located on a mountaintop, hilltop or tall building, that receives and retransmits the signals of other stations on a different channel or channels for greater range.

RTTY (radio teletype) — Narrow-band direct-printing radioteletype that uses a digital code.

Space station — An amateur station located more than 50 km above Earth's surface.

SSB (Single sideband) — A common mode of voice of Amateur Radio voice transmission.

SSTV (Slow-scan television) — An operating mode ham radio operators use to exchange still pictures from their stations.

SWL (Shortwave listener) — A person who enjoys listening to shortwave radio broadcasts or Amateur Radio conversations.

TIS (Technical Information Service) — A service of the ARRL that helps hams solve technical problems.

Transceiver — A radio transmitter and receiver combined in one unit.

Transmitter — A device that produces radio-frequency (RF) signals.

UHF (Ultra-high frequency) — The radio frequencies from 300 to 3000 MHz.

VE (Volunteer Examiner) — An Amateur Radio operator who is qualified to administer Amateur Radio licensing examinations.

VHF (Very-high frequency) — The radio frequency range from 30 to 300 MHz.

WAS (Worked All States) — An ARRL award that is earned when an Amateur Radio operator confirms two-way radio contact with other stations in all 50 US states.

Wavelength — A means of designating a frequency band, such as the 80-meter band.

Work — To contact another ham.

Contents

2.1 Introduction to Electricity
    2.1.1 Electric Charge, Voltage, and Current
    2.1.2 Electronic and Conventional Current
    2.1.3 Units of Measurement
    2.1.4 Series and Parallel Circuits
    2.1.5 Glossary — DC and Basic Electricity
2.2 Resistance and Conductance
    2.2.1 Resistance
    2.2.2 Conductance
    2.2.3 Ohm's Law
    2.2.4 Resistance of Wires
    2.2.5 Temperature Effects
    2.2.6 Resistors
2.3 Basic Circuit Principles
    2.3.1 Kirchoff's Current Law
    2.3.2 Resistors in Parallel
    2.3.3 Kirchoff's Voltage Law
    2.3.4 Resistors in Series
    2.3.5 Conductances in Series and Parallel
    2.3.6 Equivalent Circuits
    2.3.7 Voltage and Current Sources
    2.3.8 Thevenin's Theorem and Thevenin Equivalents
    2.3.9 Norton's Theorem and Norton Equivalents
2.4 Power and Energy
    2.4.1 Energy
    2.4.2 Generalized Definition of Resistance
    2.4.3 Efficiency
    2.4.4 Ohm's Law and Power Formulas
2.5 Circuit Control Components
    2.5.1 Switches
    2.5.2 Fuses and Circuit Breakers
    2.5.3 Relays and Solenoids
2.6 AC Theory and Waveforms
    2.6.1 AC In Circuits
    2.6.2 AC Waveforms
    2.6.3 Electromagnetic Energy
    2.6.4 Measuring AC Voltage, Current, and Power
    2.6.5 Glossary — AC Theory and Reactance
2.7 Capacitance and Capacitors
    2.7.1 Electrostatic Fields and Energy
    2.7.2 The Capacitor
    2.7.3 Capacitors in Series and Parallel
    2.7.4 RC Time Constant
    2.7.5 Alternating Current in Capacitors
    2.7.6 Capacitive Reactance and Susceptance
    2.7.7 Characteristics of Capacitors
    2.7.8 Capacitor Types and Uses
2.8 Inductance and Inductors
    2.8.1 Magnetic Fields and Magnetic Energy Storage
    2.8.2 Magnetic Core Properties
    2.8.3 Inductance and Direct Current
    2.8.4 Mutual Inductance and Magnetic Coupling
    2.8.5 Inductances in Series and Parallel
    2.8.6 RL Time Constant
    2.8.7 Alternating Current in Inductors
    2.8.8 Inductive Reactance and Susceptance
2.9 Working With Reactance
    2.9.1 Ohm's Law for Reactance
    2.9.2 Reactances in Series and Parallel
    2.9.3 At and Near Resonance
    2.9.4 Reactance and Complex Waveforms
2.10 Impedance
    2.10.1 Calculating Z from R and X in Series Circuits
    2.10.2 Calculating Z from R and X in Parallel Circuits
    2.10.3 Admittance
    2.10.4 More than Two Elements in Series or Parallel
    2.10.5 Equivalent Series and Parallel Circuits
    2.10.6 Ohm's Law for Impedance
    2.10.7 Reactive Power and Power Factor
2.11 Quality Factor (Q) of Components
2.12 Practical Inductors
    2.12.1 Air-core Inductors
    2.12.2 Straight-wire Inductance
    2.12.3 Iron-core Inductors
    2.12.4 Slug-tuned Inductors
    2.12.5 Powdered-Iron Toroidal Inductors
    2.12.6 Ferrite Toroidal Inductors
2.13 Resonant Circuits
    2.13.1 Series-Resonant Circuits
    2.13.2 Parallel-Resonant Circuits
2.14 Transformers
    2.14.1 Basic Transformer Principles
    2.14.2 Autotransformers
2.15 Heat Management
    2.15.1 Thermal Resistance
    2.15.2 Heat Sink Selection and Use
    2.15.3 Transistor Derating
    2.15.4 Rectifiers
    2.15.5 RF Heating
    2.15.6 Forced-Air and Water Cooling
    2.15.7 Heat Pipe Cooling
    2.15.8 Thermoelectric Cooling
    2.15.9 Temperature Compensation
    2.15.10 Thermistors
2.16 Radio Mathematics
    2.16.1 Working With Decibels
    2.16.2 Rectangular and Polar Coordinates
    2.16.3 Complex Numbers
    2.16.4 Accuracy and Significant Figures
2.17 References and Bibliography

Chapter 2

Electrical Fundamentals

Building on the work of numerous earlier contributors (most recently Roger Taylor, K9ALD's section on DC Circuits and Resistance), this chapter has been updated by Ward Silver, NØAX. Look for the section "Radio Mathematics," added at the end of this chapter, for a list of online math resources and sections on some of the mathematical techniques used in radio and electronics. Glossaries follow the sections on dc and ac theory, as well.

2.1 Introduction to Electricity

The *atom* is the primary building block of matter and is made up of a *nucleus*, containing *protons* and *neutrons*, surrounded by *electrons*. Protons have a positive electrical charge, electrons a negative charge, and neutrons have no electrical charge. An *element* (or *chemical element*) is a type of atom that has a specific number of protons, the element's *atomic number*. Each different element, such as iron, oxygen, silicon, or bromine has a distinct chemical and physical identity determined primarily by the number of protons. A *molecule* is two or more atoms bonded together and acting as a single particle.

Unless modified by chemical, mechanical, or electrical processes, all atoms are electrically neutral because they have the same number of electrons as protons. If an atom loses electrons, it has more protons than electrons and thus has a net positive charge. If an atom gains electrons, it has more electrons than protons and a net negative charge. Atoms or molecules with a positive or negative charge are called *ions*. Electrons not bound to any atom, or *free electrons*, can also be considered as ions because they have a negative charge.

2.1.1 Electric Charge, Voltage and Current

Any piece of matter that has a net positive or negative electrical charge is said to be *electrically charged*. An electrical force exists between electrically charged particles, pushing charges of the same type apart (like charges repel each other) and pulling opposite charges together (opposite charges attract). This is the *electromotive force* (or EMF), also referred to as *voltage* or *potential*. The higher the EMF's voltage, the stronger is its force on an electrical charge.

Under most conditions, the number of positive and negative charges in any volume of space is very close to balanced and so the region has no net charge. When there are extra positive ions in one region and extra negative ions (or electrons) in another region, the resulting EMF attracts the charges toward each other. The direction of the force, from the positive region to the negative region, is called its *polarity*. Because EMF results from an imbalance of charge between two regions, its voltage is always measured between two points, with positive voltage defined as being in the direction from the positively-charged to the negatively-charged region.

If there is no path by which electric charge can move in response to an EMF (called a *conducting path*), the charges cannot move together and so remain separated. If a conducting path is available, then the electrons or ions will flow along the path, neutralizing the net imbalance of charge. The movement of electrical charge is called *electric current*. Materials through which current flows easily are called *conductors*. Most metals, such as copper or aluminum are good conductors. Materials in which it is difficult for current to flow are *insulators*. *Semiconductors*, such as silicon or germanium, are materials with much poorer conductivity than metals. Semiconductors can be chemically altered to acquire properties that make them useful in solid-state devices such as diodes, transistors and integrated circuits.

Voltage differences can be created in a variety of ways. For example, chemical ions can be physically separated to form a battery. The resulting charge imbalance creates a voltage difference at the battery terminals so that if a conductor is connected to both terminals at once, electrons flow between the terminals and gradually eliminate the charge imbalance, discharging the battery's stored energy. Mechanical means such as friction (static electricity, lightning) and moving conductors in a magnetic field (generators) can also produce voltages. Devices or systems that produce voltage are called *voltage sources*.

2.1.2 Electronic and Conventional Current

Electrons move in the direction of positive voltage — this is called *electronic current*. *Conventional current* takes the other point of view — of positive charges moving in the direction of negative voltage. Conventional current was the original model for electricity and results from an arbitrary decision made by Benjamin Franklin in the 18th century when the nature of electricity and atoms was still unknown. It can be imagined as electrons flowing "backward" and is completely equivalent to electronic current.

Conventional current is used in nearly all electronic literature and is the standard used in this book. The direction of conventional current direction establishes the polarity for most electronics calculations and circuit diagrams. The arrows in the drawing symbols for transistors point in the direction of conventional current, for example.

2.1.3 Units of Measurement

To measure electrical quantities, certain definitions have been adopted. Charge is measured in *coulombs* (C) and represented by q in equations. One coulomb is equal to 6.25 × 10^{18} electrons (or protons). Current, the flow of charge, is measured in *amperes* (A) and represented by i or I in equations. One ampere represents one coulomb of charge flowing past a point in one second and so amperes can also be expressed as coulombs per second. Electromotive force is measured in *volts* (V) and represented by e, E, v, or V in equations. One volt is defined as the electromotive force between two points required to cause one ampere of current to do one *joule* (measure of energy) of work in flowing between the points. Voltage can also be expressed as joules per coulomb.

2.1.4 Series and Parallel Circuits

A *circuit* is any conducting path through which current can flow between two points of that have different voltages. An *open circuit* is a circuit in which a desired conducting path is interrupted, such as by a broken wire or a switch. A *short circuit* is a circuit in which a conducting path allows current to flow directly between the two points at different voltages.

The two fundamental types of circuits are shown in **Fig 2.1**. Part A shows a *series circuit* in which there is only one current path. The

Schematic Diagrams

The drawing in Fig 2.1 is a *schematic diagram*. Schematics are used to show the electrical connections in a circuit without requiring a drawing of the actual components or wires, called a *pictorial diagram*. Pictorials are fine for very simple circuits like these, but quickly become too detailed and complex for everyday circuits. Schematics use lines and dots to represent the conducting paths and connections between them. Individual electrical devices and electronic components are represented by *schematic symbols* such as the resistors shown here. A set of the most common schematic symbols is provided in the **Component Data and References** chapter. You will find additional information on reading and drawing schematic diagrams in the ARRL Web site Technology section at **www.arrl.org/circuit-construction**.

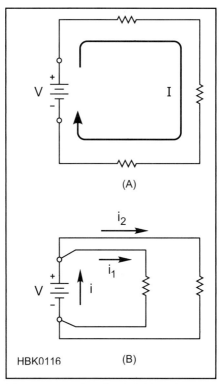

Fig 2.1 — A series circuit (A) has the same current through all components. Parallel circuits (B) apply the same voltage to all components.

current in this circuit flows from the voltage source's positive terminal (the symbol for a battery is shown with its voltage polarity shown as + and –) in the direction shown by the arrow through three *resistors* (electronic components discussed below) and back to the battery's negative terminal. Current is the same at every point in a series circuit.

Part B shows a *parallel circuit* in which there are multiple paths for the current to take. One terminal of both resistors is connected to the battery's positive terminal. The other terminal of both resistors is connected to the battery's negative terminal. Current flowing out of the battery's positive terminal divides into smaller currents that flow through the individual resistors and then recombine at the battery's negative terminal. All of the components in a parallel circuit experience the same voltage. All circuits are made up of series and parallel combinations of components and sources of voltage and current.

2.1.5 Glossary — DC and Basic Electricity

Alternating current (ac) — A flow of charged particles through a conductor, first in one direction, then in the other direction.

Ampere — A measure of flow of charged particles per unit of time. One ampere (A) represents one coulomb of charge flowing past a point in one second.

Atom — The smallest particle of matter that makes up a distinct chemical element. Atoms consist of protons and neutrons in the central region called the nucleus, with electrons surrounding the nucleus.

Circuit — Conducting path between two points of different voltage. In a *series circuit*, there is only one current path. In a *parallel circuit*, there are multiple current paths.

Conductance (G) — The reciprocal of resistance, measured in siemens (S).

Conductor — Material in which electrons or ions can move easily.

Conventional Current — Current defined as the flow of positive charges in the direction of positive to negative voltage. Conventional current flows in the opposite direction of electronic current, the flow of negative charges (electrons) from negative to positive voltage.

Coulomb — A unit of measure of a quantity of electrically charged particles. One coulomb (C) is equal to 6.25 × 10^{18} electrons.

Current (I) — The movement of electrical charge, measured in amperes and represented by i in equations.

Direct current (dc) — A flow of charged particles through a conductor in one direction only.

Electronic current — see **Conventional Current**

EMF — Electromotive Force is the term used to define the force of attraction or repulsion between electrically-charged regions. Also called *voltage* or *potential*.

Energy — Capability of doing work. It is usually measured in electrical terms as the number of watts of power consumed during a specific period of time, such as watt-seconds or kilowatt-hours.

Insulator — Material in which it is difficult for electrons or ions to move.

Ion — Atom or molecule with a positive or negative electrical charge.

Joule — Measure of a quantity of energy. One joule is defined as one newton (a measure of force) acting over a distance of one meter.

Ohm — Unit of resistance. One ohm is defined as the resistance that will allow one ampere of current when one volt of EMF is impressed across the resistance.

Polarity — The direction of EMF or voltage, from positive to negative.

Potential — See **EMF**.

Power — Power is the rate at which work is done. One watt of power is equal to one volt of EMF causing a current of one ampere through a resistor.

Resistance (R) — Opposition to current by conversion into other forms of energy, such as heat, measured in ohms (Ω).

Volt, voltage — See **EMF**.

Voltage source — Device or system that creates a voltage difference at its terminals.

2.2 Resistance and Conductance

2.2.1 Resistance

Any conductor connected to points at different voltages will allow current to pass between the points. No conductor is perfect or lossless, however, at least not at normal temperatures. The moving electrons collide with the atoms making up the conductor and lose some of their energy by causing the atoms to vibrate, which is observed externally as heat. The property of energy loss due to interactions between moving charges and the atoms of the conductor is called *resistance*. The amount of resistance to current is measured in *ohms* (Ω) and is represented by *r* or R in equations.

Suppose we have two conductors of the same size and shape, but of different materials. Because all materials have different internal structures, the amount of energy lost by current flowing through the material is also different. The material's ability to impede current flow is its *resistivity*. Numerically, the resistivity of a material is given by the resistance, in ohms, of a cube of the material measuring one centimeter on each edge. The symbol for resistivity is the Greek letter rho, ρ.

The longer a conductor's physical path, the higher the resistance of that conductor. For direct current and low-frequency alternating currents (up to a few thousand hertz) the conductor's resistance is inversely proportional to the cross-sectional area of the conductor. Given two conductors of the same material and having the same length, but differing in cross-sectional area, the one with the larger area (for example, a thicker wire or sheet) will have the lower resistance.

One of the best conductors is copper, and it is frequently convenient to compare the resistance of a material under consideration with that of a copper conductor of the same size and shape. **Table 2.1** gives the ratio of the resistivity of various conductors to the resistivity of copper.

2.2.2 Conductance

The reciprocal of resistance (1/R) is *conductance*. It is usually represented by the symbol G. A circuit having high conductance has low resistance, and vice versa. In radio work, the term is used chiefly in connection with electron-tube and field-effect transistor characteristics. The units of conductance are siemens (S). A resistance of 1 Ω has a conductance of 1 S, a resistance of 1000 Ω has a conductance of 0.001 S, and so on. A unit frequently used in regards to vacuum tubes and the field-effect transistor is the µS or one millionth of a siemens. It is the conductance of a 1-MΩ resistance. Siemens have replaced the obsolete unit *mho* (abbreviated as an upside-down Ω symbol).

2.2.3 Ohm's Law

The amount of current that will flow through a conductor when a given EMF is applied will vary with the resistance of the conductor. The lower the resistance, the greater the current for a given EMF. One ohm is defined as the amount of resistance that allows one ampere of current to flow between two points that have a potential difference of one volt. This proportional relationship is known as *Ohm's Law*:

$$R = \frac{E}{I} \qquad (1)$$

where
R = resistance in ohms,
E = potential or EMF in volts and
I = current in amperes.

Transposing the equation gives the other common expressions of Ohm's Law as:

Table 2.1
Relative Resistivity of Metals

Material	Resistivity Compared to Copper
Aluminum (pure)	1.60
Brass	3.7-4.90
Cadmium	4.40
Chromium	8.10
Copper (hard-drawn)	1.03
Copper (annealed)	1.00
Gold	1.40
Iron (pure)	5.68
Lead	12.80
Nickel	5.10
Phosphor bronze	2.8-5.40
Silver	0.94
Steel	7.6-12.70
Tin	6.70
Zinc	3.40

The Origin of Unit Names

Many units of measure carry names that honor scientists who made important discoveries in or advanced the state of scientific knowledge of electrical and radio phenomena. For example, Georg Ohm (1787-1854) discovered the relationship between current, voltage and resistance that now bears his name as Ohm's Law and as the unit of resistance, the ohm. The following table lists the most common electrical units and the scientists for whom they are named. You can find more information on these and other notable scientists in encyclopedia entries on the units that bear their names.

Electrical Units and Their Namesakes

Unit	Measures	Named for
Ampere	Current	Andree Ampere 1775-1836
Coulomb	Charge	Charles Coulomb 1736-1806
Farad	Capacitance	Michael Faraday 1791-1867
Henry	Inductance	Joseph Henry 1797-1878
Hertz	Frequency	Heinrich Hertz 1857-1894
Ohm	Resistance	George Simon Ohm 1787-1854
Watt	Power	James Watt 1736-1819
Volt	Voltage	Alessandro Volta 1745-1827

$$E = I \times R \qquad (2)$$

and

$$I = \frac{E}{R} \qquad (3)$$

All three forms of the equation are used often in electronics and radio work. You must remember that the quantities are in volts, ohms and amperes; other units cannot be used in the equations without first being converted. For example, if the current is in milliamperes you must first change it to the equivalent fraction of an ampere before substituting the value into the equations.

The following examples illustrate the use of Ohm's Law in the simple circuit of **Fig 2.2**. If 150 V is applied to a circuit and the current is measured as 2.5 A, what is the resistance of the circuit? In this case R is the unknown, so we will use equation 1:

$$R = \frac{E}{I} = \frac{150 \text{ V}}{2.5 \text{ A}} = 60 \text{ }\Omega$$

No conversion of units was necessary because the voltage and current were given in volts and amperes.

If the current through a 20,000-Ω resistance is 150 mA, what is the voltage? To find voltage, use equation 2. Convert the current from milliamperes to amperes by dividing by 1000 mA / A (or multiplying by 10^{-3} A / mA) so that 150 mA becomes 0.150 A. (Notice the conversion factor of 1000 does not limit the number of significant figures in the calculated answer.)

$$I = \frac{150 \text{ mA}}{1000 \frac{\text{mA}}{\text{A}}} = 0.150 \text{ A}$$

Then:

$$E = 0.150 \text{ A} \times 20000 \text{ }\Omega = 3000 \text{ V}$$

In a final example, how much current will flow if 250 V is applied to a 5000-Ω resistor? Since I is unknown,

$$I = \frac{E}{R} = \frac{250 \text{ V}}{5000 \text{ }\Omega} = 0.05 \text{ A}$$

It is more convenient to express the current in mA, and 0.05 A × 1000 mA / A = 50 mA.

It is important to note that Ohm's Law applies in any portion of a circuit as well as to

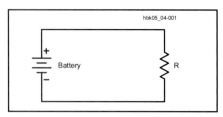

Fig 2.2 — A simple circuit consisting of a battery and a resistor.

Ohm's Law Timesaver

This simple diagram presents the mathematical equations relating voltage, current, and resistance. Cover the unkown quantity (E,I, or R) and the remaining symbols are shown as in the equation. For example, covering I shows E over R, as they would be written in the equation I=E/R.

When the current is small enough to be expressed in milliamperes, calculations are simplified if the resistance is expressed in kilohms rather than in ohms. With voltage in volts, if resistance in kilohms is substituted directly in Ohm's Law, the current will be milliamperes. Expressed as an equation: V = mA × kΩ.

the circuit as a whole. No matter how many resistors are connected together or how they are connected together, the relationship between the resistor's value, the voltage across the resistor, and the current through the resistor still follows Ohm's Law.

2.2.4 Resistance of Wires

The problem of determining the resistance of a round wire of given diameter and length — or the converse, finding a suitable size and length of wire to provide a desired amount of resistance — can easily be solved with the help of the copper wire table given in

Component Tolerance

Resistors are manufactured with a specific *nominal* value of resistance. This is the value printed on the body of the resistor or marked with stripes of colored paint. The *actual* value of resistance varies from the nominal value because of random variations in the manufacturing process. The maximum allowable amount of variation is called the *tolerance* and it is expressed in percent. For example, a 1000-Ω resistor with a tolerance of 5% could have any value of resistance between 95% and 105% of 1000 Ω; 950 to 1050 Ω. In most circuits, this small variation doesn't have much effect, but it is important to be aware of tolerance and choose the correct value (10%, 5%, 1%, or even tighter tolerance values are available for *precision components*) of tolerance for the circuit to operate properly, no matter what the actual value of resistance. All components have this same nominal-to-actual value relationship.

the chapter on **Component Data and References**. This table gives the resistance, in ohms per 1000 ft, of each standard wire size. For example, suppose you need a resistance of 3.5 Ω, and some #28 AWG wire is on hand. The wire table shows that #28 AWG wire has a resistance of 66.17 Ω / 1000 ft. Since the desired resistance is 3.5 Ω, the required length of wire is:

$$\text{Length} = \frac{R_{\text{DESIRED}}}{\frac{R_{\text{WIRE}}}{1000 \text{ ft}}} = \frac{3.5 \text{ }\Omega}{\frac{66.17 \text{ }\Omega}{1000 \text{ ft}}} \qquad (4)$$

$$= \frac{3.5 \text{ }\Omega \times 1000 \text{ ft}}{66.17 \text{ }\Omega} = 53 \text{ ft}$$

As another example, suppose that the resistance of wire in a radio's power cable must not exceed 0.05 Ω and that the length of wire required for making the connections totals 14 ft. Then:

$$\frac{R_{\text{WIRE}}}{1000 \text{ ft}} < \frac{R_{\text{MAXIMUM}}}{\text{Length}} = \frac{0.05 \text{ }\Omega}{14.0 \text{ ft}} \qquad (5)$$

$$= 5.57 \times 10^{-3} \frac{\Omega}{\text{ft}} \times \frac{1000 \text{ ft}}{1000 \text{ ft}}$$

$$\frac{R_{\text{WIRE}}}{1000 \text{ ft}} < \frac{3.57 \text{ }\Omega}{1000 \text{ ft}}$$

Find the value of R_{WIRE} / 1000 ft that is less than the calculated value. The wire table shows that #15 AWG is the smallest size having a resistance less than this value. (The resistance of #15 AWG wire is given as 3.1810 Ω / 1000 ft.) Select any wire size larger than this for the connections in your circuit, to ensure that the total wire resistance will be less than 0.05 Ω.

When the wire in question is not made of copper, the resistance values in the wire table should be multiplied by the ratios shown in Table 2.1 to obtain the resulting resistance. If the wire in the first example were made from nickel instead of copper, the length required for 3.5 Ω would be:

$$\text{Length} = \frac{R_{\text{DESIRED}}}{\frac{R_{\text{WIRE}}}{1000 \text{ ft}}}$$

$$= \frac{3.5 \text{ }\Omega}{\frac{66.17 \text{ }\Omega}{1000 \text{ ft}} \times 5.1} = \frac{3.5 \text{ }\Omega \times 1000 \text{ ft}}{66.17 \text{ }\Omega \times 5.1} \qquad (6)$$

$$\text{Length} = \frac{3500 \text{ ft}}{337.5} = 10.37 \text{ ft}$$

2.2.5 Temperature Effects

The resistance of a conductor changes with its temperature. The resistance of practically every metallic conductor increases with increasing temperature. Carbon, however, acts in the opposite way; its resistance decreases when its temperature rises. It is seldom necessary to consider temperature in making resistance calculations for amateur work. The temperature effect is important when it is necessary to maintain a constant resistance under all conditions, however. Special materials that have little or no change in resistance over a wide temperature range are used in that case.

2.2.6 Resistors

A package of material exhibiting a certain amount of resistance and made into a single unit is called a *resistor*. (See the **Component Data and References** chapter for information on resistor value marking conventions.) The size and construction of resistors having the same value of resistance in ohms may vary considerably based on how much power they are intended to dissipate, how much voltage is expected to be applied to them, and so forth (see **Fig 2.3**).

TYPES OF RESISTORS

Resistors are made in several different ways: carbon composition, metal oxide, carbon film, metal film and wirewound. In some circuits, the resistor value may be critical. In this case, precision resistors are used. These are typically wirewound or carbon-film devices whose values are carefully controlled during manufacture. In addition, special material or construction techniques may be used to provide temperature compensation, so the value does not change (or changes in a precise manner) as the resistor temperature changes.

Carbon composition resistors are simply small cylinders of carbon mixed with various binding agents to produce any desired resistance. The most common sizes of "carbon comp" resistors are ½- and ¼-W resistors. They are moderately stable from 0 to 60 °C (their resistance increases above and below this temperature range). They can absorb short overloads better than film-type resistors, but they are relatively noisy, and have relatively wide tolerances. Because carbon composition resistors tend to be affected by humidity and other environmental factors and because they are difficult to manufacture in surface-mount packages, they have largely been replaced by film-type resistors.

Metal-oxide resistors are similar to carbon composition resistors in that the resistance is supplied by a cylinder of metal oxide. Metal-oxide resistors have replaced carbon composition resistors in higher power applications because they are more stable and can operate at higher temperatures.

Wirewound resistors are made from wire, which is cut to the proper length and wound on a coil form (usually ceramic). They are capable of handling high power; their values are very stable, and they are manufactured to close tolerances.

Metal-film resistors are made by depositing a thin film of aluminum, tungsten or other metal on an insulating substrate. Their resistances are controlled by careful adjustments of the width, length and depth of the film. As a result, they have very tight tolerances. They are used extensively in surface-mount technology. As might be expected, their power handling capability is somewhat limited. They also produce very little electrical noise.

Carbon-film resistors use a film of carbon mixed with other materials instead of metal. They are not quite as stable as other film resistors and have wider tolerances than metal-film resistors, but they are still as good as (or better than) carbon composition resistors.

THERMAL CONSIDERATIONS FOR RESISTORS

Current through a resistance causes the conductor to become heated; the higher the resistance and the larger the current, the greater the amount of heat developed. Resistors intended for carrying large currents must be physically large so the heat can be radiated quickly to the surrounding air or some type of heat sinking material. If the resistor does not dissipate the heat quickly, it may get hot enough to melt or burn.

The amount of heat a resistor can safely dissipate depends on the material, surface area and design. Typical resistors used in amateur electronics (⅛ to 2-W resistors) dissipate heat primarily through the surface area of the case, with some heat also being carried away through the connecting leads. Wirewound resistors are usually used for higher power levels. Some have finned cases for better convection cooling and/or metal cases for better conductive cooling.

The major departure of resistors from ideal behavior at low-frequencies is their *temperature coefficient* (TC). (See the chapter on **RF Techniques** for a discussion of the behavior of resistors at high frequencies.) The resistivity of most materials changes with temperature, and typical TC values for resistor materials are given in **Table 2.2**. TC values are usually expressed in parts-per-million (PPM) for each degree (centigrade) change from some nominal temperature, usually room temperature (77 °F or 27 °C). A positive TC indicates an increase in resistance with increasing temperature while a negative TC indicates a decreasing resistance. For example, if a 1000-Ω resistor with a TC of +300 PPM/°C is heated to 50 °C, the *change* in resistance

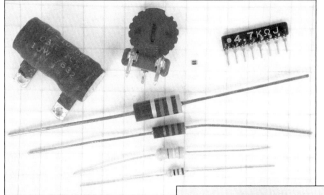

Fig 2.3 — Examples of various resistors. At the top left is a small 10-W wirewound resistor. A single in-line package (SIP) of resistors is at the top right. At the top center is a small PC-board-mount variable resistor. A tiny surface-mount (chip) resistor is also shown at the top. Below the variable resistor is a 1-W carbon composition resistor and then a ½-W composition unit. The dog-bone-shaped resistors at the bottom are ½-W and ¼-W film resistors. The ¼-inch-ruled graph paper background provides a size comparison. The inset photo shows the chip resistor with a penny for size comparison.

Table 2.2
Temperature Coefficients for Various Resistor Compositions

1 PPM = 1 part per million = 0.0001%

Type	TC (PPM/°C)
Wire wound	±(30 - 50)
Metal Film	±(100 - 200)
Carbon Film	+350 to −800
Carbon composition	±800

is $300 \times (50 - 27) = 6900$ PPM, yielding a new resistance of

$$1000\left(1 + \frac{6900}{1000000}\right) = 1006.9 \, \Omega$$

Carbon-film resistors are unique among the major resistor families because they alone have a negative temperature coefficient. They are often used to "offset" the thermal effects of the other components.

If the temperature increase is small (less than 30-40 °C), the resistance change with temperature is nondestructive — the resistor will return to normal when the temperature returns to its nominal value. Resistors that get too hot to touch, however, may be permanently damaged even if they appear normal. For this reason, be conservative when specifying power ratings for resistors. It's common to specify a resistor rated at 200% to 400% of the expected dissipation.

POTENTIOMETERS

Potentiometer (pronounced po-ten-tchee-AH-meh-tur) is a formal name for a variable resistor and the common name for these components is "pots." A typical potentiometer consists of a circular pattern of resistive material, usually a carbon compound similar to that used in carbon composition resistors, with a wiper contact on a shaft moving across the material. For higher power applications, the resistive material may be wire, wound around a core, like a wirewound resistor.

As the wiper moves along the material, more resistance is introduced between the wiper and one of the fixed contacts on the material. A potentiometer may be used to control current, voltage or resistance in a circuit. **Fig 2.4** shows several different types of potentiometers. **Fig 2.5** shows the schematic symbol for a potentiometer and how changing the position of the shaft changes the resistance between its three terminals. The figure shows a *panel pot*, designed to be mounted on an equipment panel and adjusted by an operator. The small rectangular *trimmer* potentiometers in Fig 2.5 are adjusted with a screwdriver and have wire terminals.

Typical specifications for a potentiometer include element resistance, power dissipation, voltage and current ratings, number of turns (or degrees) the shaft can rotate, type and size of shaft, mounting arrangements and resistance "taper."

Not all potentiometers have a *linear* taper. That is, the change in resistance may not be the same for a given number of degrees of shaft rotation along different portions of the resistive material. These are called *nonlinear tapers*

A typical use of a potentiometer with a nonlinear taper is as a volume control in an audio amplifier. Since the human ear has a logarithmic response to sound, a volume control must change the amplifier output much more near one end of the potentiometer than the other (for a given amount of rotation) so that the "perceived" change in volume is about the same for a similar change in the control's position. This is commonly called an *audio taper* or *log taper* as the change in resistance per degree of rotation attempts to match the response of the human ear. Tapers can be designed to match almost any desired control function for a given application. Linear and audio tapers are the most common tapers.

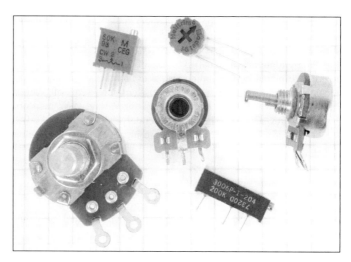

Fig 2.4 — This photo shows examples of different styles of potentiometers. The ¼-inch-ruled graph paper background provides a size comparison.

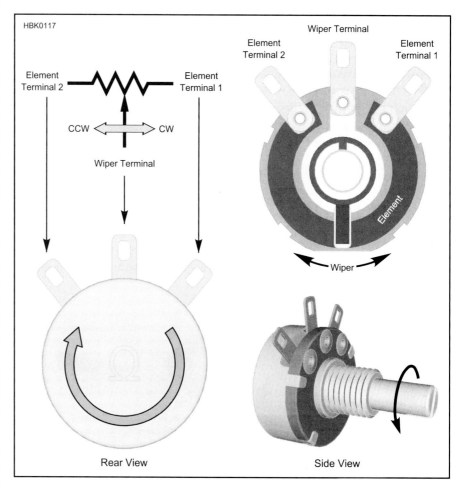

Fig 2.5 — Typical potentiometer construction and schematic symbol. Rotation on the shaft moves the wiper along the element, changing the resistance between the wiper terminal and the element terminals. Moving the wiper closer to an element terminal reduces the resistance between them.

2.3 Basic Circuit Principles

Circuits are composed of *nodes* and *branches*. A node is any point in the circuit at which current can divide between conducting paths. For example, in the parallel circuit of **Fig 2.6**, the node is represented by the schematic dot. A branch is any unique conducting path between nodes. A series of branches that make a complete current path, such as the series circuit of Fig 2.1A, is called a *loop*.

Very few actual circuits are as simple as those shown in Fig 2.1. However, all circuits, no matter how complex, are constructed of combinations of these series and parallel circuits. We will now use these simple circuits of resistors and batteries to illustrate two fundamental rules for voltage and current, known as *Kirchoff's Laws*.

2.3.1 Kirchhoff's Current Law

Kirchhoff's Current Law (KCL) states, *"The sum of all currents flowing into a node and all currents flowing out of a node is equal to zero."* KCL is stated mathematically as:

$$(I_{in1} + I_{in2} + \cdots) - (I_{out1} + I_{out2} + \cdots) = 0 \quad (7A)$$

The dots indicate that as many currents as necessary may be added.

Another way of stating KCL is that the sum of all currents flowing into a node must balance the sum of all currents flowing out of a node as shown in equation 7B:

$$(I_{in1} + I_{in2} + \cdots) = (I_{out1} + I_{out2} + \cdots) \quad (7B)$$

Equations 7A and 7B are mathematically equivalent.

KCL is illustrated by the following example. Suppose three resistors (5.0 kΩ, 20.0 kΩ and 8.0 kΩ) are connected in parallel as shown in Fig 2.6. The same voltage, 250 V, is applied to all three resistors. The current through R1 is I1, I2 is the current through R2 and I3 is the current through R3.

The current in each can be found from Ohm's Law, as shown below. For convenience, we can use resistance in kΩ, which gives current in milliamperes.

$$I1 = \frac{E}{R1} = \frac{250\,V}{5.0\,k\Omega} = 50.0\,mA$$

$$I2 = \frac{E}{R2} = \frac{250\,V}{20.0\,k\Omega} = 12.5\,mA$$

$$I3 = \frac{E}{R3} = \frac{250\,V}{8.0\,k\Omega} = 31.2\,mA$$

Notice that the branch currents are inversely proportional to the resistances. The 20,000-Ω resistor has a value four times larger than the 5000-Ω resistor, and has a current

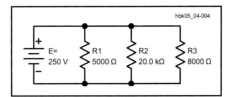

Fig 2.6 — An example of resistors in parallel.

one-quarter as large. If a resistor has a value twice as large as another, it will have half as much current through it when they are connected in parallel.

Using the balancing form of KCL (equation 7B) the current that must be supplied by the battery is therefore:

$$I_{BATT} = I1 + I2 + I3$$

$$I_{BATT} = 50.0\,mA + 12.5\,mA + 31.2\,mA$$

$$I_{BATT} = 93.7\,mA$$

2.3.2 Resistors in Parallel

In a circuit made up of resistances in parallel, the resistors can be represented as a single *equivalent* resistance that has the same value as the parallel combination of resistors. In a parallel circuit, the equivalent resistance is less than that of the lowest resistance value present. This is because the total current is always greater than the current in any individual resistor. The formula for finding the equivalent resistance of resistances in parallel is:

$$R_{EQUIV} = \cfrac{1}{\cfrac{1}{R1} + \cfrac{1}{R2} + \cfrac{1}{R3} + \cfrac{1}{R4}\cdots} \quad (8)$$

where the dots indicate that any number of parallel resistors can be combined by the same method. In the example of the previous section, the equivalent resistance is:

$$R = \cfrac{1}{\cfrac{1}{5000\,\Omega} + \cfrac{1}{20\,k\Omega} + \cfrac{1}{8000\,\Omega}} = 2.67\,k\Omega$$

The notation "//" (two slashes) is frequently used to indicate "in parallel with." Using that notation, the preceding example would be given as "5000 Ω // 20 kΩ // 8000 Ω."

For only two resistances in parallel (a very common case) the formula can be reduced to the much simpler (and easier to remember):

$$R_{EQUIV} = \frac{R1 \times R2}{R1 + R2} \quad (9)$$

Example: If a 500-Ω resistor is connected in parallel with one of 1200 Ω, what is the total resistance?

$$R = \frac{R1 \times R2}{R1 + R2} = \frac{500\,\Omega \times 1200\,\Omega}{500\,\Omega + 1200\,\Omega}$$

$$R = \frac{600000\,\Omega^2}{1700\,\Omega} = 353\,\Omega$$

Any number of parallel resistors can be combined two at a time by using equation 9 until all have been combined into a single equivalent. This is a bit easier than using equation 8 to do the conversion in a single step.

CURRENT DIVIDERS

Resistors connected in parallel form a circuit called a *resistive current divider*. For any number of resistors connected in parallel (R1, R2, R3, ... R4), the current through one of the resistors, R_n, is equal to the sum of all resistor currents multiplied by the ratio of the sum of all resistors *except* R_n to the sum of all resistors.

For example, in a circuit with three parallel resistors; R1, R2, and R3, the current through R2 is equal to:

$$I_2 = I\frac{R1 + R3}{R1 + R2 + R3}$$

where I is the total of the individual currents flowing through all resistors. If I = 100 mA, R1 = 100 Ω, R2 = 50 Ω, and R3 = 200 Ω:

$$100\,mA\,\frac{100 + 200}{100 + 50 + 200} = 85.7\,mA$$

2.3.3 Kirchhoff's Voltage Law

Kirchhoff's Voltage Law (KVL) states, *"The sum of the voltages around a closed current loop is zero."* Where KCL is somewhat intuitive, KVL is not as easy to visualize. In the circuit of **Fig 2.7**, KVL requires that the battery's voltage must be balanced exactly by the voltages that appear across the three resistors in the circuit. If it were not, the "extra" voltage would create an infinite current with no limiting resistance, just as KCL prevents charge from "building up" at a circuit node.

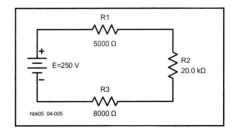

Fig 2.7 — An example of resistors in series.

KVL is stated mathematically as:

$$E_1 + E_2 + E_3 + \cdots = 0 \quad (10)$$

where each E represents a voltage encountered by current as it flows around the circuit loop.

This is best illustrated with an example. Although the current is the same in all three of the resistances in the example of Fig 2.7, the total voltage divides between them, just as current divides between resistors connected in parallel. The voltage appearing across each resistor (the *voltage drop*) can be found from Ohm's Law. (Voltage across a resistance is often referred to as a "drop" or "I-R drop" because the value of the voltage "drops" by the amount $E = I \times R$.)

For the purpose of KVL, it is common to assume that if current flows *into* the more positive terminal of a component the voltage is treated as positive in the KVL equation. If the current flows *out* of a positive terminal, the voltage is treated as negative in the KVL. Positive voltages represent components that consume or "sink" power, such as resistors. Negative voltages represent components that produce or "source" power, such as batteries. This allows the KVL equation to be written in a balancing form, as well:

$$(E_{source1} + E_{source2} + \cdots) =$$
$$(E_{sink1} + E_{sink2} + \cdots)$$

All of the voltages are treated as positive in this form, with the power sources (current flowing *out* of the more positive terminal) on one side and the power sinks (current flowing *into* the more positive terminal) on the other side.

Note that it doesn't matter what a component terminal's *absolute* voltage is with respect to ground, only which terminal of the component is more positive than the other. If one side of a resistor is at +1000 V and the other at +998 V, current flowing into the first terminal and out of the second experiences a +2 V voltage drop. Similarly, current supplied by a 9 V battery with its positive terminal at –100 V and its negative terminal at –108.5 V still counts for KVL as an 8.5 V power source. Also note that current can flow *into* a battery's positive terminal, such as during recharging, making the battery a power sink, just like a resistor.

Here's an example showing how KVL works: In Fig 2.7, if the voltage across R1 is E1, that across R2 is E2 and that across R3 is E3, then:

$$-250 + I \times R1 + I \times R2 + I \times R3 = 0$$

This equation can be simplified to:

$$-250 + I(R1 + R2 + R3) =$$
$$-250 + I(33000\ \Omega) = 0$$

Solving for I gives I = 250 / 33000 = 0.00758 A = 7.58 mA. This allows us to calculate the value of the voltage across each resistor:

E1 = I × R1 = 0.00758 A × 5000 Ω = 37.9 V

E2 = I × R2 = 0.00758 A × 20000 Ω = 152 V

E3 = I × R3 = 0.00758 A × 8000 Ω = 60.6 V

Verifying that the sum of E1, E2, and E3 does indeed equal the battery voltage of 250 V ignoring rounding errors.

$$E_{TOTAL} = E1 + E2 + E3$$
$$E_{TOTAL} = 37.9\ V + 152\ V + 60.6\ V$$
$$E_{TOTAL} = 250\ V$$

2.3.4 Resistors in Series

The previous example illustrated that in a circuit with a number of resistances connected in series, the equivalent resistance of the circuit is the sum of the individual resistances. If these are numbered R1, R2, R3 and so on, then:

$$R_{EQUIV} = R1 + R2 + R3 + R4... \quad (11)$$

Example: Suppose that three resistors are connected to a source of voltage as shown in Fig 2.7. The voltage is 250 V, R1 is 5.0 kΩ, R2 is 20.0 kΩ and R3 is 8.0 kΩ. The total resistance is then

$$R_{EQUIV} = R1 + R2 + R3$$
$$R_{EQUIV} = 5.0\ k\Omega + 20.0\ k\Omega + 8.0\ k\Omega$$
$$R_{EQUIV} = 33.0\ k\Omega$$

The current in the circuit is then

$$I = \frac{E}{R} = \frac{250\ V}{33.0\ k\Omega} = 7.58\ mA$$

VOLTAGE DIVIDERS

Notice that the voltage drop across each resistor in the KVL example is directly proportional to the resistance. The value of the 20,000 Ω resistor is four times larger than the 5000 Ω resistor, and the voltage drop across the 20,000 Ω resistor is four times larger. A resistor that has a value twice as large as another will have twice the voltage drop across it when they are connected in series.

Resistors in series without any other connections form a *resistive voltage divider*. (Other types of components can form voltage dividers, too.) The voltage across any specific resistor in the divider, R_n, is equal to the voltage across the entire string of resistors multiplied by the ratio of R_n to the sum of all resistors in the string.

For example, in the circuit of Fig 2.7, the voltage across the 5000 Ω resistor is:

$$E1 = 250\frac{5000}{5000 + 20000 + 8000} = 37.9\ V$$

This is a more convenient method than calculating the current through the resistor and using Ohm's Law.

Voltage dividers can be used as a source of voltage. As long as the device connected to the output of the divider has a much higher resistance than the resistors in the divider, there will be little effect on the divider output voltage. For example, for a voltage divider with a voltage of E = 15 V and two resistors of R1 = 5 kΩ and R2 = 10 kΩ, the voltage across R2 will be 10 V measured on a high-impedance voltmeter because the measurement draws very little current from the divider. However, if the measuring device or load across R2 draws significant current, it will increase the amount of current drawn through the divider and change the output voltage.

A good rule of thumb is that the load across a voltage divider's output should be at least ten times the value of the highest resistor in the divider to stay reasonably close to the required voltage. As the load resistance gets closer to the value of the divider resistors, the current drawn by the load affects the voltage division and causes changes from the desired value.

Potentiometers (variable resistors described previously) are often used as adjustable voltage dividers and this is how they got their name. Potential is an older name for voltage and a "potential-meter" is a device that can "meter" or adjust potential, thus potentiometer.

2.3.5 Conductances in Series and Parallel

Since conductance is the reciprocal of resistance, G = 1/R, the formulas for combining resistors in series and in parallel can be converted to use conductance by substituting 1/G for R. Substituting 1/G into equation 11 shows that conductances in series are combined this way:

$$G = \frac{1}{\frac{1}{G1} + \frac{1}{G2} + \frac{1}{G3} + \frac{1}{G4}...}$$

and two conductances in series may be combined in a manner similar to equation 9:

$$G_{EQUIV} = \frac{G1 \times G2}{G1 + G2}$$

Substituting 1/G into equation 8 shows that conductances in parallel are combined this way:

$$G_{TOTAL} = G1 + G2 + G3 + G4...$$

This also shows that when faced with a large number of parallel resistances, converting them to conductances may make the math a little easier to deal with.

2.3.6 Equivalent Circuits

A circuit may have resistances both in parallel and in series, as shown in **Fig 2.8A**. In order to analyze the behavior of such a circuit, *equivalent circuits* are created and combined by using the equations for combining resistors in series and resistors in parallel. Each separate combination of resistors, series or parallel, can be reduced to a single equivalent resistor. The resulting combinations can be reduced still further until only a single resistor remains.

The method for analyzing the circuit of Fig 2.8A is as follows: Combine R2 and R3 to create the equivalent single resistor, R_{EQ} whose value is equal to R2 and R3 in parallel.

$$R_{EQ} = \frac{R2 \times R3}{R2 + R3} = \frac{20000 \, \Omega \times 8000 \, \Omega}{20000 \, \Omega + 8000 \, \Omega}$$

$$= \frac{1.60 \times 10^8 \, \Omega^2}{28000 \, \Omega} = 5710 \, \Omega = 5.71 \, k\Omega$$

This resistance in series with R1 then forms a simple series circuit, as shown in Fig 2.8B. These two resistances can then be combined into a single equivalent resistance, R_{TOTAL}, for the entire circuit:

$$R_{TOTAL} = R1 + R_{EQ} = 5.0 \, k\Omega + 5.71 \, k\Omega$$

$$R_{TOTAL} = 10.71 \, k\Omega$$

The battery current is then:

$$I = \frac{E}{R} = \frac{250 \, V}{10.71 \, k\Omega} = 23.3 \, mA$$

The voltage drops across R1 and R_{EQ} are:

$$E1 = I \times R1 = 23.3 \, mA \times 5.0 \, k\Omega = 117 \, V$$

$$E2 = I \times R_{EQ} = 23.3 \, mA \times 5.71 \, k\Omega = 133 \, V$$

These two voltage drops total 250 V, as described by Kirchhoff's Voltage Law. E2 appears across both R2 and R3 so,

$$I2 = \frac{E2}{R2} = \frac{133 \, V}{20.0 \, k\Omega} = 6.65 \, mA$$

$$I3 = \frac{E3}{R3} = \frac{133 \, V}{8.0 \, k\Omega} = 16.6 \, mA$$

where
I2 = current through R2 and
I3 = current through R3.

The sum of I2 and I3 is equal to 23.3 mA, conforming to Kirchhoff's Current Law.

2.3.7 Voltage and Current Sources

In designing circuits and describing the behavior of electronic components, it is often useful to use *ideal sources*. The two most common types of ideal sources are the *voltage source* and the *current source*, symbols for which are shown in **Fig 2.9**. These sources are considered ideal because no matter what circuit is connected to their terminals, they continue to supply the specified amount of voltage or current. Practical voltage and current sources can approximate the behavior of an ideal source over certain ranges, but are limited in the amount of power they can supply and so under excessive load, their output will drop.

Voltage sources are defined as having zero *internal impedance*, where impedance is a more general form of resistance as described in the sections of this chapter dealing with alternating current. A short circuit across an ideal voltage source would result in the source providing an infinite amount of current. Practical voltage sources have non-zero internal impedance and this also limits the amount of power they can supply. For example, placing a short circuit across the terminals of a practical voltage source such as 1.5 V dry-cell battery may produce a current of several amperes, but the battery's internal impedance acts to limit the amount of current produced in accordance with Ohm's Law — as if the resistor in Fig 2.2 was inside of or internal to the battery.

Current sources are defined to have infinite internal impedance. This means that no matter what is connected to the terminals of an ideal current source, it will supply the same amount of current. An open circuit across the terminal of an ideal current source will result in the source generating an infinite voltage at its terminals. Practical current sources will raise their voltage until the internal power supply limits are reached and then reduce output current.

2.3.8 Thevenin's Theorem and Thevenin Equivalents

Thevenin's Voltage Theorem (usually just referred to as "Thevenin's Theorem") is a useful tool for simplifying electrical circuits or *networks* (the formal name for circuits) by allowing circuit designers to replace a circuit with a simpler equivalent circuit. Thevenin's Theorem states, *"Any two-terminal network made up of resistors and voltage or current sources can be replaced by an equivalent network made up of a single voltage source and a series resistor."*

Thevenin's Theorem can be readily applied to the circuit of Fig 2.8A, to find the current through R3. In this example, illustrated in **Fig 2.10**, the circuit is redrawn to show R1 and R2 forming a voltage divider, with R3 as the load (Fig 2.10A). The current drawn by the load (R3) is simply the voltage across R3, divided by its resistance. Unfortunately, the value of R2 affects the voltage across R3, just as the presence of R3 affects the voltage appearing across R2. Some means of separating the two is needed; hence the *Thevenin-equivalent circuit* is constructed, replacing everything connected to terminals A and B with a single voltage source (the *Thevenin-equivalent voltage*, E_{THEV}) and series resistor (the *Thevenin-equivalent resistance*, R_{TH}).

The first step of creating the Thevenin-equivalent of the circuit is to determine its *open-circuit voltage*, measured when there is no load current drawn from either terminal A or B. Without a load connected between A and B, the total current through the circuit is (from Ohm's Law):

$$I = \frac{E}{R1 + R2} \qquad (12)$$

and the voltage between terminals A and B (E_{AB}) is:

$$E_{AB} = I \times R2 \qquad (13)$$

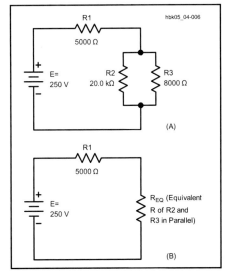

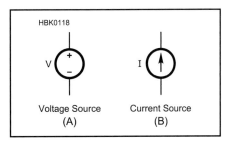

Fig 2.8 — At A, an example of resistors in series-parallel. The equivalent circuit is shown at B.

Fig 2.9 — Voltage sources (A) and current sources (B) are examples of ideal energy sources.

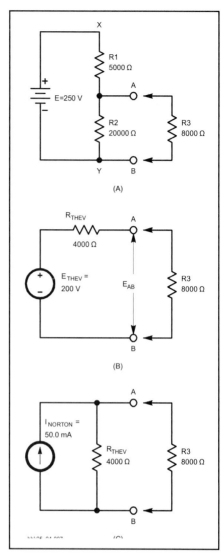

Fig 2.10 — Equivalent circuits for the circuit shown in Fig 2.8. A shows the circuit to be replaced by an equivalent circuit from the perspective of the resistor (R3 load). B shows the Thevenin-equivalent circuit, with a resistor and a voltage source in series. C shows the Norton-equivalent circuit, with a resistor and current source in parallel.

By substituting equation 12 into equation 13, we have an expression for E_{AB} in which all values are known:

$$E_{AB} = \frac{R2}{R1+R2} \times E \qquad (14)$$

Using the values in our example, this becomes:

$$E_{AB} = \frac{20.0 \text{ k}\Omega}{25.0 \text{ k}\Omega} \times 250 \text{ V} = 200 \text{ V}$$

when nothing is connected to terminals A or B. E_{THEV} is equal to E_{AB} with no current drawn.

The equivalent resistance between terminals A and B is R_{THEV}. R_{THEV} is calculated as the equivalent circuit at terminals A and B with all sources, voltage or current, replaced by their internal impedances. The ideal voltage source, by definition, has zero internal resistance and is replaced by a short circuit. The ideal current source has infinite internal impedance and is replaced by an open circuit.

Assuming the battery to be a close approximation of an ideal source, replace it with a short circuit between points X and Y in the circuit of Fig 2.10A. R1 and R2 are then effectively placed in parallel, as viewed from terminals A and B. R_{THEV} is then:

$$R_{THEV} = \frac{R1 \times R2}{R1+R2} \qquad (15)$$

$$R_{THEV} = \frac{5000 \, \Omega \times 20000 \, \Omega}{5000 \, \Omega + 20000 \, \Omega}$$

$$R_{THEV} = \frac{1.0 \times 10^8 \, \Omega^2}{25000 \, \Omega} = 4000 \, \Omega$$

This gives the Thevenin-equivalent circuit as shown in Fig 2.10B. The circuits of Figs 2.10A and 2.10B are completely equivalent from the perspective of R3, so the circuit becomes a simple series circuit.

Once R3 is connected to terminals A and B, there will be current through R_{THEV}, causing a voltage drop across R_{THEV} and reducing E_{AB}. The current through R3 is equal to

$$I3 = \frac{E_{THEV}}{R_{TOTAL}} = \frac{E_{THEV}}{R_{THEV}+R3} \qquad (16)$$

Substituting the values from our example:

$$I3 = \frac{200 \text{ V}}{4000 \, \Omega + 8000 \, \Omega} = 16.7 \text{ mA}$$

This agrees with the value calculated earlier.

The Thevenin-equivalent circuit of an ideal voltage source in series with a resistance is a good model for a real voltage source with non-zero internal resistance. Using this more realistic model, the maximum current that a real voltage source can deliver is seen to be

$$I_{sc} = \frac{E_{THEV}}{R_{THEV}}$$

and the maximum output voltage is $V_{oc} = E_{THEV}$.

Sinusoidal voltage or current sources can be modeled in much the same way, keeping in mind that the internal impedance, Z_{THEV}, for such a source may not be purely resistive, but may have a reactive component that varies with frequency.

2.3.9 Norton's Theorem and Norton Equivalents

Norton's Theorem is another method of creating an equivalent circuit. Norton's Theorem states, *"Any two-terminal network made up of resistors and current or voltage sources can be replaced by an equivalent network made up of a single current source and a parallel resistor."* Norton's Theorem is to current sources what Thevenin's Theorem is to voltage sources. In fact, the Thevenin-resistance calculated previously is also the *Norton-equivalent resistance*.

The circuit just analyzed by means of Thevenin's Theorem can be analyzed just as easily by Norton's Theorem. The equivalent Norton circuit is shown in Fig 2.10C. The short circuit current of the equivalent circuit's current source, I_{NORTON}, is the current through terminals A and B with the load (R3) replaced by a short circuit. In the case of the voltage divider shown in Fig 2.10A, the short circuit completely bypasses R2 and the current is:

$$I_{AB} = \frac{E}{R1} \qquad (17)$$

Substituting the values from our example, we have:

$$I_{AB} = \frac{E}{R1} = \frac{250 \text{ V}}{5000 \, \Omega} = 50.0 \text{ mA}$$

The resulting Norton-equivalent circuit consists of a 50.0-mA current source placed in parallel with a 4000-Ω resistor. When R3 is connected to terminals A and B, one-third of the supply current flows through R3 and the remainder through R_{THEV}. This gives a current through R3 of 16.7 mA, again agreeing with previous conclusions.

A Norton-equivalent circuit can be transformed into a Thevenin-equivalent circuit and vice versa. The equivalent resistor, R_{THEV}, is the same in both cases; it is placed in series with the voltage source in the case of a Thevenin-equivalent circuit and in parallel with the current source in the case of a Norton-equivalent circuit. The voltage for the Thevenin-equivalent source is equal to the open-circuit voltage appearing across the resistor in the Norton-equivalent circuit. The current for a Norton-equivalent source is equal to the short circuit current provided by the Thevenin source. A Norton-equivalent circuit is a good model for a real current source that has a less-than infinite internal impedance.

2.4 Power and Energy

Regardless of how voltage is generated, energy must be supplied if current is drawn from the voltage source. The energy supplied may be in the form of chemical energy or mechanical energy. This energy is measured in joules (J). One joule is defined from classical physics as the amount of energy or *work* done when a force of one newton (a measure of force) is applied to an object that is moved one meter in the direction of the force.

Power is another important concept and measures the rate at which energy is generated or used. One *watt* (W) of power is defined as the generation (or use) of one joule of energy (or work) per second.

One watt is also defined as one volt of EMF causing one ampere of current to flow through a resistance. Thus,

$$P = I \times E \qquad (18)$$

where
P = power in watts
I = current in amperes
E = EMF in volts.

(This discussion pertains only to direct current in resistive circuits. See the **AC Theory and Reactance** section of this chapter for a discussion about power in ac circuits, including reactive circuits.)

Common fractional and multiple units for power are the milliwatt (mW, one thousandth of a watt) and the kilowatt (kW, 1000 W).

Example: The plate voltage on a transmitting vacuum tube is 2000 V and the plate current is 350 mA. (The current must be changed to amperes before substitution in the formula, and so is 0.350 A.) Then:

$$P = I \times E = 2000 \text{ V} \times 0.350 \text{ A} = 700 \text{ W}$$

Power may be expressed in *horsepower* (hp) instead of watts, using the following conversion factor:

1 horsepower = 746 W

This conversion factor is especially useful if you are working with a system that converts electrical energy into mechanical energy, and vice versa, since mechanical power is often expressed in horsepower in the US. In metric countries, mechanical power is usually expressed in watts. All countries use the metric power unit of watts in electrical systems, however. The value 746 W/hp assumes lossless conversion between mechanical and electrical power; practical efficiency is taken up shortly.

2.4.1 Energy

When you buy electricity from a power company, you pay for electrical energy, not power. What you pay for is the work that the electrical energy does for you, not the rate at which that work is done. Like energy, work is equal to power multiplied by time. The common unit for measuring electrical energy is the *watt-hour* (Wh), which means that a power of one watt has been used for one hour. That is:

$$Wh = P\,t \qquad (19)$$

where
Wh = energy in watt-hours
P = power in watts
t = time in hours.

Actually, the watt-hour is a fairly small energy unit, so the power company bills you for *kilowatt-hours* (kWh) of energy used. Another energy unit that is sometimes useful is the *watt-second* (Ws), which is equal to joules.

It is important to realize, both for calculation purposes and for efficient use of power resources, a small amount of power used for a long time can eventually result in a power bill that is just as large as if a large amount of power had been used for a very short time.

A common use of energy units in radio is in specifying the energy content of a battery. Battery energy is rated in *ampere-hours* (Ah) or *milliampere-hours* (mAh). While the multiplication of amperes and hours does not result in units of energy, the calculation assumes the result is multiplied by a specified (and constant) battery voltage. For example, a rechargeable NiMH battery rated to store 2000 mAh of energy is assumed to supply that energy at a terminal voltage of 1.5 V. Thus, after converting 2000 mA to 2 A, the actual energy stored is:

$$\text{Energy} = 1.5 \text{ V} \times 2 \text{ A} \times 1 \text{ hour} = 3 \text{ Wh}$$

Another common energy unit associated with batteries is *energy density*, with units of Ah per unit of volume or weight.

One practical application of energy units is to estimate how long a radio (such as a hand-held unit) will operate from a certain battery. For example, suppose a fully charged battery stores 900 mAh of energy and that the radio draws 30 mA on receive. A simple calculation indicates that the radio will be able receive 900 mAh / 30 mA = 30 hours with this battery, assuming 100% efficiency. You shouldn't expect to get the full 900 mAh out of the battery because the battery's voltage will drop as it is discharged, usually causing the equipment it powers to shut down before the last fraction of charge is used. Any time spent transmitting will also reduce the time the battery will last. The **Power Supplies** chapter includes additional information about batteries and their charge/discharge cycles.

2.4.2 Generalized Definition of Resistance

Electrical energy is not always turned into heat. The energy used in running a motor, for example, is converted to mechanical motion. The energy supplied to a radio transmitter is largely converted into radio waves. Energy applied to a loudspeaker is changed into sound waves. In each case, the energy is converted to other forms and can be completely accounted for. None of the energy just disappears! These are examples of the Law of Conservation of Energy. When a device converts energy from one form to another, we often say it *dissipates* the energy, or power. (Power is energy divided by time.) Of course the device doesn't really "use up" the energy, or make it disappear, it just converts it to another form. Proper operation of electrical devices often requires that the power be supplied at a specific ratio of voltage to current. These features are characteristics of resistance, so it can be said that any device that "dissipates power" has a definite value of resistance.

This concept of resistance as something that absorbs power at a definite voltage-to-current ratio is very useful; it permits substituting a simple resistance for the load or power-consuming part of the device receiving power, often with considerable simplification of calculations. Of course, every electrical device has some resistance of its own in the more narrow sense, so a part of the energy supplied to it is converted to heat in that resistance even though the major part of the energy may be converted to another form.

2.4.3 Efficiency

In devices such as motors and transmitters, the objective is to convert the supplied energy (or power) into some form other than heat. In such cases, power converted to heat is considered to be a loss because it is not useful power. The efficiency of a device is the useful power output (in its converted form) divided by the power input to the device. In a transmitter, for example, the objective is to convert power from a dc source into ac power at some radio frequency. The ratio of the RF power output to the dc input is the *efficiency* (Eff or η) of the transmitter. That is:

$$\text{Eff} = \frac{P_O}{P_I} \qquad (20)$$

where
Eff = efficiency (as a decimal)
P_O = power output (W)

P_I = power input (W).

Example: If the dc input to the transmitter is 100 W, and the RF power output is 60 W, the efficiency is:

$$\text{Eff} = \frac{P_O}{P_I} = \frac{60 \text{ W}}{100 \text{ W}} = 0.6$$

Efficiency is usually expressed as a percentage — that is, it expresses what percent of the input power will be available as useful output. To calculate percent efficiency, multiply the value from equation 20 by 100%. The efficiency in the example above is 60%.

Suppose a mobile transmitter has an RF power output of 100 W with 52% efficiency at 13.8 V. The vehicle's alternator system charges the battery at a rate of 5.0 A at this voltage. Assuming an alternator efficiency of 68%, how much horsepower must the engine produce to operate the transmitter and charge the battery? Solution: To charge the battery, the alternator must produce 13.8 V × 5.0 A = 69 W. The transmitter dc input power is 100 W / 0.52 = 190 W. Therefore, the total electrical power required from the alternator is 190 + 69 = 259 W. The engine load then is:

$$P_I = \frac{P_O}{\text{Eff}} = \frac{259 \text{ W}}{0.68} = 381 \text{ W}$$

We can convert this to horsepower using the conversion factor given earlier to convert between horsepower and watts:

$$381 \text{ W} = \frac{1 \text{ horsepower}}{746 \text{ W}} = 0.51 \text{ horsepower}$$

2.4.4 Ohm's Law and Power Formulas

Electrical power in a resistance is turned into heat. The greater the power, the more rapidly the heat is generated. By substituting the Ohm's Law equivalent for E and I, the following formulas are obtained for power:

$$P = \frac{E^2}{R} \tag{21}$$

and

Ohm's Law and Power Circle

During the first semester of my *Electrical Power Technology* program, one of the first challenges issued by our dedicated instructor — Roger Crerie — to his new freshman students was to identify and develop 12 equations or formulas that could be used to determine voltage, current, resistance and power. Ohm's Law is expressed as R = E / I and it provided three of these equation forms while the basic equation relating power to current and voltage (P = I × E) accounted for another three. With six known equations, it was just a matter of applying mathematical substitution for his students to develop the remaining six. Together, these 12 equations compose the *circle* or *wheel* of voltage (E), current (I), resistance (R) and power (P) shown in **Fig 2.A1**. Just as Roger's previous students had learned at the Worcester Industrial Technical Institute (Worcester, Massachusetts), our Class of '82 now held the basic electrical formulas needed to proceed in our studies or professions. As can be seen in Fig 2.A1, we can determine any one of these four electrical quantities by knowing the value of any two others. You may want to keep this page bookmarked for your reference. You'll probably be using many of these formulas as the years go by — this has certainly been my experience. — *Dana G. Reed, W1LC*

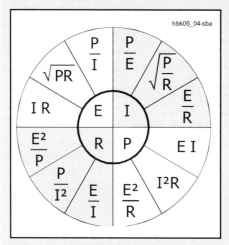

Figure 2.A1 — Electrical formulas.

$$P = I^2 \times R \tag{22}$$

These formulas are useful in power calculations when the resistance and either the current or voltage (but not both) are known.

Example: How much power will be converted to heat in a 4000-Ω resistor if the potential applied to it is 200 V? From equation 21,

$$P = \frac{E^2}{R} = \frac{40000 \text{ V}^2}{4000 \text{ Ω}} = 10.0 \text{ W}$$

As another example, suppose a current of 20 mA flows through a 300-Ω resistor. Then:

$$P = I^2 \times R = 0.020^2 \text{ A}^2 \times 300 \text{ Ω}$$

$$P = 0.00040 \text{ A}^2 \times 300 \text{ Ω}$$

$$P = 0.12 \text{ W}$$

Note that the current was changed from milliamperes to amperes before substitution in the formula.

Resistors for radio work are made in many sizes, the smallest being rated to safely operate at power levels of about $\frac{1}{16}$ W. The largest resistors commonly used in amateur equipment are rated at about 100 W. Large resistors, such as those used in dummy-load antennas, are often cooled with oil to increase their power-handling capability.

2.5 Circuit Control Components

2.5.1 Switches

Switches are used to allow or interrupt a current flowing in a particular circuit. Most switches are mechanical devices, although the same effect may be achieved with solid-state devices.

Switches come in many different forms and a wide variety of ratings. The most important ratings are the *voltage-handling* and *current-handling* capabilities. The voltage rating usually includes both the *breakdown voltage rating* and the *interrupt voltage rat-* *ing*. The breakdown rating is the maximum voltage that the switch can withstand when it is open before the voltage will arc between the switch's terminals. The interrupt voltage rating is the maximum amount of voltage that the switch can interrupt without arcing. Normally, the interrupt voltage rating is the lower value, and therefore the one given for (and printed on) the switch.

Switches typically found in the home are usually rated for 125 V ac and 15 to 20 A. Switches in cars are usually rated for 12 V dc and several amperes. The breakdown voltage rating of a switch primarily depends on the insulating material surrounding the contacts and the separation between the contacts. Plastic or phenolic material normally provides both structural support and insulation. Ceramic material may be used to provide better insulation, particularly in rotary (wafer) switches.

A switch's current rating includes both the *current-carrying capacity* and the *interrupt capability*. The current-carrying capacity of the switch depends on the contact material and size, and on the pressure exerted to keep the contacts closed. It is primarily determined from the allowable contact temperature rise. On larger ac switches and most dc switches, the interrupt capability is usually lower than the current carrying value.

Most power switches are rated for alternating current use. Because ac current goes through zero twice in each cycle, switches can successfully interrupt much higher alternating currents than direct currents without arcing. A switch that has a 10-A ac current rating may arc and damage the contacts if used to turn off more than an ampere or two of dc.

Switches are normally designated by the number of *poles* (circuits controlled) and *throws* or *positions* (circuit path choices). The simplest switch is the on-off switch, which is a single-pole, single-throw (SPST) switch as shown in **Fig 2.11A**. The off position does not direct the current to another circuit. The next step would be to change the current path to another path. This would be a single-pole, double-throw (SPDT) switch as shown in Fig 2.11B. Adding an off position would give a single-pole, double-throw, center-off switch as shown in Fig 2.11C.

Several such switches can be "ganged" to or actuated by the same mechanical activator to provide double-pole, triple-pole or even more, separate control paths all activated at once. Switches can be activated in a variety of ways. The most common methods include lever or toggle, push-button and rotary switches. Samples of these are shown in **Fig 2.12**. Most switches stay in position once set, but some are spring-loaded so they only stay in the desired position while held there. These are called *momentary* switches.

Rotary/wafer switches can provide very complex switching patterns. Several poles (separate circuits) can be included on each wafer. Many wafers may be stacked on the same shaft. Not only may many different circuits be controlled at once, but by wiring different poles/positions on different wafers together, a high degree of circuit switching logic can be developed. Such switches can select different paths as they are turned and can also "short" together successive contacts to connect numbers of components or paths.

Rotary switches can also be designed to either break one contact before making another (*break-before-make*), or to short two contacts together before disconnecting the first one (*make-before-break*) to eliminate arcing or perform certain logic functions. The two types of switches are generally not interchangeable and may cause damage if inappropriately substituted for one another during circuit construction or repair. When buying rotary switches from a surplus or flea-market vendor, check to be sure the type of switch is correct.

Microswitches are designed to be actuated by the operation of machine components, opening or closing of a door, or some other mechanical movement. Instead of a handle or button-type actuator that would be used by a human, microswitches have levers or buttons more suitable for being actuated as part of an enclosure or machine.

In choosing a switch for a particular task, consideration should be given to function, voltage and current ratings, ease of use, availability and cost. If a switch is to be operated frequently, a better-quality switch is usually less costly over the long run. If signal noise or contact corrosion is a potential problem (usually in low-current signal applications), it is best to get gold-plated contacts. Gold does not oxidize or corrode, thus providing surer contact, which can be particularly important at very low signal levels. Gold plating will not hold up under high-current-interrupt applications, however.

2.5.2 Fuses and Circuit Breakers

Fuses self-destruct to protect circuit wiring or equipment. The fuse *element* that melts or *blows* is a carefully shaped piece of soft metal, usually mounted in a cartridge of some kind. The element is designed to safely carry a given amount of current and to melt at a current value that is a certain percentage above the rated value.

The most important fuse rating is the *nominal current rating* that it will safely carry for an indefinite period without blowing. A fuse's melting current depends on the type of material, the shape of the element and the

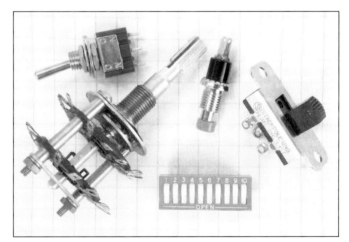

Fig 2.12 — This photo shows examples of various styles of switches. The ¼-inch-ruled graph paper background provides for size comparison.

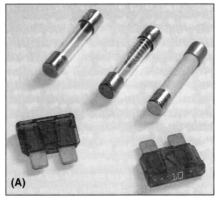

Fig 2.11 — Schematic diagrams of various types of switches. A is an SPST, B is an SPDT, and C is an SPDT switch with a center-off position.

Fig 2.13 — These photos show examples of various styles of fuses. Cartridge-type fuses (A, top) can use glass or ceramic construction. The center fuse is a slow-blow type. Automotive blade-type fuses (A, bottom) are common for low-voltage dc use. A typical home circuit breaker for ac wiring is shown at B.

heat dissipation capability of the cartridge and holder, among other factors.

Next most important are the timing characteristics, or how quickly the fuse element blows under a given current overload. Some fuses (*slow-blow*) are designed to carry an overload for a short period of time. They typically are used in motor-starting and power-supply circuits that have a large inrush current when first started. Other fuses are designed to blow very quickly to protect delicate instruments and solid-state circuits.

A fuse also has a voltage rating, both a value in volts and whether it is expected to be used in ac or dc circuits. The voltage rating is the amount of voltage an open fuse can withstand without arcing. While you should never substitute a fuse with a higher current rating than the one it replaces, you may use a fuse with a higher voltage rating.

Fig 2.13A shows typical cartridge-style cylindrical fuses likely to be encountered in ac-powered radio and test equipment. Automotive style fuses, shown in the lower half of Fig 2.13A, have become widely used in low-voltage dc power wiring of amateur stations. These are called "blade" fuses. Rated for vehicle-level voltages, automotive blade fuses should never be used in ac line-powered circuits.

Circuit breakers perform the same function as fuses — they open a circuit and interrupt current flow when an overload occurs. Instead of a melting element, circuit breakers use spring-loaded magnetic mechanisms to open a switch when excessive current is present. Once the overload has been corrected, the circuit-breaker can be reset. Circuit breakers are generally used by amateurs in home ac wiring (a typical ac circuit breaker is shown in Fig 2.13B) and in dc power supplies.

A replacement fuse or circuit breaker should have the same current rating and the same characteristics as the fuse it replaces. Never substitute a fuse with a larger current rating. You may cause permanent damage (maybe even a fire) to wiring or circuit elements by allowing larger currents to flow when there is an internal problem in equipment. (Additional discussion of fuses and circuit breakers is provided in the chapter on **Safety**.)

Fuses blow and circuit breakers open for several reasons. The most obvious reason is that a problem develops in the circuit, causing too much current to flow. In this case, the circuit problem needs to be fixed. A fuse can fail from being cycled on and off near its current rating. The repeated thermal stress causes metal fatigue and eventually the fuse blows. A fuse can also blow because of a momentary power surge, or even by rapidly turning equipment with a large inrush current on and off several times. In these cases it is only necessary to replace the fuse with the same type and value.

Panel-mount fuse holders should be wired with the hot lead of an ac power circuit (the black wire of an ac power cord) connected to the end terminal, and the ring terminal is connected to the power switch or circuit inside the chassis. This removes voltage from the fuse as it is removed from the fuse holder. This also locates the line connection at the far end of the fuse holder where it is not easily accessible.

2.5.3 Relays And Solenoids

Relays are switches controlled by an electrical signal. *Electromechanical relays* consist of a electromagnetic *coil* and a moving *armature* attracted by the coil's magnetic field when energized by current flowing in the coil. Movement of the armature pushes the switch contacts together or apart. Many sets of contacts can be connected to the same armature, allowing many circuits to be controlled by a single signal. In this manner, the signal voltage that energizes the coil can control circuits carrying large voltages and/or currents.

Relays have two positions or *states* — *energized* and *de-energized*. Sets of contacts called *normally-closed* (NC) are closed when the relay is de-energized and open when it is energized. *Normally-open* contact sets are closed when the relay is energized.

Like switches, relay contacts have breakdown voltage, interrupting, and current-carrying ratings. These are not the same as the voltage and current requirements for energizing the relay's coil. Relay contacts (and housings) may be designed for ac, dc or RF signals. The most common control voltages for relays used in amateur equipment are 12 V dc or 120 V ac. Relays with 6, 24, and 28 V dc, and 24 V ac coils are also common. **Fig 2.14** shows some typical relays found in amateur equipment.

A relay's p*ull-in voltage* is the minimum voltage at which the coil is guaranteed to cause the armature to move and change the relay's state. *Hold-in voltage* is the minimum voltage at which the relay is guaranteed to hold the armature in the energized position after the relay is activated. A relay's pull-in voltage is higher than its hold-in voltage due to magnetic hysteresis of the coil (see the section on magnetic materials below). *Current-sensing relays* activate when the current through the coil exceeds a specific value, regardless of the voltage applied to the coil. They are used when the control signal is a current rather than a voltage.

Latching relays have two coils; each moves the armature to a different position where it remains until the other coil is energized. These relays are often used in portable and low-power equipment so that the contact configuration can be maintained without the need to continuously supply power to the relay.

Reed relays have no armature. The contacts are attached to magnetic strips or "reeds" in a glass or plastic tube, surrounded by a coil. The reeds move together or apart when current is applied to the coil, opening or closing contacts. Reed relays can open and close very quickly and are often used in transmit-receive switching circuits.

Solid-state relays (SSR) use transistors in-

(A)

(B)

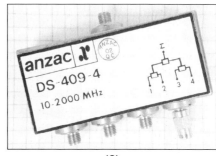

(C)

Fig 2.14 — These photos show examples of various styles and sizes of relays. Photo A shows a large reed relay, and a small reed relay in a package the size of a DIP IC. The contacts and coil can clearly be seen in the open-frame relay. Photo B shows a relay inside a plastic case. Photo C shows a four-position coaxial relay-switch combination with SMA connectors. The ¼-inch-ruled graph paper background provides a size comparison.

stead of mechanical contacts and electronic circuits instead of magnetic coils. They are designed as substitutes for electromechanical relays in power and control circuits and are not used in low-level ac or dc circuits.

Coaxial relays have an armature and contacts designed to handle RF signals. The signal path in coaxial relays maintains a specific characteristic impedance for use in RF systems. Coaxial connectors are used for the RF circuits. Coaxial relays are typically used to control antenna system configurations or to switch a transceiver between a linear amplifier and an antenna.

A *solenoid* is very similar to a relay, except that instead of the moving armature actuating switch contacts, the solenoid moves a lever or rod to actuate some mechanical device. Solenoids are not commonly used in radio equipment, but may be encountered in related systems or devices.

2.6 AC Theory and Waveforms

2.6.1 AC in Circuits

A circuit is a complete conductive path for current to flow from a source, through a load and back to the source. If the source permits the current to flow in only one direction, the current is dc or *direct current*. If the source permits the current to change direction, the current is ac or *alternating current*. **Fig 2.15** illustrates the two types of circuits. Circuit A shows the source as a battery, a typical dc source. Circuit B shows the more abstract voltage source symbol to indicate ac. In an ac circuit, both the current and the voltage reverse direction. For nearly all ac signals in electronics and radio, the reversal is *periodic*, meaning that the change in direction occurs on a regular basis. The rate of reversal may range from a few times per second to many billions per second.

Graphs of current or voltage, such as Fig 2.15, begin with a horizontal axis that represents time. The vertical axis represents the amplitude of the current or the voltage, whichever is graphed. Distance above the zero line indicates larger positive amplitude; distance below the zero line means larger negative amplitude. Positive and negative only designate the opposing directions in which current may flow in an alternating current circuit or the opposing *polarities* of an ac voltage.

If the current and voltage never change direction, then from one perspective, we have a dc circuit, even if the level of dc constantly changes. **Fig 2.16** shows a current that is always positive with respect to 0. It varies periodically in amplitude, however. Whatever the shape of the variations, the current can be called *pulsating dc*. If the current periodically reaches 0, it can be called *intermittent dc*. From another perspective, we may look at intermittent and pulsating dc as a combination of an ac and a dc current. Special circuits can separate the two currents into ac and dc *components* for separate analysis or use. There are circuits that combine ac and dc currents and voltages, as well.

2.6.2 AC Waveforms

A *waveform* is the pattern of amplitudes reached by an ac voltage or current as measured over time. The combination of ac and dc voltages and currents results in *complex* waveforms. **Fig 2.17** shows two ac waveforms fairly close in frequency and their combination. **Fig 2.18** shows two ac waveforms dissimilar in both frequency and wavelength, along with the resultant combined waveform. Note the similarities (and the differences) between the resultant waveform in Fig 2.18 and the combined ac-dc waveform in Fig 2.16.

Alternating currents may take on many useful waveforms. **Fig 2.19** shows a few that are commonly used in electronic and radio circuits. The *sine wave* is both mathematically and practically the foundation of all other forms of ac; the other forms can usually be reduced to (and even constructed from) a particular collection of sine waves. The *square wave* is vital to digital electronics. The *triangular* and *ramp waves* — the latter sometimes called a *sawtooth* waveform — are especially useful in timing circuits. The individual repeating patterns that make up these *periodic waveforms* are called *cycles*. Waveforms that do not consist of repetitive patterns are called *aperiodic* or *irregular* waveforms. Speech is

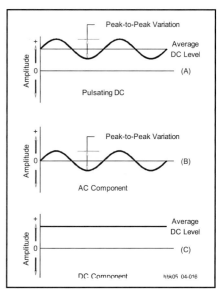

Fig 2.16 — A pulsating dc current (A) and its resolution into an ac component (B) and a dc component (C).

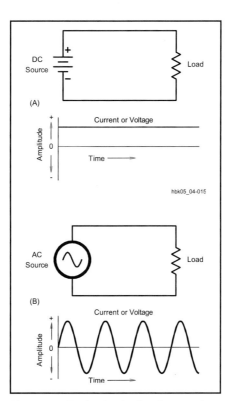

Fig 2.15 — Basic circuits for direct and alternating currents. With each circuit is a graph of the current, constant for the dc circuit, but periodically changing direction in the ac circuit.

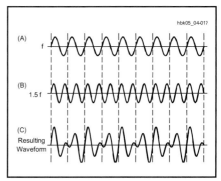

Fig 2.17 — Two ac waveforms of similar frequencies (f1 = 1.5 f2) and amplitudes form a composite wave. Note the points where the positive peaks of the two waves combine to create high composite peaks at a frequency that is the difference between f1 and f2. The beat note frequency is 1.5f − f = 0.5f and is visible in the drawing.

Electrical Fundamentals 2.15

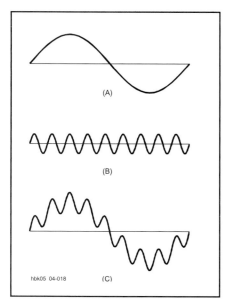

Fig 2.18 — Two ac waveforms of widely different frequencies and amplitudes form a composite wave in which one wave appears to ride upon the other.

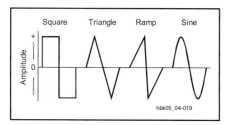

Fig 2.19 — Some common ac waveforms: square, triangle, ramp and sine.

a good example of an irregular waveform, as is noise or static.

There are numerous ways to generate alternating currents: with an ac power generator (an *alternator*), with a *transducer* (for example, a microphone) or with an electronic circuit (for example, an RF oscillator). The basis of the sine wave is circular or cyclical motion, which underlies the most usual methods of generating alternating current. The circular motion of the ac generator may be physical or mechanical, as in an alternator. Currents in the resonant circuit of an oscillator may also produce sine waves as repetitive electrical cycles without mechanical motion.

SINE WAVES AND CYCLICAL MOTION

The relationship between circular motion and the sine wave is illustrated in **Fig 2.20** as the correlation between the amplitude of an ac current (or voltage) and the relative positions of a point making circular rotations through one complete revolution of 360°. Following the height of the point above the horizontal axis, its height is zero at point 1. It rises to a maximum at a position 90° from position 1, which is position 3. At position 4, 180° from position 1, the point's height falls back to zero. Then the point's height begins to increase again, but below the horizontal axis, where it reaches a maximum at position 5.

An alternative way of visualizing the relationship between circular motion and sine waves is to use a yardstick and a light source, such as a video projector. Hold the yardstick horizontally between the light source and the wall so that it points directly at the light source. This is defined as the 0° position and the length of the yardstick's shadow is zero. Now rotate the yardstick, keeping the end farthest from the projector stationary, until the yardstick is vertical. This is the 90° position and the shadow's length is a maximum in the direction defined as positive. Continue to rotate the yardstick until it reaches horizontal again. This is the 180° position and the shadow once again has zero length. Keep rotating the yardstick in the same direction until it is vertical and the shadow is at maximum, but now below the stationary end of the yardstick. This is the 270° position and the direction of the shadow is defined to be negative. Another 90° of rotation brings the pointer back to its original position at 0° and the shadow's length to zero. If the length of the shadow is plotted against the amount of rotation of the yardstick, the result is a sine wave.

The corresponding rise and fall of a *sinusoidal* or sine wave current (or voltage) along a linear time line produces the curve accompanying the circle in Fig 2.20. The curve is sinusoidal because its amplitude varies as the sine of the angle made by the circular movement with respect to the zero position. This is often referred to as the *sine function*, written sin(angle). The sine of 90° is 1, and 90° is also the position of maximum current (along with 270°, but with the opposite polarity). The sine of 45° (point 2) is 0.707, and the value of current at point 2, the 45° position of rotation, is 0.707 times the maximum current. Both ac current and voltage sine waves vary in the same way.

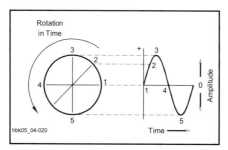

Fig 2.20 — The relationship of circular motion and the resultant graph of ac current or voltage. The curve is sinusoidal, a sine wave.

FREQUENCY AND PERIOD

With a continuously rotating generator, alternating current or voltage will pass through many equal cycles over time. This is a *periodic waveform*, composed of many identical cycles. An arbitrary point on any one cycle can be used as a marker of position on a periodic waveform. For this discussion, the positive peak of the waveform will work as an unambiguous marker. The number of times per second that the current (or voltage) reaches this positive peak in any one second is called the *frequency* of the waveform. In other words, frequency expresses the *rate* at which current (or voltage) cycles occur. The unit of frequency is *cycles per second*, or *hertz* — abbreviated Hz (after the 19th century radio-phenomena pioneer, Heinrich Hertz).

The length of any cycle in units of time is the *period* of the cycle, as measured between equivalent points on succeeding cycles. Mathematically, the period is the inverse of frequency. That is,

$$\text{Frequency (f) in Hz} = \frac{1}{\text{Period (T) in seconds}}$$
(23)

and

$$\text{Period (T) in seconds} = \frac{1}{\text{Frequency (f) in Hz}}$$
(24)

Example: What is the period of a 400-hertz ac current?

$$T = \frac{1}{f} = \frac{1}{400 \text{ Hz}} = 0.00250 \text{ s} = 2.5 \text{ ms}$$

The frequency of alternating currents used in radio circuits varies from a few hertz, or cycles per second, to thousands of millions of hertz. Likewise, the period of alternating currents amateurs use ranges from significant fractions of a second down to nanoseconds or smaller. In order to express compactly the units of frequency, time and almost everything else in electronics, a standard system of prefixes is used. In magnitudes of 1000 or 10^3, frequency is measurable in hertz, in kilohertz (1000 hertz or kHz), in megahertz (1 million hertz or MHz), gigahertz (1 billion hertz or GHz — pronounced "gee-gah" with a hard 'g' as in 'golf') and even in terahertz (1 trillion hertz or THz). For units smaller than unity, as in the measurement of period, the basic unit can be milliseconds (1 thousandth of a second or ms), microseconds (1 millionth of a second or μs), nanoseconds (1 billionth of a second or ns) and picoseconds (1 trillionth of a second or ps).

It is common for complex ac signals to contain a series of sine waves with frequencies related by integer multiples of some lowest or *fundamental* frequency. Sine waves with the higher frequency are called harmonics and are said to be *harmonically related* to the fundamental. For example, if a complex waveform is made up of sine waves with frequencies of 10, 20 and 30 kHz, 10 kHz is the fundamental and the other two are harmonics.

The uses of ac in radio circuits are many and varied. Most can be cataloged by reference to ac frequency ranges used in circuits. For example, the frequency of ac power used in the home, office and factory is ordinarily 60 Hz in the United States and Canada. In Great Britain and much of Europe, ac power is 50 Hz. Sonic and ultrasonic applications of ac run from about 20 Hz (audio) up to several MHz.

Radio circuits include both power- and sonic-frequency-range applications. Radio communication and other electronics work, however, require ac circuits capable of operation with frequencies up to the GHz range. Some of the applications include signal sources for transmitters (and for circuits inside receivers); industrial induction heating; diathermy; microwaves for cooking, radar and communication; remote control of appliances, lighting, model planes and boats and other equipment; and radio direction finding and guidance.

PHASE

When drawing the sine-wave curve of an ac voltage or current, the horizontal axis represents time. This type of drawing in which the amplitude of a waveform is shown compared to time is called the *time domain*. Events to the right take place later; events to the left occur earlier. Although time is measurable in parts of a second, it is more convenient to treat each cycle as a complete time unit divided into 360°. The conventional starting point for counting degrees in a sinusoidal waveform such as ac is the *zero point* at which the voltage or current begins the positive *half cycle*. The essential elements of an ac cycle appear in **Fig 2.21**.

The advantage of treating the ac cycle in this way is that many calculations and measurements can be taken and recorded in a manner that is independent of frequency. The positive peak voltage or current occurs at 90° into the cycle. Relative to the starting point, 90° is the *phase* of the ac at that point. Phase is the position within an ac cycle expressed in degrees or *radians*. Thus, a complete description of an ac voltage or current involves reference to three properties: frequency, amplitude and phase.

Phase relationships also permit the comparison of two ac voltages or currents at the same frequency. If the zero point of two signals with the frequency occur at the same time, there is zero phase difference between the signals and they are said to be *in phase*.

Fig 2.22 illustrates two waveforms with a constant phase difference. Since B crosses

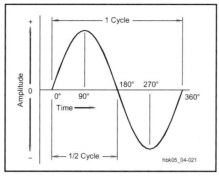

Fig 2.21 — An ac cycle is divided into 360° that are used as a measure of time or phase.

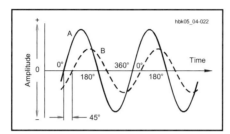

Fig 2.22 — When two waves of the same frequency start their cycles at slightly different times, the time difference or phase difference is measured in degrees. In this drawing, wave B starts 45° (one-eighth cycle) later than wave A, and so lags 45° behind A.

the zero point in the positive direction after A has already done so, there is a *phase difference* between the two waves. In the example, B *lags* A by 45°, or A *leads* B by 45°. If A and B occur in the same circuit, their composite waveform will also be a sine wave at an intermediate phase angle relative to each. Adding any number of sine waves of the same frequency always results in a sine wave at that frequency. Adding sine waves of different frequencies, as in Fig 2.17, creates a complex waveform with a *beat frquency* that is the difference between the two sine waves.

Fig 2.22 might equally apply to a voltage and a current measured in the same ac circuit. Either A or B might represent the voltage; that is, in some instances voltage will lead the current and in others voltage will lag the current.

Two important special cases appear in **Fig 2.23**. In Part A, line B lags 90° behind line A. Its cycle begins exactly one quarter cycle later than the A cycle. When one wave is passing through zero, the other just reaches its maximum value. In this example, the two sine waves are said to be *in quadrature*.

In Part B, lines A and B are 180° *out of phase*, sometimes called *anti-phase*. In this case, it does not matter which one is considered to lead or lag. Line B is always positive

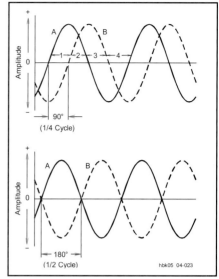

Fig 2.23 — Two important special cases of phase difference: In the upper drawing, the phase difference between A and B is 90°; in the lower drawing, the phase difference is 180°.

while line A is negative, and vice versa. If the two waveforms are of two voltages or two currents in the same circuit and if they have the same amplitude, they will cancel each other completely.

Phase also has other meanings. It is important to distinguish between *polarity*, meaning the sense in which positive and negative voltage or current relate to circuit operation, and phase, which is a function of time or position in a waveform. It is quite possible for two signals to have opposite polarities, but still be in phase, for example. In a multi-phase ac power system, "phase" refers to one of the distinct voltage waveforms generated by the utility.

Degrees and Radians

While most electronic and radio mathematics use degrees as a measure of phase, you will occasionally encounter *radians*. Radians are used because they are more convenient mathematically in certain equations and computations. Radians are used for angular measurements when the *angular frequency* (ω) is being used. There are 2π radians in a circle, just as there are 360°, so one radian = 360/2π ≈ 57.3°. Angular frequency is measured in radians/second (not revolutions/second or cycles/second) so ω = 2πf. In this *Handbook*, unless it is specifically noted otherwise, the convention will be to use degrees in all calculations that require an angle.

2.6.3 Electromagnetic Energy

Alternating currents are often loosely classified as audio frequency (AF) and radio frequency (RF). Although these designations are handy, they actually represent something other than electrical energy: They designate special forms of energy that we find useful.

Audio or *sonic* energy is the energy imparted by the mechanical movement of a medium, which can be air, metal, water or even the human body. Sound that humans can hear normally requires the movement of air between 20 Hz and 20 kHz, although the human ear loses its ability to detect the extremes of this range as we age. Some animals, such as elephants, can detect air vibrations well below 20 Hz, while others, such as dogs and cats, can detect air vibrations well above 20 kHz.

Electrical circuits do not directly produce air vibrations. Sound production requires a *transducer*, a device to transform one form of energy into another form of energy; in this case electrical energy into sonic energy. The speaker and the microphone are the most common audio transducers. There are numerous ultrasonic transducers for various applications.

RF energy occurs at frequencies for which it is practical to generate and detect *electromagnetic* or *RF waves* that exist independently of the movement of electrical charge, such as a radio signal. Like sonic energy, a transducer — an antenna — is required to convert the electrical energy in a circuit to electromagnetic waves. In a physical circuit, such as a wire, electromagnetic energy exists as both electromagnetic waves and the physical movement of electrical charge.

Electromagnetic waves have been generated and detected in many forms with frequencies from below 1 Hz to above 10^{12} GHz, including at the higher frequencies infrared, visible, and ultraviolet light, and a number of energy forms of greatest interest to physicists and astronomers. **Table 2.3** provides a brief glimpse at the total spectrum of electromagnetic energy. The *radio spectrum* is generally considered to begin around 3 kHz and end at infrared light.

Despite the close relationship between electromagnetic energy and waves, it remains important to distinguish the two. To a circuit producing or amplifying a 15-kHz alternating current, the ultimate transformation and use of the electrical energy may make no difference to the circuit's operation. By choosing the right transducer, one can produce either a sonic wave or an electromagnetic wave — or both. Such is a common problem of video monitors and switching power supplies; forces created by the ac currents cause electronic parts both to vibrate audibly *and* to radiate electromagnetic energy.

All electromagnetic energy has one thing in common: it travels, or *propagates*, at the speed of light, abbreviated c. This speed is approximately 300,000,000 (or 3×10^8) meters per second in a vacuum. Electromagnetic-energy waves have a length uniquely associated with each possible frequency. The *wavelength* (λ) is the speed of propagation divided by the frequency (f) in hertz.

$$f(Hz) = \frac{3.0 \times 10^8 \left(\frac{m}{s}\right)}{\lambda(m)} \quad (25)$$

and

$$\lambda(m) = \frac{3.0 \times 10^8 \left(\frac{m}{s}\right)}{f(Hz)} \quad (26)$$

Example: What is the frequency of an RF wave with wavelength of 80 meters?

$$f(Hz) = \frac{3.0 \times 10^8 \left(\frac{m}{s}\right)}{\lambda(m)}$$

$$= \frac{3.0 \times 10^8 \left(\frac{m}{s}\right)}{80.0 \, m}$$

$$= 3.75 \times 10^6 \, Hz$$

This is 3.750 MHz or 3750 kHz, a frequency in the middle of the ham band known as "80 meters."

A similar equation is used to calculate the wavelength of a sound wave in air, substituting the speed of sound instead of the speed of light in the numerator. The speed of propagation of the mechanical movement of air that we call sound varies considerably with air temperature and altitude. The speed of sound at sea level is about 331 m/s at 0 °C and 344 m/s at 20 °C.

To calculate the frequency of an electromagnetic wave directly in kilohertz, change the speed constant to 300,000 (3×10^5) km/s.

$$f(kHz) = \frac{3.0 \times 10^5 \left(\frac{km}{s}\right)}{\lambda(m)} \quad (27)$$

and

$$\lambda(m) = \frac{3.0 \times 10^5 \left(\frac{km}{s}\right)}{f(kHz)} \quad (28)$$

For frequencies in megahertz, change the speed constant to 300 (3×10^2) Mm/s.

$$f(MHz) = \frac{300 \left(\frac{Mm}{s}\right)}{\lambda(m)} \quad (29)$$

and

$$\lambda(m) = \frac{300 \left(\frac{Mm}{s}\right)}{f(MHz)} \quad (30)$$

Stated as it is usually remembered and used, "wavelength in meters equals 300 divided by frequency in megahertz." Assuming the proper units for the speed of light constant simplify the equation.

Example: What is the wavelength of an RF wave whose frequency is 4.0 MHz?

$$\lambda(m) = \frac{300}{f(MHz)} = \frac{300}{4.0} = 75 \, m$$

At higher frequencies, circuit elements with lengths that are a significant fraction of a wavelength can act like transducers. This property can be useful, but it can also cause problems for circuit operations. Therefore, wavelength calculations are of some importance in designing ac circuits for those frequencies.

Within the part of the electromagnetic-energy spectrum of most interest to radio applications, frequencies have been classified into groups and given names. **Table 2.4** provides a reference list of these classifications. To a significant degree, the frequencies within each group exhibit similar properties, both in circuits and as RF waves. For example, HF or high frequency waves, with frequencies from

Table 2.3
Key Regions of the Electromagnetic Energy Spectrum

Region Name	Frequency Range		
Radio frequencies	3.0×10^3 Hz	to	3.0×10^{11} Hz
Infrared	3.0×10^{11} Hz	to	4.3×10^{14} Hz
Visible light	4.3×10^{14} Hz	to	7.5×10^{14} Hz
Ultraviolet	7.5×10^{14} Hz	to	6.0×10^{16} Hz
X-rays	6.0×10^{16} Hz	to	3.0×10^{19} Hz
Gamma rays	3.0×10^{19} Hz	to	5.0×10^{20} Hz
Cosmic rays	5.0×10^{20} Hz	to	8.0×10^{21} Hz

Note: The range of radio frequencies can also be written as 3 kHz to 300 GHz

Table 2.4

Classification of the Radio Frequency Spectrum

Abbreviation	Classification	Frequency Range			
VLF	Very low frequencies	3	to	30	kHz
LF	Low frequencies	30	to	300	kHz
MF	Medium frequencies	300	to	3000	kHz
HF	High frequencies	3	to	30	MHz
VHF	Very high frequencies	30	to	300	MHz
UHF	Ultrahigh frequencies	300	to	3000	MHz
SHF	Superhigh frequencies	3	to	30	GHz
EHF	Extremely high frequencies	30	to	300	GHz

3 to 30 MHz, all exhibit *skip* or ionospheric refraction that permits regular long-range radio communications. This property also applies occasionally both to MF (medium frequency) and to VHF (very high frequency) waves, as well.

2.6.4 Measuring AC Voltage, Current and Power

Measuring the voltage or current in a dc circuit is straightforward, as **Fig 2.24A** demonstrates. Since the current flows in only one direction, the voltage and current have constant values until the resistor values are changed.

Fig 2.24B illustrates a perplexing problem encountered when measuring voltages and currents in ac circuits — the current and voltage continuously change direction and value. Which values are meaningful? How are measurements performed? In fact, there are several methods of measuring sine-wave voltage and current in ac circuits with each method providing different information about the waveform.

INSTANTANEOUS VOLTAGE AND CURRENT

By far, the most common waveform associated with ac of any frequency is the sine wave. Unless otherwise noted, it is safe to assume that measurements of ac voltage or current are of a sinusoidal waveform. **Fig 2.25** shows a sine wave representing a voltage or current of some arbitrary frequency and amplitude. The *instantaneous* voltage (or current) is the value at one instant in time. If a series of instantaneous values are plotted against time, the resulting graph will show the waveform.

In the sine wave of Fig 2.25, the instantaneous value of the waveform at any point in time is a function of three factors: the maximum value of voltage (or current) along the curve (point B, E_{max}), the frequency of the wave (f), and the time elapsed from the preceding positive-going zero crossing (t) in seconds or fractions of a second. Thus,

$$E_{inst} = E_{max} \sin (ft) \qquad (31)$$

assuming all sine calculations are done in degrees. (See the sidebar "Degrees and Radians".) If the sine calculation is done in radians, substitute $2\pi ft$ for *ft* in equation 31.

If the point's phase is known — the position along the waveform — the instantaneous voltage at that point can be calculated directly as:

$$E_{inst} = E_{max} \sin \theta \qquad (32)$$

where θ is the number of degrees of phase difference from the beginning of the cycle.

Example: What is the instantaneous value of voltage at point D in Fig 2.25, if the maximum voltage value is 120 V and point D's phase is 60.0°?

$$E_{inst} = 120 \text{ V} \times \sin 60°$$

$$= 120 \times 0.866 = 104 \text{ V}$$

PEAK AND PEAK-TO-PEAK VOLTAGE AND CURRENT

The most important of an ac waveform's instantaneous values are the maximum or *peak values* reached on each positive and negative half cycle. In Fig 2.25, points B and C represent the positive and negative peaks. Peak values (indicated by a "pk" or "p" subscript) are especially important with respect to component ratings, which the voltage or current in a circuit must not exceed without danger of component failure.

The *peak power* in an ac circuit is the

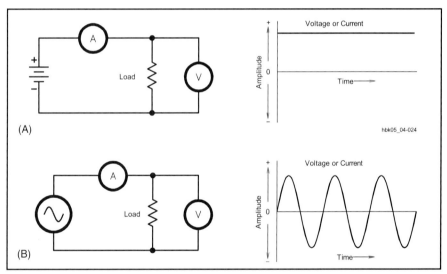

Fig 2.24 — Voltage and current measurements in dc and ac circuits.

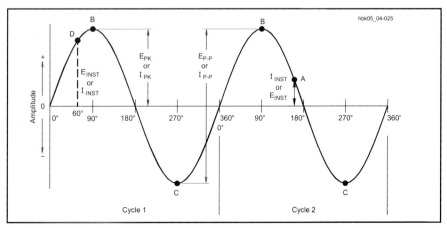

Fig 2.25 — Two cycles of a sine wave to illustrate instantaneous, peak, and peak-to-peak ac voltage and current values.

product of the peak voltage and the peak current, or

$$P_{pk} = E_{pk} \times I_{pk} \qquad (33)$$

The span from points B to C in Fig 2.25 represents the largest difference in value of the sine wave. Designated the *peak-to-peak* value (indicated by a "P-P" subscript), this span is equal to twice the peak value of the waveform. Thus, peak-to-peak voltage is:

$$E_{P-P} = 2 E_{pk} \qquad (34)$$

Amplifying devices often specify their input limits in terms of peak-to-peak voltages. Operational amplifiers, which have almost unlimited gain, often require input-level limiting to prevent the output signals from distorting if they exceed the peak-to-peak output rating of the devices.

RMS VALUES OF VOLTAGES AND CURRENTS

The *root mean square* or *RMS* values of voltage and current are the most common values encountered in electronics. Sometimes called *effective* values, the RMS value of an ac voltage or current is the value of a dc voltage or current that would cause a resistor to dissipate the same average amount of power as the ac waveform. This measurement became widely used in the early days of electrification when both ac and dc power utility power were in use. Even today, the values of the ac line voltage available from an electrical power outlet are given as RMS values. Unless otherwise specified, unlabeled ac voltage and current values found in most electronics literature are normally RMS values.

The RMS values of voltage and current get their name from the mathematical method used to derive their value relative to peak voltage and current. Start by *squaring* the individual values of all the instantaneous values of voltage or current during an entire single cycle of ac. Take the average (*mean*) of these squares (this is done by computing an integral of the waveform) and then find the square *root* of that average. This procedure produces the RMS value of voltage or current for all waveforms, sinusoidal or not. (The waveform is assumed to be periodic.)

For the remainder of this discussion, remember that we are assuming that the waveform in question is a sine wave. The simple formulas and conversion factors in this section are generally *not* true for non-sinusoidal waveforms such as square or triangle waves (see the sidebar, "Measuring Non-Sinusoidal Waveforms"). The following formulas are true *only* if the waveform is a sine wave and the circuit is *linear* — that is, raising or lowering the voltage will raise or lower the current proportionally. If those conditions are true, the following conversion factors have been computed and can be used without any additional mathematics.

For a sine wave to produce heat equivalent to a dc waveform the peak ac power required is twice the dc power. Therefore, the average ac power equivalent to a corresponding dc power is half the peak ac power.

$$P_{ave} = \frac{P_{pk}}{2} \qquad (35)$$

A sine wave's RMS voltage and current values needed to arrive at average ac power are related to their peak values by the conversion factors:

$$E_{RMS} = \frac{E_{pk}}{\sqrt{2}} = \frac{E_{pk}}{1.414} = E_{pk} \times 0.707 \qquad (36)$$

$$I_{RMS} = \frac{I_{pk}}{\sqrt{2}} = \frac{I_{pk}}{1.414} = I_{pk} \times 0.707 \qquad (37)$$

RMS voltages and currents are what is displayed by most volt and ammeters.

If the RMS voltage is the peak voltage divided by $\sqrt{2}$, then the peak voltage must be the RMS voltage multiplied by $\sqrt{2}$, or

$$E_{pk} = E_{RMS} \times 1.414 \qquad (38)$$

$$I_{pk} = I_{RMS} \times 1.414 \qquad (39)$$

Example: What is the peak voltage and the peak-to-peak voltage at the usual household ac outlet, if the RMS voltage is 120 V?

$$E_{pk} = 120 \text{ V} \times 1.414 = 170 \text{ V}$$

$$E_{P-P} = 2 \times 170 \text{ V} = 340 \text{ V}$$

In the time domain of a sine wave, the instantaneous values of voltage and current correspond to the RMS values at the 45°, 135°, 225° and 315° points along the cycle shown in **Fig 2.26**. (The sine of 45° is approximately 0.707.) The instantaneous value of voltage or current is greater than the RMS value for half the cycle and less than the RMS value for half the cycle.

Since circuit specifications will most commonly list only RMS voltage and current values, these relationships are important in finding the peak voltages or currents that will stress components.

Example: What is the peak voltage on a capacitor if the RMS voltage of a sinusoidal waveform signal across it is 300 V ac?

$$E_{pk} = 300 \text{ V} \times 1.414 = 424 \text{ V}$$

The capacitor must be able to withstand this higher voltage, plus a safety margin. (The capacitor must also be rated for ac use because of the continually reversing polarity and ac current flow.) In power supplies that convert ac to dc and use capacitive input filters, the output voltage will approach the peak value of the ac voltage rather than the RMS value. (See the **Power Supplies** chapter for more information on specifying components in this application.)

AVERAGE VALUES OF AC VOLTAGE AND CURRENT

Certain kinds of circuits respond to the *average* voltage or current (not power) of an ac waveform. Among these circuits are electrodynamic meter movements and power supplies that convert ac to dc and use heavily inductive ("choke") input filters, both of which work with the pulsating dc output of a full-wave rectifier. The average value of each ac half cycle is the *mean* of all the instantaneous values in that half cycle. (The average value of a sine wave or any symmetric ac waveform over an entire cycle is zero!) Related to the peak values of voltage and current, average values for each half-cycle

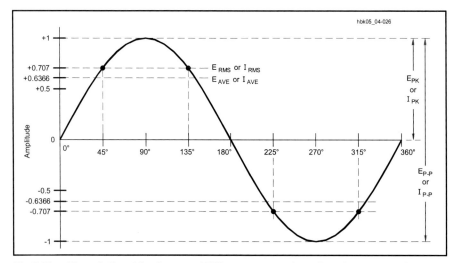

Fig 2.26 — The relationships between RMS, average, peak, and peak-to-peak values of ac voltage and current.

Measuring Nonsinusoidal Waveforms

Making measurements of ac waveforms is covered in more detail in the **Test Equipment and Measurements** chapter. However, this is a good point in the discussion to reinforce the dependence of RMS and values on the nature of the waveform being measured.

Analog meters and other types of instrumentation that display RMS values may only be calibrated for sine waves, those being the most common type of ac waveform. Using that instrumentation to accurately measure waveforms other than sine waves — such as speech, intermittent sine waves (such as CW from a transmitter), square waves, triangle waves or noise — requires the use of *calibration factors* or the measurement may not be valid.

To make calibrated, reliable measurements of the RMS or value of these waveforms requires the use of *true-RMS* instruments. These devices may use a balancing approach to a known dc value or if they are microprocessor-based, may actually perform the full root-mean-square calculation on the waveform. Be sure you know the characteristics of your test instruments if an accurate RMS value is important.

of a sine wave are $2/\pi$ (or 0.6366) times the peak value.

$$E_{ave} = 0.6366 \, E_{pk} \tag{40}$$

$$I_{ave} = 0.6366 \, I_{pk} \tag{41}$$

For convenience, **Table 2.5** summarizes the relationships between all of the common ac values. All of these relationships apply only to sine waves.

COMPLEX WAVEFORMS AND PEAK-ENVELOPE VALUES

Complex waveforms, as shown earlier in Fig 2.18, differ from sine waves. The amplitude of the peak voltage may vary significantly from one cycle to the next, for example. Therefore, other amplitude measures are required, especially for accurate measurement of voltage and power with transmitted speech or data waveforms.

An SSB waveform (either speech or data) contains an RF ac waveform with a frequency many times that of the audio-frequency ac waveform with which it is combined. Therefore, the resultant *composite* waveform appears as an amplitude envelope superimposed upon the RF waveform as illustrated by **Fig 2.27**. For a complex waveform such as this, the *peak envelope voltage* (PEV) is the maximum or peak value of voltage anywhere in the waveform.

Peak envelope voltage is used in the calculation of *peak envelope power* (PEP). The Federal Communications Commission (FCC) sets the maximum power levels for amateur transmitters in terms of peak envelope power. PEP is the *average* power supplied to the antenna transmission line by the transmitter during one RF cycle at the crest of the modulation envelope, taken under normal operating conditions. That is, the average power for the RF cycle during which PEV occurs.

Since calculation of PEP requires the average power of the cycle, and the deviation of the modulated RF waveform from a sine wave is very small, the error incurred by using the conversion factors for sine waves is insignificant. Multiply PEV by 0.707 to obtain an RMS value. Then calculate PEP by using the square of the voltage divided by the load resistance.

$$PEP = \frac{(PEV \times 0.707)^2}{R} \tag{42}$$

Example: What is the PEP of a transmitter's output with a PEV of 100 V into a 50-ohm load?

$$PEP = \frac{(100 \times 0.707)^2}{R} = \frac{(70.7)^2}{50} = 100 \, W$$

2.6.5 – Glossary — AC Theory and Reactance

Admittance (Y) — The reciprocal of impedance, measured in siemens (S).

Capacitance (C) — The ability to store electrical energy in an electrostatic field, measured in farads (F). A device with capacitance is a capacitor.

Flux density (B) — The number of magnetic-force lines per unit area, measured in gauss.

Frequency (f) — The rate of change of an ac voltage or current, measured in cycles per second, or hertz (Hz).

Fundamental — The lowest frequency in a series of sine waves whose frequencies have an integer relationship.

Harmonic — A sine wave whose frequency is an integer multiple of a fundamental frequency.

Impedance (Z) — The complex combination of resistance and reactance, measured in ohms (Ω).

Inductance (L) — The ability to store electrical energy in a magnetic field, measured in henrys (H). A device, such as a coil of wire, with inductance is an inductor.

Peak (voltage or current) — The maximum value relative to zero that an ac voltage or current attains during any cycle.

Peak-to-peak (voltage or current) — The value of the total swing of an ac voltage

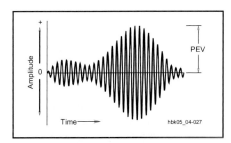

Fig 2.27 — The peak envelope voltage (PEV) for a composite waveform.

Table 2.5
Conversion Factors for Sinusoidal AC Voltage or Current

From	To	Multiply By
Peak	Peak-to-Peak	2
Peak-to-Peak	Peak	0.5
Peak	RMS	$1/\sqrt{2}$ or 0.707
RMS	Peak	$\sqrt{2}$ or 1.414
Peak-to-Peak	RMS	$1/(2 \times \sqrt{2})$ or 0.35355
RMS	Peak-to-Peak	$2 \times \sqrt{2}$ or 2.828
Peak	Average	$2/\pi$ or 0.6366
Average	Peak	$\pi/2$ or 1.5708
RMS	Average	$(2 \times \sqrt{2})/\pi$ or 0.90
Average	RMS	$\pi/(2 \times \sqrt{2})$ or 1.11

Note: These conversion factors apply only to continuous pure sine waves.

or current from its peak negative value to its peak positive value, ordinarily twice the value of the peak voltage or current.

Period (T) — The duration of one ac voltage or current cycle, measured in seconds (s).

Permeability (µ) — The ratio of the magnetic flux density of an iron, ferrite, or similar core in an electromagnet compared to the magnetic flux density of an air core, when the current through the electromagnet is held constant.

Power (P) — The rate of electrical-energy use, measured in watts (W).

Q (quality factor) — The ratio of energy stored in a reactive component (capacitor or inductor) to the energy dissipated, equal to the reactance divided by the resistance.

Reactance (X) — Opposition to alternating current by storage in an electrical field (by a capacitor) or in a magnetic field (by an inductor), measured in ohms (Ω).

Resonance — Ordinarily, the condition in an ac circuit containing both capacitive and inductive reactance in which the reactances are equal.

RMS (voltage or current) — Literally, "root mean square," the square root of the average of the squares of the instantaneous values for one cycle of a waveform. A dc voltage or current that will produce the same heating effect as the waveform. For a sine wave, the RMS value is equal to 0.707 times the peak value of ac voltage or current.

Susceptance (B) — The reciprocal of reactance, measured in siemens (S).

Time constant (τ) — The time required for the voltage in an RC circuit or the current in an RL circuit to rise from zero to approximately 63.2% of its maximum value or to fall from its maximum value 63.2% toward zero.

Toroid — Literally, any donut-shaped solid; most commonly referring to ferrite or powdered-iron cores supporting inductors and transformers.

Transducer — Any device that converts one form of energy to another; for example an antenna, which converts electrical energy to electromagnetic energy or a speaker, which converts electrical energy to sonic energy.

Transformer — A device consisting of at least two coupled inductors capable of transferring energy through mutual inductance.

2.7 Capacitance and Capacitors

It is possible to build up and hold an electrical charge in an *electrostatic field*. This phenomenon is called *capacitance*, and the devices that exhibit capacitance are called *capacitors*. (Old articles and texts use the obsolete term *condenser*.) **Fig 2.28** shows several schematic symbols for capacitors. Part A shows a fixed capacitor; one that has a single value of capacitance. Part B shows the symbol for variable capacitors; these are adjustable over a range of values. If the capacitor is of a type that is *polarized*, meaning that dc voltages must be applied with a specific polarity, the straight line in the symbol should be connected to the most positive voltage, while the curved line goes to the more negative voltage, which is often ground. For clarity, the positive terminal of a polarized capacitor symbol is usually marked with a + symbol. The symbol for *non-polarized* capacitors may be two straight lines or the + symbol may be omitted. When in doubt, consult the capacitor's specifications or the circuits parts list.

2.7.1 Electrostatic Fields and Energy

An *electrostatic field* is created wherever a voltage exists between two points, such as two opposite electric charges or regions that contain different amounts of charge. The field causes electric charges (such as electrons or ions) in the field to feel a force in the direction of the field. If the charges are not free to move, as in an insulator, they store the field's energy as *potential energy*, just as a weight held in place by a surface stores gravitational energy. If the charges are free to move, the field's stored energy is converted to *kinetic energy* of motion just as if the weight is released to fall in a gravitational field.

The field is represented by *lines of force* that show the direction of the force felt by the electric charge. Each electric charge is surrounded by an electric field. The lines of force of the field begin on the charge and extend away from charge into space. The lines of force can terminate on another charge (such as lines of force between a proton and an electron) or they can extend to infinity.

The strength of the electrostatic field is measured in *volts per meter* (V/m). Stronger fields cause the moving charges to accelerate more strongly (just as stronger gravity causes weights to fall faster) and stores more energy in fixed charges. The stronger the field in V/m, the more force an electric charge in the field will feel. The strength of the electric field diminishes with the square of the distance from its source, the electric charge.

2.7.2 The Capacitor

Suppose two flat metal plates are placed close to each other (but not touching) and are connected to a battery through a switch, as illustrated in **Fig 2.29A**. At the instant the switch is closed, electrons are attracted from the upper plate to the positive terminal of the battery, while the same quantity is repelled from the negative battery terminal and pushed into the lower plate. This imbalance of charge creates a voltage between the plates. Eventually, enough electrons move into one plate and out of the other to make the voltage between the plates the same as the battery voltage.

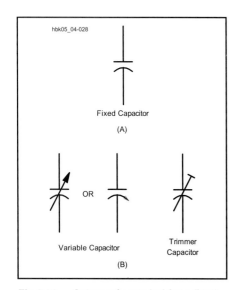

Fig 2.28 — Schematic symbol for a fixed capacitor is shown at A. The symbols for a variable capacitor are shown at B.

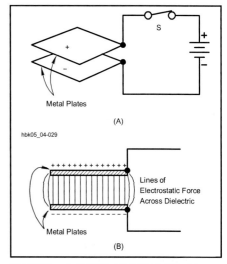

Fig 2.29 — A simple capacitor showing the basic charging arrangement at A, and the retention of the charge due to the electrostatic field at B.

At this point, the voltage between the plates opposes further movement of electrons and no further current flow occurs.

If the switch is opened after the plates have been charged in this way, the top plate is left with a deficiency of electrons and the bottom plate with an excess. Since there is no current path between the two, the plates remain *charged* despite the fact that the battery no longer is connected to a source of voltage. As illustrated in Fig 2.29B, the separated charges create an electrostatic field between the plates. The electrostatic field contains the energy that was expended by the battery in causing the electrons to flow off of or onto the plates. These two plates create a *capacitor*, a device that has the property of storing electrical energy in an electric field, a property called *capacitance*.

The amount of electric charge that is held on the capacitor plates is proportional to the applied voltage and to the capacitance of the capacitor:

$$Q = CV \qquad (43)$$

where
  Q = charge in coulombs,
  C = capacitance in farads (F), and
  V = electrical potential in volts. (The symbol E is also commonly used instead of V in this and the following equation.)

The energy stored in a capacitor is also a function of voltage and capacitance:

$$W = \frac{V^2 C}{2} \qquad (44)$$

where W = energy in joules (J) or watt-seconds.

If a wire is simultaneously touched to the two plates (short circuiting them), the voltage between the plates causes the excess electrons on the bottom plate to flow through the wire to the upper plate, restoring electrical neutrality. The plates are then *discharged*.

Fig 2.30 illustrates the voltage and current in the circuit, first, at the moment the switch is closed to charge the capacitor and, second, at the moment the shorting switch is closed to discharge the capacitor. Note that the periods of charge and discharge are very short, but that they are not zero. This finite charging and discharging time can be controlled and that will prove useful in the creation of timing circuits.

During the time the electrons are moving — that is, while the capacitor is being charged or discharged — a current flows in the circuit even though the circuit apparently is broken by the gap between the capacitor plates. The current flows only during the time of charge and discharge, however, and this time is usually very short. There is no continuous flow of direct current through a capacitor.

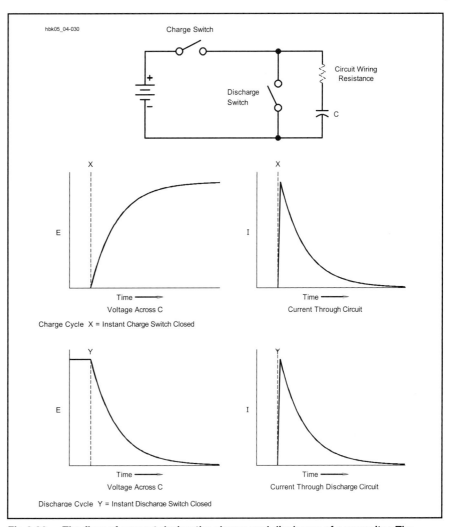

Fig 2.30 — The flow of current during the charge and discharge of a capacitor. The charging graphs assume that the charge switch is closed and the discharge switch is open. The discharging graphs assume just the opposite.

Although dc cannot pass through a capacitor, alternating current can. At the same time one plate is charged positively by the positive excursion of the alternating current, the other plate is being charged negatively at the same rate. (Remember that conventional current is shown as the flow of positive charge, equal to and opposite the actual flow of electrons.) The reverse process occurs during the second half of the cycle as the changing polarity of the applied voltage causes the flow of charge to change direction, as well. The continual flow into and out of the capacitor caused by ac voltage appears as an ac current, although with a phase difference between the voltage and current flow as described below.

UNITS OF CAPACITANCE

The basic unit of capacitance, the ability to store electrical energy in an electrostatic field, is the *farad*. This unit is generally too large for practical radio circuits, although capacitors of several farads in value are used in place of small batteries or as a power supply filter for automotive electronics. Capacitance encountered in radio and electronic circuits is usually measured in microfarads (abbreviated µF), nanofarads (abbreviated nF) or picofarads (pF). The microfarad is one millionth of a farad (10^{-6} F), the nanofarad is one thousandth of a microfarad (10^{-9} F) and the picofarad is one millionth of a microfarad (10^{-12} F). Old articles and texts use the obsolete term micromicrofarad (µµF) in place of picofarad.

CAPACITOR CONSTRUCTION

An ideal capacitor is a pair of parallel metal plates separated by an insulating or *dielectric* layer, ideally a vacuum. The capacitance of a vacuum-dielectric capacitor is given by

$$C = \frac{A \, \varepsilon_r \, \varepsilon_0}{d} \qquad (45)$$

where
  C = capacitance, in farads
  A = area of plates, in cm^2
  d = spacing of the plates in cm

Electrical Fundamentals 2.23

ε_r = dielectric constant of the insulating material

ε_0 = permittivity of free space, 8.85 × 10^{-14} F/cm.

The actual capacitance of such a parallel-plate capacitor is somewhat higher due to *end effect* caused by the electric field that exists just outside the edges of the plates.

The *larger* the plate area and the *smaller* the spacing between the plates, the *greater* the amount of energy that can be stored for a given voltage, and the *greater* the capacitance. The more general name for the capacitor's plates is *electrodes*. However, amateur radio literature generally refers to a capacitor's electrodes as plates and that is the convention in this text.

The amount of capacitance also depends on the material used as insulating material between the plates; capacitance is smallest with air or a vacuum as the insulator. Substituting other insulating materials for air may greatly increase the capacitance.

The ratio of the capacitance with a material other than a vacuum or air between the plates to the capacitance of the same capacitor with air insulation is called the *dielectric constant*, or K, of that particular insulating material. The dielectric constants of a number of materials commonly used as dielectrics in capacitors are given in **Table 2.6**. For example, if polystyrene is substituted for air in a capacitor, the capacitance will be 2.6 times greater.

In practice, capacitors often have more than two plates, with alternating plates being connected in parallel to form two sets, as shown in **Fig 2.31**. This practice makes it possible to obtain a fairly large capacitance in a small space, since several plates of smaller individual area can be stacked to form the equivalent of a single large plate of the same total area. Also, all plates except the two on the ends of the stack are exposed to plates of the other group on both sides, and so are twice as effective in increasing the capacitance.

The formula for calculating capacitance from these physical properties is:

$$C = \frac{0.2248 \, K \, A \, (n-1)}{d} \quad (46)$$

where

C = capacitance in pF,
K = dielectric constant of material between plates,
A = area of one side of one plate in square inches,
d = separation of plate surfaces in inches, and
n = number of plates.

If the area (A) is in square centimeters and the separation (d) is in centimeters, then the formula for capacitance becomes

$$C = \frac{0.0885 \, K \, A \, (n-1)}{d} \quad (47)$$

If the plates in one group do not have the same area as the plates in the other, use the area of the smaller plates.

Example: What is the capacitance of two copper plates, each 1.50 square inches in area, separated by a distance of 0.00500 inch, if the dielectric is air?

$$C = \frac{0.2248 \, K \, A \, (n-1)}{d}$$

$$C = \frac{0.2248 \times 1 \times 1.50 \, (2-1)}{0.00500}$$

$$C = 67.4 \text{ pF}$$

What is the capacitance if the dielectric is mineral oil? (See Table 2.6 for the appropriate dielectric constant.)

$$C = \frac{0.2248 \times 2.23 \times 1.50 \, (2-1)}{0.00500}$$

$$C = 150.3 \text{ pF}$$

2.7.3 Capacitors in Series and Parallel

When a number of capacitors are connected in parallel, as in **Fig 2.32A**, the total capacitance of the group is equal to the sum of the individual capacitances:

$$C_{total} = C1 + C2 + C3 + C4 + ... + C_n \quad (48)$$

When two or more capacitors are con-

Fig 2.31 — A multiple-plate capacitor. Alternate plates are connected to each other, increasing the total area available for storing charge.

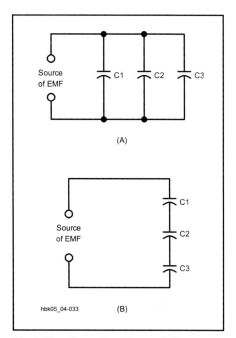

Fig 2.32 — Capacitors in parallel are shown at A, and in series at B.

Table 2.6
Relative Dielectric Constants of Common Capacitor Dielectric Materials

Material	Dielectric Constant (k)	(O)rganic or (I)norganic
Vacuum	1 (by definition)	I
Air	1.0006	I
Ruby mica	6.5 - 8.7	I
Glass (flint)	10	I
Barium titanate (class I)	5 - 450	I
Barium titanate (class II)	200 - 12000	I
Kraft paper	≈ 2.6	O
Mineral Oil	≈ 2.23	O
Castor Oil	≈ 4.7	O
Halowax	≈ 5.2	O
Chlorinated diphenyl	≈ 5.3	O
Polyisobutylene	≈ 2.2	O
Polytetrafluoroethylene	≈ 2.1	O
Polyethylene terephthalate	≈ 3	O
Polystyrene	≈ 2.6	O
Polycarbonate	≈ 3.1	O
Aluminum oxide	≈ 8.4	I
Tantalum pentoxide	≈ 28	I
Niobium oxide	≈ 40	I
Titanium dioxide	≈ 80	I

(Adapted from: Charles A. Harper, *Handbook of Components for Electronics*, p 8-7.)

nected in series, as in Fig 2.32B, the total capacitance is less than that of the smallest capacitor in the group. The rule for finding the capacitance of a number of series-connected capacitors is the same as that for finding the resistance of a number of parallel-connected resistors.

$$C_{total} = \frac{1}{\frac{1}{C1} + \frac{1}{C2} + \frac{1}{C3} + \ldots + \frac{1}{C_n}} \quad (49)$$

For only two capacitors in series, the formula becomes:

$$C_{total} = \frac{C1 \times C2}{C1 + C2} \quad (50)$$

The same units must be used throughout; that is, all capacitances must be expressed in μF, nF or pF, etc. Different units cannot be combined in the same equation.

Capacitors are often connected in parallel to obtain a larger total capacitance than is available in one unit. The voltage rating of capacitors connected in parallel is the lowest voltage rating of any of the capacitors.

When capacitors are connected in series, the applied voltage is divided between them according to Kirchhoff's Voltage Law. The situation is much the same as when resistors are in series and there is a voltage drop across each. The voltage that appears across each series-connected capacitor is inversely proportional to its capacitance, as compared with the capacitance of the whole group. (This assumes ideal capacitors.)

Example: Three capacitors having capacitances of 1, 2 and 4 μF, respectively, are connected in series as shown in **Fig 2.33**. The voltage across the entire series is 2000 V. What is the total capacitance? (Since this is a calculation using theoretical values to illustrate a technique, we will not follow the rules of significant figures for the calculations.)

$$C_{total} = \frac{1}{\frac{1}{C1} + \frac{1}{C2} + \frac{1}{C3}}$$

$$= \frac{1}{\frac{1}{1\,\mu F} + \frac{1}{2\,\mu F} + \frac{1}{4\,\mu F}}$$

$$= \frac{1}{\frac{7}{4\,\mu F}} = \frac{4\,\mu F}{7} = 0.5714\,\mu F$$

The voltage across each capacitor is proportional to the total capacitance divided by the capacitance of the capacitor in question. So the voltage across C1 is:

$$E1 = \frac{0.5714\,\mu F}{1\,\mu F} \times 2000\,V = 1143\,V$$

Similarly, the voltages across C2 and C3 are:

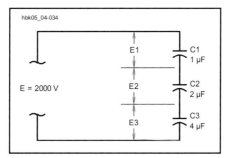

Fig 2.33 — An example of capacitors connected in series. The text shows how to find the voltage drops, E1 through E3.

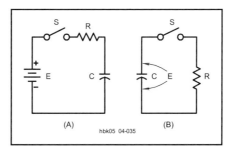

Fig 2.34 — An RC circuit. The series resistance delays the process of charging (A) and discharging (B) when the switch, S, is closed.

$$E2 = \frac{0.5714\,\mu F}{2\,\mu F} \times 2000\,V = 571\,V$$

and

$$E3 = \frac{0.5714\,\mu F}{4\,\mu F} \times 2000\,V = 286\,V$$

The sum of these three voltages equals 2000 V, the applied voltage.

Capacitors may be connected in series to enable the group to withstand a larger voltage than any individual capacitor is rated to withstand. The trade-off is a decrease in the total capacitance. As shown by the previous example, the applied voltage does not divide equally between the capacitors except when all the capacitances are precisely the same. Use care to ensure that the voltage rating of any capacitor in the group is not exceeded. If you use capacitors in series to withstand a higher voltage, you should also connect an "equalizing resistor" across each capacitor as described in the **Power Supplies** chapter.

2.7.4 RC Time Constant

Connecting a dc voltage source directly to the terminals of a capacitor charges the capacitor to the full source voltage almost instantaneously. Any resistance added to the circuit (as R in **Fig 2.34A**) limits the current, lengthening the time required for the voltage between the capacitor plates to build up to the source-voltage value. During this charging period, the current flowing from the source into the capacitor gradually decreases from its initial value. The increasing voltage stored in the capacitor's electric field offers increasing opposition to the steady source voltage.

While it is being charged, the voltage between the capacitor terminals is an exponential function of time, and is given by:

$$V(t) = E\left(1 - e^{-\frac{t}{RC}}\right) \quad (51)$$

where
  V(t) = capacitor voltage at time t,
  E = power source potential in volts,
  t = time in seconds after initiation of charging current,
  e = natural logarithmic base = 2.718,
  R = circuit resistance in ohms, and
  C = capacitance in farads.

(References that explain exponential equations, e, and other mathematical topics are found in the "Radio Math" section of this chapter.)

If t = RC, the above equation becomes:

$$V(RC) = E(1 - e^{-1}) \approx 0.632\,E \quad (52)$$

The product of R in ohms times C in farads is called the *time constant* (also called the *RC time constant*) of the circuit and is the time in seconds required to charge the capacitor to 63.2% of the applied voltage. (The lower-case Greek letter tau, τ, is often used to represent the time constant in electronics circuits.) After two time constants (t = 2τ) the capacitor charges another 63.2% of the difference between the capacitor voltage at one time constant and the applied voltage, for a total charge of 86.5%. After three time constants the capacitor reaches 95% of the applied voltage, and so on, as illustrated in the curve of **Fig 2.35A**. After five time constants, a capacitor is considered fully charged, having reached 99.24% of the applied voltage. Theoretically, the charging process is never really finished, but eventually the charging current drops to an immeasurably small value and the voltage is effectively constant.

If a charged capacitor is discharged through a resistor, as in Fig 2.34B, the same time constant applies to the decay of the capacitor voltage. A direct short circuit applied between the capacitor terminals would discharge the capacitor almost instantly. The resistor, R, limits the current, so a capacitor discharging through a resistance exhibits the same time-constant characteristics (calculated in the same way as above) as a charging capacitor. The voltage, as a function of time while the capacitor is being discharged, is given by:

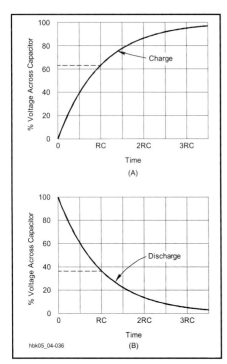

Fig 2.35 — At A, the curve shows how the voltage across a capacitor rises, with time, when charged through a resistor. The curve at B shows the way in which the voltage decreases across a capacitor when discharging through the same resistance. For practical purposes, a capacitor may be considered charged or discharged after five RC periods.

$$V(t) = E \left(e^{-\frac{t}{RC}} \right) \quad (53)$$

where t = time in seconds after initiation of discharge and E is the fully-charged capacitor voltage prior to beginning discharge.

Again, by letting t = RC, the time constant of a discharging capacitor represents a decrease in the voltage across the capacitor of about 63.2%. After five time-constants, the capacitor is considered fully discharged, since the voltage has dropped to less than 1% of the full-charge voltage. **Fig 2.35B** is a graph of the discharging capacitor voltage in terms of time constants.

Time constant calculations have many uses in radio work. The following examples are all derived from practical-circuit applications.

Example 1: A 100-µF capacitor in a high-voltage power supply is shunted by a 100-kΩ resistor. What is the minimum time before the capacitor may be considered fully discharged? Since full discharge is approximately five RC periods,

$$t = 5 \times RC = 5 \times 100 \times 10^3 \, \Omega \times 100 \times 10^{-6} \, F$$

$$= 50000 \times 10^{-3}$$

$$t = 50.0 \, s$$

Note: Although waiting almost a minute for the capacitor to discharge seems safe in this high-voltage circuit, never rely solely on capacitor-discharging resistors (often called *bleeder resistors*). Be certain the power source is removed and the capacitors are totally discharged before touching any circuit components. (See the **Power Supplies** chapter for more information on bleeder resistors.)

Example 2: Smooth CW keying without clicks requires approximately 5 ms (0.005 s) of delay in both the rising and falling edges of the waveform, relative to full charging and discharging of a capacitor in the circuit. What typical values might a builder choose for an RC delay circuit in a keyed voltage line? Since full charge and discharge require 5 RC periods,

$$RC = \frac{t}{5} = \frac{0.005 \, s}{5} = 0.001 \, s$$

Any combination of resistor and capacitor whose values, when multiplied together, equal 0.001 would do the job. A typical capacitor might be 0.05 µF. In that case, the necessary resistor would be:

$$R = \frac{0.001 \, s}{0.05 \times 10^{-6} \, F}$$

$$= 0.02 \times 10^6 \, \Omega = 20000 \, \Omega = 20 \, k\Omega$$

In practice, a builder would use the calculated value as a starting point. The final value would be selected by monitoring the waveform on an oscilloscope.

Example 3: Many modern integrated circuit (IC) devices use RC circuits to control their timing. To match their internal circuitry, they may use a specified threshold voltage as the trigger level. For example, a certain IC uses a trigger level of 0.667 of the supply voltage. What value of capacitor and resistor would be required for a 4.5-second timing period?

First we will solve equation 51 for the time constant, RC. The threshold voltage is 0.667 times the supply voltage, so we use this value for V(t).

$$V(t) = E \left(1 - e^{-\frac{t}{RC}} \right)$$

$$0.667 \, E = E \left(1 - e^{-\frac{t}{RC}} \right)$$

$$e^{-\frac{t}{RC}} = 1 - 0.667$$

$$\ln \left(e^{-\frac{t}{RC}} \right) = \ln (0.333)$$

$$-\frac{t}{RC} = -1.10$$

We want to find a capacitor and resistor combination that will produce a 4.5 s timing period, so we substitute that value for t.

$$RC = \frac{4.5 \, s}{1.10} = 4.1 \, s$$

If we select a value of 10 µF, we can solve for R.

$$R = \frac{4.1 \, s}{10 \times 10^{-6} \, F} = 0.41 \times 10^6 \, \Omega = 410 \, k\Omega$$

A 1% tolerance resistor and capacitor will give good results. You could also use a variable resistor and an accurate method of measuring time to set the circuit to a 4.5 s period.

As the examples suggest, RC circuits have numerous applications in electronics. The number of applications is growing steadily, especially with the introduction of integrated circuits controlled by part or all of a capacitor charge or discharge cycle.

2.7.5 Alternating Current in Capacitance

Whereas a capacitor in a dc circuit will appear as an open circuit except for the brief charge and discharge periods, the same capacitor in an ac circuit will both pass and limit current. A capacitor in an ac circuit does not handle electrical energy like a resistor, however. Instead of converting the energy to heat and dissipating it, capacitors store electrical energy when the applied voltage is greater than that across the capacitor and return it to the circuit when the opposite is true.

In **Fig 2.36** a sine-wave ac voltage having a maximum value of 100 V is applied to a capacitor. In the period OA, the applied voltage increases from 0 to 38, storing energy in the capacitor; at the end of this period the capacitor is charged to that voltage. In interval AB the voltage increases to 71; that is, by an additional 33 V. During this interval a smaller quantity of charge has been added than in OA, because the voltage rise during interval AB is smaller. Consequently the average current during interval AB is smaller than during OA. In the third interval, BC, the voltage rises from 71 to 92, an increase of 21 V. This is less than the voltage increase during AB, so the quantity of charge added is less; in other words, the average current during interval

---

RC Timesaver

When calculating time constants, it is handy to remember that if R is in units of MΩ and C is in units of µF, the result of R × C will be in seconds. Expressed as an equation: MΩ × µF = seconds

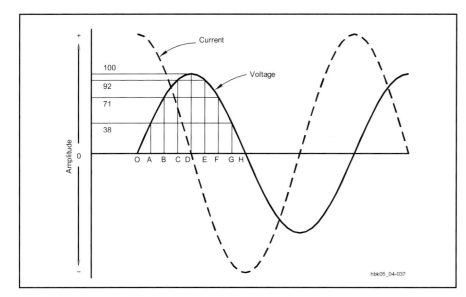

Fig 2.36 — Voltage and current phase relationships when an alternating current is applied to a capacitor.

BC is still smaller. In the fourth interval, CD, the voltage increases only 8 V; the charge added is smaller than in any preceding interval and therefore the current also is smaller.

By dividing the first quarter-cycle into a very large number of such intervals, it can be shown that the current charging the capacitor has the shape of a sine wave, just as the applied voltage does. The current is largest at the beginning of the cycle and becomes zero at the maximum value of the voltage, so there is a phase difference of 90° between the voltage and the current. During the first quarter-cycle the current is flowing in the original (positive) direction through the circuit as indicated by the dashed line in Fig 2.36, since the capacitor is being charged. The increasing capacitor voltage indicates that energy is being stored in the capacitor.

In the second quarter-cycle — that is, in the time from D to H — the voltage applied to the capacitor decreases. During this time the capacitor loses charge, returning the stored energy to the circuit. Applying the same reasoning, it is evident that the current is small in interval DE and continues to increase during each succeeding interval. The current is flowing *against* the applied voltage, however, because the capacitor is returning energy to (discharging into) the circuit. The current thus flows in the *negative* direction during this quarter-cycle.

The third and fourth quarter-cycles repeat the events of the first and second, respectively, although the polarity of the applied voltage has reversed, and so the current changes to correspond. In other words, an alternating current flows in the circuit because of the alternate charging and discharging of the capacitance. As shown in Fig 2.36, the current starts its cycle 90° before the voltage, so the current in a capacitor *leads* the applied voltage by 90°. You might find it helpful to remember the word "ICE" as a mnemonic because the current (I) in a capacitor (C) comes before voltage (E). (see the sidebar "ELI the ICE man" in the section on inductors.) We can also turn this statement around, to say the voltage in a capacitor *lags* the current by 90°.

2.7.6 Capacitive Reactance and Susceptance

The quantity of electric charge that can be placed on a capacitor is proportional to the applied voltage and the capacitance. If the applied voltage is ac, this amount of charge moves back and forth in the circuit once each cycle. Therefore, the rate of movement of charge (the current) is proportional to voltage, capacitance and frequency. Stated in another way, capacitor current is proportional to capacitance for a given applied voltage and frequency.

When the effects of capacitance and frequency are considered together, they form a quantity called *reactance* that relates voltage and current in a capacitor, similar to the role of resistance in Ohm's Law. Because the reactance is created by a capacitor, it is called *capacitive reactance*. The units for reactance are ohms, just as in the case of resistance. Although the units of reactance are ohms, there is no power dissipated in reactance. The energy stored in the capacitor during one portion of the cycle is simply returned to the circuit in the next.

The formula for calculating the reactance of a capacitor at a given frequency is:

$$X_C = \frac{1}{2 \pi f C} \quad (54)$$

where
X_C = capacitive reactance in ohms,
f = frequency in hertz,
C = capacitance in farads
π = 3.1416

Note: In many references and texts, angular frequency $\omega = 2\pi f$ is used and equation 54 would read

$$X_C = \frac{1}{\omega C}$$

Example: What is the reactance of a capacitor of 470 pF (0.000470 µF) at a frequency of 7.15 MHz?

$$X_C = \frac{1}{2 \pi f C}$$

$$= \frac{1}{2 \pi \times 7.15 \text{ MHz} \times 0.000470 \text{ µF}}$$

$$= \frac{1 \Omega}{0.0211} = 47.4 \Omega$$

Example: What is the reactance of the same capacitor, 470 pF (0.000470 µF), at a frequency of 14.29 MHz?

$$X_C = \frac{1}{2 \pi f C}$$

$$= \frac{1}{2 \pi \times 14.3 \text{ MHz} \times 0.000470 \text{ µF}}$$

$$= \frac{1 \Omega}{0.0422} = 23.7 \Omega$$

Current in a capacitor is directly related to the rate of change of the capacitor voltage. The maximum rate of change of voltage in a sine wave increases directly with the frequency, even if its peak voltage remains fixed. Therefore, the maximum current in the capacitor must also increase directly with

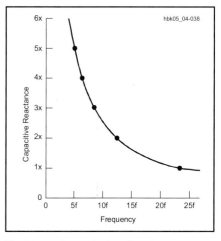

Fig 2.37 — A graph showing the general relationship of reactance to frequency for a fixed value of capacitance.

Capacitive Reactance Timesaver

The fundamental units for frequency and capacitance (hertz and farads) are too cumbersome for practical use in radio circuits. If the capacitance is specified in microfarads (μF) and the frequency is in megahertz (MHz), however, the reactance calculated from equation 54 is in units of ohms (Ω).

Table 2.7
Typical Temperature Coefficients and Leakage Resistances for Various Capacitor Constructions

Type	TC @ 20°C (PPM/°C)	DC Leakage Resistance (Ω)
Ceramic Disc	±300(NP0)	> 10 M
	+150/–1500(GP)	> 10 M
Mica	–20 to +100	> 100,000 M
Polyester	±500	> 10 M
Tantalum Electrolytic	±1500	> 10 MΩ
Small Al Electrolytic (≈ 100 μF)	–20,000	500 k - 1 M
Large Al Electrolytic (≈ 10 mF)	–100,000	10 k
Vacuum (glass)	+100	≈ ∞
Vacuum (ceramic)	+50	≈ ∞

frequency. Since, if voltage is fixed, an increase in current is equivalent to a decrease in reactance, the reactance of any capacitor decreases proportionally as the frequency increases. **Fig 2.37** illustrates the decrease in reactance of an arbitrary-value capacitor with respect to increasing frequency. The only limitation on the application of the graph is the physical construction of the capacitor, which may favor low-frequency uses or high-frequency applications.

CAPACITIVE SUSCEPTANCE

Just as conductance is sometimes the most useful way of expressing a resistance's ability to conduct current, the same is true for capacitors and ac current. This ability is called *susceptance* (abbreviated *B*). The units of susceptance are siemens (S), the same as that of conductance and admittance.

Susceptance in a capacitor is *capacitive susceptance*, abbreviated B_C. In an ideal capacitor with no losses, susceptance is simply the reciprocal of reactance. Hence,

$$B_C = \frac{1}{X_C}$$

where
X_C is the capacitive reactance, and
B_C is the capacitive susceptance.

2.7.7 Characteristics of Capacitors

The ideal capacitor does not conduct any current at dc, dissipates none of the energy stored in it, has the same value at all temperatures, and operates with any amount of voltage applied across it or ac current flowing through it. Practical capacitors deviate considerably from that ideal and have imperfections that must be considered when selecting capacitors and designing circuits that use capacitors. (The characteristics of capacitors at high frequencies is discussed in the **RF Techniques** chapter.)

LEAKAGE RESISTANCE

If we use anything other than a vacuum for the insulating layer, even air, two imperfections are created. Because there are atoms between the plates, some electrons will be available to create a current between the plates when a dc voltage is applied. The magnitude of this *leakage current* will depend on the insulator quality, and the current is usually very small. Leakage current can be modeled by a resistance R_L in parallel with the capacitance (in an ideal capacitor, R_L is infinite). **Table 2.7** shows typical dc leakage resistances for different dielectric materials. Leakage also generally increases with increasing temperature.

CAPACITOR LOSSES

When an ac current flows through the capacitor (even at low frequencies), capacitors dissipate some of the energy stored in the dielectric due to the electromagnetic properties of dielectric materials. This loss can be thought of as a resistance in series with the capacitor and it is often specified in the manufacturer's data for the capacitor as *effective (or equivalent) series resistance (ESR)*.

Loss can also be specified as the capacitor's *loss angle,* θ. (Some literature uses δ for loss angle.) Loss angle is the angle between X_C (the reactance of the capacitor without any loss) and the impedance of the capacitor (impedance is discussed later in this chapter) made up of the combination of ESR and X_C. Increasing loss increases loss angle. The loss angle is usually quite small, and is zero for an ideal capacitor.

Dissipation Factor (DF) or *loss tangent* = tan θ = ESR / X_C and is the ratio of loss resistance to reactance. The loss angle of a given capacitor is relatively constant over frequency, meaning that ESR = (tan θ) / $2\pi f C$ goes down as frequency goes up. For this reason, ESR must be specified at a given frequency.

TOLERANCE AND TEMPERATURE COEFFICIENT

As with resistors, capacitor values vary in production, and most capacitors have a tolerance rating either printed on them or listed on a data sheet. Typical capacitor tolerances and the labeling of tolerance are described in the chapter on **Component Data and References**.

Because the materials that make up a capacitor exhibit mechanical changes with temperature, capacitance also varies with temperature. This change in capacitance with temperature is the capacitor's *temperature coefficient* or *tempco (TC)*. The lower a capacitor's TC, the less its value changes with temperature.

TC is important to consider when constructing a circuit that will carry high power levels, operate in an environment far from room temperature, or must operate consistently at different temperatures. Typical temperature coefficients for several capacitor types are given in Table 2.7. (Capacitor temperature coefficient behaviors are listed in the **Component Data and References** chapter.)

VOLTAGE RATINGS AND BREAKDOWN

When voltage is applied to the plates of a capacitor, force is exerted on the atoms and molecules of the dielectric by the electrostatic field between the plates. If the voltage is high enough, the atoms of the dielectric will ionize (one or more of the electrons will be pulled away from the atom), causing a large dc current to flow discharging the capacitor. This is *dielectric breakdown,* and it is generally destructive to the capacitor because it creates punctures or defects in solid dielectrics that provide permanent low-resistance current paths between the plates. (*Self-healing* dielectrics have the ability to seal off this type of damage.) With most gas dielectrics such as air, once the voltage is removed, the arc ceases and the capacitor is ready for use again.

The *breakdown voltage* of a dielectric depends on the chemical composition and thickness of the dielectric. Breakdown voltage is not directly proportional to the thickness; doubling the thickness does not quite double the breakdown voltage. A thick dielectric must be used to withstand high voltages. Since capacitance is inversely proportional to dielectric thickness (plate spacing) for a given plate area, a high-voltage capacitor must have more plate area than a low-voltage one of the same capacitance. High-voltage, high-capacitance capacitors are therefore physically large.

Dielectric strength is specified in terms of a *dielectric withstanding voltage* (DWV), given in volts per mil (0.001 inch) at a specified temperature. Taking into account the design temperature range of a capacitor and a safety margin, manufacturers specify *dc working voltage* (dcwv) to express the maximum safe limits of dc voltage across a capacitor to prevent dielectric breakdown.

For use with ac voltages, the peak value of ac voltage should not exceed the dc working voltage, unless otherwise specified in component ratings. In other words, the RMS value of sine-wave ac waveforms should be 0.707 times the dcwv value, or lower. With many types of capacitors, further derating is required as the operating frequency increases. An additional safety margin is good practice.

Dielectric breakdown in a gas or air dielectric capacitor occurs as a spark or arc between the plates. Spark voltages are generally given with the units *kilovolts per centimeter*. For air, the spark voltage or V_s may range from more than 120 kV/cm for gaps as narrow as 0.006 cm down to 28 kV/cm for gaps as wide as 10 cm. In addition, a large number of variables enter into the actual breakdown voltage in a real situation. Among the variables are the plate shape, the gap distance, the air pressure or density, the voltage, impurities in the air (or any other dielectric material) and the nature of the external circuit (with air, for instance, the humidity affects conduction on the surface of the capacitor plate).

Dielectric breakdown occurs at a lower voltage between pointed or sharp-edged surfaces than between rounded and polished surfaces. Consequently, the breakdown voltage between metal plates of any given spacing in air can be increased by buffing the edges of the plates. If the plates are damaged so they are no longer smooth and polished, they may have to be polished or the capacitor replaced.

2.7.8 Capacitor Types and Uses

Quite a variety of capacitors are used in radio circuits, differing considerably in physical size, construction and capacitance. Some of the different types are shown in **Fig 2.38** and many other types and packages are available. (See the **Component Data and References** chapter for illustrations of capacitor types and labeling conventions.)

The dielectric determines many properties of the capacitor, although the construction of the plates strongly affects the capacitor's ac performance and some dc parameters. Various materials are used for different reasons such as working voltage and current, availability, cost, and desired capacitance range.

Fixed capacitors having a single, non-adjustable value of capacitance can also be made with metal plates and with air as the dielectric, but are usually constructed from strips of metal foil with a thin, solid or liquid dielectric sandwiched between, so that a relatively large capacitance can be obtained in a small package. Solid dielectrics commonly used in fixed capacitors are plastic films, mica, paper and special ceramics. Two typical types of fixed capacitor construction are shown in **Fig 2.39**.

For capacitors with wire leads, there are two

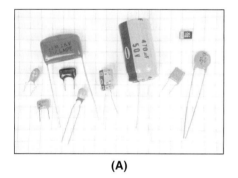

(A)

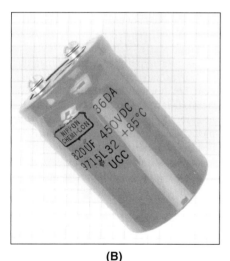

(B)

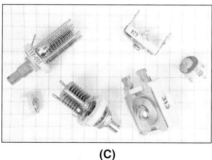

(C)

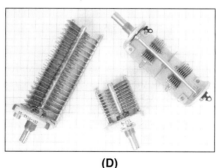

(D)

(E)

Fig 2.38 — Fixed-value capacitors are shown in parts A and B. Aluminum electrolytic capacitors are pictured near the center of photo A. The small tear-drop units to the left of center are tantalum electrolytic capacitors. The rectangular units are silvered-mica, polystyrene film and monolithic ceramic. At the right edge is a disc-ceramic capacitor and near the top right corner is a surface-mount capacitor. B shows a large "computer-grade" electrolytic. These have very low equivalent series resistance (ESR) and are often used as filter capacitors in switch-mode power supplies, and in series-strings for high-voltage supplies of RF power amplifiers. Parts C and D show a variety of variable capacitors, including air variable capacitors and mica compression units. Part E shows a vacuum variable capacitor such as is sometimes used in high-power amplifier circuits. The ¼-inch-ruled graph paper backgrounds provide size comparisons.

Electrical Fundamentals 2.29

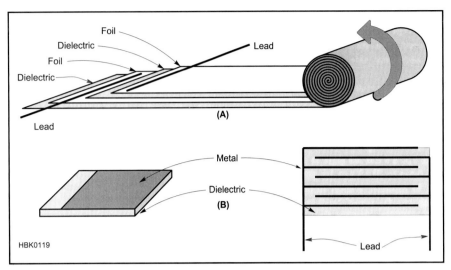

Fig 2.39 — Two common types of capacitor construction. A shows the roll method for film capacitors with axial leads. B shows the alternating layer method for ceramic capacitors. Axial leads are shown in A and radial leads in B.

basic types of lead orientation; *axial* (shown in Fig 2.39A) in which the leads are aligned with the long axis of the capacitor body and *radial* (shown in Fig 2.39B) in which the leads are at right angles to the capacitor's length or width.

Vacuum. Both fixed and variable vacuum capacitors are available. They are rated by their maximum working voltages (3 to 60 kV) and currents. Losses are specified as negligible for most applications. The high working voltage and low losses make vacuum capacitors widely used in transmitting applications. Vacuum capacitors are also unaffected by humidity, moisture, contamination, or dust, unlike air-dielectric capacitors discussed next. This allows them to be used in environments for which air-dielectric capacitors would be unsuitable.

Air. Since $K \approx 1$ for air, air-dielectric capacitors are large when compared to those of the same value using other dielectrics. Their capacitance is very stable over a wide temperature range, leakage losses are low, and therefore a high Q can be obtained. They also can withstand high voltages. Values range from a few tens to hundreds of pF.

For these reasons (and ease of construction) most variable capacitors in tuning circuits are *air-variable* capacitors made with one set of plates movable with respect to the other set to vary the area of overlap and thus the capacitance. A *transmitting-variable* capacitor has heavy plates far enough apart to withstand the high voltages and currents encountered in a transmitter. (Air variable capacitors with more closely-spaced plates are often referred to as *receiving-variables*.)

Plastic film. Capacitors with plastic film (such as polystyrene, polyethylene or Mylar) dielectrics are widely used in bypassing and coupling applications up to several megahertz. They have high leakage resistances (even at high temperatures) and low TCs. Values range from tens of pF to 1 µF. Plastic-film variable capacitors are also available.

Most film capacitors are not polarized; however, the body of the capacitor is usually marked with a color band at one end. The band indicates the terminal that is connected to the outermost plate of the capacitor. This terminal should be connected to the side of the circuit at the lower potential as a safety precaution.

Mica. The capacitance of mica capacitors is very stable with respect to time, temperature and electrical stress. Leakage and losses are very low and they are often used in transmitting equipment. Values range from 1 pF to 0.1 µF. High working voltages are possible, but they must be derated severely as operating frequency increases.

Silver-mica capacitors are made by depositing a thin layer of silver on the mica dielectric. This makes the value even more stable, but it presents the possibility of silver migration through the dielectric. The migration problem worsens with increased dc voltage, temperature and humidity. Avoid using silver-mica capacitors under such conditions. Silver-mica capacitors are often used in RF circuits requiring stable capacitor values, such as oscillators and filters.

Ceramic. Ceramic capacitors are available with values from 1 pF to 1 µF and with voltage ratings up to 1 kV. *Monolithic ceramic* capacitors are constructed from a stack of thin ceramic layers with a metal coating on one side. The layer is then compressed with alternating metal coatings connected together to form the capacitor's plates. The high dielectric constant makes these capacitors physically small for their capacitance, but their value is not as stable and their dielectric properties vary with temperature, applied voltage and operating frequency. They also exhibit piezoelectric behavior. Use them only in coupling and bypass roles. *Disc ceramic* capacitors are made similarly to monolithic ceramic capacitors but with a lower dielectric constant so they are larger and tend to have higher voltage ratings. Ceramic capacitors are useful into the VHF and UHF ranges.

Transmitting ceramic capacitors are made, like transmitting air-variables, with heavy plates and high-voltage ratings. They are relatively large, but very stable and have nearly as low losses as mica capacitors at HF.

Electrolytic. Electrolytic capacitors are constructed with plates made of aluminum-foil strips and a semi-liquid conducting chemical compound between them. They are sometimes called *aluminum electrolytics*. The actual dielectric is a very thin film of insulating material that forms on one set of plates through electrochemical action when a dc voltage is applied to the capacitor. The capacitance of an electrolytic capacitor is very large compared to capacitors having other dielectrics, because the dielectric film is so thin — much thinner than is practical with a solid dielectric. Electrolytic capacitors are available with values from approximately 1 µF to 1 F and with voltage ratings up to hundreds of volts.

Electrolytic capacitors are popular because they provide high capacitance values in small packages at a reasonable cost. Leakage resistance is comparatively low and they are polarized — there is a definite positive and negative plate, due to the chemical reaction that creates the dielectric. Internal inductance restricts aluminum-foil electrolytics to low-frequency applications such as power-supply filtering and bypassing in audio circuits. To maintain the dielectric film, electrolytic capacitors should not be used if the applied dc potential will be well below the capacitor working voltage.

A cautionary note is warranted regarding electrolytic capacitors found in older equipment, both vacuum tube and solid-state. The chemical paste in electrolytics dries out when the component is heated and with age, causing high losses and reduced capacitance. The dielectric film also disappears when the capacitor is not used for long periods. It is possible to "reform" the dielectric by applying a low voltage to an old or unused capacitor and gradually increasing the voltage. However, old electrolytics rarely perform as well as new units. To avoid expensive failures and circuit damage, it is recommended that electrolytic capacitors in old equipment be replaced if they have not been in regular use for more than ten years.

Tantalum. Related to the electrolytic capacitor, *tantalum* capacitors substitute a *slug* of extremely porous tantalum (a rare-earth metallic element) for the aluminum-foil strips as one plate. As in the electrolytic capacitor, the dielectric is an oxide film that forms on the surface of the tantalum. The slug is immersed in a liquid compound contained in a metal can that serves as the other plate. Tantalum capacitors are commonly used with values from 0.1 to several hundred µF and voltage ratings of less than 100 V. Tantalum capacitors are smaller, lighter and more stable, with less leakage and inductance than their aluminum-foil electrolytic counterparts but their cost is higher.

Paper. Paper capacitors are generally not used in new designs and are largely encountered in older equipment; capacitances from 500 pF to 50 µF are available. High working voltages are possible, but paper-dielectric capacitors have low leakage resistances and tolerances are no better than 10 to 20%.

Trimming capacitors. Small-value variable capacitors are often referred to as *trimmers* because they are used for fine-tuning or frequency adjustments, called *trimming*. Trimmers have dielectrics of Teflon, air, or ceramic and generally have values of less than 100 pF. *Compression trimmers* have higher values of up to 1000 pF and are constructed with mica dielectrics.

Oil-filled. Oil-filled capacitors use special high-strength dielectric oils to achieve voltage ratings of several kV. Values of up to 100 µF are commonly used in high-voltage applications such as high-voltage power supplies and energy storage. (See the chapter on **Power Supplies** for additional information about the use of oil-filled and electrolytic capacitors.)

2.8 Inductance and Inductors

A second way to store electrical energy is in a *magnetic field*. This phenomenon is called *inductance*, and the devices that exhibit inductance are called *inductors*. Inductance is derived from some basic underlying magnetic properties.

2.8.1 Magnetic Fields and Magnetic Energy Storage

MAGNETIC FLUX

As an electric field surrounds an electric charge, magnetic fields surround *magnets*. You are probably familiar with metallic bar, disc, or horseshoe-shaped magnets. **Fig 2.40** shows a bar magnet, but particles of matter as small as an atom can also be magnets.

Fig 2.40 also shows the magnet surrounded by lines of force called *magnetic flux*, representing a *magnetic field*. (More accurately, a *magnetostatic field*, since the field is not changing.) Similar to those of an electric field, magnetic lines of force (or *flux lines*) show the direction in which a magnet would feel a force in the field.

There is no "magnetic charge" comparable to positive and negative electric charges. All magnets and magnetic fields have a polarity, represented as *poles*, and every magnet — from atoms to bar magnets — possesses both a *north* and *south pole*. The size of the source of the magnetism makes no difference. The north pole of a magnet is defined as the one attracted to the Earth's north magnetic pole. (Confusingly, this definition means the Earth's North Magnetic Pole is magnetically a south pole!) Like conventional current, the direction of magnetic lines of force was assigned arbitrarily by early scientists as pointing *from* the magnet's north pole *to* the south pole.

An electric field is *open* — that is, its lines of force have one end on an electric charge and can extend to infinity. A magnetic field is *closed* because all magnetic lines of force form a loop passing through a magnet's north and south poles.

Magnetic fields exist around two types of materials; *permanent magnets* and *electromagnets*. Permanent magnets consist of *ferromagnetic* and *ferrimagnetic* materials whose atoms are or can be aligned so as to produce a magnetic field. Ferro- or ferrimagnetic materials are strongly attracted to magnets. They can be *magnetized*, meaning to be made magnetic, by the application of a magnetic field. Lodestone, magnetite, and ferrites are examples of ferrimagnetic materials. Iron, nickel, cobalt, Alnico alloys and other materials are ferromagnetic. Magnetic materials with high *retentivity* form permanent magnets because they retain their magnetic properties for long periods. Other materials, such as soft iron, yield temporary magnets that lose their magnetic properties rapidly.

Paramagnetic substances are very weakly attracted to a magnet and include materials such as platinum, aluminum, and oxygen. *Diamagnetic* substances, such as copper, carbon, and water, are weakly repelled by a magnet.

The second type of magnet is an electrical conductor with a current flowing through it. As shown in **Fig 2.41**, moving electrons are surrounded by a closed magnetic field, illustrated as the circular lines of force around the wire lying in planes perpendicular to the current's motion. The magnetic needle of a compass placed near a wire carrying direct current will be deflected as its poles respond to the forces created by the magnetic field around the wire.

If the wire is coiled into a *solenoid* as shown in **Fig 2.42**, the magnetic field greatly intensi-

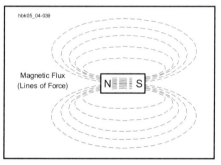

Fig 2.40 — The magnetic field and poles of a permanent magnet. The magnetic field direction is from the north to the south pole.

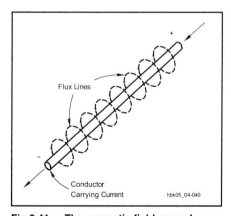

Fig 2.41 — The magnetic field around a conductor carrying an electrical current. If the thumb of your right hand points in the direction of the conventional current (plus to minus), your fingers curl in the direction of the magnetic field around the wire.

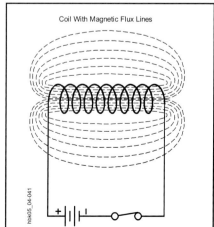

Fig 2.42 — Cross section of an inductor showing its flux lines and overall magnetic field.

Electrical Fundamentals 2.31

fies. This occurs as the magnetic fields from each successive turn in the coil add together because the current in each turn is flowing in the same direction.

Note that the resulting *electromagnet* has magnetic properties identical in principle to those of a permanent magnet, including poles and lines of force or flux. The strength of the magnetic field depends on several factors: the number and shape of turns of the coil, the magnetic properties of the materials surrounding the coil (both inside and out), the length of the coil and the amplitude of the current.

Magnetic fields and electric current have a special two-way relationship: voltage causing an electrical current (moving charges) in a conductor will produce a magnetic field and a moving magnetic field will create an electrical field (voltage) that produces current in a conductor. This is the principle behind motors and generators, converting mechanical energy into electrical energy and vice-versa.

Magnetic flux is measured in the SI units (International System of Units) of *webers* (Wb), which is a volt-second (Wb = V-s). In the *centimeter-gram-second* (*cgs*) metric system units, magnetic flux is measured in *maxwells* (Mx) and 1 Mx = 10^{-8} Wb. The volt-second is used because of the relationship described in the previous paragraph: 1 volt of electromotive force will be created in a loop of wire in which magnetic flux through the loop changes at the rate of 1 weber per second. The relationship between current and magnetic fields is one of motion and change.

MAGNETIC FLUX DENSITY

Magnetic field intensity, known as *flux density*, decreases with the square of the distance from the source, either a magnet or current. Flux density (*B*) is represented in *gauss* (G), where one gauss is equivalent to one line of force per square centimeter of area measured perpendicularly to the direction of the field (G = Mx / cm^2). The gauss is a *cgs* unit. In SI units, flux density is represented by the *tesla* (T), which is one weber per square meter (T = Wb/m^2 and 1T = 10,000 G).

The magnetizing or *magnetomotive force* ($\Im$) that produces a flux or total magnetic field is measured in *gilberts* (Gb). The SI unit of magnetomotive force is the ampere-turn, abbreviated A, just like the ampere. (1 Gb = 0.79577 A)

$$\Im = 0.4\pi N I \qquad (55)$$

where
 $\Im$ = magnetomotive strength in gilberts,
 N = number of turns in the coil creating the field,
 I = dc current in ampres in the coil, and
 π = 3.1416.

MAGNETIC FIELD STRENGTH

The magnetic field strength (*H*) measured in *oersteds* (Oe) produced by any particular magnetomotive force (measured in gilberts) is given by:

$$H = \frac{\Im}{\ell} = \frac{0.4\pi N I}{\ell} \qquad (56)$$

where
 H = magnetic field strength in oersteds, and
 ℓ = mean magnetic path length in centimeters.

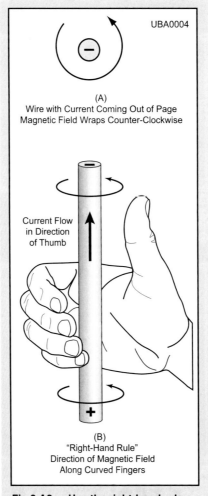

The Right-hand Rule

How do you remember which way the magnetic field around a current is pointing? Luckily, there is a simple method, called the *right-hand rule*. Make your right hand into a fist, then extend your thumb, as in **Fig 2.A2**. If your thumb is pointing in the direction of conventional current flow, then your fingers curl in the same direction as the magnetic field. (If you are dealing with electronic current, use your left hand, instead!)

Fig 2.A2 — Use the right-hand rule to determine magnetic field direction from the direction of current flow.

The *mean magnetic path length* is the average length of the lines of magnetic flux. If the inductor is wound on a closed core as shown in the next section, *l* is approximately the average of the inner and outer circumferences of the core. The SI unit of magnetic field strength is the *ampere-turn per meter*. (1 Oe = 0.0513 A/m) The *cgs* units gilbert and oersted are the units most commonly used for radio circuits.

2.8.2 Magnetic Core Properties

PERMEABILITY

The nature of the material within the coil of an electromagnet, where the lines of force are most concentrated, has the greatest effect upon the magnetic field established by the coil. All core materials are compared relatively to air. The ratio of flux density produced by a given material compared to the flux density produced by an air core is the *permeability* (μ) of the material. Air and non-magnetic materials have a permeability of one.

Suppose the coil in **Fig 2.43** is wound on an iron core having a cross-sectional area of 2 square inches. When a certain current is sent through the coil, it is found that there are 80,000 lines of force in the core. Since the area is 2 square inches, the magnetic flux density is 40,000 lines per square inch. Now suppose that the iron core is removed and the same current is maintained in the coil. Also suppose the flux density without the iron core is found to be 50 lines per square inch. The ratio of these flux densities, iron core to air, is 40,000 / 50 or 800. This ratio is the core's permeability.

Permeabilities as high as 10^6 have been attained. The three most common types of materials used in magnetic cores are these:

A. stacks of thin steel laminations (for power and audio applications, see the dis-

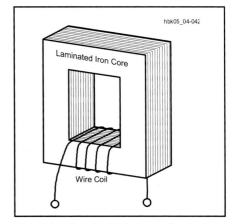

Fig 2.43 — A coil of wire wound around a laminated iron core.

Table 2.8
Properties of Some High-Permeability Materials

Material	Approximate Percent Composition					Maximum Permeability
	Fe	Ni	Co	Mo	Other	
Iron	99.91	—	—	—	—	5000
Purified Iron	99.95	—	—	—	—	180000
4% silicon-iron	96	—	—	—	4 Si	7000
45 Permalloy	54.7	45	—	—	0.3 Mn	25000
Hipernik	50	50	—	—	—	70000
78 Permalloy	21.2	78.5	—	—	0.3 Mn	100000
4-79 Permalloy	16.7	79	—	—	0.3 Mn	100000
Supermalloy	15.7	79	—	5	0.3 Mn	800000
Permendur	49.7	—	50	—	0.3 Mn	5000
2V Permendur	49	—	49	—	2 V	4500
Hiperco	64	—	34	—	2 Cr	10000
2-81 Permalloy*	17	81	—	2	—	130
Carbonyl iron*	99.9	—	—	—	—	132
Ferroxcube III**	($MnFe_2O_4$ + $ZnFe_2O_4$)		1500			

Note: all materials in sheet form except * (insulated powder) and ** (sintered powder).
(Reference: L. Ridenour, ed., *Modern Physics for the Engineer*, p 119.)

cussion on eddy currents below);

B. various ferrite compounds (for cores shaped as rods, toroids, beads and numerous other forms); and

C. powdered iron (shaped as slugs, toroids and other forms for RF inductors).

The permeability of silicon-steel power-transformer cores approaches 5000 in high-quality units. Powdered-iron cores used in RF tuned circuits range in permeability from 3 to about 35, while ferrites of nickel-zinc and manganese-zinc range from 20 to 15000. Not all materials have permeabilities higher than air. Brass has a permeability of less than one. A brass core inserted into a coil will decrease the magnetic field compared to an air core.

Table 2.8 lists some common magnetic materials, their composition and their permeabilities. Core materials are often frequency sensitive, exhibiting excessive losses outside the frequency band of intended use. (Ferrite materials are discussed separately in a later section of the chapter on **RF Techniques**.)

As a measure of the ease with which a magnetic field may be established in a material as compared with air, permeability (μ) corresponds roughly to electrical conductivity. Higher permeability means that it is easier to establish a magnetic field in the material. Permeability is given as:

$$\mu = \frac{B}{H} \quad (57)$$

where
  B is the flux density in gauss, and
  H is the magnetic field strength in oersteds.

RELUCTANCE

That a force (the magnetomotive force) is required to produce a given magnetic field strength implies that there is some opposition to be overcome. This opposition to the creation of a magnetic field is called *reluctance*. Reluctance ($\Re$) is the reciprocal of permeability and corresponds roughly to resistance in an electrical circuit. Carrying the electrical resistance analogy a bit further, the magnetic equivalent of Ohm's Law relates reluctance, magnetomotive force, and flux density: $\Re = \Im / B$.

SATURATION

Unlike electrical conductivity, which is independent of other electrical parameters, the permeability of a magnetic material varies with the flux density. At low flux densities (or with an air core), increasing the current through the coil will cause a proportionate increase in flux. This occurs because the current passing through the coil forces the atoms of the iron (or other material) to line up, just like many small compass needles. The magnetic field that results from the atomic alignment is *much* larger than that produced by the current with no core. As more and more atoms align, the magnetic flux density also increases.

At very high flux densities, increasing the current beyond a certain point may cause no appreciable change in the flux because all of the atoms are aligned. At this point, the core is said to be *saturated*. Saturation causes a rapid decrease in permeability, because it decreases the ratio of flux lines to those obtainable with the same current using an air core. **Fig 2.44** displays a typical permeability curve, showing the region of saturation. The saturation point varies with the makeup of different magnetic materials. Air and other nonmagnetic materials do not saturate.

HYSTERESIS

Retentivity in magnetic core materials is caused by atoms retaining their alignment from an applied magnetizing force. Retentivity is desirable if the goal is to create a permanent magnet. In an electronic circuit, however, the changes caused by retentivity cause the properties of the core material to depend on the history of how the magnetizing force was applied.

Fig 2.45 illustrates the change of flux density (*B*) with a changing magnetizing force (*H*). From starting point A, with no flux in the core, the flux reaches point B at the maximum magnetizing force. As the force decreases, so too does the flux, but it does not reach zero simultaneously with the force at point D. As the force continues in the opposite direction, it brings the flux density to point C. As the force decreases to zero, the flux once more lags behind. This occurs because some of the atoms in core retain their alignment, even after the external magnetizing force is removed. This creates *residual flux* that is present even

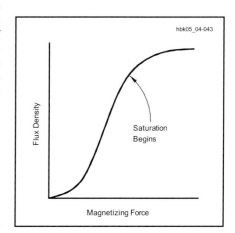

Fig 2.44 — A typical permeability curve for a magnetic core, showing the point where saturation begins.

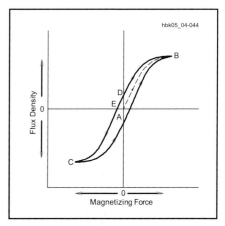

Fig 2.45 — A typical hysteresis curve for a magnetic core, showing the additional energy needed to overcome residual flux.

with no applied magnetizing force. This is the property of *hysteresis*.

In effect, a *coercive force* is necessary to reverse or overcome the residual magnetism retained by the core material. If a circuit carries a large ac current (that is, equal to or larger than saturation), the path shown in Fig 2.45 will be retraced with every cycle and the reversing force each time. The result is a power loss to the magnetic circuit, which appears as heat in the core material. Air cores are immune to hysteresis effects and losses.

2.8.3 Inductance and Direct Current

In an electrical circuit, any element whose operation is based on the transfer of energy into and out of magnetic fields is called an *inductor* for reasons to be explained shortly. **Fig 2.46** shows schematic-diagram symbols and photographs of a few representative inductors. The photograph shows an air-core inductor, a slug-tuned (variable-core) inductor with a nonmagnetic core and an inductor with a magnetic (iron) core. Inductors are often called *coils* because of their construction.

As explained above, when current flows through any conductor — even a straight wire — a magnetic field is created. The transfer of energy to the magnetic field represents work performed by the source of the voltage. Power is required for doing work, and since power is equal to current multiplied by voltage, there must be a voltage drop across the inductor while energy is being stored in the field. This voltage drop, exclusive of any voltage drop caused by resistance in the conductor, is the result of an opposing voltage created in the conductor while the magnetic field is building up to its final value. Once the field becomes constant, the *induced voltage* or *back-voltage* disappears, because no further energy is being stored. Back voltage is analogous to the opposition to current flow in a capacitor from the increasing capacitor voltage.

The induced voltage opposes the voltage of the source, preventing the current from rising rapidly when voltage is applied. **Fig 2.47A** illustrates the situation of energizing an inductor or magnetic circuit, showing the relative amplitudes of induced voltage and the delayed rise in current to its full value.

The amplitude of the induced voltage is proportional to the rate at which the current changes (and consequently, the rate at which the magnetic field changes) and to a constant associated with the inductor itself, *inductance* (*L*). (*Self-inductance* is sometimes used to distinguish between mutual inductance as described below.) The basic unit of inductance is the *henry* (abbreviated H).

Rate of Change

The symbol Δ represents change in the following variable, so that ΔI represents "change in current" and Δt "change in time". A rate of change per unit of time is often expressed in this manner. When the amount of time over which the change is measured becomes very small, the letter *d* replaces Δ in both the numerator and denominator to indicate infinitesimal changes. This notation is used in the derivation and presentation of the functions that describe the behavior of electric circuits.

$$V = L \frac{\Delta I}{\Delta t} \qquad (58)$$

where

V is the induced voltage in volts,
L is the inductance in henries, and
$\Delta I/\Delta t$ is the rate of change of the current in amperes per second.

An inductance of 1 H generates an induced voltage of one volt when the inducing current is varying at a rate of one ampere per second.

The energy stored in the magnetic field of an inductor is given by the formula:

$$W = \frac{I^2 L}{2} \qquad (59)$$

where

W = energy in joules,
I = current in amperes, and
L = inductance in henrys.

This formula corresponds to the energy-storage formula for capacitors: energy storage is a function of current squared. Inductance is proportional to the amount of energy stored in an inductor's magnetic field for a given amount of current. The magnetic field strength, H, is proportional to the number of turns in the inductor's winding, N, (see equation 56) and for a given amount of current, to the value of μ for the core. Thus, inductance is directly proportional to both N and μ.

The polarity of the induced voltage is always such as to oppose any change in the circuit current. (This is why the term "back" is used, as in back-voltage or *back-EMF* for this reason.) This means that when the current in the circuit is increasing, work is being done against the induced voltage by storing energy in the magnetic field. Likewise, if the current in the circuit tends to decrease, the stored energy of the field returns to the circuit, and adds to the energy being supplied by the voltage source. The net effect of storing and releasing energy is that inductors oppose changes in current just as capacitors oppose changes in voltage. This phenomenon tends to keep the current flowing even though the applied voltage may be decreasing or be removed entirely. Fig 2.47B illustrates the decreasing but continuing flow of current caused by the induced voltage after the source voltage is removed from the circuit.

Inductance depends on the physical configuration of the inductor. All conductors, even straight wires, have inductance. Coiling a conductor increases its inductance. In effect, the growing (or shrinking) magnetic field of each turn produces magnetic lines of force that — in their expansion (or contraction) — intercept the other turns of the coil, inducing a voltage in every other turn. (Recall the two-way relationship between a changing magnetic field and the voltage it creates in a conductor.) The mutuality of the effect, called *magnetic flux linkage* (ψ), multiplies

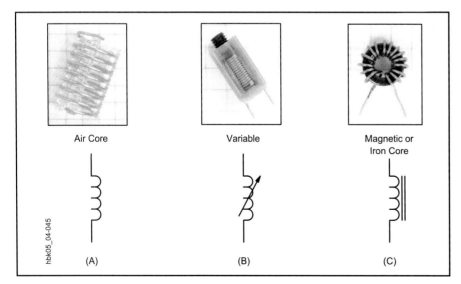

Fig 2.46 — Photos and schematic symbols for representative inductors. A, an air-core inductor; B, a variable inductor with a nonmagnetic slug and C, an inductor with a toroidal magnetic core. The ¼-inch-ruled graph paper background provides a size comparison.

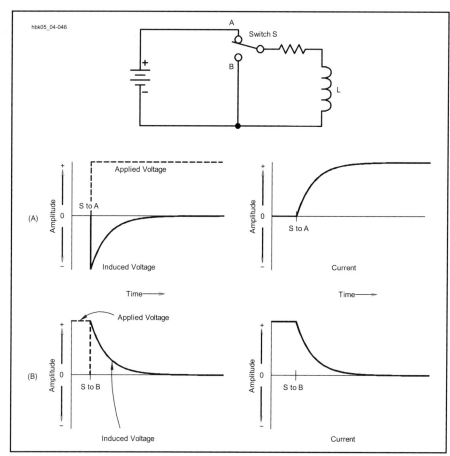

Fig 2.47 — Inductive circuit showing the generation of induced voltage and the rise of current when voltage is applied to an inductor at A, and the decay of current as the coil shorted at B.

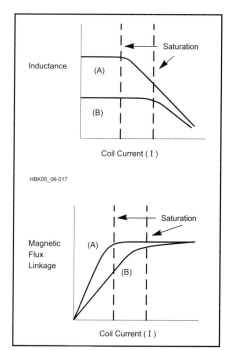

Fig 2.48 — Magnetic flux linkage and inductance plotted versus coil current for (A) a typical iron-core inductor. As the flux linkage $N\phi$ in the coil saturates, the inductance begins to decrease since inductance = flux linkage / current. The curves marked B show the effect of adding an air gap to the core. The current-handling capability has increased, but at the expense of reduced inductance.

the ability of the coiled conductor to store magnetic energy.

A coil of many turns will have more inductance than one of few turns, if both coils are otherwise physically similar. Furthermore, if an inductor is placed around a magnetic core, its inductance will increase in proportion to the permeability of that core, if the circuit current is below the point at which the core saturates.

In various aspects of radio work, inductors may take values ranging from a fraction of a nanohenry (nH) through millihenrys (mH) up to about 20 H.

EFFECTS OF SATURATION

An important concept for using inductors is that as long as the coil current remains below saturation, the inductance of the coil is essentially constant. **Fig 2.48** shows graphs of magnetic flux linkage (ψ) and inductance (L) vs. current (I) for a typical iron-core inductor both saturated and non-saturated. These quantities are related by the equation

$$\psi = N\phi = LI \qquad (60)$$

where
 ψ = the flux linkage

N = number of turns,
ϕ = flux density in webers
L = inductance in henrys, and
I = current in amperes.

In the lower graph, a line drawn from any point on the curve to the (0,0) point will show the effective inductance, $L = N\phi / I$, at that current. These results are plotted on the upper graph.

Note that below saturation, the inductance is constant because both ψ and I are increasing at a steady rate. Once the saturation current is reached, the inductance decreases because ψ does not increase anymore (except for the tiny additional magnetic field the current itself provides).

One common method of increasing the saturation current level is to cut a small air gap in the core (see **Fig 2.49**). This gap forces the flux lines to travel through air for a short distance, reducing the permeability of the core. Since the saturation flux linkage of the core is unchanged, this method works by requiring a higher current to achieve saturation. The price that is paid is a reduced inductance below saturation. The curves in Fig 2.48B show the result of an air gap added to that inductor.

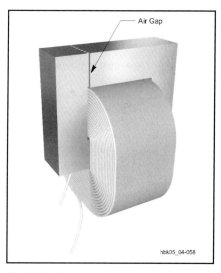

Fig 2.49 — Typical construction of a magnetic-core inductor. The air gap greatly reduces core saturation at the expense of reducing inductance. The insulating laminations between the core layers help to minimize eddy currents, as well.

Manufacturer's data sheets for magnetic cores usually specify the saturation flux density. Saturation flux density (ϕ) in gauss can be calculated for ac and dc currents from the following equations:

Electrical Fundamentals 2.35

$$\phi_{ac} = \frac{3.49\,V}{fNA}$$

$$\phi_{dc} = \frac{NIA_L}{10A}$$

where
V = RMS ac voltage
f = frequency, in MHz
N = number of turns
A = equivalent area of the magnetic path in square inches (from the data sheet)
I = dc current, in amperes, and
A_L = inductance index (also from the data sheet).

2.8.4 Mutual Inductance and Magnetic Coupling

MUTUAL INDUCTANCE

When two inductors are arranged with their axes aligned as shown in **Fig 2.50**, current flowing in through inductor 1 creates a magnetic field that intercepts inductor 2. Consequently, a voltage will be induced in inductor 2 whenever the field strength of inductor 1 is changing. This induced voltage is similar to the voltage of self-induction, but since it appears in the second inductor because of current flowing in the first, it is a mutual effect and results from the *mutual inductance* between the two inductors.

When all the flux set up by one coil intercepts all the turns of the other coil, the mutual inductance has its maximum possible value. If only a small part of the flux set up by one coil intercepts the turns of the other, the mutual inductance is relatively small. Two inductors having mutual inductance are said to be *coupled*.

The ratio of actual mutual inductance to the maximum possible value that could theoretically be obtained with two given inductors is called the *coefficient of coupling* between the inductors. It is expressed as a percentage or as a value between 0 and 1. Inductors that have nearly the maximum possible mutual inductance (coefficient = 1 or 100%) are said to be closely, or tightly, coupled. If the mutual inductance is relatively small the inductors are said to be loosely coupled. The degree of coupling depends upon the physical spacing between the inductors and how they are placed with respect to each other. Maximum coupling exists when they have a common or parallel axis and are as close together as possible (for example, one wound over the other). The coupling is least when the inductors are far apart or are placed so their axes are at right angles.

The maximum possible coefficient of coupling is closely approached when the two inductors are wound on a closed iron core. The coefficient with air-core inductors may run as high as 0.6 or 0.7 if one inductor is

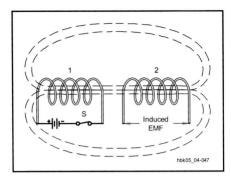

Fig 2.50 — Mutual inductance: When S is closed, current flows through coil number 1, setting up a magnetic field that induces a voltage in the turns of coil number 2.

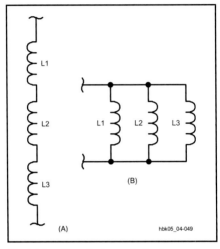

Fig 2.51 — Part A shows inductances in series, and Part B shows inductances in parallel.

wound over the other, but will be much less if the two inductors are separated. Although unity coupling is suggested by Fig 2.50, such coupling is possible only when the inductors are wound on a closed magnetic core.

Coupling between inductors can be minimized by using separate closed magnetic cores for each. Since an inductor's magnetic field is contained almost entirely in a closed core, two inductors with separate closed cores, such as the toroidal inductor in Fig 2.46 C, can be placed close together in almost any relative orientation without coupling.

UNWANTED COUPLING

The inductance of a short length of straight wire is small, but it may not be negligible. (In free-space, round wire has an inductance on the order of 1 µH/m, but this is affected by wire diameter and the total circuit's physical configuration.) Appreciable voltage may be induced in even a few inches of wire carrying ac by changing magnetic fields with a frequency on the order of 100 MHz or higher. At much lower frequencies or at dc, the inductance of the same wire might be ignored because the induced voltage would be very small.

There are many phenomena, both natural and man-made, that create sufficiently strong or rapidly-changing magnetic fields to induce voltages in conductors. Many of them create brief but intense pulses of energy called *transients* or "spikes." The magnetic fields from these transients intercept wires leading into and out of — and wires wholly within — electronic equipment, inducing unwanted voltages by mutual coupling.

Lightning is a powerful natural source of magnetically-coupled transients. Strong transients can also be generated by sudden changes in current in nearby circuits or wiring. High-speed digital signals and pulses can also induce voltages in adjacent conductors.

2.8.5 Inductances in Series and Parallel

When two or more inductors are connected in series (**Fig 2.51A**), the total inductance is equal to the sum of the individual inductances, provided that the inductors are sufficiently separated so that there is no coupling between them (see the preceding section). That is:

$$L_{total} = L1 + L2 + L3 \ldots + L_n \qquad (61)$$

If inductors are connected in parallel (Fig 2.51B), again assuming no mutual coupling, the total inductance is given by:

$$L_{total} = \frac{1}{\frac{1}{L1} + \frac{1}{L2} + \frac{1}{L3} + \ldots + \frac{1}{L_n}} \qquad (62)$$

For only two inductors in parallel, the formula becomes:

$$L_{total} = \frac{L1 \times L2}{L1 + L2} \qquad (63)$$

Thus, the rules for combining inductances in series and parallel are the same as those for resistances, assuming there is no coupling between the inductors. When there is coupling between the inductors, the formulas given above will not yield correct results.

2.8.6 RL Time Constant

As with capacitors, the time dependence of inductor current is a significant property. A comparable situation to an RC circuit exists when resistance and inductance are connected in series. In **Fig 2.52**, first consider the case in which R is zero. Closing S1 sends a current through the circuit. The instantaneous transition from no current to a finite value, however small, represents a rapid change in current, and an opposing voltage is induced in L. The value of the opposing voltage is almost equal to the applied voltage, so the resulting initial current is very small.

The opposing voltage is created by change

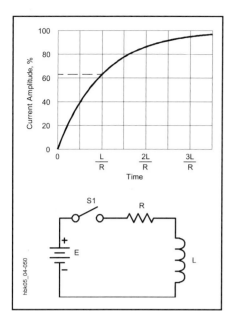

Fig 2.52 — Time constant of an RL circuit being energized.

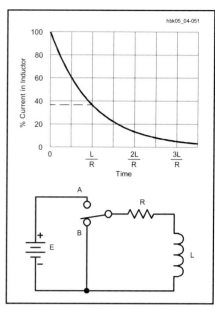

Fig 2.53 — Time constant of an RL circuit being de-energized. This is a theoretical model only, since a mechanical switch cannot change state instantaneously.

I(t) when t becomes very large; this is the *steady-state value* of I. If t = L/R, the above equation becomes:

$$V(L/R) = \frac{E}{R}(1 - e^{-1}) \approx 0.632 \frac{E}{R} \quad (65)$$

The time in seconds required for the current to build up to 63.2% of the maximum value is called the *time constant* (also the *RL time constant*), and is equal to L/R, where L is in henrys and R is in ohms. (Time constants are also discussed in the section on RC circuits above.) After each time interval equal to this constant, current increases by an additional 63.2% closer to the final value of E/R. This behavior is graphed in Fig 2.52. As is the case with capacitors, after five time constants the current is considered to have reached its maximum value. As with capacitors, we often use the lower-case Greek tau (τ) to represent the time constant.

Example: If a circuit has an inductor of 5.0 mH in series with a resistor of 10 Ω, how long will it take for the current in the circuit to reach full value after power is applied? Since achieving maximum current takes approximately five time constants,

$$t = \frac{5L}{R} = \frac{5 \times 5.0 \times 10^{-3} \text{ H}}{10 \, \Omega}$$

$$= 2.5 \times 10^{-3} \text{ seconds} = 2.5 \text{ ms}$$

Note that if the inductance is increased to 5.0 H, the required time increases by a factor of 1000 to 2.5 seconds. Since the circuit resistance didn't change, the final current is the same for both cases in this example. Increasing inductance increases the time required to reach full current.

Zero resistance would prevent the circuit from ever achieving full current. All practical inductors have some resistance in the wire making up the inductor.

An inductor cannot be discharged in the simple circuit of Fig 2.52 because the magnetic field ceases to exist or "collapses" as soon as the current ceases. Opening S1 does not leave the inductor charged in the way that a capacitor would remain charged. Energy storage in a capacitor depends on the separated charges staying in place. Energy storage in an inductor depends on the charges continuing to move as current.

The energy stored in the inductor's magnetic field attempts to return instantly to the circuit when S1 is opened. The rapidly changing (collapsing) field in the inductor causes a very large voltage to be induced across the inductor. Because the change in current is now in the opposite direction, the induced voltage also reverses polarity. This induced voltage (called *inductive kick-back*) is usually many times larger than the originally applied voltage, because the induced voltage is proportional to the rate at which the field changes.

in the inductor current and would cease to exist if the current did not continue to increase. With no resistance in the circuit, the current would increase forever, always growing just fast enough to keep the self-induced opposing voltage just below the applied voltage.

When resistance in the circuit limits the current, the opposing voltage induced in L must only equal the difference between E and the drop across R, because that is the voltage actually applied to L. This difference becomes smaller as the current approaches its final value, limited by Ohm's Law to I = E/R. Theoretically, the opposing voltage never quite disappears, and so the current never quite reaches the Ohm's Law limit. In practical terms, the difference eventually becomes insignificant, just as described above for capacitors charging to an applied voltage through a resistor.

The inductor current at any time after the switch in Fig 2.52 has been closed, can be found from:

$$I(t) = \frac{E}{R}\left(1 - e^{-\frac{tR}{L}}\right) \quad (64)$$

where
I(t) = current in amperes at time t,
E = power source potential in volts,
t = time in seconds after application of voltage,
e = natural logarithmic base = 2.718,
R = circuit resistance in ohms, and
L = inductance in henrys.

(References that explain exponential equations, e, and other mathematical topics are found in the "Radio Math" section of this chapter.) The term E/R in this equation represents the dc value of I, or the value of

The common result of opening the switch in such a circuit is that a spark or arc forms at the switch contacts during the instant the switch opens. When the inductance is large and the current in the circuit is high, large amounts of energy are released in a very short time. It is not at all unusual for the switch contacts to burn or melt under such circumstances.

The spark or arc at the opened switch can be reduced or suppressed by connecting a suitable capacitor and resistor in series across the contacts to absorb the energy non-destructively. Such an RC combination is called a *snubber network*. The current rating for a switch may be significantly reduced if it is used in an inductive circuit.

Transistor switches connected to and controlling inductors, such as relays and solenoids, also require protection from the high kick-back voltages. In most cases, a small power diode connected across the relay coil so that it does not conduct current when the inductor is energized (called a *kick-back diode*) will protect the transistor.

If the excitation is removed without breaking the circuit, as shown in **Fig 2.53**, the current will decay according to the formula:

$$I(t) = \frac{E}{R}\left(e^{-\frac{tR}{L}}\right) \quad (66)$$

where t = time in seconds after removal of the source voltage.

After one time constant the current will decay by 63.2% of its steady-state value. (It

Electrical Fundamentals   2.37

will decay to 36.8% of the steady-state value.) The graph in Fig 2.53 shows the current-decay waveform to be identical to the voltage-discharge waveform of a capacitor. Be careful about applying the terms *charge* and *discharge* to an inductive circuit, however. These terms refer to energy storage in an electric field. An inductor stores energy in a magnetic field and the usual method of referring to the process is *energize* and *de-energize* (although it is not always followed).

2.8.7 Alternating Current in Inductors

For reasons similar to those that cause a phase difference between current and voltage in a capacitor, when an alternating voltage is applied to an ideal inductance with no resistance, the current is 90° out of phase with the applied voltage. In the case of an inductor, however, the current *lags* 90° behind the voltage as shown in **Fig 2.54**, the opposite of the capacitor current-voltage relationship. (Here again, we can also say the voltage across an inductor *leads* the current by 90°.) Interpreting Fig 2.54 begins with understanding that the cause for current lag in an inductor is the opposing voltage that is induced in the inductor and that the amplitude of the opposing voltage is proportional to the rate at which the inductor current changes.

In time segment OA, when the applied voltage is at its positive maximum, the rate at which the current is changing is also the highest, a 38% change. This means that the opposing voltage is also maximum, allowing the least current to flow. In the segment AB, as a result of the decrease in the applied voltage, current changes by only 33% inducing a smaller opposing voltage. The process continues in time segments BC and CD, the latter producing only an 8% rise in current as the applied and induced opposing voltage approach zero.

In segment DE, the applied voltage changes polarity, causing current to begin to decrease, returning stored energy to the circuit from the inductor's magnetic field. As the current rate of change is now negative (decreasing) the induced opposing voltage also changes polarity. Current flow is still in the original direction (positive), but is decreasing as less energy is stored in the inductor.

As the applied voltage continues to increase negatively, the current — although still positive — continues to decrease in value, reaching zero as the applied voltage reaches its negative maximum. The energy once stored in the inductor has now been completely returned to the circuit. The negative half-cycle then continues just as the positive half-cycle.

Similarly to the capacitive circuit discussed earlier, by dividing the cycle into a large number of intervals, it can be shown that the current and voltage are both sine waves, although with a difference in phase.

ELI the ICE Man

If you have difficulty remembering the phase relationships between voltage and current with inductors and capacitors, you may find it helpful to think of the phrase, "ELI the ICE man." This will remind you that voltage across an inductor leads the current through it, because the E comes before (leads) I, with an L between them, as you read from left to right. (The letter L represents inductance.) Similarly, I comes before (leads) E with a C between them.

Compare Fig 2.54 with Fig 2.36. Whereas in a pure capacitive circuit, the current *leads* the voltage by 90°, in a pure inductive circuit, the current *lags* the voltage by 90°. These phenomena are especially important in circuits that combine inductors and capacitors. Remember that the phase difference between voltage and current in both types of circuits is a result of energy being stored and released as voltage across a capacitor and as current in an inductor.

EDDY CURRENT

Since magnetic core material is usually conductive, the changing magnetic field produced by an ac current in an inductor also induces a voltage in the core. This voltage causes a current to flow in the core. This *eddy current* (so-named because it moves in a closed path, similarly to eddy currents in water) serves no useful purpose and results in energy being dissipated as heat from the core's resistance. Eddy currents are a particular problem in inductors with iron cores. Cores made of thin strips of magnetic material, called *laminations*, are used to reduce eddy currents. (See also the section on Practical Inductors elsewhere in the chapter.)

2.8.8 Inductive Reactance and Susceptance

The amount of current that can be created in an inductor is proportional to the applied voltage but inversely proportional to the inductance because of the induced opposing voltage. If the applied voltage is ac, the rate of change of the current varies directly with the frequency and this rate of change also determines the amplitude of the induced or reverse voltage. Hence, the opposition to the flow of current increases proportionally to frequency. Stated in another way, inductor current is inversely proportional to inductance for a given applied voltage and frequency.

The combined effect of inductance and frequency is called *inductive reactance*, which — like capacitive reactance — is expressed in ohms. As with capacitive reactance, no power is dissipated in inductive reactance. The energy stored in the inductor during one portion of the cycle is returned to the circuit in the next portion.

The formula for inductive reactance is:

$$X_L = 2 \pi f L \tag{67}$$

where
X_L = inductive reactance in ohms,
f = frequency in hertz,
L = inductance in henrys, and
π = 3.1416.
(If $\omega = 2 \pi f$, then $X_L = \omega L$.)

Example: What is the reactance of an inductor having an inductance of 8.00 H at a frequency of 120 Hz?

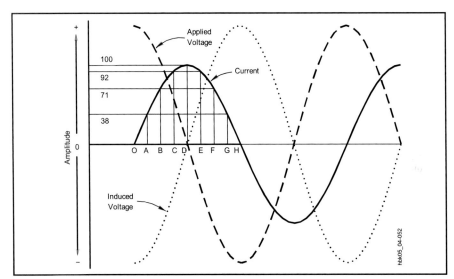

Fig 2.54 — Phase relationships between voltage and current when an alternating current is applied to an inductance.

$X_L = 2 \pi f L$
$= 6.2832 \times 120 \text{ Hz} \times 8.0 \text{ H}$
$= 6030 \, \Omega$

Example: What is the reactance of a 15.0-microhenry inductor at a frequency of 14.0 MHz?

$X_L = 2 \pi f L$
$= 6.2832 \times 14.0 \text{ MHz} \times 15.0 \, \mu\text{H}$
$= 1320 \, \Omega$

The resistance of the wire used to wind the inductor has no effect on the reactance, but simply acts as a separate resistor connected in series with the inductor.

Example: What is the reactance of the same inductor at a frequency of 7.0 MHz?

$X_L = 2 \pi f L$
$= 6.2832 \times 7.0 \text{ MHz} \times 15.0 \, \mu\text{H}$
$= 660 \, \Omega$

Inductive Reactance Timesaver

Similarly to the calculation of capacitive reactance, if inductance is specified in microhenrys (µH) and the frequency is in megahertz (MHz), the reactance calculated from equation 66 is in units of ohms (Ω). The same is true for the combination of mH and kHz.

The direct relationship between frequency and reactance in inductors, combined with the inverse relationship between reactance and frequency in the case of capacitors, will be of fundamental importance in creating resonant circuits.

INDUCTIVE SUSCEPTANCE

As a measure of the ability of an inductor to limit the flow of ac in a circuit, inductive reactance is similar to capacitive reactance in having a corresponding *susceptance*, or ability to pass ac current in a circuit. In an ideal inductor with no resistive losses — that is, no energy lost as heat — susceptance is simply the reciprocal of reactance.

$$B = \frac{1}{X_L} \qquad (68)$$

where
X_L = reactance, and
B = susceptance.

The unit of susceptance for both inductors and capacitors is the *siemens*, abbreviated S.

2.9 Working with Reactance

2.9.1 Ohm's Law for Reactance

Only ac circuits containing capacitance or inductance (or both) have reactance. Despite the fact that the voltage in such circuits is 90° out of phase with the current, circuit reactance does oppose the flow of ac current in a manner that corresponds to resistance. That is, in a capacitor or inductor, reactance is equal to the ratio of ac voltage to ac current and the equations relating voltage, current and reactance take the familiar form of Ohm's Law:

$$E = I X \qquad (69)$$

$$I = \frac{E}{X} \qquad (70)$$

$$X = \frac{E}{I} \qquad (71)$$

where
E = RMS ac voltage in volts,
I = RMS ac current in amperes, and
X = inductive or capacitive reactance in ohms.

Example: What is the voltage across a capacitor of 200 pF at 7.15 MHz, if the current through the capacitor is 50 mA?

Since the reactance of the capacitor is a function of both frequency and capacitance, first calculate the reactance:

$X_C = \dfrac{1}{2 \pi f C}$

$= \dfrac{1}{2 \times 3.1416 \times 7.15 \times 10^6 \text{ Hz} \times 200 \times 10^{-12} \text{ F}}$

$= \dfrac{10^6 \, \Omega}{8980} = 111 \, \Omega$

Next, use equation 69:

$E = I \times X_C = 0.050 \text{ A} \times 111 \, \Omega = 5.6 \text{ V}$

Example: What is the current through an 8.0-H inductor at 120 Hz, if 420 V is applied?

$X_L = 2 \pi f L$

$= 2 \times 3.1416 \times 120 \text{ Hz} \times 8.0 \text{ H}$

$= 6030 \, \Omega$

Fig 2.55 charts the reactances of capacitors from 1 pF to 100 µF, and the reactances of inductors from 0.1 µH to 10 H, for frequencies between 100 Hz and 100 MHz. Approximate values of reactance can be read or interpolated from the chart. The formulas will produce more exact values, however. (The chart can also be used to find the frequency at which an inductor and capacitor have equal reactances, creating resonance as described in the section "At and Near Resonance" below.)

Although both inductive and capacitive reactance oppose the flow of ac current, the two types of reactance differ. With capacitive reactance, the current *leads* the voltage by 90°, whereas with inductive reactance, the current *lags* the voltage by 90°. The convention for charting the two types of reactance appears in **Fig 2.56**. On this graph, inductive reactance is plotted along the +90° vertical line, while capacitive reactance is plotted along the −90° vertical line. This convention of assigning a positive value to inductive reactance and a negative value to capacitive reactance results from the mathematics used for working with impedance as described elsewhere in this chapter.

2.9.2 Reactances in Series and Parallel

If a circuit contains two reactances of the same type, whether in series or in parallel, the resulting reactance can be determined by applying the same rules as for resistances in series and in parallel. Series reactance is given by the formula

$$X_{total} = X1 + X2 + X3 \ldots + X_n \qquad (72)$$

Example: Two noninteracting inductances are in series. Each has a value of 4.0 µH, and the operating frequency is 3.8 MHz. What is the resulting reactance?

The reactance of each inductor is:

$X_L = 2 \pi f L$

$= 2 \times 3.1416 \times 3.8 \times 10^6 \text{ Hz} \times 4 \times 10^{-6} \text{ H}$

$= 96 \, \Omega$

$X_{total} = X1 + X2 = 96 \, \Omega + 96 \, \Omega = 192 \, \Omega$

We might also calculate the total reactance by first adding the inductances:

$L_{total} = L1 + L2 = 4.0 \, \mu\text{H} + 4.0 \, \mu\text{H} = 8.0 \, \mu\text{H}$

$X_{total} = 2 \pi f L$

$= 2 \times 3.1416 \times 3.8 \times 10^6 \text{ Hz} \times 8.0 \times 10^{-6} \text{ H}$

$= 191 \, \Omega$

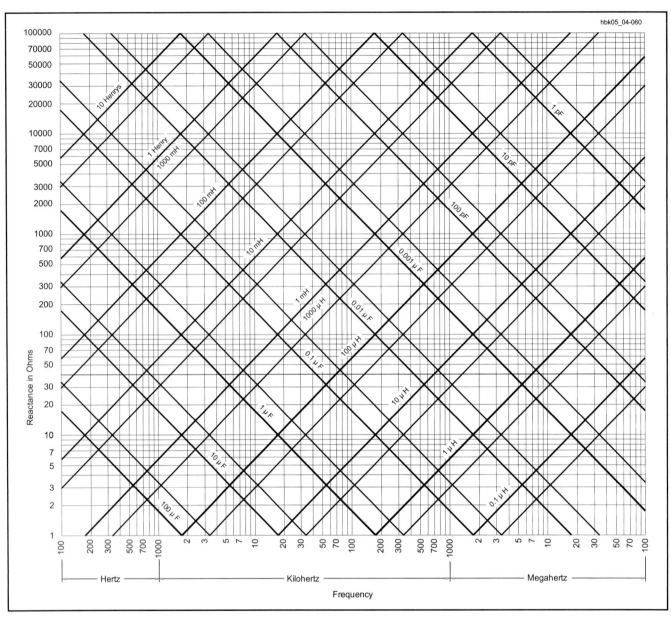

Fig 2.55 — Inductive and capacitive reactance vs frequency. Heavy lines represent multiples of 10, intermediate lines multiples of 5. For example, the light line between 10 µH and 100 µH represents 50 µH; the light line between 0.1 µF and 1 µF represents 0.5 µF, and so on. Other values can be extrapolated from the chart. For example, the reactance of 10 H at 60 Hz can be found by taking the reactance of 10 H at 600 Hz and dividing by 10 for the 10 times decrease in frequency. (Originally from Terman, *Radio Engineer's Handbook*. See references.)

(The fact that the last digit differs by one illustrates the uncertainty of the calculation caused by the limited precision of the measured values in the problem, and differences caused by rounding off the calculated values. This also shows why it is important to follow the rules for significant figures.)

Example: Two noninteracting capacitors are in series. One has a value of 10.0 pF, the other of 20.0 pF. What is the resulting reactance in a circuit operating at 28.0 MHz?

$$X_{C1} = \frac{1}{2 \pi f C}$$

$$= \frac{1}{2 \times 3.1416 \times 28.0 \times 10^6 \text{ Hz} \times 10.0 \times 10^{-12} \text{ F}}$$

$$= \frac{10^6 \, \Omega}{1760} = 568 \, \Omega$$

$$X_{C2} = \frac{1}{2 \pi f C}$$

$$= \frac{1}{2 \times 3.1416 \times 28.0 \times 10^6 \text{ Hz} \times 20.0 \times 10^{-12} \text{ F}}$$

$$= \frac{10^6 \, \Omega}{3520} = 284 \, \Omega$$

$$X_{total} = X_{C1} + X_{C2} = 568 \, \Omega + 284 \, \Omega = 852 \, \Omega$$

Alternatively, combining the series capacitors first, the total capacitance is 6.67×10^{-12} F or 6.67 pF. Then:

$$X_{total} = \frac{1}{2 \pi f C}$$

$$= \frac{1}{2 \times 3.1416 \times 28.0 \times 10^6 \text{ Hz} \times 6.67 \times 10^{-12} \text{ F}}$$

$$= \frac{10^6 \, \Omega}{1170} = 855 \, \Omega$$

(Within the uncertainty of the measured values and the rounding of values in the calculations, this is the same result as the 852 Ω we obtained with the first method.)

This example serves to remind us that *series capacitance* is not calculated in the

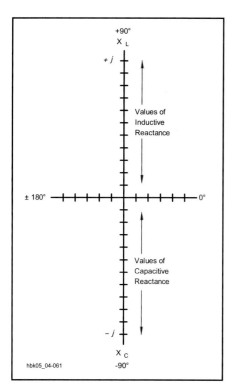

Fig 2.56 — The conventional method of plotting reactances on the vertical axis of a graph, using the upward or "plus" direction for inductive reactance and the downward or "minus" direction for capacitive reactance. The horizontal axis will be used for resistance in later examples.

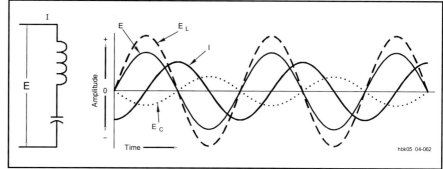

Fig 2.57 — A series circuit containing both inductive and capacitive components, together with representative voltage and current relationships.

manner used by other series resistance and inductance, but *series capacitive reactance* does follow the simple addition formula.

For reactances of the same type in parallel, the general formula is:

$$X_{total} = \frac{1}{\frac{1}{X1} + \frac{1}{X2} + \frac{1}{X3} + ... + \frac{1}{X_n}} \qquad (73)$$

or, for exactly two reactances in parallel

$$X_{total} = \frac{X1 \times X2}{X1 + X2} \qquad (74)$$

Example: Place the capacitors in the last example (10.0 pF and 20.0 pF) in parallel in the 28.0 MHz circuit. What is the resultant reactance?

$$X_{total} = \frac{X1 \times X2}{X1 + X2}$$

$$= \frac{568\,\Omega \times 284\,\Omega}{568\,\Omega + 284\,\Omega} = 189\,\Omega$$

Alternatively, two capacitors in parallel can be combined by adding their capacitances.

$$C_{total} = C_1 + C_2 = 10.0\text{ pF} + 20.0\text{ pF} = 30\text{ pF}$$

$$X_C = \frac{1}{2\pi f C}$$

$$= \frac{1}{2 \times 3.1416 \times 28.0 \times 10^6 \text{ Hz} \times 30 \times 10^{-12}\text{ F}}$$

$$= \frac{10^6\,\Omega}{5280} = 189\,\Omega$$

Example: Place the series inductors above (4.0 µH each) in parallel in a 3.8-MHz circuit. What is the resultant reactance?

$$X_{total} = \frac{X_{L1} \times X_{L2}}{X_{L1} + X_{L2}}$$

$$= \frac{96\,\Omega \times 96\,\Omega}{96\,\Omega + 96\,\Omega} = 48\,\Omega$$

Of course, a number (n) of equal reactances (or resistances) in parallel yields a reactance that is the value of one of them divided by n, or:

$$X_{total} = \frac{X}{n} = \frac{96\,\Omega}{2} = 48\,\Omega$$

All of these calculations apply only to reactances of the same type; that is, all capacitive or all inductive. Mixing types of reactances requires a different approach.

UNLIKE REACTANCES IN SERIES

When combining unlike reactances — that is, combinations of inductive and capacitive reactance — in series, it is necessary to take into account that the voltage-to-current phase relationships differ for the different types of reactance. **Fig 2.57** shows a series circuit with both types of reactance. Since the reactances are in series, the current must be the same in both. The voltage across each circuit element differs in phase, however. The voltage E_L *leads* the current by 90°, and the voltage E_C *lags* the current by 90°. Therefore, E_L and E_C have opposite polarities and cancel each other in whole or in part. The line E in Fig 2.57 approximates the resulting voltage, which is the *difference* between E_L and E_C.

Since for a constant current the reactance is directly proportional to the voltage, the net reactance is still the sum of the individual reactances as in equation 72. Because inductive reactance is considered to be positive and capacitive reactance negative, the resulting reactance can be either positive (inductive) or negative (capacitive) or even zero (no reactance).

Another common method of adding the unlike reactances is to use the absolute values of the reactances and subtract capacitive reactance from inductive reactance:

$$X_{total} = X_L - X_C \qquad (75)$$

The convention of using absolute values for the reactances and building the sense of positive and negative into the formula is the preferred method used by hams and will be used in all of the remaining formulas in this chapter. Nevertheless, before using any formulas that include reactance, determine whether this convention is followed before assuming that the absolute values are to be used.

Example: Using Fig 2.57 as a visual aid, let $X_C = 20.0\,\Omega$ and $X_L = 80.0\,\Omega$. What is the resulting reactance?

$$X_{total} = X_L - X_C$$

$$= 80.0\,\Omega - 20.0\,\Omega = +60.0\,\Omega$$

Since the result is a positive value, reactance is inductive. Had the result been a negative number, the reactance would have been capacitive.

When reactance types are mixed in a series circuit, the resulting reactance is always smaller than the larger of the two reactances. Likewise, the resulting voltage across the series combination of reactances is always smaller than the larger of the two voltages across individual reactances.

Every series circuit of mixed reactance types with more than two circuit elements can be reduced to this simple circuit by combining all the reactances into one inductive and one

Electrical Fundamentals 2.41

capacitive reactance. If the circuit has more than one capacitor or more than one inductor in the overall series string, first use the formulas given earlier to determine the total series inductance alone and the total series capacitance alone (or their respective reactances). Then combine the resulting single capacitive reactance and single inductive reactance as shown in this section.

UNLIKE REACTANCES IN PARALLEL

The situation of parallel reactances of mixed type appears in **Fig 2.58**. Since the elements are in parallel, the voltage is common to both reactive components. The current through the capacitor, I_C, *leads* the voltage by 90°, and the current through the inductor, I_L, *lags* the voltage by 90°. In this case, it is the currents that are 180° out of phase and thus cancel each other in whole or in part. The total current is the difference between the individual currents, as indicated by the line I in Fig 2.58.

Since reactance is the ratio of voltage to current, the total reactance in the circuit is:

$$X_{total} = \frac{E}{I_L - I_C} \qquad (76)$$

In the drawing, I_C is larger than I_L, and the resulting differential current retains the phase of I_C. Therefore, the overall reactance, X_{total}, is for capacitive in this case. The total reactance of the circuit will be smaller than the larger of the individual reactances, because the total current is smaller than the larger of the two individual currents.

In parallel circuits, reactance and current are inversely proportional to each other for a constant voltage and equation 74 can be used, carrying the positive and negative signs:

$$X_{total} = \frac{X_L \times (-X_C)}{X_L - X_C} = \frac{-X_L \times X_C}{X_L - X_C} \qquad (77)$$

As with the series formula for mixed reactances, follow the convention of using absolute values for the reactances, since the minus signs in the formula account for capacitive reactance being negative. If the solution yields a negative number, the resulting reactance is capacitive, and if the solution is positive, then the reactance is inductive.

Example: Using Fig 2.58 as a visual aid, place a capacitive reactance of 10.0 Ω in parallel with an inductive reactance of 40.0 Ω. What is the resulting reactance?

$$X_{total} = \frac{-X_L \times X_C}{X_L - X_C}$$

$$= \frac{-40.0\,\Omega \times 10.0\,\Omega}{40.0\,\Omega - 10.0\,\Omega}$$

$$= \frac{-400\,\Omega}{30.0\,\Omega} = -13.3\,\Omega$$

The reactance is capacitive, as indicated by the negative solution. Moreover, the resultant reactance is always smaller than the larger of the two individual reactances.

As with the case of series reactances, if each leg of a parallel circuit contains more than one reactance, first simplify each leg to a single reactance. If the reactances are of the same type in each leg, the series reactance formulas for reactances of the same type will apply. If the reactances are of different types, then use the formulas shown above for mixed series reactances to simplify the leg to a single value and type of reactance.

2.9.3 At and Near Resonance

Any series or parallel circuit in which the values of the two unlike reactances are equal is said to be *resonant*. For any given inductance or capacitance, it is theoretically possible to find a value of the opposite reactance type to produce a resonant circuit for any desired frequency.

When a series circuit like the one shown in Fig 2.57 is resonant, the voltages E_C and E_L are equal and cancel; their sum is zero. This is a *series-resonant* circuit. Since the reactance of the circuit is proportional to the sum of these voltages, the net reactance also goes to zero. Theoretically, the current, as shown in **Fig 2.59**, can become infinite. In fact, it is limited only by losses in the components and other resistances that would exist in a real circuit of this type. As the frequency of operation moves slightly off resonance and the reactances no longer cancel completely, the net reactance climbs as shown in the figure. Similarly, away from resonance the current drops to a level determined by the net reactance.

In a *parallel-resonant* circuit of the type in Fig 2.58, the current I_L and I_C are equal and cancel to zero. Since the reactance is inversely

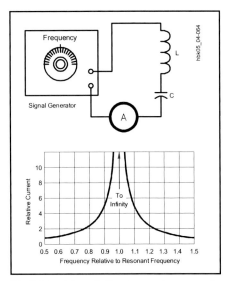

Fig 2.59 — The relative generator current with a fixed voltage in a series circuit containing inductive and capacitive reactances as the frequency approaches and departs from resonance.

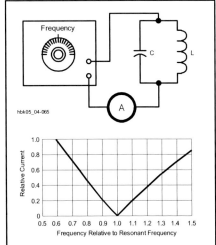

Fig 2.60 — The relative generator current with a fixed voltage in a parallel circuit containing inductive and capacitive reactances as the frequency approaches and departs from resonance. (The circulating current through the parallel inductor and capacitor is a maximum at resonance.)

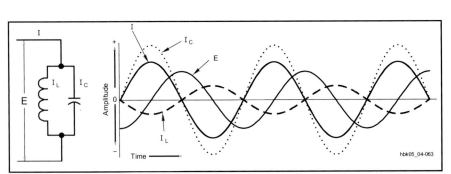

Fig 2.58 — A parallel circuit containing both inductive and capacitive components, together with representative voltage and current relationships.

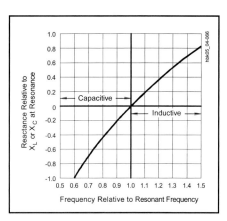

Fig 2.61 — The transition from capacitive to inductive reactance in a series-resonant circuit as the frequency passes resonance.

infinite at resonance, the opposite condition exists: above resonance, the reactance is capacitive and below resonance it is inductive, as shown in **Fig 2.62**. Of course, all graphs and calculations in this section are theoretical and presume a purely reactive circuit. Real circuits are never purely reactive; they contain some resistance that modifies their performance considerably. Real resonant circuits will be discussed later in this chapter.

2.9.4 Reactance and Complex Waveforms

All of the formulas and relationships shown in this section apply to alternating current in the form of regular sine waves. Complex wave shapes complicate the reactive situation considerably. A complex or nonsinusoidal wave can be treated as a sine wave of some fundamental frequency and a series of harmonic frequencies whose amplitudes depend on the original wave shape. When such a complex wave — or collection of sine waves — is applied to a reactive circuit, the current through the circuit will not have the same wave shape as the applied voltage. The difference results because the reactance of an inductor and capacitor depend in part on the applied frequency.

For the second-harmonic component of the complex wave, the reactance of the inductor is twice and the reactance of the capacitor is half their respective values at the fundamental frequency. A third-harmonic component produces inductive reactances that are triple and capacitive reactances that are one-third those at the fundamental frequency. Thus, the overall circuit reactance is different for each harmonic component.

The frequency sensitivity of a reactive circuit to various components of a complex wave shape creates both difficulties and opportunities. On the one hand, calculating the circuit reactance in the presence of highly variable as well as complex waveforms, such as speech, is difficult at best. On the other hand, the frequency sensitivity of reactive components and circuits lays the foundation

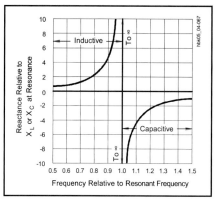

Fig 2.62 — The transition from inductive to capacitive reactance in a parallel-resonant circuit as the frequency passes resonance.

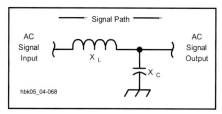

Fig 2.63 — A signal path with a series inductor and a shunt capacitor. The circuit presents different reactances to an ac signal and to its harmonics.

for filtering, that is, for separating signals of different frequencies or acting upon them differently. For example, suppose a coil is in the series path of a signal and a capacitor is connected from the signal line to ground, as represented in **Fig 2.63**. The reactance of the coil to the second harmonic of the signal will be twice that at the fundamental frequency and oppose more effectively the flow of harmonic current. Likewise, the reactance of the capacitor to the harmonic will be half that to the fundamental, allowing the harmonic an easier current path away from the signal line toward ground. See the **RF and AF Filters** chapter for detailed information on filter theory and construction.

proportional to the current, as the current approaches zero, the reactance becomes infinite. As with series circuits, component losses and other resistances in the circuit prevent the current from reaching zero. **Fig 2.60** shows the theoretical current curve near and at resonance for a purely reactive parallel-resonant circuit. Note that in both Fig 2.59 and Fig 2.60, the departure of current from the resonance value is close to, but not quite, symmetrical above and below the resonant frequency.

Example: What is the reactance of a series L-C circuit consisting of a 56.04-pF capacitor and an 8.967-µH inductor at 7.00, 7.10 and 7.20 MHz? Using the formulas from earlier in this chapter, we calculate a table of values:

Frequency (MHz)	X_L (Ω)	X_C (Ω)	X_{total} (Ω)
7.000	394.4	405.7	–11.3
7.100	400.0	400.0	0
7.200	405.7	394.4	11.3

The exercise shows the manner in which the reactance rises rapidly as the frequency moves above and below resonance. Note that in a series-resonant circuit, the reactance at frequencies below resonance is capacitive, and above resonance, it is inductive. **Fig 2.61** displays this fact graphically. In a parallel-resonant circuit, where the reactance becomes

2.10 Impedance

When a circuit contains both resistance and reactance, the combined opposition to current is called *impedance*. Symbolized by the letter Z, impedance is a more general term than either resistance or reactance. Frequently, the term is used even for circuits containing only resistance or reactance. Qualifications such as "resistive impedance" are sometimes added to indicate that a circuit has only resistance, however.

The reactance and resistance comprising an impedance may be connected either in series or in parallel, as shown in **Fig 2.64**. In these circuits, the reactance is shown as a box to indicate that it may be either inductive or capacitive. In the series circuit at A, the current is the same in both elements, with (generally) different voltages appearing across the resistance and reactance. In the parallel circuit at B, the same voltage is applied to both elements, but different currents may flow in the two branches.

In a resistance, the current is in phase with the applied voltage, while in a reactance it is 90° out of phase with the voltage. Thus, the phase relationship between current and voltage in the circuit as a whole may be anything between zero and 90°, depending on the relative amounts of resistance and reactance.

As shown in Fig 2.56 in the preceding section, reactance is graphed on the vertical (Y) axis to record the phase difference between the voltage and the current. **Fig 2.65** adds

Electrical Fundamentals 2.43

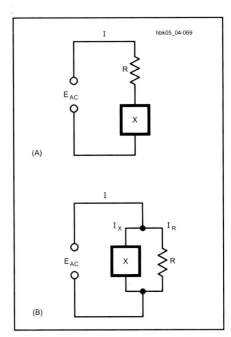

Fig 2.64 — Series and parallel circuits containing resistance and reactance.

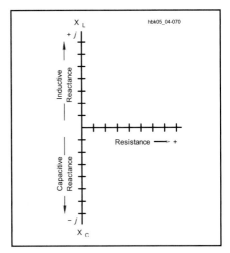

Fig 2.65 — The conventional method of charting impedances on a graph, using the vertical axis for reactance (the upward or "plus" direction for inductive reactance and the downward or "minus" direction for capacitive reactance), and using the horizontal axis for resistance.

resistance to the graph. Since the voltage is in phase with the current, resistance is recorded on the horizontal axis, using the positive or right side of the scale.

2.10.1 Calculating Z From R and X in Series Circuits

Impedance is the complex combination of resistance and reactance. Since there is a 90° phase difference between resistance and reactance (whether inductive or capacitive), simply adding the two values does not correspond to what actually happens in a circuit and will not give the correct result. Therefore, expressions like "Z = R ± X" can be misleading, because they suggest simple addition. As a result, impedance is often expressed "Z = R ± jX."

In pure mathematics, "i" indicates an imaginary number. Because i represents current in electronics, we use the letter "j" for the same mathematical operator, although there is nothing imaginary about what it represents in electronics. (References to explain imaginary numbers, rectangular coordinates, polar coordinates and how to work with them are provided in the "Radio Math" section at the end of this chapter.) With respect to resistance and reactance, the letter j is normally assigned to those figures on the vertical axis, 90° out of phase with the horizontal axis. The actual function of j is to indicate that calculating impedance from resistance and reactance requires *vector addition*.

A *vector* is a value with both magnitude and direction, such as velocity; "10 meters/sec to the north". Impedance also has a "direction" as described below. In vector addition, the result of combining two values with a 90° phase difference is a quantity different from the simple *algebraic addition* of the two values. The result will have a phase difference intermediate between 0° and 90°.

RECTANGULAR FORM OF IMPEDANCE

Because this form for impedances, Z = R ± jX, can be plotted on a graph using rectangular coordinates, this is the *rectangular form* of impedance. The rectangular coordinate system in which one axis represents real number and the other axis imaginary numbers is called the *complex plane* and impedance with both real (R) and imaginary (X) components is called *complex impedance*. Unless specifically noted otherwise, assume that "impedance" means "complex impedance" and that both R and X may be present.

Consider **Fig 2.66**, a series circuit consisting of an inductive reactance and a resistance. As given, the inductive reactance is 100 Ω and the resistance is 50 Ω. Using *rectangular coordinates*, the impedance becomes

$$Z = R + jX \quad (78)$$

where

Z = the impedance in ohms,
R = the resistance in ohms, and
X = the reactance in ohms.

In the present example,

$$Z = 50 + j100 \, \Omega$$

This point is located at the tip of the arrow drawn on the graph where the dashed lines cross.

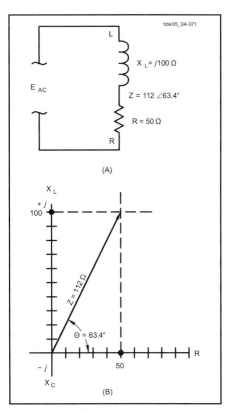

Fig 2.66 — A series circuit consisting of an inductive reactance of 100 Ω and a resistance of 50 Ω. At B, the graph plots the resistance, reactance, and impedance.

POLAR FORM OF IMPEDANCE AND PHASE ANGLE

As the graph in Fig 2.66 shows, the impedance that results from combining R and X can also be represented by a line completing a right triangle whose sides are the resistance and reactance. The point at the end of the line — the complex impedance — can be described by how far it is from the origin of the graph where the axes cross (the *magnitude of the impedance*, |Z|) and the angle made by the line with the horizontal axis representing 0° (the *phase angle* of the impedance, θ). This is the *polar form* of impedance and it is written in the form

$$Z = |Z| \angle \theta \quad (79)$$

Occasionally, θ may be given in radians. The convention in this handbook is to use degrees unless specifically noted otherwise.

The length of the hypotenuse of the right triangle represents the magnitude of the impedance and can be calculated using the formula for calculating the hypotenuse of a right triangle, in which the square of the hypotenuse equals the sum of the squares of the two sides:

$$|Z| = \sqrt{R^2 + X^2} \quad (80)$$

In this example:

$$|Z| = \sqrt{(50\,\Omega)^2 + (100\,\Omega)^2}$$

$$= \sqrt{2500\,\Omega^2 + 10000\,\Omega^2}$$

$$= \sqrt{12500\,\Omega^2} = 112\,\Omega$$

The magnitude of the impedance that results from combining 50 Ω of resistance with 100 Ω of inductive reactance is 112 Ω. From trigonometry, the tangent of the phase angle is the side opposite the angle (X) divided by the side adjacent to the angle (R), or

$$\tan\theta = \frac{X}{R} \tag{81}$$

where
X = the reactance, and
R = the resistance.

Find the angle by taking the inverse tangent, or arctan:

$$\theta = \arctan\frac{X}{R} \tag{82}$$

Calculators sometimes label the inverse tangent key as "tan^{-1}". Remember to be sure your calculator is set to use the right angular units, either degrees or radians.

In the example shown in Fig 2.66,

$$\theta = \arctan\frac{100\,\Omega}{50\,\Omega} = \arctan 2.0 = 63.4°$$

Using the information just calculated, the complex impedance in polar form is:

$$Z = 112\,\Omega\,\angle 63.4°$$

This is stated verbally as "112 ohms at an angle of 63 point 4 degrees."

POLAR TO RECTANGULAR CONVERSION

The expressions R ± jX and |Z| ∠θ both provide the same information, but in two different forms. The procedure just given permits conversion from rectangular coordinates into polar coordinates. The reverse procedure is also important. **Fig 2.67** shows an impedance composed of a capacitive reactance and a resistance. Since capacitive reactance appears as a negative value, the impedance will be at a negative phase angle, in this case, 12.0 Ω at a phase angle of − 42.0° or Z = |12.0 Ω| ∠ − 42.0°.

Remember that the impedance forms a triangle with the values of X and R from the rectangular coordinates. The reactance axis forms the side opposite the angle θ.

$$\sin\theta = \frac{\text{side opposite}}{\text{hypotenuse}} = \frac{X}{|Z|}$$

Solving this equation for reactance, we have:

$$X = |Z| \times \sin\theta \text{ (ohms)} \tag{83}$$

Likewise, the resistance forms the side adjacent to the angle.

$$\cos\theta = \frac{\text{side adjacent}}{\text{hypotenuse}} = \frac{R}{|Z|}$$

Solving for resistance, we have:

$$R = |Z| \times \cos\theta \text{ (ohms)} \tag{84}$$

Then from our example:

$$X = 12.0\,\Omega \times \sin(-42°)$$
$$= 12.0\,\Omega \times -0.669 = -8.03\,\Omega$$

$$R = 12.0\,\Omega \times \cos(-42°)$$
$$= 12.0\,\Omega \times 0.743 = 8.92\,\Omega$$

Since X is a negative value, it is plotted on the lower vertical axis, as shown in Fig 2.67, indicating capacitive reactance. In rectangular form, Z = 8.92 Ω − j8.03 Ω.

In performing impedance and related calculations with complex circuits, rectangular coordinates are most useful when formulas require the addition or subtraction of values. Polar notation is most useful for multiplying and dividing complex numbers. (See the section on "Radio Math" at the end of the chapter for references dealing with the mathematics of complex numbers.)

All of the examples shown so far in this section presume a value of reactance that contributes to the circuit impedance. Reactance is a function of frequency, however, and many impedance calculations may begin with a value of capacitance or inductance and an operating frequency. In terms of these values, equation 80 can be expressed in either of two forms, depending on whether the reactance is inductive (equation 85) or capacitive (equation 86):

$$|Z| = \sqrt{R^2 + (2\pi f L)^2} \tag{85}$$

$$|Z| = \sqrt{R^2 + \left(\frac{1}{2\pi f C}\right)^2} \tag{86}$$

Example: What is the impedance of a circuit like Fig 2.66 with a resistance of 100 Ω and a 7.00-μH inductor operating at a frequency of 7.00 MHz? Using equation 85,

$$|Z| = \sqrt{R^2 + (2\pi f L)^2}$$

$$= \sqrt{(100\,\Omega)^2 + (2\pi \times 7.0 \times 10^{-6}\,\text{H} \times 7.0 \times 10^6\,\text{Hz})^2}$$

$$= \sqrt{10000\,\Omega^2 + (308\,\Omega)^2}$$

$$= \sqrt{10000\,\Omega^2 + 94900\,\Omega^2}$$

$$= \sqrt{104900\,\Omega^2} = 323.9\,\Omega$$

Since 308 Ω is the value of inductive reactance of the 7.00-μH coil at 7.00 MHz, the phase angle calculation proceeds as given in the earlier example (equation 82):

$$\theta = \arctan\frac{X}{R} = \arctan\left(\frac{308.0\,\Omega}{100.0\,\Omega}\right)$$

$$= \arctan(3.08) = 72.0°$$

Since the reactance is inductive, the phase angle is positive.

2.10.2 Calculating Z From R and X in Parallel Circuits

In a parallel circuit containing reactance and resistance, such as shown in **Fig 2.68**, calculation of the resultant impedance from the values of R and X does not proceed by

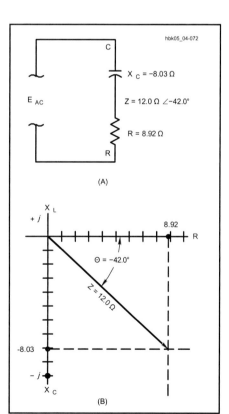

Fig 2.67 — A series circuit consisting of a capacitive reactance and a resistance: the impedance is given as 12.0 Ω at a phase angle θ of −42 degrees. At B, the graph plots the resistance, reactance, and impedance.

Electrical Fundamentals 2.45

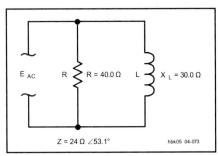

Fig 2.68 — A parallel circuit containing an inductive reactance of 30.0 Ω and a resistor of 40.0 Ω. No graph is given, since parallel impedances can not be manipulated graphically in the simple way of series impedances.

direct combination as for series circuits. The general formula for parallel circuits is:

$$|Z| = \frac{RX}{\sqrt{R^2 + X^2}} \qquad (87)$$

where the formula uses the absolute (unsigned) reactance value. The phase angle for the parallel circuit is given by:

$$\theta = \arctan\left(\frac{R}{X}\right) \qquad (88)$$

The sign of θ has the same meaning in both series and parallel circuits: if the parallel reactance is capacitive, then θ is a negative angle, and if the parallel reactance is inductive, then θ is a positive angle.

Example: An inductor with a reactance of 30.0 Ω is in parallel with a resistor of 40.0 Ω. What is the resulting impedance and phase angle?

$$|Z| = \frac{RX}{\sqrt{R^2 + X^2}} = \frac{30.0\,\Omega \times 40.0\,\Omega}{\sqrt{(30.0\,\Omega)^2 + (40.0\,\Omega)^2}}$$

$$= \frac{1200\,\Omega^2}{\sqrt{900\,\Omega^2 + 1600\,\Omega^2}} = \frac{1200\,\Omega^2}{\sqrt{2500\,\Omega^2}}$$

$$= \frac{1200\,\Omega^2}{50.0\,\Omega} = 24.0\,\Omega$$

$$\theta = \arctan\left(\frac{R}{X}\right) = \arctan\left(\frac{40.0\,\Omega}{30.0\,\Omega}\right)$$

$$\theta = \arctan(1.33) = 53.1°$$

Since the parallel reactance is inductive, the resultant angle is positive.

Example: A capacitor with a reactance of 16.0 Ω is in parallel with a resistor of 12.0 Ω. What is the resulting impedance and phase angle? (Remember that capacitive reactance is negative when used in calculations.)

$$|Z| = \frac{RX}{\sqrt{R^2 + X^2}} = \frac{-16.0\,\Omega \times 12.0\,\Omega}{\sqrt{(-16.0\,\Omega)^2 + (12.0\,\Omega)^2}}$$

$$= \frac{-192\,\Omega^2}{\sqrt{256\,\Omega^2 + 144\,\Omega^2}} = \frac{-192\,\Omega^2}{\sqrt{400\,\Omega^2}}$$

$$= \frac{-192\,\Omega^2}{20.0\,\Omega} = -9.60\,\Omega$$

$$\theta = \arctan\left(\frac{R}{X}\right) = \arctan\left(\frac{12.0\,\Omega}{-16.0\,\Omega}\right)$$

$$\theta = \arctan(-0.750) = -36.9°$$

Because the parallel reactance is capacitive and the reactance negative, the resultant phase angle is negative.

2.10.3 Admittance

Just as the inverse of resistance is conductance (G) and the inverse of reactance is susceptance (B), so, too, impedance has an inverse: admittance (Y), measured in siemens (S). Thus,

$$Y = \frac{1}{Z} \qquad (89)$$

Since resistance, reactance and impedance are inversely proportional to the current (Z = E / I), conductance, susceptance and admittance are directly proportional to current. That is,

$$Y = \frac{I}{E} \qquad (90)$$

Admittance can be expressed in rectangular and polar forms, just like impedance,

$$Y = G \pm jB = |Y|\angle\theta \qquad (91)$$

The phase angle for admittance has the same sign convention as for impedance; if the susceptance component is inductive, the phase angle is positive, and if the susceptive component is capacitive, the phase angle is negative.

One handy use for admittance is in simplifying parallel circuit impedance calculations. Similarly to the rules stated previously for combining conductance, the admittance of a parallel combination of reactance and resistance is the vector addition of susceptance and conductance. In other words, for parallel circuits:

$$|Y| = \sqrt{G^2 + B^2} \qquad (92)$$

where

|Y| = magnitude of the admittance in siemens,

G = conductance or 1 / R in siemens, and

B = susceptance or 1 / X in siemens.

Example: An inductor with a reactance of 30.0 Ω is in parallel with a resistor of 40.0 Ω. What is the resulting impedance and phase angle? The susceptance is 1 / 30.0 Ω = 0.0333 S and the conductance is 1 / 40.0 Ω = 0.0250 S.

$$Y = \sqrt{(0.0333\,S)^2 + (0.0250\,S)^2}$$

$$Y = \sqrt{0.00173\,S^2} = 0.0417\,S$$

$$Z = \frac{1}{Y} = \frac{1}{0.0417\,S} = 24.0\,\Omega$$

The phase angle in terms of conductance and susceptance is:

$$\theta = \arctan\left(\frac{B}{G}\right) \qquad (93)$$

In this example,

$$\theta = \arctan\left(\frac{0.0333\,S}{0.0250\,S}\right) = \arctan(1.33) = 53.1°$$

Again, since the reactive component is inductive, the phase angle is positive. For a capacitively reactive parallel circuit, the phase angle would have been negative. Compare these results with the example calculation of the impedance for the same circuit earlier in the section.

Conversion between resistance, reactance and impedance and conductance, susceptance and admittance is very useful in working with complex circuits and in impedance matching of antennas and transmission lines. There are many on-line calculators that can perform these operations and many programmable calculators and suites of mathematical computer software have these functions built-in. Knowing when and how to use them, however, demands some understanding of the fundamental strategies shown here.

2.10.4 More than Two Elements in Series or Parallel

When a circuit contains several resistances or several reactances in series, simplify the circuit before attempting to calculate the impedance. Resistances in series add, just as in a purely resistive circuit. Series reactances of the same kind — that is, all capacitive or all inductive — also add, just as in a purely reactive circuit. The goal is to produce a single value of resistance and a single value of reactance that can be used in the impedance calculation.

Fig 2.69 illustrates a more difficult case in which a circuit contains two different reactive elements in series, along with a series resistance. The series combination of X_C and X_L reduce to a single value using the same rules of combination discussed in the section on purely reactive components. As Fig 2.69B demonstrates, the resultant reactance is the difference between the two series reactances.

For parallel circuits with multiple resistances or multiple reactances of the same type, use the rules of parallel combination to reduce the resistive and reactive components to single elements. Where two or more reactive components of different types appear in the same circuit, they can be combined using formulas shown earlier for pure reactances. As **Fig 2.70** suggests, however, they can also be combined as susceptances. Parallel susceptances of different types add, with attention to their differing signs. The resulting single susceptance can then be combined with the conductance to arrive at the overall circuit admittance whose inverse is the final circuit impedance.

2.10.5 Equivalent Series and Parallel Circuits

The two circuits shown in Fig 2.64 are equivalent if the same current flows when a given voltage of the same frequency is applied, and if the phase angle between voltage and current is the same in both cases. It is possible, in fact, to transform any given series circuit into an equivalent parallel circuit, and vice versa.

A series RX circuit can be converted into its parallel equivalent by means of the formulas:

$$R_P = \frac{R_S^2 + X_S^2}{R_S} \tag{94}$$

$$X_P = \frac{R_S^2 + X_S^2}{X_S} \tag{95}$$

where the subscripts P and S represent the parallel- and series-equivalent values, respectively. If the parallel values are known, the equivalent series circuit can be found from:

$$R_S = \frac{R_P X_P^2}{R_P^2 + X_P^2} \tag{96}$$

and

$$X_S = \frac{R_P^2 X_P}{R_P^2 + X_P^2} \tag{97}$$

Example: Let the series circuit in Fig 2.64 have a series reactance of –50.0 Ω (indicating a capacitive reactance) and a resistance of 50.0 Ω. What are the values of the equivalent parallel circuit?

$$R_P = \frac{R_S^2 + X_S^2}{R_S} = \frac{(50.0\,\Omega)^2 + (-50.0\,\Omega)^2}{50.0\,\Omega}$$

$$= \frac{2500\,\Omega^2 + 2500\,\Omega^2}{50.0\,\Omega} = \frac{5000\,\Omega^2}{50\,\Omega} = 100\,\Omega$$

$$X_P = \frac{R_S^2 + X_S^2}{X_S} = \frac{(50.0\,\Omega)^2 + (-50.0\,\Omega)^2}{-50.0\,\Omega}$$

$$= \frac{2500\,\Omega^2 + 2500\,\Omega^2}{-50.0\,\Omega} = \frac{5000\,\Omega^2}{-50\,\Omega} = -100\,\Omega$$

In the parallel circuit of Fig 2.64, a capacitive reactance of 100 Ω and a resistance of 100 Ω to be equivalent would the series circuit.

2.10.6 Ohm's Law for Impedance

Ohm's Law applies to circuits containing impedance just as readily as to circuits having resistance or reactance only. The formulas are:

$$E = I\,Z \tag{98}$$

$$I = \frac{E}{Z} \tag{99}$$

$$Z = \frac{E}{I} \tag{100}$$

where
E = voltage in volts,
I = current in amperes, and
Z = impedance in ohms.

Z must now be understood to be a complex number, consisting of resistive and reactive components. If Z is complex, then so are E and I, with a magnitude and phase angle. The rules of complex mathematics are then applied and the variables are written in boldface type as **Z**, **E**, and **I**, or an arrow is added above them to indicate that they are complex, such as,

$$\vec{E} = \vec{I}\,\vec{Z}$$

If only the magnitude of impedance, voltage, and currents are important, however, then the magnitudes of the three variables can be combined in the familiar ways without regard to the phase angle. In this case E and I are assumed to be RMS values (or some other steady-state value such as peak, peak-to-peak, or average). **Fig 2.71** shows a simple circuit consisting of a resistance of 75.0 Ω and a reactance of 100 Ω in series. From the series-impedance formula previously given, the impedance is

$$Z = \sqrt{R^2 + X_L^2} = \sqrt{(75.0\,\Omega)^2 + (100\,\Omega)^2}$$

$$= \sqrt{5630\,\Omega^2 + 10000\,\Omega^2}$$

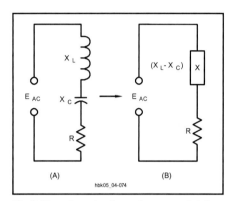

Fig 2.69 — A series impedance containing mixed capacitive and inductive reactances can be reduced to a single reactance plus resistance by combining the reactances algebraically.

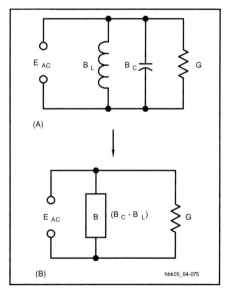

Fig 2.70 — A parallel impedance containing mixed capacitive and inductive reactances can be reduced to a single reactance plus resistance using formulas shown earlier in the chapter. By converting reactances to susceptances, as shown in A, you can combine the susceptances algebraically into a single susceptance, as shown in B.

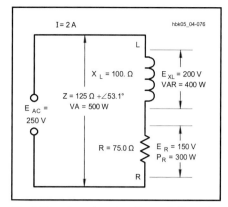

Fig 2.71 — A series circuit consisting of an inductive reactance of 100 Ω and a resistance of 75.0 Ω. Also shown is the applied voltage, voltage drops across the circuit elements, and the current.

Electrical Fundamentals 2.47

$$= \sqrt{15600 \, \Omega^2} = 125 \, \Omega$$

If the applied voltage is 250 V, then

$$I = \frac{E}{Z} = \frac{250 \text{ V}}{125 \, \Omega} = 2.0 \text{ A}$$

This current flows through both the resistance and reactance, so the voltage drops are:

$$E_R = I \, R = 2.0 \text{ A} \times 75.0 \, \Omega = 150 \text{ V}$$

$$E_{XL} = I \, X_L = 2.0 \text{ A} \times 100 \, \Omega = 200 \text{ V}$$

Illustrating one problem of working only with RMS values, the simple arithmetical sum of these two drops, 350 V, is greater than the applied voltage because the two voltages are 90° out of phase. When phase is taken into account,

$$E = \sqrt{(150 \text{ V})^2 + (200 \text{ V})^2}$$

$$= \sqrt{22500 \text{ V}^2 + 40000 \text{ V}^2}$$

$$= \sqrt{62500 \text{ V}^2} = 250 \text{ V}$$

2.10.7 Reactive Power and Power Factor

Although purely reactive circuits, whether simple or complex, show a measurable ac voltage and current, we cannot simply multiply the two together to arrive at power. Power is the rate at which energy is consumed by a circuit, and purely reactive circuits do not consume energy. The charge placed on a capacitor during part of an ac cycle is returned to the circuit during the next part of a cycle. Likewise, the energy stored in the magnetic field of an inductor returns to the circuit as the field collapses later in the ac cycle. A reactive circuit simply cycles and recycles energy into and out of the reactive components. If a purely reactive circuit were possible in reality, it would consume no energy at all.

In reactive circuits, circulation of energy accounts for seemingly odd phenomena. For example, in a series circuit with capacitance and inductance, the voltages across the components may exceed the supply voltage. That condition can exist because, while energy is being stored by the inductor, the capacitor is returning energy to the circuit from its previously charged state, and vice versa. In a parallel circuit with inductive and capacitive branches, the current circulating through the components may exceed the current drawn from the source. Again, the phenomenon occurs because the inductor's collapsing magnetic field supplies current to the capacitor, and the discharging capacitor provides current to the inductor.

To distinguish between the non-dissipated energy circulating in a purely reactive circuit and the dissipated or *real power* in a resistive circuit, the unit of *reactive power* is called the *volt-ampere reactive*, or *VAR*. The term watt is not used and sometimes reactive power is called "wattless" power. VAR has only limited use in radio circuits. Formulas similar to those for resistive power are used to calculate VAR:

$$VAR = I \times E \qquad (101)$$

$$VAR = I^2 \times X \qquad (102)$$

$$VAR = \frac{E^2}{X} \qquad (103)$$

where E and I are RMS values of voltage and current.

Real, or dissipated, power is measured in watts. *Apparent power* is the product of the voltage across and the current through an impedance. To distinguish apparent power from real power, apparent power is measured in *volt-amperes* (*VA*).

In the circuit of Fig 2.71, an applied voltage of 250 V results in a current of 2.00 A, giving an apparent power of 250 V × 2.00 A = 500 W. Only the resistance actually consumes power, however. The real power dissipated by the resistance is:

$$E = I^2 \, R = (2.0 \text{ A})^2 \times 75.0 \text{ V} = 300 \text{ W}$$

and the reactive power is:

$$VAR = I^2 \times X_L = (2.0 \text{ A})^2 \times 100 \, \Omega = 400 \text{ VA}$$

The ratio of real power to the apparent power is called the circuit's *power factor* (PF).

$$PF = \frac{P_{consumed}}{P_{apparent}} = \frac{R}{Z} \qquad (104)$$

Power factor is frequently expressed as a percentage. The power factor of a purely resistive circuit is 100% or 1, while the power factor of a pure reactance is zero. In the example of Fig 2.71 the power factor would be 300 W / 500 W = 0.600 or 60%.

Apparent power has no direct relationship to the power actually dissipated unless the power factor of the circuit is known.

$$P = \text{Apparent Power} \times \text{power factor} \qquad (105)$$

An equivalent definition of power factor is:

$$PF = \cos \theta \qquad (106)$$

where θ is the phase angle of the circuit impedance.

Since the phase angle in the example equals:

$$\theta = \arctan\left(\frac{X}{R}\right) = \arctan\left(\frac{100 \, \Omega}{75.0 \, \Omega}\right)$$

$$\theta = \arctan(1.33) = 53.1°$$

and the power factor is:

$$PF = \cos 53.1° = 0.600$$

as the earlier calculation confirms.

Since power factor is always rendered as a positive number, the value must be followed by the words "leading" or "lagging" to identify the phase of the voltage with respect to the current. Specifying the numerical power factor is not always sufficient. For example, many dc-to-ac power inverters can safely operate loads having a large net reactance of one sign but only a small reactance of the opposite sign. Hence, the final calculation of the power factor in this example would be reported as "0.600, leading."

2.11 Quality Factor (Q) of Components

Components that store energy, like capacitors and inductors, may be compared in terms of *quality factor* or *Q factor*, abbreviated *Q*. The Q of any such component is the ratio of its ability to store energy to the sum total of all energy losses within the component. In practical terms, this ratio reduces to the formula:

$$Q = \frac{X}{R} \qquad (107)$$

where

Q = quality factor (no units),
X = X_L (inductive reactance) for inductors and X_C (capacitive reactance) for capacitors (in ohms), and
R = the sum of all resistances associated with the energy losses in the component (in ohms).

The Q of capacitors is ordinarily high. Good quality ceramic capacitors and mica capacitors may have Q values of 1200 or more. Small ceramic trimmer capacitors may have Q values too small to ignore in some applications. Microwave capacitors can have poor Q values; 10 or less at 10 GHz and higher frequencies.

Inductors are subject to many types of electrical energy losses, however, such as wire resistance, core losses and skin effect. All electrical conductors have some resistance through which electrical energy is lost as heat. Moreover, inductor wire must be sized to handle the anticipated current through the inductor. Wire conductors suffer additional ac losses because alternating current tends to flow on the conductor surface due to the skin effect discussed in the chapter on **RF Techniques**. If the inductor's core is a conductive material, such as iron, ferrite, or brass, the core will introduce additional losses of energy. The specific details of these losses are discussed in connection with each type of core material.

The sum of all core losses may be depicted by showing a resistor in series with the inductor (as in Figs 2.52 and 2.53), although there is no separate component represented by the resistor symbol. As a result of inherent energy losses, inductor Q rarely, if ever, approaches capacitor Q in a circuit where both components work together. Although many circuits call for the highest Q inductor obtainable, other circuits may call for a specific Q, even a very low one.

AC Component Summary

	Resistor	Capacitor	Inductor
Basic Unit	ohm (Ω)	farad (F)	henry (H)
Units Commonly Used		microfarads (μF)	millihenrys (mH)
		picofarads (pF)	microhenrys (μH)
Time constant	(None)	RC	L/R
Voltage-Current Phase	In phase	Current leads voltage	Voltage leads current
		Voltage lags current	Current lags voltage
Resistance or Reactance	Resistance	$X_C = 1/2\pi fC$	$X_L = 2\pi fL$
Change with increasing frequency	No	Reactance decreases	Reactance increases
Q of circuit	Not defined	X_C / R	X_L / R

2.12 Practical Inductors

Various facets of radio circuits make use of inductors ranging from the tiny up to the massive. Small values of inductance, such as those inductors in **Fig 2.72A**, serve mostly in RF circuits. They may be self-supporting, air-core or air-wound inductors or the winding may be supported by nonmagnetic strips or a form. Phenolic, certain plastics and ceramics are the most common *coil forms* for air-core inductors. These inductors range in value from a few hundred μH for medium- and high-frequency circuits down to tenths of a μH at VHF and UHF.

The most common inductor in small-signal RF circuits is the *encapsulated inductor*. These components look a lot like carbon-composition or film resistors and are often marked with colored paint stripes to indicate value. (The chapter **Component Data and References** contains information on inductor color codes and marking schemes.) These inductors have values from less than 1 μH to a few mH. They cannot handle much current without saturating or over-heating.

It is possible to make solenoid inductors variable by inserting a moveable *slug* in the center of the inductor. (Slug-tuned inductors normally have a ceramic, plastic or phenolic insulating form between the conductive slug and the inductor winding.) If the slug material is magnetic, such as powdered iron, the inductance increases as the slug is moved into the center of the inductor. If the slug is brass or some other nonmagnetic material, inserting the slug will reduce the inductor's inductance.

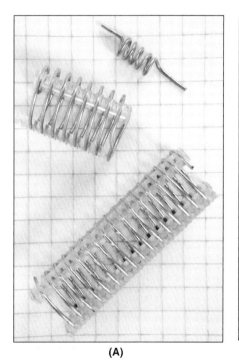

(A)

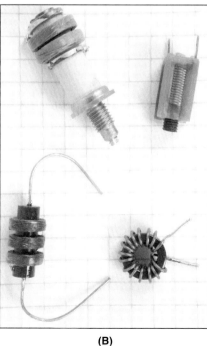

(B)

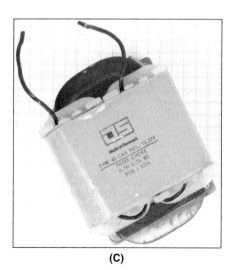

(C)

Fig 2.72 — Part A shows small-value air-wound inductors. Part B shows some inductors with values in the range of a few millihenrys and C shows a large inductor such as might be used in audio circuits or as power-supply chokes. The ¼-inch-ruled graph paper background provides a size comparison.

An alternative to air-core inductors for RF work are *toroidal* inductors (or *toroids*) wound on powdered-iron or ferrite cores. The availability of many types and sizes of powdered-iron cores has made these inductors popular for low-power fixed-value service. The toroidal shape concentrates the inductor's field nearly completely inside the inductor, eliminating the need in many cases for other forms of shielding to limit the interaction of the inductor's magnetic field with the fields of other inductors. (Ferrite core materials are discussed here and in the chapter on **RF Techniques**.)

Fig 2.72B shows samples of inductors in the millihenry (mH) range. Among these inductors are multi-section RF chokes designed to block RF currents from passing beyond them to other parts of circuits. Low-frequency radio work may also use inductors in this range of values, sometimes wound with *litz wire*. Litz wire is a special version of stranded wire, with each strand insulated from the others, and is used to minimize losses associated with skin effect.

For audio filters, toroidal inductors with values up to 100 mH are useful. Resembling powdered-iron-core RF toroids, these inductors are wound on ferrite or molybdenum-permalloy cores having much higher permeabilities.

Audio and power-supply inductors appear in Fig 2.72C. Lower values of these iron-core inductors, in the range of a few henrys, are useful as audio-frequency chokes. Larger values up to about 20 H may be found in power supplies, as choke filters, to suppress 120-Hz ripple. Although some of these inductors are open frame, most have iron covers to confine the powerful magnetic fields they produce.

Although builders and experimenters rarely construct their own capacitors, inductor fabrication is common. In fact, it is often necessary, since commercially available units may be unavailable or expensive. Even if available, they may consist of inductor stock to be trimmed to the required value. Core materials and wire for winding both solenoid and toroidal inductors are readily available. The following information includes fundamental formulas and design examples for calculating practical inductors, along with additional data on the theoretical limits in the use of some materials.

2.12.1 Air-Core Inductors

Many circuits require air-core inductors using just one layer of wire. The approximate inductance of a single-layer air-core inductor may be calculated from the simplified formula:

$$L\,(\mu H) = \frac{d^2 n^2}{18d + 40\ell} \qquad (108)$$

where
L = inductance in microhenrys,
d = inductor diameter in inches (from wire center to wire center),
ℓ = inductor length in inches, and
n = number of turns.

The notation is illustrated in **Fig 2.73**. This formula is a close approximation for inductors having a length equal to or greater than 0.4d. (Note: Inductance varies as the square of the turns. If the number of turns is doubled, the in-

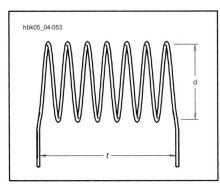

Fig 2.73 — Coil dimensions used in the inductance formula for air-core inductors.

ductance is quadrupled. This relationship is inherent in the equation, but is often overlooked. For example, to double the inductance, add additional turns equal to 1.4 times the original number of turns, or 40% more turns.)

Example: What is the inductance of an inductor if the inductor has 48 turns wound at 32 turns per inch and a diameter of ¾ inch? In this case, d = 0.75, ℓ = 48/32 = 1.5 and n = 48.

$$L\,(\mu H) = \frac{0.75^2 \times 48^2}{(18 \times 0.75) + (40 \times 1.5)}$$

$$= \frac{1300}{74} = 18\,\mu H$$

To calculate the number of turns of a single-layer inductor for a required value of inductance, the formula becomes:

$$n = \frac{\sqrt{L\,(18\,d + 40\,\ell)}}{d} \qquad (109)$$

Example: Suppose an inductance of 10.0 µH is required. The form on which the inductor is to be wound has a diameter of one inch and is long enough to accommodate an inductor of 1¼ inches. Then d = 1.00 inch, ℓ = 1.25 inches and L = 10.0. Substituting:

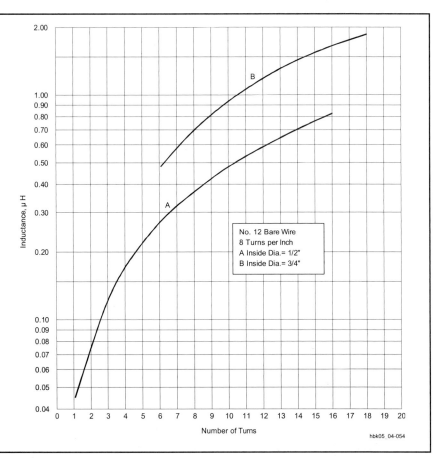

Fig 2.74 — Measured inductance of coils wound with #12 bare wire, eight turns to the inch. The values include half-inch leads.

$$n = \frac{\sqrt{10.0\,[\,(18 \times 1.0) + (40 \times 1.25)\,]}}{1}$$
$$= \sqrt{680} = 26.1 \text{ turns}$$

A 26-turn inductor would be close enough in practical work. Since the inductor will be 1.25 inches long, the number of turns per inch will be 26.1 / 1.25 = 20.9. Consulting the wire table in the **Component Data and References** chapter, we find that #17 AWG enameled wire (or anything smaller) can be used. The proper inductance is obtained by winding the required number of turns on the form and then adjusting the spacing between the turns to make a uniformly spaced inductor 1.25 inches long.

Most inductance formulas lose accuracy when applied to small inductors (such as are used in VHF work and in low-pass filters built for reducing harmonic interference to televisions) because the conductor thickness is no longer negligible in comparison with the size of the inductor. **Fig 2.74** shows the measured inductance of VHF inductors and may be used as a basis for circuit design. Two curves are given; curve A is for inductors wound to an inside diameter of ½ inch; curve B is for inductors of ¾-inch inside diameter. In both curves, the wire size is #12 AWG and the winding pitch is eight turns to the inch (⅛-inch turn spacing). The inductance values include leads ½-inch long.

Machine-wound inductors with the preset diameters and turns per inch are available in many radio stores, under the trade names of B&W Miniductor and Airdux. Information on using such coil stock is provided in the **Component Data and References** chapter to simplify the process of designing high-quality inductors for most HF applications.

Forming a wire into a solenoid increases its inductance, but also introduces distributed capacitance. Since each turn is at a slightly different ac potential, each pair of turns effectively forms a capacitor in parallel with part of the inductor. (See the chapter on **RF Techniques** for information on the effects of these and other factors that affect the behavior of the "ideal" inductors discussed in this chapter.)

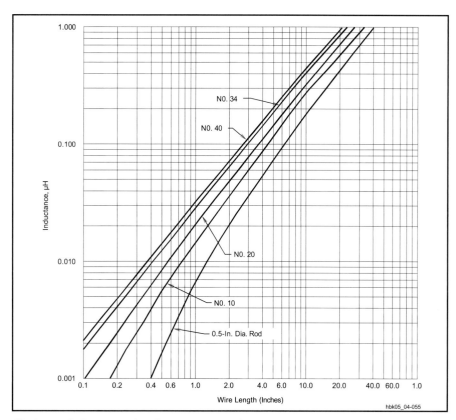

Fig 2.75 — Inductance of various conductor sizes as straight wires.

Moreover, the Q of air-core inductors is, in part, a function of the inductor shape, specifically its ratio of length to diameter. Q tends to be highest when these dimensions are nearly equal. With wire properly sized to the current carried by the inductor, and with high-caliber construction, air-core inductors can achieve Q above 200.

For a large collection of formulas useful in constructing air-core inductors of many configurations, see the "Circuit Elements" section in Terman's *Radio Engineers' Handbook*.

2.12.2 Straight-Wire Inductance

At low frequencies the inductance of a straight, round, nonmagnetic wire in free space is given by:

$$L = 0.00508\,b\left\{\left[\ln\left(\frac{2b}{a}\right)\right] - 0.75\right\} \quad (110)$$

where
L = inductance in µH,
a = wire radius in inches,
b = wire length in inches, and
ln = natural logarithm = 2.303 × common logarithm (base 10).

If the dimensions are expressed in millimeters instead of inches, the equation may still be used, except replace the 0.00508 value with 0.0002.

Skin effect reduces the inductance at VHF and above. As the frequency approaches infinity, the 0.75 constant within the brackets approaches unity. As a practical matter, skin

---

$$L = 0.0117\,b\left\{\log_{10}\left[\frac{2h}{a}\left(\frac{b + \sqrt{b^2 + a^2}}{b + \sqrt{b^2 + 4h^2}}\right)\right]\right\} + 0.00508\left(\sqrt{b^2 + 4h^2} - \sqrt{b^2 + a^2} + \frac{b}{4} - 2h + a\right)$$

where
L = inductance in µH
a = wire radius in inches
b = wire length parallel to ground plane in inches
h = wire height above ground plane in inches

Fig 2.76 — Equation for determining the inductance of a wire parallel to a ground plane, with one end grounded. If the dimensions are in millimeters, the numerical coefficients become 0.0004605 for the first term and 0.0002 for the second term.

Electrical Fundamentals 2.51

effect will not reduce the inductance by more than a few percent.

Example: What is the inductance of a wire that is 0.1575 inch in diameter and 3.9370 inches long? For the calculations, a = 0.0787 inch (radius) and b = 3.9370 inch.

$$L = 0.00508\, b \left\{ \left[\ln\left(\frac{2b}{a}\right) \right] - 0.75 \right\}$$

$$= 0.00508\,(3.9370) \times \left\{ \left[\ln\left(\frac{2 \times 3.9370}{0.0787}\right) \right] - 0.75 \right\}$$

$$= 0.020 \times [\ln(100) - 0.75]$$

$$= 0.020 \times (4.60 - 0.75)$$

$$= 0.020 \times 3.85 = 0.077\ \mu H$$

Fig 2.75 is a graph of the inductance for wires of various radii as a function of length.

A VHF or UHF tank circuit can be fabricated from a wire parallel to a ground plane, with one end grounded. A formula for the inductance of such an arrangement is given in **Fig 2.76**.

Example: What is the inductance of a wire 3.9370 inches long and 0.0787 inch in radius, suspended 1.5748 inch above a ground plane? (The inductance is measured between the free end and the ground plane, and the formula includes the inductance of the 1.5748-inch grounding link.) To demonstrate the use of the formula in Fig 2.76, begin by evaluating these quantities:

$$b + \sqrt{b^2 + a^2}$$

$$= 3.9370 + \sqrt{3.9370^2 + 0.0787^2}$$

$$= 3.9370 + 3.94 = 7.88$$

$$b + \sqrt{b^2 + 4(h^2)}$$

$$= 3.9370 + \sqrt{3.9370^2 + 4(1.5748^2)}$$

$$= 3.9370 + \sqrt{15.50 + 4(2.480)}$$

$$= 3.9370 + \sqrt{15.50 + 9.920}$$

$$= 3.9370 + 5.0418 = 8.9788$$

$$\frac{2h}{a} = \frac{2 \times 1.5748}{0.0787} = 40.0$$

$$\frac{b}{a} = \frac{3.9370}{4} = 0.098425$$

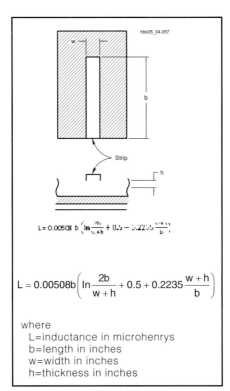

$$L = 0.00508\, b \left(\ln\frac{2b}{w+h} + 0.5 + 0.2235\frac{w+h}{b} \right)$$

where
L = inductance in microhenrys
b = length in inches
w = width in inches
h = thickness in inches

Fig 2.77 — Equation for determining the inductance of a flat strip inductor.

Substituting these values into the formula yields:

$$L = 0.0117 \times 3.9370$$

$$\left\{ \log_{10}\left[40.0 \times \left(\frac{7.88}{8.9788}\right) \right] \right\}$$

$$+ 0.00508 \times (5.0418 - 3.94 + 0.98425 -$$

$$3.1496 + 0.0787)$$

$$L = 0.0662\ \mu H$$

Another conductor configuration that is frequently used is a flat strip over a ground plane. This arrangement has lower skin-effect loss at high frequencies than round wire because it has a higher surface-area to volume ratio. The inductance of such a strip can be found from the formula in **Fig 2.77**.

2.12.3 Iron-Core Inductors

If the permeability of an iron core in an inductor is 800, then the inductance of any given air-wound inductor is increased 800 times by inserting the iron core. The inductance will be proportional to the magnetic flux through the inductor, other things being equal. The inductance of an iron-core inductor is highly dependent on the current flowing in the inductor, in contrast to an air-core inductor, where the inductance is independent of current because air does not saturate.

Iron-core inductors such as the one sketched in Fig 2.49 are used chiefly in power-supply equipment. They usually have direct current flowing through the winding, and any variation in inductance with current is usually undesirable. Inductance variations may be overcome by keeping the flux density below the saturation point of the iron. Opening the core so there is a small air gap will achieve this goal, as discussed in the earlier section on inductors. The reluctance or magnetic resistance introduced by such a gap is very large compared with that of the iron, even though the gap is only a small fraction of an inch. Therefore, the gap — rather than the iron — controls the flux density. Air gaps in iron cores reduce the inductance, but they hold the value practically constant regardless of the current magnitude.

When alternating current flows through an inductor wound on an iron core, a voltage is induced. Since iron is a conductor, eddy currents also flow in the core as discussed earlier. Eddy currents represent lost power because they flow through the resistance of the iron and generate heat. Losses caused by eddy currents can be reduced by laminating the core (cutting the core into thin strips). These strips or laminations are then insulated from each other by painting them with some insulating material such as varnish or shellac. Eddy current losses add to hysteresis losses, which are also significant in iron-core inductors.

Eddy-current and hysteresis losses in iron increase rapidly as the frequency of the alternating current increases. For this reason, ordinary iron cores can be used only at power-line and audio frequencies — up to approximately 15000 Hz. Even then, a very good grade of iron or steel is necessary for the core to perform well at the higher audio frequencies. Laminated iron cores become completely useless at radio frequencies because of eddy current and hysteresis losses.

2.12.4 Slug-Tuned Inductors

For RF work, the losses in iron cores can be reduced to a more useful level by grinding the iron into a powder and then mixing it with a binder of insulating material in such a way that the individual iron particles are insulated from each other. Using this approach, cores can be made that function satisfactorily even into the VHF range. Because a large part of the magnetic path is through a nonmagnetic material (the binder), the permeability of the powdered iron core is low compared with the values for solid iron cores used at power-line frequencies.

The slug is usually shaped in the form of a cylinder that fits inside the insulating form on which the inductor is wound. Despite the fact that the major portion of the magnetic path for the flux is in air, the slug is quite effective in increasing the inductor inductance. By pushing (or screwing) the slug in and out

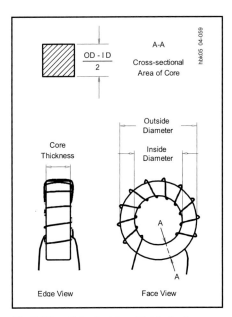

Fig 2.78 — A typical toroidal inductor wound on a powdered-iron or ferrite core. Some key physical dimensions are noted. Equally important are the core material, its permeability, its intended range of operating frequencies, and its A_L value. This is an 11-turn toroid.

of the inductor, the inductance can be varied over a considerable range.

2.12.5 Powdered-Iron Toroidal Inductors

For fixed-value inductors intended for use at HF and VHF, the powdered-iron toroidal core has become the standard in low- and medium-power circuits. **Fig 2.78** shows the general outlines of a toroidal inductor on a magnetic core.

Manufacturers offer a wide variety of core materials, or *mixes*, to create conductor cores that will perform over a desired frequency range with a reasonable permeability. Permeabilities for powdered-iron cores fall in the range of 3 to 35 for various mixes. In addition, core sizes are available in the range of 0.125-inch outside diameter (OD) up to 1.06-inch OD, with larger sizes to 5-inch OD available in certain mixes. The range of sizes permits the builder to construct single-layer inductors for almost any value using wire sized to meet the circuit current demands.

The use of powdered iron in a binder reduces core losses usually associated with iron, while the permeability of the core permits a reduction in the wire length and associated resistance in forming an inductor of a given inductance. Therefore, powdered-iron-core toroidal inductors can achieve Q well above 100, often approaching or exceeding 200 within the frequency range specified for a given core. Moreover, these inductors are considered *self-shielding* since most of the magnetic flux is within the core, a fact that simplifies circuit design and construction.

Each powdered-iron core has a value of A_L determined and published by the core manufacturer. For powdered-iron cores, A_L represents the *inductance index*, that is, the inductance in μH per 100 turns-squared of wire on the core, arranged in a single layer. The builder must select a core size capable of holding the calculated number of turns, of the required wire size, for the desired inductance. Otherwise, the inductor calculation is straightforward. To calculate the inductance of a powdered-iron toroidal inductor, when the number of turns and the core material are known, use the formula:

$$L = \frac{A_L \times N^2}{10000} \qquad (111)$$

where
L = the inductance in μH,
A_L = the inductance index in μH per 100 turns-squared, and
N = the number of turns.

Example: What is the inductance of a 60-turn inductor on a core with an A_L of 55? This A_L value was selected from manufacturer's information about a 0.8-inch OD core with an initial permeability of 10. This particular core is intended for use in the range of 2 to 30 MHz. See the **Component Data and References** chapter for more detailed data on the range of available cores.

$$L = \frac{A_L \times N^2}{10000} = \frac{55 \times 60^2}{10000}$$

$$= \frac{198000}{10000} = 19.8 \, \mu H$$

To calculate the number of turns needed for a particular inductance, use the formula:

$$N = 100\sqrt{\frac{L}{A_L}} \qquad (112)$$

Example: How many turns are needed for a 12.0-μH inductor if the A_L for the selected core is 49?

$$N = 100\sqrt{\frac{L}{A_L}} = 100\sqrt{\frac{12.0}{49}}$$

$$N = 100\sqrt{0.245} = 100 \times 0.495 = 49.5 \text{ turns}$$

If the value is critical, experimenting with 49-turn and 50-turn inductors is in order, especially since core characteristics may vary slightly from batch to batch. Count turns by each pass of the wire through the center of the core. (A straight wire through a toroidal core amounts to a one-turn inductor.) Fine adjustment of the inductance may be possible by spreading or compressing inductor turns.

The power-handling ability of toroidal cores depends on many variables, which include the cross-sectional area through the core, the core material, the numbers of turns in the inductor, the applied voltage and the operating frequency. Although powdered-iron cores can withstand dc flux densities up to 5000 gauss without saturating, ac flux densities from sine waves above certain limits can overheat cores. Manufacturers provide guideline limits for ac flux densities to avoid overheating. The limits range from 150 gauss at 1 MHz to 30 gauss at 28 MHz, although the curve is not linear. To calculate the maximum anticipated flux density for a particular inductor, use the formula:

$$B_{max} = \frac{E_{RMS} \times 10^8}{4.44 \times A_e \times N \times f} \qquad (113)$$

where
B_{max} = the maximum flux density in gauss,
E_{RMS} = the voltage across the inductor,
A_e = the cross-sectional area of the core in square centimeters,
N = the number of turns in the inductor, and
f = the operating frequency in Hz.

Example: What is the maximum ac flux density for an inductor of 15 turns if the frequency is 7.0 MHz, the RMS voltage is 25 V and the cross-sectional area of the core is 0.133 cm^2?

$$B_{max} = \frac{E_{RMS} \times 10^8}{4.44 \times A_e \times N \times f}$$

$$= \frac{25 \times 10^8}{4.44 \times 0.133 \times 15 \times 7.0 \times 10^6}$$

$$= \frac{25 \times 10^8}{62 \times 10^6} = 40 \text{ gauss}$$

Since the recommended limit for cores operated at 7 MHz is 57 gauss, this inductor is well within guidelines.

2.12.6 Ferrite Toroidal Inductors

Although nearly identical in general appearance to powdered-iron cores, ferrite cores differ in a number of important characteristics. Composed of nickel-zinc ferrites for lower permeability ranges and of manganese-zinc ferrites for higher permeabilities, these cores span a permeability range from 20 to above 10000. Nickel-zinc cores with permeabilities from 20 to 800 are useful in high-Q applications, but function more commonly in amateur applications as RF chokes. They are also useful in wide-band transformers, discussed in the chapter **RF Techniques**.

Electrical Fundamentals 2.53

Ferrite cores are often unpainted, unlike powdered-iron toroids. Ferrite toroids and rods often have sharp edges, while powdered-iron toroids usually have rounded edges.

Because of their higher permeabilities, the formulas for calculating inductance and turns require slight modification. Manufacturers list ferrite A_L values in mH per 1000 turns-squared. Thus, to calculate inductance, the formula is

$$L = \frac{A_L \times N^2}{1000000} \quad (114)$$

where
 L = the inductance in mH,
 A_L = the inductance index in mH per 1000 turns-squared, and
 N = the number of turns.

Example: What is the inductance of a 60-turn inductor on a core with an A_L of 523? (See the chapter **Component Data and References** for more detailed data on the range of available cores.)

$$L = \frac{A_L \times N^2}{1000000} = \frac{523 \times 60^2}{1000000}$$

$$= \frac{1.88 \times 10^6}{1 \times 10^6} = 1.88 \text{ mH}$$

To calculate the number of turns needed for a particular inductance, use the formula:

$$N = 1000 \sqrt{\frac{L}{A_L}} \quad (115)$$

Example: How many turns are needed for a 1.2-mH inductor if the A_L for the selected core is 150?

$$N = 1000 \sqrt{\frac{L}{A_L}} = 1000 \sqrt{\frac{1.2}{150}}$$

$$N = 1000 \sqrt{0.008} = 1000 \times 0.089 = 89 \text{ turns}$$

For inductors carrying both dc and ac currents, the upper saturation limit for most ferrites is a flux density of 2000 gauss, with power calculations identical to those used for powdered-iron cores. More detailed information is available on specific cores and manufacturers in the **Component Data and References** chapter.

2.13 Resonant Circuits

A circuit containing both an inductor and a capacitor — and therefore, both inductive and capacitive reactance — is often called a *tuned circuit* or a *resonant circuit*. For any such circuit, there is a particular frequency at which the inductive and capacitive reactances are the same, that is, $X_L = X_C$. For most purposes, this is the *resonant frequency* of the circuit. (Special considerations apply to parallel-resonant circuits; they will be covered in their own section.) At the resonant frequency — or at *resonance*, for short:

$$X_L = 2\pi f L = X_C = \frac{1}{2\pi f C}$$

By solving for f, we can find the resonant frequency of any combination of inductor and capacitor from the formula:

$$f = \frac{1}{2\pi \sqrt{LC}} \quad (116)$$

where
 f = frequency in hertz (Hz),
 L = inductance in henrys (H),
 C = capacitance in farads (F), and
 π = 3.1416.

For most high-frequency (HF) radio work, smaller units of inductance and capacitance and larger units of frequency are more convenient. The basic formula becomes:

$$f = \frac{10^3}{2\pi \sqrt{LC}} \quad (117)$$

where
 f = frequency in megahertz (MHz),
 L = inductance in microhenrys (μH),
 C = capacitance in picofarads (pF), and
 π = 3.1416.

Example: What is the resonant frequency of a circuit containing an inductor of 5.0 μH and a capacitor of 35 pF?

$$f = \frac{10^3}{2\pi \sqrt{LC}} = \frac{10^3}{6.2832 \sqrt{5.0 \times 35}}$$

$$= \frac{10^3}{83} = 12 \text{ MHz}$$

To find the matching component (inductor or capacitor) when the frequency and one component is known (capacitor or inductor) for general HF work, use the formula:

$$f^2 = \frac{1}{4\pi^2 LC} \quad (118)$$

where f, L and C are in basic units. For HF work in terms of MHz, μH and pF, the basic relationship rearranges to these handy formulas:

$$L = \frac{25330}{f^2 C} \quad (119)$$

$$C = \frac{25330}{f^2 L} \quad (120)$$

where
 f = frequency in MHz,
 L = inductance in μH, and
 C = capacitance in pF

For most radio work, these formulas will permit calculations of frequency and component values well within the limits of component tolerances.

Example: What value of capacitance is needed to create a resonant circuit at 21.1 MHz, if the inductor is 2.00 μH?

$$C = \frac{25330}{f^2 L} = \frac{25330}{(21.1^2 \times 2.0)}$$

$$= \frac{25330}{890} = 28.5 \text{ pF}$$

Fig 2.55 can also be used if an approximate answer is acceptable. From the horizontal axis, find the vertical line closest to the desired resonant frequency. Every pair of diagonals that cross on that vertical line represent a combination of inductance and capacitance that will resonate at that frequency. For example, if the desired frequency is 10 MHz, the pair of diagonals representing 5 μH and 50 pF cross quite close to that frequency. Interpolating between the given diagonals will provide more resolution — remember that all three sets of lines are spaced logarithmically.

Resonant circuits have other properties of importance, in addition to the resonant frequency, however. These include impedance, voltage drop across components in series-resonant circuits, circulating current in parallel-resonant circuits, and bandwidth. These properties determine such factors as the selectivity

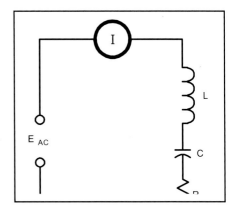

Fig 2.79 — A series circuit containing L, C, and R is resonant at the applied frequency when the reactance of C is equal to the reactance of L. The I in the circle is the schematic symbol for an ammeter.

of a tuned circuit and the component ratings for circuits handling significant amounts of power. Although the basic determination of the tuned-circuit resonant frequency ignored any resistance in the circuit, that resistance will play a vital role in the circuit's other characteristics.

2.13.1 Series-Resonant Circuits

Fig 2.79 presents a basic schematic diagram of a *series-resonant circuit*. Although most schematic diagrams of radio circuits would show only the inductor and the capacitor, resistance is always present in such circuits. The most notable resistance is associated with losses in the inductor at HF; resistive losses in the capacitor are low enough at those frequencies to be ignored. The current meter shown in the circuit is a reminder that in series circuits, the same current flows through all elements.

At resonance, the reactance of the capacitor cancels the reactance of the inductor. The voltage and current are in phase with each other, and the impedance of the circuit is determined solely by the resistance. The actual current through the circuit at resonance, and for frequencies near resonance, is determined by the formula:

$$I = \frac{E}{Z} = \frac{E}{\sqrt{R^2 + \left[2\pi f L - \frac{1}{(2\pi f C)}\right]^2}} \quad (121)$$

where all values are in basic units.

At resonance, the reactive factor in the formula is zero. (the bracketed expression under the square root symbol) As the frequency is shifted above or below the resonant frequency without altering component values, however, the reactive factor becomes significant, and the value of the current becomes smaller than at resonance. At frequencies far from resonance, the reactive components become dominant, and the resistance no longer significantly affects the current amplitude.

The exact curve created by recording the current as the frequency changes depends on the ratio of reactance to resistance. When the reactance of either the coil or capacitor is of the same order of magnitude as the resistance, the current decreases rather slowly as the frequency is moved in either direction away from resonance. Such a curve is said to be *broad*. Conversely, when the reactance is considerably larger than the resistance, the current decreases rapidly as the frequency moves away from resonance, and the circuit is said to be *sharp*. A sharp circuit will respond a great deal more readily to the resonant frequency than to

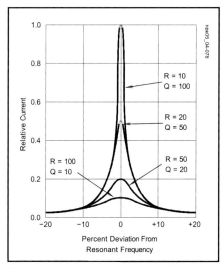

Fig 2.80 — Relative current in series-resonant circuits with various values of series resistance and Q. (An arbitrary maximum value of 1.0 represents current at resonance.) The reactance at resonance for all curves is 1000 Ω. Note that the current is hardly affected by the resistance in the circuit at frequencies more than 10% away from the resonant frequency.

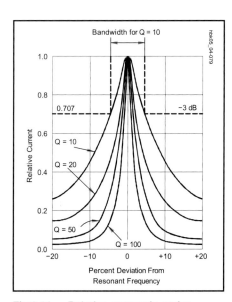

Fig 2.81 — Relative current in series-resonant circuits having different values of Q_U. The current at resonance is normalized to the same level for all curves in order to show the rate of change of decrease in current for each value of Q_U. The half-power points are shown to indicate relative bandwidth of the response for each curve. The bandwidth is indicated for a circuit with a Q_U of 10.

frequencies quite close to resonance; a broad circuit will respond almost equally well to a group or band of frequencies centered around the resonant frequency.

Both types of resonance curves are useful.

A sharp circuit gives good selectivity — the ability to respond strongly (in terms of current amplitude) at one desired frequency and to discriminate against others. A broad circuit is used when the apparatus must give about the same response over a band of frequencies, rather than at a single frequency alone.

Fig 2.80 presents a family of curves, showing the decrease in current as the frequency deviates from resonance. In each case, the inductive and capacitive reactances are assumed to be 1000 Ω. The maximum current, shown as a relative value on the graph, occurs with the lowest resistance, while the lowest peak current occurs with the highest resistance. Equally important, the rate at which the current decreases from its maximum value also changes with the ratio of reactance to resistance. It decreases most rapidly when the ratio is high and most slowly when the ratio is low.

UNLOADED Q

As noted in earlier sections of this chapter, Q is the ratio of reactance or stored energy to resistance or consumed energy. Since both terms of the ratio are measured in ohms, Q has no units and is known as the *quality factor* (and less frequently, the *figure of merit* or the *multiplying factor*). The resistive losses of the coil dominate the energy consumption in HF series-resonant circuits, so the inductor Q largely determines the resonant-circuit Q. Since this value of Q is independent of any external load to which the circuit might transfer power, it is called the *unloaded Q* or Q_U of the circuit.

Example: What is the unloaded Q of a series-resonant circuit with a loss resistance of 5 Ω and inductive and capacitive components having a reactance of 500 Ω each? With a reactance of 50 Ω each?

$$Q_{U1} = \frac{X1}{R} = \frac{500\,\Omega}{5\,\Omega} = 100$$

$$Q_{U2} = \frac{X2}{R} = \frac{50\,\Omega}{5\,\Omega} = 10$$

BANDWIDTH

Fig 2.81 is an alternative way of drawing the family of curves that relate current to frequency for a series-resonant circuit. By assuming that the peak current of each curve is the same, the rate of change of current for various values of Q_U and the associated ratios of reactance to resistance are more easily compared. From the curves, it is evident that the lower Q_U circuits pass current across a greater *bandwidth* of frequencies than the circuits with a higher Q_U. For the purpose of comparing tuned circuits, bandwidth is often defined as the frequency spread between the two frequencies at which the current ampli-

Electrical Fundamentals 2.55

tude decreases to 0.707 (or $1/\sqrt{2}$) times the maximum value. Since the power consumed by the resistance, R, is proportional to the square of the current, the power at these points is half the maximum power at resonance, assuming that R is constant for the calculations. The half-power, or –3 dB, points are marked on Fig 2.81.

For Q values of 10 or greater, the curves shown in Fig 2.81 are approximately symmetrical. On this assumption, bandwidth (BW) can be easily calculated:

$$BW = \frac{f}{Q_U} \quad (122)$$

where BW and f are in the same units, that is, in Hz, kHz or MHz.

Example: What is the bandwidth of a series-resonant circuit operating at 14 MHz with a Q_U of 100?

$$BW = \frac{f}{Q_U} = \frac{14 \text{ MHz}}{100} = 0.14 \text{ MHz} = 140 \text{ kHz}$$

The relationship between Q_U, f and BW provides a means of determining the value of circuit Q when inductor losses may be difficult to measure. By constructing the series-resonant circuit and measuring the current as the frequency varies above and below resonance, the half-power points can be determined. Then:

$$Q_U = \frac{f}{BW} \quad (123)$$

Example: What is the Q_U of a series-resonant circuit operating at 3.75 MHz, if the bandwidth is 375 kHz?

$$Q_U = \frac{f}{BW} = \frac{3.75 \text{ MHz}}{0.375 \text{ MHz}} = 10.0$$

Table 2.9 provides some simple formulas for estimating the maximum current and phase angle for various bandwidths, if both f and Q_U are known.

VOLTAGE DROP ACROSS COMPONENTS

The voltage drop across the coil and across the capacitor in a series-resonant circuit are each proportional to the reactance of the component for a given current (since E = I X). These voltages may be many times the applied voltage for a high-Q circuit. In fact, at resonance, the voltage drop is:

$$E_X = Q_U E_{AC} \quad (124)$$

where
E_X = the voltage across the reactive component,
Q_U = the circuit unloaded Q, and
E_{AC} = the applied voltage in Fig 2.79.

(Note that the voltage drop across the in-

Table 2.9
The Selectivity of Resonant Circuits

Approximate percentage of current at resonance[1] or of impedance at resonance[2]	Bandwidth (between half-power or –3 dB points on response curve)	Series circuit current phase angle (degrees)
95	f / 3Q	18.5
90	f / 2Q	26.5
70.7	f / Q	45
44.7	2f / Q	63.5
24.2	4f / Q	76
12.4	8f / Q	83

[1]For a series resonant circuit
[2]For a parallel resonant circuit

ductor is the vector sum of the voltages across the resistance and the reactance; however, for Q greater than 10, the error created by using this is not ordinarily significant.) Since the calculated value of E_X is the RMS voltage, the peak voltage will be higher by a factor of 1.414. Antenna couplers and other high-Q circuits handling significant power may experience arcing from high values of E_X, even though the source voltage to the circuit is well within component ratings.

CAPACITOR LOSSES

Although capacitor energy losses tend to be insignificant compared to inductor losses up to about 30 MHz, the losses may affect circuit Q in the VHF range. Leakage resistance, principally in the solid dielectric that forms the insulating support for the capacitor plates, is not exactly like the wire resistance losses in a coil. Instead of forming a series resistance, capacitor leakage usually forms a parallel resistance with the capacitive reactance. If the leakage resistance of a capacitor is significant enough to affect the Q of a series-resonant circuit, the parallel resistance (R_P) must be converted to an equivalent series resistance (R_S) before adding it to the inductor's resistance.

$$R_S = \frac{X_C^2}{R_P} = \frac{1}{R_P \times (2\pi f C)^2} \quad (125)$$

Example: A 10.0 pF capacitor has a leakage resistance of 10,000 Ω at 50.0 MHz. What is the equivalent series resistance?

$$R_S = \frac{1}{R_P \times (2\pi f C)^2}$$

$$= \frac{1}{1.0 \times 10^4 \times (6.283 \times 50.0 \times 10^6 \times 10.0 \times 10^{-12})^2}$$

$$= \frac{1}{1.0 \times 10^4 \times 9.87 \times 10^{-6}}$$

$$= \frac{1}{0.0987} = 10.1 \text{ }\Omega$$

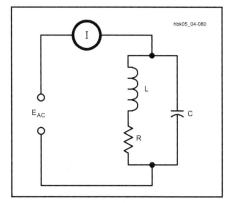

Fig 2.82 — A typical parallel-resonant circuit, with the resistance shown in series with the inductive leg of the circuit. Below a Q_U of 10, resonance definitions may lead to three separate frequencies which converge at higher Q_U levels. See text.

In calculating the impedance, current and bandwidth for a series-resonant circuit in which this capacitor might be used, the series-equivalent resistance of the unit is added to the loss resistance of the coil. Since inductor losses tend to increase with frequency because of skin effect, the combined losses in the capacitor and the inductor can seriously reduce circuit Q, without special component- and circuit-construction techniques.

2.13.2 Parallel-Resonant Circuits

Although series-resonant circuits are common, the vast majority of resonant circuits used in radio work are *parallel-resonant circuits*. **Fig 2.82** represents a typical HF parallel-resonant circuit. As is the case for series-resonant circuits, the inductor is the chief source of resistive losses, and these losses appear in series with the coil. Because current through parallel-resonant circuits is lowest at resonance, and impedance is highest, they are sometimes called *antiresonant* circuits. (You may encounter the old terms *acceptor* and *rejector* referring to series- and

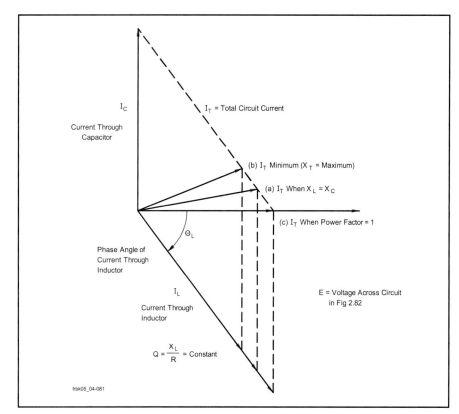

Fig 2.83 — Resonant conditions for a low- Q_U parallel circuit. Resonance may be defined as (a) $X_L = X_C$ (b) minimum current flow and maximum impedance or (c) voltage and current in phase with each other. With the circuit of Fig 2.82 and a Q_U of less than 10, these three definitions may represent three distinct frequencies.

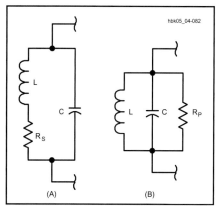

Fig 2.84 — Series and parallel equivalents when both circuits are resonant. The series resistance, R_S in A, is replaced by the parallel resistance, R_P in B, and vice versa. $R_P = X_L^2 / R_S$.

parallel-resonant circuits, respectively.)

Because the conditions in the two legs of the parallel circuit in Fig 2.82 are not the same — the resistance is in only one of the legs — all of the conditions by which series resonance is determined do not occur simultaneously in a parallel-resonant circuit. **Fig 2.83** graphically illustrates the situation by showing the currents through the two components. (Currents are drawn in the manner of complex impedances shown previously to show the phase angle for each current.) When the inductive and capacitive reactances are identical, the condition defined for series resonance is met as shown at point (a). The impedance of the inductive leg is composed of both X_L and R, which yields an impedance greater than X_C and that is not 180° out of phase with X_C. The resultant current is greater than its minimum possible value and not in phase with the voltage.

By altering the value of the inductor slightly (and holding the Q constant), a new frequency can be obtained at which the current reaches its minimum. When parallel circuits are tuned using a current meter as an indicator, this point (b) is ordinarily used as an indication of resonance. The current "dip" indicates a condition of maximum impedance and is sometimes called the *antiresonant* point or *maximum impedance resonance* to distinguish it from the condition at which $X_C = X_L$. Maximum impedance is achieved at this point by vector addition of X_C, X_L and R, however, and the result is a current somewhat out of phase with the voltage.

Point (c) in the figure represents the *unity-power-factor* resonant point. Adjusting the inductor value and hence its reactance (while holding Q constant) produces a new resonant frequency at which the resultant current is in phase with the voltage. The inductor's new value of reactance is the value required for a parallel-equivalent inductor and its parallel-equivalent resistor (calculated according to the formulas in the last section) to just cancel the capacitive reactance. The value of the parallel-equivalent inductor is always smaller than the actual inductor in series with the resistor and has a proportionally smaller reactance. (The parallel-equivalent resistor, conversely, will always be larger than the coil-loss resistor shown in series with the inductor.) The result is a resonant frequency slightly different from the one for minimum current and the one for $X_L = X_C$.

The points shown in the graph in Fig 2.83 represent only one of many possible situations, and the relative positions of the three resonant points do not hold for all possible cases. Moreover, specific circuit designs can draw some of the resonant points together, for example, compensating for the resistance of the coil by retuning the capacitor. The differences among these resonances are significant for circuit Q below 10, where the inductor's series resistance is a significant percentage of the reactance. Above a Q of 10, the three points converge to within a percent of the frequency and the differences between them can be ignored for practical calculations. Tuning for minimum current will not introduce a sufficiently large phase angle between voltage and current to create circuit difficulties.

PARALLEL CIRCUITS OF MODERATE TO HIGH Q

The resonant frequencies defined above converge in parallel-resonant circuits with Q higher than about 10. Therefore, a single set of formulas will sufficiently approximate circuit performance for accurate predictions. Indeed, above a Q of 10, the performance of a parallel circuit appears in many ways to be simply the inverse of the performance of a series-resonant circuit using the same components.

Accurate analysis of a parallel-resonant circuit requires the substitution of a parallel-equivalent resistor for the actual inductor-loss series resistor, as shown in **Fig 2.84**. Sometimes called the *dynamic resistance* of the parallel-resonant circuit, the parallel-equivalent resistor value will increase with circuit Q, that is, as the series resistance value decreases. To calculate the approximate parallel-equivalent resistance, use the formula:

$$R_P = \frac{X_L^2}{R_S} = \frac{(2\pi f L)^2}{R_S} = Q_U X_L \qquad (126)$$

Example: What is the parallel-equivalent resistance for a coil with an inductive reactance of 350 Ω and a series resistance of 5.0 Ω at resonance?

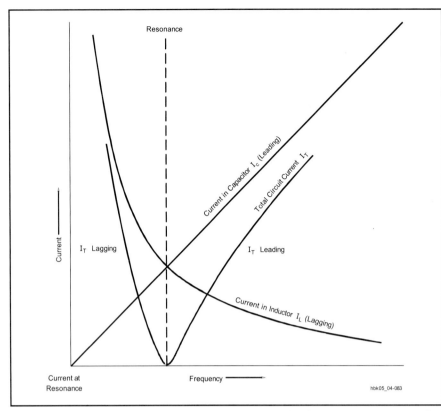

Fig 2.85 — The currents in a parallel-resonant circuit as the frequency moves through resonance. Below resonance, the current lags the voltage; above resonance the current leads the voltage. The base line represents the current level at resonance, which depends on the impedance of the circuit at that frequency.

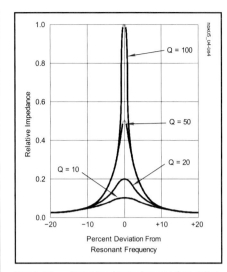

Fig 2.86 — Relative impedance of parallel-resonant circuits with different values of Q_U. The curves are similar to the series-resonant circuit current level curves of Fig 2.80. The effect of Q_U on impedance is most pronounced within 10% of the resonance frequency.

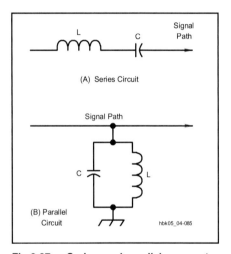

Fig 2.87 — Series- and parallel-resonant circuits configured to perform the same theoretical task: passing signals in a narrow band of frequencies along the signal path. A real design example would consider many other factors.

$$R_P = \frac{X_L^2}{R_S} = \frac{(350\,\Omega)^2}{5.0\,\Omega}$$

$$= \frac{122{,}500\,\Omega^2}{5.0\,\Omega} = 24{,}500\,\Omega$$

Since the coil Q_U remains the inductor's reactance divided by its series resistance, the coil Q_U is 70. Multiplying Q_U by the reactance also provides the approximate parallel-equivalent resistance of the coil series resistance.

At resonance, where $X_L = X_C$, R_P defines the impedance of the parallel-resonant circuit. The reactances just equal each other, leaving the voltage and current in phase with each other. In other words, the circuit shows only the parallel resistance. Therefore, equation 126 can be rewritten as:

$$Z = \frac{X_L^2}{R_S} = \frac{(2\pi f L)^2}{R_S} = Q_U X_L \quad (127)$$

In this example, the circuit impedance at resonance is 24,500 Ω.

At frequencies below resonance the current through the inductor is larger than that through the capacitor, because the reactance of the coil is smaller and that of the capacitor is larger than at resonance. There is only partial cancellation of the two reactive currents, and the total current therefore is larger than the current taken by the resistance alone. At frequencies above resonance the situation is reversed and more current flows through the capacitor than through the inductor, so the total current again increases.

The current at resonance, being determined wholly by R_P, will be small if R_P is large, and large if R_P is small. **Fig 2.85** illustrates the relative current flows through a parallel-tuned circuit as the frequency is moved from below resonance to above resonance. The base line represents the minimum current level for the particular circuit. The actual current at any frequency off resonance is simply the vector sum of the currents through the parallel equivalent resistance and through the reactive components.

To obtain the impedance of a parallel-tuned circuit either at or off the resonant frequency, apply the general formula:

$$Z = \frac{Z_C Z_L}{Z_S} \quad (128)$$

where

Z = overall circuit impedance
Z_C = impedance of the capacitive leg (usually, the reactance of the capacitor),
Z_L = impedance of the inductive leg (the vector sum of the coil's reactance and resistance), and
Z_S = series impedance of the capacitor-inductor combination as derived from the denominator of equation 121.

After using vector calculations to obtain Z_L and Z_S, converting all the values to polar form — as described earlier in this chapter — will ease the final calculation. Of course, each impedance may be derived from the resistance and the application of the basic reactance formulas on the values of the inductor and capacitor at the frequency of interest.

Since the current rises away from resonance, the parallel-resonant-circuit impedance must fall. It also becomes complex,

Table 2.10
The Performance of Parallel-Resonant Circuits
A. High- and Low-Q Circuits (in relative terms)

Characteristic	High-Q Circuit	Low-Q Circuit
Selectivity	high	low
Bandwidth	narrow	wide
Impedance	high	low
Total current	low	high
Circulating current	high	low

B. Off-Resonance Performance for Constant Values of Inductance and Capacitance

Characteristic	Above Resonance	Below Resonance
Inductive reactance	increases	decreases
Capacitive reactance	decreases	increases
Circuit resistance	unchanged*	unchanged*
Relative impedance	decreases	decreases
Total current	increases	increases
Circulating current	decreases	decreases
Circuit impedance	capacitive	inductive

*This is true for frequencies near resonance. At distant frequencies, skin effect may alter the resistive losses of the inductor.

resulting in an ever-greater phase difference between the voltage and the current. The rate at which the impedance falls is a function of Q_U. **Fig 2.86** presents a family of curves showing the impedance drop from resonance for circuit Q ranging from 10 to 100. The curve family for parallel-circuit impedance is essentially the same as the curve family for series-circuit current.

As with series-resonant circuits, the higher the Q of a parallel-tuned circuit, the sharper will be the response peak. Likewise, the lower the Q, the wider the band of frequencies to which the circuit responds. Using the half-power (–3 dB) points as a comparative measure of circuit performance, equations 122 and 123 apply equally to parallel-tuned circuits. That is, $BW = f / Q_U$ and $Q_U = f / BW$, where the resonant frequency and the bandwidth are in the same units. As a handy reminder, **Table 2.10** summarizes the performance of parallel-resonant circuits at high and low Q and above and below resonant frequency.

It is possible to use either series- or parallel-resonant circuits to do the same work in many circuits, thus giving the designer considerable flexibility. **Fig 2.87** illustrates this general principle by showing a series-resonant circuit in the signal path and a parallel-resonant circuit shunted from the signal path to ground. Assume both circuits are resonant at the same frequency, f, and have the same Q. The series-resonant circuit at A has its lowest impedance at f, permitting the maximum possible current to flow along the signal path. At all other frequencies, the impedance is greater and the current at those frequencies is less. The circuit passes the desired signal and tends to impede signals at undesired frequencies. The parallel circuit at B provides the highest impedance at resonance, f, making the signal path the lowest impedance path for the signal. At frequencies off resonance, the parallel-resonant circuit presents a lower impedance, thus presenting signals with a path to ground and away from the signal path. In theory, the effects will be the same relative to a signal current on the signal path. In actual circuit design exercises, of course, many other variables will enter the design picture to make one circuit preferable to the other.

CIRCULATING CURRENT

In a parallel-resonant circuit, the source voltage is the same for all the circuit elements. The current in each element, however, is a function of the element's reactance.

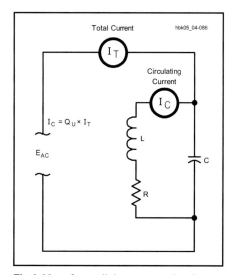

Fig 2.88 — A parallel-resonant circuit redrawn to illustrate both the total current and the circulating current.

Fig 2.88 redraws the parallel-resonant circuit to indicate the total current and the current circulating between the coil and the capacitor. The current drawn from the source may be low, because the overall circuit impedance is high. The current through the individual elements may be high, however, because there is little resistive loss as the current circulates through the inductor and capacitor. For parallel-resonant circuits with an unloaded Q of 10 or greater, this *circulating current* is approximately:

$$I_C = Q_U I_T \qquad (129)$$

where
- I_C = circulating current in A, mA or µA,
- Q_U = unloaded circuit Q, and
- I_T = total current in the same units as I_C.

Example: A parallel-resonant circuit permits an ac or RF total current of 30 mA and has a Q of 100. What is the circulating current through the elements?

$$I_X = Q_U I = 100 \times 30 \text{ mA} = 3000 \text{ mA} = 3 \text{ A}$$

Circulating currents in high-Q parallel-tuned circuits can reach a level that causes component heating and power loss. Therefore, components should be rated for the anticipated circulating currents, and not just the total current.

LOADED Q

In many resonant-circuit applications, the only power lost is that dissipated in the resistance of the circuit itself. At frequencies below 30 MHz, most of this resistance is in the coil. Within limits, increasing the number of turns in the coil increases the reactance faster than it raises the resistance, so coils for circuits in which the Q must be high are made with relatively large inductances for the frequency.

When the circuit delivers energy to a load

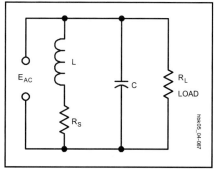

Fig 2.89 — A loaded parallel-resonant circuit, showing both the inductor-loss resistance and the load, R_L. If smaller than the inductor resistance, R_L will control the loaded Q of the circuit (Q_L).

(as in the case of the resonant circuits used in transmitters), the energy consumed in the circuit itself is usually negligible compared with that consumed by the load. The equivalent of such a circuit is shown in **Fig 2.89**, where the parallel resistor, R_L, represents the load to which power is delivered. If the power dissipated in the load is at least 10 times as great as the power lost in the inductor and capacitor, the parallel impedance of the resonant circuit itself will be so high compared with the resistance of the load that for all practical purposes the impedance of the combined circuit is equal to the load impedance. Under these conditions, the load resistance replaces the circuit impedance in calculating Q. The Q of a parallel-resonant circuit loaded by a resistive impedance is:

$$Q_L = \frac{R_L}{X} \tag{130}$$

where
 Q_L = circuit loaded Q,
 R_L = parallel load resistance in ohms, and
 X = reactance in ohms of either the inductor or the capacitor.

Example: A resistive load of 3000 Ω is connected across a resonant circuit in which the inductive and capacitive reactances are each 250 Ω. What is the circuit Q?

$$Q_L = \frac{R_L}{X} = \frac{3000\,\Omega}{250\,\Omega} = 12$$

The effective Q of a circuit loaded by a parallel resistance increases when the reactances are decreased. A circuit loaded with a relatively low resistance (a few thousand ohms) must have low-reactance elements (large capacitance and small inductance) to have reasonably high Q. Many power-handling circuits, such as the output networks of transmitters, are designed by first choosing a loaded Q for the circuit and then determining component values. See the chapter on **RF Power Amplifiers** for more details.

Parallel load resistors are sometimes added to parallel-resonant circuits to lower the circuit Q and increase the circuit bandwidth. By using a high-Q circuit and adding a parallel resistor, designers can tailor the circuit response to their needs. Since the parallel resistor consumes power, such techniques ordinarily apply to receiver and similar low-power circuits, however.

Example: Specifications call for a parallel-resonant circuit with a bandwidth of 400 kHz at 14.0 MHz. The circuit at hand has a Q_U of 70.0 and its components have reactances of 350 Ω each. What is the parallel load resistor that will increase the bandwidth to the specified value? The bandwidth of the existing circuit is:

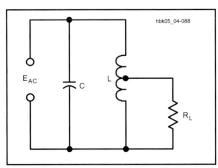

Fig 2.90 — A parallel-resonant circuit with a tapped inductor to effect an impedance match. Although the impedance presented to E_{AC} is very high, the impedance at the connection of the load, R_L, is lower.

$$BW = \frac{f}{Q_U} = \frac{14.0\,\text{MHz}}{70.0} = 0.200\,\text{MHz}$$
$$= 200\,\text{kHz}$$

The desired bandwidth, 400 kHz, requires a circuit with a Q of:

$$Q = \frac{f}{BW} = \frac{14.0\,\text{MHz}}{0.400\,\text{MHz}} = 35.0$$

Since the desired Q is half the original value, halving the resonant impedance or parallel-resistance value of the circuit is in order. The present impedance of the circuit is:

$$Z = Q_U\,X_L = 70.0 \times 350\,\Omega = 24500\,\Omega$$

The desired impedance is:

$$Z = Q_U\,X_L = 35.0 \times 350\,\Omega$$
$$= 12250\,\Omega = 12.25\,\text{k}\Omega$$

or half the present impedance.

A parallel resistor of 24,500 Ω, or the nearest lower value (to guarantee sufficient bandwidth), will produce the required reduction in Q and bandwidth increase. Although this example simplifies the situation encountered in real design cases by ignoring such factors as the shape of the band-pass curve, it illustrates the interaction of the ingredients that determine the performance of parallel-resonant circuits.

IMPEDANCE TRANSFORMATION

An important application of the parallel-resonant circuit is as an impedance matching device. Circuits and antennas often need to be connected to other circuits or feed lines that do not have the same impedance. To transfer power effectively requires a circuit that will convert or "transform" the impedances so that each connected device or system can operate properly.

Fig 2.90 shows such a situation where the source, E_{AC}, operates at a high impedance, but the load, R_L, operates at a low impedance. The technique of impedance transformation shown in the figure is to connect the parallel-resonant circuit, which has a high impedance, across the source, but connect the load across only a portion of the coil. (This is called *tapping the coil* and the connection point is a *tap*.) The coil acts as an *autotransformer*, described in the following section, with the magnetic field of the coil shared between what are effectively two coils in series, the upper coil having many turns and the lower coil fewer turns. Energy stored in the field induces larger voltages in the many-turn coil than it does in the fewer-turn coil, "stepping down" the input voltage so that energy can be extracted by the load at the required lower voltage-to-current ratio (which is impedance). The correct tap point on the coil usually has to be experimentally determined, but the technique is very effective.

When the load resistance has a very low value (say below 100 Ω) it may be connected in series in the resonant circuit (such as R_S in Fig 2.84A, for example), in which case the series L-R circuit can be transformed to an equivalent parallel L-R circuit as previously described. If the Q is at least 10, the equivalent parallel impedance is:

$$Z_R = \frac{X^2}{R_L} \tag{131}$$

where
 Z_R = resistive parallel impedance at resonance,
 X = reactance (in ohms) of either the coil or the capacitor, and
 R_L = load resistance inserted in series.

If the Q is lower than 10, the reactance will have to be adjusted somewhat — for the reasons given in the discussion of low-Q resonant circuits — to obtain a resistive impedance of the desired value.

These same techniques work in either "direction" — with a high-impedance source and low-impedance load or vice versa. Using a parallel-resonant circuit for this application does have some disadvantages. For instance, the common connection between the input and the output provides no dc isolation. Also, the common ground is sometimes troublesome with regard to ground-loop currents. Consequently, a circuit with only mutual magnetic coupling is often preferable. With the advent of ferrites, constructing impedance transformers that are both broadband and permit operation well up into the VHF portion of the spectrum has become relatively easy. The basic principles of broadband impedance transformers appear in the **RF Techniques** chapter.

2.14 Transformers

When the ac source current flows through every turn of an inductor, the generation of a counter-voltage and the storage of energy during each half cycle is said to be by virtue of self-inductance. If another inductor — not connected to the source of the original current — is positioned so the magnetic field of the first inductor intercepts the turns of the second inductor, coupling the two inductors and creating mutual inductance as described earlier, a voltage will be induced and current will flow in the second inductor. A load such as a resistor may be connected across the second inductor to consume the energy transferred magnetically from the first inductor.

Fig 2.91 illustrates a pair of coupled inductors, showing an ac energy source connected to one, called the *primary inductor*, and a load connected to the other, called the *secondary inductor*. If the inductors are wound tightly on a magnetic core so that nearly all or magnetic flux from the first inductor intersects with the turns of the second inductor, the pair is said to be *tightly coupled*. Inductors not sharing a common core and separated by a distance would be *loosely coupled*.

The signal source for the primary inductor may be household ac power lines, audio or other waveforms at low frequencies, or RF currents. The load may be a device needing power, a speaker converting electrical energy into sonic energy, an antenna using RF energy for communications or a particular circuit set up to process a signal from a preceding circuit. The uses of magnetically coupled energy in electronics are innumerable.

Mutual inductance (M) between inductors is measured in henrys. Two inductors have a mutual inductance of 1 H under the following conditions: as the primary inductor current changes at a rate of 1 A/s, the voltage across the secondary inductor is 1 V. The level of mutual inductance varies with many factors: the size and shape of the inductors, their relative positions and distance from each other, and the permeability of the inductor core material and of the space between them.

If the self-inductance values of two inductors are known (self-inductance is used in this section to distinguish it from the mutual inductance), it is possible to derive the mutual inductance by way of a simple experiment schematically represented in **Fig 2.92**. Without altering the physical setting or position of two inductors, measure the total coupled inductance, L_C, of the series-connected inductors with their windings complementing each other and again with their windings opposing each other. Since, for the two inductors, $L_C = L1 + L2 + 2M$, in the complementary case, and $L_O = L1 + L2 - 2M$ for the opposing case,

$$M = \frac{L_C - L_O}{4} \quad (132)$$

The ratio of magnetic flux set up by the secondary inductor to the flux set up by the primary inductor is a measure of the extent to which two inductors are coupled, compared to the maximum possible coupling between them. This ratio is the *coefficient of coupling* (k) and is always less than 1. If k were to equal 1, the two inductors would have the maximum possible mutual coupling. Thus:

$$M = k\sqrt{L1\,L2} \quad (133)$$

where
M = mutual inductance in henrys,
L1 and L2 = individual coupled inductors, each in henrys, and
k = the coefficient of coupling.

Using the experiment above, it is possible to solve equation 133 for k with reasonable accuracy.

Any two inductors having mutual inductance comprise a *transformer* having a *primary winding* or inductor and a *secondary winding* or inductor. The word "winding" is generally dropped so that transformers are said to have "primaries" and "secondaries." **Fig 2.93** provides a pictorial representation of a typical iron-core transformer, along with the schematic symbols for both iron-core and air-core transformers. Conventionally, the term *transformer* is most commonly applied to coupled inductors having a magnetic core material, while coupled air-wound inductors are not called by that name. They

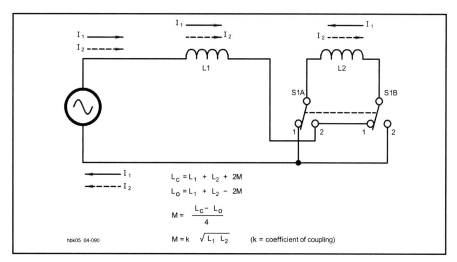

Fig 2.92 — An experimental setup for determining mutual inductance. Measure the inductance with the switch in each position and use the formula in the text to determine the mutual inductance.

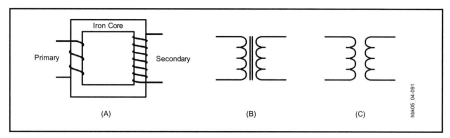

Fig 2.93 — A transformer. A is a pictorial diagram. Power is transferred from the primary coil to the secondary by means of the magnetic field. B is a schematic diagram of an iron-core transformer, and C is an air-core transformer.

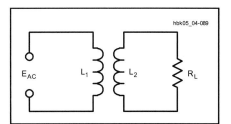

Fig 2.91 — A basic transformer: two inductors — one connected to an ac energy source, the other to a load — with coupled magnetic fields.

are still transformers, however.

We normally think of transformers as ac devices, since mutual inductance only occurs when magnetic fields are changing. A transformer connected to a dc source will exhibit mutual inductance only at the instants of closing and opening the primary circuit, or on the rising and falling edges of dc pulses, because only then does the primary winding have a changing field. There are three principle uses of transformers: to physically isolate the primary circuit from the secondary circuit, to transform voltages and currents from one level to another, and to transform circuit impedances from one level to another. These functions are not mutually exclusive and have many variations.

2.14.1 Basic Transformer Principles

The primary and secondary windings of a transformer may be wound on a core of magnetic material. The permeability of the magnetic material increases the inductance of the windings so a relatively small number of turns may be used to induce a given voltage value with a small current. A closed core having a continuous magnetic path, such as that shown in Fig 2.93, also tends to ensure that practically all of the field set up by the current in the primary winding will intercept or "cut" the turns of the secondary winding.

For power transformers and impedance-matching transformers used at audio frequencies, cores made of soft iron strips or sheets called *laminations* are most common and generally very efficient. At higher frequencies, ferrite or powdered-iron cores are more frequently used. This section deals with basic transformer operation at audio and power frequencies. RF transformer operation is discussed in the chapter on **RF Techniques**.

The following principles presume a coefficient of coupling (k) of 1, that is, a perfect transformer. The value k = 1 indicates that all the turns of both windings link with all the magnetic flux lines, so that the voltage induced per turn is the same with both windings. This condition makes the induced voltage independent of the inductance of the primary and secondary inductors. Iron-core transformers for low frequencies closely approach this ideal condition. **Fig 2.94** illustrates the conditions for transformer action.

VOLTAGE RATIO

For a given varying magnetic field, the voltage induced in an inductor within the field is proportional to the number of turns in the inductor. When the two windings of a transformer are in the same field (which is the case when both are wound on the same closed core), it follows that the induced voltages will

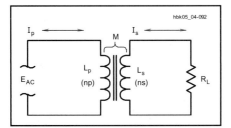

Fig 2.94 — The conditions for transformer action: two coils that exhibit mutual inductance, an ac power source, and a load. The magnetic field set up by the energy in the primary circuit transfers energy to the secondary for use by the load, resulting in a secondary voltage and current.

be proportional to the number of turns in each winding. In the primary, the induced voltage practically equals, and opposes, the applied voltage, as described earlier. Hence:

$$E_S = E_P \left(\frac{N_S}{N_P}\right) \qquad (134)$$

where
E_S = secondary voltage,
E_P = primary applied voltage,
N_S = number of turns on secondary, and
N_P = number of turns on primary.

Example: A transformer has a primary with 400 turns and a secondary with 2800 turns, and a voltage of 120 V is applied to the primary. What voltage appears across the secondary winding?

$$E_S = 120 \text{ V} \left(\frac{2800}{400}\right) = 120 \text{ V} \times 7 = 840 \text{ V}$$

(Notice that the number of turns is taken as a known value rather than a measured quantity, so they do not limit the significant figures in the calculation.) Also, if 840 V is applied to the 2800-turn winding (which then becomes the primary), the output voltage from the 400-turn winding will be 120 V.

Either winding of a transformer can be used as the primary, provided the winding has enough turns (enough inductance) to induce a voltage equal to the applied voltage without requiring an excessive current. The windings must also have insulation with a voltage rating sufficient for the voltages applied or created. Transformers are called *step-up* or *step-down* transformers depending on whether the secondary voltage is higher or lower than the primary voltage, respectively.

CURRENT OR AMPERE-TURNS RATIO

The current in the primary when no current is taken from the secondary is called the *magnetizing current* of the transformer. An ideal transformer, with no internal losses, would consume no power, since the current through the primary inductor would be 90° out of phase with the voltage. In any properly designed transformer, the power consumed by the transformer when the secondary is open (not delivering power) is only the amount necessary to overcome the losses in the iron core and in the resistance of the wire with which the primary is wound.

When power is transferred from the secondary winding to a load, the secondary current creates a magnetic field that opposes the field established by the primary current. For the induced voltage in the primary to equal the applied voltage, the original magnetizing field must be maintained. Therefore, enough additional current must flow in the primary to create a field exactly equal and opposite to the field set up by the secondary current, leaving the original magnetizing field.

In practical transformer calculations it may be assumed that the entire primary current is caused by the secondary load. This is justifiable because the magnetizing current should be very small in comparison with the primary load current at rated power output.

If the magnetic fields set up by the primary and secondary currents are to be equal, the number of ampere-turns must be equal in each winding. (See the previous discussion of magnetic fields and magnetic flux density.) Thus, primary current multiplied by the primary turns must equal the secondary current multiplied by the secondary turns.

$$I_P N_P = I_S N_S$$

and

$$I_P = I_S \left(\frac{N_S}{N_P}\right) \qquad (135)$$

where
I_P = primary current,
I_S = secondary current,
N_P = number of turns in the primary winding, and
N_S = number of turns in the secondary winding.

Example: Suppose the secondary of the transformer in the previous example is delivering a current of 0.20 A to a load. What will be the primary current?

$$I_P = 0.20 \text{ A} \left(\frac{2800}{400}\right) = 0.20 \text{ A} \times 7 = 1.4 \text{ A}$$

Although the secondary voltage is higher than the primary voltage, the secondary current is lower than the primary current, and by the same ratio. The secondary current in an ideal transformer is 180° out of phase with the primary current, since the field in the secondary just offsets the field in the primary. The phase relationship between the currents in the windings holds true no matter what the phase difference between the current and the voltage of the secondary. In fact, the phase difference, if any, between voltage and current in the secondary winding will be *reflected* back to the primary as an identical phase difference.

Power Ratio

A transformer cannot create power; it can only transfer it and change the voltage and current ratios. Hence, the power taken from the secondary cannot exceed that taken by the primary from the applied voltage source. There is always some power loss in the resistance of the windings and in the iron core, so in all practical cases the power taken from the source will exceed that taken from the secondary.

$$P_O = \eta P_I \quad (136)$$

where

P_O = power output from secondary,
P_I = power input to primary, and
η = efficiency factor.

The efficiency, η, is always less than 1 and is commonly expressed as a percentage: if η is 0.65, for instance, the efficiency is 65%.

Example: A transformer has an efficiency of 85.0% at its full-load output of 150 W. What is the power input to the primary at full secondary load?

$$P_I = \frac{P_O}{\eta} = \frac{150 \text{ W}}{0.850} = 176 \text{ W}$$

A transformer is usually designed to have the highest efficiency at the power output for which it is rated. The efficiency decreases with either lower or higher outputs. On the other hand, the losses in the transformer are relatively small at low output but increase as more power is taken. The amount of power that the transformer can handle is determined by its own losses, because these losses heat the wire and core. There is a limit to the temperature rise that can be tolerated, because too high a temperature can either melt the wire or cause the insulation to break down. A transformer can be operated at reduced output, even though the efficiency is low, because the actual loss will be low under such conditions. The full-load efficiency of small power transformers such as are used in radio receivers and transmitters usually lies between about 60 and 90%, depending on the size and design.

IMPEDANCE RATIO

In an ideal transformer — one without losses or leakage inductance (see Transformer Losses) — the primary power, $P_P = E_P I_P$, and secondary power, $P_S = E_S I_S$, are equal. The relationships between primary and secondary voltage and current are also known (equations 134 and 135). Since impedance is the ratio of voltage to current, $Z = E/I$, the impedances represented in each winding are related as follows:

$$Z_P = Z_S \left(\frac{N_P}{N_S}\right)^2 \quad (137)$$

where

Z_P = impedance at the primary terminals from the power source,
Z_S = impedance of load connected to secondary, and
N_P/N_S = turns ratio, primary to secondary.

A load of any given impedance connected to the transformer secondary will thus be transformed to a different value at the primary terminals. The impedance transformation is proportional to the square of the primary-to-secondary turns ratio. (Take care to use the primary-to-secondary turns ratio, since the secondary-to-primary ratio is more commonly used to determine the voltage transformation ratio.)

The term *looking into* is often used to mean the conditions observed from an external perspective at the terminals specified. For example, "impedance looking into" the transformer primary means the impedance measured externally to the transformer at the terminals of the primary winding.

Example: A transformer has a primary-to-secondary turns ratio of 0.6 (the primary has six-tenths as many turns as the secondary) and a load of 3000 Ω is connected to the secondary. What is the impedance looking into the primary of the transformer?

$$Z_P = 3000 \text{ Ω} \times (0.6)^2 = 3000 \text{ Ω} \times 0.36$$

$$Z_P = 1080 \text{ Ω}$$

By choosing the proper turns ratio, the impedance of a fixed load can be transformed to any desired value, within practical limits. If transformer losses can be neglected, the transformed (reflected) impedance has the same phase angle as the actual load impedance. Thus, if the load is a pure resistance, the load presented by the primary to the power source will also be a pure resistance. If the load impedance is complex, that is, if the load current and voltage are out of phase with each other, then the primary voltage and current will have the same phase angle.

Many devices or circuits require a specific value of load resistance (or impedance) for optimum operation. The impedance of the actual load that is to dissipate the power may be quite different from the impedance of the source device or circuit, so an *impedance-matching transformer* is used to change the actual load into an impedance of the desired value. The turns ratio required is:

$$\frac{N_P}{N_S} = \sqrt{\frac{Z_P}{Z_S}} \quad (138)$$

where

N_P/N_S = required turns ratio, primary to secondary,
Z_P = primary impedance required, and
Z_S = impedance of load connected to secondary.

Example: A transistor audio amplifier requires a load of 150 Ω for optimum performance, and is to be connected to a loudspeaker having an impedance of 4.0 Ω. What primary-to-secondary turns ratio is required in the coupling transformer?

$$\frac{N_P}{N_S} = \sqrt{\frac{Z_P}{Z_S}} = \frac{N_P}{N_S} \sqrt{\frac{150 \text{ Ω}}{4.0 \text{ Ω}}} = \sqrt{38} = 6.2$$

The primary therefore must have 6.2 times as many turns as the secondary.

These relationships may be used in practical circuits even though they are based on an ideal transformer. Aside from the normal design requirements of reasonably low internal losses and low leakage reactance, the only other requirement is that the primary have enough inductance to operate with low magnetizing current at the voltage applied to the primary.

The primary terminal impedance of an iron-core transformer is determined wholly by the load connected to the secondary and by the turns ratio. If the characteristics of the transformer have an appreciable effect on the impedance presented to the power source, the transformer is either poorly designed or is not suited to the voltage and frequency at which it is being used. Most transformers will operate quite well at voltages from slightly above to well below the design figure.

TRANSFORMER LOSSES

In practice, none of the formulas given so far provides truly exact results, although they afford reasonable approximations. Transformers in reality are not simply two coupled inductors, but a network of resistances and reactances, most of which appear in **Fig 2.95**. Since only the terminals numbered 1 through 4 are accessible to the user, transformer ratings and specifications take into account the additional losses created by these complexities.

In a practical transformer not all of the magnetic flux is common to both windings, although in well-designed transformers the amount of flux that cuts one winding and not the other is only a small percentage of the total flux. This *leakage flux* causes a voltage by self-induction in the winding creating the flux. The effect is the same as if a small *leakage inductance* existed independently of the main windings. Leakage inductance acts in exactly the same way as inductance inserted in series with the winding. It has, therefore, a certain reactance, depending on the amount

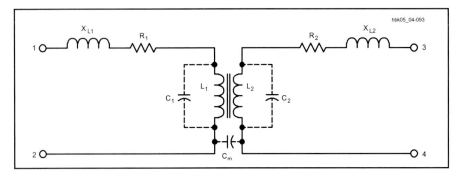

Fig 2.95 — A transformer as a network of resistances, inductances and capacitances. Only L1 and L2 contribute to the transfer of energy.

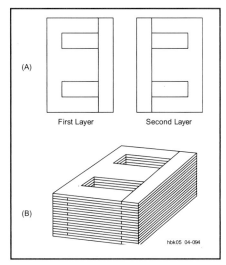

Fig 2.96 — A typical transformer iron core. The E and I pieces alternate direction in successive layers to improve the magnetic path while attenuating eddy currents in the core.

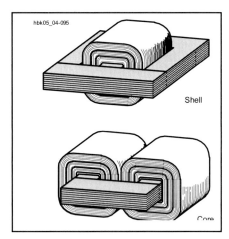

Fig 2.97 — Two common transformer constructions: shell and core.

of leakage inductance and the frequency. This reactance is called *leakage reactance* and is shown as X_{L1} and X_{L2} in Fig 2.95.

Current flowing through the leakage reactance causes a voltage drop. This voltage drop increases with increasing current; hence, it increases as more power is taken from the secondary. The resistances of the transformer windings, R1 and R2, also cause voltage drops when there is current. Although these voltage drops are not in phase with those caused by leakage reactance, together they result in a lower secondary voltage under load than is indicated by the transformer turns ratio. Thus, in a practical transformer, the greater the secondary current, the smaller the secondary terminal voltage becomes.

At ac line frequencies (50 or 60 Hz), the voltage at the secondary, with a reasonably well-designed iron-core transformer, should not drop more than about 10% from open-circuit conditions to full load. The voltage drop may be considerably more than this in a transformer operating at audio frequencies, because the leakage reactance increases with frequency.

In addition to wire resistances and leakage reactances, certain unwanted or "stray" capacitances occur in transformers. The wire forming the separate turns of the windings acts as the plates of a small capacitor, creating a capacitance between turns and between the windings. This *distributed capacitance* appears in Fig 2.95 as C1, C2, and C_M. Moreover, transformer windings can exhibit capacitance relative to nearby metal, for example, the chassis, the shield and even the core. When current flows through a winding, each turn has a slightly different voltage than its adjacent turns. This voltage causes a small current to flow in these *interwinding* and *winding-to-winding capacitances*.

Although stray capacitances are of little concern with power and audio transformers, they become important as the frequency increases. In transformers for RF use, the stray capacitance can resonate with either the leakage reactance or, at lower frequencies, with the winding reactances, L1 or L2, especially under very light or zero loads. In the frequency region around resonance, transformers no longer exhibit the properties formulated above or the impedance properties to be described below.

Iron-core transformers also experience losses within the core itself. *Hysteresis losses* include the energy required to overcome the retentivity of the core's magnetic material. Circulating currents through the core's resistance are *eddy currents*, which form part of the total core losses. These losses, which add to the required magnetizing current, are equivalent to adding a resistance in parallel with L1 in Fig 2.95.

CORE CONSTRUCTION

Audio and power transformers usually employ silicon steel as the core material. With permeabilities of 5000 or greater, these cores saturate at flux densities approaching 10^5 lines of flux per square inch of cross section. The cores consist of thin insulated laminations to break up potential eddy current paths.

Each core layer consists of an "E" and an "I" piece butted together, as represented in **Fig 2.96**. The butt point leaves a small gap. Each layer is reversed from the adjacent layers so that each gap is next to a continuous magnetic path so that the effect of the gaps is minimized. This is different from the air-gapped inductor of Fig 2.49 in which the air gap is maintained for all layers of laminations.

Two core shapes are in common use, as shown in **Fig 2.97**. In the shell type, both windings are placed on the inner leg, while in the core type the primary and secondary windings may be placed on separate legs, if desired. This is sometimes done when it is necessary to minimize capacitance between the primary and secondary, or when one of the windings must operate at very high voltage.

The number of turns required in the primary for a given applied voltage is determined by the size, shape and type of core material used, as well as the frequency. The number of turns required is inversely proportional to the cross-sectional area of the core. As a rough indication, windings of small power transformers frequently have about six to eight turns per volt on a core of 1-square-inch cross section and have a magnetic path 10 or 12 inches in length. A longer path or smaller cross section requires more turns per volt, and vice versa.

In most transformers the windings are wound in layers, with a thin sheet of treated-paper insulation between each layer. Thicker insulation is used between adjacent windings and between the first winding and the core.

SHIELDING

Because magnetic lines of force are continuous loops, shielding requires a complete path for the lines of force of the leakage flux. The high-permeability of iron cores tends to concentrate the field, but additional shielding is often needed. As depicted in **Fig 2.98**,

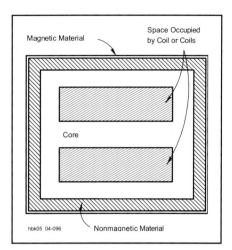

Fig 2.98 — A shielded transformer cross-section: the core plus an outer shield of magnetic material contain nearly all of the magnetic field.

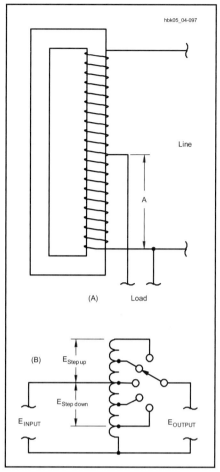

Fig 2.99 — The autotransformer is based on the transformer, but uses only one winding. The pictorial diagram at A shows the typical construction of an autotransformer. The schematic diagram at B demonstrates the use of an autotransformer to step up or step down ac voltage, usually to compensate for excessive or deficient line voltage.

enclosing the transformer in a good magnetic material can restrict virtually all of the magnetic field in the outer case. The nonmagnetic material between the case and the core creates a region of high reluctance, attenuating the field before it reaches the case.

2.14.2 Autotransformers

The transformer principle can be used with only one winding instead of two, as shown in **Fig 2.99A**. The principles that relate voltage, current and impedance to the turns ratio also apply equally well. A one-winding transformer is called an *auto-transformer*. The current in the common section (A) of the winding is the difference between the line (primary) and the load (secondary) currents, since these currents are out of phase. Hence, if the line and load currents are nearly equal, the common section of the winding may be wound with comparatively small wire. The line and load currents will be equal only when the primary (line) and secondary (load) voltages are not very different.

Autotransformers are used chiefly for boosting or reducing the power-line voltage by relatively small amounts. Fig 2.99B illustrates the principle schematically with a switched, stepped autotransformer. Continuously variable autotransformers are commercially available under a variety of trade names; Variac and Powerstat are typical examples.

Technically, tapped air-core inductors, such as the one in the network in Fig 2.90 at the close of the discussion of resonant circuits, are also autotransformers. The voltage from the tap to the bottom of the winding is less than the voltage across the entire winding. Likewise, the impedance of the tapped part of the winding is less than the impedance of the entire winding. Because in this case, leakage reactances are great and the coefficient of coupling is quite low, the relationships that are true in a perfect transformer grow quite unreliable in predicting the exact values. For this reason, tapped inductors are rarely referred to as transformers. The stepped-down situation in Fig 2.90 is better approximated — at or close to resonance — by the formula

$$R_P = \frac{R_L \, X_{COM}^2}{X_L} \qquad (139)$$

where

R_P = tuned-circuit parallel-resonant impedance,
R_L = load resistance tapped across part of the winding,
X_{COM} = reactance of the portion of the winding common to both the resonant circuit and the load tap, and
X_L = reactance of the entire winding.

The result is approximate and applies only to circuits with a Q of 10 or greater.

2.15 Heat Management

While not strictly an electrical fundamental, managing the heat generated by electronic circuits is important in nearly all types of radio equipment. Thus, the topic is included in this chapter. Information on the devices and circuits discussed in this section may be found in other chapters.

Any actual energized circuit consumes electric power because any such circuit contains components that convert electricity into other forms of energy. This dissipated power appears in many forms. For example, a loudspeaker converts electrical energy into sound, the motion of air molecules. An antenna (or a light bulb) converts electricity into electromagnetic radiation. Charging a battery converts electrical energy into chemical energy (which is then converted back to electrical energy upon discharge). But the most common transformation by far is the conversion, through some form of resistance, of electricity into heat.

Sometimes the power lost to heat serves a useful purpose — toasters and hair dryers come to mind. But most of the time, this heat represents a power loss that is to be minimized wherever possible or at least taken into account. Since all real circuits contain resistance, even those circuits (such as a loudspeaker) whose primary purpose is to convert electricity to some *other* form of energy also convert some part of their input power to heat. Often, such losses are negligible, but sometimes they are not.

If unintended heat generation becomes significant, the involved components will get warm. Problems arise when the temperature increase affects circuit operation by either

• causing the component to fail, by explosion, melting, or other catastrophic event, or, more subtly,

• causing a slight change in the properties of the component, such as through a temperature coefficient (TC).

In the first case, we can design conservatively, ensuring that components are rated to safely handle two, three or more times the maximum power we expect them to dissipate.

Electrical Fundamentals

In the second case, we can specify components with low TCs, or we can design the circuit to minimize the effect of any one component. Occasionally we even exploit temperature effects (for example, using a resistor, capacitor or diode as a temperature sensor). Let's look more closely at the two main categories of thermal effects.

Not surprisingly, heat dissipation (more correctly, the efficient removal of generated heat) becomes important in medium- to high-power circuits: power supplies, transmitting circuits and so on. While these are not the only examples where elevated temperatures and related failures are of concern, the techniques we will discuss here are applicable to all circuits.

2.15.1 Thermal Resistance

The transfer of heat energy, and thus the change in temperature, between two ends of a block of material is governed by the following heat flow equation and illustrated in **Fig 2.100**):

$$P = \frac{kA}{L}\Delta T = \frac{\Delta T}{\theta} \quad (140)$$

where
- P = power (in the form of heat) conducted between the two points
- k = *thermal conductivity*, measured in W/(m °C), of the material between the two points, which may be steel, silicon, copper, PC board material and so on
- L = length of the block
- A = area of the block, and
- ΔT = *difference* in temperature between the two points.

Thermal conductivities of various common materials at room temperature are given in **Table 2.11**.

The form of equation 140 is the same as the variation of Ohm's Law relating current flow to the ratio of a difference in potential to resistance; I = E/R. In this case, what's flowing is heat (P), the difference in potential is a temperature difference (ΔT), and what's resisting the flow of heat is the *thermal resistance*;

$$\theta = \frac{L}{kA} \quad (141)$$

with units of °C/W. (The units of resistance are equivalent to V/A.) The analogy is so apt that the same principles and methods apply to heat flow problems as circuit problems. The following correspondences hold:
- Thermal conductivity W/(m °C) ↔ Electrical conductivity (S/m).
- Thermal resistance (°C/W) ↔ Electrical resistance (Ω).
- Thermal current (heat flow) (W) ↔ Electrical current (A).

Table 2.11
Thermal Conductivities of Various Materials
Gases at 0 °C, Others at 25 °C; from *Physics*, by Halliday and Resnick, 3rd Ed.

Material	k in units of W/m°C
Aluminum	200
Brass	110
Copper	390
Lead	35
Silver	410
Steel	46
Silicon	150
Air	0.024
Glass	0.8
Wood	0.08

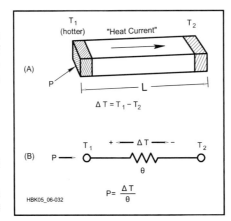

Fig 2.100 — Physical and "circuit" models for the heat-flow equation.

- Thermal potential (T) ↔ Electrical potential (V).
- Heat source ↔ Power source.

For example, calculate the temperature of a 2-inch (0.05 m) long piece of #12 copper wire at the end that is being heated by a 25 W (input power) soldering iron, and whose other end is clamped to a large metal vise (assumed to be an infinite heat sink), if the ambient temperature is 25 °C (77 °F).

First, calculate the thermal resistance of the copper wire (diameter of #12 wire is 2.052 mm, cross-sectional area is 3.31 × 10⁻⁶ m²)

$$\theta = \frac{L}{kA} = \frac{(0.05 \text{ m})}{(390 \text{ W}/(\text{m °C})) \times (3.31 \times 10^{-6} \text{ m}^2)}$$

= 38.7 °C/W

Then, rearranging the heat flow equation above yields (after assuming the heat energy actually transferred to the wire is around 10 W)

$$\Delta T = P \theta = (10 \text{ W}) \times (38.7 \text{ °C}/\text{W}) = 387 \text{ °C}$$

So the wire temperature at the hot end is 25 C + ΔT = 412 °C (or 774 °F). If this sounds a little high, remember that this is for the steady state condition, where you've been holding the iron to the wire for a long time.

From this example, you can see that things can get very hot even with application of moderate power levels. For this reason, circuits that generate sufficient heat to alter, not necessarily damage, the components must employ some method of cooling, either active or passive. Passive methods include heat sinks or careful component layout for good ventilation. Active methods include forced air (fans) or some sort of liquid cooling (in some high-power transmitters).

2.15.2 Heat Sink Selection and Use

The purpose of a heat sink is to provide a high-power component with a large surface area through which to dissipate heat. To use the models above, it provides a low thermal-resistance path to a cooler temperature, thus allowing the hot component to conduct a large "thermal current" away from itself.

Power supplies probably represent one of the most common high-power circuits amateurs are likely to encounter. Everyone has certainly noticed that power supplies get warm or even hot if not ventilated properly. Performing the thermal design for a properly cooled power supply is a very well-defined process and is a good illustration of heat-flow concepts.

This material was originally prepared by ARRL Technical Advisor Dick Jansson, KD1K, during the design of a 28-V, 10-A power supply. (The chapter **Analog Basics** discusses semiconductor diodes, rectifiers, and transistors. The **Power Supplies** chapter has more information on power supply design.) An outline of the design procedure shows the logic applied:

1. Determine the expected power dissipation (P_{in}).
2. Identify the requirements for the dissipating elements (maximum component temperature).
3. Estimate heat-sink requirements.
4. Rework the electronic device (if necessary) to meet the thermal requirements.
5. Select the heat exchanger (from heat sink data sheets).

The first step is to estimate the filtered, unregulated supply voltage under full load. Since the transformer secondary output is 32 V ac (RMS) and feeds a full-wave bridge rectifier, let's estimate 40 V as the filtered dc output at a 10-A load.

The next step is to determine our critical components and estimate their power dissipations. In a regulated power supply, the pass transistors are responsible for nearly all the power lost to heat. Under full load, and allowing for some small voltage drops in the

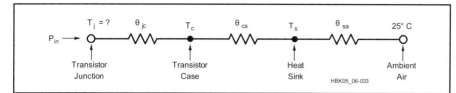

Fig 2.101 — Resistive model of thermal conduction in a power transistor and associated heat sink. See text for calculations.

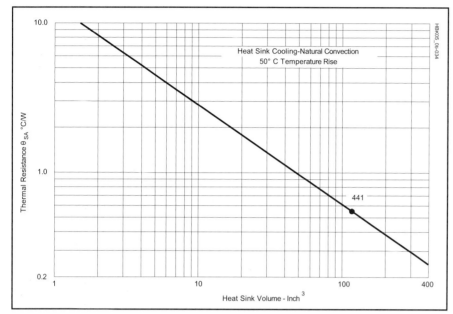

Fig 2.102 — Thermal resistance vs heat-sink volume for natural convection cooling and 50 °C temperature rise. The graph is based on engineering data from Wakefield Thermal Solutions, Inc.

power-transistor emitter circuitry, the output of the series pass transistors is about 29 V for a delivered 28 V under a 10-A load. With an unregulated input voltage of 40 V, the total energy heat dissipated in the pass transistors is (40 V – 29 V) × 10 A = 110 W. The heat sink for this power supply must be able to handle that amount of dissipation and still keep the transistor junctions below the specified safe operating temperature limits. It is a good rule of thumb to select a transistor that has a maximum power dissipation of twice the desired output power.

Now, consider the ratings of the pass transistors to be used. This supply calls for 2N3055s as pass transistors. The data sheet shows that a 2N3055 is rated for 15-A service and 115-W dissipation. But the design uses *four* in parallel. Why? Here we must look past the big, bold type at the top of the data sheet to such subtle characteristics as the junction-to-case thermal resistance, q_{jc}, and the maximum allowable junction temperature, T_j.

The 2N3055 data sheet shows $\theta_{jc} = 1.52$ °C/W, and a maximum allowable case (and junction) temperature of 220 °C. While it seems that one 2N3055 could barely, on paper at least, handle the electrical requirements — at what temperature would it operate?

To answer that, we must model the entire "thermal circuit" of operation, starting with the transistor junction on one end and ending at some point with the ambient air. A reasonable model is shown in **Fig 2.101**. The ambient air is considered here as an infinite heat sink; that is, its temperature is assumed to be a constant 25 °C (77 °F). θ_{jc} is the thermal resistance from the transistor junction to its case. θ_{cs} is the resistance of the mounting interface between the transistor case and the heat sink. θ_{sa} is the thermal resistance between the heat sink and the ambient air. In this "circuit," the generation of heat (the "thermal current source") occurs in the transistor at P_{in}.

Proper mounting of most TO-3 package power transistors such as the 2N3055 requires that they have an electrical insulator between the transistor case and the heat sink. However, this electrical insulator must at the same time exhibit a low thermal resistance. To achieve a quality mounting, use thin polyimid or mica formed washers and a suitable thermal compound to exclude air from the interstitial space. "Thermal greases" are commonly available for this function. Any silicone grease may be used, but filled silicone oils made specifically for this purpose are better.

Using such techniques, a conservatively high value for θ_{cs} is 0.50 °C/W. Lower values are possible, but the techniques needed to achieve them are expensive and not generally available to the average amateur. Furthermore, this value of θ_{cs} is already much lower than θ_{jc}, which cannot be lowered without going to a somewhat more exotic pass transistor.

Finally, we need an estimate of θ_{sa}. **Fig 2.102** shows the relationship of heat-sink volume to thermal resistance for natural-convection cooling. This relationship presumes the use of suitably spaced fins (0.35 inch or greater) and provides a "rough order-of-magnitude" value for sizing a heat sink. For a first calculation, let's assume a heat sink of roughly 6 × 4 × 2 inch (48 cubic inches). From Fig 2.102, this yields a θ_{sa} of about 1 °C/W.

Returning to Fig 2.101, we can now calculate the approximate temperature increase of a single 2N3055:

$\delta T = P\, \theta_{total}$

$= 110 \text{ W} \times (1.52 \text{ °C/W} + 0.5 \text{ °C/W} + 1.0 \text{ °C/W})$

$= 332 \text{ °C}$

Given the ambient temperature of 25 °C, this puts the junction temperature T_j of the 2N3055 at 25 + 332 = 357 °C! This is clearly too high, so let's work backward from the air end and calculate just how many transistors we need to handle the heat.

First, putting more 2N3055s in parallel means that we will have the thermal model illustrated in **Fig 2.103**, with several identical θ_{jc} and θ_{cs} in parallel, all funneled through the same θ_{as} (we have one heat sink).

Keeping in mind the physical size of the project, we could comfortably fit a heat sink of approximately 120 cubic inches (6 × 5 × 4 inches), well within the range of commercially available heat sinks. Furthermore, this application can use a heat sink where only "wire access" to the transistor connections is required. This allows the selection of a more efficient design. In contrast, RF designs require the transistor mounting surface to be completely exposed so that the PC board can be mounted close to the transistors to minimize parasitics. Looking at Fig 2.102, we see that a 120-cubic-inch heat sink yields a θ_{sa} of 0.55 °C/W. This means that the temperature of the heat sink when dissipating 110 W will be 25 °C + (110 W × 0.55 °C/W) = 85.5 °C.

Industrial experience has shown that silicon transistors suffer substantial failure when junctions are operated at highly elevated temperatures. Most commercial and military specifications will usually not permit design junction temperatures to exceed 125 °C. To arrive at a safe figure for our maximum al-

Electrical Fundamentals 2.67

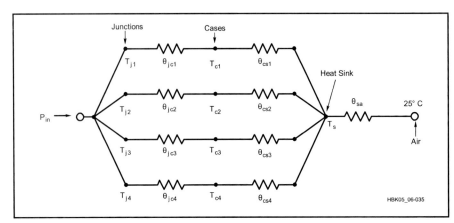

Fig 2.103 — Thermal model for multiple power transistors mounted on a common heat sink.

lowed T_j, we must consider the intended use of the power supply. If we are using it in a 100% duty-cycle transmitting application such as RTTY or FM, the circuit will be dissipating 110 W continuously. For a lighter duty-cycle load such as CW or SSB, the "key-down" temperature can be slightly higher as long as the average is less than 125 °C. In this intermittent type of service, a good conservative figure to use is $T_j = 150$ °C.

Given this scenario, the temperature rise across each transistor can be $150 - 85.5 = 64.5$ °C. Now, referencing Fig 2.103, remembering the total θ for each 2N3055 is $1.52 + 0.5 = 2.02$ °C/W, we can calculate the maximum power each 2N3055 can safely dissipate:

$$P = \frac{\delta T}{\theta} = \frac{64.5 \text{ °C}}{2.02 \text{ °C/W}} = 31.9 \text{ W}$$

Thus, for 110 W full load, we need four 2N3055s to meet the thermal requirements of the design. Now comes the big question: What is the "right" heat sink to use? We have already established its requirements: it must be capable of dissipating 110 W, and have a θ_{sa} of 0.55 °C/W (see above).

A quick consultation with several manufacturer's catalogs reveals that Wakefield Thermal Solutions, Inc. model nos. 441 and 435 heat sinks meet the needs of this application. A Thermalloy model no. 6441 is suitable as well. Data published in the catalogs of these manufacturers show that in natural-convection service, the expected temperature rise for 100 W dissipation would be just under 60 °C, an almost perfect fit for this application. Moreover, the no. 441 heat sink can easily mount four TO-3-style 2N3055 transistors as shown in **Fig 2.104**. Remember: heat sinks should be mounted with the fins and transistor mounting area vertical to promote convection cooling.

The design procedure just described is applicable to any circuit where heat buildup is a potential problem. By using the thermal-resistance model, we can easily calculate whether or not an external means of cooling is necessary, and if so, how to choose it. Aside from heat sinks, forced air cooling (fans) is another common method. In commercial transceivers, heat sinks with forced-air cooling are common.

2.15.3 Transistor Derating

Maximum ratings for power transistors are usually based on a case temperature of 25 °C. These ratings will decrease with increasing operating temperature. Manufacturer's data sheets usually specify a *derating* figure or curve that indicates how the maximum ratings change per degree rise in temperature. If such information is not available (or even if it is!), it is a good rule of thumb to select a power transistor with a maximum power dissipation of at least twice the desired output power.

2.15.4 Rectifiers

Diodes are physically quite small, and they operate at high current densities. As a result their heat-handling capabilities are somewhat limited. Normally, this is not a problem in high-voltage, low-current supplies in which rectifiers in axial-lead DO-type packages are used. (See the **Component Data and References** chapter for information on device packages.) The use of high-current (2 A or greater) rectifiers at or near their maximum ratings, however, requires some form of heat sinking. The average power dissipated by a rectifier is

$$P = I_{AVG} \times V_F \quad (142)$$

where
I_{AVG} is the average current, and
V_F is the forward voltage drop.

Average current must account for the conduction duty cycle and the forward voltage drop must be determined at the average current level.

Rectifiers intended for such high-current applications are available in a variety of packages suitable for mounting to flat surfaces. Frequently, mounting the rectifier on the main chassis (directly, or with thin mica insulating washers) will suffice. If the diode is insulated from the chassis, thin layers of thermal compound or thermal insulating washers should be used to ensure good heat conduction. Large, high-current rectifiers often require special heat sinks to maintain a safe operating temperature. Forced-air cooling is sometimes used as a further aid.

2.15.5 RF Heating

RF current often causes component heating problems where the same level of dc current may not. An example is the tank circuit of

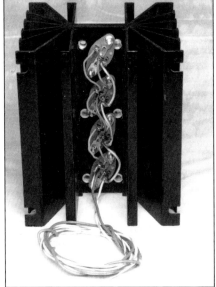

Fig 2.104 — A Wakefield 441 heat sink with four 2N3055 transistors mounted.

an RF oscillator. If several small capacitors are connected in parallel to achieve a desired capacitance, skin effect will be reduced and the total surface area available for heat dissipation will be increased, thus significantly reducing the RF heating effects as compared to a single large capacitor. This technique can be applied to any similar situation; the general idea is to divide the heating among as many components as possible.

2.15.6 Forced-Air and Water Cooling

In Amateur Radio today, forced-air cooling is most commonly found in vacuum-tube circuits or in power supplies built in small enclosures, such as those in solid-state transceivers or computers. Fans or blowers are commonly specified in cubic feet per minute (CFM). While the nomenclature and specifications differ from those used for heat sinks, the idea remains the same: to offer a low thermal resistance between the inside of the enclosure and the (ambient) exterior.

For forced air cooling, we basically use the "one resistor" thermal model of Fig 2.100. The important quantity to be determined is heat generation, P_{in}. For a power supply, this can be easily estimated as the difference between the input power, measured at the transformer primary, and the output power at full load. For variable-voltage supplies, the worst-case output condition is minimum voltage with maximum current. A discussion of forced-air cooling for vacuum tube equipment appears in the **RF Power Amplifiers** chapter.

Dust build-up is a common problem for forced-air cooling systems, even with powerful blowers and fans. If air intake grills and vents are not kept clean and free of lint and debris, air flow can be significantly reduced, leading to excessive equipment temperature and premature failure. Cleaning of air passageways should be included in regular equipment maintenance for good performance and maximum equipment life.

Water cooling systems are much less common in amateur equipment, used primarily for high duty cycle operating, such as RTTY, and at frequencies where the efficiency of the amplifier is relatively low, such as UHF and microwaves. In these situations, water cooling is used because water can absorb and transfer more than 3000 times as much heat as the same volume of air!

The main disadvantage of water cooling is that it requires pumps, hoses, and reservoirs whereas a fan or blower is all that is required for forced-air cooling. For high-voltage circuits, using water cooling also requires special insulation techniques and materials to allow water to circulate in close contact with the heat source while remaining electrically isolated.

Nevertheless, the technique can be effective. The increased availability of inexpensive materials designed for home sprinkler and other low-pressure water distribution systems make water-cooling less difficult to implement. It is recommended that the interested reader review articles and projects in the amateur literature to observe the successful implementation of water cooling systems.

2.15.7 Heat Pipe Cooling

A heat pipe is a device containing a *working fluid* in a reservoir where heat is absorbed and a channel to a second reservoir where heat is dissipated. Heat pipes work by *evaporative cooling*. The heat-absorbing reservoir is placed in thermal contact with the heat source which transfers heat to the working fluid, usually a liquid substance with a boiling point just above room temperature. The working fluid vaporizes and the resulting vapor pressure pushes the hot vapor through the channel to the cooling reservoir.

In the cooling reservoir, the working fluid gives up its heat of vaporization, returning to the fluid state. The cooled fluid then flows back through the channel to the heat-absorbing reservoir where the process is repeated.

Heat pipes require no fans or pumps — movement of the working fluid is driven entirely from the temperature difference between the two reservoirs. The higher the temperature difference between the absorbing and dissipating reservoirs, the more effective the heat pump becomes, up to the limit of the dissipating reservoir to dissipate heat.

The principle application for heat pipes is for space operations. Heat pipe applications in terrestrial applications are limited due to the gravity gradient sensitivity of heat pipes. At present, the only amateur equipment making use of heat pipes are computers and certain amplifier modules. Nevertheless, as more general-purpose products become available, this technique will become more common.

2.15.8 Thermoelectric Cooling

Thermoelectric cooling makes use of the *Peltier effect* to create heat flow across the junction of two different types of materials. This process is related to the *thermoelectric effect* by which thermocouples generate voltages based on the temperature of a similar junction. A *thermoelectric cooler* or *TEC* (also known as a *Peltier cooler*) requires only a source of dc power to cause one side of the device to cool and the other side to warm. TECs are available with different sizes and power ratings for different applications.

TECs are not available with sufficient heat transfer capabilities that they can be used in high-power applications, such as RF amplifiers. However, they can be useful in lowering the temperature of sensitive receiver circuits, such as preamplifiers used at UHF and microwave frequencies, or imaging devices, such as charge-coupled devices (CCDs). Satellites use TECs as *radiative coolers* that dissipate heat directly as thermal or infrared radiation. TECs are also found in some computing equipment where they are used to remove heat from microprocessors and other large integrated circuits.

2.15.9 Temperature Compensation

Aside from catastrophic failure, temperature changes may also adversely affect circuits if the temperature coefficient (TC) of one or more components is too large. If the resultant change is not too critical, adequate temperature stability can often be achieved simply by using higher-precision components with low TCs (such as NP0/C0G capacitors or metal-film resistors). For applications where this is impractical or impossible (such as many solid-state circuits), we can minimize temperature sensitivity by *compensation* or *matching* — using temperature coefficients to our advantage.

Compensation is accomplished in one of two ways. If we wish to keep a certain circuit quantity constant, we can interconnect pairs of components that have equal but opposite TCs. For example, a resistor with a negative TC can be placed in series with a positive TC resistor to keep the total resistance constant. Conversely, if the important point is to keep the *difference* between two quantities constant, we can use components with the *same* TC so that the pair "tracks." That is, they both change by the same amount with temperature.

An example of this is a Zener reference circuit. Since a diode is strongly affected by operating temperature, circuits that use diodes or transistors to generate stable reference voltages must use some form of temperature compensation. Since, for a constant current, a reverse-biased PN junction has a negative voltage TC while a forward-biased junction has a positive voltage TC, a good way to temperature-compensate a Zener reference diode is to place one or more forward-biased diodes in series with it.

2.15.10 Thermistors

Thermistors can be used to control temperature and improve circuit behavior or protect against excessive temperatures, hot or cold. Circuit temperature variations can affect gain, distortion or control functions like receiver AGC or transmitter ALC. Thermistors can be used in circuits that compensate for temperature changes, such as the simplified AGC system shown in **Fig 2.105** where temperature-sensitive resistors ($R_{th1} - R_{th3}$) are used to counteract the thermal changes of other resistors.

In this discussion, William Sabin, WØIYH describes how a simple temperature control circuit can be used to protect a power transistor, such as a MOSFET in an RF power amplifier, or produce a temperature-stable heater to regulate the temperature of a VFO to reduce oscillator drift.

A *thermistor* is a small bit of intrinsic (no N or P doping) metal-oxide semiconductor compound material between two wire leads. As temperature increases, the number of liberated hole/electron pairs increases exponentially, causing the resistance to decrease exponentially. You can see this in the resistance equation:

$$R(T) = R(T0)e^{-\beta(1/T0 - 1/T)} \qquad (143)$$

where T is some temperature in Kelvins and T0 is a reference temperature, usually 298 K (25°C), at which the manufacturer specifies R(T0).

The constant β is experimentally determined by measuring resistance at various temperatures and finding the value of β that best agrees with the measurements. A simple way to get an approximate value of β is to make two measurements, one at room temperature, say 25 °C (298 K) and one at 100 °C (373 K) in boiling water. Suppose the resistances are 10 kΩ and 938 Ω. Eq 1 is solved for β

$$\beta = \frac{\ln\left(\frac{R(T)}{R(T0)}\right)}{\frac{1}{T} - \frac{1}{T0}} = \frac{\ln\left(\frac{938}{1000}\right)}{\frac{1}{373} - \frac{1}{298}} = 3507 \qquad (144)$$

With the behavior of the thermistor known — either by equation or calibration table — its change in resistance can be used to create an electronic circuit whose behavior is controlled by temperature in a known fashion. Such a circuit can be used for controlling temperature or detecting specific temperatures.

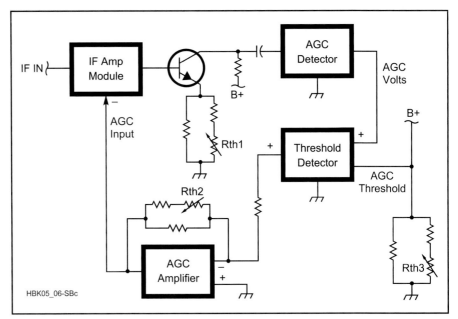

Fig 2.105 — Temperature compensation of IF amplifier and AGC circuitry. R_{th1} – R_{th3} are temperature-sensitive resistors that counteract thermal drift in gain and thresholds.

THERMISTOR-BASED TEMPERATURE CONTROLLER

For many electronics applications, the exact value of temperature is not as important as the ability to maintain that temperature. The following project shows how to use a thermistor to create a circuit that acts as a temperature-controlled switch. The circuit is then used to detect over-temperature conditions and to act as a simple on/off temperature controller.

The circuit of **Fig 2.106** uses a thermistor (R1) to detect a case temperature of about 93 °C, with an on/off operating range of about 0.3°C. A red LED warns of an over-temperature condition that requires attention. The LM339 comparator toggles when R_t = R5. Resistors R3, R4, R5 and the thermistor form a Wheatstone bridge. R3 and R4 are

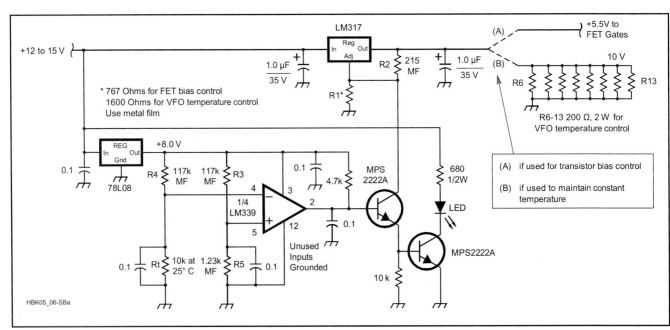

Fig 2.106 — Temperature controller for a power transistor (option A) or VFO (option B).

chosen to make the inputs to the LM339 about 0.5 V at the desired temperature. This greatly reduces self-heating of the thermistor, which could cause a substantial error in the circuit behavior.

The RadioShack Precision Thermistor (RS part number 271-110) used in this circuit is rated at 10 kΩ ±1% at 25 °C. It comes with a calibration chart from –50 °C to +110 °C that you can use to get an approximate resistance at any temperature in that range. For many temperature protection applications you don't have to know the exact temperature. You won't need an exact calibration, but you can verify that the thermistor is working properly by measuring its resistance at 20 °C (68 °F) and in boiling water (≈ 100 °C).

MOSFET POWER TRANSISTOR PROTECTOR

It is very desirable to compensate the temperature sensitivity of power transistors. In bipolar (BJT) transistors, thermal runaway occurs because the dc current gain increases as the transistors get hotter. In MOSFET transistors, thermal runaway is less likely to occur but with excessive drain dissipation or inadequate cooling, the junction temperature may increase until its maximum allowable value is exceeded.

Suppose you want to protect a power amplifier that uses a pair of MRF150 MOSFETs. You can place a thermistor on the heat sink close to the transistors so that the bias adjustment tracks the flange temperature. Another good idea is to attach a thermistor to the ceramic case of the FET with a small drop of epoxy. That way it will respond more quickly to a sudden temperature increase, possibly saving the transistor from destruction. Use the following procedure to get the desired temperature control:

The MRF150 MOSFET has a maximum allowed junction temperature of 200 °C. The thermal resistance θ_{JC} from junction to case is 0.6 °C per watt. The maximum expected dissipation of the FET in normal operation is 110 W. Select a case temperature of 93 °C. This makes the junction temperature 93 °C + (0.6 °C/W) × (110 W) = 159 °C, which is a safe 41 °C below the maximum allowed temperature.

The FET has a rating of 300 W maximum dissipation at a case temperature of 25 °C,

Fig 2.107 — Temperature-controlled VFO cabinet. ¼-inch plexiglass on the outside and ¼-inch thick insulating foam panels on the inside are used to improve temperature stability.

derated at 1.71 W per °C. At 93 °C case temperature the maximum allowed dissipation is 300 W – 1.71 W/°C × (93 °C – 25 °C) = 184 W. The safety margin at that temperature is 184 – 110 = 74 W.

A very simple way to determine the correct value of R5 is to put the thermistor in 93 °C water (let it stabilize) and adjust R5 so that the circuit toggles. At 93°C, the measured value of the thermistor should be about 1230 Ω.

To use the circuit of Fig 2.106 to control the FET bias voltage, set R1 to be 767 Ω. This value, along with R2, adjusts the LM317 voltage regulator output to 5.5V for the FET gate bias. When the thermistor heats to the set point, the LM339 comparator toggles on. This brings the FET gate voltage to a low level and completely turns off the FETs until the temperature drops about 0.3°C. Use metal film resistors throughout the circuit.

TEMPERATURE CONTROL FOR VFOS

The same circuit can be used to control the temperature of a VFO (variable frequency oscillator) with a few simple changes. Select R1 to be 1600 Ω (metal film) to adjust the LM317 for 10 V. Add resistors R6 through R13 as a heater element. Place the VFO in a thermally insulated enclosure such as shown in **Fig 2.107**. The eight 200-Ω, 2-W, metal-oxide resistors at the output of the LM317 supply about 4 W to maintain a temperature of about 33°C inside the enclosure. The resistors are placed so that heat is distributed uniformly. Five are placed near the bottom and three near the top. The thermistor is mounted in the center of the box, close to the tuned circuit and in physical contact with the oscillator ground-plane surface, using a small drop of epoxy.

Use a massive and well-insulated enclosure to slow the rate of temperature change. In one project, over a 0.1 °C range, the frequency at 5.0 MHz varied up and down ±20 Hz or less, with a period of about five minutes. Superimposed is a very slow drift of average frequency that is due to settling down of components, including possibly the thermistor. These gradual changes became negligible after a few days of "burn-in." One problem that is virtually eliminated by the constant temperature is a small but significant "retrace" effect of the cores and capacitors, and perhaps also the thermistor, where a substantial temperature transient of some kind may take from minutes to hours to recover the previous L and C values.

For a much more detailed discussion of this material, see Sabin W.E., WØIYH "Thermistors in Homebrew Projects," *QEX* Nov/Dec 2000. This article is included (with all *QT*, *QEX* and *NCJ* articles from 2000) on the *2000 ARRL Periodicals CD-ROM*. (Order No. 8209.)

2.16 Radio Mathematics

Understanding radio and electronics beyond an intuitive or verbal level requires the use of some mathematics. None of the math is more advanced than trigonometry or advanced algebra, but if you don't use math regularly, it's quite easy to forget what you learned during your education. In fact, depending on your age and education, you may not have encountered some of these topics at all.

It is well beyond the scope of this book to be a math textbook, but this section touches some of topics that most need explanation for amateurs. For introductory-level tutorials and explanations of advanced topics, a list of on-line, no-cost tutorials is presented in the sidebar, "Online Math Resources". You can browse these tutorials whenever you need them in support of the information in this reference text.

2.16.1 Working With Decibels

The *decibel* (*dB*) is a way of expressing a ratio *logarithmically*, meaning as a power of some *base* number, such as 10 or the base for *natural logarithms*, the number e ≈ 2.71828. A decibel, using the metric prefix "deci" (d) for one-tenth, is one-tenth of a *Bel* (B), a unit used in acoustics representing a ratio of 10, and named for the telephone's inventor, Alexander Graham Bell. As it turns out, the Bel was too large a ratio for common use and the deci-Bel, or decibel, became the standard unit.

In radio, the logarithmic ratio is used, but it compares signal strengths — power, voltage, or current. The decibel is more useful than a *linear* ratio because it can represent a wider scale of ratios. The numeric values encountered in radio span a very wide range and so the dB is more suited to discuss large ratios. For example, a typical receiver encounters signals that have powers differing by a factor of 100,000,000,000,000 (10^{14} or 100 trillion!). Expressed in dB, that range is 140, which is a lot easier to work with than the preceding number, even in scientific notation!

The formula for computing the decibel equivalent of a power ratio is

$$dB = 10 \log (\text{power ratio}) = 10 \log (P1/P2) \quad (145)$$

or if voltage is used

$$dB = 20 \log (\text{voltage ratio}) = 20 \log (V1/V2) \quad (146)$$

Positive values of dB mean the ratio is greater than 1 and negative values of dB indicate a ratio of less than 1. For example, if an amplifier turns a 5 W signal into a 25 W signal, that's a gain of 10 log (25 / 5) = 10 log (5) = 7 dB. On the other hand, if by adjusting a receiver's volume control the audio output signal voltage is reduced from 2 V to 0.1 V, that's a loss: 20 log (0.1 / 2) = 20 log (0.05) = –26 dB.

If you are comparing a measured power or voltage (P_M or V_M) to some reference power (P_{REF} or V_{REF}) the formulas are:

$$dB = 10 \log (P_M/P_{REF})$$

$$dB = 20 \log (V_M/V_{REF})$$

There are several commonly-used reference powers and voltages, such as 1 V or 1 mW. When a dB value uses one of them as

---

Online Math Resources

The following Web links are a compilation of on-line resources organized by topic. Other resources are available online at **www.arrl.org/tech-prep-resource-library**. Look for the Math Tutorials section.

Many of the tutorials listed below are part of the *Interactive Mathematics* Web site (**www.intmath.com**), a free, online system of tutorials. The system begins with basic number concepts and progresses all the way through introductory calculus. The lessons referenced here are those of most use to a student of radio electronics.

Basic Numbers & Formulas
Order of Operations — **www.intmath.com/Numbers/3_Order-of-operations.php**
Powers, Roots, and Radicals — **www.intmath.com/Numbers/4_Powers-roots-radicals.php**
Introduction to Scientific Notation — **www.ieer.org/clssroom/scinote.html**
Scientific Notation — **www.intmath.com/Numbers/6_Scientific-notation.php**
Ratios and Proportions — **www.intmath.com/Numbers/7_Ratio-proportion.php**
Geometric Formulas — **www.equationsheet.com/sheets/Equations-4.html**
Tables of Conversion Factors — **en.wikipedia.org/wiki/Conversion_of_units**

Metric System
Metric System Overview — **en.wikipedia.org/wiki/Metric_system**
Metric System Tutorial — **www.arrl.org/FandES/tbp/radio-lab/RLH%20Unit%203%20 Lesson%20%233.1.doc**

Conversion of Units
Metric-English — **www.m-w.com/mw/table/metricsy.htm**
Metric-English — **en.wikipedia.org/wiki/Metric_yardstick**
Conversion Factors — **oakroadsystems.com/math/convert.htm**

Fractions
Equivalent Fractions — **www.intmath.com/Factoring-fractions/5_Equivalent-fractions.php**
Multiplication and Division — **www.intmath.com/Factoring-fractions/6_Multiplication-division-fractions.php**
Adding and Subtracting — **www.intmath.com/Factoring-fractions/7_Addition-subtraction-fractions.php**
Equations Involving Fractions — **www.intmath.com/Factoring-fractions/8_Equations-involving-fractions.php**
Basic Algebra — **www.intmath.com/Basic-algebra/Basic-algebra-intro.php**

Graphs
Basic Graphs — **www.intmath.com/Functions-and-graphs/Functions-graphs-intro.php**
Polar Coordinates — **www.intmath.com/Plane-analytic-geometry/7_Polar-coordinates.php**
Exponents & Radicals — **www.intmath.com/Exponents-radicals/Exponent-radical.php**
Exponential & Logarithmic Functions — **www.intmath.com/Exponential-logarithmic-functions/Exponential-log-functionsintro.php**

Trigonometry
Basic Trig Functions — **www.intmath.com/Trigonometric-functions/Trig-functions-intro.php**
Graphs of Trig Functions — **www.intmath.com/Trigonometric-graphs/Trigo-graph-intro.php**

Miscellaneous
Complex Numbers — **www.intmath.com/Complex-numbers/imaginary-numbers-intro.php**

the references, dB is followed with a letter. Here are the most common:

- dBV means dB with respect to 1 V ($V_{REF} = 1$ V)
- dBµV means dB with respect to 1 µV ($V_{REF} = 1$ µV)
- dBm means dB with respect to 1 mW ($P_{REF} = 1$ mW)

If you are given a ratio in dB and asked to calculate the power or voltage ratio, use the following formulas:

Power ratio = antilog (dB / 10) (147)

Voltage ratio = antilog (dB / 20) (148)

Example: A power ratio of –9 dB = antilog –(9 / 10) = antilog (–0.9) = ⅛ = 0.125

Example: A voltage ratio of 32 dB = antilog (32 / 20) = antilog (1.6) = 40

Antilog is also written as $\log^{-1}$ and may be labeled that way on calculators.

CONVERTING BETWEEN DB AND PERCENTAGE

You may also have to convert back and forth between decibels and percentages. Here are the required formulas:

dB = 10 log (percentage power / 100%) (149)

dB = 20 log (percentage voltage / 100%) (150)

Percentage power =
100% × antilog (dB / 10) (151)

Percentage voltage =
100% × antilog (dB / 20) (152)

Example: A power ratio of 20% = 10 log (20% / 100%) = 10 log (0.2) = –7 dB

Example: A voltage ratio of 150% = 20 log (150% / 100%) = 20 log (1.5) = 3.52 dB

Example: –1 dB represents a percentage power = 100% × antilog (–1 / 10) = 79%

Example: 4 dB represents a percentage voltage = 100% × antilog (4 / 20) = 158%

2.16.2 Rectangular and Polar Coordinates

Graphs are drawings of what equations describe with symbols — they're both saying the same thing. Graphs are used to present a visual representation of what an equation expresses. The way in which mathematical quantities are positioned on the graph is called the *coordinate system*. *Coordinate* is another name for the numeric scales that divide the graph into regular units. The location of every point on the graph is described by a pair of *coordinates*.

The two most common coordinate systems used in radio are the *rectangular-coordinate* system shown in **Fig 2.108** (sometimes called *Cartesian coordinates*) and the *polar-coordinate* system shown in **Fig 2.109**.

The line that runs horizontally through the center of a rectangular coordinate graph is called the *X axis*. The line that runs vertically through the center of the graph is normally called the *Y axis*. Every point on a rectangular coordinate graph has two coordinates that identify its location, X and Y, also written as (X,Y). Every different pair of coordinate values describes a different point on the graph. The point at which the two axes cross — and where the numeric values on both axes are zero — is called the *origin*, written as (0,0).

In Fig 2.108, the point with coordinates (3,5) is located 3 units from the origin along

Computer Software and Calculators

Every version of the *Windows* operating system comes with a calculator program located in the Accessories program group and a number of free calculators are available on-line. Enter "online" and "calculator" into the search window of an Internet search engine for a list. If you can express your calculation as a mathematical expression, such as sin(45) or 10log(17.5), it can be entered directly into the search window at www.google.com and the Google calculator will attempt to solve for the answer.

Microsoft *Excel* (and similar spreadsheet programs) also make excellent calculators. If you are unfamiliar with the use of spreadsheets, here are some online tutorials and help programs:

homepage.cs.uri.edu/tutorials/csc101/pc/excel97/excel.html
www.baycongroup.com/el0.htm
www.bcschools.net/staff/ExcelHelp.htm

For assistance in converting units of measurement, the Web site www.onlineconversion.com is very useful. The Google online unit converter can also be used by entering the required conversion, such as "12 gauss in tesla", into the Google search window.

Decibels In Your Head

Every time you increase the power by a factor of 2 times, you have a 3-dB increase of power. Every 4 times increase of power is a 6-dB increase of power. When you increase the power by 10 times, you have a power increase of 10 dB. You can also use these same values for a decrease in power. Cut the power in half for a 3-dB loss of power. Reduce the power to ¼ the original value for a 6-dB loss in power. If you reduce the power to ¹⁄₁₀ of the original value you will have a 10-dB loss. The power-loss values are often written as negative values: –3, –6 or –10 dB. The following tables show these common decibel values and ratios.

Common dB Values and Power Ratios	
P2/P1	dB
0.1	–10
0.25	–6
0.5	–3
1	0
2	3
4	6
10	10

Common dB Values and Voltage Ratios	
V2/V1	dB
0.1	–20
0.25	–12
0.5	–6
0.707	–3
1.0	0
1.414	3
2	6
4	12
10	20

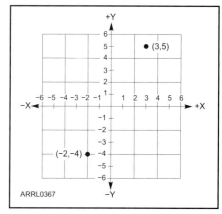

Fig 2.108 — Rectangular-coordinate graphs use a pair of axes at right angles to each other; each calibrate in numeric units. Any point on the resulting grid can be expressed in terms of its horizontal (X) and vertical (Y) values, called coordinates.

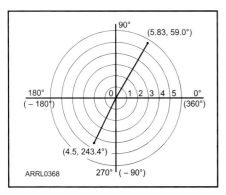

Fig 2.109 — Polar-coordinate graphs use a radius from the origin and an angle from the 0° axis to specify the location of a point. Thus, the location of any point can be specified in terms of a radius and an angle.

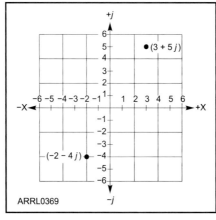

Fig 2.110 — The Y axis of a complex-coordinate graph represents the imaginary portion of complex numbers. This graph shows the same numbers as in Figure 4-1, graphed as complex numbers.

the X axis and 5 units from the origin along the Y axis. Another point at (–2,–4) is found 2 units to the left of the origin along the X axis and 4 units below the origin along the Y axis. Do not confuse the X coordinate with the X representing reactance! Reactance is usually plotted on the Y-axis, creating occasional confusion.

In the polar-coordinate system, points on the graph are described by a pair of numeric values called *polar coordinates*. In this case, a length, or *radius*, is measured from the origin, and an *angle* is measured counterclockwise from the 0° line as shown in Fig 2.109. The symbol r is used for the radius and θ for the angle. A number in polar coordinates is written r$\angle\theta$. So the two points described in the last paragraph could also be written as (5.83, $\angle$59.0°) and (4.5, $\angle$243.4°). Unlike geographic maps, 0° is always to the *right* along the X-axis and 90° at the top along the Y-axis!

Angles are specified counterclockwise from the 0° axis. A negative value in front of an angle specifies an angle measured clockwise from the 0° axis. (Angles can also be given with reference to some arbitrary line, as well.) For example, –270° is equivalent to 90°, –90° is equivalent to 270°, 0° and –360° are equivalent, and +180° and –180° are equivalent. With an angle measured clockwise from the 0° axis, the polar coordinates of the second point in Fig 2.109 would be (4.5, $\angle$–116.6°).

In some calculations the angle will be specified in *radians* (1 radian = 360 / 2π degrees = 53.1°), but you may assume that all angles are in degrees in this book. Radians are used when *angular frequency* (ω) is used instead of frequency (f). Angular frequency is measured in units of 2π radians/sec and so $\omega = 2\pi f$.

In electronics, it's common to use both the rectangular and polar-coordinate systems when dealing with impedance problems. The examples in the next few sections of this book should help you become familiar with these coordinate systems and the techniques for changing between them.

2.16.3 Complex Numbers

Mathematical equations that describe phases and angles of electrical quantities use the symbol j to represent the square root of minus one ($\sqrt{-1}$). (Mathematicians use i for the same purpose, but i is used to represent current in electronics.) The number j and any real number multiplied by j are called *imaginary numbers* because no real number is the square root of minus one. For example, 2j, 0.1j, 7j/4, and 457.6j are all imaginary numbers. Imaginary numbers are used in electronics to represent reactance, such as j13 Ω of inductive reactance or –j25 Ω of capacitive reactance.

Real and imaginary numbers can be combined by using addition or subtraction. Adding a real number to an imaginary number creates a hybrid called a *complex number*, such as 1 + j or 6 – 7j. These numbers come in very handy in radio, describing impedances, relationships between voltage and current, and many other phenomena.

If the complex number is broken up into its real and imaginary parts, those two numbers can also be used as coordinates on a graph using *complex coordinates*. This is a special type of rectangular-coordinate graph that is also referred to as the *complex plane*. By convention, the X axis coordinates represent the real number portion of the complex number and Y axis coordinates the imaginary portion. For example, the complex number 6 – 7j would have the same location as the point (6,–7) on a rectangular-coordinate graph. **Fig 2.110** shows the same points as Fig 2.108.

The number j has a number of interesting properties:

$1/j = -j$

$j^2 = -1$

$j^3 = -j$

$j^4 = 1$

Multiplication by j can also represent a phase shift or rotation of +90°. In polar coordinates $j = 1 \angle 90°$.

WORKING WITH COMPLEX NUMBERS

Complex numbers representing electrical quantities can be expressed in either rectangular form (a + jb) or polar form (r$\angle\theta$) as described above. Adding complex numbers is easiest in rectangular form:

$(a + jb) + (c + jd) = (a+c) + j(b+d)$ (153)

Multiplying and dividing complex numbers is easiest in polar form:

$a\angle\theta_1 \times b\angle\theta_2 = (a \times b) \angle(\theta_1 + \theta_2)$ (154)

and

$\dfrac{a\angle\theta_1}{b\angle\theta_2} = \left(\dfrac{a}{b}\right) \angle(\theta_1 - \theta_2)$ (155)

Converting from one form to another is useful in some kinds of calculations. For example, to calculate the value of two complex impedances in parallel you use the formula

$Z = \dfrac{Z_1 Z_2}{Z_1 + Z_2}$

To calculate the numerator ($Z_1 Z_2$) you would write the impedances in polar form. To calculate the numerator ($Z_1 + Z_2$) you would write the impedances in rectangular form. So you need to be able to convert back and forth from one form to the other. Here is the procedure:

To convert from rectangular (a+jb) to polar form (r $\angle \theta$):

$r = \sqrt{a^2 + b^2}$ and $\theta = \tan^{-1}(b/a)$ (156)

To convert from polar to rectangular form:

$a = r \cos\theta$ and $b = r \sin\theta$ (157)

Many calculators have polar-rectangular

conversion functions built-in and they are worth learning how to use. Be sure that your calculator is set to the correct units for angles, radians or degrees.

Example: Convert $3 \angle 60°$ to rectangular form:

$a = 3 \cos 60° = 3 (0.5) = 1.5$

$b = 3 \sin 60° = 3 (0.866) = 26$

$3 \angle 60° = 2.6 + j1.5$

Example: Convert $0.8 + j0.6$ to polar form:

$r = \sqrt{0.8^2 + 0.6^2} = 1$

$\theta = \tan^{-1}(0.6/0.8) = 36.8°$

$0.8 + j0.6 = 1 \angle 36.8°$

2.16.5 Accuracy and Significant Figures

The calculations you have encountered in this chapter and will find throughout this handbook follow the rules for accuracy of calculations. Accuracy is represented by the number of *significant digits* in a number. That is, the number of digits that carry numeric value information beyond order of magnitude. For example, the numbers 0.123, 1.23, 12.3, 123, and 1230 all have three significant digits.

The result of a calculation can only be as accurate as its *least* accurate measurement or known value. This is important because it is rare for measurements to be more accurate than a few percent. This limits the number of useful significant digits to two or three. Here's another example; what is the current through a 12-Ω resistor if 4.6 V is applied? Ohm's Law says I in amperes = 4.6 / 12 = 0.3833333... but because our most accurate numeric information only has two significant digits (12 and 4.6), strictly applying the significant digits calculation rule limits our answer to 0.38. One extra digit is often included, 0.383 in this case, to act as a guide in rounding off the answer.

Quite often a calculator is used, and the result of a calculation fills the numeric window. Just because the calculator shows 9 digits after the decimal point, this does not mean that is a more correct or even useful answer.

2.17 References and Bibliography

BOOKS

Alexander and Sadiku, *Fundamentals of Electric Circuits* (McGraw-Hill)
Banzhaf, W., *Understanding Basic Electronics*, 2nd ed (ARRL)
Glover, T., *Pocket Ref* (Sequoia Publishing)
Horowitz and Hill, *The Art of Electronics* (Cambridge University Press)
Kaiser, C., *The Resistor Handbook* (Saddleman Press, 1998)
Kaiser, C., *The Capacitor Handbook* (Saddleman Press, 1995)
Kaiser, C., *The Inductor Handbook* (Saddleman Press, 1996)
Kaiser, C., *The Diode Handbook* (Saddleman Press, 1999)
Kaiser, C., *The Transistor Handbook* (Saddleman Press, 1999)
Kaplan, S., *Wiley Electrical and Electronics Dictionary* (Wiley Press)
Orr, W., *Radio Handbook* (Newnes)
Terman, F., *Radio Engineer's Handbook* (McGraw-Hill)

MAGAZINE AND JOURNAL ARTICLES

ARRL Lab Staff, "Lab Notes — Capacitor Basics," *QST*, Jan 1997, pp 85-86
B. Bergeron, NU1N, "Under the Hood II: Resistors," *QST*, Nov 1993, pp 41-44
B. Bergeron, NU1N, "Under the Hood III: Capacitors," *QST*, Jan 1994, pp 45-48
B. Bergeron, NU1N, "Under the Hood IV: Inductors," *QST*, Mar 1994, pp 37-40
B. Bergeron, NU1N, "Under the Hood: Lamps, Indicators, and Displays," *QST*, Sep 1994, pp 34-37
J. McClanahan, W4JBM, "Understanding and Testing Capacitor ESR," *QST*, Sep 2003, pp 30-32
W. Silver, NØAX, "Experiment #24 — Heat Management," *QST*, Jan 2005, pp 64-65
W. Silver, NØAX, "Experiment #62 — About Resistors," *QST*, Mar 2008, pp 66-67
W. Silver, NØAX, "Experiment #63 — About Capacitors," *QST*, Apr 2008, pp 70-71
W. Silver, NØAX, "Experiment #74 — Resonant Circuits," *QST*, Mar 2009, pp 72-73
J. Smith, K8ZOA, "Carbon Composition, Carbon Film and Metal Oxide Film Resistors," *QEX*, Mar/Apr 2008, pp 46-57

Contents

3.1 Analog Signal Processing
    3.1.1 Linearity
    3.1.2 Linear Operations
    3.1.3 Nonlinear Operations

3.2 Analog Devices
    3.2.1 Terminology
    3.2.2 Gain and Transconductance
    3.2.3 Characteristic Curves
    3.2.4 Manufacturer's Data Sheets
    3.2.5 Physical Electronics of Semiconductors
    3.2.6 The PN Semiconductor Junction
    3.2.7 Junction Semiconductors
    3.2.8 Field-Effect Transistors (FET)
    3.2.9 Semiconductor Temperature Effects
    3.2.10 Safe Operating Area (SOA)

3.3 Practical Semiconductors
    3.3.1 Semiconductor Diodes
    3.3.2 Bipolar Junction Transistors (BJT)
    3.3.3 Field-Effect Transistors (FET)
    3.3.4 Optical Semiconductors
    3.3.5 Linear Integrated Circuits
    3.3.6 Comparison of Semiconductor Devices for Analog Applications

3.4 Analog Systems
    3.4.1 Transfer Functions
    3.4.2 Cascading Stages
    3.4.3 Amplifier Frequency Response
    3.4.4 Interstage Loading and Impedance Matching
    3.4.5 Noise
    3.4.6 Buffering

3.5 Amplifiers
    3.5.1 Amplifier Configurations
    3.5.2 Transistor Amplifiers
    3.5.3 Bipolar Transistor Amplifiers
    3.5.4 FET Amplifiers
    3.5.5 Buffer Amplifiers
    3.5.6 Cascaded Buffers
    3.5.7 Using the Transistor as a Switch
    3.5.8 Choosing a Transistor

3.6 Operational Amplifiers
    3.6.1 Characteristics of Practical Op-Amps
    3.6.2 Basic Op Amp Circuits

3.7 Analog-Digital Conversion
    3.7.1 Basic Conversion Properties
    3.7.2 Analog-to-Digital Converters (ADC)
    3.7.3 Digital-to-Analog Converters (DAC)
    3.7.4 Choosing a Converter

3.8 Miscellaneous Analog ICs
    3.8.1 Transistor and Driver Arrays
    3.8.2 Voltage Regulators and References
    3.8.3 Timers (Multivibrators)
    3.8.4 Analog Switches and Multiplexers
    3.8.5 Audio Output Amplifiers
    3.8.6 Temperature Sensors
    3.8.7 Radio Subsystems

3.9 Analog Glossary

3.10 References and Bibliography

Chapter 3

Analog Basics

The first section of this chapter discusses a variety of methods and techniques for working with analog signals, called signal processing. These basic functions are combined to produce useful systems that become radios, instruments, audio recorders and so on.

Subsequent sections of this chapter present several types of active components and circuits that can be used to manipulate analog signals. (An active electronic component is one that requires a power source to function, and is distinguished in this way from passive components such as resistors, capacitors and inductors that are described in the **Electrical Fundamentals** chapter.) The chapter concludes with a discussion of analog-digital conversion by which analog signals are converted into digital signals and vice versa.

New to the 2011 edition, circuit simulation files have been created for the basic transistor circuits in section 3.5. The files were created with version 4.06l of *LTSpice* from Linear Technologies, a free *SPICE*-based circuit simulator. The files can be found on the CD-ROM accompanying this book in the "Simulation Files" folder.

Material in this chapter was updated by Ward Silver, NØAX. It builds on work by Greg Lapin, N9GL, with contributions by Leonard Kay, K1NU, that appeared in previous editions. The section "Poles and Zeroes" was adapted from material previously written for the chapter on **Oscillators and Synthesizers** by David Stockton, GM4ZNX, and Fred Telewski, WA7TZY.

3.1 Analog Signal Processing

The term *analog signal* refers to voltages, currents and waves that make up ac radio and audio signals, dc measurements, even power. The essential characteristic of an analog signal is that the information or energy it carries is continuously variable. Even small variations of an analog signal affect its value or the information it carries. This stands in contrast to *digital signals* (described in the **Digital Basics** chapter) that have values only within well-defined and separate ranges called *states*. To be sure, at the fundamental level all circuits and signals are analog: Digital signals are created by designing circuits that restrict the values of analog signals to those discrete states.

Analog signal processing involves various electronic stages to perform functions on analog signals such as amplifying, filtering, modulation and demodulation. A piece of electronic equipment, such as a radio, is constructed by combining a number of these circuits. How these stages interact with each other and how they affect the signal individually and in tandem is the subject of sections later in the chapter.

3.1.1 Linearity

The premier properties of analog signals are *superposition* and *scaling*. Superposition is the property by which signals are combined, whether in a circuit, in a piece of wire, or even in air, as the sum of the individual signals. This is to say that at any one point in time, the voltage of the combined signal is the sum of the voltages of the original signals at the same time. In a *linear system* any number of signals will add in this way to give a single combined signal. (Mathematically, this is a *linear combination*.) For this reason, analog signals and components are often referred to as *linear signals* or *linear components*. A linear system whose characteristics do not change, such as a resistive voltage divider, is called *time-invariant*. If the system changes with time, it is *time-varying*. The variations may be random, intermittent (such as being adjusted by an operator) or periodic.

One of the more important features of superposition, for the purposes of signal processing, is that signals that have been combined by superposition can be separated back into the original signals. This is what allows multiple signals that have been received by an antenna to be separated back into individual signals by a receiver.

3.1.2 Linear Operations

Any operation that modifies a signal and obeys the rules of superposition and scaling is a *linear operation*. The following sections explain the basic linear operations from which linear systems are made.

AMPLIFICATION AND ATTENUATION

Amplification and *attenuation* scale signals to be larger and smaller, respectively. The operation of *scaling* is the same as multiplying the signal at each point in time by a constant value; if the constant is greater than one then the signal is amplified, if less than one then the signal is attenuated.

An *amplifier* is a circuit that increases the amplitude of a signal. Schematically, a generic amplifier is signified by a triangular symbol, its input along the left face and its output at the point on the right (see **Fig 3.1**). The linear amplifier multiplies every value of a signal by a constant value. Amplifier gain is often expressed as a multiplication factor (× 5, for example).

$$\text{Gain} = \frac{V_o}{V_i} \quad (1)$$

where V_o is the output voltage from an amplifier when an input voltage, V_i, is applied. (Gain is often expressed in decibels (dB) as explained in the chapter on **Electrical Fundamentals**.)

Certain types of amplifiers for which currents are the input and output signals have *current gain*. Most amplifiers have both voltage and current gain. *Power gain* is the ratio of output power to input power.

Ideal linear amplifiers have the same gain for all parts of a signal. Thus, a gain of 10 changes 10 V to 100 V, 1 V to 10 V and –1 V to –10 V. (Gain can also be less than one.) The ability of an amplifier to change a signal's level is limited by the amplifier's *dynamic range*, however. An amplifier's dynamic range is the range of signal levels over which the amplifier produces the required gain without distortion. Dynamic range is limited for small signals by noise, distortion and other nonlinearities.

Dynamic range is limited for large signals because an amplifier can only produce output voltages (and currents) that are within the range of its power supply. (Power-supply voltages are also called the *rails* of a circuit.) As the amplified output approaches one of the rails, the output can not exceed a given voltage near the rail and the operation of the amplifier becomes nonlinear as described below in the section on Clipping and Rectification.

Another similar limitation on amplifier linearity is called *slew rate*. Applied to an amplifier, this term describes the maximum rate at which a signal can change levels and still be accurately amplified in a particular device. *Input slew rate* is the maximum rate of change to which the amplifier can react linearly. *Output slew rate* refers to the maximum rate at which the amplifier's output circuit can change. Slew rate is an important concept, because there is

Obtaining a Frequency Response

With the computer tools such as spreadsheets, it's easy to do the calculations and make a graph of frequency response. If you don't have a spreadsheet program, then use semi-log graph paper with the linear axis used for dB or phase and the logarithmic axis for frequency. An Excel spreadsheet set up to calculate and display frequency response is available on the CD that comes with this book. You can modify it to meet your specific needs.

Follow these rules whether using a spreadsheet or graph paper:
- Measure input and output in the same units, such as volts, and use the same conventions, such as RMS or peak-to-peak.
- Measure phase shift from the input to the output. (The **Test Equipment and Measurements** chapter discusses how to make measurements of amplitude and phase.)
- Use 10 log (P_O/P_I) for power ratios and 20 log (V_O/V_I) for voltage or current.

To make measurements that are roughly equally spaced along the logarithmic frequency axis, follow the "1-2-5 rule." Dividing a range this way, for example 1-2-5-10-20-50-100-200-500 Hz, creates steps in approximately equal ratios that then appear equally spaced on a logarithmic axis.

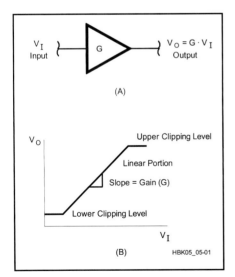

Fig 3.1 — Generic amplifier. (A) Symbol. For the linear amplifier, gain is the constant value, G, and the output voltage is equal to the input voltage times G; (B) Transfer function, input voltage along the x-axis is converted to the output voltage along the y-axis. The linear portion of the response is where the plot is diagonal; its slope is equal to the gain, G. Above and below this range are the clipping limits, where the response is not linear and the output signal is clipped.

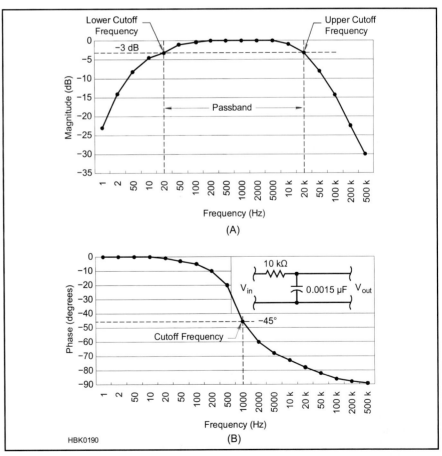

Fig 3.2 — Bode plot of (A) band-pass filter magnitude response and (B) an RC low-pass filter phase response.

a direct correlation between a signal level's rate of change and the frequency content of that signal. The amplifier's ability to react to or reproduce that rate of change affects its frequency response and dynamic range.

An *attenuator* is a circuit that reduces the amplitude of a signal. Attenuators can be constructed from passive circuits, such as the attenuators built using resistors, described in the chapter on **Test Equipment and Measurements**. Active attenuator circuits include amplifiers whose gain is less than one or circuits with adjustable resistance in the signal path, such as a PIN diode attenuator or amplifier with gain is controlled by an external voltage.

FREQUENCY RESPONSE AND BODE PLOTS

Another important characteristic of a circuit is its *frequency response*, a description of how it modifies a signal of any frequency. Frequency response can be stated in the form of a mathematical equation called a *transfer function*, but it is more conveniently presented as a graph of gain vs frequency. The ratio of output amplitude to input amplitude is often called the circuit's *magnitude* or *amplitude response*. Plotting the circuit's magnitude response in dB versus frequency on a logarithmic scale, such as in **Fig 3.2A**, is called a *Bode plot* (after Henrik Wade Bode). The combination of decibel and log-frequency scales is used because the behavior of most circuits depends on ratios of amplitude and frequency and thus appears linear on a graph in which both the vertical and horizontal scales are logarithmic.

Most circuits also affect a signal's phase along with its amplitude. This is called *phase shift*. A plot of phase shift from the circuit's input to its output is called the *phase response*, seen in Fig 3.2B. Positive phase greater than 0° indicates that the output signal *leads* the input signal, while *lagging* phase shift has a negative phase. The combination of an amplitude and phase response plot gives a good picture of what effect the circuit has on signals passing through it.

TRANSFER CHARACTERISTICS

Transfer characteristics are the ratio of an output parameter to an input parameter, such as output current divided by input current, h_{FE}. There are different families of transfer characteristics, designated by letters such as h, s, y or z. Each family compares parameters in specific ways that are useful in certain design or analysis methods. The most common transfer characteristics used in radio are the h-parameter family (used in transistor models) and the s-parameter family (used in RF design, particularly at VHF and above). See the **RF Techniques** chapter for more discussion of transfer characteristics.

POLES AND ZEROES

It's easy to imagine frequency as stretching from zero to infinity, but let's imagine that complex numbers could be used. This is pure imagination — we can't make signals at complex values (with real and imaginary components) of hertz — but if complex numbers are used for frequency, we can gain additional insight into how the circuit behaves. (See the chapter on **Electrical Fundamentals** for a discussion of complex numbers.)

This is illustrated by **Fig 3.3**. The transfer function (shown at the left of the figure) describes the behavior of the simple RC-circuit at real-world frequencies as well as imaginary frequencies whose values contain *j*. When complex numbers are used for f, the transfer function can describe the circuit's phase response, as well as amplitude. At one such frequency, $f = -j/2\pi RC$, the denominator of the transfer function is zero, and the gain is infinite! Infinite gain is a pretty amazing thing to achieve with a passive circuit — but because this can only happen at an imaginary frequency, it does not happen in the real world.

Frequencies that cause the transfer function to become infinite are called *poles*. This is shown at the bottom right of Fig 3.3 in the graph of the circuit's amplitude response for imaginary frequencies shown on the horizontal axis. (The pole causes the graph to extend up "as a pole under a tent," thus the name.) Similarly, circuits can have *zeroes* which occur at imaginary frequencies that cause the transfer function to be zero, a less imaginative name, but quite descriptive.

A circuit can also have poles and zeroes at frequencies of zero and infinity. For example, the circuit in Fig 3.3 has a zero at infinity because the capacitor's reactance is zero at infinity and the transfer function is zero, as well. If the resistor and capacitor were exchanged, so that the capacitor was in series with the output,

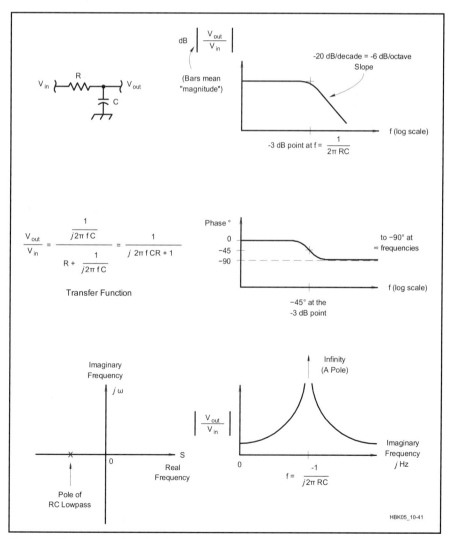

Fig 3.3 — The transfer function for a circuit describes both the magnitude and phase response of a circuit. The RC circuit shown at the upper left has a pole at $f = 2\pi RC$, the filter's –3 dB or cutoff frequency, at which the phase response is a 45° lagging phase shift. Poles cause an infinite response on the imaginary frequency axis.

then at zero frequency (dc), the output would be zero because the capacitor's reactance was infinite, creating a zero.

Complex circuits can have multiple poles or multiple zeroes at the same frequency. Poles and zeroes can also occur at frequencies that are combinations of real and imaginary numbers. The poles and zeroes of a circuit form a pattern in the complex plane that corresponds to certain types of circuit behavior. (The relationships between the pole-zero pattern and circuit behavior is beyond the scope of this book, but are covered in textbooks on circuit theory.)

What is a Pole?

Poles cause a specific change in the circuit's amplitude and phase response for real-world frequencies, even though we can't experience imaginary frequencies directly. A pole is associated with a bend in a magnitude response plot that changes the slope of the response downward with increasing frequency by 6 dB per octave (20 dB per decade; an octave is a 2:1 frequency ratio, a decade is a 10:1 frequency ratio).

There are four ways to identify the existence and frequency of a pole as shown in Fig 3.3:

1. For a downward bend in the magnitude versus frequency plot, the pole is at the –3 dB frequency for a single pole. If the bend causes a change in slope of more than 6 dB/octave, there must be multiple poles at this frequency.

2. A 90° lagging change on a phase versus frequency plot, where the lag increases with frequency. The pole is at the point of 45° added lag on the S-shaped transition. Multiple poles will add their phase lags, as above.

3. On a circuit diagram, a single pole looks like a simple RC low-pass filter. The pole is at the –3 dB frequency (f = 1/2πRC Hz). Any other circuit with the same response has a pole at the same frequency.

4. In an equation for the transfer function of a circuit, a pole is a theoretical value of frequency that would result in infinite gain. This is clearly impossible, but as the value of frequency will either be absolute zero, or will have an imaginary component, it is impossible to make an actual real-world signal at a pole frequency.

For example, comparing the amplitude responses at top and bottom of Fig 3.3 shows that the frequency of the pole is equal to the circuit's –3 dB cutoff frequency (1/2πfC) multiplied by j, which is also the frequency at which the circuit causes a –45° (lagging) phase shift from input to output.

What Is a Zero?

A zero is the complement of a pole. In math, it is a frequency at which the transfer function equation of a circuit is zero. This is not impossible in the real world (unlike the pole), so zeroes can be found at real-number frequencies as well as complex-number frequencies.

Each zero is associated with an *upward* bend of 6 dB per octave in a magnitude response. Similarly to a pole, the frequency of the zero is at the +3 dB point. Each zero is associated with a transition on a phase-versus-frequency plot that reduces the lag by 90°. The zero is at the 45° leading phase point. Multiple zeroes add their phase shifts just as poles do.

In a circuit, a zero creates gain that increases with frequency forever above the zero frequency. This requires active circuitry that would inevitably run out of gain at some frequency, which implies one or more poles up there. In real-world circuits, zeroes are usually not found by themselves, making the magnitude response go up, but rather paired with a pole of a different frequency, resulting in the magnitude response having a slope between two frequencies but flat above and below them.

Real-world circuit zeroes are only found accompanied by a greater or equal number of poles. Consider a classic RC high-pass filter, such as if the resistor and capacitor in Fig 3.3 were exchanged. The response of such a circuit increases at 6 dB per octave from 0 Hz (so there must be a zero at 0 Hz) and then levels off at 1/2πRC Hz. This leveling off is due to the presence of a pole adding its 6 dB-per-octave roll-off to cancel the 6 dB-per-octave roll-up of the zero. The transfer function for such as circuit would equal zero at zero frequency and infinity at the imaginary pole frequency.

FEEDBACK AND OSCILLATION

The *stability* of an amplifier refers to its ability to provide gain to a signal without tending to oscillate. For example, an amplifier just on the verge of oscillating is not generally considered to be "stable." If the output of an amplifier is fed back to the input, the feedback can affect the amplifier stability. If the amplified output is added to the input, the output of the sum will be larger. This larger output, in turn, is also fed back. As this process continues, the amplifier output will continue to rise until the amplifier cannot go any higher (clamps). Such *positive feedback* increases the amplifier gain, and is called *regeneration*. (The chapter on **Oscillators and Synthesizers** includes a discussion of positive feedback.)

Most practical amplifiers have some intrinsic and unavoidable feedback either as part of the circuit or in the amplifying device(s) itself. To improve the stability of an amplifier, *negative feedback* can be added to counteract any unwanted positive feedback. Negative feedback is often combined with a phase-shift *compensation* network to improve the amplifier stability.

Although negative feedback reduces amplifier or stage gain, the advantages of *stable* gain, freedom from unwanted oscillations and the reduction of distortion are often key design objectives and advantages of using negative feedback.

The design of feedback networks depends on the desired result. For amplifiers, which should not oscillate, the feedback network is customized to give the desired frequency response without loss of stability. For oscillators, the feedback network is designed to create a steady oscillation at the desired frequency.

SUMMING

In a linear system, nature does most of the work for us when it comes to adding signals; placing two signals together naturally causes them to add according to the principle of superposition. When processing signals, we would like to control the summing operation so the signals do not distort or combine in a nonlinear way. If two signals come from separate stages and they are connected together directly, the circuitry of the stages may interact, causing distortion of either or both signals.

Summing amplifiers generally use a resistor in series with each stage, so the resistors connect to the common input of the following stage. This provides some *isolation* between the output circuits of each stage. **Fig 3.4** illustrates the resistors connecting to a summing amplifier. Ideally, any time we wanted to combine signals (for example, combining an audio signal with a sub-audible tone in a 2 meter FM transmitter prior to modulating the RF signal) we could use a summing amplifier.

FILTERING

A *filter* is a common linear stage in radio equipment. Filters are characterized by their ability to selectively attenuate certain frequen-

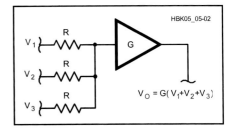

Fig 3.4 — Summing amplifier. The output voltage is equal to the sum of the input voltages times the amplifier gain, G. As long as the resistance values, R, are equal and the amplifier input impedance is much higher, the actual value of R does not affect the output signal.

cies in the filter's *stop band*, while passing or amplifying other frequencies in the *passband*. If the filter's passband extends to or near dc, it is a *low-pass* filter, and if to infinity (or at least very high frequencies for the circuitry involved), it is a *high-pass* filter. Filters that pass a range of frequencies are *band-pass* filters. *All-pass* filters are designed to affect only the phase of a signal without changing the signal amplitude. The range of frequencies between a band-pass circuit's low-pass and high-pass regions is its *mid-band*.

Fig 3.2A is the amplitude response for a typical band-pass audio filter. It shows that the input signal is passed to the output with no loss (0 dB) between 200 Hz and 5 kHz. This is the filter's *mid-band response*. Above and below those frequencies the response of the filter begins to drop. By 20 Hz and 20 kHz, the amplitude response has been reduced to one-half of (−3 dB) the mid-band response. These points are called the circuit's *cutoff* or *corner* or *half-power frequencies*. The range between the cutoff frequencies is the filter's passband. Outside the filter's passband, the amplitude response drops to 1/200th of (−23 dB) mid-band response at 1 Hz and only 1/1000th (−30 dB) at 500 kHz. The steepness of the change in response with frequency is the filter's *roll-off* and it is usually specified in dB/octave (an octave is a doubling or halving of frequency) or dB/decade (a decade is a change to 10 times or 1/10th frequency).

Fig 3.2B represents the phase response of a different filter — the simple RC low-pass filter shown at the upper right. As frequency increases, the reactance of the capacitor becomes smaller, causing most of the input signal to appear across the fixed-value resistor instead. At low frequencies, the capacitor has little effect on phase shift. As the signal frequency rises, however, there is more and more phase shift until at the cutoff frequency, there is 45° of lagging phase shift, plotted as a negative number. Phase shift then gradually approaches 90°.

Practical passive and active filters are described in the **RF and AF Filters** chapter. Filters implemented by digital computation (*digital filters*) are discussed in the chapter on **DSP and Software Radio Design**. Filters at RF may also be created by using transmission lines as described in the **Transmission Lines** chapter. All practical amplifiers are in effect either low-pass filters or band-pass filters, because their magnitude response decreases as the frequency increases beyond their gain-bandwidth products.

AMPLITUDE MODULATION/DEMODULATION

Voice and data signals can be transmitted over the air by using *amplitude modulation* (AM) to combine them with higher frequency *carrier* signals (see the **Modulation** chapter). The process of amplitude modulation can be mathematically described as the multiplication (product) of the voice signal and the carrier signal. Multiplication is a linear process — amplitude modulation by the sum of two audio signals produces the same signal as the sum of amplitude modulation by each audio signal individually. Another aspect of the linear behavior of amplitude modulation is that amplitude-modulated signals can be demodulated exactly to their original form. Amplitude demodulation is the converse of amplitude modulation, and is represented as division, also a linear operation.

In the linear model of amplitude modulation, the signal that performs the modulation (such as the audio signal in an AM transmitter) is shifted in frequency by multiplying it with the carrier. The modulated waveform is considered to be a linear function of the signal. The carrier is considered to be part of a time-varying linear system and not a second signal.

A curious trait of amplitude modulation (and demodulation) is that it can be performed nonlinearly, as well. Accurate analog multipliers (and dividers) are difficult and expensive to fabricate, so less-expensive nonlinear methods are often employed. Each nonlinear form of amplitude modulation generates the desired linear combination of signals, called *products*, in addition to other unwanted products that must then be removed. Both linear and nonlinear modulators and demodulators are discussed in the chapter on **Mixers, Modulators, and Demodulators**.

3.1.3 Nonlinear Operations

All signal processing doesn't have to be linear. Any time that we treat various signal levels differently, the operation is called *nonlinear*. This is not to say that all signals must be treated the same for a circuit to be linear. High-frequency signals are attenuated in a low-pass filter while low-frequency signals are not, yet the filter can be linear. The distinction is that the amount of attenuation at different frequencies is always the same, regardless of the amplitude of the signals passing through the filter.

What if we do not want to treat all voltage levels the same way? This is commonly desired in analog signal processing for clipping, rectification, compression, modulation and switching.

CLIPPING AND RECTIFICATION

Clipping is the process of limiting the range of signal voltages passing through a circuit (in other words, *clipping* those voltages outside the desired range from the signals). There are a number of reasons why we would like to do this. As shown in Fig 3.1, clipping is the process of limiting the positive and negative peaks of a signal. (Clipping is also called *clamping*.)

Clamping might be used to prevent a large audio signal from causing excessive deviation in an FM transmitter that would interfere with communications on adjacent channels. Clipping circuits are also used to protect sensitive inputs from excessive voltages. Clipping distorts the signal, changing it so that the original signal waveform is lost.

Another kind of clipping results in *rectification*. A *rectifier* circuit clips off all voltages of one polarity (positive or negative) and passes only voltages of the other polarity, thus changing ac to pulsating dc (see the **Power Supplies** chapter). Another use of rectification is in a *peak detection* circuit that measures the peak value of a waveform. Only one polarity of the ac voltage needs to be measured and so a rectifier clips the unwanted polarity.

LIMITING

Another type of clipping occurs when an amplifier is intentionally operated with so much gain that the input signals result in an output that is clipped at the limits of its power supply voltages (or some other designated voltages). The amplifier is said to be driven into *limiting* and an amplifier designed for this behavior is called a *limiter*. Limiters are used in FM receivers to amplify the signal until all amplitude variations in the signal are removed and the only characteristic of the original signal that remains is the frequency.

LOGARITHMIC AMPLIFICATION

It is sometimes desirable to amplify a signal logarithmically, which means amplifying low levels more than high levels. This type of amplification is often called *signal compression*. Speech compression is sometimes used in audio amplifiers that feed modulators. The voice signal is compressed into a small range of amplitudes, allowing more voice energy to be transmitted without overmodulation (see the **Modulation** chapter).

3.2 Analog Devices

There are several different kinds of components that can be used to build circuits for analog signal processing. Bipolar semiconductors, field-effect semiconductors and integrated circuits comprise a wide spectrum of active devices used in analog signal processing. (Vacuum tubes are discussed in the chapter on **RF Power Amplifiers**, their primary application in Amateur Radio.) Several different devices can perform the same function, each with its own advantages and disadvantages based on the physical characteristics of each type of device.

Understanding the specific characteristics of each device allows you to make educated decisions about which device would be best for a particular purpose when designing analog circuitry, or understanding why an existing circuit was designed in a particular way.

3.2.1 Terminology

A similar terminology is used when describing active electronic devices. The letter V or v stands for voltages and I or i for currents. Capital letters are often used to denote dc or bias values (bias is discussed later in this chapter). Lower-case often denotes instantaneous or ac values.

Voltages generally have two subscripts indicating the terminals between which the voltage is measured (V_{BE} is the dc voltage between the base and the emitter of a bipolar transistor). Currents have a single subscript indicating the terminal into which the current flows (I_C is the dc current into the collector of a bipolar transistor). If the current flows out of the device, it is generally treated as a negative value.

Resistance is designated with the letter R or r, and impedance with the letter Z or z. For example, r_{DS} is resistance between drain and source of an FET and Z_i is input impedance. For some parameters, values differ for dc and ac signals. This is indicated by using capital letters in the subscripts for dc and lower-case subscripts for ac. For example, the common-emitter dc current gain for a bipolar transistor is designated as h_{FE}, and h_{fe} is the ac current gain. (See the section on transistor amplifiers later in this chapter for a discussion of the common-emitter circuit.) Qualifiers are sometimes added to the subscripts to indicate certain operating modes of the device. SS for saturation, BR for breakdown, ON and OFF are all commonly used.

Power supply voltages have two subscripts that are the same, indicating the terminal to which the voltage is applied. V_{DD} would represent the power supply voltage applied to the drain of a field-effect transistor.

Since integrated circuits are collections of semiconductor components, the abbrevia-

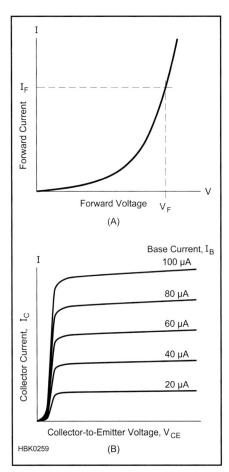

Fig 3.5 — Characteristic curves. A forward voltage vs forward current characteristic curve for a semiconductor diode is shown at (A). (B) shows a set of characteristic curves for a bipolar transistor in which the collector current vs collector-to-emitter voltage curve is plotted for five different values of base current.

tions for the type of semiconductor used also apply to the integrated circuit. For example, V_{CC} is a power supply voltage for an integrated circuit made with bipolar transistor technology in which voltage is applied to transistor collectors.

3.2.2 Gain and Transconductance

The operation of an amplifier is specified by its *gain*. Gain in this sense is defined as the change (Δ) in the output parameter divided by the corresponding change in the input parameter. If a particular device measures its input and output as currents, the gain is called a *current gain*. If the input and output are voltages, the amplifier is defined by its *voltage gain*. *Power gain* is often used, as well. Gain is technically unit-less, but is often given in V/V. Decibels are often used to specify gain, particularly power gain.

If an amplifier's input is a voltage and the output is a current, the ratio of the change in output current to the change in input voltage is called *transconductance*, g_m.

$$g_m = \frac{\Delta I_o}{\Delta V_i} \qquad (2)$$

Transconductance has the same units as conductance and admittance, Siemens (S), but is only used to describe the operation of active devices, such as transistors or vacuum tubes.

3.2.3 Characteristic Curves

Analog devices are described most completely with their *characteristic curves*. The characteristic curve is a plot of the interrelationships between two or three variables. The vertical (y) axis parameter is the output, or result of the device being operated with an input parameter on the horizontal (x) axis. Often the output is the result of two input values. The first input parameter is represented along the x-axis and the second input parameter by several curves, each for a different value.

Almost all devices of concern are nonlinear over a wide range of operating parameters. We are often interested in using a device only in the region that approximates a linear response. Characteristic curves are used to graphically describe a device's operation in both its linear and nonlinear regions.

Fig 3.5A shows the characteristic curve for a semiconductor diode with the y-axis showing the forward current, I_F, flowing through the diode and the x-axis showing forward voltage, V_F, across the diode. This curve shows the relationship between current and voltage in the diode when it is conducting current. Characteristic curves showing voltage and current in two-terminal devices such as diodes are often called *I-V curves*. Characteristic curves may include all four quadrants of operation in which both axes include positive and negative values. It is also common for different scales to be used in the different quadrants, so inspect the legend for the curves carefully.

The parameters plotted in a characteristic curve depend on how the device will be used so that the applicable design values can be obtained from the characteristic curve. The slope of the curve is often important because it relates changes in output to changes in input. To determine the slope of the curve, two closely-spaced points along that portion of the curve are selected, each defined by its location along the x and y axes. If the two points are defined by (x_1, y_1) and (x_2, y_2), the

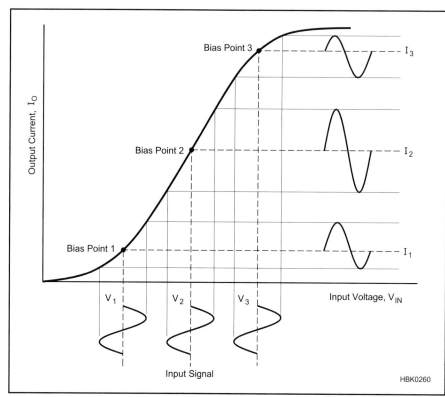

Fig 3.6 — Effect of biasing. An input signal may be reproduced linearly or nonlinearly depending on the choice of bias points.

slope, m, of the curve (which can be a gain, a resistance or a conductance, for example) is calculated as:

$$m = \frac{\Delta y}{\Delta x} = \frac{y_1 - y_2}{x_1 - x_2} \qquad (3)$$

It is important to pick points that are close together or the slope will not reflect the actual behavior of the device. A device whose characteristic curve is not a straight line will not have a linear response to inputs because the slope changes with the value of the input parameter.

For a device in which three parameters interact, such as a transistor, sets of characteristic curves can be drawn. Fig 3.5B shows a set of characteristic curves for a bipolar transistor where collector current, I_C, is shown on the y axis and collector-to-emitter voltage, V_{CE}, is shown on the x axis. Because the amount of collector current also depends on base current, I_B, the curve is repeated several times for different values of I_B. From this set of curves, an amplifier circuit using this transistor can be designed to have specific values of gain.

BIASING

The operation of an analog signal-processing device is greatly affected by which portion of the characteristic curve is used to do the processing. The device's *bias point* is its set of operating parameters when no input signal is applied. The bias point is also known as the *quiescent point* or *Q-point*. By changing the bias point, the circuit designer can affect the relationship between the input and output signal. The bias point can also be considered as a dc offset of the input signal. Devices that perform analog signal processing require appropriate input signal biasing.

As an example, consider the characteristic curve shown in **Fig 3.6**. (The exact types of device and circuit are unimportant.) The characteristic curve shows the relationship between an input voltage and an output current. Increasing input voltage results in an increase in output current so that an input signal is reproduced at the output. The characteristic curve is linear in the middle, but is quite nonlinear in its upper and lower regions.

In the circuit described by the figure, bias points are established by adding one of the three voltages, V_1, V_2 or V_3 to the input signal. Bias voltage V_1 results in an output current of I_1 when no input signal is present. This is shown as Bias Point 1 on the characteristic curve. When an input signal is applied, the input voltage varies around V_1 and the output current varies around I_1 as shown. If the dc value of the output current is subtracted, a reproduction of the input signal is the result.

If Bias Point 2 is chosen, we can see that the input voltage is reproduced as a changing output current with the same shape. In this case, the device is operating linearly. If either Bias Point 1 or Bias Point 3 is chosen, however, the shape of the output signal is distorted because the characteristic curve of the device is nonlinear in this region. Either the increasing portion of the input signal results in more variation than the decreasing portion (Bias Point 1) or vice versa (Bias Point 3). Proper biasing is crucial to ensure that a device operates linearly.

3.2.4 Manufacturer's Data Sheets

Manufacturer's data sheets list device characteristics, along with the specifics of the part type (polarity, semiconductor type), identity of the pins and leads (*pinouts*), and the typical use (such as small signal, RF, switching or power amplifier). The pin identification is important because, although common package pinouts are normally used, there are exceptions. Manufacturers may differ slightly in the values reported, but certain basic parameters are listed. Different batches of the same devices are rarely identical, so manufacturers specify the guaranteed limits for the parameters of their device. There are usually three values listed in the data sheet for each parameter: guaranteed minimum value, the guaranteed maximum value, and/or the typical value.

Another section of the data sheet lists ABSOLUTE MAXIMUM RATINGS, beyond which device damage may result. For example, the parameters listed in the ABSOLUTE MAXIMUM RATINGS section for a solid-state device are typically voltages, continuous currents, total device power dissipation (P_D) and operating- and storage-temperature ranges.

Rather than plotting the characteristic curves for each device, the manufacturer often selects key operating parameters that describe the device operation for the configurations and parameter ranges that are most commonly used. For example, a bipolar transistor data sheet might include an OPERATING PARAMETERS section. Parameters are listed in an OFF CHARACTERISTICS subsection and an ON CHARACTERISTICS subsection that describe the conduction properties of the device for dc voltages. The SMALL-SIGNAL CHARACTERISTICS section might contain a minimum Gain-Bandwidth Product (f_T or GBW), maximum output capacitance, maximum input capacitance, and the range of the transfer parameters applicable to a given device. Finally, the SWITCHING CHARACTERISTICS section might list absolute maximum ratings for Delay Time (t_d), Rise Time (t_r), Storage Time (t_s), and Fall Time (t_f). Other types of devices list characteristics important to operation of that specific device.

When selecting equivalent parts for replacement of specified devices, the data sheet provides the necessary information to tell if a given part will perform the functions of another. Lists of *cross-references* and *substitution guides* generally only specify devices that have nearly identical parameters. There are usually a large number of additional devices that can be chosen as replacements. Knowledge of the circuit requirements adds even more to the list of possible replacements. The device parameters should be compared individually to make sure that the replacement part meets or exceeds the parameter values of the original part required by the circuit. Be aware that in some applications a far superior part may fail as a replacement, however. A transistor with too much gain could easily oscillate if there were insufficient negative feedback to ensure stability.

3.2.5 Physical Electronics of Semiconductors

In a conductor, such as a metal, some of the outer, or *valence*, electrons of each atom are free to move about between atoms. These *free electrons* are the constituents of electrical current. In a good conductor, the concentration of these free electrons is very high, on the order of 10^{22} electrons/cm^3. In an insulator, nearly all the electrons are tightly held by their atoms and the concentration of free electrons is very small — on the order of 10 electrons/cm^3.

Between the classes of materials considered to be conductors and insulators is a class of elements called *semiconductors*, materials with conductivity much poorer than metals and much better than insulators. (In electronics, "semiconductor" means a device made from semiconductor elements that have been chemically manipulated as described below, leading to interesting properties that create useful applications.)

Semiconductor atoms (silicon, Si, is the most widely used) share their valence electrons in a chemical bond that holds adjacent atoms together, forming a three-dimensional *lattice* that gives the material its physical characteristics. A lattice of pure semiconductor material (one type of atom or molecule) can form a crystal, in which the lattice structure and orientation is preserved throughout the material. *Monocrystalline* or "single-crystal" is the type of material used in electronic semiconductor devices. *Polycrystalline* material is made of up many smaller crystals with their own individual lattice orientations.

Crystals of pure semiconductor material are called *intrinsic* semiconductors. When energy, generally in the form of heat, is added to a semiconductor crystal lattice, some electrons are liberated from their bonds and move freely throughout the lattice. The bond that loses an electron is then unbalanced and the space that the electron came from is referred to as a *hole*. In these materials the number of free electrons is equal to the number of holes.

Electrons from adjacent bonds can leave their positions to fill the holes, thus leaving behind a hole in their old location. As a consequence of the electron moving, two opposite movements can be said to occur: negatively charged electrons move from bond to bond in one direction and positively charged holes move from bond to bond in the opposite direction. Both of these movements represent forms of electrical current, but this is very different from the current in a conductor. While a conductor has free electrons that flow independently from the bonds of the crystalline lattice, the current in a pure semiconductor is constrained to move from bond to bond.

Impurities can be added to intrinsic semiconductors (by a process called *doping*) to enhance the formation of electrons or holes and thus improve conductivity. These materials are *extrinsic* semiconductors. Since the additional electrons and holes can move, their movement is current and they are called *carriers*. The type of carrier that predominates in the material is called the *majority carrier*. In N-type material the majority carriers are electrons and in P-type material, holes.

There are two types of impurities that can be added: a *donor impurity* with five valence electrons donates free electrons to the crystalline structure; this is called an *N-type* impurity, for the negative charge of the majority carriers. Some examples of donor impurities are antimony (Sb), phosphorus (P) and arsenic (As). N-type extrinsic semiconductors have more electrons and fewer holes than intrinsic semiconductors. *Acceptor impurities* with three valence electrons accept free electrons from the lattice, adding holes to the overall structure. These are called P-type impurities, for the positive charge of the majority carriers; some examples are boron (B), gallium (Ga) and indium (In).

It is important to note that even though N-type and P-type material have different numbers of holes and free electrons than intrinsic material, they are still electrically neutral. When an electron leaves an atom, the positively-charged atom that remains in place in the crystal lattice electrically balances the roaming free electron. Similarly, an atom gaining an electron acquires a negative charge that balances the positively-charged atom it left. At no time does the material acquire a net electrical charge, positive or negative.

Compound semiconductor material can be formed by combining equal amounts of N-type and P-type impurity materials. Some examples of this include gallium-arsenide (GaAs), gallium-phosphate (GaP) and indium-phosphide (InP). To make an N-type compound semiconductor, a slightly higher amount of N-type material is used in the mixture. A P-type compound semiconductor has a little more P-type material in the mixture.

Impurities are introduced into intrinsic semiconductors by diffusion, the same physical process that lets you smell cookies baking from several rooms away. (Molecules diffuse through air much faster than through solids.) Rates of diffusion are proportional to temperature, so semiconductors are doped with impurities at high temperature to save time. Once the doped semiconductor material is cooled, the rate of diffusion of the impurities is so low that they are essentially immobile for many years to come. If an electronic device made from a structure of N- and P-type materials is raised to a high temperature, such as by excessive current, the impurities can again migrate and the internal structure of the device may be destroyed. The maximum operating temperature for semiconductor devices is specified at a level low enough to limit additional impurity diffusion.

The conductivity of an extrinsic semiconductor depends on the charge density (in other words, the concentration of free electrons in N-type, and holes in P-type, semiconductor material). As the energy in the semiconductor increases, the charge density also increases. This is the basis of how all semiconductor devices operate: the major difference is the way in which the energy level is increased. Variations include: The *transistor*, where conductivity is altered by injecting current into the device via a wire; the *thermistor*, where the level of heat in the device is detected by its conductivity, and the *photoconductor*, where light energy that is absorbed by the semiconductor material increases the conductivity.

3.2.6 The PN Semiconductor Junction

If a piece of N-type semiconductor material is placed against a piece of P-type semiconductor material, the location at which they join is called a *PN junction*. The junction has characteristics that make it possible to develop diodes and transistors. The action of the junction is best described by a diode operating as a rectifier.

Initially, when the two types of semiconductor material are placed in contact, each type of material will have only its majority carriers: P-type will have only holes and N-type will have only free electrons. The presence of the positive charges (holes) in the P-type material attracts free electrons from the N-type material immediately across the junction. The opposite is true in the N-type material.

These attractions lead to diffusion of some of the majority carriers across the junction, which combine with and neutralize the majority carriers immediately on the other side (a process called *recombination*). As distance from the junction increases, the attraction quickly becomes too small to cause the carriers to move. The region close to the junction is then *depleted* of carriers, and so is named the *depletion region* (also the *space-charge region* or the *transition region*). The width of the depletion region is very small, on the order of 0.5 μm.

If the N-type material (the *cathode*) is placed at a more negative voltage than the P-type material (the *anode*), current will pass through the junction because electrons are attracted from the lower potential to the higher potential and holes are attracted in the opposite direction. This *forward bias* forces the majority carriers toward the junction where recombination occurs with the opposite type of majority carrier. The source of voltage supplies replacement electrons to the N-type material and removes electrons from the P-type material so that the majority carriers are continually replenished. Thus, the net effect is a *forward current* flowing through the semiconductor, across the PN junction. The *forward resistance* of a diode conducting current is typically very low and varies with the amount of forward current.

When the polarity is reversed, current does not flow because the electrons that are trying to enter the N-type material are repelled, as are the holes trying to enter the P-type material. The reverse polarity attracts majority carriers away from the junction and not towards it. Very little current flows across the PN junction — called *reverse leakage current* — in this case. Allowing only unidirectional current flow is what allows a semiconductor diode to act as rectifier.

3.2.7 Junction Semiconductors

Semiconductor devices that operate using the principles of a PN junction are called *junction semiconductors*. These devices can have one or several junctions. The properties of junction semiconductors can be tightly controlled by the characteristics of the materials used and the size and shape of the junctions.

SEMICONDUCTOR DIODES

Diodes are commonly made of silicon and occasionally germanium. Although they act similarly, they have slightly different characteristics. The *junction threshold voltage*, or *junction barrier voltage*, is the forward bias voltage (V_F) at which current begins to pass through the device. This voltage is different for the two kinds of diodes. In the diode response curve of **Fig 3.7**, V_F corresponds to the voltage at which the positive portion of the curve begins to rise sharply from the x-axis. Most silicon diodes have a junction threshold voltage of about 0.7 V, while the voltage for germanium diodes is typically 0.3 V. Reverse leakage current is much lower for silicon diodes than for germanium diodes.

The characteristic curve for a semiconductor diode junction is given by the following equation (slightly simplified) called

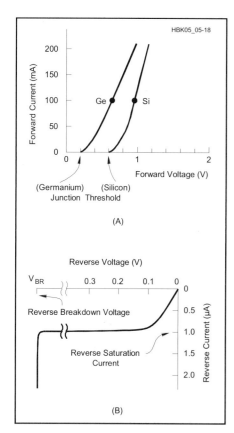

Fig 3.7 — Semiconductor diode (PN junction) characteristic curve. (A) Forward- biased (anode voltage higher than cathode) response for Germanium (Ge) and Silicon (Si) devices. Each curve breaks away from the x-axis at its junction threshold voltage. The slope of each curve is its forward resistance. (B) Reverse-biased response. Very small reverse current increases until it reaches the reverse saturation current (I_0). The reverse current increases suddenly and drastically when the reverse voltage reaches the reverse breakdown voltage, V_{BR}.

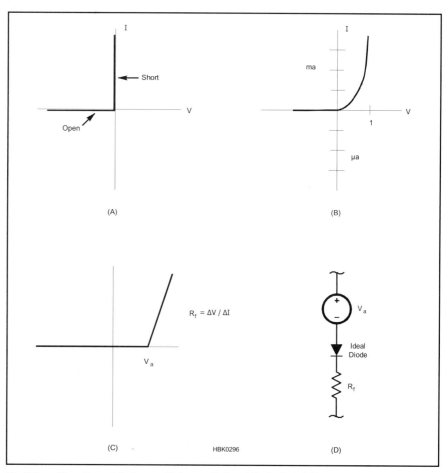

Fig 3.8 — Circuit models for rectifying switches (diodes). A: I-V curve of the ideal rectifier. B: I-V curve of a typical semiconductor diode. Note the different scales for positive and negative current. C shows a simplified diode I-V curve for dc-circuit calculations. D is an equivalent circuit for C.

the *Fundamental Diode Equation* because it describes the behavior of all semiconductor PN junctions.

$$I = I_S \left(e^{\frac{V}{\eta V_t}} \right) \qquad (4)$$

where
I = diode current
V = diode voltage
I_s = reverse-bias saturation current
V_t = kT/q, the thermal equivalent of voltage (about 25 mV at room temperature)
η = emission coefficient.

The value of I_s varies with the type of semiconductor material. η also varies from 1 to 2 with the type of material and the method of fabrication. (η is close to 2 for silicon at normal current levels, decreasing to 1 at high currents.) This curve is shown in **Fig 3.8B**.

The obvious differences between Fig 3.8A and B are that the semiconductor diode has a finite *turn-on* voltage — it requires a small but nonzero forward bias voltage before it begins conducting. Furthermore, once conducting, the diode voltage continues to increase very slowly with increasing current, unlike a true short circuit. Finally, when the applied voltage is negative, the reverse current is not exactly zero but very small (microamperes). The reverse current flow rapidly reaches a level that varies little with the reverse bias voltage. This is the *reverse-bias saturation current*, I_s.

For bias (dc) circuit calculations, a useful model for the diode that takes these two effects into account is shown by the artificial I-V curve in Fig 3.8C. This model neglects the negligible reverse bias current I_s.

When converted into an equivalent circuit, the model in Fig 3.8C yields the circuit in Fig 3.8D. The ideal voltage source V_a represents the turn-on voltage and R_f represents the effective resistance caused by the small increase in diode voltage as the diode current increases. The turn-on voltage is material-dependent: approximately 0.3 V for germanium diodes and 0.7 for silicon. R_f is typically on the order of 10 Ω, but it can vary according to the specific component. R_f can often be completely neglected in comparison to the other resistances in the circuit. This very common simplification leaves only a pure *voltage drop* for the diode model.

BIPOLAR TRANSISTOR

A bipolar transistor is formed when two PN junctions are placed next to each other. If N-type material is surrounded by P-type material, the result is a PNP transistor. Alternatively, if P-type material is in the middle of two layers of N-type material, the NPN transistor is formed (**Fig 3.9**).

Physically, we can think of the transistor as two PN junctions back-to-back, such as two diodes connected at their *anodes* (the positive terminal) for an NPN transistor or two diodes connected at their *cathodes* (the negative terminal) for a PNP transistor. The connection point is the base of the transistor. (You can't actually make a transistor this way — this is a representation for illustration only.)

A transistor conducts when the base-emitter junction is forward biased and the base-collector is reverse biased. Under these conditions, the emitter region emits majority carriers into the base region, where they become minority carriers because the materials of the emitter and base regions have opposite polarity. The excess minority carriers in the base are then attracted across the very thin

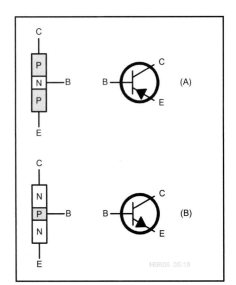

Fig 3.9 — Bipolar transistors. (A) A layer of N-type semiconductor sandwiched between two layers of P-type semiconductor makes a PNP device. The schematic symbol has three leads: collector (C), base (B) and emitter (E), with the arrow pointing in toward the base. (B) A layer of P-type semiconductor sandwiched between two layers of N-type semiconductor makes an NPN device. The schematic symbol has three leads: collector (C), base (B) and emitter (E), with the arrow pointing out away from the base.

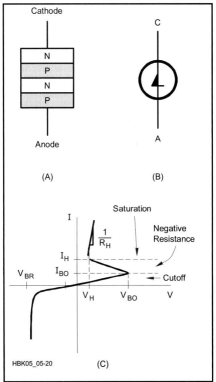

Fig 3.10 — PNPN diode. (A) Alternating layers of P-type and N-type semiconductor. (B) Schematic symbol with cathode (C) and anode (A) leads. (C) I-V curve. Reverse-biased response is the same as normal PN junction diodes. Forward biased response acts as a hysteresis switch. Resistance is very high until the bias voltage reaches V_{BO} and exceeds the cutoff current, I_{BO}. The device exhibits a negative resistance when the current increases as the bias voltage decreases until a voltage of V_H and saturation current of I_H is reached. After this, the resistance is very low, with large increases in current for small voltage increases.

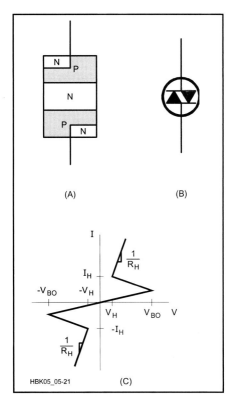

Fig 3.11 — Bilateral switch. (A) Alternating layers of P-type and N-type semiconductor. (B) Schematic symbol. (C) I-V curve. The right-hand side of the curve is identical to the PNPN diode response in Fig 3.10. The device responds identically for both forward and reverse bias so the left-hand side of the curve is symmetrical with the right-hand side.

base to the base-collector junction, where they are collected and are once again considered majority carriers before they can flow to the base terminal.

The flow of majority carriers from emitter to collector can be modified by the application of a bias current to the base terminal. If the bias current causes majority carriers to be injected into the base material (electrons flowing into an N-type base or out of a P-type base) the emitter-collector current increases. In this way, a transistor allows a small base current to control a much larger collector current.

As in a semiconductor diode, the forward biased base-emitter junction has a threshold voltage (V_{BE}) that must be exceeded before the emitter current increases. As the base-emitter current continues to increase, the point is reached at which further increases in base-emitter current cause no additional change in collector current. This is the condition of *saturation*. Conversely, when base-emitter current is reduced to the point at which collector current ceases to flow, that is the situation of *cutoff*.

THYRISTORS

Thyristors are semiconductors made with four or more alternating layers of P- and N-type semiconductor material. In a four-layer thyristor, when the anode is at a higher potential than the cathode, the first and third junctions are forward biased and the center junction reverse biased. In this state, there is little current, just as in the reverse-biased diode. The different types of thyristor have different ways in which they turn on to conduct current and in how they turn off to interrupt current flow.

PNPN Diode

The simplest thyristor is a PNPN (usually pronounced like *pinpin*) diode with three junctions (see **Fig 3.10**). As the forward bias voltage is increased, the current through the device increases slowly until the *breakover (or firing) voltage*, V_{BO}, is reached and the flow of current abruptly increases. The PNPN diode is often considered to be a switch that is off below V_{BO} and on above it.

Bilateral Diode Switch (Diac)

A semiconductor device similar to two PNPN diodes facing in opposite directions and attached in parallel is the *bilateral diode switch* or *diac*. This device has the characteristic curve of the PNPN diode for both positive and negative bias voltages. Its construction, schematic symbol and characteristic curve are shown in **Fig 3.11**.

Silicon Controlled Rectifier (Diac)

Another device with four alternate layers of P-type and N-type semiconductor is the *silicon controlled rectifier (SCR)*. (Some sources refer to an SCR as a thyristor, as well.) In addition to the connections to the outer two layers, two other terminals can be brought out for the inner two layers. The connection to the P-type material near the cathode is called the *cathode gate* and the N-type material near the anode is called the *anode gate*. In nearly all commercially available SCRs, only the cathode gate is connected (**Fig 3.12**).

Like the PNPN diode switch, the SCR is used to abruptly start conducting when the voltage exceeds a given level. By biasing the gate terminal appropriately, the breakover voltage can be adjusted.

Triac

A five-layered semiconductor whose operation is similar to a bidirectional SCR is the *triac* (**Fig 3.13**). This is also similar to a bidirectional diode switch with a bias control gate. The gate terminal of the triac can control both positive and negative breakover voltages and the devices can pass both polarities of voltage.

Thyristor Applications

The SCR is highly efficient and is used in power control applications. SCRs are available that can handle currents of greater than 100 A and voltage differentials of greater than 1000 V, yet can be switched with gate currents of less than 50 mA. Because of their high current-handling capability, SCRs are used as "crowbars" in power supply circuits, to short the output to ground and blow a fuse when an overvoltage condition exists.

SCRs and triacs are often used to control ac power sources. A sine wave with a given RMS value can be switched on and off at preset points during the cycle to decrease the RMS voltage. When conduction is delayed until after the peak (as **Fig 3.14** shows) the peak-to-peak voltage is reduced. If conduction starts before the peak, the RMS voltage is reduced, but the peak-to-peak value remains the same. This method is used to operate light dimmers and 240 V ac to 120 V ac converters. The sharp switching transients created when

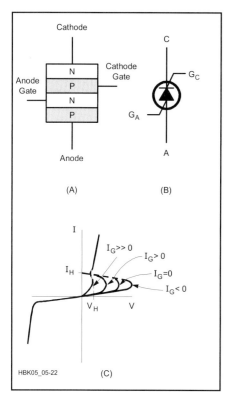

Fig 3.12 — SCR. (A) Alternating layers of P-type and N-type semiconductor. This is similar to a PNPN diode with gate terminals attached to the interior layers. (B) Schematic symbol with anode (A), cathode (C), anode gate (G_A) and cathode gate (G_C). Many devices are constructed without G_A. (C) I-V curve with different responses for various gate currents. I_G = 0 has a similar response to the PNPN diode.

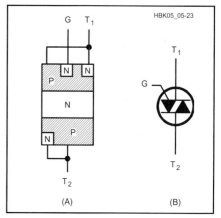

Fig 3.13 — Triac. (A) Alternating layers of P-type and N-type semiconductor. This behaves as two SCR devices facing in opposite directions with the anode of one connected to the cathode of the other and the cathode gates connected together. (B) Schematic symbol.

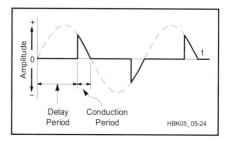

Fig 3.14 — Triac operation on sine wave. The dashed line is the original sine wave and the solid line is the portion that conducts through the triac. The relative delay and conduction period times are controlled by the amount or timing of gate current, I_G. The response of an SCR is the same as this for positive voltages (above the x-axis) and with no conduction for negative voltages.

these devices turn on are common sources of RF interference. (See the chapter on **Electromagnetic Compatibility** for information on dealing with interference from thyristors.)

3.2.8 Field-Effect Transistors (FET)

The *field-effect transistor (FET)* controls the current between two points but does so differently than the bipolar transistor. The FET operates by the effects of an electric field on the flow of electrons through a single type of semiconductor material. This is why the FET is sometimes called a *unipolar* transistor. Unlike bipolar semiconductors that can be arranged in many configurations to provide diodes, transistors, photoelectric devices, temperature sensitive devices and so on, the field effect technique is usually only used to make transistors, although FETs are also available as special-purpose diodes, for use as constant current sources.

FET devices are constructed on a *substrate* of doped semiconductor material. The channel is formed within the substrate and has the opposite polarity (a P-channel FET has N-type substrate). Most FETs are constructed with silicon.

Within the FET, current moves in a *channel* as shown in **Fig 3.15**. The channel is made of either N-type or P-type semiconductor material; an FET is specified as either an N-channel or P-channel device. Current flows from the *source* terminal (where majority carriers are injected) to the *drain* terminal (where majority carriers are removed). A *gate* terminal generates an electric field that controls the current in the channel.

In N-channel devices, the drain potential must be higher than that of the source ($V_{DS} > 0$) for electrons (the majority carriers) to flow in channel. In P-channel devices, the flow of holes requires that $V_{DS} < 0$. The polarity of the electric field that controls current in the channel is determined by the majority carriers of the channel, ordinarily positive for P-channel FETs and negative for N-channel FETs.

Variations of FET technology are based on different ways of generating the electric field. In all of these, however, electrons at the gate are used only for their charge in order to create an electric field around the channel. There is a minimal flow of electrons through the gate. This leads to a very high dc input resistance in devices that use FETs for their input circuitry. There may be quite a bit of capacitance between the gate and the other FET terminals, however, causing the input impedance to be quite low at high frequencies.

The current through an FET only has to pass through a single type of semiconductor material. Depending on the type of material and the construction of the FET, drain-source resistance when the FET is conducting ($r_{DS(ON)}$) may be anywhere from a few hundred ohms to much less than an ohm. The output impedance of devices made with FETs is generally quite low. If a gate bias voltage is added to operate the transistor near cutoff, the circuit output impedance may be much higher.

In order to achieve a higher gain-bandwidth product, other materials have been used. Gallium-arsenide (GaAs) has *electron mobility* and *drift velocity* (both are measures of how easily electrons are able to move through the crystal lattice) far higher than the standard doped silicon. Amplifiers designed with GaAsFET devices operate at much higher frequencies and with a lower noise factor at VHF and UHF than those made with silicon FETs (although silicon FETs have improved dramatically in recent years).

JFET

One of two basic types of FET, the *junction FET (JFET)* gate material is made of the opposite polarity semiconductor to the channel material (for a P-channel FET the gate is made of N-type semiconductor material). The gate-channel junction is similar to

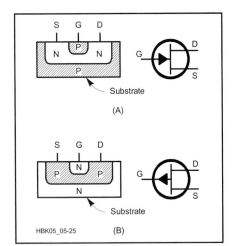

Fig 3.15 — JFET devices with terminals labeled: source (S), gate (G) and drain (D). A) Pictorial of N-type channel embedded in P-type substrate and schematic symbol. B) P-channel embedded in N-type substrate and schematic symbol.

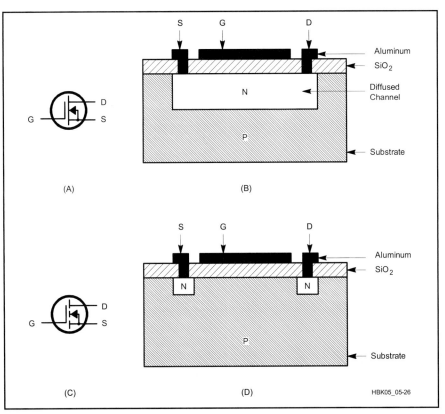

Fig 3.16 — MOSFET devices with terminals labeled: source (S), gate (G) and drain (D). N-channel devices are pictured. P-channel devices have the arrows reversed in the schematic symbols and the opposite type semiconductor material for each of the layers. (A) N-channel depletion mode device schematic symbol and (B) pictorial of P-type substrate, diffused N-type channel, SiO$_2$ insulating layer and aluminum gate region and source and drain connections. The substrate is connected to the source internally. A negative gate potential narrows the channel. (C) N-channel enhancement mode device schematic and (D) pictorial of P-type substrate, N-type source and drain wells, SiO$_2$ insulating layer and aluminum gate region and source and drain connections. Positive gate potential forms a channel between the two N-type wells.

a diode's PN junction with the gate material in direct contact with the channel. JFETs are used with the junction reverse-biased, since any current in the gate is undesirable. The reverse bias of the junction creates an electric field that "pinches" the channel. Since the magnitude of the electric field is proportional to the reverse-bias voltage, the current in the channel is reduced for higher reverse gate bias voltages. When current in the channel is completely halted by the electric field, this is called *pinch-off* and it is analogous to cutoff in a bipolar transistor. The channel in a JFET is at its maximum conductivity when the gate and source voltages are equal ($V_{GS} = 0$).

Because the gate-channel junction in a JFET is similar to a bipolar junction diode, this junction must never be forward biased; otherwise large currents will pass through the gate and into the channel. For an N-channel JFET, the gate must always be at a lower potential than the source ($V_{GS} < 0$). The prohibited condition is for $V_{GS} > 0$. For P-channel JFETs these conditions are reversed (in normal operation $V_{GS} > 0$ and the prohibited condition is for $V_{GS} < 0$).

MOSFET

Placing an insulating layer between the gate and the channel allows for a wider range of control (gate) voltages and further decreases the gate current (and thus increases the device input resistance). The insulator is typically made of an oxide (such as silicon dioxide, SiO_2). This type of device is called a *metal-oxide-semiconductor FET (MOSFET)* or *insulated-gate FET (IGFET)*.

The substrate is often connected to the source internally. The insulated gate is on the opposite side of the channel from the substrate (see **Fig 3.16**). The bias voltage on the gate terminal either attracts or repels the majority carriers of the substrate across its PN-junction with the channel. This narrows (*depletes*) or widens (*enhances*) the channel, respectively, as V_{GS} changes polarity. For example, in the N-channel enhancement-mode MOSFET, positive gate voltages with respect to the substrate and the source ($V_{GS} > 0$) repel holes from the channel into the substrate, thereby widening the channel and decreasing channel resistance. Conversely, $V_{GS} < 0$ causes holes to be attracted from the substrate, narrowing the channel and increasing the channel resistance. Once again, the polarities discussed in this example are reversed for P-channel devices. The common abbreviation for an N-channel MOSFET is *NMOS*, and for a P-channel MOSFET, *PMOS*.

Because of the insulating layer next to the gate, input resistance of a MOSFET is usually greater than 10^{12} Ω (a million megohms). Since MOSFETs can both deplete the channel, like the JFET, and also enhance it, the construction of MOSFET devices differs based on the channel size in the quiescent state, $V_{GS} = 0$.

A *depletion mode* device (also called a *normally-on MOSFET*) has a channel in the quiescent state that gets smaller as a reverse bias is applied; this device conducts current with no bias applied (see Fig 3.16A and B). An *enhancement mode* device (also called a *normally-off MOSFET*) is built without a channel and does not conduct current when $V_{GS} = 0$; increasing forward bias forms a temporary channel that conducts current (see Fig 3.16C and D).

Complementary Metal Oxide Semiconductors (CMOS)

Power dissipation in a circuit can be reduced to very small levels (on the order of a few nanowatts) by using MOSFET devices in complementary pairs (CMOS). Each amplifier is constructed of a series circuit of MOSFET devices, as in **Fig 3.17**. The gates are tied together for the input signal, as are the drains for the output signal. In saturation and cutoff, only one of the devices conducts. The current drawn by the circuit under no load is equal to the OFF leakage current of either device and the voltage drop across the pair is equal to V_{DD}, so the steady-state power used by the circuit is always equal to $V_{DD} \times I_{D(OFF)}$. Power is only consumed during the switching process, so for ac signals, power consumption is proportional to frequency.

CMOS circuitry could be built with discrete components, but the number of extra parts and the need for the complementary components to be matched has made that an unusual design technique. The low power consumption and ease of fabrication has made

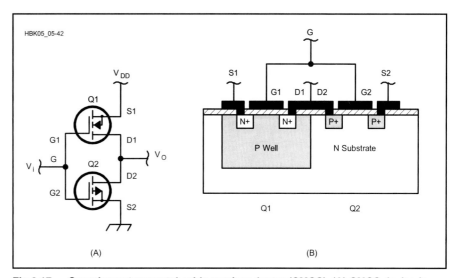

Fig 3.17 — Complementary metal oxide semiconductor (CMOS). (A) CMOS device is made from a pair of enhancement mode MOS transistors. The upper is an N-channel device, and the lower is a P-channel device. When one transistor is biased on, the other is biased off; therefore, there is minimal current from V_{DD} to ground. (B) Implementation of a CMOS pair as an integrated circuit.

CMOS the most common of all IC technologies. Although CMOS is most commonly used in digital integrated circuitry, its low power consumption has also been put to work by manufacturers of analog ICs, as well as digital ICs.

3.2.9 Semiconductor Temperature Effects

The number of excess holes and electrons in semiconductor material is increased as the temperature of a semiconductor increases. Since the conductivity of a semiconductor is related to the number of excess carriers, this also increases with temperature. With respect to resistance, semiconductors have a negative temperature coefficient. The resistance of silicon *decreases* by about 8% per °C and by about 6% per °C for germanium. Semiconductor temperature properties are the opposite of most metals, which *increase* their resistance by about 0.4% per °C. These opposing temperature characteristics permit the design of circuits with opposite temperature coefficients that cancel each other out, making a temperature insensitive circuit.

Semiconductor devices can experience an effect called *thermal runaway* as the current causes an increase in temperature. (This is primarily an issue with bipolar transistors.) The increased temperature decreases resistance and may lead to a further increase in current (depending on the circuit) that leads to an additional temperature increase. This sequence of events can continue until the semiconductor destroys itself, so circuit design must include measures that compensate for the effects of temperature.

Semiconductor Failure Caused by Heat

There are several common failure modes for semiconductors that are related to heat. The semiconductor material is connected to the outside world through metallic *bonding leads*. The point at which the lead and the semiconductor are connected is a common place for the semiconductor device to fail. As the device heats up and cools down, the materials expand and contract. The rate of expansion and contraction of semiconductor material is different from that of metal. Over many cycles of heating and cooling the bond between the semiconductor and the metal can break. Some experts have suggested that the lifetime of semiconductor equipment can be extended by leaving the devices powered on all the time, but this requires removal of the heat generated during normal operation.

A common failure mode of semiconductors is caused by the heat generated during semiconductor use. If the temperatures of the PN junctions remain at high enough levels for long enough periods of time, the impurities resume their diffusion across the PN junctions. When enough of the impurity atoms cross the depletion region, majority carrier recombination stops functioning properly and the semiconductor device fails permanently.

Excessive temperature can also cause failure anywhere in the semiconductor from heat generation within any current-carrying conductor, such as an FET channel or the bonding leads. Integrated circuits with more than one output may have power dissipation limits that depend on how many of the outputs are active at one time. The high temperature can cause localized melting or cracking of the semiconductor material, causing a permanent failure.

Another heat-driven failure mode, usually not fatal to the semiconductor, is excessive leakage current or a shift in operating point that causes the circuit to operate improperly. This is a particular problem in complex integrated circuits — analog and digital — dissipating significant amounts of heat under normal operating conditions. Computer microprocessors are a good example, often requiring their own cooling systems. Once the device cools, normal operation is usually restored.

To reduce the risk of thermal failures, the designer must comply with the limits stated in the manufacturer's data sheet, devising an adequate heat removal system. (Thermal issues are discussed in the **Electrical Fundamentals** chapter.)

3.2.10 Safe Operating Area (SOA)

Devices intended for use in circuits handling high currents or voltages are specified to have a *safe operating area* (*SOA*). This refers to the area drawn on the device's characteristic curve containing combinations of voltage and current that the device can be expected to control without damage under specific conditions. The SOA combines a number of limits — voltage, current, power, temperature and various breakdown mechanisms — in order to simplify the design of protective circuitry. The SOA is also specified to apply to specific durations of use — steady-state, long pulses, short pulses and so forth. The device may have separate SOAs for resistive and inductive loads.

You may also encounter two specialized types of SOA for turning the device on and off. *Reverse bias safe operating area* (*RBSOA*) applies when the device is turning off. *Forward bias safe operating* area (*FBSOA*) applies when turning the device on. These SOAs are used because the high rate-of-change of current and voltage places additional stresses on the semiconductor.

3.3 Practical Semiconductors

3.3.1 Semiconductor Diodes

Although many types of semiconductor diodes are available, they share many common characteristics. The different types of diodes have been developed to optimize particular characteristics for one type of application.

The diode symbol is shown in **Fig 3.18**. Forward current flows in the direction from anode to cathode, in the direction of the arrow. Reverse current flows from cathode to anode. (Current is considered to be conventional current as described in the **Electrical Fundamentals** chapter.) The anode of a semiconductor junction diode is made of P-type material and the cathode is made of N-type material, as indicated in Fig 3.18. Most diodes are marked with a band on the cathode end (Fig 3.18).

DIODE RATINGS

Five major characteristics distinguish standard junction diodes from one another: current handling capacity, maximum voltage rating, response speed, reverse leakage current and junction forward voltage. Each of these characteristics can be manipulated during manufacture to produce special purpose diodes.

Current Capacity

The ideal diode would have zero resistance

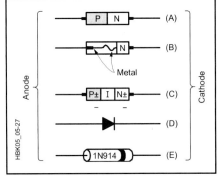

Fig 3.18 — Practical semiconductor diodes. All devices are aligned with anode on the left and cathode on the right. (A) Standard PN junction diode. (B) Point-contact or "cat's whisker" diode. (C) PIN diode formed with heavily doped P-type (P+), undoped (intrinsic) and heavily doped N-type (N+) semiconductor material. (D) Diode schematic symbol. (E) Diode package with marking stripe on the cathode end.

in the forward direction and infinite resistance in the reverse direction. This is not the case for actual devices, which behave as shown in the plot of a diode response in Fig 3.7. Note that the scales of the two graphs are drastically different. The inverse of the slope of the line (the change in voltage between two points on a straight portion of the line divided by the corresponding change in current) on the upper right is the resistance of the diode in the forward direction, R_F.

The range of voltages is small and the range of currents is large since the forward resistance is very small (in this example, about $2\,\Omega$). Nevertheless, this resistance causes heat dissipation according to $P = I_F^2 \times R_F$.

In addition, there is a forward voltage, V_F, whenever the forward current is flowing. This also results in heat dissipation as $P = I \times V_F$. In power applications where the average forward current is high, heating from forward resistance and the forward voltage drop can be significant. Since forward current determines the amount of heat dissipation, the diode's power rating is stated as a *maximum average current*. Exceeding the current rating in a diode will cause excessive heating that leads to PN junction failure as described earlier.

Peak Inverse Voltage (PIV)

In Fig 3.7, the lower left portion of the curve illustrates a much higher resistance that increases from tens of kilohms to thousands of megohms as the reverse voltage gets larger, and then decreases to near zero (a nearly vertical line) very suddenly. This sudden change occurs because the diode enters *reverse breakdown* or when the reverse voltage becomes high enough to push current across the junction. The voltage at which this occurs is the *reverse breakdown voltage*. Unless the current is so large that the diode fails from overheating, breakdown is not destructive and the diode will again behave normally when the bias is removed. The maximum reverse voltage that the diode can withstand under normal use is the *peak inverse voltage* (*PIV*) rating. A related effect is *avalanche breakdown* in which the voltage across a device is greater than its ability to control or block current flow.

Response Speed

The speed of a diode's response to a change in voltage polarity limits the frequency of ac current that the diode can rectify. The diode response in Fig 3.7 shows how that diode will act at dc. As the frequency increases, the diode may not be able to turn current on and off as fast as the changing polarity of the signal.

Diode response speed mainly depends on *charge storage* in the depletion region. When forward current is flowing, electrons and holes fill the region near the junction to recombine. When the applied voltage reverses, these excess charges move away from the junction so that no recombination can take place. As reverse bias empties the depletion region of excess charge, it begins to act like a small capacitor formed by the regions containing majority carriers on either side of the junction and the depletion region acting as the dielectric. This *junction capacitance* is inversely proportional to the width of the depletion region and directly proportional to the cross-sectional surface area of the junction.

The effect of junction capacitance is to allow current to flow for a short period after the applied voltage changes from positive to negative. To halt current flow requires that the junction capacitance be charged. Charging this capacitance takes some time; a few µs for regular rectifier diodes and a few hundred nanoseconds for *fast-recovery* diodes. This is the diode's *charge-storage time*. The amount of time required for current flow to cease is the diode's *recovery time*.

Reverse Leakage Current

Because the depletion region is very thin, reverse bias causes a small amount of reverse leakage or reverse saturation current to flow from cathode to anode. This is typically 1 µA or less until reverse breakdown voltage is reached. Silicon diodes have lower reverse leakage currents than diodes made from other materials with higher carrier mobility, such as germanium.

The reverse saturation current I_s is not constant but is affected by temperature, with higher temperatures increasing the mobility of the majority carriers so that more of them cross the depletion region for a given amount of reverse bias. For silicon diodes (and transistors) near room temperature, I_s increases by a factor of 2 every 4.8 °C. This means that for every 4.8 °C rise in temperature, either the current doubles (if the voltage across it is constant), or if the current is held constant by other resistances in the circuit, the diode voltage will *decrease* by $V_T \times \ln 2$ = 18 mV. For germanium, the current doubles every 8 °C and for gallium-arsenide (GaAs), 3.7 °C. This dependence is highly reproducible and may actually be exploited to produce temperature-measuring circuits.

While the change resulting from a rise of several degrees may be tolerable in a circuit design, that from 20 or 30 degrees may not. Therefore it's a good idea with diodes, just as with other components, to specify power ratings conservatively (2 to 4 times margin) to prevent self-heating.

While component derating does reduce self-heating effects, circuits must be designed for the expected operating environment. For example, mobile radios may face temperatures from −20° to +140 °F (−29° to 60 °C).

Forward Voltage

The amount of voltage required to cause majority carriers to enter the depletion region and recombine, creating full current flow, is called a diode's *forward voltage*, V_F. It depends on the type of material used to create the junction and the amount of current. For silicon diodes at normal currents, $V_F = 0.7$ V, and for germanium diodes, $V_F = 0.3$ V. As you saw earlier, V_F also affects power dissipation in the diode.

POINT-CONTACT DIODES

One way to decrease charge storage time in the depletion region is to form a metal-semiconductor junction for which the depletion is very thin. This can be accomplished with a *point-contact diode*, where a thin piece of aluminum wire, often called a *whisker*, is placed in contact with one face of a piece of lightly doped N-type material. In fact, the original diodes used for detecting radio signals ("cat's whisker diodes") were made with a steel wire in contact with a crystal of impure lead (galena). Point-contact diodes have high response speed, but poor PIV and current-handling ratings. The 1N34 germanium point-contact diode is the best-known example of point-contact diode still in common use.

SCHOTTKY DIODES

An improvement to point-contact diodes, the *hot-carrier diode* is similar to a point-contact diode, but with more ideal characteristics attained by using more efficient metals, such as platinum and gold, that act to lower forward resistance and increase PIV. This type of contact is known as a *Schottky barrier*, and diodes made this way are called *Schottky diodes*. The junctions of Schottky diodes, being smaller, store less charge and as a result, have shorter switching times and junction capacitances than standard PN-junction diodes. Their forward voltage is also lower, typically 0.3 to 0.4 V. In most other respects they behave similarly to PN diodes.

PIN DIODES

The PIN diode, shown in Fig 3.18C is a *slow response* diode that is capable of passing RF and microwave signals when it is forward biased. This device is constructed with a layer of intrinsic (undoped) semiconductor placed between very highly doped P-type and N-type material (called P+-type and N+-type material to indicate the extra amount of doping), creating a *PIN junction*. These devices provide very effective switches for RF signals and are often used in transmit-receive switches in transceivers and amplifiers. The majority carriers in PIN diodes have longer than normal lifetimes before recombination, resulting in a slow switching process that causes them to act more like resistors than diodes at high radio frequencies. The amount of resistance can be controlled by the amount of forward bias applied to the PIN diode and this allows them to act as current-controlled attenuators. (For additional discussion of PIN diodes and projects in which they are used, see the chapters on **Transceivers**, **RF Power Amplifiers**, and **Test Equipment and Measurements**.)

VARACTOR DIODES

Junction capacitance can be used as a circuit element by controlling the reverse bias voltage across the junction, creating a small variable capacitor. Junction capacitances are small, on the order of pF. As the reverse bias voltage on a diode increases, the width of the depletion region increases, decreasing its capacitance. A *varactor* (also known by the trade name Varicap diode) is a diode with a junction specially formulated to have a relatively large range of capacitance values for a modest range of reverse bias voltages (**Fig 3.19**).

As the reverse bias applied to a diode changes, the width of the depletion layer, and therefore the capacitance, also changes. The diode junction capacitance (C_j) under a reverse bias of V volts is given by

$$C_j = \frac{C_{j0}}{\sqrt{V_{on} - V}} \qquad (5)$$

where C_{j0} = measured capacitance with zero applied voltage.

Note that the quantity under the radical is a large *positive* quantity for reverse bias. As seen from the equation, for large reverse biases C_j is inversely proportional to the square root of the voltage.

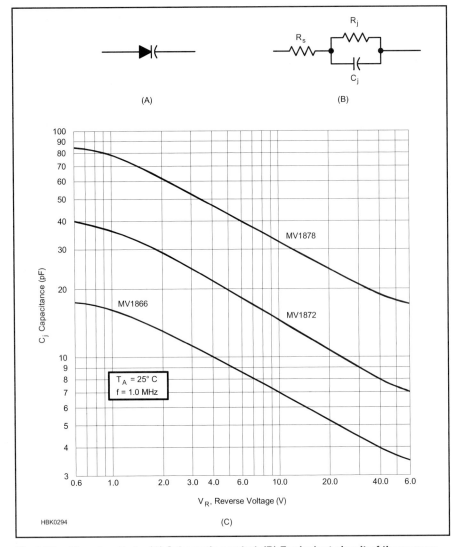

Fig 3.19 — Varactor diode. (A) Schematic symbol. (B) Equivalent circuit of the reverse biased varactor diode. R_S is the junction resistance, R_J is the leakage resistance and C_J is the junction capacitance, which is a function of the magnitude of the reverse bias voltage. (C) Plot of junction capacitance, C_J, as a function of reverse voltage, V_R, for three different varactor devices. Both axes are plotted on a logarithmic scale.

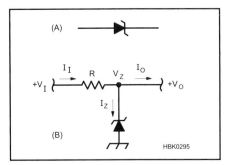

Fig 3.20 — Zener diode. (A) Schematic symbol. (B) Basic voltage regulating circuit. V_Z is the Zener reverse breakdown voltage. Above V_Z, the diode draws current until $V_I - I_I R = V_Z$. The circuit design should select R so that when the maximum current is drawn, $R < (V_I - V_Z) / I_O$. The diode should be capable of passing the same current when there is no output current drawn.

Although special forms of varactors are available from manufacturers, other types of diodes may be used as inexpensive varactor diodes, but the relationship between reverse voltage and capacitance is not always reliable.

When designing with varactor diodes, the reverse bias voltage must be absolutely free of noise since any variations in the bias voltage will cause changes in capacitance. For example, if the varactor is used to tune an oscillator, unwanted frequency shifts or instability will result if the reverse bias voltage is noisy. It is possible to frequency modulate a signal by adding the audio signal to the reverse bias on a varactor diode used in the carrier oscillator. (For examples of the use of varactors in oscillators and modulators, see the chapters on **Mixers, Modulators, and Demodulators** and **Oscillators and Synthesizers**.)

ZENER DIODES

When the PIV of a reverse-biased diode is exceeded, the diode begins to conduct current as it does when it is forward biased. This current will not destroy the diode if it is limited to less than the device's maximum allowable value. By using heavy levels of doping during manufacture, a diode's PIV can be precisely controlled to be at a specific level, called the *Zener voltage*, creating a type of voltage reference. These diodes are called Zener diodes after their inventor, American physicist Clarence Zener.

When the Zener voltage is reached, the reverse voltage across the Zener diode remains constant even as the current through it changes. With an appropriate series current-limiting resistor, the Zener diode provides an accurate voltage reference (see **Fig 3.20**).

Zener diodes are rated by their reverse-breakdown voltage and their power-handling capacity, where $P = V_Z \times I_Z$. Since the same current must always pass though the resistor to drop the source voltage down to the reference voltage, with that current divided between the Zener diode and the load, this type of power source is very wasteful of current.

The Zener diode does make an excellent and efficient voltage reference in a larger voltage regulating circuit where the load current is provided from another device whose voltage is set by the reference. (See the **Power Supplies** chapter for more information about using Zener diodes as voltage regulators.) When operating in the breakdown region, Zener diodes can be modeled as a simple voltage source.

The primary sources of error in Zener-diode-derived voltages are the variation with load current and the variation due to heat. Temperature-compensated Zener diodes are available with temperature coefficients as low as 0.0005 % per °C. If this is unacceptable, voltage reference integrated circuits based on Zener diodes have been developed that include additional circuitry to counteract temperature effects.

A variation of Zener diodes, *transient voltage suppressor* (TVS) diodes are designed to dissipate the energy in short-duration, high-voltage transients that would otherwise damage equipment or circuits. TVS diodes have large junction cross-sections so that they can handle large currents without damage. These diodes are also known as TransZorbs. Since the polarity of the transient can be positive, negative, or both, transient protection circuits can be designed with two devices connected with opposite polarities.

COMMON DIODE APPLICATIONS

Standard semiconductor diodes have many uses in analog circuitry. Several examples of

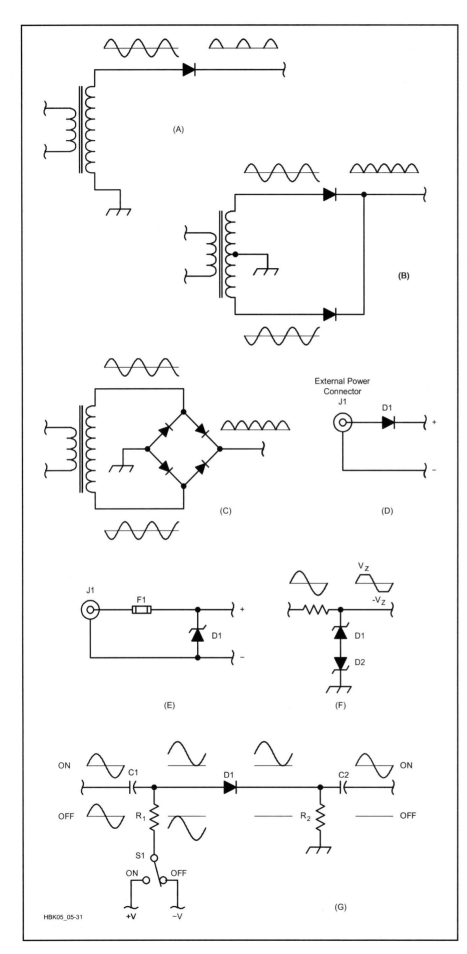

Fig 3.21 — Diode circuits. (A) Half wave rectifier circuit. Only when the ac voltage is positive does current pass through the diode. Current flows only during half of the cycle. (B) Full-wave center-tapped rectifier circuit. Center-tap on the transformer secondary is grounded and the two ends of the secondary are 180° out of phase. (C) Full-wave bridge rectifier circuit. In each half of the cycle two diodes conduct. (D) Polarity protection for external power connection. If polarity is correct, the diode will conduct and if reversed the diode will block current, protecting the circuit that is being powered. (E) Over-voltage protection circuit. If excessive voltage is applied to J1, D1 will conduct current until fuse, F1, is blown. (F) Bipolar voltage clipping circuit. In the positive portion of the cycle, D2 is forward biased, but no current is shunted to ground because D1 is reverse biased. D1 starts to conduct when the voltage exceeds the Zener breakdown voltage, and the positive peak is clipped. When the negative portion of the cycle is reached, D1 is forward biased, but no current is shunted to ground because D2 is reverse biased. When the voltage exceeds the Zener breakdown voltage of D2, it also begins to conduct, and the negative peak is clipped. (G) Diode switch. The signal is ac coupled to the diode by C1 at the input and C2 at the output. R2 provides a reference for the bias voltage. When switch S1 is in the ON position, a positive dc voltage is added to the signal so it is forward biased and is passed through the diode. When S1 is in the OFF position, the negative dc voltage added to the signal reverse biases the diode, and the signal does not get through.

diode circuits are shown in **Fig 3.21**.

The most common application of a diode is to perform rectification; that is, permitting current flow in only one direction. Power rectification converts ac current into pulsating dc current. There are three basic forms of power rectification using semiconductor diodes: half wave (1 diode), full-wave center-tapped (2 diodes) and full-wave bridge (4 diodes). These applications are shown in Fig 3.21A, B and C and are more fully described in the **Power Supplies** chapter.

The most important diode parameters to consider for power rectification are the PIV and current ratings. The peak negative voltages that are blocked by the diode must be smaller in magnitude than the PIV and the peak current through the diode when it is forward biased must be less than the maximum average forward current.

Rectification is also used at much lower current levels in modulation and demodulation and other types of analog signal processing circuits. For these applications, the diode's response speed and junction forward voltage

Analog Basics 3.17

are the most important ratings.

Diodes are also used to protect circuits against the application of voltage with the wrong polarity. In battery-powered devices a forward-biased series diode (Fig 3.21D) is often used to protect the circuitry from the user inadvertently inserting the batteries backwards. Likewise, when a circuit is powered from an external dc source, a diode is often placed in series with the power connector in the device to prevent incorrectly wired power supplies from destroying the equipment. Diodes are commonly used to protect analog meters from both reverse voltage and over voltage conditions that would destroy the delicate needle movement.

Zener diodes are sometimes used to protect low-current (a few amps) circuits from overvoltage conditions (Fig 3.21E). A reverse-biased Zener diode connected between the positive power lead and ground will conduct excessive current if its breakdown voltage is exceeded. Used in conjunction with a fuse in series with the power lead, the Zener diode will cause the fuse to blow when an overvoltage condition exists.

Diodes can be used to clip signals, similar to rectification. If the signal is appropriately biased it can be clipped at any level. Two Zener diodes placed back-to-back can be used to clip both the positive and negative peaks of a signal. (Fig 3.21F) Such an arrangement is used to convert a sine wave to an approximate square wave. If the intent is to protect a circuit against high-energy transients, the TVS diode discussed above is a better choice.

Care must be taken when using Zener diodes to process signals. The Zener diode is a relatively noisy device and can add excessive noise to the signals if operated in its breakdown region. The Zener diode is often used to intentionally generate noise, such as in the noise bridge (see the **Test Equipment and Measurements** chapter). The reverse-biased Zener diode in breakdown generates wideband (nearly white) noise levels as high as

$$\frac{2000\ \mu V}{\sqrt{Hz}}$$

The actual noise voltage developed is determined by multiplying this value by the square root of the circuit bandwidth in Hz.

Diodes are used as switches for ac coupled signals when a dc bias voltage can be added to the signal to permit or inhibit the signal from passing through the diode, see Fig 3.21G. In this case the bias voltage must be added to the ac signal and be of sufficient magnitude so that the entire envelope of the ac signal is above or below the junction barrier voltage, with respect to the cathode, to pass through the diode or inhibit the signal. Special forms of diodes, such as the PIN diode described earlier, which are capable of passing higher frequencies, are used to switch RF signals.

3.3.2 Bipolar Junction Transistors (BJT)

The bipolar transistor is a *current-controlled device* with three basic terminals; *emitter*, *collector* and *base*. The current between the emitter and the collector is controlled by the current between the base and emitter. The convention when discussing transistor operation is that the three currents into the device are positive (I_c into the collector, I_b into the base and I_e into the emitter). Kirchhoff's Current Law (see the **Electrical Fundamentals** chapter) applies to transistors just as it does to passive electrical networks: the total current entering the device must be zero. Thus, the relationship between the currents into a transistor can be generalized as

$$I_c + I_b + I_e = 0 \qquad (6)$$

which can be rearranged as necessary. For example, if we are interested in the emitter current,

$$I_e = -(I_c + I_b) \qquad (7)$$

The back-to-back diode model shown in Fig 3.9 is appropriate for visualization of transistor construction. In actual transistors, however, the relative sizes of the collector, base and emitter regions differ. A common transistor configuration that spans a distance of 3 mm between the collector and emitter contacts typically has a base region that is only 25 μm across.

The operation of the bipolar transistor is described graphically by characteristic curves as shown in **Fig 3.22**. These are similar to the I-V characteristic curves for the two-terminal devices described in the preceding sections. The parameters shown by the curves depend on the type of circuit in which they are measured, such as common emitter or common collector. The output characteristic shows a set of curves for either collector or emitter current versus collector-emitter voltage at various values of input current (either base or emitter). The input characteristic shows the voltage between the input and common terminals (such as base-emitter) versus the input current for different values of output voltage.

CURRENT GAIN

Two parameters describe the relationships between the three transistor currents at low frequencies:

$$\alpha = -\frac{\Delta I_C}{\Delta I_E} = 1 \qquad (8)$$

$$\beta = \frac{\Delta I_C}{\Delta I_B} \qquad (9)$$

The relationship between α and β is defined as

$$\alpha = -\frac{\beta}{1+\beta} \qquad (10)$$

Another designation for β is often used: h_{FE}, the *forward dc current gain*. (The "h" refers to "h parameters," a set of transfer parameters for describing a two-port network and described in more detail in the **RF Techniques** chapter.) The symbol, h_{fe}, in which the subscript is in lower case, is used for the forward current gain of ac signals.

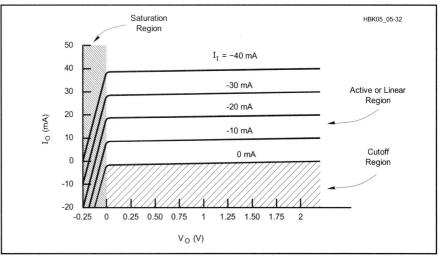

Fig 3.22 — Transistor response curve. The x-axis is the output voltage, and the y-axis is the output current. Different curves are plotted for various values of input current. The three regions of the transistor are its cutoff region, where no current flows in any terminal, its active region, where the output current is nearly independent of the output voltage and there is a linear relationship between the input current and the output current, and the saturation region, where the output current has large changes for small changes in output voltage.

OPERATING REGIONS

Current conduction between collector and emitter is described by *regions* of the transistor's characteristic curves in Fig 3.22. (References such as *common-emitter* or *common-base* refer to the configuration of the circuit in which the parameter is measured.) The transistor is in its *active* or *linear region* when the base-collector junction is reverse biased and the base-emitter junction is forward biased. The slope of the output current, I_O, versus the output voltage, V_O, is virtually flat, indicating that the output current is nearly independent of the output voltage. In this region, the output circuit of the transistor can be modeled as a constant-current source controlled by the input current. The slight slope that does exist is due to base-width modulation (known as the *Early effect*).

When both the junctions in the transistor are forward biased, the transistor is said to be in its *saturation region*. In this region, V_O is nearly zero and large changes in I_O occur for very small changes in V_O. Both junctions in the transistor are reverse-biased in the *cutoff region*. Under this condition, there is very little current in the output, only the nanoamperes or microamperes that result from the very small leakage across the input-to-output junction. Finally, if V_O is increased to very high values, avalanche breakdown begins as in a PN-junction diode and output current increases rapidly. This is the *breakdown region*, not shown in Fig 3.22.

These descriptions of junction conditions are the basis for the use of transistors. Various configurations of the transistor in circuitry make use of the properties of the junctions to serve different purposes in analog signal processing.

OPERATING PARAMETERS

A typical general-purpose bipolar-transistor data sheet lists important device specifications. Parameters listed in the ABSOLUTE MAXIMUM RATINGS section are the three junction voltages (V_{CEO}, V_{CBO} and V_{EBO}), the continuous collector current (I_C), the total device power dissipation (P_D) and the operating and storage temperature range. Exceeding any of these parameters is likely to cause the transistor to be destroyed. (The "O" in the suffixes of the junction voltages indicates that the remaining terminal is not connected, or open.)

In the OPERATING PARAMETERS section, three guaranteed minimum junction breakdown voltages are listed $V_{(BR)CEO}$, $V_{(BR)CBO}$ and $V_{(BR)EBO}$. Exceeding these voltages is likely to cause the transistor to enter avalanche breakdown, but if current is limited, permanent damage may not result.

Under ON CHARACTERISTICS are the guaranteed minimum dc current gain (β or h_{FE}), guaranteed maximum collector-emitter saturation voltage, $V_{CE(SAT)}$, and the guaranteed maximum base-emitter on voltage, $V_{BE(ON)}$. Two guaranteed maximum collector cutoff currents, I_{CEO} and I_{CBO}, are listed under OFF CHARACTERISTICS.

The next section is SMALL-SIGNAL CHARACTERISTICS, where the guaranteed minimum current gain-bandwidth product, BW or f_T, the guaranteed maximum output capacitance, C_{obo}, the guaranteed maximum input capacitance, C_{ibo}, the guaranteed range of input impedance, h_{ie}, the small-signal current gain, h_{fe}, the guaranteed maximum voltage feedback ratio, h_{re} and output admittance, h_{oe} are listed.

Finally, the SWITCHING CHARACTERISTICS section lists absolute maximum ratings for delay time, t_d, rise time, t_r, storage time, t_s and fall time, t_f.

3.3.3 Field-Effect Transistors (FET)

FET devices are controlled by the voltage level of the input rather than the input current, as in the bipolar transistor. FETs have three basic terminals, the *gate*, the *source* and the *drain*. They are analogous to bipolar transistor terminals: the gate to the base, the source to the emitter, and the drain to the collector. Symbols for the various forms of FET devices are pictured in **Fig 3.23**.

The FET gate has a very high impedance, so the input can be modeled as an open circuit. The voltage between gate and source, V_{GS}, controls the resistance of the drain-source channel, r_{DS}, and so the output of the FET is modeled as a current source, whose output current is controlled by the input voltage.

The action of the FET channel is so nearly ideal that, as long as the JFET gate does not become forward biased and inject current from the base into the channel, the drain and source currents are virtually identical. For JFETs the *gate leakage current*, I_G, is a function of V_{GS} and this is often expressed with an *input curve* (see **Fig 3.24**). The point at which there is a significant increase in I_G is called the *junction breakpoint voltage*. Because the gate of MOSFETs is insulated from the channel, gate leakage current is insignificant in these devices.

The dc channel resistance, r_{DS}, is specified in data sheets to be less than a maximum value when the device is biased on ($r_{DS(on)}$). When the gate voltage is maximum ($V_{GS} = 0$ for a JFET), $r_{DS(on)}$ is minimum. This describes the effectiveness of the device as an analog switch. Channel resistance is approximately the same for ac and dc signals until at high frequencies the capacitive reactances inherent in the FET structure become significant.

FETs also have strong similarities to vacuum tubes in that input voltage between the grid and cathode controls an output current between the plate and cathode. (See the chapter on **RF Power Amplifiers** for more information on vacuum tubes.)

FORWARD TRANSCONDUCTANCE

The change in FET drain current caused by a change in gate-to-source voltage is called *forward transconductance*, g_m.

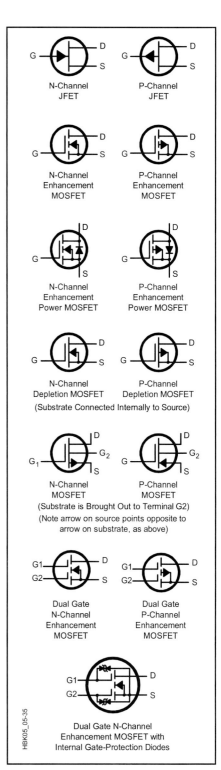

Fig 3.23—FET schematic symbols.

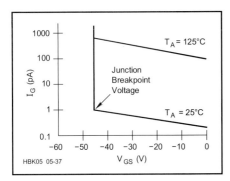

Fig 3.24 — JFET input leakage curves for common source amplifier configuration. Input voltage (V_{GS}) on the x-axis versus input current (I_G) on the y-axis, with two curves plotted for different operating temperatures, 25 °C and 125 °C. Input current increases greatly when the gate voltage exceeds the junction breakpoint voltage.

$$g_m = \frac{\Delta I_{DS}}{\Delta V_{GS}}$$

or

$$\Delta I_{DS} = g_m \Delta V_{GS} \qquad (11)$$

The input voltage, V_{GS}, is measured between the FET gate and source and drain current, I_{DS}, flows from drain to source. Analogous to a bipolar transistor's current gain, the units of transconductance are Siemens (S) because it is the ratio of current to voltage. (Both g_m and g_{fs} are used interchangeably to indicate transconductance. Some sources specify g_{fs} as the *common-source forward transconductance*. This chapter uses g_m, the most common convention in the reference literature.)

OPERATING REGIONS

The most useful relationships for FETs are the output and transconductance response characteristic curves in **Fig 3.25**. (References such as *common-source* or *common-gate* refer to the configuration of the circuit in which the parameter is measured.) Transconductance curves relate the drain current, I_D, to gate-to-source voltage, V_{GS}, at various drain-source voltages, V_{DS}. The FET's forward transconductance, g_m, is the slope of the lines in the forward transconductance curve. The same parameters are interrelated in a different way in the output characteristic, in which I_D is shown versus V_{DS} for different values of V_{GS}.

Like the bipolar transistor, FET operation can be characterized by regions. The *ohmic region* is shown at the left of the FET output characteristic curve in **Fig 3.26** where I_D is increasing nearly linearly with V_{DS} and the FET is acting like a resistance controlled by V_{GS}. As V_{DS} continues to increase, I_D saturates and becomes nearly constant. This is the FET's *saturation region* in which the channel of the FET can be modeled as a constant-current source. V_{DS} can become so large that V_{GS} no longer controls the conduction of the device and avalanche breakdown occurs as in bipolar transistors and PN-junction diodes. This is the *breakdown region*, shown in Fig 3.26 where the curves for I_D break sharply upward. If V_{GS} is less than V_P, so that transconductance is zero, the FET is in the *cutoff region*.

OPERATING PARAMETERS

A typical FET data sheet gives ABSOLUTE MAXIMUM RATINGS for V_{DS}, V_{DG}, V_{GS} and I_D, along with the usual device dissipation (P_D) and storage temperature range. Exceeding these limits usually results in destruction of the FET.

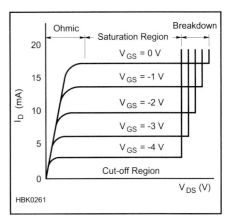

Fig 3.26 — JFET operating regions. At the left, I_D is increasing rapidly with V_{GS} and the JFET can be treated as resistance (R_{DS}) controlled by V_{GS}. In the saturation region, drain current, I_D, is relatively independent of V_{GS}. As V_{DS} increases further, avalanche breakdown begins and I_D increases rapidly.

Under OPERATING PARAMTERS the OFF CHARACTERISTICS list the gate-source breakdown voltage, $V_{GS(BR)}$, the reverse gate current, I_{GSS} and the gate-source cutoff voltage, $V_{GS(OFF)}$. Exceeding $V_{GS(BR)}$ will not permanently damage the device if current is limited. The primary ON CHARACTERISTIC parameters are the channel resistance, r_{DS}, and the zero-gate-voltage drain current (I_{DSS}). An FET's dc channel resistance, r_{DS}, is specified in data sheets to be less than a maximum value when the device is biased on ($r_{DS(on)}$). For ac signals, $r_{ds(on)}$ is not necessarily the same as $r_{DS(on)}$, but it is not very different as long as the frequency is not so high that capacitive reactance in the FET becomes significant.

The SMALL SIGNAL CHARACTERISTICS include the forward transfer admittance, y_{fs}, the output admittance, y_{os}, the static drain-source on resistance, $r_{ds(on)}$ and various capacitances such as input capacitance, C_{iss}, reverse transfer capacitance, C_{rss}, the drain-substrate capacitance, $C_{d(sub)}$. FUNCTIONAL CHARACTERISTICS include the noise figure, NF, and the common source power gain, G_{ps}.

MOSFETS

As described earlier, the MOSFET's gate is insulated from the channel by a thin layer of nonconductive oxide, doing away with any appreciable gate leakage current. Because of this isolation of the gate, MOSFETs do not need input and reverse transconductance curves. Their output curves (**Fig 3.27**) are similar to those of the JFET. The gate acts as a small capacitance between the gate and both the source and drain.

The output and transconductance curves in Fig 3.27A and 3.27B show that the depletion-mode N-channel MOSFET's transconduc-

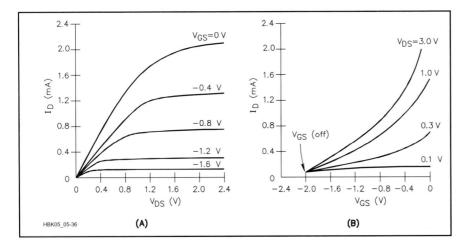

Fig 3.25 — JFET output and transconductance response curves for common source amplifier configuration. (A) Output voltage (V_{DS}) on the x-axis versus output current (I_D) on the y-axis, with different curves plotted for various values of input voltage (V_{GS}). (B) Transconductance curve with the same three variables rearranged: V_{GS} on the x-axis, I_D on the y-axis and curves plotted for different values of V_{DS}.

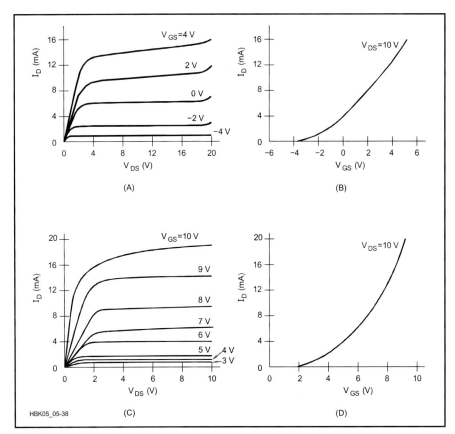

Fig 3.27 — MOSFET output [(A) and (C)] and transconductance [(B) and (D)] response curves. Plots (A) and (B) are for an N-channel depletion mode device. Note that V_{GS} varies from negative to positive values. Plots (C) and (D) are for an N-channel enhancement mode device. V_{GS} has only positive values.

tance is positive at $V_{GS} = 0$, like that of the N-channel JFET. Unlike the JFET, however, increasing V_{GS} does not forward-bias the gate-source junction and so the device can be operated with $V_{GS} > 0$.

In the enhancement-mode MOSFET, transconductance is zero at $V_{GS} = 0$. As V_{GS} is increased, the MOSFET enters the ohmic region. If V_{GS} increases further, the saturation region is reached and the MOSFET is said to be *fully-on*, with r_{DS} at its minimum value. The behavior of the enhancement-mode MOSFET is similar to that of the bipolar transistor in this regard.

The relatively flat regions in the MOSFET output curves are often used to provide a constant current source. As is plotted in these curves, the drain current, I_D, changes very little as the drain-source voltage, V_{DS}, varies in this portion of the curve. Thus, for a fixed gate-source voltage, V_{GS}, the drain current can be considered to be constant over a wide range of drain-source voltages.

Multiple gate MOSFETs are also available. Due to the insulating layer, the two gates are isolated from each other and allow two signals to control the channel simultaneously with virtually no loading of one signal by the other. A common application of this type of device is an automatic gain control (AGC) amplifier. The signal is applied to one gate and a rectified, low-pass filtered form of the output (the AGC voltage) is fed back to the other gate. Another common application is as a mixer in which the two input signals are applied to the pair of gates.

MOSFET Gate Protection

The MOSFET is constructed with a very thin layer of SiO_2 for the gate insulator. This layer is extremely thin in order to improve the transconductance of the device but this makes it susceptible to damage from high voltage levels, such as *electrostatic discharge* (ESD) from static electricity. If enough charge accumulates on the gate terminal, it can *punch through* the gate insulator and destroy it. The insulation of the gate terminal is so good that virtually none of this potential is eased by leakage of the charge into the device. While this condition makes for nearly ideal input impedance (approaching infinity), it puts the device at risk of destruction from even such seemingly innocuous electrical sources as static electrical discharges from handling.

Some MOSFET devices contain an internal Zener diode with its cathode connected to the gate and its anode to the substrate. If the voltage at the gate rises to a damaging level the Zener diode junction conducts, bleeding excess charges off to the substrate. When voltages are within normal operating limits the Zener has little effect on the signal at the gate, although it may decrease the input impedance of the MOSFET.

This solution will not work for all MOSFETs. The Zener diode must always be reverse biased to be effective. In the enhancement-mode MOSFET, $V_{GS} > 0$ for all valid uses of the part, keeping the Zener reverse biased. In depletion mode devices however, V_{GS} can be both positive and negative; when negative, a gate-protection Zener diode would be forward biased and the MOSFET gate would not be driven properly. In some depletion mode MOSFETs, back-to-back Zener diodes are used to protect the gate.

MOSFET devices are at greatest risk of damage from static electricity when they are out of circuit. Even though an electrostatic discharge is capable of delivering little energy, it can generate thousands of volts and high peak currents. When storing MOSFETs, the leads should be placed into conductive foam. When working with MOSFETs, it is a good idea to minimize static by wearing a grounded wrist strap and working on a grounded workbench or mat. A humidifier may help to decrease the static electricity in the air. Before inserting a MOSFET into a circuit board it helps to first touch the device leads with your hand and then touch the circuit board. This serves to equalize the excess charge so that little excess charge flows when the device is inserted into the circuit board.

Power MOSFETs

Power MOSFETs are designed for use as switches, with extremely low values of $r_{DS(on)}$; values of 50 milliohms (mΩ) are common. The largest devices of this type can switch tens of amps of current with V_{DS} voltage ratings of hundreds of volts. The **Component Data and References** chapter includes a table of Power FET ratings. The schematic symbol for power MOSFETs (see Fig 3.23) includes a *body diode* that allows the FET to conduct in the reverse direction, regardless of V_{GS}. This is useful in many high-power switching applications.

While the maximum ratings for current and voltage are high, the devices cannot withstand both high drain current and high drain-to-source voltage at the same time because of the power dissipated; $P = V_{DS} \times I_D$. It is important to drive the gate of a power MOSFET such that the device is fully on or fully off so that either V_{DS} or I_D is at or close to zero. When switching, the device should spend as little time as possible in the linear region where both current and voltage are nonzero because their product (P) can be substantial. This is not a big problem if switching only takes place occasionally, but if the switching is repetitive (such as in a switching power supply) care should be taken to drive the gate properly and remove excess heat from the device.

Because the gate of a power MOSFET is capacitive (up to several hundred pF for large devices), charging and discharging the gate quickly results in short current peaks of more than 100 mA. Whatever circuit is used to drive the gate of a power MOSFET must be able to handle that current level, such as an integrated circuit designed for driving the capacitive load an FET gate presents.

The gate of a power MOSFET should not be left open or connected to a high-impedance circuit. Use a pull-down or pull-up resistor connected between the gate and the appropriate power supply to ensure that the gate is placed at the right voltage when not being driven by the gate drive circuit.

GaAsFETs

FETs made from gallium-arsenide (GaAs) material are used at UHF and microwave frequencies because they have gain at these frequencies and add little noise to the signal. The reason GaAsFETs have gain at these frequencies is the high mobility of the electrons in GaAs material. Because the electrons are more mobile than in silicon, they respond to the gate-source input signal more quickly and strongly than silicon FETs, providing gain at higher frequencies (f_T is directly proportional to electron mobility). The higher electron mobility also reduces thermally-generated noise generated in the FET, making the GaAsFET especially suitable for weak-signal preamps.

Because electron mobility is always higher than hole mobility, N-type material is used in GaAsFETs to maximize high-frequency gain. Since P-type material is not used to make a gate-channel junction, a metal Schottky junction is formed by depositing metal directly on the surface of the channel. This type of device is also called a *MESFET* (metal-semiconductor field-effect transistor).

3.3.4 Optical Semiconductors

In addition to electrical energy and heat energy, light energy also affects the behavior of semiconductor materials. If a device is made to allow photons of light to strike the surface of the semiconductor material, the energy absorbed by electrons disrupts the bonds between atoms, creating free electrons and holes. This increases the conductivity of the material (*photoconductivity*). The photon can also transfer enough energy to an electron to allow it to cross a PN junction's depletion region as current flow through the semiconductor (*photoelectricity*).

PHOTOCONDUCTORS

In commercial *photoconductors* (also called *photoresistors*) the resistance can change by as much as several kilohms for a light intensity change of 100 ft-candles. The most common material used in photoconductors is cadmium sulfide (CdS), with a resistance range of more than 2 MΩ in total darkness to less than 10 Ω in bright light. Other materials used in photoconductors respond best at specific colors. Lead sulfide (PbS) is most sensitive to infrared light and selenium (Se) works best in the blue end of the visible spectrum.

PHOTODIODES

A similar effect is used in some diodes and transistors so that their operation can be controlled by light instead of electrical current biasing. These devices, shown in **Fig 3.28**, are called *photodiodes* and *phototransistors*. The flow of minority carriers across the reverse biased PN junction is increased by light falling on the doped semiconductor material. In the dark, the junction acts the same as any reverse biased PN junction, with a very low current, I_{SC}, (on the order of 10 μA) that is nearly independent of reverse voltage. The presence of light not only increases the current but also provides a resistance-like relationship (reverse current increases as reverse voltage increases). See **Fig 3.29** for the characteristic response of a photodiode. Even with no reverse voltage applied, the presence of light causes a small reverse current, as indicated by the points at which the lines in Fig 3.29 intersect the left side of the graph.

Photoconductors and photodiodes are generally used to produce light-related analog signals that require further processing. For example, a photodiode is used to detect infrared light signals from remote control devices as in Fig 3.28A. The light falling on the reverse-biased photodiode causes a change in I_{SC} that is detected as a change in output voltage.

Light falling on the phototransistor acts as base current to control a larger current between the collector and emitter. Thus the phototransistor acts as an amplifier whose input signal is light and whose output is current. Phototransistors are more sensitive to light than the other devices. Phototransistors have lots of light-to-current gain, but photodiodes normally have less noise, so they make more sensitive detectors. The phototransistor in Fig 3.28B is being used as a detector. Light falling on the phototransistor causes collector current to flow, dropping the collector voltage below the voltage at the amplifier's + input and causing a change in V_{OUT}.

PHOTOVOLTAIC CELLS

When illuminated, the reverse-biased photodiode has a reverse current caused by excess

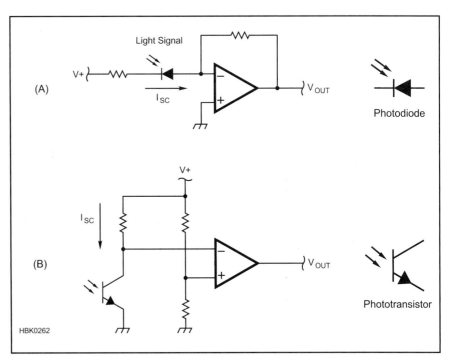

Fig 3.28 — The photodiode (A) is used to detect light. An amplifier circuit changes the variations in photodiode current to a change in output voltage. At (B), a phototransistor conducts current when its base is illuminated. This causes the voltage at the collector to change causing the amplifier's output to switch between ON and OFF.

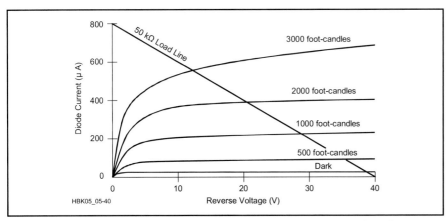

Fig 3.29 — Photodiode I-V curve. Reverse voltage is plotted on the x-axis and current through diode is plotted on the y-axis. Various response lines are plotted for different illumination. Except for the zero illumination line, the response does not pass through the origin since there is current generated at the PN junction by the light energy. A load line is shown for a 50-kΩ resistor in series with the photodiode.

minority carriers. As the reverse voltage is reduced, the potential barrier to the forward flow of majority carriers is also reduced. Since light energy leads to the generation of both majority and minority carriers, when the resistance to the flow of majority carriers is decreased these carriers form a forward current. The voltage at which the forward current equals the reverse current is called the *photovoltaic potential* of the junction. If the illuminated PN junction is not connected to a load, a voltage equal to the photovoltaic potential can be measured across it as the *terminal voltage*, V_T, or *open-circuit voltage*, V_{OC}.

Devices that use light from the sun to produce electricity in this way are called *photovoltaic (PV)* or *solar cells* or *solar batteries*. The symbol for a photovoltaic cell is shown in **Fig 3.30A**. The electrical equivalent circuit of the cell is shown in Fig 3.30B. The cell is basically a large, flat diode junction exposed to light. Metal electrodes on each side of the junction collect the current generated.

When illuminated, the cell acts like a current source, with some of the current flowing through a diode (made of the same material as the cell), a shunt resistance for leakage current and a series resistor that represents the resistance of the cell. Two quantities define the electrical characteristics of common silicon photovoltaic cells. These are an open-circuit voltage, V_{OC} of 0.5 to 0.6 V and the output *short-circuit current*, I_{SC} as above, that depends on the area of the cell exposed to light and the degree of illumination. A measure of the cell's effectiveness at converting light into current is the *conversion efficiency*. Typical silicon solar cells have a conversion efficiency of 10 to 15% although special cells with stacked junctions or using special light-absorbing materials have shown efficiencies as high as 40%.

Solar cells are primarily made from single-crystal slices of silicon, similar to diodes and transistors, but with a much greater area. *Polycrystalline silicon* and *thin-film* cells are less expensive, but have lower conversion efficiency. Technology is advancing rapidly in the field of photovoltaic energy and there are a number of different types of materials and fabrication techniques that have promise in surpassing the effectiveness of the single-junction silicon cells.

Solar cells are assembled into arrays called *solar panels*, shown in Fig 3.30C. Cells are connected in series so that the combined output voltage is a more useful voltage, such as 12 V. Several strings of cells are then connected in parallel to increase the available output current. Solar panels are available with output powers from a few watts to hundreds of watts. Note that unlike batteries, strings of solar cells can be connected directly in parallel because they act as sources of constant current instead of voltage. (More information on the use of solar panels for powering radio equipment can be found in the chapter on **Power Supplies**.)

LIGHT EMITTING DIODES AND LASER DIODES

In the photodiode, energy from light falling on the semiconductor material is absorbed to create additional electron-hole pairs. When the electrons and holes recombine, the same amount of energy is given off. In normal diodes the energy from recombination of carriers is given off as heat. In certain forms of semiconductor material, the recombination energy is given off as light with a mechanism called *electroluminescence*. Unlike the incandescent light bulb, electroluminescence is a cold (non-thermal) light source that typically operates with low voltages and currents (such as 1.5 V and 10 mA). Devices made for this purpose are called *light emitting diodes (LEDs)*. They have the advantages of low power requirements, fast switching times (on the order of 10 ns) and narrow spectra (relatively pure color).

The LED emits light when it is forward biased and excess carriers are present. As the carriers recombine, light is produced with a color that depends on the properties of the semiconductor material used. Gallium-arsenide (GaAs) generates light in the infrared region, gallium-phosphide (GaP) gives off red light when doped with oxygen or green

Fig 3.30 — A photovoltaic cell's symbol (A) is similar to a battery. Electrically, the cell can be modeled as the equivalent circuit at (B). Solar panels (C) consist of arrays of cells connected to supply power at a convenient voltage.

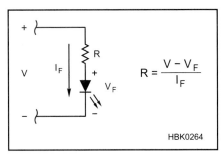

Fig 3.31 — A light-emitting diode (LED) emits light when conducting forward current. A series current-limiting resistor is used to set the current through the LED according to the equation.

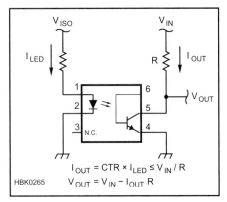

Fig 3.32 — The optoisolator consists of an LED (input) that illuminates the base of a phototransistor (output). The phototransistor then conducts current in the output circuit. CTR is the optoisolator's current transfer ratio.

light when doped with nitrogen. Orange light is attained with a mixture of GaAs and GaP (GaAsP). Silicon-carbide (SiC) creates a blue LED.

White LEDs are made by coating the inside of the LED lens with a white-light emitting phosphor and illuminating the phosphor with light from a single-color LED. White LEDs are currently approaching the cost of cold-florescent (CFL) bulbs and will eventually displace CFL technology for lighting, just as CFL is replacing the incandescent bulb.

The LED, shown in **Fig 3.31**, is very simple to use. It is connected across a voltage source (V) with a series resistor (R) that limits the current to the desired level (I_F) for the amount of light to be generated.

$$R = \frac{V - V_F}{I_F} \qquad (12)$$

where V_F is the forward voltage of the LED.

The cathode lead is connected to the lower potential, and is specially marked as shown in the chapter on **Component Data and References**. LEDs may be connected in series for additional light, with the same current flowing in all of the diodes. Diodes connected in parallel without current-limiting resistors for each diode are likely to share the current unequally, thus the series connection is preferred.

The laser diode operates by similar principles to the LED except that all of the light produced is *monochromatic* (of the same color and wavelength) and it is *coherent*, meaning that all of the light waves emitted by the device are in phase. Laser diodes generally require higher current than an LED and will not emit light until the *lasing current* level is reached. Because the light is monochromatic and coherent, laser diodes can be used for applications requiring precise illumination and modulation, such as high-speed data links, and in data storage media such as CD-ROM and DVD. LEDs are not used for high-speed or high-frequency analog modulation because of recovery time limitations, just as in regular rectifiers.

OPTOISOLATORS

An interesting combination of optoelectronic components proves very useful in many analog signal processing applications. An *optoisolator* consists of an LED optically coupled to a phototransistor, usually in an enclosed package (see **Fig 3.32**). The optoisolator, as its name suggests, isolates different circuits from each other. Typically, isolation resistance is on the order of 10^{11} Ω and isolation capacitance is less than 1 pF. Maximum voltage isolation varies from 1000 to 10,000 V ac. The most common optoisolators are available in 6-pin DIP packages.

Optoisolators are primarily used for voltage level shifting and signal isolation. Voltage level shifting allows signals (usually digital signals) to pass between circuits operating at greatly different voltages. The isolation has two purposes: to protect circuitry (and operators) from excessive voltages and to isolate noisy circuitry from noise-sensitive circuitry.

Optoisolators also cannot transfer signals with high power levels. The power rating of the LED in a 4N25 device is 120 mW. Optoisolators have a limited frequency response due to the high capacitance of the LED. A typical bandwidth for the 4N25 series is 300 kHz. Optoisolators with bandwidths of several MHz are available, but are somewhat expensive.

As an example of voltage level shifting, an optoisolator can be used to allow a low-voltage, solid-state electronic Morse code keyer to activate a vacuum-tube grid-block keying circuit that operates at a high negative voltage (typically about –100 V) but low current. No common ground is required between the two pieces of equipment.

Optoisolators can act as input protection for circuits that are exposed to high voltages or transients. For example, a short 1000-V transient that can destroy a semiconductor circuit will only saturate the LED in the optoisolator, preventing damage to the circuit. The worst that will happen is the LED in the optoisolator will be destroyed, but that is usually quite a bit less expensive than the circuit it is protecting.

Optoisolators are also useful for isolating different ground systems. The input and output signals are totally isolated from each other, even with respect to the references for each signal. A common application for optoisolators is when a computer is used to control radio equipment. The computer signal, and even its ground reference, typically contains considerable wide-band noise caused by the digital circuitry. The best way to keep this noise out of the radio is to isolate both the signal and its reference; this is easily done with an optoisolator.

The design of circuits with optoisolators is not greatly different from the design of circuits with LEDs and with transistors. On the input side, the LED is forward-biased and driven with a series current-limiting resistor whose value limits current to less than the maximum value for the device (for example, 60 mA is the maximum LED current for a 4N25). This is identical to designing with standalone LEDs.

On the output side, instead of current gain for a transistor, the optoisolator's *current transfer ratio* (*CTR*) is used. CTR is a ratio given in percent between the amount of current through the LED to the phototransistor's maximum available collector current. For example, if an optoisolator's CTR = 25%, then an LED current of 20 mA results in the output transistor being able to conduct up to 20 × 0.25 = 5 mA of current in its collector circuit.

If the optoisolator is to be used for an analog signal, the input signal must be appropriately dc shifted so that the LED is always forward biased. A phototransistor with all three leads available for connection (as in Fig 3.32) is required. The base lead is used for biasing, allowing the optical signal to create variations above and below the transistor's operating point. The collector and emitter leads are used as they would be in any transistor amplifier circuit. (There are also linear optoisolators that include built-in linearizing circuitry.) The use of linear optoisolators is not common.

FIBER OPTICS

An interesting variation on the optoisolator is the *fiber-optic* connection. Like the optoisolator, the input signal is used to drive an LED or laser diode that produces modulated light (usually light pulses). The light is transmitted in a fiber optic cable, an extruded glass fiber that efficiently carries light over long distances and around fairly sharp bends. The signal is recovered by a photo detector (photoresistor, photodiode or phototransistor). Because the fiber optic cable is nonconductive, the transmitting and receiving systems are electrically isolated.

Fiber optic cables generally have far less loss than coaxial cable transmission lines. They do not leak RF energy, nor do they pick up electrical noise. Fiber optic cables are virtually immune to electromagnetic interference! Special forms of LEDs and phototransistors are available with the appropriate optical couplers for connecting to fiber optic cables. These devices are typically designed for higher frequency operation with gigahertz bandwidth.

3.3.5 Linear Integrated Circuits

If you look inside a transistor, the actual size of the semiconductor is quite small compared to the size of the packaging. For most semiconductors, the packaging takes considerably more space than the actual semiconductor device. Thus, an obvious way to reduce the physical size of circuitry is to combine more of the circuit inside a single package.

HYBRID INTEGRATED CIRCUITS

It is easy to imagine placing several small semiconductor chips in the same package. This is known as *hybrid circuitry*, a technology in which several semiconductor chips are placed in the same package and miniature wires are connected between them to make complete circuits.

Hybrid circuits miniaturize analog electronic circuits by replacing much of the packaging that is inherent in discrete electronics. The term *discrete* refers to the use of individual components to make a circuit, each in its own package. The individual components are attached together on a small circuit board or with bonding wires.

Once widespread in electronics, hybrid ICs have largely been displaced by fully integrated devices with all the components on the same piece of semiconductor material. Manufacturers often use hybrids when small size is needed, but there is insufficient volume to justify the expense of a custom IC.

A current application for hybrid circuitry is UHF and microwave amplifiers — they are in wide use by the mobile phone industry. For example, the Motorola MW4IC915N Wideband Integrated Power Amplifier is a complete 15-W transmitting module. Its TO-272 package is only about 1 inch long by ⅜-inch wide. This particular device is designed for use between 750 and 1000 MHz and can be adapted for use on the amateur 902 MHz band. Other devices available as hybrid circuits include oscillators, signal processors, preamplifiers and so forth. Surplus hybrids can be hard to adapt to amateur use unless they are clearly

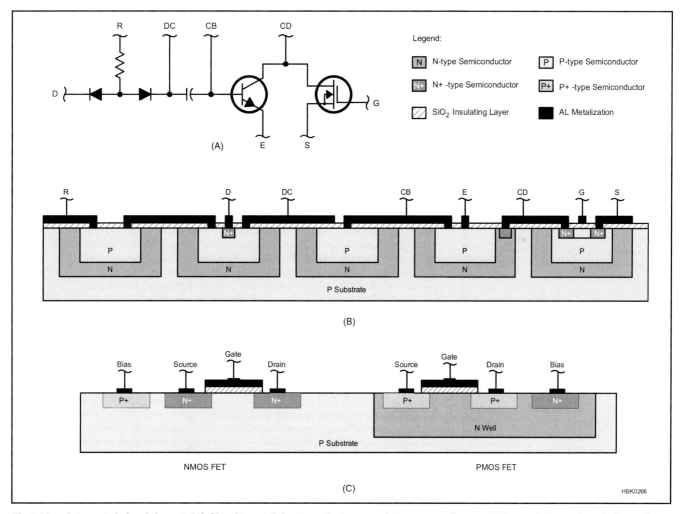

Fig 3.33 — Integrated circuit layout. (A) Circuit containing two diodes, a resistor, a capacitor, an NPN transistor and an N-channel MOSFET. Labeled leads are D for diode, R for resistor, DC for diode-capacitor, E for emitter, S for source, CD for collector-drain and G for gate. (B) Integrated circuit that is identical to circuit in (A). Same leads are labeled for comparison. Circuit is built on a P-type semiconductor substrate with N-type wells diffused into it. An insulating layer of SiO₂ is above the semiconductor and is etched away where aluminum metal contacts are made with the semiconductor. Most metal-to-semiconductor contacts are made with heavily doped N-type material (N⁺-type semiconductor).

identified with manufacturing identification such that a data sheet can be obtained.

MONOLITHIC INTEGRATED CIRCUITS

In order to build entire circuits on a single piece of semiconductor, it must be possible to fabricate resistors and capacitors, as well as transistors and diodes. Only then can the entire circuit be created on one piece of silicon called a *monolithic integrated circuit*.

An integrated circuit (IC) or "chip" is fabricated in layers. An example of a semiconductor circuit schematic and its implementation in an IC is pictured in **Fig 3.33**. The base layer of the circuit, the *substrate*, is made of P-type semiconductor material. Although less common, the polarity of the substrate can also be N-type material. Since the mobility of electrons is about three times higher than that of holes, bipolar transistors made with N-type collectors and FETs made with N-type channels are capable of higher speeds and power handling. Thus, P-type substrates are far more common. For devices with N-type substrates, all polarities in the ensuing discussion would be reversed.

Other substrates have been used, one of the most successful of which is the *silicon-on-sapphire* (SOS) construction that has been used to increase the bandwidth of integrated circuitry. Its relatively high manufacturing cost has impeded its use, however, except for the demanding military and aerospace applications.

On top of the P-type substrate is a thin layer of N-type material in which the active and passive components are built. Impurities are diffused into this layer to form the appropriate component at each location. To prevent random diffusion of impurities into the N-layer, its upper surface must be protected. This is done by covering the N-layer with a layer of silicon dioxide (SiO_2). Wherever diffusion of impurities is desired, the SiO_2 is etched away. The precision of placing the components on the semiconductor material depends mainly on the fineness of the etching. The fourth layer of an IC is made of aluminum (copper is used in some high-speed digital ICs) and is used to make the interconnections between the components.

Different components are made in a single piece of semiconductor material by first diffusing a high concentration of acceptor impurities into the layer of N-type material. This process creates P-type semiconductor — often referred to as P+-type semiconductor because of its high concentration of acceptor atoms — that isolates regions of N-type material. Each of these regions is then further processed to form single components.

A component is produced by the diffusion of a lesser concentration of acceptor atoms into the middle of each isolation region. This results in an N-type *isolation well* that contains P-type material, is surrounded on its sides by P+-type material and has P-type material (substrate) below it. The cross sectional view in Fig 3.33B illustrates the various layers. Connections to the metal layer are often made by diffusing high concentrations of donor atoms into small regions of the N-type well and the P-type material in the well. The material in these small regions is N+-type and facilitates electron flow between the metal contact and the semiconductor. In some configurations, it is necessary to connect the metal directly to the P-type material in the well.

Fabricating Resistors and Capacitors

An isolation well can be made into a resistor by making two contacts into the P-type semiconductor in the well. Resistance is inversely proportional to the cross-sectional area of the well. An alternate type of resistor that can be integrated in a semiconductor circuit is a *thin-film resistor*, where a metallic film is deposited on the SiO_2 layer, masked on its upper surface by more SiO_2 and then etched to make the desired geometry, thus adjusting the resistance.

There are two ways to form capacitors in a semiconductor. One is to make use of the PN junction between the N-type well and the P-type material that fills it. Much like a varactor diode, when this junction is reverse biased a capacitance results. Since a bias voltage is required, this type of capacitor is polarized, like an electrolytic capacitor. Nonpolarized capacitors can also be formed in an integrated circuit by using thin film technology. In this case, a very high concentration of donor ions is diffused into the well, creating an N+-type region. A thin metallic film is deposited over the SiO_2 layer covering the well and the capacitance is created between the metallic film and the well. The value of the capacitance is adjusted by varying the thickness of the SiO_2 layer and the cross-sectional size of the well. This type of thin film capacitor is also known as a metal oxide semiconductor (MOS) capacitor.

Unlike resistors and capacitors, it is very difficult to create inductors in integrated circuits. Generally, RF circuits that need inductance require external inductors to be connected to the IC. In some cases, particularly at lower frequencies, the behavior of an inductor can be mimicked by an amplifier circuit. In many cases the appropriate design of IC amplifiers can reduce or eliminate the need for external inductors.

Fabricating Diodes and Transistors

The simplest form of diode is generated by connecting to an N+-type connection point in the well for the cathode and to the P-type well material for the anode. Diodes are often converted from NPN transistor configurations. Integrated circuit diodes made this way can either short the collector to the base or leave the collector unconnected. The base contact is the anode and the emitter contact is the cathode.

Transistors are created in integrated circuitry in much the same way that they are fabricated in their discrete forms. The NPN transistor is the easiest to make since the wall of the well, made of N-type semiconductor, forms the collector, the P-type material in the well forms the base and a small region of N+-type material formed in the center of the well becomes the emitter. A PNP transistor is made by diffusing donor ions into the P-type semiconductor in the well to make a pattern with P-type material in the center (emitter) surrounded by a ring of N-type material that connects all the way down to the well material (base), and this is surrounded by another ring of P-type material (collector). This configuration results in a large base width separating the emitter and collector, causing these devices to have much lower current gain than the NPN form. This is one reason why integrated circuitry is designed to use many more NPN transistors than PNP transistors.

FETs can also be fabricated in IC form as shown in Fig 3.33C. Due to its many functional advantages, the MOSFET is the most common form used for digital ICs. MOSFETs are made in a semiconductor chip much the same way as MOS capacitors, described earlier. In addition to the signal processing advantages offered by MOSFETs over other transistors, the MOSFET device can be fabricated in 5% of the physical space required for bipolar transistors. CMOS ICs can contain 20 times more circuitry than bipolar ICs with the same chip size, making the devices more powerful and less expensive than those based on bipolar technology. CMOS is the most popular form of integrated circuit.

The final configuration of the switching circuit is CMOS as described in a previous section of this chapter. CMOS gates require two FETs, one of each form (NMOS and PMOS as shown in the figure). NMOS requires fewer processing steps, and the individual FETs have lower on-resistance than PMOS. The fabrication of NMOS FETs is the same as for individual semiconductors; P+ wells form the source and drain in a P-type substrate. A metal gate electrode is formed on top of an insulating SiO_2 layer so that the channel forms in the P-type substrate between the source and drain. For the PMOS FET, the process is similar, but begins with an N-type well in the P-type substrate.

MOSFETs fabricated in this manner also have bias (B) terminals connected to the positive power supply to prevent destructive *latch-up*. This can occur in CMOS gates because the two MOSFETs form a *parasitic SCR*. If the SCR mode is triggered and both transistors conduct at the same time, large currents can flow through the FET and destroy the IC unless power is removed. Just as discrete MOSFETs are at risk of gate destruction, IC chips made with MOSFET devices have a similar risk. They should be treated with the same care to protect them from static electricity as discrete MOSFETs.

While CMOS is the most widely used technology, integrated circuits need not be made exclusively with MOSFETs or bipolar transistors. It is common to find IC chips designed with both technologies, taking advantage of the strengths of each.

INTEGRATED CIRCUIT ADVANTAGES

The primary advantages of using integrated circuits as opposed to discrete components are the greatly decreased physical size of the circuit and improved reliability. In fact, studies show that failure rate of electronic circuitry is most closely related to the number of interconnections between components. Thus, using integrated circuits not only reduces volume, but makes the equipment more reliable.

The amount of circuitry that can be placed onto a single semiconductor chip is a function of two factors: the size of the chip and the size of the individual features that can be created on the semiconductor wafer. Since the invention of the monolithic IC in the mid-1960s, feature size limits have dropped below 100 nanometers (1/10th of a millionth of a meter) as of 2009. Currently, it is not unusual to find chips with more than one million transistors on them.

In addition to size and reliability of the ICs themselves, the relative properties of the devices on a single chip are very predictable. Since adjacent components on a semiconductor chip are made simultaneously (the entire N-type layer is grown at once; a single diffusion pass isolates all the wells and another pass fills them), the characteristics of identically formed components on a single chip of silicon are nearly identical. Even if the exact characteristics of the components are unknown, very often in analog circuit design the major concern is how components interact. For instance, push-pull amplifiers require perfectly matched transistors, and the gain of many amplifier configurations is governed by the ratio between two resistors and not their absolute values of resistance. With closely matched components on the single substrate, this type of design works very well without requiring external components to adjust or "trim" IC performance.

Integrated circuits often have an advantage over discrete circuits in their temperature behavior. The variation of performance of the components on an integrated circuit due to heat is no better than that of discrete components. While a discrete circuit may be exposed to a wide range of temperature changes, the entire semiconductor chip generally changes temperature by the same amount; there are fewer "hot spots" and "cold spots." Thus, integrated circuits can be designed to better compensate for temperature changes.

A designer of analog devices implemented with integrated circuitry has more freedom to include additional components that could improve the stability and performance of the implementation. The inclusion of components that could cause a prohibitive increase in the size, cost or complexity of a discrete circuit would have very little effect on any of these factors in an integrated circuit.

Once an integrated circuit is designed and laid out, the cost of making copies of it is very small, often only pennies per chip. Integrated circuitry is responsible for the incredible increase in performance with a corresponding decrease in price of electronics. While this trend is most obvious in digital computers, analog circuitry has also benefited from this technology.

The advent of integrated circuitry has also improved the design of high frequency circuitry, particularly the ubiquitous mobile phone and other wireless devices. One problem in the design and layout of RF equipment is the radiation and reception of spurious signals. As frequencies increase and wavelengths approach the dimensions of the wires in a circuit board, the interconnections act as efficient antennas. The dimensions of the circuitry within an IC are orders of magnitude smaller than in discrete circuitry, thus greatly decreasing this problem and permitting the processing of much higher frequencies with fewer problems of interstage interference. Another related advantage of the smaller interconnections in an IC is the lower inherent inductance of the wires, and lower stray capacitance between components and traces.

INTEGRATED CIRCUIT DISADVANTAGES

Despite the many advantages of integrated circuitry, disadvantages also exist. ICs have not replaced discrete components, even vacuum tubes, in some applications. There are some tasks that ICs cannot perform, even though the list of these continues to decrease over time as IC technology improves.

Although the high concentration of components on an IC chip is considered to be an advantage of that technology, it also leads to a major limitation. Heat generated in the individual components on the IC chip is often difficult to dissipate. Since there are so many heat generating components so close together, the heat can build up and destroy the circuitry. It is this limitation that currently causes many power amplifiers of more than 50 W output to be designed with discrete components.

Integrated circuits, despite their short interconnection lengths and lower stray inductance, do not have as high a frequency response as similar circuits built with appropriate discrete components. (There are exceptions to this generalization, of course. As described previously, monolithic microwave integrated circuits — MMICs — are available for operation to 10 GHz.) The physical architecture of an integrated circuit is the cause of this limitation. Since the substrate and the walls of the isolation wells are made of opposite types of semiconductor material, the PN junction between them must be reverse biased to prevent current from passing into the substrate. Like any other reverse-biased PN junction, a capacitance is created at the junction and this limits the frequency response of the devices on the IC. This situation has improved over the years as isolation wells have gotten smaller, thus decreasing the capacitance between the well and the substrate, and techniques have been developed to decrease the PN junction capacitance at the substrate. One such technique has been to create an N$^+$-type layer between the well and the substrate, which decreases the capacitance of the PN junction as seen by the well. As a result, analog ICs are now available with gain-bandwidth products over 1 GHz.

A major impediment to the introduction of new integrated circuits, particularly with special applications, is the very high cost of development of new designs for full custom ICs. The masking cost alone for a designed and tested integrated circuit can exceed $100,000 so these devices must sell in high volume to recoup the development costs. Over the past decade, however, IC design tools and fabrication services have become available that greatly reduce the cost of IC development by using predefined circuit layouts and functional blocks. These *application-specific integrated circuits* (*ASICs*) and *programmable gate arrays* (*PGA*) are now routinely created and used for even limited production runs. While still well beyond the reach of the individual amateur's resources, the ASIC or PGA is widely used in nearly all consumer electronics and in many pieces of radio equipment.

The drawback of the ASIC and PGA is that servicing and repairing the equipment at the integrated circuit level is almost impossible for the individual without access to the manufacturer's inventory of parts and proprietary information. Nevertheless, just as earlier amateurs moved beyond replacing individual components to diagnosing ICs, today's amateurs can troubleshoot to the module or circuit-board level, treating them as "components" in their own right.

3.3.6 Comparison of Semiconductor Devices for Analog Applications

Analog signal processing deals with changing a signal to a desired form. The three primary types of devices — bipolar transistors, field-effect transistors and integrated circuits — perform similar functions, each with specific advantages and disadvantages. The vacuum tube, once the dominant signal processing component, is relegated to high-power amplifier and display applications and is found only in the **RF Power Amplifiers** chapter of this *Handbook*. *Cathode-ray tubes* (CRTs) are covered in a supplement on the *Handbook* CD-ROM.

Bipolar transistors, when treated properly, can have virtually unlimited life spans. They are relatively small and, if they do not handle high currents, do not generate much heat. They make excellent high-frequency amplifiers. Compared to MOSFET devices they are less susceptible to damage from electrostatic discharge. RF amplifiers designed with bipolar transistors in their final amplifiers generally include circuitry to protect the transistors from the high voltages generated by reflections under high SWR conditions. Bipolar transistors and ICs, like all semiconductors, are susceptible to damage from power and lightning transients.

There are many performance advantages to FET devices, particularly MOSFETs. The extremely low gate currents allow the design of analog stages with nearly infinite input resistance. Signal distortion due to loading is minimized in this way. FETs are less expensive to fabricate in ICs and so are gradually replacing bipolar transistors in many IC applications.

The current trend in electronics is portability. Transceivers are decreasing in size and in their power requirements. Integrated circuitry has played a large part in this trend. Extremely large circuits have been designed with microscopic proportions, including combinations of analog and digital circuitry that previously required multiple devices. *Charge-coupled devices* (CCD) imaging technology has replaced vidicon tubes in video cameras and film cameras of all types. *Liquid crystal displays* (LCDs) in laptop computers and standalone computer displays have largely displaced the bulky and power-hungry cathode-ray tube (CRT), although it is still found in analog oscilloscopes and some types of video display equipment.

An important consideration in the use of analog components is the future availability of parts. At an ever increasing rate, as new components are developed to replace older technology, the older components are discontinued by the manufacturers and become unavailable for future use. ASIC technology, as mentioned earlier, brings the power of custom electronics to the radio, but also makes it nearly impossible to repair at the level of the IC, even if the problem is known. If field repair and service at the component level are to be performed, it is important to use standard ICs wherever possible. Even so, when demand for a particular component drops, a manufacturer will discontinue its production. This happens on an ever-decreasing timeline.

A further consideration is the trend toward digital signal processing and software-defined radio systems. (See the chapter on **DSP and Software Radio Design**.) More and more analog functions are being performed by microprocessors and the analog signals converted to digital at higher and higher frequencies. There will always be a need for analog circuits, but the balance point between analog and digital is shifting towards the latter. In future years, radio and test equipment will consist of a powerful, general-purpose digital signal processor, surrounded by the necessary analog circuitry to convert the signals to digital form and supply the processor with power.

3.4 Analog Systems

Many kinds of electronic equipment are developed by combining basic analog signal processing circuits, often treating them as independent functional blocks. This section describes several topics associated with building analog systems from multiple blocks. Although not all basic electronic functions are discussed here, the concepts associated with combining them can be applied generally.

An analog circuit can contain any number of discrete components (or may be implemented as an IC). Since our main concern is the effect that circuitry has on a signal, we often describe the circuit by its actions rather than by its specific components. A *black box* is a circuit that can be described entirely by the behavior of its interfaces with other blocks and circuitry. When circuits are combined in such as way as to perform sequential operations on a signal, the individual circuits are called *stages*.

The most general way of referring to a circuit is as a *network*. Two basic properties of analog networks are of principal concern: the effect that the network has on an analog signal and the interaction that the network has with the circuitry surrounding it. Interfaces between the network and the rest of the network are called *ports*.

Many analog circuits are analyzed as *two-port networks* with an input and an output port. The signal is fed into the input port, is modified inside the network and then exits from the output port. (See the chapter on **RF Techniques** for more information on two-port networks.)

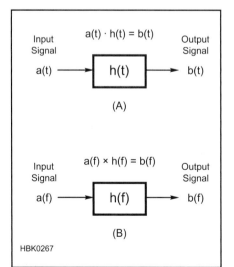

Fig 3.34 — Linear function blocks and transfer functions. The transfer function can be expressed in the time domain (A) or in the frequency domain (B). The transfer function describes how the input signal a(t) or a(f) is transformed into the output signal b(t) or b(f).

3.4.1 Transfer Functions

The specific way in which the analog circuit modifies the signal can be described mathematically as a *transfer function*. The mathematical operation that combines a signal with a transfer function is pictured symbolically in **Fig 3.34**. The transfer function, h(t) or h(f), describes the circuit's modification of the input signal in the time domain where all values are functions of time, such as a(t) or b(t), or in the frequency domain where all values are functions of frequency, such as a(f) or b(f). The mathematical operation by which h(t) operates on a(t) is called *convolution* and is represented as a dot, as in a(t) • h(t) = b(t). In the frequency domain, the transfer function multiplies the input, as in a(f) × h(f) = b(f).

While it is not necessary to understand transfer functions mathematically to work with analog circuits, it is useful to realize that they describe how a signal interacts with other signals in an electronic system. In general, the output signal of an analog system depends not only on the input signal at the same time, but also on past values of the input signal. This is a very important concept and is the basis of such essential functions as analog filtering.

3.4.2 Cascading Stages

If an analog circuit can be described with a transfer function, a combination of analog circuits can also be described similarly. This description of the combined circuits depends upon the relationship between the transfer functions of the parts and that of the combined circuits. In many cases this relationship allows us to predict the behavior of large and complex circuits from what we know about the parts of which they are made. This aids in the design and analysis of analog circuits.

When two analog circuits are cascaded (the output signal of one stage becomes the input signal to the next stage) their transfer functions are combined. The mechanism of the combination depends on the interaction between the stages. The ideal case is the functions of the stages are completely independent. In other words, when the action of a stage is unchanged, regardless of the characteristics of any stages connected to its input or output.

Just as the signal entering the first stage is modified by the action of the first transfer function, the ideal cascading of analog circuits results in changes produced only by the individual transfer functions. For any number of stages that are cascaded, the combination of their transfer functions results in a new transfer function. The signal that enters the circuit is changed by the composite transfer function to produce the signal that exits in the cascaded circuits.

While each stage in a series may use feedback within itself, feedback around more than one stage may create a function — and resultant performance — different from any of the included stages. Examples include oscillation or negative feedback.

3.4.3 Amplifier Frequency Response

At higher frequencies a typical amplifier acts as a low-pass filter, decreasing amplification with increasing frequency. Signals within a range of frequencies are amplified consistently but outside that range the amplification changes. At high gains many amplifiers work properly only over a small range of frequencies. The combination of gain and frequency response is often expressed as a *gain-bandwidth product*. For many amplifiers, gain times bandwidth is approximately constant. As gain increases, bandwidth decreases, and vice versa.

Performance at lower frequencies depends on whether the amplifier is *dc- or ac-coupled*. Coupling refers to the transfer of signals between circuits. A dc-coupled amplifier amplifies signals at all frequencies down to dc. An ac-coupled amplifier acts as a high-pass filter, decreasing amplification as the frequency decreases toward dc. Ac-coupled circuits usually use capacitors to allow ac signals to flow between stages while blocking the dc bias voltages of the circuit.

3.4.4 Interstage Loading and Impedance Matching

Every two-port network can be further defined by its input and output impedance. The input impedance is the opposition to current, as a function of frequency, seen when looking into the input port of the network. Likewise, the output impedance is similarly defined when looking back into a network through its output port.

If the transfer function of a stage changes when it is cascaded with another stage, we say that the second stage has *loaded* the first stage. This often occurs when an appreciable amount of current passes from one stage to the next. Interstage loading is related to the relative output impedance of a stage and the input impedance of the stage that is cascaded after it.

In some applications, the goal is to transfer a maximum amount of power from the output of the stage to a load connected to the output. In this case, the output impedance of the stage is *matched* or transformed to that of the load (or vice versa). This allows the stage to operate at its optimum voltage and current levels. In an RF amplifier, the impedance at the input of the transmission line feeding an antenna is transformed by means of a matching network to produce the resistance the amplifier needs in order to efficiently produce RF power.

In contrast, it is the goal of most analog signal processing circuitry to modify a signal rather than to deliver large amounts of energy. Thus, an impedance-matched condition may not be required. Instead, current between stages can be minimized by using mismatched impedances. Ideally, if the output impedance of a network is very low and the input impedance of the following stage is very high, very little current will pass between the stages, and interstage loading will be negligible.

3.4.5 Noise

Generally we are only interested in specific man-made signals. Nature allows many signals to combine, however, so the desired signal becomes combined with many other unwanted signals, both man-made and naturally occurring. The broadest definition of noise is any signal that is not the one in which we are interested. One of the goals of signal processing is to separate desired signals from noise.

One form of noise that occurs naturally and must be dealt with in low-level processing circuits is called *thermal noise*, or *Johnson noise*. Thermal noise is produced by random motion of free electrons in conductors and semiconductors. This motion increases as temperature increases, hence the name. This kind of noise is present at all frequencies and is proportional to temperature. Naturally occurring noise can be reduced either by decreasing the circuit's bandwidth or by reducing the temperature in the system. Thermal noise voltage and current vary with the circuit impedance and follow Ohm's Law. Low-noise-amplifier-design techniques are based on these relationships.

Analog signal processing stages are characterized in part by the noise they add to a signal. A distinction is made between enhancing existing noise (such as amplifying it) and adding new noise. The noise added by analog signal processing is commonly quantified by the *noise factor, f*. Noise factor is the ratio of the total output noise power (thermal noise plus noise added by the stage) to the input noise power when the termination is at the standard temperature of 290 K (17 °C). When the noise factor is expressed in dB, we often call it *noise figure, NF*. NF is calculated as:

$$NF = 10 \log \frac{P_{NO}}{A\, P_{N\,TH}} \qquad (13)$$

where

P_{NO} = total noise output power,
A = amplification gain
$P_{N\,TH}$ = input thermal noise power.

Noise factor can also be calculated as the difference between the input and output *signal-to-noise ratios* (SNR), with SNR expressed in dB.

In a system of many cascaded signal processing stages, such as a communications receiver, each stage contributes to the total noise of the system. The noise factor of the first stage dominates the noise factor of the entire system because noise added at the first stage is then multiplied by each following stage. Noise added by later stages is not multiplied to the same degree and so is a smaller contribution to the overall noise at the output.

Designers try to optimize system noise factor by using a first stage with a minimum possible noise factor and maximum possible gain. (Caution: A circuit that overloads is often as useless as one that generates too much noise.) See the **RF Techniques** chapter for a more complete discussion on noise. Circuit overload is discussed in the **Receivers** chapter.

3.4.6 Buffering

It is often necessary to isolate the stages of an analog circuit. This isolation reduces the loading, coupling and feedback between stages. It is often necessary to connect circuits that operate at different impedance levels between stages. An intervening stage, a type of amplifier called a *buffer*, is often used for this purpose.

Buffers can have high values of amplification but this is unusual. A buffer used for impedance transformation generally has a low or unity gain. In some circuits, notably power amplifiers, the desired goal is to deliver a maximum amount of power to the output device (such as a speaker or an antenna). Matching the amplifier output impedance to the output-device impedance provides maximum power transfer. A buffer amplifier may be just the circuit for this type of application. Such amplifier circuits must be carefully designed to avoid distortion. Combinations of buffer stages can also be effective at isolating the stages from each other and making impedance transformations, as well.

3.5 Amplifiers

By far, the most common type of analog circuit is the amplifier. The basic component of most electronics — the transistor — is an amplifier in which a small input signal controls a larger signal. Transistor circuits are designed to use the amplifying characteristics of transistors in order to create useful signal processing functions, regardless of whether the input signal is amplified at the output.

3.5.1 Amplifier Configurations

Amplifier configurations are described by the *common* part of the device. The word "common" is used to describe the connection of a lead directly to a reference that is used by both the input and output ports of the circuit. The most common reference is ground, but positive and negative power sources are also valid references.

The type of circuit reference used depends on the type of device (transistor [NPN or PNP] or FET [P-channel or N-channel]), which lead is chosen as common, and the range of signal levels. Once a common lead is chosen, the other two leads are used for signal input and output. Based on the biasing conditions, there is only one way to select these leads. Thus, there are three possible amplifier configurations for each type of three-lead device. (Vacuum tube amplifiers are discussed in the chapter on **RF Power Amplifiers**.)

DC power sources are usually constructed so that ac signals at the output terminals are bypassed to ground through a very low impedance. This allows the power source to be treated as an *ac ground*, even though it may be supplying dc voltages to the circuit. When a circuit is being analyzed for its ac behavior, ac grounds are usually treated as ground, since dc bias is ignored in the ac analysis. Thus, a transistor's collector can be considered the "common" part of the circuit, even though in actual operation, a dc voltage is applied to it.

Fig 3.35 shows the three basic types of bipolar transistor amplifiers: the common-base, common-emitter, and common-collector. The common terminal is shown connected to ground, although as mentioned earlier, a dc bias voltage may be present. Each type of amplifier is described in the following sections. Following the description of the amplifier, additional discussion of biasing transistors and their operation at high frequencies and for large signals is presented.

3.5.2 Transistor Amplifiers

Creating a useful transistor amplifier depends on using an appropriate model for the transistor itself, choosing the right configuration of the amplifier, using the design equations for that configuration and insuring that the amplifier operates properly at different temperatures. This section follows that sequence, first introducing simple transistor models and then extending that knowledge to the point of design guidelines for common circuits that use bipolar and FETs.

DEVICE MODELS AND CLASSES

Semiconductor circuit design is based on equivalent circuits that describe the physics of the devices. These circuits, made up of voltage and current sources and passive components such as resistors, capacitors and inductors, are called models. A complete model that describes a transistor exactly over a wide frequency range is a fairly complex circuit. As a result, simpler models are used in specific circumstances. For example, the *small-signal model* works well when the device is operated close to some nominal set of characteristics such that current and voltage interact fairly linearly. The *large-signal model* is used when the device is operated so that it enters its saturation or cut-off regions, for example.

Different frequency ranges also require different models. The *low-frequency models* used in this chapter can be used to develop circuits for dc, audio and very low RF applications. At higher frequencies, small capacitances and inductances that can be ignored at low frequencies begin to have significant effects on device behavior, such as gain or impedance. In addition, the physical structure of the device also becomes significant as gain begins to drop or phase shifts between input and output signals start to grow. In this region, *high-frequency models* are used.

Amplifiers are also grouped by their *operating class* that describes the way in which the input signal is amplified. There are several classes of analog amplifiers; A, B, AB, AB1, AB2 and C.

The analog class designators specify over how much of the input cycle the active device is conducting current. A class-A amplifier's active device conducts current for 100 percent of the input signal cycle, such as shown in Fig 3.6. A class-B amplifier conducts during one-half of the input cycle, class-AB, AB1, and AB2 some fraction between 50 and 100 percent of the input cycle, and class-C for less than 50 percent of the input signal cycle.

Digital amplifiers, in which the active device is operated as a switch that is either fully-on or fully-off, similarly to switchmode power supplies, are also grouped by classes beginning with the letter D and beyond. Each different class uses a different method of converting the switch's output waveform to the desired RF waveform.

Amplifier classes, models and their use at high-frequencies are discussed in more detail in the chapter on **RF Techniques**. In addition, the use of models for circuit simulation is discussed at length in the **Computer-Aided Circuit Design** chapter.

3.5.3 Bipolar Transistor Amplifiers

In this discussion, we will focus on simple models for bipolar transistors (BJTs). This discussion is centered on NPN BJTs but applies equally well to PNP BJTs if the bias voltage and current polarities are reversed. This section assumes the small-signal, low-frequency models for the transistors.

SMALL-SIGNAL BJT MODEL

The transistor is usually considered as a *current-controlled* device in which the base current controls the collector current:

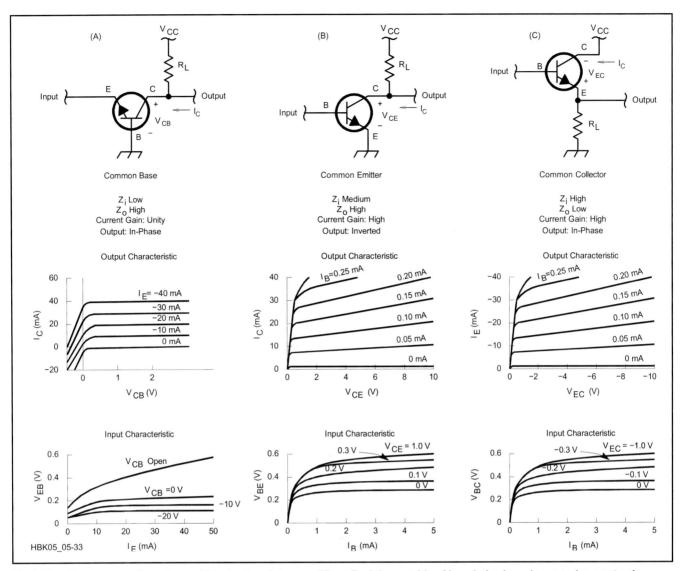

Fig 3.35 — The three configurations of bipolar transistor amplifiers. Each has a table of its relative impedance and current gain. The output characteristic curve is plotted for each, with the output voltage along the x-axis, the output current along the y-axis and various curves plotted for different values of input current. The input characteristic curve is plotted for each configuration with input current along the x-axis, input voltage along the y-axis and various curves plotted for different values of output voltage. (A) Common base configuration with input terminal at the emitter and output terminal at the collector. (B) Common emitter configuration with input terminal at the base and output terminal at the collector. (C) Common collector with input terminal at the base and output terminal at the emitter.

$$I_c = \beta I_b \quad (14)$$

where
- I_c = collector current
- I_b = base current
- β = *common-emitter current gain*, beta.

(The term "common-emitter" refers to the type of transistor circuit described below in which the transistor operates with base current as its input and collector circuit as its output.) Current is positive if it flows *into* a device terminal.

The transistor can also be treated as a voltage-controlled device in which the transistor's emitter current, I_e, is controlled by the base-emitter voltage, V_{be}:

$$I_c = I_{es}[e^{(qV_{be}/kT)} - 1] \approx I_{es}e^{(qV_{be}/kT)} \quad (15)$$

where
- q = electronic charge
- k = Boltzmann's constant
- T = temperature in degrees Kelvin (K)
- I_{es} = emitter saturation current, typically 1×10^{-13} A.

The subscripts for voltages indicate the direction of positive voltage, so that V_{be} indicates positive is from the base to the emitter. It is simpler to design circuits using the current-controlled device, but accounting for the transistor's behavior with temperature requires an understanding of the voltage-controlled model.

Transistors are usually driven by both biasing and signal voltages. Equations 14 and 15 apply to both transistor dc biasing and signal design. Both of these equations are approximations of the more complex behavior exhibited by actual transistors. Equation 15 applies to a simplification of the first *Ebers-Moll model* in Reference 1. More sophisticated models for BJTs are described by Getreu in Reference 2. Small-signal models treat only the signal components. We will consider bias later.

The next step is to use these basic equations to design circuits. We will begin with small-signal amplifier design and the limits of where the techniques can be applied. Later, we'll discuss large-signal amplifier design and the distortion that arises from operating the transistor in regions where the relationship between the input and output signals is nonlinear.

Common-Emitter Model

Fig 3.36 shows a BJT amplifier connected in the common-emitter configuration. (The emitter, shown connected to ground, is common to both the input circuit with the voltage source and the output circuit with the transistor's collector.) The performance of this circuit is adequately described by equation 14. **Fig 3.37** shows the most common of all transistor small-signal models, a controlled current source with emitter resistance.

There are two variations of the model shown in the figure. Fig 3.37B shows the base as a direct connection to the junction of a current-controlled current source ($I_c = \beta I_b$) and a resistance, r_e, the *dynamic emitter resistance* representing the change in V_{be} with I_e. This resistance also changes with emitter current:

$$r_e = \frac{kT}{qI_0} \approx \frac{26}{I_e} \qquad (16)$$

where I_e is the dc bias current in milliamperes.

The simplified approximation only applies at a typical ambient temperature of 300 K because r_e increases with temperature. In Fig 3.37A, the emitter resistance has been moved to the base connection, where it has the value $(\beta+1)r_e$. These models are electrically equivalent.

The transistor's output resistance (the Thevenin or Norton equivalent resistance between the collector and the grounded emitter) is infinite because of the current source. This is a good approximation for most silicon transistors at low frequencies (well below the transistor's gain-bandwidth product, F_T) and will be used for the design examples that follow.

As frequency increases, the capacitance inherent in BJT construction becomes significant and the *hybrid-pi model* shown in **Fig 3.38** is used, adding C_π in parallel with the input resistance. In this model the transfer parameter h_{ie} often represents the input impedance, shown here as a resistance at low frequencies.

THREE BASIC BJT AMPLIFIERS

Fig 3.39 shows a small-signal model applied to the three basic bipolar junction transistor (BJT) amplifier circuits: *common-emitter* (CE), *common-base* (CB) and *common-collector* (CC), more commonly known as the *emitter-follower* (EF). As defined earlier, the word "common" indicates that the referenced terminal is part of both the input and output circuits.

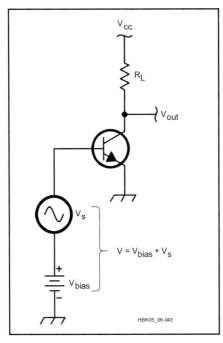

Fig 3.36 — Bipolar transistor with voltage bias and input signal.

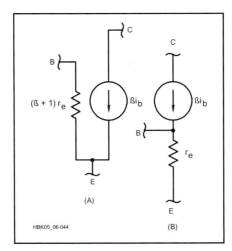

Fig 3.37 — Simplified low-frequency model for the bipolar transistor, a "beta generator with emitter resistance." $r_e = 26 / I_e$ (mA dc).

In these simple models, transistors in both the CE and CB configurations have infinite output resistance because the collector current source is in series with the output current. (The amplifier circuit's output impedance must include the effects of R_L.) The transistor connected in the EF configuration, on the other hand, has a finite output resistance because the current source is connected in parallel with the base circuit's equivalent resistance. Calculating the EF amplifier's output resistance requires including the input voltage source, V_s, and its impedance.

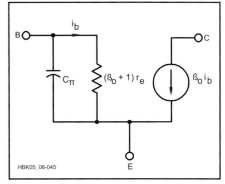

Fig 3.38 — The hybrid-pi model for the bipolar transistor.

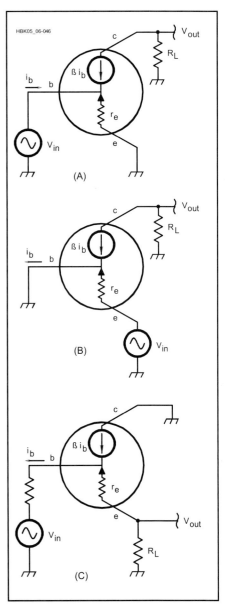

Fig 3.39 — Application of small-signal models for analysis of (A) the CE amplifier, (B) the CB and (C) the EF (CC) bipolar junction transistor amplifiers.

The three transistor amplifier configurations are shown as simple circuits in Fig 3.35. Each circuit includes the basic characteristics of the amplifier and characteristic curves for a typical transistor in each configuration. Two sets of characteristic curves are presented: one describing the input behavior and the other describing the output behavior in each amplifier configuration. The different transistor amplifier configurations have different gains, input and output impedances and phase relationships between the input and output signals.

Examining the performance needs of the amplifier (engineers refer to these as the circuit's *performance requirements*) determines which of the three circuits is appropriate. Then, once the amplifier configuration is chosen, the equations that describe the circuit's behavior are used to turn the performance requirements into actual circuit component values.

This text presents design information for the CE amplifier in some detail, then summarizes designs for the CC and CB amplifiers. Detailed design analysis for all three amplifiers is described in the texts listed in the reference section for this chapter. All of the analysis in the following sections assume the small-signal, low-frequency model and ignore the effects of the coupling capacitors. High-frequency considerations are discussed in the **RF Techniques** chapter and some advanced discussion of biasing and large signal behavior of BJT amplifiers is available on the companion CD-ROM.

LOAD LINES AND Q-POINT

The characteristic curves in Fig 3.35 show that the transistor can operate with an infinite number of combinations of current (collector, emitter and base) and voltage (collector-emitter, collector-base or emitter-collector). The particular combination at which the amplifier is operating is its *operating point*. The operating point is controlled by the selection of component values that make up the amplifier circuit so that it has the proper combination of gain, linearity and so forth. The result is that the operating point is restricted to a set of points that fall along a *load line*. The operating point with no input signal applied is the circuit's *quiescent point* or *Q-point*. As the input signal varies, the operating point moves along the load line, but returns to the Q-point when the input signal is removed.

Fig 3.40 shows the load line and Q-point for an amplifier drawn on a transistor's set of characteristic curves for the CE amplifier circuit. The two end-points of the load line correspond to transistor saturation (I_{Csat} on the I_C current axis) and cutoff (V_{CC} on the V_{CE} voltage axis).

When a transistor is in saturation, further increases in base current do not cause a further increase in collector current. In the CE amplifier, this means that V_{CE} is very close to zero and I_C is at a maximum. In the circuit of Fig 3.35B, imagine a short circuit across the collector-to-emitter so that all of V_{CC} appears across R_L. Increasing base current will not result in any additional collector current. At cutoff, base current is so small that V_{CE} is at a maximum because no collector current is flowing and further reductions in base current cause no additional increase in V_{CE}.

In this simple circuit, $V_{CE} = V_{CC} - I_C R_L$ and the relationship between I_C and V_{CE} is a straight line between saturation and cutoff. This is the circuit's *load line* and it has a slope of $R_L = (V_{CC} - V_{CE}) / I_C$. No matter what value of base current is flowing in the transistor, the resulting combination of I_C and V_{CE} will be somewhere on the load line.

With no input signal to this simple circuit, the transistor is at cutoff where $I_C = 0$ and $V_{CE} = V_{CC}$. As the input signal increases so that base current gets larger, the operating point begins to move along the load line to the left, so that I_C increases and the voltage drop across the load, $I_C R_L$, increases, reducing V_{CE}. Eventually, the input signal will cause enough base current to flow that saturation is reached, where $V_{CE} \approx 0$ (typically 0.1 to 0.3 V for silicon transistors) and $I_C \approx V_{CC} / R_L$. If R_L is made smaller, the load line will become steeper and if R_L increases, the load line's slope is reduced.

This simple circuit cannot reproduce negative input signals because the transistor is already in cutoff with no input signal. In addition, the shape and spacing of the characteristic curves show that the transistor responds nonlinearly when close to saturation and cutoff (the nonlinear regions) than it does in the middle of the curves (the linear or active region). Biasing is required so that the circuit does not operate in nonlinear regions, distorting the signal as shown in Fig 3.6.

If the circuit behaves differently for ac signals than for dc signals, a separate *ac load line* can be drawn as discussed below in the section "AC Performance" for the common-emitter amplifier. For example, in the preceding circuit, if R_L is replaced by a circuit that includes inductive or capacitive reactance, ac collector current will result in a different voltage drop across the circuit than will dc collector current. This causes the slope of the ac load line to be different than that of the dc load line.

The ac load line's slope will also vary with frequency, although it is generally treated as constant over the range of frequencies for which the circuit is designed to operate. The ac and dc load lines intersect at the circuit's Q-point because the circuit's ac and dc operation is the same if the ac input signal is zero.

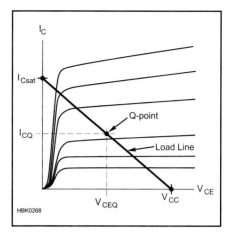

Fig 3.40 — A load line. A circuit's load line shows all of the possible operating points with the specific component values chosen. If there is no input signal, the operating point is the quiescent or Q-point.

COMMON-EMITTER AMPLIFIER

The *common-emitter amplifier* (CE) is the most common amplifier configuration of all — found in analog and digital circuits, from dc through microwaves, made of discrete components and fabricated in ICs. If you understand the CE amplifier, you've made a good start in electronics.

The CE amplifier is used when modest voltage gain is required along with an *input impedance* (the load presented to the circuit supplying the signal to be amplified) of a few hundred to a few kΩ. The current gain of the CE amplifier is the transistor's current gain, β.

The simplest practical CE amplifier circuit is shown in **Fig 3.41**. This circuit includes both coupling and biasing components. The capacitors at the input (C_{IN}) and output (C_{OUT}) block the flow of dc current to the load or to the circuit driving the amplifier. This is an ac-coupled design. These capacitors also cause the gain at very low frequencies to be

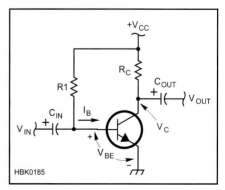

Fig 3.41 — Fixed-bias is the simplest common-emitter (CE) amplifier circuit.

reduced — gain at dc is zero, for example, because dc input current is blocked by C_{IN}. Resistor R_1 provides a path for bias current to flow into the base, offsetting the collector current from zero and establishing the Q-point for the circuit.

As the input signal swings positive, more current flows into the transistor's base through C_{IN}, causing more current to flow from the collector to emitter as shown by equation 14. This causes more voltage drop across R_L and so the voltage at the collector also drops. The reverse is true when the input signal swings negative. Thus, the output from the CE amplifier is inverted from its input.

Kirchoff's Voltage Law (KVL, see the **Electrical Fundamentals** chapter) is used to analyze the circuit. We'll start with collector circuit and treat the power supply as a voltage source.

$$V_{cc} = I_c R_c + V_{ce}$$

We can determine the circuit's voltage gain, A_V from the variation in output voltage caused by variations in input voltage. The output voltage from the circuit at the transistor collector is

$$V_c = V_{CC} - I_c R_c = V_{CC} - \beta I_B R_C \quad (17)$$

It is also necessary to determine how base current varies with input voltage. Using the transistor's equivalent circuit of Fig 3.37A,

$$I_B = \frac{V_B}{(\beta + 1) r_e}$$

so that

$$V_c = V_{CC} - V_B \frac{\beta}{\beta + 1} \times \frac{R_C}{r_e} \quad (18)$$

We can now determine the circuit's *voltage gain*, the variation in output voltage, ΔV_C, due to variations in input voltage, ΔV_B. Since V_{CC} is constant and β is much greater than 1 in our model:

$$A_V \approx -\frac{R_C}{r_e} \quad (19)$$

Because r_e is quite small (typically a few ohms, see equation 16), A_V for this circuit can be quite high.

The circuit load line's end-points are $V_{CE} = V_{CC}$ and $I_C = V_{CC} / R_C$. The circuit's Q-point is determined by the collector resistor, R_C, and resistor R_1 that causes bias current to flow into the base. To determine the Q-point, again use KVL starting at the power source and assuming that $V_{BE} = 0.7$ V for a silicon transistor's PN junction when forward-biased.

$$V_{CC} - I_B R_1 = V_B = V_{BE} = 0.7 \text{ V}$$

so

$$I_B = \frac{V_{CC} - 0.7V}{R_1} \quad (20)$$

And the Q-point is therefore

$$V_{CEQ} = V_{CC} - \beta I_B R_C \quad (21a)$$

and

$$I_{CQ} = \beta I_B \quad (21b)$$

The actual V_{BE} of silicon transistors will vary from 0.6-0.75 V, depending on the level of base current, but 0.7 V is a good compromise value and widely used in small-signal, low-frequency design. Use 0.6 V for very low-power amplifiers and 0.75 V (or more) for high-current switch circuits.

This simple *fixed-bias* circuit is a good introduction to basic amplifiers, but is not entirely practical because the bias current will change due to the change of V_{BE} with temperature, leading to thermal instability. In addition, the high voltage gain can lead to instability due to positive feedback at high frequencies.

To stabilize the dc bias, **Fig 3.42** adds R_E, a technique called *emitter degeneration* because the extra emitter resistance creates negative feedback: as base current rises, so does V_E, the voltage drop across R_E. This reduces the base-emitter voltage and lowers base current. The benefit of emitter degeneration comes from stabilizing the circuit's dc behavior with temperature, but there is a reduction in gain because of the increased resistance in the emitter circuit. Ignoring the effect of R_L for the moment,

$$A_V \approx -\frac{R_C}{R_E} \quad (22)$$

In effect, the load resistor is now split between R_C and R_E, with part of the output voltage appearing across each because the changing current flows through both resistors. While somewhat lower than with the emitter connected directly to ground, voltage gain becomes easy to control because it is the ratio of two resistances.

Biasing the CE Amplifier

Fig 3.43 adds R_1 and R_2 from a voltage divider that controls bias current by fixing the base voltage at:

$$V_B = V_{CC} \frac{R_2}{R_1 + R_2}$$

Since

$$V_B = V_{BE} + (I_B + I_C) R_E = 0.7 \text{ V} + (\beta + 1) I_B R_E$$

base current is

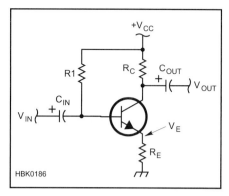

Fig 3.42 — Emitter degeneration. Adding R_E produces negative feedback to stabilize the bias point against changes due to temperature. As the bias current increases, the voltage drop across R_E also increases and causes a decrease in V_{BE}. This reduces bias current and stabilizes the operating point.

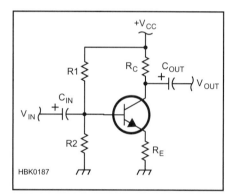

Fig 3.43 — Self-bias. R1 and R2 form a voltage divider to stabilize V_B and bias current. A good rule of thumb is for current flow through R1 and R2 to be 10 times the desired bias current. This stabilizes bias against changes in transistor parameters and component values.

$$I_B = \frac{V_B - 0.7 \text{ V}}{(\beta + 1) R_E} \quad (23a)$$

and Q-point collector current becomes for high values of β

$$I_{CQ} = \beta I_B \approx \frac{V_{CC} \frac{R_2}{R_1 + R_2} - 0.7}{R_E} \quad (23b)$$

This is referred to as *self-bias* in which the Q-point is much less sensitive to variations in temperature that affect β and V_{BE}.

A good rule-of-thumb for determining the sum of R_1 and R_2 is that the current flowing through the voltage divider, $V_{CC}/(R_1+R_2)$, should be at least 10 times the bias current, I_B. This keeps V_B relatively constant even with small changes in transistor parameters and temperature.

Q-point V_{CEQ} must now also account for the voltage drop across both R_C and R_E,

$$V_{CEQ} \approx V_{CC} - \beta I_B (R_C + R_E) \qquad (24)$$

More sophisticated techniques for designing the bias networks of bipolar transistor circuits are described in reference texts listed at the end of this chapter.

Input and Output Impedance

With R_E in the circuit, the small changes in input current, I_B, when multiplied by the transistor's current gain, β, cause a large voltage change across R_E equal to $\beta I_B R_E$. This is the same voltage drop as if I_B was flowing through a resistance equal to βR_E. Thus, the effect of β on impedance at the base is to multiply the emitter resistance, R_E by β, as well. At the transistor's base,

$$Z_B \approx (\beta + 1) R_E$$

The input source doesn't just drive the base, of course, it also has to drive the combination of R1 and R2, the biasing resistors. From an ac point of view, both R1 and R2 can be considered as connected to ac ground and they can be treated as if they were connected in parallel. When R1//R2 are considered along with the transistor base impedance, Z_B, the impedance presented to the input signal source is:

$$Z_{IN} = R1 \mathbin{/\mkern-5mu/} R2 \mathbin{/\mkern-5mu/} (\beta+1) R_E \qquad (25)$$

where // designates "in parallel with."

For both versions of the CE amplifier, the collector output impedance is high enough that

$$Z_{OUT} \approx R_C \qquad (26)$$

CE Amplifier Design Example

The general process depends on the circuit's primary performance requirements, including voltage gain, impedances, power consumption and so on. The most common situation in which a specific voltage gain is required and the circuit's Q-point has been selected based on the transistor to be used, and using the circuit of Fig 3.43, is as follows:

1) Start by determining the circuit's design constraints and assumptions: power supply $V_{CC} = 12$ V, transistor $\beta = 150$ and $V_{BE} = 0.7$ V. State the circuit's design requirements: $|A_V| = 5$, Q-point of $I_{CQ} = 4$ mA and $V_{CEQ} = 5$ V. (A $V_{CEQ} \approx \tfrac{1}{2} V_{CC}$ allows a wide swing in output voltage with the least distortion.)

2) Determine the values of R_C and R_E using equation 27: $R_C + R_E = (V_{CC} - V_{CEQ})/I_{CQ} = 1.75$ kΩ

3) $A_V = -5$, so from equation 25: $R_C = 5 R_E$, thus $6 R_C = 1.75$ kΩ and $R_E = 270$ Ω

4) Use equation 14 to determine the base bias current, $I_B = I_{CQ}/\beta = 27$ μA. By the rule of thumb, current through R_1 and R_2 =

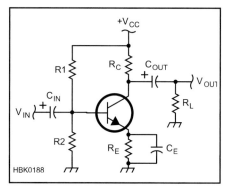

Fig 3.44 — Emitter bypass. Adding C_E allows ac currents to flow "around" R_E, returning ac gain to the value for the fixed-bias circuit while allowing R_E to stabilize the dc operating point.

$10 I_B = 270$ μA

5) Use equation 23 to find the voltage across $R_2 = V_B = V_{BE} + I_C R_E = 0.7 + 4$ mA $(0.27$ k$\Omega) = 1.8$ V. Thus, $R_2 = 1.8$ V / 270 μA = 6.7 kΩ

6) The voltage across $R_1 = V_{CC} - V_{R2} = 12 - 1.8 = 10.2$ V and $R_1 = 10.2$ V / 270 μA = 37.8 kΩ

Use the nearest standard values ($R_E = 270$ Ω, $R_1 = 39$ kΩ, $R_2 = 6.8$ kΩ) and circuit behavior will be close to that predicted.

AC Performance

To achieve high gains for ac signals while maintaining dc bias stability, the *emitter-bypass capacitor*, C_E, is added in **Fig 3.44** to provide a low impedance path for ac signals

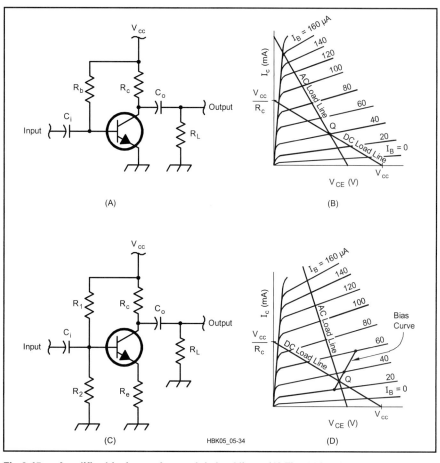

Fig 3.45 — Amplifier biasing and ac and dc load lines. (A) Fixed bias. Input signal is ac coupled through C_i. The output has a voltage that is equal to $V_{CC} - I_C \times R_C$. This signal is ac coupled to the load, R_L, through C_O. For dc signals, the entire output voltage is based on the value of R_C. For ac signals, the output voltage is based on the value of R_C in parallel with R_L. (B) Characteristic curve for the transistor amplifier pictured in (A). The slope of the dc load line is equal to $-1/R_C$. For ac signals, the slope of the ac load line is equal to $-1/(R_C \mathbin{/\mkern-5mu/} R_L)$. The quiescent-point, Q, is based on the base bias current with no input signal applied and the point where this characteristic line crosses the dc load line. The ac load line must also pass through point Q. (C) Self-bias. Similar to fixed bias circuit with the base bias resistor split into two: R1 connected to V_{CC} and R2 connected to ground. Also an emitter bias resistor, R_E, is included to compensate for changing device characteristics. (D) This is similar to the characteristic curve plotted in (B) but with an additional "bias curve" that shows how the base bias current varies as the device characteristics change with temperature. The operating point, Q, moves along this line and the load lines continue to intersect it as it changes. If C_E was added as in Fig 3.44, the slope of the ac load line would increase further.

around R_E. In addition, a more accurate formula for ac gain includes the effect of adding R_L through the dc blocking capacitor at the collector. In this circuit, the ac voltage gain is

$$A_V \approx -\frac{R_C // R_L}{r_e} \qquad (27)$$

Because of the different signal paths for ac and dc signals, the ac performance of the circuit is different than its dc performance. This is illustrated in **Fig 3.45** by the intersecting load lines labeled "AC Load Line" and "DC Load Line." The load lines intersect at the Q-point because at that point dc performance is the same as ac performance if no ac signal is present.

The equation for ac voltage gain assumes that the reactances of C_{IN}, C_{OUT}, and C_E are small enough to be neglected (less than one-tenth that of the components to which they are connected at the frequency of interest). At low frequencies, where the capacitor reactances become increasingly large, voltage gain is reduced. Neglecting C_{IN} and C_{OUT}, the low-frequency 3 dB point of the amplifier, f_L, occurs where $X_{CE} = 0.414\,r_e$,

$$f_L = \frac{2.42}{2\pi r_e C_E} \qquad (28)$$

This increases the emitter circuit impedance such that A_V is lowered to 0.707 of its mid-band value, lowering gain by 3 dB. (This ignores the effects of C_{IN} and C_{OUT}, which will also affect the low-frequency performance of the circuit.)

The ac input impedance of this version of the CE amplifier is lower because the effect of R_E on ac signals is removed by the bypass capacitor. This leaves only the internal emitter resistance, r_e, to be multiplied by the current gain,

$$Z_{IN} \approx R1 // R2 // \beta r_e \qquad (29)$$

and

$$Z_{OUT} \approx R_C \qquad (30)$$

again neglecting the reactance of the three capacitors.

The power gain, A_P, for the emitter-bypassed CE amplifier is the ratio of output power, V_O^2/Z_{OUT}, to input power, V_I^2/Z_{IN}. Since $V_O = V_I A_V$,

$$A_P = A_V^2 \frac{R1 // R2 // \beta r_e}{R_C} \qquad (31)$$

COMMON-COLLECTOR (EMITTER-FOLLOWER) AMPLIFIER

The common-collector (CC) amplifier in **Fig 3.46** is also known as the *emitter-follower* (EF) because the emitter voltage "follows" the input voltage. In fact, the amplifier has no voltage gain (voltage gain ≈ 1), but is used as a buffer amplifier to isolate sensitive circuits such as oscillators or to drive low-impedance loads, such as coaxial cables. As in the CE amplifier, the current gain of the emitter-follower is the transistor's current gain, β. It has relatively high input impedance with low output impedance and good power gain.

The collector of the transistor is connected directly to the power supply without a resistor and the output signal is created by the voltage drop across the emitter resistor. There is no 180° phase shift as seen in the CE amplifier; the output voltage follows the input signal with 0° phase shift because increases in the input signal cause increases in emitter current and the voltage drop across the emitter resistor.

The EF amplifier has high input impedance: following the same reasoning as for the CE amplifier with an unbypassed emitter resistor as described by equation 25,

$$Z_{IN} = R1 // R2 // (\beta + 1) R_E \qquad (32)$$

The impedance at the EF amplifier's output consists of the emitter resistance, R_E, in parallel with the series combination of the internal emitter resistance, r_e, the parallel combination of biasing resistors R1 and R2, and the internal impedance of the source providing the input signal. In this case, current gain acts to *reduce* the effect of the input circuit's impedance on output impedance:

$$Z_{OUT} = \left[\frac{R_S // R1 // R2}{(\beta + 1)}\right] // R_E \qquad (33)$$

In practice, with transistor β of 100 or more, $Z_{OUT} \approx R_S/\beta$. However, if a very high impedance source is used, such as an crystal microphone element or photodetector, the effects of the biasing and emitter resistors must be considered.

Because the voltage gain of the EF amplifier is unity, the power gain is simply the ratio of input impedance to output impedance,

$$A_P \approx \frac{R1 // R2 // (\beta + 1) R_E}{R_E} \qquad (34)$$

EF Amplifier Design Example

The following procedure is similar to the design procedure in the preceding section for the CE amplifier, except $A_V = 1$.

1) Start by determining the circuit's design constraints and assumptions: $V_{CC} = 12$ V (the power supply voltage), a transistor's β of 150 and $V_{BE} = 0.7$ V. State the circuit's design requirements: Q-point of $I_{CQ} = 5$ mA and $V_{CEQ} = 6$ V.

2) $R_E = (V_{CC} - V_{CEQ})/I_{CQ} = 1.2$ kΩ

3) Base current, $I_B = I_{CQ}/\beta = 33$ µA

4) Current through R1 and R2 = 10 I_B = 330 µA (10 I_B rule of thumb as with the CE amplifier)

5) Voltage across R2 = $V_{BE} + I_C R_E$ = 0.7 + 5 mA (1.2 kΩ) = 6.7 V and R2 = 6.7 V / 330 µA = 20.3 kΩ (use the standard value 22 kΩ)

6) Voltage across R1 = V_{CC} − 6.7 V = 5.3 V

7) R1 = 5.3 V / 330 µA = 16.1 kΩ (use 16 kΩ)

8) Z_{IN} = R1 // R2 // $R_E(\beta + 1) \approx 8.5$ kΩ

COMMON-BASE AMPLIFIER

The common-base (CB) amplifier of **Fig 3.47** is used where low input impedance is needed, such as for a receiver preamp with a coaxial feed line as the input signal source. Complementary to the EF amplifier, the CB amplifier has unity current gain and high output impedance.

Fig 3.47A shows the CB circuit as it is usually drawn, without the bias circuit resistors connected and with the transistor symbol turned on its side from the usual orientation so that the emitter faces the input. In order to better understand the amplifier's function, Fig 3.47B reorients the circuit in a more familiar style. We can now clearly see that the input has just moved from the base circuit to the emitter circuit.

Placing the input in the emitter circuit allows it to cause changes in the base-emitter current as for the CE and EF amplifiers, except that for the CB amplifier a positive change in input amplitude reduces base current by lowering V_{BE} and raising V_C. As a result, the CB amplifier is noninverting, just like the EF, with output and input signals in-phase.

A practical circuit for the CB amplifier is shown in **Fig 3.48**. From a dc point of view (replace the capacitors with open circuits), all of the same resistors are there as in the CE amplifier. The input capacitor, C_{IN}, allows

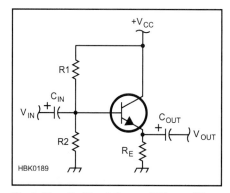

Fig 3.46 — Emitter follower (EF) amplifier. The voltage gain of the EF amplifier is unity. The amplifier has high input impedance and low output impedance, making it a good choice for use as a buffer amplifier.

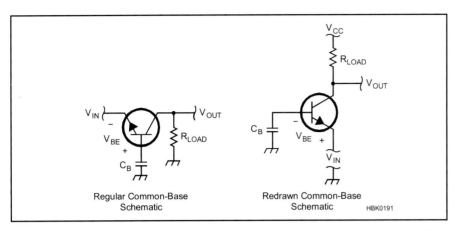

Fig 3.47 — The common-base (CB) amplifier is often drawn in an unfamiliar style (A), but is more easily understood when drawn similarly to the CE and EF amplifiers (B). The input signal to the CB amplifier is applied to the emitter instead of the base.

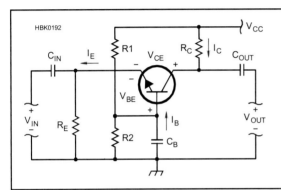

Fig 3.48 — A practical common-base (CB) amplifier. The current gain of the CB amplifier is unity. It has low input impedance and high output impedance, resulting in high voltage gain. The CB amplifier is used to amplify signals from low-impedance sources, such as coaxial cables.

the dc emitter current to bypass the ac input signal source and C_B places the base at ac ground while allowing a dc voltage for biasing. (All voltages and currents are labeled to aid in understanding the different orientation of the circuit.)

The CB amplifier's current gain,

$$A_I = \frac{i_C}{i_E} = \frac{\beta}{\beta + 1} \tag{35}$$

is relatively independent of input and output impedance, providing excellent isolation between the input and output circuits. Output impedance does not affect input impedance, allowing the CB amplifier to maintain stable input impedance, even with a changing load.

Following reasoning similar to that for the CE and EF amplifiers for the effect of current gain on R_E, we find that input impedance for the CB amp is

$$Z_{IN} = R_E // (\beta + 1) r_e \tag{36}$$

The output impedance for the CB amplifier is approximately

$$Z_{OUT} = R_C // \frac{1}{h_{oe}} \approx R_C \tag{37}$$

where h_{oe} is the transistor's collector output admittance. The reciprocal of h_{oe} is in the range of 100 kΩ at low frequencies.

Voltage gain for the CB amplifier is

$$A_V \approx \frac{R_C // R_L}{r_e} \tag{38}$$

As a result, the usual function of the CB amplifier is to convert input current from a low-impedance source into output voltage at a higher impedance.

Power gain for the CB amplifier is approximately the ratio of output to input impedance,

$$A_P \approx \frac{R_C}{R_E // (\beta + 1) r_e} \tag{39}$$

CB Amplifier Design Example

Because of its usual function as a current-to-voltage converter, the design process for the CB amplifier begins with selecting R_E and A_V, assuming that R_L is known.

1) Start by determining the circuit's design constraints and assumptions: $V_{cc} = 12$ V (the power supply voltage), a transistor's β of 150 and $V_{BE} = 0.7$ V. State the circuit's design requirements: $R_E = 50$ Ω, $R_L = 1$ kΩ, $I_{CQ} = 5$ mA, $V_{CEQ} = 6$ V.

2) Base current, $I_B = I_{CQ}/β = 33$ µA

3) Current through R1 and R2 = 10 I_B = 330 µA (10 I_B rule of thumb as with the CE amplifier)

4) Voltage across R2 = $V_{BE} + I_C R_E$ = 0.7 + 5 mA (1.2 kΩ) = 6.7 V and R2 = 6.7 V / 330 µA = 20.3 kΩ (use the standard value 22 kΩ)

5) Voltage across R1 = V_{CC} − 6.7 V = 5.3 V

6) R1 = 5.3 V / 330 µA = 16.1 kΩ (use 16 kΩ)

7) $R_C = (V_{CC} − I_{CQ} R_E − V_{CEQ}) / I_{CQ}$ = (12 − 0.25 − 5) / 5 mA = 1.35 kΩ (use 1.5 kΩ)

8) A_V = (1.5 kΩ // 1 kΩ) / (26 / I_E) = 115

3.5.4 FET Amplifiers

The field-effect transistor (FET) is widely used in radio and RF applications. There are many types of FETs, with JFETs (junction FET) and MOSFETs (metal-oxide-semiconductor FET) being the most common types. In this section we will discuss JFETs, with the understanding that the use of MOSFETs is similar. (This discussion is based on N-channel JFETs, but the same discussion applies to P-channel devices if the bias voltages and currents are reversed.)

SMALL-SIGNAL FET MODEL

While bipolar transistors are most commonly viewed as current-controlled devices, the JFET, however, is purely a voltage-controlled device — at least at low frequencies. The input gate is treated as a reverse-biased diode junction with virtually no current flow. As with the bipolar transistor amplifier circuits, the circuits in this section are very basic and more thorough treatments of FET amplifier design can be found in the references at the end of the chapter.

The operation of an N-channel JFET for both biasing and signal amplification can be characterized by the following equation:

$$I_D = I_{DSS} \left(\frac{1 − V_{SG}}{V_P} \right)^2 \tag{40}$$

where

I_{DSS} = drain saturation current
V_{SG} = the source-gate voltage
V_P = the pinch-off voltage.

I_{DSS} is the maximum current that will flow between the drain and source for a given value of drain-to-source voltage, V_{DS}. Note that the FET is a *square-law* device in which output current is proportional to the square of an input voltage. (The bipolar transistor's output current is an exponential function of input current.)

Also note that V_{SG} in this equation has the opposite sense of the bipolar transistor's V_{BE}. For this device, as V_{SG} increases (making the

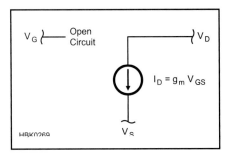

Fig 3.49 — Small-signal FET model. The FET can be modeled as a voltage-controlled current source in its saturation region. The gate is treated as an open-circuit due to the reverse-biased gate-channel junction.

source more positive than the gate), drain current decreases until at V_P the channel is completely "pinched-off" and no drain current flows at all. This equation applies only if V_{SG} is between 0 and V_P. JFETs are seldom used with the gate-to-channel diode forward-biased ($V_{SG} < 0$).

None of the terms in Equation 40 depend explicitly on temperature. Thus, the FET is relatively free of the thermal instability exhibited by the bipolar transistor. As temperature increases, the overall effect on the JFET is to reduce drain current and to stabilize the operation.

The small-signal model used for the JFET is shown in **Fig 3.49**. The drain-source channel is treated as a current source whose output is controlled by the gate-to-source voltage so that $I_D = g_m V_{GS}$. The high input impedance allows the input to be modeled as an open circuit (at low frequencies). This simplifies circuit modeling considerably as biasing of the FET gate can be done by a simple voltage divider without having to consider the effects of bias current flowing in the JFET itself.

The FET has characteristic curves as shown in Fig 3.25 that are similar to those of a bipolar transistor. The output characteristic curves are similar to those of the bipolar transistor, with the horizontal axis showing V_{DS} instead of V_{CE} and the vertical axis showing I_D instead of I_C. Load lines, both ac and dc, can be developed and drawn on the output characteristic curves in the same way as for bipolar transistors.

The set of characteristic curves in Fig 3.25 are called *transconductance response curves* and they show the relationship between input voltage (V_{GS}), output current (I_D) and output voltage (V_{DS}). The output characteristic curves show I_D and V_{DS} for different values of V_{GS} and are similar to the BJT output characteristic curve. The input characteristic curves show I_D versus V_{GS} for different values of V_{DS}.

MOSFETs act in much the same way as JFETs when used in an amplifier. They have a higher input impedance, due to the insulation between the gate and channel. The insulated gate also means that they can be operated with the polarity of V_{GS} such that a JFET's gate-channel junction would be forward biased, beyond V_P. Refer to the discussion of depletion- and enhancement-mode MOSFETs in the previous section on Practical Semiconductors.

THREE BASIC FET AMPLIFIERS

Just as for bipolar transistor amplifiers, there are three basic configurations of amplifiers using FETs; the *common-source* (CS)(corresponding to the common-emitter), *common-drain* (CD) or *source-follower* (corresponding to the emitter-follower) and the *common-gate* (CG) (corresponding to the common base). Simple circuits and design methods are presented here for each, assuming low-frequency operation and a simple, voltage-controlled current-source model for the FET. Discussion of the FET amplifier at high frequencies is available in the **RF Techniques** chapter and an advanced discussion of biasing FET amplifiers and their large-signal behavior is contained on the companion CD-ROM.

COMMON-SOURCE AMPLIFIER

The basic circuit for a common-source FET amplifier is shown in **Fig 3.50**. In the ohmic region (see the previous discussion on FET characteristics), the FET can be treated as a variable resistance as shown in **Fig 3.50A** where V_{GS} effectively varies the resistance between drain and source. However, most FET amplifiers are designed to operate in the saturation region and the model of Fig 3.49 is used in the circuit of **Fig 3.50B** in which,

$$I_D = g_m V_{GS} \quad (41)$$

where g_m is the FET's forward transconductance.

If V_O is measured at the drain terminal (just as the common-emitter output voltage is measured at the collector), then

$$\Delta V_O = -g_m \Delta V_{GS} R_D$$

The minus sign results from the output voltage decreasing as drain current and the voltage drop across R_D increases, just as in the CE amplifier. Like the CE amplifier, the input and output voltages are thus 180° out of phase. Voltage gain of the CS amplifier in terms of transconductance and the drain resistance is:

$$A_V = -g_m R_D \quad (42)$$

As long as $V_{GS} < 0$, this simple CS amplifier's input impedance at low frequencies

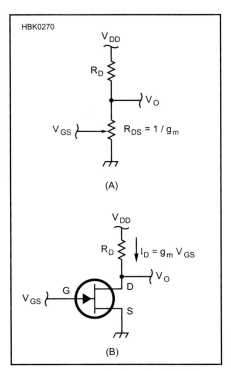

Fig 3.50 — In the ohmic region (A), the FET acts like a variable resistance, controlled by V_{GS}. In the saturation region (B), the drain-source channel of the FET can be treated like a current source with $I_D = g_m V_{GS}$.

is that of a reverse-biased diode — nearly infinite with a very small leakage current. Output impedance of the CS amplifier is approximately R_D because the FET drain-to-source channel acts like a current source with very high impedance.

$$Z_{IN} = \infty \text{ and } Z_{OUT} \approx R_D \quad (43)$$

As with the BJT, biasing is required to create a Q-point for the amplifier that allows reproduction of ac signals. The practical circuit of Fig 3.50B is used to allow control of V_{GS} bias. A load line is drawn on the JFET output characteristic curves, just as for a bipolar transistor circuit. One end point of the load line is at $V_{DS} = V_{DD}$ and the other at $I_{DS} = V_{DD} / R_D$. The Q-point for the CS amplifier at I_{DQ} and V_{DSQ} is thus determined by the dc value of V_{GS}.

The practical JFET CS amplifier shown in **Fig 3.51** uses self-biasing in which the voltage developed across the source resistor, R_S, raises V_S above ground by $I_D R_S$ volts and $V_{GS} = -I_D R_S$ since there is no dc drop across R_G. This is also called *source degeneration*. The presence of R_S changes the equation of voltage gain to

$$A_V = -\frac{g_m R_D}{1 + g_m R_S} \approx -\frac{R_D}{R_S} \quad (44)$$

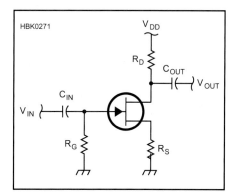

Fig 3.51 — Common-source (CS) amplifier with self-bias.

AC Performance

As with the CE bipolar transistor amplifier, a bypass capacitor can be used to increase ac gain while leaving dc bias conditions unchanged as shown in **Fig 3.52**. In the case of the CS amplifier, a *source bypass* capacitor is placed across R_S and the load, R_L, connected through a dc blocking capacitor. In this circuit voltage gain becomes:

$$A_V = -g_m (R_D // R_L) \qquad (48)$$

Assuming C_{IN} and C_{OUT} are large enough to ignore their effects, the low-frequency cut-off frequency of the amplifier, f_L, is approximately where $X_{CS} = 0.707\,(R_D // R_L)$,

$$f_L = \frac{1.414}{2\pi (R_D // R_L) C_S} \qquad (49)$$

as this reduces A_V to 0.707 of its mid-band value, resulting in a 3 dB drop in output amplitude.

The low-frequency ac input and output impedances of the CS amplifier remain

$$Z_{IN} = R_G \text{ and } Z_{OUT} \approx R_D \qquad (50)$$

COMMON-DRAIN (SOURCE-FOLLOWER) AMPLIFIER

The *common-drain* amplifier in **Fig 3.53** is also known as a *source-follower* (SF) because the voltage gain of the amplifier is unity, similar to the emitter follower (EF) bipolar transistor amplifier. The SF amplifier is used primarily as a buffer stage and to drive low-impedance loads.

At low frequencies, the input impedance of the SF amplifier remains nearly infinite. The SF amplifier's output impedance is the source resistance, R_S, in parallel with the impedance of the controlled current source, $1/g_m$.

$$Z_{OUT} = R_S // \frac{1}{g_m}$$
$$= \frac{R_S}{g_m R_S + 1} \approx \frac{1}{g_m} \text{ for } g_m R_S \gg 1 \qquad (51)$$

Design of the SF amplifier follows essentially the same process as the CS amplifier, with $R_D = 0$.

THE COMMON-GATE AMPLIFIER

The *common-gate* amplifier in **Fig 3.54** has similar properties to the bipolar transistor common-base (CB) amplifier; unity current gain, high voltage gain, low input impedance and high output impedance. (Refer to the discussion of the CB amplifier regarding placement of the input and how the circuit schematic is drawn.) It is used as a voltage amplifier, particularly for low-impedance sources, such as coaxial cable inputs.

The CG amplifier's voltage gain is

$$A_V = g_m (R_D // R_L) \qquad (52)$$

The value of R_S is obtained by substituting $V_{SG} = I_D R_S$ into Equation 40 and solving for R_S as follows:

$$R_S = \frac{-V_P}{I_{DQ}} \left(1 - \sqrt{\frac{I_{DQ}}{I_{DSS}}}\right) \qquad (45)$$

Once R_S is known, the equation for voltage gain can be used to find R_D.

The input impedance for the circuit of Fig 3.51 is essentially R_G. Since the gate of the JFET is often ac coupled to the input source through a dc blocking capacitor, C_{IN}, a value of 100 kΩ to 1 MΩ is often used for R_G to provide a path to ground for gate leakage current. If R_G is omitted in an ac-coupled JFET amplifier, a dc voltage can build up on the gate from leakage current or static electricity, affecting the channel conductivity.

$$Z_{IN} = R_G \qquad (46)$$

Because of the high impedance of the drain-source channel in the saturation region, the output impedance of the circuit is:

$$Z_{OUT} \approx R_D \qquad (47)$$

Designing the Common-Source Amplifier

The design of the CS amplifier begins with selection of a Q-point $I_{DQ} < I_{DSS}$. Because of variations in V_P and I_{DSS} from JFET to JFET, it may be necessary to select devices individually to obtain the desired performance.

1) Start by determining the circuit's design constraints and assumptions: $V_{DD} = 12$ V (the power supply voltage) and the JFET has an I_{DSS} of 35 mA and a V_P of −3.0V, typical of small-signal JFETs. State the circuit's design requirements: $|A_V| = 10$ and $I_{DQ} = 10$ mA.

2) Use equation 45 to determine $R_S = 139$ Ω

3) Since $|A_V| = 10$, $R_D = 10\,R_S = 1390$ Ω. Use standard values for $R_S = 150$ Ω and $R_D = 1.5$ kΩ.

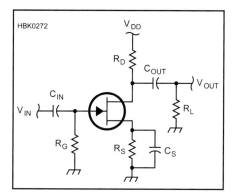

Fig 3.52 — Common-source amplifier with source bypass capacitor, C_S, to increase voltage gain without affecting the circuit's dc performance.

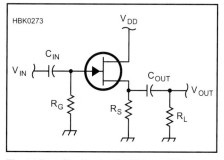

Fig 3.53 — Similar to the EF amplifier, the common-drain (CD) amplifier has a voltage gain of unity, but makes a good buffer with high input and low output impedances.

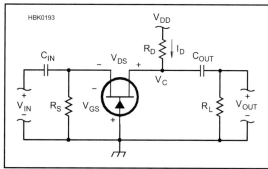

Fig 3.54 — FET common-gate (CG) amplifiers are often used as preamplifiers because of their high voltage gain and low input impedance. With the proper choice of transistor and quiescent-point current, the input impedance can match coaxial cable impedances directly.

The output impedance of the CG amplifier is very high, we must take into account the output resistance of the controlled current source, r_o. This is analogous to the appearance of h_{oe} in the equation for output impedance of the bipolar transistor CG amplifier.

$$Z_O \approx r_o (g_m R_S + 1) // R_D \quad (53)$$

The CG amplifier input impedance is approximately

$$Z_I = R_S // \frac{1}{g_m} \quad (54)$$

Because the input impedance is quite low, the cascode circuit described later in the section on buffers is often used to present a higher-impedance input to the signal source.

Occasionally, the value of R_S must be fixed in order to provide a specific value of input impedance. Solving equation 40 for I_{DQ} results in the following equation:

$$I_{DQ} = \frac{V_P}{2R_S^2 I_{DSS}} \left(V_P + \sqrt{V_P^2 - 4R_S I_{DSS} V_P} \right) - \frac{V_P}{R_S} \quad (55)$$

Designing the Common-Gate Amplifier

Follow the procedure for designing a CS amplifier, except determine the value of R_D as shown in equation 52 for voltage gain above.

1) Start by determining the circuit's design constraints and assumptions: $V_{DD} = 12$ V (the power supply voltage) and the JFET has a g_m of 15 mA/V, an I_{DSS} of 60 mA and $V_P = -6$ V. State the circuit's design requirements: $A_V = 10$, $R_L = 1$ kΩ and $R_S = 50$ Ω.

2) Solve equation 52 for R_D: $10 = 0.015 \times R_D // R_L$, so $R_D // R_L = 667$ Ω. $R_D = 667 R_L / (R_L - 667) = 2$ kΩ.

3) I_{DQ} is determined from equation 55: $I_{DQ} = 10$ mA. If I_{DQ} places the Q-point in the ohmic region, reduce A_V and repeat the calculations.

3.5.5 Buffer Amplifiers

Fig 3.55 shows common forms of buffers with low-impedance outputs: the *emitter follower* using a bipolar transistor, the *source follower* using a field-effect transistor and the *voltage follower*, using an operational amplifier. (The operational amplifier is discussed later in this chapter.) These circuits are called "followers" because the output "follows" the input very closely with approximately the same voltage and little phase shift between the input and output signals.

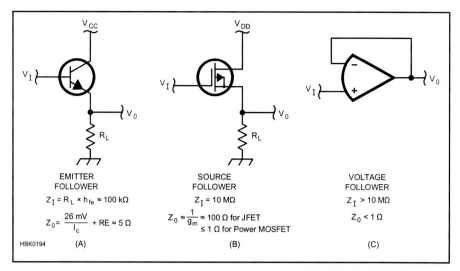

Fig 3.55 — Common buffer stages and some typical input (Z_I) and output (Z_O) impedances. (A) Emitter follower, made with an NPN bipolar transistor; (B) Source follower, made with an FET; and (C) Voltage follower, made with an operational amplifier. All of these buffers are terminated with a load resistance, R_L, and have an output voltage that is approximately equal to the input voltage (gain ≈ 1).

3.5.6 Cascaded Buffers
THE DARLINGTON PAIR

Buffer stages that are made with single active devices can be more effective if cascaded. Two types of such buffers are in common use. The *Darlington pair* is a cascade of two transistors connected as emitter followers as shown in **Fig 3.56**. The current gain of the Darlington pair is the product of the current gains for the two transistors, $\beta_1 \times \beta_2$.

What makes the Darlington pair so useful is that its input impedance is equal to the load impedance times the current gain, effectively multiplying the load impedance;

$$Z_I = Z_{LOAD} \times \beta_1 \times \beta_2 \quad (56)$$

For example, if a typical bipolar transistor has $\beta = 100$ and $Z_{LOAD} = 15$ kΩ, a pair of these transistors in the Darlington-pair configuration would have:

$$Z_I = 15 \text{ k}\Omega \times 100 \times 100 = 150 \text{ M}\Omega$$

This impedance places almost no load on the circuit connected to the Darlington pair's input. The shunt capacitance at the input of real transistors can lower the actual impedance as the frequency increases.

Drawbacks of the Darlington pair include lower bandwidth and switching speed. The extremely high dc gain makes biasing very sensitive to temperature and component tolerances. For these reasons, the circuit is usually used as a switch and not as a linear amplifier.

CASCODE AMPLIFIERS

A common-emitter amplifier followed by a common-base amplifier is called a *cascode buffer*, shown in its simplest form in **Fig 3.57**. (Biasing and dc blocking components are omitted for simplicity — replace the transistors with the practical circuits described earlier.) Cascode stages using FETs follow a

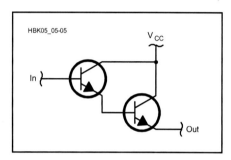

Fig 3.56 — Darlington pair made with two emitter followers. Input impedance, Z_I, is far higher than for a single transistor and output impedance, Z_O, is nearly the same as for a single transistor. DC biasing has been omitted for simplicity.

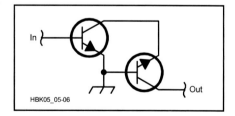

Fig 3.57 — Cascode buffer made with two NPN bipolar transistors has a medium input impedance and high output impedance. DC biasing has been omitted for simplicity.

common-source amplifier with a common-gate configuration. The input impedance and current gain of the cascode amplifier are approximately the same as those of the first stage. The output impedance of the common-base or –gate stage is much higher than that of the common-emitter or common-source amplifier. The power gain of the cascode amplifier is the product of the input stage current gain and the output stage voltage gain.

As an example, a typical cascode buffer made with BJTs has moderate input impedance (Z_{IN} = 1 kΩ), high current gain (h_{fe} = 50), and high output impedance (Z_{OUT} = 1 MΩ). Cascode amplifiers have excellent input/output isolation (very low unwanted internal feedback), resulting in high gain with good stability. Because of its excellent isolation, the cascode amplifier has little effect on external tuning components. Cascode circuits are often used in tuned amplifier designs for these reasons.

3.5.7 Using the Transistor as a Switch

When designing amplifiers, the goal was to make the transistor's output a replica of its input, requiring that the transistor stay within its linear region, conducting some current at all times. A switch circuit has completely different properties — its output current is either zero or some maximum value. **Fig 3.58** shows both a bipolar and metal-oxide semiconductor field-effect transistor (or MOSFET) switch circuit. Unlike the linear amplifier circuits, there are no bias resistors in either circuit. When using the bipolar transistor as a switch, it should operate in saturation or in cutoff. Similarly, an FET switch should be either fully-on or fully-off. The figure shows the waveforms associated with both types of switch circuits.

This discussion is written with power control in mind, such as to drive a relay or motor or lamp. The concepts, however, are equally applicable to the much lower-power circuits that control logic-level signals. The switch should behave just the same — switch between on and off quickly and completely — whether large or small.

DESIGNING SWITCHING CIRCUITS

First, select a transistor that can handle the load current and dissipate whatever power is dissipated as heat. Second, be sure that the input signal source can supply an adequate input signal to drive the transistor to the required states, both on and off. Both of these conditions must be met to insure reliable driver operation.

To choose the proper transistor, the load current and supply voltage must both be known. Supply voltage may be steady, but sometimes varies widely. For example, a car's

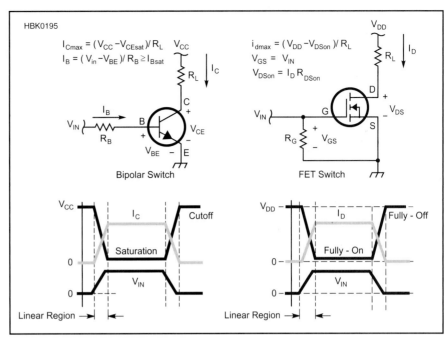

Fig 3.58 — A pair of transistor driver circuits using a bipolar transistor and a MOSFET. The input and output signals show the linear, cutoff and saturation regions.

12 V power bus may vary from 9 to 18 V, depending on battery condition. The transistor must withstand the maximum supply voltage, V_{MAX}, when off. The load resistance, R_L, must also be known. The maximum steady-state current the switch must handle is:

$$I_{MAX} = \frac{V_{MAX}}{R_L} \qquad (57)$$

If you are using a bipolar transistor, calculate how much base current is required to drive the transistor at this level of collector current. You'll need to inspect the transistor's data sheet because β decreases as collector current increases, so use a value for β specified at a collector current at or above I_{MAX}.

$$I_B = \frac{I_{MAX}}{\beta} \qquad (58)$$

Now inspect the transistor's data sheet values for V_{CEsat} and make sure that this value of I_B is sufficient to drive the transistor fully into saturation at a collector current of I_{MAX}. Increase I_B if necessary — this is I_{Bsat}. The transistor must be fully saturated to minimize heating when conducting load current.

Using the *minimum* value for the input voltage, calculate the value of R_B:

$$R_B = \frac{V_{IN(min)} - V_{BE}}{I_{Bsat}} \qquad (59)$$

The minimum value of input voltage must be used to accommodate the *worst-case* combination of circuit voltages and currents.

Designing with a MOSFET is a little easier because the manufacturer usually specifies the value V_{GS} must have for the transistor to be fully on, $V_{GS(on)}$. The MOSFET's gate, being insulated from the conducting channel, acts like a small capacitor of a few hundred pF and draws very little dc current. R_G in Fig 3.58 is required if the input voltage source does not actively drive its output to zero volts when off, such as a switch connected to a positive voltage. The MOSFET won't turn off reliably if its gate is allowed to "float." R_G pulls the gate voltage to zero when the input is open-circuited.

Power dissipation is the next design hurdle. Even if the transistors are turned completely on, they will still dissipate some heat. Just like for a resistor, for a bipolar transistor switch the power dissipation is:

$$P_D = V_{CE}I_C = V_{CE(sat)} I_{MAX}$$

where $V_{CE(sat)}$ is the collector-to-emitter voltage when the transistor is saturated.

Power dissipation in a MOSFET switch is:

$$P_D = V_{DS}I_D = R_{DS(on)} I_{MAX}^2 \qquad (60)$$

$R_{DS(on)}$ is the resistance of the channel from drain to source when the MOSFET is on. MOSFETs are available with very low on-resistance, but still dissipate a fair amount of power when driving a heavy load. The transistor's data sheet will contain $R_{DS(on)}$ specification and the V_{GS} required for it to be reached.

Power dissipation is why a switching tran-

Analog Basics 3.41

sistor needs to be kept out of its linear region. When turned completely off or on, either current through the transistor or voltage across it are low, also keeping the product of voltage and current (the power to be dissipated) low. As the waveform diagrams in Fig 3.58 show, while in the linear region, both voltage and current have significant values and so the transistor is generating heat when changing from off to on and vice versa. It's important to make the transition through the linear region quickly to keep the transistor cool.

The worst-case amount of power dissipated during each on-off transition is approximately

$$P_{transition} = \frac{1}{4} V_{MAX} I_{MAX}$$

assuming that the voltage and current increase and decrease linearly. If the circuit turns on and off at a rate of f, the total average power dissipation due to switching states is:

$$P_D = \frac{f}{2} V_{MAX} I_{MAX} \qquad (61)$$

since there are two on-off transitions per switching cycle. This power must be added to the power dissipated when the switch is conducting current.

Once you have calculated the power the switch must dissipate, you must check to see whether the transistor can withstand it. The manufacturer of the transistor will specify a *free-air dissipation* that assumes no heat-sink and room temperature air circulating freely around the transistor. This rating should be at least 50% higher than your calculated power dissipation. If not, you must either use a larger transistor or provide some means of getting rid of the heat, such as heat sink. Methods of dissipating heat are discussed in the **Electrical Fundamentals** chapter.

INDUCTIVE AND CAPACITIVE LOADS

Voltage transients for inductive loads, such as solenoids or relays can easily reach dozens of times the power supply voltage when load current is suddenly interrupted. To protect the transistor, the voltage transient must be clamped or its energy dissipated. Where switching is frequent, a series-RC *snubber* circuit (see **Fig 3.59A**) is connected across

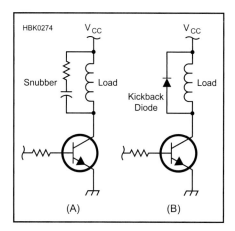

Fig 3.59 — The snubber RC circuit at (A) absorbs energy from transients with fast rise- and fall-times. At (B) a kickback diode protects the switching device when current is interrupted in the inductive load, causing a voltage transient, by conducting the energy back to the power source.

the load to dissipate the transient's energy as heat. The most common method is to employ a "kickback" diode that is reverse-biased when the load is energized as shown in Fig 3.59B. When the load current is interrupted, the diode routes the energy back to the power supply, clamping the voltage at the power supply voltage plus the diode's forward voltage drop.

Capacitive loads such as heavily filtered power inputs may temporarily act like short circuits when the load is energized or de-energized. The surge current is only limited by the internal resistance of the load capacitance. The transistor will have to handle the temporary overloads without being damaged or overheating. The usual solution is to select a transistor with an I_{MAX} rating greater to the surge current. Sometimes a small current-limiting resistor can be placed in series with the load to reduce the peak surge current at the expense of dissipating power continuously when the load is drawing current.

HIGH-SIDE AND LOW-SIDE SWITCHING

The switching circuits shown in Fig 3.58 are *low-side switches*. This means the switch is connected between the load and ground. A *high-side switch* is connected between the power source and the load. The same concerns for power dissipation apply, but the methods of driving the switch change because of the voltage of the emitter or source of the switching device will be at or near the power supply voltage when the switch is on.

To drive an NPN bipolar transistor or an N-channel MOSFET in a high-side circuit requires the switch input signal to be at least $V_{BE(sat)}$ or $V_{GS(on)}$ *above* the voltage supplied to the load. If the load expects to see the full power supply voltage, the switch input signal will have to be *greater* than the power supply voltage. A small step-up or boost dc-to-dc converter is often used to supply the extra voltage needed for the driver circuit.

One alternate method of high-side switching is to use a PNP bipolar transistor as the switching transistor. A small input transistor turns the main PNP transistor on by controlling the larger transistor's base current. Similarly, a P-channel MOSFET could also be employed with a bipolar transistor or FET acting as its driver. P-type material generally does not have the same high conductivity as N-type material and so these devices dissipate somewhat more power than N-type devices under the same load conditions.

3.5.8 Choosing a Transistor

With all the choices for transistors — Web sites and catalogs can list hundreds — selecting a suitable transistor can be intimidating. Start by determining the maximum voltage (V_{CEO} or $V_{DS(MAX)}$), current (I_{MAX}) and power dissipation ($P_{D(MAX)}$) the transistor must handle. Determine what dc current gain, β, or transconductance, g_m, is required. Then determine the highest frequency at which full gain is required and multiply it by either the voltage or current gain to obtain f_T or h_{fe}. This will reduce the number of choices dramatically.

The chapter on **Component Data and References** has tables of parameters for popular transistors that tend to be the lowest-cost and most available parts, as well. You will find that a handful of part types satisfy the majority of your building needs. Only in very special applications will you need to choose a corresponding special part.

3.6 Operational Amplifiers

An *operational amplifier*, or *op amp*, is one of the most useful linear devices. While it is possible to build an op amp with discrete components, and early versions were, the symmetry of the circuit demanded for high performance requires a close match of many components. It is more effective, and much easier, to implement as an integrated circuit.

The term "operational" comes from the op amp's origin in analog computers where it was used to implement mathematical operations.) The modern op-amp IC approaches the performance of a perfect analog circuit building block.

The op amp's performance approaches that of an ideal analog circuit building; an infinite input impedance (Z_i), a zero output impedance (Z_o) and an open loop voltage gain (A_v) of infinity. Obviously, practical op amps do not meet these specifications, but they do come closer than most other types of amplifiers. These attributes allow the circuit designer to implement many different functions with an op amp and only a few external components.

3.6.1 Characteristics of Practical Op-Amps

An op amp has three signal terminals (see **Fig 3.60**). There are two input terminals, the *noninverting input* marked with a + sign and the *inverting input* marked with a – sign. Voltages applied to the noninverting input cause the op amp output voltage to change with the same polarity.

The output of the amplifier is a single terminal with the output voltage referenced to the external circuit's reference voltage. Usually, that reference is ground, but the op amp's internal circuitry allows all voltages to *float*,

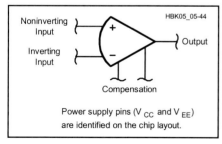

Fig 3.60 — Operational amplifier schematic symbol. The terminal marked with a + sign is the noninverting input. The terminal marked with a – sign is the inverting input. The output is to the right. On some op amps, external compensation is needed and leads are provided, pictured here below the device. Usually, the power supply leads are not shown on the op amp itself but are specified in the data sheet.

that is, to be referenced to any arbitrary voltage between the op amp's power supply voltages. The reference can be negative, ground or positive. For example, an op amp powered from a single power supply voltage amplifies just as well if the circuit reference voltage is halfway between ground and the supply voltage.

GAIN-BANDWIDTH PRODUCT AND COMPENSATION

An ideal op amp would have infinite frequency response, but just as transistors have an f_T that marks their upper frequency limit, the op amp has a *gain-bandwidth product* (GBW or BW). GBW represents the maximum product of gain and frequency available to any signal or circuit: voltage gain × frequency = GBW. If an op-amp with a GBW of 10 MHz is connected as a ×50 voltage amplifier, the maximum frequency at which that gain could be guaranteed is GBW / gain = 10 MHz / 50 = 200 kHz. GBW is an important consideration in high-performance filters and signal processing circuits whose design equations require high-gain at the frequencies over which they operate.

Older operational amplifiers, such as the LM301, have an additional two connections for *compensation*. To keep the amplifier from oscillating at very high gains it is often necessary to place a capacitor across the compensation terminals. This also decreases the frequency response of the op amp but increases its stability by making sure that the output signal can not have the right phase to create positive feedback at its inputs. Most modern op amps are internally compensated and do not have separate pins to add compensation capacitance. Additional compensation can be created by connecting a capacitor between the op amp output and the inverting input.

CMRR AND PSRR

One of the major advantages of using an op amp is its very high *common mode rejection ratio* (CMRR). *Common mode* signals are those that appear equally at all terminals. For example, if both conductors of an audio cable pick up a few tenths of a volt of 60 Hz signal from a nearby power transformer, that 60 Hz signal is a common-mode signal to whatever device the cable is connected. Since the op amp only responds to *differences* in voltage at its inputs, it can ignore or reject common mode signals. CMRR is a measure of how well the op amp rejects the common mode signal. High CMRR results from the symmetry between the circuit halves. CMRR is important when designing circuits that process low-level signals, such as microphone audio or the mV-level dc signals from sensors or thermocouples.

The rejection of power-supply imbalance is also an important op amp parameter. Shifts in power supply voltage and noise or ripple on the power supply voltages are coupled directly to the op amp's internal circuitry. The op amp's ability to ignore those disturbances is expressed by the *power supply rejection ratio* (PSRR). A high PSRR means that the op amp circuit will continue to perform well even if the power supply is imbalanced or noisy.

INPUT AND OUTPUT VOLTAGE LIMITS

The op amp is capable of accepting and amplifying signals at levels limited by the power supply voltages, also called *rails*. The difference in voltages between the two rails limits the range of signal voltages that can be processed. The voltages can be symmetrical positive and negative voltages (±12 V), a positive voltage and ground, ground and a negative voltage or any two different, stable voltages.

In most op amps the signal levels that can be handled are one or two diode forward voltage drops (0.7 V to 1.4 V) away from each rail. Thus, if an op amp has 15 V connected as its upper rail (usually denoted V+) and ground connected as its lower rail (V−), input signals can be amplified to be as high as 13.6 V and as low as 1.4 V in most amplifiers. Any values that would be amplified beyond those limits are clamped (output voltages that should be 1.4 V or less appear as 1.4 V and those that should be 13.6 V or more appear as 13.6 V). This clamping action was illustrated in Fig 3.1.

"Rail-to-rail" op amps have been developed to handle signal levels within a few tens of mV of rails (for example, the MAX406, from Maxim Integrated Products processes signals to within 10 mV of the power supply voltages). Rail-to-rail op-amps are often used in battery-powered products to allow the circuits to operate from low battery voltages for as long as possible.

INPUT BIAS AND OFFSET

The inputs of an op amp, while very high impedance, still allow some input current to flow. This is the *input bias current* and it is in the range of nA in modern op amps. Slight asymmetries in the op amp's internal circuitry result in a slight offset in the op amp's output voltage, even with the input terminals shorted together. The amount of voltage difference between the op amp's inputs required to cause the output voltage to be exactly zero is the *input offset voltage*, generally a few mV or less. Some op amps, such as the LM741, have special terminals to which a potentiometer can be connected to *null* the offset by correcting the internal imbalance. Introduction of a

Analog Basics 3.43

small dc correction voltage to the noninverting terminal is sometimes used to apply an offset voltage that counteracts the internal mismatch and centers the signal in the rail-to-rail range.

DC offset is an important consideration in op amps for two reasons. Actual op amps have a slight mismatch between the inverting and noninverting terminals that can become a substantial dc offset in the output, depending on the amplifier gain. The op amp output voltage must not be too close to the clamping limits or distortion will occur.

A TYPICAL OP AMP

As an example of typical values for these parameters, one of today's garden-variety op amps, the TL084, which contains both JFET and bipolar transistors, has a guaranteed minimum CMRR of 80 dB, an input bias current guaranteed to be below 200 pA (1 pA = 1 millionth of a µA) and a gain-bandwidth product of 3 MHz. Its input offset voltage is 3 mV. CMRR and PSRR are 86 dB, meaning that an unwanted signal or power supply imbalance of 1 V will only result in a 2.5 nV change at the op amp's output! All this for 33 cents even purchased in single quantities and there are four op-amps per package — that's a lot of performance.

3.6.2 Basic Op Amp Circuits

If a signal is connected to the input terminals of an op amp without any other circuitry attached, it will be amplified at the device's *open-loop gain* (typically 200,000 for the TL084 at dc and low frequencies, or 106 dB). This will quickly saturate the output at the power supply rails. Such large gains are rarely used. In most applications, negative feedback is used to limit the circuit gain by providing a feedback path from the output terminal to the inverting input terminal. The resulting *closed-loop gain* of the circuit depends solely on the values of the passive components used to form the loop (usually resistors and, for frequency-selective circuits, capacitors). The higher the op-amp's open-loop gain, the closer the circuit's actual gain will approach that predicted from the component values. Note that the gain of the op amp itself has not changed — it is the configuration of the external components that determines the overall gain of the circuit. Some examples of different circuit configurations that manipulate the closed-loop gain follow.

INVERTING AND NONINVERTING AMPLIFIERS

The op amp is often used in either an *inverting* or a *noninverting* amplifier circuit as shown in **Fig 3.61**. (Inversion means that the output signal is inverted from the input signal about the circuit's voltage reference as de-

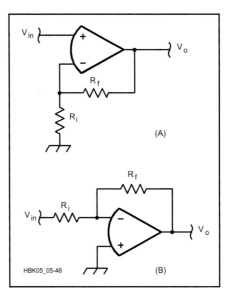

Fig 3.61 — Operational amplifier circuits. (A) Noninverting configuration. (B) Inverting configuration.

scribed below.) The amount of amplification is determined by the two resistors: the feedback resistor, R_f, and the input resistor, R_i.

In the noninverting configuration shown in Fig 3.61A, the input signal is connected to the op-amp's noninverting input. The feedback resistor is connected between the output and the inverting input terminal. The inverting input terminal is connected to R_i, which is connected to ground (or the circuit reference voltage).

This circuit illustrates how op amp circuits use negative feedback, the high open-loop gain of the op amp itself, and the high input impedance of the op amp inputs to create a stable circuit with a fixed gain. The signal applied to the noninverting input causes the output voltage of the op-amp to change with the same polarity. That is, a positive input signal causes a positive change in the op amp's output voltage. This voltage causes current to flow in the voltage divider formed by R_f and R_i. Because the current into the inverting input is so low, the current through R_f is the same as R_i.

The voltage at the *summing junction*, the connection point for the two resistors and the inverting terminal, V_{INV}, is:

$$V_{INV} = V_O \left(\frac{R_i}{R_i + R_f} \right) \quad (62)$$

The op amp's output voltage will continue to rise until the *loop error signal*, the difference in voltage between the inverting and noninverting inputs, is close to zero. At this point, the voltage at the inverting terminal is approximately equal to the voltage at the noninverting terminal, V_{in}, so that $V_{INV} = V_{in}$. Substituting in equation 62, the gain of this circuit is:

$$\frac{V_O}{V_{in}} = \left(1 + \frac{R_f}{R_i} \right) \quad (63)$$

where
V_o = the output voltage
V_{in} = the input voltage.

The higher the op amp's open-loop gain, the closer will be the voltages at the inverting and noninverting terminals when the circuit is balanced and the more closely the circuit's closed-loop gain will equal that of Equation 63. So the negative feedback creates an electronic balancing act with the op amp increasing its output voltage so that the input error signal is as small as possible.

In the inverting configuration of Fig 3.61B, the input signal (V_{in}) is connected through R_i to the inverting terminal. The feedback resistor is again connected between the inverting terminal and the output. The noninverting terminal is connected to ground (or the circuit reference voltage). In this configuration the feedback action results in the output voltage changing to whatever value is needed such that the current through R_i is balanced by an equal and opposite current through R_f. The gain of this circuit is:

$$\frac{V_O}{V_{in}} = -\frac{R_f}{R_{in}} \quad (64)$$

where V_{in} represents the voltage input to R_{in}.

For the remainder of this section, "ground" or "zero voltage" should be understood to be the circuit reference voltage. That voltage may not be "earth ground potential." For example, if a single positive supply of 12 V is used, 6 V may be used as the circuit reference voltage. The circuit reference voltage is a fixed dc voltage that can be considered to be an ac ground because of the reference source's extremely low ac impedance.

The negative sign in equation 64 indicates that the signal is inverted. For ac signals, inversion represents a 180° phase shift. The gain of the noninverting configuration can vary from a minimum of 1 to the maximum of which the op amp is capable, as indicated by A_v for dc signals, or the gain-bandwidth product for ac signals. The gain of the inverting configuration can vary from a minimum of 0 (gains from 0 to 1 attenuate the signal while gains of 1 and higher amplify the signal) to the maximum of which the device is capable.

The inverting amplifier configuration results in a special condition at the op amp's inverting input called *virtual ground*. Because the op amp's high open-loop gain drives the two inputs to be very close together, if the noninverting input is at ground potential, the inverting input will be very close to ground as well and the op amp's output will change with the input signal to maintain the inverting input at ground. Measured with a voltmeter, the input appears to be grounded, but it is merely maintained at ground potential by the action of the op amp and the feedback loop. This point in the circuit may not be connected to any other ground connection or circuit point because the resulting additional current flow will upset the balance of the circuit.

The *voltage follower* or *unity-gain buffer* circuit of **Fig 3.62** is commonly used as a buffer stage. The voltage follower has the input connected directly to the noninverting terminal and the output connected directly to the inverting terminal. This configuration has unity gain because the circuit is balanced when the output and input voltages are the same (error voltage equals zero). It also provides the maximum possible input impedance and the minimum possible output impedance of which the device is capable.

Differential and Difference Amplifier

A *differential amplifier* is a special application of an operational amplifier (see **Fig 3.63**). It amplifies the difference between two analog signals and is very useful to cancel noise under certain conditions. For instance, if an analog signal and a reference signal travel over the same cable they may pick up noise, and it is likely that both signals will have the same amount of noise. When the differential amplifier subtracts them, the signal will be unchanged but the noise will be completely removed, within the limits of the CMRR. The equation for differential amplifier operation is

$$V_O = \frac{R_f}{R_i}\left[\frac{1}{\frac{R_n}{R_g}+1}\left(\frac{R_i}{R_f}+1\right)V_n - V_i\right] \quad (65)$$

which, if the ratios R_i/R_f and R_n/R_g are equal, simplifies to:

$$V_O = \frac{R_f}{R_i}(V_n - V_1) \quad (66)$$

Note that the differential amplifier response is identical to the inverting amplifier response (equation 64) if the voltage applied to the

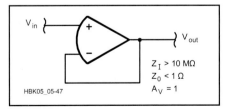

Fig 3.62 — Voltage follower. This operational amplifier circuit makes a nearly ideal buffer with a voltage gain of about one, and with extremely high input impedance and extremely low output impedance.

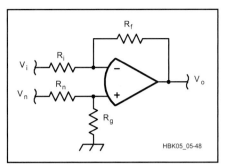

Fig 3.63 — Difference amplifier. This operational amplifier circuit amplifies the difference between the two input signals.

noninverting terminal is equal to zero. If the voltage applied to the inverting terminal (V_i) is zero, the analysis is a little more complicated but it is possible to derive the noninverting amplifier response (equation 62) from the differential amplifier response by taking into account the influence of R_n and R_g. If all four resistors have the *same* value the *difference amplifier* is created and V_O is just the difference of the two voltages.

$$V_O = V_n - V_1 \quad (67)$$

Instrumentation Amplifier

Just as the symmetry of the transistors making up an op amp leads to a device with high values of Z_i, A_v and CMRR and a low value of Z_o, a symmetric combination of op amps is used to further improve these parameters. This circuit, shown in **Fig 3.64** is called an *instrumentation amplifier*. It has three parts; each of the two inputs is connected to a noninverting buffer amplifier with a gain of 1 + R2/R1. The outputs of these buffer amplifiers are then connected to a differential amplifier with a gain of R4/R3. V2 is the circuit's inverting input and V1 the noninverting input.

The three amplifier modules are usually all part of the same integrated circuit. This means that they have essentially the same temperature and the internal transistors and resistors are very well matched. This causes the subtle gain and tracking errors caused by temperature differences and mismatched components between individual op amps to be cancelled out or dramatically reduced. In addition, the external resistors using the same designators (R2, R3, R4) are carefully matched as well, sometimes being part of a single integrated *resistor pack*. The result is a circuit with better performance than any single-amplifier circuit over a wider temperature range.

Summing Amplifier

The high input impedance of an op amp makes it ideal for use as a *summing amplifier*. In either the inverting or noninverting configuration, the single input signal can be replaced by multiple input signals that are connected together through series resistors, as shown in **Fig 3.65**. For the inverting summing amplifier, the gain of each input signal can be calculated individually using equation 64 and, because of the superposition property of linear circuits, the output is the sum of each input signal multiplied by its gain. In the noninverting configuration, the output is the gain times the weighted sum of the m different input signals:

$$V_n = V_{n1}\frac{R_{p1}}{R_1+R_{p1}} + V_{n2}\frac{R_{p2}}{R_2+R_{p2}} + \ldots$$

$$+V_{nm}\frac{R_{pm}}{R_m+R_{pm}} \quad (68)$$

where R_{pm} is the parallel resistance of all m resistors excluding R_m. For example, with three signals being summed, R_{p1} is the parallel combination of R_2 and R_3.

Comparators

A *voltage comparator* is another special

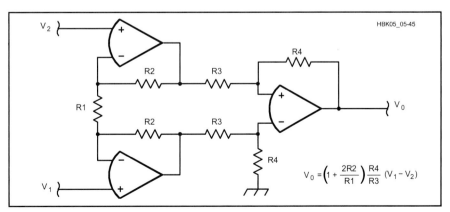

Fig 3.64 — Operational amplifiers arranged as an instrumentation amplifier. The balanced and cascaded series of op amps work together to perform differential amplification with good common-mode rejection and very high input impedance (no load resistor required) on both the inverting (V_1) and noninverting (V_2) inputs.

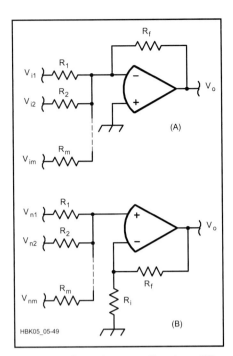

Fig 3.65 — Summing operational amplifier circuits. (A) Inverting configuration. (B) Noninverting configuration.

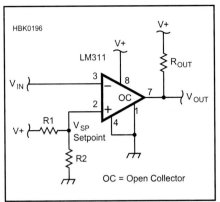

Fig 3.66 — A comparator circuit in which the output voltage is low when voltage at the inverting input is higher than the setpoint voltage at the noninverting input.

and the pull-up resistor "pulls up" the output voltage to the positive power supply voltage. When the comparator output is FALSE, the transistor is driven into the saturation and the output voltage is the transistor's $V_{CE(sat)}$.

Hysteresis

Comparator circuits also use *hysteresis* to prevent "chatter" — the output of the comparator switching rapidly back and forth when the input voltage is at or close to the setpoint voltage. There may be noise on the input signal, as shown in **Fig 3.67A**, that causes the input voltage to cross the setpoint threshold repeatedly. The rapid switching of the output can be confusing to the circuits monitoring the comparator output.

Hysteresis is a form of positive feedback form of op amp circuit, shown in **Fig 3.66**. It has two analog signals as its inputs and its output is either TRUE or FALSE depending on whether the noninverting or inverting signal voltage is higher, respectively. Thus, it "compares" the input voltages. TRUE generally corresponds to a positive output voltage and FALSE a negative or zero voltage. The circuit in Fig 3.66 uses external resistors to generate a reference voltage, called the *setpoint*, to which the input signal is compared. A comparator can also compare two variable voltages.

A standard operational amplifier can be made to act as a comparator by connecting the two input voltages to the noninverting and inverting inputs with no input or feedback resistors. If the voltage of the noninverting input is higher than that of the inverting input, the output voltage will be driven to the positive clamping limit. If the inverting input is at a higher potential than the noninverting input, the output voltage will be driven to the negative clamping limit. If the comparator is comparing an unknown voltage to a known voltage, the known voltage is called the *setpoint* and the comparator output indicates whether the unknown voltage is above or below the setpoint.

An op amp that has been intended for use as a comparator, such as the LM311, is optimized to respond quickly to the input signals. In addition, comparators often have *open-collector outputs* that use an external *pull-up* resistor, R_{OUT}, connected to a positive power supply voltage. When the comparator output is TRUE, the output transistor is turned off

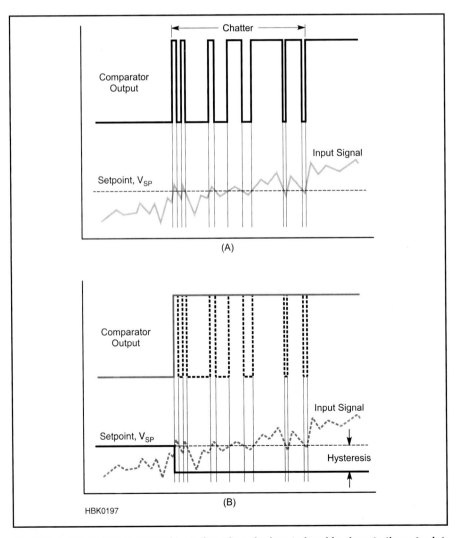

Fig 3.67 — Chatter (A) is caused by noise when the input signal is close to the setpoint. Chatter can also be caused by voltage shifts that occur when a heavy load is turned on and off. Hysteresis (B) shifts the setpoint a small amount by using positive feedback in which the output pulls the setpoint farther away from the input signal after switching.

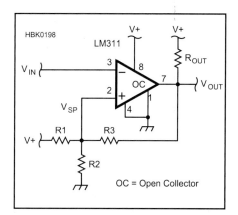

Fig 3.68 — Comparator circuit with hysteresis. R3 causes a shift in the comparator setpoint by allowing more current to flow through R1 when the comparator output is low.

that "moves" the setpoint by a few mV in the direction *opposite* to that in which the input signal crossed the setpoint threshold. As shown in Fig 3.67B, the slight shift in the setpoint tends to hold the comparator output in the new state and prevents switching back to the old state. **Fig 3.68** shows how the output of the comparator is fed back to the positive input through resistor R3, adding or subtracting a small amount of current from the divider and shifting the setpoint.

Some applications of a voltage comparator are a zero crossing detector, a signal squarer (which turns other cyclical wave forms into square waves) and a peak detector. An amateur station application: Circuits that monitor the CI-V band data output voltage from ICOM HF radios use a series of comparators to sense the level of the voltage and indicate on which band the radio is operating.

FILTERS

One of the most important type of op amp circuits is the *active filter*. Two examples of op amp filter circuits are shown in **Fig 3.69**. The simple noninverting low-pass filter in Fig 3.69A has the same response as a passive single-pole RC low-pass filter, but unlike the passive filter, the op amp filter circuit has a very high input impedance and a very low output impedance so that the filter's frequency and voltage response are relatively unaffected by the circuits connected to the filter input and output. This circuit is a low-pass filter because the reactance of the feedback capacitor decreases with frequency, requiring less output voltage to balance the voltages of the inverting and noninverting inputs.

The *multiple-feedback* circuit in Fig 3.69B results in a band-pass response while using only resistors and capacitors. This circuit is just one of many different types of active filters. Active filters are discussed in the **RF and AF Filters** chapter.

RECTIFIERS AND PEAK DETECTORS

The high open-loop gain of the op amp can also be used to simulate the I-V characteristics of an ideal diode. A *precision rectifier* circuit is shown in **Fig 3.70** along with the I-V characteristics of a real (dashed lines) and ideal (solid line) diode. The high gain of the op amp compensates for the V_F forward voltage drop of the real diode in its feedback loop with an output voltage equal to the input voltage plus V_F. Remember that the op amp's output increases until its input voltages are balanced. When the input voltage is negative, which would reverse-bias the diode, the op

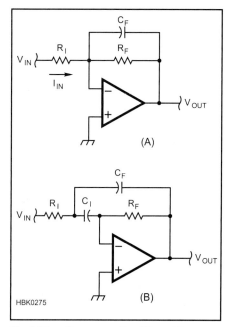

Fig 3.69 — Op amp active filters. The circuit at (A) has a low-pass response identical to an RC filter. The −3 dB frequency occurs when the reactance of C_F equals R_F. The band-pass filter at (B) is a multiple-feedback filter.

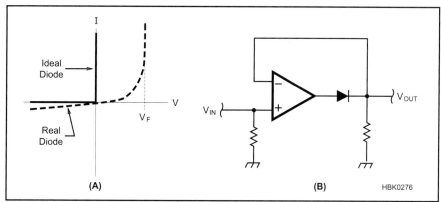

Fig 3.70 — Ideal and real diode I-V characteristics are shown at (A). The op amp precision rectifier circuit is shown at (B).

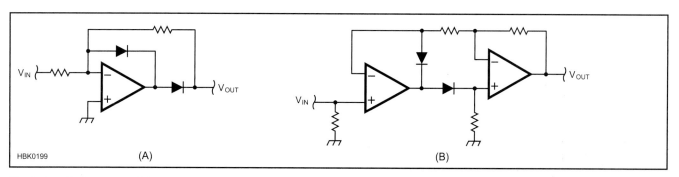

Fig 3.71 — Half-wave precision rectifier (A). The extra diode at the output of the op amp prevents the op amp from saturating on negative half-cycles and improves response time. The precision full-wave rectifier circuit at (B) reproduces both halves of the input waveform.

Analog Basics 3.47

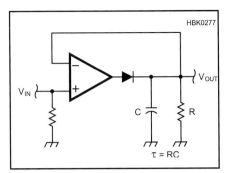

Fig 3.72 — Peak detector. Coupling a precision diode with a capacitor to store charge creates a peak detector. The capacitor will charge to the peak value of the input voltage. R discharges the capacitor with a time constant of $\tau = RC$ and can be omitted if it is desired for the output voltage to remain nearly constant.

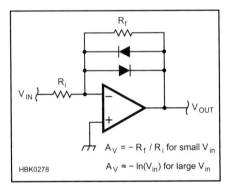

Fig 3.73 — Log amplifier. At low voltages, the gain of the circuit is $-R_f/R_i$, but as the diodes begin to conduct for higher-voltage signals, the gain changes to $-\ln(V_{in})$ in because of the diode's exponential current as described in the Fundamental Diode Equation.

amp's output can't balance the input because the diode blocks any current flow through the feedback loop. The resistor at the output holds the voltage at zero until the input voltage is positive once again. Precision half-wave and full-wave rectifier circuits are shown in **Fig 3.71** and their operation is described in many reference texts.

One application of the precision rectifier circuit useful in radio is the *peak detector*, shown in **Fig 3.72**. A precision rectifier is used to charge the output capacitor which holds the peak voltage. The output resistor sets the time constant at which the capacitor discharges. The resistor can also be replaced by a transistor acting as a switch to reset the detector. This circuit is used in AGC loops, spectrum analyzers, and other instruments that measure the peak value of ac waveforms.

LOG AMPLIFIER

There are a number of applications in radio in which it is useful for the gain of an amplifier to be higher for small input signals than for large input signals. For example, an audio compressor circuit is used to reduce the variations in a speech signal's amplitude so that the average power output of an AM or SSB transmitter is increased. A *log amplifier* circuit whose gain for large signals is proportional to the logarithm of the input signal's amplitude is shown in **Fig 3.73**. The log amp circuit is used in compressors and limiter circuits.

At signal levels that are too small to cause significant current flow through the diodes, the gain is set as in a regular inverting amplifier, $A_V = -R_f/R_i$. As the signal level increases, however, more current flows through the diode according to the Fundamental Diode Equation (equation 4) given earlier in this chapter. That means the op amp output voltage has to increase less (lower gain) to cause enough current to flow through R_i such that the input voltages balance. The larger the input voltage, the more the diode conducts and the lower the gain of the circuit. Since the diode's current is exponential in response to voltage, the gain of the circuit for large input signals is logarithmic.

Voltage-Current Converters

Another pair of useful op amp circuits convert voltage into current and current into voltage. These are frequently used to convert currents from sensors and detectors into voltages that are easier to measure. **Fig 3.74A** shows a voltage-to-current converter in which the output current is actually the current in the feedback loop. Because the op amp's high open-loop gain insures that its input voltages are equal, the current $I_{R1} = V_{IN}/R1$. Certainly, this could also be achieved with a resistor and Ohm's Law, but the op amp circuit's high input impedance means there is little interaction between the input voltage source and the output current.

Going the other way, Fig 3.74B is a current-to-voltage converter. The op amp's output will change so that the current through the feedback resistor, R1, exactly equals the input current, keeping the inverting terminal at ground potential. The output voltage, $V_O = I_{IN} R1$. Again, this could be done with just a resistor, but the op amp provides isolation between the source of input current and the output voltage. Fig 3.74C shows an application of a current-to-voltage converter in which the small currents from a photodiode are turned into voltage. This circuit can be used as a detector for amplitude modulated light pulses or waveforms.

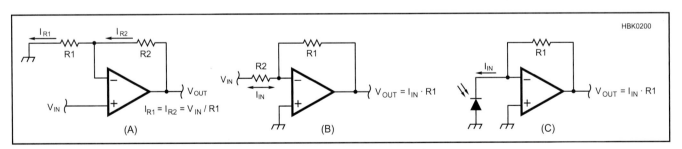

Fig 3.74 — Voltage-current converters. The current through R1 in (A) equals $V_{in}/R1$ because the op amp keeps both input terminals at approximately the same voltage. At (B), input current is balanced by the op amp, resulting in $V_{OUT} = I_{IN}R1$. Current through a photodiode (C) can be converted into a voltage in this way.

3.7 Analog-Digital Conversion

While radio signals are definitely analog entities, much of the electronics associated with radio is digital, operating on binary data representing the radio signals and the information they carry. The interface between the two worlds — analog and digital — is a key element of radio communications systems and the function is performed by analog-digital converters.

This section presents an overview of the different types of analog-digital converters and their key specifications and behaviors. The chapter on **Digital Basics** presents information on interfacing converters to digital circuitry and associated issues. The chapter on **DSP and Software Radio Design** discusses specific applications of analog-digital conversion technology in radio communications systems.

3.7.1 Basic Conversion Properties

Analog-digital conversion consists of taking data in one form, such as digital binary data or an analog ac RF waveform, and creating an equivalent representation of it in the opposite domain. Converters that create a digital representation of analog voltages or currents are called *analog-to-digital converters* (ADC), *analog/digital converters*, *A/D converters* or *A-to-D converters*. Similarly, converters that create analog voltages or currents from digital quantities are called *digital-to-analog converters* (DAC), *digital/analog converters*, *D/A converters* or *D-to-A converters*. The word "conversion" in this first section on the properties of converting information between the analog and digital domains will apply equally to analog-to-digital or digital-to-analog conversion.

Converters are typically implemented as integrated circuits that include all of the necessary interfaces and sub-systems to perform the entire conversion process. Schematic symbols for ADCs and DACs are shown in **Fig 3.75**.

Fig 3.76 shows two different representations of the same physical phenomenon; an analog voltage changing from 0 to 1 V. In the analog world, the voltage is continuous and can be represented by any real number between 0 and 1. In the digital world, the number of possible values that can represent any phenomenon is limited by the number of bits contained in each value.

In Fig 3.76, there are only four two-bit digital values 00, 01, 10, and 11, each corresponding to the analog voltage being within a specific range of voltages. If the analog voltage is anywhere in the range 0 to 0.25 V, the digital value representing the analog voltage will be 00, no matter whether the voltage is 0.0001 or 0.24999 V. The range 0.25 to 0.5 V is represented by the digital value 01, and so forth.

The process of converting a continuous range of possible values to a limited number of discrete values is called *digitization* and each discrete value is called a *code* or a *quantization code*. The number of possible codes that can represent an analog quantity is 2^N, where N is the number of bits in the code. A two-bit number can have four codes as shown in Fig 3.76, a four-bit number can have sixteen codes, an eight-bit number 256 codes, and so forth. Assuming that the smallest change in code values is one bit, that value is called the *least significant bit* (LSB) regardless of its position in the format used to represent digital numbers.

RESOLUTION AND RANGE

The *resolution* or *step size* of the conversion is the smallest change in the analog value that the conversion can represent. The *range* of the conversion is the total span of analog values that the conversion can process. The maximum value in the range is called the *full-scale* (F.S.) value. In Fig 3.76, the conversion range is 1 V. The resolution of the conversion is

$$\text{resolution} = \frac{\text{range}}{2^N}$$

In Fig 3.76, the conversion resolution is $\frac{1}{4} \times 1\,V = 0.25\,V$ in the figure. If each code had four bits instead, it would have a resolution of $\frac{1}{2^4} \times 1\,V = 0.083\,V$. Conversion range does not necessarily have zero as one end point. For example, a conversion range of 5 V may span 0 to 5 V, −5 to 0 V, 10 to 15 V, −2.5 to 2.5 V, and so on.

Analog-digital conversion can have a range that is *unipolar* or *bipolar*. Unipolar means a conversion range that is entirely positive or negative, usually referring to voltage. Bipolar means the range can take on both positive and negative values.

Confusingly, the format in which the bits are organized is also called a code. The binary code represents digital values as binary numbers with the least significant bit on the right. *Binary-coded-decimal* (BCD) is a code in which groups of four bits represent individual decimal values of 0-9. In the *hexadecimal* code, groups of four bits represent decimal values of 0-15. There are many such codes. Be careful in interpreting the word "code" to be sure the correct meaning is used or understood.

To avoid having to know the conversion range to specify resolution, *percentage resolution* is used instead.

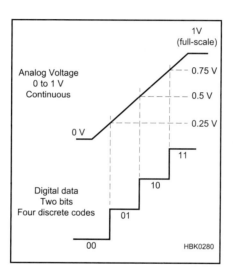

Fig 3.76 — The analog voltage varies continuously between 0 and 1 V, but the two-bit digital system only has four values to represent the analog voltage, so each can only represent a portion of the analog signal's range.

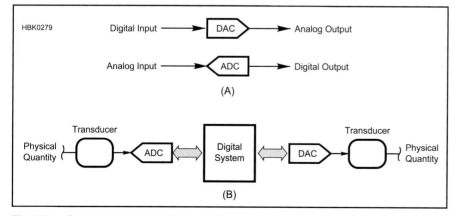

Fig 3.75 — Schematic symbols (A) for digital-to-analog converters (DAC) and analog-to-digital converters (ADC). The general block diagram of a system (B) that digitizes an analog signal, operates on it as digital data, then converts it back to analog form.

Analog Basics   3.49

$$\% \text{ resolution} = \frac{\text{resolution}}{\text{full scale}} \times 100\% \quad (69)$$

In Fig 3.76, the conversion's percentage resolution is

$$\% \text{ resolution} = \frac{0.25 \text{ V}}{1.0 \text{ V}} \times 100\% = 25\%$$

Because each code represents a range of possible analog values, the limited number of available codes creates *quantization error*. This is the maximum variation in analog values that can be represented by the same code. In Fig 3.76, any value from 0.25 through 0.50 V could be represented by the same code: 01. The quantization error in this case is 0.25 V. Quantization error can also be specified as % full-scale by substituting the value of error for resolution in equation 69.

Resolution can also be defined by the number of bits in the conversion. The higher the number of bits, the smaller the resolution as demonstrated above. Since many converters have variable ranges set by external components or voltages, referring to percent resolution or as a number of bits is preferred. The conversions between percent resolution and number of bits are as follows:

$$\% \text{ resolution} = \frac{1}{2^N} \times 100\% \quad (70)$$

and

$$N = \frac{\log\left(\frac{100\%}{\% \text{ resolution}}\right)}{\log 2} \quad (71)$$

Quantization error can also be specified as a number of least significant bits (LSB) where each bit is equivalent to the conversion's resolution.

When applied to receivers, dynamic range is more useful than percent resolution. For each additional bit of resolution, the resolution becomes two times greater, or 6.02 dB.

$$\text{Dynamic range (dB)} = N \times 6.02 \text{ dB} \quad (72)$$

For example, a 16-bit conversion has a dynamic range of $16 \times 6.02 = 96.32$ dB. This is the conversion's theoretical dynamic range with no noise present in the system. If noise is present, some number of the smallest codes will contain only noise, reducing the dynamic range available to represent a signal. The noise inherent in a particular conversion process can be specified as an analog value (mV, μA and so on) or as a number of bits, meaning the number of codes that only represent noise. For example, if a conversion has five bits of noise, any value represented by the smallest five codes is considered to be noise. The conversion's *effective number of bits* (ENOB) describes the number of bits available to contain information about the signal.

ACCURACY

A companion to resolution, *accuracy* refers to the ability of the converter to either assign the correct code to an analog value or create the true analog value from a specific code. As with resolution, it is most convenient to refer to accuracy as either a percentage of full scale or in bits. *Full-scale error* is the maximum deviation of the code's value or the analog quantity's value as a percentage of the full scale value. If a converter's accuracy is given as 0.02% F.S. and the conversion range is 5 V, the conversion can be in error by as much as $0.02\% \times 5$ V $= 1$ mV from the correct or expected value. *Offset* has the same meaning in conversion as it does in analog electronics — a consistent shift in the value of the conversion from the ideal value.

Linearity error represents the maximum deviation in step size from the ideal step size. This is also called *integral nonlinearity* (INL). In the converter of Fig 3.76, ideal step size is 0.25 V. If the linearity error for the conversion was given as 0.05% F.S., any actual step size could be in error by as much as $0.05\% \times 5$ V $= 2.5$ mV. The amount of error is based on the full-scale value, not the step-size value. *Differential nonlinearity* is a measure of how much any two adjacent step sizes deviate from the ideal step size. Errors can be represented as a number of bits, usually assumed to be least significant bits, or LSB, with one bit representing the same range as the conversion resolution.

Accuracy and resolution are particularly important in conversions for radio applications because they represent distortion in the signal being processed. An analog receiver that distorts the received signal creates undesired spurious signals that interfere with the desired signal. While the process is not exactly the same, distortion created by the signal conversion circuitry of a digital receiver also causes degradation in performance.

DISTORTION AND NOISE

Distortion and noise in a conversion are characterized by several parameters all related to linearity and accuracy. THD+N (*Total Harmonic Distortion + Noise*) is a measure of how much distortion and noise is introduced by the conversion. THD+N can be specified in percent or in dB. Smaller values are better. SINAD (*Signal to Noise and Distortion Ratio*) is related to THD+N, generally specified along with a desired signal level to show what signal level is required to achieve a certain level of SINAD or the highest signal level at which a certain level of SINAD can be maintained. *Spurious-free Dynamic Range* (SFDR) is the difference between the amplitude of the desired signal and the highest unwanted signal. SFDR is generally specified in dB and a higher number indicates better performance.

CONVERSION RATE AND BANDWIDTH

Another important parameter of the conversion is the *conversion rate* or its reciprocal, *conversion speed*. A code that represents an analog value at a specific time is called a *sample*, so conversion rate, f_S, is specified in *samples per second* (sps) and conversion speed as some period of time per sample, such as 1 μsec/s. Because of the mechanics by which conversion is performed, conversion speed can also be specified as a number of cycles of clock signal used by the digital system performing the conversion. Conversion rate then depends on the frequency of the clock. Conversion speed may be smaller than the reciprocal of conversion rate if the system controlling the conversion introduces a delay between conversions. For example, if a conversion can take place in 1 μs, but the system only performs a conversion once per ms, the conversion rate is 1 ksps, not the reciprocal of 1 μs/s = 1 Msps.

The time accuracy of the digital clock that controls the conversion process can also affect the accuracy of the conversion. An error in long-term frequency will cause all conversions to have the same frequency error. Short-term variations in clock period are called *jitter* and they add noise to the conversion.

According to the *Nyquist Sampling Theorem*, a conversion must occur at a rate twice the highest frequency present in the analog signal. This minimum rate is the *Nyquist rate* and the maximum frequency allowed in the analog signal is the *Nyquist frequency*. In this way, the converter *bandwidth* is limited to one-half the conversion rate.

Referring to the process of converting analog signals to digital samples, if a lower rate is used, called *undersampling*, false signals called *aliases* will be created in the digital representation of the input signal at frequencies related to the difference between the Nyquist sampling rate and f_S. This is called *aliasing*. Sampling faster than the Nyquist rate is called *oversampling*.

Because conversions occur at some maximum rate, there is always the possibility of signals greater than the Nyquist frequency being present in an analog signal undergoing conversion or that is being created from digital values. These signals would result in aliases and must be removed by *band-limiting filters* that remove them prior to conversion. The mechanics of the sampling process are discussed further in the chapter **DSP and Software Radio Design**.

3.7.2 Analog-to-Digital Converters (ADC)

There are a number of methods by which the conversion from an analog quantity to a set of digital samples can be performed. Each has its strong points — simplicity, speed,

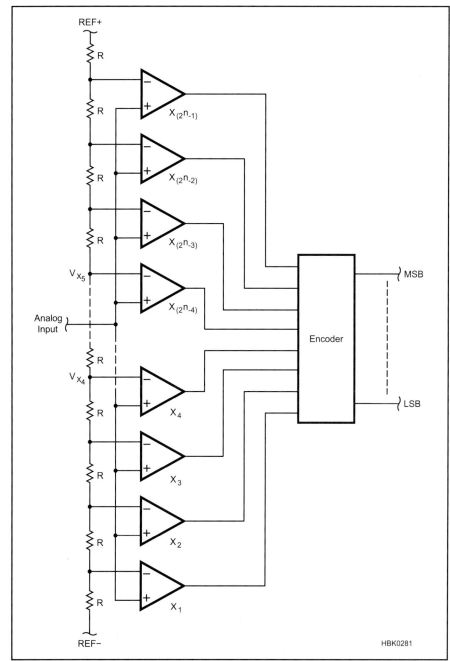

Fig 3.77 — The comparators of the flash converter are always switching state depending on the input signal's voltage. The decoder section converts the array of converter output to a single digital word.

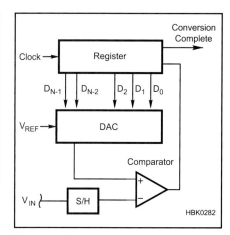

Fig 3.78 — The successive-approximation converter creates a digital word as it varies the DAC signal in order to keep the comparator's noninverting terminal close to the input voltage. A sample-and-hold circuit (S/H) holds the input signal steady while the measurement is being made.

resolution, accuracy — all affect the decision of which method to use for a particular application. In order to pick the right type of ADC, it is important to decide which of these criteria most strongly affect the performance of your application.

FLASH CONVERTER

The simplest type of ADC is the *flash converter*, shown in **Fig 3.77**, also called a *direct-conversion ADC*. It continually generates a digital representation of the analog signal at its input. The flash converter uses an array of comparators that compare the amplitude of the input signal to a set of reference voltages. There is one reference voltage for each step.

The outputs of the comparator array represent a digital value in which each bit indicates whether the input signal is greater (1) or less than (0) the reference voltage for that comparator. A digital logic *priority encoder* then converts the array of bits into a digital output code. Each successive conversion is available as quickly as the comparators can respond and the priority encoder can create the output code.

Flash converters are the fastest of all ADCs (conversion speeds can be in the ns range) but do not have high resolution because of the number of comparators and reference voltages required.

SUCCESSIVE-APPROXIMATION CONVERTER

The *successive-approximation converter* (SAC) is one of the most widely-used types of converters. As shown in **Fig 3.78**, it uses a single comparator and DAC (digital-to-analog converter) to arrive at the value of the input voltage by comparing it to successive analog values generated by the DAC. This type of converter offers a good compromise of conversion speed and resolution.

The DAC control logic begins a conversion by setting the output of the DAC to ½ of the conversion range. If the DAC output is greater than the analog input value, the output of the comparator is 0 and the most significant bit of the digital value is set to 0. The DAC output then either increases or decreases by ¼ of the range, depending on whether the value of the first comparison was 1 or 0. One test is made for each bit in digital output code and the result accumulated in a storage register. The process is then repeated, forming a series of approximations (thus the name of the converter), until a test has been made for all bits in the code.

While the digital circuitry to implement the SAC is somewhat complex, it is less expensive to build and calibrate than the array of comparators and precision resistors of the flash converter. Each conversion also takes

Analog Basics 3.51

a known and fixed number of clock cycles. The higher the number of bits in the output code, the longer the conversion takes, all other things being equal.

DUAL-SLOPE INTEGRATING CONVERTERS

The *dual-slope integrating ADC* is shown in **Fig 3.79**. It makes a conversion by measuring the time it takes for a capacitor to charge and discharge to a voltage level proportional to the analog quantity.

A constant-current source charges a capacitor (external to the converter IC) until the comparator output indicates that the capacitor voltage is equal to the analog input signal. The capacitor is then discharged by the constant-current source until the capacitor is discharged. This process is repeated continuously. The frequency of the charge-discharge cycle is determined by the values of R and C in Fig 3.79. The frequency is then measured by frequency counter circuitry and converted to the digital output code.

Dual-slope ADCs are low-cost and relatively immune to noise and temperature variations. Due to the slow speed of the conversion (measured in tens of ms) these converters are generally only used in test instruments, such as multimeters.

DELTA-ENCODED CONVERTERS

Instead of charging and discharging a capacitor from 0 V to the level of the input signal, and then back to 0 V, the *delta-encoded ADC* in **Fig 3.80** continually compares the output of a DAC to the input signal using a comparator. Whenever the signal changes, the DAC is adjusted until its output is equal to the input signal. Digital counter circuits keep track of the DAC value and generate the digital output code. Delta-encoded counters are available with wide conversion ranges and high resolution.

SIGMA-DELTA CONVERTERS

The *sigma-delta converter* also uses a DAC and a comparator in a feedback loop to generate a digital signal as shown in **Fig 3.81**. An integrator stores the sum of the input signal and the DAC output. (This is "sigma" or sum in the converter's name.) The comparator output indicates whether the integrator output is above or below the reference voltage and that signal is used to adjust the DAC's output so that the integrator output stays close to the reference voltage. (This is the "delta" in the name.) The stream of 0s and 1s from the comparator forms a high-speed digital bit stream that is digitally-filtered to form the output code.

Sigma-delta converters are used where high resolution (16 to 24 bits) is required at low sampling rates of a few kHz. The digital filtering in the converter also reduces the need for external band-limiting filters on the input signal.

ADC SUBSYSTEMS

Sample-and-Hold

ADCs that use a sequence of operations to create the digital output code must have a means of holding the input signal steady while the measurements are being made. This function is performed by the circuit of **Fig 3.82**. A high input-impedance buffer drives the external storage capacitor, C_{HOLD}, so that its voltage is the same as the input

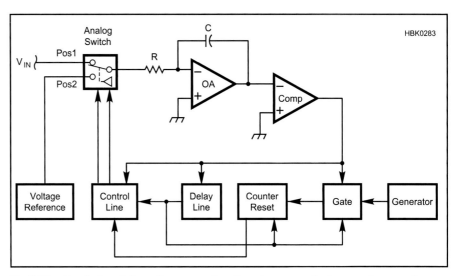

Fig 3.79 — Dual-slope integrating converter. By using a constant-current source to continually charge a capacitor to a known reference voltage then discharge it, the resulting frequency is directly proportional to the resistor value.

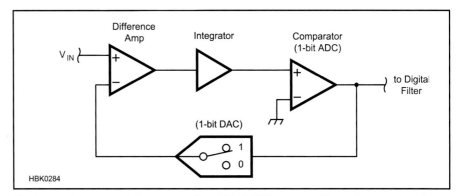

Fig 3.80 — Delta-encoded converter. The 1-bit DAC is operated in such a way that the bit stream out of the comparator represents the value of the input voltage.

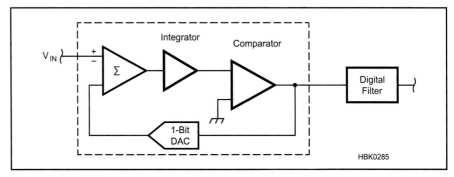

Fig 3.81 — Sigma-delta converter. Similar to the delta-encoded converter (Fig 3.80), the converter runs much faster than the output samples and uses a digital filter to derive the actual output value.

signal. Another high input-impedance buffer is used to provide a replica of the voltage on C_{HOLD} to the conversion circuitry.

When a conversion is started, a digital control signal opens the input switch, closes the output switch, and the capacitor's voltage is measured by the converter. It is important that the capacitor used for C_{HOLD} have low *leakage* so that while the measurement is being made, the voltage stays constant for the few ms required. This is of particular important in high-precision conversion.

Single-Ended and Differential Inputs

The input of most ADCs is *single-ended*, in which the input signal is measured between the input pin and a common ground. Shown in **Fig 3.83A**, this is acceptable for most applications, but if the voltage to be measured is small or is the difference between two non-zero voltages, an ADC with *differential inputs* should be used as in Fig 3.83B. Differential inputs are also useful when measuring current as the voltage across a small resistor in series with the current. In that case, neither side of the resistor is likely to be at ground, so a differential input is very useful. Differential inputs also help avoid the issue of noise contamination as discussed below.

Input Buffering and Filtering

The input impedance of most ADCs is high enough that the source of the input signal is unaffected. However, to protect the ADC input and reduce loading on the input source, an external buffer stage can be used. **Fig 3.84** shows a typical buffer arrangement with clamping diodes to protect against electrostatic discharge (ESD) and an RC-filter to prevent RF signals from affecting the input signal. In addition, to attenuate higher-frequency signals that might cause aliases, the input filters can also act as band-limiting filters.

Analog and Digital "Ground"

By definition, ADCs straddle the analog and digital domain. In principle, the signals remain separate and isolated from each other. In practice, however, voltages and currents from the analog and digital circuitry can be mixed together. This can result in the contamination of an analog signal with components of digital signals, and rarely, vice versa. Mostly, this is a problem when trying to measure small voltages in the presence of large power or RF signals.

The usual problem is that currents from high-speed digital circuitry find their way into analog signal paths and create transients and other artifacts that affect the measurement of the analog signal. Thus, it is important to have separate current paths for the two types of signals. The manufacturer of the converter will provide guidance for the proper use of the converter either in the device's data sheet or as

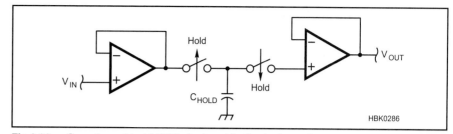

Fig 3.82 — Sample-and-hold (S/H). An input buffer isolates the sampled voltage from the input signal by charging the capacitor C_{HOLD} to that voltage with the input switch closed and the output switch open. When a measurement is being taken, the input switch is open to prevent the input signal from changing the capacitor voltage, and the output switch is closed so that the output buffer can generate a steady voltage at its output.

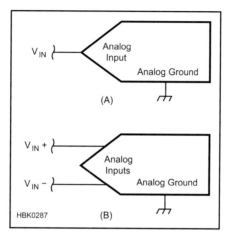

Fig 3.83 — Single-ended ADC inputs have a single active line and a ground or return line (A). Single-ended ADC input are often susceptible to noise and common mode signals or any kind of disturbance on their ground rails. In (B), the differential inputs used help the circuit "ignore" offsets and shifts in the input signal.

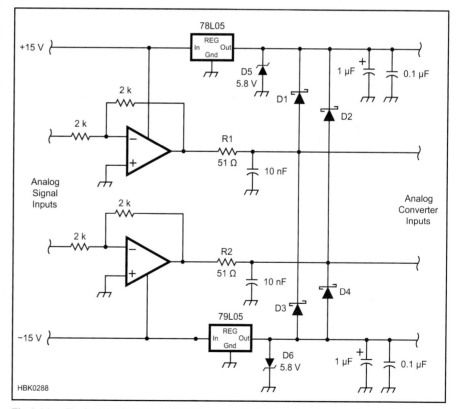

Fig 3.84 — Typical ADC input buffer-filter circuit. Unity-gain voltage followers help isolate the ADC from the input source. RC filters following the buffers act as band-limiting filters to prevent aliasing. Zener diodes are used to clamp the transient voltage and route the energy of transient into the power supply system.

Analog Basics 3.53

application notes. Look for separate pins on the converter, such as "AGND" or "DGND" that indicate how the two types of signal return paths should be connected.

3.7.3 Digital-to-Analog Converters (DAC)

Converting a digital value to an analog quantity is considerably simpler than the reverse, but there are several issues primarily associated with DACs that affect the selection of a particular converter.

As each new digital value is converted to analog, the output of the DAC makes an abrupt *step change*. Even if very small, the response of the DAC output does not respond perfectly or instantaneously. *Settling time* is the amount of time required for the DAC's output to stabilize within a certain amount of the final value. It is specified by the manufacturer and can be degraded if the load connected to the DAC is too heavy or if it is highly reactive. In these cases, using a buffer amplifier is recommended.

Monotonicity is another aspect of characterizing the DAC's accuracy. A DAC is monotonic if in increasing the digital input value linearly across the conversion range, the output of the DAC increases with every step. Because of errors in the internal conversion circuitry, it is possible for there to be some steps that are too small or too large, leading to output values that seem "out of order." These are usually quite small, but if used in a precision application, monotonicity is important.

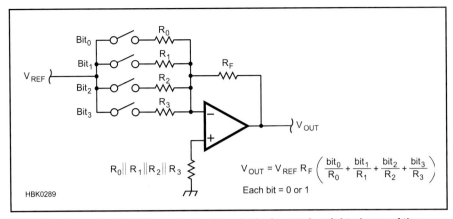

Fig 3.85 — Summing DAC. The output voltage is the inverted, weighted sum of the inputs to each summing resistor at the input. Digital data at the input controls the current into the summing resistors and thus, the output voltage.

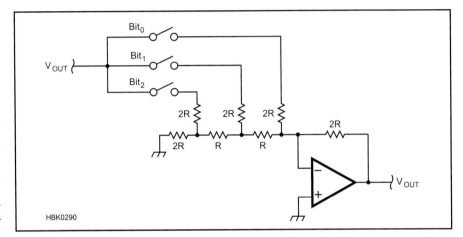

Fig 3.86 — R-2R Ladder DAC. This is the most common form of DAC because all of the resistor values are similar, making it easier to manufacture. The similarity in resistor values also means that there will be less variation of the comparator with temperature and other effects that affect all resistors similarly.

SUMMING DAC

A *summing DAC*, shown in **Fig 3.85**, is a summing amplifier with all of the inputs connected to a single reference voltage through switches. The digital value to be converted controls which switches are closed. The larger the digital value, the more switches are closed. Higher current causes the summing amplifier's output voltage to be higher, as well. Most summing DACs have a digital signal interface to hold the digital value between successive conversions, but not all, so be sure before you select a particular DAC.

The input resistors are *binary weighted* so that the summing network resistors representing the more significant bit values inject more current into the op amp's summing junction. Each resistor differs from its neighboring resistors in the amount of current it injects into the summing node by a factor of two, recreating the effect of each digital bit in the output voltage. At high resolutions, this becomes a problem because of the wide spread in resistor values — a 12-bit DAC would require a spread of 2048 between the largest and smallest resistor values. Summing DACs are generally only available with low resolution for that reason.

The *current output DAC* functions identically to the summing DAC, but does not have an op amp to convert current in the digitally-controlled resistor network to voltage. It consists only of the resistor network, so an external current-to-voltage circuit (discussed in the previous section on op amps) is required to change the current to a voltage. In some applications, the conversion to voltage is not required or it is already provided by some other circuit.

R-2R LADDER DAC

The summing and current output DACs both used binary weighted resistors to convert the binary digital value into the analog output. The practical limitation of this design is the large difference in value between the smallest and largest resistor. For example, in a 12-bit DAC, the smallest and largest resistors differ by a factor of 2048 (2^{N-1})! This can be difficult to fabricate in an IC without expensive *trimming* processes that adjust each resistor to the correct value and that maintain the powers-of-two relationship over wide temperature variations. For DACs with high-resolutions of 8 bits or more, the R-2R ladder DAC of **Fig 3.86** is a better design.

If the resistances are fairly close in value, the problems of manufacturing are greatly reduced. By using the *R-2R ladder* shown in the figure, the same method of varying current injected into an op amp circuit's summing junction can be used with resistors of only two values, R and 2R. In fact, since the op amp feedback resistor is also one of the IC resistors, the absolute value of the resistance R is unimportant, as long as the ratio of R:2R is maintained. This simplifies manufacturing greatly and is an example of IC design being based on ratios instead of absolute values. For this reason, most DACs use the R-2R ladder design and the performance differences lie mostly in their speed and accuracy.

3.7.4 Choosing a Converter

From the point of view of performance, choosing a converter, either an ADC or a DAC, comes down to resolution, accuracy and speed. Begin by determining the percent resolution or the dynamic range of the converter. Use the equations in the preceding section to determine the number of bits the converter must have. Select from converters with the next highest number of bits. For example, if you determine that you need 7 bits of resolution, use an 8-bit converter.

Next, consider accuracy. If the converter is needed for test instrumentation, you'll need to perform an *error budget* on the instrument's conversion processes, include errors in the analog circuitry. Once you have calculated percent errors, you can determine the requirements for FS error, offset error, and nonlinearities. If an ADC is going to be used for receiving applications, the spur-free dynamic range may be more important than high precision. (The chapter on **DSP and Software Radio Design** goes into more detail about the performance requirements for ADCs used in these applications.)

The remaining performance criterion is the speed and rate at which the converter can operate. Conversions should be able to be made at a minimum of twice the highest frequency of signal you wish to reproduce. (The signal is assumed to be a sinusoid. If a complex waveform is to be converted, you must account for the higher frequency signals that create the nonsinusoidal shapes.) If the converter will be running near its maximum rate, be sure that the associated digital interface, supporting circuitry, and software can support the required data rates, too!

Having established the conversion performance requirements, the next step is to consider cost, amount of associated circuitry, power requirements, and so forth. For example, a self-contained ADC is easier to use and takes up less PC board space, but is not as flexible as one that allows the designer to use an external voltage reference to set the conversion range. Many DACs with "current output" in their description are actually R-2R DACs with additional analog circuitry to provide the current output. DACs are available with both current and voltage outputs, as well. Other considerations, such as the nature of the required digital interface, as discussed in the next section, can also affect the selection of the converter.

3.8 Miscellaneous Analog ICs

The three main advantages of designing a circuit into an IC are to take advantage of the matched characteristics of its components, to make highly complex circuitry more economical, and to miniaturize the circuit and reduce power consumption. As circuits standardize and become widely used, they are often converted from discrete components to integrated circuits. Along with the op amp described earlier, there are many such classes of linear ICs.

3.8.1 Transistor and Driver Arrays

The most basic form of linear integrated circuit and one of the first to be implemented is the component array. The most common of these are the resistor, diode and transistor arrays. Though capacitor arrays are also possible, they are used less often. Component arrays usually provide space saving but this is not the major advantage of these devices. They are the least densely packed of the integrated circuits because each device requires a separate off-chip connection. While it may be possible to place over a million transistors on a single semiconductor chip, individual access to these would require a total of three million pins and this is beyond the limits of practicability. More commonly, resistor and diode arrays contain from five to 16 individual devices and transistor arrays contain from three to six individual transistors. The advantage of these arrays is the very close matching of component values within the array. In a circuit that needs matched components, the component array is often a good method of obtaining this feature. The components within an array can be internally combined for special functions, such as termination resistors, diode bridges and Darlington pair transistors. A nearly infinite number of possibilities exists for these combinations of components and many of these are available in arrays.

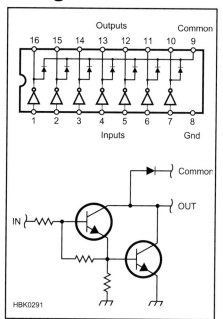

Fig 3.87 — Typical ULN2000-series driver array configuration and internal circuit. The use of driver array ICs is very popular as an interface between microprocessor or other low-power digital circuits and loads such as relays, solenoids or lamps.

Driver arrays, such as the ULN2000-series devices shown in **Fig 3.87** are very useful in creating an interface between low-power circuits such as microprocessors and higher-power loads and indicators. Each driver consists of a Darlington pair switching circuit as described earlier in this chapter. There are different versions with different types and arrangements of resistors and diodes.

Many manufacturers offer driver arrays. They are available with built-in kickback diodes to allow them to drive inductive loads, such as relays, and are heavy enough to source or sink current levels up to 1 A. (All of the drivers in the array can not operate at full load at the same time, however. Read the data sheet carefully to determine what limitations on current and power dissipation may exist.)

3.8.2 Voltage Regulators and References

One of the most popular linear ICs is the voltage regulator. There are two basic types, the three-terminal regulator and the regulator controller. Examples of both are described in the **Power Supplies** chapter.

The three-terminal regulator (input, ground, output) is a single package designed to perform all of the voltage regulation functions. The output voltage can be fixed, as in the 7800-series of regulators, or variable, as in the LM317 regulator. It contains a voltage reference, comparator circuits, current and temperature sensing protective circuits, and the main pass element. These ICs are usually contained in the same packages as power transistors and the same techniques of thermal management are used to remove excess heat.

Regulator controllers, such as the popular 723 device, contain all of the control and voltage reference circuitry, but require external components for the main pass element, current sensing, and to configure some of their control functions.

Voltage references such as the Linear Technology LT1635 are special semiconductor diodes that have a precisely controlled I-V characteristic. A buffer amplifier isolates the sensitive diode and provides a low output impedance for the voltage signal. Voltage references are used as part of power regulators and by analog-digital converter circuits.

3.8.3 Timers (Multivibrators)

A *multivibrator* is a circuit that oscillates between two states, usually with a square wave or pulse train output. The frequency of oscillation is accurately controlled with the addition of appropriate values of external resistance and capacitance. The most common multivibrator in use today is the 555 timer IC (NE555 by Signetics [now Philips] or LM555 by National Semiconductor). This very simple eight-pin device has a frequency range from less than one hertz to several hundred kilohertz. Such a device can also be used in *monostable* operation, where an input pulse generates an output pulse of a different duration, or in *a stable* or *free-running* operation, where the device oscillates continuously. Other applications of a multivibrator include a frequency divider, a delay line, a pulse width modulator and a pulse position modulator. (These can be found in the IC's data sheet or in the reference listed at the end of this chapter.)

Fig 3.88 shows the basic components of a 555. Connected between power input (V_{cc}) and ground, the three resistors labeled "R" at the top left of the figure form a *voltage divider* that divides V_{CC} into two equal steps—one at ⅔ V_{CC} and one at ⅓ V_{CC}. These serve as reference voltages for the rest of the circuit.

Connected to the reference voltages are blocks labeled *trigger comparator* and *threshold comparator*. (Comparators were discussed in a preceding section.) The trigger comparator in the 555 is wired so that its output is high whenever the trigger input is *less* than ⅓ V_{CC} and vice versa. Similarly, the threshold comparator output is high whenever the threshold input is *greater* than ⅔ V_{CC}. These two outputs control a digital *flip-flop* circuit. (Flip-flops are discussed in the **Digital Basics** chapter.)

The flip-flop output, *Q*, changes to high or low when the state of its *set* and *reset* input changes. The Q output stays high or low (it *latches* or *toggles*) until the opposite input changes. When the set input changes from low to high, Q goes low. When reset changes from low to high, Q goes high. The flip-flop ignores any other changes. An inverter makes the 555 output high when Q is low and vice versa — this makes the timer circuit easier to interface with external circuits.

The transistor connected to Q acts as a switch. When Q is high, the transistor is on and acts as a closed switch connected to ground. When Q is low, the transistor is off and the switch is open. These simple building blocks — voltage divider, comparator, flip-flop and switch — build a surprising number of useful circuits.

THE MONOSTABLE OR "ONE-SHOT" TIMER

The simplest 555 circuit is the monostable circuit. This configuration will output one fixed-length pulse when triggered by an input pulse. **Fig 3.89** shows the connections for this circuit.

Starting with capacitor C discharged, the flip-flop output, Q, is high, which keeps the discharge transistor turned on and the voltage on C below ⅔ V_{CC}. The circuit is in its stable state, waiting for a trigger pulse.

When the voltage at the trigger input drops below ⅓ V_{CC}, the trigger comparator output changes from low to high, which causes Q to toggle to the low state. This turns off the transistor (opens the switch) and allows C to begin charging toward V_{CC}.

When C reaches ⅔ V_{CC}, this causes the threshold comparator to switch its output from low to high and that resets the flip-flop. Q returns high, turning on the transistor and discharging C. The circuit has returned to its stable state. The output pulse length for the monostable configuration is:

$$T = 1.1\,R\,C_1 \qquad (73)$$

Notice that the timing is independent of the absolute value of V_{CC} — the output pulse width is the same with a 5 V supply as it is with a 15 V supply. This is because the 555 design is based on ratios and not absolute voltage levels.

THE ASTABLE MULTIVIBRATOR

The complement to the monostable circuit is the astable circuit in **Fig 3.90**. Pins 2, 6 and 7 are configured differently and timing resistor is now split into two resistors, R1 and R2.

Start from the same state as the monostable circuit, with C completely discharged. The monostable circuit requires a trigger pulse to initiate the timing cycle. In the astable circuit, the trigger input is connected directly to the capacitor, so if the capacitor is discharged, then the trigger comparator output must be high. Q is low, turning off the discharge transistor, which allows C to immediately begin charging.

C charges toward V_{CC}, but now through the combination of R1 and R2. As the capacitor voltage passes ⅔ V_{CC}, the threshold comparator output changes from low to high, resetting Q to high. This turns on the discharge transistor and the capacitor starts to discharge through R2. When the capacitor is discharged below ⅓ V_{CC}, the trigger comparator changes from high to low and the cycle begins again, automatically. This happens over and over, causing a train of pulses at the output while

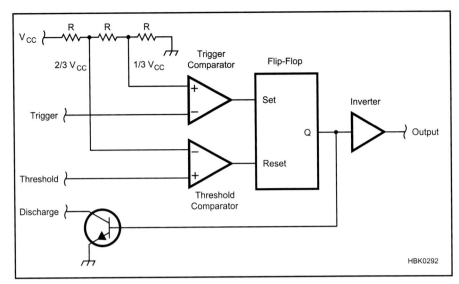

Fig 3.88 — Internal NE555 timer components. This simple array of components combine to make one of the most popular analog ICs. The 555 timer IC uses ratios of internal resistors to generate a precise voltage reference for generating time intervals based on charging and discharging a capacitor.

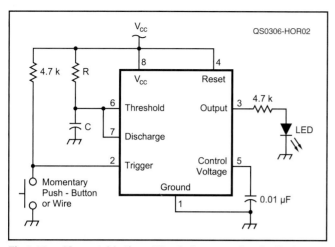

Fig 3.89 — Monostable timer. The timing capacitor is discharged until a trigger pulse initiates the charging process and turns the output on. When the capacitor has charged to 2/3 V_{CC}, the output is turned off, the capacitor is discharged and the timer awaits the next trigger pulse.

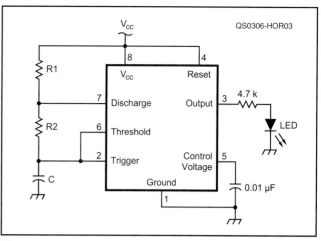

Fig 3.90 — Astable timer. If the capacitor discharge process initiates the next charge cycle, the timer will output a pulse train continuously.

C charges and discharges between ⅓ and ⅔ V_{CC} as seen in the figure.

The total time it takes for one complete cycle is the charge time, T_c, plus the discharge time, T_d:

$$T = T_c + T_d = 0.693 (R_1 + R_2) C + 0.693 R_2 C$$
$$= 0.693 (R_1 + 2R_2) C \quad (74)$$

and the output frequency is:

$$f = \frac{1}{T} = \frac{1.443}{(R_1 + 2 R_2) C} \quad (75)$$

When using the 555 in an application in or around radios, it is important to block any RF signals from the IC power supply or timing control inputs. Any unwanted signal present on these inputs, especially the Control Voltage input, will upset the timer's operation and cause it to operate improperly. The usual practice is to use a 0.01 μF bypass capacitor (shown on pin 5 in both Fig 3.89 and 3.90) to bypass ac signals such as noise or RF to ground. Abrupt changes in V_{CC} will also cause changes in timing and these may be prevented by connecting filter capacitors at the V_{CC} input to ground.

3.8.4 Analog Switches and Multiplexers

Arrays of analog switches, such as the Maxim MAX312-series, allow routing of audio through lower RF signals without mechanical switches. There are several types of switch arrays. Independent switches have isolated inputs and outputs and are turned on and off independently. Both SPST and SPDT configurations are available. Multiple switches can be wired with common control signals to implement multiple-pole configurations.

Use of analog switches at RF through microwave frequencies requires devices specifically designed for those frequencies. The Analog Devices ADG901 is a switch usable to 2.5 GHz. It absorbs the signal when off, acting as a terminating load. The ADG902 instead reflects the signal as an open circuit when off. Arrays of three switches called "tee-switches" are used when very high isolation between the input and output is required.

Multiplexers or "muxes" are arrays of SPST switches configured to act as a multiposition switch that connects one of four to sixteen input signals to a single output. Demultiplexers ("demuxes") have a single input and multiple outputs. Multiplexer ICs are available as single N-to-1 switches (the MAX4617 is an 8-to-1 mux) or as groups of N-to-1 switches (the MAX4618 is a dual 4-to-1 mux).

Crosspoint switch arrays are arranged so that any of four to sixteen signal inputs can be connected to any of four to sixteen output signal lines. The Analog Devices AD8108 is an 8-by-8 crosspoint switch with eight inputs and eight outputs. These arrays are used when it is necessary to switch multiple signal sources among multiple signal receivers. They are most commonly used in telecommunications.

All analog switches use FET technology as the switching element. To switch ac signals, most analog switches require both positive and negative voltage power supplies. An alternative is to use a single power supply voltage and ground, but bias all inputs and output at one-half the power supply voltage. This requires dc blocking capacitors in all signal paths, both input and output, and loading resistors may be required at the device outputs. The blocking capacitors can also introduce low-frequency roll-off.

The impedance of the switching element varies from a few ohms to more than 100 ohms. Check the switch data sheet to determine the limits for how much power and current the switches can handle. Switch arrays, because of the physical size of the array, can have significant coupling or *crosstalk* between signal paths. Use caution when using analog switches for high-frequency signals as coupling generally increases with frequency and may compromise the isolation required for high selectivity in receivers and other RF signal processing equipment.

3.8.5 Audio Output Amplifiers

While it is possible to use op amps as low power audio output drivers for headphones, they generally have output impedances that are too high for most audio transducers such as speakers and headphones. The LM380 series of audio driver ICs has been used in radio circuits for many years and a simple schematic for a speaker driver is shown in **Fig 3.91**.

The popularity of personal music players has resulted in the creation of many new and inexpensive audio driver ICs, such as the National Semiconductor LM4800- and LM4900-series. Drivers that operate from voltages as low as 1.5 V for battery-powered devices and up to 18 V for use in vehicles are now available.

When choosing an audio driver IC for communications audio, the most important parameters to evaluate are its power requirements and power output capabilities. An overloaded or underpowered driver will result in distortion. Driver ICs intended for music players have frequency responses well in excess

Analog Basics 3.57

of the 3000 Hz required for communications. This can lead to annoying and fatiguing hiss unless steps are taken to reduce the circuit's frequency response.

Audio power amplifiers should also be carefully decoupled from the power supply and the manufacturer may recommend specific circuit layouts to prevent oscillation or feedback. Check the device's data sheet for this information.

3.8.6 Temperature Sensors

Active temperature sensors use the temperature-dependent properties of semiconductor devices to create voltages that correspond to absolute temperature in degrees Fahrenheit (LM34) or degrees Celsius (LM35). These sensors (of which many others are available than the two examples given here) are available in small plastic packages, both leaded and surface-mount, that respond quickly to temperature changes. They are available with 1% and better accuracy, requiring only a source of voltage at very low current and ground. Complete application information is available in the manufacturer data sheets. Thermistors, a type of passive temperature sensor, are discussed in the **Electrical Fundamentals** chapter. Temperature sensors are used in radio mostly in cooling and thermal management systems.

3.8.7 Radio Subsystems

As a particular technology becomes popular, a wave of integrated circuitry is developed to service that technology and reduce its cost of production and service. A good example is the wireless telephony industry. IC manufacturers have developed a large number of devices targeting this industry; receivers, transmitters, couplers, mixers, attenuators, oscillators and so forth. In addition, other wireless technologies such as data transmission provide opportunities for manufacturers to create integrated circuits that implement radio-related functions at low cost. Space limitations prohibit a comprehensive listing of all analog special function ICs but a sampling of those that are more useful in the radio field is presented.

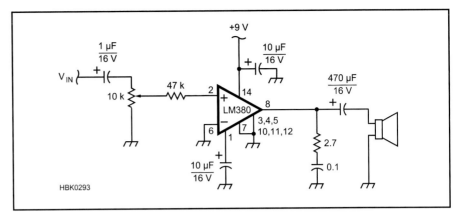

Fig 3.91 — Speaker driver. The LM380-series of audio output drivers are well-suited for low-power audio outputs, such as for headphones and small speakers. When using IC audio output drivers, be sure to refer to the manufacturer's data sheet for layout and power supply guidelines.

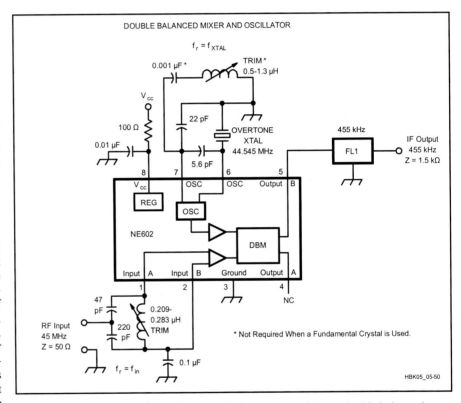

Fig 3.92 — NE/SA602 receiver subsystem. This device contains a doubly balanced mixer, a local oscillator, buffers and a voltage regulator. This application uses the NE/SA602 to convert an RF signal in a receiver to IF.

BALANCED MIXERS AND MODULATORS

The *balanced mixer* and *balanced modulator* are devices with many applications in modern radio transceivers (see the **Mixers, Modulators and Demodulators** chapter for specific circuits). Audio signals can modulate a carrier or be demodulated from the carrier with a balanced modulator. RF signals can be down-converted to intermediate frequency (IF) or IF can be up-converted to RF with a balanced mixer.

A passive balanced mixer, such as the Mini-Circuits Labs SBL-series, is made with a bridge of four matched Schottky diodes and the necessary transformers packaged in a small metal, plastic or ceramic container. Careful matching of the diodes results in good isolation between the local oscillator (LO) and the input and output signal ports.

Active balanced modulator ICs such as the NE/SA602 shown in **Fig 3.92** and the MC1496 use a "Gilbert cell" of transistors to multiply the input signals together and provide gain. These devices can perform modulation and demodulation, with many of the operating parameters controllable by the circuit designer. The SA602 includes an oscillator circuit to generate the LO, as well.

MONOLITHIC MICROWAVE INTEGRATED CIRCUIT (MMIC)

The *monolithic microwave integrated circuit* (*MMIC*) is a special type of IC that takes advantage of the small features that can be created on an IC to operate at higher frequencies

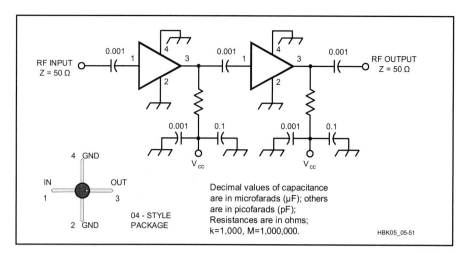

Fig 3.93 — Two MMICs in a typical circuit. Both amplifiers have both input and output impedance of 50 Ω. Low-cost MMICs have gain over a wide bandwidth, well into the GHz range. Consult the manufacturer's data sheets to find the right device for your application and appropriate component values and case styles.

and over wider bandwidths than can circuits made of discrete components. There is no formal definition of when an IC amplifier becomes an MMIC and, as the performance of IC devices improves the distinction is becoming blurred. MMIC devices typically have a fixed gain and require few external components with all biasing and power circuitry included in the IC. MMIC devices are available with operating bandwidths up to 10 GHz.

MMICs are excellent building blocks for low-power receiving and transmitting functions such as mixers, amplifiers and switches. MMICs usually have input and output impedances of 50 Ω, so they are easy to use with other equipment and circuits, not requiring any matching networks. MMICs are also quite small, typically available in a "pill" style package a few millimeters in diameter or in surface-mount packages. **Fig 3.93** shows how a pair of MMICs can be used to create high gain at UHF and above.

A good example of an MMIC is Mini-Circuit Labs' GVA-84+ amplifier. It operates from dc to 6 GHz. At low frequencies, its gain is approximately 24 dB, dropping to 12.5 dB at 6 GHz. It can develop output power of 50 to 100 mW (17 to 20 dBm) with a +5 V power supply. The noise figure is a respectable 5.5-6 dB. It has only three connections: input, ground and a combined dc power/RF output connection. The cost is about $2 in small quantities. The **Component Data and References** chapter includes a table of MMIC part numbers and data.

RECEIVER SUBSYSTEMS

High performance ICs have been designed that make up complete receivers with the addition of only a few external components. Two examples that are very similar are the Motorola MC3363 and the Philips NE627. Both of these chips have all the active RF stages necessary for a double conversion FM receiver. The MC3363 has an internal local oscillator (LO) with varactor diodes that can generate frequencies up to 200 MHz, although the rest of the circuit is capable of operating at frequencies up to 450 MHz with an external oscillator. The RF amplifier has a low noise factor and gives this chip a sensitivity of 0.3 μV. The intermediate frequency stages contain limiter amplifiers and quadrature detection. The necessary circuitry to implement receiver squelch and zero crossing detection of FSK modulation is also present. The circuit also contains received signal strength ("S-meter") circuitry (RSSI). The input and output of each stage are also brought out of the chip for versatility. The audio signal out of this chip must be appropriately amplified to drive a low-impedance speaker. This chip can be driven with a dc power source from 2 to 7 V and it draws only 3 mA with a 2 V supply.

The Philips SA627 is a newer chip than the MC3363 and has better performance characteristics even though it has essentially the same architecture. Its LO can generate frequencies up to 150 MHz and external oscillator frequencies up to 1 GHz can be used. The chip has a 4.6 dB noise figure and 0.22 μV sensitivity. The circuit can be powered with a dc voltage between 4.5 and 8 V and it draws between 5.1 and 6.7 mA. This chip is also ESD hardened so it resists damage from electrostatic discharges, such as from nearby lightning strikes.

The various stages in the receiver subsystem ICs are made available by connections on the package. There are two reasons that this is done. Filtering that is added between stages can be performed more effectively with inductors and crystal or ceramic filters, which are difficult to fabricate in integrated circuitry, so the output of one stage can be filtered externally before being fed to the next stage. It also adds to the versatility of the device. Filter frequencies can be customized for different intermediate frequencies. Stages can be used individually as well, so these devices can be made to perform direct conversion or single conversion reception or other forms of demodulation instead of FM.

TRANSMITTER SUBSYSTEMS

Single chips are available to implement FM transmitters. One implementation is the Motorola MC2831A. This chip contains a microphone preamplifier with limiting, a tone generator for CTCSS or AFSK and a frequency modulator. It has an internal voltage controlled oscillator that can be controlled with a crystal or an LC circuit. This chip also contains circuitry to check the power supply voltage and produce a warning if it falls too low. Together with an FM receiver IC, an entire transceiver can be fabricated with very few parts.

INTEGRATED POWER AMPLIFIER MODULES

A number of integrated power amplifier modules are available with power outputs up to 100 W (for a few watts of drive) and at frequency ranges from 20 MHz through 10 GHz. These modules are generally optimized for a specific type of service, such as wireless telephony base stations or mobile radios, and can be difficult to adapt to amateur bands and modes. However, some amateurs have successfully converted modules designed for wireless telephony between 700 and 1100 MHz to service in the amateur 1.2 GHz band.

If the module is a general-purpose design, manufacturer data is probably available to guide the design of a transmitter using the module. Many modules are only labeled with assembly numbers for specific products and data may be very hard to come by. In any case, it is recommended that the various amateur user's groups and online communities be enlisted for help in using these devices. Often, an amateur employed in the industry can look up data for surplus and "house label" devices and even give some guidance as to their application.

3.9 Analog Glossary

AC ground — A circuit connection point that presents a very low impedance to ac signals.

Accuracy — The ability of an analog-to-digital conversion to assign the correct code to an analog value or create the true analog value from a specific code.

Active — A device that requires power to operate.

Active region — The region in the characteristic curve of an analog device in which it is capable of processing the signal linearly.

Amplification — The process by which amplitude of a signal is increased. Gain is the amount by which the signal is amplified.

Analog signal — A signal that can have any amplitude (voltage or current) value and exists at any point in time.

Analog-to-digital converter (ADC) — Circuit (usually an IC) that generates a digital representation of an analog signal.

Anode — The element of an analog device that accepts electrons or toward which electrons flow.

Attenuation — The process of reducing the amplitude of a signal.

Avalanche breakdown — Current flow through a semiconductor device in response to an applied voltage beyond the device's ability to control or block current flow.

Base — The terminal of a bipolar transistor in which control current flows.

Beta (β) — The dc current gain of a bipolar transistor, also designated h_{FE}.

Biasing — The addition of a dc voltage or current to a signal at the input of an analog device, changing or controlling the position of the device's operating point on the characteristic curve.

Bipolar transistor — An analog device made by sandwiching a layer of doped semiconductor between two layers of the opposite type: PNP or NPN.

Black box — Circuit or equipment that is analyzed only with regards to its external behavior.

Bode plot — Graphs showing amplitude response in dB and phase response in degrees versus frequency on a logarithmic scale.

Buffer — An analog stage that prevents loading of one analog stage by another.

Carrier — (1) Free electrons and holes in semiconductor material. (2) An unmodulated component of a modulated signal.

Cascade — Placing one analog stage after another to combine their effects on the signal.

Cathode — The element of an analog device that emits electrons or from which electrons are emitted or repelled.

Characteristic curve — A plot of the relative responses of two or three analog-device parameters, usually of an output with respect to an input. (Also called *I-V* or *V-I curve*.)

Class — For analog amplifiers (Class A, B, AB, C), a categorization of the fraction of the input signal cycle during which the amplifying device is active. For digital or switching amplifiers (Class D and above), a categorization of the method by which the signal is amplified.

Clipping — A nonlinearity in amplification in which the signal's amplitude can no longer be increased, usually resulting in distortion of the waveform. (Also called *clamping* or *limiting*.)

Closed-loop gain — Amplifier gain with an external feedback circuit connected.

Collector — The terminal of a bipolar transistor from which electrons are removed.

Code — One possible digital value representing an analog quantity. (Also called *quantization code*.)

Common — A terminal shared by more than one port of a circuit or network.

Common mode — Signals that appear equally on all terminals of a signal port.

Comparator — A circuit, usually an amplifier, whose output indicates the relative amplitude of two input signals.

Compensation — The process of counteracting the effects of signals that are inadvertently fed back from the output to the input of an analog system. Compensation increases stability and prevents oscillation.

Compression — Reducing the dynamic range of a signal in order to increase the average power of the signal or prevent excessive signal levels.

Conversion efficiency — The amount of light energy converted to electrical energy by a photoelectric device, expressed in percent.

Conversion rate — The amount of time in which an analog-digital conversion can take place. *Conversion speed* is the reciprocal of conversion rate.

Coupling (ac or dc) — The type of connection between two circuits. DC coupling allows dc current to flow through the connection. AC coupling blocks dc current while allowing ac current to flow.

Cutoff frequency — Frequency at which a circuit's amplitude response is reduced to one-half its mid-band value (also called *half-power* or *corner* frequency).

Cutoff (region) — The region in the characteristic curve of an analog device in which there is no current through the device. Also called the OFF region.

Degeneration (emitter or source) — Negative feedback from the voltage drop across an emitter or source resistor in order to stabilize a circuit's bias and operating point.

Depletion mode — An FET with a channel that conducts current with zero gate-to-source voltage and whose conductivity is progressively reduced as reverse bias is applied.

Depletion region — The narrow region at a PN junction in which majority carriers have been removed. (Also called *space-charge* or *transition* region.)

Digital-to-analog converter (DAC) — Circuit (usually an IC) that creates an analog signal from a digital representation.

Diode — A two-element semiconductor with a cathode and an anode that conducts current in only one direction.

Drain — The connection at one end of a field-effect-transistor channel from which electrons are removed.

Dynamic range — The range of signal levels over which a circuit operates properly. Usually refers to the range over which signals are processed linearly.

Emitter — The terminal of a bipolar transistor into which electrons are injected.

Enhancement mode — An FET with a channel that does not conduct with zero gate-to-source voltage and whose conductivity is progressively increased as forward bias is applied.

Feedback — Routing a portion of an output signal back to the input of a circuit. Positive feedback causes the input signal to be reinforced. Negative feedback results in partial cancellation of the input signal.

Field-effect transistor (FET) — An analog device with a semiconductor channel whose width can be modified by an electric field. (Also called *unipolar transistor*.)

Forward bias — Voltage applied across a PN junction in the direction to cause current flow.

Forward voltage — The voltage required to cause forward current to flow through a PN junction.

Free electron — An electron in a semiconductor crystal lattice that is not bound to any atom.

Frequency response — A description of a circuit's gain (or other behavior) with frequency.

Gain — see *Amplification*.

Gain-bandwidth product — The relationship between amplification and frequency that defines the limits of the ability of a device to act as a linear amplifier. In many amplifiers, gain times bandwidth is approximately constant.

Gate — The control electrode of a field-effect transistor.

High-side — A switch or controlling device connecting between a power source and load.

Hole — A positively charged carrier that results when an electron is removed from an atom in a semiconductor crystal structure.

Hysteresis — In a comparator circuit, the practice of using positive feedback to shift the input setpoint in such a way as to minimize output changes when the input signal(s) are near the setpoint.

Integrated circuit (IC) — A semiconductor device in which many components, such as diodes, bipolar transistors, field-effect transistors, resistors and capacitors are fabricated to make an entire circuit.

Isolation — Eliminating or reducing electrical contact between one portion of a circuit and another or between pieces of equipment.

Junction FET (JFET) — A field-effect transistor whose gate electrode forms a PN junction with the channel.

Linearity — Processing and combining of analog signals independently of amplitude.

Load line — A line drawn through a family of characteristic curves that shows the operating points of an analog device for a given load or circuit component values.

Loading — The condition that occurs when the output behavior of a circuit is affected by the connection of another circuit to that output.

Low-side — A switch or controlling device connected between a load and ground.

Metal-oxide semiconductor (MOSFET) — A field-effect transistor whose gate is insulated from the channel by an oxide layer. (Also called *insulated gate FET* or *IGFET*)

Multivibrator — A circuit that oscillates between two states.

NMOS — N-channel MOSFET.

N-type impurity — A doping atom with an excess of valence electrons that is added to semiconductor material to act as a source of free electrons.

Network — General name for any type of circuit.

Noise — Any unwanted signal, usually random in nature.

Noise figure (NF) — A measure of the noise added to a signal by an analog processing stage, given in dB. (Also called *noise factor*.)

Open-loop gain — Gain of an amplifier with no feedback connection.

Operating point — Values of a set of circuit parameters that specify a device's operation at a particular time.

Operational amplifier (op amp) — An integrated circuit amplifier with high open-loop gain, high input impedance, and low output impedance.

Optoisolator — A device in which current in a light-emitting diode controls the operation of a phototransistor without a direct electrical connection between them.

Oscillator — A circuit whose output varies continuously and repeatedly, usually at a single frequency.

P-type impurity — A doping atom with a shortage of valence electrons that is added to semiconductor material to create an excess of holes.

Passive — A device that does not require power to operate.

Peak inverse voltage (PIV) — The highest voltage that can be tolerated by a reverse biased PN junction before current is conducted. (See also *avalanche breakdown*.)

Photoconductivity — Phenomenon in which light affects the conductivity of semiconductor material.

Photoelectricity — Phenomenon in which light causes current to flow in semiconductor material.

PMOS — P-channel MOSFET.

PN junction — The structure that forms when P-type semiconductor material is placed in contact with N-type semiconductor material.

Pole — Frequency at which a circuit's transfer function becomes infinite.

Port — A pair of terminals through which a signal is applied to or output from a circuit.

Quiescent (Q-) point — Circuit or device's operating point with no input signal applied. (Also called *bias point*.)

Pinch-off — The condition in an FET in which the channel conductivity has been reduced to zero.

Products — Signals produced as the result of a signal processing function.

Rail — Power supply voltage(s) for a circuit.

Range — The total span of analog values that can be processed by an analog-to-digital conversion.

Recombination — The process by which free electrons and holes are combined to produce current flow across a PN junction.

Recovery time — The amount of time required for carriers to be removed from a PN junction device's depletion region, halting current flow.

Rectify — Convert ac to pulsating dc.

Resolution — Smallest change in an analog value that can be represented in a conversion between analog and digital quantities. (Also called *step size*.)

Reverse bias — Voltage applied across a PN junction in the direction that does not cause current flow.

Reverse breakdown — The condition in which reverse bias across a PN junction exceeds the ability of the depletion region to block current flow. (See also **avalanche breakdown.**)

Roll-off — Change in a circuit's amplitude response per octave or decade of frequency.

Safe operating area (SOA) — The region of a device's characteristic curve in which it can operate without damage.

Sample — A code that represents the value of an analog quantity at a specific time.

Saturation (Region) — The region in the characteristic curve of an analog device in which the output signal can no longer be increased by the input signal. See **Clamping**.

Schottky barrier — A metal-to-semiconductor junction at which a depletion region is formed, similarly to a PN junction.

Semiconductor — (1) An element such as silicon with bulk conductivity between that of an insulator and a metal. (2) An electronic device whose function is created by a structure of chemically-modified semiconductor materials.

Signal-to-noise ratio (SNR) — The ratio of the strength of the desired signal to that of the unwanted signal (noise), usually expressed in dB.

Slew rate — The maximum rate at which a device can change the amplitude of its output.

Small-signal — Conditions under which the variations in circuit parameters due to the input signal are small compared to the quiescent operating point and the device is operating in its active region.

Source — The connection at one end of the channel of a field-effect transistor into which electrons are injected.

Stage — One of a series of sequential signal processing circuits or devices.

Substrate — Base layer of material on which the structure of a semiconductor device is constructed.

Superposition — Process in which two or more signals are added together linearly.

Total harmonic distortion (THD) — A measure of how much noise and distortion are introduced by a signal processing function.

Thermal runaway — The condition in which increasing device temperature increases device current in a positive feedback cycle.

Transconductance — Ratio of output current to input voltage, with units of Siemens (S).

Transfer characteristics — A set of parameters that describe how a circuit or network behaves at and between its signal interfaces.

Transfer function — A mathematical expression of how a circuit modifies an input signal.

Unipolar transistor — see ***Field-effect transistor (FET)***.

Virtual ground — Point in a circuit maintained at ground potential by the circuit without it actually being connected to ground.

Zener diode — A heavily-doped PN-junction diode with a controlled reverse breakdown voltage, used as a voltage reference or regulator.

Zero — Frequency at which a circuit's transfer function becomes zero.

3.10 References and Bibliography

REFERENCES

1. Ebers, J., and Moll, J., "Large-Signal Behavior of Junction Transistors," *Proceedings of the IRE*, 42, Dec 1954, pp 1761-1772.
2. Getreu, I., *Modeling the Bipolar Transistor* (Elsevier, New York, 1979). Also available from Tektronix, Inc, Beaverton, Oregon, in paperback form. Must be ordered as Part Number 062-2841-00.

FURTHER READING

Alexander and Sadiku, *Fundamentals of Electric Circuits* (McGraw-Hill)

Hayward, W., *Introduction to Radio Frequency Design* (ARRL, 2004)

Hayward, Campbell and Larkin, *Experimental Methods in RF Design* (ARRL, 2009)

Millman and Grabel, *Microelectronics: Digital and Analog Circuits and Systems* (McGraw-Hill, 1988)

Mims, F., *Timer, Op Amp & Optoelectronic Circuits & Projects* (Master Publishing, 2004)

Jung, W. *IC Op Amp Cookbook* (Prentice-Hall, 1986)

Hill and Horowitz, *The Art of Electronics* (Cambridge University Press, 1989)

Analog-Digital Conversion Handbook (by the staff of Analog Devices, published by Prentice-Hall)

Safe-Operating Area for Power Semiconductors (ON Semi), **www.onsemi.com/pub_link/Collateral/AN875-D.PDF**

"Selecting the Right CMOS Analog Switch", Maxim Semiconductor, **www.maxim-ic.com/appnotes.cfm/an_pk/638**

"Understanding Mixers — Terms Defined, and Measuring Performance," Mini-Circuits Labs, **www.minicircuits.com/pages/pdfs/an00009.pdf**

"AN189 — Balanced modulator/demodulator applications using the MC1496/1596" NXP Semiconductor, **www.nxp.com/acrobat_download/applicationnotes/AN189.pdf**

Hyperphysics Op-Amp Circuit Tutorials, **hyperphysics.phy-astr.gsu.edu/Hbase/Electronic/opampvar.html#c2**

Contents

4.1 Digital vs Analog

4.2 Number Systems

    4.2.1 Binary

    4.2.2 Hexadecimal

    4.2.3 Binary Coded Decimal (BCD)

    4.2.4 Conversion Techniques

4.3 Physical Representation of Binary States

    4.3.1 State Levels

    4.3.2 Transition Time

    4.3.3 Propagation Delay

4.4 Combinational Logic

    4.4.1 Boolean Algebra and the Basic Logical Operators

    4.4.2 Common Gates

    4.4.3 Additional Gates

    4.4.4 Boolean Theorems

4.5 Sequential Logic

    4.5.1 Synchronicity and Control Signals

    4.5.2 Flip-Flops

    4.5.3 Groups of Flip-Flops

    4.5.4 Multivibrators

4.6 Digital Integrated Circuits

    4.6.1 Comparing Logic Families

    4.6.2 Bipolar Logic Families

    4.6.3 Metal Oxide Semiconductor (MOS) Logic Families

    4.6.4 Interfacing Logic Families

    4.6.5 Real-World Interfacing

4.7 Microcontrollers

    4.7.1 An Overview of Microcontrollers

    4.7.2 Selecting a Microcontroller

    4.7.3 The Development Process

    4.7.4 Learning More About Microcontrollers

4.8 Personal Computer Interfacing

    4.8.1 Parallel vs Serial Signaling

    4.8.2 Data Rate

    4.8.3 Error Detection

    4.8.4 Standard Interface Buses

4.9 Glossary of Digital Electronics Terms

4.10 References and Bibliography

Chapter 4

Digital Basics

Radio amateurs have been involved with digital technology since the first spark transmitters, a form of pulse-coded transmission, were connected to an "aerial." Modern digital technology use by radio amateurs probably arrived first in CW keyers, where hams learned about flip-flops and gates to replace their semi-automatic mechanical "bug" keys.

Amateur use of digital technology echoed public use of these new abilities, starting with using the first home computers for calculations and later digital communications terminals. Today's Amateur Radio digital applications range from simple shack accessories like keyers and timers, to computer based digital modes such as PSK31 and Hellschreiber, to computer networking using TCP/IP over AX.25, computer controlled rigs, digital signal processing and software defined radios.

This chapter was written by Dale Botkin, N0XAS, building on material in previous editions by Christine Montgomery, KG0GN and Paul Danzer, N1II. It presents digital theory fundamentals and some applications of that theory in Amateur Radio. The fundamentals introduce digital mathematics, including number systems, logic devices and simple digital circuits. Next, the implementation of these simple circuits is explored in integrated circuits, their families and interfacing. Finally, some Amateur Radio applications are discussed involving digital logic, embedded microcontrollers and interfacing to personal computers.

4.1 Digital vs Analog

An analog signal can represent an infinitely variable indication of voltage, current, frequency, the position of a dial, or some other condition or value. As an example, using a potentiometer as a volume control will give you infinitely variable control over the volume of a signal. In theory, there is no limit to the difference in volume that can be produced. Though the control may be marked from 1 to 10, the actual value would have to be represented by a real number somewhere between 0 and 10. There are an infinite number of settings in between.

In its simplest form, a digital signal simply indicates the *on* or *off* state of some value or input signal. For example, the straight key you may use to key your CW transmitter (or the PTT switch of your voice transmitter) produces an on or off binary signal. In one state the transmitter produces an output signal of some sort; in the other state it does not. Another example is a simple light switch. The light is either on, or it's off. We represent these two states using 0 for off and 1 for on.

Digital electronics gets more interesting when we combine several or many simple on/off digital states to perform more complex tasks. For example, a relatively simple digital circuit can connect the antenna to either the transmitter or the receiver depending on a PTT or other keying signal. It can turn a preamp on or off depending on the state of the transmitter, mute the speaker while transmitting, and even select an antenna based on the selected frequency band. No special digital integrated circuits (chips) are needed to do any of these tasks; we can simply use bipolar transistors or MOSFETs, driven to saturation, as on/off switches. Simple circuits like this can often even be implemented with relays or diodes. The important fact is that the system is digital. There is no "almost transmitting" or "PTT switch partially pressed" state — it's either on, or it's off.

A very useful aspect of digital electronics is our ability to construct simple circuits that can maintain their on/off state indefinitely, until some event causes them to change. These *flip-flop* circuits can be used in various combinations to form *registers* that store information for later use or *counters* that count events and can be read or reset when needed. All of these circuits can be combined in ever larger groups until we finally arrive at the modern microprocessor. A microprocessor can accept input signals from many sources, follow a stored program to perform complex data storage and mathematical calculations, and produce output that we can use to do things that would be far more difficult with analog circuits.

So let's revisit our volume control example from the earlier paragraph. Let's assume we have a volume control, but it is used as an input to a digital system that will produce output at the desired level. This is

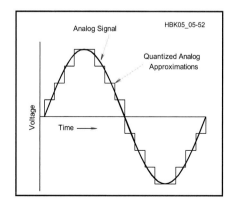

Fig 4.1 — An analog signal and its analog approximation. Note that the analog waveform has continuously varying voltage while the approximated waveform is composed of discrete steps.

quite common in modern equipment, whether it is amateur or consumer gear. Since the control is now digital, we know we can't have an infinite number of values. However, a simple on/off volume control would not be very useful. Using digital electronics, we can break the range between "off" and "fully on" into as many discrete steps as we need. With enough steps, we can give the user of the equipment an approximation of the original analog control while keeping the actual control digital.

By using coding, as discussed in the following pages, the two binary values (off and on, or 0 and 1) can represent any number of real values. **Fig 4.1** illustrates the contrast of an analog signal (in this case a sine wave) and its digital approximation. Four positive and four negative values are shown as an approximation to the sine wave, but any number of coded value steps can be used as an approximation. The more values are used to approximate the wave, the closer you can come to the actual wave form.

While the focus in this chapter will be on digital theory, many circuits and systems involve *both* digital and analog components. Often, a designer may choose between using digital technology, analog technology or a combination.

4.2 Number Systems

If you have been around computer hobbyists, some of whom are also hams, you may have seen a T-shirt or bumper sticker that reads, "There are 10 kinds of people in the world: those who understand binary, and those who don't." If this has puzzled you in the past, after reading this chapter you will be able to laugh with the rest of us.

In order to understand digital electronics, you must first understand the binary numbering system. Any number system has two distinct characteristics: a set of *symbols* (digits or numerals) and a *base* or radix. A *number* is a collection of these digits, where the left-most digit is the *most significant digit (MSD)* and the right-most digit is the *least significant digit (LSD)*. The value of this number is a weighted sum of its digits. The *weights* are determined by the system's base and the digit's position relative to the decimal point.

While these definitions may seem strange with all the technical terms, they will be more familiar when seen in a decimal system example. See **Table 4.1**. This is the "traditional" number system with which we are all familiar. In the *base-10* or decimal numbering system we use every day, the digits used are 0 through 9. The weights are powers of ten: 10^0 or 1 for the right-most column, 10^1 or 10 for the next column, 10^2 or 100 for the next and so on. Thus the number 548 represents five hundreds, four tens and eight ones. In this case, 5 is the MSD, and 8 is the LSD. Once you understand this concept, it can be applied to numbering systems using bases other than 10 such as base-2, base-8, or even base-16.

4.2.1 Binary

Binary is a *base-2* number system and therefore limited to two symbols: {0, 1}. The weight factors are now powers of 2, like 2^0, 2^1 and 2^2. For example, the decimal number, 163 and its equivalent binary number, 10100011, are shown in **Table 4.2**.

The digits of a binary number are now *bits* (short for binary digit). The MSD is the *most significant bit (MSB)* and the LSD is the *least significant bit (LSB)*. Four bits make a *nibble* (which you will occasionally see spelled *nybble*) and two nibbles, or eight bits, make a *byte*. The length of a *word* is dependent upon the hardware; it generally can consist of two or four or more bytes, but occasionally will be some other number of bits. These groupings are useful when converting to hexadecimal notation, which is explained later. It is important to remember that while everyone agrees on the meaning of a bit, a nibble (regardless of spelling) and a byte, the meaning of *word* can vary.

Counting in binary follows the same pattern we would use for decimal or any other number system. Consider the three digit binary number XXX. First fill up the right-hand column.

Binary Number	Decimal Equivalent
0000	0
0001	1

The column has been filled, and much quicker than with decimal, since there are only two values instead of 10. But just as we would with a decimal number, we now reset the right-hand column to 0, increase the next column by 1, and continue.

Table 4.1
Decimal Numbers

Example: 548

Digit = 5; Weight = 10; Position = 2

```
548  =  5(10²)    +  4(10¹)   +  8(10⁰)
     =  5(100)    +  4(10)    +  8(1)
     =  500       +  40       +  8
     =  5            4           8
        MSD                      LSD
```

Table 4.2
Decimal and Binary Number Equivalents

```
163       =  128    + 0       + 32      + 0       + 0       +0       + 2+     1       decimal
          =  1(128) + 0(64)   + 1(32)   + 0(16)   + 0(8)    +0(4)    + 1(2)   +1(1)
          =  1(2⁷)  + 0(2⁶)   + 1(2⁵)   + 0(2⁴)   + 0(2³)   +0(2²)   + 1(2¹)  +1(2⁰)

10100011  =  1        0         1         0         0         0         1       1      binary
             MSB                                                                LSB
                     |_____|       |_____|
                                Nibble                              Nibble
             |_____|
                                        Byte = 8 digits
```

0010	2
0011	3

Now the first two columns are full, so reset both back to 0 and increase the next column by 1 and continue:

0100	4
0101	5
0110	6
0111	7
1000	8
...	
1111	15

And so on.

4.2.2 Hexadecimal

The *hexadecimal*, or hex, *base-16* number system is widely used in computer systems for its ease in conversion to and from binary numbers and the fact that it is somewhat more human-friendly than long strings of 1s and 0s. A base-16 number requires 16 symbols. Since our normal mathematical number, as set up in the decimal system, has only 10 digits (0 through 9), a set of additional new symbols is required. Hex uses both numbers and characters in its set of sixteen symbols: {0, 1, 2, 3, 4, 5,6, 7, 8, 9, A, B, C, D, E, F}. Here, the letters A to F have the decimal equivalents of 10 to 15 respectively: A=10, B=11, C=12, D=13, E=14 and F=15. Again, the weights are powers of the base, such as 16^0, 16^1 and 16^2.

The four-bit binary listing in the previous paragraph shows that the individual 16 hex digits can be represented by a four-bit binary number. Since a byte is equal to eight binary digits, two hex digits provide a byte — the equivalent of 8 binary digits. Conversion from binary to hex is therefore simplified. Take a binary number, divide it into groups of four binary digits starting from the right, and convert each of the four binary digits to an individual value.

Conversion from hex to binary is equally convenient; simply replace each hex digit with its four-bit binary equivalent. As an example, the decimal number 163 is shown in Table 4.2 as binary 10100011. Divide the binary number in groups of four, so 1010 is equivalent to decimal 10 or "A" hex, and 0011 is equivalent to decimal 3, thus decimal number 163 is equivalent to hex A3.

4.2.3 Binary Coded Decimal (BCD)

The binary number system representation is the most appropriate form for fast internal computations since there is a direct mathematical relationship for every bit in the number. To interface with a human user — who usually wants to see inputs and outputs in terms of decimal numbers — other codes are more useful. The *Binary Coded Decimal (BCD)* system is a simple method for converting binary values to and from decimal for inputs and outputs for user-oriented digital systems. Back in the days when the most common method of presenting output to a user was via seven-segment LED displays, BCD was widely used. Since we now mostly use powerful microprocessors that can easily present information in decimal form, BCD is not nearly as common as it once was. You may, however, run into BCD when using or repairing older digital gear. It is also used in some chips intended for use in digital voltmeters.

In the BCD system, each decimal digit is expressed as a corresponding 4-bit binary number. In other words, the decimal digits 0 to 9 are encoded as the bit strings 0000 to 1001. To make the number easier to read, a space is left between each 4-bit group. For example, the decimal number 163 is equivalent to the BCD number 0001 0110 0011, as shown in **Table 4.3**.

The important difference between BCD and the previous number systems is that, starting with decimal 10, BCD loses the standard mathematical relationship of a weighted sum. BCD is simply a cut-off hexadecimal. Instead of using the 4-bit code strings 1010 to 1111 for decimal 10 to 15, BCD uses 0001 0000 to 0001 0101. This is one of the reasons that we have moved away from BCD.

4.2.4 Conversion Techniques

An easy way to convert a number from decimal to another number system is to do repeated division, recording the remainders in a tower just to the right. The converted number, then, is the remainders, reading up the tower. This technique is illustrated in **Table 4.4** for hexadecimal and binary conversions of the decimal number 163.

For example, to convert decimal 163 to hex, repeated divisions by 16 are performed. The first division gives 163/16 = 10 remainder 3. The remainder 3 is written in a column to the right. The second division gives 10/16 = 0 remainder 10. Since 10 decimal = A hex, A is written in the remainder column to the right. This division gave a divisor of 0 so the process is complete. Reading up the remainders column, the result is A3. The most common mistake in this technique is to forget that the Most Significant Digit ends up at the bottom.

Another technique that should be briefly mentioned can be even easier: use a calculator with a binary and/or hex mode option. Many inexpensive and readily available calculators intended for scientific and programming use will convert between number systems quite easily. In addition, calculator programs are available for all types of personal computers regardless of the operating system used.

One warning for this technique: this chapter doesn't discuss negative binary numbers. If your calculator does not give you the answer you expected, it may have interpreted the number as negative. This would happen when the number's binary form has a 1 in its MSB, such as the highest (leftmost) bit for the binary mode's default size. To avoid learning about negative binary numbers the hard way, always use a leading 0 when you enter a number in binary or hex into your calculator.

Table 4.3
Binary Coded Decimal Number Conversion

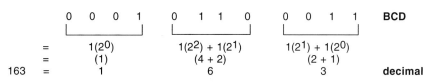

		0 0 0 1	0 1 1 0	0 0 1 1	BCD
	=	$1(2^0)$	$1(2^2) + 1(2^1)$	$1(2^1) + 1(2^0)$	
	=	(1)	(4 + 2)	(2 + 1)	
163	=	1	6	3	decimal

Table 4.4
Number System Conversions

Hex	Remainder		Binary	Remainder	
16	163		2	163		
	|10	3 LSB		|81	1 LSB
	|0	A MSB		|40	1
					20	0
					10	0
					5	0
					2	1
					1	0
				|0	1 MSB
A3 hex			1010 0011 binary		

4.3 Physical Representation of Binary States

4.3.1 State Levels

Most digital systems use the binary number system because many simple physical systems are most easily described by two state levels (0 and 1). For example, the two states may represent "on" and "off" or a "mark" and "space" in a communications transmission. In electronic systems, state levels are physically represented by voltages. A typical choice is
state 0 = 0 V
state 1 = 5 V

Since it is unrealistic to obtain these exact voltage values, a more practical choice is a range of values, such as
state 0 = 0.0 to 0.4 V
state 1 = 2.4 to 5.0 V

Fig 4.2 illustrates this representation of states by voltage levels. The undefined region between the two binary states is also known as the *transition region* or *noise margin*.

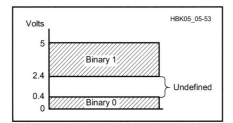

Fig 4.2 — Representation of binary states 1 and 0 by a selected range of voltage levels.

4.3.2 Transition Time

The gap in Fig 4.2, between binary 0 and binary 1, shows that a change in state does not occur instantly. There is a *transition time* between states. This transition time is a result of the time it takes to charge or discharge the stray capacitance in wires and other components because voltage cannot change instantaneously across a capacitor. (Stray inductance in the wires also has an effect because the current through an inductor can't change instantaneously.) The transition from a 0 to a 1 state is called the *rise time*, and is usually specified as the time for the pulse to rise from 10% of its final value to 90% of its final value. Similarly, the transition from a 1 to a 0 state is called the *fall time,* with a similar 10% to 90% definition. Note that these times need not be the same. **Fig 4.3A** shows an ideal signal, or *pulse*, with zero-time switching. Fig 4.3B shows a typical pulse, as it changes between states in a smooth curve.

Rise and fall times vary with the logic family used and the location in a circuit. Typical values of transition time are in the microsecond to nanosecond range. In a circuit, distributed inductances and capacitances in wires or PC-board traces may cause rise and fall times to increase as the pulse moves away from the source. One reason rise and fall times may be of interest to the radio designer is because of the possibility of generating RF noise in a digital circuit.

4.3.3 Propagation Delay

Rise and fall times only describe a relationship within a pulse. For a circuit, a pulse input into the circuit must propagate through the circuit; in other words it must pass through each component in the circuit until eventually it arrives at the circuit output. The time delay between providing an input to a circuit and seeing a response at the output is the *propagation delay* and is illustrated by **Fig 4.4**.

For modern switching logic, typical propagation delay values are in the 1 to 15 nanosecond range. (It is useful to remember that the propagation delay along a wire or printed-circuit-board trace is about 1.0 to 1.5 ns per inch.) Propagation delay is the result of cumulative transition times as well as transistor switching delays, reactive element charging times and the time for signals to travel through wires. In complex circuits, different propagation delays through different paths can cause problems when pulses must arrive somewhere at exactly the same time.

The effect of these delays on digital devices can be seen by looking at the speed of the digital pulses. Most digital devices and all PCs

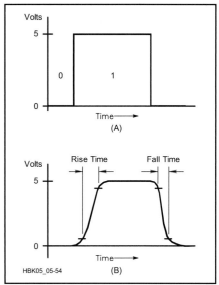

Fig 4.3 — (A) An ideal digital pulse and (B) a typical actual pulse, showing the gradual transition between states.

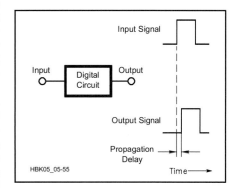

Fig 4.4 — Propagation delay in a digital circuit.

use *clock pulses*. If two pulses are supposed to arrive at a logic circuit at the same time, or very close to the same time, the path length for the two signals cannot be any different than two to three inches. This can be a very significant design problem for high-speed logic designs.

4.4 Combinational Logic

Having defined a way to use voltage levels to physically represent digital numbers, we can apply digital signal theory to design useful circuits. Digital circuits combine binary inputs to produce a desired binary output or combination of outputs. This simple combination of 0s and 1s can become very powerful, implementing everything from simple switches to powerful computers.

A digital circuit falls into one of two types: combinational logic or sequential logic. In a *combinational logic* circuit, the output depends only on the *present inputs* (if we ignore propagation delay). In contrast, in a *sequential logic* circuit, the output depends on the present inputs, the *previous sequence of inputs* and often a clock signal. Later sections of this chapter will examine some circuits built using the basics established here.

4.4.1 Boolean Algebra and the Basic Logical Operators

Combinational circuits are composed of logic gates, which perform binary operations. Logic gates manipulate binary numbers, so you need an understanding of the algebra of binary numbers to understand how logic gates operate. *Boolean algebra* is the mathematical system used to describe and design binary digital circuits. It is named after George Boole, the mathematician who developed the system. Standard algebra has a set of basic operations: addition, subtraction, multiplication and division. Similarly, Boolean algebra has a set of basic operations, called *logical operations*: NOT, AND and OR.

The function of these operators can be described by either a Boolean equation or a truth table. A Boolean *equation* describes an operator's function by representing the inputs and the operations performed on them. An equation is of the form "B = A," while an *expression* is of the form "A." In an assignment equation, the inputs and operations appear on the right and the result, or output, is assigned to the variable on the left.

A *truth table* describes an operator's function by listing all possible inputs and the corresponding outputs. Truth tables are sometimes written with Ts and Fs (for true and false) or with their respective equivalents, 1s and 0s. In company databooks (catalogs of logic devices a company manufactures), truth tables are usually written with Hs and Ls (for high and low). In the figures, 1 will mean high and 0 will mean low. This representation is called positive logic. The meaning of different logic types and why they are useful is discussed in a later section.

Each Boolean operator also has two circuit symbols associated with it. The traditional symbol — used by ARRL and other US publications — appears on top in each of the figures; for example, the triangle and bubble for the NOT function in Fig 4.7. In the traditional symbols, a small circle, or *bubble*, always represents "NOT." (This *bubble* is called a state indicator.)

Appearing just below the traditional symbol is the newer ANSI/IEEE Standard symbol. This symbol is always a square box with notations inside it. In these newer symbols, a small triangular flag represents "NOT." The new notation is an attempt to replace the detailed logic drawing of a complex function with a simpler block symbol. Adoption of the newer symbols has been spotty, and you are therefore still more likely to see the traditional symbols for basic logic functions than the ANSI/IEEE symbols.

4.4.2 Common Gates

Figs 4.5, **4.6** and **4.7** show the truth tables, Boolean algebra equations and circuit symbols for the three basic Boolean operations: AND, OR and NOT, respectively. All combinational logic functions, no matter how complex, can be described in terms of these three operators. Each truth table can be converted into words. The truth table for the two-input AND gate can be expressed as "the output

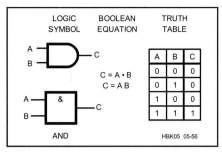

Fig 4.5 — Two-input AND gate.

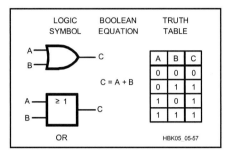

Fig 4.6 — Two-input OR gate.

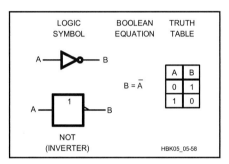

Fig 4.7 — Inverter.

C is a 1 only when the inputs are both 1s." This can be seen by examining the output column C — it remains at a 0 and becomes a 1 only when the input column A and the input column B are both 1s — the last line of the table.

The NOT operation is also called *inversion*, *negation* or *complement*. The circuit that implements this function is called an *inverter* or *inverting buffer*. The most common notation for NOT is a bar over a variable or expression. For example, NOT A is denoted $\overline{A}$. This is read as either "Not A" or "A bar." A less common notation is to denote Not A by A', which is read as "A prime." You will also see various other notations in schematic diagrams and component data sheets, such as a leading exclamation point or has symbol — !A or #A indicating "Not A."

While the inverting buffer and the noninverting buffer covered later have only one input and output, many combinational logic elements can have multiple inputs. When a combinational logic element has two or more inputs and one output, it is called a *gate*. (The term "gate" has a number of different but specific technical uses. For a clarification of the many definitions of gate, see the section on **Synchronicity and Control Signals**, later in this chapter.) For simplicity, the figures and truth tables for multiple-input elements will show the operations for only two inputs, the minimum number. Remember, though, that it is quite common to have gates with more than two inputs. A three-, four-, or eight-input gate works in the exact same manner as a two-input gate.

The output of an AND function is 1 only if *all* of the inputs are 1. Therefore, if *any* of the inputs are 0, then the output is 0. The notation for an AND is either a dot (•) between the inputs, as in C = A•B, or nothing between the inputs, as in C = AB. Read these equations as "C equals A AND B."

The OR gate detects if one or more inputs are 1. In other words, if *any* of the inputs are 1, then the output of the OR gate is 1. Since this includes the case where more than one input may be 1, the OR operation is also known as an INCLUSIVE OR. The OR operation detects if *at least one* input is 1. Only if all the inputs are 0, then the output is 0. The notation for an OR is a plus sign (+) between the inputs, as in C = A + B. Read this equation as "C equals A OR B."

4.4.3 Additional Gates

More complex logical functions are derived from combinations of the basic logical operators. These operations — NAND, NOR, XOR and the noninverter or buffer — are illustrated in **Figs 4.8** through **4.11**, respectively. As before, each is described by a truth table, Boolean algebra equation and circuit symbols. Also as before, except for the noninverter, each could have more inputs than the two illustrated.

The NAND gate (short for NOT AND) is equivalent to an AND gate followed by a NOT gate. Thus, its output is the complement of the AND output: The output is a 0 only if all the inputs are 1. If any of the inputs is 0, then the output is a 1.

The NOR gate (short for NOT OR) is equivalent to an OR gate followed by a NOT gate. Thus, its output is the complement of the OR output: If any of the inputs are 1, then the output is a 0. Only if all the inputs are 0, then the output is a 1.

The operations so far enable a designer to determine two general cases: (1) if *all* inputs have a desired state or (2) if *at least one* input has a desired state. The XOR and XNOR gates enable a designer to determine if *one and only one* input of a desired state is present.

The XOR gate (read as EXCLUSIVE OR) is a combination of an OR and a NAND gate. It has an output of 1 if one and only one of the inputs is a 1 state. The output is 0 otherwise. The symbol for XOR is ⊕. This is easy to

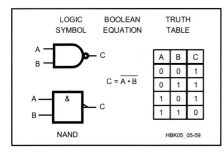

Fig 4.8 — Two-input NAND gate.

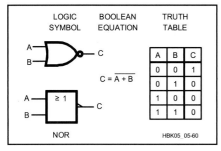

Fig 4.9 — Two-input NOR gate.

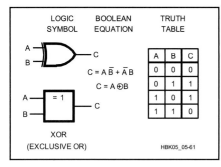

Fig 4.10 — Two-input XOR gate.

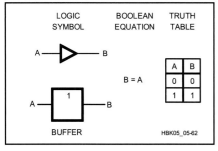

Fig 4.11 — Noninverting buffer.

remember if you think of the "+" OR symbol enclosed in an "O" for *only one*.

The XOR gate is also known as a "half adder," because in binary arithmetic it does everything but the "carry" operation. The following examples show the possible binary additions for a two-input XOR.

A	0	0	1	1
B	0	1	0	1
Sum	0	1	1	0

The XNOR gate (read as EXCLUSIVE NOR) is the complement of the XOR gate. The output is 0 if one and only one of the inputs is a 1. The output is 1 either if all inputs are 0 or more than one input is 1.

NONINVERTERS (BUFFERS)

A *noninverter*, also known as a *buffer*, *amplifier* or *driver*, at first glance does not seem to do anything. It simply receives an input and produces the same output. In reality, it is changing other properties of the signal in a useful fashion, such as amplifying the current level. While not useful for logical operations, applications of a noninverter include providing sufficient current to drive a number of gates or some other circuit such as a relay; interfacing between two logic families; obtaining a desired pulse rise time; and providing a slight delay to make pulses arrive at the proper time.

TRI-STATE GATES

Under normal circumstances, a logic element can drive or feed several other logic elements. A typical AND gate might be able to drive or feed 10 other gates. This is known as *fan-out*. However, with certain exceptions only one gate output can be connected to a single wire. If you have two possible driving sources to feed one particular wire, some logic network that probably includes a number OR gates must be used.

In many applications, including computers, data is routed internally on a set of wires called *buses*. The data on the bus can come from many circuits or drivers, and many other devices may be *listening* on the bus. To eliminate the need for the network of OR gates to drive each bus wire, a set of gates known as *tri-state* gates are used.

The symbol and truth table for a tri-state gate are shown in **Fig 4.12**. A tri-state gate can be any of the common gates previously described, but with one additional control lead. When this lead is enabled (it can be designed to allow either a 0 or a 1 to enable it) the gate operates normally, according to the truth table for that type of gate. However, when the gate is not enabled, the output goes to a high impedance, and so far as the output wire is concerned, the gate does not exist.

Each device that has to send data down a bus wire is connected to the bus wire through a tri-state gate. However, as long as only one device, through its tri-state gate, is enabled, it is as though all the other connected tri-state gates do not exist.

4.4.4 Boolean Theorems

The analysis of a circuit starts with a logic diagram and then derives a circuit description. In digital circuits, this description is in the form of a truth table or logical equation. The

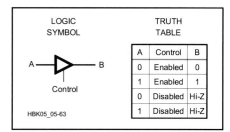

Fig 4.12 — Tri-State gate.

synthesis, or design, of a circuit goes in the reverse: starting with an informal description, determining an equation or truth table and then expanding the truth table to components that will implement the desired response. In both of these processes, we need to either simplify or expand a complex logical equation.

To manipulate an equation, we use mathematical *theorems*. Theorems are statements that have been proven to be true. The theorems of Boolean algebra are very similar to those of standard algebra, such as commutivity and associativity. Proofs of the Boolean algebra theorems can be found in an introductory digital design textbook.

BASIC THEOREMS

Table 4.5 lists the theorems for a single variable and **Table 4.6** lists the theorems for two or more variables. These tables illustrate the *principle of duality* exhibited by the Boolean theorems: Each theorem has a dual in which, after swapping all ANDs with ORs and all 1s with 0s, the statement is still true.

The tables also illustrate the *precedence* of the Boolean operations: the order in which operations are performed when not specified by parenthesis. From highest to lowest, the precedence is NOT, AND then OR. For example, the distributive law includes the expression "A + B•C." This is equivalent to "A

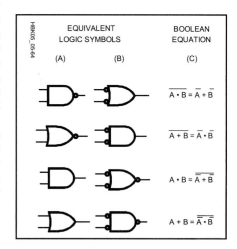

Fig 4.13 — Equivalent gates from DeMorgan's Theorem: Each gate in column A is equivalent to the opposite gate in column B. The Boolean equations in column C formally state the equivalences.

+ (B•C)." The parenthesis around (B•C) can be left out since an AND operation has higher priority than an OR operation. Precedence for Boolean algebra is similar to the convention of standard algebra: raising to a power, then multiplication, then addition.

DeMORGAN'S THEOREM

One of the most useful theorems in Boolean algebra is DeMorgan's Theorem:

$$\overline{A \cdot B} = \overline{A} + \overline{B}$$

and its dual

$$\overline{A + B} = \overline{A} \cdot \overline{B}.$$

Table 4.5
Boolean Algebra Single Variable Theorems

Identities:	$A \cdot 1 = A$	$A + 0 = A$
Null elements:	$A \cdot 0 = 0$	$A + 1 = 1$
Idempotence:	$A \cdot A = A$	$A + A = A$
Complements:	$A \cdot \overline{A} = 0$	$A + \overline{A} = 1$
Involution:	$\overline{(\overline{A})} = A$	

The truth table in **Table 4.7** proves these statements. DeMorgan's Theorem provides a way to simplify the complement of a large expression. It also enables a designer to interchange a number of equivalent gates, as shown by **Fig 4.13**.

The equivalent gates show that the duality principle works with symbols the same as it does for Boolean equations: just swap ANDs with ORs and switch the bubbles. For example, the NAND gate — an AND gate followed by an inverter bubble — becomes an OR gate preceded by two inverter bubbles. DeMorgan's Theorem is important because it means any logical function can be implemented using either inverters and AND gates or inverters and OR gates. Also, the ability to change placement of the bubbles using DeMorgan's Theorem is useful in dealing with mixed logic, to be discussed next.

POSITIVE AND NEGATIVE LOGIC

The truth tables shown in the figures in this chapter are drawn for positive logic. In *positive logic*, or *high true*, a higher voltage means true (logic 1) while a lower voltage means false (logic 0). This is also referred to as *active high*: a signal performs a named action or denotes a condition when it is "high" or 1. In *negative logic*, or *low true*, a lower voltage means true (1) and a higher voltage means false (0). An *active low* signal performs an action or denotes a condition when it is "low" or 0.

In both logic types, true = 1 and false = 0; but whether true means high or low differs. Company databooks are drawn for general truth tables: an H for high and an L for low. (Some tables also have an X for a "don't care" state.) The function of the table can differ depending on whether it is interpreted for positive logic or negative logic.

Device data sheets often show positive logic convention, or positive logic is assumed. However, a signal into an IC is represented with a bar above it, indicating that the "enable" on that wire is active low — it does *not* mean negative logic (0 V = a logical 1) is used! Similarly a bubble on the input of a logic element also usually means active low. These can be sources of confusion.

Fig 4.14 shows how a general truth table differs when interpreted for different logic types. The same truth table gives two equivalent gates: positive logic gives the function of a NAND gate while negative logic gives the function of a NOR gate.

Note that these gates correspond to the

Table 4.6
Boolean Algebra Multivariable Theorems

Commutativity:	$A \cdot B = B \cdot A$
	$A + B = B + A$
Associativity:	$(A \cdot B) \cdot C = A \cdot (B \cdot C)$
	$(A + B) + C = A + (B + C)$
Distributivity:	$(A + B) \cdot (A + C) = A + B \cdot C$
	$A \cdot B + A \cdot C = A \cdot (B + C)$
Covering:	$A \cdot (A + B) = A$
	$A + A \cdot B = A$
Combining:	$(A + B) \cdot (A + \overline{B}) = A$
	$A \cdot B + A \cdot \overline{B} = A$
Consensus:	$A \cdot B + \overline{A} \cdot C + B \cdot C = A \cdot B + \overline{A} \cdot C$
	$(A + B) \cdot (\overline{A} + C) \cdot (B + C) = (A + B) \cdot (\overline{A} + C)$
	$A + \overline{A}B = A + B$

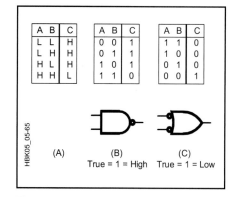

Fig 4.14 — (A) A general truth table, (B) a truth table and NAND symbol for positive logic and (C) a truth table and NOR symbol for negative logic.

Table 4.7
DeMorgan's Theorem

(A) $\overline{A \cdot B} = \overline{A} + \overline{B}$
(B) $\overline{A + B} = \overline{A} \cdot \overline{B}$
(C)

(1)	(2)	(3)	(4)	(5)	(6)	(7)	(8)	(9)	(10)
A	B	$\overline{A}$	$\overline{B}$	$A \cdot B$	$\overline{A \cdot B}$	$A + B$	$\overline{A + B}$	$\overline{A} \cdot \overline{B}$	$\overline{A} + \overline{B}$
0	0	1	1	0	1	0	1	1	1
0	1	1	0	0	1	1	0	0	1
1	0	0	1	0	1	1	0	0	1
1	1	0	0	1	0	1	0	0	0

(A) and (B) are statements of DeMorgan's Theorem. The truth table at (C) is proof of these statements: (A) is proven by the equivalence of columns 6 and 10 and (B) by columns 8 and 9.

equivalent gates from DeMorgan's Theorem. A bubble on an input or output terminal indicates an active low device. The absence of bubbles indicates an active high device.

Like the bubbles, signal names can be used to indicate logic states. These names can aid the understanding of a circuit by indicating control of an action (GO, /ENABLE) or detection of a condition (READY, /ERROR). The action or condition occurs when the signal is in its active state. When a signal is in its active state, it is called *asserted*; a signal not in its active state is called *negated* or *deasserted*.

A prefix can easily indicate a signal's active state. Active low signals are preceded by a symbol such as /, |, ! or # (for example /READY or !READY). Active low signals are also denoted by an overscore, such as $\overline{CL}$. Active high signals have no prefix or overscore. As an example, see the truth table for a flip-flop later in this chapter. Standard practice is that the signal name and input pin match (have the same active level). For example, an input with a bubble (active low) may be called /READY, while an input with no bubble (active high) is called READY. Output signal names should always match the device output pin.

In this chapter, positive logic is used unless indicated otherwise. Although using mixed logic can be confusing, it does have some advantages. Mixed logic combined with DeMorgan's Theorem can promote more effective use of available gates. Also, well-chosen signal names and placement of bubbles can promote more understandable logic diagrams.

4.5 Sequential Logic

The previous section discussed combinational logic, whose outputs depend only on the present inputs. In contrast, in *sequential logic* circuits, the new output depends not only on the present inputs but also on the present outputs. The present outputs depended on the previous inputs and outputs and those earlier outputs depended on even earlier inputs and outputs and so on. Thus, the present outputs depend on the previous *sequence of inputs* and the system has *memory*. Having the outputs become part of the new inputs is known as *feedback*.

4.5.1 Synchronicity and Control Signals

When a combinational circuit is given a set of inputs, the outputs take on the expected values after a propagation delay during which the inputs travel through the circuit to the output. In a sequential circuit, however, the travel through the circuit is more complicated. After application of the first inputs and one propagation delay, the outputs take on the resulting state; but then the outputs start trickling back through and, after a second propagation delay, new outputs appear. The same happens after a third propagation delay. With propagation delays in the nanosecond range, this cycle around the circuit is rapidly and continually generating new outputs. A user needs to know when the outputs are valid.

There are two types of sequential circuits: synchronous circuits and asynchronous circuits, which are analyzed differently for valid outputs. In *asynchronous* operation, the outputs respond to the inputs immediately after the propagation delay. To work properly, this type of circuit must eventually reach a *stable* state: the inputs and the fed back outputs result in the new outputs staying the same. When the nonfeedback inputs are changed, the feedback cycle needs to eventually reach a new stable state. Generally, the output of this type of logic is not valid until the last input has changed, and enough time has elapsed for all propagation delays to have occurred.

In *synchronous* operation, the outputs change state only at specific times. These times are determined by the presence of a particular input signal: a clock, toggle, latch or enable. Synchronicity is important because it ensures proper timing: all the inputs are present where needed when the control signal causes a change of state.

CONTROL SIGNALS

Some authors vary the meanings slightly for the different control signals. The following is a brief illustration of common uses, as well as showing uses for noun, verb and adjective. *Enabling* a circuit generally means the control signal goes to its asserted level, allowing the circuit to change state. *Latch* implies memory: a latch circuit can store a bit of information. A latch signal can cause a circuit to keep its present state indefinitely. *Gate* can have several meanings, some unrelated to synchronous control. For example, a gate can be a signal used to trigger the passage of other signals through a circuit. A gate can also be a logic circuit with two or more inputs and one output, as used earlier in this chapter. Of course, "gate" can also be one of the electrodes of an FET as described in another chapter. To *toggle* means a signal changes state, from 1 to 0 or vice versa. A *clock* signal is one that toggles at a regular rate.

Clock control is the most common method of synchronizing logic circuits, so it has some additional terms as illustrated by **Fig 4.15**. The *clock period* is the time between successive transitions in the same direction; the *clock frequency* is the reciprocal of the period. A *pulse* or *clock tick* is the first edge in a clock period, or sometimes the period itself or the first half of the period. The *duty cycle* is the percentage of time that the clock signal is at its asserted level. A common application of the use of clock pulses is to limit the input to a logic circuit such that the circuit is only enabled on one clock phase; that is the inputs occur before the clock changes to a logic 1.

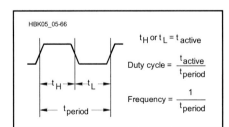

Fig 4.15 — Clock signal terms. The duty cycle would be t_H / t_{PERIOD} for an active high signal and t_L / t_{PERIOD} for an active low signal.

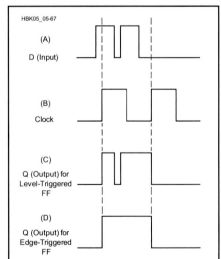

Fig 4.16 — Level-triggered vs edge-triggered for a D flip-flop: (A) input D, (B) clock input, (C) output Q for level-triggered: circuit responds whenever clock is 1. (D) output Q for edge-triggered: circuit responds only at rising edge of clock. Notice that the short negative pulse on the input D is not reproduced by the edge-triggered flip-flop.

The outputs are sampled only after this point; perhaps when the clock next changes back to a logic 0.

The reaction of a synchronous circuit to its control signal is *static* or *dynamic*. Static, *gated* or *level-triggered* control allows the circuit to change state whenever the control signal is at its active or asserted level. Dynamic, or *edge-triggered*, control allows the circuit to change state only when the control signal *changes* from unasserted to asserted. By convention, a control signal is active high if state changes occur when the signal is high or at the rising edge and active low in the opposite case. Thus, for positive logic, the convention is enable = 1 or enable goes from 0 to 1. This transition from 0 to 1 is called *positive edge-triggered* and is indicated by a small triangle inside the circuit box. A circuit responding to the opposite transition, from 1 to 0, is called *negative edge-triggered*, indicated by a bubble with the triangle.

Whether a circuit is level-triggered or edge-triggered can affect its output, as shown by **Fig 4.16**. Input D includes a very brief pulse, called a *glitch*, which may be caused by noise. The differing results at the output illustrate how noise can cause errors. We have both edge and level triggered circuits available so that we can meet the requirements of our particular design.

4.5.2 Flip-Flops

Flip-flops are the basic building blocks of sequential circuits. A *flip-flop* is a device with two stable states: the *set* state (1) and the *reset* or *cleared* state (0). The flip-flop can be placed in one or the other of the two states by applying the appropriate input. Since a common use of flip-flops is to store one bit of information, some use the term *latch* interchangeably with flip-flop. A set of latches, or flip-flops holding an n-bit number is called a *register*. While gates have special symbols, the schematic symbol for most sequential logic components is a rectangular box with the circuit name or abbreviation, the signal names and assertion bubbles. For flip-flops, the circuit name is usually omitted since the signal names are enough to indicate a flip-flop and its type. The four basic types of flip-flops are the S-R, D, T and J-K. The most common flip-flops available to Amateurs today are the J-K and D- flip-flops; the others can be synthesized if needed by utilizing these two varieties.

TRIGGERING A FLIP-FLOP

Although the S-R (Set-reset) flip flop is no longer generally available or used, it does provide insight in basic flip-flop operations and triggering. It is also not uncommon to build S-R flip-flops out of gates for jobs such as switch contact debouncing. In **Fig 4.17** the

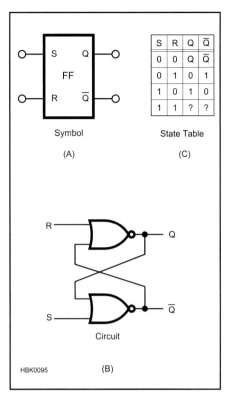

Fig 4.17 — Unclocked S-R Flip-Flop. (A) schematic symbol. (B) circuit diagram. (C) state table or truth table.

symbol for an S-R flip flop and its truth table are accompanied by a logic implementation, using NAND gates. As the truth table shows, this basic implementation requires a positive or logic 1 input on the set input to put the flip-flop in the Q or set state. Remove the input, and the flip flop stays in the Q state, which is what is expected of a flip-flop. Not until the S input receives a logic 1 input does the flip- flop change state and go to the reset or Q=0 state.

Note that the input can be a short pulse or a level; as long as it is there for some minimum duration (established by the propagation delay of the gates used), the flip-flop will respond. By contrast the clocked S-R flip-flop in **Fig 4.18** requires both a positive level to be present at either the S or R inputs and a positive clock pulse. The clock pulse is ANDed with the S or R input to trigger the flip–flop. In this case the flip-flop shown is implemented with a set of NOR gates.

A final triggering method is edge triggering. Here, instead of using the clock pulse as shown in the timing diagram of Fig 4.18, just the edge of the clock pulse is used. The edge-triggered flip-flop helps solves a problem with noise. Edge-triggering minimizes the time during which a circuit responds to its inputs: the chance of a glitch occurring during

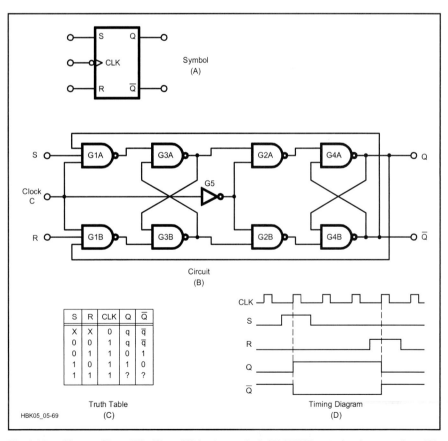

Fig 4.18 — Master-Slave Flip-Flop. (A) logic symbol. (B) NAND gate implementation. (C) truth table. (D) timing diagram.

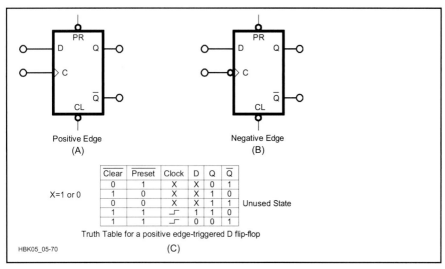

Fig 4.19 — (A & B) The D flip-flop. (C) A truth table for the positive edge-triggered D flip-flop.

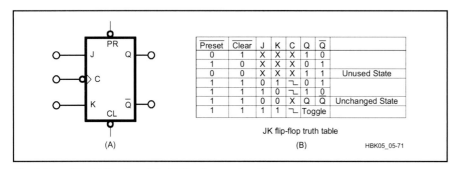

Fig 4.20 — (A) JK flip-flop. (B) JK flip-flop truth table.

the nanosecond transition of a clock pulse is remote. A side benefit of edge-triggering is that only one new output is produced per clock period. Edge-triggering is denoted by a small rising-edge or falling-edge symbol in the clock column of the flip-flop's truth table. It can also appear, instead of the clock triangle, inside the schematic symbol.

MASTER/SLAVE FLIP-FLOP

One major problem with the simple flip-flop shown up to now is the question of when is there a valid output. Suppose a flip-flop receives input that causes it to change state; at the same time the output of this flip-flop is being sampled to control some other logic element. There is a real risk here that the output will be sampled just as it is changing and thus the validity of the output is questionable.

A solution to this problem is a circuit that samples and stores its inputs before changing its outputs. Such a circuit is built by placing two flip-flops in series; both flip-flops are triggered by a common clock but an inverter on the second flip-flop's clock input causes it to be asserted only when the first flip-flop is not asserted. The action for a given clock pulse is as follows: The first, or master, flip-flop can change only when the clock is high, sampling and storing the inputs. The second, or slave, flip-flop gets its input from the master and changes when the clock is low. Hence, when the clock is 1, the input is sampled; then when the clock becomes 0, the output is generated. Note that a bubble may appear on the schematic symbol's clock input, reminding us that the output appears when the clock is asserted low. This is conventional for TTL-style J-K flip-flops, but it can be different for CMOS devices.

The master/slave method isolates output changes from input changes, eliminating the problem of series-fed circuits. It also ensures only one new output per clock period, since the slave flip-flop responds to only the single sampled input. A problem can still occur, however, because the master flip-flop can change more than once while it is asserted; thus, there is the potential for the master to sample at the wrong time. There is also the potential that either flip-flop can be affected by noise.

A master-slave, S-R clocked input flip-flop synthesized from NAND gates, Fig 4.18B, is accompanied by its logic symbol, Fig 4.18A. From the logic symbols you can tell that the output changes on a negative-going clock edge.

G3A and G3B form the master set-rest flip-flop, and G4A and G4B the slave flip flop. The input signals S and R are controlled by the positive going edge of the clock through gates G1A and G1B. G2A and G2B control the inputs into the slave flip-flop; these inputs are the outputs of the master flip-flop. Note G5 inverts the clock; thus while the positive-going edge places new data into the master flip-flop, the other edge of the clock transfers the output of the master into the slave on the following negative clock edge.

D FLIP-FLOP

In a D (data) flip-flop, the *data* input is transferred to the outputs when the flip-flop is enabled. The logic level at input D is transferred to Q when the clock is positive; the Q output retains this logic level until the next positive clock pulse (see **Fig 4.19**). The truth table summarizes this operation. If D = 1 the next clock pulse makes Q = 1. If D = 0, the next clock pulse makes Q = 0. A D flip-flop is useful to store one bit of information. A collection of D flip-flops forms a register.

J-K FLIP-FLOP

The J-K flip-flop, shown schematically in **Fig 4.20A**, has five inputs. The unit shown uses both positive active inputs (the J and K inputs) and negative active inputs (note the bubbles on the C or clock, PR or preset and CL or clear inputs). With these inputs almost any other type of flip-flop may be synthesized.

The truth table of Fig 4.20B provides an explanation. Lines (rows) 1 and 2 show the preset and clear inputs and their use. These are active low, meaning that when one (and only one) of them goes to a logic 0, the flip-flop responds, just as if it was a S-R or set-reset flip-flop. Make PR a logic 0, and leave CL a logic 1, and the flip-flop goes into the Q = 1 state (line 1). Do the reverse (line 2) – PR = 1, CL = 0 and the flip-flop goes into a Q' = 1 state. When these two inputs are used, J, K and C are marked as X or don't care, because the PR and CL inputs override them. Line 3 corresponds to the unused state of the R-S flip-flop.

Line 5 shows that if J = 1 and K = 0, the next clock transition from high to low sets Q = 1 and Q' = 0. Alternately, line 4 shows J = 0 and K = 1 sets Q = 0 and Q' = 1. Therefore if a signal is applied to J, and the inverted signal sent to K, the J-K flip-flop will mimic a D flip-flop, echoing its input.

The most unique feature of the J-K flip-flop is line 7. If both J and K are connected to a 1, then each clock 1 to 0 transition will flip or toggle the flop-flop. Thus the J-K flip-flop can be used as a T flip-flop, as in a ripple counter (see the following **Counters** section.)

Table 4.8 summarizes D and J-K flip-flops.

Table 4.8

Summary of Standard Flip-Flops

q = current state   Q = next state   X = don't care

Flip-Flop Type	Symbol	Truth Table	Characteristic Equation	Excitation Table
D	(D flip-flop symbol)	D CLK Q X ⌐ q 0 ⌐ 0 1 ⌐ 1	$Q = D \cdot CLK$	q Q D CLK 0 0 X ⌐ 0 1 1 ⌐ 1 0 0 ⌐ 1 1 X ⌐
J K	(J-K positive edge symbol)	J K CLK Q X X ⌐ q 0 0 ⌐ q 0 1 ⌐ 0 1 0 ⌐ 1 1 1 ⌐ t	$Q = (J \cdot \bar{q} + \bar{K} \cdot q) \cdot CLK$ Positive Edge Clock	q Q J K CLK 0 0 0 X ⌐ 0 1 1 X ⌐ 1 0 X 1 ⌐ 1 1 X 0 ⌐
J K	(J-K negative edge symbol)	J K CLK Q 0 0 ⌐ 0 0 1 ⌐ 0 1 0 ⌐ 1 1 1 ⌐ t	$Q = (J \cdot \bar{q} + \bar{K} \cdot q) \cdot CLK$ Negative Edge Clock	q Q J K CLK 0 0 0 X ⌐ 0 1 1 X ⌐ 1 0 X 1 ⌐ 1 1 X 0 ⌐

t: If J = K, the clock toggles the flip-flop

4.5.3 Groups of Flip-Flops

COUNTERS

Groups of flip-flops can be combined to make counters. Intuitively, a counter is a circuit that starts at state 0 and sequences up through states 1, 2, 3, to m, where m is the maximum number of states available. From state m, the next state will return the counter to 0. This describes the most common counter: the *n-bit binary counter*, with n outputs corresponding to $2^n = m$ states. Such a counter can be made from n flip-flops, as shown in **Fig 4.21**. This figure shows implementations for each of the types of synchronicity. Both circuits pass the data count from stage to stage. In the asynchronous counter, Fig 4.21A, the clock is also passed from stage to stage and the circuit is called *ripple* or *ripple-carry*.

The J-K flip-flop truth table shows that with PR (Preset) and CL (Clear) both positive, and therefore not effecting the operation, the flip-flop will toggle if J and K are tied to a logic 1. In Fig 4.21A the first stage has its J and K inputs permanently tied to a logic 1, and each succeeding stage has its J and K inputs tied to Q of the proceeding stage. This provides a direct ripple counter implementation.

Design of a synchronous counter is bit more involved. It consists of determining, for a particular count, the conditions that will make the next stage change at the same clock edge when all the stages are changing.

To illustrate this, notice the binary counting table of Fig 4.21. The right-hand column represents the lowest stage of the counter. It alternates between 1 and 0 on every line. Thus, for the first stage the J and K inputs are tied to logic 1. This provides the alternation required by the counting table.

The middle column or second stage of the counter changes state right after the lower stage is a 1 (lines C, E and G). Thus if the Q output of the lowest stage is tied to the J and K inputs of the second stage, each time the output of the lowest stage is a 1 the second stage toggles on the next clock pulse.

Finally, the third column (third stage) toggles when both the first stage and the second stage are both 1s (line D). Thus by ANDing the Q outputs of the first two stages, and then connecting them to the J and K inputs of the third stage, the third stage will toggle whenever the first two stages are 1s.

There are formal methods for determining the wiring of synchronous counters. The illustration above is one manual method that may be used to design a counter of this type.

The advantage of the synchronous counter is that at any instant, except during clock pulse transition, all counter stage outputs *are correct* and delay due to propagation through the flip flops is not a problem.

In the synchronous counter, Fig 4.21B, each stage is controlled by a common clock signal.

There are numerous variations on this first example of a counter. Most counters have the ability to *clear* the count to 0. Some counters can also *preset* to a desired count. The clear and preset control inputs are often asynchronous — they change the output state without being clocked. Counters may either count up (increment) or down (decrement). *Up/down* counters can be controlled to count in either direction. Counters can have sequences other than the standard numbers, for example a BCD counter.

Counters are also not restricted to changing state on every clock cycle. An n-bit counter that changes state only after m clock pulses is called a *divider* or *divide-by-m* counter. There are still $2^n = m$ states; however, the output after p clock pulses is now p / m. Combining different divide-by-m counters can result in almost any desired count. For example, a base 12 counter can be made from a divide-by-2 and a divide-by-6 counter; a base 10 (decade) counter consists of a divide-by-2 and a BCD divide-by-5 counter.

The outputs of these counters are binary. To produce output in decimal form, the output of a counter would be provided to a binary-to-decimal decoder chip and/or an LED display.

REGISTERS

Groups of flip-flops can be combined to make registers, usually implemented with D flip-flops. A *register* stores n bits of information, delivering that information in response to a clock pulse. Registers usually have asynchronous *set* to 1 and *clear* to 0 capabilities.

Storage Register

A storage register simply stores temporary information, for example, incoming information or intermediate results. The size is related to the basic size of information handled by a computer: 8 flip-flops for an 8-bit or *byte register* or 16 bits for a *word register*. **Fig 4.22** shows a typical circuit and schematic symbols for an 8-bit storage register.

Shift Register

Shift registers also store information and provide it in response to a clock signal, but they handle their information differently: When a clock pulse occurs, instead of each flip-flop passing its result to the output, the flip-flops pass their data to each other, up and down the row. For example, in up mode, each flip-flop receives the output of the preceding flip-flop. A data bit starting in flip-flop D0 in a left

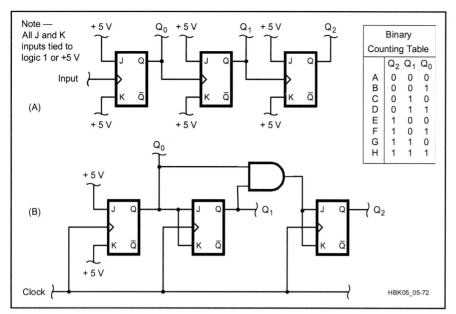

Fig 4.21 — Three-bit binary counter using J-K flip-flops: (A) asynchronous or ripple, (B) synchronous.

shifter would move to D1, then D2 and so on until it is shifted out of the register. If a 0 was input to the least significant bit, D0, on each clock pulse then, when the last data bit has been shifted out, the register contains all 0s.

Shift registers can be left shifters, right shifters or controlled to shift in either direction. The most general form, a *universal shift register*, has two control inputs for four states: Hold, Shift right, Shift left and Load. Most also have asynchronous inputs for preset, clear and parallel load. The primary use of shift registers is to convert parallel information to serial or vice versa. Additional uses for a shift register are to delay or synchronize data, and to multiply or divide a number by a factor 2^n. Data can be delayed simply by taking advantage of the Hold feature of the register control inputs. Multiplication and division with shift registers is best explained by example: Suppose a 4-bit shift register currently has the value 1000 = 8. A right shift results in the new parallel output 0100 = 4 = 8 / 2. A second right shift results in 0010 = 2 = (8 / 2) / 2. Together the 2 right shifts performed a division by 2^2. In general, shifting right n times is equivalent to dividing by 2^n. Similarly, shifting left multiplies by 2^n. This can be useful to compiler writers to make a computer program run faster.

4.5.4 Multivibrators

Multivibrators are a general type of circuit with three varieties: bistable, monostable and astable. The only truly digital multivibrator is bistable, having two stable states. The flip-flop is a *bistable multivibrator*: both of its two states are stable; it can be triggered from one stable state to the other by an external signal. The other two varieties of multivibrators are partly analog circuits and partly digital. While their output is one or more pulses, the internal operation is strictly analog.

MONOSTABLE MULTIVIBRATOR

A *monostable* or *one-shot* multivibrator has one energy-storing element in its feedback paths, resulting in one stable and one quasi-stable state. It can be switched, or *triggered*, to its quasi-stable state; then returns to the stable state after a time delay. Thus, when triggered, the one-shot multivibrator puts out a pulse of some duration, T.

A very common integrated circuit used for non-precision generation of a signal pulse is the 555 timer IC. **Fig 4.23** shows a 555 connected as a one-shot multivibrator. The one-shot is activated by a negative-going pulse between the trigger input and ground. The trigger pulse causes the output (Q) to go positive and capacitor C to charge through resistor R. When the voltage across C reaches two-thirds of V_{CC}, the capacitor is quickly discharged to ground and the output returns to 0. The output remains at logic 1 for a time determined by

$$T = 1.1\, RC$$

where:
R = resistance in ohms, and
C = capacitance in farads.

A very common, but again, non-precision application of this circuit is the generation of a delayed pulse. If there is a requirement

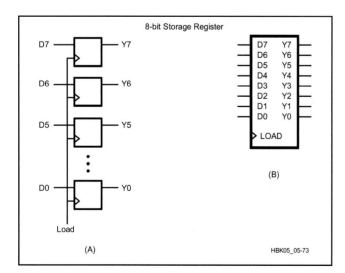

Fig 4.22 — An eight-bit storage register: (A) circuit and (B) schematic symbol.

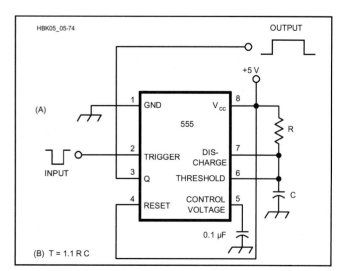

Fig 4.23 — (A) A 555 timer connected as a monostable multivibrator. (B) The equation to calculate values where T is the pulse duration in seconds, R is in ohms and C is in farads.

to generate a 50 μs pulse, but delayed from a trigger by 20 ms, two 555s might be used. The first 555, configured as an astable multivibrator, generates the 20-ms pulse, and the trailing edge of the 20-ms pulse is used to trigger a second 555 that in turn generates the 10 μs pulse. See the **Analog Basics** chapter for more information on the 555 timer and related circuits.

ASTABLE MULTIVIBRATOR

An *astable* or *free-running* multivibrator has two energy-storing elements in its feedback paths, resulting in two quasi-stable states. It continuously switches between these two states without external excitation. Thus, the astable multivibrator puts out a sequence of pulses. By properly selecting circuit components, these pulses can be of a desired frequency and width.

Fig 4.24 shows a 555 timer IC connected as an astable multivibrator. The capacitor C charges to two-thirds V_{CC} through R1 and R2 and discharges to one-third V_{CC} through R2. The ratio R1 : R2 sets the asserted high duty cycle of the pulse: t_{HIGH} / t_{PERIOD}. The output frequency is determined by:

$$f = \frac{1.46}{(R1 + 2\,R2)\,C}$$

where:
R1 and R2 are in ohms,
C is in farads and
f is in hertz.

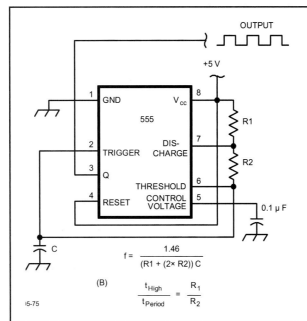

Fig 4.24 — (A) A 555 timer connected as an astable multivibrator. (B) The equations to calculate values for R1, R2 in ohms and C in farads, where f is the clock frequency in Hertz.

It may be difficult to produce a 50% duty cycle due to manufacturing tolerance for the resistors R1 and R2. One way to ensure a 50% duty cycle is to run the astable multivibrator at 2f and then divide by 2 with a toggle flip-flop.

Astable multivibrators, and the 555 integrated circuit in particular, are very often used to generate clock pulses. Although this is a very inexpensive and minimum hardware approach, the penalty is stability with temperature. Since the frequency and the pulse dimensions are set by resistors and capacitors, drift with temperature and to some extent aging of components will result in changes with time. This is no different than the problem faced by designers of L-C controlled VFOs.

4.6 Digital Integrated Circuits

Integrated circuits (ICs) are the cornerstone of digital logic devices. Modern technology has enabled electronics to become smaller and smaller in size and less and less expensive. Much of today's complex digital equipment would be impossible to build with discrete transistors and discrete components.

An IC is a miniature electronic module of components and conductors manufactured as a single unit. All you see is a ceramic or black plastic package and the silver-colored pins sticking out. Inside the package is a piece of material, usually silicon, created (fabricated) in such a way that it conducts an electric current to perform logic functions, such as a gate, flip-flop or decoder.

As each generation of ICs surpassed the previous one, they became classified according to the number of gates on a single chip. These classifications are roughly defined as:

Small-scale integration (SSI):
10 or fewer gates on a chip.

Medium-scale integration (MSI):
10-100 gates.

Large-scale integration (LSI):
100-1000 gates.

Very-large-scale integration (VLSI):
1000 or more gates.

Though SSI and MSI logic chips are still useful for building circuits to handle very simple tasks, it is more common to see them either used along with or completely replaced by programmable logic arrays and microcontrollers. In many cases you will see the smaller logic circuits referred to as "glue logic."

4.6.1 Comparing Logic Families

When selecting devices for a circuit, a designer is faced with choosing between many families and subfamilies of logic ICs. The determination of which logic subfamily is right for a specific application is based upon several desirable characteristics: logic speed, power consumption, fan-out, noise immunity and cost. From a practical viewpoint, the primary IC families available and in common use today are CMOS, with TTL a distant second place. Within these families, there are tradeoffs that can be made with respect to individual circuit capabilities, especially in the areas of speed and power consumption. Except under the most demanding circumstances, normal commercial grade tempera-

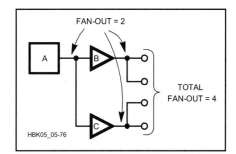

Fig 4.25 — Nonverting buffers used to increase fan-out: Gate A (fan-out = 2) is connected to two buffers, B and C, each with a fan-out of 2. Result is a total fan-out of 4.

ture rating will do for amateur service.

FAN-OUT

Fan-out is a term with which you will need to become familiar when working with TTL logic families such as 7400, 74LS or 74S. A gate output can supply only a limited amount of current. Therefore, a single output can only drive a limited number of inputs. The measure of driving ability is called fan-out, expressed as the number of inputs (of the same subfamily) that can be driven by a single output. If a logic family that is otherwise desirable does not have sufficient fan-out, consider using noninverting buffers to increase fan-out, as shown by **Fig 4.25**.

Another approach is to use a CMOS logic family. These families typically have output drivers capable of sourcing or sinking 20 to 25 mA, and input current leakage in the microampere range. Thus, fan-out is seldom a problem when using these devices.

NOISE IMMUNITY

The noise margin was illustrated in Fig 4.2. The choice of voltage levels for the binary states determines the noise margin. If the gap is too small, a spurious signal can too easily produce the wrong state. Too large a gap, however, produces longer, slower transitions and thus decreased switching speeds.

Circuit impedance also plays a part in noise immunity, particularly if the noise is from external sources such as radio transmitters. At low impedances, more energy is needed to change a given voltage level than at higher impedances.

4.6.2 Bipolar Logic Families

Two broad categories of digital logic ICs are *bipolar* and *metal-oxide semiconductor* (MOS). Numerous manufacturing techniques have been developed to fabricate each type. Each surviving, commercially available family has its particular advantages and disadvantages and has found its own special niche in the market. The designer is cautioned, however, that sometimes this niche is simply the ongoing maintenance of old products. There are still very old logic families available for reasonable prices that would be considered quite obsolete and generally not suitable for new designs.

Bipolar semiconductor ICs usually employ NPN junction transistors. (Bipolar ICs can be manufactured using PNP transistors, but NPN transistors make faster circuits.) While early bipolar logic was faster and had higher power consumption than MOS logic, the speed difference has largely disappeared as manufacturing technology has developed.

There are several families of bipolar logic devices, and within some of these families there are subfamilies. The most-used bipolar logic family is transistor-transistor logic (TTL). Another bipolar logic family, Emitter Coupled Logic (ECL), has exceptionally high speed but high power consumption.

TRANSISTOR-TRANSISTOR LOGIC (TTL)

The TTL family saw widespread acceptance through the 1960s, 1970s and 1980s because it was fast compared to early MOS and CMOS logic, and has good noise immunity. It was by far the most commonly used logic family for a couple of decades. Though TTL logic is not in widespread use today for new designs, the device numbering system devised for TTL chips survives to this day for newer technologies. You will also often see TTL, especially the later low power, higher speed TTL subfamilies, in various equipment you may use and repair.

TTL Subfamilies

The original standard TTL used bipolar transistors and "totem-pole" outputs (see **Fig 4.26A** and **B**), which were a great improvement over the earlier diode-transistor logic (DTL) and resistor-transistor logic (RTL). Still, TTL logic consumed quite a bit of power even at idle, and there were limits on how many inputs could be driven by a single output. Later versions used Schottky diodes to greatly improve switching speed, and reduced power requirements were introduced.

TTL IC identification numbers begin with either 54 or 74. The 54 prefix denotes an extended military temperature range of −55 to 125 °C, while 74 indicates a commercial temperature range of 0 to 70 °C. The next letters, in the middle of the TTL device number, indicate the TTL subfamily. Following the subfamily designation is a 2, 3 or 4-digit device-identification number. For example, a 7400 is a standard TTL NAND gate and a 74LS00 is a low-power Schottky NAND gate (The NAND gate is the workhorse TTL chip). A partial list of TTL subfamilies includes:

	74xx	standard TTL
H	74Hxx	High-speed
L	74Lxx	Low-power
S	74Sxx	Schottky
F	74Fxx	Fairchild Advanced Schottky
LS	74LSxx	Low-power Schottky
AS	74ASxx	Advanced Schottky
ALS	74ALSxx	Advanced Low-power Schottky

Each subfamily is a compromise between speed and power consumption. **Table 5.9** shows some of these characteristics. Because the speed-power product is approximately constant, less power consumption generally results in lower speed and vice versa. The advanced low power Schottky devices (ALS, F) offer both increased speed and reduced power consumption. Historically, an additional consideration to the speed-versus-power trade-off has been the cost trade-off. For the amateur, this is not nearly the factor it once was as component costs are relatively low for the newer, faster, lower powered parts.

When a TTL gate changes state, the amount of current that it draws changes rapidly. These

---

Table 4.9
TTL and CMOS Subfamily Performance Characteristics

TTL Family	Propagation Delay (ns)	Per Gate Power Consumption (mW)	Speed Power Product (pico-joules)
Standard	9	10	90
L	33	1	33
H	6	22	132
S	3	20	60
F	3	8.5	25.5
LS	9	2	18
AS	1.6	20	32
ALS	5	1.3	6.5

CMOS Family Operating with							
$4.5 < V_{CC} < 5.5$ V		f=100 kHz	f=1 MHz	f=10 MHz	f=100 kHz	f=1 MHz	f=10 MHz
HC	18	0.0625	0.6025	6.0025	1.1	10.8	108
HCT	18	0.0625	0.6025	6.0025	1.1	10.8	108
AC	5.25	0.080	0.755	7.505	0.4	3.9	39
ACT	4.75	0.080	0.755	7.505	0.4	3.6	36

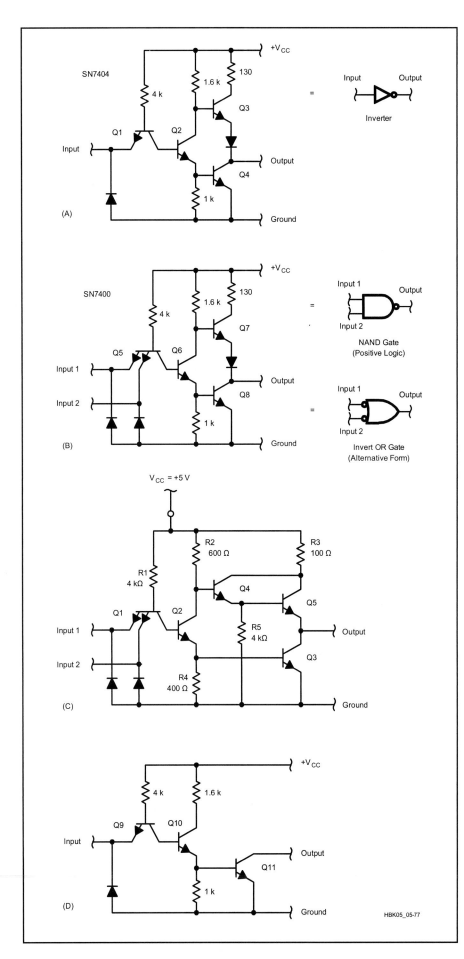

changes in current, called switching transients, appear on the power supply line and can cause false triggering of other devices. For this reason, the power bus should be adequately decoupled. For proper decoupling of TTL circuits, connect a 0.01 to 0.1 µF capacitor from V_{CC} to ground near each device to minimize the transient currents caused by device switching and magnetic coupling. These capacitors must be low-inductance, high-frequency RF capacitors (ceramic capacitors are preferred). In addition, a large-value (50 to 100 µF) capacitor should be connected from V_{CC} to ground somewhere on the board to accommodate the continually changing I_{CC} requirements of the total V_{CC} bus line. These are generally low-inductance tantalum capacitors.

Darlington and Open-Collector Outputs

Fig 4.26C and D show variations from the totem-pole configuration. They are the Darlington transistor pair and the open-collector configuration respectively.

The Darlington pair configuration replaces the single transistor Q4 with two transistors, Q4 and Q5. The effect is to provide more current-sourcing capability in the high state. This has two benefits: (1) the rise time is decreased and (2) the fan-out is increased.

Transistor(s) on the output in both the totem-pole and Darlington configurations provide active pull-up. Omitting the transistor(s) and providing an external resistor for passive pull-up gives the open-collector configuration. This configuration, unfortunately, results in slower rise time, since a relatively large external resistor must be used. The technique has some very useful applications, however: driving other devices, performing wired logic, busing and interfacing between logic devices.

Devices that need other than a 5-V supply can be driven with the open-collector output by substituting the device for the external resistor. Example devices include LEDs, relays and solenoids. Inductive devices like relay coils and solenoids need a protection diode across the coil. You must pay attention to the current ratings of open-collector outputs in such applications. You may need a switching transistor to drive some relays or other high-current loads.

Fig 4.26 — Example TTL circuits and their equivalent logic symbols: (A) an inverter and (B) a NAND gate, both with totem-pole outputs. (C) A NAND gate with a Darlington output. (D) A NAND gate with an open-collector output. (Indicated resistor values are typical. Identification of transistors is for text reference only. These are not discrete components but parts of the silicon die.)

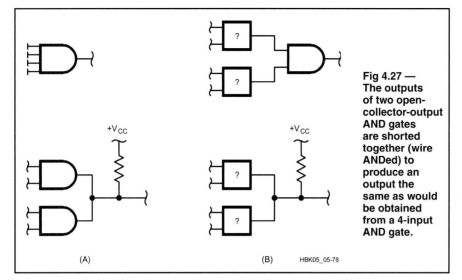

Fig 4.27 — The outputs of two open-collector-output AND gates are shorted together (wire ANDed) to produce an output the same as would be obtained from a 4-input AND gate.

Open-collector outputs can perform wired logic, rather than gated IC logic, by wire-ANDing the outputs. This can save the designer an AND gate, potentially simplifying the design. Wire-ANDed outputs are several open-collector outputs connected to a single external pull-up resistor. The overall output, then, will only be high when all pull-down transistors are OFF (all connected outputs are high), effectively performing an AND of the connected outputs. If any of the connected outputs are low, the output after the external resistor will be low. **Fig 4.27** illustrates the wire-ANDing of open-collector outputs.

The wire-ANDed concept can be applied to several devices sharing a common bus. At any time, all but one device has a high-impedance (off) output. The remaining device, enabled with control circuitry, drives the bus output.

Open-collector outputs are also useful for interfacing TTL gates to gates from other logic families. TTL outputs have a minimum high level of 2.4 V and a maximum low level of 0.4 V. When driving non-TTL circuits, a pull-up resistor (typically 2.2 kΩ) connected to the positive supply can raise the high level to 5 V. If a higher output voltage is needed, a pull-up resistor on an open-collector output can be connected to a positive supply greater than 5 V, so long as the chip output voltage and current maximums are not exceeded.

Three-State Outputs

While open-collector outputs can perform bus sharing, a more popular method is three-state output, or tristate, devices. The three states are low, high and high impedance, also called Hi-Z or *floating*. An output in the high-impedance state behaves as if it is disconnected from the circuit, except for possibly a small leakage current. Three-state devices have an additional disable input. When the enable input is active, the device provides high and low outputs just as it would normally; when enable is inactive, the device goes into its high-impedance state.

A bus is a common set of wires, usually used for data transfer. A three-state bus has several three-state outputs wired together. With control circuitry, all devices on the bus but one have outputs in the high-impedance state. The remaining device is enabled, driving the bus with high and low outputs. Care should be taken to ensure only one of the output devices can be enabled at any time, since simultaneously connected high and low outputs may result in an incorrect logic voltage. (The condition when more than one driver is enabled at the same time is called *bus contention*.) Also, the large current drain from V_{CC} to ground through the high driver to the low driver can potentially damage the circuit or produce noise pulses that can affect overall system behavior.

Unused TTL Inputs

A design may result in the need for an n-input gate when only an n + m input gate is available. In this case, the recommended solution for extraneous inputs is to give the extra inputs a constant value that won't affect the output. A low input is easily provided by connecting the input to ground. A high input can be provided with either an inverter whose input is ground or with a pull-up resistor. The pull-up resistor is preferred rather than a direct connection to power because the resistor limits the current, thus protecting the circuit from transient voltages. Usually, a 1-kΩ to 5-kΩ resistor is used; a single 1-kΩ resistor can handle up to 10 inputs.

It's important to properly handle all inputs. Design analysis would show that an unconnected, or floating TTL input is usually high but can easily be changed low by only a small amount of capacitively-coupled noise.

4.6.3 Metal Oxide Semiconductor (MOS) Logic Families

While bipolar devices use junction transistors, MOS devices use field effect transistors (FETs). MOS is characterized by simple device structure, small size (high density) and ease of fabrication. MOS circuits use the NOR gate as the workhorse chip rather than the NAND. MOS families, specifically CMOS, are used extensively in most digital devices today because of their low power consumption and high speed.

P-CHANNEL MOS (PMOS)

The first MOS devices to be fabricated were PMOS, conducting electrical current by the flow of positive charges (holes). PMOS power consumption is much lower than that of bipolar logic, but its operating speed is also lower. The only extensive use of PMOS was in calculators and watches, where low speed is acceptable and low power consumption and low cost are desirable. PMOS was replaced by NMOS, which offered substantially higher switching speeds.

N-CHANNEL MOS (NMOS)

With improved fabrication technology, NMOS became feasible and provided improved performance and TTL compatibility. The speed of NMOS is at least twice that of PMOS, since electrons rather than holes carry the current. NMOS also has greater gain than PMOS and supports greater packaging density through the use of smaller transistors. NMOS has been almost completely obsoleted by CMOS.

COMPLEMENTARY MOS (CMOS)

CMOS combines both P-channel and N-channel devices on the same substrate to achieve high noise immunity and low power consumption: less than 1 mW per gate and negligible power during standby. This accounts for the widespread use of CMOS in battery-operated equipment. The high impedance of CMOS gates makes them susceptible to electromagnetic interference, however, particularly if long traces are involved. Consider a trace ¼-wavelength long between input and output. The output is a low-impedance point, hence the trace is effectively grounded at this point. You can get high RF potentials ¼-wavelength away, which disturbs circuit operation.

A notable feature of CMOS devices is that the logic levels swing to within a few millivolts of the supply voltages. The input-switching threshold is approximately one half the supply voltage ($V_{DD} - V_{SS}$). This characteristic contributes to high noise immunity on the input signal or power supply lines. CMOS input-current drive requirements are minuscule, so the fan-out is great, at least in low-speed systems. For high-speed systems, the input capacitance increases the dynamic power dissipation and limits the fan-out.

CMOS Subfamilies

There are a large number of CMOS subfamilies available. Like TTL, the original CMOS has largely been replaced by later subfamilies using improved technologies; in turn, these will be replaced with even newer families as technology evolves. The original family, called the 4000 series, has numbers beginning with 40 or 45 followed by two or three numbers to indicate the specific device. 4000B is second generation CMOS. When introduced, this family offered low power consumption but was fairly slow and not easy to interface with TTL.

Later CMOS subfamilies offer improved performance and, in some cases, TTL compatibility. For simplicity, the later subfamilies were given numbers similar to the TTL numbering system, with the same leading numbers, 54 or 74, followed by 1 to 4 letters indicating the subfamily and as many as 5 numbers indicating the specific device. The subfamily letters usually include a "C" to distinguish them as CMOS.

Following is a description of some of the CMOS device families available. As there are a substantial number of families offered by specific manufacturer, and there are new families being introduced frequently, this information is by no means exhaustive. A check of IC suppliers' and manufacturers' Web sites will provide the designer with a complete selection of choices to meet his or her requirements.

4000	4071B	*Standard CMOS*
C	74Cxx	*CMOS versions of TTL*

Largely obsolete today, devices in the 74C subfamily are pin and functional equivalents of many of the most popular parts in the 7400 TTL family. It may be possible to replace all TTL ICs in a particular circuit with 74C-series CMOS, but this family should not be mixed with TTL in a circuit without careful design considerations. Devices in the C series are typically 50% faster than the 4000 series.

HC   74HCxx   High-speed CMOS

Devices in the 74HC subfamily have speed and drive capabilities similar to Low-power Schottky (LS) TTL but with better noise immunity and greatly reduced power consumption. High-speed refers to faster than the previous CMOS family, the 4000-series.

HCT   74HCTxx   High-Speed CMOS, TTL compatible

Devices in this subfamily were designed to interface TTL to CMOS systems. The HCT inputs recognize TTL levels, while the outputs are CMOS compatible. HCT chips are commonly used as lower powered, drop-in replacements for their pin compatible LS TTL counterparts.

AC   74ACxxxxx   Advanced CMOS

Devices in this family have reduced propagation delays, increased drive capabilities and can operate at higher speeds than standard CMOS. They are comparable to Advanced Low-power Schottky (ALS) TTL devices.

ACT   74ACTxxxxx   Advanced CMOS, TTL compatible

This subfamily combines the improved performance of the AC series with TTL-compatible inputs.

AHCT   74AHCTxxxx   Advanced High Speed CMOS, TTL Compatible

This subfamily is substantially faster than HC and HCT, but only slightly faster than ACT in switching speed. It has lower output drive capacity than AC/ACT.

New CMOS subfamilies are being introduced regularly. The current move is to lower voltage operation; where 5 V was the most common supply voltage until a few years ago, 3.3 V and lower supplies are becoming more common. Many newer microprocessors, microcontrollers and intelligent peripheral chips require 3.3 V logic.

LVC   74LVCxxxx   Low Voltage CMOS

This subfamily uses a 3.3 V supply rather than 5 V. It offers low propagation delay (under 5 ns typical), extremely low supply current, robust output drive capabilities, and 5-V TTL compatible inputs.

ALVC   74ALVCxxxx   Advanced Low Voltage CMOS

Even faster than LVC, this family offers propagation delays if under 3ns. Current demands are slightly higher, but still lower than 5V CMOS.

VCX   74VCXxxxx   Low voltage CMOS

This family operates at supply voltages of 1.8 or 3.6 V. Offering high switching speeds and low propagation delays, it does not have TTL-compatible buffered inputs.

As with TTL, each CMOS subfamily has characteristics that may make it suitable or unsuitable for a particular design. You should consult the manufacturer's data books for complete information on each subfamily being considered.

CMOS Circuits

A simplified diagram of a CMOS logic inverter is shown in **Fig 4.28**. When the input is low, the resistance of Q2 is low so a high current flows from V_{CC}. Since Q1's resistance is high, the high current flows to the output. When the input is high, the opposite occurs: Q2's resistance is low, Q1's is high and the output is low. The diodes are to protect the circuit against static charges.

Special Considerations

Some of the diodes in the input- and output-protection circuits are an inherent part of the manufacturing process. Even with the protection circuits, however, CMOS ICs are susceptible to damage from static charges. To protect against damage from static, the pins should not be inserted in Styrofoam as is sometimes done with other components. Instead, a spongy conductive foam, usually black or pink in color, is available for this purpose. Before removing a CMOS IC from its protective material, make certain that your body is grounded. Touching nearly any large metal object before handling the ICs is probably adequate to drain any static charge off your body. Some people prefer to touch a grounded metal object or to use a conductive bracelet connected to the ground terminal of a three-wire ac outlet through a 10-MΩ resistor. Since wall outlets aren't always wired properly, you should measure the voltage between the ground terminal and any metal objects you might touch. Connecting yourself to ground through a 1-MΩ to 10-MΩ resistor will limit any current that might flow through your body.

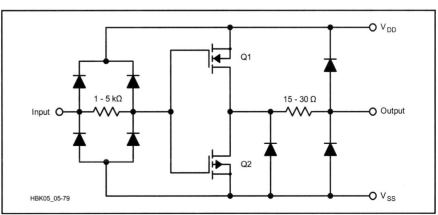

Fig 4.28 — Internal structure of a CMOS inverter.

All CMOS inputs should be tied to an input signal. A positive supply voltage or ground is suitable if a constant input is desired. Undetermined CMOS inputs, even on unused gates, may cause gate outputs to oscillate. Oscillating gates draw high current, and may overheat and self destruct.

The low power consumption of CMOS ICs made them attractive for satellite applications, but standard CMOS devices proved to be sensitive to low levels of radiation — cosmic rays, gamma rays and X rays. Later, radiation-hardened CMOS ICs, able to tolerate 10^6 rads, made them suitable for space applications. (A rad is a unit of measurement for absorbed doses of ionizing radiation, equivalent to 10^{-2} joules per kilogram.)

SUMMARY

There are many types of logic ICs, each with its own advantages and disadvantages. Regardless of the application, consult up-to-date product specification sheets and manufacturer literature when designing logic circuits. IC data sheets, application notes, databooks and more are available from IC manufacturers via their Web sites. By using a search engine and entering a few key word specifications, you will locate application notes, tutorials and a host of other information.

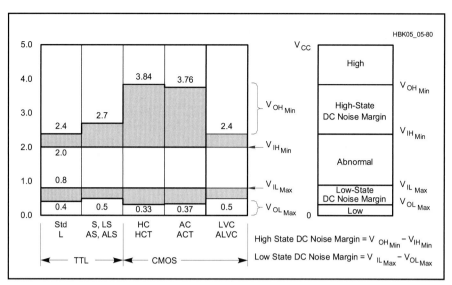

Fig 4.29 — Differences in logic levels for some TTL and CMOS families.

4.6.4 Interfacing Logic Families

Each semiconductor logic family has its own advantages in particular applications. When a design mixes ICs from different logic families, the designer must account for the differing voltage and current requirements each logic family recognizes. The designer must ensure the appropriate interface exists between the point at which one logic family ends and another begins. Knowledge of the specific input/output (I/O) characteristics of each device is necessary, and knowledge of the general internal structure is desirable to ensure reliable digital interfaces. Typical internal structures have been illustrated for each common logic family. **Fig 4.29** illustrates the logic level changes for different TTL and CMOS families. Data sheets should be consulted for manufacturer's specifications.

Often more than one conversion scheme is possible, depending on whether the designer wishes to optimize power consumption or speed. Usually one quality must be traded off for the other. The following section discusses some specific logic conversions. Where an electrical connection between two logic systems isn't possible, an optoisolator can sometimes be used.

TTL DRIVING CMOS

TTL and low-power TTL can drive 74C series CMOS directly over the commer-

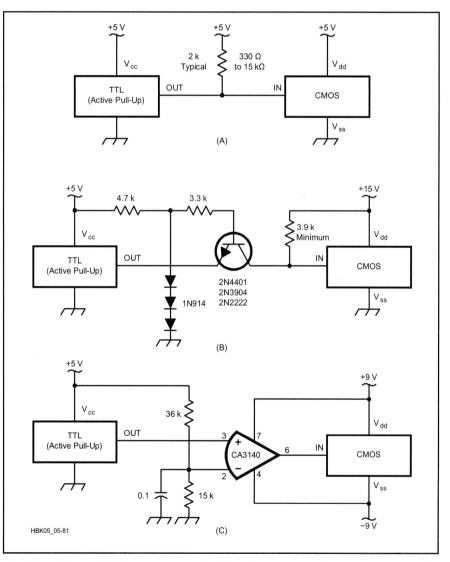

Fig 4.30 — TTL to CMOS interface circuits: (A) pull-up resistor, (B) common-base level shifter and (C) op amp configured as a comparator.

cial temperature range without an external pull-up resistor. However, they cannot drive 4000-series CMOS directly, and for HC-series devices, a pull-up resistor is recommended. The pull-up resistor, connected between the output of the TTL gate and V_{CC} as shown in **Fig 4.30A**, ensures proper operation and enough noise margin by making the high output equal to V_{DD}. Since the low output voltage will also be affected, the resistor value must be chosen with both desired high and low voltage ranges in mind. Resistor values in the range 1.5 kΩ to 4.7 kΩ should be suitable for all TTL families under worst conditions. A larger resistance reduces the maximum possible speed of the CMOS gate; a lower resistance generates a more favorable RC product but at the expense of increased power dissipation.

HCT-series and ACT-series CMOS devices were specifically designed to interface non-CMOS devices to a CMOS system. An HCT device acts as a simple buffer between the non-CMOS (usually TTL) and CMOS device and may be combined with a logic function if a suitable HCT device is available.

When the CMOS device is operating from a power supply other than +5 V, the TTL interface is more complex. One fairly simple technique uses a TTL open-collector output connected to the CMOS input, with a pull-up resistor from the CMOS input to the CMOS power supply. Another method, shown in Fig 4.30B, is a common-base level shifter. The level shifter translates a TTL output signal to a +15 V CMOS signal while preserving the full noise immunity of both gates. An excellent converter from TTL to CMOS using dual power supplies is to configure an operational amplifier as a comparator, as shown in Fig 4.30C. An FET op amp is shown because its output voltage can usually swing closer to the rails (+ and − supply voltages) than a bipolar device.

CMOS DRIVING TTL

Certain CMOS devices (including most modern 5 V powered CMOS logic families) can drive TTL loads directly. The output voltages of CMOS are compatible with the input requirements of TTL, but the input-current requirement of TTL limits the number of TTL loads that a CMOS device can drive from a single output (the fan-out).

Interfacing CMOS to TTL is a bit more complicated when the CMOS is operating at a voltage other than +5 V. One technique is shown in **Fig 4.31A**. The diode blocks the high voltage from the CMOS gate when it is in the high output state. A germanium diode is preferred because its lower forward-voltage drop provides higher noise immunity for the TTL device in the low state. The 68-kΩ resistor pulls the input high when the diode is reverse biased.

A simple resistor/Zener circuit, shown in

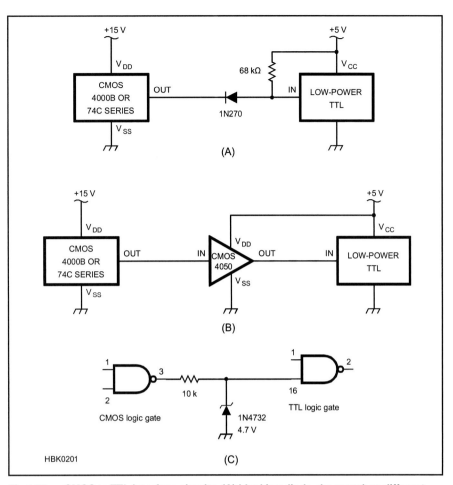

Fig 4.31 — CMOS to TTL interface circuits. (A) blocking diode chosen when different supply voltages are used. The diode is not necessary if both devices operate with a +5 V supply. (B) CMOS noninverting buffer IC. (C) Resistor/Zener circuit clamps the voltage to the TTL input at 4.7 V.

Fig. 4.31C, can also be used. This clamps the voltage to the TTL input at 4.7 V.

There are two CMOS devices specifically designed to interface CMOS to TTL when TTL is using a lower supply voltage. The CD4050 is a noninverting buffer that allows its input high voltage to exceed the supply voltage. This capability allows the CD4050 to be connected directly between the CMOS and TTL devices, as shown in Fig 4.31B. The CD4049 is an inverting buffer that has the same capabilities as the CD4050.

5 V DRIVING 3.3 V LOGIC

Low voltage logic operating with 3.3 to 3.6 V supplies is becoming more and more common. Some of these logic families have 5 V tolerant inputs and can be driven directly by TTL levels; others require buffering to keep inputs at or below the supply voltage. Output levels may or may not be sufficient to drive TTL level inputs reliably under all conditions. The use of TTL compatible buffers (74LVC, 74ALVC) on input signals is a safe way to drive these devices. Since the low logic voltages are compatible, a simple Zener clamp arrangement may also be sufficient.

3.3 V DRIVING 5 V LOGIC

Output voltages of most 3.3 V logic families are sufficient to drive the inputs of TTL-level devices. In some cases, the high level output voltage of a 3.3 V powered device may approach the lower end of the TTL level device's input range. In these cases, a pull-up resistor to 3.3 V will usually be sufficient. It is also possible to use an IC or transistor buffer.

4.6.5 Real-World Interfacing

Quite often logic circuits must either drive or be driven from non-logic sources. A very common requirement is sensing the presence or absence of a high (as compared to +5 volts) voltage or perhaps turning on or off a 120-V ac device or moving the motor in an antenna rotator. A similar problem occurs when two different units in the shack must be interfaced because induced ac voltages or ground loops can cause problems with the desired signals.

Digital Basics 4.19

A slow speed but safe way to interface such circuits is to use a relay. This provides absolute isolation between the logic circuits and the load. **Fig 4.32A** shows the correct way to provide this connection. The relay coil is selected to draw less than the available current from the driving logic circuit. The diode, most often a 1N914 or equivalent switching diode, prevents the inductive load from back-biasing the logic circuit and possibly destroying it.

It is often not possible to find a relay that meets the load requirements *and* has a coil that can be driven directly from the logic output. Fig. 4.32B shows two methods of using transistors to allow the use of higher power relays with logic gates.

Electro-optical couplers such as optoisolators and solid-state relays can also be used for this circuit interfacing. Fig 4.32C uses an optoisolator to interface two sets of logic circuits that must be kept electrically isolated, and Fig 4.32D uses a solid-state relay to control an ac line supply to a high current load. Note that this example uses a solid-state relay with internal current limiting on the input side; the LED input has an impedance of approximately 300 Ω. Some devices may need a series resistor to set the LED current; always consult the device data sheet to avoid exceeding device limits of the relay or the processor's I/O pin.

For safely using signals with voltages higher than logic levels as inputs, the same simple resistor and Zener diode circuit similar to that shown in Fig. 4.31C can be used to clamp the input voltage to an acceptable level. Care must be used to choose a resistor value that will not load the input signal unacceptably.

INTERFACING ANALOG SIGNALS

Of course not all signals in the real world are digital. It is often desirable to know the exact voltage of an analog sensor, for example, or to measure the level of something — light, water, current or some other input. Similarly, an analog output can be quite useful for controlling or indicating things that are better left in the analog domain. For this reason we have two types of converters. The analog-to-digital converter (ADC) and digital-to-analog converter (DAC) handle these tasks for us, and are often fairly easy to use. Additional information on ADCs and DACs may be found in the **Analog Basics** chapter.

ADC and DAC Interface Types

Interfaces for ADC and DAC chips are generally classed as serial or parallel. Serial interfaces can vary in speed, complexity and the number of wires required for operation. A serial interface has the advantage of requiring a small number of processor I/O pins to accommodate data of any length. Whether your ADC is providing 8, 10 or 12 bits, the same small number of I/O signals are used.

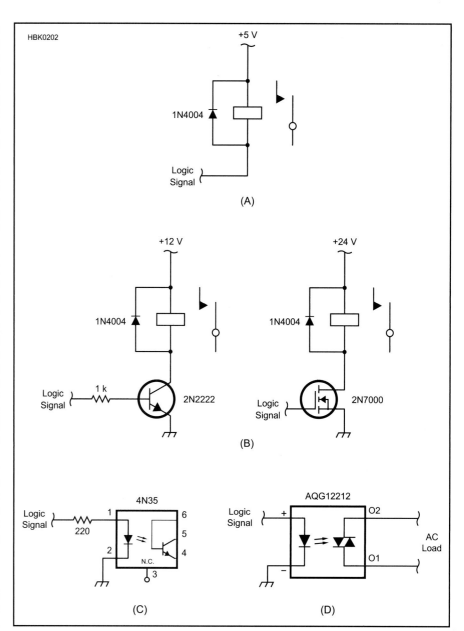

Fig 4.32— Interface circuits for logic driving real-world loads. (A) driving a relay from a logic output; (B) using a bipolar transistor or MOSFET to boost current capacity; (C) using an optoisolator for electrical isolation; (D) using a solid-state relay for switching ac loads.

Some of the most common serial interfaces used for ADC and DAC chips are as follows:

Serial Peripheral Interconnect (SPI). This four-wire, synchronous bus and protocol use a common CLK signal, plus data lines for master-to-slave (Master Out-Slave In, or MOSI) and slave-to-master (Master In-Slave Out, or MISO). Multiple devices can be used by providing each with its own Slave Select (SS) signal. Speeds can range up to 70 Mbit/s, depending on the capabilities of the master and slave devices.

Microwire. This earlier predecessor to SPI implements a half-duplex subset of SPI using the same signals.

Inter-IC Communication (I2C) bus. Originally developed by Philips, this synchronous 2-wire bus and protocol use a pair of open-drain lines, Serial Clock (SCL) and Serial Data (SDA). Communication is controlled by a master node, though there may be more than one master attached to the bus. Many peripheral devices can be attached to an I2C bus; a device address is sent by the master to initiate communication with a slave. Speeds range from 10 kbit/s (low speed) to 100 kbit/s (standard) to 400 kbit/s (high speed) and higher.

Parallel interfaces generally have eight or more data lines, plus a chip select, read/write and interrupt controls. The read and write signals may be separate, or may be a single read/write signal — low for WR, and high for RD, for example, or vice versa. The interrupt signal can be used to indicate the end of a conversion cycle to the CPU. In this way the processor can tell the converter to start a conversion, then continue processing until the conversion is complete. This can allow more processing to be done without waiting idly for the ADC or DAC to complete its conversion.

PROGRAMMING AND COMMUNICATION

Many ADC chips have multiplexed inputs that allow you to use one chip to sample and digitize more than one analog input. In these devices, there is a single converter but several inputs that can be internally switched. Additionally, it is common to be able to program the inputs as single-ended or differential. In the case of the popular ADC0832, for example, you can use its two analog input pins as two separate inputs, or as a single differential input. The ADC0838 expands this to eight single-ended or four differential inputs. In the case of the ADC0832, modes can be mixed. This allows the use of various combinations of single-ended and differential inputs, as needed.

Programming and selecting the inputs is done by sending a series of bits from the processor to the ADC at the beginning of the conversion cycle. **Fig. 4.33** shows a diagram of the signal timing used with the ADC0834 ADC. In the case of the ADC0834, a series of four bits are sent by the CPU to the ADC to select single ended or differential mode, the polarity of the input, and which of the four inputs (or two input pairs) is to be selected. Immediately following the fourth bit, the ADC starts its conversion and starts sending the resulting data to the CPU. Other chips in this family and many from different manufacturers use a similar scheme; the number of program bits send by the CPU depends on the number of inputs present.

Some chips are also configurable for data format as well. For example, a 12-bit ADC may be programmable to send a sign bit for a total of 13 data bits rather than 12. Data may also be sent LSB first or MSB first. Some chips are configurable for either parallel or serial data transfer; this selection is generally made by pulling a pin or pins high or low to select the mode. In all cases, a careful reading of the device data sheet is your best bet for successful integration of the converter into your project.

INTERFACING TO YOUR MICROCONTROLLER

Probably the simplest way to interface an external ADC to your microcontroller is by using a parallel interface. In this case, one simply manipulates the I/O pins from the CPU to start the conversion, then reads the resulting data from the data lines. This method is often less desirable than a serial connection, though, due to the limited number of I/O pins usually available. **Fig 4.34A** shows an example of a parallel interface between a 12-bit ADC (the ADS7810) and a PIC microcontroller (the PIC16F887). In this example, only eight data lines are used for twelve data bits. The HBE signal is used to gate the four

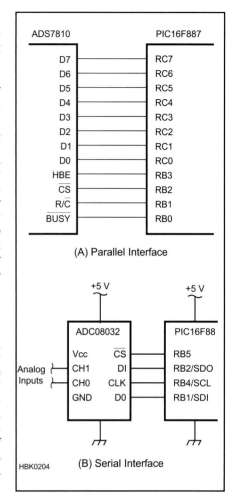

Fig 4.34 — A/D converter to microcontroller interfacing: (A) parallel; (B) serial.

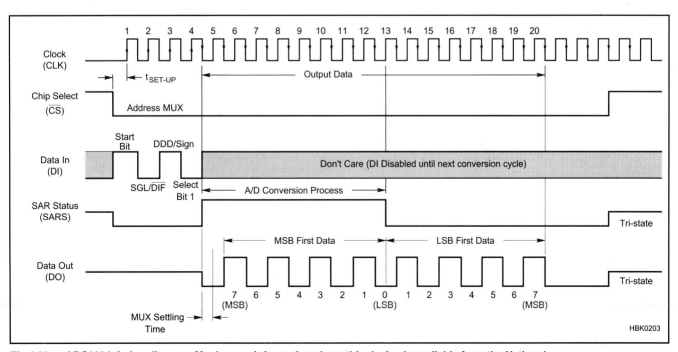

Fig 4.33 — ADC0834 timing diagram. Much more information about this device is available from the National Semiconductor datasheet.

high order bits onto the data lines after the low order bits have been read. This saves four I/O pins on the microcontroller.

Serial connections save I/O pins, allow the use of physically smaller IC packages, and can be quite easy to implement. Fig. 4.34B shows an example of a serial interface. If your CPU has a hardware SPI/Microwire interface built in, communication with the ADC can be quite simple from a program standpoint. Simply send the program word, if required, and read the resulting data — either one or two bytes, depending on the ADC resolution. If you do not have built-in hardware SPI, the program code to "bit-bang" the interface is fairly simple to implement. (See **www.dlp design.com/images/bit-bang-usb.pdf** for more information on bit-bang.)

Note that Fig. 4.34B shows separate grounds for the processor and ADC. This is required to keep switching noise and transients from affecting the accuracy of readings. A separate analog supply is also recommended. The analog supply should be well filtered using combination of smaller ceramic or monolithic, and bulk tantalum or other low-ESR capacitors. This should be done for the analog supply and reference voltage supplies. The more noise you can eliminate from the analog supply, the more reliable your readings will be.

Microcontrollers with Built-In Converters

It is not at all unusual to find microcontrollers with built-in analog-to-digital converters, complete with multiplexers, analog supply pins and V_{REF} inputs. For example, many Microchip PIC processors are equipped with 10-bit ADCs, and have 5, 8 or more pins that can be configured as multiplexed analog inputs. A few also have digital-to-analog, though this is not nearly as common. While these devices can be extremely useful, there are some limitations. Reference voltages are generally limited to the device's supply voltage. Internal CPU-generated noise may also somewhat limit the usefulness of the converter. With careful attention to device specifications and limitations, though, a built-in ADC may meet the needs of a wide range of uses.

MSI, LSI, VLSI AND HYBRID CIRCUITS

In addition to using the basic logic elements discussed in the previous sections, there are many integrated circuits available for Amateur Radio applications that include the equivalent of dozens, hundreds or many thousands of gates. It is often no harder to use one of these units than it is to use a few gates and flip-flops.

The A/D and D/A converters discussed elsewhere in this section are examples, as are microcontrollers. Hybrid (containing both analog and digital functions) integrated circuits provide other opportunities for builders. As an example, the LM3914 is an LED driver that takes an analog signal in and turns on one or more LEDs, depending on the analog voltage.

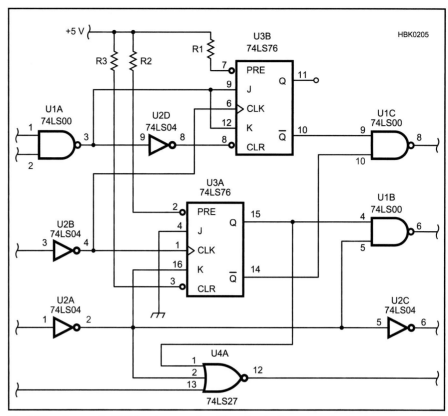

Fig 4.35 — A sample PLD implementation. The circuit shown can be implemented in a single programmable GAL logic chip.

PROGRAMMABLE LOGIC

As digital logic designs became more and more complex, the size and power consumption of their implementations grew. PLDs (*Programmable Logic Devices*) were introduced to allow the designer to produce application-specific logic without the expense and delay of fabricating custom chips. As explained earlier, the design of a logic circuit begins with a description; this description is used to determine an equation; then the equation is expanded to components. With programmable logic we can shortcut this process by implementing the equations themselves, programming them into a generic array of gates.

Several families of programmable logic devices have been introduced over the years. The earliest were Programmable Array Logic (PAL) chips, consisting of an array of gates whose inputs, interconnections and outputs could be configured by blowing internal fuses according to a specific design. PALs were followed by Generic Array Logic (GALs), which have more gates and more flexible I/O pin logic. GALs are also erasable and reprogrammable.

Later developments led to the Field Programmable Gate Array (FPGA) and other programmable logic arrays. These devices can contain anywhere from several hundred to millions of logic gates, and up to over a thousand I/O pins. The connections between gates, and thus the device's function, is determined by complex program code. PLDs can be made to replace large numbers of individual SSI, MSI and even LSI chips, up to and including entire microprocessor cores.

Shown in **Fig. 4.35** shows a control circuit implemented with discrete logic gates. In its original form it requires eight gates, two flip-flops, four chips and 58 total circuit board pins. The same function can be implemented with a single GAL16V8 chip — the smallest available — using less than half its capabilities. (These capabilities can be seen in the data sheet available from **www.latticesemi.com/ lit/docs/datasheets/pal_gal/16v8.pdf**.) The entire circuit now takes a single, 20-pin IC package. Additionally, the logic can be altered at will during troubleshooting or to change the operation of the circuit later on. As you can see, if a design requires a large number of gates it quickly becomes worthwhile to implement it using programmable logic rather than as a collection of gates implemented in SSI or MSI.

4.7 Microcontrollers

4.7.1 An Overview of Microcontrollers

Let's imagine a comparison of the average amateur transceiver of, say, 1970 to today's current state of the art. We would find that the modern equipment is far smaller, lighter, requires less power to operate, and consists of fewer "boxes" performing more functions than were thought possible when the older equipment was built. Instead of a separate transmitter and receiver set capable of CW, AM and SSB on perhaps half a dozen bands, the modern transceiver might boast CW, SSB, AM, FM and FSK on every band from 160 meters through 70 cm. In addition we are likely to find such things as a built in antenna tuner, power supply, digital display, multiple filters, digital signal processing and other features never dreamed of in 1970 — all in a physical package far smaller and lighter than a single piece of 60s or 70s vintage gear. The situation is much the same with digital electronics.

Over the past couple of decades, microcontrollers have replaced more and more functions once performed by collections of gates or programmable logic. Initially, the use of microprocessors in embedded designs was hampered by their high cost as well as by the requirement for complex and expensive support devices. The microprocessor itself was relatively expensive, and might require

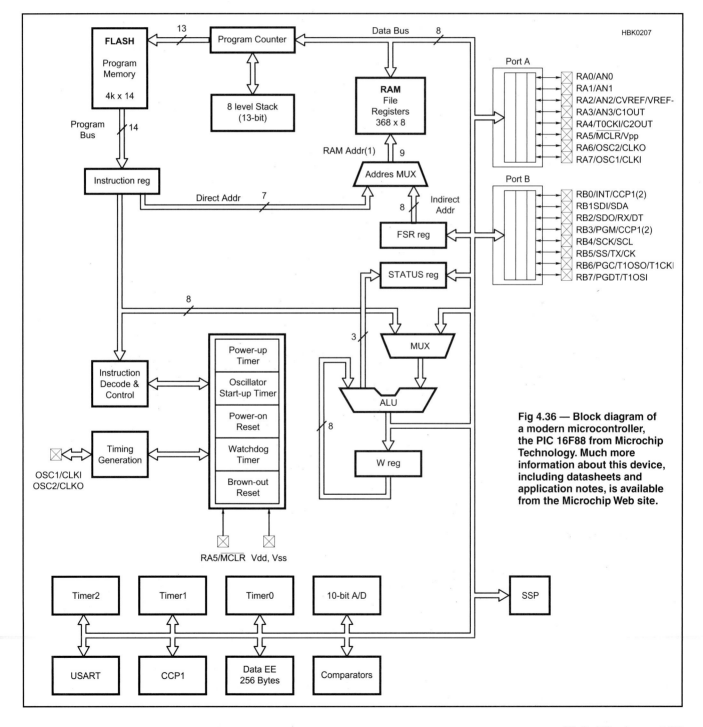

Fig 4.36 — Block diagram of a modern microcontroller, the PIC 16F88 from Microchip Technology. Much more information about this device, including datasheets and application notes, is available from the Microchip Web site.

a clock generator chip, external read-only program memory (ROM) as well as read/write data memory (RAM), I/O expanders to interface with the outside world and so forth. As the state of the art has advanced, however, more and more of these functions have been brought into the modern microcontroller.

A microcontroller can be regarded as a "computer on a chip." It usually consists of a relatively small microprocessor along with some amount of program memory, data memory, input/output ports, and often some specialized peripheral devices. An example of a fairly low-end, modern microcontroller can be seen in **Fig. 4.36**. Let's take a look at what this device provides:

- A reduced instruction set computer (RISC) microprocessor core
- 4096 words of program memory
- 368 bytes of random access data memory (RAM)
- 256 bytes of electrically erasable read-only memory (EEPROM)
- Two 2-input analog comparators
- One 10-bit ADC with 10 input channels
- A hardware module that can generate pulse-width modulated output (PWM), or measure pulse width of an input signal
- Built-in support for several different modes of serial communication
- An internal precision clock oscillator that can run at speeds from 31 kHz to 8 MHz, or can use an external crystal at up to 20 MHz
- Several internal counter/timer units, including 8- and 16-bit timer/counters
- Up to 16 I/O pins

All of these features are implemented in a single, 18 pin IC package available from a number of suppliers at less cost than a handful of SSI chips or a doughnut and a cup of hamfest coffee. You can see why most new products, and many new ham radio projects, make use of the power of these microcontrollers. A general purpose circuit can be built and its operation changed almost at will, simply by changing the program code running inside the chip.

4.7.2 Selecting a Microcontroller

There are a number of microcontroller families available from several manufactures, each with its own advantages and disadvantages. At the time of this writing, the more popular options for amateur use come from Microchip (PIC product line), Atmel (AVR, ATMega, 8051 and other product lines), Freescale (68HC11, HC08, HC16, and other product lines), and many manufacturers who are building chips based on the venerable Intel 8051 architecture. In addition to these you will see and hear of the Texas Instruments MSP430, ARM, and many others.

Designing projects using microcontrollers can range from relatively simple to quite complex. On the simpler end of the scale are several products intended to bring microcontroller based design ability to people who would otherwise not be able to use these devices.

The Parallax BASIC Stamp, for example, is a PIC microcontroller preloaded with a program that allows the user to easily write and download programs in a dialect of BASIC. The BASIC program is written in an editor program on the PC and then downloaded. Various breadboard kits are available to simplify hardware breadboarding and program downloading.

The PicAxe is another effort to bring the ability to use complex, very capable microcontrollers to those who have neither the time nor the desire to learn a low-level programming language.

In each of these cases, the user does not necessarily need to have a deep understanding of the details of the microcontroller or its operation. The circuit can be designed with attention being paid to keeping signal levels within the limits of the device. Then, one simply writes a BASIC program to perform the functions required. To be sure, there are trade-offs. The built-in firmware required to allow the user to write a program in a simple language takes up space and processor cycles, so execution speed is not as fast and programs are somewhat limited. The BASIC Stamp and PicAxe use specific variants of the PIC microcontroller, so if you need features that are not present in those chips you will need to take a different approach. Still, these can be extremely valuable tools that the average amateur can use with very little time and money invested.

There is a very broad range of microcontrollers available to meet nearly any need. Available devices include 8-bit, 16-bit and 32-bit processors with internal program memory ranging from 512 words up to 256 kbytes and more. At the low end are very low cost, physically small devices (down to 8 pin ICs) that can perform relatively simple tasks with ease. Larger, faster processors may include more memory, more I/O pins, and specialized peripherals. Some of the more common features are ADCs, DACs, pulse width modulation outputs, USB interface hardware or digital signal processing,

I/O REQUIREMENTS

Selecting a microcontroller for a new project involves evaluating your requirements and prior experience. First, evaluate the number of inputs and outputs you are likely to need. If, for example, you need to sample an analog voltage and two switch closures and generate a serial data stream, you will need one analog input and three or more digital I/O pins. Depending on memory requirements, even some of the smallest devices will fit the bill. For a more elaborate design, you may need to scan a 4×4 key switch matrix, drive an LCD module with a parallel interface, control a number of LEDs and interface to a PC through a USB port. In this case, the number of I/O pins will eliminate a large number of devices from consideration.

CPU SIZE, PERFORMANCE AND MEMORY

If you are a computer user, you are used to seeing 32- and 64-bit microprocessors, many with more than one CPU core on the same chip. In the world of embedded microcontrollers, things are a little different. Instead of handling an operating system and a large number of different programs, the microcontroller only has to execute one program and is completely dedicated to that task. Unless you require very high speed I/O, very intensive processing or will be using DSP, a simple 8-bit processor may be more than sufficient. There are dozens of ham related products that use embedded controllers — iambic keyers, APRS transponders, repeater controllers, antenna rotator controls and many other accessories that all use simple 8-bit processors with surprisingly small amounts of program memory. On the other hand, if you are designing a new rig or antenna tuner that will require DSP, chances are you may need a more robust processor.

If you are unsure of what your project will require, you may want to pick a processor family that has scalable, pin-compatible members. For example, Microchip offers processors ranging from 8 bits and 5 million instructions per second (MIPS) with 7 kbytes of program memory, to 16 bit with 48 kbytes of program memory at 30 MIPS. All are pin-compatible, so if you find you need more speed, more memory or more built-in peripherals, you can drop in a different processor without changing the hardware.

Also take into account the physical packaging. Some products are available in through-hole, dual inline (DIP) packages; others are available only in surface mount versions. This can make a difference if you lack the equipment, expertise or desire for SMT work.

SPECIALIZED PERIPHERALS

Will you need a USB interface? Analog inputs? DSP? Precision timekeeping? It may be easier to select a processor that has the features you need built in. On the other hand, it is relatively easy to add an external real-time clock, ADC or DAC, and it takes only a few I/O pins. There are several USB interface chips available that will handle communication with your PC, and present an asynchronous serial data interface to the microcontroller. This can save a lot of firmware development time and frustration.

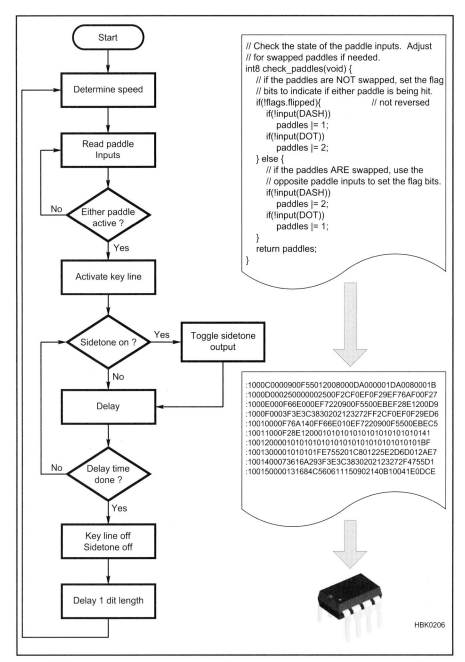

Fig 4.37 — Simplified representation of the development process. A flow chart shows the program logic. Source code is compiled to object code, and objet code is programmed into the device.

4.7.3 The Development Process

Once you have settled on a processor and worked out the general hardware configuration you are ready to begin work on the firmware. The most difficult part of using a microcontroller is, of course, writing the program code. While once a designer would sweat out the design using logic symbols and data sheets, now the tools of choice are the text editor and compiler or assembler. The level of effort is not that much different. What is different, however, is the development process itself. Where once you would need to spend hours re-wiring a circuit board to change its operation or fix a problem with your design, now you change the code, reprogram and try again.

Regardless of the microcontroller or programming language you decide to use, the development process will be roughly the same. A basic set of characteristics and features is arrived at based on your needs. You begin with a flow chart (if you are conscientious) and write your program code, commenting liberally so you can tell what you did and why you did it later on.

Fig. 4.37 shows a simplified example of a flow chart, with process and decision blocks to implement a simple Morse code keyer. To the right of the flow chart is a graphical representation of the process. First, the program source code is written. The top block shows a small block of C language source code to implement one of the function blocks. Once written, the source code is compiled into binary machine instructions (object code, shown here by a hexadecimal representation). The binary object code is programmed into the chip itself, and is ready for testing and use. These steps are covered in more detail later on.

WRITING THE PROGRAM

Once the logic of the program has been worked out, the program itself must be written. There are a number of different programming languages in use. The most common for amateur use are Assembly, BASIC and C. Which is used depends on the individual and the tools at hand. For example, if you have extensive experience writing programs in BASIC, you may want to find a BASIC compiler for your processor. If, on the other hand, you want to use only freely available tools, you may need to stick with assembly.

Assembly code is the most fundamental programming language generally used. It provides a set of human readable mnemonics for the binary machine instructions. Instead of the programmer needing to code the binary instruction, an assembly code mnemonic for that instruction is used. The source code is then processed by an assembler, which gener-

HARDWARE COST AND DEVELOPMENT TOOLS

The cost of the actual microcontroller may be a factor, or it may not. If you are building a one-off project for your own use, there is not much difference between using a $3 chip and a $30 chip with a built-in BASIC interpreter. The difference in time and effort to develop the firmware may make it well worth the extra money for the easier to use solution. If you plan to manufacture your device for a larger audience, the cost difference may lead you to make a different choice.

Take into account the cost of the software and hardware tools needed for development. Does the manufacturer offer a free set of firmware development tools, or will you have to spend extra money for an assembler or compiler? Is there a low cost development kit available, with the PC interface you need? Does the device require a simple PC interface to program, or a more expensive dedicated programmer? Do you have any previous experience with the processor family, or will it be a completely new effort for you?

ates the binary instructions needed to program the processor's internal program memory. These mnemonics are different depending on the processor used. For example, all microprocessors have an instruction to add the contents of two registers together and store the result into a register. The PIC mnemonic is ADDWF f,d where f is the source register and d is the destination. If you are using an 8051 variant, the instruction would be ADD A, r where r is the register number to be used.

Assembly language is different for each processor, and it provides very precise control over the exact operation of the program. It can, however, be quite complicated to implement some seemingly simple functions. Many processors lack machine language or assembly instructions for division and multiplication, as well as serial input/output and many other common functions. It is up to the programmer to build these functions using the available machine instructions.

Many people prefer to program in a higher level language, such as C or BASIC. This has the advantage of simplifying the programmer's job; the compiler will handle the drudgery of translating the high level instructions into machine code. As an example, **Fig 4.38** compares the task of multiplying two 8-bit numbers in Assembly (ASM), BASIC and C. You will find that C is a more common high level language than BASIC, Java or FORTH for microcontrollers.

The language you select will depend on your level of expertise, the tools available and their cost. For example, Microchip makes a fully featured development environment available for their PIC products, as do several other manufacturers — as long as you are happy writing programs in Assembly language. There are also free or low cost programs to let you write your programs in subsets of BASIC or the C programming language. If you choose an 8051 based processor, you also have the choice of several free assemblers and compilers for C, Forth and others. If funds allow, there are commercial compilers available costing anywhere from a hundred or so dollars up into the thousands, each with its own set of advantages and disadvantages.

PROGRAMMING THE CHIP

So let's assume you have picked a language, obtained an assembler or compiler and have some program code written. You now have a compiled binary program — but that binary instruction code is still stored on your PC. Now it needs to be programmed into the chip itself. This is a process you will sometimes hear referred to as "burning," a term which dates back to the very early days of programmable read-only memories. Most modern microcontrollers use reprogrammable FLASH memory, which can be erased and rewritten thousands or hundreds of thousands of times.

Depending on the microcontroller you are using, you will probably need a special programming device. The programmer is usually some simple hardware that attaches to your PC's serial, parallel or USB port. It handles the voltages and signals needed to transfer the compiled program code into the microcontroller's internal memory. The details of the interface hardware and programming software will be different for each manufacturer, and occasionally for different product families from the same manufacturer as well.

Once this process is complete, you are ready to plug the chip into your project and test it out. This is usually followed by several cycles of testing, debugging errors, rewriting program code, erasing and reprogramming before everything is working perfectly. Programmers for most microcontroller families can generally be had from a variety of sources, and for a fairly wide price range. Many can also be home brewed using simple, inexpensive circuits easily obtained from suppliers world wide. An Internet search will turn up a number of designs for programmers, as well as the PC software for loading your programs into the microcontroller.

4.7.4 Learning More About Microcontrollers

Many books and Internet resources exist for learning how to program microcontrollers. Manufacturer web sites, such as Microchip (**www.microchip.com**), Freescale (**www.freescale.com**), and Atmel (**www.atmel.com**) are a good place to start. There are also a number of message boards, forums and email lists devoted to specific processor families, such as the PICList (**www.piclist.com**); Ham-PIC (**www.njqrp.org/ham-pic/**); and **www.8052.com**. The American QRP Club has made a successful PIC Elmer course available; you can see the details at **www.amqrp.org/elmer160/**.

Several projects using microcontrollers are described in the **Station Accessories** chapter.

ASM 8x8 multiply :

```
        ;8x8 multiply routine from Microchip AN526
mpy_S   clrf H_byte
        clrf L_byte
        movlw 8
        movwf count
        movf mulcnd,W
        bcf STATUS,C ; Clear the carry bit in the status Reg.
loop    rrf mulplr, F
        btfsc STATUS,C
        addwf H_byte,Same
        rrf H_byte,Same
        rrf L_byte,Same
        decfsz count, F
        goto loop
retlw 0
```

BASIC 8x8 multiply:

```
let a = b x c
```

C 8x8 multiply:

```
a = b * c;
```

Fig 4.38 — Example of multiplication in ASM, BASIC and C.

4.8 Personal Computer Interfacing

An in-depth discussion of the personal computer is beyond the scope of this chapter. We will, however, spend some time exploring the applications of the PC in Amateur Radio and interfacing to it.

Computers are becoming more and more common in the ham shack as they become more useful to the Amateur operator. Numerous applications exist for logging, for example, with near instantaneous lookup of call signs replacing the old paper *Callbook*. Internet connectivity has led to the advent of modes such as IRLP, Echolink and WIRES-II (see the **Digital Communications** chapter) — all of which use the Internet to transport a voice signal for some part of the communication path. Hams are using computers more and more as part of software defined radio (SDR) stations. Once the domain of extremely high tech, high dollar systems, SDR has reached the point where a simple direct conversion receiver can be used along with a PC sound card to receive digital modes that may be inaudible to the human ear.

Hams are also increasingly using the personal computer to control their rigs, especially during contests. A simple serial interface can attach your transceiver to your PC, allowing your logging program to automatically retrieve and store the operating frequency and mode as a contact is logged. Simple keyboard macros can be used to switch bands, frequencies, modes, filters and antennas in less than a second, saving the operator precious time and reducing the opportunity for mistakes that may mean lost contacts or even damaged equipment.

All of these applications require some form of interfacing between the computer and other gear found in the shack or elsewhere in the station. This real-world interfacing can be quite simple, or it can get to be amazingly complex. In addition to the information presented here, several PC interface projects may be found in the **Station Accessories** chapter.

4.8.1 Parallel vs Serial Signaling

To communicate a word to someone across the room, you could hold up flash cards displaying the letters of the word. If you hold up four flash cards, each with a letter on it, all at once, then you are transmitting in parallel. If, instead, you hold up each of the flashcards one at a time, then you are transmitting in serial. *Parallel* means all the bits in a group are handled exactly at the same time. *Serial* means each of the bits is sent in turn over a single channel or wire, according to an agreed sequence. **Fig 4.39** illustrates parallel and serial signaling.

Both parallel and serial signaling are appro-

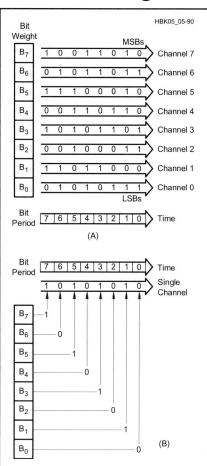

Fig 4.39 — Parallel (A) and serial (B) signaling. Parallel signaling in this example uses 8 channels and is capable of transferring 8 bits per bit period. Serial transfer only uses 1 channel and can send only 1 bit per bit period.

priate for certain circumstances. Parallel signaling is faster, since all bits are transmitted simultaneously, but each bit needs its own conductor, which can be expensive. Parallel signaling is more likely to be used for internal communications. For spanning longer distances, such as to an external device, serial signaling is more appropriate. Each bit is sent in turn, so communication is slower, but it also is less expensive, since fewer channels are needed between the devices.

Most amateur digital communications use serial transmission to minimize cost and complexity. The number of channels needed for signaling also depends on the operational mode. One channel is required per bit for simplex (one-way, from sender to receiver only) and for half-duplex (two-way communication, but only one person can talk at a time), but two channels per bit are needed for full-duplex (simultaneous communications in both directions).

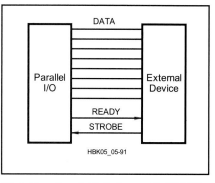

Fig 4.40 — Parallel interface with READY and STROBE handshaking lines.

PARALLEL I/O INTERFACING

Fig 4.40 shows an example of a parallel input/output interface, similar to the parallel printer ports once common in PCs. Typically, they have eight data lines and one or more handshaking lines. *Handshaking* involves a number of functions to coordinate the data transfer. For example, the READY line indicates that data are available on all 8 data lines. If only the READY line is used, however, the receiver may not be able to keep up with the data. Thus, the STROBE line is added so the receiver can determine when the transmitter is ready for the next character.

Interfacing to the parallel port is very simple and can be done in many languages and under many operating systems. It is an easy way to get a PC to act as a controller in the shack and around the home, with from 8 to 12 independent input or output wires. You can get an old computer either virtually free or for just a few dollars and use it exclusively as a controller. (New computers usually do not have a parallel port.)

By searching the Internet for a combination of "parallel port interface" and the language of your choice (for example, Visual BASIC or C), you can find detailed interfacing and programming instructions. It is a good and relatively simple place to "roll your own," compared to the protocol requirements of interfacing with the serial port or USB.

For more information, including practical circuits, see "Interfacing to the Parallel Port" by Paul Danzer, N1II, on the CD-ROM included with this book.

SERIAL I/O INTERFACING

Serial input/output interfacing is more complex than parallel, since the data must be transmitted based on an agreed sequence. For example, transmitting the 8 bits (b7, b6, ... b0) of a word includes specifying whether the least significant bit, b0, or the most significant bit, b7, is sent first. Fortunately, a number of standards have been developed to define the agreed sequence, or encoding scheme.

In order to use a serial interface between a computer and a serial device such as a modem, printer or a new transceiver, several things must be set the same on both ends of the connection. The first is the data rate, the choices for which are usually limited to specific values between 300 and 115200 bits per second for historical reasons. The next is the data format — 7 or 8 bits, and with even, odd or no parity. Finally, both devices must be using the same handshaking method. The usual choices are RTS/CTS, DSR/DTR, or XON/XOFF. The first two use extra signals (and therefore extra wires) to indicate when the transmitting device has data to send, and when the receiving device is able to accept data. In the third method, XON/XOFF, the device receiving data will send an XOFF control character to the transmitting device to indicate that it can no longer accept data. When the transmitting device sees the XOFF it stops sending until it receives an XON from the other end. Each method has its advantages and disadvantages. For example, XON/XOFF needs no extra wires, but response is not instantaneous. Both ends must have some data buffer space and logic to support the handshake.

4.8.2 Data Rate

There are a number of limitations on how fast data can be transferred: (1) The sending equipment has an upper limit on how fast it can produce a continuous stream of data. (2) The receiving equipment has an upper limit on how fast it can accept and process data. (3) The signaling channel itself has a speed limit, often based on how fast data can be sent without errors. (4) Finally, standards and the need for compatibility with other equipment may have a strong influence on the data rate.

Two ways to express data transmission rates are *baud* and *bits per second (bit/s)*. These two terms are not interchangeable: Baud describes the signaling, or symbol, rate — a measure of how fast individual signal elements *could* be transmitted through a communications system. Specifically, the baud is defined as the reciprocal of the shortest element (in seconds) in the data-encoding scheme. For example, in a system where the shortest element is 1 ms long, the maximum signaling rate would be 1000 elements per second. (Note that, since baud is measured in elements per second, the term "baud rate" is incorrect since baud is already a measure of speed, or rate.) Continuous transmission is not required, because signaling speed is based only on the shortest signaling element.

Signaling rate in baud says nothing about actual information transfer rate. The maximum information transfer rate is defined as the number of equivalent binary digits transferred per second; this is measured in bits per second.

When binary data encoding is employed, each signaling element represents one bit. Complications arise when more sophisticated data encoding schemes are used. In a quadrature- phase-shift keying (QPSK) system, a phase transition of 90° represents a level shift. There are four possible states in a QPSK system. Thus, two binary digits are required to represent the four possible states. If 1000 elements per second are transmitted in a quadriphase system where each element is represented by two bits, then the actual information rate is 2000 bit/s.

This scheme can be extended. It is possible to transmit three bits at a time using eight different phase angles (bit/s = 3 × baud). In addition, each angle can have more than one amplitude. A 9600 bit/s modem uses 12 phase angles, 4 of which have two amplitude values. This yields 16 distinct states, each represented by four binary digits. Using this technique, the information transfer rate is four times the signaling speed. This is what makes it possible to transfer data over a phone line at a rate that produces an unacceptable bandwidth using simpler binary encoding. This also makes it possible to transfer data at 2400 bit/s on 10 meters, where FCC regulations allow only 1200-baud signals.

When are transmission speed in bauds and information rate in bit/s equal? Three conditions must be met: (1) binary encoding must be used; (2) all elements used to encode characters must be equal in width; and (3) synchronous transmission at a constant rate must be employed. In all other cases, the two terms are not equivalent.

Within a given piece of equipment, it is desirable to use the highest possible data rate. When external devices are interfaced, it is normal practice to select the highest standard signaling rate at which both the sending and receiving equipment can operate.

4.8.3 Error Detection

Since data transfers are subject to errors, data transmission should include some method of detecting and correcting errors. Numerous techniques are available, each used depending on the specific circumstances, such as what types of errors are likely to be encountered. Some error detection techniques are discussed in the **Digital Modes** chapter. One of the simplest and most common techniques, parity check, is discussed here.

PARITY CHECK

Parity check provides adequate error detection for some data transfers. This method transmits a parity bit along with the data bits. In systems using odd parity, the parity bit is selected such that the number of 1 bits in the transmitted character (data bits plus parity bit) is odd. In even parity systems, the parity bit is chosen to give the character an even number of ones. For example, if the data 1101001 is to be transmitted, there are 4 (an even number) ones in the data. Thus, the parity bit should be set to 1 for odd parity (to give a total of 5 ones) or should be 0 for even parity (to maintain the even number, 4). When a character is received, the receiver checks parity by counting the 1s in the character. If the parity is correct, the data is assumed to be correct. If the parity is wrong, an error has been detected.

Parity checking only detects a small fraction of possible errors. This can be intuitively understood by noting that a randomly chosen word has a 50% chance of having even parity and a 50% chance of having odd parity. Fortunately, on relatively error-free channels, single-bit errors are the most common and parity checking will always detect a single bit in error. However, an *even* number of errors will go undetected, whereas an *odd* number of errors will be detected. Parity checking is a simple error detection strategy. Because it is easy to implement, it is frequently used. Other more complex techniques are used in commercial data transmission to both detect and correct errors.

4.8.4 Standard Interface Buses

SIGNALING LEVELS

Inside equipment and for short runs of wire between equipment, the normal practice is to use neutral keying; that is, simply to key a voltage such as +5 V on and off. In neutral keying, the off condition is considered to be 0 V. Over longer runs of wire, the line is viewed as a transmission line, with distributed inductance and capacitance. It takes longer to make the transition from 0 to 1 or vice versa because of the additional inductance and capacitance. This decreases the maximum speed at which data can be transferred on the wire and also may cause the 1s and 0s to be different lengths, called bias distortion. Also, longer lines are more likely to pick up noise, which can make it difficult for the receiver to decide exactly when the transition takes place. Because of these problems, bipolar keying is used on longer lines. Bipolar keying uses one polarity (for example +) for a logical 1 and the other (− in this example) for a 0. This means that the decision threshold at the receiver is 0 V. Any positive voltage is taken as a 1 and any negative voltage as a 0.

EIA-RS-232

The most common serial bus protocol, EIA-RS-232, addresses this issue (however, a Mark "1" is a negative voltage and a Space "0" is positive). Generally called RS-232, this protocol defines connectors and voltages between data terminal equipment (DTE) such as a PC, and data communications equipment (DCE), such as a modem or TNC.

The connector is the DB-25 (25 pin), or DB-9 (9 pin) version — though RS-232 interfaces have also been implemented using nonstandard connectors such as RJ45, 3.5 mm audio plugs and many others. Signaling voltages are defined between +3 V and +25 V for logic "0" and between −3 V and −25 V for logic "1." Although the top data rate addressed in the specification is only 20 kbit/s, speeds of up to 115 kbit/s are commonly used. Communications distances of hundreds of meters are possible at reasonable data rates.

Since neutral keying is usually used inside equipment and bipolar keying for lines leaving equipment, signals must be converted between bipolar and neutral. Discrete level shifters or op amp circuits may perform this task, or low cost specialized IC line drivers and receivers are available.

RS-422

RS-422 is a serial protocol similar to RS-232, but employing fully differential data lines. Differential data offers the important advantage that common grounds between remote units are not necessary, and an important cause of ground loops (and their associated problems) is eliminated. Available on many Apple Macintosh computers, RS-422 systems may connect to standard RS-232 modems and TNCs by building a cable that makes the following translations:

RS-422 DTE	RS-232 DCE
RXD−	RXD
TXD−	TXD
RXD+	GND
TXD+	No connection
GPi	CD

IrDA (INFRARED DATA ACCESS)

Another high-speed serial protocol is the IrDA, which is a simple, short range wireless system using infrared LEDs and detectors. Data rates up to 3 MB/s are possible between compatible units.

UNIVERSAL SERIAL BUS (USB)

USB is a computer standard for an intelligent serial data transfer protocol, and has become the standard for nearly all connections between a PC and its external peripherals. It has largely replaced serial, parallel, keyboard and mouse ports as well as SCSI and numerous other buses in consumer PCs.

In addition to its higher speed than RS-232 and parallel ports, USB offers reasonable power availability to its loads, or functions. Under certain circumstances, up to 127 hubs and functions may connect to a single computer. USB requires that each function have on-board intelligence and that it negotiate with the host for power and bandwidth allocation. USB also has the major advantage of hot-pluggablity — the PC need not reboot when new functions are added.

The USB connectors use four-conductor cable, with two bidirectional, differential data lines, power, and ground. Approximately 5 V at 100 mA is allowed per function, with up to 500 mA available if the host system has the capability. This means that relatively sophisticated devices, such as modems, small video cameras, or hand-held scanners may operate from the bus without additional power supplies. It has also become common to use USB for connecting and charging a wide range of devices such as cellular phones, cameras and GPS units. When a USB connecting cable is used, proper connections and proper flow are ensured by using a rectangular (USB "A") connector on the host and a different connector (USB "B", Mini-B or Micro-B) on the attached function.

There are currently two USB standards in general use. USB 1.1, somewhat obsolete but common in PCs just a few years old, is capable of 12 megabits per second (12 Mbit/s). USB 2.0 is the later standard and is rated up to 480 Mbit/s. Most USB 2.0 ports will allow the use of older USB 1.1 devices — that is, they are backwards compatible. However maximum cable lengths and available power to devices may be affected. Recently introduced, the new USB 3.0 specification will use additional wires and some new connectors to enable communication at up to 5 Gbit/s in its SuperSpeed configuration.

Older PCs that have just one or two USB ports can have additional ports by adding an inexpensive USB port expander, commonly called a USB hub. Some units obtain their power from the host computer's USB port, and distribute only the USB signals along with the remaining power available from the host. Others come with a small power supply that provides the normal power to each new USB port.

If the PC does not have any USB ports, an expansion card can be used to add USB ports. It is quite common to see new personal computers without any serial or parallel ports, with USB replacing these functions. If you need to use your older serial or parallel peripheral devices with these "legacy-free" PC systems, adapter cables are available to connect them to USB ports.

IEEE-1394 (FIREWIRE)

A very high speed serial protocol, IEEE-1394 (christened "FireWire" by its creator, Apple Computer), is capable of up to 400 Mbit/s of sustained transfer. It is intended for high bandwidth systems, such as live video, external hard drives, or high-speed DVD player/recorders. Up to 63 devices may daisy-chain together at once via a standard six-wire cable. Unlike USB, 1394 is peer-to-peer, meaning any device may initiate a data transfer — the PC does not have to initiate a data transfer. Similar to USB, IEEE-1394 is hot-pluggable and provides power on the cable, but the voltage may vary from 7 V to almost 40 V, and may be sourced by any device. Allowable current drain per device may reach 1 A.

ETHERNET

Just a few years ago, most common office/home networks used 10Base$N$ Ethernet protocol. 10Base2 was generally recommended for Amateur Radio installations, since it uses shielded cable (RG-58, renamed "thin coax" in this application). Additionally, no separate hub is needed as the connected computers work on a peer-to-peer basis. However, 10Base2 and 10BaseT have been almost completely replaced by newer, much faster technologies. For wired networks, 100BaseT or "fast Ethernet" has been the standard for a number of years, with low cost switches commonly available. 100BaseT systems are rated to 100 Mbit/s and use unshielded, twisted-pair Category 5 network cable. Each computer on the network connects to a central hub or switch.

More recently, 1000BaseT (gigabit Ethernet, or GigE) has become more affordable and is quickly replacing 100BaseT even at the consumer level. Gigabit Ethernet also uses Cat 5 cable, though Cat 5e or Cat 6 is often recommended. Explanation of the MBase$N$ and Category N terminology can be found in many available networking books.

Mutual interference with ham station operation is not uncommon. A 10 Mbit/s signal, if all bit positions are filled, means that unshielded wire is carrying 10 MHz and various harmonics of 10 MHz. around the shack. When digital words are going through the network, any number of frequencies may be present on the wires and the wires may be very susceptible to pick-up from HF, VHF and UHF stations. In addition, cable and DSL modems may not only transmit various frequencies (especially on VHF) but lose data when a few hundred watts on a ham band is present. Some RF in the shack that can normally be ignored can easily bring down a network. Unfortunately, it is also common for this equipment to have "noisy" switching power supplies that can wreak havoc on a ham's sensitive receiver.

WIRELESS NETWORKS

The other technology commonly found in home and office networks is IEEE 802.11 wireless. These networks use low-power, spread spectrum transceivers operating in the 2.4 and 5.2-5.8 GHz range to transfer data at nominal rates up to 54 Mbit/s and higher. Older equipment used the 801.11b protocol at

11 Mbit/s; newer gear commonly found at low prices uses 802.11g at 54 Mbit/s maximum data rate. The newer 802.11n specification is coming into use now, with data rates up to several hundred megabits per second in the 2.4 and 5 GHz ranges.

These data rates are the maximum under ideal conditions, and it is not uncommon to see links running at significantly lower speeds as the *wireless access point* (WAP) and the wireless adapter dynamically adapt to the conditions. Still, performance is adequate for most uses, and the lack of network cabling makes wireless networking attractive in many situations.

Wireless networks actually appear to be less susceptible to mutual interference, since they are not connected to long runs of unshielded wire that easily act as both transmitting and receiving antennas. The frequency bands used for wireless networks are reasonably far removed from normal ham operations at frequencies below 2.4 GHz, but a high power VHF or UHF station may interfere with such a network. Also, some of the channels used for 802.11b and 802.11g network equipment fall within the amateur spectrum allocation in the 2.4 GHz band.

DISAPPEARING "LEGACY" PORTS

As technology advances, we are seeing more of the I/O ports commonly found on older PCs disappear. It is rare to find a laptop computer with serial or parallel ports. Even FireWire is being dropped by its creator, Apple Computer, in their new machines. The original crop of keyboard and mouse ports were replaced years ago by PS/2 ports, and those are being phased out. Many of these are being dropped in favor of USB, which is truly living up to its name as a "universal" serial bus. This can create a problem for amateurs, who often have a great deal of equipment that requires serial, parallel or even joystick ports.

So what can we do to replace these ports? Fortunately, there is a fairly wide selection of ways to cope with the disappearance of our old favorites. USB to serial and USB to parallel adapters are relatively inexpensive and easy to find. Note, however, that not all of them support all of the handshaking signals we may require, and some may have delays or inadequate drivers. Be prepared to try several different brands to find one that works with your equipment. While they are becoming less common, there is also still a market for add-in PCI bus cards to add serial, parallel and other ports to desktop computers.

Finally, though USB may be rather daunting for the individual experimenter, it need not be. There are chips available that connect to a USB host and present an easy to use serial or parallel interface that can easily be adapted to your own needs. FTDI makes several different versions, both serial and parallel. These are surface-mount chips with fine-pitch pins; prototyping can become difficult. Fortunately there are also plug-in modules that include the chip, its support parts, a USB connector, and an LED or two. These modules make it easy to plug into a PC board or solderless breadboard. FTDI also makes USB to TTL serial cable; one end terminates in a USB "A" connector, the other in a six-pin single inline connector that makes it easy to plug into your own project. Device drivers for the host system (*Windows*, *Linux* or *Mac OS*) make these devices appear as standard serial ports to your applications.

SOUND CARDS

There is often confusion surrounding the use of sound card ports. While a PC sound card can be used for a variety of digital modes, these are not digital ports. The inputs and outputs associated with a sound card are purely analog — at least at the connectors. The analog signals are digitized by the sound card, but from the user's perspective all of the normal conventions for analog connections should be used. More information on using sound cards in the amateur station may be found in the **Digital Communications** chapter.

STANDARD COMPUTER CONNECTIONS

See the **Component Data and References** chapter for details on computer connector pinouts.

4.9 Glossary of Digital Electronics Terms

AND gate — A logic circuit whose output is 1 only when all of its inputs are 1.

Astable (free-running) multivibrator — A circuit that alternates between two unstable states. This circuit could be considered as an oscillator that produces square waves.

Asynchronous flip-flop — A circuit, also called a *latch*, that changes output state depending on the data inputs, without requiring a clock signal.

Binary — A base-2 number system used in digital electronics that uses the symbols 0 and 1.

Binary coded decimal (BCD) — A simple method for converting binary values to and from decimal for inputs and outputs for user-oriented digital systems. BCD was widely used in the days of 7-segment LED displays but is not common today.

Bistable multivibrator — Another name for a flip-flop circuit that has two stable output states.

Boolean algebra — The mathematical system used to describe and design binary used digital circuits, named after George Boole.

Bus — A set of wires through which data is routed internally within computers and other digital devices.

Clock — A signal that toggles at a regular rate. Clock control is the most common method of synchronizing logic circuits.

Combinational logic — A type of circuit element in which the output depends on the present inputs. (Also see **Sequential logic**.)

Complementary metal-oxide semiconductor (CMOS) — A type of construction used to make digital integrated circuits. CMOS is composed of both N-channel and P-channel MOS devices on the same chip.

Counter (divider, divide-by-n counter) — A circuit that is able to change from one state to the next each time it receives an input signal. A counter produces an output signal every time a predetermined number of input signals have been received.

Decimal — The base-10 number system we use every day that uses the symbols 0 through 9.

DeMorgan's Theorem — In Boolean algebra, a way to simplify the complement of a large expression or to enable a designer to interchange a number of equivalent gates.

Digital IC — An integrated circuit whose output is either on (1) or off (0).

Dynamic (edge-triggered) input — A control signal that allows a circuit to change state only when the control signal changes from unasserted to asserted.

Exclusive OR gate — A logic circuit whose output is 1 when either of two inputs is 1 and whose output is 0 when neither input is 1 or when both inputs are 1.

Fan-out — The ability of a logic element

to drive or feed several other logic elements.

Field Programmable Gate Array (FPGA) — A type of programmable logic array that can contain several hundred to millions of logic gates and up to 1000 or more I/O pins.

FireWire (IEEE-1394) — A very high speed serial protocol capable of up to 400 Mbit/s of sustained transfer.

Flip-flop (bistable multivibrator) — A circuit that has two stable output states, and which can change from one state to the other when the proper input signals are detected.

Gate — A combinational logic element with two or more inputs and one output. The output state depends upon the state of the inputs.

Handshaking — Functions to coordinate data transfer.

Hexadecimal — A base-16 number system widely used in computer systems that uses the 0,1,2,3,4,5,6,7,8,9,A,B,C,D,E,F.

Integrated circuit — A device composed of many bipolar or field-effect transistors manufactured on the same chip, or wafer, of silicon.

Inverter — A logic circuit with one input and one output. The output is 1 when the input is 0, and the output is 0 when the input is 1.

Latch — Another name for a bistable multivibrator (flip-flop) circuit. The term latch reminds us that this circuit serves as a memory unit, storing a bit of information.

Linear IC — An integrated circuit whose output voltage is a linear (straight line) representation of its input voltage.

Logic probe — A simple piece of test equipment used to indicate high or low logic states (voltage levels) in digital-electronic circuits.

Microcontroller — A "computer on a chip" that usually consists of a relatively small microprocessor along with some amount of program memory, data memory, input/output ports and often some specialized peripheral devices.

Monostable multivibrator (one shot) — A circuit that has one stable state. It can be forced into an unstable state for a time determined by external components, but it will revert to the stable state after that time.

NAND (NOT AND) gate — A logic circuit whose output is 0 only when both inputs are 1.

Noninverter — A logic circuit with one input and one output, and whose output state is the same as the input state (0 or 1). Sometimes called a *noninverting buffer*.

NOR (NOT OR) gate — A logic circuit whose output is 0 if either input is 1.

OR gate — A logic circuit whose output is 1 when either input is 1.

Parallel — A digital signaling method in which all the bits in a group are handled exactly at the same time.

Programmable logic device (PLD) — A device that includes a generic array of gates that can be controlled by program code. PLDs can be made to replace large numbers of individual ICs.

Propagation delay — The time delay between providing an input to a digital circuit and seeing a response at the output.

Register — A set of latches or flip-flops storing an n-bit number.

RS-232 — The most common serial bus protocol.

RS-422 — A serial protocol similar to RS-232, but employing fully differential data lines.

Serial — A digital signaling method in which each bit is sent in turn over a single channel or wire, according to an agreed sequence.

Sequential logic — A type of circuit element in which the output depends on the present inputs, the previous sequence of inputs and often a clock signal. (Also see **Combinational logic**.)

Square wave — A periodic waveform that alternates between two values, and spends an equal time at each level. It is made up of sine waves at a fundamental frequency and all odd harmonics.

Static (level-triggered, or gated) input — A control signal that allows the circuit to change state whenever the control signal is at its active or asserted level.

Synchronous flip-flop — A circuit whose output state depends on the data inputs, but that will change output state only when it detects the proper clock signal.

Transition region — The undefined region between the two binary states. Also known as the *noise margin*.

Transition time — The time it takes a digital circuit to change state. The transition from a 0 to a 1 state is called the *rise time*, and the transition from a 1 to a 0 state is called the *fall time*.

Tri-state gate — A gate with one additional control lead. When enabled, the gate operates normally; when not enabled, the output goes to a high impedance.

Truth table — A chart showing the outputs for all possible input combinations to a digital circuit.

Universal serial bus (USB) — A computer standard for an intelligent serial data transfer protocol.

4.10 References and Bibliography

DIGITAL ELECTRONICS

Bignell and Donovan, *Digital Electronics* (Delmar Learning, 2006)

Holdsworth, B., *Digital Logic Design* (Newnes, 2002)

Lancaster and Berlin, *CMDS Cookbook* (Newnes, 1997)

Tocci, Widmer and Moss, *Digital Systems: Principles and Applications* (Prentice Hall, 2006)

Tokheim, R., *Digital Electronics: Principles and Applications* (McGraw-Hill, 2008)

Logic simulation software: "Getting Started with Digital Works" www.spu.edu/cs/faculty/bbrown/circuits/howto.html

MICROPROCESSORS

Dumas, J., *Computer Architecture: Fundamentals and Principles of Computer Design* (CRC Press, 2005)

Gilmore, C., *Microprocessors: Principles and Applications* (McGraw-Hill, 1995)

Korneev and Kiselev, *Modern Microprocessors* (Charles River Press, 2004)

Tocci and Ambrosio, *Microprocessors and Microcomputers: Hardware and Software* (Prentice-Hall, 2002)

Contents

5.1 Introduction
5.2 Lumped-Element versus Distributed Characteristics
5.3 Effects of Parasitic Characteristics
    5.3.1 Parasitic Inductance
    5.3.2 Parasitic Capacitance
    5.3.3 Inductors at Radio Frequencies
    5.3.4 Skin Effect
    5.3.5 RF Heating
    5.3.6 Effect on Q
    5.3.7 Self-Resonance
    5.3.8 Dielectric Breakdown and Arcing
    5.3.9 Radiative Losses
    5.3.10 Bypassing and Decoupling
    5.3.11 Effects on Filter Performance
5.4 Ferrite Materials
    5.4.1 Ferrite Permeability and Frequency
    5.4.2 Resonances of Ferrite Cores
    5.4.3 Ferrite Series and Parallel Equivalent Circuits
    5.4.4 Type 31 Material

5.5 Semiconductor Circuits at RF
    5.5.1 The Diode at High Frequencies
    5.5.2 The Transistor at High Frequencies
    5.5.3 Amplifier Classes
    5.5.4 RF Amplifiers with Feedback
5.6 Impedance Matching Networks
    5.6.1 L Networks
    5.6.2 Pi Networks
    5.6.3 T Networks
5.7 RF Transformers
    5.7.1 Air-Core Nonresonant RF Transformers
    5.7.2 Air-Core Resonant RF Transformers
    5.7.3 Broadband Ferrite RF Transformers
    5.7.4 Transmission Line Transformers and Baluns
5.8 Noise
    5.8.1 Noise in Amplifiers
    5.8.2 Noise Figure
5.9 Two-Port Networks
    5.9.1 Two-port Parameters
    5.9.2 Return Loss
5.10 RF Design Techniques Glossary
5.11 References and Further Reading

Chapter 5

RF Techniques

This chapter is a compendium of material from ARRL publications and other sources. It assumes the reader is familiar with the concepts introduced in the **Electrical Fundamentals** and **Analog Basics** chapters. The topics and techniques discussed here are associated with the special demands of circuit design in the HF and VHF ranges. The material is collected from previous editions of this book written by Leonard Kay, K1NU; *Introduction to Radio Frequency Design* by Wes Hayward, W7ZOI; and *Experimental Methods in RF Design* by Wes Hayward, W7ZOI, Rick Campbell, KK7B, and Bob Larkin, W7PUA. Material on ferrites is drawn from publications by Jim Brown, K9YC. The editor was Ward Silver, NØAX.

5.1 Introduction

When is an inductor not an inductor? When it's a capacitor! This statement may seem odd, but it suggests the main message of this chapter. In the earlier chapter, **Electrical Fundamentals**, the basic components of electronic circuits were introduced. As you may know from experience, those simple component pictures are ideal. That is, an ideal component (or element) by definition behaves exactly like the mathematical equations that describe it, and only in that fashion. For example, the current through an ideal capacitor is equal to the capacitance times the rate of change of the voltage across it without consideration of the materials or techniques by which a real capacitor is manufactured.

It is often said that, "Parasitics are anything you don't want," meaning that the component is exhibiting some behavior that detracts from or compromises its intended use. Real components only approximate ideal components, although sometimes quite closely. Any deviation from ideal behavior a component exhibits is called *non-ideal* or *parasitic*. The important thing to realize is that *every* component has parasitic aspects that become significant when it is used in certain ways. This chapter deals with parasitic effects that are commonly encountered at radio frequencies.

Knowing to what extent and under what conditions real components cease to behave like their ideal counterparts, and what can be done to account for these behaviors, allows the circuit designer or technician to work with circuits at radio frequencies. We will explore how and why the real components behave differently from ideal components, how we can account for those differences when analyzing circuits and how to select components to minimize, or exploit, non-ideal behaviors.

5.2 Lumped-Element versus Distributed Characteristics

Most electronic circuits that we use every day are inherently and mathematically considered to be composed of *lumped elements*. That is, we assume each component acts at a single point in space, and the wires that connect these lumped elements are assumed to be perfect conductors (with zero resistance and insignificant length). This concept is illustrated in **Fig 5.1**. These assumptions are perfectly reasonable for many applications, but they have limits. Lumped element models break down when:
- Circuit impedance is so low that the small, but non-zero, resistance in the wires is important. (A significant portion of the circuit power may be lost to heat in the conductors.)
- Operating frequency is high enough that the length of the connecting wires is a significant fraction (>0.1) of the wavelength causing the propagation delay along the conductor or radiation from it to affect the circuit in which it is used.
- Transmission lines are used as conductors. (Their characteristic impedance is usually significant, and impedances connected to them are transformed as a function of the line length. See the **Transmission Lines** chapter for more information.)

Effects such as these are called *distributed*, and we talk of *distributed elements* or effects to contrast them to lumped elements.

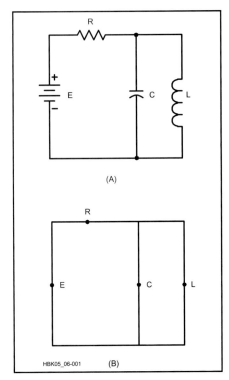

Fig 5.1 — The lumped element concept. Ideally, the circuit at A is assumed to be as shown at B, where the components are isolated points connected by perfect conductors. Many components exhibit nonideal behavior when these assumptions no longer hold.

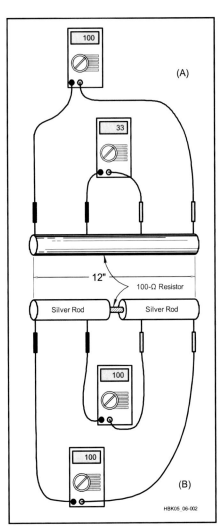

Fig 5.2 — Distributed (A) and lumped (B) resistances. See text for discussion.

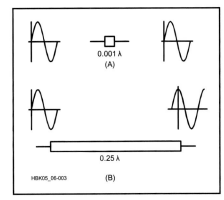

Fig 5.3 — The effects of distributed resistance on the phase of a sinusoidal current. There is no phase delay between ends of a lumped element.

To illustrate the differences between lumped and distributed elements, consider the two resistors in **Fig 5.2**, which are both 12 inches long. The resistor at A is a uniform rod of carbon. The second "resistor" B is made of two 6-inch pieces of silver rod (or other highly conductive material), with a small resistor soldered between them. Now imagine connecting the two probes of an ohmmeter to each of the two resistors, as in the figure. Starting with the probes at the far ends, as we slide the probes toward the center, the carbon rod will display a constantly decreasing resistance on the ohmmeter. This represents a distributed resistance. On the other hand, the ohmmeter connected to the other 12-inch "resistor" will display a constant resistance as long as one probe remains on each side of the small resistance and as long as we neglect the resistance of the silver rods! This represents a lumped resistance connected by perfect conductors.

Lumped elements also have the very desirable property that they introduce no phase shift resulting from propagation delay through the element. (Although combinations of lumped elements can produce phase shifts by virtue of their R, L and C properties.) Consider a lumped element that is carrying a sinusoidal current, as in **Fig 5.3A**. Since the element has negligible length, there is no phase difference in the current between the two sides of the element — *no matter how high the frequency* — precisely *because* the element length is negligible. If the physical length of the element were long, say 0.25 wavelength (0.25 λ) as shown in Fig 5.3B, the current phase would *not* be the same from end to end. In this instance, the current is delayed by 90 electrical degrees as it moves along the element. The amount of phase difference depends on the circuit's electrical length.

Because the relationship between the physical size of a circuit and the wavelength of an ac current present in the circuit will vary as the frequency of the ac signal varies, the ideas of lumped and distributed effects actually occupy two ends of a spectrum. At HF (30 MHz and below), where λ ≥ 10 m, the lumped element concept is almost always valid. In the UHF and microwave region (300 MHz and above), where λ ≤ 1 m and physical component size can represent a significant fraction of a wavelength, nearly all components and wiring exhibits distributed effects to one degree or another. From roughly 30 to 300 MHz, whether the distributed effects are significant must be considered on a case-by-case basis.

Of course, if we could make resistors, capacitors, inductors and so on, very small, we could treat them as lumped elements at much higher frequencies. For example, surface-mount components, which are manufactured in very small, leadless packages, can be used at much higher frequencies than leaded components and with fewer non-ideal effects.

It is for these reasons that circuits and equipment are often specified to work within specific frequency ranges. Outside of these ranges the designer's assumptions about the physical characteristics of the components and the methods and materials of the circuit's assembly become increasingly invalid. At frequencies sufficiently removed from the design range, circuit behavior often changes in unpredictable ways.

5.3 Effects of Parasitic Characteristics

At HF and above (where we do much of our circuit design) several other considerations become very important, in some cases dominant, in the models we use to describe our components. To understand what happens to circuits at RF we turn to a brief discussion of some electromagnetic and microwave theory concepts.

Parasitic effects due to component leads, packaging, leakage and so on are relatively common to all components. When working at frequencies where many or all of the parasitics become important, a complex but completely general model such as that in **Fig 5.4** can be used for just about any component, with the actual component placed in the box marked *. Parasitic capacitance, C_p, and leakage conductance, G_L, appear in parallel across the device, while series resistance, R_s, and parasitic inductance, L_s, appear in series with it. Package capacitance, C_{pkg}, appears as an additional capacitance in parallel across the whole device.

These small parasitics can significantly affect frequency responses of RF circuits. Either take steps to minimize or eliminate them, or use simple circuit theory to predict and anticipate changes. This maze of effects may seem overwhelming, but remember that it is very seldom necessary to consider all parasitics at all frequencies and for all applications. The **Computer-Aided Circuit Design** chapter shows how to incorporate the effect of multiple parasitics into circuit design and performance modeling.

5.3.1 Parasitic Inductance

Maxwell's equations — the basic laws of electromagnetism that govern the propagation of electromagnetic waves and the operation of all electronic components — tell us that any wire carrying a current that changes with time (one example is a sine wave) develops a changing magnetic field around it. This changing magnetic field in turn induces an opposing voltage, or *back EMF*, on the wire. The back EMF is proportional to how fast the current changes (see **Fig 5.5**).

We exploit this phenomenon when we make an inductor. The reason we typically form inductors in the shape of a coil is to concentrate the magnetic field and thereby maximize the inductance for a given physical size. However, *all* wires carrying varying currents have these inductive properties. This includes the wires we use to connect our circuits, and even the *leads* of capacitors, resistors and so on. The inductance of a straight, round, nonmagnetic wire in free space is given by:

$$L = 0.00508\, b \left[\ln\left(\frac{2b}{a}\right) - 0.75 \right] \quad (1)$$

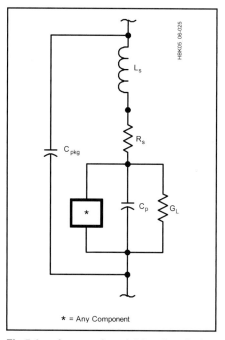

Fig 5.4 — A general model for electrical components at VHF frequencies and above. The box marked * represents the component itself. See text for discussion.

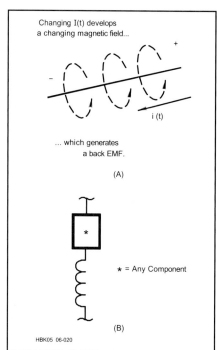

Fig 5.5 — Inductive consequences of Maxwell's equations. At A, any wire carrying a changing current develops a voltage difference along it. This can be mathematically described as an effective inductance. B adds parasitic inductance to a generic component model.

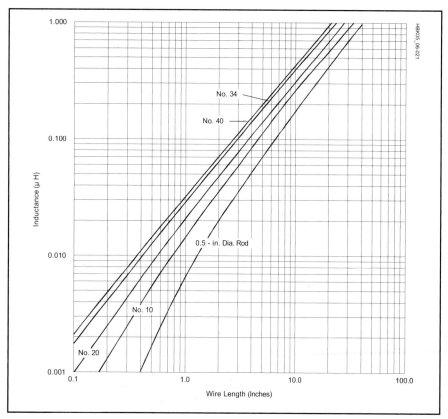

Fig 5.6 — A plot of inductance vs length for straight conductors in several wire sizes.

where
L = inductance, in µH
a = wire radius, in inches
b = wire length, in inches
ln = natural logarithm (2.303 × log₁₀)

Skin effect (discussed below) changes this formula slightly at VHF and above. As the frequency approaches infinity, the value 0.75 in the above equation increases to approach 1. This effect usually causes a change of no more than a few percent.

As an example, let's find the inductance of a typical #18 wire (diameter = 0.0403 inch and a = 0.0201) that is 4 inches long (b = 4):

$$L = 0.00508\,(4)\left[\ln\left(\frac{8}{0.0201}\right) - 0.75\right]$$

$$= 0.0203\,[5.98 - 0.75] = 0.106\ \mu H$$

Wire of this diameter has an inductance of about 25 nH per inch of length. In circuits operating at VHF and higher frequencies, including high-speed digital circuits, the inductance of component leads can become significant. (The #24 AWG wire typically used for component leads has an inductance on the order of 20 nH per inch.) At these frequencies, lead inductance can affect circuit behavior, making the circuit hard to reproduce or repair. Good design and construction practice is to minimize the effects of lead inductance by using surface-mount components or trimming the leads to be as short as possible.

The impact of reactance due to parasitic inductance is usually very small; at AF or LF, parasitic inductive reactance of most components is practically zero. To use this example, the reactance of a 0.106 µH inductor even at 10 MHz is only 6.6 Ω. **Fig 5.6** shows a graph of the inductance for wires of various gauges (radii) as a function of length. Whether the reactance is significant or not depends on the application and the frequency of use.

We can represent parasitic inductance in component models by adding an inductor of appropriate value in series with the component since the wire leads are in series with the element. This (among other reasons) is why minimizing lead lengths and interconnecting wires becomes very important when designing circuits for VHF and above.

PARASITIC INDUCTANCE IN RESISTORS

The basic construction of common resistor types is shown in **Fig 5.7**. The primary parasitic effect associated with resistors is parasitic inductance. (Some parasitic capacitance exists between the leads or electrodes due to packaging.) **Fig 5.8** shows some more accurate circuit models for resistors at low to medium frequencies. The type of resistor with the most parasitic inductance are wire-wound resistors, essentially inductors used as resistors. Their use is therefore limited to dc or low-frequency ac applications where their reactance is negligible. Remember that this inductance will also affect switching transient waveforms, even at dc, because the component will act as an RL circuit. The inductive effects of wire-wound resistors begin to become significant in the audio range above a few kHz.

As an example, consider a 1-Ω wire-wound resistor formed from 300 turns of #24 wire closely-wound in a single layer 6.3 inches long on a 0.5-inch diameter form. What is its approximate inductance? From the inductance formula for air-wound coils in the **Electrical Fundamentals** chapter:

$$L = \frac{d^2 n^2}{18d + 40l} = \frac{0.5^2 \times 300^2}{(18 \times 0.5) + (40 \times 6.3)} = 86\ \mu H$$

If we want the inductive reactance to be less than 10% of the resistor value, then this resistor cannot be used above f = 0.1 / (2π × 86 µH) = 185 Hz! Real wire-wound resistors have multiple windings layered over each other to minimize both size and parasitic inductance (by winding each layer in opposite directions, much of the inductance is canceled). If we assume a five-layer winding, the length is reduced to 1.8 inches and the inductance to approximately 17 µH, so the resistor can then be used below 937 Hz. (This has the effect of increasing the resistor's parasitic capacitance, however.)

The resistance of certain types of tubular film resistors is controlled by inscribing a spiral path through the film on the inside of the tube. This creates a small inductance that may be significant at and above the higher audio frequencies.

NON-INDUCTIVE RESISTORS

The resistors with the least amount of parasitic inductance are the bulk resistors, such as carbon-composition, metal-oxide, and ceramic resistors. These resistors are made from a single linear cylinder, tube or block of resis-

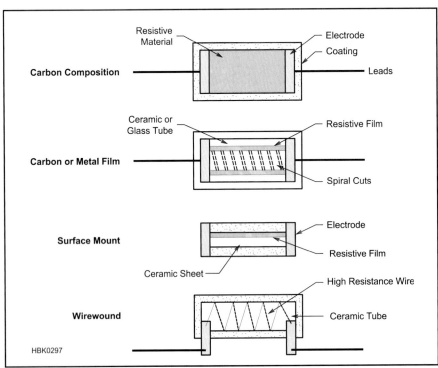

Fig 5.7 — The electrical characteristics of different resistor types are strongly affected by their construction. Reactance from parasitic inductance and capacitance strongly impacts the resistor's behavior at RF.

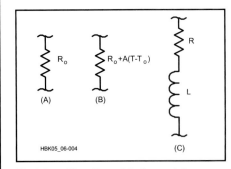

Fig 5.8 — Circuit models for resistors. The wire-wound model with associated inductance is shown at C. For designs at VHF and higher frequencies, the model at C could be used with L representing lead inductance.

tive material so that inductance is minimized. Each type of resistor has a maximum usable frequency, above which parasitic capacitance and inductance begin to become significant. Review the manufacturer's data sheet for the component to learn about its performance at high frequencies.

Some resistors advertised as "noninductive" are actually wire-wound resistors with a special winding technique that minimizes inductance. These resistors are intended for use at audio frequencies and are not suitable for use at RF. If you are not sure, ask the vendor if the resistors are suitable for use in RF circuits.

Because resistors are manufactured with an insulating coating, it can be difficult to determine their internal structure and thus estimate their parasitic inductance. In cases where a surplus or used component is to be included, it is recommended that you test the component with an impedance meter or make some other type of reactance measurement if you are unable to access the manufacturer's specifications for the resistor.

PARASITIC INDUCTANCE IN CAPACITORS

The size and shape of a capacitor's plates and the leads used to connect them to circuits create parasitic inductance, often referred to as *equivalent series inductance* (ESL) by capacitor manufacturers. **Figs 5.9** and **5.10** show reasonable models for capacitors that are good up to VHF.

Fig 5.11 shows a roll-type capacitor made of two strips of very thin metal foil and separated by a dielectric. After leads are attached to the foil strips, the sandwich is rolled up and either placed in a metal can or coated with plastic. *Radial leads* both stick out of one end of the roll and *axial leads* from both ends along the roll's axis. Because of the rolled strips, the ESL is high. Electrolytic and many types of film capacitors are made with roll construction. As a result, they are generally not useful in RF circuits.

In the stack capacitor, thin sheets of dielectric are coated on one side with a thin metal layer. A stack of the sheets is placed under pressure and heated to make a single solid unit. Metal side caps with leads attached contact the metal layers. The ESL of stack capacitors is very low and so they are useful at high frequencies. Ceramic and mica capacitors are the most common stack-style capacitor.

Parallel-plate air and vacuum capacitors used at RF have relatively low parasitic inductance, but transmitting capacitors made to withstand high voltages and current are large enough that parasitic inductance becomes significant, limiting their use to low-VHF and lower frequencies. Adjustable capacitors (air variables and compression or piston trimmer

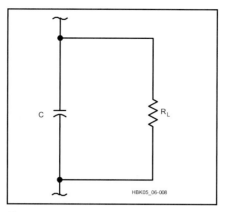

Fig 5.9 — A simple capacitor model for frequencies well below self-resonance.

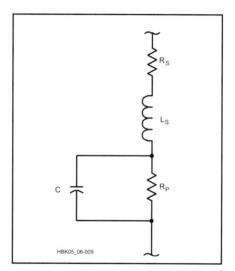

Fig 5.10 — A capacitor model for VHF and above including series resistance and distributed inductance.

capacitors) for tuning low-power circuits are much smaller and so have correspondingly lower parasitic inductance.

It is difficult for a single capacitor to work well over a very wide frequency range, so capacitors are often placed in parallel as discussed in the section below on **Bypassing and Decoupling**. It is often suggested that different types of capacitors be connected in parallel to avoid the effects of parasitic inductance at different frequencies, but without testing and careful modeling the results are often unpredictable or even counterproductive as the referenced discussion shows.

5.3.2 Parasitic Capacitance

Maxwell's equations also tell us that if the voltage between any two points changes with time, a displacement current is generated between these points as illustrated in **Fig 5.12**. This *displacement current* results from the propagation of the electromagnetic field between the two points and is not to be confused with *conduction current*, which is the movement of electrons. Displacement current is directly proportional to the rate at which the voltage is changing.

When a capacitor is connected to an ac voltage source, a steady ac current can flow because taken together, conduction current and displacement current "complete the loop" from the positive source terminal, across the plates of the capacitor, and back to the negative terminal.

In general, parasitic capacitance shows up *wherever* the voltage between two points is changing with time, because the laws of electromagnetics require a displacement current to flow. Since this phenomenon

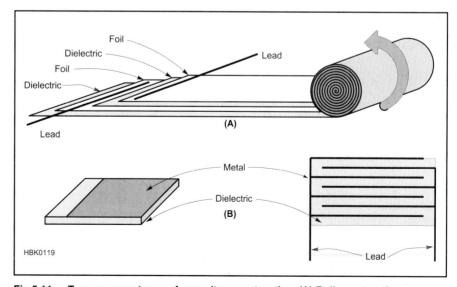

Fig 5.11 — Two common types of capacitor construction. (A) Roll construction uses two strips of foil separated by a strip of dielectric. (B) Stack construction layers dielectric material (such as ceramic or film), one side coated with metal. Leads are attached and the assembly coated with epoxy resin.

RF Techniques 5.5

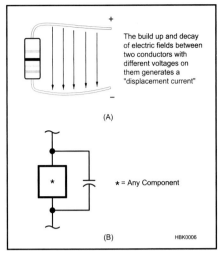

Fig 5.12 — Capacitive consequences of Maxwell's equations. A: Any changing voltage between two points, for example along a bent wire, generates a displacement current running between them. This can be mathematically described as an effective capacitance. B adds parasitic capacitance to a generic component model.

represents an *additional* current path from one point in space to another, we can add this parasitic capacitance to our component models by adding a capacitor of appropriate value in *parallel* with the component. These parasitic capacitances are typically less than 1 pF, so that below VHF they can be treated as open circuits (infinite reactances) and thus neglected.

PACKAGE CAPACITANCE

Another source of capacitance, also in the 1-pF range and therefore important only at VHF and above, is the packaging of the component itself. For example, a power transistor packaged in a TO-220 case (see **Fig 5.13**), often has either the emitter or collector connected to the metal tab itself. This introduces an extra *inter-electrode capacitance* across the junctions.

The copper traces on a PC board also create capacitance with the circuit components, to other traces, and to power and ground planes. Double-sided PC boards have a certain capacitance per square inch between the layers of copper on each side of the board. (Multi-layer PC boards have higher values of capacitance due to the smaller separation between layers.) It is possible to create capacitors by leaving unetched areas of copper on both sides of the board. The dielectric constant of inexpensive PC board materials intended for use at low-frequencies is not well-controlled, however, leading to significant variations in capacitance. For this reason, the copper on one side of a double-sided board should be completely removed under frequency-determining circuits such as VFOs.

Stray capacitance (a general term used for any "extra" capacitance that exists due to physical construction) appears in any circuit where two metal surfaces exist at different voltages. Such effects can be modeled as an extra capacitor in parallel with the given points in the circuit. A rough value can be obtained with the parallel-plate formula given in the chapter on **Electrical Fundamentals**. Similar to parasitic inductance, *any* circuit component that has wires attached to it, or is fabricated from wire, or is near or attached to metal, will have a parasitic capacitance associated with it, which again, becomes important only at RF.

Stray capacitance can be difficult to account for in circuit design because it exists *between* components and other circuit structures, depending on the physical orientation of the component. Its presence may allow signals to flow in ways that disrupt the normal operation of a circuit and may have a greater affect in a high-impedance circuit because the capacitive reactance may be a greater percentage of the circuit impedance. Also, because stray capacitance often appears in parallel with the circuit, the stray capacitor may bypass more of the desired signal at higher frequencies. Careful physical design of an RF circuit and selection of components can minimize the effects of stray capacitance.

5.3.3 Inductors at Radio Frequencies

Inductors are perhaps the component with the most significant parasitic effects. Where there are many different varieties of form for capacitors and resistors, most inductors are fundamentally similar: a coil of wire on a tubular or toroidal form. As such, they are affected by both parasitic resistance and capacitance as shown in the simple inductor model of **Fig 5.14**.

While the leakage conductance of a capacitor is usually negligible, the series resistance of an inductor often is not. This is caused by the long lengths of thin wire needed to create typical inductance values used in RF circuits and the skin effect (discussed below). Consider a typical air-core inductor, with L = 33 mH and a minimum Q of 30 measured at 2.5 MHz. This would indicate a series resistance of $R_s = 2\pi f L / Q = 17\,\Omega$ that could significantly alter the bandwidth of a circuit. The skin effect also results in parasitic resistance in the upper HF ranges and above. To minimize losses due to the skin effect, inductors at these frequencies, particularly those intended for use in transmitters and amplifiers and that are expected to carry significant currents, are often made from large-diameter wire or tubing or even flat strap to maximize the surface area for current flow. For coils with many turns for which large conductors are impractical, *Litz wire* is sometimes used. Litz wire is made of many fine insulated strands woven together, each with a diameter smaller than the skin depth at the expected frequency of use, thus presenting a larger surface area than a solid wire or normal stranded wire.

If an inductor is wound on a magnetic core, the core itself can have losses that are treated as parasitic resistance. Each type of core — iron, powdered iron or ferrite — has a frequency range over which it is designed for maximum efficiency. Outside of that range, the core may have significant losses, raising the parasitic resistance of the inductor.

Parasitic capacitance is a particular concern

Fig 5.13 — Unexpected stray capacitance. The mounting tab of TO-220 transistors is often connected to one of the device leads. Because one lead is connected to the chassis, small capacitances from the other lead to the chassis appear as additional package capacitance at the device. Similar capacitance can appear at any device with a conductive package.

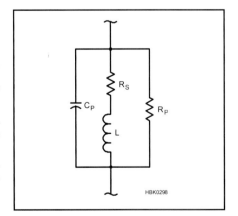

Fig 5.14 — General model for inductor with parasitic capacitance and resistance. The parasitic capacitance represents the cumulative effect of the capacitance between the different turns of wire. Parasitic resistance, R_S, depends on frequency due to the skin effect. R_P represents leakage resistance. At low frequencies, parasitic capacitance, C_P, can be neglected.

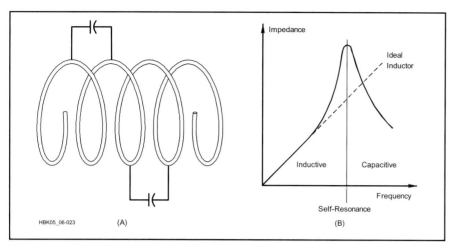
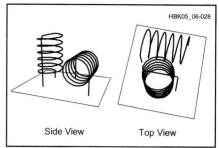

Fig 5.16 — Unshielded coils in close proximity should be mounted perpendicular to each other to minimize coupling.

Fig 5.15 — Coils exhibit distributed capacitance as explained in the text. The graph at B shows how distributed capacitance resonates with the inductance. Below resonance, the reactance is predominantly inductive which increases as frequency increases. However, above resonance, the reactance becomes predominantly capacitive which decreases as frequency increases.

for inductors because of their construction. Consider the inductor in **Fig 5.15**. If this coil has n turns, then the ac voltage between identical points of two neighboring turns is 1/n times the ac voltage across the entire coil. When this voltage changes due to an ac current passing through the coil, the effect is that of many small capacitors acting in parallel with the inductance of the coil. Thus, in addition to the capacitance resulting from the leads, inductors have higher parasitic capacitance due to their physical shape.

These various effects illustrate why inductor construction has such a large effect on performance. As an example, assume you're working on a project that requires you to wind a 5 µH inductor. Looking at the coil inductance formula in the **Electrical Fundamentals** chapter, it comes to mind that many combinations of length and diameter could yield the desired inductance. If you happen to have both 0.5 and 1-inch coil forms, why should you select one over the other? To eliminate some other variables, let's make both coils 1 inch long, close-wound, and give them 1-inch leads on each end.

Let's calculate the number of turns required for each. On a 0.5-inch-diameter form:

$$n = \frac{\sqrt{L(18d + 40l)}}{d} =$$

$$\frac{\sqrt{5[(18 \times 0.5) + (40 \times 1)]}}{0.5} = 31.3 \text{ turns}$$

This means coil 1 will be made from #20 AWG wire (29.9 turns per inch). Coil 2, on the 1-inch form, yields

$$n = \frac{\sqrt{5[(18 \times 1) + (40 \times 1)]}}{1} = 17 \text{ turns}$$

which requires #15 AWG wire in order to be close-wound.

What are the series resistances associated with each? For coil 1, the total wire length is 2 inches + (31.3 × π × 0.5) = 51 inches, which at 10.1 Ω/1000 ft gives $R_s = 0.043$ Ω at dc. Coil 2 has a total wire length of 2 inches + (17.0 × π × 1) = 55 inches, which at 3.18 Ω/1000 ft gives a dc resistance of $R_s = 0.015$ Ω, or about ⅓ that of coil 1. Furthermore, at RF, coil 1 will begin to suffer from skin effect at a frequency about 3 times lower than coil 2 because of its smaller conductor diameter. Therefore, if Q were the sole consideration, it would be better to use the larger diameter coil.

Q is not the only concern, however. Such coils are often placed in shielded enclosures. A common rule of thumb says that to prevent the enclosure from affecting the inductor, the enclosure should be at least one coil diameter from the coil on all sides. That is, 3×3×2 inches for the large coil and 1.5×1.5×1.5 for the small coil, a volume difference of over 500%.

INDUCTOR COUPLING

Mutual inductance (see the **Electrical Fundamentals** chapter) will also have an effect on the resonant frequency and Q of RF circuits. For this reason, inductors in frequency-critical circuits should always be shielded, either by constructing compartments for each circuit block or through the use of can-mounted coils. Another helpful technique to minimize coupling is to mount nearby coils with their axes perpendicular as in **Fig 5.16**.

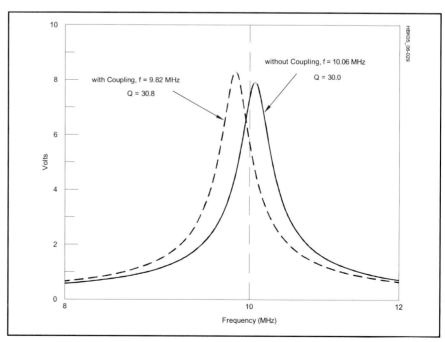

Fig 5.17 — Result of light coupling (k = 0.05) between two identical circuits of Fig 5.19A on their frequency responses.

As an example, assume we build an oscillator circuit that has both input and output resonant circuits. If we are careful to keep the two coils in these circuits uncoupled, the frequency response of either of the two circuits is that of the solid line in **Fig 5.17**.

If the two coils are coupled either through careless placement or improper shielding, the resonant frequency and Q will be affected. The dashed line in Fig 5.17 shows the frequency response that results from a coupling coefficient of k = 0.05, a reasonable value for air-wound inductors mounted perpendicularly in close proximity on a circuit chassis. Note the resonant frequency shifted from 10.06 to 9.82 MHz, or 2.4%. The Q has gone up slightly from 30.0 to 30.8 as a result of the slightly higher inductive reactance at the resonant frequency.

5.3.4 Skin Effect

The resistance of a conductor to ac is different than its value for dc. A consequence of Maxwell's equations is that thick, near-perfect conductors (such as metals) conduct ac only to a certain *skin depth*, δ, inversely proportional to the square root of the frequency of the current. Called the *skin effect*, this decreases the effective cross-section of the conductor at high frequencies and thus increases its resistance.

$$\delta = \frac{1}{\sqrt{\pi f \mu \sigma}} \qquad (2)$$

where
  μ is the conductor's permeability, and
  σ is the conducting material's conductance.

The increase in resistance caused by the skin effect is insignificant at and below audio frequencies, but beginning around 1 MHz (depending on the size of the conductor) it is so pronounced that practically all the current flows in a very thin layer near the conductor's surface. For example, at 10 MHz, the skin depth of a copper conductor is about 0.02 mm (0.00079 inch). For this reason, at RF a hollow tube and a solid rod of the same diameter and made of the same metal will have the same resistance and the RF resistance of a conductor is often much higher than its dc resistance. A rough estimate of the frequency above which a nonmagnetic wire (one made of nonferrous metal) will begin to show appreciable skin effect can be calculated from

$$f = \frac{124}{d^2} \qquad (3)$$

where
  f = frequency, in MHz
  d = diameter in mils (a mil is 0.001 inch)

Above this frequency, increase the resistance of the wire by 10× for every 2 decades of frequency (roughly 3.2× for every decade). For example, say we wish to find the RF resistance of a 2-inch length of #18 AWG copper wire at 100 MHz. From the wire tables in the **Component Data and References** chapter, we see that this wire has a dc resistance of 2 inches × 6.386 Ω/1000 ft = 1.06 mΩ. From the above formula, the frequency is found to be 124 / 40.3² = 76 kHz. Since 100 MHz is roughly three decades above this (100 kHz to 100 MHz), the RF resistance will be approximately 1.06 mΩ × 3.2³ = 1.06 mΩ × 32.8 = 34.8 mΩ. Again, values calculated in this manner are approximate and should be used qualitatively — like when you want an answer to a question such as, "Can I neglect the RF resistance of this length of connecting wire at 100 MHz?" Several useful charts regarding skin effect are available in *Reference Data for Engineers*, listed in the References section of this chapter.

Losses associated with skin effect can be reduced by increasing the surface area of the conductor carrying the RF current. Flat, solid strap and tubing are often used for that reason. In addition, because the current-carrying layer is so thin at UHF and microwave frequencies, a thin highly conductive layer at the surface of the conductor, such as silver plating, can lower resistance. Silver plating is too thin to improve HF conductivity significantly, however.

5.3.5 RF Heating

RF current often causes component heating problems where the same level of dc or low frequency ac current may not. These losses result from both the skin effect and from dielectric losses in insulating material, such as a capacitor dielectric.

An example is the tank circuit of an RF oscillator. If several small capacitors are connected in parallel to achieve a desired capacitance, skin effect will be reduced and the total surface area available for heat dissipation will be increased, thus significantly reducing the RF heating effects as compared to a single large capacitor. This technique can be applied to any similar situation; the general idea is to divide the heating among as many components as possible.

An example is shown in the circuit block in **Fig 5.18**, which is representative of the input tank circuit used in many HF VFOs. Along with L, C_{main}, C_{trim}, C1 and C3 set the oscillator frequency. Therefore, temperature effects are critical in these components. By using several capacitors in parallel, the RF current (and resultant heating, which is proportional to the square of the current) is reduced in each component. Parallel combinations are used for the feedback capacitors for the same reason.

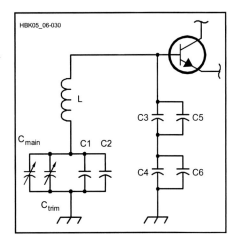

Fig 5.18 — A tank circuit of the type commonly used in VFOs. Several capacitors are used in parallel to distribute the RF current, which reduces temperature effects.

At high power levels, losses due to RF heating — either from resistive or dielectric losses — can be significant. Insulating materials may exhibit dielectric losses even though they are excellent insulators. For example, nylon plastic insulators may work very well at low frequencies, but being quite lossy at and above VHF, they are unsuitable for use in RF circuits at those frequencies. To determine whether material is suitable for use as an insulator at RF, a quick test can be made by heating the material in a microwave oven for a few seconds and measuring its temperature rise. (Take care to avoid placing parts or materials containing any metal in the microwave!) Insulating materials that heat up are lossy and thus unsuitable for use in RF circuits.

5.3.6 Effect on Q

Recall from the **Electrical Fundamentals** chapter that circuit Q, a useful figure of merit for tuned RLC circuits, can be defined in several ways:

$$Q = \frac{X_L \text{ or } X_C}{R} \qquad (4)$$

$$= \frac{\text{energy stored per cycle}}{\text{energy dissipated per cycle}}$$

Q is also related to the bandwidth of a tuned circuit's response by

$$Q = \frac{f_0}{BW_{3\,dB}} \qquad (5)$$

Parasitic inductance, capacitance and resistance can significantly alter the performance and characteristics of a tuned circuit if the design frequency is close to the self-resonant frequencies of the components.

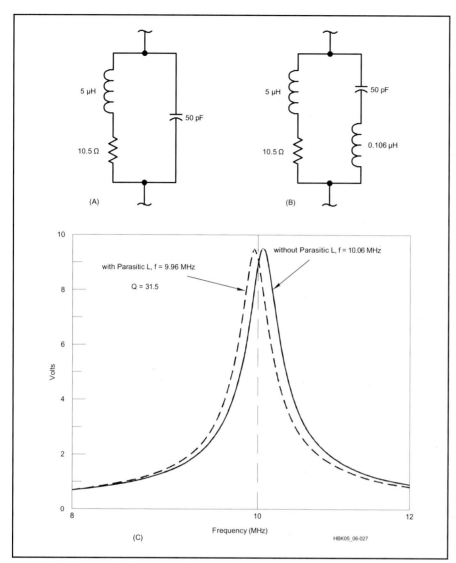

Fig 5.19 — A is a tank circuit, neglecting parasitics. B is the same circuit including L_p on capacitor. C is the frequency response curves for A and B. The solid line represents the unaltered circuit (also see Fig 5.17) while the dashed line shows the effects of adding parasitic inductance.

ics. The results are shown in the plot in Fig 5.19C, where we can see that the parasitic circuit has an f_0 of 9.96 MHz (a shift of 1%) and a Q (measured from the –3 dB points) of 31.5. For comparison, the simulation of the unaltered circuit does in fact show f_0 = 10.06 MHz and Q = 30.

5.3.7 Self-Resonance

Because of parasitic effects, a capacitor or inductor—all by itself—exhibits the properties of a resonant RLC circuit at frequencies for which the parasitic effects are significant. Figs 5.10 and 5.14 illustrate RF models for the capacitor and inductor, which are based on the general model in Fig 5.4, leaving out the packaging capacitance. Note the slight difference in configuration; the pairs C_p-R_p and L_s-R_s are in series in the capacitor but in parallel in the inductor. This is because of the different physical structure of the components.

At some sufficiently high frequency, both inductors and capacitors become *self-resonant* when the parasitic reactance cancels or equals the intended reactance, creating a series-resonant circuit. Similar to a series-resonant circuit made of discrete components, above the self-resonant frequency a capacitor will appear inductive, and an inductor will appear capacitive.

For an example, let's calculate the approximate self-resonant frequency of a 470-pF capacitor whose leads are made from #20 AWG wire (0.032-inch diameter), with a total length of 1 inch. Using equation 1, we calculate the approximate parasitic inductance

$$L\ (\mu H) = 0.00508\ (1) \left[\ln\left(\frac{2(1)}{(0.032/2)}\right) - 0.75 \right]$$

$$= 0.021\ \mu H$$

and the self-resonant frequency is roughly

$$f = \frac{1}{2\pi\sqrt{LC}} = 50.6\ \text{MHz}$$

Similarly, an inductor can also have a parallel self-resonance.

The purpose of making these calculations is to provide a feel for actual component values. They could be used as a rough design guideline, but should not be used quantitatively. Other factors such as lead orientation, shielding and so on, can alter the parasitic effects to a large extent. Large-value capacitors tend to have higher parasitic inductances (and therefore a lower self-resonant frequency) than small-value capacitors.

Self-resonance becomes critically important at VHF and UHF because the self-resonant frequency of many common components is at or below the frequency where the component will be used. In this case, either special techniques can be used to construct

As an example, consider the resonant circuit of **Fig 5.19A**, which could represent the frequency-determining resonant circuit of an oscillator. Neglecting any parasitics,

$$f_0 = \frac{1}{2\pi\sqrt{LC}} = 10.06\ \text{MHz}$$

As in many practical cases, assume the resistance arises entirely from the inductor series resistance. The data sheet for the inductor specified a minimum Q of 30, so assuming Q = 30 yields an R value of

$$\frac{X_L}{Q} = \frac{2\pi\ (10.06\ \text{MHz})\ (5\ \mu H)}{30} = 10.5\ \Omega$$

Next, let's include the parasitic inductance of the capacitor (Fig 5.19B). A reasonable assumption is that this capacitor has the same physical size as the example from the section on Parasitic Inductance for which we calculated L_s = 0.106 µH. This would give the capacitor a self-resonant frequency of 434 MHz — well above our frequency of interest. However, the added parasitic inductance does account for an extra 0.106/5.0 = 2% inductance. Since this circuit is no longer strictly series or parallel, we must convert it to an equivalent form before calculating the new f_0.

An easier and faster way is to *simulate* the altered circuit by computer. This analysis was performed on a desktop computer using *SPICE*, a standard circuit simulation program described in the **Computer-Aided Circuit Design** chapter. The voltage response of the circuit (given an input current of 1 mA) was calculated as a function of frequency for both cases, with and without parasit-

components to operate at these frequencies, by reducing the parasitic effects, or else the idea of lumped elements must be abandoned altogether in favor of microwave techniques such as striplines and waveguides.

5.3.8 Dielectric Breakdown and Arcing

Anyone who has ever seen an arc form across a transmitting capacitor's plates, seen static discharges jump across an antenna insulator or touched a metal doorknob on a dry day has experienced the effects of dielectric breakdown.

In the ideal world, we could take any two conductors and put as large a voltage as we want across them, no matter how close together they are. In the real world, there is a voltage limit (*dielectric strength*, measured in kV/cm and determined by the insulating material between the two conductors) above which the insulator will break down.

Because they are charged particles, the electrons in the atoms of a dielectric material feel an attractive force when placed in an electric field. If the field is sufficiently strong, the force will strip the electron from the atom. This electron is available to conduct current, and furthermore, it is traveling at an extremely high velocity. It is very likely that this electron will hit another atom, and free another electron. Before long, there are many stripped electrons producing a large current, forming an *arc*. When this happens, we say the dielectric has suffered *breakdown*. Arcing in RF circuits is most common in transmitters and transmission line components where high voltages are common, but it is possible anywhere two components at significantly different voltage levels are closely spaced.

If the dielectric is liquid or gas, it will repair itself when the applied voltage is removed and the molecules in the dielectric return to their normal state. A solid dielectric, however, cannot repair itself because its molecules are fixed in place and the low-resistance path created by the arc is permanent. A good example of this is a CMOS integrated circuit. When exposed to the very high voltages associated with static electricity, the electric field across the very thin gate oxide layer exceeds the dielectric strength of silicon dioxide, and the device is permanently damaged by the resulting hole created in the oxide layer.

The breakdown voltage of a dielectric layer depends on its composition and thickness (see **Table 5.1**). The variation with thickness is not linear; doubling the thickness does not quite double the breakdown voltage. Breakdown voltage is also a function of geometry: Because of electromagnetic considerations, the breakdown voltage between two conductors separated by a fixed distance is less if the surfaces are pointed or sharp-edged than if they are smooth or rounded. Therefore, a simple way to help prevent breakdown in many projects is to file and smooth the edges of conductors. (See the **Power Supplies** and **RF Power Amplifier** chapters for additional information on high voltage applications.)

Capacitors are, by nature, the component most often associated with dielectric failure. To prevent damage, the working voltage of a capacitor — and there are separate dc and ac ratings — should ideally be two or three times the expected maximum voltage in the circuit. Capacitors that are not air-insulated or have a *self-healing dielectric* should be replaced if a dielectric breakdown occurs.

Resistors and inductors also have voltage ratings associated with breakdown of their insulating coating. High-value resistors, in particular, can be bypassed by leakage current flowing along the surface of the resistor. High-voltage resistors often have elongated bodies to create a long *leakage path* to present a high resistance to leakage current. Cleaning the bodies of resistors and inductors in high-voltage circuits helps prevent arcs from forming and minimizes leakage current.

5.3.9 Radiative Losses

Any conductor placed in an electromagnetic field will have a current induced in it. We put this principle to good use when we make an antenna. The unwelcome side of this law of nature is the phrase "any conductor"; even conductors we don't intend to act as antennas will respond this way.

Fortunately, the efficiency of such "antennas" varies with conductor length. They will be of importance only if their length is a significant fraction of a wave-length. When we make an antenna, we usually choose a length on the order of $\lambda/2$. Therefore, when we *don't* want an antenna, we should be sure that the conductor length is *much less* than $\lambda/2$, no more than 0.1 λ. This will ensure a very low-efficiency antenna. This is why even unpaired 60-Hz power lines do not lose a significant fraction of the power they carry — at 60 Hz, 0.1 λ is about 300 miles!

In addition, we can use shielded cables. Such cables do allow some penetration of EM fields if the shield is not solid, but even 95% coverage is usually sufficient, especially if some sort of RF choke is used to reduce shield current.

Radiative losses and coupling can also be reduced by using twisted or parallel pairs of conductors — the fields tend to cancel. In some applications, such as audio cables, this may work better than shielding. Critical stages such as tuned circuits should be placed in shielded compartments where possible.

This argument also applies to large components — remember that a component or long wire can both radiate and receive RF energy. Measures that reduce radiative losses will also reduce unwanted RF pickup. See the **EMC** chapter for more information.

5.3.10 Bypassing and Decoupling

BYPASSING

Circuit models showing ac behavior often show ground connections that are not at dc ground. Bias voltages and currents are neglected in order to show how the circuit responds to ac signals. Rather, those points are "signal grounded" through bypass capacitors. Obtaining an effective bypass can be difficult and is often the route to design difficulty. The problem is parasitic inductance. Although we label and model parts as "capacitors," a more complete model is needed. The better model is a series RLC circuit, shown in **Fig 5.20**. Capacitance is close to the marked value while inductance is a small value that grows with component lead length. Resistance is a loss term, usually controlled by the Q of the parasitic inductor. Even a leadless SMT (surface-mount technology) component will display inductance commensurate with its dimensions. As shown in the Parasitic Inductance section, wire component leads have an inductance of about 1 nH per mm of length (20-25 nH per inch).

Bypass capacitor characteristics can be measured in the home lab with the test setup of **Fig 5.21**. **Fig 5.22** shows a test fixture with an installed 470-pF leaded capacitor. The fixture is used with a signal generator and spectrum analyzer to evaluate capacitors. Relatively long capacitor leads were required to interface to the BNC connectors, even though the

Table 5.1
Dielectric Constants and Breakdown Voltages

Material	Dielectric Constant*	Puncture Voltage**
Aisimag 196	5.7	240
Bakelite	4.4-5.4	240
Bakelite, mica filled	4.7	325-375
Cellulose acetate	3.3-3.9	250-600
Fiber	5-7.5	150-180
Formica	4.6-4.9	450
Glass, window	7.6-8	200-250
Glass, Pyrex	4.8	335
Mica, ruby	5.4	3800-5600
Mycalex	7.4	250
Paper, Royalgrey	3.0	200
Plexiglas	2.8	990
Polyethylene	2.3	1200
Polystyrene	2.6	500-700
Porcelain	5.1-5.9	40-100
Quartz, fused	3.8	1000
Steatite, low loss	5.8	150-315
Teflon	2.1	1000-2000

*At 1 MHz
**In volts per mil (0.001 in.)

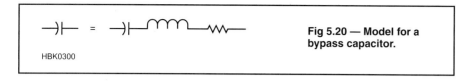

Fig 5.20 — Model for a bypass capacitor.

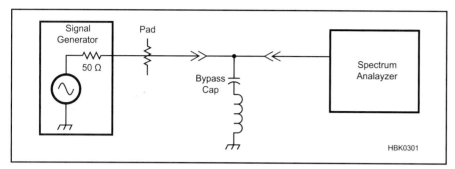

Fig 5.21 — Test set for home lab measurement of a bypass capacitor.

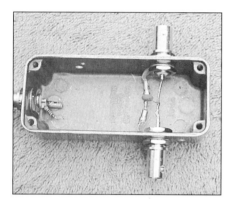

Fig 5.22 — Test fixture for measuring self-resonant frequency of capacitors.

capacitor itself was small. The signal generator was tuned over its range while examining the spectrum analyzer response, showing a minimum at the series resonant frequency. Parasitic inductance is calculated from this frequency. The C value was measured with a low-frequency LC meter. Additional test instruments and techniques are discussed in the **Test Equipment and Measurement** chapter.

The measured 470-pF capacitor is modeled as 485 pF in series with an inductance of 7.7 nH. The L is larger than we would see with shorter leads. A 0.25-inch 470-pF ceramic disk capacitor with zero lead length will show a typical inductance closer to 3 nH. The measured capacitor Q was 28 at its self-resonance of 82 MHz but is higher at lower frequency. Data from a similar measurement, but with a network analyzer is shown in **Fig 5.23**. Two 470-pF capacitors are measured, one surface mounted and the other a leaded part with 0.1-inch leads.

It is often suggested that the bandwidth for bypassing can be extended by paralleling a capacitor that works well at one frequency with another to accommodate a different part of the spectrum. Hence, paralleling the 470 pF with a 0.01-µF capacitor should extend the bypassing to lower frequencies. The calculations are shown in the plots of **Fig 5.24**. The results are terrible! While the low frequency bypassing is indeed improved, a high impedance response is created at 63 MHz. This complicated behavior is again the result of inductance. Each capacitor was assumed to have a series inductance of 7 nH. A parallel resonance is approximately formed between the L of the larger capacitor and the C of the smaller. The Smith Chart plot shows us that the impedance is nearly 50 Ω at 63 MHz. Impedance would be even higher with greater capacitor Q.

Bypassing can be improved by paralleling capacitors. However, the capacitors should be approximately identical. **Fig 5.25** shows the result of paralleling two capacitors of about the same value. They differ slightly at 390 and 560 pF, creating a hint of resonance. This appears as a small perturbation in the reactance plot and a tiny loop on the Smith Chart. These anomalies disappear as the C values become equal. Generally, paralleling is the scheme that produces the best bypassing. The ideal solution is to place an SMT capacitor on each side of a printed circuit trace or wire at a point that is to be bypassed.

Matched capacitor pairs form an effective bypass over a reasonable frequency range. Two 0.01-µF disk capacitors have a reactance magnitude less than 5 Ω from 2 to 265 MHz. A pair of the 0.1-µF capacitors was even better, producing the same bypassing impedance from 0.2 to 318 MHz. The 0.1-µF capacitors are chip-style components with attached wire leads. Even better results can be obtained with multi-layer ceramic chip capacitors. Construction with multiple layers creates an integrated paralleling.

Some applications (for example, IF amplifiers) require effective bypassing at even lower frequencies. Modern tantalum electrolytic capacitors are surprisingly effective through the RF spectrum while offering high enough C to be useful at audio. In critical applications, however, the parts should be tested to be sure of their effectiveness.

DECOUPLING

The bypass capacitor usually serves a dual role, first creating the low impedance needed to generate a "signal" ground. It also becomes part of a decoupling low-pass filter that passes dc while attenuating signals. The attenuation must function in both directions, suppressing noise in the power supply that might reach an amplifier while keeping amplifier signals from reaching the power supply. A low-pass filter is formed with alternating series and parallel component connections. A parallel bypass is followed by a series impedance, ideally a resistor.

Additional shunt elements can then be added, although this must be done with care. An inductor between shunt capacitors should have high inductance. It will resonate with the

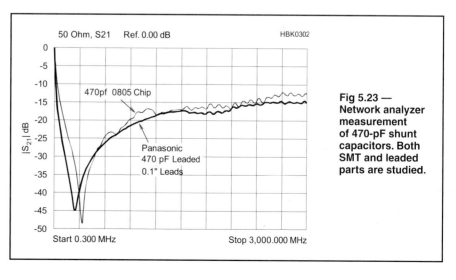

Fig 5.23 — Network analyzer measurement of 470-pF shunt capacitors. Both SMT and leaded parts are studied.

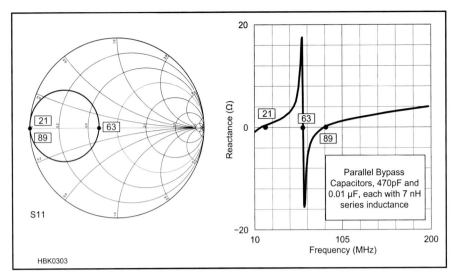

Fig 5.24 — The classic technique of paralleling bypass capacitors of two values, here 470 pF and 0.01 µF. This is a *terrible* bypass! See text.

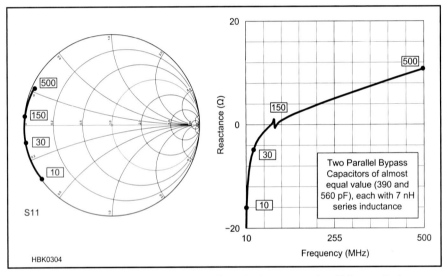

Fig 5.25 — Paralleling bypass capacitors of nearly the same value. This results in improved bypassing without complicating resonances.

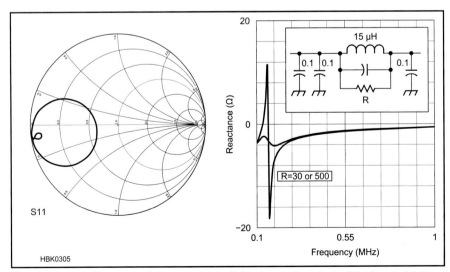

Fig 5.26 — Two different resistor values parallel a decoupling choke. The lower, 30-Ω value is more effective. See text.

shunt capacitors to create high impedances just like those that came from parasitic inductance in the bypasses. This makes it desirable to have an inductance that is high enough that any resonance is below the band of interest. But series inductors have their own problems; they have parasitic capacitance, creating their own self-resonance. As an example, a pair of typical RF chokes (RFCs) were measured (now as series elements) as described earlier. A 2.7-µH molded choke was parallel resonant at 200 MHz, indicating a parallel capacitance of 0.24 pF. The Q at 20 MHz was 52. A 15-µH molded choke was parallel resonant at 47 MHz, yielding a parallel C of 0.79 pF. This part had a Q of 44 at 8 MHz.

Large inductors can be fabricated from series connections of smaller ones. The best wideband performance will result only when all inductors in a chain have about the same value. The reasons for this (and the mathematics that describe the behavior) are identical with those for paralleling identical capacitors.

Low inductor Q is often useful in decoupling applications, encouraging us to use inductors with ferrite cores. Inductors using the Fair-Rite Type #43 material have low Q in the 4 to 10 range over the HF spectrum. One can also create low-Q circuits by paralleling a series inductor of modest Q with a resistor.

Fig 5.26 shows a decoupling network and the resulting impedance when viewed from the "bypass" end. The 15-µH RFC resonates with a 0.1-µF capacitor to destroy the bypass effect just above 0.1 MHz. A low-value parallel resistor fixes the problem.

A major reason for careful wideband bypassing and decoupling is the potential for amplifier oscillation. Instability that allows oscillations is usually suppressed by low impedance terminations. The base and collector (or gate and drain) should both "see" low impedances to ensure stability. But that must be true at all frequencies where the device can produce gain. It is never enough to merely consider the operating frequency for the amplifier. A parallel resonance in the base or collector circuits can be a disaster. When wideband bypassing is not possible, negative feedback that enhances wideband stability is often used.

Emitter bypassing is often a critical application. As demonstrated in the **Analog Basics** chapter, a few extra ohms of impedance in the emitter circuit can drastically alter amplifier performance. A parallel-resonant emitter bypass could be a profound difficulty while a series-resonant one can be especially effective.

Capacitors also appear in circuits as blocking elements. A blocking capacitor, for example, appears between stages, creating a near short circuit for ac signals while accommodating different dc voltages on the two sides. A blocking capacitor is not as critical as a bypass, for the impedances on either side will usually be higher than that of the block.

With parasitic effects having the potential to strongly affect circuit performance, the circuit designer must account for them wherever they are significant. The additional complexity of including parasitic elements can result in a design too complex to be analyzed manually. Clearly, detailed modeling is the answer to component selection and the control of parasitic effects.

5.3.11 Effects on Filter Performance

LC band-pass filters perform a critical function in determining the performance of a typical RF system such as a receiver. An input filter, usually a band pass, restricts the frequency range that the receiver must process. Transmitters use LC filters to reduce harmonic output. Audio and LF filters may also use LC elements. The LC filters we refer to in this section are narrow with a bandwidth from 1 to 20% of the center frequency. Even narrower filters are built with resonators having higher Q; the quartz crystal filter discussed in the **RF and AF Filters** chapter is used where bandwidths of less than a part per thousand are possible. The basic concepts that we examine with LC circuits will transfer to the crystal filter.

LOSSES IN FILTERS AND Q

The key elements in narrow filters are tuned circuits made from inductor-capacitor pairs, quartz crystals, or transmission line sections. These resonators share the property that they store energy, but they have losses. A chime is an example. Striking the chime with a hammer produces the waveform of **Fig 5.27**. The rate at which the amplitude decreases with time after the hammer strike is determined by the filter's Q, which is discussed in the chapter on **Analog Basics**. The higher the Q, the longer it takes for the sound to disappear. The oscillator amplitude would not decrease if it were not for the losses that expend energy stored in the resonator. The mere act of observing the oscillation will cause some energy to be dissipated.

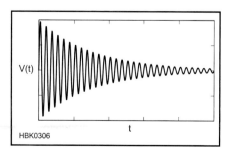

Fig 5.27 — The amplitude of a chime's ring after being struck by a hammer. Units are arbitrary.

A chime is an acoustic resonator, but the same behavior occurs in electric resonators. A pulse input to an LC circuit causes it to ring; losses cause the amplitude to diminish. The most obvious loss in an LC circuit is conductor resistance, including that in the inductor wire. This resistance is higher than the dc value owing to the skin effect, which forces high frequency current toward the conductor surface. Other losses might result from hysteresis losses in an inductor core or dielectric losses in a capacitor.

An inductor is modeled as an ideal part with a series or a parallel resistance. The resistance will depend on the Q if the inductor was part of a resonator with that quality. The two resistances are shown in **Fig 5.28**.

$$R_{Series} = \frac{2\pi f L}{Q} \quad \text{and}$$

$$R_{Parallel} = Q \times 2\pi f L \qquad (6)$$

The higher the inductor Q, the smaller the series resistance, or the larger the parallel resistance is needed to model that Q. It does not matter which configuration is used. The Q of a resonator is related to the bandwidth of the tuned circuit by

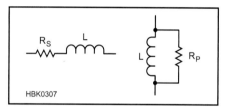

Fig 5.28 — Inductor Q may be modeled with either a series or a parallel resistance.

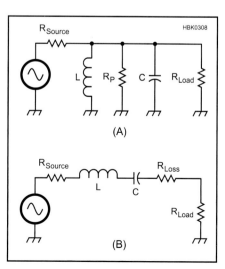

Fig 5.29 — Two simple forms of the single-tuned circuit.

$$BW = \frac{f_C}{Q} \qquad (7)$$

where f_C is the tuned circuit's center frequency. This is also the Q of an inductor in a tuned circuit if the capacitor is lossless.

The single-tuned circuit is presented in two different forms in **Fig 5.29**. In the top, a parallel tuned circuit consisting of L and C has loss modeled by three resistors. The resistor R_p is the parallel loss resistance representing the non-ideal nature of the inductor. (Another might be included to represent capacitor losses.) But the LC is here paralleled by three resistors: the source, the load and the loss element. R_p would disappear if the tuned circuit was built from perfect components. The source and load remain; they represent the source resistance that must be present if power is available and a load resistance that must be included if power is to be extracted.

Equation 6 and 7 can be applied in several ways. If the resonator is evaluated with only its intrinsic loss resistance (in either series or parallel form) the resulting Q is called the *unloaded Q*, or Q_u. If, however, the net resistance is used, which is the parallel combination of the load, the source, and the loss in the parallel tuned circuit, the resulting Q is called the *loaded Q*, Q_L. If we were working with the series tuned circuit form, the loaded Q would be related to the total series R.

Consider an example, a parallel tuned circuit (Fig 5.29A) with a 2-µH inductor tuned to 5 MHz with a 507-pF capacitor. Assume the parallel loss resistor was 12.57 kΩ. The unloaded Q calculated from equation 6 is 200. The unloaded bandwidth would be 5 MHz / 200 = 25 kHz. Assume that the source and load resistors were equal, each 2 kΩ. The net resistance paralleling the LC would then be the combination of the three resistors, 926 Ω. The loaded Q becomes 14.7 with a loaded bandwidth of 339 kHz. The loaded Q is also the filter Q, for it describes the bandwidth of the single-tuned circuit, the simplest of band-pass filters.

This filter also has *insertion loss* (IL). This is illustrated in **Fig 5.30** which shows the

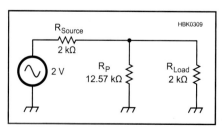

Fig 5.30 — Simplified parallel tuned circuit at resonance. The effect of loss is illustrated by removing the parallel-LC combination which at resonance has infinite impedance.

parallel-resonant LC combination removed (at resonance, the LC combination has infinite impedance), leaving only the loss resistance of 12.57 kΩ. We use an arbitrary open circuit source voltage of 2. The available power to a load is then 1 V across a resistance equaling the 2-kΩ source. If the resonator had no internal losses, this available power would be delivered to the 2-kΩ load. However, the loss R parallels the load, causing the output voltage to be 0.926 V, a bit less than the ideal 1 V. Calculation of the output power into the 2-kΩ load resistance and the available power shows that the insertion loss is 0.67 dB.

This exercise illustrates two vital points that are general for all band-pass filters. First, the bandwidth of any filter must always be larger than the unloaded bandwidth of the resonators used to build the filter. Second, any filter built from real-world components will have an insertion loss. The closer the Q of the filter approaches the unloaded resonator Q, the greater the insertion loss becomes. A parallel tuned circuit illustrated these ideas; the series tuned filter would have produced identical results. Generally, the insertion loss of a single-tuned circuit relates to loaded and unloaded Q by

$$IL(dB) = -20 \log\left(1 - \frac{Q_L}{Q_U}\right) \quad (8)$$

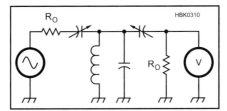

Fig 5.31 — Test setup for measuring the Q of a resonator. The source and load impedances, R_O, are assumed to be identical. The two coupling capacitors are adjusted to be equal to each other. The output signal is measured with an appropriate ac voltmeter, an oscilloscope or a spectrum analyzer.

The Q of a tuned LC circuit is easily measured with a signal generator of known output impedance (usually 50 Ω) and a sensitive detector, again with a known impedance level, often equal to that of the generator. The test configuration is shown in **Fig 5.31**. It uses equal loads and equal capacitors to couple from the terminations to the resonator. Equal capacitors, C1 and C2 guarantees that each termination contributes equally to the resonator parallel load resistance. The voltmeter across the load is calibrated in dB.

To begin measurement we remove the tuned circuit and replace it with a direct connection from generator to load. The available power delivered to the load is calculated after the voltage is measured. The resonator is then inserted between the generator and load, and the generator is tuned for a peak. The measured power is less than that available from the source, with the difference being the insertion loss for the simple filter. Capacitors C1 and C2 are adjusted until the loss is 30 dB or more. With loss this high, the intrinsic loss resistance of the resonator will dominate the loss. The generator is now tuned first to one side of the peak, and then to the other, noting the frequencies where the response is lower than the peak by 3 dB. The unloaded bandwidth, ΔF, is the difference between the two 3 dB frequencies. The unloaded Q is calculated as

$$Q_U = \frac{F}{\Delta F} \quad (9)$$

This method for Q measurement is quite universal, being effective for audio tuned circuits, simple LC RF circuits, VHF helical resonators, or microwave resonators. The form of the variable capacitors, C1 and C2, may be different for the various parts of the spectrum, but the concepts are general. Indeed, it is not even important how the coupling occurs. The Q measurement normally determines an unloaded value, but loaded values are also of interest when testing filters.

5.4 Ferrite Materials

Ferrites are ceramics consisting of various metal oxides formulated to have very high permeability. Iron, manganese, manganese zinc (MnZn), and nickel zinc (NiZn) are the most commonly used oxides. Ferrite cores and beads are available in many styles, as shown in **Fig 5.32**. Wires and cables are then passed through them or wound on them.

When ferrite surrounds a conductor, the high permeability of the material provides a much easier path for magnetic flux set up by current flow in the conductor than if the wire were surrounded only by air. The short length of wire passing through the ferrite will thus see its self-inductance "magnified" by the relative permeability of the ferrite. The ferrites used for suppression are soft ferrites — that is, they are not permanent magnets.

Recalling the definition in the **Electrical Fundamentals** chapter, *permeability* (μ) is the characteristic of a material that quantifies the ease with which it supports a magnetic field. Relative permeability is the ratio of the permeability of the material to the permeability of free space. The relative permeability of nonmagnetic materials like air, copper, and aluminum is 1, while magnetic materials have a relative permeability much greater than 1. Typical values (measured at power frequencies) for stainless steel, steel and mu-metal are on the order of 500, 1000 and 20,000 respectively. Various ferrites have values from the low tens to several thousand. The permeability of these materials changes with frequency and this affects their suitability for use as an inductor's magnetic core or as a means of providing EMI suppression.

Manufacturers vary the chemical composition (the *mix* or *material "type"*) and the dimensions of ferrites to achieve the desired electrical performance characteristics. It is important to select a mix with permeability and loss characteristics that are appropriate for the intended frequency range and application. (The **Component Data and References** chapter includes tables showing the data for different types of ferrite and powdered-iron cores and appropriate frequency ranges for each.)

Fair-Rite Products Corp. supplies most of the ferrite materials used by amateurs, and their Web site (**www.fair-rite.com**) includes extensive technical data on both the materials and the many parts made from those materials. Much can be learned

Fig 5.32 — Ferrites are made in many forms. Beads are cylinders with small center holes so that they can be slid over wires or cables. Toroidal cores are more ring-like so that the wire or cable can be passed through the center hole multiple times. Snap-on or split cores are made to be clamped over cables too large to be wound around the core or for a bead to be installed.

from the study of this data, the most extensive of any manufacturer. The Web site includes two detailed application notes on the use of ferrite materials, *How to Choose Ferrite Components for EMI Suppression* and *Use of Ferrites in Broadband Transformers*.

5.4.1 Ferrite Permeability and Frequency

Product data sheets characterize ferrite materials used as chokes by graphing their series equivalent impedance, and chokes are usually analyzed as if their equivalent circuit had only a series resistance and inductance, as shown in **Figs 5.33A** and 5.33B. (Fig 5.33 presents small-signal equivalent circuits. Issues of temperature and saturation at high power levels are not addressed in this section.) The actual equivalent circuit is closer to Fig 5.33C. The presence of both inductance and capacitance creates resonances visible in the graphs of impedance versus frequency as discussed below. One resonance is created by the pair $L_D C_D$ and the other by $L_C C_C$.

Fig 5.34A graphs the permeabilities μ'_S and μ''_S for Fair-Rite Type #61 ferrite material (Fair-Rite products are identified by a material "Type" and a number), one of the many mix choices available. This material is recommended for use in inductive applications below 25 MHz and for EMI suppression at frequencies above 200 MHz.

For the simple series equivalent circuit of Fig 5.33A and B, the permeability constant for ferrite is actually complex; $\mu = \mu'_S + j\mu''_S$. In this equation, μ'_S represents the component of permeability defining ordinary inductance, and μ''_S describes the component that affects losses in the material. You can see that up to approximately 20 MHz, μ'_S is nearly constant, meaning that an inductor wound on a core of this material will have a stable inductance. Below about 15 MHz, the chart shows that μ''_S is much smaller than μ'_S, so that losses are small, making this material good for high power inductors and transformers in this frequency range.

Above 15 MHz, μ''_S increases rapidly and so do the material's associated losses, peaking between 300 and 400 MHz. That makes the inductor very lossy at those frequencies and good for suppressing EMI by absorbing energy in the unwanted signal. **Fig 5.35** shows the manufacturer's impedance data for one "turn" of wire through a cylindrical ferrite bead (a straight wire passing through the bead) made of mix #61 with the expected peak in impedance above 200 MHz. Interestingly, X_L goes below the graph above resonance, but it isn't zero. The negative reactance is created by the capacitors in Fig 5.33C.

The Type #43 mix (see Fig 5.34B) used for the beads of **Fig 5.36** is optimized for suppres-

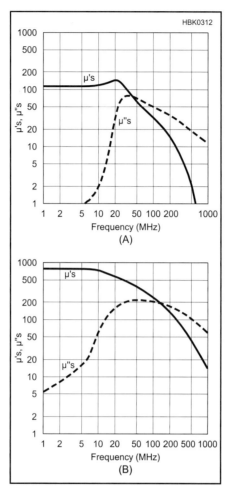

Fig 5.34 — Permeability of a typical ferrite material, Fair-Rite Type #61 (A) and of Type #41 material (B). (Based on product data published by Fair-Rite Products Corp.)

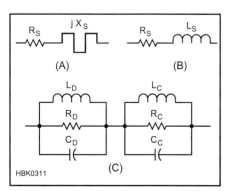

Fig 5.33 — Equivalent circuits for ferrite material. (A) shows the equivalent series circuit specified by the data sheet. (B) shows an over-simplified equivalent circuit for a ferrite choke. (C) shows a better equivalent circuit for a ferrite choke.

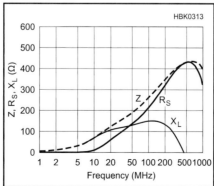

Fig 5.35 — Impedance of a bead for EMI suppression at UHF, Fair-Rite Type #61. (Based on product data published by Fair-Rite Products Corp.)

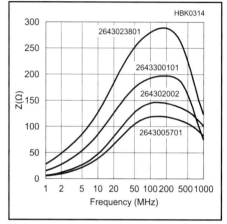

Fig 5.36 — Impedance of different length beads of Fair-Rite Type #43 material. The longer the bead, the higher the impedance. (Based on product data published by Fair-Rite Products Corp.)

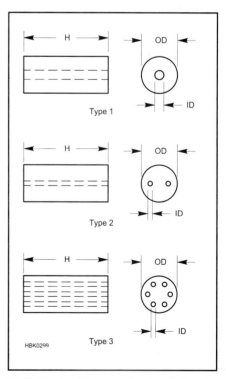

Fig 5.37 — Typical one-piece ferrite bead configurations. Component leads or cables can be inserted through beads one or more times to create inductance and loss as described in the text.

RF Techniques 5.15

Fig 5.38 — A ferrite choke consisting of multiple turns of cable wound on a 2.4-in. OD × 1.4-in. ID × 0.5-in. toroid core.

sion at VHF (30-300 MHz). Type #43 material's μ'_S is much higher than type #61, meaning that fewer turns are required to achieve the needed inductance. But μ''_S for #43 remains significant even at low frequencies, limiting its usefulness for inductor and transformer cores that must handle high power. The figure shows the impedance data for several beads of different lengths. The longer the bead, the higher the impedance. The same effect can be obtained by stringing multiple beads together on the same wire or cable.

5.4.2 Resonances of Ferrite Cores

Below resonance, the impedance of a wire passing through a ferrite cylinder is proportional to the length of the cylinder. Fig 5.36 shows the impedance of a family of beads that differ primarily in their length. There are also small differences in their cross section, which is why the resonant frequency shifts slightly as discussed below. **Fig 5.37** shows some typical ferrite bead configurations.

Like all inductors, the impedance of a ferrite choke below resonance is approximately proportional to the square of the number of turns passing through the core. **Fig 5.38** shows a multi-turn choke wound on a 2.4-inch OD × 1.4-inch ID × 0.5-inch ferrite toroid core. **Fig 5.39** is measured data for the choke in Fig 5.38 made of ferrite optimized for the VHF range (30-300 MHz). The data of **Fig 5.40** are for toroids of the same size, but wound on a material optimized for use below 2 MHz.

We'll study the $L_D C_D$ resonance first. A classic text (*Soft Ferrites, Properties and Applications*, by E. C. Snelling, published in 1969), shows that there is a dimensional resonance within the ferrite related to the velocity of propagation (V_P) within the ferrite and standing waves that are set up in the cross-sectional dimensions of the core. In general, for any given material, the smaller the core, the higher will be the frequency of this resonance, and to a first approximation, the resonant frequency will double if the core dimension is halved. In Fig 5.33C, L_D and C_D account for

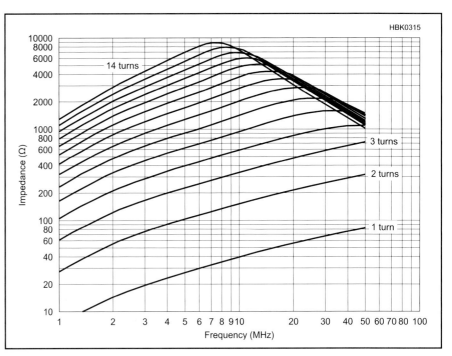

Fig 5.39 — Impedance of multi-turn choke wound on a Fair-Rite Type #43 core as shown in Fig 5.38. Type #43 material is optimized for use as a choke for the VHF range. (Measured data)

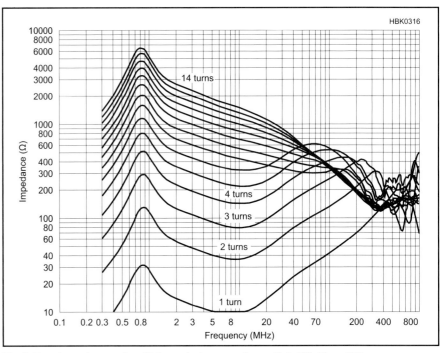

Fig 5.40 — Impedance of multi-turn choke wound on a Fair-Rite Type #78 core as shown in Fig 5.38. Type #78 material is optimized for use as a choke below 2 MHz. (Measured data)

this dimensional resonance, and R_D for losses within the ferrite. R_D is mostly due to eddy currents (and some hysteresis) in the core.

Now it's time to account for R_C, L_C and C_C. Note that there are two sets of resonances for the chokes wound around the Type #78 material (Fig 5.40), but only one set for the choke of Fig 5.39. For both materials, the upper reso-

nance starts just below 1 GHz for a single turn and moves down in frequency as the number of turns is increased. **Fig 5.41**, the reactance for the chokes of Fig 5.40, also shows both sets of resonances. That's why the equivalent circuit must include two parallel resonances!

The difference between these materials that accounts for this behavior is their chemical

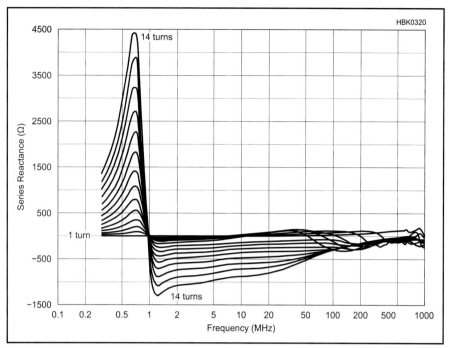

Fig 5.41 — Series reactive component of the chokes of Fig 5.40. (Measured data)

composition (the mix). Type #78 is a MnZn ferrite, while Type #43 is a NiZn ferrite. The velocity of propagation (V_P) in NiZn ferrites is roughly two orders of magnitude higher than for MnZn, and, at those higher frequencies, there is too much loss to allow the standing waves that establish dimensional resonance to exist.

To understand what's happening, we'll return to our first order equivalent circuit of a ferrite choke (Fig 5.33C). L_C, and R_C, and C_C are the inductance, resistance, (including the effect of the µ of the ferrite), and stray capacitance associated with the wire that passes through the ferrite. This resonance moves down in frequency with more turns because both L and C increase with more turns. The dimensional resonance does not move, since it depends only on the dimensions V_P of the ferrite.

What is the source of C_C if there's no "coil," only a single wire passing through a cylinder? It's the parasitic capacitance from the wire at one end of the cylinder to the wire at the other end, with the ferrite acting as the dielectric. Yes, it's a very small capacitance, but you can see the resonance it causes in the measured data and on the data sheet.

5.4.3 Ferrite Series and Parallel Equivalent Circuits

Let's talk briefly about series and parallel equivalent circuits. Many impedance analyzers express the impedance between their terminals as Z with a phase angle, and the series equivalent R_S, and X_S. They could just have easily expressed that same impedance using the parallel equivalent R_P and X_P but R_P and

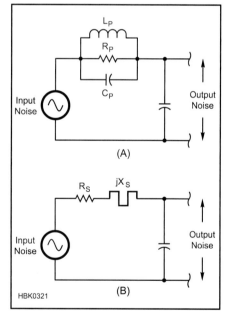

Fig 5.42 — At A, the series element of the divider is a parallel-resonant circuit. At B, the series element of the divider is the series-equivalent circuit used by ferrite data sheets.

X_P will have values that are numerically different from R_S and X_S. (See the section on series-parallel impedance transformation in the **Electrical Fundamentals** chapter.)

It is important to remember that in a series circuit, the larger value of R_S and X_S has the greatest influence, while in a parallel circuit, the smaller value of R_P and X_P is dominant.

In other words, for R_P to dominate, R_P must be small.

Both expressions of the impedance (series or parallel) are correct at any given frequency, but whether the series or parallel representation is most useful will depend on the physics of the device being measured and how that device is used in a circuit. We've just seen, for example, that the parallel equivalent circuit is a more realistic representation of a ferrite choke — the values of R_P, L_P, and C_P will come much closer to remaining constant as frequency changes than if we use the series equivalent. (R_P, L_P and C_P won't be precisely constant though, because the physical properties of all ferrites — permeability, resistivity and permittivity — all vary with frequency.)

But virtually all product data for ferrite chokes is presented as the series equivalent R_S and X_S. Why? First, because it's easy to measure and understand, second, because we tend to forget there is stray capacitance, and third because ferrite beads are most often used as chokes to reduce current in a series circuit! **Fig 5.42A** and **Fig 5.42B** are both useful representations of the voltage divider formed by a ferrite choke and a small bypass capacitor across the device input. Which we use will depend on what we know about our ferrite.

If we know R_P, L_P and C_P and they are constant over the frequency range of interest, Fig 5.42A may be more useful, because we can insert values in a circuit model and perhaps tweak the circuit. But if we have a graph of R_S and X_S vs frequency, Fig 5.42B will give us a good answer faster. Because we will most often be dealing with R_S and X_S data, the series circuit equivalents are used most often. Another reason for using R_S and X_S is that the impedance of two or more ferrite chokes in series can be computed simply by adding their R_S and X_S components, just as with any other series impedances! When you look at the data sheet plots of R_S, X_L and Z for a standard ferrite part, you are looking at the series equivalent parameters of their dominant resonance. For most MnZn materials, it is dimensional resonance, while, for most NiZn materials, it is the circuit resonance.

Values for R_P, L_P and C_P for nearly any ferrite choke can be obtained by curve-fitting the data for its parallel-resonance curve. (Self-resonance is discussed earlier in this chapter.) Because the C_P of many practical chokes is quite small and their impedance at resonance is often rather high, they are quite difficult to measure accurately, especially with reflection-based measurement systems. See the **Test Equipment and Measurements** chapter for a simple measurement method that can provide good results. More information on the use of ferrite cores and beads for EMI suppression is available in the chapter on **EMC** and in *The ARRL RFI Book*. A thorough treatment of the use of these ferrite materials

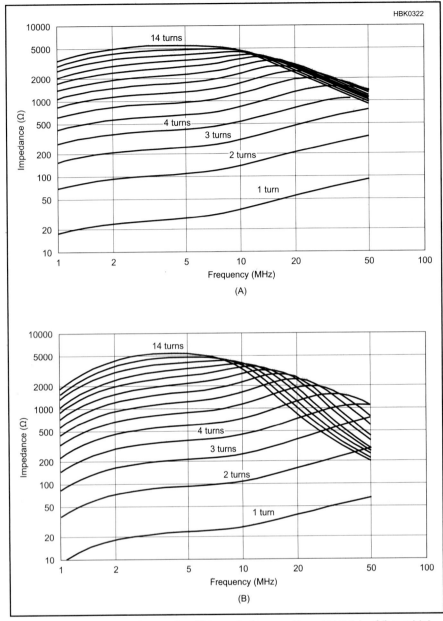

Fig 5.43 — At A, the impedance of multi-turn chokes on a Type #31 2.4-in. OD toroidal core. At B, the equivalent series resistance of the chokes of (A). (Measured data)

for EMI suppression is continued in the online publication *A Ham's Guide to RFI, Ferrites, Baluns, and Audio Interfacing* at **www.audiosystemsgroup.com/publish**. The use of ferrite beads and cores for transmitting chokes is presented in the **Transmission Lines** and **Station Accessories** chapters.

5.4.4 Type 31 Material

The relatively new Type #31 material made by Fair-Rite Products is extremely useful as a choke core, especially if some component of your problem occurs below 5 MHz. Measured data for the new material is displayed in **Figs 5.43A** and 5.43B. Compare it with **Figs 5.44A** and 5.44B, which are corresponding plots for the older Type #43 material. By comparison, Type #31 provides nearly 7 dB greater choking impedance at 2 MHz, and at least 3 dB more on 80 meters. At 10 MHz and above, the two materials are nearly equivalent, with Type #43 being about 1 dB better. If your goal is EMI suppression or a feed-line choke (a so-called "current balun"), the Type #31 material is the best all round performer to cover all HF bands, and is clearly the material of choice at 5 MHz and below. Between 5 MHz and 20 MHz, Type #43 has a slight edge (about 1 dB), and above 20 MHz they're equivalent. (Baluns are discussed more in the section below on Transmission Line Transformers and in the chapters on **Transmission Lines** and **Antennas**.)

The new Type #31 material is useful because it exhibits both of the resonances in our equivalent circuit — that is, the dimensional resonance of the core and the resonance of the choke with the lossy permeability of the core material. Below 10 MHz, these two resonances combine (in much the manner of a stagger-tuned IF) to provide significantly greater suppression bandwidth (roughly one octave, or one additional harmonically related ham band). The result is that a single choke on a Type #31 core can be made to provide very good suppression over about 8:1 frequency span, as compared to 4:1 for Type #43. Type #31 also has somewhat better temperature characteristics at HF.

5.5 Semiconductor Circuits at RF

The models used in the **Analog Basics** chapter are reasonably accurate at low frequencies, and they are of some use at RF, but more sophisticated models are required for consistent results at higher frequencies. This section notes several areas in which the simple models must be enhanced. Circuit design using models accurate at RF, particularly for large signals, is performed today using design software as described in the **Computer-Aided Circuit Design** chapter of this book. In-depth discussions of the elements of RF circuit design appear in Hayward's *Introduction to Radio Frequency Design*, an excellent text for the beginning RF designer (see the References section).

5.5.1 The Diode at High Frequencies

A DIODE AC MODEL

At high frequencies, the diode's behavior becomes less like a switch, especially for small signals that do not cause the diode's operating point to move by large amounts. In this case, the diode's small-signal dynamic behavior becomes important.

Fig 5.45 shows a simple resistor-diode circuit to which is applied a dc bias voltage plus an ac signal. Assuming that the voltage drop across the diode is 0.6 V and R_f is negligible, we can calculate the bias current to be I = (5 − 0.6) / 1 kΩ = 4.4 mA. This point is marked on the diodes I-V curve in Fig 5.45. If we draw a line tangent to this point, as shown,

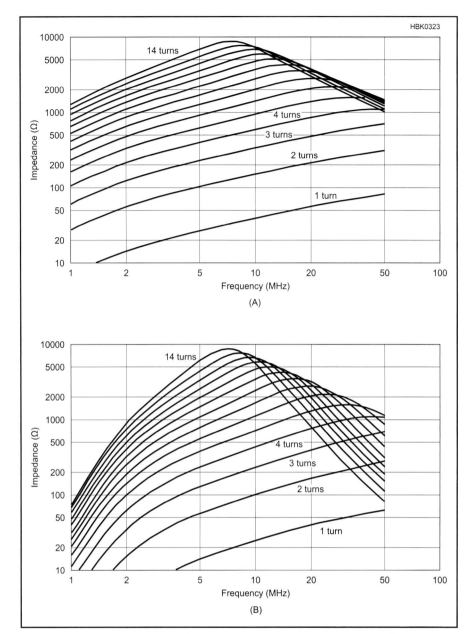

Fig 5.44 — At A, the impedance of multi-turn chokes on a Type #43 2.4-in. OD toroidal core. At B, the equivalent series resistance of the chokes of A. (Measured data).

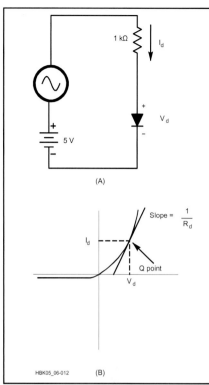

Fig 5.45 — A simple resistor-diode circuit used to illustrate dynamic resistance. The ac input voltage "sees" a diode resistance whose value is the slope of the line at the Q-point, shown in B.

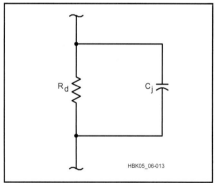

Fig 5.46 — An ac model for diodes. R_d is the dynamic resistance and C_j is the junction capacitance.

the slope of this line represents the *dynamic resistance* of the diode, R_d, experienced by a small ac signal. At room temperature, dynamic resistance is approximately

$$R_d = \frac{25}{I} \Omega \qquad (10)$$

where I is the diode current in mA. Note that this resistance changes with bias current and should not be confused with the dc forward resistance discussed in the **Analog Basics** chapter, which has a similar value but represents a different concept. **Fig 5.46** shows a low-frequency ac model for the diode, including the dynamic resistance and junction capacitance. This model should only be used when the diode's dc operating parameters can be neglected.

SWITCHING TIME

If you change the polarity of a signal applied to the ideal diode, current flow stops or starts instantaneously. Current in a real diode cannot do this, as a finite amount of time is required to move electrons and holes in or out of the diode as it changes states. Effectively, the diode junction capacitance, C_J, in Fig 5.46 must be charged or discharged. As a result, diodes have a maximum useful frequency when used in switching applications.

The operation of diode switching circuits can often be modeled by the circuit in **Fig 5.47**. The approximate switching time (in seconds) for this circuit is given by

$$t_s = \tau_p \frac{\left(\frac{V_1}{R_1}\right)}{\left(\frac{V_2}{R_2}\right)} = \tau_p \frac{I_1}{I_2} \qquad (11)$$

where τ_p is the minority carrier lifetime of the

Transistor Amplifier Design — a Practical Approach

The design of a transistor amplifier is a straightforward process. Just as you don't need a degree in mechanical engineering to drive an automobile, neither do you need detailed knowledge of semiconductor physics in order to design a transistor amplifier with predictable and repeatable properties.

This sidebar will describe how to design a small-signal "Class A" transistor amplifier, following procedures detailed in one of the best books on the subject — Solid State Design for the Radio Amateur, by Wes Hayward, W7ZOI, and Doug DeMaw, W1FB. For many years, both hams and professional engineers have used this classic ARRL book to design untold numbers of working amplifiers. This example builds on the concepts and equations presented in the **Analog Basics** chapter to develop a useful transistor RF amplifier while discussing some of the tradeoffs and considerations that designers go through in the process.

How Much Gain?

One of the simple, yet profound, observations made in Solid State Design for the Radio Amateur is that a designer should not attempt to extract every last bit of gain from a single amplifier stage. Trying to do so virtually guarantees that the circuit will be "touchy" — it may end up being more oscillator than amplifier! While engineers might debate the exact number, modern semiconductor circuits are inexpensive enough that you should try for no more than 25 dB of gain in a single stage.

For example, if you are designing a high-gain amplifier system to follow a direct-conversion receiver mixer, you will need a total of about 100 dB of audio amplification. We would recommend a conservative approach where you use four stages, each with 25 dB of gain. You might risk oscillation and instability by using only two stages, with 50 dB gain each. The component cost will not be greatly different between these approaches, but the headaches and lack of reproducibility of the "simpler" two-stage design will very likely far outweigh any small cost advantages!

Biasing the Transistor Amplifier

The first step in amplifier design is to *bias* the transistor properly. A small-signal linear amplifier is biased properly when there is current at all times. Once you have biased the stage, you can then use several simple rules of thumb to determine all the major properties of the resulting amplifier.

Solid State Design for the Radio Amateur introduces several simple transistor models. We won't get into that much detail here, except to say that the most fundamental property of a transistor is this: When there is current across the base-emitter junction, a larger current can flow across the collector-emitter junction. When the base-emitter junction is thus *forward biased*, the voltage across the base and emitter leads of a silicon transistor will be relatively constant, at 0.7 V. For most modern transistors, the dc current gain of the transistor is at least 50 to 100.

See **Fig 5.A1**, which shows a simple capacitively-coupled low-frequency amplifier suitable for use at 1 MHz. Resistors R1 and R2 form a voltage divider feeding the base of the transistor. The amount of current in the resistive voltage divider is purposely made large enough so the base current is small in comparison, thus creating a "stiff" voltage supply for the base. As stated above, the voltage at the emitter will be 0.7 V less than the base voltage for this NPN transistor. The emitter voltage V_E appears across the series combination of R4 and R5. Note that R5 is bypassed by capacitor C4 for ac current.

By Ohm's Law, the emitter current is equal to the emitter voltage V_E divided by the sum of R4 plus R5. Now, the emitter current is made up of both the base-emitter and the collector-emitter current, but since the base current is much smaller than the collector current, the amount of collector current is essentially equal to the emitter current, at $V_E / (R4 + R5)$.

Our design process starts by specifying the amount of current we want to flow in the collector, with the dc collector voltage equal to half the supply voltage. For good bias stability with temperature variation, the total emitter resistor should be at least 100 Ω for a small-signal amplifier. Let's choose a collector current of 5 mA, and use a total emitter resistance of 200 Ω, with R4 = R5 = 100 Ω each. The voltage across 200 Ω for 5 mA of current is 1.0 V. This means that the voltage at the base must be 1.0 V + 0.7 V = 1.7 V, provided by the voltage divider R1 and R2.

The dc base current requirement for a collector current of 5 mA is approximately 5 mA / 50 = 0.1 mA if the transistor's dc beta is at least 50, a safe assumption for modern transistors. To provide a "stiff" base voltage, we want the current through the voltage divider to be about five to ten times greater than the base current. For convenience then, we choose the current through R1 to be 1 mA. This is a convenient current value, because the math is simplified — we don't have to worry about decimal points for current or resistance: 1 mA × 1.8 kΩ = 1.8 V. This is very close to the 1.7 V we are seeking. We thus choose a standard value of 1.8 kΩ for R2. The voltage drop across R1 is 12 V − 1.8 V = 10.2 V. With 1 mA in R1, the necessary value is 10.2 kΩ, and we choose the closest standard value, 10 kΩ.

Let's now look at what is happening in the collector part of the circuit. The collector resistor R3 is 1 kΩ, and the 5 mA of collector current creates a 5 V drop across R3. This means that the collector dc voltage must be 12 V − 5 V = 7 V. The dc power dissipated in the transistor will be essentially all in the collector-emitter junction, and will be the collector-emitter voltage (7 V − 1 V = 6 V) times the collector current of 5 mA = 0.030 W, or 30 mW. This dissipation is well within the 0.5 W rating typical of small-signal transistors.

Now, let's calculate more accurately the result from using standard values for R1 and R2. The actual base voltage will be 12 V × [1.8 kΩ / (1.8 kΩ + 10 kΩ)] = 1.83 V, rather than 1.7 V. The resulting

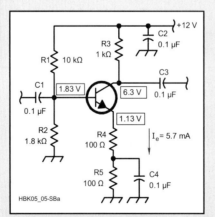

Fig 5.A1 — Example of a simple low-frequency capacitively coupled transistorized small-signal amplifier. The voltages shown are the preliminary values desired for a collector current of 5 mA. The ac voltage gain is the ratio of the collector load resistor, R3, divided by R4, the unbypassed portion of the emitter resistance.

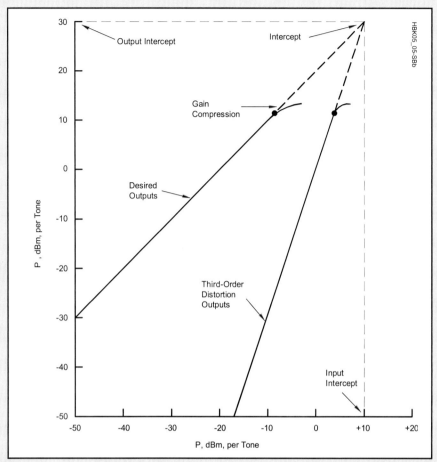

Fig 5.A2 — Output IMD (intermodulation distortion) as a function of input. In the region below the 1-dB compression point, a decrease in input level of 10 dB results in a drop of IMD products by 30 dB below the level of each output tone in a two-tone signal.

emitter voltage is 1.83 V − 0.7 V = 1.13 V, resulting in 1.13 V / 200 Ω = 5.7 mA of collector current, rather than our desired 5 mA. We are close enough — we have finished designing the bias circuitry!

Performance: Voltage Gain

Now we can analyze how our little amplifier will work. The use of the un-bypassed emitter resistor R4 provides *emitter degeneration* — a form of negative feedback to stabilize the transistor's operating point. The bottom line for us is that we can use several handy rules of thumb. The first is for the ac voltage gain of an amplifier: A_V = R3 / R4. The ac voltage gain of such an amplifier is simply the ratio of the collector load resistor and the un-bypassed resistance in the emitter circuit. In this case, the gain is R3 / R4 = 1000 / 100 = 10, which is 20 dB of voltage gain. This expression for gain is true virtually without regard for the exact kind of transistor used in the circuit, provided that we design for moderate gain in a single stage, as we have done.

Performance: Input Resistance

Another useful rule of thumb stemming from use of an un-bypassed emitter resistor is the expression for the ac input resistance: R_{IN} = beta × R4. If the ac beta at low frequencies is about 50, then the input resistance of the transistor is 50 × 100 Ω = 5000 Ω. The actual input resistance includes the shunt resistance of voltage divider R2 and R1, about 1.5 kΩ, in parallel with the transistor's input resistance. Thus the biasing resistive voltage divider essentially sets the input resistance of the amplifier.

Performance: Overload

We can accurately predict how this amplifier will perform. If we were to supply a peak positive 1 V signal to the base, the voltage at the collector will try to fall by the voltage gain of 10. However, since the dc voltage at the collector is only 7 V, it is clear that the collector voltage cannot fall 10 V. In theory, the collector voltage could fall as low as the 1.13 V dc level at the emitter. This amplifier will "run out of voltage" at a negative collector voltage swing of about 6.3 V − 1.13 V = 5.17 V, when the input voltage is 5.17 divided by the gain of 10 = 0.517 V. When a negative-going ac voltage is supplied to the base, the collector current falls, and the collector voltage will rise by the voltage gain of 10. The maximum amount of voltage possible is the 12 V supply voltage, where the transistor is cut off with no collector current. The maximum positive collector swing is from the standing collector dc voltage to the supply voltage: 12 V − 6.3 V = 5.7 V positive swing. This occurs with a peak negative input voltage of 5.7 V / 10 = 0.57 V. Our amplifier will overload rather symmetrically on both negative and positive peaks. This is no accident — we biased it to have a collector voltage halfway between ground and the supply voltage.

When the amplifier "runs out of output voltage" in either direction, another useful rule of thumb is that this is the *1 dB compression point*. This is where the amplifier just begins to depart from linearity, where it can no longer provide any more output for further input. For our amplifier, this is with a peak-to-peak output swing of approximately 5.1 V × 2 = 10.2 V, or 3.6 V RMS. The output power developed in output resistor R3 is $(3.6)^2$ / 1000 = 0.013 W = 13 mW, which is +11.1 dBm (referenced to 1 mW on 50 Ω).

At the 1 dB compression point, the third-order *IMD* (intermodulation distortion) will be roughly 25 dB below the level of each tone. **Fig 5.A2** shows a graph of output versus input levels for both the desired signal and for third-order IMD products. The rule of thumb for IMD is that if the input level is decreased by 10 dB, the IMD will decrease by 30 dB. Thus, if input is restricted to be 10 dB below the 1 dB compression point, the IMD will be 25 dB + 30 dB = 55 dB below each output tone. With very simple math we have thus designed and characterized a simple amplifier. This amplifier will be stable for both dc and ac under almost any thermal and environmental conditions conceivable. That wasn't too difficult, was it? — *R. Dean Straw, N6BV*

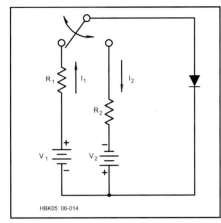

Fig 5.47 — Circuit used for computation of diode switching time.

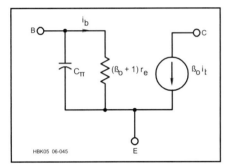

Fig 5.48 — The hybrid-pi model for the bipolar transistor.

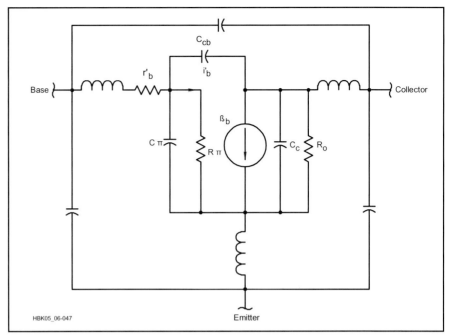

Fig 5.49 — A more refined small-signal model for the bipolar transistor. Suitable for many applications near the transistor F_T.

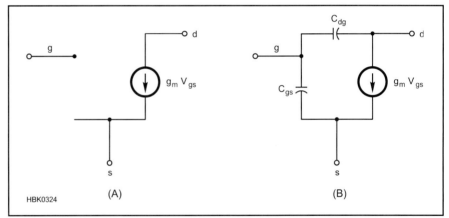

Fig 5.50 — Small-signal models for the JFET. (A) is useful at low frequency. (B) is a modification to approximate high frequency behavior.

diode, a material constant determined during manufacture, on the order of 1 ms. I_1 and I_2 are currents that flow during the switching process. The minimum time in which a diode can switch from one state to the other and back again is therefore $2\,t_s$, and thus the maximum usable switching frequency is $f_{sw}\,(Hz) = \frac{1}{2} t_s$. It is usually a good idea to stay below this by a factor of two. Diode data sheets usually give typical switching times and show the circuit used to measure them.

Note that f_{sw} depends on the forward and reverse currents, determined by I_1 and I_2 (or equivalently V_1, V_2, R_1, and R_2). Within a reasonable range, the switching time can be reduced by manipulating these currents. Of course, the maximum power that other circuit elements can handle places an upper limit on switching currents.

5.5.2 The Transistor at High Frequencies

GAIN VERSUS FREQUENCY

The circuit design equations in **Analog Basics** generally assumed that current gain in the bipolar transistor was independent of frequency. As signal frequency increases, however, current gain decreases. The low-frequency current gain, β_0, is constant through the audio spectrum, but it eventually decreases, and at some high frequency it will drop by a factor of 2 for each doubling of signal frequency. A transistor's frequency vs current gain relationship is specified by its *gain-bandwidth product*, or F_T, the frequency at which the current gain is 1. Common transistors for lower RF applications might have $\beta_0 = 100$ and $F_T = 500$ MHz. The frequency at which current gain equals β_0 is called F_b and is related to F_T by $F_b = F_T / \beta_0$.

The frequency dependence of current gain is modeled by adding a capacitor across the base resistor of **Fig 5.48**, the *hybrid-pi* model results. The capacitor's reactance should equal the low-frequency input resistance, $(\beta + 1)r_e$, at F_b. This simulates a frequency-dependent current gain.

Recognizing that transistor gain varies with frequency, it is important to know the conditions under which the simpler model is useful. Calculations show that the simple model is valid, with $\beta = F_T/F$, for frequencies well above F_b. The approximation worsens, however, as the operating frequency (f_0) approaches F_T.

SMALL-SIGNAL DESIGN AT RF

The simplified hybrid-pi model of Fig 5.48 is often suitable for non-critical designs, but **Fig 5.49** shows a better small-signal model for RF design that expands on the hybrid-pi. Consider the physical aspects of a real transistor: There is some capacitance across each of the PN junctions (C_{cb} and C_p) and capacitance from collector to emitter (C_c). There are also capacitances between the device leads (C_e, C_b and C_c). There is a resistance in each current path, emitter to base and collector. From emitter to base, there is r_π from the hybrid-pi model and $r'b$, the "base spreading" resistance. From emitter to collector is R_o, the output resistance. The leads that attach the silicon die to the external circuit present three inductances.

Manual circuit analysis with this model is best tackled with the aid of a computer and specialized software. Other methods are presented in Hayward's *Introduction to Radio Frequency Design*. Surprisingly accurate results may be obtained, even at RF, from the simple models. Simple models also give a better "feel" for device characteristics that might be obscured by the mathematics of a more rigorous treatment. Use the simplest model that describes the important features of the device and circuit at hand.

The bipolar transistor can be viewed as controlled by either voltage or current. The JFET, however, is purely a voltage controlled element, at least at low frequencies. (Other types of FETs behave similarly.) The simple model in **Fig 5.50A** is the one introduced in the **Analog Basics** chapter. Fig 5.50B contains capacitive elements which are effective in describing high-frequency behavior. Like the bipolar transistor, the JFET model will grow in complexity as more sophisticated applications are encountered.

5.5.3 Amplifier Classes

The class of operation of an amplifier is determined by the fraction of a drive cycle during which conduction occurs in the amplifying device or for switchmode devices. (Switchmode amplifiers use different criteria.) The Class A amplifier conducts for 100% of the cycle. It is characterized by constant flow of supply current, regardless of the strength of the driving signal. Most of the amplifiers we use for RF applications and many audio circuits in receivers operate in Class A.

A Class B amplifier conducts for 50% of the cycle, which is 180 degrees if we examine the circuit with regard to a driving sine wave. A Class B amplifier draws no dc current when no input signal is applied, but current begins to flow with any input, growing with the input strength. A Class B amplifier can display good *envelope linearity*, meaning that the output amplitude at the drive frequency changes linearly with the input signal. The total absence of current flow for half of the drive cycle will create harmonics of the signal drive.

A Class C amplifier is one that conducts for less than half of a cycle. No current flows without drive. Application of a small drive produces no output and no current flow. Only after a threshold is reached does the device begin to conduct and provide output. A bipolar transistor with no source of bias for the base typically operates in Class C.

The large-signal models discussed earlier are suitable for the analysis of all amplifier classes. Small-signal models are generally reserved for Class A amplifiers. The most common power amplifier class is a cross between Class A and B — the Class AB amplifier that conducts for more than half of each cycle. A Class AB amplifier at low drive levels is indistinguishable from a Class A design. However, increasing drive produces greater collector (or drain) current and greater output.

Amplifier class letter designators for vacuum tube amplifiers were augmented with a numeric subscript. A Class AB_1 amplifier operates in AB, but with no grid current flowing. A Class AB_2 amplifier's grid is driven positive with respect to the cathode and so some grid current flows. In solid-state amplifiers, which have no grids, no numeric subscripts are used.

While wide-bandwidth Class A and Class B amplifiers are common, most circuits operating in Class C and higher are tuned at the output. The tuning accomplishes two things. First, it allows different terminations to exist for different frequencies. For example, a resistive load could be presented at the drive frequency while presenting a short circuit at some or all harmonics. The second consequence of tuning is that reactive loads can be created and presented to the amplifier collector or drain. This then provides independent control of current and voltage waveforms.

While not as common as A, B, and C, Class D and E amplifiers are of increasing interest. The Class D circuit is a balanced (two transistor) switching format where the input is driven hard enough to produce square wave collector waveforms. Class E amplifiers usually use a single switching device with output tuning that allows high current to flow in the device only when the output voltage is low. Other "switching amplifier" class designators refer to the various techniques of controlling the switched currents and voltages.

Class A and AB amplifiers are capable of good envelope linearity, so they are the most common formats used in the output of SSB amplifiers. Class B and, predominantly, Class C amplifiers are used for CW and FM applications, but lack the envelope linearity needed for SSB.

Efficiency varies considerably between amplifier class. The Class A amplifier can reach a collector efficiency of 50%, but no higher, with much lower values being typical. Class AB amplifiers are capable of higher efficiency, although the wideband circuits popular in HF transceivers typically offer only 30% at full power. A Class C amplifier is capable of efficiencies approaching 100% as the conduction cycle becomes small, with common values of 50 to 75%. Both Class D and E are capable of 90% and higher efficiency. An engineering text treating power amplifier details is Krauss, Bostian, and Raab's *Solid State Radio Engineering*. A landmark paper by a Cal Tech group led by David Rutledge, KN6EK, targeted to the home experimenter, *High-Efficiency Class-E Power Amplifiers*, was presented in *QST* for May and June, 1997.

5.5.4 RF Amplifiers with Feedback

The feedback amplifier appears frequently in amateur RF circuits. This is a circuit with two forms of negative feedback with (usually) a single transistor to obtain wide bandwidth, well-controlled gain and well-controlled, stable input and output resistances. Several of these amplifiers can be cascaded to form a high gain circuit that is both stable and predictable.

The small-signal schematic for the feedback amplifier is shown in **Fig 5.51** without bias components or power supply details. The design begins with an NPN transistor biased for a stable dc current. Gain is reduced with emitter degeneration, increasing input resistance while decreasing gain. Additional feedback is then added with a parallel feedback resistor, R_f, between the collector and base. This is much like the resistor between an op-amp output and the inverting input which reduces gain and *decreases* input resistance.

Several additional circuits are presented showing practical forms of the feedback amplifier. **Fig 5.52** shows a complete circuit. The base is biased with a resistive divider from the collector. However, much of the resistor is bypassed, leaving only R_f active for actual signal feedback. Emitter degeneration is ac-coupled to the emitter. The resistor R_E dominates the degeneration since R_E is normally much smaller than the emitter bias resistor. Components that are predominantly used for biasing are marked with "B." This amplifier would normally be terminated in 50 Ω at both the input and output. The transformer has the

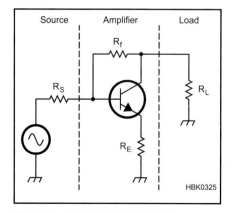

Fig 5.51 — Small-signal circuit for a feedback amplifier.

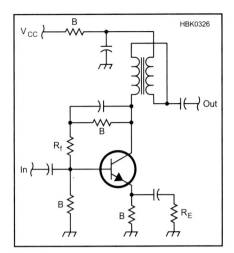

Fig 5.52 — A practical feedback amplifier. Components marked with "B" are predominantly for biasing. The 50-Ω output termination is transformed to 200 Ω at the collector. A typical RF transformer is 10 bifilar turns of #28 on a FT-37-43 ferrite toroid. The inductance of one of the two windings should have a reactance of around 250 Ω at the lowest frequency of operation.

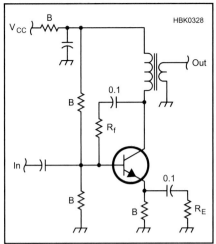

Fig 5.54 — This form of the feedback amplifier uses an arbitrary transformer. Feedback is isolated from bias components.

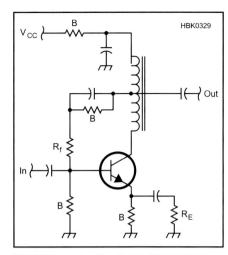

Fig 5.55 — A feedback amplifier with feedback from the output transformer tap. This is common, but can produce unstable results.

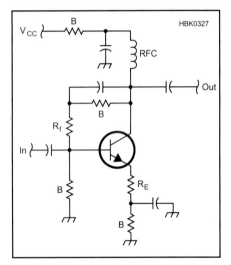

Fig 5.53 — A variation of the feedback amplifier with a 50-Ω output termination at the collector.

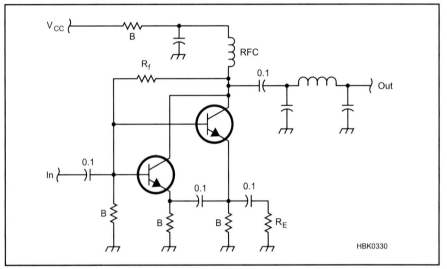

Fig 5.56 — Feedback amplifier with two parallel transistors.

effect of making the 50-Ω load "look like" a larger load value, $R_L = 200\ \Omega$ to the collector. This is a common and useful value for many HF applications.

Fig 5.53 differs from Fig 5.52 in two places. First, the collector is biased through an RF choke instead of a transformer. The collector circuit then "sees" 50 Ω when that load is connected. Second, the emitter degeneration is in series with the bias, instead of the earlier parallel connection. Either scheme works well, although the parallel configuration affords experimental flexibility with isolation between setting degeneration and biasing. Amplifiers without an output transformer are not constrained by degraded transformer performance and often offer constant or "flat" gain to several GHz.

The variation of **Fig 5.54** may well be the most general. It uses an arbitrary transformer to match the collector. Biasing is traditional and does not interact with the feedback.

Feedback is obtained directly from the output tap in the circuit of **Fig 5.55**. While this scheme is common, it is less desirable than the others, for the transformer is part of the feedback loop. This could lead to instabilities. Normally, the parallel feedback tends to stabilize the amplifiers.

The circuit of **Fig 5.56** has several features. Two transistors are used, each with a separate emitter biasing resistor. However, ac coupling causes the pair to operate as a single device with degeneration set by R_E. The parallel feedback resistor, R_f, is both a signal feedback element and part of the bias divider. This constrains the values slightly. Finally, an arbitrary output load can be presented to the composite collector through a π-type matching network. This provides some low pass filtering, but constrains the amplifier bandwidth.

5.6 Impedance Matching Networks

An impedance transforming or matching network is one that accepts power from a generator with one characteristic impedance, the source, and delivers virtually all of that power to a load at a different impedance. The simple designs in this section provide matching at only one frequency. More refined methods discussed in the reference texts can encompass a wide band of frequencies.

Both source and load impedances are likely to be complex with both real and imaginary (reactive) parts. The procedures given in this section allow the reader to design the common impedance-matching networks based on the simplification that both the source and load impedance are resistive (with no reactive component). To use these procedures with a complex load or source impedance, the usual method is to place a reactance in series with the impedance that has an equal and opposite reactance to cancel the reactive component of the impedance to be matched. Then treat the resulting purely-resistive impedance as the resistance to be transformed. For example, if a load impedance to be matched is $120 + j\,40\,\Omega$, add capacitive reactance of $-j\,40\,\Omega$ in series with the load, resulting in a load impedance of $120 + j\,0\,\Omega$ so that the following procedures can be used. The series reactance-canceling component may then be combined with the matching network's output component in some configurations.

You may wish to review the section on series-parallel impedance transformations in the **Electrical Fundamentals** chapter as those techniques form the basis of impedance-matching network design. There is additional discussion of L networks in the chapters on **Transmission Lines** and of Pi networks in the **RF Power Amplifiers** chapter. These discussions apply to more specific applications of the networks.

5.6.1 L Networks

Perhaps the most common LC impedance transforming network is the L network, so named because it uses two elements — one series element and one parallel — resembling the capital L on its side. There are eight different types of L networks as shown in **Fig 5.57**. (Types G and H are not as widely used and not covered here.) The simplified Smith Chart sketches show what range of impedance values can be transformed to the required impedance Z_0 and the path on the chart by which the transformation is achieved by the two reactances. (An introductory discussion of the Smith Chart is provided on this book's CD-ROM and a detailed treatment is available in *The ARRL Antenna Book*.) Z_{DEV} represents the output impedance of a device such as a transistor amplifier or any piece of equipment.

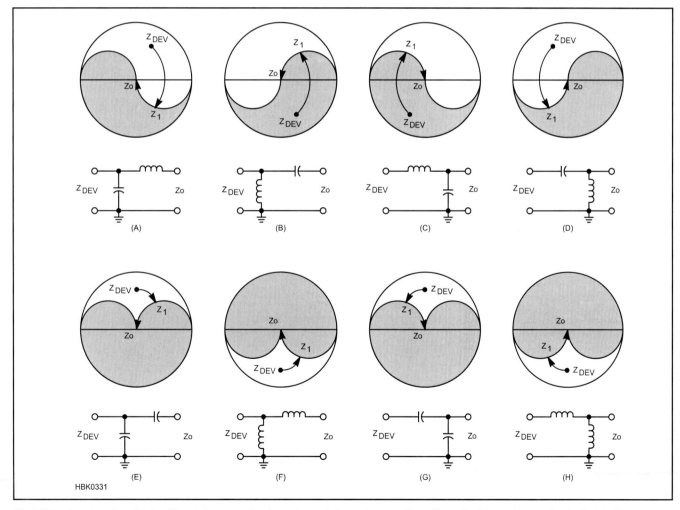

Fig 5.57 — L-networks which will match a complex impedance (shown here as Z_{DEV}, the output impedance of a device) to Z_0, a resistive source or load. Impedances within the shaded portion of the simplified Smith Chart cannot be matched by the network. Z_1 represents the impedance that is transformed from Z_{DEV} by the series element.

The process of transformation works the same in both "directions." That is, a network designed to transform Z_{DEV} to Z_0 will also transform Z_0 to Z_{DEV} if reversed. The L network (and the Pi and T networks discussed below) is *bilateral,* as are all lossless networks.

Purely from the standpoint of impedance matching, the L-network can be constructed with inductive or capacitive reactance in the series arm and the performance will be exactly the same in either case. However, the usual case in amateur circuits is to place the inductive reactance in the series arm to act as a low-pass filter. Your particular circumstances may dictate otherwise, however. For example, you may find it useful that the series-C/parallel-L network places a dc short circuit across one side of the network while blocking dc current through it.

The design procedure for L network configurations A through F in Fig 5.57 is as follows:

Given the two resistance values to be matched, connect the series arm of the circuit to the smaller of the two (R_S) and the parallel arm to the larger (R_P).

Find the ratio R_P / R_S and the L network's Q:

$$Q = \sqrt{\frac{R_P}{R_S} - 1}$$

Calculate the series reactance $X_S = Q R_S$
Calculate the parallel reactance $X_P = R_P / Q$.

Select the form of the L network that suits your needs and convert the reactances to component values:

$$C = \frac{1}{2\pi f X_c} \text{ and } L = \frac{X_L}{2\pi f}$$

5.6.2 Pi Networks

Another popular impedance matching network is the Pi network shown in **Fig 5.58**. The Pi network can be thought of as two L networks "back to back." For example, the Pi network in Fig 5.58 can be split into L networks 5.57A on the left and 5.57C on the right. The two inductors in series are combined into the single inductance of the Pi network. There are other forms of the Pi network with different configurations of inductance and capacitance, but the version shown is by far the most common in amateur circuits.

The use of two transformations allows the designer to choose Q for the Pi network, unlike the L network for which Q is determined by the ratio of the impedances to be matched. This is particularly useful for matching amplifier outputs as discussed in the **RF Power Amplifier** chapter because it allows more control of the network's frequency response and of component values. Q must be high enough that $(Q^2 + 1) > (R_1 / R_2)$. If these two quantities are equal, X_{C2} becomes infinite, meaning zero capacitance, and the Pi network reduces to the L network in Fig 5.57A. The design procedure for the Pi network in Fig 5.58 is as follows:

Determine the two resistance values to be matched, $R_1 > R_2$, and select a value for Q.

Calculate the value of the parallel reactance $X_{C1} = R_1 / Q$.

Calculate the value of the parallel reactance X_{C2}

$$X_{C2} = R_2 \sqrt{\frac{R_1 / R_2}{Q^2 + 1 - R_1 / R_2}}$$

Calculate the value of the series reactance X_L

$$X_L = \frac{Q R_1 + R_1 R_2 / X_{C2}}{Q^2 + 1}$$

Convert the reactances to component values:

$$C = \frac{1}{2\pi f X_c} \text{ and } L = \frac{X_L}{2\pi f}$$

IMPEDANCE INVERSION

The Pi network displays a useful property, that of *impedance inversion.* Consider a symmetrical Pi network designed for 50-Ω terminations on each end and a Q of 1. For any value of impedance connected at one end of the network, the value seen at the other end is the same as if the network was replaced with a quarter-wavelength of transmission line at the design frequency. This holds for all values of real and reactive impedance. For example, if $Z = 10 + j\,10$ Ω is connected to one end of the network, the impedance at the other end of the line is inverted about the network's design impedance of 50 Ω to $125 - j\,125$ Ω, just as it would be if the network were replaced by a quarter-wavelength of transmission line with a characteristic impedance of 50 Ω. The symmetrical Pi network has the easily remembered design parameters $X_C = X_L = Z_0$.

5.6.3 T Networks

The T network shown in **Fig 5.59** is especially useful for matching to relatively low impedance from 50-Ω sources with practical components and low Q. Like the Pi network, Q must be high enough that $(Q^2 + 1) > (R_1 / R_2)$. Designing the component values for this network requires the calculation of a pair of intermediate values, A and B, to make the equations more manageable.

Determine the two resistance values to be matched, $R_1 > R_2$, and select a value for Q.

Calculate the intermediate variables A and B

$$A = R_1(Q^2 + 1) \text{ and } B = \sqrt{\frac{A}{R_2} - 1}$$

Calculate the value of the input series reactance $X_{L1} = R_1 Q$.

Calculate the value of the output series reactance $X_{L2} = R_2 B$.

Calculate the value of the parallel reactance $X_C = A / (Q + B)$

Convert the reactances to component values:

$$C = \frac{1}{2\pi f X_c} \text{ and } L = \frac{X_L}{2\pi f}$$

Many amateur transmission line impedance matching circuits, ("antenna tuners") are constructed using the version of the T network in which the series components are capacitive and the parallel reactance is inductive. The circuit (for which the calculations are the same as above) is made adjustable by using variable capacitors for the series reactances and a tapped inductor for the parallel reactance. This configuration (two series variable-C, one parallel tapped-L) is easier and less expensive to construct than one with two variable-L and one variable-C. The configuration with an input series inductance and two capacitors forming the rest of the network is very useful in solid state amplifier design and covered in the reference texts listed at the end of this chapter.

The symmetrical T network with Q=1 has the same impedance inversion properties as the symmetrical Pi network discussed in the preceding section. As for the symmetrical Pi network $X_C = X_L = Z_0$.

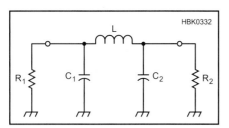

Fig 5.58 — Schematic for the Pi network, used for impedance matching.

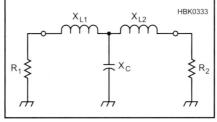

Fig 5.59 — An LCL-type T network.

5.7 RF Transformers

5.7.1 Air-Core Nonresonant RF Transformers

Air-core transformers often function as mutually coupled inductors for RF applications. They consist of a primary winding and a secondary winding in close proximity. Leakage reactances are ordinarily high, however, and the coefficient of coupling between the primary and secondary windings is low. Consequently, unlike transformers having a magnetic core, the turns ratio does not have as much significance. Instead, the voltage induced in the secondary depends on the mutual inductance.

In a very basic transformer circuit operating at radio frequencies, such as in **Fig 5.60**, the source voltage is applied to L1. R_S is the series resistance inherent in the source. By virtue of the mutual inductance, M, a voltage is induced in L2. A current flows in the secondary circuit through the reactance of L2 and the load resistance of R_L. Let X_{L2} be the reactance of L2 independent of L1, that is, independent of the effects of mutual inductance. The impedance of the secondary circuit is then:

$$Z_S = \sqrt{R_L^2 + X_{L2}^2} \qquad (12)$$

where
 Z_S = the impedance of the secondary circuit in ohms,
 R_L = the load resistance in ohms, and
 X_{L2} = the reactance of the secondary inductance in ohms.

The effect of Z_S upon the primary circuit is the same as a coupled impedance in series with L1. Fig 5.60 displays the coupled impedance (Z_P) in a dashed enclosure to indicate that it is not a new physical component. It has the same absolute value of phase angle as in the secondary impedance, but the sign of the reactance is reversed; it appears as a capacitive reactance. The value of Z_P is:

$$Z_P = \frac{(2\pi f M)^2}{Z_S} \qquad (13)$$

where
 Z_P = the impedance introduced into the primary,
 Z_S = the impedance of the secondary circuit in ohms, and
 $2\pi f M$ = the mutual reactance between the reactances of the primary and secondary coils (also designated as X_M).

5.7.2 Air-Core Resonant RF Transformers

The use of at least one resonant circuit in place of a pair of simple reactances elimi-

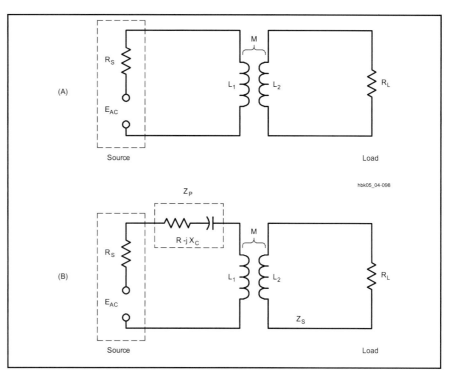

Fig 5.60 — The coupling of a complex impedance back into the primary circuit of a transformer composed of nonresonant air-core inductors.

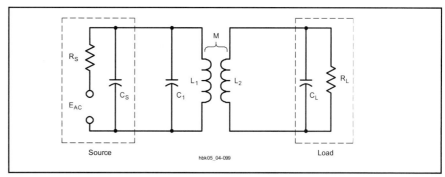

Fig 5.61 — An air-core transformer circuit consisting of a resonant primary circuit and an untuned secondary. R_S and C_S are functions of the source, while R_L and C_L are functions of the load circuit.

nates the reactance from the transformed impedance in the primary. For loaded or operating Q of at least 10, the resistances of individual components is negligible. **Fig 5.61** represents just one of many configurations in which at least one of the inductors is in a resonant circuit. The reactance coupled into the primary circuit is cancelled if the circuit is tuned to resonance while the load is connected. If the reactance of the load capacitance, C_L is at least 10 times any stray capacitance in the circuit, as is the case for low impedance loads, the value of resistance coupled to the primary is

$$R1 = \frac{X_M^2 R_L}{X_2^2 + R_L^2} \qquad (14)$$

where
 R1 = series resistance coupled into the primary circuit,
 X_M = mutual reactance,
 R_L = load resistance, and
 X_2 = reactance of the secondary inductance.

The parallel impedance of the resonant circuit is just R1 transformed from a series to a parallel value by the usual formula, $R_P = X_2 / R1$.

The higher the loaded or operating Q of the circuit, the smaller the mutual inductance required for the same power transfer. If both the primary and secondary circuits consist of resonant circuits, they can be more loosely coupled than with a single tuned circuit for the same power transfer. At the usual loaded Q of 10 or greater, these circuits are quite selective, and consequently narrowband.

Although coupling networks have to a large measure replaced RF transformer coupling that uses air-core transformers, these circuits are still useful in antenna tuning units and other circuits. For RF work, powdered-iron toroidal cores have generally replaced air-core inductors for almost all applications except where the circuit handles very high power or the coil must be very temperature stable. Slug-tuned solenoid coils for low-power circuits offer the ability to tune the circuit precisely to resonance. For either type of core, reasonably accurate calculation of impedance transformation is possible. It is often easier to experiment to find the correct values for maximum power transfer, however.

5.7.3 Broadband Ferrite RF Transformers

The design concepts and general theory of ideal transformers presented in the **Electrical Fundamentals** chapter apply also to transformers wound on ferromagnetic-core materials (ferrite and powdered iron). As is the case with stacked cores made of laminations in the classic I and E shapes, the core material has a specific permeability factor that determines the inductance of the windings versus the number of wire turns used. (See the earlier discussion on Ferrite Materials in this chapter.)

Toroidal cores are useful from a few hundred hertz well into the UHF spectrum. The principal advantage of this type of core is the self-shielding characteristic. Another feature is the compactness of a transformer or inductor. Therefore, toroidal-core transformers are excellent for use not only in dc-to-dc converters, where tape-wound steel cores are employed, but at frequencies up to at least 1000 MHz with the selection of the proper core material for the range of operating frequencies. Toroidal cores are available from micro-miniature sizes up to several inches in diameter that can handle multi-kW military and commercial powers.

One of the most common ferromagnetic transformers used in amateur circuits is the conventional broadband transformer. Broadband transformers with losses of less than 1 dB are employed in circuits that must have a uniform response over a substantial frequency range, such as a 2- to 30-MHz broadband amplifier. In applications of this sort, the reactance of the windings should be at least four times the impedance that the winding is designed to look into at the lowest design frequency.

Example: What should be the winding reactances of a transformer that has a 300-Ω primary and a 50-Ω secondary load? Relative to the 50-Ω secondary load:

$$X_S = 4\,Z_S = 4 \times 50\,\Omega = 200\,\Omega$$

and the primary winding reactance (X_P) is:

$$X_P = 4\,Z_P = 4 \times 300\,\Omega = 1200\,\Omega$$

The core-material permeability plays a vital role in designing a good broadband transformer. The effective permeability of the core must be high enough to provide ample winding reactance at the low end of the operating range. As the operating frequency is increased, the effects of the core tend to disappear until there are scarcely any core effects at the upper limit of the operating range. The limiting factors for high frequency response are distributed capacity and leakage inductance due to uncoupled flux. A high-permeability core minimizes the number of turns needed for a given reactance and therefore also minimizes the distributed capacitance at high frequencies.

Ferrite cores with a permeability of 850 are common choices for transformers used between 2 and 30 MHz. Lower frequency ranges, for example, 1 kHz to 1 MHz, may require cores with permeabilities up to 2000. Permeabilities from 40 to 125 are useful for VHF transformers. Conventional broadband transformers require resistive loads. Loads with reactive components should use appropriate networks to cancel the reactance.

The equivalent circuit in Fig 5.33 applies to any coil wound on a ferrite core, including transformer windings. (See the section on Ferrite Materials.) However, in the series-equivalent circuit, µ′S is not constant with frequency as shown in Fig 5.34A and 5.34B. Using the low-frequency value of µ′S is a useful approximation, but the effects of the parallel R and C should be included. In high-power transmitting and amplifier applications, the resistance R may dissipate some heat, leading to temperature rise in the core. Regarding C, there are at least two forms of stray capacitance between windings of a transformer; from wire-to-wire through air and from wire-to-wire through the ferrite, which acts as a dielectric material. (Ferrites with low iron content have a relative dielectric constant of approximately 10-12.)

Conventional transformers are wound in the same manner as a power transformer. Each winding is made from a separate length of wire, with one winding placed over the previous one with suitable insulation between. Unlike some transmission-line transformer designs, conventional broadband transformers provide dc isolation between the primary and secondary circuits. The high voltages encountered in high-impedance-ratio step-up transformers may require that the core be wrapped with glass electrical tape before adding the windings (as an additional protection from arcing and voltage breakdown), especially with ferrite cores that tend to have rougher edges. In addition, high voltage applications should also use wire with high-voltage insulation and a high temperature rating.

Fig 5.62 illustrates one method of transformer construction using a single toroid as the core. The primary of a step-down impedance transformer is wound to occupy the entire core, with the secondary wound over the primary. The first step in planning the winding is to select a core of the desired permeability. Convert the required reactances determined earlier into inductance values for the lowest frequency of use. To find the number of turns for each winding, use the A_L value for the selected core and the equation for determining the number of turns:

$$L = \frac{A_L \times N^2}{1000000} \quad (15)$$

where
 L = the inductance in mH
 A_L = the inductance index in mH per 1000 turns, and
 N = the number of turns.

Be certain the core can handle the power by calculating the maximum flux and comparing the result with the manufacturer's guidelines.

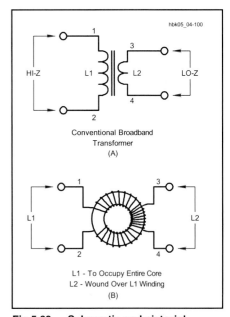

Fig 5.62 — Schematic and pictorial representation of a conventional broadband transformer wound on a toroid core. The secondary winding (L2) is wound over the primary winding (L1).

$$B_{max} = \frac{E_{RMS} \times 10^8}{4.44 \times A_e \times N \times f} \quad (16)$$

where

B_{max} = the maximum flux density in gauss
E_{RMS} = the voltage across the inductor
A_e = the cross-sectional area of the core in square centimeters
N = the number of turns in the inductor, and
f = the operating frequency in Hz.

(Both equations are from the section on ferrite toroidal inductors in the **Electrical Fundamentals** chapter and are repeated here for convenience.)

Example: Design a small broadband transformer having an impedance ratio of 16:1 for a frequency range of 2.0 to 20.0 MHz to match the output of a small-signal stage (impedance ≈ 500 Ω) to the input (impedance ≈ 32 Ω) of an amplifier.

Since the impedance of the smaller winding should be at least 4 times the lower impedance to be matched at the lowest frequency,

$$X_S = 4 \times 32 \, \Omega = 128 \, \Omega$$

The inductance of the secondary winding should be

$$L_S = \frac{X_S}{2 \pi f} = \frac{128}{6.2832 \times 2.0 \times 10^6 \, \text{Hz}}$$

$$= 0.0101 \, \text{mH}$$

Select a suitable core. For this low-power application, a ⅜ inch. ferrite core with permeability of 850 is suitable. The core has an A_L value of 420. Calculate the number of turns for the secondary.

$$N_S = 1000 \sqrt{\frac{L}{A_L}} = 1000 \sqrt{\frac{0.010}{420}}$$

$$= 4.88 \, \text{turns}$$

A 5-turn secondary winding should suffice. The primary winding derives from the impedance ratio:

$$N_P = N_S \sqrt{\frac{Z_P}{Z_S}} = 5 \sqrt{\frac{16}{1}}$$

$$= 5 \times 4 = 20 \, \text{turns}$$

This low-power application will not approach the maximum flux density limits for the core, and #28 AWG enamel wire should both fit the core and handle the currents involved.

A second style of broadband transformer

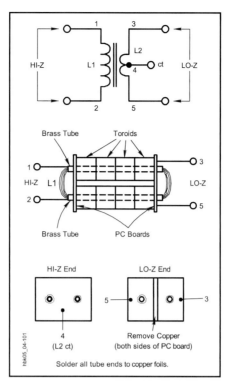

Fig 5.63 — Schematic and pictorial representation of a "binocular" style of conventional broadband transformer. This style is used frequently at the input and output ports of transistor RF amplifiers. It consists of two rows of high-permeability toroidal cores, with the winding passed through the center holes of the resulting stacks.

construction appears in **Fig 5.63**. The key elements in this transformer are the stacks of ferrite cores aligned with tubes soldered to pc-board end plates. This style of transformer is suited to high power applications, for example, at the input and output ports of transistor RF power amplifiers. Low-power versions of this transformer can be wound on "binocular" cores having pairs of parallel holes through them.

For further information on conventional transformer matching using ferromagnetic materials, see the **RF Power Amplifiers** chapter. Refer to the **Component Data and References** chapter for more detailed information on available ferrite cores. A standard reference on conventional broadband transformers using ferromagnetic materials is *Ferromagnetic Core Design and Applications Handbook* by Doug DeMaw, W1FB, published by MFJ Enterprises.

5.7.4 Transmission Line Transformers and Baluns

Conventional transformers use flux linkages to deliver energy to the output circuit. *Transmission line transformers* use transmission line modes of energy transfer between the input and the output terminals of the devices. Although toroidal versions of these transformers physically resemble toroidal conventional broadband transformers, the principles of operation differ significantly. Stray inductances and inter-winding capacitances form part of the characteristic impedance of the transmission line, largely eliminating resonances that limit high frequency response. The limiting factors for transmission line transformers include line length, deviations in the constructed line from the design value of characteristic impedance, and parasitic capacitances and inductances that are independent of the characteristic impedance of the line.

The losses in conventional transformers depend on current and include wire, eddy-current and hysteresis losses. In contrast, the losses of transmission line transformers are voltage-dependent, which make higher impedances and higher VSWR values limiting factors in design. Within design limits, the cancellation of flux in the cores of transmission line transformers permits very high efficiencies across their pass bands. Losses may be lower than 0.1 dB with the proper core choice.

Transmission-line transformers can be configured for several modes of operation, but the chief amateur use is in *baluns* (balanced-to-unbalanced transformers) and in *ununs* (unbalanced-to-unbalanced transformers). The basic principle behind a balun appears in **Fig 5.64**, a representation of the classic Guanella 1:1 balun. The input and output

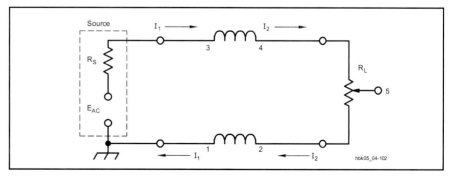

Fig 5.64 — Schematic representation of the basic Guanella "choke" balun or 1:1 transmission line transformer. The inductors are a length of two-wire transmission line. R_S is the source impedance and R_L is the load impedance.

impedances are the same, but the output is balanced about a real or virtual center point (terminal 5). If the characteristic impedance of the transmission line forming the inductors with numbered terminals equals the load impedance, then E2 will equal E1. With respect to terminal 5, the voltage at terminal 4 is E1 / 2, while the voltage at terminal 2 is –E1 / 2, resulting in a balanced output.

The small losses in properly designed baluns of this order stem from the potential gradient that exists along the length of transmission line forming the transformer. The value of this potential is –E1 / 2, and it forms a dielectric loss that can't be eliminated. Although the loss is very small in well-constructed 1:1 baluns at low impedances, the losses climb as impedances climb (as in 4:1 baluns) and as the VSWR climbs. Both conditions yield higher voltage gradients.

The inductors in the transmission-line transformer are equivalent to — and may be — coiled transmission line with a characteristic impedance equal to the load. They form a choke isolating the input from the output and attenuating undesirable currents, such as antenna current, from the remainder of the transmission line to the energy source. The result is a *current* or *choke balun*. Such baluns may take many forms: coiled transmission line, ferrite beads placed over a length of transmission line, windings on linear ferrite cores or windings on ferrite toroids.

Reconfiguring the windings of Fig 5.64 can alter the transformer operation. For example, if terminal 2 is connected to terminal 3, a positive potential gradient appears across the lengths of line, resulting in a terminal 4 potential of 2 × E1 with respect to ground. If the load is disconnected from terminal 2 and reconnected to ground, 2 × E1 appears across the load — instead of ±E1 / 2. The product of this experiment is a 4:1 impedance ratio, forming an unun. The bootstrapping effect of the new connection is applicable to many other design configurations involving multiple windings to achieve custom impedance ratios from 1:1 up to 9:1.

Balun and unun construction for the impedances of most concern to amateurs requires careful selection of the feed line used to wind the balun. Building transmission line transformers on ferrite toroids may require careful attention to wire size and spacing to approximate a 50-Ω line. Wrapping wire with polyimide tape (one or two coatings, depending upon the wire size) and then glass-taping the wires together periodically produces a reasonable 50-Ω transmission line. Ferrite cores in the permeability range of 125 to 250 are generally optimal for transformer windings, with 1.25-inch cores suitable to 300-Ω power levels and 2.4-inch cores usable to the 5 kW level. Special designs may alter the power-handling capabilities of the core sizes. For the 1:1 balun shown in Fig 5.64, 10 bifilar turns (#16 AWG wire for the smaller core and #12 AWG wire for the larger, both Thermaleze wire) yields a transformer operable from 160 to 10 meters.

Transmission-line transformers have their most obvious application to antennas, since they isolate the antenna currents from the feed line, especially where a coaxial feed line is not exactly perpendicular to the antenna. The balun prevents antenna currents from flowing on the outer surface of the coax shielding, back to the transmitting equipment. Such currents would distort the antenna radiation pattern. Appropriately designed baluns can also transform impedance values at the same time. For example, one might use a 4:1 balun to match a 12.5-Ω Yagi antenna impedance to a 50-Ω feed line. A 4:1 balun might also be used to match a 75-Ω TV antenna to 300-Ω feed line.

Interstage coupling within solid-state transmitters represents another potential for transmission-line transformers. Broadband coupling between low-impedance, but mismatched stages can benefit from the low losses of transmission-line transformers. Depending upon the losses that can be tolerated and the bandwidth needed, it is often a matter of designer choice between a transmission-line transformer and a conventional broadband transformer as the coupling device.

For further information on transmission-line transformers and their applications, see the **Transmission Lines** chapter. A chart of balun construction techniques and additional discussion is provided on this book's accompanying CD-ROM. Another reference on the subject is *Transmission Line Transformers*, by Jerry Sevick, W2FMI, published by Noble Publishing.

5.8 Noise

The ideal amplifier is linear with an output that is an exact replica of the input with the only difference being greater amplitude and a phase difference. There should be no other output signals. If two inputs are applied to an ideal linear amplifier, the result will be two outputs, each being just what would be seen if each input was applied alone, with nothing else added. Several phenomena compromise amplifiers from this ideal. They include noise, gain compression, harmonic distortion and intermodulation distortion. (Gain compression, harmonic distortion, and sources of external noise are discussed in the **Receivers** chapter.)

5.8.1 Noise in Amplifiers

Noise is a familiar corruption in an amplifier. The noise of concern is not what we most often hear coming from our HF receivers; that noise generally arises from thunderstorms somewhere in the world, or power lines somewhere in our community. Rather, we are concerned with the noise that is generated within the circuitry. The dominant component of this noise, so called thermal noise, originates from random motion of the electrons within a conductor. This noise shows up as a voltage that appears between the two conductor ends. The available power present is kTB (in watts) where k is Boltzman's constant, T is absolute temperature in Kelvin, and B is the bandwidth we use to observe the noise.

Although a power equal to kTB is available from any conductor, the related voltage is very small if the conductor is a good one. A resistor, a conductor with larger resistance, allows a larger voltage to appear, but can only provide the same amount of noise power.

Fig 5.65 shows a simple amplifier terminated in 50 Ω at both input and output. The source and load resistances generate noise. The noise generated by the output load is normally ignored during a noise analysis of the amplifier, for the circuit designer is primarily concerned with the available noise from the amplifier. The noise from the input source is increased by the amplifier gain, just as any signal would be increased. There is nothing that can be done to avoid this noise. If the amplifier available power gain is G and the available noise power from the input source is N_{IN}, the output noise will be $G \times N_{IN}$, even when the amplifier is perfect and noiseless.

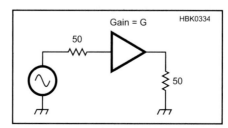

Fig 5.65 — A terminated amplifier used for noise analysis.

5.8.2 Noise Figure

A real-world amplifier will have a noise output that is even higher than the amplified input noise. The output noise is greater by a ratio that we call the *noise factor*, designated by F. The logarithmic form of noise factor is the *Noise Figure* or *NF* (dB) = 10 log (F). The two forms, algebraic ratio or dB, are used interchangeably, although the algebraic ratio is used in all of the equations that follow. The extra noise is that generated within the active device and circuit components. Noise factor is given by

$$F = \frac{N_{OUT}}{G N_{IN}} \quad (17)$$

where
  N_{OUT} is the output noise power delivered to the load,
  N_{IN} is the noise power available from the input resistance, and
  G is the available power gain of the circuit.

N_{IN} is the noise power available from the source resistance when it has a temperature of 290 K. NF is the ratio of two noise powers. The larger number (numerator) is the noise actually coming from the amplifier while the smaller (denominator = GN_{IN}) is the noise that would be coming from the amplifier if it generated no noise of its own. A perfect, noiseless amplifier would have F=1 from the equation, or converting to dB, NF=0 dB.

Gain, G, is the power gain we normally associate with an amplifier: output signal power delivered to the load, S_{OUT}, divided by S_{IN}, an input signal power. If we insert this gain ratio into the noise figure defining equation, and rearrange the terms, we obtain

$$F = \frac{S_{IN}/N_{IN}}{S_{OUT}/N_{OUT}} \quad (18)$$

This describes a combination of signal and noise. Essentially, noise figure can be interpreted to be a degradation in signal to noise ratio as we progress through the amplifier.

This equation can be rearranged to

$$F = \frac{G_{NOISE}}{G_{SIGNAL}} \quad (19)$$

where G_{NOISE} is the *noise gain*, the output noise power divided by the available input noise power. G_{SIGNAL} is the familiar signal gain used above. All forms of these equations are used in deriving some of the results we use with noise figure.

Typical NF values range from 1 to 10 dB for the amplifiers that we frequently use in RF systems. Mixers tend to have higher noise figures. Modern FET amplifiers are capable of NF as low as 0.1 to 0.2 dB at UHF with values under 1 dB even possible at 10 GHz.

We frequently ask for the noise factor of a cascade of two amplifiers. This result is

$$F_{NET} = F_1 + \frac{F_2 - 1}{G_1} \quad (20)$$

where F_1 and F_2 are noise factors for stage 1 and 2, respectively, and G_1 is the available power gain for the first stage. While the noise from both stages contributes to the net noise factor, the 2nd stage noise contribution is reduced by the gain of the first stage. Clearly, if we can calculate NF for two stages, we can perform the calculations several times and obtain the result for any number of stages.

Noise figure is a vital amplifier and receiver characteristic at VHF where external noise (from thunderstorms, for example) is low. While a low noise figure is rarely needed at lower frequencies, it becomes more important when small antennas are used. Noise figure is also a vital parameter within a receiver, for careful control of noise will allow the designer to use low gain, which keeps distortion low.

The noise power available from a resistor is kTB. A useful number to remember is that kT = –174 dBm at "room" temperature of 290 K. If the noise was observed in a receiver with a bandwidth of 3 kHz (a voice "channel"), B would be 3000 Hz and 10 log B = 34.8 dB. The noise power available from the resistor would then be –174dBm + 34.8 dB = –139.2 dBm. A receiver can be thought of as a large amplifier. If the receiver had a 10 dB noise figure, the output noise would be the same as would appear if an input noise of –139.2 dBm + 10 dB = –129.2 dBm was applied to the input of a perfect, noiseless receiver.

The related noise voltage from a resistor, R, is

$$V_N = \sqrt{4kTBR} \quad (21)$$

where k is Boltzmann's constant (1.38 × 10^{-23}), T is the resistor temperature in K, and B is bandwidth in Hz. R is the resistance in Ω. The available power, kT, is called a *spectral power density*, usually in W/Hz. The resulting voltage, V_n, is a spectral voltage density in volts-per-root-Hz. Op-amps often have noise specified in terms of an equivalent input spectral voltage density of noise. The same method is sometimes used for transistors, although noise figure is the more common parameter used to specify an RF design.

Amplifier noise figure is not always a simple constant that may be extracted from a data sheet and applied to a design. Rather, data sheet noise figure is specific to a "typical" amplifier, or more often, is the best NF one can achieve. The noise figure of a specific design then depends upon device biasing and the impedance presented to the device input.

5.9 Two-Port Networks

A *two-port network* is one with four terminals. The terminals are arranged into pairs, each being called a *port*. The general network schematic is shown in **Fig 5.66**. The input port is characterized by input voltage and current, V_1 and I_1, and the output is described by V_2 and I_2. By convention, currents into the network are usually considered positive.

Many devices of interest have three terminals rather than four. Two-port methods are used with these by choosing one terminal to be common to both input and output ports. The two-port representations of the common emitter, common base and common collector connections of the bipolar transistor are shown in **Fig 5.67**. Similar configurations may be used with FETs, vacuum tubes, ICs or passive networks.

The general concepts of two-port theory are applicable to devices with a larger number of terminals. The theory is expandable

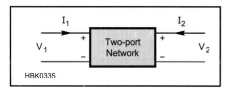

Fig 5.66 — General configuration of a two-port network. Note the voltage polarities and direction of currents.

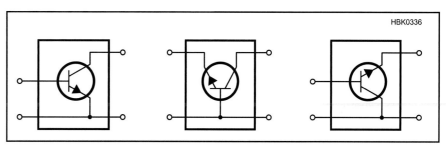

Fig 5.67 — Two-port representations of the common emitter, common base and common collector amplifiers.

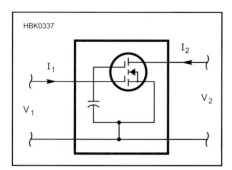

Fig 5.68 — A dual-gate MOSFET treated as a three-terminal device in a two-port network.

to any number of ports. Alternatively, the bias on some terminals can be established with attention fixed only upon two ports of a multi-element device. An example would be a dual-gate MOSFET in a common-source configuration as shown in **Fig 5.68**. The input port contain the source and gate 1 while the output port contains the source and drain leads. The fourth device terminal, gate 2, has a fixed bias potential and is treated as an ac ground. Signal currents at this terminal are ignored in the analysis.

5.9.1 Two-Port Parameters

There are four variables associated with any two-port network; two voltages and two currents. These are signal components. Any two variables may be picked as independent. The remaining variables are then dependent variables. These are expressed as an algebraic linear combination of the two independent quantities. The following overview are intended for definition purposes. A complete discussion of the use of two-port parameters can be found in the reference texts at the end of this chapter and examples of their use in RF circuit design in Hayward's *Introduction to Radio Frequency Design*.

Y AND Z PARAMETERS

Assume that the two voltages are chosen as independent variables. The two currents are then expressed as linear combinations of the voltages, $I_1 = K_a V_1 + K_b V_2$ and $I_2 = K_c V_1 + K_d V_2$. The constants of proportionality, K_a through K_d, have the dimensions of admittance. The usual representation is

$$I_1 = y_{11} V_1 + y_{12} V_2$$
$$I_2 = y_{21} V_1 + y_{22} V_2 \quad (22)$$

The independent and dependent variable sets are column vectors, leading to the equivalent matrix representation

$$\begin{pmatrix} I_1 \\ I_2 \end{pmatrix} = \begin{pmatrix} y_{11} & y_{12} \\ y_{21} & y_{22} \end{pmatrix} \begin{pmatrix} V_1 \\ V_2 \end{pmatrix} \quad (23)$$

The y matrix for a two-port network uniquely describes that network. The set of y_{11} through y_{22} are called the two-port network's *Y parameters* or *admittance parameters*. Consider the y parameters from an experimental viewpoint. The first y parameter, y_{11}, is the input admittance of the network with V_2 set to zero. Hence, it is termed the *short-circuit input admittance*. y_{21} is the *short-circuit forward transadmittance*, the reciprocal of transconductance. Similarly, if V_1 is set to zero, realized by short circuiting the input, y_{22} is the *short-circuit output admittance* and y_{12} is the *short-circuit reverse transadmittance*.

The matrix subscripts are sometimes replaced by letters. The set of y parameters can be replaced by y_i, y_r, y_f, and y_o where the subscripts indicate respectively input, reverse, forward, and output. The subscripts are sometimes modified further to indicate the connection of the device. For example, the short circuit forward transfer admittance of a common emitter amplifier would be y_{21e} or y_{fe}.

The y parameters are only one set of two-port parameters. The open-circuited Z parameters or *impedance parameters* result if the two currents are treated as independent variables

$$\begin{pmatrix} V_1 \\ V_2 \end{pmatrix} = \begin{pmatrix} z_{11} & z_{12} \\ z_{21} & z_{22} \end{pmatrix} \begin{pmatrix} I_1 \\ I_2 \end{pmatrix} \quad (24)$$

The parameter sets describe the same device; hence, they are related to each other. If equation 22 is multiplied by y_{12} and the resulting equations are subtracted, the result is the input voltage as a function of the currents

$$V_1 = \frac{I_1 y_{22} - y_{12} I_2}{y_{11} y_{22} - y_{12} y_{21}}$$

A similar procedure is used to find the output voltage as a function of the currents, leading to the general relationships

$$z_{11} = \frac{y_{22}}{\Delta y} \quad z_{12} = \frac{-y_{12}}{\Delta y}$$
$$z_{21} = \frac{-y_{21}}{\Delta y} \quad z_{22} = \frac{y_{11}}{\Delta y} \quad (25)$$

where Δy is the determinant of the y matrix, $y_{11} y_{22} - y_{12} y_{21}$. The inverse transformations, yielding the y parameters when z parameters are known, are exactly the same as those in equation 25, except that the y_{jk} and z_{jk} values are interchanged. The similarity is useful when writing transformation programs for a programmable calculator or computer.

H PARAMETERS

The *H parameters* or *hybrid parameters* are defined if the input current and output voltage are selected as independent variables

$$\begin{pmatrix} V_1 \\ I_2 \end{pmatrix} = \begin{pmatrix} h_{11} & h_{12} \\ h_{21} & h_{22} \end{pmatrix} \begin{pmatrix} I_1 \\ V_2 \end{pmatrix} \quad (26)$$

The input term, h_{11}, is an impedance, while h_{22} represents an output admittance. The forward term, h_{21}, is the ratio of the output to the input current, beta for a bipolar transistor. The reverse parameter, h_{12}, is a voltage ratio. The mixture of dimensions accounts for the "hybrid" name of the set.

SCATTERING (S) PARAMETERS

The two-port parameters presented above deal with four simple variables; input and output voltage and current at the ports. The variables are interrelated by appropriate matrices. The choice of which matrix is used depends upon which of the four variables are chosen to be independent.

There is no reason to limit the variables to simple ones. Linear combinations of the simple variables are just as valid. The more complicated variables chosen should be linearly independent and, ideally, should have some physical significance.

A transformation to other variables is certainly not new. For example, logarithmic transformations such as the dB or dBm are so common that we used them interchangeably with the fundamental quantities without even mentioning that a transformation has occurred. Such a new viewpoint can be of great utility in working with transmission line when an impedance is replaced by a reflection coefficient, $\Gamma = (Z - Z_0) / (Z + Z_0)$.

Scattering parameters or *S parameters* are nothing more than a repeat of this viewpoint. Instead of considering voltages and currents to be the fundamental variables, we use four "voltage waves." They are interrelated through an appropriate matrix of s parameters.

Fig 5.69 shows the traditional two-port network and an alternate one with voltage waves incident on and reflected from the ports. The voltage waves are defined with the letters a_1, b_1, a_2, and b_2. The a waves are considered to be incident waves on the parts and are the independent variables. The b waves are the result of reflection or "scattering" and are the dependent variables. The waves are related to voltages and currents and defined with respect to a characteristic impedance, Z_0.

The scattered waves are related to the incident ones with a set of linear equations just as the port currents were related to the port voltages with y parameters. The relating equations are

$$b_1 = S_{11} a_1 + S_{12} a_2$$
$$b_2 = S_{21} a_1 + S_{22} a_2 \quad (27)$$

or, in matrix form

$$\begin{pmatrix} b_1 \\ b_2 \end{pmatrix} = \begin{pmatrix} S_{11} & S_{12} \\ S_{21} & S_{22} \end{pmatrix} \begin{pmatrix} a_1 \\ a_2 \end{pmatrix} \quad (28)$$

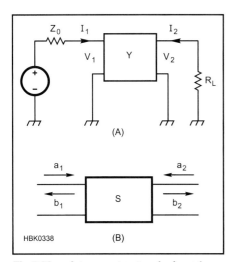

Fig 5.69 — A two-port network viewed as being driven by voltages and currents (A) or voltage waves (B). The voltages and currents are related by Y parameters while the voltage waves are related by scattering or S parameters.

Consider the meaning of S_{11}. If the incident wave at the output, a_2, set to zero, the set of equations 27 reduce to $b_1 = S_{11}a_1$ and $b_2 = S_{21}a_1$. S_{11} is the ratio of the input port reflected wave to the incident one. This reduces, using the defining equations for a_1 and b_1 to

$$S_{11} = \frac{Z - Z_0}{Z + Z_0} \qquad (29)$$

This is the *input port reflection coefficient*. Similarly, S_{21} is the voltage wave emanating from the output as the result of an incident wave at the network input. In other words, S_{21} represents a forward gain. The other two S parameters have similar significance. S_{22} is the *output reflection coefficient* when looking back into the output port of the network with the input terminated in Z_0. S_{12} is the reverse gain if the output is driven and the signal at the input port detected.

The reflection coefficient nature of S parameters makes them especially convenient for use in design and specification and even more so when displayed on a Smith Chart.

5.9.2 Return Loss

Although SWR as described in the **Transmission Lines** chapter is usually used by amateurs to describe the relationship between a transmission line's characteristic impedance and a terminating impedance, the engineering community generally finds it more convenient to use *return loss, RL,* instead.

Return loss and SWR measure the same thing — how much of the incident power, P_{INC}, in the transmission line is transferred to the load and how much is reflected by it, P_{REFL} — but state the result differently.

$$\text{Return Loss (dB)} = -10\log\left(\frac{P_{REFL}}{P_{INC}}\right) \qquad (30)$$

Because P_{REFL} is never greater than P_{FWD}, RL is always positive. The more positive RL, the less the amount of power reflected from the load compared to forward power. If all the power is transferred to the load because $Z_L = Z_0$, RL = ∞ dB. If none of the power is transferred to the load, such as at an open- or short-circuit, RL = 0 dB. (You may encounter negative values for RL in literature or data sheets. Use the absolute magnitude of these values — the negative value does not indicate power gain.)

RL can also be calculated directly from power ratios, such as dBm (decibels with respect to 1 mW) or dBW (decibels with respect to 1 watt). In this case, RL = $P_{INC} - P_{REFL}$ because the logarithm has already been taken in the conversion to dBm or dBW. (Ratios in dB are computed by subtraction, not division.) For example, if P_{INC} = 10 dBm and P_{REFL} = 0.5 dBm, RL = 10 − 0.5 = 9.5 dB. Both power measurements must have the same units (dBm, dBW, and so on) for the subtraction to yield the correct results — for example, dBW can't be subtracted from dBm directly.

Since SWR and RL measure the same thing — reflected power as a fraction of forward power — they can be converted from one to the other. Start by converting RL back to a power ratio:

$$\frac{P_{REFL}}{P_{INC}} = \log^{-1}(-0.1 \times \text{RL}) \qquad (31)$$

Now use the equation for computing SWR from forward and reflected power (see the **Transmission Lines** chapter):

$$\text{SWR} = \frac{\left[1 + \sqrt{\frac{P_{REFL}}{P_{INC}}}\right]}{\left[1 - \sqrt{\frac{P_{REFL}}{P_{INC}}}\right]} \qquad (32)$$

SWR can also be converted to RL by using the equation for power ratio in terms of SWR:

$$\frac{P_{REFL}}{P_{INC}} = \left[\frac{\text{SWR}-1}{\text{SWR}+1}\right]^2 \qquad (33)$$

Then convert to RL using equation 30.

5.10 RF Design Techniques Glossary

Arc — Current flow through an insulator due to breakdown from excessive voltage.

Balun — A device that transfers power between *bal*anced and *un*balanced systems, sometimes transforming the impedance level as well (see also **unun**).

Bead — Hollow cylinder of magnetic material through which a wire is threaded to form an inductor.

Bilateral — A network that operates or responds in the same manner regardless of the direction of current flow in the network.

Choke balun — see **current balun**.

Core — Magnetic material around which wire is wound or through which it is threaded to form an inductor.

Current balun — A balun that transfers power from an unbalanced to a balanced system by forcing current flow in the balanced system to be balanced as well (also called a **choke balun**).

Dielectric strength — The rated ability of an insulator to withstand voltage.

Distributed element — Electronic component whose effects are spread out over a significant distance, area or volume.

Dynamic resistance — The change in current in response to a small change in voltage.

Equivalent Series Inductance (ESL) — A capacitor's parasitic inductance.

Ferrite — A ferromagnetic ceramic.

Gain-bandwidth product — The frequency at which a device's gain drops unity. Below that frequency the product of the device's gain and frequency tends to be constant.

Hybrid-pi — High-frequency model for a bipolar transistor.

Impedance inversion — Dividing a characteristic impedance by the ratio of the impedance to be inverted to the characteristic impedance. For example, 25 Ω inverted about 50 Ω is 100 Ω and 200 Ω inverted about 50 Ω is 12.5 Ω.

Insertion loss (IL) — The loss inherent in a circuit due to parasitic resistance.

Inter-electrode capacitance — Capacitance between the internal elements of a semiconductor or vacuum tube.

Lumped element — Electronic component that exists at a single point.

Mix — The chemical composition of a ferrite or powdered-iron material (also called **type**).

Noise — Any unwanted signal, usually refers to signals of natural origins or random effects resulting from interfering signals.

Noise factor (F) — The amount by which noise at the output of a device is greater than that at the input multiplied by the gain of the device. A measure of how much noise is generated by a device.

Noise figure (NF) — 10 log (noise factor).

Noise gain — Circuit output noise power divided by the available input noise power. This is not always equal to signal gain, depending on the source of the noise and the location of the noise source in the circuit.

Nonideal — Behavior that deviates from that of an ideal component (see also **parasitic**).

Nonlinear — A component that acts on a signal differently depending on the signal's amplitude.

Parasitic — Unintended characteristic related to the physical structure of a component.

Permeability — The ability of a material to support a magnetic field.

Return loss (RL) — The difference in dB between forward and reflected power at a network port.

Self-resonant — Resonance of a component due to parasitic characteristics.

Simulate — Model using numerical methods, usually on a computer.

Skin effect — The property of a conductor that restricts high-frequency ac current flow to a thin layer on its surface.

Skin depth — The depth of the layer at the surface of a conductor to which ac current flow is restricted (see **skin effect**).

Spectral Power Density — The amount of power per unit of bandwidth, usually "root-Hz" or √Hz, the square root of the measurement bandwidth.

Stray — see **parasitic.**

Toroid (toroidal) — A ring-shaped continuous core.

Two-port network — A network with four terminals organized in two pairs, each pair called a port.

Two-port parameters — A set of four parameters that describe the relationship between signals at the network's two ports.

Unun — A device that transfers power between two unbalanced systems, usually performing an impedance transformation (see also **balun**).

5.11 References and Further Reading

Alley, C. and Atwood, K., *Electronic Engineering* (John Wiley & Sons, New York, 1973)

Terman, F., *Electronic and Radio Engineering* (McGraw-Hill, New York, 1955)

Carr, J., *Secrets of RF Circuit Design* (McGraw-Hill/TAB Electronics, 2000)

DeMaw, D., *Practical RF Design Manual* (MFJ Enterprises, 1997)

Hayward and DeMaw, *Solid-State Design for the Radio Amateur* (ARRL, out of print)

Hayward, Campbell, and Larkin, *Experimental Methods in Radio Frequency Design* (ARRL, 2009)

Dorf, R., Ed., *The Electrical Engineering Handbook* (CRC Press, 2006)

Hayward, W., *Introduction to RF Design* (ARRL, 1994)

Van Valkenburg, M., *Reference Data for Engineers* (Newnes, 2001)

Kaiser, C., *The Resistor Handbook* (CJ Publishing, 1994)

Kaiser, C., *The Capacitor Handbook* (CJ Publishing, 1995)

Kaiser, C., *The Inductor Handbook* (CJ Publishing, 1996)

Contents

6.1 Circuit Simulation Overview

    6.1.1 Hobby Circuit Simulation versus Circuit Simulation in Industry

    6.1.2 Real Circuits versus Simulations: The Difference

6.2 Computer-Aided Design Examples

    6.2.1 Example 1: A Two-JFET VFO and Buffer

    6.2.2 Example 2: Modeling a Phase-Lead RC Oscillator

    6.2.3 Example 3: Exploring Issues of Modeling Gain, Link Coupling and Q with a 7-MHz Filter

    6.2.4 Example 4: Checking Reality with a 10-dB Attenuator

    6.2.5 Example 5: Simulating the 40-Meter Filter in an RF-Fluent Simulator

    6.2.6 Example 6: *SPICE* and Intermodulation Modeling

    6.2.7 Circuit CAD in the Radio Amateur's Toolbox

6.3 Limitations of Simulation at RF

    6.3.1 *SPICE*-based Simulators

    6.3.2 Harmonic-Balance Simulators

    6.3.3 Contrasts in Results

    6.3.4 RF Simulation Tools

6.4 CAD for PCB Design

    6.4.1 Overview of the PCB Design Process

    6.4.2 Types of PCB Design Software

    6.4.3 Schematic Capture

    6.4.4 PCB Characteristics

    6.4.5 PCB Design Elements

    6.4.6 PCB Layout

    6.4.7 Preparation for Fabrication

6.5 References and Bibliography

Chapter 6

Computer-Aided Circuit Design

This chapter introduces Amateur Radio builders and experimenters to fundamentals of using computer-aided design (CAD) tools for RF design. These tools enable the hobbyist to harness the circuit-simulation power employed by professional electronic and RF engineers working in large-scale electronic and RF design and production for industry.

David Newkirk, W9VES, explores generic simulation tools capable of analyzing arbitrary circuit topologies. Dr Ulrich Rohde, N1UL, surveys issues associated with linear and nonlinear RF simulation and contributes three extensive papers on the accompanying CD-ROM. Dale Grover, KD8KYZ, presents a comprehensive introduction to the use of PCB design and layout software.

Software to aid design and analysis in specialized areas, for example RF filters, switchmode power supplies, transmission lines and RF power amplifiers, is covered in other chapters.

6.1 Circuit Simulation Overview

That computers are now used wherever information is generated, communicated and used is everyday knowledge. We expect, or at least can infer if we take a moment to think, that a computer was used to type and typeset these words, and to compose the pages of this book and to design its cover. Thinking a bit more deeply, we realize that computers are used in just about every phase of product design and process design and control, including the electronic devices we use at work and play. The widespread application of computers is possible because computers are, at base, generic math machines. The TV set you watch, the radio transceiver which with you talk to the world, is a math machine — math put to work in the form of a kinetic sculpture built from copper, carbon, plastic, silicon and steel. You feed radio (or audio) and ac-line energy — both of which can be readily described numerically — in. The TV or radio performs mathematical operations on those inputs, and a desired, quantifiable result comes out: a sound, a moving picture, a Field Day contact, a declaration of rook to queen's bishop three from a chess opponent in Spain.

Because radio-electronic circuitry "just does math," math can predict and analyze the action of electromagnetic signals and the radio-electronic circuitry we build to produce and process them. Program an electronic computer — which, at base, is a generic math machine — to do radio/electronics math in practically applicable ways, and you're ready to do computer-aided design (CAD) of radio and electronic circuits. (Program a computer to do radio-electronics math *in real time*, and you're ready to replace radio-electronics hardware with software, as this book's **DSP and Software Radio Design** chapter describes.)

6.1.1 Hobby Circuit Simulation versus Circuit Simulation in Industry

Professional-grade circuit simulation software exists to facilitate the construction of tightly packaged, highly integrated, no-tweaking-required modern electronics/RF products. These products work predictably well even when reproduced by automated processes in large quantities — quantities that may, with sufficient marketing success and buyer uptake, exceed millions of units.

Manufacturers of specialized *electronic design automation* (*EDA*) software serve the engineering needs of this industry. Through comprehensive CAD suites, one may proceed from graphical component-level circuit and/or IC design (*schematic capture*), through simulation of circuit and IC behavior (often using a variant of the simulator called *SPICE*, but increasingly with non-*SPICE* simulators more fluent in issues of RF and electromagnetic design), through design of PC board and IC masks suitable for driving validation, testing and production. Comprehensive EDA CAD reduces costs and speeds time to market with the help of features that can automatically modify circuits to achieve specific performance goals (*optimization*); predict effects of component tolerances and temperature on circuit behavior across large populations of copies (*Monte Carlo analysis*); and generate bills of materials (BOMs) suitable for driving purchasing and procurement at every step of the way.

Free demonstration versions are available for some EDA CAD products (see **Table 6.1**), and a subset of these are especially useful for hobby purposes. Although these *demoware* tools come to us with a large-scale-production pedigree, they are greatly (and strategically)

Table 6.1
Some Sources of Freeware/Demoware Simulation, Schematic and Layout CAD Software

Source	Address	Resource
Ansoft	www.ansoft.com	Ansoft Designer SV 2 (schematic, linear RF simulator, planar electromagnetic simulator, layout [PCB] design), more. (Students version SV2 has been discontinued in favor of an academic licensing program, but older copies of the program may still be abailable.
Cadence Design Systems	www.cadence.com	OrCAD 16 (schematic, SPICE simulator, layout [PCB] design)
CadSoft	www.cadsoft.de www.cadsoftusa.com	EAGLE schematic and layout design
gEDA	www.gpleda.org	GPLed suite of electronic design automation tools
Kicad	iut-tice.ujf-grenoble.fr/kicad	GPLed full-function schematic and layout design
Linear Technology Corp	www.linear.com	LTSpice (schematic, SPICE simulator enhanced for power-system design)

SPICE: What's in a Name?

SPICE — Simulation Program with Integrated-Circuit Emphasis — originates from the Electrical Engineering and Computer Sciences Department of the University of California at Berkeley and first appeared under its current name as SPICE1 in 1972. "SPICE", write the maintainers of the official SPICE homepage at **bwrc.eecs.berkeley.edu/Classes/icbook/SPICE/**, "is a general-purpose circuit simulation program for nonlinear dc, nonlinear transient, and linear ac analyses. Circuits may contain resistors, capacitors, inductors, mutual inductors, independent voltage and current sources, four types of dependent sources, lossless and lossy transmission lines (two separate implementations), switches, uniform distributed RC lines, and the five most common semiconductor devices: diodes, BJTs, JFETs, MESFETs, and MOSFETs."

That SPICE is "general-purpose" does not mean that its usefulness is unfocused, but rather that it is well-established as a circuit-simulation mainstay of comprehensive power. A wide, deep SPICE ecosystem exists as a result of decades of its daily use, maintenance and enhancement by industrial, academic and hobby users. Many excellent commercial versions of SPICE exist — versions that may be improved for workhorse use in particular subdisciplines of electronics, power and RF design.

Throughout this chapter we will use a particular commercial variant of the SPICE simulator: PSpice as embodied in Cadence Design System's OrCAD 16.0 EDA CAD demonstration software suite. With those particulars declared, we will usually refer to the underlying simulator merely as SPICE, expecting and intending that simulations presented in this chapter will be duplicable with any modern commercial SPICE variant. Only details associated with value-added features, such schematic and layout editing and simulation-results reporting, will vary.

CAD Software and Your Computer's Operating System

The OrCAD 16.0 and Ansoft Designer SV 2 demoware packages used in this chapter, and most other CAD products you're likely to use, are compiled to run under Microsoft Windows. So what if you want to run RF and electronics CAD software under Linux or on the Mac?

You're in luck. EAGLE schematic and layout software is available in native versions for Windows, Linux, and the Macintosh. The GPLed EDA application suite, gEDA, are primarily developed on Linux, but are intended to run under, or at least be portable to, POSIX-compliant systems in general. The GPLed schematic and layout editor Kicad runs natively under Windows and Linux, and has been tested under FreeBSD and Solaris. Further, the great strides made in the Wine translation layer (**www.winehq.com/**) allow many applications written for Windows to run well under the operating systems supported by Wine, including Linux and the Macintosh.

In your writer's experience, MicroSim DesignLab 8 (a widely distributed precursor to OrCAD 16 that can run all of the SPICE examples described in this chapter) and Ansoft Serenade SV 8.5 (a precursor to Ansoft Designer SV 2) can be run in Wine under Linux with few artifacts and their expected schematic-capture and simulation capabilities intact. Cursor handling in OrCAD 16 installs under Wine readily enough. but cursor-handling artifacts in its schematic editor, at least in the computers tried, seems to preclude its use under Wine for now. Ansoft Designer SV 2 installs but does not properly start.

All things considered, however, especially as Wine and CAD applications continue to strengthen and mature, running your favorite Windows-based applications under Wine is well worth a try.

---

feature-limited. Only a relatively few components, often representing only a subset of available component models, may be used per simulation. Monte Carlo analysis, optimization, BOM generation and similar enhancements are usually unavailable. Demoware is intended to drive software purchasing decisions and serve as college-level learning aids — learning aids in college study toward becoming electronics/RF professionals who will each day work with the unlimited, full versions of the demoware.

The radio hobbyist's circuit-simulation needs are much simpler. Most of us will build only one copy of a given design — a copy that may be lovingly tweaked and refined to our hearts' content far beyond "good enough." Many of us may build as much with the intent of learning about and exploring the behavior of circuits as achieving practical results with them. A demoware circuit simulator can powerfully accelerate such self-driven exploration and education in electronics and RF.

6.1.2 Real Circuits versus Simulations: The Difference

Having just finished stuffing and soldering the PC board for a binaural CW receiver, you find yourself in need of a 7-MHz local oscillator (LO) to drive it. The receiver designer has provided just such a circuit: a junction field-effect transistor (JFET) Hartley oscillator driving a JFET amplifier to an output power of +10 dBm. You obtain the parts, perhaps substituting components with close-enough values here and there as your junk box suggests and your experience allows. Having completed its construction, you interconnect it, the main receiver board, a power supply, headphones and an antenna. You power up the receiver, hear 40-meter signals, and —

perhaps with the aid of a frequency counter or second receiver — adjust the tuning range of your new LO to cover your favorite part of the band. Another project done, another success achieved — what could be simpler? Certainly not modeling the same circuit in circuit-simulation software — precisely because all actions in the real world are always *maximally and unimaginably complex.*

The components we use to build real circuits always operate to the full robustness of their intrinsic properties, their simplicity to our mind's eye notwithstanding — regardless of our relative ignorance of what those properties may be and how and why they operate the way they do. No real-world component operates ideally. Nothing — excepting, perhaps, events in the realm of quantum physics — is sketched in, so far as science has been able to determine. So it is that we may set out to build an amplifier only to discover at power-up that we have instead built a persistent oscillator. So it is that we may design, successfully build, and publish a circuit intended to be an oscillator only to discover that for 3 out of every 100 subsequent reader-builders it does not oscillate at all!

When we take the "just build it" approach with circuit simulation software — that is, when we expect components to "just work" as they do in real-world circuits — we may run into trouble rapidly. This can happen because the electrical and electronic components available in circuit simulators are only mathematical *models* of real-world components — because every modeled behavior of every modeled component is *only sketched in* relative to the behaviors of its real-world counterpart. Simulated component characteristics and behaviors approach those of real-world components only as closely as science may allow *and* only as closely as the simulator's designers, engineers and users need its simulations to mirror real-world behaviors to meet their design goals. Once a design works well enough, reproducibly enough, cheaply enough, it goes into production, ultimately to ship in the hundreds, thousands or millions.

Finalizing an Amateur Radio circuit design is less spectacular in comparison. Even if the designer of our figurative VFO provided a quantitative specification for its performance — +10 dBm output — we very likely will not ultimately confirm whatever numbers we have before happily putting the project into service. We're hearing signals and making radio contacts — the likely object of the exercise, after all — and so we consider our construction project successful.

Were this chapter a textbook, or part of a textbook, on computer-aided circuit design, we might begin our CAD explorations by reviewing the basics of what electronic circuits are and do, following this with a discussion of what computerized circuit simulation is and how it works. An excursion into the arcane world of active-device modeling — the construction and workings of mathematical electrical equivalents to the transistors, diodes and integrated circuits that await us at our favorite electronics suppliers and in our junk boxes — might follow. Finally, we might systematically proceed through a series of simulation examples from the basic to the more complex, progressively building our store of understood, trustworthy and applicable-to-future-work circuit-CAD concepts as we go.

But this is a chapter in a handbook, not a textbook, and following a sequence of abstract basics to concrete practice very likely does not reflect the process most of us have followed, and follow yet, in learning and using what we know about electronics and radio. More realistically, our approach is more like this: We find ourselves in need of a solution to a problem, identify one, and attempt to apply it. If it works, we move on, likely having learned little if we have not had to troubleshoot. If the solution does *not* work, we may merely abandon it and seek another, or — better, if we are open to learning — we may instead seek to understand why, with the happily revised aim of understanding what we need to understand to make the solution work. Even if we must ultimately abandon the solution as unworkable in favor of another, we do not consider our time wasted because we have further accelerated our deepening intuition by taking the initiative to understand why.

This chapter introduces circuit CAD in exactly that spirit: Especially at first, we will learn about circuit simulation through "just building," and simulating with CAD, *buildable circuits we know should work* in quantified and quantifiable ways that we also know well. If they don't work as they should, we will investigate why, fix them and review what we have learned — just as we do when building and learning from real-world hardware.

6.2 Computer-Aided Design Examples

6.2.1 Example 1: A Two-JFET VFO and Buffer

In introducing this chapter, we "just built" — in our mind's eye — a receiver LO that "just worked," as real-world projects are intended and expected to do. Building and simulating the same two-JFET design in the circuit simulator called *SPICE* transports us right to the heart of fundamental circuit-modeling problems and their solutions. If we build through graphical schematic entry — through schematic capture — its oscillator-buffer circuit exactly according to its electrical schematic, *it cannot possibly work!*

Fig 6.1 shows the circuit for real-world duplication. In it, a J310 JFET Hartley oscillator drives a J310 buffer amplifier, which drives a 50-Ω load at +10 dBm via a trifilar broadband transformer that provides an impedance step-down ratio of 9:1. Any experimenter accustomed to building and using such circuits needs more no information than that provided in the Fig 6.1 schematic and its caption to make the circuit work as expected.

Fig 6.2 shows the same circuit successfully modeled in *OrCAD 16.0*, *SPICE* based simulator software from Cadence Design Systems. To understand the differences between Fig 6.1 and Fig 6.2, we'll examine the simulation's components type by type.

RESISTORS

The designer's output power specification (+10 dBm) assumes that the VFO is connected to the 50-Ω load afforded by the mixer system in the receiver it was designed to drive. To simulate this mixer load, we have added R5.

The value of R1, specified as 1 MΩ in the original circuit, is now specified as 1E6 — scientific/engineering notation for one million. We have done this to remind ourselves that *SPICE*'s use of unit suffixes — *scale factors* in *SPICE*-speak — differs from what we are generally accustomed to seeing in electrical schematics, and that we have multiple options for specifying values numerically using integer and decimal floating-point numbers. The scale factors available in *SPICE* include:

- F — 1E–9
- G — 1E9
- K — 1E3
- M — 1E–3
- MEG — 1E6
- MIL — 25.4E–6
- N — 1E–9
- P — 1E–12
- T — 1E12
- U — 1E–6

Specifying the value of R1 as 1M would declare its value as 1 milliohm (0.001 Ω), short-circuiting the JFET's gate to common and breaking our simulation. Specifying the

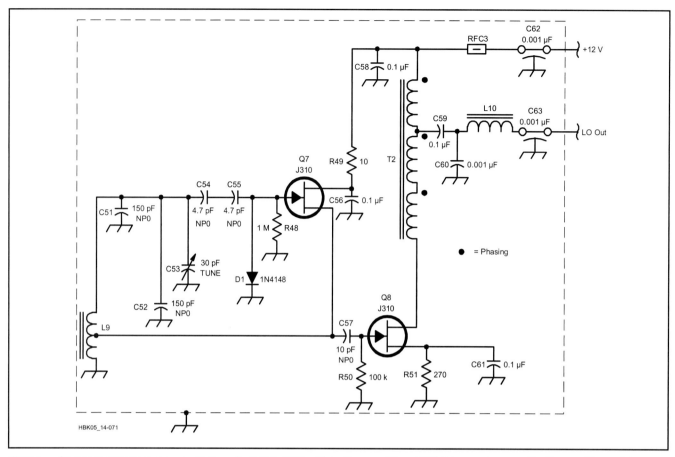

Fig 6.1 — Standard electrical representation of a 7-MHz VFO with buffer amplifier. JFET Q1 operates as a Hartley oscillator; D1, as a limiter that improves frequency stability by keeping Q1's gate voltage from going more positive than about 0.6 V; and Q2, as an amplifier that increases the oscillator output — obtained from the feedback tap on the oscillator inductor by capacitive coupling (C57) — to +10 dBm. The tapped inductor (L9), is 1.2 µH (22 turns of #28 wire on a T-30-6 toroidal core); the trifilar output transformer (T2), 10 trifilar turns of #28 wire on an Amidon FB-43-2401 ferrite bead).

Using Your Computer to Draw Schematics

The art of drawing circuit schematics predates electronic computers; the art of drawing schematics and PC-board layouts with computers predates the art and science of *simulating* circuits with computers. What if you only want to draw a circuit's schematic and design a PC board without simulating it? What computerized tools are available to *you*?

Almost any circuit-simulation program or electronic design automation (EDA) suite that uses schematic circuit capture can, within functionality limits imposed on the demoware version, serve as a first-rate schematic editor. Although demoware component library limitations usually restrict the *types* of components you can use — in CAD-speak, *place* — in a design, part-*count* limitations usually operate only at simulation time. Restrictions in physical size and layer count, and in the materials and metallizations available for substrate specification, will likely apply to whatever layout-design facilities may be available. After all, the main purpose of demoware is to let students and potential buyers taste the candy without giving away the store.

Excellent simulation-free schematic-capture and layout-design products exist, of course. The schematic style long standard in ARRL publications comes from the use of Autodesk *AutoCAD*, a fully professional product with a fully professional price. Long popular with radio amateurs and professionals alike is CadSoft *EAGLE*, a schematic-capture and layout-design product available in freeware and affordable full-version forms. You can even export *EAGLE* schematics to a *SPICE* simulator and back with Beige Bag Software's *B2 Spice*. The full-function freeware schematic and PCB-layout application *Kicad* and the EDA suite *gEDA* come to us from the open-source community.

Do-it-yourself schematic CAD can be as close as the basic drawing utility included with your computer's operating system. Some hobbyists find that cutting, copying, moving and pasting components snipped from favorite graphical schematic files into new configurations and adding new wires as graphical lines is enough for what they want to do. The connection between the world of modern EDA tools and this seemingly primitive approach to schematic creation can be as close as the operating system's clipboard: Some CAD schematic-capture programs represent circuit elements as metafile data during copying, cutting and pasting operations. Copying a schematic to the clipboard with such a program and then pasting the clipboard contents to a suitable drawing program creates a *picture* of the copied schematic — give it a try with *Windows Paint* and the schematic editor in the free demoware version of *OrCAD 16*.

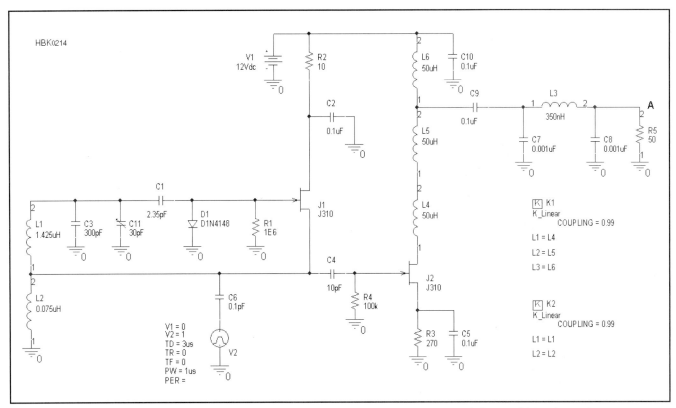

Fig 6.2 — The VFO-and-buffer circuit configured for successful *SPICE* simulation under the *OrCAD 16* CAD suite as drawn — "captured" — in the *OrCAD Capture CIS* schematic editor. To make the circuit work like the real thing, we added several components only implied in Fig 6.1, including a 12-V dc source (V1) and a 50-Ω load resistor (R5). Also new to the circuit, ac-coupled (via C6) to the oscillator JFET (J1) source, is a mysterious second voltage source (V2) — an addition without which the oscillator cannot oscillate. The numbers identifying the leads of each inductor are displayed by default in the *OrCAD* schematic editor to indicate phasing, knowledge of which is essential for properly modeling the behavior of the trifilar transformer formed by L4 through L6. We have also enabled pin-number display for R5 to help us determine the name of the circuit node — labeled A — we must probe to graph the circuit's output waveform.

value of R1 as 1MEG or 1000K would be correct alternatives. *SPICE* scale factors are case-insensitive.

Notice that *SPICE* assumes unit *dimensions* — ohms, farads, henrys and so on — from component-name context; in specifying resistance, we need not specify ohms. In parsing numbers for scale factors, *SPICE* detects only scale factors it knows and, having found one, ignores any additional letters that follow. This lets us make our schematics more human-readable by appending additional characters to values — as long as we don't confuse *SPICE* by running afoul of existing scale factors. We may therefore specify "100pF" or "2.2uF" for a capacitance rather than just "100p" or "2.2u" — a plus for schematic readability. (On the reduced readability side, however, *SPICE* requires that there be *no space* between a value and its scale factor — a limitation that stems from programming expediency and is present in many circuit-simulation programs.)

CAPACITORS

To develop the habit of keeping our simulated circuits' part counts down so we don't run up against the circuit-complexity restrictions of the *OrCAD 16.0* demoware (*all* demoware and student-version CAD packages have such limits) we have combined the seriesed and paralleled capacitor pairs in our real-world tuned circuit into single capacitances. Original C54 and C55 become C1; original C51 and C52, C3. That said, there are two reasons why we may not want to do so. If one of the objects of our simulation is to examine the voltage, current or power at the junction of C54 and C55 in the original circuit, combining them has disallowed achieving that aim. More generally, if we also intend to base a PC-board design on our captured circuit through *OrCAD 16.0 PCB Design Demo* — one of *OrCAD Capture CIS*'s sibling applications in the *OrCAD 16.0* demoware suite — we must engineer our simulation with that practical outcome in mind throughout.

INDUCTORS

Simulating the tapped coil of a Hartley oscillator immediately challenges us to learn more about our real-world circuit than we need to know to successfully build it. The original coil, 1.5 µH, consists of 22 turns of wire tapped at 5 turns, yet a tapped inductor is not available in the *SPICE* model library. Multiple approaches to simulating a tapped inductor can be used, including connecting two inductors in series (as we have done here) and proportioning their values intelligently, or basing the tuned circuit on one winding of an ideal transformer and proportioning the inductance of the secondary to simulate the tap. Either way, we must make the best educated guess we can about the inductance between the tap and ground, as its value directly affects the oscillator feedback, and hence its output.

Especially if your approach to building is more practical than theoretical, proportioning the values of the two coils in the 22:5 ratio reflected in the original's winding information might seem like a fair approximation — until we recall that a coil's inductance-versus-turns ratio is not linear. Taking that approach would give us a larger-than-life inductance value for the lower portion of the coil, resulting in more-than-realistic feedback and higher-than-realistic output. So what we have done for this simulation is calculate the inductance of 5 turns of wire on a T-30-6 core, taking the answer (0.075 µH) as the value of Fig 6.2's L2, and 1.5 µH – 0.075 µH (1.425 µH) as the value of L1, the upper portion of the coil.

To illustrate another feature of *SPICE* — and to provide one avenue for later experimentation with this simulation — we have also specified near-ideal ($K = 0.99$) coupling between L1 and L2 by means of K_Linear element K1 (in the lower right corner of Fig 6.2). *SPICE* allows us to specify coupling between any subset of inductors in a simulation, including *all* inductors in a simulation. The value of this feature in enabling greater realism in simulations of complex, cross-coupled connector, circuit-board and IC structures is profound — at the expense of requiring the realistic specification of coupling values if the power of this feature is to be realized.

Here we have coupled the two sections of our oscillator tank inductor because (1) we know that they actually *are* coupled in the real thing; (2) we want to experience specifying inductor coupling in *SPICE*; and (3) practical experiments with Hartley inductors consisting of separate toroidal cores nonetheless shows that such coupling is *not* necessary to make real Hartley oscillators work! Assuming we can get the circuit to oscillate as is, reducing the coupling between L1 and L2 would let us simulate the use of separate coils in a real-world oscillator.

Having specified coupling between the sections of the oscillator tank inductor, we are ready to welcome the similar challenge of simulating the trifilar broadband output transformer (T2) in Fig 6.1. Here, as with the tapped oscillator tank coil, no direct equivalent to this transformer topology is available in *SPICE*, but multiple alternatives can get us close enough. One option would be to use a conventional two-coil transformer, such as that available in the *OrCAD Capture CIS* component library. As with the tapped oscillator coil, however, we have decided to use three separate coupled inductors, specifying their coefficients of coupling with another K_Linear element, K2. For the inductance of each, we have drawn on our experience with the "10 multifilar turns on an FT-37-43 core"-class broadband transformers commonly used for just such applications in many ARRL RF projects, specifying an inductance (50 µH) close in value to that of a single such winding for each coil of our simulated transformer. (As a check on the intelligence of using this value, we recall that the rule of thumb for the inductance of a conventional broadband transformer winding calls for a winding reactance of at least 5 to 10 times the impedance at which the winding operates — and the reactance of 50 µH at 7 MHz equates to 2.2 kΩ, over 40 × 50 Ω.) In wiring the three inductors (L4, L5 and L6), we have also taken care to phase them properly, their **1** and **2** labels conveying the winding-sense information communicated by the phasing dots that accompany T2's windings in Fig 6.1.

Specifying for simulation the remaining inductor in Fig 6.1 — in our Fig 6.2 schematic, L3, 350 nH, one of three components in a π low-pass filter network — is straightforward enough to warrant no comment. But comment we shall, for in specifying the electrical performance of an inductor merely by setting a value for its inductance — as we have so far done for all of the inductors in our circuit — we have in no way specified its quality factor (Q). In simulation it will there act as an absolutely pure inductance — a component that cannot be built or bought. This will make a difference in how closely our simulation may approach the real thing — but how much of a difference?

All *real* inductors, capacitors, and resistors — all real components of any type — are non-ideal in many ways. For starters, as **Fig 6.3** models for a capacitor, every real L also exhibits some C and some R; every real C, some L and R; every real R, some L and C. These unwanted qualities may be termed *parasitic*, like the parasitic oscillations that sometimes occur in circuits that we want to act only as amplifiers, and in oscillators (which may simultaneously oscillate at multiples frequencies, including the frequency we intend). For experimental and proof-of-concept purposes at audio and HF radio frequencies, parasitic L, C and R can often be ignored. In oscillator and filter circuits and modeled active devices, however, and as a circuit's frequency of operation generally increases, neglecting to account for parasitic L, C and R can result in surprising performance shortfalls in real-world *and* simulated performance. In active-device modeling realistic enough to accurately simulate oscillator phase noise and amplifier phase shift and their effects on modern, phase-error-sensitive data-communication modes, device-equivalent models must even include *nonlinear* parasitic inductances and capacitances — Ls and Cs that vary as their associated voltages and currents change. As active-device operation moves from small-signal — in which the signals handled by a circuit do not significantly shift the dc bias points of its active devices — to large-signal — in which applied signals significantly shift active device dc bias and gain — the reality of *device self-heating* must be included in the device model. Examples: When amplitude stabilization occurs in an oscillator or gain reduction occurs in an amplifier as a result of voltage or current limiting or saturation.

Designers aiming for realism in simulating power circuits that include magnetic-core inductors face the additional challenge that all real magnetic cores are nonlinear. Their magnetization versus magnetic field strength (*B-H*) characteristics exhibit hysteresis. They can and will saturate (that is, fail to increase their magnetic-field strength commensurately with increasing magnetization) when overdriven. Short of saturation, the permeability of magnetic cores varies, hence changing the inductance of coils that include them, with the flow of dc through their windings. These effects can often be considered negligible in modeling ham-buildable low-power circuits (such as Fig 6.1), but designers using *SPICE* to simulate power supplies and electromechanical systems for mass fabrication and production must harness its ability to model nonlinear magnetics — a capability greatly limited in the demo version of *OrCAD 16.0*. See the **Power Supplies** chapter for more information on this topic.

So what of our non-specification of Q for the tuned-circuit inductors — and while we're at it, the tuned-circuit capacitors — in Fig 6.2? With all other factors ignored and with no steps taken to otherwise introduce realistic parasitics and losses into simulations, ideal tuned circuits will generally result in

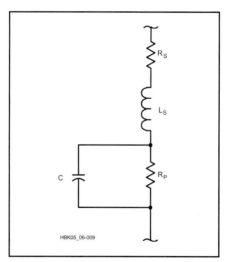

Fig 6.3 — A capacitor model that aims for improved realism at VHF and above. R_S models the net series resistance of the capacitor package; L_S, the net equivalent inductance of the structure. R_P, in parallel with the capacitance, models the effect of leakage that results in self-discharge. Intuiting the topology of this model is one thing; measuring and/or realistically calculating real-world values for R_S, L_S and R_P for application in a circuit simulator is a significant challenge. How and to what degree these parasitic characteristics may cause the electrical behavior of a capacitor to differ from the ideal depends on its role in the circuit that includes it and frequency at which the circuit operates. For simulating many ham-buildable circuits that operate below 30 MHz, the effects of component parasitic R, L and C can usually be ignored unless guidance or experience suggests otherwise.

higher-than-expected output in amplifiers and heavier clipping than we might expect (and therefore only *maybe* higher output than expected) in oscillators. Ideal transformers and *LC* filters will exhibit lower-than-expected losses and sharper-than-actual resonances. Whether this matters depends on our simulation aims. If we merely want to get an idea of the frequency response of a filter, ideal *L*s and *C*s are fine; if we want to compare the efficiencies of competing matching networks or filters or realistically model filter insertion loss, ideal *L*s and *C*s will lead us astray. So, if we want to more realistically simulate the output power of an oscillator, setting a realistic *Q* for at least its tuned-circuit *L* would seem to be a good thing.

Yet we have not yet done so in Fig 6.2. Because we are new to circuit modeling in general and this circuit in particular, and because against all odds we have chosen as our first exercise the modeling of an *LC* oscillator — and merely getting a *SPICE*-simulated oscillator to start can be enough of a challenge — we will work with ideal *L*s and *C*s for now. We will go into issues of specifying realistic *Q* later as part of our exploration into evaluating circuit gain.

SOURCES

We would expect Fig 6.2 to include a 12-V dc source, and it does (V1). Unexpectedly, however, our simulated circuit includes a second source: V2, which we have configured to generate a 1-V, 1-ns-long pulse that occurs 3 μs after simulation begins. We have included this pulse source because our simulated oscillator has a high-*Q* tuned circuit and, simulated in *SPICE*, cannot start oscillating without it.

Real oscillators need help starting, too. The difference between Fig 6.2 and its real-world equivalent is that real-world transistors in well-designed oscillators can usually get themselves started oscillating without our building in any means of kick-starting them.. A real-world *LC* oscillator can do this because its active device or devices generate internal noise, which, fed back and filtered by its tuned circuit, and amplified again and again, becomes less and less noiselike and more and more sinusoidal as it builds, until some mechanism of voltage or current limiting allows the signal to increase no further. That an oscillator can non-self-destructively reach and maintain this condition of *amplitude stabilization* is no less momentous than the occurrence of oscillation startup.

SPICE can simulate active-device noise, but only during ac circuit analysis, in which any active devices present are treated as linear — that is, as operating under small-signal conditions — before circuit behavior with ac signal input is evaluated. As an oscillator starts and then reaches amplitude stabilization, its oscillatory device(s) move from small-signal to large-signal operation. We must therefore use *SPICE*'s time-domain (also known as transient) simulation capabilities to simulate Fig 6.2 if we want to observe its output power. In time-domain simulation, *SPICE* progressively calculates the voltages at and currents through each circuit node — each point of interconnection between components — as the time steps of a simulation advance from the initial conditions at Time Zero.

Our simulation of Fig 6.2 will begin as every *SPICE* time-domain simulation begins if we do nothing to adjust the initial conditions for any of its components away from their defaults: All voltage and current sources will be at their specified levels, all capacitors will not yet be charged, and all inductors and transformers will not yet be energized. Starting such an analysis is very much like suddenly connecting a 12-V battery to the real-world circuit in Fig 6.1.

Because we want to see what happens when we "just build" a real-world circuit in a simulator, we will only mention in passing the availability of the advanced technique of explicitly specifying non-zero initial conditions of key circuit components as an aid to oscillator startup. This technique requires that we first build a high-*Q* oscillator, such as a crystal oscillator, as low-*Q*, get it going well enough to determine steady-state voltage or current values for key components, rebuild it as high-*Q*, and then re-simulate it with the steady-state values in place. That said, rather than first trying Fig 6.2 without V2 only to have it fail to start, we choose with the writer's help to magically learn from that certain failure in advance by kick-starting our simulated oscillator with a voltage pulse from V2. (We also include V2 [and C6] in Fig 6.2 from the get-go as a service to your memory: Introducing the circuit without these components and adding them later in a small, separate schematic would encourage your image-memory capabilities to snapshot a picture of a simulatable circuit *that cannot work*.)

DEVICES, DEVICE MODELS AND DEVICE PARAMETERS

In building a real circuit that uses active devices, we pull the necessary parts out of storage, solder them in, and they "just work." Like the rest of the components in the projects we build, they operate to the full robustness of the intrinsic properties of their constituents and construction regardless of our knowledge of the details. We can, and do, count on it.

Not (likely) being degreed practitioners of either general circuit-simulator mathematics or of the more specialized disciplines of device manufacturing and/or modeling, we will naturally tend to do the same when using a circuit simulator. Just as with real-world devices, once we click **Run** and simulation begins, our modeled devices will act to the fullest degree allowed by their construction. Very differently, however, *absolutely every desired behavior exhibited by a simulated device must be explicitly built into the model*, mathematical atom by mathematical atom. A real-world device always "knows" exactly what to do with whatever conditions confront it (however the resulting behavior may alarm or confound us). A simulated device can reliably simulate real-world behavior only to the extent that it has been programmed and configured to do so. A 1N4148 diode from your junk box "knows" exactly what to do when ac is applied to it, regardless of the polarity and level of the signal. Mathematically *modeling* the forward- and reverse-biased behavior of the real thing is almost like modeling two different devices. Realistically modeling the smooth transition between those modes, especially with increasing frequency, is yet another challenge.

Mathematical transistor modeling approaches the amazingly complex, especially for devices that must handle significant power at increasingly high frequencies, and especially as such devices are used in digital-communication applications where phase relationships among components of the applied signal must be maintained to keep bit error rates low. The effect of nonlinear reactances — for instance, device capacitances that vary with applied-signal level — must be taken into account if circuit simulation is to accurately predict oscillator phase noise and effects of the large-signal phenomenon known as *AM-to-PM conversion*, in which changes in signal amplitude cause shifts in signal phase. In effect, different aspects of device behavior require greatly different models — for instance, a dc model, a small-signal ac model, and a large-signal ac model. Of *SPICE*'s bipolar-junction-transistor (BJT) model, we learn from the *SPICE* web pages that "The bipolar junction transistor model in *SPICE* is an adaptation of the integral charge control model of Gummel and Poon. This modified Gummel-Poon model extends the original model to include several effects at high bias levels. The model automatically simplifies to the simpler Ebers-Moll model when certain parameters are not specified."

As an illustration of device-model complexity, **Fig 6.4** shows as a schematic the BIP linear bipolar junction transistor model from *ARRL Radio Designer*, a linear circuit simulator published by ARRL in the late 1990s. Luckily for those interested mainly in audio and relatively low-frequency RF applications, specifying values just for the parameters A (0.99 for average transistors) and RE (26 ÷ collector current in milliamperes, in which the *26* equates to 26 *millivolts*, the room-temperature value of V_T, the thermal equivalent of voltage in the transistor's semiconductor material) can suffice for good-enough-for-

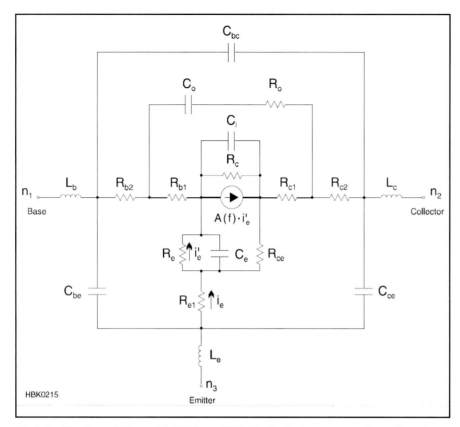

Fig 6.4 — The linear BJT model, BIP, from *ARRL Radio Designer*, a now-discontinued circuit-simulation product published by ARRL in the late 1990s. Frustratingly to users of *ARD*, real-world values for many BIP parameters could not be directly inferred from manufacturers' device datasheets. Scarcity of device parameters is much less of a problem for *SPICE* users, as the widespread use of the simulator by industry has compelled many device manufacturers to extract and publish — free for the downloading — real-world device parameter values that can be plugged directly into *SPICE*. Modern RF-fluent non-*SPICE* simulators like *Ansoft Designer SV 2* may be able to use *SPICE* device parameters directly or with a bit of parameter-renaming and value-resuffixing.

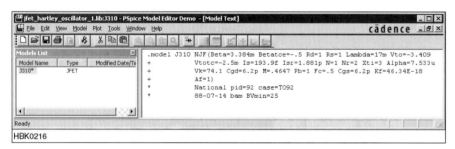

Fig 6.5 — Opening the circuit's J310 device for editing reveals that its model parameters were extracted at National Semiconductor in 1988. In the *OrCAD 16* demoware, both J310 instances in Fig 6.2 must use exactly this parameter-value set; non-demoware simulators allow separate customization of each instance of the same model. Lines that begin with asterisks are comments, ignored by the simulation engine.

students and their teachers, those who work for a semiconductor foundry that uses them, and those who produce circuit-simulation products that implement them.)

Most of us will go (and need go) no further into the arcanities of device-modeling than using *SPICE*'s JFET model for FETs like the 2N3819, J310, and MPF102, and *SPICE*'s BJT model for bipolar transistors like the 2N3904. Getting the hang of the limitations and quirks of these models may well provide challenge enough for years of modeling exploration. (We'll encounter another device-modeling option — representing devices as black boxes characterized by manufacturer-supplied network parameter datasets — later as we consider the RF-fluent simulator *Ansoft Designer SV 2*.)

Obtaining real-world-useful device-parameter values to plug into even *SPICE*'s standard diode, BJT and JFET models is critically important if we are to creating simulations that work well. *OrCAD 16.0* includes preconfigured 1N914, 1N4148, 2N2222, 2N2907A, 2N3819, 2N3904 and 2N3906 devices, among others, and these will be sufficient for many a ham-radio simulation session. To get parameters for other devices, especially RF devices but also including the J310 JFETs of Fig 6.2, we must search the Internet in general and device-manufacturer websites in particular to find the data we need. The manufacturer sites listed in **Table 6.2** will get you started.

Fig 6.5 shows the Fig 6.2 J310 parameters in *OrCAD 16*'s component editor. Because the demoware version of *OrCAD 16.0* does not allow us to edit multiple instances of a device independently of each other, the J310s in our simulation are identical.

Fig 6.6 illustrates the level of detail involved in more-accurate device modeling for VHF and UHF. The device is a California Eastern Labs NE46134, a surface-mount BJT intended to serve as a broadband linear amplifier at collector currents up to 100 mA and collector voltages up to 12.5. This manufacturer-supplied model embeds the unpackaged device chip (NE46100) within a *netlist*-based subcircuit that models parasitic reactances contributed by the transistor package, including chip-to-lead connections. At MF and HF, where the NE46134 could serve well as a strong post-mixer amplifier, modeling the device with just its basic, bare-chip characteristics (declared in the .MODEL NE46100 statement) would likely be accurate enough for many applications.

That Fig 6.6 illustrates the device parasitics using ASCII art is a side attraction. The main show — aside from the conveyance of the *SPICE* parameters of the NE46100 chip in its .MODEL statement — is the depiction of the NE46134 device as an NE46100 chip embedded in a subcircuit defined in netlist —

basic realism with the BIP model or linear BJT model equivalent to it.

Especially in the area of MOSFET and MESFET device modeling, and large-signal device modeling in general (of critical importance to designers of RF integrated circuits [RFICs] for use at microwave frequencies) *SPICE* and RF-fluent non-*SPICE* simulators include active-device models home experimenters are unlikely, even unable, to use. (Do BSIM3 and BSIM4 ring a bell? MEXTRAM? Statz, Curtice and TriQuint GaAs FET models [levels 1, 2, 3 and 6]? No points off if you can't already answer *yes* to these extra-credit questions. Several, if not most, of these models are of interest only to EE

Table 6.2
Device Parameter Sources

Source	Address	Resource
Cadence Design Systems	www.cadence.com/products/orcad/pages/downloads.aspx#cd	OrCAD-ready libraries of manufacturer-supplied device models
California Eastern Laboratories	www.cel.com	data and models NEC RF transistors
Duncan's Amp Pages	www.duncanamps.com/spice.html	SPICE models for vacuum tubes
Infineon	www.infineon.com	data and models for Siemens devices
Fairchild Semiconductor	www.fairchildsemi.com	device data and SPICE models
National Semiconductor	www.nsc.com	data and SPICE models for National op amps
NXP	www.nxp.com/models/index.html	data and models for Philips devices

The National Semiconductor J310 parameters used in simulating Fig 6.2 came from a QRP-L posting ("Re: More JFET Models") archived at **qrp.kd4ab.org/1999/990616/0098.html**.

```
* FILENAME:         NE46134.CIR
* from http://www.cel.com/pdf/models/ne46134.cir
* NEC PART NUMBER: NE46134
* LAST MODIFIED:    11/97
* BIAS CONDITIONS: Vce=5V, to 12.5V,  Ic=50mA to 100mA
* FREQ RANGE:       0.1GHz TO 2.5GHz
*
*                            CCBpkg
*                    .-------||--------.
*                    |        CCB      |
*                    |     .--||---.   |           COLLECTOR
*   BASE             |     |       |   |
*            __Tb__  |     |    /----o-o-------o----o
*   o--------|____|-o-LB---o--|Q1   |   |
*                    |      B  \   --
*                    |       -- |   --  CCE     --
*                    -- CBEpkg E o----'         --CCEpkg
*                    |          |               |
*                    |          LE              |
*                    `----------o---------------'
*                               |
*                               -
*                              | | Te
*                              | |
*                               -
*                               |
*                               o EMITTER
*
*          CCB   = 0.03 pF      LB = 1.2nH      Tb/Te:
*          CCE   = 0.5 pF       LE = 1.2nH       z=60 ohms
*          CCBpkg= 0.18pF                        l=50 mils
*          CCEpkg= 0.18pF                        a=0.0001
*          CBEpkg= 0.01pF                        f=0.9GHz
*
*              c b e
.SUBCKT NE46134/CEL 2 1 3
Q1 2 6 7 NE46100
CCB 6 2 0.03E-12
CCE 2 7 0.5E-12
LB  4 6 1.2E-9
LE  7 5 1.2E-9
TB 1 0 4 0 Z0=60 TD=9.63E-12
TE 5 0 3 0 Z0=60 TD=9.63E-12
CCBPKG 4 2 0.18E-12
CCEPKG 2 5 0.18E-12
CBEPKG 4 5 0.01E-12
.MODEL NE46100 NPN
+( IS=8.7e-16    BF=185.0      NF=0.959     VAF=30.0     IKF=0.20
+  ISE=5.70e-13  NE=1.80       BR=5.0       NR=1.0       VAR=12.4
+  IKR=0.018     ISC=1.0e-14   NC=1.95      RE=0.630     RB=6.0
+  RBM=4.0       IRB=0.004     RC=3.0       CJE=4.9e-12  VJE=0.60
+  MJE=0.450     CJC=2.50e-12  VJC=0.830    MJC=0.330    XCJC=0.20
+  CJS=0.0       VJS=0.750     MJS=0.0      FC=0.50      TF=12.9e-12
+  XTF=1.60      VTF=19.9      ITF=0.40     PTF=0.0      TR=1.70e-8
+  EG=1.11       XTB=0.0       XTI=3.0      KF=0.0       AF=1.0 )
.ENDS
*$
```

Fig 6.6 — ASCII art lives on in the depiction of package parasitics in the California Eastern Laboratories model of the NE46134 linear broadband transistor. In introducing us to the concept of subcircuits — a defined circuit chunk that, once simulated in a given simulation run, can be reused multiple times elsewhere in the simulation. This model also illustrates that SPICE, in common with other simulators, uses a network list — netlist — to communicate circuit topology, component values, and analysis instructions to its simulation engine.

```
* source JFET_HARTLEY_OSCILLATOR_1
J_J2          N012082 N01236 N01504 J310
D_D1          N00091 0 D1N4148
Kn_K2         L_L1 L_L2     0.99
C_C8          0 N02935  0.001uF
J_J1          N00193 N00091 N00753 J310
C_C7          0 N09766  0.001uF
L_L6          N19502 N02415  50uH
V_V2          N07095 0
+ PULSE 0 1 3us 0 0 1us
R_R1          0 N00091  1E6
V_V1          N02415 0 12Vdc
C_C11         0 N15520  {30pF*0.5+.001p}
R_R2          N00193 N02415  10
C_C6          N07095 N00753  0.1pF
C_C9          N09766 N19502  0.1uF
L_L1          N00753 N15520  1.425uH
C_C4          N01236 N00753  10pF
L_L4          N012082 N18891  50uH
R_R3          0 N01504  270
C_C1          N00091 N15520  2.35pF
Kn_K1         L_L4 L_L5
+ L_L6   0.99
C_C2          0 N00193  0.1uF
R_R5          0 N02935  50
C_C5          0 N01504  0.1uF
C_C10         0 N02415  0.1uF
L_L5          N18891 N19502  50uH
L_L2          0 N00753  0.075uH
R_R4          0 N01236  100k
C_C3          0 N15520  300pF
L_L3          N09766 N02935  350nH
```

Fig 6.7 — The JFET VFO circuit in *SPICE* netlist form. The N*nnnn* declarations name circuit nodes or *nets* — points of interconnection between components. Each component statement names the part's *primitive* — the leading *J* (for JFET) in *J_J1* — and identifies its instance (*J1*) to form an instance name (*J_J1*). Net numbers, generated automatically and algorithmically by the *netlister* function associated with the schematic editor, are ordered in element statements such that element *pins* — points of electrical interconnection — are connected to the correct nets as graphically depicted in the schematic. With some tedium, display of net names may be togglable in a simulator's schematic editor; net name display in *OrCAD 16.0* must be enabled one net at a time. This can be useful because netlisters occasionally make mistakes, and comparing a circuit's netlist to its schematic may aid in troubleshooting simulation errors.

network list — form. A netlist is a specialized table that names the circuit's components, specifies their electrical characteristics, and maps in text form the electrical interconnections among them. Uniquely numbered nodes or *nets* — in effect, coordinates in the connectivity space walked by the simulated circuit — serve as interconnects between components, with each component defined by a statement comprising one or more netlist lines. Statements that must span multiple lines include *continuation characters* (+) to tell the netlist parser to join them at line breaks. Asterisk (*) or other non-alphanumeric characters denote comments — informational-to-human lines to be ignored by the simulator. In Fig 6.6, header information and the ASCII-art portrayal of the device-package parasitics are *commented out* in this way.

The netlist served as the original means of circuit capture for all simulators known to the writer, including *SPICE* and the now-discontinued *ARRL Radio Designer* simulation product; *schematic* capture came later. Further reflecting *SPICE*'s pre-graphical heritage is the fact that, to this day a *SPICE* netlist may be referred to by long-time *SPICE* hands as a *SPICE deck*, as in "deck of Hollerith punch cards." In *SPICE*'s early days, circuit definitions and simulation instructions (netlist statements that begin with a period [.]) were commonly conveyed to the simulation engine in punched-paper-card form. All of the circuit simulators known to the writer still use a netlist as a means, if not *the* means, of conveying circuit topology and simulation instructions to the simulator; simulation based on schematicless user-created netlists may be possible with some.

Fig 6.7 shows the VFO circuit rendered as a partial netlist for simulation. This netlist is "partial" in the sense that it omits simulation and output commands we would expect to see in a full *SPICE* deck. At analysis time, the netlist contents we're shown by *OrCAD 16.0* are concatenated in memory with other data, including analysis setup information.

ANALYSIS SETUP

Many of us know from our smattering of learned theory that a complex waveform must be sampled at a frequency at least twice that of the highest-frequency component present if the samples taken are to be acceptably representative of reality. In doing transient (time-domain) simulations in *SPICE*, we have a similar concern: We must sample circuit behavior over time often enough to show us accurately the highest-frequency effects we want to see. Present-day *SPICE* simulators can intelligently size the maximum time step used in transient analysis to a value appropriate for useful representation of the highest-frequency fundamental source in a simulation. That said, manually setting the maximum step

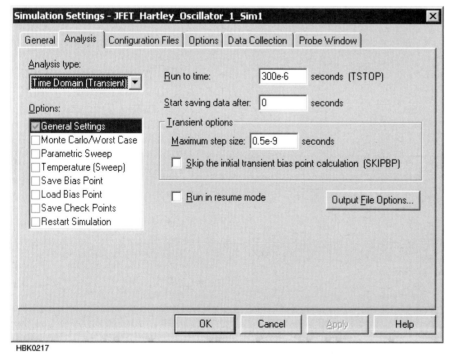

Fig 6.8 — *OrCAD 16 SPICE* setup for transient analysis of the JFET oscillator-buffer circuit. The behavior of the circuit will be progressively simulated every 0.5 ns (at most) from 0 to 300 μs.

to a smaller-than-automatic value can provide smoother-looking waveform graphs more acceptable to the eye. Smaller time steps come at the expense of longer simulation times and larger simulation-data files — issues more important in the earlier days of *SPICE* than nowadays, when gigahertz-class CPUs, gigabytes of RAM and even terabyte-class hard drives are standard. In this case, to simulate Fig 6.2 we set up a transient analysis out to 300 microseconds and a minimum step size of 0.5 ns (**Fig 6.8**). And then we click **Run**.

The *OrCAD 16.0* reporter opens when the analysis has finished. What we want to see is the circuit's output waveform across R5, the 50-Ω load we added. In reporter-terminology for this simulator, that's V(R5:2). **Fig 6.9** shows the resulting graph, rescaled a bit to center the instant of startup in the frame.

As a goal within our more general goal of getting a feel for "just building" simulated circuits as we often build them in the real world, we undertook this simulation with the aim of confirming the real-world circuit's claimed output power of +10 dBm (10 mW). That we have done, as **Fig 6.10** shows. Interestingly, the power level rises relatively slowly (and actually is still edging higher — ultimately to 9.5 mW — even after 300 μs). Does this reflect the behavior of the real thing? The same question may occur to us after we view the circuit's output spectrum, which **Fig 6.11** shows on a linear voltage scale. Shouldn't the output of our VFO be a pure sine wave? For a discussion of where such assumptions, unconscious and otherwise, may lead us, see the sidebar, "Simulation Goal or Moving Target?"

6.2.2 Example 2: Modeling a Phase-Lead RC Oscillator

The small maximum time step (0.5 ns) and long simulation period (300 μs) we set in simulating Fig 6.2 result in simulation runs that take nearly two minutes on an 868-MHz Pentium III computer with 512 MB of RAM. We chose those settings to give the circuit time to amplitude-stabilize and to allow for smoother display of waveform graphs when we zoom in. If our simulating computer has sufficient RAM and disk space to generate and handle the resulting large data file (170 MB), *OrCAD 16.0* is ready to graph simulation results from Fig 6.2 in the time it takes to start a cup of tea steeping.

As an illustration of a class of oscillators that can simulate much more rapidly, **Fig 6.12** presents an RC phase-lead circuit used as a CW sidetone generator in popular Amateur Radio transceivers of the 1970s and 1980s. In this case, the addition of a kick-start pulse is not required because charging currents in the circuit's 0.012-μF capacitors provide the stimulus for startup; **Fig 6.13** shows its startup to 40 ms.

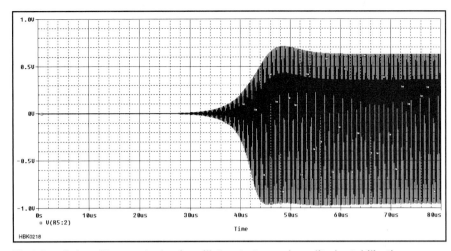

Fig 6.9 — At last: The spectacle of oscillator startup and amplitude stabilization rewards our attention to modeling detail. This graph shows the voltage at point A in Fig 6.2 — that is, the voltage, referred to ground (node 0), at node 2 of R5 [in this simulator's reporter-terminology, V(R5:2)], the circuit's 50-Ω load. The square dots scattered throughout the plot are trace markers. Note that the waveform's peak-to-peak span is not symmetrical around 0 V.

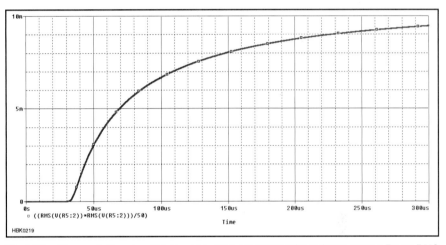

Fig 6.10 — To graph the circuit's power output, we render the voltage across R5 as RMS with the reporter's built-in RMS function, square the result by multiplying it by itself, and then divide the result by 50, the resistance of R5. By the end of the simulation, the output has reached 9 mW (9.5 dBm) — realistically close to the 10 dBm reported for the real-world circuit.

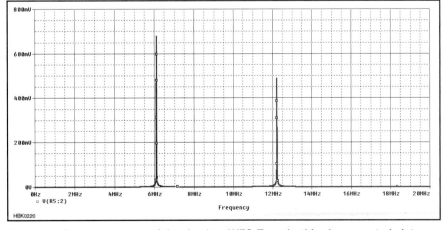

Fig 6.11 — Output spectrum of the simulated VFO. To make this clean spectral plot, we told the reporter's fast Fourier transformer (FFT) function to use only data beyond 60 microseconds after the start of simulation. Including data up through startup and stabilization data would cause the FFT to generate noise and discrete-frequency artifacts that diverge from real-life behavior.

Simulation Goal or Moving Target?

In simulating Fig 6.1, we set out to confirm the real-world circuit's output power, 10 dBm (10 mW). Fig 6.10 shows that we did just that. So, are we done? That depends.

Fig 6.A1 shows what the circuit's output waveform looks like when we zoom in on Fig 6.9 after the circuit has had time to amplitude-stabilize. It's clearly *not* a sine wave! Does our simulated circuit have a problem? What should we do next? That depends.

Circuit simulation allows us to explore circuits and what we know — and what we *think* we know — about them in a highly elastic and dynamic way. We may go into a simulation looking to find the answer to a simple question or questions, and find ourselves inventing new questions, even new modeling goals, along the way. This can bring good news and bad news. The good news is that our intuition and learning can be supercharged by whatever we may encounter. The bad news is that we can be misled by assumptions we may not even know we've been making.

An accurate RF power meter substituted for R5 in Fig 6.2 can indicate the absence or presence of RF at that point and its absolute level, and no more. The simulator's reporting function played the same role in providing us with the output-power curve shown in Fig 6.10. Whether the power-measured signal is spectrally dirty or clean, whether its frequency drifts or jumps — these characteristics, and more — cannot possibly be inferred from power measurement. Luckily — or maybe not — a simulator's reporter can tell us much more.

In real life, we would build our circuit expecting to see 10 mW output, measure that value with our RF power meter — if we even *have* a meter — connect the VFO to the receiver we built it to drive, and tune happily away. Little would we know that behind the deceptive simplicity of our measurement may lurk a signal that's other than a sine wave. (Even more arcane: The VFO output signal may be a sine wave when driving a resistive load but become non-sinusoidal when connected to the more reactive load presented by the receiver mixer's local-oscillator port.)

Perhaps the waveform in Fig 6.A1 really *does* generally reflect what real-world copies of Fig 6.1 do — but at this red-hot second we don't know enough to be sure. The description of the original circuit did not include a picture of its output waveform. A good high-frequency oscilloscope connected to the output of the real thing could tell us; a spectrum analyzer could also tell us, if a bit less directly.

Short of that, we can only intelligently speculate: Maybe the model data we used for the J310 JFET is insufficiently realistic. Maybe *SPICE*'s built-in JFET model folds up somehow as devices modeled with it move into the large-signal operation that oscillator amplitude stabilization involves. Maybe a signal like Fig 6.A1 happens whenever we diode-clamp the gate voltage of a JFET oscillator. Maybe the oscillator is overdriving the buffer amplifier. Maybe we or the author misspecified a value or left out a component. (Note to self: This is the first thing to check.) Maybe we have been unwarrantedly assuming for all of our radio-experimentation lives that unseen sources are sinusoidal until proven guilty.

Industrial-strength circuit simulators come to us as a result of engineering disciplines that seek to understand the world and predict its behavior toward the ultimate goal of achieving practical tools reproducible in quantity. As radio-hobbyist experimenter-builders, we may be just as satisfied with speculating about, and exploring of the quality and behavior of, a tool far beyond our having achieved its sufficiently practical function.

Computerized simulation can empower both approaches. The trick for us experimenters is to know when we've moved from solving a narrowly defined problem to dynamically redefining the problem such that enough is never enough. If you spend a half hour tweaking a bandpass-filter simulation for a –3-dB bandwidth of exactly 200 kHz through specification of capacitance values out to three decimal places, will you be able to achieve exactly that result with parts from your junk box? Will you even be able to know if you have? Will it even matter when you use your circuit on the air?

So what *about* that non-sinusoidality in Fig 6.A1? Is it a problem, an opportunity, or neither? Writing the rest of its story is up to you.

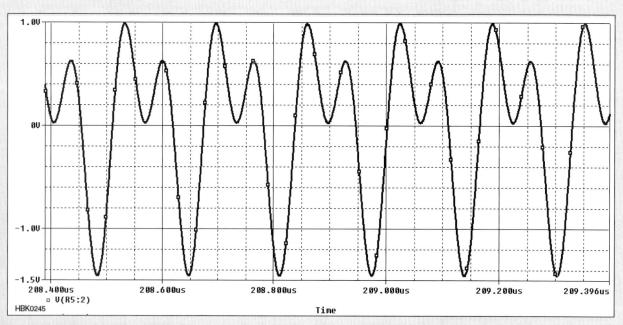

Fig 6.A1 — Zooming in on the JFET VFO's output signal confirms what the waveform asymmetry in Fig 6.9 and the second-harmonic spike in Fig 6.11 imply: that the VFO's output is not a pure sine wave. But is this a problem? That depends.

As we did with the JFET VFO example, we can use the reporter's FFT function to display the circuit's output waveform as an amplitude-vs-frequency (spectral) graph. We must do so with care, however, as the results may vary significantly with the amount and type of data used in the transform. To illustrate this, **Fig 6.14** shows the circuit's output spectrum based on analysis data from 40 ms to 500 ms, and **Fig 6.15** shows us what the FFT reports when we tell it to use only data from 480 to 500 ms. In effect, an FFT becomes surer of its results — the spectral components it reports sharpen — as we give it more data to work with; on the other hand, once the FFT has shown us all there is to see, we encounter diminishing returns — insufficient improvement in resolution as the dataset grows — if we make the simulation longer than it needs to be.

6.2.3 Example 3: Exploring Issues of Modeling Gain, Link Coupling and Q with a 7-MHz Filter

In evaluating the simulated oscillators in Figs 6.2 and 6.12, we had little difficulty in identifying the schematic junction — *node* — at which to probe the signal we wanted to evaluate. Other than working through the issue of graphing RMS power rather than peak power for our simulated VFO, we were able to graph our simulations' output without difficulty. Choosing the circuit point at which we would monitor our simulated circuits' behavior was simplified because an oscillator has only one *port* — excluding power and bias sources and control lines, only one point of interaction with the outside world.

Obtaining simulation results that we can both understand and trust becomes more complex when we simulate circuits with two or more ports. This is so because reporting the behavior of a simulated circuit is form of Q & A: Through the simulator's reporting functions we formulate a question, receiving as our answer a graph or table of circuit responses as numerical values, manipulated by such additional mathematical operations as we may specify.

From life experience we know that asking the wrong question will necessarily give us a wrong — though not necessary useless — answer. Especially if we are not electrical or electronics engineers, however, asking the wrong question of a simulator's reporting functions — its *reporter* — is deceptively easy. Our real-world experience in evaluating circuit behavior may not have prepared us for the subtleties of correctly posing even the most basic question to our simulator's reporter. The most immediate and far-reaching example of this is the evaluation of circuit gain.

The *concept* of gain as a ratio of output power, voltage or current to input power, voltage or current, perhaps expressed in decibels, is straightforward enough: Gain is at base a ratio that expresses the difference between the level of a signal at a circuit's output and the level of the same signal at the circuit's input. Sometimes we express gain numbers directly ("a voltage gain of 10"); sometimes we express gains terms of decibels ("a gain of 20 dB"). And we usually, but not always, call negative gain *loss*.

Measuring gain in real life is straightforward as well — at least gain as we are accustomed to thinking about it when we evaluate mixers, amplifiers, and filters for many ham-radio purposes. **Fig 6.16** shows a test-bench setup for measuring and displaying the gain (or loss), of a two-port circuit — a *two-port*, to put it generically — across a range of frequencies. In it, a signal generator — the tracking generator — applies a test signal to the input of the device under test (DUT), and an output-level-calibrated receiver — the spec-

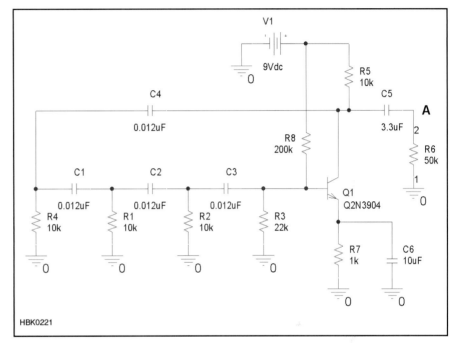

Fig 6.12 — Popular ham transceivers of the 1970s and 1980s included an RC phase-lead oscillator very much like this one as a CW sidetone generator. This modeled oscillator does not need a kick-start pulse, as the cascading disturbance of charging currents progressing through its *RC* phase-shift network is sufficient to start oscillation. Real-world builders may find that the lossiness of the circuit's feedback loop may require careful selection of the 2N3904 to ensure reliable starting.

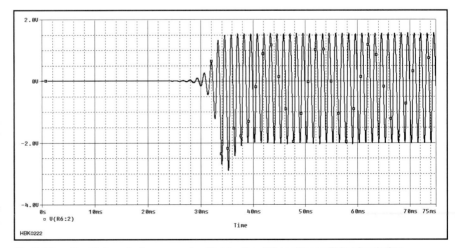

Fig 6.13 — Oscillator startup as embodied by the *RC* oscillator. This simulation (10-μs steps to 75 ms) takes only a few tens of seconds in a relatively slow computer; the Fig 6.2 simulation, *minutes*.

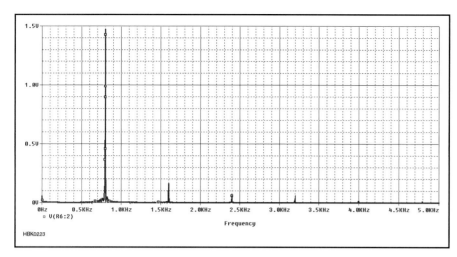

Fig 6.14 — Output spectrum of the RC phase-lead oscillator on a linear voltage as returned by the *OrCAD 16* reporter's fast Fourier transform (FFT) function. For a cleaner response — that is, to avoid displaying mathematical resultants from the circuit's rapidly changing spectral characteristics before and through startup and amplitude stabilization, and to give the FFT a larger periodic dataset to digest — we have extended the analysis to 0.5 s and excluded from the FFT analysis data between 0 and 40 ms.

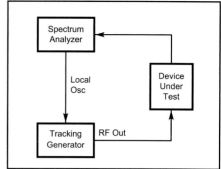

Fig 6.16 — One possible test setup for measuring the gain-versus-frequency response of a two-port device under test (DUT). This simplified diagram does not show the additional input and/or output attenuation that would be present in many actual measurement scenarios. For example, the presence of an active DUT (one expected to exhibit positive gain, as opposed to negative gain [loss]) would compel us to add attenuation between its output and the spectrum-analyzer input — not only to keep the spectrum-analyzer receiver from overloading and giving us false results, but also to protect the analyzer from damage if the DUT were to start oscillating rather than just amplifying.

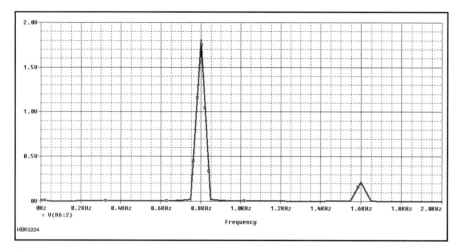

Fig 6.15 — Greatly restricting the dataset digested by the FFT degrades the amplitude-vs-frequency resolution of the report to the point of unrealism.

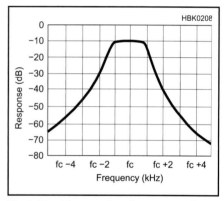

Fig 6.17 — The passband response of a crystal ladder filter as displayed by a spectrum analyzer.

trum analyzer in Fig 6.16 — receives the test signal as modified by the DUT and displays the results as an output-vs-frequency graph (**Fig 6.17**). To determine the gain or loss of the DUT at a given frequency, we'd compare that graph to the graph we get when we connect the tracking generator directly to the spectrum analyzer. When in ARRL publications we say that an RF circuit has "2 dB of loss" or "16 dB of gain," we almost always mean the type of gain derivable from measurements made by this process or its equivalent.

Type of gain? There's more than one? Yes: See the sidebar, "Defining Gain." It turns out that the gain we measure with a system like that shown in Fig 6.16 returns values that correspond to just one among multiple possible definitions of gain. In this case, our test system measures *insertion gain*: the difference between output measurements made at the load with the source connected directly to the load and with the device under test inserted between source and load. If we want to be able to compare the gain of a CAD-simulated two-port device with the gain of a counterpart real-life device measured in a test setup like Fig 6.16, we must tell our CAD program to report our circuit's insertion gain.

Using the Fig 6.16 system to measure insertion gain is straightforward because it is hard-configured to produce two-port insertion-gain measurements. Its spectrum analyzer can report only the signal level present at the load, and its TEST GENERATOR OUTPUT and SPECTRUM ANALYZER INPUT jacks enforce the insertion of the DUT at a particular point between source and load. We must merely take care not to connect the DUT backwards — unless doing so might return some other value of interest.

Reporting the insertion gain of circuit simulated in *SPICE* (even a very simple one) is fundamentally trickier because the relationships between source, DUT and load, and between signal-level metering and load, that can be safely assumed to exist in Fig 6.16 are neither predefined nor enforced in *SPICE* simulations. We must build a test-signal generator into our circuit, and we may report the signal level at one circuit node as easily as another. To explore how this complication

Defining Gain

Although the concept of gain as the ratio of the voltage, current or power at the output of a circuit to the power, voltage or current at the input of a circuit may seem to require no qualification, exactly where we measure these values in a system under test can greatly affect the gain value returned. Determining the point at which to measure circuit output is relatively straightforward: We measure output power in the load, output voltage across the load, and output current through the load. But what about input power, voltage, or current? Considering the problem only in terms of power, **Fig 6.A2** lays the groundwork for an understanding of this issue by depicting a gain-measurement setup in generic form.

Writing in NTIA Publication TR-04-410, *Gain Characterization of the RF Measurement Path*, J. Wayde Allen shows that four possible definitions can be proposed for the power gain of the 2-port in Fig 6.A2:

$$G_1 = \frac{P_{2a}}{P_{1a}} \quad (1)$$

$$G_2 = \frac{P_{2a}}{P_{1d}} \quad (2)$$

$$G_3 = \frac{P_{2d}}{P_{1a}} \quad (3)$$

$$G_4 = \frac{P_{2d}}{P_{1d}} \quad (4)$$

where *a* denotes available power and *d* denotes delivered power.

Equation 1 is commonly considered as describing the *available gain* (G_a) of the 2-port; equation 3 as describing the 2-port's *signal gain* (G_s) or transducer gain (G_t); and equation 4 as describing *power gain* (G). Further qualification of measurement conditions leads to additional, more specialized definitions for gain.

When we simulate circuits with the aim of directly comparing the results with real-world measurements, we need to simulate measurements made with a test set like that shown in Fig 6.16. To do that, we must know exactly which circuit points to probe to give a ratio that, expressed in decibels, gives the same results we'd see if we tested our simulation's real-world counterpart in the Fig 6.16 setup. Keeping Fig 6.A2 in mind,

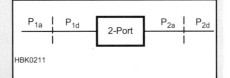

Fig 6.A2 — Generalized two-port gain evaluation in terms of available (subscript *a*) and delivered (subscript *d*) power (P) at port 1 (subscript *1*) and port 2 (subscript *2*). The power available from a source can differ from the power delivered to its load as a result of impedance mismatch between source and load. Although the subtleties of the issues involved are beyond the scope of this chapter, we must note that measuring gain becomes even more complex in the presence of sources and loads that are not purely resistive — that is, when voltage and current are not in phase.

knowing which of the equations above corresponds to the gain measured by the Fig 6.16 setup could help us determine where to probe in our simulations. (By the way, considering that throughout our example simulation discussions we have chosen to generate graphs in terms of signal voltage, here we should mention that equations 1 through 4 are just as valid for voltage and current as they are for power; it's only when we want to render G in decibels that we must remember whether to multiply the logarithm of G by 10 or 20.)

Reading further in Allen, it turns out that none of those equations will suffice, for what the Fig 6.16 setup actually measures is *insertion gain* (G_i), which, thinking in terms of power, can be defined as

$$G_i = \frac{P_d}{P_r} \quad (5)$$

where P_d is the power delivered to the load when the 2-port under test is connected between the signal generator and the load, and P_r, the reference power, is the power delivered to the load when the 2-port under test is absent and the signal generator is connected directly to the load.

Working with a spectrum analyzer to determine gain or loss involves exactly this two-step operation. What a test setup like that shown in Fig 6.16 determines is insertion gain. As we'll see in simulating a 7-MHz band-pass filter, having access to the internals of the test-signal generator lets us report a circuit's insertion gain in just *one* step with the help of a bit of math.

can affect the results we see, we'll examine a simple 7-MHz filter in *OrCAD 16.0* — that is, as simulated by *SPICE* — and in a more-RF-friendly simulator, *Ansoft Designer SV 2*.

Fig 6.18 shows the circuit, a top-coupled double-tuned-circuit (DTC) design intended to provide a 3-dB bandwidth of 200 kHz centered at 7.1 MHz. From its description, we learn that its inductors consist of 17-turn windings on T-50-6 powdered-iron cores, with 2-turn links for input and output coupling. Per the inductance-from-turns equation associated with the toroidal core properties tables in this *Handbook*'s **Component Data and References** chapter, we calculate the inductance of the resonators as 1.156 µH; of the coupling links, 0.02 µH. **Fig 6.19** shows its response as published by Hayward.

Because we want the response of our simulated filter to approach Hayward's results as closely as possible, we must somehow specify the quality factor, Q, of its resonator inductors —their *loaded* Q (Q_U). In this case, the most direct approach would be to construct the coils and measure their Q on a Q meter. Not having access to one, we do the next best thing and extrapolate from another's careful Q measurements on a similar coil. From the **Measured Toroidal Inductor Q Values** table in Zack Lau's "RF" column in March 1995 QEX, we estimate our resonators' Q as 238 based on his data for a very similar 1.165-µH coil.

Having convinced ourselves of the importance of working inductor Q into our simulation and having obtained a trustworthy Q value for our inductors, we face a more fundamental challenge: *SPICE*'s inductor (and capacitor) models afford *no* built-in means of specifying Q! Recalling that the Q of a coil can be equated to its reactance, X, divided by its (equivalent series) resistance, R_S, we realize further that R_S can be determined by dividing X by Q. For a Q of 238 in a 1.156-µH coil at 7.1 MHz, R_S works out to 0.22 Ω. It seems that all we must do to model realistic resonator-inductor Q in our simulation of Fig 6.18 is add a 0.22-Ω resistance in series with each tuned-circuit coil.

In this case, yes — but a few sentences ago we said "*equivalent* series resistance" for good reason. The less-than-infinite Q of inductors at ham-band frequencies is a result not of ohmic resistance, but ac resistance due to *skin effect*, the tendency for ac at frequencies higher than a few hundred kilohertz to flow mainly at and near the *surface* of a conductor. Adding 0.22 Ω in series with each of our resonators therefore has the unwanted side-effect of making the coils unrealistically lossy at dc and low frequencies. In this simple HF-filter-modeling case, dc accuracy doesn't matter because our coils carry no dc. If, however, we wanted to model realistic Q in a coil that also carried dc to, say, a tran-

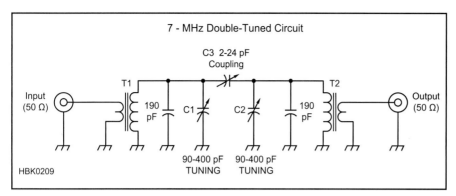

Fig 6.18 — A 7-MHz double-tuned-circuit filter as drawn for real-world builders. Each resonator consists of 17 turns of #22 wire on a T-50-6 toroidal, powdered-iron core; the coupling links consist of 3 turns of insulated wire. Per the article that described this circuit, the filter is intended to have "a 7.1-MHz center frequency and a 200-kHz bandwidth."

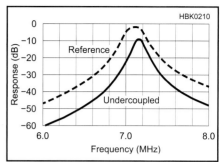

Fig 6.19 — Published response (Reference curve) of the 7-MHz filter. Careful reading reveals that these and other response graphs in the source article were generated by CAD software ("a computer generated the data for this and the other graphs in this article; experiment confirmed the data"); additional reading provides measurement results that provide real-world modeling goals: "The result: a critically coupled filter that's 178 kHz wide, with just over 2 dB of insertion loss."

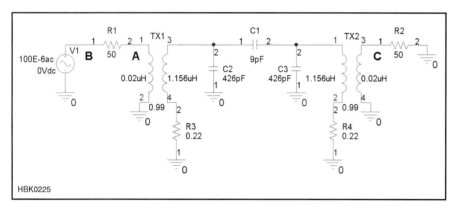

Fig 6.20 — The 7-MHz filter ready for SPICE simulation in *OrCAD 16.0 Capture CIS*. Transformer (TX) components simulate the circuit's link-coupled resonators (note the coupling coefficient, *K*, of 0.99), and 0.22-Ω resistors added in series with the tuned-circuit windings to model realistic *Q*. (No attempt is made to simulate realistic *Q* in the coupling links [0.02 μH] as real-world data is unavailable for that characteristic.) Paralleled capacitances in Fig 6.18 are represented by single values — an approach worth cultivating as a habit to stay within the component-count limitations of feature-limited demoware. Sinusoidal voltage source V1 (100 μV) and R1 serve as the test-signal generator, with R1 also serving as the filter's input termination; R2 serves as the filter load. Although intuition tells us that probing node R2:1 (labeled C) will give the circuit's output voltage, real-world test-bench experience does not immediately suggest where (point A or B?) to obtain the input-voltage value necessary for calculating the circuit's insertion gain as a real-world test setup would report it.

R1 — a loss that will vary inversely with the impedance presented by the filter and its load. This means that we always have access to the reference level we need for calculating the insertion gain of whatever two-port circuit we connect between the test generator (V1R1) and its load (R2). Because R1 = R2, we can determine the insertion gain of whatever we connect between R1 and R2 with the equation

$$\text{Gain (dB)} = 20 \log\left(2\left[\frac{V_C}{V_B}\right]\right) \quad (1)$$

where V_C is the voltage at point C of Fig 6.20 and V_B is the voltage at point B. Assuming that R1 = R2 — exactly the case in a real-world gain-measurement setup, in which the values of both will usually be 50 Ω — equation 1 returns a value of 0 when the test generator (V1R1) is directly connected to its load (R2).

Fig 6.21 shows the analysis setup settings that tell *SPICE* to do ac analysis, in which *SPICE* first determines a circuit's dc characteristics before determining its ac characteristics in response to whatever ac sources it may contain. In this case, our ac source puts out 100E–6 volts — 100 μV, a level that corresponds to a strong on-the-air signal.

Fig 6.22 graphs the analysis results — as returned by equation 1 (curve B) and (in curve A) as returned by merely expressing in decibels the ratio of the voltage across the filter output to the voltage across the filter input. Curve B reports the circuit's insertion gain as a real-world gain-measurement setup would display it. As we shall also see, Curve

sistorized power amplifier, the voltage drop across any inductor R_S added merely as a *Q*-modeling workaround might well make our simulation unacceptably inaccurate at dc. One workaround to this new problem might be to shunt the *R*-equipped *L* with a high-value ideal *L* — a dc-bypass choke — but doing so would likely introduce yet other side-effects, including parasitic resonances and reduced realism in simulating the circuit in the time domain. A better approach would be to use an inductor model that implements skin-effect-based *Q*, such as that available in *Ansoft Designer SV 2*.

Fig 6.20 shows the filter in *OrCAD Capture CIS*, ready for modeling in *SPICE*, with our *Q*-modeling resistances included as R3 and R4. Three more additional components, V1, R1 and R2, provide our test setup for measuring the circuit's insertion gain. V1 and R1 form the test-signal generator, with R1, which corresponds to the internal impedance of the generator, serving as the filter's input termination, and R2 serving as the filter's output termination and test load.

That the test-signal generator we build in treats as separate the signal source and its internal impedance gives us access to generator internals that we can't access in the real thing. Specifically, we can probe point B, the node at which the test generator's full output is available without modification by loss in the generator's internal impedance,

B reports the circuit's insertion gain as an RF-fluent circuit simulator reports it "out of the box" as one of a very important set of tools called *S parameters*. Before we meet such a simulator in the form of *Ansoft Designer SV 2*, we'll use *SPICE* to reality-check our virtual insertion-gain-measurement setup.

6.2.4 Example 4: Checking Reality with a 10-dB Attenuator

To confirm in a general way that we're on the right track with the insertion-gain probing and reporting regime we arrived at for our simulation of Fig 6.20, we'll simulate a simple circuit of known gain and frequency response. The circuit, **Fig 6.23**, consists of little more than a pi-network attenuator with its resistances configured to exhibit 10 dB of loss in a 50-Ω system. As in Fig 6.20, we add 50-Ω source and load resistors, and a voltage source as a test-signal generator. **Fig 6.24** shows the attenuator's gain: –10 dB. We have indeed built and calibrated our virtual insertion-gain test setup to duplicate the behavior of the real thing, confirming as well the validity of the results we obtained for the insertion gain of the filter in Fig 6.20.

6.2.5 Example 5: Simulating the 40-Meter Filter in an RF-Fluent Simulator

Realistically simulating Fig 6.18 in *SPICE* required the addition of resistors to set its resonators' unloaded *Q* to 238. Such workarounds are not necessary when we use a circuit simulator specialized to speak RF, as we'll illustrate by simulating Fig 6.18 in *Ansoft Designer SV 2*, the student version of *Ansoft Designer*, a linear, nonlinear, electromagnetic and system EDA CAD suite engineered for modern RF, MMIC and RFIC design. **Fig 6.25** shows the filter in the *Ansoft Designer SV 2* schematic editor. Now, instead of using ideal transformer windings, we can model the filter's resonators as inductors with realistic *Q* based on skin effect (**Fig 6.26**). But using stand-alone inductors rather than transformers presents a new challenge: How will we model the filter's input and output coupling links?

In addition to including many realistically non-ideal components (**Fig 6.27**), the *Ansoft Designer SV 2* component library includes ideal transformers — transformers with a coupling coefficient (*K*) of 1 — that can be characterized by turns ratio. The resonators in the real-world circuit consist of 17-turn coils with 2-turn links, so we can simulate the links by using transformers with turns ratios of 2:17 (0.118) and 17:2 (8.5) at the filter input and output, respectively.

Absent from our schematic are a signal

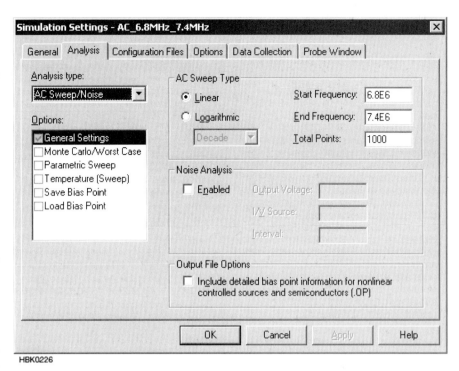

Fig 6.21 — Setup for a *SPICE* ac analysis at 1000 evenly spaced points from 6.8 MHz to 7.4 MHz.

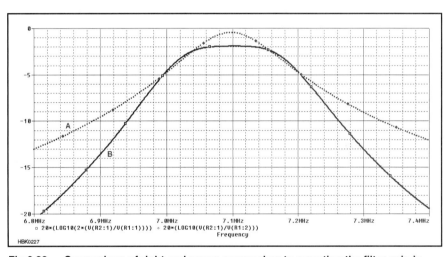

Fig 6.22 — Comparison of right and wrong approaches to reporting the filter gain in decibels. The dotted curve, A, merely plots the ratio, expressed in decibels, of the voltages at the filter output and input — points C and A — in Fig 6.20. The solid curve, B, plots values based on the ratio of the voltages at points C and B as determined by the trace definition 20*(LOG10(2*(V(R2:1)/V(R1:1)))) — the main text's Eq insertion gain calculation rendered in a form understandable by the reporter. Curve B depicts the circuit's insertion gain in decibels as the Fig 6.16 test setup would report it.

source and hardwired input and output terminations. Sources are unnecessary for linear simulation using *Ansoft Designer SV 2*. Terminations, including the 50-Ω default set within the circuit's port elements, are applied at reporting time, *after* analysis has concluded.

Fig 6.28 shows the responses available for our simulation in the *Ansoft Designer SV 2* reporter. Four-port responses are shown because the full circuit analyzed actually consists of the Fig 6.18 circuit (ports 1 and 2) and Fig 6.25 circuit (port 3 and 4) place side by side. **Fig 6.29** compares the two circuit's responses, with the responses of the Fig 6.20 circuit shown as dotted lines.

Unsurprisingly, the two modeling approaches produce slightly different responses, both of which meet the goals of the filter designer. Which one is "better" can therefore

Computer-Aided Circuit Design 6.17

Symbols, Circuits and Hierarchies

We discover something quite interesting if we happen to view the *SPICE* netlist for Fig 6.20: Its linear transformer (TX) components (**Fig 6.A3**) are actually symbols that represent subcircuits — in this case, subcircuits that consist of two inductor (L) components coupled with a K_Linear coupling component. **Fig 6.A4** displays the evidence.

Because it includes subcircuits, Fig 6.20 is a *hierarchical* circuit — a circuit with multiple levels. Although support for hierarchies is usually limited in the demoware versions of CAD software that radio amateurs are likely to encounter, subcircuits are a powerful tool for creating and simulating large-scale circuits that contain other circuits, and circuits that use multiple copies of groups of components — individual identical transistor cells used many times throughout an RFIC, switching transistors with built-in bias resistors, bias and decoupling networks — throughout a design. Creating a schematic symbol that represents a subcircuit lets you in turn add the symbolized subcircuit as a new component for subsequent point-and-click placement in schematics like any other part — just as we do whenever we place a linear transformer component in the *OrCAD 16.0* schematic editor.

The power of symbolized subcircuits also operates at analysis time: However many identical instances of a given subcircuit may be present in an analysis, the simulation engine need evaluate its structure only once, recalculating its electrical behavior as it calculates the behavior of circuits that contain it.

```
* source ZOI_40_METER_FILTER
V_V1         N07792 0 DC 0Vdc AC 100E-6ac
R_R1         N07792 N07820    50
R_R2         N04311 0    50
X_TX1        N07820 0 N01324 N00387 zoi_40m_filter_orcad16_TX1
X_TX2        N01606 N00537 N04311 0 zoi_40m_filter_orcad16_TX2
R_R3         0 N00387   0.22
R_R4         0 N00537   0.22
C_C1         N01324 N01606   9pF
C_C2         0 N01324    426pF
C_C3         0 N01606    426pF

.subckt zoi_40m_filter_orcad16_TX1 1 2 3 4
K_TX1         L1_TX1 L2_TX1 0.99
L1_TX1           1 2 0.02uH
L2_TX1           3 4 1.156uH
.ends zoi_40m_filter_orcad16_TX1

.subckt zoi_40m_filter_orcad16_TX2 1 2 3 4
K_TX2         L1_TX2 L2_TX2 0.99
L1_TX2           1 2 1.156uH
L2_TX2           3 4 0.02uH
.ends zoi_40m_filter_orcad16_TX2
```

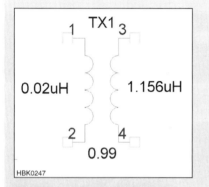

Fig 6.A3 — A linear transformer (TX) component in the *OrCAD 16.0* schematic editor. This part is actually a symbol for a subcircuit that contains three parts.

Fig 6.A4 — Inspecting the *SPICE* netlist for Fig 6.20 reveals the reality behind the TX symbol: Placing a TX element actually places two inductors (L) and a coupling element (K).

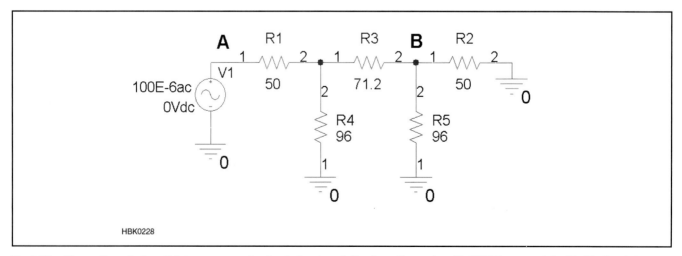

Fig 6.23 — To confirm what we think we now understand about modeling insertion gain with *SPICE*, we model a 10-dB attenuator using the same test generator and load arrangement as in Fig 6.18. Once again, we will calculate the gain of the circuit based on voltages probed between R1:1 (point A) and common (node 0) and R2:1 (point B) and common.

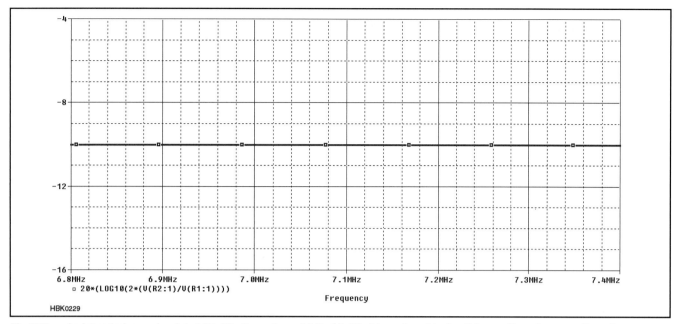

Fig 6.24 — As intended, our simulated 10-dB attenuator exhibits 10 dB of loss. In achieving this result, we have confirmed that our method of probing the circuit at R1:1 and R2:1 and common, in conjunction with reporting the circuit's gain as 20*(LOG10(2*(V(R2:1)/V(R1:1)))), validly returns the insertion gain we would measure in a real-world setup like that of Fig 6.16. In confirming this, we have also confirmed our finding of insertion gain through simulating Fig 6.20.

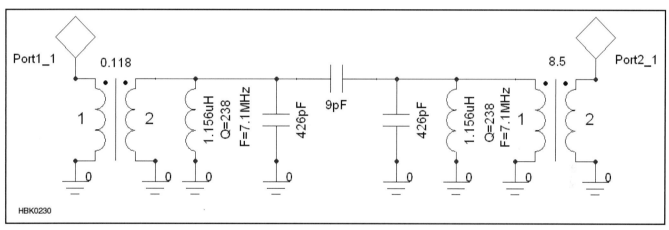

Fig 6.25 — Simulating the 7-MHz filter in *Ansoft Designer SV 2*, lets us use inductors that realistically model *Q*. Doing so then requires us to model the filter's input and output links with ideal transformers characterized in terms of turns ratio. The resonators in the real-world circuit are 17-turn toroidal coils with 2-turn links. The test generator and terminations necessary for *SPICE* analysis are absent for the reasons described in the text.

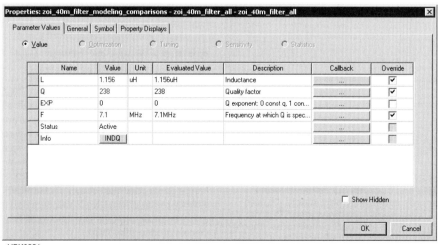

Fig 6.26 — Properties dialog for the *Ansoft Designer SV 2* INDQ component, showing settings for *Q*.

Computer-Aided Circuit Design   6.19

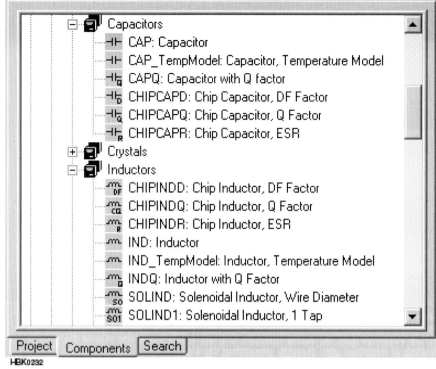

Fig 6.27 — The *Ansoft Designer SV 2* component chooser includes inductor and capacitor models that realistically model lossiness at RF.

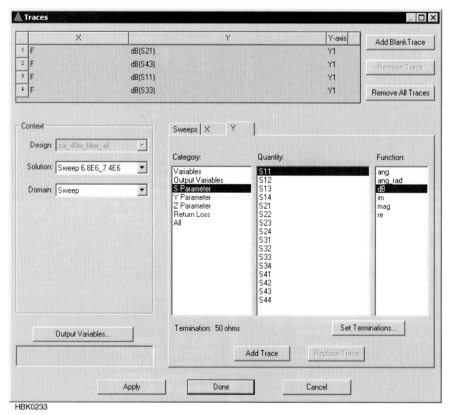

Fig 6.28 — The reporter of the *Ansoft Designer* RF-fluent linear simulator lets us evaluate circuit behavior not in terms of voltages and currents but rather in terms of *network parameters* — scattering parameters (*S*), admittance parameters (*Y*), and impedance parameters (*Z*) — and return loss. These reporter settings, used to produce the comparative graph in Fig 6.29, show responses for a four-port circuit rather than a two-port because the full analysis run included as a separate circuit copy the *Q*-from-added-resistors circuit from Fig 6.20.

be considered to be academic; discerning the difference between two real-world filters that exhibit exactly these responses would be a measurement challenge. On the air we could tell them apart only with difficulty.

Our true purpose in reevaluating the Fig 6.20 filter in *Ansoft Designer SV 2* is to illustrate the difference between RF-fluent CAD and the general-purpose *SPICE*-based CAD tools to which CAD-minded hams tend to default for circuit design. In graphing its responses in *Ansoft Designer SV2* we stand at the threshold of a class of muscular CAD tools that bring great power to hobbyists interested in RF CAD. Rather than plotting just gain, Fig 6.29 actually plots gain and *return loss*, a value of practical importance in many RF-design applications. (The higher the return loss, the more input-signal energy the filter accepts, and the less input-signal energy is reflected back to its source. For a filter of this design, we want and expect return loss to be high in its passband and low in its stopband.) Rather than merely evaluating the circuit's voltage gain, we report its response in terms of standardized *network parameters* — in this case the S (scattering) parameters S_{21} and S_{11} (for Fig 6.20, and S_{43} and S_{33} for Fig 6.25 in our two-two-ports-at-once simulation) expressed in decibels. The sidebar "S-Parameter Basics" explains more about why S parameters are important and how RF engineers use them.

The ability to handle network parameters is a profound differentiator between *SPICE* and more RF-fluent simulators. Although it's possible to derive S parameter from *SPICE* analysis through post-processing and/or the use of special subcircuits that stimulate an n-port for the purpose of more network-parameter-literate reporting, fluency in small-signal network parameters is not among *SPICE*'s simulation and reporting strengths. As we'll see in the next section, *SPICE* can only limitedly simulate intermodulation distortion, a signal-handling flaw in which multiple signals present at a circuit's input interact in circuit components — in active devices, especially — to produce output spectral components not present at a circuit's input.

6.2.6 Example 6: *SPICE* and Intermodulation Modeling

As an example of the limitations of *SPICE* for critical RF-fluent analyses, we will attempt to simulate intermodulation distortion (IMD) with *OrCAD 16.0*. IMD is of great importance to RF engineers because the span of signal levels — the *dynamic range* — we expect modern communications circuitry to handle is so wide that communication possible by means of weak legitimate signals can easily be made impossible by the weak false signals produced by IMD.

Fig 6.30 shows the circuit: the widely used

S-Parameter Basics

The tool called *S parameters* provides a standardized way of characterizing how a device behaves in response to signal energy applied to its ports — its signal inputs and outputs — usually with all of its ports terminated in identical, standard impedances (commonly, 50 Ω, resistive). A transistor, for instance, is a two-port device. By convention, the ports are labeled with numbers, Port 1 being the input and Port 2 being the output.

Signal energy applied to one port of a two-port device comes out two places: at the same port (because the device reflects some of the energy back to the generator) and at the other port. How much signal comes out relative to the applied signal tells us the device's gain (which can be negative — that is, a loss); how much signal reflects back out tells us something about the impedance match between that port's impedance and our signal generator. Determining the phase of the output or reflection signals relative to the phase of the applied signals tells us even more about the device or subcircuit under test.

Fig 6.A5 shows this idea graphically. An S parameter is a voltage ratio (commonly, but not always, expressed in decibels) annotated with two subscript numbers that indicate the ports involved. For instance, a device's forward transmission gain, S_{21}, ("S sub two one"), is the ratio of the voltage at Port 2 to the voltage applied to Port 1 — a value that must be expressed as a vector to convey the two signals' relative phase. To discuss S parameters more readily and to communicate them in tabular form, we can split each of the four basic parameters into separate components — real and imaginary parts, or, especially useful for device modeling with S parameters, magnitude and phase: MS_{11} (magnitude of input reflection) and PS_{11} (phase of input reflection); MS_{21} (magnitude of forward transmission gain) and PS_{21} (phase of forward transmission gain); MS_{12} (magnitude of reverse transmission gain) and PS_{12} (phase of reverse transmission gain); and MS_{22} (magnitude of output reflection) and PS_{22} (phase of output reflection).

From the standpoint of circuit evaluation and modeling, the great power of S parameters is that they can convey usefully realistic information about the ac behavior of a device or subcircuit through relatively few standardized numerical values. For instance, if we know the S parameters of a given transistor operating under known conditions of power- and bias-supply voltage and current, we have a very useful picture of how it looks to the outside world — a picture we can paste directly into an S-parameter-fluent linear circuit simulator. Of great value to the modeling efforts of professionals and radio amateurs alike is the fact that RF-device manufacturers commonly make their products' S parameters freely available in data formats widely used by industry-standard simulators. **Table 6.3** shows S-parameter data for the NE46134 transistor in an S-parameter format that most RF-fluent simulators can handle.

To use device S-parameter data in a simulation, we place a generic *black-box* component in our circuit and tell the simulator to read the S-parameter data when it calculates the behavior of the black box. Using black-box *n*-port parts in place of detailed mathematical device models extends the power of linear simulation for use in modeling the network responses, noise, and stability characteristics of, and developing matching networks for, devices that might otherwise not be modelable without recourse to nonlinear simulation and detailed mathematic device models configured with realistic mathematical parameter values. A significant limitation of black-box modeling is that an S-parameter dataset is static, and most accurately reflects real-world device behavior only under the conditions of voltage and current used in generating it.

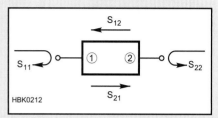

Fig 6.A5 — S parameters corresponding to input reflection, forward transmission gain, reverse transmission gain and output reflection can quite closely characterize a small-signal linear device — in this case, a two-port device. Expressing a two-port's forward transmission gain, S_{21}, in decibels returns the same number we would report for its insertion gain.

Table 6.3
Two-Port S-Parameter Data Equivalent to an NE46134 Transistor (Operating at V_{CE}=12.5 and I_C=50 mA) from the California Eastern Laboratories File *NE46134G.S2P*

```
! FILENAME:     NE46134G.S2P   VERSION: 8.0
! NEC PART NUMBER: NE46134     DATE:   07/94
! BIAS CONDITIONS: VCE=12.5V, IC= 50mA
# GHZ S MA R 50
0.050    0.432    −91.9    34.140    125.6    0.024    63.0    0.586    −56.7
0.100    0.372   −129.4    19.834    106.3    0.036    63.5    0.362    −78.4
0.200    0.348   −161.2    10.444     92.4    0.058    65.4    0.223    −97.1
0.300    0.344   −175.7     7.086     85.1    0.081    67.7    0.183   −106.7
0.400    0.343    173.5     5.332     79.3    0.104    68.1    0.170   −112.8
0.500    0.341    164.1     4.282     74.5    0.127    67.5    0.168   −116.6
0.600    0.343    157.1     3.610     70.1    0.150    66.3    0.172   −119.3
0.700    0.344    149.7     3.114     65.7    0.173    64.7    0.179   −121.0
0.800    0.349    143.8     2.755     61.8    0.196    63.1    0.188   −122.4
0.900    0.347    137.4     2.471     57.9    0.218    61.2    0.198   −123.4
1.000    0.352    132.4     2.254     54.4    0.239    59.2    0.210   −124.2
1.100    0.351    126.5     2.075     50.7    0.260    57.2    0.223   −125.0
1.200    0.353    121.2     1.927     47.1    0.280    55.0    0.236   −125.8
1.300    0.351    116.3     1.803     43.7    0.300    52.9    0.251   −126.5
1.400    0.350    111.6     1.710     40.4    0.320    50.7    0.267   −127.4
1.500    0.346    106.6     1.607     37.2    0.337    48.4    0.283   −128.0
1.600    0.343    102.0     1.523     34.0    0.354    46.2    0.300   −128.9
1.700    0.339     97.0     1.447     31.4    0.369    44.1    0.317   −130.2
1.800    0.339     93.1     1.388     29.0    0.385    42.4    0.330   −131.5
1.900    0.338     88.4     1.340     26.2    0.402    40.4    0.343   −132.1
2.000    0.336     83.3     1.295     23.5    0.418    38.2    0.357   −132.7
2.100    0.331     78.4     1.252     20.7    0.432    36.1    0.372   −133.4
2.200    0.325     73.3     1.206     18.1    0.445    34.0    0.386   −134.0
2.300    0.320     68.4     1.172     16.1    0.458    32.0    0.399   −134.7
2.400    0.318     63.0     1.140     13.6    0.471    29.9    0.411   −135.3
2.500    0.314     57.3     1.109     11.7    0.482    28.0    0.423   −135.9
```

Each of these lines conveys frequency and the S parameter magnitude and phase values $MS11$, $PS11$, $MS21$, $PS21$, $MS12$, $PS12$, $MS22$ and $PS22$. This S*n*P format (where *n* is the number of ports in the device characterized) was originally used by the EESOF Touchstone circuit simulator, and is one of several S-parameter data formats now widely used in the RF engineering community. An S*n*P file may also contain noise-modeling data.

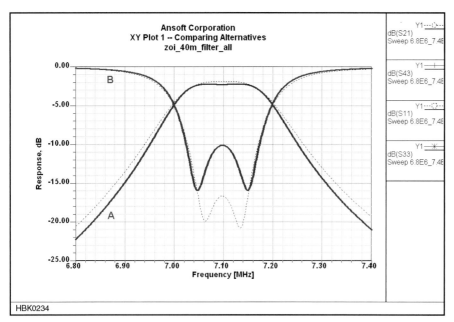

Fig 6.29 — *Ansoft Designer SV 2* report for the realistic-*L* filter showing (A) insertion gain and (B) return loss, both in decibels. (Return loss, the magnitude of S_{11}, is a positive value in dB although in this plot the Y-axis is calibrated in negative dB.) The dotted lines show the same responses for the *Q*-from-series-*R* circuit of Fig 6.20. Graphing the filter's return loss provides information about how closely the impedance of the terminated filter matches the impedance terminating the filter's input. The higher the return loss, the more closely the impedance presented by the input of the terminated filter approaches the impedance of its input termination (in this case, 50 Ω). Return Loss associated with a passive circuit, such as a filter, is always positive.

and well-characterized post-mixer feedback amplifier introduced by Hayward and Lawson in their 1981 Progressive Communications Receiver. This implementation uses the *SPICE* parameters shown in Fig 6.6 for the California Eastern Laboratories NE46134 transistor and includes two signal sources in series for the purpose of generating IMD products.

Simulating the circuit's gain vs frequency response (**Fig 6.31**) returns realistic numbers; turning on nodal voltage and current display in the schematic editor (**Fig 6.32**) confirms that the device's bias point (for a BJT, collector current) is realistic and that we are not exceeding the device's collector-to-emitter voltage rating (15).

Fig 6.33 shows the FFTed output spectrum on a linear scale. Spectral components other than those attributable to the two test signals are absent. **Fig 6.34** displays the same data on a logarithmic voltage scale, with the X-axis zoomed in on the 0- to 40-MHz span. If we know what to look for, we can see responses at frequencies attributable to harmonics of the input tones; to second-order IMD (the frequency of each tone plus and minus the frequency of the other); and third-order IMD (twice the frequency of each tone plus and minus the frequency of the other), but the graph is complicated by higher-order products — arguably good from the standpoint of realism — a high noise floor, and a rising response toward 0 Hz.

As a comparison of simulation techniques, **Fig 6.35** shows the output spectrum for the same circuit as predicted by the harmonic-balance nonlinear simulator in *Serenade SV 8.5*, the now-discontinued predecessor of *Ansoft Serenade SV 2*. Harmonic-balance simulation treats the linear and nonlinear portions of a circuit as separately solvable subsystems, analyzing the linear portion in the frequency domain and the nonlinear portion in the (steady-state) time domain. Harmonic-balance analysis offers significant speed and dynamic-range advantages over *SPICE* for circuits that include transmission lines, long time constants relative to operating frequency, and many reactive components (such as RF circuits and systems commonly contain). Alas, at this writing, harmonic-balance analysis is unavailable to hobbyists and students in free demoware form as discussed in the sidebar, "RF-fluent CAD: What We're Missing."

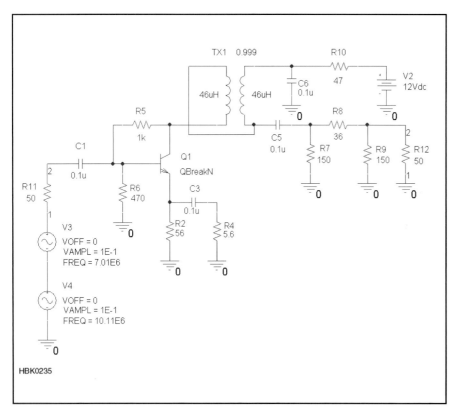

Fig 6.30 — The Progressive Communications Receiver post-mixer amplifier configured for two-tone IMD analysis in *OrCAD 16.0 SPICE*. The level of both test signals is 0.1 V — very strong, "other hams in the immediate neighborhood"-class signals. The transistor is modeled with *SPICE* data for the California Eastern Laboratories NE46134. Using a value of 47 Ω for R10 sets its collector current to 40 mA; 100 Ω, 31 mA.

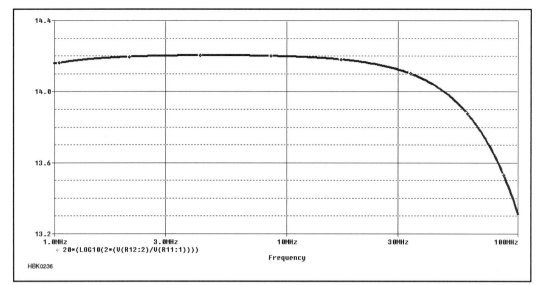

Fig 6.31 —Simulating the gain of the post-mixer amplifier with *OrCAD 16.0 SPICE* produces a realistic gain vs frequency response.

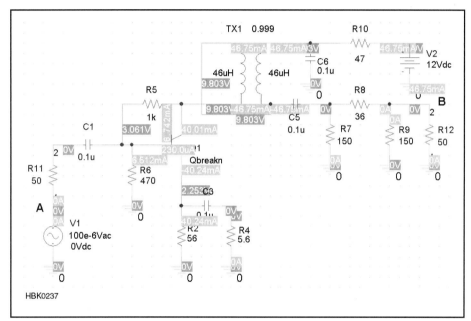

Fig 6.32 — After running at least one analysis, we can turn on voltage and current display in the schematic editor to reality-check the device bias point (for a BJT, collector current). Comparing the voltages displayed for the simulated BJT's collector, base, and emitter lets us ensure that we're not exceeding the published ratings of the real-world device.

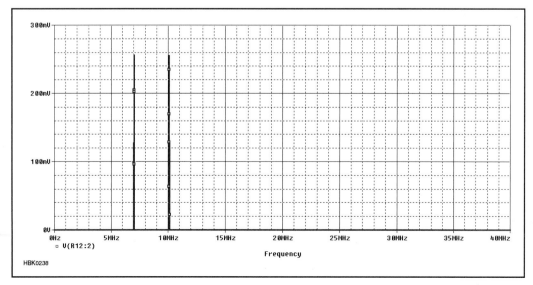

Fig 6.33 — Output spectrum of the feedback amplifier on a linear voltage scale. To generate this graph, we analyzed the Fig 6.30 circuit for 100 μs with a maximum time step of 10e–10 s, and used the *OrCAD 16.0* reporter's FFT function. Components attributable to IMD are superficially absent because of the linear *y*-axis scale.

Computer-Aided Circuit Design   6.23

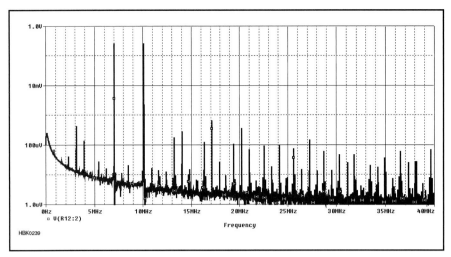

Fig 6.34 — Switching to a logarithmic voltage scale and zooming in on the 0- to 40-MHz range lets us discern spectral components at frequencies corresponding to harmonics and second- and third-order IMD products, but significant artifacts — a relatively high noise floor and a rising response toward time zero — are apparent.

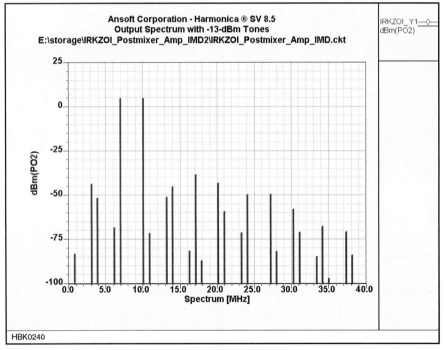

Fig 6.35 — Output power spectrum of the same circuit as predicted by the harmonic-balance nonlinear simulator in the no-longer-available Ansoft *Serenade SV 8.5*. This spectrum is simpler compared to Fig 6.34 because IM calculations were limited to fifth-order products in the student version of *Serenade 8.5*. Most striking is the large simulation dynamic range achieved without the appearance of math-noise artifacts, and the ability to report output power in decibels relative to a milliwatt (dBm). One more thing: The *SPICE* analysis for Figs 6.33 and 6.34 took over five minutes; the *Serenade SV 8.5* run that produced this graph, *under two seconds*.

Simulating Keying with a *SPICE* Behavioral Model

This chapter begins by illustrating that electronic circuits "just do math," responding to and processing electronic signals — also describable mathematically — by, in effect, performing mathematical operations on them. If, for instance, what you require in a simulation is, say, the mathematically idealized behavior of a comparator or 555 timer IC rather than detailed, step-by-time-step nodal analysis of the behaviors of its internal circuitry, you can use (after building it yourself, if necessary) a *behavioral model* of the part instead. Behavioral modeling can be used to simulate analog and digital parts.

The *OrCAD 16.0 Capture CIS* demoware component library includes quite a few behavioral models — mostly 7400-series TTL ICs, but also several analog ICs, including LF411, LM324 and 741 op amps, an LM111 comparator, and a 555 timer. As an example of what behavior models allow us to do, **Fig 6.A6** presents an analysis designed by John Seboldt, KØJD, to simulate on-off keying of a broadband feedback amplifier (Q1, QBreakN, an *OrCAD Capture CIS* "breakout" device characterized with data for an Infineon BFG135 transistor) by means of a series dc-supply switch (Q2, a 2N2907A) keyed via a 2N3904 switch by a 555 timer configured as a "ditter." So realistic is this simulation (**Fig 6.A7**) that it even models "short first dit" and backwave!

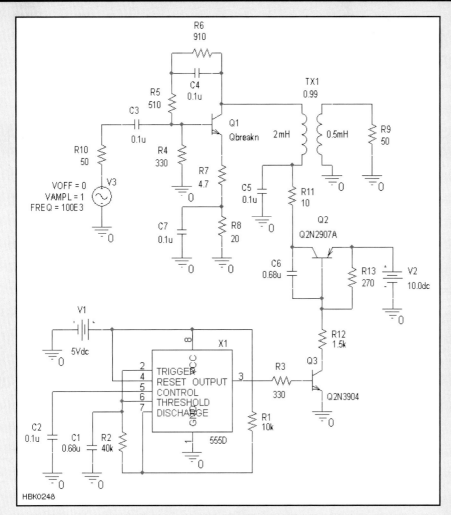

Fig 6.A6 — To simulate the keying waveform of a QRP transmitter stage in *SPICE*, John Seboldt, KØJD, built a ditter — a circuit that, powered up, sends a continuous string of Morse code dots — using a SPICE behavioral model for the 555 timer IC. The timer output drives 2N3904 and 2N2907A switches to interrupt the collector power supply of amplifier transistor QBreakN, an *OrCAD Capture CIS* "breakout" device characterized by NXP Semiconductor data for a Philips BFG135 (BFR194 chip) broadband transistor. Although a low-frequency (100-kHz) source is used to reduce the analysis time and datafile size (and we have increased the inductances in transformer TX1 appropriately relative to their values at HF), coupling and bypass capacitances are kept at their HF-appropriate values to keep RC-time-constant-related settling times consistent with the behavior of the real-world circuit at HF. To make the keyed signal's rise and fall times slower and more easily discerned in a graph, we have also increased the value of shaping capacitor C6 from Seboldt's value of 0.47 µF.

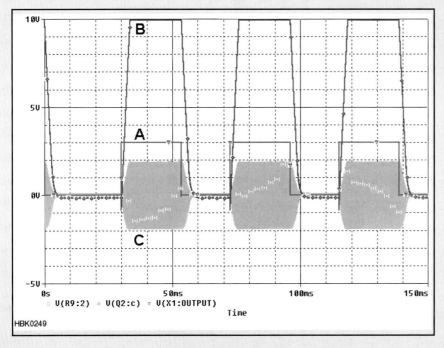

Fig 6.A7 — Results of the keying simulation, showing (A), the voltage at the 555 OUTPUT pin; (B), the keyed collector supply applied to the amplifier; and (C), the keyed amplifier output; time steps are 0.2 µs. So realistic is this simulation that it also reflects the "short first dit" problem shared by some real-world transmitters (in this case, it's an artifact of powering up the keyer and amplifier together as the analysis begins) and significant *backwave* (discernible output during key-up periods). In a real-world transmitter, we would reduce backwave to an acceptable level by keying multiple amplifier stages, a mixer, or — at the risk of degrading signal quality in additional ways — an oscillator.

Computer-Aided Circuit Design

RF-Fluent CAD: What We're Missing

SPICE-based simulators can do wonders in many classes of circuit simulation. For RF use, however, *SPICE* has significant drawbacks. For starters, *SPICE* is not RF-fluent in that it does not realistically model *physical* distributed circuit elements — microstrip, stripline, and other distributed circuit elements based on transmission lines. It cannot directly speak network parameters (*S*, *Y*, *Z* and more), stability factor and group delay. It cannot simulate component *Q* attributable to skin effect. It cannot simulate noise in nonlinear circuits, including oscillator phase noise. It cannot realistically simulate intermodulation and distortion in high-dynamic-range circuits intended to operate linearly. This also means that it cannot simulate RF mixing and intermodulation with critical accuracy.

The feature-unlimited version of *Ansoft Designer* and competing RF-fluent simulation products can do these things and more excellently — but many of these features, especially those related to nonlinear simulation, are unavailable in the student/demoware versions of these packages where such versions exist.

For awhile, from 2000 to 2005, the free demoware precursor of *Ansoft Designer SV 2*, Ansoft *Serenade SV 8.5*, brought limited use of nonlinear-simulation tools to students and experimenters. With *Serenade SV 8.5*, you could simulate mixers, and you could simulate amplifier IMD — IMD from two tones only, to be sure — to a maximum of four nonlinear ports, meaning that *Serenade SV 8.5* could simulate mixers with up to four diodes or up to two transistors (or one transistor and two diodes — you see the strategy of the limitation). See **Figs 6.A8**, **6.A9** and **6.A10**. You could simulate the conversion gain and noise figure of a mixer. Optimization was enabled. Realistic nonlinear libraries were included for several Siemens — now Infineon — parts. You could accurately predict whether or not a circuit you hoped would oscillate would *actually* oscillate, and assuming that it would, you could accurately predict its output power and frequency.

The harmonic-balance techniques used by Ansoft's nonlinear solver — and by the nonlinear solvers at the core of competing RF-fluent CAD products, such as Agilent *Advanced Design System* (ADS) — allowed you to simulate crystal oscillators as rapidly as you can simulate lower-*Q* oscillators based on LC circuits. (In *SPICE*, getting a crystal oscillator to start may be impossible without presetting current and/or voltages in key components to nonzero values, even if you try kick-starting it with a pulse as we do in this chapter's JFET VFO simulation.)

Students and CAD-minded radio amateurs alike miss the features made available in *Serenade SV 8.5* and hope that they will one day return in some form to the world of free demoware CAD software. In the meantime, if you're serious about pushing into RF CAD beyond what *SPICE* can do and someone you know has no use for their copy of *Serenade SV 8.5*, see if you talk them out of it!

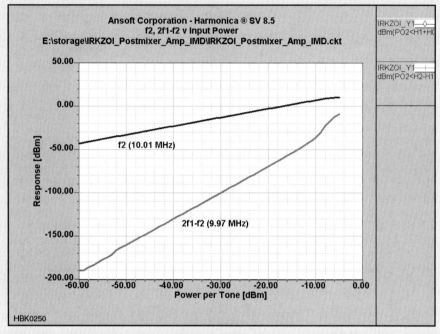

Fig 6.A8 — At this writing, the two-tone nonlinear simulation capabilities necessary to model third-order IMD were unavailable in free-demoware form. This simulation was done by Ansoft *Serenade Designer SV 8.5*, available from 2000 to 2005.

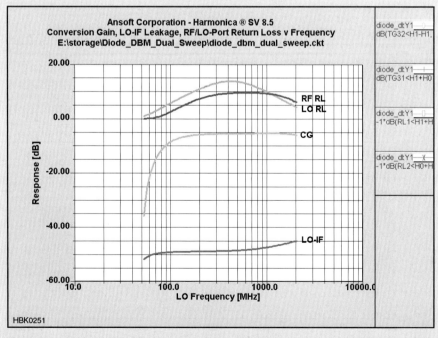

Fig 6.A9 — Professional RF circuit simulators can also simulate mixing and the small-signal characteristics of mixers, such as port return loss, conversion gain, and port-to-port isolation. (Serenade SV 8.5 *simulation*)

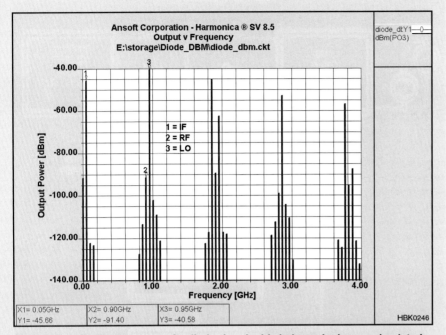

Fig 6.A10 — Output spectrum of a diode-ring doubly balanced mixer as simulated by *Serenade SV 8.5*. Note the dynamic range implicit in this graph: In a simulation that includes a local-oscillator (LO) signal at 7 dBm, we're seeing accurate values for IMD products nearly *140 dB weaker* without encountering mathematical noise — an achievement unapproachable with *SPICE*-based simulators.

6.2.7 Circuit CAD in the Radio Amateur's Toolbox

This chapter emphasizes the use of *SPICE* for circuit simulation because radio amateurs interested in circuit CAD will likely first experience it with a *SPICE*-based simulator and keep using *SPICE*. At this writing, *SPICE* is the only *nonlinear* simulator freely available to hobbyists. Radio amateurs seeking to enhance their knowledge of RF design techniques through circuit simulation will want to use an RF-fluent simulator instead of or in addition to *SPICE*, and *Ansoft Designer SV 2* can serve as a linear-simulation workhorse for this purpose. (Here we must differentiate between trialware and demoware: Although RF-fluent nonlinear simulation may be available in the feature-limited *trialware* versions of some EDA products, hobbyists need CAD capabilities that won't stop working in 30 to 90 days. We hasten to add that radio amateurs do not expect that such capabilities need be free, just affordable; the full versions of the RF-fluent simulators known to the author sell for thousands to tens of thousands of dollars.)

An expensive full-version simulator can mislead as, or more, easily than its freeware version in the absence of designer know-how teamed with *carefully and fully characterized performance data describing the actual behavior of simulatable real-world circuits*. Without tempering by experimental experience and constant comparison with real-world performance data, results obtained through CAD can lose their necessary real-world anchoring. Well-applied, however, computerized circuit simulation can greatly accelerate one's acquisition of the intuition and RF "street smarts" that make RF design an art *and* science.

6.3 Limitations of Simulation at RF

[Experienced users of circuit simulation software are wary of using any software near or outside the boundaries of circuits and parameters for which it was intended and tested. RF simulation can present just such situations, leading to software failure and unrealistic results. Introduced and summarized in this section, several detailed papers by Dr Ulrich Rohde, N1UL, exploring simulation at RF are provided on the CD-ROM accompanying this Handbook. The papers are:

• *"Using Simulation at RF" by Rohde, a survey of the issues of RF simulation and the techniques used in current modeling programs.*

• *"The Dangers of simple usage of Microwave Software" by Rohde and Hartnagel, a discussion of inaccuracies introduced by device parameter measurement and model characteristics.*

• *"Mathematical stability problems in modern non-linear simulations programs" by Rohde and Lakhe, presenting various approaches to dealing with non-linear circuit simulation. — Ed.]*

While the precise lower bound of "RF" is ill-defined, RF effects start already at about 100 kHz. This was first noticed as self-resonance of high-Q inductors for receivers. In response, Litz wire was invented in which braided copper wires were covered with cotton and then braided again to reduce self-resonance effects.

As frequencies get higher, passive elements will show the effects of parasitic elements such as lead inductance and stray capacitance. At very high frequencies, the physical dimensions of components and their interconnections reach an appreciable fraction of the signal wavelength and their RF performance can change drastically.

RF simulators fall in the categories of *SPICE*, harmonic-balance (HB) programs and EM (electromagnetic) programs. The EM simulators are more exotic programs. Two types are common, the 2D (2.5) or 2-dimensional and the full 3-dimensional versions. They are used to analyze planar circuits, including vias (connections between layers) and wraparounds (top-to-ground plane connections), and solid-shapes at RF. They go far beyond the *SPICE* concept.

6.3.1 *SPICE*-based Simulators

SPICE was originally developed for low-frequency and dc analysis. (Modern *SPICE* programs are based on *SPICE3* from University of California – Berkeley.) While doing dc, frequency, and time domain simulations

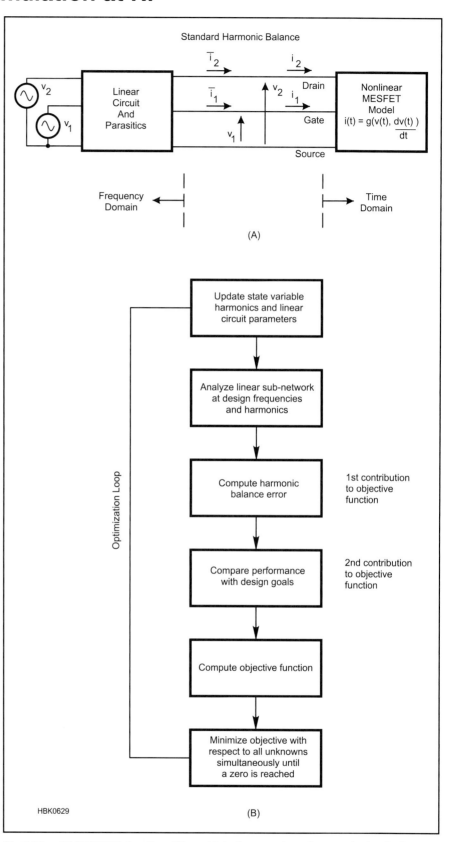

Fig 6.36 — (A) MESFET circuit partitioned into linear and nonlinear sub-circuits for harmonic-balance analysis. Applied gate and drain voltages, and relevant terminal voltages and currents, are indicated. (B) Flowchart of a general-purpose harmonic-balance design algorithm that includes optimization.

very well, *SPICE*-based simulation has some problems. The time domain calculation uses the very complex mathematics of the Newton-Raphson solution to nonlinear equations. These methods are not always stable. All kind of adjustments to the program settings may be necessary for the calculations to converge properly. Knowledge of the specifics of different types of electronic circuits can assist the user in finding an accurate solution by specifying appropriate analysis modes, options, tolerances, and suitable model parameters. For example, oscillators require certain initializations not necessary for amplifiers and bipolar transistors may need different convergence tolerances than do MOS circuits. Generally, *SPICE* finds a solution to most circuit problems. However, because of the nonlinearity of the circuit equations and a few imperfections in the analytical device models, a solution is not always guaranteed when the circuit and its specification are otherwise correct.

The next problem at RF is that the basic *SPICE* simulator uses ideal elements and some transmission line models. As we approach higher frequencies where the lumped elements turn into distributed elements and special connecting elements become necessary, the use of the standard elements ends. To complicate matters, active elements such as diodes and transistors force the designer to more complex simulators. Adding the missing component elements leads to highly complex models and problems of convergence in which the simulator gives an error advising of a numerical problem or more likely by failing to generate a solution.

SPICE also has problems with very high-Q circuits and noise analysis. Questions of the noise figure of amplifiers or phase noise of an oscillator cannot be answered by a *SPICE*-based program accurately. Noise analysis, if not based on the noise correlation matrix approach, will not be correct if the feedback capacitance (Im(Y12), the imaginary component of Y-parameter Y_{12}) is significant at the frequencies involved. Analysis of oscillators in *SPICE* does not give a reliable output frequency and some of the latest *SPICE* programs resort to some approximation calculations.

6.3.2 Harmonic-Balance Simulators

Harmonic balance (HB) analysis is performed using a spectrum of harmonically related frequencies, similar to what you would see by measuring signals on a spectrum analyzer. The fundamental frequencies are the frequencies whose integral combinations form the spectrum of harmonic frequency components used in the analysis. On a spectrum analyzer you may see a large number of signals, even if the input to your circuit is only one or two tones. The harmonic balance analysis must truncate the number of harmonically related signals so it can be analyzed on a computer.

The modern HB programs have found better solutions for handling very large numbers of transistors (>1 million transistors) and their math solutions are much more efficient, leading to major speed improvements. Memory management through the use of matrix formulations reduces the number of internal nodes and solving non-linear equations for transient analysis are some of the key factors to this success.

HB analysis performs steady-state analysis of periodically excited circuits. The circuit to be analyzed is split into linear and non-linear sub-circuits. The linear sub-circuit is analyzed in the frequency domain by using distributed models. In particular, this enables straightforward intermodulation calculations and mixer analysis. The nonlinear sub-circuit is calculated in the time domain by using non-linear models derived directly from device physics. This allows a more intuitive and logical circuit representation.

Fig 6.36A diagrams the harmonic-balance approach for a MESFET amplifier. Fig 6.36B charts a general-purpose nonlinear design al-

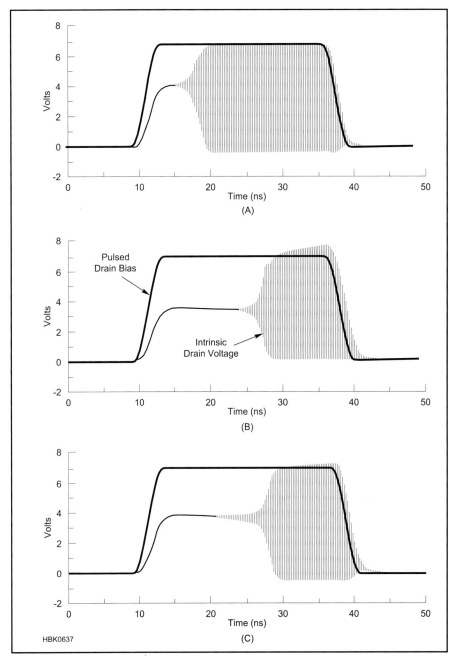

Fig 6.37 — (A) is the initial simulation of a *SPICE*-based simulator. (B) is the correct response of a pulsed microwave oscillator obtained by harmonic balance simulation using the Krylov-subspace solution. (C) is the *SPICE*-based simulation after 80 pulses of the drain voltage.

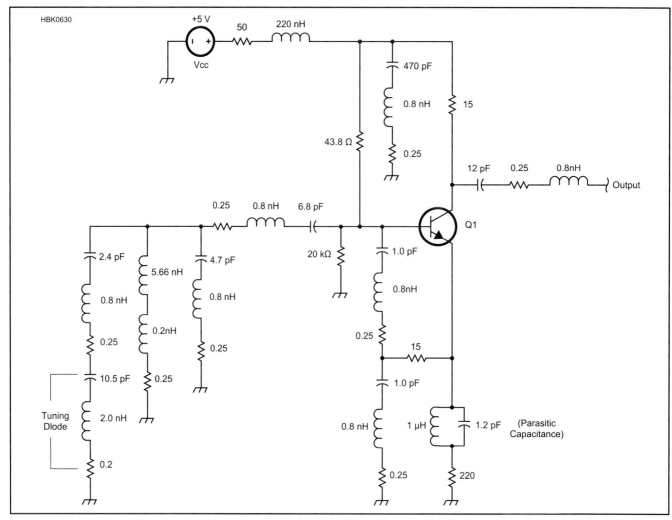

Fig 6.38 — Colpitts oscillator for 800 MHz with lumped elements modeled by their real values.

gorithm that includes optimization. Modern analysis tools that must provide accurate phase-noise calculation should be based on the principle of harmonic balance.

Analysis parameters such as Number of Harmonics specify the truncation and the set of fundamental frequencies used in the analysis. The fundamental frequencies are typically not the lowest frequencies (except in the single-tone case) nor must they be the frequencies of the excitation sources. They simply define the base frequencies upon which the complete analysis spectrum is built.

6.3.3 Contrasts in Results

The following time domain analysis is a good example of differences between *SPICE* and harmonic balance simulation. A microwave oscillator is keyed on and off and a transient analysis is performed. When using the standard *SPICE* based on *SPICE3*, the initial calculation shows an incorrect response after one iteration as seen in **Fig 6.37A**. It takes

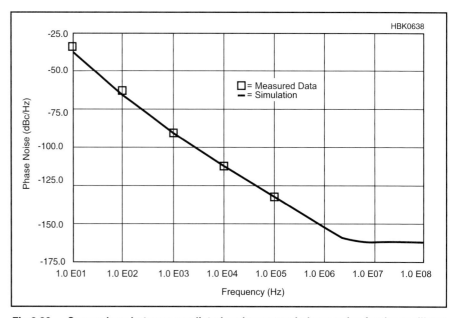

Fig 6.39 — Comparison between predicted and measured phase noise for the oscillator shown in Fig 6.38.

about 80 pulses (80th period of the pulsed drain voltage) until the simulation attains the correct answer (Fig 6.37C) of the Krylov-subspace-based harmonic balance in Fig 6.37B.

The frequencies involved need not be in the GHz range. Oscillators, in particular, can be very difficult to analyze at any frequency as shown by simulations of a low-MHz phase-shift oscillator and a 10 MHz Colpitts oscillator in the referenced papers.

Validating the harmonic-balance approach, **Fig 6.38** shows a BJT microwave oscillator entered into the schematic-capture module of a commercially available HB simulator (Ansoft Serenade 8.0); **Fig 6.39** compares this oscillator's simulated phase noise to measured data. HB analysis gives similarly accurate results for mixers.

6.3.4 RF Simulation Tools

PSPICE: This popular version of *SPICE*, available from Orcad (now Cadence, **www.cadence.com**) runs under the PC and Macintosh platforms. An evaluation version, which can handle small circuits with up to 10 transistors, is freely available, such as from **www.electronics-lab.com**. Contact Cadence for a full version or for more information. *AIM-spice* (**www.aimspice.com**) is a PC-version of *SPICE* with a revised user interface, simulation control, and with extra models. A student version can be downloaded. Table 6.1 earlier in this chapter shows other free *SPICE* offerings.

There are a number of PC-based *SPICE* programs in the $1000 range but they are designed more for switching power supplies and logic circuits optimization than RF. *ICAP4* (**www.intusoft.com/demos.htm**) and *MICROCAP9* (**www.spectrum-soft.com/index.shtm**) both have demonstration/evaluation versions available for download.

Agilent, AWR, Ansys, and Synopsis offer very modern mixed-mode CAD tools and they combine the concept of *SPICE* with the advanced technologies. These are professional-quality tools, but if one can arrange to make use of them through a friend or associate, the results are worth investing the time to learn their use.

6.4 CAD for PCB Design

[With numerous PCB design software packages available and low-quantity, low-cost PCB manufacturing services accepting orders electronically, the development of PCBs has never been easier for the amateur. As with any assembly or manufacturing process, it is important to understand the vocabulary and technology in order to achieve the desired result. Thus, this section provides a detailed description of the entire process of PCB design. — Ed.]

The primary goal of using software for printed circuit board (PCB) design is the production of so-called PCB artwork — the graphic design used to create the patterns of traces that establish connectivity on the PCB. Historically, PCB artwork was created by hand on clear film using black tape and special decals which were then photographically reduced. However, free and low-cost programs specifically for the PCB design process are now widely available. These programs not only allow the creation of artwork efficiently and accurately, but produce the required ancillary files for commercial production, exchange information with schematic capture software, produce Bills of Materials (parts lists), and even include such features as three-dimensional visualization of the finished board. While artwork files can be shared with other people for PCB production, the "source" files used by the CAD program can typically only be used by other people who share the same program.

The decision to produce a PCB must take into account the nature of the circuit itself (for example, high frequency, low noise and high current circuits require additional care). Other considerations are time available, expense, available alternatives, quantity required, ability to share and replicate the design, and non-electrical characteristics such as thermal and mechanical, as well as desired robustness.

6.4.1 Overview of the PCB Design Process

The PCB design process begins with establishing the list of components in the circuit, the connections between the components, the physical outline/size of the board, and any other physical, thermal and electrical constraints or design goals. Much of the connectivity and component information is reflected in the schematic for a circuit, so in many cases the PCB layout process begins by entering the schematic in a schematic capture program which may be integrated with the PCB CAD program or standalone. (Schematic capture is not required for PCB layout.) Once the schematic is entered, there may be other options possible such as simulating the circuit as described in the preceding sections. A clean, well-organized schematic that is easily modified is an asset regardless of the circuit production and construction methods.

With input from the schematic and other information, the board outline is created, mounting and other holes placed, the components positioned, and the pattern of traces created. Once the layout is complete, in many cases it is possible to run a design rules check — the equivalent of a "spell checker." Design rules include component connections and other information to check for problems related to connectivity and manufacturability. This step can save a great deal of time and expense by catching errors that could be fixed by hand, but would otherwise negate some of the benefits of a PCB.

The final step in the PCB layout program is to produce the collection of up to a dozen or so different files required for PCB production. In brief, the list includes the artwork for the pattern of traces, files for producing the board outline, solder masks, silk screens and holes.

The user then uploads the set of files to a PCB manufacturer. As quickly as two to three days later an envelope will be delivered with the freshly-minted boards ready for assembly! Alternatively, the user may create the board "in house" using photomechanical or other processes based on the output files from the software, as is discussed in the **Construction Techniques** chapter along with PCB assembly techniques.

6.4.2 Types of PCB Design Software

PCB software varies in features, function and cost, but for the radio amateur, the most interesting software for introductory use fall into the following categories:

1) Open-Source: PCB design software such as *GNU PCB* (**pcb.gpleda.org**, for *Linux*, Mac OS X) and *KiCad* (**kicad.sourceforge.net/wiki/index.php/Main_Page**, for *Linux*, Mac OS X, and *Windows*, includes schematic capture) are free to use and have no artificial restrictions. Support is through user-forums. Source code is available for the user to modify. *gschem* is a schematic capture sister program to *GNU PCB*.

2) Free, restricted-use/restricted-feature commercial: At least one company makes a version of their PCB and schematic software that is free to use for non-commercial purposes. Though it is restricted in number of layers (two) and maximum board size (4 × 3.2 inches), *Eagle PCB* "Light Edition"

(**www.cadsoftusa.com**, for *Linux*, Mac OS X, *Windows*) is very popular among hobbyists. Files can be shared with others; the resulting industry-standard files can be sent to nearly any PCB manufacturer. *Eagle* also contains a schematic entry program.

3) Free, restricted-output commercial: Several PCB manufacturers offer schematic and PCB software with a proprietary output format tied to their PCB manufacturing service. *PCB123* from Sunstone Circuits (**www.sunstone.com**, for *Windows*) is one such offering, including schematic capture and layout software with up to four layers and board sizes up to 12 × 18 inches (double sided). For an additional fee (per design), industry-standard files can be exported. Schematic entry is included. *Express PCB* (**www.expresspcb.com**, for *Windows*) also provides schematic capture and PCB layout capability, tied to the Express PCB board fabrication services, including the fixed-size (3.8 × 2.5 inches) Miniboard service. Advanced Circuit's proprietary *PCB Artist* software (**www.4pcb.com**, for *Windows*) includes the ability to import netlists.

4. Low-cost commercial: Eagle and many other companies offer PCB and schematic software at a range of prices from $50 to many thousands of dollars. Several versions are typically offered from each company, usually based on limitations on board size, schematic size/complexity and features such as auto-routing. Schematic entry may be included in some packages, or be a separate purchase.

PCB design software manuals and tutorials discuss the basic operation but also special keystrokes and other shortcuts that make operations such as routing traces much more efficient.

The first time designing and ordering a PCB can be daunting, so keep the initial job simple and pay attention to details (and read the instructions). When starting to use a specific software package, join a user's support group or forum if one is available. Request sample designs from other users and experiment with them to see how they are constructed and what files are required in the output data set. Once you are comfortable with the tools, you can begin on a design of your own.

6.4.3 Schematic Capture

The first step in PCB design is to create a schematic. It is possible to design a layout directly from a paper schematic, but it is much easier if the schematic is entered (or "captured") in electronic form. Schematic capture software has two outputs — the visual schematic and the component and connectivity data for subsequent PCB layout. These two separate requirements can make some operations during schematic entry more complicated than what would seem at first glance necessary. Bear in mind however, that the user is creating not only a clear graphic representation of the circuit, but of the underlying electrical connectivity.

Schematics are generally entered on a (virtual) page usually corresponding to common paper sizes — for example, 11 × 17 inches. More complicated schematics can span multiple pages, using special labels or components to indicate both visual and electrical connectivity. Often one can group logically-related elements into a module that can then be referenced as a "black box" on a higher-level schematic. For complex circuits, these features are extremely useful and make the difference between a jumbled diagram that is difficult to use and an organized, compact diagram that efficiently communicates the function and operation of the circuit.

COMPONENTS

The components (resistors, capacitors, etc) on a schematic are either selected from an existing library or created by the user and stored in a custom library. It is also possible to find components and/or additional libraries on the Internet, although each program has its own specific format.

Each component includes a great deal more than shape and pin numbers. A typical component library entry includes:

Symbol — This is the graphic representation shown on the schematic. Many components may have the same symbol (eg, the op amp symbol may be shared by many different types of op amps)

Pins — For each pin or point of electrical connection, the component model may specify the pin number, label (eg, "V_{DD}"), pin type (inverting, non-inverting) or pin functions (common).

PCB footprint — A given component may be available in a number of different packages (eg, DIP or surface mount). Many components may have the same physical footprint (eg, op amps, comparators and optoisolators could all map to the same eight-pin DIP footprint). Footprints include the electrical connections (pins) as well as mechanical mounting holes and pad sizes, and the component outline.

Value — Many components such as resistors and capacitors will have identical information except for a difference in value. All ¼-watt resistors may be instances of the same component, differing only in value and designator.

Designator — The unique reference to the component, such as R1, C7, D3. This is assigned when the component is used (often automatically and in sequence).

Source information — Part number, vendor, cost, etc. This information is for the Bill of Materials.

Components are typically placed on the schematic by opening a library and searching for the desired component. It may be tempting for the beginner to select a component that looks "about right" when faced by a long list of components in some libraries. However, even at this early stage, the physical PCB often must be taken into account. For example, either "1/8W Resistor, Axial" or "1W Resistor, Upright" will result in the same neatly drawn resistor symbol on the schematic but in the subsequent step of using the component data to create a PCB, the footprints will be dramatically different.

It is not at all uncommon to add new components to the library in the course of creating a schematic. Since many components are closely related to existing devices, the process often consists of selecting an existing schematic symbol, editing the shape and/or component data, creating a new label, and associating the part with an existing footprint. Adding a specific type of op amp is an example. This usually only needs to be done once since symbols can be saved in a personal library (and shared with others). It is usually easier to modify a part that is close to what is desired than to "build" a new part from scratch.

Component symbols can generally be rotated and flipped when placing the component instance on the schematic. Designators (R1, T34, etc) can be assigned and modified by the user although the default designators are usually selected sequentially.

CONNECTIONS

The schematic software will have a mode for making electrical connections, called "nets." For example, one might click on the "draw net" symbol then draw a line using the mouse from one pin to another pin, using intermediate mouse clicks to route the line neatly with 90° turns on a uniform grid. Internally, the software must not only draw the visual line, but recognize what electrical connectivity that connection represents. So one must click (exactly) on a component pin to start or end a line or when making a connection between two lines that intersect, explicitly indicate a net-to-net connection (often with a special "dot" component). The connections on a schematic can often be assigned additional information, such as the desired width of the trace for this connection on the PCB or a name assigned by the user, such as "input signal."

Not all connections on a schematic are drawn. To make any schematic — electronic or hand-drawn — more readable, conventions are often employed such as ground or power symbols or grouping similar connections into busses. Schematic capture software often supports these conventions. In some cases, components may be created with implicit power connections; in these cases the connections may not even be noted on the schematic but will be exported to the PCB software. However, as a general rule, software aimed at beginning PCB designers will not require the use of these advanced features.

Since it is often possible for component pins to be assigned attributes such as "power input", "output," "input," and so on, some schematic entry programs allow one to do an early design check. The program can then flag connections between two outputs, inputs that are missing connections, and so on. This is not nearly as helpful or complete as the Design Rule Check discussed below.

Free text can be placed on the schematic and there will be a text block in a corner for date, designer, version, title and the other information that identifies the schematic.

NETLISTS

Once the components are placed and connections made, the schematic may be printed and any output files for the PCB layout software produced. The connectivity and component information needed for PCB layout is captured in a *netlist* file. The flow from schematic entry to PCB may be tightly integrated, in which case the user may switch between schematic and PCB like two views of the same design (which they are). However, most schematic software will generate a separate netlist to be used by PCB layout software, whether integrated or a separate program. The netlist can also be exported to an external circuit simulation program or be used by an integrated simulator program. (See the first part of this chapter for more information on circuit simulation.)

Netlists are often human-readable text files and in most cases it is possible to create a netlist file manually. In the absence of a schematic entry program, this allows the user to take a hand-drawn schematic, extract the connectivity information, and create the netlist for the PCB program to perform design rule checks. However, a netlist is generally not required for the PCB layout software; the user will also have the option to create a PCB on-the-fly, adding components and connections as they wish.

ANNOTATION AND BILL OF MATERIALS

The important features of *forward* and *backward annotation* enter at the interface between schematic entry and PCB layout. It is not uncommon during the PCB layout process to either come across some design deficiency or realize that a change to the schematic could produce a design that would be easier to lay out. Likewise, a review of the schematic partway through the PCB layout process could reveal some needed design change. In the case of changes to the PCB (perhaps changing some pins on a connector to make routing easier), back annotation can propagate the changes "backward" to the schematic. The connectivity data will be updated; however the user may need to manually route the connection lines to neaten up the schematic. Likewise, changes to the schematic when the PCB is already (partially) routed are known as forward annotations and like the schematic, while the connectivity is updated the user will likely need to manually route the traces. Neither forward nor back annotation is necessary, but is useful in keeping the schematic and PCB consistent. In their absence, the user is strongly urged to keep the schematic and PCB up to date manually to avoid time-consuming problems later on.

Finally, the underlying data in the schematic can be used to produce a *Bill of Materials* (BOM). A BOM lists all the components of the schematic, typically ordered by reference designator(s), and may even be exportable for online ordering.

6.4.4 PCB Characteristics
PCB CONSTRUCTION

It is useful to know a little bit about PCB construction in order to make sense of the PCB design process. **Fig 6.40** shows the basic structure of a PCB and some of its design elements (discussed in later sections). The laminate material provides a stable, insulating substrate with other known characteristics (thermal, dielectric, etc). Copper is bonded to one or both sides and selectively removed (usually chemically) to leave traces and pads. The pads provide points of connection for components. Though electrical connectivity is crucial, it is important to remember that the solder and pads provide mechanical and thermal connectivity as well. Pads may be drilled for mounting *through-hole* components or left undrilled for *surface-mount* components.

A separate electrochemical process plates the inside surface of *plated-through holes* to provide connectivity between upper and lower pads. Plated-through holes whose sole purpose is to provide electrical connectivity between layers of a PCB are known as *vias*, shown in **Fig 6.41**. Since they do not need to accommodate a component lead, their hole and pad size are smaller.

While two-layer boards can mount components on either side, most PCBs will have a primary side called the *component side* upon which most of the components will be placed, and a *solder side* dominated by soldered pins and traces. Where high density is required, surface mount (and sometimes through-hole) devices are mounted on both sides, but this is considerably more complex.

Multi-layer boards are essentially a stack of two or more two-layer boards, with an insulating layer between each board. Plated-through holes make connections possible on every layer, and the laminate material is proportionally thinner so the entire multi-layer board is roughly the same thickness as a regular two-layer board. Vias that join selected, adjacent copper layers without connecting the entire stack of layers are called "buried" or "blind" vias and are typically only needed for very dense designs. Multi-layer boards provide much more flexibility in routing signals and some other benefits such as dedicated layers for power distribution and grounding, but at often substantial additional cost.

PCB MANUFACTURING SPECIFICATIONS

Unless the board is manufactured by the hobbyist, the PCB files are sent out to be manufactured by a *board house*. The most important issue for the amateur may be the pricing policies of the board house. Board size, quantity, delivery time, number of layers, number and/or density of holes, presence of solder masks and silk screens, minimum trace/separation width, type of board material, and thickness of copper will all influence pricing. One cost-saving option of the past, a single-layer board, may not be offered with low-cost, low-volume services — two layers may be the simplest option and it results in a more robust board. [Note that most ordering specifications use English units of inches and ounces. Offshore board houses may use both English and metric units, or be metric-only. English units are used here because they are the most common encountered by hobbyists. — *Ed.*]

The second issue to consider is manufacturing capabilities and ordering options. These will vary with pricing and delivery times, but include the following:

Board material and thickness — FR-4 is the most popular board material for low volume PCBs; it consists of flame-resistant woven fiberglass with epoxy binder. Typical thickness is 0.062 inch (1/16 inch), but thinner material is sometimes available. Flexible laminates are also available at greater cost and longer delivery time. Special board laminates for microwave use or high-temperature applications are also available.

Copper thickness — Expressed in ounces per square foot, typical values are 1-2 oz (1 oz corresponds to 0.0014 inch of thickness.) Other values may not be available inexpensively for small volumes. Inner layers on multi-layer boards may be thinner — check if this is important. Most board designs can assume at least 1 oz copper for double-sided boards; trace width is then varied to accommodate any high current requirements.

Layers — Two-layer boards are the most common. Because of the way PCBs are manufactured, the number of copper layers will be multiples of two. For quick-turn board houses, usually only two or four layer boards are available. PCBs with more than two layers will always be more expensive and

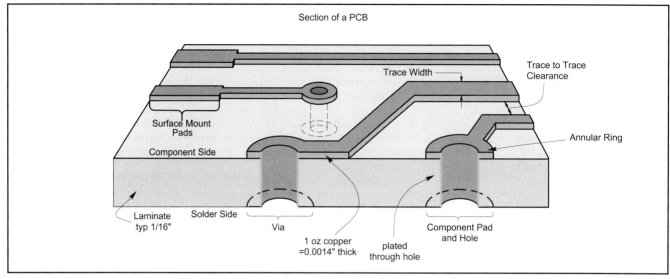

Fig 6.40 — The various elements of PCB construction and specification.

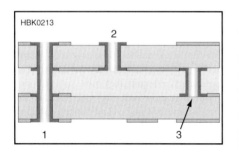

Fig 6.41 — *Via* refers to a plated-through hole that connects one board layer to another. Vias are used for signal, power, or ground connections and even for ventilation. Different via types include through-hole (1); blind (2), and buried (3).

often take longer to manufacture.

Minimum hole size, number, and density of holes — Minimum hole size will rarely be an issue, but unusual board designs with high hole density or many different hole sizes may incur additional costs. Be sure to include vias when specifying minimum hole size. Some board houses may have a specific list of drill sizes they support. Note that you can often just edit the drill file to reduce the number of different drill sizes.

Minimum trace width and clearance — Often these two numbers are close in value. Most board houses are comfortable with traces at least 0.010 inch in width, but 0.008 and 0.006 inch are often available, sometimes at a higher cost.

Minimum annular ring — A minimum amount of copper is required around each plated-through hole, since the PCB manufacturing process has variations. This may be expressed as the ratio of the pad size to hole size, but more commonly as the width of the ring.

Edge clearances — Holes, pads, and traces may not be too close to the edge of the board.

Board outline and copies — There may be options to route the outline of the board in other shapes than a rectangle, perhaps to accommodate a specific enclosure or optimize space. If multiple copies of a board are ordered, some board houses can *panelize* a PCB, duplicating it multiple times on a single larger PCB (with a reduction in cost per board). These copies may be cut apart at the board house or small tabs left to connect the boards so assembly of multiple boards can be done as a single unit.

Tin plating — Once the traces and pads have been etched and drilled, tin plating is usually applied to the exposed copper surfaces for good soldering.

Solder mask — This is a solder-resistant coating applied after tin plating to both sides of the board covering everything except the component mounting pads. It prevents molten solder from bridging the gaps between pads and traces. Solder mask is offered except by the quickest-turn services. Green is the most common color, but other colors may be available.

Silkscreen — This is the ink layer, usually white, on top of the solder mask that lays out component shapes and designators and other symbols or text. A minimum line width may be specified — if not specified, try to avoid thin lines. All but the quickest-turn services typically offer silk screening on one or both sides of the PCB.

6.4.5 PCB Design Elements

The schematic may not note the specific package of a part, nor the width or length of a connection. The PCB, being a physical object, is composed of specific instances of components (not just "a resistor," but a "¼-watt, axial-lead resistor mounted horizontally," for example) plus traces — connections between pins of components with a specific width and separation from other conductors. Before discussing the process of layout, we briefly discuss the nature of components and connections in a PCB.

COMPONENTS

A component in a PCB design is very similar to its counterpart on the schematic. **Fig 6.42** shows the PCB footprint of an opto-interrupter, including graphics and connectivity information. The footprint of a component needs to specify what the footprint is like on all applicable copper layers, any necessary holes including non-electrical mounting holes or slots, and any additional graphics such a silkscreen layer.

Take a common ¼-watt axial-lead resistor as an example. This footprint will have two pins, each associated with a pad, corresponding to the resistor's two leads. This pad will appear on both the top and bottom layers of the board, but will also have a smaller pad associated with inner layers, should there be any. The hole's size will be based on the nominal lead diameter, plus some allowance (typically 0.006 inch). The pad size will be big enough to provide a reasonable annular ring, but is usually much larger so as to allow good quality soldering. The pins will be labeled in a way that corresponds to the pin numbering on the schematic symbol (even though for this component, there is no polarity). A silkscreen layer will be defined, usually a box within which the value or designator will appear. The silkscreen layer is particularly useful for indicating orientation of parts with polarity.

More complicated parts may require additional holes which will not be associated with a schematic pin (mounting hole, for example). These are usually added to the part differently than adding a hole with a pad — in this case, the hole is desired without any annular ring or plating. The silkscreen layer may be used to outline the part above and beyond what is obvious from the pads, for example, a TO-220 power transistor laying on its back, or the plastic packaging around the opto-interrupter in Fig 6.42.

As with schematic entry, it is not uncommon to have to modify or create a new PCB footprint. Good technical drawings are often available for electronic parts; when possible the user should verify these dimensions against a real part with an inexpensive dial caliper. It is also useful to print out the finished circuit board artwork at actual size and do a quick check against any new or unusual parts.

TRACES

Traces are the other main element of PCB construction — the copper pathways that connect components electrically. PCB traces are merely planar, flat wires — they have no magical properties when compared to an equivalent thickness of copper wire. At VHF/UHF/microwave frequencies and for high-speed digital signals, PCB traces act as transmission lines and these properties need to be accounted for, and can be used to advantage, in the design.

There are few constraints on traces apart from those such as minimum width and clearance imposed by the board house. They are created by chemical etching and can take arbitrary shapes. In fact, text and symbols may be created on copper layers which may be handy if a silkscreen is not included. Traces may be of any length, vary in width, incorporate turns or curves, and so on. However, most traces will be a uniform width their entire length (a width they will likely share with other traces carrying similar signals), make neat 45° or 90° corners, and on two-layer boards have a general preference for either horizontal travel on the component side or vertical travel on the solder side.

The same considerations when building a circuit in other methods applies to PCB design, including current capacity (width of trace, thickness of copper), voltage (clearance to other signals), noise (shielding, guarding, proximity to other signals), impedance of ground and power supplies, and so on.

6.4.6 PCB Layout

With a schematic and netlist ready and all of the PCB characteristics defined, the actual layout of the PCB can begin.

BOARD SIZE AND LAYERS

The first step in PCB layout is to create the board outline to contain not only the circuit itself and any additional features such as mounting holes. For prototype or one-off designs, the board is often best made a bit larger to allow more space between components for ease in testing and debugging. (Some low-cost or freeware commercial PCB software imposes limits on board size and number of components.) The board outline may be provided in a default size that the user can modify, or the user may need to enter the outline from scratch.

As discussed above, rectangular board shapes are generally acceptable, but many board houses can accommodate more complex outlines, including curves. These outlines will be routed with reasonable accuracy and may save an assembly step if the PCB needs a cutout or odd shape to fit in a specific location.

While the software may not require deciding at the start how many layers the PCB will use, this is a decision the user should make as early as possible, since the jump from two to four or more will have a big impact on routing the traces as well as cost!

For your initial design, start with a two-layer board for a simple circuit that you have already built and tested. This will reduce the number of decisions you have to make and remove some of the unknowns from the design process.

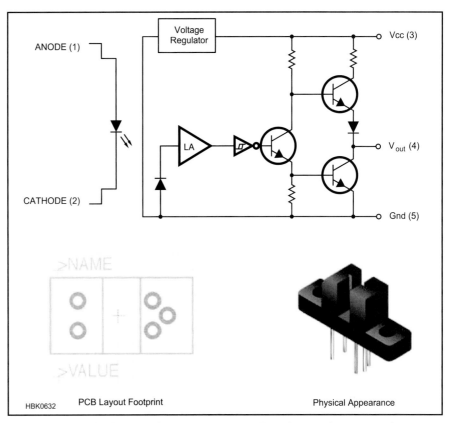

Fig 6.42 — The PCB footprint for a component, such as the opto-interrupter shown here, combines electrical connectivity as defined in the part's schematic and the part's physical attributes.

COMPONENT PLACEMENT

Good component placement is more than half the battle of PCB layout. Poor placement will require complicated routing of traces and make assembly difficult, while good placement can lead to clean, easy-to-assemble designs.

The first elements placed should be mounting holes or other fixed-location features. These are often placed using a special option selected from a palette of tools in the software rather than as parts from a library. Holes sufficient for a #4 or #6 screw are usually fine; be sure to leave room around them for the heads of the screws and nut driver or standoff below. These will be non-plated-through holes with no pad (though the board house may plate all the holes in a board, regardless).

Depending on the software and whether schematic capture was performed, the board outline may already contain the footprints of all the circuit components (sometimes stacked in a heap in one corner of the board) and the netlist will already be loaded. In this case, components may be placed by clicking and dragging the components to the desired location on the board. Most PCB programs have a "rat's-nest" option that draws a straight line for each netlist connection of a component, and this is a great aid in placement as the connections between components are apparent as the components are moved around. (See **Fig 6.43**) However, connections are shown to the nearest pin sharing that electrical connection; thus, components such as decoupling capacitors (which are often meant to be near a specific component) will show rats-nest connections to the nearest power and ground pins and not the pins the designer may have intended. These will have to be manually edited.

The PCB layout software may offer *autoplacement* in which the components are initially arranged automatically. The beginner should certainly feel free to experiment and see how well this tool performs, but it is likely not useful for the majority of designs.

PCBs need not be arranged to precisely mimic the schematic, but it is appropriate to place components in a logical flow when possible so as to minimize the length of traces in the signal path. Sensitive components may need to be isolated or shielded from other components, and grounding and decoupling attended to, just as one would do with a point-to-point soldered version.

If the PCB is being designed "on-the-fly" or using an imported netlist, components may need to be selected and placed on the board manually using the libraries of parts in the PCB software. Not all design software makes this task simple or fast — in particular, the description of component footprints may be confusing. The use of highly-condensed industry standard or non-uniform naming conventions often means the user needs to browse through the component library to see the different types of components. Resistors, diodes and capacitors seem particularly prone to a propagation of perplexing options. One solution is to open an example PCB layout and see what library element that designer used for resistors, LEDs and so on. Here, the PCB layout software directed at hobbyists may be superior in that there are fewer options than in professional programs.

During placement the user will find that different orientations of components simplify routing (for example, minimizing the number of traces that have to cross over each other or reducing the trace lengths). Components are generally rotated in increments of 90°, although free rotation may be an option. The user is strongly urged to maintain the same orientation on like devices as much as possible. Mixing the position of pin 1 on IC packages, or placing capacitors, diodes, and LEDs with random orientation invites time-consuming problems during assembly and testing that can be minimized by consistent, logical layout.

Placement and orientation of components can also affect how easily the final PCB can be assembled. Allow plenty of space for sockets, for example, and for ICs to be inserted and removed. Components with a mechanical interface such as potentiometers and switches should be positioned to allow access for adjustment. Any components such as connectors, switches, or indicators (eg, LEDs) should be positioned carefully, especially if they are to protrude through a panel. Often this will involve having the component overhang the edge of the PCB. (Beware of the required clearance between copper traces and pads to the edge of the PCB.)

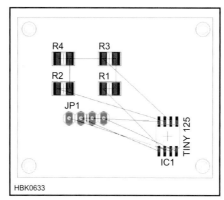

Fig 6.43 — The rat's nest view during PCB layout shows the direct connections between component pins. This helps the designer with component placement and orientation for the most convenient routing.

Components should include a silkscreen outline that shows the size of the whole component — for example, a transistor in a TO-220 package mounted flat against the PCB should have an outline that shows the mounting hole and the extent of the mounting tab. The user should also consider the clearance required by any additional hardware for mounting a component, such as nuts and bolts or heatsinks — including clearance for nut-drivers or other assembly tools.

Take care to minimize the mechanical stress on the PCB, since this can result in cracked traces, separated pads, or other problems. Utilize mounting holes or tabs when possible for components such as connectors, switches, pots. Use two-layer boards with plated-through holes even if the design can be single-layer. Component leads soldered to plated-through holes produce much stronger mechanical connections than single-layer boards in which the soldered pad is held only by the bond between copper and laminate and is easily lifted if too much heat or stress is imposed.

When prototyping a new design, add a few unconnected pads on the circuit board for extra components (eg, a 16-pin DIP, 0.4-inch spaced pads for resistors and other discrete components). Include test points and ground connections. These can be simply pads to which cut-off leads can be soldered to provide convenient test points for ground clips or to monitor signals.

Wires or cables can be directly soldered to the PCB, but this is inconvenient when swapping out boards, and is not very robust. Connectors are much preferred when possible and often provide strain relief for the wire or cable. However, if a wire is directly soldered to the PCB, the user should consider adding an unplated hole nearby just large enough to pass the wire including insulation. The wire can then be passed from the solder side through the unplated hole, then soldered into the regular plated-through hole. This provides some measure of strain relief which can be augmented with a dollop of glue if desired.

ROUTING TRACES

After placing components, mounting holes and other fixed-location features that limit component or trace placement, traces can be *routed*. That is, to complete all the connections between pins without producing short circuits.

Most PCB design programs allow components and traces to be placed on a regular grid, similarly to drawing programs. There may be two grids — a visible coarse-pitch grid, and a "snap" fine-pitch grid, to which components and other objects will be aligned when placed. It is good practice to use a 0.1- or 0.050-inch grid for component placement and to route traces on a 0.025-inch grid. While the "snap to grid" feature can usually be turned off to allow fine adjustment of placement, a board routed on a grid is likely to look cleaner and be easier to route.

The trace starts at a component pin and wends its way to any other pins to which it should be connected. Traces should start and end at the center of pads, not at the edge of a pad, so that the connection is properly recorded in the program's database. If a netlist has been loaded, most PCB software will display a rats-nest line showing a direct connection between pads. Once the route is completed, the rats-nest line for that connection disappears. The rat's-nest line is rarely the desired path for the trace and often not the correct destination. For example, when routing power traces, the user should use good design sense rather than blindly constructing a Byzantine route linking pins together in random order. For this reason, routing the power and ground early is a good practice.

High-speed, high-frequency, and low-noise circuits will require additional care in routing. In general, traces connecting digital circuits such as microprocessors and memories should not cross or be in close proximity to traces carrying analog or RF signals. Please refer to the **RF Techniques** and **Construction Techniques** chapters of this *Handbook*, and the references listed at the end of this section.

Manual routing is a core skill of PCB design, whether or not auto-routing is used. The process is generally made as simple as possible in the software, since routing will take up most of the PCB design time. A trace will be routed on the copper layer currently selected. For a single-sided board, there is only one layer for routing; for a two-layer board the component side and solder side can have traces; and for multi-layer boards additional inner layers can have traces. Often a single keystroke can change the active copper layer (sometimes automatically inserting a via if a route is in progress). The trace is drawn in straight segments and ends at the destination pin. When routing, 90° corners are normally avoided — a pair of 45° angles is the norm. **Fig 6.44** shows some sample traces.

It is good practice on a double-sided board to have one side of the board laid out with mostly horizontal traces, and the other side laid out with mostly vertical traces. A trace that needs to travel primarily vertically can do so on the side with vertical traces and use a via to move to the other side to complete the horizontal part of the route.

It is easiest during testing and debugging to route most traces on the bottom (solder) side of the board — traces on the component side often run under ICs or other components, making them hard to access or follow. It is often much clearer to connect adjacent IC or connector pins by routing a trace that leaves one pad, moves away from the IC or connector, then heads back in to the adjacent pad to connect. This makes it clear the connection on the assembled board is not a solder bridge, which a direct connection between the two pads would resemble.

It may be the case that no amount of vias or wending paths can complete a route. The one remaining tool for the PCB designer is a jumper — a wire added as a component during assembly just for the purpose of making a connection between two points on the board. Jumpers are most often required for single-sided boards; when the "jump" is rather small, uninsulated wire can be used. Jumpers are usually straight lines, and can be horizontal or vertical. Professional production PCBs use machine-insertable zero-ohm resistors as jumpers. Jumpers on double-sided boards are usually not viewed very favorably, but this is an aesthetic and efficiency issue, not a functional one.

Multi-layer boards clearly offer additional routing options, but again having some dominant routing direction (vertical or horizontal) on each layer is recommended, since mixing directions tends to cause routing problems. However, it is not uncommon to devote one or two inner layers to power and ground, rather than merely be additional layers for routing signals. This allows power and ground to be routed with minimal resistance and exposes the traces carrying interesting signals on the

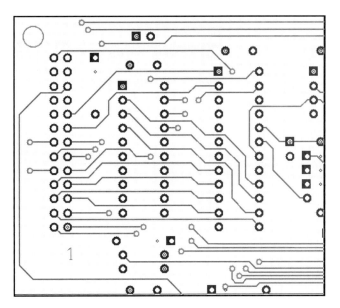

Fig 6.44 — This example shows traces on the side of the PCB for horizontal routing. Traces are routed between pins of ICs. The smaller pads are for vias to a different layer of the PCB.

component and solder sides where they are available to be probed or modified. It is very difficult to modify traces on inner layers, needless to say!

Before routing too many traces, it is helpful to run the *Design Rule Check (DRC)* on the board. (See the section on Design Rule Checking below.) Applied early and often, DRC can identify areas of concern when it is easiest to correct. For example, a given trace width may provide insufficient clearance when passing between two IC pads.

Some PCB design packages offer *auto-router* capability in which the software uses the component and connectivity data of the netlist and attempts to route the traces automatically. There are some circumstances when they save time, but view these tools with some caution. Auto-routers are good at solving the routing puzzle for a given board, but merely connecting all the pins correctly does not produce a good PCB design. Traces carrying critical signals may take "noisy" routes; components that should have short, low-resistance connections to each other may have lengthy traces instead, and so on. More sophisticated auto-routers can be provided with extensive lists of "hints" to minimize these problems. For the beginner, the time spent conveying this design information to the auto-router is likely better spent manually routing the traces.

If an auto-router is used, at a minimum, critical connections should be first routed manually. These include sensitive signals, connections whose length should be minimized, and often power and ground (for both RFI and trace width reasons). Better still is to develop a sense of what a good layout looks like (which will come with practice and analyzing well-designed boards), and learn at what stage the auto-router can be "turned loose" to finish the routing puzzle.

TRACE WIDTH AND SPACING

All traces will have some width — the width may be the default width, the last width selected, or a width provided from data in the netlist. It may be tempting to route all but the power traces using the smallest trace width available from the board house (0.008 inch or smaller), since this allows the highest density of traces and eases routing. A better design practice is to use wider traces to avoid hard-to-detect trace cracking and improve board reliability. The more common traces 0.012-inch wide can be run in parallel on a 0.025-inch grid and can pass between many pads on 0.1-inch centers. Even wider traces will make the board easier to produce "in house," though the exact process used (CNC routing, chemical etching, etc) will limit the resolution. Note

---

PCB Design and EMC

While amateur projects are rarely subject to electromagnetic compatibility (EMC) standards, using good engineering practices when designing the board still reduces unwanted RF emissions and susceptibility to RF interference. For example, proper layout of a microprocessor circuit's power and ground traces can reduce RF emissions substantially. Proper application of ground planes, bypass capacitors and especially shield connections can have a dramatic effect on RFI performance. (See the **RFI** chapter for more on RFI.) A good reference on RFI and PCB design is *Electromagnetic Compatibility Engineering*, by Henry Ott, WA2IRQ.

that it is possible to "neck down" traces where they pass between IC or connector pads — that is, the regular, thick trace is run up close to the narrow gap between the pads, passes between the pads with a narrow width, then expands back to the original width. There is little reason to use traces wider than 0.030 inch or so for most signals (see **Table 6.4**) but power and ground trace widths should be appropriate for the current.

All traces have resistance, and this resistance is a function of the cross section of the trace (width times thickness) and the length. This resistance will convert electrical power to heat. If the heat exceeds a relatively high threshold, the trace becomes a fairly expensive and difficult-to-replace fuse. The trace width should be selected such that for the worst-case expected current, heat rise is limited to some threshold, often 10 °C. In practice, power traces (especially grounds) are often made as wide as practical to reduce resistance, and they greatly exceed the width required by heat rise limits alone.

Table 6.4 summarizes maximum currents for external (component and solder side) and internal traces for some common trace widths. Internal traces (on inner layers of multi-layer boards) can carry only about half the current of external traces for the same width since the internal layers do not dissipate heat to the ambient air like external traces can. (Note that trace widths are also sometimes expressed in "mils." 1 mil = 0.001 inch; it is shorthand for "milli-inch", not millimeter!)

There is no upper bound on the effective trace width. It is common to have large areas of the board left as solid copper. These *copper fill* areas can serve as grounds, heat sinks, or may just simplify board production (especially home-made boards). It is not a good idea to place a component hole in the middle of a copper fill — the copper is a very efficient heat sink when soldering. Instead, a "wagon wheel" pattern known as a thermal relief is placed (sometimes automatically) around the solder pad, providing good electrical connectivity but reducing the heat sinking. Often, copper fill areas can be specified using a polygon and the fill will automatically flow around pads and traces in that area, but can lead to isolated pads of copper.

In practice, most boards will have only two or perhaps three different trace widths; narrow widths for signals, and a thicker width for power (usually with a healthy margin).

One final note on trace width — vias are typically one size (ie, small), but multiple vias can be used to create low-resistance connections between layers. Spacing the vias so their pads do not touch works well; the pads are then shorted on both top and bottom layers.

Voltage also figures into the routing equation, but instead of trace width, higher voltages should be met with an increased clearance between the trace and other copper. The IPC-2221 standard calls for a clearance of 0.024 inch for traces carrying 31-150 V (peak) and 0.050 inch for traces carrying 151-300 V (peak); these are external traces with no coating. (With the appropriate polymer solder mask coating, the clearances are 0.006 inch for 31-100 V and 0.016 inch for 101-300 V. Internal traces also have reduced clearance requirements.) Fully addressing the safety (and regulatory) issues around high voltage wiring is outside the scope of this brief review, however, and the reader is urged to consult UL or IPC standards.

Table 6.4
Maximum Current for 10 °C Rise, 1 oz/ft^2 Copper
Based on IPC-2221 standards (not an official IPC table)

Trace Width (inches)	Max. Current (External Trace) (A)	Max. Current (Internal Trace) (A)	Resistance (ohms/inch)
0.004	0.46	0.23	0.13
0.008	0.75	0.38	0.063
0.012	1.0	0.51	0.042
0.020	1.5	0.73	0.025
0.040	2.4	1.2	0.013
0.050	2.8	1.4	0.010
0.100	4.7	2.4	0.0051
0.200	7.8	3.9	0.0025
0.400	13	6.4	0.0013

IPC-2221 Generic Standard on Printed Circuit Design, Institute for Interconnecting and Packaging Electronic Circuits, **www.ipc.org**

SILKSCREEN AND SOLDER MASK

The silkscreen (or "silk") layer contains the text and graphics that will be silkscreened on the top of the board, shown in white in **Fig 6.45**. Components will generally have elements on the silkscreen layer that will automatically appear, such as designators and values, but other elements must be created and placed manually. Common silkscreen elements include: Circuit name, date, version, designer (and call sign), company name, power requirements (voltage, current, and fusing), labels for connections (eg, "Mic input"), warnings and cautions, labels for adjustments and switches. A solid white rectangle on the silkscreen layer can provide a good space to write a serial number or test information.

The board house will specify the minimum width for silkscreen lines, including the width of text. Text and graphics can be placed anywhere on the solder mask, but not on solder pads and holes.

Many quick-turn board houses omit the silkscreen for prototype boards. As noted earlier, many of the text elements above can be placed on the external copper layers. Component outlines are not possible since the resulting copper would short out traces, but component polarization can be noted with symbols such as a hand-made "+" made from two short traces, or a "1" from a single short trace. (Note that some component footprints follow a practice of marking the pad for pin 1 with a square pad while others are round or oval.)

The solder mask is a polymer coating that is screened onto the board before the silkscreen graphics. As shown in Fig 6.45, it covers the entire surface of the board except for pads and vias. Solder masking prevents solder bridges between pads and from pads to traces during assembly and is particularly important for production processes that use wave soldering or reflow soldering. There is one solder mask layer for the top layer and another for the bottom layer. Internal layers do not need a solder mask. Solder masking may be omitted for a prototype board, but care must be taken to keep solder from creating unwanted bridges or short circuits.

During the PCB layout process, solder mask layers are generally not shown because they do not affect connectivity. **Fig 6.46** shows a typical PCB as it appears when the PCB layout process is complete.

DESIGN RULE CHECK

If a netlist has been provided from the schematic capture program, a *design rule check* (*DRC*) can be made of the board's layout. The PCB software will apply a list of rules to the PCB, verifying that all the connections in the netlist are made, that there is sufficient clearance between all the traces, and so on. These rules can be modified based on the specific board house requirements. As stated above, it is useful to run the DRC even before all the traces have been routed — this can identify clearance or other issues that might require substantial re-routing or a different approach.

If the user has waited until all the routing is done before running the DRC, the list of violations can be daunting. However, it is often the case that many if not all of the violations represent issues that may prevent the board from operating as wished. Whenever possible, all DRC violations should be rectified before fabrication.

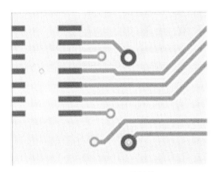
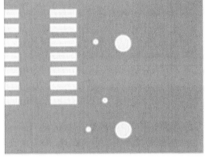
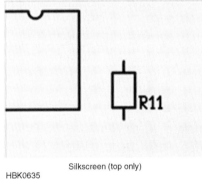

Fig 6.45 — The relationship between the layout's top copper layer with traces and pads, the solder mask that covers the copper (a separate solder mask is required for the top and bottom layers of the PCB) and the silkscreen information that shows component outlines and designators.

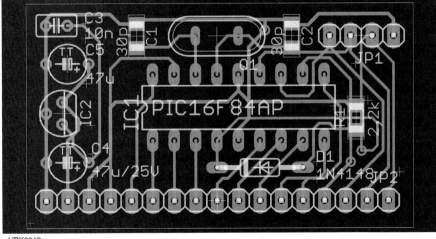

Fig 6.46 — A completed microprocessor board as it is seen in a typical PCB layout editor (*Eagle*). Solder mask layers are omitted for visibility. Traces that appear to cross each other are on different sides of the board and are in different colors in the layout software. The silkscreen layer is shown in white.

6.4.7 Preparation for Fabrication

LAYOUT REVIEW

Once the board has passed DRC, the electrical connectivity and basic requirements for manufacturability have likely been satisfied. However, the design may benefit from an additional review pass. Turn off all the layers but one copper layer and examine the traces — often simplifications in routing will be apparent without the distractions of the other copper layers. For example, a trace can be moved to avoid going between two closely-spaced pins. Densely spaced traces could be spaced farther apart. There may be opportunities to reduce vias by routing traces primarily on one layer even if that now means both vertical and horizontal travel. Repeat the exercise for all the copper layers.

Review the mechanical aspects of the board as well, including the proximity of traces to hardware. If your prototype PCB does not have a solder mask, traces that run underneath components such as crystals in a conductive case or too close to mounting hardware can form a short circuit. An insulator must be provided or the trace can be re-routed.

GENERATING OUTPUT FILES

Once the PCB design is complete, the complete set of design description files can be generated for producing the PCB. These are:

Copper layers — One file per copper layer. These are known as *Gerber files* and were text files of commands originally intended to drive a *photoplotter*. Gerber was the primary manufacturer of photoplotters, machines that moved a light source of variable width (apertures) from one location to another to draw patterns on photographic film. While photoplotters have been replaced by digital technology, the format used by Gerber has been standardized as RS-274X and is universally used except by PCB software tied to a specific manufacturer. RS-274X is related to RS-274D ("G-Code") used by machinists to program CNC machinery but is an *additive* description (essentially saying "put copper here"), rather than describing the movements of a tool to remove material. A program is thus required to translate between Gerber and G-Code if a CNC machine is used to make a PCB by mechanically removing copper.

Drill file — The file containing the coordinates and drill sizes for all the holes, plated or not. Also called the *NC* or *Excellon* file, some board houses may require a specific format for the coordinates, but these are usually available to be set as options in the PCB program. There is only one drill file for a PCB, since the holes are drilled from one side. (Exotic options such as buried vias will require more information.) Like RS-274X apertures, the drill file will generally contain a drill table.

Computer-Aided Circuit Design 6.39

Silkscreen — Also in RS-274 format. Some board houses can provide silkscreen on both sides of the board, which will require two files.

Solder mask — The solder mask file is used by the board house to create the solder mask. One file per side is required.

A Gerber preview program such as *Gerbv* (**gerbv.gpleda.org**, open source, *Linux*, Mac OS X) or *GC-Prevue* (**www.graphicode.com**, *Windows*) can be used to review the trace layout Gerber files. This is a good test — the board house will make the boards from the Gerber files, not the PCB design file. Gerber previewers can import the copper layers, silkscreen and drill files to verify they correspond and make sense.

Any of the layers can usually be printed out within the PCB program (and/or Gerber preview program) for reference and further inspection.

In addition to files for PCB production, PCB layout programs can also generate assembly diagrams, and in some cases can provide 3D views of what the assembled board will look like. These can be useful for documentation as well as verification of mechanical issues such as height clearance.

Sending the files to the PCB manufacturer or board house and ordering PCBs is explained on the manufacturing Web site or a customer service representative can walk you through the process. Some firms accept sets of files on CD-ROM and may also offer a design review service for first-time customers or on a fee basis.

6.5 References and Bibliography

J. Wayde Allen, "Gain Characterization of the RF Measurement Path," *NTIA Report TR-04-410* (Washington: US Department of Commerce: 2004). Available as **www.its.bldrdoc.gov/pub/ntia-rpt/04-410/04-410.pdf**.

W. Hayward, W7ZOI, and J. Lawson, K5IRK, "A Progressive Communications Receiver," *QST*, Nov 1981, pp 11-21. Also see *QST* Feedback: Jan 1982, p 47; Apr 1982, p 54; Oct 1982, p 41.

W. Hayward, W7ZOI, "Designing and Building Simple Crystal Filters," *QST*, Jul 1987, pp 24-29.

W. Hayward, W7ZOI, "The Double-Tuned Circuit: An Experimenter's Tutorial," *QST*, Dec 1991, pp 29-34. Included in J. Kleinman and Z. Lau, compilers, *QRP Power* (ARRL: Newington, 1991), pp 5-29 to 5-34.

W. Hayward, W7ZOI; R. Campbell, KK7B; and B. Larkin, W7PUA, *Experimental Methods in RF Design*, 2nd ed (ARRL: Newington, 2009).

P. Horowitz and W. Hill, *The Art of Electronics*, 2nd ed (Cambridge: Cambridge University Press, 1989).

NXP Semiconductor: **www.nxp.com/models/bi_models/mextram/**

Z. Lau, W1VT, "Calculating the Power Limit of Circuits with Toroids," RF, *QEX*, Mar 1995, p 24.

D. Newkirk, WJ1Z, "Modeling a Direct-Conversion Receiver's Audio Response and Gain with ARRL Radio Designer," Exploring RF, *QST*, Mar 1995 pp 76-82.

D. Newkirk, WJ1Z, "Optimizing Circuit Performance with ARRL Radio Designer," Exploring RF, *QST*, Mar 1995, pp 76-78.

D. Newkirk, WJ1Z, "Math in a Box, Transistor Modeling, and a New Meeting Place," Exploring RF, *QST*, May 1995, pp 90-92.

D. Newkirk, WJ1Z, "Transistor Modeling with ARRL Radio Designer, Part 2: Optimization Produces Realistic Transistor Simulations," Exploring RF, *QST*, Jul 1995, pp 79-81.

D. Newkirk, WJ1Z, "Transistor Modeling, Part 3: Constraints in Optimization, Two-Port Data with Noise Parameters, and Introducing ARRL Radio Designer, Exploring RF," *QST*, Sep 1995, pp 99-101.

D. Newkirk, WJ1Z, "ARRL Radio Designer as a Learning (and Just Plain Snooping) Tool," Exploring RF, *QST*, Nov 1995, pp 89-91.

D. Newkirk, WJ1Z, "An ARRL Radio Designer Voltage Probe Mystery: The 3-dB Pad that Loses 9 dB," Exploring RF, *QST*, Jan 1996, pp 79-80.

D. Newkirk, WJ1Z, "Gyrating with ARRL Radio Designer," Exploring RF, *QST*, Mar 1996, pp 67-68.

D. Newkirk, WJ1Z, "Frequently Asked Questions About ARRL Radio Designer Reports and Report Editor," Exploring RF, *QST*, May 1996, pp 78-79.

D. Newkirk, WJ1Z, "ARRL Radio Designer versus Oscillators, Part 1," Exploring RF, *QST*, Jul 1996, pp 68-69.

D. Newkirk, WJ1Z, "ARRL Radio Designer Versus Oscillators, Part 2," Exploring RF, *QST*, Sep 1996, pp 79-80.

D. Newkirk, W9VES, "Simulating Circuits and Systems with *Serenade* SV," *QST*, Jan 2001, pp 37-43.

"The Spice Page," (**bwrc.eecs.berkeley.edu/Classes/icbook/SPICE/**). The home page for *SPICE*.

Paul W. Tuinenga, "*Spice: A Guide to Circuit Simulation and Analysis Using Pspice*, 2nd ed (New York: Prentice-Hall, 1992).

Andrei Vladimirescu, *The SPICE Book* (New York: John Wiley and Sons, 1994).

PCB CAD REFERENCES

Pease, Robert A, *Troubleshooting Analog Circuits*, Butterworth-Heinemann, 1991

Analog Devices, *High Speed Design Techniques*, Analog Devices, 1996

Johnson, Howard, and Graham, Martin, *High Speed Digital Design: A Handbook of Black Magic*, Prentice Hall, 1993

Ott, Henry, *Electromagnetic Compatibility Engineering*, Wiley Press, 2009

Contents

7.1 The Need for Power Processing
7.2 AC-AC Power Conversion
    7.2.1 Fuses and Circuit Breakers
7.3 Power Transformers
    7.3.1 Volt-Ampere Rating
    7.3.2 Source Voltage and Frequency
    7.3.3 How to Evaluate an Unmarked Power Transformer
7.4 AC-DC Power Conversion
    7.4.1 Half-Wave Rectifier
    7.4.2 Full-Wave Center-Tap Rectifier
    7.4.3 Full-Wave Bridge Rectifier
    7.4.4 Comparison of Rectifier Circuits
7.5 Voltage Multipliers
    7.5.1 Half-Wave Voltage Doubler
    7.5.2 Full-Wave Voltage Doubler
    7.5.3 Voltage Tripler and Quadrupler
7.6 Current Multipliers
7.7 Rectifier Types
    7.7.1 Vacuum Tube
    7.7.2 Mercury Vapor
    7.7.3 Early Solid-State Rectifiers
    7.7.4 Semiconductor Diodes
    7.7.5 Rectifier Strings or Stacks
    7.7.6 Rectifier Ratings versus Operating Stress
    7.7.7 Rectifier Protection
7.8 Power Filtering
    7.8.1 Load Resistance
    7.8.2 Voltage Regulation
    7.8.3 Bleeder Resistors
    7.8.4 Ripple Frequency and Voltage
    7.8.5 Capacitor-Input Filters
    7.8.6 Choke-Input Filters
7.9 Power Supply Regulation
    7.9.1 Zener Diodes
    7.9.2 Linear Regulators
    7.9.3 Linear Regulator Pass Transistors
    7.9.4 Three-Terminal Voltage Regulators
7.10 "Crowbar" Protective Circuits
7.11 DC-DC Switchmode Power Conversion
    7.11.1 The Buck Converter
    7.11.2 The Boost Converter
    7.11.3 Buck-Boost and Flyback Converters
    7.11.4 The Forward Converter
    7.11.5 Parallel, Half-, and Full-Bridge Converters
    7.11.6 Building Switching Power Supplies
    7.11.7 Switchmode Control Loop Issues
7.12 High-Voltage Techniques
    7.12.1 High-Voltage Capacitors
    7.12.2 High-Voltage Bleeder Resistors
    7.12.3 High-Voltage Metering Techniques
    7.12.4 High-Voltage Transformers and Inductors
    7.12.5 Construction Techniques for High-Voltage Supplies
    7.12.6 High-Voltage Safety Considerations
7.13 Batteries
    7.13.1 Battery Capacity
    7.13.2 Chemical Hazards of Batteries
    7.13.3 Battery Types
    7.13.4 Battery Discharge Planning
    7.13.5 Battery Internal Resistance
    7.13.6 Battery Charging
    7.13.7 DC-AC Inverters
    7.13.8 Selecting a Battery for Mobile Operation
7.14 Glossary of Power Supply Terms
7.15 References and Bibliography
7.16 Power-Supply Projects
    7.16.1 Four-Output Switching Bench Supply
    7.16.2 12-V, 15-A Linear Power Supply
    7.16.3 13.8-V, 5-A Linear Power Supply
    7.16.4 High-Voltage Power Supply
    7.16.5 Micro M+ PV Charge Controller
    7.16.6 Automatic Sealed Lead-Acid Battery Charger
    7.16.7 Overvoltage Protection for AC Generators
    7.16.8 Overvoltage Crowbar Circuit

Chapter 7

Power Supplies

Our transceivers, amplifiers, accessories, computers and test equipment all require power to operate. This chapter illustrates the various techniques, components and systems used to provide power at the voltage and current levels our equipment needs. Topics range from basic transformers, rectifiers and filters to linear voltage regulation, switchmode power conversion, high voltage techniques and batteries. Expanded and updated by Rudy Severns, N6LF, with support from Chuck Mullett, KR6R, the material presented here builds on the earlier work of Ken Stuart, W3VVN, and other contributors. Alan Applegate, KØBG contributed the section on selecting batteries for mobile use.

7.1 The Need for Power Processing

Electronic equipment requires a source of electrical power. Sometimes the source can be as simple as a battery, but in many cases a battery may not be able to supply sufficient power and/or the voltages required with adequate regulation. For example, the voltage of a vehicle battery may vary from 14 V when fully charged down to 10 V or less when discharged. The lower end of the voltage range can compromise the performance of a transceiver, so it may be desirable to insert a device between the battery and the transceiver to maintain the transceiver input voltage at about 13 V as the battery voltage declines during use.

Another example is operation from commercial utility power, which may vary from 90 to 270 V ac depending on where you are in the world. In most cases the signal generating and processing equipment we use requires well-regulated dc voltages, sometimes at low voltages (3-24 V) and at other times high voltages (100-3000 V).

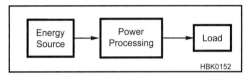

Fig 7.1 — Basic concept of power processing.

Fig 7.1 illustrates the concept of a power processing unit inserted between the energy

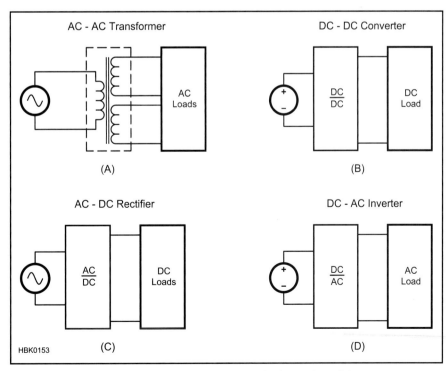

Fig 7.2 — Four power processing schemes: ac-ac, dc-dc, ac-dc and dc-ac.

source and the electronic equipment or load. The *power processor* is often referred to as the *power supply*. That's a bit misleading in that the energy "supply" actually comes from some external source (battery, utility power and so forth), which is then converted to useful forms by the power processor. Be that as it may, in practice the terms "power supply" and "power processor" are used interchangeably.

The real world is even more arbitrary. Power processors are frequently referred to as *power converters* or simply as *converters*, and we will see other terms used later in this chapter. It is usually obvious from the context of the discussion what is meant and the glossary at the end of this chapter gives some additional information.

Power conversion schemes can take the form of: ac-to-ac (usually written ac-ac), ac-dc, dc-ac and dc-dc. Examples of these schemes are given in **Fig 7.2**. Specific names may be given to each scheme: ac-dc => rectifier, dc-dc => converter and dc-ac => inverter. These are the generally recognized terms but you will see exceptions.

In the following discussion it will be assumed that the reader is familiar with the component characteristics presented in the **Electrical Fundamentals** chapter.

7.2 AC-AC Power Conversion

In most US residences, three wires are brought in from the outside electrical-service mains to the house distribution panel. In this three-wire system, one wire is neutral and should be at earth ground potential. (See the **Safety** chapter for information on electrical safety.) The neutral connection to a ground rod or electrode is usually made at the distribution panel. The voltage between the other two wires is 60-Hz ac with a potential difference of approximately 240 V RMS. Half of this voltage appears between each of these wires and the neutral, as indicated in **Fig 7.3A**. In systems of this type, the 120-V household loads are divided at the breaker panel as evenly as possible between the two sides of the power mains. Heavy appliances such as electric stoves, water heaters, central air conditioners and so forth, are designed for 240-V operation and are connected across the two ungrounded wires.

Both hot wires for 240 V circuits and the single hot wire for 120 V circuits should be protected by either a fuse or breaker. A fuse or breaker or any kind of switch should *never* be used in the neutral wire. Opening the neutral wire does not disconnect the equipment from an active or "hot" line, possibly creating a potential shock hazard between that line and earth ground.

Another word of caution should be given at this point. Since one side of the ac line is grounded (through the green or bare wire — the standard household wiring color code) to earth, all communications equipment should be reliably connected to the ac-line ground through a heavy ground braid or bus wire of #14 or heavier-gauge wire. This wire must be a separate conductor. You must not use the power-wiring neutral conductor for this safety ground. (A properly-wired 120-V outlet with a ground terminal uses one wire for the ac hot connection, one wire for the ac neutral connection and a third wire for the safety ground connection.) This provides a measure of safety for the operator in the event of accidental short or leakage of one side of the ac line to the chassis.

Remember that the antenna system is frequently bypassed to the chassis via an RF choke or tuned circuit, which could make the antenna electrically "live" with respect to earth ground and create a potentially lethal shock hazard. A *Ground Fault Circuit Interrupter* (GFCI or GFI) is also desirable for safety reasons, and should be a part of the shack's electrical power wiring.

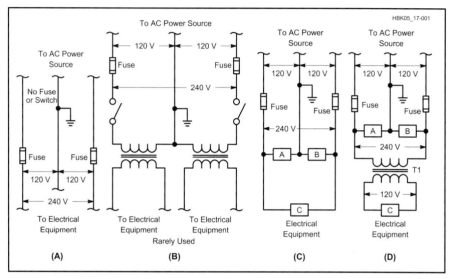

Fig 7.3 — Three-wire power-line circuits. At A, normal three-wire-line termination. No fuse should be used in the grounded (neutral) line. The ground symbol is the power company's ground, not yours! Do not connect anything other than power return wiring, including the equipment chassis, to the power neutral wire. At B, the "hot" lines each have a switch, but a switch in the neutral line would not remove voltage from either side of the line and should never be used. At C, connections for both 120 and 240-V transformers. At D, operating a 120-V plate transformer from the 240-V line to avoid light blinking. T1 is a 2:1 step-down transformer.

7.2.1 Fuses and Circuit Breakers

All transformer primary circuits should be fused properly and multiple secondary outputs should also be individually fused. To determine the approximate current rating of the fuse or circuit breaker on the line side of a power supply it is necessary to determine the total load power. This can be done by multiplying each current (in amperes) being drawn by the load or appliance, by the voltage at which the current is being drawn. In the case of linear regulated power supplies, this voltage has to be the voltage appearing at the output of the rectifiers before being applied to the regulator stage. Include the current drawn by bleeder resistors and voltage dividers. Also include filament power if the transformer is supplying vacuum tube filaments. The National Electrical Code (NEC) also specifies maximum fuse ratings based on the wire sizes used in the transformer and connections.

After multiplying the various voltages and currents, add the individual products. This is the total power drawn from the line by the supply. Then divide this power by the line voltage and add 10 to 30% to account for the inefficiency of the power supply itself. Use a fuse or circuit breaker with the nearest larger current rating. Remember that the charging of filter capacitors can create large surges of current when the supply is turned on. If fuse blowing or breaker tripping at turn on is a problem, use slow-blow fuses, which allow for high initial surge currents.

For low-power semiconductor circuits, use fast-blow fuses. As the name implies, such fuses open very quickly once the current exceeds the fuse rating by more than 10%.

7.3 Power Transformers

Numerous factors are considered to match a transformer to its intended use. Some of these parameters are:

1. Output voltage and current (volt-ampere rating).
2. Power source voltage and frequency.
3. Ambient temperature.
4. Duty cycle and temperature rise of the transformer at rated load.
5. Mechanical considerations like weight, shape and mounting.

7.3.1 Volt-Ampere Rating

In alternating-current equipment, the term *volt-ampere* (*VA*) is often used rather than the term watt. This is because ac components must handle reactive power as well as real power. If this is confusing, consider a capacitor connected directly across the secondary of a transformer. The capacitor appears as a reactance that permits current to flow, just as if the load were a resistor. The current is at a 90° phase angle, however. If we assume a perfect capacitor, there will be no heating of the capacitor, so no real power (watts) will be delivered by the transformer. The transformer must still be capable of supplying the voltage, and be able to handle the current required by the reactive load. The current in the transformer windings will heat the windings as a result of the I^2R losses in the winding resistances. The product of the voltage and current in the winding is referred to as "volt-amperes," since "watts" is reserved for the real, or dissipated, power in the load. The volt-ampere rating will always be equal to, or greater than, the power actually being drawn by the load.

The number of volt-amperes delivered by a transformer depends not only upon the dc load requirements, but also upon the type of dc output filter used (capacitor or choke input), and the type of rectifier used (full-wave center tap or full-wave bridge). With a capacitive-input filter, the heating effect in the secondary is higher because of the high peak-to-average current ratio. The volt-amperes handled by the transformer may be several times the power delivered to the load. The primary winding volt-amperes will be somewhat higher because of transformer losses. This point is treated in more detail in the section on ac-dc conversion. (See the **Electrical Fundamentals** chapter for more information on transformers and reactive power.)

7.3.2 Source Voltage and Frequency

A transformer operates by producing a magnetic field in its core and windings. The intensity of this field varies directly with the instantaneous voltage applied to the transformer primary winding. These variations, coupled to the secondary windings, produce the desired output voltage. Since the transformer appears to the source as an inductance in parallel with the (equivalent) load, the primary will appear as a short circuit if dc is applied to it. The unloaded inductance of the primary (also known as the *magnetizing inductance*) must be high enough so as not to draw an excess amount of input current at the design line frequency (normally 60 Hz in the US). This is achieved by providing a combination of sufficient turns on the primary and enough magnetic core material so that the core does not saturate during each half-cycle.

The voltage across a winding is directly related to the time rate of change of magnetic flux in the core. This relationship is expressed mathematically by $V = N\, d\Phi/dt$ as described in the section on Inductance in the **Electrical Fundamentals** chapter. The total flux in turn is expressed by $\Phi = A_e B$, where A_e is the cross-sectional area of the core and B is the flux density.

The maximum value for *flux density* (the magnetic field strength produced in the core) is limited to some percentage (< 80% for example) of the maximum flux density that the core material can stand without saturating, since in saturation the core becomes ineffective and causes the inductance of the primary to plummet to a very low level and input current to rise rapidly. Saturation causes high primary currents and extreme heating in the primary windings.

At a given voltage, 50 Hz ac creates more flux in an inductor or transformer core because the longer time period per half-cycle results in more flux and higher magnetizing current than the same transformer when excited by same 60-Hz voltage. For this reason, transformers and other electromagnetic equipment designed for 60-Hz systems must not be used on 50-Hz power systems unless specifically designed to handle the lower line frequency.

7.3.3 How to Evaluate an Unmarked Power Transformer

Hams who regularly visit hamfests frequently develop a junk box filled with used and unmarked transformers. Over time, transformer labels or markings on the coil wrappings may come off or be obscured. There is a good possibility that the transformer is still useable, but the problem is to determine what voltages and currents the transformer can supply. First consider the possibility that you may have an audio transformer or other impedance-matching device rather than a power transformer. If you aren't sure, don't connect it to ac power!

If the transformer has color-coded leads, you are in luck. There is a standard for transformer lead color-coding, as is given in the **Component Data and References** chapter. Where two colors are listed, the first one is the main color of the insulation; the second

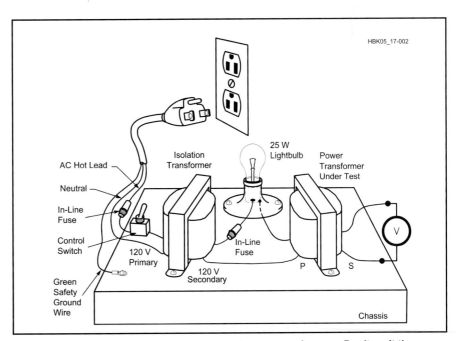

Fig 7.4 — Use a test fixture like this to test unknown transformers. Don't omit the isolation transformer, and be sure to insulate all connections before you plug into the ac mains.

is the color of the stripe.

Check the transformer windings with an ohmmeter to determine that there are no shorted (or open) windings. In particular, check for continuity between any winding and the core. If you find that a winding has been shorted to the core, do not use the transformer! The primary winding usually has a resistance higher than a filament winding and lower than a high-voltage winding.

Fig 7.4 shows that a convenient way to test the transformer is to rig a pair of test leads to an electrical plug with a 25-W household light bulb in series to limit current to safe (for the transformer) levels. For safety reasons use an isolation transformer and be sure to insulate all connections before you plug into the ac mains. Switch off the power while making or changing any connections. You can be electrocuted if the voltmeter leads or meter insulation are not rated for the transformer output voltage! If in doubt, connect the meter with the circuit turned off, then apply power while you are not in contact with the circuit. *Be careful! You are dealing with hazardous voltages!*

Connect the test leads to each winding separately. The filament/heater windings will cause the bulb to light to full brilliance because a filament winding has a very low impedance and almost all the input voltage will be across the series bulb. The high-voltage winding will cause the bulb to be extremely dim or to show no light at all because it will have a very high impedance, and the primary winding will probably cause a small glow. The bulb glows even with the secondary windings open-circuited because of the small magnetizing current in the transformer primary.

When the isolation transformer output is connected to what you think is the primary winding, measure the voltages at the low-voltage windings with an ac voltmeter. If you find voltages close to 6-V ac or 5-V ac, you know that you have identified the primary and the filament windings. Label the primary and low voltage windings.

Even with the light bulb, a transformer can be damaged by connecting ac mains power to a low-voltage or filament winding. In such a case the insulation could break down in a primary or high-voltage winding because of the high turns ratio stepping up the voltage well beyond the transformer ratings.

Connect the voltmeter to the high-voltage windings. Remember that the old TV transformers will typically supply as much as 800 V_{pk} or so across the winding, so make sure that your meter can withstand these potentials without damage and that you use the voltmeter safely.

Divide 6.3 (or 5) by the voltage you measured across the 6.3-V (or 5-V) winding in this test setup. This gives a multiplier that you can use to determine the actual no-load voltage rating of the high-voltage secondary. Simply multiply the ac voltage measured across the high-voltage winding by the multiplier.

The current rating of the windings can be determined by loading each winding with the primary connected directly (no bulb) to the ac line. Using power resistors, increase loading on each winding until its voltage drops by about 10% from the no-load figure. The current drawn by the resistors is the approximate winding load-current rating.

7.4 AC-DC Power Conversion

One of the most common power supply functions is the conversion of ac power to dc, or *rectification*. The output from the rectifier will be a combination of dc, which is the desired component, and ac *ripple* superimposed on the dc. This is an undesired but inescapable component. Since most loads cannot tolerate more than a small amount of ripple on the dc voltage, some form of filter is required. The result is that ac-dc power conversion is performed with a rectifier-filter combination as shown in **Fig 7.5**.

As we will see in the rectifier circuit examples given in the next sections, sometimes the rectifier and filter functions will be separated into two distinct parts but very often the two will be integrated. This is particularly true for voltage and current multipliers as described in the sections on multipliers later in the chapter. Even when it appears that the rectifier and filter are separate elements, there will still be a strong interaction where the design and behavior of each part depends heavily on the other. For example the current waveforms in the rectifiers and the input source are functions of the load and filter characteristics. In turn the voltage waveform applied to the filter depends on the rectifier circuit and the input source voltages. To simplify the discussion we will treat the rectifier connections and the filters separately but always keeping in mind their interdependence.

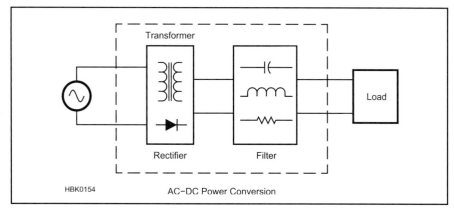

Fig 7.5 — Ac-dc power conversion with a rectifier and a filter.

The following rectifier-filter examples assume a conventional 60 Hz ac sine wave source, but these circuits are frequently used in switching converters at much higher frequencies and with square wave or quasi-square wave voltage and current waveforms. The component values may be different but the basic behavior will be very similar.

There are many different rectifier circuits or "connections" that may be used depending on the application. The following discussion provides an overview of some of the more common ones. The circuit diagrams use the symbol for a semiconductor diode, but the same circuits can be used with the older types of rectifiers that may be encountered in older equipment.

For each circuit we will show the voltage and current waveforms in the circuit for resistive, capacitive and inductive loads. The inductive and capacitive loads represent commonly used filters. We will be interested in the peak and average voltages as well as the RMS currents.

7.4.1 Half-Wave Rectifier

Fig 7.6 shows several examples of the half-wave rectifier circuit. It begins with a simple transformer with a resistive load (Fig 7.6A) and goes on to show how the output voltage and transformer current varies when a diode and filter elements are added.

Without the diode (Fig 7.6A) the output voltage (V_R) and current are just sine waves, and the RMS current in the transformer windings will be the same as the load (R) current.

Next, add a rectifier diode in series with the load (Fig 7.6B). During one half of the ac cycle, the rectifier conducts and there is current through the rectifier to the load. During the other half cycle, the rectifier is reverse-biased and there is no current (indicated by the broken line in Fig 7.6B) in R. The output voltage is pulsating dc, which is a combination of two components: an average dc value of 0.45 E_{RMS} (the voltage read by a dc voltmeter) and line-frequency ac ripple. The transformer secondary winding current is also pulsating dc. The power delivered to R is now 1/2 that for Fig 7.6A but the secondary RMS winding current in Fig 7.6B is still 0.707 times what it was in Fig 7.6A. For the same winding resistance, the winding loss, in proportion to the output power, is twice what it was in Fig 7.6A. This is an intrinsic limitation of the half-wave rectifier circuit — the RMS winding current is larger in proportion to the load power. In addition, the dc component of the secondary winding current may bias the transformer core toward saturation and increased core loss.

A filter can be used to smooth out these variations and provide a higher average dc voltage from the circuit. Because the frequency of the pulses (the ripple frequency) is low (one pulse per cycle), considerable filtering is required to provide adequately smooth dc output. For this reason the circuit is usually limited to applications where the required current is small. Parts C, D and E in Fig 7.6 show some possible capacitive and inductive filters.

As shown in Fig 7.6C and D, when a capacitor is used for filtering the output dc voltage will approach

$$V_{pk} = \sqrt{2} \times E_{RMS} \tag{1}$$

and the larger we make the filter capacitance, the smaller the ripple will be.

Unfortunately, as we make the filter capacitance larger, the diode, capacitor and transformer winding currents all become high-amplitude narrow pulses which will have a very high RMS value in proportion to the power level. These current pulses are also transmitted to the input line and inject currents at harmonics of the line frequency into the power source, which may result in interference to other equipment. Narrow high-amplitude current pulses are characteristic of capacitive-input filters in all rectifier connections when driven from voltage sources.

As shown in Fig 7.6E, it is possible to use an inductive filter instead, but a second diode (D2, sometimes called a *free-wheeling diode*) should be used. Without D2 the output voltage will get smaller as we increase the size of L to get better filtering, and the output voltage will vary greatly with load. By adding D2 we are free to make L large for small output ripple but still have reasonable voltage regulation. Currents in D1 and the winding will be approximately square waves, as indicated. This will reduce the line harmonic currents injected into the source but there will still be some.

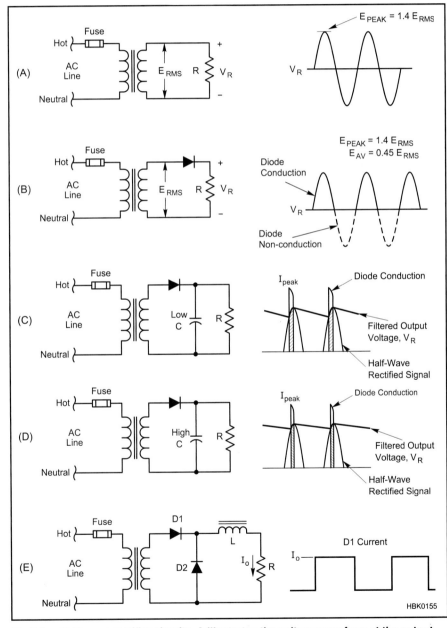

Fig 7.6 — Half-wave rectifier circuits. A illustrates the voltage waveform at the output without a rectifier. B represents the basic half-wave rectifier and the output waveform. C and D illustrate the impact of small and large filter capacitors on the output voltage and input current waveforms. E shows the effect of using an inductor filter with the half-wave rectifier. Note the addition of the shunt diode (D2) when using inductive filters with this rectifier connection.

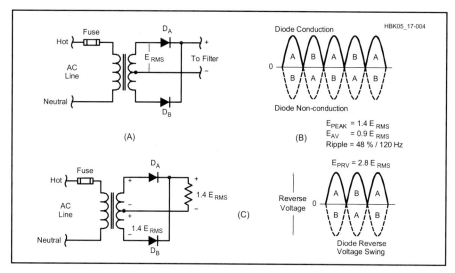

Fig 7.7 — Full-wave center-tap rectifier circuits. A illustrates the basic circuit. Diode conduction is shown at B with diodes A and B alternately conducting. The peak-inverse voltage for each diode is 2.8 E_{RMS} as depicted at C.

Peak inverse voltage (PIV) is the maximum voltage the rectifier must withstand when it isn't conducting. This varies with the load and rectifier connection. In the half-wave rectifier, with a resistive load the PIV is the peak ac voltage (1.4 × E_{RMS}); with a capacitor filter and a load drawing little or no current, the PIV can rise to 2.8 × E_{RMS}.

7.4.2 Full-Wave Center-Tap Rectifier

The full-wave center-tap rectifier circuit is shown in **Fig 7.7**. It is essentially an arrangement where the outputs of two half-wave rectifiers are combined so that both halves of the ac cycle are used to deliver power to the output. A transformer with a center-tapped secondary is required.

The average output voltage of this circuit is 0.9 × E_{RMS} of half the transformer secondary (the center-tap to one side); this is the maximum that can be obtained with a suitable choke-input filter. The peak output voltage is 1.4 × E_{RMS} of half the transformer secondary; this is the maximum voltage that can be obtained from a capacitor-input filter.

As can be seen in Fig 7.7C, the PIV impressed on each diode is independent of the type of load at the output. This is because the peak inverse voltage condition occurs when diode D_A conducts and diode D_B is not conducting. The positive and negative voltage peaks occur at precisely the same time, a condition different from that in the half-wave circuit. As the cathodes of diodes D_A and D_B reach a positive peak (1.4 E_{RMS}), the anode of diode D_B is at a negative peak, also 1.4 E_{RMS}, but in the opposite direction. The total peak inverse voltage is therefore 2.8 E_{RMS}.

Fig 7.7C shows that the ripple frequency is twice that of the half-wave rectifier (two times the line frequency). Substantially less filtering is required because of the higher ripple frequency. Since the rectifiers work alternately, each handles half of the load current. The current rating of each rectifier need be only half the total current drawn from the supply.

The problem with dc bias in the transformer core associated with the half-wave connection is largely eliminated with this circuit and the RMS current in the primary winding will also be reduced.

7.4.3 Full-Wave Bridge Rectifier

Another commonly used rectifier connection that does not require a center-tapped transformer is illustrated in **Fig 7.8**. In this arrangement, two rectifiers operate in series on each half of the cycle, one rectifier being in the lead supplying current to the load, the other being the current return lead. As shown in Figs 7.8A and B, when the top lead of the transformer secondary is positive with respect to the bottom lead, diodes D_A and D_C will conduct while diodes D_B and D_D are reverse-biased. On the next half cycle, when the top lead of the transformer is negative with respect to the bottom, diodes D_B and D_D will conduct while diodes D_A and D_C are reverse-biased.

The output voltage wave shape and ripple frequency are the same as for the full-wave center-tap connection. The average dc output voltage into a resistive load or choke-input filter is 0.9 times E_{RMS} delivered by the transformer secondary; with a capacitor filter and a light load, the maximum output voltage is 1.4 times the secondary E_{RMS} voltage.

Fig 7.8C shows the PIV to be 1.4 E_{RMS} for each diode which is half that of the full-wave center-tap connection for the same output voltage. When an alternate pair of diodes (such as D_A and D_C) is conducting, the other diodes are essentially connected in parallel (the conducting diodes are essentially short circuits) in a reverse-biased direction. The reverse stress is then 1.4 E_{RMS}. Each pair

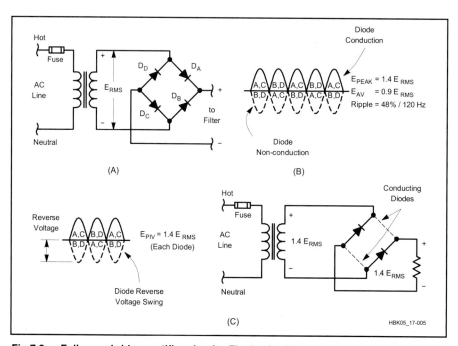

Fig 7.8 — Full-wave bridge rectifier circuits. The basic circuit is illustrated at A. Diode conduction and nonconduction times are shown at B. Diodes A and C conduct on one half of the input cycle, while diodes B and D conduct on the other. C displays the peak inverse voltage for one half cycle. Since this circuit reverse-biases two diodes essentially in parallel, 1.4 E_{RMS} is applied across each diode.

of diodes conducts on alternate half cycles, with the full load current through each diode during its conducting half cycle. Since each diode is not conducting during the other half cycle the average diode current is one-half the total load current drawn from the supply.

Compared to the half-wave and full-wave center-tapped connections, the full-bridge connection further reduces the transformer RMS winding currents. In the case of a resistive load the winding currents are the same as when the resistive load is connected directly across the secondary. The RMS winding currents will still be higher when inductive and especially capacitive filters are used because of the pulsating nature of the diode and winding currents.

7.4.4 Comparison of Rectifier Circuits

Comparing the full-wave center-tap and the full-wave bridge-rectifier connections, we can see that the center-tap has half the number of rectifiers as the bridge but these rectifiers have twice the PIV rating requirement of the bridge diodes. The diode current ratings are identical for the two circuits. The bridge makes better use of the transformer's secondary than the center-tap rectifier, since the transformer's full winding supplies power during both half cycles, while each half of the center-tap circuit's secondary provides power only during its positive half-cycle.

The full-wave center-tap rectifier is typically used in high-current, low-voltage applications because only one diode conducts at a time. This reduces the loss associated with diode conduction. In the full-wave bridge connection there are two diodes in series in conduction simultaneously, which leads to higher loss. The full-wave bridge connection is typically used for higher output voltages where this is not a serious concern. The lower diode PIV and better utilization of the transformer windings makes this connection very attractive for higher output voltages and higher powers typical of high voltage amplifier supplies.

Because of the disadvantages pointed out earlier, the half-wave configuration is rarely used in 60-Hz rectification except for bias supplies or other small loads. It does see considerable use, however, in high-frequency switching power supplies.

7.5 Voltage Multipliers

Other rectification circuits are sometimes useful, including *voltage multipliers*. These circuits function by the process of charging one or more capacitors in parallel on one half cycle of the ac waveform, and then connecting that capacitor or capacitors in series with the opposite polarity of the ac waveform on the alternate half cycle. In full-wave multipliers, this charging occurs during both half-cycles.

Voltage multipliers, particularly *voltage doublers*, find considerable use in high-voltage supplies. When a doubler is employed, the secondary winding of the power transformer need have only half the voltage that would be required for a bridge rectifier. This reduces voltage stress in the windings and decreases the transformer insulation requirements. This is not without cost, however, because the transformer-secondary *current* rating has to be correspondingly doubled for a given load current and charging of the capacitors leads to narrow high-RMS current waveforms in the transformer windings and the capacitors.

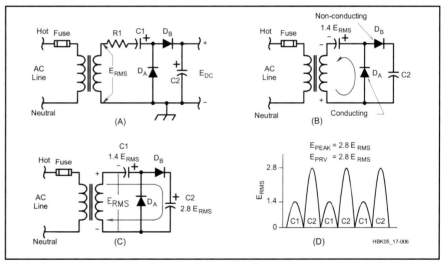

Fig 7.9 — Part A shows a half-wave voltage-doubler circuit. B displays how the first half cycle of input voltage charges C1. During the next half cycle (shown at C), capacitor C2 charges with the transformer secondary voltage plus that voltage stored in C1 from the previous half cycle. The arrows in parts B and C indicate the conventional current. D illustrates the levels to which each capacitor charges over several cycles.

7.5.1 Half-Wave Voltage Doubler

Fig 7.9 shows the circuit of a half-wave voltage doubler and illustrates the circuit operation. For clarity, assume the transformer voltage polarity at the moment the circuit is activated is that shown at Fig 7.9B. During the first negative half cycle, D_A conducts (D_B is in a nonconductive state), charging C1 to the peak rectified voltage (1.4 E_{RMS}). C1 is charged with the polarity shown in Fig 7.9B. During the positive half cycle of the secondary voltage, D_A is cut off and D_B conducts, charging capacitor C2. The amount of voltage delivered to C2 is the sum of the transformer peak secondary voltage plus the voltage stored in C1 (1.4 E_{RMS}). On the next negative half cycle, D_B is non-conducting and C2 will discharge into the load. If no load is connected across C2, the capacitors will remain charged — C1 to 1.4 E_{RMS} and C2 to 2.8 E_{RMS}. When a load is connected to the circuit output, the voltage across C2 drops during the negative half cycle and is recharged up to 2.8 E_{RMS} during the positive half cycle.

The output waveform across C2 resembles that of a half-wave rectifier circuit because C2 is pulsed once every cycle. Fig 7.9D illustrates the levels to which the two capacitors are charged throughout the cycle. In actual operation the capacitors will usually be large enough that they will discharge only partially, not all the way to zero as shown.

7.5.2 Full-Wave Voltage Doubler

Fig 7.10 shows the circuit of a full-wave voltage doubler and illustrates the circuit operation. During the positive half cycle of the transformer secondary voltage, as shown in Fig 7.10B, D_A conducts charging capacitor C1 to 1.4 E_{RMS}. D_B is not conducting at this time.

During the negative half cycle, as shown in Fig 7.10C, D_B conducts, charging capacitor C2 to 1.4 E_{RMS}, while D_A is non-conducting. The output voltage is the sum of the two capacitor voltages, which will be 2.8 E_{RMS} under no-load conditions. Fig 7.10D illustrates that each capacitor alternately receives a charge once per cycle. The effective filter capacitance is that of C1 and C2 in series, which is less than the capacitance of either C1 or C2 alone.

Resistors R1 and R2 in Fig 7.10A are used to limit the surge current through the rectifiers. Their values are based on the transformer voltage and the rectifier surge-current rating, since at the instant the power supply is turned on, the filter capacitors look like a short-circuited load. Provided the limiting resistors can withstand the surge current, their current-handling capacity is based on the maximum load current from the supply. Output voltages approaching twice the peak voltage of the transformer can be obtained with the voltage doubling circuit shown in Fig 7.10.

Fig 7.11 shows how the voltage depends upon the ratio of the series resistance to the load resistance, and the load resistance times the filter capacitance. The peak inverse voltage across each diode is 2.8 E_{RMS}. As indicated by the curves in Fig 7.11, the output voltage regulation of this doubler connection is not very good and it is not attractive for providing high voltages at high power levels.

There are better doubler connections for higher power applications, and two possibilities are shown in **Fig 7.12**. The connection in Fig 7.12A uses two bridge rectifiers in series with capacitive coupling between the ac terminals of the bridges. At the expense of more diodes, this connection will have much better output voltage regulation at higher power levels. Even better regulation can be achieved by using the connection shown in Fig 7.12B. In this example, two windings on the transformer are used. It is not essential that both windings have the same voltage, but both must be capable of providing the desired output current. In addition, the insulation of the upper winding must be adequate to accommodate the additional dc bias applied to it from the lower winding.

7.5.3 Voltage Tripler and Quadrupler

Fig 7.13A shows a voltage-tripling circuit. On one half of the ac cycle, C1 and C3 are charged to the source voltage through D1, D2 and D3. On the opposite half of the cycle, D2 conducts and C2 is charged to twice the source voltage, because it sees the transformer plus the charge in C1 as its source (D1 is cut off during this half cycle). At the same time, D3

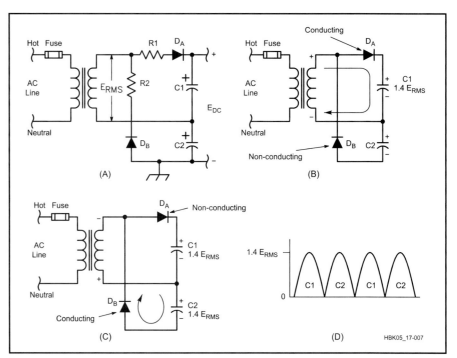

Fig 7.10 — Part A shows a full-wave voltage-doubler circuit. One-half cycle is shown at B and the next half cycle is shown at C. Each capacitor receives a charge during every input-voltage cycle. D illustrates how each capacitor is charged alternately.

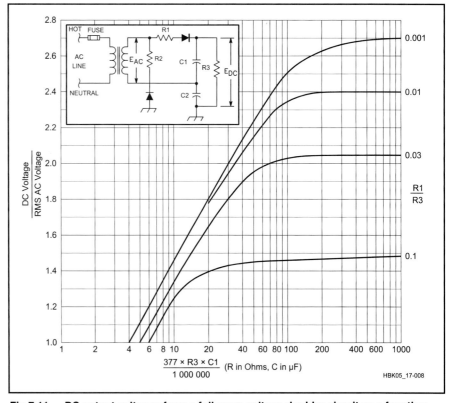

Fig 7.11 — DC output voltages from a full-wave voltage-doubler circuit as a function of the filter capacitances and load resistance. For the ratio R1 / R3 and for the R3 × C1 product, resistance is in ohms and capacitance is in microfarads. Equal resistance values for R1 and R2, and equal capacitance values for C1 and C2 are assumed. These curves are adapted from those published by Otto H. Schade in "Analysis of Rectifier Operation," *Proceedings of the I. R. E.*, July 1943.

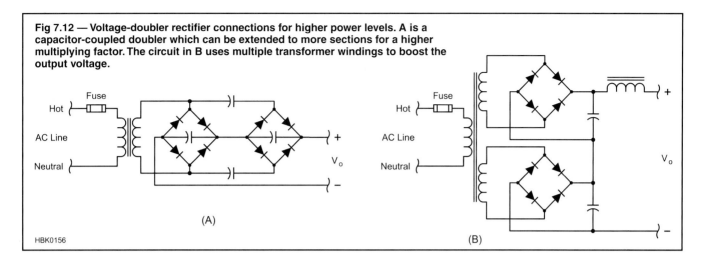

Fig 7.12 — Voltage-doubler rectifier connections for higher power levels. A is a capacitor-coupled doubler which can be extended to more sections for a higher multiplying factor. The circuit in B uses multiple transformer windings to boost the output voltage.

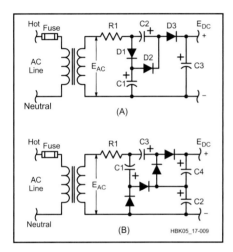

Fig 7.13 — Voltage-multiplying circuits with one side of the transformer secondary used as a common connection. A shows a voltage tripler and B shows a voltage quadrupler. Capacitances are typically 20 to 50 µF, depending on the output current demand. Capacitor dc ratings are related to E_{PEAK} (1.4 E_{RMS}):
C1 — Greater than E_{PEAK}
C2 — Greater than 2 E_{PEAK}
C3 — Greater than 3 E_{PEAK}
C4 — Greater than 2 E_{PEAK}

conducts, and with the transformer and the charge in C2 as the source, C3 is charged to three times the transformer voltage.

The voltage-quadrupling circuit of Fig 7.13B works in similar fashion. In either of the circuits of Fig 7.13, the output voltage will approach an exact multiple of the peak ac voltage when the output current drain is low and the capacitance values are large.

7.6 Current Multipliers

Just as there are voltage multiplier connections for high-voltage, low-current loads, there are current multiplier connections for low-voltage, high-current loads. An example of a current-doubler is given in **Fig 7.14A.**

To make the circuit operation easier to visualize, we can represent L1 and L2 as current sources (Fig 7.14B) which is a good approximation for steady-state operation. When terminal 1 of the secondary winding is positive with respect to terminal 2, diode D_A will be reverse-biased and therefore non-conducting. The current flows within the circuit are shown in Fig 7.14B. Note that all of the output current (I_o) flows through D_B but only half of I_o flows through the winding. At the cathode of D_B the current divides with half going to L2 and the other half to the transformer secondary. The output voltage will be one-half the voltage of the average winding voltage ($0.45 E_{RMS}$). This rectifier connection divides the voltage and multiplies the current! Because of the need for two inductors, this circuit is seldom used in line-frequency applications but it is very useful in high-frequency switching regulators with very low output voltages (<10 V) because it makes the secondary winding design easier and can improve circuit efficiency. At high frequencies, the inductors can be quite small.

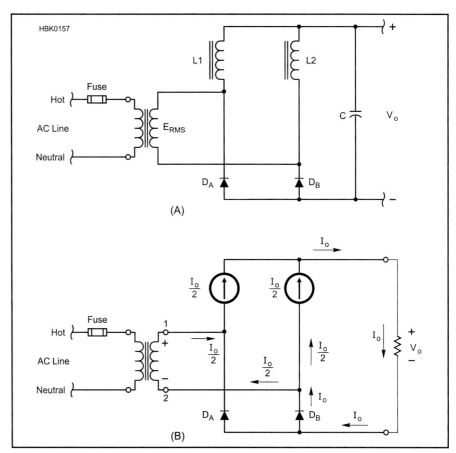

Fig 7.14 — A current doubler rectifier connection. A is the basic circuit. B illustrates current flow within the circuit.

7.7 Rectifier Types

Rectifiers have a long history beginning with mechanical rectifiers in the 1800s to today's abundant variety of semiconductor devices. While many different devices have been created for this purpose, they all have the characteristic that they block current flow in the reverse direction, withstanding substantial reverse voltage and allowing current flow in the forward direction with minimum voltage drop. The simplest rectifiers are diodes, but it is also possible to have three-terminal devices (such as a thyristor) that can be controlled to regulate the output dc in addition to providing rectification. It is also possible to use devices like MOSFETs as synchronous rectifiers with very low forward drop during conduction. This is typically done to improve efficiency for very low voltage outputs. The following is a brief description of several of the more common examples.

7.7.1 Vacuum Tube

Once the mainstay of the rectifier field, the vacuum-tube rectifier has largely been supplanted by the silicon diode, but it may be found in vintage equipment. Vacuum-tube rectifiers require filament power and are characterized by high forward voltage drop during conduction, which leads to inherently poor regulation of the output voltage. They are largely immune to ac line transients (also known as "spikes") that can destroy other rectifier types.

7.7.2 Mercury Vapor

The mercury-vapor rectifier was an improvement over the vacuum tube rectifier in that the electron stream from cathode to plate would ionize vaporized mercury in the tube and greatly reduce the forward voltage drop. Because of the lower voltage drop, the power dissipation was much lower for a given current and these tubes could carry relatively high currents. They were popular in transmitter high-voltage power supplies, some of which amateurs may still encounter.

Mercury rectifiers have to be treated with special care. When power is initially applied, the tube filament has to be turned on first to vaporize condensed mercury before high-voltage ac can be applied to the plate. This could take from one to two minutes. Also, if the tube was handled or the equipment transported, filament power would have to be applied for about a half hour to vaporize any mercury droplets that might have been shaken onto tube insulating surfaces. Mercury vapor rectifiers have mostly been replaced by silicon diodes.

7.7.3 Early Solid-State Rectifiers

Copper oxide and selenium rectifiers were the first of the solid-state rectifiers to find their way into commercial equipment. Voltage breakdown per rectifying junction was only a few volts for the copper oxide rectifiers and about 20 V for selenium. Multi-junction stacked versions of selenium rectifiers were used for higher voltage and had a relatively low forward voltage drop. Selenium rectifiers found their way into the plate supplies of test equipment and accessories that needed only a few tens of milliamperes of current at about a hundred volts. Examples include such as grid-dip meters, VTVMs and so forth.

Selenium rectifiers had a relatively low reverse resistance leading to high reverse leakage currents and were therefore inefficient. These may still be encountered in older equipment but they are usually replaced with silicon diodes. Very conveniently, a key indicator of a failed selenium rectifier is the smell of hydrogen selenide, similar to rotten eggs.

A word of caution is warranted when replacing older rectifiers such as vacuum tubes, mercury-vapor rectifiers, selenium or copper oxide. The voltage drop introduced by these rectifiers would have been included in the design of the equipment. When replaced with silicon diodes, the output voltage of the rectifier is very likely to be higher than with the original design. Care should be taken to make sure the higher voltage will not damage the filter elements or load circuits. Furthermore, mercury and selenium are toxic and should not be disposed of with regular household trash. Contact your local recycling agency for information about where to dispose of these devices.

7.7.4 Semiconductor Diodes

Rectifier diodes can be made from a number of different semiconductor materials such as germanium, silicon, silicon-carbide or gallium-arsenide, and no doubt other materials will appear in the future. The choice will depend on the application and as always cost is a factor.

Germanium diodes were the first of the solid-state semiconductor rectifiers. They have an extremely low forward voltage drop but are relatively temperature sensitive, having high reverse leakage currents at higher temperatures. They can be easily destroyed by overheating during soldering as well. Germanium diodes are no longer used as power rectifiers.

Today, silicon diodes are the primary choice for virtually all power rectifier applications. They are characterized by extremely high reverse resistance (low reverse leakage), forward drops of a volt or less and operation at junction temperatures up to 125 °C. Some multi-junction HV diodes will have forward drops of several volts, but that is still low compared to the voltage at which they are being used.

Many different types of silicon diodes are available for different applications. Silicon rectifiers fall into to two general categories: PN-junction diodes and Schottky barrier diodes (see the **Analog Basics** chapter).

Schottky diodes are the usual choice for low output voltages (<20 V) where their low forward conduction drop is critical for efficiency. For higher voltages however, the high reverse leakage of Schottky diodes is not acceptable and PN-junction diodes are normally chosen.

For 50/60 Hz applications, diodes with reverse recovery times of a microsecond or even more are suitable and very economical. For switchmode converters and inverters that regularly operate at 25 kHz and higher frequencies, fast-recovery diodes are needed. These converters typically have waveform transitions of less than 1 μs within the circuit. MOSFET power transistors often have transitions of less than 100 ns.

During the switching transitions, previously conducting diodes see a reversal of current direction. This change tends to reverse-bias those diodes, and thereby put them into an open-circuit condition. Unfortunately, as explained in the **Analog Basics** chapter, solid-state rectifiers cannot be made to cease conduction instantaneously. As a result, when the opposing diodes in a bridge rectifier or full-wave rectifier become conductive at the time the converter switches states, the diodes being turned off will actually conduct in the reverse direction for a brief time. That effectively short circuits the converter for a period of time depending on the reverse recovery characteristics of the rectifiers. This characteristic can create high current transients that stress the switching transistors and lead to increased loss and electromagnetic interference. As the switching frequency increases, more of these transitions happen each second, and more power is lost because of diode cross-conduction.

These current transients and associated losses are reduced by using *fast recovery* diodes, which are specially doped diodes designed to minimize storage time. Diodes with recovery times of 50 ns or less are available.

7.7.5 Rectifier Strings or Stacks

DIODES IN SERIES

When the PIV rating of a single diode is not sufficient for the application, similar diodes may be used in series. (Two 500 PIV diodes in series will withstand 1000 PIV and so on.) There used to be a general recommendation to place a resistor across each diode in the string to equalize the PIV drops. With modern diodes, this practice is no longer necessary.

Modern silicon rectifier diodes are constructed to have an avalanche characteristic. Simply put, this means that the diffusion process is controlled so the diode will exhibit a Zener characteristic in the reverse-biased direction before destructive breakdown of the

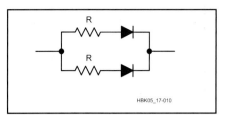

Fig 7.15 — Diodes can be connected in parallel to increase the current-handling capability of the circuit. Each diode should have a series current-equalizing resistor, with a value selected to have a drop of several tenths of a volt at the expected current.

junction can occur. This provides a measure of safety for diodes in series. A diode will go into Zener conduction before it self-destructs. If other diodes in the chain have not reached their avalanche voltages, the current through the avalanched diode will be limited to the leakage current in the other diodes. This should normally be very low. For this reason, shunting resistors are generally not needed across diodes in series rectifier strings. In fact, shunt resistors can actually create problems because they can produce a low-impedance source of damaging current to any diode that may have reached avalanche potential.

DIODES IN PARALLEL

Diodes can be placed in parallel to increase current-handling capability. Equalizing resistors should be added as shown in **Fig 7.15**. Without the resistors, one diode may take most of the current. The resistors should be selected to have a drop of several tenths of a volt at the expected peak current. A disadvantage of this form of forced current sharing will be the increase in power loss because of the added resistors.

7.7.6 Rectifier Ratings versus Operating Stress

Power supplies designed for amateur equipment use silicon rectifiers almost exclusively. These rectifiers are available in a wide range of voltage and current ratings: PIV ratings of 600 V or more and current ratings as high as 400 A are available. At 1000 PIV, the current ratings may be several amperes. It is possible to stack several units in series for higher voltages. Stacks are available commercially that will handle peak inverse voltages up to 10 kV at a load current of 1 A or more.

7.7.7 Rectifier Protection

The discussion of rectifier circuits included the peak reverse voltage seen by the rectifiers in each circuit. You will need this information to select the voltage rating of the diodes in given application. It is normal good practice to not expose the diodes to more than 75% of their rated voltage for the worst case reverse voltage. This will probably be when operating at the highest input voltage but should also take into account transients that may occur.

The important specifications of a silicon diode are:

1. PIV — the peak inverse voltage.
2. I_0 — the average dc current rating.
3. I_{REP} — the peak repetitive forward current.
4. I_{SURGE} — a non-repetitive peak half-sine wave of 8.3 ms duration (one-half cycle of 60-Hz line frequency).
5. Switching speed or reverse recovery time.
6. Power dissipation and thermal resistance.

The first two specifications appear in most catalogs. I_{REP} and I_{SURGE} are not often specified in catalogs, but they are very important. Except in some switching regulator and capacitive filter circuits, rectifier current typically flows half the time — when it does conduct, the rectifier has to pass at least twice the average direct current. With a capacitor-input filter, the rectifier conducts much less than half the time. In this case, when it does conduct, it may pass as much as 10 to 20 times the average dc current, under certain conditions.

CURRENT INRUSH

When the supply is first turned on, the filter capacitors are discharged and act like a dead short. The result can be a very heavy current surge through the diode for at least one half-cycle and sometimes more. This current transient is called I_{SURGE}. The maximum surge current rating for a diode is usually specified for a duration of one-half cycle (at 60 Hz), or about 8.3 ms. Some form of surge protection is usually necessary to protect the diodes until the filter capacitors are nearly charged, unless the diodes used have a very high surge-current rating (several hundred amperes). If a manufacturer's data sheet is not available, an educated guess about a diode's capability can be made by using these rules of thumb for silicon diodes commonly used in Amateur Radio power supplies:

Rule 1. The maximum I_{REP} rating can be assumed to be approximately four times the maximum I_0 rating, where I_0 is the average dc current rating.

Rule 2. The maximum I_{SURGE} rating can be assumed to be approximately 12 times the maximum I_0 rating. This figure should provide a reasonable safety factor. Silicon rectifiers with 750-mA dc ratings, for example, seldom have 1-cycle surge ratings of less than 15 A; some are rated up to 35 A or more. From this you can see that the rectifier should be selected on the basis of I_{SURGE} and not on I_0 ratings.

Although you can sometimes rely on the resistance of the transformer windings to provide surge-current limiting, this is seldom adequate in high-voltage power supplies. Series resistors are often installed between the secondary and the rectifier strings or in the transformer's primary circuit, but these can be a deterrent to good voltage regulation.

One way to have good surge current limiting at turn-on without affecting voltage regulation during normal operation is to have a resistor in series with the input, along with a relay across the resistor that shorts it out after 50 ms or so. This kind of arrangement is particularly important in HV supplies.

VOLTAGE TRANSIENTS

Vacuum-tube rectifiers had little problem with voltage transients on the incoming power lines. The possibility of an internal arc was of little consequence, since the heat produced was of very short duration and had little effect on the massive plate and cathode structures.

Unfortunately, such is not the case with silicon diodes. Because of their low forward voltage drop, silicon diodes create very little heat with high forward current and therefore have tiny junction areas. However, conduction in the reverse direction beyond the normal reverse recovery time (reverse avalanching) can cause junction temperatures to rise extremely rapidly with the resulting destruction of the semiconductor junction.

To protect semiconductor rectifiers from voltage transients, special surge-absorption devices are available for connection across the incoming ac bus or transformer secondary. These devices operate in a fashion similar to a Zener diode; they conduct heavily when a specific voltage level is reached. Unlike Zener diodes, however, they have the ability to absorb very high transient energy levels without damage. With the clamping level set well above the normal operating voltage range for the rectifiers, these devices normally appear as open circuits and have no effect on the power-supply circuits. When a voltage transient occurs, however, these protection devices clamp the transient and thereby prevent destruction of the rectifiers.

Transient protectors are available in three basic varieties:

1. *Silicon Zener diodes* — large junction Zeners specifically made for this purpose and available as single junction for dc (unipolar) and back-to-back junctions for ac (bipolar). These silicon protectors are available under the trade name of TransZorb from General Semiconductor Corporation and are also made by other manufacturers. They have the best transient-suppressing characteristics of the three varieties mentioned here, but are expensive and have the least energy absorbing capability per dollar of the group.

2. *Varistors* — made of a composition metal-oxide material that breaks down at a certain voltage. Metal-oxide varistors, also known as MOVs, are cheap and easily obtained, but have a higher internal resistance, which allows a greater increase in clamped voltage than the Zener variety. Varistors can also degrade with successive transients within their rated power handling limits (this is not usually a problem in the ham shack where transients are few and replacement of the varistor is easily accomplished).

Varistors usually become short-circuited when they fail. Large energy dissipation can result in device explosion. Therefore, it is a good idea to include a fuse that limits the short-circuit current through the varistor, and to protect people and circuitry from debris.

3. *Gas tube* — similar in construction to the familiar neon bulb, but designed to limit conducting voltage rise under high transient currents. Gas tubes can usually withstand the highest transient energy levels of the group. Gas tubes suffer from an ionization time problem, however. A high voltage across the tube will not immediately cause conduction. The time required for the gas to ionize and clamp the transient is inversely proportional to the level of applied voltage in excess of the device ionization voltage. As a result, the gas tube will let a little of the transient through to the equipment before it activates.

In installations where reliable equipment operation is critical, the local power is poor and transients are a major problem, the usual practice is to use a combination of protectors. Such systems consist of a varistor or Zener protector, combined with a gas-tube device. Often there is an indicator light to warn when a surge has blown out the varistor. Operationally, the solid-state device clamps the surge immediately, with the beefy gas tube firing shortly thereafter to take most of the surge from the solid-state device.

HEAT

The junction of a diode is quite small, so it must operate at a high current density. The heat-handling capability is, therefore, quite small. Normally, this is not a prime consideration in high-voltage, low-current supplies. Use of high-current rectifiers at or near their maximum ratings (usually 2-A or larger, stud-mount rectifiers) requires some form of heat sinking. Frequently, mounting the rectifier on the main chassis — directly or with thin thermal insulating washers — will suffice.

When a rectifier is directly mounted on the heatsink it is good practice to use a thin layer of thermal grease between the diode and the heat sink to assure good heat conduction. Most modern insulating thermal washers do not require the use of grease, but the older mica and other washers may benefit from a *very thin layer* of grease. Thermal grease and heat conducting insulating washers and pads are standard products available from mail-order component sellers.

Large, high-current rectifiers often require special heat sinks to maintain a safe operating temperature. Forced-air cooling from a fan is sometimes used as a further aid. Safe case temperatures are usually given in the manufacturer's data sheets and should be observed if the maximum capabilities of the diode are to be realized. See the thermal design section in the chapter on **Electrical Fundamentals** for more information.

7.8 Power Filtering

Most loads will not tolerate the ripple (an ac component) of the pulsating dc from the rectifiers. Filters are required between the rectifier and the load to reduce the ripple to a low level. As pointed out earlier, some capacitances or inductances may be inherent in the rectifier connection, reducing the ripple amplitude. In most cases, however, additional filtering is required. The design of the filter depends to a large extent on the dc voltage output, the desired voltage regulation of the power supply and the maximum load current. Power-supply filters are low-pass devices using series inductors and/or shunt capacitors.

7.8.1 Load Resistance

In discussing the performance of power-supply filters, it is sometimes convenient to characterize the load connected to the output as a resistance. This *load resistance* is equal to the output voltage divided by the total load current, including the current drawn by the bleeder resistor.

7.8.2 Voltage Regulation

In an unregulated supply, the output voltage usually decreases as more current is drawn. This happens not only because of increased voltage drops in the transformer and filter chokes, but also because the output voltage at light loads tends to soar to the peak value of the transformer voltage as a result of charging the first capacitor. Proper filter design can reduce this effect. The change in output voltage with load is called the *voltage regulation* and is expressed as a percentage.

$$\text{Percent Regulation} = \frac{(E1 - E2)}{E2} \times 100\% \quad (2)$$

where
E1 = the no-load voltage
E2 = the full-load voltage.

A steady load, such as that represented by a receiver, speech amplifier or unkeyed stages of a transmitter, does not require good (low) regulation as long as the proper voltage is obtained under load conditions. The filter capacitors must have a voltage rating safe for the highest value to which the voltage will rise when the external load is removed.

Typically the output voltage will display a larger change with long-duration changes

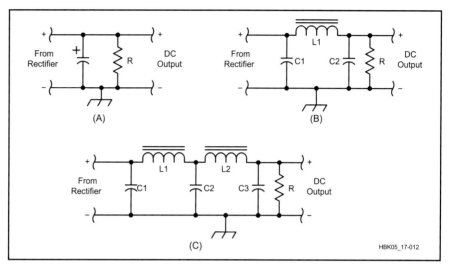

Fig 7.16 — Capacitor-input filter circuits. At A is a simple capacitor filter. B and C are single- and double-section filters, respectively.

in load resistance than with short transient changes. The reason for this is that transient load currents are supplied from energy stored in the output capacitance. The regulation with long-term changes is often called the *static regulation*, to distinguish it from the *dynamic regulation* (transient load changes). A load that varies at a syllabic or keyed rate, as represented by some audio and RF amplifiers, usually requires good dynamic regulation (<15%) if distortion products are to be held to a low level. The dynamic regulation of a power supply can be improved by increasing the output capacitance.

When essentially constant voltage is required, regardless of current variation (for stabilizing an oscillator, for example), special voltage regulating circuits described later in this chapter are used.

7.8.3 Bleeder Resistors

A *bleeder resistor* is a resistance (R) connected across the output terminals of the power supply as shown in **Fig 7.16A**. Its functions are to discharge the filter capacitors as a safety measure when the power is turned off and to improve voltage regulation by providing a minimum load resistance. When voltage regulation is not of importance, the resistance may be as high as 100 Ω per volt of output voltage. The resistance value to be used for voltage-regulating purposes is discussed in later sections. From the consideration of safety, the power rating of bleeder resistors should be as conservative as possible — having a burned-out bleeder resistor is dangerous!

7.8.4 Ripple Frequency and Voltage

The ripple at the output of the rectifier is an alternating current superimposed on a steady direct current. From this viewpoint, the filter may be considered to consist of: 1) shunt capacitors that short circuit the ac component while not interfering with the flow of the dc component; and/or 2) series chokes that readily pass dc but will impede the ac component.

The effectiveness of the filter can be expressed in terms of percent ripple, which is the ratio of the RMS value of the ripple to the dc value in terms of percentage.

$$\text{Percent Ripple (RMS)} = \frac{E1}{E2} \times 100\% \quad (3)$$

where
E1 = the RMS value of ripple voltage
E2 = the steady dc voltage.

Any frequency multiplier or amplifier supply in a CW transmitter should have less than 5% ripple. A linear amplifier can tolerate about 3% ripple on the plate voltage. Bias supplies for linear amplifiers should have less than 1% ripple. VFOs, speech amplifiers and receivers may require no greater than 0.01% ripple.

Ripple frequency refers to the frequency of the pulsations in the rectifier output waveform — the number of pulsations per second. The ripple frequency of half-wave rectifiers is the same as the line-supply frequency — 60 Hz with a 60-Hz supply. Since the output pulses are doubled with a full-wave rectifier, the ripple frequency is doubled — to 120 Hz with a 60-Hz supply.

The amount of filtering (values of inductance and capacitance) required to give adequate smoothing depends on the ripple frequency. More filtering is required as the ripple frequency is reduced. This is why the filters used for line-frequency rectification are much larger than those used in switchmode converters where the ripple frequency is often in the hundreds of kHz.

7.8.5 Capacitor-Input Filters

Typical capacitor-input filter systems are shown in Fig 7.16. The ripple can be reduced by making C1 larger, but that can lead to very large capacitances and high inrush currents at turn-on. Better ripple reduction will be obtained when moderate values for C1 are employed and LC sections are added as shown in Fig 7.16B and C.

INPUT VERSUS OUTPUT VOLTAGE

The average output voltage of a capacitor-input filter is generally poorly regulated with load-current variations. As shown earlier (Fig 7.6) the rectifier diodes conduct for only a small portion of the ac cycle to charge the filter capacitor to the peak value of the ac waveform. When the instantaneous voltage of the ac passes its peak, the diode ceases to conduct. This forces the capacitor to support the load current until the ac voltage on the opposing diode in the bridge or full wave rectifier is high enough to pick up the load and recharge the capacitor. For this reason, the peak diode currents are usually quite high.

Since the cyclic peak voltage of the capacitor-filter output is determined by the peak of the input ac waveform, the minimum voltage and, therefore, the ripple amplitude, is determined by the amount of voltage discharge, or "droop," occurring in the capacitor while it is discharging and supporting the load. Obviously, the higher the load current, the proportionately greater the discharge, and therefore the lower the average output.

There is an easy way to approximate the peak-to-peak ripple for a certain capacitor and load by assuming a constant load current. We can calculate the droop in the capacitor by using the relationship:

$$C \times E = I \times t \quad (4)$$

where
- C = the capacitance in microfarads
- E = the voltage droop, or peak-to-peak ripple voltage
- I = the load current in milliamperes
- t = the length of time in ms per cycle during which the rectifiers are not conducting, during which the filter capacitor must support the load current. For 60-Hz, full-wave rectifiers, t is about 7.5 ms.

As an example, let's assume that we need to determine the peak-to-peak ripple voltage at the dc output of a full-wave rectifier/filter combination that produces 13.8 V dc and supplies a transceiver drawing 2.0 A. The filter capacitor in the power supply is 5000 μF. Using the above relationship:

$$C \times E = I \times t \quad (5)$$

$$5000 \text{ μF} \times E = 2000 \text{ mA} \times 7.5 \text{ ms}$$

$$E = \frac{2000 \text{ mA} \times 7.5 \text{ ms}}{5000 \text{ μF}} = 3 \text{ V}_{P-P}$$

Obviously, this is too much ripple. A capacitor value of about 20,000 μF would be better suited for this application. If a linear regulator is used after this rectifier/filter combination, then it is possible to trade off higher ripple voltage against high power dissipation in the regulator. A properly designed linear regulator can reduce the ripple amplitude to a very small value.

7.8.6 Choke-Input Filters

Choke-input filters provide the benefits of greatly improved output voltage stability over varying loads and low peak-current surges in the rectifiers. On the negative side, the output voltage will be lower than that for a capacitor-input filter.

In line-frequency power supplies, choke-input filters are less popular than they once were. This change came about in part because of the high surge current capability of silicon rectifiers, but more importantly because size, weight and cost are reduced when large filter chokes are eliminated. However, choke input filters are frequently used in high-frequency switchmode converters where the chokes will be much smaller.

As long as the inductance of the choke is large enough to maintain a continuous

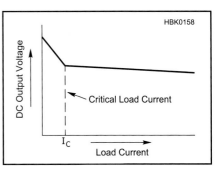

Fig 7.17 — Inductive filter output voltage regulation as a function of load current. The transition from capacitive peak charging to inductive averaging occurs at the critical load current, I_C.

current over the complete cycle of the input ac waveform, the filter output voltage will be the average value of the rectified output. The average dc value of a full-wave rectified sine wave is 0.637 times its peak voltage. Since the RMS value is 0.707 times the peak, the output of the choke input filter will be (0.637 / 0.707), or 0.9 times the RMS ac voltage.

As shown in **Fig 7.17**, there is a minimum or "critical" load current below which the choke does not provide the necessary filtering. For light loads, there may not be enough energy stored in the choke during the input waveform crest to allow continuous current over the full cycle. When this happens, the filter output voltage will rise as the filter assumes more and more of the characteristics of a capacitor-input filter. One purpose of the bleeder resistor is to keep the minimum load current above the critical value.

The value for the critical (or minimum) inductance for a given maximum value of load resistance in a single phase, full-wave

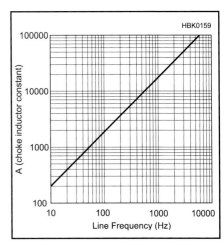

Fig 7.18 — The choke inductor constant, A, is used to solve equation 6.

rectifier with a sine wave source voltage can be approximated from:

$$L_c = \frac{R}{A} \quad (6)$$

where
R = the maximum load resistance

A = a constant obtained from **Fig 7.18**, derived from the frequency of the input current (see Reference 1).

Low values for minimum load current (high minimum load R) can lead to large values for L_c which may not be practical. Standard filter inductors typically have a relatively constant value for L as the dc current is varied but it is possible to use a *swinging choke* instead. This is an inductor which has a high inductance at low currents and much lower inductance at high currents. Using a swinging choke will usually result in a much smaller filter choke and/or better output regulation.

7.9 Power Supply Regulation

The output of a rectifier/filter system may be usable for some electronic equipment, but for today's transceivers and accessories, further measures may be necessary to provide power sufficiently clean and stable for their needs. Voltage regulators are often used to provide this additional level of conditioning.

Rectifier/filter circuits by themselves are unable to protect the equipment from the problems associated with input-power-line fluctuations, load-current variations and residual ripple voltages. Regulators can eliminate these problems, but not without costs in circuit complexity and power-conversion efficiency.

7.9.1 Zener Diodes

A Zener diode (developed by Dr Clarence Zener) can be used to maintain the voltage applied to a circuit at a practically constant value, regardless of the voltage regulation of the power supply or variations in load current.

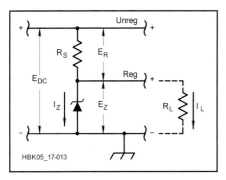

Fig 7.19 — Zener-diode voltage regulation. The voltage from a negative supply may be regulated by reversing the power-supply connections and the diode polarity.

The typical circuit is shown in **Fig 7.19**. Note that the cathode side of the diode is connected to the positive side of the supply.

Zener diodes are available in a wide variety of voltages and power ratings. The voltages range from less than 2 V to a few hundred volts, while the power ratings (power the diode can dissipate) run from less than 0.25 W to 50 W. The ability of the Zener diode to stabilize a voltage depends on the diode's conducting impedance. This can be as low as 1 Ω or less in a low-voltage, high-power diode or as high as 1000 Ω in a high-voltage, low-power diode.

The circuit in Fig 7.19 is a *shunt* regulator in that it "shunts" current through a controlling device (the Zener diode) to maintain a constant output voltage. To design a Zener shunt regulator, you must know the minimum and maximum input voltage (E_{DC}); the output voltage, which is equal to the Zener diode voltage (E_Z); and the minimum and maximum load current (I_L) through R_L. If the input voltage is variable, you must specify the maximum and minimum values for E_{DC}. As a rule of thumb, the current through the Zener should be $I_L/10$ for good regulation and must be greater than the $I_{Z(min)}$ at which the Zener diode maintains its constant voltage drop. Once these quantities are known the series resistance, R_S, can be determined:

$$R_S = \frac{E_{DC(min)} - E_Z}{1.1\, I_{L(max)}} \quad (7)$$

The power dissipation of the Zener diode, P_D, is

$$P_{DZ} = \left[\frac{E_{DC(max)} - E_Z}{R_S} - I_{L(min)}\right] E_Z \quad (8)$$

and of the series resistor, R_S,

$$P_{DR} = \frac{\left(E_{DC(max)} - E_Z\right)^2}{R_S} \quad (9)$$

It is good practice to provide a five times rated power dissipation safety margin for both the series resistor and the Zener diode. This avoids heating in the Zener and the resulting drift in voltage. High-power Zener diodes (10 W dissipation or more) will require heat-sinking as discussed in the section on Managing Heat in the **Electrical Fundamentals** chapter.

7.9.2 Linear Regulators

Linear regulators come in two varieties, *series* and *shunt*, as shown in **Fig 7.20**. The shunt regulator is simply an electronic (also called "active") version of the Zener diode. For the most part, the active shunt regulator (Fig 7.20B) is rarely used since the series regulator is a superior choice for most applications.

The series regulator (Fig 7.20A) consists of a stable voltage reference, which is usually established by a Zener diode, a transistor in series between the power source and the load (called a *series pass transistor*), and an error amplifier. In critical applications a temperature-compensated reference diode would be used instead of the Zener diode.

The output voltage is sampled by the error amplifier, which compares the output (usually scaled down by a voltage divider) to the reference. If the scaled-down output voltage becomes higher than the reference voltage, the error amplifier reduces the drive current to the pass transistor, thereby allowing the

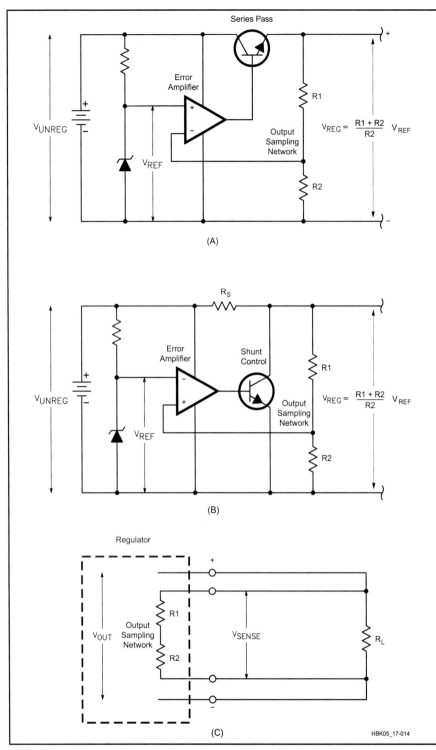

Fig 7.20 — Linear electronic voltage-regulator circuits. In these diagrams, batteries represent the unregulated input-voltage source. A transformer, rectifier and filter would serve this function in most applications. Part A shows a series regulator and Part B shows a shunt regulator. Part C shows how remote sensing overcomes poor load regulation caused by the I R drop in the connecting wires by bringing them inside the feedback loop.

output voltage to drop slightly. Conversely, if the load pulls the output voltage below the desired value, the amplifier drives the pass transistor into increased conduction.

The "stiffness" or tightness of regulation of a linear regulator depends on the gain of the error amplifier and the ratio of the output scaling resistors. In any regulator, the output is cleanest and regulation stiffest at the point where the sampling network or error amplifier is connected. If heavy load current is drawn through long leads, the voltage drop can degrade the regulation at the load. To combat this effect, the feedback connection to the error amplifier can be made directly to the load. This technique, called *remote sensing*, moves the point of best regulation to the load by bringing the connecting loads inside the feedback loop. This is shown in Fig 7.20C.

INPUT VERSUS OUTPUT VOLTAGE

In a series regulator, the pass-transistor power dissipation is directly proportional to the load current and input/output voltage differential. The series pass element can be located in either leg of the supply. Either NPN or PNP devices can be used, depending on the ground polarity of the unregulated input.

The differential between the input and output voltages is a design tradeoff. If the input voltage from the rectifiers and filter is only slightly higher than the required output voltage, there will be minimal voltage drop across the series pass transistor. A small drop results in minimal thermal dissipation and high power-supply efficiency. The supply will have less capability to provide regulated power in the event of power line brownout and other reduced line voltage conditions, however. Conversely, a higher input voltage will provide operation over a wider range of input voltage, but at the expense of increased heat dissipation.

7.9.3 Linear Regulator Pass Transistors

DARLINGTON PAIRS

A simple Zener-diode reference or IC op-amp error amplifier may not be able to source enough current to a pass transistor that must conduct heavy load current. The Darlington configuration of **Fig 7.21A** multiplies the pass-transistor beta, thereby extending the control range of the error amplifier. If the Darlington arrangement is implemented with discrete transistors, resistors across the base-emitter junctions may be necessary to prevent collector-to-base leakage currents in Q1 from being amplified and turning on the transistor pair. These resistors are contained in the envelope of a monolithic Darlington device.

When a single pass transistor is not available to handle the current required from a regulator, the current-handling capability may be increased by connecting two or more pass transistors in parallel. The circuit of Fig 7.21B shows the method of connecting these pass transistors. The resistances in the emitter leads of each transistor are necessary to equalize the currents.

TRANSISTOR RATINGS

When bipolar (NPN, PNP) power transistors are used in applications in which they are called upon to handle power on a continuous

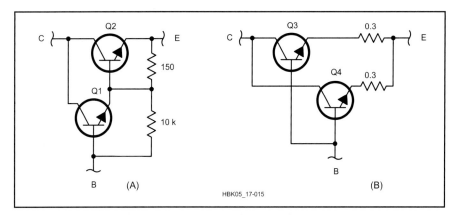

Fig 7.21 — At A, a Darlington-connected transistor pair for use as the pass element in a series-regulating circuit. At B, the method of connecting two or more transistors in parallel for high-current output. Resistances are in ohms. The circuit at A may be used for load currents from 100 mA to 5 A, and the one at B may be used for currents from 6 A to 10 A.
Q1 — Motorola MJE 340 or equivalent
Q2-Q4 — Power transistor such as 2N3055 or 2N3772

basis, rather than switching, there are four parameters that must be examined to see if any maximum limits are being exceeded. Operation of the transistor outside these limits can easily result in device failure, and these parameters must be considered during the design process.

The four limits are maximum collector current (I_C), maximum collector-emitter voltage (V_{CEO}), maximum power and *second breakdown* (I_{SB}). All four of these parameters are graphically shown on the transistor's data sheet on what is known as a *safe operating area (SOA)* graph. (see **Fig 7.22**) The first three of these limits are usually also listed prominently with the other device information, but it is often the fourth parameter — secondary breakdown — that is responsible for the "sudden death" of the power transistor after an extended operating period.

The maximum current limit of the transistor ($I_{C\ MAX}$) is usually the current limit for fusing of the bond wire connected to the emitter, rather than anything pertaining to the transistor chip itself. When this limit is exceeded, the bond wire can melt and open circuit the emitter. On the operating curve, this limit is shown as a horizontal line extending out from the Y-axis and ending at the voltage point where the constant power limit begins.

The maximum collector-emitter voltage limit of the transistor ($V_{CE\ MAX}$) is the point at which the transistor junctions can no longer withstand the voltage between collector and emitter.

With increasing collector-emitter voltage drop at maximum collector current, a point is reached where the power in the transistor will cause the junction temperature to rise to a level where the device leakage current rapidly increases and begins to dominate. In this region, the product of the voltage drop and the current would be constant and represent the maximum power (P_t) rating for the transistor; that is, as the voltage drop continues to increase, the collector current must decrease to maintain the power dissipation at a constant value.

With most transistors rated for higher voltages, a point is reached on the constant power portion of the curve whereby, with further increased voltage drop, the maximum power rating is *not* constant, but decreases as the collector to emitter voltage increases. This decrease in power handling capability continues until the maximum voltage limit is reached.

This special region is known as the *forward bias second breakdown* (FBSB) area. Reduction in the transistor's power handling capability is caused by localized heating in certain small areas of the transistor junction ("hot spots"), rather than a uniform distribution of power dissipation over the entire surface of the device.

The region of operating conditions contained within these curves is called the safe operating area, or SOA. If the transistor is always operated within these limits, it should provide reliable and continuous service for a long time.

MOSFET TRANSISTORS

The bipolar junction transistor (BJT) is rapidly being replaced by the MOSFET in new power supply designs because MOSFETs are easier to drive. The N-channel MOSFET (equivalent to the NPN bipolar) is more popular than the P-channel for pass transistor applications.

There are some considerations that should be observed when using a MOSFET as a linear regulator series pass transistor. Several volts of gate drive are needed to start conduction of the device, as opposed to less than 1 V for the BJT. MOSFETs are inherently very-high-frequency devices and will readily oscillate with stray-circuit capacitances. To prevent oscillation in the transistor and surrounding circuits, it is common practice to insert a small resistor of about 100 Ω directly in series with the gate of the series-pass transistor to reduce the gate circuit Q.

OVERCURRENT PROTECTION

Damage to a pass transistor can occur when the load current exceeds the safe amount. **Fig 7.23A** illustrates a simple current-limiter circuit that will protect Q1. All of the load current is routed through R1. A voltage difference will exist across R1; the value will depend on the exact load current at a given time. When the load current exceeds a predetermined safe value, the voltage drop across R1 will forward-bias Q2 and cause it to conduct. Because Q2 is a silicon transistor, the voltage drop across R1 must exceed 0.6 V to turn Q2 on. This being the case, R1 is chosen for a value that provides a drop of 0.6 V when the maximum safe load current is drawn. In this instance, the drop will be 0.6 V when I_L reaches 0.5 A. R2 protects the base-emitter junction of Q2 from current transients, or from destruction in the event Q1 fails under short-circuit conditions.

When Q2 turns on, some of the current through R_S flows through Q2, thereby depriving Q1 of some of its base current. This action, depending upon the amount of Q1 base current at a precise moment, cuts off Q1 conduction to some degree, thus limiting the current through it.

FOLDBACK CURRENT LIMITING

Under short-circuit conditions, a constant-

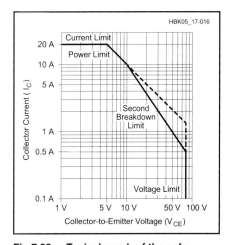

Fig 7.22 — Typical graph of the safe operating area (SOAR) of a transistor. See text for details. Safe operating conditions for specific devices may be quite different from those shown here.

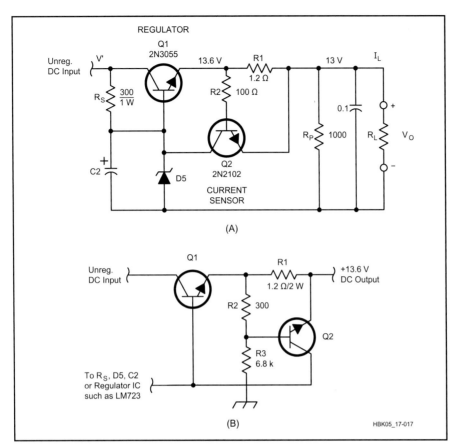

Fig 7.23 — Overload protection for a regulated supply can be implemented by addition of a current-overload-protective circuit, as shown at A. At B, the circuit has been modified to employ current-foldback limiting.

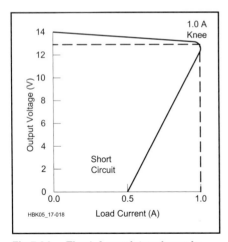

Fig 7.24 — The 1-A regulator shown in Fig 7.23B will fold back to 0.5 A under short-circuit conditions. See text.

current type current limiter must still withstand the full source voltage and limited short-circuit current simultaneously, which can impose a very high power dissipation or second breakdown stress on the series pass transistor. For example, a 12-V regulator with current limiting set for 10 A and having a source of 16 V will have a dissipation of 40 W [(16 V – 12 V) × 10 A] at the point of current limiting (knee). But its dissipation will rise to 160 W under short-circuit conditions (16 V × 10 A).

A modification of the limiter circuit can cause the regulated output current to decrease with decreasing load resistance beyond the over-current knee. With the output shorted, the output current is only a fraction of the knee current value, which protects the series-pass transistor from excessive dissipation and possible failure. Using the previous example of the 12-V, 10-A regulator, if the short-circuit current is designed to be 3 A (the knee is still 10 A), the transistor dissipation with a short circuit will be only 16 V × 3 A = 48 W.

Fig 7.23B shows how the current-limiter example given in the previous section would be modified to incorporate foldback limiting. The divider string formed by R2 and R3 provides a negative bias to the base of Q2, which prevents Q2 from turning on until this bias is overcome by the drop in R1 caused by load current. Since this hold-off bias decreases as the output voltage drops, Q2 becomes more sensitive to current through R1 with decreasing output voltage. See **Fig 7.24**.

The circuit is designed by first calculating the value of R1 for short-circuit current. For example, if 0.5 A is chosen, the value for R1 is simply 0.6 V / 0.5 A = 1.2 Ω (with the output shorted, the amount of hold-off bias supplied by R2 and R3 is very small and can be neglected). The knee current is then chosen. For this example, the selected value will be 1.0 A. The divider string is then proportioned to provide a base voltage at the knee that is just sufficient to turn on Q2 (a value of 13.6 V for 13.0 V output). With 1.0 A flowing through R1, the voltage across the divider will be 14.2 V. The voltage dropped by R2 must then be 14.2 V –13.6 V, or 0.6 V. Choosing a divider current of 2 mA, the value of R2 is then 0.6 V / 0.002 A = 300 Ω. R3 is calculated to be 13.6 V / 0.002 A = 6800 Ω.

7.9.4 Three-Terminal Voltage Regulators

The modern trend in regulators is toward the use of three-terminal devices commonly referred to as *three-terminal regulators*. Inside each regulator is a voltage reference, a high-gain error amplifier, temperature-compensated voltage sensing resistors and a pass element. Many currently available units have thermal shut-down, overvoltage protection and current foldback, making them virtually destruction-proof. It is easy to see why regulators of this sort are so popular when you consider the low price and the number of individual components they can replace.

Three-terminal regulators have connections for unregulated dc input, regulated dc output and ground, and they are available in a wide range of voltage and current ratings. Fixed-voltage regulators are available with output ratings in most common values between 5 and 28 V. Other families include devices that can be adjusted from 1.25 to 50 V.

The regulators are available in several different package styles, depending on current ratings. Low-current (100 mA) devices frequently use the plastic TO-92 and DIP-style cases. TO-220 packages are popular in the 1.5-A range, and TO-3 cases house the larger 3-A and 5-A devices. They are available in surface mount packages too.

Three-terminal regulators are available as positive or negative types. In most cases, a positive regulator is used to regulate a positive voltage and a negative regulator a negative voltage. Depending on the system ground requirements, however, each regulator type may be used to regulate the "opposite" voltage.

Fig 7.25A and B illustrate how the regulators are used in the conventional mode. Several regulators can be used with a common-input supply to deliver several voltages with a common ground. Negative regulators may be used in the same manner. If no other common supplies operate from the input

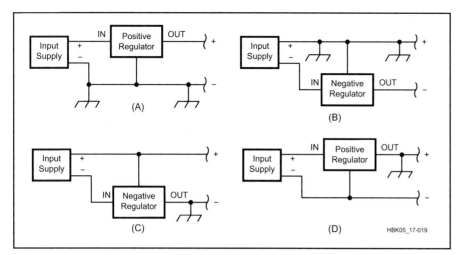

Fig 7.25 — Parts A and B illustrate the conventional manner in which three-terminal regulators are used. Parts C and D show how one polarity regulator can be used to regulate the opposite-polarity voltage.

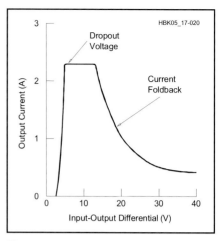

Fig 7.26 — Effects of input-output differential voltage on three-terminal regulator current.

supply to the regulator, the circuits of Fig 7.25C and D may be used to regulate positive voltages with a negative regulator and vice versa. In these configurations the input supply is floated; neither side of the input is tied to the system ground.

Manufacturers have adopted a system of family numbers to classify three-terminal regulators in terms of supply polarity, output current and regulated voltage. For example, 7805 describes a positive 5-V, 1.5-A regulator and 7905 a negative 5-V, 1.5 A unit. Depending on the manufacturer, the full part number might have a prefix such as LM, UA or MC, along with various suffixes (for example, LM7805CT or MC-7805CTG). There are many such families with widely varied ratings available from manufacturers. More information may be found in the **Component Data and References** chapter.

SPECIFYING A REGULATOR

When choosing a three-terminal regulator for a given application, the most important specifications to consider are device output voltage, output current, minimum and maximum input-to-output differential voltages, line regulation, load regulation and power dissipation. Output voltage and current requirements are determined by the load with which the supply will ultimately be used.

Input-to-output differential voltage is one of the most important three-terminal regulator specifications to consider when designing a supply. The differential value (the difference between the voltage applied to the input terminal and the voltage on the output terminal) must be within a specified range. The minimum differential value, usually about 2.5 V, is called the *dropout voltage*. If the differential value is less than the dropout voltage, no regulation will take place. Special *low dropout regulators* with lower minimum differential values are available as well. At the other end of the scale, maximum input-output differential voltage is generally about 40 V. If this differential value is exceeded, device failure may occur.

Increases in either output current or differential voltage produce proportional increases in device power consumption. By employing current foldback, as described above, some manufacturers ensure that maximum dissipation will never be exceeded in normal operation. **Fig 7.26** shows the relationship between output current, input-output differential and current limiting for a three-terminal regulator nominally rated for 1.5-A output current. Maximum output current is available with differential voltages ranging from about 2.5 V (dropout voltage) to 12 V. Above 12 V, the output current decreases, limiting the device dissipation to a safe value. If the output terminals are accidentally short circuited, the input-output differential will rise, causing current foldback, and thus preventing the power-supply components from being over stressed. This protective feature makes three-terminal regulators particularly attractive in simple power supplies.

When designing a power supply around a particular three-terminal regulator, input-output voltage characteristics of the regulator should play a major role in selecting the transformer-secondary and filter-capacitor component values. The unregulated voltage applied to the input of the three-terminal device should be higher than the dropout voltage, yet low enough that the regulator does not go into current limiting caused by an excessive differential voltage. If, for example, the regulated output voltage of the device shown in Fig 7.26 was 12 V, then unregulated input voltages of between 14.5 and 24 V would be acceptable if maximum output current is desired.

In use, all but the lowest-current regulators generally require an adequate external heat sink because they may be called on to dissipate a fair amount of power. Also, because the regulator chip contains a high-gain error amplifier, bypassing of the input and output leads is essential for stable operation.

Most manufacturers recommend bypassing the input and output directly at the leads where they protrude through the heat sink. Solid tantalum capacitors are usually recommended because of their good high-frequency capabilities.

External capacitors used with IC regulators may discharge through the IC junctions under certain circuit conditions, and high-current discharges can harm ICs. Look at the regulator data sheet to see whether protection diodes are needed, what diodes to use and how to place them in any particular application.

Adjustable Regulators

In addition to fixed-output-voltage ICs, high-current, adjustable voltage regulators are available. These ICs require little more than an external potentiometer for an adjustable output range from 5 to 24 V at up to 5 A. The unit price on these items is only a few dollars, making them ideal for test-bench power supplies. A very popular low current, adjustable output voltage three terminal regulator, the LM317, is shown in **Fig 7.27**. It develops a steady 1.25-V reference, V_{REF}, between the output and adjustment terminals. By installing R1 between these terminals, a constant current, I1, is developed, governed by the equation:

$$I1 = \frac{V_{REF}}{R1} \qquad (10)$$

Both I1 and a 100-µA error current, I2, flow

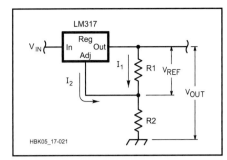

Fig 7.27 — By varying the ratio of R2 to R1 in this simple LM317 schematic diagram, a wide range of output voltages is possible. See text for details.

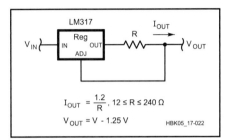

Fig 7.28 — The basic LM317 voltage regulator is converted into a constant-current source by adding only one resistor.

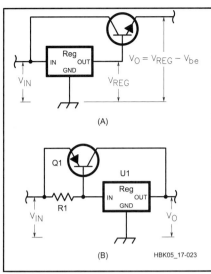

Fig 7.29 — Two methods for boosting the output-current capacity of an IC voltage regulator. Part A shows an NPN emitter follower and B shows a PNP "wraparound" configuration. Operation of these circuits is explained in the text.

through R2, resulting in output voltage V_O. V_O can be calculated using the equation:

$$V_O = V_{REF}\left(1+\frac{R2}{R1}\right) + I2 \times R2 \qquad (11)$$

Any voltage between 1.2 and 37 V may be obtained with a 40-V input by changing the ratio of R2 to R1. At lower output voltages, however, the available current will be limited by the power dissipation of the regulator.

Fig 7.28 shows one of many flexible applications for the LM317. By adding only one resistor with the regulator, the voltage regulator can be changed into a constant-current source capable of charging NiCd batteries, for example. Design equations are given in the figure. The same precautions should be taken with adjustable regulators as with the fixed-voltage units. Proper heat sinking and lead bypassing are essential for proper circuit operation.

INCREASING REGULATOR OUTPUT CURRENT

When the maximum output current from an IC voltage regulator is insufficient to operate the load, discrete power transistors may be connected to increase the current capability. **Fig 7.29** shows two methods for boosting the output current of a positive regulator, although the same techniques can be applied to negative regulators.

In A, an NPN transistor is connected as an emitter follower, multiplying the output current capacity by the transistor beta. The shortcoming of this approach is that the base-emitter junction is not inside the feed-back loop. The result is that the output voltage is reduced by the base-emitter drop, and the load regulation is degraded by variations in this drop.

The circuit at B has a PNP transistor "wrapped around" the regulator. The regulator draws current through the base-emitter junction, causing the transistor to conduct. R1 provides bias voltage for turning on Q1 so that U1 doesn't see the excess current. For example, a 6 Ω resistor will limit the current U1 sees to 100 mA. The IC output voltage is unchanged by the transistor because the collector is connected directly to the IC output (sense point). Any increase in output voltage is detected by the IC regulator, which shuts off its internal-pass transistor, and this stops the boost-transistor base current.

7.10 "Crowbar" Protective Circuits

Electronic components *do* fail from time to time. In a regulated power supply, the only component standing between an elevated dc source voltage and your transceiver is one transistor, or a group of transistors wired in parallel. If the transistor, or one of the transistors in the group, happens to short internally, your equipment could suffer lots of damage.

To safeguard the load equipment against possible overvoltage, some power-supply manufacturers include a circuit known as a *crowbar*. This circuit usually consists of a silicon-controlled rectifier (SCR) or thyristor connected directly across the output of the power supply, with an over-voltage-sensing trigger circuit tied to its gate. The SCR is large enough to take the full short-circuit output current of the supply, as if a crowbar was placed across the output terminals, thus the name.

In the event the output voltage exceeds the trigger set point, the SCR will fire, and the output is short circuited. The resulting high current in the power supply (shorted output in series with a series pass transistor failed short) will blow the power supply's line fuses. This is a protection for the supply as well as an indicator that something has malfunctioned internally. For these reasons, never replace blown fuses with ones that have a higher current rating.

An example of a crowbar overvoltage protection circuit can be found as a project at the end of this chapter. It provides basic design equations that can be adapted to a wide range of power supply applications.

7.11 DC-DC Switchmode Power Conversion

Very often the power source is dc, such as a battery or solar cell, or the output of an unregulated rectifier connected to an ac source. In most applications high conversion efficiency is desired, both to conserve energy from the source and to reduce heat dissipation in the converter. When high efficiency is needed, some form of switching circuit will be employed for dc-dc power conversion. Besides being more efficient, switching circuits are usually much smaller and lighter than conventional 60 Hz, transformer-rectifier circuits because they operate at much higher frequencies — from 25 to 400 kHz or even higher. Switching circuits go by many names; *switching regulators* and *switchmode converters* are just two of the more common names.

The possibility of achieving high conversion efficiency stems directly from the use of switches for the power conversion process, along with low-loss inductive and capacitive elements. An active switch is a device that is either ON or OFF and the state of the switch (ON or OFF) can be controlled with an external signal. The loss in the switch is always the product of the voltage across the switch and the current flowing through it ($P = E \times I$).

In the ON (conducting) state the voltage drop across the switch is small, and in the OFF state the current through the switch is small. In both cases the losses can be small relative to the power level of the converter. During transitions between ON and OFF states, however, there will simultaneously be both substantial voltage across the switch and significant current flowing through it. This results in power dissipation in the switch, called *switching loss*. This loss is minimized by making the switching transitions as short as possible. In this way even though the instantaneous power dissipation may be high, the average loss is low because of the small duty cycle of the transitions. Of course the more frequently the switch operates (higher switching frequency), the higher the average loss will be and this eventually limits the maximum operating frequency. A limitation on the lower end of the switching frequency range is that it needs to be above audible frequencies (>25 kHz).

This is quite different from the linear regulators discussed earlier in which there is always some voltage drop across the pass transistor (which is acting as a controlled variable resistor) while current is flowing through it. As a result the efficiency of a linear regulator can be very low, often 65% or less. Switchmode converters on the other hand will typically have efficiencies in the range of 85 to 95%.

Switchmode circuits can also generate radio-frequency interference (RFI) through VHF because of switching frequency harmonics and ringing induced by the rapid rise and fall times of voltage and current. In attempting to minimize the ON-OFF transition time, significant amounts of RF energy can be generated. To prevent RFI to sensitive receivers, careful bypassing, shielding and filtering of both input and output circuits is required. (RFI from switchmode or "switching" supplies is also discussed in the chapter on **Electromagnetic Capability**.)

There are literally hundreds of different switchmode circuits or "topologies" (see Reference 2) but we will only look at a few of those most commonly used by amateurs. Fortunately, the characteristics of the simpler circuits are to a large extent replicated in more complex circuits so that an understanding of the basic circuits provides an entry point to many other circuits.

The following discussion is only an introduction to the basics of switchmode converters. To really get a handle on designing these circuits the reader will have to do some additional reading. Fortunately a very large amount of useful information is freely available on-line. Many useful application notes are available from semiconductor manufacturers, and additional information can be found on the Web sites of filter capacitor and ferrite core manufacturers (see References 3 and 4). There are also numerous books on the subject, often available in libraries and used book stores.

Switchmode circuits can and have been implemented with many different kinds of switches, from mechanical vibrators to vacuum and gas tubes to semiconductors. Today however, semiconductors are the universal choice. For converters in the power range typical of amateur applications (a few watts through 2-3 kW) the most common choices of semiconductor switches would be either bipolar junction transistors (BJTs) or power MOSFETs. BJTs have a long history of use in switchmode applications, but to employ BJTs in a reliable and trouble-free circuit requires a relatively sophisticated understanding of them. While the MOSFET must also be used carefully, it is generally easier for a newcomer. The following circuit diagrams will show MOSFETs or generic switch symbols for the switches but keep in mind that all of the circuits can be implemented with other types of switches.

7.11.1 The Buck Converter

A schematic for a *buck converter* is shown in **Fig 7.30A**. This circuit is called a "buck" converter because the output voltage is always less than or equal to the input voltage. Power is supplied from the dc source (V_S) through the input filter (L1, C1) to the drain of the switch (Q1). The load (R) is connected across the output filter (L, C2).

The equivalent circuit when Q1 is ON is shown in Fig 7.30B (where the input filter

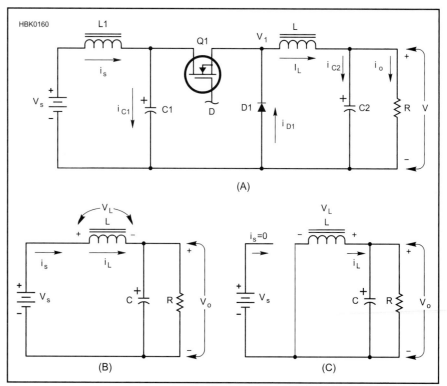

Fig 7.30 — Typical buck converter.

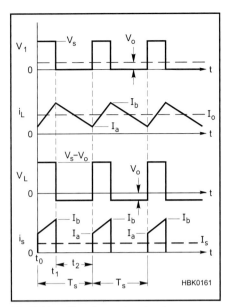

Fig 7.31 — Waveforms in a buck converter.

components are assumed to be part of V_S). V_S is connected to one end of the output inductor (L) and, because $V_O < V_S$, current flows from the source to the output delivering energy from the source to the output and also storing energy in L. At some point Q1 is turned OFF and the current flowing in L commutates (switches) to D1, as shown in Fig 7.30C. (The current in an inductor can not change instantaneously, so when Q1 turns OFF, the collapsing magnetic field in L pulls current through shunt diode D1, called a *free-wheeling diode*.) The energy in L is now being discharged into the output. This cycle is repeated at the switching frequency (f_s). The ratio of the switch ON-time to the total switching period ($T_s = t_{on} + t_{off} = 1/f_s$) is called the *duty cycle (D)*.

Typical voltage and current waveforms for a buck converter are shown in **Fig 7.31** where V_1 is the voltage at the junction of Q1, D1 and L (see Fig 7.30A).

The interval $0-t_1$ corresponds to the ON time of Q1 and $T_s - t_1$ corresponds to the OFF time ($T_s = t_1 + t_2$). From the waveforms it can be seen that the current in the inductor rises while Q1 is ON and falls while Q1 is OFF. The energy in the inductor is proportional to $LI^2/2$.

The input current (I_s) is pulsating at the switching frequency. This pulsation in the input current is the reason for the input filter (L1 and C1). We need to keep this high frequency noise (the ac component of the pulsation) out of the source. All switchmode converters require some form of filter on the input to keep switching noise out of the source. Some form of filter is also required on the output to keep the switching noise out of the load. In Fig 7.30, L and C2 form the output filter. For applications requiring low ripple it would not be unusual to see an additional stage of L-C filtering added to the output.

The voltage waveform (V_1) at the input to the output filter (L and C2) is pulse-width modulated (PWM) and V_O (the dotted line in the waveform for V_1 in Fig 7.31A) is the time average of V_1. V_O is controlled or regulated by adjusting the duty cycle of Q1 in response to input voltage or output load changes. If we increase D we increase V_O up to the point where D = 1 (the switch is ON all the time) and $V_O \approx V_S$. If we never turn Q1 on (D = 0) then $V_O = 0$. Normal operation will be somewhere between these extremes.

The inductor current waveform in Fig 7.31 does not go to zero while Q1 is OFF. In other words, not all of the energy stored in L is discharged into the output by the end of each switching cycle. This is a common mode of operation for heavier loads. It is referred to as the *continuous conduction mode* or CCM, where the conduction referred to is the current in L. There is another possibility: during the time Q1 is off all of the energy in L may be discharged and for some period of time the inductor current is zero until Q1 is turned ON again. This is referred to as the *discontinuous conduction mode* or DCM. A typical converter will operate in CCM for heavy loads but as the load is reduced, at some point the circuit operation will change to DCM. CCM is often referred to as the "heavy load" condition and DCM as the "light load" condition.

This distinction turns out to be very important because the behavior of the circuit, for both small signal and large signal, is radically different between the two modes. **Fig 7.32** shows the relationship between the duty cycle (D) of Q1 and the ratio of the output voltage to the input voltage ($M = V_O/V_S$) as a function of τ_L in Fig 7.32:

$$\tau_L = \left(\frac{L}{R}\right) f_S \qquad (12)$$

τ_L is just a convenient way to make Fig 7.32 more general by tying together the variables L, R and f_s. Smaller values for τ_L correspond to lighter loads (larger values for R). As can been seen in Fig 7.32, for very light loads the higher values for D have little effect on the output/input voltage ratio. This is basically charging of the output capacitor (C2) to the peak value of $V_1 \approx V_S$.

For CCM operation, M is a linear function of D. As we vary the load, V_O will remain relatively constant. But when we go into DCM, the relation between D and M is no longer linear and in addition is heavily dependent on the load: V_O will vary as the load is changed unless the duty cycle is varied to compensate. This kind of behavior is typical of all switching regulators, even those not directly related to the buck converter. In fact, we have seen this behavior already in the section on choke input filters in Fig 7.17 where the output voltage is close to the peak input voltage for light loads and decreases as the load increases until a point is reached (I_C) where V_O stabilizes.

In a buck converter the value for the critical inductance (L_C) can be found from:

$$L_c = R\left(\frac{1-M}{2f_S}\right) \qquad (13)$$

This looks just like equation 6 (in the earlier section on choke-input filters) with $A = 2f_s / (1 - M)$! The two phenomena are the same: peak charging versus averaging of V_1.

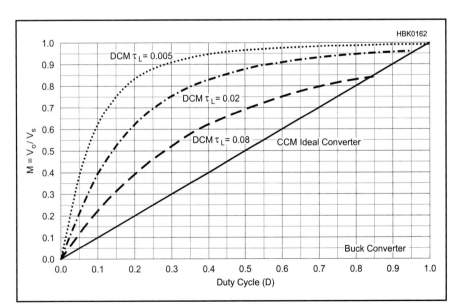

Fig 7.32 — Change in output voltage to input voltage ratio (M=VO/VS) as a function of switch duty cycle (D).

7.11.2 The Boost Converter

A *boost converter* circuit is shown in **Fig 7.33A**. This circuit is called a "boost" converter because the output voltage is always greater than or equal to the input voltage.

When Q1 is ON (Fig 7.33B), L is connected in parallel with the source (V_S) and energy is stored in it. During this time interval load energy is supplied from C. When Q1 turns OFF (Fig 7.33C), L is connected via D1 to the output and the energy in L, plus additional energy from the source, is discharged into the output. The value for L, V_S and length of time it is charged determines the energy stored in L. Again we have the possibilities that either some (CCM) or all (DCM) the energy in L is discharged during the OFF-time of Q1. The variation in the ratio of the output-to-input voltage (M) with duty cycle is shown in **Fig 7.34**.

As we saw in the buck converter, the conduction mode of L is important. In an ideal converter operating in CCM, the output voltage is substantially independent of load and $M \approx 1 / (1 - D)$. In realistic boost converters there is an important limit on the CCM value for M. In an ideal boost converter you could make the boost ratio (M) as large as you wish but in real converters the parasitic resistances associated with the components will limit the maximum value of M as indicated by the dashed line for CCM operation. The exact shape of this part of the control function will depend on the ratio of the parasitic resistance within the converter to the load resistance (see Reference 2).

There is also another very important practical effect of this limitation on the peak value for M. When the duty cycle is increased beyond the point of maximum M (this occurs at D = 0.9 in Fig 7.34), the sign of the slope of the control function changes so that the control loop will change from negative feedback to positive feedback. This can cause the converter to latch up under some load conditions if the control circuit allows D to exceed the value at maximum M. In boost converters the design of the control circuit must limit the maximum value for D so that latch-up is not possible although this may be difficult for overload conditions.

As in the case of the buck converter, in DCM the control function for M is very different from what it is in CCM and is strongly dependent on the load R. One advantage of the DCM operation is that the limitation on the maximum value for M because of parasitic circuit resistances is not nearly so pronounced. By operating in DCM it is possible to have M > 5.

An important limitation of the boost converter is that with Q1 turned OFF, you have no control over V_O. V_O will simply equal V_S. In addition, when V_S is turned on there will be an inrush current through D1, into C which cannot be controlled by Q1. In the case of the buck converter, if Q1 is kept OFF when V_S is turned on, there will be no inrush current charging the output filter capacitor (C2). In the buck converter C2 can be charged slowly by ramping up the duty cycle of Q1 during turn-on but this is not possible in the boost converter.

7.11.3 Buck-Boost and Flyback Converters

Fig 7.35 shows an example of a *buck-boost converter*. The name "buck-boost" comes from the fact that the magnitude of the output voltage can be either greater or less than the input voltage.

When S1 is ON, V_S is applied across L and energy is stored in it. During the ON time of Q1, D1 is reverse-biased and non-conducting. The output voltage (V_O) across the load (R) is supported solely from the energy stored in C. When S1 is OFF, the energy in L is discharged into C through D1. Note that the polarity of V_O is inverted from V_S: *positive V_S means negative V_O*. The relationship between V_O and V_S as a function of duty cycle is shown in **Fig 7.36**.

This graph closely resembles that for the boost converter (Fig 7.34) with one important exception: |M| begins at zero. This allows the

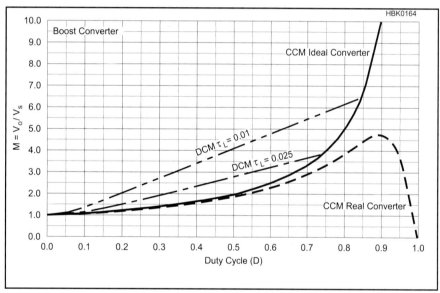

Fig 7.34 — DC control characteristic for a boost converter. M = V_O/V_S, D = duty cycle of Q1 and $\tau_L = f_s (L/R)$.

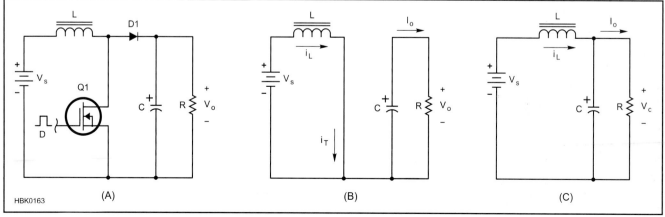

Fig 7.33 — Typical boost converter.

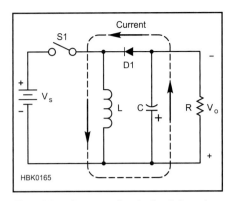

Fig 7.35 — An example of a buck-boost converter.

the transformer-coupled version of this converter, referred to as a *flyback converter*, that is employed. The relationship between the flyback and buck-boost converters is illustrated in **Fig 7.37**. In Fig 7.37A we have a standard buck-boost converter, the only change from Fig 7.35 being two parallel and equal windings on the inductor. In Fig 7.37B we remove the links at the top and bottom of the two parallel windings, converting the inductor into a transformer with primary and secondary windings. The only change is that when S1 is ON, current flows in the primary of the transformer and when S1 is OFF, the current flows in the secondary winding delivering the stored energy to the output. At this point the circuit operation is the same except that we have introduced primary-to-secondary galvanic isolation.

We are now free to change the turns ratio from 1:1 to anything we wish. We can also change the polarity of the output voltages and/or add more windings with other voltages and additional loads as shown in Fig 7.37C, a typical example of a flyback converter. These are most often used in the power range of a few watts to perhaps 200 W. For higher power levels other circuits are generally more useful.

The advantages of the flyback converter lie in its simplicity. It requires only one power switch and one diode on each of the output windings. The inductor is also the isolation transformer so you have only one magnetic

magnitude of the output voltage to be either below or above the source voltage, hence the name "buck-boost."

FLYBACK CONVERTERS

Simple buck-boost converters are occasionally used, but much more often it is

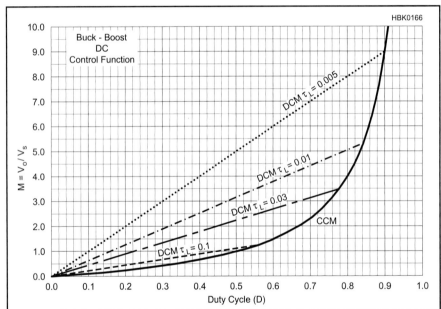

Fig 7.36 — DC control function for an ideal buck-boost converter $M = V_O/V_S$, D = duty cycle of Q1 and $\tau_L = f_s (L/R)$.

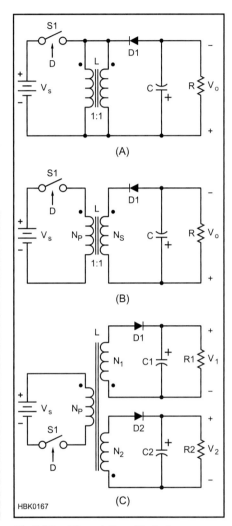

Fig 7.37 — The relationship between buck-boost and flyback converters.

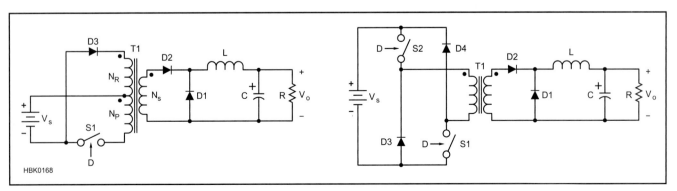

Fig 7.38 — Example of a forward converter.

7.24   Chapter 7

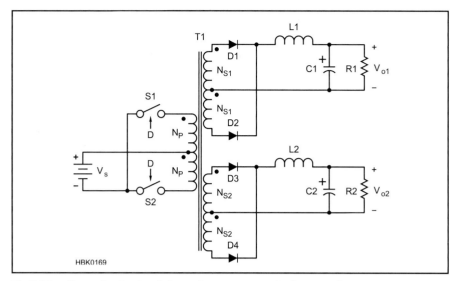

Fig 7.39 — Example of a parallel quasi-square wave dc-dc converter.

very helpful at higher line voltages. These circuits behave just like a buck converter except you can now have multiple isolated outputs with arbitrary voltages and polarities. These circuits are typically used in converters with power levels of 100 to 500 W.

7.11.5 Parallel, Half and Full-Bridge Converters

As the power level increases it becomes advantageous to use more switches in a somewhat more complex circuit. For applications where the input voltage is low (<100V) and the current high, the push-pull *quasi-square wave* circuit shown in **Fig 7.39** is often used.

S1 and S2 are switched on alternately with duty cycles <0.5. The output voltages are controlled by the duty cycle, D. The peak switch currents are equal to the input current but the peak switch voltages will be $2V_S$.

This circuit is still just a buck converter, but with an isolation transformer added that allows multiple outputs with different voltages above or below the input voltage. It would be very common to have a 5 V, high-current output and ±12 V, lower-current outputs, for example. This converter is typically used for operation from vehicular power with loads up to several hundred watts.

Operating directly from rectified ac util-

component. The disadvantage of this circuit is that both in the input and output current waveforms are pulsating. The result is that more filtering is required and the filter capacitors are exposed to high RMS currents relative to the power level.

7.11.4 The Forward Converter

The buck converter has many useful properties but it lacks input-to-output galvanic isolation, the ability to produce output voltages higher than the input voltage, and/or multiple isolated output voltages for multiple loads. We can overcome these drawbacks by inserting a transformer between the switch (S1) and the shunt diode (D1). To make this simple idea work however, we also have to add a diode in series with the output of the transformer (D2), and a third winding (N_R) with another diode (D3) to the transformer. This is done to provide a means for resetting the core (returning the magnetic flux to zero) by the end of each switching cycle. The result is the *forward converter* shown in **Fig 7.38**.

The circuit in A is the one just described. The variation in B uses two switches (S1 and S2, which switch ON and OFF *simultaneously*) instead of one but eliminates the need for a reset winding on the transformer. For the circuit in A and a given input voltage, the voltage across the switch (in the OFF state) will be equal to V_S plus the reset voltage during the OFF-time. Typically the peak switch voltage will be about $2V_S$. In the circuit shown in B the switch voltage is limited to V_S which is

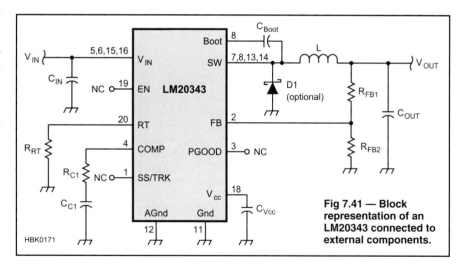

Fig 7.41 — Block representation of an LM20343 connected to external components.

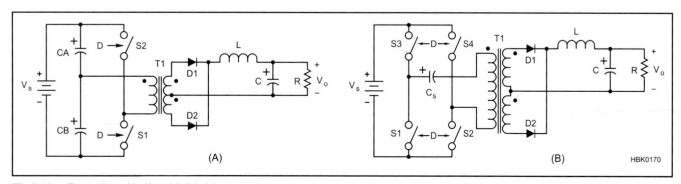

Fig 7.40 — Examples of half and full-bridge quasi-square wave dc-dc converters.

Power Supplies 7.25

ity power usually means that V_S will be 200 V or more. For these applications, **Fig 7.40** shows how the switches are configured in either a half- (Fig 7.40A) or full-bridge (Fig 7.40B) circuit.

In A, S1 and S2 switch alternately and are pulse-width modulated (PWM) to control the output. CA and CB are large capacitors that form a voltage divider with $V_S/2$ across each capacitor. The peak switch voltages will be equal to V_S but the switch currents will be $2I_S$. The peak voltage across the primary winding will be $V_S/2$. This circuit would typically be used for off-line applications with output powers of 500 W or so.

In B, S1 and S4 switch simultaneously alternating with S2 and S3 which also switch simultaneously. The output is controlled by PWM. The peak switch voltage will be equal to V_S and the peak switch current close to I_S (the peak value is a little higher due to ripple on the inductor current which is reflected back into the primary winding). The full-bridge circuit is typically used for power levels of 500 W to several kW. C_S is present in the full-bridge circuit to prevent core saturation due to any asymmetry in the primary PWM voltage waveforms.

7.11.6 Building Switching Power Supplies

Selecting a switching circuit or topology is just the first step in building a practical switching power supply. In the actual circuit you will have to sample the output voltage, provide a voltage reference against which to compare the output, include a modulator that will convert the error voltages to a variable-duty cycle signal and finally provide correct drive to the power switches. Fortunately, all of these functions can be provided with readily available integrated circuits and a few external components. These ICs typically come with extensive applications information.

Particularly for low power applications (<100 W) there are ICs that provide all the control functions *and* the power switch. In some cases a power diode is also included. **Fig 7.41** gives an example of one of these ICs. Similar IC's are made by many different manufacturers.

The IC provides most of the components for a buck converter but some external components are still needed: filter capacitors (C_{IN}, C_{OUT}, C_{VCC}), the output inductor (L), output voltage sensing (R_{FB1} and R_{FB2}) and components to set the switching frequency (R_{C1} and C_{C1}). The shunt diode (D1) is marked as

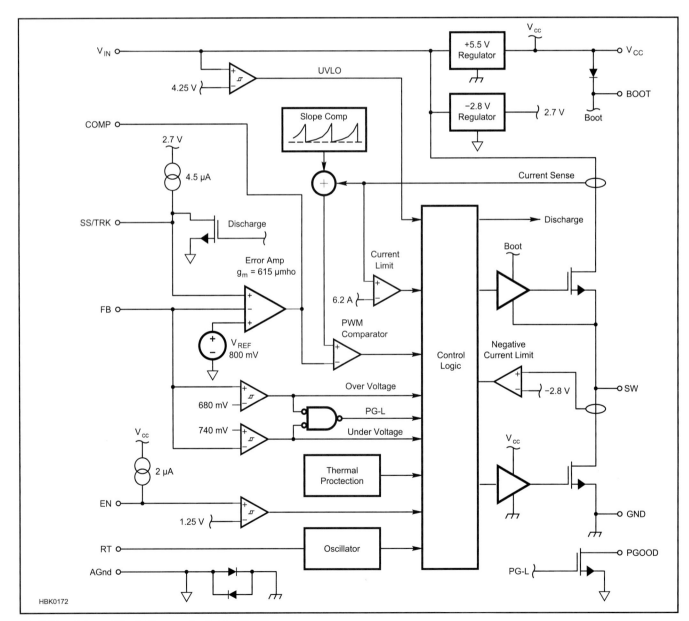

Fig 7.42 — Internal block diagram for the LM20343.

optional because there is an internal switch that can perform this function for lower powers. We can get a more detailed look at the circuitry within the IC from **Fig 7.42**.

In addition to the control functions there are also some protective functions such as input under- and overvoltage protection, over-temperature (thermal) protection, over-current limiting and slow-starting (slowly ramping up D) to limit inrush current at turn-on. ICs similar to this one are available for boost, buck-boost and flyback converters, as well as many other topologies.

For higher power applications, the power components are external to the control IC but additional features may be added. **Figs 7.43** and **7.44** are examples of typical general-purpose IC controllers. There are also ICs available that implement more complex control schemes such as current mode control and feed-forward compensation.

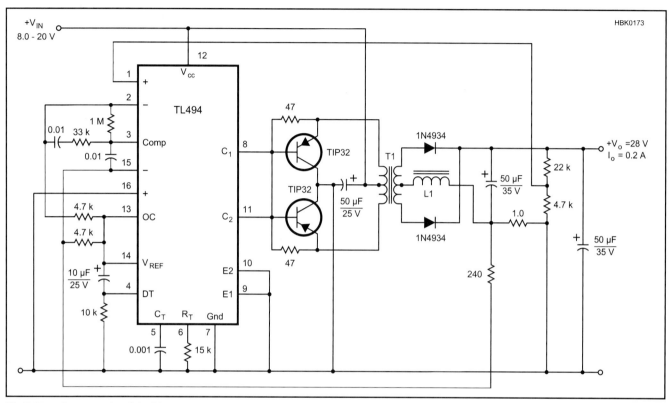

Fig 7.43 — An example of the TL494 applied to a push-pull dc-dc converter.

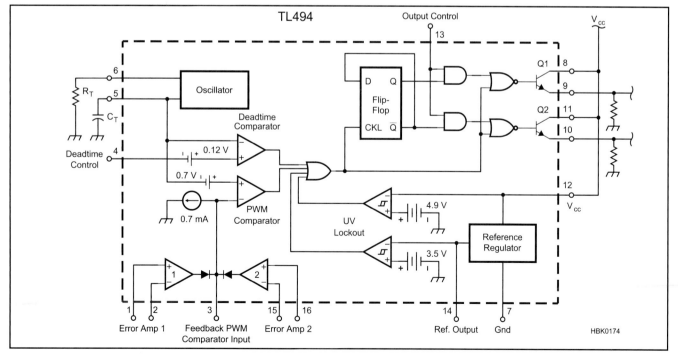

Fig 7.44 — Internal block diagram representation for the TL494.

Switchmode Converter Design Aids

By Chuck Mullett, KR6R

Today, there are hundreds of ICs on the market, from dozens of manufacturers, aimed at controlling the power conversion process. These devices have grown from their simplest forms when introduced over 20 years ago to include a long list of ancillary functions aimed at reducing the supporting circuitry and enhancing the quality and efficiency of the power conversion process. As a result, the task of designing these ICs into the final power conversion circuit has actually become more complex. To the circuit designer who is below the expert level, the task can seem daunting, indeed.

In answer to this problem, many IC manufacturers provide computer-aided design tools that greatly simplify the task of using their products. These tools take several forms, from the most rudimentary cookbook-style guides, to full-fledged circuit simulation tools like *SPICE*. The purpose of these tools is to help the designer pick appropriate resistors, capacitors and magnetic components for the design, and also to estimate the stresses on the power-handling components.

These design tools fall generally into two types: equation-based design tools and iterative circuit simulators. The differences are that the equation-based tools simply automate the basic design equations of the circuit, providing recommended choices for the components and then solving equations to compute the voltage, current and power stresses on the components. In some cases, thermal analysis is also included.

The simulators, on the other hand, use detailed mathematical models for the components and provide both dc and ac simulation of the circuits. This allows the user to see the dynamic behavior of the circuit. The simulation includes details during the switching intervals and shows rise times, parasitic effects caused by component capacitance, internal resistance and other characteristics.

These simulators use iterative mathematical processes, so they can take tens of seconds or even minutes to do the analysis, even on a modern, high-performance computer. Usually, the IC manufacturers provide fairly rudimentary versions of these simulators, capable of analyzing a basic circuit with a dozen or so components.

The free *SPICE*-based software from Linear Technology Corporation (*LTSpice*, **www.linear.com**) is capable of modeling larger circuits. More capable simulators are available from software vendors, but the cost can run from hundreds of dollars to tens of thousands.

In modern power converter design, the magnetic components — transformers, power inductors and filter inductors — can be a challenge to the designer not versed in magnetic design. As a result, the vendors of these parts usually furnish design tools created specifically to the design of these components and/or the proper choice of off-the-shelf versions.

The reader is encouraged to explore design aids available from the Web sites of manufacturers (**Table 7.A1**). The following is far from complete, but should give the reader valuable insight into the vast array of available tools at his disposal. Remember also that manufacturers of passive components such as capacitors, resistors, heat sinks and thermal hardware may also have helpful and informative design aids on their Web sites.

Table 7.A1
Switchmode Converter Design Aids

Vendor	Main Web Site	Design Tools
Fairchild Semiconductor	www.fairchildsemi.com	From the home page, select "Design Center"
Infineon	www.infineon.com	From the home page Search window, select "Website" and "Search Technical Documents", then select "Power Management IC's" from the pull-down menu. From the subsequent list, select "Application Notes"
International Rectifier	www.irf.com/indexsw.html	www.irf.com/design-center/mypower/
Intersil	www.intersil.com	www.intersil.com/design/
National Semiconductor	www.national.com/analog	www.national.com/analog/power/simple_switcher
ON Semiconductor	www.onsemi.com	onsemi.com/PowerSolutions/supportDoc.do?type=tools
Texas Instruments	www.power.ti.com	From the Power Management page, click "Tools and Software"
Linear Technology Corp	www.linear.com	Search for "LTSpice" in the home page window to access the simulation tools page.

7.11.7 Switchmode Control Loop Issues

We do not have the space in this book to explain all the control issues associated with switchmode converters, but it is important to recognize that designing stable, high performance control loops for switchmode converters is a much more complex task than it would be for a series pass-regulator or an audio amplifier. All we can do here is to alert the reader to some of the issues and suggest consulting Reference 2 with detailed explanations and examples. In addition, there are numerous books and applications notes on switchmode converter design that go into the necessary detail.

The complexities arise from the inherent behavior of switchmode circuits. This behavior is often non-intuitive and sometimes even bizarre. As we've seen already, the dc behavior differs dramatically between CCM and DCM modes of operation. This difference is even more pronounced in the small-signal control-to-output characteristics. It is very common to see poles and zeros in the control system's response (see the **Analog Basics** chapter) that move with input voltage, duty cycle and/or load. Fixed double-poles can change to moving single-poles.

Moving right-half-plane zeros (this is a zero with decreasing phase-shift instead of increasing phase-shift with frequency) are inherent in CCM operation of boost and buck-boost circuits and their derivatives. There can also be large signal instabilities such as oscillations at sub-harmonics of the switching frequency and occasionally chaotic response to large load current or line voltage changes.

This is not intended to discourage readers from working with switching converters, but to simply alert them in advance of some of the difficulties. Once a basic understanding is achieved, the design of switching converters can be very interesting and rewarding.

7.12 High-Voltage Techniques

The construction of high-voltage supplies requires special considerations in addition to the normal design and construction practices used for lower-voltage supplies. In general, the builder needs to remember that physical spacing between leads, connections, parts and the chassis must be sufficient to prevent arcing. Also, the series connection of components such as capacitor and resistor strings needs to be done with consideration for the distribution of voltage stresses across the components. High-voltages can constitute a safety hazard and great care must be taken to limit physical access to components and wiring while high potentials are present.

7.12.1 High-Voltage Capacitors

For reasons of economy and availability, electrolytic capacitors are frequently used for output filter capacitors. Because these capacitors have relatively low voltage ratings (<600 V), in HV applications it will usually be necessary to connect them in series strings to form an equivalent capacitor with the capability to withstand the higher applied voltage. Electrolytic capacitors have relatively high leakage currents (low leakage resistance) especially at higher temperatures.

To keep the voltages across the capacitors in the series string relatively constant, equal-value bypassing resistors are connected across each capacitor. These *equalizing resistors* should have a value low enough to equalize differences in capacitor leakage resistance between the capacitors, while high enough not to dissipate excessive power. Also, capacitor bodies need to be insulated from the chassis and from each other by mounting them on insulating panels, thereby preventing arcing to the chassis or other capacitors in the string. The insulated mounting for the capacitors is often plastic or other insulating material in board form. A typical example from a commercial amplifier that implements some the guidelines given in this section is shown in **Fig 7.45**.

Equalizing resistors are needed because of differences in dc leakage current between different capacitors in the series string. The data sheet for an electrolytic capacitor will usually give the dc leakage current at 20 °C in the form of an equation. For example:

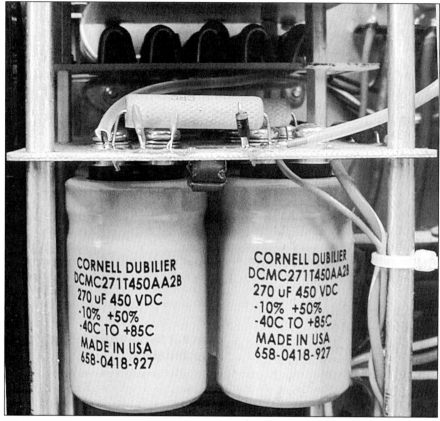

Fig 7.45 — Example of the filter capacitor/bleeder resistor installation in a high voltage power supply.

$$I_{leakage} = 3\sqrt{CV} \quad (14)$$

where
C = the value in μF
V = the working voltage
$I_{leakage}$ = leakage current in μA.

Keep in mind that the leakage current will increase as the capacitor temperature rises above 20 °C. A value of 3 to 4 times the 20 °C value would not be unusual because of normal heating.

We can use an approximation which includes allowance for heating to determine the maximum value of the divider resistors:

$$R \leq \frac{V_r - V_m}{I_{leakage}} \quad (15)$$

where
V_r = the voltage rating of the capacitor
V_m = the voltage across the capacitor during normal operation.

A typical 3000-V power supply might use eight 330-μF, 450-V capacitors in series. In that case, V_r = 450 V, V_m = 375 V and $I_{leakage}$ = 1.2 mA. This would make R = 63 kΩ or a total of 504 kΩ for the resistor string. With 3000-V across the resistor string, the total power dissipation would be 18 W or about 2.25 W per resistor. To be conservative you should use resistors rated for at least 4 W each.

The equalizing resistors may dissipate significant power and become quite warm. It is important that these resistors do not heat the capacitors they are associated with, as this will increase the capacitor leakage current. You also have to be careful that the heat from the resistors is not trapped under the plastic support panels. The best practice is to place the resistors above the capacitors and their mounting structure allowing the heat to rise unobstructed as shown in Fig 7.45.

OIL-FILLED CAPACITORS

For high voltages, oil-filled paper-dielectric capacitors are superior to electrolytics because they have lower internal impedance at high frequencies, higher leakage resistance and are available with much higher working voltages. These capacitors are available with values of several microfarads and have working voltage ratings of thousands of volts. On the other hand, they can be expensive, heavy and bulky.

Oil-filled capacitors are frequently offered for sale at flea markets at attractive prices. One caution: It is best to avoid older oil-filled capacitors because they may contain polychlorinated biphenyls (PCBs), a known cancer-causing agent. Newer capacitors have eliminated PCBs and have a notice on the case to that effect. Older oil-filled capacitors should be examined carefully for any signs of leakage. Contact with leaking oil should be avoided, with careful washing of the hands after handling. Do not dispose of any oil-filled capacitors with household trash, particularly older units. Contact your local recycling agency for information about how to dispose of them properly.

7.12.2 High-Voltage Bleeder Resistors

Bleeder resistors across the output are used to discharge the stored energy in the filter capacitors when the power supply is turned off and should be given careful consideration. These resistors provide protection against shock when the power supply is turned off and dangerous wiring is exposed. A general rule is that the bleeder should be designed to reduce the output voltage to 30 V or less within 30 seconds of turning off the power supply.

Take care to ensure that the maximum voltage rating of the resistor is not exceeded. In a typical divider string, the resistor values are high enough that the voltage across the resistor is not dissipation limited. The voltage limit is typically related to the insulation intrinsic to the resistor. Resistor maximum voltage ratings are usually given in the manufacturer's data sheet and can be found online. Two major resistor manufacturers are Ohmite Electronics (**www.ohmite.com**) and Stackpole Electronics (**www.seielect.com**).

A 2-W carbon composition resistor will have a maximum voltage rating of 500 V. As a rough estimate, larger wire-wound power resistors are typically rated at 500 V_{RMS} per inch of length — but check with the manufacturer to be sure.

The bleeder will consist of several resistors in series. Typically wire-wound power resistors are used for this application. One additional recommendation is that two separate (redundant) bleeder strings be used, to provide safety in the event one of the strings fails. When electrolytic capacitors are used, the equalizing resistors can also serve as the bleeder resistor but they should be redundant. Again, give careful attention to keeping the heat from the resistors away from the capacitors as shown in Fig 7.45.

In the example given above for calculating the equalizing resistor value, eight 330-µF capacitors in series created the equivalent of a 41-µF capacitor with a 3000-V rating. The total resistance across the capacitors was 8×63 kΩ = 504 kΩ. This gives us a time constant (R × C) of about R × C ≈ 21 seconds. To discharge the capacitors to 30 V from 3000 V would take about four time constants (about 84 seconds) which is well over a minute. To get the discharge time down to 30 seconds would require reducing the equalizing resistors to 25 kΩ each. The bleeder power dissipation would then be:

$$P = \frac{V^2}{R} = \frac{3000^2}{200,000} = 45 \text{ W} \quad (16)$$

For a 2 or 3 kV power supply, this is a reasonable value but it is still a significant amount of power and you need to make sure the resulting heat is properly managed.

7.12.3 High-Voltage Metering Techniques

Special considerations should be observed for metering of high-voltage supplies, such as the plate supplies for linear amplifiers. This is to provide safety to both personnel and to the meters themselves.

To monitor the current, it is customary to place the ammeter in the supply return (ground) line. This ensures that both meter terminals are close to ground potential. Placing the meter in the positive output line creates a hazard because the voltage on each meter terminal would be near the full high-voltage potential. Also, there is the strong possibility that an arc could occur between the wiring and coils inside the meter and the chassis of the amplifier or power supply itself. This hazardous potential cannot exist with the meter in the negative leg.

Another good safety practice is to place a low-voltage Zener diode across the terminals of the ammeter. This will bypass the meter in the event of an internal open circuit in the meter. A 1-W Zener diode will suffice since the current in the metering circuit is low.

The chapter on **RF Power Amplifiers** contains examples of how to perform current and voltage metering in high voltage supplies and amplifiers. In the past, amplifier articles have shown the meters mounted on plastic boards with stand-off insulators behind a plastic window in the amplifier or power supply front panel. While this can work it is not considered good practice today. It is usually possible to arrange the metering so that the meters are close to ground potential and may be safely mounted on the front panel of either the amplifier or the power supply.

For metering of high voltage, the builder should remember that resistors to be used in multiplier strings will often have voltage-breakdown ratings well below the total voltage being sampled. Usually, several identical series resistors will be used to reduce voltage stress across individual resistors. A basic rule of thumb is that these resistors should be limited to a maximum of 200 V, unless rated otherwise. For example, in a 2000-V power supply, the voltmeter multiplier should have a string of 10 resistors connected in series to distribute the voltage equally. The comments on resistor voltage rating in the sections on capacitor equalization and bleeder resistors apply to this case, as well.

7.12.4 High-Voltage Transformers and Inductors

Usable transformers and filter inductors can often be found at amateur flea markets but frequently these are very old units, often dating from WWII. The hermetically-sealed military components are likely to still be good, especially if they have not been used. Open-frame units, with insulation that has been directly exposed to the atmosphere and moisture for long periods of time, should be considered suspect. These units can be checked by running what is called a "hipot test" (high potential test). This involves a low current, variable output voltage power supply with a high-value resistor in series with the output. Unfortunately this equipment is seldom available to amateurs. Some motor repair shops will have an insulation tester called a "Megger" and may be willing to perform an insulation test on a transformer or inductor for you. This is not a completely definitive test but will certainly detect any gross problems with the insulation.

An alternative would be to perform the transformer tests discussed earlier with a variable autotransformer on the ac input. Slowly increase the input voltage until full line voltage is reached and let the transformer run for an hour or two while watching to see if there are any signs of failure. Doing this test in a dark room makes it easier to see any visible corona discharge, another sign of insulation problems. Because the transformer terminals will be at full voltage great care should be taken to avoid contact with the HV terminals. Some form of transparent insulating shield for the test setup is necessary.

7.12.5 Construction Techniques for High-Voltage Supplies

Layout and component arrangement in HV supplies requires some additional care beyond those for lower voltage projects. The photographs in the **RF Power Amplifiers** chapter are a good start, but there are some points which may not be obvious from the pictures.

SHARP EDGES

Sharp points, board edges and/or hardware with ragged edges can lead to localized intensification of the electric field's strength, resulting in a possible breakdown. One common offender is the component leads on the soldered side of a printed-circuit board. These are usually soldered and then cut off leaving a small but often very sharp point. The best way to handle this is to cut the lead as close to the board as practical and form a small solder ball or mound around the cut-off lead end. An example of these suggestions is given in **Fig 7.46**. The protruding component wire near the top would have a high field gradient. Below that we see a wire cut short with a rounded mound of solder covering the end of the wire.

The ends of bolts with sharp threads often protrude beyond the associated nut. One way to eliminate this is to use the dome-style nuts (cap nuts) that capture the end of the bolt or screw, forming a nicely rounded surface. An example of a cap nut is shown in Fig 7.46.

Sheet metal screws, with their needle-sharp ends, should be avoided if possible. Sheet metal screws used to close metal housings at high potential should have the screw tips inside the enclosure, which would be normal in most cases. You must also be careful of the tips of sheet metal screws that protrude on the inside of an outer enclosure. If these are in proximity to circuits at high potential, they can also lead to arcing. Keeping sheet metal screw tips well away from high voltage circuitry is the best defense.

Small pieces of sheet metal that may be part of a structure at high potential should have their edges rounded off with a file. Copper traces near the edges of circuit boards do not benefit from rounding, however. In fact, filing may make the edge sharper, ragged and more prone to breakdown. In critical areas, a small solid round wire can be soldered to the edges of a copper trace to form a rounded edge as illustrated at the right side of Fig 7.46.

INSULATORS

Some portions of the circuit may be mounted on insulators or plastic sheets. A new, clean insulator should easily withstand 10 kV per inch without creepage or breakdown across the insulating surface, but over time that surface will accumulate dirt and dust. Reducing the high-voltage stress across insulators to 5 kV per inch would be more conservative.

In theory the spacing between two smooth surfaces across an air gap can be much smaller, perhaps 20 to 30 kV per inch or even higher. But given the uncertainties of layout and voltage gradients around hardware, wiring, components and board edges, sticking with 5 kV per inch is good idea even for air gaps. This separation will seldom cause construction layout problems unless you are trying to build a very compact unit.

FUSES

Sometimes a fuse will be placed in series with a high voltage output to provide protection in case of load arcing. These fuses pose special problems. When a fuse blows, the fuse element will at least melt and perhaps even vaporize. There may be an interval when most, if not all, the output voltage appears across the fuse but the fuse has not stopped conducting. That is, there can be a sustained arc in the fuse.

Fuses have voltage ratings that are the maximum voltages across the fuse for which the arc can be expected to quench quickly. The standard 0.25 × 1.25 inch fuses used in the input ac line are typically rated for 250 V. While no doubt these ratings are conservative, this type of fuse cannot be expected to reliably clear an arc with 2 to 3 kV across the fuse and should not be used in series with a high-voltage output.

Fig 7.46 — Example of an HV board with a solder ball on a protruding lead and the use of a cap nut to terminate the end of a screw thread.

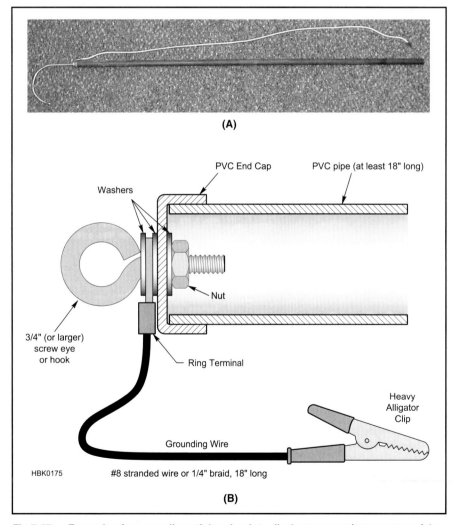

Fig 7.47 — Example of a grounding stick or hook to discharge capacitor energy safely. The end of the wire is connected to ground and the hook is touched to the capacitor terminals to discharge them. A diagram showing how to construct a grounding stick is shown at B.

Fuses with very high voltage ratings do exist however. Unfortunately, for the most part these HV fuses are for utility applications. They are physically large and not available in low current ratings. Fuses specifically rated for low currents (<2A) and several kV exist but may be very hard to locate. Even if you have a suitable fuse, care must be taken that the fuse holder can withstand the voltage. Normally high voltage fuses are mounted with end clips on an insulated board.

7.12.6 High-Voltage Safety Considerations

The voltages present in HV power supplies are potentially lethal. Every effort must be made to restrict physical access to any high potentials. A number of steps can be taken:

1) Build the power supply within a closed box, preferably a metal one.

2) Install interlocks on removable panels used for access to the interior. An interlock is usually a normally-open microswitch in series with the input power line. The microswitch is positioned so that it can be closed only when the overlying panel has been secured in place. When the panel is removed the switch moves to the open position, removing power from the supply.

3) As noted previously, the use of a bleeder resistor to rapidly discharge the capacitors is mandatory.

4) To further protect the operator when accessing the inside of a high-voltage power supply, a grounding hook like that shown in **Fig 7.47A** should be used to positively discharge each capacitor by touching the hook to each capacitor terminal. The hook is connected to the chassis via the wire shown and held by the insulated handle. It is normal practice to leave the hook in place across the capacitors while the supply is being worked on, just in case primary power is inadvertently applied. When the work is finished, remove the hook removed and replace the covers on the power supply.

Fig 7.47B shows how to construct a grounding stick of your own. A large screw eye or hook is substituted for the hook. It is important that the grounding wire be left *outside* the handle so that any hazardous voltages or currents will not be present near your hand or body.

7.13 Batteries

The availability of solid-state equipment makes it practical to use battery power under portable or emergency conditions. Handheld transceivers and instruments are obvious applications, but it's common for amateurs to power 100-W class transceivers from batteries for Field Day or emergency operation.

A battery is a group of chemical cells, usually series-connected to give some desired multiple of the *cell voltage*. The cell voltage is usually in the range of 1-4 V. Each chemical system used in the cell gives a particular nominal voltage that will vary with temperature and state of charge. This must be taken into account to make up a particular battery voltage. For example, four 1.5-V carbon-zinc cells make a 6-V battery and six 2-V lead-acid cells make a 12-V battery. The chapter on **Component Data and References** includes a table with more information about the voltage and energy capacity of common types and sizes of batteries.

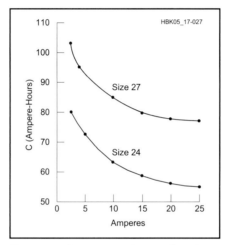

Fig 7.48 — Output capacity as a function of discharge rate for two sizes of lead-acid batteries.

7.13.1 Battery Capacity

The common rating of battery capacity is in ampere-hours (Ah) or milliampere-hours (mAh), the product of current and time. The symbol "C" is commonly used (where C is the nominal capacity in Ah). For example, C/10 would be the current continuously available for 10 hours. The value of C changes with the discharge rate and might be 110 at 2 A but only 80 at a heavier load of 20 A. **Fig 7.48** gives capacity-to-discharge rates for two standard-size vehicular lead-acid batteries. Battery capacity varies from 35 mAh for some of the small hearing-aid batteries, to more than 450 Ah for a large deep-cycle battery.

Because the current is produced by an electro-chemical reaction, cold batteries have less charge available, and some attempt to keep a battery warm before use is worthwhile. A battery may lose 70% or more of its capacity at cold extremes, but it will recover when warmer. All batteries have some tendency to freeze, but those with full charges are less susceptible. A fully charged lead-acid battery is safe to –30 °F (–34 °C) or colder. Storage batteries may be warmed somewhat by charging. Blowtorches or other open flame should never be used to heat any type of battery.

The output voltage of all batteries drops as they discharge. A practical discharge limit occurs when the load will no longer operate satisfactorily on the lower output voltage near the "discharged" point. Equipment intended for "mobile" use may be designed for an average of 13.6 V and a peak of perhaps 15 V, but may not operate well below 12 V. To obtain full use of battery charge, equipment should operate properly (if not at full power) on as little as 10.5 V with a nominal 12 to 13.6-V rating. The amount by which the battery's voltage drops depends on the type of battery and the rate of discharge. The battery manufacturer can provide information on the expected amount of voltage drop versus remaining charge at different discharge rates.

When no current is drawn from a battery, the chemical reaction essentially stops until current is required. A small amount of chemical action does continue, so stored batteries eventually will *self-discharge*. The time taken for degradation without battery use is called *shelf life*. The suggested end-of-life date is frequently marked on the side of primary batteries. Heavy-duty and industrial batteries usually have a longer shelf life.

The shelf-life of modern batteries has improved significantly, but when stored for long periods of time they will still self-discharge. One way to extend the shelf life is to keep the batteries in a refrigerator. (Do not freeze batteries as any water inside them may expand and destroy the battery.) This was standard practice with the older carbon-zinc batteries and while not as necessary with modern batteries, storage in a cool and dry place can still help extend the storage life.

7.13.2 Chemical Hazards of Batteries

In addition to the precautions given for each type of battery, the following precautions apply to all battery types. Always follow the manufacturer's advice! Extensive application information can found on manufacturer's Web sites.

Hydrogen gas escaping from storage batteries can be explosive. Keep flames or lighted tobacco products away. Use and charge batteries in well-ventilated areas to prevent hydrogen gas from building up.

No battery should be subjected to unnecessary heat, vibration or physical shock. The battery should be kept clean. Frequent inspection for leaks is a good idea. Electrolyte that has leaked or sprayed from the battery should be cleaned from all surfaces. The electrolyte is chemically active and electrically conductive, and may ruin electrical equipment. Acid may be neutralized with sodium bicarbonate (baking soda), and alkalis (found in NiCd batteries) may be neutralized with a weak acid such as vinegar. Both neutralizers will dissolve in water and should be quickly washed off. Do not let any of the neutralizer enter the battery.

Keep a record of the battery use, and include the last output voltage and, for lead-acid storage batteries, the hydrometer reading. This allows prediction of useful charge remaining, and the recharging or procuring of extra batteries, thus minimizing failure of battery power during an excursion or emergency.

Batteries can contain a number of hazardous materials such as lead, cadmium, mercury or acid, and some thought is needed for their disposal at the end of useful life. Municipal and county waste disposal sites and recycling centers will usually accept lead acid batteries because they can readily be recycled. Other types of batteries are typically not recycled and should be treated as hazardous waste. Most disposal sites and recycling centers will have occasional special programs for accepting household hazardous waste. Hardware and electronic stores may have battery recycling programs, as well. Take advantage of them.

7.13.3 Battery Types
PRIMARY BATTERIES

A *primary* battery is intended for one-time use and is then discarded. Sealed primary cells usually benefit from intermittent (rather than continuous) use. The resting period allows completion of chemical reactions needed to dispose of byproducts of the discharge.

One of the most common type of primary battery is the *alkaline* cell, in which chemical oxidation occurs during discharge. The alkaline battery has a nominal voltage of 1.5 V. Larger cells are capable of producing more mAh and less voltage drop than smaller cells. Alkaline cells have replaced the older carbon-zinc batteries that have the same nominal voltage, but substantially lower capacity.

Lithium primary batteries have a nominal voltage of about 3 V per cell and by far the best capacity, discharge, shelf-life and temperature characteristics. Their disadvantages are high cost and the fact that they cannot be readily replaced by other types in an emergency. The lithium-thionyl-chloride battery is a primary cell, and should not be recharged under any circumstances. The charging process vents hydrogen, and a catastrophic hydrogen explosion can result. Even accidental charging caused by wiring errors or a short circuit should be avoided.

Zinc-air (1.5 V), *silver oxide* (1.5 V), and *mercury* (1.4 V) batteries are very good where nearly constant voltage is desired at low currents for long time periods. Their main use is in hearing aids (zinc-air, primarily), though they may be found in other mass-produced devices such as household smoke alarms. Mercury batteries are often found as voltage references in older voltmeters.

Primary batteries should not be recharged for two reasons: It may be dangerous because of heat generated within sealed cells, and even in cases where there may be some success, both the resulting charge and life are limited. One type of alkaline battery is rechargeable, and is so marked.

SECONDARY (RECHARGEABLE) BATTERIES

A *storage* or *secondary* battery may be recharged many times. Rechargeable batteries are very popular, both as individual cells and in *battery packs* used with low-power radios, portable instruments, laptop computers and power tools. A battery pack consists of several battery cells connected in series and packaged as a single battery. In the following discussion, "battery" refers to both individual cells and battery packs.

Nickel-Cadmium

A common type of secondary battery is the nickel-cadmium (NiCd or "nye-cad"), with a nominal voltage of 1.2 V per cell. Carefully used, these are capable of 500 or more charge and discharge cycles. For best life, the NiCd battery must not be fully discharged. Where there is more than one cell in the battery, the most-discharged cell may suffer polarity reversal, resulting in a short circuit, or seal rupture. All storage batteries have discharge limits, and NiCd types should not be discharged to less than 1.0 V per cell. There is a popular belief that it is necessary to completely discharge NiCd cells in order to recharge them to full capacity. Called the "memory effect," while this was possibly true of earlier cells, this effect is not seen in modern batteries.

Caution is required in the replacement of carbon-zinc or alkaline cells by NiCd storage cells and vice versa. Eight alkaline cells will give 12 V, while 10 of the same size NiCd cells are required for the same voltage. If a 10-cell battery holder is used, the equipment should be designed for 15 V in case the alkaline units are plugged in.

Nickel-cadmium cells are not limited to "D" cells and smaller sizes. They also are available in larger varieties ranging to mammoth 1000-Ah units having carrying handles on the sides and caps on the top for adding water, similar to lead-acid types. These large cells are sold to the aircraft industry for jet-engine starting, and to the railroads for starting locomotive diesel engines. They also are used extensively for uninterruptible power supplies. Although expensive, they have very long life. Surplus cells are often available through surplus electronics dealers, and these cells often have close to their full rated capacity.

Advantages for the ham in these vented-cell batteries lie in the availability of high discharge current to the point of full discharge. Also, cell reversal is not the problem that it is in the sealed cell, since water lost through gas evolution can easily be replaced. Simply remove the cap and add distilled water. Tap water or bottled drinking water should never be added to either nickel-cadmium or lead-acid cells, since dissolved minerals in the water can hasten self-discharge and interfere with the electrochemical process.

Lead-Acid

The most widely used high-capacity rechargeable battery is the lead-acid type. In addition to high-capacity units which are universally available, small sealed "gel-cell" batteries with capacities of a few hundred mAh and up are also becoming widely available and can be very useful. In automotive service, the battery is usually expected to discharge partially at a very high rate (engine starting), and then to be recharged promptly while the alternator is also carrying the electrical load. If the conventional auto battery is allowed to discharge fully from its nominal 2 V per cell to 1.75 V per cell, fewer than 50 charge and discharge cycles may be expected, with reduced storage capacity.

The most attractive battery for extended high-power electronic applications is the so-called "deep-cycle" battery, which is intended for such uses as powering electric fishing motors and the accessories in recreational vehicles. Size 24 and 27 batteries furnish a nominal 12 V and are about the size of small and medium automotive batteries. These batteries may furnish between 1000 and 1200 Wh per charge at room temperature. When properly cared for, they may be expected to last more than 200 cycles. They often have lifting handles and screw terminals, as well as the conventional truncated-cone automotive terminals. They may also be fitted with accessories, such as plastic carrying cases, with or without built-in chargers.

"Discharged" condition for a 12-V lead-acid battery should not be less than 10.5 V. It is also good to keep a running record of hydrometer readings, but the conventional readings of 1.265 as charged and 1.100 as discharged apply only to a long, low-rate discharge. Heavy loads may discharge the battery with little reduction in the hydrometer reading.

Lead-acid batteries with liquid electrolyte usually fall into one of three classes — conventional, with filling holes and vents to permit the addition of distilled water lost from evaporation or during high-rate charge or discharge; maintenance-free, from which gas may escape but water cannot be added; and sealed. Generally, deep-cycle batteries have filling holes and vents. As in the case of the NiCd batteries, use only distilled water to top off the electrolyte.

New lead-acid batteries are usually sold "dry charged" with a separate container for the electrolyte. In a dry charged battery the chemical composition of the cell plates is the same as for a fully charged battery but the electrolyte is omitted. This allows for much longer storage time before the battery is sold or used, keeping the battery from deteriorating. New batteries should be filled with electrolyte and allowed to stand for at least half an hour before use. They should then be charged at about 15 A for 15 minutes or so. The capacity of the battery will build up slightly for the first few cycles of charge and discharge, and then have fairly constant capacity for many cycles. Slow capacity decrease may then be noticed.

Lead-acid batteries are also available with gelled electrolyte. Commonly called *gel cells*, these may be mounted in any position if sealed, but some vented types are position sensitive. Also popular are AGM (absorbed glass mat) deep-cycle batteries that are rugged, sealed, maintenance free units that do not outgas dangerous levels of hydrogen when charging.

Nickel-Metal Hydride

The nickel-metal hydride (NiMH) battery is quite similar to the NiCd, but the cadmium electrode is replaced by one made from a porous metal alloy that traps hydrogen (therefore the name metal hydride). NiMH cells do not contain any dangerous substances, while both NiCd and lead-acid cells contain quantities of toxic heavy metals. NiCd cells have been largely replaced by NiMH and other newer, less toxic, technologies in new equipment.

Many of the basic characteristics of NiMH cells are similar to NiCds. For example, the voltage is very nearly the same, they can be slow-charged from a constant-current source, and they can safely be deep-cycled.

There are also some important differences: The most attractive feature is a much higher capacity for the same cell size — often nearly twice as much as NiCd batteries! The typical AA NiMH cell has a capacity of 2000 mAh or greater, compared to the 600 to 830 mAh for the same size NiCd. The fast-charge process is different for NiMH batteries. A fast charger designed for NiCd will not correctly charge NiMH batteries, but many commercial fast chargers are designed for both types of batteries. Check the charger specifications and setting before attempting to recharge any batteries.

The internal resistance (discussed below) of NiMH cells is somewhat higher than that of NiCd cells, resulting in reduced performance at very high discharge currents. This can cause slightly reduced power output from a handheld transceiver powered by a NiMH pack, but the effect is barely noticeable and the higher capacity and resulting longer run time far outweigh this drawback. NiMH batteries outperform NiCd batteries whenever high capacity is desired, while NiCd batteries still have advantages when delivering very high peak currents, such as for power tools. At least one manufacturer warns that the self-discharge of NiMH cells is higher than for NiCd, but again, in practice this can hardly be noticed.

At the time of this writing, many cell phones and portable consumer electronics equipment use NiMH batteries, and manufacturers offer NiMH packs for Amateur Radio applications. Standard-sized NiMH cells are widely available from the major electronic parts suppliers.

Lithium-Ion

The lithium-ion battery (Li-ion) is another alternative to NiCd batteries. For the same energy storage, a Li-ion battery is about one-third the weight and one-half the volume of a NiCd. It also has a lower self-discharge rate. Typically, at room temperature, a NiCd cell will lose from 0.5 to 2% of its charge per day. The Lithium-ion cell will lose less than 0.5% per day and even this loss rate decreases after about 10% of the charge has been lost. At higher temperatures the difference is even greater in favor of Li-ion. The result is that lithium-ion cells are a much better choice for standby operation where frequent recharge is not available.

One major difference between NiCd and Li-ion cells is the cell voltage. The nominal voltage for a NiCd cell is about 1.2 V. For the Li-ion cell it is 3.6 V with a maximum cell charging voltage of 4 V. You cannot substitute Li-ion cells directly for NiCd cells. You will need one Li-ion cell for three NiCd cells. Chargers intended for NiCd batteries must not be used with Li-ion batteries, and vice-versa.

7.13.4 Battery Discharge Planning

As shown in the previous section, discharge profiles will vary with the type of battery and the characteristics of the load. Both the type of battery and the characteristics of the load need to be considered to get the maximum

Table 7.1
Battery Energy Storage Capability

Battery Type	Wh/kg	Joules/kg	Wh/liter
Lead-acid	41	146,000	100
Alkaline long-life	110	400,000	320
Carbon-zinc	36	130,000	92
NiMH	95	340,000	300
NiCd	39	140,000	140
Lithium-ion	128	460,000	230

from a given battery. Another important characteristic especially for portable operation is the energy density of a given battery. This varies widely between types. **Table 7.1** makes a comparison between several different types of batteries. The energy densities are given in terms of both weight and volume. For a fixed volume or physical size of the battery, discharge planning must also take into account the substantial differences between different types of batteries.

Transceivers usually drain a battery at two or three rates: one for receiving, one for transmit standby and one for key-down or average voice transmit. Considering just the first and last of these (assuming the transmit standby is equal to receive), average two-way communication would require the low rate 75% of the time and the high rate 25% of the time. The ratio may vary somewhat with voice. The user may calculate the percentage of battery charge used in an hour by the combination (sum) of rates. If, for example, 20% of the battery capacity is used in an hour, the battery will provide five hours of communications per charge. In most actual on-air situations, the time spent listening is much greater than that spent transmitting.

7.13.5 Battery Internal Resistance

Battery internal resistance is very important to handheld transceiver users. This is because the internal resistance is in series with the battery's output and therefore reduces the available battery voltage at the high discharge currents demanded by the transmitter. The result is reduced transmitter output power and power wasted in the cell itself by internal heating. Because of cell-construction techniques and battery chemistry, certain types of cells typically have lower internal resistance than others.

The NiCd and NiMH cells are the best battery type for high discharge current capability. The NiCd is slightly better, but NiMH battery construction is rapidly becoming competitive with NiCd. Both battery types maintain their low internal resistance throughout the *discharge curve* (a graph of output voltage versus charge remaining). **Figs 7.49** through **7.52** show typical discharge characteristics

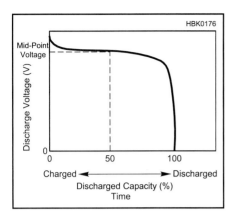

Fig 7.49 — Typical NiCd battery discharge characteristic at a constant load current.

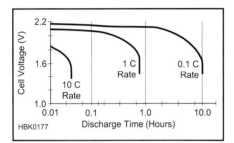

Fig 7.50 — Typical lead acid battery discharge characteristic.

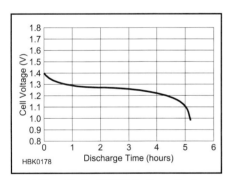

Fig 7.51 — Typical NiMH battery discharge characteristic.

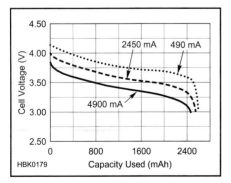

Fig 7.52 — Typical lithium-ion battery discharge characteristic at three discharge rates.

for four different types of batteries.

Alkaline primary cells have higher internal resistance than NiCd and NiMH. When these cells are used with handheld transceivers, it is not uncommon to have lower output power because of the lower battery voltage under load. Even with fresh cells, the low battery indicator may come on as the cell voltage drops more than for comparable NiCd or NiMH batteries..

The lead-acid cell, which is used in larger belt-hung battery packs and for temporary or emergency power, is pretty close to the alkaline cell for internal resistance, but only at full charge. Unlike the NiCd, the electrolyte in the lead-acid cell enters into the chemical reaction. During discharge, the specific gravity of the electrolyte gradually drops as it approaches water, and the conductivity decreases. Therefore, as the lead-acid cell approaches a discharged state, the internal resistance increases. Larger battery capacities are usually used (approximately 2 Ah) and so the effects of the higher internal resistance are consequently reduced.

The highest internal resistance is found in the common carbon-zinc flashlight cell. With the transmit current demand levels of handheld radios, these cells have limited utility although they may be useful for low-current, receive-only applications.

7.13.6 Battery Charging

Each different type of rechargeable battery must be charged according to the manufacturer's recommended charge rate and duration. Batteries often require a slight overcharge, based on the battery capacity, C. If, for instance, the specified charge rate is 0.1 C (the 10-hour rate), 12 or more hours may be needed for the charge. Once full-charge is reached, the battery charger should switch to a maintenance mode that varies with battery type. Do not attempt to charge batteries in a charger not specified for that type of battery. The result may be damage to the battery, the charger, or both. If you are attempting to charge batteries while still in a radio, using the wrong type of batteries can damage the radio, as well. Typical charge profiles for several types of batteries are shown in **Figs 7.53** through **7.56**.

CHARGING LEAD-ACID BATTERIES

Deep-cycle lead-acid cells are best charged at a slow rate, while automotive types may safely be given higher rate charges. This depends on the amount of heat generated within each cell, and cell venting to prevent pressure build-up. Some batteries have built-in temperature sensing, used to stop or reduce charging before the heat rise becomes a danger. Fast

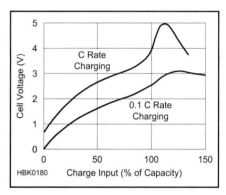

Fig 7.53 — Typical NiCd battery charge characteristic at two different rates.

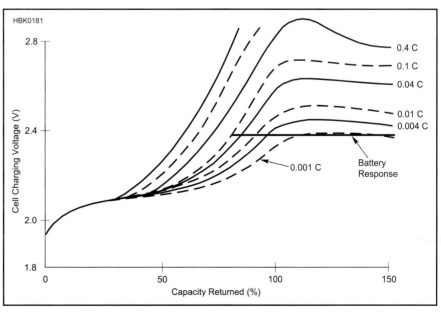

Fig 7.54 — Typical lead-acid battery charge characteristic at various rates.

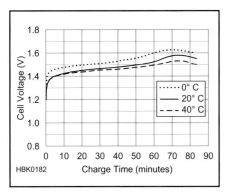

Fig 7.55 — Typical NiMH battery charge characteristic at various temperatures.

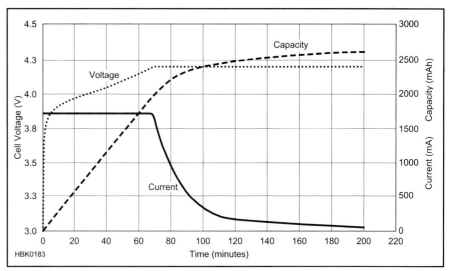

Fig 7.56 — Typical lithium-ion battery charge characteristic.

charges do not usually allow gas recombination, so some of the battery water will escape in the form of gas. If the water level falls below a certain point, acid hydrometer readings are no longer reliable. If the water level falls to the level of the battery's internal plates, permanent battery damage may result.

Gelled-electrolyte lead-acid batteries provide 2.4 V/cell when fully charged. Damage results, however, if they are overcharged. (Avoid *constant-current* or *trickle charging* unless battery voltage is monitored and charging is terminated when a full charge is reached.) Voltage-limited charging is best for these batteries. A proper charger maintains a safe charge-current level until 2.3 V/cell is reached (13.8 V for a 12-V battery). Then, the charge current is tapered off until 2.4 V/cell is reached. Once charged, the battery may be safely maintained at the *float* level, 2.3 V/cell. Thus, a 12-V gel-cell battery can be "floated" across a regulated 13.8-V system as a battery backup in the event of power failure. These batteries are readily available since they are used in household and commercial burglar/fire alarm systems as well as in uninterruptible power supplies (UPSs).

CHARGING NiCd AND NiMH BATTERIES

NiCd batteries have a flat voltage-versus-charge characteristic until full charge is reached; at this point the battery voltage rises abruptly. With further charging of the NiCd battery, the electrolyte begins to break down and oxygen gas is generated at the positive (nickel) electrode and hydrogen at the negative (cadmium) electrode.

Since the cell should be made capable of accepting an overcharge, battery manufacturers typically prevent the generation of hydrogen by increasing the capacity of the cadmium electrode. This allows the oxygen formed at the positive electrode to reach the metallic cadmium of the negative electrode and re-oxidize it. During overcharge, therefore, the cell is in equilibrium. The positive electrode is fully charged and the negative electrode less than fully charged, so oxygen evolution and recombination "wastes" the charging power being supplied.

To ensure that all cells in a NiCd battery reach a fully-charged condition, NiCd batteries should be charged by a constant current at about a 0.1-C current level. This level is about 50 mA for the AA-size cells used in most handheld radios. This is the optimum rate for most NiCd cells since 0.1 C is high enough to provide a full charge, yet it is low enough to prevent over-charge damage and provide good charge efficiency.

Although *fast-charge-rate* (three to five hours, typically) chargers are available for hand-held transceivers, they should be used with care. The current delivered by these units is capable of causing the generation of large quantities of oxygen in a fully charged cell. If the generation rate is greater than the oxygen recombination rate, pressure will build in the cell, forcing the vent to open and the oxygen to escape. This can eventually cause drying of the electrolyte, and then cell failure. The cell temperature can also rise, which can shorten cell life. To prevent overcharge from occurring, fast-rate chargers should have automatic charge-limiting circuitry that will switch or taper the charging current to a safe rate as the battery reaches a fully charged state.

Overcharging NiCds in moderation causes little loss of battery life. Continuous overcharge, however, may generate a voltage depression when the cells are later discharged. For best results, charging of NiCd cells should be terminated after 15 hours at the slow rate. Better yet, circuitry may be included in the charger to stop charging, or reduce the current to about 0.02 C when the 1.43-V-per-cell terminal voltage is reached.

Although many of the same concerns and suggestions apply to NiMH batteries, there are some important differences. After full-charge is reached, the NiMH cell electrodes discharge slightly. Many chargers use this slight drop in battery voltage as a signal that full charge has been reached. Another sign that the battery is fully charged is a rise in internal temperature. If the full charging rate is maintained, internal temperature can become dangerously high, leading to battery damage or explosion. Trickle- or float-charging is not recommended for NiMH batteries.

SOLAR CHARGING SYSTEMS

Price and availability make solar panels an attractive way to maintain the charge on your batteries. Relatively small, low-power solar arrays provide a convenient way to charge a NiCd or sealed-lead-acid battery for emergency and portable operation. This is especially popular for QRP operating and for stations on boats or used while camping. You should always connect some type of charge controller between the battery and the solar array. This will prevent overcharging the battery, and the possible resulting battery damage.

7.13.7 DC-AC Inverters

For battery-powered operation of ac-powered equipment, dc-ac inverters are used. An inverter is a dc-to-ac converter that provides 120-V ac. Inverters come with varying degrees of sophistication. The simplest type of inverter switches directly at 60 Hz to produce a square-wave output. This is no problem for lighting and other loads that don't care about the input waveform. However, some equipment will work poorly or not at all when supplied with square wave power because of the high harmonic content of the waveform.

The harmonic content of the inverter output waveform can be reduced by the simple expedient of reducing the waveform duty cycle from 50% (for the square wave) to about 40%.

For many loads, such as computers and other electronic devices, this may still not be adequate, and so many inverters use waveform shaping to approximate a sine-wave output. The simplest of these methods is a resonant inductor-capacitor filter. This adds significant weight and size to the inverter. Most modern inverters use high-frequency Pulse-width modulation (PWM) techniques to synthesize the 60 Hz sinusoidal output waveform, much like a switching power supply. See **Fig 7.57**.

Inverters are usually rated in terms of their VA or "volt-ampere product" capability although sometimes they will be rated in watts. Care is required in interpreting inverter ratings. A purely resistive load operating from a sinusoidal voltage source will have a sinusoidal current flowing in phase with the voltage. In this case, VA, the product of the voltage (V) and the current (A), will equal the actual power in watts delivered to the load so that the VA and the watt ratings are the same. Some loads, such as motors and many rectifier power supplies, will shift the phase of load current away from the source voltage or the load current will flow in short pulses as shown earlier for capacitor input filters. In these cases (which are very common), the VA product for that type of load can be much larger than the delivered power in watts. In the absence of detailed knowledge of the load characteristics it is prudent to select an inverter with a VA or wattage capability of 25% or more above the expected load.

7.13.8 Selecting a Battery for Mobile Operation

There are two basic modes of mobile operation; *in-motion*, and *stationary*. Each mode has unique power requirements and thus different battery requirements. To satisfy these needs, there are three basic battery types. It is important to understand the differences between the types and to apply them properly.

Standard vehicle batteries are referred to as *SLI* (starting, lights, ignition) or just *starter* batteries. Their primary function is to start the engine and then act as a power filter for the alternator which is the actual long-term power source. The most important rating of an SLI battery is the *cold cranking amps* (CCA) rating — the number of amps that the battery can produce at 32 °F (0 °C) for 30 seconds.

Batteries designed for repeated cycles of charging and discharging use are often called *deep-cycle* although the term is widely over-used. A true deep-cycle battery is designed to be repeatedly discharged to 20% remaining capacity. The term "deep cycle" is a misnomer, as all lead-acid batteries are considered discharged when their output voltage drops below 10.5 V at some specified current draw as outlined by the Battery Council Institute (BCI — **www.batterycouncil.org**). At

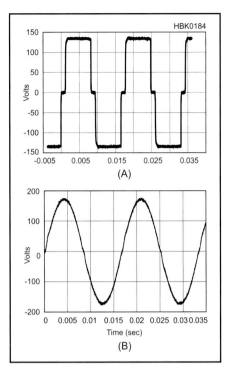

Fig 7.57 — Output waveforms of typical dc-ac inverters. At A, the output of a modified sine wave inverter. Note the stepped square waves. At B, the output of a "pure sine wave" inverter. Note the close approximation of a commercial ac sine wave.

10.5 V, a six-cell lead-acid battery is considered discharged.

Marine batteries are designed to be stored without charging for up to two years, yet still maintain enough power to start a marine engine. Contrary to common practice, they're not really designed for extended low-current power delivery. Marine batteries often have hybrid characteristics between SLI and deep-cycle batteries.

To differentiate true deep-cycle batteries from SLI and marine batteries, examine a battery's *reserve capacity* (RC). A battery's RC rating is the number of minutes that the battery can deliver 25 A while maintaining an output voltage above 10.5 V. RC batteries typically have RC ratings 20% or more higher than SLI batteries and perhaps 50% higher under ICAS (intermittent, commercial, and amateur service) conditions. A deep-cycle battery has a lower CCA rating than an SLI battery due to its internal construction that favors long-term power delivery over high-current starting loads.

With these facts in mind, we can now select the correct battery for our style of mobile operating. For in-motion operation with power outputs up to 200 W, a second trunk-mounted battery is seldom needed if the power cabling wire size is chosen correctly. (See the **Assembling a Station** chapter, section 29.2, for information on wire sizes in mobile applications.)

When an amplifier is added for higher output power levels, it is often less expensive to add a second trunk-mounted battery than to install larger cables to the main battery. In these cases, the second battery can be of almost any type, as long as it is lead-acid. The second battery should be connected in parallel to the vehicle's main SLI battery. The battery's ampere-hour rating should be close to that of the vehicle's main SLI battery.

All secondary wiring should be properly fused, as outlined in the **Assembling a Station** chapter's on mobile installations. The use of relays and circuit breakers should be avoided. Remember, should a short circuit occur, good-quality lead-acid batteries can deliver upwards of 3000 A which exceeds the break circuit ratings of most relays and circuit breakers. A better solution is a FET switch such as those made by Perfect Switch (**www.perfectswitch.com**).

Assuming the second battery is mounted inside the vehicle's passenger compartment or in the trunk, it should be an AGM type. AGM (Absorption Glass Mat) batteries do not outgas explosive hydrogen gas under normal operating conditions. Flooded (liquid electrolyte) batteries should *never* be used in an enclosed environment.

For stationary operation, select a battery with a large RC rating because it will not be continuously charged. There are two main considerations; the ampere-hour rating (Ah) and the reserve capacity rating, typically listed as C/8, C/10, or C/20, with units of hours. (C is the battery's capacity in Ah.) Dividing the Ah rating by the load amperage (8, 10 or 20 A) will give you the reserve capacity in hours, but the actual ampere-hours any given battery can deliver before the voltage reaches 10.5 V (nominal discharge level) will vary with the load, both average and peak. Heavier loads will reduce the actual ampere-hours available.

Automotive batteries are arranged in BCI group sizes (**www.rtpnet.org/teaa/bci-group.html**), from 21 through 98. Generally speaking, the larger the group size the larger the battery and the higher the Ah rating. For example, size 24 (small car) has an average rating of 40 Ah, and size 34 (large car) has an average rating of 55 Ah. Exact ratings, including their reserve capacity, are available from the manufacturers' Web sites listed below. A good rule of thumb is to select a battery as physically large as you have room for, consistent with the highest RC rating for any given Ah rating.

Batteries are heavy, and need to be properly secured inside a battery box or by using factory-supplied brackets. For example, a BIC group 34 (average SLI size) battery weighs about 55 pounds. Some battery models (such as the Optima) come supplied with mounting brackets and terminal protection covers. Even though battery boxes aren't always needed, they should be used as a safety precaution to prevent accidental contact with the terminals and can protect the battery from external items. Battery restraints should be adequate to provide 6Gs of lateral and 4Gs of vertical retention, ruling out sheet metal screws and most webbing material. Use the proper brackets!

There are three other considerations: isolating the battery electrically, recharging the battery, and output voltage regulation. Diode-based battery isolators are not all equal. Models with FET bypass switches are the preferred type because of the low voltage-drop across the FET.

If you have wired the battery in parallel with the vehicle's main SLI battery, recharging is taken care of whenever the vehicle is running. If you plan on operating in stationary mode, you'll need a separate recharging system. Most vehicle factory-installed trailer wiring systems also include a circuit for charging RV or boat "house" batteries. Check with your dealer's service personnel about these options.

Voltage regulators, commonly called "battery boosters," are almost a necessity for stationary operation. A model with a low-voltage cutoff should be used to avoid discharging the battery below 10.5 V, as discharging a lead-acid battery beyond this point drastically reduces its charge-cycle life — the number of full-charge/full-discharge cycles. (See the November 2008 *QST* Product Review column.)

For additional information on battery ratings, sizes, and configuration, visit these Web sites:
optimabatteries.com
www.exide.com
www.interstatebatteries.com
www.lifelinebatteries.com

7.14 Glossary of Power Supply Terms

Bleeder — A resistive load across the output or filter of a power supply, intended to quickly discharge stored energy once the supply is turned off.

Boost converter — A switchmode converter in which the output voltage is always greater than or equal to the input voltage.

Buck converter — A switchmode converter in which the output voltage is always less than or equal to the input voltage.

Buck-boost converter — A switchmode converter in which the magnitude of the output voltage can be either greater or less than the input voltage.

C-rate — The charging rate for a battery, expressed as a ratio of the battery's ampere-hour rating.

CCA (cold cranking amps) — A measure of a battery's ability to deliver high current to a starter motor.

Circular mils — A convenient way of expressing the cross-sectional area of a round conductor. The area of the conductor in circular mils is found by squaring its diameter in mils (thousandths of an inch), rather than squaring its radius and multiplying by pi. For example, the diameter of 10-gauge wire is 101.9 mils (0.1019 inch). Its cross-sectional area is 10380 CM, or 0.008155 square inches.

Core saturation (magnetic) — That condition whereby the magnetic flux in a transformer or inductor core is more than the core can handle. If the flux is forced beyond this point, the permeability of the core will decrease, and it will approach the permeability of air.

Crowbar — A last-ditch protection circuit included in many power supplies to protect the load equipment against failure of the regulator in the supply. The crowbar senses an overvoltage condition on the supply's output and fires a shorting device (usually an SCR) to directly short-circuit the supply's output and protect the load. This causes very high currents in the power supply, which blow the supply's input-line fuse.

Darlington transistor — A package of two transistors in one case, with the collectors tied together, and the emitter of one transistor connected to the base of the other. The effective current gain of the pair is approximately the product of the individual gains of the two devices.

DC-DC converter — A circuit for changing the voltage of a dc source to ac, transforming it to another level, and then rectifying the output to produce direct current.

Deep-cycle — A battery designed for repeated charge-discharge cycles to 20% of remaining capacity.

Equalizing resistors — Equal-value bypassing resistors placed across capacitors connected in series for use in a high-voltage power supply to keep the voltages across the capacitors in the string relatively constant.

Fast recovery rectifier — A specially doped rectifier diode designed to minimize the time necessary to halt conduction when the diode is switched from a forward-biased state to a reverse-biased state.

Flyback converter — A transformer-coupled version of the **buck-boost converter**.

Forward converter — A **buck converter** with multiple isolated outputs at different voltage levels and polarities.

Foldback current limiting — A special type of current limiting used in linear power supplies, which reduces the current through the supply's regulator to a low value under short circuited load conditions in order to protect the series pass transistor from excessive power dissipation and possible destruction.

Ground fault (circuit) interrupter (GFI or GFCI) — A safety device installed between the household power mains and equipment where there is a danger of personnel touching an earth ground while operating the equipment. The GFI senses any current flowing directly to ground and immediately switches off all power to the equipment to minimize electrical shock. GFCIs are now standard equipment in bathroom and outdoor receptacles.

Input-output differential — The voltage drop appearing across the series pass transistor in a linear voltage regulator. This term is usually stated as a minimum value, which is that voltage necessary to allow the regulator to function and conduct current. A typical figure for this drop in most three-terminal regulator ICs is about 2.5 V. In other words, a regulator that is to provide 12.5 V dc will need a source voltage of at least 15.0 V at all times to maintain regulation.

Inverter — A circuit for producing ac power from a dc source.

Li-ion — Lithium-ion, a type of rechargeable battery that is about ⅓ the weight and 1/2 the volume of a **NiCd** battery of the same capacity.

Low dropout regulator — A three-terminal regulator designed to work with a low minimum input-output differential value.

Marine — A battery designed to retain significant energy over long periods *of time without being continuously charged.*

NiCd — Nickel cadmium, a type of rechargeable battery.

NiMH — Nickel metal hydride, a type of rechargeable battery that does not contain toxic substances.

Peak inverse voltage (PIV) — The maximum reverse-biased voltage that a semiconductor is rated to handle safely. Exceeding the peak inverse rating can result in junction breakdown and device destruction.

Power converter — Another term for a power supply.

Power processor — Another term for a power supply.

Primary battery — A battery intended for one-time use and then discarded.

RC (reserve capacity) — A measure of a battery's ability to deliver current over long periods.

Regulator — A device (such as a Zener diode) or circuitry in a power supply for maintaining a constant output voltage over a range of load currents and input voltages.

Resonant converter — A form of dc-dc converter characterized by the series pass switch turning on into an effective series-resonant load. This allows a zero current condition at turn-on and turn-off. The resonant converter normally operates at frequencies between 100 kHz and 500 kHz and is very compact in size for its power handling ability.

Ripple — The residual ac left after rectification, filtration and regulation of the input power.

RMS — *Root Mean Square*. Refers to the effective value of an alternating voltage or current, corresponding to the dc voltage or current that would cause the same heating effect.

Secondary battery — A battery that may be recharged many times. Also called a *storage battery*.

Secondary breakdown — A runaway failure condition in a transistor, occurring at higher collector-emitter voltages, where hot spots occur due to (and promoting) localization of the collector current at that region of the chip.

Series pass transistor, or pass transistor — The transistor(s) that control(s) the passage of power between the unregulated dc source and the load in a regulator. In a linear regulator, the series pass transistor acts as a controlled resistor to drop the voltage to that needed by the load. In a switch-mode regulator, the series pass transistor switches between its ON and OFF states.

SLI (starter, lights, ignition) — An automotive battery designed to start the vehicle and provide power to the lighting and ignition systems.

SOA (Safe Operating Area) — The range of permissible collector current and collector-emitter voltage combinations where a transistor may be safely operated without danger of device failure.

Spike — See transient

Surge — A moderate-duration perturbation on a power line, usually lasting for hundreds of milliseconds to several seconds.

Switching regulator — Another name for a switchmode converter.

Switchmode converter — A high-efficiency switching circuit used for dc-dc power conversion. Switching circuits are usually much smaller and lighter than conventional 60 Hz, transformer-rectifier circuits because they operate at much higher frequencies — from 25 to 400 kHz or even higher.

Three-terminal regulator — A device used for voltage regulation that has three leads (terminals) and includes a voltage reference, a high-gain error amplifier, temperature-compensated voltage sensing resistors and a pass element.

Transient — A short perturbation on a power line, usually lasting for microseconds to tens of milliseconds.

Varistor — A surge suppression device used to absorb transients and spikes occurring on the power lines, thereby protecting electronic equipment plugged into that line. Frequently, the term MOV (*Metal Oxide Varistor*) is used instead.

Volt-Amperes (VA) — The product obtained by multiplying the current times the voltage in an ac circuit without regard for the phase angle between the two. This is also known as the apparent power delivered to the load as opposed to the actual or real power absorbed by the load, expressed in watts.

Voltage multiplier — A type of rectifier circuit that is arranged so as to charge a capacitor or capacitors on one half-cycle of the ac input voltage waveform, and then to connect these capacitors in series with the rectified line or other charged capacitors on the alternate half-cycle. The voltage doubler and tripler are commonly used forms of the voltage multiplier.

Voltage regulation — The change in power supply output voltage with load, expressed as a percentage.

7.15 References and Bibliography

REFERENCES

1) Landee, Davis and Albrecht, *Electronic Designer Handbook*, 2nd edition, (McGraw-Hill, 1977), page 12-9. This book can frequently be found in technical libraries and used book stores. The power supply section is well worth reading.

2) Severns and Bloom, *Modern DC-to-DC Switchmode Power Converter Circuits*, (Van Nostrand Reinhold, 1984, ISBN: 0-442-21396-4). A reprint of this book is currently available at the Power Sources Manufacturers Association Web site, **www.psma.com**.

3) Fair-Rite Web site, **www.fair-rite.com**

4) **www.mag-inc.com**

OTHER RESOURCES

M. Brown, *Power Supply Cookbook*, (Butterworth-Heinemann, 2001).

J. Fielding, ZS5JF, *Power Supply Handbook*, (RSGB, 2006).

K. Jeffrey, *Independent Energy Guide*, (Orwell Cove Press, 1995). Includes information on batteries and inverters.

7.16 Power Supply Projects

Construction of a power supply can be one of the most rewarding projects undertaken by a radio amateur. Whether it's a charger for batteries, a low-voltage, high-current monster for a new 100-W solid-state transceiver, or a high-voltage supply for a new linear amplifier, a power supply is basic to all of the radio equipment we operate and enjoy. Final testing and adjustment of most power-supply projects requires only a voltmeter, and perhaps an oscilloscope — tools commonly available to most amateurs.

General construction techniques that may be helpful in building the projects in this chapter are outlined in the **Construction Techniques** chapter. Other chapters in the *Handbook* contain basic information about the components that make up power supplies.

Safety must always be carefully considered during design and construction of any power supply. Power supplies contain potentially lethal voltages, and care must be taken to guard against accidental exposure. For example, electrical tape, insulated tubing ("spaghetti") or heat-shrink tubing is recommended for covering exposed wires, components leads, component solder terminals and tie-down points. Whenever possible, connectors used to mate the power supply to the outside world should be of an insulated type designed to prevent accidental contact.

Connectors and wire should be checked for voltage and current ratings. Always use wire with an insulation rating higher than the working voltages in the power supply. For supply voltages above 300 V, use wire with insulated rated accordingly. The **Component Data and References** chapter contains a table showing the current-carrying capability of various wire sizes. Scrimping on wire and connectors to save money could result in flashover, meltdown or fire.

All fuses and switches should be placed in the hot circuit(s) only. The neutral circuit should not be interrupted. Use of a three-wire (grounded) power connection will greatly reduce the chance of accidental shock. The proper wiring color code for 120-V circuits is: black — hot; white — neutral; and green — ground. For 240-V circuits, the second hot circuit generally uses a red wire.

POWER SUPPLY PRIMARY-CIRCUIT CONNECTOR STANDARD

The International Commission on Rules for the Approval of Electrical Equipment (CEE) standard for power-supply primary-circuit connectors for use with detachable cable assemblies is the CEE-22. The CEE-22 has been recognized by the ARRL and standards agencies of many countries. Rated for up to 250 V, 6 A at 65 °C, the CEE-22 is the most commonly used three-wire (grounded), chassis-mount primary circuit connector for electronic equipment in North America and Europe. It is often used in Japan and Australia as well.

When building a power supply requiring 6 A or less for the primary supply, a builder would do well to consider using a CEE-22 connector and an appropriate cable assembly, rather than a permanently installed line cord. Use of a detachable line cord makes replacement easy in case of damage. CEE-22 compatible cable assemblies are available with a wide variety of power plugs including most types used overseas.

Some manufacturers even supply the CEE-22 connector with a built-in line filter. These connector/filter combinations are especially useful in supplies that are operated in RF fields. They are also useful in digital equipment to minimize conducted interference to the power lines.

CEE-22 connectors are available in many styles for chassis or PC-board mounting. Some have screw terminals; others have solder terminals. Some styles even contain built-in fuse holders.

7.16.1 Four-Output Switching Bench Supply

This project by Larry Cicchinelli, K3PTO, describes the four-output bench power supply shown in **Fig 7.58** with three positive outputs and one negative output. The three positive outputs use identical switching regulator circuits that can be set independently to any voltage between 3.3 V and 20 V at up to 1 A. The fourth output is a negative regulator capable of about 250 mA. As built, the supply has two fixed outputs and two variable outputs, but any module can be built with variable output. (Construction diagrams and instructions, a complete parts list, and additional design details are included on this *Handbook's* CD-ROM.)

The only dependency among the outputs is that they are all driven by a single transformer. The transformer used is rated at 25 V and 2 A — good for 50 W. Assuming that the regulator IC being used has a 75% efficiency, a total of about 37 W is available from the power supply outputs.

One of the features of a switching regulator is that you can draw more current from the outputs than what the transformer is supplying — at a lower voltage, of course — as long as you stay within the 37 W limit and maximum current for the regulator. Most of the discussion in this article will be about the positive regulators as the negative regulator was an add-on after the original system was built.

POSITIVE REGULATOR

Fig 7.59 is the circuit for the positive regulator modules — a buck-type regulator. There are several variations of the circuit, any of which you can implement.

• L2 and C4 are optional. These two components implement a low-pass filter that will decrease high frequency noise that might otherwise appear at the output.

• The pads for R1 will accommodate a small, multi-turn potentiometer. You can insert one here or you can use the pads to connect a panel-mounted potentiometer.

• If you want a fixed output you can simply short out R1 and use R2 by itself.

• You can also insert a fixed resistor in the R1 position in the case where the calculated value is non-standard and you want to use two fixed resistors.

The formula for setting output voltage us-

Fig 7.58 — The front panel of the four-output switching supply.

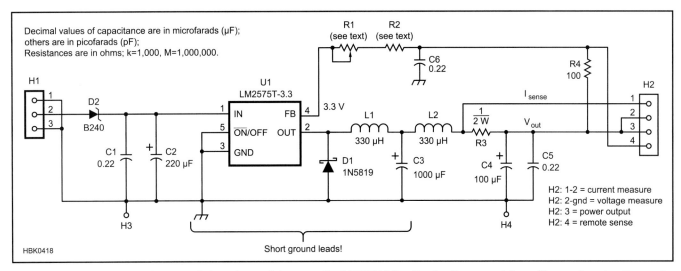

Fig 7.59 — The positive buck-type switchmode regulator uses the LM2575-3.3, a fixed-voltage regulator, with an external voltage-set resistor (R1 + R2). See text for details of the calculations needed to determine the value of R1 and R2. As noted in the text, these values are the total resistance for both parts, and can be made from one fixed resistor, one variable resistor or a combination. Some common values (R1 + R2 total) are: For a 12 V fixed supply, 7.1 kΩ; for 5 V, 1.4 kΩ; for a 3.3 to 20 V variable supply, 0-13.7 kΩ (use a 15 kΩ pot); for 5 to 15 V, 1.4-9.6 kΩ (use 1 kΩ fixed-value resistor and a 1 kΩ pot). A full parts list is included on the *Handbook* CD-ROM.

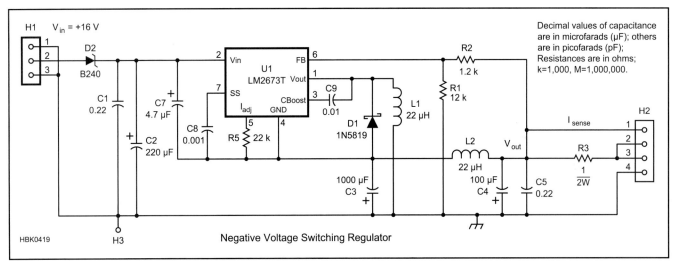

Fig 7.60 — The negative regulator uses the LM2673T in a buck-boost circuit. This circuit inverts the output voltage from the input voltage. A full parts list is included on the *Handbook* CD-ROM.

ing the 3.3 V version of the regulator is based on knowing the current (in mA) through the regulator's internal voltage divider = 3.3 V / 2.7 kΩ = 1.22 mA. The sum of R1 and R2 must cause the voltage at the regulator FB pin to equal 3.3 V. Thus, R1 + R2 in kΩ = (V_{out} – 3.3) / 1.22 and V_{out} = 1.22 (R1 + R2) + 3.3. If R1 = R2 = 0, a direct connection from the output voltage to the FB pin, the calculation results in an output of 3.3 V. The leakage current of the Error Amplifier in the regulator is somewhat less then 25 nA so it can be ignored. The values for R1 and R2 are shown above the schematic for Fig 7.59.

The only critical parts are R1 and R2 which form the voltage dividers for the regulator module. Even their values can be changed, within reason, as long as the ratios are maintained. If you want to have an accurate, fixed output voltage, select a value for R2 that is lower than the calculated value and use a potentiometer for R1 to set the voltage exactly. The value of C3 is not especially critical; however, it should be a low-ESR (equivalent series resistance) type that is intended for use in switchmode circuits.

NEGATIVE REGULATOR

The negative regulator is a buck-boost configuration — it converts a positive voltage into a negative one — see **Fig 7.60**. This design uses many of the same component values as the positive regulators except the regulator IC is an LM2673 to improve circuit stability. The author was unable to implement the current measuring circuit within a feedback loop. Several configurations introduced a significant low frequency noise component to the output voltage. There was also some 50 kHz noise present on the output, but an additional low-pass filter on the output reduced it considerably.

REMOTE SENSING

Many power supplies use remote sensing to electronically compensate for the voltage drop in the wires carrying current to the load. Even with relatively short wires, there can be significant voltage drop between the regulator and its load. There is provision for remote sensing in this circuit described in the sup-

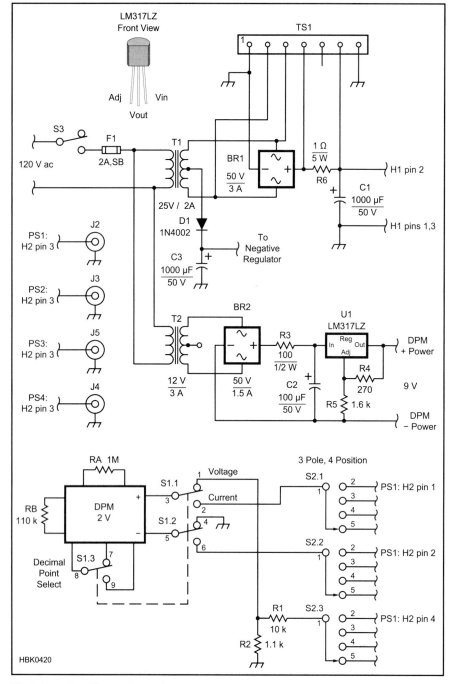

Fig 7.61 — Rectifier and metering schematic. The panel meter is switched between the four modules with a rotary switch (S2) and between voltage and current with a 3PDT toggle (S1). A separate rectifier provides power for the negative supply and a separate three-terminal regulator circuit provides power to the DPM. A full parts list is included on the *Handbook* CD-ROM.

port information for this project on the book's CD-ROM.

If you are not going to use remote sensing then you should insert a jumper in place of R4 in Fig 7.59. R4 (100 Ω) is there for protection just in case the remote sense connection is missing. If you do not want to use remote sensing you can simplify the digital panel meter (DPM) switch wiring to use a two-pole switch instead of the three-pole model listed. In this case, do not use S2.2 and connect S1.2 to the common of S2.3 instead of S2.2.

RECTIFIER CIRCUIT

Fig 7.61 shows the connections among the parts of the system: regulator boards, the digital panel meter (DPM), and rectifier circuit. The components used for the main rectifier circuit are mounted on a terminal strip. You can see the terminal strip and R6 at the top-left of **Fig 7.62**.

Barely visible at the top left, C1 is mounted underneath the terminal strip. The leads of the bridge rectifier are soldered into the holes where the terminals are riveted to the insulating strip. One of the four leads, the negative output, is soldered to a grounded terminal.

THE DIGITAL PANEL METER

Another feature of the unit is the DPM which can be switched to measure the output voltage (H2 pin 1 to ground) as well as the current draw (voltage between H2 pins 1 and 2) for each of the positive supplies. Fig 7.61 shows the 3-pole, 4-position rotary switch (S2) that selects which power supply to monitor and a 3PDT toggle switch (S1) that selects between measuring voltage and current.

In order to measure the voltage drop across the 1-Ω current sense resistors, the DPM needs either an isolated power supply or some more circuitry. This system uses an isolated power supply. A series regulator is used simply because they are somewhat easier to implement and the DPM has a very low current requirement. All components except the transformer are mounted on a piece of perforated board that you can see just below the transformer in Fig 7.62.

Since the 1.2 mA current for the feedback circuit flows through the current sense resistor it will be included in the value displayed by the DPM when current is selected.

The DPM also has a set of jumpers that allow you to set the decimal point location. As can be seen in Fig 7.61, one pole of the toggle switch selects its location.

CONSTRUCTION DETAILS

Both the DPM and the regulators use pin headers for all of the connections that come off the boards (see the parts list for details). This allows assembly of the subsystems without having to consider any attached wires. Wire lengths can be determined later, then install the mating connectors on the wires and simply push them onto the pins.

The printed circuit board (PCB) for the positive regulators contains four identical circuits. The boards could be separated, but a single board made mounting the board and heat sinking the regulator ICs easier. Regulator ICs are mounted on the bottom side instead of on the top and folded over with the flat side parallel to the PCB and farthest from the board (see **Fig 7.63**). If you fold the ICs identically, you can use their mounting holes to fasten them to the side of the enclosure. This not only is a convenient method of mounting the board it also gives the ICs a good heat sink and ground! Fig 7.62 shows the completed assembly bolted to the side of the enclosure.

The negative regulator was an addition to the positive-output system, so it is installed separately from the positive regulators on the bottom of the enclosure. If you already have a positive voltage power supply you can build the negative regulator circuit and simply connect its input to the output of your existing supply.

A caution regarding the circuit boards is in order. PC boards are available from FAR Circuits (**www.farcircuits.net**), a company that provides a lot of boards for ham-related projects. Their boards do not have plated through holes so you will have to be sure that you solder the through-hole components on *both* sides of the board.

Artwork for the PC board layout, Gerber files, and a drill file are available on the CD-ROM included with this book. The schematic capture software *DipTrace* was used in the development of this project. Source files for the schematic and PCB files are also available on the CD-ROM.

RFI

The power supply, in its aluminum enclosure, was put into a completely shielded box with an 18 inch antenna within a few inches of it. An 8-Ω, 20-W resistor was installed on the #2 output with the output voltage adjusted to +8 V. Using a calibrated EMI measurement system, the only emission that would have failed FCC Part 15 testing occurs around 88 MHz as shown in **Fig 7.64**.

7.16.2 12-V, 15-A Linear Power Supply

The supply shown in **Fig 7.65** is a linear 12-V, 15-A design by Ed Oscarson, WA1TWX, with adjustable output voltage and current limiting. Supply regulation is excellent, typically exhibiting a change of less than 20 mV from no load to 15 A. This basic design, with heftier components and additional pass transistors, can deliver over 30 A. (All numbered notes, additional circuit design information, a discussion of how to change the supply voltage and/or current ratings, construction notes, and a complete parts list are available on the CD-ROM included with this *Handbook*.)

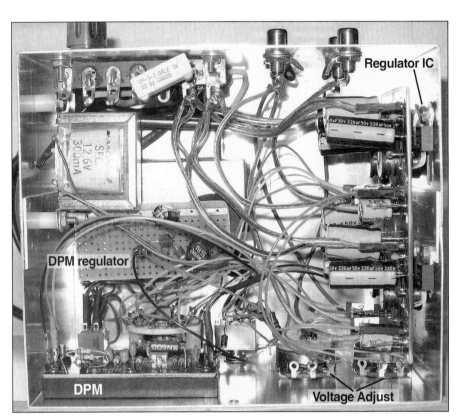

Fig 7.62 — Internal supply wiring. The positive regulator boards are attached to the enclosure wall by the mounting tabs of the regulator ICs. This provides good heat sinking for the regulators.

CIRCUIT DESCRIPTION

Fig 7.66 is the supply's schematic. The ac line input is fused by F1, switched on and off by S1 and filtered by FL1. F1 and S1 are rated at about ¼ of the output current requirement (for 15-A output, use a 4- or 5-A slow-blow fuse or a similarly rated circuit breaker). FL1 prevents any RF from the secondary or load from coupling into the power line and prevents RF on the power line from disturbing supply operation. If your ac power line is clean, and you experience no RF problems, you can eliminate FL1, but it's inexpensive insurance.

When discharged, filter capacitor C1 looks like a short circuit across the output of rectifier U2 when ac power is applied. That usually subjects the rectifier and capacitor to a large inrush current, which can damage them. Fortunately a simple and inexpensive means of inrush-current limiting is available. Keystone Carbon Company (and others) produce a line of inrush-current limiters (thermistors) for this purpose. The device (RT1) is placed in series with one of the transformer primary leads. RT1 has a current rating of 6 A,[1] and

Fig 7.63 — Regulator assemblies. The regulator ICs are bent "forward" so that the metal back of the package is exposed. A nut is epoxied on the front of the mounting tab so that a screw can be inserted through the enclosure wall to mount the regulator.

[1]See the *Handbook* CD ROM for all numbered notes and additional construction information.

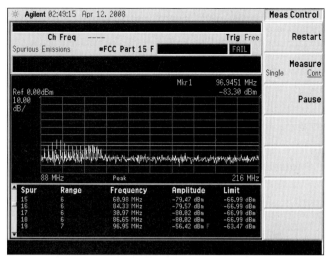

Fig 7.64 — Spectrum of the supply's RF emissions. The only frequency at which emissions exceed FCC Part 15 limits is near 88 MHz.

a cold resistance of 5 Ω. When it's hot, RT1's resistance drops to 0.11 Ω. Such a low resistance has a negligible effect on supply operation. Thermistors run *hot* so they must be mounted in free air, and away from anything that can be damaged by heat.[2,3]

The largest and most important part in the power supply is the transformer (T1). If purchased new, it can also be the most costly. Fortunately, a number of surplus dealers offer power transformers that can be used in this supply.

T1 produces 17 V ac RMS at 20 A; the center tap is not used. Bridge rectifier U2 provides full-wave rectification. Full-wave rectification reduces the ripple component of current that flows in the filter capacitor, resulting in less power dissipation in the capacitor's internal resistance. U2's voltage rating should be at least 50 V, and its current rating about 25% higher than the normal load requirement; a 2-A bridge rectifier will do. U2 is secured to the chassis (or a heat sink) because it dissipates heat.

C1 is a computer-grade electrolytic. Any capacitor value from 15,000 to 30,000 μF will suffice. This version uses a 19,000 μF, 40-V capacitor. The capacitor's voltage rating should be at least 50% higher than the expected no-load rectified dc voltage. In this supply, that voltage is 25 V, and a 40-V capacitor provides enough margin.

R5, a 75-Ω, 20-W bleeder resistor, is connected across C1's terminals to discharge the supply when no load is attached or one is removed. Any resistance value from 50 Ω to 200 Ω is fine; adjust the resistor's wattage rating appropriately.

At the terminals of C1, we have a dc voltage, but it varies widely with the load applied. When keying a CW transmitter or switching a rig from receive to full output, 5-V swings can result. The dc voltage also has an ac ripple component of up to 1.5 V under full load. Adding a solid-state regulator (U1) provides a stable output voltage even with a varying input and load.

VOLTAGE REGULATOR IC AND PASS TRANSISTORS

The LM723 used at U1 has a built-in voltage reference and sense amplifier, and a 150-mA drive output for a pass-transistor array. U1's voltage reference provides a stable point of comparison for the internal regulator circuitry. In this supply, it's connected to the non-inverting input of the voltage-sense op amp. The reference is set internally to 7.15 V, but the absolute value is not critical because an output-voltage adjustment (R12) is provided. What is important is that the voltage is stable, with a specified variation of 0.05% per 1000 hours of operation. This is more than adequate for the supply.

For the regulator to work properly, its ground reference must be at the same point as the output ground terminal. The best way to ensure this is to use the output GROUND terminal (J4) as a single-point ground for all of the supply grounds. Run wires to J4 from each component requiring a ground connection. Fig 7.66 attempts to show this graphically through the use of parallel connections to a single circuit node.

The output pass transistor array consists of a TIP112 Darlington-pair transistor (Q5) driving three 2N3055 power transistors (Q1-Q3). This two-stage design is less efficient than connecting the power transistors directly to the LM723, but Q5 can provide considerably more base current to the 2N3055s than the 150-mA maximum rating of the LM723. You can place additional 2N3055s in parallel to increase the output current capacity of the supply.

This design is not fussy about the pass transistors or the Darlington transistor used. Just ensure all of these devices have voltage ratings of at least 40 V. Q5 must have a 5-A (or greater) collector-current rating and a beta of over 100. The pass transistors should be rated for collector currents of 10 A or more, and have a beta of at least 10.[5]

Resistors R17, R18 and R19 prevent leakage current through the collector-base junction from turning on the transistor by diverting it around the base-emitter junction. When the pass transistors are hot, at the V_{CE} encountered in this design, the leakage current can be as high as 3 mA. The resulting drop across the 33-Ω resistors is 0.1 V — safely below the turn-on value for V_{BE}.

When unmatched transistors are simply connected in parallel they usually don't equally share the current.[6] By placing a low-value resistor in each transistor's emitter lead (emitter-ballasting resistors, R1-R3), equal current sharing is ensured. When a transistor with a lower voltage drop tries to pass more current, the emitter resistor's voltage drop increases,

Fig 7.65 — The completed 12-V, 15-A supply. If individual meters are not available, a DMM can be substituted.

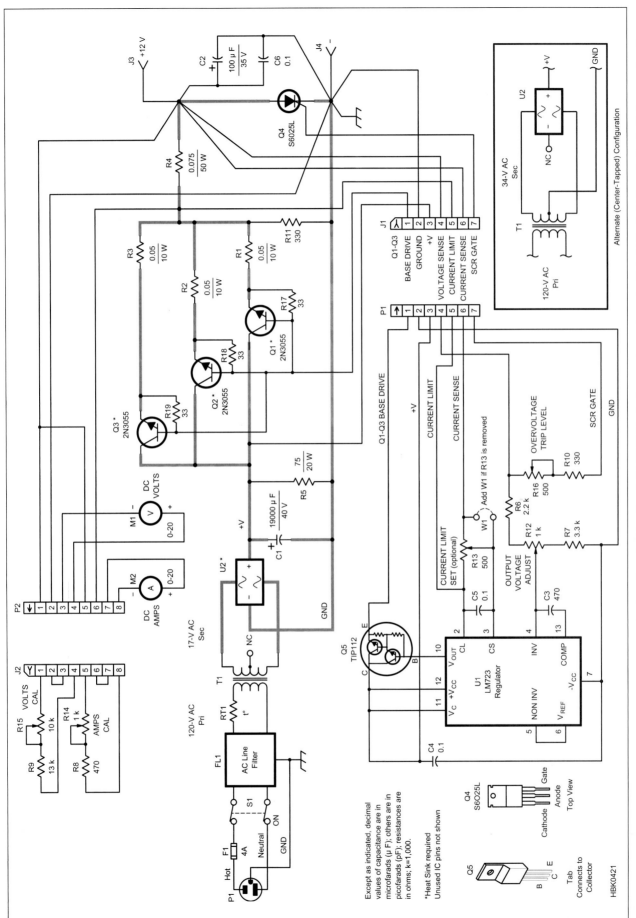

Fig 7.66 — Schematic of the power supply. A parts list may be found on the *Handbook* CD-ROM. Equivalent parts can be substituted. The bold lines indicate high-current paths that should use heavy-gauge (#10 or #12 AWG) wire. This schematic graphically shows wiring to a single-point ground; see text. The majority of the parts used in this supply are available as surplus components.

Power Supplies 7.45

allowing the other transistors to provide more current. Because the voltage-sense point is on the load side of the resistors, the transistors are forced to dynamically share the load current.

With a 5-A emitter current, 0.25 V develops across each 0.05-Ω resistor, producing 1.25 W of heat. Ideally, a resistor's power rating should be at least twice the power it's called upon to dissipate. To help the resistors dissipate the heat, mount them on a heat sink, or secure them to a metal chassis (as shown in **Fig 7.67**). You can use any resistor with a value between 0.065 and 0.1 Ω, but remember that the power dissipated is higher with higher-value resistors (10 W resistors are used here).

At the high output currents provided by this supply, the pass transistors dissipate considerable power. With a current of 5 A through each transistor-and assuming a 9-V drop across the transistor — each device dissipates 45 W. Because the 2N3055's rating is 115 W when used with a properly sized heat sink, this dissipation level shouldn't present a problem. If the supply is to be used for continuous-duty operation, increase the size of the heat sink and mount it with the fins oriented vertically to assist in air circulation.

The output-voltage sense is connected through a resistive divider to the negative input of U1. U1 uses the difference between its negative and positive inputs to control the pass transistors that in turn provide the output current. C3, a compensation capacitor, is connected between this input and a dedicated compensation pin to prevent oscillation. The output voltage is adjusted by potentiometer R12 and two fixed-value resistors, R6 and R7.[7] The voltage-sense input is connected to the supply's positive output terminal, J3.

Current sensing is done through R4, a 0.075-Ω, 50-W resistor connected between the emitter-ballasting resistors and J3. R4's power dissipation is much higher than that of R1, R2 or R3 because it sees the total output current. At 15 A, R4 dissipates 17 W. At 20 A, the dissipated power increases to 30 W.

U1 provides current limiting via two sense inputs connected across R4. Limiting takes place when the voltage across the sense inputs is greater than 0.65 V.[8] For a 15-A maximum output-current limit, this requires a 0.043-Ω resistor. By using a larger-value sense resistor and a potentiometer, you can vary the current limit. Connecting potentiometer R13 across R4 provides a current-limiting range from full limit voltage (8.7 A limit) to no limit voltage. This allows the current limit to be fine-tuned, if needed, and also permits readily available resistor values (such as the author's 0.075-Ω resistors) to be used. A current limit of 20 A is at the top end of the ammeter scale.

R20 maintains a small load of 35-40 mA depending on power supply output voltage. This reduces the effect of leakage current in

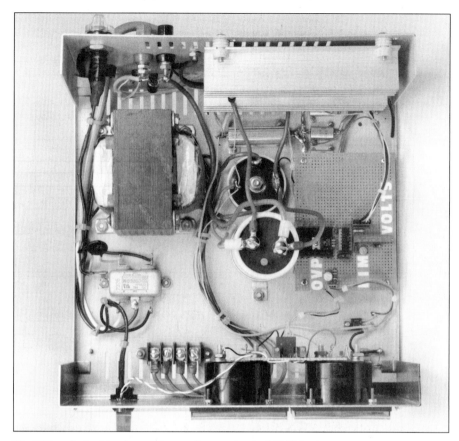

Fig 7.67 — An inside view of the neatly constructed 12-V, 15-A power supply. There's plenty of room in this cabinet to accept the components comfortably and allow hands and tools to get at them. Vents in the bottom and rear of the cabinet provide some measure of convection cooling.

the pass transistors and keeps the regulator's feedback action active, even if no external load is connected.

METERING

Voltmeter M1 is a surplus meter. R8 and potentiometer R15 provide for voltmeter calibration. If the correct fixed-value resistor is available, R15 can be omitted. The combined value of the resistor and potentiometer is determined by the full-scale current requirement of the meter used.[9]

Ammeter M2 is actually a voltmeter (also surplus) that measures the potential across R4. The positive side of M2 connects to the high side of R4. R8 and potentiometer R14 connect between the positive output terminal (J3) and the negative side of M2 to provide calibration adjustment. The values of R8 and R14 are determined by the coil-current requirements of the meter used. (Digital panel meters or a dedicated DMM can be used instead of separate analog meters.)

OUTPUT WIRING AND CROWBAR CIRCUIT

The supply output is connected to the outside world by two heavy-duty banana jacks, J3 and J4. C2, a 100-µF capacitor, is soldered directly across the terminals to prevent low-frequency oscillation. C6, a 0.1-µF capacitor, is included to shunt RF energy to ground. Heavy-gauge wire must be used for the connections between the pass transistors and J3 and between chassis ground and J4. The voltage-sense wire must connect directly to J3 and U2's ground pin must connect directly to J4 (see Fig 7.66). This provides the best output voltage regulation.

An over-voltage crowbar circuit prevents the output voltage from exceeding a preset limit. If that limit is exceeded, the output is shunted to ground until power is removed. If the current-limiting circuitry in the supply is working properly, the supply current-limits to the preset value. If the current limiting is not functioning, the crowbar causes the ac-line fuse to blow. Therefore, it's important to use the correct fuse size: 4 to 5 A for a 15-A supply.

The crowbar circuit is a simple design based on an SCR's ability to latch and conduct until the voltage source is removed. The SCR (Q4) is connected across output terminals J3 and J4. (The SCR can also be connected directly across the filter capacitor, C1, for additional protection.) R10 and potentiometer R16 in series with the Q4's gate provide a means of adjusting the trip voltage. The prototype crowbar is set to conduct at 15 V. The S6025L SCR is rated at 25 A and should be mounted on a metal chassis or heat sink. (Note: Some SCRs are isolated from their mounting tabs, others are not. The S6025L and the 65-ampere S4065J are isolated types. If the SCR you use is not isolated, use a mica washer or thermal pad to insulate it from the chassis or heat sink.)

The bold lines in Fig 7.66 indicate high-current paths that should use heavy-gauge (#10 or #12) wire. Traces that are connected to the output terminals in the schematic by individual lines should be connected directly to the terminals by individual wires. This establishes a 4-wire measurement, where the heavy wires carry the current (and have voltage drops) and the sense wires carry almost no current and therefore voltage errors are not caused by voltage drops in the wiring. If desired, the sense wire can be carried out to the load, but that may introduce noise into the sense feedback circuit, so use caution if that is done.

CONSTRUCTION

Fig 7.67 shows the inside of the prototype supply. The larger components are chassis mounted; two perf-boards contain the majority of the low-power parts of the supply. This includes the potentiometers for the regulator, meters and over-voltage adjustments. A template to make a PC board that contains all of the parts mounted on the two perf-boards is available in the **Templates** folder of this book's CD-ROM.

Q1, Q2, and Q3 must be mounted on a heat sink. The one used measures 1¼ × 6 × 3⅝ inches (HWD), the minimum size recommended. If you can find a heat sink with vertically oriented fins, so much the better. Use a small amount of heat-sink compound between the transistors and the heat sink. Because the transistor collectors are at a potential of + 25 V, they must be insulated from the heat sink, or, as in this supply, the entire heat sink can be isolated from the chassis. (It is better to use a grounded heat sink design because of the potential for short-circuits if the external heat sink is electrically "hot." — *Ed.*) If there is adequate ventilation, you can mount the heat sink inside the chassis. The three emitter-ballasting resistors and the sense resistor can be secured to the chassis rear or bottom, but they should be located near the transistors to which they are connected. Orient the components so that they can be easily soldered to the common output connections.

Also mounted on the back panel are a line-cord strain relief and fuse holder. Use a strain relief to prevent the cord from being pulled out of the chassis. (It is preferable to use a standard CEE-22 plug and socket as discussed at the beginning of this section — *Ed.*) Mount the fuse holder directly above the line cord. The output terminals, J3 and J4, are placed in the same area; use heavy-duty banana jacks or terminal blocks. If FL1 is used, mount it on the chassis. Install an insulated terminal strip near the filter to hold the inrush-current limiter,

Once all of the major components are installed, some of the wiring can be done. Wire the line cord to the fuse with the black (hot) lead at the center, and the outer ring connected to the power switch. It's important to connect the green (ground) line-cord wire to the chassis for safety. Connect the transformer secondary directly to the bridge rectifier. Use #12 AWG wire to connect the rectifier output to the filter capacitor. Use crimp-on or solder-on terminal lugs as needed, as at the filter capacitor connections. Connect C1's negative lead directly to the output GROUND terminal, J4. Connect a length of #12 AWG wire from J4 (or C1) to the chassis. J4 is the single-point ground for the rest of the system. The positive connection will be made later. Attach bleeder resistor R5 to C1.

At this point, you should test the basic dc supply. When ac power is applied, about 20 to 28 V dc should be present across C1's terminals. This potential is dependent on the transformer used, but should not exceed 30 V dc. Turn off the supply.

Next, wire the output pass transistors. If the transistors are insulated from the heat sink, use #10 AWG wire to connect the collectors together. Leave an 8- to 10-inch pigtail for later connection to C1's positive terminal. If the transistors are mounted directly to the heat sink, the pigtail can be connected to the heat sink.

Use #20 AWG wire to connect the transistor base leads together and provide a pigtail for attachment to U1. Using #12 AWG wire, connect the emitters of Q1-Q3 to their respective emitter-balancing resistors. Solder together the remaining emitter-resistor leads and use #10 AWG wire to connect them to R4. Solder the other side of R4 to J3, the positive output terminal. Connect R17, R18, and R19 directly across the transistor base and emitter terminals. Connect R11 directly from the junction of R4 and the emitter resistors to ground.

Next, attach the 100-μF (C2) and 0.1-μF capacitors (C6) across the output terminals. Keep the leads as short as possible, especially those of C6.

Once the regulator board is wired, attach its mating connector to the appropriate points on the chassis. The voltage-sense wire and ground wires must connect directly to the appropriate output terminals. Use #20 or #22 AWG wire for the voltage-sense wire and #18 AWG for the ground.

Attach the current-limit and current-sense wires directly to R4. This is essential for proper regulation and current limiting. There is little current in the wires, so use #20 or #22 AWG wire here and for the power, SCR gate and base-drive connections.

Q4 connects across J3 and J4. Q4's gate is attached to the PC board SCR gate connection at J1, pin 7. Set potentiometer R16 to its maximum resistance or disconnect the SCRs gate prior to testing the supply.

TESTING

Initial testing is done without a load. Use a 2-A fuse at F1 to protect the components in case of problems. If any of the steps do not produce the expected results, check the circuit wiring.

Connect a voltmeter to the output terminals. Turn on the supply. The voltmeter should read between 8 and 15 V. Adjust R12 for a 12-V output (13.8 V if supplying a radio). Adjust R15 for a 13.8-V reading on the front-panel meter. Turn off the supply.

Connect a 12-Ω, 20-W resistor to J3 and attach the other end through an ammeter to J4. (The ammeter must be capable of reading a current now of more than 1 A.) Turn on the supply and measure the output current which should be 1 A. Adjust R14 until ammeter M3 displays 1 A. Turn off the supply.

The next test requires a 0.5-Ω load resistor. Use a high-power-dissipation resistor. To provide additional cooling, immerse the resistor in a plastic container of clean water. Connect the resistor to the supply in place of the 12-Ω load resistor previously used. If the ammeter is left in series with the load, it must be capable of reading a current flow of at least 10 A. The front-panel ammeter may also be used to measure the current. Adjust the CURRENT LIMIT SET potentiometer (R13) to the position where the wiper is at the same end of the potentiometer as the terminal that is connected to the output side of R4. This sets the current limit to 8.7 A (if R4 is a 0.075-Ω resistor).[12] Turn on the supply. The ammeter should indicate about 9 A. If it doesn't, immediately turn off the supply. Check the wiring of the current-limiting circuit, including R13.

If the ammeter reading was okay, remove the series-connected ammeter and connect the 0.50-Ω load resistor across the output terminals. Turn on the supply and adjust CURRENT LIMIT SET pot R13 for a 20-A current indication (or the desired limit point). Turn off the supply.

At this point, the output voltage and the current limit are set. You can recalibrate M2 with a 5- or 10-A load to get better meter resolution when adjusting R14.

The following sequence assumes that the desired output voltage is 12 V, and the over-voltage trip point is 15 V. Using R12, set the output voltage to 15. Decrease the resistance of R16 until the SCR trips. When this happens, turn off the power. With the power off, adjust R12 to decrease the voltage. Turn on the supply and readjust R12 for 12 V.

Now, regulation needs to be checked. Connect a voltmeter across the output terminals with no load connected to the supply. Turn

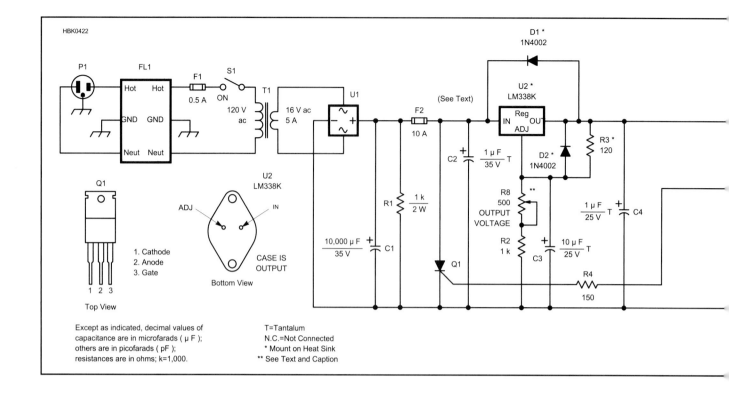

on the supply and record the output voltage. Connect a 10- or 15-A load to the supply and record the voltage. Turn off the supply. The difference between the no-load and 10- or 15-V load voltages should be less than 50 mV. (It is typically less than 20 mV on the prototype.) Higher voltage differences could be caused by the current limit being activated (if near the limit point), or by problems in the sense or single-point ground wiring.

7.16.3 13.8-V, 5-A Linear Power Supply

The power supply shown in **Fig 7.68**, designed by Ben Spencer, G4YNM, provides 13.8 V dc at 5 A, suitable for many low-power transceivers and accessories. It features time-dependent current limiting and short-circuit protection, thermal overload protection within the safe operating area of the regulator IC, and overvoltage protection for the equipment it powers.

Construction, testing and calibration are straightforward, requiring no special skills or equipment. Many of the components can be found in junk boxes, or purchased at hamfests or from mail-order suppliers.

CIRCUIT DESCRIPTION

Fig 7.69 is the power-supply schematic. Incoming ac-line current is filtered by a chassis-mounted line filter (FL1) and, after passing through the fuse (F1), is routed S1 to T1.

U1 rectifies, and C1 filters, the ac output of T1. U2 is an LM338K voltage regulator. This IC features a continuous output of 5 A, with a guaranteed peak output of 7 A, on-chip thermal and safe-operating-area protection for itself, and current limiting. U2's output voltage is set by two resistors (R2 and R3) and a trimmer potentiometer (R8), which allows for adjustment over a small range. U2's input and output are bypassed by C2, C3, and C4. D1 and D2 protect U2 against these capacitors discharging through it.

Overvoltage protection is provided by an SCR, Q1, across the regulator input. Normally, Q1 presents an open circuit, but under fault conditions, it's triggered and short-circuits the unregulated dc input to ground. This discharges C1 and blows F2, reducing the possibility of damage to any connected equipment.

U3, an overvoltage-protection IC, continuously monitors the output voltage. When the output voltage rises above a predetermined level, U3 starts charging C5. If the overvoltage duration is sufficiently long, U3 triggers Q1. This built-in delay (about 1 ms) allows short transient noise spikes on the output voltage to be safely ignored while still triggering the SCR if a true fault occurs. The monitored voltage is set by R5 and R6 and trimmer potentiometer R9, which allows for adjustment over a limited range.

D3 protects the supply from reverse-polarity discharge from connected equipment. The presence of output voltage is indicated by an LED, DS1. R7 is a current limiting resistor for DS1.

CONSTRUCTION

How you construct your supply depends on the size of the components and enclosure you use. General physical layout is not important, although there are a couple of areas that require some attention. In the unit shown in **Fig 7.70**, FL1, the fuse holders, S1, the heat sink. DS1 and the binding posts are mounted on the front and rear enclosure panels. T1 and the PC board are secured to the enclosure's bottom plate. C1's mounting clamp is attached to the rear panel. Bleeder resistor R1 is connected directly across C1's terminals. D3 is soldered directly across the output binding posts.

U2, D1, D2 and R3 are all mounted on the

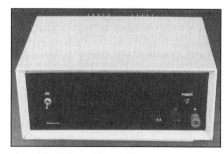

Fig 7.68 — Completed 13.8-V, 5-A power supply

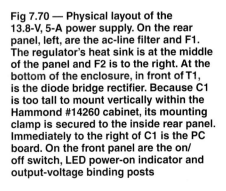

Fig 7.69 — Schematic for the 13.8-V, 5-A power supply. Unless otherwise specified, resistors are ¼-W, 5%-tolerance carbon-composition or film units. A PC board and U3 are available from FAR Circuits (www.farcircuits.net). The PC board has mounting holes and pads to allow for handling different trimmer-potentiometer footprints. A PC board template is available on the CD-ROM for this book.

C1 — 10,000 µF, 35-V electrolytic.
C2, C4 — 1 µF, 35-V tantalum.
C3 — 10 µF, 35-V tantalum.
C5 — 0.1 µF, 25-V ceramic disc.
D1-D3 — 1N4002.
DS1 — Red LED.
F2 — Fast-acting 10-A fuses; three required (see text).
F1 — Fast-acting 0.5-A fuse.
FL1 — Ac-line filter.
Q1 — BT152 400-V, 25-A SCR in TO-220A package (NTE5554)
R8 — 500 Ω, single-turn trimmer potentiometer.
R9 — 500 Ω or 1 kΩ, single-turn trimmer potentiometer.
S1 — SPST panel-mount switch.
T1 — 120-V primary, 16- to 20-V, 5-A secondary.
U1 — 100-PIV, 6-A bridge rectifier.
U2 — LM338K 5-A adjustable power regulator in a TO-3 package.
U3 — MC3423P1 overvoltage protection IC.

Misc: two panel-mount fuse holders; line cord; heat sinks for TO-3 case transistors; TO-3 mounting kit and heat-sink grease; black and red binding posts; chassis or cabinet; PC board; hardware, rubber hoods, heat-shrink tubing or electrical tape for F1 and FL1, hook-up wire.

heat sink as shown in **Fig 7.71**. It's important to keep R3 attached as closely as possible to U2's terminals to prevent instability. Use a TO-3 mounting kit and heat-conductive grease or thermal pad to electrically isolate U2 from the heat sink.

Cover all ac-input wiring (use insulated wire and heat-shrink tubing) to prevent electrical shock and route the ac wiring away from the dc wiring. The rear panel assembly is shown in **Fig 7.72**.

Mount U2, C1, and the PC board close to each other and keep the wire runs between these components as short as possible. Excessively long wire runs may lead to unpredictable behavior.

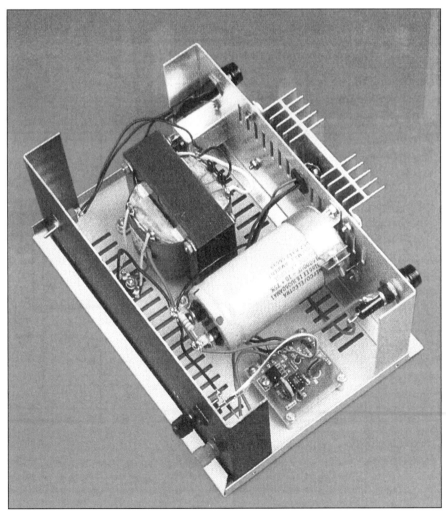

Fig 7.70 — Physical layout of the 13.8-V, 5-A power supply. On the rear panel, left, are the ac-line filter and F1. The regulator's heat sink is at the middle of the panel and F2 is to the right. At the bottom of the enclosure, in front of T1, is the diode bridge rectifier. Because C1 is too tall to mount vertically within the Hammond #14260 cabinet, its mounting clamp is secured to the inside rear panel. Immediately to the right of C1 is the PC board. On the front panel are the on/off switch, LED power-on indicator and output-voltage binding posts

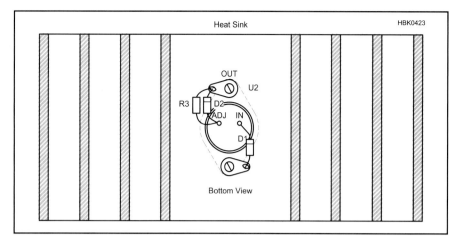

Fig 7.71 — Here's the regulator's heat-sinking arrangement, showing U2, D1, D2 and R3. U2 is electrically isolated from the heat sink by a TO-3 mounting kit.

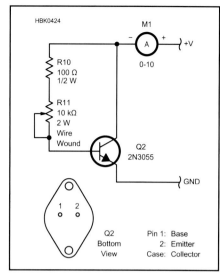

Fig 7.73 — An active resistive load to use in testing the supply. Adjustment of R11 is sensitive.

M1 — Multimeter or ammeter capable of measuring 10 A.
Q2 — 2N3055; mount on heat sink.

Fig 7.72 — A rear view of the power supply enclosure. Louvers in the enclosure bottom and on the rear panel provide convective cooling.

TEST AND CALIBRATION

An accurate multimeter covering ranges of 30 V dc and 10 A dc is required. A variable resistive load with a power rating of 100 W is also needed; this can he made using a heat-sink-mounted 2N3055 power transistor and a couple of components as shown in **Fig 7.73**.

First, set R8 (OUTPUT VOLTAGE) fully clockwise and R9 (OVERVOLTAGE) fully counterclockwise. Insert a fuse in the dc line at F2. Connect the ac line, turn on S1 and check that DS1 lights. Measure the output voltage: it should be about 12 V. Adjust R8 counterclockwise until you obtain 14.2 V output; this sets the trip voltage.

While monitoring the output voltage, gradually adjust R9 clockwise until the voltage suddenly falls to zero. This indicates that the SCR has triggered and blown F2. Disconnect the ac line cord from the wall socket. *Don't make any adjustment to R9!* Instead, adjust R8 fully clockwise.

With the ac line cord removed, check that F2 is open. Replace F2 with a new fuse (now you know why two of the three fuses are called for). Reconnect the ac line cord, and while continually monitoring the output voltage, gradually adjust R8 until Q1 again triggers at 14.2 V, blowing F2. If you find the adjustment of R9 to be too sensitive, use a 500-Ω potentiometer in its place and reduce the value of R5 (if necessary) to provide the required adjustment range.

Again disconnect the line cord from the wall socket, set R8 fully clockwise, replace F2 (there's the third fuse!) and reset R8 for 13.8 V. This completes the voltage calibration and overvoltage protection tests. The power supply is now set to 13.8 V output, with the overvoltage protection set for 14.2 V.

Adjust the variable resistive load to maximum resistance and connect it to the power supply output in series with the ammeter. Turn on the power supply and gradually adjust R10 until a current of 5 A flows. Decrease the resistance further and check that the current limits between 5.5 A and 0.5 A.

Finally, be thoroughly unpleasant and apply a short circuit via the ammeter. Check that the current-limiting feature operates correctly. The prototype limited at approximately 3.5 A. Disconnect the ac line cord and test equipment, dose up the case and your power supply is ready for service.

SUMMARY

The prototype supply powers a 25-W transmitter that continually draws 4.5 A. Under these conditions, the entire power supply gets hot, so ventilation holes were eventually drilled into the case. (On this prototype, the heat sinks are mounted inside the cabinet). The heat sink gets hot enough to scorch the skin of the unwary, so be careful!

7.16.4 High-Voltage Power Supply

This two-level, high-voltage power supply was designed and built by Dana G. Reed, W1LC. It was designed primarily for use with an RF power amplifier using a triode in class AB2 grounded-grid operation. The supply is rated at a continuous output current of 1.5 A, and will easily handle intermittent peak currents of 2 A. The 12-V control circuitry, and the low-tap setting of the plate transformer secondary, make it straightforward to adapt the design to homemade tube amplifiers.

The step-start circuit is straightforward and

ensures that the rectifier diodes are current-limited when the power supply is first turned on. A 6-kV meter is used to monitor high-voltage output.

Fig 7.74 is a schematic diagram of the bi-level supply. An ideal power supply for a high-power linear amplifier should operate from a 240-V circuit, for best line regulation. A special, hydraulic/magnetic circuit breaker also serves as the disconnect for the plate transformer primary. Don't substitute a standard circuit breaker, switch or fuses for this breaker; fuses won't operate quickly enough to protect the amplifier or power supply in case of an operating abnormality. The 100-kΩ, 3-W bleeder resistors are of stable metal-oxide film design. These resistors are wired across each of the 14 capacitors to equalize voltage drops in the series-connected bank. This choice of bleeder resistor value provides a lighter load (less than 25 W total under high-tap output) and benefits mainly the capacitor-bank filter by yielding much less heat as a result. A reasonable, but longer bleed-down time to fully discharge the capacitors results — about nine minutes after power is removed. A small fan is included to remove any excess heat from the power supply cabinet during operation.

POWER SUPPLY CONSTRUCTION

The power supply can be built in a 23½ × 10¾ × 16-inch cabinet. The plate transformer is quite heavy at 67 lbs, so use 1/8-inch aluminum for the cabinet bottom and reinforce it with aluminum angle for extra strength and stability. The capacitor bank will be sized for the specific capacitors used. This project employed ⅜-inch thick polycarbonate for reasonable mechanical stability and excellent high-voltage isolation. The full-wave bridge consists of four commercial diode block assemblies.

POWER SUPPLY OPERATION

When the front-panel breaker is turned on, a single 50-Ω, 100-W power resistor limits primary inrush current to a conservative value as the capacitor bank charges. After approximately two seconds, step-start relay K1 actuates, shorting the 50-Ω resistor and allowing full line voltage to be applied to the plate transformer. No-load output voltages under low- and high-tap settings as configured and shown in **Fig 7.74** are 3050 V and 5400 V, respectively. Full-load levels are somewhat lower, approximately 2800 V and 4900 V. If a tap-select switch is used as described in the schematic parts list, it should only be switched when the supply is off.

7.16.5 Micro M+ PV Charge Controller

The Micro M+ is an ideal photovoltaic (PV) controller for use at home or in the field. It's an easy-to-build, one-evening project even a beginner can master. This project was designed by Mike Bryce, WB8VGE. This project description is an overview of the Micro M+ controller.

An earlier charge controller called the "Micro M" (Sep 1996 *QST*) proved to be a very popular project. After feedback from readers and builders, the following changes

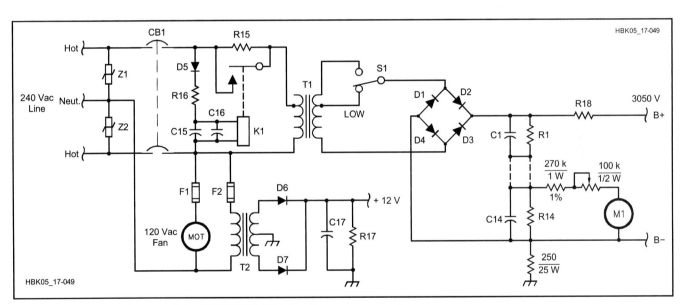

Fig 7.74 — Schematic diagram of the 3050-V/5400-V high-voltage power supply.

C1-C14 — 800 μF, 450 V electrolytic.
C15, C16 — 4700 μF, 50 V electrolytic.
C17 — 1000 μF, 50 V electrolytic.
CB1 — 20-A hydraulic/magnetic circuit breaker (Potter & Brumfield W68-X2Q12-20 or equiv). 40-A version required for commercial applications/service (Potter & Brumfield W92-X112-40).
D1-D4 — Commercial diode block assembly: K2AW HV14-1, from K2AW's Silicon Alley
D5 — 1000-PIV, 3-A, 1N5408 or equiv.
D6, D7 — 200-PIV, 3-A, 1N5402 or equiv.
F1, F2 — 0.5 A, 250 V (Littelfuse 313 Series, 3AG glass body or equiv).
K1 — DPDT power relay, 24-V dc coil; both poles of 240-V ac/25-A contacts in parallel (Potter & Brumfield PRD-11DYO-24 or equiv).
M1 — High-voltage meter, 6-kV dc full scale. (Important: Use a 1-mA or smaller meter movement to minimize parallel-resistive loading at R14. Also, select series meter-resistor and adjustment-potentiometer values to calibrate your specific meter. Values shown are for a 1-mA meter movement.)
MOT1 — Cooling fan, 119mm, 110120-V ac, 30-60 CFM, (EBM 4800Z or equiv).
R1-R14 — Bleeder resistor, 100 kΩ, 3 W, metal oxide film.
R15 — 50 Ω, 100 W.
R16 — 3.9 kΩ, 25 W.
R17 — 30 Ω, 25 W.
R18 — 20 Ω, 50 W.
S1 — Ceramic rotary, 2-pos. tap-select switch (optional). Voltage rating between tap positions should be at least 2.5 kV. Mount switch on insulated or ungrounded material such as a metal plate on standoff insulators, or an insulating plate, and use only a *nonconductive* or otherwise *electrically-isolated* shaft through the front panel for safety
T1 — High-voltage plate transformer, 220/230-V primary, 2000/3500-V, 1.5-A CCS JK secondary (Peter W. Dahl Company, Hipersil C-Core; www.harbachelectronics.com). Primary 220-V tap fed with nominal 240-V ac line voltage to obtain modest increase in specified secondary voltage levels
T2 — 120-V primary, 18-V CT, 2-A secondary (Mouser 41FJ020).
Z1-Z2 — 130-V MOV.

were made to the design, creating the Micro M+:

• Reduce the standby current at night
• Increase current handling capacity to 4 A
• Change the charging scheme to high (positive) side switching
• Improve the charging algorithm
• Keep the size as small as possible, but large enough to easily construct.

You can assemble one in about an hour. Everything mounts on one double-sided PC board. It's completely silent and makes absolutely no RFI!

The Micro M+ will handle up to 4 A of current from a solar panel. That's equal to a 75-W solar panel. (A 75-W module produces 4.4 A at 17 V. The Micro M+ can easily handle the extra 400 mA.) Standby current is less than 1 mA. All the current switching is done on the positive side, so you can connect the photovoltaic array, battery and load grounds together.

HOW IT WORKS

Fig 7.75 shows the complete Micro M+ and **Fig 7.76** shows the schematic diagram. Let's begin with the current handling part of the Micro M+. Current from the solar panel is controlled by a power MOSFET. Instead of using a common N-channel MOSFET, however, the Micro M+ uses an STMicroelectronics STP80PF55 P-channel MOSFET. This P-channel FET has a current rating of 80 A with an R_{DSon} of 0.016 Ω in a TO-220 package. Current from the solar panel is routed directly to the MOSFET source lead.

Using a P-channel MOSFET as a high-side switch (see the **Analog Basics** chapter section "Using a Transistor as a Switch") eliminates the need for a charge pump that would be required to drive an N-channel MOSFET. To turn on a P-channel MOSFET, all we have to do is pull the gate lead to ground! Since the Micro M+ does not have a charge pump, it generates no RFI!

With the P-channel MOSFET controlling the current, diode D4 — an SB520 Schottky — prevents battery current from flowing into the solar panel at night. This diode also provides reverse polarity protection to the battery in the event you connect the solar panel backwards. This protects the expensive P-channel MOSFET.

Zener diode D2, a 1N4747, protects the gate from damage due to spikes on the solar panel line. Resistor R12 pulls the gate up, ensuring the power MOSFET is off when it is supposed to be.

The Micro M+ never draws current from the battery. The solar panel provides all the power the Micro M+ needs, which means the Micro M+ goes to sleep at night. When the sun rises, the Micro M+ will start up again. As soon as the solar panel is producing enough current and voltage to start charging the battery, the Micro M+ will pass current into the battery.

To reduce the amount of stand-by current, diode D3 passes current from the solar panel to U3, the voltage regulator. U3, an LM78L08 regulator, provides a steady +8 V to the Micro M+ controller. Bypass capacitors, C6, C7 and C8 are used to keep everything happy. As long as there is power being produced by the solar panel, the Micro M+ will be awake. At sun down, the Micro M+ will go to sleep. Sleep current is on the order of less than 1 mA!

BATTERY SENSING

The battery terminal voltage is divided down to a more usable level by resistors, R1, R2 and R3. Resistor R3, a 10 kΩ trimmer, sets the state-of-charge for the Micro M+. A filter consisting of R5 and C1 helps keep the input clean from noise picked up by the wires to and from the solar panel. Diode D1 protects the op-amp input in case the battery sense line was connected backwards.

An LM358 dual op-amp is used in the Micro M+. One section (U1B) buffers the divided battery voltage before passing it along to the voltage comparator, U1A. Here the battery sense voltage is compared to the reference voltage supplied by U4. U4 is an LM336Z-5.0 precision diode. To prevent U1A from oscillating, a 10-MΩ resistor is used to eliminate any hysteresis.

As long as the voltage of the battery under charge is below the reference point, the output of U1A will be high. This saturates transistors Q1 and Q2. Q2 conducts and lights LED DS1, the CHARGING LED. Q1, also fully saturated, pulls the gate of the P channel MOSFET to ground. This effectively turns on the FET, and current flows from the solar panel into the battery via D4.

As the battery begins to take up the charge, its terminal voltage will increase. When the battery reaches the state-of-charge set point, the output of U1A goes low. With Q1 and Q2 now off, the P channel MOSFET is turned off, stopping all current into the battery. With Q2 off, the CHARGING LED goes dark.

Since we have eliminated any hysteresis in U1A, as soon as the current stops, the output of U1A pops back up high again. Why? Because the battery terminal voltage will fall back down as the charging current is removed. If left like this, the Micro M+ would sit and oscillate at the state-of-charge set point.

To prevent that from happening, the output of U1A is monitored by U2, an LM555 timer chip. As soon as the output of U1A goes low, this low trips U2. The output of U2 goes high, fully saturating transistor Q3. With Q3 turned on, it pulls the base of Q1 and Q2 low. Since both Q1 and Q2 are now deprived of base current, they remain off.

With the values shown for R15 and C2, charging current is stopped for about four seconds after the state-of-charge has been reached.

After the four second delay, Q1 and Q2 are allowed to have base drive from U1A. This lights up the charging LED and allows Q4 to pass current once more to the battery.

As soon as the battery hits the state-of-charge once more, the process is repeated. As the battery becomes fully charged, the "on" time will shorten up while the "off" time will always remain the same four seconds. In effect, a pulse of current will be sent to the battery that will shorten over time.

As a side benefit of the pulse time modulation, the Micro M+ won't go nuts if you put a large solar panel onto a small battery. The charging algorithm will always keep the off time at four seconds allowing the battery time to rest before being hit by higher current than normal for its capacity.

7.16.6 Automatic Sealed Lead-Acid Battery Charger

After experiencing premature failure of the battery in his Elecraft K2 transceiver, Bob Lewis, AA4PB, began searching for an automatic battery charger. Although this char-

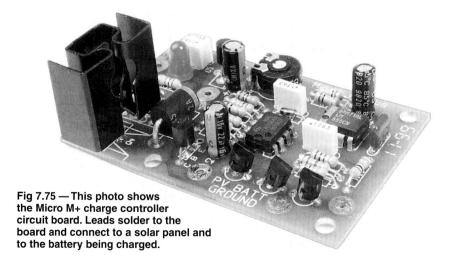

Fig 7.75 — This photo shows the Micro M+ charge controller circuit board. Leads solder to the board and connect to a solar panel and to the battery being charged.

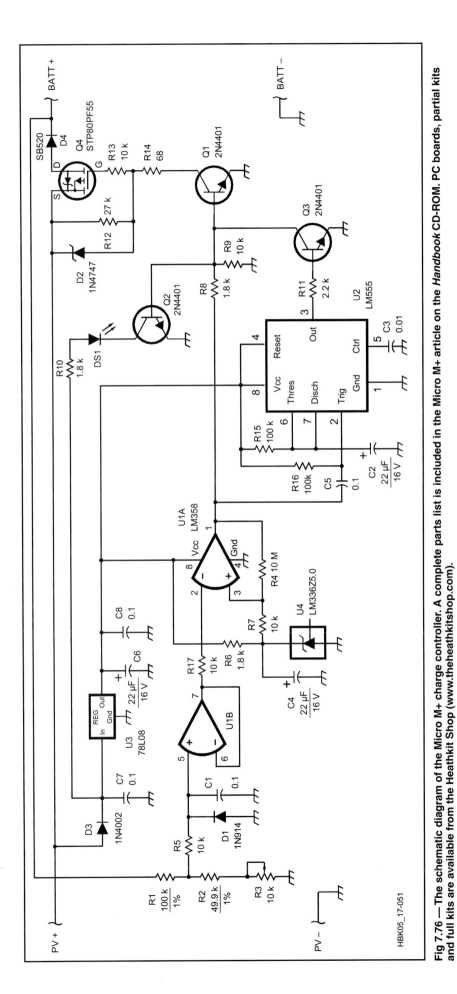

Fig 7.76 — The schematic diagram of the Micro M+ charge controller. A complete parts list is included in the Micro M+ article on the *Handbook* CD-ROM. PC boards, partial kits and full kits are available from the Heathkit Shop (www.theheathkitshop.com).

ger was designed specifically for use with the Power-Sonic PS-1229A sealed lead-acid (SLA) battery used in the Elecraft K2 transceiver, its design concepts have wide ranging applications for battery operated QRP rigs of all types. Comments pertaining to the SLA batteries and chargers apply across the board and the charger described here can be used with any similar battery.

SLAs are commonly called gel-cells because of their gelled electrolyte. As with all things, to obtain maximum service life from an SLA battery, it needs to be treated with a certain degree of care. SLA batteries must be recharged on a regular basis; they should not be undercharged or overcharged. If an SLA battery is left unused, it will gradually self-discharge.

USING A THREE-MODE CHARGER

The author first attempted to use a commercial automatic three-mode charger. However, most three-mode chargers work by sensing current and are never intended to charge a battery under load.

Three-mode chargers begin the battery charging process by applying a voltage to the battery through a 500-mA current limiter. This stage is known as *bulk-mode* charging. As the battery charges, its voltage begins to climb. When the battery voltage reaches 14.6 V the charger maintains the voltage at that level and monitors the battery charging current. This is known as the *absorption mode*, sometimes called the *overcharge mode*. By this time, the battery has achieved 85 % to 95% of its full charge. As the battery continues to charge — with the voltage held constant at 14.6 V — the charging current begins to drop.

When the charging current falls to 30 mA, the three-mode charger switches to *float mode* and lowers the applied voltage to 13.8 V. At 13.8 V, the battery becomes self-limiting, drawing only enough current to offset its normal self-discharge rate. This works great — until you attach a light load to the battery, such as a receiver. The K2 receiver normally draws about 220 mA. When the charger detects a load current above 30 mA, it's fooled into thinking that the battery needs charging, so it reverts to the absorption mode, applying 14.6 V to the battery. If left in this condition, the battery is overcharged, shortening its service life.

UC3906 IC CHARGERS

Chargers using the UC3906 SLA charge-controller IC work just like the three-mode charger described earlier except that their return from float mode to absorption mode is based on voltage rather than current. Typically, once the charger is in float mode it won't return to absorption mode until the battery voltage drops to 90% of the float-mode voltage (or about 12.4 V). Although

Power Supplies   7.53

this is an improvement over the three-mode charger, it still has the potential for overcharging a battery to which a light load is attached.

First, let's look at the situation where a UC3906-controlled charger is in absorption mode and you turn on the K2 receiver, applying a load. The battery is fully charged, but because the load is drawing 220 mA, the charging current never drops to 30 mA and the charger remains in absorption mode, thinking that it is the battery that is asking for the current. As with the three-mode charger, the battery is subject to being overcharged.

If we remove the load by turning off the K2, the current demand drops below 30 mA and the charger switches to float mode (13.8 V). When the K2 is turned on again, because the charger is able to supply the 220 mA for the receiver, the battery voltage doesn't drop, so the charger stays in float mode and all is well. However, if the transmitter is keyed (increasing the current demand), the charger can't supply the required current, so it's taken from the battery and the battery voltage begins to drop. If we un-key the transmitter before the battery voltage reaches 12.4 V, the charger stays in float mode. Now it takes much longer for the charger to supply the battery with the power used during transmit than it would have if the charger had switched to absorption mode.

Let's key the transmitter again, but this time keep it keyed until the battery voltage drops below 12.4 V. At this point, the charger switches to the absorption mode. When we un-key the transmitter, we're back to the situation where the charger is locked in absorption mode until we turn off the receiver.

WHY WORRY?

So why this concern about overcharging an SLA battery? At 13.8 V, the battery self-limits, drawing only enough current to offset its self-discharge rate (typically about 0.001 times the battery capacity, or 2.9 mA for a 2.9 Ah battery). An SLA battery can be left in this float-charge condition indefinitely without overcharging it. At 14.6 V, the battery takes more current than it needs to offset the self-discharge. Under this condition, oxygen and hydrogen are generated faster than they can be recombined, so pressure inside the battery increases. Plastic-cased SLA batteries such as the PS-1229A have a one-way vent that opens at a couple of pounds per square inch pressure (PSI) and release the gases into the atmosphere. This results in drying the gelled electrolyte and shortening the battery's service life. Both undercharging and overcharging need to be avoided to get maximum service life from the battery.

Continuing to apply 14.6 V to a 12-V SLA battery represents a relatively minor amount of overcharge and results in a gradual deterioration of the battery. Applying a potential of 16 V or excessive bulk-charging current to a small SLA battery from an uncontrolled solar panel can result in serious overcharging. Under these conditions, the overcharging can cause the battery to overheat, which causes it to draw more current and result in thermal runaway, a condition that can warp electrodes and render a battery useless in a few hours. To prevent thermal runaway, the maximum current and the maximum voltage need to be limited to the battery manufacturer's specifications.

DESIGN DECISION

To avoid the potential of overcharging a battery with an automatic charger locked up by the load, the author decided to design a charger that senses battery voltage rather than current in order to select the proper charging rate. A 500-mA current limiter sets the maximum bulk rate charge to protect the battery and the charger's internal power supply.

Like the three-mode chargers, when a battery with a low terminal voltage is first connected to the charger, a constant current of 500 mA flows to the battery. As the battery charges, its voltage begins to climb. When the battery voltage reaches 14.5 V, the charger switches off. With no charge current flowing to the battery, its voltage now begins to drop. When the current has been off for four seconds, the charger reads the battery voltage. If the potential is 13.8 V or less, the charger switches back on. If the voltage is still above 13.8 V, the charger waits until it drops to 13.8 V before turning on. The result is a series of 500-mA current pulses varying in width and duty cycle to provide an average current just high enough to maintain the battery in a fully charged condition. Because the repetition rate is very low (a maximum of one current pulse every four seconds) no RFI is generated that could be picked up by the K2 receiver. Because the K2's critical circuits are all well-regulated, slowly cycling the battery voltage between 13.8 V and 14.5 V has no ill effects on the transmitted or received signals.

As the battery continues to charge, the pulses get narrower and the time between pulses increases (a lower duty cycle). Now when the K2 receiver is turned on and begins drawing 220 mA from the battery, the battery voltage drops more quickly so the pulses widen (the duty cycle increases) to supply a higher average current to the battery and make up for that taken by the receiver. When the K2 transmitter is keyed, it draws about 2 to 3 A from the battery. Because the charger is current limited to 500 mA, it is not able to keep up with the transmitter demands. The battery voltage drops and the charger supplies a constant 500 mA. The battery voltage continues to drop as it supplies the required transmit current. When the transmitter is un-keyed, the battery voltage again begins to rise as the charger replenishes the energy used during transmit. After a short time, (depending on how long the transmitter was keyed) the battery voltage reaches 14.5 V and the pulsing begins again. The charger is now fully automatic, maintaining the battery in a charged condition and adjusting to varying load conditions.

The great thing about this charging system is that during transmit the majority of the required 2 to 3 A is taken from the battery. When you switch back to receive, the charger is able to supply the 220 mA needed to run the receiver and deliver up to 280 mA to the battery to replenish what was used during transmit. This means that the power source need only supply the average energy used over time, rather than being required to supply the peak energy needed by the transmitter. (You don't need to carry a heavy 3-A regulated power supply with your K2.) As long as you don't transmit more than about 9% of the time, this system should be able to power a K2 indefinitely.

Have you ever noticed that sometimes when your handheld radio has a low battery and you drop it into its charger you hear hum on the received signals? This charger's power supply is well filtered to ensure that there is no ripple or ac hum to get into the K2 under low battery voltage conditions.

CIRCUIT DESCRIPTION

The charger schematic is shown in **Fig 7.77**; the unit is dubbed the PCR12-500A, short for Pulsed-Charge Regulator for 12-V SLA batteries with maximum bulk charge rates of 500 mA. U1, an LM317 three-terminal voltage regulator, is used as a current limiter, voltage regulator and charge-control switch. A 15-V Zener diode (D2) sets U1 to deliver a no-load output of 16.2 V. R3 sets U1 to limit the charging current to 500mA. When Q1 is turned on by the LM555 timer (U2), the ADJ pin of U1 is pulled to ground, lowering its output voltage to 1.2 V. D4 effectively disconnects the battery by preventing battery current from flowing back into U1. A Schottky diode is used at D4 because of its low voltage drop (0.4 V).

An LM358 (U4A) operates as a voltage comparator. U5, an LM336, provides a 2.5-V reference to the positive input (pin 3) of U4. R11, R12 and R13 function as a voltage divider to supply a portion of the battery voltage to pin 2 of U4A. R13 is adjusted so that when the battery terminal voltage reaches 14.5 V, the negative input of U4A rises slightly above the 2.5-V reference and its output switches from +12 V to 0 V. When this happens, the 1-MΩ resistor (R7) causes the reference voltage to drop a little and provide some hysteresis. The battery voltage must now drop to approximately 13.8 V before U4A turns back on.

U4B is a voltage follower. It pulls the trigger input (pin 2) of U2 to 0 V, causing its output to go to 12 V. U4B's output remains at 12 V until C5 has charged through R6 (approximately four seconds) and the trigger

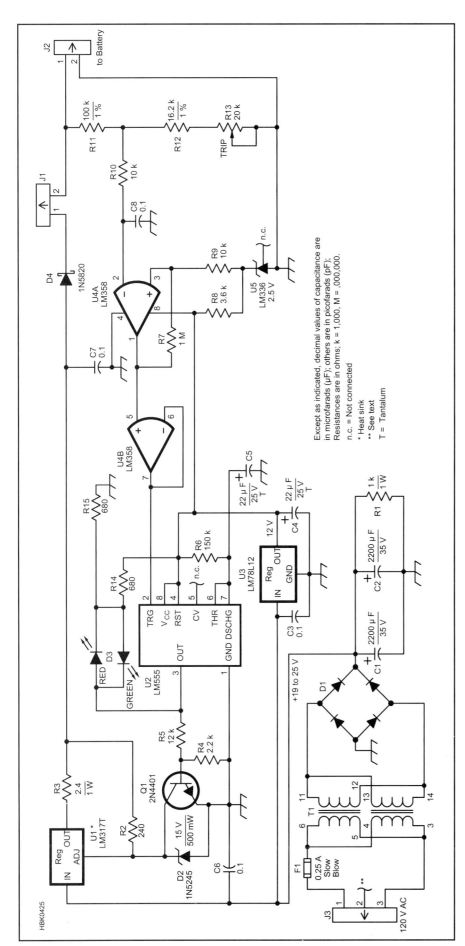

Fig 7.77 — Schematic of the SLA charger. Unless otherwise specified, resistors are ¼-W, 5%-tolerance carbon-composition or film units. Equivalent parts can be substituted; n.c. indicates no connection. A PC board is available from FAR Circuits (www.farcircuits.net).

C1, C2 — 2200 µF, 35 V electrolytic.
C3, C6, C7, C8 — 0.1 µF, 50 V metalized-film.
C4, C5 — 22 µF, 25 V tantalum.
D1 — 400 V, 4 A bridge rectifier.
D2 — 1N5245 Zener diode, 15 V, 500 mW.
D3 — Bicolor LED, red-green.
D4 — 1N5820 Schottky diode.
F1 — 0.25 A slow-blow fuse.
J1 — 2-pin header, PC mount.
J2 — 2-pin connector, PC mount.
J3 — 3-pin connector, PC mount.
Q1 — 2N4401 NPN transistor.
R13 — 20 kΩ multi-turn pot.
T1 — 15 V ac, 666 mA.
U1 — LM317T voltage regulator, TO-220 case.
U2 — LM555 timer.
U3 — LM78L12 voltage regulator, TO-92 case.
U4 — LM358 dual op amp.
U5 — LM336, 2.5 V voltage reference, TO-92 case.
Misc: PC board; TO-220 heat sink; five ¼-inch, #4-40 stand-offs; two fuse-holder clips, PC mount; two-pin shunt; two-pin connector housing; three-pin connector housing; four housing pins; enclosure

has been released by U4A sensing the battery dropping to 13.8 V or less. While the output of U2 is at 12 V, emitter/base current for Q1 flows via R5 and Q1's collector pulls U1's ADJ pin to ground, turning off the charging current.

The output of U2 also provides either +12 V or 0 V to the bicolor LED, D3. R14 and R15 form a voltage divider to provide a reference voltage to D3 such that D3 glows red when U2's output is +12 V and green when U2's output is at 0 V. When ac power is applied but U1 is switched off and not supplying current to the battery, D3 glows red. When U1 is on and supplying current to the battery, D3 is green. As the battery reaches full charge, D3 blinks green at about a four-second rate. As the battery charge increases, the on time of the green LED decreases and the off time increases. A fully charged battery may show green pulses as short as a half-second and the time between pulses may be 60 seconds or more.

T1, D1, C1, and C2 form a standard full-wave-bridge power supply providing an unregulated 20 V dc at 500 mA. U3, an LM78L12 three-terminal regulator, provides a regulated 12-V source for the control circuits.

Note that the mounting tab on U1 is not at ground potential. U1 should be mounted to a heat sink with suitable electrically insulated but thermally conductive mounting hardware to avoid short circuits. Suitable mounting hardware is included with the PC board.

Fig 7.78 — Complete SLA charger PC board.

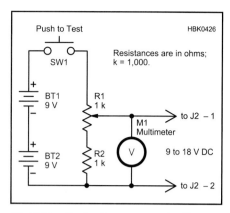

Fig 7.79 — Test voltage source for the battery charger.

OTHER BULK-CHARGE RATES

The maximum bulk-charge rate is set by the value of R3 in the series regulator circuit. The formula used to determine the value of this resistor is R (Ω) = 1200 / 1 (mA). T1 must be capable of supplying the bulk charge current and U1 must be rated to handle this current. The LM317T used here is rated for a maximum current of 1.5 A, provided it has a heat sink sufficiently large enough to dissipate the generated heat. If you increase the bulk-charge rate, you'll definitely need to increase the size of the on-board heat sink. Mounting U1 directly to the housing (be sure to use an insulator) may be a good option.

TRANSFORMER SUBSTITUTION

The transformer used for T1 is small in size and offers PC-board mounting. You can substitute any transformer rated at 15 or 16 V ac (RMS) at 500 mA or more. You may find common frame transformers to be more readily available. You can mount such a transformer to an enclosure wall and route the transformer leads to the appropriate PC-board holes.

CONSTRUCTION

There is nothing critical about building this charger. You can assemble it on a prototyping board, but a PC board and heat sink are available (**www.farcircuits.net**; see **Fig 7.78**) The specially ordered heat sink supplied with the PC board is ¼-inch higher than the one identified in the parts list and results in slightly cooler operation of U1. The remaining parts are available from Digi-Key.

Be sure to space R1 and R3 away from the board by ¼ inch or so to provide proper cooling. R13 can be a single-turn or a multi-turn pot. You'll probably find a multi-turn pot makes it easier to set the cutoff voltage to exactly 14.5 V.

R13 Adjustment

To check for proper operation and to set the trip point to 14.5 V dc, we need a test-voltage source variable from 12 to 15 V dc. A convenient means of obtaining this test voltage is to connect two 9-V transistor-radio batteries in series to supply 18 V as shown in **Fig 7.79**. Connect a 1-kΩ resistor (R2) in series with a 1-kΩ potentiometer (R1) and connect this series load across the series batteries with the fixed-value resistor to the negative lead. The voltage at the pot arm should now be adjustable from 9 to 18 V. During the following procedure, be sure to adjust the voltage with the test supply connected to the charger at 12 V because the charger loads the test-voltage supply and causes the voltage to drop a little when it's connected.

Remove the jumper at J1 and apply ac line voltage to the unit at J3. Turn R13 fully counterclockwise. D3 should glow green. Connect the test voltage to J2 and adjust R1 of Fig 7.79 for an output of 14.5 V. Slowly adjust R13 clockwise until D3 glows red. To test the circuit, wait at least four seconds, then gradually reduce the test voltage until D3 turns green. At that point, the test voltage should be approximately 13.8 V. Slowly increase the test voltage again until D3 turns red. The test voltage should now read 14.5 V. If it is not exactly 14.5 V, make a minor adjustment to R13 and try again. The aim of this adjustment is to have D3 glow red just as the test voltage reaches 14.5 V.

To test the timer functioning, remove the test voltage from J2 and set it for about 15 V. Momentarily apply the test voltage to J2. D3 should turn red for approximately four seconds, then turn green. The regulator is now calibrated and ready for operation. Remove the test voltage and ac power and install the jumper at J1.

The prototype used an 8 × 3 × 2.75-inch LMB Perf137 box (Digi-Key L171-ND) to house the charger as shown in **Fig 7.80**. An alternative enclosure is the Bud CU482A Convertabox, which measures 8×4×2 inches (available from Mouser). If you use the Convertabox, be sure to add some ventilation holes directly above the board-mounted heat sink. The LMB Perf box comes with a ventilated cover. If you are inclined to do some metal work, you could build your own enclosure using aluminum angle stock and sheet

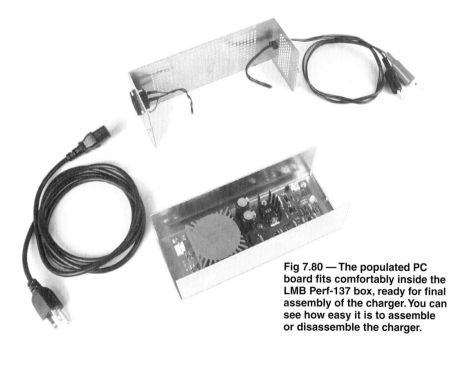

Fig 7.80 — The populated PC board fits comfortably inside the LMB Perf-137 box, ready for final assembly of the charger. You can see how easy it is to assemble or disassemble the charger.

and probably reduce the size to perhaps 8 × 3×2 inches. If you use a PC-board-mounted power transformer, watch out for potential shorts between the transformer pins (especially the 120-V ac-line pins) and the case. If you use a metal enclosure, connect the safety ground (green) wire of the ac-line cord directly to the case.

OPERATION

It is very important that this charger be connected directly to the SLA battery with no diodes, resistors or other electronics in between the two. The charger works by reading the battery voltage, so any voltage drop across an external series component results in an incorrect reading and improper charging. For example, the Elecraft K2 has internal diodes in the power-input circuit, so it's necessary to add a charging jack to the transceiver that provides a direct connection to the battery. Now the K2 can be left connected to the charger at all times and be assured that its internal battery is fully charged and ready to go at a moment's notice.

7.16.7 Overvoltage Protection for AC Generators

When using portable generators, there is always a possibility of damage to expensive equipment as a result of generator failure, especially from overvoltage. If the generator supplying power to this equipment puts out too much voltage, you run the risk of burning up power supplies or other electronic components. This project, by Jerry Paquette, WB8IOW, addresses the problem of increased voltage (not lower voltage) or surges and spikes lasting for a few microseconds.

Using a portable generator overvoltage protection device 120-V circuit and ground-fault circuit interrupter (GFCI) shown in **Fig 7.81** is good insurance. This overvoltage protection device must be used in conjunction with a GFCI at each station! (More information on GFCIs may be found in the **Safety** chapter.)

CIRCUIT DESCRIPTION

Refer to **Fig 7.82** for this description. R1

Fig 7.82 — The completed over-voltage protection PC board.

places an intentional fault on the load side of the GFCI. With the value resistor used, the fault is limited to 10 mA. (The normal tripping threshold of a GFCI is 5 mA. This current forces the GFCI to trip in just a few milliseconds. This circuit will not function at all without the use of a GFCI. A GFCI must be used at each station. If a single GFCI were used at the generator, rather than one at each location, premature tripping could occur. Several hundred feet of extension cords could have enough leakage to trip the GFCI.

You can see that the GFCI has separate lines (inputs) and loads (outputs). GFCI input terminals must be connected to the generator output. The GFCI ground must be tied to the ground of the generator. The load (computers, radios, etc) will plug into the GFCI or are wired to the load side of the GFCI. The primary of T1 is wired to the load side of the GFCI. The 12-kΩ/2-W resistor, however, must be wired to neutral on the *line* side of the GFCI in order for it to trip when used with generators that have windings isolated from ground. For safety, construct the entire unit in a single enclosure including the GFCI and its wiring. The generator connection can be made through a wired plug or using a male receptacle mounted on the enclosure.

T1 can be any 120- to 12.6-V transformer capable of delivering 100 mA or more. Mounting of this transformer varies depending on the type used. All remaining components mount on a circuit board. D1 rectifies the ac from T1 and the 100-µF capacitor filters the dc.

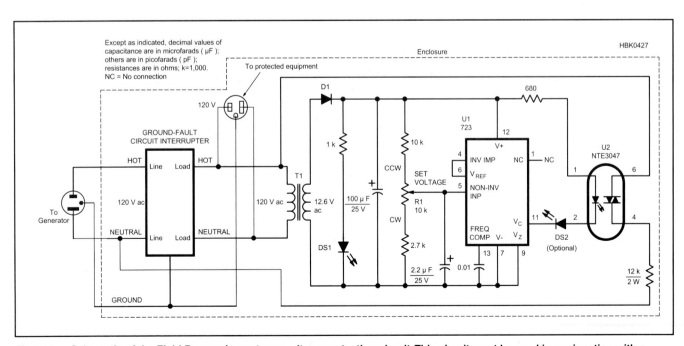

Fig 7.81 — Schematic of the Field Day equipment overvoltage-protection circuit. This circuit must be used in conjunction with a ground fault circuit interrupter (GFCI). A separate GFCI must be installed at each station. Unless otherwise specified, resistors are ¼-W, 5%-tolerance carbon-composition or film units. A PC board is available from FAR Circuits (www.farcircuits.net).

D1 — 200 PIV, 1 A diode; 1N4003 or equiv.
DS1, DS2 — Small LEDs.

R1 — 10 kΩ board-mounted, multi-turn potentiometer
T1 — 12.6 V ac transformer (see text).

U1 — 723 adjustable voltage regulator IC.
U2 — Optoisolator with TRIAC output; NTE3047 or equiv.

This voltage provides the power to the 723 voltage regulator.

Two fixed resistors and a potentiometer form the voltage-divider network supplying voltage to the LM723 input, pin 5. R1, the board-mounted potentiometer, has only three leads, but there are four pads on the circuit board, to accommodate different styles of pots. The 2.2-µF capacitor provides a slight delay, to prevent false tripping when the circuit is powered up. The 0.01-µF capacitor from pin 13 of the 723 to the negative supply bus should always be used. When the voltage at pin 5 goes higher than the reference voltage at pins 4 and 6, pin 11 goes low, turning on the trip indicator LED DS2 and the optical coupler LED. LED current is limited by the 1-kΩ resistor. The optical coupler turns on the TRIAC, which creates a 10-mA fault current between the hot wire and ground of the GFCI. DS2 will remain lit as an indicator until the 100 µF capacitor is discharged.

ADJUSTMENT

Adjustment is simple. You'll need a variable ac transformer (Powerstat or Variac). Turn R1 fully clockwise and use the variable transformer to adjust input to 130 V ac. Turn the pot counterclockwise until the GFCI trips.

7.16.8 Overvoltage Crowbar Circuit

If the regulator circuitry of a power supply should fail — for example, if a pass transistor should short — the unregulated supply voltage could appear at the output terminals. This could cause a failure in the equipment connected to the supply. The overvoltage "crowbar" circuit shown here has been shown to be effective as a last line of defense against power supply overvoltage failures. When an overvoltage condition is detected, the heavy-duty SCR is fired, becoming a short circuit — as if a crowbar (representing any over-sized conductor capable of handling the supply's short-circuit output current) had been connected across the supply, thus the name.

This circuit shown in **Fig 7.83** was originally designed for the 28-V power supply project available on the CD-ROM for this book. The use of the MC3423 overvoltage protection IC provides quicker triggering and more reliable gate drive to the SCR than comparable Zener-based circuits. Power supply

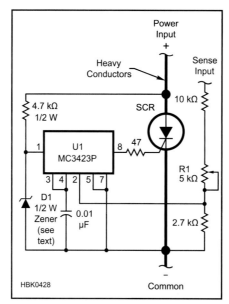

Fig 7.83 — Schematic of the over-voltage crowbar circuit. Unless otherwise specified, resistors are ¼-W, 5%-tolerance carbon-composition or film units. A PC board is available from FAR Circuits (www.farcircuits.net).

SCR — C38M stud-mount (TO-65 package).
U1 — MC3423P or NTE7172 voltage protection circuit, 8-pin DIP.
D1 — Zener diode, ½-W (see text).
R1 — 5 kΩ, PC board mount trimmer potentiometer.

builders can incorporate the overvoltage protection circuit into any dc power supply with the required component value adjustments.

The crowbar circuit is usually connected with the power input and the sense input connected together at the supply positive output and COMMON to the supply negative output. Another option is to connect power input and common across the rectifier filter capacitor and the sense input to the power supply output. In either case, the power input and common wires should be adequately sized to handle the full short-circuit current and are shown as heavy lines on the schematic.

The crowbar circuit functions as follows: the 4.7 kΩ resistor and Zener diode D1 create a supply voltage for the MC3423. U1 will function properly with a supply voltage of 4.5-40 V. Use a Zener diode with a voltage rating a few volts below that of the crowbar circuit positive-to-negative power input voltage. For example, if the crowbar is connected to a 12-V supply output, D1 voltage should be 6 to 9 V. The exact value is not critical.

U1 contains a 2.5-V reference and two comparators. When the voltage at pin 2 (sense terminal) reaches 2.5 V, the output voltage (pin 8) changes from the negative input voltage to the positive input voltage. This drives the gate of the SCR through the 47-Ω resistor. The trip voltage is set by the resistive divider across the + and − inputs:

$$V_{trip} = 2.5 \left(1 + \frac{10\ k\Omega + R}{2.7\ k\Omega}\right)$$

The application notes for the MC3423 recommend that the resistance from the sense input to the negative input be less than 10 kΩ for minimum drift, suggesting the value of 2.7 kΩ. The value of 10 kΩ for the fixed portion of the adjustable resistance is selected for $V_{trip} = 15$ V at the midpoint of the 5 kΩ potentiometer (R1) travel. For other trip voltages, the fixed resistor value should be the closest standard value to

$$R_{fixed} = 2.7\ k\Omega \left(\frac{V_{trip}}{2.5} - 1\right) - 2.5 k\Omega$$

assuming the potentiometer value remains at 5 kΩ.

When the SCR turns on, it short-circuits the inputs, causing any protective fuses or circuit breakers to open. The SCR will stay on until the current through drops below the "keep alive" threshold, at which point the SCR turns off. The SCR will stay on even if the input voltage to U1 drops below 4.5 V.

If the crowbar circuit fires due to RFI, an additional 0.01-µF capacitor should be connected from pin 2 of U1 to common and another across the D1.

More information may be found in the MC3423 datasheet and "Semiconductor Consideration for DC Power Supply Voltage Protector Circuits," ON Semi AN004E/D, both available from ON Semiconductor.

Additional Projects and Information

Additional power supply projects and supporting files are included on the *Handbook* CD-ROM.

Contents

8.1 Introduction
8.2 Analog Modulation
    8.2.1 Amplitude-Modulated Analog Signals
    8.2.2 Angle-Modulated Analog Signals
8.3 Digital Modulation
    8.3.1 Amplitude-Modulated Digital Signals
    8.3.2 Angle-Modulated Digital Signals
    8.3.3 Quadrature Modulation
    8.3.4 Multi-carrier Modulation
    8.3.5 AFSK and Other Audio-Based Modulation
    8.3.6 Spread Spectrum
    8.3.7 Filtering and Bandwidth of Digital Signals
8.4 Image Modulation
    8.4.1 Fast-Scan Television
    8.4.2 Facsimile
    8.4.3 Slow-Scan Television
8.5 Modulation Impairments
    8.5.1 Intermodulation Distortion
    8.5.2 Transmitted Bandwidth
    8.5.3 Modulation Accuracy
8.6 Modulation Glossary
8.7 References and Further Reading

Chapter 8

Modulation

Radio amateurs use a wide variety of modulation types to convey information. This chapter, written by Alan Bloom, N1AL, explores analog, digital and image modulation techniques, as well as impairments that affect our ability to communicate. A glossary of terms and suggestions for additional reading round out the chapter.

8.1 Introduction

The purpose of Amateur Radio transmissions is to send information via radio. The one possible exception to that is a beacon station that transmits an unmodulated carrier for propagation testing. In that case the only information being sent is, "Am I transmitting or not?" One can think of it as a single data bit with two states, *on* or *off*. However, in reality even a beacon station must periodically identify with the station call sign, so we can say that modulation is an essential ingredient of Amateur Radio.

To represent the information being sent, the radio signal must periodically change its state in some way that can be detected by the receiver. In the early days of radio, the only way to do that was on-off keying using Morse code. By alternating the on and off states with the proper sequence and rhythm, a pair of highly-skilled operators can exchange textual data at rates up to perhaps 60 WPM. Later, engineers figured out how to amplify the signal from a microphone and use it to vary the power of the radio signal continuously. Thus was born amplitude modulation, which allowed transmitting voices at full speaking speeds, that is, up to about 200 WPM. That led to analog modes such as television with even faster information rates. Today's digital radio systems are capable of transferring tens of megabits of information per second, equivalent to tens of millions of words per minute.

Modulation is but one component of any transmission mode. For example AM voice and NTSC television (the analog US system) both use amplitude modulation, but the transmission protocol and type of information sent are very different. Digital modes also are generally defined not only by the modulation but also by multiple layers of protocol and data coding. Of the three components of a mode (modulation, protocol and information), this chapter will concentrate on the first. The **Digital Modes** chapter covers digital transmission protocols and the **Digital Communications** chapter covers the practical aspects of operating the various digital modes.

EMISSION DESIGNATORS

We tend to think of a radio signal as being "on" a particular frequency. In reality, any modulated signal occupies a band of frequencies. The bandwidth depends on the type of modulation and the data rate. A Morse code signal can be sent within a bandwidth of a couple hundred Hz at 60 WPM, or less at lower speeds. An AM voice signal requires about 6 kHz. For high-fidelity music, more bandwidth is needed; in the United States, an FM broadcast signal occupies a 200-kHz channel. Television signals need about 6 MHz, while 802.11n, the latest generation of "WiFi" wireless LAN, uses up to 40 MHz for data transmission.

The International Telecommunication Union (ITU) has specified a system for designating radio emissions based on the bandwidth, modulation type and information to be transmitted. The emission designator begins with the bandwidth, expressed as a maximum of five numerals and one letter. The letter occupies the position of the decimal point and represents the unit of bandwidth, as follows: H = hertz, K = kilohertz, M = megahertz and G = gigahertz. The bandwidth is followed by three to five emission classification symbols, as defined in the sidebar, "Emission Classifications." The first three symbols are mandatory; the fourth and fifth symbols are supplemental. These designators are found in Appendix 1 of the ITU Radio Regulations, ITU-R Recommendation SM.1138 and in the FCC rules §2.201.

For example, the designator for a CW signal might be 150H0A1A, which means 150 Hz

Emission Classifications

Emissions are designated according to their classification and their necessary bandwidth. A minimum of three symbols is used to describe the basic characteristics of radio waves. Emissions are classified and symbolized according to the following characteristics:

I. First symbol—Type of modulation of the main carrier
II. Second symbol—Nature of signal(s) modulating the main carrier
III. Third symbol—Type of information to be transmitted
Note: A fourth and fifth symbol are provided for in the ITU Radio Regulations. Use of the fourth and fifth symbol is optional.
IV. Details of signal(s)
V. Nature of multiplexing

First symbol—type of modulation of the main carrier
(1) Emission of an unmodulated carrier .... N
(2) Emission in which the main carrier is amplitude-modulated (including cases where subcarriers are angle-modulated):
  - Double sideband .... A
  - Single sideband, full carrier .... H
  - Single sideband, reduced or variable level carrier .... R
  - Single sideband, suppressed carrier .... J
  - Independent sidebands .... B
  - Vestigial sideband C
(3) Emission in which the main carrier is angle-modulated:
  - Frequency modulation .... F
  - Phase modulation .... G
  Note: Whenever frequency modulation (F) is indicated, phase modulation (G) is also acceptable.
(4) Emission in which the main carrier is amplitude and angle-modulated either simultaneously or in a pre-established sequence .... D
(5) Emission of pulses[1]
  - Sequence of unmodulated pulses .... P
  - A sequence of pulses:
   - Modulated in amplitude .... K
   - Modulated in width/duration L
   - Modulated in position/phase .... M
  - In which the carrier is angle-modulated during the period of the pulse .... Q
  - Which is a combination of the foregoing or is produced by other means V
(6) Cases not covered above, in which an emission consists of the main carrier modulated, either simultaneously or in a pre-established sequence in a combination of two or more of the following modes: amplitude, angle, pulse .... W
(7) Cases not otherwise covered X

Second symbol—nature of signal(s) modulating the main carrier
(1) No modulating signal 0
(2) A single channel containing quantized or digital information without the use of a modulating subcarrier, excluding time-division multiplex 1
(3) A single channel containing quantized or digital information with the use of a modulating subcarrier, excluding time-division multiplex 2
(4) A single channel containing analog information .. 3
(5) Two or more channels containing quantized or digital information 7
(6) Two or more channels containing analog information 8
(7) Composite system with one or more channels containing quantized or digital information, together with one or more channels containing analog information 9
(8) Cases not otherwise covered X

Third symbol—type of information to be transmitted[2]
(1) No information transmitted .... N
(2) Telegraphy, for aural reception .... A
(3) Telegraphy, for automatic reception .... B
(4) Facsimile C
(5) Data transmission, telemetry, telecommand .... D
(6) Telephony (including sound broadcasting) .... E
(7) Television (video) .... F
(8) Combination of the above .... W
(9) Cases not otherwise covered X

Where the fourth or fifth symbol is used it shall be used as indicated below. Where the fourth or the fifth symbol is not used this should be indicated by a dash where each symbol would otherwise appear.

Fourth symbol—Details of signal(s)
(1) Two-condition code with elements of differing numbers and/or durations .... A
(2) Two-condition code with elements of the same number and duration without error-correction .... B
(3) Two-condition code with elements of the same number and duration with error-correction C
(4) Four-condition code in which each condition represents a signal element (of one or more bits) .... D
(5) Multi-condition code in which each condition represents a signal element (of one or more bits) .... E
(6) Multi-condition code in which each condition or combination of conditions represents a character .... F
(7) Sound of broadcasting quality (monophonic) ..... G
(8) Sound of broadcasting quality (stereophonic or quadraphonic) .... H
(9) Sound of commercial quality (excluding categories given in (10) and (11) below) .... J
(10) Sound of commercial quality with frequency inversion or band-splitting .... K
(11) Sound of commercial quality with separate frequency-modulated signals to control the level of demodulated signal L
(12) Monochrome .... M
(13) Color .... N
(14) Combination of the above .... W
(15) Cases not otherwise covered X

Fifth symbol—Nature of multiplexing
(1) None .... N
(2) Code-division multiplex[3] C
(3) Frequency-division multiplex .... F
(4) Time-division multiplex .... T
(5) Combination of frequency-division and time-division multiplex .... W
(6) Other types of multiplexing X

---

[1] Emissions where the main carrier is directly modulated by a signal which has been coded into quantized form (eg, pulse code modulation) should be designated under (2) or (3).
[2] In this context the word "information" does not include information of a constant unvarying nature such as is provided by standard frequency emissions, continuous wave and pulse radars, etc.
[3] This includes bandwidth expansion techniques.

bandwidth, double sideband, digital information without subcarrier, and telegraphy for aural reception. SSB would be 2K5J3E, or 2.5 kHz bandwidth, single sideband with suppressed carrier, analog information, and telephony. The designator for a PSK31 digital signal is 60H0J2B, which means 60 Hz bandwidth, single sideband with suppressed carrier, digital information using a modulating subcarrier, and telegraphy for automatic reception.

Authorized modulation modes for Amateur Radio operators depend on frequency, license class, and geographical location, as specified in the FCC regulations §97.305. Technical standards for amateur emissions are specified in §97.307. Among other things, they require that no amateur station transmission shall occupy more bandwidth than necessary for the information rate and emission type being transmitted, in accordance with good amateur practice. We will discuss the necessary bandwidth for each type of modulation as it is covered in the following sections.

8.2 Analog Modulation

While the newer digital modes are all the rage today, traditional analog modulation is still quite popular on the amateur bands. There are many reasons for that, some technical and some not. Analog equipment tends to be simpler; no computer is required. For example, a low-power CW transceiver is easy to construct and simple to understand. CW and SSB are both narrowband modes with spectral efficiency that is quite competitive with their digital equivalents. While heated arguments have been made about the relative power efficiency of analog vs digital modes, it is clear that in the hands of skilled operators CW and SSB are at least in the same ballpark with, if perhaps not better than, many digital modes.

When copying under difficult conditions, the human brain is capable of higher-level signal processing that is difficult to duplicate on a computer. For example, in a DX pileup, an operator may realize that the signal he wants is the weak one and mentally filter out all the stronger stations based on signal strength alone. Other aspects of the signal may be used for filtering as well, such as the operator's accent (on voice) or "fist" (on CW), propagation effects, the operator's speed and operating pattern, CW hum and chirp, and other factors. Different portions of the other station's call sign may be copied at different times and stitched together in the operator's brain. Missing portions of a transmission can be filled in based on context. Many operators enjoy the challenge of using their ears and brain to decipher a weak signal, an aspect of operating that is lost with the digital modes.

8.2.1 Amplitude-Modulated Analog Signals

Of the various properties of a signal that can be modulated to transmit voice information, amplitude was the first to be used. Not only are modulation and demodulation of AM signals simple in concept, but they are simple to implement as well. Since the RF output voltage of a class-C amplifier is approximately proportional to the power supply voltage, an AM modulator needs only to vary the power supply voltage in step with the audio signal, as shown in **Fig 8.1**.

To prevent overmodulation and the resulting distortion, the modulating signal should not try to reduce the power supply voltage to less than zero volts or more than twice the average voltage. The *modulation index* of an AM signal is the ratio of the peak modulation to those limits. The resulting RF signal might look like the waveform in **Fig 8.2C**.

Note that the shape of the envelope of the modulated RF signal matches the shape of the modulating signal. That suggests a possible demodulation method. A diode detector puts out a signal proportional to the envelope of the RF signal, recovering the original modulation. See **Fig 8.3**. The capacitor should be large enough that it filters out most of the RF ripple but not so large that it attenuates the higher audio frequencies.

So far we have been discussing signals as a function of time — all the graphs have time as the horizontal axis. It is equally valid

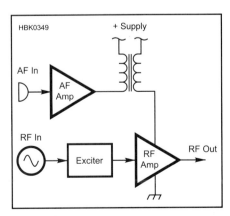

Fig 8.1 — An AM modulator. With a vacuum-tube RF power amplifier this method is known as *plate modulation* since the modulated power supply voltage is applied to the vacuum tube plate.

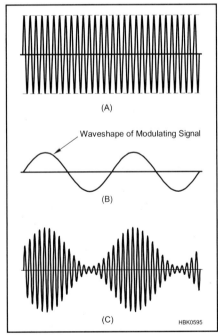

Fig 8.2 — Graphical representation of amplitude modulation. In the unmodulated carrier (A) each RF cycle has the same amplitude. When the modulating signal (B) is applied, the RF amplitude is increased or decreased according to the amplitude of the modulating signal (C). A modulation index of approximately 75% is shown. With 100% modulation the RF power would just reach zero on negative peaks of the modulating signal.

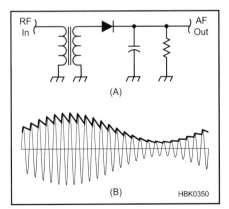

Fig 8.3 — A simple diode-type AM detector, also known as an *envelope detector* (A). The demodulated output waveform has the same shape as the envelope of the RF signal (B). In (B) the thin line is the RF signal modulated with a sine wave and the darker line is the demodulated audio frequency with some residual RF ripple.

Modulation 8.3

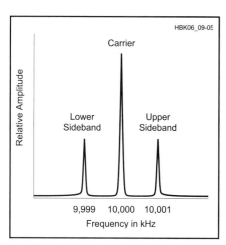

Fig 8.4 — A 10-MHz carrier that is AM-modulated by a 1-kHz sine wave.

to look in the frequency domain. It turns out that the modulation spectrum consists of a pair of sidebands above and below the RF carrier. **Fig 8.4** shows a 10-MHz signal modulated with a 1-kHz sine wave. The upper sideband is a single frequency at 10 MHz + 1 kHz = 10.001 MHz. The spectrum of the lower sideband is inverted, so it is at 10 MHz – 1 kHz = 9.999 MHz.

To see why that happens, refer to **Fig 8.5**. Simply adding the first three sine waves shown, representing a carrier and upper and lower sidebands, results in the 100%-modulated AM waveform at the bottom. Note that each sideband has half the amplitude of the carrier, which corresponds to one-quarter the power or –6 dB. Each sideband in an AM signal that is 100% modulated with a sine wave is therefore –6 dB with respect to the carrier.

If a more complex modulating signal is used, each frequency component of the modulation f_m appears at $f_c + f_m$ and $f_c - f_m$, where f_c is the carrier frequency. An AM modulator is really nothing more than a type of mixer that generates new output frequencies at the sum and difference of the carrier and modulation frequencies. The upper sideband is simply a frequency-shifted version of the modulation spectrum and the lower sideband is both frequency-shifted and inverted. For communications purposes, the audio frequency response typically extends up to about 2.5-3 kHz, resulting in a 5-6 kHz RF bandwidth.

One problem with AM is that if the amplitude of the carrier becomes attenuated for any reason, then the modulation is distorted, especially the negative-going portion near the 100%-modulation (zero power) point. This can happen due to a propagation phenomenon called *selective fading*. It occurs when the signal arrives simultaneously at the receive antenna via two or more paths, such as ground wave and sky wave. If the difference in the distance of the two paths is an odd number of half-wavelengths, then the two signals are out of phase. If the amplitudes are nearly the same, they cancel and a deep fade results. Since wavelength depends on frequency, the fading is frequency-selective. On the lower-frequency amateur bands it is possible for an AM carrier to be faded while the two sidebands are still audible.

A solution is to regenerate the carrier in the receiver. Since the carrier itself carries no information about the modulation, it is not necessary for demodulation. The transmitted signal may be a standard full-carrier AM signal or the carrier may be suppressed, resulting in *double sideband, suppressed carrier* (DSBSC). An AM detector that regenerates the carrier from the signal is known as a *synchronous detector*. Often the regenerated carrier oscillator is part of a phase-locked loop (PLL) that locks onto the incoming carrier. See **Fig 8.6**. Synchronous detectors not only reduce the effects of selective fading but also are usually more linear than diode detectors so they have less distortion. Some commercial short-wave broadcasts include a reduced, but not suppressed, carrier (DSBRC) to allow operation with PLL-type synchronous detectors.

Since the advent of single sideband in the 1960s, full-carrier double-sideband AM has become something of a niche activity in Amateur Radio. It does retain several advantages however. We have already mentioned the simplicity of the circuitry. Another advantage is that, because of the presence of the

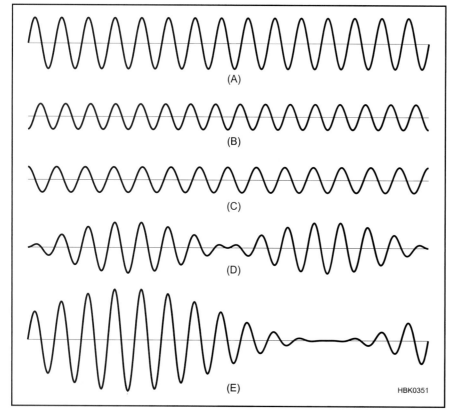

Fig 8.5 — At A is an unmodulated carrier. If the upper (B) and lower (C) sidebands are added together a double-sideband suppressed carrier (DSBSC) signal results (D). If each sideband has half the amplitude of the carrier, then the combination of the carrier with the two sidebands results in a 100%-modulated AM signal (E). Whenever the two sidebands are out of phase with the carrier, the three signals sum to zero. Whenever the two sidebands are in phase with the carrier, the resulting signal has twice the amplitude of the unmodulated carrier.

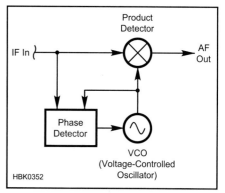

Fig 8.6 — Block diagram of a synchronous detector. The voltage-controlled oscillator (VCO) is part of a phase-locked loop that locks the oscillator to the carrier frequency of the incoming AM or DSBRC signal.

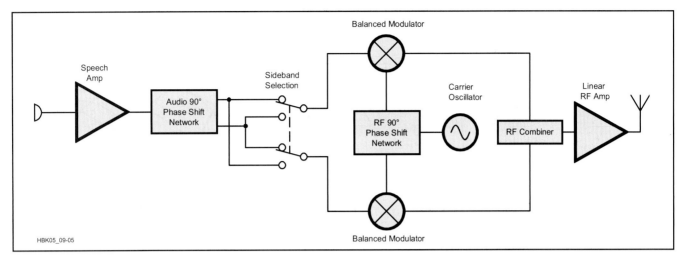

Fig 8.7 — Block diagram of a phasing-type SSB generator.

carrier, the automatic gain control system in the receiver remains engaged at all times, ensuring a constant audio level. Unlike with SSB, there is no rush of noise during every pause in speech. Also tuning is less critical than with SSB. There is no "Donald Duck" sound if the receiver is slightly mistuned. Finally, the audio quality of an AM signal is usually better than that of SSB because of the lack of a crystal filter in the transmit path and the wider filter in the receiver.

SINGLE SIDEBAND

As we have seen, since the carrier itself contains no modulation, it does not need to be transmitted, which saves at least 67% of the transmitted power. In addition, since the two sidebands carry identical information, one of them may be eliminated as well, saving half the bandwidth. The result is *single sideband, suppressed carrier* (SSBSC), which is commonly referred to simply as "SSB." If the lower sideband is eliminated, the result is called *upper sideband* (USB). If the upper sideband is eliminated, you're left with *lower sideband* (LSB). The bandwidth is a little less than half that of an equivalent AM signal. A typical SSB transmitter might have an overall audio response from 300 Hz to 2.8 kHz, resulting in 2.5 kHz bandwidth.

As with an AM synchronous detector, the carrier is regenerated in the receiver but since only one sideband is present, the synchronous detector's phase-locked loop is not possible. Instead, the detector uses a free-running *beat-frequency oscillator* (BFO). The detector itself is called a *product detector* because its output is the mathematical product of the BFO and the SSB signal. The BFO must be tuned to the same frequency as the suppressed carrier to prevent distortion of the recovered audio. In a superheterodyne receiver, that is done by carefully tuning the local oscillator (the main tuning dial) such that after conversion to the intermediate frequency, the suppressed carrier aligns with the BFO frequency.

In the transmitter, the SSB modulator must suppress both the carrier and the unwanted sideband. Carrier suppression is normally accomplished with a *balanced modulator*, a type of mixer whose output contains the sum and difference frequencies of the two input signals (the modulating signal and the carrier) but not the input signals themselves. There are several ways to eliminate the unwanted sideband, but the most common is to pass the output of the balanced mixer through a crystal filter that passes the wanted sideband while filtering out the unwanted one. This is convenient in a transceiver since the same filter can be used in the receiver by means of a transmit-receive switch.

Another method to generate single sideband is called the *phasing method*. See **Fig 8.7**. Using trigonometry, it can be shown mathematically that the sum of the signals from two balanced modulators, each fed with audio signals and RF carriers that are 90° out of phase, consists of one sideband only. The other sideband is suppressed. The output can be switched between LSB and USB simply by reversing the polarity of one of the inputs, which changes the phase by 180°. The phasing method eliminates the need for an expensive crystal filter following the modulator or, in the case of a transceiver, the need to switch the crystal filter between the transmitter and receiver sections. In addition, the audio quality is generally better because it eliminates the poor phase dispersion that is characteristic of most crystal filters. In the old days, designing and building an audio phase-shift network that accurately maintained a 90° differential over a decade-wide frequency band (300 Hz to 3000 Hz) was rather complicated. With modern DSP techniques, the task is much easier.

Compared to most other analog modulation modes, SSB has excellent power efficiency because the transmitted power is proportional to the modulating signal and no power at all is transmitted during pauses in speech. Another advantage is that the lack of a carrier results in less interference to other stations. Back in the days when AM ruled the HF amateur voice bands, operators had to put up with a cacophony of heterodynes emanating from their headphones. SSB transceivers also are well-suited for the narrowband digital modes. Because an SSB transceiver simply frequency-translates the baseband audio signal to RF in the transmitter and back to audio again in the receiver, digital modulation may be generated and detected at audio frequencies using the sound card in a personal computer.

8.2.2 Angle-Modulated Analog Signals

Angle modulation varies the phase angle of an RF signal in response to the modulating signal. In this context, "phase" means the phase of the modulated RF sine wave with respect to the unmodulated carrier. Phase modulation is one type of angle modulation, but so is frequency modulation because any change in frequency results in a change in phase. For example, the way to smoothly ramp the phase from one value to another is to change the frequency and wait. If the frequency is changed by +1 Hz, then after 1 second the phase will have changed by +360°. After 2 seconds, the phase will have changed by +720°, and so on. Change the frequency in the other direction and the phase moves in the opposite direction as well. With sine-wave modulation, the frequency and phase both vary in a sinusoidal fashion. See **Fig 8.8**.

Since any change in frequency results in a change in phase and vice versa, *frequency*

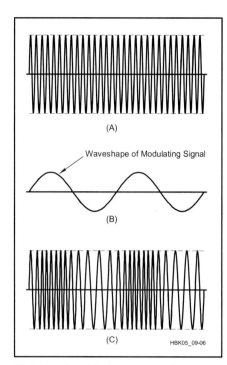

Fig 8.8 — Graphical representation of frequency modulation. In the unmodulated carrier (A) each RF cycle occupies the same amount of time. When the modulating signal (B) is applied, the radio frequency is increased or decreased according to the amplitude and polarity of the modulating signal (C).

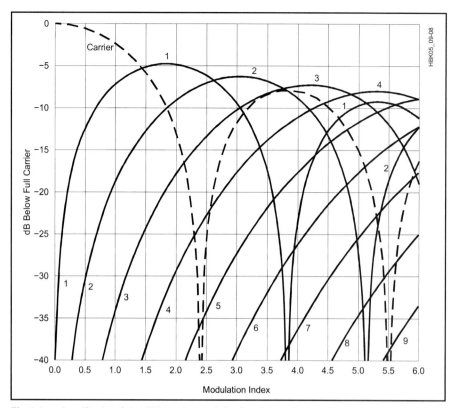

Fig 8.9 — Amplitude of the FM carrier and the first nine sidebands versus modulation index. This is a graphical representation of mathematical functions developed by F. W. Bessel. Note that the carrier completely disappears at modulation indices of 2.405 and 5.52.

modulation (FM) and *phase modulation* (PM) are fundamentally the same. The only difference is the audio frequency response. An FM transmitter with 6 dB/octave pre-emphasis of the modulating signal is indistinguishable from a PM transmitter. A PM transmitter with 6 dB/octave de-emphasis is indistinguishable from an FM transmitter. The reverse happens at the receiver. A frequency detector followed by a 6 dB/octave de-emphasis network acts like a phase detector. It is interesting to note that most VHF and UHF amateur "FM" transceivers should really be called "PM" transceivers due to the pre-emphasis and de-emphasis networks used in the transmitters and receivers respectively.

The term *frequency deviation* means the amount the RF frequency deviates from the center (carrier) frequency with a given modulating signal. The instantaneous deviation of an FM signal is proportional to the instantaneous amplitude of the modulating signal. The instantaneous deviation of a PM signal is proportional to the instantaneous *rate of change* of the modulating signal. Since the rate of change of a sine wave is proportional to its frequency as well as its amplitude, the deviation of a PM signal is proportional to both the amplitude and the frequency of the modulating signal.

Most FM and PM transmitters include some kind of audio compressor before the modulator to limit the maximum deviation. Common usage of the term *deviation* is that it refers to the maximum peak deviation allowed by the audio compressor. If the frequency swings a maximum of 5 kHz above the center frequency and 5 kHz below the center frequency, we say the deviation is 5 kHz. The *modulation index* is the ratio of the peak deviation to the highest audio frequency. The *deviation ratio* is the ratio of maximum permitted peak deviation to the maximum permitted modulating frequency.

FCC regulations limit the modulation index at the highest modulating frequency to a maximum of 1.0 for frequencies below 29 MHz. If the audio is low-pass filtered to 3 kHz, then the deviation may be no more than 3 kHz. For that reason, FM transmitters for frequencies below 29 MHz are usually set for 3 kHz deviation while FM transmitters at higher frequencies are typically set for about 5 kHz deviation. The term *narrowband FM* generally refers to deviation of no more than 3 kHz and *wideband FM* refers to deviation greater than 3 kHz.

Unlike amplitude modulation, angle modulation is nonlinear. Recall that with amplitude modulation, the shape of the RF spectrum is the same as that of the modulation spectrum, single-sided with SSB and double-sided with DSB. That is not true with angle modulation. A double-sideband AM signal with audio band-limited to 3 kHz has 6 kHz RF bandwidth. It is easy to see that an FM transmitter with the same audio characteristics but with, say, 5 kHz deviation must have a bandwidth of at least 10 kHz. And it is even worse than that. While you might think that the bandwidth equals twice the deviation, in reality the transmitted spectrum theoretically extends to infinity, although it does become vanishingly small beyond a certain point. As a rule of thumb, approximately 98% of the spectral energy is contained within the bandwidth defined by *Carson's rule*:

$$BW = 2(f_d + f_m) \qquad (1)$$

where
f_d = the peak deviation, and
f_m = the highest modulating frequency.

For example, if the deviation is 5 kHz and the audio is limited to 3 kHz, the bandwidth is approximately 2 (5 + 3) = 16 kHz.

With sine-wave modulation, the RF frequency spectrum from an angle-modulated transmitter consists of the carrier and a series of sideband pairs, each spaced by the frequency of the modulation. For example, with 3 kHz sine-wave modulation, the spectrum includes tones at ±3 kHz, ±6 kHz, ±9 kHz and so on with respect to the carrier. When the modulation index is much less than 1.0, only the first sideband pair is significant, and the spectrum looks similar to that of an AM signal (although the phases of the sidebands are different). As the modulation index is increased, more and more sidebands appear within the bandwidth predicted by Carson's rule. Unlike with AM, the carrier amplitude also changes with deviation and actually disappears altogether for certain modulation indices. Because the amplitude of an angle-modulated signal does not vary with modulation, the total power of the carrier and all sidebands is constant.

The amplitudes of the carrier and the various sidebands are described by a series of mathematical equations called Bessel functions, which are illustrated graphically in **Fig 8.9**. Note that the carrier disappears when the modulation index equals 2.405 and 5.52. That fact can be used to set the deviation of an FM transmitter. For example, to set 5 kHz deviation, connect the microphone input to a sine-wave generator set for a frequency of 5 / 2.405 = 2.079 kHz, listen to the carrier on a narrowband receiver, and adjust the deviation until the carrier disappears.

Since the amplitude carries no information, FM receivers are designed to be as insensitive to amplitude variations as possible. Because noise tends to be mostly AM in nature, that results in a quieter demodulated signal. Typically the receiver includes a *limiter*, which is a very high-gain ampli-

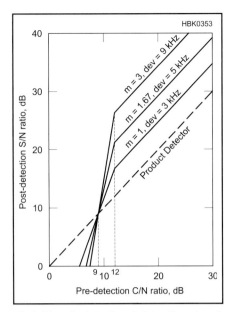

Fig 8.10 — A plot of post-detection signal-to-noise ratio (S/N) versus input carrier-to-noise ratio (C/N) for an FM detector at various modulation indices, m. For each modulation index the deviation is also noted assuming a maximum modulating frequency of 3 kHz. For comparison, the performance of an SSB product detector is also shown.

fier that causes the signal to clip, removing any amplitude variations, before being applied to the detector. Unlike with AM, as an FM signal gets stronger the volume of the demodulated audio stays the same, but the noise is reduced. Receiver sensitivity is often specified by how much the noise is suppressed for a certain input signal level. For example, if a 0.25 μV signal causes the noise to be reduced by 20 dB, then we say the receiver sensitivity is "0.25 μV for 20 dB of quieting."

The limiter also causes a phenomenon known as *capture effect*. If more than one signal is present at the same time, the limiter tends to reduce the weaker signal relative to the stronger one. We say that the stronger signal "captures" the receiver. The effect is very useful in reducing on-channel interference.

The suppression of both noise and interference is greater the wider the deviation. FM signals with wider deviation do take up more bandwidth and actually have a poorer signal-to-noise (S/N) ratio at the detector output for weak signals but have better S/N ratio and interference rejection for signal levels above a certain threshold. See **Fig 8.10**.

In addition to the noise and interference-reduction advantage, angle-modulated signals share with full-carrier AM the advantages of non-critical frequency accuracy and the continuous presence of a signal, which eases the task of the automatic gain control system in the receiver. In addition, since the signal is constant-amplitude, the transmitter does not need a linear amplifier. Class-C amplifiers may be used, which have greater power efficiency.

8.3 Digital Modulation

The first digital modulation used in Amateur Radio was on-off keying using Morse code. However, the term "digital" is commonly understood to refer to modes where the sending and receiving is done by machine. By that definition, the first popular amateur digital mode (called radioteleprinter, or RTTY) used frequency-shift keying with all the coding and decoding done by mechanical teleprinters. You still hear RTTY on the bands today, although most operators are using computers these days rather than Teletype machines. More-modern digital modes use more efficient modulation formats as well as error detection and correction to ensure reliable communications. Data can be sent much more rapidly than with Morse code or voice, with less effort, and with fewer errors. In some systems, stations operate unattended 24 hours per day, transferring data with little or no operator intervention. For point-to-point transmission of large quantities of text or other data, these modern digital modes are the obvious choice.

8.3.1 Amplitude-Modulated Digital Signals

We start this section by discussing *on-off keying* (OOK), which is a special case of *amplitude-shift keying* (ASK) and is normally sent using Morse code. For historical reasons dating from the days of spark transmitters, amateurs often refer to this as *continuous wave* (CW) even though the signal is actually keyed on and off. While most amateurs think of CW as being an analog mode, we include it here because it has technical characteristics that are more similar to other digital modes than to analog.

You can think of OOK as being the same as analog AM that is modulated with a two-level signal that switches between full power and zero power. For example, imagine that the modulation is a 10 Hz square wave, equivalent to sending a series of dits at 24 WPM. A square wave may be decomposed into a sine wave of the same frequency and a theoretically infinite succession of odd harmonics.

Recall that the lower and upper sidebands of an AM signal are simply the inverted and non-inverted spectra of the baseband modulation. The RF spectrum therefore contains sidebands at the carrier frequency ±10 Hz, ±30 Hz, ±50 Hz, and so on to plus and minus infinity. This phenomenon is called *key clicks*. Stations listening on nearby frequencies hear a click upon every key closure and opening.

To prevent interference to other stations, the modulation must be low-pass filtered, which slows down the transition times between the on and off states. See **Fig 8.11**. If the transitions are too fast, then excessive key clicks occur. If the transitions are too slow, then at high speeds the previous transition may not have finished before the next one starts, which makes the signal sound mushy and hard to read. Traditionally, filtering was done with a simple resistor-capacitor low-pass filter on the keying line, but using a transition with a raised-cosine shape allows faster transition times without excessive key

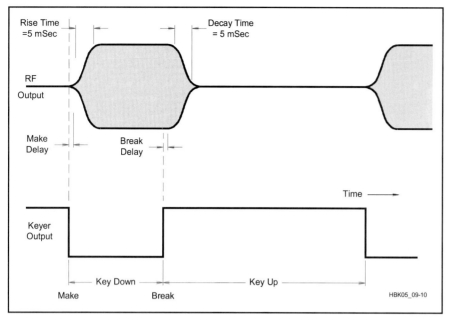

Fig 8.11 — A filtered CW keying waveform. The on-off transitions of the RF envelope should be as smooth as possible while transitioning as quickly as possible. The raised-cosine shape shown is nearly optimum in that respect.

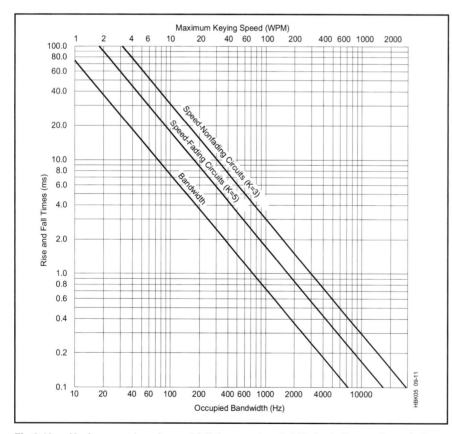

Fig 8.12 — Keying speed vs rise and fall times vs bandwidth for fading and non-fading communications circuits. For example, to optimize transmitter timing for 20 WPM on a non-fading circuit, draw a vertical line from the WPM axis to the K = 3 line. From there draw a horizontal line to the rise/fall axis (approximately 15 ms). Draw a vertical line from where the horizontal line crosses the bandwidth line and see that the bandwidth is about 50 Hz.

clicks. Some modern transceivers use DSP techniques to generate such a controlled transition shape.

The optimum transition time, and thus bandwidth, depends on keying speed. It also depends on propagation conditions. When the signal is fading, the transitions must be sharper to allow good copy. **Fig 8.12** gives recommended keying characteristics based on sending speed and propagation. As a compromise, many transmitters use a 5 ms rise and fall time. That limits the bandwidth to approximately 150 Hz while allowing good copy up to 60 WPM on non-fading channels and 35 WPM on fading channels, which covers most requirements.

With any digital system, the number of changes of state per second is called the *baud rate* or the *symbol rate*, measured in *bauds* or *symbols per second*. Sending a single Morse code dit requires two equal-length states, or symbols: *on* for the length of the dit and *off* for the space between dits. Thus, a string of dits sent at a rate of 10 dits per second has a symbol rate of 20 bauds.

Refer to **Fig 8.13**. A Morse code dash is on for three times the length of a dit. Including one symbol for the off time, the total time to send a dash is four symbols, twice the time to send a dit. For example, at a baud rate of 20 bauds, there are 10 dits per second or 5 dashes per second. The spacing between characters within a word is three symbols, two more than the normal space between dits and dashes. The spacing between words is seven symbols.

For purposes of computing sending speed, a standard word is considered to have five characters, plus the inter-word spacing. On average, that results in 50 symbols per word. From that, the speed in words per minute may be computed from the baud rate:

$$\text{WPM} = \frac{60 \, (\text{sec/min})}{50 \, (\text{symbols/word})} \times \text{bauds}$$

(2)

$$= 1.2 \times \text{bauds} = 2.4 \times \text{dits/sec}$$

Characters in Morse code do not all have the same length. Longer codes are used for characters that are used less frequently while the shortest codes are reserved for the most common characters. For example, the most common letter in the English language, E, is sent as a single dit. In that way, the average character length is reduced, resulting in a faster sending speed for a given baud rate. Such a variable-length code is known as *varicode*, a technique that has been copied in some modern digital modes.

PULSE MODULATION

Another type of digital amplitude modulation comes under the general category of *pulse modulation*. The RF signal is broken

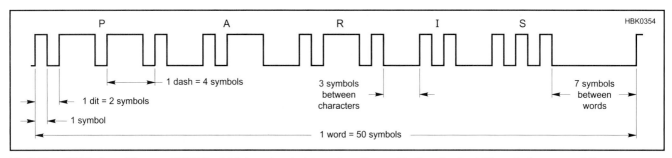

Fig 8.13 — CW timing of the word PARIS, which happens to have a length equal to the standard 50 symbols per word. By programming it multiple times into a memory keyer, the speed may be calibrated simply by counting the number of times the word is completed in one minute.

up into a series of pulses, which are usually equally-spaced in time and separated by periods of no signal. We will discuss three types of pulse modulation, PAM, PWM and PPM.

Pulse-amplitude modulation (PAM) consists of a series of pulses of varying amplitude that correspond directly to the amplitude of the modulating signal. See **Fig 8.14B**. The pulses can be positive or negative, depending on the polarity of the signal. The modulating signal can be recovered simply by low-pass filtering the pulse train. The effect of the negative pulses is to reverse the polarity of the RF signal. The result is very similar to a double-sideband, suppressed-carrier signal that is rapidly turned on and off at the pulse rate of the PAM. In other words, the DSBSC signal is periodically sampled at a certain pulse repetition rate.

For the signal to be properly represented, its highest modulating frequency must be less than the *Nyquist frequency*, which is one-half the sample rate. That condition, known as the *Nyquist criterion*, applies not only to PAM but to any digital modulation technique.

There are two variations of PAM that should be mentioned. *Natural sampling* does not hold the amplitude of each pulse constant throughout the pulse as shown in Fig 8.14B, but rather follows the shape of the analog modulation. The *single-polarity method* adds a fixed offset, or pedestal, to the modulating signal before the PAM modulator. As long as the pedestal is greater than or equal to the peak negative modulation, then the RF phase never changes and the signal is equivalent to sampled full-carrier AM rather than DSBSC.

PAM is rarely used on the amateur bands because it increases the transmitted bandwidth and adds circuit complexity with no improvement in signal-to-noise ratio for most types of noise and interference. The concept is useful, however, because PAM is similar to the signal generated by the sample-and-hold circuit that is used at the input to an analog-to-digital converter (ADC). An ADC is used in virtually every digital transmitter to convert the analog voice signal to a digital signal suitable for digital signal processing.

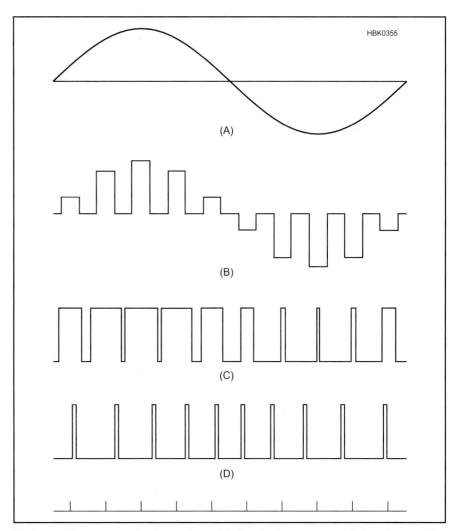

Fig 8.14 — Three types of pulse-code modulation. A sine-wave modulating signal (A) is shown at the top. Pulse amplitude modulation (B) varies the amplitude of the pulses, pulse width modulation (C) varies the pulse width, and pulse-position modulation (D) varies the pulse position, proportional to the modulating signal. The tic marks show the nominal pulse times. The RF signal is created by an AM modulator using (B), (C) or (D) as the modulating signal.

Pulse-width modulation (PWM) is a series of pulses whose width varies in proportion to the amplitude of the modulating signal. Fig 8.14C shows the pulses centered on the sample times, but in some systems the sample times may correspond to the leading or trailing pulse edges. With either method, the modulating signal can be recovered by low-pass filtering the pulse train and passing it through a coupling capacitor to remove the DC component.

Pulse-position modulation (PPM),

Fig 8.14D, varies the position, or phase, of the pulses in proportion to the amplitude of the modulating signal. With both PWM and PPM, the peak amplitude of the signal is constant. That allows the receiver to be designed to be insensitive to amplitude variations, which can result in a better post-detection signal-to-noise ratio, in a manner similar to analog angle modulation.

8.3.2 Angle-Modulated Digital Signals

Just as with analog modulation, angle-modulated digital signals have constant amplitude and vary the phase and frequency to represent the modulating signal. They share some of the same advantages as their analog counterparts. The receiver can be made relatively insensitive to amplitude noise and on-channel interference. The constant-amplitude signal simplifies the job of the receiver's automatic-gain-control system. The transmitter does not require a linear amplifier, so a more-efficient class-C amplifier may be used. However digital angle modulation also shares analog's nonlinearity so that the bandwidth is not as well-constrained as it is with linear modulation systems.

Frequency-shift keying (FSK) was the first digital angle-modulated format to come into common use. FSK can be thought of as being the same as analog FM that is modulated with a binary (two-level) signal that causes the RF signal to switch between two frequencies. It can also be thought of as equivalent to two on-off-keyed (OOK) signals on two nearby frequencies that are keyed in such a way that whenever one is on the other is off. Just as with OOK, if the transitions between states are instantaneous, then excessive bandwidth occurs — causing interference on nearby channels similar to key clicks. For that reason, the modulating signal must be low-pass filtered to slow down the speed at which the RF signal moves from one frequency to the other.

Although FSK is normally transmitted as a true constant-amplitude signal with only the frequency changing between symbols, it does not have to be received that way. The receiver can treat the signal as two OOK signals and demodulate each one separately. This is an advantage when HF propagation conditions exhibit selective fading — even if one frequency fades out completely the receiver can continue to copy the other, a form of *frequency diversity*. To take advantage of it, wide shift (850 Hz) must normally be used. With narrow-shift FSK (170 Hz), the two tones are generally too close to exhibit selective fading.

As previously discussed, selective fading is caused by the signal arriving at the receiver antenna by two or more paths simultaneously. The same phenomenon can cause another signal impairment known as *inter-symbol interference*. See **Fig 8.15**. If the difference in the two path lengths is great enough, then the signal from one path may arrive delayed by one entire symbol time with respect to the other. The receiver sees two copies of the signal that are time-shifted by one symbol. In effect, the signal interferes with itself. One solution is to slow down the baud rate so that the symbols do not overlap. It is for this reason that symbol rates employed on HF are usually no more than 50 to 100 bauds.

Multi-level FSK (MFSK) is one method to reduce the symbol rate. Unlike with conventional binary FSK, more than two shift frequencies are allowed. For example, with eight frequencies, each symbol can have eight possible states. Since three bits are required to represent eight states, three bits are transmitted per symbol. That means you get three times the data rate without increasing the baud rate. The disadvantage is a reduced signal to noise and signal to interference ratio. If the maximum deviation is the same, then the frequencies are seven times closer with 8FSK than with binary FSK and the receiver is theoretically seven times (16.9 dB) more susceptible to noise and interference. However the IF limiter stage removes most amplitude variations before the signal arrives at the FSK detector, so the actual increase in susceptibility is less for signal levels above a certain threshold.

With any type of FSK, you can theoretically make the shift as narrow as you like. The main disadvantage is that the receiver becomes more susceptible to noise and interference, as explained above. In addition, the bandwidth is not reduced as much as you might expect. Just as with analog FM, you still get sidebands whose extent depends on the symbol rate, no matter how small the deviation.

Minimum-shift keying (MSK) is FSK with a deviation that is at the minimum practical level, taking bandwidth and signal-to-noise ratio into account. That turns out to be a frequency shift from the center frequency of 0.25 times the baud rate or, using the common definition of frequency shift, a difference between the two tones of 0.5 times the baud rate. If, on a given symbol, the frequency is shifted to the upper tone, then the phase of the RF signal will change by 0.25 cycles, or 90°, during the symbol period. If the lower tone is selected, the phase shift is –90°. Thus, MSK may be regarded as either FSK with a frequency shift of 0.5 times the symbol rate, or as *differential phase-shift keying* (DPSK) with a phase shift of ±90°. The binary data causes the phase to change by either +90 or –90° from one symbol to the next.

Gaussian minimum-shift keying (GMSK) refers to MSK where the modulating signal has been filtered with a low-pass filter that has a Gaussian frequency response. As mentioned before, with any type of angle modulation the spectrum of the modulation is not duplicated at RF, but spreads out into an increased bandwidth. For that reason, there is no point in using a modulation filter with a sharp cutoff since the RF spectrum will be wider anyway. It turns out that a Gaussian filter, with its gradual transition from passband to stopband, has the optimum shape for an angle-modulated digital system. However, a Gaussian filter also has a gradual transition from symbol-to-symbol in the time domain as well. The transition is not totally completed by the time of the next symbol, which means that there is some inter-symbol interference in the transmitted signal, even in the absence of propagation impairments. With the proper choice of filter bandwidth, however, the ISI is small enough not to seriously affect performance.

Binary phase-shift keying (BPSK), often referred to simply as phase-shift keying (PSK), is included in this section because the name suggests that it is true constant-amplitude angle modulation. It is possible to implement it that way. An example is MSK which, as described above, can be considered to be differential BPSK with a ±90° phase shift. However the term BPSK is normally understood to refer to phase-shift keying with a 180° phase difference between symbols. To transition from one state to the other, the modulation filter smoothly reduces the amplitude to zero where the polarity reverses (phase changes 180°) before smoothly ramping up to full amplitude again. For that reason, BPSK as usually implemented really should not be considered to be PSK, but rather a form of amplitude-shift keying (ASK) with two modulation amplitudes, +1 and –1. Unlike true angle modulation, it is linear so that the spectrum of the modulation filter is duplicated at RF. The transmitter must use a linear amplifier to prevent distortion and excessive bandwidth similar to the splatter that results from an over-driven SSB transmitter.

8.3.3 Quadrature Modulation

Quadrature modulation encodes digital signals using a combination of amplitude

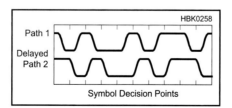

Fig 8.15 — Multipath propagation can cause inter-symbol interference (ISI). At the symbol decision points, which is where the receiver decoder samples the signal, the path 2 data is often opposing the data on path 1.

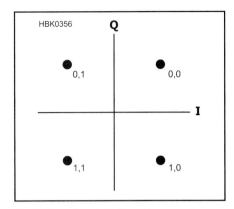

Fig 8.16 — Constellation diagram of a 4-QAM signal, also known as QPSK. The four symbol locations all have the same amplitude and have phase angles of 45°, 135°, −135° and −45°. 4-QAM is a two-bits-per-symbol format. The four symbols are selected by the four possible states of the two data bits, which can be assigned in any order. With the assignment shown, the Q value depends only on the first bit and the I value depends only on the second, an arrangement that can simplify symbol encoding.

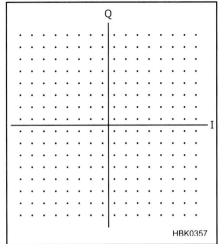

Fig 8.17 — Constellation diagram of a 256-QAM signal. Since an 8-bit number has 256 possible states, each symbol represents 8 bits of data.

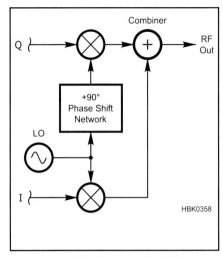

Fig 8.18 — Block diagram of an I/Q modulator. Note the similarity to Fig 8.7. By connecting an audio 90° phase-shift network to the I and Q inputs, an I/Q modulator can generate an analog SSB signal by the phasing method.

and phase modulation. With two types of modulation to work with, it is possible to cram more data bits into each modulation symbol, which allows more throughput for a given bandwidth. Modulation formats in use today have up to 8 or more bits per symbol which would be impractical for ASK, FSK or PSK alone.

Since quadrature-modulated symbols are defined by both amplitude and phase, the most common way to represent symbol states is with a polar plot, called a *constellation diagram*. See **Fig 8.16**, which shows a *four-level quadrature amplitude modulation* (4-QAM) signal, often referred to simply as QAM. The distance of each state from the origin represents the amplitude. The phase angle with respect to the +I axis represents the phase. The four states shown have phase angles of +45°, +135°, −45° and −135°. Normally, the receiver has no absolute phase information and can only detect phase differences between the states. For that reason, we could just as easily have drawn the four states directly on the I and Q axes, at 0°, +90°, 180° and −90°. Note that each of the four states has the same amplitude, differing only in phase. For that reason 4-QAM is often referred to as *quadrature phase-shift keying* (QPSK), in the same manner that two-level ASK is normally referred to as BPSK.

Since 4-QAM has four possible states, there are two data bits per symbol. However QAM is not limited to four states. **Fig 8.17** illustrates a 256-QAM signal. It takes 8 bits to represent 256 states, so 256-QAM packs 8 data bits into each symbol. The disadvantage of using lots of states is that the effect of noise and interference is worse at the receiver. Since, for the same peak power, the states of a 256-QAM signal are 15 times closer together than with 4-QAM, the receiver's decoder has to determine each symbol's location to 15 times greater accuracy. That means the ratio of peak power to noise and interference must be 20 log(15) = 23.5 dB greater for accurate decoding. That is why QAM with a large number of states is normally not used on the HF bands where fading, noise and interference are common. A more common application is digital cable television where the coaxial-cable transmission channel is much cleaner. For example, the European DVB-C standard provides for 16-QAM through 256-QAM, depending on bandwidth and data rate.

There are at least two ways to generate a QAM signal. One is to combine a phase modulator and an amplitude modulator. The phase modulator places the symbol location at the correct phase angle and the amplitude modulator adjusts the amplitude so the symbol is the correct distance from the origin. That method is seldom used, both because of the circuit complexity and because the nonlinear phase modulator makes the signal difficult to filter so as to limit the bandwidth.

A much more common method of generating QAM is with an *I/Q modulator*. See **Fig 8.18**. By using two modulators (mixers) fed with RF sine waves in quadrature (90° out of phase with each other), any amplitude and phase may be obtained by varying the amplitudes of the two modulation inputs. The input labeled I (for in-phase) moves the symbol location horizontally in the constellation diagram and the one labeled Q (for quadrature) moves it vertically. For example, to obtain an amplitude of 1 and a phase angle of −45°, set I to +0.707 and set Q to −0.707.

It is possible to generate virtually any type of modulation using an I/Q modulator. For example, to generate BPSK or on-off keying, simply disconnect the Q input and apply the modulation to I. For angle modulation, such as FM or PM, a waveform generator applies a varying signal to I and Q in such a manner as to cause the symbol to rotate at constant amplitude with the correct phase and frequency (rotation rate). Even a multi-carrier signal may be generated with a single I/Q modulator by applying the sum of a number of signals, each representing one carrier, to the I and Q inputs. The phase of each signal rotates at a rate equal to the frequency offset of its carrier from the center.

One problem with QAM is that whenever the signal trajectory between two symbol states passes through the origin, the signal amplitude momentarily goes to zero. That imposes stringent linearity requirements on the RF power amplifier, since many amplifiers exhibit their worst linearity near zero power. One solution is *offset QPSK* (OQPSK) modulation. In this case, the symbol transitions of the I and Q channels are offset by half a symbol. That is, for each symbol, the I channel changes state first then the Q channel changes half a symbol time later. That allows the symbol trajectory to sidestep around the origin, allowing use of a higher-efficiency or lower-cost power amplifier that has worse linearity.

Another solution to the zero-crossing problem is called *PI over 4 differential QPSK* (π/4 DQPSK). See **Fig 8.19**. This is actually a form of 8-PSK, where the eight symbol locations are located every 45° around a constant-amplitude circle. On any given symbol, however, only four of the symbol locations are used. The symbol location always changes by an odd multiple of 45° (π/4 radians). If the current symbol is located on the I or Q axis, then on the next symbol only the four non-axis locations are available and vice versa. As with OQPSK, that avoids transitions that pass through the origin.

Another advantage of π/4 DQPSK which is shared with other types of differential modulation is that absolute phase doesn't matter. The information is encoded only in the *difference* in the phase of successive symbols. That greatly simplifies the job of the receiver's demodulator.

The block diagram of an I/Q demodulator looks like an I/Q modulator drawn backward. See **Fig 8.20**. If the local oscillator is not tuned to exactly the same frequency as the one in the transmitter (after downconversion to the IF, assuming a superhet receiver) then the demodulated signal will rotate in the I/Q plane at a rate equal to the frequency error. Most receivers include a *carrier-recovery* circuit, which phase-locks the local oscillator to the average frequency of the incoming signal to obtain stable demodulation. While that corrects the frequency error, it does not correct the phase, which must be accounted for in some other manner such as by using differential modulation or some kind of symbol-recovery mechanism.

8.3.4 Multi-carrier Modulation

An effective method to fit more data bits into each symbol is to use more than one separately-modulated signal at a time, each on its own carrier frequency spaced an appropriate distance from the frequencies of the other carriers. An example is multi-carrier FSK. This is not to be confused with MFSK which also uses multiple frequencies, but only one at a time. With multi-carrier FSK, each carrier is present continuously and is frequency-shifted in response to a separate data stream. The total data rate equals the data rate of one carrier times the number of carriers. A disadvantage of multi-carrier FSK is that the resulting signal is no longer constant-amplitude — a linear amplifier must be used. In general, multi-carrier signals using any modulation type on each carrier tend to have high peak-to-average power ratios.

In the presence of selective fading, one or more of the carriers may disappear while the others are still present. An advantage of

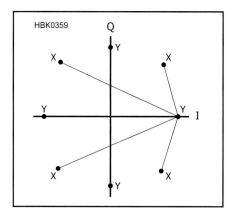

Fig 8.19 — Constellation diagram of a π/4 DQPSK signal. There are eight possible symbol locations. If the current symbol location is at one of the four positions labeled "X" then the next symbol will be at one of the four locations labeled "Y" and vice versa. That guarantees that no possible symbol trajectories (the lines between symbol locations) can ever pass through the origin.

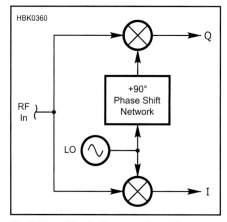

Fig 8.20 — Block diagram of an I/Q demodulator.

multi-carrier modulation is that error-correcting coding can use the unaffected carriers to reconstruct the missing data. Also, since each carrier signal is relatively narrowband, propagation conditions are essentially constant within that bandwidth. That makes it easier for the receiver to correct for other frequency-selective propagation impairments such as phase distortion. If a single-carrier signal of the same total bandwidth had been used instead, the receiver would need an adaptive equalizer to correct for the amplitude and phase variations across the transmission channel.

By using multiple carriers each with multiple-bit-per-symbol modulation it is possible to obtain quite high data rates while maintaining the low symbol rates that are required to combat the effects of multi-path propagation on the HF bands. For example, PACTOR-III achieves a raw data rate of 3600 bits per second with a 100-baud symbol rate using 18 carriers of DQPSK. (100 bauds × 2 bits/symbol × 18 carriers = 3600 bits per second.) Similarly, Clover-2000 modulation gets 3000 bits per second with a 62.5-baud symbol rate using eight carriers of 16-DPSK combined with 4-level DASK. (62.5 bauds × (4+2) bits/symbol × 8 carriers = 3000 bits per second.) Decoding is rather fragile using these complex modulation techniques, so PACTOR and Clover include means to automatically switch to simpler, more-robust modulation types as propagation conditions deteriorate.

What is the minimum carrier spacing that can be used without excessive interference between signals on adjacent frequencies? The answer depends on the symbol rate and the filtering. It turns out that it is easy to design the filtering to be insensitive to interference on frequencies that are spaced at integer multiples of the symbol rate. (See the following section on filtering and bandwidth.) For that reason, it is common to use a carrier spacing equal to the symbol rate. The carriers are said to be *orthogonal* to each other since each theoretically has zero correlation with the others.

Orthogonal frequency-division multiplexing (OFDM), sometimes called *coded OFDM* (COFDM), refers to the multiplexing of multiple data streams onto a series of such orthogonal carriers. The term usually implies a system with a large number of carriers. In that case, an efficient decoding method is to use a DSP algorithm called the *fast Fourier transform* (FFT). The FFT is the software equivalent of a hardware spectrum analyzer. It gathers a series of samples of a signal taken at regular time intervals and outputs another series of samples representing the frequency spectrum of the signal. See **Fig 8.21B**. If the length of the series of input samples equals one symbol time and if the sample rate is selected properly, then each frequency sample of the FFT output corresponds to one carrier. Each frequency sample is a complex number (containing a "real" and "imaginary" part) that represents the amplitude and phase of one of the carriers during that symbol period. Knowing the amplitude and phase of a carrier is all the information required to determine the symbol location in the I/Q diagram and thus decode the data. At the transmitter end of the circuit, an *inverse FFT* (IFFT) can be used to encode the data, that is, to convert the amplitudes and phases of each of the carriers into a series of I and Q time samples to send to the I/Q modulator. See Fig 8.21A.

One advantage of OFDM is high spectral efficiency. The carriers are spaced as closely as theoretically possible and, because of the narrow bandwidth of each carrier, the overall spectrum is very square in shape with a sharp

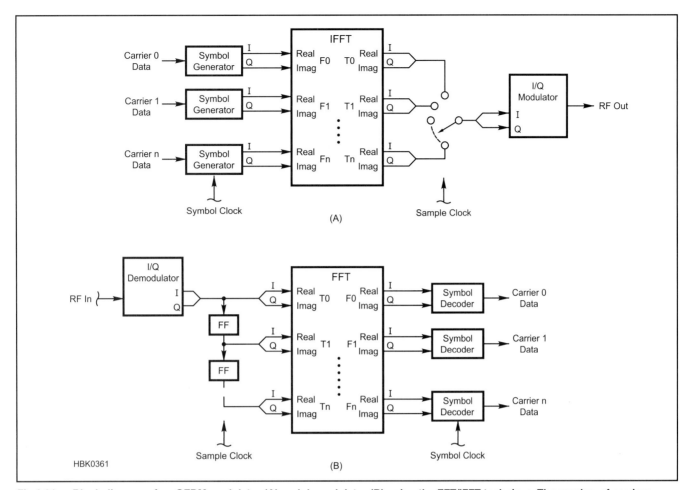

Fig 8.21 — Block diagram of an OFDM modulator (A) and demodulator (B) using the FFT/IFFT technique. The number of carriers is n, and the sample rate is n times the symbol rate. Once per symbol, all the symbol generators in the modulator are loaded with new data and the inverse fast-Fourier transform (IFFT) generates n output samples, which are selected in succession by the switch. In the demodulator, n samples are stored in a shift register (string of flip-flops) for each symbol, then the FFT generates one "frequency" output for each carrier frequency. From the amplitude and phase of each frequency, the symbol decoders can determine the symbol locations and thus the data.

drop-off at the passband edges. One disadvantage is that the receiver must be tuned very accurately to the transmitter's frequency to avoid loss of orthogonality, which causes cross-talk between the carriers.

8.3.5 AFSK and Other Audio-Based Modulation

Audio frequency-shift keying (AFSK) is the generation of radio-frequency FSK using an audio-frequency FSK signal fed into the microphone input of an SSB transmitter. Assume the SSB transmitter is tuned to 14.000 MHz, USB. If the audio signal consists of a sine wave that shifts between 2125 Hz and 2295 Hz (170-Hz frequency shift), then the RF signal is a sine wave shifting between 14.002125 and 14.002295 MHz. The frequency shift and spectral characteristics are theoretically unchanged, other than being translated 14 MHz upward in frequency. The RF signal should be indistinguishable from one generated by varying the frequency of an RF oscillator directly.

This technique works not only for FSK but also for nearly any modulation type with a bandwidth narrow enough to fit within the passband of an SSB transmitter and receiver. The most common non-voice analog modulation type to use this technique on the amateur bands is *slow-scan television* (SSTV), which uses frequency modulation for the video signal. In addition, nearly all narrowband digital signals today are generated in this manner. The audio is generated and received either from a dedicated hardware modulator/demodulator (modem) or using the sound card that is found in virtually all personal computers.

Whatever the method, it is important to ensure that unwanted interference is not caused by audio distortion or by insufficient suppression of the carrier and unwanted sideband. For example, with the AFSK tone frequencies mentioned above, 2125 and 2295 Hz, the tone harmonics cause no trouble because they fall outside of the transmitter's passband. However, some AFSK modems use 1275 and 1445 Hz (to accommodate 850-Hz shift without changing the 1275-Hz mark frequency). In that case, the second harmonics at 2550 and 2890 Hz must be suppressed since those frequencies are not well-attenuated by the transmitter. With non-constant-envelope modulation types such as QPSK or the various multi-carrier modes, it is important to set the amplitude of the audio input to the SSB transmitter below the level that activates the transmitter's automatic level control (ALC). That is because the ALC circuit itself generates distortion of signals within the bandwidth of its feedback loop.

8.3.6 Spread Spectrum

A *spread-spectrum* (SS) system is one that intentionally increases the bandwidth of a digital signal beyond that normally required by means of a special spreading code that is independent of the data sequence. There are several reasons for spreading the spectrum in that way.

Spread spectrum was first used in military systems, where the purpose was to encrypt the transmissions to make it harder for the enemy to intercept or jam them. Amateurs are not allowed to encrypt transmissions for the purpose of concealing the information, but reducing interference, intentional or otherwise, is an obvious benefit. The signal is normally spread in such a fashion that it appears like random noise to a receiver not designed to receive it, so other users of the band may not even be aware that an SS signal is present.

Another advantage to spreading the spectrum is that it can make frequency accuracy less critical. In addition, the wide bandwidth means that expensive narrow-bandwidth filters are not required in the receiver. It also provides a measure of frequency diversity. If certain frequencies are unusable because of interference or selective fading, the signal can often be reconstructed using information in the rest of the bandwidth.

There are several ways to spread the spectrum — we will cover the two most common methods below — but they all share certain characteristics. Imagine that the unspread signal occupies a 10 kHz bandwidth and it is spread by a factor of 100. The resulting SS signal is 1000 kHz (1 MHz) wide. Each 10-kHz channel contains 1/100 of the total signal power, or –20 dB. That means that any narrowband stations using one of those 10-kHz channels experience a 20 dB reduction in interference, but also are more likely to be interfered with because of the 100-times greater bandwidth of the SS station's emissions.

How is the spread-spectrum station affected by interference from narrowband stations? In effect, the SS receiver attenuates the signal received on each 10 kHz channel by 20 dB in order to obtain a full-power signal when all 100 channels are added together. That means that the interference from a narrowband station is reduced by 20 dB but, again, the interference is more likely to occur because of the 100-times greater bandwidth of the SS station's receiver.

How is the spread-spectrum station affected by interference from another SS station on the same frequency? It turns out that if the other station is using a different orthogonal spreading code then, once again, the interference reduction is 20 dB for 100-times spreading. That means that many SS stations can share the same channel without interference as long as they are all received at roughly the same signal level. Commercial mobile-telephone SS networks use an elaborate system of power control with real-time feedback to ensure that the signals from all the mobile stations arrive at the base station at approximately the same level.

That scheme works well in a one-to-many (base station to mobile stations) system architecture but would be much more difficult to implement in a typical amateur many-to-many arrangement because of the different distances and thus path losses between

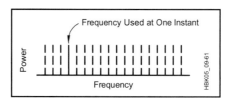

Fig 8.22 — Power versus frequency for a frequency-hopping spread spectrum signal. Emissions jump around to discrete frequencies in pseudo-random fashion. Normally the spacing of the frequencies is approximately equal to the bandwidth of the unspread signal so that the average spectrum is approximately flat.

each pair of stations in the network. On the HF bands it is not uncommon to see differences in signal levels of 80 to 90 dB or more. (For example, the difference between S1 and 40 dB over S9 is 88 dB, assuming 6 dB per S-unit.) A spread spectrum signal at S9 + 40 dB with a spreading ratio of 100 times would interfere with any other signals below about S9 + 20 dB. It works the same in the other direction as well. The SS signal would experience interference from any other stations that are more than 20 dB louder than the desired signal.

Normally, increasing the bandwidth of a transmission degrades the signal-to-noise (S/N) ratio at the receiver. A 100-times greater bandwidth contains 100 times as much noise, which causes a 20 dB reduction in S/N ratio. However SS receivers benefit from a phenomenon known as *processing gain*. Just as the receiver is insensitive to other SS signals with different orthogonal spreading codes, so it is insensitive to random noise. The improvement in S/N ratio due to processing gain is:

Processing Gain =

$$10 \times \log\left(\frac{\text{spread bandwidth}}{\text{unspread bandwidth}}\right) \text{ dB} \quad (3)$$

That is exactly equal to the reduction in S/N ratio due to the increased bandwidth. The net result is that an SS signal has neither an advantage nor a disadvantage in signal-to-noise ratio compared to the unspread version of the same signal. When someone states that, because of processing gain, an SS receiver can receive signals that are below the noise level (signals that have a negative S/N ratio), that is a true statement. However, it does not imply better S/N performance than could be obtained if the signal were not spread.

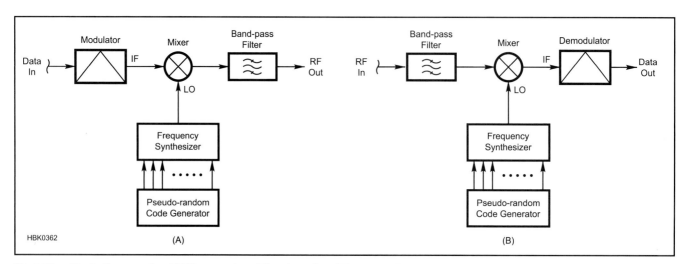

Fig 8.23 — Block diagram of an FHSS transmitter (A) and receiver (B). The receiver may be thought of as a conventional superhet with a local oscillator (LO) that is continually hopping its frequency in response to a pseudo-random code generator. The transmitter has a similar architecture to up-convert a conventionally-modulated intermediate frequency (IF) to a frequency-hopped radio frequency (RF).

FREQUENCY HOPPING SPREAD SPECTRUM

One simple way to spread the spectrum of a narrowband signal is to repetitively sweep it across the frequency range of the wider spectrum, either continuously or in a series of steps at discrete frequencies. That technique, called *chirp modulation*, can be considered a special case of *frequency-hopping spread spectrum* (FHSS), in which the narrowband signal covers the expanded spectrum by rapidly hopping back and forth from frequency to frequency in a pseudo-random manner. On average, each frequency is used the same percentage of time so that the average spectrum is flat across the bandwidth of the FHSS signal. See **Fig 8.22**.

If the receiver hops in step with the transmitter, using the same pseudo-random sequence synchronized to the one in the transmitter, then the transmitter and receiver are always tuned to the same frequency and the receiver's detector sees a continuous, non-hopped narrowband signal which can be demodulated in the normal way. We say that the signal has been *de-spread*, that is, returned to its normal narrowband form. Synchronization between the receiver and transmitter is one of the challenges in an FHSS system. If the timing of the two sequences differs by even one hop, then the receiver is always tuned to the wrong frequency, unless the same frequency happens to occur twice in succession in the pseudo-random sequence. Any signal with an unsynchronized sequence, or with a different sequence, is reduced in amplitude by the processing gain.

There are two types of FHSS based on the rate at which the frequency hops take place. *Slow-frequency hopping* refers to a hop rate slower than the baud rate. Several symbols are sent per hop. With *fast-frequency hopping*, the hop rate is faster than the baud rate. Several hops occur during each symbol. The term *chip* refers to the shortest-duration modulation state in the system. For slow-frequency hopping, that is the baud rate. For fast-frequency hopping it is the hop rate. Fast-frequency hopping can be useful in reducing the effects of multi-path propagation. If the hop period is less then the typical time delay of secondary propagation paths, then those signals are uncorrelated to the main path and are attenuated by the processing gain.

A diagram of a frequency-hopping spread spectrum system is shown in **Fig 8.23**. In both the transmitter and the receiver, a pseudo-random code generator controls a frequency synthesizer to hop between frequencies in the correct order. In this way, the narrowband signal is first spread by the transmitter, then sent over the radio channel, and finally de-spread at the receiver to obtain the original narrowband signal again. One issue

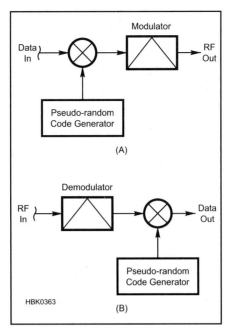

Fig 8.24 — Block diagram of a DSSS transmitter (A) and receiver (B). For BPSK modulation, the modulation has the same format as the bipolar data (±1), so the modulator and demodulator could be moved to the other side of the multiplier if desired.

with FHSS is that many synthesizers do not maintain phase coherence over successive frequency hops. That means the basic (non-spread) modulation must be a type that does not depend on phase information. That rules out PSK, QPSK and QAM. That is the reason that modulation types that do not depend on phase, such as FSK and MFSK with non-coherent detection, are frequently used as the base modulation type in FHSS systems.

DIRECT SEQUENCE SPREAD SPECTRUM

Whereas an FHSS system hops through a pseudo-random sequence of frequencies to spread the signal, a *direct-sequence spread spectrum* (DSSS) system applies the pseudo-random sequence directly to the data in order to spread the signal. See **Fig 8.24**. The binary data is considered to be in *polar* form, that is, the two possible states of each bit are represented by –1 and +1. The data bits are multiplied by a higher-bit-rate pseudo-random sequence, also in polar form, which chops the data into smaller time increments called *chips*. The ratio of the chip rate to the data's bit rate equals the ratio of the spread bandwidth to the unspread bandwidth, which is just the processing gain:

$$\text{Processing Gain} = 10 \times \log\left(\frac{\text{chip rate}}{\text{bit rate}}\right) \text{ dB} \quad (4)$$

Although it doesn't have to, DSSS normally uses a one-bit-per-symbol modulation type such as BPSK. In that case, the modulator and demodulator in Fig 8.24 would consist simply of a mixer, which multiplies the RF local oscillator by the bipolar DSSS modulating signal. Since unfiltered BPSK is constant-envelope, a nonlinear class-C power amplifier may be used for high efficiency. The demodulator in the receiver would also be a mixer, which multiplies the RF signal by a local oscillator to regenerate the DSSS modulating signal. Not shown are additional mixers and filters that would be used in a superheterodyne receiver to convert the received signal to an intermediate frequency before demodulation and decoding.

The unfiltered spectrum has the form of a sinc function. That shows up clearly in the spectrum of an actual DSSS signal in **Fig 8.25A**, which is plotted on a logarithmic scale calibrated in decibels. The humps in the response to the left and right (and additional ones not shown off scale) are not needed for communications and should be filtered out to avoid excessive occupied bandwidth. Fig 8.25B shows the DSSS signal in the presence of noise at the input to the receiver, and Fig 8.25C illustrates the improvement in signal-to-noise ratio of the de-spread narrowband signal.

CODE-DIVISION MULTIPLE ACCESS (CDMA)

As mentioned before, to receive an SS signal, the de-spreading sequence in the receiver must match the spreading sequence in the transmitter. The term *orthogonal* refers to two sequences that are coded in such a way that they are completely uncorrelated. The receiver's response to an orthogonal code is the same as to random noise, that is, it is suppressed by a factor equal to the processing gain. One can take advantage of this property to allow multiple SS stations to access the same frequency simultaneously, a technique known as *code-division multiple access* (CDMA). Each transmitter is assigned a different orthogonal code. A receiver can "tune in" any transmitter's signal by selecting the correct code for de-spreading.

If multiple stations want to be able to transmit simultaneously without using spread spectrum, they must resort to either *frequency-division multiple access* (FDMA), where each station transmits on a different frequency channel, or *time-division multiple access* (TDMA), where each transmission is broken up into short time slots which are interleaved with the time slots of the other stations. Compared to TDMA, CDMA has the advantage that it does not require an external synchronization network to make sure that different stations' time slots do not overlap. Compared to both TDMA and FDMA, CDMA has the further advantage that it experiences a gradual

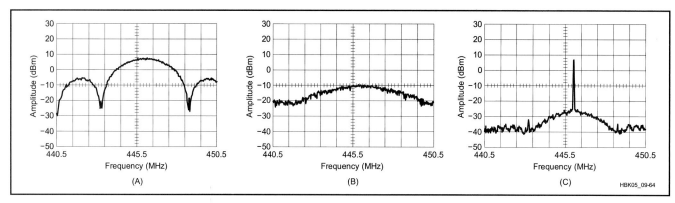

Fig 8.25 — (A) The frequency spectrum of an actual unfiltered biphase-modulated spread spectrum signal as viewed on a spectrum analyzer. In this practical system, band-pass filtering is used to confine the spread spectrum signal to the amateur band. (B) At the receiver end of the line, the filtered spread spectrum signal is apparent only as a 10-dB hump in the noise floor. (C) The signal at the output of the receiver de-spreader. The original carrier—and any modulation components that accompany it—has been recovered. The peak carrier is about 45 dB above the noise floor—more than 30 dB above the hump shown at B. (These spectrograms were made at a sweep rate of 0.1 s/division and an analyzer bandwidth of 30 kHz; the horizontal scale is 1 MHz/division.)

degradation in performance as the number of stations on the channel increases. It is relatively easy to add new users to the system. Also CDMA has inherent resistance to interference due to multi-path propagation or intentional jamming. The primary disadvantages of CDMA are the relative complexity and the necessity for accurate power-level control to make sure that unwanted signals do not exceed the level that can be rejected through processing gain.

8.3.7 Filtering and Bandwidth of Digital Signals

We have already touched on this topic in previous sections, but let us now cover it a little more systematically. The bandwidth required by a digital signal depends on the filtering of the modulation, the symbol rate and the type of modulation. For linear modulation types such as OOK, BPSK, QPSK and QAM, the bandwidth depends only on the symbol rate and the modulation filter.

As an example, an unfiltered BPSK modulating signal with alternating ones and zeroes for data (10101010...) is a square wave at one-half the symbol rate. Like any square wave, its spectrum can be broken down into a series of sine waves at the fundamental frequency (symbol rate / 2) and all the odd harmonics. If the data consists of alternating pairs of ones and pairs of zeroes (11001100...) then we have a square wave at one-fourth the symbol rate and the spectrum is a series of sine waves at one-fourth the symbol rate and all its odd harmonics. Random data contains energy at all frequencies from zero to half the symbol rate and all the odd harmonics of those frequencies. The harmonics are not needed for proper demodulation of the signal, so they can be filtered out with a low-pass filter with a cutoff frequency of one-half the symbol rate.

With random data, the shape of the unfil-

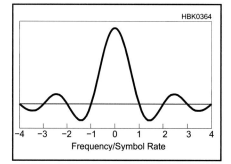

Fig 8.26 — The sinc function, which is the spectrum of an unfiltered BPSK modulating signal with random data, plotted with a linear vertical scale. The center (zero) point corresponds to the RF carrier frequency. To see what the double-sided RF spectrum looks like on a logarithmic (dB) scale, see Fig 8.25A.

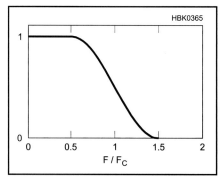

Fig 8.27 — Amplitude versus frequency for a 0.5-alpha, raised-cosine filter. The vertical scale is linear, not logarithmic as would be seen on a spectrum analyzer. The amplitude is 0.5 (–6 dB) at the cutoff frequency, Fc. The amplitude is 1.0 for frequencies less than Fc × (1 – alpha) and is 0.0 for frequencies above Fc × (1 + alpha).

tered spectrum is a sinc function,

$$\text{sinc}(f/f_s) = \frac{\sin(\pi f / f_s)}{\pi f / f_s} \quad (5)$$

where f_s is the symbol rate. See **Fig 8.26**. Note that the response is zero (minus infinity dB) whenever f is an integer multiple of the symbol rate. That is why multi-carrier modulation generally uses a carrier spacing equal to the symbol rate.

The previous discussion applies to the baseband signal, before it modulates the RF carrier. BPSK is a double-sideband-type modulation so the baseband spectrum appears above and below the RF carrier frequency, doubling the bandwidth to 2 × (symbol rate / 2) = symbol rate. In reality, since no practical filter has an infinitely-sharp cutoff, the occupied bandwidth of a BPSK signal must be somewhat greater than the symbol rate. That is also true for all other double-sideband linear modulation types, such as OOK and the various forms of QPSK and QAM.

If the low-pass filtering is not done properly, it may slow down the transition between symbols to the point where one symbol starts to run into another, causing *inter-symbol interference* (ISI). A type of filter that avoids that problem is called a *Nyquist filter*. It ensures that each symbol's contribution to the modulating signal passes through zero at the center of all other symbols, so that no ISI occurs. The most common type of Nyquist filter is the *raised-cosine* filter, so-called because the frequency response (plotted on a linear scale) in the passband-to-stopband transition region follows a raised-cosine curve. See **Fig 8.27**. The sharpness of the frequency cutoff is specified by a parameter called *alpha*. If alpha is 1.0, then the transition from passband to stopband is very gradual — it starts to roll off right at zero hertz and finally reaches zero response at two times the nominal cutoff frequency. An alpha of 0.0 specifies an ideal "brick-wall" filter that transitions instantaneously from full response to zero right at the cutoff frequency. Values in the range of 0.3 to 0.5 are common in communications systems.

Unfortunately, if any additional filter is placed before or after the Nyquist filter it destroys the anti-ISI property. In order to allow filtering in both the transmitter and the receiver many systems effectively place half the Nyquist filter in each place. Because the frequency response of each filter is the square root of the response of a Nyquist filter, they are called *root-Nyquist* filters. The *root-raised-cosine* filter is an example. While a Nyquist or root-Nyquist response theoretically could be approximated with an analog filter, they are almost always implemented as digital filters. More information on digital filters appears in the **DSP and Software Radio Design** chapter.

As mentioned before, filtering is more difficult with angle modulation because it is nonlinear. The RF spectrum is not a linear transposition of the baseband spectrum as it is with linear modes and Nyquist filtering doesn't work. Old-fashioned RTTY transmitters traditionally just used an R-C low-pass filter to slow down the transitions between mark and space. While that does not limit the bandwidth to the minimum value possible, the baud rate is low enough that the resulting bandwidth is acceptable anyway. In more modern systems, there is a tradeoff between making the filter bandwidth as narrow as possible for interference reduction and widening the bandwidth to reduce inter-symbol interference. For example, the GSM (Global System for Mobile communications) standard, used for some cellular telephone systems, uses minimum-shift keying and a Gaussian filter with a BT (bandwidth symbol-time product) of 0.3. A 0.3 Gaussian filter has a 0.3 ratio of 3-dB bandwidth to baud rate, which results in a small but acceptable amount of ISI and a moderate amount of adjacent-channel interference.

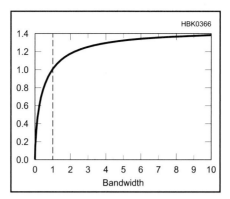

Fig 8.28 — Plot of channel capacity versus bandwidth, calculated by the Shannon-Hartley theorem. The S/N ratio has been selected to be unity at Bandwidth = 1.

CHANNEL CAPACITY

It is possible to increase the quantity of error-free data that can be transmitted over a communications channel by using an error-correcting code. That involves adding additional error-correction bits to the transmitted data. The more bits that are added, the greater the errors that can be corrected. However, the extra bits increase the data rate, which requires additional bandwidth, which increases the amount of noise. For that reason, as you add more and more error-correction bits, requiring more and more bandwidth, you eventually reach a point of limited additional return. In the 1940s, Claude Shannon worked out a formula (called the *Shannon-Hartley theorem*) for the maximum capacity possible over a communications channel, assuming a theoretically-perfect error-correction code:

$$C = B\log_2\left(1 + \frac{S}{N}\right) \text{ bits/s} \qquad (6)$$

where
  C = the net channel capacity, not including error-correcting bits,
  B = the bandwidth in Hz, and
  S/N = the signal-to-noise ratio, expressed as a power ratio.

Note that as B increases, N increases in the same proportion. **Fig 8.28** is a plot of channel capacity versus bandwidth based on the formula. As bandwidth is increased, channel capacity increases rapidly until the point where the S/N ratio drops to unity (labeled Bandwidth = 1 in the graph) after which channel capacity increases much more slowly.

8.4 Image Modulation

The following section covers the modulation aspects of amateur television and facsimile communications. More detailed information on protocols and operating standards can be found in the **Image Communications** chapter.

8.4.1 Fast-Scan Television

Amateur fast-scan television (ATV) is a wideband mode used in the amateur bands above 420 MHz. It is called "fast scan" to differentiate it from slow-scan TV.

The most popular ATV mode is based on NTSC, the same technical standard used by commercial analog television stations in the United States before most switched to digital TV in 2009. **Fig 8.29** shows the spectrum of an NTSC analog TV channel. It's basically a full-carrier, double-sideband AM signal, with filtering to partially remove the lower sideband. The partially-filtered *vestigial sideband* (VSB) extends 1.25 MHz below the carrier frequency. The channel is 6 MHz wide to accommodate the composite video

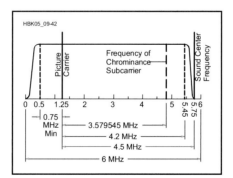

Fig 8.29 — An analog NTSC 6-MHz video channel with the video carrier 1.25 MHz up from the lower edge. The color subcarrier is at 3.58 MHz and the sound subcarrier at 4.5 MHz above the video subcarrier.

and two subcarriers, one at 3.58 MHz for the color burst and the other an FM-modulated signal at 4.5 MHz for the sound. For simplicity, amateur stations often transmit unfiltered full-DSB AM, but the normally-removed portion of the lower sideband is unused by the TV receiver. Since less than 5% of the video energy appears more than 1 MHz below the carrier, little of the transmitter power is wasted. If needed, to reduce interference to other band users, a VSB filter in the antenna line can attenuate the lower sideband color and sound subcarrier frequencies by 20-30 dB.

The video signal includes pulses to synchronize the vertical and horizontal scan oscillators in the receiver. See **Fig 8.30**. The sync pulses and the "front porch" and "back porch" areas that bracket them are "blacker than black" so that the signal is blanked during retrace. The video-to-sync ratio must remain constant throughout all of the linear amplifiers in the transmit chain as the video level from the camera changes. To maintain the sync tips at 100% of peak power, the modulator usually contains a clamp circuit that also acts as a sync stretcher to compensate for amplifier gain compression.

Given NTSC's 525 horizontal lines and its 30 frames per second scan rate, the resulting horizontal resolution bandwidth is 80 lines

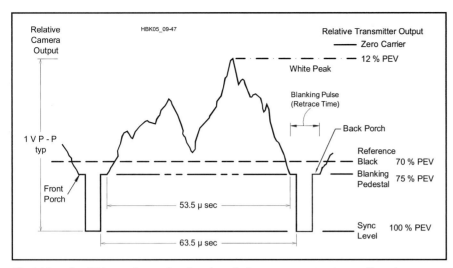
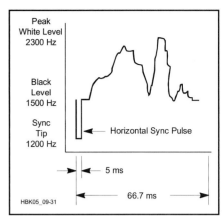

Fig 8.30 — An ATV waveform, showing the relative camera output as well as the transmitter output RF power during one horizontal line scan for black-and-white TV. (A color camera would generate a "burst" of 8 cycles at 3.58 MHz on the back porch of the blanking pedestal.) Note that "black" corresponds to a higher transmitter power than "white."

Fig 8.31 — The basic 8-second black and white transmission format developed by early SSTV experimenters. The sync pulses are "blacker than black" to blank the signal during retrace. A complete frame has 120 lines (8 seconds at 15 lines per second). Horizontal sync pulses occur at the beginning of every line; a 30-ms vertical sync pulse precedes each frame.

per MHz. Therefore, with the typical TV set's 3-dB rolloff at 3 MHz (primarily in the IF filter), up to 240 vertical black lines can be seen, corresponding to 480 pixels per line. Color bandwidth in a TV set is less than that, resulting in up to 100 color lines. (Lines of resolution should not be confused with the number of horizontal scan lines per frame.)

The PAL analog TV system in Europe used frequency modulation rather than AM. As described previously, FM achieves superior noise and interference suppression for signal levels above a certain threshold, although AM seems to work better for receiver signal levels below about 5 µV. FM ATV in the United States typically uses 4 MHz deviation with NTSC video and a 5.5-MHz sound subcarrier set to 15 dB below the video level. Using Carson's rule, the occupied bandwidth comes out to just under 20 MHz. Most available FM ATV equipment is made for the 1.2, 2.4 and 10.25-GHz bands.

Digital fast-scan TV has been explored by amateurs in Europe and the US using the commercial DVB-S satellite digital video broadcasting standard. It uses MPEG2 audio and video data compression and QPSK modulation with symbol rates up to 20 Mbauds. Much of work has been on the 23 cm band since inexpensive set-top boxes are available that cover that frequency range.

In the United States and some other countries, commercial over-the-air digital television uses a modulation type called *8-level vestigial sideband* (8-VSB), which is basically 8-level amplitude-shift keying with the lower sideband filtered out. The eight levels result in three bits per symbol, resulting in approximately 32 Mbits/second raw data rate with a 10.76 Mbaud symbol rate. The net data rate with error correction and other overhead is 19.39 Mbits/second. The DTV signal fits into the same 6 MHz channel as analog TV.

8.4.2 Facsimile

Amateur *facsimile* (fax) is a method for sending high-resolution still pictures by radio. Standard 120-line-per-minute fax uses a frequency-modulated tone between 1500 Hz (black) and 2300 Hz (white), which may be sent and received as an audio signal via any voice transmitter and receiver. Commercially, fax has been used for many years to send newspaper photos over the landline and to receive weather images from satellites.

The resolution of typical fax images greatly exceeds what can be obtained using SSTV or even conventional television. Typical images are made up of 480 to 1600 scanning lines, compared to the 240 lines in a typical SSTV image. This high resolution is achieved by slowing down the rate at which the lines are transmitted, resulting in image transmission times typically in the range of 4 to 13 minutes. For color, transmission times are longer since the image is sent three times to include the red, green and blue components.

Besides the increased resolution, the main difference with SSTV is that many fax systems use no sync pulses. Instead, they rely on the extreme accuracy of external time bases at both the transmitting and receiving ends of the circuit to maintain precise synchronization.

Modern personal computers have virtually eliminated bulky mechanical fax machines from most amateur installations. Now the incoming image can be stored in computer memory and viewed on a high-resolution computer graphics display. The use of a color display system makes it entirely practical to transmit color fax images when band conditions permit the increased transmission time.

8.4.3 Slow-Scan Television

Despite its name, so-called *slow-scan television* (SSTV) is actually a method for sending still images, like facsimile. The original monochrome analog SSTV format illustrated in **Fig 8.31** takes approximately 8 seconds to send one complete frame. The 1500 to 2300-Hz frequency-modulated audio tone resembles that of a fax signal, but is sent at a faster rate and includes pulses at 1200 Hz for synchronization.

Color may be sent using any of several methods. The first to be used was the *frame-sequential* method, in which each of the three primary colors (red, green and blue) is sent sequentially, as a complete frame. That has the disadvantage that you have to wait for the third frame to begin before colors start to become correct, and any noise or interference is three times more likely to corrupt the image and risks ruining the image registration (the overlay of the frames) and thus spoil the picture.

In the *line-sequential* method, each line is electronically scanned three times: once each for red, green and blue. Pictures scan down the screen in full color as they are received and registration problems are reduced. The Wraase SC-1 modes are examples of early line-sequential color transmission. They have a horizontal sync pulse for each of the color component scans. The major weakness of this method is that if the receiving system gets out of step, it doesn't know which scan represents which color.

Rather than sending color images with the usual RGB (red, green, blue) components, Robot Research decided to use luminance and chrominance signals for their 1200C modes. The first half or two thirds of each scan line contains the luminance information, which is a weighted average of the R, G and B components. The remainder of each line contains the chrominance signal with the color information. Existing black-and-white equipment could display the B&W-compatible image on the first part of each scan line and the rest would go off the edge of the screen. That compatibility was very beneficial when most people still had only B&W equipment.

The luminance-chrominance encoding makes more efficient use of the transmission time. A 120-line color image can be sent in 12 seconds, rather than the usual 24. Our eyes are more sensitive to details in changes of brightness than color, so the time is used more efficiently by devoting more time to luminance than chrominance. The NTSC and PAL broadcast standards also take advantage of this vision characteristic and use less bandwidth for the color part of the signal. For SSTV, luminance-chrominance encoding offers some benefits, but image quality suffers. It is acceptable for most natural images but looks bad for sharp, high-contrast edges, which are more and more common as images are altered via computer graphics. As a result, all newer modes have returned to RGB encoding.

The 1200C introduced another innovation, called *vertical interval signaling* (VIS). It encodes the transmission mode in the vertical sync interval. By using narrow FSK encoding around the sync frequency, compatibility is maintained. This new signal just looks like an extra-long vertical sync to older equipment.

The Martin and Scottie modes are essentially the same except for the timings. They have a single horizontal sync pulse for each set of RGB scans. Therefore, the receiving end can easily get back in step if synchronization is temporarily lost. Although they have horizontal sync, some implementations ignore them on receive. Instead, they rely on very accurate time bases at the transmitting and receiving stations to keep in step. The advantage of this "synchronous" strategy is that missing or corrupted sync pulses won't disturb the received image. The disadvantage is that even slight timing inaccuracies produce slanted pictures.

In the late 1980s, yet another incompatible mode was introduced. The AVT mode is different from all the rest in that it has *no horizontal sync*. It relies on very accurate oscillators at the sending and receiving stations to maintain synchronization. If the beginning-of-frame sync is missed, it's all over. There is no way to determine where a scan line begins. However, it's much harder to miss the 5-s header than the 300-ms VIS code. Redundant information is encoded 32 times and a more powerful error-detection scheme is used. It's only necessary to receive a small part of the AVT header in order to achieve synchronization. After this, noise can wipe out parts of the image, but image alignment and colors remain correct.

Digital images may be sent over Amateur Radio using any of the standard digital modulation formats that support binary file transfer. *Digital SSTV* (DSSTV) is one method of transmitting computer image files, such as JPEG or GIF, as described in an article by Ralph Taggart, WB8DQT, in the Feb 2004 issue of *QST*. This format phase-modulates a total of eight subcarriers (ranging from 590 to 2200 Hz) at intervals of 230 Hz. Each subcarrier has nine possible modulation states. This signal modulation format is known as *redundant digital file transfer* (RDFT) developed by Barry Sanderson, KB9VAK.

Most digital SSTV transmission has switched to using Digital Radio Mondiale (DRM), derived from a system developed for shortwave digital voice broadcasting. The DRM digital SSTV signal occupies the bandwidth between 350 and 2750 Hz. As many as 57 subcarriers may be sent simultaneously, all at the same level. Three pilot carriers are sent at twice the level of the other subcarriers. The subcarriers are modulated using OFDM and QAM, which were described earlier in this chapter. DRM SSTV includes several methods of error correction.

8.5 Modulation Impairments

Most of the previous discussion of the various modulation types has assumed the modulation is perfect. With analog modulation, that means the audio or video modulating signal is perfectly reproduced in the RF waveform without distortion, spurious frequencies or other unwanted artifacts. With digital modulation, the symbol timing and the locations and trajectories in the I/Q constellation are perfectly accurate. In all cases, the RF power amplifier is perfectly linear, if so required by the modulation type, and it introduces no noise or other spurious signals close to the carrier frequency.

In the real world, of course, such perfection can never be achieved. Some modulation impairments are caused by the transmitting system, some by the transmission medium through which the signal propagates, and some in the receiving system. This section will concentrate on impairments caused by the circuitry in the transmitter and, to some extent, in the receiver. Signal impairments due to propagation are covered in detail in the **Propagation of Radio Signals** chapter.

8.5.1 Intermodulation Distortion

To minimize distortion of the modulation, each stage in the signal chain must be linear, from the microphone or modem, through all the intermediate amplifiers and processors, to the modulator itself. For the linear modulation types, all the amplifiers and other stages between the modulator and the antenna must be linear as well.

If the modulation consists of a single sine wave, then nonlinearity causes only harmonic distortion, which produces new frequencies at integer multiples of the sine-wave frequency. If multiple frequencies are present in the modulation, however, then *intermodulation distortion* (IMD) products are produced. IMD occurs when a nonlinear amplifier or other device acts as a mixer, producing sum and difference frequencies of all the pairs of frequencies and their harmonics. For example if two frequencies, F1 and F2 > F1, are present, then IMD will cause spurious frequencies to appear at F1 + F2, F2 − F1, 2F1, 2F2, 2F1 − F2, 2F2 − F1, 2F2 − 2F1, 3F1, 3F2, and so on. *Odd-order* products are those that include the original frequencies an odd number of times, such as 3F1, 2F1 + F2, 3F1 − 2F2, and so on. *Even-order* products contain an even number of the original frequencies, such as 2F2, F1 + F2, 3F1 + F2, and so on. If more than two frequencies are present in the undistorted modulation, then the number of unwanted frequencies increases exponentially.

Although intermodulation distortion that occurs before modulation is fundamentally the same as IMD that occurs after the modulator (at the intermediate or radio frequency), the effects are quite different. Consider two frequency components of a modulating signal at, for example, 1000 Hz and 1200 Hz that modulate an SSB transmitter tuned to 14.000 MHz, USB. The desired RF signal has components at 14.001 and 14.0012 MHz. If the IMD occurs before modulation, then the F1 + F2 distortion product occurs at 1000 + 1200 = 2200 Hz. Since that is well within the audio passband of the SSB transmitter, there is no way to filter it out so it shows up at 14.0022 MHz at the RF output. However, if the distortion had occurred after the modulator, the F1 + F2 product would be at 14.001 + 14.0012 = 28.0022 MHz which is easily filtered out by the transmitter's low-pass filter.

That explains why RF speech processors work better than audio processors. A speech processor clips or limits the peak amplitude of the modulating signal to prevent over-modulation in the transmitter. However, the limiting process typically produces considerable intermodulation distortion. If the limiter is located after the modulator rather than before it, then it is an RF signal being clipped, rather than audio. If the RF limiter is followed by a band-pass filter then many of the distortion products are removed, resulting in a less-distorted signal.

If the distortion is perfectly symmetrical (for example equal clipping of positive and negative peaks) then only odd-order products are produced. Most distortion is not symmetrical so that both even and odd-order products appear. However, the odd-order products are of particular interest when measuring the linearity of an RF power amplifier. The reason is that even-order products that occur after the modulator occur only near harmonics of the RF frequency where they are easy to filter out. Odd-order products can fall within the desired channel, where they cause distortion of the modulation, or at nearby frequencies, where they cause interference to other stations.

The informal term for such IMD that interferes with other stations outside the desired channel is *splatter*. It becomes severe if the linear amplifier is over-driven, which causes clipping of the modulation envelope with the resulting odd-order IMD products.

The method commonly used to test the linearity of an SSB transmitter or RF power amplifier is the *two-tone test*. Two equal-amplitude audio-frequency tones are fed into the microphone input, the transmitter and/or amplifier is adjusted for the desired power level, and the output signal is observed on a spectrum analyzer. See **Fig 8.32**. The third-order products that occur near the desired signal are at 2F1 − F2 and 2F2 − F1. The fifth-order products occur at 3F1 − 2F2 and 3F2 − 2F1. If, for example, the two tones are spaced 1 kHz apart, then the two third-order products show up at 1 kHz above the high tone and 1 kHz below the low tone. The fifth-order products are at 2 kHz above and below the high and low tones respectively.

8.5.2 Transmitted Bandwidth

We have already discussed the necessary bandwidth for each of the various modulation types. The previous section explained how intermodulation distortion is one phenomenon that can cause unwanted emissions

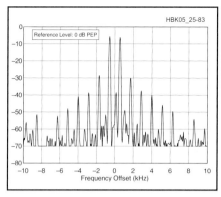

Fig 8.32 — Intermodulation products from an SSB transmitter. The two signals at the center are the desired frequencies. The frequencies of the third-order products are separated from the frequencies of the desired tones by the tone spacing. The fifth-order frequencies are separated by two times the tone spacing, and so on for higher-order products.

outside of the desired communications channel. Another is failing to properly low-pass filter the modulating signal to the minimum necessary bandwidth. That is especially a concern for linear modulation modes such as SSB, AM, OOK (CW), BPSK and QAM. For angle-modulated modes like FM and FSK, excessive bandwidth can result from simply setting the deviation too high.

There are other modulation impairments that cause emissions outside the desired bandwidth. For example, in an SSB transmitter, if the unwanted sideband is not sufficiently suppressed, the occupied bandwidth is up to twice as large as it should be. Also, an excessively strong suppressed carrier causes particularly-annoying heterodyne interference to stations tuned near that frequency. In some SSB modulators, there is an adjustment provided to optimize carrier suppression.

The carrier suppression may be degraded by a modulating signal that is too low in amplitude. For example, if the signal from the microphone to an SSB transmitter is one-tenth (−20 dB) of the proper amplitude, and if the gain of the RF amplifier stages is increased to compensate, then the carrier suppression is degraded by 20 dB.

The term *adjacent-channel power* (ACP) refers to the amount of transmitted power that falls into an adjacent communications channel above or below the desired channel. Normally the unwanted out-of channel power is worse for the immediately-adjacent channels than for those that are two channels away, the so-called *alternate* channels. ACP is normally specified as a power ratio in dB. It is measured with a spectrum analyzer that can measure the total power within the desired channel and the total power in an adjacent channel, so that the dB difference can be calculated.

The *occupied bandwidth* is the bandwidth within which a specified percentage of the total power occurs. A common percentage used is 99%. For a properly-adjusted, low-distortion transmitting system, the occupied bandwidth is determined mainly by the modulation type and filtering and, in the case of digital modulation, the symbol rate. For example, the IS-54 TDMA format that has been used in some US digital cellular networks has about 30 kHz occupied bandwidth using 24.3-kilosymbols/sec, π/4-DQPSK modulation with a 0.35-alpha root-raised-cosine filter. The GSM cellular standard requires about a 350-kHz occupied bandwidth for its 270.833-kilosymbols/sec, 0.3 Gaussian-filtered MSK signal.

8.5.3 Modulation Accuracy

For analog modes, modulation accuracy is mainly a question of maintaining the proper frequency response across the desired bandwidth with minimal distortion and unwanted signal artifacts. In-band artifacts like noise and spurious signals should not be a problem with any reasonably-well-designed system. Maintaining modulation peaks near 100% for AM signals or the proper deviation for FM signals is facilitated by an audio compressor. It can be either the type that uses a detector and an automatic-gain-control feedback loop to vary the gain in the modulation path or a clipper-type compressor that limits the peak amplitude and then filters the clipped signal to remove the harmonics and intermodulation products that result. SSB transmitters can also use audio speech compression to maintain the proper peak power level although, as explained previously, clipping of the signal before it reaches the modulator can cause unacceptable distortion unless special techniques are used.

For digital signals, there are a number of other possible sources of modulation inaccuracy. For modes that use Nyquist filtering, the cutoff frequency and filter shape must be accurate to ensure no inter-symbol interference (ISI). Fortunately, that is easy to do with digital filters. However any additional filtering in the signal path can degrade the ISI. For example, most HF digital modes use an analog SSB transceiver to up-convert the signal from audio to RF for transmission and to down-convert from RF to audio again at the receiver end. The crystal filters used in the transmitter and receiver can significantly degrade group delay flatness, especially near the edges of the filter passband. That is why most HF digital modes use a bandwidth substantially less than a typical transceiver's passband and attempt to center the signal near the crystal filters' center frequency.

Distortion can also impair the proper

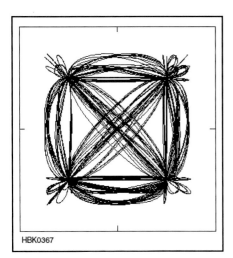

Fig 8.33 — Simulated I/Q constellation display of the trajectory of a QPSK signal over an extended period. A 0.5-alpha raised-cosine filter was used. Because it is a Nyquist filter, the trajectories pass exactly through each symbol location.

decoding of digital signals, especially formats with closely-spaced symbols such as 256-QAM. Any "flat-topping" in the final amplifier causes the symbols at the outermost corners of the constellation to be closer together than they should be. The accuracy of the symbol clock is critical in formats like JT65 that integrate the detection process over a large number of symbols. Perhaps surprisingly, the clock rate on some inexpensive computer sound cards can be off on the order of a percent, which results in a similar error in symbol rate.

The modulation accuracy of an FSK signal is normally characterized by the frequency shift at the center of each symbol time, the point at which the receiver usually makes the decision of which symbol is being received. For other digital formats, the modulation accuracy is normally characterized by the amplitude and phase at the symbol decision points. Amplitude error is typically measured as a percentage of the largest symbol amplitude. The phase error is quoted in degrees. In both cases one can specify either the average RMS value or the peak error that was detected over the measurement period. Amplitude error is the most important consideration for modulation types such as BPSK where the information is encoded as an amplitude. For a constant-envelope format like MSK, phase error is a more important metric.

For formats like QPSK and QAM, both amplitude and phase are important in determining symbol location. A measurement that includes both is called *error vector magnitude* (EVM). It is the RMS average distance between the ideal symbol location in the constellation diagram and the actual signal value at the symbol decision point, expressed as a percentage of either the RMS signal power or the maximum symbol amplitude. This is a measurement that requires specialized test equipment not generally available to home experimenters. However it is possible to estimate the effect on EVM of various design choices using computer simulations.

Previously, we have plotted constellation diagrams with the transitions between symbol locations indicated by straight lines. In an actual system, both the I and Q signals are low-pass filtered which makes the symbol transitions smoother, with no abrupt changes of direction at the symbol locations. **Fig 8.33** shows the actual symbol trajectories for a QPSK signal using a 0.5-alpha raised-cosine filter with random data. Since a raised-cosine is a type of Nyquist filter, the trajectories pass exactly through each symbol location. This is what you would see if you connected the I and Q outputs from a QPSK baseband generator to the X (horizontal) and Y (vertical) inputs of an oscilloscope. It is also what you would see in a receiver at the input to the symbol detector after the carrier and clock-recovery circuits have stabilized the signal. If the symbol trajectories were not accurate, then the dark areas at the symbol points would be less distinct and more spread out. Such a constellation diagram is also a good troubleshooting tool for other purposes, for example to check for amplitude distortion or to see if a faulty symbol encoder is missing some symbols or placing them at the wrong location.

In a system where a root-Nyquist filter is used in both the transmitter and receiver to obtain a net Nyquist response, the transmitter output will not display sharply defined symbol locations as in the figure. In that case, the measuring instrument must supply the missing root-Nyquist filter that is normally in the receiver, in order to obtain a clean display. Professional modulation analyzers normally include a means of selecting the proper matching filter.

Another way to view modulation accuracy is with an *eye diagram*. See **Fig. 8.34**. In

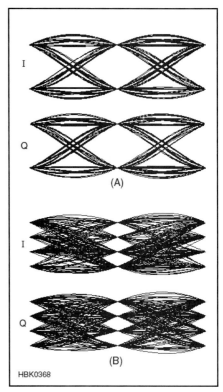

Fig 8.34 — Eye diagrams of the I and Q outputs of a QPSK generator (A) and a 16-QAM generator (B). The "eye" is the empty area between the symbol decision points, visible as the points where all the symbol trajectories come together. The bigger the "eye" the easier the signal is to decode.

this case the oscilloscope's horizontal axis is driven by its internal sweep generator, triggered by the symbol clock. The two vertical channels of the oscilloscope are connected to the I and Q outputs of the baseband generator. The "eye" is the empty area at the center of the I and Q traces in between the symbol decision points, where all the traces come together. The eye should be as "open" as possible to make the job of the receiver's symbol decoder as easy as possible. The oscilloscope would typically be set for infinite persistence, so that the worst-case excursions from the ideal symbol trajectory are recorded. QPSK has four symbol locations, one in each quadrant, such that I and Q each only have two possible values and there is only one "eye" for each. 16-QAM has 16 symbol locations with four possible locations for I and Q, which forms three "eyes."

8.6 Modulation Glossary

GENERAL

AM — Amplitude modulation.

Angle modulation — Modulation of an RF carrier by varying the phase or frequency.

Emission designator — Official ITU code to specify the bandwidth and modulation type of a radio transmission.

Frequency diversity — The use of a wideband signal to compensate for selective fading. While one band of frequencies is faded, the data can be reconstructed from other frequencies that are not faded.

FM — Frequency modulation.

IMD — Intermodulation distortion. Unwanted frequencies that occur at the sum and difference of the desired frequencies and their harmonics.

ITU — International Telecommunications Union. An agency of the United Nations that coordinates and recommends technical standards for electronic communications.

Modulation — The periodic alteration of some parameter of a carrier wave in order to transmit information.

PM — Phase modulation.

Protocol — A formal set of rules and procedures for the exchange of information within a communications network.

Selective fading — A propagation phenomenon in which closely-spaced frequencies experience markedly different fading at the same time.

Telemetry — The use of telecommunication for automatically indicating or recording measurements at a distance from the measuring instrument.

Telephony — A form of telecommunication primarily intended for the exchange of information in the form of speech.

Telegraphy — A form of telecommunication in which the transmitted information is intended to be recorded on arrival as a graphic document; the transmitted information may sometimes be presented in an alternative form or may be stored for subsequent use.

Television — A form of telecommunication for the transmission of transient images of fixed or moving objects.

ANALOG MODULATION (INCLUDING IMAGE)

ATV — Amateur fast-scan television.

BFO — Beat frequency oscillator. In an SSB or DSBSC receiver, the intermediate-frequency oscillator in the receiver that re-inserts the suppressed carrier.

Carson's rule — A rule of thumb to calculate FM bandwidth that says that 98% of the energy is typically contained within a bandwidth equal to two times the sum of the peak frequency deviation and the highest modulating frequency.

Capture effect — The tendency of the strongest signal to suppress other signals in an FM detector, which improves the signal-to-noise and signal-to-interference ratio.

Deviation ratio — The ratio of the maximum permitted peak deviation of an angle-modulated signal to the maximum permitted modulating frequency.

DSBSC — Double sideband, suppressed carrier. An AM signal in which the carrier has been removed but both sidebands remain.

Fax — Facsimile. The sending of still images by wire or by radio.

Frame — In television, one complete scanned image. On systems with interlaced scanning, there are two vertical scans per frame.

Frequency deviation — The amount the RF frequency of an angle-modulated signal deviates from the center (carrier) frequency in response to the modulating signal. The term is often understood to mean the maximum deviation available in a given system.

Limiter — A high-gain amplifier in an FM receiver that limits the peak amplitude of the signal in order to eliminate any AM component.

LSB — Lower sideband. An SSB signal with the upper sideband removed.

Martin — A series of analog SSTV formats, especially popular in Europe.

Modulation index (AM) — The ratio of the peak value of the modulation of an AM signal to the value that just causes the modulation envelope of the RF signal to reach zero on negative peaks and twice the average value on positive peaks.

Modulation index (FM) — The ratio of the peak deviation of an angle-modulated signal to the highest modulating frequency.

Narrowband FM — An FM signal with a modulation index less than or equal to 1.0.

NTSC — National Television System Committee. The analog television standard used in the US, Japan and several other countries.

PAL — Phase alteration line. The analog television standard used in many parts of Europe.

Phasing method — A method of generating an SSBSC signal that does not require a filter to remove the unwanted sideband.

Pixel — Picture element. The dots that make up images on a computer's monitor.

Product detector — A detector that multiplies a BFO signal with the received signal, typically SSB or DSBSC-modulated.

RGB — Red, green and blue. The three primary colors required to transmit a full-color image in many television and facsimile systems.

Scottie — A series of analog SSTV formats, especially popular in the US.

SSB — Single sideband. An AM signal in which one sideband has been removed. The term is usually understood to mean SSBSC.

SSBSC — Single sideband, suppressed carrier. An AM signal in which one sideband and the carrier have been removed.

SSTV — Slow scan TV, a system for sending and receiving still images.

Sync — Modulation pulses used in ATV and SSTV to synchronize the horizontal and/or vertical scanning.

Synchronous detector — A type of AM detector in which the carrier is regenerated in the receiver.

Two-tone test — A procedure for testing the IMD of an SSB transmitter. Two equal-amplitude tones are fed to the microphone input and the transmitter RF output is examined with a spectrum analyzer to determine the amplitudes of the IMD products.

USB — Upper sideband. An SSB signal with the lower sideband removed.

Vestigial sideband — The filtering of all but the bottom MHz or so of the lower sideband of an ATV or NTSC television signal.

VIS — Vertical interval signaling. Digital encoding of the transmission mode during the vertical sync interval of an SSTV frame.

Wideband FM — An FM signal with a modulation index greater than 1.0.

DIGITAL MODULATION

AFSK — Audio FSK. The use of an SSB transceiver to transmit and receive FSK using an audio-frequency modem.

ASK — Amplitude-shift keying. Digital amplitude modulation in which the

amplitude depends on the modulating code.
Baud rate — The rate at which a digital signal transitions between symbol states. Symbol rate.
Bauds — Symbols per second. Unit of baud rate or symbol rate.
Bit rate — The total number of physically transferred bits per second over a communication link. Bit rate can be used interchangeably with **baud rate** only when each modulation transition carries exactly one bit of data.
BPSK — Binary PSK. PSK with only two possible states. The term is usually understood to mean a non-constant-envelope signal in which the two states differ by 180°.
Chirp — Incidental frequency modulation of a carrier as a result of oscillator instability during keying.
CW — Continuous wave. The term used for on-off keying using Morse code.
Constellation diagram — A diagram showing the constellation of possible symbol locations on a polar plot of modulation amplitude and phase.
DBPSK — Differential BPSK.
Decision point — The point, typically in the center of a symbol time, at which the receiver decides which symbol is being sent.
Differential modulation — A modulation technique that encodes the information in the difference between subsequent symbols, rather than in the symbols themselves.
Equalization — Correction for variations in amplitude and/or phase versus frequency across a communications channel.
DQPSK — Differential QPSK.
Eye diagram — An oscilloscope measurement of a digital modulating signal with the horizontal sweep synchronized to the symbol times. With Nyquist filtering, there should be a clear separation, or "eye", in the trajectories at the symbol decision points.
EVM — Error vector magnitude. A measure of the RMS error in the symbol locations at the symbol decision points in the constellation plot of a digital signal.
FFT — Fast Fourier transform. The Fourier transform is a mathematical function that calculates the frequency spectrum of a signal. The FFT is a software algorithm that does the calculations very efficiently.
FSK — Frequency-shift keying. A form of digital frequency modulation in which the frequency deviation depends on the modulating data.
GMSK — Gaussian MSK. MSK with a Gaussian modulation filter.
I/Q modulation — Quadrature modulation implemented with an I/Q modulator, one that uses in-phase (I) and quadrature (Q) modulating signals to generate the zero-degree and 90° components of the RF signal.
ISI — Inter-symbol interference. Interference of a signal with itself, caused when energy from one symbol is delayed long enough to interfere with a subsequent symbol.
MFSK — Multi-level FSK. FSK with more than two states represented by different frequency deviations.
Modem — Modulator/demodulator. A device that generates and demodulates digital modulation signals, usually at audio frequencies. It connects between the data terminal (usually a computer) and the radio.
MSK — Minimum-shift keying. A form of FSK with a frequency shift equal to one-half the symbol rate.
Nyquist criterion — A principle that states that the sampling frequency must be greater than twice the highest frequency in the sampled signal.
Nyquist frequency — One-half the sampling frequency.
OFDM — Orthogonal frequency-division multiplexing. A transmission mode that uses multiple carriers, spaced such that modulation on each carrier is orthogonal with the others.
OOK — On-off keying. A type of ASK with only two states, on and off.
OQPSK — Offset QPSK. By offsetting in time the symbol transitions of the I and Q channels, symbol trajectories through the origin are eliminated.
Orthogonal — Refers to streams of data that are uncorrelated with each other such that there is no mutual interference.
PAM — Pulse amplitude modulation. A type of pulse modulation where the modulating signal is encoded in the pulse amplitude.
π/4 DQPSK — PI-over-four differential QPSK. A form of differential 8PSK in which the only allowed transitions are ±45 and ±135°, resulting in four allowed states per symbol.
PPM — Pulse position modulation. A type of pulse modulation where the modulating signal is encoded in the pulse position.
PSK — Phase-shift keying. A form of digital phase modulation in which the phase of the RF signal depends on the modulating code. The term often is understood to refer to BPSK.
Pulse modulation — The modulating signal is sampled at regular intervals to generate a series of modulation symbols in the form of pulses.
PWM — Pulse width modulation. A type of pulse modulation where the modulating signal is encoded in the pulse width.
QAM — Quadrature amplitude modulation. A digital modulation type in which both amplitude and phase are varied. The number that precedes it, for example 64QAM, is the number of different possible states of the amplitude and phase.
QPSK — PSK with four possible states. The term is usually understood to be equivalent to four-level QAM, in which the four states have the same amplitude and differ in phase by 90°.
Quadrature modulation — Refers to modulation using two RF carriers in phase quadrature, that is, 90° out of phase.
Symbol rate — The rate at which a digital signal transitions between different states. Baud rate.
Varicode — A coding method in which the length of each character code depends on its frequency of occurrence. It is used to optimize the ratio of characters per second to baud rate.

SPREAD SPECTRUM

CDMA — Code-division multiple access. A method of allowing several stations to use the same frequency band simultaneously by assigning each station a different orthogonal spreading code.
Chip — The shortest-duration modulation state in a SS system.
De-spreading — Conversion of a SS signal back to its narrowband equivalent by convoluting it with the spreading code.
DSSS — Direct-sequence SS. Spreading of a signal by multiplying the data stream by a higher-rate pseudo-random digital sequence.
FHSS — Frequency-hopping SS. Spreading of a signal by means of pseudo-random frequency-hopping of the unspread signal.
Processing gain — The increase in signal-to-noise ratio that occurs when an SS signal is de-spread.
SS — Spread spectrum. A system that intentionally increases the bandwidth of a digital signal by means of a special spreading code.

FILTERING AND BANDWIDTH

ACP — Adjacent channel power. The amount of transmitted power that falls into a communications channel immediately adjacent to the desired one.
Alpha — A design parameter for a Nyquist or root Nyquist filter. The smaller the alpha, the sharper the passband-

- to-stopband transition at the cutoff frequency.
- **Cutoff frequency** — The frequency at which a filter response changes from the passband to the stopband.
- **Gaussian filter** — A modulation filter with a Gaussian frequency response. The transition between passband and stopband is more gradual than with most Nyquist filters.
- **Key clicks** — Out-of-channel interference from an OOK signal caused by too-sharp transitions between the on and off states.
- **Nyquist filter** — A filter that causes no inter-symbol interference (ISI). It is so-called because the cutoff frequency is at one-half the symbol rate, the Nyquist frequency.
- **Nyquist criterion** — The rule that states that the sample rate must be greater than twice the highest frequency to be sampled.
- **Occupied bandwidth** — The bandwidth within which a specified percentage of the total power occurs, typically 99%.
- **Raised-cosine filter** — A type of Nyquist filter whose passband-to-stopband transition region has the shape of the first half-cycle of a cosine raised so that the negative peak is at zero.
- **Root Nyquist filter** — A filter that, when cascaded with another identical filter, forms a Nyquist filter. The frequency response is the square root of that of the Nyquist filter.
- **Root raised-cosine filter** — A root Nyquist filter with a frequency response that is the square root of a raised-cosine filter.
- **Shannon-Hartley theorem** — A formula that predicts the maximum channel capacity that is theoretically possible over a channel of given bandwidth and signal-to-noise ratio.
- **Sinc function** — The spectrum of a pulse or random series of pulses, equal to sin(x)/x.
- **Splatter** — Out-of-channel interference from an amplitude-modulated signal such as SSB or QPSK caused by distortion, typically in the power amplifier stages.

8.7 References and Further Reading

Agilent Technologies, "Digital Modulation in Communications Systems — An Introduction" Application Note 1298. **cp.literature.agilent.com/litweb/pdf/5965-7160E.pdf**

Costas, J. P., "Poisson, Shannon, and the Radio Amateur," *Proceedings of the IRE*, vol 47, pp 2058-2068, Dec 1959.

Ford, S., WB8IMY, *ARRL's HF Digital Handbook* (ARRL, 2008)

Ford, S., WB8IMY, *ARRL's VHF Digital Handbook* (ARRL, 2008)

Haykin, S., *Digital Communications* (Wiley, 1988)

Langner, J. WB2OSZ, "SSTV Transmission Modes," **www.comunicacio.net/digigrup/ccdd/sstv.htm**

McConnell, Bodson, and Urban, *FAX: Facsimile Technology and Systems* (Artech House 1999)

Nagle, J., K4KJ, "Diversity Reception: an Answer to High Frequency Signal Fading," *Ham Radio*, Nov 1979, pp 48-55.

Sabin, W., et al, *Single-Sideband Systems and Circuits* (McGraw Hill, 1987)

Shavkoplyas, A., VE3NEA, "CW Shaping in DSP Software," *QEX*, May/Jun 2006, pp 3-7.

Seiler, T., HB9JNX/AE4WA, et al, "Digital Amateur TeleVision (D-ATV)," proc. 2001 ARRL/TAPR Digital Communications Conference, **www.baycom.org/~tom/ham/dcc2001/datv.pdf**

Silver, H. W., NØAX, "About FM," *QST*, Jul 2004, pp 38-42.

Smith, D., KF6DX, "Distortion and Noise in OFDM Systems," *QEX*, Mar/Apr 2005, pp 57-59.

Smith, D., KF6DX, "Digital Voice: The Next New Mode?," *QST*, Jan 2002, pp 28-32.

Stanley, J., K4ERO, "Observing Selective Fading in Real Time with Dream Software," *QEX*, Jan/Feb 2007, pp 18-22.

Taggart, R., WB8DQT, "An Introduction to Amateur Television," *QST*, Apr, May and Jun 1993.

Taggart, R., WB8DQT, *Image Communications Handbook* (ARRL, 2002)

Taggart, R., WB8DQT, "Digital Slow-Scan Television," *QST*, Feb 2004, pp 47-51.

Van Valkenburg, M., et al, *Reference Data for Engineers: Radio, Electronics, Computer and Communications*, chapters 23, 24 and 25, 9th Ed. (Newnes, 2001)

Contents

9.1 How Oscillators Work
    9.1.1 Maintained Resonance
    9.1.2 Amplification
    9.1.3 Start-Up

9.2 Phase Noise
    9.2.1 Effects of Phase Noise
    9.2.2 Reciprocal Mixing
    9.2.3 A Phase Noise Demonstration
    9.2.4 Measuring Receiver Phase Noise
    9.2.5 Measuring Oscillator and Transmitter Phase Noise
    9.2.6 Low-Cost Phase Noise Testing

9.3 Oscillator Circuits and Construction
    9.3.1 LC Oscillator Circuits
    9.3.2 RC Oscillators
    9.3.3 Three High-Performance HF VFOs
    9.3.4 VFO Components and Construction
    9.3.5 Temperature Compensation
    9.3.6 Shielding and Isolation

9.4 Designing an Oscillator

9.5 Quartz Crystals in Oscillators
    9.5.1 Quartz and the Piezoelectric Effect
    9.5.2 Frequency Accuracy
    9.5.3 The Equivalent Circuit of a Crystal
    9.5.4 Crystal Oscillator Circuits
    9.5.5 VXOs
    9.5.6 Logic-Gate Crystal Oscillators

9.6 Oscillators at UHF and Above
    9.6.1 UHF Oscillators
    9.6.2 Microwave Oscillators
    9.6.3 Klystrons, Magnetrons and Traveling Wave Tubes

9.7 Frequency Synthesizers
    9.7.1 Phase-Locked Loops (PLL)
    9.7.2 A PLL Design Example
    9.7.3 PLL Measurements
    9.7.4 Common Problems in PLL Design
    9.7.5 Synthesizer ICs
    9.7.6 Improving VCO Noise Performance
    9.7.7 Microprocessor Control of Synthesizers
    9.7.8 The Role of DDS in Recent Hybrid Architectures
    9.7.9 Multi-Phase Division

9.8 Present and Future Trends in Oscillator Application

9.9 Glossary of Oscillator and Synthesizer Terms

9.10 References and Bibliography

Chapter 9

Oscillators and Synthesizers

Just say in public that oscillators are one of the most important, fundamental building blocks in radio technology and you will immediately be interrupted by someone pointing out that tuned-RF (TRF) receivers can be built without any form of oscillator at all. This is certainly true, but it shows how some things can be taken for granted. What use is any receiver without signals to receive? All intentionally transmitted signals trace back to some sort of signal generator — an oscillator or frequency synthesizer. In contrast with the TRF receivers just mentioned, a modern, all-mode, feature-laden MF/HF transceiver may contain in excess of a dozen RF oscillators and synthesizers, while a simple QRP CW transmitter may consist of nothing more than a single oscillator.

This chapter, which includes material contributed by David Stockton, GM4ZNX, Frederick J. Telewski, WA7TZY, and Jerry DeHaven, WA0ACF, explores the design, application and performance characteristics of various types of oscillators and frequency synthesizers. Ulrich Rohde, N1UL, contributed two appendices on phase noise and oscillator design which may be found on the *Handbook* CD-ROM.

In the 1980s, the main area of progress in the performance of radio equipment was the recognition of receiver intermodulation as a major limit to our ability to communicate, with the consequent development of receiver front ends with improved ability to handle large signals. So successful was this campaign that other areas of transceiver performance now required similar attention. One indication of this is any equipment review receiver dynamic range measurement qualified by a phrase like "limited by oscillator phase noise." A plot of a receiver's effective selectivity can provide another indication of work to be done: An IF filter's high-attenuation region may appear to be wider than the filter's published specifications would suggest — almost as if the filter characteristic has grown sidebands! In fact, in a way, it has: This is the result of local-oscillator (LO) or synthesizer *phase noise* spoiling the receiver's overall performance.

The sheer number of different oscillator circuits can be intimidating, but their great diversity is an illusion that evaporates once their underlying pattern is seen. Almost all RF oscillators share one fundamental principle of operation: an amplifier and a filter operate in a loop (**Fig 9.1**). There are plenty of filter types to choose from:

• LC
• Quartz crystal and other piezoelectric materials
• Transmission line (stripline, microstrip, open-wire, coax, and so on)
• Microwave cavities, YIG spheres, dielectric resonators
• Surface-acoustic-wave (SAW) devices

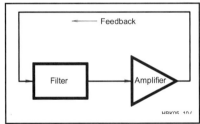

Fig 9.1 — Reduced to essentials, an oscillator consists of a filter and an amplifier operating in a feedback loop.

Should any new forms of filter be invented, it's a safe guess that they will also be applicable to oscillators. There is an equally large range of amplifiers to choose from:

• Vacuum tubes of all types
• Bipolar junction transistors
• Field effect transistors (JFET, MOSFET, GaAsFET, in all their varieties)
• Gunn diodes, tunnel diodes and other negative-resistance generators

It seems superfluous to state that anything that can amplify can be used in an oscillator, because of the well-known propensity of all prototype amplifiers to oscillate! The choice of amplifier is widened further by the option of using single- or multiple-stage amplifiers and discrete devices versus integrated circuits. Multiply all of these options with those of filter choice and the resulting set of combinations is very large, but a long way from complete. Then there are choices of how to couple the amplifier into the filter and the filter into the amplifier. And then there are choices to make in the filter section: Should it be tuned by variable capacitor, variable inductor or some form of sliding cavity or line?

Despite the number of combinations that are possible, a manageably small number of types will cover all but very special requirements. Look at an oscillator circuit and "read" it: What form of filter — *resonator* — does it use? What form of amplifier? How have the amplifier's input and output been coupled into the filter? How is the filter tuned? These are simple, easily answered questions that put oscillator types into appropriate categories and make them understandable. The questions themselves may make more sense if we understand the mechanics of oscillation, in which *resonance* plays a major role.

9.1 How Oscillators Work

9.1.1 Maintained Resonance

The pendulum, a good example of a resonator, has been known for millennia and understood for centuries. It is closely analogous to an electronic resonator, as shown in **Fig 9.2**. The weighted end of the pendulum can store energy in two different forms: The *kinetic energy* of its motion and the *potential energy* of it being raised above its rest position. As it reaches its highest point at the extreme of a swing, its velocity is zero for an instant as it reverses direction. This means that it has, at that instant, no kinetic energy, but because it is also raised above its rest position, it has some extra potential energy.

At the center of its swing, the pendulum is at its lowest point with respect to gravity and so has lost the extra potential energy. At the same time, however, it is moving at its highest speed and so has its greatest kinetic energy. Something interesting is happening: The pendulum's stored energy is continuously moving between potential and kinetic forms. Looking at the pendulum at intermediate positions shows that this movement of energy is smooth. Newton provided the keys to understanding this. It took his theory of gravity and laws of motion to explain the behavior of a simple weight swinging on the end of a length of string and calculus to perform a quantitative mathematical analysis. Experiments had shown the period of a pendulum to be very stable and predictable. Apart from side effects of air drag and friction, the length of the period should not be affected by the mass of the weight, or by the amplitude of the swing.

A pendulum can be used for timing events, but its usefulness is spoiled by the action of drag or friction, which eventually stops it. This problem was overcome by the invention of the *escapement*, a part of a clock mechanism that senses the position of the pendulum and applies a small push in the right direction and at the right time to maintain the amplitude of its swing or oscillation. The result is a mechanical oscillator: The pendulum acts as the filter, the escapement acts as the amplifier and a weight system or wound-up spring powers the escapement.

Electrical oscillators are closely analogous to the pendulum, both in operation and in development. The voltage and current in the tuned circuit — often called *tank circuit* because of its ability to store energy — both vary sinusoidally with time and are 90° out of phase. There are instants when the current is zero, so the energy stored in the inductor must be zero, but at the same time the voltage across the capacitor is at its peak, with all of the circuit's energy stored in the electric field between the capacitor's plates. There

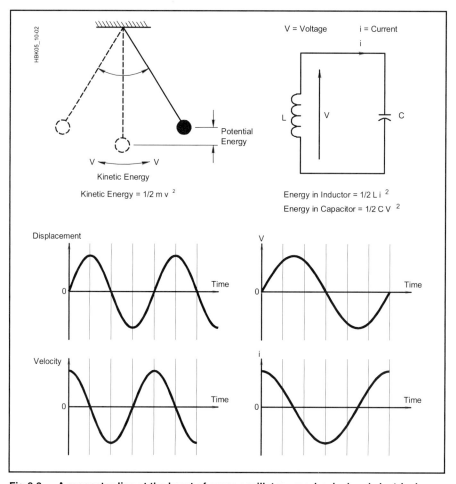

Fig 9.2 — A resonator lies at the heart of every oscillatory mechanical and electrical system. A mechanical resonator (here, a pendulum) and an electrical resonator (here, a tuned circuit consisting of L and C in parallel) share the same mechanism: the regular movement of energy between two forms — potential and kinetic in the pendulum, electric and magnetic in the tuned circuit. Both of these resonators share another trait: Any oscillations induced in them eventually die out because of losses — in the pendulum, due to drag and friction; in the tuned circuit, due to the resistance, radiation and inductance. Note that the curves corresponding to the pendulum's displacement vs velocity and the tuned circuit's voltage vs current, differ by one-quarter of a cycle, or 90°.

are also instants when the voltage is zero and the current is at a peak, with no energy in the capacitor. Then, all of the circuit's energy is stored in the inductor's magnetic field.

Just like the pendulum, the energy stored in the electrical system is swinging smoothly between two forms; electric field and magnetic field. Also like the pendulum, the tank circuit has losses. Its conductors have resistance, and the capacitor dielectric and inductor core are imperfect. Leakage of electric and magnetic fields also occurs, inducing currents in neighboring objects and just plain radiating energy off into space as radio waves. The amplitudes of the oscillating voltage and current decrease steadily as a result. Early intentional radio transmissions, such as those of Heinrich Hertz's experiments, involved abruptly dumping energy into a tuned circuit and letting it oscillate, or *ring*, as shown in **Fig 9.3**. This was done by applying a spark to the resonator. Hertz's resonator was a gapped ring, a good choice for radiating as much of the energy as possible. Although this looks very different from the LC tank of Fig 9.2, it has inductance *distributed* around its length and capacitance distributed across it and across its gap, as opposed to the *lumped* L and C values in Fig 9.2. The gapped ring therefore works just the same as the LC tank in terms of oscillating voltages and currents. Like the pendulum and the LC tank, its period, and therefore the frequency at which it oscillates, is independent of the magnitude of its excitation.

Making a longer-lasting signal with the Fig 9.3 arrangement merely involves repeating the sparks. The problem is that a truly

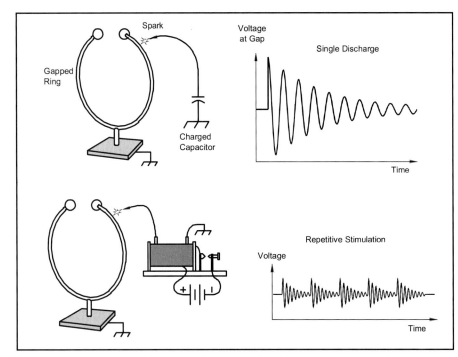

Fig 9.3 — Stimulating a resonance, 1880s style. Shock-exciting a gapped ring with high voltage from a charged capacitor causes the ring to oscillate at its resonant frequency. The result is a damped wave, each successive alternation of which is weaker than its predecessor because of resonator losses. Repetitively stimulating the ring produces trains of damped waves, but oscillation is not continuous.

continuous signal cannot be made this way. The sparks cannot be applied often enough or always timed precisely enough to guarantee that another spark re-excites the circuit at precisely the right instant. This arrangement amounts to a crude spark transmitter, variations of which served as the primary means of transmission for the first generation of radio amateurs. The use of damped waves is now entirely forbidden by international treaty because of their great impurity. Damped waves look a lot like car-ignition waveforms and sound like car-ignition interference when received.

What we need is a *continuous wave* (CW) oscillation — a smooth, sinusoidal signal of constant amplitude, without phase jumps, a "pure tone." To get it, we must add to our resonator an equivalent of the clock's escapement — a means of synchronizing the application of energy and a fast enough system to apply just enough energy every cycle to keep each cycle at the same amplitude.

9.1.2 Amplification

A sample of the tank's oscillation can be extracted, amplified and reinserted. The gain can be set to exactly compensate the tank losses and perfectly maintain the oscillation. The amplifier usually only needs to give low gain, so active devices can be used in oscillators not far below their unity-gain frequency. The amplifier's output must be lightly coupled into the tank — the aim is just to replace lost energy, not forcibly drive the tank. Similarly, the amplifier's input should not heavily load the tank. It is a good idea to think of *coupling* networks rather than *matching* networks in this application, because a matched impedance extracts the maximum available energy from a source, and this would certainly spoil an oscillator.

Fig 9.4A shows the block diagram of an oscillator. Certain conditions must be met for oscillation. The criteria that separate oscillator loops from stable loops are often attributed to Barkhausen by those aiming to produce an oscillator and to Nyquist by those aiming for amplifier stability, although they boil down to the same set of constraints for the circuit. Fig 9.4B shows the loop broken and a test signal inserted. (The loop can be broken anywhere; the amplifier input just happens to be the easiest place to do it.)

The *Barkhausen criterion* for oscillation says that at a frequency at which the phase shift around the loop is exactly zero, the net gain around the loop must equal or exceed unity (that is, one). (Later, when we design phase-locked loops, we must revisit this concept in order to check that those loops *cannot* oscillate.) Fig 9.4C shows what happens to the amplitude of an oscillator if the *loop gain* (total gain of all devices around the loop) is made a little higher or lower than one. (In many treatments of the Barkhausen criterion, the feedback block is shown as a filter. The resonator and coupling networks shown here perform the same function.)

The loop gain has to be precisely one if we want stable amplitude. Any inaccuracy will cause the amplitude to grow to clipping or shrink to zero, making the oscillator useless. Better accuracy will only slow, not stop this process. Perfect precision is clearly impossible, yet there are enough working oscillators in existence to prove that we are missing something important. In an amplifier, nonlinearity is a nuisance, leading to signal distortion and intermodulation, yet nonlinearity is what makes stable oscillation possible. All of the vacuum tubes and transistors used in oscillators tend to reduce their gain at higher signal levels. With such components, only a tiny change in gain can shift the loop's operation between amplitude growth and shrinkage. Oscillation stabilizes at that level at which the gain of the active device sets the loop gain at exactly one.

Another gain-stabilization technique involves biasing the device so that once some level is reached, the device starts to turn off over part of each cycle. At higher levels, it cuts off over more of each cycle. This effect reduces the effective gain quite strongly, stabilizing the amplitude. This badly distorts the signal (true in most common oscillator circuits) in the amplifying device, but provided the amplifier is lightly coupled to a high-Q resonant tank, the signal in the tank should not be badly distorted.

Textbooks give plenty of coverage to the frequency-determining mechanisms of oscillators, but the amplitude-determining mechanism is rarely covered. It is often not even mentioned. There is a good treatment in Clarke and Hess, *Communications Circuits: Analysis and Design*.

9.1.3 Start-Up

Perfect components don't exist, but if we could build an oscillator from them, we would naturally expect perfect performance. We would nonetheless be disappointed. We could assemble from our perfect components an oscillator that exactly met the criterion for oscillation, having slightly excessive gain that falls to the correct amount at the target operating level and so is capable of sustained, stable oscillation. But being capable of something is not the same as doing it, for there is another stable condition. If the amplifier in the loop shown in Fig 9.4A has no input signal, and is perfect, it will give no output! No signal returns to the amplifier's input via the resonator, and the result is a sustained and stable *lack* of oscillation. Something is needed to start the oscillator.

This fits the pendulum-clock analogy: A wound-up clock is stable with its pendulum at rest, yet after a push the system will sustain

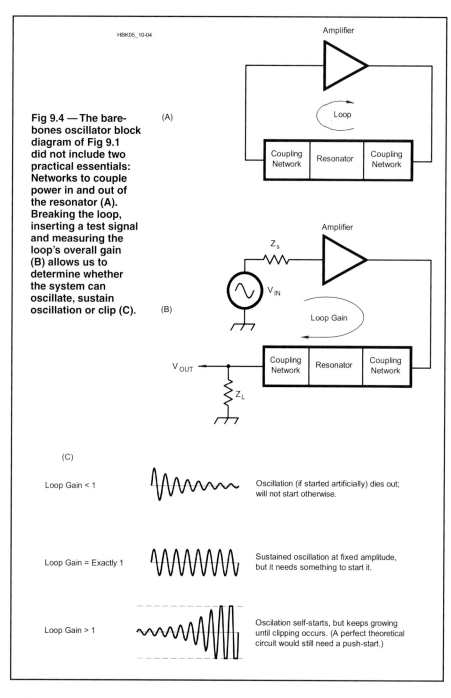

Fig 9.4 — The bare-bones oscillator block diagram of Fig 9.1 did not include two practical essentials: Networks to couple power in and out of the resonator (A). Breaking the loop, inserting a test signal and measuring the loop's overall gain (B) allows us to determine whether the system can oscillate, sustain oscillation or clip (C).

oscillation. The mechanism that drives the pendulum is similar to a Class C amplifier: It does not act unless it is driven by a signal that exceeds its threshold. An electrical oscillator based on a Class C amplifier can sometimes be kicked into action by the turn-on transient of its power supply. The risk is that this may not always happen and should some external influence stop the oscillator, it will not restart until someone notices the problem and cycles the power. This can be very inconvenient!

A real-life oscillator whose amplifier does not lose gain at low signal levels can self-start due to noise. **Fig 9.5** shows an oscillator block diagram with the amplifier's noise shown, for our convenience, as a second input that adds with the true input. The amplifier amplifies the noise. The resonator filters the output noise, and this signal returns to the amplifier input. The importance of having slightly excessive gain until the oscillator reaches operating amplitude is now obvious. If the loop gain is slightly above one, the re-circulated noise must, within the resonator's bandwidth, be larger than its original level at the input. More noise is continually summed in as a noise-like signal continuously passes around the loop, undergoing amplification and filtering as it does. The level increases, causing the gain to reduce. Eventually, it stabilizes at whatever level is necessary to make the net loop gain equal to one.

So far, so good — the oscillator is running at its proper level, but something seems very wrong. It is not making a proper sine wave; it is recirculating and filtering a noise signal. It can also be thought of as a "Q multiplier" — a controlled (high) gain amplifier, filtering a noise input and amplifying it to a set level. Narrow-band filtered noise approaches a true sine wave as the filter is narrowed to zero width. What this means is that we cannot make a true sine-wave signal — all we can do is make narrow-band filtered noise as an approximation. A high-quality, "low-noise oscillator is merely one that does tighter filtering. Even a kick-started Class-C-amplifier oscillator has noise continuously adding to its circulating signal, and so behaves similarly.

A small-signal gain greater than one is absolutely critical for reliable starting, but having too much gain can make the final operating level unstable. Some oscillators are designed around limiting amplifiers to make their operation predictable. AGC systems have also been used, with an RF detector and dc amplifier used to servo-control the amplifier gain. It is notoriously difficult to design reliable crystal oscillators that can be published or mass-produced without

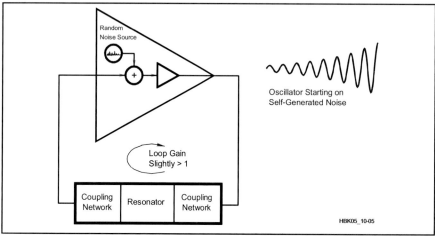

Fig 9.5 — An oscillator with noise. Real-world amplifiers, no matter how quiet, generate some internal noise; this allows real-world oscillators to self-start.

having occasional individuals refuse to start without some form of shock.

Mathematicians have been intrigued by "chaotic systems" where tiny changes in initial conditions can yield large changes in outcome. The most obvious example is meteorology, but much of the necessary math was developed in the study of oscillator start-up, because it is a case of chaotic activity in a simple system. The equations that describe oscillator start-up are similar to those used to generate many of the popular, chaotic fractal images.

9.2 Phase Noise

Viewing an oscillator as a filtered-noise generator is relatively modern. The older approach was to think of an oscillator making a true sine wave with an added, unwanted noise signal. These are just different ways of visualizing the same thing: They are equally valid views, which are used interchangeably, depending which best makes some point clear. Thinking in terms of the signal-plus-noise version, the noise surrounds the carrier, looking like sidebands and so can also be considered to be equivalent to random-noise FM and AM on the ideal sine-wave signal. This gives us a third viewpoint. These noise sidebands are called *phase noise*. If we consider the addition of a noise voltage to a sinusoidal voltage, we must take into account the phase relationship. A *phasor diagram* is the clearest way of illustrating this. **Fig 9.6A** represents a clean sine wave as a rotating vector whose length is equal to the peak amplitude and whose frequency is equal to the number of revolutions per second of its rotation. Moving things are difficult to depict on paper, so phasor diagrams are usually drawn to show the dominant signal as stationary, with other components drawn relative to this. (Tutorials on complex numbers, vectors, and phasor notation are available at **www.intmath.com**. In addition, performing an Internet search for "phasor tutorial" will turn up several excellent Web pages on the subject.)

Noise contains components at many frequencies, so its phase with respect to the dominant, theoretically pure signal — the "carrier" — is random. Its amplitude is also random. Noise can only be described in statistical terms because its voltage is constantly and randomly changing, yet it does have an average amplitude that can be expressed as an RMS value. Fig 9.6B shows noise added to the carrier phasor, with the noise represented as a region in which the sum phasor wanders randomly. The phase of the noise is uniformly random — no direction is more likely than any other — but the instantaneous magnitude

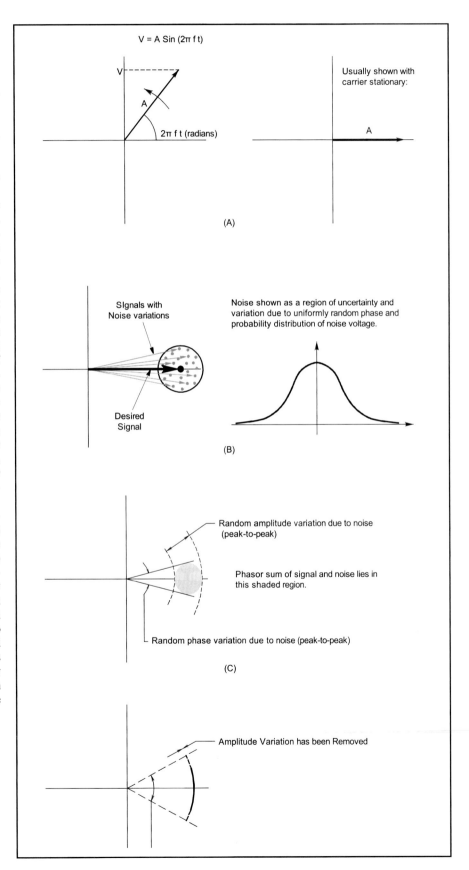

Fig 9.6 — At A, a phasor diagram of a clean (ideal) oscillator. Noise creates a region of uncertainty in the vector's length and position (B). AM noise varies the vector's length; PM noise varies the vector's relative angular position (C) Limiting a signal that contains both AM and PM noise strips off the AM and leaves the PM (D).

of the noise obeys a probability distribution like that shown, with values farther from the normal carrier frequency being progressively rarer. Fig 9.6C shows how the extremities of the noise region can be considered as extremes of phase and amplitude variation from the normal values of the carrier.

Phase modulation and frequency modulation are closely related. Phase is the integral of frequency, so phase modulation resembles frequency modulation in which deviation decreases with increasing modulating frequency. Thus, there is no need to talk of "frequency noise" because phase noise already covers it.

Fig 9.6C clearly shows AM noise as the random variation of the length of the sum phasor, yet "amplitude noise" is rarely discussed. The oscillator's amplitude control mechanism acts to reduce the AM noise by a small amount, but the main reason is that the output is often fed into some form of limiter that strips off AM components just as the limiting IF amplifier in an FM receiver removes any AM on incoming signals. The limiter can be obvious, like a circuit to convert the signal to logic levels, or it can be implicit in some other function. A diode ring mixer may be driven by a sine-wave LO of moderate power, yet this signal drives the diodes fully on and fully off, approximating square-wave switching. This is a form of limiter, and it reduces the effect of any AM on the LO. Fig 9.6D shows the result of passing a signal with noise through a limiting amplifier. For these reasons, AM noise sidebands are rarely a problem in oscillators, and so are normally ignored. There is one subtle problem to beware of, however. If a sine wave drives some form of switching circuit or limiter, and the threshold is offset from the signal's mean voltage, any level changes will affect the exact switching times and cause *phase jitter*. In this way, AM sidebands are translated into PM sidebands and pass through the limiter. This is usually called *AM-to-PM conversion* and is a classic problem of limiters.

A thorough analysis of oscillator phase noise is beyond the scope of this section. However, a detailed, state-of-the-art treatment by Dr Ulrich Rohde, N1UL, of free-running oscillators using nonlinear harmonic-balance techniques is presented as a supplement on the CD-ROM accompanying this book.

9.2.1 Effects of Phase Noise

You would be excused for thinking that phase noise is a recent discovery, but all oscillators have always produced it. Other changes have elevated an unnoticed characteristic up to the status of a serious impairment. Increased crowding and power levels on the ham bands, allied with greater expectations of receiver performance as a result of other

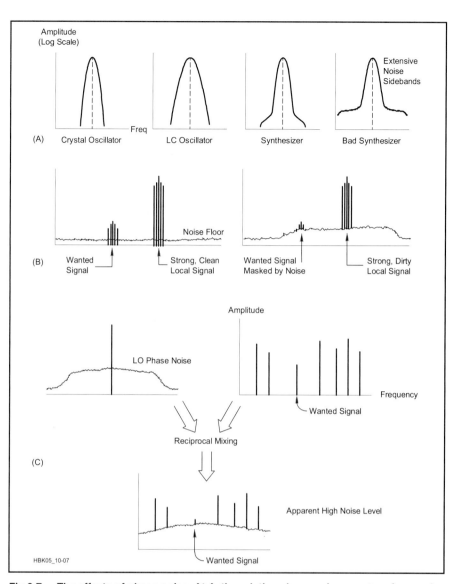

Fig 9.7 — The effects of phase noise. At A, the relative phase-noise spectra of several different signal-generation approaches. At B, how transmitted phase noise degrades the weak-signal performance even of nearby receivers with phase-quiet oscillators, raising the effective noise floor. What is perhaps most insidious about phase noise in Amateur Radio communication is that its presence in a receiver LO can allow strong, clean transmitted signals to degrade the receiver's noise floor just as if the transmitted signals were dirty. This effect, reciprocal mixing, is shown at C.

improvements, have made phase noise more noticeable, but the biggest factor has been the replacement of analog VFOs in radios by frequency synthesizers. It is a major task to develop a synthesizer that tunes in steps fine enough for SSB and CW while competing with the phase-noise performance of a reasonable-quality LC VFO. Many synthesizers have fallen far short of that target. Phase noise is worse in higher-frequency oscillators and the trend toward general-coverage, up-converting structures has required that local oscillators operate at higher and higher frequencies. **Fig 9.7A** shows sketches of the relative phase-noise performance of some oscillators. The very high Q of the quartz crystal in a crystal oscillator gives it the potential for much lower phase noise than LC oscillators. A medium-quality synthesizer has close-in phase noise performance approaching that of a crystal oscillator, while further from the carrier frequency, it degrades to the performance of a modest-quality LC oscillator. There may be a small bump in the noise spectrum at the boundary of these two zones. A bad synthesizer can have extensive noise sidebands extending over many tens (hundreds in extreme cases) of kilohertz at high levels.

Phase noise on a transmitter's local oscillators passes through its stages, is amplified and fed to the antenna along with the intentional signal. The intentional signal is thereby surrounded by a band of noise. This radiated noise exists in the same proportion to the transmitter power as the phase noise was to the oscillator power if it passes through no narrow-band filtering capable of limiting its bandwidth. This radiated noise makes for a noisier band. In bad cases, reception of nearby stations can be blocked over many tens of kilohertz above and below the frequency of the offending station. Fig 9.7B illustrates the difference between clean and dirty transmitters.

9.2.2 Reciprocal Mixing

The effects of receiver-LO phase noise are more complicated, but at least it doesn't affect other stations' reception. The process is called *reciprocal mixing*. This is an effect that occurs in all mixers, yet despite its name, reciprocal mixing is an LO, not a mixer, problem. Imagine that the outputs of two supposedly unmodulated signal generators are mixed together and the mixer output is fed into an FM receiver. The receiver produces sounds, indicating that the resultant signal is modulated nonetheless. Which signal generator is responsible? This is a trick question, of course. A moment spent thinking about how a change in the frequency of either input signal affects the output signal will show that FM or PM on either input reaches the output. The best answer to the trick question is therefore "either or both."

The modulation on the mixer output is the combined modulations of the inputs. This means that modulating a receiver's local oscillator is indistinguishable from using a clean LO and having the same modulation present on *all* incoming signals. (This is also true for AM provided that a fully linear multiplier is used as the mixer, but mixers are commonly driven into switching, which strips any AM off the LO signal. This is the chief reason why the phase component of oscillator noise is more important than any AM component.)

The word "indistinguishable" is important in the preceding paragraph. It does not mean that the incoming signals are themselves modulated, but that the signals in the receiver IF and the noise in the IF, sound exactly as if they were. What really happens is that the noise components of the LO are extra LO signals that are offset from the carrier frequency. Each of them mixes with other signals that are appropriately offset from the LO carrier into the receiver's IF. Noise is the sum of an infinite number of infinitesimal components spread over a range of frequencies, so the signals it mixes into the IF are spread into an infinite number of small replicas, all at different frequencies. This amounts to scrambling these other signals into noise. It is tedious to look at the effects of receiver LO phase noise this way. The concept of reciprocal mixing gives us an easier, alternative view that is much more digestible and produces identical results.

A poor oscillator can have significant noise sidebands extending out many tens of kilohertz on either side of its carrier. This is the same, as far as the signals in the receiver IF are concerned, as if the LO were clean and every signal entering the mixer had these noise sidebands. Not only will the wanted signal (and its noise sidebands) be received, but the noise sidebands added by the LO to signals near, but outside, the receiver's IF passband will overlap it. If the band is busy, each of many signals present will add its set of noise sidebands — and the effect is cumulative. This produces the appearance of a high background-noise level on the band. Many hams tend to accept this, blaming "conditions."

Hams now widely understand reception problems due to intermodulation, and almost everyone knows to apply RF attenuation until the signal gets no clearer. Intermodulation is a nonlinear effect, and the levels of the intermod products fall by greater amounts than the reduction in the intermodulating signals. The net result is less signal, but with the intermodulation products dropped still further. This improvement reaches a limit when more attenuation pushes the desired signal too close to the receiver's noise floor.

Reciprocal mixing is a linear process, and the mixer applies the same amount of noise "deviation" to incoming signals as that present on the LO. Therefore the ratio of noise-sideband power to signal power is the same for each signal, and the same as that on the LO. Switching in RF attenuation reduces the power of signals entering the mixer, but the reciprocal mixing process still adds the noise sidebands at the same *relative* power to each. Therefore, no reception improvement results. Other than building a quieter oscillator, the only way of improving things is to use narrow preselection to band-limit the receiver's input response and reduce the number of incoming signals subject to reciprocal mixing. This reduces the number of times the phase noise sidebands get added into the IF signal. Commercial *tracking preselectors* — selective front-end circuits that tune in step with a radio's band changes and tuning, are expensive, but one that is manually tuned would make a modest-sized home-brew project and could also help reduce intermodulation effects. When using a good receiver with a linear front end and a clean LO, amateurs accustomed to receivers with poor phase-noise performance report the impression is of a seemingly emptier band with gaps between signals — and then they begin to find readable signals in some of the gaps.

Fig 9.7C shows how a noisy oscillator affects transmission and reception. The effects on reception are worst in Europe, on 40 meters, at night. Visitors from North America, and especially Asia, are usually shocked by the levels of background noise. In ITU Region 1, the long-time Amateur Radio 40 meter allocation has been 7.0 to 7.1 MHz; above this, ultra-high-power broadcasters operate. (As of 2009, Region 1 and 2 broadcasters have begun to vacate the 7.1-7.2 MHz allocation, greatly reducing noise levels in these areas.) The front-end filters in commercial ham gear are usually fixed band-pass designs that cover the wider 40-meter allocations in the other regions. This allows huge signals to reach the mixer and mix large levels of LO phase noise into the IF. Operating co-located radios, on the same band, in a multioperator contest station, requires linear front ends, preselection and state-of-the-art phase-noise performance. Outside of amateur circles, only maritime vessel operation is more demanding, with kilowatt transmitters and receivers sharing antennas on the same mast.

9.2.3 A Phase Noise Demonstration

Healthy curiosity demands some form of demonstration so the scale of a problem can be judged "by ear" before measurements are attempted. We need to be able to measure the noise of an oscillator alone (to aid in the development of quieter ones) and we also need to be able to measure the phase noise of the oscillators in a receiver (a transmitter can be treated as an oscillator). Conveniently, a receiver contains most of the functions needed to demonstrate its own phase noise.

No mixer has perfect port-to-port isolation, and some of its local-oscillator signal leaks through into the IF. If we tune a general-coverage receiver, with its antenna disconnected, to exactly 0 Hz, the local oscillator is exactly at the IF center frequency, and the receiver acts as if it is tuned to a very strong unmodulated carrier. A typical mixer might give only 40 dB of LO isolation and have an LO drive power of at least 10 mW. If we tune away from 0 Hz, the LO carrier tunes away from the IF center and out of the passband. The apparent signal level falls. Although this moves the LO carrier out of the IF passband, some of its noise sidebands will not be, and the receiver will respond to this energy as an incoming noise signal. To the receiver operator, this sounds like a rising noise floor as the receiver is tuned toward 0 Hz. To get good noise floor at very low frequencies, some professional/military receivers, like the Racal RA1772, use very carefully balanced mixers to get as much port-to-port isolation as possible, and they also may switch a crystal notch filter into the first mixer's LO feed.

This demonstration cannot be done if the receiver tunes amateur bands only. As it is, most general-coverage radios inhibit tuning in the LF or VLF region. It could be suggested by a cynic that how low manufacturers allow you to tune is an indication of how far they

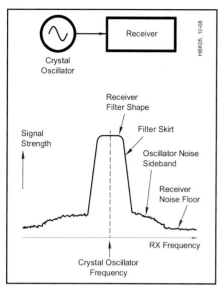

Fig 9.8 — Tuning in a strong, clean crystal-oscillator signal can allow you to hear your receiver's relative phase-noise level. Listening with a narrow CW filter switched in allows you to get better results closer to the carrier.

think their phase-noise sidebands could extend!

The majority of amateur transceivers with general-coverage receivers are programmed not to tune below 30 to 100 kHz, so means other than the "0 Hz" approach are needed to detect LO noise in these radios. Because reciprocal mixing adds the LO's sidebands to clean incoming signals, in the same proportion to the incoming carrier as they exist with respect to the LO carrier, all we need do is to apply a strong, clean signal wherever we want within the receiver's tuning range. This signal's generator must have lower phase noise than the radio being evaluated. A general-purpose signal generator is unlikely to be good enough; a crystal oscillator is needed.

It's appropriate to set the level into the receiver to about that of a strong broadcast carrier, say S9 + 40 dB. Set the receiver's mode to SSB or CW and tune around the test signal, looking for an increasing noise floor (higher hiss level) as you tune closer towards the signal, as shown in **Fig 9.8**. Switching in a narrow CW filter allows you to hear noise closer to the carrier than is possible with an SSB filter. This is also the technique used to measure a receiver's effective selectivity, and some equipment reviewers kindly publish their plots in this format. *QST* reviews, done by the ARRL Lab, often include the results of specific phase-noise measurements.

9.2.4 Measuring Receiver Phase Noise

There are several different ways of measuring phase noise, offering different tradeoffs between convenience, cost and effort. Some methods suit oscillators in isolation, others suit them in-situ (in their radios).

If you're unfamiliar with noise measurements, the units involved may seem strange. One reason for this is that a noise signal's power is spread over a frequency range, like an infinite number of infinitesimal sinusoidal components. This can be thought of as similar to painting a house. The area that a gallon of paint can cover depends on how thinly it's spread. If someone asks how much paint was used on some part of a wall, the answer would have to be in terms of paint volume per square foot. The wall can be considered to be an infinite number of points, each with an infinitesimal amount of paint applied to it. The question of what volume of paint has been applied at some specific point is unanswerable. With noise, we must work in terms of *power density*, of watts per hertz. We therefore express phase-noise level as a ratio of the carrier power to the noise's power density. Because of the large ratios involved, expression in decibels is convenient. It has been a convention to use *dBc* to mean "decibels with respect to the carrier."

For phase noise, we need to work in terms of a standard bandwidth, and 1 Hz is the obvious candidate. Even if the noise is measured in a different bandwidth, its equivalent power in 1 Hz can be easily calculated. A phase-noise level of –120 dBc in a 1-Hz bandwidth (often written as "–120 dBc/Hz") translates into each hertz of the noise having a power of 10^{-12} of the carrier power. In a bandwidth of 3 kHz, this would be 3000 times larger.

The most convenient way to measure phase noise is to buy and use a commercial phase noise test system. Such a system usually contains a state-of-the-art, low-noise frequency synthesizer and a low-frequency spectrum analyzer, as well as some special hardware. Often, a second, DSP-based spectrum analyzer is included to speed up and extend measurements very close to the carrier by using the Fast Fourier Transform (FFT). The whole system is then controlled by a computer with proprietary software. With a good system like this costing about $100,000, this is not a practical method for amateurs, although a few fortunate individuals have access to them at work. These systems are also overkill for our needs, because we are not particularly interested in determining phase-noise levels very close to and very far from the carrier.

It's possible to make respectable receiver-oscillator phase-noise measurements with less than $100 of parts and a multimeter. Although it's time-consuming, the technique is much more in keeping with the amateur spirit than using a $100k system! An ordinary multimeter will produce acceptable results; a meter capable of indicating "true RMS" ac voltages is preferable because it can give correct readings on sine waves *and* noise. **Fig 9.9** shows the setup. Measurements can only be made around the frequency of the crystal oscillator, so if more than one band is to be tested, crystals must be changed, or else a set of appropriate oscillators is needed. The oscillator should produce about +10 dBm (10 mW) and be *very* well shielded. (To this end, it's advisable to build the oscillator into a die-cast box and power it from internal batteries. A noticeable shielding improvement results even from avoiding the use of an external power switch; a reed-relay element inside the box can be positioned to connect the battery when a small permanent magnet is placed against a marked place outside the box.)

Likewise, great care must be taken with attenuator shielding. A total attenuation of around 140 dB is needed, and with so much attenuation in line, signal leakage can easily exceed the test signal that reaches the receiver. It's not necessary to be able to switch-select all 140 dB of the attenuation, nor is this desirable, as switches can leak. All of the attenuators' enclosure seams must be soldered. A pair of boxes with 30 dB of fixed attenuation each is needed to complete the set. With 140 dB of attenuation, coax cable leakage is also a problem. The only countermeasure against this is to minimize all cable lengths and to interconnect test-system modules with BNC plug-to-plug adapters (UG-491A) where possible.

Ideally, the receiver could simply be tuned across the signal from the oscillator and the response measured using its signal-strength (S) meter. Unfortunately, receiver S meters are notoriously imprecise, so an equivalent

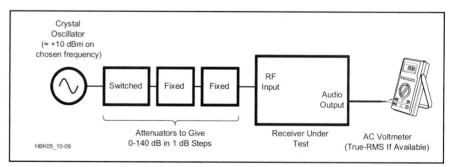

Fig 9.9 — Setup for measuring receiver-oscillator phase noise.

method is needed that does not rely on the receiver's AGC system.

The trick is not to measure the response to a fixed level signal, but to measure the changes in applied signal power needed to give a fixed response. Here is a step-by-step procedure based on that described by John Grebenkemper, KI6WX, in March and April 1988 *QST*:

1. Connect the equipment as shown in Fig 9.9, but with the crystal oscillator off. Set the step attenuator to maximum attenuation. Set the receiver for SSB or CW reception with its narrowest available IF filter selected. Switch out any internal preamplifiers or RF attenuators. Select AGC off, maximum AF and RF gain. It may be necessary to reduce the AF gain to ensure the audio amplifier is at least 10 dB below its clipping point. The ac voltmeter or an oscilloscope on the AF output can be used to monitor this.

2. To measure noise, it is important to know the bandwidth being measured. A true-RMS ac voltmeter measures the power in the noise reaching it. To calculate the noise density, we need to divide by the receiver's *noise bandwidth*. The receiver's –6-dB IF bandwidth can be used as an approximation, but purists will want to plot the top 20 dB of the receiver's bandwidth on linear scales and integrate the area under it to find the width of a rectangle of equal area and equal height. This accounts properly for the noise in the skirt regions of the overall selectivity. (The very rectangular shape of common receiver filters tends to minimize the error of just taking the approximation.)

Switch on the test oscillator and set the attenuators to give an AF output above the noise floor and below the clipping level with the receiver peaked on the signal. Tune the receiver off to each side to find the frequencies at which the AF voltage is half that at the peak. The difference between these is the receiver's –6-dB bandwidth. High accuracy is not needed: 25% error in the receiver bandwidth will only cause a 1-dB error in the final result. The receiver's published selectivity specifications will be close enough. The benefit of integration is greater if the receiver has a very rounded, low-ringing or low-order filter.

3. Retune the receiver to the peak. Switch the oscillator off and note the noise-floor voltage. Turn the oscillator back on and adjust the attenuator to give an AF output voltage 1.41 times (3 dB) larger than the noise floor voltage. This means that the noise power and the test signal power at the AF output are equal — a value that's often called the *MDS* (*minimum discernible signal*) of a receiver. Choosing a test-oscillator level at which to do this test involves compromise. Higher levels give more accurate results where the phase noise is high, but limit the lowest level of phase noise that can be measured because better receiver oscillators require a greater input signal to produce enough noise to get the chosen AF-output level. At some point, either we've taken all the attenuation out and our measurement range is limited by the test oscillator's available power, or we overload the receiver's front end, spoiling the results.

Record the receiver frequency at the peak, (f_0), the attenuator setting (A_0) and the audio output voltage (V_0). These are the carrier measurements against which all the noise measurements will be compared.

4. Now you must choose the offset frequencies — the separations from the carrier — at which you wish to make measurements. The receiver's skirt selectivity will limit how close to the carrier noise measurements can be made. (Any measurements made too close in are valid measurements of the receiver selectivity, but because the signal measured under these conditions is sinusoidal and not noise like, the corrections for noise density and noise bandwidth are not appropriate.) It is difficult to decide where the filter skirt ends and the noise begins, and what corrections to apply in the region of doubt and uncertainty. A good practical approach is to listen to the audio and tune away from the carrier until you can't distinguish a tone in the noise. The ear is superb at spotting sine tones buried in noise, so this criterion, although subjective, errs on the conservative side.

Tune the receiver to a frequency offset from f_0 by your first chosen offset and adjust the attenuators to get an audio output voltage as close as possible to V_0. Record the total attenuation, A_1 and the audio output voltage, V_1. The SSB phase noise (qualified as SSB because we're measuring the phase noise on only one side of the carrier, whereas some other methods cannot segregate between upper and lower noise sidebands and measure their sum, giving DSB phase noise) is now easy to calculate:

$$L(f) = A_1 - A_0 - 10_{\log}(BW_{noise}) \quad (1)$$

where

$L(f)$ = SSB phase noise in dBc/Hz
BW_{noise} = receiver noise bandwidth, Hz
A_0 = Attenuator setting in step three
A_1 = Attenuator setting in step four

This equation begins with the difference between the attenuation necessary to reduce the peak carrier signal to the MDS level (A_0 in step three) and the attenuation necessary to reduce the phase noise to the MDS level (A_1 in step four). Subtracting the bandwidth correction term results in the noise power per Hz of bandwidth with respect to the peak carrier signal. Note that this equation does not depend on the absolute power level of the

Table 9.1
SSB Phase Noise of ICOM IC-745 Receiver Section

Oscillator output power = –3 dBm (0.5 mW)
Receiver bandwidth (Δf) = 1.8 kHz
Audio noise voltage = –0.070 V
Audio reference voltage (V_0) = 0.105 V
Reference attenuation (A_0) = 121 dB

Offset Frequency	Attenuation (A_1) (dB)	Audio V_1 (volts)	Audio V_2 (volts)	Ratio V_2/V_1	SSB Phase Noise (kHz) (dBc/Hz)
4	35	0.102	0.122	1.20	–119
5	32	0.104	0.120	1.15	–122*
6	30	0.104	0.118	1.13	–124*
8	27	0.100	0.116	1.16	–127*
10	25	0.106	0.122	1.15	–129*
15	21	0.100	0.116	1.16	–133*
20	17	0.102	0.120	1.18	–137
25	14	0.102	0.122	1.20	–140
30	13	0.102	0.122	1.20	–141
40	10	0.104	0.124	1.19	–144
50	8	0.102	0.122	1.20	–146
60	6	0.104	0.124	1.19	–148
80	4	0.102	0.126	1.24	–150
100	3	0.102	0.126	1.24	–151
150	3	0.102	0.124	1.22	–151
200	0	0.104			–154
250	0	0.100			–154
300	0	0.98			–154
400	0	0.96			–154
500	0	0.96			–154
600	0	0.97			–154
800	0	0.96			–154
1000	0	0.96			–154

*Asterisks indicate measurements possibly affected by receiver overload (see text).

carrier signal as long as it remains constant during the test.

5. It's important to check for overload. Decrease the attenuation by 3 dB, and record the new audio output voltage, V_2. If all is well, the output voltage should increase by 22% (1.8 dB); if the receiver is operating non-linearly, the increase will be less. (An 18% increase is still acceptable for the overall accuracy we want.) Record V_2/V_1 as a check: a ratio of 1.22:1 is ideal, and anything less than 1.18:1 indicates a bad measurement.

If too many measurements are bad, you may be overdriving the receiver's AF amplifier, so try reducing the AF gain and starting again back at Step 3. If this doesn't help, reducing the RF gain and starting again at Step 3 should help if the compression is occurring late in the IF stages.

6. Repeat Steps 4 and 5 at all the other offsets you wish to measure. If measurements are made at increments of about half the receiver's bandwidth, any discrete (non-noise) spurs will be found. A noticeable tone in the audio can indicate the presence of one of these. If it is well clear of the noise, the measurement is valid, but the noise bandwidth correction should be ignored, giving a result in dBc.

Table 9.1 shows the results for an ICOM IC-745 as measured by KI6WX, and **Fig 9.10** shows this data in graphic form. His oscilla-

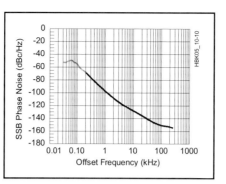

Fig 9.10 — The SSB phase noise of an ICOM IC-745 transceiver (serial number 01528) as measured by KI6WX.

tor power was only −3 dBm, which limited measurements to offsets less than 200 kHz. More power might have allowed noise measurements to lower levels, although receiver overload places a limit on this. This is not important, because the real area of interest has been thoroughly covered. When attempting phase-noise measurements at large offsets, remember that any front-end selectivity, before the first mixer, will limit the maximum offset at which LO phase-noise measurement is possible.

9.2.5 Measuring Oscillator and Transmitter Phase Noise

Measuring the composite phase noise of a receiver's LO requires a clean test oscillator. Measuring the phase noise of an incoming signal, whether from a single oscillator or an entire transmitter, requires the use of a clean receiver, with lower phase noise than the source under test. The sidebar, "Transmitter Phase-Noise Measurement in the ARRL

Transmitter Phase-Noise Measurement in the ARRL Lab

Here is a brief description of the technique used in the ARRL Lab to measure transmitter phase noise. The system consists of a phase noise test set with low-noise variable crystal oscillators for reference signals. The test set zero-beats the oscillator to the transmitter signal using phase detectors on both signals. As shown in **Fig 9.A1**, we use an attenuator after the transmitter to bring the signal down to a suitable level for the phase noise test set. The crystal oscillator output is low level, so it does not require attenuation.

The phase noise test set is an automated system run by a PC that steps through the test, setting parameters in the spectrum analyzers and reading the data back from them, in addition to performing system calibration measurements at the start of the test. The computer screen displays the transmitted phase-noise spectrum, which can be printed or saved to a file on the hard drive. To test the baseline phase noise of the system, two identical crystal oscillators are tested against each other. It is quite important to be sure that the phase noise of the reference source is lower than that of the signal under test.

A sample phase-noise plot for an amateur transceiver is shown in **Fig 9.A2**. It was produced with the test setup shown in Fig 9.A1. These plots do not necessarily reflect the phase-noise characteristics of all units of a particular model.

The reference level (the top horizontal line on the scale in

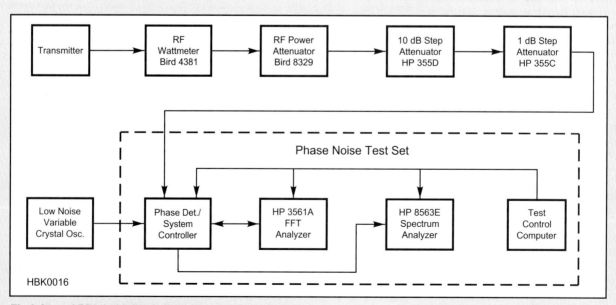

Fig 9.A1 — ARRL Lab phase-noise measurement setup.

Lab," details the method used to measure composite noise (phase noise and amplitude noise, the practical effects of which are indistinguishable on the air) for *QST* Product Reviews. Although targeted at measuring high power signals from entire transmitters, this approach can be used to measure lower-level signals simply by changing the amount of input attenuation used.

At first, this method — using a low-frequency spectrum analyzer and a low-phase-noise signal source — looks unnecessarily elaborate. A growing number of radio amateurs have acquired good-quality spectrum analyzers for their shacks since older model Tektronix and Hewlett-Packard instruments have started to appear on the surplus market at affordable prices. The obvious question is, "Why not just use one of these to view the signal and read phase-noise levels directly off the screen?" Reciprocal mixing is the problem.

Very few spectrum analyzers have clean enough local oscillators not to completely swamp the noise being measured. Phase-noise measurements involve the measurement of low-level components very close to a large carrier, and that carrier will mix the noise sidebands of the *analyzer's* LO into its IF. Some way of notching out the carrier is needed, so that the analyzer need only handle the noise sidebands. A crystal filter could be designed to do the job, but this would be expensive, and one would be needed for every different oscillator frequency to be tested. The alternative is to build a direct-conversion receiver using a clean LO like the Hewlett-Packard HP8640B signal generator and spectrum-analyze its "audio" output with an audio analyzer. This scheme mixes the carrier to dc; the LF analyzer is then ac-coupled, and this removes the carrier. The analyzer can be made very sensitive without overload or reciprocal mixing being a problem.

The remaining problem is then keeping the LO — the HP8640B in this example — at exactly the carrier frequency. 8640s are based on a shortened-UHF-cavity oscillator and can drift a little. The oscillator under test will also drift. The task is therefore to make the 8640B track the oscillator under test. For once we get something for free: The HP8640B's FM input is dc coupled, and we can use this as an electronic fine-tuning input. As a further bonus, the 8640B's FM deviation control acts as a sensitivity control for this input. We also get a phase detector for free, as the mixer output's dc component depends on the phase relationship between the 8640B and the signal under test (remember to use the dc coupled port of a diode ring mixer as the output). Taken together, the system includes everything needed to create a crude phase-locked loop that will automatically track the input signal over a small frequency range. **Fig 9.11** shows the arrangement.

The oscilloscope is not essential for operation, but it is needed to adjust the system. With the loop unlocked (8640B FM input disconnected), tune the 8640 off the signal frequency to give a beat at the mixer output. Adjust the mixer drive levels to get an undistorted sine wave on the scope. This ensures that the mixer

the plot) represents 0 dBc/Hz. Because each vertical division represents 20 dB, the plot shows the noise level between 0 dBc/Hz (the top horizontal line) and –160 dBc/Hz (the bottom horizontal line). The horizontal scale is logarithmic, with one decade per division (the first division shows noise from 100 Hz to 1000 Hz offset, whereas the last division shows noise from 100 kHz through 1 MHz offset).

What Do the Phase-Noise Plots Mean?

Although they are useful for comparing different radios, plots can also be used to calculate the amount of interference you may receive from a nearby transmitter with known phase-noise characteristics. An approximation is given by

$$A_{QRM} = NL + 10 \times \log(BW)$$

where
A_{QRM} = Interfering signal level, dBc
NL = noise level on the receive frequency, dBc
BW = receiver IF bandwidth, in Hz

For instance, if the noise level is –90 dBc/Hz and you are using a 2.5-kHz SSB filter, the approximate interfering signal will be –56 dBc. In other words, if the transmitted signal is 20 dB over S9, and each S unit is 6 dB, the interfering signal will be as strong as an S3 signal.

The measurements made in the ARRL Lab apply only to transmitted signals. It is reasonable to assume that the phase-noise characteristics of most transceivers are similar on transmit and receiver because the same oscillators are generally used in local-oscillator (LO) chain.

In some cases, the receiver may have better phase noise characteristics than the transmitter. Why the possible difference? The most obvious reason is that circuits often perform less than optimally in strong RF fields, as anyone who has experienced RFI problems can tell you. A less obvious reason results from the way that many high-dynamic-range receivers work. To get good dynamic range, a sharp crystal filter called a *roofing filter* is often placed immediately after the first mixer in the receive line. This filter removes all but a small slice of spectrum for further signal processing. If the desired filtered signal is a product of mixing an incoming signal with a noisy oscillator, signals far away from the desired one can end up in this slice. Once this slice of spectrum is obtained, however, unwanted signals cannot be reintroduced, no matter how noisy the oscillators used in further signal processing. As a result, some oscillators in receivers don't affect phase noise.

The difference between this situation and that in transmitters is that crystal filters are seldom used for reduction of phase noise in transmitting because of the high cost involved. Equipment designers have enough trouble getting smooth, click-free break-in operation in transceivers without having to worry about switching crystal filters in and out of circuits at 40-WPM keying speeds! — *Zack Lau, W1VT, and Michael Tracy, KC1SX*

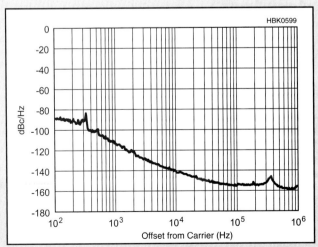

Fig 9.A2 — Sample phase noise plot for an amateur HF transceiver as published in Product Review in *QST*. This is the Elecraft K3.

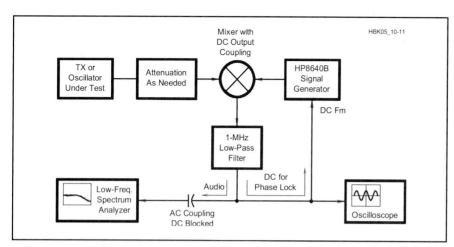

Fig 9.11 — Arrangement for measuring phase noise by directly converting the signal under test to audio. The spectrum analyzer views the signal's noise sidebands as audio; the signal's carrier, converted to dc, provides a feedback signal to phase-lock the Hewlett-Packard HP8640B signal generator to the signal under test.

is not being overdriven. While the loop is off-tuned, adjust the beat to a frequency within the range of the LF spectrum analyzer and use it to measure its level, "A_c" in dBm. This represents the carrier level and is used as the reference for the noise measurements. Connect the FM input of the signal generator, and switch on the generator's dc FM facility. Try a deviation range of 10 kHz to start with. When you tune the signal generator toward the input frequency, the scope will show the falling beat frequency until the loop jumps into lock. Then it will display a noisy dc level. Fine tune to get a mean level of 0 V. (This is a very-low-bandwidth, very-low-gain loop. Stability is not a problem; careful loop design is not needed. We actually want as slow a loop as possible; otherwise, the loop would track and cancel the slow components of the incoming signal's phase noise, preventing their measurement.)

When you first take phase-noise plots, it's a good idea to duplicate them at the generator's next lower FM-deviation range and check for any differences in the noise level in the areas of interest. Reduce the FM deviation range until you find no further improvement. Insufficient FM deviation range makes the loop's lock range narrow, reducing the amount of drift it can compensate. (It's sometimes necessary to keep gently trimming the generator's fine tune control.)

Set up the LF analyzer to show the noise. A sensitive range and 100-Hz resolution bandwidth are appropriate. Measure the noise level, "A_n" in dBm. We must now calculate the noise density that this represents. Spectrum-analyzer filters are normally *Gaussian*-shaped and bandwidth-specified at their –3-dB points. To avoid using integration to find their true-noise power bandwidth, we can reasonably assume a value of $1.2 \times BW$. A spectrum analyzer logarithmically compresses its IF signal ahead of the detectors and averaging filter. This affects the statistical distribution of noise voltage and causes the analyzer to read low by 2.5 dB. To produce the same scale as the ARRL Lab photographs prior to May 2006 *QST*, the analyzer reference level must be set to –60 dBc/Hz, which can be calculated as:

$$A_{ref} = A_c - 10 \log(1.2 \times BW) + 62.5 \text{ dBm} \quad (2)$$

where

A_{ref} = analyzer reference level, dBm
A_c = carrier amplitude, dBm

This produces a scale of –60 dBc/Hz at the top of the screen, falling to –140 dBc/Hz at the bottom. The frequency scale is 0 to 20 kHz with a resolution bandwidth (BW in the above equation) of 100 Hz. This method combines the power of *both* sidebands and so measures DSB phase noise. To calculate the equivalent SSB phase noise, subtract 3 dB for noncoherent noise (the general "hash" of phase noise) and subtract 6 dB for coherent, discrete components (that is, single-frequency spurs). This can be done by setting the reference level 3 to 6 dB higher.

9.2.6 Low-Cost Phase Noise Testing

All that expensive equipment may seem far beyond the means of the average Amateur Radio experimenter. With careful shopping and a little more effort, alternative equipment can be put together for pocket money. (All of the things needed — parts for a VXO, a surplus spectrum analyzer and so on — have been seen on sale cheap enough to total less than $100.) The HP8640B is good and versatile, but for use at one oscillator frequency, you can build a VXO for a few dollars. It will only cover one oscillator frequency, but a VXO can provide even better phase-noise performance than the 8640B. There is free software available so that you can use your PC soundcard as an LF spectrum analyzer, though you may want to add a simple preamp and some switched attenuators.

In the reference at the end of this chapter, Pontius has also demonstrated that signal-source phase-noise measurements can be accurately obtained without the aid of expensive equipment.

9.3 Oscillator Circuits and Construction

9.3.1 LC Oscillator Circuits

The LC oscillators used in radio equipment are usually designed to be *variable frequency oscillators* (VFOs). Tuning is achieved by either varying part of the capacitance of the resonator or, less commonly, by using a movable magnetic core to vary the inductance. Since the early days of radio, there has been a huge quest for the ideal, low-drift VFO. Amateurs and professionals have spent immense effort on this pursuit. A brief search of the literature reveals a large number of designs, many accompanied by claims of high stability. The quest for stability has been solved by the development of low-cost frequency synthesizers, which can give crystal-controlled stability. Synthesizers have other problems though, and the VFO still has much to offer in terms of signal cleanliness, cost and power consumption, making it attractive for home construction. No one VFO circuit has any overwhelming advantage over any other — component quality, mechanical design and the care taken in assembly are much more important.

Fig 9.12 shows three popular oscillator circuits stripped of any unessential frills so they can be more easily compared. The original Colpitts circuit (Fig 9.12A) is now often referred to as the *parallel-tuned Colpitts* because its series-tuned derivative (Fig 9.12B) has become common. All three of these circuits use an amplifier with a voltage gain less than unity, but large current gain. The N-channel JFET source follower shown appears to be the most popular choice nowadays. In the parallel-tuned Colpitts, C3 and C4 are large values, perhaps 10 times larger than typical values for C1 and C2. This means that only a small fraction of the total tank voltage is applied to the FET, and the FET can be considered to be only lightly coupled into the tank. The FET is driven by the sum of the voltages across C3 and C4, while it drives the voltage across C4 alone. This means that the tank operates as a resonant, voltage-step-up transformer compensating for the less-than-unity-voltage-gain amplifier. The resonant circuit consists of L, C1, C2, C3 and C4. The resonant frequency can be calculated by using the standard formulas for capacitors in series and parallel to find the resultant capacitance effectively connected across the inductor, L, and then use the standard formula for LC resonance:

$$f = \frac{1}{2\pi\sqrt{LC}} \qquad (3)$$

where
 f = frequency in hertz
 L = inductance in henries
 C = capacitance in farads.

For a wide tuning range, C2 must be kept small to reduce the effect of C3 and C4 swamping the variable capacitor C1. (For more information on component selection, the chapters on oscillators in *Experimental Methods for RF Design* and *Introduction to Radio Frequency Design* listed in the References provide excellent material.)

The series-tuned Colpitts circuit works in much the same way. The difference is that the variable capacitor, C1, is positioned so that it is well-protected from being swamped by the large values of C3 and C4. In fact, *small* values of C3, C4 would act to limit the tuning range. Fixed capacitance, C2, is often added across C1 to allow the tuning range to be reduced to that required, without interfering with C3 and C4, which set the amplifier coupling. The series-tuned Colpitts has a reputation for better stability than the parallel-tuned original. Note how C3 and C4 swamp the capacitances of the amplifier in both versions.

A parallel-tuned Colpitts oscillator is the subject of the detailed paper "A Design Example for an Oscillator for Best Phase Noise and Good Output Power" by Dr Ulrich Rohde, N1UL, available on the CD-ROM accompanying this book. The paper shows the design process by which both noise and power can be optimized in a simple oscillator.

The Hartley of Fig 9.12C is similar to the parallel-tuned Colpitts, but the amplifier source is connected to a tap on the tank inductance instead of the tank capacitance. A typical tap placement is 10 to 20% of the total turns up from the "cold" end of the inductor. (It's usual to refer to the lowest-signal-voltage end of an inductor as "cold" and the other, with the highest signal voltage as "hot".) C2 limits the tuning range as required; C3 is reduced to the minimum value that allows reliable starting. This is necessary because the Hartley's lack of the Colpitts' capacitive divider would otherwise couple the FET's capacitances to the tank more strongly than in the Colpitts, potentially affecting the circuit's frequency stability.

In all three circuits, there is a 1 kΩ resistor in series with the source bias choke. This resistor does a number of desirable things. It spoils the Q of the inevitable low-frequency resonance of the choke with the tank tap circuit. It reduces tuning drift due to choke impedance

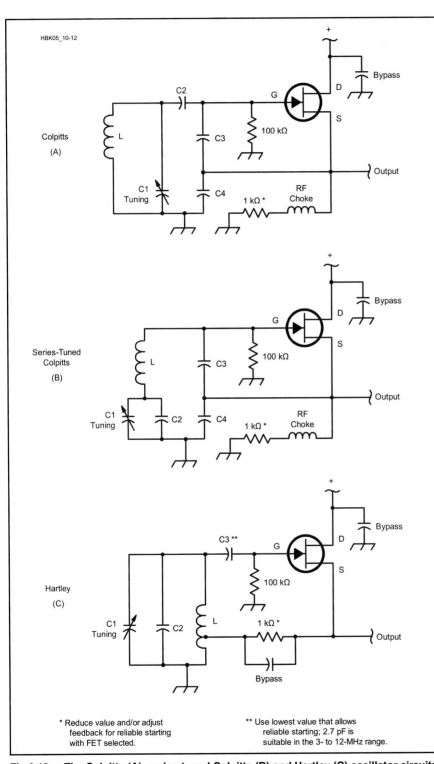

* Reduce value and/or adjust feedback for reliable starting with FET selected.

** Use lowest value that allows reliable starting; 2.7 pF is suitable in the 3- to 12-MHz range.

Fig 9.12 — The Colpitts (A), series-tuned Colpitts (B) and Hartley (C) oscillator circuits. Rules of thumb: C3 and C4 at A and B should be equal and valued such that their X_c = 45 Ω at the operating frequency; for C2 at A, X_c = 100 Ω. For best stability, use C0G or NP0 units for all capacitors associated with the FETs' gates and sources. Depending on the FET chosen, the 1-kΩ source-bias-resistor value shown may require adjustment for reliable starting.

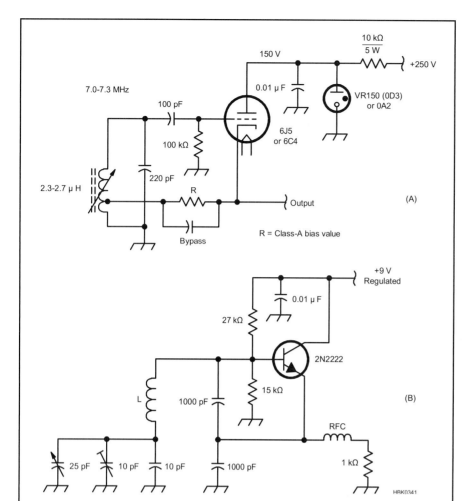

Fig 9.13 — Two more oscillator examples: at A, a triode-tube Hartley; at B, a bipolar junction transistor in a series-tuned Colpitts.

and winding capacitance variations. It also protects against spurious oscillation due to internal choke resonances. Less obviously, it acts to stabilize the loop gain of the built-in AGC action of this oscillator. Stable operating conditions act to reduce frequency drift.

Some variations of these circuits may be found with added resistors providing a dc bias to stabilize the system quiescent current. More elaborate still are variations characterized by a constant-current source providing bias. This can be driven from a separate AGC detector system to give very tight level control. The gate-to-ground clamping diode (1N914 or similar) long used by radio amateurs as a means of avoiding gate-source conduction has been shown by Ulrich Rohde, N1UL, to degrade oscillator phase-noise performance, and its use is virtually unknown in professional circles.

Fig 9.13 shows some more VFOs to illustrate the use of different devices. The triode Hartley shown includes *permeability tuning*, which has no sliding contact like that to a capacitor's rotor and can be made reasonably linear by artfully spacing the coil turns. The slow-motion drive can be done with a screw thread. The disadvantage is that special care is needed to avoid backlash and eccentric rotation of the core. If a non-rotating core is used, the slides have to be carefully designed to prevent motion in unwanted directions. The Collins Radio Company made extensive use of tube-based permeability tuners, and a semiconductor version can still be found in a number of Ten-Tec radios.

Vacuum tubes cannot run as cool as competitive semiconductor circuits, so care is needed to keep the tank circuit away from the tube heat. In many amateur and commercial vacuum-tube oscillators, oscillation drives the tube into grid current at the positive peaks, causing rectification and producing a negative grid bias. The oscillator thus runs in Class C, in which the conduction angle reduces as the signal amplitude increases until the amplitude stabilizes. As in the FET circuits of Fig 9.12, better stability and phase-noise performance can be achieved in a vacuum-tube oscillator by moving it out of true Class C — that is, by including a bypassed cathode-bias resistor (the resistance appropriate for Class A operation is a good starting value). A small number of people still build vacuum-tube radios partly to be different, partly for fun, but the semiconductor long ago achieved dominance in VFOs.

The voltage regulator (*VR*) tube shown in Fig 9.13A has a potential drawback: It is a gas-discharge device with a high striking voltage at which it starts to conduct and a lower extinguishing voltage, at which it stops conducting. Between these extremes lies a region in which decreasing voltage translates to increasing current, which implies negative resistance. When the regulator strikes, it discharges any capacitance connected across it to the extinguishing voltage. The capacitance then charges through the source resistor until the tube strikes again, and the process repeats. This *relaxation oscillation* demonstrates how negative resistance can cause oscillation. The oscillator translates the resultant sawtooth modulation of its power supply into frequency and amplitude variation. Because of VR tubes' ability to support relaxation oscillation, a traditional rule of thumb is to keep the capacitance directly connected across a VR tube to 0.1 µF or less. A value much lower than this can provide sufficient bypassing, as shown in Fig 9.13A because the dropping resistor acts as a decoupler, increasing the bypass's effectiveness.

There is a related effect called *squegging*, which can be loosely defined as oscillation on more than one frequency at a time, but which may also manifest itself as RF oscillation interrupted at an AF rate, as in a super-regenerative detector. One form of squegging occurs when an oscillator is fed from a power supply with a high source impedance. The power supply charges up the decoupling capacitor until oscillation starts. The oscillator draws current and pulls down the capacitor voltage, starving itself of power until oscillation stops. The oscillator stops drawing current, and the decoupling capacitor then recharges until oscillation restarts. The process, the low-frequency cycling of which is a form of relaxation oscillation, repeats indefinitely. The oscillator output can clearly be seen to be pulse modulated if an oscilloscope is used to view it at a suitable time-base setting. This fault is a well-known consequence of poor design in battery-powered radios. As dry cells become exhausted, their internal resistance rises quickly and circuits they power can begin to misbehave. In audio stages, such misbehavior may manifest itself in the "putt-putt" sound of the slow relaxation oscillation called *motorboating*.

Compared to the frequently used JFET, bipolar transistors, Fig 9.13B, are relatively uncommon in oscillators because their low input and output impedances are more difficult to couple into a high-Q tank without excessively loading it. Bipolar devices do tend to give better sample-to-sample amplitude uniformity for a given oscillator circuit, however, as JFETs of a given type tend to vary more in their characteristics.

9.3.2 RC Oscillators

Plenty of RC oscillators are capable of operating to several megahertz. Some of these are really constant-current-into-capacitor circuits, which are easier to make in silicon. Like the phase-shift oscillator above, the timing circuit Q is less than one, giving very poor noise performance that's unsuitable even to the least demanding radio application. One example is the oscillator section of the CD4046 phase-locked-loop IC. This oscillator has poor stability over temperature, large batch-to-batch variation and a wide variation in its voltage-to-frequency relationship. It is not recommended that this sort of oscillator be used at RF in radio systems. (The '4046 phase detector section is very useful, however, as we'll see later.) These oscillators are best suited to audio applications.

9.3.3 Three High-Performance HF VFOs

THE N1UL MODIFIED VACKAR VFO

The oscillator circuit of **Fig 9.14A** was contributed by Dr Ulrich Rohde, N1UL. It is a modified Vackar design (see the sidebar) in which a small coupling capacitor (8 pF) and voltage divider capacitor (18 pF) isolate the resonator circuit (10 μH and 50 pF tuning capacitor) from the oscillator transistor.

The oscillator transistor is followed by a buffer stage to isolate the oscillator from the load. Because the coupling between the transistor base and the resonator is fixed and light, stability of the oscillator is high across a wide tuning range from 5.5 to 6.6 MHz. Either the inductor or capacitor may be varied to tune the oscillator, but a variable capacitor is recommended as more practical and gives better performance.

Because of the oscillator transistor's large capacitors from base to ground (220 pF) and collector to ground (680 pF), the various parameters of the oscillator transistor have little practical influence on circuit performance. The widely available 2N3904 will perform well for both the oscillator and buffer transistors.

Practical resonator coil and the tuning capacitors will have a positive temperature coefficient. The 8 pF and the 18 pF capacitors should have an N150 temperature coefficient. After 1 hour, the observed drift for this circuit was less than 10 Hz / hour.

THE K7HFD LOW-NOISE DIFFERENTIAL OSCILLATOR

The other high performance oscillator example, shown in **Fig 9.15**, was designed for

The Vackar Oscillator

The original Vackar oscillator is named for Jirí Vackár, who invented the circuit in the late 1940s. The circuit's description — a refinement of the Clapp oscillator — can be found in older editions of the Radio Society of Great Britain's *Radio Communication Handbook*, with some further comments on the oscillator in RSGB's *Amateur Radio Techniques*. The circuit is also described in the Nov 1955 *QST* "Technical Correspondence" column by W9IK. The Vackar circuit optimized the Clapp oscillator for frequency stability: the oscillator transistor is isolated from the resonator, tuning does not affect the feedback coupling, and the transistor's collector output impedance is kept low so that gain is the minimum necessary to sustain oscillation.

low-noise performance by Linley Gumm, K7HFD, and appears on page 126 of ARRL's *Solid State Design for the Radio Amateur*. Despite its publication in that homebrewer's bible (out of print, but available through libraries and loans), this circuit seems to have been overlooked by many builders. It uses no unusual components and looks simple, yet it is a subtle and sophisticated circuit. It represents the antithesis of G3PDM's VFO: In the pursuit of low noise sidebands, a number of design choices have been made that will degrade the stability of frequency over temperature.

The effects of oscillator noise have already been covered, and Fig 9.7 shows the effect of

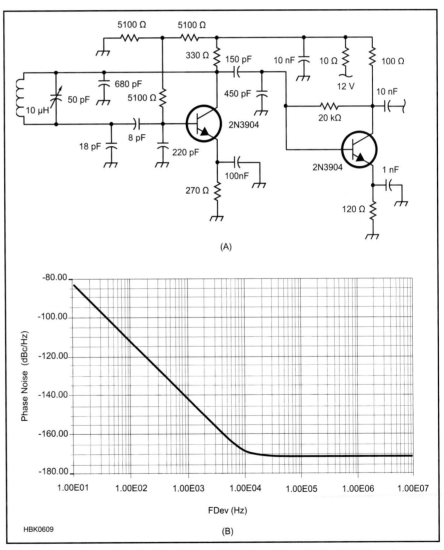

Fig 9.14A — At A, N1UL's Modified Vackar VFO is tuned from 5.5 to 6.6 MHz using the 50 pF capacitor. Tuning may be restricted to narrower ranges by placing a fixed capacitor in parallel with a smaller variable capacitor. The resonant frequency of the oscillator is determined by the 10 μH inductor and 50 pF tuning capacitor. B shows the excellent phase noise performance of the modified Vackar VFO in this *Harmonica* simulation. At 1 kHz from the carrier, noise is −144 dBc.

VFO Construction

These suggestions by G3PDM can be applied to any analog VFO circuit. Following these guidelines minimizes the effects of temperature change and mechanical vibration (microphonics) on the VFO.
- Use silvered-mica or other highly-stable capacitors for all fixed-value capacitors in the oscillator circuit. Power-supply decoupling capacitors may be any convenient type, such as ceramic or film.
- Tuning capacitors should be a high-quality component with double ball-bearings and silver-plated surfaces and contacts.
- The resonant circuit inductor should be wound on a ceramic form and solidly mounted.
- All oscillator components should be clean and attached to a solid support to minimize thermal changes and mechanical vibration.
- The enclosure should be solid and isolated from mechanical vibration.
- The power supply should be well-regulated, liberally decoupled and free of noise.
- Keep component leads short and if point-to-point wiring is employed, use heavy wire (#16 to #18 AWG).
- Single-point grounding of the oscillator components is recommended to avoid stray inductance and to minimize noise introduced from other sources. If a PCB is used, include a ground-plane.

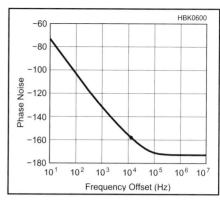

Fig 9.15 B — Modeling of the differential oscillator by Dr Ulrich Rohde, N1UL, shows its excellent phase-noise performance.

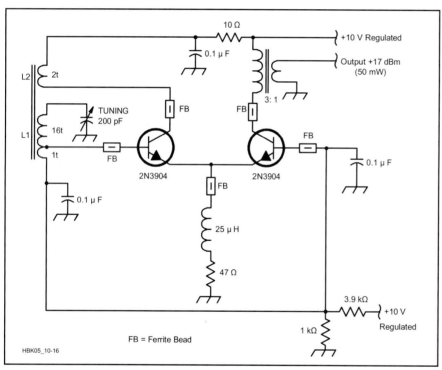

Fig 9.15 A — This low-noise oscillator design by K7HFD operates at an unusually high power level to achieve a high C/N (carrier-to-noise) ratio. Need other frequencies? See the caption of Fig 9.14 for a frequency-scaling technique.

limiting on the signal from a noisy oscillator. Because AM noise sidebands can get translated into PM noise sidebands by imperfect limiting, there is an advantage to stripping off the AM as early as possible, in the oscillator itself. An ALC system in the oscillator will counteract and cancel only the AM components within its bandwidth, but an oscillator based on a limiter will do this over a broad bandwidth. K7HFD's oscillator uses a differential pair of bipolar transistors as a limiting amplifier. The dc bias voltage at the bases and the resistor in the common emitter path to ground establishes a controlled dc bias current. The ac voltage between the bases switches this current between the two collectors. This applies a rectangular pulse of current into link winding L2, which drives the tank. The output impedance of the collector is high in both the current on and current off states. Allied with the small number of turns of the link winding, this presents a very high impedance to the tank circuit, which minimizes the damping of the tank Q. The input impedance of the limiter is also quite high and is applied across only a one-turn tap of L1, which similarly minimizes the effect on the tank Q. The input transistor base is driven into conduction only on one peak of the tank waveform. The output transformer has the inverse of the current pulse applied to it, so the output is not a low distortion sine wave, although the output harmonics will not be as extensive as simple theory would suggest because the circuit's high output impedance allows stray capacitances to attenuate high-frequency components. The low-frequency transistors used also act to reduce the harmonic power.

With an output of +17 dBm, this is a power oscillator, running with a large dc input power, so appreciable heating results that can cause temperature-induced drift. The circuit's high-power operation is a deliberate ploy to create a high signal-to-noise ratio by having as high a signal power as possible. This also reduces the problem of the oscillator's broadband noise output. The limitation on the signal level in the tank is the transistors' base-emitter-junction breakdown voltage. The circuit runs with a few volts peak-to-peak across the one-turn tap, so the full tank is running at over 50 V_{P-P}.

The single easiest way to damage a bipolar transistor is to reverse bias the base-emitter junction until it avalanches. Most devices are only rated to withstand 5 V applied this way, the current needed to do damage is small, and very little power is needed. If the avalanche current is limited to less than that needed to perform immediate destruction of the transistor, it is likely that there will be some degradation of the device, a reduction in its bandwidth and gain along with an increase in its noise. These changes are irreversible and cumulative. Small, fast signal diodes have breakdown voltages of over 30 V and less capacitance than the transistor bases, so one possible experiment would be to try the effect of adding a diode in series with the base of each transistor and running the circuit at even higher levels.

The amplitude must be controlled by the

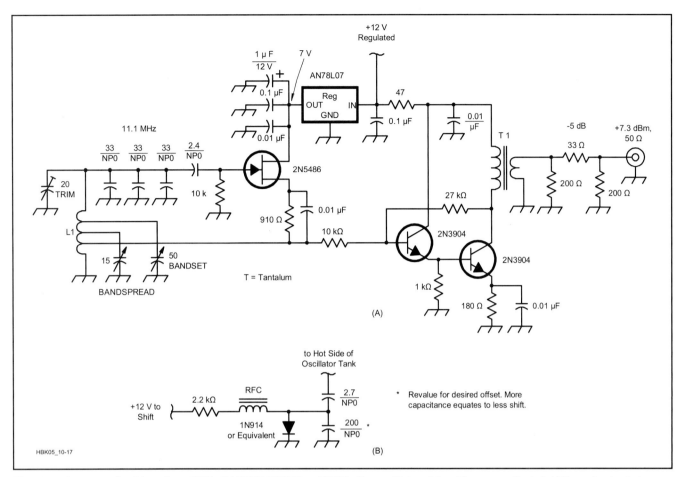

Fig 9.16 — Incorporating ideas from N1UL, KA7EXM, W7ZOI and W7EL, the oscillator at A achieves excellent stability and output at 11.1 MHz without the use of a gate-clamping diode, as well as end-running the shrinking availability of reduction drives through the use of bandset and bandspread capacitors. L1 consists of 10 turns of B & W #3041 Miniductor (#22 tinned wire, ⅝ inch in diameter, 24 turns per inch). The source tap is 2½ turns above ground; the tuning-capacitor taps are positioned as necessary for bandset and bandspread ranges required. T1's primary consists of 15 turns of #28 enameled wire on an FT-37-72 ferrite core; its secondary, 3 turns over the primary. B shows a system for adding fixed TR offset that can be applied to any LC oscillator. The RF choke consists of 20 turns of #26 enameled wire on an FT-37-43 core. Need other frequencies? See the caption of Fig 9.14 for a frequency-scaling technique.

drive current limit. The voltage on L2 must never allow the collector of the transistor driving it to go into saturation, if this happens, the transistor presents a very low impedance to L2 and badly loads the tank, wrecking the Q and the noise performance. The circuit can be checked to verify the margin from saturation by probing the hot end of L2 and the emitter with an oscilloscope. Another, less obvious, test is to vary the power-supply voltage and monitor the output power. While the circuit is under current control, there is very little change in output power, but if the supply is low enough to allow saturation, the output power will change significantly.

The use of the 2N3904 is interesting, as it is not normally associated with RF oscillators. It is a cheap, plain, general-purpose type more often used at dc or audio frequencies. There is evidence that suggests some transistors that have good noise performance *at RF* have worse noise performance at low frequencies, and that the low-frequency noise they create can modulate an oscillator, creating noise sidebands.

Experiments with low-noise audio transistors may be worthwhile, but many such devices have high junction capacitances. In the description of this circuit in *Solid State Design for the Radio Amateur*, the results of a phase-noise test made using a spectrum analyzer with a crystal filter as a preselector are given. Ten kilohertz out from the carrier, in a 3-kHz measurement bandwidth, the noise was more than 120 dB below the carrier level. This translates into better than $-120 - 10 \log (3000)$, which equals -154.8 dBc/Hz, SSB. At this offset, -140 dBc is usually considered to be excellent. This is state-of-the art performance by today's standards — in a 1977 publication.

A JFET HARTLEY VFO

Fig 9.16 shows an 11.1-MHz version of a VFO and buffer closely patterned after that used in 7-MHz transceiver designs published by Roger Hayward, KA7EXM, and Wes Hayward, W7ZOI ("The Ugly Weekender") and Roy Lewallen, W7EL ("The Optimized QRP Transceiver"). In it, a 2N5486 JFET Hartley oscillator drives the two-2N3904 buffer attributed to Lewallen. This version diverges from the originals in that its JFET uses source bias (the bypassed 910-Ω resistor) instead of a gate-clamping diode and is powered from a low-current 7-V regulator IC instead of a Zener diode and dropping resistor. The 5-dB pad sets the buffer's output to a level appropriate for "Level 7" (+7-dBm-LO) diode ring mixers.

The circuit shown was originally built with a gate-clamping diode, no source bias and a 3-dB output pad. Adjusting the oscillator bias as shown increased its output by 2 dB without degrading its frequency stability (200 to 300-Hz drift at power up, stability within ±20 Hz thereafter at a constant room temperature).

In recognition that precision mechanical tuning components are hard to obtain, the resonator uses "Bandset" and "Bandspread" variable capacitors. Terms from the early days of radio, *bandset* is for coarse-tuning and *bandspread* is for fine-tuning.

Oscillators and Synthesizers 9.17

9.3.4 VFO Components and Construction

TUNING CAPACITORS AND REDUCTION DRIVES

As most commercially made radios now use frequency synthesizers, it has become increasingly difficult to find certain key components needed to construct a good VFO. The slow-motion drives, dials, gearboxes, associated with names like Millen, National, Eddystone and Jackson are no longer available. Similarly, the most suitable silver-plated variable capacitors with ball races at both ends, although still made, are not generally marketed and are expensive. Three approaches remain: Scavenge suitable parts from old equipment; use tuning diodes instead of variable capacitors — an approach that, if uncorrected through phase locking, generally degrades stability and phase-noise performance; or use two tuning capacitors, one with a capacitance range $1/5$ to $1/10$ that of the other, in a bandset/bandspread approach.

Assembling a variable capacitor to a chassis and its reduction drive to a front panel can result in *backlash* — an annoying tuning effect in which rotating the capacitor shaft deforms the chassis and/or panel rather than tuning the capacitor. One way of minimizing this is to use the reduction drive to support the capacitor, and use the capacitor to support the oscillator circuit board.

FIXED CAPACITORS

Traditionally, silver-mica fixed capacitors have been used extensively in oscillators, but their temperature coefficient is not as low as can be achieved by other types, and some silver micas have been known to behave erratically. Polystyrene film has become a proven alternative. One warning is worth noting: Polystyrene capacitors exhibit a permanent change in value should they ever be exposed to temperatures much over 70 °C; they do not return to their old value on cooling. Particularly suitable for oscillator construction are the low-temperature-coefficient ceramic capacitors, often described as *NP0* or *C0G* types. These names are actually temperature-coefficient codes. Some ceramic capacitors are available with deliberate, controlled temperature coefficients so that they can be used to compensate for other causes of frequency drift with temperature. For example, the code N750 denotes a part with a temperature coefficient of −750 parts per million per degree Celsius. These parts are now somewhat difficult to obtain, so other methods are needed. (Values for temperature coefficients and other attributes of capacitors are presented in the **Component Data and References** chapter.)

In a Colpitts circuit, the two large-value capacitors that form the voltage divider for the active device still need careful selection. It would be tempting to use any available capacitor of the right value, because the effect of these components on the tank frequency is reduced by the proportions of the capacitance values in the circuit. This reduction is not as great as the difference between the temperature stability of an NP0 ceramic part and some of the low-cost, decoupling-quality X7R-dielectric ceramic capacitors. It's worth using low-temperature coefficient parts even in the seemingly less-critical parts of a VFO circuit — even as decouplers. Chasing the cause of temperature drift is more challenging than fun. Buy critical components like high-stability capacitors from trustworthy sources.

INDUCTORS

Ceramic coil forms can give excellent results, as can self-supporting air-wound coils (Miniductor). If you use a magnetic core, make it powdered iron, never ferrite, and support it stably. Stable VFOs have been made using toroidal cores, but again, ferrite must be avoided. Micrometals mix #6 has a low temperature coefficient and works well in conjunction with NP0 ceramic capacitors. Coil forms in other materials have to be assessed on an individual basis.

A material's temperature stability will not be apparent until you try it in an oscillator, but you can apply a quick test to identify those nonmetallic materials that are lossy enough to spoil a coil's Q. Put a sample of the coil-form material into a microwave oven along with a glass of water and cook it about 10 seconds on low power. *Do not include any metal fittings or ferromagnetic cores.* Good materials will be completely unaffected; poor ones will heat and may even melt, smoke, or burst into flame. (This operation is a fire hazard if you try more than a tiny sample of an unknown material. Observe your experiment continuously and do not leave it unattended.)

W7ZOI suggests annealing toroidal VFO coils after winding. W7EL reports achieving success with this method by boiling his coils in water and letting them cool in air.

VOLTAGE REGULATORS

VFO circuits are often run from locally regulated power supplies, usually from resistor/Zener diode combinations. Zener diodes have some idiosyncrasies that could spoil the oscillator. They are noisy, so decoupling is needed down to audio frequencies to filter this out. Zener diodes are often run at much less than their specified optimum bias current. Although this saves power, it results in a lower output voltage than intended, and the diode's impedance is much greater, increasing its sensitivity to variations in input voltage, output current and temperature. Some common Zener types may be designed to run at as much as 20 mA; check the data sheet for your diode family to find the optimum current.

True Zener diodes are low-voltage devices; above a couple of volts, so-called Zener diodes are actually avalanche types. The temperature coefficient of these diodes depends on their breakdown voltage and crosses through zero for diodes rated at about 5 V. If you intend to use nothing fancier than a common-variety Zener, designing the oscillator to run from 5 V and using a 5.1-V Zener, will give you a free advantage in voltage-versus-temperature stability. There are some diodes available with especially low temperature coefficients, usually referred to as *reference* or *temperature-compensated diodes*. These usually consist of a series pair of diodes designed to cancel each other's temperature drift. Running at 7.5 mA, the 1N829A provides 6.2 V ±5% and a temperature coefficient of just ±5 parts per million (ppm) maximum per degree Celsius. A change in bias current of 10% will shift the voltage less than 7.5 mV, but this increases rapidly for greater current variation. The 1N821A is a lower-grade version, at ±100 ppm/°C. The LM399 is a complex IC that behaves like a superb Zener at 6.95 V, ±0.3 ppm/°C. There are also precision, low-power, three-terminal regulators designed to be used as voltage references, some of which can provide enough current to run a VFO. There are comprehensive tables of all these devices between pages 334 and 337 of Horowitz and Hill, *The Art of Electronics*, 2nd ed.

OSCILLATOR DEVICES

The 2N3819 FET, a classic from the 1960s, has proven to work well in VFOs, but, like the MPF102 also long-popular with ham builders, it's manufactured to wide tolerances. Considering an oscillator's importance in receiver stability, you should not hesitate to spend a bit more on a better device. The 2N5484, 2N5485 and 2N5486 are worth considering; together, their transconductance ranges span that of the MPF102, but each is a better-controlled subset of that range. The 2N5245 is a more recent device with better-than-average noise performance that runs at low currents like the 2N3819. The 2N4416/A, also available as the plastic-cased PN4416, is a low-noise device, designed for VHF/UHF amplifier use, which has been featured in a number of good oscillators up to the VHF region. Its low internal capacitance contributes to low frequency drift. The J310 (plastic; the metal-cased U310 is similar) is another popular JFET in oscillators.

The 2N5179 (plastic, PN5179 or MPS5179) is a bipolar transistor capable of good performance in oscillators up to the top of the VHF region. Care is needed because its absolute-maximum collector-emitter voltage is only 12 V, and its collector current must not exceed 50 mA. Although these characteristics may seem to convey fragility, the 2N5179 is sufficient for circuits powered by stabilized 6-V power supplies.

VHF-UHF devices are not really necessary in LC VFOs because such circuits are rarely used above 10 MHz. Absolute frequency stability is progressively harder to achieve with increasing frequency, so free-running oscillators are rarely used to generate VHF-UHF signals for radio communication. Instead, VHF-UHF radios usually use voltage-tuned, phase-locked oscillators in some form of synthesizer. Bipolar devices like the BFR90 and MRF901, with f_Ts in the 5-GHz region and mounted in stripline packages, are needed at UHF.

Integrated circuits have not been mentioned until now because few specific RF-oscillator ICs exist. Some consumer ICs — the NE602, for example — include the active device(s) necessary for RF oscillators, but often this is no more than a single transistor fabricated on the chip. This works just as a single discrete device would, although using such on-chip devices may result in poor isolation from the rest of the circuits on the chip. There is one specialized LC-oscillator IC: the Motorola MC1648. This device has been made since the early 1970s and is a surviving member of MECL III, a long-obsolete family of emitter-coupled-logic devices. It is still used in military and commercial equipment. It is difficult to obtain, expensive, power hungry, and offers relatively low performance. Its circuitry is complex for an oscillator, with a multi-transistor limiting-amplifier cell controlled by an on-chip ALC system. The MC1648's first problem is that the ECL families use only about a 0.8 V swing between logic levels, and this same limitation applies to the signal in the oscillator tuned circuit. It is possible to improve this situation by using a tapped or transformer-coupled tank circuit to give improved Q, but risks the occurrence of the device's second problem.

Periodically, semiconductor manufacturers modernize their plants and scrap old assembly lines used to make old products. Any surviving devices then must undergo some redesign to allow their production by the new processes. One common result of this is that devices are shrunk, when possible, to fit more onto a wafer. All this increases the f_T of the transistors in the device, and such evolution has rendered today's MC1648s capable of operation at much higher frequencies than the specified 200-MHz limit. This allows higher-frequency use, but great care is needed in the layout of circuits using it to prevent spurious oscillation. A number of old designs using this part have needed reengineering because the newer parts generate spurious oscillations that the old ones didn't, using PC-board traces as parasitic tuned circuits.

The moral is that a UHF-capable device requires UHF-cognizant design and layout even if the device will be used at far lower frequencies. **Fig 9.17** shows the MC1648 in a simple circuit and with a tapped resonator. These more complex circuits have a greater risk of presenting a stray resonance within the device's operating range, risking oscillation at an unwanted frequency. This device is *not* a prime choice for an HF VFO because the physical size of the variable capacitor and the inevitable lead lengths, combined with the need to tap-couple to get sufficient Q for good noise performance, makes spurious oscillation difficult to avoid. The MC1648 is really intended for tuning-diode control in phase-locked loops operating at VHF. This difficulty is inherent in wideband devices, especially oscillator circuits connected to their tank by a single "hot" terminal, where there is simply no isolation between the amplifier's input and output paths. Any resonance in the associated circuitry can control the frequency of oscillation.

The popular SA/NE602 mixer IC has a built-in oscillator and can be found in many published circuits. This device has separate input and output pins to the tank and has proved to be quite tame. It may not have been "improved" yet (so far, it has progressed from the SA/NE602 to the SA/NE602A, the A version affording somewhat higher dynamic range than the original SA/NE602). It might be a good idea for anyone laying out a board using one to take a little extra care to keep PCB traces short in the oscillator section to build in some safety margin so that the board can be used reliably in the future. Experienced (read: "bitten") professional designers know that their designs are going to be built for possibly more than 10 years and have learned to make allowances for the progressive improvement of semiconductor manufacture.

9.3.5 Temperature Compensation

The general principle for creating a high-stability VFO is to use components with minimal temperature coefficients in circuits that are as insensitive as possible to changes in components' secondary characteristics. Even after careful minimization of the causes of temperature sensitivity, further improvement can still be desirable. The traditional method was to split one of the capacitors in the tank so that it could be apportioned between low-temperature-coefficient parts and parts with deliberate temperature dependency. Only a limited number of different, controlled temperature coefficients are available, so the proportioning between low coefficient and controlled coefficient parts was varied to "dilute" the temperature sensitivity of a part more sensitive than needed. This was a tedious process, involving much trial and error, an undertaking made more complicated by the difficulty of arranging means of heating and cooling the unit being compensated. (Hayward described such a means in December 1993 *QST*.) As commercial and military equipment have been based on frequency syn-

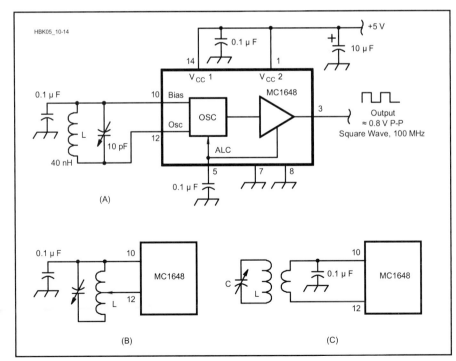

Fig 9.17 — One of the few ICs ever designed solely for oscillator service, the ECL Motorola MC1648 (A) requires careful design to avoid VHF parasitics when operating at HF. Keeping its tank Q high is another challenge; B and C show means of coupling the IC's low-impedance oscillator terminals to the tank by tapping up on the tank coil (B) or with a link (C).

thesizers for some time, supplies of capacitors with controlled temperature sensitivity are drying up. An alternative approach is needed.

A temperature-compensated crystal oscillator (TCXO) is an improved-stability version of a crystal oscillator that is used widely in industry. Instead of using controlled-temperature coefficient capacitors, most TCXOs use a network of thermistors and normal resistors to control the bias of a tuning diode. (See this chapter's Phase-Locked Loop section for information on the use of varactor diodes for oscillator tuning.) Manufacturers measure the temperature vs frequency characteristic of sample oscillators, and use a computer program to calculate the optimum normal resistor values for production. This can reliably achieve at least a tenfold improvement in stability. We are not interested in mass manufacture, but the idea of a thermistor tuning a varactor is worth stealing. The parts involved are likely to be available for a long time.

Browsing through component suppliers' catalogs shows ready availability of 4.5- to 5-kΩ-bead thermistors intended for temperature-compensation purposes, at less than a dollar each. **Fig 9.18** shows a circuit based on this form of temperature compensation. Commonly available thermistors have negative temperature coefficients, so as temperature rises, the voltage at the counterclockwise (CCW) end of R8 increases, while that at the clockwise (CW) end drops. Somewhere near the center, there is no change. Increasing the voltage on the tuning diode decreases its capacitance, so settings toward R8's CCW end simulate a negative-temperature-coefficient capacitor; toward its clockwise end, a positive-temperature-coefficient part. Choose R1 to pass 8.5 mA from whatever supply voltage is available to the 6.2-V reference diode, D1. The 1N821A/1N829A-family diode used has a very low temperature coefficient and needs 7.5 mA bias for best performance; the bridge takes 1 mA. R7 and R8 should be good-quality multi-turn trimmers. D2 and C1 need to be chosen to suit the oscillator circuit. Choose the least capacitance that provides enough compensation range. This reduces the noise added to the oscillator. (It is possible, though tedious, to solve for the differential varactor voltage with respect to R2 and R5, via differential calculus and circuit theory. The equations in Hayward's 1993 article can then be modified to accommodate the additional capacitors formed by D2 and C1.) Use a single ground point near D2 to reduce the influence of ground currents from other circuits. Use good-quality metal-film components for the circuit's fixed resistors.

The novelty of this circuit is that it was designed to have an easy and direct adjustment process. The circuit requires two adjustments, one at each of two different temperatures, and achieving them requires a stable frequency counter that can be kept far enough from the radio so that the radio, not the counter, is subjected to the temperature extremes. (Using a receiver to listen to the oscillator under test can speed the adjustments.) After connecting the counter to the oscillator to be corrected, run the radio containing the oscillator and compensator in a room-temperature, draft-free environment until the oscillator's frequency reaches its stable operating temperature rise over ambient. Lock its tuning, if possible. Adjust R7 to balance the bridge. This causes a drop of 0 V across R8, a condition you can reach by winding R8 back and forth across its range while slowly adjusting R7. When the bridge is balanced and 0 V appears across R8, adjusting R8 causes no frequency shift. When you've found this R7 setting, leave it there, set R8 to the exact center of its range and record the oscillator frequency.

Run the radio in a hot environment and allow its frequency to stabilize. Adjust R8 to restore the frequency to the recorded value. The sensitivity of the oscillator to temperature should now be significantly reduced between the temperatures at which you performed the adjustments. You will also have somewhat improved the oscillator's stability outside this range.

For best results with any temperature-compensation scheme, it's important to group all the oscillator and compensator components in the same box, avoiding differences in airflow over components. A good oscillator should not dissipate much power, so it's feasible, even advisable, to mount all of the oscillator components in an unventilated box. In the real world, temperatures change, and if the components being compensated and the components doing the compensating have different thermal time constants, a change in temperature can cause a temporary change in frequency until the slower components have caught up. One cure for this is to build the oscillator in a thick-walled metal box that's slow to heat or cool, and so dominates and reduces the possible rate of change of temperature of the circuits inside. This is sometimes called a *cold oven*.

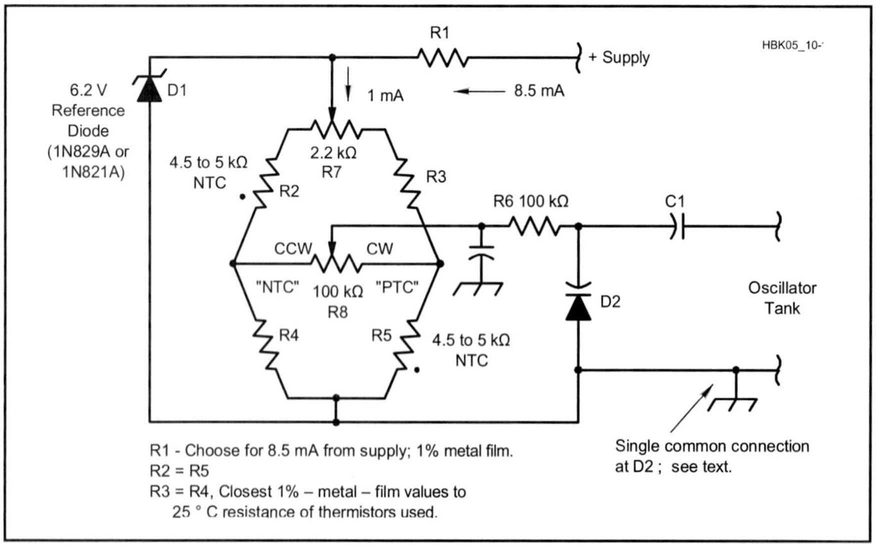

9.3.6 Shielding and Isolation

Oscillators contain inductors running at moderate power levels and so can radiate strong enough signals to cause interference with other parts of a radio, or with other radios. Oscillators are also sensitive to radiated signals. Effective shielding is therefore important. A VFO used to drive a power amplifier and antenna (to form a simple CW transmitter) can prove surprisingly difficult to shield well enough. Any leakage of the power amplifier's high-level signal back into the oscillator can affect its frequency, resulting in a poor transmitted note. If the radio gear is in the station antenna's near field, sufficient shielding may be even more difficult. The following rules of thumb continue to serve ham builders well:

• Use a complete metal box, with as few holes as possible drilled in it, with good contact around surface(s) where its lid(s) fit(s) on.

• Use feedthrough capacitors on power and control lines that pass in and out of the VFO enclosure, and on the transmitter or transceiver enclosure as well.

• Use *buffer amplifier* circuitry that amplifies the signal by the desired amount *and* provide sufficient attenuation of signal energy flowing in the reverse direction. This is known as *reverse isolation* and is a frequently

Fig 9.18 — Oscillator temperature compensation is difficult because of the scarcity of negative-temperature-coefficient capacitors. This circuit, by GM4ZNX, uses a bridge containing two identical thermistors to steer a tuning diode for drift correction. The 6.2-V Zener diode used (a 1N821A or 1N829A) is a temperature-compensated part; just any 6.2-V Zener will not do.

overlooked loophole in shielding. Figs 9.15 and 9.16 include buffer circuitry of proven performance. As another (and higher-cost) option, consider using a high-speed buffer-amplifier IC (such as the LM6321N by National Semiconductor, a part that combines the high input impedance of an op amp with the ability to drive 50-Ω loads directly up into the VHF range).

• Use a mixing-based frequency-generation scheme instead of one that operates straight through or by means of multiplication. Such a system's oscillator stages can operate on frequencies with no direct frequency relationship to its output frequency.

• Use the time-tested technique of running your VFO at a *subharmonic* of the output signal desired — say, 3.5 MHz in a 7-MHz transmitter — and *multiply* its output frequency in a suitably nonlinear stage for further amplification at the desired frequency.

9.4 Designing an Oscillator

We've covered a lot of ground about how oscillators work, their limitations and a number of interesting circuits, so the inevitable question arises of how to design one. Let's make the embarrassing confession right here at the beginning: Very few oscillators you see in published circuits or commercial equipment were designed by the equipment's designer. Almost all have been "Borrowed," "Re-used" or just plain stolen. While recycling in general is important for the environment, it means in this case that very few professional or amateur designers have ever designed an oscillator from scratch. We all have collections of circuits we've "harvested," and we adjust a few values or change a device type to produce something to suit a new project. Oscillators aren't designed, they evolve. They seem to have a life of their own. The Clarke & Hess book listed in the references contains one of the few published classical design processes.

Modern RF design is heavily reliant on simulation. Universities have shifted emphasis from hardware labs to computer simulations. Simulators work well on amplifiers, filters, modulators, mixers and so on, but not as well on oscillators. This leaves even graduate-level RF engineers rather light on the subject of oscillators.

Such a void is a back-door opportunity to oscillator analysis. Simulation has not been used much by amateurs because it is computationally intensive. Most recent PCs have plenty of power for this task, however. You may not be aware of how powerful your machine is, because things like operating systems and word processor software have become progressively inefficient. Set your computer loose on a math-intensive task like circuit simulation and see what a beast has been hiding behind the flashy graphics and zillions of type fonts.

The simulator needs to be able to handle oscillators, and it needs to be affordable on an amateur budget. High-end RF CAD packages usually contain tools for handling oscillators, but they aren't easy to use and many are afterthought additions that don't mesh with the rest of the package. Oscillators are diabolical things to simulate. Even if you can afford to shovel money at the problem, it still takes a bit of creativity to use a full "bells and whistles" simulator on an oscillator and for the results to make sense. (For detailed information on simulation, see the chapter on **Computer-Aided Circuit Design**.)

An oscillator is a simple loop with signal flowing round and round in it. We can view the signal flow as being many components adding together at some point at some time. We can think of these components as being different vintages. This component came from thermal noise X microseconds ago and has been round the loop Y times, and that component… you can see the pattern. This is hard to analyze because all components get summed, not separated. So instead of thinking of our oscillator as a single stage with everything distributed in time, let's distribute it in space as well.

The GM4ZNX "Swiss-Roll" technique is to think of a rolled-up jelly cake and unroll the oscillator circuit as illustrated in **Fig 9.19**. We break the loop open, which leaves us with a tuned amplifier circuit with an input and an output terminal. In a closed oscillator loop, the output terminal would drive the input terminal. In the unrolled circuit, the output terminal drives the input of a second stage and so on.

Instead of infinitely looping signals in a single stage, we now have a single signal flow down an infinite string of stages. This doesn't sound any better, but we only need to handle a couple of hundred stages. Think about that… we are going to simulate a 200-stage tuned amplifier! But all the stages are identical, so we only have to draw one stage and have the others built as hierarchical repeats. On a 2005 model laptop, a simulation takes several minutes, which is fine.

Affordable RF analysis programs are normally limited to linear simulations and can't handle the oscillator's amplitude control mechanism. We also need a simulator that can handle random noise at the same time so the choice of tools becomes very limited. One example of a suitable simulation package, based on *SPICE*, is available free from the Linear Technologies Web site, **www.linear.com**. (The program is called *SWCADIII* or *LTSPICE*.) It is optimized for the analysis of switch-mode power supply (SMPS) circuits. This is very good for our purposes. In an SMPS things ring at high frequencies, and slow things settle. *LTSPICE* has some trickery which allows it to easily handle RF speed

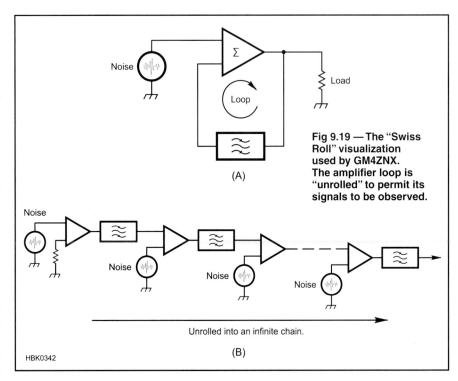

Fig 9.19 — The "Swiss Roll" visualization used by GM4ZNX. The amplifier loop is "unrolled" to permit its signals to be observed.

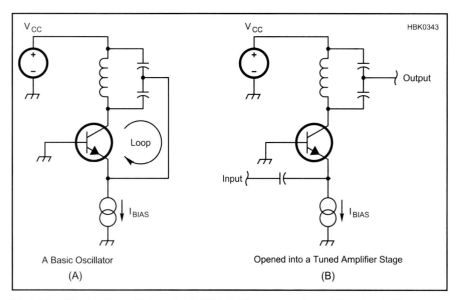

Fig 9.20 — The basic oscillator circuit (A) is built on a tuned amplifier circuit (B).

events, while running a simulation efficiently over a long enough period for control loops to settle. This is what makes it efficient for oscillators. Our oscillator circuits are run at such large signal levels that as the oscillator reaches its long term output, it will affect bias conditions, which are slow-damped by decoupling components. (Again, refer to the **Computer-Aided Circuit Design** chapter for more information on circuit simulation.)

LTSPICE can handle noise, and it can also handle hierarchical schematics. We draw our basic oscillator circuit once, and create a symbol for it. We can then use that symbol in a higher level schematic. For our worked example of an oscillator analysis, a single-stage schematic and symbol were created, then a higher level, a 10-pack of those symbols was created. The top level was made of 20 10-packs. Files for these designs are in the *Handbook's* companion CD, but to be able to set them up in *LTSPICE* you will need to read *LTSPICE*'s help files on hierarchical designs. Hierarchical schematics save the need to draw our oscillator stage 200 times. It also allows editing a parameter on one schematic to change all 200 stages at once. Without this feature, things would be tedious.

Fig 9.20 shows a basic oscillator circuit. The transistor is a bipolar type, operated in common-base mode. The resonator is an LC tank with a capacitive tap. The common-base amplifier gives less than unity current gain, low input impedance and high output impedance. There has to be an impedance transformation to make the loop have enough gain to oscillate, and the capacitive tap is it. Instead of having bias resistor networks, the stage has a voltage source to set the collector bias voltage, and a current source to set the quiescent emitter current. You can just type in your bias values without having to calculate resistor values. One drawback of *LTSPICE* is that it has few RF parts in its libraries, but a search on the Internet will find user groups and guidance on editing models for other *SPICE* variants to work with *LTSPICE*. The 2N4142 in the supplied library is used in this example. A 5 V collector bias and 0.2 mA emitter bias have been chosen as starting points.

In designing this circuit, we must create a stage gain greater than unity, and allow the operating level to rise until that gain is compressed to exactly unity. With a bipolar transistor at RF, we do not want it to go into saturation. Instead, we want it to go into cut-off before that can happen. Set the emitter current for the level you want, and then make sure there is enough collector bias voltage that the instantaneous collector voltage is always a few volts more positive than the emitter. The L/C ratio of the tank, and the capacitive tap ratio are surprisingly tolerant, the circuit will oscillate with these parameters varied over an appreciable range. If your goal is low drift, then you should go for high capacitance in the tank, to dilute the effect of strays. If your goal is low phase noise, then you should try to create as high a tank Q as possible.

Because oscillators of this type use nonlinear device operation to control amplitude, do not try to get low levels of harmonics on their output — the harmonic creation mechanism is fundamental to the oscillator. Try an oscillator with a detector and feedback loop to control its level instead. The square-law characteristic of JFETs is an excellent level-stabilizing mechanism and accounts for their popularity in oscillators. However, their wide range of I_{DSS} for any given type, can be troublesome. MMIC amplifiers with internal feedback are probably the worst choice of device, but some types can be used in special circumstances.

In the simulation circuit provided here you will find some odd things. The random number source is of uniform density, so several of them are added together to approximate Gaussian noise. The collector load resistor which represents the output of the oscillator is split into two parts, a resistor to ground on the output is needed to define the dc potential of a node between the capacitors of adjacent stages (or else *SPICE* complains) and it also loads the tank via the tap.

When you have got all the schematics in the right directories for *LTSPICE* to use them, and run the simulation, you will have an immense database of all waveforms at all nodes. You can now probe around the top level schematic. The first transistor in the chain acts as an emitter-follower so you can see its noise voltage coming out of its input if you probe there. The way the random number generators work is tied to the simulation time clock and you will see the first random number generator turn on at about 2 µs into the simulation, followed progressively by the others. This is just an oddity of the realization, but it does illustrate the summation of differently-seeded generators to approximate Gaussian distribution of noise.

Probing along, 10 stages at a time, you will see bigger and bigger noise waveforms until you start to notice the effect of the cumulative selectivity, as it starts to look like a noisy sine wave. Further along you will see the level begin to stabilize. If you zoom out to a larger time-scale, you will see that the sine wave seems to have low frequency random noise AM applied to it. As you probe further along you will see this "AM" compressed until it isn't seen, by the level control action of the oscillator and progressively narrowing bandwidth. Long before you get to the end of the 200 stages, you will have the appearance of a stable sine wave. Remember that the taps earlier in the structure represent the signal in a closed loop oscillator at times close to when it started-up. In a running oscillator, compression eventually reduces the gain of all amplification passes of all signal components. Fourier transform conversion can be used to convert these time-pictures into spectrum displays, if you want to explore further.

Among the accompanying files there is also a simulator that takes a single stage and plots its compression characteristic. It could be improved by adding earlier and later stages to make the source and load impedances more representative. The example oscillator can also have a model of a quartz crystal added in series with its output, to simulate a crystal oscillator. Explore! And have fun.

How can a lower phase noise oscillator be designed? Firstly if you use a lower noise transistor, it will take more stages of gain and filtering to bring it up to the stable output level. The greater repetition of filtering before noise reaches a significant level narrows the bandwidth of the output signal. Secondly, if you can engineer a higher-Q tank circuit, each pass through the tank has a greater narrowing effect.

The simulation files for this section are included on the CD-ROM that accompanies this book:

SwissRollDemoBasicStage.asc — (The oscillator stage. Editing this changes every stage)
SwissRollDemoBasicStage.asy — (The symbol file for use in higher level schematics)
SwissRollDemo10Pack.asc — (10 basic stages as a building block)
SwissRollDemo10Pack.ass — (The symbol for a 10 pack)
SwissRollDemoTopLevel.asc — (This is the real simulation, uses the above as components)
SwissRollDemoCompression.asc — (Crude simulation showing stage gain compression)
SwissRollDemoStageBandwidth.asc — (Crude simulation showing filter Q of one stage)

9.5 Quartz Crystals in Oscillators

Because crystals afford Q values and frequency stabilities that are orders of magnitude better than those achievable with LC circuits, fixed-frequency oscillators usually use quartz crystal resonators. Master references for frequency counters and synthesizers are always based on crystal oscillators.

So glowing is the executive summary of the crystal's reputation for stability that newcomers to radio experimentation naturally believe that the presence of a crystal in an oscillator will force oscillation at the frequency stamped on the can. This impression is usually revised after the first few experiences to the contrary! There is no sure-fire crystal oscillator circuit (although some are better than others); reading and experience soon provide a learner with plenty of anecdotes to the effect that:

• Some circuits have a reputation of being temperamental, even to the point of not always starting.

• Crystals sometimes mysteriously oscillate on unexpected frequencies.

Even crystal manufacturers have these problems, so don't be discouraged from building crystal oscillators. The occasional uncooperative oscillator is a nuisance, not a disaster, and it just needs a little individual attention. Knowing how a crystal behaves is the key to a cure.

Dr Ulrich Rohde, N1UL, has generously contributed a pair of detailed papers that discuss the crystal oscillator along with several HF and VHF designs. Both papers, "Quartz Crystal Oscillator Design" and "A Novel Grounded Base Oscillator Design for VHF/UHF Frequencies" are included on the CD-ROM accompanying this book.

9.5.1 Quartz and the Piezoelectric Effect

Quartz is a crystalline material with a regular atomic structure that can be distorted by the simple application of force. Remove the force, and the distorted structure springs back to its original form with very little energy loss. This property allows *acoustic waves* — sound — to propagate rapidly through quartz with very little attenuation, because the velocity of an acoustic wave depends on the elasticity and density (mass/volume) of the medium through which the wave travels.

If you heat a material, it expands. Heating may cause other characteristics of a material to change — such as elasticity, which affects the speed of sound in the material. In quartz, however, expansion and change in the speed of sound are very small and tend to cancel, which means that the transit time for sound to pass through a piece of quartz is very stable.

The third property of this wonder material is that it is *piezoelectric*. Apply an electric field to a piece of quartz and the crystal lattice distorts just as if a force had been applied. The electric field applies a force to electrical charges locked in the lattice structure. These charges are captive and cannot move around in the lattice as they can in a semiconductor, for quartz is an insulator. A capacitor's dielectric stores energy by creating physical distortion on an atomic or molecular scale. In a piezoelectric crystal's lattice, the distortion affects the entire structure. In some piezoelectric materials, this effect is sufficiently pronounced that special shapes can be made that bend *visibly* when a field is applied.

Consider a rod made of quartz. Any sound wave propagating along it eventually hits an end, where there is a large and abrupt change in acoustic impedance. Just as when an RF wave hits the end of an unterminated transmission line, a strong reflection occurs. The rod's other end similarly reflects the wave. At some frequency, the phase shift of a round trip will be such that waves from successive round trips exactly coincide in phase and reinforce each other, dramatically increasing the wave's amplitude. This is *resonance*.

The passage of waves in opposite directions forms a standing wave with antinodes at the rod ends. Here we encounter a seeming ambiguity: not just one, but a family of different frequencies, causes standing waves — a family fitting the pattern of 1/2, 2/3, 5/2, 7/2 and so on, wavelengths into the length of the rod. And this is the case: A quartz rod can resonate at any and all of these frequencies.

The lowest of these frequencies, where the crystal is ½-wavelength long, is called the *fundamental* mode. The others are named the third, fifth, seventh and so on, *overtones*. There is a small phase-shift error during reflection at the ends, which causes the frequencies of the overtone modes to differ slightly from odd integer multiplies of the fundamental. Thus, a crystal's third overtone is very close to, but not exactly, three times its fundamental frequency. Many people are confused by overtones and harmonics. Harmonics are additional signals at exact integer multiples of the fundamental frequency. Overtones are not signals at all; they are additional resonances that can be exploited if a circuit is configured to excite them.

The crystals we use most often resonate in the 1- to 30-MHz region and are of the *AT-cut, thickness shear* type, although these last two characteristics are rarely mentioned. A 15-MHz-fundamental crystal of this type is about 0.15 mm thick. Because of the widespread use of reprocessed war-surplus, pressure-mounted FT-243 crystals, you may think of crystals as small rectangles on the order of a half-inch in size. The crystals we

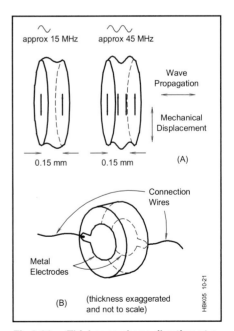

Fig 9.21 — Thickness-shear vibration at a crystal's fundamental and third overtone (A); B shows how the modern crystals commonly used by radio amateurs consist of etched quartz discs with electrodes deposited directly on the crystal surface.

commonly use today are discs, etched and/or doped to their final dimensions, with metal electrodes deposited directly on the quartz. A crystal's diameter does not directly affect its frequency; diameters of 8 to 15 mm are typical. (Quartz crystals are also discussed in the **RF and AF Filters** chapter.)

AT-cut is one of a number of possible standard designations for the orientation at which a crystal disc is sawed from the original quartz crystal. The crystal lattice atomic structure is asymmetric, and the orientation of this with respect to the faces of the disc influences the crystal's performance. *Thickness shear* is one of a number of possible orientations of the crystal's mechanical vibration with respect to the disc. In this case, the crystal vibrates perpendicularly to its thickness. This is not easy to visualize, and diagrams don't help much, but **Fig 9.21** is an attempt at illustrating this. Place a moist bathroom sponge between the palms of your hands, move one hand up and down, and you'll see thickness shear in action.

There is a limit to how thin a disc can be made, given requirements of accuracy and price. Traditionally, fundamental-mode crystals have been made up to 20 MHz, although 30 MHz is now common at a moderately raised price. Using techniques pioneered in the semiconductor industry, crystals have been made with a central region etched down to a thin membrane, surrounded by a thick ring for robustness. This approach can push fundamental resonances to over 100 MHz, but these are more lab curiosities than parts for everyday use. The easy solution for higher frequencies is to use a nice, manufacturably-thick crystal on an overtone mode. All crystals have all modes, so if you order a 28.060-MHz, third-overtone unit for a little QRP transmitter, you'll get a crystal with a fundamental resonance somewhere near 9.353333 MHz, but its manufacturer will have adjusted the thickness to plant the third overtone exactly on the ordered frequency. An accomplished manufacturer can do tricks with the flatness of the disc faces to make the wanted overtone mode a little more active and the other modes a little less active. (As some builders discover, however, this does not *guarantee* that the wanted mode is the most active!)

Quartz's piezoelectric property provides a simple way of driving the crystal electrically. Early crystals were placed between a pair of electrodes in a case. This gave amateurs the opportunity to buy surplus crystals, open them and grind them a little to reduce their thickness, thus moving them to higher frequencies. The frequency could be reduced very slightly by loading the face with extra mass, such as by blackening it with a soft pencil. Modern crystals have metal electrodes deposited directly onto their surfaces (Fig 9.21B), and such tricks no longer work.

The piezoelectric effect works both ways. Deformation of the crystal produces voltage across its electrodes, so the mechanical energy in the resonating crystal can also be extracted electrically by the same electrodes. Seen electrically, at the electrodes, the mechanical resonances look like electrical resonances. Their Q is very high. A Q of 10,000 would characterize a *poor* crystal nowadays; 100,000 is often reached by high-quality parts. For comparison, a Q of over 200 for an LC tank is considered good.

9.5.2 Frequency Accuracy

A crystal's frequency accuracy is as outstanding as its Q. Several factors determine a crystal's frequency accuracy. First, the manufacturer makes parts with certain tolerances: ±200 ppm for a low-quality crystal for use as in a microprocessor clock oscillator, ±10 ppm for a good-quality part for professional radio use. Anything much better than this starts to get expensive! A crystal's resonant frequency is influenced by the impedance presented to its terminals, and manufacturers assume that once a crystal is brought within several parts per million of the nominal frequency, its user will perform fine adjustments electrically.

Second, a crystal ages after manufacture. Aging could give increasing or decreasing frequency; whichever, a given crystal usually keeps aging in the same direction. Aging is rapid at first and then slows down. Aging is influenced by the care in polishing the surface of the crystal (time and money) and by its holder style. The cheapest holder is a soldered-together, two-part metal can with glass bead insulation for the connection pins. Soldering debris lands on the crystal and affects its frequency. Alternatively, a two-part metal can be made with flanges that are pressed together until they fuse, a process called *cold-welding*. This is much cleaner and improves aging rates roughly fivefold compared to soldered cans. An all-glass case can be made in two parts and fused together by heating in a vacuum. The vacuum raises the Q, and the cleanliness results in aging that's roughly 10 times slower than that achievable with a soldered can. The best crystal holders borrow from vacuum-tube assembly processes and have a *getter*, a highly reactive chemical substance that traps remaining gas molecules, but such crystals are used only for special purposes.

Third, temperature influences a crystal. A reasonable, professional quality part might be specified to shift not more than ±10 ppm over 0 to 70 °C. An AT-cut crystal has an *S*-shaped frequency-versus-temperature characteristic, which can be varied by slightly changing the crystal cut's orientation. **Fig 9.22** shows the general shape and the effect of changing the cut angle by only a few seconds of arc. Notice how all the curves converge at 25 °C. This is because this temperature is normally chosen as the reference for specifying a crystal. The temperature stability specification sets how accurate the manufacturer must make the cut. Better stability may be needed for a crystal used as a receiver frequency standard, frequency counter clock and so on. A crystal's temperature characteristic shows a little hysteresis. In other words, there's a bit of offset to the curve depending on whether temperature is increasing or decreasing. This is usually of no consequence except in the highest-precision circuits.

It is the temperature of the quartz that is important, and as the usual holders for crystals all give effective thermal insulation, only a couple of milliwatts dissipation by the crystal itself can be tolerated before self-heating becomes troublesome. Because such heating occurs in the quartz itself and does not come from the surrounding environment, it defeats the effects of temperature compensators and ovens.

The techniques shown earlier for VFO for temperature compensation can also be applied to crystal oscillators. An after-compensation drift of 1 ppm is routine and 0.5 ppm is good. The result is a *temperature-compensated crystal oscillator* (*TCXO*). Recently, oscillators have appeared with built-in digital thermometers, microprocessors and ROM look-up tables customized on a unit-by-unit basis to control a tuning diode via a digital-to-analog converter (DAC) for temperature compensation. These *digitally temperature-compensated oscillators* (*DTCXOs*) can reach 0.1 ppm over the temperature range. With automated production and adjustment, they promise to become the cheapest way to achieve this level of stability.

Oscillators have long been placed in temperature-controlled *ovens*, which are typically held at 80 °C. Stability of several parts per billion can be achieved over temperature, but this is a limited benefit as aging can easily dominate the accuracy. These are usually called *oven-controlled crystal oscillators* (*OCXOs*).

Fourth, the crystal is influenced by the impedance presented to it by the circuit in which it is used. This means that care is needed to make the rest of an oscillator circuit stable, in

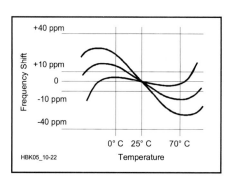

Fig 9.22 — Slight changes in a crystal cut's orientation shift its frequency-versus-temperature curve.

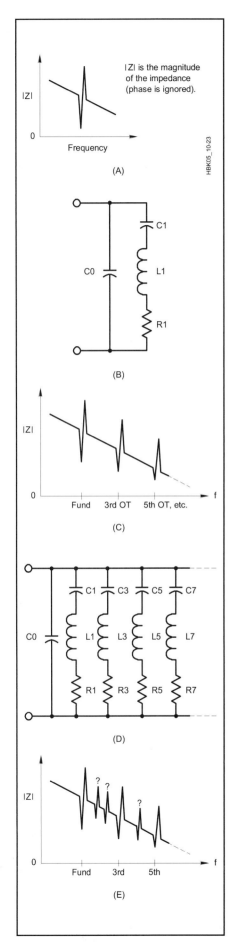

Fig 9.23 — Exploring a crystal's impedance (A) and equivalent circuit (B) through simplified diagrams. C and D extend the investigation to include overtones; E, to spurious responses not easily predictable by theory or controllable through manufacture. A crystal may oscillate on any of its resonances under the right conditions.

Table 9.2
Typical Equivalent Circuit Values for a Variety of Crystals

Crystal Type	Series L	Series C (pF)	Series R (Ω)	Shunt C (pF)
1-MHz fundamental	3.5 H	0.007	340	3.0
10-MHz fundamental	9.8 mH	0.026	7	6.3
30-MHz third overtone	14.9 mH	0.0018	27	6.2
100-MHz fifth overtone	4.28 mH	0.0006	45	7.0

terms of impedance and phase shift.

Gravity can slightly affect crystal resonance. Turning an oscillator upside down usually produces a small frequency shift, usually much less than 1 ppm; turning the oscillator back over reverses this. This effect is quantified for the highest-quality reference oscillators.

9.5.3 The Equivalent Circuit of a Crystal

Because a crystal is a passive, two-terminal device, its electrical appearance is that of an impedance that varies with frequency. **Fig 9.23A** shows a very simplified sketch of the magnitude (phase is ignored) of the impedance of a quartz crystal. The general trend of dropping impedance with increasing frequency implies capacitance across the crystal. The sharp fall to a low value resembles a series-tuned tank, and the sharp peak resembles a parallel-tuned tank. These are referred to as series and parallel resonances. Fig 9.23B shows a simple circuit that will produce this impedance characteristic. The impedance looks purely resistive at the exact centers of both resonances, and the region between them has impedance increasing with frequency, which looks inductive.

C1 (sometimes called *motional capacitance*, C_m, to distinguish it from the lumped capacitance it approximates) and L1 (*motional inductance*, L_m) create the series resonance, and as C0 and R1 are both fairly small, the impedance at the bottom of the dip is very close to R1. At parallel resonance, L1 is resonating with C1 and C0 in series, hence the higher frequency. The impedance of the parallel tank is immense, the terminals are connected to a capacitive tap, which causes them to see only a small fraction of this, which is still a very large impedance. The overtones should not be neglected, so Figs 9.23C and 9.23D include them. Each overtone has series and parallel resonances and so appears as a series tank in the equivalent circuit. C0 again provides the shifted parallel resonance.

This is still simplified, because real-life crystals have a number of spurious, unwanted modes that add yet more resonances, as shown in Fig 9.23E. These are not well controlled and may vary a lot even between crystals made to the same specification. Crystal manufacturers work hard to suppress these spurs and have evolved a number of recipes for shaping crystals to minimize them. Just where they switch from one design to another varies from manufacturer to manufacturer.

Always remember that the equivalent circuit is just a representation of crystal behavior and does not represent circuit components actually present. Its only use is as an aid in designing and analyzing circuits using crystals. **Table 9.2** lists typical equivalent-circuit values for a variety of crystals. It is impossible to build a circuit with 0.026 to 0.0006-pF capacitors; such values would simply be swamped by strays. Similarly, the inductor must have a Q that is orders of magnitude better than is practically achievable, and impossibly low stray C in its winding.

The values given in Table 9.2 are nothing more than rough guides. A crystal's frequency is tightly specified, but this still allows inductance to be traded for capacitance. A good manufacturer could hold these characteristics within a ±25% band or could vary them over a 5:1 range by special design. Similarly marked parts from different sources vary widely in motional inductance and capacitance.

Quartz is not the only material that behaves in this way, but it is the best. Resonators can be made out of lithium tantalate and a group of similar materials that have lower Q, allowing them to be *pulled* over a larger frequency range in VCXOs. Much more common, however, are ceramic resonators based on the technology of the well-known ceramic IF filters. These have much lower Q than quartz and much poorer frequency precision. They serve mainly as clock resonators for cheap microprocessor systems in which every last cent must be saved. A ceramic resonator could be used as the basis of a wide range, cheap VXO, but its frequency stability would not be as good as a good LC VFO.

9.5.4 Crystal Oscillator Circuits

(See also the papers "Quartz Crystal Oscillator Design" and "A Novel Grounded Base

Oscillator Design for VHF/UHF Frequencies" by Rohde, included on the CD-ROM accompanying this book.)

Crystal oscillator circuits are usually categorized as series- or parallel-mode types, depending on whether the crystal's low- or high-impedance resonance comes into play at the operating frequency. The series mode is now the most common; parallel-mode operation was more often used with vacuum tubes. **Fig 9.24** shows a basic series-mode oscillator. Some people would say that it is an overtone circuit, used to run a crystal on one of its overtones, but this is not necessarily true. The tank (L-C1-C2) tunes the collector of the common-base amplifier. C1 is larger than C2, so the tank is tapped in a way that transforms to a lower impedance, decreasing signal voltage, but increasing current. The current is fed back into the emitter via the crystal. The common-base stage provides a current gain of less than unity, so the transformer in the form of the tapped tank is essential to give loop gain. There are *two* tuned circuits, the obvious collector tank and the series-mode one "in" the crystal. The tank kills the amplifier's gain away from its tuned frequency, and the crystal will only pass current at the series resonant frequencies of its many modes. The tank resonance is much broader than any of the crystal's modes, so it can be thought of as the crystal setting the frequency, but the tank selecting which of the crystal's modes is active. The tank could be tuned to the crystal's fundamental, or one of its overtones.

Fundamental oscillators can be built without a tank quite successfully, but there is always the occasional one that starts up on an overtone or spurious mode. Some simple oscillators have been known to change modes while running (an effect triggered by changes in temperature or loading) or to not always start in the same mode! A series-mode oscillator should present a low impedance to the crystal at the operating frequency. In Fig 9.24, the tapped collector tank presents a transformed fraction of the 1-kΩ collector load resistor to one end of the crystal, and the emitter presents a low impedance to the other. To build a practical oscillator from this circuit, choose an inductor with a reactance of about 300 Ω at the wanted frequency and calculate C1 in series with C2 to resonate with it. Choose C1 to be 3 to 4 times larger than C2. The amplifier's quiescent ("idling") current sets the gain and hence the operating level. This is not easily calculable, but can be found by experiment. Too little quiescent current and the oscillator will not start reliably; too much and the transistor can drive itself into saturation. If an oscilloscope is available, it can be used to check the collector waveform; otherwise, some form of RF voltmeter can be used to allow the collector voltage to be set to 2 to 3 V RMS. 3.3 kΩ would be a suitable starting point for the emitter bias resistor. The transistor type is not critical; 2N2222A or 2N3904 would be fine up to 30 MHz; a 2N5179 would allow operation as an overtone oscillator to over 100 MHz (because of the low collector voltage rating of the 2N5179, a supply voltage lower than 12 V is required). The ferrite bead on the base gives some protection against parasitic oscillation at UHF.

If the crystal is shorted, this circuit should still oscillate. This gives an easy way of adjusting the tank; it is even better to temporarily replace the crystal with a small-value (tens of ohms) resistor to simulate its *equivalent series resistance* (ESR), and adjust L until the circuit oscillates close to the wanted frequency. Then restore the crystal and set the quiescent current. If a lot of these oscillators were built, it would sometimes be necessary to adjust the current individually due to the different equivalent series resistance of individual crystals. One variant of this circuit has the emitter connected directly to the C1/C2 junction, while the crystal is a decoupler for the transistor base (the existing capacitor and ferrite bead not being used). This works, but with a greater risk of parasitic oscillation.

We commonly want to trim a crystal oscillator's frequency. While off-tuning the tank a little will pull the frequency slightly, too much detuning spoils the mode control and can stop oscillation (or worse, make the circuit unreliable). The answer to this is to add

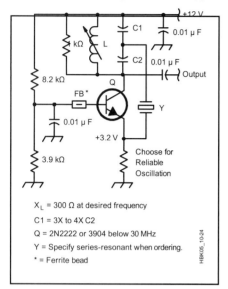

Fig 9.24 — A basic series-mode crystal oscillator. A 2N5179 can be used in this circuit if a lower supply voltage is used; see text.

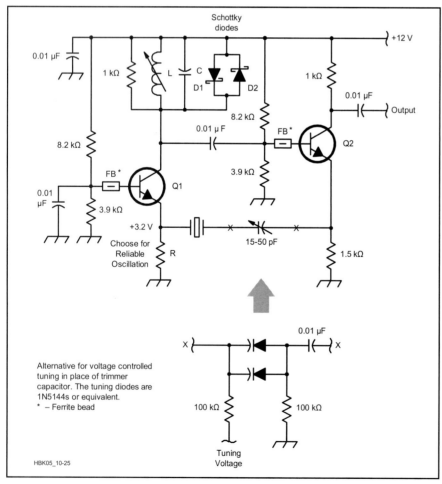

Fig 9.25 — A Butler crystal oscillator.

a trimmer capacitor, which will act as part of the equivalent series tuned circuit, in series with the crystal. This will shift the frequency in one way only, so the crystal frequency must be re-specified to allow the frequency to be varying around the required value. It is common to specify a crystal's frequency with a standard load (30 pF is commonly specified), so that the manufacturer grinds the crystal such that the series resonance of the specified mode is accurate when measured with a capacitor of this value in series. A 15- to 50-pF trimmer can be used in series with the crystal to give fine frequency adjustment. Too little capacitance can stop oscillation or prevent reliable starting. The Q of crystals is so high that marginal oscillators can take several seconds to start!

This circuit can be improved by reducing the crystal's driving impedance with an emitter follower as in **Fig 9.25**. This is the *Butler* oscillator. Again the tank controls the mode to either force the wanted overtone or protect the fundamental mode. The tank need not be tapped because Q2 provides current gain, although the circuit is sometimes seen with C split, driving Q2 from a tap. The position between the emitters offers a good, low-impedance environment to keep the crystal's in-circuit Q high. R, in the emitter of Q1, is again selected to give reliable oscillation. The circuit has been shown with a capacitive load for the crystal, to suit a unit specified for a 30-pF load. An alternative circuit to give electrical fine tuning is also shown. The diodes across the tank act as limiters to stabilize the operating amplitude and limit the power dissipated in the crystal by clipping the drive voltage to Q2. The tank

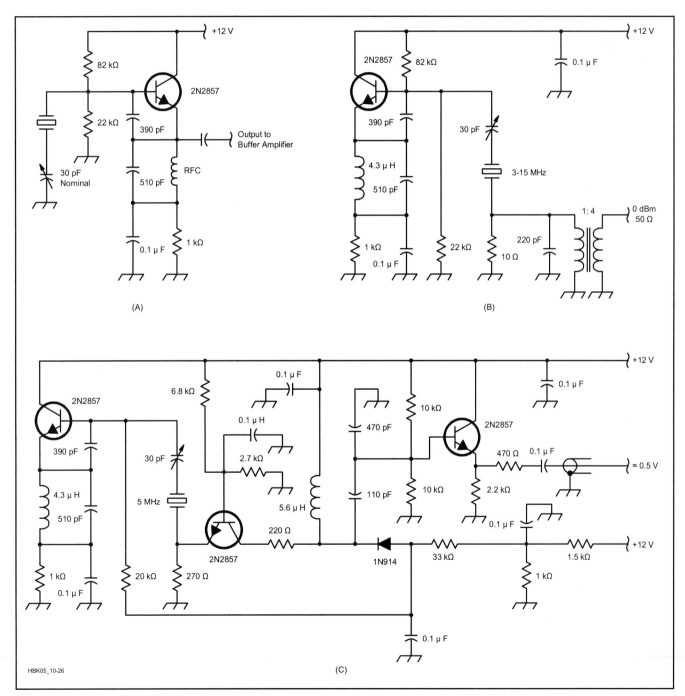

Fig 9.26 — The crystal in the series-tuned Colpitts oscillator at A operates in its series-resonant mode. B shows N1UL's low-noise version, which uses the crystal as a filter and features high harmonic suppression (from Rohde, *Microwave and Wireless Synthesizers Theory and Design*; see references). The circuit at C builds on the B version by adding a common-base output amplifier and ALC loop.

should be adjusted to peak at the operating frequency, not used to trim the frequency. The capacitance in series with the crystal is the proper frequency trimmer.

The Butler circuit works well, and has been used in critical applications to 140 MHz (seventh-overtone crystal, 2N5179 transistor). Although the component count is high, the extra parts are cheap ones. Increasing the capacitance in series with the crystal reduces the oscillation frequency but has a progressively diminishing effect. Decreasing the capacitance pulls the frequency higher, to a point at which oscillation stops; before this point is reached, start-up will become unreliable. The possible amount of adjustment, called *pulling range*, depends on the crystal; it can range from less than 10 to several hundred parts per million. Overtone crystals have much less pulling range than fundamental crystals on the same frequency; the reduction in pulling is roughly proportional to the square of the overtone number.

LOW-NOISE CRYSTAL OSCILLATORS

Fig 9.26A shows a crystal operating in its series mode in a series-tuned Colpitts circuit. Because it does not include an LC tank to prevent operation on unwanted modes, this circuit is intended for fundamental mode operation only and relies on that mode being the most active. If the crystal is ordered for 30-pF loading, the frequency trimming capacitor can be adjusted to compensate for the loading of the capacitive divider of the Colpitts circuit. An unloaded crystal without a trimmer would operate slightly off the exact series resonant frequency in order to create an inductive impedance to resonate with the divider capacitors. Ulrich Rohde, N1UL, in Fig 4-47 of his book *Digital PLL Frequency Synthesizers — Theory and Design*, published an elegant alternative method of extracting an output signal from this type of circuit, shown in Fig 9.26B. This taps off a signal from the current in the crystal itself. This can be thought of as using the crystal as a band-pass filter for the oscillator output. The RF choke in the emitter keeps the emitter bias resistor from loading the tank and degrading the Q. In this case (3-MHz operation), it has been chosen to resonate close to 3 MHz with the parallel capacitor (510 pF) as a means of forcing operation on the wanted mode. The 10-Ω resistor and the transformed load impedance will reduce the in-circuit Q of the crystal, so a further development substituted a common base amplifier for the resistor and transformer. This is shown in Fig 9.26C. The common-base amplifier is run at a large quiescent current to give a very low input impedance. Its collector is tuned to give an output with low harmonic content and an emitter follower is used to buffer this from the load. This oscillator sports a simple ALC system, in which the amplified and rectified signal is used to reduce the bias voltage on the oscillator transistor's base. This circuit is described as achieving a phase noise level of –168 dBc/Hz a few kilohertz out from the carrier. This may seem far beyond what may ever be needed, but frequency multiplication to high frequencies, whether by classic multipliers or by frequency synthesizers, multiplies the deviation of any FM/PM sidebands as well as the carrier frequency. This means that phase noise worsens by 20 dB for each tenfold multiplication of frequency. A clean crystal oscillator and a multiplier chain is still the best way of generating clean microwave signals for use with narrow-band modulation schemes.

It has already been mentioned that overtone crystals are much harder to pull than fundamental ones. This is another way of saying that overtone crystals are less influenced by their surrounding circuit, which is helpful in a frequency-standard oscillator like this one. Even though 5 MHz is in the main range of fundamental-mode crystals and this circuit will work well with them, an overtone crystal has been used. To further help stability, the power dissipated in the crystal is kept to about 50 μW. The common-base stage is effectively driven from a higher impedance than its own input impedance, under which conditions it gives a very low noise figure.

9.5.5 VXOs

Some crystal oscillators have frequency trimmers. If the trimmer is replaced by a variable capacitor as a front-panel control, we have a *variable crystal oscillator* (*VXO*): a crystal-based VFO with a narrow tuning range, but good stability and noise performance. VXOs are often used in small, simple QRP transmitters to tune a few kilohertz around common calling frequencies. Artful constructors, using optimized circuits and components, have achieved 1000-ppm tuning ranges. Poor-quality "soft" crystals are more pull-able than high-Q ones. Overtone crystals are not suited to VXOs. For frequencies beyond the usual limit for fundamental mode crystals, use a fundamental unit and frequency multipliers.

ICOM and Mizuho made some 2 meter SSB transceivers based on multiplied VXO local oscillators. This system is simple and can yield better performance than many expensive synthesized radios. SSB filters are available at 9 or 10.7 MHz, to yield sufficient image rejection with a single conversion. Choice of VXO frequency depends on whether the LO is to be above or below signal frequency and how much multiplication can be tolerated. Below 8 MHz multiplier filtering is difficult. Above 15 MHz, the tuning range per crystal narrows. A 50-200 kHz range per crystal should work with a modern front-end design feeding a good 9-MHz IF, for a contest quality 2 meter SSB receiver.

The circuit in **Fig 9.27** is a JFET VXO from Wes Hayward, W7ZOI, and Doug DeMaw, W1FB, optimized for wide-range pulling. Published in *Solid State Design for the Radio Amateur*, many have been built and its ability to pull crystals as far as possible has been proven. Ulrich Rohde, N1UL, has shown that the diode arrangement as used here to make signal-dependent negative bias for the gate confers a phase-noise disadvantage, but oscillators like this that pull crystals as far as possible need any available means to stabilize their amplitude and aid start-up. In this case, the noise penalty is worth paying. This circuit can achieve a 2000-ppm tuning range with amenable crystals. If you have some overtone crystals in your junk box whose fundamental

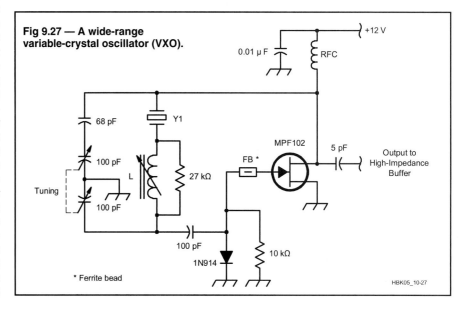

Fig 9.27 — A wide-range variable-crystal oscillator (VXO).

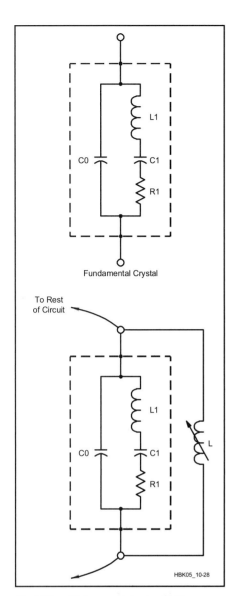

Fig 9.28 — Using an inductor to "tune out" C0 can increase a crystal oscillator's pulling range.

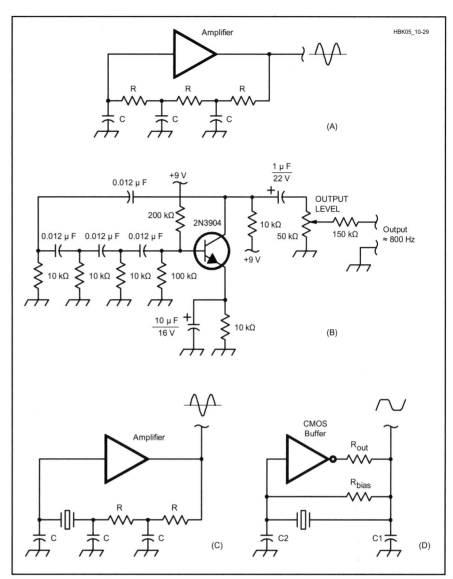

Fig 9.29 — In a phase-shift oscillator (A) based on logic gates, a chain of RC networks — three or more — provide the feedback and phase shift necessary for oscillation, at the cost of low Q and considerable loop loss. Many commercial Amateur Radio transceivers have used a phase-lead oscillator similar to that shown at B as a sidetone generator. Replacing one of the resistors in A with a crystal produces a Pierce oscillator (C), a cut-down version of which (D) has become the most common clock oscillator configuration in digital systems.

frequency is close to the wanted value, they are worth trying.

This sort of circuit doesn't necessarily stop pulling at the extremes of the possible tuning range, sometimes the range is set by the onset of undesirable behavior such as jumping mode or simply stopping oscillating. L was a 16-µH slug-tuned inductor for 10-MHz operation. It is important to minimize the stray and inter-winding capacitance of L since this dilutes the range of impedance presented to the crystal.

One trick that can be used to aid the pulling range of oscillators is to tune out the C0 of the equivalent circuit with an added inductor. **Fig 9.28** shows how. L is chosen to resonate with C0 for the individual crystal, turning it into a high-impedance parallel-tuned circuit. The Q of this circuit is orders of magnitude less than the Q of the true series resonance of the crystal, so its tuning is much broader. The value of C0 is usually just a few picofarads, so L has to be a fairly large value considering the frequency at which it is resonated. This means that L has to have low stray capacitance or else it will self-resonate at a lower frequency. The tolerance on C0 and the variations of the stray C of the inductor means that individual adjustment is needed. This technique can also work wonders in crystal ladder filters.

9.5.6 Logic-Gate Crystal Oscillators

A 180° phase-shift network and an inverting amplifier can be used to make an oscillator. A single stage RC low-pass network cannot introduce more than 90° of phase shift and only approaches that as the signal becomes infinitely attenuated. Two stages can only approach 180° and then have immense attenuation. It takes three stages to give 180° and not destroy the loop gain. **Fig 9.29A** shows the basic form of the *phase-shift* oscillator; Fig 9.29B is an example the phase-shift oscillator commonly used in commercial Amateur Radio transceivers as an audio sidetone oscillator.

The frequency-determining network of an RC oscillator has a Q of less than one, which is a massive disadvantage compared to an LC or crystal resonator. The *Pierce* crystal oscillator is a converted phase-shift oscillator, with the crystal taking the place of one series

Oscillators and Synthesizers 9.29

resistor, in Fig 9.29C. At the exact series resonance, the crystal looks resistive, so by suitable choice of the capacitor values, the circuit can be made to oscillate.

The crystal has a far steeper phase/frequency relationship than the rest of the network, so the crystal is the dominant controller of the frequency. The Pierce circuit is rarely seen in this full form. Instead, a cut-down version has become the most common circuit for crystal-clock oscillators for digital systems.

Fig 9.29D shows this minimalist Pierce, using a logic inverter as the amplifier. R_{bias} provides dc negative feedback to bias the gate into its linear region. At first sight it appears that this arrangement should not oscillate, but the crystal is a resonator (not a simple resistor) and oscillation occurs at a frequency offset from the series resonance, where the crystal appears inductive, which makes up the missing phase shift. This is one circuit that *cannot* oscillate exactly on the crystal's series resonance.

This circuit is included in many microprocessors and other digital ICs that need a clock. It is also the usual circuit inside the miniature clock oscillator cans. It is not a very reliable circuit, as operation is dependent on the crystal's equivalent series resistance and the output impedance of the logic gate, occasionally the logic device or the crystal needs to be changed to start oscillation, sometimes playing with the capacitor values is necessary. It is doubtful whether these circuits are ever designed — values (the two capacitors) seem to be arrived at by experiment. Once going, the circuit is reliable, but its drift and noise are moderate, not good — acceptable for a clock oscillator. The commercial packaged oscillators are the same, but the manufacturers have handled the production foibles on a batch-by-batch basis.

9.6 Oscillators at UHF and Above

9.6.1 UHF Oscillators

A traditional way to make signals at higher frequencies is to make a signal at a lower frequency (where oscillators are easier) and multiply it up to the wanted range. Multipliers are still one of the easiest ways of making a clean UHF/microwave signal. The design of a multiplier depends on whether the multiplication factor is an odd or even number. For odd multiplication, a Class-C biased amplifier can be used to create a series of harmonics; a filter selects the one wanted. For even multiplication factors, a full-wave-rectifier arrangement of distorting devices can be used to create a series of harmonics with strong even-order components, with a filter selecting the wanted component. At higher frequencies, diode-based passive circuits are commonly used. Oscillators using some of the LC circuits already described can be used in the VHF range. At UHF different approaches become necessary.

Fig 9.30 shows a pair of oscillators based on a resonant length of line. The first one is a return to basics: a resonator, an amplifier and a pair of coupling loops. The amplifier can be a single bipolar or FET device or one of the monolithic microwave integrated circuit (MMIC) amplifiers. The second circuit is really a Hartley, and one was made as a test oscillator for the 70 cm band from a 10-cm length of wire suspended 10 mm over an un-etched PC board as a ground plane, bent down and soldered at one end, with a trimmer at the other end. The FET was a BF981 dual-gate device used as a source follower.

No free-running oscillator will be stable enough on these bands except for use with wideband FM or video modulation and AFC at the receiver. Oscillators in this range are almost invariably tuned with tuning diodes controlled by phase-locked-loop synthesizers, which are themselves controlled by a crystal oscillator.

There is one extremely common UHF oscillator that is rarely applied intentionally. The answer to this riddle is a configuration that is sometimes deliberately built as a useful wide-tuning oscillator covering say, 500 MHz to 1 GHz — and is also the modus operandi of a very common form of spurious VHF/UHF oscillation in circuitry intended to process lower-frequency signals! This oscillator has no generally accepted name. It relies on the creation of a small negative resistance in series with a series resonant LC tank. **Fig 9.31A** shows the circuit in its simplest form.

This circuit is well-suited to construction with printed-circuit inductors. Common FR4 glass-epoxy board is lossy at these frequencies; better performance can be achieved by using the much more expensive glass-Teflon board. If you can get surplus pieces of this type of material, it has many uses at UHF and microwave, but it is difficult to use, as the adhesion between the copper and the substrate can be poor. A high-UHF transistor with a 5-GHz f_T like the BFR90 is suitable; the base inductor can be 30 mm of 1-mm trace folded into a hairpin shape (inductance, less than 10 nH).

Analyzing this circuit using a comprehensive model of the UHF transistor reveals that the emitter presents an impedance that is small, resistive and negative to the outside world. If this is large enough to more than cancel the effective series resistance of the

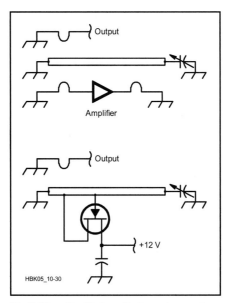

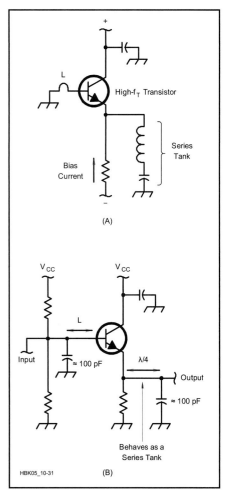

Fig 9.30 — Oscillators that use transmission-line segments as resonators. Such oscillators are more common than many of us may think, as Fig 9.31 reveals.

Fig 9.31 — High device gain at UHF and resonances in circuit board traces can result in spurious oscillations even in non-RF equipment.

emitter tank, oscillation will occur.

Fig 9.31B shows a very basic emitter-follower circuit with some capacitance to ground on both the input and output. If the capacitor shunting the input is a distance away from the transistor, the trace can look like an inductor — and small capacitors at audio and low RF can look like very good decouplers at hundreds of MHz. The length to the capacitor shunting the output will behave as a series resonator at a frequency where it is a quarter-wavelength long. This circuit is the same as in Fig 9.31A; it, too, can oscillate. (Parasitic effects are discussed in the **RF Techniques** chapter.)

The semiconductor manufacturers have steadily improved their small-signal transistors to give better gain and bandwidth so that any transistor circuit where all three electrodes find themselves decoupled to ground at UHF may oscillate at several hundred megahertz. The upshot of this is that there is no longer any branch of electronics where RF design and layout techniques can be safely ignored. A circuit must not just be designed to do what it *should* do, it must also be designed so that it cannot do what it *should not* do.

There are three ways of taming such a circuit; adding a small resistor, of perhaps 50 to 100 Ω in the collector lead, close to the transistor, or adding a similar resistor in the base lead, or by fitting a ferrite bead over the base lead under the transistor. The resistors can disturb dc conditions, depending on the circuit and its operating currents. Ferrite beads have the advantage that they can be easily added to existing equipment and have no effect at dc and low frequencies. Beware of some electrically conductive ferrite materials that can short transistor leads. If an HF oscillator uses beads to prevent any risk of spurs (Fig 9.15), the beads should be anchored with a spot of adhesive to prevent movement that can cause small frequency shifts. Ferrite beads of Fair-Rite #43 material are especially suitable for this purpose; they are specified in terms of impedance, not inductance. Ferrites at frequencies above their normal usable range become very lossy and can make a lead look not inductive, but like a few tens of ohms, resistive.

9.6.2 Microwave Oscillators
SOLID-STATE OSCILLATORS

Low-noise microwave signals are still best made by multiplying a very-low-noise HF crystal oscillator, but there are a number of oscillators that work directly at microwave frequencies. Such oscillators can be based on resonant lengths of stripline or microstrip, and are simply scaled-down versions of UHF oscillators, using microwave transistors and printed striplines on a low-loss substrate, like alumina. Techniques for printing metal traces on substrates to form filters, couplers, matching networks and so on, have been intensively developed over the past two decades. Much of the professional microwave community has moved away from waveguide and now uses low-loss coaxial cable with a solid-copper shield — semi-rigid cable — to connect circuits made flat on ceramic or Teflon based substrates. Semiconductors are often bonded on as unpackaged chips, with their bond-wire connections made directly to the traces on the substrate. At lower microwave frequencies, they may be used in standard surface-mount packages. From an Amateur Radio

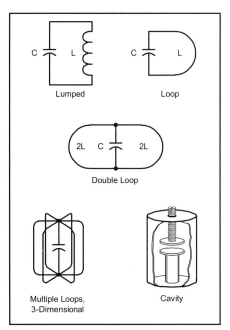

Fig 9.32 — Evolution of the cavity resonator.

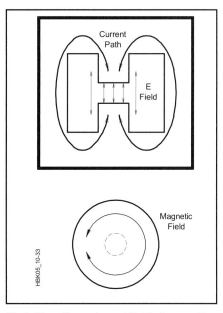

Fig 9.33 — Currents and fields in a cavity.

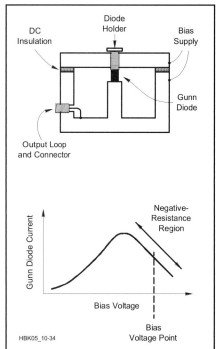

Fig 9.34 — A Gunn diode oscillator uses negative resistance and a cavity resonator to produce radio energy.

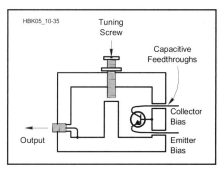

Fig 9.35 — A transistor can also directly excite a cavity resonator.

viewpoint, many of the processes involved are not feasible without access to specialized furnaces and materials. Using ordinary PC-board techniques with surface-mount components allows the construction of circuits up to 4 GHz or so. Above this, structures get smaller and accuracy becomes critical; also PC-board materials quickly become very lossy.

Older than stripline techniques and far more amenable to home construction, cavity-based oscillators can give the highest possible performance. The dielectric constant of the substrate causes stripline structures to be much smaller than they would be in free air, and the lowest-loss substrates tend to have very high dielectric constants. Air is a very low-loss dielectric with a dielectric constant of 1, so it gives high Q and does not force excessive miniaturization. **Fig 9.32** shows a series of structures used by G. R. Jessop, G6JP, to illustrate the evolution of a cavity from a tank made of lumped components. All cavities have a number of different modes of resonance, the orientation of the currents and fields are shown in **Fig 9.33**. The cavity can take different shapes, but that shown has proven to suppress unwanted modes well. The gap need not be central and is often right at the top. A screw can be fitted through the top, protruding into the gap, to adjust the frequency.

To make an oscillator out of a cavity, an amplifier is needed. Gunn and tunnel diodes have regions in their characteristics where their current *falls* with increasing bias voltage. This is negative resistance. If such a device is mounted in a loop in a cavity and bias applied, the negative resistance can more than cancel the effective loss resistance of the cavity, causing oscillation. These diodes are capable of operating at extremely high frequencies and were discovered long before transistors were developed that had any gain at microwave frequencies.

A *Gunn-diode* cavity oscillator is the basis of many of the Doppler-radar modules used to detect traffic or intruders. **Fig 9.34** shows a common configuration. The coupling loop and coax output connector could be replaced with a simple aperture to couple into waveguide or a mixer cavity. **Fig 9.35** shows a transistor cavity oscillator version using a modern microwave transistor. FET or bipolar devices can be used. The two coupling loops are completed by the capacitance of the feedthroughs.

DIELECTRIC-RESONATOR OSCILLATORS (DRO)

The *dielectric-resonator oscillator* (*DRO*) may soon become the most common microwave oscillator of all, as it is used in the downconverter of satellite TV receivers. The dielectric resonator itself is a ceramic cylinder, like a miniature hockey puck, several millimeters in diameter. The ceramic has a very high dielectric constant, so the surface (where ceramic meets air in an abrupt mismatch) reflects electromagnetic waves and makes the ceramic body act as a resonant cavity. It is mounted on a substrate and coupled to the active device of the oscillator by a stripline that runs past it. At 10 GHz, a FET made of gallium arsenide (GaAsFET), rather than silicon, is normally used. The dielectric resonator elements are made at frequencies appropriate to mass applications like satellite TV. The set-up charge to manufacture small quantities at special frequencies is likely to be prohibitive for the foreseeable future. The challenge with these devices is to devise new ways of using oscillators on industry standard frequencies. Their chief attraction is their low cost in large quantities and compatibility with microwave stripline (microstrip) techniques. Frequency stability and Q are competitive with good cavities, but are inferior to that achievable with a crystal oscillator and chain of frequency multipliers. Satellite TV downconverters need free running oscillators with less than 1 MHz of drift at 10 GHz.

YIG OSCILLATORS

The *yttrium-iron garnet* (*YIG*) oscillator was developed for a wide tuning range as a solid-state replacement for the *backward wave oscillator* (BWO), a type of vacuum tube oscillator, and many of them can be tuned over more than an octave. They are complete, packaged units that appear to be a heavy block of metal with low-frequency connections for power supplies and tuning, and an SMA connector for the RF output. The manufacturer's label usually states the tuning range and often the power supply voltages. This is very helpful because, with new units priced in the kilo-dollar range, it is important to be able to identify surplus units. The majority of YIGs are in the 2- to 18-GHz region, although units down to 500 MHz and up to 40 GHz are occasionally found. In this part of the spectrum there is no octave-tunable device that can equal their cleanliness and stability. A 3-GHz unit drifting less than 1 kHz per second gives an idea of typical stability. This seems very poor — until we realize that this is 0.33 ppm per second. Nevertheless, any YIG application involving narrow-band modulation will usually require some form of frequency stabilizer.

Good quality, but elderly, RF spectrum analyzers have found their way into the workshops of a number of dedicated constructors. A 0- to 1500-MHz analyzer usually uses a 2- to 3.5-GHz YIG as its first local oscillator, and its tuning circuits are designed around it. A reasonable understanding of YIG oscillators will help in troubleshooting and repair.

YIG spheres are resonant at a frequency controlled not only by their physical dimensions, but also by any applied magnetic field. A YIG sphere is carefully oriented within a coupling loop connected to a negative-resistance device and the whole assembly is placed between the poles of an electromagnet. Negative-resistance diodes have been used, but transistor circuits are now common. The support for the YIG sphere often contains a thermostatically controlled heater to reduce temperature sensitivity.

The first problem with a magnetically tuned oscillator is that magnetic fields, especially at low frequencies, are extremely difficult to shield; the tuning will be influenced by any local fields. Varying fields will cause frequency modulation. The magnetic core must be carefully designed to be all-enclosing in an attempt at self-shielding and then one or more nested mu-metal cans are fitted around everything. It is still important to site the unit away from obvious sources of magnetic noise, like power transformers. Cooling fans are less obvious sources of fluctuating magnetic fields, since some are 20 dB worse than a well designed 200-W 50/60-Hz transformer.

The second problem is that the oscillator's internal tuning coils need significant current from the power supply to create strong fields. This can be eased by adding a permanent magnet as a fixed "bias" field, but the bias will shift as the magnet ages. The only solution is to have a coil with many turns, hence high inductance, which will require a high supply voltage to permit rapid tuning, thus increasing the power consumption. The usual compromise is to have dual coils: One with many turns allows slow tuning over a wide range; a second with much fewer turns allows fast tuning or FM over a limited range. The main coil can have a sensitivity in the 20 MHz/mA range; and the "FM" coil, perhaps 500 kHz/mA.

The frequency/current relationship can have excellent linearity. **Fig 9.36** shows the construction of a YIG oscillator.

9.6.3 Klystrons, Magnetrons and Traveling Wave Tubes

There are a number of thermionic (vacuum-tube) devices that are used as oscillators or amplifiers at microwave frequencies. Standard vacuum tubes (see the **RF Amplifiers** chapter for an introduction to vacuum tubes) work well for frequencies up to hundreds of megahertz. At frequencies higher than this, the amount of time that it takes for the electrons to move between the cathode and the plate becomes a limiting factor. There are several special tubes designed to work at microwave frequencies, usually providing more power than can be obtained from solid-state devices.

The *klystron* tube uses the principle of velocity modulation of the electrons to avoid transit time limitations. The beam of electrons travels down a metal drift tube that has interaction gaps along its sides. RF voltages are applied to the gaps and the electric fields that they generate accelerate or decelerate the passing electrons. The relative positions of the electrons shift due to their changing velocities causing the electron density of the beam to vary. The modulation of the electron density is used to perform amplification or oscillation. Klystron tubes tend to be relatively large, with lengths ranging from 10 cm to 2 m and weights ranging from as little as 150 g to over 100 kg. Unfortunately, klystrons have relatively narrow bandwidths, and are not re-tunable by amateurs for operation on different frequencies.

The *magnetron* tube is an efficient oscillator for microwave frequencies. Magnetrons are most commonly found in microwave ovens and high-powered radar equipment. The anode

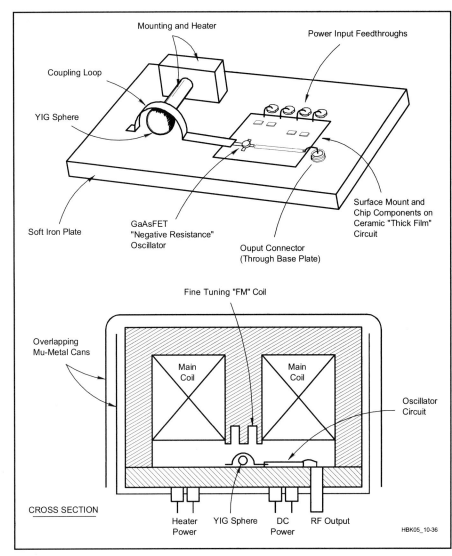

Fig 9.36 — A yttrium-iron-garnet (YIG) sphere serves as the resonator in the sweep oscillators used in many spectrum analyzers.

of a magnetron is made up of a number of coupled resonant cavities that surround the cathode. The magnetic field causes the electrons to rotate around the cathode and the energy that they give off as they approach the anode adds to the RF electric field. The RF power is obtained from the anode through a vacuum window. Magnetrons are self-oscillating with the frequency determined by the construction of their anodes; however, they can be tuned by coupling either inductance or capacitance to the resonant anode. The range of frequencies depends on how fast the tuning must be accomplished. The tube may be tuned slowly over a range of approximately 10% of the center frequency. If faster tuning is necessary, such as is required for frequency modulation, the range decreases to about 5%.

A third type of tube capable of operating in the microwave range is the *traveling wave tube* (TWT). For wide-band amplifiers in the microwave range this is the tube of choice. Either permanent magnets or electromagnets are used to focus the beam of electrons that emerges from an electron gun similar to one for a CRT display tube. The electron beam passes through a helical slow-wave structure, in which electrons are accelerated or decelerated, providing density modulation due to the applied RF signal, similar to that in the klystron. The modulated electron beam induces voltages in the helix that provides an amplified tube output whose gain is proportional to the length of the slow-wave structure. After the RF energy is extracted from the electron beam by the helix, the electrons are collected and recycled to the cathode. Traveling wave tubes can often be operated outside their designed frequencies by carefully optimizing the beam voltage.

9.7 Frequency Synthesizers

Like many of our modern technologies, the origins of frequency synthesis can be traced back to WW II. The driving force was the desire for stable, rapidly switchable and accurate frequency control technology to meet the demands of narrow-band, frequency-agile HF communications systems without resorting to large banks of switched crystals. Early synthesizers were cumbersome and expensive, and therefore their use was limited to the most sophisticated communications systems. With the help of the same technologies that have taken computers from "room-sized" to the palms of our hands, the role of frequency synthesis has gone far beyond its original purpose and has become one of the most enabling technologies in modern communications equipment.

Just about every communications device manufactured today, be it a handheld transceiver, cell phone, pager, AM/FM entertainment radio, scanner, television, HF communications equipment, or test equipment contains a synthesizer. Synthesis is the technology that allows an easy interface with both computers and microprocessors. It provides amateurs with many desirable features, such as the feel of an analog knob with 10-Hz frequency increments, accuracy and stability determined by a single precision crystal oscillator, frequency memories, and continuously variable splits. Now reduced in size to only a few small integrated circuits, frequency synthesizers have also replaced the cumbersome chains of frequency multipliers and filters in VHF, UHF and microwave equipment, giving rise to many of the highly portable communications devices we use today. Frequency synthesis has also had a major impact in lowering the cost of modern equipment, as well as reducing manufacturing complexity.

Frequency synthesizers have been categorized in two general types, *direct synthesizers* (not to be confused with direct digital synthesis) and *indirect synthesizers*. The architecture of the direct types consists mainly of multipliers, dividers, mixers, filters and copious amounts of shielding. They are also cumbersome and expensive, and have all but disappeared from the market. The indirect varieties make use of programmable dividers and phase-locked loops (PLL), thereby considerably reducing their complexity. There are a number of variations on the indirect theme. They include programmable division, variable or dual modulus division, and fractional-N. Other techniques that have been developed since the direct/indirect classification are *direct digital synthesis* (DDS) and *rate multiplier synthesis*. Synthesizers used in equipment today are usually a hybrid of

An Introduction to Phase-Locked Loops

By Jerry DeHaven, WA0ACF

Phase locked loops (PLLs) are used in many applications from tone decoders to stabilizing the pictures in your television set, to demodulating your local FM station or 2-m repeater. They are used for recovering weak signals from deep space satellites and digging out noisy instrumentation signals. Perhaps the most common usage for PLLs in Amateur Radio is to combine the variability of an LC oscillator with the long term stability and accuracy of a crystal oscillator in PLL frequency synthesizers. Strictly speaking a PLL is not necessarily a frequency synthesizer and a frequency synthesizer is not necessarily a PLL, although the terms are used interchangeably.

PLL Block Diagram

The block diagram in **Fig 9.A3** shows a basic *control loop*. An example of a control loop is the furnace or air conditioning system in your house or the cruise control in your car. You input a desired output and the control loop causes the system output to change to and remain at that desired output (called a *setpoint*) until you change the setpoint. Control loops are characterized by an input, an output and a feedback mechanism as shown in the simple control loop diagram in Fig 9.A3. The setpoint in this general feedback loop is called the *reference*.

In the case of a heating system in your house, the reference would be the temperature that you set at the thermostat. The feedback element would be the temperature sensor inside the thermostat. The thermostat performs the comparison function as well. The generator would be the furnace which is turned on or off depending on the output of the comparison stage. In general terms, you can see that the reference input is controlling the output of the generator.

Like other control loops, the design of a PLL is based on feedback, comparison and correction as shown in Fig 9.A3. In this section we will focus on the concepts of using a phase locked loop to generate (synthesize) one or more frequencies based on a single reference frequency.

In a typical PLL frequency synthesizer there are six basic functional blocks as shown in **Fig 9.A4**. The PLL frequency synthesizer block diagram is only slightly more complicated than the simple control loop diagram, so the similarity should be evident. After a brief description of the function of each block we will describe the operational behavior of a PLL frequency synthesizer.

In Fig 9.A4, the general control loop is implemented with a bit more detail.

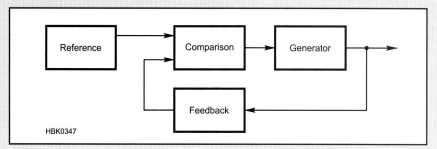

Fig 9.A3 — Simple control loop. The Comparison block creates a control signal for the Generator by comparing the output of the Reference and Feedback blocks.

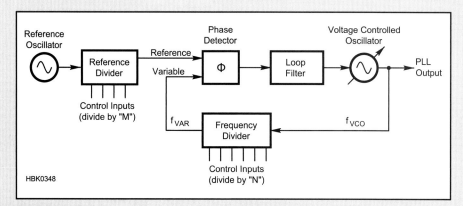

Fig 9.A4 — PLL frequency synthesizer block diagram. The PLL is locked when the frequency PLL Output / N is the same as the Reference Oscillator / M.

The Reference block is composed of the Reference Oscillator and the Reference Divider. The Comparison block consists of the Phase Detector and Loop Filter. The Generator is replaced by the Voltage-Controlled Oscillator (or VCO) and the Feedback block is replaced by the Frequency Divider. The two signals being compared are both frequencies; the reference frequency and variable frequency signals. The output of the Loop Filter is made up of dc and low frequency ac components that act to change the VCO frequency. The output of the PLL is an integer multiple of the reference frequency (the output of the Reference Divider). It is probably not obvious why, but read on to find out!

The Reference Oscillator — The reference oscillator is usually a crystal oscillator with special attention paid to thermal stability and low mechanical and electronic noise. The main function of the reference oscillator is to generate a stable frequency for the PLL. Let us make a distinction between the reference oscillator frequency and the reference frequency. The output of the reference oscillator is at the crystal frequency, say 5.000 MHz. The choice of the actual crystal frequency depends on the PLL design, the availability of another oscillator in the radio, the avoidance of spurious responses in the receiver, and so on.

The *reference frequency* is the output of the reference (crystal) oscillator divided by the integer M to a relatively low frequency, say 10 kHz. In this case, 10 kHz is the reference frequency. The reference frequency equals the step size between the PLL output frequencies. In this simple example the crystal frequency can be any frequency that is an integer multiple of 10 kHz and within the operating range of the Reference Divider.

The Phase Detector — There are many types of phase detectors but for now let us just consider a basic mixer, or a multiplier. The phase detector has two inputs and one output. In its simplest form, the output of the phase detector is a dc voltage proportional to the phase difference between the two inputs. In practice, a phase detector can be built using a diode double-balanced mixer or an active multiplier.

The output of the mixer consists of products at many frequencies both from multiplication and from nonlinearities. In this application, the desired output is the low frequency and dc terms so all of the RF products are terminated in a load and the low frequency and dc output is passed to the PLL loop filter.

When the PLL is locked — meaning that the output frequency is the desired multiple of the reference frequency — the phase detector output is a steady dc voltage somewhere between ground and the PLL power supply voltage. When the PLL is commanded to change to another frequency, the phase detector output will be a complex, time-varying signal that gradually settles in on its final value.

The Loop Filter — The loop filter is a low-pass filter that filters the output of the phase detector. The loop filter can be a simple resistor-capacitor (RC) low-pass filter or an active circuit built with bipolar transistors or operational amplifiers. The cutoff frequency of the low pass filter is on the order of 100 Hz to 10 kHz. Although the frequencies involved are dc and low audio frequencies the design of the loop filter is critical for good reference suppression and low noise performance of the PLL frequency synthesizer.

The output of the loop filter is a dc voltage that "steers" the following stage, the Voltage Controlled Oscillator. The dc value can vary over a substantial portion of the PLL power supply. For example, if the loop filter runs off a 9-V power supply you may expect to see the output voltage range from about 2 V to 7 V depending on the VCO frequency range.

The Voltage Controlled Oscillator (VCO) — The VCO is probably the most critical stage in the PLL frequency synthesizer. This stage is the subject of many conflicting design goals. The VCO must be electrically tunable over the desired frequency range with low noise and very good mechanical stability. Ideally, the output frequency of the VCO will be directly proportional to the input control voltage. A key characteristic is the *VCO constant* or *tuning gain*, usually expressed in MHz per volt (MHz/V). The VCO has one input, the dc voltage from the loop filter, and usually two outputs; one for the PLL output, the other driving the feedback stage — the frequency divider.

The Frequency Divider — The VCO frequency divider that acts as the feedback block is a programmable frequency divider or counter. The division ratio N is set either by thumbwheel switches, diode arrays or a microprocessor. The input to the divider is the VCO output frequency. The output of the divider is the VCO output frequency divided by N.

When the PLL is unlocked, such as when the PLL is first turned on, or commanded to change to another frequency, the frequency divider's output frequency will vary in a nonlinear manner until the PLL locks. Under locked conditions, the divider's output frequency (VCO Output / N) is the same as the reference frequency (Reference Oscillator / M).

The function of the frequency divider as the feedback element is easier to understand with an example. Let's assume that the desired PLL output frequency is 14.000 MHz to 14.300 MHz and it should change in steps of 10 kHz (0.010 MHz). A step size of 10 kHz means that the reference frequency must be 10 kHz. There are 31 output frequencies available — one at a time, not simultaneously. (Don't forget to count 14.000 MHz.) Let us arbitrarily choose the reference oscillator as a 5.000 MHz crystal oscillator. To generate the 10 kHz reference frequency, the reference divider must divide by M = 500 = 5.000 MHz / 0.010 MHz. (Use the same units, don't divide MHz by kHz.)

To generate a matching 10 kHz signal from the VCO output frequency, the Frequency Divider (the feedback stage), must be programmable to divide by N = 1400 to 1430 = 14.000 to 14.300 MHz / 0.010 MHz. The reference divisor (M = 500) is fixed, but the Frequency Divider stage needs to be programmable to divide by N = 1400 to 1430 in steps of 1. In this way, the output frequency of the PLL is compared and locked to the stable, crystal-generated value of the reference frequency.

PLL Start-up Operation

Let's visualize the PLL start-up from power on to steady state, visualizing in sequence what each part of the PLL is doing. Some portions of the loop will act quickly, in microseconds; other parts will react more slowly, in tens or hundreds of milliseconds.

The reference oscillator and its dividers will probably start oscillating and stabilize within tens of microseconds. The reference oscillator divider provides the reference frequency to one input of the phase detector.

The VCO will take a bit longer to come up to operation because of extensive power supply filtering with high-value capacitors. The VCO and the programmable dividers will probably reach full output within a few hundred microseconds although the VCO will not yet be oscillating at the correct frequency.

At this point the phase detector has two inputs but they are at different frequencies. The mixer action of the phase detector produces a beat note at its output. The beat note, which could be tens or hundreds of kHz, is low-pass filtered by the PLL loop filter. That low-pass filter action averages the beat note and applies a complicated time-varying ac/dc voltage to the VCO input.

Now the VCO can begin responding to the control voltage applied to its input. An important design consideration is that the filtered VCO control signal must steer the VCO in the correct direction. If the polarity of the control signal is incorrect, it will steer the VCO *away* from the right frequency and the loop will never lock. Assuming correct design, the control voltage will begin steering the VCO in such a direction that the VCO divider frequency will now get closer to the reference frequency.

With the phase detector inputs a little bit closer in frequency, the beat note will be lower in frequency and the low pass filtered average of the phase detector output will change to a new level. That new control voltage will continue to steer the VCO in the right direction even closer to the "right" frequency. With each successive imaginary trip around the loop you can visualize that eventually a certain value of VCO control voltage will be reached at which the VCO output frequency, when divided by the frequency divider, will produce zero frequency difference at the inputs to the phase detector. In this state, the loop is *locked*. Once running, the range over which a PLL can detect and lock on to a signal is its *capture range*.

PLL Steady-state Operation

When the divided-down VCO frequency matches the reference frequency, the PLL will be in its steady-state condition. Since the comparison stage is a phase detector, not just a frequency mixer, the control signal from the loop filter to the VCO will act in such a way that maintains a constant relationship between the *phase* of the reference frequency and the frequency divider output. For example, the loop might stabilize with a 90° difference between the inputs to the phase detector. As long as the phase difference is constant, the reference frequency and the output of the frequency divider (and by extension, the VCO output) must also be the same.

As the divisor of the frequency divider is changed, the loop's control action will keep the divided-down VCO output phase locked to the reference frequency, but with a phase difference that gets closer to 0 or 180°, depending on which direction the input frequency changes. The range over which the phase difference at the phase detector's input varies between 0° and 180° is the widest range over which the PLL can keep the input and VCO signals locked together. This is called the loop's *lock range*.

If the divisor of the frequency divider is changed even further, the output of the loop filter will actually start to move in the opposite direction and the loop will no longer be able to keep the VCO output locked to the reference frequency and the loop is said to be *out of lock* or *lost lock*.

Loop Bandwidth

Whether you picture a control loop such as the thermostat in your house, the speed control in your car or a PLL frequency synthesizer, they all have "bandwidth," that is, they can only respond at a certain speed. Your house will take several minutes to heat up or cool down, your car will need a few seconds to respond to a hill if you are using the speed control. Likewise, the PLL will need a finite time to stabilize, several milliseconds or more depending on the design.

This brings up the concept of *loop bandwidth*. The basic idea is that disturbances outside the loop have a low-pass characteristic up to the loop bandwidth, and that disturbances inside the loop have a high-pass characteristic up to the loop bandwidth. The loop bandwidth is determined by the phase detector gain, the loop filter gain and cutoff frequency, the VCO gain and the divider ratio. Loop bandwidth and the phase response of the PLL determine the stability and transient response of the PLL.

PLLs and Noise

Whether the PLL frequency synthesizer output is being used to control a receiver or a transmitter frequency, the spectral cleanliness of the output is important. Since the loop filter output voltage goes directly to the VCO, any noise or spurious frequencies on the loop filter output will modulate the VCO causing FM sidebands. On your transmitted signal, the sidebands are heard as noise by adjacent stations. On receive, they can mix with other signals, resulting in a higher noise floor in the receiver.

A common path for noise to contaminate the loop filter output is via the power supply. This type of noise tends to be audible in your receiver or transmitter audio. Use a well-regulated supply to minimize 120-Hz power supply ripple. Use a dedicated low-power regulator separate from the receiver audio stages so you do not couple audio variations from the speaker amplifier into the loop filter output. If you have a computer / sound card interface, be careful not to share power supplies or route digital signals near the loop filter or VCO.

A more complicated spectral concern is the amount of reference frequency feed-through to the VCO. This type of spurious output is above the audio range (typically a few kHz) so it may not be heard but it will be noticed, for example, in degraded receiver adjacent channel rejection. In the example above, Fig 9.A4, the reference frequency is 10 kHz. High amounts of reference feed-through on the VCO control input will result in FM sidebands on the VCO output.

Loop filter design and optimization is a complicated subject with many tradeoffs. This subject is covered in detail in the PLL references, but the basic tradeoff is between loop bandwidth (which affects the speed at which a PLL can change frequency and lock) and reference rejection. Improved phase detectors can be used which will minimize the amount of reference feed-through for the loop filter to deal with. Another cause of high reference sidebands is inadequate buffering between the VCO output and the frequency divider stage.

One indication of PLL design difficulty is the spread of the VCO range expressed either in percent or in octaves. It is fairly difficult to make a VCO that can tune over one octave, a two-to-one frequency spread. Typically, the greater the spread of the VCO tuning range, the worse the noise performance gets. The conflicting design requirements can sometimes be conquered by trading off complexity, cost, size, and power consumption.

A common source of noise in the VCO is microphonics. Very small variations in capacitance can frequency modulate the VCO and impose audible modulation on the VCO output. Those microphonic products will be heard on your transmit audio or they will be superimposed on your receiver audio output. An air-wound inductor used in a VCO can move slightly relative to a shield can under vibration, for example. A common technique to minimize microphonics is to cover the sensitive parts with wax, Q-Dope, epoxy or a silicone sealer. As with the loop filter use a well-filtered, dedicated regulator for the VCO to minimize modulation by conducted electrical noise.

Types of PLLs

One of the disadvantages of a single loop PLL frequency synthesizer is that the reference frequency determines the step size of the output frequency. For two-way radio and other systems with fixed channel spacing that is generally not a problem. For applications like amateur radio CW and SSB operation, tuning steps smaller than 10 Hz is almost a necessity. Like so many design requirements there are trade-offs with making the reference frequency smaller and smaller. Most evident is that the loop filter bandwidth would have to keep getting lower and lower to preserve spectral cleanliness and loop stability. Because low frequency filters take longer to settle, a single-loop PLL with a 10 Hz loop bandwidth would take an unacceptably long time to settle to the next frequency.

Creative circuit architectures came up involving two or more PLLs operating together to achieve finer tuning increments with only moderate complexity. The main portion of this chapter discusses these multi-loop PLL frequency synthesizers. Once you understand the basics of a single loop PLL you will be able to learn and understand the operation of a multi-loop PLL.

---

the indirect technique and some of the latter developments. There are entire textbooks devoted to treating these various methods in depth, and as such it is not practical to discuss each one in detail in this handbook. We will focus on the indirect approach, based on phase-locked loops. This approach came to dominate frequency synthesis long ago, and is still dominant today.

9.7.1 Phase-Locked Loops (PLL)

To understand the indirect synthesizer, we need to understand the *phase-locked loop* (PLL). The sidebar "An Introduction to Phase-Locked Loops" provides an overview of this technology. This section presents a detailed discussion of design and performance topics of the PLL and PLL synthesizers and the circuits used to construct them. The digital logic circuits and terms included in this discussion are covered in the **Digital Basics** chapter.

The principle of the phase-locked loop (PLL) synthesizer is very simple. An oscillator can be built to cover the required frequency range, so what is needed is a system to keep it tuned to the desired frequency. This is done by continuously comparing the phase of the oscillator to a stable reference, such as a crystal oscillator and using the result to steer the tuning. If the oscillator frequency is too high, the phase of its output will start to lead that of the reference by an increasing amount. This is detected and is used to decrease the frequency of the oscillator. Too low a frequency causes an increasing phase lag, which is detected and used to increase the frequency of the oscillator. In this way, the oscillator is locked into a fixed-phase relationship with the reference, which means that their frequencies are held exactly equal.

The oscillator is now under control, but is locked to a fixed frequency. **Fig 9.37** shows the next step. The phase detector does not simply compare the oscillator frequency with the reference oscillator; both signals have now been passed through frequency dividers. Advances in digital integrated circuits have made frequency dividers up to microwave frequencies (over 10 GHz) commonplace.

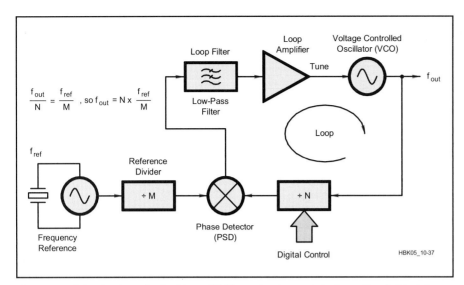

Fig 9.37 — A basic phase-locked-loop (PLL) synthesizer acts to keep the divided-down signal from its voltage-controlled oscillator (VCO) phase-locked to the divided-down signal from its reference oscillator. Fine tuning steps are therefore possible without the complication of direct synthesis.

The divider on the reference path has a fixed division factor, but that in the VCO path is programmable, the factor being entered digitally as a (usually) binary word. The phase detector is now operating at a lower frequency, which is a submultiple of both the reference and output frequencies.

The phase detector, via the loop amplifier, steers the oscillator to keep both its inputs equal in frequency. The reference frequency, F_{ref}, divided by M, is equal to the output frequency divided by N. The output frequency equals the reference frequency × N/M. N is programmable (and an integer), so this synthesizer is capable of tuning in steps of F_{ref}/M.

As an example, to make a 2 meter FM radio covering 144 to 148 MHz with a 10.7-MHz IF, we need a local oscillator covering 154.7 to 158.7 MHz. If the channel spacing is 20 kHz, then F_{ref}/M is 20 kHz, so N is 7735 to 7935. There is a free choice of F_{ref} or M, but division by a round binary number is easiest. Crystals are readily available for 10.24 MHz, and one of these (divided by 512) will give 20 kHz at low cost. ICs are readily available containing most of the circuitry necessary for the reference oscillator (less the crystal), with the programmable divider and the phase detector.

LOOP FREQUENCY RESPONSE

The phase detector in this example compares the relative timing of digital pulses at 20 kHz. Inevitably, the phase detector output will contain strong components at 20 kHz and its harmonics in addition to the desired "steering" signal. The loop filter must reject these unwanted signals; otherwise the loop amplifier will amplify them and apply them to the VCO, generating unwanted FM. No filtering can be perfect and the VCO is very sensitive, so most synthesizers have measurable sidebands spaced at the phase detector operating frequency (and harmonics) away from the carrier. These are called *reference-frequency sidebands*. (Exactly which is the reference frequency is ambiguous: Do we mean the frequency of the reference oscillator, or the frequency applied to the reference input of the phase detector? You must look carefully at context whenever "reference frequency" is mentioned.)

The loop filter is not usually built as a single block of circuitry, it is often made up of three sub-circuits: Some components directly at the output of the phase detector, a shaped frequency response in the loop amplifier, and more components between the loop amplifier and the VCO.

The PLL is a feedback control loop, although the "signal" around its loop is represented by frequency in some places, by phase in others and by voltage in others. Like any feedback amplifier, there is a risk of instability and oscillations, which can be seen as massive FM on the output and a strong ac signal on the tuning line to the VCO. The loop's frequency response has to be designed to prevent this by controlling filtering and gain.

The following example illustrates the effect of frequency response on loop action: Imagine that we shift the reference oscillator by 1 Hz. The reference applied to the phase detector would shift 1/M Hz, and the loop would respond by shifting the output frequency to produce a matching shift at the other phase detector port, so the output is shifted by N/M Hz.

Imagine now that we apply a very small amount of FM to the reference oscillator. The amount of deviation will be amplified by N/M — but this is only true for low modulating frequencies. If the modulating frequency is increased, the VCO has to change frequency at a higher rate. As the modulating frequency continues to increase, eventually roll-off of the low-pass loop filter frequency response starts to reduce the gain around the loop, the loop ceases to track the modulation and the deviation at the output falls. This is referred to as the *closed-loop frequency response* or *closed-loop bandwidth*. The design of the loop filtering and the gain around the loop sets the closed-loop performance.

Poorly designed loops can have poor closed-loop responses. For example, the gain can have large peaks above N/M at some reference modulation frequencies, indicating marginal stability and excessively amplifying any noise at those offsets from the carrier.

STEP SIZE TRADEOFFS

In a single-loop synthesizer, there is a trade-off between how small the step size can be versus the performance of the loop. A loop with a very small step size requires the phase detector to operate at a very low frequency, so the loop bandwidth has to be kept very, very low to keep the reference-frequency sidebands low. Also, the reference oscillator usually has much better phase-noise performance than the VCO, and (within the loop bandwidth) the loop acts to oppose the low-frequency components of the VCO phase-noise sidebands. This very useful cleanup activity is lost when loops have to be narrowed in bandwidth to allow narrow steps.

Low-bandwidth loops are slow to respond — they exhibit overly long *settling times* — to the changes in N necessary to change channels. The tradeoffs between step size and loop performance make it difficult to obtain small step sizes from a phase-locked loop.

For a number of reasons, single-loop synthesizers are okay for the 2 meter FM radio discussed earlier: Channel spacing (the step size) in this band is not too small, FM is more tolerant of phase noise than other modes, 2 meter FM is rarely used for weak-signal DXing, and the channelization involved does away with the desirability of simulating fast, smooth, continuous tuning. None of these excuses apply to MF/HF radios, however, and ways to circumvent the problems are needed.

A clue was given earlier, by carefully referring to *single*-loop synthesizers. It's possible to use *several* PLLs, or for that matter, some of the other mentioned techniques such as DDS, fractional-N, variable- or dual-modulus etc, as components in a larger structure, dividing the frequency of one loop and adding it to the frequency of another that has not been divided. This represents a form of hybrid between the old direct synthesizer and the PLL. A form of PLL containing a mixer in place of the programmable divider can be used to perform the frequency addition.

Let us continue our discussion of phase-locked loop synthesizers by examining the role of each of the component pieces of the system. They are, the VCO, the dividers, prescalers, the phase detector and the loop-compensation amplifier.

VOLTAGE-CONTROLLED OSCILLATORS

Voltage-controlled oscillators are commonly referred to as *VCOs*. (Voltage-tuned oscillator, VTO, more accurately describes the circuitry most commonly used in VCOs, but tradition is tradition! An exception to this is the YIG oscillator, described earlier as sometimes used in UHF and microwave PLLs, which is current-tuned.) In all the oscillators described so far, except for permeability tuning, the frequency is controlled or trimmed by a variable capacitor. These are modified for voltage tuning by using a tuning diode. The oscillator section of this chapter shows the schematic symbol and construction detail of a voltage-controlled or *varactor* (tuning) diode. (see **Fig 9.38**)

The VCO circuit of **Fig 9.39** is representative of a VCO implemented with discrete components. (Most PLL ICs include a VCO with all necessary frequency control elements except the tank circuit inductor.) The circuit is a Colpitts oscillator whose tank capacitance has been replaced by the back-to-back tuning diodes. The control voltage applied through the 10-kΩ resistor is assumed to be decoupled and bypassed by components not shown on the schematic.

When this oscillator is used in a PLL, the PLL adjusts the tuning voltage to completely cancel any drift in the oscillator (provided it does not drift further than it can be tuned) regardless of its cause, so no special care is needed to compensate a PLL VCO for temperature effects. Adjusting the inductor core will not change the frequency; the PLL will adjust the tuning voltage to hold the oscillator on frequency. This adjustment is important, though. It's set so that the tuning voltage neither gets too close to the maximum available, nor too low for acceptable Q, as the PLL is tuned across its full range.

In a high-performance, low noise synthesizer, the VCO may be replaced by a bank of switched VCOs, each covering a section of the total range needed, each VCO having better Q and lower noise because the lossy diodes constitute only a smaller part of the total tank capacity than they would in a single, full-range oscillator. Another method uses tuning diodes for only part of the tuning range and switches in other capacitors (or inductor taps) to extend the range. The performance of wide-range VCOs can be improved by using a large number of varactors in parallel in a high-C, low-L tank.

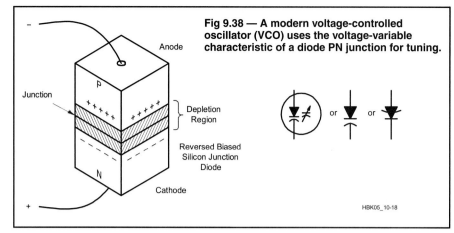

Fig 9.38 — A modern voltage-controlled oscillator (VCO) uses the voltage-variable characteristic of a diode PN junction for tuning.

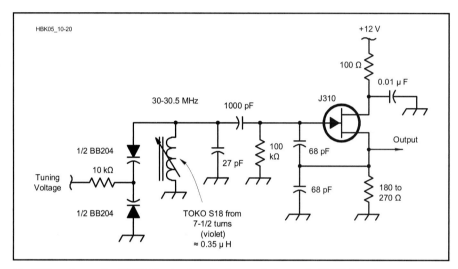

Fig 9.39 — A practical VCO. The tuning diodes are halves of a BB204 dual, common-cathode tuning diode (capacitance per section at 3 V, 39 pF) or equivalent. The ECG617, NTE617 and MV104 are suitable dual-diode substitutes, or use pairs of 1N5451s (39 pF at 4 V) or MV2109s (33 pF at 4 V).

PROGRAMMABLE DIVIDERS

Designing your own programmable divider is now a thing of the past, as complete systems have been made as single ICs for over 10 years. The only remaining reason to do so is when an ultra-low-noise synthesizer is being designed.

It may seem strange to think of a digital counter as a source of random noise, but digital circuits have propagation delays caused by analog effects, such as the charging of stray capacitances. Thermal noise in the components of a digital circuit or signal artifacts coupled from nearby circuitry or the power supply, when added to the signal voltage will slightly change the times at which thresholds are crossed. As a result, the output of a digital circuit will have picked up some timing *jitter*, which is noise in time instead of in amplitude. Jitter in one or both of the signals applied to a phase detector can be viewed as low-level random phase modulation. Within the loop bandwidth, the phase detector steers the VCO so as to oppose and cancel it, so phase jitter or noise is applied to the VCO in order to cancel out the jitter added by the divider. This makes the VCO noisier.

The maximum frequency of operation of the programmable divider shown as part of **Fig 9.40** is limited by the minimum time needed to reliably detect that the division cycle has completed and then reload the counter. (A programmable divider is a type of counter circuit that outputs one pulse for every pre-programmed number of input pulses. See the **Digital Basics** chapter for more information on counters.) Timing limits are determined by the speed of available logic. Noise performance of the high-speed CMOS logic normally used in programmable dividers is good. For ultra-low-noise dividers, high-speed ECL devices are sometimes used.

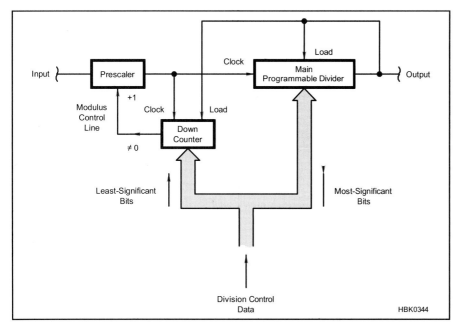

Fig 9.40 — The function of a dual-modulus prescaler. The counter is reloaded with N when the count reaches 0 or 1, depending on the sequencer action.

The synchronous reset cycle forces the entire divider to be made of equally fast logic. Just adding a fast *prescaler* — a fixed-ratio divider ahead of the programmable one — will increase the maximum frequency that can be used, but it also scales up the loop step size. Equal division has to be added to the reference input of the phase detector to restore the step size, reducing the phase detector's operating frequency. The loop bandwidth has to be reduced to restore the filtering needed to suppress reference frequency sidebands. This makes the loop much slower to change frequency and degrades the noise performance.

VARIABLE OR DUAL-MODULUS PRESCALERS

A plain programmable divider for use at VHF would need to be built from ECL devices. It would be expensive, hot and power-hungry. The idea of a fast front end ahead of a CMOS programmable divider would be perfect if these problems could be circumvented. Consider a fast divide-by-10 prescaler ahead of a programmable divider where division by 947 is required. If the main divider is set to divide by 94, the overall division ratio is 940. The prescaler goes through its cycle 94 times and the main divider goes through its cycle once, for every output pulse. If the prescaler is changed to divide by 11 for 7 of its cycles for every cycle of the entire divider system, the overall division ratio is now $[(7 \times 11) + (87 \times 10)] = 947$. At the cost of a more elaborate prescaler and the addition of a slower programmable counter to control it, this prescaler does not multiply the step size and avoids all the problems of fixed prescaling.

Fig 9.40 shows the general block diagram of a dual-modulus prescaled divider. The down counter controls the modulus of the prescaler. The numerical example, just given, used decimal arithmetic for ease of understanding, although binary is used by the logic itself.

Each cycle of the system begins with the last output pulse having loaded the frequency control word, shown as "Division Control Data" in Fig 9.40, into both the main divider and the prescaler controller. If the division control data's least significant bits loaded into the prescaler controller are not zero, the prescaler is set to divide by 1 greater than its normal ratio. Each cycle of the prescaler clocks the down counter. Eventually, it reaches zero, and two things happen: The counter is designed to hold (stop counting) at zero (and it will remain held until it is next reloaded) and the prescaler is switched back to its normal ratio until the next reload. One way of visualizing this is to think of the prescaler as just being a divider of its normal ratio, but with the ability to "steal" a number of input pulses controlled by the data loaded into its companion down counter. Note that a dual-modulus prescaler system has a *minimum* division ratio, needed to ensure there are enough cycles of the prescaler to allow enough input pulses to be stolen.

Because the technique is widely used, dual-modulus prescaler ICs are widely used and widely available. Devices for use to a few hundred megahertz are cheap, and devices in the 2.5-GHz region are commonly available. Common prescaler IC division ratio pairs are: 8-9, 10-11, 16-17, 32-33, 64-65 and so on. Many ICs containing programmable dividers are available in versions with and without built-in prescaler controllers.

PHASE DETECTORS

A *phase detector* (PSD) produces an output voltage that depends on the phase relationship between its two input signals. If two signals, *in phase on exactly the same frequency*, are mixed together in a conventional diode-ring mixer with a dc-coupled output port, one of the products is direct current (0 Hertz). If the phase relationship between the signals changes, the mixer's dc output voltage changes. With both signals in phase, the output is at its most positive; with the signals 180° out of phase, the output is at its most negative. When the phase difference is 90° (the signals are said to be *in quadrature*), the output is 0 V.

Applying sinusoidal signals to a phase detector causes the detector's output voltage to vary sinusoidally with phase angle, as in **Fig 9.41A**. This nonlinearity is not a problem, as the loop is usually arranged to run with phase differences close to 90°. What might seem to be a more serious complication is that the detector's phase-voltage characteristic repeats every 180°, not 360°. Two possible input phase differences can therefore produce a given output voltage. This turns out not to be a problem, because the two identical voltage points lie on opposing slopes on the detector's output-voltage curve. One direction of slope gives positive feedback (making the loop unstable and driving the VCO away from what otherwise would be the lock angle) over to the other slope, to the true and stable lock condition.

MD108, SBL-1 and other diode mixers can be used as phase detectors, as can active mixers like the MC1496. Mixer manufacturers make some parts (such as the Mini-Circuits Labs RPD-1) that are specially optimized for phase-detector service. All these devices can make excellent, low-noise phase detectors but are not commonly used in ham equipment. A high-speed sample-and-hold circuit, based on a Schottky-diode bridge, can form a very low-noise phase detector and is sometimes used in specialized instruments. This is just a variant on the basic mixer; it produces a similar result, as shown in Fig 9.41B.

The most commonly used simple phase detector is just a single exclusive-OR (XOR) logic gate. This circuit gives a logic 1 output only when *one* of its inputs is at logic 1; if both inputs are the same, the output is a logic 0. If inputs A and B in Fig 9.41C are almost in phase, the output will be low most of the time, and its average filtered value will be close to the logic 0 level. If A and B are almost in opposite phase, the output will be high most of the time, and its average voltage will be close to logic 1. This circuit is very similar to the mixer. In fact, the internal circuit of ECL XOR gates is the same transistor circuit found

Oscillators and Synthesizers 9.39

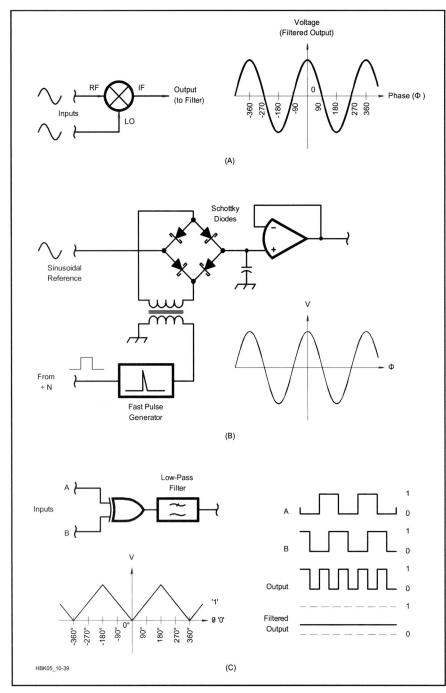

Fig 9.41 — Simple phase detectors: a mixer (A), a sampler (B) and an exclusive-OR gate (C).

in the MC1496 and similar mixers, with some added level shifting. Like the other simple phase detectors, it produces a cyclic output, but because of the square-wave input signals, produces a triangular output signal. To achieve this circuit's full output-voltage range, it's important that the reference and VCO signals applied operate at a 50% duty cycle.

PHASE-FREQUENCY DETECTORS

All the simple phase detectors described so far are really specialized mixers. If the loop is out of lock and the VCO is far off frequency, such phase detectors give a high-frequency output midway between zero and maximum. This provides no information to steer the VCO towards lock, so the loop remains unlocked. Various solutions to this problem, such as crude relaxation oscillators that start up to sweep the VCO tuning until lock is acquired, or laboriously adjusted "pretune" systems in which a DAC, driven from the divider control data, is used to coarse tune the VCO to within locking range, have been used in the past. Many of these solutions have been superseded, although pretune systems are still used in synthesizers that must change frequency very rapidly.

The *phase-frequency detector* is the usual solution to lock-acquisition problems. It behaves like a simple phase detector over an extended phase range, but its output remains at maximum or minimum and does not repetitively cycle with larger and larger phase errors. Because its output voltage stays high or low, depending on which input is higher in frequency, this PSD can steer a loop towards lock from anywhere in its tuning range. **Fig 9.42** shows the internal logic of the phase-frequency detector in the CD4046 PLL chip. (The CD4046 also contains an XOR PSD). When the phase of one input leads that of the other, one of the output MOSFETs is pulsed on with a duty cycle proportional to the phase difference. Which MOSFET receives drive depends on which input is leading. If both inputs are in phase, the output will include small pulses due to noise, but their effects on the average output voltage will cancel. If one signal is at a higher frequency than the other, its phase will lead by an increasing amount, and the detector's output will be held close to either V_{DD} or ground, depending on which input signal is higher in frequency. To get a usable voltage output, the MOSFET outputs can be terminated in a high-value resistor to $V_{DD}/2$, but the CD4046's output stage was really designed to drive current pulses into a capacitive load, with pulses of one polarity charging the capacitor and pulses of the other discharging it. The capacitor then integrates the pulses. Simple phase detectors are normally used to lock their inputs in quadrature (90° difference), but phase-frequency detectors are used to lock their input signals in phase (0° difference).

Traditional textbooks and logic-design courses give extensive coverage to avoiding *race conditions*, timing hazards caused by near-simultaneous signals racing through parallel circuit paths to control a single output. Avoiding such situations is important in many circuits because the outcome is strongly dependent on slight differences in gate speed. The phase-frequency detector is the one circuit whose entire function depends on its built-in ability to *make* both its inputs race. Consequently, it's not easy to home-brew a phase-frequency detector from ordinary logic parts. Fortunately, they are available in IC form, usually combined with other functions. The MC4044 is an old stand-alone TTL phase-frequency detector; the MC12040 is an ECL derivative, and the much faster and compatible MCH 12140 can be used to over 600 MHz. The Hittite HMC403S8G can be used to 1.3 GHz. CMOS versions can be found in the CD4046 PLL chip and in almost all current divider-PLL synthesizer chips.

The race-condition hazard does cause one problem in phase-frequency detectors:

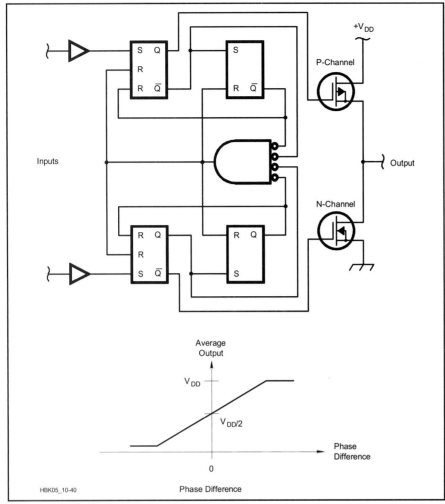

Fig 9.42 — Input signals very far off frequency can confuse a simple phase-detector; a phase-frequency detector solves this problem.

A device's delays and noise, rather than its input signals, control its phase-voltage characteristic in the zero-phase-difference region. This degrades the loop's noise performance and makes its phase-to-voltage coefficient uncertain and variable in a small region, a "dead zone" — unfortunately, in the detector's normal operating range! It's therefore normal to bias operation slightly away from *exact* phase equality to avoid this problem. Fortunately, the newer, faster phase detectors like the MCH 12140 and HMC403S8G tend to minimize this "dead zone" problem.

THE LOOP COMPENSATION AMPLIFIER

Phase-locked loops have acquired a troublesome reputation for a variety of reasons. In the past, a number of commercially made radios have included poorly designed synthesizers that produced excessive noise sidebands, or wouldn't lock reliably. Some un-producible designs have been published for home construction that could not be made to work. Many experimenters have toiled over an unstable loop, desperately trying anything to get it to lock stably. So the PLL has earned a shady reputation.

Luckily, the proliferation of synthesis techniques and PLLs in contemporary equipment have lead to a number of excellent integrated circuits, as well as fine application notes to support them.

The *first* task is to take action to ensure that all previously discussed components of the loop are functioning properly and stably before attempting to close the loop. This means that the oscillators, dividers and amplifiers must all be unconditionally stable. Any instability or "squegging" in these components must be dealt with prior to attempting to close the loop.

The *second* task is to produce a closed-loop characteristic that best suits the application. Herein is where the greatest difficulty frequently lies. The selection of the closed-loop bandwidth and phase margin or peaking is usually one of great compromise and thought. The process frequently begins with an educated approximation that is then evaluated and modified. There are many factors to consider, including the degree of reference frequency suppression, switching speed, noise, microphonics and modulation, if any. Once the optimum loop characteristic is established, the next problem is to maintain it with respect to variations of the division resulting from the synthesis, and also gain variations associated with the nonlinear tuning curve of the VCO. While not always required in amateur applications, loops in sophisticated synthesizers frequently employ programmable multiplying DACs to maintain constant loop bandwidths and phase margin. In an example later in this chapter, we will actually describe the process for establishing the loop bandwidth for a simple synthesizer.

The *third* task is to design a loop compensation amplifier (loop filter) that will ensure stable operation when the loop is closed. The design of this compensation network or filter requires knowledge of the VCO gain and linearity, the phase-detector gain, and the effective division ratio and its variation in the loop. The foregoing requirements imply the necessity for measurement and calculation to have any reasonable hope of producing a successful outcome.

Note carefully that we are designing a *loop*. Trial and error, intuitive component choice, or "reusing" a loop amplifier design from a different synthesizer may lead to a low probability of success. All oscillators are loops, and any loop will oscillate when certain conditions are met. Look again at the RC phase-shift oscillator from which the "Pierce" crystal oscillator is derived (Fig 9.29A). Our PLL is a loop, containing an amplifier and a number of RC sections. It has all the parts needed to make an oscillator. If a loop is unstable and oscillates, there will be an oscillatory voltage superimposed on the VCO tuning voltage. This will produce massive unwanted FM on the output of the PLL, which is absolutely undesirable.

Before you turn to a less demanding chapter, consider this: It's one of nature's jokes that easy procedures often hide behind a terror-inducing facade. The math you need to design a good PLL may look weird when you see it for the first time and seem to involve some obviously impossible concepts, but it ends in a very simple procedure that allows you to calculate the response and stability of a loop. Professional designers must handle these things daily, and because they like differential equations no more than anyone else does, they use a graphical method based on the work of the mathematician, Laplace, to generate PLL designs. (We'll leave proofs in the textbooks, where they belong! Incidentally, if you can get the math required for PLL design under control, as a free gift you will also be ready to do modern filter synthesis and design beam antennas, since the math they require is essentially the same.)

The easy solutions to PLL problems can only be performed during the design phase. We must deliberately design loops with sufficient stability safety margins so that all foreseeable variations in component tolerances, from part-to-part, over temperature and across the tuning range, cannot take the loop anywhere near the threshold of oscillation. More than this, it is important to have adequate stability margin because loops with lesser margins will exhibit amplified phase-noise sidebands, and that is another PLL problem that can be designed out.

At the beginning of this chapter (following the text cite of Fig 9.4A), the criterion for stability/oscillation of a loop was mentioned: Barkhausen's for oscillation or Nyquist's for stability. This time Harry Nyquist is our hero.

A PLL is a negative feedback system, with the feedback opposing the input, this opposition or inversion around the loop amounts to a 180° phase shift. This is the frequency-*independent* phase shift round the loop, but there are also frequency-*dependent* shifts that will add in. These will inevitably give an increasing, lagging phase with increasing frequency. Eventually an extra 180° phase shift will have occurred, giving a total of 360° around the loop at some frequency. We are in trouble if the gain around the loop has not dropped below unity (0 dB) by this frequency.

Note that we are not concerned only with the loop operating frequency. Consider a 30-MHz low-pass filter passing a 21-MHz signal. Just because there is no signal at 30 MHz, our filter is no less a 30-MHz circuit. Here we use the concept of frequency to describe "what would happen *if* a signal was applied at that frequency."

The next sections use the graphical concepts of poles and zeros to calculate the gain and phase response around a loop. The graphical concepts are quite useful in that they provide insight into the effect of various component choices on the loop performance. (Poles and zeroes are discussed in the **Analog Basics** chapter.)

Poles and Zeroes in the Loop Amplifier

The loop-amplifier circuit used in the example loop has a blocking capacitor in its feedback path. This means there is no dc feedback. At higher frequencies the reactance of this capacitor falls, increasing the feedback and so reducing the gain. This low-pass filter is an *integrator*. The gain is immense at 0 Hz (dc) and rolls off at 6 dB per octave, pointing to a pole at 0 Hz. This is true whatever the value of the series resistance feeding the signal into the amplifier inverting input and whatever the value of the feedback capacitor. The values of these components scale

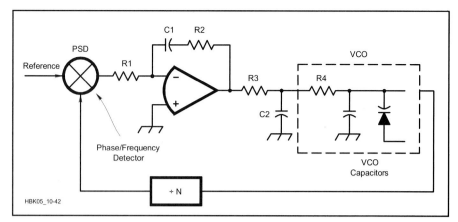

Fig 9.43 — A common loop-amplifier/filter arrangement.

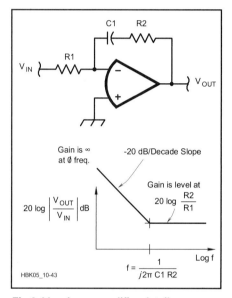

Fig 9.44 — Loop-amplifier detail.

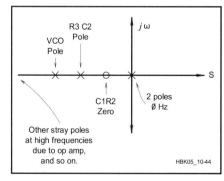

Fig 9.45 — The loop's resulting pole-zero "constellation."

the frequency response of the gain rather than shape it. The roll-off of the integrator gain is fixed at –6 dB per octave, but these components allow us to move it to achieve some wanted gain at some wanted frequency.

At high frequencies, the feedback capacitor in an ordinary integrator will have very low reactance, giving the circuit very low gain. In our loop amplifier, there is a resistor in series with the capacitor. This limits how low the gain can go, in other words the integrator's downward (with increasing frequency) slope is leveled off. This resistor and capacitor create a zero, and its +6 dB per octave slope cancels the –6 dB per octave slope created by the pole at 0 Hz.

RECIPE FOR A PLL POLE-ZERO DIAGRAM

We can choose a well-tried loop-amplifier circuit and calculate values for its components to make it suit our loop. **Fig 9.43** shows a common synthesizer loop arrangement. The op amp operates as an integrator, with R2 added to level off its falling gain. An integrator converts a dc voltage at its input to a ramping voltage at its output. Greater input voltages yield faster ramps. Reversing the input polarity reverses the direction of the ramp.

The system's phase-frequency detector (connected to work in the right sense) steers the integrator to ramp in the direction that tunes the VCO toward lock. As the VCO approaches lock, the phase detector reduces the drive to the integrator, and the ramping output slows and settles on the right voltage to give the exact, locked output frequency. Once lock is achieved, the phase detector outputs short pulses that "nudge" the integrator to keep the divided VCO frequency exactly locked in phase with the reference. Now we'll take a look around the entire loop and find the poles and zeros of the circuits.

The integrator produces a –20 dB per decade roll off from 0 Hz, so it has a pole at 0 Hz. It also includes R2 to cancel this slope, which is the same as adding a rising slope that exactly offsets the falling one. This implies a zero at $f = 1/(2\pi C1 R2)$ Hz, as **Fig 9.44** shows. R3 and C2 make another pole at $f = 1/(2\pi C2 R3)$ Hz. A VCO usually includes a series resistor that conveys the control voltage to the tuning diode, which also is loaded by various capacitors. This creates another pole.

The VCO generates frequency, while the phase detector responds to phase. Phase is the integral of frequency, so together they act as another integrator and add another pole at 0 Hz.

What about the frequency divider? The frequency divider's effect on loop response is that of a simple attenuator. Imagine a signal generator that is switched between 10 and 11 MHz on alternate Tuesdays. Imagine that we divide its output by 10. The output of the divider will change between 1 and 1.1 MHz, still on alternate Tuesdays. The divider divides the deviation of the frequency (or phase) modulated signal, but it cannot affect the modulating frequency.

A possible test signal passing around the example loop is in the form of frequency modulation as it passes through the divider, so it is simply attenuated. A divide-by-N circuit will give $20 \log_{10}(N)$ dB of attenuation (reduction of loop gain). This completes the loop. We can plot all this information as shown in **Fig 9.45**, and the pattern of poles and zeroes is sometimes referred to as a *pole-zero constellation*.

OPEN-LOOP GAIN AND PHASE

Just putting together the characteristics of the blocks around the loop, without allowing for the loop itself, gives us the system's *open-loop* characteristic, which is all we really need.

What Is Stability? — Here "open-loop response" means the gain around the loop that would be experienced by a signal at some frequency *if* such a signal were inserted. We do not insert such signals in the actual circuit, but the concept of frequency-dependent gain is still valid. We need to ensure that there is insufficient gain, at *all* frequencies, to ensure that the loop cannot create a signal and begin oscillating. More than this, we want a good safety margin to allow for component variations and because loops that are *close* to instability perform poorly.

The loop gain and phase are doing interesting things on our plots at frequencies in the AF part of the spectrum. This does not mean that there is visible operation at those frequencies. Imagine a bad, unstable loop. It will have a little too much loop gain at some frequency, and unavoidable noise will build up until a strong signal is created. The amplitude will increase until nonlinearities limit it. An oscilloscope view of the tuning voltage input to the VCO will show a big signal, often in the hertz to tens of kilohertz region, and often large enough to drive the loop amplifier to the limits of its output swing, close to its supply voltages. As this big signal is applied to the tuning voltage input to the VCO, it will modulate the VCO frequency across a wide range. The output will look like that of a sweep generator on a spectrum analyzer. What is wrong? How can we fix this loop? Well, the problem may be excessive loop gain, or improper loop time constants. Rather than work directly with the loop time constants, it is far easier to work with the pole and zero frequencies of the loop response. It's just a different view of the same things.

Now imagine a good, stable loop, with an adequate stability safety margin. There will be some activity around the loop: the PSD (phase detector) will demodulate the phase noise of the VCO and feed the demodulated noise through the loop amplifier in such a way as to cancel the phase noise. This is how a good PLL should give *less* phase noise than the VCO alone would suggest. In a good loop, this noise will be such a small voltage that a 'scope will not show it. In fact, connecting test equipment in an attempt to measure it can usually *add* more noise than is normally there.

Finally, imagine a poor loop that is just barely stable. With an inadequate safety margin, the action will be like that of a Q multiplier or a regenerative receiver close to the point of oscillation. There will be an amplified noise peak at some frequency. This spoils the effect of the phase detector trying to combat the VCO phase noise and gives the opposite effect. The output spectrum will show prominent regions of exaggerated phase noise, as the excess noise frequency modulates the VCO.

Now, let's use the pole-zero diagram as a graphic tool to find the system open-loop gain and phase. As we do so, we need to keep in mind that the frequency we have been discussing in designing a loop response is the frequency of a theoretical test signal passing around the loop. In a real PLL, the loop signal exists *in two forms*: Between the output of the phase detector and the input of the VCO it is a sinusoidal voltage, and elsewhere it is a sinusoidal modulation of the VCO frequency. The VCO translates the signal from voltage to frequency.

With our loop's pole-zero diagram in hand, we can pick a frequency at which we want to know the system's open-loop gain and phase. We plot this value on the graph's vertical $j\omega$ axis and draw lines between it and each of the poles and zeros, as shown in **Fig 9.46**. Next, we measure the lengths of the lines and the "angles of elevation" of the lines. The loop gain is proportional to the product of the lengths of all the lines to zeros *divided by* the product of all the lengths of the lines to the poles. The phase shift around the loop (lagging phase equates to a positive phase shift) is equal to the sum of the pole angles minus the sum of the zero angles. We can

---

Complex Frequency

We are accustomed to thinking of frequency as a real number — so many cycles per second — but frequency can also be a complex number, s, with both a real part, designated by σ, and an imaginary part, designated by $j\omega$. (ω is also equal to $2\pi f$.) The resulting complex frequency is written as $s = \sigma + j\omega$. Complex frequency is used in Laplace transforms, a mathematical technique used for circuit and signal analysis. (Thorough treatments of the application of complex frequency can be found in college-level textbooks on circuit and signal analysis.)

The graphs in Fig 9.44 and 9.45 plot s on a pair of real and imaginary axes. This is also called the *s-plane*. Each pole and zero has a complex frequency that is plotted on the s-plane just as complex numbers are plotted in Cartesian coordinates. The graph of the poles and zeroes is called a *pole-zero map* or *pole-zero diagram*.

When complex frequency is used, a sinusoidal signal is described by Ae^{st}, where A is the amplitude of the signal and t is time. Because s is complex, $Ae^{st} = Ae^{(\sigma+j\omega)t} = A(e^{\sigma t})(e^{j\omega t})$. The two exponential terms describe independent characteristics of the signal. The second term, $e^{j\omega t}$, is the sine wave with which we are familiar and that has frequency f, where $f = \omega/2\pi$. The first term, $e^{\sigma t}$, represents the rate at which the signal increases or decreases. If σ is negative, the exponential term decreases with time and the signal gets smaller and smaller. If σ is positive, the signal gets larger and larger. If $\sigma = 0$, the exponential term equals 1, a constant, and the signal amplitude stays the same.

Complex frequency is very useful in describing a circuit's stability. If the response to an input signal is at a frequency on the right-hand side of the s-plane for which $\sigma > 0$, the system is *unstable* and the output signal will get larger until it is limited by the circuit's power supply or some other mechanism. If the response is on the left-hand side of the s-plane, the system is *stable* and the response to the input signal will eventually die out. The larger the absolute value of σ, the faster the response changes. If the response is precisely on the $j\omega$ axis where $\sigma = 0$, the response will persist indefinitely.

The practical effects of complex frequency can be experienced in a narrow CW crystal or LC filter. The poles of such a filter are just to the left of the $j\omega$ axis, so the input signal causes the filters to "ring", or output a damped sine wave along with the desired signal. Similarly, the complex frequency of an oscillator's output at power-up must have $\sigma > 0$ or the oscillation would never start! The output amplitude continues to grow until limiting takes place, reducing gain until $\sigma = 1$ for a steady output.

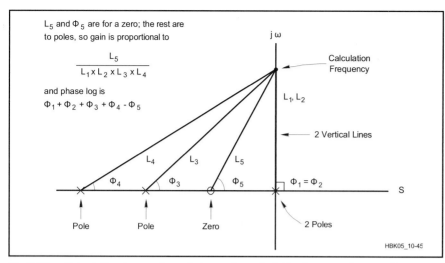

Fig 9.46 — Calculating loop gain and phase characteristics from a pole-zero diagram.

repeat this calculation for a number of different frequencies and draw graphs of the loop gain and phase versus frequency.

All the lines to poles and zeros are hypotenuses of right triangles, so we can use Pythagoras's rule and the tangents of angles to eliminate the need for scale drawings. Much tedious calculation is involved because we need to repeat the whole business for each point on our open-loop response plots. This much tedious calculation is an ideal application for a computer.

The procedure we've followed so far gives only *proportional* changes in loop gain, so we need to calculate the loop gain's *absolute* value at some (chosen to be easy) frequency and then relate everything to this. Let's choose 1 Hz as our reference. (Note that it's usual to express angles in radians, not degrees, in these calculations and that this normally renders frequency in peculiar units of *radians per second*. We can keep frequencies in hertz if we remember to include factors of 2π in the right places. A frequency of 1 Hz = 2π radians/second, because 2π radians = 360°.)

We must then calculate the loop's proportional gain at 1 Hz from the pole-zero diagram, so that the constant of proportion can be found. For starters, we need a reasonable estimate of the VCO's *voltage-to-frequency gain* — how much it changes frequency per unit change of tuning voltage. As we are primarily interested in stability, we can just take this number as the slope, in Hertz per volt, at the steepest part of voltage-versus-frequency tuning characteristic, which is usually at the low-frequency end of the VCO tuning range. (You can characterize a VCO's voltage-versus-frequency gain by varying the bias on its tuning diode with an adjustable power supply and measuring its tuning characteristic with a voltmeter and frequency counter.)

The loop divide-by-N stage divides our theoretical modulation — the tuning corrections provided through the phase detector, loop amplifier and the VCO tuning diode — by its programmed ratio. The worst case for stability occurs at the divider's lowest N value, where the divider's "attenuation" is least. The divider's voltage-versus-frequency gain is therefore 1/N, which, in decibels, equates to –20 log (N).

The change from frequency to phase has a voltage gain of one at the frequency of 1 radian/second, $1/(2\pi)$, which equates to –16 dB, at 1 Hz. The phase detector will have a specified "gain" in volts per radian. To finish off, we then calculate the gain of the loop amplifier, including its feedback network, at 1 Hz.

There is no need even to draw the pole-zero diagram. **Fig 9.47** gives the equations needed to compute the gain and phase of a system with up to four poles and one zero. They can be extended to more singularities and put into a simple computer program. Alternatively, you can type them into your favorite spreadsheet and get printed plots. There are numerous on-line calculators to help, such as those found at **www.circuitsage.com/pll.html**.

Fig 9.48 shows the sort of gain- and phase-versus-frequency plots obtained from a "recipe" loop design. We want to know where the loop's phase shift equals, or becomes more negative than, –180°. Added to the –180° shift inherent in the loop's feedback (the polarity of which must be negative to ensure that the phase detector drives the VCO toward lock instead of away from it), this will be the point at which the loop itself oscillates — *if* any loop gain remains at this point. The game is to position the poles and zero, and set the unit frequency gain, so that the loop gain falls to

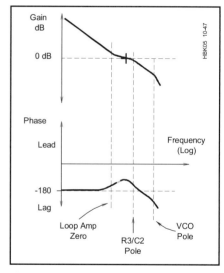

Fig 9.48 — An open-loop gain/phase plot.

$$\text{Open loop gain (dB)} = 20 \log \left[\frac{\sqrt{P_1^2+1}\sqrt{P_2^2+1}\sqrt{P_3^2+1}\sqrt{P_4^2+1}\sqrt{Z^2+f^2}}{\sqrt{P_1^2+f^2}\sqrt{P_2^2+f^2}\sqrt{P_3^2+f^2}\sqrt{P_4^2+f^2}\sqrt{Z^2+1}} \times 10^{\frac{\text{unity freq gain}}{20}} \right]$$

$$\text{Phase (lead)} = \tan^{-1}\left(\frac{f}{Z}\right) - \tan^{-1}\left(\frac{f}{P_2}\right) - \tan^{-1}\left(\frac{f}{P_3}\right) - \tan^{-1}\left(\frac{f}{P_4}\right)$$

f is the frequency of the point to be characterized. P_1, P_2, P_3, P_4 and Z are the frequencies of the poles and zero (all in the same units)

Fig 9.47 — Pole-zero frequency-response equations capable of handling up to four poles and one zero.

below 0 dB before the –180° line is crossed.

At the low-frequency end of the Fig 9.48 plots, the two poles have had their effect, so the gain falls at 40 dB/decade, and the phase remains just infinitesimally on the safe side of –180°. The next influence is the zero, which throttles the gain's fall back to 20 dB/decade and peels the phase line away from the –180° line. The zero would eventually bring the phase back to 90° of lag, but the next pole bends it back before that and returns the gain slope back to a fall of 20 dB/decade. The last pole, that attributable to the VCO, pushes the phase over the –180° line.

It's essential that there not be a pole between the 0-Hz pair and the zero, or else the phase will cross the line early. In this example, the R3-C2 pole and the zero do all the work in setting this critical portion of the loop's response. Their frequency spacing should create a phase bump of 30 to 45°, and their particular frequency positions are constrained as a compromise between sufficient loop bandwidth and sufficient suppression of reference-frequency sidebands. We know the frequency of the loop amplifier zero (that contributed by R1 and C1), so all we need to do now is design the loop amplifier to exhibit a frequency response such that the open-loop gain passes through 0 dB at a frequency close to that of the phase-plot bump.

This loop-design description may have been hard to follow, but a loop amplifier can be designed in under 30 minutes with a little practice. The high output impedance of commonly used CMOS phase detectors favors loop amplifiers based on FET-input op amps and allows the use of high-impedance RC networks (large capacitors can therefore be avoided in their design). LF356 and TL071 op amps have proven successful in loop-amplifier service and have noise characteristics suited to this environment.

To sum up, the recipe gives a proven pole zero pattern (listed in order of increasing frequency):

1. Two poles at 0 Hz (one is the integrator, the other is implicit in all PLLs)
2. A zero controlled by the RC series in the loop amplifier feedback path
3. A simple RC low-pass pole
4. A second RC low-pass pole formed by the resistor driving the capacitive load of the tuning voltage inputs of the VCO

To design a loop: Design the VCO, divider, phase detector and find their coefficients to get the loop gain at unit frequency (1 Hz), then, keeping the poles and zeros in the above order, move them about until you get the same 45° phase bump at a frequency close to your desired bandwidth. Work out how much loop-amplifier unit-frequency gain is needed to shift the gain-frequency-frequency plot so that 0-dB loop gain occurs at the center of the phase bump. Then calculate R and C values to position the poles and zero and the feedback-RC values to get the required loop amplifier gain at 1 Hz.

Design tips: You can scale the example loop to other frequencies: Just take the reactance values at the zero and pole frequencies and use them to scale the components, you get the nice phase plot bump at a scaled frequency. Even so, you still must do the loop-gain design.

Don't forget to do this for your lowest division factor, as this is usually the least stable condition because there is less attenuation. Also, VCOs are usually most sensitive at the low end of their tuning range.

NOISE IN PHASE-LOCKED LOOPS

Differences in Q usually make the phase-noise sidebands of a loop's reference oscillator much smaller than those of the VCO. Within its loop bandwidth, a PLL acts to correct the phase-noise components of its VCO and impose those of the reference. Dividing the reference oscillator to produce the reference signal applied to the phase detector also divides the deviation of the reference oscillator's phase-noise sidebands, translating to a 20-dB reduction in phase noise per decade of division, a factor of $20 \log (M)$ dB, where M is the reference divisor. Offsetting this, within its loop bandwidth the PLL acts as a *frequency multiplier*, and this multiplies the deviation, again by 20 dB per decade, a factor of $20 \log (N)$ dB, where N is the loop divider's divisor. Overall, the reference sidebands are increased by $20 \log (N/M)$ dB. Noise in the dividers is, in effect, present at the phase detector input, and so is increased by $20 \log (N)$ dB. Similarly, op-amp noise can be calculated into an equivalent phase value at the input to the phase detector, and this can be increased by $20 \log (N)$ to arrive at the effect it has on the output.

Phase noise can be introduced into a PLL by other means. Any amplifier stages between the VCO and the circuits that follow it (such as the loop divider) will contribute some noise, as will microphonic effects in loop- and reference-filter components (such as those due to the piezoelectric properties of ceramic capacitors and the crystal filters sometimes used for reference-oscillator filtering). Noise on the power supply to the system's active components can modulate the loop. The fundamental and harmonics of the system's ac line supply can be coupled into the VCO directly or by means of ground loops.

Fig 9.49 shows the general shape of the PLL's phase noise output. The dashed curve shows the VCO's noise performance when unlocked; the solid curve shows how much locking the VCO to a cleaner reference improves its noise performance. The two noise "bumps" are a classic characteristic of a phase-locked loop. If the loop is poorly designed and has a low stability margin, the bumps may be exaggerated — a sign of noise amplification due to an overly peaky loop response.

Exaggerated noise bumps can also occur if the loop bandwidth is less than optimum. Increasing the loop bandwidth in such a case would widen the band over which the PLL acts, allowing it to do a better job of purifying the VCO — but this might cause other problems in a loop that's deliberately bandwidth-limited for better suppression of reference-frequency sidebands. Immense loop bandwidth is not desirable, either: Farther away from the carrier, the VCO may be so quiet on its own that widening the loop would make it *worse*.

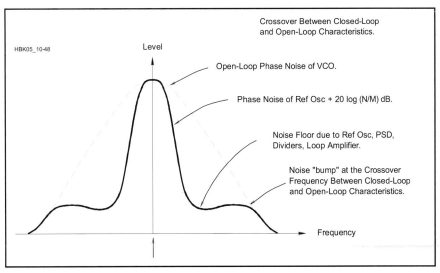

Fig 9.49 — A PLL's open- and closed-loop phase-noise characteristics. The noise bumps at the crossover between open- and closed-loop characteristics are typical of PLLs; the severity of the bumps reflects the quality of the system's design.

9.7.2 A PLL Design Example

In this section, we will explore the application of the design principles previously covered, as well as some of the tradeoffs required in a practical design. We will also cover measurement techniques designed to give the builder confidence that the loop design goals have been met. Finally, we will cover some common troubleshooting problems.

As our design example, a synthesized local oscillator chain for a 10-GHz transverter will be considered. **Fig 9.50** is a simplified block diagram of a 10-GHz converter. This example was chosen as it represents a departure from the traditional multi-stage multiply-and-filter approach. It permits realization of the oscillator system with two very simple loops and minimal RF hardware. It is also representative of what is achievable with current hardware, and can fit in a space of 2 to 3 square inches. This example is intended as a vehicle to explore the loop design aspects and is not offered as a "construction project." The additional detail required would be beyond the scope of this chapter.

In this example, two synthesized frequencies, 10 GHz and 340 MHz, are required. Since 10.368 GHz is one of the popular traffic frequencies, we will mix this with the 10-GHz LO to produce an IF of 368 MHz. The 368-MHz IF signal will be subsequently mixed with a 340-MHz LO to produce a 28-MHz final IF, which can be fed into the 10 meter input of any amateur transceiver. We will focus our attention on the design of the 10-GHz synthesizer only. Once this is done, the same principles may be applied to the 340-MHz section. Our goal is to attempt to design a low-noise LO system (this implies minimum division) with a loop reference oscillator that is an integer multiple of 10 MHz. Using this technique will allow the entire system to be locked at a later time to a 10-MHz standard for precise frequency control.

Ideally, we would like to start with a low-noise 100-MHz crystal reference and a 10-GHz oscillator divided by 100. It is here that we are already confronted with our first practical design tradeoff. We are considering using a line of microwave integrated circuits made by Hittite Microwave. These include a selection of prescalers operating to 12 GHz, a 5-bit counter that will run to 2.2 GHz and a phase/frequency detector that will run up to 1.3 GHz. The 5-bit counter presents the first problem. Our ideal scheme would be to use a divide-by-4 prescaler and enter the 5-bit counter at 2.5 GHz, subsequently dividing by 25 to 100 MHz. The problem is that the 5-bit counter is only rated to 2.2 GHz and not 2.5 GHz. This forces us to use a divide-by-8 prescaler and enter the 5-bit divider at 1.25 GHz. This is well within the range of the

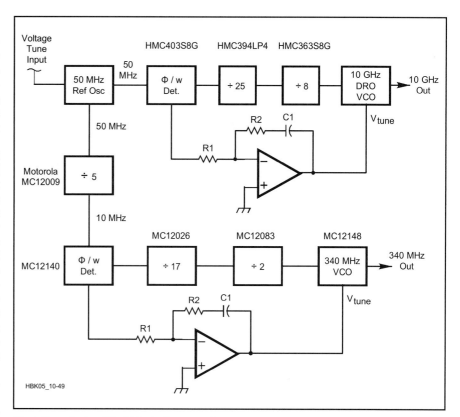

Fig 9.50 — A simplified block diagram of a local oscillator, for a 10-GHz converter.

divider, but we can no longer use an integer division (ie, 1.25 GHz/12.5 = 100 MHz) to get to 100 MHz. The next easiest option is to reduce the reference frequency to 50 MHz and to let the 5-bit divider run as a divide-by-25, giving us a total division ratio of 200.

Having already faced our first design tradeoff, a number of additional aspects must be considered to minimize the conflicts in the design:

First, this is a *static* synthesizer — that is, it will not be required to change frequency during operation. As a result, switching time considerations are irrelevant, as are the implications that the switching time would have had on the loop bandwidth. The effects of variable division ratios are also eliminated, as well as any problems associated with nonlinear tuning of the VCOs.

Second, the reference frequencies (50 MHz and 10 MHz) are very large with respect to any practically desirable loop bandwidths. This makes the requirement of the loop filter to eliminate reference sidebands quite easy to achieve. In fact, reference sideband suppression will more likely be a function of board layout and shielding effectiveness rather than suppression in the loop filter. This also allows placement of the reference suppression poles far out in the pole zero constellation and well outside the loop bandwidth at about 10 × BW3 (the 3-dB bandwidth of the loop). This placement of the reference suppression poles will give us the option of using a "type 2, 2nd-order" loop approximation, which will greatly simplify the calculation process.

Third, based on all of the foregoing tradeoffs, the loop bandwidth can be chosen almost exclusively on the basis of noise performance.

Before we can proceed with the loop calculations we need some additional information: specifically VCO gain, VCO noise performance, divider noise performance, phase detector gain, phase detector noise performance and finally reference noise performance. The phase detector and divider information is available from specification sheets. Now we need to select the 50-MHz reference and the 10-GHz VCO. An excellent choice for a low noise reference is the one described by John Stephensen on page 13 in Nov/Dec 1999 *QEX*. The noise performance of this VCXO is in the order of –160 dBc/Hz at 10-kHz offset at the fundamental frequency. For the 10-GHz VCO, there are many possibilities, including cavity oscillators, YIG oscillators and dielectric resonator oscillators. Always looking for parts that are easily available, useable and economical, salvaging a dielectric resonator from a Ku band LNB (see Jan-Feb 2002 *QEX*, p 3) appears promising. These high-Q oscillators can be fitted with a varactor and tuned over a limited range with good results. The tuning sensitivity of our dielectric resonator VCO is about 10 MHz per volt and

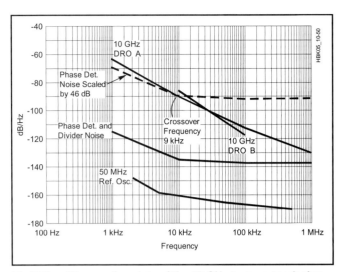

Fig 9.51 — Phase noise plots of the 10-GHz transverter design example.

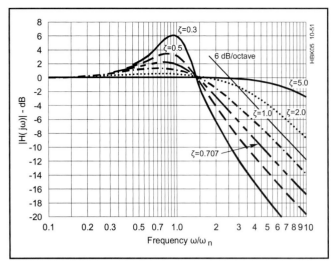

Fig 9.52 — Closed-loop frequency response of a second-order loop as a function of the loop's natural frequency (ω_n) and damping factor (ζ). (Based on Gardner, *Phaselock Techniques*. See references.)

the phase noise at 10-kHz offset is –87dBc/Hz. We are now in a position to plot the phase noise of the reference, the dividers, the division effect (46 dB) and the VCO. The phase noise plots are shown in **Fig 9.51**. The noise of the reference is uniformly about 25 dB below the divider and phase detector noise floor. This means that the noise of the phase detector and dividers will be the dominant contribution within the loop bandwidth. The cross-over point is at approximately 9 kHz. At frequencies lower than 9 kHz, the loop will reduce the noise of the VCO (eg, at 1 kHz the reduction will be about 20 dB). At frequencies above 9 kHz, the natural noise roll off of the VCO will dominate. At this point, we can also see the effect of having chosen a reference frequency of 50 MHz. Had we been able to use a 100-MHz reference instead of 50 MHz and reduce the division factor from 200 to 100, we could have put the loop bandwidth at about 30 kHz and picked up an additional 6-dB improvement in close-in noise performance. Nevertheless, even with the 6-dB penalty, this will be quite a respectable 9-GHz LO. It should be apparent that a fair amount of thought and effort was required to be able to determine an applicable bandwidth for this loop.

Now that we have some estimate of the noise performance and the required loop bandwidth, it is time to design the loop. The second consideration cited above enabling the application of the "type-2, 2nd order" approximation is now going to pay off. The math for this loop is really quite simple. There are two main variables, the *natural frequency* of the loop, ω_n, in radians/second, and the *damping factor*, ζ, which is dimensionless.

The closed-loop bandwidth of the loop is a function of ω_n and ζ. **Fig 9.52** shows loop response as a function of ω_n and ζ. Values of ζ less than 0.5 ("underdamped") are not desirable, as they tend towards instability and poor phase margin. A value for ζ of 0.707 is referred to as "critically damped" and is favored in applications where settling time is critical (see Gardner). For our application, we will choose a ζ of 1.0. This will give acceptable phase margin and further simplify the math. For a ζ of 1, $\omega_n = (6.28 \times BW3) / 2.48$. Substituting the 9-kHz loop bandwidth for BW3 yields 22790 radians per second for ω_n. Our loop filter will take the basic form shown in Fig 9.44. We can compute values for R1, R2 and C1 as follows.

$$R1 = \frac{K_{PD} \times K_{VCO}}{N \times \omega_n^2 \times C1} \quad (4)$$

where

K_{PD} is the phase detector gain, 1/6.28 V/radian in this case,
K_{VCO} is the VCO gain in radians/Hz, 6.28×10^7 radians/V in this case,
N is the division factor, 200 in this case, and
C1 is the feedback capacitor value in farads.

To proceed, we need to start with an estimate for C1. Practically speaking, since odd values of capacitors are more difficult to obtain, let us try a value for C1 of 0.01 µF.

$$R2 = \frac{2 \times \zeta}{\omega_n^2 \times C1} \quad (5)$$

R1 computes as 9626 Ω and R2 is 8775 Ω. Since we will be using a phase/frequency detector with differential outputs, we will have to modify the circuit of Fig 9.44 to become a differential amplifier. We will also add a passive input filter to keep the inputs of the op amp from being stressed by the very fast and short pulses emanating from the phase/frequency detector. We will also add a passive "hash filter" at the output of the op amp to limit the amount of out of band noise delivered to the VCO tuning port. Both of these filters will be designed for a cutoff frequency of 90 kHz (i.e. 10 times the 9-kHz closed-loop bandwidth, as cited above in the second consideration. Both of these filters will also aid in the rejection of reference frequency sidebands. For the input filter, we simply divide the input resistor in half and add a capacitor to ground. The value of this capacitor is determined as follows:

$$C2 = \frac{4}{62.8 \times BW3 \times R1} = 735 \text{ pF} \quad (6)$$

Using 680 or 750 pF will be adequate. The output filter is also quite simple, with one minor stipulation. Op amps can exhibit stability problems when driving a capacitive load through too low a resistance. One very safe way around this is not to use a value for R3 lower than the recommended minimum impedance for full output of the amplifier. For many amplifiers, this value is around 600 Ω. For the output filter, we compute C3 as follows, for R3 = 600 Ω:

$$C3 = \frac{1}{62.8 \times BW3 \times R3} = 2950 \text{ pF} \quad (7)$$

Using 3000 pF will be adequate. This completes the computations for the loop compensation amplifier. The complete amplifier is shown in **Fig 9.53**.

9.7.3 PLL Measurements

One of the first things we need to measure

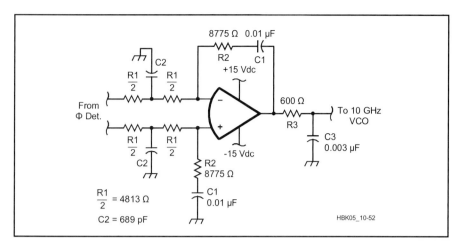

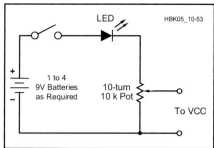

Fig 9.53 — A complete loop-compensation amplifier for the design example 10-GHz transverter.

Fig 9.54 — Clean, variable dc voltage source used to measure VCO gain in a PLL design.

when designing a PLL is the VCO gain. The tools we will need include a voltmeter, some kind of frequency measuring device like a receiver or frequency counter and a clean source of variable dc voltage. The setup in **Fig 9.54** containing one or more 9-V batteries and a 10-turn, 10-kΩ pot will do nicely. One simply varies the voltage some amount and then records the associated frequency change of the VCO. The gain of the VCO is then the change in f divided by the change in voltage (Hz/V). The phase detector gain constant can usually be found on the specification sheets of the components selected.

Measuring closed-loop bandwidth is slightly more complicated. The way it is commonly done in the laboratory is to replace the reference oscillator with a DCFM-able (ie, a signal generator whose FM port is dc coupled) signal generator and then feed the tracking generator output of a low-frequency spectrum analyzer into the DCFM-able generator while observing the spectrum of the tuning voltage on the spectrum analyzer (see **Fig 9.55**). While this approach is quite straightforward, few amateurs have access to or can afford the test equipment to do this. Today however, thanks to the PC sound card-based spectrum analyzer and tracking generator programs, amateurs can measure the closed-loop response of loops that are less than 20 kHz in bandwidth for significantly less than a king's ransom in test equipment! Here's how:

One approach is to build the reference oscillator with some "built-in test equipment" (BITE) already in the design. This BITE takes the form of some means of DC-FMing the reference oscillator. This is one of the things that makes John Stephenson's oscillator (previously mentioned) attractive. The oscillator includes a varactor input for the purpose of phase locking it to a high-stability low-frequency standard oscillator. This same varactor tuning input takes the place of the aforementioned DCFM-able signal generator.

Now all we need to complete our test setup is some batteries. First, we need to make sure the input of our sound card is ac-coupled (when connected to the VCO tuning port, there will be a dc component present and the sound card may not like this) and that the ac coupling is flat down to at least 10% of the natural frequency of the loop. If the sound card is not ac-coupled, a capacitor will be needed at the input to block the dc on the tuning voltage.

The next step is to determine the tuning sensitivity of the reference oscillator. This is done in the same manner as was done for the VCO described above. Once this has been established, we will set the audio oscillator for an output voltage that will produce between 100 Hz and 1 kHz of deviation. We will also need to establish that the output of the sound card is dc-coupled so that we can place a small battery in series with the audio generator and the tuning port of the reference oscillator. We can provide a dc return on the output by putting a resistive 10-dB attenuator of the appropriate impedance on the output of the sound card and passing the varactor bias through the attenuator to ground. This is required to bias the varactors and also assure that the audio voltage will not drive the varactors into conduction. We can now connect the ac-coupled input of the sound card to the VCO tuning voltage and turn on the loop.

Do not be surprised to see signal components at multiples of the power line frequency, as well as the signal from the audio oscillator. The presence of line frequency components is an indication that the loop is doing its job and removing these components from the spectrum of the VCO. We can now "sweep" the audio oscillator and, by plotting the amplitude of the audio oscillator's response, determine the closed-loop bandwidth of the PLL. In the case of the loop described above, (ω_n = 22790, ζ = 1.0) we should expect to see

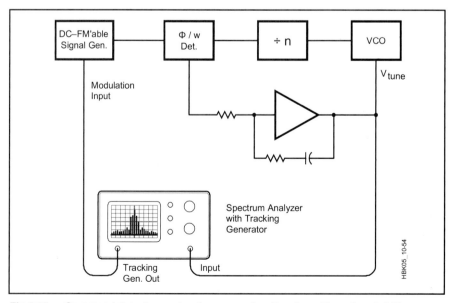

Fig 9.55 — Common laboratory setup for measuring the closed-loop bandwidth of loops.

slightly over 1 dB of peaking at 0.7 ω_n and the 3-dB down point should fall at 2.5 ω_n. In any event, if the peaking exceeds 3 dB, the loop phase margin is growing dangerously small and steps should be taken to improve it.

9.7.4 Common Problems in PLL Design

Here are some frequently encountered problems in PLL designs:

• The outputs of the phase detector are inverted. This results in the loop going to one or the other rail. The loop cannot possibly lock in this condition. Solution: Swap the phase detector outputs.

• The loop cannot comply with the tuning voltage requirements of the VCO. If the loop runs out of tuning voltage before the required voltage for a lock is reached, the locked condition is not possible. Solution: Re-center the VCO at a lower tuning voltage or increase the rail voltages on the op amp.

• The loop is very noisy and the tuning voltage is very low. The tuning voltage on the varactor diodes should not drop below the RF voltage swing in the oscillator tank circuit. Solution: Adjust the VCO so that the loop locks with a higher tuning voltage.

9.7.5 Synthesizer ICs

In the earlier section of this chapter on oscillators and synthesizers, we have explored oscillators and how they work along with modern simulation techniques. In the synthesis section, we have covered the basics of the PLL and its components and how to compute a proper loop. The subject of loop measurements and trouble shooting has also been covered. Some of the more common but traditional architectures have also been examined.

The following section is not intended to be project oriented, but rather is designed to expose the reader to additional concepts that are too complex to be fully explored here. The reader, being made aware of these ideas, may wish to examine them in detail using references at the end of this chapter.

In this section, we will explore several areas:

• Using commercially available synthesizer chips
• Improving VCO performance
• The microprocessor in a house keeping function
• The role of DDS in more recent hybrid architectures
• Multi-phase division

There have been many synthesizer chips introduced in recent years for cellular and WiFi applications. One might reasonably ask if any of these devices are well suited to amateur applications. The chips conform to a fairly strict applications profile. They are designed for minimum power consumption at battery potentials of 3 to 5 V. Applications using these chips strive for a minimum of external components and a maximum of on-chip components. The communication environment is typically one of sending a few tens of megabits for distances of less than 3 miles (5 km). Some are even designated as "low noise" with respect to other chips of their ilk, but not really with respect to most amateur radio requirements.

What this means is that these devices are not the type of synthesizer that one would directly use in amateur applications. For many reasons, the phase noise while adequate for their intended application, is not good enough for many HF, VHF and UHF applications without additional measures being taken. Despite all this, they do have some interesting properties and capabilities that can be exploited with additional components. Some properties that can be exploited are programmable charge pump current, a rich set of division options for creating multiple-loop synthesizers and typically low power consumption.

One of the easiest ways to improve the overall performance is to follow one of these chips with additional division. Consider the following example. One of these chips could be used as a 500 to 550 MHz synthesizer, followed with two cascaded decade dividers for a total division of 100. This division would reduce the phase noise profile by 40 dB making a reasonable quiet synthesizer for the 5 to 5.5 MHz range. The step size would also be reduced by a factor of 100 making the spacing required in the 500 to 550 MHz range equal to 1 kHz for a step size of 10 Hz at 5 MHz. This would make a good local oscillator for a simple traditional radio that covers 80 and 20 meters using both mixer products against a 9 MHz SSB generator.

There are also other techniques that can prove helpful. Decoupling the VCO from the chip will permit one to avoid much of the "on chip" noise that VCOs are susceptible to at the expense of some more complexity. If the VCO is implemented externally, there is an opportunity to design it with increased operating Q and thereby improving the overall phase noise performance. An external VCO also opens other opportunities.

As mentioned earlier, these chips are designed for operation at 3 to 5 V V_{CC}, typically using a internal charge pump to generate the operating voltage. This limits the tuning voltage swing to V_{CC}. One must be able to fit the entire tuning range voltage and the ac voltage excursion in the VCO tank circuit within the V_{CC} range. This problem could be overcome with the addition of voltage gain in an external operational amplifier, running at a higher voltage. This would allow the signal-to-noise ratio of the oscillator to be improved by increasing the tank voltage swing and still having adequate voltage range to perform the tuning function. These are just a few examples of how the performance of these devices could be improved for amateur applications.

9.7.6 Improving VCO Noise Performance

It is tempting to regard VCO design as a matter of coming up with a suitable oscillator topology with a variable capacitor, simply replacing the variable capacitor with a suitable varactor diode and applying a tuning voltage to the diode. Unfortunately, things are more complex. There is the matter of applying the tuning voltage to the diode without significantly disturbing the oscillator performance. There is also an issue of not introducing parasitic oscillation with the varactor circuit.

Next, as mentioned earlier, one would not like the tuning voltage to drop below the voltage swing in the oscillator tank. If this is allowed to occur, the tuning diodes will go into conduction and the oscillator noise will get worse. A good first approach is to use varactor diodes back to back as shown in Fig 9.39. This will allow the tank voltage swing to be developed across two diodes instead of one, as well as allow for a more balanced loading of the oscillator's tank. The semiconductor industry has realized this and there are a number of varactors available prepackaged in this configuration today.

IMPROVING THE OPERATING Q OF THE VCO

Many times the Q of passive capacitors available for use in an oscillator tank will exceed the Q of available varactors. One way of reducing the influence of the varactor is to use only the amount of varactor capacity required to tune the oscillator over its desired range. The balance of the capacity is supplied to the tank in the form of a higher Q fixed capacitor. This has the advantage of not requiring the varactor to supply all the capacity needed to make the circuit function, and will often allow for the use of a lower capacity varactor.

Lower capacity varactors typically exhibit higher Q values than their larger capacity counterparts. This concept can be extended by splitting the oscillator tuning range, say 70 MHz to 98 MHz (for a typical lower-side up-converter receiver design popular for the last three decades), into multiple bands. For this example let us consider four oscillators, each with 7 MHz of tuning range. The varactor Q effects can further be swamped by high-Q fixed capacitors. This can be further im-proved through the use of a "segment-tuned VCO" discussed below. A secondary but important benefit of this is to reduce the effective tuning gain (MHz/volt) of the oscillators, making them less susceptible to other noise voltage sources in the synthesizer loop. These noise sources can come from a variety of places including, but not limited to, varactor leakage current, varactor tuning drive impedance and output noise of the driving operational amplifier or charge pump.

Segment-tuned VCOs provide the designer with additional benefits, but with them come additional challenges. By segmented we mean that circuit elements, which could be both inductance and capacitance, are selected for each range that the VCO is expected to tune. **Fig 9.56** shows the frequency range-switching section of a typical VCO in which several diode switches are used to alter the total tank circuit inductance in small steps. Thus, the segments create a type of "coarse tuning" for the VCO and the output of the loop filter performs "fine tuning." Usually the component values are arranged in some sort of binary tree. In some advanced designs, 64 or even 128 sub-bands are available from the VCO. This represents the ultimate in reducing the undesirable effects of the varactor.

Many times the segmented-VCO concept is applied to an oscillator design that is required to cover several octaves in frequency. For example, this would be the case with a single-IF radio with the IF at about 8 MHz. The VCO would be required to cover 8 MHz to 48 MHz for 100 kHz to 50 MHz coverage of a typical transceiver. While segmentation will allow one to deal with the tuning of a multi-octave oscillator, segmentation also imposes some additional constraints.

Recalling the earlier section in this chapter on oscillators, the condition for oscillation is an oscillator loop gain of 1 at a phase angle of 0 degrees. Over several octaves, the gain of the oscillating device (typically a transistor) will vary, decreasing as operating frequency increases. While conventional limiting in the oscillator circuit will deal with some of this, it is not good practice to let the normal limiting process handle the entire gain variation. This becomes the role of automatic gain control (AGC) in multi-octave oscillator designs. Application of AGC will allow for the maintenance of the oscillation criteria, a uniform output and good starting characteristics.

Finally, a properly designed segmented VCO can improve synthesizer switching speed. The segmentation allows the designer to "pre-steer" the system and reduce the time required for the integrator to slew the VCO to the desired frequency for lock.

9.7.7 Microprocessor Control of Synthesizers

So far we have examined some interesting components that share one interesting trait, in that they require direction in order to work

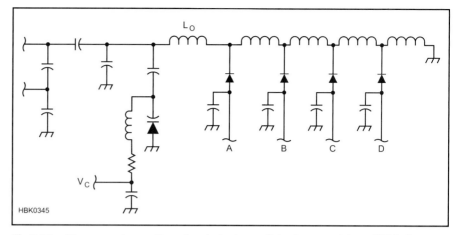

Fig 9.56 — The resonator portion of an inductor-switched segment-tuned VCO. A varactor diode provides tuning over a small range. Diode switches will alter the total inductance, changing the frequency in larger steps. Inductor-switched, capacitor-switched, and combinations are all used in VCO designs. (From Hayward, *Introduction to Radio Frequency Design*, Chapter 7. See references.)

in a system. This direction or "housekeeping" is best often left to a microprocessor and some software. One can differentiate this from DDS or DSP in that the microprocessor does not operate directly on some digital representation of the signal, but rather directs a number of components to perform the task.

The microprocessor may be tasked with selecting the appropriate frequency divisor values, depending on the dial or keyboard input, for the synthesizer. It may also select the appropriate charge pump current (loop gain) as well as pre-steer the system to the proper VCO segment for operation. There are also gain linearizations and offsets that may be supplied with the help of multiplying digital-to-analog converters.

9.7.8 The Role of DDS in Recent Hybrid Architectures

While DSP and DDS (Direct Digital Synthesis) techniques are improving rapidly and are capable of producing a good and economical medium performance transceiver, they are not as yet capable of "competition grade" performance. Present techniques still require some sort of analog/digital hybrid to achieve this. (DSP and DDS are discussed in the **DSP and Software Radio Design** chapter.)

One method of incorporating DDS is to use a hybrid system in which the DDS is allowed to supply the least significant digits and to sum that DDS output into a conventional divide-by-N loop that supplies the most significant digits. This method is employed in a number of commercial signal generators as well as some of the more modern transceivers. Keep in mind that the industry is still looking for cost-to-performance advantages, and that this architecture may be more complex than other methods.

For many years, designers have recognized that there is a trade off between stability and the best phase noise performance in an oscillator. Stable, frequency-standard-grade oscillators operate their crystals at low drive levels to diminish the effects of aging on the crystal. By necessity, this reduces the signal-to-noise ratio in the oscillator and consequently the phase noise performance. Designers have also come to realize that much of the noise performance can be recaptured by passing the output of the stable low-drive configuration through an additional crystal filter and removing significant amounts of the phase noise sidebands.

Designers have also realized that if one could move the reference in frequency or "rubber reference", which is normally contradictory to the idea set forth in the last paragraph, they could supply the frequency deviation required to fill in the least significant digits. The big problem is how to achieve this in a mathematically determinable manner. The benefits of as such a system would be great, as it would yield one of the ultimate goals of synthesis, i.e., the true single-loop synthesizer without additional mixing.

This application has recently become fertile ground for DDS techniques. A DDS is clocked from a stable reference oscillator. The output of the DDS is only allowed to deviate enough, through its programming, as is required to fill in the least significant digits of the synthesis. The output of the DDS is then passed through a narrow crystal filter which removes the artifacts of the DDS as well as enhancing the phase noise. The output of the DDS through the filter is then applied to a conventional PLL IC. The PLL IC then directs a highly segmented VCO to achieve the desired output. The beauty of this system is that it is entirely mathematically determinate and may be executed under the direction of a microprocessor as a housekeeper. The performance of such a system can be competition-grade, while less than a Euro-card size with excellent economy. This technique is used in some of the very latest transceiver designs.

9.7.9 Multi-Phase Division

So far, we have discussed a variety of improvements that can be made to synthesizer components and architectures. One of the limiting factors is the division itself, especially for large divisors. It is relatively common to accept that the signal-to-noise ratio of the divider is as it is and there is not much that can be done about it. *Multi-phase division* (see the patent by Telewski and Drucker in the listed references) can in fact be used to improve the phase noise performance of conventionally available dividers and phase detectors. In its simplest form, the output of the VCO is sent through independent, but identical, dividers and recombined after the phase detector. This is illustrated in the block diagram of **Fig 9.57**.

The systems works in that the noise generated in the individual divider chains is not correlated, and therefore sums as the root of the squares or in quadrature. This yields a 3-dB increase in noise. The signals, on the other hand, sum by linear superposition and effectively increase by 6 dB, giving a 3-dB advantage in signal-to-noise ratio for each time the divider system is doubled. One might reasonably argue that this is a fair amount of effort for improvement of 3 dB, but remember that multiple dividers can easily and economically be integrated in one higher-performance chip. It would not be unreasonable to see this technique employed in newer chips, yielding substantially better division performance.

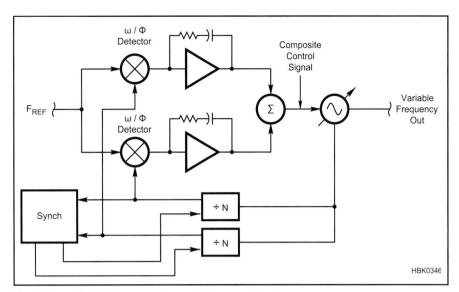

Fig 9.57 — A block diagram of a multi-phase synthesizer.

9.8 Present and Future Trends in Oscillator Application

In this chapter a wide variety of oscillator types and frequency-synthesis schemes are discussed. Which techniques are the most important today and which ones will likely become important in the future?

As we follow radio technology from the invention of the vacuum tube, we can see a continuous evolution in the types of oscillators deemed most useful at any given time. Broadly speaking, communications systems started out with LC oscillators, and when the need for greater frequency stability arose, designers moved to crystal oscillators. As the need to vary frequency became apparent, low-drift VFOs were developed to replace the crystals. As higher frequencies began to be exploited, stability again became the dominant problem. Multi-conversion systems that used low-frequency stable VFOs with crystal oscillators were developed to establish the desired high frequency stability. Those designs gave way to highly-stable frequency-synthesis techniques more amenable to digital control for variability and programmability. As more and more of the radio becomes digital and processing power increases, direct digital synthesis (DDS) is found even in entry-level equipment.

When we look inside today's transceivers, we usually see only two types of oscillators. The first is a temperature-compensated crystal oscillator (TCXO), whose main purpose is to set the frequency calibration of the transceiver. The second type is a DDS VFO interfaced directly to the controlling microprocessor. In the previous generation of equipment, the second type of oscillator was the low phase-noise, voltage-controlled oscillator (VCO) used in synthesizer PLLs.

Yesteryear's mechanically tuned VFOs have fallen by the wayside, with the exception perhaps being their use in low-power, all-analog radios. Even this could change in the near future, as very low power consumption synthesizers are already employed in cell phones and handhelds.

As we look to the future, we can see several emerging trends. Contemporary radios are now employing DSP at the IF. As A-D/D-A technology offers 16 bits of resolution at sample rates exceeding 60 MHz, 1.8 to 30-MHz transceivers are available that are implemented entirely by DSP. Their only remaining analog RF sections are input filtering and gain compensation, and the output power amplifier and filtering. All of the traditional local oscillator synthesis hardware, IF filters, IF amplifiers, mixers and demodulation are done in a DSP engine that resides between an A-D/D-A converter pair. For the frequency ranges covered by the A-D/D-A pair, the synthesis process is substantially "hidden." This technology is discussed at length in this *Handbook*'s chapter on **DSP and Software Radio Design**.

What comes next? On the design horizon in labs around the world are Cognitive Radio and the rapid adoption of digital modes and protocols, even in Amateur Radio. What kinds of oscillators will be required for this architecture and what parameters will be most important?

Cognitive Radio, the ability of the radio to configure itself on-the-fly to adapt to the requirements of its user, will demand oscillators that are highly frequency-agile and that can be adapted to all kinds of currently-unknown digital protocols and modulations. Efforts to share spectrum between competing services — just as amateurs have done manually for a century — will place new demands on spectral purity, as well. The use of digital modes, particularly for spread-spectrum and high-speed data transmission, will put a premium on frequency accuracy and stability. Both types of oscillators will be integrated into the digital "radio-on-a-chip" architectures being developed in the wireless telephony industry.

The extreme (in today's terms) requirements for accuracy, stability, and noise may cause even the high-stability TCXO to join other designs of the past. GPS technology can lock terrestrial time-bases to the master rubidium frequency standards carried by each satellite and synchronized to the world-wide network. As this capability is being relentlessly squeezed into consumer-grade devices, Amateur Radio will naturally adopt this inexpensive source of technology, just as it adopted the components manufactured for broadcast radios 80 years ago.

Having discussed the technology for HF transceivers, what about VHF, UHF and microwave amateur equipment? It is reasonable to assume that the direct DSP approach could be extended up in frequency as the A-D/D-A technologies improve in speed, however, there are some limiting factors regarding phase noise performance. The method of using a transverter and a high-performance DSP-based HF transceiver is likely to remain the practical option for some time to come.

If the transverter's oscillators are phase-locked to the HF transceiver's GPS-derived time base, the result would be frequency synthesizers with low phase noise and greater accuracy than the mixer-based up- and down-conversion designs in use today. The low-noise VCO will be an important element in these designs. These UHF and microwave VCOs will continue to take advantage of low-noise, highly-sensitive technology developed for wireless telephony and networking technology and direct satellite broadcast services.

The only thing that stays constant in oscillator and synthesizer design and application is that it changes constantly! As each generation of technology matures, its replacement is already being tested in laboratory prototypes. Amateur Radio will continue to adopt — and even advance — state-of-the art technology in the quest for better performance.

9.9 Glossary of Oscillator and Synthesizer Terms

Buffer — A circuit that amplifies the output of a circuit while isolating it from the load.

Bypass — Create a low ac impedance to ground at a point in the circuit.

Cavity — A hollow structure used as an electrical resonator.

Closed-loop — Operation under the control of a feedback loop (see also **open-loop**).

Coupling — The transfer of energy between circuits or structures.

Damping (factor) — the characteristics of the decay in a system's response to an input signal. The **damping factor, ζ,** is a numeric value specifying the degree of damping. An **underdamped** system alternately overshoots and undershoots the eventual steady-state output. An **overdamped** system approaches the steady-state output gradually, without overshoot. A **critically-damped** system approaches the steady-state output as quickly as possible without overshoot.

dBc — Decibels with respect to a carrier level.

DC-FM — control of a signal generator's output frequency by a dc voltage.

Decouple — To provide isolation between circuits, usually by means of filtering.

Direct digital synthesis (DDS) — Generation of signals by using counters and accumulators to create an output waveform.

Distributed — Circuit elements that are inherent properties of an extended structure, such as a transmission line.

ESR — Equivalent series resistance.

Free-running — Oscillating without any form of external control.

Fundamental — Lowest frequency of natural vibration or oscillation.

Integrator — A low-pass filter whose output is approximately the integral of the input signal.

Intermodulation — Generation of distortion products from two signals interacting in a nonlinear medium, device, or connection.

Isolation — Preventing signal flow between two circuits or systems. **Reverse isolation** refers to signal flow against the desired signal path.

Jitter (phase jitter) — Random variations of a signal in time, usually refers to random variations in the transition time of digital signals between states.

Linearization — Creation of a linear amplification or frequency characteristic through corrections supplied by an external system.

Loop gain — The total gain applied to a signal traveling around a feedback control loop.

Lumped (element) — Circuit elements whose electrical functions are concentrated at one point in the form of an electronic component.

Match — Equal values of impedance.

Modulus — The number of states of a digital counter or divider.

Motional capacitance (inductance) — The electrical effect of a crystal's mechanical properties, modeled as a capacitance (inductance).

Natural frequency (ω_n) — Frequency at which a system oscillates without any external control.

Noise bandwidth — The width of an ideal rectangular filter that would pass the same noise power from white noise as the filter being compared (also called **equivalent noise bandwidth**).

Open-loop — Operation without controlling feedback.

Oscillation — Repetitive mechanical motion or electrical activity created by the application of positive feedback.

Overtones — Vibration or oscillation at frequencies above the **fundamental**, usually harmonically related to the fundamental.

Permeability tuning — Varying the permeability of the core of an inductor used to control an oscillator's frequency.

Phase-lock — Maintain two signals in a fixed phase relationship by means of a control system.

Phase noise — Random variations of a signal in time, expressed as variations in phase of a sinusoidal signal.

Phasor — Representation of a sinusoidal signal as an amplitude and phase.

Power density — Amount of power per unit of frequency, usually specified as dBc/Hz or as RMS voltage/$\sqrt{Hz}$.

Prescaler — A frequency divider used to reduce the frequency of an input signal for processing by slower circuitry.

Preselector — Filters applied at a receiver's input to reject out-of-band signals.

Pull — Change the frequency at which a crystal oscillates by changing reactance of the circuit in which it is installed.

Quadrature — A 90° phase difference maintained between two signals.

Reciprocal mixing — Noise in a mixer's output due to the LO's noise sidebands mixing with those of the desired signal.

Relaxation oscillation — Oscillation produced by a cycle of gradual accumulation of energy followed by its sudden release.
Resonator — Circuit or structure whose resonance acts as a filter.
Simulation — Calculate a circuit's behavior based on mathematical models of the components.
Spurious (spur) — A signal at an undesired frequency, usually unrelated to the frequency of a desired frequency.
Squegg — Chaotic or random jumps in an oscillator's amplitude and/or frequency.
Static (synthesizer) — A synthesizer designed to output a signal whose frequency does not change or that is not changed frequently.
Synthesis (frequency) — The generation of variable-frequency signals by means of nonlinear combination and filtering (direct synthesis) or by using phase-lock or phase-control techniques (indirect synthesis).
TCXO — Temperature-compensated crystal oscillator. A **digitally temperature-compensated oscillator (DTCXO)** is controlled by a microcontroller or computer to maintain a constant frequency. **Oven-controlled crystal oscillators (OCXO)** are placed in a heated enclosure to maintain a constant temperature and frequency.
Temperature coefficient (tempco) — The amount of change in a component's value per degree of change in temperature.
Temperature compensation — Causing a circuit's behavior to change with temperature in such as way as to oppose and cancel the change with temperature of some temperature-sensitive component, such as a crystal.
Varactor (Varicap) — Reverse-biased diode used as a tunable capacitor.
VCO — Voltage-controlled oscillator (also called **voltage-tuned oscillator**).
VFO — Variable-frequency oscillator.
VXO — Variable crystal oscillator, whose frequency is adjustable around that of the crystal.

9.10 References and Bibliography

Clarke and Hess, *Communications Circuits; Analysis and Design* (Addison-Wesley, 1971; ISBN 0-201-01040-2). Wide coverage of transistor circuit design, including techniques suited to the design of integrated circuits. Its age shows, but it is especially valuable for its good mathematical treatment of oscillator circuits, covering both frequency- and amplitude-determining mechanisms. Look for a copy at a university library or initiate an interlibrary loan.

J. Grebenkemper, "Phase Noise and Its Effects on Amateur Communications," Part 1, *QST*, Mar 1988, pp 14-20; Part 2, Apr 1988 pp 22-25. Also see Feedback, *QST*, May 1988, p 44.

W. Hayward and D. DeMaw, *Solid State Design for the Radio Amateur* (Newington, CT: ARRL, 1986). Now out of print, this is dated but a good source of RF design ideas.

W. Hayward, W7ZOI, R. Campbell, KK7B, and B. Larkin, W7PUA, *Experimental Methods in RF Design* (Newington, CT: ARRL, 2003). A good source of RF design ideas, with good explanation of the reasoning behind design decisions.

W. Hayward, W7ZOI, *Introduction to Radio Frequency Design* (Newington, CT: ARRL, 1994). In-depth treatments of circuits and techniques used at RF.

"Spectrum Analysis…Random Noise Measurements," Hewlett-Packard Application Note 150-4. Source of correction factors for noise measurements using spectrum analyzers. HP notes should be available through local HP dealers.

B. Parzen, A. Ballato. *Design of Crystal and Other Harmonic Oscillators* (Wiley-Interscience, 1983, ISBN 0-471-08819-6). This book shows a thorough treatment of modern crystal oscillators including the optimum design. It is also the only book that looks at the base and collector limiting effect. The second author is affiliated with the US Army Electronics Technology & Devices Laboratory, Fort Monmouth, New Jersey.

U. Rohde, *Microwave and Wireless Synthesizers Theory and Design* (John Wiley and Sons, 1997, ISBN 0-471-52019-5). This book contains the textbook-standard mathematical analyses of frequency synthesizers combined with unusually good insight into what makes a better synthesizer and *a lot* of practical circuits to entertain serious constructors. A good place to look for low-noise circuits and techniques.

U. Rohde, D. Newkirk, *RF/Microwave Circuit Design for Wireless Applications* (John Wiley and Sons, 2000, ISBN 0-471-29818-2) While giving a deep insight into circuit design for wireless applications, Chapter 5 (RF/Wireless Oscillators) shows a complete treatment of both discrete and integrated circuit based oscillators. Chapter 6 (Wireless Synthesizers) gives more insight in synthesizers and large list of useful references.

U. Rohde, J. Whitaker, *Communications Receivers DSP, Software Radios, and Design*, 3rd ed., (McGraw-Hill, 2001, ISBN 0-07-136121-9). While mostly dedicated to communication receivers, this book has a useful chapter on Frequency Control and Local Oscillators, specifically on Fractional Division Synthesizers.

G. Vendelin, A. Pavio, U. Rohde, *Microwave Circuit Design Using Linear and Nonlinear Techniques*, 2nd ed., (Wiley-Interscience, 2005, ISBN 0-471-414479-4). This is a general handbook on microwave circuit design. It covers (in 200 pages) a large variety of oscillator design problems including microwave applications. It provides a step-by-step design guide for both linear and nonlinear oscillator designs.

U. Rohde, A. Poddar, G. Boeck, *The Design of Modern Microwave Oscillators for Wireless Applications Theory and Optimization* (Wiley-Interscience, 2005, ISBN 0-471-72342-8). This is the latest and most advanced textbook on oscillator design for high frequency/microwave application. It shows in detail how to design and optimize high frequency and microwave VCOs. It also contains a thorough analysis of the phase noise as generated in oscillators. While highly mathematical, the appendix gives numerical solutions that can be easily applied to various designs. It covers both bipolar transistors and FETs.

U. Rohde, "All About Phase Noise in Oscillators," Part 1, *QEX*, Dec 1993 pp 3-6; Part 2, *QEX*, Jan 1994 pp 9-16; Part 3, *QEX*, Feb 1994, pp 15-24.

U. Rohde, "Key Components of Modern

Receiver Design," Part 1, *QST*, May 1994, pp 29-32; Part 2, *QST*, June 1994, pp 27-31; Part 3, *QST*, Jul 1994, pp 42-45. Includes discussion of phase-noise reduction techniques in synthesizers and oscillators.

U. Rohde, "A High-Performance Hybrid Frequency Synthesizer," *QST*, Mar 1995, pp 30-38.

E.W. Pappenfus. W. Bruene and E.O. Schoenike, *Single Sideband Circuits and Systems*, (McGraw-Hill, 1964). A book on HF SSB transmitters, receivers and accoutrements by Rockwell-Collins staff. Contains chapters on synthesizers and frequency standards. The frequency synthesizer chapter predates the rise of the DDS and other recent techniques, but good information about the effects of synthesizer performance on communications is spread throughout the book.

Motorola Applications Note AN-551, "Tuning Diode Design Techniques," included in Motorola RF devices data books. A concise explanation of varactor diodes, their characteristics and use.

I. Keyser, "An Easy to Set Up Amateur Band Synthesizer," *RADCOM* (RSGB), Dec 1993, pp 33-36.

B. E. Pontius, "Measurement of Signal Source Phase Noise with Low-Cost Equipment," *QEX*, May-Jun 1998, pp 38-49.

V. Manassewitsch, *Frequency Synthesizers Theory and Design* (Wiley-Interscience, 1987, ISBN 0 471-01116-9).

Frederick J Telewski and Eric R Drucker, "Noise Reduction Method and Apparatus for Phase-locked Loops", U.S. Patent 5,216,387.

David Stockton, "Polyphase noise-shaping fractional-N frequency synthesizer", U.S. Patent 6,509,800.

F. Gardner, *Phaselock Techniques*, (John Wiley and Sons, 1966, ISBN 0-471-29156-0).

INTERESTING MODERN JOURNAL PUBLICATIONS

Oscillator References

C.-C. Wei, H.-C. Chiu, and W.-S. Feng, "An Ultra-Wideband CMOS VCO with 3-5 GHz Tuning Range," IEEE, *International Workshop on Radio-Frequency Integration Technology*, Nov 2005, pp 87-90.

A, P. S. (Paul) Khanna, "Microwave Oscillators: The State of The Technology", *Microwave Journal*, Apr 2006, pp. 22-42.

M-S Yim, K. K. O., "Switched Resonators and Their Applications in a Dual Band Monolithic CMOS LC Tuned VCO," *IEEE MTT*, Vol. 54, Jan 2006, pp 74-81.

A. D. Berny, A. M. Niknejad, and R. G. Meyer, "A 1.8 GHz LC VCO with 1.3 GHz Tuning Range and Digital Amplitude Calibration," *IEEE Journal of SSC*, vol. 40. no. 4, 2005, pp 909-917.

U.L. Rohde, "A New Efficient Method of Designing Low Noise Microwave Oscillators," Dr.-Ing. Dissertation, TU- Berlin, Germany, 12 Feb 2004.

N. Nomura, M. Itagaki, and Y. Aoyagi, "Small packaged VCSO for 10 Gbit Ethernet Application", *IEEE IUFFCS*, 2004, pp. 418-421.

A. K. Poddar, "A Novel Approach for Designing Integrated Ultra Low Noise Microwave Wideband Voltage-Controlled Oscillators," Dr.-Ing. Dissertation, TU- Berlin, Germany, 14 Dec 2004.

Sheng Sun and Lei Zhu, "Guided-Wave Characteristics of Periodically Nonuniform Coupled Microstrip Lines- Even and Odd Modes," *IEEE Transaction on MTT*, Vol. 53, No. 4, Apr 2005, pp. 1221-1227.

A. Ravi, B. R. Carlton, G. Banerjee, K. Soumyanath, "A 1.4V, 2.4/5.2 GHz, 90 nm CMOS system in a Package Transreceiver for Next Generation WLAN," *IEEE Symp. on VLSI Tech.*, 2005.

A. Chenakin, "Phase Noise Reduction in Microwave Oscillators", *Microwave Journal*, Oct 2009, pp 124-140.

Synthesizer References

R. E. Best, Phase-Locked Loops: Design, Simulation, and Applications, 5th ed., McGraw-Hill, 2003.

H. Arora, N. Klemmer, J. C. Morizio, and P. D. Wolf, "Enhanced Phase Noise Modeling of Fractional-N Frequency Synthesizer," IEEE Trans. Circuits and Systems-I: Regular Papers, vol 52, Feb 2005, pp 379-395.

A. Natrajan, A. Komijani, and A. Hajimiri, "A Fully 24-GHz Phased-Array Transmitter in CMOS," IEEE Journal of SSC, vol 40, no 12, Dec 2005, pp 2502-2514.

C. Lam, B. Razavi, "A 2.6/5.2 GHz Frequency Synthesizer in 0.4um CMOS Technology," IEEE JSSC, vol 35, May 2000, pp 788-79.

W. Rahajandraibe, L. Zaid, V. C. de Beaupre, and G. Bas, "Frequency Synthesizer and FSK Modulator For IEEE 802.15.4 Based Applications," 2007 RFIC Symp. Digest, pp 229-232.

C. Sandner, A. Wiesbauer, "A 3 GHz to 7 GHz Fast-Hopping Frequency Synthesizer For UWB," Proc. Int. Workshop on Ultra Wideband Systems 2004, pp 405-409.

J. Lee, "A 3-to-8 GHz Fast-Hopping Frequency Synthesizer in 0.18-um CMOS Technology," IEEE Journal of SSC, vol 41, no 3, Mar 2006, pp 566-573.

R. van de Beek. D. Leenaerts, G. van der Weide, "A Fast-Hopping Single-PLL 3-band UWB Synthesizer in 0.25um SiGe BiCMOS," Proc. European Solid-State Circuit Conf. 2005, pp 173-176.

R. B. Staszewski and others, "All Digital PLL and Transmitter for Mobile Phones," IEEE J. Solid-State Circuits, vol 40, no 12, Dec 2005, pp 2469-2482.

Contents

10.1 The Mechanism of Mixers and Mixing
    10.1.1 What is a Mixer?
    10.1.2 Putting Multiplication to Work

10.2 Mixers and Amplitude Modulation
    10.2.1 Overmodulation
    10.2.2 Using AM to Send Morse Code
    10.2.3 The Many Faces of Amplitude Modulation
    10.2.4 Mixers and AM Demodulation

10.3 Mixers and Angle Modulation
    10.3.1 Angle Modulation Sidebands
    10.3.2 Angle Modulators
    10.3.3 Mixers and Angle Demodulation

10.4 Putting Mixers, Modulators and Demodulators to Work
    10.4.1 Dynamic Range: Compression, Intermodulation and More
    10.4.2 Intercept Point

10.5 A Survey of Common Mixer Types
    10.5.1 Gain-Controlled Analog Amplifiers As Mixers
    10.5.2 Switching Mixers
    10.5.3 The Diode Double-Balanced Mixer: A Basic Building Block
    10.5.4 Active Mixers — Transistors as Switching Elements
    10.5.5 The Tayloe Mixer
    10.5.6 The NE602/SA602/SA612: A Popular Gilbert Cell Mixer
    10.5.7 An MC-1496P Balanced Modulator
    10.5.8 An Experimental High-Performance Mixer

10.6 References and Bibliography

Chapter 10

Mixers, Modulators and Demodulators

At base, radio communication involves translating information into radio form, letting it travel for a time as a radio signal, and translating it back again. Translating information into radio form entails the process we call modulation, and demodulation is its reverse. One way or another, every transmitter used for radio communication, from the simplest to the most complex, includes a means of modulation; one way or another, every receiver used for radio communication, from the simplest to the most complex, includes a means of demodulation.

Modulation involves varying one or both of a radio signal's basic characteristics — amplitude and frequency (or phase) — to convey information. A circuit, stage or piece of hardware that modulates is called a modulator.

Demodulation involves reconstructing the transmitted information from the changing characteristic(s) of a modulated radio wave. A circuit, stage or piece of hardware that demodulates is called a demodulator.

Many radio transmitters, receivers and transceivers also contain mixers — circuits, stages or pieces of hardware that combine two or more signals to produce additional signals at sums of and differences between the original frequencies.

This chapter, by David Newkirk, W9VES, and Rick Karlquist, N6RK, examines mixers, modulators and demodulators. Related information may be found in the **Modulation** chapter, and in the chapters on **Receivers** and **Transmitters**. Quadrature modulation implemented with an I/Q modulator, one that uses in-phase (I) and quadrature (Q) modulating signals to generate the 0° and 90° components of the RF signal, is covered in the chapters on **Modulation** and **DSP and Software Radio Design**.

Amateur Radio textbooks have traditionally handled mixers separately from modulators and demodulators, and modulators separately from demodulators. This chapter examines mixers, modulators and demodulators together because the job they do is essentially the same. Modulators and demodulators translate information into radio form and back again; mixers translate one frequency to others and back again. All of these translation processes can be thought of as forms of frequency translation or frequency shifting — the function traditionally ascribed to mixers. We'll therefore begin our investigation by examining what a mixer is (and isn't), and what a mixer does.

10.1 The Mechanism of Mixers and Mixing

10.1.1 What is a Mixer?

Mixer is a traditional radio term for a circuit that shifts one signal's frequency up or down by combining it with another signal. The word *mixer* is also used to refer to a device used to blend multiple audio inputs together for recording, broadcast or sound reinforcement. These two mixer types differ in one very important way: A radio mixer makes new frequencies out of the frequencies put into it, and an audio mixer does not.

MIXING VERSUS ADDING

Radio mixers might be more accurately called "frequency mixers" to distinguish them from devices such as "microphone mixers," which are really just signal *combiners*, *summers* or *adders*. In their most basic, ideal forms, both devices have two inputs and one output. The combiner simply *adds* the instantaneous voltages of the two signals together to produce the output at each point in time (**Fig 10.1**). The mixer, on the other hand, *multiplies* the instantaneous voltages of the two signals together to produce its output signal from instant to instant (**Fig 10.2**). Comparing the output spectra of the combiner and mixer, we see that the combiner's output contains only the frequencies of the two inputs, and nothing else, while the mixer's output contains *new* frequencies. Because it combines one energy with another, this process is sometimes called *heterodyning*, from the Greek words for *other* and *power*. The sidebar, "Mixer Math: Mixing as Multiplication," describes this process mathematically.

The key principle of a radio mixer is that in mixing multiple signal voltages together, *it adds and subtracts their frequencies to produce new frequencies.* (In the field of signal processing, this process, *multiplication in the time domain*, is recognized as equivalent to the process of *convolution in the frequency domain*. Those interested in this alternative approach to describing the generation of new frequencies through mixing can find more information about it in the many textbooks available on this subject.)

The difference between the mixer we've been describing and any mixer, modulator or demodulator that you'll ever use is that it's ideal. We put in two signals and got just two signals out. *Real* mixers, modulators and demodulators, on the other hand, also produce *distortion* products that make their output spectra "dirtier" or "less clean," as well as putting out some energy at input-signal frequencies and their harmonics. Much of the art and science of making good use of multiplication in mixing, modulation and demodulation goes

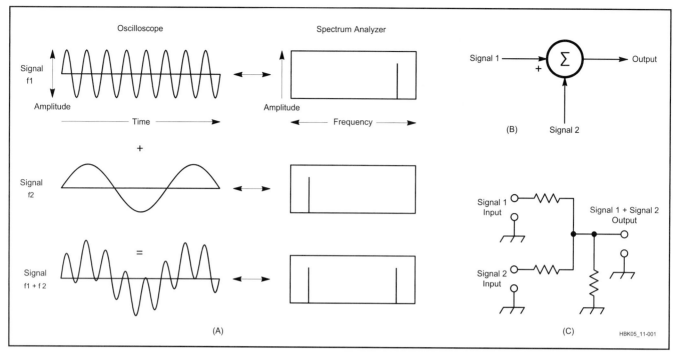

Fig 10.1 — *Adding or summing* two sine waves of different frequencies (f1 and f2) combines their amplitudes without affecting their frequencies. Viewed with an *oscilloscope* (a real-time graph of amplitude versus time), adding two signals appears as a simple superimposition of one signal on the other. Viewed with a *spectrum analyzer* (a real-time graph of signal amplitude versus frequency), adding two signals just sums their spectra. The signals merely coexist on a single cable or wire. All frequencies that go into the adder come out of the adder, and no new signals are generated. Drawing B, a block diagram of a summing circuit, emphasizes the stage's mathematical operation rather than showing circuit components. Drawing C shows a simple summing circuit, such as might be used to combine signals from two microphones. In audio work, a circuit like this is often called a mixer — but it does not perform the same function as an RF mixer.

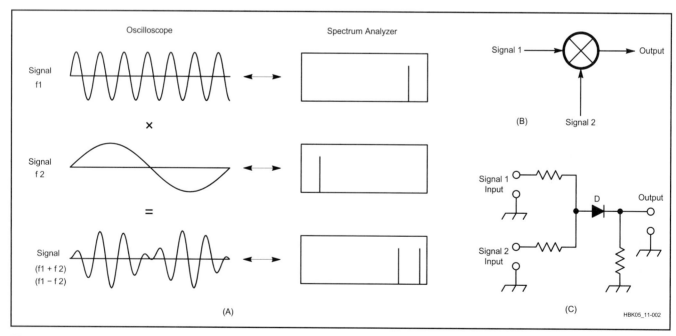

Fig 10.2 — *Multiplying* two sine waves of different frequencies produces a new output spectrum. Viewed with an oscilloscope, the result of multiplying two signals is a composite wave that seems to have little in common with its components. A spectrum-analyzer view of the same wave reveals why: The original signals disappear entirely and are replaced by two new signals — at the *sum* and *difference* of the original signals' frequencies. Drawing B diagrams a multiplier, known in radio work as a mixer. The *X* emphasizes the stage's mathematical operation. (The circled *X* is only one of several symbols you may see used to represent mixers in block diagrams, as Fig 10.3 explains.) Drawing C shows a very simple multiplier circuit. The diode, D, does the mixing. Because this circuit does other mathematical functions and adds them to the sum and difference products, its output is more complex than f1 + f2 and f1 – f2, but these can be extracted from the output by filtering.

Mixer Math: Mixing as Multiplication

Since a mixer works by means of multiplication, a bit of math can show us how they work. To begin with, we need to represent the two signals we'll mix, A and B, mathematically. Signal A's instantaneous amplitude equals

$$A_a \sin 2\pi f_a t \tag{1}$$

in which A is peak amplitude, f is frequency, and t is time. Likewise, B's instantaneous amplitude equals

$$A_b \sin 2\pi f_b t \tag{2}$$

Since our goal is to show that multiplying two signals generates sum and difference frequencies, we can simplify these signal definitions by assuming that the peak amplitude of each is 1. The equation for Signal A then becomes

$$a(t) = A \sin(2\pi f_a t) \tag{3}$$

and the equation for Signal B becomes

$$b(t) = B \sin(2\pi f_b t) \tag{4}$$

Each of these equations represents a sine wave and includes a subscript letter to help us keep track of where the signals go.

Merely combining Signal A and Signal B by letting them travel on the same wire develops nothing new:

$$a(t) + b(t) = A \sin(2\pi f_a t) + B \sin(2\pi f_b t) \tag{5}$$

As simple as equation 5 may seem, we include it to highlight the fact that multiplying two signals is a quite different story. From trigonometry, we know that multiplying the sines of two variables can be expanded according to the relationship

$$\sin x \sin y = \frac{1}{2}[\cos(x-y) - \cos(x+y)] \tag{6}$$

Conveniently, Signals A and B are both sinusoidal, so we can use equation 6 to determine what happens when we multiply Signal A by Signal B. In our case, x = $2\pi f_a t$ and y = $2\pi f_b t$, so plugging them into equation 6 gives us

$$a(t) \times b(t) = \frac{AB}{2} \cos(2\pi[f_a - f_b]t) - \frac{AB}{2} \cos(2\pi[f_a + f_b]t) \tag{7}$$

Now we see two momentous results: a sine wave at the frequency *difference* between Signal A and Signal B $2\pi(f_a - f_b)t$, and a sine wave at the frequency *sum* of Signal A and Signal B $2\pi(f_a + f_b)t$. (The products are cosine waves, but since equivalent sine and cosine waves differ only by a phase shift of 90°, both are called *sine waves* by convention.)

This is the basic process by which we translate information into radio form and translate it back again. If we want to transmit a 1-kHz audio tone by radio, we can feed it into one of our mixer's inputs and feed an RF signal — say, 5995 kHz — into the mixer's other input. The result is two radio signals: one at 5994 kHz (5995 – 1) and another at 5996 kHz (5995 + 1). We have achieved modulation.

Converting these two radio signals back to audio is just as straightforward. All we do is feed them into one input of another mixer, and feed a 5995-kHz signal into the mixer's other input. Result: a 1-kHz tone. We have achieved demodulation; we have communicated by radio.

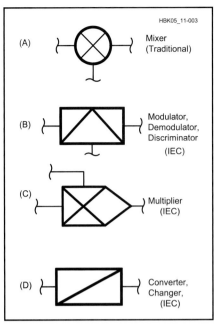

Fig 10.3 — We commonly symbolize mixers with a circled *X* (A) out of tradition, but other standards sometimes prevail (B, C and D). Although the converter/changer symbol (D) can conceivably be used to indicate frequency changing through mixing, the three-terminal symbols are arguably better for this job because they convey the idea of two signal sources resulting in a new frequency. (*IEC* stands for *International Electrotechnical Commission*.)

10.1.2 Putting Multiplication to Work

Piecing together a coherent picture of how multiplication works in radio communication isn't made any easier by the fact that traditional terms applied to a given multiplication approach and its products may vary with their application. If, for instance, you're familiar with standard textbook approaches to mixers, modulators and demodulators, you may be wondering why we didn't begin by working out the math involved by examining *amplitude modulation*, also known as *AM*. "Why not tell them about the *carrier* and how to get rid of it in a *balanced modulator*?" A transmitter enthusiast may ask "Why didn't you mention *sidebands* and how we conserve spectrum space and power by getting rid of one and putting all of our power into the other?" A student of radio receivers, on the other hand, expects any discussion of the same underlying multiplication issues to touch on the topics of *LO feedthrough*, *mixer balance* (*single* or *double*?), *image rejection* and so on.

You likely expect this book to spend some time talking to you about these things, so it will. But *this* radio-amateur-oriented discussion of mixers, modulators and demodulators will take a look at their common underlying mechanism *before* turning you loose on practical mixer, modulator and demodulator circuits. Then you'll be able to tell the forest from the trees. **Fig 10.3** shows the block symbol for a traditional mixer along with several IEC symbols for other functions mixers may perform.

It turns out that the mechanism underlying multiplication, mixing, modulation and demodulation is a pretty straightforward thing: Any circuit structure that *nonlinearly distorts* ac waveforms acts as a multiplier to some degree

into minimizing these unwanted multiplication products (or their effects) and making multipliers do their frequency translations as efficiently as possible.

NONLINEAR DISTORTION?

The phrase *nonlinear distortion* sounds redundant, but isn't. Distortion, an externally imposed change in a waveform, can be linear; that is, it can occur independently of signal amplitude. Consider a radio receiver front-end filter that passes only signals between 6 and 8 MHz. It does this by *linearly distorting* the single complex waveform corresponding to the wide RF spectrum present at the radio's antenna terminals, reducing the amplitudes of frequency components below 6 MHz and above 8 MHz relative to those between 6 and 8 MHz. (Considering multiple signals on a wire as one complex waveform is just as valid, and sometimes handier, than considering them as separate signals. In this case, it's a bit easier to think of distortion as something that happens to a waveform rather than something that happens to separate signals relative to each other. It would be just as valid — and certainly more in keeping with the consensus view — to say merely that the filter attenuates signals at frequencies below 6 MHz and above 8 MHz.) The filter's output waveform certainly differs from its input waveform; the waveform has been distorted. But because this distortion occurs independently of signal level or polarity, the distortion is linear. No new frequency components are created; only the amplitude relationships among the wave's existing frequency components are altered. This is *amplitude* or *frequency* distortion, and all filters do it or they wouldn't be filters.

Phase or *delay distortion*, also linear, causes a complex signal's various component frequencies to be delayed by different amounts of time, depending on their frequency but independently of their amplitude. No new frequency components occur, and amplitude relationships among existing frequency components are not altered. Phase distortion occurs to some degree in all real filters.

The waveform of a non-sinusoidal signal can be changed by passing it through a circuit that has only linear distortion, but only *nonlinear distortion* can change the waveform of a simple sine wave. It can also produce an output signal whose output waveform changes as a function of the input amplitude, something not possible with linear distortion. Nonlinear circuits often distort excessively with overly strong signals, but the distortion can be a complex function of the input level.

Nonlinear distortion may take the form of *harmonic distortion*, in which integer multiples of input frequencies occur, or *intermodulation distortion (IMD)*, in which different components multiply to make new ones.

Any departure from absolute linearity results in some form of nonlinear distortion, and this distortion can work for us or against us. Any amplifier, including a so-called linear amplifier, distorts nonlinearly to some degree; any device or circuit that distorts

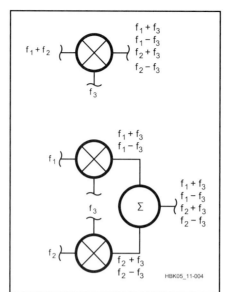

Fig 10.4 — Feeding two signals into one input of a mixer results in the same output as if f_1 and f_2 are each first mixed with f_3 in two separate mixers, and the outputs of these mixers are combined.

nonlinearly can work as a mixer, modulator, demodulator or frequency multiplier. An amplifier optimized for linear operation will nonetheless mix, but inefficiently; an amplifier biased for nonlinear amplification may be practically linear over a given tiny portion of its input-signal range. The trick is to use careful design and component selection to maximize nonlinear distortion when we want it (as in a mixer), and minimize it when we don't. Once we've decided to maximize nonlinear distortion, the trick is to minimize the distortion products we don't want, and maximize the products we want.

KEEPING UNWANTED DISTORTION PRODUCTS DOWN

Ideally, a mixer multiplies the signal at one of its inputs by the signal at its other input, but does not multiply a signal at the same input by itself, or multiple signals at the same input by themselves or by each other. (Multiplying a signal by itself — squaring it — generates harmonic distortion [specifically, *second-harmonic* distortion] by adding the signal's frequency to itself per equation 7. Simultaneously squaring two or more signals generates simultaneous harmonic and intermodulation distortion, as we'll see later when we explore how a diode demodulates AM.)

Consider what happens when a mixer must handle signals at two different frequencies (we'll call them f_1 and f_2) applied to its first input, and a signal at a third frequency (f_3) applied to its other input. Ideally, a mixer multiplies f_1 by f_3 and f_2 by f_3, but does not multiply f_1 and f_2 by each other. This produces output at the sum and difference of f_1 and f_3, and the sum and difference of f_2 and f_3, but *not* the sum and difference of f_1 and f_2. **Fig 10.4** shows that feeding two signals into one input of a mixer results in the same output as if f_1 and f_2 are each first mixed with f^3 in two separate mixers, and the outputs of these mixers are combined. This shows that a mixer, even though constructed with nonlinearly distorting components, actually behaves as a *linear frequency shifter*. Traditionally, we refer to this process as mixing and to its outputs as *mixing products*, but we may also call it frequency *conversion*, referring to a device or circuit that does it as a *converter*, and to its outputs as *conversion products*. If a mixer produces an output frequency that is higher than the input frequency, it is called an upconverter; if the output frequency is lower than the input, a downconverter.

Real mixers, however, at best act only as *reasonably* linear frequency shifters, generating some unwanted IMD products — spurious signals, or *spurs* — as they go. Receivers are especially sensitive to unwanted mixer IMD because the signal-level spread over which they must operate without generating unwanted IMD is often 90 dB or more, and includes infinitesimally weak signals in its span. In a receiver, IMD products so tiny that you'd never notice them in a transmitted signal can easily obliterate weak signals. This is why receiver designers apply so much effort to achieving "high dynamic range."

The degree to which a given mixer, modulator or demodulator circuit produces unwanted IMD is often *the* reason why we use it, or don't use it, instead of another circuit that does its wanted-IMD job as well or even better.

OTHER MIXER OUTPUTS

In addition to desired sum-and-difference products and unwanted IMD products, real mixers also put out some energy at their input frequencies. Some mixer implementations may *suppress* these outputs — that is, reduce one or both of their input signals by a factor of 100 to 1,000,000, or 20 to 60 dB. This is good because it helps keep input signals at the desired mixer-output sum or difference frequency from showing up at the IF terminal — an effect reflected in a receiver's *IF rejection* specification. Some mixer types, especially those used in the vacuum-tube era, suppress their input-signal outputs very little or not at all.

Input-signal suppression is part of an overall picture called *port-to-port isolation*. Mixer input and output connections are traditionally called *ports*. By tradition, the port to which we apply the shifting signal is the *local-oscillator (LO)* port. By convention, the signal or signals to be frequency-shifted are applied to the *RF (radio frequency)* port, and the frequency-shifted (product) signal or signals emerge at the *IF (intermediate frequency)* port. This illustrates the function of a mixer in a receiver: Since it is often impractical to achieve the desired gain and filtering at the incoming signal's frequency (at RF), a mixer is used to translate the incoming RF signal to an intermediate frequency (the IF), where gain and filtering can be applied. The IF maybe be either lower or higher than the incoming RF signal. In a transmitter, the modulated signal may be created at an IF, and then translated in frequency by a mixer to the operating frequency.

Some mixers are *bilateral*; that is, their RF and IF ports can be interchanged, depending on the application. Diode-based mixers are usually bilateral. Many mixers are not bilateral (*unilateral*); the popular SA602/612 Gilbert cell IC mixer is an example of this.

It's generally a good idea to keep a mixer's input signals from appearing at its output port because they represent energy that we'd rather not pass on to subsequent circuitry. It therefore follows that it's usually a good idea to keep a mixer's LO-port energy from appearing at its RF port, or its RF-port energy from making it through to the IF port. But there are some notable exceptions.

10.2 Mixers and Amplitude Modulation

Now that we've just discussed what a fine thing it is to have a mixer that doesn't let its input signals through to its output port, we can explore a mixing approach that outputs one of its input signals so strongly that the fed-through signal's amplitude at least equals the combined amplitudes of the system's sum and difference products! This system full-carrier, *amplitude modulation*, is the oldest means of translating information into radio form and back again. It's a frequency-shifting system in which the original unmodulated signal, traditionally called the *carrier*, emerges from the mixer along with the sum and difference products, traditionally called *sidebands*. The sidebar, "Mixer Math: Amplitude Modulation," describes this process mathematically.

10.2.1 Overmodulation

Since the information we transmit using AM shows up entirely as energy in its sidebands, it follows that the more energetic we make the sidebands, the more information energy will be available for an AM receiver to "recover" when it demodulates the signal. Even in an ideal modulator, there's a practical limit to how strong we can make an AM signal's sidebands relative to its carrier, however. Beyond that limit, we severely distort the waveform we want to translate into radio form.

We reach AM's distortion-free modulation limit when the sum of the sidebands and carrier at the modulator output *just reaches zero* at the modulating wave-form's most negative peak (**Fig 10.5**). We call this condition *100% modulation*, and it occurs when *m* in equation 8 equals 1. (We enumerate *modulation percentage* in values from 0 to 100%. The lower the number, the less information energy is in the sidebands. You may also see modulation enumerated in terms of a *modulation factor* from 0 to 1, which directly equals *m*; a modulation factor of 1 is the same as 100% modulation.) Equation 9 shows that each sideband's voltage is half that of the carrier. Power varies as the square of voltage, so the power in each sideband of a 100%-modulated signal is therefore $(\frac{1}{2})^2$ times, or ¼, that of the carrier. A transmitter capable of 100% modulation when operating at a carrier power of 100 W therefore puts out a 150-W signal at 100% modulation, 50 W of which is attributable to the sidebands.

(The *peak* envelope power [PEP] output of a double-sideband, full-carrier AM transmitter at 100% modulation is four times its carrier PEP. This is why our solid-state, "100-W" MF/HF transceivers are usually rated for no more than about 25 W carrier output at 100% amplitude modulation.)

Mixer Math: Amplitude Modulation

We can easily make the carrier pop out of our mixer along with the sidebands merely by building enough *dc level shift* into the information we want to mix so that its waveform never goes negative. Back at equations 1 and 2, we decided to keep our mixer math relatively simple by setting the peak voltage of our mixer's input signals directly equal to their sine values. Each input signal's peak voltage therefore varies between +1 and –1, so all we need to do to keep our modulating- signal term (provided with a subscript *m* to reflect its role as the modulating or information waveform) from going negative is add 1 to it. Identifying the carrier term with a subscript *c*, we can write

$$\text{AM signal} = (1 + m \sin 2\pi f_m t) \sin 2\pi f_c t \qquad (8)$$

Notice that the modulation ($2\pi f_m t$) term has company in the form of a coefficient, *m*. This variable expresses the modulating signal's varying amplitude — variations that ultimately result in amplitude modulation. Expanding equation 8 according to equation 6 gives us

$$\text{AM signal} = \sin 2\pi f_c t + \frac{1}{2} m \cos (2\pi f_c - 2\pi f_m) t - \frac{1}{2} m \cos (2\pi f_c + 2\pi f_m) t \qquad (9)$$

The modulator's output now includes the carrier ($\sin 2\pi f_c t$) in addition to sum and difference products that vary in strength according to m. According to the conventions of talking about modulation, we call the sum product, which comes out at a frequency higher than that of the carrier, the *upper sideband (USB)*, and the difference product, which comes out a frequency lower than that of the carrier, the *lower sideband (LSB)*. We have achieved amplitude modulation.

Why We Call It Amplitude Modulation

We call the modulation process described in equation 8 *amplitude modulation* because the complex waveform consisting of the sum of the sidebands and carrier varies with the information signal's magnitude (m). Concepts long used to illustrate AM's mechanism may mislead us into thinking that the *carrier* varies in strength with modulation, but careful study of equation 9 shows that this doesn't happen. The carrier, $\sin 2\pi f_c t$, goes into the modulator — we're in the modulation business now, so it's fitting to use the term *modulator* instead of *mixer* — as a sinusoid with an unvarying maximum value of |1|. The modulator multiplies the carrier by the dc level (+1) that we added to the information signal (m sin $2\pi f_m t$). Multiplying $\sin 2\pi f_c t$ by 1 merely returns $\sin 2\pi f_c t$. We have proven that the carrier's amplitude does not vary as a result of amplitude modulation — a fact that makes sense when we realize that in simple AM receivers the carrier serves as an AGC control signal and (during "diode" detection) provides, at the correct power level and frequency, the "LO" signal necessary to heterodyne the sidebands back to baseband.

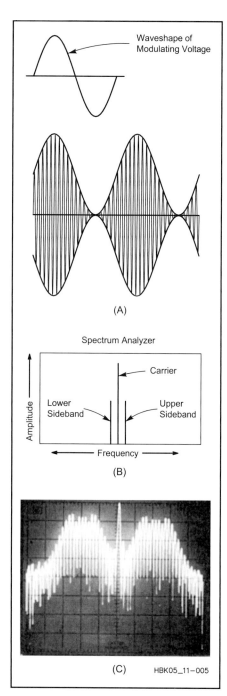

Fig 10.5 — Graphed in terms of amplitude versus time (A), the *envelope* of a properly modulated AM signal exactly mirrors the shape of its modulating waveform, which is a sine wave in this example. This AM signal is modulated as fully as it can be — 100% — because its envelope *just* hits zero on the modulating wave's negative peaks. Graphing the same AM signal in terms of amplitude versus frequency (B) reveals its three spectral components: Carrier, upper sideband and lower sideband. B shows sidebands as single-frequency components because the modulating waveform is a sine wave. With a complex modulating waveform, the modulator's sum and difference products really do show up as *bands* on either side of the carrier (C).

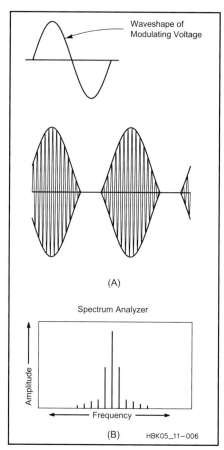

Fig 10.6 — Negative-going overmodulation of an AM transmitter results in a modulation envelope (A) that doesn't faithfully mirror the modulating waveform. This distortion creates additional sideband components that broaden the transmitted signal (B). Positive-going modulation beyond 100% is used by some AM broadcasters in conjunction with negative-peak limiting to increase "talk power" without causing negative overmodulation.

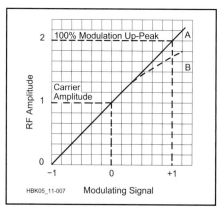

Fig 10.7 — An ideal AM transmitter exhibits a straight-line relationship (A) between its instantaneous envelope amplitude and the instantaneous amplitude of its modulating signal. Distortion, and thus an unnecessarily wide signal, results if the transmitter cannot respond linearly across the modulating signal's full amplitude range.

One-hundred-percent negative modulation is a brick-wall limit because an amplitude modulator can't reduce its output to less than zero. Trying to increase negative modulation beyond the 100% point results in *over-modulation* (**Fig 10.6**), in which the modulation envelope no longer mirrors the shape of the modulating wave (Fig 10.6A). A negatively overmodulated wave contains more energy than it did at 100% modulation, but some of the added energy now exists as *harmonics of the modulating waveform* (Fig 10.6B). This distortion makes the modulated signal take up more spectrum space than it needs. In voice operation, overmodulation commonly happens only on syllabic peaks, making the distortion products sound like transient noise we refer to as *splatter*.

MODULATION LINEARITY

If we increase an amplitude modulator's modulating-signal input by a given percentage, we expect a proportional modulation increase in the modulated signal. We expect good *modulation linearity*. Suboptimal amplitude modulator design may not allow this, however. Above some modulation percentage, a modulator may fail to increase modulation in proportion to an increase in its input signal (**Fig 10.7**). Distortion, and thus an unnecessarily wide signal, results.

10.2.2 Using AM to Send Morse Code

Fig 10.8A closely resembles what we see when a properly adjusted CW transmitter sends a string of dots. Keying a carrier on and off produces a wave that varies in amplitude and has double (upper and lower) sidebands that vary in spectral composition according to the duration and envelope shape of the on-off transitions. The emission mode we call *CW* is therefore a form of AM. The concepts of modulation percentage and overmodulation are usually not applied to generating an on-off-keyed Morse signal, however. This is related to how we copy CW by ear, and the fact that, in CW radio communication, we usually don't translate the received signal all the way back into its original pre-modulator (*baseband*) form, as a closer look at the process reveals.

In CW transmission, we usually open and close a keying line to make dc transitions that turn the transmitted carrier on and off. See Fig 10.8B. CW reception usually does not entirely reverse this process, however. Instead of demodulating a CW signal all the way back to its baseband self — a shifting dc level — we want the presences and absences of its carrier to create long and short audio tones. Because the carrier is RF and not AF, we must mix it with a locally generated RF signal — from a *beat-frequency oscillator*

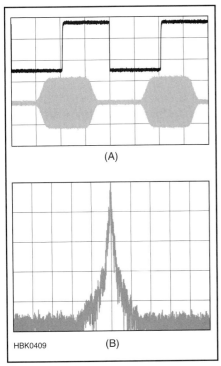

Fig 10.8 — Telegraphy by on-off-keying a carrier is also a form of AM, called *CW* (short for *continuous wave*) for reasons of tradition. *Waveshaping* in a CW transmitter often causes a CW signal's RF envelope (lower trace in the amplitude-versus-time display at A) to contain less harmonic energy than the abrupt transitions of its key closure waveform (upper trace in A) suggest should be the case. B, an amplitude-versus-frequency display, shows that even a properly shaped CW signal has many sideband components.

(BFO) — that's close enough in frequency to produce a difference signal at AF (this BFO can, of course, also be inserted at an IF stage). What goes into our transmitter as shifting dc comes out of our receiver as thump-delimited tone bursts of dot and dash duration. We have achieved CW communication.

It so happens that we always need to hear one or more harmonics of the fundamental keying waveform for the code to sound sufficiently crisp. If the transmitted signal will be subject to fading caused by varying propagation — a safe assumption for any long-distance radio communication — we can *harden* our keying by making the transmitter's output rise and fall more quickly. This puts more energy into keying sidebands and makes the signal more copiable in the presence of fading — in particular, *selective fading*, which linearly distorts a modulated signal's complex waveform and randomly changes the sidebands' strength and phase relative to the carrier and each other. The appropriate keying hardness also depends on the keying speed. The faster the keying in WPM, the faster the on-off times — the harder the keying — must be for the signal to remain ear- and machine-readable through noise and fading.

Instead of thinking of this process in terms of modulation percentage, we just ensure that a CW transmitter produces sufficient keying-sideband energy for solid reception. Practical CW transmitters ususally do not do their keying with a modulator stage as such. Instead, one or more stages are turned on and off to modulate the carrier with Morse, with rise and fall times set by R and C values associated with the stages' keying and/or power supply lines. A transmitter's CW *waveshaping* is therefore usually hardwired to values appropriate for reasonably high-speed sending (35 to 55 WPM or so) in the presence of fading. However, some transceivers allow the user to vary keying hardness at will as a menu option. Rise and fall times of 1 to 5 ms are common; 5-ms rise and fall times equate to a keying speed of 36 WPM in the presence of fading and 60 WPM if fading is absent.

The faster a CW transmitter's output changes between zero and maximum, the more bandwidth its carrier and sidebands occupy. See Fig 10.8B. Making a CW signal's keying too hard is therefore spectrum-wasteful and unneighborly because it makes the signal wider than it needs to be. Keying sidebands that are stronger and wider than necessary are traditionally called *clicks* because of what they sound like on the air. A more detailed discussion of keying waveforms appears in the **Transmitters** chapter of this *Handbook*.

10.2.3 The Many Faces of Amplitude Modulation

We've so far examined mixers, multipliers and modulators that produce complex output signals of two types. One, the action of which equation 7 expresses, produces only the frequency sum and frequency difference between its input signals. The other, the amplitude modulator characterized by equations 8 and 9, produces carrier output in addition to the frequency sum of and frequency difference between its input signals. Exploring the AM process led us to a discussion of on-off-keyed CW, which is also a form of AM.

Amplitude modulation is nothing more and nothing less than varying an output signal's amplitude according to a varying voltage or current. All of the output signal types mentioned above are forms of amplitude modulation, and there are others. Their names and applications depend on whether the resulting signal contains a carrier or not, and both sidebands or not. Here's a brief overview of AM-signal types, what they're called, and some of the jobs you may find them doing:

• *Double-sideband (DSB), full-carrier AM* is often called just *AM*, and often what's meant when radio folk talk about just *AM*. (When the subject is broadcasting, *AM* can also refer to broadcasters operating in the 535- to 1705-kHz region, generically called *the AM band* or *the broadcast band* or *medium wave*. These broadcasters used only double-sideband, full-carrier AM for many years, but many now use combinations of amplitude modulation and *angle modulation*, explored later in this chapter, to transmit stereophonic sound in analog and digital form.) Equations 8 and 9 express what goes on in generating this signal type. What we call *CW* — Morse code done by turning a carrier on and off — is a form of DSB, full-carrier AM.

• *Double-sideband, suppressed-carrier AM* is what comes out of a circuit that does what equation 7 expresses — a sum (upper sideband), a difference (lower sideband) and no carrier. We didn't call its sum and difference outputs upper and lower sidebands earlier in equation 7's neighborhood, but we'd do so in a transmitting application. In a transmitter, we call a circuit that suppresses the carrier while generating upper and lower sidebands a *balanced modulator*, and we quantify its *carrier suppression*, which is always less than infinite. In a receiver, we call such a circuit a *balanced mixer*, which may be *single-balanced* (if it lets either its RF signal or its LO [carrier] signal through to its output) or *double-balanced* (if it suppresses both its input signal and LO/carrier in its output), and we quantify its *LO suppression* and *port-to-port isolation*, which are always less than infinite. (Mixers [and amplifiers] that afford no balance whatsoever are sometimes said to be *single-ended*.) Sometimes, DSB suppressed-carrier AM is called just *DSB*.

• *Vestigial sideband (VSB), full-carrier AM* is like the DSB variety with one sideband partially filtered away for bandwidth reduction. Some amateur television stations use VSB AM, and it was used by commercial television systems that transmitted analog AM video in the pre-digital TV days.

• *Single-sideband, suppressed-carrier AM* is what you get when you generate a DSB, suppressed carrier AM signal and suppress one sideband with filtering or phasing. We usually call this signal type just *single sideband (SSB)* or, as appropriate, *upper sideband (USB)* or *lower sideband (LSB)*. In a modulator or demodulator system, the *unwanted sideband* — that is, the sum or difference signal we don't want — may be called just that, or it may be called the *opposite sideband*, and we refer to a system's *sideband rejection* as a measure of how well the opposite sideband is suppressed. In receiver mixers not used for demodulation and transmitter mixers not used for modulation, the unwanted sum or difference signal, or the input signal that produces the unwanted sum or difference, is the *image*, and we refer to a system's *image rejection*.

A pair of mixers specially configured to suppress either the sum or the difference output is an *image-reject mixer (IRM)*. In receiver demodulators, the unwanted sum or difference signal may just be called the opposite sideband, or it may be called the *audio image*. A receiver capable of rejecting the opposite sideband or audio image is said to be capable of *single-signal* reception.

• *Single-sideband, full-carrier AM* is akin to full-carrier DSB with one sideband missing. Commercial and military communicators may call it *AM equivalent (AME)* or *compatible AM (CAM)* — *compatible* because it can be usefully demodulated in AM and SSB receivers and because it occupies about the same amount of spectrum space as SSB.)

• *Independent sideband (ISB) AM* consists of an upper sideband and a lower sideband containing different information (a carrier of some level may also be present). Radio amateurs sometimes use ISB to transmit simultaneous slow-scan-television and voice information; international broadcasters sometimes use it for point-to-point audio feeds as a backup to satellite links.

10.2.4 Mixers and AM Demodulation

Translating information from radio form back into its original form — demodulation — is also traditionally called *detection*. If the information signal we want to detect consists merely of a baseband signal frequency-shifted into the radio realm, almost any low-distortion frequency-shifter that works according to equation 7 can do the job acceptably well.

Sometimes we recover a radio signal's information by shifting the signal right back to its original form with no intermediate frequency shifts. This process is called *direct conversion*. More commonly, we first convert a received signal to an *intermediate frequency* so we can amplify, filter and level-control it prior to detection. This is *superheterodyne* reception, and most modern radio receivers work in this way. Whatever the receiver type, however, the received signal ultimately makes its way to one last mixer or demodulator that completes the final translation of information back into audio, video, or into a signal form suitable for device control or computer processing. In this last translation, the incoming signal is converted back to recovered-information form by mixing it with one last RF signal. In heterodyne or *product* detection, that final frequency-shifting signal comes from a BFO. The incoming-signal energy goes into one mixer input port, BFO energy goes into the other, and audio (or whatever form the desired information takes) results.

If the incoming signal is full-carrier AM and we don't need to hear the carrier as a tone,

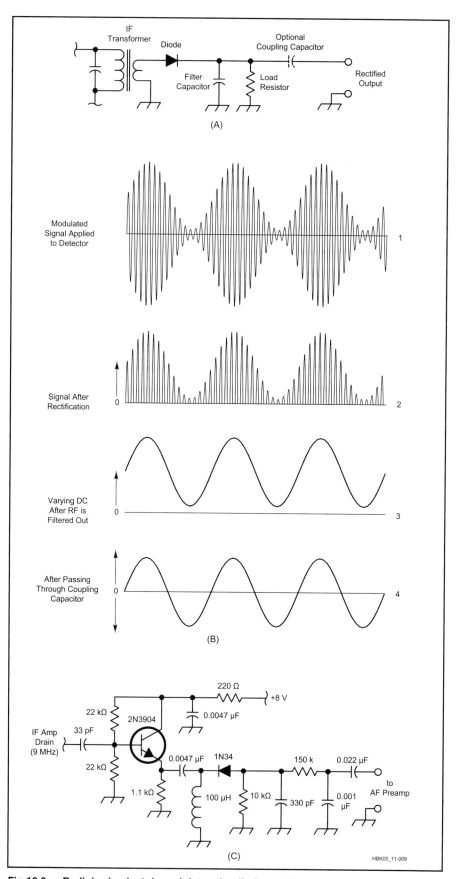

Fig 10.9 — Radio's simplest demodulator, the diode rectifier (A), demodulates an AM signal by acting as a switch that multiplies the carrier and sidebands to produce frequency sums and differences, two of which sum into a replica of the original modulation (B). Modern receivers often use an emitter follower to provide low-impedance drive for their diode detectors (C).

we can modify this process somewhat, if we want. We can use the carrier itself to provide the heterodyning energy in a process called *envelope detection*.

ENVELOPE DETECTION AND FULL-CARRIER AM

Fig 10.5 graphically represents how a full-carrier AM signal's *modulation envelope* corresponds to the shape of the modulating wave. If we can derive from the modulated signal a voltage that varies according to the modulation envelope, we will have successfully recovered the information present in the sidebands. This process is called envelope detection, and we can achieve it by doing nothing more complicated than half-wave-rectifying the modulated signal with a diode (**Fig 10.9**).

That a diode demodulates an AM signal by allowing its carrier to multiply with its sidebands may jar those long accustomed to seeing diode detection ascribed merely to "rectification." But a diode is certainly nonlinear. It passes current in only one direction, and its output voltage is (within limits) proportional to the square of its input voltage. These nonlinearities allow it to multiply.

Exploring this mathematically is tedious with full-carrier AM because the process squares three summed components (carrier, lower sideband and upper sideband). Rather than fill the better part of a page with algebra, we'll instead characterize the outcome verbally: In "just rectifying" a DSB, full-carrier AM signal, a diode detector produces

• Direct current (the result of rectifying the carrier);
• A second harmonic of the carrier;
• A second harmonic of the lower sideband;
• A second harmonic of the upper sideband;
• Two difference-frequency outputs (upper sideband minus carrier, carrier minus lower sideband), each of which is equivalent to the modulating wave-form's frequency, and both of which sum to produce the recovered information signal; and
• A second harmonic of the modulating waveform (the frequency difference between the two sidebands).

Three of these products are RF. Low-pass filtering, sometimes little more than a simple RC network, can remove the RF products from the detector output. A capacitor in series with the detector output line can block the carrier-derived dc component. That done, only two signals remain: the recovered modulation and, at a lower level, its second harmonic — in other words, second-harmonic distortion of the desired information signal.

10.3 Mixers and Angle Modulation

Amplitude modulation served as our first means of translating information into radio form because it could be implemented as simply as turning an electric noise generator on and off. (A spark transmitter consisted of little more than this.) By the 1930s, we had begun experimenting with translating information into radio form and back again by modulating a radio wave's angular velocity (frequency or phase) instead of its overall amplitude. The result of this process is *frequency modulation (FM)* or *phase modulation (PM)*, both of which are often grouped under the name *angle modulation* because of their underlying principle.

A change in a carrier's frequency or phase for the purpose of modulation is called *deviation*. An FM signal deviates according to the amplitude of its modulating waveform, independently of the modulating waveform's frequency; the higher the modulating wave's amplitude, the greater the deviation. A PM signal deviates according to the amplitude *and frequency* of its modulating waveform; the higher the modulating wave's amplitude *and/or frequency*, the greater the deviation. See the sidebar, "Mixer Math: Angle Modulation" for a numerical description of these processes.

10.3.1 Angle Modulation Sidebands

Although angle modulation produces un-

> **Mixer Math: Angle Modulation**
>
> An angle-modulated signal can be mathematically represented as
>
> $$f_c(t) = \cos(2\pi f_c t + m \sin(2\pi f_m t))$$
> $$= \cos(2\pi f_c t)\cos(m \sin(2\pi f_m t)) - \sin(2\pi f_c t)\sin(m \sin(2\pi f_m t)) \qquad (10)$$
>
> In equation 10, we see the carrier frequency ($2\pi f_c t$) and modulating signal (sin $2\pi f_m t$) as in the equation for AM (equation 8). We again see the modulating signal associated with a coefficient, m, which relates to degree of modulation. (In the AM equation, *m* is the modulation factor; in the angle-modulation equation, *m* is the *modulation index* and, for FM, equals the deviation divided by the modulating frequency.) We see that angle-modulation occurs as the cosine of the sum of the carrier frequency ($2\pi f_c t$) and the modulating signal (sin $2\pi f_m t$) times the modulation index (*m*). In its expanded form, we see the appearance of sidebands above and below the carrier frequency.
>
> Angle modulation is a multiplicative process, so, like AM, it creates sidebands on both sides of the carrier. Unlike AM, however, angle modulation creates an *infinite* number of sidebands on either side of the carrier! This occurs as a direct result of modulating the carrier's angular velocity, to which its frequency and phase directly relate. If we continuously vary a wave's angular velocity according to another periodic wave's cyclical amplitude variations, the rate at which the modulated wave repeats *its* cycle — its frequency — passes through an infinite number of values. (How many individual amplitude points are there in one cycle of the modulating wave? An infinite number. How many corresponding discrete frequency or phase values does the corresponding angle-modulated wave pass through as the modulating signal completes a cycle? An infinite number!) In AM, the carrier frequency stays at one value, so AM produces two sidebands — the sum of its carrier's unchanging frequency value and the modulating frequency, and the difference between the carrier's unchanging frequency value and the modulating frequency. In angle modulation, the modulating wave shifts the frequency or phase of the carrier through an infinite number of different frequency or phase values, resulting in an infinite number of sum and difference products.

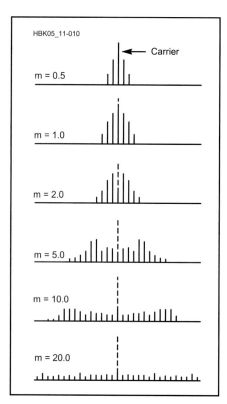

Fig 10.10 — Angle-modulation produces a carrier and an infinite number of upper and lower sidebands spaced from the average ("resting," unmodulated) carrier frequency by integer multiples of the modulating frequency. (This drawing is a simplification because it only shows relatively strong, close-in sideband pairs; space constraints prevent us from extending it to infinity.) The relative amplitudes of the sideband pairs and carrier vary with modulation index, *m*.

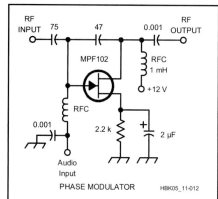

Fig 10.12 — A series reactance modulator acts as a variable shunt around a reactance — in this case, a 47-pF capacitor — through which the carrier passes.

countable sum and difference products, most of them are vanishingly weak in practical systems. They emerge from the modulator spaced from the average ("resting," unmodulated) carrier frequency by integer multiples of the modulating frequency (**Fig 10.10**). The strength of the sidebands relative to the carrier, and the strength and phase of the carrier itself, vary with the degree of modulation — the modulation index. (The *overall* amplitude of an angle-modulated signal does not change with modulation, however; when energy goes out of the carrier, it shows up in the sidebands, and vice versa.) In practice, we operate angle-modulated transmitters at modulation indexes that make all but a few of their infinite sidebands small in amplitude.

(A mathematical tool called *Bessel functions* helps determine the relative strength of the carrier and sidebands according to modulation index. The **Modulation** chapter includes a graph to illustrate this relationship.) Selectivity in transmitter and receiver circuitry further modify this relationship, especially for sidebands far away from the carrier.

10.3.2 Angle Modulators

If you vary a reactance in or associated with an oscillator's frequency-determining element(s), you vary the oscillator's frequency. If you vary the tuning of a tuned circuit through which a signal passes, you vary the signal's phase. A circuit that does this is called a *reactance modulator*, and can be little more than a tuning diode or two connected to a tuned circuit in an oscillator or amplifier (**Fig 10.11**). Varying a reactance through which the signal passes (**Fig 10.12**) is another way of doing the same thing.

The difference between FM and PM depends solely on how, and not how much, deviation occurs. A modulator that causes deviation in proportion to the modulating wave's amplitude and frequency is a phase modulator. A modulator that causes deviation only in proportion to the modulating signal's amplitude is a frequency modulator.

INCREASING DEVIATION BY FREQUENCY MULTIPLICATION

Maintaining modulation linearity is just as important in angle modulation as it is in AM, because unwanted distortion is always our enemy. A given angle-modulator circuit can frequency- or phase-shift a carrier only so much before the shift stops occurring in strict proportion to the amplitude (or, in PM, the amplitude and frequency) of the modulating signal.

If we want more deviation than an angle modulator can linearly achieve, we can operate the modulator at a suitable sub-harmonic — submultiple — of the desired frequency, and process the modulated signal through a series of *frequency multipliers* to bring it up to the desired frequency. The deviation also increases by the overall multiplication factor, relieving the modulator of having to do it all directly. A given FM or PM radio design may achieve its final output frequency through a combination of mixing (frequency shift, no deviation change) and frequency multiplication (frequency shift *and* deviation change).

THE TRUTH ABOUT "TRUE FM"

Something we covered a bit earlier bears closer study:

"An FM signal deviates according to the amplitude of its modulating waveform, independently of the modulating wave-form's frequency; the higher the modulating wave's

Fig 10.11 — One or more tuning diodes can serve as the variable reactance in a reactance modulator. This HF reactance modulator circuit uses two diodes in series to ensure that the tuned circuit's RF-voltage swing cannot bias the diodes into conduction. D1 and D2 are "30-volt" tuning diodes that exhibit a capacitance of 22 pF at a bias voltage of 4. The BIAS control sets the point on the diode's voltage-versus-capacitance characteristic around which the modulating waveform swings.

amplitude, the greater the deviation. A PM signal deviates according to the amplitude *and frequency* of its modulating waveform; the higher the modulating wave's amplitude *and/or frequency*, the greater the deviation."

The practical upshot of this excerpt is that we can use a phase modulator to generate FM. All we need to do is run a PM transmitter's modulating signal through a low-pass filter that (ideally) halves the signal's amplitude for each doubling of frequency (a reduction of "6 dB per octave," as we sometimes see such responses characterized) to compensate for its phase modulator's "more deviation with higher frequencies" characteristic. The result is an FM, not PM, signal. FM achieved with a phase modulator is sometimes called *indirect FM* as opposed to the *direct FM* we get from a frequency modulator.

We sometimes see claims that one piece of gear is better than another solely because it generates "true FM" as opposed to indirect FM. We can debunk such claims by keeping in mind that direct and indirect FM *sound exactly alike in a receiver* when done correctly.

CONVEYING DC LEVELS WITH ANGLE MODULATION

Depending on the nature of the modulation source, there *is* a practical difference between a frequency modulator and a phase modulator. Answering two questions can tell us whether this difference matters: Does our modulating signal contain a dc level or not? If so, do we need to accurately preserve that dc level through our radio communication link for successful communication? If both answers are *yes*, we must choose our hardware and/or information-encoding approach carefully, because a frequency modulator can convey dc-level shifts in its modulating waveform, while a phase modulator, which responds only to instantaneous changes in frequency and phase, cannot.

Consider what happens when we want to frequency-modulate a phase-locked-loop-synthesized transmitted signal. **Fig 10.13** shows the block diagram of a PLL frequency modulator. Normally, we modulate a PLL's VCO because it's the easy thing to do. As long as our modulating frequency results in frequency excursions too fast for the PLL to follow and correct — that is, as long as our modulating frequency is outside the PLL's *loop bandwidth* — we achieve the FM we seek. Trying to modulate a dc level by pushing the VCO to a particular frequency and holding it there fails, however, because a PLL's loop response includes dc. The loop, therefore, detects the modulation's dc component as a correctable error and "fixes" it. FMing a PLL's VCO therefore can't buy us the dc response "true FM" is supposed to allow.

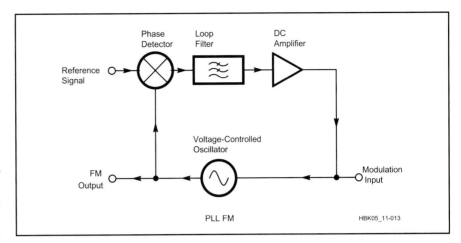

Fig 10.13 — Frequency modulation, PLL-style.

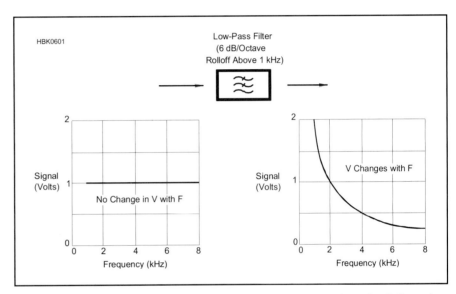

Fig 10.14 — Frequency-sweeping a constant-amplitude signal and passing it through a low-pass filter results in an output signal that varies in amplitude with frequency. This is the principle behind the angle-demodulation process called *frequency discrimination*.

We *can* dc-modulate a PLL modulator, but we must do so by modulating the frequency of the loop *reference*. The PLL then adjusts the VCO to adapt to the changed reference, and our dc level gets through. In this case, the modulating frequency must be *within* the loop bandwidth — which dc certainly is — or the VCO won't be corrected to track the shift.

10.3.3 Mixers and Angle Demodulation

With the awesome prospect of generating an infinite number of sidebands still fresh in our minds, we may be a bit disappointed to learn that we commonly demodulate angle modulation by doing little more than turning it into AM and then envelope- or product-detecting it! But this is what happens in many of our FM receivers and transceivers, and we can get a handle on this process by realizing that a form of angle-modulation-to-AM conversion begins quite early in an angle-modulated signal's life because of linear distortion of the modulation by amplitude-linear circuitry — something that happens to angle-modulated signals, it turns out, in any linear circuit that doesn't have an amplitude-versus-frequency response that's utterly flat out to infinity.

Think of what happens, for example, when we sweep a constant-amplitude signal up in frequency — say, from 1 kHz to 8 kHz — and pass it through a 6-dB-per-octave filter (**Fig 10.14**). The filter's rolloff causes the output signal's amplitude to decrease as frequency increases. Now imagine that we linearly sweep our constant-amplitude signal *back and forth* between 1 kHz and 8 kHz at a constant rate of 3 kHz per second. The filter's output *amplitude* now varies cyclically over time as the input signal's *frequency* varies cyclically over time. Right before our eyes, a frequency change turns into an amplitude change. The process of converting angle modulation to amplitude modulation has begun.

This is what happens whenever an angle-modulated signal passes through circuitry with an amplitude-versus-frequency response that isn't flat out to infinity. As the signal deviates across the frequency-response curves of whatever circuitry passes it, its angle modulation is, to some degree, converted to AM — a form of crosstalk between the two modulation types, if we wish to look at it that way. (Variations in system phase linearity also cause distortion and FM-to-AM conversion, because the sidebands do not have the proper phase relationship with respect to each other and with respect to the carrier.)

All we need to do to put this effect to practical use is develop a circuit that does this frequency-to-amplitude conversion linearly across the frequency span of the modulated signal's deviation. Then we envelope-demodulate the resulting AM, and we have achieved angle demodulation.

Fig 10.15 shows such a circuit — a *discriminator* — and the sort of amplitude-versus-frequency response we expect from it. It's actually possible to use an AM receiver to recover understandable audio from a narrow angle-modulated signal by "off-tuning" the signal so its deviation rides up and down on one side of the receiver's IF selectivity curve. This *slope detection* process served as an early, suboptimal form of frequency discrimination in receivers not designed for FM. It is always worth trying as a last-resort-class means of receiving narrowband FM with an AM receiver.

QUADRATURE DETECTION

It's also possible to demodulate an angle-modulated signal merely by multiplying it with a time-delayed copy of itself in a double-balanced mixer as shown in **Fig 10.16**; the sidebar, "Mixer Math: Quadrature Demodulation," explains the process numerically.

An ideal quadrature detector puts out 0 V dc when no modulation is present (with the carrier at f_c). The output of a real quadrature detector may include a small *dc offset* that requires compensation. If we need the detector's response all the way down to dc, we've got it; if not, we can put a suitable

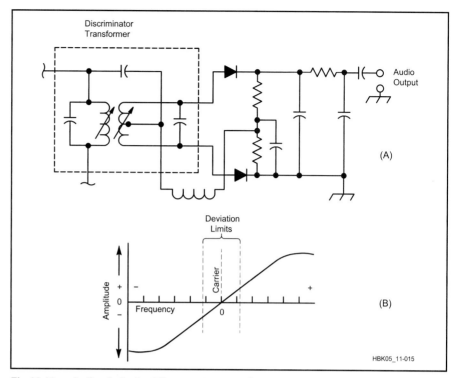

Fig 10.15 — A *discriminator* (A) converts an angle-modulated signal's deviation into an amplitude variation (B) and envelope-detects the resulting AM signal. For undistorted demodulation, the discriminator's amplitude-versus-frequency characteristic must be linear across the input signal's deviation. A *crystal discriminator* uses a crystal as part of its frequency-selective circuitry.

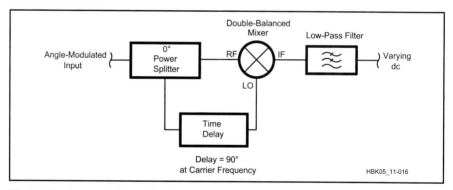

Fig 10.16 — In *quadrature detection*, an angle-modulated signal multiplies with a time-delayed copy of itself to produce a dc voltage that varies with the amplitude and polarity of its phase or frequency excursions away from the carrier frequency. A practical quadrature detector can be as simple as a 0° power splitter (that is, a power splitter with in-phase outputs), a diode double-balanced mixer, a length of coaxial cable ¼-λ (electrical) long at the carrier frequency, and a bit of low-pass filtering to remove the detector output's RF components. IC quadrature detectors achieve their time delay with one or more resistor-loaded tuned circuits (Fig 10.17).

blocking capacitor in the output line for ac-only coupling.

Quadrature detection is more common than frequency discrimination nowadays because it doesn't require a special discriminator transformer or resonator, and because the necessary balanced-detector circuitry can easily be implemented in IC structures along with limiters and other receiver circuitry. The NXP Semiconductor SA604A FM IF IC is

Mixer Math: Quadrature Demodulation

Demodulating an angle-modulated signal merely by multiplying it with a time-delayed copy of itself in a double-balanced mixer results in quadrature demodulation (Fig 10.16). To illustrate this mathematically, for simplicity's sake, we'll represent the mixer's RF input signal as just a sine wave with an amplitude, A:

$$A \sin(2\pi ft) \tag{11}$$

and its time-delayed twin, fed to the mixer's LO input, as a sine wave with an amplitude, A, and a time delay of d:

$$A \sin[2\pi f(t+d)] \tag{12}$$

Setting this special mixing arrangement into motion, we see

$$A \sin(2\pi ft) \times A \sin(2\pi ft + d)$$

$$= \frac{A^2}{2}\cos(2\pi fd) - \frac{A^2}{2}\cos(2\pi fd)\cos(2 \times 2\pi ft) + \frac{A^2}{2}\sin(2\pi fd)\sin(2 \times 2\pi ft) \tag{13}$$

Two of the three outputs — the second and third terms — emerge at twice the input frequency; in practice, we're not interested in these, and filter them out. The remaining term — the one we're after — varies in amplitude and sign according to how far and in what direction the carrier shifts away from its resting or center frequency (at which the time delay, d, causes the mixer's RF and LO inputs to be exactly 90° out of phase — in *quadrature* — with each other). We can examine this effect by replacing f in equations 11 and 12 with the sum term $f_c + f_s$, where f_c is the center frequency and f_s is the frequency shift. A 90° time delay is the same as a quarter cycle of f_c, so we can restate d as

$$d = \frac{1}{4f_c} \tag{14}$$

The first term of the detector's output then becomes

$$\frac{A^2}{2}\cos(2\pi(f_c + f_s)d)$$

$$= \frac{A^2}{2}\cos\left(2\pi(f_c + f_s)\frac{1}{4f_c}\right)$$

$$= \frac{A^2}{2}\cos\left(\frac{\pi}{2} + \frac{\pi f_s}{2f_c}\right) \tag{15}$$

When f_s is zero (that is, when the carrier is at its center frequency), this reduces to

$$\frac{A^2}{2}\cos\left(\frac{\pi}{2}\right) = 0 \tag{16}$$

As the input signal shifts higher in frequency than f_c, the detector puts out a positive dc voltage that increases with the shift. When the input signal shifts lower in frequency than f_c, the detector puts out a negative dc voltage that increases with the shift. The detector therefore recovers the input signal's frequency or phase modulation as an amplitude-varying dc voltage that shifts in sign as f_s varies around f_c — in other words, as ac. We have demodulated FM by means of quadrature detection.

---

one example of this; **Fig 10.17** shows another, the Freescale Semiconductor (formerly Motorola) MC3359 (equivalent, NTE860).

PLL ANGLE DEMODULATION

Back at Fig 10.14, we saw how a phase-locked loop can be used as an angle modulator. A PLL also makes a fine angle *de*modulator. Applying an angle-modulated signal to a PLL keeps its phase detector and VCO hustling to maintain loop lock through the input signal's angle variations. The loop's error voltage therefore tracks the input signal's modulation, and its variations mirror the modulation signal. Turning the loop's varying dc error voltage into audio is just a blocking capacitor away.

Although we can't convey a dc level by directly modulating the VCO in a PLL angle modulator, a PLL demodulator can respond down to dc quite nicely. A constant frequency offset from f_c (a dc component) simply causes a PLL demodulator to swing its VCO over to the new input frequency, resulting in a proportional dc offset on the VCO control-voltage line. Another way of looking at the difference between a PLL angle modulator and a PLL angle demodulator is that a PLL demodulator works with a varying reference signal (the input signal), while a PLL angle modulator generally doesn't.

AMPLITUDE LIMITING REQUIRED

By now, it's almost household knowledge that FM radio communication systems are superior to AM in their ability to suppress and ignore static, manmade electrical noise and (through a characteristic called *capture effect*) co-channel signals sufficiently weaker than the desired signal. AM-noise immunity and capture effect are not intrinsic to angle modulation, however; they must be designed into the angle-modulation receiver in the form of signal amplitude *limiting*.

If we note the progress of A from the left to the right side of the equal sign in equation 13, we realize that the amplitude of a quadrature detector's input signal affects the amplitude of a quadrature detector's three output signals. A quadrature detector therefore responds to AM, and so does a frequency discriminator. To achieve FM's stored noise immunity, then, these angle demodulators must be preceded by *limiter* circuitry that removes all amplitude variations from the incoming signal.

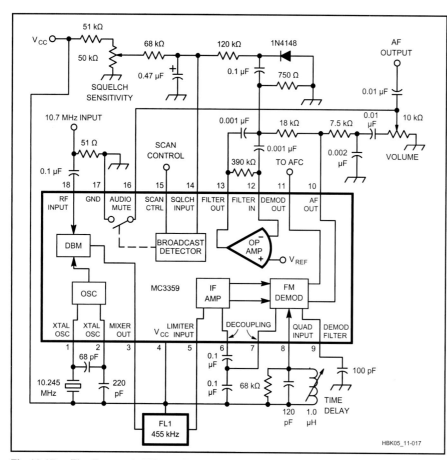

Fig 10.17 — The Freescale MC3359/NTE680 is one of many FM subsystem ICs that include limiter and quadrature-detection circuitry. The TIME DELAY coil is adjusted for minimum recovered-audio distortion.

10.4 Putting Mixers, Modulators and Demodulators to Work

Variations on relatively few mixer types provide mixing, modulation, and demodulation functions in the majority of Amateur Radio receivers, transmitters, and transceivers. Comparing their behaviors and specifications begins with a grounding in key mixer characteristics related to *dynamic range* — the span of signal strengths a circuit can handle without generating false signals that can interfere with communication.

10.4.1 Dynamic Range: Compression, Intermodulation and More

The output of a linear stage — including a mixer, which we want to act as a linear frequency shifter — tracks its input signal decibel by decibel, every 1-dB change in its input signal(s) corresponding to an identical 1-dB output change. This is the stage's *first-order* response.

Because no device is perfectly linear, however, two or more signals applied to it generate sum and difference frequencies. These IMD products occur at frequencies and amplitudes that depend on the order of the IMD response as follows:

• *Second-order* IMD products change 2 dB for every decibel of input-signal change (this figure assumes that the IMD comes from equal-level input signals), and appear at frequencies that result from the simple addition and subtraction of input-signal frequencies. For example, assuming that its input bandwidth is sufficient to pass them, an amplifier subjected to signals at 6 and 8 MHz will produce second-order IMD products at 2 MHz

Putting Mixer Math to Work through DSP

In describing the idealized behavior of mixing, modulation and demodulation circuits and processes through mathematics, this chapter's Mixer Math sidebars reflect a profound underlying reality: If what a mixer, modulator or demodulator does can be described with math, then *all a mixer, modulator or demodulator does is math*. This is true not only of mixers, modulators and demodulators, but of *all* electrical and electronic circuitry, analog and digital. An electrical or electronic circuit is merely a math machine — a machine that, for much of the history of human use of radio and electronics, has necessarily been hardwired to be capable of solving only one equation or set or class of equations. When you build an electronic circuit, interconnecting its resistors, capacitors, inductors, active devices and other components, you build out the signal-processing equation it will solve, atom by mathematical atom. Adjusting the controls of such a purpose-built math machine, every switch you throw or control modifies the equation it solves, and therefore the work it performs.

No matter how painstakingly and elegantly engineered and constructed such a practical realization of math may be, however, it is only so operationally flexible. If you want to significantly alter the equation(s) it solves, you must change its circuitry by replacing and/or reconnecting components and/or by adjusting whatever controls you may have included to tweak its operation on the fly.

The **Computer-Aided Circuit Design** chapter of this *Handbook* also characterizes electrical and electronic circuits as math machines. It does so for the purpose of exploring the use of personal computers — which are generic, arbitrarily and readily programmable and reprogrammable math machines — as engineering tools capable of simulating, through mathematical modeling, the electrical and electronic behavior of real-world circuits. The next logical expansion of the use of electronic computers in radio is to move beyond just *simulating* the mathematical signal-processing functions of circuits to actually *performing those functions in real time through mathematics applied in real time* — as physical circuitry has done for decades and still does, but better, and more flexibly, repeatably, and modifiably. Such real-time applied-math techniques, well-established and increasingly widely applied as *digital signal processing* (DSP) and fundamental to the design, evolution, and growing power of *software-defined radios* (SDRs), are likely already in use, or will soon be, in a radio you own. You can learn more about DSP and its techniques, applications and limitations in this *Handbook*'s **DSP and Software Radio Design** chapter.

(8–6) and 14 MHz (8+6).

• *Third-order* IMD products change 3 dB for every decibel of input-signal change (this also assumes equal-level input signals), and appear at frequencies corresponding to the sums and differences of twice one signal's frequency plus or minus the frequency of another. Assuming that its input bandwidth is sufficient to pass them, an amplifier subjected to signals at 14.02 MHz (f_1) and 14.04 MHz (f_2) produces third-order IMD products at 14.00 ($2f_1 - f_2$), 14.06 ($2f_2 - f_1$), 42.08 ($2f_1 + f_2$) and 42.10 ($2f_2 + f_1$) MHz. The subtractive products (the 14.02 and 14.04-MHz products in this example) are close to the desired signal and can cause significant interference. Thus, third-order-IMD performance is of great importance in receiver mixers and RF amplifiers.

It can be seen that the IMD order determines how rapidly IMD products change level per unit change of input level. *N*th- order IMD products therefore change by *n* dB for every decibel of input-level change.

IMD products at orders higher than three can and do occur in communication systems, but their magnitudes are usually much smaller than the lower-order products. The second- and third-order products are most important in receiver front ends. In transmitters, third- and higher-odd-order products are important because they widen the transmitted signal.

10.4.2 Intercept Point

The second type of dynamic range concerns the receiver's *intercept point*, sometimes simply referred to as *intercept*. Intercept point is typically measured (or calculated by extrapolation of the test results) by applying two or three signals to the antenna input, tuning the receiver to count the number of resulting spurious responses, and measuring their level relative to the input signal.

Because a device's IMD products increase more rapidly than its desired output as the input level rises, it might seem that steadily increasing the level of multiple signals applied to an amplifier would eventually result in equal desired-signal and IMD levels at the amplifier output. Real devices are incapable of doing this, however. At some point, every device *overloads*, and changes in its output level no longer equally track changes at its input. The device is then said to be operating in *compression*; the point at which its first-order response deviates from linearity by 1 dB is its *1-dB compression point*. Pushing the process to its limit ultimately leads to *saturation*, at which point input-signal increases no longer increase the output level.

The power level at which a device's second-order IMD products would equal its first-order output (a point that must be extrapolated because most likely the device is in compression by this point) is its *second-order intercept point*. Likewise, its *third-order intercept point* is the power level at which third-order responses would equal the desired signal. **Fig 10.18** graphs these relationships.

Input filtering can improve second-order intercept point; device nonlinearities determine the third, fifth and higher-odd-number intercept points. In preamplifiers and active mixers, third-order intercept point is directly related to dc input power; in passive switching mixers, to the local-oscillator power applied.

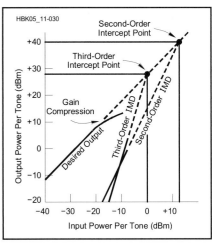

Fig 10.18 — A linear stage's output tracks its input decibel by decibel on a 1:1 slope — its *first-order* response. *Second-order intermodulation distortion (IMD) products* produced by two equal-level input signals ("tones") rise on a 2:1 slope — 2 dB for every 1 dB of input increase. *Third-order IMD products* likewise increase 3 dB for every 1 dB of increase in two equal tones. For each IMD order *n*, there is a corresponding *intercept point* IP_n at which the stage's first-order and *n*th order products are equal in amplitude. The first-order output of real amplifiers and mixers falls off (the device overloads and goes into *compression*) before IMD products can intercept it, but intercept point is nonetheless a useful, valid concept for comparing radio system performance. The higher an amplifier or mixer's intercept point, the stronger the input signals it can handle without overloading. The input and output powers shown are for purposes of example; every device exhibits its own particular IMD profile. (After W. Hayward, *Introduction to Radio Frequency Design*, Fig 6.17)

Testing and Calculating Intermodulation Distortion in Receivers

Second and third-order IMD can be measured using the setup of **Fig 10.A1**. The outputs of two signal generators are combined in a 3-dB hybrid coupler. Such couplers are available from various companies, and can be homemade. The 3-dB coupler should have low loss and should itself produce negligible IMD. The signal generators are adjusted to provide a known signal level at the output of the 3-dB coupler, say, –20 dBm for each of the two signals. This combined signal is then fed through a calibrated variable attenuator to the device under test. The shielding of the cables used in this system is important: At least 90 dB of isolation should exist between the high-level signal at the input of the attenuator and the low-level signal delivered to the receiver.

The measurement procedure is simple: adjust the variable attenuator to produce a signal of known level at the frequency of the expected IMD product ($f_1 \pm f_2$ for second-order, $2f_1 - f_2$ or $2f_2 - f_1$ for third-order IMD).

To do this, of course, you have to figure out what equivalent input signal level at the receiver's operating frequency corresponds to the level of the IMD product you are seeing. There are several ways of doing this. One way — the way used by the ARRL Lab in their receiver tests — uses the minimum discernible signal. This is defined as the signal level that produces a 3-dB increase in the receiver audio output power. That is, you measure the receiver output level with no input signal, then insert a signal at the operating frequency and adjust the level of this input signal until the output power is 3 dB greater than the no-signal power. Then, when doing the IMD measurement, you adjust the attenuator of Fig 10.A1 to cause a 3-dB increase in receiver output. The level of the IMD product is then the same as the MDS level you measured.

There are several things I dislike about doing the measurement this way. The problem is that you have to measure noise power. This can be difficult. First, you need an RMS voltmeter or audio power meter to do it at all. Second, the measurement varies with time (it's noise!), making it difficult to nail down a number. And third, there is the question of the audio response of the receiver; its noise output may not be flat across the output spectrum. So I prefer to measure, instead of MDS, a higher reference level. I use the receiver's S meter as a reference. I first determine the input signal level it takes to get an S1 reading. Then, in the IMD measurement, I adjust the attenuator to again give an S1 reading. The level of the IMD product signal is now equal to the level I measured at S1. Note that this technique gives a different IMD level value than the MDS technique. That's OK, though. What we are trying to determine is the *difference* between the level of the signals applied to the receiver input and the level of the IMD product. Our calculations will give the same result whether we measure the IMD product at the MDS level, the S1 level or some other level.

An easy way to make the reference measurement is with the setup of Fig 10.A1. You'll have to switch in a lot of attenuation (make sure you have an attenuator with enough range), but doing it this way keeps all of the possible variations in the measurement fairly constant. And this way, the difference between the reference level and the input level needed to produce the desired IMD product signal level is simply the difference in attenuator settings between the reference and IMD measurements.

Calculating Intercept Points

Once we know the levels of the signals applied to the receiver input and the level of the IMD product, we can easily calculate the intercept point using the following equation:

$$IP_n = \frac{n \times P_A - P_{IM_n}}{n - 1} \tag{A}$$

Here, n is the order, P_A is the receiver input power (of one of the input signals), P_{IM_n} is the power of the IMD product signal,

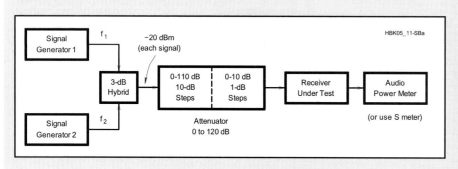

Fig 10.A1 — Test setup for measurement of IMD performance. Both signal generators should be types such as HP 608, HP 8640, or Rohde & Schwarz SMDU, with phase-noise performance of –140 dBc/Hz or better at 20 kHz from the signal frequency.

10.5 A Survey of Common Mixer Types

Depending on the application, frequency-mixer implementations may vary from the extremely simple to the complex. For example, a simple half-wave rectifier (a signal diode, such as a 1N34 [germanium] or a 1N914 [silicon]) can do the job (as we saw at Fig 10.9 in our discussion of full-carrier AM demodulation); this is an example of a *switching mixer*, in which mixing occurs as one signal — in this case, the carrier, which in effect turns the diode on and off as its polarity reverses — interrupts the transmission of another (in demodulation of full-carrier AM, the sideband[s]).

A switch can be thought of as an amplifier toggled between two gain states, off and on, by a control signal. It turns out that a binary amplifier is not necessary; *any* device that can be gain-varied in accordance with the amplitude of a control signal can serve as a frequency mixer.

10.5.1 Gain-Controlled Analog Amplifiers As Mixers

Fig 10.19 shows two analog-amplifier mixing arrangements commonly encoun-

and IP_n is the nth-order intercept point. All powers should be in dBm. For second and third-order IMD, equation A results in the equations:

$$IP_2 = \frac{2 \times P_A - P_{IM_2}}{2-1} \qquad (B)$$

$$IP_3 = \frac{3 \times P_A - P_{IM_3}}{3-1} \qquad (C)$$

You can measure higher-order intercept points, too.

Example Measurements

To get a feel for this process, it's useful to consider some actual measured values.

The first example is a Rohde & Schwarz model EK085 receiver with digital preselection. For measuring second-order IMD, signals at 6.00 and 8.01 MHz, at –20 dBm each, were applied at the input of the attenuator. The difference in attenuator settings between the reference measurement and the level needed to produce the desired IMD product signal level was found to be 125 dB. The calculation of the second-order IP is then:

$$IP_2 = \frac{2(-20\text{ dBm}) - (-20\text{ dBm} - 125\text{ dB})}{2-1}$$

$$= -40\text{ dBm} + 20\text{ dBm} + 125\text{ dB} = +105\text{ dB}$$

For IP_3, we set the signal generators for 0 dBm at the attenuator input, using frequencies of 14.00 and 14.01 MHz. The difference in attenuator settings between the reference and IMD measurements was 80 dB, so:

$$IP_3 = \frac{3(0\text{ dBm}) - (0\text{ dBm} - 80\text{ dB})}{3-1}$$

$$= \frac{0\text{ dBm} + 80\text{ dB}}{2} = +40\text{ dBm}$$

We also measured the IP_3 of a Yaesu FT-1000D at the same frequencies, using attenuator-input levels of –10 dBm. A difference in attenuator readings of 80 dB resulted in the calculation:

$$IP_3 = \frac{3(-10\text{ dBm}) - (-10\text{ dBm} - 80\text{ dB})}{3-1}$$

$$= \frac{-30\text{ dBm} + 10\text{ dBm} + 80\text{ dB}}{2}$$

$$= \frac{-20\text{ dBm} + 80\text{ dB}}{2}$$

$$= +30\text{ dBm}$$

Synthesizer Requirements

To be able to make use of high third-order intercept points at these close-in spacings requires a low-noise LO synthesizer. You can estimate the required noise performance of the synthesizer for a given IP_3 value. First, calculate the value of receiver input power that would cause the IMD product to just come out of the noise floor, by solving equation A for P_A, then take the difference between the calculated value of P_A and the noise floor to find the dynamic range. Doing so gives the equation:

$$ID_3 = \frac{2}{3}(IP_3 + P_{min}) \qquad (D)$$

Where ID_3 is the third-order IMD dynamic range in dB and P_{min} is the noise floor in dBm. Knowing the receiver bandwidth, BW (2400 Hz in this case) and noise figure, NF (8 dB) allows us to calculate the noise floor, P_{min}:

$$P_{min} = -174\text{ dBm} + 10\log(BW) + NF$$

$$= -174\text{ dBm} + 10\log(2400) + 8$$

$$= -132\text{ dBm}$$

The synthesizer noise should not exceed the noise floor when an input signal is present that just causes an IMD product signal at the noise floor level. This will be accomplished if the synthesizer noise is less than:

$$ID + 10\log(BW) = 114.7\text{ dB} + 10\log(2400)$$

$$= 148.5\text{ dBc/Hz}$$

in the passband of the receiver. Such synthesizers hardly exist.
— Dr Ulrich L. Rohde, N1UL

---

tered in vacuum-tube-based vintage radio gear. In one (Fig 10.19A), the RF and LO signals are applied to the control grid of a pentode amplifier; in the second, the RF and LO signals are applied to different grids of a tube specifically designed to act as a mixer. RF, LO, and IF (RF and LO sum and difference frequencies) energy is present at the output of both. In both circuits, a double-tuned transformer selects the desired output frequency and provides some reduction of unwanted components. The pentagrid mixer circuit (Fig 10.19A) provides better isolation between its RF and LO inputs because of its separate RF and LO-injection grids.

Fig 10.20 shows a simple JFET front end that consists of an AGCed amplifier (2N4393) and 2N3819 or 2N5454 mixer with source LO injection. As in the tube circuits of Fig 10.19, RF, LO, and IF spectral components are all present in the output of this simple circuit.

Mixers based on FET cascode amplifiers (**Fig 10.21**) can provide excellent performance, and were common (in dual-gate MOSFET form) across at least two generations of Amateur Radio gear. As with the circuits presented in Figs 10.19 and 10.20,

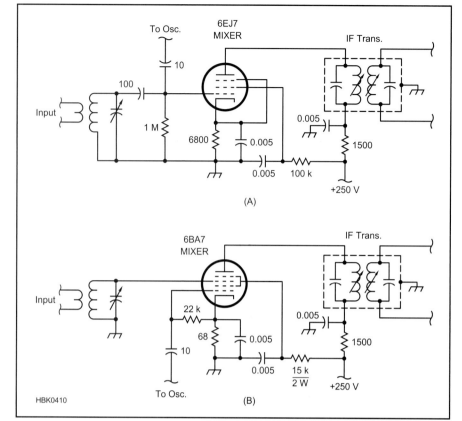

Fig 10.19 — Vacuum-tube amplifiers as frequency mixers. At A, RF and LO signals are applied to the control grid of a 6EJ7 pentode; at B, a RF and LO signals are applied to grids 1 and 3, respectively, of a 6BA7 pentagrid mixer. The 6BA7 and its relatives and similars (6A8, 6K8, 6SA7, 6SB7, and 6BE6, among others), designed specifically for frequency-conversion use, can be configured to generate their own LO signal; such a self-oscillating tube mixer is commonly called a *converter*.

RF, LO, and IF spectral components are present in the output of these mixers, and filtering is necessary to select the desired component and reduce unwanted ones.

10.5.2 Switching Mixers

Most modern radio mixers act more like fast analog switches than analog multipliers. In using a mixer as a fast switching device, we apply a square wave to its LO input with a square wave rather than a sine wave, and feed sine waves, audio, or other complex signals to the mixer's RF input. The RF port serves as the mixer's "linear" input, and therefore must preferably exhibit low intermodulation and harmonic distortion. Feeding a ±1-V square wave into the LO input alternately multiplies the linear input by +1 or –1. Multiplying the RF-port signal by +1 just transfers it to the output with no change. Multiplying the RF-port signal by –1 does the same thing, except that the signal inverts (flips 180° in phase). The LO port need not exhibit low intermodulation and harmonic distortion; all it has to do is preserve the fast rise and fall times of the switching signal.

REVERSING-SWITCH MIXERS

We can multiply a signal by a square wave without using an analog multiplier at all. All we need is a pair of balun transformers and four diodes (**Fig 10.22A**).

With no LO energy applied to the circuit, none of its diodes conduct. RF-port energy (1)

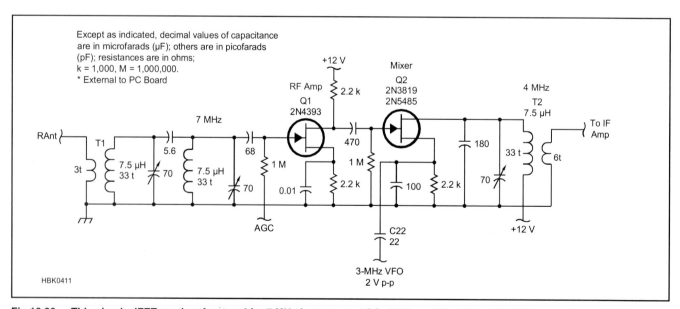

Fig 10.20 — This simple JFET receiver front end for 7 MHz features an AGCed RF amplifier (Q1, a 2N4393) and gain-controlled-amplifier mixer (Q2, a 2N3819 or 2N5485). LO voltage applied to the source of Q2 varies the stage's gain to produce sum and difference outputs. T2 selects the product (4 MHz). Any JFET with a pinchoff voltage of –1 V is suitable for Q1; the point labeled AGC can be grounded to operate the RF stage at full gain. The inductors consist of no. 22 enameled wire wound on plastic bobbins used to hold the lower thread in "New Home" brand sewing machines; if toroidal equivalents are substituted, the turns ratios in T1 and T2 should be preserved. This circuit was described in "Build the No-Excuses QRP Transceiver," December 2000 *QST*.

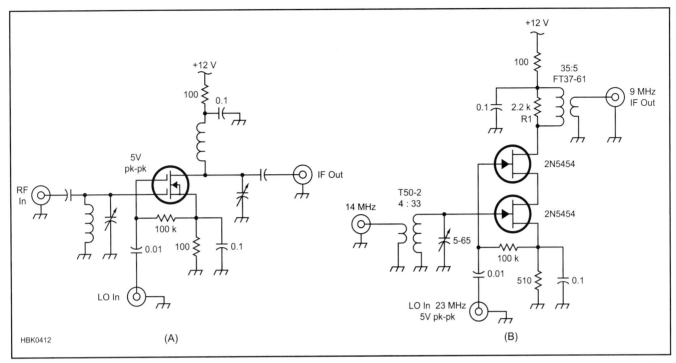

Fig 10.21 — Cascode FET amplifiers used as mixers, as presented by Hayward, Campbell and Larkin in *Experimental Methods in RF Design*. RF is applied to the gate of one device, and LO and a small positive bias are applied to the gate of the other; commercial applications of the MOSFET version commonly include positive bias on the RF-port gate. A dual-gate MOSFET is shown at A; suitable devices include the 3N187, 3N211, 40673 and BF998. (Internally, a dual-gate MOSFET consists of two single-gate MOSFETs in cascode.) Applying AGC rather than this circuit's combination of LO and source-derived bias to gate 2 of the MOSFET would turn this arrangement into one of the AGCed IF amplifiers used in commercial ham gear for over 40 years. Two JFETs in cascode are shown at B; suitable devices include the 2N4416, 2N3819, 2N5400-series parts, J310/U310 and MPF102, among others. R1 loads the circuit to eliminate oscillation at the RF-port resonance; it may be increased in value or eliminated altogether if the circuit is stable without it. Ratios accompanying the transformers at B convey the number of turns. Per *EMRFD*, the JFET circuit shown, biased for 3.4 mA at 12 V, operates with a conversion gain of 8 dB; a noise figure of 10 dB; and a third-order intercept point of +5 dBm.

can't make it to the LO port because there's no direct connection between the secondaries of T1 and T2, and (2) doesn't produce IF output because T2's secondary balance results in energy cancellation at its center tap, and because no complete IF-energy circuit exists through T2's secondary with both of its ends disconnected from ground.

Applying a square wave to the LO port biases the diodes so that, 50% of the time, D1 and D2 are on and D3 and D4 are reverse-biased off. This unbalances T2's secondary by leaving its upper wire floating and connecting its lower wire to ground through T1's secondary and center tap. With T2's secondary unbalanced, RF-port energy emerges from the IF port.

The other 50% of the time, D3 and D4 are on and D1 and D2 are reverse-biased off. This unbalances T2's secondary by leaving its lower wire floating, and connects its upper wire to ground through T1's secondary and center tap. With T2's secondary unbalanced, RF-port energy again emerges from the IF port — shifted 180° relative to the first case because T2's active secondary wires are now, in effect, transposed relative to its primary.

A reversing switch mixer's output spectrum is the same as the output spectrum of a multiplier fed with a square wave. This can be analyzed by thinking of the square wave in terms of its Fourier series equivalent, which consists of the sum of sine waves at the square wave frequency and all of its odd harmonics. The amplitude of the equivalent series' fundamental sine wave is $4/\pi$ times (2.1 dB greater than) the amplitude of the square wave. The amplitude of each harmonic is inversely proportional to its harmonic number, so the third harmonic is only $\frac{1}{3}$ as strong as the fundamental (9.5 dB below the fundamental), the 5th harmonic is only $\frac{1}{5}$ as strong (14 dB below the fundamental) and so on. The input signal mixes with each harmonic separately from the others, as if each harmonic were driving its own separate mixer, just as we illustrated with two sine waves in Fig 10.4. Normally, the harmonic outputs are so widely removed from the desired output frequency that they are easily filtered out, so a reversing-switch mixer is just as good as a sine wave driven analog multiplier for most practical purposes, and usually better — for radio purposes — in terms of dynamic range and noise.

An additional difference between multiplier and switching mixers is that the signal flow in a switching mixer is reversible (that is, bilateral). It really only has one dedicated input (the LO input). The other terminals can be thought of as I/O (input/output) ports, since either one can be the input as long as the other is the output.

CONVERSION LOSS IN SWITCHING MIXERS

Fig 10.22B shows a perfect *multiplier* mixer. That is, the output is the product of the input signal and the LO. The LO is a perfect square wave. Its peak amplitude is ±1.0 V and its frequency is 8 MHz. Fig 10.22C shows the output waveform (the product of two inputs) for an input signal whose value is 0 dBm and whose frequency is 2 MHz. Notice that for each transition of the square-wave LO, the sine-wave output waveform polarity reverses. There are 16 transitions during the interval shown, at each zero-crossing point of the output waveform. Fig 10.22D shows the mixer output spectrum. The principle components are at 6 MHz and 10 MHz, which are the sum and difference of the signal and LO frequencies. The amplitude of each of these is −3.9 dBm. Numerous other pairs of output frequencies occur that are also spaced 4 MHz

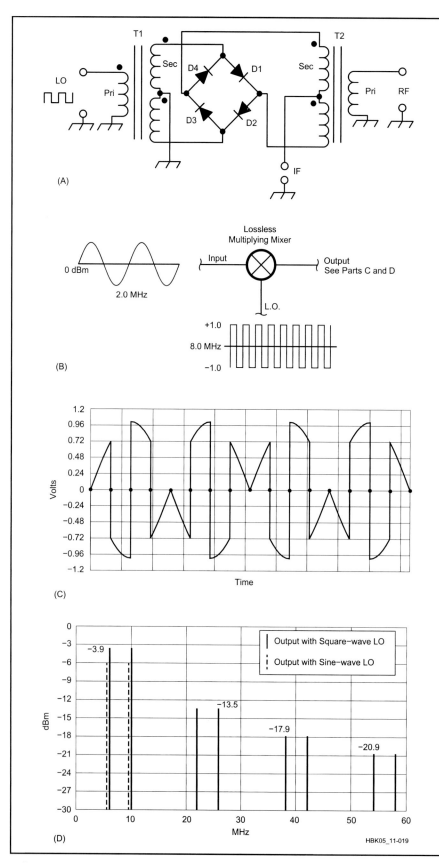

Fig 10.22 — Part A shows a general-purpose diode *reversing-switch* mixer. This mixer uses a square-wave LO and a sine-wave input signal. The text describes its action. Part B is an ideal *multiplier* mixer. The square-wave LO and a sine-wave input signal produce the output waveform shown in part C. The solid lines of part D show the output spectrum with the square-wave LO. The dashed lines show the output spectrum with a sine-wave LO.

apart and centered at 24 MHz, 40 MHz and 56 MHz and higher odd harmonics of 8 MHz. The ones shown are at –13.5 dBm, –17.9 dBm and –20.9 dBm. Because the mixer is lossless, the sum of all of the outputs must be exactly equal to the value of the input signal. As explained previously, this output spectrum can also be understood in terms of each of the odd-harmonic components of the square-wave LO operating independently.

If the mixer switched losslessly, such as in Fig 10.22A, with diodes that are perfect switches, the results would be mathematically identical to the above example. The diodes would commutate the input signal exactly as shown in Fig 10.22C.

Now consider the perfect multiplier mixer of Fig 10.22B with an LO that is a perfect sine wave with a peak amplitude of ±1.0 V. In this case the dashed lines of Fig 10.22D show that only two output frequencies are present, at 6 MHz and 10 MHz (see also Fig 10.2). Each component now has a –6 dBm level. The product of the 0 dBm sine-wave input at one frequency and the ±1.0V sine-wave LO at another frequency (see equation 6 in this chapter) is the –3 dBm total output.

These examples illustrate the difference between the square-wave LO and the sine-wave LO, for a perfect multiplier. For the same peak value of both LO waves, the square-wave LO delivers 2.1 dB more output at 6 MHz and 10 MHz than the sine-wave LO. An actual diode mixer such as Fig 10.22A behaves more like a switching mixer. Its sine-wave LO waveform is considerably flattened by interaction between the diodes and the LO generator, so that it looks somewhat like a square wave. The diodes have nonlinearities, junction voltages, capacitances, resistances and imperfect parameter matching. (See the **RF Techniques** chapter.) Also, "re-mixing" of a diode mixer's output with the LO and the input is a complicated possibility. The practical end result is that diode double-balanced mixers have a conversion loss, from input to each of the two major output frequencies, in the neighborhood of 5 to 6 dB. (Conversion loss is discussed in a later section.)

10.5.3 The Diode Double-Balanced Mixer: A Basic Building Block

The diode *double-balanced mixer* (DBM) is standard in many commercial, military and amateur applications because of its excellent balance and high dynamic range. DBMs can serve as mixers (including image-reject types), modulators (including single- and double-sideband, phase, biphase, and quadrature-phase types) and demodulators, limiters, attenuators, switches, phase detectors and frequency doublers. In some of these applications, they work in conjunction with power dividers, combiners and hybrids.

THE BASIC DBM CIRCUIT

We have already seen the basic diode DBM circuit (Fig 10.22A). In its simplest form, a DBM contains two or more unbalanced-to-balanced transformers and a Schottky-diode ring consisting of 4 × n diodes, where n is the number of diodes in each leg of the ring. Each leg commonly consists of up to four diodes.

As we've seen, the degree to which a mixer is *balanced* depends on whether either, neither or both of its input signals (RF and LO) emerge from the IF port along with mixing products. An unbalanced mixer suppresses neither its RF nor its LO; both are present at its IF port. A single-balanced mixer suppresses its RF or LO, but not both. A double-balanced mixer suppresses its RF *and* LO inputs. Diode and transformer uniformity in the Fig 10.22 circuit results in equal LO potentials at the center taps of T1 and T2. The LO potential at T1's secondary center tap is zero (ground); therefore, the LO potential at the IF port is zero.

Balance in T2's secondary likewise results in an RF null at the IF port. The RF potential between the IF port and ground is therefore zero — except when the DBM's switching diodes operate, of course!

The Fig 10.22 circuit normally also affords high RF-IF isolation because its balanced diode switching precludes direct connections between T1 and T2. A diode DBM can be used as a current-controlled switch or attenuator by applying dc to its IF port, albeit with some distortion. This causes opposing diodes (D2 and D4, for instance) to conduct to a degree that depends on the current magnitude, connecting T1 to T2.

TRIPLE-BALANCED MIXERS

The triple-balanced mixer shown in **Fig 10.23** (sometimes called a "double double-balanced mixer") is an extension of the single-diode-ring mixer. The diode rings are fed by power-splitting baluns at the RF and LO ports. An additional balun is added at the IF output. The circuit's primary advantage is that the IF output signal is balanced and isolated from the RF and LO ports over a large bandwidth — commercial mixer IF ranges of 0.5 to 10 GHz are typical.

It has higher signal-handling capability and dynamic range (a 1-dB compression point within 3 to 4 dB below LO signal levels) and lower intermodulation levels (by 10 dB or more) than a single-ring mixer. The triple-balanced mixer is used when a very wide IF range is required.

Adding the balancing transformer in the IF output path increases IF-to-LO and IF-to-RF isolation. This makes the conversion process much less sensitive to IF impedance mis-

Testing Mixer Performance

In order to make proper tests on mixers using signal generators, a hybrid coupler with at least 40 dB of isolation between the two input ports and an attenuator are required. The test set-up provided by DeMaw in *QST*, Jan 1981, shown in **Fig 10.B1** is ideal for this. Two signal generators operating near 14 MHz are combined in the hybrid coupler, then isolated from the mixer under test (MUT) by a variable attenuator. The LO is supplied by a VFO covering 5.0-5.5 MHz and applied to the MUT through another variable attenuator. The output is isolated with another attenuator, amplified and applied to a spectrum analyzer for analysis.

Attenuation should be sufficient to provide isolation (minimum of 6 to 10 dB required) and to result in signal levels to the mixer under test (MUT) appropriate for the required testing and as suitable for the particular mixer device.

The 2N5109 amplifier shown may not be sufficient for extremely high intercept point tests as this stage may no longer be transparent (operate linearly) at high signal levels. For stability tests, it is recommended to have a reactive network at the output of the mixer for the sole purpose of checking whether the mixer can become unstable.

The two 14 MHz oscillators must have extremely low harmonic content and very low noise sidebands. A convenient oscillator circuit is provided in **Fig 10.B2**, based on a 1975 *Electronic Design* article. — *Dr Ulrich L. Rohde, N1UL*

DeMaw, W1FB, and Collins, AD0W, "Modern Receiver Mixers for High Dynamic Range," *QST*, Jan 1981, p 19.
U. Rohde, "Crystal Oscillator Provides Low Noise," *Electronic Design*, Oct 11, 1975.

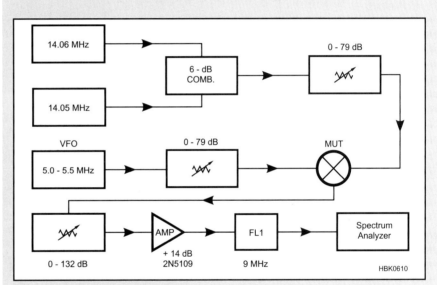

Fig 10.B1 — The equipment setup for measuring mixer performance at HF.

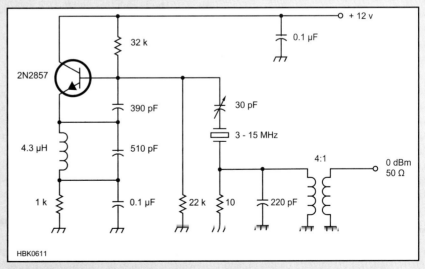

Fig 10.B2 — A low-noise VXO circuit for driving the LO port of the mixer under test in Figure 10.B1.

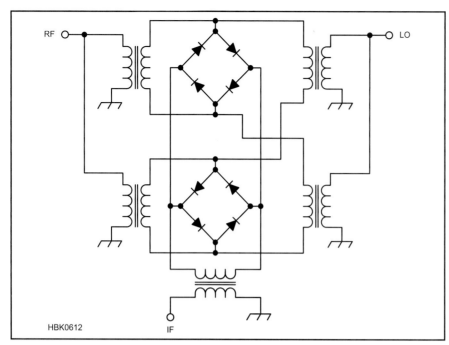

Fig 10.23 — The triple-balanced mixer uses a pair of diode rings and adds an additional balancing transformer to the IF port.

matches. Since the IF port is isolated from the RF and LO ports, the three frequency ranges (RF, LO and IF) can overlap. A disadvantage of IF transformer coupling is that a dc (or low-frequency) IF output is not available, so the triple-balanced mixer cannot be used for direct-conversion receivers.

DIODE DBM COMPONENTS

Commercially manufactured diode DBMs generally consist of a supporting base, a diode ring, two or more ferrite-core transformers commonly wound with two or three twisted-pair wires, encapsulating material, an enclosure.

Diodes

Hot-carrier (Schottky) diodes are the devices of choice for diode-DBM rings because of their low ON resistance, although ham-built DBMs for non-critical MF/HF use commonly use switching diodes like the 1N914 or 1N4148. The forward voltage drop, V_f, across each diode in the ring determines the mixer's optimum local-oscillator drive level. Depending on the forward voltage drop of each of its diodes and the number of diodes in each ring leg, a diode DBM will often be specified by the optimum LO drive level in dBm (typical values are 0, 3, 7, 10, 13, 17, 23 or 27). As a rule of thumb, the LO signal must be 20 dB stronger than the RF and IF signals for proper operation. This ensures that the LO signal, rather than the RF or IF signals, switches the mixer's diodes on and off — a critical factor in minimizing IMD and maximizing dynamic range.

Transformers

From the DBM schematic shown in Fig 10.22, it's clear that the LO and RF transformers are unbalanced on the input side and balanced on the diode side. The diode ends of the balanced ports are 180° out of phase throughout the frequency range of interest. This property causes signal cancellations that result in higher port-to-port isolation. **Fig 10.24A** plots LO-RF and LO-IF isolation versus frequency for Synergy Microwave's CLP-403 DBM, which is specified for +7 dBm LO drive level. Isolations on the order of 70 dB occur at the lower end of the band as a direct result of the balance among the four diode-ring legs and the RF phasing of the balanced ports.

As we learned in our discussion of generic switching mixers, transformer efficiency plays an important role in determining a mixer's conversion loss and drive-level requirement. Core loss, copper loss and impedance mismatch all contribute to transformer losses. Ferrite in toroidal, bead, balun (multi-hole) or rod form can serve as DBM transformer cores. Radio amateurs commonly use Fair-Rite Mix 43 ferrite (μ = 950).

RF transformers combine lumped and distributed capacitance and inductance. The interwinding capacitance and characteristic impedance of a transformer's twisted wires sets the transformer's high-frequency response. The core's μ and size, and the number of winding turns, determine the transformer's lower frequency limit. Covering a specific frequency range requires a compromise in the number of turns used with a given core. Increasing a transformer's core size and number of turns improves its low-frequency response. Cores may be stacked to meet low-frequency performance specs.

Inexpensive mixers operating up to 2 GHz most commonly use twisted trifilar (three-wire) windings made of a wire size between #36 and #32. The number of twists per unit length of wire determines a winding's characteristic impedance. Twisted wires are analogous to transmission lines. The transmission-line effect predominates at the higher end of a transformer's frequency range.

DIODE DBMS IN PRACTICE

Important DBM specifications include conversion loss and amplitude flatness across the required IF bandwidth; variation of conversion loss with input frequency; variation of conversion loss with LO drive, 1-dB compression point; LO-RF, LO-IF and RF-IF isolation; intermodulation products; noise figure (usually within 1 dB of conversion loss); port SWR; and dc offset, which is directly related to isolation among the RF, LO and IF ports. Most of these parameters also apply to other mixer types.

Conversion Loss

Fig 10.24B shows conversion loss versus intermediate frequency in a typical DBM. The curves show conversion loss for two fixed RF-port signals, one at 100 kHz and

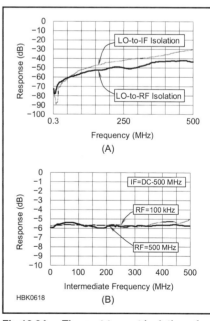

Fig 10.24 — The port-to-port isolation of a diode DBM depends on how well its diodes match and how well its transformers are balanced. (A) shows LO-IF and LO-RF isolation versus frequency and (B) shows conversion loss for a typical diode DBM, the Synergy Microwave CLP-403 mixer. In (B), LO driver level is +7 dBm.

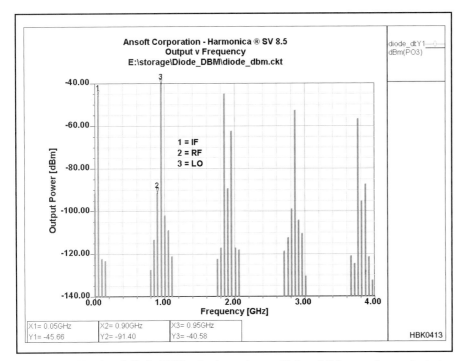

Fig 10.25— Simulated diode-DBM output spectrum with four LO harmonics evaluated. Note that the desired output products (the highest two products, RF − LO and RF + LO) emerge at a level 5 to 6 dB below the mixer's RF input (−40 dBm). This indicates a mixer conversion loss of 5 to 6 dB. (*Serenade SV8.5* simulation.)

the another at 500 MHz, while varying the LO frequency from 100 kHz to 500 MHz.

Fig 10.25 graphs a diode DBM's simulated output spectrum. Note that the RF input (900 MHz) is −40 dBm and the desired IF output (51 MHz, the frequency difference between the RF and LO signals) is −46 dBm, implying a conversion loss of 6 dB. Very nearly the same value (5 dB) applies to the sum of both signals (RF + LO). We minimize a diode DBM's conversion loss, noise figure and intermodulation by keeping its LO drive high enough to switch its diodes on fully and rapidly. Increasing a mixer's LO level beyond that sufficient to turn its switching devices all the way on merely makes them dissipate more LO power without further improving performance.

Insufficient LO drive results in increased noise figure and conversion loss. IMD also increases because RF-port signals have a greater chance to control the mixer diodes when the LO level is too low.

APPLYING DIODE DBMS

At first glance, applying a diode DBM is easy: We feed the signal(s) we want to frequency-shift (at or below the maximum level called for in the mixer's specifications, such as −10 dBm for the Mini-Circuits SBL-1 and TUF-3, and Synergy Microwave S-1, popular 7 dBm LO power parts) to the DBM's RF port, feed the frequency-shifting signal (at the proper level) to the LO port, and extract the sum and difference products from the mixer's IF port.

There's more to it than that, however, because diode DBMs (along with most other modern mixer types) are *termination-sensitive*. That is, their ports — particularly their IF (output) ports — must be resistively terminated with the proper impedance (commonly 50 Ω, resistive). A wideband, resistive output termination is particularly critical if a mixer is to achieve its maximum dynamic range in receiving applications. Such a load can be achieved by:

• Terminating the mixer in a 50-Ω resistor or attenuator pad (a technique usually avoided in receiving applications because it directly degrades system noise figure);

• Terminating the mixer with a low-noise, high-dynamic-range *post-mixer amplifier* designed to exhibit a wideband resistive input impedance; or

• Terminating the mixer in a *diplexer*, a frequency-sensitive signal splitter that appears as a two-terminal resistive load at its input while resistively dissipating unwanted outputs and passing desired outputs through to subsequent circuitry.

Termination-insensitive mixers are available, but this label can be misleading. Some termination-insensitive mixers are nothing more than a termination-sensitive mixer packaged with an integral post-mixer amplifier. True termination-insensitive mixers are less common and considerably more elaborate. Amateur builders will more likely use one of the many excellent termination-sensitive mixers available in connection with a diplexer, post-mixer amplifier or both.

Fig 10.26 shows one diplexer implementation. In this approach, L1 and C1 form a series-tuned circuit, resonant at the desired IF, that presents low impedance between the diplexer's input and output terminals at the IF. The high-impedance parallel-tuned circuit formed by L2 and C2 also resonates at the desired IF, keeping desired energy out of the diplexer's 50-Ω load resistor, R1.

The preceding example is called a *bandpass diplexer*. **Fig 10.27** shows another type: a *high-pass/low-pass diplexer* in which each inductor and capacitor has a reactance of 70.7 Ω at the 3-dB cutoff frequency. It can be used after a "difference" mixer (a mixer in which the IF is the difference between the signal frequency and LO) if the desired IF and its image frequency are far enough apart so that the image power is "dumped" into the network's 51-Ω resistor. (For a "summing" mixer — a mixer in which the IF is the sum of the desired signal and LO — interchange the 50-Ω idler load resistor and the diplexer's "50-Ω Amplifier" connection.) Richard Weinreich, KØUVU, and R. W. Carroll described this circuit in November 1968 *QST* as one of several absorptive TVI filters.

Fig 10.28 shows a BJT post-mixer amplifier design made popular by Wes Hayward, W7ZOI, and John Lawson, K5IRK. RF feedback (via the 1-kΩ resistor) and emitter degeneration (the ac-coupled 5.6-Ω emitter resistor) work together to keep the stage's input impedance near 50 Ω and uniformly resistive across a wide bandwidth. Performance comparable to the Fig 10.28 circuit can be obtained at MF and HF by using paralleled 2N3904s as shown in **Fig 10.29**.

AMPLITUDE MODULATION WITH A DBM

We can generate DSB, suppressed-carrier AM with a DBM by feeding the carrier to its RF port and the modulating signal to the IF port. This is a classical *balanced modulator*, and the result — sidebands at radio frequencies corresponding to the carrier signal plus audio and the RF signal minus audio — emerges from the DBM's LO port. If we also want to transmit some carrier along with the sidebands, we can dc-bias the IF port (with a current of 10 to 20 mA) to upset the mixer's balance and keep its diodes from turning all the way off. (This technique is sometimes used for generating CW with a balanced modulator otherwise intended to generate DSB as part of an SSB-generation process.) **Fig 10.30** shows a more elegant approach to generating full-carrier AM with a DBM.

As we saw earlier when considering the many faces of AM, two DBMs, used in con-

Fig 10.26 — A diplexer resistively terminates energy at unwanted frequencies while passing energy at desired frequencies. This band-pass diplexer (A) uses a series-tuned circuit as a selective pass element, while a high-C parallel-tuned circuit keeps the network's terminating resistor R1 from dissipating desired-frequency energy. Computer simulation of the diplexer's response with *ARRL Radio Designer 1.0* characterizes the diplexer's insertion loss and good input match from 8.8 to 9.2 MHz (B) and from 1 to 100 MHz (C); and the real and imaginary components of the diplexer's input impedance from 8.8 to 9.2 MHz with a 50-Ω load at the diplexer's output terminal (D). The high-C, low-L nature of the L2-C2 circuit requires that C2 be minimally inductive; a 10,000-pF chip capacitor is recommended. This diplexer was described by Ulrich L. Rohde and T. T. N. Bucher in *Communications Receivers: Principles and Design*.

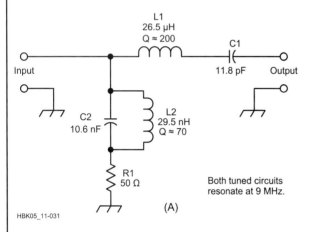
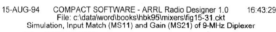
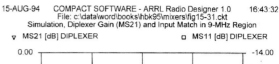
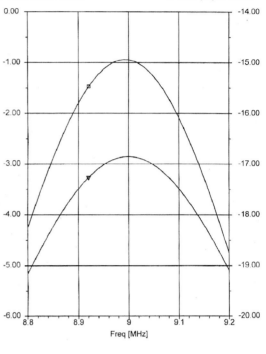

(A)

(B)

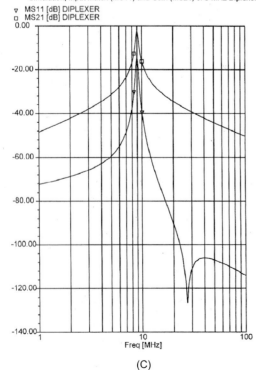

(C)

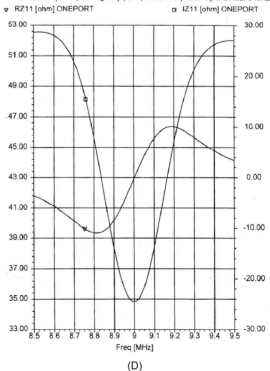

(D)

junction with carrier and audio phasing, can be used to generate SSB, suppressed-carrier AM. Likewise, two DBMs can be used with RF and LO phasing as an image-reject mixer.

PHASE DETECTION WITH A DBM

As we saw in our exploration of quadrature detection, applying two signals of equal frequency to a DBM's LO and RF ports produces an IF-port dc output proportional to the cosine of the signals' phase difference (**Fig 10.31**). This assumes that the DBM has a dc-coupled IF port, of course. If it doesn't — and some DBMs don't — phase-detector operation is out. Any dc output offset introduces error into this process, so critical phase-detection applications use low-offset DBMs optimized for this service.

BIPHASE-SHIFT KEYING MODULATION WITH A DBM

Back in our discussion of square-wave mixing, we saw how multiplying a switching mixer's linear input with a square wave causes a 180° phase shift during the negative part of the square wave's cycle. As **Fig 10.32** shows, we can use this effect to produce *biphase-shift keying (BPSK)*, a digital system that conveys data by means of carrier phase reversals. A related system, *quadrature phase-shift keying (QPSK)* uses two DBMs and phasing to convey data by phase-shifting a carrier in 90° increments.

10.5.4 Active Mixers — Transistors as Switching Elements

We've covered diode DBMs in depth because their home-buildability, high performance and suitability for direct connection into 50-Ω systems makes them attractive to Amateur Radio builders. The abundant availability of high-quality manufactured diode mixers at reasonable prices makes them

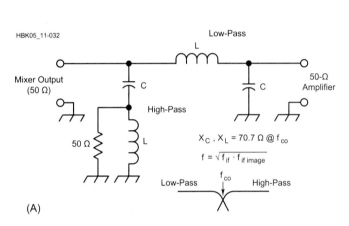

Fig 10.27 — All of the inductors and capacitors in this high-pass/low-pass diplexer (A) exhibit a reactance of 70.7 Ω at its tuned circuits' 3-dB cutoff frequency (the geometric mean of the IF and IF image). B and C show *ARRL Radio Designer* simulations of this circuit configured for use in a receiver that converts 7 MHz to 3.984 MHz using a 10.984-MHz LO. The IF image is at 17.984 MHz, giving a 3-dB cutoff frequency of 8.465 MHz. The inductor values used in the simulation were therefore 1.33 μH (Q = 200 at 25.2 MHz); the capacitors, 265 pF (Q = 1000). This drawing shows idler load and "50-Ω Amplifier" connections suitable for a receiver in which the IF image falls at a frequency *above* the desired IF. For applications in which the IF image falls below the desired IF, interchange the 50-Ω idler load resistor and the diplexer's "50-Ω Amplifier" connection so the idler load terminates the diplexer low-pass filter and the 50-Ω amplifier terminates the high-pass filter.

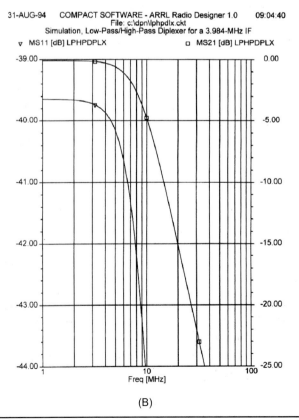

(B)

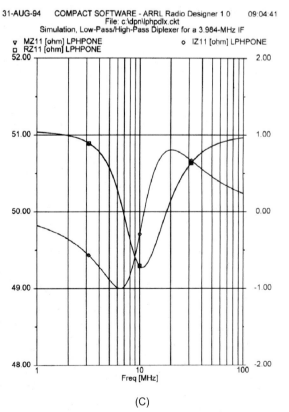

(C)

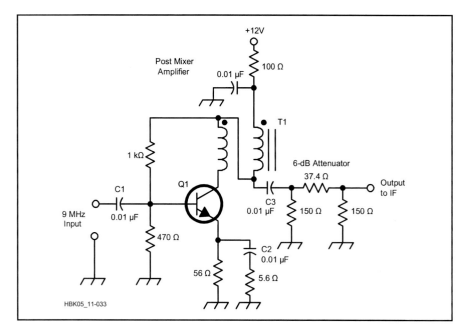

Fig 10.28 — The post-mixer amplifier from Hayward and Lawson's Progressive Communications Receiver (November 1981 *QST*). This amplifier's gain, including the 6-dB loss of the attenuator pad, is about 16 dB; its noise figure, 4 to 5 dB; its output intercept, 30 dBm. The 6-dB attenuator is essential if a crystal filter follows the amplifier; the pad isolates the amplifier from the filter's highly reactive input impedance. This circuit's input match to 50 Ω below 4 MHz can be improved by replacing 0.01-µF capacitors C1, C2 and C3 with low-inductance 0.1-µF units (chip capacitors are preferable). Q1 — TO-39 CATV-type bipolar transistor, f_T = 1 GHz or greater. 2N3866, 2N5109, 2SC1252, 2SC1365 or MRF586 suitable. Use a small heat sink on this transistor. T1 — Broadband ferrite transformer, ≈42 µH per winding: 10 bifilar turns of #28 enameled wire on an FT 37-43 core.

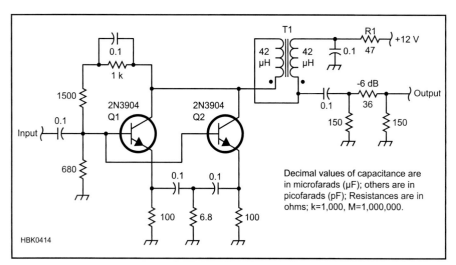

Fig 10.29 — At MF and HF, paralleled 2N3904 BJTs can provide performance comparable to that of the Fig 10.28 circuit with sufficient attention paid to device standing current, here set at ≈30 mA for the pair. The value of decoupling resistor R1 is critical in that small changes in its value cause a relatively large change in the 2N3904s' bias point. This circuit is part of the EZ-90 Receiver, described by Hayward, Campbell and Larkin in *Experimental Methods in RF Design*.

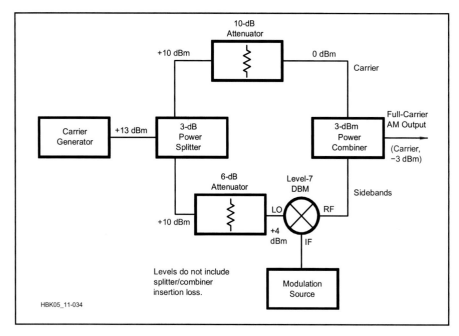

Fig 10.30 — Generating full-carrier AM with a diode DBM. A practical modulator using this technique is described in *Experimental Methods in RF Design*.

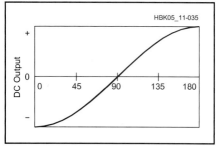

Fig 10.31 — A phase detector's dc output is the cosine of the phase difference between its input and reference signals.

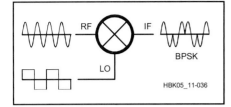

Fig 10.32 — Mixing a carrier with a square wave generates biphase-shift keying (BPSK), in which the carrier phase is shifted 180° for data transmission. In practice, as in this drawing, the carrier and data signals are phase-coherent so the mixer switches only at carrier zero crossings.

excellent candidates for home construction projects. Although diode DBMs are common in telecommunications as a whole, their conversion loss and relatively high LO power requirement have usually driven the manufacturers of high-performance MF/HF Amateur Radio receivers and transceivers to other solutions. Those solutions have generally involved single- or double-balanced FET mixers — MOSFETs in the late 1970s and early 1980s, JFETs from the early 1980s to date. A comprehensive paper that explores the differences between various forms of active mixers, "Performance Capabilities of Active Mixers," by Ulrich Rohde, N1UL, is included on the CD-ROM accompanying this *Handbook*.

Many of the JFET designs are variations of a single-balanced mixer circuit introduced to *QST* readers in 1970! **Fig 10.33** shows the circuit as it was presented by William Sabin in "The Solid-State Receiver," *QST*, July 1970. Two 2N4416 JFETs operate in a common-source configuration, with push-pull RF input and parallel LO drive. **Fig 10.34** shows a similar circuit as implemented in the ICOM IC-765 transceiver. In this version, the JFETs (2SK125s) operate in common-gate, with the LO applied across a 220-Ω resistor between the gates and ground.

Current state of the art for active mixers in the HF through GHz range replaces discrete device designs with integrated designs such as the Analog Devices AD8342 (**www.analog.com**). Using an IC greatly improves matching of the active devices, improving circuit balance. The AD8342 has a conversion gain of 3.7 dB, a noise figure of 12.2 dB, and an input IP3 of 22.7 dBm. The device operates with a single-voltage power supply and is well-suited to interface with digital hardware, such as for SDR applications. Reference circuits for applications at HF and VHF/UHF are provided in the device's datasheet.

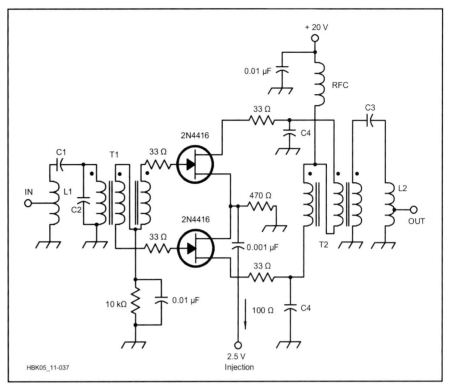

Fig 10.33 — Two 2N4416 JFETs provide high dynamic range in this mixer circuit from Sabin, *QST*, July 1970. L1, C1 and C2 form the input tuned circuit; L2, C3 and C4 tune the mixer output to the IF. The trifilar input and output transformers are broadband transmission-line types.

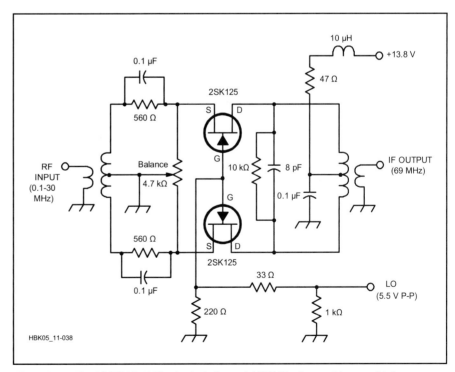

Fig 10.34 — The ICOM IC-765's single-balanced 2SK125 mixer achieves a high dynamic range (per *QST* Product Review, an IP_3 of 10.5 dBm at 14 MHz with preamp off). The first receive mixer in many commercial Amateur Radio transceiver designs of the 1980s and 1990s used a pair of 2SK125s or similar JFETs in much this way.

10.5.5 The Tayloe Mixer

[The following description of the Tayloe Product Detector (a.k.a. — the Tayloe Mixer) is adapted from the July 2002 QEX article, "Software-Defined Radio For the Masses, Part 1" by Gerald Youngblood AC5OG. — Ed.]

The beauty of the Tayloe detector (see references by Tayloe) is found in both its design elegance and its exceptional performance. In its simplest form, you can build a complete quadrature downconverter with only three or four ICs (less the local oscillator) at a cost of less than $10.

Fig 10.35 illustrates a single-balanced version of the Tayloe detector. It can be visualized as a four-position rotary switch revolving at a rate equal to the carrier frequency. The 50-Ω antenna impedance is connected to the rotor and each of the four switch positions is connected to a *sampling capacitor*. Since the switch rotor is turning at exactly the RF carrier frequency, each capacitor will track the carrier's amplitude for exactly one-quarter of the cycle and will then hold its value for the remainder of the cycle. The rotating switch will therefore sample the signal at 0°, 90°, 180° and 270°, respectively.

As shown in **Fig 10.36**, the 50-Ω impedance of the antenna and the sampling capacitors form an R-C low-pass filter during the period when each respective switch is turned on. Therefore, each sample represents the integral or average voltage of the signal during its respective one-quarter cycle. When the switch is off, each sampling capacitor will hold its value until the next revolution. If the RF carrier and the rotating frequency were exactly in phase, the output of each capacitor will be a dc level equal to the average value of the sample.

If we differentially sum outputs of the 0° and 180° sampling capacitors with an op amp (see Fig 10.35), the output would be a dc voltage equal to two times the value of the individually sampled values when the switch rotation frequency equals the carrier frequency. Imagine, 6 dB of noise-free gain! The same would be true for the 90° and 270° capacitors as well. The 0°/180° summation forms the *I* channel and the 90°/270° summation forms the *Q* channel of a quadrature downconversion. (See the **Modulation** chapter for more information on I/Q modulation.)

As we shift the frequency of the carrier away from the sampling frequency, the values of the inverting phases will no longer be dc levels. The output frequency will vary according to the "beat" or difference frequency between the carrier and the switch-rotation frequency to provide an accurate representation of all the signal components converted to baseband.

Fig 10.37 is the schematic for a simple, single-balanced Tayloe detector. It consists

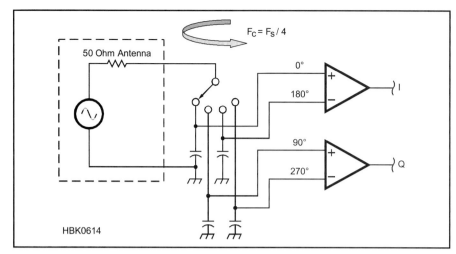

Fig 10.35 — Tayloe detector: The switch rotates at the carrier frequency so that each capacitor samples the signal once each revolution. The 0° and 180° capacitors differentially sum to provide the in-phase (I) signal and the 90° and 270° capacitors sum to provide the quadrature (Q) signal.

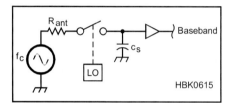

Fig 10.36 — Track-and-hold sampling circuit: Each of the four sampling capacitors in the Tayloe detector form an RC track-and-hold circuit. When the switch is on, the capacitor will charge to the average value of the carrier during its respective one-quarter cycle. During the remaining three-quarters cycle, it will hold its charge. The local-oscillator frequency is equal to the carrier frequency so that the output will be at baseband.

of a PI5V331, 1:4 FET demultiplexer (an analog switch) that switches the signal to each of the four sampling capacitors. The 74AC74 dual flip-flop is connected as a divide-by-four Johnson counter to provide the two-phase clock to the demultiplexer chip. The outputs of the sampling capacitors are differentially summed through the two LT1115 ultra-low-noise op amps to form the *I* and *Q* outputs, respectively.

Note that the impedance of the antenna forms the input resistance for the op-amp gain as shown in Equation 17. This impedance may vary significantly with the actual antenna. In a practical receiver, a buffer amplifier should be used to stabilize and control the impedance presented to the mixer.

Since the duty cycle of each switch is 25%, the effective resistance in the RC network is the antenna impedance multiplied by four in the op-amp gain formula:

$$G = \frac{R_f}{4R_{ant}} \quad (17)$$

For example, with a feedback resistance, R_f, of 3.3 kΩ and antenna impedance, R_{ant}, of 50 Ω, the resulting gain of the input stage is:

$$G = \frac{3300}{4 \times 50} = 16.5$$

The Tayloe detector may also be analyzed as a *digital commutating filter* (see reference by Kossor). This means that it operates as a very-high-Q tracking filter, where Equation 18 determines the bandwidth and n is the number of sampling capacitors, R_{ant} is the antenna impedance and C_s is the value of the individual sampling capacitors. Equation 19 determines the Q_{det} of the filter, where f_c is the center frequency and BW_{det} is the bandwidth of the filter.

$$BW_{det} = \frac{1}{\pi\, n\, R_{ant}\, C_s} \quad (18)$$

$$Q_{det} = \frac{f_c}{BW_{det}} \quad (19)$$

By example, if we assume the sampling capacitor to be 0.27 µF and the antenna impedance to be 50 Ω, then BW and Q at an operating frequency of 14.001 MHz are computed as follows:

$$BW_{det} = \frac{1}{\pi \times 4 \times 50 \times (2.7 \times 10^{-7})} = 5895\ Hz$$

$$Q_{det} = \frac{14.001 \times 10^{-6}}{5895} = 2375 \quad (20)$$

The real payoff in the Tayloe detector is its performance. It has been stated that the

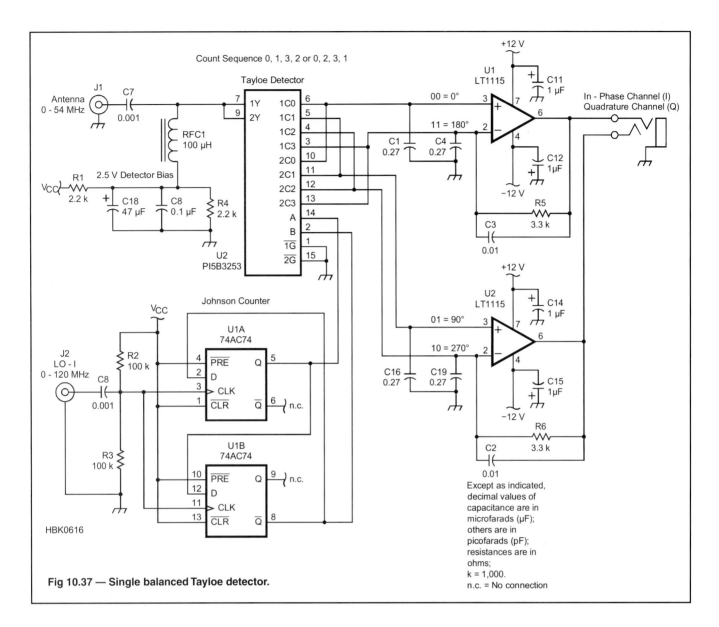

Fig 10.37 — Single balanced Tayloe detector.

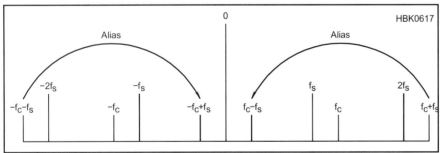

Fig 10.38 — Alias summing on Tayloe detector output: Since the Tayloe detector samples the signal, the sum frequency ($f_c + f_s$) and its image ($-f_c - f_s$) are located at the first alias frequency. The alias signals sum with the baseband signals to eliminate the mixing product loss associated with traditional mixers. In a typical mixer, the sum frequency energy is lost through filtering thereby increasing the noise figure of the device.

ideal commutating mixer has a minimum conversion loss (which equates to noise figure — see the **RF Techniques** chapter) of 3.9 dB. Typical high-level diode mixers have a conversion loss of 6-7 dB and noise figures 1 dB higher than the loss. The Tayloe detector has less than 1 dB of conversion loss, remarkably. How can this be? The reason is that it is not really a mixer but a sampling detector in the form of a quadrature track-and-hold circuit. This means that the design adheres to discrete-time sampling theory, which, while similar to mixing, has its own unique characteristics. Because a track and hold actually holds the signal value between samples, the signal output never goes to zero. (See the **DSP and Software Radio Design** chapter for more on sampling theory.)

This is where aliasing can actually be used to our benefit. Since each switch and capacitor in the Tayloe detector actually samples the RF signal once each cycle, it will respond to alias frequencies as well as those within the Nyquist frequency range. In a traditional direct-conversion receiver, the local-oscillator frequency is set to the carrier frequency so that the difference frequency, or IF, is at 0 Hz and the sum frequency is at two times the carrier frequency. We normally remove the sum frequency through low-pass filter-

Mixers, Modulators and Demodulators 10.29

ing, resulting in conversion loss and a corresponding increase in noise figure. In the Tayloe detector, the sum frequency resides at the first alias frequency as shown in **Fig 10.38**. Remember that an alias is a real signal and will appear in the output as if it were a baseband signal. Therefore, the alias adds to the baseband signal for a theoretically lossless detector. In real life, there is a slight loss, usually less than 1 dB, due to the resistance of the switch and aperture loss due to imperfect switching times.

10.5.6 The NE602/SA602/SA612: A Popular Gilbert Cell Mixer

Introduced as the Philips NE602 in the mid-1980s, the NXP SA602/SA612 mixer-oscillator IC has become greatly popular with amateur experimenters for receive mixers, transmit mixers and balanced modulators. The SA602/612's mixer is a *Gilbert cell* multiplier. **Fig 10.39** shows its equivalent circuit. A Gilbert cell consists of two differential transistor pairs whose bias current is controlled by one of the input signals. The other signal drives the differential pairs' bases, but only after being "predistorted" in a diode circuit. (This circuit distorts the signal equally and oppositely

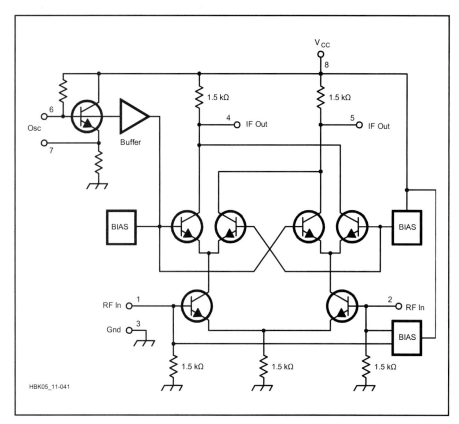

Fig 10.39 — The SA602/612's equivalent circuit reveals its Gilbert-cell heritage.

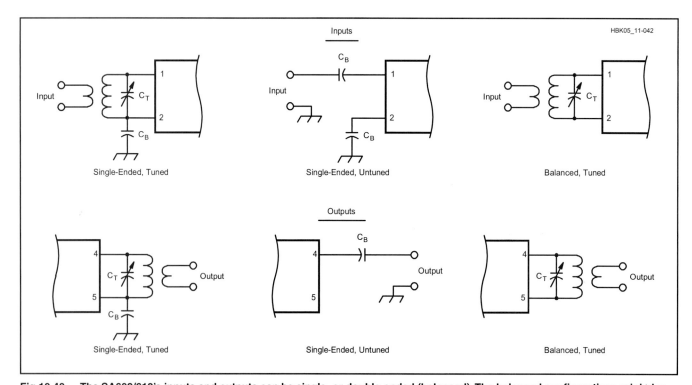

Fig 10.40 — The SA602/612's inputs and outputs can be single- or double-ended (balanced). The balanced configurations minimize second-order IMD and harmonic distortion, and unwanted envelope detection in direct-conversion service. C_T tunes its inductor to resonance; C_B is a bypass or dc-blocking capacitor. The arrangements pictured don't show all the possible input/output configurations; for instance, a center-tapped broadband transformer can be used to achieve a balanced, untuned input or output.

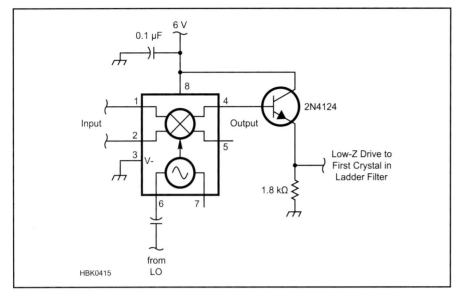

Fig 10.41 — An NPN transistor at the output of an SA602/612 mixer provides power gain and low-impedance drive for a 4.914-MHz crystal filter. A low-reactance coupling capacitor can be added between the emitter and the circuitry it drives if dc blocking is necessary. *(Circuit from the Elecraft KX1 transceiver courtesy of Wayne Burdick, N6KR)*

to the inherent distortion of the differential pair.) The resulting output signal is an accurate multiplication of the input voltages.

SA602/612A VARIANTS

The SA602/612 began life as the NE602/SA602. SA-prefixed 602/612 parts are specified for use over a wider temperature range than their NE-prefixed equivalents. Parts without the A suffix have a slightly lower IP_3 specification than their A counterparts. The pinout-identical NE612A and SA612A cost less than their 602 equivalents as a result of wider tolerances. All variants of this popular part should work satisfactorily in most "NE602" experimenter projects. The same mixer/oscillator topology, modified for slightly higher dynamic range at the expense of somewhat less mixer gain, is also available in the mixer/oscillator/FM IF chips NE/SA605 (input IP_3 typically –10 dBm) and NE/SA615 (input IP_3 typically –13 dBm).

SA602/612 USAGE NOTES

The SA602/612's typical current drain is 2.4 mA; its supply voltage range is 4.5 to 8.0 V. Its inputs (RF) and outputs (IF) can be single- or double-ended (balanced) according to design requirements (**Fig 10.40**). The equivalent ac impedance of each input is approximately 1.5 kΩ in parallel with 3 pF; each output's resistance is 1.5 kΩ. **Fig 10.41** shows the use of an NPN transistor at the SA602/612 output to obtain low-impedance drive for a crystal filter; **Fig 10.42** shows how AGC can be applied to an SA602/612.

The SA602/612 mixer can typically handle signals up to 500 MHz. At 45 MHz, its noise figure is typically 5.0 dB; its typical conversion gain, 18 dB. Note that in contrast to the diode-based mixers described earlier, which have conversion *loss*, most Gilbert-cell mixers have conversion *gain*. Considering the SA602/612's low current drain, its input IP_3 (measured at 45 MHz with 60-kHz spacing) is usefully good at –15 dBm. Factoring in the mixer's conversion gain results in an equivalent output IP_3 of about 3 dBm.

The SA602/612's on-board oscillator can operate up to 200 MHz in LC and crystal-controlled configurations (**Fig 10.43** shows three possibilities). Alternatively, energy from an external LO can be applied to the chip's pin 6 via a dc blocking capacitor. At least 200 mV P-P of external LO drive is required for proper mixer operation.

The SA602/612 was intended to be used as the second mixer in double-conversion FM cellular radios, in which the first IF is typically 45 MHz, and the second IF is typically 455 kHz. Such a receiver's second mixer can be relatively weak in terms of dynamic range because of the adjacent-signal protection afforded by the high selectivity of the first-IF filter preceding it. When

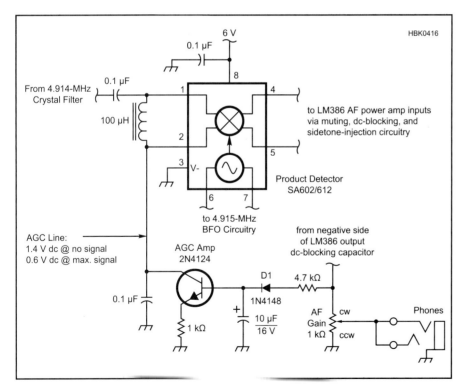

Fig 10.42 — SA602/612 product detector AGC from the Elecraft KX1 transceiver. Designed by Wayne Burdick, N6KR, this circuit first appeared in the Wilderness Radio SST transceiver with an LED used at D1 for simultaneous signal indication and rectification. The selectivity provided by the crystal filter preceding the detector works to mitigate the effects of increasing detector distortion with gain reduction. *(Circuit courtesy of Wayne Burdick, N6KR, and Bob Dyer, K6KK)*

Mixers, Modulators and Demodulators 10.31

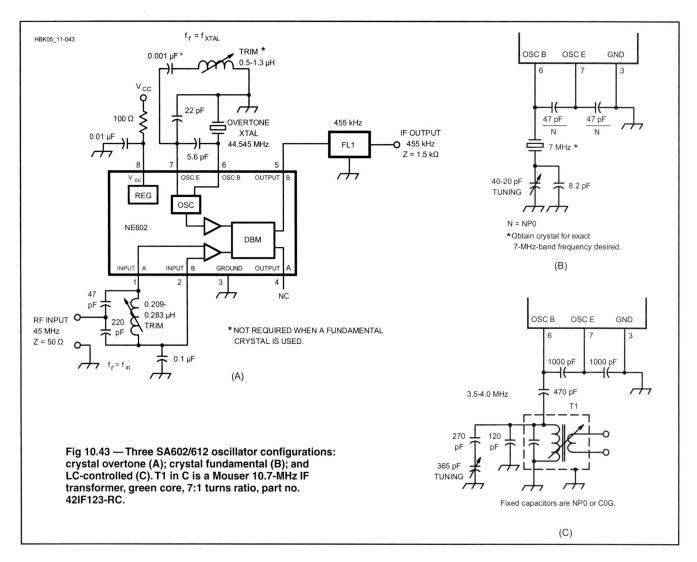

Fig 10.43 — Three SA602/612 oscillator configurations: crystal overtone (A); crystal fundamental (B); and LC-controlled (C). T1 in C is a Mouser 10.7-MHz IF transformer, green core, 7:1 turns ratio, part no. 42IF123-RC.

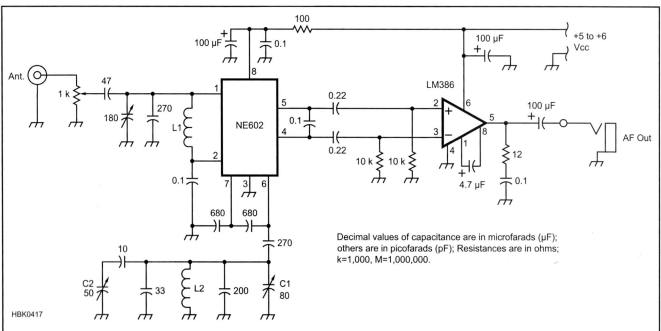

Fig 10.44 — A 7-MHz direct-conversion receiver based on the NE602/SA602/612. Equipped with a stage or two of audio filtering and a means of muting during transmit periods, such a receiver is entirely sufficient for basic Amateur Radio communication at MF and HF. L1 and L2 are 1.2 µH. This receiver is described in greater detail in *Experimental Methods in RF Design*.

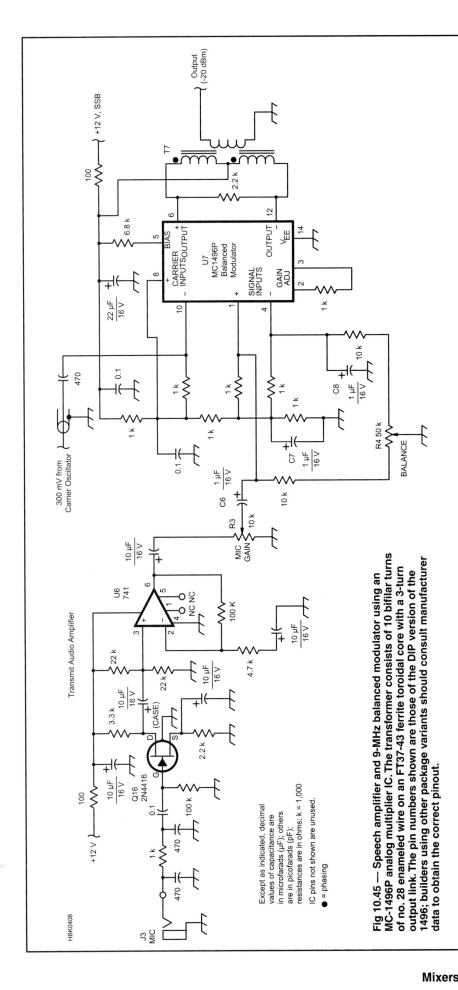

Fig 10.45 — Speech amplifier and 9-MHz balanced modulator using an MC-1496P analog multiplier IC. The transformer consists of 10 bifilar turns of no. 28 enameled wire on an FT37-43 ferrite toroidal core with a 3-turn output link. The pin numbers shown are those of the DIP version of the 1496; builders using other package variants should consult manufacturer data to obtain the correct pinout.

used as a first mixer, the SA602/612 can provide a two-tone third-order dynamic range between 80 and 90 dB, but this figure is greatly diminished if a preamplifier is used ahead of the SA602/612 to improve the system's noise figure.

When the SA602/612 is used as a second mixer, the sum of the gains preceding it should not exceed about 10 dB. An SA602/612 can serve as low-distortion (THD <1%) product detector if overload is avoided through the use of AGC and appropriate attenuation between the '602/612 and the IF strip that drives it.

The SA602/612 is generally not a good choice for VHF and higher-frequency mixers because of its input noise and diminishing IMD performance at high frequencies. There are applications, however, where 6-dB noise figure and 60- to 70-dB dynamic range performance is adequate. If your target specifications exceed these numbers, you should consider other mixers at VHF and up.

Fig 10.44 shows the schematic of a complete 7-MHz direct-conversion receiver based on the SA602/612 and the widely used LM386 AF power amplifier IC. Such simple product-detector-based receivers sometimes suffer from incidental envelope detection, which causes audio from strong, full-carrier-AM shortwave or mediumwave broadcast stations to be audible regardless of where the receiver LO is tuned. RF attenuation and/or band-limiting the receiver input with a double- or triple-tuned-circuit filter can usually reduce this effect to inaudibility.

10.5.7 An MC1496P Balanced Modulator

Although it predates the SA602/612, Freescale's MC1496 Gilbert cell multiplier remains a viable option for product detection and balanced modulator service. **Fig 10.45** shows an MC1496-based balanced modulator that is capable of a carrier suppression greater than 50 dB. Per its description in Hayward, Campbell, and Larkin's *Experimental Methods in RF Design*, its output with audio drive should be kept to about –20 dBm with this circuit. LO drive should be 200 to 500 mV P-P.

10.5.8 An Experimental High-Performance Mixer

Examining the state of the art we find that the best receive-mixer dynamic ranges are achieved with quads of RF MOSFETs operating as passive switches, with no drain voltage applied. The best of these techniques involves following a receiver's first mixer with a diplexer and low-loss roofing crystal filter, rather than terminating the mixer in a strong wideband amplifier.

Colin Horrabin's (G3SBI) experimenta-

Mixers, Modulators and Demodulators 10.33

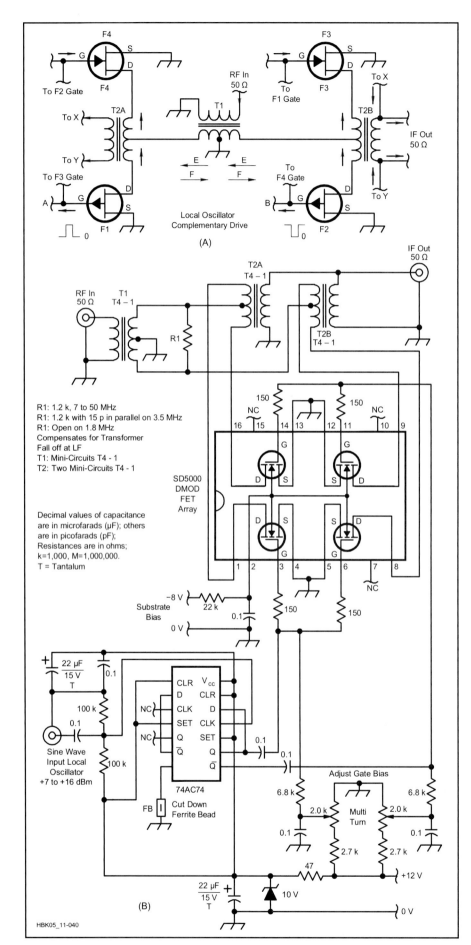

Fig 10.46 — A shows the operation of the H-mode switching mixer developed by Colin Horrabin, G3SBI. B shows the actual mixer circuit implemented with the Linear Integrated Systems SD5000 DMOS FET quad switch IC and a 74AC74 flip-flop.

tions with variations of an original high-performance mixer circuit by Jacob Mahkinson, N6NWP, led to the development of a new mixer configuration, called an H-mode mixer. This name comes from the signal path through the circuit. (See **Fig 10.46A**.) Horrabin is a professional scientist/engineer at the Science and Engineering Research Council's Darebury Laboratory, which has supported his investigative work on the H-mode switched-FET mixer, and consequently holds intellectual title to the new mixer. This does not prevent readers from taking the development further or using the information presented here.

Inputs A and B are complementary square-wave inputs derived from the sine-wave local oscillator at twice the required square-wave frequency. If A is ON, then FETs F1 and F3 are ON and F2 and F4 are OFF. The direction of the RF signal across T1 is given by the E arrows. When B is ON, FETs F2 and F4 are ON and F1 and F3 are OFF. The direction of the RF signal across T1 reverses, as shown by the F arrows.

This is still the action of a switching mixer, but now the source terminal of each FET switch is grounded, so that the RF signal switched by the FET cannot modulate the gate voltage. In this configuration the transformers are important: T1 is a Mini-Circuits type T4-1 and T2 is a pair of these same transformers with their primaries connected in parallel.

10.6 References and Bibliography

The following mixer references include books, journals and other publications as well as websites. They are arranged by publication date.

BOOKS

E. E. Bucher, *The Wireless Experimenter's Manual* (New York: Wireless Press, Inc., 1920).

F. E. Terman, *Radio Engineers' Handbook* (New York: McGraw-Hill, 1943).

R. W. Landee, D. C. Davis and A. P. Albrecht, *Electronic Designer's Handbook*, 2nd ed, Section 22 (New York: McGraw-Hill, 1957).

A. A. M. Saleh, *Theory of Resistive Mixers* (Cambridge, MA: MIT Press, 1971).

L.G. Giacoletto, *Electronics Designers' Handbook*, 2nd ed (New York: McGraw-Hill 1977).

P. Horowitz, W. Hill, *The Art of Electronics*, 2nd ed (New York: Cambridge University Press, 1989).

RF/IF Designer's Handbook (Brooklyn, NY: Scientific Components, 1992).

RF Communications Handbook (Philips Components-Signetics, 1992).

W. Hayward, *Introduction to Radio Frequency Design* (Newington, CT: ARRL, 1994).

L. E. Larson, *RF and Microwave Circuit Design for Wireless Communication*, (Artech House, 1996).

S. A. Maas, The RF and Microwave Circuit Design Cookbook (Artech House, 1998).

U. L Rohde, D. P. Newkirk, *RF/Microwave Circuit Design for Wireless Applications* (New York: John Wiley & Sons, 2000). (Large number of useful references in the mixer chapter.)

U. L. Rohde, J. Whitaker, *Communications Receivers: DSP, Software Radios, and Design*, 3rd ed (New York: McGraw-Hill, 2001. (Large number of useful references in the mixer chapter.)

H. I. Fujishiro, Y. Ogawa, T. Hamada and T. Kimura, *Electronic Letters*, Vol 37, Iss 7, Mar 2001, pp 435-436. See IET Digital Library (www.ietdl.org).

S. A. Maas, *Nonlinear Microwave and RF Circuits*, 2nd ed (Artech House, 2003).

S. A. Maas, Noise in Linear and Nonlinear Circuits (Artech House, 2005).

G. D. Vendelin, A. M. Pavio and U. L. Rohde, *Microwave Circuit Design Using Linear and Nonlinear Techniques* (New York: John Wiley & Sons, 2005). (Large number of useful references in the mixer chapter.)

Synergy Microwave Corporation Product Handbook (Paterson, NJ: Synergy Microwave Corporation, 2007).

JOURNALS AND OTHER PUBLICATIONS

R. Weinreich and R. W. Carroll, "Absorptive Filter for TV Harmonics," *QST*, Nov 1968, pp 20-25.

B. Gilbert, "A Precise Four-Quadrant Multiplier with Sub Nanosecond Response," *IEEE Journ Solid-State Circuits*, Vol 3, 1968, pp 364-373.

W. Sabin, "The Solid-State Receiver," *QST*, Jul 1970, pp 35-43.

G. Begemann, A. Hecht, "The Conversion Gain and Stability of MESFET Gate Mixers" 9th EMC, 1979, pp 316-319.

P. Will, "Broadband Double Balanced Mixer Having Improved Termination Insensitivity Characteristics", Adams-Russell Co, Waltham, MA, US Patent No 4,224,572, Sep 23, 1980.

C. Tsironis, R. Meierer, R. Stahlmann, "Dual-Gate MESFET Mixers," *IEEE Trans MTT*, Vol 32, No 3, 04-1984.

B. Henderson and J. Cook, "Image-Reject and Single-Sideband Mixers", Watkins-Johnson Tech Note, Vol 12 No 3, 1985.

H. Statz, P. Newman, I. Smith, R. Pucel and H. Haus, *IEEE Trans Electron Devices* 34 (1987) 160.

J. Dillon, "The Neophyte Receiver," *QST*, Feb 1988, pp 14-18.

B. Henderson, "Mixers in Microwave Systems (Part 1)", Watkins-Johnson Tech Note, Vol 17 No 1, 1990, Watkins-Johnson.

B. Henderson, "Mixers in Microwave Systems (Part 2)", Watkins-Johnson Tech Note, Vol 17 No 2, 1990, Watkins-Johnson.

R. Zavrel, "Using the NE602," Technical Correspondence, *QST*, May 1990, pp 38-39.

D. Kazdan "What's a Mixer?" *QST*, Aug 1992, pp 39-42.

L. Richey, "W1AW at the Flick of a Switch," *QST*, Feb 1993, pp 56-57.

J. Makhinson, "High Dynamic Range MF/HF Receiver Front End," *QST*, Feb 1993, pp 23-28. Also see Feedback, *QST*, Jun 1993, p 73.

J. Vermusvuori, "A Synchronous Detector for AM Transmisions," *QST*, pp 28-33. Uses SA602/612 mixers as phase-locked and quasi-synchronous product detectors.

P. Hawker, Ed., "Super-Linear HF Receiver Front Ends," Technical Topics, *Radio Communication,* Sep 1993, pp 54-56.

S. Joshi, "Taking the Mystery Out of Double-Balanced Mixers," *QST*, Dec 1993, pp 32-36.

U. L. Rohde, "Key Components of Modern Receiver Design," Part 1, *QST*, May 1994, pp 29-32; Part 2, *QST*, Jun 1994, pp 27-31; Part 3, *QST*, Jul 1994, 42-45.

U. L. Rohde, "Testing and Calculating Intermodulation Distortion in Receivers," *QEX,* Jul 1994, pp 3-4.

B. Gilbert, "Demystifying the Mixer," self-published monograph, 1994.

S. Raman, F. Rucky, and G. M. Rebeiz, "A High Performance W-band Uniplanar Subharmonic Mixer," *IEEE Trans MTT*, Vol 45, pp 955-962, Jun 1997.

M. J. Roberts, S. Iezekiel and C. M. Snowdan, "A Compact Subharmonically Pumped MMIC Self Oscillating Mixer for 77 GHz Applications," *IEEE MTT-S Digest*, 1998, pp 1435-1438.

M. Kossor, WA2EBY, "A Digital Commutating Filter," *QEX*, May/Jun 1999, pp 3-8.

K. S. Ang, A. H. Baree, S. Nam and I. D. Robertson, "A Millimeter-Wave Monolithic Sub-Harmonically Pumped Resistive Mixer," *Proc APMC*, Vol 2, 1999, pp 222-225.

C. H. Lee, S. Han, J. Laskar, "GaAs MESFET Dual-Gate Mixer with Active Filter Design for Ku-Band Application," *IEEE MTT-S Digest*, 1999, pp 841-844.

H. Darabi and A. A. Abidi, "Noise in RF-CMOS Mixers: A Simple Physical Model," *IEEE JSSC*, Vol 35, Jan 2000, pp 15-25.

P. S. Tsenes, G. E. Stratakos, N. K. Uzunoglu, M. Lagadas, G. Deligeorgis, "An X-band MMIC Down-Converter," WOCSDICE 2000, pp X.7-X.8

M. T. Terrovitis and R. G. Meyer, "Intermodulation Distortion in Current-Commutating CMOS Mixers," *IEEE Journ of SSC*, Vol 35, No 10, October 2000.

M. W. Chapman and S. Raman, "A 60 GHz Uniplanar MMIC 4X Subharmonic Mixer," *IEEE MTT-S Int Microwave Symp Dig*, 2001, pp 95–98, 2001.

M. Sironen, Y. Qian and T. Itoh, "A Subharmonic Self-Oscillating Mixer with Integrated Antenna for 60 GHz Wireless Applications," *IEEE Trans MTT*, 2001, pp 442–450

D. Tayloe, N7VE, "Letters to the Editor, Notes on 'Ideal' Commutating Mixers (Nov/Dec 1999)," *QEX*, March/April 2001, p 61.

K. Kanaya, K. Kawakami, T. Hisaka, T. Ishikawa and S. Sakmoto, "A 94 GHz High Performance Quadruple Subharmonic Mixer MMIC," *IEEE MTT-S Int Microwave Symp Dig*, 2002, pp 1249–1252.

J. Kim, Y. Kwon, "Intermodulation Analysis of Dual-Gate FET Mixers," *IEEE Trans MTT*, Vol 50, No 6, Jun 2002.

B. Matinpour, N. Lal, J. Laskar, R. E. Leoni and C. S. Whelan, "A Ka Band Subharmonic Down-Converter in a GaAs Metamorphic HEMT Process," *IEEE MTT-S Int Microwave Symp Dig*, 2001, pp 1337–1339.

T. S. Kang, S. D. Lee, B. H. Lee, S. D. Kim, H. C. Park, H. M. Park and J. K. Rhee, *Journ Korean Phys Soc 41*, 2002, 533.

D. An, B. H. Lee, Y. S. Chae, H. C. Park, H. M. Park and J. K. Rhee, *Journ Korean Phys Soc 41*, 2002, 1013.

D. Metzger, "Build the 'No Excuses' Transceiver," *QST*, Dec 2002, pp 28–34.

W. S. Sul, S. D. Kim, H. M. Park and J. K. Rhee, *Jpn Journ Appl Phys*, Vol 42, 2003, pp 7189-7193 (see **jjap.ipap.jp/link?JJAP/42/7189/**).

H. Morkner, S. Kumar and M. Vice, "An 18-45 GHz Double-Balanced Mixer with Integrated LO Amplifier and Unique Suspended Broadside-Coupled Balun," *2003 Gallium Arsenide Integrated Circuit Symp*, Nov 2003, pp 267-270.

D. Tayloe, N7VE, "A Low-noise, High-performance Zero IF Quadrature Detector/Preamplifier", March 2003, *RF Design*.

D. Tayloe, N7VE, US Patent 6,230,000, "A Product Detector and Method Therefore".

M.-F, Lei, P.-S, Wu, T.-W. Huang, H. Wang, "Design and Analysis of a Miniature W-band MMIC Sub Harmonically Pumped Resistive Mixer," *2004 IEEE MTT-S Int Microwave Symp*, Vol 1, Jun 2004, pp 235-238.

Ping-Chun Yeh, Wei-Cheng Liu and Hwann-Kaeo Chiou, "Compact 28-GHz Subharmonically Pumped Resistive Mixer MMIC Using a Lumped-Element High-Pass/Band-Pass Balun," *IEEE Microwave and Wireless Component Letters*, Vol 15, No 2, Feb 2005.

J. Browne, "Wideband Mixers Hit High Intercept Points," *Microwave & RF Journal*, Sep 2005, pp 98-104.

Chien-Chang H. and Wei-Ting C., "Multi-Band/Multi-Mode Current Folded Up-Convert Mixer Design," *Proc APMC 2006*, Japan.

X. Wang, M. Chen and O. B. Lubecke, "0.25 um CMOS Resistive Sub threshold Mixer," *Proc APMC 2006*, Japan.

Ro-Min Weng, Jing-Chyi Wang and Hung-Che Wei, "A 1V 2.4 GHz Down Conversion Folded Mixer", *IEEE APCCAS*, Dec 4-7, 2006, Singapore, pp 1476-1478.

B. M. Motlagh, S. E. Gunnarsson, M. Ferndahl and H. Zirath, "Fully Integrated 60-GHz Single-Ended Resistive Mixer in 90-nm CMOS Technology," *IEEE Microwave and Wireless Component Letters*, Vol 16, No 1, Jan 2006.

F. Ellinger, "26.5-30-GHz Resistive Mixer in 90-nm VLSI SOI CMOS Technology With High Linearity for WLAN" *IEEE Trans MTT*, Vol 53, No 8, Aug 2005.

U. L. Rohde and A. K. Poddar, "High Intercept Point Broadband, Cost Effective and Power Efficient Passive Reflection FET DBM," *EuMIC Symposium*, 10-15 Sep 2006, UK.

U. L. Rohde and A. K. Poddar, "A Unified Method of Designing Ultra-Wideband, Power-Efficient, and High IP3 Reconfigurable Passive FET Mixers," *IEEE/ICUWB*, Sep 24-27, 2006, MA, USA.

U. L. Rohde and A. K. Poddar, "Low Cost, Power-Efficient Reconfigurable Passive FET Mixers", 20th *IEEE CCECE* 2007, 22-26 Apr 2007, British Columbia, Canada.

INTERESTING WEB SITES FOR MIXERS

www.analog.com/en/content/0,2886,770%255F848%255F67465,00.html

www.maximic.com/solutions/base_stations/index.mvp?CMP=3758&gclid=CIzn1J_lm40CFSgRGgodD2Go6g

www.pulsarmicrowave.com/products/mixers/mixers.htm

www.synergymwave.com/products/mixers/?LocTitle=Mixers

news.thomasnet.com/fullstory/507272

www.wj.com/products/HighInterceptFETMixers.aspx

Contents

11.1 Introduction
11.2 Filter Basics
    11.2.1 Filter Magnitude Response
    11.2.2 Filter Order
    11.2.3 Filter Families
    11.2.4 Group Delay
    11.2.5 Transient Response
    11.2.6 Filter Family Summary
11.3 Lumped-Element Filters
    11.3.1 Low-Pass Filters
    11.3.2 High-Pass Filters
    11.3.3 Low-Pass to Band-Pass Transformation
    11.3.4 High-Pass to Band-Stop Transformation
    11.3.5 Refinements in Band-Pass Design
    11.3.6 Effect of Component Q
    11.3.7 Side Effects of Passband Ripple
11.4 Filter Design Examples
    11.4.1 Normalized Values
    11.4.2 Filter Family Selection
11.5 Active Audio Filters
    11.5.1 SCAF Filters
    11.5.2 Active RC Filters
    11.5.3 Active Filter Responses
    11.5.4 Using Active Filter Design Tools

11.6 Quartz Crystal Filters
    11.6.1 Crystal Filter Parameters
    11.6.2 Crystal Filter Design
    11.6.3 Crystal Characterization
    11..6.4 Crystal Filter Evaluation
11.7 SAW Filters
11.8 Transmission Line Filters
    11.8.1 Stripline and Microstrip Filters
    11.8.2 Transmission Line Band-Pass Filters
    11.8.3 Quarter-Wave Transmission Line Filters
    11.8.4 Emulating LC Filters with Transmission Line Filters
11.9 Helical Resonators
    11.9.1 Helical Resonator Design
    11.9.2 Helical Filter Construction
    11.9.3 Helical Resonator Tuning
    11.9.4 Helical Resonator Insertion Loss
    11.9.5 Coupling Helical Resonators
11.10 Use of Filters at VHF and UHF
11.11 Filter Projects
    11.11.1 Crystal Ladder Filter for SSB
    11.11.2 Broadcast-Band Rejection Filter
    11.11.3 Wave Trap for Broadcast Stations
    11.11.4 Optimized Harmonic Transmitting Filters
    11.11.5 Diplexer Filter
    11.11.6 High-Performance, Low-Cost 1.8 to 54 MHz Low-Pass Filter
11.12 Filter Glossary
11.13 References and Bibliography

Chapter 11

RF and AF Filters

This chapter discusses the most common types of filters used by radio amateurs. The sections describing basic concepts, lumped element filters and some design examples were initially prepared by Jim Tonne, W4ENE, and updated by Ward Silver, NØAX. Material on active filters was provided by Dan Tayloe, N7VE, and the section on crystal filters was developed by Dave Gordon-Smith, G3UUR.

11.1 Introduction

Electrical filters are circuits used to process signals based on their frequency. For example, most filters are used to pass signals of certain frequencies and reject others. The electronics industry has advanced to its current level in large part because of the successful use of filters. Filters are used in receivers so that the listener can hear only the desired signal; other signals are rejected. Filters are used in transmitters to pass only one signal and reject those that might interfere with other spectrum users. **Table 11.1** shows the usual signal bandwidths for several signal types.

The simplified SSB receiver shown in **Fig 11.1** illustrates the use of several common types of filters. A *preselector* filter is placed between the antenna and the first mixer to pass all frequencies within a given amateur band with low loss. Strong out-of-band signals from broadcast, commercial or military stations are rejected to prevent them from overloading the first mixer. The preselector filter is almost always built with *lumped-element* or "LC" technology.

An intermediate frequency (IF) filter is placed between the first and second mixers to pass only the desired signal. Finally, an audio filter is placed somewhere between the detector and the speaker rejects unwanted products of detection, power supply hum and noise. The audio

Table 11.1
Typical Filter Bandwidths for Typical Signals

Source	Required Bandwidth
Fast-scan analog television (ATV)	4.5 MHz
Broadcast-quality speech and music	15 kHz (from 20 Hz to 15 kHz)
Communications-quality speech	3 kHz (from 300 Hz to 3 kHz)
Slow-scan television (SSTV)	3 kHz (from 300 Hz to 3 kHz)
HF RTTY	250 to 1000 Hz (depends on frequency shift)
Radiotelegraphy (Morse code, CW)	200 to 500 Hz
PSK31 digital modulation	100 Hz

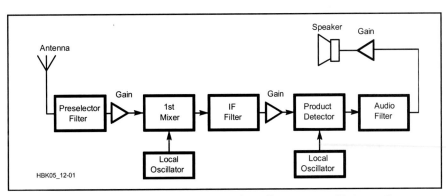

Fig 11.1 — Single-band SSB receiver. At least three filters are used between the antenna and speaker.

filter is often implemented with *active filter* technology.

The complementary SSB transmitter block diagram is shown in **Fig 11.2**. A similar array of filters appears in reverse order. First, an audio filter between the microphone and the balanced mixer rejects out-of-band audio signals such as 60 Hz hum. Since the balanced mixer generates both lower and upper sidebands, an IF filter is placed at the mixer output to pass only the desired lower (or upper) sideband. Finally, a filter at the output of the transmit mixer passes only signals within the amateur band in use rejects unwanted frequencies generated by the mixer to prevent them from being amplified and transmitted. Filters at the transmitter output attenuate the harmonics of the transmitted signals.

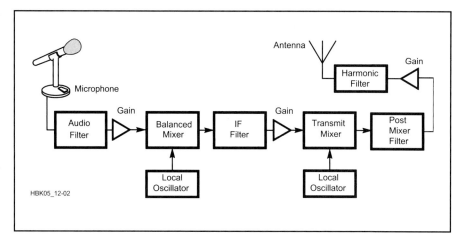

Fig 11.2 — Single-band SSB transmitter. At least three filters are needed to ensure a clean transmitted signal.

11.2 Filter Basics

A common type of filter, the *band-pass*, passes signals in a range of frequencies — the *passband* — while rejecting signals outside that range — the *stop band*. To pass signals from dc up to some *cutoff frequency* at which output power is halved (reduced by 3 dB) we would use a *low-pass* filter. To pass signals above a cutoff frequency (also called the "3-dB frequency") we would use a *high-pass* filter. Similarly, to pass signals within a range of frequencies we would use a *band-pass* filter. To pass signals at all frequencies *except* those within a specified range requires a *band-stop* or *notch filter*. The region between the passband and the stop band is logically called the *transition region*. The attenuation for signal frequencies in the stop band far from the cutoff frequency is called the *ultimate attenuation*.

Many of the basic terms used to describe filters are defined in the chapters on **Electrical Fundamentals** and **Analog Basics**. In addition, a glossary is provided at the end of this chapter.

11.2.1 Filter Magnitude Responses

Fig 11.3 illustrates the *magnitude response* of a low-pass filter. Signals lower than the cutoff frequency (3 MHz in this case) are passed with some small amount of attenuation while signals higher than that frequency are attenuated. The degree of attenuation is dependent on several variables, filter complexity being a major factor.

Of the graphs in this chapter that show a filter's magnitude response, the vertical axes are labeled "Transmission (dB)" with 0 dB at the top of the axis and negative values increasing towards the x-axis. Increasingly negative values of transmission in dB are the same as

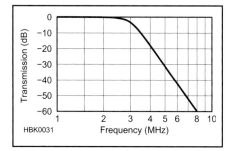

Fig 11.3 — Example of a low-pass magnitude response.

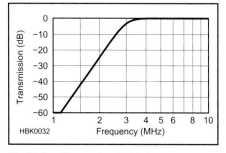

Fig 11.4 — Example of a high-pass magnitude response.

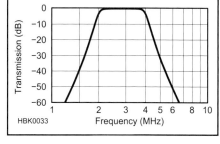

Fig 11.5 — Example of a band-pass magnitude response.

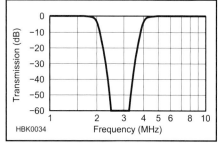

Fig 11.6 — Example of a band-stop magnitude response.

increasingly positive values of attenuation in dB. For example, –40 dB transmission is the same as 40 dB of attenuation.

Fig 11.4 illustrates the magnitude response of a high-pass filter. Signals above the 3 MHz cutoff frequency are passed with minimum attenuation while signals below that frequency are attenuated. Again, the degree of attenuation is dependent on several variables.

Fig 11.5 illustrates the magnitude response of a band-pass filter. Signals within the band-pass range (between the lower and upper cutoff frequencies) are passed with minimum attenuation while signals outside that range are attenuated. In this example the filter was designed with cutoff frequencies of 2 MHz and 4 MHz, for a passband width of 2 MHz.

Fig 11.6 illustrates the magnitude response of a band-stop filter. Signals within the band-stop range are attenuated while all other signals are passed with minimum attenuation. A notch filter is a type of band-stop filter with a narrow stop-band in which the attenuation

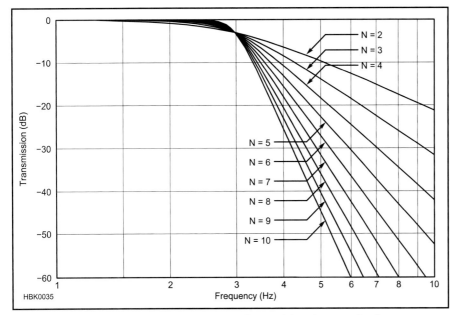

Fig 11.7 — Magnitude responses for low-pass filters with orders ranging from very simple (N=2) to very complex (N=10).

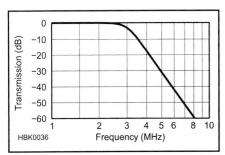

Fig 11.8 — Filters from the Butterworth family exhibit flat magnitude response in the passband.

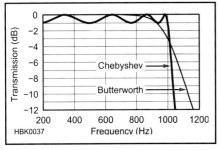

Fig 11.9 — This plot compares the response of a Chebyshev filter with a 1-dB ripple bandwidth of 1000 Hz and a Butterworth filter 3-dB bandwidth of 1000 Hz.

is a maximum at a single frequency.

An *ideal filter* — a low-pass filter, for example — would pass all frequencies up to some point with no attenuation at all and totally reject everything beyond that point. This is known as a *brick wall* response because the filter's passband and stop band are *flat*, meaning no attenuation, and the rolloff in the transition region is infinitely steep. The magnitude response of such a filter would be drawn as a flat line representing 0 dB attenuation up to the cutoff frequency that then abruptly changes at the cutoff frequency to a flat line at infinite attenuation throughout the stop band to infinite frequency.

11.2.2 Filter Order

The steepness of the descent from the passband to the region of attenuation — the stop band — is dependent on the complexity of the filter, properly called the *order*. In a lumped element filter made from inductors and capacitors, the order is determined by the number of separate energy-storing elements (either L or C) in the filter. For example, an RC filter with one resistor and one capacitor has an order of 1. A notch filter made of a single series-LC circuit has an order of 2. The definition of order for active and digital signal processing filters can be more complex, but the general understanding remains valid that the order of a filter determines how rapidly its response can change with frequency.

For example, **Fig 11.7** shows the magnitude response of one type of low-pass filter with orders varying from very simple ("N=2") to the more complex ("N=10"). In each case on this plot the 3-dB point (the filter's cutoff or corner frequency) remained the same, at 3 Hz.

As frequency increases through the transition region from the passband into the stop band, the attenuation increases. *Rolloff*, the rate of change of the attenuation for frequencies above cutoff, is expressed in terms of dB of change per octave of frequency. Rolloff for this type of filter design is always 6.02 times the order, regardless of the circuit used to fabricate the circuit. As the order goes up, the rate of change also goes up. A filter's response with large values of rolloff is called "sharp" or "steep" and "soft" or "shallow" if rolloff is low.

11.2.3 Filter Families

There is no single "best" way to develop the parts values for a filter. Instead we have to decide on some traits and then choose the most appropriate *family* for our design. Different families have different traits, and filter families are commonly named after the mathematician or engineer responsible for defining their behavior mathematically.

Real filters respond more gradually and with greater variation than ideal filters. Two primary traits of the most importance to amateurs are used to describe the behavior of filter families: *ripple* (variations in the magnitude response within the passband and stop band) and rolloff. Different families of filters have different degrees of ripple and rolloff (and other characteristics, such as phase response).

With a flat magnitude response (and so no ripples in the passband) we have what is known as a *Butterworth* family design. This family also goes by the name of *Maximally Flat Magnitude* or *Maximally Flat Gain*. An example of the magnitude response of a filter from the Butterworth family is shown in both **Fig 11.7** and **Fig 11.8**. (Most filter discussions are based on low-pass filters because the same concepts are easily extended to other types of filters and much of the mathematics behind filters is equivalent for the various types of frequency responses.)

By allowing magnitude response ripples in the passband, we can get a somewhat steeper rolloff from the passband into the stop band, particularly just beyond the cutoff frequency. A family that does this is the *Chebyshev* family. **Fig 11.9** illustrates how allowing magnitude ripple in the passband provides a sharper filter rolloff. This plot compares the 1-dB ripple Chebyshev with the no-ripple Butterworth filter down to 12 dB of attenuation.

Fig 11.9 also illustrates the usual definition of bandwidth for a low-pass Butterworth filter — the filter's 3-dB or cutoff frequency. For a low-pass Chebyshev filter, *ripple bandwidth* is used — the frequency range over which the filter's passband ripple is no greater than the specified limit. For example, the ripple bandwidth of a low-pass Chebyshev filter designed to have 1 dB of passband ripple is the highest frequency at which attenuation is 1 dB or less. The 3-dB bandwidth of the frequency of the filter will be somewhat greater.

A Butterworth filter is defined by specify-

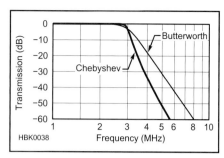

Fig 11.10 — A Chebyshev filter (0.2 dB passband ripple) allows a sharper cutoff than a Butterworth design with no passband ripple.

ing the order and bandwidth. The Chebyshev filter is defined by specifying the order, the ripple bandwidth, and the amount of passband ripple. In Fig 11.9, the Chebyshev filter has 1 dB of ripple; its ripple bandwidth is 1000 Hz. The Butterworth filter has a 3-dB bandwidth also of 1000 Hz. Some filter textbooks use the 3-dB point to define Chebyshev filters; most use the ripple bandwidth as illustrated here. The schematics (if you ignore parts values) of those two families are identical.

Even small amounts of ripple can be beneficial in terms of increasing a filter's rolloff. **Fig 11.10** compares a Butterworth filter (with the narrow line plot, no ripple in the passband) with a Chebyshev filter (wide line plot, 0.2 dB of ripple in the passband) down to 60 dB of attenuation. (For this comparison the cutoff frequencies at 3 dB of attenuation are the same for each filter.) Even that small amount of ripple in the Chebyshev filter passband allows a noticeably steeper rolloff between the passband and the stop band. As the passband ripple specification is increased, the steepness of the transition from passband to stop band increases, compared to the Butterworth family although the rolloff of the two families eventually becomes equal.

For Chebyshev filters, when the value of the passband ripple is changed, the magnitude response in the stop band region also changes. **Fig 11.11** compares the stop band response of Chebyshev filters with passband ripple ranging from 0.01 to 1 dB.

It is possible to obtain even steeper rolloff into the stop band by adding "traps" whose frequencies are carefully calculated. The resonant frequencies of those traps are in the stop band region and are set to yield best performance. When this is done, we have a *Cauer* family design (also called the *elliptic-function* design). The rolloff of the Cauer filter from the passband into the stop band is the steepest of all analog filter types provided that the behavior in the passband is uniform (either no ripple or a uniform amount of ripple). **Fig 11.12** shows the response of the Chebyshev and Cauer designs for comparison.

The Cauer filter is defined by specifying the order, the ripple bandwidth, and the passband ripple, just as for the Chebyshev. Again, an alternative bandwidth definition is to use the 3-dB frequency instead of the ripple bandwidth. The Cauer family requires one more specification: the *stop band frequency* and/or the *stop band depth*. The stop band frequency is the lowest frequency of a null or

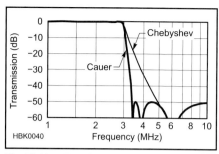

Fig 11.12 — The Cauer family has an even steeper rolloff from the passband into the stop band than the Chebyshev family. Note that the ultimate attenuation in the stop band is a design parameter rather than ever-increasing as is the case with other families.

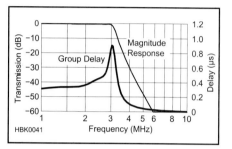

Fig 11.13 — Magnitude response and group delay of a Chebyshev low-pass filter.

notch in the stop band. The stop band depth is the minimum amount of attenuation allowed in the stop band. In Fig 11.12, the stop band frequency is about 3.5 MHz and the stop band depth is 50 dB. For this comparison both designs have the same passband ripple value of 0.2 dB.

A downside to the Cauer filter is that the ultimate attenuation is some chosen value rather than ever increasing, as is the case with the other families. In the far stop band region (where frequency is much greater than the cutoff frequency) the rolloff of Cauer filters ultimately reaches 6 or 12 dB per octave, depending on the order. Odd-ordered Cauer filters have an ultimate rolloff rate of about 6 dB per octave while the even-ordered versions have an ultimate rolloff rate of about 12 dB per octave.

11.2.4 Group Delay

Another trait that can influence the choice of which filter family to use is the way the transit time of signals through the filter varies with frequency. This is known as *group delay*. ("Group" refers to a group of waves of similar frequency and phase moving through a media, in this case, the filter.) The wide line (lower plot) in **Fig 11.13** illustrates the group delay characteristics of a Chebyshev low-pass filter, while the upper plot (nar-

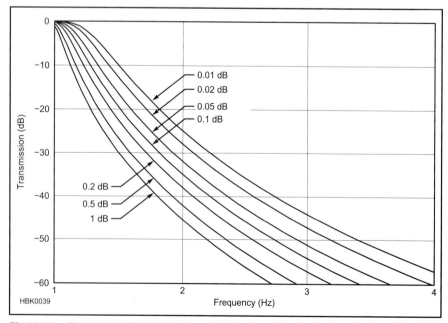

Fig 11.11 — These stop band response plots illustrate the Chebyshev family with various values of passband ripple. These plots are for a seventh-order low-pass design with ripple values from 0.01 to 1 dB. Ignoring the effects in the passband of high ripple values, increasing the ripple will allow somewhat steeper rolloff into the stop band area, and better ultimate attenuation in the stop band.

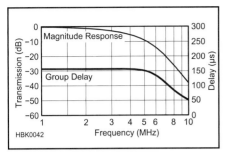

Fig 11.14 — Magnitude response and group delay of a Bessel low-pass filter.

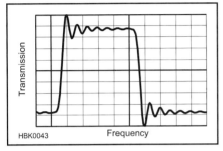

Fig 11.15 — Transient response of a Chebyshev low-pass filter.

row line) shows the magnitude response. As shown in Fig 11.13, the group delay of components near the cutoff frequency becomes quite large when compared to that of components at lower and higher frequencies. This is a result of the phase shift of the filter's transmission being nonlinear with frequency; it is usually greater near the cutoff frequency. By delaying signals at different frequencies different amounts, the signal components are "smeared" in time. This causes distortion of the signal and can seriously disrupt high-speed data signals.

If a uniform group delay for signals throughout the passband is needed, then the *Bessel* filter family should be selected. The Bessel filter can be used as a delay-line or time-delay element although the gentle rolloff of the magnitude response may need to be taken into account. The magnitude and delay characteristics for the Bessel family are shown in **Fig 11.14**. Comparing the Bessel filter's response to that of the Butterworth in Fig 11.8 shows the difference in roll off.

A downside of Bessel family filters is that the rolloff characteristic is quite poor; they are not very good as a magnitude-response-shaping filter. The Bessel family is characterized largely by its constant group delay in the passband (for a low- or high-pass filter) shown as the bottom plot in Fig 11.14. The *constant-delay* characteristic of the Bessel extends into the stop band. The Bessel filter bandwidth is commonly defined by its 3-dB point (as with the Butterworth).

Because the Bessel filter is used when phase response is important, it is often characterized by the frequency at which a specific amount of phase shift occurs, usually one radian. (One radian is equal to $360/2\pi = 57.3°$.) Since the delay causes the output signal to lag behind the input signal, the filter is specified by its *one-radian lag frequency*.

11.2.5 Transient Response

Some applications require that a signal with sharp rising and falling edges (such as a digital data waveform) applied to the input of a low-pass design have a minimum *overshoot* or *ringing* as seen at the filter's output. Overshoot (and undershoot) occurs when a signal exceeds (falls below) the final amplitude temporarily before settling at its final value. Ringing is a repeated sequence of overshoot and undershoot. If a sharp-cutoff filter is used in such an application, overshoot or ringing will occur. The appearance of a signal such as a square wave with sharp rising and falling edges as it exits from the filter may be as shown in **Fig 11.15**. The scales for both the X- and the Y-axes would depend on the frequency and magnitude of the waveform.

The square wave used in this discussion as a test waveform is composed of a fundamental and an endless series of odd harmonics. If harmonics only up to a certain order are used to create the square wave — that is, if the square wave is passed through a sharp-cutoff low-pass filter that attenuates higher frequencies — then that waveform will have the overshoot or ringing as shown as it exits from the filter.

All of the traits mentioned so far in this chapter apply to a filter regardless of how it is implemented, whether it is fabricated using *passive* lumped-element inductors and capacitors or using op amps with resistors and capacitors (an *active* filter). The traits are general descriptions of filter behavior and can be applied to any type of filter technology.

11.2.6 Filter Family Summary

Selecting a filter family is one of the first steps in filter design. To make that choice easier, the following list of filter family attributes is provided:

• Butterworth — No ripple in passband, smooth transition region, shallow rolloff for a given filter order, high ultimate attenuation, smooth group delay change across transition region

• Chebyshev — Some passband ripple, abrupt transition region, steep rolloff for a given filter order, peak in group delay near cutoff frequency, high ultimate attenuation

• Cauer (or Elliptical-function) — Some passband ripple, abrupt transition region, steepest roll off, ripple in stop band due to traps, ultimate attenuation smaller than Butterworth and Chebyshev, group delay peaks near cutoff frequency and in stop band

• Bessel — No ripple in passband, smooth transition region, constant group delay in passband, shallow rolloff for a given filter order, smooth changes in group delay, high ultimate attenuation

All characterizations such as "steep" and "abrupt" are relative with respect to filter designs from other families with similar orders. Other factors, such as number of components, sensitivity to component value and so on may need to be considered when selecting a filter family for a specific application.

11.3 Lumped-Element Filters

This part of the chapter deals with passive LC filters fabricated using discrete inductors and capacitors (which gives rise to their name, *lumped-element*). We will begin with a discussion of basic low-pass filters and then generalize to other types of filters.

11.3.1 Low-Pass Filters

A very basic LC filter built using inductors and capacitors is shown in **Fig 11.16**. In the first case (Fig 11.16A), less power is delivered to the load at higher frequencies because the reactance of the inductor in series with the load increases as the test frequency increases. The voltage appearing at the load goes down as the frequency increases. This configuration would pass direct current (dc) and reject higher frequencies, and so it would be a low-pass filter.

With the second case (Fig 11.16B), less power is delivered to the load at higher frequencies because the reactance of the capacitor in parallel with the load decreases as the test frequency increases. Again, the voltage appearing at the load goes down as the frequency increases and so this, too, would be called a low-pass filter. In the real world, combinations of both series and parallel components are used to form a low-pass filter.

A high-pass filter can be made using the opposite configuration — series capacitors and

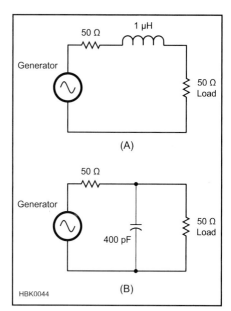

Fig 11.16 — A basic low-pass filter can be formed using a series inductor (A) or a shunt capacitor (B).

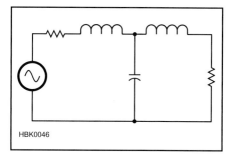

Fig 11.18 — A third-order inductor-input low-pass filter.

shunt inductors. And a band-pass (or band-stop) filter can be made using pairs of series and parallel tuned circuits. These filters, made from alternating LC elements or LC tuned circuits, are called *ladder filters*.

We spoke of filter order, or complexity, earlier in this chapter. **Fig 11.17** illustrates *capacitor-input* low-pass filters with orders of 3, 4 and 5. Remember that the order corresponds to the number of energy-storing elements. For example, the third-order filter in Fig 11.17A has three energy-storing elements (two capacitors and one inductor),

while the fifth-order design in Fig 11.17C has five elements total (three capacitors, two inductors). For comparison, a third-order filter is illustrated in **Fig 11.18**. The filters in Fig 11.17 are *capacitor-input* filters because a capacitor is connected directly across the input source. The filter in Fig 11.18 is an *inductor-input* filter.

As mentioned previously, the Cauer family has traps (series or parallel tuned circuits) carefully added to produce dips or notches (properly called *zeros*) in the stop band. Schematics for the Cauer versions of capacitor-input low-pass filters with orders of 3, 4 and 5 are shown in **Fig 11.19**. The capacitors in parallel with the series inductors create the notches at calculated frequencies to allow the Cauer filter to be implemented.

The capacitor-input and the inductor-input versions of a given low-pass design have identical characteristics for their magnitude, phase and time responses, but they differ in the impedance seen looking into the filter. The capacitor-input filter has low impedance in the stop band while the inductor-input filter has high impedance in the stop band.

11.3.2 High-Pass Filters

A high-pass filter passes signals above its cutoff frequency and attenuates those below. Simple high-pass equivalents of the filters in Fig 11.16 are shown in **Fig 11.20**.

The reactance of the series capacitor in Fig 11.20A increases as the test frequency is lowered and so at lower frequencies there will be less power delivered to the load. Similarly, the reactance of the shunt inductor in Fig 11.20B decreases at lower frequency, with the same effect. Similarly to the low-pass filter designs presented in Fig 11.17, a high-pass filter in a real world design would typically use both series and shunt components but with the positions of inductors and capacitors exchanged.

An example of a high-pass filter application would be a broadcast-reject filter designed to pass amateur-band signals in the range of 3.5 MHz and above while rejecting broadcast signals at 1.7 MHz and below. A high-pass filter with a design cutoff of 2 MHz is illustrated in **Fig 11.21**. It can be implemented as a capacitor-input (Fig 11.21A) or inductor-input (Fig 11.21B) design. In each case, signals above the cutoff are passed with minimum attenuation while signals below the cutoff are attenuated, in a manner similar to the action of a low-pass filter.

11.3.3 Low-Pass to Band-Pass Transformation

A band-pass filter is defined in part by a *bandwidth* and a *center frequency*. (An alternative method is to specify a lower and an upper cutoff frequency.) A low-pass design such as the one shown in Fig 11.17A can be converted to a band-pass filter by resonating each of the elements at the center frequency. **Fig 11.22** shows a third-order low-pass filter with a design bandwidth of 2 MHz for use in a 50-Ω system. It should be mentioned that the next several designs and the various manipulations were done using a computer; other design methods will be shown later in the chapter.

If the shunt elements are now resonated with a parallel component, and if the series elements are resonated with a series component, the result is a band-pass filter as shown in **Fig 11.23**. The series inductor value and the shunt capacitor values are the same as those for the original low-pass design. Those compo-

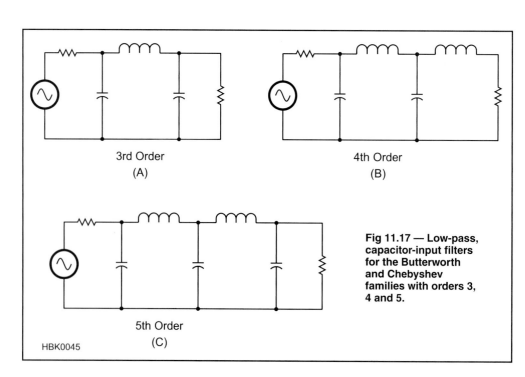

Fig 11.17 — Low-pass, capacitor-input filters for the Butterworth and Chebyshev families with orders 3, 4 and 5.

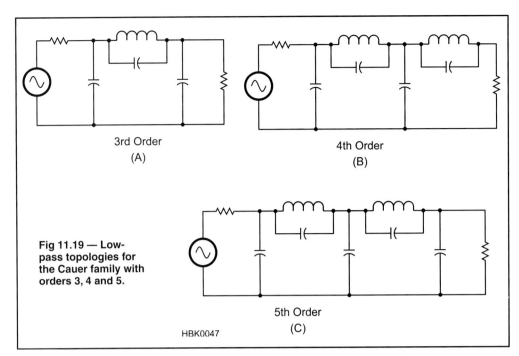

Fig 11.19 — Low-pass topologies for the Cauer family with orders 3, 4 and 5.

nents have been resonated at the center frequency for the filter (2.828 MHz in this case).

Now we can compare the magnitude response of the original low-pass filter with the band-pass design; the two responses are shown in **Fig 11.24**. The magnitude response of the low-pass version of this filter is down 3 dB at 2 MHz. The band-pass response is down 3 dB at 2 MHz and 4 MHz (the difference from the high side to the low side is 2 MHz). The response of the low-pass design is down 33 dB at 7 MHz while the response of the band-pass version is down 33 dB at 1 and 8 MHz (the difference from high side to low side is 7 MHz).

11.3.4 High-Pass to Band-Stop Transformation

Just as a low-pass filter can be transformed to a band-pass type, a high-pass filter can be transformed into a band-stop (also called a *band-reject*) filter. The procedures for doing this are similar in nature to those of the transformation from low-pass to band-pass. As with the band-pass filter example, to transform a high-pass to a band-stop we need to specify a center frequency. The bandwidth of a band-stop filter is measured between the frequencies at which the magnitude response drops 3 dB in the transition region into the stop band.

Fig 11.25 shows how to convert the 2 MHz capacitor-input high-pass filter of Fig 11.21A to a band-stop filter centered at 2.828 MHz. The original high-pass components are resonated at the chosen center frequency to form a band-stop filter. Either a capacitor-input high-pass or an inductor-input high-pass may be transformed in this manner. In this case the series capacitor values and the shunt inductor values for the band-stop are the same as those for the high-pass. The series elements are resonated with an element in parallel with

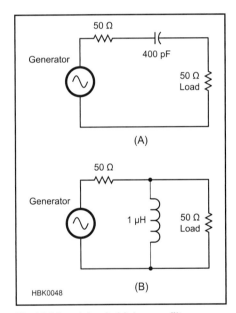

Fig 11.20 — A basic high-pass filter can be formed using a series capacitor (A) or a shunt inductor (B).

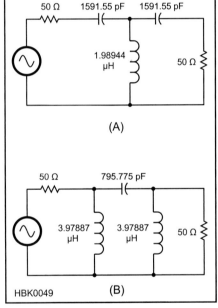

Fig 11.21 — Capacitor-input (A) and inductor-input (B) high-pass filters. Both designs have a 2 MHz cutoff.

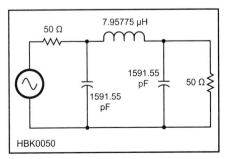

Fig 11.22 — Third-order low-pass filter with a bandwidth of 2 MHz in a 50 Ω system.

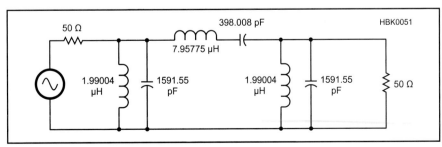

Fig 11.23 — The low-pass filter of Fig 11.22 can be transformed to a band-pass filter by resonating the shunt capacitors with a parallel inductor and resonating the series inductor with a series capacitor. Bandwidth is 2 MHz and the center frequency is 2.828 MHz.

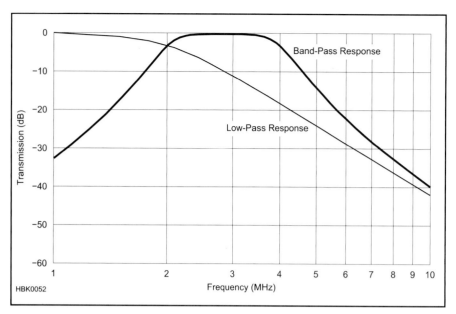

Fig 11.24 — Comparison of the response of the low-pass filter of Fig 11.22 with the band-pass design of Fig 11.23.

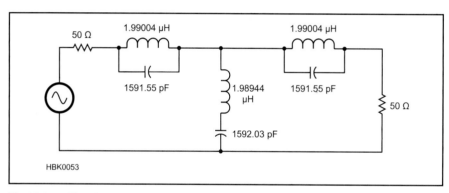

Fig 11.25 — The high-pass filter of Fig 11.21A can be converted to a band-stop filter with a bandwidth of 2 MHz, centered at 2.828 MHz.

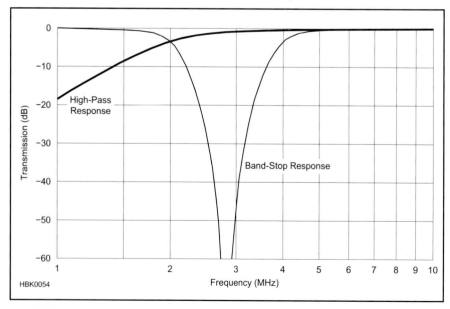

Fig 11.26 — Comparison of the response of the high-pass filter of Fig 11.21A with the band-stop design of Fig 11.25.

them. Similarly, the shunt elements are also resonated with an element in series with them. In each case the pair resonate at the center frequency of the band-stop. **Fig 11.26** shows the responses of both the original high-pass and the resulting band-stop filter.

11.3.5 Refinements in Band-Pass Design

The method of transforming a low-pass filter to a band-pass filter described previously is simple, but it has a major drawback. The resulting component values are often awkward for building a narrowband band-pass filter or inappropriate for the frequencies involved. This can be addressed through additional transformation techniques or by changing the filter to a different *topology* (the general organization of the filter circuit elements and how they are connected).

The following discussion shows an example of band-pass filter design, starting with a low-pass filter and then applying various techniques to create a filter with appropriate component values and improved response characteristics. These transformations are tedious and are best done using filter design software. Some manual design methods are presented later in this chapter.

Fig 11.27 shows a band-pass filter centered at 3 MHz with a width of 100 kHz that was designed by resonating the elements of a 3-MHz low-pass ladder-type filter as previously described. This filter is impractical for several reasons. The shunt capacitors (0.0318 µF) will probably have poor characteristics at the center frequency of 3 MHz because of the series inductance likely to be present in a practical capacitor of that value. Similarly, because of parasitic capacitance, the series inductor (159 µH) will certainly have a parallel self-resonance that is quite likely to alter the filter's response. For a narrowband band-pass filter, this method of transformation from low-pass to band-pass is not practical. (See the **RF Techniques** chapter for a discussion of parasitic effects.)

A better approach to narrowband band-pass filter design may be found by changing the filter topology. There are many different filter topologies, such as the LC "ladder" filters discussed in the preceding sections. One such topology is the *nodal-capacitor-coupled* design and **Fig 11.28** shows how the filter in Fig 11.27 can be redesigned using such an approach.

Some component values are improved, but the shunt capacitor values are still quite large for the center frequency of 3 MHz. In this case, the technique of *impedance scaling* offers some improvement. The filter was originally designed for a 50-Ω system. By scaling the filter impedance upwards from

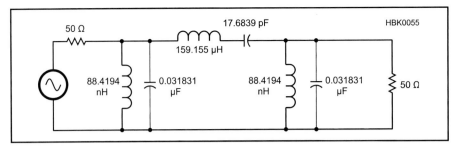

Fig 11.27 — This band-pass filter, centered at 3 MHz with a bandwidth of 100 kHz, was designed using a simple transformation. The resulting component values make the design impractical, as discussed in the text.

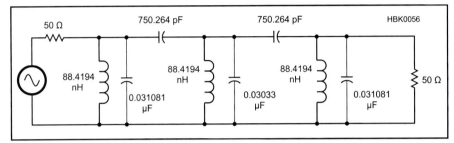

Fig 11.28 — A narrowband band-pass filter using a nodal-capacitor-coupled design improves component values somewhat.

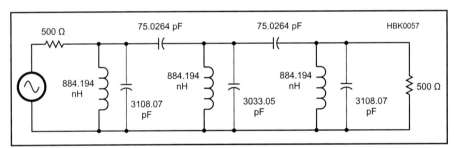

Fig 11.29 — The filter in Fig 11.28 scaled to an impedance of 500 Ω.

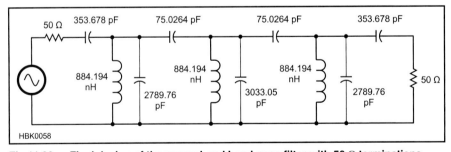

Fig 11.30 — Final design of the narrowband band-pass filter with 50 Ω terminations.

Try *Elsie* for "Filter Design"!

The *ARRL Handbook CD* at the back of this book includes *Elsie*, a *Windows* program for design and analysis of lumped-element LC filters. In addition to providing parts values for filters with various topologies from various families, tools are included to assist with practical construction. *Elsie* is provided courtesy of Jim Tonne, W4ENE, who wrote the introductory section of this chapter.

50 to 500 Ω, the reactances of all elements are multiplied by a factor of 500 / 50 = 10. (Capacitors will get smaller and inductors larger.) The 0.03-μF capacitors will decrease in value to only 3000 pF, a much better value for a center frequency of 3 MHz. The shunt inductor values and the nodal coupling capacitor values (about 75 pF in this case) are also realistic. The 500 Ω version of the filter in Fig 11.28 is shown in **Fig 11.29**.

From a component-value viewpoint this is a better design, but it must be terminated in 500 Ω at each end. To use this filter in a 50-Ω system, impedance matching components must be added as shown in **Fig 11.30**. This topology is typical of a radio receiver front-end preselector or anywhere else that a narrow — in terms of percentage of the center frequency — filter is needed.

When the nodal capacitor-coupled band-pass topology is used for a filter whose bandwidth is wide (generally, a bandwidth 20% or more of the center frequency, f_0, resulting in a filter Q = f_0 / BW less than 5), the attenuation at frequencies above the center frequency will be less than below the center frequency. This characteristic should be taken into account when attenuation of harmonics of signals in the passband is of concern. The filter shown in **Fig 11.31A** is designed to pass the amateur 75 meter band and suffers from this defect. By going to the *nodal inductor-coupled* topology we can correct this problem. The resulting design is shown in Fig 11.31B. Another way to accomplish the task is by going to a *mesh capacitor-coupled* design as shown in Fig 11.31C. The last two designs have identical magnitude responses.

Two other topologies are shown for comparison. The design in Fig 11.31B has only three capacitors (a minimum-capacitor design) while the design in Fig 11.31C has only three inductors (a minimum-inductor design).

The magnitude response of the capacitor-coupled design in Fig 11.31A is shown in **Fig 11.32** as the solid line. The designs in Fig 11.31B and C have magnitude responses as shown in Fig 11.32 as a dashed line. Note that the rolloff on the high-frequency side is steeper for the designs in Fig 11.31B and C. This will be the case for those filters with relatively wide (percentage-wise) designs. Such a design would be useful where harmonics of a signal are to be especially attenuated. It might also improve image rejection in a receiver RF stage where high-side injection of the local oscillator is used.

11.3.6 Effect of Component Q

When components with less-than-ideal characteristics are used to fabricate a filter, the performance will also be less than ideal. One such item to be concerned about is component "Q." As described in the **Electrical Fundamentals** chapter, Q is a measure of the loss in an inductor or capacitor, as determined by its resistive component. Q is the ratio of

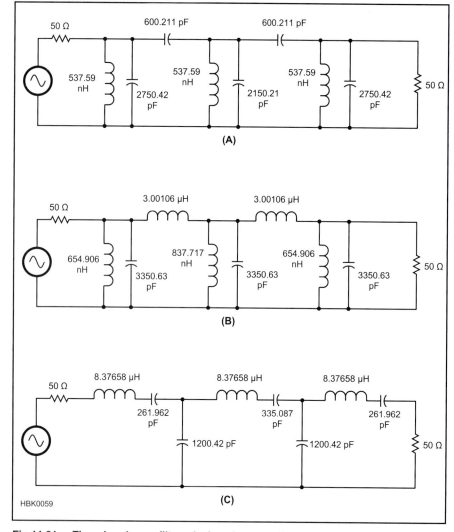

Fig 11.31 — Three band-pass filters designed to pass the 75-meter amateur band. The nodal-capacitor-coupled topology is shown at A, nodal-inductor-coupled at B, and mesh capacitor-coupled at C.

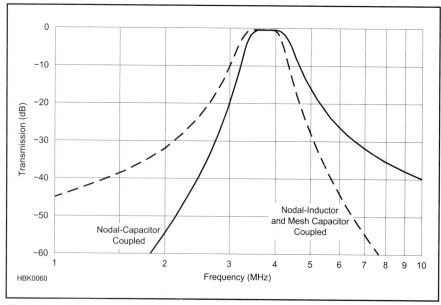

Fig 11.32 — Responses of the filters in Fig 11.31.

component's reactance to the loss resistance and is specified at a given test frequency. The loss resistance referred to here includes not only the value as measured by an ohmmeter but includes all sources of loss, such as skin effect, dielectric heating and so on.

The Q values for capacitors are usually greater than 500 and may reach a few thousand. Q values for inductors seldom reach 500 and may be as low as 20 or even worse for miniaturized parts. A good toroidal inductor can have a Q value in the vicinity of 250 to 400.

Q values can affect both the *insertion loss* of signals passing through the filter and the steepness of the filter's rolloff. Band-pass filters (and especially narrowband band-pass filters) are more vulnerable to this problem than low-pass and high-pass filters. **Fig 11.33** shows the effect of finite values of inductor Q values on the response of a low-pass filter. The Q values for each plot are as shown.

Inadequate component Q values introduce loss and more importantly they compromise the filter's response at cutoff, especially problematic in the case of narrowband band-pass filters. **Fig 11.34** illustrates the effect of finite inductor Q values on a narrowband band-pass filter. In the case of a band-pass filter, the Q values required to support a given response shape are much higher than those required for the low-pass or high-pass filter (by the ratio of center to width). Capacitor Q values are generally much high than inductor Q values and so contribute far less to this effect.

In general, component-value adjustment will not be able to fully compensate for inadequate component Q values. However, if the filter is deliberately mismatched (by changing the input and/or output terminations) then a limited amount of *response-shape* correc-

Table 11.2
Passband Ripple, VSWR and Return Loss

Passband Ripple (dB)	VSWR	Return Loss (dB)
0.0005	1.022	39.38
0.001	1.031	36.37
0.002	1.044	33.36
0.005	1.07	29.39
0.01	1.101	26.38
0.02	1.145	23.37
0.05	1.24	19.41
0.1	1.355	16.42
0.2	1.539	13.46
0.5	1.984	9.636
1	2.66	6.868

Note: As the passband ripple specification is changed so do the other items. Conversely, to get a particular value of VSWR or return loss, use this table to find the passband ripple that should be used to design the filter.

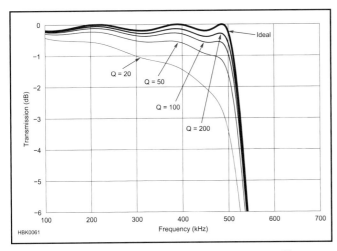

Fig 11.33 — Effect of inductor Q values on a low-pass filter.

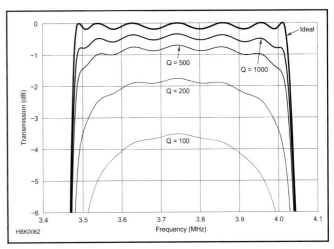

Fig 11.34 — Effect of inductor Q values on a narrowband band-pass filter.

tion can sometimes be achieved by network component value optimization ("tweaking"). The loss caused by Q problems (at dc in the case of a low-pass or at the center frequency in the case of a band-pass) may increase if such correction is attempted.

11.3.7 Side Effects of Passband Ripple

Especially in RF applications it is desirable to design a filter such that the impedance seen looking into the input side remains fairly constant over the passband. Increasing the value of passband ripple increases the rate of descent from the passband into the stop band, giving a sharper cutoff. But it also degrades the uniformity of the impedance across the passband as seen looking into the input of the filter. This may be shown in terms of VSWR or return loss; those are simply different ways of stating the same effect. (*Return loss* is explained in the **RF Techniques** chapter.)

Designs using a low value of passband ripple are preferred for RF work. Audio-frequency applications are generally not as critical, and so higher ripple values (up to about 0.2 dB) may be used in audio work.

Table 11.2 shows the values of VSWR and return loss for various values of passband ripple. Note that lower values for passband ripple yield better values for VSWR and return loss. Specifying a filter to have a passband ripple value of 0.01 dB will result in that filter's VSWR figure to be about 1.1:1. Or restated, the return loss will be about 26 dB. These values will be a function of frequency and at some test frequencies may be much better.

11.4 Filter Design Examples

Each topology of filter has its own mathematical techniques to determine the filter components, given the frequencies and attenuations that define the filter behavior. These techniques are very precise, involving the use of tables and scaling techniques as discussed in the following sections. With the availability of filter design software, such as *Elsie* and *SVC Filter Designer* (both available on the CD-ROM included with this book) filters using different topologies can be designed easily, however. (Additional programs and design resources are listed in the References section at the end of this chapter.)

Whether designing a filter manually or through the use of software, the design process begins with developing a set of filter *performance requirements* for:

• Type of response — low-pass, high-pass, band-pass, band-stop, or all-pass.

• Passband ripple — none or some maximum amount.

• Cutoff frequency (or frequencies) or bandwidth.

• Steepness of rolloff in transition region.

• Minimum stop-band attenuation and stop-band depth (optional).

• Group delay (optional).

The requirements shown as optional may not be important for simple filters. The next step is to identify the filter families that can meet those requirements. For example, if no ripple is allowed in the passband, the Butterworth or Bessel filter families would be suitable. If a very steep rolloff is required, try the Chebyshev or Cauer family.

After entering the basic requirements into the filter design software and observing the results of the calculation, the design is likely to require some adjustment. Component values can be unrealistic or the required order may be too high for practical implementation. At this point, your experience with filter design helps guide the choices of what changes to make — a different filter family, raising or lowering some of your requirements and so forth. The key is to experiment and observe to build understanding of filters.

Filter design and analysis programs show the responses of the resulting network in graphic form. This makes it easy to compare your filter to your requirements, making design adjustments quickly. Many design packages also allow the selection of nearest-5% values, tuning and other useful features. For example, the program *SVC Filter Designer* (SVC is *standard value component*) automatically selects the nearest 5% capacitor values. It selects the nearest 5% inductor values as an

> ### SVC Filter Design Software
>
> The *ARRL Handbook CD* at the back of this book includes *SVC Filter Designer*, a *Windows* program for the design of lumped-element high-pass and low-pass filters by Jim Tonne, W4ENE. The software shows ideal values and also the nearest 5% values for capacitors and inductors. It also analyzes those filters and shows the deviation of key responses from ideal when those 5% values are used.

Table 11.3
Butterworth Normalized Values

Order	G(1)	G(2)	G(3)	G(4)	G(5)	G(6)	G(7)	G(8)	G(9)	G(10)	G(11)	R_{load}
3	1	2	1									1
4	0.7654	1.848	1.848	0.7654								1
5	0.618	1.618	2	1.618	0.618							1
6	0.5176	1.414	1.932	1.932	1.414	0.5176						1
7	0.445	1.247	1.802	2	1.802	1.247	0.445					1
8	0.3902	1.111	1.663	1.962	1.962	1.663	1.111	0.3902				1
9	0.3473	1	1.532	1.879	2	1.879	1.532	1	0.3473			1
10	0.3129	0.908	1.414	1.782	1.975	1.975	1.782	1.414	0.908	0.3129		1
11	0.2846	0.8308	1.31	1.683	1.919	2	1.919	1.683	1.31	0.8308	0.2846	1

Table 11.4
Bessel Normalized Values

Order	G(1)	G(2)	G(3)	G(4)	G(5)	G(6)	G(7)	G(8)	G(9)	G(10)	R_{load}
3	2.203	0.9705	0.3374								1
4	2.24	1.082	0.6725	0.2334							1
5	2.258	1.111	0.804	0.5072	0.1743						1
6	2.265	1.113	0.8538	0.6392	0.4002	0.1365					1
7	2.266	1.105	0.869	0.702	0.5249	0.3259	0.1106				1
8	2.266	1.096	0.8695	0.7303	0.5936	0.4409	0.2719	0.0919			1
9	2.265	1.086	0.8639	0.7407	0.6306	0.5108	0.377	0.2313	0.078		1
10	2.264	1.078	0.8561	0.742	0.6493	0.5528	0.4454	0.327	0.1998	0.0672	1

Table 11.5
Chebyshev Normalized Values
Passband ripple 0.01 dB

Order	G(1)	G(2)	G(3)	G(4)	G(5)	G(6)	G(7)	G(8)	G(9)	G(10)	G(11)	R_{load}
3	0.6292	0.9703	0.6292									1
4	0.7129	1.2	1.321	0.6476								0.9085
5	0.7563	1.305	1.577	1.305	0.7563							1
6	0.7814	1.36	1.69	1.535	1.497	0.7098						0.9085
7	0.7969	1.392	1.748	1.633	1.748	1.392	0.7969					1
8	0.8073	1.413	1.782	1.683	1.853	1.619	1.555	0.7334				0.9085
9	0.8145	1.427	1.804	1.713	1.906	1.713	1.804	1.427	0.8145			1
10	0.8196	1.437	1.819	1.731	1.936	1.759	1.906	1.653	1.582	0.7446		0.9085
11	0.8235	1.444	1.83	1.744	1.955	1.786	1.955	1.744	1.83	1.444	0.8235	1

Table 11.6
Chebyshev Normalized Values
Passband ripple 0.044 dB

Order	G(1)	G(2)	G(3)	G(4)	G(5)	G(6)	G(7)	G(8)	G(9)	G(10)	G(11)	R_{load}
3	0.855	1.104	0.855									1
4	0.9347	1.293	1.581	0.7641								0.8175
5	0.9747	1.372	1.805	1.372	0.9747							1
6	0.9972	1.413	1.896	1.55	1.729	0.8153						0.8175
7	1.011	1.437	1.943	1.621	1.943	1.437	1.011					1
8	1.02	1.452	1.969	1.657	2.026	1.61	1.776	0.8341				0.8175
9	1.027	1.462	1.986	1.677	2.067	1.677	1.986	1.462	1.027			1
10	1.031	1.469	1.998	1.69	2.091	1.709	2.067	1.633	1.797	0.8431		0.8175
11	1.035	1.474	2.006	1.698	2.105	1.727	2.105	1.698	2.006	1.474	1.035	1

Table 11.7
Chebyshev Normalized Values
Passband ripple 0.2 dB

Order	G(1)	G(2)	G(3)	G(4)	G(5)	G(6)	G(7)	G(8)	G(9)	G(10)	G(11)	R_{load}
3	1.228	1.153	1.228									1
4	1.303	1.284	1.976	0.8468								0.65
5	1.339	1.337	2.166	1.337	1.339							1
6	1.36	1.363	2.239	1.456	2.097	0.8838						0.65
7	1.372	1.378	2.276	1.5	2.276	1.378	1.372					1
8	1.38	1.388	2.296	1.522	2.341	1.493	2.135	0.8972				0.65
9	1.386	1.394	2.309	1.534	2.373	1.534	2.309	1.394	1.386			1
10	1.39	1.398	2.318	1.542	2.39	1.554	2.372	1.507	2.151	0.9035		0.65
11	1.393	1.402	2.324	1.547	2.401	1.565	2.401	1.547	2.324	1.402	1.393	1

Table 11.8
Cauer Normalized Values
Passband ripple 0.01 dB, Stop band depth 30 dB

Order	G(1)	G(2)	H(2)	G(3)	G(4)	H(4)	G(5)	G(6)	H(6)	G(7)	R_{load}	F_{stop}
3	0.5835	0.8848	0.06895	0.5835							1	3.524
4	0.4844	0.954	0.1799	1.109	0.6392	0					1	2.215
5	0.6173	1.106	0.1913	1.264	0.7084	0.6533	0.3403				1	1.418
6	0.4158	0.936	0.3958	1.074	0.7642	0.7979	0.9396	0.7292	0		1	1.252
7	0.6095	1.117	0.2497	1.027	0.5503	1.459	0.8809	0.5513	1.175	0.1594	1	1.105

Table 11.9
Cauer Normalized Values
Passband ripple 0.01 dB, Stop band depth 40 dB

Order	G(1)	G(2)	H(2)	G(3)	G(4)	H(4)	G(5)	G(6)	H(6)	G(7)	R_{load}	F_{stop}
3	0.6074	0.9299	0.03092	0.6074							1	5.12
4	0.5463	1.057	0.09503	1.144	0.634	0					1	2.885
5	0.6678	1.179	0.1179	1.359	0.9065	0.3574	0.4904				1	1.687
6	0.5089	1.063	0.2549	1.208	0.992	0.4761	1.066	0.7275	0		1	1.416
7	0.6636	1.197	0.173	1.186	0.7766	0.8845	1.04	0.7415	0.7093	0.3321	1	1.191

Table 11.10
Cauer Normalized Values
Passband ripple 0.01 dB, Stop band depth 50 dB

Order	G(1)	G(2)	H(2)	G(3)	G(4)	H(4)	G(5)	G(6)	H(6)	G(7)	R_{load}	F_{stop}
3	0.6194	0.9518	0.01415	0.6194							1	7.47
4	0.5813	1.116	0.05176	1.165	0.6309	0					1	3.797
5	0.7007	1.225	0.07357	1.432	1.045	0.2096	0.5872				1	2.045
6	0.5739	1.156	0.168	1.317	1.167	0.3023	1.157	0.7255	0		1	1.634
7	0.7019	1.252	0.122	1.319	0.9711	0.5823	1.192	0.899	0.4614	0.4575	1	1.309

Table 11.11
Cauer Normalized Values
Passband ripple 0.01 dB, Stop band depth 60 dB

Order	G(1)	G(2)	H(2)	G(3)	G(4)	H(4)	G(5)	G(6)	H(6)	G(7)	R_{load}	F_{stop}
3	0.6246	0.9617	0.00652	0.6246							1	10.94
4	0.6011	1.15	0.02862	1.178	0.629	0					1	5.025
5	0.7209	1.254	0.04608	1.482	1.137	0.1269	0.6486				1	2.514
6	0.6191	1.222	0.1121	1.399	1.296	0.1978	1.223	0.7238	0		1	1.913
7	0.7283	1.291	0.08673	1.425	1.131	0.3982	1.322	1.024	0.312	0.5489	1	1.463

Table 11.12
Cauer Normalized Values
Passband ripple 0.044 dB, Stop band depth 30 dB

Order	G(1)	G(2)	H(2)	G(3)	G(4)	H(4)	G(5)	G(6)	H(6)	G(7)	R_{load}	F_{stop}
3	0.7873	0.9891	0.09924	0.7873							1	2.788
4	0.6172	1.046	0.2284	1.239	0.8101	0					1	1.883
5	0.794	1.119	0.2452	1.353	0.6862	0.8268	0.4694				1	1.286
6	0.5391	0.9646	0.4632	1.064	0.7337	0.9564	0.9908	0.8958	0		1	1.171
7	0.7808	1.11	0.3034	1.058	0.4585	1.892	0.8492	0.5271	1.379	0.2962	1	1.065

Table 11.13
Cauer Normalized Values
Passband ripple 0.044 dB, Stop band depth 40 dB

Order	G(1)	G(2)	H(2)	G(3)	G(4)	H(4)	G(5)	G(6)	H(6)	G(7)	R_{load}	F_{stop}
3	0.8229	1.05	0.04455	0.8229							1	4.02
4	0.6958	1.176	0.1209	1.287	0.8064	0					1	2.429
5	0.8597	1.211	0.1509	1.491	0.9058	0.4527	0.6448				1	1.504
6	0.6471	1.112	0.299	1.227	0.9893	0.5657	1.126	0.8992	0		1	1.304
7	0.8479	1.204	0.2085	1.242	0.6828	1.118	1.045	0.7251	0.833	0.4797	1	1.132

Table 11.14
Cauer Normalized Values
Passband ripple 0.044 dB, Stop band depth 50 dB

Order	G(1)	G(2)	H(2)	G(3)	G(4)	H(4)	G(5)	G(6)	H(6)	G(7)	R_{load}	F_{stop}
3	0.8401	1.079	0.02038	0.8401							1	5.85
4	0.7411	1.253	0.06596	1.316	0.8039	0					1	3.178
5	0.9012	1.269	0.094	1.594	1.064	0.2658	0.761				1	1.803
6	0.724	1.22	0.1977	1.362	1.192	0.3584	1.227	0.9004	0		1	1.487
7	0.8928	1.27	0.1462	1.401	0.8857	0.7262	1.232	0.8927	0.5426	0.6158	1	1.229

Table 11.15
Cauer Normalized Values
Passband ripple 0.044 dB, Stop band depth 60 dB

Order	G(1)	G(2)	H(2)	G(3)	G(4)	H(4)	G(5)	G(6)	H(6)	G(7)	R_{load}	F_{stop}
3	0.8481	1.093	0.0094	0.8481							1	8.553
4	0.767	1.297	0.0365	1.333	0.8025	0					1	4.192
5	0.9275	1.307	0.05888	1.667	1.172	0.1613	0.8371				1	2.199
6	0.7782	1.298	0.1323	1.465	1.346	0.2346	1.301	0.9008	0		1	1.725
7	0.9275	1.317	0.104	1.531	1.056	0.4957	1.395	1.027	0.3689	0.7207	1	1.359

Table 11.16
Cauer Normalized Values
Passband ripple 0.2 dB, Stop band depth 30 dB

Order	G(1)	G(2)	H(2)	G(3)	G(4)	H(4)	G(5)	G(6)	H(6)	G(7)	R_{load}	F_{stop}
3	1.116	0.9996	0.1584	1.116							1	2.207
4	0.8176	1.087	0.3037	1.335	1.065	0					1	1.61
5	1.084	1.031	0.3443	1.482	0.5907	1.153	0.6819				1	1.18
6	0.7319	0.9415	0.5675	1.024	0.6677	1.186	1.002	1.149	0		1	1.106
7	1.065	1.01	0.4035	1.124	0.3345	2.756	0.8246	0.451	1.785	0.5138	1	1.036

Table 11.17
Cauer Normalized Values
Passband ripple 0.2 dB, Stop band depth 40 dB

Order	G(1)	G(2)	H(2)	G(3)	G(4)	H(4)	G(5)	G(6)	H(6)	G(7)	R_{load}	F_{stop}
3	1.174	1.08	0.07104	1.174							1	3.147
4	0.9224	1.254	0.1607	1.399	1.067	0					1	2.047
5	1.175	1.14	0.2106	1.686	0.8175	0.6264	0.8974				1	1.351
6	0.8638	1.107	0.3667	1.221	0.9522	0.6869	1.139	1.164	0		1	1.21
7	1.156	1.116	0.2739	1.35	0.5366	1.558	1.073	0.6422	1.069	0.7205	1	1.084

Table 11.18
Cauer Normalized Values
Passband ripple 0.2 dB, Stop band depth 50 dB

Order	G(1)	G(2)	H(2)	G(3)	G(4)	H(4)	G(5)	G(6)	H(6)	G(7)	R_{load}	F_{stop}
3	1.203	1.118	0.03253	1.203							1	4.554
4	0.9842	1.355	0.08773	1.438	1.067	0					1	2.654
5	1.233	1.211	0.1309	1.842	0.9885	0.3676	1.047				1	1.595
6	0.9599	1.23	0.2433	1.387	1.19	0.4317	1.246	1.173	0		1	1.36
7	1.213	1.192	0.1904	1.548	0.7322	0.9884	1.313	0.8105	0.694	0.8771	1	1.161

Table 11.19
Cauer Normalized Values
Passband ripple 0.2 dB, Stop band depth 60 dB

Order	G(1)	G(2)	H(2)	G(3)	G(4)	H(4)	G(5)	G(6)	H(6)	G(7)	R_{load}	F_{stop}
3	1.216	1.137	0.01501	1.216							1	6.64
4	1.02	1.413	0.04859	1.46	1.067	0					1	3.484
5	1.272	1.256	0.08207	1.953	1.108	0.2237	1.149				1	1.924
6	1.029	1.32	0.1634	1.517	1.375	0.2818	1.328	1.178	0		1	1.56
7	1.257	1.244	0.1348	1.716	0.9025	0.6672	1.527	0.9477	0.4718	1.001	1	1.269

option. It also shows the resulting response degradations (which may be minor).

Although filter-design software has greatly simplified the design of lumped-element filters, going through the manual design process, which is based on the use of tables of component values, will give some insight into how to make better use the software to develop and refine a filter.

11.4.1 Normalized Values

The equations used to design filters are quite complex, so to ease the design process, tables of component values have been developed for different families of filters. Developing a set of tables for all possible filter requirements would be impractical, so *tables of normalized values* are used based on a low-pass filter with a bandwidth of 1 radian per second and with 1-Ω input termination impedance. Performance requirements for the desired filter are *normalized* to 1 rad/sec bandwidth and 1-Ω terminations and the tables used to determine the various component values for those frequencies and impedance. That filter is then transformed to the desired response if necessary — high-pass, band-pass, etc. Then the filter is converted to the desired frequency and impedance by *denormalizing* or *scaling* the component values with another set of transformations. In this way, a (relatively) small set of tables can be used to design filters of any type as you'll see in the following examples.

Summarizing, the table-based design process consists of:
- Determining performance requirements.
- Normalizing the requirements to a 1-Hz, 1-Ω, low-pass response.
- Selecting a family and order.
- Obtaining normalized component values from the tables.
- Transforming the filter to the desired response.
- Denormalizing the component values for frequency and impedance.

11.4.2 Filter Family Selection

The tables presented here are for low-pass filters from different families (Butterworth, Bessel, Chebyshev and Cauer), with various specifications for passband ripple and so on. In the normalized tables, capacitor values are in farads (F) and inductor values are in henries (H). These are converted to actual component values through the process of denormalization described below.

The Butterworth family (described by normalized component values in **Table 11.3**) is used when there should be no ripple at all in the passband, so that the response is to be as flat as possible. This trait is particularly apparent near dc for the low-pass response, at the center frequency for a band-pass response, and at infinity for the high-pass response. The resulting magnitude response will also have a relatively gentle transition from the passband into the stop band.

When a low-pass filter with constant group delay throughout the passband is desired then the Bessel family, whose normalized component values are shown in **Table 11.4**, should be used.

The Chebyshev family is used when a sharper cutoff is desired for a given number of components and where at least a small amount of ripple is allowable in the passband. The Chebyshev filter tables are broken into groups according to logically chosen values of passband ripple. The tables presented here are for passband ripple values of 0.01 dB (**Table 11.5**, for critical RF work), 0.044 dB (**Table 11.6**, an intermediate value offering 20 dB of return loss or 1.2:1 VSWR) and 0.2 dB (**Table 11.7**, for audio and less-critical work where a steeper rolloff into the stop band is of primary concern). Some published tables show figures for passband ripple values less than 0.01 dB. These are difficult to implement because of the tight component tolerances required to achieve the expected responses.

When steepness of rolloff from passband into stop band is the item of greatest importance, then the Cauer filter family is used. Cauer filters involve a more complicated set of choices. In addition to selecting a passband ripple, the designer must also assign a stop band depth (or stop band frequency). Some of the items interact; they can't all be selected arbitrarily.

The most likely combinations of items to be chosen for Cauer filters are presented in **Tables 11.8** to **11.19**. These tables are for passband ripple values of 0.01, 0.044 and 0.2 dB (the same values that were chosen for the Chebyshev family) and stop band depths of 30, 40, 50 and 60 dB. The frequency at which the attenuation first reaches the design stop band depth value is shown in the final column, labeled F_{stop}. As with the Chebyshev filters, some published tables show ripple values of less than 0.01 dB but such designs are difficult to implement in practice because of the tight tolerances required on all of the components.

DESIGN EXAMPLE — CHEBYSHEV LOW-PASS

Here is an example of using normalized-value tables to design a low-pass Chebyshev filter with 0.01 dB of passband ripple, a bandwidth of 4.2 MHz, a rolloff requirement of 25 dB of attenuation one octave above cutoff, and an input termination of 50 Ω. **Fig 11.35** illustrates the attenuation to be expected for various orders (N) of a Chebyshev low-pass filter with 0.01 dB of passband ripple. The next step is to determine the lowest order that can meet the requirements for roll off. Based on the rolloff requirement, a fifth-order filter is the lowest order filter for which the attenuation curve is below 25 dB at twice the normalized cutoff frequency of 1. (If 14 dB of attenuation one octave above the cutoff frequency had been required, a fourth-order filter would suffice.)

Fig 11.36 shows the schematic of a fifth-order low-pass filter that will meet the requirements. (If a higher- or lower-order filter is required, add or subtract elements, beginning with G(1) at the filter input, with capacitors as the parallel or shunt elements and inductors as the series elements.)

Next, refer to Table 11.5 to obtain the normalized component values for the 1-Ω and 1 radian/second low-pass filter. Choosing the table row with component values for Order = 5, G(1) is 0.7563, G(2) is 1.305, G(3) is 1.577, G(4) is 1.305 and G(5) is 0.7563. The last value, R_{load}, is used to calculate the output termination.

The equations to calculate the actual component values are shown in Fig 11.36A. Ω_c

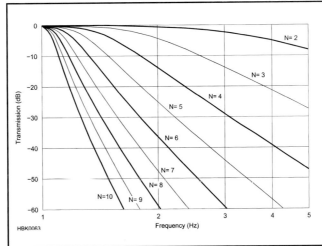

Fig 11.35 — This graph shows transmission for Chebyshev low-pass filters with 0.01 dB passband ripple filters with orders from 2 through 10. For a given order, higher passband ripple values will have higher attenuation (less transmission) at a given frequency while Butterworth filters will have less attenuation at a given frequency.

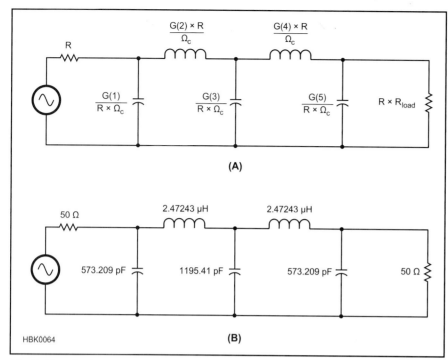

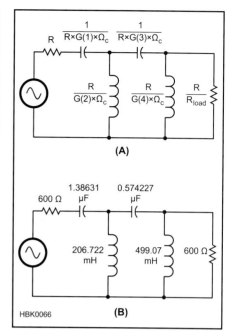

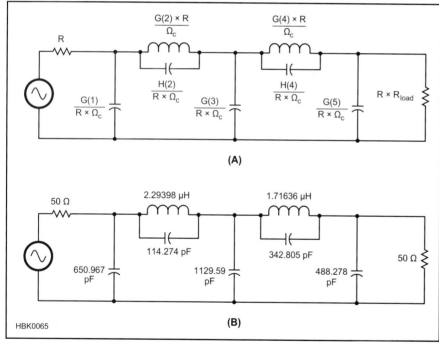

Fig 11.36 — Design example for a Chebyshev low-pass filter using the normalized filter tables. The topology and design equations are shown at A, with the resulting calculated parts values at B.

Fig 11.37 — Design example for a Cauer low-pass filter using the normalized filter tables. The topology and design equations are shown at A, with the resulting calculated parts values at B.

Fig 11.38 — Design example for a Butterworth high-pass filter using the normalized filter tables. The topology and design equations are shown at A, with the resulting calculated parts values at B. For a Cauer family high-pass, trap capacitors would be inserted in series with the shunt inductors. Their value is calculated by the same expression as the series capacitors, but using the $H(n)$ values from the Cauer tables.

column under R_{load}.) After denormalizing, the resulting component values are shown in Fig 11.36B. At this point in a table-based design process, the closest available standard value or adjustable components would be used to fabricate the actual circuit.

DESIGN EXAMPLE — CAUER LOW-PASS

The design of a Cauer low-pass filter is very similar to the design of a Chebyshev, but it has added capacitors across the series inductors to form the traps that create the stop band notches. Graphs showing the stop band performance or loss of the various Cauer filters in the tables are impractical because of the large number of options to be chosen.

This example illustrates the design of a fifth-order Cauer capacitor-input low-pass filter with 0.044 dB passband ripple and 40-dB stop band depth. The input termination will be 50 Ω and the ripple bandwidth will again be 4.2 MHz.

Fig 11.37 shows the filter schematic. The first step is to obtain from Table 11.13 the normalized 1-Ω and 1 radian/second component values. Choosing the table row of component values for Order = 5, G(1) is 0.8597, G(2) is 1.211, G(3) is 1.491, G(4) is 0.9058 and G(5)

is the denormalizing factor used to scale the normalized filter component values to the desired frequency

$$\Omega_c = 2\pi\, F_{width} \qquad (1)$$

In this case, $\Omega_c = 2 \times 3.1416 \times 4.2 = 26.389$. R is the filter's input termination, in this case

50 Ω. The termination on the output side of the filter will be the input termination value multiplied by R_{load}. In this case, $R_{load} = 1$, so the input and output terminations are equal. (Chebyshev *even-ordered* low-pass and high-pass filters have an output termination different from the input, as shown in the last

is 0.6448. In addition, the "trap" capacitor values must be retrieved. The trap capacitors are shown as "H" values. H(2) is 0.1509 and H(4) is 0.4527. The output termination is 1 and so is the same as the input termination.

The filter schematic with the denormalization equations appears in Fig 11.37A. Ω_c and R have the same definitions as in the previous example. The resulting actual component values for the required frequency and impedance are shown in Fig 11.37B.

DESIGN EXAMPLE — HIGH-PASS BUTTERWORTH

The design of a high-pass filter will now be illustrated. Fig 11.35 can be used to estimate the response of a high-pass design just as it was for a low-pass. The only manipulation that needs to be done is to use the *reciprocal* of frequency when estimating the magnitude response. For example, if the attenuation requirement for the high-pass filter is for 10 dB of attenuation at 0.2 times the cutoff frequency, then when using the low-pass filter response graphs, look up the attenuation at 1/0.2 = 5 times the cutoff frequency, instead.

For this example, we will design a fourth-order capacitor-input Butterworth high-pass response with the 3-dB cutoff frequency at 250 Hz and system impedance of 600 Ω. The filter schematic is shown in **Fig 11.38**. (Remember that even-order ladder filters can have either a series or parallel element at the input.)

The normalized values for the low-pass response are taken from Table 11.3, and the design equations are shown in Fig 11.38A. From the table row of component values for Order = 4, the value for G(1) is 0.7654, G(2) is 1.848, G(3) is 1.848 and G(4) is 0.7654.

Ω_c is computed the same way as in the previous examples. For the high-pass configuration, the output termination is based on the *reciprocal* of R_{load}. In this case, the table shows that the output termination is 1, so it is the same as the input. (Unlike the Chebyshev family, the output termination for the Butterworth family is always the same as the input termination.) Fig 11.38B shows the resulting denormalized component values for the actual filter.

DESIGN EXAMPLE — WIDE BAND-PASS

The design of a band-pass filter of appreciable percentage bandwidth will be looked at next. For the purposes of this discussion, "appreciable" means a bandwidth of 20% of the center frequency or higher (a filter Q of 5 or less).

Fig 11.39 shows this example, a third-order shunt-input Chebyshev type with 0.2 dB of passband ripple with an input termination of 75 Ω. The center frequency is

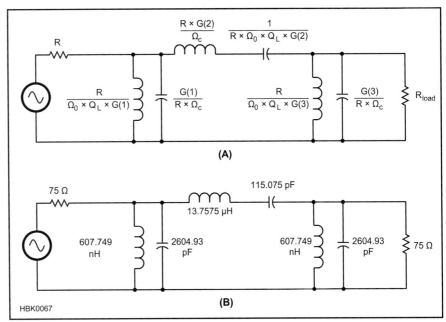

Fig 11.39 — Design example for a Chebyshev wide band-pass filter using the normalized filter tables. The topology and design equations are shown at A, with the resulting calculated parts values at B.

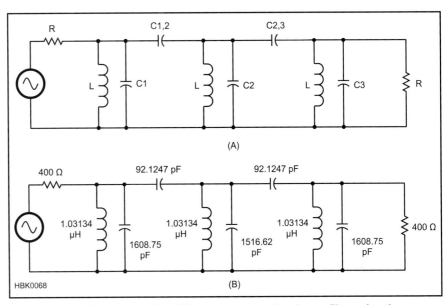

Fig 11.40 — Design example for a Chebyshev narrow band-pass filter using the normalized filter tables. The topology and design equations are shown at A, with the resulting calculated parts values at B.

to be 4 MHz and the ripple bandwidth is to be 1 MHz. Again, Fig 11.35 can be used to determine the required filter order, bearing in mind that it is precise for passband ripple of 0.01 dB.

The normalized values for the low-pass are taken from Table 11.7. Choosing the table row of component values for Order = 3, the value for G(1) is 1.228, G(2) is 1.153 and G(3) is 1.228.

In the equations used to calculate the actual parts values, Q_L is the loaded Q. This is the ratio of F_{center} to F_{width}. Note that in this topology both Ω_c and Ω_o are used. Ω_c is defined as in the previous examples, while

$$\Omega_o = 2\pi F_{center} \qquad (2)$$

The filter schematic with the denormalizing equations appears in Fig 11.39A, and calculated values for the actual filter are shown in Fig 11.39B.

DESIGN EXAMPLE —

NARROW BAND-PASS

Narrow band-pass filters involve more calculations than wide band-pass types. The filter topology used in this discussion is shown in **Fig 11.40** — a set of shunt-connected parallel-tuned resonators coupled by relatively small-value coupling capacitors.

This example will be for a band-pass filter with a bandwidth of 200 kHz, centered at 3.8 MHz, using the Chebyshev family with a passband ripple of 0.044 dB. The order will be three, and the normalized values are found in Table 11.6.

As in the wide band-pass example, the equations used to calculate the actual parts values use Q_L (the loaded Q), which is the ratio of F_{center} to F_{width}. Once again both Ω_c and Ω_o are used as before The same inductor, L, is used for each resonator. The value of that inductor is given by:

$$L = \frac{R}{\Omega_o \times Q_L \times G(1)} \quad (3)$$

Each resonator has a basic tuning capacitor whose value is given by:

$$C_{basic} = \frac{G(1)}{R \times \Omega_c} \quad (4)$$

Next calculate the inter-resonator coupling capacitors:

$$C1,2 = \frac{G(1)}{R \times \Omega_o} \sqrt{\frac{1}{G(1) \times G(2)}} \quad (5)$$

$$C2,3 = \frac{G(1)}{R \times \Omega_o} \sqrt{\frac{1}{G(2) \times G(3)}} \quad (6)$$

The actual shunt tuning capacitors for each resonator are then the basic tuning capacitor minus the coupling capacitors on each side, as shown here:

C1 = C_{basic} − C1,2
C2 = C_{basic} − C1,2 − C2,3
C3 = C_{basic} − C2,3

The completed design is shown in Fig 11.40B. It should be evident how to extend this procedure to higher-order filters.

11.5 Active Audio Filters

Below RF, in what is broadly referred to as the "audio" range between a few Hz and a few hundred kHz, designers have several choices of filter technology.
- Passive LC
- Digital Signal Processing (DSP)
- Switched-Capacitor Audio Filter (SCAF)
- Active RC

LC audio filters are not used much in current designs except in high-power audio applications, such as speaker crossover networks, and are not covered here. LC filters were once popular as external audio filters for CW reception. These designs tended to be large and bulky, often using large surplus 44 or 88 mH core inductors. (A classic *ARRL Handbook* project to construct a passive LC CW filter can be found on the CD-ROM included with this book.) LC filters used in very low-level receiver applications can be very compact, but these tend to suffer from relatively high insertion loss, which reduces the receiver sensitivity (if no preamp is used ahead of the filter) or its large-signal dynamic range (if a loss-compensating preamp is used ahead of the filter). In addition, LC filters used in low-level receiver applications tend to pick up 60/120 Hz hum from stray magnetic fields from ac power transformers. Even "self shielded" inductors can produce noticeable ac hum pick up when followed

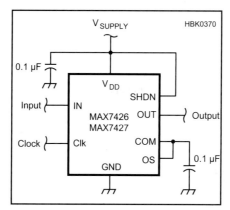

Fig 11.41 — Simple SCAF low pass filter, taken from the MAX7426 data sheet.

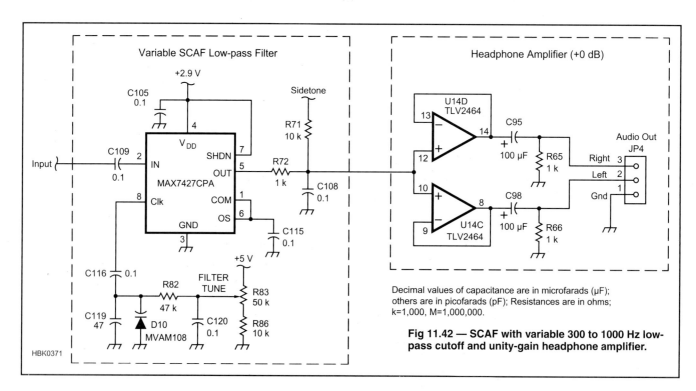

Fig 11.42 — SCAF with variable 300 to 1000 Hz low-pass cutoff and unity-gain headphone amplifier.

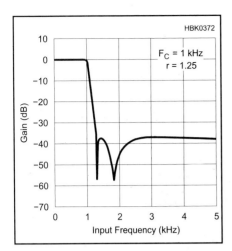

Fig 11.43 — Frequency response of a low-pass filter using the MAX7426 set to a 1 kHz cutoff frequency.

by 90 to 120 dB of audio gain!

DSP filtering can yield filters that are superior to analog filters. (See the **DSP and Software Radio Design** chapter.) High signal-level DSP-based external audio filters such as the popular Timewave DSP-9 (**www.timewave.com**) and MFJ Enterprises MFJ-784B (**www.mfjenterprises.com**) have been available for some time. In addition, if the DSP is coupled to a very high-performance audio A/D converter such as an Asahi-Kasei AK5394A, DSP filtering can provide excellent filtering for even very low-level audio signals.

The main drawback to DSP processing is that of expense. A low-end DSP external audio filter costs at least $100, while a DSP with a high-end A/D converter (such as an audio sound card front-end to a PC) can run in the $150 to $400+ price range, which does not include the cost of the host PC.

11.5.1 SCAF Filters

Simple SCAF filter designs can produce extremely effective low-power audio filters. The implementation of these filters in practical IC form usually involves small-value capacitors and high-value resistors. The use of high-value resistors tends to generate enough noise to make these filters unsuitable for very low-level audio processing such as the front end of a receiver audio chain. Thus, SCAF filters tend to be limited to filtering at the output end of the audio chain — headphone or speaker level audio applications, an area in which they very much excel. The cutoff frequency of a SCAF filter is set by an external clock signal. Thus, this class of filter naturally lends itself to use in variable frequency filters. For simplicity, excellent frequency selectivity and relatively low cost, SCAF filters are highly recommended when additional audio filtering is desired on the output of an existing receiver.

An example of a very simple, but highly effective SCAF low-pass filter IC is the Maxim MAX7426 in **Fig 11.41.** The filter's cutoff frequency can be set by placing an appropriately sized capacitor across the clock oscillator inputs because the part can generate its own internal clock signal. For example, connecting a 180 pF capacitor across the MAX7426's clock inputs will produce a 1 kHz low-pass filter. The 3-V version of this chip was used in the NC2030 QRP transceiver along with a MVAM108 varactor diode to create a low-pass filter that was tunable from 300 Hz to 1 kHz.

A portion of that schematic is shown in **Fig 11.42.** The SCAF low-pass filter is followed by a 3-V unity-gain headphone amplifier (U14C and U14D) as the filter IC itself cannot directly drive headphones. The low-pass cutoff frequency is tuned by using R83 to vary the voltage applied across the MVAM108 varactor diode D10, thus changing the capacitance across the clock input (pin 8) of the chip. C116 was used to isolate the dc voltage across the varactor diode from the bias voltage on the

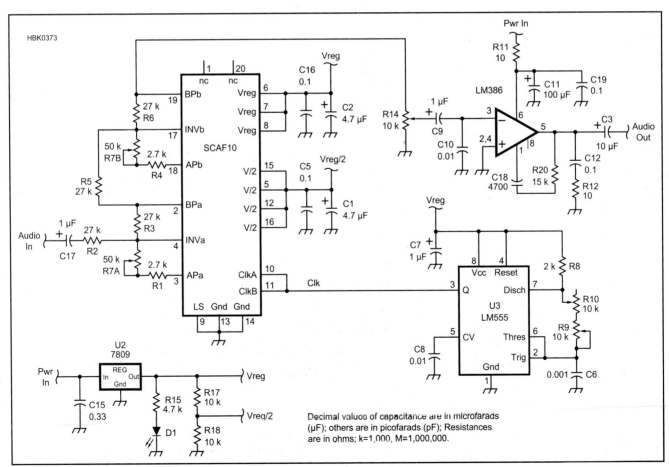

Fig 11.44 — Example of a simple SCAF band-pass filter from a New England QRP Club kit.

clock input line. The frequency response of such a filter is shown in **Fig 11.43**.

As can be seen, this is an extremely sharp filter, producing almost 40 dB of attenuation very close to the cut off frequency — not too bad for an inexpensive part. The slightly more expensive MAX7403 gives an even steeper 80 dB cutoff. Both parts are specified for an 80 dB signal-to-noise ratio based on an assumed 4-V_{P-P} signal. Sensitive modern in-ear type headphones require only 20 mV_{P-P} to produce a fairly loud signal. A 20 mV signal is 46 dB below 4 V_{P-P}, so at such headphone levels, the noise floor is actually only 34 dB below the 4 V_{P-P} signal — fairly quiet, but is a lot less noise margin than one would tend to think given the 80 dB specification.

Again, with a noise floor only 34 dB below typical headphone levels, SCAF filters such as these are fine at the end of the receiver chain, but are not useful as an audio filter at very low signal levels early on in a receiver.

Fig 11.44 is an example of a slightly more complex SCAF band-pass filter. This filter features both a variable center frequency (450 Hz to 1000 Hz) and a variable bandwidth (90 Hz to 1500 Hz). This very popular filter was designed by the New England QRP Club (www.newenglandqrp.org) and is currently offered in kit form.

This filter features both sections of the SCAF10 IC configured as identical band-pass filters. The bandwidth (Q) is adjustable via R7A and R7B, while the center frequency is adjustable by R10 and the trimmer R9 as these resistors set the frequency of the LM555 clock generator.

As in the previous SCAF filter example, this filter also includes an audio amplifier (LM386) for driving either an external speaker or headphones as none of the SCAF filter ICs are capable of driving headphones directly.

11.5.2 Active RC Filters

Active RC filters based on op amp circuits can be used in either high-level audio output or very low-level direct-conversion receiver front-end filtering applications and thus are extremely flexible. Unlike LC filters, active RC filters can provide both filtering and gain at the same time, eliminating the relatively high insertion loss of a physically small, sharp LC filter. In addition, an active RC filter is not susceptible to the same ac hum pickup in low signal level applications as LC filters.

Active filters can be designed for gain and they offer excellent stage-to-stage isolation. The circuits require only resistors and capacitors, avoiding the limitations associated with inductors. By using gain and feedback, filter Q is controllable to a degree unavailable to passive LC filters. Despite the advantages, there are also some limitations. They require power, and performance may be limited by the op amp's finite input and output levels, gain and bandwidth. Active filters that drive speakers or other heavy loads usually employ an audio output amplifier circuit to boost the output power after a low-power filter stage.

A particular advantage of active RC filters for use in receivers is that they can be capable of extremely low noise operation, allowing the filtering of extremely small signals. At the same time they are capable of handling extremely large signals. Op amps are often capable of using ±18-V dual supply voltages or 36-V single supply voltages allowing the construction of a very high performance audio filter/amplifier chain that can handle signals up to 33 V_{P-P}. The ability to provide gain while also providing filtering provides a lot of flexibility in managing the sensitivity of a receiver audio chain.

The main disadvantage of active RC filters is their relatively high parts count compared to other filter types. In addition, active RC filters tend to be fixed-frequency designs unlike SCAF filters whose frequency can be moved simply by changing the clock frequency that drives the SCAF IC.

11.5.3 Active Filter Responses

Active filters can implement any of the passive LC filter responses described in the preceding section: low-pass, high-pass, band-pass, band-stop and all-pass. Filter family responses such as Butterworth, Chebyshev, Bessel and Cauer (elliptic) can be realized. All of the same family characteristics apply equally to passive and active filters and will not be repeated here. (Op amps are discussed in the **Analog Basics** chapter.)

Fig 11.45 presents circuits for first-order low-pass (Fig 11.45A) and high-pass (Fig 11.45B) active filters. The frequency response of these two circuits is the same as a parallel and series RC circuit, respectively, except that these two circuits can have a passband gain greater than unity. Rolloff is shallow at 6 dB/octave. The two responses can be combined to form a simple band-pass filter (Fig 11.45C). This combination cannot produce sharp band-pass filters because of the shallow rolloff.

To achieve high-order responses with steeper rolloff and narrower bandwidths, more complex circuits are required in which combinations of capacitors and resistors create *poles* and *zeroes* in the frequency response. (Poles and zeros are described in the **Analog Basics** chapter.) The various filter response families are created by different combinations of additional poles and zeroes. There are a variety of circuits that can be configured to implement the equations that describe the various families of filter responses. The most common circuits are *Sallen-Key* and *multiple-feedback*, but there are numerous other choices.

There are many types of active filters — this section presents some commonly used circuits as examples. A book on filter design (see the References section) will present more choices and how to develop designs based on the different circuit and filter family types.

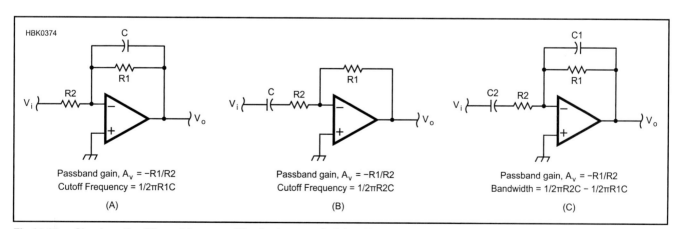

Fig 11.45 — Simple active filters. A low-pass filter is shown at A; B is a high-pass filter. The circuit at C combines the low- and high-pass filters into a wide band-pass filter.

SECOND-ORDER ACTIVE FILTERS

Fig 11.46 shows circuits for four second-order filters: low-pass (Fig 11.46A), high-pass (Fig 11.46B), band-pass (Fig 11.46C) and band-reject or notch (Fig 11.46D). Sequences of these filters are used to create higher-order circuits by connecting them in series. Two second-order filter stages create a fourth-order filter, and so forth.

The low-pass and high-pass filters use the Sallen-Key circuit. Note that the high-pass circuit is just the low-pass circuit with the positions of R1-R2 and C1-C2 exchanged. R3 and R4 are used to control gain in the low- and high-pass configurations. The band-pass filter is a multiple-feedback design. The notch filter is based on the twin-T circuit. All of the filter design equations and tables will result in a Butterworth family response. (For the Chebyshev and other filter responses, consult the references listed at the end of the chapter.)

The circuits in Fig 11.46 assume the op amp is operating from a balanced, bipolar power supply, such as ±12 V. If a single supply is used (such as +12 V and ground), the circuit must have a dc offset added and blocking capacitors between filter sections to prevent the dc offset from causing the op amp to saturate. The references listed at the end of the chapter provide more detailed information on single-supply circuit design.

Avoid electrolytic or tantalum capacitors as frequency-determining components in active filter design. These capacitors are best used for bypassing and power filtering as their tolerance is generally quite low, they have significant parasitic effects, and are usually polarized. Very small values of capacitance (less than 100 pF) can be affected by stray capacitance to other circuit components and wiring. High-order and high-Q filters require close attention to component tolerance and temperature coefficients, as well.

SECOND-ORDER ACTIVE FILTER DESIGN PROCEDURES

The following simple procedures are used to design filters based on the schematics in Fig 11.46. Equations and a design example are provided in the figure.

Low- and High-Pass Filter Design

To design a low- or high-pass filter using the circuits in Fig 11.46, start by determining your performance requirements for filter order (2, 4, 6, or 8), gain (K) and cutoff frequency (f_C). Calculate $\omega_C = 2\pi f_C$ and C_2 or C as required.

Table 11.20 provides design coefficients to create the Butterworth response from successive Sallen-Key low- and high-pass stages. A different coefficient is used for each stage. Obtain design coefficient "a" from Table 11.20. Calculate the remaining component values from the equations provided.

Band-Pass Filter Design

To design a band-pass filter as shown in Fig 11.46C, begin by determining the filter's required Q, gain, and center frequency, f_0. Choose a value for C and solve for the resistor values. Very high or very low values of resistance (above 1 MΩ or lower than 10 Ω) should be avoided. Change the value of C until suitable values are obtained. High gain and Q may be difficult to obtain in the same stage with reasonable component values. Consider a separate stage for additional gain or to narrow the filter bandwidth.

Band-Reject (Notch) Filter Design

Band-reject filter design begins with the selection of center frequency, f_0, and Q. Calculate the value of K. Choose a value for C1 that is approximately 10 μF / f_0. This determines the values of C2 and C3 as shown. Calculate the value for R1 that results in the desired value for f_0. This determines the values of R2 and R3 as shown. Select a convenient value for R4 + R5 that does not load the output amplifier. Calculate R4 and R5 from the value of K.

The depth of the notch depends on how closely the values of the components match the design values. The use of 1%-tolerance resistors is recommended and, if possible, matched values of capacitance. If all identical components are used, two capacitors can be paralleled to create C3 and two resistors in parallel create R3. This helps to minimize thermal drift.

11.5.4 Using Active Filter Design Tools

While the simple active filter examples presented above can be designed manually, more sophisticated circuits are more easily designed using filter-design software. Follow the same general approach to determining the filter's performance requirements and then the filter family as was presented in the section on LC filters. You can then enter the values or make the necessary selections for the design software. Once a basic design has been calculated, you can then "tweak" the design performance, use standard value components and make other adjustments. The design example presented below shows how a real analog design is assembled by understanding the performance requirements and then using design software to experiment for a "best" configuration.

The next several sections will make use of Texas Instrument's "freeware" filter design software, *FilterPro*. This package is extremely useful in designing active RC filters. This package allows filter parameters to be adjusted so as to "tweak" the design close to standard component values. (Either go to **www.ti.com** and search for "FilterPro" or go directly to **focus.ti.com/docs/toolsw/folders/print/filterpro.html**.) The reader is encouraged to follow along and experiment with *FilterPro* as a means of becoming familiar with the software so that it can be used for other filter design tasks.

More free online tools, such as National Semiconductor's excellent "Webench" suite of Web-based design software (**webench.national.com**) and online filter design calculators can also be used. Even if a manual design process is used, using a software tool to double-check the results is a good way to verify the design before building the circuit.

DESIGN EXAMPLE: 750-HZ HIGH-PERFORMANCE DIRECT CONVERSION RECEIVER FILTER

This example illustrates the design of the front-end filter for a high-performance direct-conversion I-Q phasing CW receiver. Along with the design of the filter, there will be some discussion of how much gain to assign to each circuit in a sequence of stages. This is included to illustrate some of the processes by which performance requirements for a circuit are established.

A unity-gain filter with a fixed-gain block in front of it means that the fixed-gain block will have no out-of band signal rejection and thus may be subject to overload due to strong out-of-band signals. Conversely, a unity-gain filter followed by a fixed-gain block could degrade signal sensitivity due to the noise internal to the unity-gain filter section. A filter with distributed gain is better than either of these two situations as the signals can be both

Table 11.20
Factor "a" for Low- and High-Pass Filters in Fig 11.46

No. of Stages	Stage 1	Stage 2	Stage 3	Stage 4
1	1.414	–	–	–
2	0.765	1.848	–	–
3	0.518	1.414	1.932	–
4	0.390	1.111	1.663	1.962

These values are truncated from those of Appendix C of Williams, *Electronic Filter Design Handbook*, for even-order Butterworth filters

filtered and amplified stage by stage.

The input signal for this example will be assumed to be straight from the detector with a 100-Ω output impedance and may be as large as 4 V_{P-P} (+16 dBm). This filter is assumed to operate from 12-V supplies so that up to 8 V_{P-P} can easily be handled by commonly available op amps such as a LM5532. (Unless an op amp is capable of rail-to-rail output, its output voltage can approach no closer than 2 V from either the positive or negative supply rails, thus 12 − 2 − 2 = 8 V of total output swing.)

A typical receiver requires approximately 80 to 90 dB of total gain for adequate headphone output levels. Speaker-level output usually requires an additional 20 dB of gain. If the volume control is placed too early in the gain chain (at the antenna input, for example), the audio stage will always be running at maximum gain and the result will be a high level of unwanted internally generated receiver hiss, noise that is not affected by the volume control. On the other hand, if the volume control is too late in the gain chain (such as at the receiver output), all the signals being received will be amplified by the maximum receiver gain of 80 to 90 dB. This can cause strong signals in the passband to saturate the audio chain, causing unwanted distortion unless the receiver uses AGC to reduce the gain.

A reasonable compromise is to place the volume control roughly halfway along the chain of gain stages. Thus, if 80 dB of total gain is desired, roughly 40 dB would occur before the volume control and 40 dB after. This keeps the gain after the volume control low enough that unwanted receiver hiss will

Unless otherwise specified, values of R are in ohms, C is in farads, F in hertz and ω in radians per second. Calculations shown here were performed on a scientific calculator.

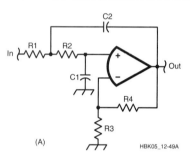
(A)

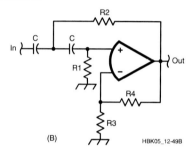
(B)

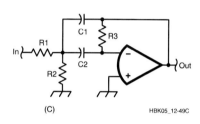
(C)

Low-Pass Filter

$$C_1 \leq \frac{\left[a^2 + 4(K-1)\right]C_2}{4}$$

$$R_1 = \frac{2}{\left[aC_2 + \sqrt{\left[a^2 + 4(K-1)\right]C_2^2 - 4C_1C_2}\right]\omega_C}$$

$$R_2 = \frac{1}{C_1 C_2 R_1 \omega_C^2}$$

$$R_3 = \frac{K(R_1 + R_2)}{K - 1} \quad (K > 1)$$

$$R_4 = K(R_1 + R_2)$$

where
K = gain
f_c = −3 dB cutoff frequency
$\omega_c = 2\pi f_c$
C_2 = a standard value near $10/f_c$ (in μF)
Note: For unity gain, short R4 and omit R3.
Example:
  a = 1.414 (see table, one stage)
  K = 2
  f = 2700 Hz
  ω_c = 16,964.6 rad/sec
  C_2 = 0.0033 μF
  C1 ≤ 0.00495 μF (use 0.0050 μF)
  R1 ≤ 25,265.2 Ω (use 24 kΩ)
  R2 = 8,420.1 Ω (use 8.2 kΩ)
  R3 = 67,370.6 Ω (use 68 kΩ)
  R4 = 67,370.6 Ω (use 68 kΩ)

High-Pass Filter

$$R_1 = \frac{4}{\left[a + \sqrt{a^2 + 8(K-1)}\right]\omega_C C}$$

$$R_2 = \frac{1}{\omega_C^2 C^2 R_1}$$

$$R_3 = \frac{KR_1}{K - 1} \quad (K > 1)$$

$$R_4 = KR_1$$

where
K = gain
f_c = −3 dB cutoff frequency
$\omega_c = 2\pi f_c$
C = a standard value near $10/f_c$ (in μF)

Note: For unity gain, short R4 and omit R3.

Example:
  a = 0.765 (see table, first of two stages)
  K = 4
  f = 250 Hz
  ω_c = 1570.8 rad/sec
  C = 0.04 μF (use 0.039 μF)
  R1 = 11,123.2 Ω (use 11 kΩ)
  R2 = 22,722 Ω (use 22 kΩ)
  R3 = 14,830.9 Ω (use 15 kΩ)
  R4 = 44,492.8 Ω (use 47 kΩ)

Band-Pass Filter

Pick K, Q, $\omega_o = 2\pi f_c$
where f_c = center freq.
Choose C
Then

$$R1 = \frac{Q}{K_0 \omega_0 C}$$

$$R2 = \frac{Q}{(2Q^2 - K_0)\omega_0 C}$$

$$R3 = \frac{2Q}{\omega_0 C}$$

Example:
  K = 2, f_o = 800 Hz, Q = 5 and C = 0.022 μF
  R1 = 22.6 kΩ (use 22 kΩ)
  R2 = 942 Ω (use 910 Ω)
  R3 = 90.4 kΩ (use 91 kΩ)

Fig 11.46 — Equations for designing a low-pass RC active audio filter are given at A. B, C and D show design information for high-pass, band-pass and band-reject filters, respectively. All of these filters will exhibit a Butterworth response. Values of K and Q should be less than 10. See Table 11.20 for values of "a".

be largely eliminated when the volume control is turned all the way down as long as a low-noise amplifier chain is used. This implies that the first 40 dB of gain will be before the volume control, which in this design will be rolled into the active RC filter.

The design objectives at this point are:

1) To provide 40 dB of gain in the desired 750 Hz passband, but 0 dB of gain at 2 kHz and higher.

2) At every stage in the filter, the signal at 2 kHz should not be allowed to exceed 8 V_{P-P} for a 4-V_{P-P} input. This means gain at 2 kHz should be 20 log (8/4) or 6 dB or less out of each stage.

3) The filter should be designed to minimize the impact on receiver sensitivity. It should not add unnecessary noise that would mask weak signals or be susceptible to overload from strong signals.

The first and second goals are attempts to ensure that no signal out of the detector at 2 kHz or higher will overload the filter section. This enables the creation of a very high performance direct-conversion receiver. The second goal is to allow that receiver to have high sensitivity which in turn has noise implications on the filter design.

Audio Filter Q Implications

When using the *FilterPro* software, Q must be specified for each filter section. In experimenting with the filter types of the Bessel, Butterworth, and Chebyshev and observing the frequency responses, it can be quickly seen that higher Q is associated with sharper filter frequency rolloff. From this observation, the conclusion could be drawn that high Q in an active RC filter is a good thing. This conclusion is not entirely true.

From a receiver design perspective, the goal is to reject undesired signals to the highest degree possible. An ideal 750-Hz low-pass filter would pass all signals at and below 750 Hz while completely eliminating all signals 751 Hz and higher. A high-order active RC filter with high-Q filter sections comes closest to this ideal. However, this sharp frequency rolloff does not come without a price.

The first problem with high-Q filter sections is ringing. **Fig 11.47** was generated using *FilterPro* for a 5th-order Chebyshev filter with 1 dB of ripple, a cutoff frequency of 750 Hz, and 40 dB of gain. This filter provides 20 dB of attenuation at 2 kHz exceeding our design goal of 0 dB. However, notice in particular the sharp peak filter group delay response at the 750 Hz cutoff frequency. This sharp group delay peak is associated with audio ringing.

The effect of ringing in a filter is much the same as ringing a bell. Strike a bell with a hammer, and the bell "rings" at a certain frequency. Likewise, when noisy, static-filled band noise hits a high-Q filter such as the one shown above, this impulse noise tends to produce an audible "ring" sound at the frequency of the delay spike. The effect of the filter ringing is that it actually creates audible interference that interferes with and can mask the desired signal.

It should be noted that simple crystal ladder filters used in many simple superheterodyne or "superhet" receivers have a band-pass characteristic. Thus, there is both a high and a low band-pass edge where group delay peaks occur. That means the typical narrow 300- to 500 Hz-wide CW crystal filter tends to ring badly at both a high (top end of the bandpass response) and a low frequency (bottom end of the band-pass response) at the same

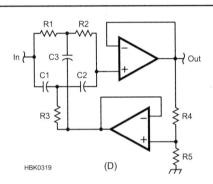

Band-Reject Filter

$$F_0 = \frac{1}{2\pi R1 C1}$$

$$K = 1 - \frac{1}{4Q}$$

$$R \gg (1-K)R1$$

where

$$C1 = C2 = \frac{C3}{2} = \frac{10\,\mu F}{f_0}$$

R1 = R2 = 2R3
R4 = (1 − K)R
R5 = K × R

Example:
f_0 = 500 Hz, Q=10
K = 0.975
C1 = C2 = 0.02 µF (or use 0.022 µF)
C3 = 0.04 µF (or use 0.044 µF)
R1 = R2 = 15.92 kΩ (use 15 kΩ)
R3 = 7.96 kΩ (use 8.2 kΩ)
R ≫ 1 kΩ
R4 = 25 Ω (use 24 Ω)
R5 = 975 Ω (use 910 Ω)

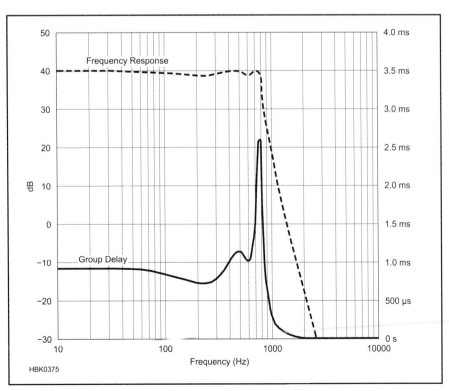

Fig 11.47 — Frequency response and group delay of 5th order Chebyshev, 1 dB passband ripple, 40 dB of gain.

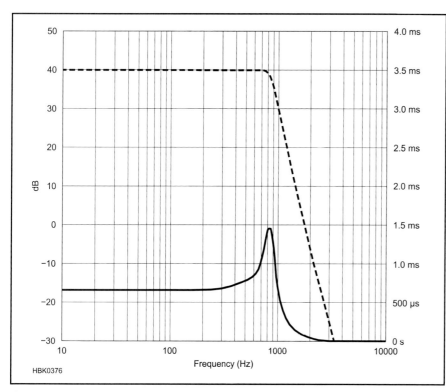

Fig 11.48 — Frequency response (dashed line) and group delay (solid line) of 5th order Chebyshev, 0.06 dB passband ripple.

time, which makes the ringing audio artifacts twice as bad.

The second problem with high-Q filters is an effect that is not at all obvious. It is simply the fact that our ears do not like them. This is caused by both the high phase and delay variations near the edge of the filter. Something in our brain "notices" these variations and objects to them. To our ears, a filter than has lower phase and delay variations "sounds" better, even outside the ringing issue.

In practice, the problems associated with both ringing and phase or delay variations can be reduced by limiting the highest Q filter section to a maximum Q of around 3. When using a design tool like *FilterPro*, this can be done by selecting a filter type of Chebyshev, and then reducing the specified allowed ripple dB value until the Q is reduced to roughly 3. For a 5th-order Chebyshev, this means reducing the allowable ripple from 1 dB to 0.06 dB. The group delay and frequency response of such a filter are shown in **Fig 11.48**.

Notice that the delay peak (the lower line) is now both smaller in amplitude and much broader than that in the previous example, which is exactly what we are after. Notice also that the frequency rolloff (the top curve) is not as sharp either, but it is still 8 dB better than our target of 0 dB at 2 kHz and above.

What does this mean? In active RC filter design, there is a tradeoff between simple and useful vs. more complex and better sounding. A higher-Q active RC filter may require only three filter sections while a lower-Q, better-sounding active RC filter with similar rolloff characteristics may require four filter sections. It is very valuable to realize that a tradeoff is possible and that an active RC filter can be built which is both sharp and sounds good (low phase/delay variations) at the same time.

Noise Implications

Resistors create noise that can mask the small signals we are trying to filter. If this filter is being used at the high-signal end of an audio chain, high-value resistors can be selected without any real harm. Resistance values as high as 1 MΩ can be useful in allowing the selection of small value capacitors which are readily and cheaply available in 2% or 5% tolerance values. However, if this filter is to be used in the front end of a receiver chain, the resistor values need to be much lower. (Receiver noise is also discussed in the **Receivers** chapter.)

A 50-Ω resistor creates about $0.85 \text{ nV}/\sqrt{\text{Hz}}$ of noise. Think of this as the noise generated by a 50-Ω antenna system. This noise voltage varies with the square root of the resistance change. Thus, a 1-MΩ resistor produces $0.85\sqrt{1,000,000/50}$ or $120 \text{ nV}/\sqrt{\text{Hz}}$ of noise. Thus, using a 1-MΩ resistor in the first stage of an active RC filter would reduce the sensitivity of an ideal receiver by 20 log (120/0.85) or 43 dB which is not good for receiver sensitivity.

If each stage of the active RC filter has gain, the effect of that gain is to lessen the impact of the noise contributed by the resistors in each succeeding stage. For example, if we want 40 dB (100× voltage gain) in three filter stages, then if the gain is evenly distributed across all the stages, each stage will have a gain of the cube root of 100 or a gain of 4.6 in each of the three stages. In this case, an input signal at $0.85 \text{ nV}/\sqrt{\text{Hz}}$ into the first stage will be 4.6× larger ($3.9 \text{ nV}/\sqrt{\text{Hz}}$) into the second stage and 4.6× larger ($18.3 \text{ nV}/\sqrt{\text{Hz}}$) into the third stage. One goal would be to use resistors than generate at most half the noise voltage of the desired signal. Thus the second stage ($3.9 \text{ nV}/\sqrt{\text{Hz}}$) should have resistors no larger than $3.9 = 0.85\sqrt{x/50}$ or $50 \times (3.9/0.85) \times (3.9/0.85) = 1052$ Ω. It should be no problem in the second and third stages to restrict resistor values to the 500-1000-Ω range in order to minimize their noise contribution.

FILTER DESIGN AND COMPONENT VALUE OPTIMIZATION

With an understanding of the gain and component values, we can use *FilterPro* to design our filter. We want a gain of around 4.6× per stage, resistors in the under-1000 Ω range, a filter Q of 3 or less, and a cutoff frequency of 750 Hz since this is a CW filter. The initial *FilterPro* screen appears as in **Fig 11.49** with a default Passband setting of Low-Pass.

Note the filter frequency response in the graph. Although we have not yet added any gain to any of the stages (we want to add 40 dB or 100×), we can see that the attenuation at 2 kHz is only 20 dB. If we were to add 40 dB of gain, we would still have 20 dB of gain at 2 kHz. Thus we need a sharper filter than a three-pole Butterworth.

We will now change the Filter type to "Chebychev" (one of several spellings), and the Cutoff Frequency to 750 Hz. Under Components, change the resistor setting to "Exact."

In the particular example of a direct-conversion receiver post-detector low-pass filter, it is best to use an odd number of poles so that the odd one-pole section can be readily configured as a receiver post detector pre-amplifier stage. (One-pole low- and high-pass filters were introduced in Fig 11.45.) Experimenting with *FilterPro*, it can be seen that even numbered pole filters produce more complex stages. Thus, when a three-pole filter is not good enough, the next step up should be a five-pole filter in this particular application.

When using a five-pole Chebyshev filter with a 750 Hz cutoff frequency and an initial value for R1 of 1 kΩ (remember, we want resistances of 1 kΩ or below), *FilterPro* gives the result shown in **Fig 11.50**.

Notice that at 2 kHz, the filter attenuation is now 60 dB. When we create 40 dB of gain

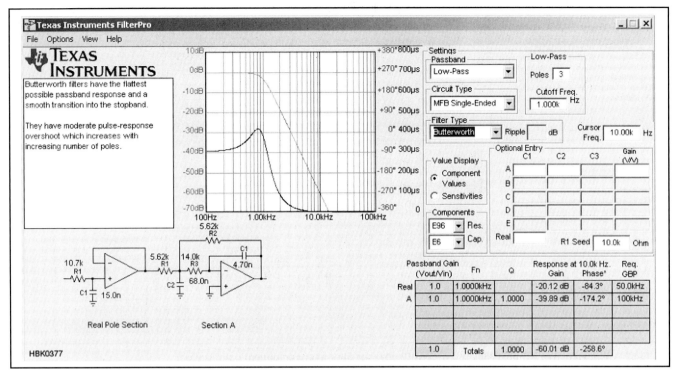

Fig 11.49 — *FilterPro* startup view.

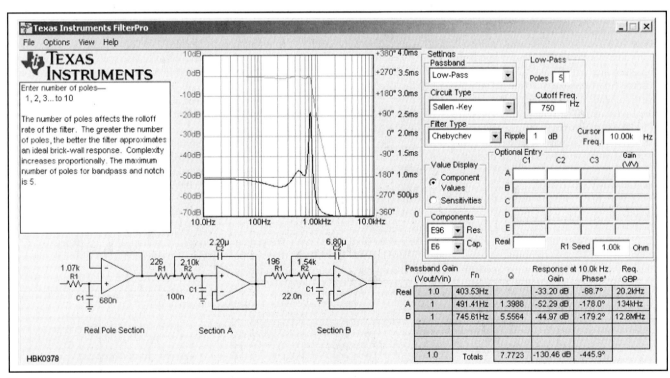

Fig 11.50 — First pass at designing a 5-pole 750-Hz low-pass Chebyshev filter.

in these filter stages, the filter attenuation will still be 20 dB better than needed. However, in the bottom right hand corner, the Q of each section is given, and the highest Q (section B) is 5.55 — higher than the desired maximum Q of 3. The Q of these stages can be adjusted by changing the passband ripple specification, which is set to 1 dB by default for the Chebyshev response. Manually lower the allowable ripple until the Q goes below 3. Experimentally you will find that lowering the passband ripple from 1 dB down to 0.05 dB produces the desired Q of a bit less than 3.

Next set the stages up for the proper gain. You will notice in Fig 11.50 that Circuit Type has been changed to Sallen-Key. This configuration tends to work a bit better as properly biasing the first stage to half the supply voltage will also dc bias all the stages after it and this configuration tends to produce values that are easier to work with. Earlier it was calculated that each stage needed a gain of 4.6× (4.6 × 4.6 × 4.6 = ~100×). This is not

RF and AF Filters 11.25

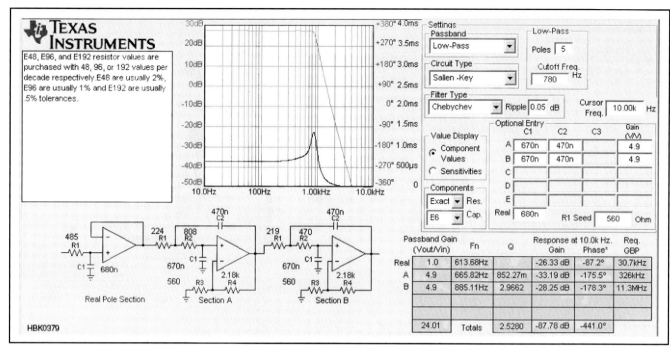

Fig 11.51 — Final result — a 5-pole 780-Hz low-pass Chebyshev filter after component optimization.

a precise value. We want to use component values that are close to standard values. Thus we can play with the gain per stage (staying close to 4.6×), and set the "R1 seed" value to obtain values for R3 and R4 close to standard values, then play with C1 and C2 and perhaps adjust the cutoff frequency a bit to get values close to standard values for the other resistors in the two stages. The final result is shown in **Fig 11.51** — a 5-pole, 780-Hz low-pass filter. The software makes the process of "cut and try" much easier than a manual design!

The result comes about from tweaking the cutoff frequency a bit (750 Hz to 780 Hz) and playing with the capacitor values. Capacitor values of 0.67 µF were used instead of 0.68 µF as that value is expensive and hard to find. The 0.67 µF caps are composed of one more-available 0.47 µF capacitor and two very commonly available 0.1 µF capacitors in parallel. This allows the main two sections (sec-

tion A and B) of the filter to use a total of four 0.47 µF and four 0.1 µF capacitors. In optimizing for component values, section B with higher Q is most sensitive to component value, so section B was optimized to get very close to 220 and 470 Ω (standard values), while the lower-Q section will not be as close when using 220 and 820 Ω.

A note about the gain-setting resistors R3 and R4: The goal was to keep resistors under 1 kΩ in these two stages, but R4 is 2.2 kΩ. As far as noise contributions are concerned, R4 and R3 "look" like they are in parallel to the input of the op amps. Thus, together they look like 1/(1/560 + 1/2180) or 445 Ω, which is indeed much less than 1 kΩ.

This still leaves the one-pole "real pole section" to be configured with the proper gain. There are two approaches that can be taken. The simplest circuit is presented in **Fig 11.52** with R1 = 485 Ω and C1 = 680 nF.

The input connects to C4 and the output is taken from the output pin of U5.

To match the receiver detector output impedance of 100 Ω, we scale R1 from 485 to 100 Ω. We must also scale C1 up by the same amount (4.85×) to make C1 = 3.3 µF. R1 can actually be eliminated as a separate component, replacing it with the 100-Ω output impedance of the detector. If R1 is eliminated, C1 should be moved to the input side of C4. C1 should be a ceramic type capacitor, not an electrolytic.

If the detector produces its own 1.5-V bias from the 3-V supply (like a Tayloe detector), the bias components R4, R5, R6 and C3 can also be eliminated along with the dc isolation capacitor, C4. U5 is shown as an LT6231 low-noise 3-V op amp. If a less expensive 12 V device is used, such as an LM5532, the bias

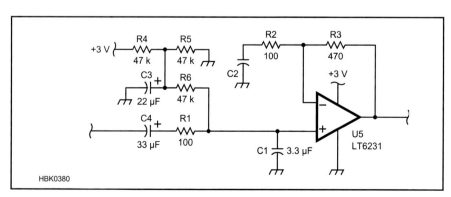

Fig 11.52 — Active filter circuit that implements the "Real Pole" section with biasing and ac decoupling components (see text).

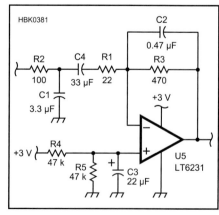

Fig 11.53 — Alternate "Real Pole" circuit with higher performance than Fig 11.52.

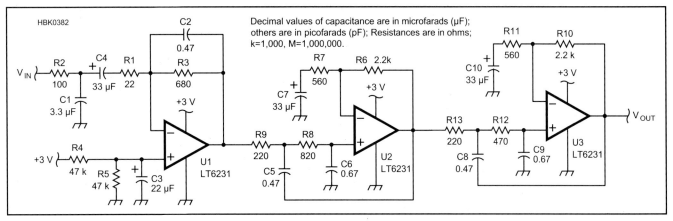

Fig 11.54 — Schematic of the complete filter, including dc biasing components.

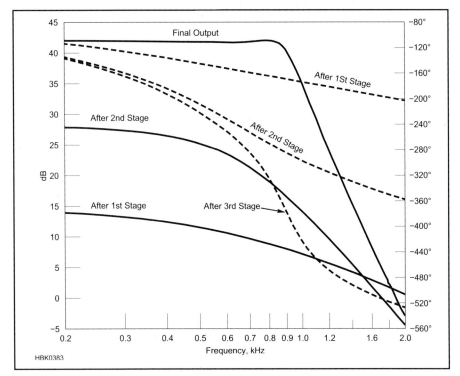

Fig 11.55 — Frequency response of all stages added one at a time. Dashed lines show phase response.

network will be used to set the bias voltage to 6 V (1/2 the power-supply voltage) and all these parts will be needed.

Another higher-performance variation of the "real pole" section is shown in **Fig 11.53**. Again, R2 can be eliminated if the output impedance of the detector is 100 Ω. In this configuration, C1 and C2 work together to provide rolloff. This configuration provides around 22 dB of attenuation at 10 kHz compared to 12 dB in the first implementation above. The gain of the stage is now R3/(R1+R2) or about 5.5×. Both circuits were simulated using *LTSpice* (free circuit simulation software from Linear Technologies at **www.linear.com/software**), and R3 and C2 were selected (using trial and error) to provide a similar gain peak (13.7 dB vs 13.5 dB) and a similar gain at the filter cutoff frequency of 780 Hz. (8.6 dB vs 9.2 dB).

Fig 11.54 is the complete five-pole filter with all necessary biasing and dc isolation, using standard value parts. If a 12-V op amp is used, the 3-V supply voltages shown above will be replaced with 12-V supplies. Remember that the 670 nF capacitors are composed of one 0.47 µF and two common 0.1 µF capacitors in parallel.

The resulting frequency plot (as modeled in *LT Spice*) is shown in **Fig 11.55**. The lowest gain curve is the frequency response of the initial "real pole" section. The gain almost reaches 0 dB by 2 kHz as desired. The next highest gain curve is the frequency response at the output of the second stage (the net response of the first and second stages). Although the total gain is higher, the gain still drops below 0 dB before 2 kHz. The highest gain line shows the total filter response and shows a slight rise near the cutoff frequency. The gain is almost 42 dB (125× voltage gain), close to the desired 40 dB target gain and also drops below 0 dB of gain before 2 kHz.

With this filter placed just after the receiver detector, no signal out of the detector (4 to 5 V_{P-P}) at 2 kHz or higher (just 1.22 kHz above the filter cutoff of 780 Hz) is capable of overloading the front end of the receiver.

Stage Order

In the example above, the stages were ordered from lowest Q (the real section) to the highest Q (section B). This order gives the best protection from inter-stage overload, but is not necessarily the best order for best receiver sensitivity. Fig 11.55 was generated showing the net frequency response adding one stage at a time. However, **Fig 11.56** shows the frequency response of each of the three stages separately.

Notice that the section responses labeled "Real Pole" and "Section A (Q=0.85)" both lose a lot of gain approaching the filter edge at 780 Hz. This loss of gain means that the resistor noise will have more impact than expected, in effect reducing the receiver sensitivity somewhat. Notice that the highest-Q section, labeled "Section B (Q = 3)" actually has a gain peak near the band edge. A peak like this will be present in any filter stage with a Q higher than 1. The higher the Q, the higher and sharper this peak will be.

The "real pole" section has a simpler configuration that makes it easy to use as the first pre-amp stage of the filter, so it comes first. However, moving Section B from the last stage to the second stage will help overcome the stage gain reduction of the first stage and provide for better receiver sensitivity overall as shown in **Fig 11.57**.

When the three stages are ganged in this order — Real Pole, Section B, Section A — the signal out of the second section is higher near the filter cutoff of 780 Hz, allowing for better low-noise performance. As can be seen in the final output curve, the stage ordering does not change the overall filter frequency response.

A drawback of this configuration could be that the second stage of the filter might now be more susceptible to overload (due to the gain peak) from large signals nearer the edge of the filter than before. This is not really the case in this filter, however. Since the filter was designed with roughly equal gain in each stage, an overload to the second stage is automatically also an overload to the last stage, since it applies additional gain to the second filter section output. However, if this filter had been designed with unity gain per section, the ordering of the stages (and stage overload) could be more of a concern.

In summary, for best ultra-low-noise results (which is only important with very small signals) the stages should be ordered:
• Highest Q sections to lowest Q sections
• If odd order, odd Real Section goes first

For best signal overload protection within the filter bandpass (which is only important if there is little or no gain per section), the sections should be ordered from lowest Q (a real pole section) first (if any) to the highest Q section.

If the filter has many sections and has little or no gain per stage, an optimum balance between good low noise results and good internal passband filter overload protection might be to alternate the sections between the lowest Q and the highest Q sections. For example if a 9-pole filter has a real section, and four other filter sections with Qs of 0.6, 0.9, 1.8, and 5.7, the best section order might be: Real Pole section, followed by Q=5.7, Q=0.6, Q=1.8, and lastly Q=0.9. Simulating and studying the combined frequency response using a program like *LTSpice* is very useful in understanding what is going on.

ALL-PASS ACTIVE FILTERS

A filter that is often seen in phasing-type receivers is the all-pass filter. This filter passes signals of all frequencies without affecting their amplitudes, but creates a controlled phase shift that varies with frequency. This phase shift is used in quadrature direct-conversion receivers for creating single frequency reception by creating a 90° phase difference used to cancel out an unwanted sideband. (See the **Receivers** and the **DSP and Software Radio Design** chapters for more information on direct-conversion receivers.)

The circuit of **Fig 11.58** takes the quadrature I and Q outputs of a direct-conversion quadrature detector (like a Tayloe detector) and adds two all-pass filters designed to create 90° of phase difference between the top and bottom two-stage all-pass sections. U1 and U2 form one phase delay section while U3 and U4 form the other. The 90° difference in phase between these two unity gain

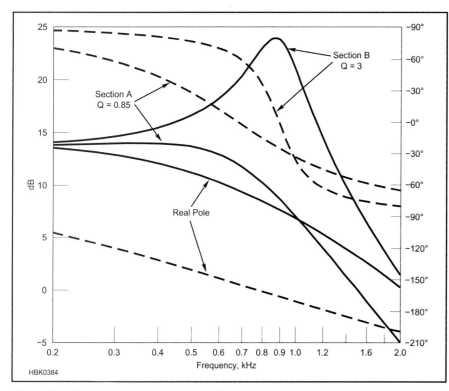

Fig 11.56 — Frequency response of each filter stage shown separately. Dashed lines show phase response.

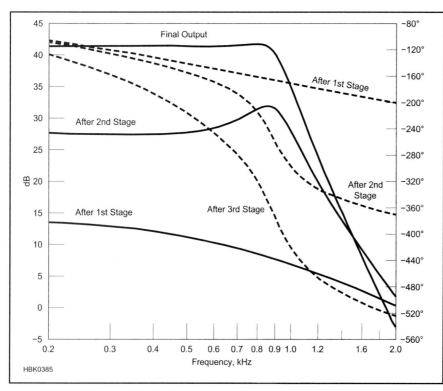

Fig 11.57 — Cumulative frequency response with the high-Q stage in the middle. Dashed lines show phase response.

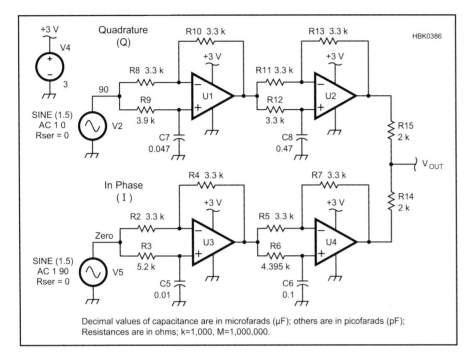

Fig 11.58 — Two all-pass filters create SSB direct-conversion receiver audio from 300 to 1000 Hz.

all-pass sections is shown in **Fig 11.59**.

In a direct-conversion quadrature receiver, the I and Q outputs are 90° apart from each other and contain the audio from both sidebands. For one of the sideband signals, the I output is 90° ahead of the Q output and for the other sideband signal, Q is 90° ahead of I. Adding in an additional 90° of delay will cause I and Q to both have the same delay on one sideband (−90 + 90 = 0) and 180° of phase difference (90 + 90 = 180) on the other sideband. Since the output is taken via the sum of R14 and R15, a 180° difference will cause the signals from one sideband to cancel, while a 0° difference will allow the signals of the other sideband to add together.

The 90° difference holds well only over a limited range, so suppression of the opposite sideband gets worse at the high and low end of the "sweet spot" where the signals are precisely 90° apart and suppression of the unwanted sideband is best. When adjusted properly, opposite sideband suppression can be excellent over a limited range as shown in Fig **11.60** This was a phasing section designed for CW covering 300 Hz to 1 kHz with well over 50 dB of opposite sideband rejection. A 300-Hz high-pass filter and a 1000-Hz low-pass filter can be used to attenuate the high and low frequency ranges in which sideband suppression drops below 50 dB.

It is difficult to hand-pick components to the 0.1% precision needed to get 60 dB of suppression. Small trimmer resistors can be placed in series with fixed values at R3 and R6 in order to allow the filter to be tuned. The book *Experimental Methods in RF Design* discusses all-pass phasing sections in great detail. The program *QuadNet* for designing and analyzing active quadrative networks is included on the *Handbook* CD-ROM.

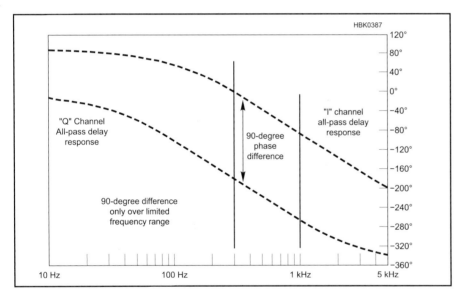

Fig 11.59 — Differential phase delay for two all-pass phasing sections.

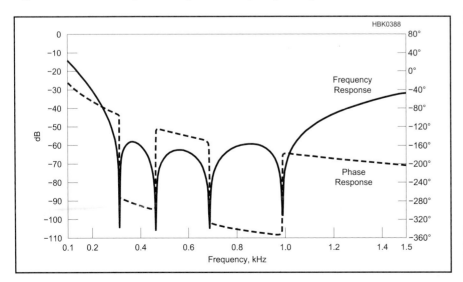

Fig 11.60 — More than 50 dB of opposite sideband suppression is obtained by using the differential phase shift between two all-pass phasing stages.

RF and AF Filters 11.29

11.6 Quartz Crystal Filters

Inductor Q values effectively limit the minimum bandwidths that can be achieved with LC band-pass filters. Higher-Q circuit elements, such as quartz crystal, PLZT ceramic, or mechanical resonators, are required to extend these limits. Quartz crystals (actually slices of quartz crystals) offer the highest Q and best stability with time and temperature of all available resonators. They are manufactured for a wide range of frequencies from audio to VHF, using *cuts* (orientation of the slice with respect to the crystal's axes of symmetry) that suit the frequency and application of the resonator. The AT-cut is favored for operation at HF and VHF, whereas other types (DT, SL and so on) are more convenient for use at lower frequencies.

Each crystal plate has several *modes* of mechanical vibration, which can be excited electrically through the piezoelectric effect, but generally resonators are designed to maximize their response at one particular frequency. Consequently, for filter design, quartz crystal resonators are modeled using the simplified equivalent circuit shown in **Fig 11.61**. L_m, C_m and r_m represent the *motional parameters* of the resonator with r_m being the loss resistance, which is also known as the *equivalent series resistance*, or ESR. C_o is a combination of the static capacitance formed between the two metal electrodes with the quartz as dielectric (ε_r = 4.54 for AT-cut crystals) and some additional capacitance introduced by the metal case, base and mount. There is a physical relationship between C_m and the static capacitance formed by the resonator electrodes, but, unfortunately, the added holder capacitance masks this direct relationship causing C_o/C_m to vary from 200 to over 500. However, for modern fundamental AT-cut crystals between 1 and 30 MHz, their values of C_m are typically between 0.003 and 0.03 pF.

The *motional inductance* of a crystal resonator is essentially the same whether it is operated on the *fundamental* or one of its *overtones* (vibration at a higher frequency, often harmonically related to the fundamental frequency). However, the motional capacitance varies between fundamental and overtone as $1/n^2$, where n is the order of the overtone.

An important parameter for crystal filter design is the *unloaded Q* at the resonant frequency of the series arm, f_s, usually denoted by Q_U.

$$Q_u = 2\pi f_s L_m / r_m \qquad (7)$$

Q_U is very high, usually exceeding 20,000, even for VHF overtone crystals, making it possible to design quartz crystal band-pass filters with a tremendously wide range of bandwidths and center frequencies.

The basic filtering action of a crystal can be seen from **Fig 11.62**, which shows a plot of attenuation versus frequency for the test circuit shown in the inset. The series arm of the crystal equivalent circuit forms a series-tuned circuit, which passes signals with little attenuation at its resonant frequency, f_s, but appears inductive above this frequency and parallel resonates with C_o at f_p to produce a deep notch in transmission. The difference between f_s and f_p is known as the *pole-zero separation* (PZ), and is dependent on C_m/C_o as well as f_s. Further information on quartz crystal theory and operation can be found in the **Oscillators and Synthesizers** chapter and in "Introduction to Quartz Crystal

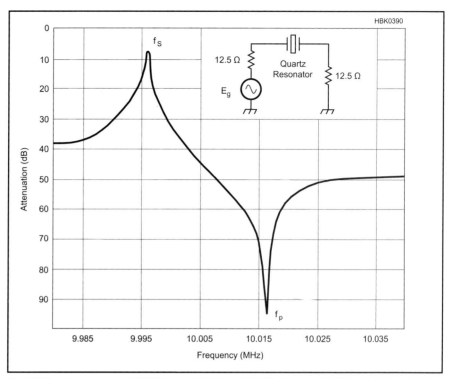

Fig 11.62 — Response of 10 MHz crystal (C_m = 0.0134 pF, L_m = 18.92 mH, ESR = 34 Ω) in a series test circuit (see inset) showing peak of transmission (lowest attenuation) at the series resonant frequency, f_s, and a null (maximum attenuation) caused by the parallel resonance due to C_o at f_p.

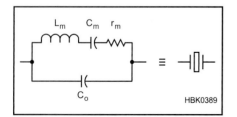

Fig 11.61 — Equivalent circuit of a quartz crystal and its circuit symbol.

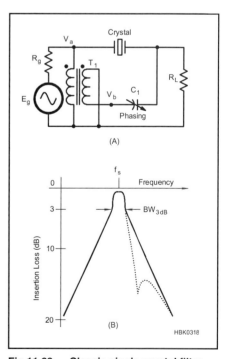

Fig 11.63 — Classic single-crystal filter in A has the response shown in B. The phasing capacitor can be adjusted to balance out C_o (solid line), or set to a lesser or greater value to create a movable null to one side, or other, of the passband (dotted line).

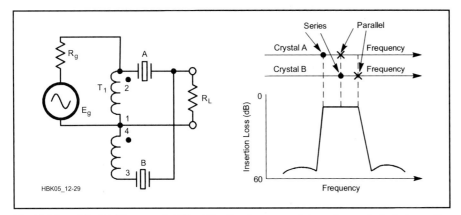

Fig 11.64 — A half-lattice crystal filter. The C_o of one crystal can be made to balance the C_o of the other, or C_o across the higher crystal can be deliberately increased to create nulls on either side of the passband.

electrically coupled crystal resonators.

Multi-pole filters can be built by coupling duals together using external capacitors, the C_o from each dual input and output being absorbed in the coupling capacitors or terminations, or by including more acoustically coupled resonators in the electrode structure on a single quartz plate. Though not common, 3-pole and 4-pole monolithic filters housed in standard, single crystal holders are available from some manufacturers.

11.6.1 Crystal Filter Parameters

An ideal band-pass filter would pass the desired signal with no loss and completely attenuate everything else. Practically, it is not possible to achieve such a response with a finite number of resonators, and approximations to this ideal have to be accepted. Greater stop-band attenuation and steeper sides can be achieved if more and more crystals are used, and the response gets nearer to the ideal brick wall shape. This feature of a filter is expressed as a *shape factor*, which specifies the ratio of the bandwidth at an attenuation of 60 dB to the bandwidth at 6 dB — both of these levels being taken relative to the actual passband peak to eliminate insertion loss from the calculation.

An ideal brick wall filter would have a shape factor of 1, and practical filters have shape factors that depend on the number of crystals used in the design and the type of response chosen for the passband. A 1 dB Chebyshev design, for example, can typically produce shape factors that vary from about 4.1 for a 4-pole to 1.5 for a 10-pole, but the actual figures obtained in practice are very much dependent on Q_u and how much greater it is than the filter $Q = f_o / BW$, where f_o is the center frequency and BW the 3dB bandwidth. The ratio of Q_u to f_o / BW is often quoted as Q_o, or q_o, and this, along with the order and type of response, determines the insertion loss of a crystal filter. Q_o also determines how closely the passband follows the design response, and how much the passband ripple is smoothed out and the edges of the response rounded off by crystal loss.

Commercial filter manufacturers usually choose the Chebyshev *equiripple* design because it gives the best compromise between passband response and the steepness of the sides. As a result, 1 dB-ripple Chebyshev designs are the de facto standard for speech bandwidths. Tolerances in component values and crystal frequencies can cause the ripple in the passband to exceed 1 dB, so often the maximum ripple is specified as 2 dB even though the target ripple is lower.

Insertion loss, the loss through the filter in its passband, also varies with the type of response and increases as the order of the filter

Unit Design," by Bottom [Ref 1]. Poles and zeroes are discussed in the **Analog Basics** chapter.

A simple crystal filter developed in the 1930s is shown in **Fig 11.63A**. The voltage-reversing transformer T1 was usually an IF transformer, but nowadays could be a bifilar winding on a ferrite core. Voltages V_a and V_b have equal magnitude but 180° phase difference. When $C1 = C_o$, the effect of C_o will disappear and a well-behaved single resonance will occur as shown in Fig 11.63B. However, if C1 is adjusted to unbalance the circuit, a movable null of increasing depth away from the passband is created, as shown by the dotted curve, which can be set to either side of the passband to attenuate interfering adjacent signals. This "crystal gate" type of filter, operating at 455 kHz, was present in many high-quality amateur communications receivers from the 1930s thru to the 1960s. When the filter was switched into the receiver IF amplifier, the bandwidth was reduced to a few hundred Hz for CW reception. The design could also be used to good effect at frequencies up to 1.7 MHz with increased bandwidth.

The *half-lattice filter* shown in **Fig 11.64** offers an improvement in performance over a single-crystal filter. The quartz crystal static capacitances, C_o, cancel each other. The remaining series-resonant arms, if offset in frequency and terminated properly, will produce an approximate 2-pole Butterworth or Chebyshev response. The crystal spacing for simple Chebyshev designs is usually around two-thirds of the bandwidth. Half-lattice filter sections can be cascaded to produce composite filters with multiple poles. Many of the older commercial filters are coupled half-lattice types using 4, 6 or 8 crystals, and this is still the favored technology for some crystal filters at lower frequencies. Reference 11 discusses the computer design of half-lattice filters.

Fig 11.65 — A dual monolithic crystal filter has two sets of electrodes acoustically coupled to provide a two-pole response in a single crystal unit. The center lead is connected to the case and grounded in normal operation.

Most commercial HF and VHF crystal filters produced today use dual monolithic structures as their resonant elements. These *monolithic* (Greek for "one stone") crystal filters consist of a single quartz plate onto which two sets of metal electrodes have been deposited, physically separated to control the acoustic (mechanical) coupling between them. An example of a dual monolithic filter (2-pole) is shown in **Fig 11.65**. Monolithic crystal filters are available with center frequencies from 9 MHz to well over 120 MHz.

In effect, the dual monolithic structure behaves just like a pair of coupled crystal resonators. There is a subtle difference, however, because the static capacitance in dual monolithic filters ("duals") appears across the input and output terminations and doesn't produce a null above the passband, as it would for two

increases, just as for LC filters. The insertion loss for a given order and bandwidth is higher for high-ripple Chebyshev designs than it is for low-ripple designs, and Butterworth designs have lower insertion loss than any Chebyshev type for equivalent order.

Passband amplitude response and shape factor are important parameters for assessing the performance of filters used for speech communications. However, group delay is also important for data and narrow-band CW reception. Differential group delay (different group delays for signals of different frequencies) can cause signal distortion on data signals if the variations are greater than the data recovery system's automatic equalizer can handle. Ringing can be an annoying problem when using very narrow CW filters, and group delay must be minimized to reduce this effect.

When narrow bandwidth filters are being considered, shape factor has to be sacrificed to reduce group delay and its associated ringing problems. It's no good having a narrow Chebyshev design with a shape factor of 2 if the filter produces unacceptable ringing. Both Bessel and *linear phase* (equiripple 0.05°) responses have practically constant, low group delay across the entire passband and well beyond on either side, making both types good choices for narrow CW or specialized data use. They also have the great advantage of offering the lowest possible insertion loss of all the types of response currently in use, which is important when Q_o is low, as it often is for very narrow bandwidth filters. The insertion loss of the Bessel design is marginally lower than that of the linear phase, but the latter has a superior shape factor giving it the best balance of low group delay and good selectivity. A 6-pole linear phase (equiripple 0.05°) design has a shape factor of 3.39, whereas a 6-pole Bessel has 3.96.

11.6.2 Crystal Filter Design

A wide variety of crystals are produced for use with microprocessors and other digital integrated circuits. They are offered in several case styles, but the most common are HC-49/U and HC-49/US. Crystal resonators in the HC-49/U style case are fabricated on 8 mm diameter quartz discs, whereas those in the low-profile HC-49/US cases are fabricated on 8 mm by 2 mm strips of quartz. At any given frequency, C_m will be lower for HC-49/US crystals because the active area is smaller than it is in the larger HC-49/U crystals.

Both types of crystals are cheap and have relatively small frequency spreads (the variation in resonant frequency of a batch of crystals), making them ideal for use in the LSB ladder configuration suggested by Dishal (Ref 2) — see **Fig 11.66**. This arrangement requires the motional inductances of all the

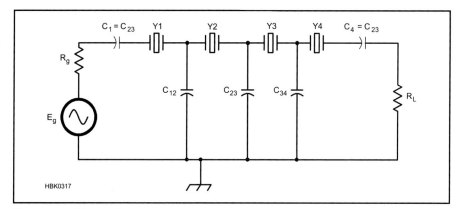

Fig 11.66 — Dishal LSB crystal-ladder filter configuration. Crystals must have identical motional inductances, and the coupling capacitors and termination resistors are selected according to the bandwidth and type of passband response required.

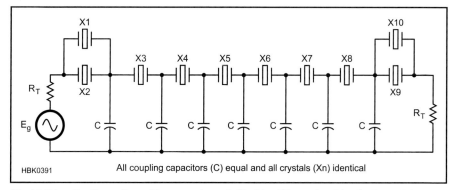

Fig 11.67 — Configuration of improved crystal ladder filter using identical crystals and equal coupling capacitor values. The parallel resonator end-sections (PRES) can provide excellent passband responses, giving either quasi-equiripple (QER) or minimum-loss (PRESML) responses with just a change in termination resistance.

crystals to be identical, and each loop in isolation (crystal and coupling capacitors either side) to be resonant at the same frequency. Series capacitors to trim individual crystals are needed to achieve this in some of the more advanced designs, where production frequency spreads are not sufficient to satisfy the latter requirement. References 3 through 8 contain design information on Dishal LSB crystal ladder filters.

The *min-loss* or *minimum-loss* form of Cohn ladder filter, where $C_{12} = C_{23} = C_{34}$, has become very popular in recent years because it's so simple to design and build. However, it suffers from the drawback that the ripple in its passband response increases dramatically with increasing order, and ringing can be a real problem at bandwidths below 500 Hz for Cohn min-loss filters of 6th-order or higher. The ripple may not be a problem in most narrow filters because it's smoothed out almost completely by loss, but the ringing can be tiring. For wider bandwidths, where the ratio of Q_u to f_o / BW is much greater, the ripple is not smoothed out, and is very evident.

One way around the problem of ripple, without sacrificing simplicity, is to use an arrangement originally designed for variable bandwidth applications by Dave Gordon-Smith, G3UUR, shown in **Fig 11.67**. It was devised to make the mesh frequencies track together as the amount of coupling varies the bandwidth, and only requires variable resistance terminations ganged to the variable coupling capacitors to give excellent responses at any setting of the bandwidth. These can be optimized for minimum loss at minimum bandwidth and almost equal ripple at maximum bandwidth, making it an ideal alternative for fixed bandwidth speech applications to the Cohn min-loss. Two crystals are used in parallel to halve the motional inductance and double the motional capacitance of the resonators in the end sections. Although the two additional crystals do not increase the order of the filter by two, they do reduce the passband ripple substantially while maintaining the simplicity of design and construction offered by the min-loss filter.

In addition, the group delay of the *parallel-resonator-end-section* (PRES) configuration is less than that of the min-loss. The coupling capacitors are still all equal, and the filter can be terminated to achieve

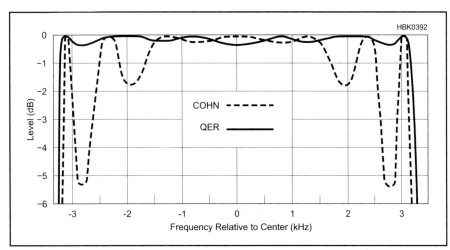

Fig 11.68 — Comparison of 8-pole Cohn min-loss passband response with that of the quasi-equiripple (QER) type. Note the almost equal ripple in the passband of the QER response.

Table 11.21
3 dB-down k & q Values for Quasi-Equiripple (QER) Ladder Filters

Order	q	k_{12}	k_{23}	Shape Factor	Max Ripple (dB)
4	0.9442	0.7660	0.5417	4.56	0.002
5	1.0316	0.7625	0.5391	3.02	0.018
6	1.0808	0.7560	0.5346	2.31	0.09
7	1.1876	0.7459	0.5275	1.90	0.16
8	1.2532	0.7394	0.5228	1.66	0.31
9	1.3439	0.7335	0.5187	1.50	0.42
10	1.4115	0.7294	0.5158	1.40	0.60
11	1.4955	0.7261	0.5134	1.33	0.72
12	1.5506	0.7235	0.5116	1.28	0.90

Table 11.22
Typical Parameters for AT-Cut Quartz Resonators

Freq (MHz)	Mode n	rs (Ω)	Cp (pF)	Cs (pF)	L (mH)	Q_U
1.0	1	260	3.4	0.0085	2900	72,000
5.0	1	40	3.8	0.011	100	72,000
10.0	1	8	3.5	0.018	14	109,000
20	1	15	4.5	0.020	3.1	26,000
30	3	30	4.0	0.002	14	87,000
75	3	25	4.0	0.002	2.3	43,000
110	5	60	2.7	0.0004	5.0	57,000
150	5	65	3.5	0.0006	1.9	27,000
200	7	100	3.5	0.0004	2.1	26,000

Courtesy of Piezo Crystal Co, Carlisle, Pennsylvania

a *quasi-equiripple response* (QER) so that its passband resembles that of a Chebyshev design, or minimum loss (PRESML) with a response like that of the min-loss. **Fig 11.68** shows the Cohn min-loss and QER passband responses with infinite crystal Q_u for comparison. Values of k and q for QER filters from 4 to 12 poles are given in **Table 11.21**, along with the maximum ripple and shape factor for each order.

The coupling capacitor value for any bandwidth can be determined from k_{23} using equation 8. The end-section resonators formed by the two parallel crystals have twice the effective motional capacitance of the inner resonators, and since k_{12} is always 1.414 times k_{23} for the QER design, the value of C_{12} will be the same as calculated for C_{23} and all the other coupling capacitors, C, in the design.

$$C = \frac{f_0 C_m}{BW\, k_{23}} \quad (8)$$

The termination resistance, R_T, must be calculated using half the motional inductance of a single crystal, as illustrated in equation 9, where L_m is the motional inductance of one of the parallel crystals used in the end-sections.

$$R_T = 2\pi\, BW\, L_m / 2q = \pi\, BW\, L_m / q \quad (9)$$

11.6.3 Crystal Characterization

Simple crystal filters can be constructed using cut-and-try methods, but sometimes results can be very disappointing. The only sure way to guarantee good results is to fully characterize the crystals beforehand, so that only those most suitable are used in a design appropriate for the application. When crystal parameters are known, computer modeling can be used to assess the effect of Q_u and C_o on bandwidth, before proceeding to the construction phase. Along with *Elsie* design program on the CD-ROM, *AADE Filter Design and Analysis* (www.aade.com) provides free filter modeling. **Table 11.22** lists typical parameters for the common AT-cut quartz crystals.

Crystal characterization can be performed with very modest or very advanced test equipment, the only difference being the accuracy of the results. The *zero phase* method for measuring C_m used in industry can be implemented by amateurs if a dual-channel oscilloscope is available to substitute as a phase detector. (Ref 7 gives details of this method.) However, many successful crystal filter constructors have achieved excellent results with home-built test equipment. Ref 6 describes a simple switched-capacitor method of measuring C_m originated by G3UUR, which only requires a frequency counter in addition to constructing an oscillator. Accurate LC meters that can measure down to 0.01 pF and 1 nH can now be constructed using PIC technology. C_m can usually be estimated to within 10% by using equation 10 and a value for C_o measured on one of these instruments.

$$C_m(pF) = \frac{C_0 - 0.95}{175} \quad (10)$$

Commercial LC meters with amazingly good specifications are also available at moderate prices if a PIC LC meter seems too ambitious a project for home construction at this stage.

Values of Q_u for crystals can vary considerably, even within the same batch, and the ratio of the best to the worst, excluding dead ones, can be as high as 6 for cheap mass-produced crystals. The ratio can still be as high as 2 for batches of high-quality crystals.

Q_u needs to be known for each crystal to weed out the poor ones and more accurately model the filter performance prior to construction. The simplest means of doing this, if the switched-capacitor method (Ref 6) is used for determining C_m and a sufficient quantity

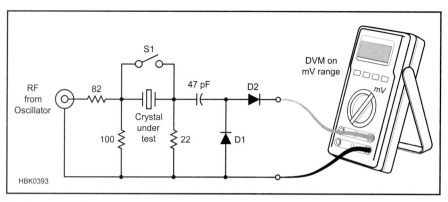

Fig 11.69 — A simple jig for measuring crystal ESR. The jig may be driven by a signal generator, or a modified crystal test oscillator, and the signal through the crystal detected on a DMM using the mV dc range. Resistors are ⅛ or ¼ W, 5%. D1 and D2 are small signal germanium or Schottky barrier diodes. S1 is a miniature pushbutton switch.

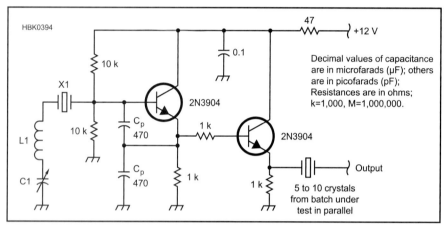

Fig 11.70 — Modified crystal test oscillator suitable for driving the ESR measuring jig in Fig 11.69. L1 can be an RF choke, and in combination with C1 must be capable of swinging the oscillator frequency about 1 kHz above and below the series-resonant frequency of the crystal under test.

of crystals has been purchased, is to drive the jig shown in **Fig 11.69** with the switched-capacitor oscillator converted to a variable crystal oscillator as shown in **Fig 11.70.**

The harmonic output of the original oscillator is quite high and can cause significant errors, so it must be filtered at its output with a simple series clean-up circuit comprising 5 to 10 parallel crystals from the batch being measured. This filter is very effective at removing the harmonics, but must have a bandwidth several times greater than the crystal under test, and five or more crystals in parallel should ensure this.

The choice of C1 and L1 depend on the frequency and tolerance of the crystals, but 0.05% of L_m is a good starting point for L1, and C1 can be anything from 75 pF to 250 pF, depending on what's available. It's imperative that the oscillator frequency can swing above and below the series resonant frequency of the test crystals, which may be 1 kHz below the marked frequency at 4 MHz and could be as much as 3 or 4 kHz below it at 12 MHz, so that a peak in the dc output voltage can be discerned.

The jig can be built on a 2 × 1.5 inch piece of PCB material, with two 4 mm banana plugs soldered to the board so that it can plug directly into the DMM. A narrow strip of copper around the +V terminal needs to be removed to isolate it, then the jig can be built in the "ugly" or "dead-bug" style on the remaining copper ground plane. The push-button shorting switch, S_1, can be mounted on another piece of PCB material about 0.5 × 0.75 inch soldered at 90° to the main board.

When measuring crystal ESR, the oscillator frequency should be adjusted to peak the reading on the DMM with switch S1 open, and the voltage noted. Switch S1 should then be closed to short out the crystal, and a second reading taken. The pairs of dc voltages obtained with this jig can be calibrated for r_m using metal-film or carbon-film resistors in place of the crystal under test. A table can be drawn up of voltage pairs for resistor values from, say, 10 to 100 Ω with S1 both open and closed to compare with the results from the crystal measurements. Don't be tempted to use the dc voltage readings to work out r_m directly, because they will not be linearly related to the RF signal voltages at the mV levels used in the jig.

11.6.4 Crystal Filter Evaluation

The simplest means of assessing the performance of a crystal filter is to temporarily install it in a finished transceiver, or receiver, and use a strong on-air signal, or locally generated carrier, to run through the filter passband and down either side to see if there are any anomalies. Provided that the filter crystals have been carefully characterized in the first instance, and computer modeling has shown that the design is close to what's required, this may be all that's required to confirm a successful project.

More elaborate checks on both the passband and stop band can be made if a direct digital synthesis (DDS) signal generator, or vector network analyzer (VNA), is available. A test set-up for evaluating the response of a filter using a DDS generator requires an oscilloscope to display the response, whereas a VNA controlled via the USB port of a PC can display the response on the PC screen with suitable software. (See the N2PK Web site — **n2pk.com**.) In addition, it can measure the phase and work out the group delay of the filter. Construction of a VNA is a very worthwhile project, and in addition to its many other applications it can be used to characterize the crystals prior to making the filter as well as evaluating filter performance after completion.

11.7 SAW Filters

The resonators in a monolithic crystal filter are coupled together by bulk acoustic waves. These acoustic waves are generated and propagated in the interior of a quartz plate. It is also possible to launch, by an appropriate transducer, acoustic waves that propagate only along the surface of the quartz plate. These are called "surface-acoustic-waves" because they do not appreciably penetrate the interior of the plate.

A *surface-acoustic-wave* (SAW) filter consists of thin aluminum electrodes, or fingers, deposited on the surface of a piezoelectric substrate as shown in **Fig 11.71**. Lithium Niobate ($LiNbO_3$) is usually favored over quartz because it yields less insertion loss. The electrodes make up the filter's transducers. RF voltage is applied to the input transducer and generates electric fields between the fingers. The piezoelectric material vibrates in response, launching an acoustic wave along the surface. When the wave reaches the output transducer it produces an electric field between the fingers. This field generates a voltage across the load resistor.

Since both input and output transducers are not entirely unidirectional, some acoustic power is lost in the acoustic absorbers located behind each transducer. This lost acoustic power produces a mid-band electrical insertion loss typically greater than 10 dB. The SAW filter frequency response is determined by the choice of substrate material and finger pattern. The finger spacing, (usually one-quarter wavelength) determines the filter center frequency. Center frequencies are available from 20 to 1000 MHz. The number and length of fingers determines the filter loaded Q and shape factor.

Loaded Qs are available from 2 to 100, with a shape factor of 1.5 (equivalent to a dozen poles). Thus the SAW filter can be made broadband much like the LC filters that it replaces. The advantage is substantially reduced volume and possibly lower cost. SAW filter research was driven by military needs for exotic amplitude-response and time-delay requirements. Low-cost SAW filters are presently found in television IF amplifiers where high mid-band loss can be tolerated.

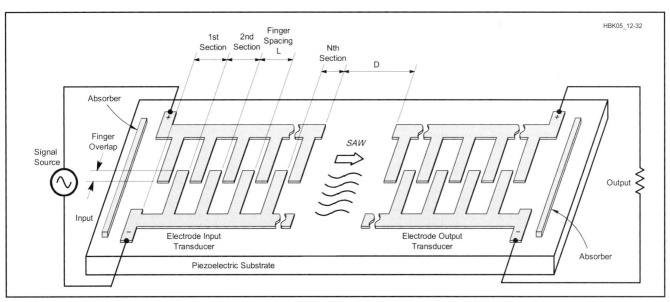

Fig 11.71 — The *interdigitated* transducer, on the left, launches SAW energy to a similar transducer on the right (see text).

11.8 Transmission Line Filters

LC filter calculations are based on the assumption that the reactances are *lumped*—that the physical dimensions of the components are considerably less than the operating wavelength. In such cases the unavoidable inter-turn parasitic capacitance associated with inductors and the unavoidable series parasitic inductance associated with capacitors are neglected as being secondary effects. If careful attention is paid to circuit layout and miniature components are used, lumped LC filter technology can be used up to perhaps 1 GHz.

Replacing lumped reactances with selected short sections of *Transverse Electromagnetic Mode* (TEM) transmission lines results in transmission line filters. (In TEM the electric and magnetic fields associated with a transmission line are at right angles (transverse) to the direction of wave propagation.) Coaxial cable, stripline and microstrip are examples of TEM components. Waveguides and waveguide resonators are not TEM components.

Coaxial cable transmission line filters are often used at HF and VHF frequencies. Stripline and microstrip transmission-line filters predominate from 500 MHz to 10 GHz. In addition they are often used down to 50 MHz when narrowband ($Q_L > 10$) band-pass filtering is required. In this application they exhibit considerably lower loss than their LC counterparts and are useful at frequencies where coaxial transmission lines are too lossy. A detailed treatment of the use of coaxial cable to form transmission line filters is presented in the **Transmission Lines** chapter. This section focuses on stripline and microstrip filters used at VHF and above.

11.8.1 Stripline and Microstrip Filters

Fig 11.72 shows three popular transmission lines used in transmission line filters. The circular coaxial transmission line (*coax*) shown in Fig 11.72A consists of two concentric metal cylinders separated by dielectric (insulating) material. The first transmission-line filters were built from sections of coaxial line. Their mechanical fabrication is expensive and it is difficult to provide electrical coupling between line sections.

Fabrication difficulties are reduced by the use of shielded strip transmission line (*stripline*) shown in Fig 11.72B. The outer conductor of stripline consists of two flat parallel metal plates (ground planes) and the inner conductor is a thin metal strip. Sometimes the inner conductor is a round metal rod. The dielectric between ground planes and strip can be air or a low-loss plastic such as polyethylene. The outer conductors (ground planes or shields) are separated from each other by distance b.

Striplines can be easily coupled together by locating the strips near each other as shown in Fig 11.72B. Stripline Z_0 vs width (w) is plotted in **Fig 11.73**. Air-dielectric stripline technology is best for low bandwidth ($Q_L > 20$) band-pass filters.

The most popular transmission line at UHF and microwave is *microstrip* (unshielded stripline), shown in Fig 11.72C. It can be fabricated with standard printed-circuit processes and is the least expensive configuration. In microstrip the outer conductor is a single flat metal ground-plane. The inner conductor is a thin metal strip separated from the ground-plane by a solid dielectric substrate. Typical substrates are 0.062-inch G-10 fiberglass ($\varepsilon = 4.5$) for the 50-MHz to 1-GHz frequency range and 0.031-inch Teflon ($\varepsilon = 2.3$) for frequencies above 1 GHz. Unfortunately, microstrip has the most loss of the three types of transmission line; therefore it is not suitable for narrow, high-Q, band-pass filters.

Conductor separation must be minimized or radiation from the line and unwanted coupling to adjacent circuits may become problems. Microstrip characteristic impedance and the effective dielectric constant (ε) are shown in **Fig 11.74**. Unlike coax and stripline, the effective dielectric constant is less than that of the substrate since a portion of the electromagnetic wave propagating along the microstrip travels in the air above the substrate.

The characteristic impedance for stripline and microstrip-lines that results in the lowest loss is not 75 Ω as it is for coax. Loss decreases as line width increases, which leads to clumsy, large structures. Therefore, to conserve space, filter sections are often constructed from 50-Ω stripline or microstrip stubs even though the loss at that characteristic impedance is not a minimum for that type of transmission line.

11.8.2 Transmission Line Band-Pass Filters

Band-pass filters can also be constructed from transmission line stubs. (See the **Transmission Lines** chapter for information on stub behavior and their use as filters at HF and VHF.) At VHF the stubs can be considerably shorter than a quarter-wavelength (¼-λ), yielding a compact filter structure with less mid-band loss than its LC counterpart. The single-stage 146-MHz stripline band-pass filter shown in **Fig 11.75** is an example. This filter consists of a single inductive 50-Ω strip-line stub mounted into a 2 × 5 × 7-inch aluminum box. The stub is resonated at 146 MHz with the "APC" variable capacitor, C1. Coupling to the 50-Ω generator and load is provided by the coupling capacitors

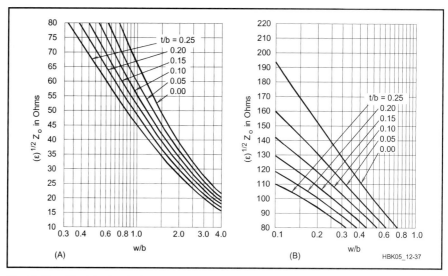

Fig 11.73 — The Z_0 of stripline varies with w, b and t (conductor thickness). See Fig 11.72B. The conductor thickness is t and the plots are normalized in terms of t/b.

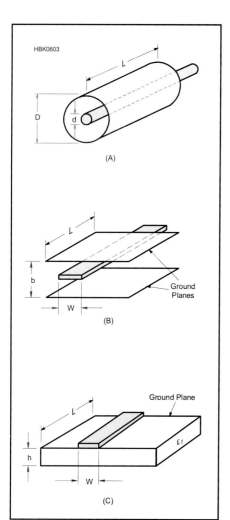

Fig 11.72 — Transmission lines. A: Coaxial line. B: Coupled stripline, which has two ground planes. C: Microstrip has only one ground plane.

Z_0 Ω	$\varepsilon=1$ (AIR) W/h	$\varepsilon=2.3$ (RT/Duroid) W/h	$\sqrt{\varepsilon_e}$	$\varepsilon=4.5$ (G-10) W/h	$\sqrt{\varepsilon_e}$
25	12.5	7.6	1.4	4.9	2.0
50	5.0	3.1	1.36	1.8	1.85
75	2.7	1.6	1.35	0.78	1.8
100	1.7	0.84	1.35	0.39	1.75
	$\sqrt{\varepsilon}=1$				

Fig 11.74 — Microstrip parameters (after H. Wheeler, *IEEE Transactions on MTT*, March 1965, p 132). ε_e is the effective ε.

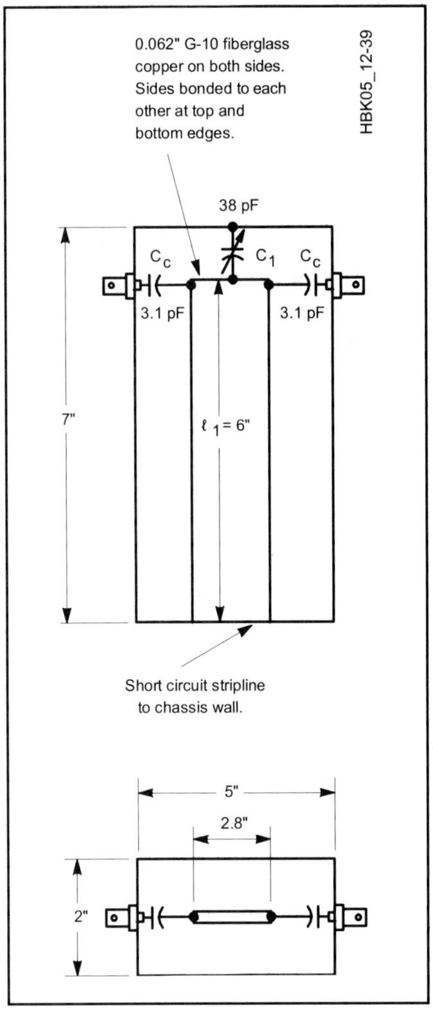

Fig 11.75 — This 146-MHz stripline band-pass filter has been measured to have a Q_L of 63 and a loss of approximately 1 dB.

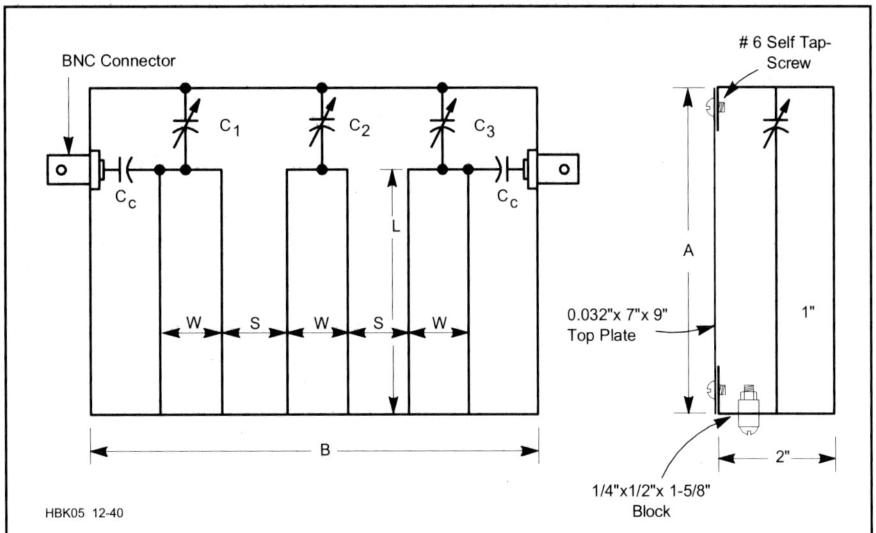

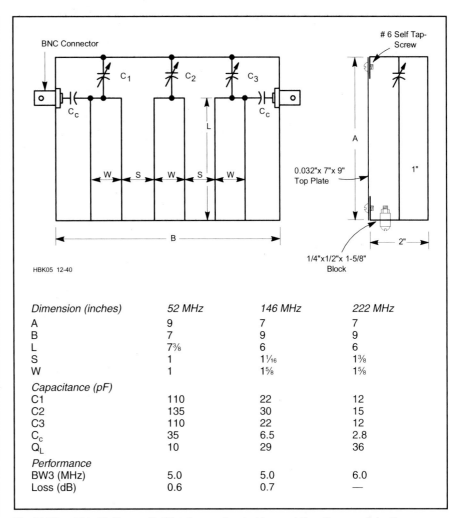

Dimension (inches)	*52 MHz*	*146 MHz*	*222 MHz*
A	9	7	7
B	7	9	9
L	7⅜	6	6
S	1	1 1/16	1⅜
W	1	1⅝	1⅝
Capacitance (pF)			
C1	110	22	12
C2	135	30	15
C3	110	22	12
C_C	35	6.5	2.8
Q_L	10	29	36
Performance			
BW3 (MHz)	5.0	5.0	6.0
Loss (dB)	0.6	0.7	—

Fig 11.76 — This Butterworth filter is constructed in combline. It was originally discussed by R. Fisher in December 1968 *QST*.

C_C. The measured performance of this filter is: f_o = 146 MHz, BW = 2.3 MHz (Q_L = 63) and mid-band loss = 1 dB.

Single-stage stripline filters can be coupled together to yield multistage filters. One method uses the capacitor coupled band-pass filter synthesis technique to design a 3-pole filter. Another method allows closely spaced inductive stubs to magnetically couple to each other. When the coupled stubs are grounded on the same side of the filter housing, the structure is called a "combline filter." Three examples of combline band-pass filters are shown in **Fig 11.76**. These filters are constructed in 2 × 7 × 9-inch aluminum boxes.

11.8.3 Quarter-Wave Transmission Line Filters

The reactance of a ¼-λ shorted-stub is infinite, as discussed in the **Transmission Lines** chapter. Thus, a ¼-λ shorted stub behaves like a parallel-resonant LC circuit. Proper input and output coupling to a ¼-λ resonator yields a practical band-pass filter. Closely spaced ¼-λ resonators will couple together to form a multistage band-pass filter. When the resonators are grounded on opposite walls of the filter housing, the structure is called an *interdigital filter* because the resonators look like interlaced fingers. Two examples of 3-pole UHF interdigital filters are shown in **Fig 11.77**. Design graphs for round-rod interdigital filters are given in Ref 9. The ¼-λ resonators may be tuned by physically changing their lengths or by tuning the screw opposite each rod.

If the short-circuited ends of two ¼-λ resonators are connected to each other, the resulting ½-λ stub will remain in resonance, even when the connection to the ground-plane is removed. Such a floating ½-λ microstrip line, when bent into a U-shape, is called a *hairpin* resonator. Closely coupled hairpin resonators can be arranged to form multistage band-pass filters. Microstrip hairpin band-pass filters are popular above 1 GHz because they can be easily fabricated using photo-etching techniques. No connection to the ground-plane is required.

11.8.4 Emulating LC Filters with Transmission Line Filters

Low-pass and high-pass transmission-line filters are usually built from short sections of transmission lines (stubs) that emulate lumped LC reactances. Sometimes low-loss lumped capacitors are mixed with transmission line inductors to form a hybrid component filter. For example, consider the 720-MHz, 3-pole microstrip low-pass filter shown in **Fig 11.78A** that emulates the LC filter shown in Fig 11.78B. C1 and C3 are replaced with 50-Ω open-circuit shunt stubs ℓ_C long. L2 is replaced with a short section of 100-Ω line ℓ_L long. The LC filter, Fig 11.78B, was designed for f_c = 720 MHz. Such a filter could be connected between a 432-MHz transmitter and antenna to reduce harmonic and spurious emissions. A reactance chart

Fig 11.77 — These 3-pole Butterworth filters for 432 MHz (shown at A, 8.6 MHz bandwidth, 1.4 dB passband loss) and 1296 MHz (shown at B, 110 MHz bandwidth, 0.4 dB passband loss) are constructed as interdigitated filters. The material is from R. E. Fisher, March 1968 *QST*.

shows that X_C is 50 Ω, and the inductor reactance is 100 Ω at f_c. The microstrip version is constructed on G-10 fiberglass 0.062-inch thick, with ε = 4.5. Then, from Fig 11.74, w is 0.11 inch and $\ell_C = 0.125\ \lambda_g$ for the 50-Ω capacitive stubs. Also, from Fig 11.74, w is 0.024 inch and ℓ_L is 0.125 ℓ_g for the 100-Ω inductive line. The inductive line length is approximate because the far end is not a short circuit. ℓ_g is 300/(720 × 1.75) = 0.238 m, or 9.37 inches Thus ℓ_C is 1.1 inch and ℓ_L is 1.1 inches

This microstrip filter exhibits about 20 dB of attenuation at 1296 MHz. Its response rises again, however, around 3 GHz. This is because the fixed-length transmission line stubs change in terms of wavelength as the frequency rises. This particular filter was designed to eliminate third-harmonic energy near 1296 MHz from a 432-MHz transmitter and does a better job in this application than the Butterworth filter in Fig 11.77 which has spurious responses in the 1296-MHz band.

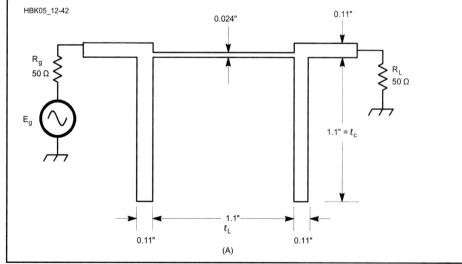

Fig 11.78 — A microstrip 3-pole emulated-Butterworth low-pass filter with a cutoff frequency of 720 MHz. A: Microstrip version built with G-10 fiberglass board (ε = 4.5, h = 0.062 inches). B: Lumped LC version of the same filter. To construct this filter with lumped elements very small values of L and C must be used and stray capacitance and inductance would have to be reduced to a tiny fraction of the component values.

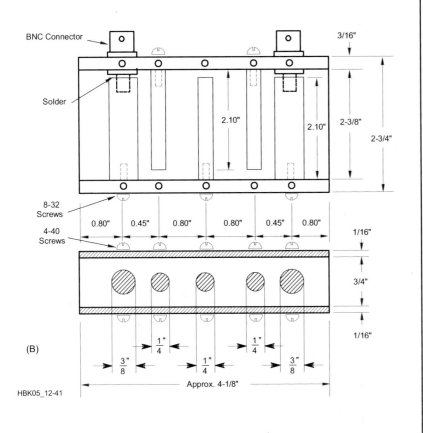

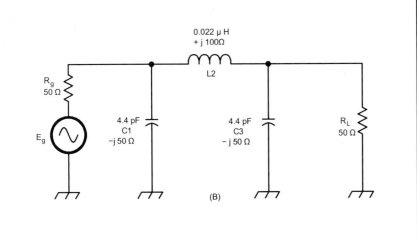

11.9 Helical Resonators

Ever-increasing occupancy of the radio spectrum brings with it a parade of receiver overload and spurious responses. Overload problems can be minimized by using high-dynamic-range receiving techniques, but spurious responses (such as the image frequency) must be filtered out before mixing occurs. Conventional tuned circuits cannot provide the selectivity necessary to eliminate the plethora of signals found in most urban and many suburban environments. Other filtering techniques must be used.

Helical resonators are usually a better choice than ¼-λ cavities on 50, 144 and 222 MHz to eliminate these unwanted inputs because they are smaller and easier to build than coaxial cavity resonators, although their Q is not as high as that of cavities. In the frequency range from 30 to 100 MHz it is difficult to build high-Q inductors, and coaxial cavities are very large. In this frequency range the helical resonator is an excellent choice. At 50 MHz for example, a capacitively tuned, ¼-λ coaxial cavity with an unloaded Q of 3000 would be about 4 inches in diameter and nearly 5 ft long. On the other hand, a helical resonator with the same unloaded Q is about 8.5 inches in diameter and 11.3 inches long. Even at 432 MHz, where coaxial cavities are common, the use of helical resonators results in substantial size reductions.

The helical resonator was described by the late Jim Fisk, W1HR, in a June 1976 *QST* article. [Ref 10] The resonator is described as a coil surrounded by a shield, but it is actually a shielded, resonant section of helically wound transmission line with relatively high characteristic impedance and low propagation velocity along the axis of the helix. The electrical length is about 94% of an axial ¼-λ or 84.6°. One lead of the helical winding is connected directly to the shield and the other end is open circuited as shown in **Fig 11.79**. Although the shield may be any shape, only round and square shields will be considered here.

11.9.1 Helical Resonator Design

The unloaded Q of a helical resonator is determined primarily by the size of the shield. For a round resonator with a copper coil on a low-loss form, mounted in a copper shield, the unloaded Q is given by

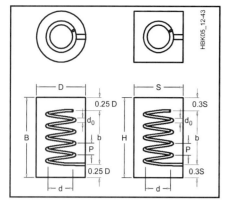

Fig 11.79 — Dimensions of round and square helical resonators. The diameter, D (or side, S) is determined by the desired unloaded Q. Other dimensions are expressed in terms of D or S (see text).

$$Q_U = 50\, D\, \sqrt{f_0} \qquad (11)$$

where
 D = inside diameter of the shield, in inches
 f_o = frequency, in MHz.

D is assumed to be 1.2 times the width of one side for square shield cans. This formula includes the effects of losses and imperfections in practical materials. It yields

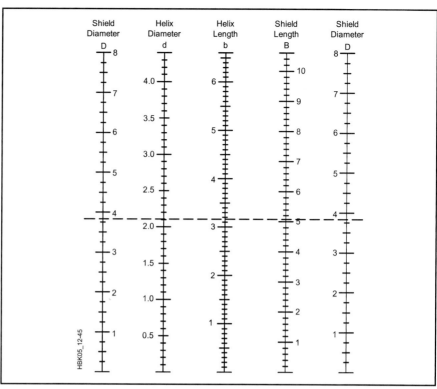

Fig 11.81 — The helical resonator is scaled from this design nomograph. Starting with the shield diameter, the helix diameter, d, helix length, b, and shield length, B, can be determined with this graph. The example shown has a shield diameter of 3.8 inches. This requires a helix mean diameter of 2.1 inches, helix length of 3.1 inches, and shield length of 5 inches.

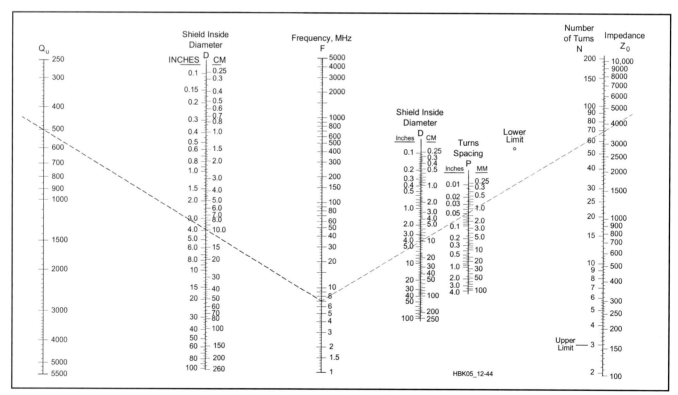

Fig 11.80 — The design nomograph for round helical resonators starts by selecting Q_U and the required shield diameter. A line is drawn connecting these two values and extended to the frequency scale (example here is for a shield of about 3.8 inches and Q_U of 500 at 7 MHz). Finally the number of turns, N, winding pitch, P, and characteristic impedance, Z_0, are determined by drawing a line from the frequency scale through selected shield diameter (but this time to the scale on the right-hand side. For the example shown, the dashed line shows P ≈ 0.047 inch, N = 70 turns, and Z_n = 3600 Ω).

values of unloaded Q that are easily attained in practice. Silver plating the shield and coil increases the unloaded Q by about 3% over that predicted by the equation. At VHF and UHF, however, it is more practical to increase the shield size slightly (that is, increase the selected Q_U by about 3% before making the calculation). The fringing capacitance at the open-circuit end of the helix is about 0.15D pF (that is, approximately 0.3 pF for a shield 2 inches in diameter). Once the required shield size has been determined, the total number of turns, N, winding pitch, P and characteristic impedance, Z_0, for round and square helical resonators with air dielectric between the helix and shield, are given by:

$$N = \frac{1908}{f_0 D} \quad (12A)$$

$$P = \frac{f_0 D^2}{2312} \quad (12B)$$

$$Z_0 = \frac{99,000}{f_0 D} \quad (12C)$$

$$N = \frac{1590}{f_0 S} \quad (12D)$$

$$P = \frac{f_0 S^2}{1606} \quad (12E)$$

$$Z_0 = \frac{82,500}{f_0 S} \quad (12F)$$

In these equations, dimensions D and S are in inches and f_0 is in megahertz. The design nomograph for round helical resonators in **Fig 11.80** is based on these formulas.

Although there are many variables to consider when designing helical resonators, certain ratios of shield size to length and coil diameter to length, provide optimum results. For helix diameter, d = 0.55 D or d = 0.66 S. For helix length, b = 0.825D or b = 0.99S. For shield length, B = 1.325 D and H = 1.60 S.

Design of filter dimensions can be done using the nomographs in this section or with computer software. The program *Helical* for designing and analyzing helical filters is included on this book's accompanying CD-ROM. Use of the nomographs is described in the following paragraphs.

Fig 11.81 simplifies calculation of these dimensions. Note that these ratios result in a helix with a length 1.5 times its diameter, the condition for maximum Q. The shield is about 60% longer than the helix — although it can be made longer — to completely contain the electric field at the top of the helix and the magnetic field at the bottom.

The winding pitch, P, is used primarily to determine the required conductor size. Adjust the length of the coil to that given by the equations during construction. Conduc-

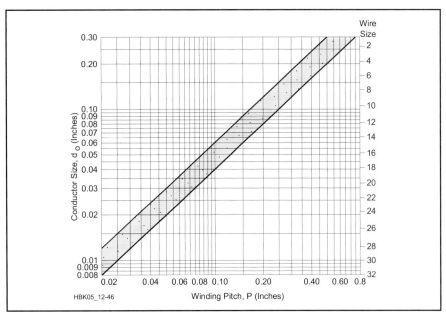

Fig 11.82 — This chart provides the design information of helix conductor size vs winding pitch, P. For example, a winding pitch of 0.047 inch results in a conductor diameter between 0.019 and 0.028 inch (#22 or #24 AWG).

tor size ranges from 0.4 P to 0.6 P for both round and square resonators and are plotted graphically in **Fig 11.82**.

Obviously, an area exists (in terms of frequency and unloaded Q) where the designer must make a choice between a conventional cavity (or lumped LC circuit) and a helical resonator. The choice is affected by physical shape at higher frequencies. Cavities are long and relatively small in diameter, while the length of a helical resonator is not much greater than its diameter. A second consideration is that point where the winding pitch, P, is less than the radius of the helix (otherwise the structure tends to be non-helical). This condition occurs when the helix has fewer than three turns (the "upper limit" on the design nomograph of Fig 11.80).

11.9.2 Helical Filter Construction

The shield should not have any seams parallel to the helix axis to obtain as high an unloaded Q as possible. This is usually not a problem with round resonators because large-diameter copper tubing is used for the shield, but square resonators require at least one seam and usually more. The effect on unloaded Q is minimized if the seam is silver soldered carefully from one end to the other.

Results are best when little or no dielectric is used inside the shield. This is usually no problem at VHF and UHF because the conductors are large enough that a supporting coil form is not required. The lower end of the helix should be soldered to the nearest point on the inside of the shield.

Although the external field is minimized by the use of top and bottom shield covers, the top and bottom of the shield may be left open with negligible effect on frequency or unloaded Q. Covers, if provided, should make electrical contact with the shield. In those resonators where the helix is connected to the bottom cover, that cover must be soldered solidly to the shield to minimize losses.

11.9.3 Helical Resonator Tuning

A carefully built helical resonator designed from the nomograph of Fig 11.80 will resonate very close to the design frequency. Slightly compress or expand the helix to adjust resonance over a small range. If the helix is made slightly longer than that called for in Fig 11.81, the resonator can be tuned by pruning the open end of the coil. However, neither of these methods is recommended for wide frequency excursions because any major deviation in helix length will degrade the unloaded Q of the resonator.

Most helical resonators are tuned by means of a brass tuning screw or high-quality air-variable capacitor across the open end of the helix. Piston capacitors also work well, but the Q of the tuning capacitor should ideally be several times the unloaded Q of the resonator. Varactor diodes have sometimes been used where remote tuning is required, but varactors can generate unwanted harmonics and other spurious signals if they are excited by strong, nearby signals.

When a helical resonator is to be tuned by a variable capacitor, the shield size is based on the chosen unloaded Q at the operating frequency. Then the number of turns, N and the winding pitch, P, are based on resonance at 1.5 f_0. Tune the resonator to the desired operating frequency, f_0.

11.9.4 Helical Resonator Insertion Loss

The insertion loss (dissipation loss), I_L, in decibels, of all single-resonator circuits is given by

$$I_L = 20 \log_{10} \left(\frac{1}{1 - \frac{Q_L}{Q_U}} \right) \quad (13)$$

where
Q_L = loaded Q
Q_U = unloaded Q

This is plotted in **Fig 11.83**. For the most practical cases ($Q_L > 5$), this can be closely approximated by $I_L \approx 9.0 \, (Q_L/Q_U)$ dB. The selection of Q_L for a tuned circuit is dictated primarily by the required selectivity of the circuit. However, to keep dissipation loss to 0.5 dB or less (as is the case for low-noise VHF receivers), the unloaded Q must be at least 18 times the Q_L.

11.9.5 Coupling Helical Resonators

Signals are coupled into and out of helical resonators with inductive loops at the bottom of the helix, direct taps on the coil or a combination of both. Although the correct tap point can be calculated easily, coupling by loops and probes must be determined experimentally.

The input and output coupling is often pro-

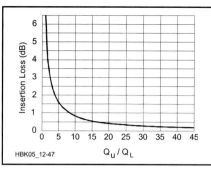

Fig 11.83 — The ratio of loaded (Q_L) to unloaded (Q_U) Q determines the insertion loss of a tuned resonant circuit.

vided by probes when only one resonator is used. The probes are positioned on opposite sides of the resonator for maximum isolation. When coupling loops are used, the plane of the loop should be perpendicular to the axis of the helix and separated a small distance from the bottom of the coil. For resonators with only a few turns, the plane of the loop can be tilted slightly so it is parallel with the slope of the adjacent conductor.

Helical resonators with inductive coupling (loops) exhibit more attenuation to signals above the resonant frequency (as compared to attenuation below resonance), whereas resonators with capacitive coupling (probes) exhibit more attenuation below the passband, as shown for a typical 432-MHz resonator in **Fig 11.84**. Consider this characteristic when choosing a coupling method. The passband can be made more symmetrical by using a combination of coupling methods (inductive input and capacitive output, for example).

If more than one helical resonator is re-

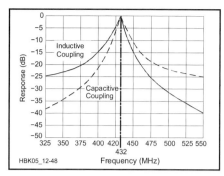

Fig 11.84 — This response curve for a single-resonator 432-MHz filter shows the effects of capacitive and inductive input/output coupling. The response curve can be made symmetrical on each side of resonance by combining the two methods (inductive input and capacitive output, or vice versa).

quired to obtain a desired band-pass characteristic, adjacent resonators may be coupled through apertures in the shield wall between the two resonators. Unfortunately, the size and location of the aperture must be found empirically, so this method of coupling is not very practical unless you're building a large number of identical units.

Since the loaded Q of a resonator is determined by the external loading, this must be considered when selecting a tap (or position of a loop or probe). The ratio of this external loading, R_b, to the characteristic impedance, Z_0, for a ¼-λ resonator is calculated from:

$$K = \frac{R_b}{Z_0} = 0.785 \left(\frac{1}{Q_L} - \frac{1}{Q_U} \right) \quad (14)$$

11.10 Use of Filters at VHF and UHF

Even when filters are designed and built properly, they may be rendered totally ineffective if not installed properly. Leakage around a filter can be quite high at VHF and UHF, where wavelengths are short. Proper attention to shielding and good grounding is mandatory for minimum leakage. Poor coaxial cable shield connection into and out of the filter is one of the greatest offenders with regard to filter leakage. Proper dc-lead bypassing throughout the receiving system is good practice, especially at VHF and above. Ferrite beads placed over the dc leads may help to reduce leakage. Proper filter termination is required to minimize loss.

Most VHF RF amplifiers optimized for noise figure do not have a 50-Ω input impedance. As a result, any filter attached to the input of an RF amplifier optimized for noise figure will not be properly terminated and filter loss may rise substantially. As this loss is directly added to the RF amplifier noise figure, carefully choose and place filters in the receiver.

11.11 Filter Projects

The filter projects to follow are by no means the only filter projects in this book. Filters for specific applications may be found in other chapters of this *Handbook*. Receiver input filters, transmitter filters, inter-stage filters and others can be extracted from the various projects and built for other applications. Since filters are a first line of defense against *electromagnetic interference* (EMI) problems, additional filter projects appear in the **EMC** chapter.

11.11.1 Crystal Ladder Filter for SSB

One of the great advantages of building your own crystal filter is the wide choice of crystal frequencies currently available. This allows the filter's center frequency to be chosen to fit more conveniently between adjacent amateur bands than those commercially available on 9 and 10.7 MHz do.

The 8.5 MHz series crystals used in this project were chosen in preference to ones on 9 MHz to balance the post-mixer filtering requirements on the 30- and 40-m bands, and reduce the spurious emission caused by the second harmonic of the IF when operating on the 17-m band. The crystals have the equivalent circuit shown in **Fig 11.85** and were obtained from Digi-Key (part number X418-ND).

The crystal-to-case capacitance becomes part of the coupling capacitance when the case is grounded, as it should be for best ultimate attenuation. The motional capacitance and inductance can vary by as much as ±5% from crystal to crystal, although the resonant frequency only varies by ±30 ppm. The ESR (equivalent series resistance) exhibits the most variation with mass-produced crystals, and in the case of these 8.5 MHz crystals can be anything from under 25 Ω to over 125 Ω.

Since the frequency variation has a spread of more than twice what can be tolerated in a 2.4 kHz SSB filter design, and the ESR is so variable, individual crystals need to be checked and selected for frequency matching and ESR. In order to allow plenty for selection, 30 crystals were purchased for the prototype. A frequency counter capable of making measurements with 10 Hz resolution, a DMM with a dc mV range and two simple self-constructed test circuits are required for crystal selection.

The project filter shown in **Fig 11.86** is based on a 7-pole QER (quasi-equiripple) design that has a shape factor of 1.96 and offers simplicity, flexibility, and a great passband. The 39 pF coupling capacitors can be silver mica or low-k disc ceramic types with a tolerance of ±2%. The termination resistance shown in the diagram has been reduced to allow for the loss in the end crystals (roughly 36 Ω for each parallel pair). The total termination resistance should be 335 Ω, theoretically, and the actual value used should be adjusted to reflect the effective ESR of your end pairs.

If more than nine suitable crystals are available from the batch tested, the order of the filter can be increased without changing the value of the coupling capacitors. The bandwidth will change very little — by less than +1% per unit increment in order. The termination resistance will need to be reduced as the order is increased, however, by the ratio of the q1 values given in Table 11.21. Increasing the order will improve the shape factor and further reduce adjacent channel interference, but also increase the passband ripple.

Effectively, each crystal in the middle section of the filter has a load of around 20.25 pF because of the 39 pF capacitors and 1.5 pF crystal-to-case capacitance on either side. The end crystal pairs are also shifted up in frequency as if they have the same load. Therefore, for best matching, all crystals should be checked in an oscillator with this load capacitance. The oscillator for this test is shown in **Fig 11.87**, and it will be seen that it shares many common parts with the VXO circuit used for ESR measurements shown in Fig 11.70. The 22 pF input capacitor and two 470 pF feedback capacitors in series present about 20 pF to the crystal under test. A transistor socket can be used as a quick means of connecting the crystals in circuit,

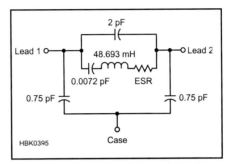

Fig 11.85 — Real electrical equivalent circuit of an ECS 8.5MHz crystal with its metal case grounded.

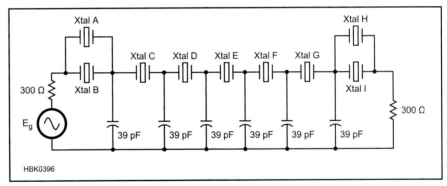

Fig 11.86 — 7-Pole 2.4 kHz QER ladder filter using ECS 8.5 MHz crystals (ECS-85-S-4) from Digi-Key (X418-ND).

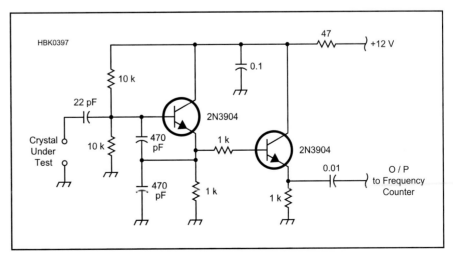

Fig 11.87 — Test oscillator with 20 pF capacitive load for matching crystal frequencies.

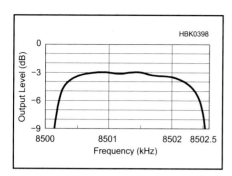

Fig 11.88 — Passband response of prototype 8.5 MHz QER crystal ladder filter with 3 dB insertion loss.

and their 20 pF frequencies and ESR values recorded, the best selection strategy is to look over the figures for a group of nine crystals that have a spread of frequencies of less than 10% of the 2.4 kHz bandwidth. They can then be considered for position in the filter on the basis of their ESR values. The ones with the lowest values should be selected for the middle positions, with the lowest of all as the central crystal.

The crystals with the highest ESR values should be paired up for use as the end parallel crystals because their loss can be absorbed in the terminations. Try to use a pair of crystals with similar values of overall ESR if you want the two terminating resistors to have the same value after subtracting the effective loss resistance of each pair of end crystals from the required theoretical termination resistance of 335 Ω. **Table 11.23** shows how the nine crystals chosen from the batch of 30 obtained for the prototype were selected for position to obtain the passband curve shown in **Fig 11.88**. It can be seen that the spread of frequencies for a load of 20 pF in this case was 215 Hz, and the values of ESR varied from 27 Ω (best) to 108 Ω (worst). The prototype bandwidth (–6 dB) came out at 2.373 kHz with 39 pF coupling capacitors that were all approximately 1% high of their nominal value. A random selection of ±2% capacitors should produce a bandwidth of between 2.35 and 2.45 kHz. If a wider bandwidth is required, 33 pF coupling can be used instead of 39 pF, with the theoretical termination resistance increased to 393 Ω. This should provide a bandwidth of around 2.8 kHz.

The motional capacitance of the ECS 9 MHz series crystals available from Digi-Key (part number X419-ND) should be only slightly higher than that of the 8.5 MHz crystals, so they could be used in this design with a corresponding increase in bandwidth — probably

Table 11.23
Measured parameters for the 9 crystals selected for use in the prototype 7-pole 2.4 kHz SSB filter.

Xtal	Freq (20pF)	ESR (Ω)	Q
A	8501.342 kHz	87	30k
B	8501.441 kHz	60	43k
C	8501.482 kHz	33	78k
D	8501.557 kHz	32	81k
E	8501.530 kHz	27	97k
F	8501.372 kHz	28	93k
G	8501.411 kHz	52	50k
H	8501.477 kHz	108	24k
I	8501.400 kHz	54	48k

around 300 Hz, making the overall filter bandwidth at 9 MHz about 2.7 kHz (±60 Hz with 2% tolerance, 39 pF capacitors).

Construction can be a matter of availability and ingenuity. Reclaimed filter cans from unwanted wide-bandwidth commercial crystal filters could be utilized if they can be picked up cheaply enough. Otherwise, inverted crystals can be soldered side by side, and in line, to a base plate made of a piece of PCB material, copper, or brass sheet as shown in **Fig 11.89**. The coupling capacitors can then be soldered between the crystal leads and the base plate, or crystal cases, depending on which is more convenient. Excellent ultimate attenuation (>120dB) can be achieved using direct wiring and good grounding like this, particularly if lead lengths are kept as short as possible and if additional shielding is added at each end of the filter to prevent active circuitry at either end of the filter from coupling to and leaking signal around it. A case made from pieces of the same material can be soldered around the base plate to form a fully shielded filter unit with feed-thru insulators for the input and output connections.

11.11.2 Broadcast-Band Rejection Filter

Inadequate front-end selectivity or poorly performing RF amplifier and mixer stages often result in unwanted cross-talk and overloading from adjacent commercial or amateur stations. The filter shown is inserted between the antenna and receiver. It attenuates the out-of-band signals from broadcast stations but passes signals of interest (1.8 to 30 MHz) with little or no attenuation.

The high signal strength of local broadcast stations requires that the stop-band attenuation of the high-pass filter also be high. This filter provides about 60 dB of stop-band attenuation with less than 1 dB of attenuation above 1.8 MHz. The filter input and output ports match 50 Ω with a maximum SWR of 1.353:1 (reflection coefficient = 0.15). A 10-element filter yields adequate stop-band

rather than soldering and de-soldering each one in turn. If test crystals are soldered, adequate time should be allowed for them to cool so that their frequency stabilizes before a reading is taken.

The crystals need to be numbered in sequence with a permanent marker to identify them, and their 20 pF load frequencies recorded in a table or spreadsheet as they are measured. When this is complete, the oscillator can then be converted to a VXO, as shown in Fig 11.70, and using this and the jig in Fig 11.69 the ESR of each crystal can be assessed by comparison with metal or carbon film resistors that give the same output readings on the DMM.

The variable capacitor used in the VXO to check the crystals for the prototype filter was a polyvaricon type with maximum capacitance of 262 pF and the inductor was a 10 μH miniature RF choke. If smaller values of variable capacitor are used the inductor value will need to be increased in order to ensure the frequency swing is great enough to tune thru the peak of the lowest frequency crystal in the batch.

Once all the crystals have been checked

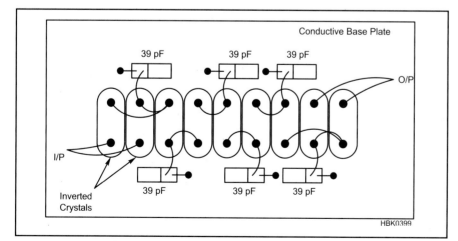

Fig 11.89 — Suggested construction arrangement using inverted crystals with their cases soldered to a conductive base plate and direct point-to-point wiring using only component leads.

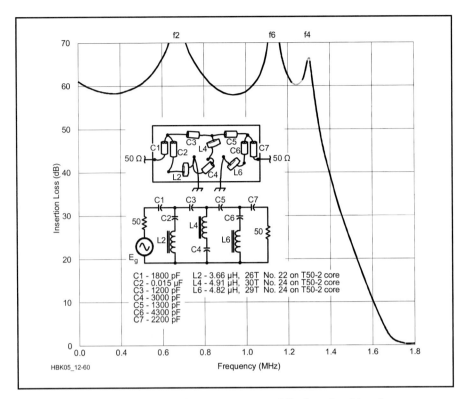

Fig 11.90 — Schematic, layout and response curve of the broadcast band rejection filter.

Construction of the filter is shown in **Fig 11.91**. Use Panasonic P-series polypropylene (type ECQ-P) capacitors. These capacitors are available through Digi-Key and other suppliers. The powdered-iron T50-2 toroidal cores are available through Amidon, Palomar Engineers and others.

For a 3.4-MHz cutoff frequency, divide the L and C values by 2. (This effectively doubles the frequency-label values in Fig 11.90.) For the 80-m version, L2 through L6 should be 20 to 25 turns each, wound on T50-6 cores. The actual turns required may vary one or two from the calculated values. Parallel-connect capacitors as needed to achieve the nonstandard capacitor values required for this filter.

FILTER PERFORMANCE

The measured filter performance is shown in Fig 11.90. The stop-band attenuation is more than 58 dB. The measured cutoff frequency (less than 1 dB attenuation) is under 1.8 MHz. The measured passband loss is less than 0.8 dB from 1.8 to 10 MHz. Between 10 and 100 MHz, the insertion loss of the filter gradually increases to 2 dB. Input impedance was measured between 1.7 and 4.2 MHz. Over the range tested, the input impedance of the filter remained within the 37- to 67.7-Ω input-impedance window (equivalent to a maximum SWR of 1.353:1).

Fig 11.91 — The filter fits easily in a 2 × 2 × 5-inch enclosure. The version in the photo was built on a piece of perfboard.

attenuation and a reasonable rate of attenuation rise. The design uses only standard-value capacitors.

BUILDING THE FILTER

The filter parts layout, schematic diagram, response curve and component values are shown in **Fig 11.90**. The standard capacitor values listed are within 2.8% of the design values. If the attenuation peaks (f2, f4 and f6) do not fall at 0.677, 1.293 and 1.111 MHz, tune the series-resonant circuits by slightly squeezing or separating the inductor windings.

11.11.3 Wave Trap for Broadcast Stations

Nearby medium-wave broadcast stations can sometimes cause interference to HF receivers over a broad range of frequencies. A wave trap can catch the unwanted frequencies and keep them out of your receiver.

CIRCUIT DESCRIPTION

The way the circuit works is quite simple. Referring to **Fig 11.92**, you can see that it consists essentially of only two components, a coil L1 and a variable capacitor C1. This series-tuned circuit is connected in parallel with the antenna circuit of the receiver. The characteristic of a series-tuned circuit is that the coil and capacitor have a very low impedance (resistance) to frequencies very close to the frequency to which the circuit is tuned. All other frequencies are almost unaffected. If the circuit is tuned to 1530 kHz, for example, the signals from a broadcast station on that frequency will flow through the filter to ground, rather than go on into the receiver. All other frequencies will pass straight into the receiver. In this way, any interference caused in the receiver by the station on 1530 kHz is significantly reduced.

CONSTRUCTION

This is a series-tuned circuit that is adjustable from about 540 kHz to 1600 kHz. It is

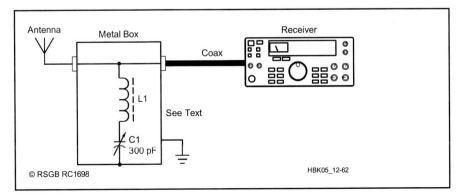

Fig 11.92 — The wave trap consists of a series tuned circuit, which 'shunts' signals on an unwanted frequency to ground.

RF and AF Filters 11.45

built into a metal box, **Fig 11.93**, to shield it from other unwanted signals and is connected as shown in Fig 11.92. To make the inductor, first make a *former* by winding two layers of paper on the ferrite rod. Fix this in place with black electrical tape. Next, lay one end of the wire for the coil on top of the former, leaving about an inch of wire protruding beyond the end of the ferrite rod. Use several turns of electrical tape to secure the wire to the former. Now, wind the coil along the former, making sure the turns are in a single layer and close together. Leave an inch or so of wire free at the end of the coil. Once again, use a couple of turns of electrical tape to secure the wire to the former. Finally, remove half an inch of enamel from each end of the wire.

Alternatively, if you have an old AM transistor radio, a suitable coil can usually be recovered already wound on a ferrite rod. Ignore any small coupling coils. Drill the box to take the components, then fit them in and solder together as shown in **Fig 11.94**. Make sure the lid of the box is fixed securely in place, or the wave trap's performance will be adversely affected by pick-up on the components.

CONNECTION AND ADJUSTMENT

Connect the wave trap between the antenna and the receiver, then tune C1 until the interference from the offending broadcast station is a minimum. You may not be able to eliminate interference completely, but this handy little device should reduce it enough to listen to the amateur bands. Let's say you live near an AM transmitter on 1530 kHz, and the signals break through on your 1.8-MHz receiver. By tuning the trap to 1530 kHz, the problem is greatly reduced. If you have problems from more than one broadcast station, the problem needs a more complex solution.

11.11.4 Optimized Harmonic Transmitting Filters

Low-pass filters should be placed at the output of transmitters to ensure that they meet the various regulatory agency requirements for harmonic suppression. These are commonly designed to pass a single amateur band and provide attenuation at harmonics of that band sufficient to meet the requirements. The material presented here by Jim Tonne, W4ENE, is based on material originally published in the September/October 1998 issue of *QEX*. The basic approach is to use a computer to optimize the performance in the passband (a single amateur band) while simultaneously maximizing the attenuation at the second and third harmonic of that same band. When this is done, the higher harmonics will also be well within spec.

The schematic of this filter along with parts values for the 3.5 to 4.0 MHz amateur band is shown in **Fig 11.95**. The responses of that filter are shown in **Fig 11.96**.

Component values for the 160-meter through the 6-meter amateur bands are shown in **Table 11.24**. The capacitors are shown in pF and the inductors in µH. The capacitors are the nearest 5% values; both the nearest 5% and the exact inductor values are shown.

Using the nearest-5% inductor values will result in satisfactory operation. If the construction method is such that exact-value (adjustable) inductors can be used then the "Exact" values are preferred. These values were obtained from the program *SVC Filter Designer* which is on the *ARRL Handbook CD* included with this book.

11.11.5 Diplexer Filter

This section, covering diplexer filters, was written by William E. Sabin, WØIYH. The diplexer is helpful in certain applications, such as frequency mixer terminations.

Fig 11.93 — The wave trap can be roughly calibrated to indicate the frequency to which it is tuned.

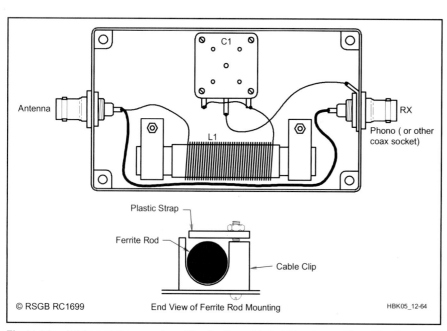

Fig 11.94 — Wiring of the wave trap. The ferrite rod is held in place with cable clips.
C1 — 300 pF polyvaricon variable.
L1 — 80 turns of 30 SWG enameled wire, wound on a ferrite rod.
Associated items: Case (die-cast box), knobs to suit, connectors to suit, nuts and bolts, plastic cable clips.

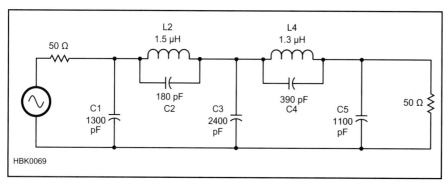

Fig 11.95 — Optimized low-pass filter. This design is for the 80 meter amateur band. It is similar to a Cauer design but the parts values have been optimized as described in the text and in the Sep/Oct 1998 issue of *QEX*.

(The terms "diplexer" and "duplexer" are often confused. A duplexer is a device such as a circulator that allows a transmitter and a receiver to use the same antenna *without* the use of filters. Diplexers use filters so that the signal frequencies must be far apart, such as on different bands. Diplexers have a constant filter-input resistance that extends to the stop band as well as the passband. Ordinary filters that become highly reactive or have an open or short-circuit input impedance outside the passband may degrade performance of the devices to which they are attached. (For example, impedances far from 50 Ω outside the operating frequency range may cause an amplifier to develop parasitic oscillations.)

Fig 11.97 shows a *normalized* prototype 5-element, 0.1-dB Chebyshev low-pass/high-pass (LP/HP) filter. This idealized filter is driven by a voltage generator with zero internal resistance, has load resistors of 1.0 Ω and a cutoff frequency of 1.0 radian per second (0.1592 Hz). The LP prototype values are taken from standard filter tables.[1] The first element is a series inductor. The HP prototype is found by:

a) replacing the series L (LP) with a series C (HP) whose value is 1/L, and

b) replacing the shunt C (LP) with a shunt L (HP) whose value is 1/C.

For the Chebyshev filter, the return loss is improved several dB by multiplying the prototype LP values by an experimentally derived number, K, and dividing the HP values by the same K. You can calculate the LP values in henrys and farads for a 50-Ω RF application with the following formulas:

$$L_{LP} = \frac{KL_{P(LP)} R}{2\pi f_{CO}} \; ; \; C_{LP} = \frac{KC_{P(LP)}}{2\pi f_{CO} R}$$

where

$L_{P(LP)}$ and $C_{P(LP)}$ are LP prototype values

K = 1.005 (in this specific example)

R = 50 Ω

f_{CO} = the cutoff (−3-dB response) frequency in Hz.

For the HP segment:

$$L_{HP} = \frac{L_{P(HP)} R}{2\pi f_{CO} K} \; ; \; C_{HP} = \frac{C_{P(HP)}}{2\pi f_{CO} KR}$$

where $L_{P(HP)}$ and $C_{P(HP)}$ are HP prototype values.

Fig 11.98 shows the LP and HP responses of a diplexer filter for the 80 meter band. The following items are to be noted:

• The 3 dB responses of the LP and HP meet at 5.45 MHz.

• The input impedance is close to 50 Ω at all frequencies, as indicated by the high value of return loss (SWR <1.07:1).

• At and near 5.45 MHz, the LP input reactance and the HP input reactance are conjugates; therefore, they cancel and produce an almost perfect 50-Ω input resistance in that region.

• Because of the way the diplexer filter is derived from synthesis procedures, the transfer characteristic of the filter is mostly independent of the actual value of the amplifier dynamic output impedance.[2] This is a useful feature, since the RF power amplifier output impedance is usually not known or specified.

• The 80 meter band is well within the

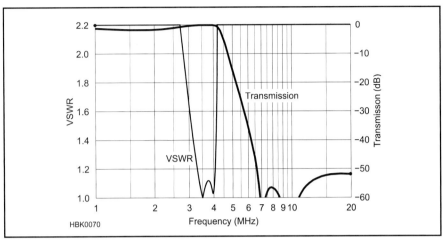

Fig 11.96 — Responses of the filter shown in Fig 11.95. Note the low values of SWR from 3.5 to 4 MHz. At the same time the harmonics are attenuated to meet regulations. Responses for the other amateur bands are very similar except for the frequency scaling.

Table 11.24
Values for the Optimized Harmonic Filters

Band (meters)	C1 (pF)	L2, 5% (μH)	L2, Exact (μH)	C2 (pF)	C3 (pF)	L4, 5% (μH)	L4, Exact (μH)	C4 (pF)	C5 (pF)
160	2400	3.0	2.88	360	4700	2.4	2.46	820	2200
80	1300	1.5	1.437	180	2400	1.3	1.29	390	1100
60	910	1.0	1.029	120	1600	0.91	0.8897	270	750
40	680	0.75	0.7834	91	1300	0.62	0.6305	220	560
30	470	0.56	0.5626	68	910	0.47	0.4652	160	430
20	330	0.39	0.3805	47	620	0.33	0.3163	110	300
17	270	0.30	0.3063	36	510	0.27	0.2617	82	240
15	220	0.27	0.2615	30	430	0.22	0.2245	68	200
12	200	0.24	0.241	27	390	0.20	0.2042	62	180
10	180	0.20	0.2063	24	330	0.18	0.1721	56	150
6	91	0.11	0.108	13	180	0.091	0.0911	30	82

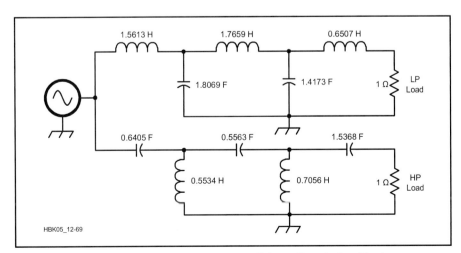

Fig 11.97 — Low-pass and high-pass prototype diplexer filter design. The low-pass portion is at the top, and the high-pass at the bottom of the drawing. See text.

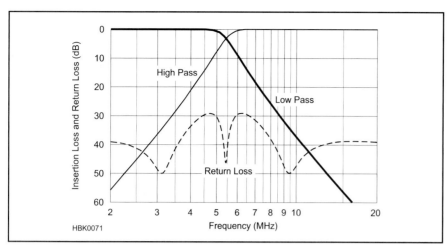

Fig 11.98 — Response for the low-pass and high-pass portions of the 80-meter diplexer filter. Also shown is the return loss of the filter.

LP response.

- The HP response is down more than 20 dB at 4 MHz.
- The second harmonic of 3.5 MHz is down only 18 dB at 7.0 MHz. Because the second harmonic attenuation of the LP is not great, it is necessary that the amplifier itself be a well-balanced push-pull design that greatly rejects the second harmonic. In practice this is not a difficult task.
- The third harmonic of 3.5 MHz is down almost 40 dB at 10.5 MHz.

Fig 11.99A shows the unfiltered output of a solid-state push-pull power amplifier for the 80 meter band. In the figure you can see that:

- The second harmonic has been suppressed by a proper push-pull design.
- The third harmonic is typically only 15 dB or less below the fundamental.

The amplifier output goes through our diplexer filter. The desired output comes from the LP side, and is shown in Fig 11.99B. In it we see that:

- The fundamental is attenuated only about 0.2 dB.
- The LP has some harmonic content; however, the attenuation exceeds FCC requirements for a 100-W amplifier.

Fig 11.99C shows the HP output of the diplexer that terminates in the HP load or *dump* resistor. A small amount of the fundamental frequency (about 1%) is also lost in this resistor. Within the 3.5 to 4.0 MHz band, the filter input resistance is almost exactly the correct 50-Ω load resistance value. This is because power that would otherwise be *reflected* back to the amplifier is absorbed in the dump resistor.

Solid state power amplifiers tend to have stability problems that can be difficult to debug.[3] These problems may be evidenced by level changes in: load impedance, drive, gate or base bias, supply voltage, etc. Problems may arise from:

- The reactance of the low-pass filter outside the desired passband. This is especially true for transistors that are designed for high-frequency operation.
- Self-resonance of a series inductor at some high frequency.
- A stop band impedance that causes voltage, current and impedance reflections back to the amplifier, creating instabilities within the output transistors.

Intermodulation performance can also be degraded by these reflections. The strong third harmonic is especially bothersome for these problems.

The diplexer filter is an approach that can greatly simplify the design process, especially for the amateur with limited PA-design experience and with limited home-lab facilities. For these reasons, the amateur homebrew enthusiast may want to consider this solution, despite its slightly greater parts count and expense.

The diplexer is a good technique for narrowband applications such as the HF amateur bands.[4] From Fig 11.98, we see that if the signal frequency is moved beyond 4.0 MHz the amount of desired signal lost in the dump resistor becomes large. For signal frequencies below 3.5 MHz the harmonic reduction may be inadequate. A single filter will not suffice for all the HF amateur bands.

This treatment provides you with the information to calculate your own filters. A *QEX* article has detailed instructions for building and testing a set of six filters for a 120-W amplifier. These filters cover all nine of the MF/HF amateur bands.[5]

You can use this technique for other filters such as Bessel, Butterworth, linear phase, Chebyshev 0.5, 1.0, etc.[6] However, the diplexer idea does *not* apply to the elliptic function types.

The diplexer approach is a resource that can be used in any application where a constant value of filter input resistance over a wide range of passband and stop band frequen-

Diplexer Filter Design Software

The *ARRL Handbook CD* at the back of this book includes *DiplexerDesigner*, a *Windows* program for design and analysis of diplexer filters by Jim Tonne, W4ENE. The software is interactive and plots performance changes as various parameters are altered.

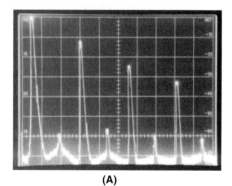
(A)

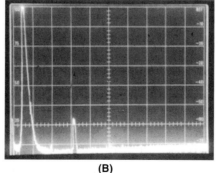
(B)

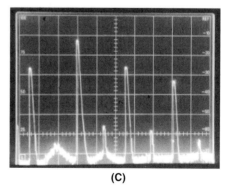
(C)

Fig 11.99 — At A, the output spectrum of a push-pull 80 meter amplifier. At B, the spectrum after passing through the low-pass filter. At C, the spectrum after passing through the high-pass filter.

cies is desirable for some reason. Computer modeling is an ideal way to finalize the design before the actual construction. The coil dimensions and the dump resistor wattage need to be determined from a consideration of the power levels involved, as illustrated in Fig 11.97.

Another significant application of the diplexer is for elimination of EMI, RFI and TVI energy. Instead of being reflected and very possibly escaping by some other route, the unwanted energy is dissipated in the dump resistor.[7]

Notes

[1] Williams, A. and Taylor, F., *Electronic Filter Design Handbook*, any edition, McGraw-Hill.
[2] Storer, J.E., *Passive Network Synthesis*, McGraw-Hill 1957, pp 168-170. This book shows that the input resistance is ideally constant in the passband and the stop band and that the filter transfer characteristic is ideally independent of the generator impedance.
[3] Sabin, W. and Schoenike, E., *HF Radio Systems and Circuits*, Chapter 12, Noble Publishing, 1998. Also the previous edition of this book, *Single-Sideband Systems and Circuits*, McGraw-Hill, 1987 or 1995.
[4] Dye, N. and Granberg, H., *Radio Frequency Transistors, Principles and Applications*, Butterworth-Heinemann, 1993, p 151.
[5] Sabin, W.E. WØIYH, "Diplexer Filters for the HF MOSFET Power Amplifier," *QEX*, Jul/Aug, 1999. Also check ARRLWeb at **www.arrl.org/qex**.
[6] See note 1. *Electronic Filter Design Handbook* has LP prototype values for various filter types, and for complexities from 2 to 10 components.
[7] Weinrich, R. and Carroll, R.W., "Absorptive Filters for TV Harmonics," *QST*, Nov 1968, pp 10-25.

Fig 11.100 — The 1.8 to 54 MHz low-pass filter is housed in a die-cast box.

Table 11.25
Low-Pass Filter Parts List

Qty	Description
1	Miniature brass strip, 1 × 12 in., 0.032 in. thick (C3)
1	Miniature brass strip, 2 × 12 in., 0.064 in. thick (C1, C2)
5 ft	1/8 inch diameter soft copper tubing
4	1/4 × 20 × 1/2 in. long hex head bolt
4	Plastic spacer or washer, 0.5 in. OD, 0.25 in. ID, 0.0625 in. thick
6	1/4 × 20 hex nut with integral tooth lock washer
1	1/4 × 20 × 4 in. long bolt
1	1/4 × 20 threaded nut insert, PEM nut, or "Nutsert"
1	1 × 0.375 in. diameter nylon spacer. ID smaller than 0.25 in. (used for C3 plunger).
4	Nylon spacer, 0.875 in. OD, 0.25 to 0.34 in. ID, approx. 0.065 in. or greater thickness (used to attach brass capacitor plates).

Aluminum diecast enclosure is available from Jameco Electronics (**www.jameco.com**) part no. 11973. The box dimensions are 7.5 × 4.3 × 2.4 in. The 0.03125 in. thick Teflon sheet is available from McMaster-Carr Supply Co (**www.mcmaster.com**), item #8545K21 is available as a 12 × 12 in. sheet.

11.11.6 High-Performance, Low-Cost 1.8 to 54 MHz Low-Pass Filter

The low-pass filter shown in **Fig 11.100** offers low insertion loss, mechanical simplicity, easy construction and operation on all amateur bands from 160 through 6 meters. Originally built as an accessory filter for a 1500-W 6-meter amplifier, the filter easily handles legal limit power. No complicated test equipment is necessary for alignment. It was originally described by Bill Jones, K8CU, in November 2002 *QST*.

Although primarily intended for coverage of the 6-meter band, this filter has low insertion loss and presents excellent SWR characteristics for all HF bands. Although harmonic attenuation at low VHF frequencies near TV channels 2, 3 and 4 does not compare to filters designed only for HF operation, the use of this filter on HF is a bonus to 6-meter operators who also use the HF bands. Six-meter operators may easily tune this filter for low insertion loss and SWR in any favorite band segment, including the higher frequency FM portion of the band.

ELECTRICAL DESIGN

The software tool used to design this low-pass filter is *Elsie* by Jim Tonne, W4ENE, which may be found on the CD included with this book. The *Elsie* format data file for this filter, DC54.lct, may be downloaded for your own evaluation from the author's Web site at **www.realhamradio.com**.

Fig 11.101 is a schematic diagram of the filter. The use of low self-inductance capacitors with Teflon dielectric easily allows legal limit high power operation and aids in the ultimate stop band attenuation of this filter. Capacitors with essentially zero lead length will not introduce significant series inductance that upsets filter operation. This filter also uses a trap that greatly attenuates second harmonic frequencies of the 6-meter band. The parts list for the filter is given in **Table 11.25**.

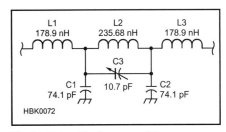

Fig 11.101 — The low-pass filter schematic.
C1, C2 — 74.1 pF. 2 × 2.65 inch brass plate sandwiched with 0.03125 inch thick Teflon sheet. The metal enclosure is the remaining grounded terminal of this capacitor.
C3 — See text and Fig 11.102.
L1, L3 — 178.9 nH. Wind with 1/8 inch OD soft copper tubing, 3.5 turns, 0.75 inch diameter form, 0.625 inch long, 1/4 inch lead length for soldering to brass plate. The length of the other lead to RF connector as required.
L2 — 235.68 nH. Wind with 1/8 inch OD soft copper tubing, 5 turns, 0.75 inch diameter form, 1.75 inches long. Leave 1/4 inch lead length for soldering.

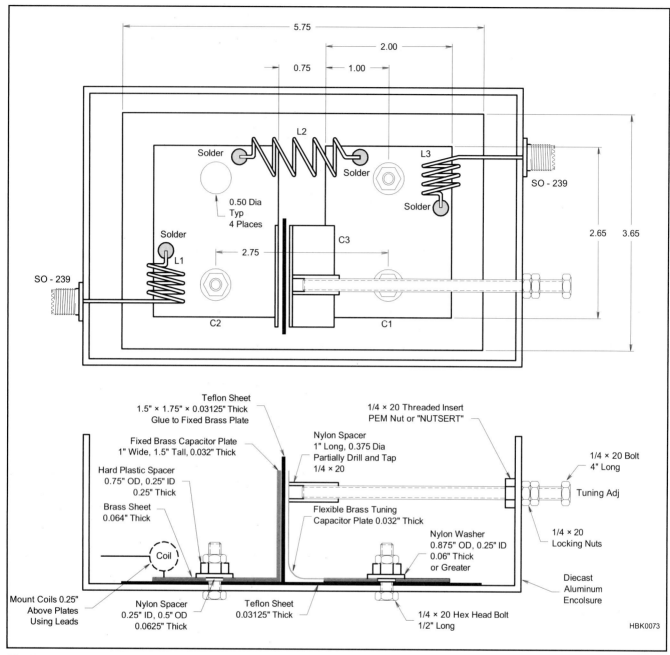

Fig 11.102 — Assembly drawing for the low-pass filter. All dimensions are in inches. See Table 11.24 for a complete parts list.

MECHANICAL DETAILS

A detailed mechanical drawing is shown in **Fig 11.102**. One design goal of this filter was easy tuning with modest home test equipment. To realize this, build the coils carefully. The homemade coils solder directly to the top surface of the brass capacitor plates. Capacitors C1 and C2 are made using a brass to Teflon to aluminum case sandwich. C3 is made from two pieces of 0.032-inch thick brass plate and a Teflon insulator. The filter inductors are mounted at right angles to each other to help maintain good stop band attenuation.

The *Elsie* software tool will calculate the details of each inductor. Inductors L1 and L3 are designed with a half turn winding. This allows short connections to the brass capacitor plate and the RF connectors mounted on the enclosure walls. The coils are physically spaced with ¼-inch lead lengths, and then soldered to the brass plates.

Many of the parts required to make this filter are available at hardware stores. In particular, the 1 and 2 inch wide brass strips (sold as hobby or miniature brass), ⅛ inch diameter soft copper tubing, nuts, bolts, and nylon spacers and washers are commonly available at low cost. It is important that the specified 0.03125-inch thickness of Teflon be used since another size will result in a different capacitor value. If you have another Teflon thickness available, you will need to calculate the specific capacitor values depending upon the new thickness and brass plate sizes.

The opaque white Teflon used here has a dielectric constant of 2.1. The clear varieties typically have lower values and will result in different capacitance for the same size brass plates. The capacitance will decrease if the assembly bolts are loose, so be sure to have the bolts tightened. Also, use the 0.064-inch thick brass plate for the bolted down capacitors. When under compression, the thinner brass size used for the variable capacitor tends to flex more and doesn't fit as flat to the Teflon.

A separate Teflon sheet, also used in the variable capacitor, is glued to the stationary brass plate. This insulator is used to prevent a short circuit in case the tuning screw is tightened too much. Teflon is extremely slick, and doesn't glue well unless chemically prepared. One way to get acceptable glue joint performance between the brass support plate and the insulator is to scuff the Teflon and brass surfaces with sandpaper. The intention is to increase the available surface area as much as possible, and provide more places for the glue to fasten to. Glue the Teflon in place with a bead of RTV or epoxy. After drying, the Teflon sheet can be intentionally peeled from the brass plate, but it appears to hold reasonably well. Special Teflon that has been treated to allow good adhesion is available, but the expense isn't justified for this simple application. This Teflon variable capacitor insulator sheet measures 1.5 inches wide by 1.75 inches tall and is larger than the two brass plates. This gives an outside edge insulation some safety margin.

CALCULATING CAPACITANCE

The 0.064-inch thick brass capacitor plates have two 0.5-inch holes in them for the mounting bolts and washers. The surface area of each hole is πR^2, so the two holes combined have a total surface area of 0.3925 square inch. The brass plate size is 2 inches by 2.65 inches. This equals 5.3 square inches of surface area. Subtracting the area of the two holes gives a total surface area of 4.9 square inches. The formula for capacitance is

$$C = 0.2248 \, (kA/d)$$

where

C = capacitance in pF
K = dielectric constant of Teflon
A = surface area of one plate in square inches
d = thickness of insulator

The dielectric constant of the Teflon used here is 2.1, and the thickness used is 0.03125 inch. The calculated capacitance of each plate equals 74.1 pF. Measured values agree closely with this number. When built as described, the capacitor plates measured between 2% and 2.5% of the calculated value. This is acceptable for a practical filter.

The brass sheet material acts like a large heat sink, so an adequate soldering iron is required. A large chisel point 125-W iron will work well. The soldering heat does not affect the Teflon material. However, beware of the temptation to use a small propane torch. Two bolts in each capacitor hold the Teflon sheet and brass plates firmly together. The bolts are insulated from the brass plates by nylon spacers the same thickness as the brass. The nylon plunger for the tuning capacitor needs to be drilled and tapped to accept the ¼ × 20 thread of the adjustment bolt. A threaded insert or PEM nut in the enclosure provides support for the tuning bolt.

ADJUSTMENT

C3 is shunted across coil L2. This coil and capacitor combination acts like a tunable trap for second harmonic frequencies when operating in the 6-meter band. After soldering into place, the flexible tuning plate of this capacitor is simply bent toward the adjustment screw. Brass of this thickness has a definite spring effect. Just bend the plate well toward the tuning screw, and then tighten the tuning bolt inward. This will result in a stable variable capacitor.

SIX-METER ALIGNMENT PROCEDURE

If you wish to use this filter from 1.8 to 30 MHz only (no 6 meter operation), C3 adjustment is not critical and does not affect HF SWR performance. Simply set the C3 plates 0.1 inch apart and disregard the following steps. However, don't eliminate the capacitor entirely. The software predicts degraded VHF response with it missing. For use on the HF bands only, the tuning screw and associated nylon plunger may be omitted.

Normally, tuning this filter would be a challenge, since three variables (with two interacting) are involved (L1, L2 and C3). I realized that the *Elsie* software "Tune" mode held the answer. After studying what the software predicted, I generated this tuning procedure. My very first attempt to exactly tune this filter was successful, and was completed in just a few minutes. This method was predicted by software and then confirmed in practice. A common variable SWR analyzer is required. These steps may seem complicated, but are actually pretty straightforward once you get a feel for it. Read first before you start adjusting.

Step One

Adjust C3 until the top plate spacing is about 0.1 inch apart. Using an SWR analyzer, search for a very low SWR null anywhere in the vicinity of about 45 to 60 MHz. If a low SWR value (near 1:1) can be found, even though the frequency of the low SWR isn't where you want it, proceed to step two. Otherwise, adjust the input coil L1 by expanding or compressing the turns until a low SWR can be obtained anywhere in 45 to 60 MHz range. If you have a way to measure the notch response at 100.2 MHz, proceed to step two. Otherwise, proceed to step three.

Step Two

Now apply a 100.2 MHz signal to the filter input. Adjust C3 until the 6-meter second harmonic at 100.2 MHz is nulled on the filter output. Hook up the SWR analyzer again,

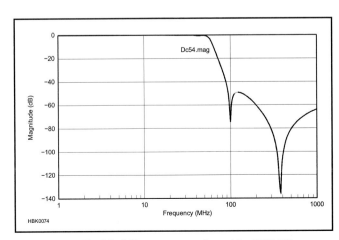

Fig 11.103 — Modeled filter response from 1 to 1000 MHz.

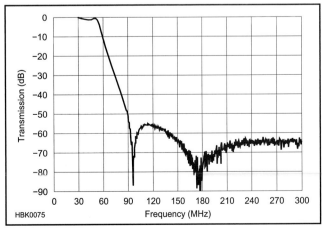

Fig 11.104 — Measured filter response from 300 kHz to 300 MHz.

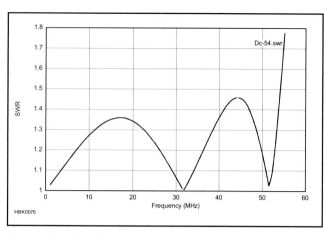

Fig 11.105 — The filter SWR from 1 to 55 MHz.

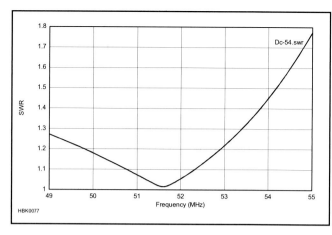

Fig 11.106 — Six meter filter SWR.

and sweep the 6-meter band with the SWR analyzer. If the low SWR frequency is too low, adjust L2 for less inductance (expand turns farther apart), and then readjust the variable capacitor to bring the notch back on frequency. Continue these iterations until the SWR null is where you want, and the notch frequency is correctly set.

If the desired SWR low spot is too high in frequency, adjust L2 for more inductance (compress the coil turns), and then readjust the variable capacitor for the second harmonic notch. Continue this until both the low SWR frequency location and the notch null are set where you want.

You may need to unsolder one end of coil L2 to allow the adjustment for a longer or shorter coil length as you expand or compress turns. Just solder the end again after you make your length correction. Note that you will probably need to install the enclosure lid during the very final tuning steps.

I was able to reduce the second harmonic into the noise floor of an IFR-1200S spectrum display, but the lid needed to be installed. The lid also interacts with the variable capacitor. Once the SWR and the notch frequency are set, the tuning process is complete and the filter is optimally adjusted. Do not perform step three below.

Step Three

This step is only performed if you don't have a way to generate the 100.2 MHz input signal, and then detect a null on the filter's output terminal. C3 will become your SWR adjustment to move the SWR null spot to the portion of the 6-meter band you desire. If you run out of adjustment range (C3 turned all the way in), just compress the L2 turns together and try again. If C3 is backed completely off, just expand the L2 turns and try again.

After your SWR is set, you are finished. Although the second harmonic notch probably isn't exactly on frequency, you will still have good (but not optimum) suppression since the notch is very deep.

PERFORMANCE DISCUSSION

Assuming the 6-meter SWR is set to a low value for a favorite part of the band, the worst case calculated forward filter loss is about 0.18 dB. The forward loss is better in the HF bands, with a calculated loss of only 0.05 dB from 1.8 through 30 MHz. The filter cutoff frequency is about 56 MHz, and the filter response drops sharply above this. There are parasitic capacitors on coils L1 and L3. These are also included in this filter analysis. The calculated self-capacitance of each coil is almost 1 pF. These small capacitors are included on the schematic and are also included in the software for the model. These capacitors occur naturally, so do not solder a 1-pF capacitor across each of the end coils in this filter. The capacitors have the effect of placing additional notches somewhere in the UHF region. The calculated self-resonant frequency of L1 and L3 is about 365 MHz.

Fig 11.103 shows the calculated filter response from 1 to 1000 MHz. The impressive notch near 365 MHz is because of these inherent stray capacitances across each of the coils. Slight variations in each coil will make slightly different tuned traps. This will introduce a stagger-tuned effect that results in a broader notch.

These exact capacitance values are hard to predict because of variations in home made coil dimensions and exact placement of each coil inside the enclosure. The best way to determine their effect is to physically measure the UHF response of this filter. Using low self-inductance capacitors in a VHF filter helps to take advantage of predicted filter attenuation at extended stop-band frequencies.

The measured response of the completed filter is shown in **Fig 11.104**. It corresponds closely with the modeled response. The SWR across the HF bands and 6 meters is shown in **Fig 11.105**. **Fig 11.106** shows only the 6-meter band SWR.

11.12 Filter Glossary

Active filter — A filter that uses active (powered) devices to implement its function.

All-pass — Filter response in which the magnitude response does not change with frequency, but the phase response does change with frequency.

Amplitude response — see **magnitude response**

Band-pass — Filter response in which signals are passed in a range of frequencies and rejected outside that range.

Band-stop — Filter response in which signals are rejected in a range of frequencies and passed outside that range (also called *band-reject* or *notch* filter).

Bandwidth — Range of frequencies over which signals are passed (low-pass, high-pass, band-pass) or rejected (band-stop).

Brick wall response — An ideal filter response in which signals are either passed with no attenuation or attenuated completely.

Cutoff frequency — Frequency at which a filter's output is 3 dB below its passband output (also called *corner frequency* or *3 dB frequency*).

Decade — A ratio of 10 in frequency.

Equiripple — Equalized ripple in a filter's magnitude response across the passband, stop band, or both.

Flat — Refers to a filter's magnitude response that is constant across a range of frequencies.

Group delay — The transit time of signals through a filter.

High-pass — Filter response in which signals above the cutoff frequency are passed and rejected at lower frequencies.

Ideal filter — Filter that passes signals without loss or attenuates them completely. An ideal filter has no transition regions. (see also **brick wall response**).

Insertion loss — The loss incurred by signals in a filter's passband.

Low-pass — Filter response in which signals below the cutoff frequency are passed and rejected at higher frequencies.

Lumped element — Discrete inductors and capacitors; a filter made from discrete inductors and capacitors.

Magnitude response — Graph of a filter's output amplitude versus frequency.

Microstrip — A type of transmission line made from a strip of metal separated from a ground plane by a layer of insulating material, such as on a printed-circuit board..

Normalize — The technique of converting numeric values to their ratio with respect to some reference value. (To denormalize is to reverse the normalization, converting the ratios back to the original values.)

Notch filter — see **band-stop filter**.

Octave — A ratio of two in frequency (see also **decade**).

Overshoot — The condition in which the output of a circuit, in responding to a change in its input, temporarily exceeds the steady-state value that the input should cause.

Overtone — Vibration mode at a higher frequency than the fundamental mode, usually harmonically related.

Passband — The range of frequencies passed by a filter.

Passive filter — A filter that does not require power to perform its function (see also **lumped element**).

Phase response — Graph of a the difference in angular units (degrees or radians) between a filter's input and output versus frequency.

Radian — Unit of angular measurement equal to $1/2\pi$ of a circle, equal to $360/2\pi$ degrees

Ringing — The condition in which the output of a circuit, in responding to a change in its input, exhibits a damped alternating sequence of exceeding and falling below the steady-state value that the input should cause before settling at the steady-state value.

Ripple — A regular variation with frequency in a filter's magnitude response.

Rolloff — The rate of change in a filter's magnitude response in the transition region and stop band.

Scaling — Changing a filter's impedance or frequency characteristics through multiplication or division by a constant.

Shape factor — The ratio of a filter's bandwidth between the points at which its magnitude response is 6 dB and 60 dB below the response in the filter's passband.

Stop band — The range of frequencies that are rejected by a filter.

Stripline — A transmission line consisting of a metal strip suspended between two ground planes.

TEM — Transverse electromagnetic mode in which the electric and magnetic fields of electromagnetic energy are aligned perpendicularly to the direction of motion.

Topology — The arrangement of connections of components in the filter. For example, "capacitor-input" and "inductor-input" are two different topologies.

Transition region — Range of frequencies between a filter's passband and stop band.

11.13 References and Bibliography

REFERENCES

1. Virgil E. Bottom, "Introduction to Quartz Crystal Unit Design," Van Nostrand Reinhold Company, 1982, ISBN 0-442-26201-9.
2. M. Dishal, "Modern Network Theory Design of Single Sideband Crystal Ladder Filters," *Proc IEEE*, Vol 53, No 9, Sep 1965.
3. J. A. Hardcastle, G3JIR, "Ladder Crystal Filter Design," *QST*, Nov 1980, p 20.
4. W. Hayward, W7ZOI, "A Unified Approach to the Design of Ladder Crystal Filters," *QST*, May 1982, p 21.
5. J. Makhinson, N6NWP, "Designing and Building High-Performance Crystal Ladder Filters," *QEX*, Jan 1995, pp 3-17.
6. W. Hayward, W7ZOI, "Refinements in Crystal Ladder Filter Design," *QEX*, June 1995, pp 16-21.
7. J.A. Hardcastle, G3JIR, "Computer-Aided Ladder Crystal Filter Design," *Radio Communication*, May 1983.
8. John Pivnichny, N2DCH, "Ladder Crystal Filters," MFJ Publishing Co, Inc.
9. W. S. Metcalf, "Graphs Speed Interdigitated Filter Design," *Microwaves*, Feb 1967.
10. J. Fisk W1DTY, "Helical-Resonator Design Techniques," *QST*, June 1976, pp 11-14.

FOR FURTHER READING

General Filter Design

Applied Radio Labs, Design Note DN004, "Group Delay Explanations and Applications," **www.radiolab.com.au**

S. Butterworth, "On the Theory of Filter Amplifiers," *Experimental Wireless and*

Wireless Engineer, Oct 1930, pp 536-541.

Laplace Transforms: P. Chirlian, *Basic Network Theory*, McGraw Hill, 1969.

S. Darlington, "Synthesis of Reactance 4-Poles Which Produce Prescribed Insertion Loss Characteristics," *Journal of Mathematics and Physics*, Sep 1939, pp 257-353.

M. Dishal, "Top Coupled Band-pass Filters," *IT&T Handbook*, 4th edition, American Book, Inc, 1956, p 216.

Phillip R. Geffe, *Simplified Modern Filter Design* (New York: John F. Rider, 1963, OCLC: 2470870).

Fourier Transforms: *Reference Data for Engineers*, Chapter 7, 7th edition, Howard Sams, 1985.

Randall W. Rhea, *HF Filter Design and Computer Simulation* (New York: McGraw-Hill, ISBN 0070520550, OCLC 32013486).

Cauer Elliptic Filters: R. Saal, *The Design of Filters Using the Catalogue of Normalized Low-Pass Filters*,(German; brief introduction in English) (AEH Telefunken, 1968, OCLC 13988270). Also *Reference Data for Radio Engineers*, pp 9-5 to 9-11.

W. E. Sabin, WØIYH, "Designing Narrow Band-Pass Filters with a BASIC Program," *QST*, May 1983, pp 23-29.

L. Weinberg, "Network Design by use of Modern Synthesis Techniques and Tables," *Proceedings of the National Electronics Conference*, vol 12, 1956.

Donald R.J. White, *A Handbook on Electrical Filters (Synthesis, Design and Applications)* (White Electromagnetics, 1980, ISBN 0932263070).

A. B. Williams, *Electronic Filter Design Handbook* (New York: McGraw-Hill, 1981).

O. Zobel, "Theory and Design of Electric Wave Filters," *Bell System Technical Journal*, Jan 1923.

A.I. Zverev, *Handbook of Filter Synthesis* (New York: John Wiley and Sons, 1967, ISBN 0471986801, OCLC 972252).

Crystal, Helical, and Interdigital Filter Design

R. Fisher, W2CQH, "Combline VHF Bandpass Filters," *QST*, Dec 1968, p 44.

R. Fisher, W2CQH, "Interdigital Bandpass Filters for Amateur VHF/UHF Applications," *QST*, Mar 1968, p 32.

U. L. Rohde, DJ2LR, "Crystal Filter Design with Small Computers" *QST*, May 1981, p 18.

Practical Filter Designs

P. Antoniazzi, IW2ACD and M. Arecco, IK2WAQ, "Easy Microwave Filters Using Waveguides and Cavities," *QEX*, Sep/Oct 2006, pp 37-42.

B. Bartlett, VK4UW and J. Loftus, VK4EMM, "Band-Pass Filters for Contesting," *National Contest Journal*, Jan 2000, pp 11-15.

T. Cefalo Jr, WA1SPI, "Diplexers, Some Practical Applications," *Communications Quarterly*, Fall 1997, pp 19-24.

P.R. Cope, W2GOM/7, "The Twin-T Filter," *QEX*, July/Aug 1998, pp 45-49.

D. Gordon-Smith, G3UUR, "Seventh-Order Unequal-Ripple Low-pass Filter Design," *QEX*, Nov/Dec 2006, pp 31-34.

W. Hayward, W7ZOI, "Extending the Double-Tuned Circuit to Three Resonators," *QEX*, Mar/Apr 1998, pp 41-46.

F. Heemstra, KT3J, "A Double-Tuned Active Filter with Interactive Coupling," *QEX*, Mar/Apr 1999, pp 25-29.

D. Jansson, WD4FAB and E. Wetherhold, W3NQN, "High-Pass Filters to Combat TVI," *QEX*, Feb 1987, pp 7-8 and 13.

Z. Lau, W1VT, "A Narrow 80-Meter Band-Pass Filter," *QEX*, Sept/Oct 1998, p 57.

R. Lumachi, WB2CQM. "How to Silverplate RF Tank Circuits," *73 Amateur Radio Today*, Dec 1997, pp 18-23.

T. Moliere, DL7AV, "Band-Reject Filters for Multi-Multi Contest Operations," *CQ Contest*, Feb 1996, pp 14-22.

W. Rahe, DC8NR, "High Performance Audio Speech Low-Pass and CW Band-Pass Filters in SVL Design," *QEX*, Jul/Aug 2007, pp 31-39.

W. Sabin, WØIYH, "Diplexer Filters for the HF MOSFET Power Amplifier," *QEX*, July/Aug 1999, pp 20-26.

W. Sabin, WØIYH, "Narrow Band-Pass Filters for HF," *QEX*, Sept/Oct 2000, pp 13-17.

J. Tonne, WB6BLD, "Harmonic Filters, Improved," *QEX*, Sept/Oct 1998, pp 50-53.

E. Wetherhold, W3NQN, "Modern Design of a CW Filter Using 88 and 44-mH Surplus Inductors," *QST*, Dec 1980, pp14-19. See also Feedback in Jan 1981 *QST*, p 43.

E. Wetherhold, W3NQN, "Band-Stop Filters for Attenuating High-Level Broadcast-Band Signals," *QEX*, Nov 1995, pp 3-12.

E. Wetherhold, W3NQN, " CW and SSB Audio Filters Using 88-mH Inductors," *QEX*-82, pp 3-10, Dec 1988; *Radio Handbook*, 23rd edition, W. Orr, editor, p 13-4, Howard W. Sams and Co., 1987; and, "A CW Filter for the Radio Amateur Newcomer," *Radio Communication* (RSGB) Jan 1985, pp 26-31.

E. Wetherhold, W3NQN, "Clean Up Your Signals with Band-Pass Filters," Parts 1 and 2, *QST*, May and June 1998, pp 44-48 and 39-42.

E. Wetherhold, W3NQN, "Second-Harmonic-Optimized Low-Pass Filters," *QST*, Feb 1999, pp 44-46.

E. Wetherhold, W3NQN, "Receiver Band-Pass Filters Having Maximum Attenuation in Adjacent Bands," *QEX*, July/Aug 1999, pp 27-33.

Contents

12.1 Introduction
    12.1.1 Characterizing Receivers
    12.1.2 Elements of Early Receivers
12.2 Basics of Heterodyne Receivers
    12.2.1 The Direct Conversion Receiver
    Project: A Rock-Bending Receiver For 7 MHz
12.3 The Superheterodyne Receiver
    12.3.1 Superhet Bandwidth
    12.3.2 Selection of IF Frequency for a Superhet
    Project: The MicroR2 — An Easy to Build SSB or CW Receiver
12.4 Superhet Receiver Design Details
    12.4.1 Receiver Sensitivity
    12.4.2 Receiver Image Rejection
    12.4.3 Receiver Dynamic Range
12.5 Control and Processing Outside the Primary Signal Path
    12.5.1 Automatic Gain Control (AGC)
    12.5.2 Audio-Derived AGC
    12.5.3 AGC Circuits

12.6 Pulse Noise Reduction
    12.6.1 The Noise Limiter
    12.6.2 The Noise Blanker
    12.6.3 Operating Noise Limiters and Blankers
    12.6.4 DSP Noise Reduction
12.7 VHF and UHF Receivers
    12.7.1 FM Receivers
    12.7.2 FM Receiver Weak-Signal Performance
    12.7.3 FM Receiver ICs
    12.7.4 VHF Receive Converters
12.8 UHF Techniques
    12.8.1 UHF Construction
    12.8.2 UHF Design Aids
    12.8.3 A 902 to 928 MHz (33-cm) Receiver
    12.8.4 Microwave Receivers
12.9 References and Bibliography

Chapter 12

Receivers

The major subsystems of a radio receiving system are the antenna, the receiver and the information processor. The antenna's task is to provide a transition from an electromagnetic wave in space to an electrical signal that can be conducted on wires. The receiver has the job of retrieving the information content from a *particular* ac signal coming from the antenna and presenting it in a useful format to the processor for use. The processor typically is an operator, but can also be an automated system. When you consider that most "processors" require signals in the range of volts (to drive an operator's speaker or headphones, or even the input of an A/D converter), and the particular signal of interest arrives from the antenna at a level of mere microvolts, the basic function of the receiver is to amplify the desired signal by a factor of a million.

One of the largest challenges of the receiver is the processing of that *particular* signal. In a real world environment, the antenna will often deliver a multitude of signals with a wide range of amplitudes. The biggest job of the receiver is often to extract the information content of the desired signal from a within large group of signals, with many perhaps much stronger in amplitude than the one we want.

This chapter was written by Joel Hallas, W1ZR, including updating of existing material from previous editions of this book.

12.1 Introduction

This first section defines some important receiver concepts that will be used throughout the chapter.

12.1.1 Characterizing Receivers

As we discuss receivers we will need to characterize their performance, and often their performance limitations, using certain key parameters. The most commonly encountered are as follows:

Sensitivity — This parameter is a measure of how weak a signal the receiver can extract information from. This generally is expressed at a particular signal-to-noise ratio (SNR) since noise is generally the limiting factor. A typical specification might be: "Sensitivity: 1 µV for 10 dB SNR with 3 kHz bandwidth." The bandwidth is stated because the amount (or power) of the noise, the denominator of the SNR fraction, increases directly with bandwidth. Generally the noise parameter refers to the noise generated within the receiver, often less than the noise that arrives with the signal from the antenna.

Selectivity — Selectivity is just the bandwidth discussed above. This is important because to a first order it identifies the receiver's ability to separate stations. With a perfectly sharp filter in an ideal receiver, stations within the bandwidth will be heard, while those outside it won't be detected. The selectivity thus describes how closely spaced adjacent channels can be. With a perfect 3 kHz bandwidth selectivity, and signals restricted to a 3 kHz bandwidth at the transmitter, a different station can be assigned every 3 kHz across the spectrum. In a less than ideal situation, it is usually necessary to include a *guard band* between channels.

Note that the word *channel* is used here in its generic form, meaning the amount of spectrum occupied by a signal, and not defining a fixed frequency such as an AM broadcast channel. A CW channel is about 300 Hz wide, a SSB channel about 2.5-3 kHz wide, and so forth. "Adjacent channel" refers to spectrum immediately higher or lower in frequency.

Dynamic Range — In the case of all real receivers, there is a range of signals that a receiver can respond to. This is referred to as dynamic range and, as will be discussed in more detail, can be established based on a number of different criteria. In its most basic form, dynamic range is the range in amplitude of signals that can be usefully received, typically from as low as the receiver's noise level, or *noise floor*, to a level at which stages overload in some way.

The type and severity of the overload is often part of the specification. A straightforward example might be a 130 dB dynamic range with less than 3% distortion. The nature of the distortion will determine the observed phenomenon. If the weakest and strongest signals are both on the same channel, for example, we would not expect to be able to process the weaker of the two. However, the more interesting case would be with the strong signal in an adjacent channel. In an ideal receiver, we would never notice that the adjacent signal was there. In a real receiver with a finite dynamic range or nonideal selectivity, there will be some level of adjacent channel signal or signals that will interfere with reception of the weaker on-channel signal.

The parameters described above are often the key performance parameters, but in many cases there are others that are important to specify. Examples are audio output power, power consumption, size, weight, control capabilities and so forth.

12.1.2 Elements of Early Receivers

A modern receiver is typically composed of multiple stages. The circuits are typically ones that have been presented in other sections of this book combined in ways to provide improved performance of the various key parameters we have discussed.

THE DETECTOR

The primary function of a receiver takes place in a *detector*, the circuit that extracts the information from the modulated RF signal that arrives from the antenna. As shown in **Fig 12.1**, the simplest receiver consists of only a detector. The earliest receivers were constructed in just that way, starting with some unusual electromechanical marvels from Marconi's days. The crystal detector, in the form of the crystal set, was the first radio experienced by youngsters from the 1930s to the 1980s and was the mainstay of commercial users until vacuum tubes became available. While even simpler configurations are possible, the circuit of **Fig 12.2** is typical.

The key performance parameters of this receiver are easy to describe. The sensitivity is the signal level required at the antenna input to create an audible signal as a consequence of the diode acting as a square-law detector. The selectivity is that determined by the loaded Q of the single-tuned resonant circuit while the dynamic range is from the sensitivity to the point that the headphone diaphragms hit their limits, or the diode opens due to high current.

While this receiver looks like it would be rather limited in performance (and is), it does work and was even the basis for most microwave radar receivers from WWII and for some years thereafter. That was the origin of the venerable 1N34 diode. An example of a crystal set used as the receiver in an on-off keyed data communications system is shown in **Fig 12.3**. The diode detector is still used as the detector for full carrier amplitude modulated signals in more complex modern receivers.

Following early crystal receivers were various forms of detectors that used vacuum tubes, and later transistors, to provide gain in the signal path. An oscillating detector, called a *regenerative* detector, provided high sensitivity and improved selectivity. It could also be used for the reception of CW signals by increasing the gain to the point of oscillation and tuning it near the frequency of the received signal. This is the precursor to the modern product detector to be discussed in a following section.

AMPLIFIERS

The next element added to the detector

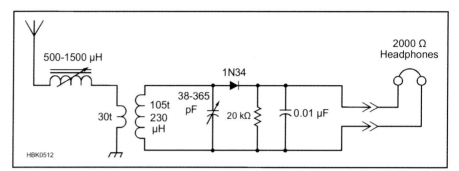

Fig 12.2 — Circuit of a typical "modern" crystal set with a semiconductor diode as the crystal detector.

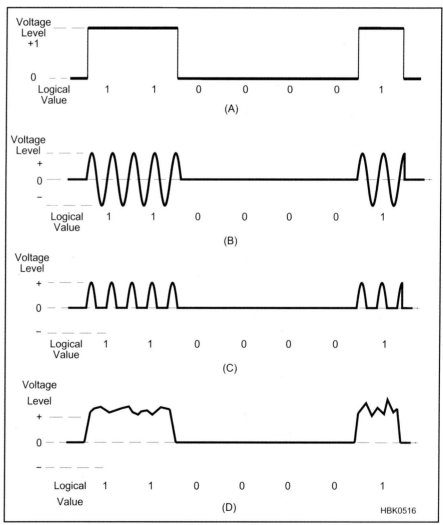

Fig 12.3 — The signals in a basic on-off keyed data system using a crystal receiver. At (A), voltage pulses corresponding to the ASCII representation of the lower case letter "a." At (B), The signals at the output of an on-off keyed radio transmitter. At (C), received pulses at output of detector in the absence of any smoothing. At (D) pulse output to processor following smoothing by shunt capacitor.

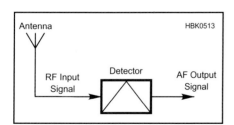

Fig 12.1 — The minimal receiver is just a detector.

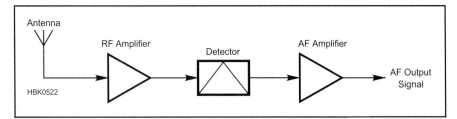

Fig 12.4 — Detector with RF and AF amplifiers to improve sensitivity, selectivity and power output.

was the amplifier. The first amplifiers only provided bandwidth capable of amplifying audio frequency signals and thus were inserted following the detector. In that position, they offered no improvement in sensitivity, which was still limited by the diode; however, they appeared to, because signals were now louder. It was now possible to have more than one person listen to a received signal through the use of a loudspeaker.

As shown in **Fig 12.4**, it wasn't long before vacuum tube and circuit design technology improved to the point that amplifiers could be used at RF as well as AF. The addition of RF amplifier(s) provided two performance improvements. First, the level of signals into the detector could be increased, providing an improvement in sensitivity. Second, if each amplifier were coupled using tuned resonant circuits, as was the practice, the selectivity could be significantly improved.

A receiver with one or more tuned RF amplifier stages was called a *tuned radio frequency*, or TRF receiver. Some had as many as three or four stages, initially with separate controls, and later with up to a five-gang variable capacitor for station selection. Needless to say, a certain amount of skill was needed to adjust each stage so it would properly *track* as it was tuned over the frequency range. There were other challenges with the TRF. With the high gain of the multiple stages all tuned to the same frequency, it was not trivial to avoid oscillation. Another concern was the selectivity provided over the tuning range. If the selectivity of the five tuned circuits could provide a bandwidth of 2% of the tuned frequency, that would result in 30 kHz at the top of the broadcast band (around 1500 kHz) but only 10 kHz at the bottom (500 kHz). Top-end receivers of the day used various methods to automatically maintain similar bandwidth across the range.

12.2 Basics of Heterodyne Receivers

A receiver design that avoids many of the issues inherent in a TRF receiver is called a *heterodyne*, or often *superheterodyne* (*superhet* for short) receiver. The superhet combines the input signal with a locally generated signal in a nonlinear device called a *mixer* to result in the sum and difference frequencies as shown in **Fig 12.5**. The receiver may be designed so the output signal is anything from dc (a so-called *direct conversion* receiver) to any frequency above or below either of the two frequencies. The major benefit is that most of the gain, bandwidth setting and processing are performed at a single frequency.

By changing the frequency of the local oscillator, the operator shifts the input fre-

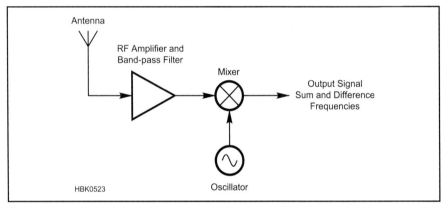

Fig 12.5 — Basic architecture of a heterodyne direct conversion receiver.

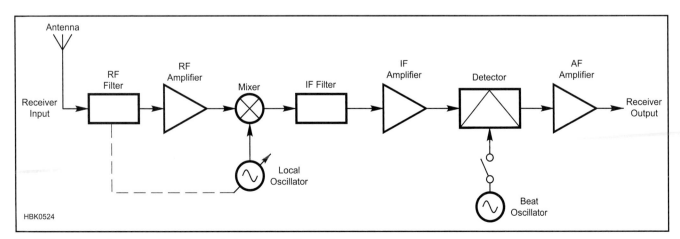

Fig 12.6 — Elements of a traditional superheterodyne radio receiver

quency that is *translated* to the output, along with all its modulated information. In most receivers the mixer output frequency is designed to be an RF signal, either the sum or difference — the other being filtered out at this point. This output frequency is called an *intermediate frequency* or *IF*. The IF amplifier system can be designed to provide the selectivity and other desired characteristics centered at a single fixed frequency — much easier to process than the variable arrangement of a TRF set.

A block diagram of a typical superhet is shown in **Fig 12.6**. In traditional form, the RF filter is used to limit the input frequency range to those frequencies that include only the desired sum or difference but not the other — the so-called *image* frequency. The dotted line represents the fact that in receivers with a wide tuning range, the input filter is often tracked along with the local oscillator. While this is similar to one of the issues about TRF receivers, note that generally there are fewer tuned circuits involved, making it easier to accomplish. The IF filter is traditionally used to establish operating selectivity — that required by the information bandwidth.

For reception of suppressed carrier single-sideband voice (SSB) or on-off or frequency-shift keyed (FSK) signals, a second *beat frequency oscillator* or *BFO* is employed to provide an audible voice, an audio tone or tones at the output for operator or FSK processing. This is the same as a heterodyne mixer with an output centered at dc, although the IF filter is usually designed to remove one of the output products.

In many instances, it is not possible to achieve all the receiver design goals with a single-conversion receiver and multiple conversion steps are taken. Traditionally, the first conversion is tasked with removing the RF image signals, while the second allows processing of the IF signal to provide the information based IF processing.

Modern receivers using digital signal processing for operating bandwidth and information detection often have an additional conversion step to a final IF at a frequency in the tens of kHz. This is established by the maximum sampling rate of a finite frequency response analog-to-digital converter (ADC). Advances in the art have resulted in ADC and processor speeds sufficient that the ADC has been moving closer and closer to the RF frequency resulting in fewer, and in some cases, no required conversions.

12.2.1 The Direct Conversion Receiver

As noted in the introduction, the heterodyne process can occur at a number of different points in the receiver. The simplest form of heterodyne receiver is called a *direct conversion* receiver because it performs the translation directly from the signal frequency to the audio output. It is, in effect, just the BFO and detector of the general superhet shown in Fig 12.6. In this case, the detector is often preceded by an RF amplifier with a typical complete receiver shown in **Fig 12.7**. Such a receiver can be very simple to construct, yet can be quite effective — especially for an ultra-compact low-power consumption oriented portable station.

The basic function of a mixer is to multiply two sinusoidal signals and generate two new signals with frequencies that are the sum and

---

Nonlinear Signal Combinations

Although a mixer is often thought of as nonlinear, it is neither necessary nor desirable for a mixer to be nonlinear. An ideal mixer is one that linearly multiplies the LO voltage by the signal voltage, creating two products at the sum and difference frequencies and only those two products. From the signal's perspective, it is a perfectly linear but time-varying device. Ideally a mixer should be as linear as possible.

If a signal is applied to a nonlinear device, however, the output will not be just a copy of the input, but can be described as the following infinite series of output signal products:

$$V_{OUT} = K_0 + K_1 \times V_{IN} + K_2 \times V_{IN}^2 + K_3 \times V_{IN}^3 + \ldots + K_N \times V_{IN}^N \quad \text{(A)}$$

What happens if the input V_{IN} consists of two sinusoids at F_1 and F_2, or $A \times [\sin(2\pi_1) \times t]$ and $B \times [\sin(2\pi F_2) \times t]$? Begin by simplifying the notation to use angular frequency in radians/second ($2\pi F = \omega$). Thus V_{IN} becomes $A\sin\omega_1 t$ and $B\sin\omega_2 t$ and equation A becomes:

$$V_{OUT} = K_0 + K_1 \times (A\sin\omega_1 t + B\sin\omega_2 t) + K_2 \times (A\sin\omega_1 t + B\sin\omega_2 t)^2 +$$
$$K_3 \times (A\sin\omega_1 t + B\sin\omega_2 t)^3 + \ldots + K_N \times (A\sin\omega_1 t + B\sin\omega_2 t)^N \quad \text{(B)}$$

The zero-order term, K_0, represents a dc component and the first-order term, $K_1 \times (A\sin\omega_1 t + B\sin\omega_2 t)$, is just a constant times the input signals. The second-order term is the most interesting for our purposes. Performing the squaring operation, we end up with:

$$\text{Second order term} = [K_2 A^2 \sin^2\omega_1 t + 2K_2 AB(\sin\omega_1 t \times \sin\omega_2 t) + K_2 B^2 \sin^2\omega_2 t] \quad \text{(C)}$$

Using the trigonometric identity (see Reference 1):

$$\sin á \sin â = \tfrac{1}{2} \{\cos(á - â) - \cos(á + â)\}$$

the product term becomes:

$$K_2 AB \times [\cos(\omega_1 - \omega_2)t - \cos(\omega_1 + \omega_2)t] \quad \text{(D)}$$

Here are the products at the sum and difference frequency of the input signals! The signals, originally sinusoids are now cosinusoids, signifying a phase shift. These signals, however, are just two of the many products created by the nonlinear action of the circuit, represented by the higher-order terms in the original series.

In the output of a mixer or amplifier, those unwanted signals create noise and interference and must be minimized or filtered out. This nonlinear process is responsible for the distortion and intermodulation products generated by amplifiers operated nonlinearly in receivers and transmitters.

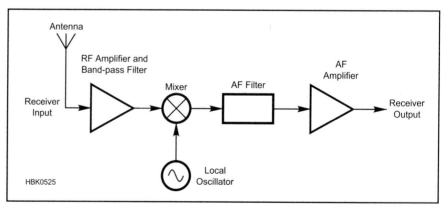

Fig 12.7 — Block diagram of a direct conversion receiver.

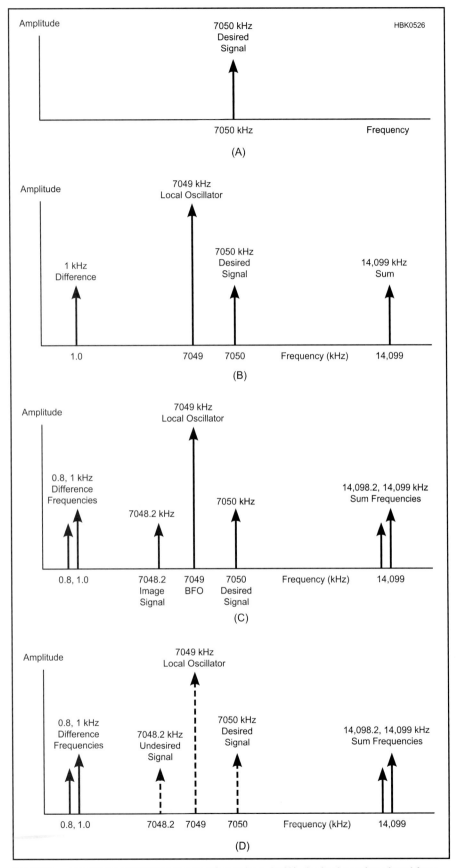

Fig 12.8 — Frequency relationships in DC receiver. At (A), desired receive signal from antenna, 7050 kHz on-off keyed carrier. At (B), internal local oscillator and receive frequency relationships. At (C), frequency relationships of mixer/detector products (not to scale). At (D), sum and difference outputs from double balanced mixer (not to scale). Note the balanced out mixer inputs that cancel at the output (dashed lines).

difference of the two original signals. This function can be performed by a linear multiplier, a switch that turns one input signal on and off at the frequency of the other input signal, or a nonlinear circuit such as a diode. (The output of a nonlinear circuit is made up of an infinite series of products, all different combinations of the two input signals, as described in the sidebar on Nonlinear Signal Combinations.) Much more information about the theory, operation and application of mixers may be found in the **Mixers, Modulators and Demodulators** chapter.

Fig 12.8 shows the progression of the spectrum of an on-off keyed CW signal through such a receiver based on the relationships described above. In 12.8C, we include an undesired image signal on the other side of the local oscillator that also shows up in the output of the receiver. Note each of the desired and undesired responses that occur as outputs of the mixer.

Some mixers are designed to be balanced. A *balanced* mixer is designed to cancel one of the input signals at the output while a *double-balanced* mixer cancels both. A double-balanced mixer simplifies the output filtering job as shown in Fig 12.8D. We will see these again in transmitters in the next chapter.

Products generated by nonlinearities in the mixing process (see the sidebar on Nonlinear Mixing) are heard as intermodulation distortion signals that we will discuss later. Note that the nonlinearities also allow mixing with unwanted signals near multiples of local oscillator frequency. These signals, such as those from TV or FM broadcast stations, must be eliminated in the filtering before the mixer since their audio output will be right in the desired passband on the output of the mixer.

Project: A Rock-Bending Receiver for 7 MHz

There are many direct conversion receiver (DC) construction articles in the amateur literature. Home builders considering construction of a DC receiver should read Chapter 8 in *Experimental Methods in RF Design*, which outlines many of the pitfalls and design limitations of such receivers, as well as a providing suggestions for making a successful DC receiver.[2]

This DC receiver design by Randy Henderson, WI5W, is presented here as an example. This receiver represents about as simple a receiver as can be constructed that will offer reasonable performance.[3] The user of such a DC receiver will appreciate its simplicity and the nice sounding response of CW and SSB signals that it receives, but only if there aren't many signals sharing the band.

A major shortcoming of the DC receiver architecture will soon become evident — it has no rejection of *image* signals. Looking again

Digital Processing and Software Defined Radios

In a day in which even a toaster might include a processor (and perhaps its own IP address!), most receivers and transceivers have processors — so what's the big deal? I think we need to make a distinction between a software-*controlled* radio and a software-*defined* radio.

Software-Controlled Radio

Most radios — or toasters for that matter — fall into the controlled category. In this case the configuration is pretty much fixed. No matter how much you change the software, you can't make that toaster into a radio — or anything besides a toaster. The toaster is limited or constrained by its physical configuration to make toast. The processor, along with associated sensors and control features, may be able to make perfect toast — just don't try to get it to do too much else.

In a similar manner, the processor in a software-controlled radio is used to sense control positions and cause logic elements to react in predictable ways. The processor may be used to count and display frequency, or form code characters at the speed you like. No matter what you do to the software controlling it, the radio is limited by its physical architecture to do what the designers thought you'd want on the day they built it.

Software-Defined Radio

As we'll discuss, there may be a range of definitions — subject to some controversy — on what constitutes a software-defined radio (SDR) in the Amateur Radio world. The FCC has defined the SDR concept in terms of their commercial certification process as:

"…a radio that includes a transmitter in which the operating parameters of the transmitter, including the frequency range, modulation type or conducted output power can be altered by making a change in software without making any hardware changes."

The FCC expects this to yield streamlined equipment authorization procedures by allowing "manufacturers to develop reconfigurable transceivers that can be multi-service, multi-standard, multi-mode and multi-band…."[1] In this context, they are envisioning radios that can be modified *at the factory* by using different software to meet different requirements. While they allow for field changes, their focus is different from ours.

SDR in the Amateur World

In the amateur environment, we are particularly interested in radios that can be changed through software by the end user or operator to meet their needs or to take advantages of newly developed capabilities.

The *ideal* SDR would thus have a minimum of physical constraints. On the receive side, the antenna would be connected to an analog-to-digital converter that would sample the entire radio spectrum. The digitized signal would enter a processor that could be programmed to analyze and decode any form of modulation or encoding and present the result as sights and sounds on the output side of the processor.

So What's Wrong With This Picture?

Not surprisingly, our utopian SDR is much easier to imagine than to construct. As a practical matter, the digital signal processor for most amateur work is a PC, which has some constraints that don't allow us to do quite what we want since it was not intended for use as a radio signal processor. Still, for a few hundred dollars, it is possible to purchase a PC that gets us fairly close.

The key to amateur SDR operation with a PC is the sound card. This card, or sometimes an external interface device, can accept an analog signal and convert it to digital data samples for processing. Advanced SDRs such as the FlexRadio FLEX-5000 have the analog-to-digital conversion function built in, usually providing higher performance than a sound card ADC. The software will determine the type of processing and the nature of the signals we can deal with. It can also take the results of processing and convert them into an analog signal. This sounds like just what we are looking for to make an SDR — and it is. Such an SDR in receive mode would consist of the blocks in **Fig 12.A1**. We do have a few significant limitations, however.

First, the analog-to-digital converter has a specific rate at which it samples the analog signal.[2] The signal frequency response that it can deal with, without going to extraordinary means, is limited to no higher than half the sampling rate.[3] For most sound cards, the sampling rate is 192 kHz or less, limiting the received analog signal to a frequency of 96 kHz (and sometimes only 48 kHz). Some kinds of dual-channel processing allow a response as high as the sampling rate.

Second, the available signal levels are not quite what we would want. And of course, we don't have amateur bands in the dc-to-96 kHz region of the spectrum. Thus we are faced with the need to insert some external processing functions outside the PC. These will be used, at a minimum, to translate the frequency range we wish to use to one that the sound card can deal with on the receive side, and make sure the signal levels are compatible.

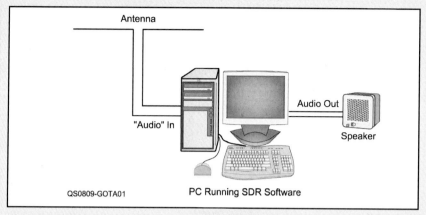

Fig 12.A1 — Conceptual block diagram of an ideal software defined radio (SDR) receiver.

Fig 12.A2 — The FlexRadio Flex-5000A SDR transceiver. A blank front panel SDR.

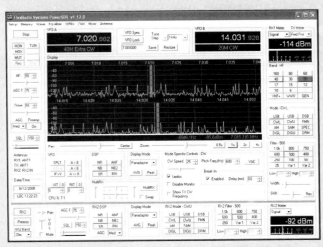

Fig 12.A3 — One version of the main operating screen of PowerSDR operating software.

Fig 12.A4 — The Elecraft K3. It looks like a traditional amateur transceiver, but is actually an SDR.

SDRs usually require a baseband signal with two versions of the input signal separated by 90° of phase. We have previously discussed how such signals can be used to process a signal to remove one sideband. By having the two versions of the signal 90° apart, they are in *phase quadrature* and referred to as the I (in-phase) and Q (in-quadrature) channels. Signal processing algorithms then combine the signals in various ways to demodulate it and recover the information. (If you'd like to hear what the I and Q outputs sound like for yourself, you can build "The Binaural Receiver" by Rick Campbell, KK7B, from the instructions on the CD-ROM included with this book.)

These topics are covered in detail in the **DSP and Software Radio Design** chapter.

So How Does it All Come Together?

The SDR designer, as with all designers, is faced with a trade-off. The equipment external to the PC required to make it do what we want may also limit the choices we can make by software change in the PC. The more hardware features we build in, the fewer choices we may have. In addition to PC software, there is often *firmware*, hard wired instruction in the box outside the PC. This has resulted in two general approaches in SDR.

The "Blank Front Panel" Architecture

Radios marketed as SDRs tend to be of this type. The classic is the FlexRadio Flex-5000A, reviewed in *QST* for July 2008.[4] As shown in **Fig 12.A2**, the front panel has no controls. All control functions are accessible via the soft buttons on the *PowerSDR* software's computer screen (**Fig 12.A3**), or via computer-connected pointing and knob devices.

There are a number of other radios using a similar configuration, and useable with the same open source *PowerSDR* software freely available from FlexRadio under the GNU Public License. These include the modular High Performance SDR system available from Tucson Amateur Packet Radio (TAPR), and some other very low cost, but no longer available SDRs such as the Firefly and SoftRock 40 (although derivatives of the SoftRock 40 are still available).[5] Some firmware for these radios is generally so-called "open source," meaning that if you are a programmer, you can view and modify the source code and thus not only upgrade, but also make the radio do what you want.

The "Looks Like a Radio" Approach

Many current radios are actually built as SDRs. Some, transceivers such as the Elecraft K3 (see **Fig 12.A4**), ICOM IC-7800, Ten-Tec Orion and Yaesu FT-2000, for example, are provided with a mechanism to allow an easy end-user upgrade to new firmware revisions. These radios look like most any other pre-SDR radio in that they have front panels with knobs and dials. Unless you looked at all the revisions to the operating instructions you wouldn't know that they were field-reconfigurable.

Another distinction between the groups is that most of the firmware for the radios in this group is proprietary with revisions available only from the manufacturer, at least as of this writing. That isn't to say that a solid programmer couldn't and perhaps hasn't developed custom software for one of these, but if it's happened, it hasn't happened often.

While all radios in this group are primarily designed to operate without an external computer, they all can be computer-controlled using aftermarket software, available from multiple developers. While this software can make them feel a bit like the radios in the other group, the operating parameter ranges are all set by the radio's internal operating firmware.

So What's it All Mean?

The blank front panel type generally has the most flexibility in operation, since they are not constrained by the physical buttons and knobs on the front panel. The more traditional looking SDR versions may take advantage of the hardware constraints that limit some operating choices to gain improved performance, but a look at the specs will indicate that it isn't always the case. Some blank panel SDRs offer top shelf performance. — *Joel R. Hallas, W1ZR*

Notes

[1] FCC Report and Order 01-264, released Sep 14, 2001.
[2] J. Taylor, K1RFD, "Product Review — Computer Sound Cards for Amateur Radio," *QST*, May 2007, pp 63-70. Available at **www.arrl.org/product-review**.
[3] H. Nyquist, "Certain Topics in Telegraph Transmission Theory," *Transactions of the AIEE*, Vol 47, pp 617-644, Apr 1928. Reprinted as classic paper in *Proceedings of the IEEE*, Vol 90, No. 2, Feb 2002.
[4] R. Lindquist, N1RL, "Product Review — The FlexRadio Flex-5000A," *QST*, Jul 2008, pp 39-45. Available on the ARRLWeb at **www.arrl.org/product-review**.
[5] Contact Softrock developer Tony Parks, KB9YIG, at **raparks@ctcisp.com** to see if a new version is available.

at Fig 12.8B, while the 7049 kHz oscillator will translate the desired 7050 kHz signal to an audio tone of 1 kHz, it will equally well translate a signal at 7048 kHz to the same frequency. This is called an *image* and results in interference. The same kind of thing happens while receiving SSB; the entire channel on the other side of the BFO is received as well.

Building a stable oscillator is often the most challenging part of a simple receiver. This one uses a tunable crystal-controlled oscillator that is both stable and easy to reproduce. All of its parts are readily available from multiple sources and the fixed value capacitors and resistors are common components available from many electronics parts suppliers.

THE CIRCUIT

This receiver works by mixing two radio-frequency signals together. One of them is the signal you want to hear, and the other is generated by an oscillator circuit (Q1 and associated components) in the receiver. In **Fig 12.9**, mixer U1 puts out sums and differences of these signals and their harmonics. We don't use the sum of the original frequencies, which comes out of the mixer in the vicinity of 14 MHz. Instead, we use the frequency *difference* between the incoming signal and the receiver's oscillator — a signal in the audio range if the incoming signal and oscillator frequencies are close enough to each other. This signal is filtered in U2, and amplified in U2 and U3. An audio transducer (a speaker or headphones) converts U3's electrical output to audio.

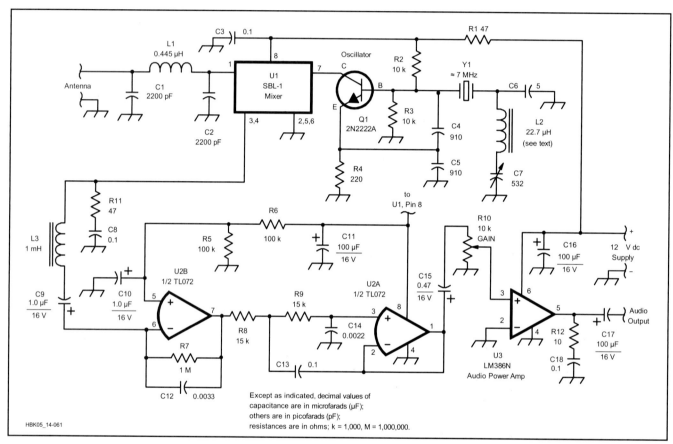

Fig 12.9 — An SBL-1 mixer (U1, which contains two small RF transformers and a Schottky-diode quad), a TL072 dual op-amp IC (U2) and an LM386 low-voltage audio power amplifier IC (U3) do much of the Rock-Bending Receiver's magic. Q1, a variable crystal oscillator (VXO), generates a low-power radio signal that shifts incoming signals down to the audio range for amplification in U2 and U3. All of the circuit's resistors are ¼ W, 5% tolerance types; the circuit's polarized capacitors are 16 V electrolytics, except C10, which can be rated as low as 10 V. The 0.1 µF capacitors are monolithic or disc ceramics rated at 16 V or higher.

C1, C2 — Ceramic or mica, 10% tolerance.
C4, C5, and C6 — Polystyrene, dipped silver mica, or C0G (formerly NP0) ceramic, 10% tolerance.
C7 — Dual gang broadcast variable capacitor (14-380 pF per section), ¼ inch dia shaft, available as #BC13380 from Ocean State Electronics. A rubber equipment foot serves as a knob. (Any variable capacitor with a maximum capacitance of 350 to 600 pF can be substituted; the wider the capacitance range, the better.)
C12, C13, C14 — 10% tolerance. For SSB, change C12, C13 and C14 to 0.001 µF.
U2 — TL072CN or TL082CN dual JFET op amp.
L1 — Four turns of AWG #18 wire on ¾ inch PVC pipe form. Actual pipe OD is 0.85 inch. The coil's length is about 0.65 inch; adjust turns spacing for maximum signal strength. Tack the turns in place with cyanoacrylate adhesive, coil dope or RTV sealant. (As a substitute, wind 8 turns of #18 wire around 75% of the circumference of a T-50-2 powdered-iron core. Once you've soldered the coil in place and have the receiver working, expand and compress the coil's turns to peak incoming signals, and then cement the winding in place.)
L2 — Approximately 22.7 µH; consists of one or more encapsulated RF chokes in series (two 10-µH chokes [Mouser #43HH105 suitable] and one 2.7-µH choke [Mouser #43HH276 suitable] used by author). See text.
L3 — 1 mH RF choke. As a substitute, wind 34 turns of #30 enameled wire around an FT-37-72 ferrite core.
Q1 — 2N2222, PN2222 or similar small-signal, silicon NPN transistor.
R10 — 5 or 10 kΩ audio-taper control (RadioShack No. 271-215 or 271-1721 suitable).
U1 — Mini-Circuits SBL-1 mixer.
Y1 — 7 MHz fundamental-mode quartz crystal. Ocean State Electronics carries 7030, 7035, 7040, 7045, 7110 and 7125 kHz units.
PC boards for this project are available from FAR Circuits.

How the Rock Bender Bends Rocks

The oscillator is a tunable crystal oscillator — a variable crystal oscillator, or *VXO*. Moving the oscillation frequency of a crystal like this is often called *pulling*. Because crystals consist of precisely sized pieces of quartz, crystals have long been called *rocks* in ham slang — and receivers, transmitters and transceivers that can't be tuned around due to crystal frequency control have been said to be *rockbound*. Widening this rockbound receiver's tuning range with crystal pulling made *rock bending* seem just as appropriate!

L2's value determines the degree of pulling available. Using FT-243 style crystals and larger L2 values, the oscillator reliably tunes from the frequency marked on the holder to about 50 kHz below that point with larger L2 values. (In the author's receiver a 25 kHz tuning range was achieved.) The oscillator's frequency stability is very good.

Inductor L2 and the crystal, Y1, have more effect on the oscillator than any other components. Breaking up L2 into two or three series-connected components often works better than using one RF choke. (The author used three molded RF chokes in series — two 10 µH chokes and one 2.7 µH unit.) Making L2's value too large makes the oscillator stop.

The author tested several crystals at Y1. Those in FT-243 and HC-6-style holders seemed more than happy to react to adjustment of C7 (TUNING). Crystals in the smaller HC-18 metal holders need more inductance at L2 to obtain the same tuning range. One tiny HC-45 unit from International Crystals needed 59 µH to eke out a mere 15 kHz of tuning range.

Input Filter and Mixer

C1, L1, and C2 form the receiver's input filter. They act as a peaked *low-pass* network to keep the mixer, U1, from responding to signals higher in frequency than the 40-meter band. (This is a good idea because it keeps us from hearing video buzz from local television transmitters, and signals that might mix with harmonics of the receiver's VXO.) U1, a Mini-Circuits SBL-1, is a passive diode-ring mixer. Diode-ring mixers usually perform better if the output is terminated properly. R11 and C8 provide a resistive termination at RF without disturbing U2A's gain or noise figure.

Audio Amplifier and Filter

U2B amplifies the audio signal from U1. U2A serves as an active low-pass filter. The values of C12, C13 and C14 are appropriate for listening to CW signals. If you want SSB stations to sound better, make the changes shown in the caption for Fig 12.9.

U3, an LM386 audio power amplifier IC, serves as the receiver's audio output stage. The audio signal at U3's output is much more

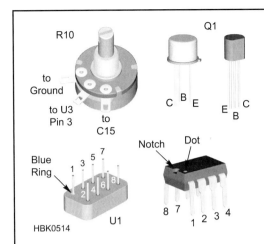

Fig 12.10 — Two Q1-case styles are shown because plastic or metal transistors will work equally well for Q1. If you build your Rock-Bending Receiver using a prefab PC board, you should mount the ICs in 8-pin mini-DIP sockets rather than just soldering the ICs to the board.

Fig 12.11 — Ground-plane construction, PC-board construction — either approach can produce the same good Rock Bending Receiver performance.

powerful than a weak signal at the receiver's input, so don't run the speaker/earphone leads near the circuit board. Doing so may cause a squeal from audio oscillation at high volume settings.

CONSTRUCTION

If you're already an accomplished builder, you know that this project can be built using a number of construction techniques, so have at it! If you're new to building, you should consider building the Rock-Bending Receiver on a printed circuit (PC) board. (The parts list tells where you can buy one ready-made.) See **Fig 12.10** for details on the physical layout of several important components used in the receiver. The receiver can be constructed either using a PC board or by using ground-plane (a.k.a "ugly") construction. **Fig 12.11** shows a photo of the receiver built on a ground-plane formed by an unetched piece of PC board material.

If you use the PC board available from FAR Circuits or a homemade double-sided circuit board based on the PC pattern on the *Handbook* CD, you'll notice that it has more holes than it needs to. The extra holes (indicated in the part-placement diagram with square pads) allow you to connect its ground plane to the ground traces on its foil side. (Doing so reduces the inductance of some of the board's ground paths.) Pass a short length of bare wire (a clipped-off component lead is fine) into each of these holes and solder on both sides. Some of the circuit's components (C1, C2 and others) have grounded leads accessible on both sides of the board. Solder these leads on both sides of the board.

Another important thing to do if you use a homemade double-sided PC board is to countersink the ground plane to clear all ungrounded holes. (Countersinking clears copper away from the holes so components won't short-circuit to the ground plane.) A ¼-inch diameter drill bit works well for this. Attach a control knob to the bit's shank and you can safely use the bit as a manual countersinking tool. If you countersink your board in a drill press, set it to about 300 rpm or less, and use very light pressure on the feed handle.

Mounting the receiver in a metal box or cabinet is a good idea. Plastic enclosures can't shield the TUNING capacitor from the presence of your hand, which may slightly affect the receiver tuning. You don't have to completely enclose the receiver — a flat aluminum panel screwed to a wooden base is an acceptable alternative. The panel supports the tuning capacitor, GAIN control and your choice of audio connector. The base can support the circuit board and antenna connector.

CHECKOUT

Before connecting the receiver to a power source, thoroughly inspect your work to spot obvious problems like solder bridges, incorrectly inserted components or incorrectly wired connections. Using the schematic (and PC-board layout if you built your receiver on a PC board), recheck every component and connection one at a time. If you have a digital voltmeter (DVM), use it to measure the resistance between ground and everything that should be grounded. This includes things like pin 4 of U2 and U3, pins 2, 5, 6 of U1, and the rotor of C7.

If the grounded connections seem all right, check some supply-side connections with the meter. The connection between pin 6 of U3 and the positive power-supply lead should show less than 1 Ω of resistance. The resistance between the supply lead and pin 8 of U1 should be about 47 Ω because of R1.

If everything seems okay, you can apply power to the receiver. The receiver will work with supply voltages as low as 6 V and as high as 13.5 V, but it's best to stay within the 9 to 12 V range. When first testing your receiver, use a current-limited power supply (set its limiting between 150 and 200 mA) or put a 150 mA fuse in the connection between the receiver and its power source. Once you're sure that everything is working as it should, you can remove the fuse or turn off the current limiting.

If you don't hear any signals with the antenna connected, you may have to do some troubleshooting. Don't worry; you can do it with very little equipment.

TROUBLE?

The first clue to look for is noise. With the GAIN control set to maximum, you should hear a faint rushing sound in the speaker or headphones. If not, you can use a small metallic tool and your body as a sort of test-signal generator. (If you have any doubt about the safety of your power supply, power the Rock-Bending Receiver from a battery during this test.) Turn the GAIN control to maximum. Grasp the metallic part of a screwdriver, needle or whatever in your fingers, and use the tool to touch pin 3 of U3. If you hear a loud scratchy popping sound, that stage is working. If not, then something directly related to U3 is the problem.

You can use this technique at U2 (pin 3, then pin 6) and all the way to the antenna. If you hear loud pops when touching either end of L3 but not the antenna connector, the oscillator is probably not working. You can check for oscillator activity by putting the receiver near a friend's transceiver (both must be in the same room) and listening for the VXO. Be sure to adjust the TUNING control through its range when checking the oscillator.

The dc voltage at Q1's base (measured without the RF probe) should be about half the supply voltage. If Q1's collector voltage is about equal to the supply voltage, and Q1's base voltage is about half that value, Q1 is probably okay. Reducing the value of L2 may be necessary to make some crystals oscillate.

OPERATION

Although the Rock-Bending Receiver uses only a handful of parts and its features are limited, it performs surprisingly well. Based on tests done with a Hewlett-Packard HP 606A signal generator, the receiver's minimum discernible signal (by ear) appears to be 0.3 µV. The author could easily copy 1 µV signals with his version of the Rock-Bending Receiver.

Although most HF-active hams use transceivers, there are advantages in using separate receivers and transmitters. This is especially true if you are trying to assemble a simple home-built station.

12.3 The Superheterodyne Receiver

The *superheterodyne*, or *superhet* for short, uses the principles of the direct conversion heterodyne receiver above — at least twice. In a superhet, a local oscillator and mixer are used to translate the received signal to an *intermediate frequency* or *IF* rather than directly to audio. This provides an opportunity for additional amplification and processing. Then a second mixer is used as in the DC receiver to detect the IF signal, translating it to audio. The configuration was shown previously in Fig 12.6.

In a typical configuration, the local oscillator (LO) and RF amplifier stages are adjusted so that as the LO is changed in frequency, the RF amplifier is also tuned to the appropriate frequency to receive the desired station. An example may help. Let's pick a common IF frequency used in an AM broadcast radio, 455 kHz. Now if we want to listen to a 600 kHz broadcast station, the RF stage should be set to amplify the 600 kHz signal and the LO should be set to 600 + 455 kHz or 1055 kHz.[4] The 600 kHz signal, along with any audio information it contains, is translated to the IF frequency and is amplified. It is then detected, just as if it were a TRF DC receiver at 455 kHz.

Note that to detect standard AM signals, the second oscillator, usually called a *beat frequency oscillator* or *BFO*, is turned off since the AM station provides its own carrier signal over the air. Receivers designed only for standard AM reception, the typical "table or kitchen radio," generally don't have a BFO at all.

It's not clear yet that we've gained anything by doing this; so let's look at another example. If we decide to change from listening to the station at 600 kHz and want to listen to another station at, say, 1560 kHz, we can tune the single dial of our superhet to 1560 kHz. With the appropriate ganged and tracked tuning capacitors, the RF stage is tuned to 1560 kHz, and the LO is set to 1560 + 455 or 2010 kHz and now that station is translated to our 455 kHz IF. Note that the bulk of our amplification can take place at the 455 kHz IF frequency, so not as many stages must be tuned each time we change to a new frequency. Note also that with the superheterodyne configuration the selectivity (the ability to separate stations) occurs primarily in the intermediate-frequency (IF) stages and is thus the same no matter what frequencies we choose to listen to. The superhet design has thus eliminated the major limitations of the TRF at the cost of two additional building blocks.

It should be no surprise then that the superhet, in various flavors that we will discuss, has become the primary receiver architecture in use today. The concept was introduced by Major Edwin Armstrong,[5] a US Army artillery officer, just as WW I was coming to a close. The superhet quickly gained popularity and, following the typical patent battles of the times, became the standard of a generation of vacuum-tube broadcast receivers. These were found in virtually every US home from the late 1920s through the 1960s, when they were slowly replaced by transistor sets, but still of superhet design.

While the superheterodyne receiver is still the standard in many applications — Amateur Radio, broadcast radio, televisions, microwave, and radar, to name but a few — it may be surprising to amateurs to learn that in terms of sheer numbers, the direct-conversion receiver is far and away the most widely used technology. More than one billion wireless telephone handsets — "cell phones" — are manufactured every year and more than 90% of them employ direct-conversion receivers!

12.3.1 Superhet Bandwidth

Now we will discuss the bandwidth requirements of different operating modes and how that affects superhet design. One advantage of a superhet is that the operating bandwidth can be established by the IF stages, and further limited by the audio system. It is thus independent of the RF frequency to which the receiver is tuned. It should not be surprising that the detailed design of a superhet receiver is dependent on the nature of the signal being received. We will briefly discuss the most commonly received modulation types and the bandwidth implications of each below. (Each modulation type is discussed in more detail in the **Modulation** or **Digital Modes** chapters.)

AMPLITUDE MODULATION (AM)

As shown in Fig 12.8, multiplying (in other words, modulating) a carrier with a single tone results in the tone being translated to frequencies of the sum and difference of the two. Thus, if a transmitter were to multiply a 600 Hz tone by a 600 kHz carrier signal, we would generate additional new frequencies at 599.4 and 600.6 kHz. If instead we were to modulate the 600 kHz carrier signal with a band of frequencies corresponding to (telephone company "toll quality") human speech of 300 to 3300 Hz, we would have a pair of bands of information carrying waveforms extending from 596.7 to 603.3 kHz, as shown in **Fig 12.12**. These bands are called *sidebands*, and some form of these is present in any AM signal that is carrying information.

Note that the total bandwidth of this AM voice signal is twice the highest frequency transmitted, or 6600 Hz. If we choose to transmit speech and limited music, we might allow modulating frequencies up to 5000 Hz, resulting in a bandwidth of 10,000 Hz or 10 kHz. This is the standard channel spacing that commercial AM broadcasters use in the US. In actual use, the adjacent channels are generally geographically separated, so broadcasters can extend some energy into the next channels for improved fidelity. We would refer to this as a *narrow-bandwidth* mode.

What does this say about the bandwidth needed for our receiver? If we want to receive the full information content transmitted by a US AM broadcast station, then we need to set the bandwidth to at least 10 kHz. What if our receiver has a narrower bandwidth? Well, we will lose the higher frequency components of the transmitted signal — perhaps ending up with a radio suitable for voice but not very good at reproducing music.

On the other hand, what is the impact of having too wide a bandwidth in our receiver? In that case, we will be able to receive the full transmitted spectrum but we will also receive some of the adjacent channel information. This will sound like interference and reduce the quality of what we are receiving. If there are no adjacent channel stations, we will get any additional noise from the additional bandwidth and minimal additional information. The general rule is that the received bandwidth should be matched to the bandwidth of the signal we are trying to receive to maximize SNR and minimize interference.

As the receiver bandwidth is reduced, intelligibility suffers, although the SNR is improved. With the carrier centered in the receiver bandwidth, most voices are difficult to understand at bandwidths less than around 4 kHz. In cases of heavy interference, full carrier AM can be received as if it were SSB, as described below, with the carrier inserted at the receiver, and the receiver tuned to whichever sideband has the least interference.

SINGLE-SIDEBAND SUPPRESSED-CARRIER MODULATION (SSB)

The standard commercial AM format is very convenient for receiver design, since the carrier needed to demodulate the received sidebands is sent along with the sidebands. Applying the total received signal to a detector or multiplier, allows the audio to be recovered without having to worry about any of the finer points we will discuss later. This is very cost effective in a broadcast environment in which there are many inexpensive receivers and only a relatively few expensive transmitters.

In looking at Fig 12.12, you might have noticed that both sidebands carry the same information, and are thus redundant. In addition, the carrier itself conveys no information. It is thus possible to transmit a *single sideband* and *no carrier*, as shown in **Fig 12.13**, relying on the BFO (beat frequency oscillator) in the receiver to provide a signal with which to multiply the sideband in order to provide demodulated audio output. The implications in the receiver are that the bandwidth can be slightly less than half that required for double sideband AM (DSB). There must be an additional mechanism to carefully replace the missing carrier within the receiver. This is the function of the BFO, which must be at just exactly the right frequency. If the frequency is improperly set, even by a few Hz, a baritone can come out sounding like a soprano and vice versa!

This makes a requirement for a much more stable receiver design with a much finer tuning system — a more expensive proposition. An alternate is to transmit a reduced level carrier and have the receiver lock on to the weak carrier, usually called a *pilot* carrier. Note that the pilot carrier need not be of sufficient amplitude to demodulate the signal, just enough to allow a BFO to lock to it. These alternatives are effective, but tend to make SSB receivers expensive, complex and most appropriate for the case in where a small number of receivers are listening to a single transmitter, as is the case of two-way communication.

Note that the bandwidth required to effectively demodulate an SSB signal is actually less than half that required for the AM signal because the area around the AM carrier need not be received. Thus the toll quality spectrum (a term carried over from long-distance telephone systems) of 300 to 3300 Hz can be received in a bandwidth of 3000 not 3300 Hz. Early SSB receivers typically used a bandwidth of around 3 kHz, but with the heavy interference frequently found in the amateur bands, it is more common for amateurs to

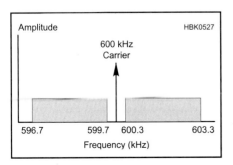

Fig 12.12 — Spectrum of sidebands of an AM voice signal sent on a 600 kHz carrier.

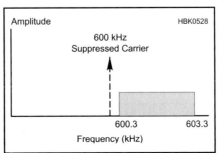

Fig 12.13 — Spectrum of single sideband AM voice signal sent adjacent to a suppressed 600 kHz carrier.

use bandwidths of 1.8 to 2.4 kHz with the corresponding loss of some of the higher consonant sounds.

RADIOTELEGRAPHY (CW)

We have described radiotelegraphy as being transmitted by "on-off keying of a carrier." You might think that since a carrier takes up just a single frequency, the receive bandwidth needed should be almost zero. This is only true if the carrier is never turned on and off. In the telegraphy case, it will be turned on and off quite rapidly. The rise and fall of the carrier results in sidebands extending out from the carrier for some distance, and they must be received in order to reconstruct the signal in the receiver.

A rule of thumb is to consider the rise and fall time as about 10% of the pulse width and the bandwidth as the reciprocal of the quickest of rise or fall time. This results in a bandwidth requirement of about 50 to 200 Hz for the usual radiotelegraph transmission rates. Another way to visualize this is with the bandwidth being set by a *high-Q* tuned circuit. Such a circuit will continue to "ring" after the input pulse is gone. Thus too narrow a bandwidth will actually "fill in" between the code elements and act like a "no bandwidth" full period carrier.

DATA COMMUNICATION

(This is a short overview of receivers and data communications. See the **Modulation, Digital Modes** and **Digital Communications** chapters for more in-depth treatment.)

The Baudot code (used for teletype communications) and ASCII code — two popular digital communications codes used by amateurs — are constructed with sequences of ON-OFF elements or bits as shown in **Figs 12.14** and **12.15**. The state of each bit — ON or OFF — is represented by a signal at one of two distinct frequencies: one designated *mark* and one designated *space*. (These states are named after the early Morse tape readers that placed a pen *mark* on a paper tape when the key was down and made a *space* when the key was up.) This is referred to as *frequency shift keying* (FSK). The transmitter frequency shifts back and forth with each character's individual elements.

Amateur Radio operators typically use a 170 Hz separation between the mark and space frequencies, depending on the data rate and local convention, although 850 Hz is sometimes used.[6] The minimum bandwidth required to recover the data is somewhat greater than twice the spacing between the tones. Note that the tones can be generated by directly shifting the carrier frequency (direct FSK), or by using a pair of 170 Hz spaced audio tones applied to the audio input of an SSB transmitter (audio FSK or AFSK). Direct FSK and AFSK sound the same to a receiver.

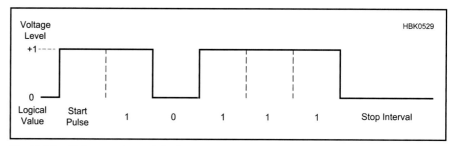

Fig 12.14 — Voltage pulses of the letter "A" in Baudot code, with start and stop pulses.

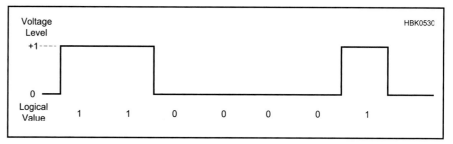

Fig 12.15 — Voltage pulses corresponding to the lower case letter "a" in ASCII code.

Note that if the standard audio tones of 2125 Hz (mark) and 2295 Hz (space) are used, they fit within the bandwidth of a voice channel and thus the facilities of a voice transmitter and receiver can be employed without any additional processing needed outside the radio equipment. Alternately, the receiver can employ detectors for each frequency and provide an output directly to a computer.

If the receiver can shift its BFO frequency appropriately, the two tones can be received through a filter designed for CW reception with a bandwidth of about 300 Hz or wider. Some receivers provide such a narrow filter with the center frequency shifted midway between the tones (2210 Hz) to avoid the need for retuning. The most advanced receivers provide a separate filter for mark and space frequencies, thus maximizing interference rejection and signal-to-noise ratio (SNR). Using a pair of tones for FSK or AFSK results in a maximum data rate of about 1200 bit/s over a high-quality voice channel.

Phase shift keying (PSK) can also be used to transmit bit sequences, requiring good frequency stability to maintain the required time synchronization to detect shifts in phase. If the channel has a high SNR, as is often the case at VHF and higher, telephone network data-modem techniques can be used.

At HF, the signal is subjected to phase and amplitude distortion as it travels. Noise is also substantially higher on the HF bands. Under these conditions, modulation and demodulation techniques designed for "wireline" connections become unusable at bit rates of more than a few hundred bps. As a result, amateurs have begun adopting and developing state of the art digital modulation techniques. These include the use of multiple carriers (MFSK, Clover, PACTOR III, etc.), multiple amplitudes and phase shifts (QAM and QPSK techniques), and advanced error detection and correction methods to achieve a data throughput as high as 3600 bits per second (bps) over a voice-bandwidth channel. (Spread-spectrum techniques are also being adopted on the UHF bands, but are beyond the scope of this discussion.)

The bandwidth required for data communications can be as low as 100 Hz for PSK31 to 1 kHz or more for the faster speeds of PACTOR III and Clover. Beyond having sufficient bandwidth for the data signal, the primary requirements for receivers used for data communications are linear amplitude and phase response over the bandwidth of the data signal. The receiver must also have excellent frequency stability to avoid drift and frequency resolution to enable the receiver filters to be set on frequency.

FREQUENCY MODULATION (FM)

Another popular voice mode is frequency modulation or FM. FM can be found in a number of variations depending on purpose. In Amateur Radio and commercial mobile communication use on the shortwave bands, it is universally *narrow band FM* or *NBFM*. In NBFM, the frequency deviation is limited to around the maximum modulating frequency, typically 3 kHz. The bandwidth requirements at the receiver can be approximated by $2 \times (D + M)$, where D is the deviation and M is the maximum modulating frequency.

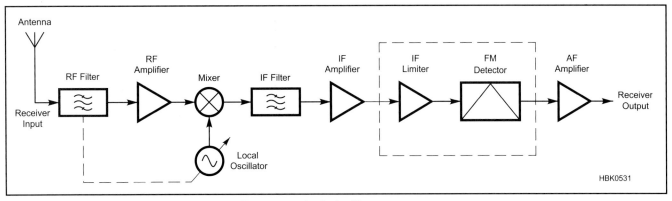

Fig 12.16 — Block diagram of an FM superhet. Changes are in dashed box.

Thus 3 kHz deviation and a maximum voice frequency of 3 kHz results in a bandwidth of 12 kHz, not far beyond the requirements for broadcast AM. (Additional signal components extend beyond this bandwidth, but are not required for voice communications.)

In contrast, broadcast or *wide-band FM* or *WBFM* occupies a channel width of 150 kHz. Originally, this provided for a higher modulation index, even with 15 kHz audio that resulted in an improved SNR. However, with multiple channel stereo and sub-channels all in the same allocated bandwidth the deviation is around the maximum transmitted signal bandwidth.

In the US, FCC amateur rules limit wideband FM use to frequencies above 29 MHz. Some, but not all, HF communication receivers provide for FM reception. For proper FM reception, two changes are required in the receiver architecture as shown within the dashed line in **Fig 12.16**. The fundamental change is that the detector must recover information from the frequency variations of the input signal. The most common type is called a *discriminator*. The discriminator does not require the BFO, so that is turned off, or eliminated in a dedicated FM receiver. Since amplitude variations convey no information in FM, they are generally eliminated by a *limiter*. The limiter is a high-gain IF amplifier stage that clips the positive and negative peaks of signals above a certain threshold. Since most noise of natural origins is amplitude modulated, the limiting process also strips away noise from the signal.

SUMMARY OF RECEIVER BANDWIDTH REQUIREMENTS

We now have briefly discussed the typical operating modes expected to be encountered by an HF communications receiver. These are tabulated in **Table 12.1** and will be used as design requirements as we develop the various receiver architectures.

12.3.2 Selection of IF Frequency for a Superhet

Now that we have established the range of bandwidths that our receiver will need to pass, we are in a position to discuss the selection of the IF frequency at which those bandwidths will be established.

For many years, the operating bandwidth of a receiver was established by multiple tuned interstage transformers in the IF portion of the receiver. All things being equal, the lower the IF frequency the narrower the selectivity; the higher the IF frequency the wider the passband. The actual bandwidth was determined by the details of the design. The number and quality of the tuned circuits, and the amount of coupling between them had a major impact, not only on the nominal bandwidth, but also the slope of the response curve. **Table 12.2** shows the selectivity of some typical early IF arrangements based on tuned circuits. The typical single IF stage had two tuned transformers, one on each side. Adding each additional stage typically added two tuned circuits, with advanced models sometimes having additional tuned circuits between stages.

The usual AM broadcast receiver, and low end "communications" receiver of the vacuum tube day, has a single IF stage at 455 kHz, and as can be seen, this provided reasonable selectivity for use in areas that did not have signals on adjacent channels.

It was relatively easy for a designer to increase the bandwidth for wider bandwidth modes, either by stagger tuning the transformers to slightly different frequencies across the band, or even by using resistive loading to reduce the circuit Q. Decreasing the bandwidth for narrower modes was a bit trickier and early communications receivers often provided a single crystal filter that could be used to provide a narrow band pass for CW operation.

IF IMAGE RESPONSE

As noted earlier, a superhet with a single local oscillator or LO and specified IF can receive two frequencies, selected by the tuning of the RF stage. For example, using a receiver with an IF of 455 kHz to listen to a desired signal at 7000 kHz can use an LO of 7455 kHz. However, the receiver will also receive a signal at 455 kHz above the LO frequency, or 7910 kHz. This undesired signal frequency, located at twice the IF frequency from the desired signal, is called an *image*.

Images will be separated from the desired frequency by twice the IF and must be filtered

Table 12.1
Typical Communications Bandwidths for Various Operating Modes

Mode	Bandwidth (kHz)
FM Voice	15
AM Broadcast	10
AM Voice	4-6.6
SSB Voice	1.8-3
RTTY (850 Hz shift)	0.3-1.0
CW	0.1-0.5

Table 12.2
Selectivity of Early Superhet IF Amplifier Arrangements

IF Amplifier	Response		
	−6 dB	−20 dB	−40 dB
One stage 50 kHz (iron core)	0.8	1.4	2.8
One stage 455 kHz (air core)	8.7	17.8	32.3
One stage 455 kHz (iron core)	4.3	10.3	20.4
Two stages 455 kHz (iron core)	2.9	6.4	10.8
Two stages 1600 kHz (iron core)	11.0	16.6	27.4

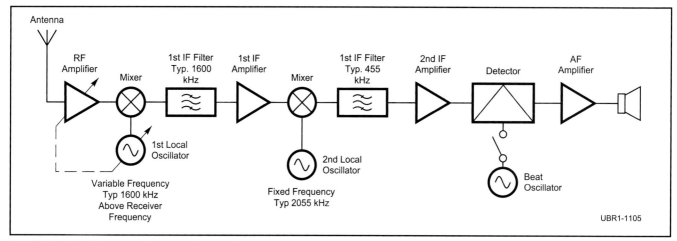

Fig 12.17 — Double-conversion superhet receiver. Early type.

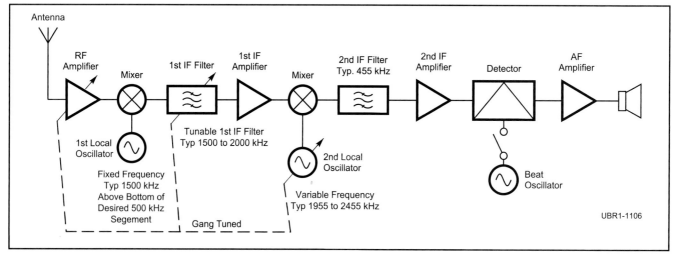

Fig 12.18 — Double-conversion superhet receiver. Collins system.

out ahead of the associated mixer by the ratio of desired image rejection. For a given IF, this gets more difficult as the received frequency is increased. For example, with a 455 kHz single conversion system tuned to 1 MHz, the image will be at 1.91 MHz, almost a 2:1 frequency ratio and relatively easy to reject with a filter. The same receiver tuned to 30 MHz, would have an image at 30.91 MHz, a much more difficult filtering problem

The popular 455 kHz IF frequency generally started to have serious image problems at receiving frequencies around 10 MHz and higher. Early top-of-the-line receivers extended this range somewhat by having multiple gang-tuned RF stages ahead of the mixer to reduce the image, but the nonlinear responses of the additional gain stages ahead of the mixer often resulted in additional unwanted mixer products that could be as bad as the image.

While an image that fell on an occupied channel was obviously troubling — it's rarely desirable to receive two signals at the same time — it is a problem even if the image frequency is clear of signals. This is because the atmospheric noise in the image bandwidth is added to the noise of the desired channel, as well as any internally-generated noise in the RF amplifier stage. If the image response is at the same level as the desired signal response, there will be a 3 dB reduction in SNR.

REDUCING IMAGES WITH A HIGHER IF FREQUENCY

An obvious solution to the RF image response is to raise the IF frequency high enough so that signals at twice the IF frequency from the desired signal are sufficiently attenuated by the tuned circuit(s) ahead of the mixer. This can easily be done, with most selections in the area of 5 to 10% of the highest receiving frequency (1.5 to 3 MHz for a receiver that covers the 3-30 MHz HF band). The concept is used at higher frequencies as well. The FM broadcast band (150 kHz wide channels over 87.9 to 108 MHz in the US) is generally received on superhet receivers with an IF of 10.7 MHz, which places all image frequencies outside the FM band, eliminating interference from other FM stations.

The use of higher-frequency tuned circuits for IF selectivity works well for the 150 kHz wide FM broadcast channels, but not so well for the relatively narrow channels encountered on HF or lower, or even for many V/UHF narrowband services. Fortunately, there are three solutions that were commonly used to resolve this problem.

Double Conversion

In the 1950s, a new technique called *double conversion* was defined to solve the image problem. Rather than having to decide between a high IF frequency for good image rejection, or a low IF frequency for narrow channel selectivity, some bright soul decided to do *both*! As shown in **Fig 12.17**, a conversion of the desired signal to a relatively high IF is followed by a second conversion to a lower IF to set the selectivity. This arrangement solved the image problem nicely,

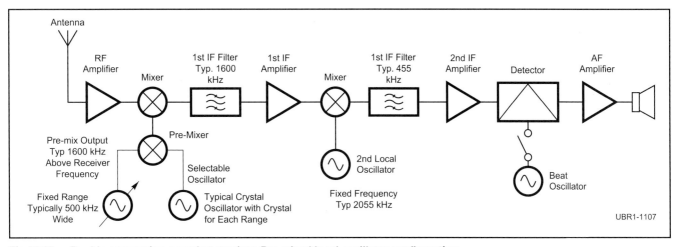

Fig 12.19 — Double-conversion superhet receiver. Pre-mixed local oscillator configuration.

while the rest of the receiver could be pretty much kept the same as before. A number of manufacturers offered revised receivers in the fifties that added an additional conversion stage, often just above 10 or 15 MHz, regions that had poor image rejection in the original design. Note that major changes were not required in the receiver, and some looked just like their single conversion predecessors from the outside. The extra stage was shoehorned in, with just a retuning of the first local oscillator required.

Some manufacturers decided that if two conversions were good, three must be even better. These designs often started in the same place, but converted the 455 kHz second IF down to a third IF, often in the 50 to 100 kHz range, for even sharper selectivity. This approach lasted until crystal lattice filters became available that outperformed the low frequency LC circuits and avoided the need for the additional conversion stage.

In the 1940s, a visionary radio pioneer, Arthur Collins, WØCXX, founder of Collins Radio in the 1930s, came up with another approach to double conversion. One problem with earlier MF and HF receivers was that they used a standard LC (first) oscillator using a variable capacitor of the type used in broadcast receivers. As noted earlier, these capacitors typically had a 9:1 capacitance range resulting in a 3:1 frequency range. The typical tuning arrangement for multi-band receivers was .5 to 1.5, 1.5 to 4.5, 4.5 to 13.5, and 13.5 to 30 MHz. This covered from the broadcast band to the top of the HF range in four bands.[7] With this arrangement, the tuning rate and dial calibration marks got less and less precise as higher bands were selected making tuning difficult.

The Collins system (see **Fig 12.18**) switched the variable oscillator to the second position and used crystal-controlled oscillators as the first oscillator. Although many more bands were required (30 for the famous 51J series that covered 0.5 to 30.5 MHz) each tuned with exactly the same tuning rate. Collins went a step further and designed an inductance-tuned oscillator (usually called a permeability-tuned oscillator or PTO) that was linear throughout its range. These oscillators could be tuned to the nearest 1 kHz from beginning to end, avoiding the tuning uncertainty of other receiver types. For this to work properly, synchronized or gang tuning is required between the oscillator, first IF variable filter and RF stages. This was no problem for the engineers at Collins Radio, but their equipment brought a premium price.

A third approach to double conversion was a mix of the first two (no pun intended). The *pre-mixed* arrangement, see **Fig 12.19**, uses a single variable oscillator range, as with the Collins, but does the mixing outside of the signal path. This avoids the need for variable tuning at the first IF, allowing a tighter *roofing filter* but trades it for the need for filtering at the output of the pre-mixer (not shown).

High Frequency Crystal Lattice Filters

Just as double conversion receivers were settling into becoming the "way things were done," the use of piezoelectric crystals as elements of band-pass filters became feasible. While crystal lattice filters at 455 kHz were popular following WW2 because of availability of large stocks of surplus crystals at that range, they didn't address the image response issue, just provided better shape factors at the same IF frequency. Starting in the 1950s, crystal lattice filter technology moved higher in the MF region and into HF.

Commercial filters at CW and SSB appropriate bandwidths became available at center frequencies up to perhaps 10 MHz. Now a single-conversion receiver with an IF in the HF range could provide both high image rejection and needed channel selectivity. **Fig 12.20** is an example of a single-conversion superhet using simple filters at 1500 kHz. While the filters shown are actually buildable at low cost, multiple-section filters with much better performance can be purchased or constructed to make this a top performing receiver. Still, the circuit shown will demonstrate the concepts and can be reproduced at low cost. Other IF frequencies can be used, perhaps depending on crystal or filter availability.

The Image-Rejecting Mixer

The third technique for reduction of image response in receivers is not as commonly encountered in HF receivers as the preceding designs, but it deserves mention because it has some very significant applications. The image-canceling, or more accurately *image-rejecting* mixer requires phase-shift networks, as shown in **Fig 12.21**. Frequency F_1 represents the input frequency while F_2 is that of the local oscillator. Note that the two 90° phase shifts are at different frequencies. While conceptually straightforward, the phase shift network following mixer one is at a fixed center frequency corresponding to the IF, while the phase shift network at frequency two must provide the required phase shift as the local oscillator tunes across the band.

As usual, the devil is in the details. If the local oscillator is required to tune over a limited fractional frequency range, this is a very feasible approach. On the other hand, maintaining a 90° phase shift over a wide range can be tricky.

The good news is that this approach provides image reduction that is independent of, and in addition to, any other mechanisms employed toward that end. This method can be found in a number of places in some receivers.

- It is the only way to provide "single signal" reception with a direct conversion receiver, effectively reducing the audio image. This can make the DC receiver a very good performer, although the added complexity is not always warranted in many DC applications..

- This option is frequently found in microwave receivers in which sufficiently selective RF filtering can be difficult to obtain. Since they often operate on fixed frequencies, maintaining the required phase shift can be straightforward.
- It is found in advanced receivers that are trying to achieve optimum performance. Even with a high first IF frequency, additional image rejection can be provided.
- In transmitters, the same system is called the *phasing method* of SSB generation. The same blocks run "backwards," with one of the phase shift networks applied to the speech band, can be used to cancel one sideband. More about this technique will be presented in the **Transmitters** chapter. The following is the description of a direct conversion receiver that uses an image rejecting mixer as a product detector for single sideband or single signal CW reception.

Project: The MicroR2 — An Easy to Build SSB or CW Receiver

Throughout the 100-year history of Amateur Radio, the receiver has been the basic element of the amateur station. This single

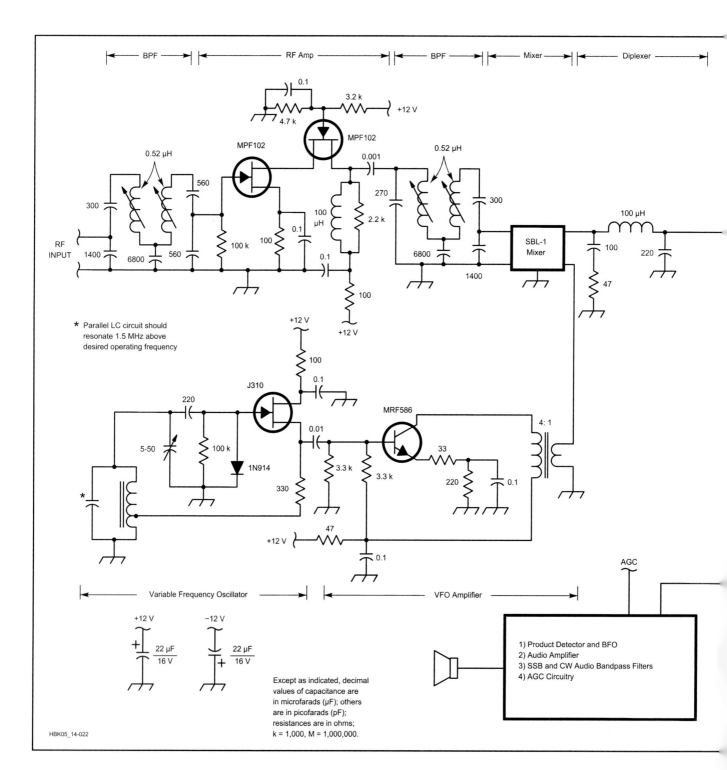

PC board receiver, originally described by Rick Campbell, KK7B, captures some of the elegance of the classic all-analog projects of the past while achieving the performance needed for operation in 21st-century band conditions across a selected segment of a single band.

A major advantage of a station with a separate receiver and transmitter is that each can be used to test the other. There is a particularly elegant simplicity to the combination of a crystal controlled transmitter and simple receiver. The receiver doesn't need a well-calibrated dial, because the transmitter is used to spot the frequency. The transmitter doesn't need sidetone, VFO offset or other circuitry because the crystal determines the transmitted frequency and the receiver is used to monitor the transmit note off the air. Even the receiver frequency stability may be relaxed, because there is no on-the-air penalty for touching up receiver tuning during a contact.

Let's dive into the design and construction of a simple receiver (**Fig 12.22**) that may be used as a companion for an SSB or CW transmitter on 40 meters, or as a tunable IF for higher bands. **Fig 12.23** is the block diagram of this little receiver, which can be built in a few evenings. The block diagram is similar to the simple direct conversion receivers of the early 1970s, but there the similarity ends. This radio takes advantage of 40 years of

Fig 12.20 — Schematic of representative superhet using crystal lattice filters to achieve selectivity with a high IF frequency.

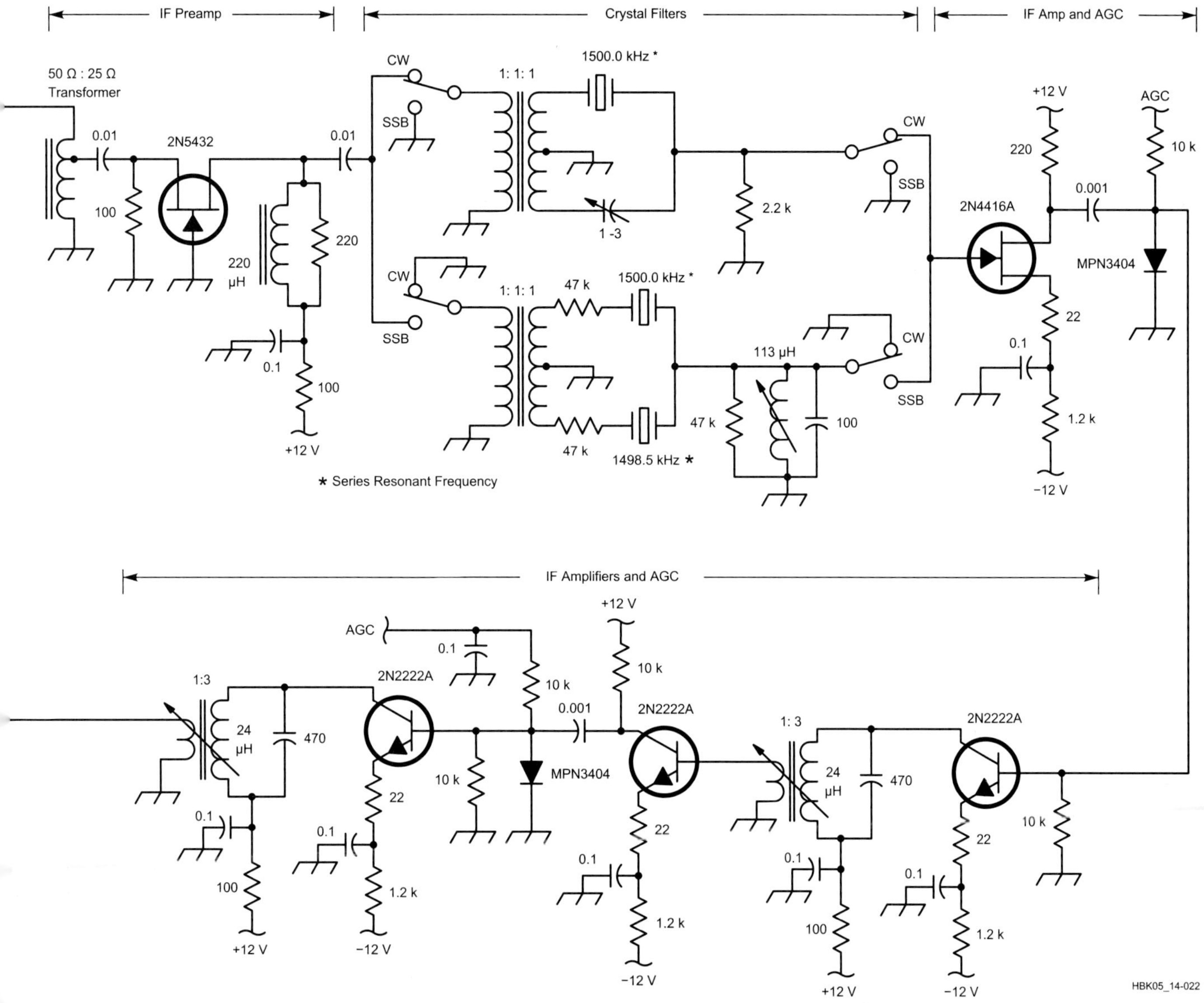

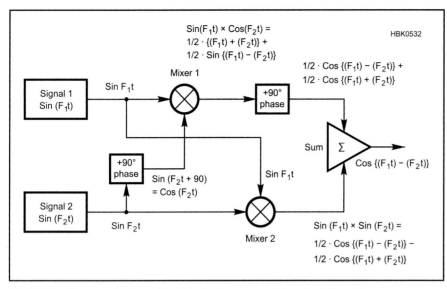

Fig 12.21 — Image rejecting mixer, block diagram and signal relationships.

Fig 12.22 — Two versions of the MicroR2. Hard to beat a big vernier dial!

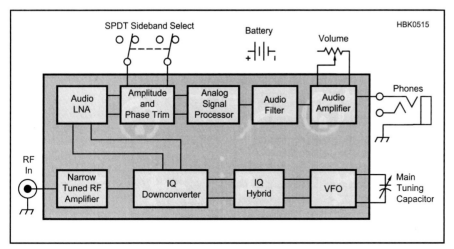

Fig 12.23 — Block diagram of the MicroR2 SSB/CW receiver.

evolution of direct conversion HF receivers by some very talented designers.

"SINGLE SIGNAL" RECEPTION MAKES ALL THE DIFFERENCE

One major difference from early direct conversion receivers is the use of phasing to eliminate the opposite sideband. The receiver in Fig 12.23 is a "single-signal" receiver that only responds to signals on one side of zero beat. Basic receivers in the '50s and '60s would hear every CW signal at two places on the dial, on either side of zero beat. As you tuned through a CW signal, first you heard a high-pitched tone, and then progressively lower audio frequencies until the individual beats became audible, and then finally all the way down to zero cycles per second: zero beat. As you tuned further you heard the pitch rise back up until the high frequency CW signal became inaudible again at some high pitch.

It's not too terrible to hear the desired signal at two places on the dial — but having twice as much interference and noise is inconvenient at best. Most modern radios, even simple ones, include a simple crystal or mechanical filter for single-signal IF selectivity. Since a direct conversion receiver has no IF, selectivity is obtained by a combination of audio filtering and an image-rejecting mixer. Chapter 9 of *Experimental Methods in RF Design* (*EMRFD*) has a complete discussion of image-reject or "phasing" receivers, and an overview of both conventional superheterodyne and phasing receivers can be found elsewhere in this chapter.

BLOCK BY BLOCK DESCRIPTION

A brief description of each block in Fig 12.23 and the schematic in **Fig 12.24** follows. (Additional detail and rationale for the circuits is provided in *EMRFD*.)

The 50-Ω RF input connects to a narrow tuned-RF amplifier. The first circuit elements are a low-pass filter to prevent strong signals above the receiver tuning range — primarily local TV and FM broadcast stations — from arriving at the mixers where they would be down-converted to audio by harmonics of the local oscillator. The common-gate JFET low-noise amplifier (LNA) sets the noise figure of the receiver and provides some gain, but even more importantly, it isolates the mixers from the antenna. Isolation prevents the impedance at the RF input from affecting the opposite sideband suppression, and reverse isolation prevents local oscillator leakage from radiating out the antenna and causing interference to others as well as tunable hum.

The tuned drain circuit provides additional selectivity and greatly improves dynamic range for signals more than a few hundred kHz from the desired signal. Lifting the source of the JFET allows a convenient mute

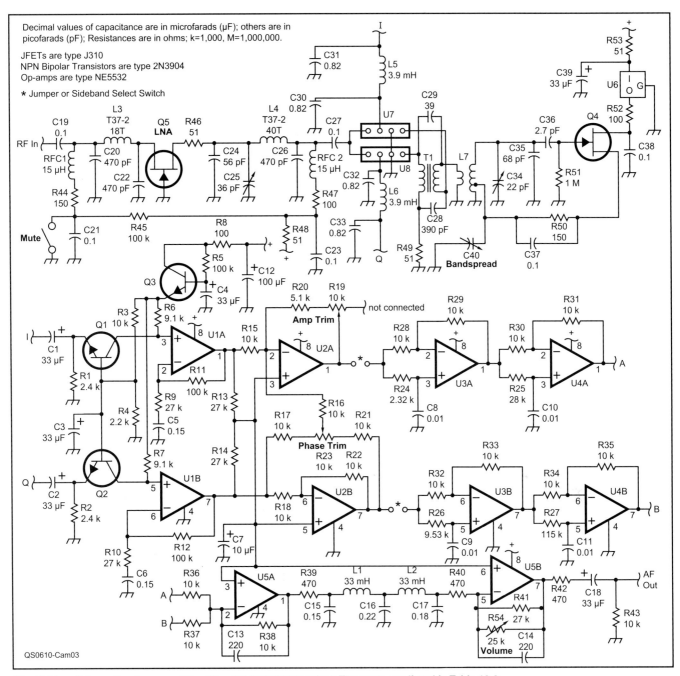

Fig 12.24 — Schematic diagram of the MicroR2 SSB/CW receiver. The parts are listed in Table 12.3.

control — changing the LNA from a 10 dB amplifier to a 40 dB attenuator.

Following the LNA are the I-Q (short for "in-phase" and "quadrature," the labels we use for the two signal paths in a phasing system) down-converter and local oscillator (LO) I-Q hybrid. Many of the more subtle difficulties of phasing direct conversion receivers are avoided by a compact, symmetrical layout and direct connection of all the mixer ports to the appropriate circuit elements without the use of transmission lines. The audio low-pass filters ensure that only audio exits the down-converter block, and set the close-in dynamic range of the receiver. The inductors in the low-pass filters will pick up magnetic hum from nearby power transformers — more on this topic later.

To the right of the I-Q hybrid is a basic JFET Hartley VFO. In this application, it is stripped of all the usual accoutrements like buffer amplifiers and receiver incremental tuning (RIT). The drain resistor and diodes in the mixers set the operating level. This simplified configuration, with a link on the VFO inductor directly driving the twisted-wire hybrid, works very well over a 5% bandwidth (350 kHz at 7 MHz) when the quadrature hybrid is optimized at the center frequency and all voltages except the LO are less than a few millivolts.[8]

The top half of the block diagram operates entirely at audio. The left block is a matched pair of audio LNAs with a noise figure of about 5 dB and near 50 Ω input impedance, to properly terminate the low-pass filters at the I-Q mixer IF ports. These are directly coupled into a circuit that performs the mathematical operations needed to remove amplitude and phase errors from the I and Q signal paths. This is a classic use of operational amplifiers to do basic math.

The next block is a pair of second-order all-pass networks. Use of second order networks limits the opposite sideband suppres-

Receivers 12.19

Table 12.3
MicroR2 Parts List
Parts for CW and SSB Version

C1-C4, C18, C39 — 33 µF, 16 V electrolytic.
C5, C6 — 0.15 µF, 5% polyester.
C7 — 10 µF, 16 V electrolytic.
C8-C11 — 0.01 µF, polyester matched to 1%.
C12 — 100 µF, 16 V electrolytic.
C13, C14 — 220 pF, NP0 ceramic.
C15 — 0.15 µF, 5% polyester, see Note A.
C16 — 0.22 µF, 5% polyester, see Note A.
C17 — 0.18 µF, 5% polyester, see Note A.
C19, C21, C23, C27, C37, C38 — 0.1 µF, 5% polyester.
C20, C22, C26 — 470 pF, NP0 ceramic.
C24 — 56 pF, NP0 ceramic.
C25 — 3-36 pF, poly trimmer capacitor.
C28 — 390 pF, NP0 ceramic.
C29 — 39 pF, NP0 ceramic on back of board, see Note B.
C30-C33 — 0.82 µF, 5% polyester, see Note C.

C34 — 2.2-22 pF, poly trimmer capacitor.
C35 — 68 pF, NP0 ceramic.
C36 — 2.7 pF 5% NP0 1000V ceramic.
C40 — Variable capacitor, see Note D.
L1, L2 — 33 mH inductor, see Note A.
L3 — 18t #28 T37-2, see Note E.
L4 — 40t #30 T37-2, see Note E.
L5, L6 — 3.9 mH inductor, see Note C.
L7 — 36t #28 T50-6, tap at 8 turns with a 2 turn link, see Note E.
Q1-Q3 — 2N3904.
Q4, Q5 — J310.
R1, R2 — 2.4 kΩ.
R3, R15-R18, R21, R22, R38, R43 — 10 kΩ.
R4 — 2.2 kΩ.
R5, R11, R12, R45 — 100 kΩ.
R6, R7 — 9.1 kΩ.
R8, R47, R52 — 100 Ω.
R9, R10, R13, R14, R41 — 27 kΩ.

R19, R23 — 10 kΩ trimpot.
R20 — 5.1 kΩ.
R24 — 2.32 kΩ, 1% see Note A.
R25 — 28.0 kΩ, 1% see Note A.
R26 — 9.53 kΩ, 1% see Note A.
R27 — 115 kΩ 1% see Note A.
R28-R37 — 10.0 kΩ 1%.
R39, R40, R42 — 470 Ω.
R44, R50 — 150 Ω.
R46, R48, R49, R53 — 51 Ω.
R51 — 1 MΩ.
R54 — 25 kΩ volume control.
RFC1, RFC2 — 15 µH molded RF choke.
T1 — 17 t two colors #28 bifilar T37-2, see Note E.
U1-U5 — NE5532 or equivalent dual low-noise high-output op-amp.
U6 — LM7806 or equivalent 6 V three terminal regulator.
U7, U8 — Mini-Circuits TUF-3 diode ring mixer.

Note A: Make the following parts substitutions for a CW only version:
R24 — 4.75 kΩ, 1%.
R25 — 41.2 kΩ, 1%.
R26 — 16.9 kΩ, 1%.
R27 — 147 kΩ, 1%.
C15 — 0.47 µF, 5% polyester.
C16 — 0.68 µF, 5% polyester.
C17 — 0.56 µF, 5% polyester.
L1, L2 — 100 mH inductor.

Note B: The total reactance of the parallel combination of C28 and C29 plus the capacitance between the windings of T1 is $-j50\ \Omega$ at the center of the tuning range. Placing most of the capacitance at one end is a different but equivalent arrangement to the quadrature hybrid we often use with equal capacitors. C29 is only needed if there is no standard value for C28 within a few % of the required value. C29 is tack soldered to the pads provided on the back of the PC board, and may be a surface mount component if desired.

Note C: To sacrifice close-in dynamic range and selectivity for reduced 60 Hz hum pickup, make the following substitutions, as illustrated in Fig 12.26. This modification is recommended if the MicroR2 must be used near 60 Hz power transformers.
C30, C32 — 0.10 µF 5% polyester.
C31, C33 — Not used.
L5, L6 — Not used. Place wire jumper between pads.

Note D: Capacitor C40 is the tuning capacitor for the receiver. The MicroR2 was specifically designed to use an off-board tuning capacitor to provide mounting flexibility, and to encourage the substitution of whatever high-quality dual bearing capacitor and reduction drive may be in the individual builder's junk box. Any variable capacitor may be used, but values around 100 pF provide a little more than 100 kHz of tuning range on the 40 meter band, which is a practical maximum. The MicroR2 needs to be realigned (RF amplifier peaked and the amplitude and phase trimpots readjusted) when making frequency excursions of more than about ±50 kHz on the 40 meter band. To prevent tunable hum and other common ills of direct conversion receivers, the MicroR2 PC board and tuning capacitor should be in a shielded enclosure. See Chapter 8 of *EMRFD* for a complete discussion.

Note E: L3, L4, L7 and T1 are listed as number of turns on the specified core rather than a specific inductance. For those who wish to study the design with a calculator, simulator and inductance meter, L3 should be about $+j100\ \Omega$ at about 8 MHz (not particularly critical). L4 and L7 should be about $+j250\ \Omega$ at mid band. Each winding of T1 should be $+j50\ \Omega$ at mid band.

---

sion to 37 dB, which sounds exceptionally clean with multiple CW signals and static crashes in the passband.

It is straightforward to improve the opposite sideband suppression by 10 dB or so, but doing so requires much more circuitry, not just a third-order all-pass network. Relaxing the opposite sideband suppression to a level comparable with the Drake 2B permitted cutting the parts count in half, and the construction and alignment time by a factor of 10.

The final blocks are the summer, audio low-pass filter and headphone amplifier. These are conventional, near duplicates of those used 10 years ago. When the board is finished it is an operating receiver. The completed circuitry is shown in **Fig 12.25**.

ALIGNMENT

The receiver PC board may be aligned on the bench without connecting the bandspread capacitor. There are only four adjustments, and the complete alignment takes just a few minutes with just an antenna, the crystal oscillator in the companion transmitter (or a test oscillator) and headphones. First, center the trimming potentiometers using the dots.

Fig 12.25 — Photo of the inside of the receiver. The PC board and parts are available from Kanga US (www.kangaus.com).

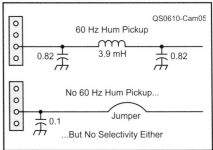

Fig 12.26 — Modified circuit to minimize hum.

Then slowly tune the VFO variable capacitor until the crystal oscillator is heard. The signal should be louder on one side of zero beat. Peak the LNA tuning carefully for the loudest signal. Then tune to the weaker side of zero beat and null the signal with the amplitude and phase trim pots. That's all there is.

If the adjustments don't work as expected or there is no opposite sideband suppression, look for a construction error. Don't forget to jumper the MUTE connection on the LNA — it is a "ground- to-receive" terminal just like simple receivers from the '60s. Once the receiver PC board is working on the bench, mount it in a box, attach the bandspread capacitor (with two wires — the ground connection needs to be soldered to the PC board too). Reset the VFO tuning capacitor for the desired tuning range, and do a final alignment in the center of the desired frequency range.

The simple on-board VFO is stable enough for casual listening and tuning around the band, but it can be greatly improved by eliminating the film trimmer capacitor that sets the VFO tuning range. It may be replaced by an air trimmer, a piston trimmer, a small compression trimmer, or a combination of small fixed NP0 ceramic capacitors experimentally selected to obtain the correct tuning range. The holes in the PC board are large enough to easily remove the film trimmer capacitor with solder wick. The VFO stability may be improved still further by following the thermal procedure described by W7ZOI in *EMRFD*.

PERFORMANCE

Is it working? I don't hear anything! Most of us are familiar with receivers with a lot of extra gain and noise, and AGC systems that turn the gain all the way up when there are no signals in the passband. Imagine instead a perfect receiver with no internal noise. With no antenna connected, you would hear absolutely nothing, just like a CD player with no disk. The MicroR2 is far from perfect — but it may be much quieter than your usual 40 meter receiver, and it probably has less gain. It is designed to drive good headphones. Weak signals will be weak and clear and strong signals strong and clear over the entire 70 dB in-channel dynamic range — just like a CD player. This receiver rewards listening skill and a good antenna — don't expect to hear much on a couple feet of wire in a noisy room. The measured performance is shown in **Table 12.4**.

Now, about that hum. One reason this receiver sounds so good is that the selectivity is established right at the mixer IF ports, before any audio gain. Then, after much of the system gain, a second low-pass filter provides additional selectivity. Distributed selectivity is a classic technique. Earlier designs included aggressive high-pass filtering to suppress frequencies below 300 Hz. That works well and has other advantages — but the audio sounds a bit thin. K7XNK suggested trying a receiver with the low frequency response wide open. It sounds great, but 60 Hz hum is audible if the receiver is within about a meter of a power transformer. Any closer and the hum gets loud. This makes it difficult to measure receiver performance in the lab, as there is a power transformer in every piece of test equipment.

Hum pickup may be eliminated by modifying the circuit as shown in **Fig 12.26**. This hurts performance in the presence of strong off-channel signals, but in some applications — an instrumentation receiver on the bench — it is a good trade-off. The receiver sounds the same either way. For battery operation in the hills, or on a picnic table in the backyard, hum pickup isn't a problem.

12.3.3 Multiple-Conversion in the Digital Age

Each of the design choices described above are still found in current receiver architectures, but some recent advances have added a few new twists. While advances in

Table 12.4
MicroR2 Receiver Performance

Parameter	Measured in the ARRL Lab
Frequency coverage:	7.0-7.1 or 7.2-7.3 MHz.
Power requirement:	50 mA, tested at 13.8 V.
Modes of operation:	SSB, CW.
SSB/CW sensitivity:	−131 dBm.
Opposite sideband rejection:	1000 Hz, 42 dB
Two-tone, 3rd-order IMD dynamic range:	20 kHz spacing, 81 dB
Third-order intercept:	20 kHz spacing, −5 dBm.

Third-order intercept points were determined using equivalent S5 reference.

microminiaturization of all circuit elements have had a radical change in the dimensions of communication equipment, perhaps most significant technology impacts on architecture come in two areas — the application of digital signal processing and direct digital synthesis frequency generation. Both are discussed in detail in the chapter on **DSP and Software Radio Design**. An overview is presented here.

Digital signal processing provides a level of bandwidth setting filter performance not practical with other technologies. While much better than most low frequency IF LC bandwidth filters, the very good crystal or mechanical bandwidth filters in amateur gear are not very close to the rectangular shaped frequency response of an ideal filter, but rather have skirts with a 6 to 60 dB response of perhaps 1.4 to 1. That means if we select an SSB filter with a nominal (6 dB) bandwidth of 2400 Hz, the width at 60 dB down will typically be 2400 × 1.4 or 3360 Hz. Thus a signal in the next channel that is 60 dB stronger than the signal we are trying to copy (as often happens) will have energy just as strong as our desired signal.

DSP filtering approaches the ideal response. **Fig 12.27** shows the ARRL Lab measured response of a DSP bandwidth filter with a 6 dB bandwidth of 2400 Hz. Note how rapidly the skirts drop to the noise level.

In addition, while analog filtering generally requires a separate filter assembly for each desired bandwidth, DSP filtering is adjustable — often in steps as narrow as 50 Hz — in both bandwidth and center frequency. In addition to bandwidth filtering, the same DSP can often provide digital noise reduction and digital notch filtering to remove interference from fixed frequency carriers.

Early DSP filters operated in the audio frequency range, based on the frequency response limitations of the analog-to-digital converter (ADC) that provides the digitization for further processing. The limitation was based on the Nyquist sampling theorem (AT&T, 1924) that established that a periodic signal could be sampled and then reconstructed from its samples if sampled at least twice during each cycle of its highest frequency component. Early ADCs were limited to sampling rates less than about 50 kHz. Such a DSP filter could be added to the audio signal at the output of any receiver and work its magic on the signals in the audio stream. Unfortunately, at that time ADC and DSP that worked at typical receiver IF frequencies were either unavailable, or cost prohibitive for amateur use.

Receiver designers quickly resurrected the earlier triple conversion receiver architecture—now with IF frequencies even lower, in the 12-20 kHz range, so that DSP could

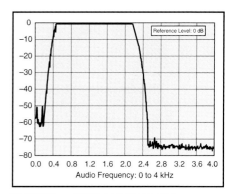

Fig 12.27 — ARRL Lab measured response of an aftermarket 2400 Hz DSP bandwidth filter.

be employed in the IF rather than audio section of receivers. This allowed the receiver to eliminate interfering signals before the application of automatic gain control, avoiding gain reduction in the presence of off-channel signals. As time and technology have advanced, the clocking speed and input bandwidth of ADCs continues to increase while prices continue to fall, allowing the third IF frequency to move toward the second. Soon the additional conversion step will no longer be required, and we will soon see high-performance HF receivers that apply signals at the input frequency directly to the ADC.

12.4 Superhet Receiver Design Details

The previous sections have provided an overview of the major architectural issues involved in receiver design. In this section, we will cover more of the details for the superheterodyne receiver constructed with a hybrid of analog electronics and DSP functions applied at an IF stage, the primary receiver design architecture used in Amateur Radio today.

12.4.1 Receiver Sensitivity

The sensitivity of a receiver is a measure of the lowest power input signal that can be received with a specified signal-to-noise ratio. In the early days of radio, this was a very important parameter and designers tried to achieve the maximum practical sensitivity. In recent years, device and design technology have improved to the point that other parameters may be of higher importance, particularly in the HF region and below.

THE RELATIONSHIP OF NOISE TO SENSITIVITY

Noise level is as important as signal level in determining sensitivity. Before beginning this section, review the discussion of noise in the **RF Techniques** chapter. The most important noise parameters affecting receiver sensitivity are noise bandwidth, noise figure, noise factor and noise temperature.

Since received noise power is directly proportional to receiver bandwidth, any specification of sensitivity must be made for a particular noise bandwidth. For DSP receivers with extremely steep filter skirts, receiver bandwidth is approximately the same as the filter or operating bandwidth. For other filter types, noise bandwidth is somewhat larger than the filter's 6 dB response bandwidth.

The relationship of noise bandwidth to noise power is one of the reasons that narrow bandwidth modes, such as CW, have a significant signal-to-noise advantage over modes with wider bandwidth, such as voice, assuming the receiver bandwidth is the minimum necessary to receive the signal. For example, compared to a 2400-Hz SSB filter bandwidth, a CW signal received in a 200 Hz bandwidth will have a 2400/200 = 12 = 10.8 dB advantage in received noise power. That is the same difference as an increase in transmitter from 100 to 1200 W.

Sources of Noise

Any electrical component will generate a certain amount of noise due to random electron motion. Any gain stages after the internal noise source will amplify the noise along with the signal. Thus a receiver with no signal input source will have a certain amount of noise generated and amplified within the receiver itself.

Upon connecting an antenna to a receiver, there will be introduction of any noise external to the receiver that is on the received frequency. The usual sources and their properties are described below. **Table 12.5** presents typical levels of external noise in a 10 kHz bandwidth present in the environment from different sources.

Atmospheric noise. This is noise generated within our atmosphere due to natural phenomena. The principal cause is lightning, acting very much like an early spark transmitter and sending wideband signals great distances. All points on the earth receive this noise, but it is much stronger in some regions than others depending on the amount of lo-

Table 12.5
Typical Noise Levels (Into the Receiver) and Their Source, by Frequency

Frequency range	Dominant noise sources	Typical level ($\mu v/m$)*
LF 30 to 300 kHz	atmospheric	150
MF 300 to 3000 kHz	atmospheric/man-made	70
Low HF 3 to 10 MHz	man-made/atmospheric	20
High HF 10 to 30 MHz	man-made/thermal	10
VHF 30 to 300 MHz	thermal/galactic	0.3
UHF 300 to 3000 MHz	galactic/ thermal	0.2

*The level assumes a 10 kHz bandwidth. Data from *Reference Data for Engineers*, 4th Ed, p 273, Fig 1.

cal lightning activity. This source is usually the strongest noise source in the LF range and may dominate well into the HF region, depending on the other noises in the region. The level of atmospheric noise tends to drop off by around 50 dB every time the frequency is increased by a factor of 10. This source usually drops in importance by the top of the HF range (30 MHz).

Man-made noise. This source acts in a similar manner to atmospheric noise, although it is more dependent on local activity rather than geography and weather. The sources tend to be rotating and other kinds of electrical machinery as well as gasoline engine ignition systems and some types of lighting. All things being equal, this source, on average, drops off by about 20 dB every time the frequency is increased by a factor of ten. The higher frequency is due to the sparks having faster rise times than lightning. The effect tends to be comparable to atmospheric noise in the broadcast band, less at lower frequencies and a bit more at HF.

Galactic Noise. This is noise generated by the radiation from heavenly bodies outside our atmosphere. Of course, while this is noise to *communicators*, it is the desired signal for *radio-astronomers*. This noise source is a major factor at VHF and UHF and is quite dependent on exactly where you point an antenna (antennas for those ranges tend to be small and are often *pointable*). It also happens that the Earth turns and sometimes moves an antenna into a position where it inadvertently is aimed at a noisy area of the galaxy. If the sun, not surprisingly the strongest signal in our solar system, appears behind a communications satellite, communications is generally disrupted until the sun is out of the antenna's receiving pattern. Galactic noise occurs on HF, as well: Noise from the planet Jupiter can be heard on the 15 meter band under quiet conditions, for example.

Thermal Noise. Unlike the previous noise sources, this one comes from our equipment. All atomic particles have electrons that move within their structures. This motion results in very small currents that generate small amounts of wideband signals. While each particle's radiation is small, the cumulative effect of all particles becomes significant as the previous sources roll off with increasing frequency. The reason that this effect is called *thermal* noise is because the electron motion increases with the particle's temperature. In fact the noise strength is directly proportional to the temperature, if measured in terms of absolute zero (0 K). For example, if we increase the temperature from 270 to 280 K, that represents an increase in noise power of 10/270, 0.037, or about 0.16 dB. Some extremely sensitive microwave receivers use cryogenically-cooled front-end amplifiers to provide large reductions in thermal noise.

Oscillator Phase Noise. As noted in the **Oscillators and Synthesizers** chapter, real oscillators do not produce sinusoids that are as pure as we might like. A receiver local oscillator, depending on the design, will have phase-noise sidebands that extend out on either side of the nominal carrier frequency at hopefully low amplitudes. Any such noise will be transferred to the received signal and create mixing products from signals on adjacent channels through the mixer. A good receiver will be designed to have phase noise that is well below the level of other internally generated noise.

Noise Power and Sensitivity

There are a number of related measures that can be used to specify the amount of noise that is generated within a receiver. If that noise approaches, or is within perhaps 10 dB of the amount of external noise received, then it must be carefully considered and becomes a major design parameter. If the internally generated noise is less than perhaps 10 dB below that expected from the environment, efforts to minimize internal noise are generally not beneficial and can, in some cases, be counterproductive.

While the total noise in a receiving system is, as discussed, proportional to bandwidth, the noise generating elements are generally not. Thus it is useful to be able to specify the internal noise of a system in a way that is independent of bandwidth. It is important to note that even though such a specification is useful, the actual noise is still directly proportional to bandwidth and any bandwidth beyond that needed to receive signal information will result in reduced SNR.

To evaluate the effect of noise power on sensitivity, we will use the following equations from the discussion of noise in the **RF Techniques** chapter:

$$F = 1 + \frac{T_E}{T_0} \quad (1)$$

where
F is the noise factor,
T_E is the equivalent noise temperature and
T_0 is the standard noise temperature, usually 290 K.

$$N_i = k \times B \times T_E \quad (2)$$

where
N_i is the equivalent noise power in watts at the input of a perfect receiver that would result in the same noise output,
k is Boltzman's constant, 1.38×10^{-23} Joules/kelvin, and
B is the system noise bandwidth in Hertz.

Expressing N_i in dBm is more convenient and is calculated as:

$$dBm_i = -198.6 + (10 \times \log_{10} B) + (10 \times \log_{10} T_E) \quad (3)$$

If N_i is greater than the noise generated

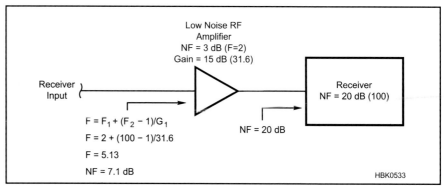

Fig 12.28 — The effect of adding a low noise preamplifier in front of a noisy receiver system.

internally by the receiver, the receiver's sensitivity is limited by the external noise. This is usually the case for HF receivers where atmospheric and man-made noise are much stronger than the receiver's internal noise floor. If N_i is within, perhaps, 10 dB greater than the receiver's internal noise, then the effect of the receiver's internal circuits on overall system sensitivity must be taken into account.

Noise Figure of Cascaded Stages

Often we are faced with the requirement of determining the noise figure of a system of multiple stages. In general adding an amplification stage between the antenna and the rest of the system will reduce the equivalent noise figure of the system by the amount of gain of the stage but adds in the noise of the added stage directly. The formula for determining the resultant noise factor is:

$$F = F_1 + \frac{F_2 - 1}{G_1} \qquad (4)$$

where
F_1 is the noise factor of the stage closest to the input,
F_2 is the noise factor of the balance of the system, and
G_1 is the gain of the stage closest to the input.

Note that these are noise *factors* not noise *figures*. The change to noise figure is easily computed at the end, if desired.

This is commonly encountered through the addition of a *low noise preamplifier* ahead of a noisy receiver as shown in **Fig 12.28**. As shown in the figure for fairly typical values, while the addition of a low noise preamplifier does reduce the noise figure, as can be observed, the amplifier gain and noise figure of the rest of the receiver can make a big difference.

In many cases elements of a receiving system exhibit loss rather than gain. Since they reduce the desired signal, while not changing the noise of following stages, they increase the noise figure at their input by an amount equal to the loss. This is the reason that VHF low noise preamps are often antenna mounted. If the coax between the antenna and radio has a loss of 1 dB, and the preamp has a noise figure of 1 dB, the resulting noise figure with the preamp at the radio will be 2 dB, while if at the antenna, the noise added by the coax will be reduced by the gain of the preamp, resulting in a significant improvement in received SNR at VHF and above. **Fig 12.29** shows a typical example.

To determine the system noise factor of multiple cascaded stages, Equation 4 can be extended as follows:

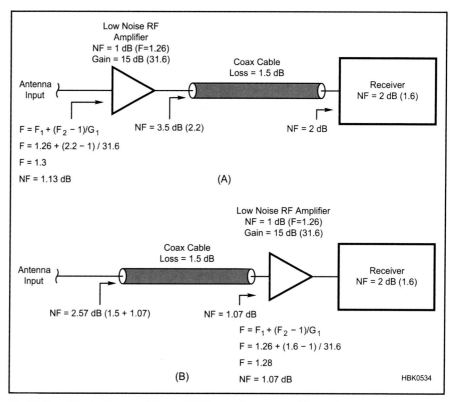

Fig 12.29 — Illustration of adding a VHF low noise preamplifier at the antenna (A) compared to one at the receiver system input (B).

$$F = F_1 + \frac{F_2 - 1}{G_1} + \frac{F_3 - 1}{G_1 G_2} + \ldots \qquad (5)$$
$$+ \frac{F_N - 1}{G_1 G_2 \ldots G_{N-1}}$$

where F_N and G_N are the noise factor and gain of the Nth stage.

This can proceed from the antenna input to the last linear stage, stopping at the point of detection or analog-to-digital conversion, whichever occurs first.

Why Do We Care?

A receiver designer needs to know how strong the signals are to establish the range of signals the receiver will be required to handle. One may compare the equivalent noise power determined in Equation 3 with the expected external noise to determine whether the overall receiver SNR will be determined by external or internal noise. A reasonable design objective is the have the internal noise be less than perhaps 10 to 20 dB below the expected noise. As noted above, this is related closely to the frequency of signals we want to receive. Any additional sensitivity will not provide a noticeable benefit to SNR, and may result in reduced dynamic range, as will be discussed in the next section.

For frequencies at which external noise sources are strongest, the noise power (and signal power) will also be a function of the antenna design. In such cases, the signal-to-noise ratio can be improved by using an antenna that picks up more signal and less noise, such as with a directional antenna that can reject noise from directions other than that of the signal.

12.4.2 Receiver Image Rejection

The major design choice in determining image response is selection of the first and any following IF frequencies. This is important in receiving weak signals on bands where strong signals are present. If the desired signal is near the noise, a signal at an image frequency could easily be 100 dB stronger, and thus to avoid interference, an image rejection of 110 dB would be needed. While some receivers meet that target, the receiver sections of most current amateur transceivers are in the 70 to 100 dB range.

Receiver Architectures for Image Rejection

As noted earlier, crystal filter technology has allowed down-conversion HF receivers to use a higher single IF than had early receivers, typically in the 4-10 MHz range. With a 10 MHz IF and an LO above the signal frequency, a 30 MHz signal would have an image at 50 MHz. This makes filtering relatively straightforward, although many receiver IF frequencies tend to be at the lower end of the above range. Still, as will be discussed in the next section, they have other advantages.

Many current HF receivers (or receiver sections of transceivers) have elected to employ an upconverting architecture. They typically have an IF in the VHF range, perhaps 60 to 70 MHz. This makes HF image rejection easy. A 30 MHz signal with a 60 MHz IF will have an image at 150 MHz. Not only is it five times the signal frequency, but signals in this range (other than perhaps the occasional taxicab) tend to be weaker than some undesired HF signals. Receivers with this architecture have image responses at the upper end of the above range, often with the image rejected by a relatively simple low-pass filter with a cut off at the top of the receiver range.

Another advantage of this architecture is that the local oscillator can cover a feasible continuous range, making it convenient for a general coverage receiver. For example, with a 60 MHz IF, a receiver designed for LF through HF would need an LO covering 60.03 to 90 MHz, a 1.5 to 1 range, easily provided by a number of synthesizer technologies, as described in the **Oscillators and Synthesizers** chapter.

The typical upconverting receiver uses multiple conversions to move signals to frequencies at which operating bandwidth can be established. While crystal filters in the VHF range used by receivers with upconverting IFs have become available with bandwidths as narrow as around 3 kHz, they do not yet achieve the shape factor of similar bandwidth filters at MF and HF. Thus, these are commonly used as *roofing filters*, discussed in the next section, prior to a conversion to one or more lower IF frequencies at which the operating bandwidth is established. **Fig 12.30** is a block diagram of a typical upconverting receiver using DSP for setting the operating bandwidth.

12.4.3 Receiver Dynamic Range

Dynamic range can be defined as the ratio of the strongest to the weakest signal that a system, in this case our receiver, can respond to linearly. Table 12.3 gives us an idea of how small a signal we might want to receive. The designer must create a receiver that will handle signals from below the noise floor to as strong as the closest nearby transmitter can generate. Most receivers have a specified (or sometimes not) highest input power that can be tolerated, representing the other end of the spectrum. Usually the maximum power specified is the power at which the receiver will not be damaged, while a somewhat lower power level is generally the highest that the receiver can operate at without overload and the accompanying degradation of quality of reception of the desired signal.

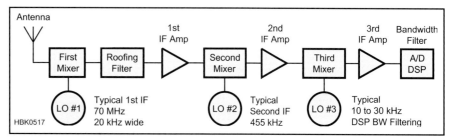

Fig 12.30 — Block diagram of a typical upconverting receiver using DSP for operating bandwidth (BW) determination.

ON-CHANNEL DYNAMIC RANGE

The signal you wish to listen to can range from the strongest to the weakest, sometimes changing rapidly with conditions, or in a situation with multiple stations such as a net. While a slow change in signal level can be handled with manual gain controls, rapid changes require automatic systems to avoid overload and operator, or neighbor, discomfort.

This is a problem that has been long solved with automatic gain control (AGC) systems. These systems are described in a further section of the chapter, but it is worth pointing out that the measurement and gain control points need to be applied carefully to the most appropriate portions of the receiver to maintain optimum performance. If all control is applied to early stages, the SNR for strong stations may suffer, while if applied in later stages, overload of early stages may occur in the presence of strong stations. Thus, gain control has to be designed into the receiver distributed from the input to the detector.

The next two sections illustrate a frequent limitation of receiver performance — dynamic range between the reception of a weak signal in the presence of one or more strong signals outside of the channel.

BLOCKING GAIN COMPRESSION

A very strong signal outside the channel bandwidth can cause a number of problems that limit receiver performance. *Blocking gain compression* (or "blocking") occurs when strong signals overload the receiver's high gain amplifiers and reduce its ability to amplify weak signals.

While listening to a weak signal, all stages are at maximum gain. If the weak signal were at a level of S-0, a strong signal could be at S-9 + 60 dB. Using the Collins standard of S-9 equal to 50 µV, and each S-unit a change of 6 dB, the receiver's front-end stages would be receiving a 0.1 µV signal and a 50,000 µV into the front end at the same time. After a few stages of full gain amplification, the stronger signal causes amplifier clipping, which reduces the gain and thus the amplitude of the weaker signal. Of course a perfectly linear receiver would amplify each equally until the undesired signal is eliminated at the operating bandwidth setting stage.

The ratio in dB between the strongest signal that a receiver can amplify linearly, with no more than 1 dB of gain reduction, and the receiver's noise floor in a specified bandwidth is called the receiver's *blocking dynamic range* or *BDR* or the *compression-free dynamic range* or *CFDR*. In an analog superhet, BDR is established by the linear regions of the IF amplifiers and mixers. If the receiver employs DSP, the range of the analog-to-digital converter usually establishes the receiver's BDR.

INTERMODULATION DYNAMIC RANGE

Blocking dynamic range is the straightforward response of a receiver to a single strong interfering signal outside the operating passband. In amateur operation, we often have more than one interferer. While such signals contribute to the blocking gain compression in the same manner as a single signal described above, multiple signals also result in a potentially more serious problem resulting from *intermodulation products*.

If we look again at Equation B in the earlier sidebar on Nonlinear Signal Combinations, we note that there are an infinite number of higher order terms. In general, the coefficients of these terms are progressively lower in amplitude, but they are still greater than zero. Of primary interest is third-order term, $K_3 \times V_{IN}^3$, when considering V_{IN} as the sum of two interfering signals (f_1 and f_2) near our desired signal (f_0) and within the first IF passband, but outside the operating bandwidth.

$$\begin{aligned}V_{OUT} &= K_3 \times [A\sin(f_1)t + B\sin(f_2)t]^3 \quad (6)\\ &= K_3 \times \{ A^3\sin^3(f_1)t + 3A^2B\,[\sin^2(f_1)t \times \sin(f_2)t] + 3AB^2\,[\sin(f_1)t \times \sin^2(f_2)t]\\ &\quad + B^3\sin^3(f_2)t \}\end{aligned}$$

The cubic terms in Equation 6 (the first and last terms) result in products at three times the frequency and can be ignored in this discussion. Using trigonometric identities to reduce the remaining $\sin^2$ terms and the subsequent cos( )sin( ) products (see Note 1) reveal individual intermodulation (IM) products, recognizing that the signals have cross-modulated each other due to the nonlinear action of the circuit. IM products have frequencies that are *linear combinations* of the input signal frequencies, written as $n(f_1) \pm m(f_2)$, where n and m are integer values. The entire group of products that result from intermodulation are broadly referred to as *intermodulation distortion* or *IMD*. The ratio in dB between the amplitude of the interfering signals, f_1 and f_2, and the resulting IM products is called the *intermodulation ratio*.

If *all* of the higher-order terms in the original equation are considered, n and m can take on any integer value. If the sum of n and m is odd, (2 and 1, or 3 and 2, or 3 and 4, etc.) the result is products that have frequencies near our desired signal, for example, $2(f_1) - 1(f_2)$. Those are called *odd-order* products. Odd-order products have frequencies close enough to those of the original signals that they can cause interference to the desired signal. If the sum of n and m is three, those are *third-order IM products* or *third-order IMD*. For fifth-order IMD, the sum of n and m is five, and so forth. The higher the order of the IM products, the smaller their amplitude, so our main concern is with third-order IMD.

If the two interfering signal have frequencies of $f_0 + \Delta$, and $f_0 + 2\Delta$, where Δ is some offset frequency, we have for the third-order term:

$$V_{OUT} = K_3 \times \left[A\sin(f_0 + \Delta)t + B\sin(f_0 + 2\Delta)t \right]^3 \quad (7)$$

A good example would be interfering signals with offsets of 2 kHz and 2 × 2 kHz or 4 kHz from the desired frequency, a common situation on the amateur bands.

Discarding the cubic terms and applying the necessary trigonometric identities shows that a product can be produced from this combination of interfering frequencies that has a frequency of exactly f_0 — the same frequency as the desired signal! (The higher-order terms of Equation B can also produce products at f_0, but their amplitude is usually well below those of the third-order products.)

Thus we have two interfering signals that are not within our operating bandwidth so we don't hear either by themselves. Yet they combine in a nonlinear circuit and produce a signal exactly on top of our desired signal. If the interfering signals are within the passband of our first IF and are strong enough the IM product will be heard.

As the strength of the interfering signals increases, so does that of the resulting intermodulation products. For every dB of increase in the interfering signals, the third-order IM products increase by approximately 3 dB. Fifth-order IM increases by 5 dB for every dB increase in the interfering signals, and so forth. Our primary concern, however, is with the third-order products because they are the strongest and cause the most interference.

INTERCEPT POINT

Intercept point describes the IMD performance of an individual stage or a complete receiver. For example, third-order IM products increase at the rate of 3 dB for every 1-dB increase in the level of each of the interfering input signals. (ideally, but not always exactly true) As the input levels increase, the distortion products seen at the output on a spectrum analyzer could catch up to, and equal, the level of the two desired signals if the receiver did not begin to exhibit blocking as discussed earlier.

The input level at which this occurs is the *input intercept point*. **Fig 12.31** shows the concept graphically, and also derives from the geometry an equation that relates signal level, distortion and intercept point. The intercept point of the most interest in receiver evaluation is that for third-order IM products and is called the *third-order intercept point* or IP_3. A similar process is used to get a

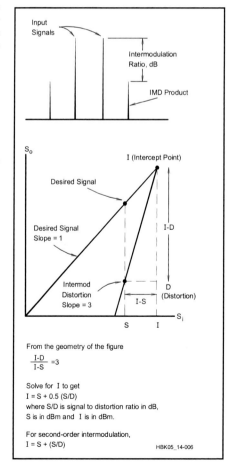

Fig 12.31 — Graphical representation of the third-order intercept concept.

second-order intercept point for second-order IMD. A higher IP_3 means that third-order IM products will be weaker for specific input signal strengths and the operator will experience less interference from IM products from strong adjacent signals.

These formulas are very useful in designing radio systems and circuits. If the input intercept point (dBm) and the gain of the stage (dB) are added the result is an output intercept point (dBm). Receivers are specified by input intercept point, referring distortion back to the receive antenna input. Intercept point is a major performance limiting parameter of receivers used in high density contest or DX operations. Keep in mind that we have been discussing this as an effect of two signals, one that is Δ away from our operating frequency and another at twice Δ. In real life, we may be trying to copy a weak signal at f_0, and have other signals at $f_0 \pm 500$, 750, 1000, 1250...5000 Hz. There will be many combinations that produce products at or near our weak signal's frequency.

Note that the products don't need to end up exactly on top of the desired signal to cause a problem; they just need to be within the operating bandwidth. So far we have been talking about steady carriers, such as would be encountered during CW operation with interference from nearby CW stations. SSB or other wider bandwidth modes with spectrum distributed across a few kHz will have signal components that go in and out of a relationship that results in on-channel interference from IMD. This manifests itself as a time-varying synthetic noise floor, composed of all the resulting products across the channel. The difference in this low level "noise" can be dramatic between different receivers.

SPURIOUS-FREE DYNAMIC RANGE

IM products increase with the amplitude of the interfering signals that cause them and at some point become detectable above the receiver's noise floor. The ratio of the strength of the interfering signals to the noise floor, in dB, is the receiver's *spurious-free dynamic range* or *SFDR*. This is the range of signal strengths over which the receiver does not produce any detectable spurious products. SFDR can be specified for a specific order of IM products; for example, SFDR3 is the SFDR measured for third-order IM products only. The bandwidth for which the receiver's noise floor is measured must also be specified, since smaller bandwidths will result in a lower noise floor.

RECEIVER DESIGN FOR DYNAMIC RANGE

As noted previously, there are a number of possible architectural choices for an amateur receiver. In the past, the receivers with the best close-in third order intermodulation distor-

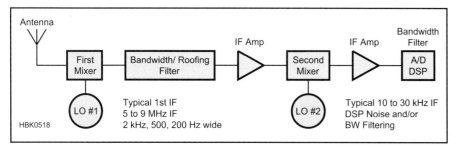

Fig 12.32 — Block diagram of a downconverting amateur band receiver with HF IF filtering to eliminate near channel interference.

tion and maximum blocking dynamic range were amateur-band-only receivers, such as the primary receiver in the Ten-Tec Orion series, the receiver in the Elecraft K2, the earlier Ten-Tec Omni VI and both receivers in the Elecraft K3. A look at a typical block diagram, as shown in **Fig 12.32**, makes it easy to see why. The problems resulting from strong unwanted signals near a desired one are minimized if the unwanted signals are kept out of the places in the receiver where they can be amplified even more and cause the nonlinear effects that we try to avoid.

Note that in Fig 12.32, the only place where the desired and undesired signals all coexist is before the first mixer. If the first mixer and any RF preamp stages have sufficient strong-signal handling capability, the undesired signals will be eliminated in the filter immediately behind the first mixer. This HF crystal filter is generally switchable to support desired bandwidths as narrow as 200 Hz. The later amplifier, mixer and DSP circuits only have to deal with the signal we want.[9]

Now look at a typical modern general coverage receiver as shown previously in Fig 12.30. In this arrangement, a single digital synthesizer, perhaps covering from 70 to 100 MHz, shifts the incoming signal(s) to a VHF IF, often near 70 MHz. A roofing filter at 70 MHz follows the first mixer. This arrangement offers simplified local oscillator (LO) design and the possibility of excellent image rejection. Unfortunately, crystal filter technology has only recently been able to produce narrow filters for 70 MHz, and so far they have much wider skirts than the crystal filters used in Fig 12.32.

Many receivers and transceivers set this filter bandwidth wider than any operating bandwidth and use DSP filtering much later in the signal chain to set the final operating bandwidth. For a receiver that will receive FM and AM, as well as SSB and CW, that usually means a roofing filter with a bandwidth of around 20 kHz.[10] With this arrangement, all signals in that 20 kHz bandwidth pass all the way through IF amplifiers and mixers and into the A/D converter before we attempt to eliminate them using DSP filters. By that time they have had an opportunity to generate intermodulation products and cause the blocking and IMD problems that we are trying to eliminate.

A hybrid architecture has recently appeared in the Ten-Tec Omni VII transceiver that effectively combines the two technologies. The first IF has a 20 kHz wide roofing filter at 70 MHz, followed by selectable steep skirted 455 kHz Collins mechanical filters at the second IF and then DSP filters at the third IF. The topology is shown in **Fig 12.33**. Careful attention to gain distribution among the stages between the filters maintains desired sensitivity, but not so high that the undesired products have a chance to become a serious problem. With bandwidths of 20, 6 and 2.5 kHz supplied, and 500 and 300 Hz as accessories, the undesired close-in signals are eliminated before they have an opportunity to cause serious trouble in the DSP stages that follow.

WHAT KIND OF PERFORMANCE CAN WE EXPECT?

As we would expect, the blocking and IMD dynamic range (IMD DR) performance of a receiver will depend on a combination of the early stage filtering, the linearity of the mixers and amplifiers, and the dynamic range of any ADC used for DSP. Product reviews of receivers and the receiver sections of transceivers published in *QST* now provide the measured dynamic range in the presence of interfering signals with spacings of 2, 5 and 20 kHz. (Details of the test procedures used are given in the **Test Equipment and Measurements** chapter.) At 20 kHz spacing, the interfering signal is usually outside of the roofing filter bandwidth of any of the above architectures.

A look at recent receiver measurements indicates that receivers have IMD dynamic range with 2 kHz spacing results in the following ranges:

Upconverting with VHF IF (Fig 12.30): 60 to 80 dB

Downconverting with HF IF (Fig 12.32): 75 to 102 dB

Hybrid distributed architecture (Fig 12.33), with the Omni VII the only sample, as this is written: 82 dB

Let's take an example of what this would mean. If we are listening to a signal at S-3, for signals to generate a third-order IMD product at the same level in a receiver with a dynamic range of 60 dB, the $f_0 + \Delta$ and $f_0 + 2\Delta$ signals would have to have a combined power equal to S-9 +27 dB, or each at S-9 +24 dB. This is not unusual on today's amateur bands. On the other hand, if we had an IMD dynamic range of 102 dB, the interfering signals would have to be at S-9 +66 dB, much less likely. How much dynamic range you need depends in large measure on the kind of operating you do, how much gain your receiving antennas have and how close is the proximity of the nearest station that operates on the same bands as you. Keep in mind also that it is often difficult to tell whether or not you are suffering from IMD — it just sounds like there are many more signals than are really there.

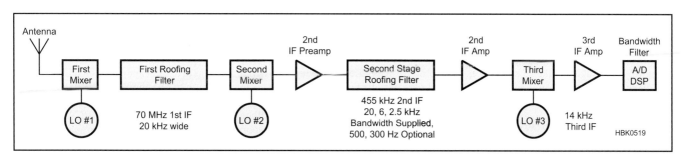

Fig 12.33 — Block diagram of a upconverting multiple conversion receiver with distributed roofing filter architecture.

12.5 Control and Processing Outside the Primary Signal Path

Discussion to date has been about processes that occur within the primary path for signals between antenna input and transducer or system output. There are a number of control and processing subsystems that occur outside of that path that contribute to the features and performance of receiving systems.

12.5.1 Automatic Gain Control (AGC)

The amplitude of the desired signal at each point in the receiver is generally controlled by AGC, although manual control is usually provided as well. Each stage has a distortion versus signal level characteristic that must be known, and the stage input level must not become excessive. The signal being received has a certain signal-to-distortion ratio that must not be degraded too much by the receiver. For example, if an SSB signal has −30 dB distortion products the receiver should have −40 dB quality. A correct AGC design ensures that each stage gets the right input level. It is often necessary to redesign some stages in order to accomplish this.

THE AGC LOOP

Fig 12.34A shows a typical AGC loop that is often used in amateur receivers. The AGC is applied to the stages through RF decoupling circuits that prevent the stages from interacting with each other. The AGC amplifier helps to provide enough AGC loop gain so that the gain-control characteristic of Fig 12.34B is achieved. The AGC action does not begin until a certain level, called the AGC *threshold*, is reached. The THRESHOLD VOLTS input in Fig 12.34A serves this purpose. After that level is exceeded, the audio level slowly increases. The audio rise beyond the threshold value is usually in the 5 to 10 dB range. Too much or too little audio rise are both undesirable for most operators.

As an option, the AGC to the RF amplifier is held off, or "delayed," by the 0.6 V forward drop of the diode so that the RF gain does not start to decrease until larger signals appear. This prevents a premature increase of the receiver noise figure. Also, a time constant of one or two seconds after this diode helps keep the RF gain steady for the short term.

Fig 12.35 is a typical plot of the signal levels at the various stages of a certain ham band receiver. Each stage has the proper level and a 115 dB change in input level produces a 10 dB change in audio level. A manual gain control could produce the same effect.

AGC TIME CONSTANTS

In Fig 12.34, following the precision rectifier, R1 and C1 set an *attack* time, to prevent excessively fast application of AGC. One or two milliseconds is a good value for the R1 × C1 product. If the antenna signal suddenly disappears, the AGC loop is opened because the precision rectifier stops conducting. C1 then discharges through R2 and the C1 × R2 product can be in the range of 100 to 200 ms. At some point the rectifier again becomes active, and the loop is closed again.

An optional modification of this behavior is the *hang AGC* circuit. If we make R2 × C1 much longer, say 3 seconds or more, the AGC voltage remains almost constant until the R5, C2 circuit decays with a switch selectable time constant of 100 to 1000 ms. At that time R3 quickly discharges C1 and full receiver gain is quickly restored. This type of control is appreciated by many operators because of the lack of AGC *pumping* due to modulation, rapid fading and other sudden signal level changes.

AGC PUMPING

AGC pumping can occur in receivers in which the AGC measurement point is located ahead of the stages that determine operating bandwidth, such as adding external audio DSP filtering to a receiver. If the weak signal is the only signal within the first IF passband, the AGC will cause the receiver to be at maximum gain and optimum SNR. If an interfering signal is within the first IF passband, but outside the audio DSP filter's passband, we won't hear the interfering signal, but it will enter the AGC system and reduce the gain so we might not hear our desired weak signal.

AGC LOOP RESPONSE PROBLEMS

If the various stages have the property that each 1 V change in AGC voltage changes the gain by a constant amount (in dB), the AGC loop is said to be *log-linear* and regular feedback principles can be used to analyze and design the loop. But there are some difficulties that complicate this textbook model. One has already been mentioned, that when the signal is rapidly decreasing the loop becomes open and the various capacitors discharge in an open loop manner. As the signal is increasing beyond the threshold, or if it is decreasing slowly enough, the feedback theory applies more accurately. In SSB and CW receivers rapid changes are the rule and not the exception. It is important that the AGC loop not overshoot or ring when the signal level rises past the threshold. [The idea is to design the ALC loop to be stable when the loop is closed. It obviously won't oscillate when open (during decay time). But the loop must have good transient response when the loop goes from open to closed state.]

Another problem involves the narrow band-pass IF filter. The group delay of this filter constitutes a time lag in the loop that can make loop stabilization difficult. Moreover, these filters nearly always have much greater group delay at the edges of the passband, so that loop problems are aggravated at these frequencies. Overshoots and undershoots, called *gulping*, are very common. Compensation networks that advance the phase of the feedback help to offset these group delays. The design problem arises because some of the AGC is applied before the filter and some after the filter. It is a good idea to put as much fast AGC as possible after the filter and use a slower decaying AGC ahead of the filter. The delay diode and RC in Fig 12.34A are helpful in that respect. Complex AGC designs using two or more compensated loops are also in the literature. If a second cascaded narrow filter is used in the IF it is usually a lot easier to leave the second or *downstream* filter out of the AGC loop at the risk of allowing AGC pumping as described in the preceding section.

Another problem is that the control characteristic is often not log-linear. For example, dual-gate MOSFETs tend to have much larger dB/V at large values of gain reduction. Many IC amplifiers have the same problem. The result is that large signals cause instability because of excessive loop gain. There are variable gain op amps and other ICs available that are intended for gain control loops.

Audio frequency components on the AGC bus can cause problems because the amplifier gains are modulated by the audio and distort the desired signal. A hang AGC circuit can reduce or eliminate this problem.

Finally, if we try to reduce the audio rise to a very low value, the required loop gain becomes very large, and stability problems become very difficult. It is much better to accept a 5 to 10 dB variation of audio output.

Because many parameters are involved and many of them are not strictly log-linear, it is best to achieve good AGC performance through an initial design effort and finalize the design experimentally. Use a signal generator, attenuator and a signal pulser (2 ms rise and fall times, adjustable pulse rate and duration) at the antenna and a synchronized oscilloscope to look at the IF envelope. Tweak the time constants and AGC distribution by means of resistor and capacitor decade boxes. Be sure to test throughout the passband of each filter. The final result should be a smooth and pleasant sounding SSB/CW response, even with maximum RF gain and strong signals. Patience and experience are helpful.

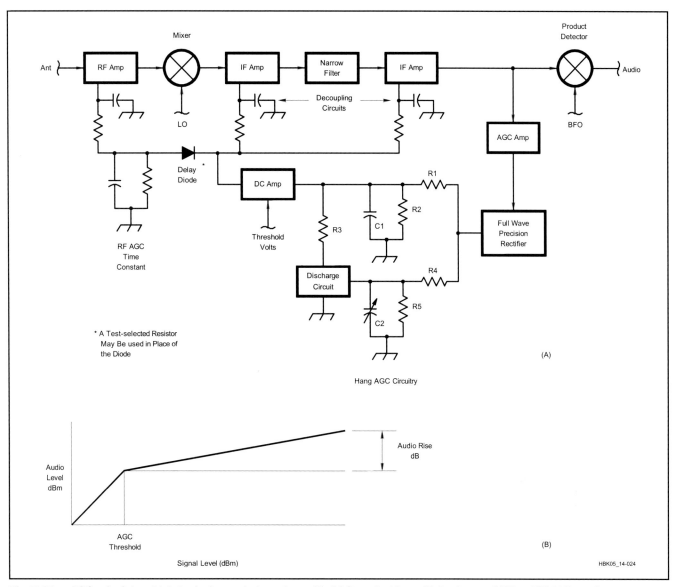

Fig 12.34 — AGC principles. At A: typical superhet receiver with AGC applied to multiple stages of RF and IF. At B: audio output as a function of antenna signal level.

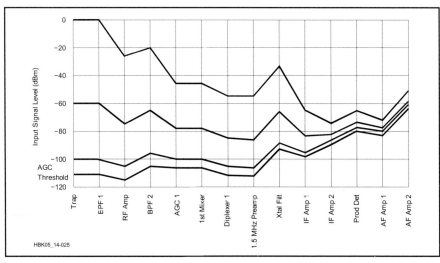

Fig 12.35 — Gain control of a ham-band receiver using AGC. A manual gain control could produce the same result.

12.5.2 Audio-Derived AGC

Some receivers, especially direct-conversion types, use audio-derived AGC. There is a problem with this approach also. At low audio frequencies the AGC can be slow to develop. That is, low-frequency audio sine waves take a long time to reach their peaks. During this time the RF/IF/AF stages can be over-driven. If the RF and IF gains are kept at a low level this problem can be reduced. Also, attenuating low audio frequencies prior to the first audio amplifier should help. With audio AGC, it is important to avoid so-called "charge pump" rectifiers or other slow-responding circuits that require multiple cycles to pump up the AGC voltage. Instead, use a peak-detecting circuit that responds accurately on the first positive or negative transition.

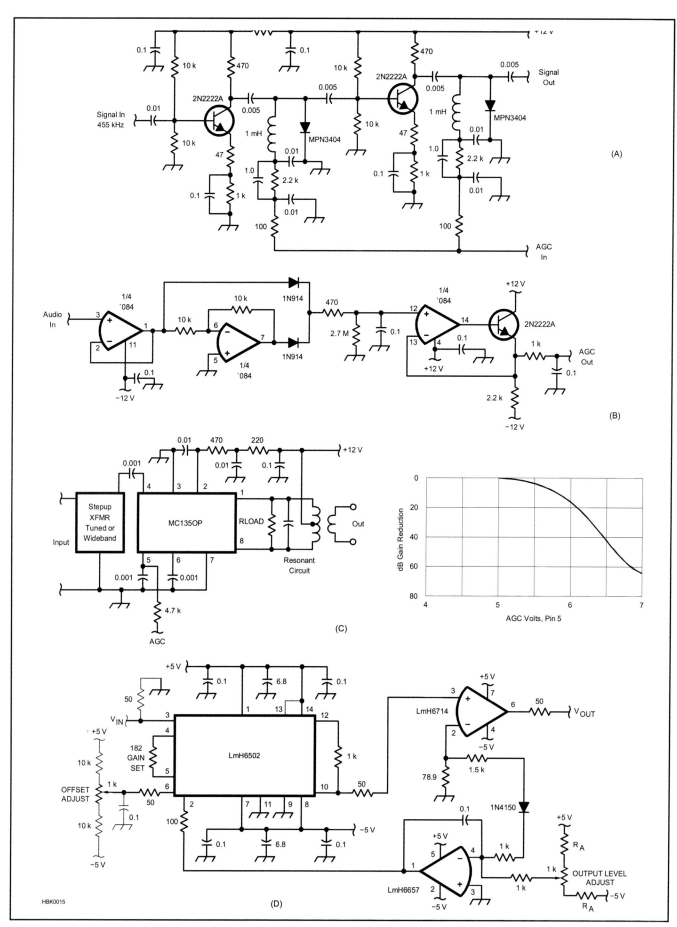

Fig 12.36 — Some gain controllable amplifiers and a rectifier suitable for audio derived AGC.

12.5.3 AGC Circuits

Fig 12.36 shows some gain controllable circuits. Fig 12.36A shows a two-stage 455-kHz IF amplifier with PIN diode gain control. This circuit is a simplified adaptation from a production receiver, the Collins 651S. The IF amplifier section shown is preceded and followed by selectivity circuits and additional gain stages with AGC. The 1.0 µF capacitors aid in loop compensation. The favorable thing about this approach is that the transistors remain biased at their optimum operating point. Right at the point at which the diodes start to conduct, a small increase in IMD may be noticed, but that goes away as diode current increases slightly. Two or more diodes can be used in series, if this is a problem (it very seldom is). Another solution is to use a PIN diode that is more suitable for a such a low-frequency IF. Look for a device with $\tau > 10 / (2\pi f)$ where τ is the minority carrier lifetime in µs and f is the frequency in MHz.

Fig 12.36B is an audio derived AGC circuit using a full-wave rectifier that responds to positive or negative excursions of the audio signal. The RC circuit follows the audio closely.

Fig 12.36C shows a typical circuit for the MC1350P RF/IF amplifier. The graph of gain control versus AGC voltage shows the change in dB/V. If the control is limited to the first 20 dB of gain reduction this chip should be favorable for good AGC transient response and good IMD performance. Use multiple low-gain stages rather than a single high-gain stage for these reasons. The gain control within the MC1350P is accomplished by diverting signal current from the first amplifier stage into a *current sink*. This is also known as the *Gilbert cell multiplier* architecture.

Another chip of this type is the NE/SA5209. This type of approach is simpler to implement than discrete circuit approaches, such as dual-gate MOSFETs that are now being replaced by IC designs.

Fig 12.36D shows the high end performance National Semiconductor LMH6502MA (14-pin DIP plastic package) voltage controlled amplifier. It is specially designed for accurate log-linear AGC from 0 to 40 dB with respect to a preset maximum voltage gain from 6 to 40 dB. Its ±3 dB bandwidth is 130 MHz. It is an excellent IF amplifier for high performance receiver or transmitter projects.

Additional info on voltage-controlled amplifier ICs can be found on the Analog Devices web site (**www.analog.com**). Search the site for Tutorial MT-073, which describes the operation of various types of gain-controlled amplifiers with numerous product examples.

12.6 Pulse Noise Reduction

A major problem for those listening to receivers has historically been local impulse noise. For HF and VHF receivers it is often from the sparks of internal combustion engine spark plugs, electric fence chargers, light dimmers and many other similar devices that put out short duration wide band signals. In the UHF and microwave region, radar systems can cause similar problems. There have been three general methods of attempting to deal with such noise over the years, some more successful than others. We will briefly describe the approaches.

12.6.1 The Noise Limiter

The first device used in an attempt to limit impulse noise was called a *noise limiter* or *clipper* circuit.[11] This circuit would *clip* or limit noise (or signal) peaks that exceeded a preset limit. The idea was to have the limit set to about as loud as you wanted to hear anything and nothing louder would get through. This was helpful in eliminating the loudest part of impulse noise or even nearby lightning crashes, but it had two problems. First it didn't eliminate the noise, it just reduced the loudness; second, it also reduced the loudness of loud non-noise signals and in the process distorted them considerably.

The second problem was fixed fairly early, with the advent of the *automatic* noise limiter or ANL.[12] The ANL automatically set the clipping threshold to that of a loud signal. It thus would adjust itself as the loudness of signals you listened to changed with time. An ANL was fairly easy to implement and became standard equipment on amateur receivers from the late 1930s on.

12.6.2 The Noise Blanker

It turned out that improvements in receiver selectivity over the 1950s and beyond, while improving the ability to reduce random noise, actually made receiver response to impulse noise worse. The reason for this is that a very short duration pulse will actually be lengthened while going through a narrow filter.[13] The noise limiters described so far were all connected at the output of the IF amplifiers and thus the noise had passed most of the selectivity before the limiter and had been widened by the receiver filters. In addition, modern receivers include *automatic gain control* (AGC), a system that reduces the receiver gain in the presence of strong signals to avoid overload of both receiver circuits and ears. In SSB receivers, since signals vary in strength as someone talks, the usual AGC responds quickly to reduce the gain of a strong signal and then slowly increases it if the signal is no longer there. This means that a strong noise pulse may reduce the receiver gain for much longer than it lasts.

The solution — a *noise blanker*. A noise blanker is almost a separate wideband receiver. It gets its input from an early stage in the receiver before much selectivity or AGC happens. It amplifies the wideband signal and detects the narrow noise pulses. The still-narrow noise pulses are used to shut off the receiver at a point ahead of the selectivity and AGC, thus keeping the noise from getting to the parts of the receiver at which the pulses would be extended.

A well-designed noise blanker can be very effective. Instead of just keeping the noise at the level of the signal as a noise limiter does, the noise blanker can actually *eliminate* the noise. If the pulses are narrow enough, the loss of desired signal during the time the receiver is disabled is not noticeable and the noise may seem to disappear entirely.

In addition to an ON/OFF switch, many noise blanker designs include a control labeled THRESHOLD. The THRESHOLD control adjusts the level of noise that will blank the receiver. If it is set for too low a level, it will blank on signal peaks as well as noise, resulting in distortion of the signal. The usual approach is to turn on the blanker, then adjust the THRESHOLD control until the noise is just blanked. Don't forget to turn it off when the noise goes away.

Noise blankers can also create problems. The wide-band receiver circuit that detects the noise pulses detects any signals in that bandwidth. If such a signal is strong and has sharp peaks (as voice and CW signals do), the noise blanker will treat them as noise pulses and shuts down the receiver accordingly. This causes tremendous distortion and can make it sound as if the strong signal to which the noise blanker is responding is generating spurious signals that cause the distortion. Before you assume that the strong signal is causing problems, turn the noise blanker on and off to check. When the band is full of strong signals, a noise blanker may cause more problems than it solves.

12.6.3 Operating Noise Limiters and Blankers

Many current receivers include both a noise limiter and a noise blanker. In some cases the noise blanker is an extra cost option. If your receiver has both, they will have separate controls and it is worthwhile to try them both. There are times at which one will work better than the other, and other times when it goes the other way, depending on the characteristics of the noise. There are other times when both work better than either. In any case, they can make listening a lot more pleasant — just remember to turn them off when you don't need them since either type can cause some distortion, especially on strong signals that should otherwise be easy to listen to.

OTHER TECHNIQUES

The previous techniques have been around for quite a few years and represented the most commonly available techniques to reduce impulse noise. There are a few other solutions as well. Note that we haven't been talking about reducing interference here. By interference, we mean another intended signal encroaching on the channel we want to listen to. There are a number of techniques to reduce interference, and some also can help with impulse noise.

Many times impulse noise is coming from a particular direction. If so, by using a directional antenna, we can adjust the direction for minimum noise. When we think about directional antennas, the giant HF Yagi springs to mind. For receiving purposes, especially on the lower bands such as 160, 80 and 40 meters (where the impulse noise often seems the worst), a small indoor or outdoor receiving loop antenna can be very effective at eliminating either interfering stations or noise (both if they happen to be in the same direction).[14]

Another technique that can be used to eliminate either interference or noise is to obtain a copy of the noise (or interference) that is 180° out of phase from the one you are receiving. By adjusting the amplitude to match the incoming signal, the signal can be cancelled at the input to the receiver. Several available commercial units perform this task.

Digital signal processing, described in the next section, is another multifunction system that can help with all kinds of noise.

12.6.4 DSP Noise Reduction

The previous techniques have been around for quite a few years and represented about all we could do about impulse noise prior to the current "digital age." Now processing capability seems to be in everything from toasters to artillery shells — why wouldn't we expect to find processing in our radios, too? Modern radios use digital processing in many ways — setting frequency, display functions, memories and, import in this discussion, digital signal processing or DSP. (The **DSP and Software Radio Design** chapter covers these topics thoroughly.)

DSP noise reduction can actually look at the statistics of the signal and noise and figure out which is which and then reduce the noise significantly. It can't quite eliminate the noise, and it needs enough of the signal to figure out what's happening, so it won't work if the signal is far below the noise. Many DSP systems "color" the resulting audio to a degree. Nonetheless, they do improve the SNR of a signal in random or impulse noise. It's always worth experimenting with the radio's features to find out.

12.7 VHF and UHF Receivers

Most of the basic ideas presented in previous section apply equally well in receivers that are intended for the VHF and UHF bands. This section will focus on the difference between VHF/UHF and HF receivers.

12.7.1 FM Receivers

Narrow-band frequency modulation (NBFM) is the most common mode used on VHF and UHF. **Fig 12.37A** is a block diagram of an FM receiver for the VHF/UHF amateur bands.

FRONT END

A low-noise front end is desirable because of the decreasing atmospheric noise level at these frequencies and also because portable gear often uses short rod antennas at ground level. Nonetheless, the possibilities for gain compression and harmonic IMD, multi-tone IMD and cross modulation are also substantial. Therefore dynamic range is an important design consideration, especially if large, high-gain antennas are used. FM limiting should not occur until after the crystal filter. Because of the high occupancy of the VHF/UHF spectrum by powerful broadcast transmitters and nearby two-way radio services, front-end preselection is desirable, so that a low noise figure can be achieved economically within the amateur band.

DOWN-CONVERSION

Down-conversion to the final IF can occur in one or two stages. Favorite IFs are in the 5 to 10 MHz region, but at the higher frequencies rejection of the image 10 to 20 MHz away can be difficult, requiring considerable preselection. At the higher frequencies an intermediate IF in the 30 to 50 MHz region is a better choice. Fig 12.37A shows dual down-conversion.

IF FILTERS

The customary peak frequency deviation in amateur FM on frequencies above 29 MHz is about 5 kHz and the audio speech band extends to 3 kHz. This defines a maximum modulation index (defined as the deviation ratio) of $5/3 = 1.67$. An inspection of the Bessel functions that describe the resulting FM signal shows that this condition confines most of the 300 to 3000 Hz speech information sidebands within a 15 kHz or so bandwidth. Using filters of this bandwidth, channel separations of 20 or 25 kHz are achievable.

Many amateur FM transceivers are channelized in steps that can vary from 1 to 25 kHz. For low distortion of the audio output (after FM detection), this filter should have good phase linearity across the bandwidth. This would seem to preclude filters with very steep descent outside the passband, which tend to have very nonlinear phase near the band edges. But since the amount of energy in the higher speech frequencies is naturally less, the actual distortion due to this effect may be acceptable for speech purposes. The normal practice is to apply pre-emphasis to the higher speech frequencies at the transmitter and de-emphasis compensates at the receiver.

LIMITING

After the filter, hard limiting of the IF is needed to remove any amplitude modulation components. In a high-quality receiver, special attention is given to any nonlinear phase shift that might result from the limiter circuit design. This is especially important in data receivers in which phase response must be controlled. In amateur receivers for speech it may be less important. Also, the "ratio detector" (see the **Mixers, Modulators and Demodulators** chapter) largely eliminates the need for a limiter stage, although the limiter approach is probably still preferred.

FM DETECTION

The discussion of this subject is deferred to the **Mixers, Modulators and Demodulators** chapter. Quadrature detection is used in some popular FM multistage ICs. An example receiver chip will be presented later.

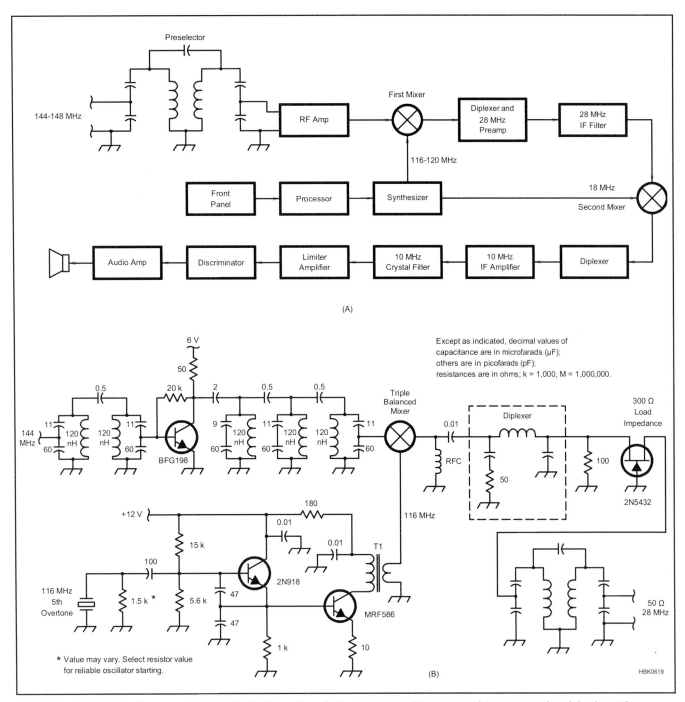

Fig 12.37 — At A, block diagram of a typical VHF FM receiver. At B, a 2 meter to 10 meter receive converter (partial schematic; some power supply connections omitted.)

12.7.2 FM Receiver Weak-Signal Performance

The noise bandwidth of the IF filter is not much greater than twice the audio bandwidth of the speech modulation, less than it would be in wideband FM. Therefore such things as capture effect, the threshold effect and the noise quieting effect so familiar to wideband FM are still operational, but somewhat less so, in FM. For FM receivers, sensitivity is specified in terms of a SINAD (see the **Test Equipment and Measurements** chapter) ratio of 12 dB. Typical values are –110 to –125 dBm, depending on the low-noise RF pre-amplification that often can be selected or deselected (in strong signal environments).

LO PHASE NOISE

In an FM receiver, LO phase noise superimposes phase modulation, and therefore frequency modulation, onto the desired signal. This reduces the ultimate signal-to-noise ratio within the passband. This effect is called "incidental FM (IFM)." The power density of IFM (W/Hz) is proportional to the phase noise power density (W/Hz) multiplied by the square of the modulating frequency (the familiar parabolic effect in FM). If the receiver uses high-frequency de-emphasis at the audio output (–6 dB per octave from 300 to 3000 Hz, a common practice), the IFM level at higher audio frequencies can be reduced. Ordinarily, as the signal increases the noise would be "quieted" (that is, "captured") in an FM receiver, but in this case the signal and the phase noise riding "piggy back" on the signal increase in the same proportion.[15] IFM is not a significant problem in modern FM radios, but phase noise can become a concern for adjacent-channel interference.

As the signal becomes large the signal-to-noise ratio therefore approaches some final value. A similar ultimate SNR effect occurs in SSB receivers. On the other hand, a perfect AM receiver tends to suppress LO phase noise.[16]

12.7.3 FM Receiver ICs

A wide variety of special ICs for communications-bandwidth FM receivers are available. Many of these were designed for "cordless" or cellular telephone applications and are widely used. **Fig 12.38** shows some popular versions for a 50 MHz FM receiver. One is an RF amplifier chip (NE/SA5204A) for 50 Ω input to 50 Ω output with 20 dB of gain. The second chip (NE/SA602A) is a front-end device with an RF amplifier, mixer and LO. The third is an IF amplifier, limiter and quadrature FM detector (NE/SA604A) that also has a very useful RSSI (logarithmic received signal strength indicator) output and also a "mute" function. The fourth is the LM386, a widely used audio-amplifier chip. Another FM receiver chip, complete in one package, is the MC3371P.

The NE/SA5204A plus the two tuned circuits help to improve image rejection. An alternative would be a single double-tuned filter with some loss of noise figure. The Mini-Circuits MAR/ERA series of MMIC amplifiers are excellent devices also. The crystal filters restrict the noise bandwidth as well as the signal bandwidth. A cascade of two low-cost filters is suggested by the vendors. Half-lattice filters at 10 MHz are shown, but a wide variety of alternatives, such as ladder networks, are possible.

Another recent IC is the MC13135, which features double conversion and two IF amplifier frequencies. This allows more gain on a single chip with less of the cross coupling that can degrade stability. This desirable feature of multiple down-conversion was mentioned previously in this chapter.

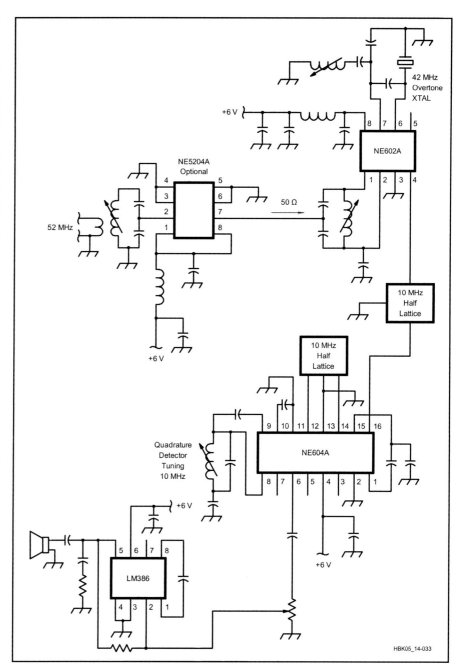

Fig 12.38 — The NE/SA5204A, NE/SA602A, NE/SA604A NBFM ICs and the LM386 audio amplifier in a typical amateur application for 50 MHz.

The diagram in Fig 12.38 is (intentionally) only a general outline that shows how chips can be combined to build complete equipment. The design details and specific parts values can be learned from a careful study of the data sheets and application notes provided by the IC vendors. Amateur designers should learn how to use these data sheets and other information such as application notes available (usually for free) from the manufacturers or on the Web.

12.7.4 VHF Receive Converters

Rather than building an entire transceiver for VHF SSB and CW, one approach is to use a receive converter. A receive converter (also called a *downconverter*) takes VHF signals and converts them to an HF band for reception using existing receiver or transceiver as a tunable IF.

Although many commercial transceivers cover the VHF bands (either multiband, multimode VHF/UHF transceivers, or HF+VHF transceivers), receive converters are sometimes preferred for demanding applications because they may be used with high-performance HF transceivers. Receive converters are often packaged with a companion transmit converter and control circuitry to make a *transverter*.

A typical 2 meter downconverter uses an IF of 28-30 MHz. Signals on 2 meters are amplified by a low-noise front-end before mixing with a 116 MHz LO. Fig 12.37B shows the schematic for a high-performance converter. The front-end design was contributed by Dr Ulrich Rohde, N1UL, who recommends a triple-balanced mixer such as the Synergy CVP206 or SLD-K5M.

The diplexer filter at the mixer output selects the difference product: (144 to 146 MHz) – 116 = (28 to 30 MHz). A common-base buffer amplifier (the 2N5432 FET) and tuned filter form the input to the 10 meter receiver. (N1UL suggests that using an IF of 21 MHz and an LO at 165 MHz would avoid interference problems with 222 MHz band signals.) For additional oscillator designs, refer to the papers on oscillators by N1UL in the Chapter 10 folder of the CD-ROM that accompanies this *Handbook*.

Based on the Philips BFG198 8 GHz transistor, the 20 dB gain front-end amplifier is optimized for noise figure (NF is approximately 2.6 dB), not for input impedance. The output circuit is optimized for best selectivity. The transistor bias is designed for dc stability at $I_C = 30$ mA and $V_C = 6$ V. Both of the transistor's emitter terminals should be grounded to prevent oscillation. NF might be improved with a higher performance transistor, such as a GaAs FET, but stability problems are often encountered with FET designs in this application.

If a mast-mounted preamplifier is used to improve the system noise figure, an attenuator should be available to prevent overload. Simulation predicts the circuit to have an IP3 figure of at least +25 dBm at 145 MHz with an I_C of 30 mA and a terminating impedance of 50 Ω.

12.8 UHF Techniques

The ultra high frequency spectrum comprises the range from 300 MHz to 3 GHz. All of the basic principles of radio system design and circuit design that have been discussed so far apply as well in this range, but the higher frequencies require some special thinking about the methods of circuit design and the devices that are used.

12.8.1 UHF Construction

Modern receiver designs make use of highly miniaturized monolithic microwave ICs (MMICs). Among these are the Avago MODAMP and the Mini Circuits MAR and MAV/ERA lines. They come in a wide variety of gains, intercepts and noise figures for frequency ranges from dc to well into the GHz range. (See the **Component Data and References** chapter for information on available parts.)

Fig 12.39 shows the schematic diagram and the physical construction of a typical RF circuit at 430 MHz. It is a GaAsFET preamplifier intended for low noise SSB/CW, moonbounce or satellite reception. The construction uses ceramic chip capacitors, small helical inductors and a stripline surface-mount GaAsFET, all mounted on a G10 (two layers of copper) glass-epoxy PC board. The very short length of interconnection leads is typical. The bottom of the PC board is a ground plane. At this frequency, lumped components are still feasible, while microstrip circuitry tends to be rather large.

At higher frequencies, microstrip methods become more desirable in most cases because of their smaller dimensions. However, the advent of tiny chip capacitors and chip resistors has extended the frequency range of discrete components. For example, the literature shows methods of building LC filters at as high as 2 GHz or more, using chip capacitors and tiny helical inductors. Amplifier and mixer circuits operate at well into the GHz range using these types of components on controlled-dielectric PC board material such as Duroid or on ceramic substrates.

Current designs emphasize simplicity of construction and adjustment. The use of printed-circuit microstrip filters that require little or no adjustment, along with IC or MMIC devices, or discrete transistors, in precise PC-board layouts that have been carefully worked out, make it much easier to "get going" on the higher frequencies.

12.8.2 UHF Design Aids

Circuit design and evaluation at the higher frequencies usually require some kind of minimal lab facilities, such as a signal generator, a calibrated noise generator and, hopefully, some kind of simple (or surplus) spectrum analyzer. This is true because circuit behavior and stability depend on a number of factors that are difficult to "guess at," and intuition is often unreliable. The ideal instrument is a vector network analyzer with all of the attachments (such as an S parameter measuring setup), but very few amateurs can afford this.

Another very desirable thing would be a circuit design and analysis program for the personal computer. Software packages created especially for UHF and microwave circuit design are available. They tend to be somewhat expensive, but worthwhile for a serious designer. Inexpensive *SPICE* programs are a good compromise. See the chapter on **Computer-Aided Circuit Design** for information on these tools.

12.8.3 A 902 to 928 MHz (33-cm) Receiver

This 902-MHz downconverter is a fairly typical example of receiver design methods for the 500 to 3000 MHz range, in which down conversion to an existing HF receiver (or 2-meter multimode receiver) is the most convenient and cost-effective approach for most amateurs. At higher frequencies a double down-conversion with a first IF of 200 MHz or so, to improve image rejection, might be necessary. Usually, though, the presence of strong signals at image frequencies is less likely. Image-reducing mixers plus down-conversion to 28 MHz is also coming into use, when strong interfering signals are not likely at the image frequency.

Fig 12.40A is the block diagram of the 902-MHz down converting receiver. A cavity resonator at the antenna input provides high selectivity with low loss. The first RF amplifier is a GaAsFET. Two additional 902 MHz band-pass microstrip filters and a second RF amplifier transistor provide more gain and image rejection (at RF – 56 MHz) for the

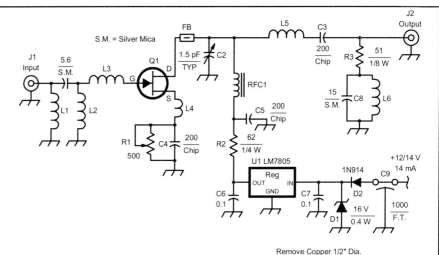
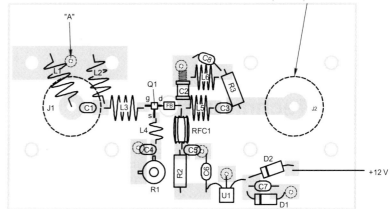
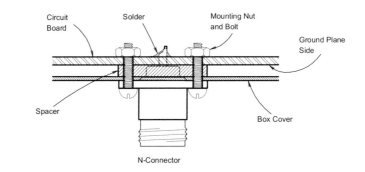

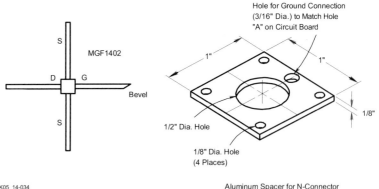

Fig 12.39 — GaAsFET preamplifier schematic and construction details for 430 MHz. Illustrates circuit, parts layout and construction techniques suitable for 430-MHz frequency range.

C1 — 5.6 pF silver-mica or same as C2.
C2 — 0.6 to 6 pF ceramic piston trimmer (Johanson 5700 series or equiv).
C3, C4, C5 — 200 pF ceramic chip.
C6, C7 — 0.1 µF disc ceramic, 50 V or greater.
C8 — 15 pF silver-mica.
C9 — 500 to 1000 pF feedthrough.
D1 — 16 to 30 V, 500 mW Zener (1N966B or equiv).
D2 — 1N914, 1N4148 or any diode with ratings of at least 25 PIV at 50 mA or greater.
J1, J2 — Female chassis-mount Type-N connectors, PTFE dielectric (UG-58 or equiv).
L1, L2 — 3t, #24 tinned wire, 0.110-inch ID spaced 1 wire dia.
L3 — 5t, #24 tinned wire, 3/16-inch ID, spaced 1 wire dia. or closer. Slightly larger diameter (0.010 inch) may be required with some FETs.
L4, L6 — 1t #24 tinned wire, 1/8-inch ID.
L5 — 4t #24 tinned wire, 1/8-inch ID, spaced 1 wire dia.
Q1 — Mitsubishi MGF1402.
R1 — 200 or 500-Ω Cermet potentiometer (initially set to midrange).
R2 — 62 Ω, 1/4-W.
R3 — 51 Ω, 1/8-W carbon composition resistor, 5% tolerance.
RFC1 — 5t #26 enameled wire on a ferrite bead.
U1 — 5 V, 100-mA 3 terminal regulator (LM78L05 or equiv. TO-92 package).

mixer. The output is at 28.0 MHz so that an HF receiver can be used as a tunable IF/demodulator stage.

CUMULATIVE NOISE FIGURE

Fig 12.40B shows the cumulative noise figure (NF) of the signal path, including the 28 MHz receiver. The 1.5 dB cumulative NF of the input cavity and first RF-amplifier combination, considered by itself, is degraded to 1.9 dB by the rest of the system following the first RF amplifier. The NF values of the various components for this example are reasonable, but may vary somewhat for actual hardware. Also, losses prior to the input such as transmission line losses (very important) are not included. They would be part of the complete receive system analysis, however. It is common practice to place a low noise preamp outdoors, right at the antenna, to overcome coax loss (and to permit use of less expensive coax).

LOCAL OSCILLATOR (LO) DESIGN

The +7-dBm LO at 874 to 900 MHz is derived from a set of crystal oscillators and frequency multipliers, separated by band-pass filters. These filters prevent a wide assortment of spurious frequencies from appearing at the mixer LO port. They also enhance the ability of the doubler stage to generate the second harmonic. That is, they have a very low impedance at the input frequency, thereby causing a large current to flow at the fundamental frequency. This increases the nonlinearity of the circuit, which increases the second-harmonic component. The higher filter impedance at the second harmonic produces a large harmonic output.

For very narrow-bandwidth use, such as EME, the crystal oscillators are often oven controlled or otherwise temperature compensated. The entire LO chain must be of low-noise design and the mixer should have good isolation from LO port to RF port (to minimize noise transfer from LO to RF).

A phase-locked loop using GHz range prescalers (as shown in Fig 12.40C) is an alternative to the multiplier chain. The divide-by-N block is a simplification; in practice, an auxiliary dual-modulus divider in a "swallow count" loop would be involved in this segment. The cascaded 902 MHz band-pass filters in the signal path should attenuate any image frequency noise (at RF–56 MHz) that might degrade the mixer noise figure.

12.8.4 Microwave Receivers

The world above 3 GHz is a vast territory with a special and complex technology and an "art form" that are well beyond the scope of this chapter. We will scratch the surface by describing a specific receiver for the 10 GHz frequency range and point out some of the important special features that are unique to this frequency range.

A 10 GHZ PREAMPLIFIER

Fig 12.41B is a schematic and parts list, Fig 12.41C is a PC board parts layout and Fig 12.41A is a photograph of a 10 GHz preamp, designed by Senior ARRL Lab Engineer Zack Lau, W1VT. With very careful design and packaging techniques a noise figure approaching the 1 to 1.5 dB range was achieved. This depends on an accurate 50-Ω generator impedance and noise matching the input using a microwave circuit-design program such as *Touchstone* or *Harmonica*. Note that microstrip capacitors, inductors and transmission-line segments are used almost exclusively. The circuit is built on a 15-mil Duroid PC board. In general, this kind of performance requires some elegant measurement equipment that few amateurs have. On the other hand, preamp noise figures in the 2 to 4-dB range are much easier to get (with simple test equipment) and are often satisfactory for amateur terrestrial communication.

Articles written by those with expertise and the necessary lab facilities almost always include PC board patterns, parts lists and detailed instructions that are easily duplicated by readers. Microwave ham clubs and their publications are a good way to get started in

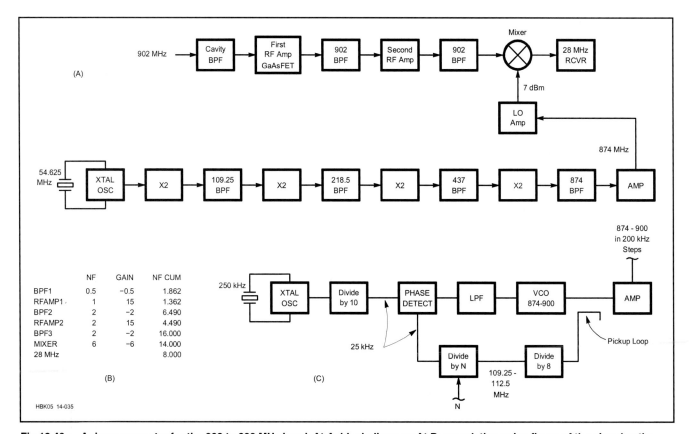

Fig 12.40 — A down converter for the 902 to 928 MHz band. At A: block diagram; At B: cumulative noise figure of the signal path; At C: alternative LO multiplier using a phase locked loop.

microwave amateur technology.

Because of the frequencies involved, dimensions of microstrip circuitry must be very accurate. Dimensional stability and dielectric constant reliability of the boards must be very good.

System Performance

At microwaves, an estimation of system performance can often be performed using known data about the signal path terrain, atmosphere, transmitter and receivers systems. In the present context of receiver design we wish to establish an approximate goal for the receiver system, including the antenna and transmission line. **Fig 12.42** shows a simplified example of how this works.

A more detailed analysis includes terrain variations, refraction effects, the Earth's curvature, diffraction effects and interactions with the atmosphere's chemical constituents and temperature gradients.

In microwave work, where very low noise levels and low noise figures are encountered, experimenters like to use the "effective noise temperature" concept, rather than noise factor. The relationship between the two is given by

$$T_E = 290(F - 1) \qquad (8)$$

where the noise factor $F = 10^{NF/10}$ and NF is the noise figure in dB.

T_E is a measure, in terms of temperature, of the "excess noise" of a component (such as an amplifier). A resistor at T_E would have the same available noise power as the device (referred to the device's input) specified by T_E. For a passive device (such as a lossy transmission line or filter) that introduces no noise of its own, T_E is zero and G is a number less than one equal to the power loss of the device. The cascade of noise temperatures is similar to the formula for cascaded noise factors.

$$T_S = T_G + TE_1 + \frac{TE_1}{G_1} + \frac{TE_3}{G_1 G_2} + \frac{TE_4}{G_1 G_2 G_3} + ... \qquad (9)$$

where T_S is the system noise temperature (including the generator, which may be an antenna) and T_G is the noise temperature of the generator or the field of view of the antenna, usually assumed 290 °K for terrestrial communications.

The 290 number in the formulas for T_E is the standard ambient temperature (in kelvins) at which the noise factor of a two-port transducer is defined and measured, according to an IEEE recommendation. So those formulas relate a noise factor F, measured at 290 K, to the temperature T_E. In general, though, it is perfectly correct to say that the ratio $(S_I/N_I)/(S_O/N_O)$ can be thought of as the ratio of total system output noise to that system output noise attributed to the "generator" alone, regardless of the temperature of the equipment or the nature of the generator, which may be an antenna at some arbitrary temperature, for example. This ratio is, in fact, a special "system noise factor (or figure), F_S" that need not be tied to any particular temperature such as 290 K. (Note that regular noise factor (or figure) does depend on reference temperature.) The use of the F_S notation avoids any confusion. As the example of Fig 12.42 shows, the value of this system noise factor F_S is just the ratio of the total system temperature to the antenna temperature.

Having calculated a system noise temperature, the receive system noise floor (that is, the antenna input level of a signal that would exactly equal system noise, both observed

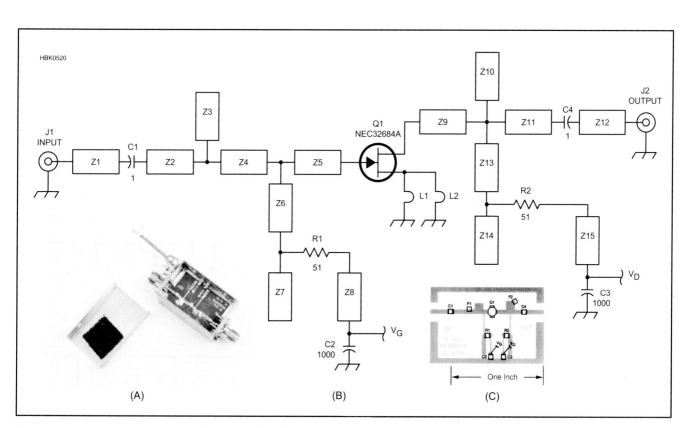

Fig 12.41 — At A a low-noise preamplifier for 10 GHz, illustrating the methods used at microwaves. At B: schematic. At C: PC board layout. Use 15-mil 5880 Duroid, dielectric constant of 2.2 and a dissipation factor of 0.0011. A template of the PC board is available on the CD-ROM included with this book.

C1, C4 — 1 pF ATC 100 A chip capacitors. C1 must be very low loss.
C2, C3 — 1000 pF chip capacitors. (Not critical.) The ones from Mini Circuits work fine.
F1, F2 — Pieces of copper foil used to tune the preamp.
J1, J2 — SMA jacks. Ideally these should be microstrip launchers. The pin should be flush against the board.
L1, L2 — The 15 mil lead length going through the board to the ground plane.
R1, R2 — 51 Ω chip resistors.
Z1-Z15 — Microstrip lines etched on the PC board.

Analysis of a 10.368 GHz communication link with SSB modulation:

Free space path loss (FSPL) over a 50-mile line-of-sight path (S) at F = 10.368 GHz:
FSPL = 36.6 (dB) + 20 log F (MHz) + 20 log S (Mi) = 36.6 + 80.3 + 34 = 150.9 dB.

Effective isotropic radiated power (EIRP) from transmitter:
EIRP (dBm) = P_{XMIT} (dBm) + Antenna Gain (dBi)

The antenna is a 2-ft diameter (D) dish whose gain GA (dBi) is:
GA = 7.0 + 20 log D (ft) + 20 log F (GHz) = 7.0 + 6.0 + 20.32 = 33.3 dBi

Assume a transmission-line loss L_T, of 3 dB
The transmitter power P_T = 0.5 (mW PEP) = –3 (dBm PEP)
$P_{XMIT} = P_T$ (dBm PEP) – L_T (dB) = (–3) – (3) = –6 (dBm PEP)
EIRP = $P_{XMIT} + G_A$ = –6 + 33.3 = 27.3 (dBm PEP)
Using these numbers the received signal level is:
P_{RCVD} = EIRP (dBm) – Path loss (dB) = 27.3 (dBm PEP) – 150.9 (dB) = –123.6 (dBm PEP)
Add to this a receive antenna gain of 17 dB. The received signal is then P_{RCVD} = –123.6 +17 = –106.6 dBm

Now find the receiver's ability to receive the signal:
The antenna noise temperature T_A is 200 K. The receiver noise figure NF_R is 6 dB (FR=3.98, noise temperature T_R = 864.5 K) and its noise bandwidth (B) is 2400 Hz. The feedline loss L_L is 3 dB (F = 2.00, noise temperature T_L = 288.6 K). The system noise temperature is:
$T_S = T_A + T_L + (L_L)(T_R)$
T_S = 200 + 288.6 + (2.0) (864.5) = 2217.6 K
$N_S = kT_SB$ = 1.38 × 10^{-23} × 2217.6 × 2400 = 7.34 × 10^{-17} W = –131.3 dBm
This indicates that the PEP signal is –106.6 –(–131.3) = 24.7 dB above the noise level. However, because the average power of speech, using a speech processor, is about 8 dB less than PEP, the average signal power is about 16.7 dB above the noise level.

To find the system noise factor F_S we note that the system noise is proportional to the system temperature T_S and the "generator" (antenna) noise is proportional to the antenna temperature T_A. Using the idea of a "system noise factor":

$F_S = T_S / T_A$ = 2217.6 / 200 = 11.09 = 10.45 dB.

If the antenna temperature were 290 K the system noise figure would be 9.0 dB, which is precisely the sum of receiver and receiver coax noise figures (6.0 + 3.0).

Fig 12.42 — Example of a 10-GHz system performance calculation. Noise temperature and noise factor of the receiver are considered in detail.

at the receiver output) associated with that temperature is:

$$N = k\, T_S\, B_N \quad (10)$$

where
$k = 1.38 \times 10^{-23}$ (Boltzmann's constant)
and
B_N = noise bandwidth

The system noise figure F_S is indicated in the example also. It is higher than the sum of the receiver and coax noise figures.

The example includes a loss of 3 dB in the receiver transmission line. The formula for T_S in the example shows that this loss has a double effect on the system noise temperature, once in the second term (288.6) and again in the third term (2.0). If the receiver (or high-gain preamp with a 6 dB NF) were mounted at the antenna, the receive-system noise temperature would be reduced to 1064.5 K and a system noise figure, F_S, of 7.26 dB, a very substantial improvement. Thus, it is the common practice to mount a preamp at the antenna, although transmission line loss must still be included in system noise figure calculations.

MICROWAVE RECEIVER FOR 10 GHZ

Here is a good example of amateur techniques for the 10 GHz band. The intended use for the radio is narrowband CW and SSB work, which requires extremely good frequency stability in the LO. Here, we will discuss the receiver circuit.

Block Diagram

Fig 12.43 is a block diagram of the receiver. Here are some important facets of the design.

1) The antenna should have sufficient gain. At 10 GHz, gains of 30 dBi are not difficult to get, as the example of Fig 12.42 demonstrates. A 4-foot dish might be difficult to aim, however.
2) For best results a very low-noise preamp at the antenna reduces loss of system sensitivity when antenna temperature is low. For example, if the antenna temperature at a quiet direction of the sky is 50 K and the receiver noise figure is 4 dB (due in part to transmission-line loss), the system temperature is 488 K for a system noise figure of 4.3 dB. If the receiver noise figure is reduced to 1.5 dB by adding a preamp at the antenna the system temperature is reduced to 170 K for a system noise figure of 2.0 dB, which is a very big improvement.
3) After two stages of RF amplification using GaAsFETs, a probe-coupled cavity resonator attenuates noise at the mixer's image frequency, which is 10.368 – 0.288 = 10.080 GHz. An image reduction of 15 to 20 dB is enough to prevent image frequency noise generated by the RF amplifiers from affecting the mixer's noise figure.
4) The single-balanced diode mixer uses a "rat-race" 180° hybrid. Each terminal of the ring is ¼ wavelength (90°) from its closest neighbors. So the anodes of the two diodes are 180° (½ wavelength) apart with respect to the LO port, but in-phase with respect to the RF port. The inductors (L1, L2) connected to ground present a low impedance at the IF frequency. The mixer microstrip circuit is carefully "tweaked" to improve system performance. Use the better mixer in the transmitter.
5) The crystal oscillator is a fifth-overtone Butler circuit that is capable of high stability. The crystal frequency error and drift are multiplied 96 times (10.224/0.1065), so for

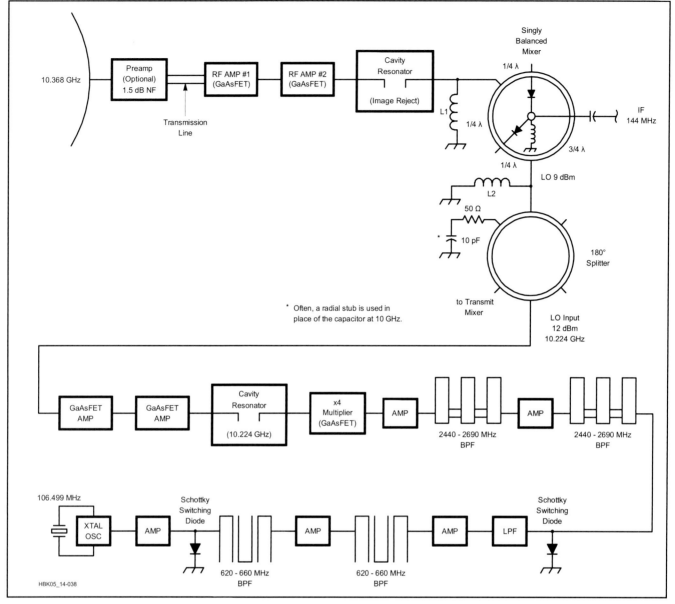

Fig 12.43 — A block diagram of the microwave receiver discussed in the text.

narrowband SSB or CW work it may be difficult to get on (and stay on) the "calling frequency" at 10.368 GHz. One acceptable (not perfect) solution might be to count the 106.5 MHz with a frequency counter whose internal clock is constantly compared with WWV. Adjust to 106.5 MHz as required. At times there may be a small Doppler shift on the WWV signal. It may be necessary to switch to a different WWV frequency, or WWV's signals may not be strong enough. Surplus frequency standards of high quality are sometimes available. Many operators just "tune" over the expected range of uncertainty.

6) The frequency multiplier chain has numerous band-pass filters to "purify" the harmonics by reducing various frequency components that might affect the signal path and cause spurious responses. The final filter is a tuned cavity resonator that reduces spurs from previous stages. Oscillator phase noise amplitude is multiplied by 96 also, so the oscillator must have very good short-term stability to prevent contamination of the desired signal.

7) A second hybrid splitter provides an LO output for the transmitter section of the radio. The 50-Ω resistor improves isolation between the two output ports. A two-part *QST* article is recommended reading for this very interesting project, which provides a fairly straightforward (but not extremely simple) way to get started on 10 GHz.[17]

12.9 References and Bibliography

[1] *CRC Standard Mathematical Tables*, CRC Press, Cleveland, OH, 1974, p 231.

[2] W. Hayward, W7ZOI, R. Campbell, KK7B, and B. Larkin, W7PUA, *Experimental Methods in RF Design*. (ARRL, Newington, 2009)

[3] R. Henderson, WI5W, "A Rock Bending Receiver for 7 MHz," *QST*, Aug 1995, pp 22-25.

[4] Note that we could have also set the local oscillator to 600 – 455 kHz or 145 kHz. By setting it to the sum, we reduce the relative range that the oscillator must tune. To cover the 500 to 1500 kHz with the difference, our LO would have to cover from 45 to 1045, a 23:1 range. Using the sum requires LO coverage from 955 to 1955 kHz, a range of about 2:1 — much easier to implement.

[5] He is the same Armstrong who invented frequency modulation (FM) some years later and who held many radio patents between WW I and WW II.

[6] The Baudot code is still the standard employed in wire-line teletype (TTY) keyboard terminals used by hearing-impaired people.

[7] A notable exception was the Hammarlund Manufacturing Corporation whose receivers used a 2:1 tuning range. The typical bands were .5 to 1, 1 to 2, 2 to 4, 4 to 8, 8 to 16 and 16 to 32 MHz. By covering the range in six rather than four bands, the tuning was noticeably more precise, especially across the higher bands.

[8] R. Fisher, W2CQH, "Twisted-Wire Quadrature Hybrid Directional Couplers," *QST*, Jan 1978, pp 21-24.

[9] For additional discussion of the issues, see J. Hallas, "International Radio Roofing Filters for the Yaesu FT-1000 MP Series Transceivers," Product Review, *QST*, Feb 2005, pp 80-81.

[10] Exceptions with general coverage receivers and switchable VHF roofing filters include the ICOM IC-7800 and Yaesu FTdx9000 and FT-2000 transceivers.

[11] H. Robinson, W3LW, "Audio Output Limiters for Improving the Signal-to-Noise Ratio in Reception", *QST*, Feb 1936, pp 27-29.

[12] J. Dickert, "A New Automatic Noise Limiter", *QST*, Nov 1938, pp 19-21.

[13] You can demonstrate this in your receiver if it has a narrow filter. Find a band with heavy impulse noise and switch between wide and narrow filters. If your narrow filter is 500 Hz or less, I'll bet the noise is more prominent with the narrow filter.

[14] R. D. Straw, Ed., *The ARRL Antenna Book*, 20th Edition (Newington 2003), pp 5-24 to 5-26.

[15] J. Grebenkemper, KI6WX, "Phase Noise and its Effects on Amateur Communications," *QST*, Mar and Apr 1988.

[16] W. Sabin, "Envelope Detection and AM Noise-Figure Measurement," *RF Design*, Nov 1988, p 29.

[17] Z. Lau, W1VT, "Home-Brewing a 10-GHz SSB/CW Transverter," *QST*, May and Jun 1993.

BIBLIOGRAPHY

Rohde, Whittaker, and Bucher, *Communications Receivers*, McGraw-Hill, 1997.

Rohde and Whittaker, *Communications Receivers: DSP, Software Radios, and Design*, 3rd Edition, McGraw-Hill Professional, 2000.

McClanning and Vito, *Radio Receiver Design*, Noble Publishing Corporation, 2001.

Contents

13.1 Introduction

13.2 Early Transmitter Architectures

    13.2.1 Alternator Transmitters

13.3 Modulation Types and Methods Applied to Transmitter Design

    13.3.1 Data, Telegraph or Pulse Transmitters

    13.3.2 A Two-stage HF CW Transmitter

    13.3.3 Amplitude-Modulated full-Carrier Voice Transmission

    13.3.4 Single-Sideband Suppressed-Carrier Transmission

    13.3.5 Angle-Modulated Transmitters

    13.3.6 A Superhet SSB/CW Transmitter

    13.3.7 CW Mode Operation

    13.3.8 Wideband Noise

    13.3.9 Automatic Level Control (ALC)

Project: The MicroT2 — A Compact Single-Band SSB Transmitter

Project: The MKII — An Updated Universal QRP Transmitter

13.4 Modern Baseband Processing

    13.4.1 Digital Signal Processing for Signal Generation

13.5 Increasing Transmitter Power

    13.5.1 Types of Power Amplifiers

    13.5.2 Linear Amplifiers

    13.5.3 Nonlinear Amplifiers

    13.5.4 Hybrid Amplifiers

13.6 Transmitter Performance and Measurement

    13.6.1 Power Measurement

    13.6.2 Frequency Calibration and Stability

    13.6.3 Transmitter Spurious Responses

    13.6.4 Harmonics of the Transmitter Frequency

    13.6.5 Key Clicks and Waveshaping

    13.6.6 Intermodulation Distortion

13.7 References and Bibliography

Chapter 13

Transmitters

Transmitters are the companion to the receivers discussed in the previous chapter. Many of the concepts presented here build on other sections of this book that detail modulation, mixing, filters and other related concepts. Amplifiers for power levels above 100 W are covered in the **RF Power Amplifiers** chapter. **DSP and Software Radio Design** has more information on digital techniques. Transmitters may contain hazardous voltages, and at higher power levels RF exposure issues must be considered — review the **Safety** chapter for more information. Material from previous editions was updated and new material contributed by Joel R. Hallas, W1ZR.

13.1 Introduction

Transmitters have what would appear to be a rather straightforward job — generate an ac signal and modify it in some way to carry information. In the early days, it was found that making an electrical spark in a circuit that was connected to an antenna could be received by various devices connected to another antenna at some distance. It didn't take folks very long to figure out that if the sparks were sent in patterns corresponding to Morse or other telegraphy codes, information could be sent distances without connecting wires. Wireless communication was born!

Technology quickly advanced through a number of variations of spark and arc transmitters and a wide variety of ingenious receiving devices until wireless telegraphy became a serious business involved with monitoring the safety of ships at sea in ways that hadn't been possible before. See **Fig 13.1** for a view of an early shipboard wireless station. The systems of the

Fig 13.1 — Marconi shipboard radiotelegraph station from 1900. (*Photo courtesy of Marconi Company, PLC*)

day were more electromechanical than what we would consider electronic today, culminating in very high speed rotary alternators that could generate high power ac directly. An RF alternator was an electromechanical alternating current generator that produced ac at radio frequencies.

Transmitter technology advanced in a parallel process similar to that of the technology of receivers. While transmitters are composed of many of the same named blocks as those used in receivers, it's important to keep in mind that they may not be the same size. An RF amplifier in a receiver may deal with amplifying picowatts while one in a transmitter may output up to megawatts. While the circuits may even look similar, the size of the components, especially cooling systems and power supplies, may differ significantly in scale. Still, many of the same principles apply.

13.2 Early Transmitter Architectures

13.2.1 Alternator Transmitters

A very basic alternator might consist of a coil with a core of *soft iron* (pure molecular iron that will magnetize and demagnetize easily) rotating past two poles of a magnet inside a round iron mounting as in **Fig 13.2**. As one end of the rotating iron cored coil passes the north magnetic pole, a magnetic field is built up in it. As a result of the magnetic field increasing and then decreasing in the rotor coil's core, one half of an ac cycle is induced in its coil. As it continues to rotate and passes the south pole, an opposite half cycle of ac is induced in it. To produce 60 Hz ac, the coil would have to rotate 60 times per second, or 3600 r/min, a very fast rotation. The two ends of the rotating coil are connected to two *slip rings* on the shaft that rotates the coil. There are two brushes of fixed carbon or other material that make a constant contact with the rotating slip rings. Any ac voltage that is generated is available at these two brushes. By using multiple rings and brushes, multiple cycles of ac will be produced with each rotation.

Better systems were used to generate lower frequency RF ac. Ernst Alexanderson, a prodigious Swedish/American electrical engineer and inventor, and Canadian inventor Reginald Fessenden first produced a lower power 60 kHz RF alternator in 1906 with which they could transmit telegraphy as well as voice and music. Later the higher powered *Alexanderson alternators* used frequencies closer to 20 kHz. Other popular alternators of the period were the German Joly-Arco and the Goldschmidt, which was somewhat similar to the Alexanderson.[1]

The final form of the Alexanderson alternator uses a rotary toothed disc, called an *inductor*. This inductor is a large round, flat, soft-iron disk often with teeth ground out at its edge as shown in **Fig 13.3**. The many teeth of the inductor are rotated rapidly by a constant speed ac motor between the N and S poles of double wound field pole electromagnets mounted one tooth-width apart all around the inside of the machine. As an iron tooth is passing between the N and S ends of the dc-excited electromagnet field poles it provides a better path for the N to S magnetic lines of force. As a result the magnetism in the field pole builds up, but collapses as the tooth moves on.

As the next inductor tooth passes the field pole it develops another magnetic field build up and collapse. These varying magnetic fields induce ac cycles in all of the secondary RF output field-pole coils. Since there are as many field poles around the inside of the machine as teeth on the inductor, ac voltages are developed in the field pole ac pick-up coils all at the same time and are all inductively coupled to a coil/antenna/ground circuit that radiates the RF waves. The speed of the motor rotating the inductor and the number of teeth determines the output frequency.[2]

13.2.2 Vacuum Tube Transmitters

By the early 1920s, large vacuum tubes were developed that could be used in high powered transmitters. Most of the big alternators were replaced by less massive high power tube transmitters. In 1920, KDKA in Pittsburgh, Pennsylvania became the first federally licensed commercial AM broadcast station in the country, operating on 1020 kHz, initially using a 100 W vacuum tube transmitter.

OSCILLATOR TRANSMITTERS

Amateur transmitters also quickly moved from spark and arc to vacuum tube sets. Be-

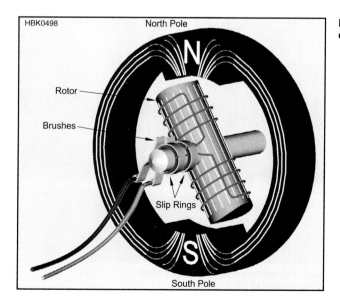

Fig 13.2 — Basic alternator consisting of a soft-iron cored coil rotating past two poles of a magnet.

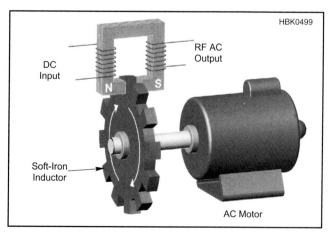

Fig 13.3 — High frequency alternator using an inductor disc.

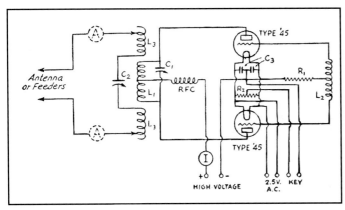

Fig 13.4 — Transmitter schematic as shown in an early *QST* magazine.

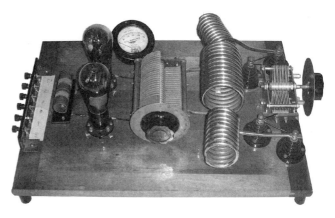

Fig 13.5 — Photo of a transmitter built from the schematic in Fig 13.4.

cause the vacuum tube oscillators generated a single-frequency RF signal directly and continuously, transmitters using them were called *continuous wave* or *CW* transmitters. While generally having less nominal power than the spark transmitters of the day, more power was actually provided to the antenna by CW transmitters. Signals were of a much narrower bandwidth, allowing more usable channels once receiver capabilities caught up. The early tube transmitters were generally single-stage self-excited oscillators coupled directly to an antenna as shown in **Figs 13.4** and **13.5**.

While a vast improvement over early technology, such transmitters had their limitations. Note that the antenna was connected directly to the tuned circuit that controlled the frequency. This meant that every time the antenna moved in the wind, the resultant change in loading shifted the transmit frequency. If the cat walked by and was lucky enough not to be electrocuted by the exposed high voltage wiring, the frequency moved even on a calm day. The frequency would often shift with each key closure as the power supply voltage sagged with the additional current drain, resulting in signals with characteristic chirps and whoops.

MASTER OSCILLATOR-POWER AMPLIFIER TRANSMITTERS

All of these issues were at least partly addressed by adding a power amplifier between the oscillator and the antenna. Morse code transmission was accomplished by using the key to turn the power amplifier stage on and off. If properly designed and adjusted and powered by a solid power supply, the oscillator could be left on between the transmitted dots and dashes. This stabilized the transmitted signal's frequency and the antenna was well-isolated from the frequency determining components. This architecture was termed a *master oscillator-power amplifier* (MOPA) configuration and remained popular in WWII military radios and amateur equipment into the 1950s. The popular ARC-5 series military aircraft HF transmitters were of MOPA design using a single-triode variable-tuned oscillator and a parallel pair of tetrode amplifier tubes. While the military ran these at more conservative ratings, amateurs used them to provide up to 100 W output, all from a box the size of a small loaf of bread. Two examples are shown in **Fig 13.6**.

CRYSTAL-CONTROLLED TRANSMITTERS

The transmitters discussed so far operated on frequencies under the control of variable-tuned LC circuits. This provided a certain amount of flexibility, but had the down side of making the operating frequency uncertain. Stability over time, supply voltage and temperature was not as good as might be desired.

Properties of crystal structures were studied at least as far back as the 18th century, but the first major application of the piezoelectric effect that converts crystal motion to voltage and back occurred during WWI. The French developed a piezoelectric ultrasonic transducer that was used as the transmitter in an acoustical submarine detection system. Between WWI and WWII, the use of resonant wafers of quartz crystal as frequency determining elements in oscillators became feasible and packaged crystals became readily available. Crystal frequency control was particularly suitable for radios that operated on assigned fixed channels, common to most services other than Amateur Radio. Still, crystal-controlled transmitters were popular in amateur service, with most stations having a large collection of crystals that could be selected to change frequency. The now-discontinued entry-level Novice class license for many years

Fig 13.6 — A pair of 7-9.1 MHz ARC-5 transmitters from WW2. On the left the US Army Air Forces version (BC-459), on the right a Navy equivalent (T-22, ARC-5). These were available for typically $5 in "new in box" condition after WW2.

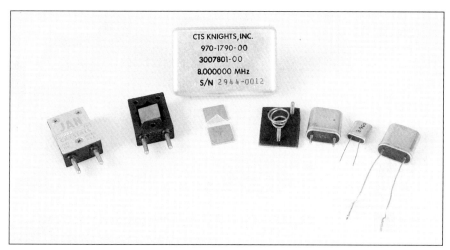

Fig 13.7 — Sample of typical frequency control crystals for amateur use. The units on the right are modern sealed types. On the far left is a WWII era crystal with a disassembled holder in the center.

required crystal control of the transmitter. **Fig 13.7** shows a selection of popular crystal types.

MULTIBAND TRANSMITTERS

The transmitters discussed so far generally operated over a single band or frequency range. In order to operate on multiple bands, a number of approaches were used. The early amateur bands were selected to be harmonically related in order to avoid interference to other services from spurious harmonic signals. The amateur bands of the early years were 160, 80, 40, 20, 10, 5 and 2½ meters.

Frequency Multiplier Transmitters

A popular approach to early AM and CW transmitters designed for multiple bands was to use frequency multiplier stages to raise the frequency to a desired harmonic. A frequency multiplier was just an amplifier operated in class C with the output tuned circuit tuned to a multiple of the input frequency. The plate current pulses would cause the output tank circuit to ring at the selected harmonic, filling in the cycles between the pulses. If the Q of the tank circuit was high enough, harmonics through the third or fourth could be generated with acceptable distortion. Additional tuned stages would reinforce the desired harmonic and reject the other signals.

A block diagram of a typical period transmitter for 80, 40 and 20 meters is shown in **Fig 13.8**. Note that a particular advantage of this configuration is that a single crystal at say 3.505 MHz can be used at 7.010 and 14.020 MHz. This arrangement was quite common in transmitters designed for frequency modulation in which not only the frequency, but also the deviation, is multiplied at each step.

While the figure shows cascaded doublers, some used circuits switchable between doublers, triplers and quadruplers to avoid the need for additional stages. Note that a limitation of this arrangement is that any drift or instability is also multiplied, making it necessary that particular attention is paid to oscillator stability. The tuning rate is also different on each band.

Heterodyne Transmitters

Another transmitter configuration looks and sounds a lot like the architecture of a modern receiver, but backwards — the heterodyne transmitter. A block diagram of a representative transmitter of that type is shown in **Fig 13.9**. This is the basis for the transmitter section of most SSB transmitters as well as modern SSB transceivers described in the **Transceivers** chapter. There are some clear advantages to this approach. In the early days, it was possible to have both oscillators running continuously with CW keying applied to the mixer stage. This provided the highest frequency stability of any configuration discussed so far.

Another plus compared to the multiplying transmitter was that the tuning rate of the variable frequency oscillator (VFO) was the same on each band. This allows more precise tuning adjustment on the higher frequency ranges. As with heterodyne receivers, multiple conversions are encountered, particularly in radios working into the VHF and UHF regions and those with DSP waveform generation. Any images resulting from the conversion process must be eliminated by the filtering between the mixer and power amplifier stages, with similar constraints to the image issues of receivers.

Some early heterodyne SSB transmitters

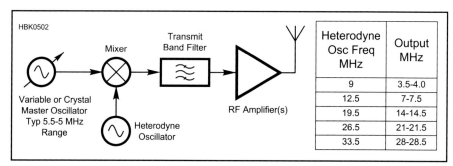

Fig 13.9 — Block diagram of typical heterodyne type transmitter.

Heterodyne Osc Freq MHz	Output MHz
9	3.5-4.0
12.5	7-7.5
19.5	14-14.5
26.5	21-21.5
33.5	28-28.5

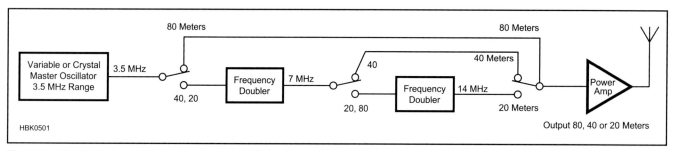

Fig 13.8 — Block diagram of typical frequency multiplier transmitter.

took advantage of the fact that the 9 MHz heterodyne oscillator frequency could actually be used on both 80/75 and 20 meters with the same 5.0 to 5.5 MHz frequency range in the VFO. The difference product at 9 MHz — the VFO frequency (4 to 3.5 MHz) was used for the lower band while the sum product at 9 MHz + the VFO frequency (14 to 14.5 MHz) was used for 20 meters. This was a popular dual-band scheme in the early days of SSB voice. One drawback was that the VFO tuned in the opposite direction on each band. For many hams of the day, it was a small price for the simplicity.

13.3 Modulation Types and Methods Applied to Transmitter Design

13.3.1 Data, Morse Code or Pulse Transmitters

The simplest transmitter consists of an oscillator generating a signal at the frequency we want to transmit. If the oscillator is connected to an antenna, the signal will propagate outward and be picked up by any receivers within range. Such a transmitter will carry little information, except perhaps for its location — it could serve as a rudimentary *beacon*, for direction finding or radiolocation, although real beacons generally transmit identification data. It also indicates whether or not it is turned on, perhaps useful as part of an alarm system.

To actually transmit information, we must *modulate* the transmitter. The modulation process, covered in detail in the **Modulation** chapter, involves changing one or more of the signal parameters to apply the information content. This must be done in such a way that the information can be extracted at the receiver. As previously noted, the parameters available for modulation are:

Frequency — this is the number of cycles the signal makes per second.

Amplitude — although the amplitude, or strength, of a sinusoid is constantly changing with time, we can express the amplitude by the maximum value that it reaches.

Phase — the phase of a sinusoid is a measure of when a sinusoid starts compared to another sinusoid of the same frequency.

We could use any of the above parameters to modulate a simple transmitter with pulse type information, but the easiest to visualize is probably amplitude modulation. If we were to just turn the transmitter on and off, with it on for binary "one" and off for a "zero" we could surely send Morse code or other types of pulse-coded data. This type of modulation is called "On-Off-Keying" or "OOK" for short.

Some care is needed in how we implement such a function. Note that if we performed the obvious step of just removing and turning on the power supply, we might be surprised to find that it takes too much time for the voltage to rise sufficiently at the oscillator to actually turn it on at the time we make the connection. Similarly, we might be surprised to find that when we turn off the power we would still be transmitting for some time after the switch is turned. These finite intervals are referred to as *rise* and *fall times* and generally depend on the time constants of filter and switching circuits in the transmitter.

There is an easy to visualize analogy with using light to transmit code signals. If you've ever seen a WWII movie showing a Navy Morse operator sending data between ships with a lantern, you may have wondered why they used a special mechanical shutter device instead of turning the light bulb on and off.[3] One reason is that the bulb continues to radiate for a period while the filament is still hot, even after the power is off.

A simple low power HF CW oscillator transmitter is shown schematically in **Fig 13.10**. By using a crystal-controlled oscillator, we gain frequency accuracy and stability, but lose frequency agility.

Note that in this design the key essentially just turns the power on and off; however, the line bypass capacitor value was chosen so that it would not delay the rise or fall at reasonable keying speed. This transmitter would actually work and generate a few mW of RF power at the crystal frequency, but it has a number of limitations in common with the early vacuum tube oscillator transmitters of the 1920s. These designs were still a major improvement over the spark transmitters that preceded them.

First, the oscillator is dependent on the environment. Changes in the antenna, due to wind, for example, could change the load and cause the frequency to shift. Second, the oscillator generates a signal at the crystal fundamental frequency, but also at harmonics (multiples) of the frequency. Third, every time the key is closed, the oscillator must start up, taking a finite time to stabilize at its rest frequency. Such behavior is often described as "chirp", since a signal that changes frequency when it is keyed sounds like a bird's chirp in the receiver.

13.3.2 A Two-Stage HF CW Transmitter

By adding an amplifier stage to our little transmitter, we can address the first problem. While we're at it, we will add filtering to solve the second problem. We will defer dealing with the third problem, as discussed below, but could if we needed to.

Fig 13.11 is the circuit of a much better, but still simple radiotelegraph transmitter. This circuit uses an oscillator similar to that of Fig 13.10 followed by a stage of amplification. The amplifier increases the output power to about ⅓ W. Note that the output circuit includes a filter to reduce any harmonic output below the levels required by the US Federal Communications Commission (FCC).

A simple oscillator transmitter can be useful, and is employed in special circumstances, particularly in the microwave region. An advantage of separating the oscillator from the device providing the output power is that, for applications requiring frequency stability, an oscillator that doesn't have to deliver

Fig 13.10 — A simple solid-state low-power HF crystal controlled oscillator-transmitter. The unspecified tuned circuit values are resonant at the crystal frequency.

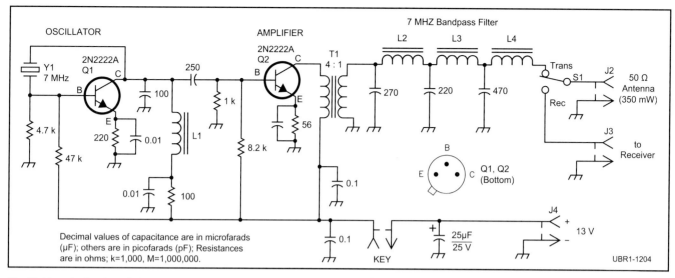

Fig 13.11 — A simple solid-state low-power HF crystal controlled oscillator-transmitter. The unspecified tuned circuit values are resonant at the crystal frequency.

Fig 13.12 — Photo of a Tuna Tin 2, built into an actual tuna can.

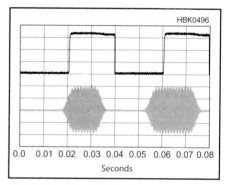

Fig 13.13 — The CW keying waveform of a transmitter with good spectrum control. The top trace is the key closure, with the start of the first contact closure on the left edge at 60 WPM using full break-in. Below that is the nicely rounded RF envelope.

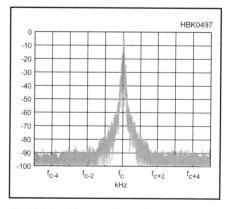

Fig 13.14 — The resultant signal spectrum from the keying shown in Fig 13.13. Note that the signal amplitude is about 80 dB down at a spacing of ±1 kHz, with a floor of −90 dB over the 10 kHz shown.

much power is more easily stabilized. This is an example of the master oscillator-power amplifier or MOPA, discussed in the previous section.

While ⅓ W doesn't sound like a lot of power (and isn't!), this transmitter has been built by many Amateur Radio operators and used for worldwide communication on the 7 MHz (40 meter) amateur band. The original model was built into a tuna fish can as shown in **Fig 13.12**.[4] This has become a classic miniature amateur transmitter, popularly known as the "Tuna Tin 2."

REAL WORLD CW KEYING

In the **Modulation** chapter, the importance of shaping the time envelope of the keying pulse of an on-off keyed transmitter is discussed. There are serious ramifications of not paying close attention to this design parameter. The optimum shape of a transmitter envelope should approach the form of a sinusoid raised to a power with a tradeoff between occupied bandwidth and overlap between the successive pulses. This can be accomplished either through filtering of the pulse waveform before modulation, in a linear transmitter, or through direct generation of the pulse shape using DSP.

The differences between well-designed and poor pulse shaping can perhaps be best described by looking at some results. The following figures are from recent *QST* product reviews of commercial multimode 100 W HF transceivers.

Fig 13.13 shows the CW keying waveform of a transmitter with good spectrum control. The top trace is the key closure, with the start of the first contact closure on the left edge at 60 WPM using full break-in. Below it is the nicely rounded RF envelope. **Fig 13.14** shows the resultant signal spectrum. Note that the signal amplitude is about 80 dB down at a spacing of 1 kHz, with a floor of −90 dB over the 10 kHz shown. **Figs 13.15** and **13.16** are similar data taken from a different manufacturer's transceiver. Note the sharp corners of the RF envelope, as well as the time it takes for the first "dit" to be developed. The resulting spectrum is not even down 40 dB at 1 kHz and shows a floor that doesn't quite make −60 dB over the 10 kHz range. It's easy to see the problems that the latter transmitter will cause to receivers trying to listen to a weak signal near its operating frequency. The unwanted components of the signal are heard on adjacent

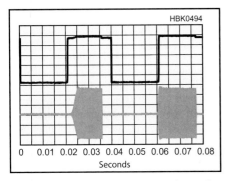

Fig 13.15 — The CW keying waveform of a transmitter with poor spectrum control. The top trace is the key closure, with the start of the first contact closure on the left edge at 60 WPM using full break-in. Note the sharp corners of the RF envelope that result in excessive bandwidth products.

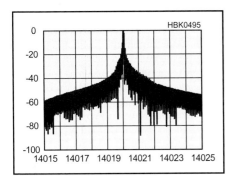

Fig 13.16 — The resultant signal spectrum from the keying shown in Fig 13.15. The resulting spectrum is not even down 40 dB at ±1 kHz and shows a floor that doesn't quite make –60 dB below the carrier across the 10 kHz.

channels as sharp clicks when the signal is turned on and off, called *key clicks*.

Note that even the best-shaped keying waveform in a linear transmitter will become sharp with a wide spectrum if it is used to drive a stage such as an external power amplifier beyond its linear range. This generally results in clipping or limiting with subsequent removal of the rounded corners on the envelope. Trying to get the last few dB of power out of a transmitter can often result in this sort of unintended signal impairment.

13.3.3 Amplitude-Modulated Full-Carrier Voice Transmission

While the telegraph key in the transmitters of the previous section can be considered a *modulator* of sorts, we usually reserve that term for a somewhat more sophisticated system that adds information to the transmitted signal. As noted earlier, there are three signal parameters that can be used to modulate a radio signal and they all can be used in various ways to add voice (or other information) to a transmitted signal.

One way to add voice to a radio signal is to first convert the analog signal to digital data and then transmit it as "ones" and "zeros". This can be done even using the simple telegraph transmitters of the last section. This is a technique frequently employed for some applications using data applied to our "pulse" transmitter described earlier. Here we will talk about the more direct application of the analog voice signal to a radio signal.

A popular form of voice amplitude modulation is called *high-level amplitude modulation*. It is generated by mixing (or modulating) an RF carrier with an audio signal. **Fig 13.17** shows the conceptual view of this. **Fig 13.18** is a more detailed view of how such a voice transmitter would actually be implemented. The upper portion is the RF channel, and you can think of the previously described Tuna Tin Two transmitter as a transmitter that we could use, after shifting the frequency into the voice portion of the band. The lower portion is the audio frequency or AF channel, usually called the *modulator*, and is nothing more than an audio amplifier designed to be fed from a microphone and with an output designed to match the anode or collector impedance of the final RF amplifier stage.

The output power of the modulator is applied in series with the dc supply of the output stage (only) of the RF channel of the transmitter. The level of the voice peaks needs to be just enough to vary the supply to the RF amplifier collector between zero volts, on negative peaks, and twice the normal supply voltage on positive voice peaks. This usually requires an AF amplifier with about half the average power output as the dc input power (product of dc collector or plate voltage times the current) of the final RF amplifier stage.

The output signal, called *full-carrier double-sideband AM*, occupies a frequency spectrum as shown in **Fig 13.19**. The spectrum

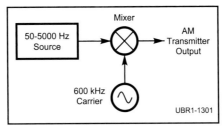

Fig 13.17 — Block diagram of a conceptual AM transmitter

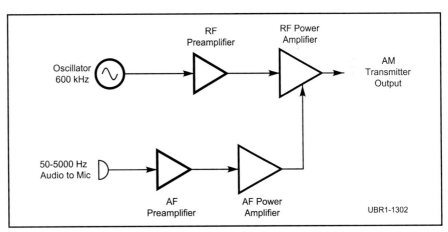

Fig 13.18 — Block diagram of a 600 kHz AM broadcast transmitter.

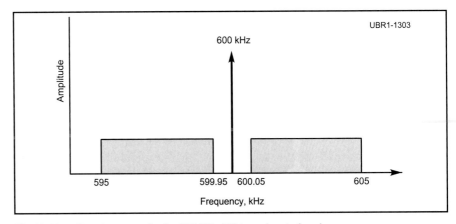

Fig 13.19 — Spectrum mask of a 600 kHz AM broadcast signal.

shown would be that of a standard broadcast station with an audio passband from 50 Hz to 5 kHz. Note that the resulting channel width is twice the highest audio frequency transmitted. If the audio bandwidth were limited to typical "telephone quality speech" of 300 to 3300 Hz, the resulting bandwidth would be reduced to 6.6 kHz. Note also that while a perfect multiplication process would result in just two sidebands and no carrier, this implementation actually provides the sum of the carrier and the sidebands from the product terms. (See the **Receivers** chapter for the mathematical description of signal multiplication.)

Full-carrier double-sideband AM is used in fewer and fewer applications. The spectral and power efficiency are significantly lower than single sideband (SSB), and the equipment becomes quite costly as power is increased. The primary application is in broadcasting — largely because AM transmissions can be received on the simplest and least expensive of receivers. With a single transmitter and thousands of receivers, the overall system cost may be less and the audience larger than for systems that use more efficient modulation techniques. While the PEP output of an AM transmitter is four times the carrier power, none of the carrier power is necessary to carry the information, as we will discuss in the next section.

13.3.4 Single-Sideband Suppressed-Carrier Transmission

The two sidebands of a standard AM transmitter carry (reversed) copies of the same information, and the carrier carries essentially no information. We can more efficiently transmit the information with just one of the sidebands and no carrier. In so doing, we use somewhat less than half the bandwidth, a scarce resource, and also consume much less transmitter power by not transmitting the carrier and the second sideband.

SINGLE-SIDEBAND — THE FILTER METHOD

The block diagram of a simple single-sideband suppressed-carrier (SSB) transmitter is shown in **Fig 13.20**. This transmitter uses a balanced mixer as a *balanced modulator* to generate a double sideband suppressed carrier signal without a carrier. (See the **Mixers, Modulators, and Demodulators** chapter for more information on these circuits.) That signal is then sent through a filter designed to pass just one (either one, by agreement with the receiving station) of the sidebands.[5] Depending on whether the sideband above or below the carrier frequency is selected, the signal is called *upper sideband* (*USB*) or *lower sideband* (*LSB*), respectively. The resulting SSB signal is amplified to the desired power level and we have an SSB transmitter.

While a transmitter of the type in Fig 13.20 with all processing at the desired transmit frequency will work, the configuration is not often used. Instead the carrier oscillator and sideband filter are often at an intermediate frequency that is heterodyned to the operating frequency as shown in **Fig 13.21**. The reason is that the sideband filter is a complex narrow-band filter and most manufacturers would rather not have to supply a new filter design every time a transmitter is ordered for a new frequency. Many SSB transmitters can operate on different bands as well, so this avoids the cost of additional mixers, oscillators and expensive filters.

Note that the block diagram of our SSB transmitter bears a striking resemblance to the diagram of a superheterodyne receiver as shown in the **Receivers** chapter, except that the signal path is reversed to begin with information and produce an RF signal. The same kind of image rejection requirements for intermediate frequency selection that were design constraints for the superhet receiver applies here as well.

SINGLE-SIDEBAND — THE PHASING METHOD

Most current transmitters use the method of SSB generation shown in Fig 13.20 and discussed in the previous section to generate the SSB signal. That is the *filter method*, but really occurs in two steps — first a balanced modulator is used to generate sidebands and eliminate the carrier, then a filter is used to eliminate the undesired sideband, and often to improve carrier suppression as well.

The *phasing method* of SSB generation is exactly the same as the image-rejecting mixer described in the **Receivers** chapter. This uses two balanced modulators and a phase-shift network for both the audio and RF carrier signals to produce the upper sideband signal

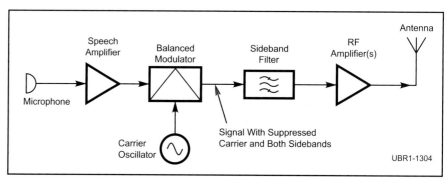

Fig 13.20 — Block diagram of a filter type single sideband suppressed carrier (SSB) transmitter.

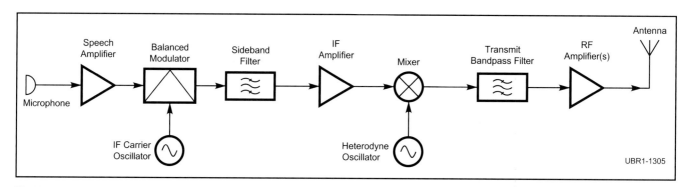

Fig 13.21 — Block diagram of a heterodyne filter-type SSB transmitter for multiple frequency operation.

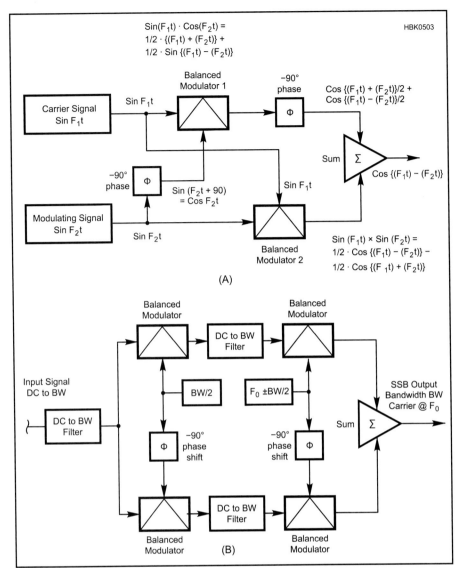

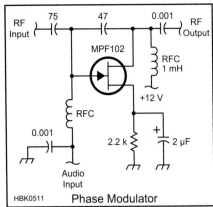

Fig 13.23 — Simple FET phase modulator circuit.

Fig 13.22 — Block diagrams of phasing type SSB transmitters for single frequency operation.

as shown in **Fig 13.22A**. By a shift in the sign of either of the phase-shift networks, the opposite sideband can be generated. This method trades a few phase-shift networks and an extra balanced modulator for the sharp sideband filter of the filter method. While it looks deceptively simple, a limitation is in the construction of a phase-shift network that will have a constant 90° phase shift over the whole audio range. Errors in phase shift result in less than full carrier and sideband suppression. Nonetheless, there have been some successful examples offered over the years.

SINGLE-SIDEBAND — THE WEAVER METHOD

Taking the phasing method one step further, the Weaver method solves the problem of requiring phase-shift networks that must be aligned across the entire audio range. Instead, the Weaver method, shown in Fig 13.22B, first mixes one copy of the message (shown with a bandwidth of dc to BW Hz) with an

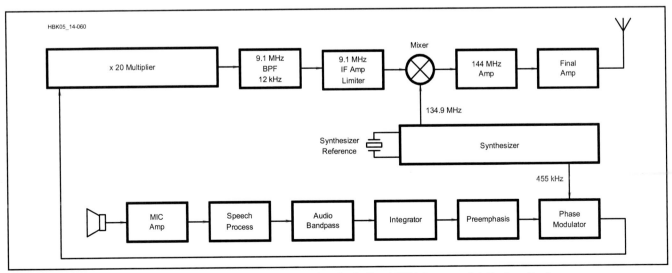

Fig 13.24 — Block diagram of a VHF/UHF NBFM transmitter using the indirect FM (phase modulation) method.

Transmitters 13.9

in-band signal at BW/2 Hz and another copy with a signal at BW/2 Hz that is phase-shifted by −90°. Instead of phase-shifting the message, only the signal at BW/2 Hz must be phase-shifted — a much simpler task!

The outputs of each balanced modulator is filtered, leaving only components from dc to BW/2. These signals are then input to a second pair of balanced modulators with a more conventional LO signal at the carrier frequency, f_0, offset by +BW/2 for USB and −BW/2 for LSB. The output of the balanced modulators is summed to produce the final SSB signal.

The Weaver method is difficult to implement in analog circuitry, but is well-suited to digital signal processing systems. As more transmitters incorporate DSP technology, the Weaver method will become common.

13.3.5 Angle-Modulated Transmitters

Transmitters using frequency modulation (FM) or phase modulation (PM) are generally grouped into the category of *angle modulation* since the resulting signals are often indistinguishable. An instantaneous change in either frequency or phase can create identical signals, even though the method of modulating the signal is somewhat different. To generate an FM signal, we need an oscillator whose frequency can be changed by the modulating signal.

We can make use of an oscillator whose frequency can be changed by a "tuning voltage". If we apply a voice signal to the TUNING VOLTAGE connection point, we will change the frequency with the amplitude and frequency of the applied modulating signal, resulting in an FM signal.

The phase of a signal can be varied by changing the values of an R-C phase-shift network. One way to accomplish phase modulation is to have an active element shift the phase and generate a PM signal. In **Fig 13.23**, the current through the field-effect transistor is varied with the applied modulating signal, varying the phase shift at the stage's output. Because the effective load on the stage is changed, the carrier is also amplitude-modulated and must be run through an FM receiver-type limiter in order to remove the amplitude variations.

FREQUENCY MODULATION TRANSMITTER DESIGN

Frequency modulation is widely used as the voice mode on VHF for repeater and other point-to-point communications. **Fig 13.24** shows the phase-modulation method, also known as *indirect FM*, as used in many FM transmitters. It is the most widely used approach to FM. Phase modulation is performed at a low frequency, say 455 kHz. Prior to the phase modulator, speech filtering and processing perform four functions:

1. Convert phase modulation to frequency modulation (see below).

2. Apply pre-emphasis (high-pass filtering) to the speech audio higher speech frequencies for improved signal-to-noise ratio after de-emphasis (low-pass filtering) of the received audio.

3. Perform speech processing to emphasize the weaker speech components.

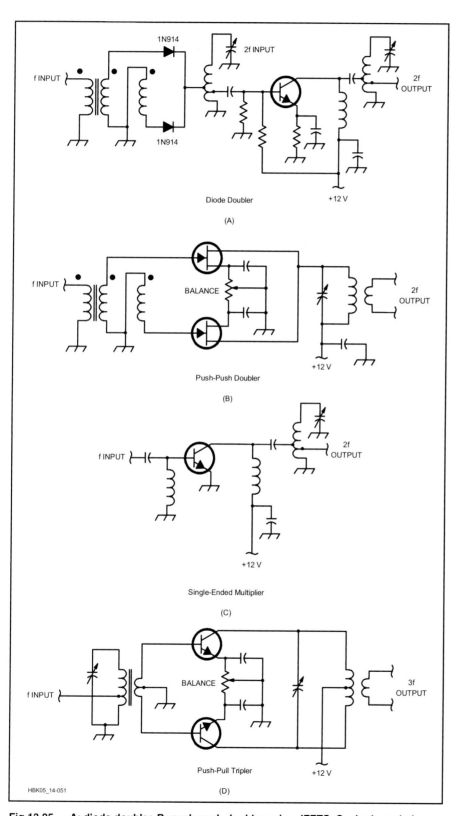

Fig 13.25 — A: diode doubler. B: push-push doubler using JFETS. C: single-ended multiplier using a BJT. D: push-pull tripler using BJTs.

4. Compensate for the microphone's frequency response and possibly also the operator's voice characteristics.

Multiplier stages then move the signal to some desired higher IF; also multiply the frequency deviation to the desired final value. If the FM deviation generated in the 455 kHz modulator is 250 Hz, the deviation at 9.1 MHz is 20 × 250, or 5 kHz.

Frequency Multipliers

Frequency multipliers are frequently used in FM transmitters as a way to increase the deviation along with the carrier frequency. They are composed of devices that exhibit high levels of harmonic distortion, usually an undesired output product. In this case the desired harmonic is selected and enhanced through filtering. The following examples show the way this can be done, both with amplifiers and with passive diode circuits.

A passive multiplier using diodes is shown in **Fig 13.25A**. The full-wave rectifier circuit can be recognized, except that the dc component is shorted to ground. If the fundamental frequency ac input is 1.0 V_{RMS} the second harmonic is 0.42 V_{RMS} or 8 dB below the input, including some small diode losses. This value is found by calculating the Fourier series coefficients for the full-wave-rectified sine wave, as shown in many textbooks.

Transistor and vacuum-tube frequency multipliers operate on the following principle: if a sine wave input causes the plate/collector/drain current to be distorted (not a sine wave) then harmonics of the input are generated. If an output resonant circuit is tuned to a harmonic, the output at the harmonic is emphasized and other frequencies are attenuated. For a particular harmonic the current pulse should be distorted in a way that maximizes that harmonic. For example, for a doubler the current pulse should look like a half-wave rectified sine wave (180° of conduction). A transistor with Class B bias would be a good choice. For a tripler, use 120° of conduction (Class C).

An FET, biased at a certain point, is very nearly a *square-law* device as described in the **Analog Basics** chapter. That is, the drain-current change is proportional to the square of the gate-voltage change. It is then an efficient frequency doubler that also de-emphasizes the fundamental.

A push-push doubler is shown in Fig 13.25B. The FETs are biased in the square-law region and the BALANCE potentiometer minimizes the fundamental frequency. Note that the gates are in push-pull and the drains are in parallel. This causes second harmonics to add in-phase at the output and fundamental components to cancel.

Fig 13.25C shows an example of a bipolar-transistor doubler. The efficiency of a doubler of this type is typically 50%, that of a tripler 33% and of a quadrupler 25%. Harmonics other than the one to which the output tank is tuned will appear in the output unless effective band-pass filtering is applied. The collector tap on L1 is placed at the point that offers the best compromise between power output and spectral purity.

A push-pull tripler is shown in Fig 13.25D. The input and output are both push-pull. The balance potentiometer minimizes even harmonics. Note that the transistors have no bias voltage in the base circuit; this places the transistors in Class C for efficient third-harmonic production. Choose an input drive level that maximizes harmonic output.

The step recovery diode (SRD) shown in **Fig 13.26A** is an excellent device for harmonic generation, especially at microwave frequencies. The basic idea of the SRD is as follows: When the diode is forward conducting, a charge is stored in the diode's diffusion capacitance; and if the diode is quickly reverse-biased, the stored charge is very suddenly released into an LC harmonic-tuned circuit. The circuit is also called a "comb generator" because of the large number of harmonics that are generated. (The spectral display looks like a comb.) A phase-locked loop (PLL) can then lock onto the desired harmonic.

A varactor diode can also be used as a multiplier. Fig 13.26B shows an example. This circuit depends on the fact that the capacitance of a varactor changes with the instantaneous value of the voltage across it, in this case the RF excitation voltage. This is a nonlinear process that generates harmonic currents through the diode. Power levels up to 25 W can be generated in this manner.

Following frequency multiplication, a second conversion to the final output frequency is performed. Prior to this final translation, IF band-pass filtering is performed in order to minimize adjacent channel interference that might be caused by excessive frequency deviation. This filter needs good phase linearity to assure that the FM sidebands maintain the correct phase relationships. If this is not done, an AM component is introduced to the signal, which can cause nonlinear distortion problems in the PA stages. The final frequency translation retains a constant value of FM deviation for any value of the output signal frequency.

The IF/RF amplifiers can be nonlinear Class C amplifiers because the signal in each amplifier contains, at any one instant, only a single value of instantaneous frequency and not multiple simultaneous frequencies whose relationship must be preserved as in SSB. These amplifiers are not sources of IMD, so they need not be "linear." The sidebands that appear in the output are a result only of the FM process. (The spectrum of an FM signal is described by Bessel functions.)

In phase modulation, the frequency deviation is directly proportional to the frequency of the audio signal. (In FM, the deviation is proportional to the audio signal's amplitude.) To make deviation independent of the audio frequency, an audio-frequency response that rolls off at 6 dB per octave is needed. An op-amp low-pass circuit in the audio amplifier accomplishes this function. This process converts phase modulation to frequency modulation.

In addition, audio speech processing helps to maintain a constant value of speech ampli-

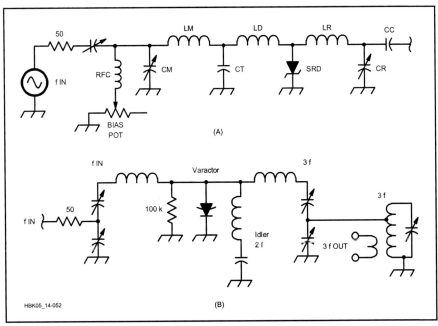

Fig 13.26 — Diode frequency multipliers. A: step-recovery diode multiplier. B: varactor diode multiplier.

tude, and therefore constant IF deviation, with respect to audio speech levels. Preemphasis of speech frequencies (a 6 dB per octave high-pass response from 300 to 3000 Hz) is commonly used to improve the signal-to-noise ratio at the receive end. Analysis shows that this is especially effective in FM systems when the corresponding de-emphasis (complementary low-pass response) is used at the receiver.[6] By increasing the amplitude of the higher audio frequencies before transmission and then reducing them in the receiver, high-frequency audio noise from the demodulation process is also reduced, resulting in a "flat" audio response with lower hiss and high-frequency noise.

An IF limiter stage may be used to ensure that any amplitude changes that are created during the modulation process are removed. The indirect-FM method allows complete frequency synthesis to be used in all the transmitter local oscillators (LOs), so that the channelization of the output frequency is very accurate. The IF and RF amplifier stages are operated in a highly efficient Class-C mode, which is helpful in portable equipment operating on small internal batteries.

FM is more tolerant of frequency misalignments between the transmitter and receiver than is SSB. In commercial SSB communication systems, this problem is solved by transmitting a *pilot carrier* with an amplitude 10 or 12 dB below the full PEP output level. The receiver is then phase-locked to this pilot carrier. The pilot carrier is also used for squelch and AGC purposes. A short-duration "memory" feature in the receiver bridges across brief pilot-carrier dropouts, caused by multipath nulls.

In a "direct FM" transmitter, a high-frequency (say, 9 MHz or so) crystal oscillator is frequency-modulated by varying the voltage on a varactor diode. The audio is pre-emphasized and processed ahead of the frequency modulator as for indirect-FM.

13.3.6 A Superhet SSB/CW Transmitter

The modern linear transmitting chain makes use of the concepts presented previously. We will now go through them in more detail. The same kind of mixing schemes, IF frequencies and IF filters that are used for superhet receivers can be, and very often are, used for a transmitter. In the following discussion, "in-band" refers to signal frequencies within the bandwidth of the desired signal. For example, for an upper sideband voice signal with a carrier frequency of 14.200 MHz, frequencies of approximately 14.2003 to 14.203 would be considered in-band. Out-of-band refers to frequencies outside this range, such as on adjacent channels.

For this example we will use the commonly-encountered dual-conversion scheme with the SSB generation at 455 kHz, a conversion to a 70 MHz IF and then a final conversion to the HF operating frequency. **Fig 13.27** is a block diagram of the system under consideration. Let's discuss the various elements in detail, starting at the microphone.

MICROPHONES

A microphone (mic) is a transducer that converts sound waves into electrical signals. For communications quality speech, its frequency response should be as flat as possible from around 200 to above 3500 Hz. Response peaks in the microphone can increase the peak to average ratio of speech, which then degrades (increases) the peak to average ratio of the transmitted signal. Some transmitters use "speech processing", which is essentially a specialized form of speech amplification either at audio or IF/RF. Since most microphones pick up a lot of background ambient noises, the output of the transmitter due to background noise pickup in the absence of speech may be as much as 20 dB greater than

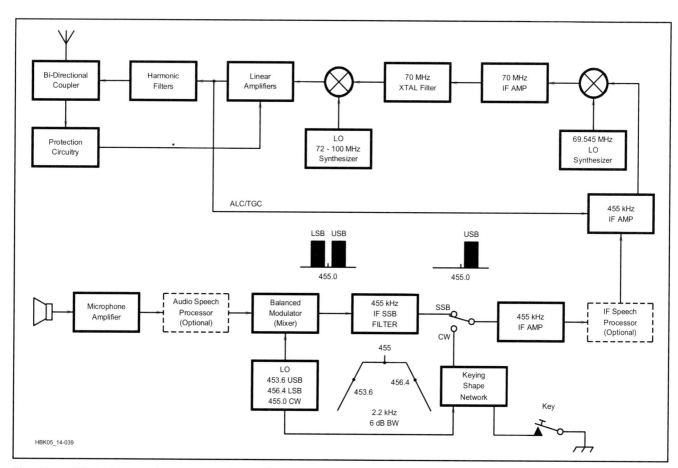

Fig 13.27 — Block diagram of an upconversion SSB/CW transmitter.

without speech processing. A *noise canceling* microphone is recommended to reduce this background pickup if there is much background noise. Microphone output levels vary, depending on the microphone type. Typical amateur microphones produce about 10 to 100 mV output levels.

Ceramic

Ceramic mics have high output impedances but low level outputs. They require a high-resistance load (usually about 50 kΩ) for flat frequency response and lose low-frequency response as this resistance is reduced (electrically, the mic "looks like" a small capacitor). These mics vary widely in quality, so a cheap mic is not a good bargain because of its effect on the transmitted power level and generally poor speech quality.

Dynamic

A dynamic mic resembles a small loudspeaker, with an impedance of about 600 Ω and an output of about 12 mV on voice peaks. Early dynamic mics (designed for vacuum tube transmitters) included a built-in transformer to transform the impedance to 100 kΩ suitable for high input impedance speech amplifiers. Currently available dynamic mics provide the output directly, although the transformers are available for connection outside the mic. Dynamic mics are widely used by amateurs.

Electret

Electret mics use a piece of special insulator material, such as Teflon, that contains a "trapped" polarization charge (Q) at its surfaces to create a capacitance (C). Sound waves modulate the capacitance of the material and cause a voltage change according to ΔV/V = –ΔC/C. For small changes in capacitance the change in voltage is almost linear. These mics have been greatly improved in recent years and are used in most cellular handsets.

The output level of the electret is fairly low and an integrated preamp is generally included in the mic cartridge. A voltage of about 4 V dc is required to power the preamp, and some commercial transceivers provide this voltage at the mic connector. The dc voltage must be blocked by a coupling capacitor if a dynamic mic element is to be used with a transceiver that supplies power to the microphone. The dynamic mic is unlikely to be damaged by the applied voltage, but the usual symptom is very low audio output with a muffled sound.

In recent years, micro electro-mechanical systems (MEMS) technology has been applied to microphones. The sound pressure is applied to a small transducer integrated on a silicon chip, which creates a mechanical motion in response to the force of the sound waves. This motion is converted into an electrical signal, usually by the varying capacitance of the transducer. Mass production of MEMS microphones is being driven by the cellular handset market.

MICROPHONE AMPLIFIERS

The balanced modulator and (or) the audio speech processor need a certain optimum level, which can be in the range of 0.3 to 0.6 V ac into perhaps 1 kΩ to 10 kΩ. Excess noise generated within the microphone amplifier should be minimized, especially if speech processing is used. The circuit in **Fig 13.28** uses a low-noise BiFET op amp. The 620 Ω resistor is selected for a low impedance microphone, and switched out of the circuit for high-impedance mics. The amplifier gain is set by the 100 kΩ potentiometer.

It is also a good idea to experiment with the low-and high-frequency responses of the mic amplifier to compensate for the frequency response of the mic and the voice of the operator.

SPEECH PROCESSING

The output of the speech amplifier can be applied directly to the balanced modulator. The resulting signal will be reproducible with the maximum fidelity and dynamic range available for the bandwidth provided. The communications efficiency of the system will depend on the characteristics of the particular voice used. The usual peak to average ratio is such that the average transmitted power is on the order of 5% of the peak power.

For communications use, it is generally beneficial to increase the average-to-peak ratio by distorting the speech waveform in a measured way. If one were to merely increase the amplifier gain, a higher average power would be obtained. The problem is that clipping, heavy distortion and likely spurious signals beyond legal limits would be generated. By carefully modifying the speech waveform before application to the balanced modulator, a synthetic waveform with a higher average-to-peak waveform can be generated that will retain most of the individual voice characteristics and avoid spurious signals. We will discuss two such techniques of *speech processing* below and another as we move into the RF circuitry.

Audio Speech Clipping

If the audio signal from the microphone amplifier is further amplified, say by as much as 12 dB, and then if the peaks are clipped (sometimes called *slicing* or *limiting*) by 12 dB by a speech clipper, the output peak value is the same as before the clipper, but the average value is increased considerably. The resulting signal contains harmonics and IMD but the speech intelligibility, especially in a white-noise background, is improved by 5 or 6 dB.

The clipped waveform frequently tends to have a square-wave appearance, especially on voice peaks. It is then band-pass filtered to remove frequencies below 300 and above 3000 Hz. The filtering of this signal can create a "repeaking" effect. That is, the peak value tends to increase noticeably above its clipped value.

An SSB generator responds poorly to a square-wave audio signal, creating significant peaks in the RF envelope. (This is described mathematically as the Hilbert Transform effect.) These peaks cause out-of-band splatter in transmitter's linear output power amplifier unless Automatic Level Control (ALC, to be discussed later) cuts back on the RF gain. The peaks increase the peak-to-average ratio and the ALC reduces the average SSB power output, thereby reducing some of the benefit of the speech processing. The square-wave

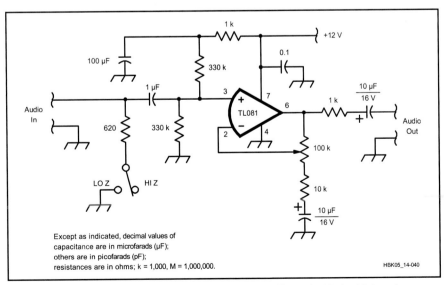

Fig 13.28 — Schematic diagram of a microphone amplifier suitable for high or low impedance microphones.

effect is also reduced by band-pass filtering (300 to 3000 Hz) the input to the clipper as well as the output.

Fig 13.29 is a circuit for a simple audio speech clipper. A CLIP LEVEL potentiometer before the clipper controls the amount of clipping and an OUTPUT LEVEL potentiometer controls the drive level to the balanced modulator. The correct adjustment of these potentiometers is done with a two-tone audio input or by talking into the microphone, rather than driving with a single tone, because single tones don't exhibit the repeaking effect.

Audio Speech Compression

Although it is desirable to keep the voice level as high as possible, it is difficult to maintain constant voice intensity when speaking into the microphone. To overcome this variable output level, it is possible to use an automatic gain control that follows the average variations in speech amplitude. This can be done by rectifying and filtering some of the audio output and applying the resultant dc to a control terminal in an early stage of the amplifier.

If an audio AGC circuit derives control voltage from the output signal, the system is a closed loop. If short attack time is necessary, the rectifier-filter bandwidth must be opened up to allow syllabic modulation of the control voltage. This allows some of the voice frequency signal to enter the control terminal, causing distortion and instability. Because the syllabic frequency and speech-tone frequencies have relatively small separation, the simpler feedback AGC systems compromise fidelity for fast response.

Problems with loop dynamics in audio AGC can be side-stepped by eliminating the loop and using a forward-acting system. The control voltage is derived from the input of the amplifier, rather than from the output. Eliminating the feedback loop allows unconditional stability, but the trade-off between response time and fidelity remains. Care must be taken to avoid excessive gain between the signal input and the control voltage output. Otherwise the transfer characteristic can reverse; that is, an increase in input level can cause a decrease in output. A simple forward-acting compressor is shown in Fig 13.30.

BALANCED MODULATORS

A balanced modulator is a mixer. A more complete discussion of balanced modulator design was provided in the **Mixers, Modulators, and Demodulators** chapter. Briefly, the IF frequency LO (455 kHz in the example of Fig 13.25) translates the audio frequencies up to a pair of IF frequencies, the LO plus the audio frequency and the LO minus the audio frequency. The balance from the LO port to the IF output causes the LO frequency to be suppressed by 30 to 40 dB. Adjustments are provided to improve the LO null.

The filter method of SSB generation uses an IF band-pass filter to pass one of the sidebands and block the other. In Fig 13.27 the filter is centered at 455.0 kHz. The LO is offset to 453.6 kHz or 456.4 kHz so that the upper sideband or the lower sideband (respectively) can pass through the filter. This creates a problem for the other LOs in the radio, because they must now be properly offset so that the final transmit output's carrier (suppressed) frequency coincides with the frequency readout on the front panel of the radio.

Various schemes have been used to create the necessary LO offsets. One method uses two crystals for the 69.545 MHz LO that can be selected. In synthesized radios the programming of the microprocessor controls the various LOs. Some synthesized radios use two IF filters at two different frequencies, one for USB and one for LSB, and a 455.0 kHz LO, as shown in Fig 13.27. These radios can be designed to transmit two independent sidebands (ISB) resulting in two separate channels in the spectrum space of the usual AM channel.

In times past, balanced modulators using diodes, balancing potentiometers and numerous components were used. These days it doesn't make sense to use this approach. ICs and packaged diode mixers do a much better job and are less expensive. The most widely known modulator IC, the MC1496, has been around for more than 25 years and is still one of the best and least expensive. **Fig 13.31** is a typical balanced modulator circuit using the MC1496.

The data sheets for balanced modulators and mixers specify the maximum level of audio for a given LO level. Higher audio levels create excessive IMD. The IF filter following the modulator removes higher-order IMD products that are outside its passband but the in-band IMD products should be at least 40 dB below each of two equal test tones. Speech clipping (AF or IF) can degrade this to 10 dB or so, but in the absence of speech processing the signal should be clean, in-band.

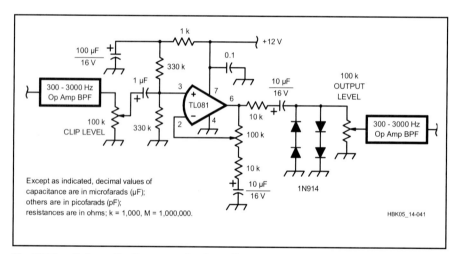

Fig 13.29 — Schematic diagram a simple audio speech clipper.

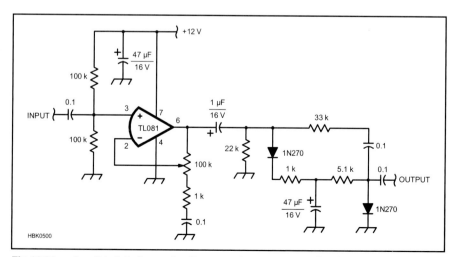

Fig 13.30 — A solid state forward acting speech compressor circuit.

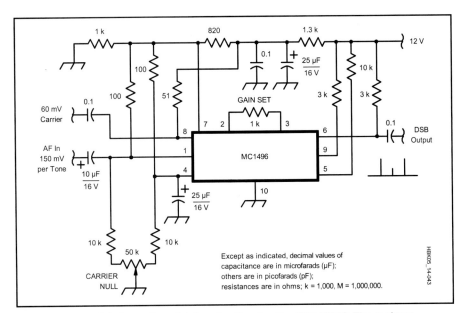

Fig 13.31 — An IC balanced modulator circuit using the MC1496 IC. The resistor between pins 2 and 3 sets the subsystem gain.

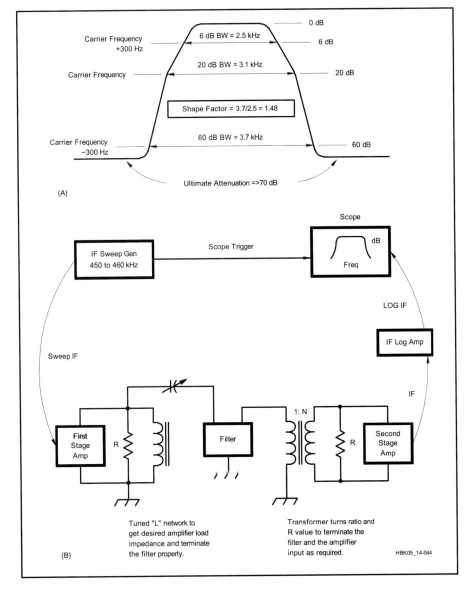

Fig 13.32 — At (A), desired response of a SSB IF filter. At (B), one method of terminating a mechanical filter that allows easy and accurate tuning adjustment and also a possible test setup for performing the adjustments.

IF FILTERS

The desired IF filter response is shown in **Fig 13.32A**. The reduction of the carrier frequency is augmented by the filter response. It is common to specify that the filter response be down 20 dB at the carrier frequency. Rejection of the opposite sideband should (hopefully) be 60 dB, starting at 300 Hz below the carrier frequency, which is the 300-Hz point on the opposite sideband. The ultimate attenuation should be at least 70 dB. This would represent a very good specification for a high quality transmitter. The filter passband should be as flat as possible (with passband ripple less than 1 dB or so).

Special filters, designated as USB or LSB, are designed with a steeper roll-off on the carrier frequency side, in order to improve rejection of the carrier and opposite sideband. Mechanical filters are available that do this. Crystal-ladder filters (see the **RF and AF Filters** chapter) are frequently called "single-sideband" filters because they also have this property. The steep skirt can be on the low side or the high side, depending on whether the crystals are across the signal path or in series with the signal path, respectively.

Filters require special attention to their terminations. The networks that interface the filter with surrounding circuits should be accurate and stable over temperature. They should be easy to adjust. One very good way to adjust them is to build a narrow-band sweep generator and look at the output IF envelope with a logarithmic amplifier, as indicated in Fig 13.32B. There are three goals:

• The driver stage must see the desired load impedance.

• The stage after the filter must see the desired source (generator) impedance.

• The filter must be properly terminated at both ends.

Lack of any of these conditions will result in loss of specified filter response. Fig 13.32B shows two typical approaches. This kind of setup is a very good way to make sure the filters and other circuitry are working properly.

Finally, overdriven filters (such as crystal or mechanical filters) can become nonlinear and generate distortion. Thus it is necessary to stay within the manufacturer's specifications. Magnetic core materials used in the tuning networks must be sufficiently linear at the signal levels encountered. They should be tested for IMD separately.

Transmitters 13.15

IF SPEECH CLIPPER

Audio clipper speech processors generate a considerable amount of in-band harmonics and IMD (involving different simultaneously occurring speech frequencies). The total distortion detracts somewhat from speech intelligibility. Other problems were mentioned in the earlier section on speech processing. IF clippers overcome most of these problems, especially the Hilbert Transform problem.[7]

Fig 13.33A is a schematic diagram of a 455 kHz IF clipper using high-frequency op-amps. 20 dB of gain precedes the diode clippers. A second amplifier establishes the desired output level. The clipping produces a wide band of IMD products close to the IF frequency. Harmonics of the IF frequency are easily rejected by subsequent selectivity. "Close-in" IMD distortion products are band-limited by the 2.5 kHz wide IF filter so that out-of-band splatter is eliminated. The in-band IMD products are at least 10 dB below the speech tones.

Fig 13.33B shows a block diagram of an adaptation of the above system to an audio in-audio out configuration that can be inserted into the mic input of any transmitter to provide the benefits of RF speech processing. These are sometimes offered as aftermarket accessories.

Fig 13.34 shows oscilloscope pictures of an IF clipped two-tone signal at various levels of clipping.[8] The level of clipping in a radio can be estimated by comparing with these photos. Listening tests verify that the IMD does not sound nearly as bad as harmonic distortion. In fact, processed speech sounds

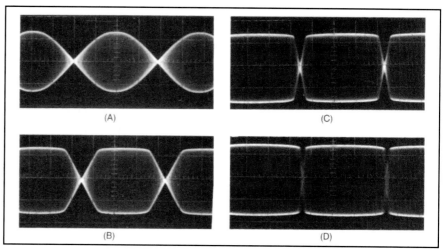

Fig 13.34 — Two-tone envelope patterns with various degrees of RF clipping. All envelope patterns are formed using tones of 600 and 1000 Hz. At A, clipping threshold; B, 5 dB of clipping; C, 10 dB of clipping; D, 15 dB of clipping (from).

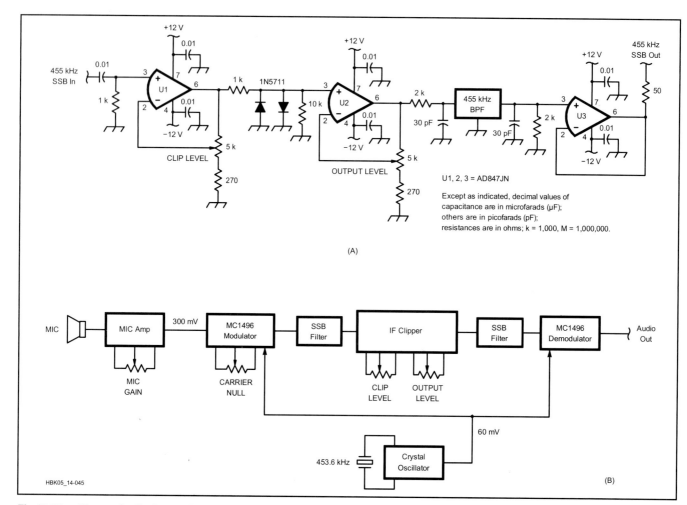

Fig 13.33 — IF speech clipping. At (A), schematic diagram of a 455 kHz IF clipper using high-frequency op amps. At (B) block diagram of an adaptation of the above system to an audio in-audio out configuration.

relatively clean and crisp. Tests also verify that speech intelligibility in a noise background is improved by 8 dB.

The repeaking effect from band-pass filtering the clipped IF signal occurs, and must be accounted for when adjusting the output level. A two-tone audio test signal or a speech signal should be used. The ALC circuitry (discussed later) will reduce the IF gain to prevent splattering in the power amplifiers. If the IF filter is of high quality and if subsequent amplifiers are clean, the transmitted signal is of very high quality and is very effective in noisy situations and often also in pile-ups.

The extra IF gain implies that the IF signal entering the clipper must be free of noise, hum and spurious products. The cleanup filter also helps reduce the carrier frequency, which is outside the passband.

An electrically identical approach to the IF clipper can be achieved at audio frequencies. If the audio signal is translated to, say 455 kHz, processed as described and translated back to audio, all the desirable effects of IF clipping are retained. This output then plugs into the transmitter's microphone jack. Fig 13.33B shows the basic method. The mic amplifier and the MC1496 circuits have been previously shown and the clipper circuit can be the same as in Fig 13.33A.

The interesting operating principle in all of these examples is that the characteristics of the IF clipped (or equivalent) speech signal do not change during frequency translation, even if translated down to audio and then back up to IF in a balanced modulator.

IF Linearity and Noise

Fig 13.27 indicates that after the last SSB filter, whether it is just after the SSB modulator or after the IF clipper, subsequent BPFs are considerably wider. For example, the 70 MHz crystal filter may be 15 to 30 kHz wide. This means that there is a "gray region" in the transmitter in which out-of-band IMD that is generated in the IF amplifiers and mixers can cause adjacent-channel interference.

A possible exception, not shown in Fig 13.27, is that there may be an intermediate IF in the 10 MHz region that also contains a narrow filter.

The implication is that special attention must be paid to the linearity of these circuits. It's the designer's job to make sure that distortion in this gray area is much less than distortion generated by the PA and also less than the phase noise generated by the final mixer. Recall also that the total IMD generated in the exciter stages is the resultant of several amplifier and mixer stages in cascade; therefore, each element in the chain must have at least 40 to 50 dB IMD quality. The various drive levels should be chosen to guarantee this. This requirement for multistage linearity is one of the main technical and cost burdens of the SSB mode.

Of interest also in the gray region are additive white, thermal and excess noises originating in the first IF amplifier after the SSB filter and highly magnified on their way to the output. This noise can be comparable to the phase noise level if the phase noise is low, as it is in a high-quality radio. Recall also that phase noise is at its worst on modulation peaks, but additive noise may be (and often is) present even when there is no modulation. This is a frequent problem in co-located transmitting and receiving environments. Many transmitter designs do not have the benefit of the narrow filter at 70 MHz, so the amplified noise can extend over a much wider frequency range.

13.3.7 CW Mode Operation

Radiotelegraph or CW operation can be easily obtained from the transmitter architecture design shown in Fig 13.27. For CW operation, a carrier is generated at the center of the SSB filter passband. There are two ways to make this carrier available. One way is to unbalance the balanced modulator so that the LO can pass through. Each kind of balanced modulator circuit has its own method of doing this. The approach chosen in Fig 13.27 is to go around the modulator and the SSB filter.

A shaping network controls the envelope of the IF signal to accomplish two things: control the shape of the Morse code character in a way that limits wideband spectrum emissions that can cause interference, and make the Morse code signal easy and pleasant to copy.

RF ENVELOPE SHAPING

On-off keying (CW) is a special kind of low-level amplitude modulation (a low signal-level stage is turned on and off). It is special because the sideband power is subtracted from the carrier power, and not provided by a separate "modulator" circuit, as in high-level AM. It creates a spectrum around the carrier frequency whose amplitude and bandwidth are influenced by the rates of signal amplitude rise and fall and by the curvature of the keyed waveform. Refer to the discussion of keying speed, rise and fall times, and bandwidth in the **Modulation** chapter and the earlier discussion in this chapter for some information about this issue.

Now look at **Fig 13.35**. The vertical axis is labeled Rise and Fall Times (ms). For a rise/fall time of 6 ms (between the 10% and 90%

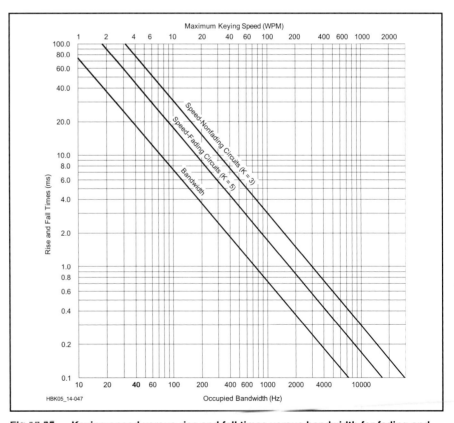

Fig 13.35 — Keying speed versus rise and fall times versus bandwidth for fading and nonfading communications circuits. For example, for transmitter output waveform rise and fall times of approximately 6 ms, draw a horizontal line from 6.0 ms on the rise and fall times scale to the bandwidth line. Then draw a vertical line to the occupied bandwidth scale at the bottom of the graph. In this case the bandwidth is about 130 Hz. Also extend the 6.0 ms horizontal line to the K = 3 line for a nonfading circuit. Finally draw a vertical line from the K = 3 line to the WPM axis. The 6 ms rise and fall time should be suitable for keying speeds up to about 50 WPM in this example.

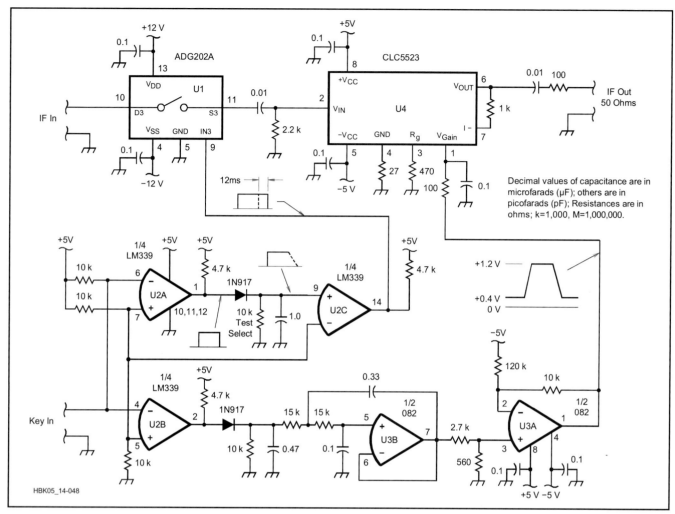

Fig 13.36 — This schematic diagram shows a CW waveshaping and keying circuit suitable for use with an SSB/CW transmitter such as is shown in Fig 13.27.

values) go horizontally to the line marked Bandwidth. A −20 dB bandwidth of roughly 120 Hz is indicated on the lower horizontal axis. Continuing to the K = 5 and K = 3 lines, the upper horizontal axis suggests code speeds of 30 WPM and 50 WPM respectively.

These code speeds can be accommodated by the rise and fall times displayed on the vertical axis. For code speeds greater than these the Morse code characters become "soft" sounding and difficult to copy, especially under less-than-ideal propagation conditions.

For a narrow spectrum and freedom from adjacent channel interference, a further requirement is that the spectrum must fall off very rapidly beyond the −20 dB bandwidth indicated in Fig 13.35. A sensitive narrowband CW receiver that is tuned to an adjacent channel that is only 1 or 2 kHz away can detect keying sidebands that are 80 to 100 dB below the key-down level of a strong CW signal. An additional consideration is that during key-up a residual signal, called "backwave" should not be noticeable in a nearby receiver.

A backwave level at least 90 dB below the key-down carrier is a desirable goal.

Fig 13.36 is the schematic of one waveshaping circuit that has been used successfully. A Sallen-Key third-order op amp low-pass filter (0.1 dB Chebyshev response) shapes the keying waveform, produces the rate of rise and fall and also softens the leading and trailing corners just the right amount. The key closure activates the CMOS switch, U1, which turns on the 455-kHz IF signal. At the key-up time, the input to the waveshaping filter is turned off, but the IF signal switch remains closed for an additional 12 ms.

The keying waveform is applied to the gain control pin of a CLC5523 amplifier IC. This device, like nearly all gain-control amplifiers, has a *logarithmic* control of gain; therefore some experimental "tweaking" of the capacitor values was used to get the result shown in **Fig 13.37A**. The top trace shows the on/off operation of the IF switch, U1. The signal is turned on shortly before the rise of the keying pulse begins and remains on for about 12 ms after the keying pulse is turned off, so that the waveform falls smoothly to a very low value. The result is an excellent spectrum and an almost complete absence of backwave. Compare this to the factory transmitter waveshapes shown in Fig 13.13 and 13.15. The bottom trace shows the resulting keyed RF output waveshape. It has an excellent spectrum, as verified by critical listening tests. The thumps and clicks that are found in some CW transmitters are virtually absent. The rise and fall intervals have a waveshape that is approximately a cosine. Spread-spectrum frequency-hop waveforms have used this approach to minimize wideband interference.

Fig 13.37B is an accurate *SPICE* simulation of the wave shaping circuit output before the signal is processed by the CLC5523 amplifier. To assist in adjusting the circuit, create a steady stream of 40 ms dots that can be seen on an RF oscilloscope that is looking at the final PA output envelope. It is important to make sure that the excellent waveshape is not degraded on its way to the transmitter output. Single-sideband linear power amplifiers are

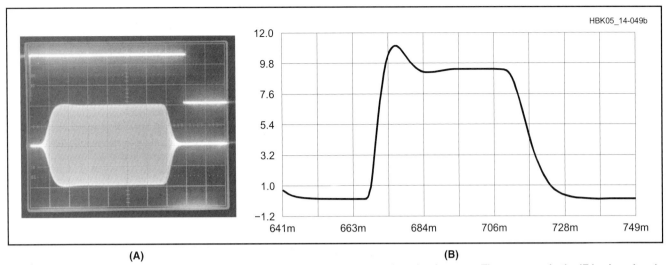

Fig 13.37 — At (A) is the oscilloscope display of the CW waveshaping and keying circuit output. The top trace is the IF keying signal applied to S1 of Fig 13.36. The bottom trace is the transmitter output RF spectrum. At (B) is a *SPICE* simulation of the waveshaping network. When this signal is applied to the logarithmic control characteristic of the CLC5523 amplifier, the RF envelope is modified slightly to the form shown in A.

well suited for a CW transmitter, but they must stay within their linear range, and the backwave problem must be resolved.

When evaluating the spectrum of an incoming CW signal during on-the-air operations, a poor receiver design can contribute problems caused by its vulnerability to a strong but clean adjacent channel signal. Clicks, thumps, front end overload, reciprocal mixing, etc can be created in the receiver. It is important to put the blame where it really belongs.

For additional information see "A 455 kHz IF Signal Processor for SSB/CW," William Sabin, WØIYH, *QEX*, Mar/Apr 2002, pp 11-16.

13.3.8 Wideband Noise

In the block diagram of Fig 13.27 the last mixer and the amplifiers after it are wideband circuits that are limited only by the harmonic filters and by any selectivity that may be in the antenna system. Wide-band phase noise transferred onto the transmitted modulation by the last LO can extend over a wide frequency range, therefore LO (almost always a synthesizer of some kind) cleanliness is always a matter of great concern.[9]

The amplifiers after this mixer are also sources of wide-band "white" or additive noise. This noise can be transmitted even during times when there is no modulation, and it can be a source of local interference. To reduce this noise, use a high-level mixer with as much signal output as possible, and make the noise figure of the first amplifier stage after the mixer as low as possible.

Transmitters that are used in close proximity to receivers, such as on ships, are always designed to control wideband emissions of both additive noise and phase noise, referred to as "composite" noise.

TRANSMIT MIXER SPURIOUS SIGNALS

The last IF and the last mixer LO in Fig 13.27 are selected so that, as much as possible, harmonic IMD products are far enough away from the operating frequency that they fall outside the passband of the low-pass filters and are highly attenuated. This is difficult to accomplish over the transmitter's entire frequency range. It helps to use a high-level mixer and a low enough signal level to minimize those products that are unavoidable. Low-order crossovers that cannot be sufficiently reduced are unacceptable, however; the designer must go back to the drawing board.

13.3.9 Automatic Level Control (ALC)

The purpose of ALC is to prevent the various stages in the transmitter from being overdriven. Over-drive can generate too much out-of-band distortion or cause excessive power dissipation, either in the amplifiers or in the power supply. ALC does this by sampling the peak amplitude of the modulation (the envelope variations) of the output signal and then developing a dc gain-control voltage that is applied to an early amplifier stage, as suggested in Fig 13.27.

ALC is usually derived from the last stage in a transmitter. This ensures that this last stage will be protected from overload. However, other stages prior to the last stage may not be as well protected; they may generate excessive distortion. It is possible to derive a composite ALC from more than one stage in a way that would prevent this problem. But designers usually prefer to design earlier stages conservatively enough so that, given a temperature range and component tolerances, the last stage can be the one source of ALC. The gain control is applied to an early stage so that all stages are aided by the gain reduction.

SPEECH PROCESSING WITH ALC

A fast response to the leading edge of the modulation is needed to prevent a transient overload. After a strong peak, the control voltage is "remembered" for some time as the voltage in a capacitor. This voltage then decays partially through a resistor between peaks. An effective practice provides two capacitors and two time constants. One capacitor decays quickly with a time constant of, say 100 ms, the other with a time constant of several seconds. With this arrangement a small amount of speech processing, about 1 or 2 dB, can be obtained. (Explanation: The dB of improvement mentioned has to do with the improvement in speech intelligibility in a random noise background. This improvement is equivalent to what could be achieved if the transmit power were increased that same number of dB).

The gain rises a little between peaks so that weaker speech components are enhanced. But immediately after a peak it takes a while for the enhancement to take place, so weak components right after a strong peak are not enhanced very much. **Fig 13.38A** shows a complete ALC circuit that performs speech processing.

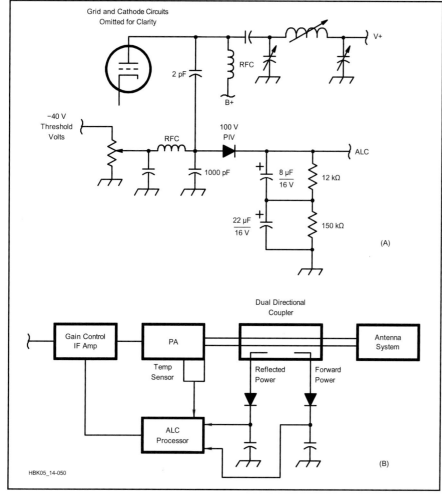

Fig 13.38 — At (A), an ALC circuit with speech processing capability. At (B), protection method for a solid-state transmitter.

While it is a bit more involved to generate an SSB signal than a CW signal, we greatly simplify the task if all the necessary circuitry is on a single PC board exciter module. Once we have a high-quality low-level SSB signal, a 5 or 500 W SSB transmitter is as easy to build as a 5 or 500 W CW transmitter. Simple transmitters are delightful, but relaxed standards are not. The MicroT2 is designed to be clean, stable and reliable, exceed FCC Part 97 requirements, and sound good, too. A more complete description of the circuitry in this transmitter can be found in *Experimental Methods in RF Design*[10] (EMRFD) of which the project's designer was a co-author.

THE CASE FOR CRYSTAL CONTROL

A jumper on the PC board makes it easy to use an external VFO for frequency control if you wish. For some applications the narrow tuning range offered by the on-board VXO is sufficient, and the secure knowledge that the transmitter is actually on frequency and stable is a real virtue. **Fig 13.39** shows just one example — a simple battery-operated transmitter intended to keep in touch with the folks back home on an HF net frequency. The crystal oscillator in the transmitter functions as a calibrator for the receiver, so that the receiver may be tuned around the band and then easily reset to the net frequency. The current drain during receive is only 60 mA, because the entire transmitter is turned off except when actually transmitting.

ALC IN SOLID-STATE POWER AMPLIFIERS

Fig 13.38B shows how a dual directional coupler can be used to provide ALC for a solid-state power amplifier (PA). The basic idea is to protect the PA transistors from excessive SWR and dissipation by monitoring both the forward power and the reflected power.

TRANSMIT GAIN CONTROL (TGC)

This is a widely used feature in commercial and military equipment. A calibrated "tune-up" test carrier of a certain known level is applied to the transmitter. The output carrier level is sampled, using a diode detector. The resulting dc voltage is used to set the gain of a low-level stage. This control voltage is digitized and stored in memory so that it is semi-permanent. A new voltage may be generated and stored after each frequency change, or the stored value may be used. A test signal is also used to do automatic antenna tuning. A dummy load is used to set the level and a low-level signal (a few mW) is used for the antenna tune-up.

Project: The MicroT2 — A Compact Single-Band SSB Transmitter

As an example of an SSB transmitter including many aspects of design covered heretofore, we present the MicroT2, a simple SSB transmitter that generates a high-quality USB or LSB signal on any single band from 1.8 MHz to 50 MHz. Rick Campbell, KK7B, developed the MicroT2 as a companion to the MicroR2 receiver project described in the **Receivers** chapter.

Fig 13.39 — This 40 meter version of the MicroT2 uses the on-board VXO. The black box on top is the 0.5 W amplifier.

Table 13.1
MicroT2 Parts List

Parts for 40 meter version. See Note A.

C1, C7, C32, C34 — 10 µF, 16 V electrolytic capacitor.
C2-C4, C6, C17, C18, C22, C23, C33, C35-C39 — 0.1 µF, 5% polyester capacitor.
C5 — 0.22 µF, 5% polyester capacitor.
C8-C13 — 0.01 µF polyester capacitor, matched to 1%.
C14 — 100 µF, 16 V electrolytic capacitor.
C15, C16 — 33 µF, 16 V electrolytic capacitor.
C19 — 470 pF NP0 ceramic capacitor.
C20 — 56 pF NP0 ceramic capacitor.
C21 — 3-36 pF poly trimmer capacitor.
C24, C30, C31 — 390 pF, NP0 ceramic capacitor.
C25 — 39 pF, NP0 ceramic on back of board capacitor.
C26-C29 — 0.82 µF, 5% polyester capacitor.
C40, C41 — 220 pF, NP0 ceramic capacitor.
C42 — 36 pF, VXO variable off board capacitor; see Note C.
C43 — 0.01 µF ceramic capacitor.
D1 — 4.7 V Zener.
D2 — 1N4148 diode.
L1, L2 — 33 mH inductor.
L3 — 40 turns #30 enameled wire on T37-2 toroid core; see Note D.
L4, L5 — 3.9 mH inductor.
L6 — 22 turns #28 enameled wire on T30-2 toroid core; see Note D.
Mx1, Mx2 — Mini-Circuits TUF-3 diode ring mixer.
Q1-Q5 — 2N3904 transistor.
Q6 — J310 field effect transistor.
RFC1, RFC2 — 15 µH molded RF choke.
R1, R2, R8, R9, R28-R31, R34, R35, R41, R42, R49, R50 — 10 kΩ resistor.
R3 — 4.7 kΩ resistor.
R4 — 33 kΩ resistor.
R5, R6 — 470 Ω resistor.
R7 — 2.7 kΩ; audio gain select resistor; see Note E.
R10 — 1.5 kΩ, 1% resistor.
R11, R13-R15, R17, R19-R21, R23, R25-R27 — 10.0 kΩ, 1% resistor.
R12 — 5.23 kΩ, 1% resistor.
R16 — 11.3 kΩ, 1% resistor.
R18 — 23.2 kΩ, 1% resistor.
R22 — 51.1 kΩ, 1% resistor.
R24 — 178 kΩ, 1% resistor.
R32 — 5.1 kΩ resistor.
R33, R36 — 10 kΩ trimpot resistor.
R38, R40, R53 — 680 Ω resistor.
R37, R39, R66 — 10 Ω resistor.
R43, R44, R46, R48, R54, R62, R65, R67 — 51 Ω resistor.
R45 — 100 kΩ resistor.
R47 — 100 Ω resistor.
R51, R60 — 33 Ω resistor.
R52 — 22 Ω resistor.
R55 — 3.9 kΩ resistor.
R56, R58, R63 — 270 Ω resistor.
R57 — 18 Ω resistor.
R59, R61 — 150 Ω resistor.
R64 — 1 kΩ resistor.
T1 — 17 turns two colors #28 enameled wire bifilar wound on T37-2 toroid core; see Note D.
T2 — 5 turns #28 enameled wire trifilar wound on FT23-43 toroid core; see Note D.
T3 — 7 turns #28 enameled wire bifilar wound on FT23-43 toroid core; see Note D.
U1-U5 — NE5532 or equivalent dual low-noise high-output op-amp.
U6 — LM7806 or equivalent 6 V three terminal regulator.

Note A: C19, 20, 24, 25, 30, 31, 40, 41; L3, L6 and T1 values are for operation in the 40 meter band.
Note B: The total reactance of the parallel combination of C24 and C25 plus the capacitance between the windings of T1 is $-j50$ Ω at the center of the tuning range. Placing most of the capacitance at one end is a different but equivalent arrangement of the quadrature hybrid we often use with equal capacitors. C25 is only needed if there is no standard value for C24 within a few percent of the required value. C25 is tack soldered to the pads provided on the back of the PC board, and may be a surface mount component if desired.
Note C: Capacitor C42 is the VXO tuning capacitor for the exciter.
Note D: L3, L6, and T1, T2 and T3 are listed as number of turns on the specified core rather than a specific inductance. For those who wish to study the design with a calculator, simulator and inductance meter, L3 should be about $+j300$ Ω at 7.2 MHz, L6 should be $+j100$ Ω at 7.2 MHz and each winding of T1 should be $j50$ Ω at 7.2 MHz. T2 and T3 are noncritical broadband transformers with about 40 µH total inductance using the specified number of turns.
Note E: Resistor R7 sets the audio signal processor gain. If the gain is set too high, intermodulation distortion products generated in the diode ring modulators will be objectionable. Select R7 for a peak exciter output level no greater than 0 dBm.

EXCITER BLOCK DIAGRAM

Fig 13.40 is the block diagram of the circuitry on the PC board and **Fig 13.41** is a detailed schematic. It is a complete VXO SSB exciter on a 2.5 × 3.8 inch circuit board with 1 mW (0 dBm) peak output. The exciter uses the phasing method of SSB generation, which makes it easy to operate on different frequencies. In a phasing SSB exciter, two identical signals with a 90° phase difference are generated and then combined so that one sideband adds and the other subtracts. The signal quality from this exciter is not merely adequate for the HF amateur bands — it is exceptional.

RF Circuitry

The frequency generator consists of a VXO, buffer amplifier and quadrature hybrid. It has a lot of parts: three transistors; a voltage regulator IC, a Zener diode, four toroids and many resistors and capacitors. There are no adjustments. You just build it and it works. The frequency stability is better than that of most commercial radios, even when portable.

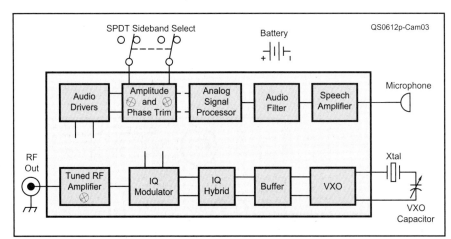

Fig 13.40 — Block diagram of the circuitry on the PC board.

There are simpler VXO circuits, but this one is excellent. It provides 0 dBm sine wave output, draws 4 mA, and has virtually no start-up drift — about 2 Hz at 7 MHz. (It makes a lovely keyed CW generator too.) The three-resistor attenuator between the output of the VXO and input of the buffer amplifier is a convenient place to insert a frequency multiplier or externally generated VFO. Just leave off the top resistor to break the path. Note that the crystal and variable capacitor are mounted off the PC board. Frequency stability is determined by the temperature of the crystal. Slip a foam packing bead over the crystal and support it by its leads between the PC board and VXO capacitor. A front-panel-mounted crystal socket looks very classy, but noise picked up by the crystal body gets into the oscillator, and the signal radiated by the crystal sounds like one with poor carrier suppression. If you use a crystal socket, put it inside the case or behind a shield door.

The buffer amplifier provides a 50 Ω source of broadband drive to the quadrature hybrid and isolates the frequency generator from impedance variations at the mixer local oscillator (LO) ports. The simple arrangement used in the MicroR2 receiver doesn't work here, because high-level audio into the diode ring mixers modulates not just the signal at the RF port, but the impedance at the LO and IF ports as well. Experiments confirm that a directly connected VFO experiences severe frequency pulling on voice peaks. The gain of the buffer is set to provide +7 dBm drive to each mixer with an input level of 0 dBm. It draws some current and the transistor gets a little warm — but only while transmitting.

The quadrature hybrid is a venerable circuit first described by Reed Fisher in *QST*.[11] It is the lumped element equivalent of a pair of tightly coupled quarter-wavelength transmission lines. The total capacitance does not need to be symmetrical between the two ends of the inductor, as is commonly shown. It may all be at one end or divided unequally. When driven from a 50 Ω source and terminated in a 51 Ω resistor and two mixer LO ports, the 90° phase difference is nearly perfect across a wide band and the amplitude difference between the two outputs is within 0.5 dB across a 10% bandwidth — more than enough bandwidth to cover the usual SSB portion of any amateur band.

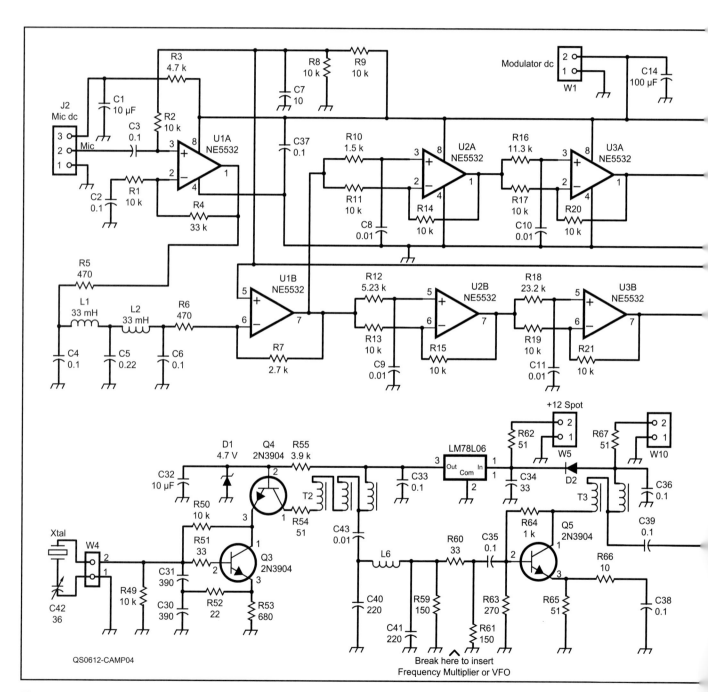

Fig 13.41 — Schematic diagram of the 40 meter version of the MicroT2. See Table 13.1 for parts values. A kit version including parts and PC board is available from Kanga US (www.kangaus.com).

A pair of Mini-Circuits TUF-3 mixers serves as the I and Q balanced modulators. These provide good carrier suppression without adjustment, and reasonable output at a modest distortion level. The carrier suppression may be improved by soldering the metal cans of the TUF-3 mixers directly to the PC board ground. IM products in the opposite sideband are more than 30 dB down at the exciter output. The aggressive low-pass filtering right at the mixer IF ports prevents wideband noise and harmonic distortion in the audio stages from contributing to the I and Q modulation. Energy outside the modulation bandwidth is then just intermodulation distortion in the mixers and linear amplifiers. Sideband suppression will be more than 40 dB if the four 0.82 µF capacitors and two 3.9 mH inductors are matched to within 1%. The resultant carrier suppression is greater than 50 dB on any of the HF bands.

The exciter's output RF amplifier uses a common-gate JFET. This stage provides a broadband resistive termination to the mixer RF port IQ summing junction to isolate it from variations in impedance at the exciter output. The RF amplifier has relatively low gain, good harmonic suppression and a very clean 0 dBm SSB signal at the output. This is an appropriate level to drive a linear amplifier, balanced mixer or transverter. It is also a low enough level that it is easy to adjust the exciter with simple equipment. The exciter output signal meets FCC regulations for direct connection to an antenna for flea power experiments.

The transmitter shown here adds the simple two transistor circuit in **Fig 13.42** for 0.5 W PEP output. The seventh order Chebyshev low-pass filter on the output is noncritical and assures a clean signal that easily meets FCC regulations.

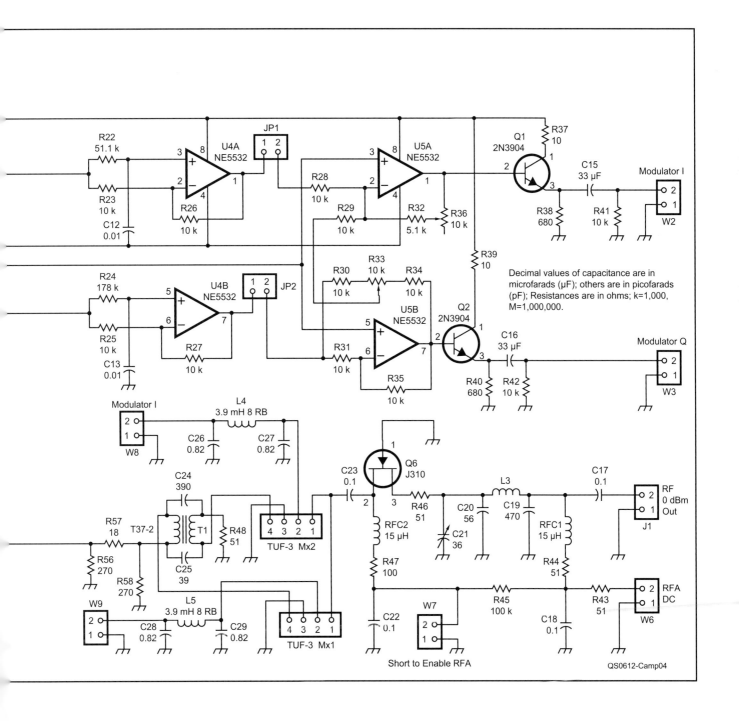

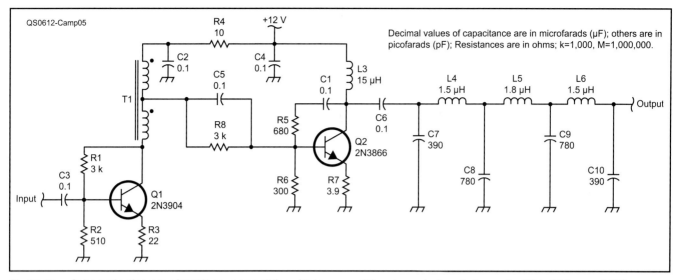

Fig 13.42 — Schematic diagram of 0.5 W power amplifier with low-pass filter shown in the photographs. T1 is 10 turns of #28 enameled wire bifilar wound on an FT37-43 toroid core.

Audio Section

The top half of the PC board contains all of the audio circuitry. There are no PC board traces connecting the two halves, and it is okay to cut the board into separate functional blocks for packaging flexibility, or to use the audio and RF portions of the circuitry in other projects.

The speech amplifier drives a passive low-pass filter using two series inductors and three shunt capacitors. The combination of this low-pass filter and the mixer IF port filters limits speech frequencies to just over 3 kHz for natural sounding speech and good spectral purity. There is no ripple in the audio passband.

The position of the sideband select switch in the signal path allows switching without readjusting the amplitude and phase trimmers. For most applications, one sideband will be used exclusively, and the sideband switch may be replaced by a pair of jumpers on the PC board. If that results in the wrong sideband, reverse the connections between the audio driver transistors and mixers.

The I and Q Class A audio drivers are emitter followers directly driven by the amplitude and phase trim op-amps. Emitter followers were used successfully in the original T2 and work well.[12] An emitter follower can only source current, so it must be biased with more than the negative peak current required by the mixer IF ports.

Q1 and Q2 each have a quiescent current of more than 10 mA. For a receive application, a different approach would save operating current. Since the exciter draws only a fraction of the total transmitter current in most applications and is turned completely off during receive, use of clean Class A emitter followers to drive the mixer IF ports is a good trade-off. Another option is connecting the feedback resistor to the emitter rather than the base of each follower. A circuit simulator shows a tiny bit less distortion — but also some potential high frequency instability with that connection. This design sticks with the proven, conservative approach, and drops the distortion still further by burning a little more current in each of the transistors.

ADJUSTING THE SSB EXCITER

There are only three adjustments on the SSB generator PC board: RF AMPLIFIER TUNING, AMPLITUDE TRIM and PHASE TRIM. Each adjustment may be set once and then left alone. Adjust the SSB exciter by ear using a wideband audio noise source — such as a spare SSB receiver tuned to noise. Plug a cable into the "noise receiver" headphone jack and connect to the exciter microphone input, starting with the volume all the way down. With about 60 dB of attenuation on the RF output of the exciter, feed it directly into another receiver with selectable sidebands. Turn off the receiver AGC, if possible, and reduce the RF GAIN so that the peak exciter signal does not move the S-meter. Tune the receiver to zero beat on carrier leakage and then slowly turn up the volume on the noise source until the exciter noise output is a strong signal in the receiver.[13]

Switch back and forth between upper and lower sideband on the receiver and confirm that one sideband is much stronger than the other.[14] Switch to the desired sideband (if it is the wrong one, reverse the connections between the I and Q modulators and audio drivers) and peak the RF amplifier capacitor. Then switch to the opposite sideband and adjust the amplitude and phase trimpots for zero noise output. Alternate between the two trimpots, as these two adjustments become increasingly critical as each one approaches zero. Once adjusted, the energy in the opposite sideband will be more than 40 dB below the energy in the desired sideband. At that level, intermodulation products in the opposite sideband will dominate, and all you will hear are the familiar unintelligible pops and clicks on voice peaks common to any clean SSB signal driving a practical linear amplifier.

OTHER MODES WITH THE MICROT2

This exciter is very similar to the one in a 1958 Central Electronics 20A SSB exciter. If you look at the front panel of the 20A you see a mode switch marked USB, LSB, DSBsc, AM, NBPM, CW. The back of the switch has just a bunch of wires and resistors used to unbalance a modulator, turn off the I or Q audio channel, etc. We could get all of those modes out of this exciter, too, by inserting a switch in the wires connecting the I and Q modulator outputs to the I and Q mixer IF inputs.

Connecting either the I or Q signal path (but not both) will result in DSB. Connecting only the I (or Q) path and inserting a little carrier by connecting a 4.7 kΩ resistor from that side to the +12 V supply will generate AM. Connecting only the I path and inserting a 4.7 kΩ resistor from the Q mixer IF port to +12 V will generate narrowband phase modulation (NBPM). Disconnecting both the I and Q paths, or simply turning off the power to the AF section, and connecting 4.7 kΩ resistors from either or both mixer IF ports will generate CW. For CW, key both the voltage on the mixer IF ports and the RF amplifier enable line. **Fig 13.43** is a good circuit to accomplish this.

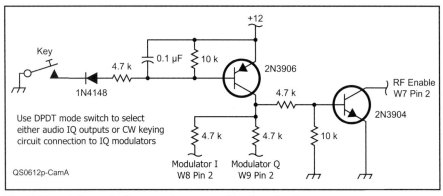

Fig 13.43 — Schematic diagram of suggested keying circuit.

SO, HOW DOES IT SOUND?

Try evaluating SSB exciters this way: First, translate the frequency to some very quiet spot in the HF spectrum using a crystal oscillator and mixer. Then put enough attenuators before and after the mixer to obtain a peak SSB output level of around –60 dBm — weak enough to not overload the receiver but strong enough to hear off-channel garbage another 60 or 70 dB below the peak transmitter output. Connect a CD player with some good acoustic folk music — lots of voice, guitar transients, perhaps a mandolin but no drums — to the exciter microphone input. Then connect the frequency converted exciter output directly into the input of a good receiver and tune it in. Don't connect it to an antenna — amateurs are not permitted to transmit music!

Any receiver with selectable sidebands and a manual RF gain control will do. Turn the receiver AGC off and manually reduce the gain so the receiver noise floor is well below the peak signal level. If your receiver is lacking in audio fidelity, run the receiver "line out" into a stereo amplifier. Play the CD through the SSB exciter, through the attenuators and frequency translator, into the receiver, and back into the stereo and out the speakers. It's an acid test, and this exciter sounds pretty good — better than most AM broadcast stations, and even some badly adjusted FM stations. Friends who hear you on the air will say, "Wow, it sounds exactly like you!"

Project: The MkII — An Updated Universal QRP Transmitter

A frequently duplicated project in the now out-of-print book *Solid State Design for the Radio Amateur*[15] was a universal QRP transmitter. This was a simple two-stage, crystal-controlled, single-band circuit with an output of about 1.5 W. The no frills design used manual transmit-receive (TR) switching. It operated on a single frequency with no provision for frequency shift. The simplicity prompted many builders to pick this QRP rig as a first solid state project.

The design simplicity compromised performance. A keyed crystal controlled oscillator often produces chirps, clicks or even delayed starting. The single pi-section output network allowed too much harmonic energy to reach the antenna, and the relatively low output of 1.5 W may seem inadequate to a first time builder.

A THREE-STAGE TRANSMITTER

Wes Hayward, W7ZOI, updated the design to the MKII (**Fig 13.44**). The circuit, shown in **Fig 13.45**, develops an output of 4 W on any single band within the HF spectrum, if provided with 12 V dc. Q1 is a crystal controlled oscillator that functions with either fundamental or overtone mode crystals. It operates at relatively low power to minimize stress to some of the miniature crystals now available. The stage has a measured output at point X of +12 dBm (16 mW) on all bands. This is applied to drive control R17 to set final transmitter output.

A three stage design provides an easy way to obtain very clean keying. Shaped dc is applied to driver Q2 through a keying switch and integrator, Q4.[16] A secondary keying switch, Q5, applies dc to the oscillator Q1. This is a time-sequence scheme in which the oscillator remains on for a short period (about 100 ms) after the key is released. The keyed waveform is shown in **Fig 13.46**.

The semiconductor basis for this transmitter is an inexpensive Panasonic 2SC5739. This part, with typical F_T of 180 MHz, is specified for switching applications, making it ideal as a class C amplifier. The transistor is conveniently housed in a plastic TO-220 package with no exposed metal. This allows it to be bolted to a heat sink with none of the insulating hardware required with many power transistors. A 2 × 4 inch scrap of circuit board served as both a heat sink and as a ground plane for the circuitry.

Another 2SC5739 serves as the driver, Q2. This circuit is a feedback amplifier with RF feedback resistors that double to bias the transistor.[17] Driver output up to 300 mW is available at point Y. Ferrite transformer T2 moves the 200 Ω output impedance seen looking into the Q2 collector to 50 Ω. The maximum output power of this stage can be changed with different R20 values. Higher stage current, obtained with lower R20 values, is needed on the higher bands. The 2SC5739 needs only to be bolted to the circuit board for heat sinking.

The Q3 power amplifier input is matched with transformer T3. The nominal 50 Ω of the driver is transformed to 12 Ω by T3.

The original design started with a simple L network output circuit at the Q3 collector followed by a third-order elliptic low-pass section to enhance harmonic suppression.[18] C5 is a moderately high reactance capacitor at the collector to bypass VHF components. This L network presented a load resistance of 18 Ω to the Q3 collector, the value needed for the desired 4 W output. But this circuit displayed instabilities when either the drive power or the supply voltage was varied. The output amplifier sometimes even showed a

Fig 13.44 — The MKII QRP transmitter includes VXO frequency control, TR switching and a sidetone generator.

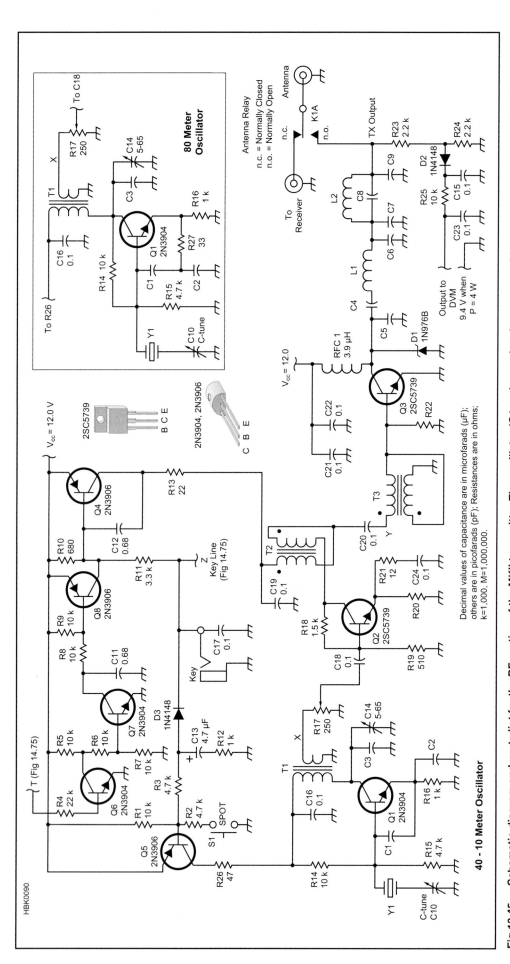

Fig 13.45 — Schematic diagram and parts list for the RF portion of the MKII transmitter. The oscillator (Q1 and associated components) in the main drawing is for the 40-10 m bands, while a modified version for 80 m is shown in the inset (see text). Fixed resistors are ¼ W, 5% carbon film unless otherwise noted. A kit of component parts is available from Kanga US (www.kangaus.com). TR switching is performed with a relay and additional circuitry (Fig 13.48).

C1-C9 — See Table 13.2, all 50 V ceramic or mica.
C10 — VXO control to provide some frequency adjustment around the crystal frequency. Use what you have in your junk box, although smaller capacitance values provide a wider tuning range. The prototype uses a small 2 to 19 pF trimmer. See text.
C11, 12 — 0.68 µF, 50 V metal film or Mylar.
C13 — 4.7 µF, 25 V electrolytic.
C14 — 5-65 pF, compression or plastic dielectric trimmer.
C15-24 — 0.1 µF, 50 V ceramic.
D1 — 1N976B, 43 V Zener diode.
K1A — See Fig 13.48.
L1, L2 — See Table 13.2.

Q1, Q6, Q7 — 2N3904, NPN silicon small signal transistor.
Q2, Q3 — 2SC5739 NPN silicon switching power transistor.
Q4, Q5, Q8 — 2N3906, PNP silicon small signal transistor.
R17 — 250 Ω, potentiometer (a 500 Ω potentiometer in parallel with 270 Ω fixed resistor can be substituted).
R20, R22 — See Table 13.2, carbon film.

RFC1 — 3.9 µH, 0.5 A molded RF choke. In place of a manufactured product, a T68-2 toroid wound with 26 turns of #22 enameled wire can be used.
T1 — See Table 13.2
T2 — 10 bifilar turns #28 enameled wire on FT-37-43 or FB-43-2401 ferrite toroid core.
T3 — 7 bifilar turns #22 enameled wire on FT-37-43 or FB-43-2401 ferrite toroid core.

Table 13.2
Band Specific Components of the MKII Transmitter

Band MHz	T1 turns-turns	C1 pF	C2 pF	C3 pF	R20 Ω	R22 Ω	L1 nH, turns wire, core	L2 nH, turns wire, core	C4 pF	C5 pF	C6 pF	C7 pF	C8 pF	C9 pF
3.5	51t-3t #26, T68-2	270	270	82	33	18	3000, 26t #28, T37-2	1750, 20t #28, T37-2	1000	390	1000	1000	300	1000
7	32t-4t #28, T50-6	390	100	82	33	33	1750, 19t #26, T37-2	890, 14t #22, T37-2	470	200	560	470	150	470
10.1	32t-4t #28, T50-6	390	100	0	33	33	1213, 19t #28, T37-6	617, 13t #28, T37-6	330	120	390	330	100	330
14	32t-4t #28, T50-6	390	100	0	33	33	875, 16t #28, T37-6	445, 11t #28, T37-6	220	100	270	220	75	220
18.1	20t-3t #28, T37-6	100	33	0	33	33	680, 14t #28, T37-6	346, 9t #28, T37-6	180	75	220	180	56	180
21	20t-3t #28, T37-6	100	33	0	18	33	583, 12t #28, T37-6	297, 9t #28, T37-6	150	62	180	150	50	150
24.9	20t-3t #28, T37-6	33	18	0	18	33	490, 11t #28, T37-6	249, 8t #28, T37-6	133	56	150	133	43	133
28	20t-3t #28, T37-6	33	18	0	18	33	438, 10t #28, T37-6	223, 7t #28, T37-6	120	47	140	120	39	120

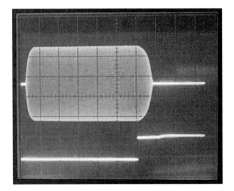

Fig 13.46 — Keyed waveform. The lower trace is the keyer input, which triggered the oscilloscope in this measurement. The horizontal time scale is 5 ms/div.

above could be observed with either an oscilloscope or a spectrum analyzer and were one of the more interesting subtleties of this project. The oscilloscope waveform looked like amplitude modulation. In the more extreme cases, every other RF cycle had a different amplitude that showed up as a half frequency component in the spectrum analyzer. The amplitude modulation appeared as unwanted sidebands in the spectrum display for the "moderately robust" instabilities. (Never assume that designing even a casual QRP rig will offer no development excitement!)

The output spectrum of this transmitter was examined with V_{CC} set to 12.0 V and the drive control set for an output of 4 W. The third harmonic output is –58 dBc and the others >70 dB down.

The author breadboarded the oscillator and buffer section for all HF amateur bands from 3.5 to 28 MHz.[20] The power amplifier circuit has been built at 3.5, 7, 14 and 21 MHz. The crystals, obtained from Kanga US (**www.kangaus.com**), were fundamental

divide-by-two characteristic. The original L network was modified with the original inductor replaced with an LC combination, C4 and L1. The new series element has the same reactance at the operating frequency as the original L network inductor. This narrow band modification provided stability on all bands. The components for the various bands are listed in **Table 13.2**.

The inductance values shown in Table 13.2 are those calculated for the networks, but the number of turns is slightly lower than the calculated value. After the inductors were wound, they were measured with a digital LC meter.[19] Turns were compressed to obtain the desired L value. Eliminate this step if an instrument is not available.

The divide-by-two oscillations mentioned

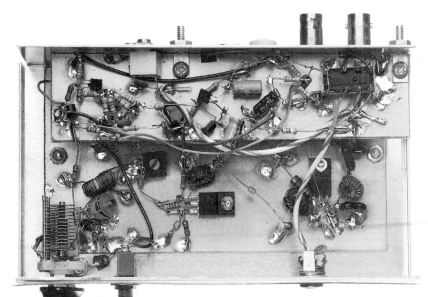

Fig 13.47 — The transmitter packaged in a 2 × 3 × 6 inch LMB #138 box. The basic RF circuitry is on the larger board. TR control is on the smaller board along the top.

mode units through 21 MHz, and third overtone above. The breadboard was built on two scraps of circuit board. Q1 and Q2 were on one with Q2 bolted to the board to serve as a heat sink. The second board had Q3 bolted to it, also serving as a heat sink.

After the breadboarding work was done, the circuits were moved to an available 2 × 3 × 6 inch box, an LMB #138. A new circuit board scrap was used, but most of the circuitry was moved intact from the breadboard. A diode detector was added to aid tune-up. The final RF board is shown in **Fig 13.47**.

CONSTRUCTION NOTES

Since publication in *QST*, some builders have encountered questions or difficulties. This section addresses those difficulties and further information may be found in *QST*.[21]

VXO Capacitor Grounding

The VXO capacitor, C10 of Fig 13.45, is mounted on the front panel of the transmitter rather than the circuit board. Grounding of the capacitor has been reported to be critical. A lead, ideally a short one, should go from the variable capacitor to the ground foil near the oscillator stage, Q1. In one of the transmitters built, the builder had merely attached the variable capacitor to the panel and relied on the ground connection that held the board to the box. This was, unfortunately, close to the power amplifier. The result was that the crystal oscillator would not always come on when the SPOT button was pushed. Adding a cleaner grounding wire solved the problem. The prototype uses a ground lug on the chassis very close to variable capacitor C10 and soldered directly to the PC foil right next to Q1.

Oscillator Changes

Some builders of the 40-meter version reported difficulty with tuning C14, the variable capacitor that tunes the collector circuit of the oscillator. The variable capacitor was too close to minimum C and a well defined peak was not always found. Of greater significance, tuning C14 to some values could allow the circuit to oscillate without crystal control of the frequency, producing oscillation in the 6.4-6.9 MHz region. Solutions to both problems are simple. First, change C3 from 100 to either 82 pF to remove the tuning ambiguity. (Removing a turn or two from the high L winding on T1 will accomplish the same end.) Second, adding C1 at 390 pF to the 40 meter circuit produces an oscillator that is always crystal controlled for any tuning of C14. Further experimentation with higher frequency versions revealed that the oscillators were generally well behaved but undesired modes could be found with extreme tuning of C14. Adding C1 to the circuit when it was initially absent always fixed this problem. The corrected component values are shown in Table 13.2.

Oscillation without crystal control was also observed with a misadjusted 80 meter circuit. Increasing the value of C1 helped but did not completely remove the problem. Analysis showed that the 80 meter oscillator starting gain was higher than was available on the higher-frequency bands. The gain was high enough that an instability was observed when the crystal was removed and the circuit was driven as an amplifier with a signal generator. Amplifier output jumped as the generator frequency was tuned.

Common "fixes" for amplifier instability include loading and the application of

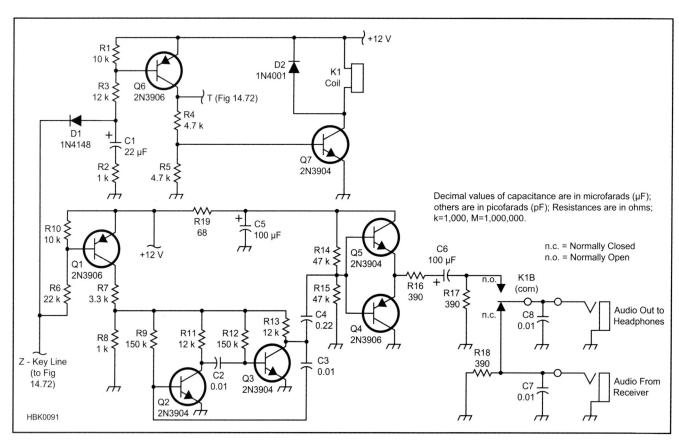

Fig 13.48 — Detailed schematic diagram and parts list for transmit-receive control section and sidetone generator of the universal QRP transmitter. Resistors are ¼ W, 5% carbon film. A kit of component parts is available from KangaUS (www.kangaus.com).

C1 — 22 µF, 25 V electrolytic.
C2, C3, C7, C8 — 0.01 µF, 50 V ceramic.
C4 — 0.22 µF, 50 V ceramic.

C5, C6 — 100 µF, 25 V electrolytic.
K1 — DPDT 12 V coil relay. An NAIS DS2Y-S-DC12, 700 Ω, 4 ms relay was used in this example.

Q1, Q4, Q6 — 2N3906, PNP silicon small signal transistor.
Q2, Q3, Q5, Q7 — 2N3904, NPN silicon small signal transistor.

negative feedback. Increased loading through adjustment of R17 helped, but this eliminated the ability to adjust overall transmitter output with this control. So, negative feedback was tried. The resulting oscillator circuit is shown in the inset of Fig 13.45. Emitter degeneration is added in the form of a 33 Ω resistor (R27), while parallel feedback is realized by moving the 10 kΩ base bias resistor (R14) from the bypass capacitor to the Q1 collector. The circuit, as shown, would not oscillate without crystal control. Care is still required for initial adjustment (with a receiver or spectrum analyzer) to avoid crystal controlled oscillation on a crystal spurious resonance. For example, one oscillator tested achieved crystal controlled operation at 3.8 MHz with a crystal built for operation at 3.56 MHz. Crystal spurious modes of this sort are found in virtually all crystals and should not be regarded as a crystal problem.

TRANSMIT-RECEIVE (TR) SWITCHING

Numerous schemes, generally part of a transceiver, are popular for switching an antenna between transmitter and receiver functions. When carefully refined, full-break-in keying becomes possible, an interesting option for transceivers. But these schemes tend to get in the way when one is developing both simple receivers and transmitters, perhaps as separate projects. A simple relay based TR scheme is then preferred and is presented here. In this system, the TR relay not only switches the antenna from the receiver to the transmitter, but disconnects the headphones from the receiver and attaches them to a sidetone oscillator that is keyed with the transmitter.

The circuitry that does most of the switching is shown in **Fig 13.48**. Line Z connects to the key. A key closure discharges capacitor C1. R2, the 1 kΩ resistor in series with C1, prevents a spark at the key. Of greater import, it also does not allow us to "ask" that the capacitor be discharged instantaneously, a common request in similar published circuits. Key closure causes Q6 to saturate, causing Q7 to also saturate, turning the relay on. The relay picked for this example has a 700 Ω, 12 V coil with a measured 4 ms pull-in time.

Relay contacts B switch, the audio line. R17 and 18 suppress clicks related to switching. A depressed key turns on PNP switch Q1, which then turns on the sidetone multivibrator, Q2 and Q3. The resulting audio is routed to switching amplifier Q4 and Q5. Although the common bases are biased to half of the supply voltage, emitter bias does not allow any static dc current to flow. The only current that flows is that related to the sidetone signal during key down intervals. Changing the value of R16 allows the audio volume to be adjusted, to compensate for the particular low-impedance headphones used.

There is an additional interface between Figs 13.48 and 13.45. Recall that Q4 of Fig 13.45 keys buffer Q2 while Q5 provides a time sequence control to oscillator Q1. Additional circuitry uses Q6, Q7 and Q8, and related parts. Under static key up conditions, Q7 is saturated, which keeps C11 discharged. Saturated Q7 also keeps PNP transistor Q8 saturated. This closed switch is across the emitter-base junction of Q4. Hence, pressing the key will start relay timing and will allow the oscillator to come on, but will not allow immediate keying of Q2 through Q4. Key closure causes Q6 in Fig 13.48 to saturate causing point T to become positive. This saturates Q6 of Fig 13.45 which turns Q7 off, allowing C11 to charge. When C11 has charged high enough, Q8 is no longer saturated and Q4 can begin its integrator action to key Q2.

This hold-off addition has solved a problem of a loud click, yielding a transmitter that is a pleasure to use. There is still a flaw resulting in the initial CW character being shortened. The result is that an I sent at 40 WPM and faster comes out as an E. Further refinement of timing component values should resolve this. The TR system circuitry is built on a narrow scrap of circuit board that is then bolted to the transmitter rear panel.

What's Next?

This has been an interesting project from many viewpoints. The resulting transmitter, which is usually used with the S7C receiver from *Experimental Methods in RF Design*,[22] is a lot of fun to use and surprisingly effective in spite of its crystal control. Primitive simplicity continues to have its place in Amateur Radio. Also, the development was more exciting than expected. The observed instabilities were interesting, as were the subtleties of the control system. Perhaps we should not approach simple CW systems with a completely casual attitude, for they continue to offer education and enlightenment.

There are clearly numerous refinements available for this transmitter. The addition of an adjustable reactance in series with the crystal will allow its frequency to move more. Try just a small variable capacitor. Two or more similar crystals in parallel form a "super VXO" topology for even greater tuning range. Higher power supply voltage will produce greater output power — over 10 W on the test bench. The transmitter could certainly be moved down to 160 meters for the top band DXer looking for QRP sport. It is not certain that the 2SC5739 will allow operation as high the 6 meter band. The transmitter could easily be converted to a modest power direct conversion transceiver using, for example, the Micromountaineer scheme offered in *QST*.[23]

13.4 Modern Baseband Processing

The term *baseband* refers to the signal or signals that comprise the information content at their natural frequency. For a communications audio signal, it would be a spectrum typically extending from 300 to 3300 Hz. Many transmitter architectures are designed to process and transmit a spectral range rather than any particular type of information. For example, the typical transmitter that shifts the modulating spectrum to occupy a single sideband adjacent to a suppressed carrier — our usual SSB transmitter, is just as happy to handle voice, modem tones or the two tones from an RTTY converter. The transmitter performs a linear operation to shift the input spectrum, the baseband signal, to the output frequency independent of the form or information content of the input spectrum.

This approach has a number of advantages for the transmitter designer and manufacturer. The baseband spectrum width, amplitude and dynamic range are inputs to the design process. The designer can thus focus on establishing the system between the baseband and the antenna port. This leaves the design of the baseband processing subsystem to perhaps another department or another company. Similarly the baseband processing equipment design may have multiple applications. Its output can be plugged into cable systems, HF transmitters or microwave systems as long as they support the required bandwidth, amplitude and dynamic range.

13.4.1 Digital Signal Processing for Signal Generation

The transmitter architecture that was described in Fig 13.27 was based on the classical analog approach to waveform generation and modulation with information content. Many current transmitters have replaced the early analog signal processing stages with a digital

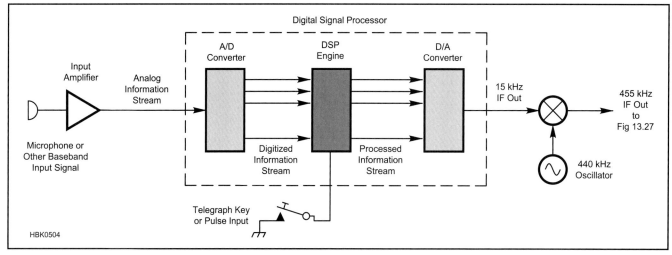

Fig 13.49 — Block diagram of a 15 kHz DSP-based baseband signal processor and IF generator.

signal processor (DSP). (Additional discussion is provided in the **DSP and Software Radio Design** chapter.)

Fig 13.49 provides a simplified block diagram of such a DSP based subsystem. The 455 kHz output would connect up to the 455 kHz amplifier stage in Fig 13.25, in place of the SSB generation and low IF subsystem that occupies the lower half of Fig 13.27.

Note that a DSP engine can generate virtually any desired waveform or modulating envelope desired by using different firmware. Thus the transmitted bandwidth, voice processing, carrier rejection or insertion, sideband selection pulse shaping and control functionality can be established with different firmware. This is a great benefit to manufacturers who can provide a single hardware platform that is usable for multiple services.

13.5 Increasing Transmitter Power

The functions described so far that process input data and information and result in a signal on the desired output radio frequency generally occur at a low level. The one exception is full-carrier AM, in which the modulation is classically applied to the final amplification stage. More modern linear transmitter systems generate AM in the same way as SSB at low levels, typically between 1 mW and 1 W.

13.5.1 Types of Power Amplifiers

The **RF Power Amplifiers** chapter provides a detailed view of power amplifiers; however, we will take a quick peek here to set the stage for the following discussions. Amplifiers use dc power applied to active devices in order to increase the power or level of signals. As will all real devices, they introduce some distortion in the process, and are generally limited by the level of distortion products. Power amplifiers can be constructed using either solid-state devices or vacuum tubes as the active device. At higher powers, typically above a few hundred watts, vacuum tubes are more frequently found, although there is a clear trend toward solid state at all amateur power levels.

Independent of the device, amplifiers are divided into classes based on the fraction of the input cycle over which they conduct. A sinusoidal output signal is provided either by the *flywheel* action of a resonant circuit or by other devices contributing in turn. The usual amplifier classes are summarized in **Table 13.3**. Moving from Class A toward Class C, the amplifiers become progressively less linear but more efficient. The amplifiers with a YES in the LINEAR column, thus are not all equally linear, however A, AB or Class B amplifiers can be suitable for operation in a linear transmitter chain. Class C amplifiers can be used only for amplification of signals that do not have modulation information contained in the amplitude, other than on-off keyed signals. Thus class C amplifiers are useful for amplification of sinusoids, CW, FM or as the nonlinear stage at which high-level AM modulation is employed.

In handheld digital cellular transceivers and base stations, the power amplifier must operate in a linear fashion to meet spectral purity requirements for the complex digital modulation schemes used. Since linear amplifiers are generally not very efficient, this is a major contributor to the energy consumption. In an effort to extend battery life in portable units, and reduce wasted energy in fixed equipment, considerable research is underway to use nonlinear amplifiers to provide linear amplification. Such techniques include pre-distortion of the low-level signal, polar modulation, envelope elimination and restoration, Cartesian-loop feedback and others.

13.5.2 Linear Amplifiers

While transmitters at power levels of 1 mW to 1 W have been successfully used

Table 13.3
Summary of Characteristics of Power Amplifier Classes
Values are Typical

Class	Conduction	Linear	Efficiency
A	360°	Yes	30%
AB	270°	Yes	55%
B	180°	Yes	65%
C	90°	No	74%

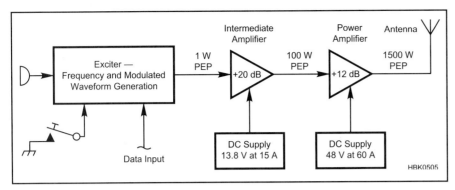

Fig 13.50 — Block diagram of a linear transmitter chain with multiple amplifier stages.

for communication across many portions of the spectrum, most communications systems operate with more success at higher powers. The low level stage is usually referred to as an exciter, while higher power is provided by one or more linear amplifier stages as shown in **Fig 13.50**.

The power levels shown at the various points in Fig 13.50 are fairly typical for a high powered amateur station. The 1500 W PEP output represents the legal limit for US amateurs in most bands (200 W PEP on 30 meters and 50 W ERP_D on the 60 meter channels are notable exceptions). The first amplifier block may contain more than one stage, while the final output amplifier is often composed of multiple parallel active devices.

Typical power supply requirements for the amplifier stages are noted for a number of reasons. First, while power is rarely an issue at the exciter level, it often is a significant issue at the power levels shown for the amplifiers. The power supplies represent a large portion of the cost and weight of the system as the power increases. Some manufacturers are beginning to use switching-type power supplies for high-power amplifiers, resulting in a major reduction in size and weight.

Note also that a gross amplifier efficiency of about 50% is assumed for the amplifiers, taking into account ancillary subsystems as well as the inefficiency of the active devices in linear mode. The 50% that doesn't result in actual RF output is radiated as heat from the amplifier and must be removed from the amplifier as it is generated to avoid component damage. This represents another cost and weight factor that increases rapidly with power level.

The voltages shown for the supplies are those typical of modern solid state amplifiers. While virtually all commercial equipment now includes solid state amplifiers at the 100 W level, vacuum tube active devices are frequently found at higher levels, although the trend is clearly moving toward solid state. Vacuum tubes typically operate at voltages in the 2 to 4 kV range, requiring stringent measures be taken to avoid arcing across components. In addition, vacuum tube amplifiers typically dissipate up to 100 W of filament power that must be added to the power supply and heat dissipation planning.

13.5.3 Nonlinear Amplifiers

Nonlinear transmitters are somewhat different in architecture than the linear systems discussed previously. The configuration of a high-level AM modulated transmitter is shown in **Fig 13.51**. Note that none of the upper RF stages (the "RF chain") needs to be particularly linear. The final stage must be nonlinear to have the modulation applied. Thus the RF stages can be the more power-efficient Class C amplifiers if desired.

There are some observations to be made here. Note that the RF chain is putting out the full carrier power whenever in transmit mode, requiring a 100% duty cycle for power and amplifier components, unlike the SSB systems discussed previously. This imposes a considerable weight and cost burden on the power supply system. Note also that the PEP output of a 100% modulated AM system is equal to four times the carrier power.

The typical arrangement to increase the power of such a system is to add not only an RF amplifier stage capable of handling the desired power, but also to add additional audio power amplification to fully modulate the final RF stage. For 100% high-level plate modulation, an audio power equal to half the dc input power (plate voltage times plate current of a vacuum tube amplifier) needs to be provided. This arrangement is shown in **Fig 13.52**. In the example shown, the lower level audio stages are provided by those of the previous 50 W transmitter, now serving as an exciter for the power amplifier and as a driver for the modulating stage. This was frequently provided for in some transmitters of the AM era, notably the popular E. F. Johnson Ranger series, which provided special taps on its modulation transformer for use as a driver for higher-power systems.

It is worth mentioning that in those days the FCC US amateur power limit was expressed in terms of dc *input* to the final stage and was limited to 1000 W, rather than the 1500 W PEP *output* now specified. A fully modulated 1000 W dc input AM transmitter would likely have a carrier output of 750 W or 3000 W PEP — 3 dB above our current limit. If you end up with that classic Collins KW-1 transmitter, throttle it back to make it last and stay out of trouble!

13.5.4 Hybrid Amplifiers

Another alternative that is convenient with current equipment is to use an AM transmitter with a linear amplifier. This can be successful if the relationship that

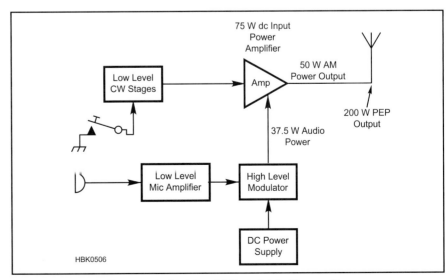

Fig 13.51 — Block diagram of a high level AM modulated transmitter.

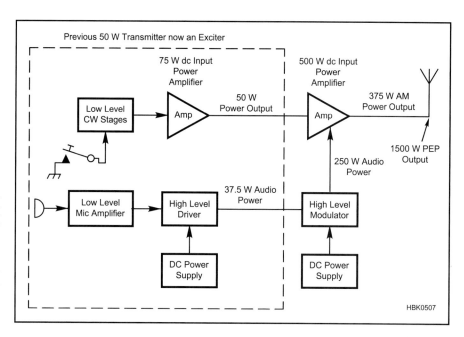

Fig 13.52 — Block diagram of a high level AM modulated transmitter with added output stage.

PEP = 4 × Carrier Power is maintained. **Fig 13.53** shows a 1500 W PEP output linear amplifier following a typical 50 W AM transmitter. In this example, the amplifier would be adjusted to provide a 375 W carrier output with no modulation applied to the exciter. During voice peaks the output seen on a special PEP meter, or using an oscilloscope should be 1500 W PEP.

Note that during AM operation, the amplifier is producing a higher average power than it would without the carrier being present, as in SSB mode. The duty cycle specification of the amplifier should be checked to be sure it can handle the heavier load. If the amplifier has an RTTY rating, it should be safe to run an AM carrier at 66% of the RTTY output, following the required on and off time intervals.

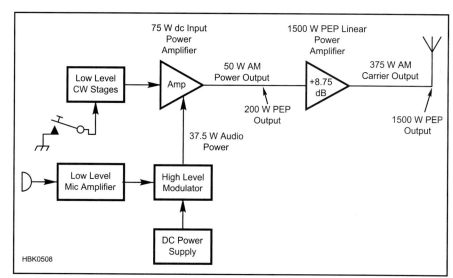

Fig 13.53 — Block diagram of a hybrid nonlinear/linear AM transmission system.

13.6 Transmitter Performance and Measurement

Throughout this chapter, we have discussed various transmitter design and performance parameters that are important to operation. Here we will mention some of the key parameters that should be evaluated before actually putting a transmitter on the air and periodically thereafter.

13.6.1 Power Measurement

Some parameters are easy to define and measure. These include CW power output — power meters with adequate accuracy are available for reasonable prices. Some also can sample and hold to measure the PEP output while driven with complex waveforms. This kind of measurement should be made frequently to ensure that transmitters are not being driven beyond their ratings — a sure invitation to the generation of distortion products. It is also important for those with high power stations that their operations are kept within the FCC power limits.

A calibrated high frequency oscilloscope can also be used to determine peak power and will also indicate some forms of distortion, such as clipping.

Similarly, it is not too difficult to measure the dc input power by measuring power supply voltage and current, although measurement of current requires some thought with high current solid state transmitters to avoid voltage drop. The combination of the above measurements can provide a check on transmitter efficiency.

13.6.2 Frequency Calibration and Stability

Whether crystal controlled or variable, a transmitter carrier frequency is a key parameter to keep under observation, both to be sure that you are in an allowed band segment and so other stations will know where to look for you. Frequency counters are a popular way to measure this parameter and are available at reasonable cost. Be aware that they are no more accurate than their clock reference, and that should be derived from, or compared to, a standard of known accuracy periodically.

Once long term frequency calibration is established, one can gain insight into transmitter stability by letting the counter settle for some time, perhaps an hour, and then measuring the transmitter frequency on turn on and tracking it every few minutes until it stabilizes. It is interesting to know how far and in what direction it shifts during start up, and how long it takes to stabilize. Even after it stabilizes, it is interesting to plot its frequency every few minutes to find out how close to its nominal frequency it stays. At this point you are likely measuring both transmitter and counter frequency shift, so a third "tie breaker" system is convenient. In the absence of a counter, a stable and accurately-calibrated communications receiver can be used for this measurement.

13.6.3 Transmitter Spurious Responses

The ideal linear transmitter will put out a signal on its intended frequency that is an exact frequency translated model of its baseband input spectrum. In the case of a CW or pulse transmitter, it will represent the minimum bandwidth necessary to generate the pulse with minimum overlap between pulses. A linear amplifier, based on the meaning of the word produces an exact replica of the input signal and its spectrum at a higher level. Any other signals that occur at the output of a transmitter are *spurious* responses.

Spurious responses can occur in transmitters in a number of ways and manifest themselves in the form of various symptoms.

FCC REGULATIONS RELATED TO SPURIOUS EMISSIONS

The following is extracted from FCC Part 97, the rules related to the Amateur Radio Service as applied to HF transmitters. There are similar requirements for other portions of the spectrum and all can be found at **www.access.gpo.gov/nara/cfr/waisidx_08/47cfr97_08.html**, downloadable as text or in *Adobe Acrobat* format.

Subpart D — Technical Standards, Sec. 97.307 Emission standards.

(a) No amateur station transmission shall occupy more bandwidth than necessary for the information rate and emission type being transmitted, in accordance with good amateur practice.

(b) Emissions resulting from modulation must be confined to the band or segment available to the control operator. Emissions outside the necessary bandwidth must not cause splatter or keyclick interference to operations on adjacent frequencies.

(c) All spurious emissions from a station transmitter must be reduced to the greatest extent practicable. If any spurious emission, including chassis or power line radiation, causes harmful interference to the reception of another radio station, the licensee of the interfering amateur station is required to take steps to eliminate the interference, in accordance with good engineering practice.

(d) For transmitters installed after January 1, 2003, the mean power of any spurious emission from a station transmitter or external RF power amplifier transmitting on a frequency below 30 MHz must be at least 43 dB below the mean power of the fundamental emission. For transmitters installed on or before January 1, 2003, the mean power of any spurious emission from a station transmitter or external RF power amplifier transmitting on a frequency below 30 MHz must not exceed 50 mW and must be at least 40 dB below the mean power of the fundamental emission. For a transmitter of mean power less than 5 W installed on or before January 1, 2003, the attenuation must be at least 30 dB. A transmitter built before April 15, 1977, or first marketed before January 1, 1978, is exempt from this requirement.

13.6.4 Harmonics of the Transmitter Frequency

Signals at multiples of the transmitter frequency are perhaps the most common forms of distortion produced by amplifiers. We discussed frequency multipliers earlier, and this is the same effect, except that for an amplifier, the transmitter is supposed to select the output at the fundamental frequency and attenuate harmonics in its output network or filter.

Aging or failure of filter components can result in additional harmonic energy at the output. Similarly, overdriving an amplifier stage can result in excessive harmonic content such that the filtering can't reduce it to a low enough level to be acceptable. Unfortunately, as with most forms of spurious signals, the fundamental signal is not noticeably changed and all seems well for receivers listening in band.

Harmonics can be best measured using a wideband spectrum analyzer that measures up to perhaps five times the operating frequency — harmonics tend to get smaller as the products get to higher orders. Alternately, a communications receiver with a calibrated S-meter can be used to measure harmonics if its response is constant over the frequency range. Note that harmonics of transmitters built or placed in operation after Jan 1, 2003 must meet the –43 dB from the carrier output level standard if over 5 W output.

13.6.5 Key Clicks and Waveshaping

Earlier in the chapter, as part of the discussion of design criteria, we discussed the importance of waveshaping in order to generate a clean and narrow CW signal. You may want to look again at Fig 13.14 through 13.16 to get an appreciation for the significance of this measurement.

A typical keying spectrum test setup is shown in **Fig 13.54**. For this test an automatic keyer is used to drive the transmitter KEY input. If the transmitter includes an internal keyer, a simple switch closure on the dot side will suffice. Generally the keyer is run at high speed (the ARRL Lab uses 60 WPM) and the spectrum is observed. A transmitter with good keying sidebands will exhibit a spectrum similar to that of Fig 13.14. If it looks more like that of 13.16, it needs improvement in the shape of the transmitted waveform.

As with many other measurements, a high-quality communications receiver with a calibrated S-meter, or an accurate step attenuator, can be used to take the measurements if a spectrum analyzer is not available.

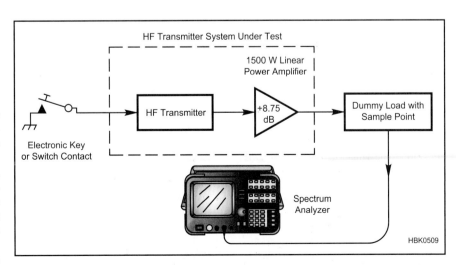

Fig 13.54 — Typical transmit keying spectrum test setup.

13.6.6 Intermodulation Distortion

The measurements described to this point represent data resulting from a single carrier frequency. Any kind of complex modulating signal contains information on multiple frequencies, and as with receivers, nonlinearities in amplifiers and mixers can result in intermodulation products. While a continuous information spectrum, such as a voice channel, has components that appear over time throughout the channel, most measurements are taken using two discrete frequencies within the voice channel.

A typical transmit IMD setup is shown in **Fig 13.55**. Note that both frequencies selected must be within the transmit audio passband and should not be harmonically related to avoid confusing results. The results of a typical test are shown graphically in **Fig 13.56**. Note that the two primary signals resulting from the tone inputs are set to equal amplitudes at 3 dB below the rated output to result in the total power at full output. The successive intermodulation products occur on either side of the generated tones. Note that the third order products are 26 dB down and are outside of the channel bandwidth, while higher products are in channels further from the transmitted signal and all products would interfere with other communications.

Of course, in a real operation, neither the tones nor the IM products would be discrete, rather they would manifest themselves as the "buckshot" type noise encountered frequently on crowded bands. It is worth noting that receiver overload or improper operation of noise blankers can result in distortion with similar characteristics to spurious transmitter output signals. Be sure your receiver is operating properly before assuming that a neighboring signal is the source of interference.

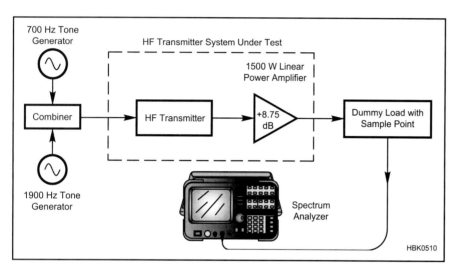

Fig 13.55 — Typical transmit IMD test setup using two-tone generators.

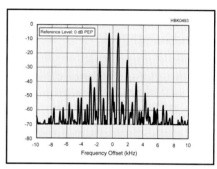

Fig 13.56 — Typical transmit IMD test results as seen in the ARRL Lab.

13.7 References and Bibliography

[1]R. Shrader, W6BNB, "When Radio Transmitters Were Machines," *QST*, Jan 2009, pp 36-38.

[2]The only remaining Alexanderson alternator transmitter is the Grimeton VLF transmitter (**en.wikipedia.org/wiki/Grimeton_VLF_transmitter**) that operates on 17.2 kHz.

[3]The Aldis lamp (**en.wikipedia.org/wiki/Signal_lamp**) was used on British navy vessels in the late 1800s and is still in use on naval vessels as it allows signaling between ships and from ship to shore without breaking radio silence.

[4]E. Hare, "The Tuna Tin Two Today," *QST*, Mar 2000, pp 37-40.

[5]Amateur practice is to use USB above 10 MHz and LSB on lower frequencies. The exception is 60 meter channels, on which amateurs are required to use USB.

[6]M. Schwartz, *Information Transmission, Modulation and Noise*, third edition, McGraw-Hill, 1980.

[7]W. Sabin and E. Schoenike, Editors, Single-Sideband Systems and Circuits, McGraw-Hill, 1987.

[8]W. Sabin, "RF Clippers for SSB," *QST*, Jul 1967, pp 13-18.

[9]J. Grebenkemper, KI6WX, "Phase Noise and its Effect on Amateur Communications," *QST*, Mar, Apr 1988.

[10]W. Hayward, W7ZOI, R. Campbell, KK7B, and B. Larkin, W7PUA, *Experimental Methods in RF Design*, ARRL, 2003.

[11]R. Fisher, W2CQH, "Twisted-Wire Quadrature Hybrid Directional Couplers," *QST*, Jan 1978, pp 21-24.

[12]R. Campbell, KK7B, "A Multimode Phasing Exciter For 1 to 500 MHz," *QST*, Apr 1993, pp 27-32.

[13]Some amateur receivers have inadequate shielding, and may overload on signal leakage from the exciter crystal oscillator — in that case, use a simple frequency converter to translate the exciter output to a different frequency. The converter may also be used to translate the frequency all the way down to 10 kHz as discussed on *EMRFD* page 7.35 to analyze the exciter signal using a computer sound card and audio spectrum analyzer software.

[14]Even a poorly adjusted phasing single sideband generator will have significant opposite sideband rejection. If it has little or none, stop and find the construction error. With over 100 parts on a PC board, it's easy to make an error — a lead left unsoldered or swapped components.

[15]W. Hayward, W7ZOI, and D. DeMaw, W1FB (SK), *Solid State Design for the Radio Amateur*, ARRL, 1977, pp 26-27.

[16]See Note 10, p 6.63.

[17]See Note 10, pp 2.24-2.28, for information on feedback amplifiers.

[18]See Note 10, p 3.6, for low-pass filter designs, 3.29 for matching network designs.

[19]See **www.aade.com** for details. A homebrew alternative is offered in Note 10, Chapter 7. Also see Carver, "The LC Tester," *Communications Quarterly*, Winter 1993, pp 19-27.

[20]See Note 10, p 1.2, and Hayward and Hayward, "The Ugly Weekender," *QST*, Aug 1981, p 18.

[21]W. Hayward, W7ZOI, "Crystal Oscillator Experiments," Technical Correspondence, *QST*, Jul 2006, pp 65-66.

[22]See Note 2, p 12.16.

[23]Hayward and White, "The Micro-mountaineer Revisited," *QST*, Jul 2000, pp 28-33.

Contents

14.1 The Transceiver Appears

14.2 Early SSB Transceiver Architectures
    14.2.1 The Collins System
    14.2.2 Other Early Transceivers
    14.2.3 Limitations of the Early Transceivers

14.3 Modern Transceiver Architecture and Capabilities
    14.3.1 Evolution of Transceiver Architectures
    14.3.2 Features of Current Transceivers

14.4 Selecting an HF Transceiver
    14.4.1 Entry Level Transceivers
    14.4.2 Portable and Mobile Transceivers
    14.4.3 Mid-Range Transceivers
    14.4.4 Upper Mid-Range Transceivers
    14.4.5 Top-Drawer Transceivers
    14.4.6 Making a Selection

14.5 Transceivers With VHF/UHF Coverage
    14.5.1 Using an HF Transceiver at VHF/UHF

14.6 Transceiver Control and Interconnection
    14.6.1 Automated Transceive
    14.6.2 Break-In Operation
    14.6.3 Push-To-Talk for Voice Operators
    14.6.4 Voice-Operated Transmit-Receive Switching (VOX)
    14.6.5 TR Switching With a Linear Amplifier

14.7 Transceiver Projects
    14.7.1 Transceiver Kit
    14.7.2 *Project*: The TAK-40 SSB CW Transceiver
    14.7.3 *Project*: A Homebrew High Performance HF Transceiver — the HBR-2000

14.8 References

Chapter 14

Transceivers

Receivers and transmitters were covered in the previous two chapters. Putting them together creates a transceiver, the heart of the modern Amateur Radio station. This chapter will present some history and perspective on transceiver development, transceiver architecture and information on selecting a commercial transceiver. Then details of control and interconnection circuits needed to integrate a receiver and transmitter into a transceiver are presented. There are two transceiver projects at the end, and more projects and related information may be found on the *Handbook* CD-ROM. This chapter was updated and new material contributed by Joel R. Hallas, W1ZR.

14.1 The Transceiver Appears

In the beginning of the 20th century, from the days of spark transmitters through the early days of single-sideband (SSB) voice operation in the 1950s, the typical Amateur Radio station included a separate transmitter and receiver. In the mid- to late-1950s, it became clear that, unlike the earlier AM and CW radios, SSB receivers and transmitters used many of the same functional blocks and could be combined into a single unit sharing the common parts. This concept became well established with the 1957 introduction of the Collins Radio KWM-2 SSB and CW transceiver, shown in **Fig 14.1**.

Transceivers offered a number of key advantages to SSB operation. Perhaps the key operational benefit was that a single variable frequency oscillator (VFO) tuned both transmitter and receiver simultaneously. Thus, having tuned in another station properly, your transmit signal would be tuned for them to hear your signal properly as well — avoiding the need to make a separate operation of trying to *zero-beat* your transmitter to the receive frequency. Because they were not designed for full-carrier AM operation, SSB transceivers didn't require the heavy transformers of the typical AM transmitters of the day, making them compact and lightweight compared to their AM brethren.

The KWM-2 transceiver shared the appearance of the Collins 75S-1 and 32V-1 receiver and transmitter respectively, fitting into the same compact cabinet as either one of the separate units. Considerable flexibility was lost in the combined unit, however. The separate receiver and transmitter could be inter-cabled to transceive as a package, while still allowing split-frequency operation. CW operation was better supported in the separate units and multiple selectivity choices were provided in the separate receiver. Nonetheless, the compact single-unit transceiver offered excellent SSB operation in a single compact enclosure. A separate VFO was provided to allow the KWM-2 to use

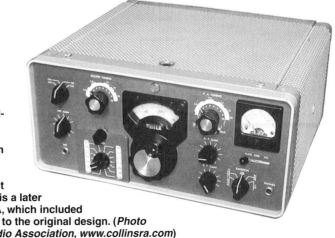

Fig 14.1 — The 1957 vintage Collins KWM-2 transceiver, circa 1957, was one of the first in the move from separate receivers and transmitters to transceivers. The unit shown in this photo is a later version, the KWM-2A, which included some improvements to the original design. (*Photo courtesy Collins Radio Association, www.collinsra.com*)

separate transmit and receive frequencies for split-frequency operation, but this somewhat negated the single-unit benefit.

The successful Collins equipment was quickly followed by compact transceivers from R. L. Drake and Heathkit — two other very popular early manufacturers of the period. Their vacuum tube equipment carried on well into the 1970s, after which they (and many other manufacturers) moved to hybrid and finally all solid-state designs.

Since then, advanced design techniques, microminiaturization and competition have resulted in a range of available HF SSB transceivers ranging in price from around $500 to over $10,000 with a wide range of choices in features and performance. It is safe to say that, for the radio amateur, there is no longer anything to be given up by selecting a transceiver rather than a separate transmitter and receiver.

14.2 Early SSB Transceiver Architectures

Since the SSB transmitter and receiver are so similar, it is possible to combine them into a single package sharing many functions. In many transceivers, the sideband filter, some amplifiers, and other filters are shared between transmit and receive modes through the use of extensive switching. A simple example is illustrated in **Fig 14.2** in which common oscillators are used for transmitting and receiving. An advantage of this configuration in a two-way system is that the radios at both ends are set to transmit and receive on the same frequency at all times. This is particularly important for SSB operation in which a tuning error of even 50 Hz can be noticeable.

14.2.1 The Collins System

The KWM-2 had an additional conversion stage and the single-sideband filter was shared between transmit and receive. The suppressed-carrier oscillator frequencies were above and below the 455 kHz center frequency of the 2.1 kHz bandwidth mechanical filter — 453.65 kHz for lower sideband and 456.35 kHz for upper sideband. Note that the carrier frequency is 300 Hz outside each filter edge. This provides a voice bandwidth of 300 Hz to 2400 Hz, adequate for communication purposes, if not quite telephone "toll quality." By placing the carrier frequency on the response skirt of the filter, the carrier suppression of the balanced modulator is enhanced by that amount of attenuation.

The first conversion stage used a variable oscillator with a tuning range of approximately 2.7 to 2.5 MHz (shifted depending on sideband selected) resulting in a variable intermediate frequency in the range of 3.155 to 2.955 MHz. The next conversion stage employed a choice of 14 crystal-controlled oscillator frequencies covering individual 200 kHz-wide bands over frequencies ranging from 3.4 to 5.0 and 6.5 to 30.0 MHz. (That was increased to 28 oscillator frequencies in the "A" model shown in Fig 14.1) This arrangement allowed an easy adaptation to use of the radio by other services merely by changing crystals. Note that 14 bands, each 200 kHz wide, in the basic unit did not quite cover all segments of even the 1957 amateur bands.

The HF circuits were all tuned simultaneously with the VFO using a mechanical slug rack that moved up and down with the tuning. This architecture is similar to that used in the earlier 75A, 51J and 32V series AM equipment and 75A-4 SSB receiver.

14.2.2 Other Early Transceivers

Drake and Heathkit provided transceivers that were functionally similar to the KWM-2, although they differed in some details in their design. A fundamental difference was that while Collins used its 455 kHz mechanical filter as a sideband filter, both Drake and Heathkit used crystal lattice filters in the low HF range for carrier and sideband suppression. Drake transceivers (but interestingly, not their separate units) used a single carrier frequency and switched between two filters for sideband selection — one on each side of the carrier frequency. Transceivers from both manufacturers were quite popular since they offered similar performance to the KWM-2 at a lower price.

14.2.3 Limitations of the Early Transceivers

FREQUENCY AGILITY

The first generation of SSB transceivers was quite usable for voice communication in which both ends of the link were on the same frequency. This is quite useful for net operations and casual conversations, but in many cases operators need the capability to operate split, with each operator transmitting on a different frequency. Sometimes the two stations are in different countries with different portions of the band allocated for voice operation, or sometimes a sought-after DX station might want to transmit on a different frequency so those calling would not interfere with the DX station's transmissions. The

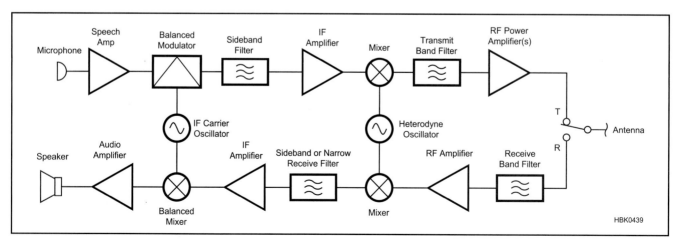

Fig 14.2 — Block diagram of a simple SSB transceiver sharing oscillator frequencies.

early transceivers could not do this.

Manufacturers eventually met this need with a second VFO for many models. This would allow transmission on one frequency and reception on another. Most were able to switch to a reverse mode (swapping transmit and receive VFOs) to simplify setting each frequency. While this was a workable solution, the need for an additional enclosure negated some of the benefits of having an "all in one" box.

Later versions began to offer *receive incremental tuning* (RIT) that allowed the receive frequency to be adjusted independently from the transmit frequency. (RIT is also called "clarify" by some manufacturers.) In some transceivers, RIT allowed sufficient separation of receive and transmit frequencies to be useful for limited split-frequency operation.

All the early transceivers were of the "downconverting" design (from the point of view of the receiver side). Multiple-conversion was employed to provide both a constant tuning rate, as defined in the Collins system (see the **Receivers** chapter), and sufficient image rejection. A limitation of this approach is that it makes *general-coverage receive* (continuous coverage from below 3 to 30 MHz or more) more complicated.

INTERFERENCE REJECTION

Early SSB transceivers generally were equipped with a sideband selection filter that provided appropriate selectivity for reception of SSB signals as well as *single-signal* CW reception if too many nearby channels weren't in use. Those with separate receivers generally enjoyed multiple filter bandwidths for different modes, as well as special filters to notch out interfering carriers.

OPERATING MODES

Most early SSB transceivers made some accommodation to allow CW operation, but it was not always totally satisfactory. The general design architecture of the early transceivers was actually almost ideal for CW — the frequency stability and tuning rate of an SSB transceiver were very close to optimum for CW. In addition, the duty cycle of CW was a good match for transmitter power amplifiers and high-voltage power supplies designed for SSB.

The problem was that some designers seemed to add CW as an afterthought. The worst offender was probably the KWM-2, which generated CW by keying a 1750 Hz audio oscillator and sending its signal through the audio stages.[1] This arrangement resulted in a much higher than usual receive tone at the far end, often causing the receive operator to change frequency to provide for comfortable listening. If both operators in a contact were using KWM-2s (which unfortunately had no RIT), they would often chase each other across the band as each operator retuned for a more appropriate CW note.

Other radios generally injected an actual CW carrier into the transmitter on a frequency that would fall comfortably within the receiving filter's passband at an appropriate tone for comfortable listening. Initially, no additional selectivity was provided. Many radios eventually offered an additional filter, or a switchable position for an optional narrow bandwidth filter, typically 500 Hz, that could be used to provide a bandwidth appropriate for CW operation.

Other aspects of operation were not often optimum for CW. Transmit-receive switching tended to share the voice-operated delay circuitry used for SSB rather than provide timing appropriate for CW. Little attention was paid to shaping of the transmit waveform, nor were the choices offered for automatic gain control (AGC) time constants appropriate for CW.

Most early CW/SSB transmitters did not provide any facility for AM, FM or digital modes. These were the days before PCs and sound cards, so support of digital modes (RTTY) was not supported until later models added dedicated keying circuits that shifted the transmitter frequency to provide frequency-shift keying (FSK).

14.3 Modern Transceiver Architecture and Capabilities

With each generation of transceivers, in what has become a highly competitive marketplace, additional features were added as technology marched on. The deficiencies of early transceivers, in comparison to separate receivers and transmitters, quickly disappeared to the point that 100 W (or higher power) class transceivers exceeded the features of the best separate receivers and transmitters of the past.

The current generation of transceivers was designed using modern solid-state components that permit abundant functionality in a small enclosure. The designers and manufacturers have taken advantage of the possibilities of integrated electronics and microprocessors, incorporating many more functions into a transceiver than could have been envisioned in the early days. This is largely a function of improved technology becoming available at reduced cost.

Separate receivers and transmitters could be built with similar features and performance, but the required duplication of subsystems would make each unit cost about the same as a transceiver. Several manufacturers do make stand-alone receivers using components from transceivers, but they are generally aimed toward different markets such as shortwave listeners or military/commercial users.

14.3.1 Evolution of Transceiver Architectures

Getting from the early transceiver designs to what is currently available occurred in a number of steps, generally corresponding to availability of technology at prices that could be included in competitive radios that amateurs could afford, as well as applied in support of commercial and military product lines. Advances in transceiver technology can be divided up in a number of ways, the following discussion being one view of the march of technology as applied to amateur transceivers.

THE MOVE TO SOLID STATE

Early solid-state devices became available first as low-speed, low-power diodes and transistors. These began to creep into otherwise vacuum-tube-based transceiver designs right from the beginning. The move of major subsystems to solid state awaited the development of high-frequency transistors that could serve in receiver and low-level transmitter stages with comparable performance to the highly developed vacuum tube circuits of the day and with greatly reduced power consumption and heat.

This occurred in the early 1970s with so-called *hybrid* transceiver designs in which all active circuits with the exception of the transmit driver and final power amplifier stages were solid state. Such radios were still entirely analog in design, and required a power supply that delivered both low and high voltage dc for the different devices. These radios offered small benefits in comparison to the mostly vacuum tube counterparts: The hybrid designs were only slightly smaller, generated slightly less heat, and perhaps had less warm up drift. They still required manual tuning of the transmitter output circuits.

In the late 1970s, both R. L. Drake and Ten-Tec offered 200 W dc power input (about 130 W PEP output, in today's terms) transceivers that were entirely solid state. The transceivers were still analog, with mechanical VFO tuning, although the Drake TR-7 included a digital frequency counter along with its mechanical dial display. The major advance of the solid-state transmitter was that no tuning was required at the output of the final transmit amplifier. This allowed instant frequency changing and even band-switching without retuning — if the antenna offered a low SWR at all the frequencies involved. The vacuum tube transceiver had a bit of an advantage here in that it could compensate for a wider level of antenna system mismatch with its manual tuning controls. Still, an external antenna tuner could be used to compensate for a mismatched antenna at the cost of some additional controls and desk space. These radios generated much less heat than their tube counterparts and could run from a single 12-V dc supply. This meant no additional hardware for portable or mobile operation, if a 12-V source were available.

DIGITAL TECHNOLOGY ARRIVES

The first application of note to utilize digital technology was frequency display, as mentioned above. In the next (1980s) generation of transceivers, virtually all included some form of digital display, sometimes as an option. Next, digital control logic was used to allow *soft* controls, such as electronically-switched PIN diode attenuators or filter selection, instead of hard-wired switches and rotary potentiometers.

As digital logic components became smaller and less expensive, they moved into more and more areas such as frequency synthesis, allowing wider tuning ranges and electronically simulating the dual external VFOs of the earlier era through the use of dual memories for a digitally controlled frequency generator. Variable analog oscillators tuned by a variable capacitor or inductor were replaced by programmable digital dividers in phase-locked loops (PLLs) and a single stable reference oscillator. PLL-based frequency synthesis is significantly more stable than analog frequency generation.

Other early features included memory registers to allow saving and recalling frequencies and operating modes and direct keypad entry of frequency. As technology continued to evolve, direct digital synthesis (DDS) replaced the PLL, offering faster frequency switching and eliminating the synthesizer design tradeoff of phase noise versus lock time. Soon embedded processor chips were controlling most aspects of transceiver operation.

DIGITAL SIGNAL PROCESSING

A major leap in capability occurred with the availability of reasonably priced digital signal processing (DSP) integrated circuits around the beginning of the 21st century. While the previous digital processing functions made operation convenient and allowed for an unprecedented level of automation, today DSP is applied directly in the signal path. The analog radio signals are sampled and converted into digital form many thousands of times a second in an *analog-to-digital converter* (ADC). The digital samples are processed at high speed using DSP algorithms designed to accomplish a number of benefits and then returned to analog form in a *digital-to-analog converter* (DAC) as if they never changed. (A complete discussion of this technology is provided in the **DSP and Software Radio Design** chapter.)

While in digital mode, a number of functions can be applied. Note that not all DSP systems provide all these functions, depending on their design objectives. Also while almost all current HF transceivers provide DSP functions inside the box, it is also possible to purchase external DSP processors that can be added to older transceivers or receivers, with significant limitations as noted below.

• *DSP selectivity* — DSP can do a marvelous job of setting the selectivity or bandwidth. If you like to compare curves, note how steep the skirts of **Figs 14.3** and **14.4** are compared to the response of analog filters. This was the selectivity of an outboard audio DSP filter made by SGC and reviewed in the January 2004 issue of *QST*. It has far sharper skirts than any pre-DSP analog filter. While this is most important for interference rejection, sharp selectivity will also reduce random noise. More recent internal IF-based DSP filters are sometimes even sharper and often provide for adjustable skirt slopes so signals don't suddenly appear (and disappear) while tuning. DSP filters are also immune to the inaccuracies of analog filters, which are subject to drift over time and temperature as capacitor and inductor values shift or crystals age.

• *DSP notch filtering* — A steady carrier signal, either from an operator adjusting an antenna tuner, or the carrier of a shortwave broadcast station on a shared band, is more interference than noise, but perhaps even more annoying. DSP can do an incredible job of eliminating narrow-band steady carriers, and can even track carriers that drift slowly — something that would require manual readjustment of an analog notch filter.

• *DSP noise reduction* — Analog filtering and notching has been used for years, although current DSP technology can generally do it better. On the other hand, DSP can also do something that analog systems can't. DSP noise reduction can actually look at the statistics of the signal and noise and figure out which is which. The DSP can then reduce the noise significantly. It can't quite eliminate the noise, and it needs enough of the signal to figure out what's happening, so it won't work if the signal is far below the noise. DSP may also help with some kinds of impulse noise, but the impulse noise may look like a signal to the DSP and not get reduced, depending on the software. It's always worth trying the feature to find out.

The first DSP systems and aftermarket DSP processors, operated at audio frequencies. As processors and sampling speeds got faster and faster, DSPs were able to operate at intermediate radio frequencies within a radio. Initially the frequencies were in or just above the audio range in the tens of kilohertz, although this is rapidly advancing with higher-speed technology.

DSP at the IF allows the processing of signals before they get into the AGC system, a significant advantage. Otherwise, a signal that gets through the radio's early filtering but isn't "heard" by the outboard DSP filter can cause the AGC to reduce the gain of the receiver, reducing the level of both the desired and undesired signals.

UPCONVERTING RECEIVER ARCHITECTURE

A significant change in the architecture of receivers and transceivers became popular in the 1980s and, with a few notable exceptions became almost universal in commercial products over the following decade. A limitation of the architecture shown in Fig 14.2 is that it is not trivial to provide operation on frequencies near the first IF. The typical transceiver designer selected a first IF frequency away

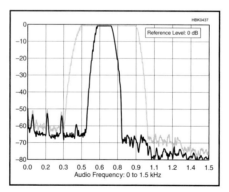

Fig 14.3 — The selectivity provided by a DSP processor is far closer to rectangular perfection than an analog filter. This shows narrow and wide CW bandwidths.

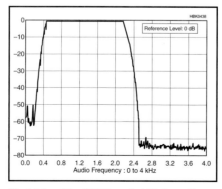

Fig 14.4 — The SSB bandwidth from the same DSP filter as in Fig 14.3.

from the desired operating frequencies and proceeded on that basis.

A Move to Flexibility

Amateur bands at 30, 17 and 24 meters were approved at the 1979 ITU World Administrative Radio Conference (WARC) and opened to hams over the following years. The difficulties of managing image rejection on the new bands and the desire for continuous receiver coverage of LF, MF and HF bands (called general-coverage receive as noted earlier) required a change in receiver conversion architecture.

The solution was to move to the upconverting architecture shown in **Fig 14.5**. By selecting a first IF well above the highest receive frequency, the first local oscillator can cover the entire receive range without any gaps. With the 70 MHz IF shown, the full range from 0 to 30 MHz can be covered by an LO covering 70 to 100 MHz, less than a 1.5:1 range, making it easy to implement with modern PLL or DDS technology. Note that the high IF makes image rejection very easy and, rather than the usual tuned bandpass front end, we can use more universal low-pass filtering. The low-pass filter is generally shared with the transmit side and designed with octave cutoff frequencies to reduce transmitter harmonic content. A typical set of HF transceiver low-pass filter cut-off frequencies would be 1, 2, 4, 8, 16 and 32 MHz.

This architecture offers significant benefits. By merely changing the control system programming, any frequency range or ranges can be provided with no change to the architecture or hardware implementation. Unlike the more traditional transceiver architecture (Fig 14.2), continuous receive frequency coverage over the range is actually easier to provide than to not provide, offering a marketing advantage for those who like to also do shortwave or broadcast listening.

An Achille's Heel

Almost every design choice offers both advantages and disadvantages, and the generally beneficial upconverting architecture is no exception. The primary limitation is not in the design itself, but rather in the characteristics of crystal filters in the low VHF range combined with the manufacturer's desire to maintain maximum flexibility. Early upconverting radios had a first IF bandwidth of typically 15 to 20 kHz. This allowed signals with narrow to wide bandwidth to pass through this portion of the receiver with the operating bandwidth set by the DSP filter, typically in the third IF stages, accommodating FM, AM and CW as well as SSB.

The result was that very strong signals within the first IF passband would travel all the way through the receiver's gain and mixing stages until they reached the ADC of the DSP system. The hope was that signals outside the operating DSP bandwidth, generally narrower than the initial first IF roofing filter bandwidth, would be eliminated in the DSP. Unfortunately, if the signals were strong enough they could exceed the dynamic range of the ADC. This could result in both reduction in receiver gain for the desired signal (blocking) or, if there were multiple signals, third-order intermodulation distortion (IMD) products *within* the desired DSP bandwidth due to two or more signals *outside* the DSP bandwidth. These effects can be a major limitation in a band crowded with strong signals.

In the architecture of Fig 14.2, the first IF bandwidth can be set as narrow as HF crystal filters can be made, typically down to 200 Hz, eliminating the problem but also eliminating continuous coverage below 30 MHz. Narrower VHF crystal filters have started to become available, and some top-end upconverting radios now offer selectable VHF first IF filters at 10, 6 and 3 kHz bandwidths, although their skirts are considerably wider than good HF filters. It wasn't long ago that such filters were unheard of, but they keep getting better and better, so perhaps it will become a moot point in the future.

Distributed Roofing Filters

The architectures discussed so far have applied narrow filtering at either the front (Fig 14.2) or the back (Fig 14.5) of the receiver chain. One manufacturer observed that there is a middle ground that could be used to maintain the near-in dynamic performance almost as well as radios with an HF first IF, yet maintain the advantages of upconverting. The Ten-Tec Omni VII has spread the filter-

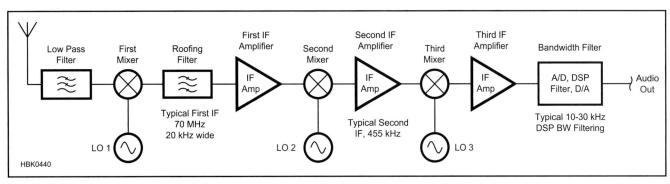

Fig 14.5 — Simplified block diagram of upconverting general coverage transceiver, receiver section shown.

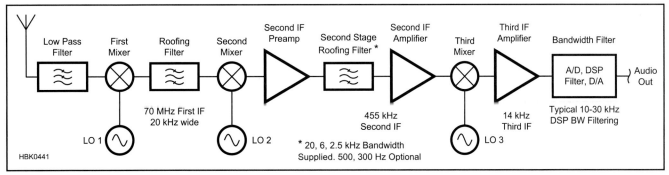

Fig 14.6 — Simplified block diagram of Ten-Tec's distributed roofing filter architecture.

ing to all three IFs as shown in **Fig 14.6**. The Omni VII achieves good dynamic receiver performance through an architecture that is worth a few words of explanation.

The problems resulting from strong unwanted signals near a desired one are minimized if the unwanted signals are kept out of stages in the receiver where they can cause the nonlinear effects that we try to avoid. Note that in Fig 14.2, the only place where the desired and undesired signals all coexist is in the first mixer. In top performing transceivers, the "sideband filter," shown on the receive side in Fig 14.2, can be switched to a much narrower sharp filter for narrow bandwidth modes such as CW or data. If the first mixer has sufficient strong-signal handling capability, the undesired signals will be eliminated in the filter immediately behind the first mixer. The later amplifier, mixer and DSP circuits only have to deal with the desired signal.

The Omni VII has effectively combined two technologies. The first IF has a 20-kHz-wide roofing filter at 70 MHz, plus selectable steep-skirted 455 kHz Collins mechanical filters at the second IF and then DSP filters at the third IF. Careful attention to the gain in stages between the filters maintains desired sensitivity, but not so much sensitivity that the undesired signals have a chance to become a serious problem. With bandwidths of 20, 6 and 2.5 kHz supplied, and 500 and 300 Hz as accessories, the undesired close-in signals are eliminated before they have an opportunity to cause serious trouble in the DSP, usually a key limitation for dynamic range. Some representative ARRL Lab receiver test results are shown in **Table 14.1**. The downconverting data shown is based on *QST* Product Review data on the Elecraft K3; the 20 kHz upconverting data from the ICOM IC-756 ProIII; the 3 kHz upconverting data from the ICOM IC-7700 and the distributed data from the Ten-Tec Omni VII.[2-5] It should be noted that there are significant price differences between the transceivers, but they are representative of the difference in the performance that one can expect from the distinct architectures in 2010.

14.3.2 Features of Current Transceivers

Current 100-W (or higher) transceivers tend to include the following features. Some features tend to appear in transceivers of different price ranges, thus the features have been placed in generic groups based generally on price class.

BASIC FEATURES

Even the lowest price entry-level transceivers of every generation have features never even considered in earlier higher-end units. The following is not an exhaustive list but should provide an indication of the benefits of some of the more popular features.

• A general coverage receiver — Typical coverage is from below the medium-wave broadcast band (sometimes with reduced sensitivity) to the top of HF, sometimes well into VHF, depending on transceiver frequency coverage. While this is useful to allow listening to international broadcast stations or other services, it may be even more useful as an adjunct to an amateur's workbench as a test instrument.

• Wide transmit frequency coverage — While this section has so far addressed "HF" transceivers, it is important to note that all current models extend downward to MF (160 meters) and most extend into VHF as well. Many provide coverage through 6 meters, and some go further into VHF and even UHF. In some cases the additional coverage may obviate the need for one or more additional radios in the station, making for a more compact, and sometimes less expensive, station.

• Multiple modes of operation — Most current transceivers offer SSB, CW, AM and FM modulation and detection. Some offer provision for digital modes, including both modulation and demodulation. Depending on frequency coverage, some "HF+" transceivers can also serve as VHF FM transceivers, offering standard repeater frequency offsets, squelch and tone encoding.

• Choices of selectivity — All current generation transceivers offer DSP-based variable selectivity. In lower-cost models it tends to be a small but adequate number of discrete choices, while in higher priced units selectivity is often continuously variable in 50 or 100 Hz steps. Advanced units allow shaping of the passband as well as high- and low-frequency rolloff adjustment. Some offer a dual-passband filter, designed to separately pass each tone of an FSK signal and not the frequencies between them.

• Dual VFO operation — All current designs offer instant selection of either of two operating frequencies to allow split-frequency operation. This allows the setting of one frequency for transmit and another for receive. A single switch or button can be used to toggle between the two so that the transmit frequency can be monitored before transmitting.

• DSP impulse, noise and carrier reduction — Most DSP-equipped transceivers provide for the reduction of impulse noise such as that from ignition systems or appliances.

• CW keyer — Most current model transceivers include the capability to connect keyer paddles for sending CW via an internal keyer. Some also include memories to allow one-button transmission of short messages.

• Removable front panel — Transceivers designed for mobile operation often have a removable front panel. This allows a compact control head to be mounted near the operator, while the radio's main chassis can be located in the trunk or other out-of-the-way spot.

FEATURES FOUND IN ADVANCED TRANSCEIVERS

More advanced transceivers tend to have additional features for their increased price. Most are larger than the basic units so that the front panels can accommodate additional controls and displays to support the extra features.

• Multiple roofing filter bandwidth — Most basic transceivers incorporate a single roofing filter with sufficient width to permit use on all modes. A typical value is 15 to 20 kHz. Improved close-in dynamic range performance can be obtained by having narrower roofing filters for narrow bandwidth modes as discussed previously and as described in the **Receivers** chapter.

• Adjustable noise blanking — Some transceivers include a separate wideband receive channel designed to detect and blank the main receiver in the presence of impulse noise. This capability is often independent of the similar DSP-based function. One or the other may be found to be better at eliminating noise from particular sources, while sometimes using both is even more helpful. Some transceivers offer two separate methods

---

Table 14.1
Comparison of Dynamic Performance of Samples of Three Architectures

Performance Data	Downconverting (500 Hz)	Upconverting (20 kHz)	Upconverting (3 kHz)	Distributed
BDR (20/5/2 kHz) dB	142/140/139	122/102/98	133/120/108	137/134/134
IMD DR (20/5/2 kHz) dB	106/105/103	102/78/70	108/99/87	91/84/82

*BDR = Blocking Dynamic Range
** IMD DR = Intermodulation Distortion Dynamic Range

of noise blanking, each suited for different common types of impulse noise.

• Choices of AGC parameters — Most transceivers provide automatic AGC settings for different operating modes. Some advanced transceivers provide additional choices to allow optimization for different kinds of operation and different ambient noise levels. For example, to maximize sensitivity for weak signals, the AGC operating threshold should generally be set above the external noise level. This will vary depending on ionospheric conditions. Other parameters include the slope of the AGC curve and hold and attack times.

• Frequency memories — Advanced (and some basic) transceivers include the capability to store operating frequencies and other settings in memory. "Band-stacking registers" are multiple memories for each band that are typically accessed by pressing the band selection button multiple times. Typically they will store two or three frequencies per band, along with mode, filter and other settings, in a last-in, first-out register stack.

• Dual receive — Some higher-end transceivers include two independent receive channels. Dual receive channels can be useful to monitor two frequencies simultaneously. For example, you could listen simultaneously to transmit and receive frequencies being used by a DX station operating in split mode.

There are some subtle differences between the *sub-receiver* capabilities of different model transceivers. In some transceivers, the sub-receiver can only be used in the same band as the main receiver. This is not a limitation for normal split operation, but it does not permit monitoring a different band for openings or other activity. While some transceivers offer a duplicate receiver system for the sub-receiver, others have sub-receivers with lower dynamic range performance or fewer filter options than the main receiver.

Some sub-receivers have independent antenna input connections that allow operation with two different antennas. If the tuning of the two receivers can be locked together, that can provide a kind of diversity operation to reduce fading. Typically that might entail polarization diversity (one horizontal antenna, one vertical); space diversity with two antennas separated by a wavelength or so; or arrival angle diversity, using antennas at different heights.

• Panadapter — While a visual view of the received spectrum has been possible for decades with a station accessory called a *panadapter*, these devices were not too common in amateur stations until manufacturers started building display screens into transceivers. Many current transceivers offer panadapters in different forms. Often they offer views of signals on a band, adjustable over a range from the immediate vicinity of the operating frequency to the whole band. Some transceiv-

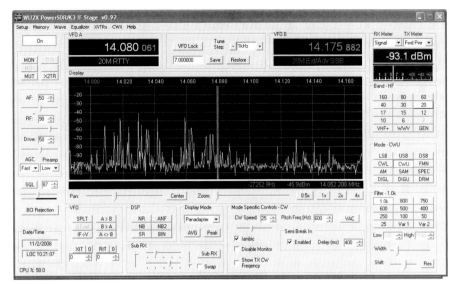

Fig 14.7 — Wideband view of the CW portion of 20 meters during a contest, using the K3 with LP-Pan software defined IQ panadapter.

ers with dual receivers provide either dual panadapters or the ability to use one on the second receiver to keep track of band openings while communicating on the main band. In some cases, the receive circuitry is shared with the panadapter so that the audio is disabled during the time of the scan.

A *real-time* panadapter scans the frequency segment and repeats the scan over and over, giving an appearance of continuous updates. Thus newly arrived signals will show up during each scan. Some panadapters offer a single-pass or *one-shot* functionality in which a button press initiates a single scan over the selected range. The display retains the same snapshot of band activity until the next time the button is depressed. While this can provide a sample of band activity, it does not provide the same level of information as a real-time display.

Some transceivers provide a wideband IF output ahead of roofing filters so that an external panadapter can be used to monitor the received band. **Fig 14.7** shows the display of a TelePost LP-Pan IQ Panadapter connected to the output of an Elecraft K-3 transceiver. This system provides a highly capable real time spectrum analysis capability as well as interactive operation with the transceiver. The wideband IF output can also be used by other external signal processing equipment.

• Higher power — Transceivers are available at many power levels, from the very compact and portable low power (QRP) transceivers to medium power output sets. While we have been focusing on the usual 100 W PEP output transceivers that are the most popular, some transceivers provide 200 W PEP output or more.

• AF equalization — Long a capability of amateurs interested in broadcast-quality voice signals, newer transceivers provide a DSP-based audio equalization capability similar to professional studio control consoles. This is often available in both transmit and receive. In receive mode, it functions largely as a higher resolution tone control. In transmit mode, equalization can be used to shape audio response to provide higher fidelity. Or it can be used to focus the response to best punch through a DX pileup or compensate for individual voice or microphone response.

• Memory voice recorder — Sometimes called a *voice keyer* because of the similarity to a CW memory keyer, this function allows you to record and play back short messages. Voice keyers use digital technology and store the voice signal into memory. Different radios have different numbers of such memories, and some also can record received audio for playback on the air or to record a contact.

• RTTY and PSK31 decode — Some transceivers, generally at the upper end of the price spectrum, include the capability to decode RTTY and PSK31 transmissions on the display screen, eliminating the need for a separate PC and sound card. There usually is a limited amount of screen space for decoded messages, but it is usually plenty for contest exchanges or occasional contacts. A few transceivers, notably recent ICOM transceivers, provide connections for a computer keyboard to allow sending without an attached PC. The Elecraft K3 offers the ability to use the CW keyer or to send information as CW; the radio then converts the CW to RTTY or PSK31 for transmission.

• Flexible configuration — Two recent transceivers allow the purchaser to custom-order the configuration of features. The Elecraft K3 and Yaesu FTdx9000 Contest transceivers are each available with multiple feature sets spanning around a 2:1 price range, depending on options selected.

14.4 Selecting an HF Transceiver

As with most major purchases, there are a wide range of choices from the most basic functionality to more bells and whistles than most operators will ever use. As might be expected, advanced features and performance generally go hand-in-hand with advanced prices.

This section considers "all-band" transceivers with 100 W or more output. Although we call them "HF" transceivers, all cover 160 meters (which is a medium frequency — MF — band). Many also cover 6 meters, some include 2 meters, and a few feature 70 cm and higher bands. Such wide frequency coverage is a plus for the Technician operators who may upgrade in the future, or for anyone who wants to explore those bands without resorting to multiple radios. All radios discussed here include a general coverage receiver that tunes at least from the AM broadcast band to 30 MHz, so they can also be used as entertainment or shortwave listening receivers as well.

The information presented here is current as of mid-2009, but both products and prices can change dynamically in such a competitive marketplace. Check dealer and manufacturer Web sites, *QST* ads and Product Reviews as you get close to making a choice.

There are also some nice transceivers available as kits, as well as some that operate at lower power levels. While any of these can provide lots of fun, a new operator may not be ready for the challenge of low power operation or serious kit building.

CATEGORIES

Transceivers tend to be grouped into a number of categories based on price points and physical configuration. The price is the typical "street price" — the price they sell for at dealers — not the "list price." Prices are approximate, as of mid-2009, and subject to change (up or down). The categories break down as follows:

Entry-level — Desktop transceivers that cost from around $500 to $1000. They offer features and performance that cost much more only a few years ago.

Mid-range — Desktop transceivers that cost from around $1000 to $2000.

Upper mid-range — Desktop transceivers that cost from around $2000 to $4000.

Top drawer — Transceivers costing around $4000 amd up.

Portable and Mobile — Radios that straddle the entry to mid-range prices, but are specifically designed for portable and mobile operation.

Unless otherwise noted, the 100 W radios described are designed to operate from a nominal 13.8 V dc power supply, usually at around

Fig 14.8 — The ICOM IC-718 desktop HF transceiver.

Fig 14.9 — The Yaesu FT-450 desktop HF and 6 meter transceiver.

20 to 22 A. The radios at higher power levels tend to have special power supply configurations, often requiring operation from ac mains. Most manufacturers offer a "matching" power supply, or other general purpose power supplies meeting specifications may be used.

One of the most active areas of transceiver development is in the application of digital signal processing and software-defined radios. These technologies are given their own chapter of this book for that reason, as well as a long sidebar on the subject in the **Receivers** chapter. It is worth continuing to educate yourself on these technologies as they are displacing a significant number of analog technologies.

Obviously, the assignment of a transceiver to a particular class is somewhat arbitrary and depends on personal tastes and operating requirements. Take the following lists as general assessments and don't be afraid to make your own rankings! In addition, new models and features are continuously introduced. Watch for products in *QST* and on the manufacturer and dealer Web sites.

14.4.1 Entry Level Transceivers

As of mid-2009, there are only four radios in full production that fall in this category, the ICOM IC-718 and IC-7200 and the Yaesu FT-450 and the FT-897D. The IC-7200 and FT-897D are marketed as portable units, so they are also shown in that category. Radios in the portable and mobile category that fall in this price range may be worth checking out, even if you plan on operating from home.

The IC-718, shown in **Fig 14.8**, is available at the lowest price of any new 100 W HF radio in any category, about $570 at this writing. The IC-718 offers a lot of bang for the buck, but it has been around for a while so it doesn't have all the latest digital technology or cover 60 or 6 meters. This radio was produced before IF DSP was common throughout the industry, but it includes an audio DSP add-on module that offers digital noise reduction and a digital notch filter and provided as standard equipment.

The operating bandwidth is set via discrete physical filters. A 6 kHz filter for AM and a 2.4 kHz filter for SSB are provided. An additional slot is provided for one of two optional CW bandwidth filters (250 or 500 Hz; $180 each), or for a narrower (1.9 kHz) or wider (3.3 kHz) bandwidth SSB filter at about $200. While they can be installed later, the filters do need to be soldered to the PC board, a bit of a tricky operation even for an old timer, so if you want one, you might want to have it installed by your dealer. Aftermarket filters are also available.

The IC-718 does not include an antenna tuner, but provision is made for one of two ICOM external tuners — the AT-180 for coax fed antennas with an SWR of 3:1 or less ($370), or the AH-4 for wire antennas ($260). Aftermarket tuners are also available from a number of manufacturers.

The IC-718 is generally easy to operate and includes all the basic controls and capabilities to get you on the air on HF. If you need to add features through available options, the price will go up considerably, bringing it more in line with some of the other choices.

The Yaesu FT-450, shown in **Fig 14.9**, is

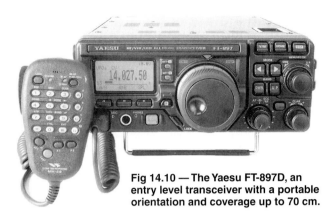

Fig 14.10 — The Yaesu FT-897D, an entry level transceiver with a portable orientation and coverage up to 70 cm.

Fig 14.12 — Kenwood TS-480HX, the only transceiver in this category with a 200 W transmitter.

Fig 14.11 — The ICOM IC-7200 desktop HF and 6 meter transceiver — similar in size to the IC-718, but with an outdoor focus.

Fig 14.13 — The ICOM IC-746PRO offers operation up to 2 meters with full DSP features, but doesn't include the 60 meter SSB channels.

a fairly recent addition to the Yaesu line of current generation transceivers and is at the lower end of the price range. The FT-450 includes the 60 meter channels and 6 meters. It also provides for FM in addition to SSB, AM, CW and data operation. It is very much like its bigger and more expensive siblings in that it's built around a common DSP architecture that provides multiple operating bandwidths to cover each mode. There are three fixed bandwidth choices each for CW, SSB and AM. These are all built in, so an FT-450 ends up costing about the same as an IC-718 with an optional filter. The FT-450 can be purchased with an internal antenna tuner for an additional $100.

The FT-897D (see **Fig 14.10**) and IC-7200 (see **Fig 14.11**) are in a way opposites to the previous transceivers of the same brands. That is, the IC-7200 is a newly designed, DSP based, radio sharing some features of the current crop of ICOM radios but packaged in many ways similar to the IC-718. The FT-897D, on the other hand, is of the same generation as the IC-718, in that it uses analog IF filters and provides audio-based DSP.

Both transceivers are marketed as portable units, with the IC-7200 emphasizing physical ruggedness. The FT-897D offers a number of options for portable use including a rechargeable internal battery pack that can make the transceiver self-contained as a low-powered transceiver.

Table 14.2
Transceivers in the Entry and Portable/Mobile Categories
All 100 W transceivers, except as noted

Entry Level Model	Street Price	DSP	60 Meters	V/UHF	Ant Tuner
ICOM IC-718	$570	NB, NF	No	No	No
ICOM IC-7200*	$1050	IF	Yes	6 Meters	No
Yaesu FT-450*	$650	IF	Yes	6 Meters	$100
Yaesu FT-897D*	$800	AF	Yes	6, 2 m; 70 cm	No
Portable/Mobile					
ICOM IC-706MKIIG	$920	NB, NF	No	6, 2 m; 70 cm	No
ICOM IC-7000	$1200	IF	Yes	6, 2 m; 70 cm	No
ICOM IC-7200*	$1050	IF	Yes	6 Meters	No
Kenwood TS-480SAT	$989	AF	Yes	6 Meters	Yes
Kenwood TS-480HX**	$1050	AF	Yes	6 Meters	No
Kenwood TS-2000B***	$1300	IF	Yes	6, 2 m; 70 cm	Yes
Yaesu FT-857D	$680	AF	Yes	6, 2 m; 70 cm	No
Yaesu FT-897D*	$800	AF	Yes	6, 2 m; 70 cm	No

*Single unit radio, others have separable control head.
**200 W output; requires one 40 A, or two 20 A power supplies.
***Includes second receiver for AM or FM reception.

Fig 14.14 — Elecraft K3 HF and 6 meter transceiver. The flexibility of selectable options for this top performing radio gets it listed in three price categories.

Fig 14.15 — FlexRadio Flex-3000 software-defined radio. This radio operates with the same *PowerSDR* software as the Flex-5000 series, and provides excellent performance in a small box. (*Photo courtesy FlexRadio Systems*)

Fig 14.16 — Kenwood TS-2000. This transceiver operates on MF, HF, VHF and UHF into microwaves with an option, providing broad frequency coverage in a single package.

Fig 14.17 — The Ten-Tec Jupiter HF transceiver provides solid basic HF coverage in a modern DSP driven configuration.

14.4.2 Portable and Mobile Transceivers

There are many choices in this category. These are 100 W (or more) HF transceivers that usually include one or more VHF/UHF bands. They have many of the features of larger radios, but they are compact and designed for the tight cockpit of modern vehicles or for easy transportation to a portable location.

There is no reason these radios can't be operated from a home station as well. The usual trade-off is that they have smaller front panels with fewer and smaller controls. They often make up for the missing controls with more programmable *menus* that some operators may find restricting. Still, they may be perfect for a compact home radio station, and can be moved to a vehicle as well.

There are too many radios in this category to discuss separately, so we'll highlight some of the differences in **Table 14.2**. Note that all, except for the portable or field oriented FT-897D and ICOM IC-7200, have a removable front panel that is designed to be mounted in the front of a vehicle while the body of the radio can be mounted in the rear. All can be combined together in some way to operate as a single unit for home or field use. Perhaps notable in this group, the Kenwood TS-480 (**Fig 14.12**) offers a choice of models with either an internal antenna tuner or a 200 W output transmitter, the only one in this price category. The Kenwood TS-2000B is a version of the TS-2000 that incorporates a detachable front panel.

Some of these radios can operate into the VHF and UHF range. Not only do they operate FM there, but they can also operate SSB, CW and even AM, making them much more versatile than the usual VHF FM-only mobile setup. Some will even allow reception of wideband FM broadcast signals, but none will be confused with a high fidelity audio system.

14.4.3 Mid-Range Transceivers

What do you get if you dig a bit deeper into the checkbook? Generally, you get a somewhat larger radio with easier to grasp controls, more features — or more choices within a feature type, such as more operating bandwidths to choose from. All are relatively recent designs with IF DSP, allowing a wide range of operating bandwidths. All, except for the otherwise wide-frequency-range IC-746PRO (see **Fig 14.13**), offer operation on the five 60 meter channels available in the US.

You also may get better receiver performance, perhaps one of the key elements that separate the radios at the higher ranges. In this case we are talking about the ability to receive a weak signal within a kHz or two of a strong one — called *close-in dynamic range*. The higher the number the better, and as noted there is quite a range. While this is only one of the many parameters evaluated in the ARRL

Table 14.3
Transceivers in the Mid Range Category
All are 100 W transceivers

Model	Street Price	DSP	60 Meters	V/UHF	Tuner	IMD DR (2 kHz)
Elecraft K3/100F	$2090*	IF	Yes	6 Meters	$330	103 dB
FlexRadio Flex-3000	$1599	IF	Yes	6 Meters	No	95 dB
ICOM IC-746PRO	$1600	IF	No	6, 2 Meters	Yes	70 dB
Kenwood TS-2000**	$1400	IF	No	6, 2 m; 70 cm	Yes	57 dB
Kenwood TS-2000X**	$1870	IF	No	6, 2 m; 70, 23 cm	Yes	57 dB
Ten-Tec Jupiter	$1595	IF	Yes	No	$299	63 dB
Yaesu FT-950	$1500	IF	Yes	6 Meters	Yes	71 dB

*Base assembled 100 W unit. IMD DR measured with optional 400 Hz roofing filter. Kit version also available.
**Includes a second receiver for simultaneous AM or FM reception only.

Fig 14.18 — The Yaesu FT-950 HF and 6 meter transceiver provides most of the FT-2000 functionality except the second receiver.

Fig 14.19 — The Ten-Tec Omni VII offers very good receiver performance and a single-pass panadapter.

Fig 14.20 — The ICOM IC-9100H transceiver offers coverage of 1.8 through 440 MHz with an option for 1.3 GHz operation. (Icom America photo)

Fig 14.21 — The Yaesu FT-2000 offers a second HF SSB receiver as standard equipment. A 200-W version (the D model) is available.

Lab for product reviews, some believe it is the key in a crowded operating environment such as contesting or DX chasing. When the band is crowded with very strong signals operating nearby, you can experience interference *generated in your receiver* by mixing products from those nearby signals. The higher the close-in dynamic range, the less likely you are to experience internally generated interference to stations you are trying to hear. (Interference from IMD products is discussed in the **Receivers** chapter and in the **Test Equipment and Measurements** chapter.)

The Elecraft K3 (**Fig 14.14**) provides top performance in this category in any price range, and shows up in almost all because of its configuration flexibility. A new addition to this grouping is the FlexRadio FLEX-3000 (**Fig 14.15**) software-defined radio. This transceiver is very much like its larger sibling, the FLEX-5000, but is in a compact enclosure that makes it a good match for a laptop PC. The FLEX-3000 offers excellent receiver performance and the epitome of flexibility. Although it doesn't have the space for the second receiver and antenna options of its larger brother, it will look the same on the PC screen and likely sound the same to the far end.

Again, due to space limitations, we have summarized some of the key features and parameters in **Table 14.3**. Some of the equipment needs a bit of additional explanation. Note that none of the radios in Table 14.2 include a fully-capable second HF receiver, but the TS-2000 (**Fig 14.16**) does have a second receiver mainly for VHF FM use — perhaps handy for some who wish to monitor their local repeater while operating HF. In the upper middle range some do have independent second receivers so that you can listen to signals on two frequencies — one in each ear, if you wish — handy while operating on split frequencies and a popular option with DX chasers. Other popular radios in this category include the Ten-Tec Jupiter, recently upgraded with a new display and shown in **Fig 14.17**, and the Yaesu FT-950 (**Fig 14.18**) sharing many features of the mid-range FT-2000, but with a single receiver channel.

14.4.4 Upper Mid-Range Transceivers

Transceivers in the upper mid-range group offer a number of choices between different desirable features. Key parameters are noted in **Table 14.4**. Some offer more or different features, while others make a push toward higher performance. At the top of the next bracket, manufacturers try to provide everything, while here you need to look carefully and decide what is most important.

The Ten-Tec Omni VII (**Fig 14.19**) was discussed previously as having a

Table 14.4
Transceivers in the Upper Middle Range Category

All are 100 W transceivers except as noted. All have IF DSP and 60 meter coverage.

Model	Street Price	2nd Rcvr	V/UHF	Tuner	IMD DR (2 kHz)
Elecraft K3/100F	$2690*	Yes	6 Meters	$330	103 dB
FlexRadio Flex-5000A	$2519**	$699	6 Meters	$299	99 dB
ICOM IC-7600	$3899	DW***	6 Meters	Yes	88 dB
ICOM IC-9100	TBD	No	6, 2, 70 cm†	Yes	N/A††
Ten-Tec Omni VII	$2695	No	6 Meters	$300	82 dB
Yaesu FT-2000‡	$2250	Yes	6 Meters	Yes	69 dB

*Assembled with dual receiver with 2700 Hz roofing filter in each receiver. IMD measured with optional 400 Hz roofing filter ($120), many other options are available to a maximum equipped price of $4747 not including available VHF and UHF transverters. Options may be added at any time.
**Requires a PC for operation.
***Dual watch — combines reception of two signals in same band into same audio channel.
†23 cm available as an option.
††Not tested in ARRL Lab at press time, check *QST* Product Reviews at **www.arrl.org**.
‡Includes internal power supply. A 200 W version, the FT-2000D, is available for $2825 including external supply.

Fig 14.22 — The FlexRadio 5000A, a software-defined radio that provides excellent performance, flexibility and growth potential through its computer control capability and optional second receiver and tuner.

Fig 14.23 — The display of a FlexRadio FLEX-5000A connected to a PC running *PowerSDR* software.

Fig 14.24 — The ICOM IC-7600 transceiver provides improved receiver performance through included 3, 6 and 15 kHz roofing filters. (*Photo courtesy Icom America*)

Fig 14.25 — The Ten-Tec Orion II offers excellent receive dynamic range through very narrow HF roofing filters in its main receiver.

Fig 14.26 — The Yaesu FTDX5000 is one of a family of top-performing 200 W transceivers expandable with a 300 Hz roofing filter and multiple display options. The unit on top is the companion SM-5000 station monitor.

Fig 14.27 — The ICOM IC-7800 provides excellent performance and an extra-flexible display system.

unique distributed roofing filter architecture that provided very good receiver dynamic performance while still offering a general coverage receiver along with a single-scan panadapter display.[5] The ICOM IC-9100 (**Fig 14.20**) has the capability to serve as both an MF/HF transceiver as well as a multimode VHF/UHF radio with an option for the 1.2 GHz band.

The Yaesu FT-2000 (**Fig 14.21**) is the only transceiver in this group that comes equipped with a second receiver and two audio channels.[14] The second receiver is an analog design that seems quite good, although its operating bandwidth is set by optional crystal lattice filters. The FlexRadio 5000A (**Fig 14.22**) is a software-defined radio that requires a separate PC to operate it.[6] Controls and displays are implemented via the PC as shown in **Fig 14.23**. FlexRadio also offers a version with an internal dedicated computer, described in the top drawer listing. Either can make use of an optional independent identical second receiver.

The ICOM IC-7600[13] (**Fig 14.24**) fills the slot of the previously available and popular IC-756PROIII[3], but offers additional dynamic performance due to a choice of 3, 6 and 15 kHz roofing filters, supporting all modes.

The Elecraft K3 shows up in both middle categories and in the top drawer as well, since it is modular in design and can be purchased in different configurations.[2] There are many

Table 14.5
Transceivers in the Top-Drawer Category
All have IF DSP and 60 meter coverage.

Model	Street Price	2nd Rcvr	Power	V/UHF	Tuner	IMD DR (2 kHz)
Elecraft K3/100F*	$4747*	Yes	100 W	6 Meters	Yes	103 dB
FlexRadio Flex-5000C	$4769**	$699	100 W	6 Meters	$299	99 dB
ICOM IC-7700	$6899	No	200 W	6 Meters	Yes	87 dB
ICOM IC-7800	$10,990	Yes	200 W	6 Meters	Yes	86 dB
Ten-Tec Orion II	$3995	Yes	100 W	No	$300	94 dB
Yaesu FTdx5000	$5700	Yes	200 W	6 Meters	Yes	103 dB
Yaesu FTdx9000 C	$4700	$1950***	200 W	6 Meters	Yes	79 dB
Yaesu FTdx9000D	$9699	Yes	200 W	6 Meters	Yes	87 dB
Yaesu FTdx9000MP	$10,699	Yes	400 W	6 Meters	Yes	85 dB

*Assembled including all options, not including available VHF and UHF transverters. Options may be added at any time.
**Includes built in PC.
***Must be specified at time of transceiver order, Other options are also available at time of purchase.

more options which some may find desirable, but the listed prices are for a fully assembled, basic 100 W transceiver in Table 14.2 with a single receiver and in Table 14.3 with a dual receiver. The K3 is also offered in kit form at significant savings.

14.4.5 Top Drawer Transceivers

Transceivers at the very top of the price range are available from a number of manufacturers. These transceivers span a considerable variation of prices, but from $4000 and up it probably doesn't make too much sense to subdivide the list. Buying decisions are driven by subtle differences in features or the desire for optimum receiver performance, a key issue with many contest and DX-focused operators.

The Ten-Tec Orion II (**Fig 14.25**) has been a mainstay of this category, with its excellent main channel receive performance and smooth CW break-in operation.[15] It offers the narrowest of first IF roofing filters in its amateur band-only main receiver and general coverage in its sub-receiver.

Among many families of transceivers, there is some flexibility here in terms of hardware. The previously discussed Elecraft K3[2] epitomizes flexibility through its extensive options list, as well as top-shelf receiver performance. It is definitely a radio that can grow with the operator's needs! Yaesu offers two families of radio systems in the Yaesu FTDX5000 series (**Fig 14.26**) with a top performing receiver and the FTDX9000.[16] Both have base units at the 200 W level and both offer available additional display options. The FTDX9000 also offers a 400 W (MP suffix) version. At ICOM's top end, the IC-7700 is basically a top of the line IC-7800 (**Fig 14.27**) without a second receiver — not something that is needed for every application.[4]

14.4.6 Software Defined Radios

We described some HF transceiver choices as "software-defined radios" or SDR. It may be worth a short digression to discuss this topic in the context of available equipment. Software-defined radio architecture and design is covered in the **DSP and Software Radio Design** chapter.

As we'll discuss, there are a range of definitions — subject to some controversy — on what constitutes an SDR in the Amateur Radio world. The FCC has defined the SDR concept in terms of their commercial certification process as:

"…a radio that includes a transmitter in which the operating parameters of the transmitter, including the frequency range, modulation type or conducted output power can be altered by making a change in software without making any hardware changes."

The FCC expects this to yield streamlined equipment authorization procedures by allowing "manufacturers to develop reconfigurable transceivers that can be multi-service, multi-standard, multi-mode and multi-band…."[18] In this context, they are envisioning radios that can be modified at the factory by using different software to meet different requirements. While they allow for field changes, the FCC's focus is different from ours.

SDR IN THE AMATEUR WORLD

In the amateur environment, we are particularly interested in radios that can be changed through software by the end user or operator to meet their needs or to take advantages of newly developed capabilities.

The ideal SDR would thus have a minimum of physical constraints. On the receive side, the antenna would be connected to an analog-to-digital converter that would sample the entire radio spectrum. The digitized signal would enter a processor that could be programmed to analyze and decode any form of modulation or encoding and present the result as sights and sounds on the output side of the processor.

On the transmit side, the processor would accept any form of information content, convert to digital if needed, process it into a waveform for transmission and send out a complex waveform conveying the information as an RF signal on the appropriate frequency or frequencies, at the desired power level to transmit from the antenna.

Not surprisingly, our utopian SDR is much easier to imagine than to construct. As a practical matter, our usual PC has some constraints that don't allow us to do quite what we want. Still, for a few hundred dollars, it is possible to purchase a PC that gets us fairly close.

The key to amateur SDR operation with a PC is the sound card.[19] This card, or sometimes an external interface device, can accept an analog signal and convert it to a digital one for processing. Advanced SDRs such as those from FlexRadio have the functionality built in. The software will determine the type of processing and the nature of the signals we can deal with. It can also take the results of processing and convert them into an analog signal. This sounds like just what we are looking for to make an SDR — and it is. Such an SDR in receive mode would consist of the blocks in **Fig 14.28**. We do have a few significant limitations:

- For most sound cards the sampling rate is 192 kHz or less, limiting the received analog signal to a frequency of 96 kHz. Some kinds of dual channel processing allow a response as high as the sampling rate. (Sampling rate limitations, also referred to as Nyquist rate[20], require sampling the incoming signal at twice the highest frequency in the sampled signal.)
- Most sound cards do not have the sensitivity required in a radio receiver and on the transmit side (see **Fig 14.29**) can only output 100 mW or less.

Thus we are faced with the need to insert some external processing functions outside the PC. These will be used, at a minimum, to translate the frequency range we wish to use to one that the sound card can deal with on the receive side. On the transmit side, we will need to translate the frequency range up to the desired portion of the radio spectrum and increase the signal to our desired transmit power level.

The SDR designer, as with all designers, is faced with a trade off. The equipment external to the PC required to make it do what we want

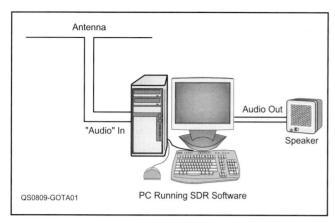

Fig 14.28 — Conceptual block diagram of an ideal software-defined radio (SDR) receiver.

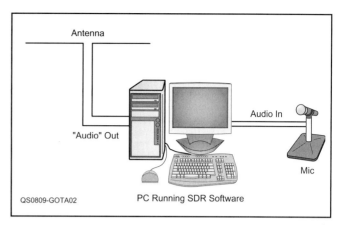

Fig 14.29 — Conceptual block diagram of the transmit side of an ideal SDR transceiver.

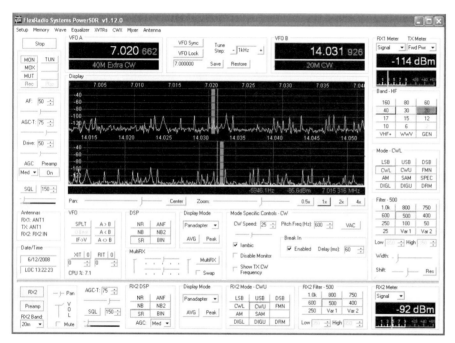

Fig 14.30 — One version of the main operating screen of PowerSDR operating software.

IC-7800, Ten-Tec Orion and Yaesu FT-2000, for example, are provided with a mechanism to allow an easy end-user upgrade to new firmware revisions. These radios look like most any other pre-SDR radio in that they have front panels with knobs and dials. Unless you looked at all the revisions to the operating instructions you wouldn't know that they were field-reconfigurable.

Another distinction between the groups is that most of the firmware for the radios in this group is proprietary with revisions available only from the manufacturer, at least as of this writing. That isn't to say that a solid programmer couldn't and perhaps hasn't developed custom software for one of these radios, but it hasn't happened often.

While all radios in this group are primarily designed to operate without an external computer, they all can be computer controlled using aftermarket software, available from multiple developers. While this software can make them feel a bit like the radios in the other group, the operating parameter ranges are all set by the radio's internal operating firmware.

Differences in Performance

The blank front panel architecture radios generally have the most flexibility in operation, since they are not constrained by the physical buttons and knobs on the front panel. The more traditional-looking versions with physical controls and displays may take advantage of those hardware constraints to gain improved performance at the expense of operating flexibility, but a look at the specs will indicate that it isn't always the case. Some blank front panel SDRs offer top shelf performance.

There will always be some who prefer the more traditional radios and are happy to have it configured and to let it stay the way they like it. Others, especially those who enjoy computers as well as radio, will prefer a transceiver that might get better with the

may also limit the choices we can make by software change in the PC. The more hardware features we build in, the fewer choices we may have. In addition to PC software, there is often firmware, hard wired instructions in the box outside the PC. This has resulted in two general approaches in SDR.

The "Blank Front Panel" Architecture

Radios marketed as SDRs tend to be of this type. The classic is the FlexRadio Flex-5000A, reviewed in *QST* for July 2008. The front panel has no controls and all control functions are implemented by soft buttons on the *PowerSDR* software's computer screen (**Fig 14.30**), or via pointing and knob devices that connect to the computer.

There are a number of similarly configured radios useable with the same open source *PowerSDR* software freely available from FlexRadio under the GNU Public License. These include the modular High Performance SDR system available through Tucson Amateur Packet Radio (TAPR)[21], and some other very low cost, but no longer available SDRs such as the Firefly[22] and Softrock-40[23]. (Source code for software called "open source" can be viewed and modified. Thus, not only can you upgrade the radio, but also change functionality to make the radio do what you want.)

The "Looks Like a Radio" Approach

Many current radios are actually built as SDRs. Some, such as the Elecraft K3, ICOM

next generation of computers, sound cards, or software. What's really nice is that we can go whichever direction we choose!

14.4.7 Making a Selection

You know the facts — how can you choose? Looking at tables and product reviews provides a great start, but just as with finding the perfect life partner, the numbers don't tell it all. With radios there's also an element of love at first sight, tempered by the way the radio feels to you as you operate it — ergonomics. Each manufacturer has a different philosophy for structuring the controls and menus, and you may find a great preference for one person-to-machine interface over another.

If you have an opportunity, try out the radios you are thinking of buying. If you are in a local club, find out who uses radios that you are considering and seek an invitation to come over and try them out. Most hams love to show off their stations. Perhaps you have a nearby dealer who has some demo setups, or you can find some at a larger hamfest, or at Field Day. If you're in the northeast, drop by ARRL Headquarters and be a guest operator at W1AW — we have many radios available to try. Nothing beats a test drive!

14.5 Transceivers With VHF/UHF Coverage

Transceivers discussed so far include a wide range of features and operational capability. While some transceivers offer capability into the microwave region, most amateurs start with a transceiver that operates over the MF to HF frequency range, often extending into VHF at 6 meters. It is also true that most amateurs have at least a passing acquaintance with higher VHF bands, perhaps 2 meter or 70 cm FM communications through local repeaters.

At some point, many amateurs have heard enough about satellite communication or exotic beyond-line-of-sight propagation such as troposcatter, sporadic-E layer communication or moonbounce to want to try these activities. They take place using SSB, CW and narrow bandwidth digital modes, so a VHF/UHF FM transceiver won't do. Two current full size HF transceivers extend operation well into the VHF and UHF regions, The Kenwood TS-2000[17] and ICOM IC-9100H not only offer all modes of operation from 160 meters to 70 cm, but each also offers an internal option for the 23 cm (1240 to 1300 MHz) microwave amateur band.

14.5.1 Using an HF Transceiver at VHF/UHF

If you have the HF transceiver you want and would like to try the higher VHF and UHF bands, a viable option is a VHF or UHF transverter. A transverter essentially adds an additional conversion stage, along with pre- and post-amplification, to translate receive and transmit frequencies to a new range.

At VHF, UHF and microwave frequencies, transverters that interact with factory-made transceivers in the HF or VHF range are common and are often home-built. These units convert the transceiver transmit signal up to a higher frequency and convert the VHF/UHF receive frequency down to the transceiver receive frequency. The configuration of a 2-meter transverter is shown in **Fig 14.31**.

For microwave frequencies, it's common for transverters to have a 144-MHz IF for connection to a multimode 2-meter transceiver.

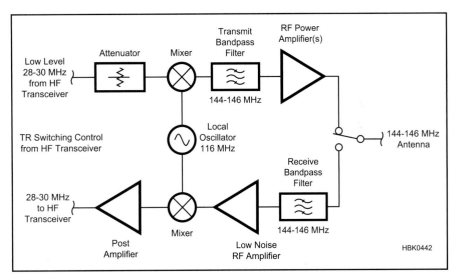

Fig 14.31 — Block diagram of a basic 2-meter to 10-meter transverter.

Fig 14.32 — The Down East Microwave 144-28HP transverter turns an HF transceiver into a 2-meter all-mode transceiver with sensitive receiver and 60 W output.

Use of a higher IF makes image filtering easier. Sometimes transverters use two stages of conversion — microwave to 2 meters, and then 2 meters to 10 meters for use with an HF transceiver.

The resulting performance and signal quality at the higher frequencies are enhanced by the frequency stability, bandwidth filters and signal processing capabilities of the transceiver. A transverter makes stable SSB and CW operation feasible on bands from 144 through 10 GHz and higher.

TRANSVERTER DESIGN

The methods of individual circuit design for a transverter are not much different from methods that have already been described. The most informative approach would be to

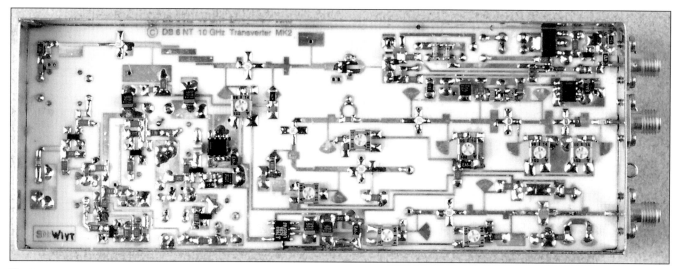

Fig 14.33 — Inside view of the Kuhne Electronics (DB6NT) MKU 10 GHz transverter kit.

Table 14.6
Key Performance and Operational Specifications of Transverters Measured in the ARRL Lab

Model	Band	Receive Gain (dB)	Noise Figure (dB)	Image Rejection (dB)	Output Power (W)
Down East 144-28HP	2 meters	18	1.0	101	60
Elecraft K144XV2*	2 meters	25	1.0	106	10
Elecraft XV144	2 meters	25	1.0	70	20
Kuhne-Electronics MKU	10 GHz	20	1.2	not measured	3
SSB Electronic LT2S MkII	2 meters	21	1.0	not measured	20

*Internal option for K3 transceiver.

study carefully an actual project description.

The interface between the transceiver and transverter requires some careful planning. For example, the transceiver power output must be compatible with the transverter's input requirements. This may require an attenuator (or an amplifier) or some modifications to a particular transverter or transceiver. There is no standard level among transceiver and transverter brands, so check to see that your HF transceiver has a low-level transmitter output. Also important: a dedicated receiver input for transverter use, as well as some provision for TR switching.

The Elecraft K3 is an example of an HF transceiver with well-thought-out transverter provisions. It has dedicated, separate transverter input and output ports that are available on an optional interface board. In addition, the band switch directly supports transverters just as if they were bands in the transceiver. The frequency display shows the VHF or UHF frequency directly for up to nine transverters. The appropriate IF frequency is set up for each band, typically 10 or 6 meters, but any others also can be used. An offset is applied to the frequency calibration to compensate for any error in the local oscillator frequency of the selected transverter and a control signal is sent to select the transverter depending on band selected.

The receive converter gain must not be so large that the transceiver front-end is overdriven, causing intermodulation and blocking. On the other hand, the transverter gain must be high enough and its noise figure low enough so that the overall system noise figure is within a dB or so of the transverter's own noise figure. The formulas in the **Receivers** chapter for cascaded noise figures should be used during the design process to assure good system performance. The transceiver's performance should be either known or measured to assist in this effort.

AVAILABLE TRANSVERTERS

If building a transverter from scratch is not for you, a number of manufacturers produce assembled or kit transverters for many VHF through microwave bands. Transverters that have been reviewed in *QST* include the following, with key ARRL Lab results summarized in **Table 14.6**:

Down East Microwave makes transverters from 6 meters to 3 cm. The ARRL reviewed their 144-28HP 2 meter transverter in *QST*, see **Fig 14.32**.[7]

Elecraft makes transverter kits from 6 meters to 70 cm including 1.25 meters. The ARRL reviewed their XV144 2 meter transverter kit in *QST*.[8]

Kuhne Electronics (DB6NT) makes transverters from 2 meters to 3 cm, except for 1.25 meters. The ARRL reviewed their MKU 10 GHz transverter in *QST*, see **Fig 14.33**.[9]

SSB Electronic offers transverters through the microwave range. ARRL reviewed their LT2S MkII in *QST*.[10]

14.6 Transceiver Control and Interconnection

In the early days of Amateur Radio, there was a logical sequence to the process of switching from receive to transmit. We used separate receivers and transmitters with minimal automation. Upon hearing the distant station say "over" or send "K", the process would begin:
- Switch the receiver from receive to standby.
- Switch the antenna from the receiver to the transmitter.
- Switch the transmitter from standby to transmit.

To go the other way you went through the sequence in reverse, with the switches changed the opposite way. If the sequence were not followed, you risked damage to your ears, equipment or both.

14.6.1 Automated Transceivers

The better equipment of the mid-1950s provided capability to allow for the automation of this process. The Johnson Viking II transmitter, for example, included a socket on the rear panel with 120 V ac applied whenever the standby/transmit switch was moved to the transmit position. By connecting this socket to an antenna relay, the antenna could be switched from the receiver to the transmitter each time the transmitter was switched to transmit. Note that many coaxial relays of the day provided a pair of *auxiliary* contacts that were actuated whenever the relay was energized. These contacts could be used to disable or *mute* the receiver upon switching to transmit. With this arrangement, the station was set up for "one-switch operation," a goal of most early amateurs!

This arrangement worked for both voice and CW operation. Note that in those days, most amateurs used full-carrier AM for voice, rather than the modern suppressed-carrier single-sideband (SSB). With AM operation, the carrier was switched on and the transmitter started putting out power as soon as the operator switched to transmit, whether or not he or she was talking. With a properly adjusted SSB transmitter, there is virtually no signal transmitted except when the operator is actually speaking into the microphone (or the dog barks!). In a similar manner, most CW transmitters only transmit when the key is depressed, so some operators would leave the transmitter turned on and count on the key to determine when a signal would be transmitted.

14.6.2 Break-In CW Operation

Some top-flight CW operators were not quite satisfied with "one-switch operation," but would prefer to just transmit when the key is down and receive when it's up. A simple way that early amateurs accomplished this was to use separate antennas for transmit and receive (to avoid the requirement to switch) and run low enough power so that the transmitter would not overload the receiver.

If the receiver recovered fast enough to hear distant signals between transmitted dits, the configuration was called "break-in keying." This allowed the sending operator to listen to the frequency while transmitting and thus take appropriate action if there was interference. The distant operator could also hit the key a few times to alert the transmitting station that information needed to be repeated. The Q signal QSK (meaning "I can hear between dits") is often used to designate "full break-in" operation.

Fortunately technology has solved this problem, and QSK operators aren't limited to running low power and separate transmit and receive antennas. Most current 100 W class HF transceivers use high-speed relays (with the relay actually following the CW keying) or solid-state PIN diodes to implement full break-in CW. Some RF power amplifiers use high-speed vacuum relays for the TR switching function.

The term "semi-break-in" is used to designate a CW switching system in which closing the key initiates transmission, but switching back to receive happens between words, not between individual dits. Some operators find this less distracting than full break-in, and it is easier to implement with less-expensive relays for the TR switching.

14.6.3 Push-To-Talk for Voice Operators

Another advance in amateur station switching followed longstanding practices of aircraft and mobile voice operators who had other things to contend with besides radio switches. Microphones in those services included built-in switches to activate TR switching. Called push-to-talk (PTT), this function is perhaps the most self-explanatory description in our acronym studded environment.

Relays controlled the various switching functions when the operator pressed the PTT switch. Some top-of-the-line transmitters of the period included at least some of the relays internally and had a socket designed for PTT microphones. **Fig 14.34** is a view of the ubiquitous Astatic D-104 microphone with PTT stand, produced from the 1930s to 2004, and still popular at flea markets and auction sites. PTT operation allowed the operator to be out of reach of the radio equipment while operating, permitting "easy chair" operation for the first time.

Modern transceivers include some form of PTT (or "one switch operation"). Relays, diodes, transistors and other components seamlessly handle myriad transmit-receive changeover functions inside the transceiver. Most transceivers have additional provisions for manually activating PTT via a front-panel switch. And many have one or more jacks for external PTT control via foot switches, computer interfaces or other devices.

14.6.4 Voice-Operated Transmit-Receive Switching (VOX)

How about break-in for voice operators? SSB operation enabled the development of voice operated transmit/receive switching, or VOX. During VOX operation, speaking into the microphone causes the station to switch from receive to transmit; a pause in speaking results in switching back to receive mode. Although VOX technology can work with AM or FM, rapidly turning the carrier signal on and off to follow speech does not provide the smooth operation possible with SSB. (During SSB transmission, no carrier or signal is sent while the operator is silent.)

VOX OPERATION

VOX is built into current HF SSB transceivers. In most, but not all, cases they also provide for PTT operation, with switches

Fig 14.34 — A classic Astatic D-104 mic with PTT stand.

or menu settings to switch among the various control methods. Some operators prefer VOX, some prefer PTT and some switch back and forth depending on the operating environment.

VOX controls are often considered to be in the "set and forget" category and thus may be controlled by a software meu or by controls on the rear panel, under the top lid or behind an access panel. The following sections discuss the operation and adjustment of radio controls associated with VOX operation. Check your transceiver's operating manual for the specifics for your radio.

Before adjusting your radio's VOX controls, it's important to understand how your particular mic operates. If it has no PTT switch, you can go on to the next section! Some mics with PTT switches turn off the audio signal if the PTT switch is released, while some just open the control contacts. If your mic does the former, you will need to lock the PTT switch closed, have a different mic for VOX or possibly modify the internal mic connections to make it operate with the VOX. If no audio is provided to the VOX control circuit, it will never activate. If the mic came with your radio, or from its manufacturer, you can probably find out in the radio or mic manual.

VOX Gain

Fig 14.35 shows some typical transceiver VOX controls. The VOX GAIN setting determines how loud speech must be to initiate switchover, called "tripping the VOX." With a dummy load on the radio, experiment with the setting and see what happens. You should be able to advance it so far that it switches with your breathing. That is obviously too sensitive or you will have to hold your breath while receiving! If not sensitive enough, it may cause the transmitter to switch off during softly spoken syllables. Notice that the setting depends on how close you are to the microphone, as well as how loud you talk. A headset-type microphone (a "boom set") has an advantage here in that you can set the microphone distance the same every time you use it.

The optimum setting is one that switches to transmit whenever you start talking, but isn't so sensitive that it switches when the mic picks up other sounds such as a cooling fan turning on or normal household noises.

VOX Delay

As soon as you stop talking, the radio can switch back to receive. Generally, if that happens too quickly, it will switch back and forth between syllables causing a lot of extra and distracting relay clatter. The VOX DELAY control determines how long the radio stays in the transmit position once you stop talking. If set too short, it can be annoying. If set too long, you may find that you miss a response to a question because the other station started talking while you were still waiting to switch over.

You may find that different delay settings work well for different types of operation. For example, in a contest the responses come quickly and a short delay is good. For casual conversation, longer delays may be appropriate. Again, experiment with these settings with your radio connected to a dummy load.

Anti-VOX

This is a control with a name that may mystify you at first glance! While you are receiving, your loudspeaker is also talking to your mic — and tripping your VOX — even if you aren't! Early VOX users often needed to use headphones to avoid this problem. Someone finally figured out that if a sample of the speaker's audio signal were fed back to the mic input, out-of-phase and at the appropriate amplitude, the signal from the speaker could be cancelled out and would not cause the VOX circuit to activate the transmitter. The ANTI-VOX (called ANTI-TRIP in the photo) controls the amplitude of the sampled speaker audio, while the phase is set by the transceiver design.

As you tune in signals on your receiver with the audio output going to the speaker, you may find that the VOX triggers from time to time. This will depend on how far you turn up the volume, which way the speaker is pointed and how far it is from the mic. You should be able to set the ANTI-VOX so that the speaker doesn't trip the VOX during normal operation.

Generally, setting ANTI-VOX to higher values allows the speaker audio to be louder without activating the VOX circuit. Keep in mind that once you find a good setting, it may need to be changed if you relocate your mic or speaker. With most radios, you should find a spot to set the speaker, mic and ANTI-VOX so that the speaker can be used without difficulty.

14.6.5 TR Switching With a Linear Amplifier

Virtually every amateur HF transceiver includes a rear panel jack called something along the lines of KEY OUT intended to provide a contact closure while in transmit mode. This jack is intended to connect to a corresponding jack on a linear amplifier called something like KEY IN. Check the transceiver and amplifier manuals to find out what they are called on your units. A diagram of the proper cabling to connect the transceiver and amplifier will be provided in the manual.

ENSURING AMPLIFIER-TRANSCEIVER COMPATIBILITY

While you're there, check the ratings to find out how much voltage and current

Fig 14.35 — The function of VOX controls is described in the text. They require adjustment for different types of operating, so front-panel knobs make the most convenient control arrangement. In some radios, VOX settings are adjusted through the menu system.

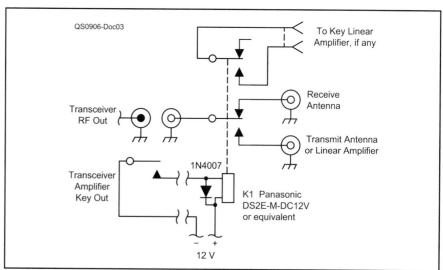

Fig 14.36 — Schematic of an external box that allows a modern transceiver to key a linear amplifier with TR switch voltage or current requirements that exceed the transceiver's ratings. As a bonus, it can also be used to allow reception from a low-noise receiving antenna.

the transceiver can safely switch. In earlier days, this switching was usually accomplished by relay contacts. More modern radios tend to use solid-state devices to perform the function. Although recent amplifiers are usually compatible with the switching capabilities of current transceivers, the voltage and/or current required to switch the relays in an older linear amplifier may exceed the ratings. Fortunately, it is pretty easy to find out what your amplifier requires, and almost as easy to fix if it's not compatible with your radio.

If your amplifier manual doesn't say what the switching voltage is, you can find out with a multimeter or DMM. Set the meter to read voltage in a range safe for 250 V dc or higher. Connect the positive meter probe to amplifier key jack's center conductor, and connect the negative meter probe to the chassis ground (or other key jack terminal if it's not grounded). This will tell you what the open circuit voltage is on the amplifier key jack. You may need to try a lower voltage range, or an ac range, or switch the probes (if the key line is a negative dc voltage) to get a reading.

Now set the meter to read current. Start with a range that can read 1 A dc, and with the leads connected as before, you should hear the amplifier relay close and observe the current needed to operate the TR relay or circuit. Adjust the meter range, if needed, to get an accurate reading.

These two levels, voltage and current, are what the transceiver will be asked to switch. If *either* reading is higher than the transceiver specification, do not connect the transceiver and amplifier together. Doing so will likely damage your transceiver. You will need a simple interface circuit to handle the amplifier's switching voltage and current.

The simple, low-cost relay circuit shown in **Fig 14.36** can be used to switch an older amplifier with a modern transceiver. It offers an added benefit: Another potential use of the transceiver KEY OUT jack is to switch to a separate low-noise receive antenna on the lower bands. While most high-end transceivers have a separate receive-only antenna connection built in, many transceivers don't. If you don't need one of the extra functions, just leave off those wires.

14.7 Transceiver Projects

There are many transceiver designs available, aimed mainly at the advanced builder. We will provide two. One describes the construction of a 5-W single-band SSB/CW transceiver with many features. The other is the description of the design and construction process that resulted in a 100 W transceiver for all HF bands with performance (confirmed in the ARRL Lab) that is as good as some of the best commercial products. Construction details for these projects and support files are available on the ARRL Web site. Additional projects may be found on the *Handbook* CD-ROM.

Fig 14.37 — MFJ Cub QRP transceiver built from a kit.

14.7.1 Transceiver Kits

One of the simplest kits to build, is also one of the least expensive. The ARRL offers the MFJ Enterprises (**www.mfjenterprises.com**) Cub 40 meter transceiver kit (**Figs 14.37** and **14.38**) bundled with *ARRL's Low Power Communication — The Art and Science of QRP* for less than $100 as we write this.[14] Other kits are available from many manufacturers.

Ten-Tec, maker of some of the best HF transceivers, offers single band CW QRP kits for less than $100 for your choice of 80, 40, 30 or 20 meters. Elecraft offers a range of kits from tiny travel radios to the K2 high performance multiband CW and SSB transceiver as well as the top performing K3 semi-kit (mechanical assembly only, no soldering required). The K2 and K3 can both start out as 10 W models and be upgraded to 100 W, if desired. DZkit, the company founded by Brian Wood, WØDZ to produce Heathkit-style radio kits offers modular transmitter and receiver kit sections that can be combined to

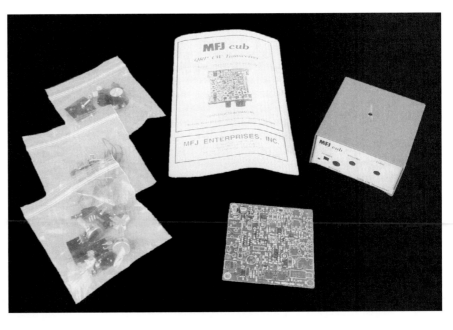

Fig 14.38 — MFJ Cub QRP transceiver kit showing parts supplied.

form a high performance transceiver.

If your interest is strictly QRP transceiver kits, there are many small and specialty manufacturers. The QRP Amateur Radio Club, Inc. (QRP ARCI) maintains a manufacturer list on their Web site at **www.qrparci.org**, available in the "QRP Links" section of the site.

14.7.2 *Project:* The TAK-40 SSB CW Transceiver

Jim Veatch, WA2EUJ, was one of two winners of the first ARRL *Homebrew Challenge* with his TAK-40 transceiver shown in **Fig 14.39**. This challenge was to build a homebrew 5 W minimum output voice and CW transceiver using all new parts for under $50. The resulting radio had to meet FCC spurious signal requirements. The submitted radios were evaluated as to operational features and capabilities by a panel of ARRL technical staff. Jim's transceiver met the requirements for a transceiver that required a connected PC for setup — to load the PIC controller. This information was presented in May 2008 *QST* in an article provided by Jim as a challenge requirement.

The following is a list of the criteria for the Homebrew Challenge and a brief description of how the TAK-40 meets the requirements:

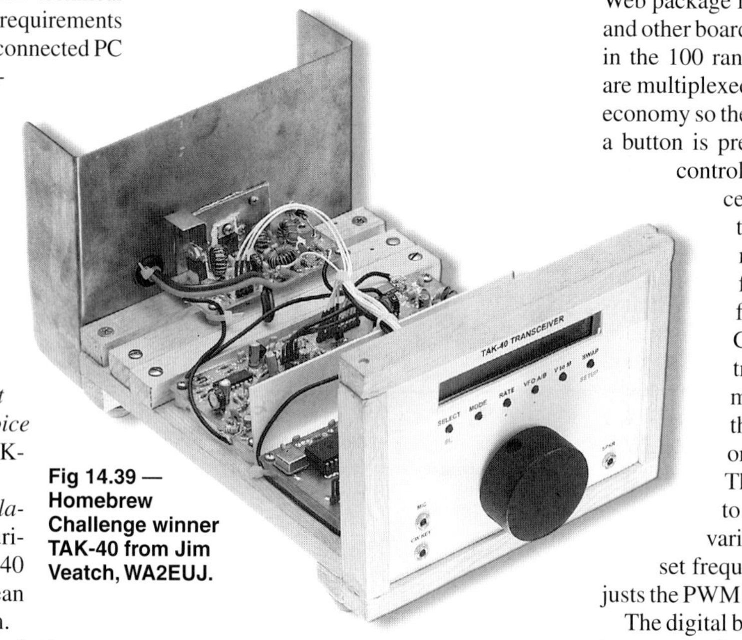

• *The station must include a transmitter and receiver that can operate on the CW and voice segments of 40 meters.* The TAK-40 covers 7.0 to 7.3 MHz.

• *It must meet all FCC regulations for spectral purity.* All spurious emissions from the TAK-40 are at least 43 dB below the mean power of fundamental emission.

• *It must have a power output of at least 5 W PEP.* The TAK-40 generates at least 5 W PEP for voice and CW modes. The ALC can be set as high as 7 W if desired.

• *It can be constructed using ordinary hand tools.* Construction of the TAK-40 uses all leaded components and assembly only requires hand tools, soldering iron and an electric drill (helpful but not strictly necessary).

• *It must be capable of operation on both voice and CW.* The TAK-40 operates USB and LSB as well as CW. USB was included to allow the TAK-40 to easily operate in digital modes such as PSK31 using a PC and sound card.

• *Parts must be readily available either from local retailers or by mail order. No "flea market specials" allowed.* The TAK-40 is constructed from materials available from DigiKey, Mouser, Jameco and Amidon.

• *Any test equipment other than a multimeter or radio receiver must either be constructed as part of the project or purchased as part of the budget.* The TAK-40 only requires a multimeter for construction, and extensive built-in setup functions in the software include a frequency counter to align the oscillators and a programmable voltage source for controlling the oscillators.

• *Equipment need only operate on a single band, 40 meters. Multiband operation is acceptable and encouraged.* The TAK-40 operates on the 40 meter band.

• *The total cost of all parts, except for power supply, mic, key, headphones or speaker, and usual supplies such as wire, nuts and bolts, tape, antenna, solder or glue must be less than $50.* The total cost of the parts required to build the TAK-40 was $49.50 at the time of the contest judging.

The TAK-40 also includes some features that make it very smooth to operate.

Fig 14.39 — Homebrew Challenge winner TAK-40 from Jim Veatch, WA2EUJ.

• Automatic Gain Control — regulates the audio output for strong and weak signals.

• S-Meter — simplifies signal reports.

• Digital frequency readout — reads the operating frequency to 100 Hz.

• Dual Tuning Rates — FAST for scanning the band and SLOW for fine tuning.

• Speech Processor — get the most from the 5 W output.

• Automatic Level Control — prevents overdriving the transmitter.

• Transmit power meter — displays approximate power output.

• Boot loader — accepts firmware updates via a computer (cable and level converter optional).

CIRCUIT DESCRIPTION

The TAK-40 transceiver is designed to be constructed as four modules:

• Digital section and front panel
• Variable frequency oscillator (VFO)
• Intermediate frequency (IF) board
• Power amplifier (PA)

The overall design is a classic superheterodyne with a 4-MHz IF and a 3- to 3.3-MHz VFO. The same IF chain is used for transmitting and receiving by switching the oscillator signals between the two mixers. **Fig 14.40** shows the block diagram of the TAK-40 transceiver. Each board is described below with a detailed parts list on the *QST* binaries Web site.[12]

Digital Board

The digital board contains the microprocessor, front panel controls, liquid crystal display (LCD), the digital to analog converter for the VFO, the beat frequency oscillator (BFO) and the oscillator switching matrix. See the Web package for detailed schematics of this and other boards. Components are numbered in the 100 range. The front panel switches are multiplexed on the LCD control lines for economy so the display will not update when a button is pressed. The BFO is a voltage controlled oscillator (VFO) using a ceramic resonator (Y101) as a tuned circuit. The pulse width modulator (PWM) output from the microprocessor is filtered (R124, C114, R131, C108) and used as the control voltage for the BFO. The microprocessor (U105) varies the BFO frequency for upper or lower sideband modulation. The microprocessor is also used to stabilize the BFO, if the BFO varies more than 10 Hz from the set frequency; the microprocessor adjusts the PWM to correct the BFO frequency.

The digital board also contains the switching matrix for the VFO and BFO (U106, U107). The NE-612 mixers on the IF board work nicely when driven with square waves and aren't sensitive to duty cycle. One section of each tri-state buffer (74HC125s) is used to convert the output of the VFO and BFO to a square wave. The remaining sections control which oscillator goes to which mixer and which oscillator is applied to the frequency counter. The counter counts the VFO then the BFO and adds the result to calculate the operating frequency. The 20 MHz oscillator (OSC101) that runs the microprocessor is only accurate to 100 PPM, so the frequency displayed may be as much a 1 kHz off. Don't operate within 1 kHz of the band/segment edge just to be sure. A commercial transceiver or frequency counter can be used to calibrate your operating frequency.

The digital-to-analog converter (DAC) (U103) used to drive the VFO is a Microchip

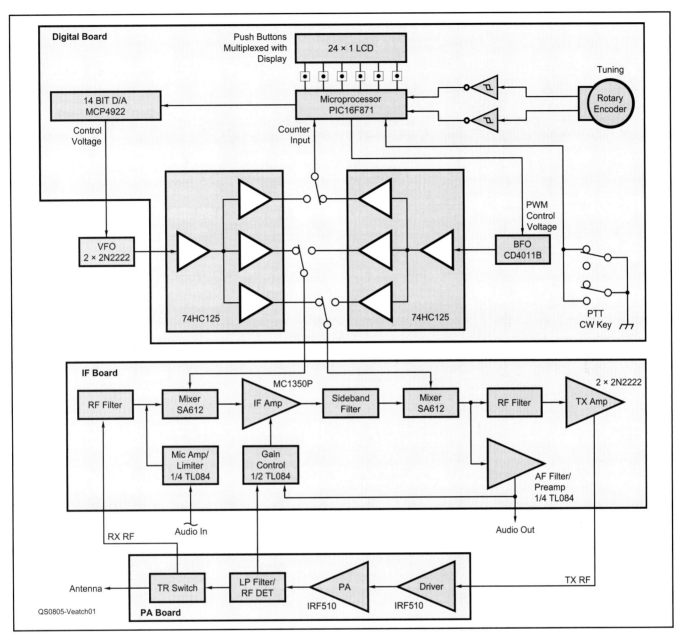

Fig 14.40 — Block diagram of TAK-40, homebrew challenge winner. This radio included many more features and capabilities than we expected to find in a $50 radio! Complete schematics, parts lists and firmware can be found at www.arrl.org/QST-in-depth. Look in the 2008 section for QS0508Veatch.zip.

MCP4922 dual 12-bit DAC. The outputs of each converter are coupled with an 8:1 resistive divider (R116, R117) effectively creating a 15-bit DAC. Since the band is split into two 150 kHz sections this results in approximately 10 Hz steps. Tuning is not linear, so frequency steps at the bottom of each band are slightly larger than steps at the top.

IF Board

In receive mode, RF is filtered in an impedance matching RF filter (C254, L208, C240, C235, L207, C234, C226), applied to the first mixer and mixed (U201) with the VFO signal to result in a 4 MHz IF. This signal is filtered in a 6-element crystal ladder filter (Y201-Y206) with a bandwidth of a bit more than 2 kHz. The signal is then amplified by an MC1350P IF amplifier (U202). The amplified signal is applied to the second mixer (U203) and mixed with the BFO signal to produce the receive audio.

The audio is filtered and amplified to drive a set of computer speakers. Audio is also used to generate an AGC signal for the IF amp and the S-meter. The audio-derived AGC pops and clicks a bit but it's an improvement over manual gain control. There is a provision for adding a manual gain control (it didn't fit into the budget). All audio and AGC functions are handled by a quad op-amp (U204).

In transmit mode, microphone audio is amplified and limited (soft-clipped) then applied to the first mixer and mixed with the BFO to create a 4 MHz IF signal. The IF is filtered, amplified (just as with the received signals) and sent to the second mixer to be mixed with the VFO to create an RF signal. The RF signal filter is identical to the receive-input filter. This is followed by two stages of amplification to bring the transmit signal to $2 V_{P-P}$. There is a manual transmit gain control setting and the RF detector on the PA board reduces the IF gain for automatic level control (ALC) on transmit.

A diplexer is used to route the transmit and receive signals into and out of the mixers so no switching is necessary. Switches are used to mute the receive audio during transmit and a diode RF switch is used at the input to the receive filter to protect the input during transmit. Parts on the IF board are numbered in the 200 range.

VFO Board

The VFO circuit is a straightforward Colpitts oscillator (Q301) with an emitter follower buffer (Q302). Tuning is achieved by varying the reverse bias on an MV209 (D302) varactor diode. A second MV209 (D301) is used to switch between the lower and upper 150 kHz sections of the band. There are no expensive trimmer caps to set the range. Tuning is achieved by winding too many turns on the inductor (L301) the removing turns until the correct tuning range is achieved. The VFO drifts a bit for the first ½ hour after power is applied but eventually settles down. Parts on the VFO board are numbered in the 300 range.

Power Amplifier Board

The PA board includes the driver (Q401) and PA (Q402). Both stages use an IRF-510 MOSFET, which is overkill for the driver but very inexpensive. The gate voltage is pulled down during receive to reduce current draw and heat. The PA is biased class A and can produce 7 to 8 W output. The RF detector (R405, R406, D403 and C413) simply measures the RF voltage at the output so it's only accurate into a 50-Ω load. No protection is provided for excessive SWR conditions, thus it is possible to damage the PA transistor with prolonged operation into a poorly matched antenna system. A tuned TR switch (C412, D404, D405 and L405) isolate the receiver input during transmissions. Parts on the PA board are numbered in the 400 range.

Chassis

Fortunately the TAK-40 requires relatively little chassis wiring. A small harness for the LCD and push-button switches, cable for the rotary encoder audio in and out and key line wiring are all that is required for the front panel. The IF board connects to the VFO, IF and PA boards for control and metering. Two RF lines run between the IF and PA boards.

CONSTRUCTION

The best way to build this radio would be to buy the printed circuit boards (PCB) but it won't fit in the $50 budget. Files included in the Web *QST* binaries package that can be sent to **www.expresspcb.com** and they will send you two complete sets of boards for just over $100. (If you order more, the pre-set price decreases.) By ordering the PCBs you would complete the radio sooner than if you built it using any other technique and with a higher probability of success. To build the TAK-40 for less than $50 we'll have to resort to more creative techniques. Perfboard is expensive! "Dead bug" or "ugly" style construction is messy and difficult to rework/troubleshoot so a different approach is used in the prototype.

Print out the mechanical files for each board. Each drawing includes a parts placement, hole position, top copper and bottom copper drawing. Cut out the hole drawing and stick it to the copper side of the copper clad PCB using glue stick or print it on a self-ticking label. Make sure that the printer is printing a 1:1 size ratio using the dimensions of the board shown on the drawings. (Print out a test copy and check it against the actual parts to be sure.) Mark each hole with a center punch (hammer and a small, sharp nail works fine) then remove the drawing and drill all of the holes. Then refer to the top copper drawing and mark every hole that does not connect to the ground plane with a Sharpie pen. Next touch each hole with a larger diameter drill bit to remove the copper but don't go all the way through.

Using the parts placement drawing, the bottom copper drawing and the schematic, build the board. Take care not to short the non-grounded component leads to the copper and directly solder component leads that need to be grounded. The technique results in a good ground, short signal runs and solid mounting.

The prototype is built on a wooden frame and the front panel printed on photo paper in an inkjet printer. The tuning knob was made by using a hole-saw to make a circular wooden slug, drilling and tapping the sides for set screws and cutting off 6-32 screws to use as set screws. The encoder is made from rebuilding a potentiometer with the guts of a wheel mouse (see binaries package). The author mounted the VFO board in an Altoids tin for three reasons; mechanical stability, electrical shielding and it's not really a cool home-brew radio unless part of it is in a food container.

Separate the inductors L101, L201, L202, L204, L207, L208 and L405 from the PCB by ¼ inch because close proximity of the copper ground plane seems to detune the tuned circuits. Scrape the copper from under the toroidal inductors L301, L401, L402, L403 and L404. RadioShack sells a pack of magnet wire that includes #22 and #26 enameled wire. To make bifilar windings, twist two conductors using a clamp on one end and a drill tool to twist the wire. It's very important to get 8 to 10 twists per inch in the wire before it goes on the core. The driver (Q401) doesn't need a heat sink but the final transistor (Q402) needs about 30 in^2 of aluminum or copper attached to the heat sink tab. The prototype used copper flashing but aluminum cake pans, soda cans or anything you can find to spread the heat will work.

ADJUSTMENTS

After completion of the digital board and front panel, the microprocessor can be powered up and the BFO aligned. Carefully recheck all connections looking for shorts and wiring errors and apply power. The display should show a frequency around 4 MHz. Powering the TAK-40 while holding the SWAP/SETUP button places the TAK-40 in setup mode. Repeatedly pressing SWAP/SETUP toggles between the five setup modes. Here is a summary of the setup modes in the order they appear:

LSB BFO Setup

The left portion of the display shows the frequency of the BFO at 100 Hz and the right portion shows the BFO setting at 10 Hz resolution. The main tuning knob adjusts the BFO setting (right display). Pressing SELECT stores the setting and updates the BFO. The microprocessor stabilizes the BFO frequency by counting the frequency with 10 Hz resolution and adjusting the BFO as necessary. USB/CW BFO setup is the same as the LSB setup, but adjusts the setting for USB and CW modes.

VFO A Range Test

The left portion of the display shows the VFO frequency and the main tuning knob adjusts the VFO frequency. This is useful when adjusting the VFO circuit and verifying the tuning range. VFO B Range Test is the same as VFO A test but displays the upper frequency range.

BFO Range Test

The right portion of the display shows the BFO frequency and the main tuning knob adjusts the BFO frequency. This is useful for setting VR102 to make sure the BFO tuning range is 3.995 to 4.005 MHz.

Once the digital board is working properly, assemble, inspect and connect the VFO board to the digital board. Adjust the number of turns on L301 for 3.000 to 3.150 MHz in VFO A test #3 and 3.150 to 3.300 MHz in VFO B test #4.

FINAL ASSEMBLY

Build the IF board and wire it to the MIC and SPEAKER jacks and the digital board. Connect a set of amplified computer speakers and a 40 meter antenna to the RF INPUT port of the IF board and you should be able to receive signals.

Build the PA board and connect it to the IF board and digital boards, follow the alignment procedure and you're almost ready to operate.

Final Tune Up

There are five potentiometers to adjust to align the TAK-40 (VR101 is not used):

VR-102 — sets the BFO range use setup

mode 5 above to display frequency of the BFO and rotate the MAIN TUNING knob CW until the frequency stops increasing. Set VR102 for a BFO frequency of 4.006 MHz. Rotate the MAIN TUNING knob counterclockwise until the BFO stops decreasing and verify that the BFO tunes below 3.995 MHz.

VR201 — Sets the AGC threshold. With no signal applied to the TAK-40 adjust VR201 for 2.5 VDC at pin 4 of the microprocessor (U105).

VR401 — Sets the PA bias. Adjust for 600 mA current draw LSB mode key down, VR 203 set to minimum.

VR203 — Sets the transmit drive level. Set for 7 W (3.7 V dc at pin 7 of U105) into 50 Ω on CW mode with VR202 set to minimum (wiper toward R227).

VR202 — Sets the ALC. Adjust to reduce CW output to 6 W (3.4 V dc at pin 7 of U105) into 50 Ω.

OPERATION

Operating this radio is a breeze, the receiver is not super sensitive but it seems relatively impervious to strong signals. A rule of thumb is if the noise level increases when the antenna is plugged in, the receiver is sensitive enough given the current operating environment. With a GAP Triton on the roof of a Baltimore row house the TAK-40 receiver works just fine. Don't scoff at 5 W either. A 5-W transceiver is 13 dB below a 100-W unit, so if you hear a signal from a 100 W transmitter that's 20 or 30 dB above the noise, the other operator will hear you just fine.

On the air most operators can't believe that it's only 5 W. The author worked 15 states on LSB in about a two-week period. Lots of phone operators use more than 100 W, but you can work then as well and they are usually excited about working a QRP station especially one that is homebrew. CW is even easier. The radio should give good results on PSK31, as well. Don't expect to sit on a frequency, call CQ and rake in the DX, but practice, patience, good operating skills and lots of listening, will be rewarded with plenty of ham radio action.

Controls

Here is a brief summary of the front panel controls and what they do. The switches are multiplexed with the LCD lines so if you hold down a switch the display won't update. Normal operation resumes when the switch is released. It's also possible that pressing a switch may corrupt an important bit if display data just cycle the power and the LCD will recover.

MAIN TUNING knob — Used to adjust the frequency, can be programmed for left or right hand operation by swapping the A and B encoder lines.

SELECT — Used in setup mode, also for future expansion (CW keyer, RIT, PBT) and other functions if the software developer ever gets going. Holding the SELECT button down during start-up puts the TAK-40 in boot loader mode ready to accept new firmware.

MODE — Selects LSB, USB or CW. Current setting retained following power off.

RATE — Selects fast or slow tuning speeds, defaults to slow on power up.

VFO A/B — Selects 7.0 to 7.15 MHz range or 7.15 to 7.3 MHz range.

V to M — Stores the current frequency in memory.

SWAP — Swaps the current and memory frequencies. Holding SWAP during power up places the TAK-40 in setup mode.

All circuitry used in the TAK-40 was designed specifically for use in the TAK-40. The author looked at many designs on the Internet and in printed sources but no circuits were taken directly from any specific source. The most valuable tools were manufacturers' data sheets, *The ARRL Handbook* and the Internet.

14.7.3 *Project:* A Homebrew High Performance HF Transceiver — the HBR-2000

Markus Hansen, VE7CA, shows us that it's still possible to roll your own full feature HF transceiver — and get competitive performance! This project is a condensed version of an article that appeared in March 2006 *QST* and should provide an inspiration to all homebrewers. (The full original article is available in PDF format on the CD-ROM included with this book.)

Have you ever dreamed of building an Amateur Radio transceiver? Have you thought how good it would feel to say, "The rig here is homebrew"? The author is the proud owner of a homebrew, high performance, 100 W HF transceiver named the HBR-2000 (see **Fig 14.41**).

The secret to being able to successfully build a major project such as this is to divide it into many small modules, as indicated in the block diagram (**Fig 14.42**). Each module represents a part of the whole, with each module built and tested before starting on the next. To choose the actual circuits that were to be built into each module, the author searched past issues of *QST*, *QEX*, *The ARRL Handbook* and publications dedicated to home brewing such as *Experimental Methods in RF Design*, published by ARRL. It's import to learn by reading, building a circuit and then taking measurements. After you build a particular circuit and measure the voltage at different points, you begin to understand how that particular circuit works.

Do not go on to the next module until finishing the previous module, including testing it to make sure it worked as expected. After building a module, follow the same process to decide on the circuit and build the next module. Then connect the two modules together and check to make sure that, when combined together, they performed as expected. It is really that simple, one step at a time. Anyone who has had some building experience can build a receiver and transmitter using this procedure.

For a receiver, begin by building the audio amplifier and product detector module, then the BFO circuit, testing each separately and then the growing receiver, step by step. From there, build the IF/AGC module, then the VFO and the heterodyne LO system, then the receiver mixer and on and on until reaching the antenna. At that point you have a functioning receiver. It is a thrilling day when you hook up an antenna to the receiver and tune across the Amateur Radio bands listening to signals emanating from the speaker.

After you build a particular module, you don't want RF from outside sources to get into the modules you build and you don't want RF signals produced inside the modules to travel to other parts of the receiver, other than through shielded coaxial lines. The reason that you don't want RF floating around the receiver is that stray RF can produce unwanted birdies in the receiver, adversely affect the AGC system or cause other subtle forms of mischief. To prevent this from happening enclose each module in a separate RF tight

Fig 14.41 — Head on view of the HBR-2000. This looks a lot like a commercial transceiver.

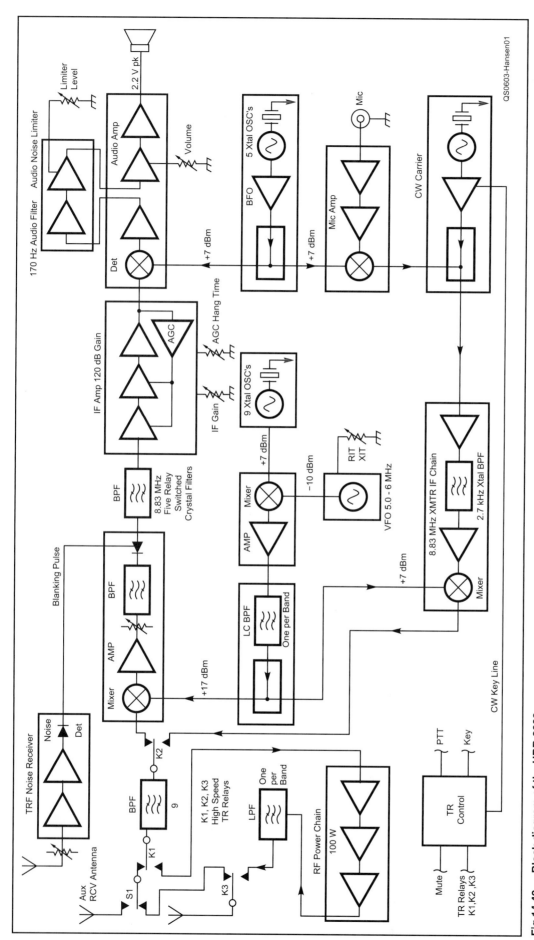

Fig 14.42 — Block diagram of the HBR-2000.

Fig 14.43 — Sample of a box made with PC board with a sheet brass cover overlapping all four sides. Note the BNC connectors and feed thru insulator.

box and used coax for all the RF lines with BNC or phono connectors on each end. All dc and control lines are connected via feed thru insulators.

Modules are enclosed in boxes made from unetched copper clad material. For the covers, cut sheet brass ½ inch wider and longer than the size of the PC box. Then lay the box on top of the brass and center the box so that there is about ¼ inch overlap around the perimeter of the box and draw a line around the box with a felt pen. Then cut the corners out with tin snips and bend the edges of the brass cover over in a vise. By drilling small holes around the perimeter of the box, inserting wires through the holes, soldering the wires to the inside of the box and to the overlapping edges, you produce an RF-tight enclosure. See **Fig 14.43** for an example of this technique.

TYPES OF CONSTRUCTION

The audio board and IF board are etched PC boards purchased from Far Circuits. The construction method the author learned to appreciate the most is "ugly construction," discussed in the **Circuit Construction** chapter. Each module is designed for an input and output impedance of 50 Ω except for the audio output (8 Ω for speaker connection). Thus 50-Ω coax cable with BNC connectors can be employed to connect the RF paths between the different modules. The concept is shown in **Fig 14.44**.

Fig 14.44 — The inside of the HBR-2000. All the boxes are connected together with 50 Ω coax and BNC connectors for the RF runs. All dc power and control leads, in and out of the boxes, are through feed through insulators.

MEASUREMENTS AND TEST EQUIPMENT

Making meaningful and accurate measurements is a major part of producing a successful project. You must make measurements as you progress or you have no way of knowing whether a module is performing according to design specifications.

One of the specifications chosen for the HBR-2000 was that it should have a very strong front end. To accomplish this, the design uses a mixer that requires that the LO port be fed with +17 dBm at 50 Ω. Having the test equipment to measure these parameters confirms the expected results.

To help in the construction of this project, the author purchased surplus test instruments including an oscilloscope and signal generator. He also built test equipment such as crystal-controlled, very low power, oscillators and attenuators to measure receiver sensitivity, and high-power oscillators and a combiner to measure receiver blocking and dynamic range. He also built a spectrum analyzer, which turned out to be one of the most useful instruments because the HBR-2000 has 19 band-pass filters and 22 low-pass filters. (This instrument was described in August and September 1998 *QST* by Wes Hayward, W7ZOI, and Terry White, K7TAU.) Later he built the RF power meter featured in June 2001 *QST*, authored by W7ZOI and Bob Larkin, W7PUA. These instruments allow adjustment and measurement of the performance of each module built. A good selection of test equipment can be affordable. The author's lab is shown in **Fig 14.45**.

OTHER CIRCUITS

Here are some of the sources used for deciding on the circuits for the other modules in the HBR-2000 design. The low-distortion audio module is from the "R1 High Performance Direct Conversion Receiver" by Rick Campbell, KK7B, in August 1993 *QST*. The VFO design is from June *QST* ("Build a

Fig 14.45 — The VE7CA work shop shows some of the home built and surplus test equipment used in the construction and evaluation of the HBR-2000.

Fig 14.46 — The 100-W amplifier board.

Fig 14.47 — The 100-W amplifier, 10 low-pass filters and the power supplies are located in a separate enclosure.

Universal VFO" by Doug DeMaw, W1FB). The first mixer, post-mixer amplifier and crystal heterodyne oscillator design was taken from "A Progressive Communication Receiver" design by Wes Hayward and John Lawson, K5IRK, which appeared in November 1981 *QST* and was featured in *The ARRL Handbook* for many years. This is a classic radio article with many good circuit ideas.

The receiver input RF band-pass filter and diplexer designs along with the noise blanker and 100 W amplifier circuits were taken from the three-part series beginning in the May/June 2000 issue of *QEX* titled The "ATR-2000: A Homemade, High-Performance HF Transceiver" by John Stephenson, KD6OZH. The power amplifier and output circuit filters are in a shielded sub-enclosure as shown in **Fig 14.46** and **Fig 14.47**.

The BFO and power supply circuits were lifted right out of the *ARRL Handbook*. The transmitter portion of the transceiver consists of combinations of various circuits found in *Experimental Method in RF Design*.

RECEIVER SPECIFICATIONS

Some readers may question whether an amateur can build a competitive grade contest class transceiver from scratch. For the skeptics, the actual measured performance of the HBR-2000, as confirmed in the ARRL Lab is shown in **Table 14.7**.

Receiver sensitivity measurements on all bands are within ±0.5 dB of −130 dBm. All measurements were made with an IF filter bandwidth of 400 Hz. Test oscillators are two, separately boxed, crystal oscillators, low pass filtered and designed for a 50-Ω output impedance. MDS measurements were made with a HP-8640B signal generator and a true reading RMS voltmeter across the speaker output:

The receiver specifications were made

Table 14.7
HBR-2000 Test Measurements

Spacing:	20 kHz	5 kHz	2 kHz
Two-tone blocking dynamic range:	>126 dB	124 dB	122 dB
Third-order IMD dynamic range:	103.5 dB	102.5 dB	93.0 dB
Third-order intercept:	25.5 dBm	24.0 dBm	14.5 dBm

Image rejection all bands:	>135 dBm
Receive to transmit time:	8 ms (incl 2 ms click filter)
CW, full QSK transmit to receive time:	17 ms (30 WPM = 20 ms dot)

Fig 14.48 — VE7CA's partly home brew Amateur Radio station.

following ARRL procedures as outlined in the ARRL "Lab Test Procedures Manual" at www.arrl.org/product-review. Making accurate receiver measurements is not a trivial matter and should be approached with the understanding of the limitations of the test equipment being used and thorough knowledge of the subject.

You may notice that the author made no attempt to miniaturize the HBR-2000. With large knobs and large labeling, he is able to operate the transceiver without the need for reading glasses. When you build your own equipment, you get to decide the front panel layout, what size knobs to use and where they should be located. That is a real bonus! See **Fig 14.48**.

Why not plan to build your dream station? If you haven't built any Amateur Radio equipment, begin with a small project. As you gain experience you will eventually have the confidence to build more complex equipment. Then someday you too can say with a smile, "my rig here is homebrew."

14.8 References

[1] R. Bitzer, WB2ZKW, "Modifying the Collins KWM-2 for Serious CW Operation," QST, Jan 2008, pp 30-34.

[2] J. Hallas, W1ZR, "Elecraft K3/100 HF and 6 Meter Transceiver," Product Review, QST, Jan 2009, pp 43-49. (Product Reviews are available on the ARRLWeb at www.arrl.org/product-review.)

[3] R. Lindquist, N1RL, "ICOM IC-756 ProIII HF and 6 Meter Transceiver," Product Review, QST, Mar 2005, pp 56-59.

[4] R. Lindquist, WW3DE, "ICOM IC-7700 HF and 6 Meter Transceiver," Product Review, QST, Oct 2008, pp 41-47.

[5] J. Hallas, W1ZR, "Ten-Tec Omni VII HF and 6 Meter Transceiver," Product Review, QST, Jul 2009, pp 58-64.

[6] R. Lindquist, N1RL, "The FlexRadio FLEX-5000A," Product Review, QST, Jul 2008, pp 39-45.

[7] G. Zimmerman, W3ZZ, "Down East Microwave 144-28HP 2 Meter Transverter," Product Review, QST, Jan 2006, pp 66-70.

[8] G. Zimmerman, W3ZZ, "Elecraft XV144 2 Meter Transverter Kit," Product Review, QST, Jan 2006, pp 66-70.

[9] Z. Lau, W1VT, "DB6NT MKU 10 GHz Transverter Kit," Product Review, QST, May 2001, pp 78-80.

[10] M. Wilson, K1RO, "SSB Electronic LT2S MkII 2-Meter Transverter," Product Review, QST, Nov 1997, pp 62-63.

[11] R. Arland, K7SZ, *ARRL's Low Power Communication — The Art and Science of QRP* bundled with an MFJ Cub transceiver kit available from your ARRL dealer or the ARRL Bookstore, ARRL order no. 102YK. Telephone 860-594-0355 or foll-free in the US 888-277-5289, www.arrl.org/publications-online-store; pubsales@arrl.org.

[12] www.arrl.org/QST-in-depth. Look in the 2008 section for file **QS0508Veatch.zip**.

[13] R. Lindquist, WW3DE, "ICOM IC-7600 HF and 6 Meter Transceiver," Product Review, QST, Nov 2009, pp 54-59.

[14] N. Fusaro, W3IZ, "Yaesu FT-2000D HF and 6 Meter Transceiver," Product Review, QST, Oct 2007, pp 69-75.

[15] R. Lindquist, N1RL, "Ten Tec Orion II HF Transceiver Model 566," Product Review, QST, Sep 2006, pp 59-64.

[16] R. Lindquist, N1RL, "Yaesu FTDX9000 Contest HF and 6 Meter Transceiver," Product Review, QST, Mar 2006, pp 61-66.

[17] R. Schetgen, KU7G, "The Kenwood TS-2000 All-Mode Multiband Transceiver," Product Review, QST, Jul 2001, pp 74-81.

[18] FCC Report and Order 01-264, released Sep 14, 2001.

[19] J. Taylor, K1RFD, "Product Review — Computer Sound Cards for Amateur Radio," QST, May 2007, pp 63-70. Available on the ARRL Web site.

[20] H. Nyquist, "Certain Topics in Telegraph Transmission Theory," Transactions of the AIEE, Vol 47, pp 617-644, Apr 1928. Reprinted as classic paper in Proceedings of the IEEE, Vol 90, No. 2, Feb 2002.

[21] Tucson Amateur Packet Radio, www.tapr.org.

[22] Hendricks QRP Kits, Firefly by Dan Tayloe, N7VE, www.qrpkits.com/firefly.html. The kit is sold out but the user manuals are downloadable.

[23] American QRP Club, Softrock-40, www.amqrp.org/kits/softrock40. The kit is sold out but design information is available at the referenced URL.

Contents

15.1 Introduction
15.2 Typical DSP System Block Diagram
    15.2.1 Data Converters
    15.2.2 DSP
15.3 Digital Signals
    15.3.1 Sampling — Digitization in Time
    15.3.2 Quantization — Digitization in Amplitude
15.4 Digital Filters
    15.4.1 FIR Filters
    15.4.2 IIR Filters
    15.4.3 Adaptive Filters
15.5 Miscellaneous DSP Algorithms
    15.5.1 Sine Wave Generation
    15.5.2 Tone Decoder
15.6 Analytic Signals and Modulation
    15.6.1 I/Q Modulation and Demodulation
    15.6.2 SSB Using I/Q Modulators and Demodulators
15.7 Software-Defined Radios (SDR)
    15.7.1 SDR Hardware
    15.7.2 SDR Software
15.8 Notes and Bibliography
15.9 Glossary

Chapter 15

DSP and Software Radio Design

In recent years, digital signal processing (DSP) technology has progressed to the point where it is an integral part of our radio equipment. DSP is rapidly replacing hardware circuits with software, offering amateurs flexibility and features only dreamed of in the past. This chapter, by Alan Bloom, N1AL, explores DSP and its use in radio design. DSP projects and additional background and support materials may be found on the *Handbook* CD.

15.1 Introduction

Digital signal processing (DSP) has been around a long time. The essential theory was developed by mathematicians such as Newton, Gauss and Fourier in the 17th, 18th and 19th centuries. It was not until the latter half of the 20th century, however, that digital computers became available that could do the calculations fast enough to process signals in real time. Today DSP is important in many fields, such as seismology, acoustics, radar, medical imaging, nuclear engineering, audio and video processing, as well as speech and data communications.

In all those systems, the idea is to process a digitized signal so as to extract information from it or to control its characteristics in some way. For example, an EKG monitor in a hospital extracts the essential characteristics of the signal from the patient's heart for display on a screen. A digital communications receiver uses DSP to filter and demodulate the received RF signal before sending it to the speaker or headphones. In some systems, the signal to be processed may have more than one dimension. An example is image data, which requires two-dimensional processing. Similarly, the controller for an electrically-steerable antenna array uses multi-dimensional DSP techniques to determine the amplitude and phase of the RF signal in each of the antenna elements. A CT scanner analyzes X-ray data in three dimensions to determine the internal structures of a human body.

SOFTWARE-DEFINED RADIO

The concept of a *software-defined radio* (SDR) became popular in the 1990s. By then, DSP technology had developed to the point that it was possible to implement almost all the signal-processing functions of a transceiver using inexpensive programmable digital hardware. The frequency, bandwidth, modulation, filtering and other characteristics can be changed under software control, rather than being fixed by the hardware design as in a conventional radio. Adding a new modulation type or a new improved filter design is a simple matter of downloading new software. In addition, with the same hardware design, a single radio can have several different modulation modes.

SDR is appealing to regulatory bodies such as the FCC because it makes possible a communications system called *cognitive radio* in which multiple radio services can share the same frequency spectrum.[1] Each node in a wireless network is programmed to dynamically change its transmission or reception characteristics to avoid interference with other users. In this way, services that in the past enjoyed fixed frequency allocations but that only use their channels a small percentage of the time can share their spectrum with other wireless users with minimal interference.

DSP ADVANTAGES

Digital signal processing has the reputation of being more complicated than the analog circuitry that it replaces. In reality, once the analog signal has been converted into the digital domain, complicated functions can be implemented in software much more simply than would be possible with analog components. For example, the traditional "phasing" method of generating an SSB signal without an expensive crystal filter requires various mixers, oscillators, filters and a wide-band audio-frequency phase-shift network built with a network of high-precision resistors and capacitors. To implement the same function in a DSP system requires adding one additional subroutine to the software program — no additional hardware is needed.

There are many features that are straightforward with DSP techniques but would be difficult or impractical to implement with analog circuitry. A few examples drawn just from the communications field are imageless mixing, noise reduction, OFDM modulation and adaptive channel equalization. Digital signals can have much more dynamic range than analog signals, limited only by the number of bits used to represent the signal. For example, it is easy to add an extra 20 or 30 dB of headroom to the intermediate signal processing stages to ensure that there is no measurable degradation of the signal, something that might be difficult or impossible with analog circuitry. Replacing analog circuitry with software algorithms eliminates the problems of nonlinearity and drift of component values with time and temperature. The programmable nature of most DSP systems means you can make the equivalent of circuit modifications without having to unsolder any components.

DSP LIMITATIONS

Despite its many advantages, we don't mean to imply that DSP is best in all situations. High-power and very high-frequency signals are still the domain of analog circuitry. Where simplicity and low power consumption are primary goals, a DSP solution may not be the best choice. For example, a simple CW receiver that draws a few milliamps from the power supply can be built with two or three analog ICs and a handful of discrete components. In many high-performance systems, the performance of the analog-to-digital converter (ADC or A/D converter) and digital-to-analog converter (DAC or D/A converter) are the limiting factors. That is why, even with the latest generation of affordable ADC technology, it is still possible to obtain better blocking dynamic range in an HF receiver using a hybrid analog-digital system rather than going all-digital by routing the RF input directly to an ADC.

The plan of this chapter is first to discuss the overall hardware and software structure of a DSP system, including general information on factors to be considered when designing at the system level. Then we will cover the basic theory of digital signals, with emphasis on topics relevant to radio communications. Following that is a section on digital filters and another section that describes several other miscellaneous DSP applications. The concept of analytic signals (negative frequencies and all that) is important for understanding software-defined radios, so we cover that before getting into SDRs themselves. The final two sections, on SDR hardware and software, use many of the concepts explained in previous sections to show how a radio can be built with most of the signal processing done digitally. For the fullest understanding of this chapter the reader should have a basic familiarity of the topics covered in the **Electrical Fundamentals**, **Analog Basics** and **Digital Basics** chapters as well as some high-school trigonometry.

15.2 Typical DSP System Block Diagram

A typical DSP system is conceptually very simple. It consists of only three sections, as illustrated in **Fig 15.1**. An ADC at the input converts an analog signal into a series of digital numbers that represent snapshots of the signal at a series of equally spaced sample times. The digital signal processor itself does some kind of calculations on that digital signal to generate a new stream of numbers at its output. A DAC then converts those numbers back into analog form.

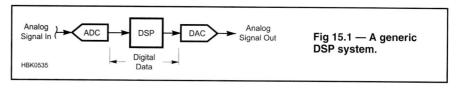

Fig 15.1 — A generic DSP system.

Some DSP systems may not have all three components. For example, a DSP-based audio-frequency generator does not need an ADC. Similarly, there is no need for a DAC in a measurement system that monitors some sensor output, processes the signal and stores the result in a computer file or displays it on a digital readout.

The term "DSP" is normally understood to imply processing that occurs in real time, at least in some sense. For example, an RF or microwave signal analyzer might include a DSP co-processor that processes chunks of sampled data in batch mode for display a fraction of a second later. However, a computer program that analyzes historical sunspot data or stock prices normally would not be called "digital signal processing" even though the types of calculations might be very similar.

15.2.1 Data Converters

In this chapter we will discuss only briefly several aspects of ADC and DAC specifications and performance that directly affect design decisions at the system level. The **Analog Basics** chapter has additional details that must be considered when doing an actual circuit design.

The first requirement when selecting a DAC or ADC is that it be able to handle the required *sample rate*. For communications-quality voice, a sample rate on the order of 8000 samples per second (8 ksps) should be adequate. For high-quality music, sample rates are typically an order of magnitude higher and for processing wideband RF signals, you'll need data converters with sample rates in the megasamples per second (Msps) range. In many systems the input and output sample rates are different. For example, a software-defined receiver might sample the input RF signal at 100 Msps while the output audio DAC is running at only 8 ksps.

The *resolution* of a data converter is expressed as the number of bits in the data words. For example, an 8-bit ADC can only represent the sampled analog signal as one of $2^8 = 256$ possible numbers. The smallest signal that it can resolve is therefore $1/256$ of full scale. Even with an ideal, error-free ADC, the *quantization error* is up to ±½ of one least-significant bit (LSB) of the digital word, or ±$1/512$ of full scale with 8-bit resolution. Similarly, a DAC can only generate the analog signal to within ±½ LSB of the desired value. Later in the chapter we will discuss how to determine the required sample rate and resolution for a given application.

Another important data converter specification is the *spurious-free dynamic range* (SFDR). This is the ratio, normally expressed in dB, between a (usually) full-scale sine wave and the worst-case spurious signal. While higher-resolution ADCs and DACs tend to have better SFDR, that is not guaranteed. Devices that are intended for signal-processing applications normally specify the SFDR on the data sheet.

While sample rate, resolution and SFDR are the principal selection criteria for data converters in a DSP system, other parameters such as signal-to-noise ratio, harmonic and intermodulation distortion, full-power bandwidth, and aperture delay jitter can also affect performance. Of course, basic specifications such as power requirements, interface type (serial or parallel) and cost also determine a device's suitability for a particular application. As with any electronic component, it is very important to read and fully understand the data sheet.

15.2.2 DSP

The term *digital signal processor* (DSP) is commonly understood to mean a special-purpose microprocessor with an architecture that has been optimized for signal processing. And indeed, in many systems the box labeled "DSP" in Fig 15.1 is such a device. A microprocessor has the advantage of flexibility because it can easily be re-programmed. Even with a single program, it can perform many completely-different tasks at different points in the code. On-chip hardware resources such as multipliers and other computational units are used efficiently because they are shared among various processes.

That is also the Achilles' heel of programmable DSPs. Any hardware resource that is shared among various processes can be used by only one process at a time. That can create bottlenecks that limit the maximum computation speed. Some DSP chips include multiple computational units or multiple *cores* (basically multiple copies of the entire processor) that can be used in parallel to speed up processing.

DIGITAL SIGNAL PROCESSING WITHOUT A "DSP"

Another way to speed up processing is to move all or part of the computations from the programmable DSP to an *application-specific integrated circuit* (ASIC), which has an architecture that has been optimized to perform some specific DSP function. For example, *direct digital synthesis* (DDS) frequency synthesizer chips are available that run at rates that would be impossible with a microprocessor-type DSP.

You could also design your own application-specific circuitry using a PC board full of discrete logic devices. Nowadays, however, it is more common to do that with a *programmable-logic device* (PLD). This is an IC that includes many general-purpose logic elements, but the connections between the elements are undefined when the device is manufactured. The user defines those connections by programming the device to perform whatever function is required. PLDs come in a wide variety of types, described by an alphabet soup of acronyms.

Programmable-array logic (PAL), *programmable logic array* (PLA), and *generic array logic* (GAL) devices are relatively simple arrays of AND gates, OR gates, inverters and latches. They are often used as "glue logic" to replace the miscellaneous discrete logic ICs that would otherwise be used to interface various larger digital devices on a circuit board. These are sometimes grouped under the general category of *small PLD* (SPLD). A *complex PLD* (CPLD) is similar but bigger, often consisting of an array of PALs with programmable interconnections between them.

A *field-programmable gate array* (FPGA) is bigger yet, with up to millions of gates per device. An FPGA includes an array of *complex logic blocks* (CLB), each of which includes some programmable logic, often implemented with a RAM *look-up table* (LUT), and output registers. *Input/output blocks* (IOB) also contain registers and can be configured as input, output, or bi-directional interfaces to the IC pins. The interconnections between blocks are much more flexible and complicated than in CPLDs. Some FPGAs also include higher-level circuit blocks such as general-purpose RAM, dozens or hundreds of hardware multipliers, and even entire on-chip microprocessors.

Some of the more inexpensive PLDs are *one-time programmable* (OTP), meaning you have to throw the old device away if you want to change the programming. Other devices are re-programmable or even *in-circuit programmable* (ICP) which allows changing the internal circuit configuration after the device has been soldered onto the PC board, typically under the control of an on-board microprocessor. That offers the best of both worlds, with speed nearly as fast as an ASIC but retaining many of the benefits of the reprogrammability of a microprocessor-type DSP. Most large FPGAs store their programming in *volatile memory*, which is RAM that must be re-loaded every time power is applied, typically by a ROM located on the same circuit board. Some FPGAs have programmable ROM on-chip.

Programming a PLD is quite different from programming a microprocessor. A microprocessor performs its operations sequentially — only one operation can be performed at a time. Writing a PLD program is more like designing a circuit. Different parts of the circuit can be doing different things at the same time. Special *hardware-description languages* (HDL) have been devised for programming the more complicated parts such as ASICs and FPGAs. The two most common industry-standard HDLs are Verilog and VHDL. (The arguments about which is "best" approach the religious fervor of the *Windows* vs *Linux* wars!) There is also a version of the C++ programming language called SystemC that includes a series of libraries that extend the language to include HDL functions. It is popular with some designers because it allows simulation and hardware description using the same software tool.

Despite the speed advantage of FPGAs, most amateurs use microprocessor-type devices for their DSP designs, supplemented with off-the-shelf ASICs where necessary. The primary reason is that the design process for an FPGA is quite complicated, involving obtaining and learning to use several sophisticated software tools. The steps involved in programming an FPGA are:

1. Simulate the design at a high abstraction level to prove the algorithms.
2. Generate the HDL code, either manually or using some tool.
3. Simulate and test the HDL program.
4. Synthesize the gate-level netlist.
5. Verify the netlist.
6. Perform a timing analysis.
7. Modify the design if necessary to meet timing constraints.
8. "Place and route" the chip design.
9. Program and test the part.

Table 15.1
PLD Manufacturers

Company	Devices	URL	Notes
Achronix	FPGA	www.achronix.com	High-speed FPGAs
Actel	FPGA	www.actel.com	Mixed-signal flash-based FPGAs
Altera	CPLD, FPGA, ASIC	www.altera.com	One of the two big FPGA vendors
Atmel	SPLD, CPLD, FPGA, ASIC	www.atmel.com	Fine-grain-reprogrammable FPGAs with AVR microprocessors on chip
Cypress Semiconductor	SPLD, CPLD	www.cypress.com	
Lattice Semiconductor	SPLD, CPLD, FPGA	www.latticesemi.com	Leading supplier of flash-based nonvolatile FPGAs
SiliconBlue	FPGA	www.siliconbluetech.com	Low-power FPGAs
Texas Instruments	SPLD, ASIC	www.ti.com	
Xilinx	CPLD, FPGA	www.xilinx.com	One of the two big FPGA vendors

Many of the software tools needed to perform those steps are quite expensive, although some manufacturers do offer free proprietary software for their own devices. Some principal manufacturers are listed in **Table 15.1**.

MICROPROCESSOR-TYPE DSP CHIPS

In contrast with designing an FPGA, programming a DSP chip is relatively easy. C compilers are available for most devices, so you don't have to learn assembly language. Typically you include a connector on your circuit board into which is plugged an *in-circuit programmer* (ICP), which is connected to a PC via a serial or USB cable. The software is written and compiled on the PC and then downloaded to the DSP. The same hardware often also includes an *in-circuit debugger* (ICD) so that the program can be debugged on the actual circuitry used in the design. The combination of the editor, compiler, programmer, debugger, simulator and related software is called an *integrated development environment* (IDE).

Until recently you had to use an *in-circuit emulator* (ICE), which is a device that plugs into the circuit board in place of the microprocessor. The ICE provides sophisticated debugging tools that function while the emulator runs the user's software on the target device at full speed. Nowadays, however, it is more common to use the ICD function that is built into many DSP chips and which provides most of the functions of a full-fledged ICE. It is much cheaper and does not require using a socket for the microprocessor chip.

The architecture of a digital signal processor shares some similarities to that of a general-purpose microprocessor but also differs in important respects. For example, DSPs generally don't spend much of their lives handling large computer files, so they tend to have a smaller memory address space than processors intended to be used in computers. On the other hand, the memory they do have is often built into the DSP chip itself to improve speed and to reduce pin count by eliminating the external address and data bus.

Most microprocessors use the traditional *Von Neumann architecture* in which the program and data are stored in the same memory space. However, most DSPs use a *Harvard architecture*, which means that data and program are stored in separate memories. That speeds up the processor because it can be reading the next program instruction at the same time as it is reading or writing data in response to the previous instruction. Some DSPs have two data memories so they can read and/or write two data words at the same time. Most devices actually use a modified Harvard architecture by providing some (typically slower and less convenient) method for the processor to read and write data to program memory.

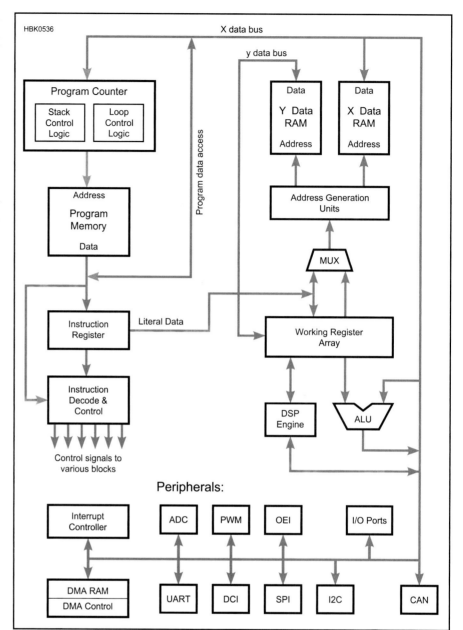

Fig 15.2 — Simplified block diagram of a dsPIC processor.

Probably the key difference between general-purpose and digital-signal processors is in the computational core, often called the *arithmetic logic unit* (ALU). The ALU in a traditional microprocessor only performs integer addition, subtraction and bitwise logic operations such as AND, OR, one-bit shifting and so on. More-complicated calculations, such as multiplication, division and operations with floating-point numbers, are done in software routines that exercise the simple resources of the ALU multiple times to generate the more-complicated results.

In contrast, a DSP has special hardware to perform many of these operations much faster. For example, the *multiplier-accumulator* (MAC) multiplies two numbers and adds (accumulates) the product with the previous results in a single step. Many common DSP algorithms involve the sum of a large number of products, so nearly all DSPs include this function. **Fig 15.2** is a simplified block diagram of the dsPIC series from Microchip. Its architecture is basically that of a general-purpose microcontroller to which has been added a DSP engine, which includes a MAC, a barrel shifter and other DSP features. It does use a modified Harvard architecture with two data memories that can be simultaneously accessed.

A *floating-point* number is the binary equivalent of scientific notation. Recall that the decimal integer 123000 is expressed as 1.23×10^5 in scientific notation. It is common practice to place the decimal point after the first non-zero digit and indicate how many digits the decimal point must be moved by the *exponent* of ten, 5 in this case. The 1.23 part is called the *mantissa*. In a computer, base-2 binary numbers are used in place of the base-10 decimal numbers used in scientific notation. The *binary point* (equivalent to the decimal point in a decimal number) is assumed to be to the left of the first non-zero bit. For example the binary number 00110100 when converted to a 16-bit floating point number would have an 11-bit mantissa of 11010000000 (with the binary point assumed to be to the left of the first "1") and a 5-bit exponent of 00110 (decimal +6).

A floating point number can represent a signal with much more dynamic range than an integer number with the same number of bits. For example, a 16-bit signed integer can vary from −32768 to +32767. The difference between the smallest (1) and largest signal that can be represented is $20 \log(65535) = 96$ dB. If the 16 bits are divided into an 11-bit mantissa and 5-bit exponent, the available range is $20 \log(2048) = 66$ dB from the mantissa and $20 \log(2^{32}) = 193$ dB from the exponent for a total of 259 dB. The disadvantage is that the mantissa has less resolution, potentially increasing noise and distortion. Normally floating-point numbers are at least 32 bits wide to mitigate that effect.

Some DSPs can process floating-point numbers directly in hardware. Fixed-point DSPs can also handle floating-point numbers, but it must be done in software. The additional dynamic range afforded by floating-point processing is normally not needed for radio communications signals since the dynamic range of radio signals is typically less than can be handled by the 16-bit data words used by most integer DSPs. Using integer arithmetic saves the additional cost of a floating-point processor or the additional computational overhead of floating-point calculations on a fixed-point device. However, it requires careful attention to detail on the part of the programmer to make sure the signal can never exceed the maximum integer value or get so weak that the signal-to-noise ratio is degraded. If cost or computation time is not an issue, it is much easier to program in floating point since dynamic range issues can be ignored for most computations.

The term *pipeline* refers to the ability of a microprocessor to perform portions of several instructions at the same time. The sequence of operations required to perform an instruction is broken down into steps. Since each step is performed by a different chunk of hardware, different chunks can be working on different instructions at the same time. Most DSPs have at least a simple form of pipelining in which the next instruction is being fetched while the previous instruction is being executed. Some DSPs can do a multiply-accumulate while the next two multiplicands are being read from memory and the previous accumulated result is being stored so that the entire operation can occur in a single clock cycle. MACs per second is a common figure of merit for measuring DSP speed. For conventional microprocessors, a more common figure of merit is millions of instructions per second (MIPS) or floating-point operations per second (FLOPS).

Many DSPs have a sophisticated address generation unit that can automatically increment one or more data memory pointers so that repetitive calculations can step through memory without the processor having to calculate the addresses. *Zero-overhead looping* is the ability to automatically jump the address pointer back to the beginning of the array when it reaches the end. That saves several microprocessor instructions per loop that normally would be required to check the current address and jump when it reaches a predetermined value.

While most DSPs do not include a full hardware divider, some do include special instructions and hardware to speed up division calculations. A *barrel shifter* is another common DSP feature. It allows shifting a data word a specified number of bits in a single clock cycle. *Direct memory access* (DMA) refers to special hardware that can automatically transfer data between memory and various peripheral devices or ports without processor overhead.

DSP IN EMBEDDED SYSTEMS

An *embedded system* is a device that is not a computer but nevertheless has a micropro-

Table 15.2
Manufacturers of Embedded DSPs

Company	Family	Data Bits	Speed MMACs	Nr. of Cores	ROM (bytes)	RAM (bytes)	Notes
Analog Devices	ADSP-21xx	16	75-160	1	12k-144k	8k-112k	Easy assembly language
www.analog.com	SHARC	32/40 fp	300-900	1	2-4M	0.5-5M	Runs fixed or floating point
	Blackfin	16/32	400-2400	1-2	External	53k-328k	Many on-chip peripherals
Cirrus Logic	CS48xxxx	32	150	1		96k	Audio applications
www.cirrus.com	CS49xxxx	32	300	2	512k	296k-328k	Audio applications
Freescale	DSP568xx	16	32-120	1	2k-576k	2k-128k	Also a microcontroller
www.freescale.com	DSP563xx	24	80-275	1	External	576k	
	StarCore	16	1000-48,000	1-6	External	0-1436k	
Microchip	dsPIC	16	30-40	1	6k-256k	256-32k	Also a microcontroller
www.microchip.com							Free IDE software
Texas Instruments	C5000	16	50-600	1	8k-256k	0-1280k	
www.ti.com	C6000	16/64 fp	300-24,000	1-3	0-384k	32k-3072	Fixed or floating point ver.
Zilog	Z89xxx	16	20	1	4k-8k	512	
www.zilog.com							

cessor or DSP chip embedded somewhere in its circuitry. Examples are microwave ovens, automobiles, mobile telephones and software-defined radios. DSPs intended for embedded systems often include a wide array of on-chip peripherals such as various kinds of timers, multiple hardware interrupts, serial ports of various types, a real-time clock, pulse-width modulators, optical encoder interfaces, A/D and D/A converters and lots of general-purpose digital I/O pins. Some DSPs not only include lots of peripherals but in addition have architectures that are well-suited for general-purpose control applications as well as digital signal processing.

Table 15.2 lists some manufacturers of DSP chips targeted to embedded systems. It should be mentioned that microprocessors intended for personal computers made by Intel and AMD also include extensive DSP capability. However, they are large, complicated, power-hungry ICs that are not often used in embedded applications.

When selecting a DSP device for a new design, often the available development environment is more important than the characteristics of the device itself. Microchip's dsPIC family of DSPs was chosen for the examples in this chapter because their integrated development environment is extensive and easy to use and the IDE software is available for free download from their Web site.[2] The processor instruction set is a superset of the PIC24 family of general-purpose microcontrollers, with which many hams are already familiar. The company offers a line of low-cost evaluation boards and starter kits as well as an inexpensive in-circuit debugger, the ICD 2. The free IDE software includes a simulator that can run dsPIC software on a PC (at a much slower rate, of course), so that you can experiment with DSP algorithms before buying any hardware.

The Microchip DSP family is limited to 40 million instructions per second. In a system with, say, a 40 kHz sample rate, 1000 instructions per sample are available which should be plenty if the calculations are not too complex. However if the sample rate is 1 MHz, then you get only 40 instructions per sample, which likely would be insufficient.

If more horsepower is required, you'll need to select a processor from a different manufacturer. Look for one with a well-integrated suite of development software that is powerful and easy to use. Also check out the cost and availability of development hardware such as evaluation kits, programmers and debuggers. Once those requirements are met, then you can move on to selecting a specific device with the performance and features required for your application. It can be helpful at the beginning of a project to first write some of the key software routines and test them on a simulator to estimate execution times, in order to determine how powerful a processor is needed.

When estimating execution time, don't forget to include the effect of interrupts. Most DSP systems require real-time response and make extensive use of interrupts to ensure that certain events happen at the correct times. Although this is hidden from the programmer's view when programming in C, the interrupt service routines contain quite a bit of overhead each time they are called (to save the processor state when responding and to recall the state just before returning from the interrupt). Sometimes an interrupt may be called more often than you expect, which can eat up processor cycles and so increase the execution time of other unrelated routines.

In the past, may embedded systems were written in assembly language so save memory and increase processing speed. Many early microprocessors and DSPs did not have enough memory to support a high-level language. Today, most processors have sufficient memory and processing speed to support a C kernel and library without difficulty. For anything but the simplest of programs, it is not only faster and easier to develop software in C but it is easier to support and maintain as well, especially if people other than the original programmer might become involved. Far more people know the C programming language than any particular processor's assembly language. It is true that the version of C used on a DSP chip is usually modified from standard ANSI C to support specific hardware features, but it would still be far easier to learn for a programmer familiar with writing C code on a PC or on a different DSP.

A common technique is first to write the entire application in C. Then, if execution time is not acceptable, analyze the system to determine in which software routines the bottlenecks are occurring. Those routines can then be re-written in assembly language. Having an already-working version written in C (even if too slow) can be helpful in testing and troubleshooting the equivalent assembly language.

15.3 Digital Signals

Digital signals differ from analog signals in two ways. One is that they are digitized in time, a process called *sampling*. The other is that they are digitized in amplitude, a process called *quantization*. Sampling and quantization affect the digitized signal in different ways so the following sections will consider their effects separately.

15.3.1 Sampling — Digitization in Time

Sampling is the process of measuring a signal at discrete points of time and recording the measured values. An example from history is recording the number of sunspots. If an observer goes out at noon every day and writes down the number of observed sunspots, then that data can be used to plot sunspot number versus time. In this case, we say the *sample rate* is one sample per day. The data can then be analyzed in various ways to determine short and long-term trends. After recording only a few months of data it will quickly become apparent that sunspot number has a marked periodicity — the numbers tend to repeat every 27 days (which happens to be the rotation rate of the sun as seen from earth).

What if, instead of taking a reading once a day, the readings were taken only once per month? With a 30-day sample period, the 27-day periodicity would likely be impossible to see. Clearly, the sample rate must be at least some minimum value to accurately represent the measured signal. Based on earlier work by Harry Nyquist, Claude Shannon proved in 1948 that in order to sample a signal without loss of information, the sample rate must be greater than the *Nyquist rate*, which is two times the bandwidth of the signal. In other words, the bandwidth must be less than the *Nyquist frequency*, which is one-half the sample rate. This is known as the *Nyquist sampling criterion*.

That simple rule has some profound implications. If all the frequency components of a signal are contained within a bandwidth of

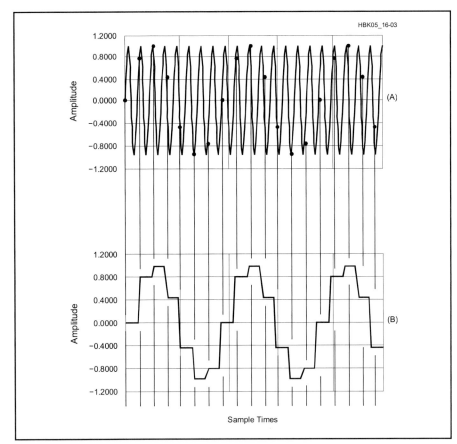

Fig 15.3 — Undersampled sine wave (A). Samples aliased to a lower frequency (B).

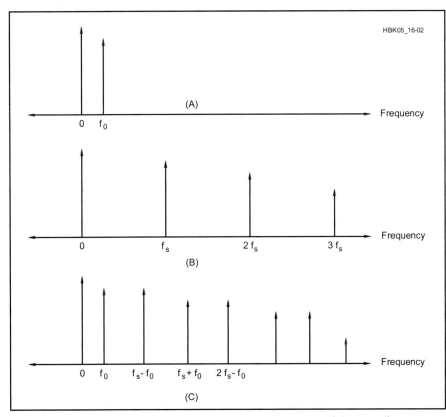

Fig 15.4 — Spectrum of an analog sine wave (A). The spectrum of the sampling function, including all harmonics (B). The spectrum of the sampled sine wave (C).

B Hz, then sampling at a rate greater than 2B samples per second is sufficient to represent the signal with 100% accuracy and with no loss of information. It is theoretically possible to convert the samples back to an analog signal that is exactly identical to the original.

Of course, a real-world digital system measures those samples with only a finite number of bits of resolution, with consequences that we will investigate in the section on quantization that follows. In addition, sampling theory assumes that there is absolutely no signal energy outside the specified bandwidth; in other words the stopband attenuation is infinity dB. Any residual signal in the stopband shows up as distortion or noise in the sampled signal.

To simplify the discussion, let's think about sampling a signal of a single frequency (a sine wave). **Fig 15.3** illustrates what happens if the sample rate is too low. As shown, the sample rate is approximately ⅞ the sine-wave frequency. You can see that the sampled signal has a period about 8 times greater than the period of the sine wave, or ⅛ the frequency. The samples are the same as if the analog signal had been a sine wave of ⅛ the actual frequency.

That is an example of a general principle. If the sample rate is too low, the sampled signal will be *aliased* to a frequency equal to the difference between the actual frequency of the analog signal and the sample rate. In the above example, the alias frequency f_o is

$$f_o = f_{sig} - f_s = \left(1 - \frac{7}{8}\right) f_{sig} = \left(\frac{1}{8}\right) f_{sig}$$

where f_{sig} is the frequency of the signal before sampling and f_s is the sample rate.

If the analog signal's frequency is even higher, then it aliases relative to whichever harmonic of the sample rate is closest. **Fig 15.4C** shows all the signal frequencies that alias to a frequency of f_o, calculated from the equation

$$f_o = \left| f_{sig} - N f_s \right|$$

where N is the harmonic number. One way to think of it is that a sampler is a harmonic mixer. The sampled signal (equivalent to the mixer output) contains the sum and difference frequencies of the input signal and all the harmonics of the sample frequency.

To avoid aliasing, most systems use an *anti-aliasing filter* before the sampler, as shown in **Fig 15.5**. For a baseband signal (one that extends to zero Hz), the anti-aliasing filter is a low-pass filter whose stopband extends from the Nyquist frequency to infinity. Of course, practical filters do not transition instantaneously from the passband to the stopband, so the bandwidth of the passband must be somewhat less than half the sample rate.

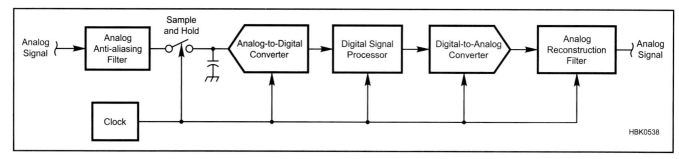

Fig 15.5 — A more complete block diagram of a DSP system.

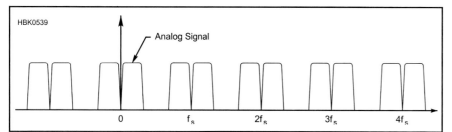

Fig 15.6 — An ideal sampled signal repeats the spectrum of the analog signal at all harmonics of the sample rate, f_s.

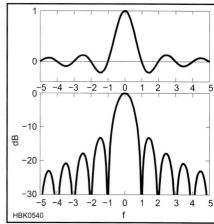

Fig 15.7 — The sinc function, where the horizontal axis is frequency normalized to the sample rate. At the bottom is the same function in decibels.

It is actually possible to accurately sample signals above the Nyquist frequency so long as their bandwidth does not violate the Nyquist criterion, a process called *undersampling* or *harmonic sampling*. Imagine an LSB signal at 455 kHz with a bandwidth of 3 kHz that is being sampled at a 48 kHz rate. The 455 kHz signal mixes with the ninth harmonic of the sample rate at 432 kHz, resulting in a sampled signal with its suppressed carrier at 455 − 432 = 23 kHz and extending 3 kHz below that to 20 kHz. So long as the incoming signal has no significant energy below 432 kHz or above 456 kHz [432 + (48/2)] kHz no unwanted aliasing occurs.

With harmonic sampling, the anti-alias filter must be a band-pass type. In the previous example, you'd probably need to use a crystal or mechanical filter in order to have a sufficiently sharp transition from the top edge of the passband slightly below 455 kHz to the stopband edge at 456 kHz.

Fig 15.3 shows each sample being held at a constant value for the duration of one sample period. However, sampling theory actually assumes that the sample is only valid at the instant the signal is sampled; it is zero or undefined at all other times. A series of such infinitely-narrow impulses has harmonics all the way to infinite frequency. Each harmonic has the same amplitude and is modulated by the signal being sampled. See **Fig 15.6**. When a digitized signal is converted back to analog form, unwanted harmonics must be filtered out by a *reconstruction filter* as shown in Fig 15.5. This is similar to the anti-aliasing filter used at the input in that its bandwidth should be no greater than one-half the sample rate. It is a low-pass filter for a baseband signal and a band-pass filter for an undersampled signal.

Most DACs actually do hold each sample value for the entire sample period. This is called *zero-order hold* and results in a frequency response in the shape of a sinc function

$$\text{sinc}(f) = \frac{\sin(\pi f)}{\pi f}$$

where f is normalized to the sample rate, f = frequency / sample rate.

The graph of the sinc function in **Fig 15.7** shows both positive and negative frequencies for reasons explained in the Analytic Signals section. Note that the logarithmic frequency response has notches at the sample rate and all of its harmonics. If the signal bandwidth is much less than the Nyquist frequency, then most of the signal at the harmonics falls near the notch frequencies, easing the task of the reconstruction filter. If the signal bandwidth is small enough (sample rate is high enough), the harmonics are almost completely notched out and a reconstruction filter may not even be required.

The sin(πf)/πf frequency response also affects the passband. For example if the passband extends to sample rate / 4 (f = 0.25), then the response is

$$20 \log \frac{\sin(\pi \cdot 0.25)}{\pi \cdot 0.25} = -0.9 \text{ dB}$$

at the top edge of the passband. At the Nyquist frequency, (f = 0.5), the error is 3.9 dB. If the signal bandwidth is a large proportion of the Nyquist frequency, then some kind of digital or analog compensation filter may be required to correct for the high-frequency rolloff.

DECIMATION AND INTERPOLATION

The term *decimation* simply means reducing the sample rate. For example to decimate by two, simply eliminate every second sample. That works fine as long as the signal bandwidth satisfies the Nyquist criterion at the lower, output sample rate. If the analog anti-aliasing filter is not narrow enough, then a digital anti-aliasing filter in the DSP can be used to reduce the bandwidth to the necessary value. This must be done *before* decimation to satisfy the Nyquist criterion.

If you need to decimate by a large amount, then the digital anti-aliasing filter must have a very small bandwidth compared to the sample rate. As we will see later, a digital filter with a small bandwidth is computationally intensive. For this reason, large decimation factors

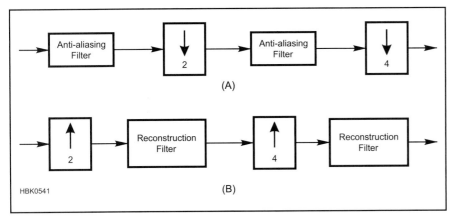

Fig 15.8 — Decimation (A) and interpolation (B). The arrow's direction indicates decimation (down) or interpolation (up) and the number is the factor.

are normally accomplished in multiple steps, as shown in **Fig 15.8A**. The first decimation is by a small factor, typically 2, so that the first anti-alias filter can be as simple as possible. The second decimation stage then does not have to decimate by such a large factor, simplifying its task. In addition, since it is running at only half the input sample rate it has more time to do its calculations. Generally it is most efficient to decimate by the smallest factor in the first stage, a larger factor in the second, and the largest factors in the third and any subsequent stages. The larger the total decimation factor, the greater the number of stages is appropriate but more than three stages is uncommon.

Interpolation means increasing the sample rate. One way to do that is simply to insert additional zero-value samples, a process called *zero-stuffing*. For example, to interpolate by a factor of three, insert two zero-value samples after each input sample. That works, but may not give the results you expect. Recall that a sampled signal has additional copies of the baseband signal at all harmonics of the sample rate. All of those harmonics remain in the resampled signal, even though the sample rate is now higher. To eliminate them, the signal must be filtered after interpolation. After filtering, there is signal only at baseband and around the harmonics of the interpolated (higher-frequency) sample rate. It's as if the analog signal had been sampled at the higher rate to begin with.

Just as with decimation, interpolation by a large factor is best done in stages, as shown in Fig 15.8B. In this case, the stage running at the lowest sample rate (again the first stage) is the one with the lowest interpolation factor.

Zero-stuffing followed by filtering is not the only way to interpolate. Really what you are trying to do is to fill in between the lower-rate samples with additional samples that "connect the dots" in as smooth a manner as possible. It can be shown that that is mathematically equivalent to zero-stuffing and filtering. For example, if instead of inserting zero-value samples you instead simply repeat the last input sample, you have a situation similar to the zero-order hold of a DAC output. It is equivalent to zero-stuffing followed by a low-pass filter with a frequency response of $\sin(\pi f)/\pi f$. If you do a straight-line interpolation between input samples (a "first-order" interpolation), it turns out that it is equivalent to a low pass filter with a frequency response of $[\sin(\pi f)/\pi f]^2$, which has a sharper cutoff and better stop-band rejection than a zero-order interpolation. Higher-order interpolations have smoother responses in the time domain which translate to better filter responses in the frequency domain.

So far we have only covered decimation and interpolation by integer factors. It is also possible to change the sample rate by a non-integer factor, which is called *resampling* or *multi-rate* conversion. For example, if you want to increase the sample rate by a factor of ⁴⁄₃, simply interpolate by 4 and then decimate by 3. That method can become impractical for some resample ratios. For example, to convert an audio file recorded from a computer sound card at 48 kHz to the 44.1 kHz required by a compact disc, the resample ratio is 44,100 / 48,000 = 147 / 160. After interpolation by 147, the 44.1 kHz input file is sampled at 6.4827 MHz, which would result in excessive processing overhead.

In addition, the interpolation/decimation method only works for resample ratios that are rational numbers (the ratio of two integers). To resample by an irrational number, a different method is required. The technique is as follows. For each output sample, first determine the two nearest input samples. Calculate the coefficients of the Nth-order equation that describes the trajectory between the two input samples. Knowing the trajectory between the input samples and the output sample's relative position between them, the value of the output sample can be calculated from the equation.

15.3.2 Quantization — Digitization in Amplitude

While sampling (digitization in time) theoretically causes no loss of signal information, quantization (digitization in amplitude) always does. For example, an 8-bit signed number can represent a signal as a value from –128 to +127. For each sample, the A/D converter assigns whichever number in that range is closest to the analog signal at that instant. If a particular sample has a value of 10, there is no way to tell if the original signal was 9.5, 10.5 or somewhere in between. That information has been lost forever.

When quantizing a complex signal such as speech, this error shows up as noise, called *quantization noise*. See **Fig 15.9**. The error is random — it is equally likely to be anywhere in the range of –1/2 to +1/2 of a single step of the ADC. We say that the maximum error is one-half of one *least-significant bit* (LSB). It can be shown mathematically that a series of uniformly-distributed random numbers between +0.5 LSB and –0.5 LSB has an RMS value of

$$\frac{LSB}{\sqrt{12}}$$

which is –10.79 dB less than one LSB. Each time you add one bit to the data word, the number of LSBs in the range doubles, which means each LSB gets two times smaller reducing the noise by 6.02 dB. A full-scale sine wave has an RMS power –3.01 dB from a full-scale dc voltage. Combining that information results in the following equation for signal-to-noise ratio in decibels for a data word of width b bits:

$$SNR = 1.76 + 6.02b \text{ dB}$$

For example, with 8-bit data, SNR = 49.9 dB. An ideal 16-bit ADC would achieve a 98.1 dB signal-to-noise ratio. Of course, real-world devices are never perfect so actual performance would be somewhat less.

One critical point that is sometimes overlooked is that quantization noise is spread over the entire bandwidth from zero Hz to the sample rate. If you are digitizing a 3 kHz audio channel with a 48 ksps sampler, only a fraction of the noise power is within the channel. For that reason, the effective signal-to-noise ratio depends not only on the number of bits but also the sample rate, f_s, and the signal bandwidth, B:

$$SNR_{eff} = SNR + 10\log\left(\frac{f_s}{2B}\right) \text{ dB}$$

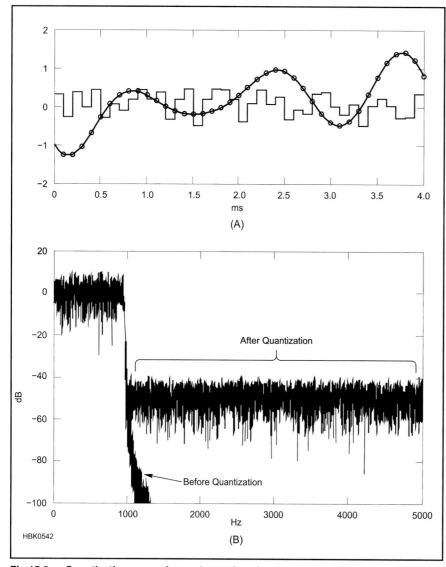

Fig 15.9 — Quantization error of a random noise signal that has been band-limited to 1 kHz to simulate an audio signal. (A) The sampler resolution is 8 bits and the sample rate is 10 kHz. Sample values are indicated by circles. Also shown is the quantization error, in units of LSB. Below is the frequency spectrum of the signal before and after quantization. (B)

The reason for the factor of two in the denominator is that the bandwidth of a positive-frequency scalar signal should be compared to the Nyquist bandwidth, $f_s/2$. When filtering a complex signal (one that contains I and Q parts), the 2B in the denominator should be replaced by B.

When choosing an A/D converter don't forget that the effective SNR depends on the sample rate. As an example, let's compare an Analog Devices AD9235 12-bit, 65 Msps ADC to an AD7653, which is a 16-bit 100 ksps ADC from the same manufacturer. Assume a 10 kHz signal bandwidth.

An ideal 12-bit ADC has a SNR of 1.76 + 6.02 × 12 = 74.0 dB. The AD9235's performance is not far from the ideal; its SNR is specified at 70.5 dB at its 65 Msps maximum sample rate. In a 10 kHz bandwidth, the effective SNR is 70.5 + 10log(65,000/20) = 105.6 dB.

An ideal 16-bit ADC has an SNR of 98.1 dB. The AD7653 is specified at 86 dB. The effective SNR is 86 + 10log(100/20) = 93 dB.

So the 12-bit ADC with 70.5 dB SNR is actually 12.6 dB better than the 16-bit device with 86 dB SNR! Even an ideal 16-bit, 100 ksps ADC would only have an effective SNR of 98.1 + 10log(100/20) = 105.1 dB, still worse than the actual performance of the AD9235 when measured with the same bandwidth. Note that to actually realize 105.6 dB of dynamic range the signal from the ADC would need to be filtered to a 10 kHz bandwidth while increasing the bits of data resolution.

Oversampling is the name given to the technique of using a higher-than-necessary sample rate in order to achieve an improved S/N ratio. Don't forget that when the high-sample-rate signal is decimated the data words must have enough bits to support the higher dynamic range at the lower sample rate. As a rule of thumb, the quantization noise should be at least 10 dB less than the signal noise in order not to significantly degrade the SNR. In the AD9235 example, assuming a 100 kHz output sample rate, about 18 bits would be required: 1.76 + 6.02 × 18 + 10log(100/20) = 117.1 dB, which is 11.5 dB better than the 105.6 dB dynamic range of the ADC in a 10 kHz bandwidth.

Most ADCs and DACs used in high-fidelity audio systems use an extreme form of oversampling, where the internal converter may oversample by a rate of 128 or 256 times, but with very low resolution (in some cases just a 1-bit ADC!). In addition, such converters use a technique called noise shaping to push most of the quantization noise to frequencies near the sample rate, and reduce it in the audio spectrum. The noise is then removed in the decimation filter.

Although quantization error manifests itself as noise when digitizing a complex non-periodic signal, it can show up as discrete spurious frequencies when digitizing a periodic signal. **Fig 15.10** illustrates a 1 kHz sine wave sampled with 8-bit resolution at a 9.5 kHz rate. On average there are 9.5 samples per cycle of the sine wave so that the sampling error repeats every second cycle. That 500-Hz periodicity in the error signal causes a spurious signal at 500 Hz and harmonics. As the signal frequency is changed, the spurs move around in a complicated manner that depends on the ratio of sample rate to signal frequency. In real-world ADCs, nonlinearities in the transfer function can also create spurious signals that vary unpredictably as a function of the signal amplitude, especially at low signal levels.

In many applications, broadband noise is preferable to spurious signals on discrete frequencies. The solution is to add *dithering*. Essentially this involves adding a small amount of noise, typically on the order of an LSB or two, in order to randomize the quantization error. Some DACs have dithering capability built in to improve the SFDR, even though it does degrade the SNR slightly. Dithering is also useful in cases where the input signal to an ADC is smaller than one LSB. Even though the signal would be well above the noise level after narrow-band filtering, it cannot be detected if the ADC input is always below one LSB. In many systems there is sufficient noise at the input, both from input amplifiers as well as from the ADC itself, to cause natural dithering.

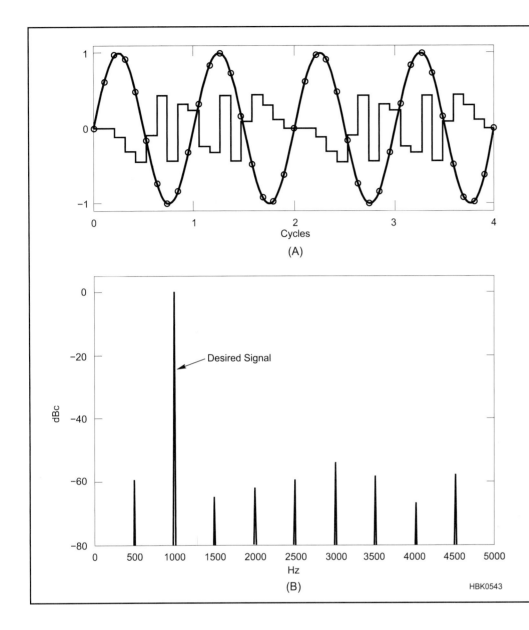

Fig 15.10 — Quantization error of a 1 kHz sine wave sampled at 9.5 kHz with 8-bit resolution (A). Sample values are indicated by circles. Also shown is the quantization error, in units of LSB. Below is the frequency spectrum, showing the spurious frequencies caused by the quantization. (B)

15.4 Digital Filters

As radio amateurs, most of us are well-acquainted with the concept of frequency. We know, for example, that a pure sine wave consists of a single frequency, which is inversely proportional to the wavelength. If the sine wave is distorted, additional harmonic frequencies appear at integer multiples of the fundamental. For example, a square wave consists of sine waves at the fundamental frequency and all the odd harmonics. In general, any periodic waveform can be decomposed into a combination of sine waves at various phase angles with frequencies that are integer multiples of the repetition rate of the waveform.

Even a non-periodic waveform can be decomposed into sine waves, although in this case they are not harmonically-related. For example a single pulse of width τ seconds has a frequency spectrum proportional to sinc(fτ) = sin(πfτ)/(πfτ). You can think of this as an infinite number of sine waves spaced infinitely closely together with amplitudes that trace out that spectral shape. It is interesting to note that if τ is decreased, the value of f must increase by the same factor for any given value of sin(πfτ)/(πfτ). In other words, the narrower the pulse the wider the spectrum. Of course that applies to sine waves and other periodic waveforms as well — the smaller the wavelength the higher the frequency. In general, anything that makes the signal "skinnier" in the time domain makes it "fatter" in the frequency domain and vice versa.

As the pulse becomes narrower and narrower, the frequency spectrum spreads out more and more. In the limit, if the pulse is made infinitely narrow, the spectrum becomes flat from zero hertz to infinity. An infinitely-narrow pulse is called an *impulse* and is a very useful concept because of its flat frequency spectrum. If you feed an impulse into the input of a filter, the signal that comes out, the *impulse response*, has a frequency spectrum equal to the frequency response of the filter. One way to design a filter is to determine the impulse response that corresponds to the desired frequency spectrum and then design the filter to have that impulse response. That method is ideally suited for designing FIR filters.

15.4.1 FIR Filters

A *finite impulse response* (FIR) filter is a filter whose impulse response is finite, ending in some fixed time. Note that analog filters have an infinite impulse response — the output theoretically rings forever. Even a simple R-C low-pass filter's output dies exponentially toward zero but theoretically never quite reaches it. In contrast, an FIR filter's impulse response becomes exactly zero at some time after receiving the impulse and stays zero forever (or at least until another impulse comes along).

Given that you have somehow figured out the desired impulse response, how would you design a digital filter to have that response? The obvious method would be to pre-calculate a table of impulse response values, sampled at the sample rate. These are called the filter *coefficients*. When an impulse of a certain amplitude is received, you multiply that amplitude by the first entry in the coefficient table and send the result to the output. At the next sample time, multiply the impulse by the second entry, and so on until you have used up all the entries in the table.

A circuit to do that is shown in **Fig 15.11**. The input signal is stored in a shift register. Each block labeled "Delay" represents a delay of one sample time. At each sample time, the signal is shifted one register to the right. Each register feeds a multiplier and the other input to the multiplier comes from one of the coefficient table entries. All the multiplier outputs are added together. Since the input is assumed to be a single impulse, at any given time all the shift registers contain zero except one, which is multiplied by the appropriate table entry and sent to the output.

We've just designed an FIR filter! By using a shift register with a separate multiplier for each tap, the filter works for continuous signals as well as for impulses. Since this is a linear system, the continuing signal is affected by the filter the same as an individual impulse.

It should be obvious from the diagram how to implement an FIR filter in software. You set up two buffers in memory, one for the filter coefficients and one for the data. The length of each buffer is the number of filter taps. (A *tap* is the combination of one filter coefficient, one shift register and one multiplier/accumulator.) Each time a new data value is received, it is stored in the next available position in the data buffer and the accumulator is set to zero. Next, a software loop is executed a number of times equal to the number of taps. During each loop, pointers to the two buffers are incremented, the next coefficient is multiplied by the next data value and the result is added to the current accumulator value. After the last loop, the accumulator contents are the output value. Normally the buffers are implemented as *circular buffers* — when the address pointer gets to the end it is reset back to the beginning.

Now you can see why a hardware multiplier-accumulator (MAC) is such an important feature of a DSP chip. Each tap of the FIR filter involves one multiplication and one addition. With a 1000-tap FIR filter, 1000 multiplications and 1000 additions must be performed during each sample time. Being able to do each MAC in a single clock cycle saves a lot of processing time.

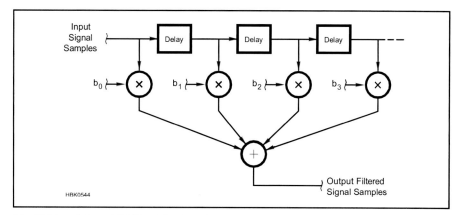

Fig 15.11 — A 4-tap FIR filter. The b_n values are the filter coefficients.

An FIR filter is a hardware or software implementation of the mathematical operation called *convolution*. We say that the filter *convolves* the input signal with the impulse response of the filter. It turns out that convolution in the time domain is mathematically equivalent to multiplication in the frequency domain. That means that the frequency spectrum of the output equals the frequency spectrum of the input times the frequency spectrum of the filter. Expressed in decibels, the output spectrum equals the input spectrum plus the filter frequency response, all in dB. If at some frequency the input signal is +3 dB and the filter is –10 dB compared to some reference, then the output signal will be 3 – 10 = –7 dB at that frequency.

An FIR filter whose bandwidth is very small compared to the sample rate requires a long impulse response with lot of taps. This is another consequence of the "skinny" versus "fat" relationship between the frequency and time domains. If the filter is narrow in the frequency domain, then its impulse response is wide. Actually, if you want the frequency response to go all the way to zero (minus infinity dB) throughout the stop band, then the impulse response theoretically becomes infinitely wide. Since we're designing a *finite* impulse response filter we have to truncate the impulse response at some point to get it to fit in the coefficient table. When you do that, however, you no longer have infinite attenuation in the stopband. The more heavily you truncate (the narrower the impulse response) the worse the stopband attenuation and the more ripple you get in the passband. Assuming optimum design techniques for selecting coefficients, you can estimate the minimum length L of the impulse response from the following equation:

$$L = 1 - \frac{10\log(\delta_1 \delta_2) - 15}{14\left(\frac{f_T}{f_s}\right)} \text{ taps}$$

where

δ_1 and δ_2 = the passband and stopband ripple expressed as a fraction

f_T = the transition bandwidth (frequency difference between passband and stopband edges)

f_s = the sample rate.

For example, for a low-pass filter with a passband that extends up to 3 kHz, a stopband that starts at 4 kHz ($f_T = 4 - 3 = 1$ kHz), $f_s = 10$ kHz sample rate, ±0.1 dB passband ripple ($\delta_1 = 10^{0.1/20} - 1 = 0.0116$), and 60 dB stopband rejection ($\delta_2 = 10^{-60/20} = 0.001$), we get

$$L = 1 - \frac{10\log(0.0116 \times 0.001) - 15}{14\left(\frac{1}{10}\right)}$$

$$= 1 - \frac{-49.4 - 15}{1.4} = 47 \text{ taps}$$

Overflow is a potential problem when doing the calculations for an FIR filter. Multiplying two N-bit numbers results in a product with 2N bits, so space must be provided in the accumulator to accommodate that. Although the final result normally will be scaled and truncated back to N bits, it is best to carry through all the intermediate results with full resolution in order not to lose any dynamic range. In addition, the sum of all the taps can be a number with more than 2N bits. For example, if the filter width is 256 taps, then if all coefficients and data are at full scale, the final result could theoretically be 256 times larger, requiring an extra 8 bits in the accumulator. We say "theoretically" because normally most of the filter coefficients are much less than full scale and it is highly unlikely that all 256 data values would ever simultaneously be full-scale values of the correct polarity to cause overflow. The dsPIC processors use 16-bit multipliers with 32-bit results and a 40-bit accumulator, which should handle any reasonable circumstances.

After all taps have been calculated, the final result must be retrieved from the accumulator. Since the accumulator has much more resolution than the processor's data words, normally the result is truncated and scaled to fit. It is up to the circuit designer or programmer to scale by the correct value to avoid overflow. The worst case is when each data value in the shift register is full-scale — positive when it is multiplying a positive coefficient and negative for negative coefficients. That way, all taps add to the maximum value. To calculate the worst-case accumulator amplitude, simply add the absolute values of all the coefficients. However, that normally gives an unrealistically pessimistic value because statistically it is extremely unlikely that such a high peak will ever be reached. For a low-pass filter, a better estimate is to calculate the gain for a dc signal and add a few percent safety margin. The dc gain is just the sum of all the coefficients (not the absolute values). For a band-pass filter, add the sum of all the coefficients multiplied by a sine wave at the center frequency.

CALCULATING FIR FILTER COEFFICIENTS

So far we have ignored the question of how to determine the filter coefficients. For an ideal "brick-wall" low-pass filter, the answer turns out to be pretty simple. A "brick-wall" low-pass filter is one that has a constant response from zero hertz up to the cutoff frequency and zero response above. Its impulse response is proportional to the sinc function:

$$C(n) = C_o \, \text{sinc}(2Bn) = \frac{\sin(2\pi Bn)}{2\pi Bn}$$

where C(n) are the filter coefficients, n is the sample number with n = 0 at the center of the impulse response, C_o is a constant, and B is the single-sided bandwidth normalized to the sample rate, B = bandwidth / sample rate.

It is interesting that this has the same form as the frequency response of a pulse, as was shown in Fig 15.7. That is because a brick-wall response in the frequency domain has the same shape as a pulse in the time domain. A pulse in one domain transforms to a sinc function in the other. This is an example of the general principle that the transformation between time and frequency domains is symmetrical. We will discuss this more later, in the section on Fourier transforms.

Normally, the filter coefficients are set up with the peak of the sinc function, sinc(0), at the center of the coefficient table so that there is an equal amount of "tail" on both sides. That points up the principle problem with this method of determining filter coefficients. Theoretically, the sinc function extends from minus infinity to plus infinity.

Table 15.3
Routine for dsPIC Processor to Calculate Filter Coefficients

```
// Calculate FIR filter coefficients
// using the windowed-sinc method
void set_coef (
 double sample_rate;
 double bandwidth;)
{
extern int c[FIR_LEN]; // Coefficient array
int i; // Coefficient index
double ph; // Phase in radians
double coef; // Filter coefficient
int coef_int; // Digitized coefficient
double bw_ratio; // Normalized bandwidth

bw_ratio = 2 * bandwidth / sample_rate;
for (i = 0; i < (FIR_LEN/2); i++) {
// Brick-wall filter:
ph = PI * (i + 0.5) * bw_ratio;
coef = sin(ph) / ph;
// Hann window:
ph = PI * (i + 0.5) / (FIR_LEN/2);
coef *= (1 + cos(ph)) / 2;
// Convert from floating point to int:
coef *= 1 << (COEF_WIDTH - 1);
coef_int = (int)coef;
// Symmetrical impulse response:
c[i + FIR_LEN/2] = coef_int;
c[FIR_LEN/2 - 1 - i] = coef_int;
 }
}
```

Abruptly terminating the tails causes the frequency response to differ from an ideal brick-wall filter. There is ripple in the passband and non-zero response in the stopband, as shown in the graph in the upper right of **Fig 15.12**. This is mainly caused by the abruptness of the truncation. In effect, all the coefficients outside the limits of the coefficient table have been set to zero. The passband and stopband response can be improved by tapering the edges of the impulse response instead of abruptly transitioning to zero.

The process of tapering the edges of the impulse response is called *windowing*. The impulse response is multiplied by a window, a series of coefficients that smoothly taper to zero at the edges. For example, a rectangular window is equivalent to no window at all. Many different window shapes have been developed over the years — at one time it seemed that every doctoral candidate in the field of signal processing did their dissertation on some new window. Each window has its advantages and disadvantages. A window that transitions slowly and smoothly to zero has excellent passband and stopband response but a wide transition band. A window that has a wider center portion and then transitions more abruptly to zero at the edges has a narrower transition band but poorer passband and stopband response. The equations for the windows in Fig 15.12 are included in a sidebar.

The routine shown in **Table 15.3** is written for a dsPIC processor so it can be used to calculate filter coefficients "on the fly" as the operator adjusts a bandwidth control. The same code should also work using a generic C compiler on a PC so the coefficients could be downloaded into an FIR filter implemented in hardware.

The windowed-sinc method works pretty well for a simple low-pass filter, but what if some more-complicated spectral shape is desired? The same general design approach still applies. You determine the desired spectral shape, transform it to the time domain using an inverse Fourier transform, and apply a window. There is lots of (often free) software available that can calculate the fast Fourier transform (FFT) and inverse fast Fourier transform (IFFT). So the technique is to generate the desired spectral shape, transform to the time domain with the IFFT, and multiply the resulting impulse response by the desired window. Then you can transform back to the frequency domain with the FFT to see how the window affected the result. If the result is not

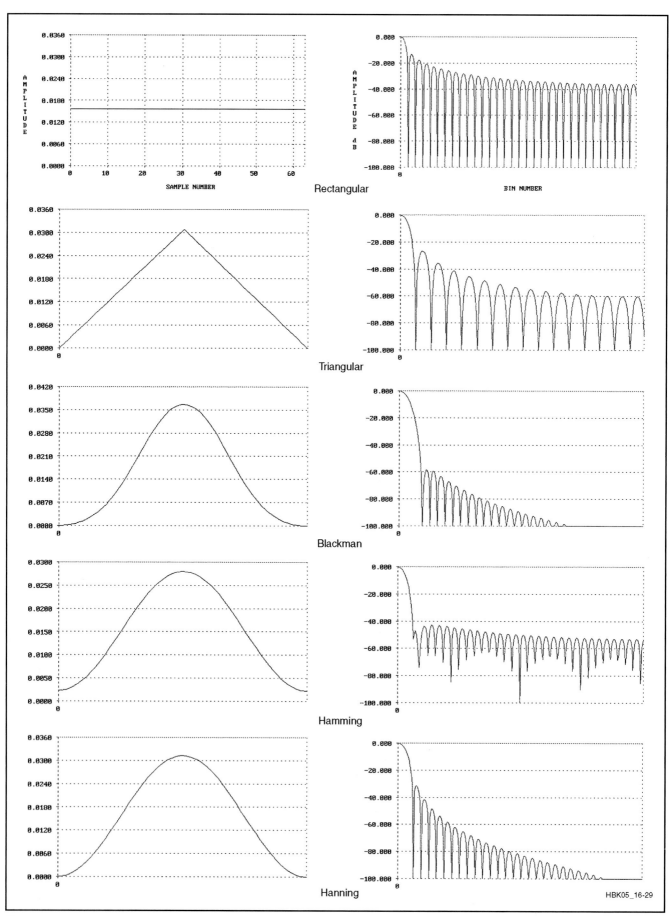

Fig 15.12 — Various window functions and their Fourier transforms.

Equations for Window Functions

For each window function, the center of the response is considered to be at time t = 0 and the width of the impulse is L. Each window is 1.0 when t = 0 and 0.0 when the |t| > L/2.

Rectangular:

$w(t) = 1.0$

Triangular (Bartlett):

$w(t) = 2\left(\dfrac{L/2 - |t|}{L}\right)$

Blackman:

$w(t) = 0.42 + 0.5\cos\left(\dfrac{2\pi t}{L}\right)$
$+ 0.08\cos\left(\dfrac{4\pi t}{L}\right)$

Hamming:

$w(t) = 0.54 + 0.46\cos\left(\dfrac{2\pi t}{L}\right)$

Hanning (Hann):

$w(t) = 0.5 + 0.5\cos\left(\dfrac{2\pi t}{L}\right)$

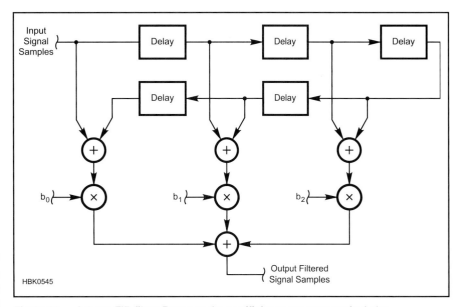

Fig 15.13 — A 6-tap FIR filter. Because the coefficients are symmetrical, the symmetrical taps may be combined before multiplication.

satisfactory, you can either choose a different window or modify the original spectral shape and go through the process again.

Windowing methods are useful because they are simple to program and the resulting software routines execute quickly. For example, you can include a bandwidth knob on your DSP filter and calculate filter coefficients "on the fly" as the user turns the knob. However, while the filter performance that results is pretty good, it is not "optimum" in the sense that it does not have minimum passband ripple and maximum stopband attenuation for a given number of filter coefficients. For that, you need what is known as an *equal-ripple*, or *Chebyshev* filter. The calculations to determine Chebyshev filter coefficients are more complicated and time-consuming. For that reason, the coefficients are normally calculated in advance on a PC and stored in DSP program memory for retrieval as needed.

Engineers have not had much success in devising a mathematical algorithm to calculate the Chebyshev coefficients directly, but in 1972 Thomas Parks and James McClellan figured out a method to do it iteratively. The *Parks-McClellan algorithm* is supported in most modern filter-design software, including a number of programs available for free download on the Web. Typically you enter the sample rate, the passband and stopband frequency ranges, the passband ripple and the stopband attenuation. The software then determines the required number of filter coefficients, calculates them and displays a plot of the resulting filter frequency response.

Filter design software typically presents the filter coefficients as floating-point numbers to the full accuracy of the computer. You will need to scale the values and truncate the resolution to the word size of your filter implementation. Truncation of filter coefficients affects the frequency response of the filter but does not add noise in the same manner as truncating the signal data.

As you look at impulse responses for various FIR filters calculated by various methods you soon realize that most of them are symmetrical. If the center of the impulse response is considered to be at time zero, then the value at time t equals the value at time −t for all t. If you know in advance that the filter coefficients are symmetrical, you can take advantage of that in the filter design. By re-arranging the adders and multipliers, the number of multipliers can be reduced by a factor of two, as shown in **Fig 15.13**. This trick is less useful in a software implementation of an FIR filter because the number of additions is the same and many DSPs take the same amount of time to do an addition as a multiply-accumulate.

In addition to the computational benefit, a symmetrical impulse response also has the advantage that it is *linear phase*. The time delay through such a filter is one-half the length of the filter for all frequencies. For example, for a 1000-tap filter running at 10 kHz the delay is 500/10,000 = 0.05 second. Since the time delay is constant for all frequencies, the phase delay is directly proportional to the frequency. For example, if the phase delay at 20 Hz is one cycle (0.05 second) it is ten cycles at 200 Hz (still 0.05 second). Linear phase delay is important with digital modulation signals to avoid distortion and intersymbol interference. It is also desirable with analog modulation where it can result in more natural-sounding audio. All analog filters are non-linear-phase; the phase distortion tends to be worse the more abrupt the transition between passband and stopband. That is why an SSB signal sounds unnatural after being filtered by a crystal filter with a small shape factor even though the passband ripple may be small and distortion minimal.

A band-pass filter can be constructed from a low-pass filter simply by multiplying the impulse response by a sine wave at the desired center frequency. This can be done before or after windowing. The linear-phase property is retained but with reference to the center frequency of the filter, that is, the phase shift is proportional to the difference in frequency from the center frequency. The frequency response is a double-sided version of the low-pass response with the zero-hertz point of the low-pass filter shifted to the frequency of the sine wave.

DSP and Software Radio Design   15.15

15.4.2 IIR Filters

An *infinite impulse response* (IIR) filter is a filter whose impulse response is infinite. After an impulse is applied to the input, theoretically the output never goes to zero and stays there. In practice, of course, the signal eventually does decay until it is below the noise level (analog filter) or less than one LSB (digital filter).

Unlike a symmetrical FIR filter, an IIR filter is not generally linear-phase. The delay through the filter is not the same for all frequencies. Also, IIR filters tend to be harder to design than FIR filters. On the other hand, many fewer adders and multipliers are typically required to achieve the same passband and stop band ripple in a given filter, so IIR filters are often used where computations must be minimized.

All analog filters have an infinite impulse response. For a digital filter to be IIR it must have feedback. That means a delayed copy of some internal computation is applied to an earlier stage in the computation. A simple but useful example of an IIR filter is the exponential decay circuit in **Fig 15.14**. In the absence of a signal at the input, the output on the next clock cycle is always $(1-\delta)$ times the current output. The time constant (the time for the output to die to $1/e = 36.8\%$ of the initial value) is very nearly

$$\tau = f_s \left(\frac{1}{\delta} - \frac{1}{2} \right)$$

where f_s is the sample rate. The circuit is the digital equivalent of a capacitor with a resistor in parallel and might be useful for example in a digital automatic gain control circuit.

One issue with IIR filters is resolution. Because of the feedback, the number of bits of resolution required for intermediate computations can be much greater than at the input or output. In the previous example, δ is very small for very long time constants. When the value in the register falls below a certain level the multiplication by $(1-\delta)$ will no longer be accurate unless the bit width is increased. In practice, the increased resolution required with IIR filters often cancels out part of the savings in the number of circuit elements.

Another issue with IIR filters is stability. Because of the feedback it is possible for the filter to oscillate if care is not taken in the design. Stability can also be affected by non-linearity at low signal levels. A circuit that is stable with large signals may oscillate with small signals due to the round-off error in certain calculations, which causes faint tones to appear when strong signals are not present. This is known as an unstable *limit cycle*. These issues are part of the reason that IIR filters have a reputation for being hard to design.

Design techniques for IIR filters mostly involve first designing an analog filter using any of the standard techniques and then transforming the design from the analog to the digital domain. The *impulse-invariant* method attempts to duplicate the filter response directly by making the digital impulse response equal the impulse response of the equivalent analog filter. It works fairly well for low-pass filters with bandwidths much less than the sample rate. Its problem is that it tries to duplicate the frequency response all the way to infinity hertz, but that violates the Nyquist criterion resulting in a folding back of the high-frequency response down into low frequencies. It is similar to the aliasing that occurs in a DSP system when the input signal to be sampled is not band-limited below the Nyquist frequency.

The *bilinear transform* method gets around that problem by distorting the frequency axis such that infinity hertz in the analog domain becomes sample rate / 2 in the digital domain. Low frequencies are fairly accurate, but high frequencies are squeezed together more and more the closer you get to the Nyquist fre-

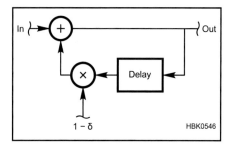

Fig 15.14 — An exponential decay circuit.

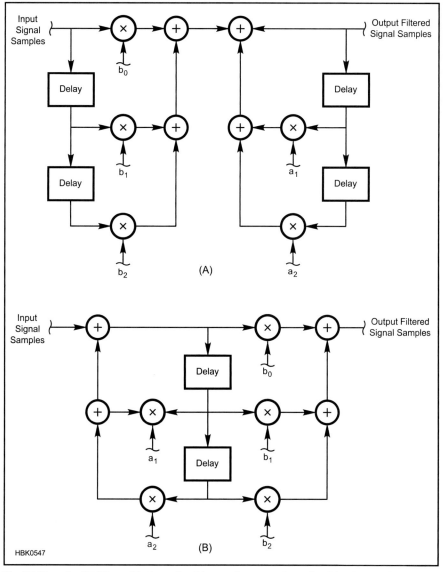

Fig 15.15 — An IIR filter with three feed-forward taps and two feed-back taps. Direct form I (A) and the equivalent direct form II (B).

quency. It avoids the aliasing problem at the expense of a change in the spectrum shape, especially at the high-frequency end. For example, when designing a low-pass filter it may be necessary to change the cutoff frequency to compensate. Again, the method works best for filters with passband frequencies much less than the sample rate.

In general, the output of an IIR filter is a combination of the current and previous input values (feed-forward) and previous output values (feed-back). **Fig 15.15A** shows the so-called *direct form I* of an IIR filter. The b_i coefficients represent feed-forward and the a_i coefficients feed-back. For example the previous value of the y output is multiplied by a_1, the second previous value is multiplied by a_2, and so on. Because the filter is linear, it doesn't matter whether the feed forward or feed back stage is performed first. By reversing the order, the number of shift registers is reduced as in Fig 15.15B. There are other equivalent topologies as well. The mathematics for generating the a_i and b_i coefficients for both the impulse-invariant and bilinear transform methods is fairly involved, but fortunately some filter design programs can handle IIR as well as FIR filters.

15.4.3 Adaptive Filters

An *adaptive filter* is one that automatically adjusts its filter coefficients under the control of some algorithm. This is often done in situations where the filter characteristics are not known in advance. For example, an adaptive channel equalizer corrects for the non-flatness in the amplitude and phase spectrum of a communications channel due to multipath

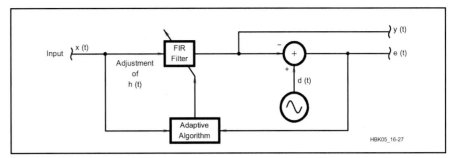

Fig 15.16 — An adaptive filter.

propagation. Typically, the transmitting station periodically sends a known sequence of data, known as a *training sequence*, which is used by the receiver to determine the channel characteristics and adjust its filter coefficients accordingly.

Another example is an automatic notch filter. An algorithm determines the frequency of an interfering tone and automatically adjusts the notch frequency to remove the tone. Noise cancellation is another application. It can be thought of as the opposite of a notch filter. In this case, all the sine-wave tones in the input signal are considered to be desired and the filter coefficients are configured to enhance them. That method works not only for CW signals but for voice as well since the human voice consists largely of discrete frequencies.

A generic block diagram of an adaptive filter is shown in **Fig 15.16**. The variable filter is typically an FIR type with coefficients calculated by the update algorithm. By some means, an estimate of the desired, unimpaired signal, d, is generated and compared to the filter output y. The difference between y and d is the error, e, which is used by the update algorithm to modify the filter coefficients to improve the accuracy of y. The algorithm is capable of acting as a noise-reduction filter and a notch filter simultaneously. Assuming d is in the form of a pure tone (sine wave), the tone is simultaneously optimized in the y output and minimized in the e output.

A common algorithm for minimizing the error signal is called *least mean squares* (LMS). The LMS algorithm includes a performance parameter, μ, which can be adjusted between 0 and 1 to control the tradeoff between adjustment speed and accuracy. A value near 1 results in fast convergence but the convergence is not very accurate. For better accuracy at the cost of slower adjustment, lower the value of μ. Some implementations adjust μ on the fly, using a large value at first to get faster lock-in when the error is large then a smaller value after convergence to reduce the error. That works as long as the signal characteristics are not changing too rapidly.

15.5 Miscellaneous DSP Algorithms

15.5.1 Sine Wave Generation

There are a number of techniques available for building a digital sine-wave generator, either in hardware or in software. One obvious idea is to make a digital oscillator, analogous to a conventional analog oscillator. Simply design a band-pass filter at the desired frequency and include positive feedback around it with a loop gain of unity. The problem is that, because of round-off error in the digital calculations, it is difficult to get the loop gain to be *exactly* 1.0. If it is slightly less, then the oscillation will eventually die out. If it is slightly greater, then the oscillation will gradually increase in amplitude until it exceeds the maximum signal that the digital circuitry can handle. There are two techniques to handle this problem. One idea is to include an automatic gain-control circuit to detect the amplitude, low-pass filter it and feed the result to a multiplier in the feedback path to control the gain. Another idea is to intentionally set the loop gain slightly greater than 1.0 and include a clipping stage in the feedback path that limits the peak amplitude in a controlled manner.

With both of those techniques there is a tradeoff between distortion and start-up time. It can take many oscillation cycles for the amplitude to stabilize. If the loop gain is increased or the AGC time constant is reduced to improve the start-up time, worse distortion results.

Probably the most common technique for generating sine waves is the *numerically-controlled oscillator* (NCO), or *direct digital synthesizer* (DDS) as shown in **Fig 15.17**.

At each clock cycle, the phase accumulator increments the phase by an amount equal to 360° times f / f_s, where f is the sine-wave frequency and f_s is the sample rate. The current phase value is used as a pointer to the proper address in a sine-wave lookup table. As the phase increases, the pointer moves through the look-up table, tracing out the sine-wave amplitude. Different frequencies can be obtained by changing the step size in the phase accumulator.

The lookup table size is a factor of two and the accumulator output is scaled such that the maximum count, corresponding to 360°, accesses the final entry in the table. When the phase passes 360° it automatically jumps back to zero as required. For good frequency resolution, the word size of the phase accu-

Fourier Transform

The *Fourier transform* is the software equivalent of a hardware spectrum analyzer. It takes in a signal in the time domain and outputs a signal in the frequency domain that shows the spectral content of the input signal. The Fourier transform works on both periodic and non-periodic signals, but since the periodic case is easier to explain we will start with that.

A periodic signal is one that repeats every τ seconds, where τ is the period. That means that the signal can consist only of frequencies whose sinusoidal waveforms have an integer number of cycles in τ seconds. In other words, the signal is made up of sinusoids that are at the frequency $1/\tau$ and its harmonics. Fourier's idea was that you can determine if a frequency is present by multiplying the waveform by a sinusoid of that frequency and integrating the result. The result of the integration yields the amplitude of that harmonic. If the integration yields zero, then that frequency is not present.

To see how that works, look at **Fig 15-A1**. For the purpose of discussion, assume the signal to be tested consists of a single tone at the second harmonic as shown at (A). The first test frequency is the fundamental, shown at (B). When you multiply the two together you get the waveform at (C). Integrating that signal gives its average value, which is zero. However if you multiply the test signal by a sine wave at the second harmonic (D), the resulting waveform (E) has a large dc offset so the integration yields a large non-zero value. It turns out that all harmonics other than the second yield a zero result. That is, the second harmonic is *orthogonal* to all the others.

If the test waveform included more than one frequency, each of those frequencies would yield a non-zero result when tested with the equivalent-frequency sine wave. The presence of additional frequencies does not disturb the tests for other frequencies since they are all orthogonal with each other.

You may have noticed that this method only works if the test sine wave is in phase with the one in the signal. If they are 90° out of phase, the integration yields zero. The Fourier transform therefore multiplies the signal by both a sine wave and a cosine wave at each frequency. The results of the two tests then yield both the amplitude and phase of that frequency component of the signal using the equations

$$A = \sqrt{a^2 + b^2}$$

and

$$\varphi = \arctan\left(\frac{b}{a}\right)$$

where A is the amplitude, φ is the phase, a is the cosine amplitude and b is the sine amplitude.

If one period of the signal contains, say, 256 samples, then testing a single frequency requires multiplying the signal by the sine wave and by the cosine wave 256 times and adding the results 256 times as well, for a total of 512 multiplications and additions. There are 128 frequencies that must be tested, since the 128th harmonic is at the Nyquist frequency. The total number of calculations is therefore $512 \times 128 = 256^2$ multiplications and additions. That is a general result. For any sample size, n, calculating the digital Fourier transform requires n^2 multiply-accumulates.

The FFT

The number of calculations grows rapidly with sample size. Calculating the Fourier transform on 1024 samples requires over a million multiply-accumulates. However, you may notice that there is some redundancy in the calculations. When testing the second harmonic, for example, each of the two cycles of the test sine wave is identical. It would be possible to pre-add signal data from the first and second halves of the sequence and then just multiply once by a single cycle of the test sine wave. Also, the first quarter cycle of a sine wave is just a mirror-image of the second quarter cycle and the first half is just the negative of the second half.

In 1965, J. W. Cooley and John W. Tukey published an algorithm that takes advantage of all the symmetries inherent in the Fourier transform to speed up the calculations. The *Cooley-Tukey algorithm*, usually just called the *fast Fourier transform* (FFT), makes the number of calculations proportional to $n\log_2(n)$ instead of n^2. For a 1024-point FFT, the calculation time is proportional to $1024\log_2(1024) = 10,240$ instead of $1024^2 = 1,048,576$, more than a 100-times improvement.

You'll notice that sample sizes are usually a power of two, such as $2^7 = 128$, $2^8 = 256$ and $2^9 = 512$. That is because the FFT algorithm is most efficient with sequences of such sizes. The algorithm uses a process called *radix-2 decimation in time*, that is, it first breaks the data into two chunks of equal size, then breaks each of those chunks into two still-smaller chunks of equal size, and so on. It is possible to squeeze even a little more efficiency out of the algorithm with a *radix-4* FFT which is based on decimation by four instead of by two. That is why you often see sample sizes that are powers of four, such as $4^3 = 64$, $4^4 = 256$ and $4^5 = 1024$. Other variations on the algorithm include decimation in frequency rather than time, mixed-radix FFTs that use different decimation factors at different stages in the calculation, and in-place calculation that puts the results into the same storage buffer as the input data, saving memory. The latter method causes the order of the output data to be scrambled by bit-reversing the address words. For example, address 01010000 becomes 00001010.

Non-periodic signals

So far we have assumed that the signal to be transformed is periodic, so that there is an integer number of cycles of each sine wave harmonic in the sequence. With a non-periodic signal, that is not necessarily so. The various frequencies in the signal are not exact harmonics of $1/\tau$ and are no longer

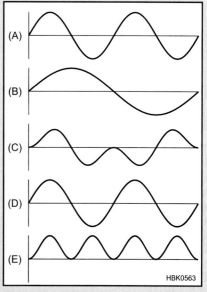

Fig 15-A1 — Signal to be tested for frequency content (A). Fundamental test frequency (B). Product of signal and fundamental (C). Second harmonic test frequency (D). Product of signal and second harmonic (E).

orthogonal to the test frequencies. The result is *spectral leakage*; a single frequency in the signal may give a non-zero result when tested at a number of different harmonic frequencies. In **Fig 15-A2**, (B) illustrates the FFT of a single sine wave at a non-harmonic frequency. You can see that the spurious response extends quite far from the actual frequency.

Those far-out spurious responses are primarily caused by the abrupt termination of the signal at the edges of the sequence. The spectrum can be cleaned up considerably by tapering the edges with a window, in a manner similar to windowing FIR filter coefficients as previously described. In fact, the same windows work for both. Fig 15-A2 part C illustrates the result of applying a Hamming window to the signal in (A) and the resulting improved spectrum is shown at (D). Just as with FIR filters, different windows excel in different areas. Windows with a gradual transition to zero at the edges do a better job of suppressing spurious responses but smear adjacent spectral lines, analogous to using a wider resolution bandwidth in an analog spectrum analyzer. Windows with a fatter center section and a more abrupt transition to zero at the edges have less smearing but worse spurious responses.

While it is interesting and instructional to write your own FFT from scratch, most programmers don't bother to try to re-invent the wheel. Many implementations have been published on the Web and in books and articles. Most of the software development systems offered by DSP vendors include an FFT library routine, which runs faster than anything you are likely to come up with on your own.

Fig 15-A2 — Illustrating the use of windowing to minimize spectral leakage, the figures show (A) a cosine waveform not at a harmonic frequency, (B) the resulting unwindowed power spectrum, (C) the same cosine waveform with a Hamming window, and (D) the much narrower power spectrum of the windowed waveform.

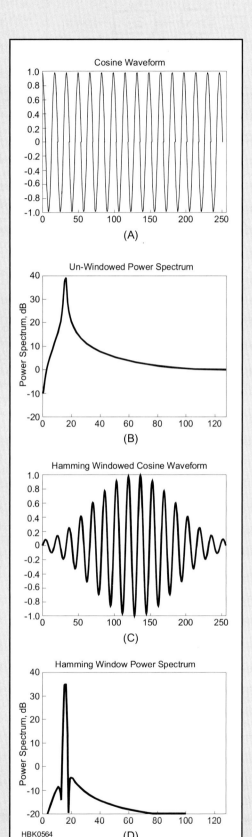

mulator, 32 in this example, is normally much greater than the address width of the look-up table, p. The less-significant accumulator bits are not used in the address. For example, with a 100 MHz clock rate and a 32-bit phase accumulator the frequency resolution is $100 \times 10^6 / 2^{32} = 0.023$ Hz.

The address width of the look-up table determines the number of table entries and thus the phase accuracy of the samples. The table size can be reduced by a factor of four by including only the first quarter-cycle of the sine wave in the table. The other three quadrants can be covered by modifying the look-up address appropriately and by negating the output when required under the control of some additional logic. For example, the method to transform quadrant three to quadrant one is illustrated in **Fig 15.18**. Another technique to improve waveform accuracy without increasing table size is to interpolate between the entries. Instead of using the look-up table output directly, the output value is calculated by interpolating between the two nearest table entries using either a straight-line interpolation as shown in **Fig 15.19** or a higher-order curve fit.

Taking that last idea to an extreme, it is possible to generate a good approximation to one quadrant of a sine wave using a fifth-order interpolation between the zero and 90° points. Assuming that the phase x has been scaled so that x = 1.0 corresponds to 90°, the sine formula is

$$\sin(x) = Ax + Bx^2 + Cx^3 + Dx^4 + Ex^5$$

where the coefficients are A = 3.140625, B = 0.02026367, C = −5.325196, D = 0.5446778 and E = 1.800293.

With those coefficients, which come from an old Analog Devices DSP manual, all the harmonics of the sine wave are more than 100 dB below the carrier.[3] With a 16-bit integer DSP the performance would be limited only by the roundoff error. The formula can be reformulated as

$$\sin(x) = (A + (B + (C + (D + Ex)x)x)x)x$$

which reduces the number of multiplications required from 15 to 5, as shown in **Fig 15.20**.

15.5.2 Tone Decoder

Tone decoders have a number of applications in Amateur Radio. A Morse code reader might use a tone decoder to determine the on and off states of the incoming CW signal. A sub-audible tone detector in a VHF FM receiver is another application. A DTMF decoder needs to detect two tones simultane-

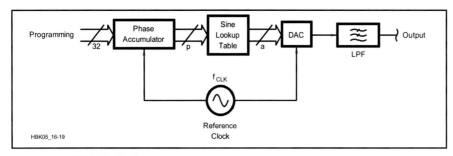

Fig 15.17 — DDS block diagram.

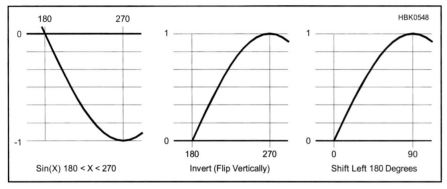

Fig 15.18 — The values of sin(x) between 180 and 270° are the same as those between 0 and 90°, after the curve has been flipped vertically and shifted 180°.

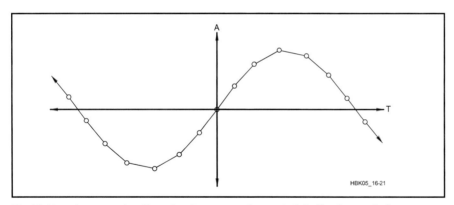

Fig 15.19 — An approximation of a sine wave using a straight-line interpolation between lookup table entries.

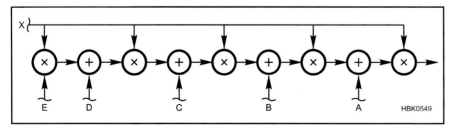

Fig 15.20 — Method to calculate a fifth-order interpolation between lookup table entries.

ously to determine which of the Touch-Tone keys has been pressed.

An FFT is one type of tone decoder. A 1024-point FFT simultaneously decodes 512 frequencies up to one-half the sample rate using a very efficient algorithm. You can control the detection bandwidth by choosing the number of points in the FFT; the spacing of the frequency samples is just the sample rate divided by the number of FFT points. However, in many applications you don't need that much resolution. For example, a DTMF signal consists of two tones, each of which can be one of four frequencies, for a total of 8 possible frequencies. Instead of performing a complete FFT, you can simply convolve sinusoidal waves of each of those 8 frequencies with the sequence of incoming samples to detect energy at those 8 frequencies. This will take less computation than a full FFT whenever the number of test frequencies is less than the base-2 logarithm of the number of points in the sequence. Since $\log_2(1024)$ is 10, decoding 8 frequencies separately would be more efficient than a 1024-point FFT. Spectral leakage is just as much of a problem with this method as with an FFT, so you still need to window the sequence of samples before performing the convolution.

If only a single frequency needs to be detected it might make more sense to mimic analog techniques and use a band-pass digital filter followed by an amplitude detector. That would have the advantage that the passband and stopband characteristics of the filter could be much more precisely controlled than with an FFT. In addition, the output is updated continuously instead of in "batch mode" after each batch of samples is collected and processed.

15.6 Analytic Signals and Modulation

In the area of modulation, the topic that seems to give people the most trouble is the concept of negative frequency. What in the world is meant by that? Consider a single-frequency signal oscillating at ω radians per second. (Recall that ω = 2πf, where f is frequency in Hz.) Let's represent the signal by a cosine wave with a peak amplitude of 1.0, x(t) = cos(ωt), where t is time. Changing the sign of the frequency is equivalent to running time backwards because (−ω)t = ω(−t). By examining **Fig 15.21A** you can see that, because a cosine wave is symmetrical about the time t = 0 point, a negative frequency results in exactly the same signal. That is, as you may remember from high-school trigonometry, cos(−ωt) = cos(ωt). If, for example, you add a positive-frequency cosine wave to its negative-frequency twin, you get the same signal with twice the amplitude.

That assumes that the phase of the signal is such that it reaches a peak at t = 0. What if instead we had a sine wave, which is zero at t = 0? From Fig 15.21B you can see that running time backwards results in a reversal of polarity, sin(−ωt) = −sin(ωt). If you add positive and negative-frequency sine waves of the same frequency and amplitude, they cancel, resulting in zero net signal.

A sinusoidal wave of any arbitrary amplitude and phase may be represented by the weighted sum of a sine and cosine wave:

x(t) = I cos(ωt) + Q sin(ωt)

For computational purposes, it is convenient to consider the *in-phase* (*I*) and *quadrature* (*Q*) components separately. Since the I and Q components are 90° out of phase in the time domain, they are often plotted on a polar graph at a 90° angle from each other. See **Fig 15.22**. For example if Q = 0, then as time increases the signal oscillates along the I (horizontal) axis, tracing out the path back and forth between I = +1 and I = −1 in a sinusoidal fashion. Conversely, if I = 0, then the signal oscillates along the Q axis.

What if both I and Q are non-zero, for example I = Q = 1? Recall that the cosine and sine are 90° out of phase. When t = 0, cos(ωt) = 1 and sin(ωt) = 0. A quarter cycle later, cos(ωt) = 0 and sin(ωt) = 1. Comparing Fig 15.22 with Fig 15.21 it should not be hard to convince yourself that the signal is tracing out a circle in the counter-clockwise direction.

What about negative frequency? Again, it should not be hard to convince yourself that changing ω to −ω results in a signal that circles the origin in the clockwise direction. If you combine equal-amplitude signals of opposite frequency, the sine portions cancel out and you are left with a simple cosine wave of twice the amplitude:

$$x(t) = [\cos(\omega t) + \sin(\omega t)]$$
$$+ [\cos(-\omega t) + \sin(-\omega t)]$$
$$= 2\cos(\omega t)$$

You can see that graphically in **Fig 15.23**. Imagine the two vectors rotating in opposite directions. If you mentally add them by placing the tail of one vector on the head of the other, as shown by the dotted line, the result always lies on the I axis and oscillates between +2 and −2.

That is why we say that a single scalar sinusoidal signal, cos(ωt), actually contains two frequencies, +ω and −ω. It also offers a logical explanation of why a mixer or modulator produces the sum and difference of the frequencies of the two inputs. For example, an AM modulator produces sidebands at the carrier frequency plus and minus the modulating frequency precisely because those positive and negative frequencies are actually already present in the modulating signal.

For many purposes, it is useful to separate the portion of the signal that specifies the amplitude and phase (I and Q) from the oscillating part (sin(ωt) and cos(ωt)). For mathematical convenience, the I/Q part is represented by a complex number, x = I + *j*Q. The oscillating part is also a complex number $e^{-j\omega t} = \cos(\omega t) - j\sin(\omega t)$. (Don't worry if you don't know where that equation comes from — concentrate on the part to the right of the equals sign.[4]) In the equations, $j = \sqrt{-1}$. Of course, −1 does not have a real square root (any real number multiplied by itself is positive) so *j*, or any real number multiplied by *j*, is called an *imaginary number*. A number with both real and imaginary parts is called a *complex number*. The total *analytic signal* is a complex number equal to

$$x(t) = xe^{-j\omega t} = (I + jQ)(\cos(\omega t) - j\sin(\omega t))$$

In the above equation, the cos(ωt) − sin(ωt) portion generally represents an RF carrier, with ω being the carrier frequency (a positive or negative value). The I + *j*Q part is the modulation. The *scalar* value of a modulated signal (what you would measure with an oscilloscope) is just the real part of the analytic signal. Using the fact that $j^2 = -1$,

$$\text{Re}[x(t)] = \text{Re}[(I + jQ)(\cos(\omega t) - j\sin(\omega t))]$$

$$\text{Re}[x(t)] = \text{Re}\begin{bmatrix} (I\cos(\omega t) + Q\sin(\omega t)) \\ + j(Q\cos(\omega t) - I\sin(\omega t)) \end{bmatrix}$$

$$\text{Re}[x(t)] = I\cos(\omega t) + Q\sin(\omega t)$$

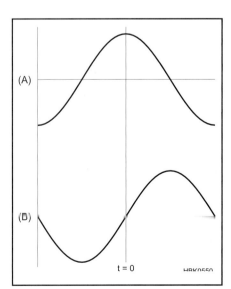

Fig 15.21 — Cosine wave (A) and sine wave (B).

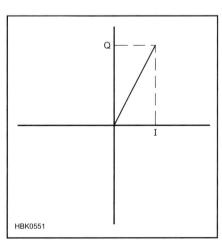

Fig 15.22 — In-phase (I) and quadrature (Q) portions of a signal.

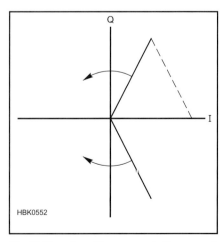

Fig 15.23 — A real frequency is the sum of a positive and negative analytic frequency.

Note that if the modulation (I and Q) varies with time, the above equation assumes that the modulated signal does not overlap zero Hz. That is, I and Q have no frequency components greater than ω.

Normally the I/Q diagram shows only I and Q (the modulation) and not the oscillating part. We call such a representation a *phasor diagram*. The I/Q vector represents the *difference* in phase and amplitude of the carrier signal compared to the unmodulated carrier. For example, if the I/Q vector is at 90°, that means the carrier has been phase-shifted by 90° from what it otherwise would have been. If the I/Q vector is rotating counter-clockwise 10 times per second, then the carrier frequency has been increased by 10 Hz.

It is worth noting that the modulation can be specified either by the in-phase and quadrature (I and Q) values as shown or alternatively by the amplitude and phase. The amplitude is the length of the I/Q vector in the phasor diagram,

$$A = \sqrt{I^2 + Q^2}$$

The phase is the angle of the vector with respect to the +I axis,

$$\varphi = \arctan\left[\frac{Q}{I}\right]$$

An alternative expression for the modulated analytic signal using amplitude and phase is

$$x(t) = Ae^{-j(\varphi + \omega t)}$$
$$= A\left[\cos(\varphi + \omega t) + j\sin(\varphi + \omega t)\right]$$

and for the scalar signal

$$\text{Re}[x(t)] = A\cos(\varphi + \omega t)$$

One final comment. So far we have been looking at signals that consist of a single sinusoidal frequency. In any linear system, anything that is true for a single frequency is also true for a combination of many frequencies. Each frequency is affected by the system as though the others were not present. Since any complicated signal can be broken down into a (perhaps large) number of single-frequency sinusoids, all our previous conclusions apply to multi-frequency signals as well.

15.6.1 I/Q Modulation and Demodulation

An *I/Q modulator* is just a device that controls the amplitude and phase of an RF signal directly from the in-phase (I) and quadrature (Q) components. See **Fig 15.24A**. An *I/Q demodulator* is basically the same circuit in reverse. It puts out I and Q signals that represent the in-phase and quadrature components of the incoming RF signal. See Fig 15.24B. Assuming the demodulator's local oscillator is on the same frequency and is in phase with the carrier of the signal being received then the I/Q output of the receiver's demodulator is theoretically identical to the I/Q input at the transmitter end.

I/Q modulators and demodulators can be built with analog components. The LO could be a transistor oscillator and the 90° phase-shift network could be implemented with coils and capacitors. The circles with the multiplication symbol would be double-balanced mixers. Not shown in the diagram are trim adjustments to balance the amplitude between the I and Q channels and to adjust the phase shift as close as possible to 90°.

No analog circuit is perfect, however. If the 90° phase-shift network is not exactly 90° or the amplitudes of the I and Q channels are not perfectly balanced, you don't get perfect opposite-sideband rejection. The modulator output includes a little bit of signal on the unwanted sideband and the I/Q signal from the demodulator includes a small signal rotating in the wrong direction. If there is a small dc offset in the amplifiers feeding the modulator's I/Q inputs, that shows up as carrier feedthrough. On receive, a dc offset makes the demodulator think there is a small signal at a constant amplitude and phase angle that is always there even when no actual signal is being received. Nor is analog circuitry distortion-free, especially the mixers. Intermodulation distortion shows up as out-of-channel "splatter" on transmit and unwanted out-of channel responses on receive.

All those problems can be avoided by going digital. If the analog I/Q inputs to the modulator are converted to streams of digital numbers with a pair of ADCs, then the mixers, oscillator, phase-shift network and summer can all be digital. In many systems, the I and Q signals are also generated digitally, so that the digital output signal has perfect unwanted sideband rejection, no carrier feedthrough and no distortion within the dynamic range afforded by the number of bits in the data words. A similar argument holds for a digital demodulator. If the incoming RF signal is first digitized with an ADC, then the demodulation can be done digitally without any of the artifacts caused by imperfections in the analog circuitry.

You can think of an I/Q modulator as a device that converts the analytic signal $I + jQ$ into a scalar signal at some RF frequency. The spectrum of the I/Q signal, both positive and negative frequencies, is translated upward in frequency so that it is centered on the carrier frequency. Thinking in terms of the phasor diagram, any components of the I/Q signal that are rotating counter-clockwise appear above the carrier frequency and clockwise components appear below.

15.6.2 SSB Using I/Q Modulators and Demodulators

As an example of how this works, let's walk through the process of generating an upper-sideband signal using an I/Q modulator. See **Fig 15.25**. We'll first describe the mathematics in the following paragraph and then give the equivalent explanation using the phasor diagram.

The modulating signal is a sine wave at a frequency of ω_m radians per second ($\omega_m / 2\pi$ cycles per second). Because ω_m is a positive frequency the signals applied to the I/Q inputs are $I(t) = \cos(\omega_m t)$ and $Q(t) = \sin(\omega_m t)$. Assume the modulating frequency ω_m is much less than the RF frequency ω. The analytic signal is

$$x(t) = \left[\cos(\omega_m t) + j\sin(\omega_m t)\right]$$
$$\times \left[\cos(\omega t) - j\sin(\omega t)\right]$$

so that the real, scalar signal that appears at the modulator output is

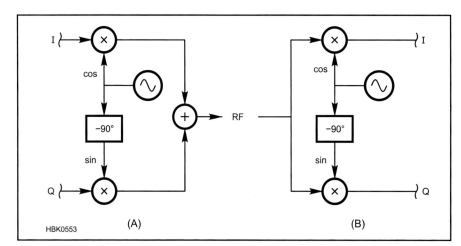

Fig 15.24 — I/Q modulator (A) and demodulator (B).

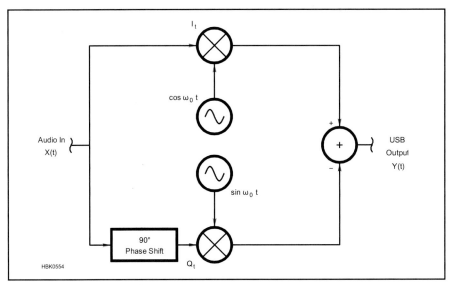

Fig 15.25 — Generating a USB signal with an I/Q modulator.

$$\text{Re}[x(t)] = \cos(\omega_m t)\cos(\omega t)$$
$$+ \sin(\omega_m t)\sin(\omega t)$$

At the moment when t = 0, then $\cos(\omega_m t) = 1$ and $\sin(\omega_m t) = 0$, so the real signal is just $\cos(\omega t)$, the RF signal with zero phase. One quarter of a modulation cycle later $\omega_m t = \pi/2$, so $\cos(\omega_m t) = 0$ and $\sin(\omega_m t) = 1$, and the real signal is now $\sin(\omega t)$, the RF signal with a phase of $+\pi/2$, or +90°. Every quarter cycle of the modulating signal, the RF phase, increases by 90°. That means that the RF phase increases by one full cycle for every cycle of the modulation, which is another way of saying the frequency has shifted by ω_m. We have an upper sideband at a frequency of $\omega + \omega_m$.

On the phasor diagram, the I/Q signal is rotating counterclockwise at a frequency of ω_m radians per second. As it rotates it is increasing the phase of the RF signal at the same rate, which causes the frequency to increase by ω_m radians per second. To cause the phasor to rotate in the opposite direction, you could change the polarity of either I or Q or you could swap the I and Q inputs. In that case you would have a lower sideband.

For that to work, the baseband signals applied to the I and Q inputs must be 90° out of phase. That's not hard to do for a single sine wave, but to generate a voice SSB signal, all frequencies in the audio range must be simultaneously phase-shifted by 90° without changing their amplitudes. To do that with analog components requires a broadband phase-shift network consisting of an array of precision resistors and capacitors and a number of operational amplifiers.

THE HILBERT TRANSFORMER

To do that with DSP requires a *Hilbert transformer*, an FIR filter with a constant 90° phase shift at all frequencies. Recall that a symmetrical FIR filter has a constant *delay* at all frequencies. That means that the phase shift is not constant — it increases linearly with frequency. It turns out that with an *anti-symmetrical* filter, in which the top half of the coefficients are the negative of the mirror image of the lower half, the phase shift is 90° at all frequencies, which is exactly what we need to generate an SSB signal.

The Hilbert transformer is connected in series with either the I or Q input, depending on whether USB or LSB is desired. Just as with any FIR filter, a Hilbert transformer has a delay equal to half its length, so an equal delay must be included in the other I/Q channel as shown in **Fig 15.26**. It is possible to combine the Hilbert transformer with the normal FIR filter that may be needed anyway to filter the baseband signal. The other I/Q channel then simply uses a similar filter with the same delay but without the 90° phase shift.

Because the RF output of the modulator is normally at a much higher frequency than the audio signal, it is customary to use a higher sample rate for the output signal than for the input. The FIR filters can still run at the lower rate to save processing time, and their output is then upsampled to a higher rate with an interpolator. It is convenient to use an output sample rate that is exactly four times the carrier frequency because each sample advances the RF phase by exactly 90°. The sequence of values for the sine wave is 0, 1, 0 and –1. To generate the 90° phase shift for the cosine wave, simply start the sequence at the second sample: 1, 0, –1, 0. The complete block diagram is shown in **Fig 15.27**.

A Hilbert transformer may also be used in an SSB demodulator at the receiver end of the communications system. It is basically the same block diagram drawn backwards, as illustrated in **Fig 15.28**.

Amateurs who have been in the hobby for many years may recognize this as the "phasing method" of SSB generation. It was popular when SSB first became common on the amateur bands back in the 1950s because suitable crystal filters were expensive or difficult to obtain.[5] The phasing method had the reputation of producing signals with excellent-quality audio, no doubt due to the lack of the phase distortion caused by crystal filters.

It is important to note that an ideal Hilbert transformer is impossible to construct because it theoretically has an infinitely-long impulse response. However, with a sufficiently-long impulse response, the accuracy is much better than an analog phase-shift network. Just as with an analog network, the frequency passband must be limited both at

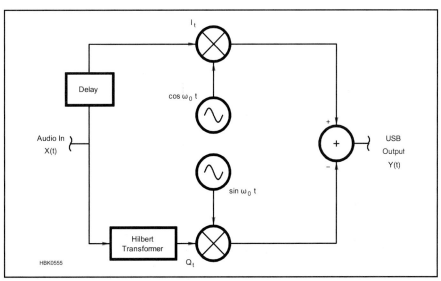

Fig 15.26 — Generating a non-sinusoidal USB signal with an I/Q modulator.

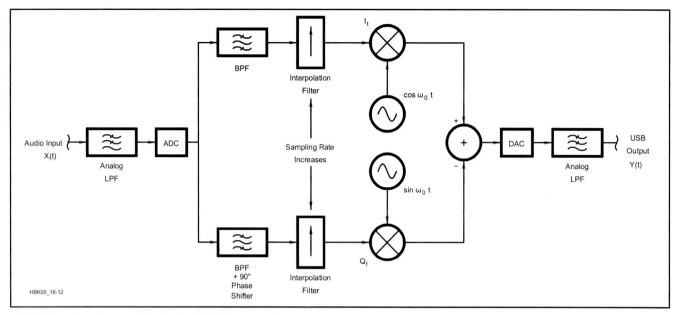

Fig 15.27 — Block diagram of a digital SSB modulator.

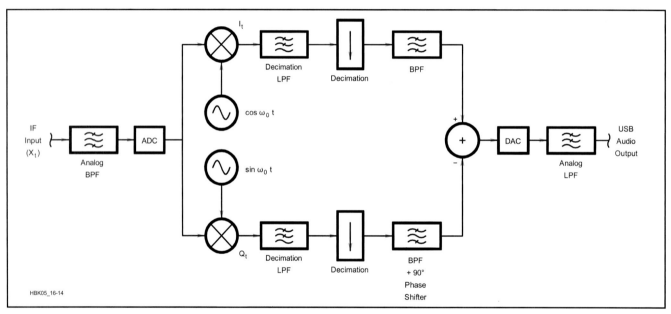

Fig 15.28 — Block diagram of a digital SSB demodulator.

the low end as well as the high end. That is, the audio must be band-pass filtered before the 90° phase shift. Actually, the filtering and phase shifting can be combined into one operation using the following method.

First design a low-pass FIR filter with a bandwidth one-half the desired audio bandwidth. For example, if the desired passband is 300 to 2700 Hz, the low-pass filter bandwidth should be (2700 − 300)/2 = 1200 Hz. Then multiply the impulse response coefficients with a sine wave of a frequency equal to the center frequency of the desired passband, (2700 + 300)/2 = 1500 Hz in this case. That results in a band-pass filter with the desired 300 − 2700 Hz response. By using sine waves 90° out of phase for the I and Q channels, you end up with two band-pass filters with the same amplitude response and delay but a 90° phase difference at all frequencies. Multiply by a cosine for zero phase and by a sine for a 90° phase shift.

Old timers may notice that this bears a striking resemblance to the Weaver method, the so-called "third method" of SSB generation, that was used back in the late 1950s to eliminate the need for a wide-band audio phase-shift network.[6,7] It is almost as if there is no such thing as truly new technology, just old ideas coming back with new terminology! Analog modulators and demodulators using the phasing and Weaver methods are covered in the **Transmitters** and **Mixers, Modulators and Demodulators** chapters.

USES FOR I/Q MODULATORS AND DEMODULATORS

While I/Q modulators and demodulators can be used for analog modes such as SSB, they really shine when used with digital modulation modes. The **Modulation** chapter shows how the modulation states of the various digital formats map to positions in the phasor diagram, what is called a *constellation diagram*. The transmitter can generate the correct modulation states simply by placing the correct values on the I and Q inputs to the I/Q modulator. In the receiver, the filtered I and Q values are sampled at the symbol decision times to determine which modulation state they most closely match.

I/Q modulators and demodulators can also be used as so-called *imageless mixers*. A normal mixer with inputs at f_1 and f_2 produces outputs at $f_1, f_2, f_1 + f_2$, and $f_1 - f_2$. A balanced mixer eliminates the f_1 and f_2 terms but both the sum and difference terms remain, even though normally only one is desired. By feeding an RF instead of AF signal into the input of an SSB modulator, we can choose the sum or difference frequency in the same way as choosing the upper or lower sideband. If the input signal is a sine wave, the Hilbert transformer can be replaced by a simple 90° phase shifter. Similarly, a mixer with the same architecture as an SSB demodulator can be used to downconvert an RF signal to IF with zero image response. Analog imageless mixers are covered in the **Receivers** chapter. They are sometimes used in microwave receivers and transmitters where it is difficult to build filters narrow enough to reject the image response, but they typically only achieve image rejection in the 20-30 dB range. With a digital imageless mixer, the image rejection is "perfect" within the dynamic range of the bit resolution.

15.7 Software-Defined Radios (SDR)

There has been much, sometimes heated, discussion about the precise definition of a *software-defined radio* (SDR). Most feel that, at minimum, an SDR must implement in software at least some of the functions that are normally done in hardware. Others feel that a radio doesn't count as an SDR unless nearly all the signal-processing functions, from the input mixer to the audio output (for the receiver) and from the microphone ADC to the power amplifier input (for the transmitter), are done in software. Others add the requirement that the software must be reconfigurable by downloading new code, preferably open-source. For our purposes we will use a rather loose definition and consider any signal-processing function done in software to fall under the general category of SDR.

Some SDRs use a personal computer to do the computational heavy lifting and external hardware to convert the transmitted and received RF signals to lower-frequency signals that the computer's sound card can handle. Some SDRs avoid the use of the sound card by including their own audio codec and transferring the data to the PC via a USB port. Modern PCs provide a lot of computational power for the buck and are getting cheaper and more powerful all the time. They also come with a large color display, a keyboard for easy data entry, and a large memory and hard disk, which allows running logging programs and other software while simultaneously doing the signal processing required by the SDR.

Other SDRs look more like conventional analog radios with everything contained in one box, which makes for a neater, more compact installation. The signal processing is done with one or more embedded DSPs. For those who prefer a knob and button user interface, this is much preferred to having to use a mouse. Especially for contesting and competitive DXing, it is much faster to have a separate control for each critical function rather than having to select from pull-down menus. In addition SDRs of this type often have some performance advantages over PC-based SDRs, as we shall see.

Either method offers all the most-important advantages of applying DSP techniques to signal processing. The channel filter can have a much better *shape factor* (the ratio between the width of the passband and the frequency difference of the stopband edges). FIR filters are linear phase and have less ringing than analog filters of the same bandwidth and shape factor. Once the signal is in the digital domain all the fancy digital signal processing algorithms can be applied such as automatic notch filters, adaptive channel equalization, noise reduction, noise blanking, and feed-forward automatic gain control. Correcting bugs, improving performance or adding new features is as simple as downloading new software.

15.7.1 SDR Hardware

The transition between analog and digital signals can occur at any of several places in the signal chain between the antenna and the human interface. Back in 1992, Dave Hershberger W9GR designed an audio-frequency DSP filter based on the TMS320C10, one of the earliest practical DSP chips available.[8] This was an external unit that plugged into the headphone jack of a receiver and included FIR filters with various bandwidths, an automatic multi-frequency notch filter, and an adaptive noise filter. The advantage of doing the DSP at AF is that it can easily be added to an unmodified analog radio as in **Fig 15.29**. It is the technique used today to implement many digital modulation modes using the sound card of a PC connected to the audio input and output of a conventional transceiver.

A related technique is to downconvert a slice of the radio spectrum to baseband audio using a technique similar to the direct-conversion receivers popular with simple low-power CW transceivers. This idea was pioneered by Gerald Youngblood, AC5OG (now K5SDR), with the SDR-1000 transceiver, which he described in a series of *QEX* articles in 2002-2003.[9] The receiver block diagram is shown in Fig **15.30**. It uses a unique I/Q demodulator designed by Dan Tayloe, N7VE, to convert the RF frequency directly to baseband I and Q signals, which are fed to the stereo input of a PC's sound card, represented by the low-pass filters and A/D converters in the figure.[10] Software in the PC does all the signal pro-

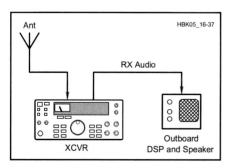

Fig 15.29 — An outboard DSP processor.

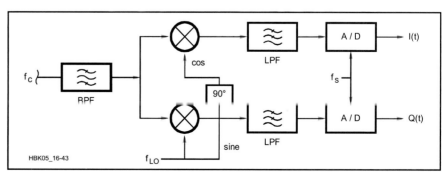

Fig 15.30 — Block diagram of K5SDR's direct-conversion software-defined receiver.

cessing and demodulation. The transmitter is the same block diagram in reverse, with an I/Q modulator converting an I/Q signal from the sound card up to the RF frequency where it is filtered and amplified to the final power level.

The sound card method manages to achieve reasonable performance with simple inexpensive hardware. Once the signal is digitized by the A/D converters in the sound card, the powerful DSP capability of the PC can do amazing things with it. The software for the SDR-1000 is open-source and available for free download on the Web.[11] In addition to implementing conventional transceiver functions such as several types of detector, variable-bandwidth filters, software AGC, an S-meter and speech compression, the software includes some extra goodies such as an automatic notch filter, noise reduction, and a panadapter spectrum display.

The simple hardware does impose some performance limitations. Because of imperfections in the analog downconverter, unwanted-sideband rejection is not perfect. This is called "image rejection" in the SDR-1000 literature. On the panadapter display, strong signals show up weakly on the opposite side of the display, equally-spaced from the center. Dc offset in the analog circuitry causes a spurious signal to appear at the center of the bandwidth. To prevent an unwanted tone from appearing in the audio output, the software demodulator is tuned slightly off frequency, but that means interference at the image frequency can cause problems because of the imperfect image rejection. The dynamic range depends on the sound card performance as well as the RF hardware. Some newer SDRs include an integrated audio codec optimized for the application so that the PC's sound card is not needed.

Among the integrated, one-box, software-defined radios, the most common place to perform the analog-digital transition is at an intermediate frequency. In the receiver, placing the ADC after a crystal IF filter improves the *blocking dynamic range* (BDR) for interfering signals that fall outside the crystal filter bandwidth. BDR is the ratio, expressed in dB, between the noise level (normally assuming a 500 Hz bandwidth) and an interfering signal strong enough to cause 1 dB gain reduction of the desired signal. With careful design, a receiver with such an architecture can achieve up to about 140 dB of BDR. The third-order dynamic range is similar to what can be achieved with a conventional analog architecture since the circuitry up to the crystal filter is the same.

Another advantage of the IF-based approach compared to sampling right at the final RF frequency is that the ADC does not have to run at such a high sample rate. In fact, because the crystal filter acts as a high-perfor-

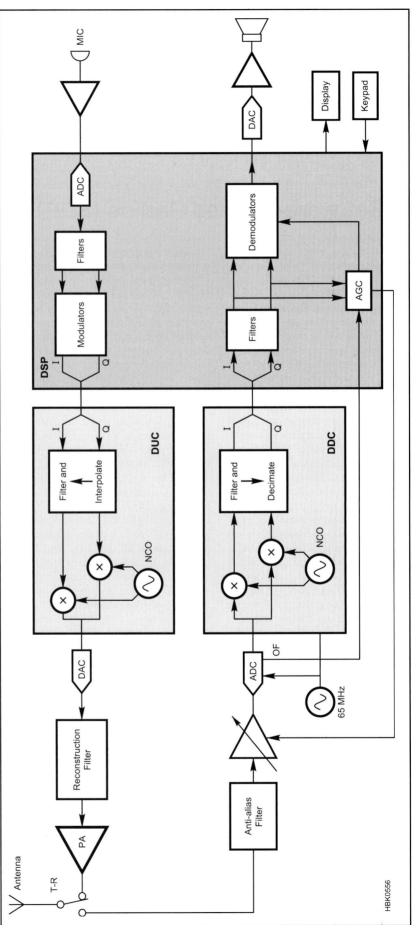

Fig 15.31 — An SDR transceiver that samples directly at the RF frequency.

mance, narrow-bandwidth anti-aliasing filter, undersampling is possible. With bandwidths of a few kHz or less, sample rates in the 10s of kHz can be used even though the center frequency of the IF signal is much higher, so long as the ADC's sample-and-hold circuit has sufficient bandwidth. Another common approach is to add a conventional analog mixer after the crystal filter to heterodyne the signal down to a second, much lower IF in the 10-20 kHz range, which is then sampled by an inexpensive, low-sample-rate ADC in the normal fashion. IF-based SDRs tend to have the highest overall performance at the expense of additional complexity.

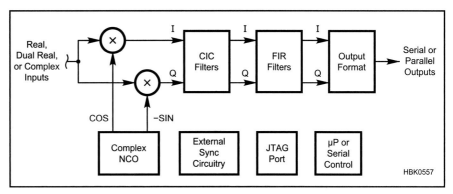

Fig 15.32 — The Analog Devices AD6620 is a digital downconverter (DDC) IC. The CIC filters and FIR filter are all decimating types.

SAMPLING AT RF

The ultimate SDR architecture is to transition between the analog and digital domains right at the frequency to be transmitted or received. In the receiver, the only remaining analog components in the signal chain are a wide-band anti-aliasing filter and an amplifier to improve the noise figure of the ADC. See **Fig 15.31**. The local oscillator, mixer, IF filters, AGC, demodulators and other circuitry are all replaced by digital hardware and software. It has only been fairly recently that low-cost high-speed ADCs have become available with specifications good enough to allow reasonable performance in a communications receiver. Today it is possible to achieve blocking dynamic range in the low 120s of dB. That is not as good as the best analog radios but is comparable to some medium-priced models currently available on the Amateur Radio market.

Third-order dynamic range is not a meaningful specification for this type of radio because it assumes that distortion products increase 3 dB for each 1 dB increase in signal level, which is not true for an ADC. The level of the distortion products in an ADC tends to be more-or-less independent of signal level until the signal peak exceeds full scale, at which point the distortion spikes up dramatically. Compared to a conventional analog mixer, ADCs tend to give very good results with a two-tone test but don't do as well when simultaneously handling a large number of signals, which results in a high peak-to-average ratio. It is important to read the data sheet carefully and note the test conditions for the distortion measurements.

There are definite advantages to sampling at RF. For one thing, it saves a lot of analog circuitry. Even if the ADC is fairly expensive the radio may be end up being cheaper because of the reduced component count. Performance is improved in some areas. For example, image rejection is no longer a worry, as long as the anti-aliasing filter is doing its job. The dynamic range theoretically does not depend on signal spacing—close-in dynamic range is often better than with a conventional architecture that uses a wide roofing filter. With no crystal filters in the signal chain, the entire system is completely linear-phase which can improve the quality of both analog and digital signals after demodulation.

The biggest challenge with RF sampling is what to do with the torrent of high-speed data coming out of the receiver ADC and how to generate transmit data fast enough to keep up with the DAC. To cover the 0-30 MHz HF range without aliasing requires a sample rate of at least 65 or 70 MHz. That is much faster than a typical microprocessor-type DSP can handle. The local oscillator, mixer and decimator or interpolator must be implemented in digital hardware so that the DSP can send and receive data at a more-reasonable sample rate. Analog Devices makes a series of *digital downconverters* (DDC) which perform those functions and output a lower-sample-rate digital I/Q signal to the DSP.[12] See **Fig 15.32**. It would also be possible to implement your own DDC in an FPGA. The same company also makes *digital upconverters* (DUC) that do the same conversion in reverse for the transmitter. Some of their DUCs even include the capability to

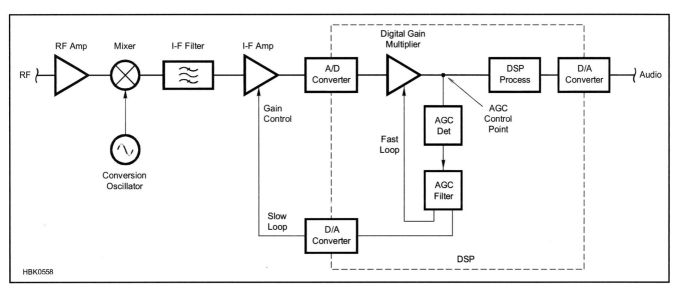

Fig 15.33 — DSP-based feedback type of AGC showing a combination of analog and digital gain-control points.

encode several digital modulation formats such as GMSK, QPSK and π/4 DQPSK.

DIGITAL AGC

In the transmitter portion of a software-defined radio, dynamic range is generally not a problem because the transmitted signal always has approximately the same power level. Even if power control is implemented digitally, a 1-100 W adjustment range only adds 20 dB to the dynamic range. The receiver is another story. Assuming 6 dB per S-unit, the difference between S1 and 60 dB over S9 is 108 dB. Considering peak power rather than average, the actual range is much greater than that. While AGC implemented in software can be quite effective in regulating the signal level at the speaker output, it does nothing to prevent the ADC from being overloaded on signal peaks. For that, some kind of hardware AGC is needed in the signal path ahead of the A/D converter. This device could be a switched attenuator or variable-gain amplifier as illustrated in **Fig 15.33**. It normally runs at full gain and only attenuates the incoming signal when very strong peak signal levels are encountered. It could be totally self-contained with its own threshold detector or it can be controlled as shown by the DSP, which activates the hardware AGC whenever it detects ADC overflow.

One issue with most AGC systems is response time. The level detector is normally placed after the gain-control stage. Because of delays in the feedback loop, by the time the AGC circuit detects an over-range condition it is already too late to reduce the gain without overshoot. One of the advantages of digital AGC is that it is easy to use feed-forward rather than feed-back control. In **Fig 15.34**, the gain control stage is placed after the level detector. A small delay is included in the signal path so that the AGC circuit can reduce the gain just before the large signal arrives at the gain multiplier. With proper design, that totally eliminates overshoot and makes for a very smooth-operating AGC.

THE LOCAL OSCILLATOR

With direct-RF sampling, the digital local oscillator is normally implemented with a direct digital synthesizer, operating totally in digital hardware. DDS operation was explained previously in the sine-wave generation section. A separate DDS chip with a built-in DAC is sometimes used in IF-sampled SDRs as well as in some analog radios. One advantage that a DDS oscillator has over a phase-locked loop (PLL) synthesizer is very fast frequency changing. That can be important in transceivers that use the same local oscillator for both the receiver and transmitter. If the transmitter and receiver are tuned to different frequencies, each time the rig is keyed the LO frequency must settle at its new value before a signal is transmitted.

The phase noise of the DDS clock is just as important as the phase noise of the local oscillator in a conventional radio. Phase noise shows up as broadband noise that gradually

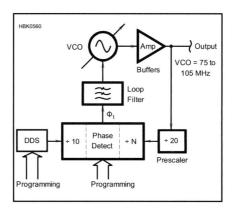

Fig 15.35 — A hybrid DDS/PLL synthesizer.

diminishes the farther you get from the oscillator frequency. In a receiver, phase noise causes a phenomenon called *reciprocal mixing*, in which a strong off-channel signal mixes with off-channel phase noise to cause a noise-modulated spurious signal to appear in the receiver passband. In many receivers, dynamic range measurements are phase-noise-limited because the spurious response due to reciprocal mixing is louder than the distortion products. One way to reduce the phase noise from the DDS is to use a conventional PLL synthesizer to generate a signal with large frequency steps and combine it with a DDS synthesizer to obtain the fine-grained frequency resolution, as suggested in **Fig 15.35**. In this way, you get the phase noise of the DDS and PLL within the loop bandwidth of the PLL and the phase noise of the VCO outside that bandwidth.

One advantage a PLL has over a DDS oscillator is lower spurious signal levels. A DDS with a wideband spurious-free dynamic range (SFDR) specification of 60 dB would be better than most, but that could cause spurious responses in the receiver only 60 dB down. The hybrid PLL/DDS technique can suppress these spurs as well.

15.7.2 SDR Software

When designing DSP software, it is sometimes surprising how much of your intuition about analog circuits and systems transfers directly over to the field of digital signal processing. The main difference is that you need to forget much of what you have learned about the imperfections of analog circuitry. For example, a multiplier is the DSP equivalent of an ideal double-balanced mixer. The multiplier output contains frequencies only at the sum and difference of the two input signals. There is no intermodulation distortion to create spurious frequencies.

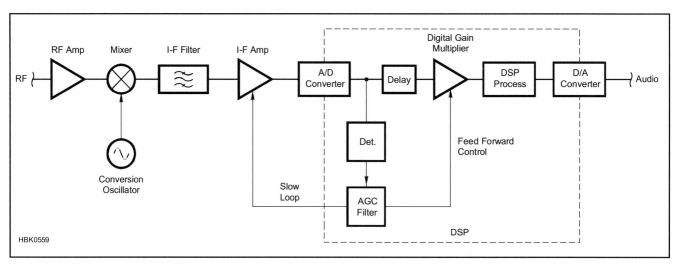

Fig 15.34 — DSP-based AGC with analog feedback and digital feed forward control.

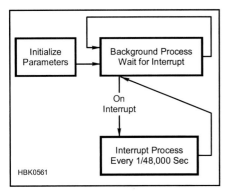

Fig 15.36 — Main program flow of a typical DSP program.

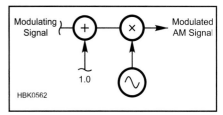

Fig 15.37 — A digital AM modulator.

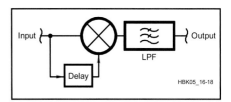

Fig 15.38 — A digital quadrature detector.

Multiplication by a constant is equivalent to an amplifier or attenuator, but with no dc offset and with a very precisely-set gain that does not drift with time or temperature. A distortionless AM "diode" detector is just a software routine that forces the signal to zero whenever it is negative. To build an ideal full-wave diode detector, just take the absolute value of the signal.

When you design an analog circuit you have to take into account all the things that don't appear on the schematic. For example your crystal filter may show a beautiful frequency response in your filter design program, but in the actual circuit the passband is skewed because of the input impedance of the post-amplifier and the stopband response is degraded by signals leaking around the filter due to the PC board layout. With software, what you see in the simulation is what you get (assuming you did the calculations correctly!)

SOFTWARE ARCHITECTURE

Incoming data to the DSP from A/D converters or other devices is normally handled in an *interrupt service routine* (ISR) that is called automatically whenever new data is ready. The above AM and AGC detection code could be included right in the ISR, but it is almost always better to limit the ISR function to the minimum necessary to service the hardware that calls the interrupt and do all the signal processing elsewhere. For example, **Table 15.4** shows an ISR that inputs 16-bit I and Q data coming in on the dsPIC serial data communications interface (DCI). The first line is a secret incantation that defines this function to be the interrupt service routine for the DCI interface.

As you can see, there is minimal functionality in the ISR itself. All the heavy computational lifting is done in processes that run in the background and are interrupted periodically when new data is available. The data_flag variable is a semaphore to signal the signal-processing routine that new data is ready. **Fig 15.36** illustrates the basic program architecture of some DSP projects for the Analog Devices EZ-Kit Lite DSP development board described in the ARRL publication *Experimental Methods in RF Design*. That book, by the way, is an excellent source for practical "how-to" information on designing DSP projects.

SOFTWARE MODULATORS AND DEMODULATORS

In this chapter we've already covered many of the algorithms needed for a software-defined radio. For example, we know how to make I/Q modulators and demodulators and use them to build an SSB modulator and detector. Let's say we want our software-defined transceiver to operate on AM voice as well. How do you make an AM modulator and demodulator?

The modulator is easy. Simply add a constant value, representing the carrier, to the audio signal and multiply the result by a sine wave at the carrier frequency, as shown in **Fig 15.37**.

Demodulation is almost as easy. We could just simulate a full-wave rectifier by taking the absolute value of the signal, as mentioned previously, and low-pass filter the result to remove the RF energy. If the signal to be demodulated is complex, with I and Q components, then instead of absolute value we take the magnitude

$$A = \sqrt{I^2 + Q^2}$$

The dc bias can be removed by adding a "series blocking capacitor" — a high-pass filter with a suitable cut-off frequency.

A little more elegant way to do it would be to include the AM detector as part of the AGC loop. In the C code snippet shown in **Table 15.5**, the variable "carrier" is the average AM carrier level. It is passed to another subroutine to control the gain.

Note that no "series capacitor" is needed since the audio signal is computed by subtracting the average historical value, carrier, from the magnitude of the current I/Q signal, am. A small fraction of its value is added to the historical value so that the AGC tracks the average AM carrier level. AGC speed is controlled by that fraction. Dividing by $2^{10} = 1024$ gives a time constant of about 1024 clock cycles.

Another type of detector we haven't discussed yet is for frequency modulation. For a scalar signal, the *quadrature detector* shown in **Fig 15.38** is one elegant solution. This is the same circuit whose analog equivalent is used today in millions of FM receivers around the world. In the digital implementation, the delay block is a FIFO buffer constructed from a series of shift registers. Multiplying the signal by a delayed

Table 15.4
Interrupt Service Routine

```
void __attribute__((auto_psv, interrupt)) _DCIInterrupt(void)
{
extern int i, q, data_flag;

// Clear interrupt flag:
DDC_DCI_IFS &= DDC_DCI_IMASK;
// Input data is 2's complement:
i = RXBUF0;
q = RXBUF1;
data_flag = 1;
return;
}
```

Table 15.5
AM Detector

```
static long int carrier;
long int am;
int i, q, signal;

/* Code that generates i and q omitted */
am = (long int)sqrt((long int)i*i + (long int)q*q);
signal = am - carrier;
// Divide signal by 2^10:
carrier += signal >> 10;
// Audio output to DAC via SPI bus:
SPI1BUF = signal;
```

version of itself gives an output with a co-sinusoidal response versus frequency. The response crosses zero whenever the carrier frequency is $f = N/(4\tau)$, where N is an odd integer and the delay in seconds is $\tau = n/f_s$, where n is the number of samples of delay and f_s is the sample frquency. As the carrier deviates above and below the zero-crossing frequency the output varies above and below zero, just what we want for an FM detector.

For an I/Q signal, probably the most straightforward FM detector is a phase detector followed by a differentiator to remove the 6 dB per octave rolloff caused by the phase detector. The phase is just

$$\varphi = \arctan\left(\frac{Q}{I}\right)$$

You have to be a little careful since there is a 180° phase ambiguity in the arctangent function. For example,

$$\arctan\left(\frac{1}{1}\right) = \arctan\left(\frac{-1}{-1}\right)$$

Software will have to check which quadrant of the phasor diagram the I/Q signal is in and add 180° when necessary. If there is no arctan function in the library, one can be constructed using a look-up table. Frequency is the derivative of the phase. A differentiator is nothing more than a subtractor that takes the difference between successive samples.

$$f = \frac{\varphi_n - \varphi_{n-1}}{2\pi\, f_s}$$

where n is the sample number and f_s is the sample rate. It is important to make sure that the difference equation functions properly around 360°. If the phase variable is scaled so that 360° equals the difference between the highest and lowest representable numbers, then standard two's complement subtraction should roll over to the right value at the 360° / 0° transition. Another thing to watch out for is that the derivative of the phase may be a rather small signal, so it might be necessary to carry through all the calculations using long integers or floating point numbers.

OTHER SOFTWARE FUNCTIONS

A *carrier-locked loop* is a circuit that automatically tunes a receiver so that it is centered on the carrier of the incoming signal. One way to achieve that is to make the receiver local oscillator controllable by a frequency detector. For example if the local oscillator (LO) in the receiver were an analog voltage-controlled oscillator (VCO), the output of the FM detector described above could be applied to a DAC that generates an error voltage to pull the VCO on frequency. If the LO were an NCO or other digitally-controlled synthesizer, then the error signal could be used to control the frequency digitally. An even more elegant way to do it is to leave the LO alone and tune the frequency of the I/Q signal directly. Conceptually, you determine the amplitude

$$A = \sqrt{I^2 + Q^2}$$

and phase

$$\varphi = \arctan\left(\frac{Q}{I}\right)$$

of the signal, add or subtract the phase error from φ to keep its average value from changing and then convert back to $I = A\cos(\phi)$ and $Q = A\sin(\phi)$ again for further processing. This is easy to do with a signal that does not change phase, such as AM phone. For an FM or PM signal, considerable averaging must be done of the error signal so that it represents the average phase of the carrier rather than the instantaneous phase of the modulation.

Speech processing is a function that lends itself well to digital signal processing. The human voice has a high peak-to-average power ratio, typically on the order 15 dB. That means, that without processing, a 100-W PEP SSB transmitter is only putting out about 3 W average! Most SSB transmitters do have an *automatic level control* (ALC) circuit that can reduce the peak-to-average ratio by 3-6 dB, but that still means your 100-W transmitter is only putting out 6-12W on average.

The problem is that if the ALC setting is too aggressive, considerable distortion of the audio can result. A transmitter's ALC circuit operates much like the AGC in a receiver. It can do a fair job of keeping the peak power from overdriving the amplifier but it can do little to reduce the short-term power variations between speech syllables. With digital processing, it is fairly easy to use feed forward gain control rather than feed back, in a manner similar to the AGC system illustrated in Fig 15.34.

The gold standard of speech processing of SSB signals is RF clipping. By clipping at radio frequency instead of at audio, many of the distortion products fall outside the passband where they can be filtered out, by a crystal filter in an analog radio or by a digital band-pass filter in an SDR. "RF" clipping doesn't actually have to be done at a high RF frequency. An IF of a few kHz is sufficient, so long as the center frequency is greater than about twice the audio bandwidth. That can all be done in software and then the signal can be converted back to baseband audio if desired.

Developing DSP software is a wonderful homebrew activity for the Radio Amateur. As electronic devices have become smaller and smaller and more and more sophisticated it has become harder and harder to get a soldering iron on the tiny pins of surface-mount ICs. Software development allows hobbyists to experiment to their heart's content with no danger that an expensive piece of electronics will be destroyed by one moment of clumsiness. With nothing more than a PC and some free software, the enthusiast can while away hours exploring the fascinating world of digital signal processing and software radios.

15.8 Glossary

Adaptive filter — A filter whose coefficients can be changed automatically.

Analog-to-digital converter (ADC) — A device that samples an analog signal and outputs a digital number representing the amplitude of the signal.

Analytic signal — A representation of the phase and amplitude of a signal (often in the form of the in-phase and quadrature components), without explicitly including the oscillating part (the carrier).

Anti-aliasing filter — A band-limiting filter placed before a sampler to make sure the incoming signal obeys the Nyquist criterion.

Anti-symmetrical — A function that is anti-symmetrical about point x=0 has the property that $f(x) = -f(-x)$.

Application-specific integrated circuit (ASIC) — A non-programmable IC that is designed for a particular application.

Arithmetic logic unit (ALU) — The portion of a microprocessor that performs basic arithmetic and logical operations.

Automatic gain control (AGC) — The circuit in a receiver that keeps the signal level approximately constant.

Automatic level control (ALC) — The circuit in a transmitter that keeps the peak signal level approximately constant.

Barrel shifter — A circuit in a microprocessor that can bit-shift a number by multiple bits at one time.

Baseband — The low-frequency portion of a signal. This is typically the modulation.

Bilinear transform — A design technique for IIR filters in which the frequency axis is transformed to prevent the filter bandwidth from violating the Nyquist criterion.

Binary point — The symbol that separates the integer part from the fractional part of a binary number.

Blocking dynamic range (BDR) — The difference between the noise level (usually in a 500-Hz bandwidth) and the signal level that causes a 1 dB in the level of a weaker signal.

Carrier-locked loop — A feedback control loop to automatically tune a receiver or demodulator to the frequency of a received carrier.

Chebyshev filter — A filter with equal ripple in the passband, stopband or both.

Circular buffers — A buffer in which the final entry is considered to be adjacent to the first.

Cognitive radio — A radio system in which a wireless node automatically changes its transmission or reception parameters to avoid interference with other nodes.

Complex number — A number that contains real and imaginary parts.

Complex PLD (CPLD) — A programmable logic device that is more complex than a small PLD, such as a PAL, but with a similar architecture.

Constellation diagram — A phasor diagram showing the locations of all the modulation states of a digital modulation type.

Convolution — A mathematical operation that modifies a sequence of numbers with another sequence of numbers so as to produce a third sequence with a different frequency spectrum or other desired characteristic. An FIR filter is a convolution engine.

Cooley-Tukey algorithm — Another name for the fast Fourier transform.

Decimation — Reduction of sample rate by an integer factor.

Decimation in time — The division of a sequence of numbers into successively smaller sub-sequences in order to facilitate calculations such as the Fourier transform.

Digital downconverters (DDC) — A device that translates a band of frequencies to baseband, typically at a lower sample rate.

Digital signal processing — The processing of sequences of digital numbers that represent signals.

Digital signal processor (DSP) — A device to do digital signal processing. The term normally is understood to refer to a microprocessor-type device with special capabilities for signal processing.

Digital-to-analog converter (DAC) — A device that converts digital numbers to an analog signal with an amplitude proportional to the digital numbers.

Digital upconverters (DUC) — A device that frequency-translates a baseband signal to a higher frequency, typically at a higher sample rate.

Direct digital synthesis (DDS) — The generation of a periodic waveform by directly calculating the values of the waveform samples.

Direct form — A circuit topology for an IIR filter that corresponds directly to the standard filter equation.

Direct memory access (DMA) — The ability of a microprocessor chip to transfer data between memory and some other device without the necessity to execute any processor instructions.

Dithering — Randomly varying the amplitude or phase of a signal in order to overcome quantization effects.

Embedded system — A system that includes a microprocessor for purposes other than general-purpose computing.

Equal-ripple filter — A filter in which the variation in passband or stopband response is constant across the band.

Exponent — The number of digits that the radix point must be moved to represent a number.

Fast Fourier transform (FFT) — An algorithm that can calculate the discrete Fourier transform with an execution time proportional to $n\log(n)$, instead of n^2 as is required by the straight-forward application of the Fourier transform equation.

Field-programmable gate array (FPGA) — An IC that contains a large array of complex logic blocks whose function and connections can be re-programmed in the field.

Filter coefficient — One of a series of numbers that define the transfer function of a filter.

Finite impulse response (FIR) — An impulse response that is zero for all time that is greater than some finite amount from the time of the impulse.

Floating-point — Refers to a number whose value is represented by a mantissa and an exponent.

Fourier transform — A mathematical operation that derives the frequency spectrum of a time-domain signal.

Hardware-description languages (HDL) — A computer language to specify the circuitry of a digital device or system.

Harmonic sampling — The use of a sample rate that is less than twice the highest frequency of the signal to be sampled. The sample rate must be greater than two times the bandwidth of the signal.

Harvard architecture — A computer architecture in which the program and data are stored in separate memories.

Hilbert transformer — A filter that creates a constant 90° phase difference over a band of frequencies.

Imageless mixer — A mixer in which the output contains only the sum or difference of the two input frequencies, but not both.

Imaginary number — A real number multiplied by the square root of minus one.

Impulse — A pulse of finite energy with a width that approaches zero.

Impulse-invariant — A design technique for IIR filters in which the impulse response is the same as the impulse response of a certain analog filter.

Impulse response — The response versus time of a filter to an impulse.

In-circuit emulator (ICE) — A device that emulates the operation of a microprocessor while providing debugging tools to the operator. The ICE normally plugs into an IC socket that normally holds the microprocessor.

In-circuit debugger (ICD) — A device that uses debugging features built into the microprocessor so that it can be tested while in the circuit.

In-circuit programmable (ICP) — A programmable IC that can be programmed while it is connected to the application circuit.

In-circuit programmer (ICP) — A device to facilitate programming of programmable ICs while they are connected to the application circuit.

In-phase (I) — The portion of a radio signal that is in phase with a reference carrier.

Infinite impulse response (IIR) — An impulse response that theoretically never goes to zero and stays there.

Integrated development environment (IDE) — An integrated collection of software and hardware tools for developing a microprocessor project.

I/Q demodulator — A device to derive the in-phase and quadrature portions of an oscillating signal.

I/Q modulator — A device to generate an oscillating signal with specified in-phase and quadrature parts.

Interpolation — Increasing the sample rate by an integer factor.

Interrupt service routine (ISR) — A software subroutine that is called automatically when the main routine is interrupted by some event.

Least mean squares (LMS) — An algorithm for adaptive filters that minimizes the mean square error of a signal.

Least-significant bit (LSB) — When used as a measurement unit, the size of the smallest step of a digital number.

Limit cycle — A non-linear oscillation in an IIR filter.

Linear phase — Refers to a system in which the delay is constant at all frequencies, which means that the phase is linear with frequency.

Mantissa — The decimal or binary part of a logarithm or floating-point number.

Multiplier-accumulator (MAC) — A device that can multiply two numbers and add the result to a previous result all in one operation.

Multi-rate — Refers to a system with more than one sample rate.

Numerically-controlled oscillator (NCO) — An oscillator that synthesizes the output frequency from a fixed timebase. A DDS oscillator.

Nyquist criterion — The requirement that the sample rate must be at least twice the bandwidth of the signal.

Nyquist frequency — One half the sample rate.

Nyquist rate — Twice the signal bandwidth.

One-time programmable (OTP) — A programmable device that may not be re-programmed.

Orthogonal — Perpendicular. In analogy with the mathematics of perpendicular geometrical vectors, the term is used in communications to refer to two signals that produce zero when convolved.

Oversampling — Use of a sample rate higher than required by the Nyquist criterion in order to improve the signal-to-noise ratio.

Parks-McClellan algorithm — An optimized design technique for equal-ripple filters.

Phasor diagram — A polar plot of the magnitude of the in-phase and quadrature components of a signal.

Pipeline — An arrangement of computational units in a microprocessor or other digital device so that different units can be working on different instructions or signal samples at the same time.

Programmable-array logic (PAL) — A type of small PLD that consists of an array of AND gates, OR gates, inverters and latches.

Programmable-logic device (PLD) — A device with many logic elements whose connections are not defined at manufacturer but must be programmed.

Quadrature (Q) — The portion of a radio signal that is 90° out of phase with a reference carrier.

Quadrature detector — An FM detector that multiplies the signal by a delayed version of the same signal.

Quantization — The representation of a continuous analog signal by a number with a finite number of bits.

Quantization error — The difference in amplitude between an analog signal and its digital samples.

Quantization noise — Noise caused by random quantization error.

Radix – The base of a number system. Binary is radix 2 and decimal is radix 10.

Radix point — The symbol that separates the integer part from the fractional part of a number.

Reciprocal mixing — A spurious response in a receiver to an off-channel signal caused by local oscillator phase noise at the same frequency offset as the interference.

Reconstruction filter — A filter located after a digital-to-analog conversion or interpolation to filter out sampling spurs.

Resampling — Changing the sample rate by a non-integer ratio.

Resolution — The number of bits required to represent a digital number from its smallest to its largest value.

Sample rate — The rate at which samples are generated, processed or output from a system.

Sampling — The process of measuring and recording a signal at discrete points of time.

Software-defined radio (SDR) — A transmitter and/or receiver whose principal signal processing functions are defined by software.

Spectral leakage — In a Fourier transform, the indication of frequencies that are not actually present in the signal due to inadequate windowing.

Tap — One processing block, consisting of a coefficient memory, signal register, multiplier and adder, of an FIR filter.

Training sequence — A sequence of one or more known symbols transmitted for the purpose of training the adaptive filter in a receiver.

Undersampling — Harmonic sampling.

Volatile memory — A memory that requires the presence of power supply voltage for data retention.

Von Neumann architecture — A computer architecture that includes a processing unit and a single separate read/write memory to hold both program and data.

Windowing — Tapering the edges of a data sequence so that the samples do not transition abruptly to zero. This avoids passband and stopband ripple in an FIR filter and spectral leakage in a Fourier transform.

Zero-order hold — Holding of a data value for the entire sample period.

Zero-overhead looping — The ability of a microprocessor to automatically jump from the end of a memory block back to the beginning without additional instructions.

Zero-stuffing — Interpolation by inserting zero-valued samples in between the original samples.

15.9 References and Bibliography

REFERENCES

1. The FCC report and order authorizing software-defined radios in the commercial service is available at **www.fcc.gov/Bureaus/Engineering_Technology/Orders/2001/fcc01264.pdf**.
2. The Microchip IDE software can be downloaded free at **www.microchip.com**.
3. Mar, chapter 4. [See bibliography]
4. This is Euler's famous formula, eix = cos(x) + $j$sin(x), named after Leonhard Euler, an 18th-century mathematician. The factor e ≈ 2.718 is the base of the natural logarithm and $j = \sqrt{-1}$ is the imaginary unit.
5. Blanchard, R., W6UYG, "Sugar-Coated Single Sideband," *QST*, Oct 1952, p 38ff.
6. Weaver, D.K., Jr, "A Third Method of Generation and Detection of Single-Sideband Signals" *Proc. IRE*, Dec. 1956.
7. Wright, Jr., H., W1PNB, "The Third Method of S.S.B.," *QST*, Sep 1957, pp 11-15.
8. Hershberger, D., W9GR, "Low-Cost Digital Signal Processing for the Radio Amateur," *QST*, Sep 1992, pp 43-51.
9. Youngblood. (See the Articles section below.)
10. Tayloe, D., N7VE, "Notes on 'Ideal' Commutating Mixers (Nov/Dec 1999)" Letters to the Editor, *QEX*, Mar 2001, p 61.
11. *PowerSDR* software is available on the Flex Radio Web site, **www.flex-radio.com**.
12. Analog Devices has lots of good free tutorial literature available for download on their Web site, **www.analog.com**.

BIBLIOGRAPHY

(Key: **D** = disk included, **A** = disk available, **F** = filter design software)

DSP Software Tools

Alkin, O., *PC-DSP*, Prentice Hall, Englewood Cliffs, NJ, 1990 (**DF**).
Kamas, A. and Lee, E., *Digital Signal Processing Experiments*, Prentice Hall, Englewood Cliffs, NJ, 1989 (**DF**).
Momentum Data Systems, Inc., *QEDesign*, Costa Mesa, CA, 1990 (**DF**).
Stearns, S. D. and David, R. A., *Signal Processing Algorithms in FORTRAN and C*, Prentice Hall, Englewood Cliffs, NJ, 1993 (**DF**).

Textbooks

Frerking, M. E., *Digital Signal Processing in Communication Systems*, Van Nostrand Reinhold, New York, NY, 1994.
Hayward, Wes, W7ZOI et al, *Experimental Methods in RF Design*, chapter 10 "DSP Components" and chapter 11 "DSP Applications in Communications," ARRL, 2009 (**D**).
Ifeachor, E. and Jervis, B., *Digital Signal Processing: A Practical Approach*, Addison-Wesley, 1993 (**AF**).
Madisetti, V. K. and Williams, D. B., Editors, *The Digital Signal Processing Handbook*, CRC Press, Boca Raton, FL, 1998 (**D**).
Mar, Amy (Editor), *Digital Signal Processing Applications Using the ADSP-2100 Family, Volume I*, Prentice Hall 1990. While oriented to the ADSP-2100, there is much good general information on DSP algorithms. Both volume I and II are available for free download on the Analog Devices Web site, **www.analog.com**.
Oppenheim, A. V. and Schafer, R. W., *Digital Signal Processing*, Prentice Hall, Englewood Cliffs, NJ, 1975.
Parks, T. W. and Burrus, C.S., *Digital Filter Design*, John Wiley and Sons, New York, NY, 1987.
Proakis, J. G. and Manolakis, D., *Digital Signal Processing*, Macmillan, New York, NY, 1988.
Proakis, J. G., Rader, C. M., et. al., *Advanced Digital Signal Processing*, Macmillan, New York, NY, 1992.
Rabiner, L. R. and Schafer, R. W., *Digital Processing of Speech Signals*, Prentice Hall, Englewood Cliffs, NJ, 1978.
Rabiner, L. R. and Gold, B., *Theory and Application of Digital Signal Processing*, Prentice Hall, Englewood Cliffs, NJ, 1975.
Sabin, W. E. and Schoenike, E. O., Eds., *HF Radio Systems and Circuits*, rev. 2nd ed., Noble Publishing Corp, Norcross, GA, 1998.
Smith, D., *Digital Signal Processing Technology: Essentials of the Communications Revolution*, ARRL, 2001.
Widrow, B. and Stearns, S. D., *Adaptive Signal Processing*, Prentice Hall, Englewood Cliffs, NJ, 1985.
Zverev, A. I., *Handbook of Filter Synthesis*, John Wiley and Sons, New York, NY, 1967.

Articles

Albert, J. and Torgrim, W., "Developing Software for DSP," *QEX*, Mar 1994, pp 3-6.
Anderson, P. T., "A Simple SSB Receiver Using a Digital Down-Converter," *QEX*, Mar, 1994, pp 17-23.
Anderson, P. T., "A Faster and Better ADC for the DDC-Based Receiver," *QEX*, Sep/Oct 1998, pp 30-32.
Applebaum, S. P., "Adaptive arrays," *IEEE Transactions Antennas and Propagation*, Vol. PGAP-24, pp 585-598, September, 1976.
Åsbrink, L., "Linrad: New Possibilities for the Communications Experimenter," *QEX*, Part 1, Nov/Dec 2002; Part 2, Jan/Feb 2003; Part 3 May/Jun 2003.
Ash, J. et al., "DSP Voice Frequency Compandor for Use in RF Communications," *QEX*, Jul 1994, pp 5-10.
Beals, K., "A 10-GHz Remote-Control System for HF Transceivers," *QEX*, Mar/Apr 1999, pp 9-15.
Bloom, J., "Measuring SINAD Using DSP," *QEX*, Jun 1993, pp 9-18.
Bloom, J., "Negative Frequencies and Complex Signals," *QEX*, Sep 1994.
Bloom, J., "Correlation of Sampled Signals," *QEX*, Feb 1996.
Brannon, B., "Basics of Digital Receiver Design," *QEX*, Sep/Oct 1999, pp 36-44.
Cahn, H., "Direct Digital Synthesis — An Intuitive Introduction," *QST*, Aug 1994, pp 30-32.
Cercas, F. A. B., Tomlinson, M. and Albuquerque, A. A., "Designing With Digital Frequency Synthesizers," *Proceedings of RF Expo East*, 1990.
de Carle, B., "A Receiver Spectral Display Using DSP," *QST* Jan 1992, pp 23-29.
Dick, R., "Tune SSB Automatically," *QEX*, Jan/Feb 1999, pp 9-18.
Dobranski, L., "The Need for Applications Programming Interfaces (APIs) in Amateur Radio," *QEX*, Jan/Feb 1999, pp 19-21.
Emerson, D., "Digital Processing of Weak Signals Buried in Noise," *QEX*, Jan 1994, pp 17-25.
Forrer, J., "Programming a DSP Sound Card for Amateur Radio," *QEX*, Aug 1994.
Forrer, J., KC7WW, "A DSP-Based Audio Signal Processor," *QEX*, Sep 1996.
Forrer, J., KC7WW, "Using the Motorola DSP56002EVM for Amateur Radio DSP Projects," *QEX*, Aug 1995.
Gradijan, S., WB5KIA, "Build a Super Transceiver — Software for Software Controllable Radios," *QEX*, Sept/Oct 2004, pp 30-34.
Green, R., "The Bedford Receiver: A New Approach," *QEX*, Sep/Oct, 1999, pp 9-23.
Hale, B., "An Introduction to Digital Signal Processing," *QST*, Sep 1992, pp 43-51.

Hershberger, D., W9GR, "DSP — An Intuitive Approach," *QST*, Feb 1996.

Kossor, M., "A Digital Commutating Filter," *QEX*, May/Jun 1999, pp 3-8.

Larkin, B., W7PUA, "The DSP-10: An All-Mode 2-Meter Transceiver Using a DSP IF and PC-Controlled Front Panel, Part 1," *QST*, Sep 1999, pp 33-41.

Larkin, B., W7PUA, "The DSP-10: An All-Mode 2-Meter Transceiver Using a DSP IF and PC-Controlled Front Panel, Part 2," *QST*, Oct 1999, pp 34-40.

Larkin, B., W7PUA, "The DSP-10: An All-Mode 2-Meter Transceiver Using a DSP IF and PC-Controlled Front Panel, Part 3," *QST*, Nov 1999, pp 42-45.

Mack, Ray, W5IFS, "SCR: Simplified," *QEX*, Jan/Feb 2009, pp 53-56 and Mar/Apr 2009, pp 41-44.

Morrison, F., "The Magic of Digital Filters," *QEX*, Feb 1993, pp 3-8.

Olsen, R., "Digital Signal Processing for the Experimenter," *QST*, Nov 1994, pp 22-27.

Reyer, S. and Herschberger, D., "Using the LMS Algorithm for QRM and QRN Reduction," *QEX*, Sep 1992, pp 3-8.

Rohde, D., "A Low-Distortion Receiver Front End for Direct-Conversion and DSP Receivers," *QEX*, Mar/Apr 1999, pp 30-33.

Runge, C., *Z. Math. Physik*, Vol 48, 1903; also Vol 53, 1905.

Scarlett, J., "A High-Performance Digital Transceiver Design," *QEX*, Part 1, Jul/Aug 2002; Part 2, Mar/Apr 2003.

Smith, D., "Introduction to Adaptive Beamforming," *QEX*, Nov/Dec 2000.

Smith, D., "Signals, Samples and Stuff: A DSP Tutorial, Parts 1-4," *QEX*, Mar/Apr-Sep/Oct, 1998.

Stephensen, J., KD6OZH, Software Defined Radios for Digital Communications," *QEX*, Nov/Dec 2004, pp 23-35. [Open-source platform]

Ulbing, Sam, "Surface-Mount Technology — You Can Work With It!," Parts 1-4, *QST*, Apr-Jul 1999.

Ward, R., "Basic Digital Filters," *QEX*, Aug 1993, pp 7-8.

Wiseman, J., KE3QG, "A Complete DSP Design Example Using FIR Filters," *QEX*, Jul 1996.

Youngblood, G., "A Software-Defined Radio for the Masses," *QEX*; Part 1, Jul/Aug 2002; Part 2, Sep/Oct 2002; Part 3, Nov/Dec 2002; Part 4, Mar/Apr 2003.

SDR and DSP Receiver Testing

Hakanson, E., "Understanding SDRs and their RF Test Requirements," Anritsu Application Note. Available online from **www.us.anritsu.com/downloads/files/11410-00403.pdf**.

Kundert, K. "Accurate and Rapid Measurement of IP2 and IP3," Designer's Guide Consulting, 2002-2006. Available online from **www.designers-guide.org/Analysis/intercept-point.pdf**.

MacLeod, J.R.; Beach, M.A.; Warr, P.A.; Nesimoglu, T., "Software Defined Radio Receiver Test-Bed," IEEE Vehicular Tech Conf, Fall 2001, VTS 54th, Vol 3, pp 565-1569. Available online from **ieeexplore.ieee.org/Xplore/login.jsp?url=/iel5/7588/20685/00956461.pdf**.

R. Sierra, Rhode & Schwarz, "Challenges In Testing Software Defined Radios," SDR Forum Sandiego Workshop, 2007.

Sirmans, D. and Urell, B., "Digital Receiver Test Results," Next Generation Weather Radar Program. Available online from **www.roc.noaa.gov/eng/docs/digital%20receiver%20test%20results.pdf**.

"Testing and Troubleshooting Digital RF Communications Receiver Designs," Agilent Technologies Application Note 1314. Available online from **cp.literature.agilent.com/litweb/pdf/5968-3579E.pdf**.

Contents

16.1 Digital "Modes"
    16.1.1 Symbols, Baud, Bits and Bandwidth
    16.1.2 Error Detection and Correction
    16.1.3 Data Representations
    16.1.4 Compression Techniques
    16.1.5 Compression vs. Encryption

16.2 Unstructured Digital Modes
    16.2.1 Radioteletype (RTTY)
    16.2.2 PSK31
    16.2.3 MFSK16
    16.2.4 DominoEX
    16.2.5 THROB
    16.2.6 MT63
    16.2.7 Olivia

16.3 Fuzzy Modes
    16.3.1 Facsimile (fax)
    16.3.2 Slow-Scan TV (SSTV)
    16.3.3 Hellschreiber, Feld-Hell or Hell

16.4 Structured Digital Modes
    16.4.1 FSK441
    16.4.2 JT6M
    16.4.3 JT65
    16.4.4 WSPR
    16.4.5 HF Digital Voice
    16.4.6 ALE

16.5 Networking Modes
    16.5.1 OSI Networking Model
    16.5.2 Connected and Connectionless Protocols
    16.5.3 The Terminal Node Controller (TNC)
    16.5.4 PACTOR-I
    16.5.5 PACTOR-II
    16.5.6 PACTOR-III
    16.5.7 G-TOR
    16.5.8 CLOVER-II
    16.5.9 CLOVER-2000
    16.5.10 WINMOR
    16.5.11 Packet Radio
    16.5.12 APRS
    16.5.13 Winlink 2000
    16.5.14 D-STAR
    16.5.15 P25

16.6 Glossary

16.7 References and Bibliography

Chapter 16

Digital Modes

The first Amateur Radio transmissions were digital using Morse code. Computers have now become so common, households often have several and there are a large number of other digital information appliances. E-mail and "texting" are displacing traditional telephone communications and Amateur Radio operators are creating new digital modes to carry this traffic over the air. Sometimes it seems nearly every month a new mode or variation appears to solve a particular need. Many of these modes can be entirely implemented in software for use with a computer sound card, making tools for experimentation and implementation available worldwide via the Internet.

This chapter will focus on the protocols for transferring various data types. The material was written or updated by Scott Honaker, N7SS, with support from Kok Chen, W7AY, on the sections regarding RTTY, PSK and MFSK. Modulation methods are covered in the **Modulation** chapter and the **Digital Communications** chapter will discuss the practical considerations of operating using these modes. This chapter addresses the process by which data is encoded, compressed and error checked, packaged and exchanged.

There is a broad array of digital modes to service various needs with more coming. The most basic modes simply encode text data for transmission in a keyboard-to-keyboard chat-type environment. These modes may or may not include any mechanism for error detection or correction. The second class of modes are generally more robust and support more sophisticated data types by structuring the data sent and including additional error-correction information to properly reconstruct the data at the receiving end. The third class of modes discussed will be networking modes with protocols often the same or similar to versions used on the Internet and computer networks.

16.1 Digital "Modes"

The ITU uses *Emission Designators* to define a "mode" as demonstrated in the **Modulation** chapter. These designators include the bandwidth, modulation type and information being sent. This system works well to describe the physical characteristics of the modulation, but digital modes create some ambiguity because the type of information sent could be text, image or even the audio of a CW session. As an example, an FM data transmission of 20K0F3D could transmit spoken audio (like FM 20K0F3E or 2K5J3E) or a CW signal (like 150H0A1A). These designators don't help identify the type of data supported by a particular mode, the speed that data can be sent, if it's error-corrected, or how well it might perform in hostile band conditions. Digital modes have more characteristics that define them and there are often many variations on a single mode that are optimized for different conditions. We'll need to look at the specifics of these unique characteristics to be able to determine which digital modes offer the best combination of features for any given application.

16.1.1 Symbols, Baud, Bits and Bandwidth

The basic performance measure of a digital mode is the *data rate*. This can be measured a number of ways and is often confused. Each change of state on a transmission medium defines a *symbol* and the *symbol rate* is also known as *baud*. (While commonly used, "baud rate" is redundant because "baud" is already defined as the rate of symbols/second.) Modulating a carrier increases the frequency range, or bandwidth, it occupies. The FCC currently limits digital modes by symbol rate on the various bands as an indirect (but easily measurable) means of controlling bandwidth.

The *bit rate* is the product of the *symbol rate* and the number of bits encoded in each symbol. In a simple two-state system like an RS-232 interface, the bit rate will be the same as baud. More complex waveforms can represent more than two states with a single symbol so the bit rate will be higher than the baud. For each additional bit encoded in a symbol, the number of states of the carrier doubles. This makes each state less distinct from the others, which in turn makes it more difficult for the receiver to detect each symbol correctly in the presence of noise. A V.34 modem may transmit symbols at a baud rate of 3420 baud and each symbol can represent up to 10 discrete states or bits, resulting in a *gross (or raw) bit rate* of 3420 baud × 10 or 34,200 bits per second (bit/s). Framing bits and other overhead reduce the *net bit rate* to 33,800 bit/s.

Bits per second is abbreviated here as *bit/s* for clarity but is also often seen as *bps*. Bits per second is useful when looking at the protocol but is less helpful determining how long it

Table 16.1
Data Rate Symbols and Multipliers

Name	Symbol	Multiplier
kilobit per second	kbit/s or kbps	1000 or 10^3
Megabit per second	Mbit/s or Mbps	1,000,000 or 10^6
Gigabit per second	Gbit/s or Gbps	1,000,000,000 or 10^9

takes to transmit a specific size file because the number of bits consumed by overhead is often unknown. A more useful measure for calculating transmission times is *bytes per second* or *Bps* (note the capitalization). Although there are only eight bits per byte, with the addition of start and stop bits, the difference between bps and Bps is often tenfold. Since the net bit rate takes the fixed overhead into account, Bytes per second can be calculated as $bps_{net}/8$. Higher data rates can be expressed with their metric multipliers as shown in **Table 16.1**.

Digital modes constantly balance the relationship between symbol rate, bit rate, bandwidth and the effect of noise. The Shannon-Hartley theorem demonstrates the maximum channel capacity in the presence of Gaussian white noise and was discussed in the **Modulation** chapter in the Channel Capacity section. This theorem describes how an increased symbol rate will require an increase in bandwidth and how a reduced signal-to-noise ratio (SNR) will reduce the potential *throughput* of the channel.

16.1.2 Error Detection and Correction

Voice modes require the operator to manually request a repeat of any information required but not understood. Using proper phonetics makes the information more easily understood but takes longer to transmit. If 100% accuracy is required, it may be necessary for the receiver to repeat the entire message back to the sender for verification. Computers can't necessarily distinguish between valuable and unnecessary data or identify likely errors but they offer other options to detect and correct errors.

ERROR DETECTION

The first requirement of any accurate system is to be able to detect when an error has occurred. The simplest method is *parity*. With 7-bit ASCII data, it was common to transmit an additional 8th parity bit to each character. The parity bit was added to make the total number of 1 bits odd or even. The binary representation for an ASCII letter Z is 1011010. Sent as seven bits with even parity, the parity bit would be 0 because there are already an even number of 1 bits and the result would be 01011010. The limitation of parity is that it only works with an odd number of bit inversions. If the last two bits were flipped to 01011001 (the ASCII letter Y), it would still pass the parity check because it still has an even number of bits. Parity is also rarely used on 8-bit data so it cannot be used when transferring binary data files.

Checksum is a method similar to the "check" value in an NTS message. It is generally a single byte (8-bit) value appended to the end of a packet or frame of data. It is calculated by adding all the values in the packet and taking the least significant (most unique) byte. This is a simple operation for even basic processors to perform quickly but can also be easily mislead. If two errors occur in the packet of equal amounts in the opposite direction (A becomes B and Z becomes Y), the checksum value will still be accurate and the packet will be accepted as error-free.

Cyclic redundancy check (CRC) is similar to checksum but uses a more sophisticated formula for calculating the check value of a packet. The formula most closely resembles long division, where the quotient is thrown away and the remainder is used. It is also common for CRC values to be more than a single byte, making the value more unique and likely to identify an error. Although other error detection systems are currently in use, CRC is the most common.

ERROR CORRECTION

There are two basic ways to design a protocol for an error correcting system: *automatic repeat request (ARQ)* and *forward error correction (FEC)*. With ARQ the transmitter sends the data packet with an error detection code, which the receiver uses to check for errors. The receiver will request retransmission of packets with errors or sends an acknowledgement (ACK) of correctly received data, and the transmitter re-sends anything not acknowledged within a reasonable period of time.

With forward error correction (FEC), the transmitter encodes the data with an *error-correcting code (ECC)* and sends the encoded message. The receiver is not required to send any messages back to the transmitter. The receiver decodes what it receives into the "most likely" data. The codes are designed so that it would take an "unreasonable" amount of noise to trick the receiver into misinterpreting the data. It is possible to combine the two, so that minor errors are corrected without retransmission, and major errors are detected and a retransmission requested. The combination is called *hybrid automatic repeat-request (hybrid ARQ)*.

There are many error correcting code (ECC) algorithms available. Extended Golay coding is used on blocks of ALE data, for example, as described in the section below on G-TOR. In addition to the ability to detect and correct errors in the data packets, the modulation scheme allows sending multiple data streams and interleaving the data in such a way that a noise burst will disrupt the data at different points.

16.1.3 Data Representations

When comparing digital modes, it is important to understand how the data is conveyed. There are inherent limitations in any method chosen. PSK31 might seem a good choice for sending data over HF links because it performs well, but it was only designed for text (not 8-bit data) and has no inherent error correction. It is certainly possible to use this modulation scheme to send 8-bit data and add error correction to create a new mode. This would maintain the weak signal performance but the speed will suffer from the increased overhead. Similarly, a digital photo sent via analog SSTV software may only take two minutes to send, but over VHF packet it could take 10 minutes, despite the higher speed of a packet system. This doesn't mean SSTV is more efficient. Analog SSTV systems generally transmit lower resolution images with no error correction. Over good local links, the VHF packet system will be able to deliver perfect images faster or of higher quality.

TEXT REPRESENTATIONS

Morse code is well known as an early code used to send text data over a wire, then over the air. Each letter/number or symbol is represented with a varying length code with the more common letters having shorter codes. This early *varicode* system is very efficient and minimizes the number of state changes required to send a message.

The Baudot code (pronounced "bawd-OH") was invented by Émile Baudot and is the predecessor to the character set currently known more accurately as International Telegraph Alphabet No 2 (ITA2). This code is used for radioteletype communications and contains five bits with start and stop pulses. This only allows for 2^5 or 32 possible characters to be sent, which is not enough for all 26 letters plus numbers and characters. To resolve this, ITA2 uses a LTRS code to select a table of upper case (only) letters and a FIGS code to select a table of numbers, punctuation and special symbols. The code is defined in the ITA2 codes table on the CD included with this book.

Early computers used a wide variety of alphabetic codes until the early 1960s until the advent of the American Standard Code for Information Interchange or ASCII (pronounced "ESS-key"). At that point many computers standardized on this character set to allow simple transfer of data between machines. ASCII is a 7-bit code which allows for 2^7 or 128 characters. It was designed without the *control characters* used by Baudot for more reliable transmissions and the letters appear in English alphabetical order for easy sorting. The code can be reduced to only six bits and still carry numbers and uppercase letters. Current FCC regulations provide that amateur use of ASCII conform to ASCII as defined in ANSI standard X3.4-1977. The international counterparts are ISO 646-1983 and International Alphabet No. 5 (IA5) as published in ITU-T Recommendation V.3. A table of ASCII characters is presented as "ASCII Character Set" on the CD included with this book.

ASCII has been modified and initially expanded to eight bits, allowing the addition of foreign characters or line segments. The different extended versions were often referred to as *code pages*. The IBM PC supported code page 437 which offers line segments, and English *Windows* natively supports code page 1252 with additional foreign characters and symbols. All of these extended code pages include the same first 128 ASCII characters for backward compatibility. In the early 1990s efforts were made to support more languages directly and *Unicode* was created. Unicode generally requires 16 bits per character and can represent nearly any language.

More recent schemes use varicode, where the most common characters are given shorter codes (see **en.wikipedia.org/wiki/Prefix_code**). Varicode is used in PSK31 and MFSK to reduce the number of bits in a message. Although the PSK31 varicode contains all 128 ASCII characters, lower-case letters contain fewer bits and can be sent more quickly. Tables of PSK and MFSK varicode characters are included on the CD included with this book.

IMAGE DATA REPRESENTATIONS

Images are generally broken into two basic types, *raster* and *vector*. Raster or *bitmap* images are simply rows of colored points known as *pixels* (picture elements). Vector images are a set of drawing primitives or instructions. These instructions define shapes, placement, size and color. Similar coding is used with plotters to command the pens to create the desired image.

Bitmap images can be stored at various *color depths* or bits per pixel indicating how many colors can be represented. Common color depths are 1-bit (2 colors), 4-bit (16 colors), 8-bit (256 colors), 16-bit (65,536 colors also called *high color*) and 24-bit (16 million also called *true color*). True color is most commonly used with digital cameras and conveniently provides eight bits of resolution each for the red, green and blue colors. Newer scanners and other systems will often generate 30-bit (2^{30} colors) and 36-bit (2^{36} colors). Images with 4- and 8-bit color can store more accurate images than might be obvious because they include a *palette* where each of their 16 or 256 colors respectively are chosen from a palette of 16 million. This palette is stored in the file which works well for simple images. The GIF format only supports 256 colors (8-bit) with a palette and lossless compression.

Digital photographs are raster images at a specific resolution and color depth. A typical low resolution digital camera image would be 640×480 pixels with 24-bit color. The 24-bit color indicated for each pixel requires three bytes to store (24 bits / 8 bits per byte) and can represent one of a possible 2^{24} or 16,777,216 colors with eight bits each for red, green and blue intensities. The raw (uncompressed) storage requirement for this image would be 640×480x3 or 921,600 bytes. This relatively low resolution image would require significant time to transmit over a slow link.

Vector images are generally created with drawing or CAD packages and can offer significant detail while remaining quite small. Because vector files are simply drawn at the desired resolution, there is no degradation of the image if the size is changed and the storage requirements remain the same at any resolution. Typical drawing primitives include lines and polylines, polygons, circles and ellipses, Bézier curves and text. Even computer font technologies such as TrueType create each letter from Bézier curves allowing for flawless scaling to any size and resolution on a screen or printer.

Raster images can be resized by dropping or adding pixels or changing the color depth. The 640×480×24-bit color image mentioned above could be reduced to 320×240×16-bit color with a raw size of 153,600 bytes — a significant saving over the 921,600 byte original. If the image is intended for screen display and doesn't require significant detail, that size may be appropriate. If the image is printed full sized on a typical 300 dpi (dots per inch) printer, each pixel in the photo will explode to nearly 100 dots on the printer and appear very blocky or pixilated.

AUDIO DATA REPRESENTATIONS

Like images, audio can be stored as a sampled waveform or in some type of primitive format. Storing a sampled waveform is the most versatile but can also require substantial storage capacity. MIDI (musical instrument digital interface, pronounced "MID-ee") is a common music format that stores instrument, note, tempo and intensity information as a musical score. There are also voice coding techniques that store speech as *allophones* (basic human speech sounds).

As with images, storage of primitives can save storage space (and transmission times) but they are not as rich as a high quality sampled waveform. Unfortunately, the 44,100 Hz 16-bit sample rate of an audio compact disc (CD) requires 176,400 bytes to store each second and 10,584,000 bytes for each minute of stereo audio.

The Nyquist-Shannon sampling theorem states that perfect reconstruction of a signal is possible when the sampling frequency is at least twice the maximum frequency of the signal being sampled. With its 44,100 Hz sample rate, CD audio is limited to a maximum frequency response of 22,050 Hz. If only voice-quality is desired, the sample rate can easily be dropped to 8000 Hz providing a maximum 4000 Hz frequency response. (See the **DSP and Software Radio Design** chapter for more information on sampling.)

The *bit depth* (number of bits used to represent each sample) of an audio signal will determine the theoretical dynamic range or signal-to-noise ratio (SNR). This is expressed with the formula SNR = (1.761 + 6.0206 × bits) dB. A dynamic range of 40 dB is adequate for the perception of human speech. **Table 16.2** compares the audio quality of various formats with the storage (or transmission) requirements.

VIDEO DATA REPRESENTATIONS

The most basic video format simply stores a series of images for playback. Video can place huge demand on storage and bandwidth because 30 frames per second is a common rate for smooth-appearing video. Rates as low as 15 frames per second can still be considered "full-motion" but will appear jerky and even this rate requires substantial storage. Primitive formats are less common for video

Table 16.2
Typical Audio Formats

Audio Format	Bits per Sample	Dynamic Range	Maximum Frequency	kbytes per Minute
44.1 kHz stereo	16	98	22.050 kHz	10,584
22 kHz mono	16	98	11 kHz	2,640
8 kHz mono	8	50	4 kHz	960

but animation systems like Adobe *Flash* are often found on the Web.

16.1.4 Compression Techniques

There are a large variety of compression algorithms available to reduce the size of data stored or sent over a transmission medium. Many techniques are targeted at specific types of data (text, audio, images or video). These compression techniques can be broken into two major categories: *lossless compression* and *lossy compression*. Lossless algorithms are important for compressing data that must arrive perfectly intact but offer a smaller compression ratio than lossy techniques. Programs, documents, databases, spreadsheets would all be corrupted and made worthless if a lossy compression technique were used.

Images, audio and video are good candidates for lossy compression schemes with their substantially higher compression rates. As the name implies, a lossy compression scheme deliberately omits or simplifies that data to be able to represent it efficiently. The human eye and ear can easily interpolate missing information and it simply appears to be of lower quality.

Audio and video compression are often implemented as *codecs* (from *co*der-*dec*oder). These codecs provide the real-time encoding or decoding of the audio or video stream. Many codecs implement a proprietary compression scheme and have licensing requirements. The codecs can be implemented as operating system or application plug-ins or even in silicon, as with the P25 IMBE and D-STAR AMBE codecs.

LOSSLESS COMPRESSION TECHNIQUES

One of the earliest lossless compression schemes is known as *Huffman coding*. Huffman coding creates a tree of commonly used data values and gives the most common values a lower bit count. Varicode is based on this mechanism. In 1984, Terry Welch released code with improvements to a scheme from Abraham Lempel and Jacob Ziv, commonly referred to as *LZW* (Lempel-Ziv-Welch). The Lempel-Ziv algorithm and variants are the basis for most current compression programs and is used in the GIF and optionally TIFF graphic formats. LZW operates similarly to Huffman coding but with greater efficiency.

The actual amount of compression achieved will depend on how redundant the data is and the size of the data being compressed. Large files will achieve greater compression rates because the common data combinations will be seen more frequently. Simple text and documents can often see 25% compression rates. Spreadsheets and databases generally consist of many empty cells and can often achieve nearly 50% compression. Graphic and video compression will vary greatly depending on the complexity of the image. Simple images with solid backgrounds will compress well where complex images with little recurring data will see little benefit. Similarly, music will compress poorly but spoken audio with little background noise may see reasonable compression rates.

Run length encoding (*RLE*) is a very simple scheme supported on bitmap graphics (*Windows* BMP files). Each value in the file is a color value and "run length" specifying how many of the next pixels will be that color. It works well on simple files but can make a file larger if the image is too complex.

It is important to note that compressing a previously compressed file will often yield a larger file because it simply creates more overhead in the file. There is occasionally some minor benefit if two different compression algorithms are used. It is not possible to compress the same file repeatedly and expect any significant benefit. Modern compression software does offer the additional benefit of being able to compress groups of files or even whole directory structures into a single file for transmission.

The compression mechanisms mentioned above allow files to be compressed prior to transmission but there are also mechanisms that allow near real-time compression of the transmitted data. V.92 modems implement a LZJH adaptive compression scheme called V.44 that can average 15% better throughput over the wire. Winlink uses a compression scheme called B2F and sees an average 44% improvement in performance since most Winlink data is uncompressed previously. Many online services offer "Web Accelerators" that also compress data going over the wire to achieve better performance. There is a slight delay (latency) as a result of this compression but over a slow link, the additional latency is minimal compared to the performance gain.

LOSSY COMPRESSION SCHEMES

Lossy compression schemes depend on the human brain to "recover" or simply ignore the missing data. Since audio, image and video data have unique characteristics as perceived by the brain, each of these data types have unique compression algorithms. New compression schemes are developed constantly to achieve high compression rates while maintaining the highest quality. Often a particular file format actually supports multiple compression schemes or is available in different versions as better methods are developed. In-depth discussion of each of these algorithms is beyond the scope of this book but there are some important issues to consider when looking at these compression methods.

Lossy Audio Compression

Audio compression is based on the psychoacoustic model that describes which parts of a digital audio signal can be removed or substantially reduced without affecting the perceived quality of the sound. Lossy audio compression schemes can typically reduce the size of the file 10- or 12-fold with little loss in quality. The most common formats currently are MP3, WMA, AAC, Ogg Vorbis and ATRAC.

Lossy Image Compression

The Joint Photographic Experts Group developed the JPEG format (pronounced "JAY-peg") in 1992 and it has become the primary format used for lossy compression of digital images. It is a scalable compression scheme allowing the user to determine the level of compression/quality of the image. Compression rates of 10-fold are common with good quality and can be over 100-fold at substantially reduced quality. The JPEG format tends to enhance edges and substantially compress fields of similar color. It is not well suited when multiple edits will be required because each copy will have generational loss and therefore reduced quality.

Lossy Video Compression

Because of the massive amount of data required for video compression it is almost always distributed with a lossy compression

Table 16.3
Audio and Video Bit Rates

Audio
8 kbit/s	Telephone quality audio (using speech codecs)
32 kbit/s	AM broadcast quality (using MP3)
96 kbit/s	FM broadcast quality (using MP3)
224-320 kbit/s	Near-CD quality, indistinguishable to the average listener (using MP3)

Video
16 kbit/s	Videophone quality (minimum for "talking head")
128-384 kbit/s	Business videoconference system quality
1.25 Mbit/s	VCD (video compact disk) quality
5 Mbit/s	DVD quality
10.5 Mbit/s	Actual DVD video bit rate

scheme. Lossless compression is only used when editing to eliminate generational loss. Video support on a computer is generally implemented with a codec that allows encoding and decoding the video stream. Video files are containers that can often support more than one video format and the specific format information is contained in the file. When the file is opened, this format information is read and the appropriate codec is activated and fed into the data stream to be decoded.

The most common current codecs are H.261 (video conferencing), MPEG-1 (used in video CDs), MPEG-2 (DVD Video), H.263 (video conferencing), MPEG-4, DivX, Xvid, H.264, Sorenson 3 (used by *QuickTime*), WMV (Microsoft), VC-1, RealVideo (RealNetworks) and Cinepak.

The basic mechanism of video compression is to encode a high quality "key-frame" that could be a JPEG image as a starting image. Successive video frames or "inter-frames" contain only changes to the previous frame. After some number of frames, it is necessary to use another key-frame and start over with inter-frames. The resolution of the images, frame rate and compression quality determine the size of the video file.

BIT RATE COMPARISON

Table 16.3 provides an indication of minimum bit rates required to transmit audio and video such that the average listener would not perceive them significantly worse than the standard shown.

16.1.5 Compression vs. Encryption

There is some confusion about compression being a form of encryption. It is true that a text file after compression can no longer be read unless uncompressed with the appropriate algorithm. In the United States the FCC defines encryption in part 97.113 as "messages encoded for the purpose of obscuring their meaning." Compressing a file with ZIP or RAR (common file compression methods) and transmitting it over the air is simply an efficient use of spectrum and time and is not intended to "obscure its meaning."

As amateur digital modes interact more with Internet-based services, the issue arises because many of these services utilize encryption of various types. Banks and other retailers may encrypt their entire transactions to insure confidentiality of personal data. Other systems as benign as e-mail may simply encrypt passwords to properly authenticate users. The FCC has offered no additional guidance on these issues.

16.2 Unstructured Digital Modes

The first group of modes we'll examine are generally considered "sound card modes" for keyboard-to-keyboard communications. Because each of these modes is optimized for a specific purpose by blending multiple features, they often defy simple categorization.

16.2.1 Radioteletype (RTTY)

Radioteletype (*RTTY*) consists of a *frequency shift keyed* (*FSK*) signal that is modulated between two carrier frequencies, called the *mark* frequency and the *space* frequency. The protocol for amateur RTTY calls for the mark carrier frequency to be the higher of the two in the RF spectrum. The difference between the mark and space frequencies is called the *FSK shift*, usually 170 Hz for an amateur RTTY signal.

At the conventional data speed of 60 WPM, binary information modulates the FSK signal at 22 ms per bit, or equivalent to 45.45 baud. Characters are encoded into binary 5-bit Baudot coded data. Each character is individually synchronized by adding a start bit before the 5-bit code and by appending the code with a stop bit. The start bit has the same duration as the data bits, but the stop bit can be anywhere between 22 to 44 ms in duration. The stop bit is transmitted as a mark carrier, and the RTTY signal "rests" at this state until a new character comes along. If the number of stop bits is set to two, the RTTY signal will send a minimum of 44 ms of mark carrier before the next start bit is sent. A start bit is sent as a space carrier. A zero in the Baudot code is sent as a space signal and a one is sent as a mark signal. **Fig 16.1** shows the character D sent by RTTY.

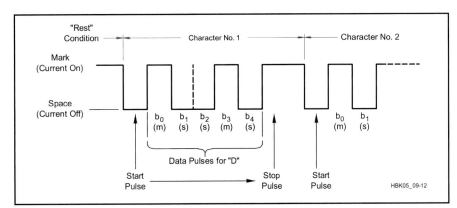

Fig 16.1 — The character "D" sent via RTTY.

BAUDOT CODE

The Baudot code (see the *Handbook* CD) is a 5-bit code; thus, it is capable of encoding only 32 unique characters. Since the combination of alphabets, decimal numbers and common punctuations exceeds 32, the Baudot code is created as two sets of tables. One table is called the LTRS Shift, and consists mainly of alphabetic characters. The second table is called the FIGS Shift, and consists mainly of decimal numerals and punctuation marks. Two unique Baudot characters called LTRS and FIGS are used by the sender to command the decoder to switch between these two tables.

Mechanical teletypewriter keyboards have two keys to send the LTRS and FIGS characters. The two keys behave much like the caps lock key on a modern typewriter keyboard. Instead of locking the keyboard of a typewriter to upper case shift or lower case shift, the two teletypewriter keys lock the state of the teletypewriter into the LTRS table or the FIGS table. LTRS and FIGS, among some other characters such as the space character, appear in both LTRS and FIGS tables so that you can send LTRS, FIGS and shift no matter which table the encoder is using.

To send the letter Q, you need to first make sure that the decoder is currently using the LTRS table, and then send the Baudot codeword for Q, 17 (hexadecimal or "hex" value). If the same hex 17 is received when the decoder is in the FIGS shift, the number 1 will be decoded instead of the Q. Modern software does away with the need for the operator to manually send the LTRS and FIGS codes. When the operator sends a character from the LTRS table, the software first checks to make sure that the previous character had also used a character from the LTRS table; if not, the software will first send a LTRS character.

Noise can often cause the LTRS or FIGS character to be incorrectly received. This will cause subsequent characters to decode into the wrong Baudot character, until a correct LTRS or FIGS is received (see also USOS below). Instead of asking that the message be repeated by the sender, a trick that many RTTY operators use is to observe that on a standard QWERTY keyboard, the Q key is below the 1 key in the row above it, the W key is below the 2 key, and so on. Q and 1 happen to share the same Baudot code; W and 2 share the same Baudot code, and so forth. Given this visual aid, a printed UE can easily be interpreted by context to 73 and TOO interpreted as 599.

If the sender uses one stop bit, an RTTY character consists of a total of seven bits after adding the start and stop bits. If the sender uses 1.5 stop bits, each RTTY character has a total length of 7.5 bits. The least significant bit of the Baudot code follows the start bit of a character.

INVERTED SIGNALS

There are times when the sender does not comply with the RTTY standard and reverses the mark and space order in the spectrum. This is often called an "inverted" signal or "reverse shift." Most RTTY modulators and demodulators have a provision to reverse the shift of an inverted signal.

DEMODULATION AND DECODING

The most common way to decode an RTTY signal is to use a single sideband (SSB) receiver to first translate the two FSK carriers into two audio tones. If the carriers are 170 Hz apart, the two audio tones (called the *tone pair*) will also be 170 Hz apart. The RTTY demodulator, in the form of a terminal unit (TU) or a software modem (modulator-demodulator), then works to discriminate between the two audio tones. Some packet TNCs (terminal node controllers) can be made to function as RTTY demodulators, but often they do not work as well under poor signal-to-noise conditions because their filters are matched to packet radio FSK shifts and baud rate instead of to RTTY shift and baud rate.

As long as the tone pair separation is 170 Hz, the frequencies of the two audio tones can be quite arbitrary. Many TUs and modems are constructed to handle a range of tone pairs. If reception uses a lower sideband (LSB) receiver, the higher mark carrier will become the lower of the two audio tones. If upper sideband (USB) is used, the mark carrier will remain the higher of the tone pair. It is common to use 2125 Hz as the mark tone and 2295 Hz as a space tone. Since the mark tone (2125 Hz) is lower in frequency, the receiver will be set to LSB. In general, modem tone pairs can be "reversed" (see the Inverted Signals section), so either an LSB or a USB receiver can be used. Moreover, the tone pairs of many modems, especially software modems, can be moved to place them where narrowband filters are available on the receiver.

In the past, audio FSK demodulators were built using high-Q audio filters followed by a "slicer" to determine if the signal from the mark filter is stronger or weaker than the signal that comes out of the space filter. To counter selective fading, where ionospheric propagation can cause the mark carrier to become stronger or weaker than the space carrier, the slicer's threshold can be designed to adapt to the imbalance. Once "sliced" into a bi-level signal, the binary stream is passed to the decoder where start bit detection circuitry determines where to extract the 5-bit character data. That data is then passed to the Baudot decoder, which uses the current LTRS or FIGS state to determine the decoded character. The mark and space transmitted carriers do not overlap, although this can occur after they pass through certain HF propagation conditions. Sophisticated demodulators can account for this distortion.

A software modem performs the same functions as a hardware terminal unit, except that the software modems can apply more sophisticated mathematics that would be too expensive to implement in hardware. Modern desktop computers have more than enough processing speed to implement an RTTY demodulator. Software modems first convert the audio signal into a sequence of binary numbers using an analog-to-digital converter which is usually part of an audio chip set either on the motherboard or sound card. Everything from that point on uses numerical algorithms to implement the demodulation processes.

FSK VS AFSK MODULATION

An RTTY signal is usually generated as an F1B or F2B emission. F1B is implemented by directly shifting an RF carrier between the two (mark and space) frequencies. This method of generating FSK is often called *direct FSK*, or *true FSK*, or simply FSK. F2B is implemented by shifting between two audio tones, instead of two RF carriers. The resultant audio is then sent to an SSB transmitter to become two RF carriers. This method of first generating an audio FSK signal and then modulating an SSB transmitter to achieve the same FSK spectrum is usually called AFSK (audio frequency shift keying).

AFSK can be generated by using either an upper sideband (USB) transmitter or a lower sideband (LSB) transmitter. With a USB transmitter, the mark tone must be the *higher* of the two audio tones in the audio FSK signal. The USB modulator will then place the corresponding mark carrier at the higher of the two FSK carrier frequencies. When LSB transmission is used, the mark tone must be the *lower* of the two audio tones in the audio FSK signal. The LSB modulator will then place the corresponding mark carrier at the higher of the two FSK carrier frequencies. As when receiving, the actual audio tones are of no importance. The important part of AFSK is to have the two audio tones separated by 170 Hz, and have the pair properly flipped to match the choice of USB or LSB transmission. (See the **Modulation** chapter for more information on USB and LSB modulation and the relationship between modulating frequency and transmitted signal frequency.)

When using AFSK with older transceivers, it is wise to choose a high tone pair so that harmonic by-products fall outside the passband of the transmitter. Because of this, a popular tone pair is 2125 Hz/2295 Hz. Most transceivers will pass both tones and also have good suppression of the harmonics of the two tones. Not all transceivers that have an FSK input are FSK transmitters. Some transceivers will take the FSK keying input and modulate an internal AFSK generator, which is then used to modulate an SSB transmitter. In this case, the transmitter is really operating as an F2B emitter. This mode of operation is often called "keyed AFSK."

"SPOTTING" AN RTTY SIGNAL

By convention, RTTY signals are identified by the frequency of the mark carrier on the RF spectrum. "Spotting" the suppressed carrier frequency dial of an SSB receiver is useless for someone else unless they also know whether the spotter is using upper or lower sideband and what tone pair the spotter's demodulator is using. The mark and space carriers are the only two constants, so the amateur RTTY standard is to spot the frequency of the mark carrier.

DIDDLE CHARACTERS

In between the stop bit of a preceding character and the start bit of the next character, the RTTY signal stays at the mark frequency. When the RTTY decoder is in this "rest" state, a mark-to-space transition tells a decoder that the start of a new character has arrived. Noise that causes a start bit to be misidentified can cause the RTTY decoder to fall out of sync. After losing sync, the decoder will use subsequent data bits to help it identify the location of the next potential start bit.

Since all mark-to-space transitions are potential locations of the leading edge of a start bit, this can cause multiple characters to be incorrectly decoded until proper synchronization is again achieved. This "character slippage" can be minimized somewhat by not allowing the RTTY signal to rest for longer than its stop bit duration. An idle or *diddle character* (so called because of the sound of the demodulated audio from an idle RTTY signal sending the idle characters) is inserted immediately after a stop bit when the operator is not actively typing. The idle character is a non-printing character from the Baudot set and most often the LTRS character is used. Baudot encodes a LTRS as five bits of all ones making it particularly useful when the decoder is recovering from a misidentified start bit.

An RTTY diddle is also useful when there is selective fading. Good RTTY demodulators counter selective fading by measuring the amplitudes of the mark and space signals and automatically adjusting the decoding threshold when making the decision of whether a mark or a space is being received. If a station does not transmit diddles and has been idle for a period of time, the receiver will have no idea if selective fading has affected the space frequency. By transmitting a diddle, the RTTY demodulator is ensured of a measurement of the strength of the space carrier during each character period.

UNSHIFT-ON-SPACE (USOS)

Since the Baudot code aliases characters (for example, Q is encoded to the same 5-bit code as 1) using the LTRS and FIGS Baudot shift to steer the decoder, decoding could turn into gibberish if the Baudot shift characters are altered by noise. For this reason, many amateurs use a protocol called *unshift-on-space* (USOS). Under this protocol, both the sender and the receiver agree that the Baudot character set is always shifted to the LTRS state after a space character is received. In a stream of text that includes space characters, this provides additional, implicit, Baudot shifts.

Not everyone uses USOS. When used with messages that have mostly numbers and spaces, the use of USOS causes extra FIGS characters to be sent. A decoder that complies with USOS will not properly decode an RTTY stream that does not have USOS set. Likewise, a decoder that has USOS turned off will not properly decode an RTTY stream that has USOS turned on.

OTHER FSK SHIFTS AND RTTY BAUD RATE

The most commonly used FSK shift in amateur RTTY is 170 Hz. However, on rare occasions stations can be found using 425 and 850 Hz shifts. The wider FSK shifts are especially useful in the presence of selective fading since they provide better frequency diversity than 170 Hz.

Because HF packet radio uses 200 Hz shifts, some TNCs use 200 Hz as the FSK shift for RTTY. Although they are mostly compatible with the 170 Hz shift protocol, under poor signal to noise ratio conditions these demodulators will produce more error hits than a demodulator that is designed for 170 Hz shift. Likewise, a signal that is transmitted using 170 Hz shift will not be optimally copied by a demodulator that is designed for a 200 Hz shift.

To conserve spectrum space, amateurs have experimented with narrower FSK shifts, down to 22.5 Hz. At 22.5 Hz, optimal demodulators are designed as *minimal shift keyed* (*MSK*) instead of frequency shift keyed (FSK) demodulators.

SOME PRACTICAL CHARACTERISTICS

When the demodulator is properly implemented, RTTY can be very resilient against certain HF fading conditions, namely when selective fading causes only one of the two FSK carriers to fade while the other carrier remains strong. However, RTTY is still susceptible to "flat fading" (where the mark and space channels both fade at the same instant). There is neither an error correction scheme nor an interleaver (a method of rearranging — interleaving — the distribution of bits to make errors easier to correct) that can make an RTTY decoder print through a flat and deep fade. The lack of a data interleaver, however, also makes RTTY a very interactive mode. There is practically no latency when compared to a mode such as MFSK16, where the interleaver causes a latency of over 120 bit durations before incoming data can even be decoded. This makes RTTY attractive to operating styles that have short exchanges with rapid turnarounds, such as in contests.

Although RTTY is not as "sensitive" as PSK31 (when there is no multipath, PSK31 has a lower error rate than RTTY when the same amount of power is used) it is not affected by phase distortion that can render even a strong PSK31 signal from being copied. When HF propagation conditions deteriorate, RTTY can often function as long as sufficient power is used. Tuning is also moderately uncritical with RTTY. When the signal-to-noise ratio is good, RTTY tuning can be off by 50 Hz and still print well.

16.2.2 PSK31

PSK31 is a family of modes that uses differentially encoded varicode (see the next section), envelope-shaped phase shift keying. BPSK31, or binary PSK31, operates at 31.25 bit/s (one bit every 32 ms). QPSK31, or quadrature PSK31, operates at 31.25 baud. Each symbol consists of four possible quadrature phase change (or *dibits*) at the signaling rate of one dibit every 32 ms. QPSK31 sends a phase change symbol of 0°, 90°, 180° or 270° every 32 ms.

Characters that are typed from the keyboard are first encoded into variable-length varicode binary digits. With BPSK31, the varicode bits directly modulate the PSK31 modulator, causing a 180° phase change if the varicode bit is a 0 and keeping a constant phase if the varicode bit is a 1. With QPSK31, the varicode bits first pass through two convolution encoders to create a sequence of bit pairs (dibits). Each dibit is then used to shift the QPSK31 modulator into one of four different phase changes.

PSK63 is a double-clock-rate version of a PSK31 signal, operating at the rate of one symbol every 16 ms (62.5 symbols per second). PSK125 is a PSK31 signal clocked at four times the rate, with one symbol every 8 ms (125 symbols per second). Although PSK63 and PSK125 are both in use, including the binary and quadrature forms, the most popular PSK31 variant remains BPSK31.

Most implementations of PSK31 first generate an audio PSK31 signal. The audio signal is then used to modulate an SSB transmitter. Since BPSK31 is based upon phase reversals, it can be used with either upper sideband (USB) or lower sideband (LSB) systems. With QPSK31 however, the 90° and 270° phase shifts have to be swapped when using LSB transmitters or receivers.

PSK31 VARICODE

Characters that are sent in PSK31 are encoded as varicode, a variable length prefix code as varicode. As described earlier, characters that occur more frequently in English text, such as spaces and the lower-case e, are encoded into a fewer number of bits than characters that are less frequent in English, such as the character q and the upper case E.

PSK31 varicode characters always end with two bits of zeros. A space character is sent as a one followed by the two zeros (100); a lower case e is sent as two ones, again terminated by the two bits of zero (1100). None of the varicode code words contain two consecutive zero bits. Because of this, the PSK31 decoder can uniquely identify the boundary between characters. A special "character" in PSK31 is the idle code, which consists of nothing but the two prefix bits. A long pause at the keyboard is encoded into a string of even numbers of zeros. The start of a new character is signaled by the first non-zero bit received after at least two consecutive zeros.

CONVOLUTION CODE

As described earlier, QPSK31 encodes the varicode stream with two convolution encoders to form the dibits that are used to modulate the QPSK31 generator. Both convolution encoders are fourth-order polynomials, shown in **Fig 16.2**.

The varicode data is first inverted before the bits are given to the convolution encoders. The first polynomial generates the most significant bit and the second polynomial generates the least significant bit of the dibit. Since we are working with binary numbers, the GF(2) sums shown in the above figure are exclusive-OR functions and the delay elements x1, x2, x3 and x4 are stages of binary shift registers. As each inverted varicode bit is available, it is clocked into the shift register and the two bits of the dibit are computed from the shift register taps.

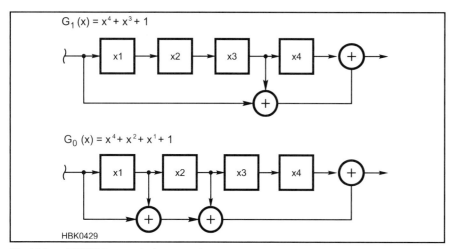

Fig 16.2 — QPSK31 convolution encoders.

Table 16.4
QPSK31 Modulation

Dibit	Phase Change
00	0°
01	90°
10	180°
11	270°

MODULATION

PSK31 uses both differential phase shift modulation and envelope modulation to maintain its narrow-band characteristics. The most common way to generate an envelope-shaped PSK31 signal is to start with baseband in-phase (I) and quadrature (Q) signals. Both I and Q signals settle at either a value of +1 or a value of −1. When no phase transition is needed between symbols, the I and Q signals remain constant (at their original +1 or −1 values). See the **Modulation** chapter for information on creating I and Q signals.

To encode a 180° transition between symbols, both I and Q signals are slewed with a cosinusoidal envelope. If the in-phase signal had a value of +1 during the previous symbol, it is slewed to −1 cosinusoidally. If the in-phase signal had a value of −1 in the previous symbols, it is slewed cosinusoidally to +1. The quadrature signal behaves in a likewise manner. To encode a 90° phase shift in QPSK31, only the in-phase signal is slewed between the two symbols, the quadrature signal remains constant between the two symbols. To encode a 270° phase shift in QPSK31, only the quadrature signal is slewed between the two symbols; the in-phase signal remains constant between the two symbols.

The envelope of the real signal remains constant if there is no phase change. When the signal makes a 180° phase transition, the amplitude of a PSK31 signal will drop to zero in between the two symbols. The actual phase reversal occurs when the signal amplitude is zero, when there is no signal energy. During the 90° and 270° phase shifts between symbols of QPSK31, the amplitude of the signal does not reach zero. It dips to only half the peak power in between the two symbols.

To provide a changing envelope when the operator is idle at the keyboard, a zero in the varicode (remember that an idle varicode consists of two 0 bits) is encoded as a 180° phase change between two BPSK31 symbols. A 1 in the varicode is encoded as no phase change from one BPSK31 symbol to the next symbol. This changing envelope allows the receiver to extract bit timing information even when the sender is idle. Bit clock recovery is implemented by using a comb filter on the envelope of the PSK31 signal.

The convolution code that is used by QPSK31 converts a constant idle stream of zeros into a stream of repeated 10 dibits. To produce the same constant phase change during idle periods as BPSK31, the QPSK31 10 dibit is chosen to represent the 180° phase shift modulating term. **Table 16.4** shows the QPSK31 modulation.

To produce bit clocks during idle, combined with the particular convolution code that was chosen for QPSK31, results in a slightly suboptimal (non-Gray code) encoding of the four dibits. At the end of a transmission, PSK31 stops all modulation for a short period and transmits a short unmodulated carrier. The intended function is as a squelch mechanism.

DEMODULATION AND DECODING

Many techniques are available to decode differential PSK, where a reference phase is not present. Okunev has a good presentation of the methods (reference: Yuri Okunev, *Phase and Phase-difference Modulation in Digital Communications* (1997, Artech House), ISBN 0-89006-937-9).

As mentioned earlier, there is sufficient amplitude information to extract the bit clock from a PSK31 signal. The output of a differential-phase demodulator is an estimate of the phase angle difference between the centers of one symbol and the previous one. With BPSK31, the output can be compared to a threshold to determine if a phase reversal or a non-reversal is more likely. The decoder then looks for two phase reversals in a row, followed by a non-reversal, to determine the beginning of a new character. The bits are gathered until two phase reversals are again seen and the accumulated bits are decoded into one of the characters in the varicode table.

QPSK31 decoding is more involved. As in the BPSK31 case, the phase difference demodulator estimates the phase change from one bit to another. However, one cannot simply invert the convolution function to derive the data dibits. Various techniques exist to decode the measured phase angles into dibits. The *Viterbi algorithm* is a relatively simple algorithm for the convolution polynomials used in QPSK31. The estimated phase angles can first be fixed to one of the four quadrature angles before the angles are submitted to the Viterbi algorithm. This is called a hard-decision *Viterbi decoder*.

A soft-decision Viterbi decoder (that is not that much more complex to construct) usually gives better results. The soft-decision decoder uses arbitrary phase angles and some measure of how "far" an angle is away from one of the four quadrature angles. Error correction occurs within the trellis that implements the Viterbi algorithm. References to convolution code, trellis and the Viterbi algorithm can be found at **en.wikipedia.org/wiki/Convolutional_code**.

16.2.3 MFSK16

MFSK16 uses M-ary FSK modulation with 16 "tones" (also known as 16-FSK modulation), where only a single tone is present at any instant. MFSK16 has a crest factor of 1, with no wave shaping performed on the data bits. The tone centers of MFSK16 are separated by 15.625 Hz. Data switches at a rate of 15.625 baud (one symbol change every 64 ms).

Characters are first encoded into the MFSK16 varicode, creating a constant bit stream whose rate is one bit every 32 ms. This results in a similar, although not identical, character rate as PSK31. The difference is due to the different varicode tables that are used by the two modes.

This bit stream is clocked into a pair of 6th-order convolution encoders. The result is a pair of bits (dibit) for each varicode bit, at a rate of one dibit every 32 ms. Consecutive pairs of dibits are next combined into sets of four bits (*quadbit* or *nibble*). Each nibble, at the rate of one per 64 ms, is then passed through an interleaver. The interleaver produces one nibble for each nibble that is sent to it. Each nibble from the interleaver is then Gray-coded and the resultant 4-bit Gray code is the ordered index of each 16 tones that are separated by 15.625 Hz. The result is a 16-FSK signal with a rate of one symbol per

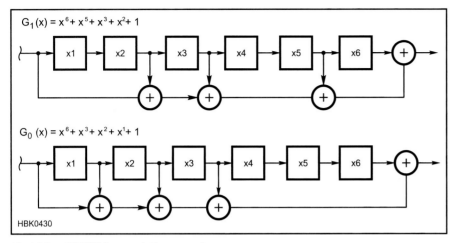

Fig 16.3 — MFSK16 convolution encoders.

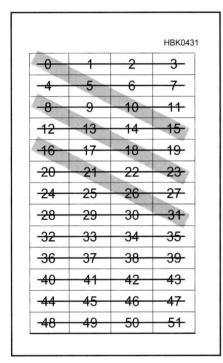

Fig 16.4 — IZ8BLY interleaver.

64 ms. Since the symbol time is the reciprocal of the tone separation, a 16-point fast Fourier transform (FFT) can be conveniently used as an MFSK16 modulator.

MFSK16 VARICODE

Although the varicode table that is used by MFSK16 (see the *Handbook* CD) is not the same as the one used by PSK31, they share similar characteristics. Please refer to the PSK31 section of this chapter for a more detailed description of varicode. Unlike PSK31 varicode, MSK16 varicode encodings can contain two or more consecutive zero bits, as long as the consecutive zeros are at the tail of a code word. Character boundaries are determined when two or more consecutive zeros are followed by a one.

CONVOLUTION CODE

As described earlier, MFSK16 encodes the varicode stream with two convolution encoders to form the dibits that are passed on to the interleaver. Both convolution encoders are sixth order polynomial, shown in **Fig 16.3**. The first polynomial generates the most significant bit and the second polynomial generates the least significant bit of the dibit. Since we are working with binary numbers, the GF(2) sums shown in the above figure are exclusive OR functions and the delay elements x1, x2, x3, x4, x5 and x6 are stages of binary shift registers. As each varicode bit is available, it is clocked into the shift register and the two bits of the dibit are computed from the shift register taps.

INTERLEAVER

HF fading channels tend to generate "burst" errors that are clumped together. An interleaver is used to permute the errors temporally so that they appear as uncorrelated errors to the convolution decoder. MFSK16 uses 10 concatenated stages of an IZ8BLY Diagonal Interleaver. More information on the interleaver can be found at **www.qsl.net/zl1bpu/MFSK/Interleaver.htm**. **Fig 16.4** illustrates how a single IZ8BLY interleaver spreads a sequence of bits.

The bits enter the IZ8BLY interleaver in the order 0, 1, 2, 3, 4,... and are passed to the output in the order 0, 5, 10, 15, 8, 13, ... (shown in the diagonal boxes). In MFSK16, the output of one interleaver is sent to the input of a second interleaver, for a total of 10 such stages. Each stage spreads the bits out over a longer time frame.

This concatenated 10-stage interpolator is equivalent to a single interpolator that is 123 bits long. **Fig 16.5** shows the structure of the single interpolator and demonstrates how four consecutive input bits are spread evenly over 123 time periods.

An error burst can be seen to be spread over a duration of almost two seconds. This gives MFSK16 the capability to correct errors over deep and long fades. While it is good for correcting errors, the delay through the long interleaver also causes a long decoding latency.

GRAY CODE

The Gray code creates a condition where

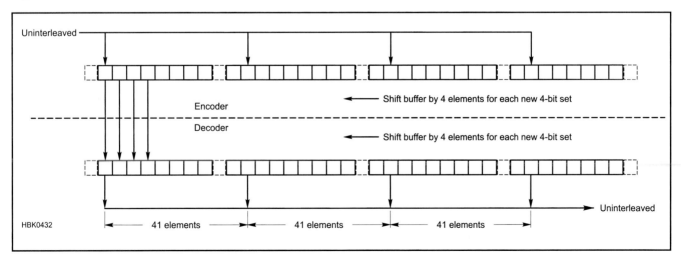

Fig 16.5 — Bit spreading through interpolation.

tones that are next to one another are also different by a smaller *Hamming* distance. This optimizes the error correction process at the receiver. References to Gray code and Hamming distance can be found at **en.wikipedia.org/wiki/Gray_code** and **en.wikipedia.org/wiki/Hamming_distance**.

DEMODULATION AND DECODING

A 16-point FFT can be used to implement a set of matched filters for demodulating an MFSK16 signal once the input waveform is properly time aligned so that each transform is performed on an integral symbol and the signal is tuned so that the MFSK16 tones are perfectly centered in the FFT bins. A reference to a matched filter can be found at **en.wikipedia.org/wiki/Matched_filter**.

The 16 output bins from an FFT demodulator can first be converted to the "best" 4-bit index of an FFT frequency bin, or they can be converted to a vector of four numerical values representing four "soft" bits. The four bits are then passed through an MFSK16 de-interleaver. In the case of "soft decoding," the de-interleaver would contain numerical values rather than a 0 or 1 bit.

The output of the de-interleaver is passed into a convolution decoder. The Gray code makes sure that adjacent FFT bins also have the lowest Hamming distance; i.e., the most likely error is also associated with the closest FFT bin. The hard or the soft Viterbi Algorithm can be used to decode and correct errors.

16.2.4 DominoEX

DominoEX is a digital mode with MFSK (multi-frequency shift keying) designed for simplex chats on HF by Murray Greenman, ZL1BPU. It was designed to be easier to use and tune than other similar modes, offer low latency for contesting or other quick exchange situations, offer reliable copy down into the noise floor, and work well as an NVIS (near-vertical incidence skywave, see the **Propagation of Radio Signals** chapter) mode for emergency communications.

Generally MFSK requires a high degree of tuning accuracy and frequency stability and can be susceptible to multipath distortion. DominoEX specifically addresses these issues. To avoid tuning issues, IFK (*incremental frequency keying*) is used. With IFK, the data is represented not by the frequency of each tone, but by the frequency difference between one tone and the next. It also uses offset incremental keying to reduce inter-symbol interference caused by multipath reception. These techniques provide a tuning tolerance of 200 Hz and a drift of 200 Hz/minute. DominoEX also features an optional FEC mode that increases latency but provides communications over even more difficult channels. More information can be found online at **www.qsl.net/zl1bpu/MFSK/DEX.htm**.

DominoEX uses M-ary FSK modulation with 18 tones in which only a single tone is present at any instant. Information is sent as a sequence of separate 4-bit ("nibble") symbols. The value of each nibble is represented during transmission as the position of the single tone.

The position of the tone is computed as the difference of the current nibble value from the nibble value of the previously transmitted symbol. In addition, a constant offset of 2 is applied to this difference. Because there are 18 tones, any possible 4-bit value between 0 and 15 can be represented, including the offset of 2.

The additional offset of 2 tone positions ensures that a transmitted tone is separated from the previously transmitted tone by at least two tone positions. It is thus impossible for two sequential symbols to result in the same tone being transmitted for two sequential tone periods. This means sequential tones will always be different by at least two positions, an important consideration in maintaining sync.

This minimum separation of successive tones of incremental frequency keying (IFK) in DominoEX reduces the inter-symbol distortion that results from a pulse being temporally smeared when passing through an HF channel. The double-tone spacing of DominoEX modes (see **Table 16.5**) further reduces inter-symbol distortion caused by frequency smearing.

Incremental frequency keying allows the DominoEX nibbles to immediately decode without having to wait for the absolute tone to be identified. With MFSK16, a tone cannot be uniquely identified until the lowest and highest of the 16 tones have passed through the receiver. This contributes to the decoding latency of MFSK16. There is no such latency with IFK.

Since IFK depends upon frequency differences and not absolute tone frequencies, DominoEX tolerates tuning errors and drifting signals without requiring any additional automatic frequency tracking algorithms.

Like MFSK16, the DominoEX signal is not wave-shaped and has constant output power. The baud rates for DominoEX are shown in Table 16.5. The tone spacings for DominoEX 11, DominoEX 16 and DominoEX 16 have the same values as their baud rates. The tone spacings for DominoEX 8, DominoEX 5 and DominoEX 4 are twice the value of their baud rates.

Unlike PSK31 and MFSK16, characters in DominoEX are encoded into varicode nibbles instead of encoding into varicode bits. The DominoEX varicode table can be found in the CD-ROM included with this book.

BEACON MESSAGE

Instead of transmitting an idle varicode symbol when there is no keyboard activity, DominoEX transmits a "beacon" message from an alternate set of varicode (SECVAR columns in the DominoEX varicode table). This user-supplied repeating beacon message is displayed at the receiving station when the sending station is not actively sending the primary message. On average, the character rate of the beacon channel is about half of the character rate of the primary channel.

FORWARD ERROR CORRECTION (FEC)

When FEC is turned off, DominoEX has very low decoding latency, providing an interactive quality that approaches RTTY and PSK31. The first character is decoded virtually instantly by the receiver after it is transmitted. Because of that, FEC is not usually used even though it is available in most software that implements DominoEX.

DominoEX FEC is similar to the FEC that is used in MFSK16. When FEC is on, each 4-bit IFK symbol that is decoded by the receiver is split into two di-bits. The dibits enter an identical convolution coder to that used by MFSK16. However, instead of a 10-stage IZ8BLY interleaver (see Fig 16.4), only 4 cascaded stages of the basic 4-bit interleaver are present in DominoEX. In the presence of long duration fading, the performance of the shortened interleaver is moderately poor when used with DominoEX 16 and DominoEX 22. However, the interleaver is quite efficient in countering fading when FEC is used with DominoEX 4, DominoEX 5 and DominoEX 8, with their longer symbol periods.

Since DominoEX works in dibit units rather than nibble units when FEC is turned on, it also switches to using the same binary varicode used by MFSK16 instead of using the nibble-based varicode. DominoEX does not implement Gray code as is used by MFSK16 FEC.

Table 16.5
Comparison of DominoEX Modes

Mode	Baud/(sec)	BW (Hz)	Tones	Speed (WPM)	FEC (WPM)	Tone Spacing
DominoEX 4	3.90625	173	18	~25	~12	Baud rate ×2
DominoEX 5	5.3833	244	18	~31	~16	Baud rate ×2
DominoEX 8	7.8125	346	18	~50	~25	Baud rate ×2
DominoEX 11	10.766	262	18	~70	~35	Baud rate ×1
DominoEX 16	15.625	355	18	~100	~50	Baud rate ×1
DominoEX 22	21.533	524	18	~140	~70	Baud rate ×1

Even without FEC, DominoEX works well under many HF propagation conditions, including the ITU NVIS ("Mid-latitude Disturbed") propagation profile. However, there are conditions where DominoEX is not usable unless FEC is switched on, specifically the CCIR Flutter and ITU High Latitude Moderate Conditions profiles. DominoEX modes, especially those with tone spacings that are twice the baud rate, are very robust even under these extreme conditions once FEC is switched on.

DominoEX performance charts (character error rates versus signal-to-noise ratios) are included as the HTML document "DominoEX Performance" in the CD-ROM accompanying this book and on-line at homepage. mac.com/chen/Technical/DominoEX/Measurements/index.html.

CHIP64/128

Chip64 and Chip128 modes were released in 2004 by Antonino Porcino, IZ8BLY. The modes were tested on the air by IZ8BLY, Murray Greenman, ZL1BPU, (who also contributed in the design of the system), Chris Gerber, HB9BDM, and Manfred Salzwedel, OH/DK4ZC. According to IZ8BLY, "The design of this new digital mode served to introduce the spread spectrum technology among radio amateurs by providing a communication tool to experiment with. Its purpose was to prove that it's possible to take advantage of the spread spectrum techniques even on the HF channels, making the communication possible under conditions where traditional narrowband modes fail.

"Among the different possible implementations of spread spectrum, Chip64 uses the so-called Direct Sequence Spread Spectrum (DSSS). In a DSSS transmission, the low speed signal containing the data bits to be transmitted is mixed (multiplied) with a greatly higher speed signal called code. The result of this mixing operation, called s*preading*, is a high-speed bit stream which is then transmitted as a normal DBPSK. Indeed, a DSSS signal looks like nothing else than wideband BPSK.

"The system proved to be efficient and we found it comparable to the other modern digital modes. Being totally different in its architecture, it shows better performance during certain circumstances, while in others it shows no actual gain. In particular, it performs better under multipath where normal BPSK can't track arriving symbols, but in quiet environments it doesn't show any improvement over plain BPSK. This is expected because of the losses that occur due to the imperfect autocorrelation of the codes.

Chip64 has a total data rate of 37.5 bit/s and the more robust Chip128 is 21.09 bit/s. Both use the same varicode used by MFSK16. The software can be downloaded from **xoomer. virgilio.it/aporcino/Chip64/index.htm** and more information is available at **www.arrl. org/technical-characteristics**. Spread spectrum is discussed in the **Modulation** chapter, as well.

16.2.5 THROB

Throb is an experimental mode written by Lionel Sear, G3PPT, and gets the name from the "throbbing" sound it makes on the air. It uses either single tones or pairs from a possible nine tones spaced 8 or 16 Hz apart, resulting in a bandwidth of 72 or 144 Hz, respectively. It has three transmission speeds — 1, 2 and 4 throbs/s — resulting in data rates of 10, 20 and 40 WPM, respectively. The 1 and 2 throb/s speeds use a tone spacing of 8 Hz for a 72 Hz bandwidth and the 4-throb/s speed uses a spacing of 16 Hz for a 144 Hz bandwidth. It is implemented as a standalone application or included in a multimode package such as *MixW* (**www.mixw.net**).

16.2.6 MT63

MT63 is a mode developed by Pawel Jalocha, SP9VRC. MT63 is very complex with wide bandwidth, low speed and very high noise immunity. By using 64 different modulated tones, MT63 includes a large amount of extra data in the transmission of each character, so that the receiving equipment can work out, with no ambiguity, which character was sent, even if 25% of the character is obliterated. MT63 also features a secondary channel that operates simultaneously with the main channel that can be used for an ID or beacon.

MT63 likely has the most extensive error correction and can be quite processor intensive. It uses a Walsh function that spreads the data bits of each 7-bit ASCII character across all 64 of the tones of the signal spectrum and simultaneously repeats the information over a period of 64 symbols within any one tone. This coding takes several seconds. The combination of time domain (temporal) and frequency domain (spectral) interleaving results in superb impulse noise rejection. At the same time, in the frequency domain, significant portions of the signal can be masked by unwanted noise or other transmissions without any noticeable effect on successful reception.

On each of the 64 tones, the transmission data rate is fairly slow, which suits the nature of ionospheric disturbances. Despite the low data rate, good text speed is maintained because the text is sent on many tones at once. The system runs at several different speeds, which can be chosen to suit conditions but 100 WPM is typical of the MT63-1K mode. Although the 1 kHz bandwidth mode is typical, MT63 can also run at 500 Hz and 2 kHz bandwidth where the tone spacing and baud rate are halved or doubled and the throughput is halved or doubled, respectively.

Tuning of MT63 modes is not critical. This is because the mode can use FEC techniques to examine different combinations of the 64 tones that calculate the correct location within the spectrum. As an example, MT63-1K will still work if the decoder is off-frequency by as much as 100 Hz. MT63-2K requires even less precision and can tolerate an error of 250 Hz.

The incredible noise immunity comes at a price beyond the large bandwidth required. There is a large latency caused by the error correction and interleaving process. Quick-turnaround QSOs are not possible because there is a several second delay between typing the last character and it being transmitted.

Without confirming each transmission with some type of ARQ mode, there is no more robust digital mode than MT63. The mode was evaluated and recommended for Navy MARS message handling. The evaluation is published on the Navy MARS Web site (**www.navymars.org**), along with other information on this mode.

16.2.7 Olivia

Olivia is an MFSK-based protocol designed to work in difficult (low signal-to-noise ratio plus multipath propagation) conditions on the HF bands. The signal can still be copied accurately at 10 dB below the noise floor. Olivia was developed in 2003 by Pawel Jalocha, SP9VRC, and performs well for digital data transfer with white noise, fading and multipath, polar path flutter and auroral conditions.

Olivia transmits a stream of 7-bit ASCII characters. The characters are sent in blocks of five with each block requiring two seconds to transmit. This results in an effective data rate of 2.5 characters/second or 150 characters/minute. A transmission bandwidth of 1000 Hz and the baud rate of 31.25 MFSK tones/second, also known as *Olivia 1000/32*, is the most common. To adapt to different propagation conditions, the number of tones and the bandwidth can be changed and the time and frequency parameters are proportionally scaled. The number of tones can be 2, 4, 8, 16, 32, 64, 128 or 256 and the bandwidth can be 125, 250, 500, 1000 or 2000 Hz.

The Olivia is constructed of two layers: the lower, modulation and FEC code layer is a classical MFSK while the higher layer is an FEC code based on Walsh functions. More detail on Walsh functions is available online at **en.wikipedia.org/wiki/Walsh_function**.

Assuming Olivia 1000/32 is being used, in the first layer the orthogonal functions are cosine functions, with 32 different tones. Since only one of those 32 tones is being sent at a time, the demodulator measures the amplitudes of all the 32 possible tones and identifies the tone with the highest amplitude. In the second layer every ASCII character is encoded as one of 64 possible Walsh functions. The receiver again measures the amplitudes for all 64 vectors and selects the greatest as the true value.

To avoid simple transmitted patterns (like a constant tone) and to minimize the chance for a false lock at the synchronizer, the characters encoded into the Walsh function pass through a scrambler and interleaver. The receiver synchronizes automatically by searching through time and frequency offsets for a matching pattern.

More information can be found online at **n1su.com/olivia** and Olivia is supported in a number of digital multimode packages such as *MixW*, *MultiPSK* and *Ham Radio Deluxe*.

16.3 Fuzzy Modes

There is a group of modes referred to as "fuzzy modes" because although they are machine generated and decoded, they are designed to be human-read. These include facsimile (fax), slow-scan TV (SSTV) and Hellschreiber.

16.3.1 Facsimile (fax)

Facsimile was developed as a mechanically transmitted technology where the source material was placed on a spinning drum and scanned line by line into an electrical signal which would be transmitted by wire or over the air. It is important that the receiving station have their drum spinning at the correct speed in order to correctly recreate the image. A value known as the *index of cooperation* (*IOC*) must also be known to decode a transmission. IOC governs the image resolution and is the product of the total line length and the number of lines per unit length divided by π. Most fax transmissions are sent with LPM (RPM) at 120 and an IOC of 576.

Facsimile is generally transmitted in single sideband with a tone of 1500 Hz representing black and 2300 Hz representing white. The *automatic picture transmission* (*APT*) format is used by most terrestrial weather facsimile stations and geostationary weather satellites. It features a start tone that triggers the receiving system, originally used to allow the receiving drum to come up to speed. It also includes a phasing signal with a periodic pulse that synchronizes the receiver so the image appears centered on the page. A stop tone, optionally followed by black, indicates the end of the transmission. The APT format is shown in **Table 16.6**.

Stations with Russian equipment sometimes use RPM 60 or 90 and sometimes an IOC of 288. Photofax transmissions such as those from North Korea use RPM 60 and an IOC 352 with gray tones, and satellite rebroadcast use also RPM 120 IOC 576, with gray tones (4 or more bit depth). For software decoding of weather fax images it is best to decode with Black and White (2-bit depth).

16.3.2 Slow-Scan TV (SSTV)

Slow-Scan TV or SSTV is similar to facsimile where a single image is converted to individual scanned lines and those lines sent as variable tones between 1500 and 2300 Hz. Modern systems use computer software and a sound card to generate and receive the required tones. (Some SSTV communication uses purely digital protocols, such as WinDRM described later, in which the picture content is sent as digital data and not directly represented in the modulation scheme.)

There are a number of different SSTV "modes" that define image resolution and color scheme. A color image takes about 2 minutes to transmit, depending on mode. Some black and white modes can transmit an image in under 10 seconds. More information about SSTV may be found in the **Image Communications** chapter.

16.3.3 Hellschreiber, Feld-Hell or Hell

Hellschreiber is a facsimile-based mode developed by Rudolph Hell in the 1920s. The name is German and means "bright writer" or "light writer" and is a pun on the inventor's name. In Hellschreiber, text is transmitted by dividing each column into seven pixels and transmitting them sequentially starting at the lowest pixel. Black pixels are transmitted as a signal and white as silence at 122.5 bit/s (about a 35 WPM text rate). Originally the text was printed on continuous rolls of paper so the message could be any length.

Even though each pixel is only transmitted once, they are printed twice, one below the other. This compensates for slight timing errors in the equipment that causes the text to slant. If properly in sync, the text will appear as two identical rows, one below the other or a line of text in the middle with chopped lines top and bottom. Regardless of the slant, it is always possible to read one copy of the text. Since the text is read visually, it can be sent in nearly any language and tends to look like an old dot matrix printer. More information can be found online at **www.qsl.net/zl1bpu/FUZZY/Feld.htm** and Randy, K7AGE, has a great introduction to Hellschreiber available on YouTube at **www.youtube.com/watch?v=yR-EmyEBVqA**.

Table 16.6
Facsimile Automatic Picture Format

Signal	Duration	IOC576	IOC288	Remarks
Start Tone	5 s	300 Hz	675 Hz	200Hz for color fax modes
Phasing Signal	30 s			White line interrupted by black pulse
Image	Variable	1200 lines	600 lines	At 120 LPM
Stop Tone	5 s	450 Hz	450 Hz	
Black	10 s			

16.4 Structured Digital Modes

This group of digital modes has more structured data. This provides more robust data connections and better weak signal performance or more sophisticated data. Each of these modes bundles data into packets or blocks that can be transmitted and error checked at the receive end.

16.4.1 FSK441

FSK441 was implemented in the *WSJT* software by Joe Taylor, K1JT, in 2001. FSK441 was designed for meteor scatter using four-tone frequency shift keying at a rate of 441 baud and was given the technical name FSK441. Meteor scatter presents an unusual problem because the signals are reflected by the ionization trails left by meteors that often last less than a second. The signals are weak, of short duration with widely varying signal strength and include some amount of Doppler shift. Since these meteors enter the atmosphere constantly, this mode can be used any time of day or year to make contacts on VHF between 500 and 1100 miles with a single large Yagi and 100 W.

FSK441 uses a 43-character alphabet compatible with the earlier PUA43 mode by Bob Larkin, W7PUA. This alphabet includes letters, numbers and six special characters and is shown in the **Table 16.7**. These characters are encoded by the four tones at 882, 1323, 1764 and 2205 Hz and are designated 0-3 in the table below. For example, the letter T has the code 210 and is transmitted by sending sequentially the tones at 1764, 1323 and 882 Hz. Since the modulation rate is specified as 441 baud, or 441 bit/s, the character transmission rate is 441/3 = 147 characters per second. At this speed, a ping lasting 0.1 s can convey 15 characters of text. Four signals (000, 111, 222 and 333) generate single solid tones that are easily decoded and are reserved for the standard meteor scatter messages — R26, R27, RRR and 73.

When receiving FSK441, the receiver time and frequency need to be synchronized precisely. The computer clock must be set as accurately as possible (ideally to WWV or other time standard) before attempting to transmit or receive. When the receiver hears a burst of signal the decoder attempts to identify the correct frequency shift and align the message. For reasons of transmission efficiency, no special synchronizing information is embedded in an FSK441 message. Instead, the proper synchronization is established from the message content itself, making use of the facts that (a) three-bit sequences starting with 3 are never used, and (b) the space character is coded as 033, as shown in FSK441 character code table. Messages sent by *WSJT* always contain at least one trailing space and the software will insert one if necessary.

Timed message sequences are a must, and according to the procedures used by common consent in North America, the westernmost station transmits first in each minute for 30 seconds, followed by 30 seconds of receiving. *WSJT* messages are also limited to 28 characters which are plenty to send call sign and signal report information but clearly does not make *WSJT* a good ragchew mode.

The meteor scatter procedures are also well-defined and supported directly by templates in the *WSJT* software. Signal reports are conventionally sent as two-digit numbers. The first digit characterizes the lengths of pings being received, on a 1-5 scale, and the second estimates their strength on a 6-9 scale. The most common signal reports are 26 for weak pings and 27 for stronger ones, but under good conditions reports such as 38 and higher are sometimes used. Whatever signal report is sent to the QSO partner, it is important that it not be changed. You never know when pings will successfully convey fragments of your message to the other end of the path, and you want the received information to be consistent.

More information can be found at the *WSJT* Web site in the user manual and other articles at **pulsar.princeton.edu/~joe/K1JT**.

16.4.2 JT6M

JT6M is another *WSJT* mode optimized for meteor and ionospheric scatter on 6 meters. Like FSK441, JT6M uses 30 second periods for transmission and reception to look for enhancements produced by short-lived meteor trail ionization. It will also account for Doppler shift of ± 400 Hz. JT6M includes some improvements that allow working signals many dB weaker than FSK441.

JT6M uses 44-tone FSK with a synchronizing tone and 43 possible data tones — one for each character in the supported alphanumeric set, the same set used for FSK441. The sync tone is at 1076.66 Hz, and the 43 other possible tones are spaced at intervals of 11025/512 = 21.53 Hz up to 2002.59 Hz. Transmitted symbols are keyed at a rate of 21.53 baud, so each one lasts for 1/21.53 = 0.04644 s. Every 3rd symbol is the sync tone, and each sync symbol is followed by two data symbols. The transmission rate of user data is therefore (2/3)×21.53 = 14.4 characters per second. The transmitted signal sounds a bit like piccolo music.

Contacts in FSK441 and JT6M are often scheduled online on Ping Jockey at **www.pingjockey.net**.

16.4.3 JT65

JT65 is another *WSJT*-supported mode designed to optimize Earth-Moon-Earth (EME) contacts on the VHF bands, and conforms efficiently to the established standards and procedures for such QSOs. It also performs well for weak signal VHF/UHF, and for HF skywave propagation. JT65 includes error-correcting features that make it very robust, even with signals much too weak to be heard. With extended transmission duration and three decoders, JT65 can reliably decode 24 to 30 dB *below* the noise floor.

JT65 does not transmit messages character by character, as done in Morse code. Instead, whole messages are translated into unique strings of 72 bits, and from those into sequences of 63 six-bit symbols. These symbols are transmitted over a radio channel; some of them may arrive intact, while others are corrupted by noise. If enough of the symbols are correct, the full 72-bit compressed message can be recovered exactly. The decoded bits are then translated back into the human-readable message that was sent. The coding scheme and robust FEC assure that messages are never received in fragments. Message components cannot be mistaken for one another, and call signs are never displayed with a few characters missing or incorrect. There is no chance for the letter O or R in a call sign to be confused with a signal report or an acknowledgment.

JT65 uses 60-second transmit-receive sequences and carefully structured messages. Standard messages are compressed so that two call signs and a grid locator can be transmitted with just 71 bits. A 72nd bit serves as a flag to indicate that the message consists of arbitrary text (up to 13 characters) instead of call signs and a grid locator. Special formats allow other information such as call sign prefixes (for example, ZA/PA2CHR) or numeri-

Table 16.7
FSK441 Character Codes

Character	Tones	Character	Tones
1	001	H	120
2	002	I	121
3	003	J	122
4	010	K	123
5	011	L	130
6	012	M	131
7	013	N	132
8	020	O	133
9	021	P	200
.	022	Q	201
,	023	R	202
?	030	S	203
/	031	T	210
#	032	U	211
space	033	V	212
$	100	W	213
A	101	X	220
B	102	Y	221
C	103	0	223
D	110	E	230
F	112	Z	231
G	113		

cal signal reports (in dB) to be substituted for the grid locator.

The aim of source encoding is to compress the common messages used for EME QSOs into a minimum fixed number of bits. After being compressed into 72 bits, a JT65 message is augmented with 306 uniquely defined error-correcting bits. The FEC coding rate is such that each message is transmitted with a "redundancy ratio" of 5.25. With a good error-correcting code, however, the resulting performance and sensitivity are far superior to those obtainable with simple five times message repetition. The high level of redundancy means that JT65 copes extremely well with QSB. Signals that are discernible to the software for as little as 10 to 15 seconds in a transmission can still yield perfect copy.

JT65 requires tight synchronization of time and frequency between transmitter and receiver. Each transmission is divided into 126 contiguous time intervals or symbols, each lasting 0.372 s. Within each interval the waveform is a constant-amplitude sinusoid at one of 65 pre-defined frequencies. Half of the channel symbols are devoted to a pseudo-random synchronizing vector interleaved with the encoded information symbols. The sync vector allows calibration of relative time and frequency offsets between transmitter and receiver. A transmission nominally begins at t = 1 s after the start of a UTC minute and finishes at t = 47.8 s.

WSJT does its JT65 decoding in three phases: a soft-decision *Reed-Solomon* decoder, the deep search decoder and the decoder for shorthand messages. In circumstances involving birdies, atmospherics, or other interference, operator interaction is an essential part of the decoding process. The operator can enable a "Zap" function to excise birdies, a "Clip" function to suppress broadband noise spikes and a "Freeze" feature to limit the frequency range searched for a sync tone. Using these aids and the program's graphical and numerical displays appropriately, the operator is well equipped to recognize and discard any spurious output from the decoder.

The JT65 procedures are also well-defined and supported directly by templates in the *WSJT* software. More information can be found at the *WSJT* Web site in the user manual and other articles at **pulsar.princeton. edu/~joe/K1JT**.

16.4.4 WSPR

WSPR (pronounced "whisper") implements a protocol designed for probing potential propagation paths with low-power transmissions. Each transmission carries a station's call sign, Maidenhead grid locator, and transmitter power in dBm. The program can decode signals with S/N as low as −28 dB in a 2500 Hz bandwidth. Stations with Internet access can automatically upload their reception reports to a central database called WSPRnet, which includes a mapping facility.

WSPR implements a protocol similar to JT65 called MEPT_JT (Manned Experimental Propagation Tests, by K1JT). In receive mode the program looks for all detectable MEPT_JT signals in a 200-Hz passband, decodes them, and displays the results. If nothing is decoded, nothing will be printed. In T/R mode the program alternates in a randomized way between transmit and receive sequences. Like JT65, MEPT_JT includes very efficient data compression and strong forward error correction. Received messages are nearly always exactly the same as the transmitted message, or else they are left blank.

Basic specifications of the MEPT_JT mode are as follows:
• Transmitted message: call sign + 4-character-locator + dBm Example: "K1JT FN20 30"
• Message length after lossless compression: 28 bits for call sign, 15 for locator, 7 for power level; 50 bits total.
• Forward error correction (FEC): Long-constraint convolutional code, K =32, r = 1/2.
• Number of channel symbols: nsym = (50 + K – 1) × 2 = 162.
• Keying rate: 12000/8192 = 1.46 baud.
• Modulation: Continuous phase 4-FSK. Tone separation 1.46 Hz.
• Synchronization: 162-bit pseudo-random sync vector.
• Data structure: Each channel symbol conveys one sync bit and one data bit.
• Duration of transmission: (162 × 8192)/12000 = 110.6 s
• Occupied bandwidth: About 6 Hz
• Minimum S/N for reception: Around −27 dB on the *WSJT* scale (2500 Hz reference bandwidth).

The current version and documentation of *WSPR* can be found at **www.physics. princeton.edu/pulsar/K1JT/wspr.html** and WSPRnet propagation data is available at **wsprnet.org**.

16.4.5 HF Digital Voice

AOR

In 2004, AOR Corporation introduced its HF digital voice and data modem, the AR9800. Digital voice offers a quality similar to FM with no background noise or fading as long as the signal can be properly decoded. The AR9800 can alternatively transmit binary files and images. AOR later released the AR9000 which is compatible with the AR9800 but less expensive and only supports the HF digital voice mode.

The AR9800 uses a protocol developed by Charles Brain, G4GUO. The protocol uses the AMBE (Advanced Multi-Band Excitation) codec from DVSI Inc. to carry voice. It uses 2400 bits/s for voice data with an additional 1200 bits/s for Forward Error Correction for a total 3600 bits/s data stream. The protocol is detailed below:
• Bandwidth: 300-2500 Hz, 36 carriers
• Symbol Rate: 20 ms (50 baud)
• Guard interval: 4 ms
• Tone steps: 62.5 Hz
• Modulation method: 36 carriers: DQPSK (3.6K)
• AFC: ±125 Hz
• Error correction: Voice: Golay and Hamming
• Video/Data: Convolution and Reed-Solomon
• Header: 1 s; 3 tones plus BPSK training pattern for synchronization
• Digital voice: DVSI AMBE2020 coder, decoder
• Signal detection: Automatic Digital detect, Automatic switching between analog mode and digital mode
• Video Compression: AOR original adaptive JPEG

AOR has more information online at **www. aorusa.com/ard9800.html** and Amateur Radio Video News featured a number of digital voice modes including the AOR devices available on DVD at **www.arvideonews. com/dv/index.html**.

WinDRM

Digital Radio Mondiale (*DRM*) — not to be confused with Digital Rights Management — is a set of commercial digital audio broadcasting technologies designed to work over the bands currently used for AM broadcasting, particularly shortwave. DRM can fit more channels than AM, at higher quality, into a given amount of bandwidth, using various MPEG-4 codecs optimized for voice or music or both. As a digital modulation, DRM supports data transmission in addition to the audio channels. The modulation used by DRM is COFDM (coded orthogonal frequency division multiplex) with each carrier coded using QAM (quadrature amplitude modulation). The use of multiple carriers in COFDM provides a reasonably robust signal on HF.

Two members of Darmstadt University of Technology in Germany, Volker Fisher and Alexander Kurpiers, developed an open-source software program to decode commercial DRM called *Dream*. Typically, the DRM commercial shortwave broadcasts use 5 kHz, 10 kHz or 20 kHz of bandwidth (monaural or stereo) and require large bandwidths in receivers.

Shortly after, Francesco Lanza, HB9TLK, modified the *Dream* software to use the DRM technology concepts in a program that only used 2.5 kHz of bandwidth for ham radio use. His software, named *WinDRM*, includes file transfer functionality and is available for download at **n1su.com/windrm/**.

FDMDV

FDMDV was written by Francesco Lanza, HB9TLK, based on ideas by Peter Martinez, G3PLX. FDMDV (Frequency Division Multiplex Digital Voice) is based on 15 carriers using the 1400 LPC (Linear Predictive Coding) codec. High power in each carrier combined with a narrow 1.125 kHz bandwidth provides robustness with fast sync for a near SSB experience. FDMDV is not derived from DRM technology.

FDMDV features:
- 50 baud 14 QPSK voice data
- 1 Center BPSK carrier with 2x power for Auto Tuning and frame indication.
- 1.125 kHz spectrum bandwidth with 75 Hz carrier spacing
- 1450 bit/s data rate with 1400 bit/s open source LPC codec and 50 bit/s text
- Adjustable squelch
- TX ALC boost average power while reducing the peak power
- No FEC for fast synchronization.
- 48000 sample rate/16-20 bit/AC97 sound card compatible
- F6CTE's RSID for identifying and synchronizing FDM signals

The FDMDV software and documentation are available for download at **n1su.com/fdmdv**. More technical details can be found at **www.arrl.org/technical-characteristics**.

16.4.6 ALE

Automatic link establishment (ALE) was created as a series of protocols for government users to simplify HF communications. The protocol provides a mechanism to analyze signal quality on various channels/bands and choose the best option. The purpose is to provide a reliable rapid method of calling and connecting during constantly changing HF ionospheric propagation, reception interference and shared spectrum use of busy or congested HF channels. It also supports text messages with a very robust protocol that can get through even if no voice-quality channel can be found.

Each radio ALE station uses a call sign or address in the ALE controller. When not actively in communication with another station, each HF SSB transceiver constantly scans through a list of frequencies, listening for its call sign. It decodes calls and soundings sent by other stations, using the bit error rate to store a quality score for that frequency and sender call sign.

To reach a specific station, the caller simply enters the call sign, just like dialing a phone number. The ALE controller selects the best available frequency and sends out brief digital selective calling signals containing the call signs. When the distant scanning station detects the first few characters of its call sign, it stops scanning and stays on that frequency. The two stations' ALE controllers automatically handshake to confirm that a link of sufficient quality is established and they are ready to communicate.

When successfully linked, the receiving station which was muted will typically emit an audible alarm and visual alert for the receiving operator of the incoming call. It also indicates the call sign of the linked station. The operators then can talk in a regular conversation. At the conclusion of the QSO, one of the stations sends a disconnect signal to the other station, and they each return their ALE stations to the scanning mode. Some military / commercial HF transceivers are available with ALE available internally. Amateur Radio operators commonly use the PCALE soundcard software ALE controller, interfaced to a ham transceiver via rig control cable and multi-frequency antenna.

The ALE waveform is designed to be compatible with the audio passband of a standard SSB radio. It has a robust waveform for reliability during poor path conditions. It consists of 8-ary frequency-shift keying (FSK) modulation with eight orthogonal tones, a single tone for a symbol. These tones represent three bits of data, with least significant bit to the right, as shown in **Table 16.8**.

The tones are transmitted at a rate of 125 tones per second, 8 ms per tone. The resultant transmitted bit rate is 375 bit/s. The basic ALE word consists of 24 bits of information. Details can be found in Federal Standard 1045, Detailed Requirements at **www.its.bldrdoc.gov/fs-1054a/ 45-detr.htm**.

It would require a lot of time for the radio to go through the sequence of calling a station on every possible frequency to establish a link. Time can be decreased by using a "smarter" way of predictive or synchronized linking. With *Link Quality Analysis* (LQA), an ALE system uses periodic sounding and linking signals between other stations in the network to stay in touch and to predict which channel is likely to support a connection to the desired station at any given time. Various stations may be operating on different channels, and this enables the stations to find and use a common open channel.

The PCALE software developed by Charles Brain, G4GUO, is available for download at **hflink.com/software**. Much more ALE information and real-time data is available online at **hflink.com**.

Table 16.8
ALE Tones

Frequency	Data
750 Hz	000
1000 Hz	001
1250 Hz	011
1500 Hz	010
1750 Hz	110
2000 Hz	111
2250 Hz	101
2500 Hz	100

16.5 Networking Modes

The modes described in this section operate using features and functions associated with computer-to-computer networking. Even though communication using these modes may not involve the creation of a network, the modes are referred to as "networking modes" because of their structure. In cases such as Winlink 2000 and D-STAR, the most common use is to implement a networked system and those features are described along with the modes and protocols used to implement communications within the network.

16.5.1 OSI Networking Model

The Open Systems Interconnection Model or *OSI Model* is an abstract description for computer network protocol design. It defines seven different *layers* or functions performed by a *protocol stack*. In the OSI model, the highest level is closest to the user and the lowest is closest to the hardware required to transport the data (network card and wire or radio). The seven layers are described in **Table 16.9**. The modes examined previously implemented the protocols as a *monolithic stack* where all the functions are performed inside a single piece of code. The modes described in this section implement network-

Table 16.9
OSI Seven Layer Networking Model

Layer	Description
7 — Application Layer	End-user program or "application" that uses the network
6 — Presentation Layer	The format of data after transfer (code conversion, encryption)
5 — Session Layer	Manages the transfer process
4 — Transport Layer	Provides reliable data transfer to the upper layers
3 — Network Layer	Controls data routing
2 — Data Link Layer	Provides error detection and flow control
1 — Physical Layer	Signal used on the medium—voltage, current, frequency, etc.

Table 16.10
Networking Protocols in the OSI Model

Layer	Examples	IP Protocol Suite
7 — Application		NNTP, DNS, FTP, Gopher, HTTP, DHCP, SMTP, SNMP, TELNET
6 — Presentation	ASCII, EBCDIC, MIDI, MPEG	MIME, SSL
5 — Session	Named Pipes, NetBIOS, Half Duplex, Full Duplex, Simplex	Sockets, Session establishment in TCP
4 — Transport		TCP, UDP
3 — Network	AX.25	IP, IPsec, ICMP, IGMP
2 — Data Link	802.3 (Ethernet), 802.11a/b/g/n MAC/LLC, ATM, FDDI, Frame Relay, HDLC, Token Ring, ARP (maps layer 3 to layer 2 address)	PPP, SLIP, PPTP, L2TP
1 — Physical	RS-232, T1, 10BASE-T, 100BASE-TX, POTS, DSL, 802.11a/b/g/n, Soundcard, TNC, Radio	

ing features in a more modular fashion. This allows greater flexibility (and complexity) when mixing features.

As a data packet moves through these layers the header or preamble is removed and any required action performed before the data is passed to the next layer, much like peeling away layers of an onion until just the basic clean data is left. The OSI model does not define any interfaces between layers; it is just a conceptual model of the functions required. Real-world protocols rarely implement each layer individually and often span multiple layers.

This description is by no means exhaustive and more information can be found online and in every networking textbook. **Table 16.10** shows the placement of commonly recognized protocols within the OSI layered structure.

16.5.2 Connected and Connectionless Protocols

The protocols discussed to this point have been *connectionless* meaning they don't establish a connection with a specific machine for the purpose of transferring data. Even with packetized modes like FSK441 with a destination call sign specified, the packet is transmitted and it's up to the user to identify they are the intended recipient. In a packet-switched network, connectionless mode transmission is a transmission in which each packet is prepended with a header containing a destination address to allow delivery of the packet without the aid of additional instructions. A packet transmitted in a connectionless mode is frequently called a *datagram*.

In *connection-oriented protocols*, the stations about to exchange data need to first declare to each other they want to "establish a connection". A connection is sometimes defined as a logical relationship between the peers exchanging data. Connected protocols can use a method called *automatic repeat request* (*ARQ*) to insure accurate delivery of packets using *acknowledgements* and *timeouts*. This allows the detection and correction of corrupted packets, misdelivery, duplication, or out-of-sequence delivery of the packets.

Connectionless modes can have error correction and detection included by a higher layer of the protocol but they have no mechanism to request a correction. An advantage of connectionless mode over connection-oriented mode is that it has a low data overhead. It also allows for *multicast* and *broadcast* (net-type) operations, which may save even more network resources when the same data needs to be transmitted to several recipients. In contrast, a connected mode is always *unicast* (point-to-point).

Another drawback of the connectionless mode is that no optimizations are possible when sending several frames between the same two peers. By establishing a connection at the beginning of such a data exchange the components (routers, bridges) along the network path would be able to pre-compute (and hence cache) routing-related information, avoiding re-computation for every packet.

Many network modes incorporate both types of protocol for different purposes. In the Internet TCP/IP protocol, TCP is a connection-oriented transport protocol where UDP is connectionless.

16.5.3 The Terminal Node Controller (TNC)

While a *terminal node controller* (*TNC*) is nominally an OSI Physical layer device, the internal firmware often implements a protocol such as PACTOR that handles all the routing, and error correction through the transport layer. This greatly simplifies the coding of any protocol or application that uses these devices.

A TNC is actually a computer that contains the protocols implemented in firmware and a *modem* (modulator/demodulator). The TNC generally connects to a PC as a serial or USB device on one side and to the radio with appropriate audio and PTT cables on the other. Most of the newer rigs have dedicated data connections available that feature audio lines with fixed levels that are unaffected by settings in the radio. These jacks make swapping mike cables unnecessary when switching between voice and digital modes. Bypassing internal audio processing circuitry eliminates a number of issues that can cause problems with digital modes and makes the use of digital modes more reproducible/reliable by eliminating a number of variables when configuring equipment. These same data jacks are recommended when using a computer sound card.

Although many of the modes discussed can use a computer sound card to generate the required modulation and a separate mechanism to support push-to-talk (PTT), a TNC offers some advantages:

- TNC hardware can be used with any computer platform.
- A computer of nearly any vintage/performance level can be used.
- Data transmission/reception is unaffected by computer interruptions from virus checkers or other "inits."
- Initialization settings are held internal to the TNC and can easily be reset as needed — once working, they stay working.
- Virtually eliminates the computer as a problem/failure point.
- Offers features independent of the computer (digipeat, BBS, APRS beacon, telemetry, weather beacon, and so forth).

The majority of TNCs are designed for 300 or 1200-bit/s packet and implement the Bell 103 or Bell 202 modulation respectively. A *multimode communications processor* (*MCP*) or *multi-protocol controller* (*MPC*) may offer the capability to operate RTTY, CW, AMTOR, PACTOR, G-TOR, Clover, fax, SSTV and other modes in addition to packet. Some of these modes are only available in TNC hardware because the real-time operating system in the TNC provides a more reliable platform to implement the mode and it also helps protect proprietary intellectual property.

16.5.4 PACTOR-I

PACTOR, now often referred to as PACTOR-I, is an HF radio transmission system developed by German amateurs Hans-Peter Helfert, DL6MAA, and Ulrich Strate, DF4KV. It was designed to overcome the shortcomings of AMTOR and packet radio. It performs well under both weak-signal and

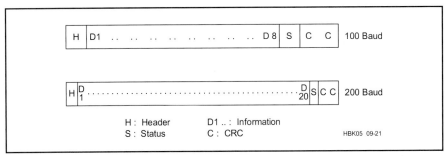

Fig 16.6 — PACTOR packet structure.

Table 16.11
PACTOR Timing

Object	Length (seconds)
Packet	0.96 (200 baud: 192 bits; 100 baud: 96 bits)
CS receive time	0.29
Control signals	0.12 (12 bits at 10 ms each)
Propagation delay	0.17
Cycle	1.25

Table 16.12
PACTOR Control Signals

Code	Chars (hex)	Function
CS1	4D5	Normal acknowledge
CS2	AB2	Normal acknowledge
CS3	34B	Break-in (forms header of first packet from RX to TX)
CS4	D2C	Speed change request

All control signals are sent only from RX to TX

Table 16.13
PACTOR Initial Contact

Master Initiating Contact

Size (bytes)	1	8	6
Content	/Header	/SLAVECAL	/SLAVECAL/
Speed (baud)	100	100	200

Slave Response
The receiving station detects a call, determines mark/space polarity, and decodes 100 baud and 200-bd call signs. It uses the two call signs to determine if it is being called and the quality of the communication path. The possible responses are:

First call sign does not match slave's call sign (Master not calling this slave)	none
Only first call sign matches slave's call sign (Master calling this slave, poor communications)	CS1
First and second call signs both match the slaves (good circuit, request speed change to 200 baud)	CS4

high-noise conditions. PACTOR-I has been overtaken by PACTOR-II and PACTOR-III but remains in use.

TRANSMISSION FORMATS

All packets have the basic structure shown in **Fig 16.6**, and their timing is as shown in **Table 16.11**.

• Header: Contains a fixed bit pattern to simplify repeat requests, synchronization and monitoring. The header is also important for the Memory ARQ function. In each packet that carries new information, the bit pattern is inverted.

• Data: Any binary information. The format is specified in the status word. Current choices are 8-bit ASCII or 7-bit ASCII (with Huffman encoding). Characters are not broken across packets. ASCII RS (hex 1E) is used as an IDLE character in both formats.

• Status word: See Table 16.11
• CRC: The CRC is calculated according to the CCITT standard, for the data, status and CRC.

The PACTOR acknowledgment signals are shown in **Table 16.12**. Each of the signals is 12-bits long. The characters differ in pairs in eight bits (Hamming offset) so that the chance of confusion is reduced. If the CS is not correctly received, the TX reacts by repeating the last packet. The request status can be uniquely recognized by the 2-bit packet number so that wasteful transmissions of pure RQ blocks are unnecessary.

The receiver pause between two blocks is 0.29 s. After deducting the CS lengths, 0.17 s remains for switching and propagation delays so that there is adequate reserve for DX operation.

CONTACT FLOW

In the listen mode, the receiver scans any received packets for a CRC match. This method uses a lot of computer processing resources, but it's flexible.

A station seeking contacts transmits CQ packets in an FEC mode, without pauses for acknowledgment between packets. The transmit time length, number of repetitions and speed are the transmit operator's choice. (This mode is also suitable for bulletins and other group traffic.) Once a listening station has copied the call, the listener assumes the TX station role and initiates a contact. Thus, the station sending CQ initially takes the RX station role. The contact begins as shown in **Table 16.13**.

With good conditions, PACTOR's normal signaling rate is 200 baud, but the system automatically changes from 200 to 100 baud and back, as conditions demand. In addition, Huffman coding can further increase the throughput by a factor of 1.7. There is no loss of synchronization speed changes; only one packet is repeated.

When the RX receives a bad 200-baud packet, it can acknowledge with CS4. TX immediately assembles the previous packet in 100-baud format and sends it. Thus, one packet is repeated in a change from 200 to 100 baud.

The RX can acknowledge a good 100-baud packet with CS4. TX immediately switches to 200 baud and sends the next packet. There is no packet repeat in an upward speed change.

The RX station can become the TX station by sending a special change-over packet in response to a valid packet. RX sends CS3 as the first section of the changeover packet. This immediately changes the TX station to RX mode to read the data in that packet and responds with CS1 and CS3 (acknowledge) or CS2 (reject).

PACTOR provides a sure end-of-contact procedure. TX initiates the end of contact by sending a special packet with the QRT bit set in the status word and the call of the RX station in byte-reverse order at 100 baud. The RX station responds with a final CS.

16.5.5 PACTOR-II

This is a significant improvement over PACTOR-I, yet it is fully compatible with the older mode. PACTOR-II uses 16PSK to transfer up to 800 bit/s at a 100 baud rate. This keeps the bandwidth less than 500 Hz.

PACTOR-II uses digital signal processing (DSP) with Nyquist waveforms, Huffman and Markov compression and powerful Viterbi decoding to increase transfer rate and sensitivity into the noise level. The effective transfer rate of text is over 1200 bit/s. Features of PACTOR II include:
• Frequency agility — it can automatically adjust or lock two signals together over a ±100 Hz window.
• Powerful data reconstruction based upon computer power — with over 2 Mbyte of available memory.
• Cross correlation — applies analog Memory ARQ to acknowledgment frames and headers.
• Soft decision making — Uses artificial intelligence (AI), as well as digital information received to determine frame validity.
• Extended data block length — when transferring large files under good conditions, the data length is doubled to increase the transfer rate.
• Automatic recognition of PACTOR-I, PACTOR-II and so on, with automatic mode switching.
• Intermodulation products are canceled by the coding system.
• Two long-path modes extend frame timing for long-path terrestrial and satellite propagation paths.

This is a fast, robust mode that has excellent coding gain as well. PACTOR-II stations acknowledge each received transmission block. PACTOR-II employs computer logic as well as received data to reassemble defective data blocks into good frames. This reduces the number of transmissions and increases the throughput of the data.

16.5.6 PACTOR-III

PACTOR-III is a software upgrade for existing PACTOR-II modems that provides a data transmission mode for improved speed and robustness. Both the transmitting and receiving stations must support PACTOR-III for end-to-end communications using this mode.

PACTOR-III's maximum uncompressed speed is 2722 bit/s. Using online compression, up to 5.2 kbit/s is achievable. This requires an audio passband from 400 Hz to 2600 Hz (for PACTOR-III speed level 6). On an average channel, PACTOR-III is more than three times faster than PACTOR-II. On good channels, the effective throughput ratio between PACTOR-III and PACTOR-II can exceed five. PACTOR-III is also slightly more robust than PACTOR-II at their lower SNR edges.

The ITU emission designator for PACTOR-III is 2K20J2D. Because PACTOR-III builds on PACTOR-II, most specifications like frame length and frame structure are adopted from PACTOR-II. The only significant difference is PACTOR III's multi-tone waveform that uses up to 18 carriers while PACTOR-II uses only two carriers. PACTOR-III's carriers are located in a 120 Hz grid and modulated with 100 symbols per second DBPSK or DQPSK. Channel coding is also adopted from PACTOR-II's Punctured Convolutional Coding.

PACTOR-III Link Establishment

The calling modem uses the PACTOR-I FSK connect frame for compatibility. When the called modem answers, the modems negotiate to the highest level of which both modems are capable. If one modem is only capable of PACTOR-II, then the 500 Hz PACTOR-II mode is used for the session. With the MYLevel (MYL) command a user may limit a modem's highest mode. For example, a user may set MYL to 1 and only a PACTOR-I connection will be made, set to 2 and PACTOR-I and II connections are available, set to 3 and PACTOR-I through III connections are enabled. The default MYL is set to 2 with the current firmware and with PACTOR-III firmware it will be set to 3. If a user is only allowed to occupy a 500 Hz channel, MYL can be set to 2 and the modem will stay in its PACTOR-II mode.

The PACTOR-III Protocol Specification is available on the Web at **www.scs-ptc.com/pactor.html**. More information can also be found online at **www.arrl.org/technical-characteristics** or in *ARRL's HF Digital Handbook* by Steve Ford, WB8IMY.

16.5.7 G-TOR

This brief description has been adapted from "A Hybrid ARQ Protocol for Narrow Bandwidth HF Data Communication" by Glenn Prescott, WB0SKX, Phil Anderson, W0XI, Mike Huslig, KB0NYK, and Karl Medcalf, WK5M (May 1994 *QEX*).

G-TOR is short for Golay-TOR, an innovation of Kantronics. It was inspired by HF automatic link establishment (ALE) concepts and is structured to be compatible with ALE. The purpose of the G-TOR protocol is to provide an improved digital radio communication capability for the HF bands. The key features of G-TOR are:
• Standard FSK tone pairs (mark and space)
• Link-quality-based signaling rate: 300, 200 or 100 baud
• 2.4-s transmission cycle
• Low overhead within data frames
• Huffman data compression — two types, on demand
• Embedded run-length data compression
• Golay forward-error-correction coding
• Full-frame data interleaving
• CRC error detection with hybrid ARQ
• Error-tolerant "Fuzzy" acknowledgments.

Since one of the objectives of this protocol is ease of implementation in existing TNCs, the modulation format consists of standard tone pairs (FSK), operating at 300, 200 or 100 baud, depending upon channel conditions. G-TOR initiates contacts and sends ACKs only at 100 baud. The G-TOR waveform consists of two phase-continuous tones (BFSK), spaced 200 Hz apart (mark = 1600 Hz, space = 1800 Hz); however, the system can still operate at the familiar 170 Hz shift (mark = 2125 Hz, space = 2295 Hz), or with any other convenient tone pairs. The optimum spacing for 300-baud transmission is 300 Hz, but you trade some performance for a narrower bandwidth.

Each transmission consists of a synchronous ARQ 1.92-s frame and a 0.48-s interval for propagation and ACK transmissions (2.4 s cycles). All advanced protocol features are implemented in the signal-processing software.

Data compression is used to remove redundancy from source data. Therefore, fewer bits are needed to convey any given message. This increases data throughput and decreases transmission time — valuable features for HF. G-TOR uses run-length encoding and two types of Huffman coding during normal text transmissions. Run-length encoding is used when more than two repetitions of an 8-bit character are sent. It provides an especially large savings in total transmission time when repeated characters are being transferred.

The Huffman code works best when the statistics of the data are known. G-TOR applies Huffman A coding with the upper- and lower-case character set, and Huffman B coding with upper-case-only text. Either type of Huffman code reduces the average number of bits sent per character. In some situations, however, there is no benefit from Huffman coding. The encoding process is then disabled. This decision is made on a frame-by-frame basis by the information sending station.

The real power of G-TOR resides in the properties of the (24, 12) extended Golay error-correcting code, which permits correction of up to three random errors in three received bytes. The (24, 12) extended Golay code is a half-rate error-correcting code: Each 12 data bits are translated into an additional 12 parity bits (24 bits total). Further, the code can be implemented to produce separate input-data and parity-bit frames.

The extended Golay code is used for G-TOR because the encoder and decoder are simple to implement in software. Also, Golay code has mathematical properties that make it an ideal choice for short-cycle synchronous communication. More information can also be found online at **www.arrl.org/technical-characteristics** or in *ARRL's HF Digital Handbook* by Steve Ford, WB8IMY.

16.5.8 CLOVER-II

The desire to send data via HF radio at high data rates and the problem encountered when using AX.25 packet radio on HF radio led Ray Petit, W7GHM, to develop a unique modulation waveform and data transfer protocol that is now called CLOVER-II. Bill Henry, K9GWT, supplied this description of the CLOVER-II system.

CLOVER modulation is characterized by the following key parameters:

• Very low base symbol rate: 31.25 symbols/second (all modes).

• Time-sequence of amplitude-shaped pulses in a very narrow frequency spectrum.

• Occupied bandwidth = 500 Hz at 50 dB below peak output level.

• Differential modulation between pulses.

• Multilevel modulation.

The low base symbol rate is very resistant to multipath distortion because the time between modulation transitions is much longer than even the worst-case time-smearing caused by summing of multipath signals. By using a time-sequence of tone pulses, Dolph-Chebychev "windowing" of the modulating signal and differential modulation, the total occupied bandwidth of a CLOVER-II signal is held to 500 Hz.

Multilevel tone, phase and amplitude modulation gives CLOVER a large selection of data modes that may be used (see **Table 16.14**). The adaptive ARQ mode of CLOVER senses current ionospheric conditions and automatically adjusts the modulation mode to produce maximum data throughput. When using the Fast bias setting, ARQ throughput automatically varies from 11.6 byte/s to 70 byte/s.

The CLOVER-II waveform uses four tone pulses that are spaced in frequency by 125 Hz. The time and frequency domain

Table 16.14
CLOVER-II Modulation Modes
As presently implemented, CLOVER-II supports a total of seven different modulation formats: five using PSM and two using a combination of PSM and ASM (Amplitude Shift Modulation).

Name Rate	Description	In-Block Data
16P4A	16 PSM, 4-ASM	750 bps
16PSM	16 PSM	500 bps
8P2A	8 PSM, 2-ASM	500 bps
8PSM	8 PSM	375 bps
QPSM	4 PSM	250 bps
BPSM	Binary PSM	125 bps
2DPSM	2-Channel Diversity BPSM	62.5 bps

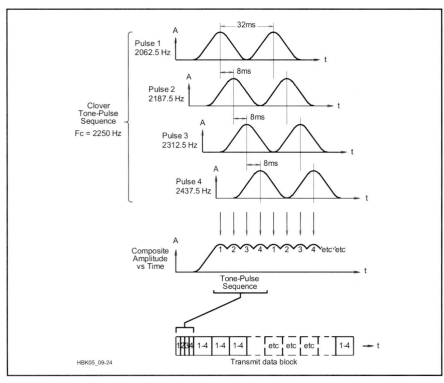

Fig 16.7 — Amplitude vs time plots for CLOVER-II's four-tone waveform.

characteristics of CLOVER modulation are shown in **Figs 16.7, 16.8** and **16.9**. The time-domain shape of each tone pulse is intentionally shaped to produce a very compact frequency spectrum. The four tone pulses are spaced in time and then combined to produce the composite output shown. Unlike other modulation schemes, the CLOVER modulation spectrum is the same for all modulation modes.

Data is modulated on a CLOVER-II signal by varying the phase and/or amplitude of the tone pulses. Further, all data modulation is differential on the same tone pulse — data is represented by the phase (or amplitude) difference from one pulse to the next. For example, when binary phase modulation is used, a data change from 0 to 1 may be represented by a change in the phase of tone pulse one by 180° between the first and second occurrence of that pulse. Further, the phase state is changed only while the pulse amplitude is zero. Therefore, the wide frequency spectra normally associated with PSK of a continuous carrier is avoided. This is true for all CLOVER-II modulation formats. The term *phase-shift modulation* (PSM) is used when describing CLOVER modes to emphasize this distinction.

CLOVER-II has four "coder efficiency" options: 60%, 75%, 90% and 100% ("efficiency" being the approximate ratio of real data bytes to total bytes sent). 60% efficiency corrects the most errors but has the lowest net data throughput. 100% efficiency turns the encoder off and has the highest throughput but fixes no errors. There is therefore a tradeoff between raw data throughput versus the number of errors that can be corrected without resorting to retransmission of the entire data block.

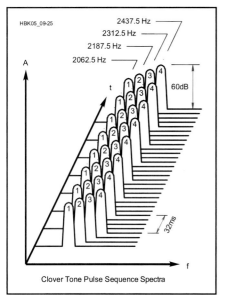

Fig 16.8 — A frequency-domain plot of a CLOVER-II waveform.

Table 16.15
Data Bytes Transmitted Per Block

Block Size	Reed-Solomon Encoder Efficiency			
	60%	75%	90%	100%
17	8	10	12	14
51	28	36	42	48
85	48	60	74	82
255	150	188	226	252

Note that while the In Block Data Rate numbers listed in the table go as high as 750 bit/s, overhead reduces the net throughput or overall efficiency of a CLOVER transmission. The FEC coder efficiency setting and protocol requirements of FEC and ARQ modes add overhead and reduce the net efficiency. **Tables 16.15** and **16.16** detail the relationships between block size, coder efficiency, data bytes per block and correctable byte errors per block.

With seven different modulation formats, four data-block lengths (17, 51, 85 or 255 bytes) and four Reed-Solomon coder efficiencies (60%, 75%, 90% and 100%), there are 112 (7 × 4 × 4) different waveform modes that could be used to send data via CLOVER. Once all of the determining factors are considered, however, there are eight different waveform combinations that are actually used for FEC and/or ARQ modes.

16.5.9 CLOVER-2000

CLOVER-2000 is a faster version of CLOVER (about four times faster) that uses eight tone pulses, each of which is 250 Hz wide, spaced at 250 Hz centers, contained within the 2 kHz bandwidth between 500 and 2500 Hz. The eight tone pulses are sequential, with only one tone being present at any instant and each tone lasting 2 ms. Each frame consists of eight tone pulses lasting a total of 16 ms, so the base modulation rate of a CLOVER-2000 signal is always 62.5 symbols per second (regardless of the type of modulation being used). CLOVER-2000's maximum raw data rate is 3000 bit/s.

Allowing for overhead, CLOVER-2000 can deliver error-corrected data over a standard HF SSB radio channel at up to 1994 bit/s, or 249 characters (8-bit bytes) per second. These are the uncompressed data rates; the maximum throughput is typically doubled for plain text if compression is used. The effective data throughput rate of CLOVER-2000 can be even higher when binary file transfer mode is used with data compression.

The binary file transfer protocol used by HAL Communications operates with a terminal program explained in the HAL E2004 engineering document. Data compression algorithms tend to be context sensitive — compression that works well for one mode (say, text), may not work well for other data forms (graphics, for example). The HAL terminal program uses the PK-WARE compression algorithm, which has proved to be a good general-purpose compressor for most computer files and programs. Other algorithms may be more efficient for some data formats, particularly for compression of graphic image files and digitized voice data. The HAL Communications CLOVER-2000 modems can be operated with other data compression algorithms in the users' computers.

CLOVER-2000 is similar to the previous version of CLOVER, including the transmission protocols and Reed-Solomon error detection and correction algorithm. The original descriptions of the CLOVER Control Block (CCB) and Error Correction Block (ECB) still apply for CLOVER-2000, except for the higher data rates inherent to CLOVER-2000. Just like CLOVER, all data sent via CLOVER-2000 is encoded as 8-bit data bytes and the error-correction coding and modulation formatting processes are transparent to the data stream — every bit of source data is delivered to the receiving terminal without modification.

Control characters and special "escape sequences" are not required or used by CLOVER-2000. Compressed or encrypted data may therefore be sent without the need to insert (and filter) additional control characters and without concern for data integrity. Five different types of modulation may be used in the ARQ mode — BPSM (Binary Phase Shift Modulation), QPSM (Quadrature PSM), 8PSM (8-level PSM), 8P2A (8PSM + 2-level Amplitude-Shift Modulation) and 16P4A (16 PSM plus 4 ASM).

The same five types of modulation used in

Table 16.16
Correctable Byte Errors Per Block

Block Size	Reed-Solomon Encoder Efficiency			
	60%	75%	90%	100%
17	1	1	0	0
51	9	5	2	0
85	16	10	3	0
255	50	31	12	0

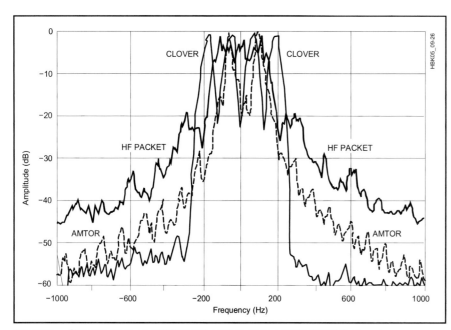

Fig 16.9 — Spectra plots of AMTOR, HF packet-radio and CLOVER-II signals.

ARQ mode are also available in Broadcast (FEC) mode, with the addition of 2-Channel Diversity BPSM (2DPSM). Each CCB is sent using 2DPSM modulation, 17-byte block size and 60% bias. The maximum ARQ data throughput varies from 336 bit/s for BPSM to 1992 bit/s for 16P4A modulation. BPSM is most useful for weak and badly distorted data signals, while the highest format (16P4A) needs extremely good channels, with high SNRs and almost no multipath.

Most ARQ protocols designed for use with HF radio systems can send data in only one direction at a time. CLOVER-2000 does not need an OVER command; data may flow in either direction at any time. The CLOVER ARQ time frame automatically adjusts to match the data volume sent in either or both directions. When first linked, both sides of the ARQ link exchange information using six bytes of the CCB. When one station has a large volume of data buffered and ready to send, ARQ mode automatically shifts to an expanded time frame during which one or more 255 byte data blocks are sent.

If the second station also has a large volume of data buffered and ready to send, its half of the ARQ frame is also expanded. Either or both stations will shift back to CCB level when all buffered data has been sent. This feature provides the benefit of full-duplex data transfer but requires use of only simplex frequencies and half-duplex radio equipment. This two-way feature of CLOVER can also provide a back-channel "order-wire" capability. Communications may be maintained in this chat mode at 55 WPM, which is more than adequate for real-time keyboard-to-keyboard communications.

More information can also be found at **www.arrl.org/technical-characteristics** or in *ARRL's HF Digital Handbook* by Steve Ford, WB8IMY.

16.5.10 WINMOR

While the various PACTOR modes currently dominate and generally represent the best available performance HF ARQ protocols suitable for digital messaging, PC sound cards with appropriate DSP software can now begin to approach PACTOR performance. The WINMOR (Winlink Message Over Radio) protocol is an outgrowth of the work SCAMP (Sound Card Amateur Message Protocol) by Rick Muething, KN6KB. SCAMP put an ARQ "wrapper" around Barry Sanderson's RDFT (Redundant Digital File Transfer) then integrated SCAMP into a Client and Server for access to the Winlink message system. (More on Winlink in a later section.) SCAMP worked well on good channels but suffered from the following issues:

• The RDFT batch-oriented DLLs were slow and required frame pipelining, increasing complexity and overhead.

• RDFT only changed the RS encoding on its 8PSK multi carrier waveform to achieve a 3:1 range in speed/robustness which is not enough.

• RDFT was inefficient in Partial Frame recovery (no memory ARQ).

• RDFT was a 2.4 kHz mode and limited to narrow HF sub bands.

• SCAMP's simple multi-tone ACK/NAK did not carry session ID info, increasing chances of fatal cross session contamination.

WINMOR is an ARQ mode generated from the ground up to address the limitations of SCAMP/RDFT and leverage what was learned. Today, a viable message system (with the need for compression and binary attachments) requires true "error-free" delivery of binary data. To achieve this there must be some "back channel" or ARQ so the receiving station can notify the sender of lost or damaged data and request retransmission or repair. **Table 16.17** outlines the guidelines used in the development of WINMOR.

Perhaps the most challenging of these requirements are:

• The ability to quickly tune, lock and acquire the signal which is necessary for practical length ARQ cycles in the 2-6 s range.

• The ability to automatically adapt the modulation scheme to changing channel conditions. An excellent example of this is Pactor III's extremely wide range of speed/robustness (18:1) and is one reason it is such an effective mode in both good and poor channel conditions.

The most recent development effort has focused on 62.5 baud BPSK, QPSK and 16QAM and 31.25-baud 4FSK using 1 (200 Hz), 3 (500 Hz) and 15 (2000 Hz). With carriers spaced at twice the symbol rate. These appear to offer high throughput and robustness especially when combined with multi-level FEC coding.

WINMOR uses several mechanisms for error recovery and redundancy.

1) FEC data encoding currently using:
• 4,8 Extended Hamming Dmin = 4 (used in ACK and Frame ID)
• 16-bit CRC for data verification
Two-level Reed-Solomon (RS) FEC for data:
• First level Weak FEC, for example RS 140,116 (corrects 12 errors)
• Second level Strong FEC, for example RS 254,116 (corrects 69 errors)

2) Selective ARQ. Each carrier's data contains a Packet Sequence Number (PSN). The ACK independently acknowledges each PSN so only carriers with failed PSNs get repeated. The software manages all the PSN accounting and re-sequencing.

3) Memory ARQ. The analog phase and amplitude of each demodulated symbol is saved for summation (phasor averaging) over multiple frames. Summation is cleared and restarted if max count reached. Reed-Solomon FEC error decoding done after summation.

4) Multiple Carrier Assignment (MCA). The same PSN can be assigned to multiple carriers (allows tradeoff of throughput for robustness). Provides an automatic mechanism for frequency redundancy and protection from interference on some carriers.

5) Dynamic threshold adjustment (used on QAM modes) helps compensate for fading which would render QAM modes poor in fading channels.

In trying to anticipate how WINMOR might be integrated into applications they came up with a "Virtual TNC" concept. This essentially allows an application to integrate the WINMOR protocol by simply treating the WINMOR code as just another TNC and writing a driver for that TNC. Like all TNCs

---

Table 16.17
WINMOR Development Guidelines

Absolute Requirements	Desirable Requirements
Work with standard HF (SSB) radios	Modest CPU & OS demands
Accommodate Automatic Connections	Bandwidth options (200, 500, 3000 Hz)
Error-free transmission and confirmation	Work with most sound cards/interfaces
Fast Lock for practical length ARQ cycles	Good bit/s/Hz performance ~ P2 goal
Auto adapt to a wide range of changing channel conditions	Efficient modulation and demodulation for acceptable ARQ latency
Must support true transparent binary to allow attachments and compression	Selective ARQ & memory ARQ to maximize throughput and robustness.
Must use loosely synchronous ARQ timing to accommodate OS and DSP demands	Near Pactor ARQ efficiency (~70% of raw theoretical throughput)

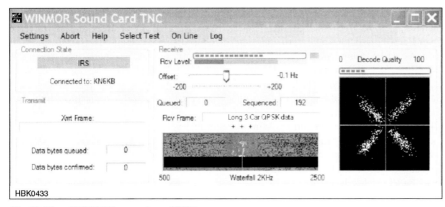

Fig 16.10 — WINMOR sound card TNC screen.

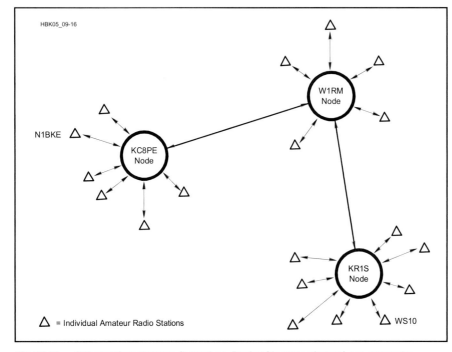

Fig 16.11 — DX spotting clusters (based on *PacketCluster* software) are networks comprised of individual nodes and stations with an interest in DXing and contesting. In this example, N1BKE is connected to the KC8PE node. If he finds a DX station on the air, he'll post a notice — otherwise known as a spot — which the KC8PE node distributes to all its local stations. In addition, KC8PE passes the information along to the W1RM node. W1RM distributes the information and then passes it to the KR1S node, which does the same. Eventually, WS1O — who is connected to the KR1S node — sees the spot on his screen. Depending on the size of the network, WS1O will receive the information within minutes after it was posted by N1BKE. Many such networks can also be found on the Web or are accessible by the use of *TELNET* software.

there are some (<10) parameters to set up: call sign, timing info, sound card, keying mechanism, etc. A sample image of the virtual TNC appears in **Fig 16.10**.

The WINMOR software DLL can even be made to appear as a physical TNC by "wrapping" the DLL with code that accesses it through a virtual serial port or a TCP/IP port. Like a physical TNC WINMOR has a "front panel" with flashing lights. But since operation is automatic with no front panel user interaction required the WINMOR TNC can be visible or hidden.

WINMOR looks promising and the testing to date confirms:

• Sound card ARQ is possible with a modern CPU and OS while making acceptable CPU processing demands. (CPU Loading of < 20% on a 1.5 GHz Celeron/Win XP)

• Throughput and robustness can be adjusted automatically to cover a wide range of bandwidth needs and channel conditions. (10:1 bandwidth range, 57:1 throughput range)

• ARQ throughput in excess of 0.5 bit/s/Hz is possible in fair to good channels (0.68 - 0.82 bit/s/Hz measured)

• Good ARQ efficiency — 70-75%

• Throughput is currently competitive with P2 and P3 and significantly better than P1

More information about WINMOR will be available as it develops at **www.winlink.org/WINMOR**.

16.5.11 Packet Radio

Amateur packet radio began in Canada after the Canadian Department of Communications permitted amateurs to use the mode in 1978. The FCC permitted amateur packet radio in the US in 1980. In the first half of the 1980s, packet radio was the habitat of a small group of experimenters who did not mind communicating with a limited number of potential fellow packet communicators. In the second half of the decade, packet radio took off as the experimenters built a network that increased the potential number of packet stations that could intercommunicate and thus attracted tens of thousands of communicators who wanted to take advantage of this potential.

Packet radio provides error-free data transfer via the AX.25 protocol. The receiving station receives information exactly as the transmitting station sends it, so you do not waste time deciphering communication errors caused by interference or changes in propagation. This simplifies the effort required to build other protocols or applications on top of packet technology since much of the data transfer work is handled seamlessly by the TNC in the lower layers. There are a number of other advantages to the packet radio design:

• Packet uses time efficiently, since packet bulletin-board systems (PBBSs) permit packet operators to store information for later retrieval by other amateurs.

• Packet uses the radio spectrum efficiently, since one radio channel may be used for multiple communications simultaneously, or one radio channel may be used to interconnect a number of packet stations to form a "cluster" that provides for the distribution of information to all of the clustered stations. The DX PacketCluster nodes are typical examples (see **Fig 16.11**). Each local channel may be connected to other local channels to form a network that affords interstate and international data communications. This network can be used by interlinked packet bulletin-board systems to transfer information, messages and third party traffic via HF, VHF, UHF, satellite and the Internet.

• Packet uses other stations efficiently, since any packet-radio station can use one or more other packet-radio stations to relay data to its intended destination.

• Packet uses current station transmitting and receiving equipment efficiently, since the same equipment used for voice communications may be used for packet communications. The outlay for the additional equipment necessary to add packet capability to a voice station may be less than $100.

DIGIPEATERS

A *digipeater* is a packet-radio station capable of recognizing and selectively repeating packet frames. An equivalent term used in the network industry is *bridge*. Virtually any TNC can be used as a single-port digipeater, because the digipeater function is included in the AX.25 Level 2 protocol firmware. The digipeater function is handy when you need a relay and no node is available, or for on-the-air testing. Digipeaters are used extensively with APRS (Automatic Packet Reporting System) covered below.

TCP/IP

Despite its name, TCP/IP (Transmission Control Protocol/Internet Protocol) is more than two protocols; it's actually a set of several protocols. TCP/IP provides a standardized set of protocols familiar to many and compatible with existing network technologies and applications. TCP/IP has a unique solution for busy networks. Rather than transmitting packets at randomly determined intervals, TCP/IP stations automatically adapt to network delays as they occur. As network throughput slows down, active TCP/IP stations sense the change and lengthen their transmission delays accordingly. As the network speeds up, the TCP/IP stations shorten their delays to match the pace. This kind of intelligent network sharing virtually guarantees that all packets will reach their destinations with the greatest efficiency the network can provide.

With TCP/IP's adaptive networking scheme, you can chat using the TELNET protocol with a ham in a distant city and rest assured that you're not overburdening the system. Your packets simply join the constantly moving "freeway" of data. They might slow down in heavy traffic, but they will reach their destination eventually. This adaptive system is used for all TCP/IP packets, no matter what they contain.

TCP/IP excels when it comes to transferring files from one station to another. By using the TCP/IP File Transfer Protocol (FTP), you can connect to another station and transfer computer files — including software. As might be imagined, transferring large files can take time. With TCP/IP, however, you can still send and receive mail (using the SMTP protocol) or talk to another ham while the transfer is taking place.

When you attempt to contact another station using TCP/IP, all network routing is performed automatically according to the TCP/IP address of the station you're trying to reach. In fact, TCP/IP networks are transparent to the average user. To operate TCP/IP, all you need is a computer, a 2 meter FM transceiver and a TNC with KISS (keep it simple, stupid) capability. As you might guess, the heart of your TCP/IP setup is software. The TCP/IP software set was written by Phil Karn, KA9Q, and is called NOSNET or just NOS for short. There are dozens of NOS derivatives available today. All are based on the original NOSNET. NOS takes care of all TCP/IP functions, using your "KISSable" TNC to communicate with the outside world. The only other item necessary is your own IP address in Network 44, termed AMPRNet (AMateur Packet Radio Network). Individual IP Address Coordinators assign addresses to new TCP/IP users in the 44.x.x.x subnet based on physical location worldwide. Your local coordinator can be found on AMPR.org on the list at **noh.ucsd.edu/~brian/amprnets.txt**.

More packet information can be found in the annual ARRL/TAPR Digital Communications Conference proceedings and on the TAPR (Tucson Amateur Packet Radio) Web site at **www.tapr.org**. TAPR also maintains the AX.25 protocol specification at **www.tapr.org/pub_ax25.html**.

16.5.12 APRS

APRS (**aprs.org**) was developed by Bob Bruninga, WB4APR, for tracking and digital communications with mobile GPS equipped stations with two-way radio. APRS is different from regular packet in four ways:

• Integration of maps and other data displays to organize and display data
• By using a one-to-many protocol to update everyone in real time
• By using generic digipeating so that prior knowledge of the network is not required
• Provides worldwide transparent Internet backbone, linking everyone worldwide

APRS turns packet radio into a real-time tactical communications and display system for emergencies, public service applications and global communications. Normal packet radio has shown usefulness in passing bulk message traffic (e-mail) from point to point. It has been difficult to apply conventional packet to real time events where information has a very short life time and needs to get to everyone.

The APRS network consists of five types of APRS stations:

• Basic — Full transmit and receive capability
• Tracker — Transmit-only device (portable or weather station)
• RELAY — Provides basic digipeating
• WIDE — Dedicated digipeaters with specific coverage areas
• IGate — Internet gateways to repeat APRS packets to servers on the Internet

APRS stations are generally set to beacon their position and other information at frequent intervals. Fixed stations don't require a GPS and should beacon infrequently to avoid crowding the channel with needless reports. Moving mobile stations can receive their coordinates from a GPS and should report more frequently. More sophisticated APRS devices support "smart beaconing" and "corner pinning" where a station will beacon more frequently when it moves faster or automatically beacon when changing direction more than a predefined angle. An APRS beacon will generally contain a call sign with *Secondary Station ID* (*SSID*), position report (lat/long), heading, speed, altitude, display icon type, status text and routing information. It may also contain antenna/power information, weather data or short message data.

Call signs generally include an SSID that historically provided information about the type of station. For example, a call sign of N7SS-9 would generally indicate a mobile station. Since there are several systems in use, the SSID information is not definitive. Recently a number of dedicated APRS radios have appeared that feature an internal TNC and APRS software. These radios include a method of displaying and entering messages without requiring the use of a computer meaning their user can directly respond to a message sent. There is value in knowing someone can respond and it is recommended they use an SSID of -7. Tactical call signs can also be used if the assigned call sign is included as part of the status text.

The APRS packet includes PATH information that determines how many times it should be digipeated. A local group should be able to offer guidance on how the PATH should be set to not overload the local network. Listening to squawking signals on the national APRS frequency of 144.390 MHz will provide some idea how busy the channel is. In much of the country the APRS network is well developed and a beacon doesn't need more than a couple hops to cover a wide area and find an IGate.

APRS SOFTWARE

There is APRS software available for most any platform with varying levels of support. Some of the most common are DOSAPRS (DOS), WinAPRS / APRS+SA / APRSPoint / UI-View (*Windows*), MacAPRS (Macintosh), PocketAPRS (Palm devices), APRSce (*Windows* CE devices) and Xastir / X-APRS (*Linux*). Because APRS beacons find their way to the Internet via IGates, it is also possible to track station or view the network online. There is a good network view available at **APRS.fi** and individual station information can be found at **map.findu.com/call sign**. The Findu.com site also offers historical weather data, a message display and a large list of other queries.

APRS INTEGRATION WITH OTHER TECHNOLOGIES

APRS is very similar to the *AIS (Automatic Identification System)* used to track ships in coastal waters. The AIS transponder equipment is becoming more common in commercial ships and private vessels. The AIS information can be displayed simultaneously with APRS data on the APRS.fi Web site. More information about AIS can be found at **en.wikipedia.org/wiki/Automatic_Identification_System**.

APRS position reports are also available from GPS-equipped D-STAR radios operating through a D-STAR gateway (more on D-STAR in a later section). The D-STAR DPRS position reports are received and converted on the D-STAR gateway, so they can be sent to the APRS-IS server. The DPRS reports are unique because they are sent simultaneously with the voice data and the station can be seen moving in real-time if the operator is talking. Although the DPRS reports are available to on the Internet servers, they will not appear on the local RF network unless specifically routed by an APRS IGate sysop.

Keith Sproul, WU2Z, operates an e-mail gateway that allows APRS messages to be converted to short e-mail or text messages. When entering a message, E-MAIL is used for the destination address. The message body must then contain the actual e-mail address followed by a space and very short message. If the message is picked up by an IGate it will be sent to the WU2Z mail server and out to the recipient, with a confirmation APRS message. For dozens of other messaging options, check online at **aprs.org/aprs-messaging.html**.

There is connectivity between APRS and the Winlink system via APRSLink. APRSLink monitors all APRS traffic gated to the Internet, worldwide, and watches for special commands that allow APRS users to:

• Read short e-mail messages sent to their call sign@winlink.org

• Send short e-mail messages to any valid e-mail address or Winlink 2000 user

• Perform e-mail related maintenance (see commands below)

• Be notified of pending Winlink e-mail via APRS message

• Query APRSLink for information on the closest Winlink RMS packet station

Attention must be paid to the APRS traffic generated when using these features but they can be very handy. Details on the APRSLink system are available at **www.winlink.org/aprslink**.

16.5.13 Winlink 2000

Winlink 2000 or *WL2K* is a worldwide radio messaging system that takes advantage of the Internet where possible. It does this in order to allow the end-user more radio spectrum on the crowded spectrum. The system provides radio interconnection services including: e-mail with attachments, position reporting, graphic and text weather bulletins, emergency / disaster relief communications, and message relay. The PACTOR-I, II, and III protocols are used on HF, and AX.25 Packet, D-STAR and 802.11 are used on VHF/UHF.

Winlink 2000 has been assisting the maritime and RV community around the clock for many years. More recently there has been an increasing interest in emergency communications (emcomm), and the Winlink 2000 development team has responded by adding features and functions that make the system more reliable, flexible and redundant. The role of Winlink 2000 in emergency communications is to supplement existing methodologies to add another tool in the toolkit of the volunteer services deploying emergency communications in their communities.

The Winlink 2000 system is a "star" based network containing five mirror-image, redundant Common Message Servers (CMS) located in San Diego (USA), Washington DC (USA), Vienna (Austria), Halifax (Canada) and Perth (Australia). These ensure that the system will remain in operation should any piece of the Internet become inoperative. Each Radio Message Server node (RMS) is tied together as would be the ends of a spoke on a wheel with the hub function performed by the Common Message Servers. Traffic goes in and out between the CMS and the Internet e-mail recipient, and between the end users and the Radio Message Server (RMS) gateways. Multiple radio-to-radio addresses may be mixed with radio-to-Internet e-mail addresses, allowing complete flexibility.

Because each Radio Message Server gateway is a mirror image of the next, it does not matter which station is used. Each can provide over 700 text-based or graphic weather products, and each can relay the user's position to a Web-based view of reporting users and interoperates with the APRS system.

One of the most important objectives in the eyes of the Winlink development team was to reduce the use of the HF spectrum to only that required to exchange messages with a user, and to do that at full "machine" speeds. The HF spectrum is very crowded, and limiting the forwarding of messages between WL2K RMS stations to the Internet, a great deal of radio air time is eliminated, making the time and spectrum available to individual users either for message handling or for other operations. A number of the RMS PACTOR servers found on HF restrict their protocols to PACTOR II (400 to 800 bit/s) and PACTOR III (1400 to 3600 bit/s.) In doing so, these RMS stations typically have a much higher ratio of traffic minutes and message counts to connect times than do the RMS stations that also receive the slower PACTOR I (100 to 200 bps) protocol. In other words, the amount of traffic that is passed with an Express station is much greater for an equivalent amount of connect time with approximately the same number of connections. On average, this translates to a PACTOR I station downloading an 80,000 byte file in approximately 80 minutes while on PACTOR III, the same download takes approximately six minutes.

The RMS servers provide endpoint connectivity to users via HF or VHF and can be found worldwide. When connecting via the Internet, users can connect directly to the CMS server operational closest to them via TELNET. Formerly, the full-featured message servers on HF were known as PMBOs (Participating Mail Box Offices) and there was a packet radio to TELNET bridge component for VHF/UHF called Telpac. These have been replaced with the newer RMS PACTOR (HF) and RMS Packet (VHF/UHF) software although the older terms are still in use. Over the air, PACTOR is used on HF, AX.25 packet on VHF/UHF and both employ the B2F compressed binary format to maximize transmission efficiency. A diagram of the Winlink 2000 architecture is shown in **Fig 16.12**.

CLIENT ACCESS TO WINLINK

The primary purpose of the Winlink 2000 network system is to assist the mobile or remotely located user and to provide emergency e-mail capabilities to community agencies. Because of this, WL2K supports a clean, simple interface to the Internet SMTP e-mail system. Any message sent or received may include multiple recipients and multiple binary attachments. The radio user's e-mail address, however, must be known to the system as a radio user or the message will be rejected. This simple Internet interface protocol has an added benefit in case of an emergency where local services are interrupted and the system must be used by non-Amateur groups as an alternative to normal SMTP e-mail.

Connecting to any one of the WL2K publicly-used RMS Packet stations via HF or the specialized non-public emcomm RMS stations, can immediately and automatically connect a local amateur station to the Internet for emergency traffic. Using a standard SMTP e-mail client, the *Paclink* mini-e-mail server can replace a network of computers (behind a

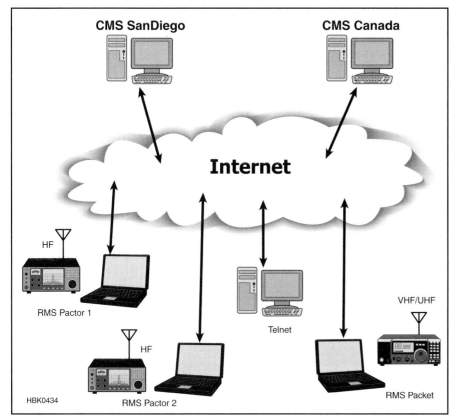

Fig 16.12 — Winlink 2000 system.

router) as a transparent substitute for normal SMTP mail. WL2K uses no external source for sending or receiving Internet e-mail. It is a stand-alone function which interacts directly with the Internet rather than through any external Internet service provider.

Airmail is independently developed, distributed and supported by Jim Corenman, KE6RK. It is the oldest and most widely used program for sending and receiving messages using the WL2K system. Airmail may be used for HF Pactor, VHF/UHF Packet, and for TELNET connections over any TCP/IP medium including the Internet and high-speed radio media, such as D-STAR and HSMM. Once connected to a WL2K station, message transfer is completely automatic. On the ham bands, Airmail can transfer messages automatically with any station supporting the BBS or F6FBB protocols, such as Winlink 2000, F6FBB, MSYS and other Airmail stations. When used with WL2K, Airmail also contains position reporting capabilities, and a propagation prediction program to determine which of the participating Winlink stations will work from anywhere on Earth. Airmail also contains a limited mail server that allows it to host e-mail independently from the Winlink network.

To obtain a copy of Airmail, including the installation and operating instructions, download the program from the Airmail Web page at **www.siriuscyber.net/ham** and support is available from the Airmail user group at **groups.yahoo.com/group/airmail2000**.

When an RF connection is not available (or necessary) but Web access is available, Winlink 2000 messages can be retrieved or sent via a Web interface. The Web browser access is limited to text-based messages without the use of bulletins or file attachments. Password-protected access to the Winlink mailbox is available through the Winlink page at **www.winlink.org/webmail**. There is also a terminal mode for interactive keyboard commands, allowing a terminal rather than computer-based software to connect to an RMS via RF. Because of the inefficient use of airtime, this method is discouraged but may be used for the listing and deletion of messages only.

Airmail provides a super-fast replica of WL2K radio operations while directly connected through the Internet to one of the CMS servers. This method of obtaining messages over the Internet allows multiple attachments, catalog bulletins, and all other Winlink 2000 services normally available over radio channels, but at Internet speeds. In order to use this service, a user must currently be listed as a radio user, and obtain the connection information for the closest CMS server. Both Paclink and Airmail support the TELNET client service. This operation allows regular use of the system with the same software configuration at high speed, without using RF bandwidth and provides a full-featured mechanism to access the system where no RF connection is available. An Emergency Operations Center with Internet access can use the TELNET path with RF as a fallback with just a minor change in the software. Similarly, an RV or marine user can use the software from an Internet café or home via TELNET and switch the software back to the TNC when mobile, with no change in functionality.

WINLINK FEATURES

To address the needs of mobile users for near real-time data, WL2K uses an "on-demand" bulletin distribution mechanism. (Note that such bulletins are not the same as traditional AX.25 Packet Bulletins.) Users must first select requested bulletins from an available "catalog" list managed in Airmail. When bulletin requests are received by an RMS station, a fresh locally cached copy of the requested bulletin is delivered. If no fresh locally-cached version is available, the RMS accesses the Internet and finds the bulletin which is then downloaded to the RMS and then sent to the user. The global catalog currently includes over 700 available weather, propagation, and information bulletins, including, instructions for using the system, World news, and piracy reports. All WL2K RMS stations support a single global catalog which ensures users can access any bulletin from any RMS. Bulletins can contain basic text, graphic fax or satellite images, binary or encoded files like GRIB or WMO weather reports. Local processing is used to re-process images to sizes suitable for HF Pactor transmission. The system prevents bulletin duplication and automatically purges obsolete time-sensitive weather bulletins and replaces them with the current version.

The system also has the ability to contain bulletins with attachment information which is local to each RMS. This is especially useful for the non-public emcomm RMS which may house valuable procedural information pertinent to complex information or instruction needed by specific agencies in any community emergency.

Multiple binary or text-based file attachments of any type or number may be attached to a message by simply selecting the file to be sent from a *Windows* selection dialog in the user's HF AirMail or the Winlink 2000 VHF/UHF Paclink server with a standard SMTP e-mail client. E-mail message attachments sent through the Winlink 2000 system must be limited in size. Users are provided an option to allow this limit to be determined. When using the default B2F format, the protocol chosen by the user usually determines the file size of an attachment. A user may also turn off the ability to receive file attachments. Certain file attachment types are blocked from the

Digital Modes 16.25

system for the protection of the user from virus attacks.

The WL2K network administrator may post notices that are delivered to all individual WL2K users as a private message. This is a valuable tool for notifying users of system changes, outages, software upgrades, emergencies, etc.

The integrated nature of the system makes possible other services beyond just simple messaging. The bulletin services mentioned above is beyond normal messaging, but WL2K also provides rapid position reporting from anywhere in the world. This facility is interconnected with the APRS, ShipTrak, and YotReps networks. It supports weather reporting from cruising yachts at sea and an interconnection with the YotReps network which is used by government forecasters for weather observations in parts of the world where no others are available. It also allows the maritime user to participate in the National Weather Service's NOAA MAROB a voluntary marine observation reporting program.

To ensure equitable access to the system individual users are assigned daily time limits on HF frequencies by RMS sysops. The default time per any 24-hour period is 30 minutes, however, the user may request more time from the RMS sysop should it be needed. The time limit is individual to each RMS station. Utilization of the PACTOR-II and PACTOR-III protocols are a great timesaver, allowing the user up to 18 times the volume of messages over that of PACTOR-I for the same period of time.

The system has a number of other secondary features to help keep it healthy. Extensive traffic reports are collected, the state of individual RMS stations are monitored and reported if it becomes inactive and daily backups are performed automatically at all RMS stations as well as the Common Message Servers to ensure the system integrity. Security is ensured through the vigorous updating of virus definitions and automatic virus screening for all Internet mail and files. The system has the ability to block any user by both radio (by frequency band) and Internet (by e-mail address) to prevent abuse of the system. Spam is controlled through the use of a secure "acceptance list" or "white list" methodology. More information about the Winlink 2000 system is available on the Winlink Web site at **www.winlink.org**.

16.5.14 D-STAR

D-STAR (Digital Smart Technologies for Amateur Radio) is a digital voice and data protocol specification developed as the result of research by the Japan Amateur Radio League (JARL) to investigate digital technologies for Amateur Radio in 2001. While there are other digital on-air technologies being used by amateurs that were developed for other services, D-STAR is one of the first on-air protocols to be widely deployed and sold by a major radio manufacturer that is designed specifically for amateur service use. D-STAR transfers both voice and data via a data stream over the 2 meter (VHF), 70 cm (UHF), and 23 cm (1.2 GHz) Amateur Radio bands either simplex or via repeater.

One of the interesting features about the D-STAR protocol is the fact the system uses Amateur Radio call signs not only as an identifier, but for signal routing. In the most common configuration, the most vital part of a D-STAR system is the gateway server, which networks a single system into a D-STAR network via a *trust server*. The trust server provides a central, master database to look up users and their associated system. Allowing Amateur Radio operators to respond to calls made to them, regardless of their location on the D-STAR Network. While almost all documentation references the Internet as the connection point for a network, any IP network connectivity will work, depending on signal latency. Additionally, D-STAR can provide a zero infrastructure support system utilizing point-to-point "backbone" 10 GHz connections. Currently, the global D-STAR trust server is maintained by a group of dedicated D-STAR enthusiasts from Dallas, Texas — the Texas Interconnect Team.

The D-STAR protocol specifies two modes; Digital Voice (DV) and Digital Data (DD). In the protocol, the DV mode provides both voice and low speed data channel on 2 meters, 70 cm and 23 cm over a 4800-bit/s data stream. In the protocol, the DV mode uses a data rate of 4800 bit/s. This data stream is broken down to three main packages; voice, forward error correction (FEC) and data. The largest portion of the data stream is the voice package, which is a total of 3600 bits/s with 1200 bits/s dedicated to forward error correction, leaving 1200 bit/s for data. This additional data contains various data flags as well as the data header, leaving about 950 bit/s available for either GPS or serial data. This portion of the data stream does not provide any type of error correction, which has been overcome by implementing error correction in the application software.

While there are various techniques of encoding and transporting a DV signal, the focus of D-STAR's design was the most efficient way to conserve RF spectrum. While D-STAR's "advertised" occupied bandwidth is 6.25 kHz, tests reveal a band plan of 10 kHz spacing is adequate to incorporate the D-STAR signal as well as provide space for channel guards.

In addition to DV mode, the D-STAR protocol outlines the high speed Digital Data (DD) mode. This higher speed data, 128 kbit/s, is available only on the 23 cm band because it requires an advertised 130 kHz bandwidth, only available at 23 cm in world-wide band plans. Unlike the DV mode repeaters, the DD mode module operates as an "access point" operating in half duplex, switching quickly on a single channel.

As with the DV mode, there is a portion of the data stream used for signal identification with the data header as well as various system flags and other D-STAR related items. Once this portion of the data stream is taken into consideration, the 128 kbit/s is reduced to approximately 100 kbit/s — still more than double a dial-up connection speed with significant range. Another consideration is the data rate specified at 128 kbit/s is the gross data rate. Therefore, the system developers are challenged by the area coverage/potential user issue. Meaning the higher the elevation of the system, the more potential users and the slower the system will become as all the users split the data bandwidth. Finally, there is an issue from the days of packet radio. While technically, the opportunity for "hidden transmitter" issues does exist and collisions do occur, the T/R switching is very fast and this effect is handled by TCP/IP as it is for WiFi access points.

The simplex channel eliminates the need for duplexers at a repeater site if only the DD mode system is installed. It is still recommended to have filtering, such as a band-pass filter, in place to reduce possible interference from other digital sources close to the 23 cm band as well as reduce RF overload from nearby RF sources. While some DD Mode system owners would like more sensitivity or more output power (10 W), at the time of print, no manufacturer has developed pre-amps or RF power amplifiers with an adequate T/R switching time to boost the signals.

Radios currently providing DV mode data service use a serial port for low-speed data (1200 bit/s), while the DD mode radio offers a standard Ethernet connection for high-speed (128 kbit/s) connections, to allow easy interfacing with computer equipment. The DD-mode Ethernet jack allows two radios to act as an Ethernet bridge without any special software support required. This allows standard file sharing, FTP, TELNET, HTTP/Web browsing, IRC chat or even Remote Desktop Connections to function as if connected by wire.

In a Gateway configuration, ALL users must be registered in the network. This provides the DD mode sysop a layer of authorization, meaning that if someone wants to use a DD-mode system, and they have not received authorization to use the gateway, their DD-mode access will be denied. Any gateway registered user, on the common network, can use any DD-mode system, even if the registration was not made on that system. While we

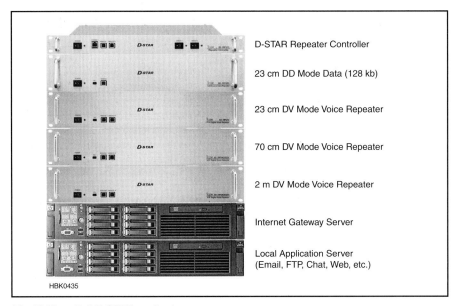

Fig 16.13 — Full D-STAR system.

are not able to use encryption in the Amateur Radio service, security can be implemented in standard software or consumer routers and firewalls.

A D-STAR repeater system consists of at least one RF module and a controller. While any combination of RF modules can be installed, typically a full system includes the three voice modules (2 m, 70 cm and 23 cm) and the 23 cm DD mode module shown in **Fig 16.13**. A computer with dual Ethernet ports, running the Gateway software is required for Internet access to the global network. An additional server, as shown in the diagram, can be incorporated for local hosting of e-mail, chat, FTP, Web and other services. In a D-STAR system installation, the standard repeater components (cavities, isolators, antennas, etc.) are not shown but are required as with any analog system. Some groups have removed analog gear and replaced it with D-STAR components on the same frequency with no additional work beyond connecting the power and feed lines. More information about D-STAR systems may be found in the **Repeaters** and **Digital Communications** chapters.

USER-CREATED FEATURES AND TOOLS

D-STAR has already benefitted from a strong user community that has been developing applications, products and upgrades at a rapid pace. The *DPlus* gateway add-on provides a number of new features, such as an echo test, voice-mail, simulcast to all modules, a linking feature, playback of voice or text messages and more. There are also a number of "D-STAR Reflectors" around the world that act like conference bridges for connecting multiple repeater systems.

One of the other projects that came from the Open D-STAR project is the DV Dongle by Robin Cutshaw, AA4RC. The DV Dongle contains the DSVI AMBE codec chip and a USB interface to allow a computer user to talk with other D-STAR voice users much the way Echolink users can connect to the system with a PC. The DV Tool software is written in Java to provide cross-platform support. The DV Dongle is available at ham radio dealers and the latest software and manual can be downloaded from **www.opendstar.org/tools**.

Pete Loveall, AE5PL, created a DPRS-to-APRS gateway that moves D-STAR GPS data to the APRS-IS network. Although it doesn't natively appear on the APRS RF network (unless IGated specifically), it does appear in all the online APRS tools. DPRS has been implemented in software with DStarmon and the DPRS-specific version of javAPRSSrvr, DPRS Interface. DPRS has been implemented in hardware with the µSmartDigi manufactured by Rich Painter, ABØVO. The µSmartDigi offers a compact, portable TNC designed to act as a gateway for position packets between a D-STAR digital network and a conventional analog APRS RF network via a D-STAR radio and a conventional FM radio. More DPRS and µSmartDigi information is available at **www.aprs-is.net/images/D-PRS.pdf** and **www.usmartdigi.com** respectively.

Satoshi Yasuda, 7M3TJZ/AD6GZ, created a D-STAR node adapter or "hot spot" that connects to a standard simplex FM radio and allows a D-STAR radio user to access the D-STAR network. The system requires the FM rig be accurately configured for a clean signal and loses the D-STAR benefit of only consuming 6.25 kHz of bandwidth. It does allow D-STAR users access to the network via RF where there is Internet access and no repeater infrastructure available. More information can be found online at **d-star.dyndns.org/rig.html.en**.

Brian Roode, NJ6N, wrote one of the earliest end-user applications for D-STAR. His d*Chat is a *Windows*-based keyboard to keyboard communication application that uses the D-STAR DV mode data capability. d*Chat is used to enable text-based communication between multiple stations on a simplex frequency or through a repeater. More information can be found online at **nj6n.com/dstar/dstar_chat.html**.

Dan Smith, KK7DS developed D-RATS with emcomm users in mind. In addition to chat, D-RATS supports file/photo downloads bulletins, forms, e-mail and a sophisticated mapping capability. It also has a "repeater" functionality that allows sharing a D-STAR radio data stream over a LAN. The repeater feature is interesting in an environment where the radio would be physically separated from the software user, like in an emergency operations center. Dan has current developments and downloads available at **www.d-rats.com/wiki**.

D-STAR TV was developed by John Brown, GM7HHB, and provides still images over D-STAR DV mode and streaming video over DD mode. The limited bandwidth of DV mode data and the lack of error correction mean the images take a while and may have some corruption (just like SSTV). The DD mode has much more bandwidth and can support live video. More information is available online at **www.dstartv.com**.

This is just a sample of the development taking place and there will undoubtedly be more applications coming as the user base increases. The two Web sites with the most current D-STAR information are **www.dstarinfo.com** and **www.dstarusers.org**.

16.5.15 P25

Project 25 (P25) or APCO-25 is a suite of standards for digital radio communications for use by public safety agencies in North America. P25 was established by Association of Public-Safety Communications Officials (APCO) to address the need for common digital public safety radio communications standards for first responders and Homeland Security/Emergency Response professionals. In this regard, P25 fills the same role as the European Tetra protocol, although not interoperable with it.

P25 has been developed in phases with

Phase 1 completed in 1995. P25 Phase 1 radio systems operate in 12.5 kHz analog, digital or mixed mode. Phase 1 radios use Continuous 4-level FM (C4FM) modulation for digital transmissions at 4800 baud and two bits per symbol, yielding 9600 bit/s total channel throughput. In the case of data transmission, data packets basically consist of a header, containing overhead information, followed by data. In the case of digitized voice transmission, after the transmission of a header containing error protected overhead information, 2400 bit/s is devoted to periodically repeating the overhead information needed to allow for late entry (or the missed reception of the header).

After evaluating several candidates, Project 25 selected the IMBE (Improved MultiBand Excitation) vocoder from DVSI for phase 1, operating at 4400 bits/s. An additional 2800 bits/s of forward error correction is added for error correction of the digitized voice. An 88.9-bit/s low-speed data channel is provided in the digitized voice frame structure and several forms of encryption are supported.

P25 Phase 2 requires a further bandwidth reduction to 6.25 kHz and operates TDMA (Time Division Multiple Access) with support for trunking systems. It will use the newer AMBE (Advanced MultiBand Excitation) codec from DVSI to accommodate the bit rate reduction to 4800 bit/s. Phase 2 radios will still offer backward compatibility with Phase 1. Phase 3 work has started and will support access to the new 700 MHz band in the US.

P25 radios offer a number of features of interest to public service agencies such as an emergency mode, optional text messaging, over the air programming/deactivation. More P25 information can be found at **www.apco911.org/frequency/project25/information.html#documents** and **www.tiaonline.org/standards/technology/project_25**.

16.6 Glossary

ACK — Acknowledgment, the control signal sent to indicate the correct receipt of a transmission block.

Address — A character or group of characters that identifies a source or destination.

AFSK — Audio frequency-shift keying.

ALE — Automatic link establishment.

Algorithm — Numerical method or process.

APCO — Association of Public Safety Communications Officials.

ARQ — Automatic Repeat reQuest, an error-sending station, after transmitting a data block, awaits a reply (ACK or NAK) to determine whether to repeat the last block or proceed to the next.

ASCII — American National Standard Code for Information Interchange, a code consisting of seven information bits.

AX.25 — Amateur packet-radio link-layer protocol.

Baud — A unit of signaling speed equal to the number of discrete conditions or events per second. (If the duration of a pulse is 20 ms, the signaling rate is 50 baud or the reciprocal of 0.02, abbreviated Bd).

Baudot code — A coded character set in which five bits represent one character. Used in the US to refer to ITA2.

Bell 103 — A 300-baud full-duplex modem using 200-Hz-shift FSK of tones centered at 1170 and 2125 Hz.

Bell 202 — A 1200-baud modem standard with 1200-Hz mark, 2200-Hz space, used for VHF FM packet radio.

BER — Bit error rate.

BERT — Bit-error-rate test.

Bit stuffing — Insertion and deletion of 0s in a frame to preclude accidental occurrences of flags other than at the beginning and end of frames.

Bit — Binary digit, a single symbol, in binary terms either a one or zero.

Bit/s or *bps* — Bits per second.

Bitmap — See **raster.**

Bit rate — Rate at which bits are transmitted in bits/s or bps. **Gross** (or **raw**) **bit rate** includes all bits transmitted, regardless of purpose. **Net bit rate** (also called **throughput**) only includes bits that represent data.

BLER — Block error rate.

BLERT — Block-error-rate test.

BPSK — Binary phase-shift keying in which there are two combinations of phase used to represent data symbols.

Byte — A group of bits, usually eight.

Cache — To store data or packets in anticipation of future use, thus improving routing or delivery performance.

Channel — Medium through which data is transmitted.

Checksum — Data representing the sum of all character values in a packet or message.

CLOVER — Trade name of digital communications system developed by Hal Communications.

COFDM — Coded Orthogonal Frequency Division Multiplex, OFDM plus coding to provide error correction and noise immunity.

Code — Method of representing data.

Codec — Algorithm for compressing and decompressing data.

Collision — A condition that occurs when two or more transmissions occur at the same time and cause interference to the intended receivers.

Compression — Method of reducing the amount of data required to represent a signal or data set. **Decompression** is the method of reversing the compression process.

Constellation — A set of points that represent the various combinations of phase and amplitude in a QAM or other complex modulation scheme.

Contention — A condition on a communications channel that occurs when two or more stations try to transmit at the same time.

Control characters — Data values with special meanings in a protocol, used to cause specific functions to be performed.

Control field — An 8-bit pattern in an HDLC frame containing commands or responses, and sequence numbers.

Convolution — The process of combining or comparing signals based on their behavior over time.

Cyclic Redundancy Check (CRC) — The result of a calculation representing all character values in a packet or message. The result of the CRC is sent with a transmission block. The receiving station uses the received CRC to check transmitted data integrity.

Datagram — A data packet in a connectionless protocol.

Data stream — Flow of information, either over the air or in a network.

DBPSK — Differential binary phase-shift keying.

Dibit — A two-bit combination.

DQPSK — Differential quadrature phase-shift keying.

Domino — A conversational HF digital mode similar in some respects to MFSK16.

DRM — Digital Radio Mondiale. A consortium of broadcasters, manufacturers, research and governmental organizations which developed a system for digital sound broadcasting in bands between 100 kHz and 30 MHz.

Emission — A signal transmitted over the air.

Encoding — Changing data into a form represented by a particular code.

Encryption — Process of using codes and encoding in an effort to obscure the meaning of a transmitted message. **Decryption** reverses the encryption process to recover the original data.

Error Correcting Code (ECC) — Code used to repair transmission errors.

Eye pattern — An oscilloscope display in the shape of one or more eyes for observing the shape of a serial digital stream and any impairments.

Fast Fourier Transform (FFT) — An algorithm that produces the spectrum of a signal from the set of sampled value of the waveform.

FEC — Forward error correction.

Frame — Data set transmitted as one package or set.

Facsimile (fax) — A form of telegraphy for the transmission of fixed images, with or without half-tones, with a view to their reproduction in a permanent form.

FCS — Frame check sequence. (see also **CRC**)

FDM — Frequency division multiplexing.

FDMA — Frequency division multiple access.

FEC — Forward error correction, an error-control technique in which the transmitted data is sufficiently redundant to permit the receiving station to correct some errors.

FSK — Frequency-shift keying.

Gray code — A code that minimizes the number of bits that change between sequential numeric values.

G-TOR — A digital communications system developed by Kantronics.

HDLC — High-level data link control procedures as specified in ISO 3309.

Hellschreiber — A facsimile system for transmitting text.

Host — As used in packet radio, a computer with applications programs accessible by remote stations.

IA5 — International Alphabet, designating a specific set of characters as an ITU standard.

Information field — Any sequence of bits containing the intelligence to be conveyed.

Interleave — Combine more than one data stream into a single stream or alter the data stream in such a way as to optimize it for the modulation and channel characteristics being used.

IP — Internet Protocol, a network protocol used to route information between addresses on the Internet.

ISO — International Organization for Standardization.

ITU — International Telecommunication Union, a specialized agency of the United Nations. (See www.itu.int.)

ITU-T — Telecommunication Standardization Sector of the ITU, formerly CCITT.

Jitter — Unwanted variations in timing or phase in a digital signal.

Layer — In communications protocols, one of the strata or levels in a reference model.

Least significant bit (LSB) — The bit in a byte or word that represents the smallest value.

Level 1 — Physical layer of the OSI reference model.

Level 2 — Link layer of the OSI reference model.

Level 3 — Network layer of the OSI reference model.

Level 4 — Transport layer of the OSI reference model.

Level 5 — Session layer of the OSI reference model.

Level 6 — Presentation layer of the OSI reference model.

Level 7 — Application layer of the OSI reference model.

Lossless (compression) — Method of compression that results in an exact copy of the original data

Lossy (compression) — Method of compression in which some of the original data is lost

MFSK16 — A multi-frequency shift communications system

Modem — Modulator-demodulator, a device that connects between a data terminal and communication line (or radio). Also called **data set**.

Most significant bit (MSB) — The bit in a byte or word that represents the greatest value.

Multicast — A protocol designed to distribute packets of data to many users without communications between the user and data source.

MSK — Frequency-shift keying where the shift in Hz is equal to half the signaling rate in bit/s.

MT63 — A keyboard-to-keyboard mode similar to PSK31 and RTTY.

Nibble — A four-bit quantity. Half a byte.

Node — A point within a network, usually where two or more links come together, performing switching, routine and concentrating functions.

OFDM — Orthogonal Frequency Division Multiplex. A method of using spaced subcarriers that are phased in such a way as to reduce the interference between them.

OSI-RM — Open Systems Interconnection Reference Model specified in ISO 7498 and ITU-T Recommendation X.200.

PACTOR — Trade name of digital communications protocols offered by Special Communications Systems GmbH & Co KG (SCS).

Packet — 1) Radio: communication using the AX.25 protocol. 2) Data: transmitted data structure for a particular protocol (see also **frame.**)

Parity (parity check) — Number of bits with a particular value in a specific data element, such as a byte or word or packet. Parity can be odd or even. A **parity bit** contains the information about the element parity.

Pixel — Abbreviation for "picture element."

Primitive — An instruction for creating a signal or data set, such as in a drawing or for speech.

Project 25 — Digital voice system developed for APCO, also known as P25.

Protocol — A formal set of rules and procedures for the exchange of information.

PSK — Phase-shift keying.

PSK31 — A narrow-band digital communications system developed by Peter Martinez, G3PLX.

QAM — Quadrature Amplitude Modulation. A method of simultaneous phase and amplitude modulation. The number that precedes it, for example, 64QAM, indicates the number of discrete stages in each symbol.

QPSK — Quadrature phase-shift keying in which there are four different combinations of signal phase that represent symbols.

Raster (image) — Images represented as individual data elements, called pixels.

Router — A network packet switch. In packet radio, a network level relay station capable of routing packets.

RTTY — Radioteletype.

Sample — Convert an analog signal to a set of digital values.

Shift — 1) The difference between mark and space frequencies in an FSK or AFSK signal. 2) To change between character sets, such as between LTRS and FIGS in RTTY.

SSID — Secondary station identifier. In AX.25 link-layer protocol, a multipurpose octet to identify several packet radio stations operating under the same call sign.

State — Particular combination of signal attributes used to represent data, such as amplitude or phase.

Start (stop) bit — Symbol used to synchronize receiving equipment at the

beginning (end) of a data byte.
Symbol — Specific state or change in state of a transmission representing a particular signaling event. **Symbol rate** is the number of symbols transmitted per unit of time (see also **baud**).
TAPR — Tucson Amateur Packet Radio Corporation, a nonprofit organization involved in digital mode development.
TDM — Time division multiplexing.
TDMA — Time division multiple access.
Throb — A multi-frequency shift mode like MFSK16.
TNC — Terminal node controller, a device that assembles and disassembles packets (frames).
Trellis — The set of allowed combination of signal states that represent data. (see also **Viterbi coding** and **constellation**.)
Turnaround time — The time required to reverse the direction of a half-duplex circuit, required by propagation, modem reversal and transmit-receive switching time of transceiver.
Unicast — A protocol in which information is exchanged between a single pair of points on a network.
Varicode — A code in which the different data values are represented by codes with different lengths.
Vector — Image represented as a collection of drawing instructions.
Word — Set of bits larger than a byte.

16.7 References and Bibliography

A. Bombardiere, K2MO, "A Quick-Start Guide to ALE400 ARQ FAE," *QST*, Jun 2010, pp 34-36.

B.P. Lathi, *Modern Digital and Analog Communication Systems*, Oxford University Press, 1998

B. Sklar, *Digital Communications, Fundamentals and Applications*, Prentice Hall, 1988

S. Ford, WB8IMY, *HF Digital Handbook*, 4th edition, ARRL, 2007

S. Ford, WB8IMY, *VHF Digital Handbook*, ARRL, 2008

Contents

17.1 High Power, Who Needs It?
17.2 Types of Power Amplifiers
    17.2.1 Why a "Linear" Amplifier?
    17.2.2 Solid State vs Vacuum Tubes
17.3 Vacuum Tube Basics
    17.3.1 Thermionic Emission
    17.3.2 Components of a Vacuum Tube
    17.3.3 Tube Nomenclature
    17.3.4 Tube Mounting and Cooling Methods
    17.3.5 Vacuum Tube Configurations
    17.3.6 Classes of Operation in Tube Amplifiers
    17.3.7 Understanding Tube Operation
17.4 Tank Circuits
    17.4.1 Tank Circuit Q
    17.4.2 Tank Circuit Efficiency
    17.4.3 Tank Output Circuits
17.5 Transmitting Device Ratings
    17.5.1 Plate, Screen and Grid Dissipation
    17.5.2 Tank Circuit Components
    17.5.3 Other Components
17.6 Sources of Operating Voltages
    17.6.1 Tube Filament or Heater Voltage
    17.6.2 Vacuum-Tube Plate Voltage
    17.6.3 Grid Bias
    17.6.4 Screen Voltage For Tubes
17.7 Tube Amplifier Cooling
    17.7.1 Blower Specifications
    17.7.2 Cooling Design Example
    17.7.3 Other Considerations
17.8 Amplifier Stabilization
    17.8.1 Amplifier Neutralization
    17.8.2 VHF and UHF Parasitic Oscillations
    17.8.3 Reduction of Distortion

17.9 Design Example: A High Power Vacuum Tube HF Amplifier
    17.9.1 Tube Capabilities
    17.9.2 Input and Output Circuits
    17.9.3 Filament Supply
    17.9.4 Plate Choke and DC Blocking
    17.9.5 Tank Circuit Design
    17.9.6 Checking Operation
17.10 Solid-State Amplifiers
    17.10.1 Solid State vs Vacuum Tubes
    17.10.2 Classes of Operation
    17.10.3 Modeling Transistors
    17.10.4 Impedance Transformation — "Matching Networks"
    17.10.5 Combiners and Splitters
    17.10.6 Amplifier Stabilization
17.11 A New 250-W Broadband Linear Amplifier
    17.11.1 The 1.8 to 55 MHz PA — Detailed Description
    17.11.2 Control and Protection
    17.11.3 Low-Pass Filter
    17.11.4 Power Supply
    17.11.5 Amplifier Cooling
    17.11.6 Metering and Calibration
    17.11.7 Chassis
    17.11.8 Performance
17.12 Tube Amplifier Projects
    Project: 3CX1500D7 RF Linear Amplifier
    Project: A 6-Meter Kilowatt Amplifier
17.13 References and Bibliography

Chapter 17

RF Power Amplifiers

Amateur Radio operators typically use a very wide range of transmitted power. With the milliwatts used by the QRP operator or the full legal power limit of 1.5 kW, amateur activities are successfully and enjoyably carried out by the amateur fraternity. This chapter covers RF power amplifiers beyond the 100-150 W level of the typical transceiver. The sections on tube-type amplifiers were prepared by John Stanley, K4ERO. Later sections on solid-state amplifiers were contributed by Richard Frey, K4XU. Roger Halstead, K8RI, contributed material on amplifier tuning and the use of surplus components in amplifier construction.

17.1 High Power, Who Needs It?

There are certain activities where higher power levels are recommended, almost required, for the greatest success. A good example is contest operation. While there are some outstanding operators who choose lower power as a sort of handicap and enjoy being competitive in spite of that disadvantage, sooner or later, the high-power stations have the biggest scores.

Another operator likely to run the legal limit is the avid DXer, although it is quite possible to work a lot of DX with a more modest station. There's plenty of DX activity on 40, 80 and 160 meters — where power becomes more important. On 160 meters, it's a real challenge to work 100 countries without the benefit of an amplifier. On 20 through 10 meters, power is not as important when skywave propagation permits working the world with a few watts. For skywave signals, power is much less important than band conditions and patient monitoring. On 10, 6 and 2 meter ground wave paths, power determines range.

Antennas are probably more important than raw power in both DX and contest operation in that the antenna helps both on receive and on transmit. Operating skill is more important than either.

Another useful and important area where higher power may be needed is in net operations. The net control station needs to be heard by all potential net participants, some of whom may be hindered by a noisy location or limited antenna options or operating mobile. For general operations, rag chewing and casual contacts, the need for high power will depend on many factors and can best be learned by experience. As a quick rule of thumb, if you hear a lot of stations that don't seem to hear you, you probably need to run more power. If you operate on bands where noise levels are high (160 and 75/80 meters) or at times when signals are weak then you may find that running the legal limit makes operations more enjoyable. On the other hand, many stations will find that they can be heard fine with the standard 100 W transmitter or even with lower power.

Power requirements also depend on the mode being used. Some digital modes, such as PSK31, work very well with surprisingly low power. CW is more power efficient than SSB voice. Least effective is full carrier AM, which is still used by vintage equipment lovers. Once you have determined that higher power will enhance your operations, you should study the material in this chapter no matter whether you plan to build or buy your amplifier.

Note that many power amplifiers are capable of exceeding the legal limit, just as most automobiles are capable of exceeding posted speed limits. That does not mean that every operator with an amplifier capable of more than 1.5 kW output is a scofflaw. Longer tube life and a cleaner signal result from running an amplifier below its maximum rating. FCC rules call for an accurate way to determine output power, especially when you are running close to the limit.

Safety First!

YOU CAN BE KILLED by coming in contact with the high voltages inside a commercial or homebrew RF amplifier. Please don't take foolish chances. Remember that *you cannot go wrong* by treating each amplifier as potentially lethal! For a more thorough treatment of this all-important subject, please review the applicable sections of the *Safety* chapter in this *Handbook*.

17.2 Types of Power Amplifiers

Power amplifiers are categorized by their power level, intended frequencies of operation, device type, class of operation and circuit configuration. Within each of these categories there are almost always two or more options available. Choosing the most appropriate set of options from all those available is the fundamental concept of design.

17.2.1 Why a "Linear" Amplifier?

The amplifiers commonly used by amateurs for increasing their transmitted power are often referred to as "linears," rather than amplifiers or linear amplifiers. What does this mean and why is it important?

The active device in amplifiers, either tube or transistor, is like a switch. In addition to the "on" and "off" states of a true switch, the active device has intermediate conditions where it presents a finite value of resistance, neither zero nor infinity. As discussed in more detail in the **RF Techniques** chapter, active devices may be operated in various *classes of operation*. Class A operation never turns the device full on or off; it is always somewhere in between. Class B turns the device fully off for about half the time, but never fully on. Class C turns the device off for about 66% of the time, and almost achieves the fully on condition. Class D switches as quickly as possible between the on and off conditions. Other letters have been assigned to various rapid switching methods that try to do what Class D does, only better. Class E and beyond use special techniques to enable the device to make the switching transition as quickly as possible.

During the operating cycle, the highest efficiency is achieved when the active device spends most of its time in the on or off condition and the least in the resistive condition. For this reason, efficiency increases as we go from Class A to B to C to D.

A *linear amplifier* is one that produces an output signal that is identical to the input signal, except that it is stronger. Not all amplifiers do this. Linear amplifiers use Class A, AB or B operation. They are used for modes such as SSB where it is critical that the output be a close reproduction of the input.

The Class C amplifiers used for FM transmitters are *not* linear. A Class C amplifier, properly filtered to remove harmonics, reproduces the frequencies present in the input signal, but the *envelope* of the signal is distorted or even flattened completely. (See the **Modulation** chapter for more information on waveforms, envelopes and other signal characteristics.)

An FM signal has a constant amplitude, so it carries no information in the envelope. A CW signal does carry information in the amplitude variations. Only the on and off states must be preserved, so a Class C amplifier retains the information content of a CW signal. However, modern CW transmitters carefully shape the pulses so that key clicks are reduced to the minimum practical value. A Class C amplifier will distort the pulse shape and make the key clicks worse. Therefore, except for FM, a linear amplifier is recommended for all amateur transmission modes.

Some digital modes, such as RTTY using FSK, are a form of FM and can also use a nonlinear Class C, D or E amplifier. If these signals are not clean, however, a Class C amplifier may make them worse. Also, Class C or even D and E can be used for very slow CW, for very simple low-power CW transmitters or on uncrowded bands where slightly worse key clicks are not so serious. After all, Class C was used for many years with CW operation.

Class of operation as it relates to tube-type amplifier design is discussed in more detail in a later section of this chapter.

ACHIEVING LINEAR AMPLIFICATION

How is linear amplification achieved? Transistors and tubes are capable of being operated in a linear mode by restricting the input signal to values that fall on the linear portion of the curve that relates the input and output power of the device. Improper bias and excessive drive power are the two most common causes of distortion in linear amplifiers. All linear amplifiers can be improperly biased or overdriven, regardless of the power level or whether transistors or tubes are used.

Fig 17.1 shows a simple circuit capable of operating in a linear manner. For linear operation, the bias on the base of the transistors must be such that the circuit begins to produce an output signal even with very small input signal values. As shown in **Fig 17.2**, when properly adjusted for linear operation, the amplifier's output signal faithfully tracks the input signal. Without proper bias on the transistors, there will be no output signal until the input voltage goes above a threshold. As shown in Fig 17.2B, the improperly adjusted amplifier suddenly switches on and produces output when the input signal reaches 0.5 V.

Some tubes are designed for *zero bias* operation. This means that an optimum bias current is inherent in the design of the tube when it is operated with the correct plate voltage. Other types of tubes and all transistors must have bias applied with circuits made for that purpose.

All amplifiers have a limit to the amount of power they can produce, even if the input power is very large. When the output power runs up against this upper limit (that is, when additional drive power results in no more output power), *flat topping* occurs and the output is distorted, as shown in **Fig 17.3**. Many amplifiers use *automatic level control* (ALC) circuits to provide feedback between the amplifier and the companion transceiver or transmitter.

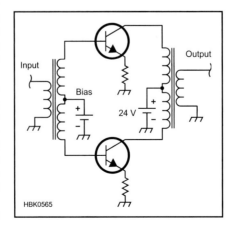

Fig 17.1 — This simple circuit can operate in a linear manner if properly biased.

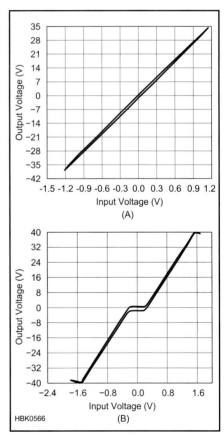

Fig 17.2 — Input versus output signals from an amplifier, as observed with the X-Y display on an oscilloscope. At A, an amplifier with proper bias and input voltage. At B, the same amplifier with improper bias and high input voltage.

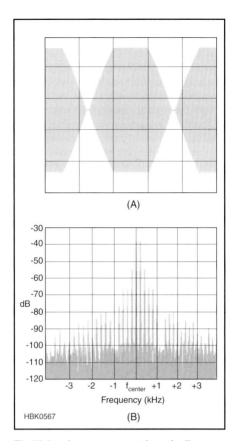

Fig 17.3 — Improper operation of a linear using a two-tone test will show peak clipping on an oscilloscope (A) and the presence of additional frequencies on a spectrum analyzer (B). When these patterns appear, your "linear" has become a "nonlinear" amplifier. Users of adjacent channels will not be happy. More information on transmitter testing may be found in the Test Equipment and Measurements chapter.

When adjusted properly, ALC will control the transmitter output power, preventing the worst effects of overdriving the amplifier. Even with ALC, however, overdriving can occur.

17.2.2 Solid State vs Vacuum Tubes

With the exception of high-power amplifiers, nearly all items of amateur equipment manufactured commercially today use solid-state (semiconductor) devices exclusively. Semiconductor diodes, transistors and integrated circuits (ICs) offer several advantages in designing and fabricating equipment. Solid-state equipment is smaller, offers broadband (no-tune-up) operation, and is easily manufactured using PC boards and automated (lower cost) processes.

Based on all these facts, it might seem that there would be no place for vacuum tubes in a solid-state world. Transistors and ICs do have significant limitations, however, especially in a practical sense. Individual present-day transistors cannot generally handle the combination of current and voltage needed nor can they safely dispose of the amount of heat dissipated for RF amplification to high power levels. Pairs of transistors, or even pairs of pairs, are usually employed in practical power amplifier designs at the 100 W level and beyond. Sometimes various techniques of power combination from multiple amplifiers must be used.

Tube amplifiers can be more economical to build for a given output power. Vacuum tubes operate satisfactorily at surface temperatures as high as 150-200 °C, so they may be cooled by simply blowing sufficient ambient air past or through their relative large cooling surfaces. The very small cooling surfaces of power transistors should be held to 75-100 °C to avoid drastically shortening their life expectancy. Thus, assuming worst-case 50 °C ambient air temperature, the large cooling surface of a vacuum tube can be allowed to rise 100-150 °C above ambient, while the small surface of a transistor must not be allowed to rise more than about 50 °C.

Furthermore, RF power transistors are much less tolerant of electrical abuse than are most vacuum tubes. An overvoltage spike lasting only microseconds can — and is likely to — destroy RF power transistors. A comparable spike is unlikely to have any effect on a tube. So the important message is this: designing with RF power transistors demands caution to ensure that adequate thermal and electrical protection is provided.

Even if one ignores the challenge of the RF portions of a high-power solid-state amplifier, there is the dc power supply to consider. A solid-state amplifier capable of delivering 1 kW of RF output might require regulated (and transient-free) 50 V at more than 40 A. Developing that much current can be challenging. A vacuum tube amplifier at the same power level might require 2000 to 3000 V, unregulated, at less than 1 A.

At the kilowatt level, the vacuum tube is still a viable option for amateur constructors because of its cost-effectiveness and ease of equipment design. Because tube amplifiers and solid-state amplifiers are quite different in many ways, we shall treat them in different sections of this chapter. Also, the author of the solid-state section presents a slightly different perspective on the tube-vs-solid-state discussion.

17.3 Vacuum Tube Basics

The term *vacuum tube* describes the physical construction of the devices, which are usually tubular and have a vacuum inside. The British call them electron valves which describes the operation of the devices, since they control the flow of electrons, like a water valve controls the flow of water.

17.3.1 Thermionic Emission

Metals are electrical conductors because the electrons in them readily move from one atom to the next under the influence of an electrical field. It is also possible to cause the electrons to be emitted into space if enough energy is added to them. Heat is one way of adding energy to metal atoms, and the resulting flow of electrons into space is called *thermionic emission*. As each electron leaves the metal surface, it is replaced by another provided there is an electrical connection from outside the tube to the heated metal.

In a vacuum tube, the emitted electrons hover around the surface of the metal unless acted upon by an electric field. If a positively charged conductor is placed nearby, the electrons are drawn through the vacuum and arrive at that conductor, thus providing a continuous flow of current through the vacuum tube.

17.3.2 Components of a Vacuum Tube

A basic vacuum tube contains at least two parts: a *cathode* and a *plate*. The electrons are emitted from the *cathode*. The cathode can be *directly heated* by passing a large dc current through it, or it can be located adjacent to a heating element (*indirectly heated*). Although ac currents can also be used to directly heat cathodes, if any of the ac voltage mixes with the signal, ac hum will be introduced into the output. If the ac heater supply voltage can be obtained from a center tapped transformer, and the center tap is connected to the signal ground, hum can be minimized.

The difficulty of producing thermionic emission varies with the metal used for the cathode, and is called the "work function" of that metal. An ideal cathode would be made of a metal with a low work function that can sustain high temperatures without melting. Pure tungsten was used in early tubes as it could be heated to a very high temperature. Later it was learned that a very thin layer of thorium greatly increased the emission. Oxides of metals with low work functions were also developed. In modern tubes, thoriated-tungsten is used for the

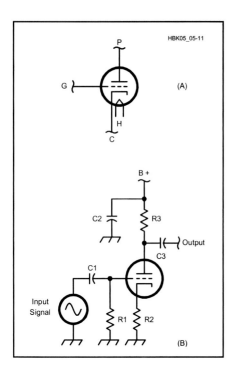

Fig 17.4 — Vacuum tube triode. (A) Schematic symbol detailing heater (H), cathode (C), grid (G) and plate (P). (B) Audio amplifier circuit using a triode. C1 and C3 are dc blocking capacitors for the input and output signals to isolate the grid and plate bias voltages. C2 is a bypass filter capacitor to decrease noise in the plate bias voltage, B+. R1 is the grid bias resistor, R2 is the cathode bias resistor and R3 is the plate bias resistor. Note that although the cathode and grid bias voltages are positive with respect to ground, they are still negative with respect to the plate.

higher power tubes and oxide-coated metals are commonly used at lower power levels.

Filament voltage is important to the proper operation of a tube. If too low, the emission will not be sufficient. If too high, the useful life of the tube will be greatly shortened. It is important to know which type of cathode is being used. Oxide-coated cathodes can be run at 5 to 10% under their rated voltage with an increase in their life, provided performance is satisfactory. They should never be run above the nominal value. Thoriated-tungsten tubes can also gain operational life by running with reduced filament voltage, and as they approach the end of their useful life, can be run at higher than the nominal value as necessary to maintain emission. Carelessly running either type at higher than nominal voltage when new is sure to lead to shorter than normal life with no performance advantage.

Every vacuum tube needs a receptor for the emitted electrons. After moving though the vacuum, the electrons are absorbed by the *plate,* also called the *anode*. This two-element tube — anode and cathode — is called a *diode*. The diode tube is similar to a semiconductor diode: it allows current to pass in only one direction. If the plate goes negative relative to the cathode, current cannot flow because electrons are not emitted from the plate. Years ago, in the days before semiconductors, tube diodes were used as rectifiers.

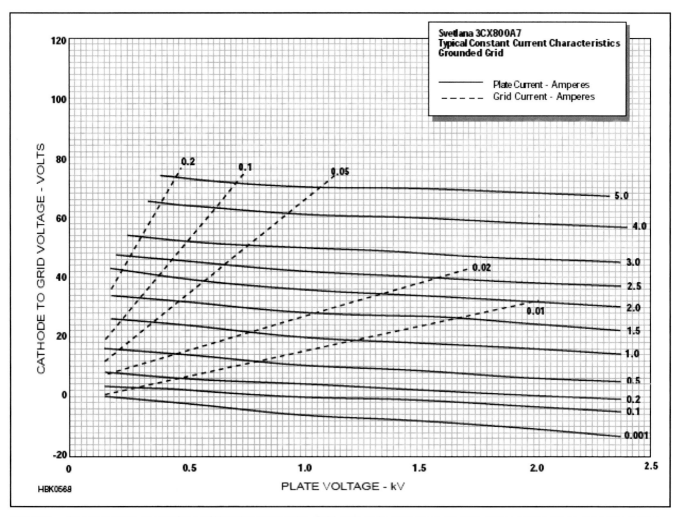

Fig 17.5 — Characteristic curves for a 3CX800A7 triode tube. Grid voltage is plotted on the left, plate voltage along the bottom. The solid lines are plate current, the dashed lines are grid current. This graph is typical of characteristic curves shown in this chapter and used with the *TubeCalculator* program described in the text and included on the *Handbook* CD.

TRIODES

To amplify signals, a vacuum tube must also contain a control *grid*. This name comes from its physical construction. The grid is a mesh of wires located between the cathode and the plate. Electrons from the cathode pass between the grid wires on their way to the plate. The electrical field that is set up by the voltage on these wires affects the electron flow from cathode to plate. A negative grid voltage repels electrons, blocking their flow to the plate. A positive grid voltage enhances the flow of electrons to the plate. Vacuum tubes containing a cathode, a grid and a plate are called *triode* tubes (*tri* for three components). See **Fig 17.4**.

The input impedance of a vacuum tube amplifier is directly related to the grid current. Grid current varies with grid voltage, increasing as the voltage becomes more positive. When the grid voltage is negative, no grid current flows and the input impedance of a tube is nearly infinite. When the grid is driven positive, it draws current and thus presents a lower input impedance, and requires significant drive power. The load placed across the plate of the tube strongly affects its output power and efficiency. An important part of tube design involves determining the optimum load resistance. These parameters are plotted as *characteristic curves* and are used to aid the design process. **Fig 17.5** shows an example.

Since the elements within the vacuum tube are conductors that are separated by an insulating vacuum, the tube is very similar to a capacitor. The capacitance between the cathode and grid, between the grid and plate, and between the cathode and plate can be large enough to affect the operation of the amplifier at high frequencies. These capacitances, which are usually on the order of a few picofarads, can limit the frequency response of an amplifier and can also provide signal feedback paths that may lead to unwanted oscillation. Neutralizing circuits are sometimes used to prevent such oscillations. Techniques for neutralization are presented later in this chapter.

TETRODES

The grid-to-plate capacitance is the chief source of unwanted signal feedback. Therefore tubes were developed with a second grid, called a *screen grid*, inserted between the original grid (now called a *control grid*) and the plate. Such tubes are called tetrodes (having four elements). See **Fig 17.6**. This second grid is usually tied to RF ground and acts as a screen between the grid and the plate, thus preventing energy from feeding back, which could cause instability. Like the control grid, the screen grid is made of a wire mesh and electrons pass through the spaces between the wires to get to the plate.

The screen grid carries a high positive voltage with respect to the cathode, and its proximity to the control grid produces a strong electric field that enhances the attraction of electrons from the cathode. The gain of a tetrode increases sharply as the screen voltage is increased. The electrons accelerate toward the screen grid and most of them pass through the spaces and continue to accelerate until they reach the plate. In large tubes this is aided by careful alignment of the screen wires with the grid wires. The effect of the screen can also be seen in the flattening of the tube curves. Since the screen shields the grid from the plate, the plate current vs plate voltages becomes almost flat, for a given screen and grid voltage. **Fig 17.7** shows characteristic curves for a typical tetrode, and curves for many more tube types are included on the *Handbook* CD.

A special form of tetrode concentrates the electrons flowing between the cathode and the plate into a tight beam. The decreased electron-beam area increases the efficiency of the tube. *Beam tetrodes* permit higher plate currents with lower plate voltages and large power outputs with smaller grid driving power. The 6146 is an example of a beam power tube.

PENTODES

Another unwanted effect in vacuum tubes is the so-called *secondary emission*. The electrons flowing within the tube can have so much energy that they are capable of dislodging electrons from the metal atoms in the grids and plate. Secondary emission can cause a grid, especially the screen grid, to lose more electrons than it absorbs. Thus while a screen usually draws current from its supply, it occasionally pushes current into the supply. Screen supplies must be able to absorb as well as supply current.

A third grid, called the *suppressor grid*, can be added between the screen grid and the plate. This overcomes the effects of secondary emission in tetrodes. A vacuum tube with three grids is called a *pentode* (penta for five components). The suppressor grid is connected to a low voltage, often to the cathode.

17.3.3 Tube Nomenclature

Vacuum tubes are constructed with their elements (cathode, grid, plate) encased in an envelope to maintain the vacuum. Tubes with glass envelopes, such as the classic transmit-

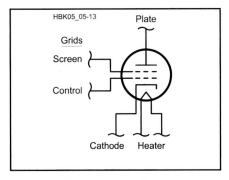

Fig 17.6 — Vacuum tube tetrode. Schematic symbol detailing heater (H), cathode (C), the two grids: control and screen, and the plate (P).

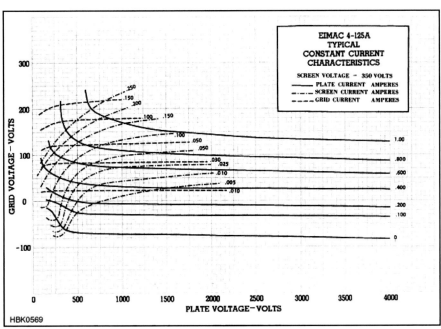

Fig 17.7 — Characteristic curves for a 4-125 tetrode tube.

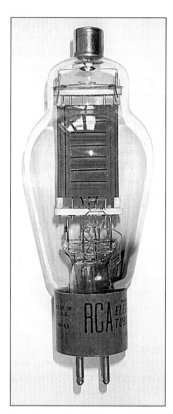

Fig 17.8 — The RCA 811A is an example of a transmitting triode with a glass envelope. (Photo courtesy the Virtual Valve Museum, www.tubecollector.org.)

Fig 17.9 — Modern power tubes, such as this 4CX1000A tetrode, tend to use metal and ceramic construction.

Because of their long history, vacuum tube types do not all follow a single logical system of identification. Many smaller tubes types begin with an indication of the filament voltage, such as the 6AU6 or 12AT7. Other tubes such as the 811 (Fig 17.8) and 6146 were assigned numbers in a more or less chronological order, much as transistors are today. Some glass envelope power tubes follow a numbering system that indicates number of tube elements and plate dissipation — 3-500Z and 4-1000A are two common examples in amateur circles.

Some power tubes follow the 3CX and 4CX numbering system. The first number indicates a triode (3) or tetrode (4) and the C indicates ceramic/metal construction. The X indicates cooling type: X for air, W for water and V for vapor cooling. The cooling type is followed by the plate dissipation. (The tubes that amateurs use typically have three or four numbers indicating plate dissipation; those used in commercial and broadcast service can have much higher numbers.) Thus a 4CX250 is a ceramic, air cooled 250 W tetrode. A 3CX1200 is a ceramic, air cooled, 1200 W triode. Often these tubes have additional characters following the plate dissipation to indicate special features or an upgraded design. For example, a 4CX250R is a special version of the 4CX250 designed for AB1 linear operation. A 4CX1500B is an updated version of the 4CX1500A.

During the heyday of tube technology, some tube types were developed with several tubes in the same glass envelope, such as the 12AX7 (a dual triode). Except for a very few devices used in the specialty audio market, tubes of this type are no longer made.

17.3.4 Tube Mounting and Cooling Methods

Most tubes mount in some kind of socket so that they can be easily replaced when they reach the end of their useful life. Connections to the tube elements are typically made through pins on the base. The pins are arranged or keyed so that the tube can be inserted into the socket only one way and are sized and spaced to handle the operating voltages and currents involved. Tubes generally use a standard base or socket, although a great many different bases developed over the years. Tube data sheets show pinouts for the various tube elements, just like data sheets for ICs and transistors. See the **Component Data and References** chapter for base diagrams of some popular transmitting tubes.

For transmitting tubes, a common arrangement is for filament and grid connections to be made through pins in the main base, while the plate connection is made through a large pin or post at the top of the tube for easier connection to the high voltage supply and tank circuit. This construction is evident in the examples shown in Figs 17.8 and 17.9. To reduce stray reactances, in some older glass tubes the grid used a separate connection.

Heat dissipation from the plate is one of the major limiting factors for vacuum tube power amplifiers. Most early vacuum tubes were encased in glass, and heat passed through it as infrared radiation. If more cooling was needed, air was simply blown over the outside of the glass. Modern ceramic tubes suitable for powers up to 5 kW are usually cooled by forcing air directly through an external anode. These tubes require a special socket that allows free flow of air. The large external anode and cooling fins may be seen in the example in Fig 17.9. Conduction through an insulating block and water cooling are other options, though they are not often seen in amateur equipment. Practical amplifier cooling methods are discussed in detail later in this chapter.

17.3.5 Vacuum Tube Configurations

Just as the case with solid-state devices described in the **Analog Basics** chapter, any of the elements of the vacuum tube can be common to both input and output. A common plate connection — called a cathode follower and similar to the emitter follower — was once used to reduce output impedance (current gain) with little loss of voltage. This application is virtually obsolete.

Most modern tube applications use either the common cathode or the common grid connection. Fig 17.4B shows the common cathode connection, which gives both current and voltage gain. The common grid (often

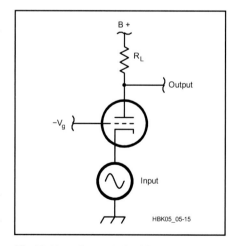

Fig 17.10 — Grounded grid amplifier schematic. The input signal is connected to the cathode, the grid is biased to the appropriate operating point by a dc bias voltage, $-V_G$, and the output voltage is obtained by the voltage drop through R_L that is developed by the plate current, I_P.

ting tube shown in **Fig 17.8**, are most familiar. Over time, manufacturers started exploring other, more rugged, methods for making high power transmitting tubes. Modern power tubes tend to be made of metal parts separated by ceramic insulating sections (**Fig 17.9**).

called *grounded grid*) connection shown in **Fig 17.10** gives only voltage gain. Thus the common cathode connection is capable of 20 to 30 dB of gain in a single stage, whereas the grounded grid connection typically gives 10 to 15 dB of gain.

The input impedance of a grounded grid stage is low, typically less than a few hundred ohms. The input impedance of a grounded cathode stage is much higher.

17.3.6 Classes of Operation in Tube Amplifiers

Class of operation was discussed briefly in the previous section describing the need for linear operation of RF power amplifiers for most Amateur Radio modes except FM. The class of operation of an amplifier stage is defined by its conduction angle, the angular portion of each RF drive cycle, in degrees, during which plate current flows. The conduction angle is determined by the bias on the device, and to a lesser extent on the drive level. These, in turn, determine the amplifier's efficiency, linearity and operating impedances. Refer to **Fig 17.11** for the following discussion.

Class A is defined as operation where plate current is always flowing. For a sine wave this means during 360° of the wave. Class A has the best linearity, but poor efficiency.

Class B is when the bias is set so that the tube is cut off for negative input signals, but current flows when the input signal is positive. Thus for a sine wave, conduction occurs during ½ of the cycle, or 180°. Class B is linear only when two devices operate in push-pull, so as to provide the missing half of the wave, or when a tuned circuit is present to restore the missing half by "flywheel" action (discussed later in this chapter).

Class AB is defined as operation that falls between Class A and Class B. For a sine wave, the conduction angle will be more than 180°, but less than 360°. In practice Class AB amplifiers usually fall within the gray area shown in the center area of the graph. Like Class B, a push pull connection or a tuned circuit are needed for linear operation. Class AB is less efficient that class B, but better than Class A. The linearity is better than class B but worse than class A.

Class AB vacuum tube amplifiers are further defined as class AB1 or AB2. In class AB1, the grid is not driven positive, so no grid current flows. Virtually no drive power is required. In Class AB2 the grid is driven positive at times with respect to the cathode and some grid current flows. Drive power and output both increase as compared to AB1. Most linear amplifiers used in the Amateur service operate Class AB2, although for greater linearity some operate Class AB1 or even Class A.

Class C is when conduction angle is less than 180° — typically 120° to 160° for vacuum tube amplifiers or within the gray area to the left in Fig 17.11. The tube is biased well beyond cutoff when no drive signal is applied. Output current flows only during positive crests in the drive cycle, so it consists of relatively narrow pulses at the drive frequency. Efficiency is high, but nonlinear operation results. Class C amplifiers always use tuned circuits at the input and output. Attempts to achieve extreme efficiency with very narrow pulses (small conduction angles) require very high drive power, so a point of diminishing returns is eventually reached.

Classes D through H use various switched mode techniques. These are used almost exclusively with solid state circuits.

17.3.7 Understanding Tube Operation

Vacuum tubes have complex current transfer characteristics, and each class of operation produces different RMS values of RF current through the load impedance. As described earlier, tube manufacturers provide characteristic curves that show how the tube behaves as operating parameters (such as plate and grid voltage and current) vary. See Figs 17.5 and 17.7 for examples of characteristic curves for two different transmitting tubes. The use of tube curves provides the best way to gain insight into the characteristics of a given tube. Before designing with a tube, get a set of these curves and study them thoroughly.

Because of the complexity and interaction among the various parameters, computer-aided design (CAD) software is useful in analyzing tube operation. One such program, *TubeCalculator*, is included on the *Handbook* CD, along with curves for many popular transmitting tubes. This software makes it easier to do analysis of a given operation with the tube you have chosen. **Fig 17.12** shows a *TubeCalculator* screen with an example of "constant current" curves for a typical tube used in high power amplifiers and tables showing values for the various operating parameters. The curves shown have grid voltage on the vertical axis and plate voltage on the horizontal axis. Older tubes and some newer ones use a slightly different format in which plate voltage is plotted on the horizontal axis and plate current on the vertical. *TubeCalculator* allows analysis using either type.

The tube is the heart of any amplifier. Using the software to arrive at the desired operating parameters is a major step toward understanding and designing an amplifier. The second most important part of the design is the tuning components. Before they can be designed, the required plate load resistance must be determined and *TubeCalculator* will do that. In addition it will give insight into what happens when a tube is under driven or over driven, when the bias is wrong, or when the load resistance incorrect.

If a tube is to be used other than with nominal voltages and currents, analysis using the tube curves is the only solution short of trial

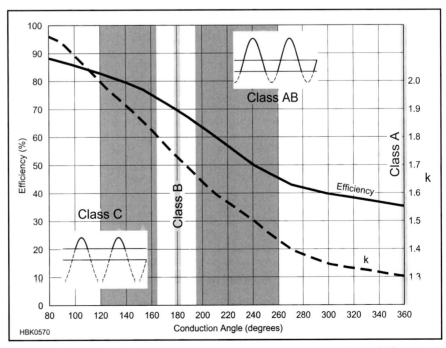

Fig 17.11 — Efficiency and K for various classes of operation. Read the solid line to determine efficiency. Read the dashed line for K, which is a constant used to calculate the plate load required, as explained in the text.

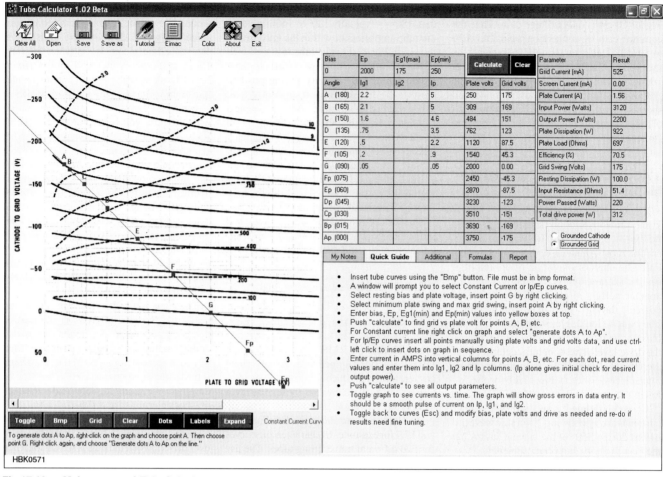

Fig 17.12 — Main screen of *TubeCalculator* program (included on the *Handbook* CD). A characteristic curve plot is loaded in the window on the left side of the screen.

and error. Trial and error is not a good idea because of the high voltages and currents found in high power amplifiers. It's best to conduct your analysis using curves and software, or else stick to operating the tube very close to the voltages, currents, drive levels and load values specified by the manufacturer.

ANALYZING OPERATING PARAMETERS

Characteristic curves allow a detailed look at tube operation as voltages and currents vary. For example, you can quickly see how much negative grid voltage is required to set the plate current to any desired value, depending also on the plate voltage and screen voltages. You can also see how much grid voltage is needed to drive the plate current to the maximum desired value.

With RF power amplifiers, both the grid voltage and the plate voltage will be sinusoidal and will be 180° out of phase. With constant current curves, an operating line can be drawn that will trace out every point of the operating cycle. This will be a straight line connecting two points. One point will be at the intersection of the peak plate voltage and the peak negative grid voltage. The other point will be at the intersection of the peak positive grid voltage and the minimum plate voltage.

If we plot the plate current along this line

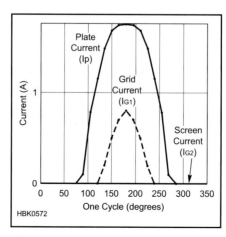

Fig 17.13 — Class AB2 plate and grid current over one cycle as plotted by the *TubeCalculator* program. This plot is for a triode, so there is no curve for screen (G2) current.

as a function of time, it will be seen that it changes in a nonlinear fashion; the exact shape of which depends on the class of operation. For class AB2 operation, which is the most commonly used in linear RF power amplifiers, it will look something like the plot shown in **Fig 17.13**. This rather complicated waveform is not easily evaluated using simple formulas, but it can be analyzed by taking the current values vs time and applying averaging techniques. This method was developed by Chaffee in 1936 and popularized by Eimac, the company that developed many of the power tubes used today.

With *TubeCalculator*, data can be extracted from the curves and can be converted into many useful operating parameters such as input power, output power, grid power required, grid and plate dissipation and required load resistance.

MANUAL METHODS FOR TUBE PARAMETER SELECTION

For those not wishing to use a computer for design, most tube manufacturers will supply a table of typical operating values. A summary of some of this information is available in the **Component Data and References** chapter. These values have already been determined both from the tube characteristics and actual operational tests. As long as your proposed operation is close to the typical values in terms of voltages and currents, the typical values can provide you the desired load resistance and expected output power and efficiency. The typical operating parameters may also include a suggested optimum load resistance. If not, we can use well established "rules of thumb."

The optimum load resistance for vacuum-tube amplifiers can be approximated by the ratio of the dc plate voltage to the dc plate current at maximum signal, divided by a constant appropriate to each class of operation. The load resistance, in turn, determines the maximum power output and efficiency the amplifier can provide. The approximate value for tube load resistance is

$$R_L = \frac{V_p}{K \times I_p} \quad (1)$$

where

R_L = the appropriate load resistance, in ohms
V_P = the dc plate potential, in V
I_P = the dc plate current, in A
K = a constant that approximates the RMS current to dc current ratio appropriate for each class. For the different classes of operation: Class A, $K \approx 1.3$; Class AB, $K \approx 1.5\text{-}1.7$; Class B, $K \approx 1.57\text{-}1.8$; Class C, $K \approx 2$. The way in which K varies for different conduction angles is shown in Fig 17.11 (right scale).

Once we determine the optimum load resistance value for the tube(s) to be used we are ready to design the output networks for the amplifier. After tube selection, this is the most important part of the total design.

17.4 Tank Circuits

Usually we want to drive a transmission line, typically 50 Ω, with the output of our amplifier. An output network is used to transform that 50 Ω impedance to the optimum load resistance for the tube. This transformation is accomplished by resonant output networks which also serve to reduce harmonics to a suitable level. The **Electrical Fundamentals** chapter of this *Handbook* gives a detailed analysis of the operation of resonant circuits. We summarize here only the most important points.

Resonant circuits have the ability to store energy. Capacitors store electrical energy in the electric field between their plates; inductors store energy in the magnetic field induced by the coil winding. These circuits are referred to as *tank circuits* since they act as storage "tanks" for RF energy. This energy is continuously passed back and forth between the inductive storage and the capacitive storage. It can be shown mathematically that the "alternating" current and voltage produced by this process are sinusoidal in waveform with a frequency of

$$f = \frac{1}{2\pi\sqrt{LC}} \quad (2)$$

which, of course, is the resonant frequency of the tank circuit.

17.4.1 Tank Circuit Q

In order to quantify the ability of a tank circuit to store energy, a quality factor, Q, is defined. Q is the ratio of energy stored in a system during one complete RF cycle to energy lost.

$$Q = 2\pi \frac{W_S}{W_L} \quad (3)$$

where
W_S = is the energy stored
W_L = the energy lost to heat and the load

The Flywheel Effect

The operation of a resonant tank is sometimes referred to as the "flywheel" effect. A flywheel does illustrate certain functions of a resonant tank, but a flywheel alone is nonresonant; that is, it has no preferred frequency of operation. A better representation of a resonant tank is found in the *balance wheel* as used in old fashioned mechanical watches and small clocks.

A weighted wheel rotates back and forth, being returned to its center position by a spiral spring, sometimes called a hairspring. The balance wheel stores energy as inertia and is analogous to an inductor. The hairspring also stores energy and is similar in operation to a capacitor. The spring has a adjustment for the frequency of operation or resonance (1 Hz). An escapement mechanism gives the wheel a small kick with each tick of the watch to keep it going. For the purpose of our illustration, imagine energy from the balance wheel being used to drive the clock hands, which is not exactly how real clocks work. Most of the energy would go into the gears that drive the hands. Some energy would be wasted in heat due to friction.

A plot of the radial velocity of the balance wheel will give a sine function. It thus converts the pulses from the escapement into smooth sinusoidal motion. In a similar way, pulses of current in the plate of a tube are smoothed into a sine wave in the "tank" circuit, which is so called because it stores electrical energy, just as a tank stores water.

The plate voltage is a sine wave, even though the plate current is made up of pulses. The frequency at which pulses are added must match the natural resonant frequency of the tank which is adjusted with the "plate tune" capacitor.

Efficient operation of the tank itself occurs when the losses are small compared to the energy transferred to the output. In a similar manner, lowering the losses in a balance wheel by using jeweled bearings makes the watch more efficient. The amount of energy stored in the balance wheel should also be kept as low as practical since excess "oscillation" of the wheel wastes energy to the air and bearings. Likewise, the "circulating currents" in the tank must be limited to reduce heating from the inevitable losses in the components.

Fig 17.A1 — A balance wheel and hairspring in an old clock illustrate the flywheel effect in tank circuits.

A load connected to a tank circuit has exactly the same effect on tank operation as circuit losses. Both consume energy. It just happens that energy consumed by circuit losses becomes heat rather than useful output. When energy is coupled out of the tank circuit into a load, the loaded Q (Q_L) is:

$$Q_L = \frac{X}{R_{Loss} + R_{Load}} \quad (4)$$

where R_{Load} is the load resistance. Energy dissipated in R_{Loss} is wasted as heat. And X represents the reactance of the inductor or the capacitor, assumed to be equal at resonance. Ideally, all the tank circuit energy should be delivered to R_{Load}. This implies that R_{Loss} should be as small as possible.

17.4.2 Tank Circuit Efficiency

The efficiency of a tank circuit is the ratio of power delivered to the load resistance (R_{Load}) to the total power dissipated by losses (R_{Load} and R_{Loss}) in the tank circuit. Within the tank circuit, R_{Load} and R_{Loss} are effectively in series and the circulating current flows through both. The power dissipated by each is proportional to its resistance. The loaded tank efficiency can, therefore, be defined as

$$\text{Tank Efficiency} = \frac{R_{Load}}{R_{Load} + R_{Loss}} \times 100 \quad (5)$$

where efficiency is stated as a percentage. The loaded tank efficiency can also be expressed as

$$\text{Tank Efficiency} = \left(1 - \frac{Q_L}{Q_U}\right) \times 100 \quad (6)$$

where

Q_L = the tank circuit loaded Q, and
Q_U = the unloaded Q of the tank circuit.

For practical circuits, Q_U is very nearly the Q of the coil with switches, capacitors and parasitic suppressors making a smaller contribution. It follows, then, that tank efficiency can be maximized by keeping Q_L low which keeps the circulating current low and the I^2R losses down. Q_U should be maximized for best efficiency; this means keeping the circuit losses low. With a typical Q_L of 10, about 10% of the stored energy is transferred to the load in each cycle. This energy is replaced by energy supplied by the tube. It is interesting to contemplate that in a typical amplifier which uses a Q_L of 10 and passes 1.5 kW from the tube to the output, the plate tank is storing about 15 kW of RF energy. This is why component selection is very important, not only for low loss, but to resist the high voltages and currents.

Resonant circuits are always used in the plate circuit. When the grid is used as the input (common cathode), both matching and a tuned circuit may be used or else a low impedance load is connected from grid to ground with a broad matching transformer or network. The "loaded grid" reduces gain, but improves stability. In grounded grid operation, a tuned circuit may not be needed in the cathode circuit, as the input Z may be close to 50 Ω, but a tuned network may improve the match and usually improves the linearity. These resonant circuits help to ensure that the voltages on grid and plate are sine waves. This wave-shaping effect is the same thing as harmonic rejection. The reinforcing of the fundamental frequency and rejection of the harmonics is a form of filtering or selectivity.

The amount of harmonic suppression is dependent upon circuit loaded Q_L, so a dilemma exists for the amplifier designer. A low Q_L is desirable for best tank efficiency, but yields poorer harmonic suppression. High Q_L keeps amplifier harmonic levels lower at the expense of some tank efficiency. At HF, a compromise value of Q_L can usually be chosen such that tank efficiency remains high and harmonic suppression is also reasonable. At higher frequencies, tank Q_L is not always readily controllable, due to unavoidable stray reactances in the circuit. Unloaded Q_U can always be maximized, however, regardless of frequency, by keeping circuit losses low.

17.4.3 Tank Output Circuits

THE PI NETWORK

The pi network with the capacitors to ground and the inductor in series is commonly used for tube type amplifier matching. This acts like a low pass filter, which is helpful for getting rid of harmonics. Harmonic suppression of a pi network is a function of the impedance transformation ratio and the Q_L of the circuit. Second-harmonic attenuation is approximately 35 dB for a load impedance of 2000 Ω in a pi network with a Q_L of 10. In addition to the low pass effect of the pi network, at the tube plate the third harmonic is already typically 10 dB lower and the fourth approximately 7 dB below that. A typical pi network as used in the output circuit of a tube amplifier is shown in **Fig 17.14**. The **RF and AF Filters** chapter describes harmonic filters that can also greatly reduce harmonics. These are typically not switched but left in the circuit at all times. With such a filter, the requirements for reducing the harmonics on the higher bands with the amplifier pi network is greatly reduced.

The formulas for calculating the component values for a pi network, for those who wish to use them, are included on the CD that accompanies this *Handbook*, along with tabular data for finished designs. The input variables are desired plate load resistance, output impedance to be matched (usually 50 Ω), the desired loaded Q (typically 10 or 12) and the frequency. With these inputs one can calculate the values of the components C1, L1 and C2. These components are usually referred to as the plate tune capacitor, the tank inductor and the loading capacitor. In a multi-frequency amplifier, the coil inductance is changed with a band switch and the capacitors are adjusted to the correct value for the band in question.

Tank circuit component values are most easily found using computer software. The program *PI-EL Design* by Jim Tonne, W4ENE, is included on the *Handbook CD* and is illustrated here. With this software all of the components for a pi or pi-L network (described in the next section) can be quickly calculated. Since there are so many possible variables, especially with a pi-L network, it is impractical to publish graphical or tabular data to cover all cases. Therefore, the use of this software is highly recommended for those designing output networks. The software allows many "what-if" possibilities to be quickly checked and an optimum design found.

THE PI-L NETWORK

There are some advantages in using an additional inductor in the output network, effectively changing it from a pi network to a pi-L network as shown in the bottom right corner of **Fig 17.15**. The harmonic rejection is increased, as shown in **Fig 17.16**, and the

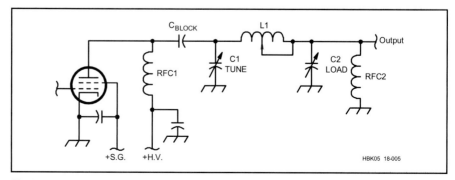

Fig 17.14 — A pi matching network used at the output of a tetrode power amplifier. RFC2 is used for protective purposes in the event C_{BLOCK} fails.

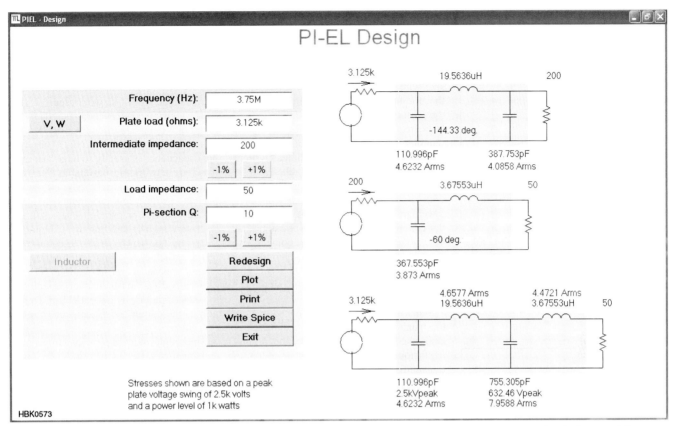

Fig 17.15— PI-EL Design software, included on the *Handbook* CD, may be used to design pi and pi-L networks. A pi network is shown at top right and a pi-L network at bottom right.

component values may become more convenient. Alternatively, the Q can be reduced to lower losses while retaining the same harmonic rejection as the simple pi. This can reduce maximum required tuning capacitance to more easily achievable values.

With a pi-L design, there are many more options than with the simple pi network. This is because the intermediate impedance can take on any value we wish to assign between the output impedance (usually 50 Ω) and the desired plate resistance. This intermediate impedance need not be the same for each frequency, providing the possibility of further optimizing the design. For that reason, it is especially desirable that the software be used instead of using the chart values when a pi-L is contemplated. Using this software, one can quickly determine component values for the required load resistance and Q values for any frequency as well as plotting the harmonic rejection values. Even the voltage and current ratings of the components are calculated.

Further analysis can be done using various versions of *SPICE*. A popular spice version is *LTspice* available from Linear Technologies and downloadable for free on their Web site, www.linear.com. The *PI-EL Design* software mentioned above generates files for *LTspice* automatically. *PI-EL Design* assumes that the

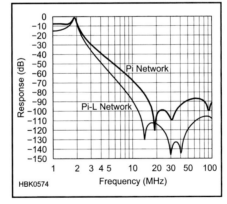

Fig 17.16 — Relative harmonic rejection of pi and pi-L circuits

blocking capacitor has negligible reactance and the RF choke has infinite reactance. There are times when these assumptions may not be valid. The effect of these components and changes to compensate for them can easily be evaluated using *LTspice*. It can also evaluate the effects of parasitic and stray effects, which all components have. The deep nulls in the response curves in Fig 17.16 are caused by the stray capacitance in the inductors. These can be used to advantage but, if not understood, can also lead to unexpected results.

See the **RF Techniques** chapter for more information.

MANUAL METHODS FOR TANK DESIGN

For those who wish to try designing a pi network without a computer, pi designs in chart form are provided. These charts (**Fig 17.17** to **Fig 17.19**) give typical values for the pi network components for various bands and desired plate load resistance. For each value of load resistance the component values can be read for each band. For bands not shown, an approximate value can be reached by interpolation between the bands shown. It's not necessary to be able to read these values to high precision. In practice, unaccounted-for stray capacitance and inductance will likely make the calculated values only an approximate starting point. For those desiring greater precision, this data is available in tabular form on the *Handbook* CD along with the formulas for calculating them.

Several things become obvious from these charts. The required capacitance is reduced and the required inductance increased when a higher load resistance is used. This means that an amplifier with higher plate voltage and lower plate current will require smaller

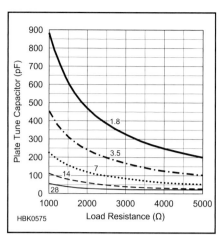

Fig 17.17 — Plate tuning capacitor values for various bands and values of load resistance. Figs 17.17 through 17.19 may be used for manual design of a tank circuit, as explained in the text.

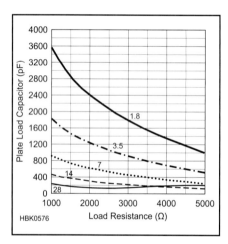

Fig 17.18 — Plate loading capacitor values for various bands and values of load resistance.

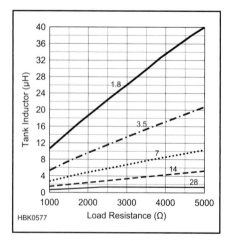

Fig 17.19 — Tank inductor values for various bands and values of load resistance.

capacitor values and larger inductors. The capacitors will, of course, also have to withstand higher voltages, so their physical size may or may not be any smaller. The inductors will have less current in them so a smaller size wire may be used. It will be obvious that for an amplifier covering many bands, the most challenging parts of the design are at the frequency extremes. The 1.8 MHz band requires the most inductance and capacitance and the 28 MHz band requires the least. Often, the output capacitance of the tube plus the minimum value of the plate capacitor along with assorted stray capacitance will put a lower limit on the effective plate capacitance that can be achieved. These charts assume that value to be 25 pF and the other components are adjusted to account for that minimum value. An inevitable trade-off here is that the Q of the tank will be higher than optimum on the highest bands.

When tuning to the lower frequency bands, the maximum value of the capacitor, especially the loading capacitor, will be a limiting factor. Sometimes, fixed mica or ceramic transmitting rated capacitors will be switched in parallel with the variable capacitor to reach the total value required.

COMPONENT SELECTION FOR THE PI-L NETWORK

For those wishing to use manual methods to design a pi-L network, there are look-up tables on the CD along with the mathematical formulas from which they are derived. **Table 17.1** is an abbreviated version that shows the general trends. Values shown are for 1.8 MHz. Other bands can be approximated by dividing all component values by the frequency ratio. For example, for 18 MHz, divide all component values by 10. For 3.6 MHz, divide values by two, and so on.

For a given Q_L, the pi-L circuit has better harmonic rejection than the pi circuit. This allows the designer to use a lower Q_L, resulting in lower losses and lower capacitor values, which is an advantage on the lower frequencies. However, the inductor values will be higher, and the lower capacitor values may be unachievable at higher frequencies. With the pi-L circuit there are many more variables to work with and, thus, one can try many different possibilities to make the circuit work within the limits of the components that are available.

PROBLEMS AT VHF AND HIGHER FREQUENCIES

As the size of a circuit approaches about 5% of a wavelength, components begin to seriously depart from the pure inductance or capacitance we assume them to be. Inductors begin to act like transmission lines. Capacitors often exhibit values far different from their marked values because of stray internal reactances and lead inductance. Therefore, tuned circuits are frequently fabricated in the form of striplines or other transmission lines in order to circumvent the problem of building "pure" inductances and capacitances. The choice of components is often more significant than the type of network used.

The high impedances encountered in VHF tube-amplifier plate circuits are not easily matched with typical networks. Tube output capacitance is usually so large that most matching networks are unsuitable. The usual practice is to resonate the tube output capacitance with a low-loss inductance connected in series or parallel. The result can be a very high-Q tank circuit. Component losses must be kept to an absolute minimum in order to achieve reasonable tank efficiency. Output impedance transformation is usually performed by a link inductively coupled to the tank circuit or by a parallel transformation of the output resistance using a series capacitor.

Since high values of plate load impedance call for low values of plate tuning capacitance, one might be tempted to add additional tubes in parallel to reduce the required load impedance. This only adds to the stray plate capacitance, and the potential for parasitic oscillations is increased. For these reasons, tubes in parallel are seldom used at VHF. Push-pull circuits offer some advantages, but with modern compact ceramic tube types, most VHF amplifiers use a single tube along with distributed type tuned networks. Other approaches are discussed later in this chapter.

"COLD TUNING" AN AMPLIFIER

Because of the high voltage and current involved, as well as the danger of damaging an expensive tube or other component, it is prudent to "cold tune" an amplifier before applying power to it. This can actually be done early in the construction as soon as tank components and the tube are in place. Cold tuning requires some test equipment, but is not difficult or time consuming. Only if you have a problem in getting the tuning right will it take much time, but that is exactly the case in which you would not want to turn on the power without having discovered that there is a problem. With cold tuning, you can also add and remove additional components, such as the RF choke, and see how much it affects the tuning.

There are always stray capacitances and inductances in larger sized equipment, so there is a good chance that your carefully designed circuits may not be quite right. Even with commercial equipment, you may want to become aware of the limitations of the tuning ranges and the approximate settings for the dials for each band. The equipment manual may provide this information, but what if the

Table 17.1
Pi-L Values for 1.8 MHz

Plate Load Resistance (Ω)	Plate Tune Capacitor (pF)	Plate Inductor (μH)	Plate Load Capacitor (pF)	Output Inductor (μH)	Intermediate Resistance (Ω)	Loaded Q (Q_L)	Harmonic Attenuation (3rd/5th, dB)
1000	883	10.5	3550	None (pi)	None (pi)	12	–30/–42
1000	740	15	2370	7.7	200	12	–38/–54
1000	499	21.5	1810	7.7	200	8	–35/–51
2200	376	26.4	1938	7.7	200	12	–39/–55
2200	255	37.7	1498	7.7	200	8	–36/–52
5000	180	50.4	1558	7.7	200	12	–39/–55
5000	124	71	1210	7.7	200	8	–36/–53
5000	114	83	928	11.7	400	8	–39/–55

amplifier you have is a bit out of calibration? In all of these cases, cold tests provide cheap insurance against damage caused by bad tuning and, at the same time, give you practice in setting up.

The first step is to ensure that the equipment is truly cold by removing the power plug from the wall. Since you will be working around the high voltage circuits, you may want to remove fuses or otherwise ensure that power cannot come on. In a well-designed amplifier there will be interlocks that prevent turn on and, perhaps, also short out the plate voltage. If these are in a place that affects the RF circuits, they may have to be temporarily removed. Just be sure to put them back when done.

The adjustment of the plate circuit components is the most important. The easiest way to check those is to attach a resistor across the tube from plate to ground. The resistor should be the same value as the design load resistance and must be noninductive. Several series resistors in the 500 Ω range, either carbon film or the older carbon composition type will work, but *not* wire wound. The tube must remain in its socket and all the normal connections to it should be in place. Covers should be installed, at least for the final tests, since they may affect tuning.

Connect a test instrument to the 50 Ω out-

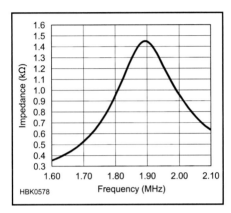

Fig 17.20 — Impedance as would be measured at the plate of an amplifier tube with the output network tuned to 1.9 MHz and the output terminated in its proper load.

put. This can be an antenna bridge or other 50 Ω measuring device. Examples of suitable test equipment would be a vector network analyzer, preferably with an impedance step up transformer, a vector impedance meter or an RX meter, such as the Boonton 250A. If the wattage of the resistor across the plate can stand it, it can even be a low power transmitter.

Select the correct band and tune the plate and load capacitors so that the instrument at the output connector shows 50 Ω or a low SWR at the 50 Ω point. Since this type of circuit is bilateral, you now have settings that will be the same ones you need to transform the 50 Ω load to look like the resistive load you used at the tube plate for the test. If you have an instrument capable of measuring relatively high impedances at RF frequencies, you can terminate the output with 50 Ω and measure the impedance on the plate. It will look something like **Fig 17.20**.

A similar test will work for the input, although it is sometimes more difficult to know what the input impedance at the tube will be. In fact, it will change somewhat with drive level so a low power cold test may not give a full picture. However, just like the case of the plate circuit, you can put your best estimate of the input impedance at the grid using a noninductive resistor. Then, tune the input circuits for a match.

An old timer's method for tuning these circuits, especially the plate circuit, is to use a dip meter. These are getting pretty hard to find these days and, in any case, only show resonance. You won't know if the transformation ratio between plate and output is correct, but it is better than nothing. At least, if you can't get a dip at the proper frequency, you will know that something is definitely wrong.

17.5 Transmitting Device Ratings

17.5.1 Plate, Screen and Grid Dissipation

The ultimate factor limiting the power-handling capability of a tube is often (but not always) its maximum plate dissipation rating. This is the measure of how many watts of heat the tube can safely dissipate, if it is cooled properly, without exceeding critical temperatures. Excessive temperature can damage or destroy internal tube components or vacuum seals — resulting in tube failure. The same tube may have different voltage, current and power ratings depending on the conditions under which it is operated, but its safe temperature ratings must not be exceeded in any case! Important cooling considerations are discussed in more detail later in this chapter.

The efficiency of a power amplifier may range from approximately 25% to 85%, depending on its operating class, adjustment and circuit losses. The efficiency indicates how much of the dc power supplied to the stage is converted to useful RF output power; the rest is dissipated as heat, mostly by the plate. The *TubeCalculator* program will calculate the dissipation of the plate as one of its outputs. Otherwise, it can be determined by multiplying the plate voltage (V) times the plate current (A) and subtracting the output power (W).

For a class AB amplifier, the resting dissipation should also be noted, since with no RF input, *all* of the dc power is dissipated in the plate. Multiply plate voltage times the resting plate current to find this resting dissipation value. Screen dissipation is simply screen voltage times screen current. Grid dissipation is a bit more complicated since some of the power into the grid goes into the bias supply, some is passed through to the output (when grounded grid is used) and some is dissipated in the grid. Some tubes have very fragile grids and cannot be run with any grid current at all.

Almost all vacuum-tube power amplifiers in amateur service today operate as linear amplifiers (Class AB or B) with efficiencies of approximately 50% to 65%. That means that a useful power output of approximately 1 to 2 times the plate dissipation generally can be achieved. This requires, of course, that the tube is cooled enough to realize its maximum plate dissipation rating and that no other tube rating, such as maximum plate current or grid dissipation, is exceeded.

Type of modulation and duty cycle also influence how much output power can be achieved for a given tube dissipation. Some types of operation are less efficient than others, meaning that the tube must dissipate more heat. Some forms of modulation, such as CW or SSB, are intermittent in nature, causing less average heating than modulation formats in which there is continuous transmission (RTTY or FM, for example).

Power-tube manufacturers use two different rating systems to allow for the variations in service. CCS (Continuous Commercial Service) is the more conservative rating and is used for specifying tubes that are in constant use at full power. The second rating system is based on intermittent, low-duty-cycle operation, and is known as ICAS (Intermittent Commercial and Amateur Service). ICAS ratings are normally used by commercial manufacturers and individual amateurs who wish to obtain maximum power output consistent with reasonable tube life in CW and SSB service. CCS ratings should be used for FM, RTTY and SSTV applications. (Plate power transformers for amateur service are also rated in CCS and ICAS terms.).

MAXIMUM RATINGS

Tube manufacturers publish sets of maximum values for the tubes they produce. No maximum rated value should ever be exceeded. As an example, a tube might have a maximum plate-voltage rating of 2500 V, a maximum plate-current rating of 500 mA, and a maximum plate dissipation rating of 350 W. Although the plate voltage and current ratings might seem to imply a safe power input of 2500 V × 500 mA = 1250 W, this is true only if the dissipation rating will not be exceeded. If the tube is used in class AB2 with an expected efficiency of 60%, the maximum safe dc power input is

$$P_{IN} = \frac{100 P_D}{100 - N_D} = \frac{100 \times 350}{100 - 60} = 875 \text{ W}$$

17.5.2 Tank Circuit Components

CAPACITOR RATINGS

The tank capacitor in a high-power amplifier should be chosen with sufficient spacing between plates to preclude high-voltage breakdown. The peak RF voltage present across a properly loaded tank circuit, without modulation, may be taken conservatively as being equal to the dc plate voltage. If the dc supply voltage also appears across the tank capacitor, this must be added to the peak RF voltage, making the total peak voltage twice the dc supply voltage. At the higher voltages, it is usually desirable to design the tank circuit so that the dc supply voltages do not appear across the tank capacitor, thereby allowing the use of a smaller capacitor with less plate spacing. Capacitor manufacturers usually rate their products in terms of the peak voltage between plates. Typical plate spacings are given in **Table 17.2**.

Output tank capacitors should be mounted as close to the tube as possible to allow short low inductance leads to the plate. Especially at the higher frequencies, where minimum circuit capacitance becomes important, the capacitor should be mounted with its stator plates well spaced from the chassis or other shielding. In circuits in which the rotor must be insulated from ground, the capacitor should be mounted on ceramic insulators of a size commensurate with the plate voltage involved and — most important of all, from the viewpoint of safety to the operator — a well-insulated coupling should be used between the capacitor shaft and the knob. The section of the shaft attached to the control knob should be well grounded. This can be done conveniently by means of a metal shaft bushing at the panel.

COIL RATINGS

Tank coils should be mounted at least half their diameter away from shielding or other large metal surfaces, such as blower housings, to prevent a marked loss in Q. Except perhaps at 24 and 28 MHz, it is not essential that the coil be mounted extremely close to the tank capacitor. Leads up to 6 or 8 inches are permissible. It is more important to keep the tank capacitor, as well as other components, out of the immediate field of the coil.

The principal practical considerations in designing a tank coil usually are to select a conductor size and coil shape that will fit into available space and handle the required power without excessive heating. Excessive power loss as such is not necessarily the worst hazard in using too-small a conductor. It is not uncommon for the heat generated to actually unsolder joints in the tank circuit and lead to physical damage or failure. For this reason it's extremely important, especially at power levels above a few hundred watts, to ensure that all electrical joints in the tank circuit are secured mechanically as well as soldered.

Table 17.3 shows recommended conductor sizes for amplifier tank coils, assuming loaded tank circuit Q of 15 or less on the 24 and 30 MHz bands and 8 to 12 on the lower frequency bands. In the case of input circuits for screen-grid tubes where driving power is quite small, loss is relatively unimportant and almost any physically convenient wire

Table 17.2
Typical Tank-Capacitor Plate Spacings

Spacing Inches	Peak Voltage	Spacing Inches	Peak Voltage	Spacing Inches	Peak Voltage
0.015	1000	0.07	3000	0.175	7000
0.02	1200	0.08	3500	0.25	9000
0.03	1500	0.125	4500	0.35	11000
0.05	2000	0.15	6000	0.5	13000

Table 17.3
Copper Conductor Sizes for Transmitting Coils for Tube Transmitters

Power Output (W)	Band (MHz)	Minimum Conductor Size
1500	1.8-3.5	10
	7-14	8 or 1/8"
	18-28	6 or 3/16"
500	1.8-3.5	12
	7-14	10
	18-28	8 or 1/8"
150	1.8-3.5	16
	7-14	12
	18-28	10

*Whole numbers are AWG; fractions of inches are tubing ODs.

size and coil shape is adequate.

The conductor sizes in Table 17.3 are based on experience in continuous-duty amateur CW, SSB and RTTY service and assume that the coils are located in a reasonably well ventilated enclosure. If the tank area is not well ventilated and/or if significant tube heat is transferred to the coils, it is good practice to increase AWG wire sizes by two (for example, change from #12 to #10) and tubing sizes by 1/16 inch.

Larger conductors than required for current handling are often used to maximize unloaded Q, particularly at higher frequencies. Where skin depth effects increase losses, the greater surface area of large diameter conductors can be beneficial. Small-diameter copper tubing, up to 3/8 inch outer diameter, can be used successfully for tank coils up through the lower VHF range. Copper tubing in sizes suitable for constructing high-power coils is generally available in 50-ft rolls from plumbing and refrigeration equipment suppliers. Silver-plating the tubing may further reduce losses. This is especially true as the tubing ages and oxidizes. Silver oxide is a much better conductor than copper oxide, so silver-plated tank coils maintain their low-loss characteristics even after years of use. (There is some debate in amateur circles about the benefits of silver plating.).

At VHF and above, tank circuit inductances do not necessarily resemble the familiar coil. The inductances required to resonate tank circuits of reasonable Q at these higher frequencies are small enough that only strip lines or sections of transmission line are practical. Since these are constructed from sheet metal or large diameter tubing, current-handling capabilities normally are not a relevant factor.

17.5.3 Other Components
RF CHOKES

The characteristics of any RF choke vary with frequency. At low frequencies the choke presents a nearly pure inductance. At some higher frequency it takes on high impedance characteristics resembling those of a parallel-resonant circuit. At a still higher frequency it goes through a series-resonant condition, where the impedance is lowest — generally much too low to perform satisfactorily as a shunt-feed plate choke. As frequency increases further, the pattern of alternating parallel and series resonances repeats. Between resonances, the choke will show widely varying amounts of inductive or capacitive reactance.

In most high-power amplifiers, the choke is directly in parallel with the tank circuit, and is subject to the full tank RF voltage. See **Fig 17.21A**. If the choke does not present a sufficiently high impedance, enough power will be absorbed by the choke to burn it out. To avoid this, the choke must have a sufficiently high reactance to be effective at the lowest frequency (at least equal to the plate load resistance) and yet have no series resonances near any of the higher frequency bands. A resonant-choke failure in a high-power amplifier can be very dramatic and damaging!

Thus, any choke intended for shunt-feed use should be carefully investigated. The best way would be to measure its reactance to ground with an impedance measuring instrument. If the dip meter is used, the choke must be shorted end-to-end with a direct, heavy braid or strap. Because nearby metallic objects affect the resonances, it should be mounted in its intended position, but disconnected from the rest of the circuit. A dip meter coupled an inch or two away from one end of the choke nearly always will show a deep, sharp dip at the lowest series-resonant frequency and shallower dips at higher series resonances.

Any choke to be used in an amplifier for the 1.8 to 28 MHz bands requires careful (or at least lucky!) design to perform well on all amateur bands within that range. Most simply put, the challenge is to achieve sufficient inductance that the choke doesn't "cancel" a large part of tuning capacitance at 1.8 MHz. At the same time, try to position all its series resonances where they can do no harm. In general, close wind enough #20 to #24 magnet wire to provide about 135 µH inductance on a 3/4 to 1-inch diameter cylindrical form of ceramic, Teflon or fiberglass. This gives a reactance of 1500 Ω at 1.8 MHz and yet yields a first series resonance in the vicinity of 25 MHz. Before the advent of the 24 MHz band this worked fine. But trying to "squeeze" the resonance into the narrow gaps between the 21, 24 and/or 28-MHz bands is quite risky unless sophisticated instrumentation is available. If the number of turns on the choke is selected to place its first series resonance at 23.2 MHz, midway between 21.45 and 24.89 MHz, the choke impedance will typically be high enough for satisfactory operation on the 21, 24 and 28 MHz bands. The choke's first series resonance should be measured very carefully as described above using a dip meter and calibrated receiver or RF impedance bridge, with the choke mounted in place on the chassis.

Investigations with a vector impedance meter have shown that "trick" designs, such as using several shorter windings spaced along the form, show little if any improvement in choke resonance characteristics. Some commercial amplifiers circumvent the problem by band switching the RF choke. Using a larger diameter (1 to 1.5 inches) form does move the first series resonance somewhat higher for a given value of basic inductance. Beyond that, it is probably easiest for an all-band amplifier to add or subtract enough turns to move the first resonance to about 35 MHz and settle for a little less than optimum reactance on 1.8 MHz.

However, there are other alternatives. If one is willing to switch the choke when changing bands, it is possible to have enough inductance for 1.8 to 10 MHz, with series resonances well above 15 MHz. Then for 14 MHz and above, a smaller choke is used which has its resonances well above 30 MHz. Providing an extra pole on the band switch is, of course, the trade-off. This switch must withstand the full plate voltage. Switches suitable for changing bands for the pi network would handle this fine.

Another approach is to feed the high-voltage dc through the main tank inductor, putting the RF choke at the loading capacitor,

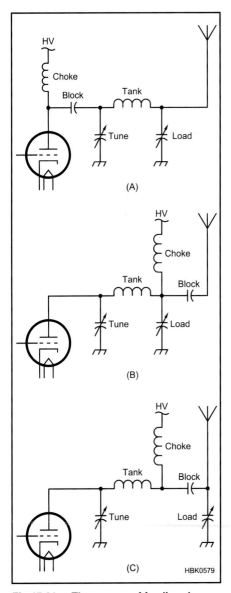

Fig 17.21 — Three ways of feeding dc to a tube via an RF choke. See text for a discussion of the tradeoffs.

instead of at the tube. (See Fig 17.21B) This puts a much lower RF voltage on the choke and, thus, not as much reactance is required for satisfactory rejection of the RF voltage. However, this puts both dc and RF voltages on the plate and loading capacitors which may be beyond their ratings. The blocking capacitor can be put before the loading capacitor, as in Fig 17.21C. This removes the dc from the loading capacitor, which typically has a lower voltage rating than the plate capacitor, but puts high current in the blocker.

Yet another method involves using hollow tubing for the plate tank and passing the dc lead through it. This lowers the RF voltage on the choke without putting dc voltage on the tuning components. This method works best for higher power transmitters where the tuning inductor can be made of 1/8 inch or larger copper tubing.

BLOCKING CAPACITORS

A series capacitor is usually used at the input of the amplifier output circuit. Its purpose is to block dc from appearing on matching circuit components of the antenna. As mentioned in the section on tank capacitors, output-circuit voltage requirements are considerably reduced when only RF voltage is present.

To provide a margin of safety, the voltage rating for a blocking capacitor should be at least 25% to 50% greater than the dc voltage applied. A large safety margin is desirable, since blocking capacitor failure can bring catastrophic results. The worse case is when dc is applied to the output of the transmitter and even to the antenna, with potentially fatal results. Often an RF choke is placed from the RF output jack to ground as a safety backup. A shorted blocker will blow the power supply fuse.

To avoid affecting the amplifier's tuning and matching characteristics, the blocking capacitor should have a low impedance at all operating frequencies. If it presents more than 5% of the plate load resistance, the pi components should be adjusted to compensate. Use of a *SPICE* analysis provides a useful way to see what adjustments might be required to maintain the desired match.

The capacitor also must be capable of handling, without overheating or significantly changing value, the substantial RF current that flows through it. This current usually is greatest at the highest frequency of operation where tube output capacitance constitutes a significant part of the total tank capacitance. A significant portion of circulating tank current, therefore, flows through the blocking capacitor. When using the connection of the RF choke shown in Fig 17.21C, the entire circulating current must be accommodated.

Transmitting capacitors are rated by their manufacturers in terms of their RF current-carrying capacity at various frequencies. Below a couple hundred watts at the high frequencies, ordinary disc ceramic capacitors of suitable voltage rating work well in high-impedance tube amplifier output circuits. Some larger disk capacitors rated at 5 to 8 kV also work well for higher power levels at HF. For example, two inexpensive Centralab type DD-602 discs (0.002 µF, 6 kV) in parallel have proved to be a reliable blocking capacitor for 1.5-kW amplifiers operating at plate voltages to about 2.5 kV. At very high power and voltage levels and at VHF, ceramic "doorknob" transmitting capacitors are needed for their low losses and high current handling capabilities. When in doubt, adding additional capacitors in parallel is cheap insurance against blocking capacitor failure and also reduces the impedance. So-called "TV doorknobs" may break down at high RF current levels and should be avoided.

The very high values of Q_L found in many VHF and UHF tube-type amplifier tank circuits often require custom fabrication of the blocking capacitor. This can usually be accommodated through the use of a Teflon "sandwich" capacitor. Here, the blocking capacitor is formed from two parallel plates separated by a thin layer of Teflon. This capacitor often is part of the tank circuit itself, forming a very low-loss blocking capacitor. Teflon is rated for a minimum breakdown voltage of 2000 V per mil of thickness, so voltage breakdown should not be a factor in any practically realized circuit. The capacitance formed from such a Teflon sandwich can be calculated from the information presented elsewhere in this *Handbook* (use a dielectric constant of 2.1 for Teflon). In order to prevent any potential irregularities caused by dielectric thickness variations (including air gaps), Dow-Corning DC-4 silicone grease should be evenly applied to both sides of the Teflon dielectric. This grease has properties similar to Teflon, and will fill in any surface irregularities that might cause problems.

17.6 Sources of Operating Voltages

17.6.1 Tube Filament or Heater Voltage

A power vacuum tube can use either a directly heated filament or an indirectly heated cathode. The filament voltage for either type should be held within 5% of rated voltage. Because of internal tube heating at UHF and higher, the manufacturers' filament voltage rating often is reduced at these higher frequencies. The derated filament voltages should be followed carefully to maximize tube life.

Series dropping resistors may be required in the filament circuit to attain the correct voltage. Adding resistance in series will also reduce the inrush current when the tube is turned on. Cold tungsten has much lower resistance than when hot. Circuits are available that both limit the inrush current at turn on and also regulate the voltage against changes in line voltage.

The voltage should be measured with a true RMS meter at the filament pins of the tube socket while the amplifier is running. The filament choke and interconnecting wiring all have voltage drops associated with them. The high current drawn by a power-tube heater circuit causes substantial voltage drops to occur across even small resistances. Also, make sure that the plate power drawn from the power line does not cause the filament voltage to drop below the proper value when plate power is applied.

Thoriated filaments lose emission when the tube is overloaded appreciably. If the overload has not been too prolonged, emission, sometimes, may be restored by operating the filament at rated voltage, with all other voltages removed, for a period of 30 to 60 minutes. Alternatively, you might try operating the tube at 20% above rated filament voltage for five to ten minutes.

17.6.2 Vacuum-Tube Plate Voltage

DC plate voltage for the operation of RF amplifiers is most often obtained from a transformer-rectifier-filter system (see the **Power Supplies** chapter) designed to deliver the required plate voltage at the required current. It is not unusual for a power tube to arc over internally (generally from the plate to the screen or control grid) once or twice, especially soon after it is first placed into service. The flashover by itself is not normally dangerous to the tube, provided that instantaneous maximum plate current to the tube is held to a safe value and the high-voltage plate supply is shut off very quickly.

A good protective measure against this is the inclusion of a high-wattage power resistor in series with the plate high-voltage circuit. The value of the resistor, in ohms, should be approximately 10 to 15 times the no-load plate voltage in KV. This will limit peak fault current to 67 to 100 A. The series resistor should be rated for 25 or 50-W power dissipation; vitreous enamel coated wire-wound resistors have been found to be capable of handling repeated momentary fault-current surges without damage. Aluminum-cased resistors (Dale) are not recommended for this application. Each resistor also must be large enough to safely handle the maximum value of normal plate current; the wattage rating required may be calculated from $P = I^2R$. If the total filter capacitance exceeds 25 µF, it is a good idea to use 50-W resistors in any case. Even at high plate-current levels, the addition of the resistors does little to affect the dynamic regulation of the plate supply.

Since tube (or other high-voltage circuit) arcs are not necessarily self-extinguishing, a fast-acting plate overcurrent relay or primary circuit breaker is also recommended to quickly shut off ac power to the HV supply when an arc begins. Using this protective system, a mild HV flashover may go undetected, while a more severe one will remove ac power from the HV supply. (The cooling blower should remain energized, however, since the tube may be hot when the HV is removed due to an arc.) If effective protection is not provided, however, a "normal" flashover, even in a new tube, is likely to damage or destroy the tube, and also frequently destroys the rectifiers in the power supply as well as the plate RF choke. A power tube that flashes over more than about 3 to 5 times in a period of several months likely is defective and will have to be replaced before long.

17.6.3 Grid Bias

The grid bias for a linear amplifier should be highly filtered and well regulated. Any ripple or other voltage change in the bias circuit modulates the amplifier. This causes hum and/or distortion to appear on the signal. Since most linear amplifiers draw only small amounts of grid current, these bias-supply requirements are not difficult to achieve.

Fixed bias for class AB1 tetrode and pentode amplifiers is usually obtained from a variable-voltage regulated supply. Voltage adjustment allows setting bias level to give the desired resting plate current. **Fig 17.22A** shows a simple Zener-diode-regulated bias supply. The dropping resistor is chosen to allow approximately 10 mA of Zener current. Bias is then reasonably well regulated for all drive conditions up to 2 or 3 mA of grid current. The potentiometer allows bias to be adjusted between Zener and approximately 10 V higher. This range is usually adequate to allow for variations in the characteristics of different tubes. Under standby conditions, when it is desirable to cut off the tube entirely, the Zener ground return is interrupted so the full bias supply voltage is applied to the grid.

In Fig 17.22B and C, bias is obtained from the voltage drop across a Zener diode in the cathode (or filament center-tap) lead. Operating bias is obtained by the voltage drop across D1 as a result of plate (and screen) current flow. The diode voltage drop effectively raises the cathode potential relative to the grid. The grid is, therefore, negative with respect to the cathode by the Zener voltage of the diode. The Zener-diode wattage rating should be twice the product of the maximum cathode current times the rated Zener voltage. Therefore, a tube requiring 15 V of bias with a maximum cathode current of 100 mA would dissipate 1.5 W in the Zener diode. To allow a suitable safety factor, the diode rating should be 3 W or more. The circuit of Fig 17.22C illustrates how D1 would be used with a cathode driven (grounded grid) amplifier as opposed to the grid driven example at B.

In all cases, the Zener diode should be bypassed by a 0.01-µF capacitor of suitable voltage. Current flow through any type of diode generates shot noise. If not bypassed, this noise would modulate the amplified signal, causing distortion in the amplifier output.

17.6.4 Screen Voltage For Tubes

Power tetrode screen current varies widely with both excitation and loading. The current may be either positive or negative, depending on tube characteristics and amplifier operating conditions. In a linear amplifier, the screen voltage should be well regulated for all values of screen current. The power output from a tetrode is very sensitive to screen voltage, and any dynamic change in the screen potential can cause distorted output. Zener diodes are commonly used for screen regulation.

Fig 17.23 shows a typical example of a regulated screen supply for a power tetrode amplifier. The voltage from a fixed dc supply is dropped to the Zener stack voltage by the current-limiting resistor. A screen bleeder resistor is connected in parallel with the Zener stack to allow for the negative screen current developed under certain tube operating conditions. Bleeder current is chosen to be roughly 10 to 20 mA greater than the expected maximum negative screen current, so that screen voltage is regulated for all values of current between maximum negative screen current and maximum positive screen current. For external-anode tubes in the 4CX250 family, a typical screen bleed-

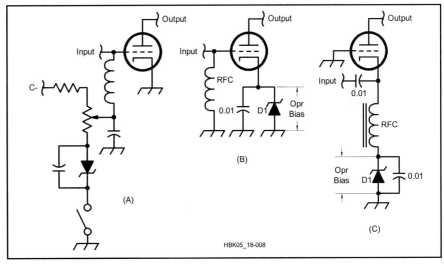

Fig 17.22 — Various techniques for providing operating bias with tube amplifiers.

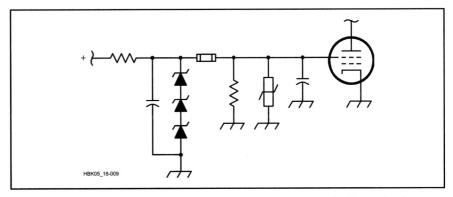

Fig 17.23 — A Zener-regulated screen supply for use with a tetrode. Protection is provided by a fuse and a varistor.

er current value would be 20 mA. For the 4CX1000 family, a screen-bleeder current of 70 mA is required.

Screen voltage should never be applied to a tetrode unless plate voltage and load also are applied; otherwise, the screen will act like an anode and will draw excessive current. Perhaps the best way to insure this is to include logic circuits that will not allow the screen supply to turn on until it senses plate voltage. Supplying the screen through a series-dropping resistor from the plate supply affords a measure of protection, since the screen voltage only appears when there is plate voltage. Alternatively, a fuse can be placed between the regulator and the bleeder resistor. The fuse should not be installed between the bleeder resistor and the tube because the tube should never be operated without a load on the screen. Without a load, the screen potential tends to rise to the anode voltage. Any screen bypass capacitors or other associated circuits are likely be damaged by this high voltage.

In Fig 17.23, a varistor is connected from screen to ground. If, because of some circuit failure, the screen voltage should rise substantially above its nominal level, the varistor will conduct and clamp the screen voltage to a low level. If necessary to protect the varistor or screen dropping resistors, a fuse or overcurrent relay may be used to shut off the screen supply so that power is interrupted before any damage occurs. The varistor voltage should be approximately 30% to 50% higher than normal screen voltage.

17.7 Tube Amplifier Cooling

Vacuum tubes must be operated within the temperature range specified by the manufacturer if long tube life is to be achieved. Tubes having glass envelopes and rated at up to 25-W plate dissipation may be used without forced-air cooling if the design allows a reasonable amount of convection cooling. If a perforated metal enclosure is used, and a ring of ¼ to ⅜-inch-diameter holes is placed around the tube socket, normal convective airflow can be relied on to remove excess heat at room temperatures.

For tubes with greater plate dissipation ratings, and even for very small tubes operated close to maximum rated dissipation, forced-air cooling with a fan or blower is needed. Most manufacturers rate tube-cooling requirements for continuous-duty operation. Their literature will indicate the required volume of airflow, in cubic feet per minute (CFM), at some particular back pressure. Often, this data is given for several different values of plate dissipation, ambient air temperature and even altitude above sea level.

One extremely important consideration is often overlooked by power-amplifier designers and users alike: a tube's plate dissipation rating is only its maximum potential capability. The power that it can actually dissipate safely depends directly on the cooling provided. The actual power capability of virtually all tubes used in high-power amplifiers for amateur service depends on the volume of air forced through the tube's cooling structure.

17.7.1 Blower Specifications

This requirement usually is given in terms of cubic feet of air per minute (CFM), delivered into a back pressure, representing the resistance of the tube cooler to air flow, stated in inches of water. Both the CFM of airflow required and the pressure needed to force it through the cooling system are determined by ambient air temperature and altitude (air density), as well as by the amount of heat to be dissipated. The cooling fan or blower must be capable of delivering the specified airflow into the corresponding back pressure. As a result of basic air flow and heat transfer principles, the volume of airflow required through the tube cooler increases considerably faster than the plate dissipation, and back pressure increases even faster than airflow.

Table 17.4
Specifications of Some Popular Tubes, Sockets and Chimneys

Tube	CFM	Back Pressure (inches)	Socket	Chimney
3-500Z	13	0.13	SK-400, SK-410	SK-416
3CX800A7	19	0.50	SK-1900	SK-1906
3CX1200A7	31	0.45	SK-410	SK-436
3CX1200Z7	42	0.30	SK-410	—
3CX1500/8877	35	0.41	SK-2200, SK-2210	SK-2216
4-400A/8438	14	0.25	SK-400, SK-410	SK-406
4-1000A/8166	20	0.60	SK-500, SK-510	SK-506
4CX250R/7850	6.4	0.59	SK602A, SK-610, SK-610A, SK-611, SK-612, SK-620, SK-620A, SK-621, SK-630	
4CX400/8874	8.6	0.37	SK1900	SK606
4CX400A	8	0.20	SK2A	—
4CX800A	20	0.50	SK1A	—
4CX1000A/8168	25	0.20	SK-800B, SK-810B, SK-890B	SK-806
4CX1500B/8660	34	0.60	SK-800B, SK-1900	SK-806
4CX1600B	36	0.40	SK3A	CH-1600B

These values are for sea-level elevation. For locations well above sea level (5000 ft/1500 m, for example), add an additional 20% to the figure listed.

Table 17.5
Blower Performance Specifications

Wheel Dia	Wheel Width	RPM	Free Air CFM	CFM for Back Pressure (inches)					Stock No.
				0.1	0.2	0.3	0.4	0.5	
2"	1"	3340	12	9	6	—	—	—	1TDN2
2¹⁵⁄₁₆"	1½"	3388	53	52	50	47	41	23	1TDN5
3"	1⅞"	3036	50	48	44	39	32	18	1TDN7
3"	1⅞"	3010	89	85	78	74	66	58	1TDP1
3¹⁵⁄₁₆"	1¹⁵⁄₁₆"	3016	75	71	68	66	61	56	1TDP3
3¾"	1⅞"	2860	131	127	119	118	112	105	1TDP5

Representative sample of Dayton squirrel cage blowers. More information and other models available from Grainger Industrial Supply (www.grainger.com).

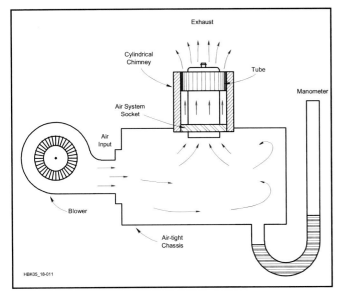

Fig 17.24 — Air is forced into the chassis by the blower and exits through the tube socket. The manometer is used to measure system back pressure, which is an important factor in determining the proper size blower.

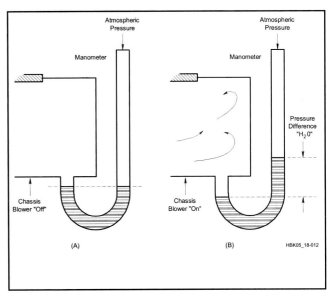

Fig 17.25 — At A the blower is "off" and the water will seek its own level in the manometer. At B the blower is "on" and the amount of back pressure in terms of inches of water can be measured as indicated.

In addition, blower air output decreases with increasing back pressure until, at the blower's so-called "cutoff pressure," actual air delivery is zero. Larger and/or faster-rotating blowers are required to deliver larger volumes of air at higher back pressure.

Values of CFM and back pressure required to realize maximum rated plate dissipation for some of the more popular tubes, sockets and chimneys (with 25 °C ambient air and at sea level) are given in **Table 17.4**. Back pressure is specified in inches of water and can be measured easily in an operational air system as indicated in **Figs 17.24** and **17.25**. The pressure differential between the air passage and atmospheric pressure is measured with a device called a *manometer*. A manometer is nothing more than a piece of clear tubing, open at both ends and fashioned in the shape of a "U." The manometer is temporarily connected to the chassis and is removed after the measurements are completed. As shown in the diagrams, a small amount of water is placed in the tube. At Fig 17.25A, the blower is "off" and the water seeks its own level, because the air pressure (ordinary atmospheric pressure) is the same at both ends of the manometer tube. At B, the blower is "on" (socket, tube and chimney in place) and the pressure difference, in terms of inches of water, is measured. For most applications, a standard ruler used for measurement will yield sufficiently accurate results.

Table 17.5 gives the performance specifications for a few of the many Dayton blowers, which are available through Grainger Industrial Supply (**www.grainger.com**). Other blowers having wheel diameters, widths and rotational speeds similar to any in Table 17.5 likely will have similar flow and back pressure characteristics. If in doubt about specifications, consult the manufacturer. Tube temperature under actual operating conditions is the ultimate criterion for cooling adequacy and may be determined using special crayons or lacquers that melt and change appearance at specific temperatures. The setup of Fig 17.25, however, nearly always gives sufficiently accurate information.

17.7.2 Cooling Design Example

As an example, consider the cooling design of a linear amplifier to use one 3CX800A7 tube to operate near sea level with the air temperature not above 25 °C. The tube, running 1150-W dc input, easily delivers 750-W continuous output, resulting in 400-W plate dissipation ($P_{DIS} = P_{IN} - P_{OUT}$). According to the manufacturer's data, adequate tube cooling at 400 W P_D requires at least 6 CFM of air at 0.09 inches of water back pressure. In Table 17.5, a Dayton no. 1TDN2 will do the job with a good margin of safety.

If the same single tube were to be operated at 2.3 kW dc input to deliver 1.5 kW output (substantially exceeding its maximum electrical ratings!), P_{IN} would be about 2300 W and $P_D \approx 800$ W. The minimum cooling air required would be about 19 CFM at 0.5 inches of water pressure — doubling P_{DIS}, more than tripling the CFM of air flow required and increasing back pressure requirements on the blower by a factor of 5.5!

However, two 3CX800A7 tubes are needed to deliver 1.5 kW of continuous maximum legal output power in any case. Each tube will operate under the same conditions as in the single-tube example above, dissipating 400 W. The total cooling air requirement for the two tubes is, therefore, 12 CFM at about 0.09 inches of water, only two-thirds as much air volume and one-fifth the back pressure required by a single tube. While this may seem surprising, the reason lies in the previously mentioned fact that both the airflow required by a tube and the resultant back pressure increase much more rapidly than P_D of the tube. Blower air delivery capability, conversely, decreases as back pressure is increased. Thus, a Dayton 1TDN2 blower can cool two 3CX800A7 tubes dissipating 800 W total, but a much larger (and probably noisier) no. 1TDN7 would be required to handle the same power with a single tube.

In summary, three very important considerations to remember are these:

• A tube's actual safe plate dissipation capability is totally dependent on the amount of cooling air forced through its cooling system. Any air-cooled power tube's maximum plate dissipation rating is meaningless unless the specified amount of cooling air is supplied.

• Two tubes will always safely dissipate a given power with a significantly smaller (and quieter) blower than is required to dissipate the same power with a single tube of the same type. A corollary is that a given blower can virtually always dissipate more power when cooling two tubes than when cooling a single tube of the same type.

• Blowers vary greatly in their ability to deliver air against back pressure so blower selection should not be taken lightly.

17.7.3 Other Considerations

A common method for directing the flow of air around a tube involves the use of a pressurized chassis. This system is shown in Fig 17.24. A blower attached to the chassis forces air around the tube base, often through holes in its socket. A chimney is used to guide air leaving the base area around the tube envelope or anode cooler, preventing it from dispersing and concentrating the flow for maximum cooling.

A less conventional approach that offers a significant advantage in certain situations is shown in **Fig 17.26**. Here the anode compartment is pressurized by the blower. A special chimney is installed between the anode heat exchanger and an exhaust hole in the compartment cover. When the blower pressurizes the anode compartment, there are two parallel paths for airflow: through the anode and its chimney, and through the air system socket. Dissipation, and hence cooling air required, generally is much greater for the anode than for the tube base. Because high-volume anode airflow need not be forced through restrictive air channels in the base area, back pressure may be very significantly reduced with certain tubes and sockets. Only airflow actually needed is bled through the base area. Blower back pressure requirements may sometimes be reduced by nearly half through this approach.

Table 17.4 also contains the part numbers for air-system sockets and chimneys available for use with the tubes that are listed. The builder should investigate which of the sockets listed for the 4CX250R, 4CX300A, 4CX1000A and 4CX1600A best fit the circuit needs. Some of the sockets have certain tube elements grounded internally through the socket. Others have elements bypassed to ground through capacitors that are integral parts of the sockets.

Depending on your design philosophy and tube sources, some compromises in the cooling system may be appropriate. For example, if glass tubes are available inexpensively as broadcast pulls, a shorter life span may be acceptable. In such a case, an increase of convenience and a reduction in cost, noise, and complexity can be had by using a pair of "muffin" fans. One fan may be used for the filament seals and one for the anode seal, dispensing with a blower and air-system socket and chimney. The airflow with this scheme is not as uniform as with the use of a chimney. The tube envelope mounted in a cross flow has flow stagnation points and low heat transfer in certain regions of the envelope. These points become hotter than the rest of the envelope. The use of multiple fans to disturb the cross airflow can significantly reduce this problem. Many amateurs have used this cooling method successfully in low-duty-cycle CW and SSB operation but it is not recommended for AM, SSTV or RTTY service.

The true test of the effectiveness of a forced air cooling system is the amount of heat carried away from the tube by the air stream. The power dissipated can be calculated from the airflow temperatures. The dissipated power is

$$P_D = 0.543 \, Q_A \, (T_2 - T_1) \qquad (7)$$

where

P_D = the dissipated power, in W
Q_A = the air flow, in CFM (cubic feet per minute).
T_1 = the inlet air temperature, °C (normally quite close to room temperature).
T_2 = the amplifier exhaust temperature, °C.

The exhaust temperature can be measured with a cooking thermometer at the air outlet. The thermometer should not be placed inside the anode compartment because of the high voltage present.

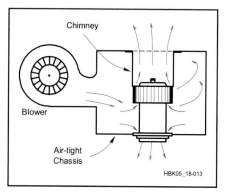

Fig 17.26 — Anode compartment pressurization may be more efficient than grid compartment pressurization. Hot air exits upwards through the tube anode and through the chimney. Cool air also goes down through the tube socket to cool tube's pins and the socket itself.

17.8 Amplifier Stabilization

Purity of emissions and the useful life (or even survival) of a tube depend heavily on stability during operation. Oscillations can occur at the operating frequency, or far from it, because of undesired positive feedback in the amplifier. Unchecked, these oscillations pollute the RF spectrum and can lead to over-dissipation and subsequent failure. Each type of oscillation has its own cause and its own cure.

17.8.1 Amplifier Neutralization

An RF amplifier, especially a linear amplifier, can easily become an oscillator at various frequencies. When the amplifier is operating, the power at the output side is large. If a fraction of that power finds it way back to the input and is in the proper phase, it can be re-amplified, repeatedly, leading to oscillation. An understanding of this process can be had by studying the sections on feedback and oscillation in the **Analog Basics** chapter. Feedback that is self-reinforcing is called "positive" feedback, even though its effects are undesirable. Even when the positive feedback is insufficient to cause actual oscillation, its presence can lead to excessive distortion and strange effects on the tuning of the amplifier and it, therefore, should be eliminated or at least reduced. The deliberate use of "negative" feedback in amplifiers to increase linearity is discussed briefly elsewhere in this chapter.

The power at the output of an amplifier will couple back to the input of the amplifier through any path it can find. It is a good practice to isolate the input and output circuits of an amplifier in separate shielded compartments. Wires passing between the two compartments should be bypassed to ground if possible. This prevents feedback via paths external to the tube.

However, energy can also pass back through the tube itself. To prevent this, a process called neutralization can be used. Neutralization seeks to prevent or to cancel out any transfer of energy from the plate of the tube back to its input, which will be either the grid or the cathode. An effective way to neutralize a tube is to provide a grounded shield between the input and the output. In the grounded grid connection, the grid itself serves this purpose. For best neutralization, the grid should be connected through a low inductance conductor to a point that is at RF ground. Ceramic external tube types may have multiple low inductance leads to ground to enhance the shielding effect. Older glass type tubes may have significant inductance inside the tube and in the socket, and this will limit the effectiveness of the shielding effect of the grid. Thus, using a grounded grid circuit with those tube types does not rule out the need for further efforts at neutralization, especially at the higher HF frequencies.

When tetrodes are used in a grounded cathode configuration, the screen grid acts as an RF shield between the grid and plate. Special tube sockets are provided that provide a very low inductance connection to RF ground. These reduce the feed through from plate to grid to a very small amount, making the effective grid-to-plate capacitance a tiny fraction of one picofarad. If in doubt about amplifiers that will work over a large frequency range, use a network analyzer or impedance measuring instrument to verify how well grounded a "grounded" grid or screen really is. If at some frequencies the impedance is more than an ohm or two, a different grounding configuration may be needed.

For amplifiers to be used at only a single frequency, a series resonant circuit can be used at either the screen or grid to provide nearly perfect bypassing to ground. Typical values for a 50 MHz amplifier are shown in **Fig 17.27**.

For some tubes, at a certain frequency, the lead inductance to ground can just cancel the grid-to-plate capacitance. Due to this effect, some tube and socket combinations have a naturally self neutralizing frequency based on the values of screen inductance and grid-to-plate capacitance. For example, the "self neutralizing frequency" of a 4-1000 is about 30 MHz. This effect has been utilized in some VHF amplifiers.

BRIDGE NEUTRALIZATION

When the shielding effect of a grid or screen bypassed to ground proves insufficient, other circuits must be devised to cancel out the remaining effect of the grid-to-plate or grid-to-cathode capacitance. These, in effect, add an additional path for negative feedback that will combine with the undesired positive feedback and cancel it. The most commonly used circuit is the "bridge neutralization" circuit shown in **Fig 17.28**. This method gets its name from the fact that the four important capacitance values can be redrawn as a bridge circuit, as shown in **Fig 17.29**. Clearly when the bridge is properly balanced, there is no transfer of energy from the plate to the grid tanks. Note that four different capacitors are part of the bridge. C_{gp} is characteristic of the chosen tube, somewhat affected by the screen or grid bypass mentioned earlier. The other components must be chosen properly so that bridge balance is achieved. C1 is the neutralizing capacitor. Its value should be adjustable to the point where

$$\frac{C1}{C3} = \frac{C_{gp}}{C_{IN}} \qquad (8)$$

where

C_{gp} = tube grid-plate capacitance
C_{IN} = tube input capacitance

The tube input capacitance must include all strays directly across the tube. C3 is not simply a bypass capacitor on the ground side of the grid tank, but rather a critical part of the bridge. Hence, it must provide a stable value of capacitance. Sometimes, simple bypass capacitors are of a type which change their value drastically with temperature. These are not suitable in this application.

Neutralization adjustment is accomplished by applying energy to the output of the amplifier, and measuring the power fed through the input. Conversely, the power may be fed to the input and the output power measured. *with the power off*, the neutralization capacitor C1 is adjusted for minimum feed through, while keeping the output tuning circuit and the input tuning (if used) at the point of maximum response. Since the bridge neutralization circuit is essentially broad band, it will work over a range of frequencies. Usually, it is adjusted at the highest anticipated frequency of operation, where the adjustment is most critical.

BROADBAND TRANSFORMER

Another neutralizing method is shown in **Fig 17.30**, where a broadband transformer provides the needed out of phase signal. C4 is adjusted so that the proper amount of negative feedback is applied to the input to just cancel the feedback via the cathode to plate capacitance. Though many 811A amplifiers have been built without this neutralization, its use makes tuning smoother on the higher bands. This circuit was featured in June 1961 *QST* and then appeared in the *RCA Transmitting Tube Handbook*. Amplifiers featuring this basic circuit are still being manufactured in 2009 and are a popular seller. Many thousands of hams have built such amplifiers as well.

An alternate method of achieving stable operation is to load the grid of a grounded cathode circuit with a low value of resistance. A convenient value is 50 Ω as it provides a match for the driver. This approach reduces the power being fed back to the grid from the output to a low enough level that good stability is achieved. However, the amplifier

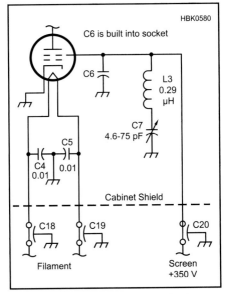

Fig 17.27 — A series-resonant circuit can be used to provide nearly perfect screen or grid bypassing to ground. This example is from a single-band 50 MHz amplifier.

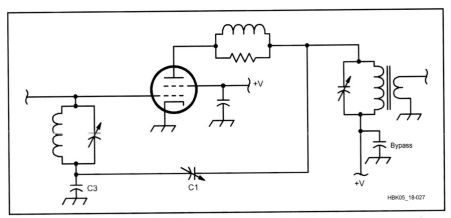

Fig 17.28 — A neutralization circuit uses C1 to cancel the effect of the tube internal capacitance.

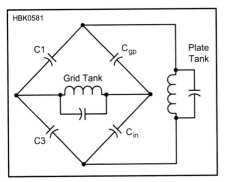

Fig 17.29 — The "bridge neutralization" circuit of Fig 17.28 redrawn to show the capacitance values.

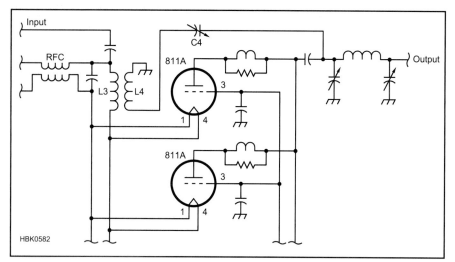

Fig 17.30 — In this neutralizing method, a broadband transformer (L3, L4) provides the needed out-of-phase signal. L3 is 6 turns of #14 wire closewound, ½ in. dia. L4 is 5 turns of insulated wire over L3. C4 is 6 pF with 0.06-in. spacing. This circuit was originally featured in June 1961 *QST* and is still found in modern amplifiers using 811A tubes.

gain will be much less than without the grid load. Also, unlike the grounded grid circuit, where much of the power applied to the input feeds through to the output, with this "loaded grid" approach, the input power is lost in the load, which must be able to dissipate such power. Distortion may be low in that the driver stage sees a very constant load. In addition, no tuning of the input is required.

17.8.2 VHF and UHF Parasitic Oscillations

RF power amplifier circuits contain parasitic reactances that have the potential to cause so-called parasitic oscillations at frequencies far above the normal operating frequency. Nearly all vacuum-tube amplifiers designed for operation in the 1.8 to 29.7-MHz frequency range exhibit tendencies to oscillate somewhere in the VHF-UHF range — generally between about 75 and 250 MHz depending on the type and size of tube. A typical parasitic resonant circuit is shown in **Fig 17.31**. Stray inductance between the tube plate and the output tuning capacitor forms a high-Q resonant circuit with the tube's C_{OUT}. C_{OUT} normally is much smaller (higher X_C) than any of the other circuit capacitances shown. The tube's C_{IN} and the tuning capacitor C_{TUNE} essentially act as bypass capacitors, while the various chokes and tank inductances shown have high reactances at VHF. Thus, the values of these components have little influence on the parasitic resonant frequency.

Oscillation is possible because the VHF resonant circuit is an inherently high-Q parallel-resonant tank that is not coupled to the external load. The load resistance at the plate is very high and thus, the voltage gain at the parasitic frequency can be quite high, leading to oscillation. The parasitic frequency, f_r, is approximately:

$$f_r = \frac{1000}{2\pi\sqrt{L_P C_{OUT}}} \quad (9)$$

where
- f_r = parasitic resonant frequency in MHz
- L_P = total stray inductance between tube plate and ground via the plate tuning capacitor (including tube internal plate lead) in µH.
- C_{OUT} = tube output capacitance in pF.

In a well-designed HF amplifier, L_P might be in the area of 0.2 µH and C_{OUT} for an 8877 is about 10 pF. Using these figures, the equation above yields a potential parasitic resonant frequency of

$$f_r = \frac{1000}{2\pi\sqrt{0.2 \times 10}} = 112.5 \text{ MHz}$$

For a smaller tube, such as the 3CX800A7 with C_{OUT} of 6 pF, f_r = 145 MHz. Circuit details affect f_r somewhat, but these results do, in fact, correspond closely to actual parasitic oscillations experienced with these tube types. VHF-UHF parasitic oscillations can be prevented (*not* just minimized!) by reducing the loaded Q of the parasitic resonant circuit so that gain at its resonant frequency is insufficient to support oscillation. This is possible with any common tube, and it is especially

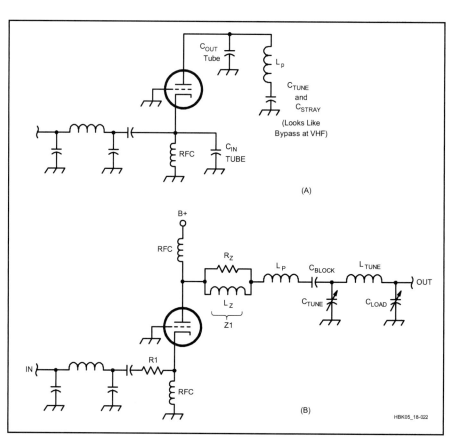

Fig 17.31 — At A, typical VHF/UHF parasitic resonance in plate circuit. The HF tuning inductor in the pi network looks like an RF choke at VHF/UHF. The tube's output capacitance and series stray inductance combine with the pi-network tuning capacitance and stray circuit capacitance to create a VHF/UHF pi network, presenting a very high impedance to the plate, increasing its gain at VHF/UHF. At B, Z1 lowers the Q and therefore gain at parasitic frequency.

easy with modern external-anode tubes like the 8877, 3CX800A7 and 4CX800A.

PARASITIC SUPPRESSORS

Z1 of Fig 17.31B is a parasitic suppressor. Its purpose is to add loss to the parasitic circuit and reduce its Q enough to prevent oscillation. This must be accomplished without significantly affecting normal operation. L_z should be just large enough to constitute a significant part of the total parasitic tank inductance (originally represented by L_P), and located right at the tube plate terminal(s). If L_z is made quite lossy, it will reduce the Q of the parasitic circuit as desired.

The inductance and construction of L_z depend substantially on the type of tube used. Popular glass tubes like the 3-500Z and 4-1000A have internal plate leads made of wire. This significantly increases L_P when compared to external-anode tubes. Consequently, L_z for these large glass tubes usually must be larger in order to constitute an adequate portion of the total value of L_P. Typically a coil of 3 to 5 turns of #10 wire, 0.25 to 0.5 inches in diameter and about 0.5 to 1 inches long is sufficient. For the 8877 and similar tubes it usually is convenient to form a "horseshoe" in the strap used to make the plate connection. A "U" about 1-inch wide and 0.75 to 1 inch deep usually is sufficient. In either case, L_z carries the full operating-frequency plate current; at the higher frequencies this often includes a substantial amount of circulating tank current, and L_z must be husky enough to handle it without overheating even at 29 MHz. Fig **17.32** shows a typical parasitic suppressor.

Regardless of the form of L_z, loss may be introduced as required by shunting L_z with one or more suitable non-inductive resistors. In high-power amplifiers, two composition or metal film resistors, each 100 Ω, 2 W, connected in parallel across L_z usually are adequate. For amplifiers up to perhaps 500 W a single 47-Ω, 2-W resistor may suffice. The resistance and power capability required to prevent VHF/UHF parasitic oscillations, while not overheating as a result of normal plate circuit current flow, depend on circuit parameters. Operating-frequency voltage drop across L_z is greatest at higher frequencies, so it is important to use the minimum necessary value of L_z in order to minimize power dissipation in R_z.

The parasitic suppressors described above very often will work without modification, but in some cases it will be necessary to experiment with both L_z and R_z to find a suitable combination. Some designers use nichrome or other resistance wire for L_z.

In exceptionally difficult cases, particularly when using glass tetrodes or pentodes, additional parasitic suppression may be attained by connecting a low value resistor (about 10 to 15 Ω) in series with the tube input, near the tube socket. This is illustrated by R1 of Fig 17.31B. If the tube has a relatively low input impedance, as is typical of grounded-grid amplifiers and some grounded-cathode tubes with large C_{IN}, R1 may dissipate a significant portion of the total drive power.

Fig 17.32 — Typical parasitic suppressor.

TESTING TUBE AMPLIFIERS FOR VHF-UHF PARASITIC OSCILLATIONS

Every high-power amplifier should be tested, before being placed in service, to insure that it is free of parasitic oscillations. For this test, nothing is connected to either the RF input or output terminals, and the band switch is first set to the lowest-frequency range. If the input is tuned and can be band switched separately, it should be set to the highest-frequency band. The amplifier control system should provide monitoring for both grid current and plate current, as well as a relay, circuit breaker or fast-acting fuse to quickly shut off high voltage in the event of excessive plate current. To further protect the tube grid, it is a good idea to temporarily insert in series with the grid current return line a resistor of approximately 1000 Ω to prevent grid current from soaring in the event a vigorous parasitic oscillation breaks out during initial testing.

Apply filament and bias voltages to the amplifier, leaving plate voltage off and/or cutoff bias applied until any specified tube warm-up time has elapsed. Then apply the lowest available plate voltage and switch the amplifier to transmit. Some idling plate current should flow. If it does not, it may be necessary to increase plate voltage to normal or to reduce bias so that at least 100 mA or so does flow. Grid current should be zero. Vary the plate tuning capacitor slowly from maximum capacitance to minimum, watching closely for any grid current or change in plate current, either of which would indicate a parasitic oscillation. If a tunable input network is used, its capacitor (the one closest to the tube if a pi circuit) should be varied from one extreme to the other in small increments, tuning the output plate capacitor at each step to search for signs of oscillation. If at any time either the grid or plate current increases to a large value, shut off plate voltage immediately to avoid damage! If moderate grid current or changes in plate current are observed, the frequency of oscillation can be determined by loosely coupling an RF absorption meter or a spectrum analyzer to the plate area. It will then be necessary to experiment with parasitic suppression measures until no signs of oscillation can be detected under any conditions. This process should be repeated using each band switch position.

When no sign of oscillation can be found, increase the plate voltage to its normal operating value and calculate plate dissipation (idling plate current times plate voltage). If dissipation is at least half of, but not more than its maximum safe value, repeat the previous tests. If plate dissipation is much less than half of maximum safe value, it is desirable (but not absolutely essential) to reduce bias until it is. If no sign of oscillation is detected, the temporary grid resistor should be removed and the amplifier is ready for normal operation.

LOW-FREQUENCY PARASITIC OSCILLATIONS

The possibility of self-oscillations at frequencies lower than VHF is significantly lower than in solid-state amplifiers. Tube amplifiers will usually operate stably as long as the input-to-output isolation is greater than the stage gain. Proper shielding and dc-power-lead bypassing essentially eliminate feedback paths, except for those through the tube itself.

On rare occasions, tube-type amplifiers will oscillate at frequencies in the range of about 50 to 500 kHz. This is most likely with high-gain tetrodes using shunt feed of dc voltages to both grid and plate through RF chokes. If the resonant frequency of the grid RF choke and its associated coupling capacitor occurs close to that of the plate choke and its blocking capacitor, conditions may support a tuned-plate tuned-grid oscillation. For example, using typical values of 1 mH and 1000 pF, the expected parasitic frequency would be around 160 kHz.

Make sure that there is no low-impedance, low-frequency return path to ground through inductors in the input matching networks in series with the low impedances reflected by a transceiver output transformer. Usually, oscillation can be prevented by changing choke or capacitor values to insure that the input resonant frequency is much lower than that of the output.

17.8.3 Reduction of Distortion

As mentioned previously, a common cause of distortion in amplifiers is over drive (flat topping). The use of automatic level control (ALC) is a practical way of reducing the ill effects of flat topping while still being assured of having a strong signal. This circuit detects the voltage applied to the input of the amplifier. Other circuits are based on detecting the onset of grid current flow. In either case, when the threshold is reached, the ALC circuit applies a negative voltage to the ALC input of the transceiver and forces it to cut back on the driving power, thus keeping the output power within set limits. Most transceivers also apply an ALC signal from their own output stage, so the ALC signal from the amplifier will add to or work in parallel with that. See **Fig 17.33** for a representative circuit.

Some tube types have inherently lower distortion than others. Selection of a tube specifically designed for linear amplifier service, and operating it within the recommended voltage and current ranges is a good start. In addition, the use of tuned circuits in the input circuits when running class AB2 will help by maintaining a proper load on the driver stages over the entire 360° cycle, rather then letting the load change as the tube begins to draw grid current. Another way to accomplish this is with the "loaded grid," the use of a rather low value of resistance from the grid to cathode. Thus, when grid current flows, the change in impedance is less drastic, having been swamped by the resistive load.

For applications where the highest linearity is desired, operating class A will greatly reduce distortion but at a high cost in efficiency. Some solid-state amateur transceivers have provision for such operation. The use of negative feedback is another way of greatly reducing distortion. High efficiency is maintained, but there is a loss of overall gain. Often, two stages of gain are used and the feedback applied around both stages. In this way, gain can be as high as desired, and both stages are compensated for any inherent nonlinearities. Amplifiers using RF negative feedback can achieve values of intermodulation distortion (IMD) as much as 20 dB lower than amplifiers without feedback.

It must be remembered that distortion tends to be a cumulative problem, with each nonlinear part of the transmission chain adding its part. It is not worthwhile to have a super clean transceiver if it is followed by an amplifier with poor linearity. In the same way, a very good linear will look bad if the transceiver driving it is poor. It is even possible to have a clean signal out of your amplifier, but have it spoiled by a ferrite core tuner inductor or balun that is saturated.

Distortion in a linear amplifier is usually measured with a spectrum analyzer while transmitting a two tone test. If the spectrum analyzer input is overloaded, this can also produce apparent distortion in the amplifier. Reducing the level so that the analyzer is not clipping the input signal is necessary to see the true distortion in the amplifier chain. Use of test equipment for various types of measurements is covered in the **Test Equipment and Measurements** chapter.

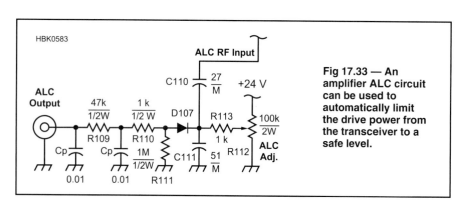

Fig 17.33 — An amplifier ALC circuit can be used to automatically limit the drive power from the transceiver to a safe level.

17.9 Design Example: A High Power Vacuum Tube HF Amplifier

Most popular HF transceivers produce approximately 100 W output. The EIMAC 8877 can deliver 1500 W output for approximately 60 W of drive when used in a grounded grid circuit. Grounded-grid operation is usually the easiest tube amplifier circuit to implement. Its input impedance is relatively low, often close to 50 Ω. Input/output shielding provided by the grid and negative feedback inherent in the grounded-grid circuit configuration reduce the likelihood of amplifier instability and provide excellent linearity without critical adjustments. Fewer supply voltages are needed in this configuration compared to others, often just high-voltage dc for the plate and low-voltage ac for the filament.

17.9.1 Tube Capabilities

The first step in the amplifier design process is to verify that the tube is actually capable of producing the desired results while remaining within manufacturer's ratings. The plate dissipation expected during normal operation of the amplifier is computed first. Since the amplifier will be used for SSB, a class of operation producing linear amplification must be used. Class AB2 provides a very good compromise between linearity and good efficiency, with effective efficiency typically exceeding 60%. Given that efficiency, an input power of 2500 W is needed to produce the desired 1500 W output. Operated under these conditions, the tube will dissipate about 1000 W — well within the manufacturer's specifications, provided adequate cooling airflow is supplied.

The grid in modern high-mu triodes is a relatively delicate structure, closely spaced to the cathode and carefully aligned to achieve high gain and excellent linearity. To avoid shortening tube life or even destruction of the tube, the specified maximum grid dissipation must not be exceeded for more than a few milliseconds under any conditions. For a given power output, the use of higher plate voltages tends to result in lower grid dissipation. It is important to use a plate voltage that is high enough to result in safe grid current levels at maximum output. In addition to maximum ratings, manufacturers' data sheets often provide one or more sets of "typical operation" parameters. This makes it even easier for the builder to achieve optimum results.

According to typical operation data, the 8877, operating at 3500 V, can produce 2075 W of RF output with excellent linearity and 64 W of drive. Operating at 2700 V it can deliver 1085 W with 40 W of drive. To some extent, the ease and cost of constructing a high-power amplifier, as well as its ultimate reliability, are enhanced by using the lowest plate voltage that will yield completely satisfactory performance. Working with various load lines on the characteristic curves shows that the 8877 can comfortably deliver 1.5 kW output with a 3100 V plate supply and 50 to 55 W of drive. Achieving 1640 W output power at this plate voltage requires 800 mA of plate current — well within the 8877's maximum rating of 1.0 A.

17.9.2 Input and Output Circuits

The next step in the design process is to calculate the optimum plate load resistance at this plate voltage and current for Class AB2 operation and design an appropriate output-

Parameter	Result
Grid Current (mA)	34.2
Screen Current (mA)	0.00
Plate Current (A)	0.800
Input Power (Watts)	2470
Output Power (Watts)	1640
Plate Dissipation (W)	831
Plate Load (Ohms)	2210
Efficiency (%)	66.3
Grid Swing (Volts)	80.0
Resting Dissipation (W)	620
Input Resistance (Ohms)	62.5
Power Passed (Watts)	48.8
Total drive power (W)	51.6

Fig 17.35 — Typical operating parameters for the 8877 triode used in the design example detailed in the text.

matching network. Using the operating line shown in **Fig 17.34**, the *TubeCalculator* program gives these values (**Fig 17.35**). The simple formula for load resistance shown earlier in this chapter (equation 1) gives a similar value, about 2200 Ω.

Several different output networks might be used to transform the nominal 50-Ω resistance of the actual load to the 2200-Ω load resistance required by the 8877, but experience shows that pi and pi-L networks are most practical. Each can provide reasonable harmonic attenuation, is relatively easy to build mechanically and uses readily available components. The pi-L gives significantly greater harmonic attenuation than the pi and usually is the better choice — at least in areas where there is any potential for TVI or cross-band interference. In a multiband amplifier, the extra cost of using a pi-L network is the "L" inductor and its associated band switch section.

The input impedance of a grounded-grid 8877 is typically on the order of 50 to 55 Ω, shunted by input capacitance of about 38 pF. While this average impedance is close enough to 50 Ω to provide negligible input SWR, the instantaneous value varies greatly over the drive cycle — that is, it is nonlinear. This nonlinear impedance is reflected back as a nonlinear load impedance at the exciter output, resulting in increased intermodulation distortion, reduced output power, and often meaningless exciter SWR meter indications. In addition, the tube's parallel input capacitance, as well as parasitic circuit reactances, often are significant enough at 28 MHz to create significant SWR.

A tank circuit at the amplifier input can solve both of these problems by tuning out the stray reactances and stabilizing (linearizing) the tube input impedance through its flywheel effect. The input tank should have a loaded Q of at least two for good results. A Q of five results in a further small improvement in linearity and distortion, but at the cost of a narrower operating bandwidth. Using the *PI-EL Design* software, one can quickly determine values for C1, L1 and C2 for the various bands as well as the bandwidth that various values of Q will provide. Since the 3.5 to 4 MHz band is the widest in terms of percentage bandwidth, using a lower Q for that band seems wise. If we wish to cover two bands with the same switch position, for example the 24 and 28 MHz bands, that will also call for a lower Q. For the 40, 30 and 20 meter bands alone, a rather high Q would work fine. Remember to subtract the input capacitance of the tube from the calculated value of C2.

Fig 17.36 illustrates these input and output networks applied in the amplifier circuit. The schematic shows the major components in the amplifier RF section, but with bandswitching and cathode dc return circuits omitted for clarity. C1 and C2 and L1 form the input pi network. C3 is a blocking capacitor to isolate the exciter from the cathode dc potential. Note that when the tube's average input resistance is close to 50 Ω, as in the case of the 8877, a simple parallel-resonant tank often can successfully perform the tuning and flywheel functions, since no impedance transformation is necessary. In this case, it is important to minimize stray lead inductance

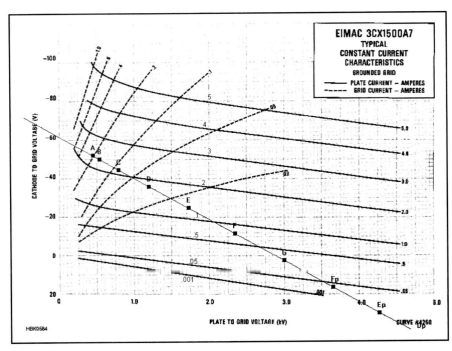

Fig 17.34 — Characteristic curves for the 8877 triode used in the design example detailed in the text.

Tuning Your Vacuum-Tube Amplifier

Today most hams are used to a wide variety of amplifiers using triodes that are tuned for maximum power output. Unfortunately, not all amplifiers or tubes are rugged enough for this approach. For example, amplifiers that use tetrodes employ a different tuning procedure in which tuning for maximum power may result in a destroyed control grid or screen grid! Procedures vary depending not only on the ratings of the tetrode, but the voltages as well. When tuning any amplifier, monitor grid currents closely and do not exceed the specified maximum current as those are the easiest elements of the tube to damage.

For commercial amplifiers, "Read the Manual" as the manufacturer's directions may not follow the following procedures exactly and can be quite different in some cases. In all cases, the last tuning adjustments should be made at full power output, not at reduced power, because the characteristics of the tube change with different power levels. Operating the amplifier at high power after tuning at low power can result in spurious emissions or over-stressing the output network components.

Begin by making sure you have all band-switching controls set properly. If the amplifier TUNE and LOAD controls (sometimes referred to as "Plate Tune" and "Output", respectively) have recommended settings on a particular band, start at those settings. If your amplifier can be set to a TUNE mode, do so. Set the initial amount of drive (input power) from the exciter — read the amplifier manual or check the tube's specifications if a manual is not available. The exciter output should be one-half or more of full power so that the exciter's ALC systems function properly.

Tuning a triode-based, grounded-grid amplifier is the simplest: Tune for maximum output power without exceeding the tube ratings, particularly grid current, or the legal output power limit. The typical procedure is, while monitoring grid current, to increase drive until the plate current equals about one-quarter to one-half of the target current (depending on the tube and grid bias) while monitoring the output on a wattmeter or the internal power meter. Adjust the TUNE control then advance the LOAD control for maximum output. Repeat the sequence of peaking TUNE then increasing LOAD until no more output power can be obtained without exceeding the ratings for the tube or the legal power limit. If necessary, increase drive and re-peak both the TUNE and LOAD controls.

Operating somewhat differently, a grid-driven tetrode (or pentode) amplifier operating near its designed output power uses the TUNE control for peaking output power and the LOAD control for increasing, but not exceeding, the maximum allowable screen current. Generally, the first part of tetrode amplifier tuning is the same as for a triode amplifier with both the TUNE and LOAD controls adjusted for maximum power output while monitoring screen and control grid current. After the initial tuning step, the LOAD control is used to peak the screen current. Maximum power should coincide with maximum screen current. The screen current is just a better indicator. As with the triode amplifier, if drive needs to be increased, readjust the TUNE and LOAD controls.

For both triode and tetrode amplifiers, once the procedures above have been completed, try moving the TUNE control a small amount higher or lower and re-peaking the LOAD control. Also known as "rocking" the TUNE control, this small variation can find settings with a few percent more output power or better efficiency.

Once tuning has been completed, it is a good idea to mark the settings of the TUNE and LOAD controls for each band. This reduces the amount of time for on-the-air or dummy-load tuning, reducing stress to the tube and interference to other stations. Usually, a quick "fine-tune" adjustment is all that is required for maximum output. The set of markings also serves as a diagnostic tool, should the settings for maximum power suddenly shift. This indicates a change in the antenna system, such as a failing connector or antenna. — *Roger Halstead, K8RI*

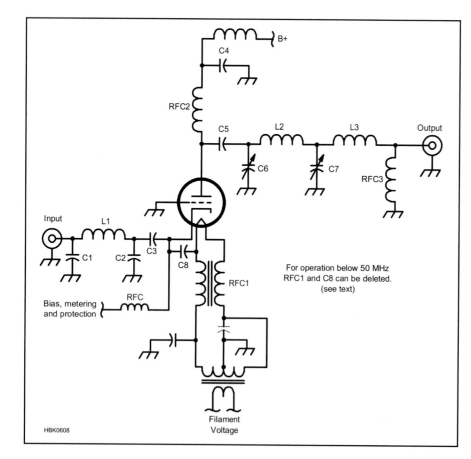

between the tank and tube to avoid undesired impedance transformation.

17.9.3 Filament Supply

The filament or "heater" in indirectly heated tubes such as the 8877 must be very close to the cathode to heat the cathode efficiently. A capacitance of several picofarads exists between the two. Particularly at very high frequencies, where these few picofarads represent a relatively low reactance, RF drive intended for the cathode can be capacitively coupled to the lossy filament and dissipated as heat. To avoid this, above about 50 MHz, the filament must be kept at a high RF impedance above ground. The high impedance (represented by choke RFC1 in Fig 17.36) minimizes RF current flow in the filament circuit so that RF dissipated in the filament becomes negligible. The choke's low-frequency resistance should be kept to a minimum to lessen voltage drops in the high-current filament circuit.

The choke most commonly used in this application is a pair of heavy-gauge insu-

Fig 17.36 — A simplified schematic of a grounded-grid amplifier using a pi network input and pi-L network output.

lated wires, bifilar-wound over a ferrite rod. The ferrite core raises the inductive reactance throughout the HF region so that a minimum of wire is needed, keeping filament-circuit voltage drops low. The bifilar winding technique assures that both filament terminals are at the same RF potential.

Below 30 MHz, the use of such a choke seldom is necessary or beneficial, but actually can introduce another potential problem. Common values of cathode-to-heater capacitance and heater-choke inductance often are series resonant in the 1.8 to 29.7 MHz HF range. A capacitance of 5 pF and an inductance of 50 µH, for example, resonate at 10.0 MHz; the actual components are just as likely to resonate near 7 or 14 MHz. At resonance, the circuit constitutes a relatively low impedance shunt from cathode to ground, which affects input impedance and sucks out drive signal. An unintended resonance like this near any operating frequency usually increases input SWR and decreases gain on that one particular band. While aggravating, the problem rarely completely disables or damages the amplifier, and so is seldom pursued or identified.

Fortunately, the entire problem is easily avoided — below 30 MHz the heater choke can be deleted. At VHF-UHF, or wherever a heater isolation choke is used for any reason, the resonance can be moved below the lowest operating frequency by connecting a sufficiently large capacitance (about 1000 pF) between the tube cathode and one side of the heater. It is good practice also to connect a similar capacitor between the heater terminals. It also would be good practice in designing other VHF/UHF amplifiers, such as those using 3CX800A7 tubes, unless the builder can ensure that the actual series resonance is well outside of the operating frequency range.

17.9.4 Plate Choke and DC Blocking

Plate voltage is supplied to the tube through RFC2. C5 is the plate blocking capacitor. The output pi-L network consists of tuning capacitor C6, loading capacitor C7, pi coil L2, and output L coil L3. RFC3 is a high-inductance RF choke placed at the output for safety purposes. Its value, usually 100 µH to 2 mH, is high enough so that it appears as an open circuit across the output connector for RF. However, should the plate blocking capacitor fail and allow high voltage onto the output matching network, RFC3 would short the dc to ground and blow the power-supply fuse or breaker. This prevents dangerous high voltage from appearing on the feed line or antenna. It also prevents electrostatic charge — from the antenna or from blocking capacitor leakage — from building up on the tank capacitors and causing periodic dc discharge arcs to ground. If such a dc discharge occurs while the amplifier is transmitting, it can trigger a potentially damaging RF arc.

17.9.5 Tank Circuit Design

The output pi-L network must transform the nominal 50-Ω amplifier load to a pure resistance of 2200 Ω. We previously calculated that the 8877 tube's plate must see 2200 Ω for optimum performance. In practice, real antenna loads are seldom purely resistive or exactly 50 Ω; they often exhibit SWRs of 2:1 or greater on some frequencies. It's desirable that the amplifier output network be able to transform any complex load impedance corresponding to an SWR up to about 2:1 into a resistance of 2200 Ω. The network also must compensate for tube C_{OUT} and other stray plate-circuit reactances, such as those of interconnecting leads and the plate RF choke. These reactances, shown in **Fig 17.37**, must be taken into account when designing the matching networks. Because the values of most stray reactances are not accurately known, the most satisfactory approach is to estimate them, and then allow sufficient flexibility in the matching network to accommodate modest errors.

Fig 17.37 shows the principal reactances in the amplifier circuit. C_{OUT} is the actual tube output capacitance of 10 pF plus the stray capacitance between its anode and the enclosure metalwork. This stray C varies with layout; we will approximate it as 5 pF, so C_{OUT} is roughly 15 pF. L_{OUT} is the stray inductance of leads from the tube plate to the tuning capacitor (internal to the tube as well as external circuit wiring.) External-anode tubes like the 8877 have essentially no internal plate leads, so L_{OUT} is almost entirely external. It seldom exceeds about 0.3 µH and is not very significant below 30 MHz. L_{CHOKE} is the reactance presented by the plate choke, which usually is significant only below 7 MHz. C_{STRAY} represents the combined stray capacitances to ground of the tuning capacitor stator and of interconnecting RF plate circuit leads. In a well-constructed, carefully thought out power amplifier, C_{STRAY} can be estimated to be approximately 10 pF. Remaining components C_{TUNE}, C_{LOAD}, and the two tuning inductors, form the pi-L network proper.

The values for the output network components can be calculated using the *PI-EL Design* software, taken from the graphical charts in Figs 17.17 to 17.19, or from tables included on the *Handbook* CD. For pi networks, a Q of 12 is a good compromise between harmonic suppression and circuit losses. In practice, it often is most realistic and practical with both pi and pi-L output networks to accept somewhat higher Q values on the highest HF frequencies — perhaps as large as 18 or even 20 at 28 MHz. When using a pi-L on the 1.8 and 3.5 MHz bands, it often is desirable to choose a moderately lower Q, perhaps 8 to 10, to permit using a more reasonably-sized plate tuning capacitor.

CIRCUIT REACTANCES

The calculated output network values must be adjusted to allow for circuit reactances outside the pi-L proper. First, low-frequency component values should be examined. At 3.5 MHz, assuming that total tuning capacitance C1 is 140 pF, we know that three other stray reactances are directly in parallel with C_{TUNE} (assuming that L_{OUT} is negligible at the operating frequency as it should be). The tube's internal and external plate capacitance to ground, C_{OUT}, is about 15 pF. Strays in the RF circuit, C_{STRAY}, are roughly 10 pF.

The impedance of the plate choke, X_{CHOKE}, is also in parallel with C_{TUNE}. Plate chokes with self-resonance characteristics suitable for use in amateur HF amplifiers typically have inductances of about 90 µH. At 3.5 MHz this is an inductive reactance of +1979 Ω. This appears in parallel with the tuning capacitance, effectively canceling an equal value of capacitive reactance. At 3.5 MHz, an X_C of 1979 Ω corresponds to 23 pF of capacitance — the amount by which tuning capacitor C_{TUNE} must be increased at 3.5 MHz to compensate for the effect of the plate choke.

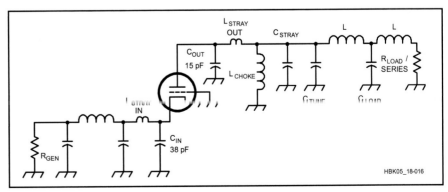

Fig 17.37 — The effective reactances for the amplifier in Fig 17.36.

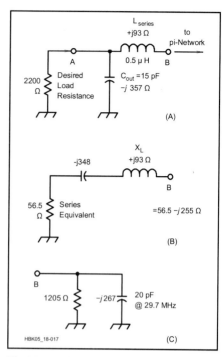

Fig 17.38 — The effect of adding a small inductor in series with the tube plate to aid matching at high frequencies. See text for details.

If the pi-L network requires an effective capacitance of 140 pF at its input at 3.5 MHz, subtracting the 25 pF provided by C_{OUT} and C_{STRAY} and adding the 23 pF canceled by X_{CHOKE}, the actual value of C_{TUNE} must be $140 - 25 + 23 = 138$ pF. It is good practice to provide at least 10% extra capacitance range to allow matching loads having SWRs up to 2:1. So, if 3.5 MHz is the lower frequency limit of the amplifier, a variable tuning capacitor with a maximum value of at least 150 to 160 pF should be used.

PERFORMANCE AT HIGH FREQUENCIES

Component values for the high end of the amplifier frequency range also must be examined, for this is where the most losses will occur. At 29.7 MHz we can assume a minimum pi-L input capacitance of 35 pF. Since C_{OUT} and C_{STRAY} contribute 25 pF, C_{TUNE} must have a minimum value no greater than 10 pF. A problem exists, because this value is not readily achievable with a 150 to 160-pF air variable capacitor suitable for operation with a 3100 V plate supply. Such a capacitor typically has a minimum capacitance of 25 to 30 pF. Usually, little or nothing can be done to reduce the tube's C_{OUT} or the circuit C_{STRAY}, and in fact the estimates of these may even be a little low. If 1.8 MHz capability is desired, the maximum tuning capacitance will be at least 200 to 250 pF, making the minimum-capacitance problem at 29.7 MHz even more severe.

There are three potential solutions to this dilemma. We could accept the actual minimum value of pi-L input capacitance, around 50 to 55 pF, realizing that this will raise the pi-L network's loaded Q to about 32. This results in very large values of circulating tank current. To avoid damage to tank components — particularly the band switch and pi inductor — from heat due to I^2R losses, it will be necessary to either use oversize components or reduce power on the highest-frequency bands. Neither option is appealing.

A second potential solution is to reduce the minimum capacitance provided by C_{TUNE}. We could use a vacuum variable capacitor with a 300-pF maximum and a 5-pF minimum capacitance. These are rated at 5 to 15 kV, and are readily available. This reduces the minimum effective circuit capacitance to 30 pF, allowing use of the pi-L values for a Q of 12 on all bands from 1.8 through 29.7 MHz. While brand new vacuum variables are quite expensive, suitable models are widely available in the surplus and used markets for prices not much higher than the cost of a new air variable. A most important caveat in purchasing a vacuum capacitor is to ensure that its vacuum seal is intact and that it is not damaged in any way. The best way to accomplish this is to "hi-pot" test the capacitor throughout its range, using a dc or peak ac test voltage of 1.5 to 2 times the amplifier plate supply voltage. For all-band amplifiers using plate voltages in excess of about 2500 V, the initial expense and effort of securing and using a vacuum-variable input tuning capacitor often is well repaid in efficient and reliable operation of the amplifier.

A third possibility is the use of an additional inductance connected in series between the tube and the tuning capacitor. In conjunction with C_{OUT} of the tube, the added inductor acts as an L network to transform the impedance at the input of the pi-L network up to the 2200-Ω load resistance needed by the tube. This is shown in **Fig 17.38A**. Since the impedance at the input of the main pi-L matching network is reduced, the loaded Q for the total capacitance actually in the circuit is lower. With lower Q, the circulating RF currents are lower, and thus tank losses are lower.

C_{OUT} in Fig 17.38 is the output capacitance of the tube, including stray C from the anode to metal enclosure. X_L is the additional series inductance to be added. As determined previously, the impedance seen by the tube anode must be a 2200 Ω resistance for best linearity and efficiency, and we have estimated C+ of the tube as 15 pF. If the network consisting of C_{OUT} and X_L is terminated at A by 2200 Ω, we can calculate the equivalent impedance at point B, the input to the pi-L network, for various values of series X_L. The pi-L network must then transform the nominal 50-Ω load at the transmitter output to this equivalent impedance.

IMPEDANCE TRANSFORMATIONS

We work backwards from the plate of the tube towards the C_{TUNE} capacitor. First, calculate the series-equivalent impedance of the parallel combination of the desired 2200-Ω plate load and the tube X_{OUT} (15 pF at 29.7 MHz = $-j\,357$ Ω). The series-equivalent impedance of this parallel combination is $56.5 - j\,348$ Ω, as shown in Fig 17.38B. Now suppose we use a 0.5 µH inductor, having an impedance of $+j\,93$ Ω at 29.7 MHz, as the series inductance X_L. The resulting series-equivalent impedance is $56.5 - j348 + j93$, or $56.5 - j255$ Ω. Converting back to the parallel equivalent gives the network of Fig 17.38C: 1205 Ω resistance in parallel with $-j267$ Ω, or 20 pF at 29.7 MHz. The pi-L tuning network must now transform the 50-Ω load to a resistive load of 1205 Ω at B, and absorb the shunt capacitance of 20 pF.

Using the *PI-EL Design* software or pi network formulas on the *Handbook* CD, R1 = 1205 Ω and Q = 15 at 29.7 MHz, yields a required total capacitance of about 67 pF at 29.7 MHz. Note that for the same loaded Q for a 2200-Ω load line without the series inductor, the capacitance was about 36 pF. When the 20 pF of transformed input capacitance is subtracted from the 67 pF total needed, the amount of capacitance is 47 pF. If the minimum capacitance in C_{TUNE} is 25 pF and the stray capacitance is 10 pF, then there is a margin of $47 - 35 = 12$ pF beyond the minimum capacitance for handling SWRs greater than 1:1 at the load.

The series inductor should be a high-Q coil wound from copper tubing to keep losses low. This inductor has a decreasing, yet significant effect, on progressively lower frequencies. A similar calculation to the above should be made on each band to determine the transformed equivalent plate impedance, before calculating the network values. The impedance-transformation effect of the additional inductor decreases rapidly with decreasing frequency. Below 21 MHz, it usually may be ignored and pi-L network values calculated for R1 = 2200 Ω.

The nominal 90-µH plate choke remains in parallel with C_{TUNE}. It is rarely possible to calculate the impedance of a real HF plate choke at frequencies higher than about 5 MHz because of self-resonances. However, as mentioned previously, the choke's reactance should be sufficiently high that the calculations are not seriously affected if the choke's first series-resonance is at 23.2 MHz.

This amplifier is made operational on multiple bands by changing the values of inductance at L2 and L3 for different bands. The usual practice is to use inductors for the lowest operating frequency, and short out part of each inductor with a switch, as necessary, to provide the inductance needed for each individual band. Wiring to the switch and the switch itself add stray inductance and capacitance to the circuit. To minimize these effects at the higher frequencies, the unswitched 10-m L2 should be placed closest to the high-impedance end of the network at C6. Stray capacitance associated with the switch then is effectively in parallel with C7, where the impedance level is around 300 Ω. The effects of stray capacitance are relatively insignificant at this low impedance level. This configuration also minimizes the peak RF voltage that the switch insulation must withstand.

17.9.6 Checking Operation

After the input and output networks are designed, cold tuning as described earlier will confirm that all of the tuned circuits are working properly. These tests are well worth doing before any power is applied to the amplifier. The band switch itself will have significant inductance especially on the higher frequencies. To determine the proper taps for the various bands on the tank inductor, start with the highest frequency and verify that the calculated number of turns gives the frequency range desired, moving the tap as needed to allow for the inductance of the band switch. As each tap is located, it should be securely wired with strap or braid and the process repeated for successively lower bands. Once the cold tuning looks good, proceed to the tests for parasitics.

You should now be ready to apply full power to the amplifier and see how it performs.

17.10 Solid-State Amplifiers

17.10.1 Solid State vs Vacuum Tubes

Solid-state amplifiers have become the norm in transceivers over the past 20 years, but their use in external high power amplifiers has not. The primary reason is economic. It is more expensive to generate a kilowatt or more with transistors because they are smaller and have less dissipation capability. This is changing. The increasing demand for auto-tuning amplifiers, an expensive undertaking with tube amplifiers, is raising the demand for solid-state amplifiers and lowering their price in the process. A number of solid-state Amateur Radio amplifiers in the 500-1000 W power class are available. **Fig 17.39** shows one of many available examples.

Vacuum tubes will not be around forever. Commercial broadcasting and industrial RF plasma, the principal commercial users of vacuum tubes, have gone solid state for all new designs. The higher acquisition cost of solid-state amplifiers is offset by their lower operating cost, which leads to a lower lifetime cost. The remaining commercial demand is the replacement market for failed tubes in existing equipment. Lower demand for vacuum tubes reduces factory volumes, driving up costs. Many tube manufacturers have gone out of business or have been absorbed by their competitors. Most of the tube types available today are not manufactured by the same company that designed them. Tube design is a withering art because there is so little demand for it.

On the bright side, new and more powerful transistors are being designed all the time. There is actually a commercially available 1500-W solid-state amplifier that uses only two transistors — Tokyo Hy-Power's HL-2.5KFX.

17.10.2 Classes of Operation

This topic applies to transistors as well as tubes, and it was covered earlier in this chapter and also in the **RF Techniques** chapter. In communications amplifiers, Class A is used mainly for driver stages where linearity is desired and efficiency is not a concern.

Class B is usually passed over in favor of the more linear Class AB. Class AB offers increased linearity, mainly less crossover distortion, for a very small (perhaps 1 or 2%) reduction in efficiency. It is the most commonly used class of operation for linear power amplifiers that must cover a wide range of frequencies. Broadband solid-state Class AB amplifiers typically achieve 50 to 60% efficiency.

Class C is used where efficiency is important and linearity and bandwidth or harmonics are not. FM transmitters are the most com-

Fig 17.39 — Solid-state amateur power amplifiers such as this 1000-W class Tokyo Hy-Power HL-1.5KFX are much more common today than they were just a decade ago.

mon application in communications. Single-band tuned amplifiers can be as much as 80% efficient. However, in a single-ended amplifier they require a tank circuit. Class C amplifiers are *not* suitable for on-off keyed modes like CW without extensive pre-distortion of the driving signal to prevent key clicks.

Class D and E are most efficient, up to 95% in practical circuit applications. But these both require a tuned tank circuit to achieve this efficiency. They are not linear; the output is essentially either on or off. They can be quite effectively used for linear amplification by the process of EER — envelope elimination and restoration — but it is always in a single-band circuit. On-off keying can be employed if the power supply is keyed with a properly shaped envelope. EER is difficult to do well and requires very complex circuitry. Without careful system design, EER results in poor SSB performance.

17.10.3 Modeling Transistors

The design method using performance curves that was detailed earlier in this chapter is more applicable to vacuum tube amplifier design than solid state. The most common method used with solid state is electronic design analysis (EDA, also called computer-aided design, or CAD) using electronic models. *SPICE* or S-parameter models are available for some high power transistors, and simple amplifiers can be readily designed with the aid of an appropriate analysis program. (See the **Computer-Aided Circuit Design** chapter for more information on *SPICE* and related modeling techniques.).

Full-featured circuit design and analysis programs are expensive, and the resulting designs are only as good as the accuracy of the models used. A complicating factor is that any design relies heavily on models for all the passive components in the circuit. While passive part models can be obtained for some commercial components, many others — such as ferrite-loaded transformers — must be designed also. It is not unusual for the electronic design to take much longer and cost more than the benefits are worth.

As detailed in the **Computer-Aided Circuit Design** chapter, many of the EDA vendors offer free or inexpensive "student versions" of their products. These are fully capable up to a certain level of circuit complexity. Although they usually are not big enough to analyze a whole amplifier, student versions are still useful for looking at parts of it.

Electronic design is very useful for getting the circuit design "in the ballpark." The design will be close enough that it will work when built, and any necessary fine tuning can be done easily once it is constructed. Computer modeling is very useful for evaluating the stresses on the circuit's passive components so they can be properly sized. Another very helpful use of CAD is in the development of the output filters. W4ENE's *SVC Filter Designer* program included with the *Handbook* CD is exceptional in this regard.

17.10.4 Impedance Transformation — "Matching Networks"

Aside from the supply voltage, there is little difference between the operation of a tube amplifier and a transistor amplifier. Each amplifies the input signal, and each will only work into a specific load impedance. In a tube amplifier, the proper plate load impedance is provided by an adjustable pi or pi-L plate tuning network, which also transforms the impedance down to 50 Ω.

A single-transistor amplifier can be made in the same way, and in fact most single-band VHF amplifier "bricks" are. A tuned matching network provides the proper load impedance for the transistor and transforms it to 50 Ω. The major difference is that the proper load impedance for a transistor, at any reasonable amount of power, is much *lower* than 50 Ω. For vacuum tubes it is much *higher* than 50 Ω.

BROADBAND TRANSFORMERS

Broadband transformers are often used in matching to the input impedance or optimum load impedance in a power amplifier. Multi-octave power amplifier performance can be achieved by appropriate application of these transformers. The input and output transformers are two of the most critical components in a broadband amplifier. Amplifier efficiency, gain flatness, input SWR, and even linearity all are affected by transformer design and application.

There are two basic RF transformer types, as described in the **RF Techniques** and **Transmission Lines** chapters: the conventional transformer and the transmission-line transformer. More information on RF transformers is included on the *Handbook* CD as well.

The conventional transformer is wound much the same way as a power transformer. Primary and secondary windings are wound around a high-permeability core, usually made from a ferrite or powdered-iron material. Coupling between the secondary and primary is made as tight as possible to minimize leakage inductance. At low frequencies, the coupling between windings is predominantly magnetic. As the frequency rises, core permeability decreases and leakage inductance increases; transformer losses increase as well.

Typical examples of conventional transformers are shown in **Fig 17.40**. In Fig 17.40A, the primary winding consists of brass or copper tubes inserted into ferrite sleeves. The tubes are shorted together at one end by a piece of copper-clad circuit board material, forming a single turn loop. The secondary winding is threaded through the tubes. Since the low-impedance winding is only a single turn, the transformation ratio is limited to the squares of integers — 1, 4, 9, 16 and so on.

The lowest effective transformer frequency is determined by the inductance of the one-turn winding. It should have a reactance, at the lowest frequency of intended operation, at least four times greater than the impedance it is connected to. The coupling coefficient between the two windings is a function of the primary tube diameter and its length, and the diameter and insulation thickness of the wire used in the high-impedance winding. High impedance ratios, greater than 36:1, should use large-diameter secondary windings. Miniature coaxial cable (using only the braid as the conductor) works well. Another use for coaxial cable braid is illustrated in Fig 17.40B. Instead of using tubing for the primary winding, the secondary winding is threaded through copper braid. Performance of the two units is almost identical.

The cores used must be large enough so the

(A)

(B)

Fig 17.40 — The two methods of constructing the transformers outlined in the text. At A, the one-turn loop is made from brass tubing; at B, a piece of coaxial cable braid is used for the loop.

core material will not saturate at the power level applied to the transformer. Core saturation can cause permanent changes to the core permeability, as well as overheating. Transformer nonlinearity also develops at core saturation. Harmonics and other distortion products are produced, clearly an undesirable situation. Multiple cores can be used to increase the power capabilities of the transformer.

Transmission-line transformers are similar to conventional transformers, but can be used over wider frequency ranges. In a conventional transformer, high-frequency performance deterioration is caused primarily by leakage inductance, which rises with frequency. In a transmission-line transformer, the windings are arranged so there is tight capacitive coupling between the two. A high coupling coefficient is maintained up to considerably higher frequencies than with conventional transformers.

The upper frequency limit of the transmission line transformer is limited by the length of the lines. As the lines approach ¼ wavelength, they start to exhibit resonant line effects and the transformer action becomes erratic.

MATCHING NETWORKS AND TRANSFORMERS

The typical tube amplifier tank circuit is an impedance transforming network in a pi or pi-L configuration. With reasonable loaded Q, it also functions as a low-pass filter to reduce the output signal harmonic levels below FCC minimums.

If a transistor amplifier uses a broadband transformer, it must be followed by a separate low-pass filter to achieve FCC harmonic suppression compliance. This is one reason broadband transistor amplifiers are operated in push-pull pairs. The balance between the circuit halves naturally discriminates against even harmonics, making the filtering job easier, especially for the second harmonic. The push-pull configuration provides double the power output when using two transistors, with very little increase in circuit complexity or component count. Push-pull pairs are also easier to match. The input and output impedance of a push-pull stage is twice that of a single-ended stage. The impedance is low, and raising it usually makes the matching task easier

The transistor's low-impedance operation provides the opportunity to use a simple broadband transformer to provide the transformation needed from the transistor's load impedance up to 50 Ω. This low impedance also swamps out the effects of the device's output capacitance and, with some ferrite loading on the transformer, the amplifier can be made to operate over a very wide bandwidth without adjustment. This is not possible with tubes.

On the other hand, a tube amplifier with its variable output network can be adjusted for the actual output load impedance. The transistor amplifier with its fixed output network cannot be adjusted and is therefore much less forgiving of load variations away from 50 Ω. Circuits are needed to protect the transistor from damage by mismatched loads. These protection circuits generally operate "behind the scenes" without any operator intervention. They are essential for the survival of any transistor amplifier operating in the real world.

CALCULATING PROPER LOAD IMPEDANCE

The proper load impedance for a single transistor (or a tube for that matter) is defined by $R = E^2/P$. Converting this from RMS to peak voltage, the formula changes to $R = E^2/2P$. If two devices are used in push-pull (with twice the power and twice the impedance) the formula becomes $R = 2E^2/P$.

There is a constraint. Transformer impedance ratios are constrained to the square of their turns ratios to integer values, 1, 4, 9, 16 and so on. Real-world transformers quickly lose their bandwidth at ratios larger than 25:1 due to stray capacitance and leakage inductance. A design solution must be found which uses one of these ratios. We will use the ubiquitous 100-W, 12-V transceiver power amplifier as an example. Using the push-pull formula, the required load impedance is $2 \times 12.5^2/100 = 3.125$ Ω. The required transformer ratio is $50/3.125 = 16$, which gives a turns ratio of 4:1.

Transistor manufacturers, recognizing the broadband transformer constraint, have developed devices that operate effectively using automotive and military battery voltages and practical transformer ratios: 65 W devices for 12 V operation and 150 W devices for 48 V. Bigger 50-V devices have been designed (for example, the MRF154) that will put out 600 W, but practical transformer constraints limit 50-V push-pull output power to 900 W. These device and transformer limitations on output power can be overcome by combining the outputs of several PA modules.

17.10.5 Combiners and Splitters

With some exceptions, practical solid state amplifiers have an upper power limit of about 500 W. This is a constraint imposed by the available devices and, to some extent, the ability to cool them. As devices are made more powerful by increasing the area of silicon die, the power density can become so high that only water cooling can provide adequate heat removal. Large devices also have large parasitic capacitances that make securing a broadband match over several octaves very difficult. By building a basic amplifier cell or "brick" and then combining several cells together, transmitter output powers are only limited by the complexity. Combiners and splitters have losses and add cost, so there are practical limits.

Broadband combiners usually take the form of an N-way 0° hybrid followed by an N:1 impedance transformer. The square ratio rule applies here too because the output impedance of a broadband combiner with N input ports is 50/N Ω — so combiners are usually 2, 4, 9 and 16-way. The higher the ratio, the lower the bandwidth will be.

Many types of combiners have been developed. The most common is the 4-way. It is easy to construct and has very good bandwidth. Most of the commercial "1-kW" broadband amplifiers use a 4-way combiner to sum the output of four 300-W push-pull modules operating on 48 V. Every combiner has loss. It may only be a few percent, but this represents a considerable amount of heat and loss of efficiency for a kilowatt output. This is a case where 4×300 does not make 1200.

The combiner approach to make a 1-kW solid-state amplifier uses a large number of individual parts. A comparable 1-kW tube amplifier requires relatively few. This makes the high-powered solid-state unit more expensive and potentially less reliable.

There is an alternative. If we had higher voltage transistors, we could use the same output transformer network configuration to get more output because power rises with the square of the operating voltage. There are high voltage transistors that can operate on 200 V or more. The problem is that these transistors must be capable of handling the corresponding higher power dissipation. The downside of making bigger, more powerful devices is an increase in parasitic capacitance. These bigger transistors become harder to drive and to match over broad bandwidths. However, the circuit simplicity and elimination of the combiner and its losses makes the high voltage approach quite attractive.

17.10.6 Amplifier Stabilization

Purity of emissions and the useful life (or even survival) of the active devices in a tube or transistor circuit depend heavily on stability during operation. Oscillations can occur at the operating frequency, or far from it, because of undesired positive feedback in the amplifier. Unchecked, these oscillations pollute the RF spectrum and can lead to tube or transistor over-dissipation and subsequent failure. Each type of oscillation has its own cause and its own cure.

In a linear amplifier, the input and output circuits operate on the same frequency. Un-

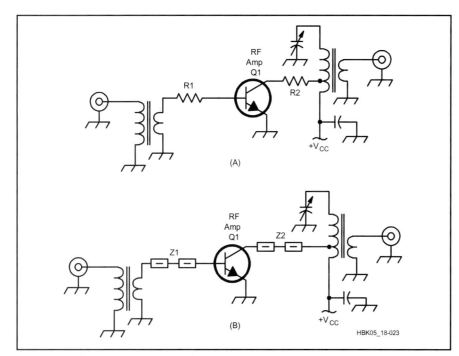

Fig 17.41 — Suppression methods for VHF and UHF parasitics in solid-state amplifiers.

less the coupling between these two circuits is kept to a small enough value, sufficient energy from the output may be coupled in phase back to the input to cause the amplifier to oscillate. Care should be used in arranging components and wiring of the two circuits so that there will be negligible opportunity for coupling external to the tube or transistor itself. A high degree of shielding between input and output circuits usually is required. All RF leads should be kept as short as possible and particular attention should be paid to the RF return paths from input and output tank circuits to emitter or cathode.

In general, the best arrangement using a tube is one in which the input and output circuits are on opposite sides of the chassis. Individual shielded compartments for the input and output circuitry add to the isolation. Transistor circuits are somewhat more forgiving, since all the impedances are relatively low. However, the high currents found on most amplifier circuit boards can easily couple into unintended circuits. Proper layout, the use of double-sided circuit boards (with one side used as a ground plane and low-inductance ground return), and heavy doses of bypassing on the dc supply lines often are sufficient to prevent many solid-state amplifiers from oscillating.

PARASITIC OSCILLATIONS

In low-power solid-state amplifiers, parasitic oscillations can be prevented by using a small amount of resistance in series with the base or collector lead, as shown in Fig 17.41A. The value of R1 or R2 typically should be between 10 and 22 Ω. The use of both resistors is seldom necessary, but an empirical determination must be made. R1 or R2 should be located as close to the transistor as practical.

At power levels in excess of approximately 0.5 W, the technique of parasitic suppression shown in Fig 17.41B is effective. The voltage drop across a resistor would be prohibitive at the higher power levels, so one or more ferrite beads placed over connecting leads can be substituted (Z1 and Z2). A bead permeability of 125 presents a high impedance at VHF and above without affecting HF performance. The beads need not be used at both circuit locations. Generally, the terminal carrying the least current is the best place for these suppression devices. This suggests that the resistor or ferrite beads should be connected in the base lead of the transistor.

C3 of **Fig 17.42** can be added to some power amplifiers to dampen VHF/UHF parasitic oscillations. The capacitor should be low in reactance at VHF and UHF, but must present a high reactance at the operating frequency. The exact value selected will depend upon the collector impedance. A reasonable estimate is to use an X_C of 10 times the collector impedance at the operating frequency. Silver-mica or ceramic chip capacitors are suggested for this application. An additional advantage is the resultant bypassing action for VHF and UHF harmonic energy in the collector circuit. C3 should be placed as close to the collector terminal as possible, using short leads.

The effects of C3 in a broadband amplifier are relatively insignificant at the operating frequency. However, when a narrow-band collector network is used, the added capacitance of C3 must be absorbed into the network design in the same manner as the C_{OUT} of the transistor.

LOW-FREQUENCY PARASITIC OSCILLATIONS

Bipolar transistors exhibit a rising gain characteristic as the operating frequency is lowered. To preclude low-frequency instabilities because of the high gain, shunt and degenerative feedback are often used. In the regions where low-frequency self-oscillations are most likely to occur, the feedback increases by nature of the feedback network, reducing the amplifier gain. In the circuit of Fig 17.42, C1 and R3 provide negative feedback, which increases progressively as the frequency is lowered. The network has a small effect at the desired operating frequency but has a pronounced effect at the lower frequencies. The values for C1 and R3 are usually chosen experimentally. C1 will usually be between 220 pF and 0.0015 μF for HF-band amplifiers while R3 may be a value from 51 to 5600 Ω.

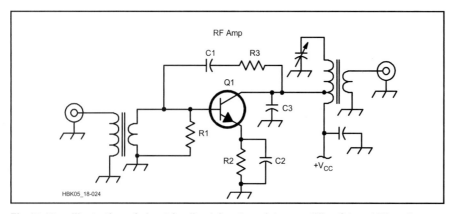

Fig 17.42 — Illustration of shunt feedback in a transistor amplifier. C1 and R3 make up the feedback network.

R2 of Fig 17.42 develops emitter degeneration at low frequencies. The bypass capacitor, C2, is chosen for adequate RF bypassing at the intended operating frequency. The impedance of C2 rises progressively as the frequency is lowered, thereby increasing the degenerative feedback caused by R2. This lowers the amplifier gain. R2 in a power stage is seldom greater than 10 Ω, and may be as low as 1 Ω. It is important to consider that under some operating and layout conditions R2 can cause instability. This form of feedback should be used only in those circuits in which unconditional stability can be achieved.

R1 of Fig 17.42 is useful in swamping the input of an amplifier. This reduces the chance for low-frequency self oscillations, but has an effect on amplifier performance in the desired operating range. Values from 3 to 27 Ω are typical. When connected in shunt with the normally low base impedance of a power amplifier, the resistors lower the effective device input impedance slightly. R1 should be located as close to the transistor base terminal as possible, and the connecting leads must be kept short to minimize stray reactances. The use of two resistors in parallel reduces the amount of inductive reactance introduced compared to a single resistor.

17.11 A New 250-W Broadband Linear Amplifier

The amplifier described here and shown in **Fig 17.43** is neither revolutionary nor daring. It uses commonly available parts — no special parts and no flea market specials. It is based on well-proven commercial designs and "best design practices" acquired over the past 30 years as solid-state technology has matured. This section will serve as a detailed design example for solid-state power amplifiers as well as a practical project that you can build.

A block diagram for the project is given in **Fig 17.44**. The amplifier is built on three PC boards — a PA module, a low-pass filter assembly, and a board for control, protection and metering circuitry. The *Handbook* CD includes *ExpressPCB* files for these boards, and the artwork can be used to have boards made in small quantities (see **www.expresspcb.com** for details).

The basic PA configuration has been in the *Handbook* for several years. It is intended to be a "ham-proof" external amplifier for QRP transceivers that put out 15 W or less. It is designed for a gain of 30×, or 15 dB. Drive power of less than 10 W will provide 250 W output from 1.8 through 50 MHz. The amplifier provides exceptionally linear performance, necessary for high quality SSB and PSK modes, and is rugged enough to withstand the most rigorous contest environment.

Amplifier design tends to focus on the RF section, but a successful stand-alone solid-state amplifier is equally dependent on its control system. The control requirements for a tube amplifier are well known, while those for solid-state amplifiers are not. This is mostly because the functions of a solid-state amplifier's control system are generally transparent to the user. Parameters are monitored and protection is applied without any operator intervention. This must be. While tubes are fairly forgiving of abuse, semiconductors can heat so quickly that intervention *must* be automatic or they can be destroyed.

Transistors are sensitive to heat, so cooling and temperature compensation are critical to a successful design. Transistors require a heat sink. Power amplifier tubes have large surface areas and are cooled by air blown on or through them. Transistors are small.

Fig 17.43 — This 250-W amplifier for 160 through 6 meters provides a detailed design example as well as a practical project. Additional photos and information about the interior layout may be found on the *Handbook* CD-ROM.

Mounting them on a heat sink increases their thermal mass and provides a much larger surface area so the heat dissipated in the devices can be removed either by convection or forced air. The thermal design of an amplifier is just as important as the electrical design. More information on thermal design may be found in the **Analog Basics** chapter.

Silicon's thermal coefficient causes the bias current to increase as the device heats up if the bias source is fixed. The increased current causes even more heating and can lead to thermal runaway. For stable Class AB linear operation, the gate bias for a MOSFET or bipolar transistor must track the temperature of the device. The control circuit typically uses another silicon device such as a diode thermally coupled to the amplifier heat sink near the transistor to sense the temperature and adjust the bias to maintain a constant bias current.

Transistor power amplifiers are designed to operate into 50 Ω. Operation into a VSWR other than 1:1 will cause an increase in device dissipation and other stress. The success of the solid-state transceiver is due to its integrated PA protection system. The temperature of the heatsink, the load VSWR, the output power and the supply current are all monitored by the control system. If any of these exceed their threshold limits, the RF drive is reduced by the transceiver's ALC system.

An external solid-state PA protection system must perform the same functions, but the driver's ALC circuit is not always available so other means must be used to protect the PA. This is usually accomplished simply by taking the amplifier out of the circuit. An indicator then tells the operator which condition caused the fault so appropriate action can be taken. Access to the driver's ALC system would make this protection task more automatic, smoother and less troublesome, but no two transceiver models have the same ALC characteristic. This makes the design of a universal ALC interface more difficult.

17.11.1 The 1.8 to 55 MHz PA — Detailed Description

Fig 17.45 shows the power amplifier (PA) schematic. Two Microsemi VRF151 MOSFETs are used in this amplifier. The circuit topology is a 4:1 transmission line transformer type, rather than a "tube and sleeve" type common in many PA designs and discussed earlier. This style offers more bandwidth, necessary to provide performance on 6 meters. Typical gain is 15 dB; 10-W drive will easily provide 250-W output with a 48-V supply. There is a lot of latitude in this design. It can be operated on an unregulated supply. As long as the maximum unloaded voltage does not exceed 65 V, the transistors will not be overstressed. Other devices such as the MRF151, SD2931 or BLF177 would probably also work but have not been tested. They will require a regulated power supply, however.

FEEDBACK — TWO KINDS

The amplifier's gain is controlled by two

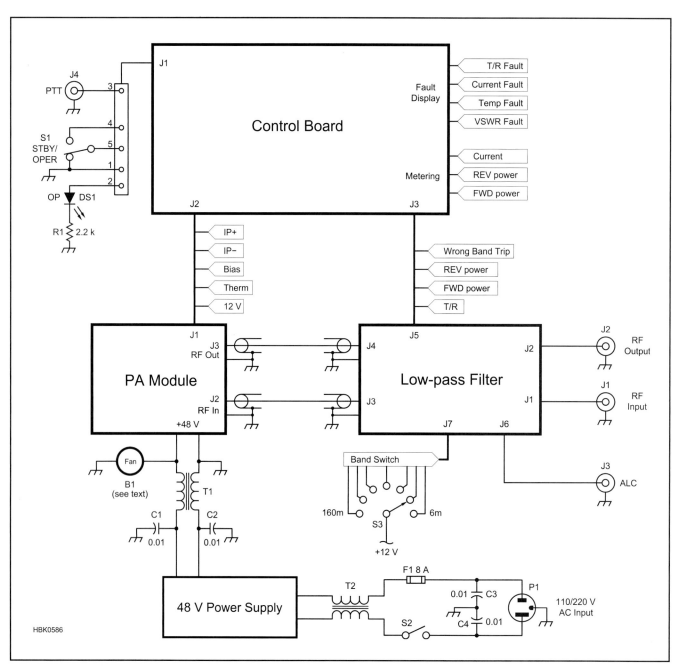

Fig 17.44 — Block diagram of the 250-W solid-state amplifier. It is built on three PC boards, which are described in the text and accompanying diagrams. T1 and T2 are 2 turns through a Fair-Rite 2643540002 ferrite bead if needed to suppress RFI from the switching power supply.

kinds of feedback. Shunt feedback (from drain to gate) is provided by the link on T2 through resistors R5 and R6. It tends to lower the input impedance but it also helps to keep the gain constant over frequency and improve the linearity. Series feedback is provided by the 0.05 Ω of resistance in each source. This increases the input impedance, cuts the gain by 3 dB, and most importantly, it has a huge effect on the linearity.

Without any feedback at all, the amplifier would have more than 30 dB gain (×1000) at some frequencies, tending to make it unstable — prone to parasitic oscillation. And the linearity would be terrible, –20 dB or so IMD products. It would also be very sensitive to load changes. The input SWR is 1.2:1 on 160 meters and rises to 1.5:1 on 6 meters. The amplifier's gain is 15 dB ±0.5 dB over the same frequency range.

PA INPUT

On the input side, T1 is a sleeve-and-tube RF transformer. Its small size and tight coupling make it suitable for very wide bandwidth operation. When properly compensated, this type is able to provide a match to a 12.5-Ω load over a very wide frequency range from 1 to 100 MHz. The problem is that the input impedance of the two transistors is not a flat resistive load.

The gate of a MOSFET is essentially a high-Q capacitor. A voltage greater than its threshold (V_{th}) applied between the gate and source will control the conductivity of the drain-to-source path. No power can be dis-

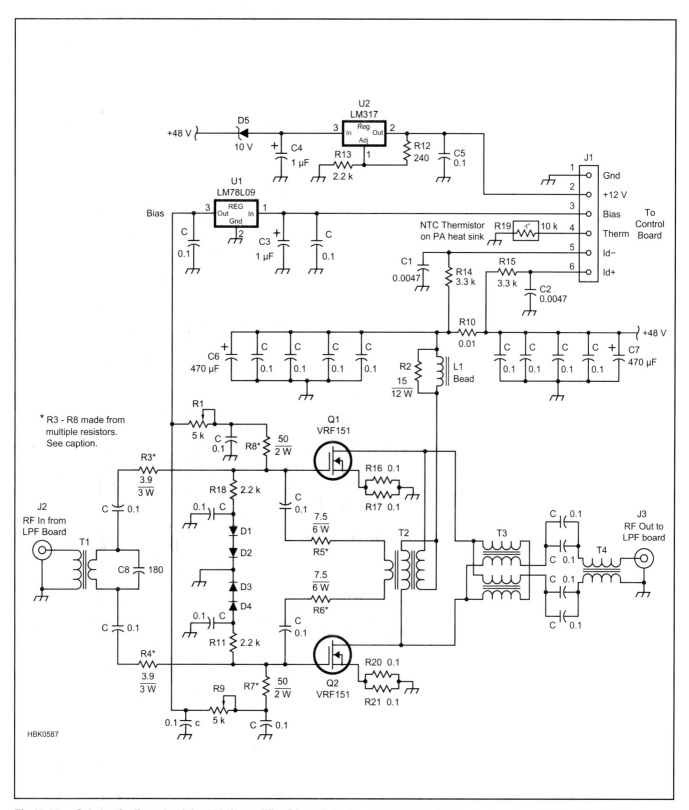

Fig 17.45 — Schematic diagram of the 250-W amplifier PA module. A complete parts list may be found on the *Handbook* CD.

C — 0.1 µF, 100 V X7R 1206 SMT.
L1 — 2t on Fair-Rite 2643480102 ferrite bead.
R2 — 15 Ω, 12 W.
R3, R4 — Three 12 Ω, 2 W SMT resistors in parallel.
R5, R6 — Two 15 Ω, 3 W resistors in parallel.

R7, R8 — Two 100 Ω, 1 W SMT resistors in parallel.
R19 — 10 kΩ, 5% NTC thermistor, SMT (DigiKey 541-1150-1-ND).
T1 — 2t #22 wire on CCI RF400-0 core (or two Fair-Rite 2643006302 cores).
T2 — Primary, 1t #22 wire; secondary 8t #22 wire bifilar wound, on Fair-Rite 5961004901 core.

T3 — 2 × 3t 25 Ω coax on Fair-Rite 28610100002 core (see text).
T4 — 3t RG-188 coax on Fair-Rite 2643665802 core (on cable to LPF board, not on PC board).
Q1, Q2 — Microsemi VRF151 MOSFET.

sipated in a purely reactive element, so we cannot match to a capacitor. Fortunately, all real reactive elements have losses and it is this loss that we would match to in a MOSFET gate for single-band operation. But that is no good here because we want a broadband match.

There are several ways to match a MOSFET over a broad frequency range. The most common is to swamp the gate capacitors with resistors. If the resistor value is lower than the impedance of the transistor gate capacitance, it dominates what the transformer sees as a load at the lower end of the bandwidth. The gate capacitor impedance decreases at higher frequency so some series resistance is added, R3 and R4, so that there is always a minimum real part to the load on the secondary of T1.

For a 1:1 input SWR, T1 wants to be terminated by a 12.5 Ω load. More than half of this is provided by R3 and R4. The rest is through shunt loads R7 and R8, plus the impedance of the output passed through the feedback network. T1 has no center tap so a balanced load is forced by the action of resistors R7 and R8. Having "soft" center taps on both the input and output absorbs any differences between transistors and greatly improves the network's RF balance. This in turn improves the cancellation of even harmonics at the output. The gate impedance is raised by the effect of the source resistors, further improving the match.

DC FEED TRANSFORMER

In addition to providing the link for the feedback, T2 also acts as the dc feed choke. It is wound with two parallel bifilar #22 wires and a single turn for the feedback. At dc, the current flows in opposite directions through each #22 wire so the net current is zero and the core does not saturate with dc. At RF, the choke with its ferrite core provides at least 50 Ω of inductive reactance making it essentially invisible to the RF signals across it. As the feedback transformer it provides a $\frac{1}{16}$ sample of the drain-drain voltage to the gate feedback loop.

It is not often mentioned in the literature, but the dc feed transformer T2 acts as a 180° hybrid combiner (**Fig 17.46**). A hybrid combiner has four ports: two inputs, the sum and the difference port. The sum of the two input signals appears on the sum port and the difference between them (differences due to voltage or phase) appears at the difference port. In this case T2 is terminated with a 1:4 balanced transformer that brings the output impedance up to 50 Ω. The sum port is across the whole secondary. The two input ports are between each end of the secondary and ground. The difference port is between the center tap and ground.

No two transistors or their circuit layouts are exactly equal. In a push-pull circuit

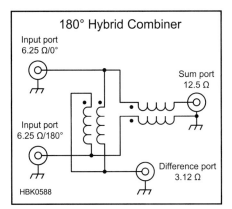

Fig 17.46 — Dc feed transformer T2 acts as a 180° hybrid combiner.

these slight differences gives rise to imbalance between sides which gives rise to even harmonics at the output. By not placing the usual heavy RF bypassing at the center (the difference port) of the bifilar winding in T2 the difference between the two drain voltages shows up across L1 and is dissipated in the parallel resistor R2. This enables the amplifier to achieve better than 40 dB suppression of even harmonics, helps the efficiency, and greatly improves its stability when driving mismatched loads.

THE PA OUTPUT TRANSFORMERS

The output circuit of the PA uses a 4:1 impedance transformer. We have two VRF151 transistors that will each put out 150 W if properly cooled. That means 300 W output power for the pair operating on 48 V. The classic formula relating drain-to-drain load impedance to the supply and output power is

$$R_L = \frac{2 V_{dd}^2}{P_O} \qquad (10)$$

For 48 V this gives 15.36 Ω. But the drain cannot swing all the way to zero because of the finite on-resistance of the MOSFET. The actual available swing is typically 90% of the supply voltage so a more practical formula is

$$R_L = \frac{2 (0.9 V_{dd})^2}{P_O} \qquad (11)$$

A 4:1 transformer gives 12.5 Ω. This is a nice load for these devices.

The 4:1 impedance ratio is performed by a simple transmission line transformer. The impedance of the coax used should be the geometric mean between the two impedances to be matched:

$$Z_0 = \sqrt{50 \times 12.5} = 25 \, \Omega$$

Coax with this characteristic impedance is hard to find and expensive when you do. A very close approximation is provided by using #22 shielded 600-V Teflon wire such as Belden #83305. The inner and outer diameters of the conductors and the dielectric properties (the e_r) of the Teflon combine to produce a coaxial cable with a characteristic impedance (Z_0) of 25 Ω.

These two coax lines are connected in parallel on the drain end and in series on the 50 Ω output end. Any voltage on the input is put in series at the output, giving a 2× voltage ratio or a 4× impedance ratio, exactly what we want. If coiled up in separate coils, the transformer will work without any ferrite. The two coils cannot be allowed to couple so they cannot be on the same form.

In order to get a wider frequency response, the inductance of these coax coils is increased by winding them on a ferrite core. The core used here has two holes, so independent coils can be wound on each side without any coupling between them and it makes a nice neat package. Separate cores would work just as well. The ferrite core is type 61 material with a permeability of $\mu_i = 125$. It is a binocular bead but could be replaced by two 0.5 dia × 1-inch long sleeves of the same material. The ferrite "load" on the coax makes its outside shield a high impedance from end to end, while inside the shield the coax maintains its 25 Ω characteristic impedance between center and shield. As long as the coax lines of T3 are wound as two non-coupled coils, the amplifier will operate from 21 to 80 MHz without any ferrite at all.

The advantage of the transmission line type of RF transformer is that it does not have the leakage reactance that plagues the tube-and-sleeve type of transformer used on many solid-state amplifiers. Simply stated, a parasitic leakage inductance is introduced in series with the primary due to incomplete coupling of the flux between the primary and secondary winding. This increases the apparent impedance of the low-Z side of the transformer as the frequency increases. The output impedance of the transistors decreases with frequency — a double hit of mismatch that causes the gain to drop off quickly.

There is one disadvantage to the transmission line transformer. It has a balanced input and output. Sometimes designers will ground one side of the output and rely on the ferrite loading to decouple the ground side of the output. This has a negative effect on the balance of the amplifier, and on those even harmonics we want to minimize. It also doubles the flux stress in the ferrite causing it to heat more. The solution is T4, a simple current balun — four passes of the 50 Ω output line through a toroid of type 61 ferrite. With T4 in place, a balanced load on the output of T3 is maintained, the even harmonics are suppressed, and the efficiency is 5 to 10% better on most bands.

PA LAYOUT

The PA board (**Fig 17.47**) was designed

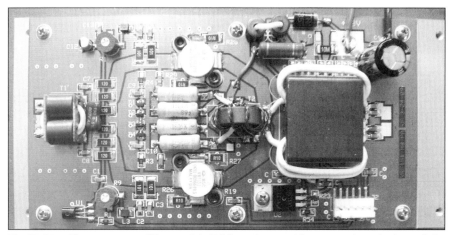

Fig 17.47 — The power amplifier module PC board assembly mounted on the heat sink.

with all parts mounted on the top surface — no through-hole parts at all. The back side of the PC board is a continuous ground plane and is mounted directly to the heatsink without spacers.

17.11.2 Control and Protection

The control board appears far more complicated than the PA but in reality, it is just a few analog and logic ICs. The various control and protection circuits are shown in **Figs 17.48** and the LED displays and drivers (also on the control board) are shown in **Fig 17.49**. This circuitry monitors several parameters, displays them, and if necessary, puts the amplifier into standby if one of them goes out of range. This control system could be used on any amplifier. All solid-state amplifiers need similar protection. The amplifier is protected for:

1. Over temperature, by a thermistor on the heatsink and setting a limit.
2. Over current, by measuring the PA current and setting a maximum limit.
3. Over SWR, by monitoring the reflected power and setting a maximum limit.
4. Selection of a low-pass filter lower than the frequency in use.

Each of these fault trips results in lighting an error LED and forcing the amplifier into the standby position, out of the RF path. There is also an ALC level detector that generates a negative-going feedback voltage for the driver when the RF drive goes above the level corresponding to maximum power. If this PA were part of a transceiver, the several faults described above would generate inputs into the ALC system and turn back the drive rather than taking it off the air. We do not always have that luxury so the best course is to take it off line until the cause can be fixed.

THERMAL COMPENSATION

MOSFETs are sensitive to temperature. If a fixed bias is used on the gate to set the quiescent bias at 100 mA when the device is cold, the current will increase as the device heats up. In some devices, it will cause thermal runaway. The hotter it gets, the more current it draws, causing even more heating and so on. The solution is to sense the temperature of the device and reduce the gate bias as it heats up. The VRF151 is relatively insensitive compared to similar high power RF MOSFETS.

The compensation system is quite simple, effective and foolproof. It relies on the thermal characteristic of silicon diodes that as a diode heats up, the forward voltage across it goes down approximately 2.4 mV/°C. Two diodes in series are used at the bottom of each gate voltage divider (D1-D4). Mounted on the PA board, they heat up along with the transistors and reduce the gate voltage by a proportional amount. The 100 mA of bias at 25 °C is less than 150 mA at 200 °C. This compensation system may not work as well if other MOSFET types are substituted because they have different thermal coefficients of V_{th} and may require more aggressive thermal compensation. Gate bias voltage is provided via a PTT-activated 9-V regulator.

A simple way to check or adjust compensation is to place the amplifier module, board and heatsink, on mother's electric frying pan. Set it to 212 °F (100 °C) and monitor the drain current. It should stay within 150% of the cold setting. Be patient. This takes a while because the response time of this arrangement is quite long. Under some operating conditions, it is possible for the transistor to get very hot before the heat travels to the diodes. For this reason, we have additional means to protect the transistors.

OVER-CURRENT PROTECTION

One simple protection method is to limit the power into the PA module by limiting the maximum supply current. The drain current is sensed across shunt resistor R10. The sense voltage is amplified by U2A and sent to the current meter. It also goes to comparator U2B where it is monitored and compared to the limit voltage set by R13. If it exceeds this limit, the comparator trips the fault latch U1, lights the OC LED, and opens the PTT.

OVER-TEMPERATURE PROTECTION

The heat sink temperature is monitored directly by a negative temperature coefficient thermistor, R19, mounted on the PA assembly. It is the lower half of a voltage divider sensed by U3A and fed to comparator U3B. When the thermal sense voltage exceeds the limit set by R14, the U1 fault latch is tripped and the PTT line opened. Any time the fault latch is tripped it lights an LED to indicate the cause of the fault so the operator can address the problem. Cycling the OPERATE - STANDBY switch resets the fault latch and restores normal operation.

VSWR PROTECTION

The amplifier is protected for high VSWR. The amplifier is designed to operate into 50 Ω. It acts somewhat like a constant voltage source. It will dutifully try to put the same voltage across whatever load it is given, even a short. The SWR bridge produces a voltage proportional to both the output power and the relative mismatch between the load and 50 Ω.

T1 on the LPF board is the heart of a dual directional coupler that provides both the VSWR detection and the forward power monitoring. It is really two transformers on one two-hole core, Fair-Rite # 2843010402. Each has 17 turns of #28 AWG as the secondary and a single "turn" of #22 AWG Teflon hookup wire as the primary.

T2 is a single directional coupler that looks at the power reflected from the low-pass filter. Its secondary has 20 turns of #28 AWG wire, and the primary is a single pass of #22 AWG wire. In order to minimize the impedance bump it places in the RF signal path between the PA and the LPF, the core is mounted in a small window cut in the board. This allows the toroid to be mounted so the primary wire can go straight through the center of the toroid. Detailed photos of all the transformers are contained on the accompanying CD.

Several load conditions can produce the same value of indicated VSWR. At one mismatch load condition the amp might be trying to put out way too much power. At a different reflection coefficient, it might see a very high reflected voltage. This could raise the peak voltage on the drains past the voltage breakdown limit of the MOSFET. The voltage from the detector on the reflected power port of the SWR bridge is brought to the meter and to a comparator that will trip the fault latch when the voltage is past a limit. Again the PTT is opened. The operator can either lower the SWR by improving load match or

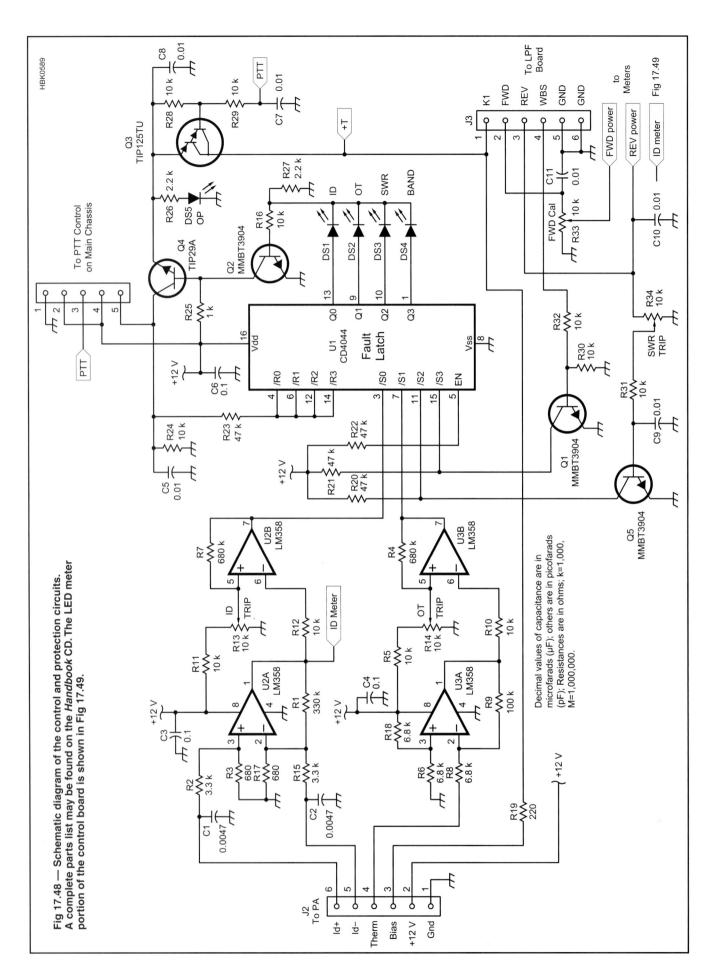

Fig 17.48 — Schematic diagram of the control and protection circuits. A complete parts list may be found on the *Handbook* CD. The LED meter portion of the control board is shown in Fig 17.49.

17.38   Chapter 17

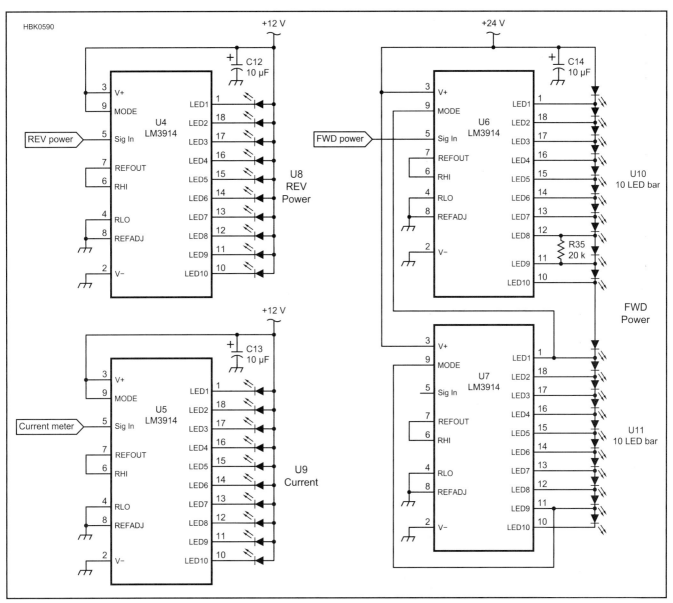

Fig 17.49 — Schematic for the LED meters, which are also located on the control board with the circuitry shown in Fig 17.48. A complete parts list may be found on the *Handbook* CD.

reduce the output power to limit the reflected power from the bad match. The amplifier is happy either way.

Transformer T1 on the LPF board is the heart of the directional coupler used for VSWR protection and power metering. It consists of identical transformers wound on each side of a two-hole ferrite core. The secondary is formed from 17 turns of #28 AWG enameled wire. The primary is a single pass of #22 AWG insulated hookup wire though the hole. A picture of the transformer is contained in the companion CD.

BAND FAULT PROTECTION

One problem unique to external broadband amplifiers is that they generally do not care what frequency they amplify, but their low-

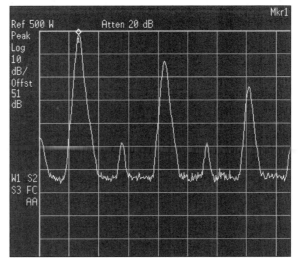

Fig 17.50 — The output of the 250 W amplifier without a low-pass filter. The even harmonics are suppressed by more than 40 dB. The odd harmonics are not suppressed by the push-pull balance and must be attenuated in the LPF. Harmonics after the LPF are all greater than 60 dB down.

RF Power Amplifiers 17.39

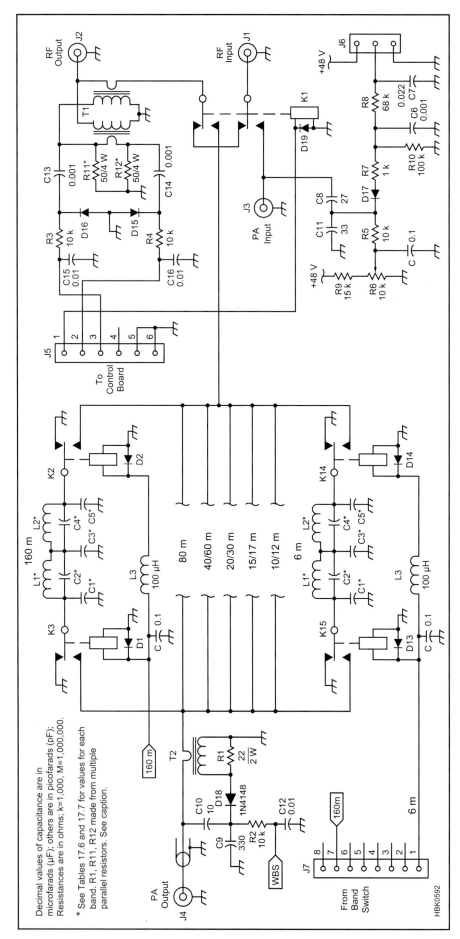

pass filters on the output certainly do. The problem arises when the operator forgets to change the band switch on the amplifier when moving to a higher band. All the power from the amp is reflected back from the LPF and the amplifier is distressed — all that power with no load. The solution is to place a reflected power sensor between the PA and LPF. A monitor will see this condition and trip the fault latch for Wrong Band Selection.

T2 is single directional coupler used for reflected power only. The transformer is mounted in a small window cut in the board that allows the primary wire to go straight through the center of the toroid. This coupler normally sees all the reflected harmonic power from the LPF. The maximum harmonic power from the PA output is down 13 dB from the fundamental. See Fig. 17.50. When the wrong filter is selected, all of the power is reflected by the filter so the threshold is set to detect this 20× (13 dB) difference.

17.11.3 Low-Pass Filter

For amateur use, FCC §97.307(d) requires that harmonics be suppressed at least 43 dB for operation below 30 MHz, and at least 60 dB on 6 meter and higher frequency bands. The output signal from a broadband amplifier itself contains harmonics and needs to have a separate filter to meet the FCC's requirements for harmonic suppression. **Fig 17.50** shows the harmonic output of this amplifier without any low-pass filtering.

In a well designed solid-state push-pull amplifier, the second harmonic is 30 to 40 dB below the fundamental if the balance is good, but the third harmonic is only down 13 dB. This means the low-pass filter needs to supply about 10-13 dB of attenuation at the 2nd harmonic and 30 dB or more to the rest. On 6 meters the filter must provide at least 35 dB attenuation at the 2nd harmonic and 50 dB to the rest. None of this is difficult for a properly designed low-pass filter. In a commercial application, all harmonics must be at least 60 dB down. This usually requires

Fig 17.51 — Schematic for the low-pass filter board. The filters are described in detail in the text, and values are shown in Tables 17.6 and 17.7. This board also contains the TR relays, directional couplers and ALC circuitry. A complete parts list may be found on the *Handbook* CD.

K1 — DPDT, 12 V coil, 8 A contacts (Potter & Brumfield RTE24012F).
K2-K15 — SPDT, 12 V coil, 10 A contacts (Omron G5LA-14-12DC).
R1 — Two 47 W, 2 W resistors in parallel.
R11, R12 — Two 100 W, 2 W resistors in parallel.
T1 — Dual transformer (see text).
T2 — Secondary: 20t #28 AWG on Fair-Rite 5961000201 toroid. Primary: 1t #22 AWG (see text).

a more complicated filter, especially if continuous coverage is desired.

The largest board in the amplifier, shown schematically in **Fig 17.51**, contains the low-pass filters, TR antenna relay, SWR bridge and the ALC detector. Amateur bands are harmonically related, so a filter for 40 meters will not do anything to reduce the second harmonic on 80 meters. If an amplifier is going to cover 160 to 10 meters, it will need at least five filters. With 30, 17 and 12 meters, the number of filters is usually increased by one to provide better suppression of harmonics. When 6 meters is added another filter is required, and it must be able to bring all 6-meter harmonics to –60 dB. This requires a more complicated filter.

FILTER DESIGN

The HF band filters can easily meet their requirements with simple five-element, 0.044-dB ripple Cauer filters. These filters use an elliptic topology. The nulls can be arranged to provide specific treatment of the third harmonic. Also, the insertion loss is the lowest of the several common filter types. The *SVC Filter Designer* software provided on the *Handbook* CD was used to design the LPFs used here. (See the **RF and AF Filters** chapter for more information on Cauer and other filter types.).

Low-pass filters are precision-tuned circuits. If the values are not right, the filter will have high loss and/or high VSWR in the passband. The calculated capacitor values are rounded to the closest 5% standard values. This is done by the *SVC Filter Designer* program. In a 5th order Cauer filter, there are two parallel resonant circuits that set the nulls in the response. If the exact calculated C value is not used, its paired L must be adjusted so the desired null still hits at the proper frequency. **Table 17.6** gives all the LPF capacitor and inductor values, the corner frequency (F_c), and the frequencies of the nulls (F1 and F2). L1-C2 resonate at F1 and L2-C4 resonate at F2.

LPF Inductors

Inductor winding details are given in **Table 17.7**. The low-frequency coils are wound on Micrometals T80-2 powdered iron toroid cores. For 20/30 and 15/17 meters the mix is changed to T80-6. Use of toroid cores keeps the Q of the coils high, makes physically smaller coils and provides magnetic shielding. This construction also helps to prevent the various sections of the filter from "talking to each other" and causing "suck-outs" in the passband or "lumps" in the stop band.

The 10/12 and 6 meter coils are self-supporting air-wound types with no cores which gives the highest possible Q. Coil adjustment is done by compressing or spreading turns on the cores. On the high bands, the coils are 5% high when wound tight. Spreading the coils slightly brings them to the proper value.

As mentioned before, it is important that the nulls in the Cauer filter response occur at the right frequency. Since the capacitor values have been rounded to the closest 5% values, the value of each parallel inductor has to be tweaked to set the null on the proper frequency. This is easy to do with a network analyzer but rather difficult for the home builder because the nulls are at various frequencies up to 154 MHz.

LPF Capacitors

Selecting capacitors for the LPF is a bit trickier. At 300 W of RF, the requirement is 123 V RMS or 174 V peak. A 500-V capacitor will easily handle this. We are also looking for RF current handling capability. At 300 W of RF, the requirement is 2.5 A RMS into 50 Ω. If a capacitor carrying this current has an equivalent series resistance (ESR) of 0.4 Ω it will dissipate 2.5 W. Harmonic currents can increase the heating dramatically.

Ceramic capacitors come in several grades of dielectric quality — X7R, Z5U and NP0 (or C0G). The NP0 is called a "Class 1" dielectric (see the **Component Data and References** chapter for more on capacitor characteristics). NP0 (C0G) capacitors are more expensive and harder to find, but they are the only type suitable for use in RF power filters. The best way to make capacitors for a PA low-pass filter is to use several in parallel. This spreads the RF current across several units and permits obtaining odd values by combining standard value parts.

RF capacitors are very difficult to find, and all the required values are never in the dealer's stock. Several capacitor manufacturers were tried and most quoted 6 to 14 week lead times and had minimum order quantities of $100

Table 17.6
Low Pass Filters
5th order Cauer
Ap = 0.044, As = 40 dB

Band (m)	Fc (MHz)	C1 (pF)	L1 (µH)	C2 (pF)	F1 (MHz)	C3 (pF)	L2 (µH)	C4 (pF)	F2 (MHz)	C5 (pF)
6	57.4	36	0.1571	6.8	154	75	0.1245	20	100.85	27
10/12	30.9	68	0.307	12	82.9	150	0.2387	36	54.3	51
15/17	22.265	120	0.424	22	52.09	220	0.338	62	34.771	91
20/30	15.053	180	0.619	33	35.217	330	0.458	100	23.508	130
40/60	8.837	300	1.058	56	20.674	560	0.831	160	13.801	240
80	4.778	560	2.027	100	11.177	1000	1.517	300	7.461	430
160	2.243	1200	4.181	220	5.248	2200	3.329	620	3.503	910

Table 17.7
Low Pass Filter Inductor Winding Details

Band (m)	L1 (nH)	Core Type	Wire Size (AWG)	No. of Turns	Inside Dia. (in.)	L2 (nH)	Core Type	Wire Size (AWG)	No. of Turns	Inside Dia. (in.)
6	157	—	16	5	0.312	124	—	16	4	0.33
10/12	307	—	16	7	0.35	238	—	16	3	0.33
15/17	424	T80-6	18	10	—	338	T80-6	18	9	—
20/30	619	T80-6	18	12	—	458	T80-6	18	10	—
40/60	1058	T80-2	18	14	—	831	T80-2	18	12	—
80	2027	T80-2	20	19	—	1516	T80-2	20	17	—
160	4180	T80-2	20	28	—	3329	T80-2	20	24	—

All of the cores are wound with enameled copper wire. The size is as large as will fit to maximize coil Q.

to $500 of each value! The best ceramic RF capacitors are made by ATC. Fortunately they also had most of the values in stock and in reasonable minimum quantities. Several values were paralleled when a particular value was not available. They are quite expensive, about $3 each for 500 V units below 200 pF and $8 each for the larger units. The filter set uses 29 different values. A commercial amplifier manufacturer would have to make a large capital investment to stock an LPF production line, another barrier for a solid-state legal-limit amplifier.

An alternative capacitor solution would be to use 500-V silver-mica capacitors. These are suitable for all but the 10-meter and 6-meter filters where a single unit's RF current rating might be exceeded and the lead parasitics changes the net value considerably. While SMT silver-micas are available, they are also hard to find in all the values needed. The leaded versions are far more common and are made by several companies. The downside is that the LPF board was laid out for surface-mount capacitors, all mounted on the bottom side of the board. Using leaded through-hole parts will require laying out a new LPF board and it will have to be larger.

With the capacitors used, the filter is suitable for a full kW output in a system with an SWR protection circuit as implemented in this amplifier. The coils might need to be implemented on the next larger size core.

LPF Construction

The LPF board, shown in **Fig 17.52**, has a continuous ground plane on the top side. All the coils are mounted through the board. As noted, all the capacitors are leadless SMT types mounted on the bottom side. Plated-thru vias complete the circuit to the top side ground. This arrangement with the coils on the top and capacitors on the bottom provides the best space efficiency and gives repeatable filter performance. Repeatability is not so much a concern for the home builder, but critical for commercial equipment.

To prevent coupling between bands, the filters are not in order across the board. On the two high band filters, the axis of each coil is perpendicular to the next to prevent coupling between them. Finally, each filter is shorted out by the selection relays when not in use, which helps prevent "sneak paths" around the selected filter.

The filter selection relays are used in pairs. Each is rated for 10 A, overkill for this application. They are quite inexpensive, around $1 each in quantities of 25, and have very low loss and parasitic inductance. The 10/12 meter and 6 meter filter values were tweaked for lowest loss and best efficiency after the whole filter was built. Computer modeling of the stray capacitances and inductances proved tricky and in the end it was better to temporarily replace the end capacitors with trimmers and tune for best performance to find the final values and then replace them with fixed units.

ALC DETECTOR

There is also an automatic level control (ALC) detector on the LPF board. ALC provides a feedback signal to the exciter to prevent overdriving the amplifier, and it was once a very common connection between the exciter and amplifier. Now that most solid-state transceivers have an output power control, it is used less often but still useful.

ALC interfacing is difficult because there is no standard ALC system specification among different transceiver brands and models. The detector here is a generic circuit with median values that generates a negative-going ALC signal. D17 rectifies the drive signal after a detection threshold, set by R69, has been exceeded. The attack time and filtering time constants are set by the RC network following the diode. Some adjustment of the R/C values may be required to work with a particular model of transceiver because of loop dynamic stability issues.

17.11.4 Power Supply

The PA requires 48 V at 10 A peak. It does not have to be regulated provided the following conditions are met. The maximum no-load voltage of 55 V is preferred, and it cannot exceed 60 V under any condition to provide a suitable safety margin. It must have at least 10,000 µF of filtering to prevent noticeable hum modulation. A simple power transformer feeding a 20-A bridge rectifier and a filter capacitor is sufficient. It may require a step-start circuit to ease the high current surge on the line power switch. See the **Power Supplies** chapter for more information.

The alternative is to use a switching supply. Supplies rated for 48 V and 500-600 W supplies regularly go for less than $100 at online auction. They are small and light. You may have to deal with RFI issues on some of the cheaper models, though. This usually just requires a ferrite bead and additional bypass capacitors on both the ac input lines and the dc output lines as well.

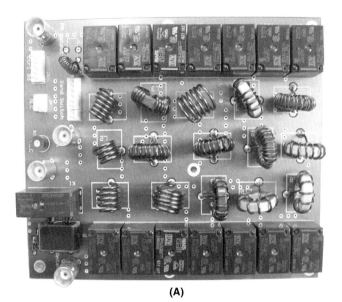

(A)

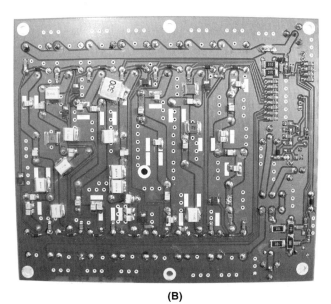

(B)

Fig 17.52 — The assembled low-pass filter board with the coils, relays and other components mounted on the top side (A) and the LPF capacitors mounted on the bottom (B). The prototype used available capacitors — two sizes of ATC ceramic RF chip capacitors, a metal clad mica and three unencapsulated SMT mica units. All are serviceable, but the ATC ceramics work best.

Power output varies with the square of supply voltage. The amplifier will put out 75 W if run from 25 V. The gain changes very little, just the output power. This squared relationship makes it more sensitive to ripple modulation, so make sure that the power supply, no matter how it is made, has reasonable filtering. It does not need to be current-limited since the PA control has this built-in. It does need a suitable fuse, however. A 15-A low-voltage fuse will do nicely.

The power supply used in the test amplifier was purchased on eBay. It was called "500 W 48 V 10.4 A Switching Power Supply for Radio" and was less than $100 including shipping. It is 8.5-inch long, 2-inch high and 5.54-inch wide. It was the best value of all power supply options investigated. However, as in all compromise decisions, it was not without a downside as will be described later.

17.11.5 Amplifier Cooling

Cooling is the single most important requirement for a transistor's reliability. Tubes can get hot and angry, but they are made from high-melting-point materials so they can stand the heat. Transistors cannot. They will melt. The active area of a VRF151 die is 0.034 square inch, about the area inside the letter P on a computer keyboard. It must dissipate up to 150 W when key down. The heat dissipated in the transistor die is conducted through to the transistor's package base. Here it is conducted to the heatsink which, because of its large surface area, can effectively couple the dissipated heat into the surrounding air.

Fig 17.53 shows the VRF151s on the PC board and heatsink. There is no substitute for ensuring a good mechanical fit before mounting the transistors. The heatsink surface must be as flat as possible and the transistors must also be flat. A sheet of 600 grit sandpaper on a glass plate makes a good way to check the flatness of high power metal-backed transistors. A couple of light strokes across the paper will reveal any high spots on the bottom of the device. Keep rubbing until the back is all the same color. The plating on the back is not necessary once it is mounted — bare copper is fine.

Similarly, the heatsink must also be flat. Extruded aluminum heat sinks are notoriously "lumpy." A careful swipe across the width of the sink with a large flat file, perpendicular to the direction of the fins, will reveal any ridges and valleys. File it until they disappear. Be careful to keep the file clean or you can do more harm than good. In commercial practice, extruded heat sinks are usually milled flat before use.

The interface between the transistor and the heatsink is always greased with thermal heatsink compound. Thermal grease is a suspension of zinc oxide powder in silicone or mineral oil. It is very similar to the white sun screen paint used to protect your nose at the beach. Here it is the oil that does the work. The zinc oxide is simply a filler to keep the oil from running away. Oil is not a good conductor of heat but is much better than air. The thermal grease is used to fill any microscopic gaps and scratches between the sink and the transistor. Use only as much grease as needed to fill the gaps. Spread an even thin coat of grease on the bottom of the transistor with a knife blade then put the part on the sink. Before putting in the screws, wiggle the part around on the sink to force out any trapped air and to make the layer of grease is as thin as possible. You should be able to feel the sink grab the part as you move it. It does not want to "float" on the grease!

THERMAL DESIGN

The VRF151's maximum junction temperature rating is 200 °C, but device lifetime is seriously degraded at this temperature. Industry design standards typically aim for 150 °C absolute maximum, and typical operating temperature about 130 °C. The heat sink chosen for this amplifier is a 7-inch length of aluminum extrusion 3.25 inches wide and 1 inch high with nine longitudinal fins. This provides a lot of surface area, but in order to remove the 250 W of heat dissipated when the amplifier is running at maximum CW power, the heat sink must have air forced through it. It is impossible to dissipate this much heat by convection alone. The forced air cooling keeps the junction temperature below 150 °C at all times. The combination of a fairly large heat sink area and a relatively small fan allows the amplifier to "keep its cool" without making a lot of noise.

A single 3-inch 24-V dc fan (47 CFM) provides pressurized air across the sink. The closed chassis forms a pressurized plenum. The air comes in through the fan but can only leave the chassis after traveling down the length of the heat sink and out the 3 × 1 inch opening in the rear panel. Ducting the air this way makes very effective use of the fin area on the heat sink.

The original prototype of the amplifier had two smaller cooling fans mounted on the rear panel. When the amplifier was put in its place on the operating table, it was overheating because both the air intake and exhaust were on the same end of the chassis. It was sucking in its own hot air. A new chassis was bent up and this time the fans were mounted on one side of the cover. This eliminated the self-heating problem and allowed a much cleaner layout of the rear panel. It provided enough space to mount the fuse holder outside rather than inside. The final version uses a single high capacity 24-V fan. A 24-V 5-W Zener diode is placed in series to run it from the 48-V supply. A 48-V fan would be a better choice, but a suitable unit was not available from distributor stock, only by special order.

Note of caution: The amplifier cannot be run at power without the top cover in place because the chassis must be pressurized in order to force the cooling air thorough the fins of the heat sink.

The fan runs at full speed all the time. It would be more "operator friendly" to run the fan at a very low speed to start off with and use the temperature sensing circuit to increase the fan speed when the heat sink reaches a moderate temperature, say 100 °F. A second threshold point would run the fan at full speed when the sink temperature reaches 120 °F. This is a project for the next version. As it is, the fan does not require any controls. It will continue to run after the supply is switched off until it bleeds down the filter capacitors.

Fig 17.53 — It's important to make sure that the VRF151 power transistors mount flush to the heat sink, without gaps or air pockets that would impede heat transfer (see text for details). Once the transistors are mounted, a piece of wire is used as a lead forming device to bend the transistor leads down to the PC board pads for soldering.

17.11.6 Metering and Calibration

The cost of quality moving-needle meters has gone beyond reasonable, in part because digital panel meters have become more popular. However, you need to be able to assess the health of the amplifier at a glance and digital meters are not good for this. LED bar graphs are very inexpensive and are used for metering in this amplifier. They are easy to read and because of their low cost we can have one for each parameter being measured. Output power is displayed on a 20 LED string (made from two 10 LED bars). There is a "hang" built in so the SSB voice peaks can be seen easily. The drain current and reflected power are each shown on 10 LED bars. The output of the thermistor on the heatsink is also available for display by changing a jumper or incorporating a selector switch.

Absolute meter calibration is not so important in an amplifier like this. As long as none of the displays overflow in normal operation, they will serve as a way to easily monitor the operation of the amplifier.

Calibration of the meters will require using external standards. An ammeter in series with the power supply will be enough to calibrate the current meter. It should be set to display 15 A at full scale. R1 is adjusted to light bar #10 at 15 A. It would work out exactly right if all the resistors were 1% values. Forward power similarly requires a calibrated power meter on the output. R33 should be adjusted for 250 W when 18 bars of the 20 are lit. The top two LEDs are red and the others green so the operator can readily tell when the drive is too high.

Calibrating the reflected power is a little more difficult. The simplest method requires a calibrated external wattmeter. Run 30 W of RF power from another source backward through the amplifier, from antenna to input, *with the operate switch in the standby position*. Set R34 so the SWR LED trips at this point. This corresponds to an SWR of 2:1 at 250 W output. The amplifier should be stable and totally reliable up to this mismatch. If your load has a higher VSWR than 2:1, you can back off on the drive and continue to operate. As long as the reflected power is less than 30 W, the amplifier is happy.

Protecting the amplifier from damage requires a combination of careful attention to the operating conditions and reliance on the automatic limits in the control system.

17.11.7 Chassis

As seen in **Fig 17.54**, the chassis is a double clamshell configuration made from aluminum sheet stock. The top cover is 0.031-inch thick and the chassis bottom is 0.062-inch thick. This provides both adequate strength and workability with simple hand tools. Aluminum 12 × 24 inch sheets are available from several suppliers. They were sheared to size and bent into U-shaped parts at a local sheet metal shop.

The power supply and the fans are mounted on the cover. It overhangs the chassis base by 1/16 inch on all sides. The top is attached by 4-40 threaded L-brackets, DigiKey part 612K-ND, that are riveted to the main chassis. A drawing of the chassis layout and front and rear panels is included on the *Handbook* CD.

17.11.8 Performance

When the design was nearly complete, it became apparent that things were not coming out right. There was plenty of cooling, but the available power supply is a 500-W 48-V unit that goes into sudden current limiting when the load exceeds 10.4 A. The low-pass filters have about 0.2 dB loss, or about 5%. If the output was increased to make up for the filter loss, the power supply would limit on some bands causing severe distortion on SSB and CW.

The maximum for the PA design itself is 300 W. Increasing its output past 300 W to make up for filter loss quickly degrades the IMD performance. It needs some headroom. So, in very un-amateur fashion, this amplifier is conservatively rated at 250 W output. This provides a clean signal and plenty of margin for wrong antenna selection, disconnected feed lines, and all the other things that can kill amplifiers that are run too close to their limit.

The PA will provide 250 W PEP for sideband or PSK and 250 W CW. The design goal for this PA was to make it reliable and at least as good as any competitive transceiver. The harmonics are –60 dB on HF and –70 dB on 6 meters. Transmit IMD is >38 dB down from either tone as shown in **Fig 17.55**.

Parasitics are not usually a problem in broadband amplifiers because of the feedback used. The prototype was tested into a 3:1 SWR load at all phase angles without breaking into parasitic oscillation anywhere.

This design project was undertaken to provide a design example for a practical solid-state amplifier for the 2010 *Handbook*. It is a continuing project.

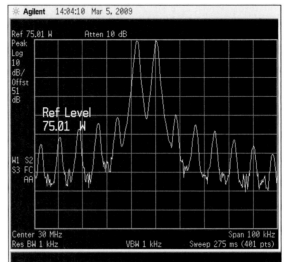

Fig 17.55 — Output of the 250 W amplifier during two-tone IMD testing. All IMD products are better than 38 dB down from either tone.

Fig 17.54 — The control and display circuitry mounts on a PC board that attaches to the front panel. LED bar graphs show forward and reflected power and current.

17.12 Tube Amplifier Projects

Project: 3CX1500D7 RF Linear Amplifier

The following describes a 10-to-160-meter RF linear amplifier that uses the new compact Eimac 3CX1500D7 metal ceramic triode. It was designed and constructed by Jerry Pittenger, K8RA.

The amplifier features instant-on operation and provides a solid 1500 W RF output with less than 100 W drive. Specifications for this rugged tube include 1500-W anode dissipation, 50-W grid dissipation and plate voltages up to 6000 V. A matching 4000-V power supply is included. The amplifier can be easily duplicated and provides full output in key-down service with no time constraints in any mode. **Fig 17.56** shows the RF deck and power supply cabinets.

DESIGN OVERVIEW

The Eimac 3CX1500D7 was designed as a compact, but heavy-duty, alternative to the popular lineup of a pair of 3-500Z tubes. It has a 5-V/30-A filament and a maximum plate dissipation of 1500 W, compared to the 1000-W dissipation for a pair of 3-500Zs. The 3CX1500D7 uses the popular Eimac SK410 socket and requires forced air through the anode for cooling. The amplifier uses a conventional grounded-grid design with an adjustable grid-trip protection circuit. See the RF Deck schematic in **Fig 17.57**.

Output impedance matching is accomplished using a pi-L tank circuit for good harmonic suppression. The 10 to 40-meter coils are hand wound from copper tubing, and they are silver plated for efficiency. Toroids are used for the 80- and 160-meter coils for compactness. The amplifier incorporates a heavy-duty shorting-type band-switch. Vacuum variable capacitors are used for pi-L tuning and loading.

A unique feature of this amplifier is the use of a commercial computer-controlled input network module from LDG Electronics (**www.ldgelectronics.com**). This greatly simplifies the amplifier design by eliminating the need for complex ganged switches and sometimes frustrating setup adjustments. [Unfortunately this module has been discontinued so the builder will need to develop an alternative input circuit such as the AT-100AMP kit from **www.w4rt.com/LDG/AT-100AMP.htm**. — *Ed.*]

An adjustable ALC circuit is also included to control excess drive power. The amplifier metering circuits allow simultaneous monitoring of plate current, grid current, and a choice of RF output, plate voltage or filament voltage.

The blower was sized to allow full 1500-W plate dissipation (65 cfm at 0.45 inches H_2O

Fig 17.56—At the top, front panel view of RF Deck and Power Supply for 3CX1500D7 amplifier. At bottom, rear view of RF Deck and Power Supply.

hydrostatic backpressure). The design provides for blower mounting on the rear of the RF deck or optionally in a remote location to reduce ambient blower noise in the shack. The flange on the socket for connecting an air hose was ground off for better air flow. (This is not necessary.).

The power supply is built in a separate cabinet with casters and is connected to the RF deck using a 6-conductor control cable, with a separate high voltage (HV) cable. The power transformer has multiple primary taps (220/230/240 V ac) and multiple secondary taps (2300/2700/3100 V ac). No-load HV ranges can be selected from 3200 to 4600 V dc using different primary-secondary combinations. The amplifier is designed to run at 4000 V dc under load to maintain a reasonable plate resistance and component size. A step-start circuit is included to protect against current surge at turn on that can damage the diode bank. The power supply schematic is shown in **Fig 17.58** and a photo of the inside of the power supply is shown in **Fig 17.59**.

Both +12-V and +24-V regulated power supplies are included in the power supply. The +12 V is required for the computer-controlled input network and +24 V is needed for the output vacuum relay. The input and output relays are time sequenced to avoid amplifier drive without a 50-Ω load. Relay actuation from the exciter uses a low-voltage/low-current circuit to accommodate the amplifier switching constraints imposed by many new solid-state radios.

Much thought was put into the physical appearance of the amplifier. The goal was to obtain a unit that looks commercial and that would look good sitting on the operating table. To accomplish the desired look, commercial cabinets were used. Not only does this help obtain a professional look but it eliminates a large amount of the metal work required in construction. Careful attention was taken making custom meter scales and cabinet labeling. The results are evident in the pictures provided.

GENERAL CONSTRUCTION NOTES

The amplifier was constructed using basic shop tools and does not require access to a sophisticated metal shop or electronics test bench. Basic tools included a band saw, a jig saw capable of cutting thin aluminum sheet, a drill press and common hand tools. Some skill in using tools is needed to obtain good results and insure safety, but most people can accomplish this project with careful planning and diligence.

Metal work can be a laborious activity. Building cabinets is an art within itself. This part of the project can be greatly simplified by using commercial cabinets. However, commercial cabinets are expensive (~$250 each) and could be a place where some dollars could be saved.

The amplifier is built in modules. This breaks the project into logical steps and facilitates testing the circuits along the way. For example, modules include the HV power supply, LV power supply, input network, control circuits, tank circuit and wattmeter. Each module can be tested prior to being integrated into the amplifier.

The project also made extensive use of computer tools in the design stage. The basic layout of all major components was done using the *Visio* diagramming software package. The printed-circuit boards

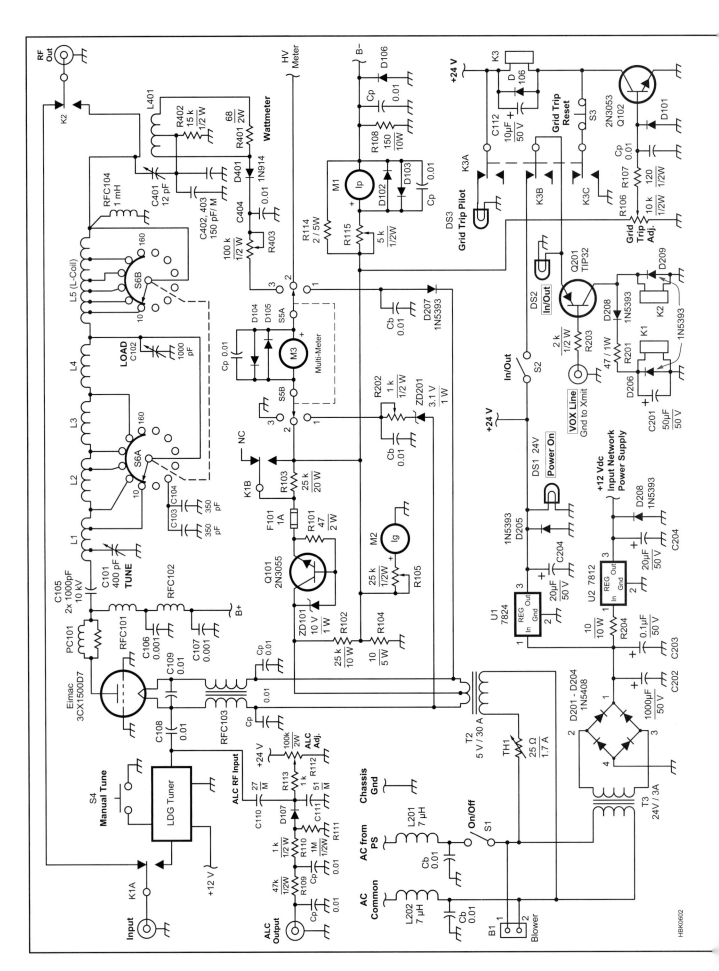

Fig 17.57 — Schematic of the RF deck and control circuitry.
B1 — Dayton 4C763 squirrel-cage blower.
Cb, Cp — 0.01 μF, 1 kV disc ceramic.
C101 — 400 pF, 10 kV Jennings vacuum variable, UCSL-400.
C102—1000 pF, 5 kV Jennings vacuum variable, UCSL-1000.
C103, C104—350 pF, 5 kV ceramic doorknob.
C105—two parallel 1000 pF, 10 kV ceramic doorknob capacitors (Ukrainian mfg).
C106, C107—0.001 μF, 7.5 kV disc ceramic.
C108, C109—0.01 μF, 3 kV transmitting mica (1 kV disc ceramics can be used).
C401—12 pF piston trimmer.
C403, C403—150 pF silver mica.
D101, D107, D205-D209—1N5393 (200 V, 1.5 A).
D102-D106, D201-D204—1N5408 (1 kV, 3 A).
K1—4PDT, 24 V dc KHP style (gold contacts).
K2—SPST vacuum relay, Kilovac H8/S4.
K3—4PDT, 24 V dc KHP style (gold contacts).
L1-L5—See Table 17.8.
L201, L202—Line chokes, 7 μH.
L401—24 t #22 enamel wire, center tapped on T50-6 core.
M1-M3—Simpson Designer Series, Model 523, 1 mA movement.
PC101—2 t ¾-inch diameter × 2-inch long, ½-inch brass strap with two 150 Ω, 2 W non-inductive carbon resistors in parallel.
Q101—2N3055 TO-220 case on heat sink.
Q102—2N3053 TO-18 case.
R103—25 kΩ, 25 W wire-wound.
R104—10 Ω, 5 W.
R108—150 Ω, 10 W wire-wound.
R112—100 kΩ, 2 W potentiometer
R403—100 kΩ, 0.5 W trim pot.
RFC101—90 μH, 3 A Plate Choke, Peter W. Dahl p/n CKRF000100, www.pwdahl.com.
RFC102—14 t #18 enamel wire wound on 100 Ω, 2 W resistor.
RFC103—Bifilar 30 A filament choke, Peter W. Dahl p/n CKRF000080, www.pwdahl.com.
RFC104—1 mH, 300 mA RF choke.
S1-S2—Alco 164TL5 DPDT switch (only SPST contacts are used), www.alliedelec.com/.
S3-S4—Alco 164TL2 momentary DPDT (only SPST contacts are used; S3 wired as normally closed, S4 as normally open), www.alliedelec.com/.
S5—2 pole, 3 position rotary switch.
S6—RadioSwitch model 86, double-pole 12-position (30° indexing) with 6-finger wiper on each deck, p/n R862R1130001, www.multi-tech-industries.com.
T2—5 V, 30 A center-tapped transformer, Peter W. Dahl EI-150 × 1.5 core, primary 115/230 V ac, www.pwdahl.com.
TH1—Thermistor, Thermometrics CL-200 (Mouser 527-CL200).
ZD101—10 V, 1 W Zener 1N4740A.
ZD201—3.1 V, 1 W Zener.
Other parts:
Cabinet—Buckeye Shapeform DSC-1054-16 (10×17×16-inch H×W×D), www.buckeyeshapeform.com.
Chimney (Teflon)—A. Howell, KB8JCY, PO Box 5842, Youngstown, OH 44504.
LDG Tuner—AT-100AMP autotuner, see text.
Tube socket—Eimac SK-410.

were designed using a free layout program called *ExpressPCB* (**www.expresspcb.com**). Masks were developed and the iron-on transfer technique was used to transfer the traces to copper-clad board. The boards were then etched with excellent results. The layout underneath the RF Deck is shown in **Fig 17.60A** and the top side of the RF Deck is shown in Fig 17.60B.

Meter scales were made using an excellent piece of software called *Meter* by Jim Tonne, W4ENE (**www.tonnesoftware.com**). Also, K8RA wrote an *Excel* spreadsheet to calculate the pi-L tank parameters. A copy of the spreadsheet, *Meter Basic* software and *ExpressPCB* files for the PC boards are all included on the CD-ROM that accompanies this book.

Although using computer tools simplifies the design step, all design work can be done without the use of a computer. Be creative and use the tools and resources at hand! There are many different ways to construct this design. The key secret is diligence and not compromising until it is done right. Note that the tank coils in this amplifier were wound at least three times, the inside side panels were cut twice and many printed circuit boards ended in the trash before acceptable boards were fabricated.

Since this project was built, Peter W. Dahl has discontinued business, Dahl transformers are now available from Harbach Electronics (**www.harbachelectronics.com**). Part numbers may have changed, and some parts used in this project (Fig 17.57 and 17.58) may no longer be available.

CABINET METAL WORK

By purchasing commercial cabinets, metal work required was minimized but not eliminated. The power supply components are very heavy. The transformer weighs about 70 pounds by itself. Therefore the base plate of the power supply cabinet needed to be reinforced. The original base plate for the cabinet was not used. One-eighth-inch plate was purchased from a local aluminum scrap company. Two pieces were sandwiched to provide a ¼-inch plate. Of course ¼-inch material could have been used but it was not available at the time of purchase.

The plate can be cut on a metal band saw using a guide or on a radial arm saw. Metal blades are readily available from Sears for both saws. If using a radial arm saw, multiple passes are required, lowering the blade slightly with each pass. Be sure to wear eye protection because the metal chips fly. The edges were then cleaned and straightened using a 4-inch belt grinder. If a belt sander is not available, a large file will work.

The two metal plates were held together with the mounting bolts on the four casters. The power supply base plate exactly matches the original base plate and fastened to the cabinet using the original tapped screw holes. All the heavy components are mounted on the base plate. The power supply must always be handled by lifting the base plate, since the cabinet does not have the structural integrity to bear the weight by itself.

The RF deck needed both a chassis plate and a front sub-panel. See Fig 17.60B. The sub-panel is used to mount the load and tune capacitors, the bandswitch and also provides RF shielding for the meters. Side plates were needed because of the cabinet configuration. The side plates, chassis plate and sub-panel all use 1/16-inch aluminum plate. After the side plates are cut and mounted to the cabinet sides, the chassis plate and front sub-panel are mounted using ½-inch aluminum angle to join the edges.

Cutting holes can often be a challenge. If a drill bit is the correct size, drilling a hole is easy, of course. But large-size round holes and square holes can be a challenge. This was especially true in this project since the front and rear panels are ⅛-inch aluminum plate.

The large meter holes can be cut using a hole saw on a drill press. For odd sizes, a "fly cutter" can be used. Fly cutters are available from Sears but a special warning is in order. These devices work well but are extremely dangerous. Make sure the cutting bit and the placement into the drill chuck are secure.

Large square holes are required for the turn counters. Mark the square hole to be cut. Drill a hole in each corner. The hole must be at least the size of the saw blade if a jigsaw is used to finish the hole. Note that the jigsaw must have a removable straight blade. If a metal-cutting jigsaw is not available, a series of small holes can be drilled in a straight line on all four sides and the edges smoothed with a file. Almost any hole can be custom cut by making a hole the approximate size and finishing it to the exact dimension with a file. It is slow and laborious but it works. When using a file on panels, be very careful that the file does not slip out of the hole and put an undesired scratch in the panel!

Once panel holes are cut, carefully label the panels before mounting the components. Dry transfers are used on both the power supply and the RF deck. Dry transfers of all sizes and fonts are available at graphics art stores and hobby shops. The author has found that hobby shops carry an excellent selection of dry transfers in the model railroad section.

RF DECK CONTROL CIRCUITS

The control circuits in the amplifier are not complex due to the simplicity of the grounded-grid design and the instant-on capability of the 3CX1500D7 tube. 120 V ac is routed from the HV power supply to the RF Deck in the 6-conductor control cable. When the on/off switch (S1) is pressed,

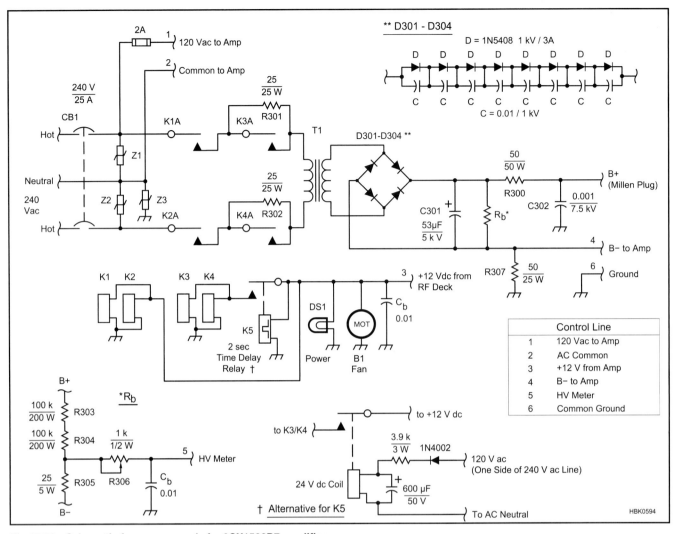

Fig 17.58—Schematic for power supply for 3CX1500D7 amplifier.

B1—12 V dc brushless fan, 2¼ inch (Mouser 432-31432).
DS1—12 V dc pilot lamp (Alco 164-TZ).
Cb—0.01 µF, 1 kV disc ceramic bypass capacitor.
C301—53 µF, 5 kV oil-filled.
CB1—2 pole 25 A, 240 V circuit breaker.
K1-K4—SPST solid-state relay 240 V ac, 25 A line voltage with 12 V dc input (the author used surplus Crydom relays but a readily available substitute is the Tyco/P&B SSR-240D25R).

K5—2 second trimpot time delay relay (surplus Bourns 3900H-1-125 or equiv; or FTD-12N03 3 second glass timer relay from Surplus Sales of Nebraska). Note: This part may be difficult to find. You can build the equivalent from a 24 V dc SPST relay, diode, resistor and capacitor as shown in the drawing inset. This technique is borrowed from the K6GT HV supply shown later in this chapter (Fig 17.63).

R303, R304—100 kΩ, 200 W wirewound resistor).
T1—Plate transformer, primary 220/230/240 V; secondary 2300/2700/3100 V at 1.5 A CCS (Peter W. Dahl Co P/N Pittenger PT-3100).
Z1-Z3—130 V MOV.
Cabinet—Buckeye Shapeform DSC-1204-16 (12×18×16 inches HWD), www.buckeyeshapeform.com.

120 V ac is sent to the primary of the low voltage transformer (T3) and the filament transformer (T2). The surge current to the filament of the tube is suppressed by the thermistor (TH1) in one leg of the filament transformer primary. These are excellent current limiting devices that have a resistance of approximately 25 Ω cold but decrease to less then 1 Ω as they heat. Keep the thermistor in open air away from other components since they are designed to run hot.

The low voltage supply provides regulated +12 V dc and +28 V dc. The voltages are regulated using simple three-terminal regulators. Pilot lights are included in each push button switch, S1-S4, and a power indicator on the HV power supply. When the low-voltage power supply first comes on, +12 V dc is directed through the control cable back to the HV supply. High voltage is applied immediately to the instant-on tube. Therefore the amplifier is turned on and ready to go instantly—You don't have to listen to your friends working that rare one for three or four tense minutes while you wait for your amplifier to time in!

The amplifier is switched in and out of the circuit using a 4PDT KHP style relay (K1) for the input and a SPDT vacuum relay (K2) for the output. It is important to select the timing constants for the input relay (C201 and R201) so the input relay closes a few milliseconds after the vacuum relay. This avoids hot switching the output, which could fuse the vacuum-relay contacts. This is a balancing act since the brief time the input relay is open will present an open circuit to the exciter. Many modern radios now have exciter-timing circuits that close the amplifier relay circuit a few milliseconds before RF is transmitted.

It is recommended that timing components for the input relay be located in a place where they can be easily changed. Another approach

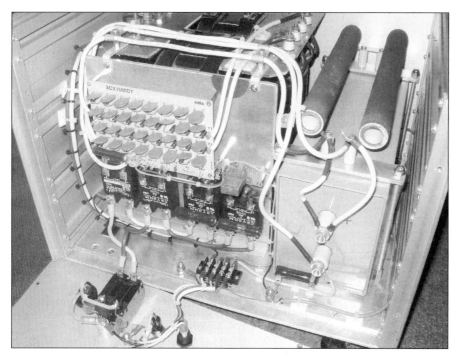

Fig 17.59—Inside view of the power supply, showing rectifier stack, control relays and HV filter capacitor with bleeder resistors. The heavy-duty Peter W. Dahl transformer is at the upper left in this photo.

is to build a breadboard circuit that feeds the relay coils in parallel but places the contacts in series. Feed a low voltage through the contacts of the two relays and monitor the timing with a dual-trace oscilloscope. This technique allows precise timing of contact closure as the two relays work together. Note that different relays will need different timing-circuit component values. A set of contacts on input relay, K1, is used to short across bias resistor, R103. The resistor biases the tube to cutoff in standby.

Approximately +10 V bias is provided to the center tap of the filament transformer to limit the idle current of the tube to approximately 125 mA. The bias is developed using the three components D101, R101 and Q101. These components could be replaced with a single 10-V/50-W Zener diode. However, 50-W zeners are expensive and they are difficult to obtain. Using the circuit shown, the bias is provided by a common NPN transistor (Q101) and a one-watt zener (D101) you can obtain from RadioShack.

TUBE PROTECTION CIRCUIT

The main protection for the tube is a plate-current surge resistor and a grid-trip circuit. The current surge resistor (R308, 50 Ω/50 W) is in series with the B+ line and acts as a fuse should excessive current be drawn from the HV power supply. Ohm's law says that up to 1-A plate current can be drawn through the resistor and still stay within the 50-W rating of R308. However, let's assume a problem occurs and 5 A flows through the resistor. Resistor R3 must now dissipate 1250 W. The resistor will quickly fail and will shut down the HV to the 3CX1500D7 tube.

Q102 is a grid-trip circuit that snaps the amplifier offline if the grid current exceeds 400 mA. The grid current is drawn through the 10-Ω resistor (R104) connected between the B– line and chassis ground. The current creates a voltage across R104 that is fed to the grid-trip adjustment potentiometer, R106. Q102 is turned on when the base voltage reaches 0.6 V and actuates the grid-trip relay K3. K3 contacts break the +28 V dc input and output relay lines (K3B), locks the relay closed (K3C) and extinguishes the pilot bulb (K3A) of the GRID-TRIP RESET normally closed push-button switch (S3) located on the front panel. Pushing the GRID TRIP RESET switch (S3) breaks the current path for the grid-trip relay K3 and resets the relay. The reason the grid trip was actuated should be determined prior to attempting to use the amplifier again. Usually, this is caused by improper setting of the load capacitor or transmitting into the wrong antenna.

INPUT NETWORK

As mentioned before, this amplifier uses a unique concept for the input-matching network, getting rid of a switched network mechanically ganged to the main band-switch. Not only can such a switching arrangement be awkward mechanically, but obtaining a reasonable network Q and a low SWR over an entire band can be difficult.

Thus the author decided to use a commercial automatic tuner integrated into the RF deck (see Fig 17.60A). The tuner is made by LDG and is based on their popular AT-100 Pro Autotuner, but was supplied without the front panel or rear panel connections and switches. This application is simple but elegant. The unit automatically initiates a retune if the input SWR exceeds approximately 1.5:1. The tuning cycle takes three to five seconds to execute. But retuning does not happen often because the tuner has over 4000 memories and remembers the settings for different frequency ranges. As the amplifier is used on each band, the tuner *learns* and stores settings into the memory. When switching bands, it only takes milliseconds to retrieve the data from memory and actuate the correct tuner relays.

Integration of the tuning network requires connections for RF input and output, +12 V and ground. RF input goes to the center of T1 and ground goes to J2 (clearly marked on the board). RF output goes to J3 and ground goes to J6. The +12-V dc connection is the larger of the three holes at J10 (next to L10). The other two holes are grounds for dc connections.

A momentary contact switch (S4) is mounted on the front panel to provide manual control of the tuner. A normally open contact on S4 is connected to the input pin J9 (next to L12) and ground. (The pin is marked as the ring for the connector that is not installed.) The correct hole is on the C56 side. If the switch is pushed for less than ½ second, the tuner alternates between bypass and in-line modes. If S4 is pressed between 0.5 to 2.5 seconds, it does a memory tune from the stored data tables. If S4 is pressed for more than 2.5 seconds with RF applied, it skips the memory access, retunes and stores the new settings into the memory table. The manual retune function is seldom, if ever, used.

The tuner works perfectly and it really simplified the input-network design and construction. The SWR never exceeds 1.5:1 (typically it is 1.2:1). LDG provides the unit as a commercial product to amplifier builders.

PI-L NETWORK

A pi-L network is used to insure good harmonic attenuation. The pi-L circuit is actually a pi-network, followed by an L-network that provides additional harmonic attenuation. The L-section transforms the load of 50 Ω up to an intermediate resistance of 300 Ω. The pi section then transforms 300 Ω up to the desired plate load resistance of 3100 Ω. The plate load is calculated using the following formula:

$$R_L = \frac{E_p}{1.7 \times I_p} = \frac{4000}{1.7 \times 0.750} = 3137 \Omega$$

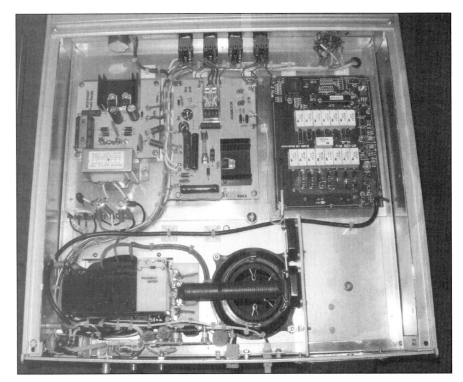

Fig 17.60—At the top, under the chassis of the RF Deck. The autotuner used as the input network for this amplifier is at the upper right. At bottom, view of the Pi-L output network in the RF Deck.

higher value of loaded Q than optimum. The solution is two fold. First, connect the plate-tune capacitor (C101) one turn into the 10-meter coil. This actually forms an L-pi-L circuit. Second, accept a higher value of loaded Q so that the variable capacitor can still be tuned. **Table 17.8** shows the loaded Q finally used for each band setting. The disadvantages of higher loaded Qs are high circulating currents in the tank circuit and the need to retune during excursions across the higher-frequency bands. This amplifier works fine on all bands, delivering a solid 1500 W output even on 10 meters.

Another pi-L tank circuit design constraint in this amplifier is the bandswitch. Many amplifiers use a single-pole, 12-position, non-shorting switch. Although this type of switch is easier to find, it can be problematic because high voltages are generated that could result in arcing in the bandswitch—usually from the wiper to the high frequency taps. You should use a switch with a multiple-finger wiper (see Fig 17.57) that shorts out lower-frequency coil taps not being used. For example, when the amplifier is used on 20 meters, the 40-, 80- and 160-meter taps are shorted to the wiper.

However, shorting switches only allow for six connections with 30° indexing. The common shorting wiper consumes 180° of switch deck on 160 meters. This results in having to design the 10/12-meter and 15/17-meter bands to use single taps for each frequency pair. Again, this is accomplished by adjusting the loaded Q for each band so that shared bands so they require nearly the same inductance. From Table 17.8, the same band switch position is shared on the 10/12-meter bands (1.4 µH) and the 15/18-meter bands (2.2 µH).

In actual construction of the tank circuit, it is very useful to have access to both a capacitance meter and an inductance meter. The author used an Elenco LCM-1950 meter that measures both capacitance and inductance and is available for under $100 (**www.elenco.com**). With the tune and load capacitors mounted and connected to calibrated knobs or turns counters, make a table of capacitance verses knob settings. This is useful to estimate the initial setting for each band during setup and test. Also, measure the inductance of each coil turn to determine initial coil taps for each band. On this amplifier, only the 10-meter tap had to be adjusted from the predetermined settings.

As mentioned above, the pi-L tank circuit was designed for 3100-Ω plate-load resistance. Such a high plate resistance demands higher inductance values to obtain reasonable tank circuit Qs. Table 17.8 shows that 160 meters requires 42 µH. If air-wound coils were used exclusively, the coils would require many turns and would take up a lot of cabinet space. To maintain a reasonable physical coil size, therefore, toroidal coils were used for

A nominal Q of 12 is used for the network. But as with most RF amplifiers, the capacitance needed for the higher-frequency bands is less than is physically possible using variable capacitors. For Q = 12, the tune capacitance (C101) for 10 meters is 14 pF. Using a vacuum variable capacitor for C1 helps because the minimum achievable capacitance (12 pF) is substantially less than with an air variable. But the tune capacitance is the sum of variable C101, 7.1 pF for the output capacitance of the 3CX1500D7 tube and any stray capacitance resulting from the physical layout of the amplifier.

The minimum obtainable capacitance is thus on the order of 30 pF, which yields a

Table 17.8
Pi-L Component Values

Frequency (MHz)	C1 (pF)	C2 pF	L1 μH	L2 μH	Q
1.850	211	1262	44.3	9.6	12
3.700	105	631	22.2	4.8	12
7.150	65	364	9.7	2.5	14
14.150	33	184	4.9	1.26	14
18.100	45	208	2.23	0.98	23
21.200	33	159	2.21	0.84	20
24.900	36	161	1.48	0.71	25
28.250	29	133	1.43	0.63	23

Tank Circuit Coils

Coil	Band	Inductance	Construction
L1	10/12-15/17 m	2.3 μH	7½ t, ¼-in. copper tube, 2-in. ID silver-plated 10/12-m tap @ 3½ t 15/17-m tap @ 7½ t
L2	20-40 m	7.4 μH	19 t, 3/16-in. copper tube, 2-in. ID silver plated 20 tap @ 8 t 40 tap @ 19 t
L3	80 m	12.4 μH	17 t on 3×T225-2 cores, #10 Teflon silver wire
L4	160 m	22.0 μH	23 t on 3×T300-2 cores, #10 Teflon silver wire
L5	L-Coil	9.6 μH	19 t on 2×T225-2 cores, #12 tinned wire w/Teflon sleeve 10/12-m tap @ 2 t 15/17-m tap @ 4 t 20-m tap @ 5 t 40-m tap @ 7 t 80-m tap @ 12 t 160-m tap @ 19 t

80 (L3) and 160 (L4) meters in addition to the output coil (L5) (see Fig 17.60B). You should use substantial core material for high-power operation to avoid core heating. Core sizes were increased by using multiple cores taped together. Each ferrite core is wrapped with three layers of high temperature fiberglass tape, available from RF Parts (**www.rfparts.com**). Teflon-insulated #10 wire was wound to obtain the desired inductance in L3 and L4. Both coils are mounted on ceramic standoffs and held in place with Teflon blocks.

The output coil is wound on a pair of T225-2 cores using #12 tinned wire covered with a Teflon sleeve. Taps onto the coil are made by carefully trimming a small ⅛-inch space from the Teflon sleeve on the inner edge of the core facing the bandswitch. Taps are then made from the back section of the 2-pole bandswitch using #12 tinned wire. The proper placement of each tap is determined by first winding #12 insulated wire around the core. A small slit is carved into each turn and the inductance was measured. The copper wire is removed and the final Teflon-covered #12 tinned wire is wound onto the core. Using the output L-coil (L102) design values in Table 17.8, permanent taps were made.

Note that the taps for the output coil are not extremely critical. Select the closest turn to the value needed. The output coil is mounted on the back of the bandswitch on one of the switch wafer screws using a threaded 1-inch diameter Teflon rod. The Teflon rod holds the position of the coil. The weight is carried by the wire taps from the coil to the bandswitch contacts. Table 17.8 also gives the inductance and construction instructions for each coil.

L1 and L2 (10-40 meters) are silver plated. They were wound using a 2-inch aluminum pipe as a form. Clean the copper tubing with #0000 steel wool prior to winding. Wind the copper tubing close spaced on the pipe. Leave plenty of pigtail on each end of the coil. The ends can be trimmed to fit the mounting positions precisely. After winding the desired number of turns, plug the ends by closing the tube ends with a hammer, spread the coil windings and rinse the coil in acetone to remove any oil. Allow a few minutes to dry. The coil is now ready to plate.

Go to any photo shop and beg/buy a gallon of used photographic fixer solution. Note that used fixer solution has silver remnant. The more the solution has been used, the more the silver content. The coils can be silver plated by dipping the clean coil into the solution. Do not leave the coil in the solution too long or it will turn black. A thin but bright silver coat will be deposited on the copper tube. This is called *flash plating*. After dipping the coil into the solution, immediately rinse in a bath of clean water and blow dry under pressure with an air compressor, heat gun or hair dryer. If a thicker silver coat is desired, electroplating is necessary, a subject beyond the scope of this article.

A #10 lug is crimped and soldered onto the end of each coil and used to mount the coil. The L2 coil is mounted using a Teflon block that is held in place to the front sub-panel with small screws. The block is carefully drilled with 3/16-inch holes the desired spacing of the coil about ¾-inch from one edge. The block is sawed down through the holes creating two matching blocks. At each end of the block a hole is drilled and tapped (6-32 tap). The silver plated coil is sandwiched between the two blocks for secure support. The tapped screws serve as the connecting points for the ends of the coil.

METERING

The amplifier uses three separate meters to simultaneously monitor plate current, grid current and a choice of plate voltage, power output or filament voltage. Each meter is identical with a 1-mA full-scale movement. As mentioned previously, the custom scales for each movement were designed using the *Meter Basic* software from Tonne Software. This allows up to three scales on each meter. Scales can be designed as either linear or log and the number of major and minor tick marks can be specified. Each scale can be labeled using different font size and color. The author printed the scales using a color inkjet printer onto glossy photo paper. The scales were carefully cut to match the meter faceplate and glued into place using a thin coat of adhesive.

Plate current is measured by M1 in series in the B– line using a current divider (R114 and R115) as shown in Fig 17.58. Adjust R114 to obtain full scale with 1-A of plate current. The meter was calibrated prior to installation using a low-voltage power supply with adjustable current limiting in series with an accurate digital meter.

M2 monitors grid current by measuring the voltage drop created by grid current flow through the 10-Ω resistor, R104. Connecting a voltage source (ie, small variable power supply) across R104 and measuring the actual current flow with an external meter provides a way to set the calibration pot, R105.

M3 is a multimeter that reads HV, RF power or filament voltage. The metering circuit is selected using a 2P3T rotary switch (S5). The HV metering circuit is in the HV power supply and fed to the RF deck through the control cable. The filament-voltage detect circuit is shown on the control circuit diagram (Fig 17.57: D207, D201 and R202). Adjust R202 for the proper reading on Meter M3. The 3.1-V zener (D201) expands the meter scale for more precise reading.

The RF wattmeter circuit is also shown in Fig 17.57. Only forward power is measured and potentiometer R403 is used for calibration. The wattmeter is not a precise instrument but gives a relative output reading. It is adequate for peaking power output when tuning. The meter provides good accuracy through 40 meters and then begins to read lower on the higher-frequency bands. This is due to the simplicity of the circuit and the toroid used. Quite honestly, don't expect much accuracy from this wattmeter.

HV POWER SUPPLY

The matching HV Power Supply (Fig 17.58) provides approximately 4000 V under load. It uses a full-wave bridge rectifier and is filtered using a single 53 µF/5000-V oil filled capacitor (C301). Whenever the HV supply is plugged into the 240-V line, live 120 V ac is routed to the RF deck through the control cable. The 120-V ac line is obtained from L1 and neutral of the 240-V ac line. The neutral line is isolated from ground for safety.

Actuating the on/off switch S1 on the front panel of the RF deck provides ac power to the low-voltage power supply. In turn, +12 V is returned to the power supply through the control cable and routed to a pair of solid state power relays (K1, K2). Also, +12 V is routed to a timer relay that provides a two-second delay in applying +12 V to the second pair of solid-state relays (K3, K4). During the two-second delay, each leg of the 240-V ac primary voltage is routed through a 25-Ω resistor (R301, R302) to reduce the current surge when charging the filter capacitor, C301.

HV is metered at the bottom of the two series 100,000-Ω bleeder resistors (R303, R304). A current divider is created using a small potentiometer (R306) in parallel with a 25-Ω/5-W resistor (R305). The current divider is in series with the bleeder resistors and tied to the B– line. R306 is set to allow 1 mA of current to flow to the HV meter located on the front panel of the RF deck with 5000 V HV dc. The potentiometer R306 and the paralleled fixed resistor R305 need handle only a small amount of power, since the voltage and current flow is quite small at this point in the circuit.

The HV cable between the RF Deck and power supply is made from a length of automotive-ignition cable that has a #20 wire and 60,000-V insulation. Be sure to get a solid-wire center conductor and not the resistive carbon material. Also use high-quality HV connectors that are intended for such an application. Millen HV connectors (50001) were used in this amplifier. Coax boots intended for coaxial cable are used on each connector for added insulation and physical strength. The mounting holes for the Millen connectors are oversized and plastic screws were used for safety.

TUNING AND OPERATION

The amplifier is very easy to tune after the initial settings of the tune (C101) and load (C102) controls are determined. The correct settings are determined with a plate current of 700 mA with a corresponding grid current of 200 mA. The turn counters provide excellent resetability once the proper settings have been found initially. Required drive power is about 75 W for 1500 W output.

Thanks to CPI Eimac Division, LDG Electronics, Peter W Dahl Co, and MTI Inc (Radio Switch) for their support in this project.

Project: A 6-Meter Kilowatt Amplifier

The Svetlana 4CX1600B tube has attracted a lot of attention because of its potent capabilities and relatively low cost. Because of its high gain and its large anode dissipation capabilities, the tube has relatively large input and output capacitances—85 pF at the input and 12 pF at the output. Stray capacitance of about 10 pF must be added in as well. On bands lower than 50 MHz, these capacitances can be dealt with satisfactorily with a broadband 50-Ω input resistor and conventional output tuning circuitry.

See the article by George Daughters, K6GT, "The Sunnyvale/Saint Petersburg Kilowatt-Plus" in 2005 and earlier Handbooks and included in the Templates section of the *Handbook CD* for details on suitable control and power-supply circuitry. This 6-meter amplifier uses the same basic design as K6GT's, except for modified input and output circuits in the RF deck. See **Fig 17.61**, a photograph of the front panel of the 6-meter amplifier built by the late Dick Stevens, W1QWJ.

On the 50-MHz band the tube's high input capacitance must be tuned out. The author used a T network so that the input impedance looks like a nonreactive 50 Ω to the transceiver. To keep the output tuning network's loaded Q low enough for efficient power generation, he used a 1.5 to 46 pF Jennings CHV1-45-5S vacuum-variable capacitor, in a Pi-L configuration to keep harmonics low. You should use a quarter-wave shorted coaxial stub in parallel with the output RF connector to make absolutely sure that the second harmonic is reduced well below the FCC specification limits.

To guarantee stability, the author had to make sure the screen grid was kept as close as possible to RF ground. This allows the screen to do its job "screening"—this minimizes the capacitance between the control grid and the anode. He used the Svetlana SK-3A socket, which includes a built-in screen bypass capacitor, and augmented that with a 50-MHz series-tuned circuit to ground. In addition, to prevent VHF parasitics, he used a parasitic suppressor in the anode circuit.

Unlike the K6GT HF amplifier, this 6-meter amplifier uses no cathode degeneration. The author wanted maximum stable power gain, with less drive power needed on 6 meters. He left the SK-3A socket in stock form, with the cathode directly grounded. This amplifier requires about 25 W of drive power to produce full output.

Fig 17.62 is a schematic of the RF deck. The control and power supply circuitry are basically the same as that used in the K6GT HF amplifier, except that plate current is monitored with a meter in series with the B– lead, since the cathode in this amplifier is grounded directly. The K6GT power supply is modified by inserting a 250-Ω, 25-W power resistor to ground in place of the direct ground connection. See **Fig 17.63**. In Fig 17.62, C1 blocks grid-bias dc voltage from appearing at the transceiver, while L1, L2 and C2 make up the T-network that tunes out the input capacitance of V1. R1 is a non-reactive 50-Ω 50-W resistor.

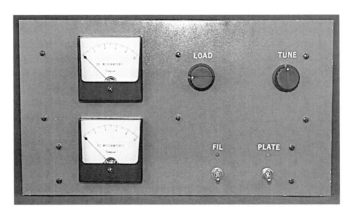

Fig 17.61—Photo of the front panel of the 6-meter 4CX1600B amplifier.

17.52   Chapter 17

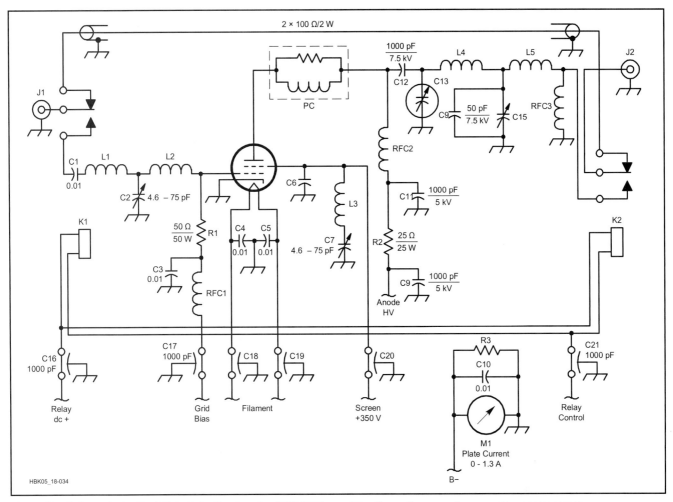

Fig 17.62—Schematic for the RF deck for the 6-meter 4CX1600B amplifier. Capacitors are disc ceramic unless noted.

C2, C7—4.6-75 pF, 500-V air-variable trimmer capacitor, APC style.
C6—Screen bypass capacitor, built into SK-3A socket.
C13—1-45 pF, 5 kV, Jennings CHV1-45-5S vacuum-variable capacitor.
C14—50 pF, 7.5 kV, NP0 ceramic doorknob capacitor.
C15—4-102 pF, 1100V, HFA-100A type air-variable capacitor.
C16, 17, 18, 19, 20, 21—1000 pF, 1 kV feedthrough capacitors.

L1—11 turns, #16, ⅜-inch diameter, 1-inch long.
L2—9 turns #16, ⅜-inch diameter, close-wound.
L3—8 turns #16, ⅜-inch diameter, ⅞-inch long.
L4—¼-inch copper tubing, 4½ turns, 1¼ inches diameter, 4½ inches long.
L5—5 turns #14, ½-inch diameter, 1⅜ inches long.
M1—0-1.3 A meter, with homemade shunt resistor, R3, across 0-10 mA movement meter.
PC—Parasitic suppressor, 2 turns #14, ½-inch diameter, shunted by two 100-Ω, 2-W carbon composition resistors in parallel.
RFC1—10 µH, grid-bias choke.
RFC2—Plate choke, 40 turns #20, ½-inch diameter, close-wound.
RFC3—Safety choke, 20 turns #20, ⅜ inch diameter.

C6 is the built-in screen bypass capacitor in the SK-3A socket, while L3 and C7 make up the series-tuned screen bypass circuit. RFC3 is a safety choke, in case blocking capacitor C12 should break down and short, which would otherwise place high voltage at the output connector.

CONSTRUCTION

Like the K6GT amplifier, this amplifier is constructed in two parts: an RF deck and a power supply. Two aluminum chassis boxes bolted together and mounted to a front panel are used to make the RF deck.

Fig 17.64 shows the 4CX1600B tube and the 6-meter output tank circuit.

Fig 17.65 shows the underside of the RF deck, with the input circuitry shown in more detail in **Fig 17.66**. The 50-Ω, 50-W noninductive power resistor is shown in Fig 17.66. Note that the tuning adjustment for the input circuit is accessed from the rear of the RF deck.

AMPLIFIER ADJUSTMENT

The tune-up adjustments can be done without power applied to the amplifier and with the top and bottom covers removed. You can use readily available test instruments: an MFJ-259 SWR Analyzer and a VTVM with RF probe.

1. Activate the antenna changeover relay, either mechanically or by applying control voltage to it. Connect a 2700-Ω, ½-W carbon composition resistor from anode to ground using short leads. Connect the SWR analyzer, tuned to 50 MHz, to the output connector. Adjust plate tuning and loading controls for a 1:1 SWR. You are using the Pi-L network in reverse this way.
2. Now, connect the MFJ-259 to the input connector and adjust the input

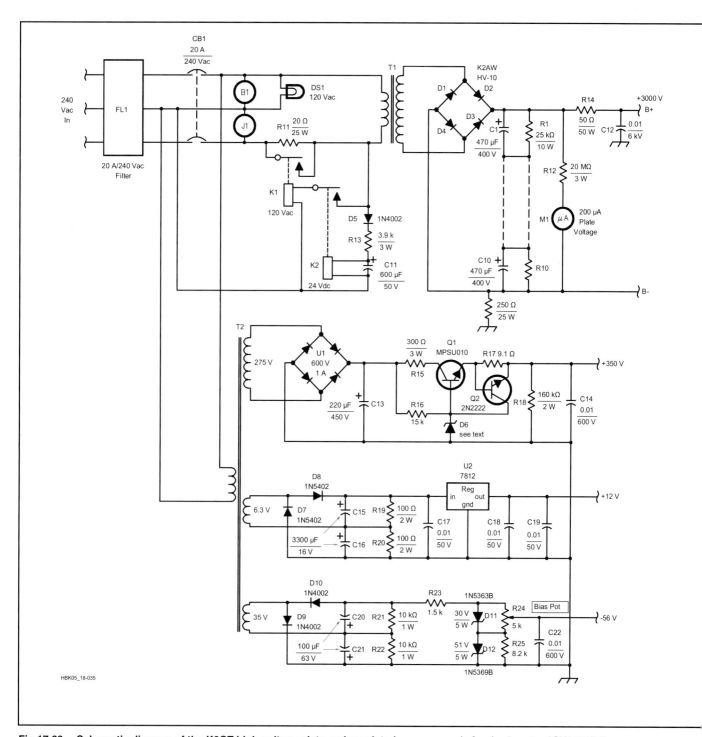

Fig 17.63— Schematic diagram of the K6GT high-voltage plate and regulated screen supply for the 6 meter 4CX1600B linear amplifier. K1, K2 and associated circuitry provide a "step-start" characteristic to limit the power-on surge of charging current for the filter capacitors. Resistors are ½ W unless noted. Capacitors are disc ceramic unless noted and those marked with a + are electrolytic.

T-network for a 1:1 SWR. Some spreading of the turns of the inductor may be required.

3. Disconnect the Pi-L output network from the tube's anode, leaving the 2700-Ω carbon composition resistor from the anode still connected. Connect the RF probe of the VTVM to the anode and run your exciter at low power into the amplifier's input connector. Tune the screen series-tuned bypass circuit for a distinct dip on the VTVM. The dip will be sharp and the VTVM reading should go to zero.

4. Now, disconnect the 2700-Ω carbon resistor from the anode and replace the covers. Connect the power supply and control circuitry. When you apply power to the amplifier, you should find that only a slight tweaking of the output controls will be needed for final adjustment.

B1 — Muffin fan (Rotron SU2A1 or similar).
C1-C10 — Filter capacitors; 470 µF, 400 V electrolytic.
C11 — 600 µF, 50 V electrolytic.
C12 — 0.01 µF, 6 kV disc ceramic.
C13 — 220 µF, 450 V electrolytic.
C14, C22 — 0.01 µF, 600 V disc ceramic.
C15, C16 — 3300 µF, 16 V electrolytic.
C17, C18, C19 — 0.01 µF, 50 V disc ceramic.
C20, C21 — 100 µF, 63 V electrolytic.
CB1 — two pole 20 A, 240 V ac circuit breaker.
D1-D4 — K2AW's HV-10 rectifier diodes.
D5, D9, D10 — 1N4002.
D6 — Zener diodes, three 1N4764A and one 1N5369B to total approximately 350 V dc.
D7, D8 — 1N5402.
D11 — Zener diode, 1N5363B (30 V, 5 W).
D12 — Zener diode, 1N5369B (51 V, 5 W).
DS1 — 120 V ac indicator lamp (red).
FL1 — 240 V ac, 20 A EMI filter.
K1 — 120 V ac DPDT relay; both poles of 240 V ac/15 A contacts in parallel.
K2 — 24 V dc relay; 120 V ac, 5 A contacts.
M1 — 200 µA meter movement.
Q1 — MPSU010.
Q2 — 2N2222.
R1-R10 — Bleeder resistors; 25 kΩ, 10 W.
R11 — 20 Ω, 25 W.
R12 — 20 MΩ, 3 W (Caddock MX430).
R13 — 3.9 kΩ, 3 W.
R14 — 50 Ω, 50 W mounted on standoff insulators.
R15 — 300 Ω, 3 W.
R18 — 160 kΩ, 2 W.
R19, R20 — 100 Ω, 2 W.
R21, R22 — 10 kΩ, 1 W.
R24 — 5 kΩ potentiometer; sets control grid bias for desired no-signal cathode current.
T1 — Plate transformer (Peter W. Dahl No. RRL-002).
T2 — Power transformer, 120 V / 275 V at 6 A, 6.3 V at 2 A, 35 V at 0.15 A.
U1 — 50 V, 1 A rectifier bridge.
U2 — 7812, +12 V IC voltage regulator.

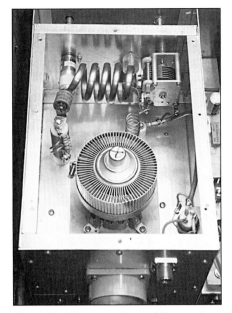

Fig 17.64—Close-up photo of the anode tank circuit for 6-meter kW amplifier. The air-cooling chimney has been removed in this photo.

Fig 17.66—Close-up photo of the input circuitry for the 6-meter kW amplifier. Input tuning capacitor C2 is adjusted from the rear panel during operation, if necessary. The series-tuning capacitor C7 used to thoroughly ground the screen for RF is shown at the lower right. It is adjusted through a normally plugged hole in the rear panel during initial adjustment only.

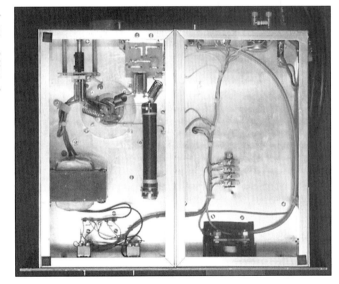

Fig 17.65—Underneath the 6-meter kW amplifier RF deck, showing on the left the tube socket and input circuitry.

Using "Surplus" Parts for Your Amplifier

First-time builders of power amplifiers soon discover that buying all new electronic components from a parts list is costly. Manufacturers realize significant savings by purchasing components in bulk, but the builder of a single unit will pay top dollar for each new part thus negating any savings in the labor costs. While some impressive equipment has been built using all new parts, pride of design and workmanship are usually the goals, not cost savings.

The way to break this economic barrier is to use surplus parts. These are items left over when a project is finished, a design changes, or a war ends. Sometimes, costly parts can be recovered from relatively new but obsolete equipment. Alert dealers locate sources of surplus parts, buy them at auction, often for pennies on the dollar, and make them available in stores, online or at hamfests.

Parts can become available when a project is abandoned for which parts have been gathered. Many hams maintain a "junk box" of parts against future needs, only to find it has grown beyond a manageable size. This may contain a mixture of new, used and NOS parts (new old stock, meaning unused but stored for many years).

When buying electronic parts from other than a trusted supplier of new stock, some precautions are in order. Before shopping the surplus shelves or making an on-line purchase, do some research. You are leaving behind the security of buying a specific part that the designer of the amplifier has tested in actual operation. One has to be sure not only that the part is sound, but that it is suited for the intended purpose. This effort is justified by the anticipated savings of up to 80%. Using "odd" parts may make your amplifier a "one of a kind" project, since the next builder won't be able to obtain the same items. Your amplifier may not be reproducible, but it can still be an object of pride and usefulness.

Surplus vacuum tubes may be of military origin. In the 50s and 60s WWII tubes powered many ham amplifiers. Tubes made in Russia in past decades are now available[1]. Many hams have used these tubes for new designs and even as replacements for hard-to-find or more expensive tube types in existing amplifiers. Since tubes have a limited life, one may wish to buy a spare or two before committing to using a surplus type, since future availability is always a concern. Broadcast or medical "pulls" or "pull-outs" provide a source of good used tubes. For highest equipment reliability, these tubes are replaced on a scheduled basis rather than at failure. They will have less life than a new tube, but can be so cheap that they are still a good deal.

When buying used tubes at a hamfest, a quick ohmmeter check is important. Between the filament pins there should be only a few ohms, with high resistance between all other connections. Signs of obvious overheating, such as discoloration or warping, are a concern and call at the least for a reduced price. There is no 100% guarantee that a tube is good apart from trying it in a circuit, since tube testers usually will not test large power tubes. Experience with on-line suppliers of NOS tubes indicates that a small percentage will be duds. If your supplier guarantees them you should be willing to pay a bit more. Some dealers will test or even match the tubes for parallel operation.

Finding sockets for odd tubes may be a challenge. Several enterprising hams supply sockets for popular types, such as the Russian GI7B[2], or you can sometimes bolt the tubes into place with straps and screws.

Using vacuum capacitors really adds class and value to an amplifier. Check the vacuum by setting the capacitor to minimum capacity and watching to see if the vacuum is sufficiently strong to pull the plates all the way in when the screw or hand pressure is released. Weak vacuum means the capacitor will not withstand its rated voltage. With the plates fully meshed, check for a short circuit, which indicates damaged plates. Be sure that the adjusting screw moves freely. Lubrication won't always free up a sticky screw. Check that the voltage and current ratings are suitable for your application. The voltage rating is usually printed on the capacitor. The current rating will be on a factory data sheet or use physical size as an indication.

Vacuum TR (transmit-receive) relays work well in an amplifier and often switch quickly enough to be used for full-QSK (full break-in) operation. If you shop carefully, a surplus relay can cost little more than a much-inferior open-frame type. Visual inspection of the contacts and an ohmmeter check on the coil will generally insure that the relay is good. If the relay's coil voltage is 24 V or some value other than what you wanted, remember that a small power supply can be added cheaply. Be aware that the amplifier keying circuits of most transceivers are not rated to switch the large coil currents of surplus relays, particularly those that operate at 24 V or higher. An outboard switching interface will be needed for these relays. Older non-vacuum relays also have long switching times that, if not accounted for in the amplifier design, may cause "hot-switching" and contact damage.

Inductors can usually be evaluated by inspection since hidden faults are rare. The current rating is determined by conductor size. With formulas found in the **Electrical Fundamentals** chapter, you can determine the inductance. Carry a hand calculator when shopping.

Choose RF switches based on physical size and inspection. Avoid badly arced-over or pitted contacts or insulators.

For power transformers, finding the exact required voltage may be difficult. You can use extra low-voltage windings to buck or boost the primary voltage. External transformers can also be used

17.13 References and Bibliography

TUBE TYPE AMPLIFIER SECTION

ARRL members can download many *QST* articles on amplifiers from: **www.arrl.org/QST**.

All About the GI-7B, K4POZ, **gi7b.com**

Badger, G., W6TC, "The 811A: Grandfather of the Zero Bias Revolution," **www.arrl.org/qst**

Care and Feeding of Power Grid Tubes, **www.cpii.com/library.cfm/9**

EIMAC tube performance computer, **www.cpii.com/library.cfm/9**

Grammer, G., "How to Run Your Linear," *QST,* Nov 1962, pp 11-14.

Knadle, R., "A Strip-line Kilowatt Amplifier 432 MHz," *QST,* Apr 1972, pp 49-55.

LTSPICE, download from Linear Technology , **www.linear.com/designtools/software/ltspice.jsp**

Meade, "A High-Performance 50-MHz Amplifier," *QST,* Sep 1975, pp 34-38.

Meade, "A 2-KW PEP Amplifier for 144 MHz," *QST,* Dec 1973, pp 34-38.

Measures, R., "Improved Parasitic Suppression for Modern Amplifier Tubes," *QST,* Oct 1988, pp 36-38, 66, 89.

Neutralizing Amplifiers, **www.w8ji.com/neutralizing_amplifier.htm**

Orr, *Radio Handbook*, 22nd Ed (Howard W. Sams & Co, Inc, 1981).

in an autotransformer connection for adjustment. Use of a variable autotransformer (Variac or similar) can adapt a transformer to your needs, but the extra winding and core losses may cause voltage sag under load. Amplifiers normally use a separate transformer for supplying the tube filaments as filaments are often turned on before the high-voltage supply. Therefore, filament windings on a plate transformer are not needed. Also, with either center-tapped, full bridge or voltage doubler circuits, you have three options for ac secondary voltage.

Transformers should be rated for 50 or 60 Hz operation, not 400 Hz as used in aircraft. Check that the current and overall VA (volt-amp) ratings are adequate. Use an ohmmeter to detect shorts or open windings, and to be sure that you understand the terminal connections. Check each winding to ground and to each other. Reject transformers with charred insulation. Use your nose to detect burned smells from internal overloads. A rusty core can be wire brushed and spray painted, but might indicate moisture has gotten to the windings. Test the primary with 120 or 240 V ac, if possible. High current or loud buzzing can indicate shorted turns. Be very careful when measuring the secondary windings as the high voltage can easily destroy your multimeter, not to mention endanger your life! (See the **Power Supplies** chapter for a method of testing transformers safely.)

Blowers can be found on otherwise worthless equipment, such as a rack of old tube gear. It may be worth buying the whole rack just to get a nice blower. The pressure and volume ratings of a blower may not be known, but apart from noise, there is no disadvantage in over-sizing your blower. An over-sized blower can be slowed down by reducing the voltage or throttled back with baffles. Remember that if the voltage rating is strange, providing an odd voltage may be cheaper than buying a more expensive blower. Avoid 400 Hz aircraft blowers. For blowers needing an external capacitor, try to obtain the capacitor with the blower.

Racks and chassis components can often be refurbished for an amplifier. Standard 19-inch rack-type cabinets are widely available surplus and at hamfests. Even new, they are available at modest cost and have good fit and finish. Filling the many panel holes and painting or adding a new front panel can make them look like new, while saving a lot of expensive and time consuming metal work.[3]
— Roger Halstead, K8RI

[1]www.nd2x.net/base-1.html
[2]k4poz.com/gi7b/index.html
[3]W Yoshida, KH6WZ, "Recycling Old Cabinets and Chassis Boxes," *QST*, Jul 2008, p 30.

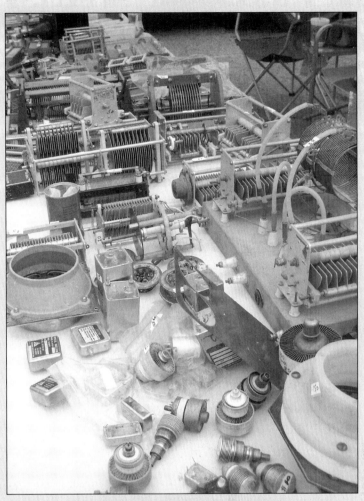

Photo A — A hamfest is often a good source of used, surplus, and even new parts for building amplifiers and other high-power RF equipment, such as impedance matching units or full-power filters. Follow the cautions in this sidebar when evaluating parts and pieces.

Pappenfus, Bruene and Schoenike, *Single Sideband Principles and Circuits* (McGraw-Hill, 1964).

Peck, Floyd K., "A Compact High-Power Linear," *QST*, Jun 1961, pp 11-14.

Potter and Fich, *Theory of Networks and Lines* (Prentice-Hall, 1963).

RCA Transmitting Tubes Technical Manual TT-5, Radio Corporation of America, 1962.

Reference Data for Radio Engineers, ITT, Howard & Sams Co, Inc.

Simpson, *Introductory Electronics for Scientists and Engineers* (Allyn and Bacon, 1975).

Terman, *Electronic and Radio Engineering* (McGraw-Hill).

Wingfield, E., "New and Improved Formulas for the Design of Pi and Pi-L Networks," *QST*, Aug 1983, pp 23-29. (Feedback, *QST*, Jan 1984, p 49.).

Wingfield, E., "A Note on Pi-L Networks," *QEX*, Dec 1983, pp 5-9.

SOLID-STATE AMPLIFIER SECTION

Abulet, Mihai, *RF Power Amplifiers* (Noble Publishing, 2001, ISBN 1884932126).

Blinchikoff, Zverev, *Filtering in the Time and Frequency Domains* (Noble Publishing, 2001, ISBN: 1884932177).

Bowick, C. Ajluni, and J. Blyler, *RF Circuit Design,* 2nd ed. (Newnes, 2007, ISBN-10: 0750685182).

Classic Works in RF Engineering : Combiners, Couplers, Transformers and Magnetic Materials Edited by Myer, Walker, Raab, Trask (Artech House, 2005, ISBN-13: 9781580530569).

Cripps, *Advanced Techniques in RF Power Amplifier Design* (Artech House, 2002, ISBN-10: 1580532829).

Davis and Agarwal, *Radio Frequency Circuit Design* (Artech House, 2001, ISBN: 0471350524).

Dye, Norman. and Helge Granberg, *Radio Frequency Transistors — Principles and Practical Applications,* (Butterworth-Heinemann, 1993, ISBN: 0750690593; 2nd ed, 2001, ISBN 0750672811).

Frenzel, Louis E., *RF Power for Industrial Applications,* (Prentice Hall, 2003, ISBN 0-1309-6577-4).

Grebennikov and Sokal, *Switchmode RF Power Amplifiers,* (Newnes, 2007, ISBN: 075067962X).

Grebennikov, *RF and Microwave Power Amplifier Design,* (McGraw-Hill, 2004, ISBN: 0071444939).

Kraus, Bostian and Raab, *Solid State Radio Engineering,* (Wiley, 1980, ISBN 04710301X).

Oxner, E., *Power FETs and Their Applications,* (Prentice-Hall, 1982, ISBN 0136869238).

Raghavan, Srirattana, Laskar *Modeling and Design Techniques for RF Power Amplifiers,* (Wiley-IEEE Press, 2008, ISBN: 0471717460).

Rutledge, David B., *The Electronics of Radio,* (Cambridge University Press, 1999, ISBN-13: 9780521646451).

Sabin, Schoenike, et al, *Single Sideband Systems and Circuits,* (McGraw-Hill, 1993, ISBN 0079120385).

Sevick, J., *Transmission Line Transformers,* 4th ed (Noble Publishing, 2001, ISBN: 1884932185).

Shirvani, A. and Wooley, B.A., *Design and Control of RF Power Amplifiers,* (Springer; 2003, ISBN-10: 1402075626).

Vizmuller, *RF Design Guide: Systems, Circuits, and Equations,* (Artech House, 1995, ISBN: 0890067546).

Zverev, A. I., *Handbook of Filter Synthesis,* (Wiley, 1970, ISBN: 0471986801).

Articles

Davis and Rutledge, "Industrial Class-E Power Amplifiers With Low-cost Power MOSFETs and Sine-wave Drive" Conf. Papers for RF Design '97, Santa Clara, CA, Sep 1997, pp 283-297.

Franke and Noorani, "Lumped-Constant Line Stretcher For Testing Power Amplifier Stability," *RF Design,* Mar/Apr 1983.

Frey, *A 50 MHz, 250W Amplifier using Push-Pull ARF448A/B,* APT Application Note APT9702.

Granberg, H., "Broadband Transformers and Power Combining Techniques for RF," AN-749, Motorola Semiconductor Products Inc.

Hilbers, A.H., *Design of HF Wideband Power Transformers,* Philips Semiconductors, ECO 6907, Mar 1998.

Hilbers, A.H., *Design of HF Wideband Power Transformers Part II,* Philips Semiconductors, ECO 7213, Mar 1998.

Hilbers, A.H., *Power Amplifier Design,* Philips Semiconductors, Application Note, Mar 1998.

Trask, C., "Designing Wide-band Transformers for HF and VHF Power Amplifiers," *QEX,* Mar/Apr 2005, pp 3-15.

Microwave — Cellular Applications

Cripps, *RF Power Amplifiers for Wireless Communication,* (Artech House, 1999, ISBN: 1596930187).

Groe and L. E. Larson, *CDMA Mobile Radio Design,* (Artech House, 2000, ISBN: 1580530591).

Pozar, *Microwave Engineering,* 3rd ed. (Wiley, 2004, ISBN: 0471448788).

Raghavan, Srirattana, Laskar *Modeling and Design Techniques for RF Power Amplifiers,* (Wiley-IEEE Press, 2008, ISBN: 0471717460).

van Nee and Prasad, *OFDM for Wireless Multimedia Communications,* (Artech House, 2000, ISBN: 0890065306).

York and Popovic, *Active and Quasi-Optical Arrays for Solid-State Power Combining,* (Wiley, 1997, ISBN: 0471146145).

Contents

18.1 A Brief History
18.2 Repeater Overview
    18.2.1 Types of Repeaters
    18.2.2 Advantages of Using a Repeater
    18.2.3 Repeaters and the FCC
18.3 FM Voice Repeaters
    18.3.1 FM Repeater Operation
    18.3.2 Home and Mobile Equipment
    18.3.3 Coded Squelch and Tones
    18.3.4 Frequency Coordination and Band Plans
    18.3.5 Linked Repeaters

18.4 Glossary of FM and Repeater Terminology
18.5 D-STAR Repeater Systems
    18.5.1 D-STAR Station IDs
    18.5.2 Station Routing
    18.5.3 Enhancing D-STAR Operation with *DPlus*
    18.5.4 D-STAR Repeater Hardware
18.6 References and Further Reading

Chapter 18

Repeaters

For decades, FM has been a dominant mode of Amateur Radio operation. FM and repeaters fill the VHF and UHF bands, and most hams have at least one handheld or mobile FM radio. Thousands of repeaters throughout the country provide reliable communication for amateurs operating from portable, mobile and home stations. Now, Amateur Radio is beginning to see the emergence of fully digital voice modulation, though analog is expected to dominate for the foreseeable future. The first part of this chapter, by Paul M. Danzer, N1II, with contributions by Gary Pearce, KN4AQ, describes FM voice repeaters. Later sections, contributed by Jim McClellan, N5MIJ, and Pete Loveall, AE5PL, of the Texas Interconnect Team, cover more recent developments in D-STAR digital voice and data repeaters.

18.1 A Brief History

Few hams today don't operate at least some VHF/UHF FM, and for many hams, FM *is* Amateur Radio. That wasn't always the case.

Until the late 1960s, the VHF and UHF Amateur Radio bands were home to a relatively small number of highly skilled operators who used CW and SSB for long distance communication and propagation experiments. This operation used just a small fraction of the spectrum available at 50 MHz and above. A somewhat larger number of hams enjoyed low power, local operation with AM transceivers on 6 and 2 meters. Our spectrum was underutilized, while public safety and commercial VHF/UHF two-way operation, using FM and repeaters, was expanding rapidly.

The business band grew so rapidly in the early 1960s that the FCC had to create new channels by cutting the existing channels in half. Almost overnight, a generation of tube-type, crystal-controlled FM equipment had to be replaced with radios that met the new channel requirements. Surplus radios fell into the hands of hams for pennies on the original dollar. This equipment was designed to operate around 150 MHz and 450 MHz, just above the 2-meter and 70-cm amateur bands. It was fairly easy to order new crystals and retune them to operate inside the ham bands. Hams who worked in the two-way radio industry led the way, retuning radios and building repeaters that extended coverage. Other hams quickly followed, attracted by the noise-free clarity of FM audio, the inexpensive equipment, and the chance to do something different.

That initial era didn't last long. The surplus commercial equipment was cheap, but it was physically large, ran hot and consumed lots of power. By the early 1970s, manufacturers recognized an untapped market and began building solid-state equipment specifically for the Amateur Radio FM market. The frequency synthesizer, perfected in the mid-1970s, eliminated the need for crystals. The stage was set for an explosion that changed the face of Amateur Radio. Manufacturers have added plenty of new features to equipment over the years, but the basic FM operating mode remains the same.

In the 1980s, hams experimenting with data communications began modulating their FM radios with tones. *Packet radio* spawned a new system of *digipeaters* (digital repeaters).

Digitized audio has been popular since audio compact discs (CDs) were introduced in the 1980s. In the 1990s, technology advanced enough to reduce the bandwidth needed for digital audio, especially voice, to be carried over the Internet and narrowband radio circuits. The first digital-voice public safety radio systems (called APCO-25) appeared, and a variety of Internet voice systems for conferencing and telephone-like use were developed.

Hams are using VHF/UHF digital voice technology, too. The Japan Amateur Radio League (JARL) developed a true ham radio standard called D-STAR, a networked VHF/UHF repeater system for digital voice and data that is beginning to make inroads around the world. Hams are also using surplus first generation APCO-25 radios and have adapted Internet voice systems for accessing and linking repeaters.

18.2 Repeater Overview

Amateurs learned long ago that they could get much better use from their mobile and portable radios by using an automated relay station called a *repeater*. Home stations benefit as well — not all hams are located near the highest point in town or have access to a tall tower. A repeater, whose basic idea is shown in **Fig 18.1**, can be more readily located where the antenna system is as high as possible and can therefore cover a much greater area.

18.2.1 Types of Repeaters

The most popular and well-known type of amateur repeater is an FM voice system on the 144 or 440 MHz bands. Amateurs operate many repeaters on the 29, 50, 222, 902 and 1240 MHz bands as well, but 2 meters and 70 cm are the most popular. Tens of thousands of hams use mobile and handheld radios for casual ragchewing, emergency communications, public service activities or just staying in touch with their regular group of friends during the daily commute.

FM is still the mode of choice for voice repeaters, as it was in commercial service, since it provides a high degree of immunity to mobile noise pulses. Operations are *channelized* — all stations operate on the same transmit frequency and receive on the same receive frequency. In addition, since the repeater receives signals from mobile or fixed stations and retransmits these signals simultaneously, the transmit and receive frequencies are different, or *split*. Direct contact between two or more stations that listen and transmit on the same frequency is called operating *simplex*.

Individuals, clubs, emergency communications groups and other organizations all sponsor repeaters. Anyone with a valid amateur license for the band can establish a repeater in conformance with the FCC rules. No one owns specific repeater frequencies, but nearly all repeaters are *coordinated* to minimize repeater-to-repeater interference. Frequency coordination and interference are discussed later in this chapter. Operational aspects are covered in more detail in The ARRL Operating Manual.

BEYOND FM VOICE

In addition to FM voice, there are several other types of ham radio repeaters.

ATV (amateur TV) — ATV repeaters are used to relay wideband television signals in the 70 cm and higher bands. They are used to extend coverage areas, just like voice repeaters. ATV repeaters are often set up for *crossband* operation, with the input on one band (say, 23 cm) and the output on another (say, 70 cm). More information on ATV repeaters may be found in the **Image Communications** chapter.

AM and SSB — There is no reason to limit repeaters to FM. There are a number of other modulation-type repeaters, some experimental and some long-established.

Digital voice — Repeaters for D-STAR

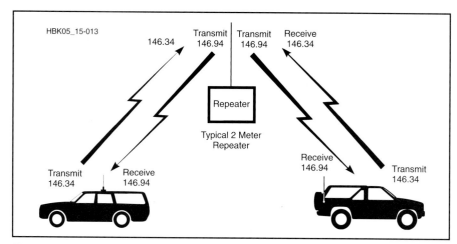

Fig 18.1 — Typical 2-m repeater, showing mobile-to-mobile communication through a repeater station. Usually located on a hill or tall building, the repeater amplifies and retransmits the received signal on a different frequency.

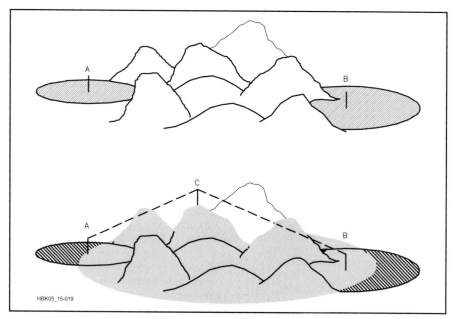

Fig 18.2 — In the upper diagram, stations A and B cannot communicate because their mutual coverage is limited by the mountains between them. In the lower diagram, stations A and B can communicate because the coverage of each station falls within the coverage of repeater C, which is on a mountaintop.

Fig 18.3 — In the Rocky Mountain west, handheld transceivers can often cover great distances, thanks to repeaters located atop high mountains. *(photo courtesy KC0ETU and K0NR)*

digital voice and data signals are discussed later in this chapter. There are also some P-25 repeaters operating in the Amateur service (see the **Digital Communications** chapter for additional information.)

Digipeaters — Digital repeaters used primarily for packet communications, including APRS (the Amateur Packet/position Reporting System). They can use a single channel (single port) or several channels (multi-port) on one or more VHF and UHF bands. See the **Digital Modes** and **Digital Communications** chapters for details of these systems.

Multi-channel (wideband) — Amateur satellites are best-known examples. Wide bandwidth (perhaps 50 to 200 kHz) is selected to be received and transmitted so all signals in bandwidth are heard by the satellite (repeater) and retransmitted, usually on a different VHF or UHF band. See the **Space Communications** chapter for details.

18.2.2 Advantages of Using a Repeater

When we use the term *repeater* we are almost always talking about transmitters and receivers on VHF or higher bands, where radio-wave propagation is normally line of sight. Don't take "line of sight" too literally. VHF/UHF radio signals do refract beyond what you can actually see on the horizon, but the phrase is a useful description. (See the **Propagation of Radio Signals** chapter for more information on these terms.)

We know that the effective range of VHF and UHF signals is related to the height of each antenna. Since repeaters can usually be located at high points, one great advantage of repeaters is the extension of coverage area from low-powered mobile and portable transceivers.

Fig 18.2 illustrates the effect of using a repeater in areas with hills or mountains. The same effect is found in metropolitan areas, where buildings provide the primary blocking structures.

Siting repeaters at high points can also have disadvantages. When two nearby repeaters use the same frequencies, your transceiver might be able to receive both. But since it operates FM, the *capture effect* usually ensures that the stronger signal will capture your receiver and the weaker signal will not be heard — at least as long as the stronger repeater is in use.

It is also simpler to provide a very sensitive receiver, a good antenna system, and a slightly higher power transmitter at just one location — the repeater — than at each mobile, portable or home location. A superior repeater system compensates for the low power (5 W or less), and small, inefficient antennas that many hams use to operate through them. The repeater maintains the range or coverage we want, despite our equipment deficiencies. If both the handheld transceiver and the repeater are at high elevations, for example, communication is possible over great distances, despite the low output power and inefficient antenna of the transceiver (see **Fig 18.3**).

Repeaters also provide a convenient meeting place for hams with a common interest. It's usually geographic — your town, or your club. A few repeaters are dedicated to a particular interest such as DX or passing traffic. Operation is channelized, and usually in any area you can find out which channel — or repeater — to pick to ragchew, get highway information or whatever your need or interest is. The conventional wisdom is that you don't have to tune around and call CQ to make a contact, as on the HF bands. Simply call on a repeater frequency — if someone is there and they want to talk, they will answer you. But with a few dozen repeaters covering almost any medium size town, you probably use the scan function in your radio to seek activity.

EMERGENCY OPERATIONS

When there is a weather-related emergency or a disaster (or one is threatening), most repeaters in the affected area immediately spring to life. Emergency operation and traffic always take priority over other ham activities, and many repeaters are equipped with emergency power sources just for these occasions. See **Fig 18.4**.

Almost all Amateur Radio emergency organizations use repeaters to take advantage of their extended range, uniformly good coverage and visibility. Most repeaters are well known — everyone active in an area with suitable equipment knows the local repeater frequencies.

18.2.3 Repeaters and the FCC

The law in the United States changes over time to adapt to new technology and changing times. Since the early 1980s, the trend has been toward deregulation, or more accurately in the case of radio amateurs, self-regulation. Hams have established band plans, calling frequencies, digital protocols and rules that promote efficient communication and interchange of information.

Originally, repeaters were licensed separately with detailed applications and control rules. Repeater users were forbidden to use their equipment in any way that could be interpreted as commercial. In some cases, even calling a friend at an office where the receptionist answered with the company name was interpreted as a problem.

The rules have changed, and now most nonprofit groups and public service events can be supported and businesses can be called — as long as the participating radio amateurs are not earning a living from this specific activity.

We can expect this trend to continue. For the latest rules and how to interpret them, see *QST* and the ARRL Web site, **www.arrl.org**.

Fig 18.4 — During disasters, repeaters over a wide area are used solely for emergency-related communication until the danger to life and property is past. *(photo courtesy WA9TZL)*

18.3 FM Voice Repeaters

Repeaters normally contain at least the sections shown in **Fig 18.5**. Repeaters consist of a receiver and transmitter plus a couple more special devices. One is a *controller* that routes the audio between the receiver and transmitter, keys the transmitter and provides remote control for the repeater licensee or designated control operators.

The second device is the *duplexer* that lets the repeater transmit and receive on the same antenna. A high power transmitter and a sensitive receiver, operating in close proximity within the same band, using the same antenna, present a serious technical challenge. You might think the transmitter would just blow the receiver away. But the duplexer keeps the transmit energy out of the receiver with a series of tuned circuits. Without a duplexer, the receiver and transmitter would need separate antennas, and those antennas would need to be 100 or more feet apart on a tower. Some repeaters do just that, but most use duplexers. A 2 meter duplexer is about the size of a two-drawer filing cabinet. See **Fig 18.6**.

Receiver, transmitter, controller, and duplexer: the basic components of most repeaters. After this, the sky is the limit on imagination. As an example, a remote receiver site can be used to extend coverage (**Fig 18.7**).

The two sites can be linked either by telephone ("hard wire") or a VHF or UHF link. Once you have one remote receiver site it is natural to consider a second site to better hear those "weak mobiles" on the other side of town (**Fig 18.8**). Some of the stations using the repeater are on 2 meters while others are on 440? Just link the two repeaters! (**Fig 18.9**).

For even greater flexibility, you can add an auxiliary receiver, perhaps for a NOAA weather channel (**Fig 18.10**).

The list goes on and on. Perhaps that is why so many hams have put up repeaters.

18.3.1 FM Repeater Operation

There are almost as many operating procedures in use on repeaters as there are repeaters. Only by listening can you determine the customary procedures on a particular machine. A number of common operating techniques are found on many repeaters, however.

One such common technique is the transmission of *courtesy tones*. Suppose several stations are talking in rotation — one following another. The repeater detects the end of a transmission of one user, waits a few seconds, and then transmits a short tone or beep. The next station in the rotation waits until the beep before transmitting, thus giving any other station wanting to join in a brief period to transmit their call sign. Thus the term *courtesy tone* — you are politely pausing to allow other stations to join in the conversation.

Another common repeater feature that encourages polite operation is the *repeater timer*. A 3-minute timer is actually designed to comply with an FCC rule for remotely controlled stations, but in practice the timer serves a more social function. Since repeater operation is channelized — allowing many stations to use the same frequency — it is polite to keep your transmissions short. If

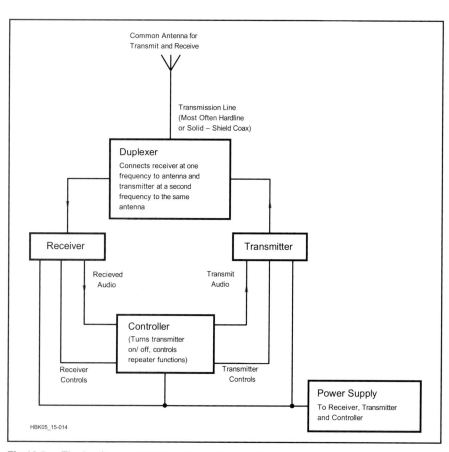

Fig 18.5 — The basic components of a repeater station. In the early days of repeaters, many were home-built. Today, most are commercial, and are far more complex than this diagram suggests.

Fig 18.6 — The W4RNC 2-meter repeater includes the repeater receiver, transmitter and controller in the rack. The large object underneath is the duplexer.
(photo courtesy KN4AQ)

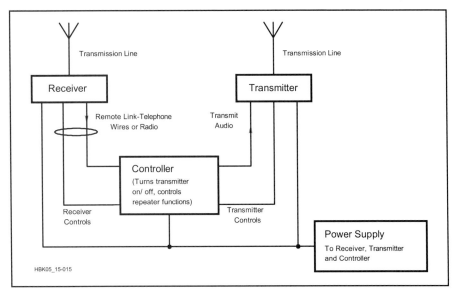

Fig 18.7 — Separating the transmitter from the receiver can extend the repeater's coverage area. The remote receiver can be located on a different building or hill, or consist of a second antenna at a different height on the tower.

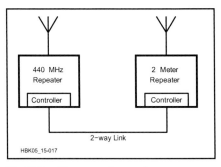

Fig 18.9 — Two repeaters using different bands can be linked for added convenience.

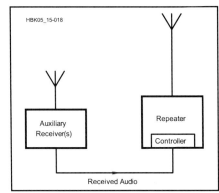

Fig 18.10 — For even greater flexibility, you can add an auxiliary receiver.

you forget this little politeness many repeaters simply cut off your transmission after 2 or 3 minutes of continuous talking. After the repeater "times out," the timer is reset and the repeater is ready for the next transmission.

A general rule, in fact law — both internationally and in areas regulated by the FCC — is that emergency transmissions always have priority. These are defined as relating to life, safety and property damage. Many repeaters are voluntarily set up to give mobile stations priority, at least in checking onto the

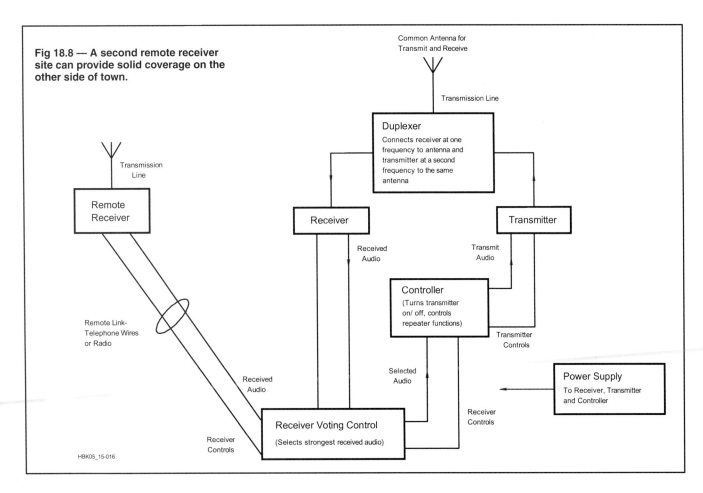

Fig 18.8 — A second remote receiver site can provide solid coverage on the other side of town.

Repeaters 18.5

repeater. If there is going to be a problem requiring help, the request will usually come from a mobile station. This is particularly true during rush hours; some repeater owners request that fixed stations refrain from using the repeater during these hours. Since fixed stations usually have the advantages of fixed antennas and higher power, they can operate simplex more easily. This frees the repeater for mobile stations that need it.

A chart of suggested operating priorities is given in **Fig 18.11**. Many but not all repeaters conform to this concept, so it can be used as a general guideline.

The figure includes a suggested priority control for *closed repeaters*. These are repeaters whose owners wish, for any number of reasons, not to have them listed as available for general use. Often they require transmission of a *subaudible* or *CTCSS* tone (discussed later). Not all repeaters requiring a CTCSS tone are closed — many open repeaters use tones to minimize interference among machines in adjacent areas using the same frequency pair. Other closed repeaters require the transmission of a special tone sequence to turn on. It is desirable that all repeaters, including generally closed repeaters, be made available at least long enough for the presence of emergency information to be made known.

Repeaters have many uses. In some areas they are commonly used for formal traffic nets, replacing or supplementing HF nets. In other areas they are used with tone alerting for severe weather nets. Even when a particular repeater is generally used for ragchewing it can be linked for a special purpose. As an example, an ARRL volunteer official may hold a periodic section meeting across her state, with linked repeaters allowing both announcements and questions directed back to her.

One of the most common and important uses of a repeater is to aid visiting hams. Since repeaters are listed in the *ARRL Repeater Directory* and other directories, hams traveling across the country with mobile or handheld radios often check into local repeaters asking for travel route, restaurant or lodging information. Others just come on the repeater to say hello to the local group. In most areas courtesy prevails — the visitor is given priority to say hello or get the needed help.

Detailed information on repeater operating techniques is included in a full chapter of the *ARRL Operating Manual*.

18.3.2 Home and Mobile Equipment

There are many options available in equipment used on repeaters. A number of these options are shown in **Fig 18.12**.

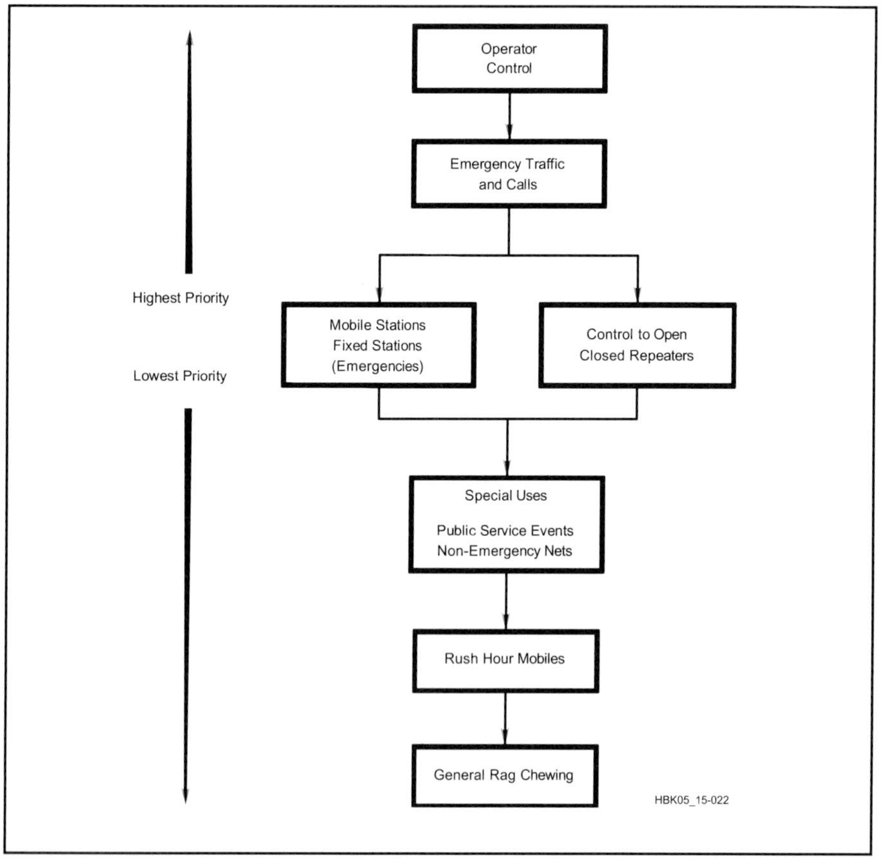

HANDHELD TRANSCEIVERS

A basic handheld radio with 500 mW to 5 W output can be mounted in an automobile with or without an external power amplifier.

Several types of antennas can be used in the handheld mode. The smallest and most convenient is a rubber flex antenna, known as a "rubber duckie," a helically wound antenna encased in a flexible tube. Unfortunately, to obtain the small size the use of a wire helix or coil produces a very low efficiency.

A quarter-wave whip, which is about 19 inches long for the 2-meter band, is a good choice for enhanced performance. The rig and your hand act as a ground plane and a reasonably efficient result is obtained. A longer antenna, consisting of several electrical quarter-wave sections in series, is also commercially available. Although this antenna usually produces extended coverage, the mechanical strain of 30 or more inches of antenna mounted on the radio's antenna connector can cause problems. After several months, the strain may require replacement of the connector.

The latest handheld radios are supplied with lithium-ion (Li-ion) batteries. These high-capacity batteries are lightweight and allow operation for much longer periods than the classic NiCd battery pack. Charging is accomplished either with a "quick" charger in an hour or less or with a trickle charger overnight. See the **Power Supplies** chapter for more information on batteries.

Power levels higher than 7 W may cause a safety problem on handheld units, since the antenna is usually close to the operator's head and eyes. See the **Safety** chapter for more information.

For mobile operation, a power cord plugs into the vehicle cigarette lighter. In addition, commercially available "brick" amplifiers can be used to raise the output power level of the handheld radio to 50 W or more.

MOBILE EQUIPMENT

Compact mobile transceivers operate from 11-15 V dc and generally offer several transmit power levels up to about 50 W. They can operate on one or more bands. Most common are the single band and dual-band transceivers. "Dual-band" usually means 2 meter and 70 cm, but other combinations are available, as are radios that cover three or more bands.

Mobile antennas range from the quick and easy magnetic mount to "drill through the car roof" assemblies. The four general classes of mobile antennas shown in the center section of Fig 18.12 are the most popular choices. Before experimenting with antennas for your vehicle, there are some precautions to be taken.

Fig 18.11 — The chart shows recommended repeater operating priorities. Note that, in general, priority goes to mobile stations.

Through-the-glass antennas: Rather than trying to get the information from your dealer or car manufacturer, test any such antenna first using masking tape or some other temporary technique to hold the antenna in place. Some windshields are metallicized for defrosting, tinting and car radio reception. Having this metal in the way of your through-the-glass antenna will seriously decrease its efficiency.

Magnet-mount antennas are convenient, but only if your car has a metal roof or trunk. The metal also serves as the ground plane.

Through-the-roof antenna mounting: Most hams are squeamish about drilling a hole in their car roof, unless they intend to keep the car for a long time. This mounting method provides the best efficiency, however, since the (metal) roof serves as a ground plane. Before you drill, carefully plan and measure how you intend to get the antenna cable down under the interior car headliner to the radio. Be especially careful of side-curtain air bags. Commercial two-way shops can install antennas, usually for a reasonable price.

Trunk lip and clip-on antennas: These antennas are good compromises. They are usually easy to mount and they perform acceptably. Cable routing must be planned. If you are going to run more than a few watts, do not mount the antenna close to one of the car windows — a significant portion of the radiated power may enter the car interior.

More information on mobile equipment may be found in the **Assembling a Station** and **Antennas** chapters.

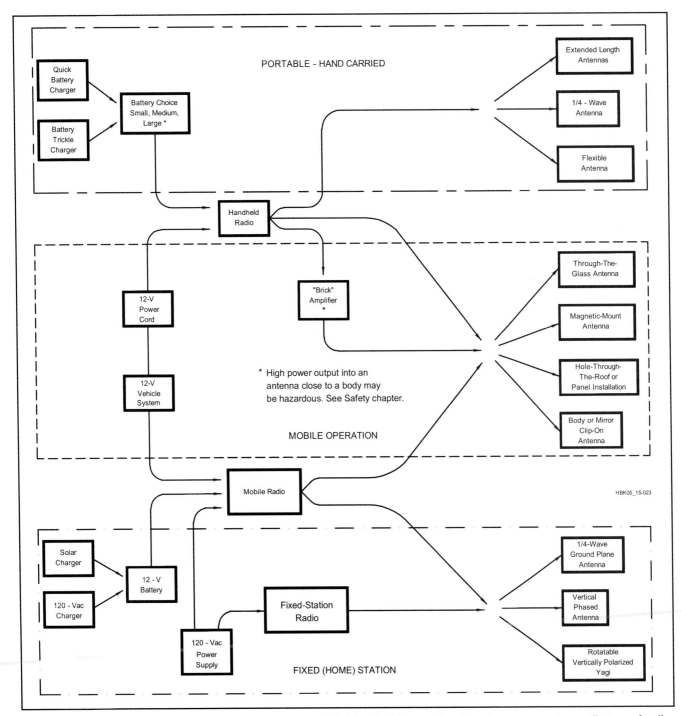

Fig 18.12 — Equipment choices for use with repeaters are varied. A handheld transceiver is perhaps the most versatile type of radio, as it can be operated from home, from a vehicle and from a mountaintop.

HOME STATION EQUIPMENT

Most "base station" FM radios are actually mobile rigs, powered either from rechargeable batteries or ac-operated power supplies. Use of batteries has the advantage of providing back-up communications ability in the event of a power interruption. Some home station transceivers offer operation on the HF/VHF or HF/VHF/UHF bands, or multi-mode (FM/SSB/CW) operation on several VHF/UHF bands. Using them means that you will not be able to monitor a local FM frequency while operating HF.

The general choice of fixed-location antennas is also shown in Fig 18.12. Most hams use an omnidirectional vertical, but if you are in an area between two repeaters on the same channel, a rotatable Yagi may let you pick which repeater you will use without interfering with the other repeater. Vertical polarization is the universal custom, since it is easiest to accomplish in a mobile installation. VHF/UHF SSB operation is customarily horizontally polarized. An operator with a radio that does both has a tough choice, as there is a serious performance hit between stations using cross-polarized antennas.

Both commercial and homemade ¼-λ and larger antennas are popular for home use. A number of these are shown in the **Antennas** chapter. Generally speaking, ¼-λ sections may be stacked up to provide more gain on any band. As you do so, however, more and more power is concentrated toward the horizon. This may be desirable if you live in a flat area. See **Fig 18.13**.

18.3.3 Coded Squelch and Tones

Squelch is the circuit in FM radios that turns off the loud rush of noise with no signal present. Most of the time, hams use *noise squelch*, also called *carrier squelch*, a squelch circuit that lets any signal at all come through. But there are ways to be more selective about what signal gets to your speaker or keys up your repeater. That's generically known as *coded squelch*, and more than half of the repeaters on the air require you to send coded squelch to be able to use the repeater.

CTCSS

The most common form of coded squelch has the generic name *CTCSS* (Continuous Tone Coded Squelch System), but is better know by Motorola's trade name PL (Private Line), or just the nickname "tone." Taken from the commercial services, subaudible tones are not generally used to keep others from using a repeater but rather are a method of minimizing interference from users of the same repeater frequency. CTCSS tones are sine-wave audio tones between 67 and 250 Hz, that are added to the transmit audio at a fairly low level. They are *sub*audible only because your receiver's audio circuit is supposed to filter them out. A receiver with CTCSS will remain silent to all traffic on a channel unless the transmitting station is sending the correct tone. Then the receiver sends the transmitted audio to its speaker.

For example, in **Fig 18.14** a mobile station on hill A is nominally within the normal coverage area of the Jonestown repeater (146.16/76). The Smithtown repeater, also on the same frequency pair, usually cannot hear stations 150 miles away. The mobile is on a

Fig 18.13 — As with all line-of-sight communications, terrain plays an important role in how your signal gets out.

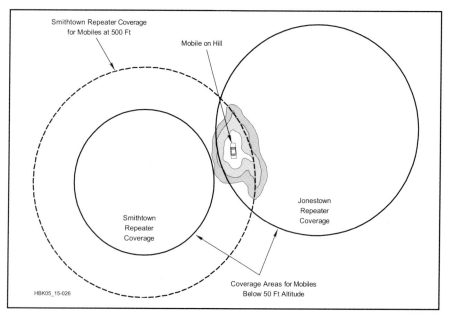

Fig 18.14 — When two repeaters operate on the same frequencies, a well-situated operator can key up both repeaters simultaneously. A directional antenna may help.

Table 18.1
CTCSS Tone Frequencies

The purpose of CTCSS is to reduce cochannel interference during band openings. CTCSS equipped repeaters would respond only to signals having the CTCSS tone required for that repeater. These repeaters would not respond to weak distant signals on their inputs and correspondingly not transmit and repeat to add to the congestion. The standard ANSI/EIA frequency codes, in hertz, are as follows:

67.0	114.8	177.3
69.3	118.8	179.9
71.9	123.0	183.5
74.4	127.3	186.2
77.0	131.8	189.9
79.7	136.5	192.8
82.5	141.3	199.5
85.4	146.2	203.5
88.5	151.4	206.5
91.5	156.7	210.7
94.8	159.8	218.1
97.4	162.2	225.7
100.0	165.5	229.1
103.5	167.9	233.6
107.2	171.3	241.8
110.9	173.8	250.3
		254.1

hill and so he is in the coverage area of both Jonestown and Smithtown. Whenever the mobile transmits both repeaters hear him.

The common solution to this problem, assuming it happens often enough, is to equip the Smithtown repeater with a CTCSS decoder and require all users of the repeater to transmit a CTCSS tone to access the repeater. Thus, the mobile station on the hill does not come through the Smithtown repeater, since he is not transmitting the required CTCSS tone.

Table 18.1 shows the available CTCSS tones. Most radios built since the early 1980s have a CTCSS encoder built in, and most radios built since the early 1990s also have a CTCSS *decoder* built in. Newer radios have a "tone scan" feature that will hunt the tone, *if* the repeater is sending tone. Most repeaters that require tone also transmit their tone, but they don't have to. Listings in the *ARRL Repeater Directory* include the CTCSS tone required, if any.

If your local repeater sends a CTCSS tone, you can use your decoder to monitor just that repeater, and avoid hearing the co-channel neighbor, intermod or the annoying fizzes of nearby consumer electronics. Radios typically store CTCSS frequency and mode in their memory channels.

DIGITAL-CODED SQUELCH (DCS)

A newer form of coded squelch is called *DCS* (Digital-Coded Squelch). DCS appeared in commercial service because CTCSS didn't provide enough tones for the many users. DCS adds another 100 or so code options. DCS started showing up in Amateur Radio transceivers several years ago and is now a standard feature in most new radios. Open repeaters generally still use CTCSS rather than DCS, since many older radios still in use don't have DCS.

DTMF

In the days before widespread use of cell phones, one of the most attractive features of repeaters was the availability of autopatch services that allowed a mobile or portable station to use a standard telephone DTMF (dual-tone multi-frequency, or Touch-Tone) key pad to connect the repeater to the local telephone line and make outgoing calls.

Although autopatches see less use today, DTMF key pads are still used for sending control signals. DTMF can also be used as a form of squelch, to turn a receiver on, though it's more often used to control various functions like autopatch and talking S-meters. Some repeaters that require CTCSS have a DTMF "override" that puts the repeater into carrier-squelch mode for a few minutes if you send the proper digits. Other applications for DTMF tones include controlling linked repeaters, described in a later section.

Table 18.2
Standard Telephone (DTMF) Tones

	Low Tone Group		High Tone Group	
	1209 Hz	1336 Hz	1477 Hz	1633 Hz
697 Hz	1	2	3	A
770 Hz	4	5	6	B
852 Hz	7	8	9	C
941 Hz	*	0	#	D

Table 18.3
Standard Frequency Offsets for Repeaters

Band	Offset
29 MHz	−100 kHz
52 MHz	Varies by region −500 kHz, −1 MHz, −1.7 MHz
144 MHz	+ or −600 kHz
222 MHz	−1.6 MHz
440 MHz	+ or −5 MHz
902 MHz	12 MHz
1240 MHz	12 MHz

Table 18.2 shows the DTMF tones. Some keyboards provide the standard 12 sets of tones corresponding to the digits 0 through 9 and the special signs # and *. Others include the full set of 16 pairs, providing special keys A through D. The tones are arranged in two groups, usually called the low tones and high tones. Two tones, one from each group, are required to define a key or digit. For example, pressing 5 will generate a 770-Hz tone and a 1336-Hz tone simultaneously.

The standards used by the telephone company require the amplitudes of these two tones to have a certain relationship.

18.3.4 Frequency Coordination and Band Plans

Since repeater operation is channelized, with many stations sharing the same frequency pairs, the amateur community has formed coordinating groups to help minimize conflicts between repeaters and among repeaters and other modes. Over the years, the VHF bands have been divided into repeater and nonrepeater sub-bands. These frequency-coordination groups maintain lists of available frequency pairs in their areas (although in most urban areas, there are no "available" pairs on 2 meters, and 70 cm pairs are becoming scarce). A complete list of frequency coordinators, band plans and repeater pairs is included in the *ARRL Repeater Directory*.

Each VHF and UHF repeater band has been subdivided into repeater and nonrepeater channels. In addition, each band has a specific *offset* — the difference between the transmit frequency and the receive frequency for the repeater. While most repeaters use these standard offsets, others use "oddball splits." These nonstandard repeaters are generally also coordinated through the local frequency coordinator. **Table 18.3** shows the standard frequency offsets for each repeater band.

FM repeater action isn't confined to the VHF and UHF bands. There are a large handful of repeaters on 10 meters around the US and the world. "Wideband" FM is permitted only above 29.0 MHz, and there are four band-plan repeater channels (outputs are 29.62, 29.64, 29.66 and 29.68 MHz), plus the simplex channel 29.60 MHz. Repeaters on 10 meters use a 100 kHz offset, so the corresponding inputs are 29.52, 29.54, 29.56 and 29.58 MHz. During band openings, you can key up a repeater thousands of miles away, but that also creates the potential for interference generated when multiple repeaters are keyed up at the same time. CTCSS would help reduce the problem, but not many 10-m repeater owners use it and too many leave their machines on "carrier access."

18.3.5 Linked Repeaters

Most repeaters are standalone devices, providing their individual pool of coverage and that's it. But a significant number of repeaters are linked — connected to one or more other repeaters. Those other repeaters can be on other bands at the same location, or they can be in other locations, or both. Linked repeaters let users communicate between different bands and across wider geographic areas than they can on a single repeater. **Fig 18.15** shows an example.

There are many ways to link repeaters. Two-meter and 70 cm repeaters on the same tower can just be wired together, or they may even share the same controller. Repeaters within a hundred miles or so of each other can use a radio link — separate link transmitters and receivers at each repeater, with antennas pointed at each other. Multiple repeaters, each still about 100 miles apart, can "daisy-chain" their links to cover even wider territory. There are a few linked repeater systems in the country that cover several states with dozens of repeaters, but most radio-linked repeater systems have more modest ambitions, covering just part of one or two states.

Repeater linking via the Internet has creat-

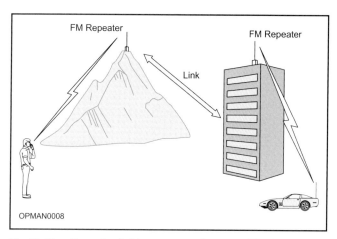

Fig 18.15 — Repeater linking can greatly expand VHF/UHF communication distances. Repeater links are commonly made via dedicated radio hardware or via the Internet.

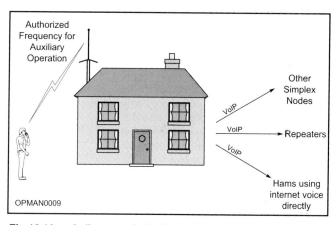

Fig 18.16 — A diagram of a VoIP simplex node. If a control operator is not physically present at the station location and the node is functioning with wireless remote control, the control link must follow the rules for *auxiliary* operation.

ed the ability to tie repeaters together around the world and in nearly unlimited number. We'll talk more about Internet linking in the next section.

There are several ways linked repeaters can be operated, coming under the categories of *full-time* and *on demand*. Full-time linked repeaters operate just as the name implies — all the repeaters in a linked network are connected all the time. If you key up one of them, you're heard on all of them, and you can talk to anyone on any of the other repeaters on the network at any time. You don't have to do anything special to activate the network, since it's always there.

In an on-demand system, the linked repeaters remain isolated unless you take some action, usually by sending a code by DTMF digits, to connect them. Your DTMF sequence may activate all the repeaters on the network, or the system may let you address just one specific repeater, somewhat like dialing a telephone. When you're finished, another DTMF code drops the link, or a timer may handle that chore when the repeaters are no longer in use.

INTERNET LINKING

The Internet and *Voice over Internet Protocol* (VoIP) has expanded repeater linking exponentially, making worldwide communication through a local repeater commonplace. Two Internet linking systems, IRLP and EchoLink, have reached critical mass in the US and are available almost everywhere. A brief overview is included here; more information may be found in the **Digital Communications** chapter.

IRLP

The Internet Radio Linking Project (IRLP) is the most "radio" based linking system. User access is only via radio, using either simplex or repeater stations, while linking is done using VoIP on the Internet. An IRLP system operator establishes a *node* by interfacing his radio equipment to a computer with an Internet connection, and then running IRLP software. Once that's set up, repeater users send DTMF tones to make connections, either directly to other individual repeater or simplex nodes (**Fig 18.16**), or to *reflectors* — servers that tie multiple nodes together as one big party line.

The direct connections work like on-demand linked repeaters. You dial in the node number you want to connect to and access code (if required), and you are connected to the distant repeater. Once connected, everyone on both ends can communicate. When finished, another DTMF sequence takes the link down. Someone from a distant repeater can make a connection to you as well.

Reflectors work like a hybrid between on-demand and full-time linked repeaters. You can connect your local repeater to a reflector and leave it there all day, or you can connect for a special purpose (like a net), and drop it when the event is over.

EchoLink

EchoLink requires a PC with sound card and appropriate software. It allows repeater connections like IRLP, and it has Conference Servers, similar to IRLP reflectors that permit multiple connections. The big difference is that EchoLink allows individuals to connect to the network from their computers, without using a radio.

The EchoLink conference servers all have more or less specific functions. Some are just regional gathering places, while some are region, topic or activity based (SKYWARN and National Hurricane Center Nets, Jamboree on the Air, and so on).

You can connect your EchoLink-enabled computer to your base station radio fairly easily through a sound card, and create an on-air node. Don't pipe EchoLink to a local repeater without permission from the repeater owner, though.

If you decide to create a full-time link from a computer to a repeater, consider using a dedicated UHF link frequency rather than just a base station on the repeater input. This applies to IRLP connections as well. Of course, the Internet is infrastructure dependent, and both power and Internet access can be interrupted during storms or other disasters.
f a voice transmission.

18.4 D-STAR Repeater Systems

D-STAR is a digital protocol developed by the Japan Amateur Radio League (JARL) that takes Amateur Radio into the 21st century. D-STAR is a bit-streaming protocol able to encapsulate voice and low speed data (DV) or higher speed Ethernet data (DD). Because the protocol is entirely digital (GMSK modulation is used for amateur station transceivers and the respective repeaters), signaling is carried entirely out-of-band (ie, control codes are a separate part of the data stream from the voice or Ethernet information). When looking at the D-STAR repeater systems, it is important to keep that aspect in mind. Amateurs are more familiar in-band signaling (DTMF tones, for example) in the analog world. Additional information on D-STAR may be found in the **Digital Communications** chapter.

18.4.1 D-STAR Station IDs

The D-STAR specification defines a protocol that can be used for simplex communication or repeater operations. When used simplex, the D-STAR radios function similarly to analog radios with the addition to enable selective listening based on station ID (call sign and a character or space). To operate through a repeater, you must know the repeater's station ID as well as the frequency the repeater is on. In this case, consider the repeater's station ID equivalent to a unique CTCSS tone in the analog world.

To talk to someone via D-STAR, you need to set your radio with four station IDs:
• your station ID (MYCALL)
• their station ID (URCALL)
• your local repeater station ID (RPT1)
• your local gateway station ID (RPT2). MYCALL is always set to your call sign, followed by a character or space. RPT1 is set to your local repeater's station ID (the repeater call sign, followed by a character or space).

If you want to talk locally through the repeater:
• Set URCALL to CQCQCQ.
• Set RPT2 to your local gateway station ID (this isn't needed for local communication, but allows some new network functions to operate, and has become the default recommendation).

The power of the D-STAR protocol becomes evident when D-STAR *gateways* are implemented providing interconnectivity between repeater systems. This interconnectivity is the same whether you are using voice or Ethernet data.

To talk to someone elsewhere in the D-STAR world, beyond your local area:
• Set URCALL to the other station's ID.
• Set RPT2 to your local gateway station ID.

The RPT2 setting is very important. You don't need to know where the other station is. You simply tell your local repeater to send your bit stream (everything is digital) to the local gateway so that gateway can determine where to send it next. The local gateway looks at URCALL (remember, this is part of the bit stream) and determines where that station was heard last. It then sends the bit stream on to that remote gateway, which looks at URCALL again to determine which repeater at the far end should transmit the bit stream.

Sounds complex? Yes, the implementation can be complex but the user is shielded from all of this by simply setting the four station IDs. **Fig 18.17** shows an example.

18.4.2 Station Routing

There are several different ways to set the destination station ID using URCALL:
• If URCALL is set to CQCQCQ, this means "don't go any farther."
• If URCALL is set to the remote station's ID and RPT2 is set to the local gateway, this means "gateway, send my bit stream to be transmitted out the repeater that the remote station was last heard on."
• If URCALL is set to/followed by a remote *repeater* station ID, and RPT2 is set to the local gateway, this means "gateway, send my bit stream to be transmitted out the repeater designated."

Some people mistakenly call this "call sign routing." In fact, this is "station routing" because you are specifying the station you want to hear your bit stream (voice or Ethernet data). This is not source routing because the source station is only defining what station they want to talk to; it is up to the gateways to determine routing (very similar to Internet routers).

You can specify a destination but you have no guarantee that:

1) The designated station is on the air to receive your transmission.

2) The repeater that the designated station is using is not busy.

3) Other factors will not prevent your bit stream from reaching the designated station.

Station routing is similar to your address book. In your address book you may have a work number, home number, cell number, fax

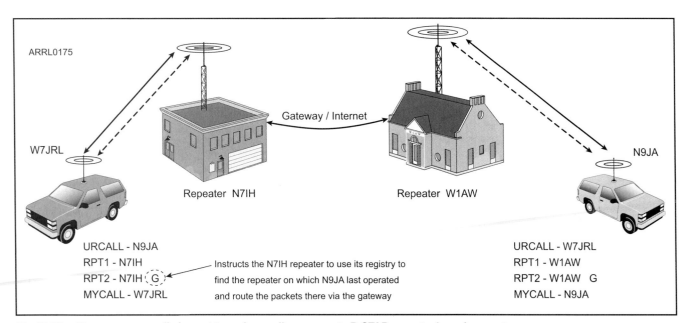

Fig 18.17 — The necessary call sign set to make a call on a remote D-STAR repeater by using a gateway.

D-STAR Network Overview

The D-STAR specification defines the repeater controller/gateway communications and defines the general D-STAR network architecture. The diagram shown here as **Fig 18.A1** is taken from the English translation of the D-STAR specification:

The Comp. IP and Own IP are shown for reference if this was a DD communications. As they do not change and are not passed as part of the D-STAR protocol, they can safely be ignored for the purposes of the following explanation.

Headers 1 through 4 are W$1QQQ calling W$1WWW. Headers 5 through 8 are W$1WWW calling W$1QQQ. Note that "Own Callsign" and "Companion Call sign" are never altered in either sequence. The "Destination Repeater Call sign" and the "Departure Repeater Callsign" are changed between the gateways. This is so the receiving gateway and repeater controller know which repeater to send the bit stream to. It also makes it easy to create a "One Touch" response as ICOM has done by simply placing the received "Own Callsign" in the transmitted "Companion Callsign", the received "Destination Repeater Callsign" in the transmitted "Departure Repeater Callsign", and the received "Departure Repeater Callsign" in the transmitted "Destination Repeater Callsign".

Use of the "special" character "/" at the beginning of a call sign indicates that the transmission is to be routed to the repeater specified immediately following the slash. For instance, entering "/K5TIT B" in the "Companion Callsign" would cause the transmission to be routed to the "K5TIT B" repeater for broadcast. Using the above example, W$1QQQ would put "/W$1SSS" in the "Companion Callsign" for the same sequence 1 through 4 to occur. At the W$1VVV gateway, however, the "/W$1SSS" in the "Companion Callsign" would be changed to "CQCQCQ". All stations within range of W$1SSS would see the transmission as originating from W$1QQQ and going to CQCQCQ just like that station was local (but the "Departure Repeater Callsign" would be "W$1VVV G" and the "Destination Repeater Callsign" would be "W$1SSS"). Replying would still be done the same way as before since the received "Companion Callsign" is ignored when programming the radio to reply.

Every "terminal" (station) has an IP address assigned to it for DD purposes. The address is assigned from the 10.0.0.0/8 address range. The D-STAR gateway is always 10.0.0.2. The router to the Internet is always 10.0.0.1. The addresses 10.0.0.3-31 are reserved for local-to-the-gateway (not routable) use. What this makes possible is the ability to send Ethernet packets to another station by only knowing that terminal's IP address and the remote station can directly respond based solely on IP address. This is because the gateway software can correlate IP address with call sign and ID. This makes it possible to route DD Ethernet packets based on the "Companion Callsign" or based on IP address with "Companion Callsign" set to "CQCQCQ". — *Pete Loveall, AE5PL*

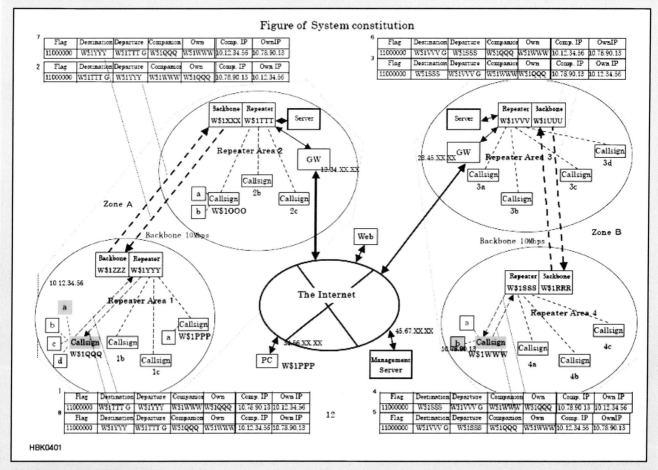

Fig 18.A1 — A D-STAR system overview.

Fig 18.18 — A full rack of D-STAR equipment on the bench of Jim McClellan, N5MIJ. Top to bottom: ICOM IP-RPC2 controller, ID-RP2V 1.2-GHz voice repeater, ID-RP4000 440-MHz voice and data repeater, a blank panel and an ID-RP2000 146-MHz voice and data repeater.

number and email address for a person. How do you know which one to contact them at? Maybe the email address is best if you want to send them data (equivalent to the station ID for their Ethernet data radio). Maybe the cell phone number if that is what they normally have with them (equivalent to their hand-held transceiver station ID).

How do I know if they heard me? When they talk (or send Ethernet data) back to you. D-STAR is connectionless. Therefore, there is no equivalence in D-STAR to repeater linking as we think of it in the analog world. However, there have been independent implementations of linking of repeaters similar to the linking we see with IRLP (see the section on *DPlus*).

In all cases of D-STAR repeater use, all digital voice (DV) signals with a proper RPT1 are always repeated so everyone hears your transmission through the repeater, regardless of the other settings. Ethernet data (DD) signals are not repeated on frequency because the "repeater" is actually a half-duplex Ethernet bridge operating on a single frequency.

For more details, see the sidebar, "D-STAR Network Overview."

18.4.3 Enhancing D-STAR Operation with *DPlus*

Because D-STAR is a true digital protocol, repeaters have no need for decoding the audio transmitted as bits from each radio. As mentioned previously, this requires all signaling to be out-of-band (with regard to the voice or Ethernet data).

Applications can be built to work with this out-of-band signaling to implement enhancements to the base product without modifying those products. One of these applications is *DPlus*, software that runs on the D-STAR Gateway computer at the repeater site. It provides many capabilities and more are being added as the software develops.

A key concept is the capability to link repeaters either *directly* (everything heard on one repeater is heard on another) or *indirectly* through a *reflector* (everything sent to a reflector is reflected back out to all linked repeaters). A reflector is a special version of *DPlus* that runs on a standalone computer that is not part of a repeater system. Linking and unlinking a repeater is done by altering the contents of URCALL according to the *DPlus* documentation. (These features continue to evolve, so specific operating commands are not covered here.)

As of mid-2009, *DPlus* is solely focused on the digital voice (DV) part of D-STAR, not the Ethernet data (DD) part. As such, there is not a way to directly link two DD "repeaters." Because DD is Ethernet bridging, however, full TCP support is available, allowing each individual station to make connections as needed to fit their requirements.

For a station to make use of a linked repeater, the station must have URCALL set to CQCQCQ and RPT2 set to the local gateway. If RPT2 is not set to the local gateway, *DPlus* running on the local gateway computer will never see the bit stream and therefore not be able to forward the bit stream to the linked repeater or reflector. This is why setting RPT2 for your local gateway should be your default setting.

If your radio is set for automatic low-speed data transmission, remember that low-speed data is carried as part of the digital voice bit stream and is not multiplexed. Therefore, any low-speed data transmission will block the frequency for the time the transmission is occurring. Caution: if you are using a linked repeater or if you have set URCALL to something other than CQCQCQ, your bit stream will be seen by all stations that are on the other end of that transmission. A reflector could have over 100 other repeaters listening to your transmission.

18.4.4 D-STAR Repeater Hardware

D-STAR repeaters are a bit different from the FM repeaters with which we're all familiar. A quick comparison will help to illustrate.

A complete FM repeater consists of at least three identifiable components. A receiver receives the original signal and demodulates it. The demodulated audio is routed to a controller, where it is mixed with other audio. The resulting combined audio signal is routed back to one or more transmitters. At least one additional source of audio is present in the controller, as it is a legal requirement to ID the repeater transmitter correctly. A well-constructed system includes validation that the levels and frequency response of the processed audio are consistent and true to the originally transmitted signals.

A D-STAR repeater's block diagram looks very similar, but functions very differently. A receiver receives the original signal and demodulates it. That signal is passed to a controller, which then drives one or more

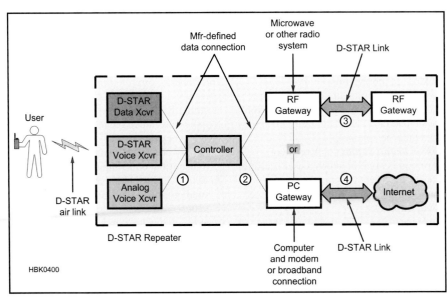

Fig 18.19 — Internal and external connections of a D-STAR repeater stack.

transmitters. The most significant difference is that there is no audio involved! D-STAR is a digital protocol. All required manipulation of information is performed in the digital domain, including the necessary ID functions. Most existing D-STAR repeaters do not contain the vocoders necessary to recover audio information, so there exist no local speaker or microphone. There is also no level-setting to consider with D-STAR.

D-STAR REPEATER OPTIONS

There is much discussion about "homebrew" D-STAR repeaters. The two most common approaches are to modify an existing FM repeater to pass the digital signal, or to wire two radios back-to-back. Both approaches provide the desired extended RF coverage, but fail to accurately process the digital signal. Thus both approaches fall short in either functional capability or in legality of the transmitted signal.

It is relatively simple to modify an existing FM repeater to pass the digital signal. The limitations of this approach are that there can be no additional capability added (for example, a D-STAR Gateway), since the digital signal is never decoded. Additionally, this method lacks the ability to encapsulate the ID for the repeater transmitter into the transmitted data.

It is also very simple to wire two D-STAR radios back-to-back, such that the incoming signals are retransmitted. Again, the functionality is limited by the inability to process the entire data stream. This approach presents an additional consideration, as the radio used as a transmitter does not process the data stream as is done in the D-STAR repeater system, and the ID of the originating transmitter is lost, replaced by the ID of the "repeater" transmitter.

Current commercially produced D-STAR repeaters are designed to be used across a broad frequency range. They do not have some of the tight front-end filtering provided by our familiar FM repeaters, so repeater builders must provide that front-end filtering externally. Installing additional band-pass filters between the antenna and the repeater will significantly improve the performance of the system. This is true for both digital voice (DV) repeaters and digital data (DD) access points.

Following good engineering practices will ensure good performance of the system. A properly constructed and installed D-STAR repeater can exhibit performance improvements of 15% or more in range, as compared to a comparably constructed FM repeater. This performance gain comes from the combination of the forward error correction (FEC) contained in the transmitted signal, and the fact that the radiated power is contained within a narrower bandwidth.

D-STAR is an exciting new system for the amateur community, providing significant opportunities for us to develop applications and implementations using capabilities we've never had before. We truly are limited only by our imaginations. The growth of the world-wide D-STAR network illustrates the level of interest in the technology by both amateurs and our served agencies. Amateurs are once again developing at the leading edge of technology. What we do with the new tools is up to us. How will you use D-STAR?

18.5 Glossary of FM and Repeater Terminology

Access code — One or more numbers and/or symbols that are keyed into the repeater with a DTMF tone pad to activate a repeater function, such as an autopatch.

Autopatch — A device that interfaces a repeater to the telephone system to permit repeater users to make telephone calls. Often just called a "patch."

Carrier-operated relay (COR) — A device that causes the repeater to transmit in response to a received signal. Solid state versions may be called COS (carrier-operated switch).

Channel — The pair of frequencies (input and output) used by a repeater.

Channel step — The difference (in kHz) between FM channels. The common steps are 15 and 20 kHz for 2-meter repeaters, 20 kHz for 222 MHz repeaters, and 25 kHz for 440 MHz repeaters. Closer spacing is beginning to be used in some congested areas.

Closed repeater — A repeater whose access is limited to a select group (see *open repeater*).

Control operator — The Amateur Radio operator who is designated to "control" the operation of the repeater, as required by FCC regulations.

Courtesy tone — An audible indication that a repeater user may go ahead and transmit.

Coverage — The geographic area within which the repeater provides communications.

Crossband — A repeater with its input on one band and output on another.

CTCSS — Abbreviation for continuous tone-controlled squelch system, a subaudible tone sent with an FM voice transmission to access a repeater.

DCS — Digital Coded Squelch. A newer version of CTCSS that uses a subaudible digital code instead of an analog tone to selectively open a receiver's squelch.

Digipeater — A packet radio (digital) repeater, usually using store-and-forward on a single frequency.

DTMF — Abbreviation for *dual-tone multifrequency*, commonly called Touch Tone, the series of tones generated from a keypad on a ham radio transceiver (or a regular telephone).

Duplex or full duplex — A mode of communication in which a user transmits on one frequency and receives on another frequency simultaneously (see *half duplex*).

Duplexer — A device that allows the repeater transmitter and receiver to use the same antenna simultaneously.

Frequency coordinator — An individual or group responsible for assigning frequencies to new repeaters without causing interference to existing repeaters.

Full quieting — A received signal that contains no noise.

Half duplex — A mode of communication in which a user transmits at one time and receives at another time.

Handheld — A small, lightweight portable transceiver small enough to be carried easily.

Hang time — A few seconds of repeater carrier following a user transmission that allows others who want to access the repeater a chance to do so; the *courtesy beep* sounds during the hang time.

Input frequency — The frequency of the repeater's receiver (and your transceiver's transmitter).

Intermod — *Intermodulation distortion (IMD)*, the unwanted mixing of two strong RF signals that causes a signal to be transmitted on an unintended frequency.

Key up — To turn on a repeater by transmitting on its input frequency.

Li-Ion — Lithium-Ion battery. Longer life, smaller and lighter than Ni-Cd, Li-Ion batteries are becoming more popular for use with handheld radios.

Machine — A repeater system.

Mag mount — Magnetic mount, an antenna with a magnetic base that permits quick installation and removal from a motor vehicle or other metal surface.

NiCd — A nickel-cadmium battery that may be recharged many times; often used to power portable transceivers. Pronounced *NYE-cad*.

NiMH — Nickel-metal-hydride battery; rechargeable, offers more capacity and lighter weight than an NiCd battery. Often used to power portable transceivers.

Offset — the spacing between a repeater's input and output frequencies.

Open repeater — a repeater whose access is not limited.

Output frequency — the frequency of the repeater's transmitter (and your transceiver's receiver).

Over — A word used to indicate the end of a voice transmission.

Repeater Directory — An annual ARRL publication that lists repeaters in the US, Canada and other areas.

Separation — The difference (in kHz) between a repeater's transmitter and receiver frequencies, also called the *offset*, or *split*. Repeaters that use unusual separations, such as 1 MHz on 2 m, are sometimes said to have "oddball splits."

Simplex — A mode of communication in which users transmit and receive on the same frequency.

Squelch tail — The noise burst heard in a receiver that follows the end of an FM transmission.

Time-out — To cause the repeater or a repeater function to turn off because you have transmitted for too long.

Timer — A device that measures the length of each transmission and causes the repeater or a repeater function to turn off after a transmission has exceeded a certain length.

Tone pad — An array of 12 or 16 numbered keys that generate the standard telephone dual-tone multifrequency (*DTMF*) dialing signals. Resembles a standard telephone keypad. (see *autopatch*).

18.6 References and Bibliography

D-STAR Users Web site — **www.dstarusers.org**

Ford, S., WB8IMY, *ARRL's VHF Digital Handbook* (ARRL, 2008). Chapter 4 and Appendix B cover D-STAR.

Japan Amateur Radio League, *Translated D-STAR System Specification*, **www.jarl.com/d-star**

Loveall, P., AE5PL, "D-STAR Uncovered," *QEX*, Nov/Dec 2008, pp 50-56.

Pearce, G., KN4AQ, "ICOM IC-92AD Dual-Band Hand Held Transceiver," Product Review, *QST*, Sep 2008, pp 39-43.

Pearce, G., KN4AQ, "Operating D-STAR," *QST*, Sep 2007, pp 30-33.

Texas Interconnect Team Web site — **www.k5tit.org**

The Repeater Builder's Technical Information Page — **www.repeaterbuilder.com**

Wilson, M., K1RO, ed., *The ARRL Operating Manual*, 9th ed. (ARRL: 2007). Chapter 2 covers FM, Repeaters, Digital Voice and Data.

Contents

19.1 Fundamentals of Radio Waves
- 19.1.1 Velocity
- 19.1.2 Free Space Attenuation and Absorption
- 19.1.3 Refraction
- 19.1.4 Scattering
- 19.1.5 Reflection
- 19.1.6 Knife-Edge Diffraction
- 19.1.7 Ground Wave

19.2 Sky-Wave Propagation and the Sun
- 19.2.1 Structure of the Earth's Atmosphere
- 19.2.2 The Ionosphere
- 19.2.3 Ionospheric Refraction
- 19.2.4 Maximum and Lowest Usable Frequencies
- 19.2.5 NVIS Propagation
- 19.2.6 Ionospheric Fading
- 19.2.7 Polarization at HF
- 19.2.8 The 11-Year Solar Cycle
- 19.2.9 The Sun's 27-Day Rotation
- 19.2.10 Disturbances to Propagation
- 19.2.11 D-Layer Propagation
- 19.2.12 E-Layer Propagation
- 19.2.13 F-Layer Propagation
- 19.2.14 Emerging Theories of HF and VHF Propagation

19.3 MUF Predictions
- 19.3.1 MUF Forecasts
- 19.3.2 Statistical Nature of Propagation Predictions
- 19.3.3 Direct Observation
- 19.3.4 WWV and WWVH
- 19.3.5 Beacons
- 19.3.6 Space Weather Information

19.4 Propagation in the Troposphere
- 19.4.1 Line of Sight
- 19.4.2 Tropospheric Scatter
- 19.4.3 Refraction and Ducting in the Troposphere
- 19.4.4 Tropospheric Fading

19.5 VHF/UHF Mobile Propagation
- 19.5.1 Rayleigh Fading
- 19.5.2 Multipath Propagation
- 19.5.3 Effect on the Receiver

19.6 Propagation for Space Communications
- 19.6.1 Faraday Rotation
- 19.6.2 Scintillation
- 19.6.3 Earth-Moon-Earth
- 19.6.4 Satellites

19.7 Noise and Propagation
- 19.7.1 Man-Made Noise
- 19.7.2 Lightning
- 19.7.3 Precipitation Static and Corona Discharge
- 19.7.4 Cosmic Sources

19.8 Glossary of Radio Propagation Terms

19.9 Further Reading

Chapter 19

Propagation of Radio Signals

Radio waves, like light waves — and all other forms of electromagnetic radiation, normally travel in straight lines. Obviously this does not happen all the time, because long-distance communication depends on radio waves traveling beyond the horizon. How radio waves propagate in other than straight-line paths is a complicated subject, but one that need not be a mystery. This chapter, by Emil Pocock, W3EP, with updates by Carl Luetzelschwab, K9LA, provides basic understanding of the principles of electromagnetic radiation, the structure of the Earth's atmosphere and solar-terrestrial interactions necessary for a working knowledge of radio propagation. The section on VHF/UHF mobile propagation was contributed by Alan Bloom, N1AL. More detailed discussions and the underlying mathematics of radio propagation physics can be found in the references listed at the end of this chapter.

19.1 Fundamentals of Radio Waves

Radio belongs to a family of electromagnetic radiation that includes infrared (radiation heat), visible light, ultraviolet, X-rays and the even shorter-wavelength gamma and cosmic rays. Radio has the longest wavelength and thus the lowest frequency of this group. See **Table 19.1**.

Electromagnetic waves are composed of an inter-related electric and magnetic field. The electric and magnetic components are oriented at right angles to each other and are also perpendicular to the direction of travel. The polarization of a radio wave is usually designated the same as the orientation of its electric field. This relationship can be visualized in **Fig 19.1**. Unlike sound waves or ocean waves, electromagnetic waves need no propagating medium, such as air or water. This property enables electromagnetic waves to travel through the vacuum of space.

19.1.1 Velocity

Radio waves, like all forms of electromagnetic radiation, travel nearly 300,000 km (186,400 mi) per second in a vacuum. Radio waves travel more slowly through any other medium. The decrease in speed through the atmosphere is so slight that it is usually ignored, but sometimes even this small difference is significant. The speed of a radio wave in a piece of wire, by contrast, is about 95% that in free space, and the speed can be even slower in other media.

The speed of a radio wave is always the product of wavelength and frequency, whatever the medium. That relationship can be stated simply as:

$$c = f \lambda$$

where
  c = speed in meters/second
  f = frequency in hertz
  λ = wavelength in meters

The *wavelength* (λ) of any radio frequency can be determined from this simple formula by rearranging the above equation to $\lambda = c/f$. For example, in free space the wavelength of a 30-MHz radio signal is thus 10 meters. A simplified equation in metric units is λ in meters = 300 divided by the frequency in MHz. Alternately in English units, λ in feet = 984 divided by the frequency in MHz.

Wavelength decreases in other media because the propagating speed is slower. In a piece

Table 19.1

The Electromagnetic Spectrum

Radiation	Frequency	Wavelength
X-ray	3×10^5 THz and higher	10 Å and shorter
Ultraviolet	800 THz - 3×10^5 THz	4000 - 10 Å
Visible light	400 THz - 800 THz	8000 - 4000 Å
Infrared	300 GHz - 400 THz	1 mm - 0.0008 mm
Radio	10 kHz - 300 GHz	30,000 km - 1 mm

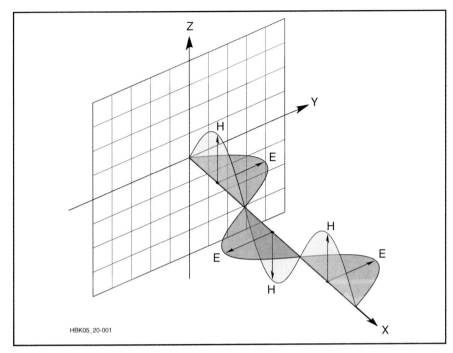

Fig 19.1 — Electric and magnetic field components of the electromagnetic wave. The polarization of a radio wave is the same direction as the plane of its electric field.

of wire, the wavelength of a 30-MHz signal shortens to about 9.5 meters. This factor must be taken into consideration in antenna designs, in transmission line designs, and in other applications.

19.1.2 Free Space Attenuation and Absorption

The intensity of a radio wave decreases as it travels. There are two mechanisms by which the intensity decreases: free space attenuation and absorption.

Free-space attenuation results from the dispersal (spherical spreading) of radio energy from its source. See **Fig 19.2**. Attenuation grows rapidly with distance because signal intensity decreases with the square of the distance traveled. If the distance between transmitter and receiver is increased from 1 km to 10 km (0.6 to 6 mi), the signal will be only one-hundredth as strong. Attenuation increases with frequency as well. Free space attenuation (path loss) can be expressed as

$$L_{fs} = 32.45 + 20 \log d + 20 \log f$$

where
L_{fs} = free space path loss in dB
d = distance in km
f = frequency in MHz

Free-space attenuation is a major factor governing signal strength, but radio signals undergo a variety of other losses as well. Energy is lost to *absorption* when radio waves travel through media other than a vacuum. Radio waves propagate through the atmosphere or solid material (like a wire) by exciting electrons, which then reradiate energy at the same frequency. This process is not perfectly efficient, so some radio energy is transformed into heat and retained by the medium. The amount of radio energy lost in this way depends on the characteristics of the medium and on the frequency. Attenuation in the atmosphere is minor from 10 MHz to 3 GHz, but at higher frequencies, absorption due to water vapor and oxygen can be high.

Radio energy is also lost during refraction, diffraction and reflection — the very phenomena that allow long-distance propagation. Indeed, any form of useful propagation is accompanied by attenuation. This may vary from the slight losses encountered by refraction from sporadic E clouds near the maximum usable frequency, to the more considerable losses involved with tropospheric forward *scatter* (not enough ionization for refraction or reflection, but enough to send weak electromagnetic waves off into varied directions) or D Layer absorption in the lower HF bands. These topics will be covered later. In many circumstances, total losses can become so great that radio signals become too weak for communication (they are below the sensitivity of a receiver).

19.1.3 Refraction

Electromagnetic waves travel in straight lines until they are deflected by something. Radio waves are *refracted*, or bent, slightly when traveling from one medium to another. Radio waves behave no differently from other familiar forms of electromagnetic radiation in this regard. The apparent bending of a pencil partially immersed in a glass of water demonstrates this principle quite dramatically.

Refraction is caused by a change in the velocity of a wave when it crosses the boundary between one propagating medium and another. If this transition is made at an angle, one portion of the wavefront slows down (or speeds up) before the other, thus bending the wave slightly. This is shown schematically in **Fig 19.3**.

The amount of bending increases with the ratio of the *refractive indices* of the two media. Refractive index is simply the velocity of a radio wave in free space divided by its velocity in the medium. The refractive properties of air may be calculated from temperature, moisture and atmospheric pressure. The index of refraction of air, at a very wide range of frequencies, may be calculated from:

$$N = \frac{77.6 \, p}{T} + \frac{3.73 \times 10^5 \, e}{T^2}$$

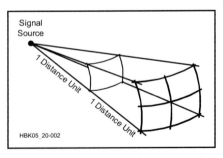

Fig 19.2 — Radio energy disperses as the square of the distance from its source. For the change of one distance unit shown the signal's power per unit of area is only one quarter as strong. Each spherical section has the same surface area.

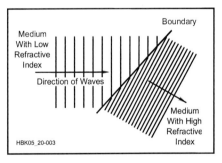

Fig 19.3 — Radio waves are refracted as they pass at an angle between dissimilar media. The lines represent the crests of a moving wave front and the distance between them is the wavelength. The direction of the wave changes because one end of the wave slows down before the other as it crosses the boundary between the two media. The wavelength is simultaneously shortened, but the wave frequency (number of crests that pass a certain point in a given unit of time) remains constant.

where
- N = index of refraction, N units (number of millionths by which the index of refraction exceeds 1.0)
- p = atmospheric pressure, millibars (mb)
- e = partial pressure of water vapor, millibars
- T = temperature, Kelvins

The refraction of radio signals is a function of the change in the index of refraction with altitude. N varies between 290 and 400 at the Earth's surface and normally diminishes with altitude at the rate of 40 N units per kilometer within the first few kilometers.

Radio waves are commonly refracted when they travel through different layers of the atmosphere, whether the highly charged ionospheric layers roughly 100 km (60 mi) and higher, or the weather-sensitive area near the Earth's surface. When the ratio of the refractive indices of two media is great enough, radio waves can be reflected, just like light waves striking a mirror. The Earth is a rather lossy reflector, but a metal surface works well if it is several wavelengths in diameter.

19.1.4 Scattering

The direction of radio waves can also be altered through *scattering*. The effect seen by a beam of light attempting to penetrate fog is a good example of light-wave scattering. Even on a clear night, a highly directional searchlight is visible due to a small amount of atmospheric scattering perpendicular to the beam. Radio waves are similarly scattered when they encounter randomly arranged objects of wavelength size or smaller, such as masses of electrons or water droplets. When the density of scattering objects becomes great enough, they behave more like a propagating medium with a characteristic refractive index.

If the scattering objects are arranged in some alignment or order, scattering takes place only at certain angles. A rainbow provides a good analogy for *field-aligned scattering* of light waves. The arc of a rainbow can be seen only at a precise angle away from the sun, while the colors result from the variance in scattering across the light-wave frequency range. Ionospheric electrons can be field-aligned by magnetic forces in auroras and under other unusual circumstances. Scattering in such cases is best perpendicular to the Earth's magnetic field lines.

19.1.5 Reflection

At amateur frequencies above 30 MHz, reflections from a variety of large objects, such as water towers, buildings, airplanes, mountains and the like, can provide a useful means of extending over-the-horizon paths several hundred km. Two stations need only beam toward a common reflector, whether stationary or moving.

Maximum range is limited by the radio line-of-sight distance of both stations to the reflector and by reflector size and shape. The reflectors must be many wavelengths in size and ideally have flat surfaces. Large airplanes make fair reflectors and may provide the best opportunity for long-distance contacts. The calculated limit for airplane reflections is 900 km (560 mi), assuming the largest jets fly no higher than 12,000 meters (40,000 ft), but actual airplane reflection contacts are likely to be considerably shorter.

19.1.6 Knife-Edge Diffraction

Radio waves can also pass behind solid objects with sharp upper edges, such as a mountain range, by *knife-edge diffraction*. This is a common natural phenomenon that affects light, sound, radio and other coherent waves, but it is difficult to comprehend. **Fig 19.4** depicts radio signals approaching an idealized knife-edge. The portion of the radio waves that strike the base of the knife-edge is entirely blocked, while that portion passing several wavelengths above the edge travel on relatively unaffected. It might seem at first glance that a knife-edge as large as a mountain, for example, would completely prevent radio signals from appearing on the other side but that is not quite true. Something quite unexpected happens to radio signals that pass just over a knife-edge.

Normally, radio signals along a wave front interfere with each other continuously as they propagate through unobstructed space, but the overall result is a uniformly expanding wave. When a portion of the wave front is blocked by a knife-edge, the resulting interference pattern is no longer uniform. This can be understood by visualizing the radio signals right at the knife-edge as if they constituted a new and separate transmitting point, but in-phase with the source wave at that point. The signals adjacent to the knife-edge still interact with signals passing above the edge, but they cannot interact with signals that have been obstructed below the edge. The resulting *interference pattern* no longer creates a uniformly expanding wave front, but rather appears as a pattern of alternating strong and weak bands of waves that spread in a nearly 180° arc behind the knife-edge.

The crest of a range of hills or mountains 50 to 100 wavelengths long can produce knife-edge diffraction at UHF and microwave frequencies. Hillcrests that are clearly defined and free of trees, buildings and other clutter make the best knife-edges, but even rounded hills may serve as a diffracting edge. Alternating bands of strong and weak signals, corresponding to the interference pattern, will

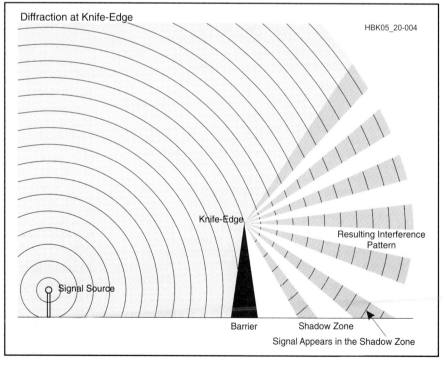

Fig 19.4 — VHF and UHF radio waves, light and other waves are diffracted around the sharp edge of a solid object that is large in terms of wavelengths. Diffraction results from interference between waves right at the knife-edge and those that are passing above it. Some signals appear behind the knife-edge as a consequence of the interference pattern. Hills or mountains can serve as natural knife-edges at radio frequencies.

Propagation Summary, by Band

OUR ONLY MEDIUM FREQUENCY (MF) BAND

1.8-2.0 MHz (160 meters)

Top band, as it is sometimes called, suffers from daytime D layer absorption. Daytime communication is limited to ground-wave coverage and a single E hop out to about 1500 km for well equipped stations (running the full legal limit, a quarter-wave vertical with a good ground system, and a low noise receiving environment). At night, the D layer quickly disappears and worldwide 160 meter communication becomes possible via F_2 layer skip and ducting. Atmospheric and man-made noise limits propagation. Tropical and mid latitude thunderstorms cause high levels of static in summer, making winter evenings the best time to work DX at 1.8 MHz. A proper choice of receiving antenna (Beverage, 4-square, small loop) can often significantly reduce the amount of received noise to improve the signal-to-noise ratio.

HIGH FREQUENCY (HF) BANDS (3-30 MHz)

A wide variety of propagation modes are useful on the HF bands. The lowest two bands in this range share many daytime characteristics with 160 meters. The transition between bands primarily useful at night or during the day appears around 10 MHz. Most long-distance contacts are made via F_2 layer skip. Above 21 MHz, more exotic propagation, including TE, sporadic E, aurora and meteor scatter, begins to be practical.

3.5-4.0 MHz (80 meters for the lower end, 75 meters for the higher end)

The lowest HF band is similar to 160 meters in many respects. Daytime absorption is significant, but not quite as extreme as at 1.8 MHz. At night, signals are often propagated halfway around the world. As at 1.8 MHz, atmospheric noise is a nuisance, making winter the most attractive season for the 80/75 meter DXer.

5.3-5.4 MHz (60 meters)

The distance covered during daytime propagation will fall in between that achievable on the 80 meter and 40 meter bands. At night, worldwide propagation is possible in spite of the relatively low power limit. Signal strengths will typically be higher than on 80 meters but not as high as on 40 meters.

7.0-7.3 MHz (40 meters)

The popular 40 meter band has a clearly defined skip zone during the day due to insufficient ionization to refract high angles. D layer absorption is not as severe as on the lower bands, so short-distance skip via the E and F layers is possible. During the day, a typical station can cover a radius of approximately 800 km (500 mi). At night, reliable worldwide communication via F_2 is common on the 40 meter band.

Atmospheric noise is much less troublesome than on 160 and 80 meters, and 40 meter DX signals are often of sufficient strength to override even high-level summer static. For these reasons, 40 meters is the lowest-frequency amateur band considered reliable for DX communication in all seasons. Even during the lowest point in the solar cycle, 40 meters may be open for worldwide DX throughout the night.

10.1-10.15 MHz (30 meters)

The 30 meter band is unique because it shares characteristics of both daytime and nighttime bands. D layer absorption is not a significant factor. Communication up to 3000 km (1900 mi) is typical during the daytime, and this extends halfway around the world via all-darkness paths. The band is generally open via F_2 on a 24-hour basis, but during a solar minimum, the MUF on some DX paths may drop below 10 MHz at night. Under these conditions, 30 meters adopts the characteristics of the daytime bands at 14 MHz and higher. The 30 meter band shows the least variation in conditions over the 11-year solar cycle, thus making it generally useful for long-distance communication anytime.

14.0-14.35 MHz (20 meters)

The 20 meter band is traditionally regarded as the amateurs' primary long-haul DX favorite. Regardless of the 11-year solar cycle, 20 meters can be depended on for at least a few hours of worldwide F_2 propagation during the day. During solar-maximum periods, 20 meters will often stay open to distant locations throughout the night. Skip distance is usually appreciable and is always present to some degree. Daytime E layer propagation may be detected along very short paths. Atmospheric noise is not a serious consideration, even in the summer. Because of its popularity, 20 meters tends to be very congested during the daylight hours.

18.068-18.168 MHz (17 meters)

The 17 meter band is similar to the 20 meter band in many respects, but the effects of fluctuating solar activity on F_2 propagation are more pronounced. During the years of high solar activity, 17 meters is reliable for daytime and early-evening long-range communication, often lasting well after sunset. During moderate years, the band may open only during sunlight hours and close shortly after sunset. At solar minimum, 17 meters will open to middle and equatorial latitudes, but only for short periods during midday on north-south paths.

21.0-21.45 MHz (15 meters)

The 15 meter band has long been considered a prime DX band during solar cycle maxima, but it is sensitive to changing solar activity. During peak years, 15 meters is reliable for daytime F_2 layer DXing and will often stay open well into the night. During periods of moderate solar activity, 15 meters is basically a daytime-only band, closing shortly after sunset. During solar minimum periods, 15 meters may not open at all except for infrequent north-south transequatorial circuits. Sporadic E is observed occasionally in early summer and midwinter, although this is not common and the effects are not as pronounced as on the higher frequencies.

24.89-24.99 MHz (12 meters)

This band offers propagation that combines the best of the 10 and 15 meter bands. Although 12 meters is primarily a daytime band during low and moderate sunspot years, it may stay open well after sunset during the solar maximum. During years of moderate solar activity, 12 meters opens to the low and middle latitudes during the daytime hours, but it seldom remains open after sunset. Periods of low solar activity seldom cause this band to go completely dead, except at higher latitudes. Occasional daytime openings, especially in the lower latitudes, are likely over north-south paths. The main sporadic E season on 24 MHz lasts from late spring through summer and short openings may be observed in mid-winter.

28.0-29.7 MHz (10 meters)

The 10 meter band is well known for extreme variations in characteristics and a variety of propagation modes. During solar maxima, long-distance F_2 propagation is so efficient that very low power can produce strong signals halfway around the globe. DX is abundant with modest equipment. Under these conditions, the band is usually open from sunrise to a few hours past sunset. During periods of moderate solar activity, 10 meters usually opens only to low and transequatorial latitudes around noon. During the solar minimum, there may be no F_2 propagation at any time during the day or night.

Sporadic E is fairly common on 10 m, especially May through August, although it may appear at any time. Short skip, as sporadic E is sometimes called on the HF bands, has little relation to the solar cycle and occurs regardless of F layer conditions. It provides single-hop communication from 300 to 2300 km (190 to 1400 mi) and multiple-hop opportunities of 4500 km (2800 mi) and farther.

Ten meters is a transitional band in that it also shares some of the propagation modes more characteristic of VHF. Meteor scatter, aurora, auroral E and transequatorial propagation provide the means of making contacts out to 2300 km (1400 mi) and farther, but these modes often go unnoticed at 28 MHz. Techniques similar to those used at VHF can be very effective on 10 meters, as signals are usually stronger and more persistent. These exotic modes can be more fully exploited, especially during the solar minimum when F_2 DXing has waned.

VERY HIGH FREQUENCY (VHF) BANDS (30-300 MHz)

A wide variety of propagation modes are useful in the VHF range. F layer skip appears on 50 MHz during solar cycle peaks. Sporadic E and several other E layer phenomena are most effective in the VHF range. Still other forms of VHF ionospheric propagation, such as field-aligned irregularities (FAI) and transequatorial propagation (TE), are rarely observed at VHF. Tropospheric propagation, which is not a factor at HF, becomes increasingly important above 50 MHz.

50-54 MHz (6 meters)

The lowest amateur VHF band shares many of the characteristics of both lower and higher frequencies. In the absence of any favorable ionospheric propagation conditions, well-equipped 50-MHz stations work regularly over a radius of 300 km (190 mi) via troposcatter, depending on terrain, power, receiver capabilities and antenna. Weak-signal troposcatter allows the best stations to make 500-km (310-mi) contacts nearly any time. Weather effects may extend the normal range by a few hundred km, especially during the summer months, but true tropospheric ducting is rare.

During the peak of the 11-year sunspot cycle (especially during the winter months), worldwide 50-MHz DX is possible via the F_2 layer during daylight hours. F_2 backscatter provides an additional propagation mode for contacts as far as 4000 km (2500 mi) when the MUF is just below 50 MHz. TE paths as long as 8000 km (5000 mi) across the magnetic equator are common around the spring and fall equinoxes of peak solar cycle years.

Sporadic E is probably the most common and certainly the most popular form of propagation on the 6 meter band. Single-hop E-skip openings may last many hours for contacts from 600 to 2300 km (370 to 1400 mi), primarily during the spring and early summer. Multiple-hop E_s provides transcontinental contacts several times a year, and contacts between the US and South America, Europe and Japan via multiple-hop E-skip occur nearly every summer.

Other types of E layer ionospheric propagation make 6 meters an exciting band. Maximum distances of about 2300 km (1400 mi) are typical for all types of E layer modes. Propagation via FAI often provides additional hours of contacts immediately following sporadic E events. Auroral propagation often makes its appearance in late afternoon when the geomagnetic field is disturbed. Closely related auroral E propagation may extend the 6 meter range to 4000 km (2500 mi) and sometimes farther across the northern states and Canada, usually after midnight. Meteor scatter provides brief contacts during the early morning hours, especially during one of the dozen or so prominent annual meteor showers.

144-148 MHz (2 meters)

Ionospheric effects are significantly reduced at 144 MHz, but they are far from absent. F layer propagation is unknown except for TE, which is responsible for the current 144-MHz terrestrial DX record of nearly 8000 km (5000 mi). Sporadic E occurs as high as 144 MHz less than a tenth as often as at 50 MHz, but the usual maximum single-hop distance is the same, about 2300 km (1400 mi). Multiple-hop sporadic E contacts greater than 3000 km (1900 mi) have occurred from time to time across the continental US, as well as across Southern Europe.

Auroral propagation is quite similar to that found at 50 MHz, except that signals are weaker and more Doppler-distorted. Auroral E contacts are rare. Meteor-scatter contacts are limited primarily to the periods of the great annual meteor showers and require much patience and operating skill. Contacts have been made via FAI on 144 MHz, but its potential has not been fully explored.

Tropospheric effects improve with increasing frequency, and 144 MHz is the lowest VHF band at which weather plays an important propagation role. Weather-induced enhancements may extend the normal 300- to 600-km (190- to 370-mi) range of well-equipped stations to 800 km (500 mi) and more, especially during the summer and early fall. Tropospheric ducting extends this range to 2000 km (1200 mi) and farther over the continent and at least to 4000 km (2500 mi) over some well-known all-water paths, such as that between California and Hawaii.

222-225 MHz (135 cm)

The 135-cm band shares many characteristics with the 2 meter band. The normal working range of 222-MHz stations is nearly as far as comparably equipped 144-MHz stations. The 135-cm band is slightly more sensitive to tropospheric effects, but ionospheric modes are more difficult to use. Auroral and meteor-scatter signals are somewhat weaker than at 144 MHz, and sporadic E contacts on 222 MHz are extremely rare. FAI and TE may also be well within the possibilities of 222 MHz, but reports of these modes on the 135-cm band are uncommon. Increased activity on 222 MHz will eventually reveal the extent of the propagation modes on the highest of the amateur VHF bands.

ULTRA-HIGH FREQUENCY (UHF) BANDS (300-3000 MHz) AND HIGHER

Tropospheric propagation dominates the bands at UHF and higher, although some forms of E layer propagation are still useful at 432 MHz. Above 10 GHz, atmospheric attenuation increasingly becomes the limiting factor over long-distance paths. Reflections from airplanes, mountains and other stationary objects may be useful adjuncts to propagation at 432 MHz and higher.

420-450 MHz (70 cm)

The lowest amateur UHF band marks the highest frequency on which ionospheric propagation is commonly observed. Auroral signals are weaker and more Doppler distorted; the range is usually less than at 144 or 222 MHz. Meteor scatter is much more difficult than on the lower bands, because bursts are significantly weaker and of much shorter duration. Although sporadic E and FAI are unknown as high as 432 MHz and probably impossible, TE may be possible.

Well-equipped 432-MHz stations can expect to work over a radius of at least 300 km (190 mi) in the absence of any propagation enhancement. Tropospheric refraction is more pronounced at 432 MHz and provides the most frequent and useful means of extended-range contacts. Tropospheric ducting supports contacts of 1500 km (930 mi) and farther over land. The current 432-MHz terrestrial DX record of more than 4000 km (2500 mi) was accomplished by ducting over water.

902-928 MHz (33 cm) and Higher

Ionospheric modes of propagation are nearly unknown in the bands above 902 MHz. Auroral scatter may be just within amateur capabilities at 902 MHz, but signal levels will be well below those at 432 MHz. Doppler shift and distortion will be considerable, and the signal bandwidth may be quite wide. No other ionospheric propagation modes are likely, although high-powered research radars have received echoes from auroras and meteors as high as 3 GHz.

Almost all extended-distance work in the UHF and microwave bands is accomplished with the aid of tropospheric enhancement. The frequencies above 902 MHz are very sensitive to changes in the weather. Tropospheric ducting occurs more frequently than in the VHF bands and the potential range is similar. At 1296 MHz, 2000-km (1200-mi) continental paths and 4000-km (2500-mi) paths between California and Hawaii have been spanned many times. Contacts of 1000 km (620 mi) have been made on all bands through 10 GHz in the US and over 1600 km (1000 mi) across the Mediterranean Sea. Well-equipped 903- and 1296-MHz stations can work reliably up to 300 km (190 mi), but normal working ranges generally shorten with increasing frequency.

Other tropospheric effects become evident in the GHz bands. Evaporation inversions, which form over very warm bodies of water, are usable at 3.3 GHz and higher. It is also possible to complete paths by scattering from rain, snow and hail in the lower GHz bands. Above 10 GHz, attenuation caused by atmospheric water vapor and oxygen become the most significant limiting factors in long-distance communication.

---

appear on the surface of the Earth behind the mountain, known as the *shadow zone*. The phenomenon is generally reciprocal, so that two-way communication can be established under optimal conditions. Knife-edge diffraction can make it possible to complete paths of 100 km or more that might otherwise be entirely obstructed by mountains or seemingly impossible terrain.

19.1.7 Ground Wave

A *ground wave* is the result of a special form of diffraction that primarily affects longer-wavelength vertically polarized radio waves. It is most apparent in the 80 and 160 meter amateur bands, where practical ground-wave distances may extend beyond 200 km (120 mi). It is also the primary mechanism used by AM broadcast stations in the medium-wave bands. The term ground wave is often mistakenly applied to any short-distance communication, but the actual mechanism is unique to the longer-wave bands.

Radio waves are bent slightly as they pass over a sharp edge, but the effect extends to edges that are considerably rounded. At medium and long wavelengths, the curvature of the Earth looks like a rounded edge. Bending results when the lower part of the wave front loses energy due to currents induced in the ground. This slows down the lower part of the wave, causing the entire wave to tilt forward slightly. This tilting follows the curvature of the Earth, thus allowing low- and medium-wave radio signals to propagate over distances well beyond line of sight.

Ground wave is most useful during the day at 1.8 and 3.5 MHz, when D layer absorption makes skywave propagation more difficult. Vertically polarized antennas with excellent ground systems provide the best results. Ground-wave losses are reduced considerably over saltwater and are worst over dry and rocky land.

19.2 Sky-Wave Propagation and the Sun

The Earth's atmosphere is composed primarily of nitrogen (78%), oxygen (21%) and argon (1%), with smaller amounts of a dozen other gases. Water vapor can account for as much as 5% of the atmosphere under certain conditions. This ratio of gases is maintained until an altitude of about 80 km (50 mi), when the mix begins to change. At the highest levels, helium and hydrogen predominate.

Solar radiation acts directly or indirectly on all levels of the atmosphere. Adjacent to the surface of the Earth, solar warming controls all aspects of the weather, powering wind, rain and other familiar phenomena. *Solar ultraviolet (UV) radiation* creates small concentrations of ozone (O_3) molecules between 10 and 50 km (6 and 30 mi). Most UV radiation is absorbed by this process and never reaches the Earth.

At even higher altitudes, UV and X-ray radiation partially ionize atmospheric gases. Electrons freed from gas atoms eventually recombine with positive ions to recreate neutral gas atoms, but this takes some time. In the low-pressure environment at the highest altitudes, atoms are spaced far apart and the gases may remain ionized for many hours. At lower altitudes, recombination happens rather quickly, and only constant radiation can keep any appreciable portion of the gas ionized.

19.2.1 Structure of the Earth's Atmosphere

The atmosphere, which reaches to more than 600 km (370 mi) altitude, is usually divided into a number of regions based on a transitioning characteristic of the atmosphere — like temperature. For propagation purposes, the important regions are shown in **Fig 19.5**. The weather-producing *troposphere* lies between the surface and an average altitude of 10 km (6 mi). Between 10 and 50 km (6 and 30 mi) are the *stratosphere* and the embedded *ozonosphere*, where ultraviolet absorbing ozone reaches its highest concentrations. About 99% of atmospheric gases are contained within these two lowest regions.

Above 50 km to about 600 km (370 mi) is the *ionosphere*, notable for its effects on radio propagation. At these altitudes, atomic oxygen, molecular oxygen, molecular nitrogen, and nitric oxide predominate under very low pressure and are the important species to consider for propagation. High-energy solar UV and X-ray radiation ionize these constituents, creating a broad region where ions are created in relative abundance. The ionosphere is subdivided into distinctive D, E and F regions.

The *magnetosphere* begins around 600 km (370 mi) and extends as far as 160,000 km (100,000 mi) into space. The predominant component of atmospheric gases gradually shifts from atomic oxygen, to helium and finally to hydrogen at the highest levels. The lighter gases may reach escape velocity or be swept off the atmosphere by the *solar wind* (electrically charged particles emitted by the sun and traveling through space). At about 3200 and 16,000 km (2000 and 9900 mi, respectively), the Earth's magnetic field traps energetic electrons and protons in two bands, known as the *Van Allen belts*. These have only a minor effect on terrestrial radio propagation.

19.2.2 The Ionosphere

The ionosphere plays a basic role in long-distance communications in all the amateur bands from 1.8 MHz to 30 MHz. The effects of the ionosphere are less apparent at the very high frequencies (30-300 MHz), but they persist at least through 432 MHz. As early as 1902, Oliver Heaviside and Arthur E. Kennelly independently suggested the existence of a layer in the upper atmosphere that could account for the long-distance radio transmissions made the previous year by Guglielmo Marconi and others. Edward Appleton confirmed the existence of the Kennelly-Heaviside layer in publications beginning in 1925 and used the letter E on his diagrams to designate the strength of the electric field of the waves that were apparently reflected from the layer he measured.

In late 1927 Appleton reported the existence of an additional layer in the "ionosphere." (Robert Watson-Watt coined the term "ionosphere", but it wasn't commonly used until 1932.) The additional higher-altitude layer was named "F" and subsequently another was termed "D." For a time during the 1930s, a "C" layer was proposed, and later discarded. Appleton was reluctant to alter this arbitrary nomenclature for fear of discovering yet other lower layers, so it has stuck to the present day. The basic physics of ionospheric propagation was largely worked out by the 1930s, yet both amateur and professional experimenters made further discoveries during the 1930s, 1940s and 1950s. Sporadic E, aurora, meteor scatter and several types of field-aligned scattering were among additional ionospheric phenomena that required explanation.

Although the term "layer" is used in this chapter, this could lead to the erroneous assumption that the ionosphere consists of distinct thin sheets separated by emptiness in between. This is not so, and as we'll see later in the chapter the ionosphere is a continuous electronic density versus altitude, with definite peaks and inflections points that define the D, E, and F regions. Studies have shown that the height of the various layers may also vary by latitude. Research into ionospheric physics is ongoing in an attempt to prove the existing theories and better understand the mechanism involved.

Fig 19.5 — Regions of the lower atmosphere and the ionosphere.

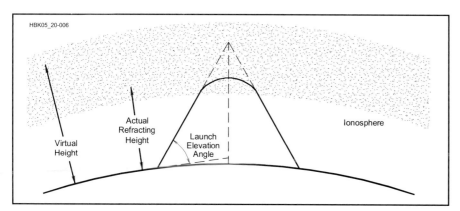

Fig 19.6 — Gradual refraction in the ionosphere allows radio signals to be propagated long distances. It is often convenient to imagine the process as a reflection with an imaginary reflection point at some virtual height above the actual refracting region. The other figures in this chapter show ray paths as equivalent reflections, but you should keep in mind that the actual process is a gradual refraction.

19.2.3 Ionospheric Refraction

The refractive index of an ionospheric layer decreases from a value of 1.00 as the density of free-moving electrons increases (this is opposite from the refractive index in the troposphere since the ionosphere is a dispersive medium). In the densest regions of the F layer, that density can reach a trillion electrons per cubic meter (10^{12} e/m³). Even at this high level, radio waves are refracted gradually over a considerable vertical distance, usually amounting to tens of km. Radio waves become useful for terrestrial propagation only when they are refracted enough to bring them back to Earth. See **Fig 19.6**.

Although refraction is the primary mechanism of ionospheric propagation, it is usually more convenient to think of the process as a reflection. The *virtual height* of an ionospheric layer is the equivalent altitude of a reflection that would produce the same effect as the actual refraction. The virtual height of any ionospheric layer can be determined using

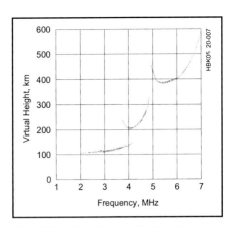

Fig 19.7 — Simplified vertical incidence ionogram showing echoes returned from the E, F_1 and F_2 layers. The critical frequencies of each layer (4.1, 4.8 and 6.8 MHz) can be read directly from the ionogram scale.

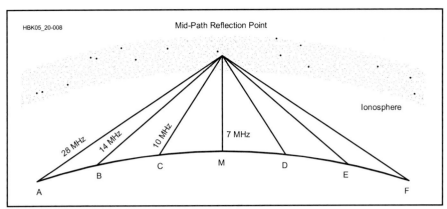

Fig 19.8 — The relationships between critical frequency, maximum usable frequency (MUF) and skip zone can be visualized in this simplified, hypothetical case. The critical frequency is 7 MHz, allowing frequencies below this to be used for short-distance ionospheric communication by stations in the vicinity of point M. These stations cannot communicate by the ionosphere at 14 MHz. Stations at points B and E (and beyond) can communicate because signals at this frequency are refracted back to Earth because they encounter the ionosphere at an oblique angle of incidence. At greater distances, higher frequencies can be used because the MUF is higher at the larger angles of incidence (low launch angles). In this figure, the MUF for the path between points A and F, with a small launch angle, is shown to be 28 MHz. Each pair of stations can communicate at frequencies at or below the MUF of the path between them, but not below the LUF — see text.

an ionospheric sounder, or *ionosonde*, a sort of vertically oriented radar. The ionosonde sends pulses that sweep over a wide frequency range, generally from 2 MHz to 20 MHz or higher, straight up into the ionosphere. The frequencies of any echoes are recorded against time and then plotted as distance on an *ionogram*. **Fig 19.7** depicts a simple ionogram. Real-time ionograms can be found online at the Digital Ionogram Database, sponsored by the University of Massachusetts Lowell (**um-lcar.uml.edu/DIDBase/**). For an extensive discussion of ionogram interpretation, download *UAG-23A: URSI Handbook of Ionogram Interpretation and Reduction* from the Australian IPS (Ionospheric Prediction Service) Web site at **www.ips.gov.au/IPSHosted/INAG/uag_23a/uag_23a.html**.

The highest frequency that returns echoes from the E and F regions at vertical incidence is known as the *vertical incidence* or *critical frequency*. (There is a D region critical frequency, but its value is well below 1 MHz and thus has minimal impact on our Amateur Radio bands.) The critical frequency is a function of ion density. The higher the ionization at a particular altitude, the higher becomes the critical frequency. Strictly speaking, the critical frequency is the term applicable to the peak electron density of a region. Physicists call any electron density in any part of the ionosphere a *plasma frequency*, because technically gases in the ionosphere are in a plasma, or partially ionized state. F layer critical frequencies commonly range from about 1 MHz to as high as 15 MHz.

19.2.4 Maximum and Lowest Usable Frequencies

When the frequency of a vertically incident signal is raised above the critical frequency of an ionospheric layer, that portion of the ionosphere is unable to refract the signal back to Earth. However, a signal above the critical frequency may be returned to Earth if it enters the layer at an *oblique angle*, rather than at vertical incidence. This is fortunate because it permits two widely separated stations to communicate on significantly higher frequencies than the critical frequency. See **Fig 19.8**.

The highest frequency supported by the ionosphere between two stations is the *maximum usable frequency* (MUF) for that path. If the separation between the stations is increased, a still higher frequency can be supported at lower launch angles. The MUF for this longer path is higher than the MUF for the shorter path because more refraction can occur for electromagnetic waves that encounter the ionosphere at more oblique angles. When the distance is increased to the maximum one-hop distance, the launch angle of the signals between the two stations is zero (that is, the ray path is tangential to the Earth at the two stations) and the MUF for this path is the highest that can be supported by that layer of the ionosphere at that location. This maximum distance is about 4000 km (2500 mi) for the F_2 layer and about 2000 km (1250 mi) for the E layer. See **Fig 19.9**.

The MUF is a function of path, time of day, season, location, solar UV and X-ray radiation levels and ionospheric disturbances. For vertically incident waves, the MUF is the same as the critical frequency. For path lengths at the limit of one-hop propagation,

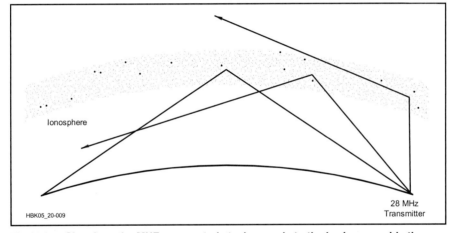

Fig 19.9 — Signals at the MUF propagated at a low angle to the horizon provide the longest possible one-hop distances. In this example, 28-MHz signals entering the ionosphere at higher angles are not refracted enough to bring them back to Earth.

Table 19.2

Maximum Usable Frequency Factors (M-factors) for 2000 km E Hops and 3000 km F Hops

Layer	Maximum Critical Frequency (MHz)	M-factor	Useful Operating Frequencies (MHz)
F_2	15.0	3.3-4.0	1-60
F_1*	5.5	4.0	10-20
E*	4.0	4.8	5-20
Es	30.0	5.3	20-160
D*	Not observed	—	None

*Daylight only

the MUF can be several times the critical frequency. The ratio between the MUF and the critical frequency is known as the *M-factor*.

The M-factor can be estimated using simple geometry in a spherical model of the Earth-ionosphere system. The angle of incidence on the ionosphere of an electromagnetic wave launched from the ground depends on the launch angle and the height of the ionospheric region. Due to the spherical geometry of the Earth-ionosphere system, the angle of incidence on the ionosphere does not approach zero as the launch angle approaches zero — it is limited to approximately 19° and 11° for the F_2 and E regions, respectively, which then limits the M-factor to approximately 3 and 5 for these two regions (from the equation MUF = 1 / the sine of the angle of incidence on the ionosphere). See **Table 19.2** for typical M-factors of the various regions.

The term *skip zone* is closely related to MUF. When two stations are unable to communicate with each other on a particular frequency because the ionosphere is unable to refract the signal enough from one to the other through the required angle — that is, the operating frequency is above the MUF — the stations are said to be in the skip zone for that frequency. Stations within the skip zone may be able to work each other on a lower frequency, or by ground wave or other mechanisms if they are close enough. There is no skip zone at frequencies below the MUF.

The MUF at any time on a particular path is just that — the *maximum* usable frequency. Frequencies below the MUF will also propagate along the path, but ionospheric absorption and noise at the receiving location (due to man-made noise and/or noise from local or distant thunderstorms) may make the received signal-to-noise ratio too low to be usable. In this case, the frequency is said to be below the *lowest usable frequency* (LUF). This occurs most frequently below 10 MHz, where atmospheric and man-made noises are most troublesome.

The LUF can be lowered somewhat by the use of high power and directive antennas, or through the use of communication modes that permit reduced receiver bandwidth or are less demanding of SNR — CW or PSK31 instead of SSB, for example. This is not true of the MUF, which is limited by the physics of ionospheric refraction, no matter how high your transmitter power or how narrow your receiver bandwidth. The LUF can be higher than the MUF. This is a common occurrence on 160 meters during the day due to too much absorption; another scenario would be too much noise on the higher bands due to thunderstorm activity or man-made noise. When the LUF is higher than the MUF, there is no frequency that supports communication on the particular path at that time.

19.2.5 NVIS Propagation

In the previous section, the statement was made that *stations within the skip zone may be able to work each other at a lower frequency, or by ground wave if they are close enough.* This statement summarizes the purpose of *Near Vertical Incidence Skywave* (*NVIS*) propagation — to bridge the gap between where ground wave is too weak and where the skip zone ends. By going to lower frequencies, communications can be maintained over these relatively short distances.

Propagation over short distances means high elevation angles — this is not DXing in which lower angles in general are most effective. For example, a path from San Francisco to Sacramento is 75 miles, and requires an average elevation angle of 78 degrees. To radiate maximum energy at these higher angles, relatively low height antennas need to be used.

In the December 2005 issue of *QST*, Dean Straw N6BV used the *VOACAP* propagation prediction program to analyze a variety of NVIS paths centered on San Francisco. His analysis showed area coverage maps (signal strength contours versus distance from the transmitter) and elevation patterns of antennas at various height. His analysis allowed him to formulate a very nice summary: *As a rule-of-thumb, for ham band NVIS, I would recommend that 40 meters be used during the day; 80 meters during the night.*

19.2.6 Ionospheric Fading

HF signal strengths typically rise and fall over periods of a few seconds to several minutes, and rarely hold at a constant level for very long. Fading is generally caused by the interaction of several radio waves from the same source arriving along different propagation paths. Waves that arrive in-phase combine to produce a stronger signal, while those out-of-phase cause destructive interference and lower net signal strength. Short-term variations in ionospheric conditions may change individual path lengths or signal strengths enough to cause fading. Even signals that arrive primarily over a single path may vary as the propagating medium changes. Fading may be most notable at sunrise and sunset, especially near the MUF, when the ionosphere undergoes dramatic transformations. Other ionospheric traumas, such as auroras and geomagnetic storms, also produce severe forms of HF fading.

19.2.7 Polarization at HF

Although the ionosphere varies on a short-term basis and results in somewhat random polarization, there is more order to polarization than realized. The ionosphere is immersed in the Earth's magnetic field, and the result of an electromagnetic wave propagating through an ionized medium (called a plasma, which is what our ionosphere is) immersed in a magnetic field is the propagation of two characteristic waves: the ordinary wave (O-wave) and the extraordinary wave (X-wave). Ordinary and extraordinary are terms borrowed from the science of optics. These two waves are generally elliptically polarized (the tip of the polarization vector traces out an ellipse), and are orthogonal to each other.

Upon entering the ionosphere, and depending on the location and the heading, a linearly polarized wave (from our horizontal or vertical antenna) will either couple all of its energy into the O-wave, all of its energy into the X-wave, or divide its energy into the O-wave and X-wave. Our horizontal antenna will couple best into one characteristic wave, and our vertical antenna will couple best into the other characteristic wave. The same coupling issue is present when the characteristic waves exit the ionosphere. The bottom line is that on the HF bands of 80 meters and higher, both characteristic waves generally propagate with similar absorption, and thus the use of a vertical or horizontal antenna is not too critical with respect to polarization. One of the characteristic waves will couple into whichever antenna is being used.

On 160 meters, though, the X-wave is heavily absorbed, leaving the O-wave. In general, vertical polarization is best for those of us in North America and at high northern latitudes. However, 160 meter operators have observed that under some conditions, horizontally polarized antennas outperform vertically polarized antennas.

19.2.8 The 11-Year Solar Cycle

The density of ionospheric layers depends on the amount of solar radiation reaching the Earth, but solar radiation is not constant. Variations result from daily and seasonal motions of the Earth, the sun's own 27-day rotation and the 11-year cycle of solar activity. One visual indicator of both the sun's rotation and the solar cycle is the periodic appearance of dark spots on the sun, which have been observed continuously since the mid-18th century. On average, the number of *sunspots* reaches a maximum every 10.7 years, but the period has varied between 7 and 17 years. Cycle 19 peaked in 1958, with a smoothed sunspot number of 201, the highest recorded to date. **Fig 19.10** shows average monthly sunspot numbers for the past five cycles.

Sunspots are cooler areas on the sun's surface associated with high magnetic activity. Active regions adjacent to sunspot groups, called *plages*, are capable of producing great flares and sustained bursts of radiation in the radio through X-ray spectrum. During the peak of the 11-year solar cycle, average solar radiation increases along with the number of flares and sunspots. The ionosphere becomes more intensely ionized as a consequence, resulting in higher critical frequencies, particularly in the F_2 layer. The possibilities for long-distance communications are considerably improved during solar maxima, especially in the higher-frequency bands.

One key to forecasting F layer critical frequencies, and thus long-distance propagation, is the intensity of ionizing UV and X-ray radiation. Until the advent of satellites, UV and X-ray radiation could not be measured directly, because they were almost entirely absorbed in the upper atmosphere during the process of ionization). The sunspot number provided the most convenient approximation of general solar activity. The sunspot number is not a simple count of the number of visual spots, but rather the result of a complicated formula that takes into consideration size, number and grouping. The smoothed sunspot number (a running average of monthly mean sunspot numbers from 6 months before to 6 months after the desired month) varies from near zero during the solar-cycle minimum to over 200.

Another method of gauging solar activity is the *solar flux*, which is a measure of the intensity of 2800-MHz (10.7-cm) radio noise coming from the sun (10^{-22} W m^{-2} Hz^{-1}). The smoothed 2800-MHz radio flux correlates well with the intensity of ionizing UV and X-ray radiation and provides a convenient alternative to sunspot numbers. Solar flux values commonly vary on a scale of 60-300 and can be related to sunspot numbers, as shown in **Fig 19.11**. The Dominion Radio Astrophysical Observatory, Penticton, British Columbia, measures the 2800-MHz solar flux daily at local noon. Radio station WWV broadcasts the latest solar-flux index at 18 minutes after each hour; WWVH does the same at 45 minutes after the hour. Solar flux and other useful data is also available online at **www.swpc.noaa.gov**.

The Penticton solar flux is employed in a wide variety of other applications. Daily, weekly, monthly and even 13-month smoothed average solar flux readings are commonly used in propagation predictions (but be aware that our HF propagation predictions are based on the correlation between smoothed solar flux or smoothed sunspot number and monthly median ionospheric parameters — more on this later in the chapter).

High flux values generally result in higher MUFs, but the actual procedures for predicting the MUF at any given hour and path is quite complicated. Solar flux is not the sole determinant, as the angle of the sun to the Earth, season, time of day, exact location of the radio path and other factors must all be taken into account. MUF forecasting a few days or months ahead involves additional variables and even more uncertainties.

TRENDS IN SOLAR CYCLES

If one looks at all 23 recorded solar cycles,

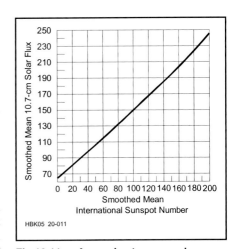

Fig 19.11 — Approximate conversion between solar flux and sunspot number. Note that this is for smoothed values of Solar Flux and Sunspot Numbers.

three characteristics stand out. First, we see a cyclic nature to the maximum smoothed sunspot numbers. Second, we've been through three high cycle periods of 50 years or so (consisting of several solar cycles) and two low cycle periods of 50 years or so (again, consisting of several solar cycles), and we appear to be headed into a third low cycle period. Third, we've lived through the highest period of high cycle activity, which has allowed excellent worldwide propagation on the higher frequency bands and provided great enjoyment for radio amateurs.

If we look at solar cycles prior to recorded history through various proxies for solar activity (for example, carbon-14 in trees and beryllium-10 in ice cores), we'll see extended periods of very low solar activity referred to as Grand Minima (for example, the Maunder Minimum between 1640 and 1710 AD). It's likely that we'll again enter one of these extended periods, but trying to predict when this will happen is, at best, a wild guess.

With respect to solar minimum periods between solar cycles, historical data shows a great variation. The average length of solar minimum, for example defined as the number of months in which the smoothed sunspot number is below 20, is around 37 months. Using this definition, the shortest minimum was 17 months (between Cycles 1 and 2) and the longest minimum was 96 months (between Cycles 5 and 6). The minimum period between Cycle 23 and Cycle 24 looks like it will end up at around the average value of 37 months. Interestingly, up until the minimum between Cycle 23 and Cycle 24, in our lifetimes we have experienced minimum periods of approximately 24 months — much shorter than the average. This leads us to believe this solar minimum period is unusual. But histori-

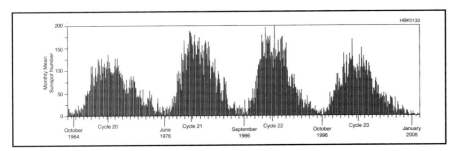

Fig 19.10 — Average monthly sunspot numbers for Solar Cycles 19 to 23 (through January 2009).

Status of the Present Solar Minimum and Cycle 24

This information was prepared in March 2010.

Smoothed sunspot data through August 2009 strongly suggests the longest solar minimum period of our lifetimes is over. The plot in **Fig 19A1** compares the present solar minimum period and the previous five solar minimum periods.

The previous five solar minimum periods were below a smoothed sunspot number of 20 for only about two years. This is what we came to expect for this current solar minimum period. Unfortunately that did not happen, much to the dismay of many solar scientists and Amateur Radio operators.

Using the predictions at **www.swpc.noaa.gov/ftpdir/weekly/Predict.txt**, it appears that this solar minimum will be below a smoothed sunspot number of 20 for just over four years. Interestingly, when compared to *all* solar minimum periods, this solar minimum period is not that unusual as the average time a solar minimum period spent below a smoothed sunspot number of 20 is just over three years.

Cycle 24 is predicted to rise to a maximum smoothed sunspot number of around 90 in early 2013 as seen at **www.swpc.noaa.gov/SolarCycle**. This corresponds to a smoothed 10.7 cm solar flux of around 140. If this prediction is accurate, Cycle 24 will be a slightly lower-than-average solar cycle.

Since the smoothed sunspot number (or smoothed 10.7 cm solar flux) is a heavily averaged measurement, it is six months behind the current month (that's why the latest smoothed data is August 2009 even though we have monthly sunspot data for February 2010). **Fig 19A2** shows monthly mean (average) sunspot data, which includes data from January 2006 through February 2010.

Note the overlap of Cycle 23 sunspots and Cycle 24 sunspots for about 16 months (January 2008 through April 2009). This overlap is typical of a solar minimum period. More importantly, Cycle 24 sunspots continue to ascend in the monthly data. This of course translates to a continued ascent in the smoothed data.

The current projections mean propagation on our higher HF bands (15, 12, and 10 meters) will be similar to, or slightly poorer than, propagation on these bands as experienced during Cycle 23's solar maximum period (1999 to 2003). It should be noted that several scientists are still predicting a Cycle 24 being 30 to 50% higher than Cycle 23. Because of our limited data (only 23 samples from 23 solar cycles) and our lack of understanding of solar processes, only time will tell which predictions are the most accurate. — *Carl Luetzelschwab, K9LA*

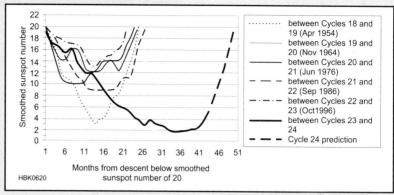

Fig 19A1 — The present solar minimum compared to the previous five solar minumums.

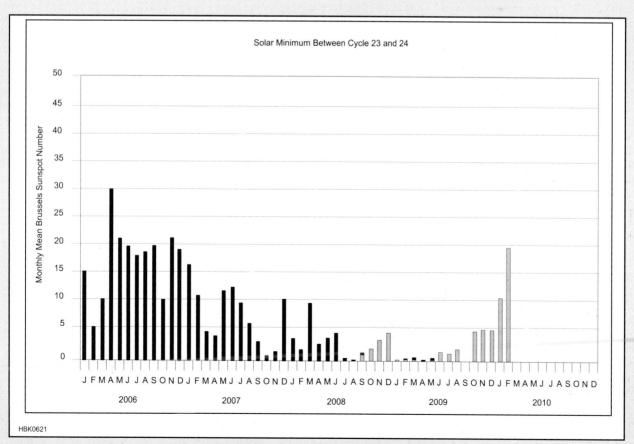

Fig 19A2 — Monthly mean (average) sunspot data January 2006 through February 2010.

cal data, with its great amount of variation, says otherwise.

19.2.9 The Sun's 27-Day Rotation

Sunspot observations also reveal that the sun rotates on its own axis. The sun is composed of extremely hot gases and does not turn uniformly (it is essentially a fluid). At the equator, the period is just over 25 days, but it approaches 35 days at the poles. Sunspots that affect the Earth's ionosphere, which appear almost entirely within 35° of the sun's equator, take about 26 days for one rotation. After taking into account the Earth's movement around the sun, the apparent period of solar rotation is about 27 days.

Active regions must face the Earth in the proper orientation to have an impact on the ionosphere. They may face the Earth only once before rotating out of view, but they often persist for several solar rotations. The net effect is that solar activity often appears in 27-day cycles corresponding to the sun's rotation, even though the active regions themselves may last for several solar rotations.

19.2.10 Disturbances to Propagation

Like a campfire that occasionally spits out a flaming ember, our sun sometimes erupts spasmodically — but on a much grander scale than a summer campfire here on Earth. After all, any event that violently releases as much as 10 billion tons of solar material traveling up to four and a half million miles per hour or releases large amounts of electromagnetic radiation at extremely short wavelengths has to be considered pretty impressive!

Following the lead of the Space Weather Prediction Center, there are three types of disturbances to propagation: geomagnetic storms (designated G), solar radiation storms (designated S) and radio blackouts (designated R). For more details and the scaling associated with these designators, visit **www.swpc.noaa.gov/NOAAscales/**.

GEOMAGNETIC STORMS

Geomagnetic storms are generally caused by *coronal mass ejections* and high speed wind streams from *coronal holes*. A *coronal mass ejection* (CME) originates in the sun's outer atmosphere, its corona. With several sophisticated satellites launched in the mid 1990s, we have gained powerful new tools to monitor the intricacies of solar activity. Using the latest satellite technology (and also some re-engineered earthbound instruments), scientists have observed many CMEs, greatly expanding our knowledge about them. Previously, the only direct observations we had of coronal activity were during solar eclipses — and eclipses don't occur very often. CMEs are observed with an instrument called a *coronagraph*, which has an occulting disk to block out the main portion of the sun. In essence a coronagraph creates an artificial eclipse.

A *coronal hole* is a region on the sun where the magnetic field is open to the interplanetary magnetic field (IMF) and ionized particles can escape into the solar wind. Normally the solar wind blows at 400 km per second. During CMEs and coronal holes, solar wind speeds can increase to 2000 km per second.

Coronal holes and coronal mass ejections that are Earth-directed (these are also called full halo CMEs, as the explosion surrounds the occulting disk of a coronagraph) concurrent with the IMF oriented in a southerly direction result in the most disturbance to propagation. It usually takes up to a couple days for a CME or the effects of a coronal hole to reach and impact the Earth's ionosphere, so this generally gives us ample warning of the impending disturbance.

SOLAR RADIATION STORMS AND RADIO BLACKOUTS

Solar radiation storms and radio blackouts are caused by large *solar flares*. When a large solar flare erupts from the sun's surface, it can launch out into space a wide spectrum of electromagnetic energy. Since electromagnetic energy travels at the speed of light, the first indication of a solar flare reaches the Earth in about eight minutes. A large flare shows up as an increase in visible brightness near a sunspot group, accompanied by increases in UV and X-ray radiation and high levels of noise in the VHF radio bands. It is the X-ray radiation that results in radio blackouts on the daytime side of the Earth due to increased D region absorption, and this is called a *sudden ionospheric disturbance* (SID). The lower frequencies are affected for the longest period. In extreme cases, nearly all background noise will be gone as well. SIDs may last up to an hour, after which ionospheric conditions return to normal.

A large solar flare can also release matter into space, mainly in the form of very energetic protons. These cause solar radiation storms, whereby increased absorption in the polar cap (that area inside the auroral oval) degrades over-the-pole paths. This is called a *polar cap absorption* (PCA) event. A PCA event may last for days, dramatically affecting transpolar HF propagation. An interesting fact with respect to PCAs is that they do not necessarily affect the northern and southern polar regions similarly. Thus if the short path between two points is degraded over one pole, the long path may still be available over the other pole.

At one time, scientists believed that solar flares and CMEs were causally related, but now they recognize that many CMEs occur without an accompanying flare. And while many flares do result in an ejection of some solar material, many do not. It now seems clear that flares don't cause CMEs and vice versa.

Since geomagnetic storms have the most impact to propagation, it is instructive to understand their occurrence statistics. The number of geomagnetic storms varies considerably from year to year, but peak geomagnetic activity follows the peak of solar activity. See **Fig 19.12**. Also, geomagnetic activity affects the ionosphere mostly in the equinox months (March and September).

MONITORING GEOMAGNETIC ACTIVITY

Geomagnetic activity is monitored by devices known as *magnetometers*. These may be as simple as a magnetic compass rigged to record its movements. Small variations in the geomagnetic field are scaled to two measures known as the K and A indices. The *K* index provides an indication of magnetic activity on a finite logarithmic scale of 0-9, and it is updated every three hours. Very quiet conditions are reported as 0 or 1, while geomagnetic storm levels begin at 4. See **Table 19.3** for the latest NOAA descriptions of geomagnetic storms.

A worldwide network of magnetometers constantly monitors the Earth's magnetic field, because the Earth's magnetic field varies with location. K indices that indicate average planetary conditions are indicated as K_p. Daily geomagnetic conditions are also summarized by the open-ended linear A index, which corresponds roughly to the cumulative K index values (it's the daily average of the eight K indices after converting the K indices to a linear scale). The A index commonly varies between 0 and 30 during quiet to active conditions, and up to 100 or higher during geomagnetic storms.

At 18 minutes past the hour, radio stations

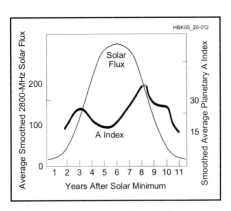

Fig 19.12 — Geomagnetic activity (measured as the A index) also follows an 11-year cycle. Average values over the past few cycles show that geomagnetic activity peaks before and after the peak of solar flux.

Table 19.3
Geomagnetic Storms

Typical Kp	Description	Days per Solar Cycle
9	Extreme	4
8	Severe	60
7	Strong	130
6	Moderate	360
5	Minor	900

WWV and WWVH broadcast the latest solar flux number, the average planetary A Index and the latest mid latitude K Index. In addition, they broadcast a descriptive account of the condition of the geomagnetic field and a forecast for the next three hours. You should keep in mind that the A Index is a description of what happened yesterday. Strictly speaking, the K Index is valid only for mid latitudes. However, the trend of the K Index is very important for propagation analysis and forecasting. A rising K foretells worsening HF propagation conditions, particularly for transpolar paths. At the same time, a rising K alerts VHF operators to the possibility of enhanced auroral activity, particularly when the K Index rises above 3. Another source of useful information about solar disturbances is **www.spaceweather.com**.

19.2.11 D Layer Propagation

The *D layer* is the lowest region of the ionosphere, situated between 55 and 90 km (30 and 60 mi). See **Fig 19.13**. It is ionized primarily by the strong emission of solar hydrogen at 121.5 nanometers and short X-rays, both of which penetrate through the upper atmosphere to ionize nitric oxide and all other constituents, respectively. The D layer exists only during daylight, because constant radiation is needed to replenish ions that quickly recombine into neutral molecules. The D layer abruptly disappears at night so far as amateur MF and HF signals are concerned. D layer ionization varies a small amount over the solar cycle. It is unsuitable as a refracting medium for HF radio signals (but it is very important for VLF signals).

DAYTIME D LAYER ABSORPTION

Nevertheless, the D layer plays an important role in HF communications. During daylight hours, radio energy as high as 5 MHz is effectively absorbed by the D layer, severely limiting the range of daytime 1.8- and 3.5-MHz signals. Signals at 7 MHz and 10 MHz pass through the D layer and on to the E and F layers only at relatively high angles. Low-angle waves, which must travel a much longer distance through the D layer, are subject to greater absorption. As the frequency increases above 10 MHz, radio waves pass through the D layer with increasing ease (less absorption).

NIGHTTIME D LAYER

D layer ionization falls 100-fold as soon as the sun sets and the source of ionizing radiation is removed. Low-band HF signals are then free to pass through to the E layer (also greatly diminished at night) and on to the F layer, where the MUF is almost always high enough to propagate 1.8- and 3.5-MHz signals half way around the world. Long-distance propagation at 7 and 10 MHz generally improves at night as well, because absorption is less and low-angle waves are able to reach the F layer.

D LAYER IONOSPHERIC FORWARD SCATTER

Radio signals in the 25-100 MHz range can be scattered by ionospheric irregularities, turbulence and stratification in the D and lower reaches of the E layers. Signals propagated by ionospheric forward scatter undergo very high losses, so signals are apt to be very weak. Typical scatter distances at 50 MHz are 800-1500 km (500-930 mi). This is not a common mode of propagation, but under certain conditions, ionospheric forward scatter can be very useful.

Ionospheric forward scatter is best during daylight hours from 10 AM to 2 PM local time, when the sun is highest in the sky and D layer ionization peaks. It is worst at night. Scattering may be marginally more effective during the summer and during the solar cycle maximum due to somewhat higher D layer ionization. The maximum path length of less than 2000 km (1200 mi) is limited by the height of the scattering region, which is centered about 70 km (40 mi). Ionospheric scatter signals are typically weak, fluttery and near the noise level. Ionization from meteors sometimes temporarily raises signals well out of the noise for up to a few seconds at a time.

This mode may find its greatest use when all other forms of propagation are absent, primarily because ionospheric scatter signals are so weak. For best results at 28 and 50 MHz, a 3-element Yagi or larger, several hundred watts of power and a sensitive receiver are required. The paths are direct (as opposed to skewed, which implies a path not following a great circle path). CW is preferred, although under optimal conditions ionospheric scatter signals may be consistent enough to support SSB communications. Scattering is not efficient below 25 MHz. The very best-equipped pairs of 144-MHz stations may also be able to complete ionospheric scatter contacts.

19.2.12 E Layer Propagation

The *E layer* lies between 90 and 150 km (60 and 90 mi) altitude, but a narrower region centered at 95 to 120 km (60 to 70 mi) is more important for radio propagation. In the E layer nitric oxide and molecular oxygen are ionized by short-wavelength UV and long-wavelength X-ray radiation. The normal E layer exists primarily during daylight hours, because like the D layer, it requires a constant source of ionizing radiation. Recombination is not as fast as in the denser D layer and absorption is much less. The E layer has a daytime critical frequency that varies between 3 and 4 MHz with the solar cycle. At night, the normal E layer decays to a minimum critical frequency of about 0.5 MHz, which is still enough to refract low elevation angle 1.8-MHz signals.

DAYTIME E LAYER

The E layer plays a small role in propagating HF signals but can be a major factor limiting propagation during daytime hours. Its usual critical frequency of 3 to 4 MHz, with a maximum M-factor of about 4.8, suggests that single-hop E layer skip might be useful between 5 and 20 MHz at distances up to 2300 km (1400 mi). In practice this is not the case, because the potential for E layer skip is severely limited by D layer absorption. Signals radiated at low angles at 7 and 10 MHz, which might be useful for the longest-distance contacts, are largely absorbed by the D layer. Only high-angle signals pass through the D layer at these frequencies, but high-angle E layer skip is typically limited to 1200 km (750 mi) or so. Signals at 14 MHz penetrate the D layer at lower angles at the cost of some absorption, but the casual operator may not be able to distinguish between signals propagated by the E layer or higher-angle F layer propagation.

An astonishing variety of other propagation modes find their home in the E layer, and this

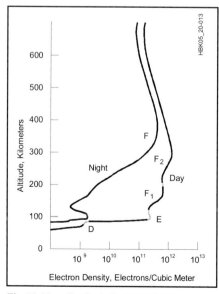

Fig 19.13 — Typical electron densities for the various ionospheric regions.

perhaps more than makes up for its ordinary limitations. Each of these other modes — sporadic E, field-aligned irregularities, aurora, auroral E and meteor scatter — are related forms of E layer propagation with unique characteristics. They are primarily useful only on the highest HF and lower VHF bands.

Sporadic E

Short skip, long familiar on the 10 meter band during the summer months, affects the VHF bands as high as 222 MHz. *Sporadic E* (E_s), as this phenomenon is properly called, commonly propagates 28, 50 and 144-MHz radio signals between 500 and 2300 km (300 and 1400 mi). Signals are apt to be exceedingly strong, allowing even modest stations to make E_s contacts. At 21 MHz, the skip distance may only be a few hundred km. During the most intense E_s events, skip may shorten to less than 200 km (120 mi) on the 10 meter band and disappear entirely on 15 meters. Multiple-hop E_s has supported contacts up to 10,000 km (6200 mi) on 28 and 50 MHz and more than 3000 km (1900 mi) on 144 MHz. The first confirmed 220-MHz E_s contact was made in June 1987, but such contacts are likely to remain very rare.

Sporadic E at mid latitudes (roughly 15° to 45°) may occur at any time, but it is most common in the Northern Hemisphere during May, June and July, with a less-intense season at the end of December and early January. Its appearance is independent of the solar cycle. Sporadic E propagation is most likely to occur from 9 AM to noon local time and again early in the evening between 5 PM and 8 PM. Mid latitude E_s events may last only a few minutes or can persist for many hours. In contrast, sporadic E is an almost constant feature of the polar regions at night and the equatorial belt during the day.

Efforts to predict mid latitude E_s have not been successful, probably because its causes are complex and not well understood. Studies have demonstrated that thin and unusually dense patches of ionization in the E layer, between 100 and 110 km (60 and 70 mi) altitude and 10 to 100 km (6 to 60 mi) in extent, are responsible for most E_s reflections. Sporadic E clouds may form suddenly, move quickly from their birthplace, and dissipate within a few hours. Professional studies have recently focused on the role of heavy metal ions, probably of meteoric origin, and wind shears as two key factors in creating the dense patchy regions of E layer ionization.

Sporadic E clouds exhibit an MUF that can rise from 28 MHz through the 50-MHz band and higher in just a few minutes. When the skip distance on 28 MHz is as short as 400 or 500 km (250 or 310 mi), it is an indication that the MUF has reached 50 MHz for longer paths at low launch angles. Contacts at the maximum one-hop sporadic E distance, about 2300 km (1400 mi), should then be possible at 50 MHz. *E-skip* (yet another term for sporadic E) contacts as short as 700 km (435 mi) on 50 MHz, in turn, may indicate that 144-MHz contacts in the 2300-km (1400 mi) range can be completed. See **Fig 19.14**. Sporadic E openings occur about a tenth as often at 144 MHz as they do at 50 MHz and for much shorter periods.

Sporadic E can also have a detrimental effect on HF propagation by masking the F_2 layer from below. HF signals may be prevented from reaching the higher levels of the ionosphere and the possibilities of long F_2 skip. Reflections from the tops of sporadic E clouds can also have a masking effect, but they may also lengthen the F_2 propagation path with a top-side intermediate hop that never reaches the Earth.

E LAYER FIELD-ALIGNED IRREGULARITIES

Amateurs have experimented with a little-known scattering mode known as *field-aligned irregularities* (FAI) at 50 and 144 MHz since 1978. FAI commonly appears directly after sporadic E events and may persist for several hours. Oblique-angle scattering becomes possible when electrons are compressed together due to the action of high-velocity ionospheric acoustic (sound) waves. The resulting irregularities in the distribution of free electrons are aligned parallel to the Earth's magnetic field, in something like moving vertical rods. A similar process of electron field-alignment takes place during radio aurora, making the two phenomena quite similar.

Most reports suggest that 8 PM to midnight may be the most productive time for FAI. Stations attempting FAI contacts point their antennas toward a common scattering region that corresponds to an active or recent E_s reflection point. The best direction must be probed experimentally, for the result is rarely along the great-circle path. Stations in south Florida, for example, have completed 144-MHz FAI contacts with north Texas when participating stations were beamed toward a common scattering region over northern Alabama.

FAI-propagated signals are weak and fluttery, reminiscent of aurora signals. Doppler shifts of as much as 3 kHz have been observed in some tests. Stations running as little as 100 W and a single Yagi should be able to complete FAI contacts during the most favorable times, but higher power and larger antennas may yield better results. Contacts

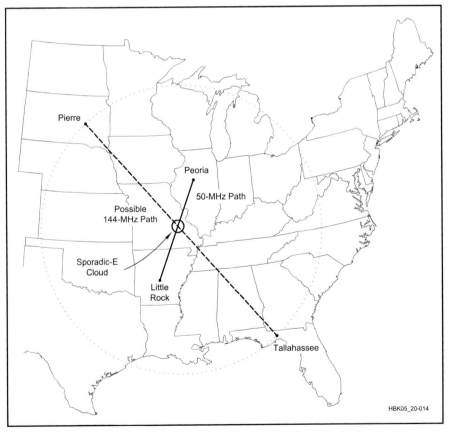

Fig 19.14 — 50 MHz sporadic E contacts of 700 km (435 mi) or shorter (such as between Peoria and Little Rock) indicate that the MUF on longer paths is above 144 MHz. Using the same sporadic E region reflecting point, 144-MHz contacts of 2200 km (1400 mi), such as between Pierre and Tallahassee, should be possible.

have been made on 50 and 144 MHz, and 222-MHz FAI seems probable as well. Expected maximum distances should be similar to other forms of E layer propagation, or about 2300 km (1400 mi).

AURORA

Radar signals as high as 3000 MHz have been scattered by the *aurora borealis* or northern lights (*aurora australis* in the Southern Hemisphere), but amateur aurora contacts are common only from 28 through 432 MHz. By pointing directional antennas generally north toward the center of aurora activity, oblique paths between stations up to 2300 km (1400 mi) apart can be completed. See **Fig 19.15**. High power and large antennas are not necessary. Stations with small Yagis and as little as 10 W output have used auroras on frequencies as high as 432 MHz, but contacts at 902 MHz and higher are exceedingly rare. Auroral propagation works just as well in the Southern Hemisphere, in which case antennas must be pointed south.

The appearance of auroras is closely linked to solar activity. During massive geomagnetic storms, low energy particles that were trapped in the magnetosphere flow into the atmosphere near the polar regions, where they ionize the gases of the E layer and higher. This unusual ionization produces spectacular visual auroral displays, which often spread southward into the mid latitudes. Higher energy electrons get down to lower altitudes to cause auroral ionization (via collisions with neutral particles) in the E layer, and this scatters radio signals in the VHF and UHF ranges.

In addition to scattering radio signals, auroras have other effects on worldwide radio propagation. Communication below 20 MHz is disrupted in high latitudes, primarily by absorption, and is especially noticeable over polar and near-polar paths. Signals on the AM broadcast band through the 40 meter band late in the afternoon may become weak and watery. The 20 meter band may close down altogether. Satellite operators have also noticed that 144-MHz downlink signals are often weak and distorted when satellites pass near the polar regions. At the same time, the MUF in equatorial regions may temporarily rise dramatically, including an enhancement of transequatorial paths at frequencies as high as 50 MHz.

Auroras occur most often around the spring and fall equinoxes (March-April and September-October), but auroras may appear in any month. Aurora activity generally peaks about two years before and after solar cycle maximum (the "before" generally due to CMEs and the "after" generally due to high-speed wind streams from coronal holes). Radio aurora activity is usually heard first in late afternoon and may reappear later in the evening. Auroras may be anticipated by following the A and K indices reports on WWV. A K index of five or greater and an A index of at least 30 are indications that a geomagnetic storm is in progress and an aurora likely. The probability, intensity and southerly extent of auroras increase as the two index numbers rise. Stations north of 42° latitude in North America experience many auroral openings each year, while those in the Gulf Coast states may hear auroral signals no more than once a year, if that often.

Aurora-scattered signals are easy to identify. On 28- and 50-MHz SSB, signals sound very distorted and somewhat wider than normal; at 144 MHz and above, the distortion may be so severe that only CW is useful. Auroral CW signals have a distinctive note variously described as a buzz, hiss or mushy sound. This characteristic auroral signal is due to Doppler broadening, caused by the movement of electrons within the aurora. An additional Doppler shift of 1 kHz or more may be evident at 144 MHz and several kilohertz at 432 MHz. This second Doppler shift is the result of massive electrical currents that sweep electrons toward the sun side of the Earth during magnetic storms. Doppler shift and distortion increase with higher frequencies, while signal strength dramatically decreases.

It is not necessary to see an aurora to make auroral contacts. Useful auroras may be 500-1000 km (310-620 mi) away and below the visual horizon. Antennas should be pointed generally north and then probed east and west to peak signals, because auroral ionization is field aligned. This means that for any pair of stations, there is an optimal direction for aurora scatter. Offsets from north are usually greatest when the aurora is closest and often provide the longest contacts. There may be some advantage to antennas that can be elevated, especially when auroras are high in the sky.

AURORAL E

Radio auroras may evolve into a propagation mode known as *auroral E* at 28, 50 and rarely 144 MHz. Auroral E is essentially sporadic E in the auroral zone. Doppler distortion disappears and signals take on the characteristics of sporadic E. The most effective antenna headings shift dramatically away from oblique aurora paths to direct great-circle bearings. The usual maximum distance is 2300 km (1400 mi), typical for E layer modes, but 28- and 50-MHz auroral E contacts of 5000 km (3100 mi) are sometimes made across Canada and the northern US, apparently using two hops. Contacts at 50 MHz between Alaska and the east coasts of Canada and the northern US have been completed this way. Transatlantic 50-MHz auroral E paths are also likely.

Typically, 28- and 50-MHz auroral E appears across the northern third of the US and southern Canada when aurora activity is diminishing. This usually happens after midnight on the eastern end of the path. Auroral E signals sometimes have a slightly hollow sound to them and build slowly in strength over an hour or two, but otherwise they are indistinguishable from sporadic E. Auroral E paths are almost always east-west oriented, perhaps because there are few stations at very northern latitudes to take advantage of this propagation. A common auroral E path on 28 MHz and 21 MHz is from North America

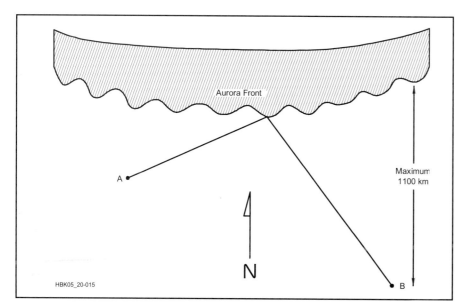

Fig 19.15 — Point antennas generally north to make oblique long-distance contacts on 28 through 432 MHz via aurora scattering. Optimal antenna headings may shift considerably to the east or west depending on the location of the aurora. This necessitates moving your antenna azimuth direction for best propagation, and continuing to do so as the aurora progresses.

in the late afternoon hours in the autumn months to the Scandinavian countries. This path is open when normal F layer propagation is not possible — even to more southern locations in Europe.

Auroral E may also appear while especially intense auroras are still in progress, as happened during the great aurora of March 1989. On that occasion, 50-MHz propagation shifted from Doppler-distorted aurora paths to clear-sounding auroral E over a period of a few minutes. Many 6 meter operators as far south as Florida and Southern California made single- and double-hop auroral E contacts across the country. At about the same time, the MUF reached 144 MHz for stations west of the Great Lakes to the Northeast, the first time auroral E had been reported so high in frequency.

METEOR SCATTER

Contacts between 800 and 2300 km (500 and 1400 mi) can be made at 28 through 432 MHz via reflections from the ionized trails left by meteors as they travel through the atmosphere. The kinetic energy of meteors no larger than grains of rice are sufficient to ionize a column of air 20 km (12 mi) long in the E layer. The particle itself evaporates and never reaches the ground, but the ionized column may persist for a few seconds to a minute or more before it dissipates. This is enough time to make very brief contacts by reflections from the ionized trails. Millions of meteors enter the Earth's atmosphere every day, but few have the required size, speed and orientation to the Earth to make them useful for meteor-scatter propagation.

Radio signals in the 30- to 100-MHz range are reflected best by meteor trails, making the 50-MHz band prime for meteor-scatter work. The early morning hours around dawn are usually the most productive, because the morning side of the Earth faces in the direction of the planet's orbit around the sun. The relative velocity of meteors that head toward the Earth's morning side are thus increased by up to 30 km/sec, which is the average rotational speed of the Earth in orbit. See **Fig 19.16**. The maximum velocity of meteors in orbit around the sun is 42 km/sec. Thus when the relative velocity of the Earth is considered, most meteors must enter the Earth's atmosphere somewhere between 12 and 72 km/sec.

Meteor contacts ranging from a second or two to more than a minute can be made nearly any morning at 28 or 50 MHz. Meteor-scatter contacts at 144 MHz and higher are more difficult because reflected signal strength and duration drop sharply with increasing frequency. A meteor trail that provides 30 seconds of communication at 50 MHz will last only a few seconds at 144 MHz, and less than a second at 432 MHz.

Meteor scatter opportunities are some-

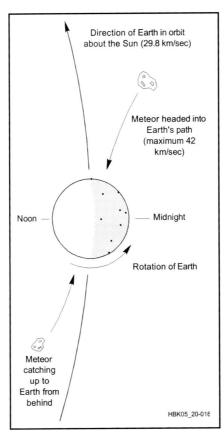

Fig 19.16 — The relative velocity of meteors that meet the Earth head-on is increased by the rotational velocity of the Earth in orbit. Fast meteors strike the morning side of the Earth because their velocity adds to the Earth's rotational velocity, while the relative velocity of meteors that "catch up from behind" is reduced.

what better during July and August for North America because the average number of meteors entering the Earth's atmosphere peaks during those months. The best times are during one of the great annual *meteor showers*, when the number of useful meteors may increase ten-fold over the normal rate of five to ten per hour. See **Table 19.4**. A meteor shower occurs when the Earth passes through a relatively dense stream of particles, thought to be the remnants of a comet in orbit around the sun. The most-productive showers are relatively consistent from year to year,

Table 19.4
Major Annual Meteor Showers

Name	Peak Dates	Approximate Rate (meteors/hour)
Quadrantids	Jan 3	50
Arietids	Jun 7-8	60
Perseids	Aug 11-13	80
Orionids	Oct 20-22	20
Geminids	Dec 12-13	60

although several can occasionally produce great storms.

Because meteors provide only fleeting moments of communication even during one of the great meteor showers, special operating techniques are often used to increase the chances of completing a contact. Prearranged schedules between two stations establish times, frequencies and precise operating standards. Usually, each station transmits on alternate 15-second periods until enough information is pieced together a bit at a time to confirm contact. High-speed Morse code of several hundred words per minute, generated and slowed down by special computer programs, can make effective use of very short meteor bursts. Non-scheduled random meteor contacts are common on 50 MHz and 144 MHz, but short transmissions and alert operating habits are required.

It is helpful to run several hundred watts to a single Yagi, but meteor-scatter can be used by modest stations under optimal conditions. During the best showers, a few watts and a small directional antenna are sufficient at 28 or 50 MHz. At 144 MHz, at least 100 W output and a long Yagi are needed for consistent results. Proportionately higher power is required for 222 and 432 MHz even under the best conditions.

A technique allowing even smaller VHF/UHF stations to take advantage of meteor scatter and other modes (ionospheric scatter and even moonbounce) is available in software called *WSJT*. This software uses the sound card in a PC and a variety of modulation and coding techniques to allow reception of signals up to 10 dB below your noise level. For more information, visit **www.physics.princeton.edu/pulsar/K1JT/**.

19.2.13 F Layer Propagation

The region of the *F layers*, from 150 km (90 mi) to over 400 km (250 mi) altitude, is by far the most important for long-distance HF communications. F-region oxygen atoms are ionized primarily by ultraviolet radiation. During the day, ionization reaches maxima in two distinct layers. The F_1 layer forms between 150 and 250 km (90 and 160 mi), is most prevalent in the summer months, and disappears at night. The F_2 layer extends above 250 km (160 mi), with a peak of ionization around 300 km (190 mi). At night, F-region ionization collapses into one broad layer at 300-400 km (190-250 mi) altitude. Ions recombine very slowly at these altitudes, because atmospheric density is relatively low. Maximum ionization levels change significantly with time of day, season and year of the solar cycle.

F1 LAYER

The daytime F_1 layer is not important to HF communication. It exists only during daylight hours and is largely absent in winter. Radio signals below 10 MHz are not likely to reach the F_1 layer, because they are either absorbed by the D layer or refracted by the E layer. Signals higher than 20 MHz that pass through both of the lower ionospheric regions are likely to pass through the F_1 layer as well, because the F_1 MUF rarely rises above 20 MHz. Absorption diminishes the strength of any signals that continue through to the F_2 layer during the day. Some useful F_1 layer refraction may take place between 10 and 20 MHz during summer days, yielding paths as long as 3000 km (1900 mi), but these would be practically indistinguishable from F_2 skip.

F2 AND NIGHTTIME F LAYERS

The F_2 layer forms between 250 and 400 km (160 and 250 mi) during the daytime and persists throughout the night as a single consolidated F region 50 km (30 mi) higher in altitude. Typical ion densities are the highest of any ionospheric layer, with the possible exception of some unusual E layer phenomenon. In contrast to the other ionospheric layers, F_2 ionization varies considerably with time of day, season and position in the solar cycle, but it is never altogether absent. These two characteristics make the F_2 layer the most important for long-distance HF communications.

The F_2 layer MUF is nearly a direct function of ultraviolet (UV) solar radiation, which in turn closely follows the solar cycle. During the lowest years of the cycle, the daytime MUF may climb above 14 MHz for only a few hours a day. In contrast, the MUF may rise beyond 50 MHz during peak years and stay above 14 MHz throughout the night. The virtual height of the F_2 region averages 330 km (210 mi), but varies between 200 and 400 km (120 and 250 mi). Maximum one-hop distance is about 4000 km (2500 mi). Near-vertical incidence skywave propagation just below the critical frequency provides reliable coverage out to 200-300 km (120-190 mi) with no skip zone. It is most often observed on 7 MHz during the day.

The extremely high-angle *Pedersen Ray* can create effective single-hop paths of 5000 to 12,000 km under certain conditions, but most operators will not be able to distinguish Pedersen-Ray paths from normal F layer propagation. A Pederson Ray is a path that follows the contour of the Earth near the height of the maximum F_2 region electron density, and requires a fairly stable ionosphere. Pedersen-Ray paths are most evident over high-latitude east-west paths at frequencies near the MUF. They appear most often about noon local time at mid-path when

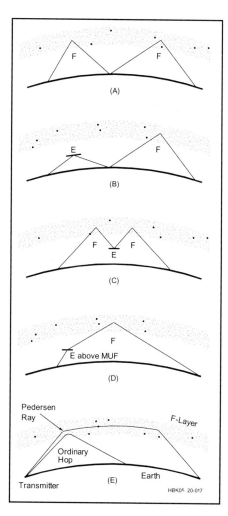

Fig 19.17 — Multihop paths can take many different configurations, including a mixture of E and F layer hops. (A) Two F layer hops. Five or more consecutive F layer hops are possible. (B) An E layer hookup to the F layer. (C) A top-side E layer reflection can shorten the distance of two F layer hops. (D) Refraction in the E layer above the MUF is insufficient to return the signal to Earth, but it can go on to be refracted in the F layer. (E) The Pedersen Ray, which originates from a signal launched at a relatively high angle above the horizon into the E or F region, may result in a single-hop path, 5000 km (3100 mi) or more. This is considerably further than the normal 4000-km (2500 mi) maximum F-region single-hop distance, where the signal is launched at a very low takeoff angle. The Pedersen Ray can easily be disrupted by any sort of ionospheric gradient or irregularity. Not shown in this figure is a chordal hop — since chordal hops are most prevalent in the equatorial ionosphere, refer to Fig 19.21 and Fig 19.22.

the geomagnetic field is very quiet. Pedersen-Ray propagation may be responsible for 50 MHz paths between the US Northeast and Western Europe, for example, when ordinary MUF analysis could not explain the 5000-km contacts. See part E in **Fig 19.17**.

At any given location on Earth, in general both F_2 layer ionization and MUF at that point build rapidly at sunrise, usually reach a maximum in the afternoon, and then decrease to a minimum prior to sunrise. Depending on the season, the MUF is generally highest within 20° of the equator and lower toward the poles. For this reason, transequatorial paths may be open at a particular frequency when all other paths are closed

In contrast to all the other ionospheric layers, daytime ionization in the winter F_2 layer averages four times the level of the summer at the same period in the solar cycle, doubling the MUF. This so-called *winter anomaly* is caused by a seasonal increase in the ratio of atoms to molecules at F_2 layer heights (atoms are instrumental in the production of electrons, whereas molecules are instrumental in the loss of electrons). Winter daytime F_2 conditions are much superior to those in summer, because the MUF is much higher.

MULTIHOP F LAYER PROPAGATION

Most HF communication beyond 4000 km (2500 mi) takes place via multiple ionospheric hops. Radio signals are reflected from the Earth back toward space for additional ionospheric refractions. A series of ionospheric refractions and terrestrial reflections commonly create paths half-way around the Earth. Each hop involves additional attenuation and absorption, so the longest-distance signals tend to be the weakest. Even so, it is possible for signals to be propagated completely around the world and arrive back at their originating point. Multiple reflections within the F layer may at times bypass ground reflections altogether, creating what are known as *chordal hops*, with lower total attenuation. It takes a radio signal about 0.15 second to make an around-the-world trip.

Multihop paths can take on many different configurations, as shown in the examples of Fig 19.17. E layer (especially sporadic E) and F layer hops may be mixed. In practice, multi-hop signals arrive via many different paths, which often increases the problems of fading. Analyzing multi-hop paths is complicated by the effects of D- and E layer absorption, possible reflections from the tops of sporadic E layers, disruptions in the auroral zone and other phenomena.

In general, when a band is opening and closing as the MUF changes, extremely low elevation angles are dictated (less than 5°). During the main part of the opening, the elevation angle is higher — generally in the range of 5° to 20°. This is why stations with extremely high antennas (which have higher gain at lower takeoff angles) perform better at band openings and closings.

F LAYER LONG PATH

Most HF communication takes place along the shortest great-circle path between two stations. Short-path propagation is always less than 20,000 km (12,000 mi) — halfway around the Earth. Nevertheless, it may be possible at times to make the same contact in exactly the opposite direction via the *long path*. The long-path distance will be 40,000 km (25,000 mi) minus the short-path length. Signal strength via the long path is usually considerably less than the more direct short-path. When both paths are open simultaneously, there may be a distinctive sort of echo on received signals. The time interval of the echo represents the difference between the short-path and long-path distances.

Sometimes there is a great advantage to using the long path when it is open, because signals can be stronger and fading less troublesome, or because fewer interfering signals lie along the path between the stations. There are times when the short path may be closed or disrupted by E layer *blanketing* (when the E-region ionization is high enough to keep waves from penetrating, regardless of elevation angle), D layer absorption or F layer gaps, especially when operating just below the MUF. Long paths that predominantly cross the night side of the Earth, for example, are sometimes useful because they generally avoid blanketing and absorption problems. Daylight-side long paths may take advantage of higher F layer MUFs that occur over the sunlit portions of the Earth.

F LAYER GRAY-LINE

Gray-line paths can be considered a special form of long-path propagation that take into account the unusual ionospheric configuration along the twilight region between night and day. The gray line, as the twilight region is sometimes called, extends completely around the world. It is not precisely a line, for the distinction between daylight and darkness is a gradual transition due to atmospheric scattering. On one side, the gray line heralds sunrise and the beginning of a new day; on the opposite side, it marks sunset and the end of the day.

The ionosphere undergoes a significant transformation between night and day. As day begins, the highly absorbent D and E layers are recreated, while the F layer MUF rises from its pre-dawn minimum. At the end of the day, the D and E layers quickly disappear, while the F layer MUF continues its slow decline from late afternoon. For a brief period just along the gray-line transition, the D and E layers are not well formed, yet the F_2 MUF usually remains higher than 5 MHz. This provides a special opportunity for stations at 1.8 and 3.5 MHz.

Normally, long-distance communication

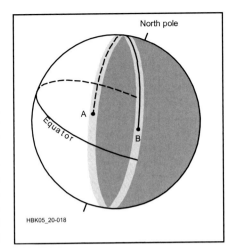

Fig 19.18 — The gray line encircles the Earth, but the tilt at the equator to the poles varies over 46° with the seasons. Long-distance contacts can often be made halfway around the Earth along the gray line, even as low as 1.8 and 3.5 MHz. The strength of the signals, characteristic of gray-line propagation, indicates that multiple Earth-ionosphere hops are not the only mode of propagation, since losses in many such hops would be very great. Chordal hops, where the signals are confined to the ionosphere for at least part of the journey, may be involved.

on the lowest two amateur bands can take place only via all-darkness paths because of daytime D layer absorption. The gray-line propagation path, in contrast, extends completely around the world. See **Fig 19.18**. This unusual situation lasts less than an hour at sunrise and sunset when the D layer is largely absent, and may support contacts that are difficult or impossible at other times.

The gray line generally runs north-south, but it varies by 23° either side of true north as measured at the equator over the course of the year. This variation is caused by the tilt in the Earth's axis. The gray line is exactly north-south through the poles at the equinoxes (March 21 and September 21) and is at its 23° extremes on June 21 and December 20. Over a one-year period, the gray line crosses a 46° sector of the Earth north and south of the equator, providing optimum paths to slightly different parts of the world each day. Many commonly available computer programs plot the gray line on a flat map or globe.

A Web based application to calculate worldwide sunrise and sunset times is available at **aa.usno.navy.mil/data/docs/RS_OneDay.html**. An alternative method of determining sunrise and sunset times, along with seeing great circle paths on the same plot, is W6EL's *W6ELProp* software, available free at **www.qsl.net/w6elprop**. For an online gray line map, visit **dx.qsl.net/propagation/greyline.html**.

F LAYER BACKSCATTER AND SIDESCATTER

Special forms of F layer scattering can create unusual paths within the skip zone. *Backscatter* and *sidescatter* signals are usually observed just below the MUF for the direct path and allow communications not normally possible by other means. Stations using backscatter point their antennas toward a common scattering region at the one-hop distance, rather than toward each other. Backscattered signals are generally weak and have a characteristic hollow sound. Useful communication distances range from 100 km (60 mi) to the normal one-hop distance of 4000 km (2500 mi).

Backscatter and sidescatter are closely related and the terminology does not precisely distinguish between the two. Backscatter usually refers to single-hop signals that have been scattered by the Earth or the ocean at some distant point back toward the transmitting station. Two stations spaced a few hundred km apart can often communicate via a backscatter path near the MUF. See **Fig 19.19**.

Sidescatter usually refers to a circuit that

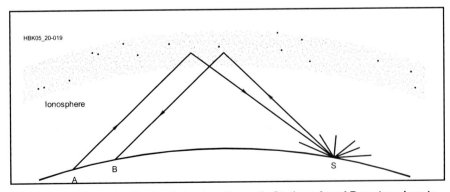

Fig 19.19 — Schematic of a simple backscatter path. Stations A and B are too close to make contact via normal F layer ionospheric refraction. Signals scattered back from a distant point on the Earth's surface (S), often the ocean, may be accessible to both and create a backscatter circuit.

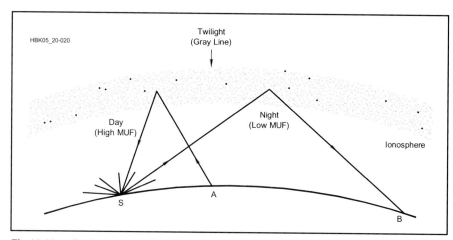

Fig 19.20 — Backscatter path across the gray line. Stations A and B are too close to make contact via normal ionospheric refraction, but may hear each other's signals scattered from point S. Station A makes use of a high-angle refraction on the day side of the gray line, where the MUF is high. Station B makes use of a night-time refraction, with a lower MUF and lower angle of propagation. Note that station A points away from B to complete the circuit.

is oblique to the normal great-circle path and is sometimes referred to as *skew path*. Two stations can make use of a common side-scattering region well off the direct path, often toward the south. European and North American stations sometimes complete 28-MHz contacts via a scattering region over Africa. US and Finnish 50-MHz operators observed a similar effect early one morning in November 1989 when they made contact by beaming off the coast of West Africa.

When backscattered signals cross an area where there is a sharp gradient in ionospheric density, such as between night and day, the path may take on a different geometry, as shown in **Fig 19.20**. In this case, stations can communicate because backscattered signals return via the day side ionosphere on a shorter hop than the night side. This is possible because the dayside MUF is higher and thus the skip distance shorter. The net effect is to create a backscatter path between two stations within the normal skip zone.

TRANSEQUATORIAL PROPAGATION

Discovered in 1947, *transequatorial propagation* (commonly abbreviated TE) supports propagation between 5000 and 8000 km (3100 and 5000 mi) across the magnetic equator from 28 MHz to as high as 432 MHz. Stations attempting TE contacts must be nearly equidistant from the geomagnetic equator. Many contacts have been made at 50 and 144 MHz between Europe and South Africa, Japan and Australia and the Caribbean region and South America. Fewer contacts have been made on the 222-MHz band, and TE signals have been heard at 432 MHz.

Unfortunately for most continental US stations, the *geomagnetic equator* dips south of the geographic equator in the Western Hemisphere, as shown in **Fig 19.21**, making only the most southerly portions of Florida and Texas within TE range. TE contacts from the southeastern part of the country may be possible with Argentina, Chile and even South Africa.

Transequatorial propagation peaks between 5 PM and 10 PM during the spring and fall equinoxes, especially during the peak years of the solar cycle. The lowest probability is during the summer. Quiet geomagnetic conditions are required for TE to form. High power and large antennas are not required to work TE, as VHF stations with 100 W and single long Yagis have been successful.

TE propagation depends on bulges of intense F_2 layer ionization on both sides of the geomagnetic equator. This field-aligned ionization forms shortly after sunset via a process known as the *fountain effect* in an

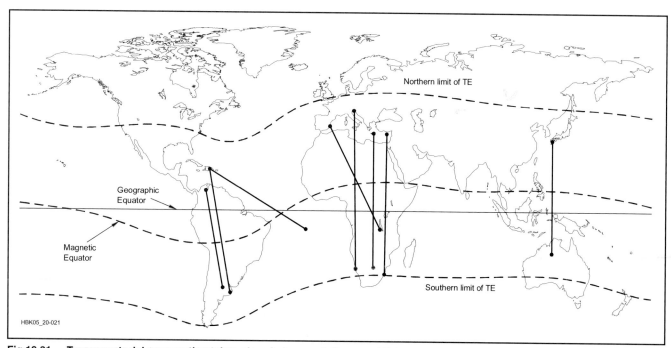

Fig 19.21 — Transequatorial propagation takes place between stations equidistant across the geomagnetic equator. Distances up to 8000 km (5000 mi) are possible on 28 through 432 MHz. Note the geomagnetic equator is considerably south of the geographic equator in the Western Hemisphere.

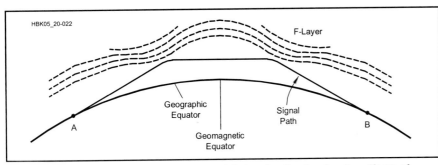

Fig 19.22 — Cross-section of a transequatorial signal path, showing the effects of ionospheric bulging and a double refraction above the normal MUF.

area 100-200 km (60-120 mi) north and south of the geomagnetic equator and 500-3000 km (310-1900 mi) wide. It moves west with the setting sun. The MUF may increase to twice its normal level 15° either side of the geomagnetic equator. See **Fig 19.22**.

19.2.14 Emerging Theories of HF and VHF Propagation

Although much is known about the ionosphere and propagation, there are still many instances of unusual propagation that cannot be easily explained with our textbook knowledge. It is encouraging to note that Amateur Radio operators continue to be at the forefront of trying to understand these unusual modes. Several recent studies have included:

1) The tie between polar mesospheric summer echoes and 6 meter propagation in "Polar Mesospheric Summer Echoes," *WorldRadio On-Line*, Feb 2009, **www.cq-amateur-radio.com**. These summer echoes are also called SSSP (summer solstice short path), a mode that Japanese amateurs have noted.

2) Theta aurora and 6 meter propagation across the polar cap in "More Alaska to EU on 6 m," **mysite.verizon.net/k9la/**.

3) The impact of galactic cosmic rays on 160 meter *ducting* in "A Theory on the Role of Galactic Cosmic Rays in 160-Meter Propagation," *CQ*, Nov 2008. *Ducting* refers to an electromagnetic wave successively refracting between a lower boundary and an upper boundary (for example, between the top of the E region and the bottom of the F region for 160 meters at night)

19.3 MUF Predictions

F layer MUF prediction is key to forecasting HF communications paths at particular frequencies, dates and times, but forecasting is complicated by several variables. Solar radiation varies over the course of the day, season, year and solar cycle. These regular intervals provide the main basis for prediction, yet recurrence is far from reliable. In addition, forecasts are predicated on a quiet geomagnetic field, but the condition of the Earth's magnetic field is most difficult to predict weeks or months ahead. For professional users of HF communications, uncertainty is a nuisance for maintaining reliable communications paths, while for many amateurs it provides an aura of mystery and chance that adds to the fun of DXing. Nevertheless, many amateurs want to know what to expect on the HF bands to make best use of available on-the-air time, plan contest strategy, ensure successful net operations or engage in other activities.

19.3.1 MUF Forecasts

Long-range forecasts several months ahead provide only the most general form of prediction. A series of 48 charts on the mem-

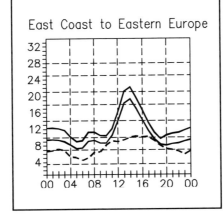

Fig 19.23 — Propagation prediction chart for East Coast to Eastern Europe from the ARRL Web members-only site for April 2001. An average 2800-MHz (10.7-cm) solar flux of 159 was assumed for the month. On 10% of the days, the highest frequency propagated is predicted to be at least as high as the uppermost curve (the Highest Possible Frequency, or HPF, approximately 33 MHz), and for 50% of the days as high as the middle curve, the MUF. The lowest curve shows the Lowest Usable Frequency (LUF) for a 1500-W CW transmitter.

bers-only ARRL Web site (**www.arrl.org/propagation**), similar to **Fig 19.23**, forecast average propagation for a one-month period over specific paths. The charts assume a single average solar flux value for the entire month and they assume that the geomagnetic field is undisturbed.

The uppermost curve in Fig 19.23 shows the highest frequency that will be propagated on at least 10% of the days in the month. The given values might be exceeded considerably on a few rare days. On at least half the days, propagation should be possible on frequencies as high as the middle curve. Propagation will exceed the lowest curve on at least 90% of the days. The exact MUF on any particular day cannot be determined from these statistical charts, but you can determine when you should start monitoring a band to see if propagation actually does occur that day — particularly at frequencies above 14 MHz.

Short-range forecasts of a few days ahead are marginally more reliable than long-range forecasts, because underlying solar indices and geomagnetic conditions can be anticipated with greater confidence. The tendency for solar disturbances to recur at 27-day intervals may enhance short-term forecasts. Daily forecasts may not be any better, as the ionosphere does not instantly react to small changes in solar flux (or sunspot number) and geomagnetic field indices. Regardless of these limitations, it is always good to know the current solar and geophysical data, as well as understanding warnings provided by observations of the sun in the visual to X-ray range.

The CD-ROM bundled with *The ARRL Antenna Book* contains even more detailed propagation-prediction tables from more than 150 locations around the world for six levels of solar activity, for the 12 months of the year. Again, keep in mind that these long-range forecasts assume quiet geomagnetic conditions. Real-time MUF forecasts are also available in a variety of text and graphical forms on the Web. Forecasts can also be made by individuals using one of several popular programs for personal computers, including *ASAPS*, *VOACAP*, *W6ELProp* and *WinCAP Wizard*. (See the sidebar, "MUF Prediction on the Home Computer.")

19.3.2 Statistical Nature of Propagation Predictions

The model of the ionosphere in propagation prediction software is a monthly median model — thus the results (usually MUF and signal strength) are statistical over a month's time frame. The median aspect means that the predicted value will actually occur on at least half the days of the month — but unfortunately it's tough to tell which are the "better" days and which are the "worse" days.

This monthly median model came about due to the efforts of scientists to correlate what the ionosphere was doing to what the sun was doing. The best correlation was between smoothed sunspot numbers and monthly median ionospheric parameters. Thus our propagation prediction software was never intended to give daily predictions. One good way to achieve better short-term predictions is to use the concept of an effective sunspot number — SSNe.

EFFECTIVE SUNSPOT NUMBER (SSNe)

The concept of SSNe is quite simple. The F_2 layer critical frequency data from worldwide ionosondes is monitored, and then the sunspot number in an F_2 layer model of the ionosphere is varied until it provides a best fit to the monitored data. The current SSNe comes from Northwest Research Associates in Tucson (AZ), and can be found at **www.nwra-az.com/spawx/ssne.html**. By plugging the current SSNe into your favorite propagation prediction program, you will have a better (but not perfect) picture of what the ionosphere is doing "now."

19.3.3 Direct Observation

Propagation conditions can be determined directly by listening to the HF bands. The simplest method is to tune higher in frequency until no more long-distance stations are heard. This point is roughly just above the MUF to anywhere in the world at that moment. The highest usable amateur band would be the next lowest one. If HF stations seem to disappear around 23 MHz, for example, the 15-meter band at 21 MHz might make a good choice for DXing. By carefully noting station locations as well, the MUF in various directions can also be determined quickly.

The shortwave broadcast bands (see **Table 19.5**) are most convenient for MUF browsing, because there are many high-powered stations on regular schedules. Take care to ensure that programming is actually transmitted from the originating country. A Radio Moscow or BBC program, for example, may be relayed to a transmitter outside Russia or England for retransmission. An excellent guide to shortwave broadcast stations is the *World Radio TV Handbook*, available through the ARRL.

19.3.4 WWV and WWVH

The standard time stations WWV (Fort Collins, Colorado) and WWVH (Kauai, Hawaii), which transmit on 2.5, 5, 10, 15 and 20 MHz, are also popular for propagation monitoring. They transmit 24 hours a day. Daily monitoring of these stations for signal strength and quality can quickly provide a good basic indication of propagation conditions. In addition, each hour they broadcast the geomagnetic A and K indices, the 2800-MHz (10.7-cm) solar flux, and a short forecast of conditions for the next day. These are heard on WWV at 18 minutes past each hour and on WWVH at 45 minutes after the hour. The same information, along with a lot more space weather information, is available at various Web sites, such as **www.swpc.noaa.gov** or **www.spaceweather.com**. The K index is updated every three hours, while the A index and solar flux are updated after 2100 UTC. These data are useful for making predictions on home computers, especially when averaged over several days of solar flux observations.

19.3.5 Beacons

Automated *beacons* in the higher amateur bands can also be useful adjuncts to propagation watching. Beacons are ideal for this purpose because most are designed to transmit 24 hours a day.

One of the best organized beacon systems is the International Beacon Project, sponsored by the Northern California DX Foundation (NCDXF) and International Amateur Radio Union (IARU). The beacons operate 24 hours a day at 14.100, 18.110, 21.150, 24.930 and 28.200 MHz. Eighteen beacons on five continents transmit in successive 10-second intervals (each beacon transmits once every 3 minutes). More on this system can be found at the Northern California DX Foundation website **www.ncdxf.org/beacons.html**.

A list of many 28-MHz beacons can be found on the 10-10 International Web site, **www.ten-ten.org** (look for the beacons link). Beacons often include location as part of their automated message, and many can be located from their call sign. Thus, even casual scanning of beacon subbands can be useful. **Table 19.6** provides the frequencies where beacons useful to HF propagation are most commonly placed.

There are also many beacons on VHF and higher bands. "G3USF's Worldwide List of 50 MHz Beacons" may be found at **www.keele.ac.uk/depts/por/50.htm**. Information on North American beacons on 144 MHz and up is maintained by Ron Klimas, WZ1V, at **www.newsvhf.con/beacons2.html**.

Table 19.5
Shortwave Broadcasting Bands

Frequency (MHz)	Band (meters)
2.300-2.495	120
3.200-3.400	90
3.900-4.000	75
4.750-5.060	60
5.959-6.200	49
7.100-7.300	41
9.500-9.900	31
11.650-12.050	25
13.600-13.800	22
15.100-15.600	19
17.550-17.900	16
21.450-21.850	13
25.600-26.100	11

Table 19.6
Popular Beacon Frequencies

(MHz)	Comments
14.100, 18.110, 21.150, 24.930, 28.200	Northern California DX Foundation beacons
28.2-28.3	Several dozen beacons worldwide
50.0-50.1	Most US beacons are within 50.06-50.08 MHz
70.03-70.13	Beacons in England, Ireland, Gibraltar and Cyprus

MUF Prediction on the Home Computer

Like predicting the weather, predicting HF propagation — even with the best computer software available — is not an exact science. The processes occurring as a signal is propagated from one point on the Earth to another are enormously complicated and subject to an incredible number of variables. Experience and knowledge of propagation conditions (as related to solar activity, especially unusual solar activity, such as flares or coronal mass ejections) are needed when you actually get on the air to check out the bands. Keep in mind, too, that ordinary computer programs are written mainly to calculate propagation for great-circle paths via the F layer. Scatter, skew-path, auroral and other such propagation modes may provide contacts when computer predictions indicate no contacts are possible.

What follows is some brief information about commercially available propagation-prediction programs for the IBM PC and compatible computers. See **Table 19.A**. These programs generally allow predictions from 3 to 30 MHz. Unfortunately prediction programs are not available on 160 meters (because of an incomplete understanding of the lower ionosphere and ducting mechanisms that contribute to propagation on that band) and on 6 meters (because of an incomplete understanding of openings when the MUF is not predicted to be high enough). As a general guideline, you should look for 160 meter openings when the path to your target station is in darkness and around sunrise/sunset, and you should look for 6 meter F_2 openings in the daytime during winter months near solar maximum.

You may find an article that appeared in the Summer 2008 issue of *CQ VHF* to be helpful with your 6 meter endeavors: "Predicting 6-meter F2 Propagation," by Carl Luetzelschwab, K9LA.

ASAPS Version 5.4

An agency of the Australian government has developed the *ASAPS* program, which stands for Advanced Stand-Alone Prediction System. It rivals *IONCAP* (see below) in its analysis capability and in its prediction accuracy. It is a *Windows* program that interacts reasonably well with the user, once you become accustomed to the acronyms used. If you change transmit power levels, antennas and other parameters, you can see the new results almost instantly without further menu entries. Available from IPS Radio and Space Services. See **www.ips.gov.au/**.

IONCAP and VOACAP

IONCAP, short for Ionospheric Communications Analysis and Prediction, was written by an agency of the US government and has been under development for about 30 years in one form or another. The *IONCAP* program has a well-deserved reputation for being difficult to use, since it came from the world of Fortran punch cards and mainframe computers.

VOACAP is a version of *IONCAP* adapted to Voice of America predictions, but this one includes a sophisticated *Windows* interface. The Voice of America (VOA) started work on *VOACAP* in the early 1990s and continued for several years before funding ran out. The program was maintained by a single, dedicated computer scientist, Greg Hand, at NTIA/ITS (Institute for Telecommunication Sciences), an agency of the US Department of Commerce in Boulder, Colorado. Greg Hand is now retired, but considerable documentation can be found at the website referenced below, along with links to other *VOACAP*-related Web sites. Although *VOACAP* is not specifically designed for amateurs (and thus doesn't include some features that amateurs are fond of, such as entry of locations by ham-radio call signs and multiple receiving antennas), it is available for free by downloading from **elbert.its.bldrdoc.gov/hf.html**.

W6ELProp, Version 2.70

In 2001, W6EL ported his well known *MINIPROP PLUS* program into the *Windows* world. It uses the same Fricker-based F2 region computation engine as its predecessor. (This method was developed by Raymond Fricker of the BBC.) *W6ELProp* has a highly intuitive, ham-friendly user interface. It produces detailed output tables, along with a number of useful charts and maps, including the unique and useful "frequency map," which shows the global MUFs from a given transmitting location for a particular month/day/time and solar-activity level. *W6ELProp* is available for free by downloading from **www.qsl.net/w6elprop**.

PropLab Pro, Version 3.0

PropLab Pro by Solar Terrestrial Dispatch represents the high end of propagation-prediction programs. It is the only com-

19.3.6 Space Weather Information

Living in the space age results in a wealth of information about the sun-Earth environment. Not only do we have measurements available, but we also have extensive tutorials available on what is being measured, how it is measured, and its general impact to propagation.

SPACE WEATHER SATELLITES

Two of the most beneficial satellites for propagation information are the Advanced Composition Explorer (ACE) and the Solar and Heliospheric Observatory (SOHO).

ACE is on a line from the Earth to the sun, and is about 1 million miles from the Earth. Its primary purpose is to measure solar wind characteristics. For example, it measures the direction of the IMF (interplanetary magnetic field), the solar wind speed, and the dynamic solar wind pressure. More detailed information about ACE can be found at **www.srl.caltech.edu/ACE/**.

SOHO moves around the sun in step with Earth — it always faces the sun. Its primary purpose is to study the sun. For example, it views coronal mass ejections, and in conjunction with the Michelson Doppler Imager (MDI), allows scientists to "see" sunspots on the far side of the sun. More detailed information about SOHO and MDI can be found at **sohowww.nascom.nasa.gov** and **soi.stanford.edu**, respectively.

The above satellite data, plus a lot more, can be found at **swpc.noaa.gov** (ACE data) and at **spaceweather.com** (SOHO data)

mercial program presently available that can do complete 3D ray tracing through the ionosphere, even taking complex geomagnetic effects and electron-neutral collisions (what causes absorption) into account. The number of computations is huge, especially in the full-blown 3D mode and operation can be slow and tedious. The *Windows*-based user interface is easier to use than previous versions, which were complex and difficult to learn. It is fascinating to see exactly how a signal can bend off-azimuth or how it can split into the ordinary and extraordinary waves. It is best considered a propagation analysis tool, as opposed to a propagation prediction program. See **www.spacew.com/proplab**.

Table 19.A

Features and Attributes of Several Currently Available Propagation Prediction Programs

	ASAPS V. 5.4	VOACAP Windows	W6ELProp V. 2.70	PropLab Pro 3.0
User Friendliness	Good	Good	Good	Good
Operating System	Windows	Windows	Windows	Windows
Uses K index	No	No	Yes	Yes
User library of QTHs	Yes/Map	Yes	Yes	No
Bearings, distances	Yes	Yes	Yes	Yes
MUF calculation	Yes	Yes	Yes	Yes
LUF calculation	Yes	Yes	No	Yes
Wave angle calculation	Yes	Yes	Yes	Yes
Vary minimum wave angle	Yes	Yes	Yes	Yes
Path regions and hops	Yes	Yes	Yes	Yes
Multipath effects	Yes	Yes	No	Yes
Path probability	Yes	Yes	Yes	Yes
Signal strengths	Yes	Yes	Yes	Yes
S/N ratios	Yes	Yes	Yes	Yes
Long path calculation	Yes	Yes	Yes	No
Antenna selection	Yes	Yes	Indirectly	Yes
Vary antenna height	Yes	Yes	Indirectly	Yes
Vary ground characteristics	Yes	Yes	No	No
Vary transmit power	Yes	Yes	Indirectly	Yes
Graphic displays	Yes	Yes	Yes	2D/3D
UT-day graphs	Yes	Yes	Yes	Yes
Area Mapping	Yes	Yes	Yes	Yes
Documentation	Yes	Online	Yes	Yes
Price class	$AUD350[1]	free[2]	free[3]	$240[4]

Prices are for early 2010 and are subject to change.

[1] *ASAPS* for North America. See **www.ips.gov.au**
[2] *VOACAP* available at **elbert.its.bldrdoc.gov/hf.html**
[3] *W6ELProp*, see **www.qsl.net/w6elprop**
[4] *PropLab Pro*, see **www.spacew.com/proplab/**

19.4 Propagation in the Troposphere

All radio communication involves propagation through the troposphere for at least part of the signal path. Radio waves traveling through the lowest part of the atmosphere are subject to refraction, scattering and other phenomena, much like ionospheric effects. Tropospheric conditions are rarely significant below 30 MHz, but they are very important at 50 MHz and higher. Much of the long-distance work on the VHF, UHF and microwave bands depends on some form of tropospheric propagation. Instead of watching solar activity and geomagnetic indices, those who use tropospheric propagation are much more concerned about the weather.

19.4.1 Line of Sight

At one time it was thought that communications in the VHF range and higher would be restricted to line-of-sight paths. Although this has not proven to be the case even in the microwave region, the concept of line of sight is still useful in understanding tropospheric propagation. In the vacuum of space or in a completely homogeneous medium, radio waves do travel essentially in straight lines, but these conditions are almost never met in terrestrial propagation.

Radio waves traveling through the troposphere are ordinarily refracted slightly earthward. The normal drop in temperature, pressure and water-vapor content with increasing altitude change the index of refraction of the atmosphere enough to cause refraction. Under average conditions, radio waves are refracted toward Earth enough to make the horizon appear 1.15 times farther away than the visual horizon. Under unusual conditions, tropospheric refraction may extend this range significantly.

A simple formula can be used to estimate the distance to the radio horizon under average conditions:

$$d = \sqrt{2h}$$

where
d = distance to the radio horizon, miles
h = height above average terrain, ft.

and

$d = \sqrt{17h}$

where
d = distance to the radio horizon, km
h = height above average terrain, m.

The distance to the radio horizon for an antenna 30 meters (98 ft) above average terrain is thus 22.6 km (14 mi). A station on top of a 1000-meter (3280-ft) mountain has a radio horizon of 130 km (80 mi).

ATMOSPHERIC ABSORPTION

Atmospheric gases, most notably oxygen and water vapor, absorb radio signals, but neither is a significant factor below 10 GHz. See **Fig 19.24**. Attenuation from rain becomes important at 3.3 GHz, where signals passing through 20 km (12 mi) of heavy showers incur an additional 0.2 dB loss. That same rain would impose 12 dB additional loss at 10 GHz and losses continue to increase with frequency. Heavy fog is similarly a problem only at 5.6 GHz and above.

19.4.2 Tropospheric Scatter

Contacts beyond the radio horizon out to a working distance of 100 to 500 km (60 to 310 mi), depending on frequency, equipment and local geography, are made every day without the aid of obvious propagation enhancement. At 1.8 and 3.5 MHz, local communication is due mostly to ground wave. At higher frequencies, especially in the VHF range and above, the primary mechanism is scattering in the troposphere, or *troposcatter*.

Most amateurs are unaware that they use troposcatter even though it plays an essential role in most local communication. Radio signals through the VHF range are scattered primarily by wave-length sized gradients in the index of refraction of the lower atmosphere due to turbulence, along with changes in temperature. Radio signals in the microwave region can also be scattered by rain, snow, fog, clouds and dust. That tiny part that is scattered forward and toward the Earth creates the over-the-horizon paths. Troposcatter path losses are considerable and increase with frequency.

The maximum distance that can be linked via troposcatter is limited by the height of a scattering volume common to two stations, shown schematically in **Fig 19.25**. The highest altitude for which scattering is efficient at amateur power levels is about 10 km (6 mi). An application of the distance-to-the-horizon formula yields 800 km (500 mi) as the limit for troposcatter paths, but typical maxima are about half that. Tropospheric scatter varies little with season or time of

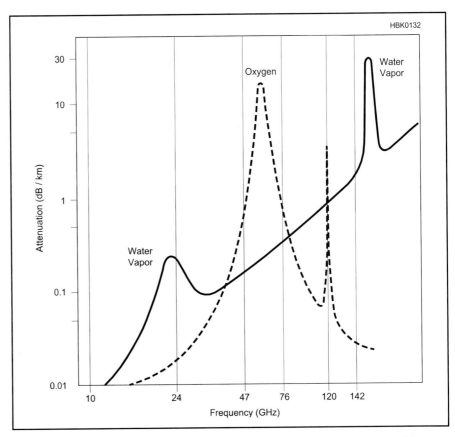

Fig 19.24 — Attenuation caused by oxygen and water vapor at 10 grams per cubic meter (equivalent to 40% humidity at 25 °C). There is little attenuation caused by atmospheric gases at 10 GHz and lower. Note that the 24 GHz band lies near the center of a peak of water vapor absorption and the 120 GHz band is near a peak of oxygen absorption.

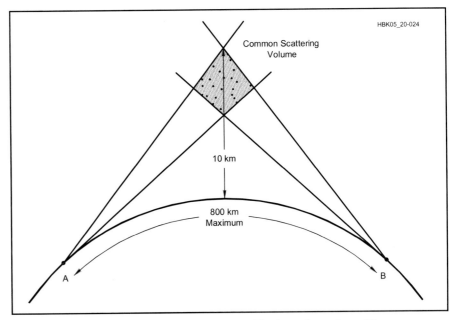

Fig 19.25 — Tropospheric-scatter path geometry. The lower boundary of the common scattering volume is limited by the take-off angle of both stations. The upper boundary of 10 km (6 mi) altitude is the limit of efficient scattering in the troposphere. Signal strength increases with the scattering volume.

day, but it is difficult to assess the effect of weather on troposcatter alone. Variations in tropospheric refraction, which is very sensitive to the weather, probably account for most of the observed day-to-day differences in troposcatter signal strength.

Troposcatter does not require special operating techniques or equipment, as it is used unwittingly all the time. In the absence of all other forms of propagation, especially at VHF and above, the usual working range is essentially the maximum troposcatter distance. Ordinary working range increases most dramatically with antenna height, because that lowers the take-off angle to the horizon. Working range increases less quickly with antenna gain and transmitter power. For this reason, a mountaintop is the choice location for extending ordinary troposcatter working distances.

RAIN SCATTER IN THE TROPOSPHERE

Scatter from raindrops is a special case of troposcatter practical in the 3.3 to 24 GHz range. Stations simply point their antennas toward a common area of rain. A certain portion of radio energy is scattered by the raindrops, making possible over-the-horizon or obstructed-path contacts, even with low power. The theoretical range for rain scatter is as great as 600 km (370 mi), but the experience of amateurs in the microwave bands suggests that expected distances are less than 200 km (120 mi). Snow and hail make less efficient scattering media unless the ice particles are partially melted. Smoke and dust particles are too small for extraordinary scattering, even in the microwave bands.

19.4.3 Refraction and Ducting in the Troposphere

Radio waves are refracted by natural gradients in the index of refraction of air with altitude, due to changes in temperature, humidity and pressure. Refraction under standard atmospheric conditions extends the radio horizon somewhat beyond the visual line of sight. Favorable weather conditions further enhance normal tropospheric refraction, lengthening the useful VHF and UHF range by several hundred kilometers and increasing signal strength. Higher frequencies are more sensitive to refraction, so its effects may be observed in the microwave bands before they are apparent at lower frequencies.

Ducting takes place when refraction is so great that radio waves are bent back to the surface of the Earth. When tropospheric ducting conditions exist over a wide geographic area, signals may remain very strong over distances of 1500 km (930 mi) or more. Ducting results from the gradient created by a sharp increase in temperature with altitude, quite the opposite of normal atmospheric conditions. A simultaneous drop in humidity contributes to increased refractivity.

Normally the temperature steadily decreases with altitude, but at times there is a small portion of the troposphere in which the temperature increases and then again begins decreasing normally. This is called an *inversion*. Useful temperature inversions form between 250 and 2000 meters (800-6500 ft) above ground. The elevated inversion and the Earth's surface act something like the boundaries of a natural open-ended waveguide. Radio waves of the right frequency range caught inside the duct will be propagated for long distances with relatively low losses. Several common weather conditions can create temperature inversions.

RADIATION INVERSIONS IN THE TROPOSPHERE

Radiation inversions are probably the most common and widespread of the various weather conditions that affect propagation. Radiation inversions form only over land after sunset as a result of progressive cooling of the air near the Earth's surface. As the Earth cools by radiating heat into space, the air just above the ground is cooled in turn. At higher altitudes, the air remains relatively warmer, thus creating the inversion. A typical radiation-inversion temperature profile is shown in **Fig 19.26A**.

The cooling process may continue through the evening and predawn hours, creating inversions that extend as high as 500 meters (1500 ft). Deep radiation inversions are most common during clear, calm, summer evenings. They are more distinct in dry climates, in valleys and over open ground. Their formation is inhibited by wind, wet ground and cloud cover. Although radiation inversions are common and widespread, they are rarely strong enough to cause true ducting. The enhanced conditions so often observed after sunset during the summer are usually a result of this mild kind of inversion.

HIGH-PRESSURE WEATHER SYSTEMS

Large, sluggish, high-pressure systems (or *anticyclones*) create the most dramatic and widespread tropospheric ducts due to *subsidence*. Subsidence inversions in high-pressure systems are created by air that is sinking. As air descends, it is compressed and heated. Layers of warmer air — temperature inversions — often form between 500 and 3000 meters (1500-10,000 ft) altitude, as shown in Fig 19.26B. Ducts usually intensify during the evening and early morning hours, when surface temperatures drop and suppress the tendency for daytime ground-warmed air to rise. In the Northern Hemisphere, the longest and strongest radio paths usually lie to the south of high-pressure centers. See **Fig 19.27**.

Sluggish high-pressure systems likely to contain strong temperature inversions are common in late summer over the eastern half of the US. They generally move southeastward out of Canada and linger for days over the Midwest, providing many hours of extended propagation. The southeastern part of the country and the lower Midwest experience the most high-pressure openings; the upper

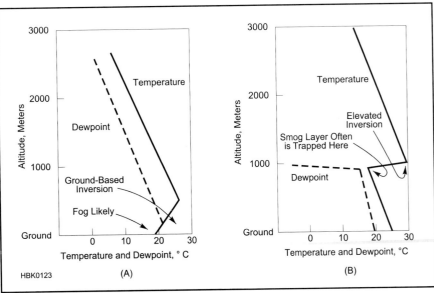

Fig 19.26 — Temperature and dew point profile of an early-morning radiation inversion is shown at A. Fog may form near the ground. The midday surface temperature would be at least 30 °C. At B, temperature and humidity profile across an elevated duct at 1000-m altitude. Such inversions typically form in summertime high-pressure systems. Note the air is very dry in the inversion.

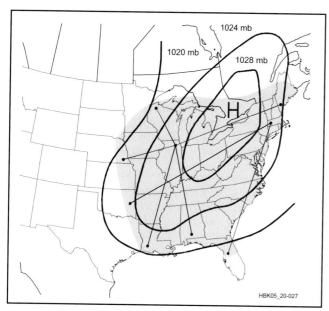

Fig 19.27 — This surface weather map shows the eastern US was dominated by a sprawling high-pressure system. The shaded portion shows the area in which ducting conditions existed on 144 through 1296 MHz and higher.

Fig 19.28 — This surface weather map shows a typical spring wave cyclone over the southeastern quarter of the US. The shaded portion shows where ducting conditions existed.

Midwest and East Coast somewhat less frequently; the western mountain regions rarely.

Semi-permanent high-pressure systems, which are nearly constant climatic features in certain parts of the world, sustain the longest and most exciting ducting paths. The Eastern Pacific High, which migrates northward off the coast of California during the summer, has been responsible for the longest ducting paths reported to date. Countless contacts in the 4000-km (2500 mi) range have been made from 144 MHz through 5.6 GHz between California and Hawaii. The *Bermuda High* is a nearly permanent feature of the Caribbean area, but during the summer it moves north and often covers the southeastern US. It has supported contacts in excess of 2800 km (1700 mi) from Florida and the Carolinas to the West Indies, but its full potential has not been exploited. Other semi-permanent highs lie in the Indian Ocean, the western Pacific and off the coast of western Africa.

WAVE CYCLONE

The *wave cyclone* is a more dynamic weather system that usually appears during the spring over the middle part of the American continent. The wave begins as a disturbance along a boundary between cooler northern and warmer southern air masses. Southwest of the disturbance, a cold front forms and moves rapidly eastward, while a warm front moves slowly northward on the eastward side. When the wave is in its open position, as shown in **Fig 19.28**, north-south radio paths 1500 km (930 mi) and longer may be possible in the area to the east of the cold front and south of the warm front, known as the warm sector. East-west paths nearly as long may also open in the southerly parts of the warm sector.

Wave cyclones are rarely productive for more than a day in any given place, because the eastward-moving cold front eventually closes off the warm sector. Wave-cyclone temperature inversions are created by a southwesterly flow of warm, dry air above 1000 meters (3200 ft) that covers relatively cooler and moister Gulf air flowing northward near the Earth's surface. Successive waves spaced two or three days apart may form along the same frontal boundary.

WARM FRONTS AND COLD FRONTS

Warm fronts and cold fronts sometimes bring enhanced tropospheric conditions, but rarely true ducting. A warm front marks the surface boundary between a mass of warm air flowing over an area of relatively cooler and more stationary air. Inversion conditions may be stable enough several hundred kilometers ahead of the warm front to create extraordinary paths.

A cold front marks the surface boundary between a mass of cool air that is wedging itself under more stationary warm air. The warmer air is pushed aloft in a narrow band behind the cold front, creating a strong but highly unstable temperature inversion. The best chance for enhancement occurs parallel to and behind the passing cold front.

OTHER CONDITIONS ASSOCIATED WITH DUCTS

Certain kinds of wind may also create useful inversions. The *Chinook* wind that blows off the eastern slopes of the Rockies can flood the Great Plains with warm and very dry air, primarily in the springtime. If the ground is cool or snow-covered, a strong inversion can extend as far as Canada to Texas and east to the Mississippi River. Similar kinds of *foehn* winds, as these mountain breezes are called, can be found in the Alps, Caucasus Mountains and other places.

The *land breeze* is a light, steady, cool wind that commonly blows up to 50 km (30 mi) inland from the oceans, although the distance may be greater in some circumstances. Land breezes develop after sunset on clear summer evenings. The land cools more quickly than the adjacent ocean. Air cooled over the land flows near the surface of the Earth toward the ocean to displace relatively warmer air that is rising. See **Fig 19.29**. The warmer ocean air, in turn, travels at 200-300 meters (600-1000 ft) altitude to replace the cool surface air. The land-sea circulation of cool air near the ground and warm air aloft creates a mild inversion that may remain for hours. Land-breeze inversions often bring enhanced conditions and occasionally allow contacts in excess of 800 km (500 mi) along coastal areas.

In southern Europe, a hot, dry wind known as the *sirocco* sometimes blows northward from the Sahara Desert over relatively cooler and moister Mediterranean air. Sirocco inversions can be very strong and extend from Israel and Lebanon westward past the Straits of Gibraltar. Sirocco-type inversions are probably responsible for record-breaking microwave contacts in excess of 1500 km (930 mi) across the Mediterranean.

MARINE BOUNDARY LAYER EFFECTS

Over warm water, such as the Caribbean and other tropical seas, *evaporation inversions*

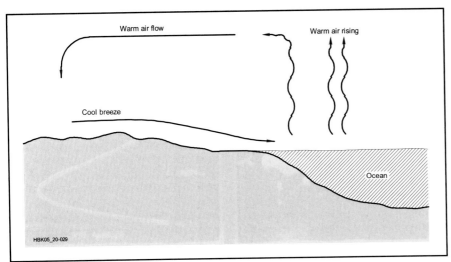

Fig 19.29 — Land-breeze convection along a coast after sunset creates a temperature inversion over the land.

may create ducts that are useful in the microwave region between 3.3 and 24 GHz. This inversion depends on a sharp drop in water-vapor content rather than on an increase in temperature to create ducting conditions. Air just above the surface of water at least 30°C is saturated because of evaporation. Humidity drops significantly within 3 to 10 meters (10 to 30 ft) altitude, creating a very shallow but stable duct. Losses due to water vapor absorption may be intolerable at the highest ducting frequencies, but breezes may raise the effective height of the inversion and open the duct to longer wavelengths. Stations must be set up right on the beaches to ensure being inside an evaporation inversion.

TROPOSPHERIC DUCTING FORECASTS

To help decide if weather conditions may support tropospheric ducting, visit "William Hepburn's Worldwide Tropospheric Ducting Forecasts" at **www.dxinfocentre.com/tropo.html**. The principal content on this Web site is the forecast maps — 6-day preview as well as 42-hour preview (in 6-hour increments) maps for virtually every region of the world.

The purpose of these maps is to display potential duct paths for VHF, UHF and microwave signals, indicated by the color shading on the maps using the Hepburn Tropo Index. This index indicates the degree of tropospheric bending forecast to occur over a particular area. It is an indication of the overall tropospheric radio signal strength on a linear scale from 0 to 10 (with 10 representing an "extremely intense opening" for propagation by ducting).

Also shown on the maps are predicted "unstable signal areas" where weather conditions could potentially disrupt signal paths and cause unusual and sometimes rapid variations in signal strengths

19.4.4 Tropospheric Fading

Because the atmosphere is not homogeneous, radio signals are refracted slightly as they encounter variations in density and humidity. A signal may arrive at the receiver by two or more different paths simultaneously, causing addition or partial cancellation depending on the relative phases and amplitudes of the paths. As the wind blows, the paths change, which causes the net amplitude of the received signal to vary slowly, a process known as *scintillation fading*. This is the same process that causes the stars to twinkle in the visible portion of the electromagnetic spectrum. The higher the frequency and the longer the path length, the more pronounced the effect. While scintillation is generally not significant for short-range VHF/UHF repeater work, it can be the main limitation on power budgets for microwave point-to-point links.

The distribution of signal amplitude is similar to that of log-normal fading of a VHF mobile signal, as described below, but the bandwidth is typically less (that is, the fading rate is slower). Theoretically, the fading bandwidth is 1.426 times the Fresnel frequency F_F, given by

$$F_F = v \sqrt{\frac{1}{2 \pi \lambda R}}$$

where
- v = wind velocity (meters/second),
- λ = RF wavelength (meters), and
- R = path length (meters).

As an example, let v = 10 m/s (22 mph), R = 1000 meters (0.6 miles), and F = 10 GHz (λ = 0.03 meter). The bandwidth calculates to approximately 1 Hz, which implies a fading rate on the order of once per second. At the other end of the spectrum, if we let v = 0.5 m/s (1.1 mph), R = 100 km (62 miles), and F = 430 MHz (0.7 meter), we get a fading bandwidth of approximately 0.001 Hz, or one fade about every 15 minutes.

Fast-flutter fading at 28 MHz and above is often the result of an airplane that temporarily creates a second propagation path. Flutter results as the phase relationship between the direct signal and that reflected by the airplane changes with the airplane's movement.

19.5 VHF/UHF Mobile Propagation

Most amateurs are aware that radio signals in free space obey the inverse-square law: the received signal power is inversely proportional to the square of the distance between the transmitting and receiving antennas. That law applies *only* if the transmitting and receiving antennas have an obstruction-free radio path between them.

Imagine two operators using hand-held 144 MHz radios, each standing on a mountaintop so that they have a direct line of sight. How far apart can they be and still maintain reliable communications? Assume 5 W transmitter power, 0 dBi antenna gain (2.15 dB worse than a dipole), 5 dB receiver noise figure, 10 dB S/N ratio and 12 kHz receiver bandwidth.

Ask experienced VHF mobile operators that question and you'll generally get guesses in the range of 20 to 30 miles because their experience tells them that's about the most you can expect when not communicating through a repeater. The correct answer, however, is over 9500 km (5938 miles) based on the parameters given in the previous paragraph! This also explains how some operators have been able to work the Amateur Radio station on the International Space Station using only a handheld transceiver.

The discrepancy is explained by the fact that the line-of-sight scenario is not realistic for a mobile station located close to ground level. At distances greater than a few miles, there usually is no line of sight — the signal is reflected at least once on its journey from transmitter to receiver. As a result, path loss is typically proportional to distance to the third or fourth power, not the second power as the inverse-square law implies.

19.5.1 Rayleigh Fading

Not only is the signal reflected, but it usually arrives at the receiver by several different paths simultaneously. See **Fig 19.30**. Because the length of each path is different, the signals are not in phase. If the signals on two paths happen to be 180° out of phase and at the same amplitude, they will cancel. If they are in phase then their amplitudes will add. As the mobile station moves about, the phases of the various paths vary in a random fashion. However they tend to be uncorrelated over distances greater than λ/4 or so, which is about 20 inches on the 2 meter band. That is why if the repeater you are listening to drops out when you are stopped at a traffic light, you can often get it back again by creeping forward a few inches.

Because there are typically dozens of paths, it is rare for their amplitudes and phases to be such that they all cancel perfectly. Fades of 20 to 30 dB or more are common. The range of signal strengths has a *Rayleigh* distribu-

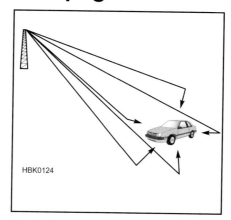

Fig 19.30 — At distances greater than a few miles, there normally is no line-of-sight path to the mobile station. Each path experiences at least one reflection. (Because the radio paths are reciprocal, propagation is the same for both receiving and transmitting.)

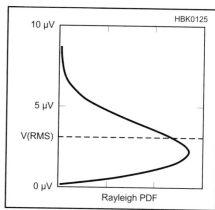

Fig 19.31 — The horizontal axis is the Rayleigh probability density function (PDF), which is the relative probability of the different signal levels shown on the vertical axis. For this graph, an RMS value of 3.16 µV has been selected to represent a typical average signal level in a marginal-coverage area.

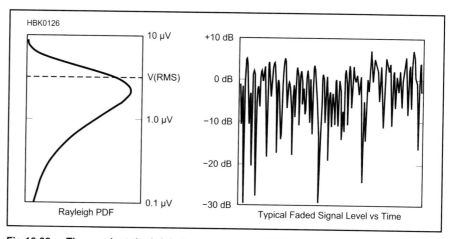

Fig 19.32 — The graph at the left is the same as Fig 19.31 except that the vertical axis is logarithmic (proportional to dB). At the right is a typical Rayleigh-faded signal of the same RMS voltage level plotted using the same vertical scale.

tion, named after the physicist/mathematician who first derived the mathematical formula. That is why the phenomenon is called *Rayleigh fading*. **Fig 19.31** shows the relative probability of various signal strengths. **Fig 19.32** is the same graph plotted on a logarithmic (dB) scale, along with a typical plot of signal strength versus time as the mobile station moves down the road.

The closer a reflecting object is to the antenna the smaller is the path loss. A reflector that is close to the transmitter or receiver antenna gives a much stronger signal than one located halfway between. Even a weak reflector, such as a tree branch or telephone pole, is significant if it is close to the mobile station. Because there are many such close-in reflectors, many rays arrive from all directions.

Rays arriving from in front of a forward-moving vehicle experience a positive Doppler frequency shift and rays from the rear have a negative Doppler shift. Those from the sides are somewhere in-between, proportional to the cosine of the angle of arrival. The received signal is the sum of all those rays, which results in *Doppler spreading* of the signal as illustrated in **Fig 19.33**. At normal vehicle speeds on the 2 meter band, the Doppler spread is only plus and minus 10 or 15 Hz, calculated from

Doppler frequency = $F_c \, v \, / \, c$

where

F_c = the carrier frequency
v = the vehicle speed
c = the speed of light

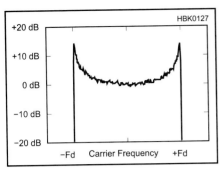

Fig 19.33 — Rayleigh fading spectrum of a CW (unmodulated) signal. The maximum deviation (F_d) from the center frequency is typically less than 10-15 Hz on the 144 MHz band.

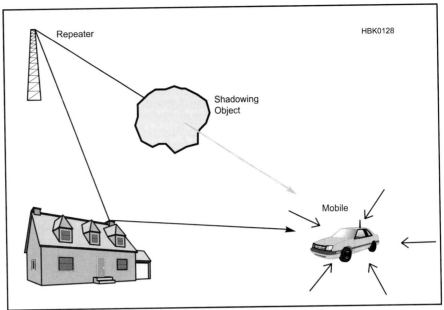

Fig 19.34 — When multipath propagation occurs, each main path is typically Rayleigh-faded by multiple reflectors close to the mobile station.

Use the same units for v and c. On an FM voice signal the only effect is a slight distortion of the audio, but Doppler spreading can severely affect digital signals, as will be discussed later.

19.5.2 Multipath Propagation

In addition to scattering by local reflectors, it is not uncommon also to have more than one main radio path caused by strong reflectors, such as large metal buildings located some distance away. See **Fig 19.34**. Each main path typically does not reach the mobile station directly, but is separately Rayleigh-faded by the local reflectors.

As the mobile station moves around, the shadowing of various paths by intervening hills, buildings and other objects causes the average signal level to fade in and out, but at a much slower rate than Rayleigh fading. This is called *shadowing* or *slow fading*. It is also called *log-normal fading* because the distribution of average signal levels tends to follow a log-normal curve. That means that the *logarithm* of the signal level (on a dB scale, if you will) has a *normal* distribution (the famous bell-shaped curve). This effect typically causes the average signal level on each path to vary plus and minus 10-20 dB (at two standard deviations) from the mean value. This is in addition to the signal variation due to Rayleigh fading.

19.5.3 Effect on the Receiver

Radio direction finding (RDF) enthusiasts have noticed that their RDF receivers usually do not give stable or accurate indications unless located on a hilltop or other location with a clear line of sight to the transmitter because of fading and reflections. The best technique is to record the bearing to the hidden transmitter from a high location clear of nearby reflecting objects, then drive in that direction and take the next reading from another (hopefully closer) hilltop. Only when close to the transmitter will the RDF equipment typically give good readings while the vehicle is in motion.

Under weak-signal conditions Rayleigh fading causes the signal to drop out periodically even though the average signal level would be high enough to maintain reliable communications if there were no fading. *Picket fencing*, as such rapid periodic dropout is called, is a common occurrence when traveling in a weak-signal area at highway speeds. With analog modulation, normally the only solution is to stop at a location where the signal is strong. Moving the vehicle forward or backward a few inches is often enough to change an unreadable signal to solid copy.

Another possible solution is to employ *diversity reception*. If two (or more) mobile receiver antennas are spaced a half-wavelength or more apart, their Rayleigh fading will be almost entirely uncorrelated. That means it is relatively rare for both to experience a deep fade at the same time. The receiving system must have circuitry to determine which of the antennas has a stronger signal at any given time and to automatically combine the signals using some scheme that minimizes the probability of signal dropout. (One engineering text that has a fairly readable discussion of fading is *Cellular Radio Performance Engineering* by Asha Mehrotra; see the "Further Reading" section at the end of this chapter.)

One low-tech scheme is to use stereo headphones with each channel connected to a separate receiver and antenna. (The Ten-Tec Orion operating manual, downloadable from **www.tentec.com**, has a nice discussion of diversity reception.) That method works better with linear modulation (AM, SSB, CW) than with FM because of the noise burst that occurs when the FM signal drops out.

Diversity antennas can also be used at a repeater site. The conditions are different because most repeater antennas are located in the clear with few local reflectors. The diversity antennas must be located much farther apart, typically on the order of 10 to 20 wavelengths, for the fading to be uncorrelated.

FADING AND DIGITAL SIGNALS

Digitally-modulated signals, such as those used in the cellular telephone industry, use several techniques to combat Rayleigh fading. One is *error-correcting coding*, which adds redundant bits to the transmitted signal in a special way such that the receiver can "fill in" missing bits to obtain error-free reception. That technique does not work if the signal drops out completely for longer than a few consecutive bits. The solution often employed is called *interleaving*. This takes data bits that represent points close together in time in the original voice signal and shuffles them into several different time slots for transmission. That way, a single brief dropout does not affect all the bits for that time period. Instead,

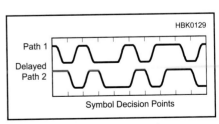

Fig 19.35 — Multipath propagation can cause inter-symbol interference (ISI). At the symbol decision points, where the receiver decoder samples the signal, the path 2 data is often opposing the data on path 1.

the lost data are scattered over several different time periods. Since only a few bits are missing at any one point, the error-correcting decoder in the receiver has enough information to reconstruct the missing information.

As the signal gradually gets weaker, the error correction in a digital receiver produces nearly-perfect voice quality until the signal gets too weak for the decoder to handle. At that point reception stops abruptly and the call is dropped. With an analog signal, reception gets scratchier as the signal gets weaker; this gives advance warning before the signal drops out completely.

The main paths of a multipath-faded signal can differ in length by several miles. A 10 km (6 mile) difference in path length results in a difference in propagation delay of over 30 μs. While that is not noticeable on an FM voice signal, it can wreak havoc with digital signals by causing *intersymbol interference (ISI)*. See **Fig 19.35**. If the delay difference is on the order of one symbol, the receiver sees adjacent symbols superimposed. In effect, the signal interferes with itself.

One solution is to use an *equalizer*. This is a special type of digital filter that filters the demodulated signal to remove the delay differences so that multiple paths become time-aligned. Since the path characteristics change as the mobile station moves around, a special *training sequence* is sent periodically so that the receiver can re-optimize the filter coefficients based on the known symbols in the training sequence.

ISI is an even bigger problem at HF than at VHF/UHF because the path lengths are much greater. It is not unusual to have path length differences up to 3000 km (1800 miles), which correspond to propagation delay differences of up to 10 ms. That is why digital modes on HF with symbol periods of less than about 10 to 20 ms (that is, with symbol rates greater than 50 to 100 baud) are not very practical without some method of equalization.

There is much more detailed information available about mobile radio propagation, but unfortunately most of it is written at an engineering level. For example, Agilent Technologies manufactures a fading simulator that operates in conjunction with their E4438C ESG signal generator. The simulator software may be downloaded for free from **www.agilent.com** — search for the product number N5115A. Although a license must be purchased to make it actually work with a signal generator, it is instructive to play with the user interface, and there is quite a bit of tutorial material on fading in the help file.

For those who would like to explore the subject further, a *Mathcad* file with equations and explanatory text related to Rayleigh fading is available at **www.arrl.org/qst-in-depth**. Look in the 2006 section for Bloom0806.zip. The file was used to generate some of the graphics in this section. For those without access to *Mathcad*, a read-only PDF version of the file is available in the same directory.

19.6 Propagation for Space Communications

Communication of all sorts into space has become increasingly important. Amateurs confront extraterrestrial propagation when accessing satellite repeaters or using the moon as a reflector. (More information on these modes may be found in the chapter on **Space Communications**.) Special propagation problems arise from signals that travel from the Earth through the ionosphere (or a substantial portion of it) and back again. Tropospheric and ionospheric phenomena, so useful for terrestrial paths, are unwanted and serve only as a nuisance for space communication. A phenomenon known as *Faraday rotation* may change the polarization of radio waves traveling through the ionosphere, presenting special problems to receiving weak signals. Cosmic noise also becomes an important factor when antennas are intentionally pointed into space.

19.6.1 Faraday Rotation

Magnetic and electrical forces rotate the polarization of radio waves passing through the ionosphere. For example, signals that leave the Earth as horizontally polarized and return after a reflection from the moon may not arrive with the same polarization. Additional path loss can occur when polarization is shifted by 90°, resulting in fading of the received signal.

Faraday rotation is difficult to predict and its effects change over time and with operating frequency. At 144 MHz, the polarization of space waves may shift back into alignment with the antenna within a few minutes, so often just waiting can solve the Faraday problem. At 432 MHz, it may take half an hour or longer for the polarization to become realigned. Use of circular polarization completely eliminates this problem, but creates a new one for EME paths. The sense of circularly polarized signals is reversed with reflection, so two complete antenna systems are normally required, one with left-hand and one with right-hand polarization.

19.6.2 Scintillation

Extraterrestrial signals experience scintillation fading as they travel through the lower atmosphere, as described above in the *Tropospheric Fading* section. However, they also experience scintillation fading as they traverse the ionosphere. The rule of thumb is that the ionosphere dominates below about 2 GHz and the atmosphere is usually more significant above 2 GHz. Scintillation in the ionosphere is more complex than in the troposphere because, unlike the troposphere, the ionosphere is highly anisotropic and irregularities tend to be aligned with the Earth's magnetic field lines.

Ionospheric scintillation varies with geographical location, time of day, the 11-year

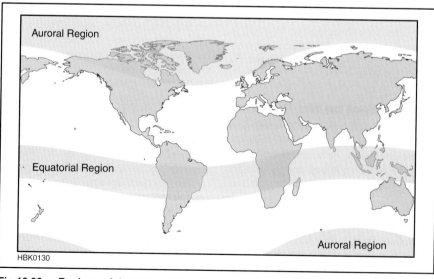

Fig 19.36 — Regions of the world (shaded) where scintillation fading effects can be especially severe.

sunspot cycle, and the presence of geomagnetic storms. **Fig 19.36** shows the worldwide areas where scintillation occurs. The most intense and frequent disturbances are located within about 20 degrees of the Earth's magnetic equator. Fading depth in this region can be as much as 20 to 30 dB. Scintillation fading also is common within about 30 degrees of the magnetic poles, although fade depths are generally no more than 10 dB or so. At mid latitudes ionospheric scintillation is rarely an issue. Longitude is important as well. In South America, Africa and India, the frequency of occurrence is significantly lower in the southern hemisphere winter (May-August) than during the summer, while the pattern is exactly opposite in the central Pacific.

In the polar regions disturbances occur at any time of day, but in the equatorial region they tend to start an hour or so after sunset and last a few hours, from perhaps 1900-2400 local time at the equinox. Fading events are sometimes intermittent, starting abruptly and ending abruptly some minutes later. The signal paths to slow-moving sources, such as the moon or geostationary satellites, tend to fade about once or twice a second. While not much work has been done on characterizing ionospheric scintillation fading on low earth orbit (LEO) satellites, the fade rate should be more than an order of magnitude faster since the orbital period is typically on the order of 1.5 hours, much less than the Earth's 24-hour rotational period.

As might be expected, sunspots also are important. Ionospheric scintillation is much more severe and frequent near the peak of the 11-year solar cycle. Geomagnetic storms also have a major effect, principally in the polar regions.

19.6.3 Earth-Moon-Earth

Amateurs have used the moon as a reflector on the VHF and UHF bands since 1960. Maximum allowable power and large antennas, along with the best receivers, are normally required to overcome the extreme free-space and reflection losses involved in Earth-Moon-Earth (EME) paths. More modest stations make EME contacts by scheduling operating times when the moon is at perigee on the horizon. The moon, which presents a target only one-half degree wide, reflects only 7% of the radio signals that reach it. Techniques have to be designed to cope with Faraday rotation, cosmic noise, Doppler shift (due to the moon's movements) and other difficulties. In spite of the problems involved, hundreds (and possibly thousands) of amateur stations have made contacts via the moon on all bands from 50 MHz to 10 GHz. The techniques of EME communication are discussed in the chapter on **Space Communications**.

19.6.4 Satellites

Accessing amateur satellites generally does not involve huge investments in antennas and equipment, yet station design does have to take into account special challenges of space propagation. Free-space loss is a primary consideration, but it is manageable when satellites are only a few hundred kilometers distant. Free-space path losses to satellites in high-Earth orbits are considerably greater, and appropriately larger antennas and higher powers are needed.

Satellite frequencies below 30 MHz can be troublesome. Ionospheric absorption and refraction may prevent signals from reaching space, especially to satellites at very low elevations. In addition, man-made and natural sources of noise are high. VHF and especially UHF are largely immune from these effects, but free-space path losses are greater. Problems related to polarization, including Faraday rotation, intentional or accidental satellite tumbling and the orientation of a satellite's antenna in relation to terrestrial antennas, are largely overcome by using circularly polarized antennas. More on using satellites can be found in the chapter on **Space Communications**.

19.7 Noise and Propagation

Noise simply consists of unwanted radio signals that interfere with desired communications. In some instances, noise imposes the practical limit on the lowest usable frequencies. Noise may be classified by its sources: man-made, terrestrial and cosmic. Interference from other transmitting stations on adjacent frequencies is not usually considered noise and may be controlled, to some degree anyway, by careful station design or by simply moving to a frequency further away from the interference.

19.7.1 Man-Made Noise

Many unintentional radio emissions result from man-made sources. Broadband radio signals are produced whenever there is a spark, such as in contact switches, electric motors, gasoline engine spark plugs and faulty electrical connections. Household appliances, such as fluorescent lamps, microwave ovens, lamp dimmers and anything containing an electric motor may all produce undesirable broadband radio energy. Devices of all sorts, especially computers and anything controlled by microprocessors, television receivers and many other electronics also emit radio signals that may be perceived as noise well into the UHF range. In many cases, these sources are local and can be controlled with proper measures. See the **EMC / Direction Finding** chapter.

High-voltage transmission lines and associated equipment, including transformers, switches and lightning arresters, can generate high-level radio signals over a wide area, especially if they are corroded or improperly maintained. Transmission lines may act as efficient antennas at some frequencies, adding to the noise problem. Certain kinds of street lighting, neon signs and industrial equipment also contribute their share of noise.

Fig 19.37 shows typical noise levels versus frequency, in terms of power, for various

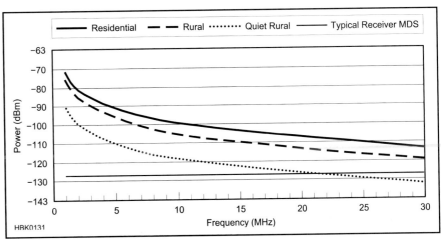

Fig 19.37 — Typical noise levels versus frequency for various environments. (Man-made noise in a 500-Hz bandwidth, from Rec. ITU-R P.372.7, *Radio Noise*)

noise environments. Note that man-made noise prevails below roughly 14 MHz, and it, rather than the sensitivity of your receiver (here depicted as the MDS — minimum discernible receiver), determines how weak a signal you can hear.

19.7.2 Lightning

Static is a common term given to the ear-splitting crashes of noise commonly heard on nearly all radio frequencies, although it is most severe on the lowest frequency bands. Atmospheric static is primarily caused by lightning and other natural electrical discharges. Static may result from close-by thunderstorms, but most static originates with tropical storms. Like any radio signals, lightning-produced static may be propagated over long distances by the ionosphere. Thus static is generally higher during the summer, when there are more nearby thunderstorms, and at night, when radio propagation generally improves. Static is often the limiting factor on 1.8 and 3.5 MHz, making winter a more favorable time for using these frequencies. (Note that the quiet "winter" months in the Southern Hemisphere are June through August.)

19.7.3 Precipitation Static and Corona Discharge

Precipitation static is an almost continuous hash-type noise that often accompanies various kinds of precipitation, including snowfall. Precipitation static is caused by rain drops, snowflakes or even wind-blown dust, transferring a small electrical charge on contact with an antenna. Electrical fields under thunderstorms are sufficient to place many objects such as trees, hair and antennas, into corona discharge. *Corona noise* may sound like a harsh crackling in the radio — building in intensity, abruptly ending, and then building again, in cycles of a few seconds to as long as a minute. A corona charge on an antenna may build to some critical level and then discharge in the atmosphere with an audible pop before recharging. Precipitation static and corona discharge can be a nuisance from LF to well into the VHF range.

19.7.4 Cosmic Sources

The sun, distant stars, galaxies and other cosmic features all contribute radio noise well into the gigahertz range. These *cosmic sources* are perceived primarily as a more-or-less constant background noise at HF. In the VHF range and higher, specific sources of cosmic noise can be identified and may be a limiting factor in terrestrial and space communications. The sun is by far the greatest source of radio noise, but its effects are largely absent at night. The center of our own galaxy is nearly as noisy as the sun. Galactic noise is especially noticeable when high-gain VHF and UHF antennas, such as may be used for satellite or EME communications, are pointed toward the center of the Milky Way. Other star clusters and galaxies are also radio hotspots in the sky. Finally, there is a much lower cosmic background noise that seems to cover the entire sky.

19.8 Glossary of Radio Propagation Terms

A index — An open-ended linear index that corresponds roughly to the cumulative **K index** values (it's the daily average of the eight K indices after converting the K indices to a linear scale). The A index commonly varies between 0 and 30 during quiet to active conditions, and up to 100 and higher during geomagnetic storms.

Absorption — The dissipation of the energy of a radio wave as it travels through a medium such as the ionosphere.

Antipode — Locations directly opposite each other on a globe.

Atmosphere — The mass of air surrounding the Earth. Radio signals travel through the atmosphere and different conditions in the atmosphere affect how those signal travel or propagate.

Aurora — A disturbance of the atmosphere at high latitudes resulting from an interaction between electrically charged particles from the sun and the magnetic field of the Earth. *Auroral propagation* occurs when HF through UHF signals are reflected from the aurora to reach another station.

Auroral E — Sporadic E in the auroral zone.

Backscatter — Single-hop signals that have been scattered by the Earth or the ocean at some distant point back toward the transmitting station.

Beacon station — A station that transmits continuously, allowing other stations to assess propagation to and from the location of the beacon station.

Coronal hole — A region on the sun where the magnetic field is open to the interplanetary magnetic field (IMF) and ionized particles can escape into the solar wind.

Critical angle — The largest angle at which a radio wave of a specified frequency can be returned to Earth by the ionosphere.

Critical frequency — The highest frequency that returns echoes from the E and F regions at vertical incidence.

D region — The lowest region of the ionosphere. The D region (or layer) contributes very little to short-wave radio propagation. It absorbs energy from radio waves as they pass through it. This absorption has a significant effect on signals below about 7.5 MHz during daylight.

Diffraction — Bending of waves by an edge or corner.

E region — The second lowest ionospheric region, the E region (or layer) exists only during the day. Under certain conditions, it may refract radio waves enough to return them to Earth.

Earth-Moon-Earth (EME) or *Moonbounce* — A method of communicating with other stations by reflecting radio signals off the Moon's surface.

Electromagnetic wave — A wave of energy composed of an electric and magnetic field.

Equinoxes — One of two points in the orbit of the Earth around the Sun at which the Earth crosses a horizontal plane extending through the equator of the Sun. The vernal equinox marks the beginning of spring and the autumnal equinox marks the beginning of autumn.

Faraday rotation — A rotation of the polarization of radio waves when the waves travel through the ionized magnetic field of the ionosphere.

F region — A combination of the two highest ionospheric regions (or layers), the F1 and F2 regions. The F region refracts radio waves and returns them to Earth. Its height varies greatly depending on the time of day, season of the year and amount of sunspot activity.

Field — A region of space in which energy is stored and through which electrical and magnetic forces act.

Field-aligned irregularities (FAI) — A propagation mechanism observed at 50 and 144 MHz that occurs when irregularities in the distribution of free electrons in the ionosphere are aligned parallel to the Earth's magnetic field.

Free-space attenuation — The dissipation of the energy of a radio wave that results from the dispersal (spherical spreading) of radio energy from its source.

Gray-line — A special form of **long-path**

propagation that takes into account the unusual ionospheric configuration along the twilight region between night and day.

Ground-wave propagation — The method by which radio waves travel along the Earth's surface.

High frequency (HF) — The term used for the frequency range between 3 MHz and 30 MHz. The amateur HF bands are where you are most likely to make long-distance (worldwide) contacts.

Ionosphere — A region of electrically charged (ionized) gases high in the atmosphere. The ionosphere bends radio waves as they travel through it, returning them to Earth. Also see **sky-wave propagation**.

K index — A geomagnetic-field measurement that is updated every three hours at Boulder, Colorado. Changes in the K index can be used to indicate HF propagation conditions. Rising values generally indicate disturbed conditions while falling values indicate improving conditions.

Line-of-sight propagation — The term used to describe VHF and UHF propagation in a straight line directly from one station to another.

Long-path propagation — Propagation between two points on the Earth's surface that follows a path along the great circle between them, but in a direction opposite from the shortest distance between them.

Lowest usable frequency (LUF) — The frequency below the **maximum usable frequency** (MUF) at which ionospheric absorption and noise at the receiving location make the received signal-to-noise ratio too low to be usable.

M-factor — The ratio between the **maximum usable frequency** (MUF) and the critical frequency.

Maximum usable frequency (MUF) — The highest-frequency radio signal that will reach a particular destination using **sky-wave propagation**, or *skip*. The MUF may vary for radio signals sent to different destinations.

Meteor-scatter communication — A method of radio communication that uses the ionized trail of a meteor that burned up in the Earth's atmosphere to reflect radio signals back to Earth.

Microwave — Radio waves or signals with frequencies greater than 1000 MHz (1 GHz). This is not a strict definition, just a conventional way of referring to those frequencies.

Moonbounce — A common name for EME communication in which signals are bounced off the Moon before being received.

Multihop propagation — Long-distance radio propagation using several skips or hops between the Earth and the ionosphere.

Multipath — A fading effect caused by the transmitted signal traveling to the receiving station over more than one path.

Near Vertical Incidence Skywave (NVIS) propagation — A propagation mechanism that allows stations located within the **skip zone** but too far apart for ground wave propagation to maintain communications by going to a lower frequency.

Path loss — The total signal loss between transmitting and receiving stations relative to the total radiated signal energy.

Pedersen ray — A high-angle radio wave that penetrates deeper into the F region of the ionosphere, so the wave is bent less than a lower-angle wave, and thus for some distance parallels the Earth's surface in the F region, returning to Earth at a distance farther than normally expected for single-hop propagation.

Polarization — The orientation of the electrical-field of a radio wave. An antenna that is parallel to the surface of the earth, such as a dipole, produces horizontally polarized waves. One that is perpendicular to the earth's surface, such as a quarter-wave vertical, produces vertically polarized waves. An antenna that has both horizontal and vertical polarization is said to be circularly polarized.

Propagation — The process by which radio waves travel.

Radio frequency (RF) signals — RF signals are generally considered to be any electrical signals with a frequency higher than 20,000 Hz, up to 300 GHz.

Radio horizon — The position at which a direct wave radiated from an antenna becomes tangent to the surface of the Earth. Note that as the wave continues past the horizon, the wave gets higher and higher above the surface.

Radiation inversion — A weather condition that affects VHF and above propagation. Radiation inversions form only over land after sunset as a result of progressive cooling of the air near the Earth's surface.

Rain scatter — A special case of tropospheric scatter practical in the 3.3 to 24 GHz range that is caused by scatter from raindrops.

Reflection — Signals that travel by **line-of-sight propagation** are reflected by large objects like buildings.

Refraction — Bending waves by changing the velocity of propagation. Radio waves refract as they travel through the ionosphere. If the radio waves refract enough they will return to Earth. This is the basis for long-distance communication on the HF bands.

Scattering — Radio wave propagation by means of multiple reflections in the layers of the atmosphere or from an obstruction. Scatter propagation also occurs in the ionosphere when there is not enough ionization for refraction or reflection, but enough to send weak electromagnetic waves off into varied directions.

Scintillation fading — Fading that occurs when a signal arrives at the receiver by two or more different paths simultaneously, causing addition or partial cancellation depending on the relative phases and amplitudes of the paths.

Selective fading — A variation of radio-wave intensity that changes over small frequency changes. It may be caused by changes in the medium through which the wave is traveling or changes in transmission path, among other things.

Short path — The shorter of the two great circle paths between two stations.

Skip — Propagation by means of ionospheric reflection. Traversing the distance to the ionosphere and back to the ground is called a *hop*.

Skip zone — A ring-shaped area of poor radio communication, too distant for ground waves and too close for sky waves.

Sky-wave propagation — The method by which radio waves travel through the ionosphere and back to Earth. Sometimes called *skip*, sky-wave propagation has a far greater range than **line-of-sight** and **ground-wave propagation**.

Solar cycle — The 10.7 year period of variation in solar activity.

Solar flare — An eruption on the surface of the sun that launches a wide spectrum of electromagnetic energy into space, disrupting communications on Earth.

Solar wind — Electrically charged particles emitted by the Sun and traveling through space. Variations in the solar wind may have a sudden impact on radio communications when they arrive at the atmosphere of the Earth.

Sporadic E — A form of enhanced radio-wave propagation that occurs when radio signals are reflected from small, dense ionization patches in the E region of the ionosphere. Sporadic E is observed on the 15, 10, 6 and 2-meter bands, and occasionally on the 1.25-meter band.

Sunspot cycle — The number of **sunspots** increases and decreases in a predictable

cycle that lasts about 11 years.
Sunspots — Dark spots on the surface of the sun. When there are few sunspots, long-distance radio propagation is poor on the higher-frequency bands. When there are many sunspots, long-distance HF propagation improves.
Temperature inversion — A condition in the atmosphere in which a region of cool air is trapped beneath warmer air.
Transequatorial propagation — A form of F layer ionospheric propagation, in which signals of higher frequency than the expected MUF are propagated across the Earth's magnetic equator.
Troposphere — The region in Earth's atmosphere just above the Earth's surface and below the ionosphere.
Tropospheric bending — When radio waves are bent in the troposphere, they return to Earth farther away than the visible horizon.
Tropospheric ducting — A type of VHF propagation that can occur when warm air overruns cold air (a temperature inversion).
Tropospheric scatter — A method of radio communication at VHF and above propagation mechanism that takes advantage of scattering in the **troposphere** to allow contacts beyond the radio horizon out to a working distance of 100 to 500 km (60 to 310 mi), depending on frequency.
Ultra high frequency (UHF) — The term used for the frequency range between 300 MHz and 3000 MHz (3 GHz). Technician licensees have full privileges on all Amateur UHF bands.
Very high frequency (VHF) — The term used for the frequency range between 30 MHz and 300 MHz. Technician licensees have full privileges on all Amateur VHF bands.
Visible horizon — The most distant point one can see by line of sight.
WWV/WWVH — Radio stations run by the US NIST (National Institute of Standards and Technology) to provide accurate time and frequencies.

19.9 References and Bibliography

A. Barter, G8ATD, ed., *International Microwave Handbook*, 2nd edition (Potters Bar: RSGB, 2008). Includes a chapter on microwave propagation.

A. Barter, G8ATD, ed., *VHF/UHF Handbook*, 2nd edition (Potters Bar: RSGB, 2008). Includes a chapter on VHF/UHF propagation.

B. R. Bean and E. J. Dutton, *Radio Meteorology* (New York: Dover, 1968).

K. Davies, *Ionospheric Radio* (London: Peter Peregrinus, 1989). Excellent though highly technical text on propagation.

R. D. Hunsucker, J.K. Hargreaves, *The High Latitude Ionosphere and Its Effect on Radio Propagation* (Cambridge University Press, 2003). Highly technical, but with an excellent chapter on fundamental physics of the ionosphere.

G. Jacobs, T. Cohen, R. Rose, *The New Shortwave Propagation Handbook*, CQ Communications, Inc. (Hicksville, NY: CQ Communications, 1995)

C. Luetzelschwab, K9LA, **mysite.verizon.net/k9la**. Many articles about the ionosphere and propagation as they relate to 160 meters, HF, VHF, contesting and more.

L. F. McNamara, *Radio Amateur's Guide to the Ionosphere* (Malabar, Florida: Krieger Publishing Company, 1994). Excellent, quite-readable text on HF propagation.

A. Mehrotra, *Cellular Radio Performance Engineering* (Artech House, 1994). See Chapter 4 "Propagation" and the introduction to Chapter 3 "Cellular Environment." Diversity reception techniques are covered in Chapter 8.

D. Straw, N6BV, "What's the Deal About 'NVIS'?," *QST*, Dec 2005.

Contents

20.1 Transmission Line Basics
    20.1.1 Fundamentals
20.1.2 Matched and Mismatched Lines
    20.1.3 Reflection Coefficient and SWR
    20.1.4 Losses in Transmission Lines
20.2 Choosing a Transmission Line
20.3 The Transmission Line as Impedance Transformer
    20.3.1 Transmission Line Stubs
    20.3.2 Transmission Line Stubs as Filters
    20.3.3 Project: A Field Day Stub Assembly
20.4 Matching Impedances in the Antenna System
    20.4.1 Conjugate Matching
    20.4.2 Impedance Matching Networks
    20.4.3 Matching Antenna Impedance at the Antenna
    20.4.4 Matching the Line to the Transmitter
    20.4.5 Adjusting Antenna Tuners
    20.4.6 Myths About SWR
20.5 Baluns and Transmission-Line Transformers
    20.5.1 Quarter-wave Baluns
    20.5.2 Transmission Line Transformers
    20.5.3 Coiled-Coax Choke Baluns
    20.5.4 Transmitting Ferrite Choke Baluns
20.6 Using Transmission Lines in Digital Circuits
20.7 Waveguides
    20.7.1 Evolution of a Waveguide
    20.7.2 Modes of Waveguide Propagation
    20.7.3 Waveguide Dimensions
    20.7.4 Coupling to a Waveguide
20.8 Glossary of Transmission Line Terms
20.9 References and Bibliography

Chapter 20

Transmission Lines

RF power is rarely generated right where it will be used. A transmitter and the antenna it feeds are a good example. The most effective antenna installation is outdoors and clear of ground and energy-absorbing structures. The transmitter, however, is most conveniently installed indoors, where it is out of the weather and is readily accessible. A *transmission line* is used to convey RF energy from the transmitter to the antenna. A transmission line should transport the RF from the source to its destination with as little loss as possible. This chapter, written by Dean Straw, N6BV, and updated by George Cutsogeorge, W2VJN, explores transmission line theory and applications. Jim Brown, K9YC, contributed updated material on transmitting choke baluns.

20.1 Transmission Line Basics

There are three main types of transmission lines used by radio amateurs: coaxial, open-wire and waveguide. The most common type is the *coaxial* line, usually called *coax*, shown in various forms in **Fig 20.1**. Coax is made up of a center conductor, which may be either stranded or solid wire, surrounded by a concentric outer conductor with a *dielectric* center insulator between the conductors. The outer conductor may be braided shield wire or a metallic sheath. A flexible aluminum foil or a second braided shield is employed in some coax to improve shielding over that obtainable from a standard woven shield braid. If the outer conductor is made of solid aluminum or copper, the coax is referred to as *hardline*.

The second type of transmission line uses parallel conductors, side by side, rather than the concentric ones used in coax. Typical examples of such *open-wire* lines are 300-Ω TV ribbon line or *twin-lead* and 450-Ω ladder line (sometimes called *window line*), also shown in Fig 20.1. Although open-wire lines are enjoying a sort of renaissance in recent years because of their inherently lower losses in simple multiband antenna systems, coaxial cables are far more prevalent because they are much more convenient to use.

The third major type of transmission line is the *waveguide*. While open-wire and coaxial lines are used from power-line frequencies to well into the microwave region, waveguides are used at microwave frequencies only. Waveguides will be covered at the end of this chapter.

20.1.1 Fundamentals

In either coaxial or open-wire line, currents flowing in the two conductors travel in opposite directions as shown in Figs 20.1E and 20.1I. If the physical spacing between the two parallel conductors in an open-wire line, S, is small in terms of wavelength, the phase difference between the currents will be very close to 180°. If the two currents also have equal amplitudes, the field generated by each conductor will cancel that generated by the other, and the line will not radiate energy, even if it is many wavelengths long.

The equality of amplitude and 180° phase difference of the currents in each conductor in an open-wire line determine the degree of radiation cancellation. If the currents are for some reason unequal, or if the phase difference is not 180°, the line will radiate energy. How such imbalances occur and to what degree they can cause problems will be covered in more detail later.

In contrast to an open-wire line, the outer conductor in a coaxial line acts as a shield, confining RF energy within the line as shown in Fig 20.1E. Because of *skin effect* (see the **RF Techniques** chapter), current flowing in the outer conductor of a coax does so on the inner surface of the outer conductor. The fields generated by the currents flowing on the outer surface of the inner conductor and on the inner surface of the outer conductor cancel each other out, just as they do in open-wire line.

VELOCITY FACTOR

In free space, electrical waves travel at the speed of light, or 299,792,458 meters per second. Converting to feet per second yields 983,569,082. The length of a wave in space may be related to frequency as wavelength = λ = velocity/frequency. Thus, the wavelength of a

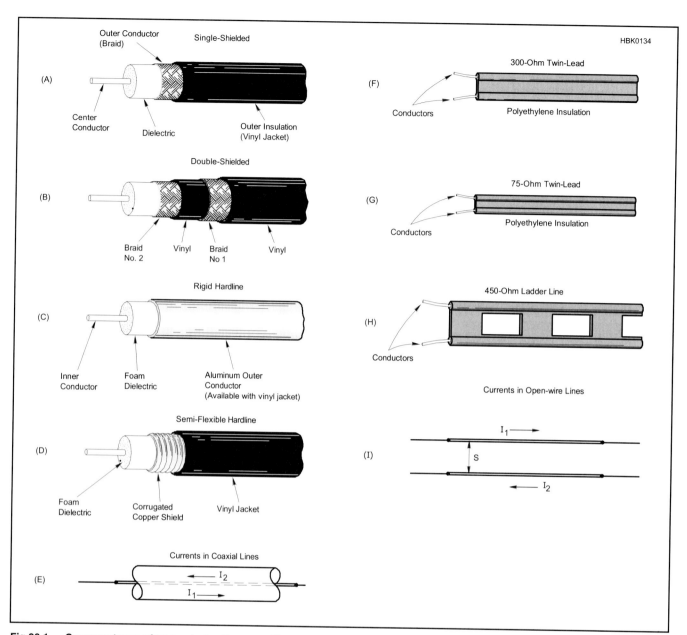

Fig 20.1 — Common types of transmission lines used by amateurs. Coaxial cable, or "coax," has a center conductor surrounded by insulation. The second conductor, called the shield, cover the insulation and is, in turn, covered by the plastic outer jacket. Various types are shown at A, B, C and D. The currents in coaxial cable flow on the outside of the center conductor and the inside of the outer shield (E). Open-wire line (F, G and H) has two parallel conductors separated by insulation. In open-wire line, the current flows in opposite directions on each wire (I).

1 Hz signal is 983,569,082 ft. Changing to a more useful expression gives:

$$\lambda = \frac{983.6}{f} \quad (1)$$

where
λ = wavelength, in ft
f = frequency in MHz.

Thus, at 14 MHz the wavelength is 70.25 ft.

Wavelength (λ) may also be expressed in electrical degrees. A full wavelength is 360°, ½ λ is 180°, ¼ λ is 90°, and so forth.

Waves travel slower than the speed of light in any medium denser than a vacuum or free space. A transmission line may have an insulator which slows the wave travel down. The actual velocity of the wave is a function of the dielectric characteristic of that insulator. We can express the variation of velocity as the *velocity factor* for that particular type of dielectric — the fraction of the wave's velocity of propagation in the transmission line compared to that in free space. The velocity factor is related to the dielectric constant of the material in use.

$$VF = \frac{1}{\sqrt{\varepsilon}} \quad (2)$$

where
VF = velocity factor
ε = dielectric constant.

So the wavelength in a real transmission line becomes:

$$\lambda = \frac{983.6}{f} VF \quad (3)$$

As an example, many coax cables use polyethylene dielectric over the center conductor as the insulation. The dielectric constant for polyethylene is 2.3, so the VF is 0.66. Thus, wavelength in the cable is about two-thirds as long as a free-space wavelength.

The VF and other characteristics of many types of lines, both coax and twin lead, are shown in the table "Nominal Characteristics of Commonly used Transmission Lines" in the **Component Data and References** chapter.

There are differences in VF from batch to batch of transmission line because there are some variations in dielectric constant during the manufacturing processes. When high accuracy is required, it is best to actually measure VF by using an antenna analyzer to measure the resonant frequency of a length of cable. (The antenna analyzer's user manual will describe the procedure.)

CHARACTERISTIC IMPEDANCE

A perfectly lossless transmission line may be represented by a whole series of small inductors and capacitors connected in an infinitely long line, as shown in **Fig 20.2**. (We first consider this special case because we need not consider how the line is terminated at its end, since there is no end.)

Each inductor in Fig 20.2 represents the inductance of a very short section of one wire and each capacitor represents the capacitance between two such short sections. The inductance and capacitance values per unit of line depend on the size of the conductors and the spacing between them. Each series inductor acts to limit the rate at which current can charge the following shunt capacitor, and in so doing establishes a very important property of a transmission line: its *surge impedance*, more commonly known as its *characteristic impedance*. This is usually abbreviated as Z_0.

$$Z_0 \approx \sqrt{\frac{L}{C}}$$

where L and C are the inductance and capacitance per unit length of line.

The characteristic impedance of an air-insulated parallel-conductor line (see Fig 20.1I), neglecting the effect of the insulating spacers, is given by

$$Z_0 = 276 \log_{10} \frac{2S}{d} \quad (4)$$

where
Z_0 = characteristic impedance
S = center-to-center distance between conductors
d = diameter of conductor (in same units as S).

The characteristic impedance of an air-insulated coaxial line is given by

$$Z_0 = 138 \log_{10} \left(\frac{b}{a}\right) \quad (5)$$

where
Z_0 = characteristic impedance
b = inside diameter of outer conductors
a = outside diameter of inner conductor (in same units as b).

It does not matter what units are used for S, d, a or b, as long as they are the *same* units. A line with closely spaced, large conductors will have a low characteristic impedance, while one with widely spaced, small conductors will have a relatively high characteristic impedance. Practical open-wire lines exhibit characteristic impedances ranging from about 200 to 800 Ω, while coax cables have Z_0 values between 25 to 100 Ω. Except in special instances, coax used in amateur radio has an impedance of 50 or 75 Ω.

All practical transmission lines exhibit some power loss. These losses occur in the resistance that is inherent in the conductors that make up the line, and from leakage currents flowing in the dielectric material between the conductors. We'll next consider what happens when a real transmission line, which is not infinitely long, is terminated in an actual load impedance, such as an antenna.

20.1.2 Matched and Mismatched Lines

Real transmission lines do not extend to infinity, but have a definite length. In use they are connected to, or *terminate* in, a load (such as an antenna), as illustrated in **Fig 20.3A**. If the load is a pure resistance whose value equals the characteristic impedance of the line, the line is said to be *matched*. To current traveling along the line, such a load at the end of the line acts as though it were still more transmission line of the same characteristic impedance. In a matched transmission line, energy travels outward along the line from the source until it reaches the load, where it is completely absorbed (or radiated if the load is an antenna).

MISMATCHED LINES

Assume now that the line in Fig 20.3B is terminated in an impedance Z_a which is not equal to Z_0 of the transmission line. The line is now a *mismatched* line. RF energy reaching the end of a mismatched line will not be fully absorbed by the load impedance. Instead, part of the energy will be reflected back toward the source. The amount of reflected versus absorbed energy depends on the degree of mismatch between the characteristic impedance of the line and the load impedance connected to its end.

The reason why energy is reflected at a discontinuity of impedance on a transmission line can best be understood by examining some limiting cases. First, consider the rather extreme case where the line is shorted at its end. Energy flowing to the load will encounter the short at the end, and the voltage at that point

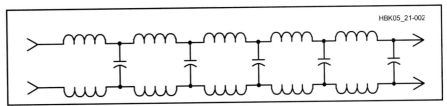

Fig 20.2 — Equivalent of an infinitely long lossless transmission line using lumped circuit constants.

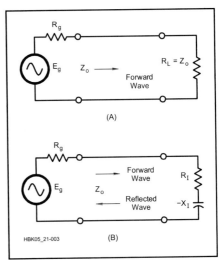

Fig 20.3 — At A the coaxial transmission line is terminated with resistance equal to its Z_0. All power is absorbed in the load. At B, coaxial line is shown terminated in an impedance consisting of a resistance and a capacitive reactance. This is a mismatched line, and a reflected wave will be returned back down the line toward the generator. The reflected wave adds to the forward wave, producing a standing wave on the line. The amount of reflection depends on the difference between the load impedance and the characteristic impedance of the transmission line.

will go to zero, while the current will rise to a maximum. Since the current can't develop any power in a dead short, the energy will all be reflected back toward the source generator.

If the short at the end of the line is replaced with an open circuit, the opposite will happen. Here the voltage will rise to maximum, and the current will by definition go to zero. The phase will reverse, and all energy will be reflected back towards the source. By the way, if this sounds to you like what happens at the end of a half-wavelength dipole antenna, you are quite correct. However, in the case of an antenna, energy traveling along the antenna is lost by radiation on purpose, whereas a good transmission line will lose little energy to radiation because of field cancellation between the two conductors.

For load impedances falling between the extremes of short- and open-circuit, the phase and amplitude of the reflected wave will vary. The amount of energy reflected and the amount of energy absorbed in the load will depend on the difference between the characteristic impedance of the line and the impedance of the load at its end.

What actually happens to the energy reflected back down the line? This energy will encounter another impedance discontinuity, this time at the generator. Reflected energy flows back and forth between the mismatches at the source and load. After a few such journeys, the reflected wave diminishes to nothing, partly as a result of finite losses in the line, but mainly because of absorption at the load. In fact, if the load is an antenna, such absorption at the load is desirable, since the energy is actually radiated by the antenna.

If a continuous RF voltage is applied to the terminals of a transmission line, the voltage at any point along the line will consist of a vector sum of voltages, the composite of waves traveling toward the load and waves traveling back toward the source generator. The sum of the waves traveling toward the load is called the *forward* or *incident* wave, while the sum of the waves traveling toward the generator is called the *reflected wave*.

20.1.3 Reflection Coefficient and SWR

In a mismatched transmission line, the ratio of the voltage in the reflected wave at any one point on the line to the voltage in the forward wave at that same point is defined as the *voltage reflection coefficient*. This has the same value as the current reflection coefficient. The reflection coefficient is a complex quantity (that is, having both amplitude and phase) and is generally designated by the Greek letter ρ (rho), or sometimes in the professional literature as Γ (Gamma). The relationship between R_l (the load resistance), X_l (the load reactance), Z_0 (the line characteristic impedance, whose real part is R_0 and whose reactive part is X_0) and the complex reflection coefficient ρ is

$$\rho = \frac{Z_l - Z_0}{Z_l + Z_0} = \frac{(R_l \pm jX_l) - (R_0 \pm jX_0)}{(R_l \pm jX_l) + (R_0 \pm jX_0)} \quad (6)$$

For most transmission lines the characteristic impedance Z_0 is almost completely resistive, meaning that $Z_0 = R_0$ and $X_0 \cong 0$. The magnitude of the complex reflection coefficient in equation 6 then simplifies to:

$$|\rho| = \sqrt{\frac{(R_l - R_0)^2 + X_l^2}{(R_l + R_0)^2 + X_l^2}} \quad (7)$$

For example, if the characteristic impedance of a coaxial line is 50 Ω and the load impedance is 120 Ω in series with a capacitive reactance of −90 Ω, the magnitude of the reflection coefficient is

$$|\rho| = \sqrt{\frac{(120 - 50)^2 + (-90)^2}{(120 + 50)^2 + (-90)^2}} = 0.593$$

Note that if R_l in equation 6 is equal to R_0 and X_l is 0, the reflection coefficient, ρ, is 0. This represents a matched condition, where all the energy in the incident wave is transferred to the load. On the other hand, if R_l is 0, meaning that the load is a short circuit and has no real resistive part, the reflection coefficient is 1.0, regardless of the value of R_0. This means that all the forward power is reflected since the load is completely reactive.

The concept of reflection is often shown in terms of the *return loss* (RL), which is given in dB and is equal to 20 times the log of the reciprocal of the reflection coefficient.

$$RL(dB) = -10\log\left(\frac{P_r}{P_f}\right) = -20\log(\rho) \quad (8)$$

In the example above, the return loss is 20 log (1/0.593) = 4.5 dB.

If there are no reflections from the load, the voltage distribution along the line is constant or *flat*. A line operating under these conditions is called either a *matched* or a *flat* line. If reflections do exist, a voltage *standing-wave* pattern will result from the interaction of the forward and reflected waves along the line. For a lossless transmission line at least ¼-λ long, the ratio of the maximum peak voltage anywhere on the line to the minimum value anywhere along the line is defined as the *voltage standing-wave ratio*, or VSWR. (The line must be ¼ λ or longer for the true maximum and minimum to be created.) Reflections from the load also produce a standing-wave pattern of currents flowing in the line. The ratio of maximum to minimum current, or ISWR, is identical to the VSWR in a given line.

In amateur literature, the abbreviation SWR is commonly used for standing-wave ratio, as the results are identical when taken from proper measurements of either current or voltage. Since SWR is a ratio of maximum to minimum, it can never be less than one-to-one. In other words, a perfectly flat line has an SWR of 1:1. The SWR is related to the magnitude of the complex reflection coefficient and vice versa by

$$SWR = \frac{1 + |\rho|}{1 - |\rho|} \quad (9A)$$

and

$$|\rho| = \frac{SWR - 1}{SWR + 1} \quad (9B)$$

The definitions in equations 8 and 9 are valid for any line length and for lines which are lossy, not just lossless lines longer than ¼ λ at the frequency in use. Very often the load impedance is not exactly known, since an antenna usually terminates a transmission line, and the antenna impedance may be influenced by a host of factors, including its height above ground, end effects from insulators, and the effects of nearby conductors. We may also express the reflection coefficient in terms of forward and reflected power, quantities which can be easily measured using a directional RF wattmeter. The reflection coefficient and SWR may be computed as

$$|\rho| = \sqrt{\frac{P_r}{P_f}} \quad (10A)$$

and

$$SWR = \frac{1 + \sqrt{\frac{P_r}{P_f}}}{1 - \sqrt{\frac{P_r}{P_f}}} \quad (10B)$$

where
P_r = power in the reflected wave
P_f = power in the forward wave.

If a line is not matched (SWR > 1:1) the difference between the forward and reflected powers measured at any point on the line is the net power going toward the load from that point. The forward power measured with a directional wattmeter (often referred to as a reflected power meter or *reflectometer*) on a mismatched line will thus always appear greater than the forward power measured on a flat line with a 1:1 SWR.

The software program *TLW*, written by Dean Straw, N6BV, and included on the *ARRL Antenna Book* CD solves these complex equations. This should come as a big relief for most radio amateurs. The characteristics of many common types of transmission lines are included in the software so that real antenna matching problems may be easily solved. Detailed instructions on using the program are included with it. The various examples in this chapter have been solved with *TLW*.

Equation 6 is also solved in a normalized form with a graphical method called the Smith Chart. Matching problems can be handled much easier than solving complex arithmetic equations by using this polar chart. More detailed information about using the Smith Chart is included in *The ARRL Antenna Book*. Many references to Smith Charts and their use may be found on the Web.

20.1.4 Losses in Transmission Lines

A real transmission line exhibits a certain amount of loss, caused by the resistance of the conductors used in the line and by dielectric losses in the line's insulators. The *matched-line loss* for a particular type and length of transmission line, operated at a particular frequency, is the loss when the line is terminated in a resistance equal to its characteristic impedance. The loss in a line is lowest when it is operated as a matched line.

Line losses increase when SWR is greater than 1:1. Each time energy flows from the generator toward the load, or is reflected at the load and travels back toward the generator, a certain amount will be lost along the line. The net effect of standing waves on a transmission line is to increase the average value of current and voltage, compared to the matched-line case. An increase in current raises I^2R (ohmic) losses in the conductors, and an increase in RF voltage increases E^2/R losses in the dielectric. Line loss rises with frequency, since the conductor resistance is related to skin effect, and also because dielectric losses rise with frequency.

Matched-line loss (ML) is stated in decibels per hundred feet at a particular frequency. The matched-line loss per hundred feet versus frequency for a number of common types of lines, both coaxial and open-wire balanced types, is shown graphically and as a table in the **Component Data and References** chapter. For example, RG-213 coax cable has a matched-line loss of 2.5 dB/100 ft at 100 MHz. Thus, 45 ft of this cable feeding a 50-Ω load at 100 MHz would have a loss of

$$\text{Matched line loss} = \frac{2.5 \text{ dB}}{100 \text{ ft}} \times 45 \text{ ft}$$

$$= 1.13 \text{ dB}$$

If a line is not matched, standing waves will cause additional loss beyond the inherent matched-line loss for that line.

Total Mismatched Line Loss (dB)

$$= 10 \log \left[\frac{a^2 - |\rho^2|}{a\left(1 - |\rho^2|\right)} \right] \quad (11)$$

where
  $a = 10^{ML/10}$
  ML = the line's matched loss in dB.

For most types of line and for modest values of SWR, the additional line loss due to SWR is of little concern. As the line's loss increases or at higher frequencies, the total line loss (the sum of matched-line loss and additional loss due to SWR) can be surprisingly high at high values of SWR.

Because of losses in a transmission line, the measured SWR at the input of the line is lower than the SWR measured at the load end of the line. This does *not* mean that the load is absorbing any more power. Line loss absorbs power as it travels to the load and again on its way back to the generator, so the difference between the generator output power and the power returning from the load is higher than for a lossless line. Thus, P_r/P_f is smaller than at the load and so is the measured SWR.

For example, RG-213 solid-dielectric coax cable exhibits a matched-line loss at 28 MHz of 1.14 dB per 100 ft. A 250-ft length of this cable has a matched-line loss of 1.14 × 250/100 = 2.86 dB. Assume that we measure the SWR at the load as 6:1, the total mismatched line loss from equation 11 is 5.32 dB.

The additional loss due to the 6:1 SWR at 28 MHz is 5.32 – 2.86 = 2.46 dB. The SWR at the input of the 250-ft line is only 2.2:1, because line loss has masked the true magnitude of SWR (6:1) at the load end of the line.

The losses increase if coax with a larger matched-line loss is used under the same conditions. For example, RG-58A coaxial cable is about one-half the diameter of RG-213, and it has a matched-line loss of 2.81 dB/100 ft at 28 MHz. A 250-ft length of RG-58A has a total matched-line loss of 7.0 dB. With a 6:1 SWR at the load, the additional loss due to SWR is 3.0 dB, for a total loss of 10.0 dB. The additional cable loss due to the mismatch reduces the SWR measured at the input of the line to 1.33:1. An unsuspecting operator measuring the SWR at his transmitter might well believe that everything is just fine, when in truth only about 10% of the transmitter power is getting to the antenna! Be suspicious of very low SWR readings for an antenna fed with a long length of coaxial cable, especially if the SWR remains low across a wide frequency range. Most antennas have narrow SWR bandwidths, and the SWR *should* change across a band.

On the other hand, if expensive ⅞-inch diameter 50-Ω hardline cable is used at 28 MHz, the matched-line loss is only 0.19 dB/100 ft. For 250 ft of this Hardline, the matched-line loss is 0.475 dB, and the additional loss due to a 6:1 SWR is 0.793 dB. Thus, the total loss is 1.27 dB.

At the upper end of the HF spectrum, when the transmitter and antenna are separated by a long transmission line, the use of bargain coax may prove to be a very poor cost-saving strategy. Adding a 1500-W linear amplifier (providing 8.7 dB of gain over a 200 W transmitter), to offset the loss in RG-58A compared to hardline, would cost a great deal more than higher-quality coax. Furthermore, no *transmitting* amplifier can boost *receiver* sensitivity — loss in the line has the same effect as putting an attenuator in front of the receiver.

At the lower end of the HF spectrum, say 3.5 MHz, the amount of loss in common coax lines is less of a problem for the range of SWR

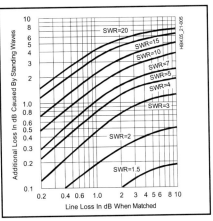

Fig 20.4 — Increase in line loss because of standing waves (SWR measured at the load). To determine the total loss in decibels in a line having an SWR greater than 1, first determine the matched-line loss for the particular type of line, length and frequency, on the assumption that the line is perfectly matched. For example, Belden 9913 has a matched-line loss of 0.49 dB/100 ft at 14 MHz. Locate 0.49 dB on the horizontal axis. For an SWR of 5:1, move up to the curve corresponding to this SWR. The increase in loss due to SWR is 0.66 dB beyond the matched line loss.

values typical on this band. For example, consider an 80 meter dipole cut for the middle of the band at 3.75 MHz. It exhibits an SWR of about 6:1 at the 3.5 and 4.0 MHz ends of the band. At 3.5 MHz, 250 ft of RG-58A small-diameter coax has an additional loss of 2.1 dB for this SWR, giving a total line loss of 4.0 dB. If larger-diameter RG-213 coax is used instead, the additional loss due to SWR is 1.3 dB, for a total loss of 2.2 dB. This is an acceptable level of loss for most 80-meter operators.

The loss situation gets dramatically worse as the frequency increases into the VHF and UHF regions. At 146 MHz, the total loss in 250 ft of RG-58A with a 6:1 SWR at the load is 21.4 dB, 10.1 dB for RG-213A, and 2.7 dB for ⅞-inch, 50-Ω hardline. At VHF and UHF, a low SWR is essential to keep line losses low, even for the best coaxial cable. The length of transmission line must be kept as short as practical at these frequencies.

The effect of SWR on line loss is shown graphically in **Fig 20.4**. The horizontal axis is the attenuation, in decibels, of the line when perfectly matched. The vertical axis gives the additional attenuation due to SWR. If long coaxial cable transmission lines are necessary, the matched loss of the coax used should be kept as low as possible, meaning that the highest-quality, largest-diameter cable should be used.

20.2 Choosing a Transmission Line

It is no accident that coaxial cable has become as popular as it has since it was first widely used during WWII. Coax is mechanically much easier to use than open-wire line. Because of the excellent shielding afforded by its outer shield, coax can be run up a metal tower leg, taped together with numerous other cables, with virtually no interaction or crosstalk between the cables. At the top of a tower, coax can be used with a rotatable Yagi or quad antenna without worrying about shorting or twisting the conductors, which might happen with an open-wire line.

Class 2 PVC (P2) noncontaminating outer jackets are designed for long-life outdoor installations. Class 1 PVC (P1) outer jackets are not recommended for outdoor installations. (See the table of coaxial cables in the **Component Data and References** chapter.) Coax can be buried underground, especially if it is run in plastic piping (with suitable drain holes) so that ground water and soil chemicals cannot easily deteriorate the cable. A cable with an outer jacket of polyethylene (PE) rather than polyvinyl chloride (PVC) is recommended for direct-bury installations.

Open-wire line must be carefully spaced away from nearby conductors, by at least several times the spacing between its conductors, to minimize possible electrical imbalances between the two parallel conductors. Such imbalances lead to line radiation and extra losses. One popular type of open-wire line is called *ladder line* because the insulators used to separate the two parallel, uninsulated conductors of the line resemble the steps of a ladder. Long lengths of ladder line can twist together in the wind and short together if not properly supported.

MULTIBAND OPERATION WITH OPEN WIRE LINE

Despite the mechanical difficulties associated with open-wire line, there are some compelling reasons for its use, especially in simple multiband antenna systems. Every antenna system, no matter what its physical form, exhibits a definite value of impedance at the point where the transmission line is connected. Although the input impedance of an antenna system is seldom known exactly, it is often possible to make a close estimate of its value with computer modeling software. As an example, **Table 20.1** lists the computed characteristics versus frequency for a multiband, 100-ft long center-fed dipole, placed 50 ft above average ground. These values were computed using *EZNEC-3*. A nonresonant 100-ft length was chosen as an illustration of a practical size that many radio amateurs could fit into their backyards, although nothing in particular recommends this antenna over other forms. It is merely used as an example.

Examine Table 20.1 carefully in the following discussion. Columns three and four show the SWR on a 50-Ω RG-213 coaxial transmission line directly connected to the antenna, followed by the total loss in 100 ft of this cable. The impedance for this nonresonant, 100-ft long antenna varies over a very wide range for the nine operating frequencies. The SWR on a 50-Ω coax connected directly to this antenna would be *extremely* high on some frequencies, particularly at 1.8 MHz, where the antenna is highly capacitive because it is very short of resonance. The loss in 100 ft of RG-213 at 1.8 MHz is a staggering 26 dB.

Contrast this to the loss in 100 ft of 450-Ω open-wire line. Here, the loss at 1.8 MHz is 8.8 dB. While 8.8 dB of loss is not particularly desirable, it is about 17 dB better than the coax! Note that the RG-213 coax exhibits a good deal of loss on almost all the bands due to mismatch. Only on 14 MHz does the loss drop down to 0.9 dB, where the antenna is just past ½-λ resonance. From 3.8 to 28.4 MHz the open-wire line has a maximum loss of only 0.6 dB.

Columns six and seven in Table 20.1 list the maximum RMS voltage for 1500 W of RF power on the 50-Ω coax and on the 450-Ω open-wire line. The maximum RMS voltage for 1500 W on the open-wire line is extremely high, at 10,950 V at 1.8 MHz. The voltage for a 100-W transmitter would be reduced by a ratio of

$$\sqrt{\frac{1500}{100}} = 3.87 : 1$$

This is 2829 V, still high enough to cause arcing in many antenna tuners, although it only occurs at specific points that are multiples of ½ λ from the load. In practice, the lower voltages present along the transmission line are within the operating range of most tuners although you should remain aware that high voltages may be present along the line at some points.

Table 20.1
Modeled Data for a 100-ft Flat-Top Antenna

Freq (MHz)	Antenna Impedance (Ω)	Input VSWR RG-213 Coax	Loss of 100 ft RG-213 Coax (dB)	Loss of 100 ft 450-Ω Line (dB)	Max Voltage RG-213 Coax at 1500 W	Max Voltage 450-Ω Line at 1500 W
1.8	4.18 −j 1590	33.7	26.0	8.8	1507	10950
3.8	37.5 −j 354	16.7	5.7	0.5	1177	3231
7.1	447 +j 956	12.3	5.9	0.2	985	2001
10.1	2010 −j 2970	12.1	10.1	0.6	967	2911
14.1	87.6 −j 156	1.6	0.9	0.3	344	1632
18.1	1800 +j 1470	7.7	6.8	0.3	753	1600
21.1	461 −j 1250	4.6	3.2	0.1	585	828
24.9	155 +j 150	3.6	2.6	0.2	516	1328
28.4	2590 +j 772	6.7	9.4	0.5	703	1950

Notes
1) Antenna is a 100 ft long, 50 ft high, center-fed dipole over average ground, using coaxial (RG-213) or open-wire transmission lines. Each transmission line is 100 ft long.
2) Antenna impedance computed using *EZNEC-3* computer program using 499 segments and with the Real Ground model.
3) Note the extremely reactive impedance levels at many frequencies, but especially at 1.8 MHz. If this antenna is fed directly with RG-213 coax, the losses are unacceptably large on 160 meters, and undesirably high on most other bands also.
4) The RF voltage at 1.8 MHz for high-power operation with open-wire line is extremely high also, and would probably result in arcing either on the line itself, or more likely in the antenna tuner.

In general, such a nonresonant antenna is a proven, practical multiband radiator when fed with 450-Ω open-wire ladder line connected to an antenna tuner. A longer antenna would be preferable for more efficient 160 meter operation, even with open-wire line. The tuner and the line itself must be capable of handling the high RF voltages and currents involved for high-power operation. On the other hand, if such a multiband antenna is fed directly with coaxial cable, the losses on most frequencies are prohibitive. Coax is most suitable for antennas whose resonant feed-point impedances are close to the characteristic impedance of the feed line.

20.3 The Transmission Line as Impedance Transformer

If the complex mechanics of reflections, SWR and line losses are put aside momentarily, a transmission line can very simply be considered as an impedance transformer. A certain value of load impedance, consisting of a resistance and reactance, at the end of the line is transformed into another value of impedance at the input of the line. The amount of transformation is determined by the electrical length of the line, its characteristic impedance, and by the losses inherent in the line. The input impedance of a real, lossy transmission line is computed using the following equation

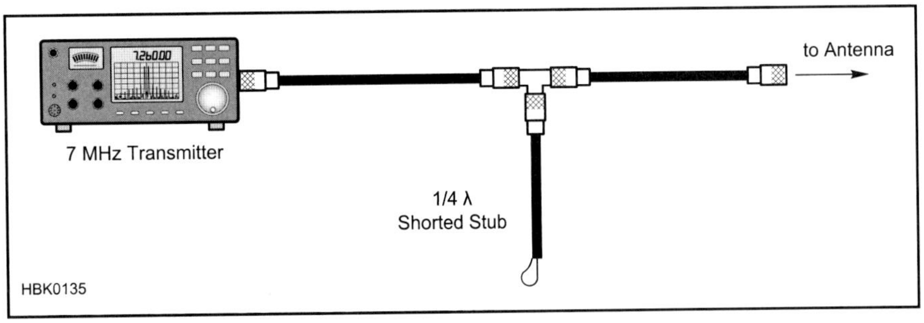

Fig 20.5 — Method of attaching a stub to a feed line.

$$Z_{in} = Z_0 \times \frac{Z_L \cosh(\eta\ell) + Z_0 \sinh(\eta\ell)}{Z_L \sinh(\eta\ell) + Z_0 \cosh(\eta\ell)} \quad (12)$$

where

Z_{in} = complex impedance at input of line = $R_{in} \pm j\,X_{in}$

Z_L = complex load impedance at end of line = $R_l \pm j\,X_l$

Z_0 = characteristic impedance of line = $R_0 \pm j\,X_0$

η = *complex loss coefficient* = $\alpha + j\,\beta$

α = matched line loss *attenuation constant*, in nepers/unit length (1 *neper* = 8.688 dB, so multiply line loss in dB per unit length by 8.688)

β = *phase constant* of line in radians/unit length (multiply electrical length in degrees by 2π radians/360 degrees)

ℓ = electrical length of line in same units of length as used for α.

Solving this equation manually is tedious, since it incorporates hyperbolic cosines and sines of the complex loss coefficient, but it may be solved using a traditional paper Smith Chart or a computer program. *The ARRL Antenna Book* has a chapter detailing the use of the Smith Chart. *TLW* software performs this transformation, but without Smith Chart graphics.

20.3.1 Transmission Line Stubs

The impedance-transformation properties of a transmission line are useful in a number of applications. If the terminating resistance is zero (that is, a short) at the end of a low-loss transmission line which is less than ¼-λ long, the input impedance consists of a reactance, which is given by a simplification of equation 12.

$$X_{in} \cong Z_0 \tan \ell \quad (13)$$

If the line termination is an open circuit, the input reactance is given by

$$X_{in} = Z_0 \cot \ell \quad (14)$$

The input of a short (less than ¼-λ) length of line with a short circuit as a terminating load appears as an inductance, while an open-circuited line appears as a capacitance. This is a useful property of a transmission line, since it can be used as a low-loss inductor or capacitor in matching networks. Such lines are often referred to as *stubs*.

A line that is electrically ¼-λ long is a special kind of a stub. When a ¼-λ line is short circuited at its load end, it presents an open circuit at its input. Conversely, a ¼-λ line with an open circuit at its load end presents a short circuit at its input. Such a line inverts the impedance of a short or an open circuit at the frequency for which the line is ¼-λ long. This is also true for frequencies that are odd multiples of the ¼-λ frequency. However, for frequencies where the length of the line is ½-λ, or integer multiples thereof, the line will duplicate the termination at its end.

20.3.2 Transmission Line Stubs as Filters

The impedance transformation properties of stubs can be put to use as filters. For example, if a shorted line is cut to be ¼-λ long at 7.1 MHz, the impedance looking into the input of the cable will be an open circuit. The line will have no effect if placed in parallel with a transmitter's output terminals. However, at twice the *fundamental* frequency, 14.2 MHz, that same line is now ½-λ, and the line looks like a short circuit. The line, often dubbed a *quarter-wave stub* in this application, will act as a trap for not only the second harmonic, but also for higher even-order harmonics, such as the fourth or sixth harmonics.

This filtering action is extremely useful in multitransmitter situations, such as Field Day, emergency operations centers, portable communications facilities and multioperator contest stations. Transmission line stubs can operate at high power where lumped-constant filters would be expensive. Using stub filters reduces noise, harmonics and strong fundamental signals from the closely spaced antennas that cause overload and interference to receivers.

Quarter-wave stubs made of good-quality coax, such as RG-213, offer a convenient way to lower transmitter harmonic levels. Despite the fact that the exact amount of harmonic attenuation depends on the impedance (often unknown) into which they are working at the harmonic frequency, a quarter-wave stub will typically yield 20 to 25 dB of attenuation of the second harmonic when placed directly at the output of a transmitter feeding common amateur antennas.

Because different manufacturing runs of coax will have slightly different velocity factors, a quarter-wave stub is usually cut a little longer than calculated, and then carefully pruned by snipping off short pieces, while using an antenna analyzer to monitor the response at the fundamental frequency. Because the end of the coax is an open circuit while pieces are being snipped away, the input of a ¼-λ line will show a short circuit exactly at the fundamental frequency. Once the coax has been pruned to frequency, a short jumper is soldered across the end, and the response at the second harmonic frequency is measured. **Fig 20.5** shows how to connect a shorted stub to a transmission line and **Fig 20.6** shows a typical frequency response.

The shorted quarter-wave stub shows low

loss at 7 MHz and at 21 MHz where it is ¾-λ long. It nulls 14 and 28 MHz. This is useful for reducing the even harmonics of a 7 MHz transmitter. It can be used for a 21 MHz transmitter as well, and will reduce any spurious emissions such as phase noise and wideband noise which might cause interference to receivers operating on 14 or 28 MHz.

The open-circuited quarter-wave stub has a low impedance at the fundamental frequency, so it must be used at two times the frequency for which it is cut. For example, a quarter-wave open stub cut for 3.5 MHz will present a high impedance at 7 MHz where it is ½-λ long. It will present a high impedance at those frequencies where it is a multiple of ½-λ, or 7, 14 and 28 MHz. It would be connected in the same manner as Fig 20.5 shows, and the frequency plot is shown in **Fig 20.7**.

This open stub can protect a receiver operating on 7, 14, 21 or 28 MHz from interference by a 3.5 MHz transmitter. It also has nulls at 10.5, 17.5 and 24.5 MHz — the 3rd, 5th and 7th harmonics. The length of a quarter-wave stub may be calculated as follows:

$$L_e = \frac{VF \times 983.6}{4f} \qquad (15)$$

where

L_e = length in ft

VF = propagation constant for the coax in use

f = frequency in MHz.

For the special case of RG-213 (and any similar cable with VF = 0.66), equation 15 can be simplified to:

$$L_e = \frac{163.5}{f} \qquad (16)$$

where

L_e = length in ft

f = frequency in MHz.

Table 20.2 solves this equation for the major contesting bands where stubs are often used. The third column shows how much of the stub to cut off if the desired frequency is 100 kHz higher in frequency. For example: To cut a stub for 14.250 MHz, reduce the overall length shown by 2.5 × 1 inches, or 2.5 inches. There is some variation in dielectric constant of coaxial cable from batch to batch or manufacturer to manufacturer, so it is always best to measure the stub's fundamental resonance before proceeding.

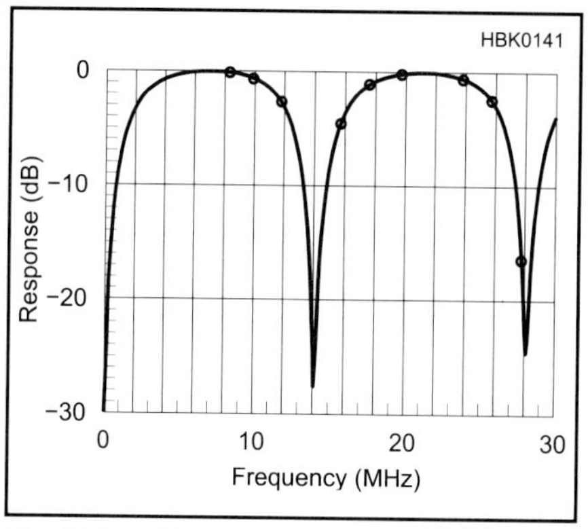

Fig 20.6 — Frequency response with a shorted stub.

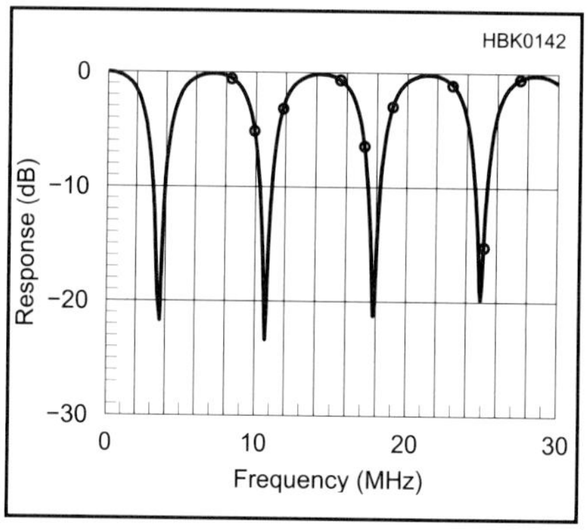

Fig 20.7 — Frequency response with an open stub.

Table 20.2

Quarter-Wave Stub Lengths for the HF Contesting Bands

Freq (MHz)	*Length (L_e)**	*Cut off per 100 kHz*
1.8	90 ft, 10 in	57⅜ in
3.5	46 ft, 9 in	15½ in
7.0	23 ft, 4 in	4 in
14.0	11 ft, 8 in	1 in
21.0	7 ft, 9 in	7⁄16 in
28.0	5 ft, 10 in	¼ in

*Lengths shown are for RG-213 and any similar cable, assuming a 0.66 velocity factor (L_e = 163.5/f). See text for other cables.

CONNECTING STUBS

Stubs are usually connected in the antenna feed line close to the transmitter. They may also be connected on the antenna side of a switch used to select different antennas. Some small differences in the null depth may occur for different positions.

To connect a stub to the transmission line it is necessary to insert a coaxial T (as shown in Fig 20.5). If a female-male-female T is used, the male can connect directly to the transmitter while the antenna line and the stub connect to the two females. It should be noted that the T inserts a small additional length in series with the stub that lowers the resonant frequency. The additional length for an Amphenol UHF T is about ⅜ inch. This length is negligible at 1.8 and 3.5 MHz, but on the higher bands it should not be ignored.

MEASURING STUBS WITH ONE-PORT METERS

Many of the common measuring instruments used by amateurs are one-port devices, meaning they have one connector at which the measurement — typically VSWR — is made. Probably the most popular instrument for this type of work is the antenna analyzer, available from a number of manufacturers.

To test a stub using an antenna analyzer, connect the stub to the meter by itself and tune the meter for a minimum impedance value, ignoring the VSWR setting. It is almost impossible to get an accurate reading on the higher HF bands, particularly with open stubs. For example, when a quarter-wave open stub cut for 20 meters was nulled on an MFJ-259 SWR analyzer, the frequency measured 14.650 MHz, with a very broad null. A recheck with a professional-quality network analyzer measured 14.018 MHz. (Resolution on the network analyzer is about ±5 kHz.) Running the same test on a quarter-wave shorted stub gave a measurement of 28.320 MHz on the MFJ-259 and 28.398 MHz on the network analyzer. (These inaccuracies are typical of amateur instrumentation and are meant to illustrate the difficulties of using inexpensive instruments for sensitive measurements.)

Other one-port instruments that measure phase can be used to get a more accurate reading. The additional length added by the required T must be accounted for. If the measurement is made without the T and then with the T, the average value will be close to correct.

MEASURING STUBS WITH TWO-PORT INSTRUMENTS

A two-port measurement is made with a signal generator and a separate detector. A T connector is attached to the generator with the stub connected to one side. The other side is connected to a cable of any length that goes to the detector. The detector should present a 50-Ω load to the cable. This is how a network analyzer is configured, and it is similar to how the stub is connected in actual use. If the generator is accurately calibrated, the measurement can be very good. There are a number of ways to do this without buying an expensive piece of lab equipment.

An antenna analyzer can be used as the signal generator. Measurements will be quite accurate if the detector has 30 to 40 dB dynamic range. Two setups were tested by the author for accuracy. The first used a digital voltmeter (DVM) with a diode detector. (A germanium diode must be used for the best dynamic range.) Tests on open and shorted stubs at 14 MHz returned readings within 20 kHz of the network analyzer. Another test was run using an oscilloscope as the detector with a 50-Ω load on the input. This test produced results that were essentially the same as the network analyzer.

A noise generator can be used in combination with a receiver as the detector. (An inexpensive noise generator kit is available from Elecraft, **www.elecraft.com**.) Set the receiver for 2-3 kHz bandwidth and turn off the AGC. An ac voltmeter connected to the audio output of the receiver will serve as a null detector. The noise level into the receiver without the stub connected should be just at or below the limiting level. With the stub connected, the noise level in the null should drop by 25 or 30 dB. Connect the UHF T to the noise generator using any necessary adapters. Connect the stub to one side of the T and connect the receiver to the other side with a short cable. Tune the receiver around the expected null frequency. After locating the null, snip off pieces of cable until the null moves to the desired frequency. Accuracy with this method is within 20 or 30 kHz of the network analyzer readings on 14 MHz stubs.

STUB COMBINATIONS

A single stub will give 20 to 30 dB attenuation in the null. If more attenuation is needed, two or more similar stubs can be combined. Best results will be obtained if a short coupling cable is used to connect the two stubs rather than connecting them directly in parallel. The stubs may be cut to the same frequency for maximum attenuation, or to two slightly different frequencies such as the CW and SSB frequencies in one band. Open and shorted stubs can be combined together to attenuate higher harmonics as well as lower frequency bands.

An interesting combination is the parallel connection of two $1/8$-λ stubs, one open and the other shorted. The shorted stub will act as an inductor and the open stub as a capacitor. Their reactance will be equal and opposite, forming a resonant circuit. The null depth with this arrangement will be a bit better than a single quarter-wave shorted stub. This presents some possibilities when combinations of stubs are used in a band switching system.

Table 20.3
Stub Selector Operation
See Fig 20.8 for circuit details.

Relay K1 Position	Relay K2 Position	Bands Passed (meters)	Bands Nulled (meters)
Open	Open	All	None
Energized	Energized	80	40, 20, 15, 10
Energized	Open	40, 15	20, 10
Open	Energized	20, 10	40, 15

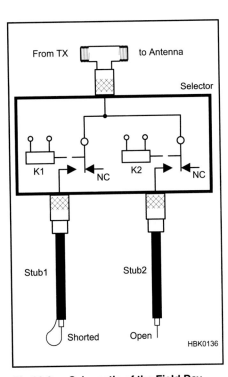

Fig 20.8 — Schematic of the Field Day stub switching relay control box. Table 20.3 shows which relays should be closed for the desired operating band.

20.3.3 Project: A Field Day Stub Assembly

Fig 20.8 shows a simple stub arrangement that can be useful in a two-transmitter Field Day station. The stubs reduce out-of-band noise produced by the transmitters that would cause interference to the other stations — a common Field Day problem where the stations are quite close together. This noise can not be filtered out at the receiver and must be removed at the transmitter. One stub assembly would be connected to each transmitter output and manually switched for the appropriate band.

Two stubs are connected as shown. The two-relay selector box can be switched in four ways. Stub 1 is a shorted quarter-wave 40-meter stub. Stub 2 is an open quarter-wave 40-meter stub. Operation is as shown in **Table 20.3**.

The stubs must be cut and tuned while connected to the selector relays. RG-213 may be used for any amateur power level and will provide 25 to 30 dB reduction in the nulls. For power levels under 500 W or so, RG-8X may be used. It will provide a few dB less reduction in the nulls because of its slightly higher loss than RG-213.

20.4 Matching Impedances in the Antenna System

Only in a few special cases is the antenna impedance the exact value needed to match a practical transmission line. In all other cases, it is necessary either to operate with a mismatch and accept the SWR that results, or else to bring about a match between the line and the antenna.

When transmission lines are used with a transmitter, the most common load is an antenna. When a transmission line is connected between an antenna and a receiver, the receiver input circuit is the load, not the antenna, because the power taken from a passing wave is delivered to the receiver.

Whatever the application, the conditions existing at the load, and *only* the load, determine the reflection coefficient, and hence the SWR, on the line. If the load is purely resistive and equal to the characteristic impedance of the line, there will be no standing waves. If the load is not purely resistive, or is not equal to the line Z_0, there will be standing waves. No adjustments can be made at the input end of the line to change the SWR at the load. Neither is the SWR affected by changing the line length, except as previously described when the SWR at the input of a lossy line is masked by the attenuation of the line.

20.4.1 Conjugate Matching

Technical literature sometimes uses the term *conjugate match* to describe the condition where the impedance seen looking toward the load from any point on the line is the complex conjugate of the impedance seen looking toward the source. This means that the resistive and reactive magnitudes of the impedances are the same, but that the reactances have opposite signs. For example, the complex conjugate of $20 + j\,75$ is $20 - j\,75$. The complex conjugate of a purely resistive impedance, such as $50 + j\,0\,\Omega$, is the same impedance, $50 + j\,0\,\Omega$. A conjugate match is necessary to achieve the maximum power gain possible from most signal sources.

For example, if 50 ft of RG-213 is terminated in a $72 - j\,34\,\Omega$ antenna impedance, the transmission line transforms the impedance to $35.9 - j\,21.9\,\Omega$ at its input. (The *TLW* program is used to calculate the impedance at the line input.) To create a conjugate match at the line input, a matching network would have to present an impedance of $35.9 + j\,21.9\,\Omega$. The system would then become resonant, since the $\pm j\,21.9\,\Omega$ reactances would cancel, leaving $35.9 + j\,0\,\Omega$.

A conjugate match is not the same as transforming one impedance to another, such as from $35.9 - j0\,\Omega$ to $50 + j0\,\Omega$. An additional impedance transformation network would be required for that step.

Conjugate matching is often used for small-signal amplifiers, such as preamps at VHF and above, to obtain the best power gain. The situation with high-power amplifiers is complex and there is considerable discussion as to whether conjugate matching delivers the highest efficiency, gain and power output. Nevertheless, conjugate matching is the model most often applied to impedance matching in antenna systems.

20.4.2 Impedance Matching Networks

When all of the components of an antenna system — the transmitter, feed line, and antenna — have the same impedance, all of the power generated by the transmitter is transferred to the antenna and SWR is 1:1. This is rarely the case, however, as antenna feed point impedances vary widely with frequency and design. This requires some method of *impedance matching* between the various antenna system components.

Many amateurs use an *impedence-matching unit or "antenna tuner"* between their transmitter and the transmission line feeding the antenna. (This is described in a following section.) The antenna tuner's function is to transform the impedance, whatever it is, at the transmitter end of the transmission line into the 50 Ω required by their transmitter. Remember that the use of an antenna tuner at the transmitter does *not tune the antenna*, reduce SWR on the feed line or reduce feed line losses!

Some matching networks are built directly into the antenna (for example, the gamma and beta matches) and these are discussed in the chapter on **Antennas** and in *The ARRL Antenna Book*. Impedance matching networks made of fixed or adjustable components can also be used at the antenna and are particularly useful for antennas that operate on a single band.

Remember, however, that impedance can be transformed anywhere in the antenna system to match any other desired impedance. A variety of techniques can be used as described in the following sections, depending on the circumstances.

An electronic circuit designed to convert impedance values is called an *impedance matching network*. The most common impedance matching network circuits for use in systems that use coax cable are:

1) The low-pass L-network.
2) The high-pass L-network.
3) The low-pass Pi network.
4) The high-pass T-network.

Basic schematics for each of the circuits are shown in **Fig 20.9**. Properties of the circuits are shown in **Table 20.4**. As shown in Table 20.4, the L-networks can be reversed if matching does not occur in one direction. L-networks are the most common for single-band antenna matching. The component in parallel is the *shunt* component, so the L-networks with the shunt capacitor or inductor at the input (Figs 20.9A and 20.9C) are *shunt-input* networks and the others are *series-input* networks.

Impedance matching circuits can use fixed-value components for just one band when a particular antenna has an impedance that is too high or low, or they can be made to be adjustable when matching is needed on several bands, such as for matching a dipole antenna fed with open-wire line.

Additional material by Bill Sabin, WØIYH, on matching networks can be found on the CD-ROM accompanying this book along with his program *MATCH*.

DESIGNING AN L-NETWORK

The L-network, shown in Fig 20.9A through 20.9D, only requires two components and is a particularly good choice of matching network for single-band antennas. The L-network is easy to construct so that it can be mounted at or near the feed point of the antenna, resulting in 1:1 SWR on the transmission line to the shack. (Note that L-networks as well as Pi- and T-networks can easily be designed with the *TLW* software.)

To design an L-network, both the source and load impedances must be known. Let us assume that the source impedance, R_S, will be 50 Ω, representing the transmission line to the transmitter, and that the load is an arbitrary value, R_L.

First, determine the circuit Q.

$$Q^2 + 1 = \frac{R_L}{50} \qquad (17A)$$

or

$$Q = \sqrt{\frac{R_L}{50} - 1} \qquad (17B)$$

Next, select the type of L-network you want from Fig 20.9. Note that the parallel component is always connected to the higher of the two impedances, source or load. Your choice should take into account whether either

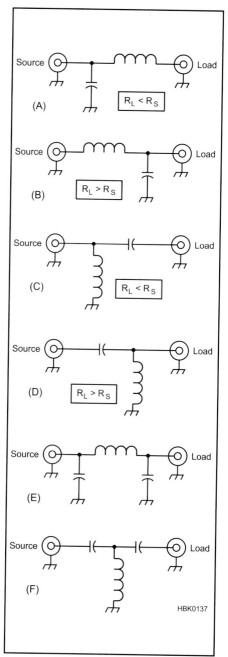

Fig 20.9 — Matching network variations. A through D show L-networks. E is a Pi-network, equivalent to a pair of L-networks sharing a common series inductor. F is a T-network, equivalent to a pair of L-networks sharing a common parallel inductor.

Table 20.4
Network Performance

Fig 20.9 Section	Circuit Type	Match Higher or Lower?	Harmonic Attenuation?
(A)	Low-pass L network	Lower	Fair to good
(B)	Reverse Low-pass L network	Higher	Fair to good
(C)	High-pass L network	Lower	No
(D)	Reverse high-pass L network	Higher	No
(E)	Low-pass Pi network	Lower and Higher	Good
(F)	High-pass T network	Lower and higher	No

the source or load require a dc ground (parallel or shunt-L) and whether it is necessary to have a dc path through the network, such as to power a remote antenna switch or other such device (parallel- or shunt-C). Once you have selected a network, calculate the values of X_L and X_C:

$$X_L = Q R_S \quad (18)$$

and

$$X_C = \frac{R_L}{Q} \quad (19)$$

As an example, we will design an L-network to match a 300-Ω antenna (R_L) to a 50-Ω transmission line (R_S). $R_L > R_S$ so we can select either Fig 20.9B or Fig 20.9D. The network in B is a low-pass network and will attenuate harmonics, so that is the usual choice.

$$Q = \sqrt{\frac{300}{50} - 1} = 2.236$$

$$X_L = 50 \times 2.236 = 112 \, \Omega$$

$$X_C = \frac{300}{2.236} = 134 \, \Omega$$

If the network is being designed to operate at 7 MHz, the actual component values are:

$$L = \frac{X_L}{2\pi f} = 2.54 \, \mu H$$

$$C = \frac{1}{2\pi f \, X_C} = 170 \, pF$$

The components could be fixed-value or adjustable.

The larger the ratio of the impedances to be transformed, the higher Q becomes. High values of Q (10 or more) may result in impractically high or low component values. In this case, it may be easier to design the matching network as a pair of L-networks back-to-back that accomplish the match in two steps. Select an intermediate value of impedance, R_{INT}, the geometric mean between R_L and the source impedance:

$$R_{INT} = \sqrt{R_L R_S}$$

Construct one L-network that transforms R_L to R_{INT} and a second L-network that transforms R_{INT} to R_S.

20.4.3 Matching Antenna Impedance at the Antenna

This section describes methods by which a network can be installed at or near the antenna feed point to provide matching to a transmission line. Having the matching system at the antenna rather than down in the shack at the end of a long transmission line does seem intuitively desirable, but it is not always very practical, especially in multiband antennas as discussed below.

RESONATING THE ANTENNA

If a highly reactive antenna can be tuned to resonance, even without special efforts to make the resistive portion equal to the line's characteristic impedance, the resulting SWR is often low enough to minimize additional line loss due to SWR. For example, the multiband 100-ft long flat-top antenna in Table 20.1 has a feed point impedance of 4.18 $-j$ 1590 Ω at 1.8 MHz. Assume that the antenna reactance is tuned out with a network consisting of two symmetrical inductors whose reactance is $+j$ 1590/2 = j 795 Ω each, with a Q of 200. These inductors are 70.29 μH coils in series with inherent loss resistances of 795/200 = 3.98 Ω. The total series resistance is thus 4.18 + 2 × (3.98) = 12.1 Ω. If placed in series with each leg of the antenna at the feed point as in **Fig 20.10**, the antenna reactance and inductor reactance cancel out, leaving a purely resistive impedance at the antenna feed point.

If this tuned system is fed with 50-Ω coaxial cable, the SWR is 50/12.1 = 4.13:1, and the loss in 100 ft of RG-213 cable would be 0.48 dB. The antenna's radiation efficiency is the ratio of the antenna's radiation resistance (4.18 Ω) to the total feed point resistance including the matching coils (12.1 Ω), so efficiency is 4.18/12.1 = 34.6% which is equivalent to 4.6 dB of loss compared to a 100% efficient antenna. Adding the 0.48 dB of loss in the line yields an overall system loss of 5.1 dB. Compare this to the loss of 26 dB if the RG-213 coax is used to feed the antenna directly, without any matching at the antenna. The use of moderately high-Q resonating inductors has yielded almost 21 dB of "gain" (that is, less loss) compared to the case without the inductors. The drawback of course is that the antenna is now resonant on only one frequency, but it certainly is a lot more efficient on that one frequency!

THE QUARTER-WAVE TRANSFORMER OR "Q" SECTION

The range of impedances presented to the transmission line is usually relatively small on a typical amateur antenna, such as a dipole or a Yagi when it is operated close to resonance. In such antenna systems, the impedance-transforming properties of a ¼-λ section of transmission line are often utilized to match the transmission line at the antenna.

Fig 20.11 shows one example of this technique to feed an array of stacked Yagis on a single tower. Each antenna is resonant and is fed in parallel with the other Yagis, using equal lengths of coax to each antenna called *phasing lines*. A stacked array is used to produce not only gain, but also a wide vertical elevation pattern, suitable for coverage of a broad geographic area. (See *The ARRL Antenna Book* for details about Yagi stacking.)

The feed point impedance of two 50-Ω Yagis fed with equal lengths of feed line connected in parallel is 25 Ω (50/2 Ω); three in parallel yield 16.7 Ω; four in parallel yield 12.5 Ω. The nominal SWR for a stack of four Yagis is 4:1 (50/12.5). This level of SWR does not cause excessive line loss, provided that low-loss coax feed line is used. However, many station designers want to be able to select, using relays, any individual antenna in the array, without having the load seen by the transmitter change. (Perhaps they might

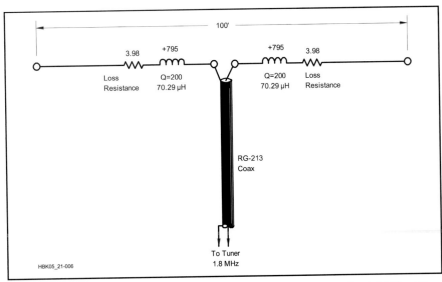

Fig 20.10 — The efficiency of the dipole in Table 20.1 can be improved at 1.8 MHz with a pair of inductors inserted symmetrically at the feed point. Each inductor is assumed to have a Q of 200. By resonating the dipole in this fashion the system efficiency, when fed with RG-213 coax, is about 21 dB better than using this same antenna without the resonator. The disadvantage is that the formerly multiband antenna can only be used on a single band.

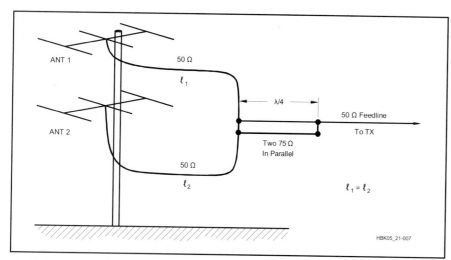

Fig 20.11 — Array of two stacked Yagis, illustrating use of ¼-λ matching sections. At the junction of the two equal lengths of 50-Ω feed line the impedance is 25 Ω. This is transformed back to 50 Ω by the two paralleled 75-Ω, ¼-λ lines, which together make a net characteristic impedance of 37.5 Ω. This is close to the 35.4 Ω value computed by the formula.

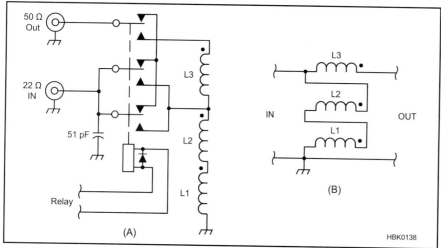

Fig 20.12 — Schematic for the impedance matching transformer described in the text. The complete schematic is shown at A. The physical positioning of the windings is shown at B.

wish to turn one antenna in the stack in a different direction and use it by itself.) If the load changes, the amplifier must be retuned, an inconvenience at best.

If the antenna impedance and the characteristic impedance of a feed line to be matched are known, the characteristic impedance needed for a quarter-wave matching section of low-loss cable is expressed by another simplification of equation 12.

$$Z = \sqrt{Z_1 Z_0} \qquad (20)$$

where
Z = characteristic impedance needed for matching section
Z_1 = antenna impedance
Z_0 = characteristic impedance of the line to which it is to be matched.

Such a matching section is called a *synchronous quarter-wave transformer* or a *quarter-wave transformer*. (Synchronous because the match is only achieved at which the length of the matching section is exactly ¼-λ long.)

Example: To match a 50-Ω line to a Yagi stack consisting of two antennas fed in parallel to produce a 25-Ω load, the quarter-wave matching section would require a characteristic impedance of

$$Z = \sqrt{50 \times 25} = 35.4 \ \Omega$$

A transmission line with a characteristic impedance of 35 Ω could be closely approximated by connecting two equal ¼-λ sections of 75-Ω cable (such as RG-11A) in parallel to yield the equivalent of a 37.5-Ω cable.

Three Yagis fed in parallel would require a ¼-λ transformer made using a cable having a characteristic impedance of

$$Z = \sqrt{16.7 \times 25} = 28.9 \ \Omega$$

This is approximated by using a ¼-λ section of 50-Ω cable in parallel with a ¼-λ section of 75-Ω cable, yielding a net impedance of 30 Ω, quite close to the desired 28.9 Ω. Four Yagis fed in parallel would require a ¼-λ transformer made up using cable with a characteristic impedance of 25 Ω, easily created by using two 50-Ω cables in parallel.

The 100-ft flat-top example in the previous section with the two resonating coils has an impedance of 12 Ω at the feed point. Two RG-58A cables, each ¼-λ long at 1.8 MHz (90 ft) can be connected in parallel to feed this antenna. An additional 10-ft length of RG-213 can make up the required 100 ft. The match will be almost perfect. The disadvantage of this system is that it limits the operation to one band, but the overall efficiency will be quite good.

Another use of ¼-λ transformers is in matching the impedance of full-wave loop antennas to 50-Ω coax. For example, the driven element of a quad antenna or a full-wave 40 meter loop has an impedance of 100 to 150 Ω. Using a ¼-λ transformer made from 62, 75 or 93-Ω coaxial cable would lower the line SWR to a level where losses were insignificant.

This use of ¼-λ transformers is limited to one band at a time. Additional ¼-λ lines need to be switched in to change bands.

MATCHING TRANSFORMERS

There is another matching technique that uses wide-band toroidal transformers. Transformers can be made that operate over very wide frequency ranges and that will match various impedances.

A very simple matching transformer consists of three windings connected in series as shown in **Fig 20.12A**. The physical arrangement of the three windings is shown in Fig 20.12B. This arrangement gives the best bandwidth. **Fig 20.13** shows a picture of this type of transformer. An IN/OUT relay is included with the transformer. One relay pole switches the 50-Ω input port while two poles in parallel switch the 22-Ω port. Three 14-inch lengths of #14 AWG wire are taped together so they lie flat on the core. A #61-mix toroid core 2.4-inch in diameter will handle full legal power.

The impedance ratio of this design is 1:2.25 or 22.22-to-50 Ω. This ratio turns out to work well for two or three 50-Ω antennas in parallel. Two in parallel will give an SWR of 25/22.22 or 1.125:1. Three in parallel give an SWR of 22.22/16.67 or 1.33. The unit shown in Fig 20.13 has an SWR bandwidth of 1.5 MHz to more than 30 MHz. The 51 pF capacitor is connected at the low impedance side to ground and tunes out some inductive reactance.

This is a good way to stack two or three triband antennas. If they have the same length feed lines and they all point the same way, their patterns will add and some gain will result. However, they don't even need to be on the same tower or pointed in the same direction or fed with the same length lines to have some benefit. Even dissimilar antennas can sometimes show a benefit when connected together in this fashion.

20.4.4 Matching the Line to the Transmitter

So far we have been concerned mainly with the measures needed to achieve acceptable amounts of loss and a low SWR when real coax lines are connected to real antennas. Not only is feed line loss minimized when the SWR is kept within reasonable bounds, but also the transmitter is able to deliver its rated output power, at its rated level of distortion, when connected to the load resistance it was designed to drive.

Most modern amateur transmitters use broadband, untuned solid-state final amplifiers designed to work into a 50-Ω load. Such a transmitter very often utilizes built-in protection circuitry that automatically reduces output power if the SWR rises to more than about 2:1. Protective circuits are needed because many solid-state devices will willingly and almost instantly destroy themselves attempting to deliver power into low-impedance loads. Solid-state devices are a lot less forgiving than vacuum tube amplifiers, which can survive momentary overloads without being destroyed instantly. Pi networks used in vacuum-tube amplifiers typically have the ability to match a surprisingly wide range of impedances on a transmission line. (See the **RF Power Amplifiers** chapter.)

Besides the rather limited option of using only inherently low-SWR antennas to ensure that the transmitter sees the load for which it was designed, an *impedance-matching unit* or *antenna tuner* ("tuner" for short) can be used. The function of an antenna tuner is to transform the impedance at the input end of the transmission line, whatever that may be, to the 50-Ω value required by the transmitter for best performance.

Do not forget that a tuner does *not* alter the SWR on the transmission line between the tuner and the antenna; it only adjusts the impedance at the transmitter end of the feed line to the value for which the transmitter was designed. Other names for antenna tuners include *transmatch*, *impedance matcher*, *matchbox* or *antenna coupler*. Since the SWR on the transmission line between the

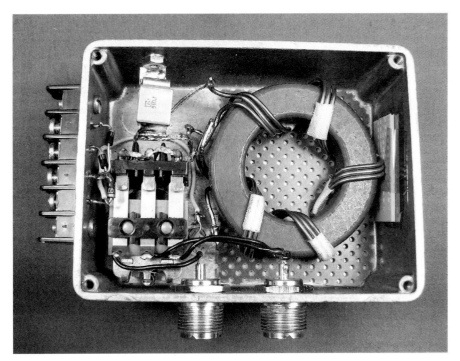

Fig 20.13 — The completed impedance matching transformer assembly.

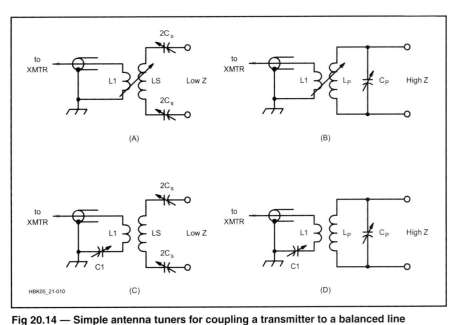

Fig 20.14 — Simple antenna tuners for coupling a transmitter to a balanced line presenting a load different from the transmitter's design load impedance, usually 50 Ω. A and B, respectively, are series and parallel tuned circuits using variable inductive coupling between coils. C and D are similar but use fixed inductive coupling and a variable series capacitor, C1. A series tuned circuit works well with a low-impedance load; the parallel circuit is better with high impedance loads (several hundred ohms or more).

antenna and the output of the antenna tuner is rarely 1:1, some loss in the feed line due to the mismatch is unavoidable, even though the SWR on the short length of line between the tuner and the transmitter is 1:1.

If separate feed lines are used for different bands, the tuner can be inserted in one feed line, tuned for best VSWR, and left at that setting. If a particular antenna has a minimum VSWR in the CW portion of a band and operation in the SSB end is desired, the tuner can be used for matching and switched out when not needed. Multiband operation generally requires retuning for each band in use.

Antenna tuners for use with balanced or open-wire feed lines include a balun or link-coupling circuit as seen in **Fig 20.14**. This allows a transmitter's unbalanced coaxial

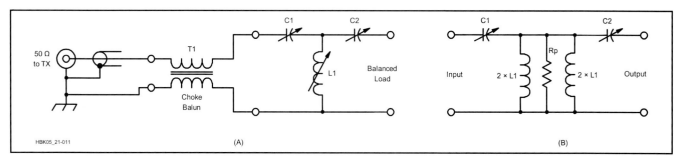

Fig 20.15 — Antenna tuner network in T configuration. This network has become popular because it has the capability of matching a wide range of impedances. At A, the balun transformer at the input of the antenna tuner preserves balance when feeding a balanced transmission line. At B, the T configuration is shown as two L networks back to back. (in the L network version, the two ½ L1 inductors are assumed to be adjustable with identical values).

output to be connected to the balanced feed line. A fully-balanced tuner has a symmetrical internal circuit with a tuner circuit for each side of the feed line and the balun at the input to the tuner where the impedance is close to 50 Ω. Most antenna tuners are unbalanced, however, with a balun located at the output of the impedance matching network, connected directly to the balanced feed line. At very high or very low impedances, the balun's power rating may be exceeded at high transmitted power levels.

Automatic antenna tuners use a microprocessor to adjust the value of the internal components. Some models sense high values of SWR and retune automatically, while others require the operator to initiate a tuning operation. Automatic tuners are available for low- and high-power operation and generally handle the same values of impedance as their manually-adjusted counterparts.

Some solid-state transmitters incorporate (usually at extra cost) automatic antenna tuners so that they too can cope with practical antennas and transmission lines. The range of impedances that can be matched by the built-in tuners is typically rather limited, however, especially at lower frequencies. Most built-in tuners specify a maximum SWR of 3:1 that can be transformed to 1:1.

THE T-NETWORK

Over the years, radio amateurs have derived a number of circuits for use as tuners. The most common form of antenna tuner in recent years is some variation of a T-network, as shown in **Fig 20.15A**. Note that the choke or current balun can be used at the input or output of the tuner to match parallel lines.

The T-network can be visualized as being two L-networks back to front, where the common element has been conceptually broken down into two inductors in parallel (see Fig 20.15B). The L-network connected to the load transforms the output impedance $R_a \pm jX_a$ into its parallel equivalent by means of the series output capacitor C2. The first L-network then transforms the parallel equivalent back into the series equivalent and resonates the reactance with the input series capacitor C1.

Note that the equivalent parallel resistance R_p across the shunt inductor can be a very large value for highly reactive loads, meaning that the voltage developed at this point can be very high. For example, assume that the load impedance at 3.8 MHz presented to the antenna tuner is $Z_a = 20 - j\,1000$. If C2 is 300 pF, then the equivalent parallel resistance across L1 is 66,326 Ω. If 1500 W appears across this parallel resistance, a peak voltage of 14,106 V is produced, a very substantial level indeed. Highly reactive loads can produce very high voltages across components in a tuner.

The ARRL computer program *TLW* calculates and shows graphically the antenna-tuner values for operator selected antenna impedances transformed through lengths of various types of practical transmission lines. The **Station Accessories** chapter includes antenna tuner projects, and *The ARRL Antenna Book* contains detailed information on tuner design and construction.

ANTENNA TUNER LOCATION

The tuner is usually located near the transmitter in order to adjust it for different bands or antennas. If a tuner is in use for one particular band and does not need to be adjusted once set up for minimum VSWR, it can be placed in a weatherproof enclosure near the antenna. Some automatic tuners are designed to be installed at the antenna, for example. For some situations, placing the tuner at the base of a tower can be particularly effective and eliminates having to climb the tower to perform maintenance on the tuner.

It is useful to consider the performance of the entire antenna system when deciding where to install the antenna tuner and what types of feed line to use in order to minimize system losses. Here is an example, using the program *TLW*. Let's assume a flat-top antenna 50 ft high and 100 ft long and not resonant on any amateur band. As extreme examples, we will use 3.8 and 28.4 MHz with 200 ft of transmission line. There are many ways to configure this system, but three examples are shown in **Fig 20.16**.

Example 1 in Fig 20.16A shows a 200-ft run of RG-213 going to a 1:1 balun that feeds the antenna. A tuner in the shack reduces the VSWR for proper matching in the transmitter. Example 2 shows a similar arrangement using 300-Ω transmitting twin lead. Example 3 shows a 50-ft run of 300-Ω line dropping straight down to a tuner near the ground and 150 ft of RG-213 going to the shack. **Table 20.5** summarizes the losses and the tuner values required.

Some interesting conclusions can be drawn. First, direct feeding this antenna with coax through a balun is very lossy — a poor solution. If the flat-top was ½-λ long — a resonant half-wavelength dipole — direct coax feed would be a good method. In the second example, direct feed with 300-Ω low-loss line does not always give the lowest loss. The combination method in Example 3 provides the best solution.

There are other considerations as well. Hanging a balun at the antenna adds stress to the wires, but can be avoided. Example 3

Table 20.5
Tuner Settings and Performance

Example (Fig 20.16)	Frequency (MHz)	Tuner Type	L (µH)	C (pF)	Total Loss (dB)
1	3.8	Rev L	1.46	2308	8.53
	28.4	Rev L	0.13	180.9	12.3
2	3.8	L	14.7	46	2.74
	28.4	L	0.36	15.6	3.52
3	3.8	L	11.37	332	1.81
	28.4	L	0.54	94.0	2.95

has some additional advantages. It feeds the antenna in a symmetrical arrangement which is best to reduce noise pickup on the shield of the feed line. The shorter feed line will not weigh down the antenna as much. The coax back to the shack can be buried or laid on the ground and it is perfectly matched. Burial of the cable will also prevent any currents from being induced on the coax shield. Once in the shack, the tuner is adjusted for minimum SWR per the manufacturer's instructions.

20.4.5 Adjusting Antenna Tuners

The process of adjusting an antenna tuner is described here and results in minimum SWR to the transmitter and also minimizes power losses in the tuner circuitry. If you have a commercial tuner and have the user's manual, the manufacturer will likely provide a method of adjustment that you should follow, including initial settings.

If you do not have a user's manual, first open the tuner and determine the circuit for the tuner. The most common circuit for commercial tuners is the high-pass T-network shown in Fig 20.9F. To adjust this type of tuner:

1) Set the series capacitors to maximum value. This may not correspond to the highest number on the control scale — verify that the capacitor's plates are fully meshed.

2) Set the inductor to maximum value. This corresponds to placing a switch tap or roller inductor contact so that it is electrically closest to circuit ground.

3) If you have an antenna analyzer, connect it to the TRANSMITTER connector of the tuner. Otherwise, connect the transceiver and tune it to the desired frequency, but do not transmit.

In the following step, it is important to verify that you hear a peak in received noise before transmitting significant power through the tuner. Tuners can sometimes be adjusted to present a low SWR to the transmitter while coupling little energy to the output. Transmitting into a tuner in this configuration can damage the tuner's components.

4) Adjust the inductor throughout its range, watching the antenna analyzer for a dip in the SWR or listen for a peak in the received noise. Return the inductor to the setting for lowest SWR or highest received noise.

    4a) If no SWR minimum or noise peak is detected, reduce the value of the capacitor closest to the transmitter in steps of about 20% and repeat.

    4b) If still no SWR minimum or noise peak is detected, return the input capacitor to maximum value and reduce the output capacitor value in steps of about 20%.

    4c) If still no SWR minimum or noise peak is detected, return the output capacitor to maximum value and reduce both input and output capacitors in 20% steps.

5) Once a combination of settings is found with a definite SWR minimum or noise peak:

    5a) If you are using an antenna analyzer, make small adjustments to find the combination of settings that produce minimum SWR with the maximum value of input and output capacitance.

    5b) If you do not have an antenna analyzer, set the transmitter output power

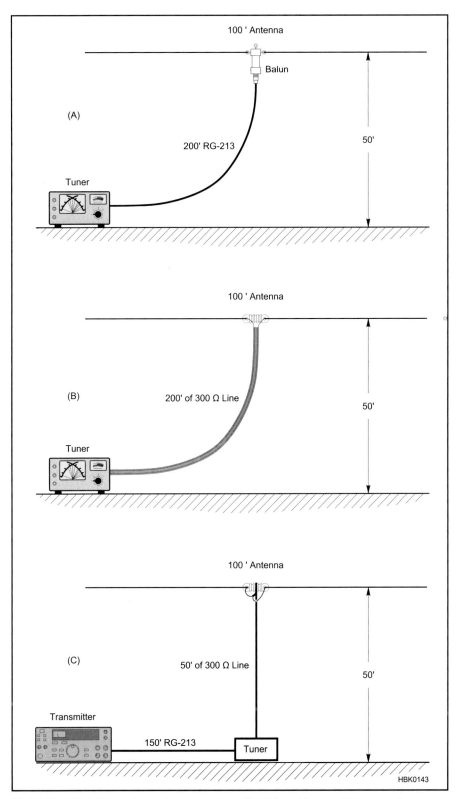

Fig 20.16 — Variations of an antenna system with different losses. The examples are discussed in the text.

to about 10 W, ensure that you won't cause interference, identify with your call sign, and transmit a steady carrier by making the same adjustments as in step 5a.

5c) For certain impedances, the tuner may not be able to reduce the SWR to an acceptable value. In this case, try adding feed line at the output of the tuner from $\frac{1}{8}$- to $\frac{1}{2}$-λ electrical wavelengths long. This will not change the feed line SWR, but it may transform the impedance to a value more suitable for the tuner components.

In general, for any type of tuner, begin with the maximum reactance to ground (maximum inductance or minimum capacitance) and the minimum series reactance between the source and load (minimum inductance or maximum capacitance). The configuration that produces the minimum SWR with maximum reactance to ground and minimum series reactance will generally have the highest efficiency and broadest tuning bandwidth.

To reduce on-the-air tune-up time, record the settings of the tuner for each antenna and band of operation. If the tuner requires readjustment across the band, record the settings of the tuner at several frequencies across the band. Print out the results and keep it near the tuner — this will allow you to adjust the tuner quickly with only a short transmission to check or fine tune the settings. This also serves as a diagnostic, since changes in the setting indicate a change in the antenna system.

20.4.6 Myths About SWR

This is a good point to stop and mention that there are some enduring and quite misleading myths in Amateur Radio concerning SWR.

- Despite some claims to the contrary, a high SWR *does not by itself* cause RFI, or TVI or telephone interference. While it is true that an antenna located close to such devices can cause overload and interference, the SWR on the feed line to that antenna has nothing to do with it, providing of course that the tuner, feed line or connectors are not arcing. The antenna is merely doing its job, which is to radiate. The transmission line is doing its job, which is to convey power from the transmitter to the radiator.
- A second myth, often stated in the same breath as the first one above, is that a high SWR will cause excessive radiation from a transmission line. SWR has nothing to do with excessive radiation from a line. *Imbalances* in feed lines cause radiation, but such imbalances are not related to SWR. An asymmetric arrangement of a transmission line and antenna can result in current being induced on the transmission line — on the shield of coax or as an imbalance of currents in an open-wire line. This current will radiate just as if it was on an antenna. A *choke balun* is used on coaxial feed lines to reduce these currents as described in the section on baluns later in this chapter.
- A third and perhaps even more prevalent myth is that you can't "get out" if the SWR on your transmission line is higher than 1.5:1, or 2:1 or some other such arbitrary figure. On the HF bands, if you use reasonable lengths of good coaxial cable (or even better yet, open-wire line), the truth is that you need not be overly concerned if the SWR at the load is kept below about 6:1. This sounds pretty radical to some amateurs who have heard horror story after horror story about SWR. The fact is that if you can load up your transmitter without any arcing inside, or if you use a tuner to make sure your transmitter is operating into its rated load resistance, you can enjoy a very effective station, using antennas with feed lines having high values of SWR on them. For example, a 450-Ω open-wire line connected to the multiband dipole shown in Table 20.1 would have a 19:1 SWR on it at 3.8 MHz. Yet time and again this antenna has proven to be a great performer at many installations.

Fortunately or unfortunately, SWR is one of the few antenna and transmission-line parameters easily measured by the average radio amateur. Ease of measurement does not mean that a low SWR should become an end in itself! The hours spent pruning an antenna so that the SWR is reduced from 1.5:1 down to 1.3:1 could be used in far more rewarding ways — making contacts, for example, or studying transmission-line theory.

20.5 Baluns and Transmission-Line Transformers

Center-fed dipoles and loops are *balanced*, meaning that they are electrically and physically symmetrical with respect to the feed point. A balanced antenna may be fed by a balanced feeder system to preserve this symmetry, thereby avoiding difficulties with unbalanced currents on the line and undesirable radiation from the transmission line itself. Line radiation can be prevented by a number of devices which detune or *decouple* the line, greatly reducing currents induced onto the feed line from the signal radiated by the antenna.

Many amateurs use center-fed dipoles or Yagis, fed with unbalanced coaxial line. Some method should be used for connecting the line to the antenna without upsetting the symmetry of the antenna itself. This requires a circuit that will isolate the balanced load from the unbalanced line, while still providing efficient power transfer. Devices for doing this are called *baluns* (a contraction for "balanced to unbalanced"). A balanced antenna fed with balanced line, such as two-wire ladder line, will maintain its inherent balance, so long as external causes of unbalance are avoided. However, even they will require some sort of balun at the transmitter, since modern transmitters have unbalanced (coaxial) outputs.

If a balanced antenna is fed at the center by a coaxial feed line without a balun, as indicated in **Fig 20.17A**, the inherent symmetry and balance is upset because one side of the $\frac{1}{2}$-radiator is connected to the shield while the other is connected to the inner conductor. On the side connected to the shield, current can be diverted from flowing into the antenna, and instead can flow away from the antenna on the outside of the coaxial shield. The field thus set up cannot be canceled by the field from the inner conductor because the fields inside the cable cannot escape through the shielding of the outer conductor. Hence currents flowing on the outside of the line will be responsible for some radiation from the line, just as if they were flowing on an antenna.

This is a good point at which to say that striving for perfect balance in a line and antenna system is not always absolutely mandatory. For example, if a nonresonant, center-fed dipole is fed with open-wire line and a tuner for multiband operation, the most desirable radiation pattern for general-purpose communication is actually an omnidirectional pattern. A certain amount of feed-line radiation might actually help fill in otherwise undesirable nulls in the azimuthal pattern of the antenna itself. Furthermore, the radiation pattern of a coaxial-fed dipole that is only a few tenths of a wavelength off the ground (50 ft high on the 80 meter band, for example) is not very directional anyway, because of its severe interaction with the ground.

Purists may cry out in dismay, but there are many thousands of coaxial-fed dipoles in daily use worldwide that perform very effectively without the benefit of a balun. See **Fig 20.18A** for a worst-case comparison between a dipole with and without a balun at its feed point. This is with a 1-λ feed line slanted downward 45° under one side of the antenna. *Common-mode currents* are conducted and induced onto the outside of the shield of the feed line, which in turn radiates. The amount of pattern distortion is not particularly severe for a dipole. It is debatable whether the bother and expense of installing a balun for such an antenna is worthwhile.

Some form of balun should be used to preserve the pattern of an antenna that is

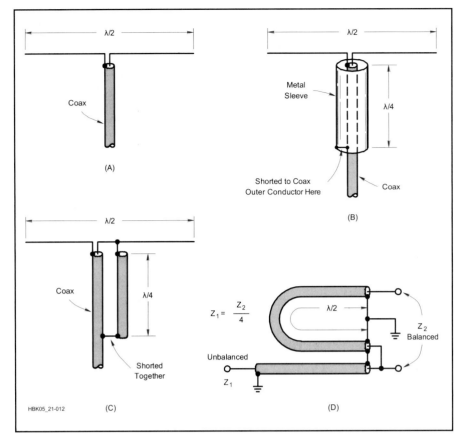

Fig 20.17 — Quarter-wavelength baluns. Radiator with coaxial feed (A) and methods of preventing unbalanced currents from flowing on the outside of the transmission line (B and C). The ½ λ-phasing section shown at D is used for coupling to an unbalanced circuit when a 4:1 impedance ratio is desired or can be accepted.

purposely designed to be highly directional, such as a Yagi or a quad. Fig 20.18B shows the distortion that can result from common-mode currents conducted and radiated back onto the feed line for a 5-element Yagi. This antenna has purposely been designed for an excellent pattern but the common-mode currents seriously distort the rearward pattern and reduce the forward gain as well. A balun is highly desirable in this case.

Choke or current baluns force equal and opposite currents to flow in the load (the antenna) by creating a high common-mode impedance to currents that are equal in both conductors or that flow on the outside of coaxial cable shields, such as those induced by the antenna's radiated field. The result of using a current balun is that currents coupled back onto the transmission line from the antenna are effectively reduced, or "choked off," even if the antenna is not perfectly balanced. Choke baluns are particularly useful for feeding asymmetrical antennas with unbalanced coax line. The common-mode impedance of the choke balun varies with frequency, but the line's differential-mode impedance is unaffected.

Reducing common-mode current on a feed line also reduces:
- Radiation from the feed line that can distort an antenna's radiation pattern
- Radiation from the feed line that can cause RFI to nearby devices
- RF current in the shack and on power-line wiring

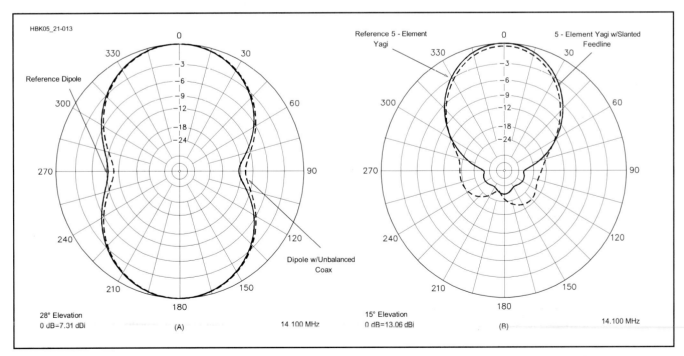

Fig 20.18 — At A, computer-generated azimuthal responses for two λ/2 dipoles placed 0.71-λ high over typical ground. The solid line is for a dipole with no feed line. The dashed line is for an antenna with its feed line slanted 45° down to ground. Current induced on the outer braid of the 1-λ-long coax by its asymmetry with respect to the antenna causes the pattern distortion. At B, azimuthal response for two 5-element 20-meter Yagis placed 0.71 λ over average ground. Again, the solid line is for a Yagi without a feed line and the dashed line is for an antenna with a 45° slanted, 1-λ long feed line. The distortion in the radiated pattern is now clearly more serious than for a simple dipole. A balun is needed at the feed point, and most likely point, preferably ¼ λ from the feed point, to suppress the common-mode currents and restore the pattern.

Baluns, Chokes, and Transformers

The term "balun" applies to any device that transfers differential-mode signals between a balanced (*bal-*) system and an unbalanced (*un-*) system while maintaining symmetrical energy distribution at the terminals of the balanced system. The term only applies to the function of energy transfer, not to how the device is constructed. It doesn't matter whether the balanced-unbalanced transition is made through transmission line structures, flux-coupled transformers, or simply by blocking unbalanced current flow. A common-mode *choke balun*, for example, performs the balun function by putting impedance in the path of common-mode currents and is therefore a balun.

A *current balun* forces symmetrical current at the balanced terminals. This is of particular importance in feeding antennas, since antenna currents determine the antenna's radiation pattern. A *voltage balun* forces symmetrical voltages at the balanced terminals. Voltage baluns are less effective in causing equal currents at their balanced terminals, such as at an antenna's feed point.

An *impedance transformer* may or may not perform the balun function. Impedance transformation (changing the ratio of voltage and current) is not required of a balun nor is it prohibited. There are balanced-to-balanced impedance transformers (transformers with isolated primary and secondary windings, for example) just as there are unbalanced-to-unbalanced impedance transformers (autotransformer and transmission-line designs). A *transmission-line transformer* is a device that performs the function of power transfer (with or without impedance transformation) by utilizing the characteristics of transmission lines.

Multiple devices are often combined in a single package called a "balun." For example, a "4:1 current balun" is a 1:1 current balun in series with a 4:1 impedance transformer or voltage balun. Other names for baluns are common, such as "line isolator" for a choke balun. Baluns are often referred to by their construction — "bead balun," "coiled-coax balun," "sleeve balun," and so forth. What is important is to separate the function (power transfer between balanced and unbalanced systems) from the construction.

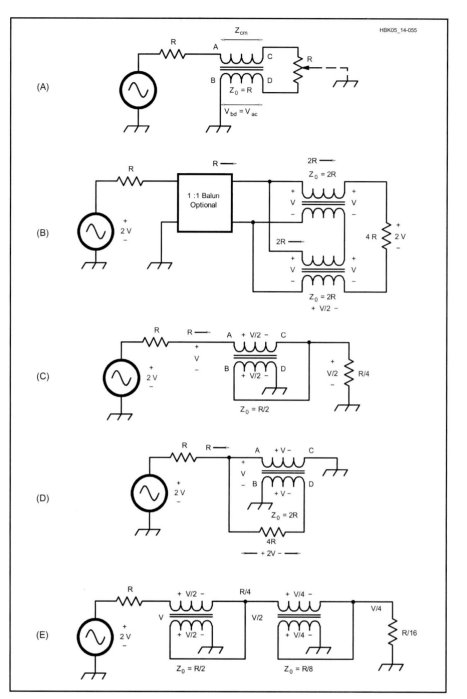

Fig 20.19 — (A) Basic current or choke balun. (B) Guanella 1:4 transformer. (C) Ruthroff 4:1 unbalanced transformer. (D) Ruthroff 1:4 balanced transformer. (E) Ruthroff 16:1 unbalanced transformer.

• Coupling of noise currents on the feed line to receiving antennas
• Currents from noise sources coupling to the feed line
• Coupling between different antennas via their feed lines

A single choke balun at the antenna feed point may not be sufficient to reduce common-mode current everywhere along a long feed line. If common-mode current on the line far from the antenna feed point is a problem, additional choke baluns can be placed at approximately ¼-λ intervals along the line. This breaks up the line electrically into segments too short to act as effective antennas. The chokes in this case function similarly to insulators used to divide tower guy wires into non-resonant lengths.

20.5.1 Quarter-Wave Baluns

Fig 20.17B shows a balun arrangement known as a *bazooka*, which uses a sleeve over the transmission line. The sleeve, together with the outside portion of the outer coax conductor, forms a shorted ¼-λ section of transmission line. The impedance looking into the open end of such a section is very high, so the end of the outer conductor of the coaxial line is effectively isolated from the part of the line below the sleeve. The length is an electrical ¼-λ, and because of the velocity factor may be physically shorter if the insulation between the sleeve and the line is not air. The bazooka has no effect on antenna impedance at the frequency where the ¼-λ sleeve is resonant. However, the sleeve adds inductive shunt reactance at frequencies lower, and capacitive shunt reactance at frequencies higher than the ¼-λ resonant frequency. The bazooka is mostly used at VHF, where its physical size does not present a major problem.

Another method that gives an equivalent effect is shown at Fig 20.17C. Since the voltages at the antenna terminals are equal and opposite (with reference to ground), equal and opposite currents flow on the surfaces of the line and second conductor. Beyond the shorting point, in the direction of the transmitter, these currents combine to cancel out each other. The balancing section acts like an open circuit to the antenna, since it is a ¼-λ parallel-conductor line shorted at the far end, and thus has no effect on normal antenna operation. This is not essential to the line-balancing function of the device, however, and baluns of this type are sometimes made shorter than ¼-λ to provide a shunt inductive reactance required in certain matching systems (such as the hairpin match described in the **Antennas** chapter).

Fig 20.17D shows a third balun, in which equal and opposite voltages, balanced to ground, are taken from the inner conductors of the main transmission line and a ½-λ phasing section. Since the voltages at the balanced end are in series while the voltages at the unbalanced end are in parallel, there is a 4:1 step-down in impedance from the balanced to the unbalanced side. This arrangement is useful for coupling between a 300-Ω balanced line and a 75-Ω unbalanced coaxial line.

20.5.2 Transmission Line Transformers

The basic transmission line transformer, from which other transformers are derived, is the 1:1 *choke balun* or *current balun*, shown in **Fig 20.19A**. To construct this type of balun, a length of coaxial cable or a pair of close-spaced, parallel wires forming a transmission line are wrapped around a ferrite rod or toroid or inserted through a number of beads. (The coiled feed line choke balun is discussed in the next section.) For the HF bands, use type 75 or type 31 material. Type 43 is used on the VHF bands. The Z_0 of the line should equal the load resistance, R.

Because of the ferrite, a high impedance exists between points A and C and a virtually identical impedance between B and D. This is true for parallel wire lines and it is also true for coax. The ferrite affects the A to C impedance of the coax inner conductor and the B to D impedance of the outer braid equally.

The conductors (two wires or coax braid and center-wire) are tightly coupled by electromagnetic fields and therefore constitute a good conventional transformer with a turns ratio of 1:1. The voltage from A to C is equal to and in-phase with that from B to D. These are called the *common-mode voltages* (*CM*).

A common-mode (CM) current is one that has the same value and direction in both wires (or on the shield and center conductor). Because of the ferrite, the CM current encounters a high impedance that acts to reduce (choke) the current. The normal *differential-mode* (*DM*) signal does not encounter this CM impedance because the electromagnetic fields due to equal and opposite currents in the two conductors cancel each other at the ferrite, so the magnetic flux in the ferrite is virtually zero. (See the section on Transmitting Ferrite Choke Baluns.)

The main idea of the transmission line transformer is that although the CM impedance may be very large, the DM signal is virtually unopposed, especially if the line length is a small fraction of a wavelength. But it is very important to keep in mind that the common-mode voltage across the ferrite winding that is due to this current is efficiently coupled to the center wire by conventional transformer action, as mentioned before and easily verified. This equality of CM voltages, and also CM impedances, reduces the *conversion* of a CM signal to an *undesired* DM signal that can interfere with the *desired* DM signal in both transmitters and receivers.

The CM current, multiplied by the CM impedance due to the ferrite, produces a CM voltage. The CM impedance has L and C

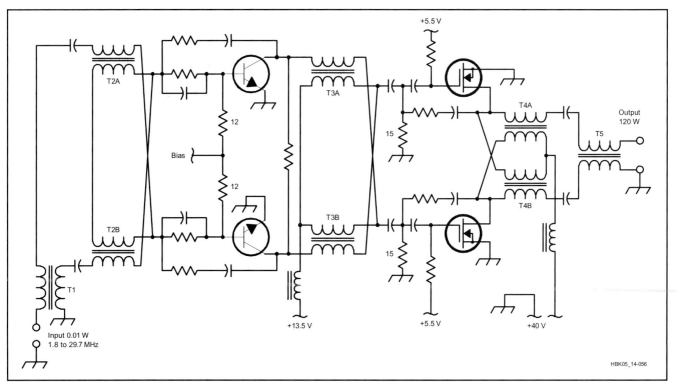

Fig 20.20 — This illustrates how transmission-line transformers can be used in a push-pull power amplifier.

reactance and also R. So L, C and R cause a broad parallel self-resonance at some frequency. The R component also produces some dissipation (heat) in the ferrite. This dissipation is an excellent way to dispose of a small amount of unwanted CM power.

Because of the high CM impedance, the two output wires of the balun in Fig 20.19A have a high impedance with respect to, and are therefore "isolated" from, the generator. This feature is very useful because now any point of R at the output can be grounded. In a well-designed balun circuit almost all of the current in one conductor returns to the generator through the other conductor, despite this ground connection. Note also that the ground connection introduces some CM voltage across the balun cores and this has to be taken into account. This CM voltage is a maximum if point C is grounded. If point D is grounded and if all "ground" connections are at the same potential, which they often are not, the CM voltage is zero and the balun may no longer be needed. In a coax balun the return current flows on the inside surface of the braid.

We now look briefly at a transmission line transformer that is based on the choke balun. Fig 20.19B shows two identical choke baluns whose inputs are in parallel and whose outputs are in series. The output voltage amplitude of each balun is identical to the common input, so the two outputs add in-phase (equal time delay) to produce twice the input voltage. It is the high CM impedance that makes this voltage addition possible. If the power remains constant the load current must be one-half the generator current, and the load resistor is 2V/0.5I = 4V/I = 4R.

THE GUANELLA TRANSFORMER

The CM voltage in each balun is V/2, so there is some flux in the cores. The right side floats. This is named the *Guanella* transformer. If Z_0 of the lines equals 2R and if the load is a pure resistance of 4R then the input resistance R is independent of line length. If the lines are exactly one-quarter wavelength, then $Z_{IN} = (2R)^2 / Z_L$, an impedance inverter, where Z_{IN} and Z_L are complex. The quality of balance can often be improved by inserting a 1:1 balun (Fig 20.19A) at the left end so that both ends of the 1:4 transformer are floating and a ground is at the far left side as shown. The Guanella transformer can also be operated from a grounded right end to a floating left end. The 1:1 balun at the left then allows a grounded far left end.

THE RUTHROFF TRANSFORMER

Fig 20.19C is the *Ruthroff* transformer in which the input voltage V is divided into two equal in-phase voltages AC and BD (they are tightly coupled), so the output is V/2. And because power is constant, $I_{OUT} = 2I_{IN}$ and the load is R/4. There is a CM voltage V/2 between A and C and between B and D, so in normal operation the core is not free of magnetic flux. The input and output both return to ground so it can also be operated from right to left for a 1:4 impedance stepup.

The Ruthroff transformer is often used as an amplifier interstage transformer, for example between 200 Ω and 50 Ω. To maintain low attenuation the line length should be much less than one-fourth wavelength at the highest frequency of operation, and its Z_0 should be R/2. A balanced version is shown in Fig 20.19D, where the CM voltage is V, not V/2, and transmission is from left-to-right only. Because of the greater flux in the cores, no different than a conventional transformer, this is not a preferred approach, although it could be used with air wound coils (for example in antenna tuner circuits) to couple 75-Ω unbalanced to 300-Ω balanced. The tuner circuit could then transform 75 Ω to 50 Ω.

POWER AMPLIFIER AND COMBINER USE

Fig 20.20 illustrates, in skeleton form, how transmission-line transformers can be used in a push-pull solid state power amplifier. The idea is to maintain highly balanced stages so that each transistor shares equally in the amplification in each stage. The balance also minimizes even-order harmonics so that low-pass filtering of the output is made much easier. In the diagram, T1 and T5 are current (choke) baluns that convert a grounded connection at one end to a balanced (floating) connection at the other end, with a high impedance to ground at both wires. T2 transforms the 50-Ω generator to the 12.5 Ω (4:1 impedance) input impedance of the first stage. T3 performs a similar step-down transformation from the collectors of the first stage to the gates of the second stage. The MOSFETs require a low impedance from gate to ground. The drains of the output stage require an impedance step up from 12.5 Ω to 50 Ω, performed by T4. Note how the choke baluns and the transformers collaborate to maintain a high degree of balance throughout the amplifier. Note also the various feedback and loading networks that help keep the amplifier frequency response flat.

Quite often the performance of a single stage can be greatly improved by combining two identical modules. Because the input power is split evenly between the two modules the drive source power can be twice as great and the output power will also be twice as great. In transmitters, especially, this often works better than a single transistor with twice the power rating. Or, for the same drive and output power, each module need supply only one-half as much power, which usually means better distortion performance. Often, the total number of stages can be reduced in this manner, with resulting cost savings. If the combining is performed properly, using hybrid transformers, the modules interact with each other much less, which can avoid certain problems. These are the system-design implications of module combining.

Three methods are commonly used to combine modules: parallel (0°), push-pull (180°) and quadrature (90°). In RF circuit design, the combining is often done with special types of "hybrid" transformers called *splitters* and *combiners*. These are both the same type of transformer that can perform either function. The splitter is at the input, the combiner at the output. We will only touch very briefly on these topics in this chapter and suggest that the reader consult the **RF Power Amplifiers** chapter and the very considerable literature for a deeper understanding and for techniques used at different frequency ranges.

Fig 20.21 illustrates one example of each of the three basic types. In a 0° hybrid splitter at the input the tight coupling between the two windings forces the voltages at A and B to be equal in amplitude and also equal in phase if the two modules are identical. The 2R resistor between points A and B greatly reduces the transfer of power between A and B via the transformer, but only if the generator resistance is closely equal to R. The output combiner separates the two outputs C and D from each other in the same manner, if the output load is equal to R, as shown. No power is lost in the 2R resistor if the module output levels are identical.

APPLICATIONS OF TRANSMISSION-LINE TRANSFORMERS

There are many transformer schemes that use the basic ideas of Fig 20.19. Several of them, with their toroid winding instructions, are shown in **Fig 20.22**. Two of the most commonly used devices are the 1:1 current balun and 4:1 impedance transformer wound on toroid cores as shown in **Fig 20.23**.

Because of space limitations, for a comprehensive treatment we suggest Jerry Sevick's books *Transmission Line Transformers* and *Building and Using Baluns and Ununs*. For applications in solid-state RF power amplifiers, see Sabin and Schoenike, *HF Radio Systems and Circuits*, Chapter 12.

20.5.3 Coiled-Coaxial Choke Baluns

The simplest construction method for a 1:1 choke balun made from coaxial feed line is simply to wind a portion of the cable into a coil (see **Fig 20.24**), creating an inductor from the shield's outer surface. This type of choke balun is simple, cheap and reduces common-mode current. Currents on the outside of the shield encounter the coil's impedance, while currents on the inside are unaffected.

A scramble-wound flat coil (like a coil of rope) shows a broad resonance that easily covers three octaves, making it reasonably

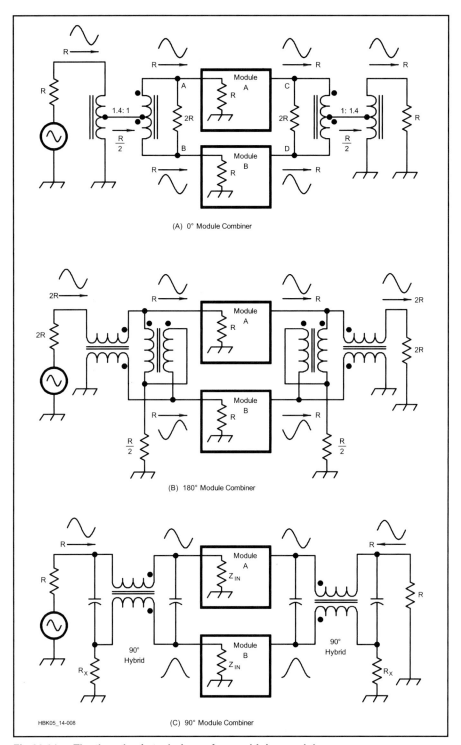

Fig 20.21 — The three basic techniques for combining modules.

Table 20.6
Coiled-Coax Choke Baluns

Wind the indicated length of coaxial feed line into a coil (like a coil of rope) and secure with electrical tape. (Diameter 6-8 inches.)

The balun is most effective when the coil is near the antenna.

Lengths and diameter are not critical.

Single Band (Very Effective)

Freq (MHz)	RG-213, RG-8	RG-58
3.5	22 ft, 8 turns	20 ft, 6-8 turns
7	22 ft, 10 turns	15 ft, 6 turns
10	12 ft, 10 turns	10 ft, 7 turns
14	10 ft, 4 turns	8 ft, 8 turns
21	8 ft, 6-8 turns	6 ft, 8 turns
28	6 ft, 6-8 turns	4 ft, 6-8 turns

Multiple Band

Freq (MHz)	RG-8, 58, 59, 8X, 213
3.5-30	10 ft, 7 turns
3.5-10	18 ft, 9-10 turns
1.8-3.5	40 ft, 20 turns
14-30	8 ft, 6-7 turns

trolled) by winding the cable as a single-layer solenoid around a section of plastic pipe, an empty bottle or other suitable cylinder. **Fig 20.25** shows how to make this type of choke balun. A coil diameter of about 5 inches is reasonable. This type of construction reduces the stray capacitance between the ends of the coil.

For both types of coiled-coaxial chokes, use cable with solid insulation, not foamed, to minimize migration of the center conductor through the insulation toward the shield. The diameter of the coil should be at least 10 times the cable diameter to avoid mechanically stressing the cable.

20.5.4 Transmitting Ferrite Choke Baluns

A ferrite choke is simply a very low-Q parallel-resonant circuit tuned to the frequency where the choke should be effective. Passing a conductor through most ferrite cores (that is, one turn) produces a resonance around 150 MHz. By choosing a suitable core material, size and shape, and by adding multiple turns and varying their spacing, the choke can be "tuned" (optimized) for the required frequency range.

Transmitting chokes differ from other common-mode chokes because they must be designed to work well when the line they are choking carries high power. They must also be physically larger so that the bend radius of the coax is large enough that the line is not deformed. Excellent common-mode chokes having very high power handling capability can be formed simply by winding multiple turns of coax through a sufficiently large ferrite core or multiple cores. (Chokes made by

effective over the entire HF range. If particular problems are encountered on a single band, a coil that is resonant on that band may be added. The choke baluns described in **Table 20.6** were constructed to have a high impedance at the indicated frequencies as measured with an impedance meter. This construction technique is not effective with open-wire or twin-lead line because of coupling between adjacent turns.

The inductor formed by the coaxial cable's shield is self-resonant due to the distributed capacitance between the turns of the coil. The self-resonant frequency can be found by using a dip meter. Leave the ends of the choke open, couple the coil to the dip meter, and tune for a dip. This is the parallel resonant frequency and the impedance will be very high.

The distributed capacitance of a flat-coil choke balun can be reduced (or at least con-

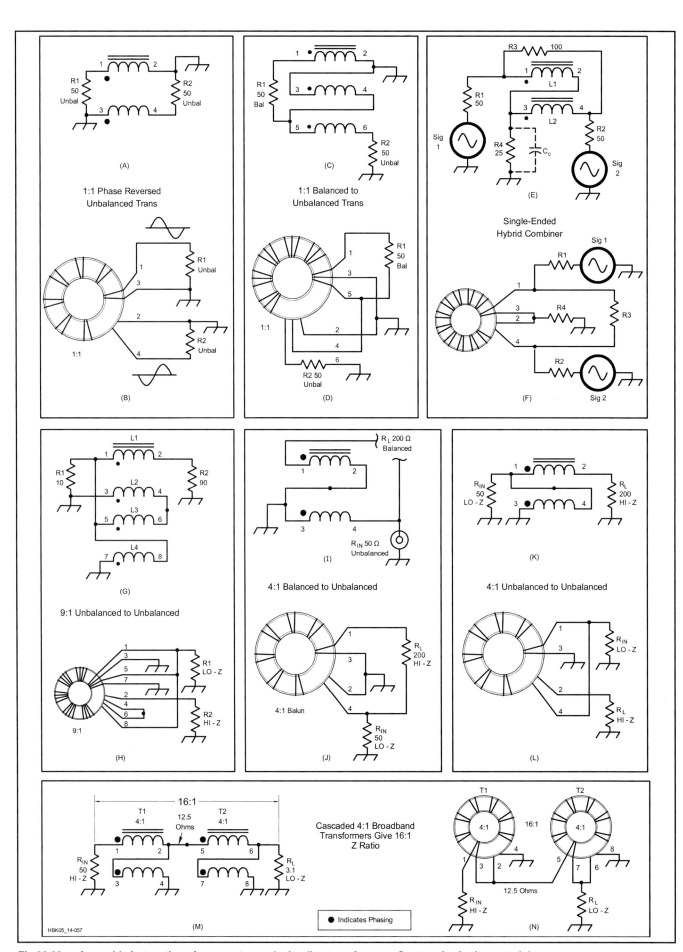

Fig 20.22 — Assembly instructions for some transmission-line transformers. See text for ferrite material type.

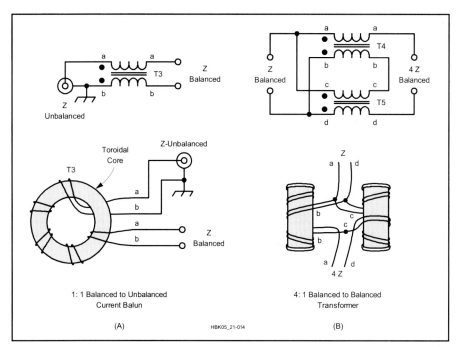

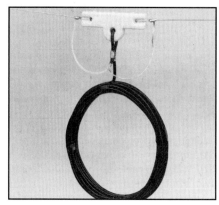

Fig 20.24 — RF choke formed by coiling the feed line at the point of connection to the antenna. The inductance of the choke isolates the antenna from the outer surface of the feed line.

Fig 20.23 — Broadband baluns. (A) 1:1 current balun and (B) Guanella 4:1 impedance transformer wound on two cores, which are separated. Use 12 bifilar turns of #14 AWG enameled wire, wound on 2.4-inch OD cores for A and B. Distribute bifilar turns evenly around core. See text for ferrite material type.

winding coaxial cable on ferrite cores will be referred to as "wound-coax chokes" to distinguish them from the coiled-coax chokes of the preceding section.)

CHOKES ON TRANSMISSION LINES

A transmission line can be wound around a ferrite core to form a common-mode choke. If the line is coax, all of the magnetic flux associated with differential mode current is confined to the dielectric (the insulating material between the center conductor and the shield). The external ferrite core carries only flux associated with common-mode current.

If the line is made up of parallel wires (a bifilar winding), a significant fraction of the flux associated with differential current will leak outside the line to the ferrite core. Leakage flux can exceed 30% of the total flux for even the most tightly-spaced bifilar winding. In addition to this leakage flux, the core will also carry the flux associated with common-mode current.

When a transformer (as opposed to a choke) is wound on a magnetic core, all of the field associated with current in the windings is carried by the core. Similarly, all forms of voltage baluns require all of the transmitted power to couple to the ferrite core. Depending on the characteristics of the core, this can result in considerable heating and power loss. Only a few ferrite core materials have loss characteristics suitable for use as the cores of high power RF transformers. Type 61 material has reasonably low dissipation below about 10 MHz, but its loss tangent rises rapidly above that frequency. The loss tangent of type 67 material makes it useful in high power transformers to around 30 MHz.

Leakage flux, corresponding to 30-40% of the transmitter power, causes heating in the ferrite core and attenuates the transmitted signal by a dB or so. At high power levels, temperature rise in the core also changes its magnetic properties, and in the extreme case, can result in the core temporarily losing its magnetic properties. A flux level high enough to make the core hot is also likely to saturate the core, producing distortion (harmonics, splatter, clicks).

Flux produced by common-mode current can also heat the core — if there is enough common-mode current. Dissipated power is equal to I^2R, so it can be made very small by making the common-mode impedance so large that the common-mode current is very small.

DESIGN CRITERIA

It can be shown mathematically, and experience confirms, that wound-coax chokes having a resistive impedance at the transmit frequency of at least 5000 Ω and wound with RG-8 or RG-11-size cable on five toroids are conservatively rated for 1500 W under high duty-cycle conditions, such as contesting or digital mode operation. While chokes wound with smaller coax (RG-6, RG-8X, RG-59, RG-58 size) are conservatively rated for dissipation in the ferrite core, the voltage and current ratings of those smaller cables suggests a somewhat lower limit on their power handling. Since the chokes see only the common-mode voltage, the only effect of high SWR on power handling of wound-coax chokes is the peaks of differential current and voltage along the line established by the mismatch.

Experience shows that 5000 Ω is also a good design goal to prevent RFI, noise coupling and pattern distortion. While 500-1000 Ω has long been accepted as sufficient to prevent pattern distortion, W1HIS has correctly observed that radiation and noise coupling from the feed line should be viewed as a form of pattern distortion that fills in the nulls of a directional antenna, reducing its ability to reject noise and interference.

Chokes used to break up a feed line into segments too short to interact with another antenna should have a choking impedance on the order of 1000 Ω to prevent interaction with simple antennas. A value closer to 5000 Ω may be needed if the effects of common-mode current on the feed line are filling the null of directional antenna.

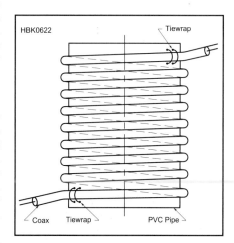

Fig 20.25 — Winding a coaxial choke balun as a single-layer solenoid may increase impedance and self-resonant frequency compared to a flat-coil choke.

BUILDING WOUND-COAX FERRITE CHOKES

Coaxial chokes should be wound with a bend radius sufficiently large that the coax is not deformed. When a line is deformed, the spacing between the center conductor and the shield varies, so voltage breakdown and heating are more likely to occur. Deformation also causes a discontinuity in the impedance; the resulting reflections may cause some waveform distortion and increased loss at VHF and UHF. (Coaxial cable has a specified "minimum bend radius".)

Chokes wound with any large diameter cable have more stray capacitance than those wound with small diameter wire. There are two sources of stray capacitance in a ferrite choke: the capacitance from end-to-end and from turn-to-turn via the core; and the capacitance from turn-to-turn via the air dielectric. Both sources of capacitance are increased by increased conductor size, so stray capacitance will be greater with larger coax. Turn-to-turn capacitance is also increased by larger diameter turns.

At low frequencies, most of the inductance in a ferrite choke results from coupling to the core, but some is the result of flux outside the core. At higher frequencies, the core has less permeability, and the flux outside the core makes a greater contribution.

The most useful cores for wound-coax chokes are the 2.4-inch OD, 1.4-inch ID toroid of type 31 or 43 material, and the 1-inch ID × 1.125-inch long clamp-on of type 31 material. Seven turns of RG-8 or RG-11 size cable easily fit through these toroids with no connector attached, and four turns fit with a PL-259 attached. Four turns of most RG-8 or RG-11 size cable fit within the 1-inch ID clamp-on. The toroids will accept at least 14 turns of most RG-6, RG-8X or RG-59 size cables.

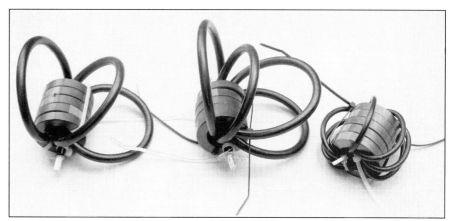

Fig 20.26 — Typical transmitting wound-coax common-mode chokes suitable for use on the HF ham bands.

PRACTICAL CHOKES

Fig 20.26 shows typical wound-coax chokes suitable for use on the HF ham bands. **Fig 20.27**, **Fig 20.28**, and **Fig 20.29** are graphs of the magnitude of the impedance for HF transmitting chokes of various sizes. Fourteen close-spaced, 3-inch diameter turns of RG-58 size cable on a #31 toroid is a very effective 300-W choke for the 160 and 80 meter bands.

Table 20.7 summarizes designs that meet the 5000-Ω criteria for the 160 through 6 meter ham bands and several practical transmitting choke designs that are "tuned" or optimized for ranges of frequencies. The table entries refer to the specific cores in the preceding paragraph. If you construct the chokes using toroids, remember to make the diameter of the turns large enough to avoid deformation of the coaxial cable. Space turns evenly around the toroid to minimize inter-turn capacitance.

USING FERRITE BEADS

The early "current baluns" developed by Walt Maxwell, W2DU, formed by stringing multiple beads in series on a length of coax to obtain the desired choking impedance, are really common-mode chokes. Maxwell's designs utilized 50 very small beads of type 73 material as shown in **Fig 20.30**. Product data sheets show that a single type 73 bead has a very low-Q resonance around 20 MHz, and has a predominantly resistive impedance of 10-20 Ω on all HF ham bands. Stringing 50 beads in series simply multiplies the impedance of one bead by 50, so the W2DU "current balun" has a choking impedance of 500-1000 Ω, and because it is strongly resistive, any resonance with the feed line is minimal.

This is a fairly good design for moderate power levels, but suitable beads are too small to fit most coax. A specialty coaxial cable

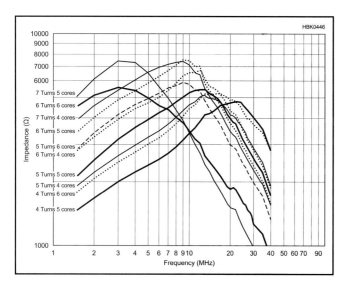

Fig 20.27 — Impedance versus frequency for HF wound-coax transmitting chokes using 2.4-inch toroid cores of #31 material with RG-8X coax.

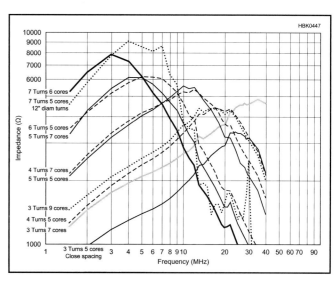

Fig 20.28 — Impedance versus frequency for HF wound-coax transmitting chokes using toroid cores of #31 material with RG-8 coax. Turns are 5-inch diameter and wide-spaced unless noted.

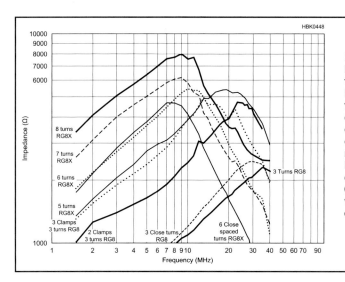

Fig 20.29 — Impedance versus frequency for HF wound-coax transmitting chokes wound on big clamp-on cores of #31 material with RG-8X or RG-8 coax. Turns are 6-inch diameter, wide-spaced except as noted.

Fig 20.30 — W2DU bead balun consisting of 50 FB-73-2401 ferrite beads over a length of RG-303 coax. See text for details.

Table 20.8
Combination Ferrite and Coaxial Coil

Freq (MHz)	7 ft, 4 turns of RG-8X	1 Core	2 Cores
1.8	—	—	520 Ω
3.5	—	660	1.4 kΩ
7	—	1.6 kΩ	3.2 kΩ
14	560 Ω	1.1 kΩ	1.4 kΩ
21	42 kΩ	500 Ω	670 Ω
28	470 Ω	—	—

Measured Impedance

Table 20.7
Transmitting Choke Designs

Freq Band(s) (MHz)	Mix	RG-8, RG-11 Turns	Cores	RG-6, RG-8X, RG-58, RG-59 Turns	Cores
1.8, 3.8	#31	7	5 toroids	7	5 toroids
				8	Big clamp-on
3.5-7		6	5 toroids	7	4 toroids
				8	Big clamp-on
10.1	#31 or #43	5	5 toroids	8	Big clamp-on
				6	4 toroids
7-14		5	5 toroids	8	Big clamp-on
14		5	4 toroids	8	2 toroids
		4	6 toroids	5-6	Big clamp-on
21		4	5 toroids	4	5 toroids
		4	6 toroids	5	Big clamp-on
28		4	5 toroids	4	5 toroids
				5	Big clamp-on
7-28, 10.1-28 or 14-28	#31 or #43	Use two chokes in series: #1 — 4 turns on 5 toroids #2 — 3 turns on 5 toroids		Use two chokes in series: #1 — 6 turns on a big clamp-on #2 — 5 turns on a big clamp-on	
14-28		Two 4-turn chokes, each w/one big clamp-on		4 turns on 6 toroids, or 5 turns on a big clamp-on	
50		Two 3-turn chokes, each w/one big clamp-on			

Notes: Chokes for 1.8, 3.5 and 7 MHz should have closely spaced turns. Chokes for 14-28 MHz should have widely spaced turns. Turn diameter is not critical, but 6 inches is good.

such as RG-303 must be used for high-power applications. Even with high-power coax, the choking impedance is often insufficient to limit current to a low enough value to prevent overheating. Equally important — the lower choking impedance is much less effective at rejecting noise and preventing the filling of nulls in a radiation pattern.

Newer "bead balun" designs use type 31 and 43 beads, which are resonant around 150 MHz, are inductive below resonance, and have only a few tens of ohms of strongly inductive impedance on the HF bands. Even with 20 of the type 31 or 43 beads in the string, the choke is still resonant around 150 MHz, is much less effective than a wound coaxial ferrite choke, and is still inductive on the HF bands (so it will be ineffective at frequencies where it resonates with the line).

Joe Reisert, W1JR, introduced the first coaxial chokes wound on ferrite toroids. He used low-loss cores, typically type 61 or 67 material. **Fig 20.31** shows that these high-Q chokes are quite effective in the narrow frequency range near their resonance. However, the resonance is quite difficult to measure and it is so narrow that it typically covers only one or two ham bands. Away from resonance, the choke becomes far less effective, as choking impedance falls rapidly and its reactive component resonates with the line.

Air-wound coaxial chokes are less effective than bead baluns. Their equivalent circuit is also a simple high-Q parallel resonance and they must be used below resonance. They are simple, inexpensive and unlikely to overheat. Choking impedance is purely inductive and not very great, reducing their effectiveness. Effectiveness is further reduced when the inductance resonates with the line at frequencies where the line impedance is capacitive and there is almost no resistance to damp the resonance.

Adding ferrite cores to a coiled-coax balun is a way to increase their effectiveness. The resistive component of the ferrite impedance damps the resonance of the coil and increases its useful bandwidth. The combinations of ferrite and coil baluns shown in **Table 20.8** demonstrate this very effectively. Eight feet of RG-8X in a 5-turn coil is a great balun for 21 MHz, but it is not particularly effective on other bands. If one type 43 core (Fair-Rite 2643167851) is inserted in the same coil of coax, the balun can be used from 3.5–21 MHz. If two of these cores are spaced a

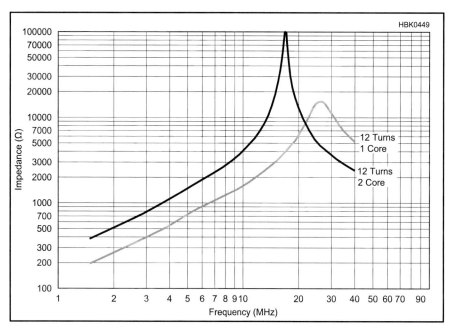

Fig 20.31 — Impedance versus frequency for HF wound-coax transmitting chokes wound with RG-142 coax on toroid cores of #61 material. For the 1-core choke: R = 15.6 kΩ, L = 25 µH, C = 1.4 pF, Q = 3.7. For the 2-core choke: R = 101 kΩ, L = 47 µH, C = 1.9 pF, Q = 20.

Fig 20.32 — Choke balun that includes both a coiled cable and ferrite beads at each end of the cable.

few inches apart on the coil as in **Fig 20.32**, the balun is more effective from 1.8 to 7 MHz and usable to 21 MHz. If type 31 material was used (the Fair-Rite 2631101902 is a similar core), low-frequency performance would be even better. The 20-turn, multiple-band, 1.8-3.5 MHz coiled-coax balun in Table 20.6 weighs 1 pound, 7 ounces. The single ferrite core combination balun weighs 6.5 ounces and the two-core version weighs 9.5 ounces.

MEASURING FERRITE CHOKE IMPEDANCE

A ferrite RF choke creates a parallel resonant circuit from inductance and resistance coupled from the core and stray capacitance resulting from interaction of the conductor that forms the choke with the permittivity of the core. If the choke is made by winding turns on a core (as opposed to single-turn bead chokes) the inter-turn capacitance also becomes part of the choke's circuit.

These chokes are very difficult to measure for two fundamental reasons. First, the stray capacitance forming the parallel resonance is quite small, typically 0.4 to 5 pF, which is often less than the stray capacitance of the test equipment used to measure it. Second, most RF impedance instrumentation measures the reflection coefficient (see the section Reflection Coefficient and SWR) in a 50-Ω circuit.

As a result, reflection-based measurements have increasingly poor accuracy when the unknown impedance is more than about three times the characteristic impedance of the analyzer, because the value of the unknown is computed by differencing analyzer data. When the differences are small, as they are for high impedances measured this way, even very small errors in the raw data cause very large errors in the computed result. While the software used with reflection-based systems use calibration and computation methods to remove systemic errors such as fixture capacitance from the measurement, these methods have generally poor accuracy when the impedance being measured is in the range of typical ferrite chokes.

The key to accurate measurement of high impedance ferrite chokes is to set up the choke as the series element, Z_X, of a voltage divider. Impedance is then measured using a well-calibrated voltmeter to read the voltage across a well-calibrated resistor that acts as the voltage divider's load resistor, R_{LOAD}. The fundamental assumption of this measurement method is that the unknown impedance is much higher than the impedance of both the generator and the load resistor.

The RF generator driving the high impedance of the voltage divider must be terminated by its calibration impedance because the generator's output voltage, V_{GEN}, is calibrated only when working into its calibration impedance. An RF spectrum analyzer with its own internal termination resistor can serve as both the voltmeter and the load. Alternatively, a simple RF voltmeter or scope can be used, with the calibrated load impedance being provided by a termination resistor of known value in the frequency range of the measurement.

With the ferrite choke in place, obtain values for the voltage across the load resistor (V_{LOAD}) and the generator in frequency increments of about 5% over the range of interest, recording the data in a spreadsheet. If multiple chokes are being measured, use the same frequencies for all chokes so that data can be plotted and compared. Using the spreadsheet, solve the voltage divider equation backward to find the unknown impedance.

$$|Z_X| = R_{LOAD} [V_{GEN} / V_{LOAD}]$$

Plot the data as a graph of impedance (on the vertical axis) vs frequency (on the horizontal axis). Scale both axes to display logarithmically.

Obtaining R, L, and C Values

This method yields the magnitude of the impedance, $|Z_X|$, but no phase information. Accuracy is greatest for large values of unknown impedance (worst case 1% for

5000 Ω, 10% for 500 Ω). Accuracy can be further improved by correcting for variations in the loading of the generator by the test circuit. Alternatively, voltage at the generator output can be measured with the unknown connected and used as V_{GEN}. The voltmeter must be unterminated for this measurement.

In a second spreadsheet worksheet, create a new table that computes the magnitude of the impedance of a parallel resonant circuit for the same range of frequencies as your choke measurements. (The required equations can be found in the section Parallel Circuits of Moderate to High Q of the **Electrical Fundamentals** chapter.) Set up the spreadsheet to compute resonant frequency and Q from manually-entered values for R, L and C. The spreadsheet should also compute and plot impedance of the same range of frequencies as the measurements and with the same plotted scale as the measurements.

1) Enter a value for R equal to the resonant peak of the measured impedance.

2) Pick a point on the resonance curve below the resonant frequency with approximately one-third of the impedance at resonance and compute L for that value of inductive reactance.

3) Enter a value for C that produces the same resonant frequency of the measurement.

4) If necessary, adjust the values of L and C until the computed curve most closely matches the measured curve.

The resulting values for R, L, and C form the equivalent circuit for the choke. The values can then be used in circuit modeling software (*NEC*, *SPICE*) to predict the behavior of circuits using ferrite chokes.

Accuracy

This setup can be constructed so that its stray capacitance is small, but it won't be zero. A first approximation of the stray capacitance can be obtained by substituting for the unknown a noninductive resistor whose resistance is in the same general range as the chokes being measured, then varying the frequency of the generator to find the –3 dB point where $X_C = R$. This test for one typical setup yielded a stray capacitance value of 0.4 pF. A thin-film surface-mount or chip resistor will have the lowest stray reactances. If a surface-mount resistor is not available, use a ¼-watt carbon composition leaded resistor with leads trimmed to the minimum amount necessary to make the connections.

Since the measured curve includes stray capacitance, the actual capacitance of the choke will be slightly less than the computed value. If you have determined the value of stray capacitance for your test setup, subtract it from the computed value to get the actual capacitance. You can also use this corrected value in the theoretical circuit to see how the choke will actually behave in a circuit — that is, without the stray capacitance of your test setup. You won't see the change in your measured data, only in the theoretical RLC equivalent.

Dual Resonances

In NiZn materials (#61, #43), there is only circuit resonance, but MnZn materials (#77, #78, #31) have both circuit resonance and dimensional resonance. (See the **RF Techniques** chapter for a discussion of ferrite resonances.) The dimensional resonance of #77 and #78 material is rather high-Q and clearly defined, so R, L, and C values can often be computed for both resonances. This is not practical with chokes wound on #31 cores because the dimensional resonance occurs below 5 MHz, is very low-Q, is poorly defined, and blends with the circuit resonance to broaden the impedance curve. The result is a dual-sloped resonance curve — that is, curve fitting will produce somewhat different values of R, L, and C when matching the low-frequency slope and high frequency slope. When using these values in a circuit model, use the values that most closely match the behavior of the choke in the frequency range of interest.

20.6 Using Transmission Lines in Digital Circuits

The performance of digital logic families covers a wide range of signal transition times. The signal rise and fall times are most important when considering how to construct a circuit. The operating frequency of a circuit is not the primary consideration. A circuit that uses high-speed logic yet runs only at a few kHz can be difficult to tame if long point-to-point wiring is used.

If the path between two points has a delay of more than ⅙ the logic family rise time, some form of transmission line should be considered. We know that waves propagate at 300 million meters per second in air and at 0.66 times as fast in common coax cable. So, in about 5 ns a wave will travel 1 meter or in 1 ns it will travel 0.2 m.

Consider a logic family which has 2 ns rise and fall time. Using the rule mentioned above, if the path length exceeds 0.066 meter or about 2.6 inches we need to use a transmission line. Another way to look at it is the approximate equivalent analog bandwidth is:

$$BW = \frac{0.35}{\text{rise time}} = \frac{0.35}{2 \text{ ns}} = 175 \text{ MHz} \quad (21)$$

If we were building an analog circuit that operated at 175 MHz, we would have to keep the wire lengths down to a fraction of an inch. So, even if our logic's clock is running at a few kHz, we still need to use these short wire lengths. But, suppose we are building a nontrivial circuit that has a number of gates and other digital blocks to interconnect. In order to reach several ICs from the clock's source, we will need to run wires over several inches in length.

It is possible to build a high speed circuit in breadboard style if small coax cable is used for interconnections. However, if a PC board is designed with *microstrip* transmission line interconnections, success is more likely. **Fig 20.33** shows the way microstrip transmission line is made. Typical dimensions are shown for ¹⁄₁₆-inch thick FR4 material and 50-Ω line.

There are several ways that the actual circuit can be configured to assure that the desired signal reaches the receiving device input. To avoid multiple reflections that would distort the signals and possibly cause false triggering, the line should either be matched at the load end or the source end. We know that matching at the load end will completely absorb the signal so that there are no reflections. However, the signal level will be reduced because of voltage division with the source impedance in the sending gate. The dc levels will also shift because of the load, reducing the logic noise immunity. A considerable amount of power can be dissipated in the load, which might overload the source gate particularly if 50-Ω line is in use. It is possible to put a capacitor in series with the load resistor, but only if the waveform duty cycle is near 50%. If not, the dc average voltage will reduce the noise immunity of the receiving gate.

A better method is to match the transmis-

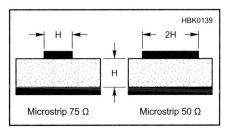

Fig 20.33 — Microstrip transmission lines. The approximate geometries to produce 75 Ω (A) and 50 Ω (B) microstrip lines with FR-4 PC board material are shown. This technique is used at UHF and microwave frequencies.

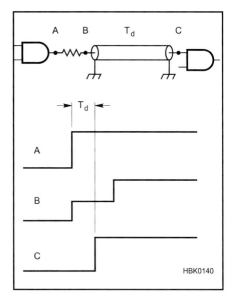

Fig 20.34 — Reflections in a transmission line cause stepping in the leading edge of a digital pulse at the generator (A). By adding a resistor in series with the gate output, a step is generated at the input to the transmission line (B), but a full-voltage step is created at the high input impedance of the receiving gate.

sion line at the signal source as in **Fig 20.34**. A resistor is added in series with the output of the sending gate, raising the gate output impedance to match Z_0 of the transmission line, which is connected to the resistor at B and has an arbitrary length T_D. No load is required on the receiving end of the transmission line, which is assumed to be connected to a gate with an input impedance much higher than Z_0.

When the sending output goes high at point A, generating the leading edge of a pulse with voltage V, the load on the output resistor is equal to the Z_0 of the transmission line so a voltage divider is formed and the voltage at point B initially goes to ½ V. The pulse travels down the transmission line and after T_D it completely reflects off the open end at the receiving gate. The voltage at C reaches V since the direct signal and the reflected signal add together. When the reflected edge of the pulse returns to point B after a round trip time of $2T_D$, the voltage level at B increases to V.

The receiving end can be terminated in Z_0 if a pair of resistors, each equal to $2 \times Z_0$, are connected from the positive power supply to ground at point C. The transmission line and gate input are connected to the resistor junction. The main problem with this method is the steady-state current required by the resistors. Some logic gates may not have adequate current output to drive this load.

20.7 Waveguides

A waveguide is a hollow conducting tube through which microwave energy is transmitted in the form of electromagnetic waves. The tube does not carry a current in the same sense that the wires of a two-conductor line do. Instead, it is a boundary that confines the waves to the enclosed space. Skin effect on the inside walls of the waveguide confines electromagnetic energy inside the guide in much the same manner that the shield of a coaxial cable confines energy within the coax. Microwave energy is injected at one end (either through capacitive or inductive coupling or by radiation) and is received at the other end. The waveguide merely confines the energy of the fields, which are propagated through it to the receiving end by means of reflections off its inner walls.

20.7.1 Evolution of a Waveguide

Suppose an open-wire line is used to carry UHF energy from a generator to a load. If the line has any appreciable length, it must be well-insulated from the supports in order to avoid high losses. Since high-quality insulators are difficult to make for microwave frequencies, it is logical to support the transmission line with quarter-wave stubs, shorted at the far end. The open end of such a stub presents an infinite impedance to the transmission line, provided that the shorted stub is non-reactive. However, the shorting link has finite length and, therefore, some inductance. This inductance can be nullified by making the RF current flow on the surface of a plate rather than through a thin wire. If the plate is large enough, it will prevent the magnetic lines of force from encircling the RF current.

An infinite number of these quarter-wave stubs may be connected in parallel without affecting the standing waves of voltage and current. The transmission line may be supported from the top as well as the bottom, and when infinitely many supports are added, they form the walls of a waveguide at its *cutoff frequency*. **Fig 20.35** illustrates how a rectangular waveguide evolves from a two-wire parallel transmission line. This simplified analysis also shows why the cutoff dimension is ½ λ.

While the operation of waveguides is usually described in terms of fields, current does flow on the inside walls, just as fields exist between the current-carrying conductors of a two-wire transmission line. At the waveguide

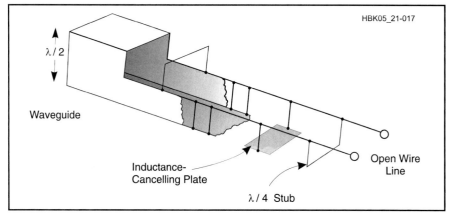

Fig 20.35 — At its cutoff frequency a rectangular waveguide can be analyzed as a parallel two-conductor transmission line supported from top and bottom by an infinite number of quarter-wave stubs.

Table 20.9
Wavelength Formulas for Waveguide

	Rectangular	Circular
Cut-off wavelength	2X	3.41R
Longest wavelength transmitted with little attenuation	1.6X	3.2R
Shortest wavelength before next mode becomes possible	1.1X	2.8R

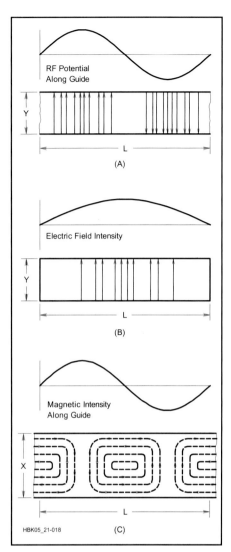

Fig 20.36 — Field distribution in a rectangular waveguide. The TE$_{1,0}$ mode of propagation is depicted.

cutoff frequency, the current is concentrated in the center of the walls, and disperses toward the floor and ceiling as the frequency increases.

Analysis of waveguide operation is based on the assumption that the guide material is a perfect conductor of electricity. Typical distributions of electric and magnetic fields in a rectangular guide are shown in **Fig 20.36**. The intensity of the electric field is greatest (as indicated by closer spacing of the lines of force in Fig 20.35B) at the center along the X dimension and diminishes to zero at the end walls. Zero field intensity is a necessary condition at the end walls, since the existence of any electric field parallel to any wall at the surface would cause an infinite current to flow in a perfect conductor, an impossible situation.

20.7.2 Modes of Waveguide Propagation

Fig 20.36 represents a relatively simple distribution of the electric and magnetic fields. An infinite number of ways exist in which the fields can arrange themselves in a guide, as long as there is no upper limit to the frequency to be transmitted. Each field configuration is called a *mode*. All modes may be separated into two general groups. One group, designated *TM* (transverse magnetic), has the magnetic field entirely crosswise to the direction of propagation, but has a component of electric field in the propagation direction. The other type, designated *TE* (transverse electric) has the electric field entirely crosswise to the direction of propagation, but has a component of magnetic field in the direction of propagation. TM waves are sometimes called *E-waves*, and TE waves are sometimes called *H-waves*. The TM and TE designations are preferred, however.

The particular mode of transmission is identified by the group letters followed by subscript numbers; for example TE$_{1,1}$, TM$_{1,1}$ and so on. The number of possible modes increases with frequency for a given size of guide. There is only one possible mode (called the *dominant mode*) for the lowest frequency that can be transmitted. The dominant mode is the one normally used in practical applications.

20.7.3 Waveguide Dimensions

In rectangular guides the critical dimension (shown as X in Fig 20.36C) must be more than one-half wavelength at the lowest frequency to be transmitted. In practice, the Y dimension is usually made about equal to ½ X to avoid the possibility of operation at other than the dominant mode. Cross-sectional shapes other than rectangles can be used; the most important of those is the circular pipe.

Table 20.9 gives dominant-mode wavelength formulas for rectangular and circular guides. X is the width of a rectangular guide, and R is the radius of a circular guide.

20.7.4 Coupling to Waveguides

Energy may be introduced into or extracted from a waveguide or resonator by means of either the electric or magnetic field. The energy transfer frequently takes place through a coaxial line. Two methods for coupling are shown in **Fig 20.37**. The probe at A is simply a short extension of the inner conductor of the feed coaxial line, oriented so that it is parallel to the electric lines of force. The loop shown at B is arranged to enclose some of the magnetic lines of force. The point at which maximum coupling will be obtained depends on the particular mode of propagation in the guide or cavity; the coupling will be maximum when the coupling device is in the most intense field.

Coupling can be varied by rotating the probe or loop through 90°. When the probe is perpendicular to the electric lines the coupling will be minimum; similarly, when the plane of the loop is parallel to the magnetic lines, the coupling will be minimum. See *The ARRL Antenna Book* for more information on waveguides.

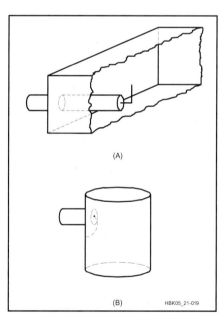

Fig 20.37 — Coupling to waveguide and resonators. The probe at A is an extension of the inner conductor of coax line. At B an extension of the coax inner conductor is grounded to the waveguide to form a coupling loop.

20.8 Glossary of Transmission Line Terms

Antenna tuner — A device that matches the antenna system input impedance to the transmitter, receiver or transceiver output impedance. Also called an *antenna-matching network, impedance matcher, transmatch, ATU, matchbox*.

Balanced line — A symmetrical two-conductor feed line that has uniform voltage and current distribution along its length.

Balun — Contraction of "balanced to unbalanced." A device to couple a balanced load to an unbalanced feed line or device, or vice versa. May be in the form of a choke balun, or a transformer that provides a specific impedance transformation (including 1:1). Often used in antenna systems to interface a coaxial transmission line to the feed point of a balanced antenna, such as a dipole.

Characteristic impedance — The ratio of voltage to current in a matched feed line, it is determined by the physical geometry and materials used to construct the feed line. Also known as *surge impedance* since it represents the impedance electromagnetic energy encounters when entering a feed line.

Choke balun — A balun that prevents current from flowing on the outside of a coaxial cable shield when connected to a balanced load, such as an antenna.

Coax — See **coaxial cable**.

Coaxial cable — Transmission lines that have the outer shield (solid or braided) concentric with the same axis as the inner or center conductor. The insulating material can be a gas (air or nitrogen) or a solid or foam insulating material.

Common-mode current — Current that flows equally and in phase on all conductors of a feed line or multiconductor cable.

Conductor — A metal body such as tubing, rod or wire that permits current to travel continuously along its length.

Conjugate match — Creating a purely resistive impedance by connecting an impedance with an equal-and-opposite reactive component.

Current balun — see **Choke balun**.

Decibel — A logarithmic power ratio, abbreviated dB. May also represent a voltage or current ratio if the voltages or currents are measured across (or through) identical impedances. Suffixes to the abbreviation indicate references: dBi, isotropic radiator; dBm, milliwatt; dBW, watt.

Dielectrics — Various insulating materials used in antenna systems, such as found in insulators and transmission lines.

Dielectric constant (k) — Relative figure of merit for an insulating material used as a dielectric. This property determines how much electric energy can be stored in a unit volume of the material per volt of applied potential.

Electric field — An electric field exists in a region of space if an electrically charged object placed in the region is subjected to an electrical force.

Electromagnetic wave — A wave of energy composed of an electric and magnetic field.

Feed line — See **transmission line**.

Feed point — The point at which a feed line is electrically connected to an antenna.

Feed point impedance — The ratio of RF voltage to current at the feed point of an antenna.

Ferrite — A ceramic material with magnetic properties.

Hardline — Coaxial cable with a solid metal outer conductor to reduce losses compared to flexible cables. Hardline may or may not be flexible.

Impedance match — To adjust impedances to be equal or the case in which two impedances are equal. Usually refers to the point at which a feed line is connected to an antenna or to transmitting equipment. If the impedances are different, that is a **mismatch**.

Impedance matcher — See **Antenna tuner**.

Impedance matching (circuit) — A circuit that transforms impedance from one value to another. Adjustable impedance matching circuits are used at the output of transmitters and amplifiers to allow maximum power output over a wide range of load impedances.

Impedance transformer — A transformer designed specifically for transforming impedances in RF equipment.

L network — A combination of two reactive components used to transform or match impedances. One component is connected in series between the source and load and the other shunted across either the source or the load. Most L networks have one inductor and one capacitor, but two-inductor and two-capacitor configurations are also used.

Ladder line — see **Open-wire line**.

Lambda (λ) — Greek symbol used to represent wavelength.

Line loss — The power dissipated by a transmission line as heat, usually expressed in decibels.

Load — (noun) The component, antenna, or circuit to which power is delivered; (verb) To apply a load to a circuit or a transmission line.

Loading — The process of a transferring power from its source to a load. The effect a load has on a power source.

Magnetic field — A region through which a magnetic force will act on a magnetic object.

Matched-line loss — The line loss in a feed line terminated by a load equal to its characteristic impedance.

Matching — The process of effecting an impedance match between two electrical circuits of unlike impedance. One example is matching a transmission line to the feed point of an antenna. Maximum power transfer to the load (antenna system) will occur when a matched condition exists.

Microstrip — A transmission line made from a strip of printed-circuit board conductor above a ground plane, used primarily at UHF and microwave frequencies.

Open-wire line — Parallel-conductor feed line with parallel insulators at regular intervals to maintain the line spacing. The dielectric is principally air, making it a low-loss type of line. Also known as *ladder line* or *window line*.

Output impedance — The equivalent impedance of a signal source.

Parallel-conductor line — A type of transmission line that uses two parallel wires spaced from each other by insulating material. Also known as *open-wire, ladder* or *window line*.

Phasing lines — Sections of transmission line that are used to ensure the correct phase relationship between the elements of a driven array, or between bays of an array of antennas. Also used to effect impedance transformations while maintaining the desired phase.

Q section — Term used in reference to transmission-line matching transformers and phasing lines.

Reflection coefficient (ρ) — The ratio of the reflected voltage at a given point on a transmission line to the incident voltage at the same point. The reflection coefficient is also equal to the ratio of reflected and incident currents. The Greek letter rho (ρ) is used to represent reflection coefficient.

Reflectometer — see **SWR bridge**

Resonance — (1) The condition in which a system's natural response and the frequency of an applied or emitted signal are the same. (2) The frequency at which a circuit's capacitive and inductive reactances are equal and cancel.

Resonant frequency — The frequency

at which the maximum response of a circuit occurs. In an antenna, the resonant frequency is one at which the feed point impedance is purely resistive.

Return loss — The absolute value of the ratio in dB of the power reflected from a load to the power delivered to the load.

Rise time — The time it takes for a waveform to reach a maximum value.

Series-input network — A network such as a filter or impedance matching circuit in which the input current flows through a component in series with the input.

Shunt-input network — A network such as a filter or impedance matching circuit with a component connected directly across the input.

Skin effect — The phenomenon in which ac current at high frequencies flows in a thin layer near the surface of a conductor.

Smith Chart — A coordinate system developed by Phillip Smith to represent complex impedances graphically. This chart makes it easy to perform calculations involving antenna and transmission-line impedances and SWR.

Standing-wave ratio (SWR) — Sometimes called voltage standing-wave ratio (VSWR). A measure of the impedance match between a feed line's characteristic impedance and the attached load (usually an antenna). VSWR is the ratio of maximum voltage to minimum voltage along the feed line, or of antenna impedance to feed line impedance.

Stacking — The technique of placing similar directive antennas atop or beside one another, forming a "stacked array." Stacking provides more gain or directivity than a single antenna.

Stub — A section of transmission line used to perform impedance matching or filtering.

Surge impedance — see **Characteristic impedance**.

SWR — see **Standing-wave ratio**.

SWR bridge — Device for measuring SWR in a transmission line. Also known as an SWR meter or reflectometer.

TE mode — Transverse electric field mode. Condition in a waveguide in which the E-field component of the traveling electromagnetic energy is oriented perpendicular to (transverse) the direction the energy is traveling in the waveguide.

TM mode — Transverse magnetic field mode. Condition in a waveguide in which the H-field (magnetic field) component of the traveling electromagnetic energy is oriented perpendicular to (transverse) the direction the energy is traveling in the waveguide.

Transmatch — See **Antenna tuner.**

Transmission line — The wires or cable used to connect a transmitter or receiver to an antenna. Also called **feed line**.

Twin-lead — Parallel-conductor transmission line in which both conductors are completely embedded in continuous strip of insulating material.

Unbalanced line — Feed line with one conductor at dc ground potential, such as coaxial cable.

Universal stub system — A matching network consisting of a pair of transmission line stubs that can transform any impedance to any other impedance.

Velocity factor (velocity of propagation) — The speed at which an electromagnetic wave will travel through a material or feed line stated as a fraction of the speed of the wave in free space (where the wave would have its maximum velocity).

VSWR — Voltage standing-wave ratio. See **SWR**.

Waveguide — A hollow conductor through which electromagnetic energy flows. Usually used at UHF and microwave frequencies instead of coaxial cable.

Window line – see **Open-wire line.**

20.9 References and Bibliography

H.W. Johnson and M. Graham, *High Speed Digital Design* (Prentice Hall, 1993).

F. Regier, "Series-Section Transmission Line Impedance Matching," *QST,* Jul 1978, pp 14-16.

J. Sevick, W2FMI, *Transmission Line Transformers*, 4th Edition (Noble Publishing, 2001).

P. Smith, *Electronic Applications of the Smith Chart* (Noble Publishing, 1995).

R.D. Straw, Ed., *The ARRL Antenna Book*, 21st Edition (Newington: ARRL, 2007). Chapters 24 through 28 include material on transmission lines and related topics.

F. Witt, "Baluns in the Real (and Complex) World," *The ARRL Antenna Compendium Vol 5* (Newington: ARRL, 1996).

The *ARRL UHF/Microwave Experimenter's Manual*, (Newington: ARRL, 2000). Chapters 5 and 6 address transmission lines and impedance matching. (This book is out of print).

C. Counselman, W1HIS, "Common-mode Chokes", **www.yccc.org/Articles/W1HIS/CommonModeChokesW1HIS2006Apr06.pdf**.

ONLINE RESOURCES

A Collection of Smith Chart References — **sss-mag.com/smith.html**

A 100 W Z-Match tuner — **freespace.virgin.net/geoff.cottrell/Z_match.htm**

Ferrite and Powdered Iron Cores — **www.fair-rite.com**, **www.amidoncorp.com** and **www.cwsbytemark.com**

R. Lewellan. W7EL, "Baluns: What They Do and How They Do It," 1985, **www.eznec.com/Amateur/Articles/Baluns.pdf**

Microwave oriented downloads — **www.microwaves101.com**

Table of transmission line properties — **hf-antenna.com/trans/**

Times Microwave catalog — **www.timesmicrowave.com**

Transmission line loss factors — **www.microwaves101.com/encyclopedia/transmission_loss.cfm**

Transmission line matching with the Smith chart — **www.odyseus.nildram.co.uk/RFMicrowave_Theory_Files/SmithChartPart2.pdf**

Transmission line transformer theory — **www.bytemark.com/products/tlttheory.htm**

Using Ferrites and Baluns — **www.audiosystemsgroup.com/RFI-Ham.pdf**

Z-Match tuner description —**members.optushome.com.au/vk6ysf/vk6ysf/zmatch.htm**

Contents

21.1 Antenna Basics
    21.1.1 Directivity and Gain
    21.1.2 Antenna Polarization
    21.1.3 Current and Voltage Distribution
    21.1.4 Impedance
    21.1.5 Impedance and Height Above Ground
    21.1.6 Antenna Bandwidth
    21.1.7 Effects of Conductor Diameter
    21.1.8 Radiation Patterns
    21.1.9 Elevation Angle
    21.1.10 Imperfect Ground

21.2 Dipoles and the Half-Wave Antenna
    21.2.1 Radiation Characteristics
    21.2.2 Feed Methods
    21.2.3 Baluns
    21.2.4 Building Dipoles and Other Wire Antennas
    21.2.5 Dipole Orientation
    21.2.6 Inverted-V Dipole
    21.2.7 Sloping Dipole
    21.2.8 Shortened Dipoles
    21.2.9 Half-Wave Vertical Dipole (HVD)
    21.2.10 Folded Dipoles
    21.2.11 Multiband Dipole Systems
    21.2.12 NVIS Antennas
    Project: Multiband Center-Fed Dipole
    Project: 40-15 Meter Dual-Band Dipole
    Project: W4RNL Inverted-U Antenna
    Project: Two W8NX Multiband, Coax-Trap Dipoles
    Project: Extended Double-Zepp for 17 Meters

21.3 Vertical (Ground-Plane) Antennas
    21.3.1 Ground Systems
    21.3.2 Full-Size Vertical Antennas
    21.3.3 Physically Short Verticals
    Project: Top-Loaded Low-Band Antenna
    21.3.4 Cables and Control Wires on Towers
    21.3.5 Multiband Trap Verticals

21.4 T and Inverted-L Antennas

21.5 Slopers and Vertical Dipoles
    21.5.1 The Half-Sloper Antenna
    Project: Half-Wave Vertical Dipole (HVD)
    Project: Compact Vertical Dipole (CVD)
    Project: All-Wire 30 Meter CVD

21.6 Yagi Antennas
    21.6.1 Parasitic Excitation
    21.6.2 Yagi Gain, Front-to-Back Ratio, and SWR
    21.6.3 Two-Element Beams
    21.6.4 Three-Element Beams
    21.6.5 Construction of Yagi Antennas
    Project: Family of Computer-Optimized HF Yagis

21.7 Quad and Loop Antennas
    Project: Five-Band, Two-Element HF Quad
    Project: Wire Quad for 40 Meters
    21.7.1 Loop Antennas
    Project: Vertical Loop Antenna for 28 MHz
    Project: Multiband Horizontal Loop Antenna

21.8 HF Mobile Antennas
    21.8.1 Simple Whips
    21.8.2 Coil-Loaded Whips
    21.8.3 Base vs Center vs Continuous Loading
    21.8.4 Remotely Controlled HF Mobile Antennas
    21.8.5 Ground Losses
    21.8.6 Antenna Mounting
    21.8.7 Mobile HF Antenna Matching
    21.8.8 Remotely-Tuned Antenna Controllers
    21.8.9 Efficiency
    Project: Mounts for Remotely-Tuned Antennas
    Project: Retuning a CB Whip Antenna

21.9 VHF/UHF Mobile Antennas
    21.9.1 VHF/UHF Antenna Mounts

21.10 VHF/UHF Antennas
    21.10.1 Gain
    21.10.2 Radiation Pattern
    21.10.3 Height Gain
    21.10.4 Physical Size
    21.10.5 Polarization
    21.10.6 Circular Polarization
    21.10.7 Transmission Lines
    21.10.8 Impedance Matching
    21.10.9 Baluns and Impedance transformers
    Project: Simple, Portable Ground-Plane Antenna
    Project: Dual-Band Antenna for 146/446 MHz

21.11 VHF/UHF Yagis
    21.11.1 Stacking Yagis
    Project: Three and Five-Element Yagis for 6 Meters
    Project: Medium-Gain 2 Meter Yagi
    Project: Cheap Yagis by WA5VJB

21.12 Direction-Finding Antennas
    21.12.1 RDF Antennas for HF Bands
    21.12.2 Methods for VHF/UHF RDF
    21.12.3 Direction-Finding Techniques

21.13 Glossary

21.14 References and Bibliography

Chapter 21

Antennas

In the world of radio, the antenna is "where the rubber meets the road!" With antennas so fundamental to communication, it is important that the amateur have a basic understanding of their function. That understanding enables effective selection and application of basic designs to whatever communications task is at hand. In addition, the amateur is then equipped to engage in one of the most active areas of amateur experimentation, antenna design. The goal of this chapter is to define and illustrate the fundamentals of antennas and provide a selection of basic designs; simple verticals and dipoles, quads and Yagi beams, and other antennas. The reader will find additional in-depth coverage of these and other topics in the *ARRL Antenna Book* and other references provided. This chapter was originally written by Chuck Hutchinson, K8CH, and has been updated by Ward Silver, NØAX. Alan Applegate, KØBG, contributed the sections on mobile antennas. The section on Radio Direction Finding Antennas was written by Joe Moell, KØOV.

21.1 Antenna Basics

This section covers a range of topics that are fundamental to understanding how antennas work and defines several key terms. (A glossary is included at the end of the chapter.) While the discussion in this section uses the dipole as the primary example, the concepts apply to all antennas.

21.1.1 Directivity and Gain

All antennas, even the simplest types, exhibit directive effects in that the intensity of radiation is not the same in all directions from the antenna. This property of radiating more strongly in some directions than in others is called the *directivity* of the antenna. Directivity is the same for receiving as transmitting.

The directive pattern of an antenna at a given frequency is determined by the size and shape of the antenna, and on its position and orientation relative to the Earth and any other reflecting or absorbing surfaces.

The more an antenna's directivity is enhanced in a particular direction, the greater the *gain* of the antenna. This is a result of the radiated energy being concentrated in some directions at the expense of others. Similarly, gain describes the ability of the antenna to receive signals preferentially from certain directions. Gain does not create additional power beyond that delivered by the feed line — it only focuses that energy.

Gain is usually expressed in decibels, and is always stated with reference to a *standard* antenna — usually a dipole or an *isotropic radiator*. An isotropic radiator is a theoretical antenna that would, if placed in the center of an imaginary sphere, evenly illuminate that sphere with radiation. The isotropic radiator is an unambiguous standard, and for that reason frequently used as the comparison for gain measurements.

When the reference for gain is the isotropic radiator in free space, gain is expressed in dBi. When the standard is a dipole, also located in *free space*, gain is expressed in dBd. Because the dipole has some gain (2.1 dB) in its favorite direction with respect to the isotropic antenna (see the next section on the dipole antenna), the dipole's gain can be expressed as 2.1 dBi. Gain in dBi can be converted to dBd by subtracting 2.1 dB and from dBd to dBi by adding 2.1 dB.

Gain also takes losses in the antenna or surrounding environment into account. For example, if a practical dipole antenna's wire element dissipated 0.5 dB of the transmitter power as heat, that specific dipole's gain with respect to an isotropic antenna would be 2.1 − 0.5 = 1.6 dBi.

21.1.2 Antenna Polarization

An electromagnetic wave has two components: an electric field and a magnetic field at right angles to each other. For most antennas, the field of primary interest is the electric, or *E-field*. The magnetic field is called the *H-field*. (The abbreviations E- and H- come from Maxwell's equations that describe electromagnetic waves.) By convention, the orientation of the E-field is the reference for determining the electromagnetic wave's *polarization*. The E-field of an electromagnetic wave can be oriented in any direction, so orientation with respect to the Earth's surface is the usual frame of reference. The wave's polarization can be vertical, horizontal, some intermediate angle, or even circular.

Antennas are considered to have polarization, too, determined by the orientation of the E-field of the electromagnetic field radiated by the antenna. Because the E-field of the radiated wave is parallel to the direction of current flow in the antenna's elements, the polarization of the wave and the orientation of the antenna elements is usually the same. For example, the E-field radiated by an antenna with linear elements is parallel to those elements, so that the polarization of the radiated wave is the same as the orientation of the elements. (This is somewhat over-simplified and additional considerations apply for elements that are not linear.) Thus a radiator that is parallel to the earth radiates a horizontally polarized wave, while a vertical antenna radiates a vertically polarized wave. If a wire antenna is slanted, it radiates waves with an E-field that has both vertical and horizontal components.

Antennas function symmetrically — a received signal will create the strongest antenna current when the antenna's elements are parallel to the E-field of the incoming wave just as the radiated wave's E-field will be strongest parallel to current in the antenna's radiating elements. This also means that for the strongest received signal, the antenna elements should have the same polarization as that of the incoming wave. Misalignment of the receiving antenna's elements with the passing wave's E-field reduces the amount of signal received. This is called *cross-polarization*. When the polarizations of antenna and wave are at right angles, very little antenna current is created by the incoming signal.

For best results in line-of-sight communications, antennas at both ends of the circuit should have the same polarization. However, it is not essential for both stations to use the same antenna polarity for ionospheric propagation or sky wave (see the **Propagation** chapter). This is because the radiated wave is bent and rotated considerably during its travel through the ionosphere. At the far end of the communications path the wave may be horizontal, vertical or somewhere in between at any given instant. For that reason, the main consideration for a good DX antenna is a low angle of radiation rather than the polarization.

Most HF-band antennas are either vertically or horizontally polarized. Although circular polarization is possible, just as it is at VHF and UHF, it is seldom used at HF. While most amateur antenna installations use the Earth's surface as their frame of reference, in cases such as satellite communication or EME the terms "vertical" and "horizontal" have no meaning with respect to polarization.

21.1.3 Current and Voltage Distribution

When power is fed to an antenna, the current and voltage vary along its length. The current is minimum at the ends, regardless of the antenna's length. The current does not actually reach zero at the current minima, because of capacitance at the antenna ends. Insulators, loops at the antenna ends, and support wires all contribute to this capacitance, which is also called the *end effect*. The opposite is true of the RF voltage. That is, there is a voltage maximum at the ends.

In the case of a half-wave antenna there is a current maximum at the center and a voltage minimum at the center as illustrated in **Fig 21.1**. The voltage and current in this case are 90° out of phase. The pattern of alternating current and voltage maxima a quarter-wavelength apart repeats every half-wavelength along a linear antenna as shown in Fig 21.1B. The phase of the current and voltage are inverted in each successive half-wavelength section.

The voltage is not zero at its minimum because of the resistance of the antenna, which consists of both the RF resistance of the wire (ohmic loss resistance) and the *radiation resistance*. The radiation resistance is the equivalent resistance that would dissipate the power the antenna radiates, with a current flowing in it equal to the antenna current at a current maximum. Radiation resistance represents the work done by the electrons in the antenna in transferring the energy from the signal source to the radiated electromagnetic wave. The loss resistance of a half-wave antenna is ordinarily small, compared with the radiation resistance, and can usually be neglected for practical purposes except in electrically small antennas, such as mobile HF antennas.

21.1.4 Impedance

The *impedance* at a given point in the antenna is determined by the ratio of the voltage to the current at that point. For example, if there were 100 V and 1.4 A of RF current at a specified point in an antenna and if they were in phase, the impedance would be approximately 71 Ω. The antenna's *feed point impedance* is the impedance at the point where the feed line is attached. If the feed point location changes, so does the feed point impedance.

Antenna impedance may be either resistive or complex (that is, containing resistance and reactance). The impedance of a *resonant* antenna is purely resistive anywhere on the antenna, no matter what value that impedance may be. For example, the impedance of a resonant half-wave dipole may be low at the center of the antenna and high at the ends, but it is purely resistive in all cases, even though its magnitude changes.

The feed point impedance is important in determining the appropriate method of matching the impedance of the antenna and the transmission line. The effects of mismatched antenna and feed line impedances are described in detail in the **Transmission Lines** chapter of this book. Some mistakenly believe that a mismatch, however small, is a serious matter. This is not true. The

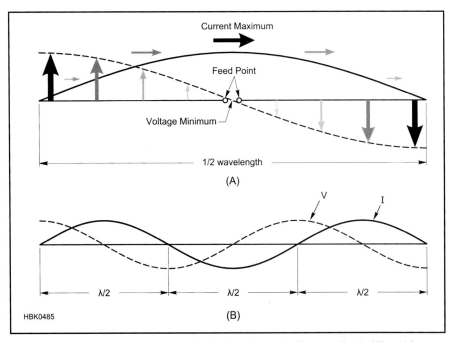

Fig 21.1 — The current and voltage distribution along a half-wave dipole (A) and for an antenna made from a series of half-wave dipoles (B).

significance of a perfect match becomes more pronounced only at VHF and higher, where feed line losses are a major factor. Minor mismatches at HF are rarely significant.

21.1.5 Impedance and Height Above Ground

The feed point impedance of an antenna varies with height above ground because of the effects of energy reflected from and absorbed by the ground. For example, a ½-λ (or half-wave) center-fed dipole will have a feed point impedance of approximately 75 Ω in free space far from ground, but **Fig 21.2** shows that only at certain electrical heights above ground will the feed point impedance be 75 Ω. The feed point impedance will vary from very low when the antenna is close to the ground to a maximum of nearly 100 Ω at 0.34-λ above ground, varying between ±5 Ω as the antenna is raised farther. The 75-Ω feed point impedance is most likely to be realized in a practical installation when the horizontal dipole is approximately ½, ¾ or 1 wavelength above ground. This is why few amateur λ/2-dipoles exhibit a center-fed feed point impedance of 75 Ω, even though they may be resonant.

Fig 21.2 compares the effects of perfect ground and typical soil at low antenna heights. The effect of height on the radiation resistance of a horizontal half-wave antenna is not drastic so long as the height of the antenna is greater than 0.2 λ. Below this height, while decreasing rapidly to zero over perfectly conducting ground, the resistance decreases less rapidly with height over actual lossy ground. At lower heights the resistance stops decreasing at around 0.15 λ, and thereafter increases as height decreases further. The reason for the increasing resistance is that more and more energy from the antenna is absorbed by the earth as the height drops below ¼ λ, seen as an increase in feed point impedance.

21.1.6 Antenna Bandwidth

The *bandwidth* of an antenna refers generally to the range of frequencies over which the antenna can be used to obtain a specified level of performance. The bandwidth can be specified in units of frequency (MHz or kHz) or as a percentage of the antenna's design frequency.

Popular amateur usage of the term antenna bandwidth most often refers to the 2:1 SWR bandwidth, such as, "The 2:1 *SWR bandwidth* is 3.5 to 3.8 MHz" or "The antenna has a 10% SWR bandwidth" or "On 20 meters, the antenna has an SWR bandwidth of 200 kHz." Other specific bandwidth terms are also used, such as the *gain bandwidth* (the bandwidth over which gain is greater than a specified level) and the *front-to-back ratio bandwidth* (the bandwidth over which front-to-back ratio is greater than a specified level).

As operating frequency is lowered, an equivalent bandwidth in percentage becomes narrower in terms of frequency range in kHz or MHz. For example, a 5% bandwidth at 21 MHz is 1.05 MHz (more than wide enough to cover the whole band) but at 3.75 MHz only 187.5 kHz! Because of the wide percentage bandwidth of the lower frequency bands 160 meters is 10.5% wide, 80 meters is 3.4% wide) it is difficult to design an antenna with a bandwidth sufficient to include the whole band.

It is important to recognize that SWR bandwidth does not always relate directly to gain bandwidth. Depending on the amount of feed line loss, an 80 meter dipole with a relatively narrow 2:1 SWR bandwidth can still radiate a good signal at each end of the band, provided that an antenna tuner is used to allow the transmitter to load properly. Broadbanding techniques, such as fanning the far ends of a dipole to simulate a conical type of dipole, can help broaden the SWR bandwidth.

21.1.7 Effects of Conductor Diameter

The impedance and resonant frequency of an antenna also depend on the diameter of the conductors that make up its elements in relation to the wavelength. As diameter of a conductor increases, its capacitance per unit length increases and inductance per unit length decreases. This has the net effect of lowering the frequency at which the antenna element is resonant, as illustrated in **Fig 21.3**. The larger the conductor diameter in terms of wavelength, the smaller its *length-to-diameter ratio (l/d)* and the lower the frequency at

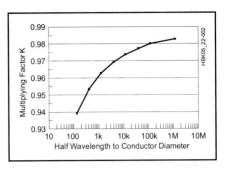

Fig 21.3 — Effect of antenna diameter on length for half-wavelength resonance, shown as a multiplying factor, K, to be applied to the free-space, half-wavelength equation.

which a specific length of that conductor is ½-wavelength long electrically, in free space.

$$l/d = \frac{\lambda/2}{d} = \frac{300}{2f \times d} \quad (1)$$

where f is in MHz and d is in meters. For example, a ½-wavelength dipole for 7.2 MHz made from #12 AWG wire (0.081-inch dia) has an l/d ratio of

$$l/d = \frac{300}{2f \times d} = \frac{300}{2 \times 7.2 \times \frac{0.081\,in}{39.37\,in/m}} = 10,126$$

The effect of l/d is accounted for by the factor K which is based on l/d. From Fig 21.3 an l/d ratio of 10,126 corresponds to K ≈ 0.975, so the resonant length of that ½-wave dipole would be 0.975 × (300 / 2f) = 20.31 meters instead of the free-space 20.83 meters.

Most wire antennas at HF have l/d ratios in the range of 2500 to 25,000 with K = 0.97 to 0.98. The value of K is taken into account in the classic formula for ½-wave dipole length, 468/f (in MHz). If K = 1, the formula would be 492/f (in MHz). (This is discussed further in the following section on Dipoles and the Half-Wave Antenna.)

For single-wire HF antennas, the effects of ground and antenna construction make a precise accounting for K unnecessary in practice. At and above VHF, the effects of l/d ratio can be of some importance, since the wavelength is small.

Since the radiation resistance is affected relatively little by l/d ratio, the decreased L/C ratio causes the Q of the antenna to decrease. This means that the change in antenna impedance with frequency will be less, increasing the antenna's SWR bandwidth. This is often used to advantage on the lower HF bands by using multiple conductors in a cage or fan to decrease the l/d ratio.

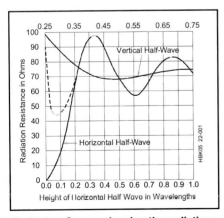

Fig 21.2 — Curves showing the radiation resistance of vertical and horizontal half-wavelength dipoles at various heights above ground. The broken-line portion of the curve for a horizontal dipole shows the resistance over *average* real earth, the solid line for perfectly conducting ground.

21.1.8 Radiation Patterns

Radiation patterns are graphic representations of an antenna's directivity. Two examples are given in **Figs 21.4** and **21.5**. Shown in polar coordinates (see the math references in the **Electrical Fundamentals** chapter for information about polar coordinates), the angular scale shows direction and the scale from the center of the plot to the outer ring, calibrated in dB, shows the relative strength of the antenna's radiated signal (gain) at each angle. A line is plotted showing the antenna's relative gain (transmitting and receiving) at each angle. The antenna is located at the exact center of the plot with its orientation specified separately.

The pattern is composed of *nulls* (angles at which a gain minimum occurs) and *lobes* (a range of angles in which a gain maximum occurs). The *main lobe* is the lobe with the highest amplitude unless noted otherwise and unless several plots are being compared, the peak amplitude of the main lobe is placed at the outer ring as a 0 dB reference point. The peak of the main lobe can be located at any angle. All other lobes are *side lobes* which can be at any angle, including to the rear of the antenna.

Fig 21.4 is an *azimuthal* or *azimuth pattern* that shows the antenna's gain in all horizontal directions (azimuths) around the antenna. As with a map, 0° is at the top and bearing angle increases clockwise. (This is different from polar plots generated for mathematical functions in which 0° is at the right and angle increases counter-clockwise.)

Fig 21.5 is an *elevation pattern* that shows the antenna's gain at all vertical angles. In this case, the horizon at 0° is located to both sides of the antenna and the zenith (directly overhead) at 90°. The plot shown in Fig 21.5 assumes a ground plane (drawn from 0° to 0°) but in free-space, the plot would include the missing semicircle with –90° at the bottom. Without the ground reference, the term "elevation" has little meaning, however.

You'll also encounter E-plane and H-plane radiation patterns. These show the antenna's radiation pattern in the plane parallel to the E-field or H-field of the antenna. It's important to remember that the E-plane and H-plane do not have a fixed relationship to the Earth's surface. For example, the E-plane pattern from a horizontal dipole is an azimuthal pattern, but if the same dipole is oriented vertically, the E-plane pattern becomes an elevation pattern.

Antenna radiation patterns can also be plotted on rectangular coordinates with gain on the vertical axis in dB and angle on the horizontal axis as shown in **Fig 21.6**. This is particularly useful when several antennas are being compared. Multiple patterns in polar coordinates can be difficult to read, particularly close to the center of the plot.

The amplitude scale of antenna patterns is almost always in dB. The scale rings can be calibrated in several ways. The most common is for the outer ring to represent the peak amplitude of the antenna's strongest lobe as 0 dB. All other points on the pattern represent *relative gain* to the peak gain. The antenna's *absolute gain* with respect to an isotropic (dBi) antenna or dipole (dBd) is printed as a label somewhere near the pattern. If several antenna radiation patterns are shown on the same plot for comparison, the pattern with the largest gain value is usually assigned the role of 0 dB reference.

The gain amplitude scale is usually divided in one of two ways. One common division is to have rings at 0, –3, –6, –12, –18, and –24 dB. This makes it easy to see where the gain has fallen to one-half of the reference or peak value (–3 dB), one-quarter (–6 dB), one-sixteenth (–12 dB), and so on. Another popular division of the amplitude scale is 0, –10, –20, –30, and –40 dB with intermediate rings or tick marks to show the –2, –4, –6, and –8 dB levels. You will encounter a number of variations on these basic scales.

RADIATION PATTERN MEASUREMENTS

Given the basic radiation pattern and scales, it becomes easy to define several useful measurements or metrics by which antennas are compared, using their azimuthal patterns. Next to gain, the most commonly-used metric for directional antennas is the *front-to-back ratio (F/B)* or just "front-to-back." This is the difference in dB between the antenna's gain in the specified "forward" direction and in the opposite or "back" direction. The front-to-back ratio of the antenna in Fig 21.4 is about 11 dB. *Front-to-side ratio* is also used and is the difference between the antenna's "forward" gain and gain at right angles to the

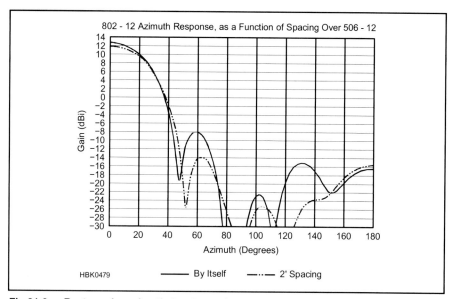

Fig 21.6 — Rectangular azimuthal pattern of an 8-element 2 meter Yagi beam antenna by itself and with another identical antenna stacked two feet above it. This example shows how a rectangular plot allows easier comparison of antenna patterns away from the main lobe.

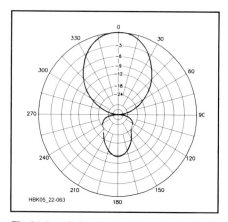

Fig 21.4 — Azimuthal pattern of a typical three-element Yagi beam antenna in free space. The Yagi's boom is along the 0° to 180° axis.

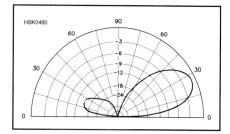

Fig 21.5 — Elevation pattern of a 3-element Yagi beam antenna placed ½-λ above perfect ground.

forward direction. This assumes the radiation pattern is symmetric and is of most use to antennas such as Yagis and quads that have elements arranged in parallel planes. The front-to-side ratio of the antenna in Fig 21.4 is more than 30 dB. Because the antenna's rear-ward pattern can have large amplitude variations, the *front-to-rear ratio* is sometimes used. Front-to-rear uses the average of rear-ward gain over a specified angle, usually the 180° semicircle opposite the direction of the antenna's maximum gain, instead of a single gain figure at precisely 180° from the forward direction.

The antenna's *beamwidth* is the angle over which the antenna's main lobe gain is within 3 dB of the peak gain. Stated another way, the beamwidth is the angle between the directions at which the antenna's gain is –3 dB. In Fig 21.4, the antenna's main lobe beamwidth is about 54°, since the pattern crosses the –3 dB gain scale approximately 27° to either side of the peak direction. Antenna patterns with comparatively small beamwidths are referred to as "sharp" or "narrow."

An antenna with an azimuthal pattern that shows equal gain in all directions is called *omnidirectional*. This is not the same as an isotropic antenna that has equal gain in all directions, both vertical both horizontal.

21.1.9 Elevation Angle

For long-distance HF communication, the (vertical) *elevation angle* of maximum radiation, or *radiation angle*, is of considerable importance. You will want to erect your antenna so that its strongest radiation occurs at vertical angles resulting in the best performance at the distances over which you want to communicate. In general, the greater the height of a horizontally polarized antenna, the stronger its gain will be at lower vertical angles. **Fig 21.7** shows this effect at work in horizontal dipole antennas. (See the **Propagation** chapter and the *ARRL Antenna Book* for more information about how to determine the best elevation angles for communication.)

Since low radiation angles usually are most effective for long distance communications, this generally means that horizontal antennas should be high — higher is usually better. (The optimum angle for intercontinental contacts on the HF bands is generally 15° or lower.) Experience shows that satisfactory results can be attained on the bands above

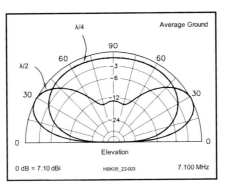

Fig 21.7 — Elevation patterns for two 40 meter dipoles over average ground (conductivity of 5 mS/m and dielectric constant of 13) at ¼-λ (33 ft) and ½-λ (66 ft) heights. The higher dipole has a peak gain of 7.1 dBi at an elevation angle of about 26°, while the lower dipole has more response at high elevation angles.

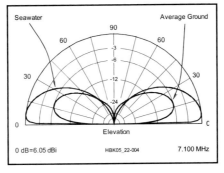

Fig 21.8 — Elevation patterns for a vertical dipole over sea water compared to average ground. In each case the center of the dipole is just over ¼-λ high. The low-angle response is greatly degraded over average ground compared to sea water, which is virtually a perfect ground.

14 MHz with antenna heights between 40 and 70 feet.

Higher vertical angles can be useful for medium to short-range communications. For example, elevation angles between 20° and 65° are useful on the 40 and 80 meter bands over the roughly 550-mile path between Cleveland and Boston. Even higher angles may be useful on shorter paths when using these lower HF frequencies. A 75 meter dipole between 30 and 70 ft high works well for ranges out to several hundred miles.

For even shorter-range communications centered on your location, such as for emergency communications and regional nets, a very low antenna is used, generating its strongest radiation straight up. This is referred to as *Near-Vertical Incidence Skywave* (*NVIS*) communication. The antenna should be less than ¼-λ above ground and the frequency used should be below the ionosphere's critical frequency so that the signal is completely reflected back toward the ground over a wide area.

Azimuthal patterns must also specify at what elevation angle the antenna gain is measured or calculated. While an azimuthal pattern may be in the plane of the antenna (an elevation angle of 0°), for antennas located above ground, the gain will vary strongly with elevation angle.

21.1.10 Imperfect Ground

Earth conducts, but is far from being a perfect conductor. This influences the radiation pattern of the antennas that we use. The effect is most pronounced at high vertical angles (the ones most important for short-range communications and least important for long-distance communications) for horizontal antennas. The consequences for vertical antennas are greatest at low angles, and are quite dramatic as can be clearly seen in **Fig 21.8**, where the elevation pattern for a 40 meter vertical half-wave dipole located over average ground is compared to one located over saltwater. At 10° elevation, the saltwater antenna has about 7 dB more gain than its landlocked counterpart.

An HF vertical antenna may work very well for a ham living in an area with rich soil. Ground of this type has very good conductivity. By contrast, a ham living where the soil is rocky or in a desert area may not be satisfied with the performance of a vertical HF antenna over such poorly conducting ground.

When evaluating or comparing antennas, it is also important to include the effects of ground on antenna gain. Depending on height above ground and the qualities of the ground, reflections can increase antenna gain by up to 6 dB. Because the actual installation of the antenna is unlikely to duplicate the environment in which the gain with reflections is claimed or measured, it is preferable to rely on free-space gain measurements that are independent of reflecting surfaces.

21.2 Dipoles and the Half-Wave Antenna

A fundamental form of antenna is a wire whose length is half the transmitting wavelength. It is the unit from which many more complex forms of antennas are constructed and is known as a *dipole antenna*. (The name di- meaning *two* and -pole meaning *electrical terminal* comes from the antenna having two distinct regions of electrical polarity as shown in Fig 21.1.) A dipole is resonant when it is electrically ½-λ long so that the current and voltage in the antenna are exactly 90° out of phase as shown in Fig 21.1.

The actual length of a resonant ½-λ antenna will not be exactly equal to the half wavelength of the radio wave of that frequency in free space, but depends on the thickness of the conductor in relation to the wavelength as shown in Fig 21.3. An additional shortening effect occurs with wire antennas supported by insulators at the ends because of current flow through the capacitance at the wire ends due to the end effect. Interaction with the ground and any nearby conductors also affects the resonant length of the physical antenna.

The following formula is sufficiently accurate for dipoles below 10 MHz at heights of ⅛ to ¼ λ and made of common wire sizes. To calculate the length of a half-wave antenna in feet,

$$\text{Length (ft)} = \frac{492 \times 0.95}{f\,(\text{MHz})} = \frac{468}{f\,(\text{MHz})} \quad (2)$$

Example: A half-wave antenna for 7150 kHz (7.15 MHz) is 468/7.15 = 65.5 ft, or 65 ft 6 inches.

For antennas at higher frequencies and/or higher above ground, a denominator value of 485 to 490 is more useful. In any case, be sure to include additional wire for attaching to insulators and be prepared to adjust the length of the antenna once installed in its intended position.

Above 30 MHz use the following formulas, particularly for antennas constructed from rod or tubing. K is taken from Fig 21.3.

$$\text{Length (ft)} = \frac{492 \times K}{f\,(\text{MHz})} \quad (3)$$

$$\text{Length (in)} = \frac{5904 \times K}{f\,(\text{MHz})} \quad (4)$$

Example: Find the length of a half-wave antenna at 50.1 MHz, if the antenna is made of ½-inch-diameter tubing. At 50.1 MHz, a half wavelength in space is

$$\frac{492}{50.1} = 9.82 \text{ ft}$$

The ratio of half wavelength to conductor diameter (changing wavelength to inches) is

$$\frac{(9.82 \text{ ft} \times 12 \text{ in/ft})}{0.5 \text{ in}} = 235.7$$

From Fig 21.3, K = 0.945 for this ratio. The length of the antenna, from equation 3 is

$$\frac{492 \times 0.945}{50.1} = 9.28 \text{ ft}$$

or 9 ft 3⅜ inches. The answer is obtained directly in inches by substitution in equation 4

$$\frac{5904 \times 0.945}{50.1} = 111.4 \text{ in}$$

Regardless of the formula used to calculate the length of the half-wave antenna, the effects of ground and conductive objects within a wavelength or so of the antenna usually make it necessary to adjust the installed length in order to obtain the lowest SWR at the desired frequency. Use of antenna modeling software may provide a more accurate initial length than a single formula.

The value of the SWR indicates the quality of the match between the impedance of the antenna and the feed line. If the lowest SWR obtainable is too high, an impedance-matching network may be used, as described in the **Transmission Lines** chapter. (High SWR may cause modern transmitters with solid-state power amplifiers to reduce power output as a protective measure for the output transistors.)

21.2.1 Radiation Characteristics

The radiation pattern of a dipole antenna in free space is strongest at right angles to

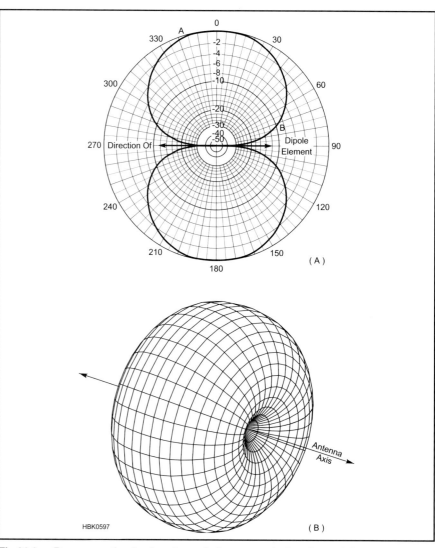

Fig 21.9 — Response of a dipole antenna in free space in the plane of the antenna with the antenna oriented along the 90° to 270° axis (A). The full three-dimensional pattern of the dipole is shown at (B). The pattern at A is a cross-section of the three-dimensional pattern taken perpendicularly to the axis of the antenna.

the wire as shown in **Fig 21.9**, a free-space radiation pattern. In an actual installation, the figure-8 pattern is less directive due to reflections from ground and other conducting surfaces. As the dipole is raised to ½-λ or greater above ground, nulls off the ends of the dipole become more pronounced. Sloping the antenna above ground and coupling to the feed line tend to distort the pattern slightly.

As a horizontal antenna is brought closer to ground, the elevation pattern peaks at a higher elevation angle as shown in Fig 21.7. **Fig 21.10** illustrates what happens to the directional pattern as antenna height changes. Fig 21.10C shows that there is significant radiation off the ends of a low horizontal dipole. For the ½-λ height (solid line), the radiation off the ends is only 7.6 dB lower than that in the broadside direction.

Fig 21.10 also shows that for short-range communication that depends on high vertical angles of radiation (NVIS communications), a dipole can be too high. For these applications, the dipole should be installed at or below ¼-λ so that the antenna radiates strongly at high vertical angles and with little horizontal directivity.

A classic dipole antenna is ½-λ long and fed at the center. The low feed point impedance at the dipole's resonant frequency, f_0, and its odd harmonics results in a low SWR when fed with coaxial cable feed lines. The feed point impedance and resulting SWR with coaxial cable will be high at even harmonics and other frequencies.

When fed with ladder line and a wide-range impedance-matching unit, such an antenna can be used on nearly any frequency, including non-resonant frequencies. (An example of such an antenna system is presented as a project farther along in this section.)

21.2.2 Feed Methods

The feed line is attached directly to the dipole, generally at the center, where an insulator separates the antenna's conductor into two sections. This is the antenna's *feed point*. One conductor of the feed line is attached to each section. **Figs 21.11** and **21.12** show how the two types of feed lines are attached. There are numerous variations, of course. You can make your own insulators from plastic or ceramic and there are many commercial insulators available, including some with built-in coax connectors for the feed point.

A dipole can be fed (feed line attached) anywhere along its length, although the impedance of the antenna will vary as discussed earlier. One common variation is the *off-center-fed* (*OCF*) dipole where the feed point is offset from center by some amount and an impedance transformer used to match the resulting moderately-high impedance to that of coaxial cable. Another variation, shown in Fig 21.12B, is the *end-fed Zepp*, named for its original application as an antenna deployed from Zeppelin airships. The feed point impedance of a "Zepp" is quite high, requiring open-wire feed line and impedance matching techniques to deliver power effectively.

Either *coaxial cable* ("coax") or *open-wire* transmission line or feed line is used to connect the transmitter and antenna. There are pro's and con's for each type of feed line. Coax is the common choice because it is readily available, its characteristic impedance is close to that of a center-fed dipole, and it may be easily routed through or along walls and among other cables. Where a very long feed line is required or the antenna is to be used at frequencies for which the feed point impedance is high, coax's increased RF loss and low working voltage (compared to that of open-wire line) make it a poor choice. Refer to the **Transmission Lines** and **Component Data and References** chapters for information that will help you evaluate the RF loss of coaxial cable at different lengths and SWR.

Respect coax's power-handling ratings. Cables such as RG-58 and RG-59 are suitable for power levels up to 300 W with low SWR. RG-8X cable can handle higher power and there are number of variations of this type of cable. For legal-limit power or moderate SWR, use the larger diameter cable types, such as RG-8 or RG-213, that are 0.4 inches in diameter or larger. Subminiature cables, such as RG-174, are useful for very short lengths at low power levels, but the high RF losses associated with these cables make them unsuitable for most uses as antenna feed lines.

The most common open-wire transmission lines are *ladder line* (also known as *window line*) and *twin-lead*. Since the conductors are

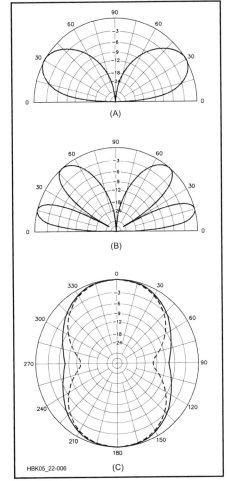

Fig 21.10 — At A, the elevation response pattern of a dipole antenna placed ½-λ above a perfectly conducting ground. At B, the pattern for the same antenna when raised to 1-λ. For both A and B, the conductor is coming out of the paper at a right angle. C shows the azimuth patterns of the dipole for the two heights at the most-favored elevation angle, the solid-line plot for the ½-λ height at an elevation angle of 30°, and the broken-line plot for the 1-λ height at an elevation angle of 15°. The conductor in C lies along the 90° to 270° axis.

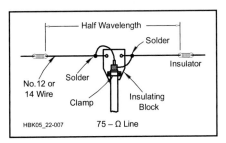

Fig 21.11 — Method of attaching feed line to the center of a dipole antenna. A plastic block is used as a center insulator. The coax is held in place by a clamp. A balun is often used to feed dipoles or other balanced antennas to ensure that the radiation pattern is not distorted. See text for explanation.

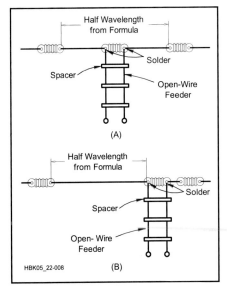

Fig 21.12 — Center-fed multiband *Zepp* antenna at A and an end-fed Zepp at B. See also Fig 21.15 for connection details.

not shielded, two-wire lines are affected by their environment. Use standoffs and insulators to keep the line several inches from structures or other conductors. Ladder line has very low loss (twin-lead has a little more), and it can stand very high voltages (created by high SWR) as long as the insulators are clean. Twin-lead can be used at power levels up to 300 W and ladder line to the full legal power limit.

The characteristic impedance of open-wire line varies from 300 Ω for twin-lead to 450 to 600 Ω for most ladder and window line. When used with ½-λ dipoles, the resulting moderate to high SWR requires an impedance-matching unit at the transmitter. The low RF losses of open-wire lines make this an acceptable situation on the HF bands.

21.2.3 Baluns

Open-wire transmission lines and center-fed dipole antennas are *balanced*, that is, each conductor or section has the same impedance to earth ground. This is different from *unbalanced* coaxial cable, in which the shield is generally connected to an earth ground at some point, generally at the transmitter. To use balanced open-wire transmission lines with unbalanced equipment — most amateur equipment is unbalanced — a *balun* is required to make the transition between the balanced and unbalanced parts of the antenna system. "Balun" is an abbreviation of "balanced-to-unbalanced," the function of the device — it allows power to be transferred between the balanced and unbalanced portions of an antenna system in either direction. The most common application of baluns is to connect an unbalanced feed line to a balanced antenna.

Because dipoles are balanced, a balun is often used at the feed point when a dipole is fed with coax. Due to the skin effect discussed in the **RF Techniques** chapter, the inside and outside of the coaxial cable shield are separate conductors at RF. This "third conductor" of a coaxial cable unbalances the symmetry of the dipole antenna when the coax is connected directly to the dipole as shown in Fig 21.11. As a result, RF current can flow on the outside of the cable shield to the enclosures of station equipment connected to the cable.

Shield currents can impair the function of instruments connected to the line (such as SWR meters and SWR-protection circuits in the transmitter). The shield current also produces some feed line radiation, which changes the antenna radiation pattern, and allows objects near the cable to affect the antenna-system performance.

The consequences may be negligible: A slight skewing of the antenna pattern usually goes unnoticed. Or, they may be significant: False SWR readings may cause the transmitter to reduce power unnecessarily; radiating coax near a TV feed line may cause strong local interference from overload. Therefore, it is better to eliminate feed line radiation whenever possible, and a balun should be used at any transition between balanced and unbalanced systems. Even so, balanced or unbalanced systems without a balun often operate with no apparent problems. For temporary or emergency stations, do not let the lack of a balun deter you from operating.

A balun can be constructed in a number of ways: the simplest being to coil several turns of coaxial cable at the antenna feed point. This creates inductance on the outer surface of the cable (the inner surface of the shield and the center conductor are unaffected) and the resulting reactance opposes RF current flow. There are other methods, such as the use of ferrite beads and cores, that are discussed in the **Transmission Lines** chapter.

21.2.4 Building Dipoles and Other Wire Antennas

The purpose of this section is to offer information on the actual physical construction of wire antennas. Because the dipole, in one of its configurations, is probably the most common amateur wire antenna, it is used in the following examples. The techniques described here, however, enhance the reliability and safety of all wire antennas.

WIRE

Choosing the right type of wire for the project at hand is the key to a successful antenna — the kind that works well and stays up through a winter ice storm or a gusty spring wind storm. What gauge of wire to use is the first question to settle; the answer depends on strength, ease of handling, cost, availability and visibility. Generally, antennas that are expected to support their own weight, plus the weight of the feed line should be made from #12 AWG wire. Horizontal dipoles, Zepps, some long wires and the like fall into this category. Antennas supported in the center, such as inverted-V dipoles and delta loops, may be made from lighter material, such as #14 AWG wire — the minimum size called for in the National Electrical Code.

The type of wire to be used is the next important decision. The wire specifications table in the **Component Data and References** chapter shows popular wire styles and sizes. The strongest wire suitable for antenna service is *copper-clad steel*, also known as *Copperweld*. The copper coating is necessary for RF service because steel is a relatively poor conductor. Practically all of the RF current is confined to the copper coating because of *skin effect*. Copper-clad steel is outstanding for permanent installations, but it can be difficult to work with because of the stiffness of the steel core.

Solid-copper wire, either hard-drawn or soft-drawn, is another popular material. Easier to handle than copper-clad steel, solid copper is available in a wide range of sizes. It is generally more expensive however, because it is all copper. Soft-drawn tends to stretch under tension, so periodic pruning of the antenna may be necessary in some cases. Enamel-coated *magnet-wire* is a suitable choice for experimental antennas because it is easy to manage, and the coating protects the wire from the weather. Although it stretches under tension, the wire may be pre-stretched before final installation and adjustment. A local electric motor rebuilder might be a good source for magnet wire.

Hook-up wire, speaker wire or even ac lamp cord are suitable for temporary installations. Almost any copper wire may be used, as long as it is strong enough for the demands of the installation.

Aluminum wire can be used for antennas, but is not as strong as copper or steel for the same diameter and soldering it to feed lines requires special techniques. Galvanized and steel wire, such as that used for electric fences, is inexpensive, but it is a much poorer conductor at RF than copper and should be avoided.

Kinking, which severely weakens wire, is a potential problem when handling any solid conductor. When uncoiling solid wire of any type — copper, steel, or aluminum — take care to unroll the wire or untangle it without pulling on a kink to straighten it. A kink is actually a very sharp twist in the wire and the wire will break at such a twist when flexed, such as from vibration in the wind.

Solid wire also tends to fail at connection or attachment points at which part of the wire is rigidly clamped. The repeated flexing from wind and other vibrations eventually causes metal fatigue and the wire breaks. Stranded wire is preferred for antennas that will be subjected to a lot of vibration and flexing. If stranded wire is not suitable, use a heavier gauge of solid wire to compensate.

Insulated vs Bare Wire

Losses are the same (in the HF region at least) whether the antenna wire is insulated or bare. If insulated wire is used, a 3 to 5% shortening from the length calculated for a bare wire is required to obtain resonance at the desired frequency. This is caused by the increased distributed capacitance resulting from the dielectric constant of the plastic insulating material. The actual length for resonance must be determined experimentally by pruning and measuring because the dielectric constant of the insulating material varies from wire to wire. Wires that might come into contact with humans or animals should be insulated to reduce the chance of shock or burns.

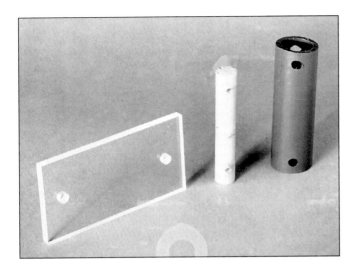

Fig 21.13 — Some ideas for homemade antenna insulators.

Fig 21.14 — Some homemade dipole center insulators. The one in the center includes a built-in SO-239 connector. Others are designed for direct connection to the feed line.

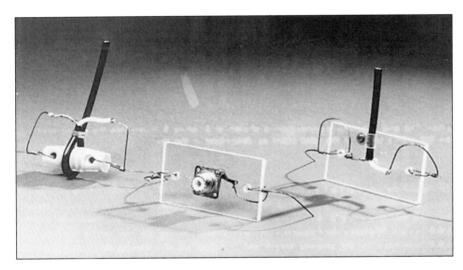

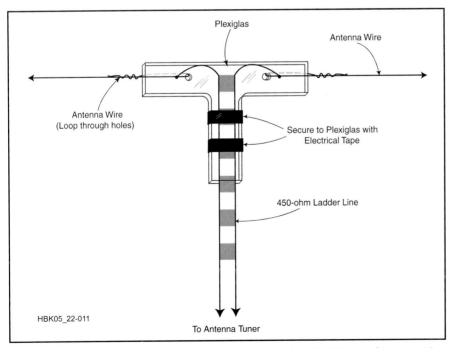

Fig 21.15 — A piece of cut Plexiglas can be used as a center insulator and to support a ladder-line feeder. The Plexiglas acts to reduce the flexing of the wires where they connect to the antenna.

INSULATORS

Wire antennas must be insulated at the ends. Commercially available insulators are made from ceramic, glass or plastic. Insulators are available from many Amateur Radio dealers. RadioShack and local hardware stores are other possible sources.

Acceptable homemade insulators may be fashioned from a variety of material including (but not limited to) acrylic sheet or rod, PVC tubing, wood, fiberglass rod or even stiff plastic from a discarded container. **Fig 21.13** shows some homemade insulators. Ceramic or glass insulators will usually outlast the wire, so they are highly recommended for a safe, reliable, permanent installation. Other materials may tear under stress or break down in the presence of sunlight. Many types of plastic do not weather well.

Most wire antennas require an insulator at the feed point. Although there are many ways to connect the feed line, there are a few things to keep in mind. If you feed your antenna with coaxial cable, you have two choices. You can install an SO-239 connector on the center insulator and use a PL-259 on the end of your coax, or you can separate the center conductor from the braid and connect the feed line directly to the antenna wire. Although it costs less to connect direct, the use of connectors offers several advantages.

Coaxial cable braid acts as a wick to soak up water. If you do not adequately seal the antenna end of the feed line, water will find its way into the braid. Water in the feed line will lead to contamination, rendering the coax useless long before its normal lifetime is up. It is not uncommon for water to drip from the end of the coax inside the shack after a year or so of service if the antenna connection is not properly waterproofed. Use of a PL-259/SO-239 combination (or other connector of your choice) makes the task of waterproofing connections much easier. Another advantage to using the PL-259/SO-239 combination is that feed line replacement is much easier, should that become necessary or desirable.

Whether you use coaxial cable, ladder line, or twin lead to feed your antenna, an often-overlooked consideration is the mechanical strength of the connection. Wire antennas and feed lines tend to move a lot in the breeze, and unless the feed line is attached securely, the connection will weaken with time. The resulting failure can range from a frustrating intermittent electrical connection to a complete separation of feed line and antenna. **Fig 21.14** illustrates several different ways of attaching the feed line to the antenna. An idea for supporting ladder line is shown in **Fig 21.15**.

PUTTING IT TOGETHER

Fig 21.16 shows details of antenna construction. Although a dipole is used for the examples, the techniques illustrated here apply to

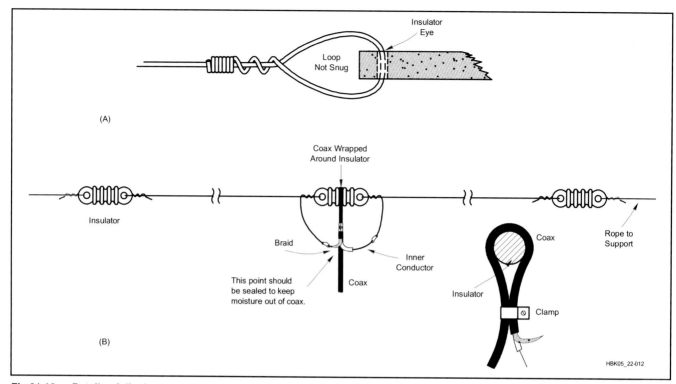

Fig 21.16 — Details of dipole antenna construction. The end insulator connection is shown at A, while B illustrates the completed antenna. This is a balanced antenna and is often fed with a balun at the feed point as described in the text.

Table 21.1
Dipole Dimensions for Amateur Bands

Freq (MHz)	Overall Length ft	in	Leg Length ft	in
28.4	16	6	8	3
24.9	18	9½	9	4¾
21.1	22	2	11	1
18.1	25	10	12	11
14.1	33	2	16	7
10.1	46	4	23	2
7.1	65	10	32	11
5.37	87	2	43	7
3.6	130	0	65	0

any type of wire antenna. **Table 21.1** shows dipole lengths for the amateur HF bands. These lengths do not include the extra wire required to attach the wire to the insulator as shown in Fig 21.16. Determine the extra amount of wire required by experimenting with the insulator you intend to use. Add twice this amount of wire to the leg lengths in Table 21.1, one extra length for each insulator.

Most dipoles require a little pruning to reach the desired resonant frequency due to the effects of ground and nearby conducting objects and surfaces. So, cut the wire to result in a constructed length 2 to 3% longer than the calculated or table length and record the constructed length with all insulators attached. (The constructed length is measured between the ends of the loops at each end of the wire.) Next, raise the dipole to the working height and find the frequency at which minimum SWR occurs. Multiply the frequency of the SWR minimum by the antenna length and divide the result by the desired f_0. The result is the finished length; trim both ends equally to reach that length and you're done. For example, if you want the SWR minimum to occur at 14.1 MHz and the first attempt with a constructed length of 33.8 ft results in an SWR minimum at 13.9 MHz, the final length for the antenna is

$$\frac{13.9}{14.1} \times 33.8 = 33.3 \text{ ft}$$

In determining how well your antenna will work over the long term, how well you put the pieces together is second only to the ultimate strength of the materials used. Even the smallest details, such as how you connect the wire to the insulators (Fig 21.16A), contribute significantly to antenna longevity. By using plenty of wire at the insulator and wrapping it tightly, you will decrease the possibility of the wire pulling loose in the wind. There is no need to solder the wire once it is wrapped. There is no electrical connection here, only mechanical. The high heat needed for soldering can anneal the wire, significantly weakening it at the solder point.

Similarly, the feed line connection at the center insulator should be made to the antenna wires after they have been secured to the insulator (Fig 21.16B). This way, you will be assured of a good electrical connection between the antenna and feed line without compromising the mechanical strength. Do a good job of soldering the antenna and feed line connections. Use a heavy iron or a torch, and be sure to clean the materials thoroughly before starting the job. If possible, solder the connections at a workbench, where the best possible joints may be made. Poorly soldered or unsoldered connections will become headaches as the wire oxidizes and the electrical integrity degrades with time. Besides degrading your antenna performance, poorly made joints can even be a cause of TVI because of rectification. Spray the connections with a UV-resistant acrylic coating for waterproofing.

So that the antenna stays up after installation, keep it away from tree branches and other objects that might rub or fall on the antenna. If the supports for the antenna move in the wind, such as trees, leave enough slack in the antenna that it is not pulled overly tight in normal winds. Other options are to use pulleys and counterweights to allow the antenna supports to flex without pulling on the antenna. (This and other installation topics are covered in the *ARRL Antenna Book*.)

If made from the right materials and installed in the clear, the dipole should give years of maintenance-free service. As you build your antenna, keep in mind that if you get it right the first time, you won't have to do it again for a long time.

21.2.5 Dipole Orientation

Dipole antennas need not be installed horizontally and in a straight line. They are generally tolerant of bending, sloping or drooping. Bent dipoles may be used where antenna space is at a premium. **Fig 21.17** shows a couple of possibilities; there are many more. Bending distorts the radiation pattern somewhat and may affect the impedance as well, but compromises may be acceptable when the situation demands them. When an antenna bends back on itself (as in Fig 21.17B) some of the signal is canceled; avoid this if possible. Remember that dipole antennas are RF conductors. For safety's sake, mount all antennas away from conductors (especially power lines), combustibles and well beyond the reach of passersby.

21.2.6 Inverted-V Dipole

An *inverted-V* dipole is supported at the center with a single support, such as a tree or mast. While *V* describes the shape of this antenna, this antenna should not be confused with long-wire horizontal-V antennas, which are highly directive.

The inverted-V's radiation pattern and feed point impedance depend on the *apex angle* between the legs: As the apex angle decreases, so does feed point impedance, and the radiation pattern becomes less directive. At apex angles below 90°, the antenna efficiency begins to decrease, as well.

The proximity of ground to the antenna ends will lower the resonant frequency of the antenna so that a dipole may have to be shortened in the inverted-V configuration. Losses in the ground increase when the antenna ends are close to the ground. Keeping the ends eight feet or higher above ground reduces ground loss and also prevents humans and animals from coming in contact with the antenna.

Remember that antenna current produces the radiated signal, and current is maximum at the dipole center. Therefore, performance is best when the central area of the antenna is high and clear of nearby objects.

21.2.7 Sloping Dipole

A sloping dipole is shown in **Fig 21.18**. This antenna is often used to favor one direction (the *forward direction* in the figure). With a non-conducting support and poor ground, signals off the back are weaker than those off the front. With a non-conducting mast and good ground, the response is omnidirectional. There is no gain in any direction with a non-conducting mast.

A conductive support such as a tower acts as a parasitic element. (So does the coax shield, unless it is routed at 90° from the antenna.) The parasitic effects vary with ground

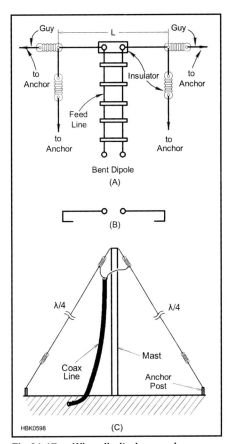

Fig 21.17 — When limited space is available for a dipole antenna, the ends can be bent downward as shown at A, or back on the radiator as shown at B. The inverted-V at C can be erected with the ends bent parallel with the ground when the available supporting structure is not high enough.

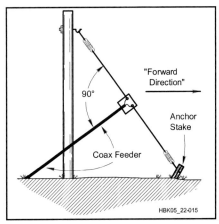

Fig 21.18 — Example of a sloping ½-λ dipole, or *full sloper*. On the lower HF bands, maximum radiation over poor to average earth is off the sides and in the *forward direction* as indicated, if a non-conductive support is used. A metal support will alter this pattern by acting as a parasitic element. How it alters the pattern is a complex issue depending on the electrical height of the mast, what other antennas are located on the mast and on the configuration of any guy wires.

quality, support height and other conductors on the support (such as a beam at the top or other wire antennas). With such variables, performance is very difficult to predict.

Losses increase as the antenna ends approach the support or the ground, so the same cautions about the height of the antenna ends applies as for the inverted-V antenna. To prevent feed line radiation, route the coax away from the feed point at 90° from the antenna as far as possible.

21.2.8 Shortened Dipoles

Inductive loading increases the electrical length of a conductor without increasing its physical length. Therefore, we can build physically-short dipole antennas by placing inductors in the antenna. These are called *loaded dipoles*, and *The ARRL Antenna Book* shows how to design them. There are some trade-offs involved: Inductively loaded antennas are less efficient and have narrower bandwidths than full-size antennas. Generally they should not be shortened more than 50%.

21.2.9 Half-Wave Vertical Dipole (HVD)

Unlike its horizontal counterpart, which has a figure-8 pattern, the azimuthal pattern of a vertical dipole is omnidirectional. In other words, it looks like a circle. Look again at Figs 21.7 and 21.8 and note the comparison between horizontal and vertical dipole elevation patterns. These two figures illustrate the fact that performance of a horizontal dipole depends to a great extent on its height above ground. By contrast, *half-wave vertical dipole* (HVD) performance is highly dependent on ground conductivity and dielectric constant.

After looking at these figures, you might easily conclude that there is no advantage to an HVD. Is that really the case? Experiments at K8CH between 2001 and 2003 showed that the HVD mounted above average ground works well for long-distance (DX) contacts. Two antennas were used in the trials. The first was a 15 meter HVD with its base 14 ft above ground (feed point at 25 ft). The second (reference) antenna was a 40 meter inverted-V modified to operate with low SWR on 15 meters. The reference antenna's feed point was at 29 ft and the ends drooped slightly for an apex angle of 160°. Signals from outside North America were usually stronger on the HVD. Antenna modeling revealed the reasons for this behavior.

Fig 21.19 shows the elevation patterns for the vertical dipole and for the reference dipole at a pattern peak and at a null. The vertical dipole does not look impressive, does it? The large lobe in the HVD pattern at 48° is

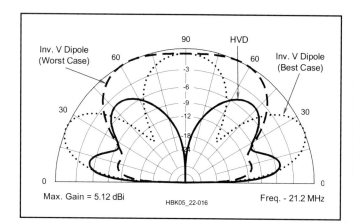

Fig 21.19 — Elevation patterns for the HVD (solid line) and the inverted-V comparison antenna in its best case (dotted line) and worst case (dashed line).

feed point impedance of a single-wire dipole is multiplied by the square of the number of conductors in the antenna. In this case, there are two conductors in the antenna, so the feed point impedance is $2^2 = 4$ times that of a single-wire dipole. (A three-wire folded dipole would have a nine times higher feed point impedance and so forth.)

A common use of the folded dipole is to raise the feed point impedance of the antenna to present a better impedance match to high impedance feed line. For example, if a very long feed line to a dipole is required, open-wire feed line would be used. By raising the dipole's feed point impedance, the SWR on the open-wire line is reduced from that of a single-wire dipole fed with open-wire feed line.

A variation of the folded dipole called the *twin-folded terminated dipole* (TFTD) adds a resistor in the top conductor. Values of 300 to 600 Ω are used. The function of the resistor is to act as a *swamping* load, reducing the higher feed point impedances over a wide frequency range. A TFTD ½-λ long at 80 meters can be constructed to cover the entire 2-30 MHz range with SWR of 3:1 or less. The resistor dissipates some of the transmitter power (more than 50% at some frequencies!), but the improvement in SWR allows a coaxial feed line to be used without an impedance-matching unit. The increased convenience and installation outweigh the reduction in radiated signal. TFTD antennas are popular for emergency communications and where only a single HF antenna can be installed and high performance is not required.

caused by the antenna being elevated 14 ft above ground. This lobe will shrink at lower heights.

Recall that the optimum elevation angles for long-distance contacts fall below 15°. Azimuthal patterns for the HVD at an elevation angle of 10° are shown in **Fig 21.20** and for 3° in **Fig 21.21**. You can clearly see in the patterns the DX potential of an HVD.

Another advantage of the HVD is its radiation resistance at low heights. Look back at Fig 21.2 at the curve for the vertical half-wave antenna. With its base just above ground, the HVD will have a radiation resistance of over 90 Ω. That can easily be turned to an advantage. Capacitive loading will lower the radiation resistance *and* shorten the antenna. It is possible to make a loaded vertical dipole that is half the height of an HVD and that has a good SWR when fed with 50-Ω coax.

21.2.10 Folded Dipoles

Fig 21.22 shows a *folded dipole* constructed from open-wire transmission line. The dipole is made from a ½-λ section of open-wire line with the two conductors connected together at each end of the antenna. The top conductor of the open-wire length is continuous from end to end. The lower conductor, however, is cut in the middle and the feed line attached at that point. Open-wire transmission line is then used to connect the transmitter.

A folded dipole has exactly the same gain and radiation pattern as a single-wire dipole. However, because of the mutual coupling between the upper and lower conductors, the

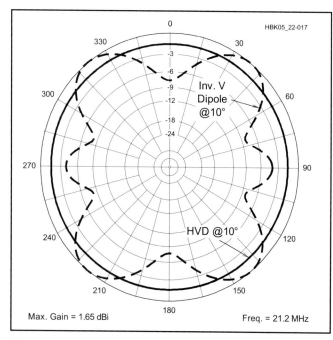

Fig 21.20 — Azimuth patterns at 10° elevation for the HVD (solid line) and inverted-V (dashed line).

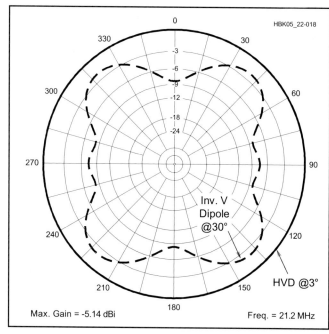

Fig 21.21 — Azimuth patterns at 3° elevation for the HVD (solid line) and the inverted-V (dashed line).

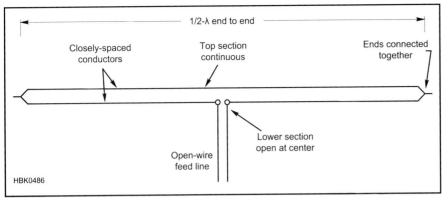

Fig 21.22 — The folded dipole is constructed from open-wire transmission line with the ends connected together. The close proximity of the two conductors and the resulting coupling act as an impedance transformer to raise the feed point impedance over that of a single-wire dipole by the square of the number of conductors used.

21.2.11 Multiband Dipole Systems

There are several ways to construct coax-fed multiband dipole systems. These techniques apply to dipoles of all orientations. Each method requires a little more work than a single dipole, but the materials don't cost much.

PARALLEL DIPOLES

Parallel dipoles as shown in **Fig 21.23** are a simple and convenient answer. Center-fed dipoles have low feed point impedances near f_0 and its odd harmonics, and high impedances at other frequencies. This lets us construct simple multiband systems that automatically select the appropriate antenna. Consider a 50-Ω resistor connected in parallel with a 5-kΩ resistor. A generator connected across the two resistors will see 49.5 Ω, and 99% of the current will flow through the 50-Ω resistor. When resonant and non-resonant antennas are connected in parallel, the same result occurs: The non-resonant antenna has a high impedance, so little current flows in it and it has little effect on the total feed point impedance. Thus, we can connect several dipoles together at the feed point, and power naturally flows to the resonant antenna.

There are some limits, however. Wires in close proximity tend to couple due to mutual inductance. In parallel dipoles, this means that the resonant length of the shorter dipoles lengthens a few percent. Shorter antennas don't affect longer ones much, so adjust for resonance in order from longest to shortest. Mutual inductance also reduces the bandwidth of shorter dipoles, so an impedance-matching unit may be needed to achieve an acceptable SWR across all bands covered. These effects can be reduced by spreading the ends of the dipoles apart.

Also, the power-distribution mechanism requires that only one of the parallel dipoles is near resonance on any amateur band. Separate dipoles for 80 and 30 meters should not be connected in parallel because the higher band is near an odd harmonic of the lower band ($80/3 \approx 30$) and center-fed dipoles have low impedance near odd harmonics. (The 40 and 15 meter bands have a similar relationship.) This means that you must either accept the performance of the low-band antenna operating on a harmonic or erect a separate antenna for those odd-harmonic bands. For example, four parallel-connected dipoles cut for 80, 40, 20 and 10 meters (fed by a single impedance-matching unit and coaxial cable) work reasonably on all HF bands from 80 through 10 meters.

TRAP DIPOLES

Trap dipoles (also called "trapped dipoles") provide multiband operation from a coax-fed single-wire dipole. **Fig 21.24** shows a two-band trap antenna. A trap consists of inductance and capacitance in parallel with a resonant frequency on the higher of the two bands of operation. The high impedance of the trap at its resonant frequency effectively disconnects the wire beyond the trap. Thus, on the higher of the two bands of operation at which traps are resonant, only the portion of the antenna between the traps is active.

Above resonance, the trap presents a capacitive reactance. Below resonance, the trap is inductive. On the lower of the two bands of operation, then, the inductive reactance of the trap acts as a loading coil to create a shortened or loaded dipole with the wire beyond the trap.

Traps may be constructed from coiled sec-

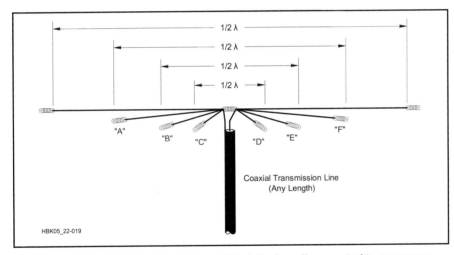

Fig 21.23 — Multiband antenna using paralleled dipoles, all connected to a common 50 or 75-Ω coax line. The ½-λ dimensions may be either for the centers of the various bands or selected for favorite frequencies in each band. The length of a ½-λ in feet is 468/frequency in MHz, but because of interaction among the various elements, some pruning for resonance may be needed on each band. See text.

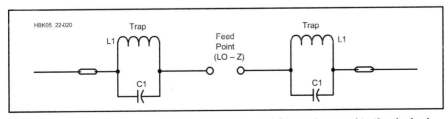

Fig 21.24 — Example of a trap dipole antenna. L1 and C1 can be tuned to the desired frequency by means of a grid-dip meter or SWR analyzer *before* they are installed in the antenna.

Antenna Modeling by Computer

Modern computer programs have made it a *lot* easier for a ham to evaluate antenna performance. The elevation plots for the 135-ft long center-fed dipole were generated using a sophisticated computer program known as *NEC*, short for "Numerical Electromagnetics Code." *NEC* is a general-purpose antenna modeling program, capable of modeling almost any antenna type, from the simplest dipole to extremely complex antenna designs. Various mainframe versions of *NEC* have been under continuous development by US government researchers for several decades.

But because it is a general-purpose program, *NEC* can be very slow when modeling some antennas — such as long-boom, multi-element Yagis. There are other, specialized programs that work on Yagis much faster than *NEC*. Indeed, *NEC* has developed a reputation for being accurate (if properly applied!), but decidedly difficult to learn and use. A number of commercial software developers have risen to the challenge and created more *user-friendly* versions. Check the ads in *QST*.

NEC uses a *Method of Moments* algorithm. The mathematics behind this algorithm are pretty formidable to most hams, but the basic principle is simple. An antenna is broken down into a set of straight-line wire *segments*. The fields resulting from the current in each segment and from the mutual interaction between segments are vector-summed in the far field to create azimuth and elevation-plane patterns.

The most difficult part of using a *NEC*-type of modeling program is setting up the antenna's geometry — you must condition yourself to think in three-dimensional coordinates. Each end point of a wire is represented by three numbers: an x, y and z coordinate. An example should help sort things out. See **Fig 21.A1**, showing a *model* for a 135-foot center-fed dipole, made of #14 wire placed 50 ft above flat ground. This antenna is modeled as a single, straight wire.

For convenience, ground is located at the *origin* of the coordinate system, at (0, 0, 0) feet, directly under the center of the dipole. The dipole runs parallel to, and above, the y-axis. Above the origin, at a height of 50 feet, is the dipole's feed point. The *wingspread* of the dipole goes toward the left (that is, in the *negative y* direction) one-half the overall length, or −67.5 ft. Toward the right, it goes +67.5 ft. The x dimension of our dipole is zero. The dipole's ends are thus represented by two points, whose coordinates are: (0, −67.5, 50) and (0, 67.5, 50) ft. The thickness of the antenna is the diameter of the wire, #14 gauge.

To run the program you must specify the number of segments into which the dipole is divided for the method-of-moments analysis. The guideline for setting the number of segments is to use at least 10 segments per half-wavelength. In Fig 21.A1, our dipole has been divided into 11 segments for 80 meter operation. The use of 11 segments, an odd rather than an even number such as 10, places the dipole's feed point (the *source* in *NEC*-parlance) right at the antenna's center and at the center of segment number six.

Since we intend to use our 135-foot long dipole on all HF amateur bands, the number of segments used actually should vary with frequency. The penalty for using more segments in a program like *NEC* is that the program slows down roughly as the square of the segments—double the number and the speed drops to a fourth. However, using too few segments will introduce inaccuracies, particularly in computing the feed point impedance. The commercial versions of *NEC* handle such nitty-gritty details automatically.

Let's get a little more complicated and specify the 135-ft dipole, configured as an inverted-V. Here, as shown in **Fig 21.A2**, you must specify *two* wires. The two wires join at the top, (0, 0, 50) ft. Now the specification of the source becomes more complicated. The easiest way is to specify two sources, one on each end segment at the junction of the two wires. If you are using the *native* version of *NEC*, you may have to go back to your high-school trigonometry book to figure out how to specify the end points of our droopy dipole, with its 120° included angle. Fig 21.A2 shows the details, along with the trig equations needed.

So, you see that antenna modeling isn't entirely a cut-and-dried procedure. The commercial programs do their best to hide some of the more unwieldy parts of *NEC*, but there's still some art mixed in with the science. And as always, there are trade-offs to be made — segments versus speed, for example.

However, once you do figure out exactly how to use them, computer models are wonderful tools. They can help you while away a dreary winter's day, designing antennas on-screen — without having to risk life and limb climbing an ice-covered tower. And in a relatively short time a computer model can run hundreds, or even thousands, of simulations as you seek to optimize an antenna for a particular parameter. Doesn't that sound better than trying to optimally tweak an antenna by means of a thousand cut-and-try measurements, all the while hanging precariously from your tower by a climbing belt?!
— *R. Dean Straw, N6BV*

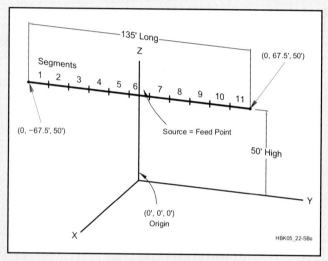

Fig 21.A1 — Model of a 135-ft center-fed dipole.

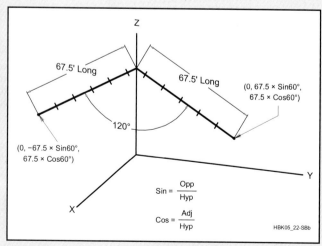

Fig 21.A2 — Model of the 135-ft dipole configured as an inverted V.

tions of coax or from discrete inductors and capacitors. (Traps are also available commercially.) Choose capacitors (C1 in the figure) that are rated for high current and voltage. Mica transmitting capacitors are good. Ceramic transmitting capacitors may work, but their values may change with temperature. Use large wire for the inductors to reduce loss. Any reactance (X_L and X_C) above 100 Ω (at f_0) will work, but bandwidth increases with reactance (up to several thousand ohms). Check trap resonance before installation. This can be done with a grid-dip meter and a receiver or with an SWR analyzer or impedance bridge.

To construct a trap antenna, build a dipole for the higher band of operation and connect the pre-tuned traps to its ends. It is fairly complicated to calculate the additional wire needed for each band, so just add enough wire to make the antenna ½-λ long on the lower band of operation, pruning it as necessary. Because the inductance in each trap reduces the physical length needed for resonance, the finished antenna will be shorter than a simple ½-λ dipole on the lower band.

21.2.12 NVIS Antennas

The use of very low dipole antennas that radiate at very high elevation angles has become popular in emergency communications ("emcomm") systems. This works at low frequencies (7 MHz and below) that are lower than the ionosphere's critical frequency — the highest frequency for which a signal traveling vertically will be reflected. (See the **Propagation** chapter.) The most common band for NVIS communication is 75 meters because the critical frequency is almost always above 4 MHz.

No special antenna construction techniques are required for NVIS antennas — just build a ½-λ dipole and install it at a height of 0.25-λ or below. 0.1-λ is often cited as the optimum height for NVIS antennas, but it is not critical.

At these low heights, the dipole's resonant frequency will be reduced because of the effects of ground. Shortening the antenna will restore the desired resonant frequency, although feed point impedance will drop.

Project: Multiband Center-Fed Dipole

An 80 meter dipole fed with ladder line is a versatile antenna. If you add a wide-range matching network, you have a low-cost antenna system that works well across the entire HF spectrum, and even 6 meters. Countless hams have used one of these in single-antenna stations and for Field Day operations.

Fig 21.25A shows a typical installation for such an antenna. The inverted-V configu-

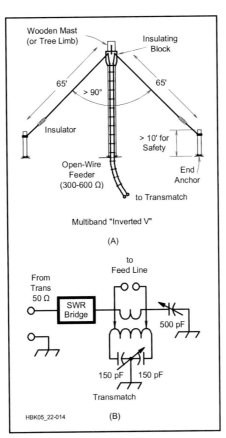

Fig 21.25 — At A, details for an inverted-V fed with open-wire line for multiband HF operation. An impedance-matching unit is shown at B, suitable for matching the antenna to the transmitter over a wide frequency range. The included angle between the two legs should be greater than 90° for best performance.

ration is shown, lowering the total antenna length to 130 ft from the 135 ft used if the entire antenna is horizontal. Either configuration will work well. Fig 21.25B shows the schematic of an impedance-matching unit or "antenna tuner" that you can build yourself. You can also use a balanced impedance-matching unit with a balun between it and the transmitter. Many amateurs are successful in using unbalanced impedance-matching units with a balun at either the output or the input of the tuner. Don't be afraid to experiment!

This configuration is popular with other lengths for the antenna:
- 105 ft — 80 through 10 meters
- 88 ft — 80 through 10 meters
- 44 ft — 40 through 10 meters

The next lower band may also be covered if the impedance-matching unit has sufficient range, although the adjustment will be fairly sharp. Six meter coverage is possible, but depends on the station layout, length of feed line, and impedance-matching unit abilities. Again, don't be afraid to experiment!

For best results place the antenna as high as you can, and keep the antenna and lad-

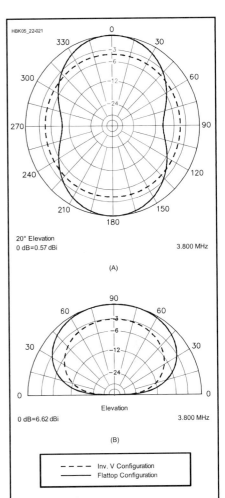

Fig 21.26 — Patterns on 80 meters for 135-ft, center-fed dipole erected as a horizontal dipole at 50 ft, and as an inverted-V with the center at 50 ft and the ends at 10 ft. The azimuth pattern is shown at A, where the conductor lies in the 90° to 270° plane. The elevation pattern is shown at B, where the conductor comes out of paper at a right angle. At the fundamental frequency the patterns are not markedly different.

der line clear of metal and other conductive objects. Despite significant SWR on some bands, the open-wire feed line keeps system losses low as described in the **Transmission Lines** chapter.

ARRL staff analyzed a 135-ft dipole at 50 ft above typical ground and compared that to an inverted-V with the center at 50 ft, and the ends at 10 ft. The results show that on the 80 meter band, it won't make much difference which configuration you choose. (See **Fig 21.26**.) The inverted-V exhibits additional losses because of its proximity to ground.

Fig 21.27 shows a comparison between a 20 meter flat-top dipole and the 135-ft flat-top dipole when both are placed at 50 ft above ground. At a 10° elevation angle, the 135-ft dipole has a gain advantage. This advantage comes at the cost of two deep, but narrow,

Antennas 21.15

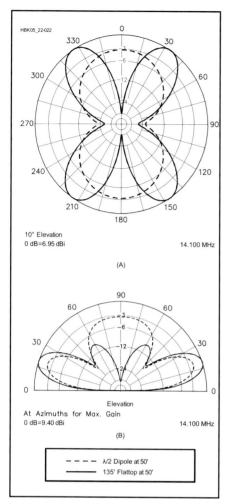

Fig 21.27 — Patterns on 20 meters comparing a standard ½-λ dipole and a multiband 135-ft dipole. Both are mounted horizontally at 50 ft. The azimuth pattern is shown at A, where conductors lie in the 90° to 270° plane. The elevation pattern is shown at B. The longer antenna has four azimuthal lobes, centered at 35°, 145°, 215°, and 325°. Each is about 2 dB stronger than the main lobes of the ½-λ dipole. The elevation pattern of the 135-ft dipole is for one of the four maximum-gain azimuth lobes, while the elevation pattern for the ½-λ dipole is for the 0° azimuthal point.

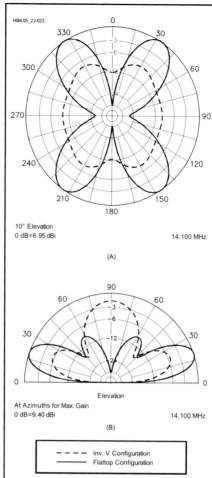

Fig 21.28 — Patterns on 20 meters for two 135-ft dipoles. One is mounted horizontally as a flat-top and the other as an inverted-V with 120° included angle between the two legs. The azimuth pattern is shown at A, and the elevation pattern is shown at B. The inverted-V has about 6 dB less gain at the peak azimuths, but has a more uniform, almost omnidirectional, azimuthal pattern. In the elevation plane, the inverted-V has a fat lobe overhead, making it a somewhat better antenna for local communication, but not quite so good for DX contacts at low elevation angles.

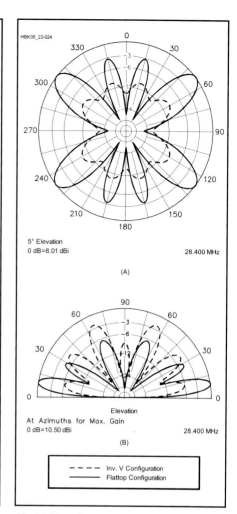

Fig 21.29 — Patterns on 10 meters for 135-ft dipole mounted horizontally and as an inverted-V, as in Fig 21.28. The azimuth pattern is shown at A, and the elevation pattern is shown at B. Once again, the inverted-V conFiguration yields a more omnidirectional pattern, but at the expense of almost 8 dB less gain than the flat-top configuration at its strongest lobes.

nulls that are broadside to the wire.

Fig 21.28 compares the 135-ft dipole to the inverted-V configuration of the same antenna on 14.1 MHz. Notice that the inverted-V pattern is essentially omnidirectional. That comes at the cost of gain, which is less than that for a horizontal flat-top dipole.

As expected, patterns become more complicated at 28.4 MHz. As you can see in **Fig 21.29**, the inverted-V has the advantage of a pattern with slight nulls, but with reduced gain compared to the flat-top configuration.

Installed horizontally, or as an inverted-V, the 135-ft center-fed dipole is a simple antenna that works well from 3.5 to 30 MHz (and on 1.8 MHz if the impedance-matching unit has sufficient range). If extremely high SWR or evidence of RF on objects at the operating position ("RF in the shack") is encountered, change the feed line length by adding or subtracting ⅛-λ at the problem frequency. A few such adjustments should yield a workable solution.

Project: 40-15 Meter Dual-Band Dipole

As mentioned earlier, dipoles have harmonic resonances at odd multiples of their fundamental resonances. Because 21 MHz is the third harmonic of 7 MHz, 7-MHz dipoles are harmonically resonant in the popular ham band at 21 MHz. This is attractive because it allows you to install a 40 meter dipole, feed it with coax, and use it without an antenna tuner on both 40 and 15 meters.

But there's a catch: The third harmonic resonance is actually higher than three times the fundamental resonant frequency. This is because there is no end effect in the center portion of the antenna where there are no insulators.

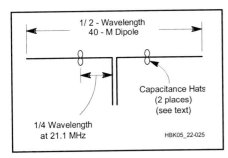

Fig 21.30 — Figure-8-shaped capacitance hats made and placed as described in the text, can make a 40 meter dipole resonate anywhere in the 15 meter band.

An easy fix for this, as shown in **Fig 21.30**, is to add capacitive loading to the antenna about ¼-λ wavelength (at 21.2 MHz) away from the feed point in both halves of the dipole. Known as *capacitance hats*, the simple loading wires shown lower the antenna's resonant frequency on 15 meters without substantially affecting resonance on 40 meters. This scheme can also be used to build a dipole that can be used on 80 and 30 meters and on 75 and 10 meters. (A project for a 75 and 10 meter dipole is included on the CD-ROM included with this book.)

Measure, cut and adjust the dipole to resonance at the desired 40 meter frequency. Then, cut two 2-ft-long pieces of stiff wire (such as #12 or #14 AWG house wire) and solder the ends of each one together to form two loops. Twist the loops in the middle to form figure-8s, and strip and solder the wires where they cross. Install these capacitance hats on the dipole by stripping the antenna wire (if necessary) and soldering the hats to the dipole about a third of the way out from the feed point (placement isn't critical) on each wire. To resonate the antenna on 15 meters, adjust the loop shapes (*not while you're transmitting!*) until the SWR is acceptable in the desired segment of the 15 meter band. Conversely, you can move the hats back and forth along the antenna until the desired SWR is achieved and then solder the hats to the antenna.

Project: W4RNL Inverted-U Antenna

This simple rotatable dipole was designed and built by L. B. Cebik, W4RNL, for use during the ARRL Field Day. For this and other portable operations we look for three antenna characteristics: simplicity, small size, and light weight. Complex assemblies increase the number of things that can go wrong. Large antennas are difficult to transport and sometimes do not fit the space available. Heavy antennas require heavy support structures, so the overall weight seems to increase exponentially with every added pound of antenna.

Today, a number of light-weight collapsible masts are available. When properly guyed, some will support antennas in the 5-10 pound range. Most are suitable for 10 meter tubular dipoles and allow the user to hand-rotate the antenna. Extend the range of the antenna to cover 20-10 meters, and you put these 20-30-foot masts to even better use. The inverted-U meets this need.

THE BASIC IDEA OF THE INVERTED-U

A dipole's highest current occurs within the first half of the distance from the feed point to the outer tips. Therefore, very little performance is lost if the outer end sections are bent. The W4RNL inverted-U starts with a 10 meter tubular dipole. You add extensions for 12, 15, 17 or 20 meters to cover those bands.

You only need enough space to erect a 10 meter rotatable dipole. The extensions hang down. **Fig 21.31** shows the relative proportions of the antenna on all bands from 10 to 20 meters. The 20 meter extensions are the length of half the 10 meter dipole. Therefore, safety dictates an antenna height of at least 20 feet to keep the tips above 10 feet high. At any power level, the ends of a dipole have high RF voltages, and we must keep them out of contact with human body parts.

Not much signal strength is lost by drooping up to half the overall element length straight down. What is lost in bidirectional gain shows up in decreased side-nulls. **Fig 21.32** shows the free-space E-plane (azimuth) patterns of the inverted-U with a 10 meter horizontal section. There is an undetectable decrease in gain between the 10 meter and 15 meter versions. The 20 meter version shows a little over a half-dB gain decrease and a signal increase off the antenna ends.

The real limitation of an inverted-U is a function of the height of the antenna above ground. With the feed point at 20 ft above ground, we obtain the elevation patterns

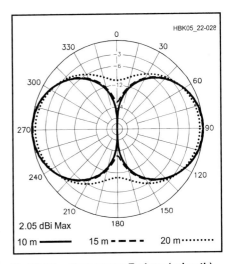

Fig 21.32 — Free-space E-plane (azimuth) patterns of the inverted-U for 10, 15, and 20 meters, showing the pattern changes with increasingly longer vertical end sections.

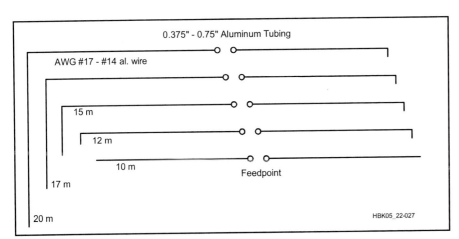

Fig 21.31 — The general outline of the inverted-U field dipole for 20 through 10 meters. Note that the vertical end extension wires apply to both ends of the main 10 meter dipole, which is constant for all bands.

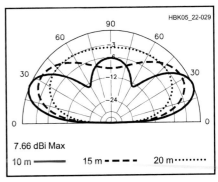

Fig 21.33 — Elevation patterns of the inverted-U for 10, 15, and 20 meters, with the antenna feed point 20 ft above average ground. Much of the decreased gain and higher elevation angle of the pattern at the lowest frequencies is due to its ever-lower height as a fraction of a wavelength.

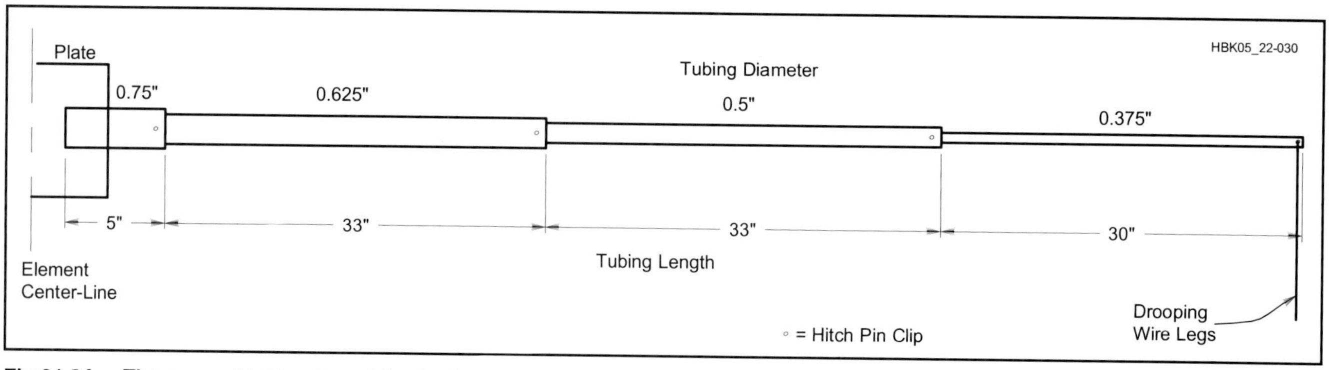

Fig 21.34 — The general tubing layout for the inverted-U for each half element. The opposite side of the dipole is a mirror image of the one shown.

shown in **Fig 21.33**. The 10 meter pattern is typical for a dipole that is about ⅝ λ above ground. On 15, the antenna is only 0.45 λ high, with a resulting increase in the overall elevation angle of the signal and a reduction in gain. At 20 meters, the angle grows still higher, and the signal strength diminishes as the antenna height drops to under 0.3 λ. Nevertheless, the signal is certainly usable. A full-size dipole at 20 meters would show only a little more gain, and the elevation angle would be similar to that of the invert U, despite the difference in antenna shape. If we raise the inverted-U to 40 feet, the 20 meter performance would be very similar to that shown by the 10 meter elevation plot in Fig 21.29.

The feed point impedance of the inverted-U remains well within acceptable limits for virtually all equipment, even at 20 feet above ground. Also, the SWR curves are very broad, reducing the criticalness of finding exact dimensions, even for special field conditions.

BUILDING AN INVERTED-U

Approach the construction of an inverted-U in 3 steps: 1) the tubing arrangement, 2) the center hub and feed point assembly, and 3) the drooping extensions. A parts list appears in **Table 21.2**.

The Aluminum Tubing Dipole for 10 Meters

The aluminum tubing dipole consists of three longer sections of tubing and a short section mounted permanently to the feed point plate, as shown in **Fig 21.34**. Let's consider each half of the element separately. Counting from the center of the plate — the feed point — the element extends five inches using ¾-inch aluminum tubing. Then we have two 33-inch exposed tubing sections, with an additional three inches of tubing overlap per section. These sections are ⅝- and ½-inch diameter, respectively. The outer section is 30 inches long exposed (with at least a three-inch overlap) and consists of ⅜-inch diameter tubing.

Since the ⅝- and ½-inch sections are 36 inches long, you can make the outer ⅜-inch section the same overall length and use more overlap, or you can cut the tubing to 33 inches and use the 3-inch overlap. Three inches of overlap is sufficient to ensure a strong junction, and it minimizes excess weight. However, when not in use, the three outer tubing sections will nest inside each other for storage, and a 36-inch length for the outer section is a bit more convenient to un-nest for assembly. Keep the end hitch pin on the ⅜ inch tubing as an easy way of pulling it into final position. You may use the readily available 6063-T832 aluminum tubing that nests well and has a long history of antenna service.

The only construction operation that you need to perform on the tubing is to drill a hole at about the center of each junction to pass a hitch pin clip. Obtain hitch pin clips (also called hairpin cotter pin clips in some literature) that fit snugly over the tubing. One size will generally handle about two or three tubing sizes. This antenna uses 3⁄32 (pin diameter) by 2⅝-inch long clips for the ¾- to ⅝-inch and the ⅝- to ½-inch junctions, with 3⁄32 by 1⅝-inch pins for the ½- to ⅜-inch junction and for the final hitch pin clip at the outer end of the antenna. Drill the ⅛-inch diameter holes for the clips with the adjacent tubes in position relative to each other. Tape the junction temporarily for the drilling. Carefully de-burr the holes so that the tubing slides easily when nested.

The hitch pin clip junctions, shown in **Fig 21.35**, hold the element sections in position. Actual electrical contact between sections is made by the overlapping portions of the tube. Due to the effects of weather, junctions of this type are not suitable for a permanent installation, but are completely satisfactory for short-term use. Good electrical contact requires clean, dry aluminum surfaces, so do not use any type of lubricant to assist the nesting and un-nesting of the tubes. Instead, clean both the inner and outer surfaces of the tubes before and after each use.

Hitch pin clips are fairly large and harder to lose in the grass of a field site than most

Table 21.2
Parts List for the Inverted-U

Amount	*Item*	*Comments*
6'	0.375" OD aluminum tubing	2 - 3' pieces
6'	0.5" OD aluminum tubing	2 - 3' pieces
6'	0.625" OD aluminum tubing	2 - 3' pieces
10"	0.75" OD aluminum tubing	2 - 5" pieces
4"	0.5" nominal (⅝" OD) CPVC	
50'	Aluminum wire AWG #17	
8	Hitch pin clips	Sized to fit tubing junctions.
1	4" by 4" by ¼" Lexan plate	Other materials suitable.
2	SS U-bolts	Sized to fit support mast
2	Sets SS #8/10 1.5" bolt, nut, washers	SS = stainless steel
2	Sets SS #8 1" bolt, nut, washers	
2	Sets SS #8 0.5" bolt, nut, washers	
1	Coax connector bracket, 1⁄16" aluminum	See text for dimensions and shape
1	Female coax connector	
2	Solder lugs, #8 holes	
2	Short pieces copper wire	From coax connector to solder lugs

Note: 6063-T832 aluminum tubing is preferred and can be obtained from such outlets as Texas Towers (**www.texastowers.com**). Lexan (polycarbonate) is available from such sources as McMasters-Carr (**www.mcmasters.com**), as are the hitch pin clips (if not locally available). Other items should be available from local home centers and radio parts stores.

Fig 21.35 — A close-up of the element mounting plate assembly, including the hitch pin clips used to secure the next section of tubing.

inch lengths of ¾-inch aluminum tubing, is just above the centerline of the plate (to allow room for the coax fitting below). ½-inch nominal CPVC has an outside diameter of about ⅝ inch and makes a snug fit inside the ¾-inch tubing. The CPVC aligns the two aluminum tubes in a straight line and allows for a small (about ½ inch) gap between them. When centered between the two tubes, the CPVC is the same width as the plate. A pair of 1.5-inch #8 or #10 stainless steel bolts — with washers and a nut — secures the element to the plate.

Note in the sketch that you may insert the ⅝-inch tube as far into the ¾-inch tube as it will go and be assured of a three-inch overlap. Drill all hitch pin clip holes perpendicular to the plate. Although this alignment is not critical to the junctions of the tubes, it is important to the outer ends of the tubes when you use the antenna below 10 meters.

Mount a single-hole female UHF connector on a bracket made from a scrap of 1/16-inch-thick L-stock that is 1 inch on a side. Drill the UHF mounting hole first, before cutting the L-stock to length and trimming part of the mounting side. Then drill two holes for ½-inch long #8 stainless steel bolts about 1 inch apart, for a total length of L-stock of about 1.5 inches. The reason for the wide strip is to place the bolt heads for the bracket outside the area where the mast will meet the plate on the back side. Note in Fig 21.36 that the bracket nuts are on the bracket-side of the main plate, while the heads face the mast. The bracket-to-plate mounting edge of the bracket needs to be only about ¾ inch wide, so you may trim that side of the L-stock accordingly.

nuts and bolts. To avoid losing the clips, attach a short colorful ribbon to the loop end of each clip.

Each half element is 101 inches long, for a total 10 meter dipole element length of 202 inches (16 ft 10 inches). Length is not critical within about one inch, so you may pre-assemble the dipole using the listed dimensions. However, if you wish a more precisely tuned element, tape the outer section in position and test the dipole on your mast at the height that you will use. Adjust the length of the outer tubing segments equally at both ends for the best SWR curve on the lower 1 MHz of the 10 meter band. Even though the impedance will be above 50 Ω throughout the band, you should easily obtain an SWR curve under 2:1 that covers the entire band segment.

The Center Hub: Mounting and Feed Point Assembly

Construct the plate for mounting the element and the mast from a 4 × 4 × ¼-inch-thick scrap of polycarbonate (trade name Lexan), as shown in **Fig 21.36**. You may use other materials so long as they will handle the element weight and stand up to field conditions.

At the top and bottom of the plate are holes for the U-bolts that fit around the mast. Since masts may vary in diameter at the top, size your U-bolts and their holes to suit the mast.

The element center, consisting of two five-

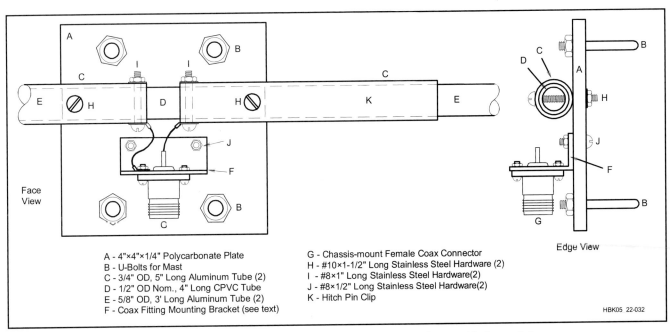

A - 4"×4"×1/4" Polycarbonate Plate
B - U-Bolts for Mast
C - 3/4" OD, 5" Long Aluminum Tube (2)
D - 1/2" OD Nom., 4" Long CPVC Tube
E - 5/8" OD, 3' Long Aluminum Tube (2)
F - Coax Fitting Mounting Bracket (see text)
G - Chassis-mount Female Coax Connector
H - #10×1-1/2" Long Stainless Steel Hardware (2)
I - #8×1" Long Stainless Steel Hardware (2)
J - #8×1/2" Long Stainless Steel Hardware (2)
K - Hitch Pin Clip

Fig 21.36 — The element and feed point mounting plate, with details of the construction used in the prototype.

Table 21.3
Inverted-U Drooping Wire Lengths

Band (meters)	Wire Length (inches)
10	n/a
12	15.9
15	37.4
17	62.0
20	108.0

Note: The wire length for the drooping ends is measured from the end of the tubular dipole to the tip for #17 AWG wire. Little change in length occurs as a function of the change in wire size. However, a few inches of additional wire length is required for attachment to the element.

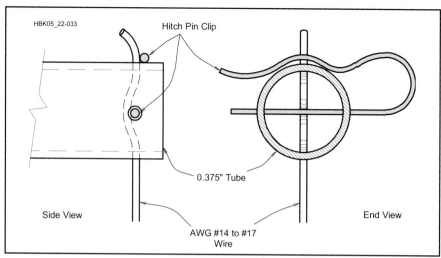

Fig 21.37 — A simple method of clamping the end wires to the ⅜-inch tube end using a hitch pin clip.

With the element center sections and the bracket in place, drill two holes for one-inch long #8 stainless steel bolts at right angles to the mounting bolts and as close as feasible to the edges of the tubing at the gap. These bolts have solder lugs attached for short leads to the coax fitting. Solder lugs do not come in stainless steel, so you should check these junctions before and after each use for any corrosion that may require replacement.

With all hardware in place, the hub unit is about $4 \times 10 \times 1$ inch (plus U-bolts). It will remain a single unit from this point onward, so that your only field assembly requirements will be to extend tubing sections and install hitch pin clips. You are now ready to perform the initial 10 meter resonance tests on your field mast.

The Drooping Extensions for 12-20 Meters

The drooping end sections consist of aluminum wire. Copper is usable, but aluminum is lighter and quite satisfactory for this application. **Table 21.3** lists the approximate lengths of each extension *below* the element. Add three to five inches of wire — less for 12 meters, more for 20 meters — to each length listed.

Common #17 AWG aluminum electric fencing wire works well. Fence wire is stiffer than most wires of similar diameter, and it is cheap. Stiffness is the more important property, since you do not want the lower ends of the wire to wave excessively in the breeze, potentially changing the feed point properties of the antenna while it is in use.

When stored, the lengths of wire extensions for 12 and 15 meters can be laid out without any bends. However, the longer extensions for 17 and 20 meters will require some coiling or folding to fit the same space as the tubing when nested. Fold or coil the wire around any kind of small spindle that has at least a two-inch diameter (larger is better). This measure prevents the wire from crimping and eventually breaking. Murphy dictates that a wire will break in the middle of an operating session. So carry some spare wire for replacement ends. All together, the ends require about 50 ft of wire.

Fig 21.37 shows the simple mounting scheme for the end wires. Push the straight wires through a pair of holes aligned vertically to the earth and bend the top portion slightly. To clamp the wire, insert a hitch pin clip though holes parallel to the ground, pushing the wire slightly to one side to reach the far hole in the tube. The double bend holds the wire securely (for a short-term field operation), but allows the wire to be pulled out when the session is over or to change bands.

Add a few inches to the lengths given in Table 21.3 as an initial guide for each band. Test the lengths and prune the wires until you obtain a smooth SWR curve below 2:1 at the ends of each band. Since an inverted-U antenna is full length, the SWR curves will be rather broad and suffer none of the narrow bandwidths associated with inductively loaded elements. **Fig 21.38** shows typical SWR curves for each band to guide your expectations.

You should not require much, if any, adjustment once you have found satisfactory lengths for each band. So you can mark the wire when you finish your initial test adjustments. However, leave enough excess so that you can adjust the lengths in the field.

Do not be too finicky about your SWR curves. An initial test and possibly one adjustment should be all that you need to arrive at an SWR value that is satisfactory for your equipment. Spending half of your operating time adjusting the elements for as near to a 1:1 SWR curve as possible will rob you of valuable contacts without changing your signal strength is any manner that is detectable.

Changing bands is a simple matter. Remove the ends for the band you are using and install the ends for the new band. An SWR check and possibly one more adjustment of the end lengths will put you back on the air.

FINAL NOTES

The inverted-U dipole with interchangeable end pieces provides a compact field antenna. All of the parts fit in a 3-ft long bag. A drawstring bag works very well. **Fig 21.39** shows the parts in their travel form. When assembled and mounted at least 20 feet up (higher is even better), the antenna will compete with just about any other dipole mounted at the same height. But the inverted-U is lighter than most dipoles at frequencies lower than 10 meters. It also rotates easily by hand — assuming that you can rotate the mast by hand. Being able to broadside the dipole to your target station gives the inverted-U a strong advantage over a fixed wire dipole.

With a dipole having drooping ends, safety is very important. Do not use the antenna unless the wire ends for 20 meters are higher than any person can touch when the antenna is in use. Even with QRP power levels, the RF voltage on the wire ends can be dangerous. With the antenna at 20 feet at its center, the ends should be at least 10 feet above ground.

Equally important is the maintenance that you give the antenna before and after each use. Be sure that the aluminum tubing is clean — both inside and out — when you nest and un-nest the sections. Grit can freeze the sections together, and dirty tubing can prevent good electrical continuity. Carry a few extra hitch pin clips in the package to be sure you have spares in case you lose one.

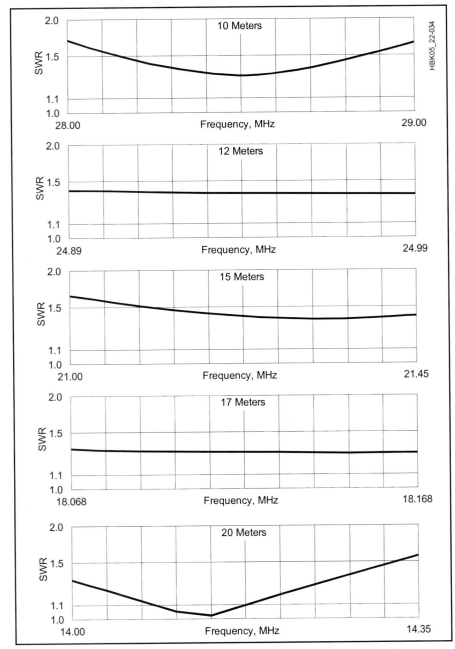

Fig 21.38 — Typical 50-Ω SWR curves for the inverted-U antenna at a feed point height of 20 ft.

Fig 21.39 — The entire inverted-U antenna parts collection in semi-nested form, with its carrying bag. The tools stored with the antenna include a wrench to tighten the U-bolts for the mast-to-plate mount and a pair of pliers to help remove end wires from the tubing. The pliers have a wire-cutting feature to help replace a broken end wire. A pair of locking pliers makes a good removable handle for turning the mast. The combination of the locking and regular pliers helps to uncoil the wire extensions for any band; give them a couple of sharp tugs to straighten the wire.

Scaling Up the Inverted-U

To inverted-U's rotatable dipole can be scaled up by as much as a factor of three with correspondingly heavier mounting plate and tubing diameters. This results in a rotatable dipole that can be used as low as 30 meters with excellent performance. A full-sized dipole is a very efficient antenna and will hold its own with two-element Yagis at comparable heights.

Suitable tubing and mounting hardware can be obtained by scavenging pieces from old tri-band HF Yagis that have been damaged or taken out of service. Metal boom-to-mast plates can be used if the antenna elements are insulated by enclosing them in a piece of exterior plastic electrical conduit whose inside diameter is a close match to the element's outside diameter. Cut a ¼-inch slot along the length of the conduit so that it can be compressed around the element by the U-bolt or muffler clamp.

Project: Two W8NX Multiband, Coax-Trap Dipoles

Over the last 60 or 70 years, amateurs have used many kinds of multiband antennas to cover the traditional HF bands. The availability of the 30, 17 and 12 meter bands has expanded our need for multiband antenna coverage.

Two different antennas are described here. The first covers the traditional 80, 40, 20, 15 and 10 meter bands, and the second covers 80, 40, 17 and 12 meters. Each uses the same type of W8NX trap — connected for different modes of operation — and a pair of short capacitive stubs to enhance coverage. The W8NX coaxial-cable traps have two different modes: a high- and a low-impedance mode. The inner-conductor windings and shield windings of the traps are connected in series for both modes. However, either the low- or high-impedance point can be used as the trap's output terminal. For low-impedance trap operation, only the center conductor turns of the trap windings are used. For high-impedance operation, all turns are used, in the conventional manner for a trap. The short stubs on each antenna are strategically sized and located to permit more flexibility in adjusting the resonant frequencies of the antenna.

80, 40, 20, 15 AND 10 M DIPOLE

Fig 21.40 shows the configuration of the 80, 40, 20, 15 and 10 meter antenna. The radiating elements are made of #14 AWG stranded copper wire. The element lengths are the wire span lengths in feet. These lengths do not include the lengths of the pigtails at the balun, traps and insulators. The 32.3-ft-long inner 40 meter segments are measured from the eyelet of the input balun to the tension-relief hole in the trap coil form. The 4.9-ft segment length is measured from the tension-relief hole in the trap to the 6-ft stub. The 16.1-ft outer-segment span is measured from the stub to the eyelet of the end insulator.

The coaxial-cable traps are wound on PVC pipe coil forms and use the low-impedance output connection. The stubs are 6-ft lengths of ⅛-inch stiffened aluminum or copper rod hanging perpendicular to the radiating elements. The first inch of their length is bent 90° to permit attachment to the radiating elements by large-diameter copper crimp connectors. Ordinary #14 AWG wire may be used for the stubs, but it has a tendency to curl up and may tangle unless weighed down at the end. You should feed the antenna with 75-Ω coax cable using a good 1:1 balun.

This antenna may be thought of as a modified W3DZZ antenna (see the References) due to the addition of the capacitive stubs. The length and location of the stub give the antenna designer two extra degrees of freedom to place the resonant frequencies within the amateur bands. This additional flexibility is particularly helpful to bring the 15 and 10 meter resonant frequencies to more desirable locations in these bands. The actual 10 meter resonant frequency of the original W3DZZ antenna is somewhat above 30 MHz, pretty remote from the more desirable low frequency end of 10 meters.

80, 40, 17 AND 12 M DIPOLE

Fig 21.41 shows the configuration of the 80, 40, 17 and 12 meter antenna. Notice that the capacitive stubs are attached immediately outboard after the traps and are 6.5 ft long, ½-ft longer than those used in the other antenna. The traps are the same as those of the other antenna, but are connected for the high-impedance parallel-resonant output mode. Since only four bands are covered by this antenna, it is easier to fine tune it to precisely the desired frequency on all bands. The 12.4-ft tips can be pruned to a particular 17 meter frequency with little effect on the 12 meter frequency. The stub lengths can be pruned to a particular 12 meter frequency with little effect on the 17 meter frequency. Both such pruning adjustments slightly alter the 80 meter resonant frequency. However, the bandwidths of the antennas are so broad on 17 and 12 meters that little need for such pruning exists. The 40 meter frequency is nearly independent of adjustments to the capacitive stubs and outer radiating tip elements. Like the first antennas, this dipole is fed with a 75-Ω balun and feed line.

Fig 21.42 shows the schematic diagram

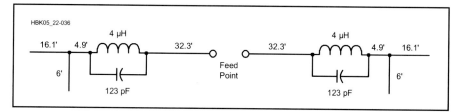

Fig 21.40 — A W8NX multiband dipole for 80, 40, 20, 15 and 10 meters. The values shown (123 pF and 4 μH) for the coaxial-cable traps are for parallel resonance at 7.15 MHz. The low-impedance output of each trap is used for this antenna.

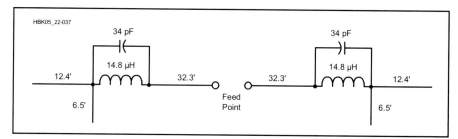

Fig 21.41 — A W8NX multiband dipole for 80, 40, 17 and 12 meters. For this antenna, the high-impedance output is used on each trap. The resonant frequency of the traps is 7.15 MHz.

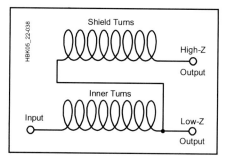

Fig 21.42 — Schematic for the W8NX coaxial-cable trap. RG-59 is wound on a 2⅜-inch OD PVC pipe.

of the traps. It illustrates the difference between the low and high-impedance modes of the traps. Notice that the high-impedance terminal is the output configuration used in most conventional trap applications. The low-impedance connection is made across only the inner conductor turns, corresponding to one-half of the total turns of the trap. This mode steps the trap's impedance down to approximately one-fourth of that of the high-impedance level. This is what allows a single trap design to be used for two different multiband antennas.

Fig 21.43 is a drawing of a cross-section of the coax trap shown through the long axis of the trap. Notice that the traps are conventional coaxial-cable traps, except for the

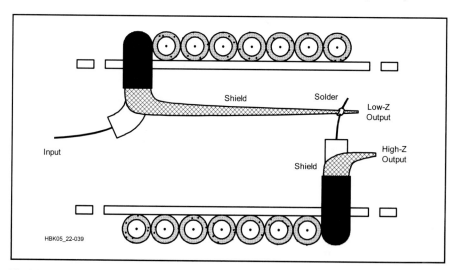

Fig 21.43 — Construction details of the W8NX coaxial-cable trap.

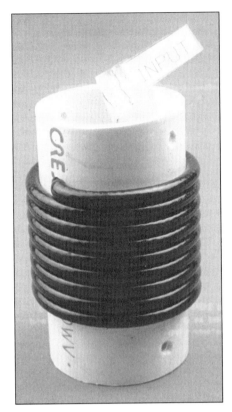

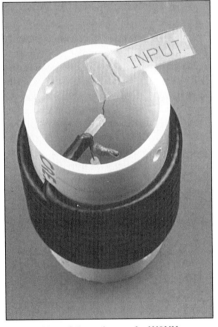

Fig 21.44 — Other views of a W8NX coax-cable trap.

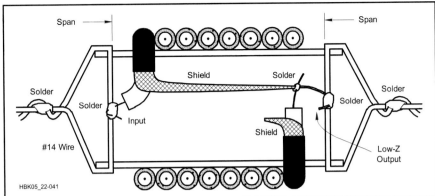

Fig 21.45 — Additional construction details for the W8NX coaxial-cable trap.

added low-impedance output terminal. The traps are 8¾ close-spaced turns of RG-59 (Belden 8241) on a 2⅜-inch-OD PVC pipe (schedule 40 pipe with a two-inch ID) coil form. The forms are 4⅛ inches long. Trap resonant frequency is very sensitive to the outer diameter of the coil form, so check it carefully. Unfortunately, not all PVC pipe is made with the same wall thickness. The trap frequencies should be checked with a dip meter and general-coverage receiver and adjusted to within 50 kHz of the 7150 kHz resonant frequency before installation. One inch is left over at each end of the coil forms to allow for the coax feed-through holes and holes for tension-relief attachment of the antenna radiating elements to the traps. Be sure to seal the ends of the trap coax cable with RTV sealant to prevent moisture from entering the coaxial cable.

Also, be sure that you connect the 32.3-ft wire element at the start of the inner conductor winding of the trap. This avoids detuning the antenna by the stray capacitance of the coaxial-cable shield. The trap output terminal (which has the shield stray capacitance) should be at the outboard side of the trap. Reversing the input and output terminals of the trap will lower the 40 meter frequency by approximately 50 kHz, but there will be negligible effect on the other bands.

Fig 21.44 shows a coaxial-cable trap. Further details of the trap installation are shown in **Fig 21.45**. This drawing applies specifically to the 80, 40, 20, 15 and 10 meter antenna, which uses the low-impedance trap connections. Notice the lengths of the trap pigtails: three to four inches at each terminal of the trap. If you use a different arrangement, you must modify the span lengths accordingly. All connections can be made using crimp connectors rather than by soldering. Using a crimping tool instead of a soldering iron allows easier access to the trap's interior.

PERFORMANCE

The performance of both antennas has been very satisfactory. W8NX uses the 80, 40, 17 and 12 meter version because it covers 17 and 12 meters. (He has a tri-band Yagi for 20, 15 and 10 meters.) The radiation pattern on 17 meters is that of ½-wave dipole. On 12 meters, the pattern is that of a ⅝-wave dipole. At his location in Akron, Ohio, the antenna runs essentially east and west. It is installed as an inverted-V, 40 ft high at the center, with a 120° included angle between the legs. Since the stubs are very short, they radiate little power and make only minor contributions to the radiation patterns. In theory, the pattern has four major lobes on 17 meters, with maxima to the northeast, southeast, southwest and northwest. These provide low-angle radiation into Europe, Africa, South Pacific, Japan and Alaska. A narrow pair of minor broadside lobes provides north and south coverage into Central America, South America and the polar regions.

There are four major lobes on 12 meters, giving nearly end-fire radiation and good low-angle east and west coverage. There are also three pairs of very narrow, nearly broadside, minor lobes on 12 meters, down about 6 dB from the major end-fire lobes. On 80 and 40 meters, the antenna has the usual figure-8 patterns of a half-wave-length dipole.

Both antennas function as electrical half-wave dipoles on 80 and 40 meters with a low SWR. They both function as odd-harmonic current-fed dipoles on their other operating frequencies, with higher, but still acceptable, SWR. The presence of the stubs can either raise or lower the input impedance of the antenna from those of the usual third and fifth harmonic dipoles. Again W8NX recommends that 75-Ω, rather than 50-Ω, feed line be used because of the generally higher input impedances at the harmonic operating frequencies of the antennas.

The SWR curves of both antennas were carefully measured using a 75 to 50-Ω transformer from Palomar Engineers inserted at the junction of the 75-Ω coax feed line and a 50-Ω SWR bridge. The transformer is required for accurate SWR measurement if a 50-Ω SWR bridge is used with a 75-Ω line. Most 50-Ω rigs operate satisfactorily with a 75-Ω line, although this requires different tuning and load settings in the final output stage of the rig or antenna tuner. The author uses the 75 to 50-Ω transformer only when making SWR measurements and at low power levels. The transformer is rated

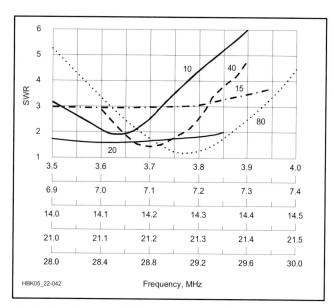

Fig 21.46 — Measured SWR curves for an 80, 40, 20, 15 and 10 meter antenna, installed as an inverted-V with 40-ft apex and 120° included angle between legs.

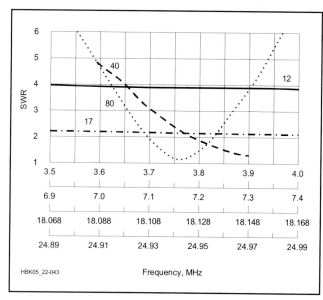

Fig 21.47 — Measured SWR curves for an 80, 40, 17 and 12 meter antenna, installed as an inverted-V with 40-ft apex and 120° included angle between legs.

for 100 W, and when he runs his 1-kW PEP linear amplifier the transformer is taken out of the line.

Fig 21.46 gives the SWR curves of the 80, 40, 20, 15 and 10 meter antenna. Minimum SWR is nearly 1:1 on 80, 1.5:1 on 40, 1.6:1 on 20, and 1.5:1 on 10 meters. The minimum SWR is slightly below 3:1 on 15 meters. On 15 meters, the stub capacitive reactance combines with the inductive reactance of the outer segment of the antenna to produce a resonant rise that raises the antenna input resistance to about 220 Ω, higher than that of the usual ½-wavelength dipole. An antenna tuner may be required on this band to keep a solid-state final output stage happy under these load conditions.

Fig 21.47 shows the SWR curves of the 80, 40, 17 and 12 meter antenna. Notice the excellent 80 meter performance with a nearly unity minimum SWR in the middle of the band. The performance approaches that of a full-size 80 meter wire dipole. The short stubs and the low-inductance traps shorten the antenna somewhat on 80 meters. Also observe the good 17 meter performance, with the SWR being only a little above 2:1 across the band.

But notice the 12 meter SWR curve of this antenna, which shows 4:1 SWR across the band. The antenna input resistance approaches 300 Ω on this band because the capacitive reactance of the stubs combines with the inductive reactance of the outer antenna segments to give resonant rises in impedance. These are reflected back to the input terminals. These stub-induced resonant impedance rises are similar to those on the other antenna on 15 meters, but are even more pronounced.

Too much concern must not be given to SWR on the feed line. Even if the SWR is as high as 9:1 *no destructively high voltages will exist on the transmission line*. Recall that transmission-line voltages increase as the square root of the SWR in the line. Thus, 1 kW of RF power in 75-Ω line corresponds to 274 V line voltage for a 1:1 SWR. Raising the SWR to 9:1 merely triples the maximum voltage that the line must withstand to 822 V. This voltage is well below the 3700-V rating of RG-11, or the 1700-V rating of RG-59, the two most popular 75-Ω coax lines. Voltage breakdown in the traps is also very unlikely. As will be pointed out later, the operating power levels of these antennas are limited by RF power dissipation in the traps, not trap voltage breakdown or feed line SWR.

TRAP LOSSES AND POWER RATING

Table 21.4 presents the results of trap Q measurements and extrapolation by a two-frequency method to higher frequencies above resonance. W8NX employed an old, but recently calibrated, Boonton Q meter for the measurements. Extrapolation to higher-frequency bands assumes that trap resistance losses rise with skin effect according to the square root of frequency, and that trap dielectric losses rise directly with frequency. Systematic measurement errors are not increased

Table 21.4
Trap Q

Frequency (MHz)	3.8	7.15	14.18	18.1	21.3	24.9	28.6
High Z out (Ω)	101	124	139	165	73	179	186
Low Z out (Ω)	83	103	125	137	44	149	*155*

Table 21.5
Trap Loss Analysis: 80, 40, 20, 15, 10 Meter Antenna

Frequency (MHz)	3.8	7.15	14.18	21.3	28.6
Radiation Efficiency (%)	96.4	70.8	99.4	99.9	100.0
Trap Losses (dB)	0.16	1.5	0.02	0.01	0.003

Table 21.6
Trap Loss Analysis: 80, 40, 17, 12 Meter Antenna

Frequency (MHz)	3.8	7.15	18.1	24.9
Radiation Efficiency (%)	89.5	90.5	99.3	99.8
Trap Losses (dB)	0.5	0.4	0.03	0.006

by frequency extrapolation. However, random measurement errors increase in magnitude with upward frequency extrapolation. Results are believed to be accurate within 4% on 80 and 40 meters, but only within 10 to 15% at 10 meters. Trap Q is shown at both the high- and low-impedance trap terminals. The Q at the low-impedance output terminals is 15 to 20% lower than the Q at the high-impedance output terminals.

W8NX computer-analyzed trap losses for both antennas in free space. Antenna-input resistances at resonance were first calculated, assuming lossless, infinite-Q traps. They were again calculated using the Q values in Table 21.4. The radiation efficiencies were also converted into equivalent trap losses in decibels. **Table 21.5** summarizes the trap-loss analysis for the 80, 40, 20, 15 and 10 meter antenna and **Table 21.6** for the 80, 40, 17 and 12 meter antenna.

The loss analysis shows radiation efficiencies of 90% or more for both antennas on all bands except for the 80, 40, 20, 15 and 10 meter antenna when used on 40 meters. Here, the radiation efficiency falls to 70.8%. A 1-kW power level at 90% radiation efficiency corresponds to 50-W dissipation per trap. In W8NX's experience, this is the trap's survival limit for extended key-down operation. SSB power levels of 1 kW PEP would dissipate 25 W or less in each trap. This is well within the dissipation capability of the traps.

When the 80, 40, 20, 15 and 10 meter antenna is operated on 40 meters, the radiation efficiency of 70.8% corresponds to a dissipation of 146 W in each trap when 1 kW is delivered to the antenna. This is sure to burn out the traps — even if sustained for only a short time. Thus, the power should be limited to less than 300 W when this antenna is operated on 40 meters under prolonged key-down conditions. A 50% CW duty cycle would correspond to a 600-W power limit for normal 40 meter CW operation. Likewise, a 50% duty cycle for 40 meter SSB corresponds to a 600-W PEP power limit for the antenna.

The author knows of no analysis where the burnout wattage rating of traps has been rigorously determined. Operating experience seems to be the best way to determine trap burn-out ratings. In his own experience with these antennas, he's had no traps burn out, even though he operated the 80, 40, 20, 15 and 10 meter antenna on the critical 40 meter band using his AL-80A linear amplifier at 600-W PEP output. He did not make a continuous, key-down, CW operating tests at full power purposely trying to destroy the traps!

Some hams may suggest using a different type of coaxial cable for the traps. The dc resistance of 40.7 Ω per 1000 feet of RG-59 coax seems rather high. However, W8NX has found no coax other than RG-59 that has the necessary inductance-to-capacitance ratio to create the trap characteristic reactance required for the 80, 40, 20, 15 and 10 meter antenna. Conventional traps with wide-spaced, open-air inductors and appropriate fixed-value capacitors could be substituted for the coax traps, but the convenience, weatherproof configuration and ease of fabrication of coaxial-cable traps is hard to beat.

Project: Extended Double-Zepp For 17 Meters

Although the Extended Double-Zepp (EDZ) antenna shown in **Fig 21.48** has several attractive features, it is rarely used by hams, perhaps out of concern over the Zepp's high feed point impedance. The antenna's overall length is 1.28 λ and its pattern is bidirectional broadside to the antenna. The SWR of the antenna is low enough near the design frequency that it can be fed with coax and an impedance-matching unit or open-wire line can be used for wider range and multiband use. This project describes an EDZ for 17 meters.

The Zepp antenna (a half-wave dipole, fed at one end) was introduced earlier in this section. The Zepp can be modified in two ways. The first is to double the length of the antenna and feed it in the middle, making a *double-Zepp*. This creates a one-wavelength dipole, with the expected high feed point impedance and about 1.6 dBd gain. A ½-λ center-fed dipole operated on its second harmonic is effectively a double-Zepp. The second modification is to extend the double-Zepp to be 0.64-λ (close to ⅝-λ) long on each side of the feed point. The feed point is then no longer at a high-impedance point on the antenna. This creates the extended, double-Zepp. (The EDZ is described in more detail in the *ARRL Antenna Book*.)

The overall length of the EDZ is calculated as follows:

$$\frac{984}{f\,(MHz)} \times 1.28 = \text{length in feet} \quad (5)$$

Using this formula, an 18.1 MHz EDZ is 69.6 feet (69 feet, 7 inches.) long. The EDZ has 3 dBd of gain in a figure-8 pattern of two major lobes broadside to the antenna and four minor lobes at smaller angles to the axis of the antenna. The feed point impedance is

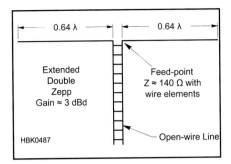

Fig 21.48 — The Extended Double-Zepp antenna consists of two 0.64-λ sections placed end to end and fed in the middle. The high-impedance points of the antenna have been moved away from the feed point, lowering feed point impedance. The antenna has gain of approximately 3 dBd broadside to the antenna.

approximately 140 Ω.

The EDZ is useful at lower frequencies, as well. On 20 meters, the 17-meter EDZ is just slightly longer than the double-Zepp, with 1.6 to 2 dBd of gain and a rather high feed point impedance of several hundred ohms. On 40 meters, the antenna is a slightly long ½-λ dipole. If your antenna tuner has sufficient range, the antenna can also serve as a shortened dipole for 75/80 meters. At these lower frequencies, the antenna's radiation pattern is a single lobe, broadside to the antenna.

On higher frequencies, the pattern continues to split into more lobes. For example, on 15 meters, there are four lobes at approximately 45° from the antenna axis. On 10 meters, where the antenna is approximately two full-wavelengths long, the pattern is similar, with the lobes a bit closer to the antenna axis and smaller lobes beginning to appear.

Some hams use a 4:1 impedance transformer to reduce the feed point impedance and improve SWR as the operating frequency moves away from the design frequency. This works best if the antenna is to be used on a single band. However, if the antenna is to be used on multiple bands, a better solution is to use open-wire feed line and an antenna tuner. If you wish to operate on a frequency at which the feed point impedance is high, use a feed line length near an odd multiple of a quarter-wavelength long, presenting a lower impedance to your antenna tuner that may be easier to match.

(This project is based on a "Hints and Kinks" item by Bob Baird, W7CSD, from the January 1992 issue of *QST*.)

21.3 Vertical (Ground-Plane) Antennas

One of the more popular amateur antennas is the *vertical*. It usually refers to a single radiating element erected vertically over the ground. A typical vertical is an electrical ¼-λ long and is constructed of wire or tubing. The vertical antenna is more accurately named the *ground plane* because it uses a conductive surface (the ground plane) to create a path for return currents, effectively creating the "missing half" of a ½-λ antenna. Another name for this type of antenna is the *monopole* (sometimes *unipole*).

The ground plane can be a solid, conducting surface, such as a vehicle body for a VHF/UHF mobile antenna. At HF, this is impractical and systems of *ground radials* are used; wires laid out on the ground radially from the base of the antenna. One conductor of the feed line is attached to the vertical radiating element of the antenna and the remaining conductor attached to the ground plane.

Single vertical antennas are omnidirectional radiators. This can be beneficial or detrimental, depending on the situation. On transmission there are no nulls in any direction, unlike most horizontal antennas. However, QRM on receive can't be nulled out from the directions that are not of interest unless multiple verticals are used in an array.

Ground-plane antennas need not be mounted vertically. A ground-plane antenna can operate in any orientation as long as the ground plane is perpendicular to the radiating element. Other considerations, such as minimizing cross-polarization between stations, may require a specific mounting orientation though. In addition, due to the size of HF antennas, mounting them vertically is usually the most practical solution.

A vertical antenna can be mounted at the Earth's surface, in which case it is a *ground-mounted vertical*. The ground plane is then constructed on the surface of the ground. A vertical antenna and the associated ground plane can also be installed above the ground. This often reduces ground losses, but it is more difficult to install the necessary number of radials. *Ground-independent* verticals are often mounted well above the ground because their operation does not rely on a ground plane.

21.3.1 Ground Systems

When compared to horizontal antennas, verticals also suffer more acutely from two main types of losses — *ground return losses* for currents in the near field, and *far-field ground losses*. Ground losses in the near field can be minimized by using many ground radials. This is covered in the sidebar, "Optimum Ground Systems for Vertical Antennas."

Far-field losses are highly dependent on

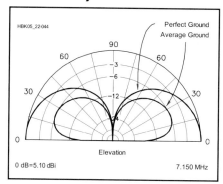

Fig 21.49 — Elevation patterns for two quarter-wave vertical antennas over different ground. One vertical is placed over perfect ground, and the other is placed over average ground. The far-field response at low elevation angles is greatly affected by the quality of the ground — as far as 100 λ away from the vertical antenna.

the conductivity and dielectric constant of the earth around the antenna, extending out as far as 100 wavelengths from the base of the antenna. There is very little that someone can do to change the character of the ground that far away — other than moving to a small island surrounded by saltwater! Far-field losses greatly affect low-angle radiation, causing the radiation patterns of practical vertical antennas to fall far short of theoretical patterns over *perfect ground*, often seen in classical texts.

Fig 21.49 shows the elevation pattern response for two different 40 meter quarter-wave verticals. One is placed over a theoretical infinitely large, infinitely conducting ground. The second is placed over an extensive radial system over average soil, having a conductivity of 5 mS/m and a dielectric constant of 13. This sort of soil is typical of heavy clay found in pastoral regions of the US mid-Atlantic states. At a 10° elevation angle, the real antenna losses are almost 6 dB compared to the theoretical one; at 20° the difference is about 3 dB. See *The ARRL Antenna Book* chapter on the effects of ground for further details.

While real verticals over real ground are not a magic method to achieve low-angle radiation, cost versus performance and ease of installation are incentives that inspire many antenna builders. For use on the lower frequency amateur bands — notably 160 and 80 meters — it is not always practical to erect a full-size vertical. At 1.8 MHz, a full-sized quarter-wave vertical is 130 ft high. In such instances it is often necessary to accept a shorter radiating element and use some form of loading.

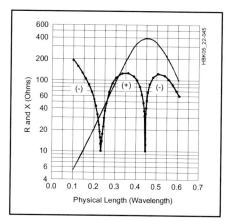

Fig 21.50 — Radiation resistance (solid curve) and reactance (dotted curve) of vertical antennas as a function of their electrical height.

Fig 21.50 provides curves for the physical height of verticals in wavelength versus radiation resistance and reactance. Although the plots are based on perfectly conducting ground, they show general trends for installations where many radials have been laid out to make a ground screen. As the radiator is made shorter, the radiation resistance decreases — with 6 Ω being typical for a 0.1-λ high antenna. The lower the radiation resistance, the more the antenna efficiency depends on ground conductivity and the effectiveness of the ground screen. Also, the bandwidth decreases markedly as the length is reduced toward the left of the scale in Fig 21.50. It can be difficult to develop suitable matching networks when radiation resistance is very low.

Generally a large number of shorter radials results in a better ground system than a few longer ones. For example, eight ¼-λ radials are preferred over four ¼-λ radials. Optimum radial lengths are described in the sidebar.

The conductor size of the radials is not especially significant. Wire gauges from #4 to #20 AWG have been used successfully by amateurs. Copper wire is preferred, but where soil is low in acid (or alkali), aluminum wire can be used. The wires may be bare or insulated, and they can be laid on the earth's surface or buried a few inches below ground. Insulated wires will have greater longevity by virtue of reduced corrosion and dissolution from soil chemicals.

When property dimensions do not allow a classic installation of equally spaced radial wires, they can be placed on the ground as space permits. They may run away from the antenna in only one or two compass directions. They may be bent to fit on your property. Hardware cloth and chicken wire are also quite effective, although the galvanizing must be of high-quality to prevent rapid rusting.

A single ground rod, or group of them bonded together, is seldom as effective as a collection of random-length radial wires.

All radial wires should be connected together at the base of the vertical antenna. The electrical bond needs to be of low resistance. Best results will be obtained when the wires are soldered together at the junction point. When a grounded vertical is used, the ground wires should be affixed securely to the base of the driven element.

Ground return losses are lower when vertical antennas and their radials are elevated above ground, a point that is well-known by those using ground plane antennas on their roofs. Even on 160 or 80 meters, effective vertical antenna systems can be made with as few as four ¼ λ long radials elevated 10 to 20 feet off the ground.

21.3.2 Full-Size Vertical Antennas

When it is practical to erect a full-size ¼-λ vertical antenna, the forms shown in **Fig 21.51** are worthy of consideration. The example at A is the well-known *vertical ground plane*. The ground system consists of four above-ground radial wires. The length of the driven element and ¼-λ radials is derived from the standard equation

$$L \text{ (ft)} = \frac{234}{f \text{ (MHz)}} \qquad (6)$$

With four equidistant radial wires drooped at approximately 30° (Fig 21.51A), the feed point impedance is roughly 50 Ω. When the radials are at right angles to the radiator (Fig 21.51B) the impedance approaches 36 Ω.

Besides minimizing ground return losses, another major advantage in this type of vertical antenna over a ground-mounted type is that the system can be elevated well above nearby conductive objects (power lines, trees, buildings and so on). When drooping radials are used, they can also serve as guy wires for the mast that supports the antenna. The coax shield braid is connected to the radials, and the center conductor to the driven element.

The *Marconi* vertical antenna shown in Fig 21.51C is the classic form taken by a ground-mounted vertical. It can be grounded at the base and *shunt fed*, or it can be isolated from ground, as shown, and *series fed*. As always, this vertical antenna depends on an effective ground system for efficient performance. If a perfect ground were located below the antenna, the feed impedance would be near 36 Ω. In a practical case, owing to imperfect ground, the impedance is more likely to be in the vicinity of 50 Ω.

Vertical antennas can be longer than ¼-λ, too. ⅜-λ, ½-λ, and ⅝-λ verticals can all be used with good results, although none will present a 50-Ω feed point impedance at the base. Non-resonant lengths have become popular for the same reasons as non-resonant horizontal antennas; when fed with low-loss feed line and a wide-range impedance matching unit, the antenna can be used on multiple bands. Various matching networks, described in the **Transmission Lines** chapter, can be employed. Antenna lengths above ½-λ are not recommended because the radiation pattern begins to break up into more than one lobe, developing a null at the horizon at 1-λ.

A gamma-match feed system for a grounded ¼-λ vertical is presented in Fig 21.51D. (The gamma match is also discussed in the **Transmission Lines** chapter.) Some rules of thumb for arriving at workable gamma-arm and capacitor dimensions are to make the rod length 0.04 to 0.05 λ, its diameter ⅓ to

Optimum Ground Systems for Vertical Antennas

A frequent question brought up by old-timers and newcomers alike is: "So, how many ground radials do I *really* need for my vertical antenna?" Most hams have heard the old standby tales about radials, such as "if a few are good, more must be better" or "lots of short radials are better than a few long ones."

John Stanley, K4ERO, eloquently summarized a study he did of the professional literature on this subject in his article "Optimum Ground Systems for Vertical Antennas" in December 1976 *QST*. His approach was to present the data in a sort of "cost-benefit" style in **Table 21.A**, reproduced here. John somewhat wryly created a new figure of merit—the total amount of wire needed for various radial configurations. This is expressed in terms of wavelengths of total radial wire.

Table 21.A
Optimum Ground-System Configurations

Configuration Designation	A	B	C	D	E	F
Number of radials	16	24	36	60	90	120
Length of each radial in wavelengths	0.1	0.125	0.15	0.2	0.25	0.4
Spacing of radials in degrees	22.5	15	10	6	4	3
Total length of radial wire installed, in wavelengths	1.6	3	5.4	12	22.5	48
Power loss in dB at low angles with a quarter-wave radiating element	3	2	1.5	1	0.5	0*
Feed point impedance in ohms with a quarter-wave radiating element	52	46	43	40	37	35

Note: Configuration designations are indicated only for text reference.

*Reference. The loss of this configuration is negligible compared to a perfectly conducting ground.

The results almost jumping out of this table are:
- If you can only install 16 radials (Case A), they needn't be very long — 0.1 λ is sufficient. You'll use 1.6 λ of radial wire in total, which is about 450 feet at 3.5 MHz.
- If you have the luxury of laying down 120 radials (Case F), they should be 0.4 λ long, and you'll gain about 3 dB over the 16-radial case. You'll also use 48 λ of total wire—For 80 meters, that would be about 13,500 feet!
- If you can't put out 120 radials, but can install 36 radials that are 0.15 λ long (Case C), you'll lose only 1.5 dB compared to the optimal Case F. You'll also use 5.4 λ of total wire, or 1,500 feet at 3.5 MHz.
- A 50-Ω SWR of 1:1 isn't necessarily a good thing — the worst-case ground system in Case A has the lowest SWR.

Table 21.A represents the case for "Average" quality soil, and it is valid for radial wires either laid on the ground or buried several inches in the ground. Note that such ground-mounted radials are detuned because of their proximity to that ground and hence don't have to be the classical quarter-wavelength that they need to be were they in "free space."

In his article John also made the point that ground-radial losses would only be significant on transmit, since the atmospheric noise on the amateur bands below 30 MHz is attenuated by ground losses, just as actual signals would be. This limits the ultimate signal-to-noise ratio in receiving.

So, there you have the tradeoffs — the loss in transmitted signal compared to the cost (and effort) needed to install more radial wires. You take your pick.

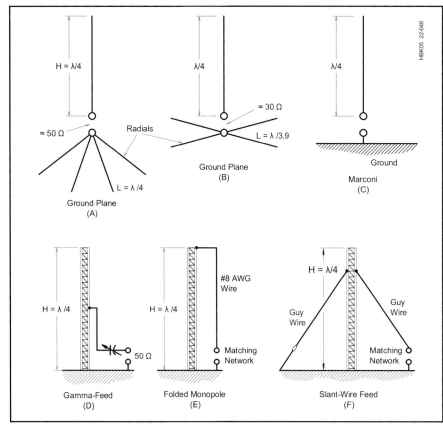

Fig 21.51 — Various types of vertical antennas.

21.3.3 Physically Short Verticals

A group of short vertical radiators is presented in **Fig 21.52**. Illustrations A and B are for top and center loading. A capacitance hat is shown in each example. The hat should be as large as practical to increase the radiation resistance of the antenna and improve the bandwidth. The wire in the loading coil is chosen for the largest gauge consistent with ease of winding and coil-form size. The larger wire diameters will reduce the resistive (I^2R) losses in the system. The coil-form material should have a medium or high dielectric constant. Phenolic or fiberglass tubing is entirely adequate.

A base-loaded vertical is shown at C of Fig 21.52. The primary limitation is that the high current portion of the vertical exists in the coil rather than the driven element. With center loading, the portion of the antenna below the coil carries high current, and in the top-loaded version the entire vertical element carries high current. Since the high-current part of the antenna is responsible for most of the radiating, base loading is the least effective of the three methods. The radiation resistance of the coil-loaded antennas shown is usually less than 16 Ω.

A method for using guy wires to top load a short vertical is illustrated in Fig 21.52D. This system works well with gamma feed. The loading wires are trimmed to provide an electrical quarter wavelength for the overall system. This method of loading will result in a higher radiation resistance and greater bandwidth than the systems shown at A through C. If an HF or VHF array is at the top of the tower, it will simply contribute to the top loading.

A three-wire monopole is shown in Fig 21.52E. Two #8 AWG drop wires are connected to the top of the tower and brought to ground level. The wires can be spaced any convenient distance from the tower — normally 12 to 30 inches from the sides. C1 is adjusted for best SWR. This type of vertical has a fairly narrow bandwidth, but because C1 can be motor driven and controlled from the operating position, frequency changes can be accomplished easily. This technique will not be suitable for matching to 50-Ω line unless the tower is less than an electrical quarter wavelength high.

A different method for top loading is shown in Fig 21.52F. Barry Boothe, W9UCW, described this method in December 1974 *QST*. An extension is used at the top of the tower to create an electrical quarter-wavelength vertical. L1 is a loading coil with sufficient inductance to provide antenna resonance. This type of antenna lends itself to operation on 160 meters.

½ that of the driven element and the center-to-center spacing between the gamma arm and the driven element roughly 0.007 λ. The capacitance of C1 at a 50-Ω matched condition will be about 7 pF per meter of wavelength. The absolute value of C1 will depend on whether the vertical is resonant and on the precise value of the radiation resistance. For best results, make the radiator approximately 3% shorter than the resonant length.

Amateur antenna towers lend themselves to use as shunt-fed verticals, even though an HF-band beam antenna is usually mounted on the tower. The overall system should be close to resonance at the desired operating frequency if a gamma feed is used. The HF-band beam will contribute somewhat to *top loading* of the tower. The natural resonance of such a system can be checked by dropping a #12 or #14 AWG wire from the top of the tower (connecting it to the tower top) to form a folded monopole (Fig 21.51E). A four- or five-turn link can be inserted between the lower end of the drop wire and the ground system. A dip meter is then inserted in the link to determine the resonant frequency.

If the tower is equipped with guy wires, they should be broken up with strain insulators to prevent unwanted loading of the vertical. In such cases where the tower and beam antennas are not able to provide ¼-λ resonance, portions of the top guy wires can be used as top-loading capacitance. Experiment with the guy-wire lengths (using the dip-meter technique) while determining the proper dimensions.

A folded-monopole is depicted in Fig 21.51E. This system has the advantage of increased feed point impedance. Furthermore, an impedance-matching unit can be connected between the bottom of the drop wire and the ground system to permit operation on more than one band. For example, if the tower is resonant on 80 meters, it can be used as shown on 160 and 40 meters with reasonable results, even though it is not electrically long enough on 160 to act as a full-size antenna. The drop wire need not be a specific distance from the tower, but you might try spacings between 12 and 30 inches.

The method of feed shown at Fig 21.51F is commonly referred to as *slant-wire feed*. The guy wires and the tower combine to provide quarter-wave resonance. A matching network is placed between the lower end of one guy wire and ground and adjusted for an SWR of 1:1. It does not matter at which level on the tower the guy wires are connected, assuming that the impedance-matching unit is capable of effecting a match to 50 Ω.

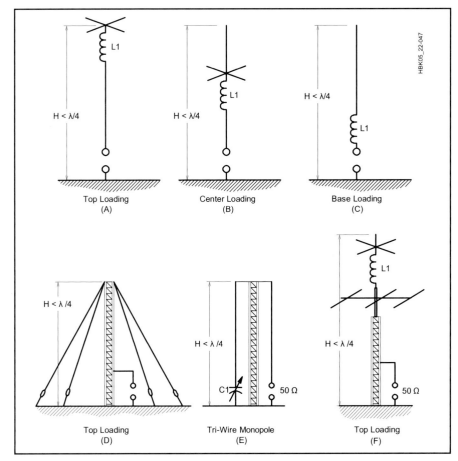

Fig 21.52 — Vertical antennas that are less than one-quarter wavelength in height.

Fig 21.53 — W9SR's short vertical uses top loading and a capacitance hat with a 10-ft TV mast to make a compact antenna for 160, 80 or 40 meters.

Project: Top-Loaded Low-Band Antenna

The short, top-loaded vertical antenna described here by Dick Stroud, W9SR, is of interest to hams with limited space or who are portable operators. It has been used on 40, 80 and 160 meters. The antenna uses a single 10-ft TV mast section for the vertical radiator, along with a capacitance hat, loading coil and short top mast. The overall height is less than 15 ft, as seen in **Fig 21.53**. The capacitance hat and loading coil assembly can be used with longer vertical radiators with changes to the coil inductance.

CAPACITANCE HAT

The capacitance hat consists of a hub that mounts to the top mast above the coil and six elements made from aluminum rod. The machining, drilling and tapping of the hub assembly can be done by nearly any machine shop if you don't have the facilities. Be sure to use stainless steel hardware throughout. (Thanks to Fred Gantzer, W0AWD, for building the original hub.)

The hub is made of two pieces of ½-inch thick aluminum as shown in **Fig 21.54**. The two pieces (Fig 21.55A and B) are bolted together to form the hub, which slides over the 1⅛-inch top mast. It is held in place with three 10-32 screws, as shown in Fig 21.54C. The six elements of the capacitance hat are made from 3/16-inch aluminum rod, each 4.5 ft long. These are held in place with 6-32 screws.

LOADING COIL

The coil form is made from fiberglass tubing available from Small Parts, Inc. A 6-inch length of 1.25-inch OD fiberglass tubing (part no. LFT-125/16-30) is centered over a 10-inch length of 1-inch OD tubing (part no. LFT-125/20-30). The tubes telescope tightly and it may be necessary to lightly sand the smaller tube for a smooth fit.

The loading coil is wound on the 1.25-inch OD fiberglass tube, as shown in Fig 21.54D.

After the coil is optimized, it is covered with a length of shrink sleeving for weather protection. Coil winding information in the drawing is for use with a 10-ft mast.

The bottom section of the coil assembly fits directly into the tapered upper section of the TV mast and the exposed upper fiberglass section of the coil assembly fits into a 2.5-ft long, 1.125-inch OD, aluminum upper mast. Stainless 8-32 screws join the pieces together and also provide a connection point for the loading coil wires. If you use a painted steel mast, be sure to remove paint from the connection point, and then weather-seal it after adjustment is complete.

The capacitance hat hub slides over the upper mast and is held in place with three screws. The hub is about 6 inches above the coil, but the location can be moved to change the resonance of the antenna slightly.

The completed capacitance hat and loading coil assembly are shown in **Fig 21.55**. A dab of Glyptol (exterior varnish or Loctite also works) locks the screws in place once adjustments have been made and the antenna is ready for installation.

The coaxial feed line attaches to the bottom of the TV mast. Again, use an 8-32 screw for attachment and clean any paint from the metal. To match the antenna to a 50-Ω transmission line, a small parallel (shunt) inductance is needed at the base. The inductor is air-wound with #16 AWG wire (see **Table 21.7**).

Table 21.7
Shunt Inductor Winding Details

Band	Turns
160 m	10 turns #16 AWG, spaced ⅛ inch
80 m	8.75 turns #16 AWG, spaced ⅛ inch
40 m	7.5 turns #16 AWG, spaced ⅛ inch

Note: All inductors are air wound, 1.75 inch ID. Dimensions shown are for use with 10 ft mast.

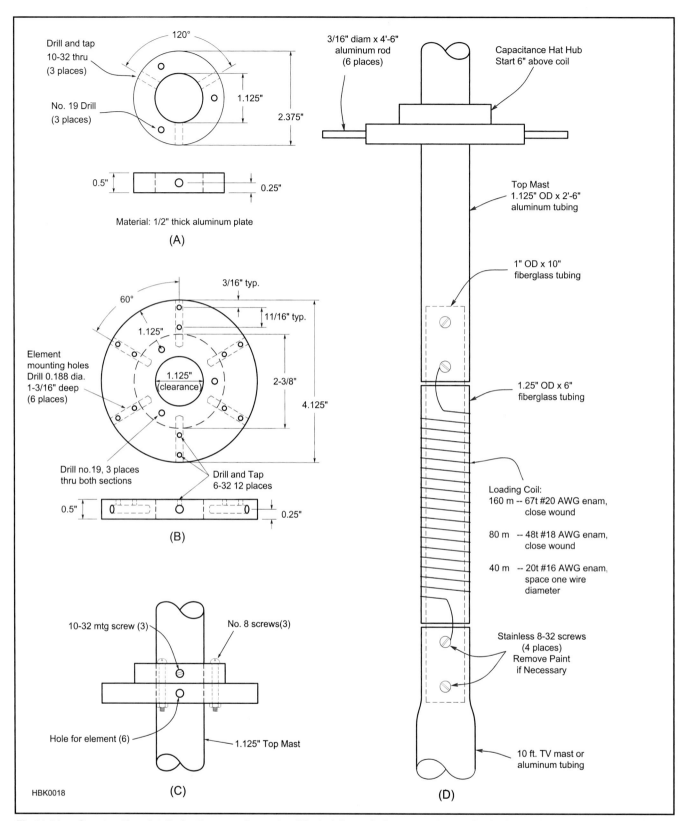

Fig 21.54 — Construction details for the capacitance hat. The hub is made from two pieces of ½-inch aluminum plate (A and B). It's attached to the mast as shown in C. The six elements are made from 4.5-ft lengths of 3/16-inch aluminum rod. Loading coil details are shown in D.

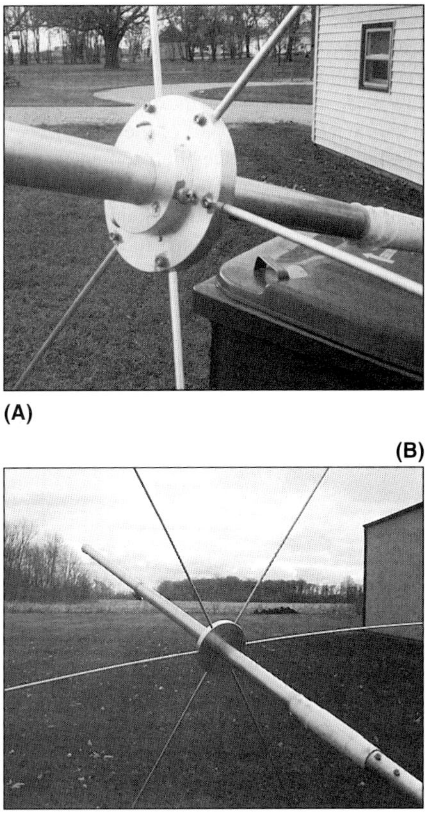

Fig 21.55 — The completed capacitance hat assembly is shown at A, and B shows a view of the upper mast assembly with the loading coil and capacitance hat.

MOUNTING THE ANTENNA

A glass beverage bottle serves well as the base insulator, with the neck of the bottle fitting snugly into the TV mast. To support the base insulator, drill a hole large enough to accept the base of the bottle in the center of a 2 × 6 board about 14 inches long. Nail a piece of ¼-inch plywood over the bottom to keep the bottle from slipping through. To keep the base support board from moving around, drill a couple of holes and secure it to the ground with stakes.

The antenna is top-heavy and will need to be guyed. A simple insulated guy ring can be made from a 2-inch PVC coupling and placed on the mast just below the loading coil. The PVC is locked to the mast with three ¼-20 bolts. They are 1-inch long and have nuts on the inside and outside of the PVC. Three ¼-inch holes are drilled for the guy lines, made from lengths of $^3/_{16}$-inch non-absorbent rope.

OPERATION

On-air results with a 10-ft mast have been very good, even with low power. The ground system for the early tests was nothing more than an 8-ft ground rod hammered into Hoosier soil. With this setup, and using about 90 W, many DX contacts were made over one week's time and stateside contacts were plentiful.

There is plenty of room to experiment however. Performance could be improved by using an extended radial system or raised and insulated radials. (Expect much better performance over average ground with a system of radials) Two, or even three, mast sections could be used with additional guys and proper loading coil inductance. If you use multiple masts, be sure to make a good electrical connection at the joints. The upper assembly is now permanently used to top load a 60-ft pole for transmitting on 160 meters.

21.3.4 Cables and Control Wires on Towers

Most vertical antennas of the type shown in Fig 21.51 and 21.52C-E consist of towers, usually with HF or VHF beam antennas at the top. The rotator control wires and the coaxial feeders to the top of the tower will not affect antenna performance adversely. In fact, they become a part of the composite antenna. To prevent unwanted RF currents from following the wires into the shack, simply dress them close to the tower legs and bring them to ground level. (Running the cables inside the tower works even better.) This decouples the wires at RF. The wires should then be routed along the earth surface (or buried underground) to the operating position. It is not necessary to use bypass capacitors or RF chokes in the rotator control leads if this is done, even when maximum legal power is employed.

21.3.5 Multiband Trap Verticals

The two-band trap vertical antenna of **Fig 21.56** operates in much the same manner as a trap dipole or trap Yagi. The notable difference is that the vertical is one-half of a dipole. The radial system (in-ground or above-ground) functions as a ground plane for the antenna, and provides an equivalent for the missing half of the dipole. Once again, the more effective the ground system, the better will be the antenna performance.

Trap verticals usually are designed to work as ¼-λ radiators. The portion of the antenna below the trap is adjusted as a ¼-λ radiator at the higher proposed operating frequency. That is, a 20/15 meter trap vertical would be a resonant quarter wavelength at 15 meter from the feed point to the bottom of the trap. The trap and that portion of the antenna above the trap (plus the 15 meter section below the trap) constitute the complete antenna during 20 meter operation. But because the trap is in the circuit, the overall physical length of the vertical antenna will be slightly less than that of a single-band, full-size 20 meter vertical.

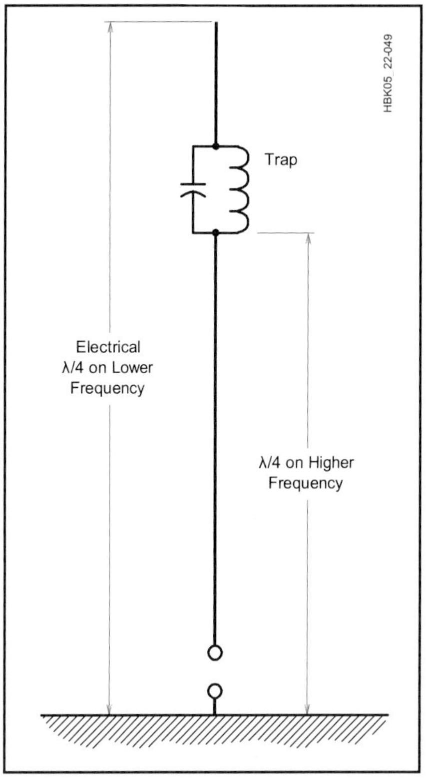

Fig 21.56 — A two-band trap vertical antenna. The trap should be resonated by itself as a parallel resonant circuit at the center of the operating range for the higher frequency band. The reactance of either the inductor or the capacitor range from 100 to 300 Ω. At the lower frequency the trap will act as a loading inductor, adding electrical length to the total antenna.

Recent "ground-independent" multiband vertical antennas also have traps, but are designed to be electrically longer than ¼-λ. A common electrical length, is ⅜ λ, for example. The "traps" in these antennas are generally not parallel-LC circuits as described above. A variety of techniques are used with both parallel-LC traps and short resonant structures similar to stubs being used to change the antenna's electrical length at different frequencies.

TRAPS

The trap functions as the name implies: the high impedance of the parallel resonant circuit "traps" the 15 meter energy and confines it to the part of the antenna below the trap. (See the **Electrical Fundamentals** chapter for more information on resonant LC circuits.) During 20 meter operation it allows the RF energy to reach all of the antenna. The trap in this example is tuned as a parallel resonant circuit to 21 MHz. At this frequency it electrically disconnects the top section of the vertical from the lower section because it presents a high impedance at 21 MHz, blocking 21 MHz current. Generally, the trap inductor and capacitor have a reactance of 100 to 300 Ω. Within that range it is not critical.

The trap is built and adjusted separately from the antenna. It should be resonated at the center of the portion of the band to be operated. Thus, if one's favorite part of the 15 meter band is between 21.0 and 21.1 MHz, the trap should be tuned to 21.05 MHz.

Resonance is checked by using a dip meter and detecting the dipper signal in a calibrated receiver. An SWR analyzer can also be used. Once the trap is adjusted it can be installed in the antenna, and no further adjustment will be required. It is easy, however, to be misled after the system is assembled: Attempts to check the trap resonance in the antenna will suggest that the trap has moved much lower in frequency (approximately 5 MHz lower in a 20/15 meter vertical). This is because the trap is now part of the overall antenna, and the resultant resonance is that of the total antenna. Measure the trap's resonant frequency separately from the rest of the antenna.

Multiband operation is quite practical by using the appropriate number of traps and tubing sections. The construction and adjustment procedure is the same, regardless of the number of bands covered. The highest frequency trap is always closest to the feed end of the antenna, and the lowest frequency trap is always the farthest from the feed point. As the operating frequency is progressively lowered, more traps and more tubing sections become a functional part of the antenna.

Traps should be weatherproofed to prevent moisture from detuning them. Several coatings of high dielectric compound, such as polystyrene Q Dope or Liquid Electrical Tape, are effective. Alternatively, a protective sleeve of heat-shrink tubing can be applied to the coil after completion. The coil form for the trap should be of high insulating quality and be rugged enough to sustain stress during periods of wind.

21.4 T and Inverted-L Antennas

This section covers variations on the vertical antenna. **Fig 21.57** shows a flat-top T vertical. The T is basically a shortened ¼-λ vertical with the flat-top T section acting as capacitive loading to length the antenna electrically. Dimension H should be as large as possible (up to ¼-λ) for best results. The horizontal section, L, is adjusted to a length that provides resonance. Maximum radiation is polarized vertically despite the horizontal top-loading wire because current in each horizontal half creates out-of-phase radiation that cancels.

A variation of the T antenna is depicted in **Fig 21.58**. This antenna is commonly referred to as an *inverted-L* and is basically a ⁵⁄₁₆-λ vertical bent in the middle so that the top section runs parallel to the ground. Similarly to the T antenna, the vertical section should be as long as possible. L is then added to provide an electrical ⁵⁄₁₆-λ overall.

Because the horizontal section does carry some current, there will be some horizontally-polarized radiation at high angles. This is often considered desirable because it provides local and regional coverage. The horizontal section need not be perfectly horizontal — sloping the wire at a shallow angle from horizontal does not greatly affect antenna performance. This allows the inverted-L to be constructed with a single vertical support.

A sidearm or a length of line attached to a tower can be used to support the vertical section of the T or inverted-L antenna. (Keep the vertical section of the antennas as far from the tower as is practical. Certain combinations of tower height and top loading can create a resonance that interacts severely with the antennas — a 70-ft tower and a 5-element Yagi, for example.)

Both the T and inverted-L antennas are ground-plane antennas and require a good ground system to be effective. If the T or inverted-L are used with a very good ground system, the feed-point impedance will approach 35-40 Ω so that the SWR approaches 1.4:1.

The inverted-L is constructed longer than resonance as illustrated in Fig 21.58 so that the feed point resistance increases to 50 Ω plus some inductive reactance due to the extra length. A series capacitor at the feed point then cancels the reactance, leaving a 50-Ω impedance suitable for direct connection by coaxial cable.

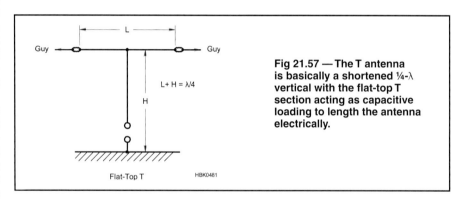

Fig 21.57 — The T antenna is basically a shortened ¼-λ vertical with the flat-top T section acting as capacitive loading to length the antenna electrically.

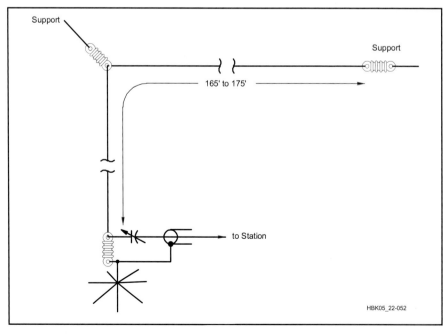

Fig 21.58 — The inverted-L antenna designed for the 1.8 MHz band. Overall wire length is 165 to 175 ft. The variable capacitor has a maximum capacitance of 500 to 800 pF.

21.5 Slopers and Vertical Dipoles

21.5.1 Half-Sloper Antenna

Many hams have had excellent results with *half-sloper* antennas, while others have not had such luck. Investigations by ARRL Technical Advisor John S. Belrose, VE2CV, have brought some insight to the situation through computer modeling and antenna-range tests. The following is taken from VE2CV's Technical Correspondence in Feb 1991 *QST*, pp 39 and 40. Essentially, the half-sloper is a top-fed vertical antenna that uses the structure at the top of the tower plane (such as a grounded Yagi antenna) as a ground plane and the tower acts as a reflector.

For half-slopers, the input impedance, the resonant length of the sloping wire and the antenna pattern all depend on the tower height, the angle (between the sloper and tower) the type of Yagi and the Yagi orientation. Here are several configurations extracted from VE2CV's work:

At 160 meters — use a 40 meter beam on top of a 95-ft tower with a 55° sloper apex angle. The radiation pattern varies little with Yagi type. The pattern is slightly cardioid with about 8 dB front-to-back ratio at a 25° takeoff angle (see Fig 21.59B and C). Input impedance is about 50 Ω.

At 80 meters — use a 20 meter beam on top of a 50-ft tower with a 55° sloper apex angle. The radiation pattern and input imped-

Table 21.8
HVD Dimensions
Length using 0.875-inch aluminum tubing

MHz	Feet	Inches
18.11	33	11
21.2	22	0
24.94	18	9
28.4	16	5

These lengths should divided by two to determine the length of the dipole legs

ance are similar to those of the 160 meter half-sloper.

At 40 meters — use a 20 meter beam on top of a 50-ft tower with a 55° sloper apex angle. The radiation pattern and impedance depend strongly on the azimuth orientation of the Yagi. Impedance varies from 76 to 127 Ω depending on Yagi direction.

Project: Half-Wave Vertical Dipole (HVD)

Chuck Hutchinson, K8CH, describes a 15 meter vertical dipole (HVD) that he built for the ARRL book, *Simple and Fun Antennas for Hams*. The performance of this antenna, with its base at 14 ft, compares favorably with a horizontal dipole at 30 ft when making intercontinental QSOs.

CONSTRUCTION OF A 15 METER HVD

The 15 meter HVD consists of four 6-ft lengths of 0.875-inch aluminum tube with 0.058 wall thickness. In addition there are two 1-ft lengths of 0.75-inch tubing for splices, and two one-foot lengths of 0.75-inch fiberglass rod for insulators. See **Table 21.8** for dimensions.

Start by cutting off 1 foot from a 6-ft length of 0.875-inch tubing. Next, insert six inches of one of the 1-foot-long 0.75-inch tubes into the machine-cut end of your tubing and fasten the tubes together. Now, slide an end of a 6-ft length of 0.875 tube over the protruding end of the 0.75 tube and fasten them together. Repeat this procedure with the remaining 0.875-inch tubing.

You should now have two 11-ft-long elements. As you can see in **Fig 21.60**, K8CH was temporarily out of aluminum pop rivets, so he used sheet metal screws. Either will work fine, but pop rivets can easily be drilled out and the antenna disassembled if you ever want to make changes.

Because hand-made cuts are not perfectly square, put those element ends at the center of the antenna. Slip these cut ends over the

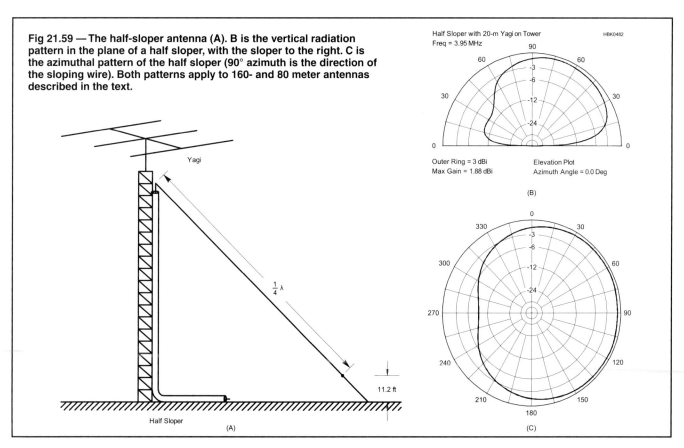

Fig 21.59 — The half-sloper antenna (A). B is the vertical radiation pattern in the plane of a half sloper, with the sloper to the right. C is the azimuthal pattern of the half sloper (90° azimuth is the direction of the sloping wire). Both patterns apply to 160- and 80 meter antennas described in the text.

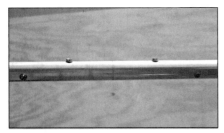

Fig 21.60 — Element splice uses a 1-ft length of 0.75-inch tubing inserted into the 0.875-inch sections to join them together. Self-tapping sheet-metal screws are used in this photo, but aluminum pop rivets or machine screws with washers and nuts can be used.

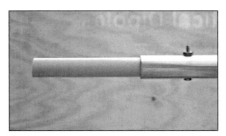

Fig 21.62 — The HVD base insulator is a 1-ft length of 0.75-inch fiberglass rod.

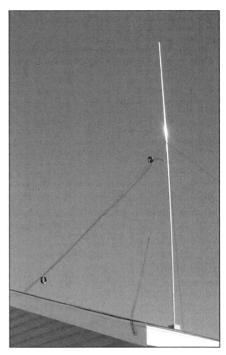

Fig 21.65 — The HVD installed at K8CH. An eye screw that is used for securing one of the guy lines is visible in the foreground. You can also see the two choke baluns that are used in the feed system (see text).

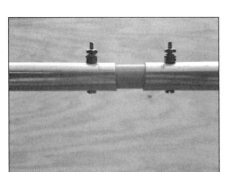

Fig 21.61 — The center insulator of the 15 meter HVD is a 1-ft length of 0.75-inch fiberglass rod. Insulator and elements have been drilled to accept #8 hardware.

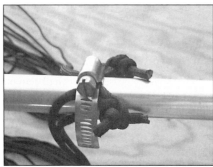

Fig 21.63 — Guys are made of Dacron line that is attached to the HVD by a stainless-steel worm-screw-type hose clamp. A self-tapping sheet-metal screw (not visible in the photo) prevents the clamp from sliding down the antenna.

ends of a 1-ft length of 0.75-inch fiberglass rod. This rod serves as the center insulator. Leave about a 1-inch gap at the center. Drill aluminum and fiberglass for #8 hardware as shown in **Fig 21.61**.

Now, slip half of the remaining 1-ft length of 0.75-inch fiberglass rod into one end the dipole. (This end will be the bottom end or base.) Drill and secure with #8 hardware. See **Fig 21.62**.

The final step is to secure the guy wires to your vertical. You can see how K8CH did that in **Fig 21.63**. Start by drilling a pilot hole and then drive a sheet metal screw into the antenna about a foot above the center. The purpose of that screw is to prevent the clamp and guys from sliding down the antenna.

The guys are clean lengths of 3/16-inch Dacron line. (The Dacron serves a dual purpose: it supports the antenna vertically, and it acts as an insulator.) Tie secure knots into the guy ends and secure these knotted ends to the antenna with a stainless-steel worm-screw-type hose clamp. Take care to not over tighten the clamps. You don't want the clamp to slip (the knots and the sheet-metal screw will help), but you especially don't want to cut your guy lines. Your antenna is ready for installation.

Fig 21.64 — At K8CH, the HVD base insulator sits in this saddle-shaped wooden fixture. This was photo was taken before the fixture was painted—a necessary step to protect against the weather.

INSTALLATION

Installation requires two things. First, a place to sit or mount the base insulator. Second, you need anchors for the support guys.

K8CH used a piece of 2 × 6 lumber to make a socket to hold the HVD base securely in place. He drilled a ¾-inch-deep hole with a ¾-inch spade bit. A couple of pieces of 2 × 2 lumber at the ends of the base form a saddle which nicely straddles the ridge at the peak of his garage roof. You can see this in **Fig 21.64**. The dimensions are not critical. Paint your base to protect it from the weather.

BALUN

This antenna needs a common-mode choke balun to ensure that stray RF doesn't flow on the shield of the coax. (See the **Transmission Lines** chapter for more information on choke baluns.) Unlike a horizontal dipole, don't consider it an option to omit the common-mode choke when building and installing an HVD.

You can use 8 ft of the RG-213 feed line wound into 7 turns for a balun. Secure the turns together with electrical tape so that each turn lies parallel with the next turn, forming a solenoid coil. Secure the feed line and balun to one of the guy lines with UV-resistant cable ties.

Because the feed line slants away from the antenna, you'll want to do *all* that you can to eliminate common-mode currents from the feed line. For that reason, make another balun about 11.5 ft from the first one. This balun also consists of 8 ft of the RG-213 feed line wound into 7 turns. See **Fig 21.65** for a photo of the installed antenna.

Project: Compact Vertical Dipole (CVD)

An HVD for 20 meters will be about 33 ft tall, and for 30 meters, it will be around 46 ft tall. Even the 20 meter version can prove to be a mechanical challenge. The compact vertical dipole (CVD), designed by Chuck Hutchinson, K8CH, uses capacitance loading to shorten the antenna. Starting with the 15 meter HVD described in the previous project, Chuck added capacitance loading wires to lower the resonance to 30 meters. Later, he shortened the wires to move resonance to the 20 meter band. This project describes those two CVDs.

PERFORMANCE ISSUES

Shortened antennas frequently suffer reduced performance caused by the shortening. A dipole that is less than ½-λ in length is a compromise antenna. The issue becomes how much is lost in the compromise. In this case there are two areas of primary interest, radiation efficiency and SWR bandwidth.

Table 21.9
CVD Loading Wires
Length using #14 AWG insulated copper wire

Band	Feet	Inches	
30 m	6	0	Top & Bottom
20 m	4	2¼	Top
20 m	3	½	Bottom

Radiation Efficiency

Capacitance loading at the dipole ends is the most efficient method of shortening the antenna. Current distribution in the high-current center of the antenna remains virtually unchanged. Since radiation is related directly to current, this is the most desirable form of loading. Computer modeling shows that radiation from a 30 meter CVD is only 0.66 dB less than that from a full-size 30 meter HVD when both have their bases 8 ft above ground. The angle of maximum radiation shifts up a bit for the CVD. Not a bad compromise when you consider that the CVD is 22-ft long compared to the approximately 46-ft length of the HVD.

SWR and SWR Bandwidth

Shortened antennas usually have lower radiation resistance and less SWR bandwidth than the full-size versions. The amount of change in the radiation resistance is related to the amount and type of loading (shortening), being lower with shorter the antennas. This can be a benefit in the case of a shortened vertical dipole. In Fig 21.2 you can see that vertical dipoles have a fairly high radiation resistance. With the dipole's lower end ⅛-λ above ground, the radiation resistance is roughly 80 Ω. In this case, a shorter antenna can have a better SWR when fed with 50-Ω coax.

SWR bandwidth tends to be wide for vertical dipoles in general. A properly designed CVD for 7 MHz or higher should give you good SWR (1.5:1 or better) across the entire band!

As you can see, in theory the CVD provides excellent performance in a compact package. Experience confirms the theory.

CONSTRUCTION

To convert the K8CH 15 meter HVD to 20 or 30 meters, you'll need to add four loading wires at the top and four more at the bottom of the HVD. The lengths are shown in **Table 21.9**. The upper wires droop at a 45° angle and the lower wires run horizontally. The antenna is supported by four guy lines. See **Fig 21.66**. You can connect the wires to the vertical portion with #8 hardware. Crimp and solder terminals on the wire ends to make connections easier. The technique is illustrated in **Fig 21.67**.

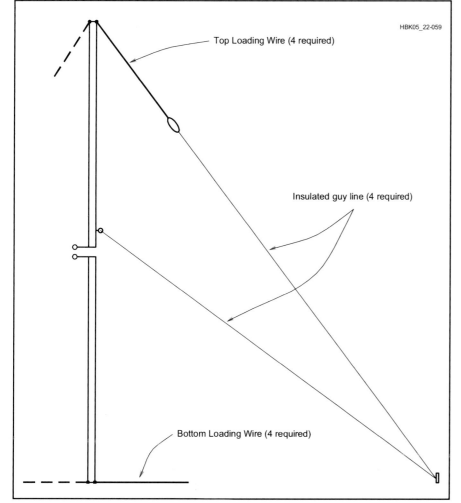

Fig 21.66 — The CVD consists of a vertical dipole and loading wires. Only one set of the four loading wires and only one guy line is shown in this drawing. See text for details.

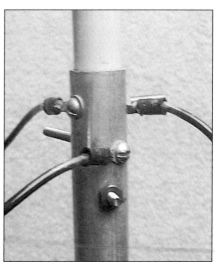

Fig 21.67 — CVD loading wires can be attached using #8 hardware. Crimp and solder terminals on the wire ends to make connections easier.

The upper loading wires can be extended with insulated line and used for additional guying. The lower wires are extended with insulated line and fasten to the guy lines so that the lower wires run horizontally.

Prune the lower wires for best SWR across the band of interest. The K8CH CVD has its base at 14 ft. This antenna has an SWR of less than 1.2:1 on 30 meters and less than 1.3:1 across the entire 20 meter band.

EXPERIENCE

The 30 meter CVD was compared to a ground-mounted quarter-wave vertical and a horizontal dipole at 30 ft. In tests, the CVD was always the superior antenna. Every DX station that was called by K8CH responded with one or two calls. What more could you ask for?

Later, the CVD loading wires were shortened for operation on 20 meters. Once again the results were very encouraging. Many contest QSOs were entered in the log using this antenna.

Finally, a late winter ice storm deposited about ¾-inch of radial ice on the antenna, loading wires and guys. The antenna would probably have survived had it not been for the sustained 45 mph winds that followed. The upper loading wires and their guy lines were not heavy enough to support the load and the antenna bent and broke. This combination of ice and wind is very unusual.

Project: All-Wire 30 Meter CVD

If you have a tree or other support that will support the upper end of a CVD at 32 ft above the ground, you might want to consider an all-wire version of the 30 meter CVD. The vertical is 24 ft long and it will have an SWR of less than 1.1:1 across the band. The four loading wires at top and bottom are each 5 ft, 2 inches long.

The configuration is shown in **Fig 21.68**. As with any vertical dipole, you'll need to use a balun between the feed line and the antenna.

Alternatively you can use two loading wires at the top and two at the bottom. In this case each of the loading wires is 8 ft, 7.5 inches long.

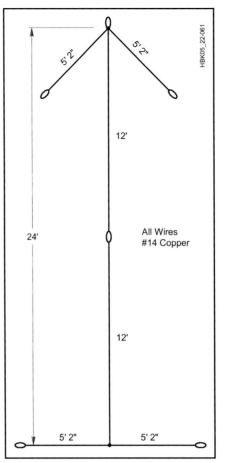

Fig 21.68 — The all-wire 30 meter CVD consists of a vertical dipole and loading wires. It can be made entirely with #14 AWG wire. Support lines have been omitted for simplicity. See text for details.

21.6 Yagi Antennas

Most antennas described earlier in this chapter have unity gain compared to a dipole, or just slightly more. For the purpose of obtaining gain and directivity it is convenient to use a Yagi-Uda *beam* antenna. The former is commonly called a *Yagi*. There are other forms of directive antennas, but the Yagi is by far the most popular used by amateurs. (For more information on phased arrays and other types of directive antennas, see the *ARRL Antenna Book*.)

Most operators prefer to erect these antennas for horizontal polarization, but they can be used as vertically polarized antennas merely by rotating the elements by 90°. In effect, the beam antenna is turned on its side for vertical polarization. The number of elements used will depend on the gain desired and the limits of the supporting structure. At HF, many amateurs obtain satisfactory results with only two elements in a beam antenna, while others have four or five elements operating on a single amateur band, called a *monoband beam*. On VHF and above, Yagis with many elements are common, particularly for simplex communication without repeaters. For fixed point-to-point communications, such as repeater links, Yagis with three or four elements are more common.

Regardless of the number of elements used, the height-above-ground considerations discussed earlier for dipole antennas remain valid with respect to the angle of radiation. This is demonstrated in **Fig 21.69** at A and B where a comparison of radiation characteristics is given for a 3-element Yagi at one-half and one wavelength above average ground. It can be seen that the higher antenna (Fig 21.69B) has a main lobe that is more favorable for DX work (roughly 15°) than the lobe of the lower antenna in Fig 21.69A (approximately 30°). The pattern at B shows that some useful high-angle radiation exists also, and the higher lobe is suitable for short-skip contacts when propagation conditions dictate the need.

The azimuth pattern for the same antenna is provided in **Fig 21.70**. (This is a free-space pattern, so the pattern is taken in the plane of the antenna. Remember that azimuth

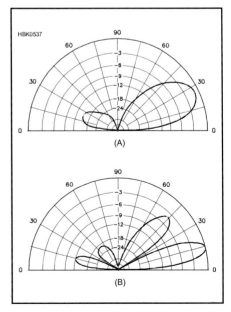

Fig 21.69 — Elevation-plane response of a 3-element Yagi placed ½ λ above perfect ground at A and the same antenna spaced 1 λ above ground at B.

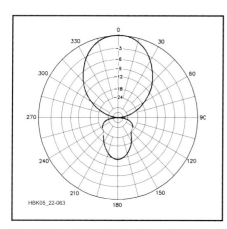

Fig 21.70 — Azimuthal pattern of a typical three-element Yagi in free space. The Yagi's boom is along the 0° to 180° axis.

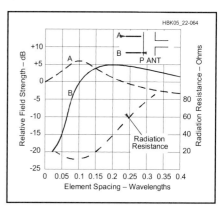

Fig 21.71 — Gain vs element spacing for a two-element Yagi, having one driven and one parasitic element. The reference point, 0 dB, is the field strength from a half-wave antenna alone.

patterns taken over a reflecting surface must also specify the elevation angle at which the pattern was measured or calculated.) Most of the power is concentrated in the main lobe at 0° azimuth. The lobe directly behind the main lobe at 180° is often called the *back lobe* or *rear lobe*. The front-to-back ratio (F/B) of this antenna is just less than 12 dB — the peak power difference, in decibels, between the main lobe at 0° and the rearward lobe at 180°. It is infrequent that two 3-element Yagis with different element spacing and tuning will yield the same lobe patterns. The patterns also change with frequency of operation. The pattern of Fig 21.70 is shown only for illustrative purposes.

21.6.1 Parasitic Excitation

In a Yagi antenna only one element (the *driven element*) is connected to the feed line. The additional elements are *coupled* to the driven element because they are so close. (Element-to-element spacing in a Yagi antenna is generally on the order of $1/10$-$1/8$ wavelength.) This *mutual coupling* results in currents being induced in the non-driven elements from the radiated field of the driven element. These elements are called *parasitic elements* and the Yagi antenna is therefore a *parasitic array*. (An antenna in which multiple elements all receive power from the transmitter is called a *driven array*.) The currents induced in the parasitic elements also result in radiated fields, just as if the current were the result of power from a feed line. This is called *re-radiation*. The combination of the field radiated by the driven element, the fields from the parasitic elements, and the physical spacing of the elements results in the fields having the proper phase relationship so as to focus the radiated energy in the desired direction and reject it in other directions.

The parasitic element is called a *director* when it reinforces radiation along a line pointing to it from the driven element, and a *reflector* in the opposite case. Whether the parasitic element is a director or reflector depends on the parasitic element tuning, which is usually adjusted by changing its length. The structure on which the elements are mounted is called the *boom* of the antenna.

21.6.2 Yagi Gain, Front-to-Back Ratio and SWR

The gain of a Yagi antenna with parasitic elements varies with the spacing and tuning of the elements. Element tuning is a function of length, diameter and *taper schedule* (the steps in length and diameter) if the element is constructed with telescoping tubing. For any given number of elements and the spacing between them, there is a tuning condition that will result in maximum gain. However, the maximum front-to-back ratio seldom, if ever, occurs at the same condition that gives maximum forward gain. The impedance of the driven element in a parasitic array, and thus the SWR, also varies with the tuning and spacing.

It is important to remember that all these parameters change as the operating frequency is varied. For example, if you operate both the CW and phone portions of the 20 meter band with a Yagi antenna, you probably will want an antenna that *spreads out* the performance over most of the band. Such designs typically must sacrifice a little gain in order to achieve good F/B and SWR performance across the band.

Gain and F/B performance generally improve with the number of elements. In Yagi antennas with more than three elements (a driven element and one director and reflector), the additional elements are added as directors, since little additional benefit is obtained from multiple reflectors. Wider spacing also improves gain and F/B up to a certain point, depending on a number of factors, beyond which performance begins to fall. Optimizing element spacing is a complex problem and no single spacing satisfies all design requirements. For the lower HF bands, the size of the antenna quickly becomes impractical for truly *optimal* designs, and compromise is necessary.

21.6.3 Two-Element Beams

A two-element beam is useful — especially where space or other considerations prevent the use of a three-element, or larger, beam. The general practice is to tune the parasitic element as a reflector and space it about 0.15 λ from the driven element, although some successful antennas have been built with 0.1-λ spacing and director tuning.

Gain vs element spacing for a two-element antenna is given in **Fig 21.71** for the special case where the parasitic element is resonant. It is indicative of the performance to be expected under maximum-gain tuning conditions. Changing the tuning of the driven element in a Yagi or quad will not materially affect the gain or F/R. Thus, only the spacing and the tuning of the single parasitic element have any effect on the performance of a 2-element Yagi.

In Fig 21.71, the greatest gain is in the direction A (in which the parasitic element is acting as a director) at spacings of less than 0.14 λ, and in direction B (in which the parasitic element is a reflector) at greater spacings. The front-to-back ratio is the difference in decibels between curves A and B. The figure also shows variation in radiation resistance of the driven element.

These curves are for the special case of a self-resonant parasitic element, but are representative of how a two-element Yagi works. At most spacings the gain as a reflector can be increased by slight lengthening of the parasitic element; the gain as a director can be increased by shortening. This also improves the front-to-rear ratio.

Most two-element Yagi designs achieve a compromise F/R of about 10 dB, together with an acceptable SWR and gain across a frequency band with a percentage bandwidth less than about 4%.

21.6.4 Three-Element Beams

A theoretical investigation of the three-element case (director, driven element and reflector) has indicated a maximum gain of about 9.7 dBi (7.6 dBd). A number of experimental investigations have shown that the spacing between the driven element and reflector for maximum gain is in the region of 0.15 to 0.25 λ. With 0.2-λ reflector spacing, **Fig 21.72** shows that the gain variation with director spacing is not especially critical.

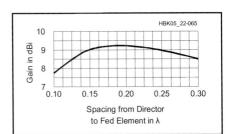

Fig 21.72 — General relationship of gain of three-element Yagi vs director spacing, the reflector being fixed at 0.2 λ. This antenna is tuned for maximum forward gain.

Also, the overall length of the array (boom length in the case of a rotatable antenna) can be anywhere between 0.35 and 0.45 λ with no appreciable difference in the maximum gain obtainable.

If maximum gain is desired, wide spacing of both elements is beneficial because adjustment of tuning or element length is less critical and the input resistance of the driven element is generally higher than with close spacing. A higher input resistance improves the efficiency of the antenna and makes a greater bandwidth possible. However, a total antenna length, director to reflector, of more than 0.3 λ at frequencies of the order of 14 MHz introduces difficulty from a construction standpoint. Lengths of 0.25 to 0.3 λ are therefore used frequently for this band, even though they are less than optimum from the viewpoint of maximum gain.

In general, Yagi antenna gain drops off less rapidly when the reflector length is increased beyond the optimum value than it does for a corresponding decrease below the optimum value. The opposite is true of a director. It is therefore advisable to err, if necessary, on the long side for a reflector and on the short side for a director. This also tends to make the antenna performance less dependent on the exact frequency at which it is operated. An increase above the design frequency has the same effect as increasing the length of both parasitic elements, while a decrease in frequency has the same effect as shortening both elements. By making the director slightly short and the reflector slightly long, there will be a greater spread between the upper and lower frequencies at which the gain starts to show a rapid decrease.

21.6.5 Construction of Yagi Antennas

Most beams and verticals are made from sections of aluminum tubing. Compromise beams have been fashioned from less-expensive materials such as electrical conduit (steel) or bamboo poles wrapped with conductive tape or aluminum foil. The steel conduit is heavy, is a poor conductor and is subject to rust. Similarly, bamboo with conducting material attached to it may deteriorate rapidly in the weather. Given the drawbacks of alternative materials, aluminum tubing (or rod for VHF and UHF Yagis) is far and away the best choice for antenna construction.

For reference, **Table 21.10** details the standard sizes of aluminum tubing, available in many metropolitan areas. Dealers may be found in the Yellow Pages under *Aluminum*. Tubing usually comes in 12-ft lengths, although 20-ft lengths are available in some sizes. Your aluminum dealer will probably also sell aluminum plate in various thicknesses needed for boom-to-mast and boom-to-element connections. Distributors of antenna towers and masts often sell aluminum tubing, as well.

Aluminum is rated according to its hardness. The most common material used in antenna construction is grade 6061-T6. This material is relatively strong and has good workability. In addition, it will bend without taking a *set*, an advantage in antenna applications where the pieces are constantly flexing in the wind. The softer grades (5051, 3003 and so on) will bend much more easily, while harder grades (7075 and so on) are more brittle.

Wall thickness is of primary concern when selecting tubing. It is of utmost importance that the tubing fits snugly where the element sections join. Sloppy joints will make a mechanically unstable antenna. The magic wall thickness is 0.058 inch. For example (from Table 21.10), 1-inch outside diameter (OD) tubing with a 0.058-inch wall has an inside diameter (ID) of 0.884 inch. The next smaller size of tubing, ⅞ inch, has an OD of 0.875 inch. The 0.009-inch difference provides just the right amount of clearance for a snug fit.

Fig 21.73 shows several methods of fastening antenna element sections together.

The slot and hose clamp method shown at the upper left is probably the best for joints where adjustments are needed. Generally, one adjustable joint on each side of the element is sufficient to tune the antenna — usually the tips at each end of an element are made adjustable. Stainless steel hose clamps (beware — some "stainless steel" models do not have a stainless screw and will rust) are recommended for longest antenna life.

The remaining photos show possible fastening methods for joints that are not adjustable. At the upper right, machine screws and nuts hold the elements in place. At the lower left, sheet metal screws are used. At the lower right, rivets secure the tubing. If the antenna is to be assembled permanently, rivets are the best choice. Once in place, they are permanent. They will never work free, regardless of vibration or wind. If aluminum rivets with aluminum mandrels are employed, they will never rust. Also, being aluminum, there is no danger of corrosion from interaction between dissimilar metals. If the antenna is to be disassembled and moved periodically, either machine or sheet metal screws will work. If machine screws are used, however, take precautions to keep the nuts from vibrating free. Use of lock washers, lock nuts and flexible adhesive such as silicone bathtub sealant will keep the hardware in place. For portable or temporary use, such as Field Day, rivets may be held in place with electrical tape and removed when the operation is finished.

Use of a conductive grease at the element joints is essential for long life. Left untreated, the aluminum surfaces will oxidize in the weather, resulting in a poor connection. Some trade names for this conductive grease are Penetrox, Noalox and Dow Corning Molykote 41. Many electrical supply houses carry these products.

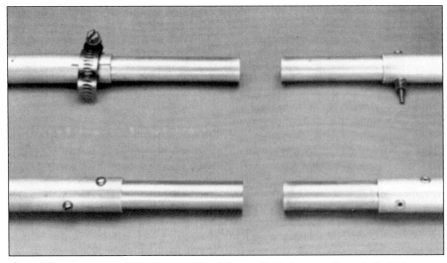

Fig 21.73 — Some methods of connecting telescoping tubing sections to build beam elements. See text for a discussion of each method.

Table 21.10
Standard Sizes of Aluminum Tubing
6061-T6 (61S-T6) Round Aluminum Tube in 12-ft Lengths

OD (in)	Wall Thickness (in)	stubs ga	ID (in)	Approx Weight (lb) per ft	per length	OD (in)	Wall Thickness (in)	stubs ga	ID (in)	Approx Weight (lb) per ft	per length
3/16	0.035	no. 20	0.117	0.019	0.228	1 1/8	0.035	no. 20	1.055	0.139	1.668
	0.049	no. 18	0.089	0.025	0.330		0.058	no. 17	1.009	0.228	2.736
1/4	0.035	no. 20	0.180	0.027	0.324	1 1/4	0.035	no. 20	1.180	0.155	1.860
	0.049	no. 18	0.152	0.036	0.432		0.049	no. 18	1.152	0.210	2.520
	0.058	no. 17	0.134	0.041	0.492		0.058	no. 17	1.134	0.256	3.072
5/16	0.035	no. 20	0.242	0.036	0.432		0.065	no. 16	1.120	0.284	3.408
	0.049	no. 18	0.214	0.047	0.564		0.083	no. 14	1.084	0.357	4.284
	0.058	no. 17	0.196	0.055	0.660	1 3/8	0.035	no. 20	1.305	0.173	2.076
3/8	0.035	no. 20	0.305	0.043	0.516		0.058	no. 17	1.259	0.282	3.384
	0.049	no. 18	0.277	0.060	0.720	1 1/2	0.035	no. 20	1.430	0.180	2.160
	0.058	no. 17	0.259	0.068	0.816		0.049	no. 18	1.402	0.260	3.120
	0.065	no. 16	0.245	0.074	0.888		0.058	no. 17	1.384	0.309	3.708
7/16	0.035	no. 20	0.367	0.051	0.612		0.065	no. 16	1.370	0.344	4.128
	0.049	no. 18	0.339	0.070	0.840		0.083	no. 14	1.334	0.434	5.208
	0.065	no. 16	0.307	0.089	1.068		*0.125	1/8"	1.250	0.630	7.416
1/2	0.028	no. 22	0.444	0.049	0.588		*0.250	1/4"	1.000	1.150	14.823
	0.035	no. 20	0.430	0.059	0.708	1 5/8	0.035	no. 20	1.555	0.206	2.472
	0.049	no. 18	0.402	0.082	0.948		0.058	no. 17	1.509	0.336	4.032
	0.058	no. 17	0.384	0.095	1.040	1 3/4	0.058	no. 17	1.634	0.363	4.356
	0.065	no. 16	0.370	0.107	1.284		0.083	no. 14	1.584	0.510	6.120
5/8	0.028	no. 22	0.569	0.061	0.732	1 7/8	0.508	no. 17	1.759	0.389	4.668
	0.035	no. 20	0.555	0.075	0.900	2	0.049	no. 18	1.902	0.350	4.200
	0.049	no. 18	0.527	0.106	1.272		0.065	no. 16	1.870	0.450	5.400
	0.058	no. 17	0.509	0.121	1.452		0.083	no. 14	1.834	0.590	7.080
	0.065	no. 16	0.495	0.137	1.644		*0.125	1/8"	1.750	0.870	9.960
3/4	0.035	no. 20	0.680	0.091	1.092		*0.250	1/4"	1.500	1.620	19.920
	0.049	no. 18	0.652	0.125	1.500	2 1/4	0.049	no. 18	2.152	0.398	4.776
	0.058	no. 17	0.634	0.148	1.776		0.065	no. 16	2.120	0.520	6.240
	0.065	no. 16	0.620	0.160	1.920		0.083	no. 14	2.084	0.660	7.920
	0.083	no. 14	0.584	0.204	2.448	2 1/2	0.065	no. 16	2.370	0.587	7.044
7/8	0.035	no. 20	0.805	0.108	1.308		0.083	no. 14	2.334	0.740	8.880
	0.049	no. 18	0.777	0.151	1.810		*0.125	1/8"	2.250	1.100	12.720
	0.058	no. 17	0.759	0.175	2.100		*0.250	1/4"	2.000	2.080	25.440
	0.065	no. 16	0.745	0.199	2.399	3	0.065	no. 16	2.870	0.710	8.520
1	0.035	no. 20	0.930	0.123	1.467		*0.125	1/8"	2.700	1.330	15.600
	0.049	no. 18	0.902	0.170	2.040		*0.250	1/4"	2.500	2.540	31.200
	0.058	no. 17	0.884	0.202	2.424
	0.065	no. 16	0.870	0.220	2.640
	0.083	no. 14	0.834	0.281	3.372

*These sizes are extruded; all other sizes are drawn tubes. Shown here are standard sizes of aluminum tubing that are stocked by most aluminum suppliers or distributors in the United States and Canada.

DRIVEN ELEMENT

The ARRL recommends *plumbers delight* construction, in which all elements are mounted directly on, and grounded to, the boom. This puts the entire array at dc ground potential, affording better lightning protection. A gamma- or T-match section can be used for matching the feed line to the array.

An alternative method is to insulate the driven element from the boom, but use a *hairpin* or *beta match*, the center point of which is electrically neutral and can be attached directly to the boom, restoring the dc ground for the driven element.

Direct feed designs in which the feed point impedance of the driven element is close to 50 Ω, requiring no impedance matching structure, presents some issues. First, a current or choke balun should be used (see the **Transmission Lines** chapter) to prevent the outer surface of the feed line shield from interacting with the antenna directly or by picking up the radiated signal. Such interaction can degrade the antenna's radiation pattern, especially by compromising signal rejection to the side and rear. Second, the driven element must be insulated from the boom, requiring some additional mechanical complexity.

BOOM MATERIAL

The boom size for a rotatable Yagi or quad should be selected to provide stability to the entire system. The best diameter for the boom depends on several factors, but mostly the element weight, number of elements and overall length. Two-inch-diameter booms should not

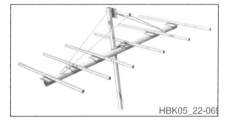

Fig 21.74 — A long boom needs both vertical and horizontal support. The crossbar mounted above the boom can support a double truss, which will help keep the antenna in position.

be made any longer than 24 ft unless additional support is given to reduce both vertical and horizontal bending forces. Suitable reinforcement for a long 2-inch boom can

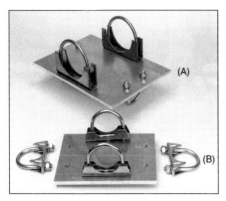

Fig 21.75 — The boom-to-element plate at A uses muffler-clamp-type U-bolts and saddles to secure the round tubing to the flat plate. The boom-to-mast plate at B is similar to the boom-to-element plate. The main difference is the size of materials used.

Table 21.11
10 Meter Optimized Yagi Designs

	Spacing Between Elements (in)	Seg 1 Length (in)	Seg 2 Length (in)	Seg 3 Length (in)	Midband Gain F/R
310-08					
Refl	0	24	18	66.750	7.2 dBi
DE	36	24	18	57.625	22.9 dB
Dir 1	54	24	18	53.125	
410-14					
Refl	0	24	18	64.875	8.4 dBi
DE	36	24	18	58.625	30.9 dB
Dir 1	36	24	18	57.000	
Dir 2	90	24	18	47.750	
510-24					
Refl	0	24	18	65.625	10.3 dBi
DE	36	24	18	58.000	25.9 dB
Dir 1	36	24	18	57.125	
Dir 2	99	24	18	55.000	
Dir 3	111	24	18	50.750	

Note: For all antennas, the tube diameters are: Seg 1=0.750 inch, Seg 2=0.625 inch, Seg 3=0.500 inch.

Table 21.12
12 Meter Optimized Yagi Designs

	Spacing Between Elements (in)	Seg 1 Length (in)	Seg 2 Length (in)	Seg 3 Length (in)	Midband Gain F/R
312-10					
Refl	0	36	18	69.000	7.5 dBi
DE	40	36	18	59.125	24.8 dB
Dir 1	74	36	18	54.000	
412-15					
Refl	0	36	18	66.875	8.5 dBi
DE	46	36	18	60.625	27.8 dB
Dir 1	46	36	18	58.625	
Dir 2	82	36	18	50.875	
512-20					
Refl	0	36	18	69.750	9.5 dBi
DE	46	36	18	61.750	24.9 dB
Dir 1	46	36	18	60.500	
Dir 2	48	36	18	55.500	
Dir 3	94	36	18	54.625	

Note: For all antennas, the tube diameters are: Seg 1 = 0.750 inch, Seg 2 = 0.625 inch, Seg 3 = 0.500 inch.

consist of a truss or a truss and lateral support, as shown in **Fig 21.74**.

A boom length of 24 ft is about the point where a three-inch diameter begins to be very worthwhile. This dimension provides a considerable amount of improvement in overall mechanical stability as well as increased clamping surface area for element hardware. The latter is extremely important to prevent rotation of elements around the boom if heavy icing is commonplace. Pinning an element to the boom with a large bolt helps in this regard. On smaller diameter booms, however, the elements sometimes work loose and tend to elongate the pinning holes in both the element and the boom. After some time the elements shift their positions slightly (sometimes from day to day) and give a ragged appearance to the system, even though this may not harm the electrical performance.

A 3-inch-diameter boom with a wall thickness of 0.065 inch is very satisfactory for antennas up to about a five-element, 20 meter array that is spaced on a 40-ft boom. A truss is recommended for any boom longer than 24 ft. One possible source for large boom material is irrigation tubing sold at farm supply houses.

PUTTING IT TOGETHER

Once you assemble the boom and elements, the next step is to fasten the elements to the boom securely and then fasten the boom to the mast or supporting structure using mounting plates as shown in **Fig 21.75**. Be sure to leave plenty of material on either side of the U-bolt holes on the element-to-boom mounting plates. The U-bolts selected should be a snug fit for the tubing. If possible, buy muffler-clamp U-bolts that come with saddles.

The *boom-to-mast* plate shown in Fig 21.75B is similar to the *boom-to-element* plate in 21.75A. The size of the plate and number of U-bolts used will depend on the size of the antenna. Generally, antennas for the bands up through 20 meters require only two U-bolts each for the mast and boom. Longer antennas for 15 and 20 meters (35-ft booms and up) and most 40 meter beams should have four U-bolts each for the boom and mast because of the torque that the long booms and elements exert as the antennas move in the wind. When tightening the U-bolts, be careful not to crush the tubing. Once the wall begins to collapse, the connection begins to weaken. Many aluminum suppliers sell ¼-inch or ⅜-inch thick plates just right for this application. Often they will shear pieces to the correct size on request. As with tubing, the relatively hard 6061-T6 grade is a good choice for mounting plates.

The antenna should be put together with good-quality hardware. Stainless steel is best for long life. Rust will attack plated steel hardware after a short while, making nuts difficult, if not impossible, to remove. If stainless muffler clamps are not available, the next best thing is to have them plated. If you can't get them plated, then at least paint them with a good zinc-chromate primer and a finish coat or two. Good-quality hardware is more expensive initially, but if you do it

Table 21.13
15 Meter Optimized Yagi Designs

	Spacing Between Elements (in)	Seg 1 Length (in)	Seg 2 Length (in)	Seg 3 Length (in)	Seg 4 Length (in)	Midband Gain F/R
315-12						
Refl	0	30	36	18	61.375	7.6 dBi
DE	48	30	36	18	49.625	25.5 dB
Dir 1	92	30	36	18	43.500	
415-18						
Refl	0	30	36	18	59.750	8.3 dBi
DE	56	30	36	18	50.875	31.2 dB
Dir 1	56	30	36	18	48.000	
Dir 2	98	30	36	18	36.625	
515-24						
Refl	0	30	36	18	62.000	9.4 dBi
DE	48	30	36	18	52.375	25.8 dB
Dir 1	48	30	36	18	47.875	
Dir 2	52	30	36	18	47.000	
Dir 3	134	30	36	18	41.000	

Note: For all antennas, the tube diameters (in inches) are: Seg 1 = 0.875, Seg 2 = 0.750, Seg 3 = 0.625, Seg 4 = 0.500.

Table 21.14
17 Meter Optimized Yagi Designs

	Spacing Between Elements (in)	Seg 1 Length (in)	Seg 2 Length (in)	Seg 3 Length (in)	Seg 4 Length (in)	Seg 5 Length (in)	Midband Gain F/R
317-14							
Refl	0	24	24	36	24	60.125	8.1 dBi
DE	65	24	24	36	24	52.625	24.3 dB
Dir 1	97	24	24	36	24	48.500	
417-20							
Refl	0	24	24	36	24	61.500	8.5 dBi
DE	48	24	24	36	24	54.250	27.7 dB
Dir 1	48	24	24	36	24	52.625	
Dir 2	138	24	24	36	24	40.500	

Note: For all antennas, tube diameters (inches) are: Seg 1=1.000, Seg 2=0.875, Seg 3=0.750, Seg 4=0.625, Seg 5=0.500.

Table 21.15
20 Meter Optimized Yagi Designs

	Spacing Between Elements (in.)	Seg 1 Length (in.)	Seg 2 Length (in.)	Seg 3 Length (in.)	Seg 4 Length (in.)	Seg 5 Length (in.)	Seg 6 Length (in.)	Midband Gain F/R
320-16								
Refl	0	48	24	20	42	20	69.625	7.3 dBi
DE	80	48	24	20	42	20	51.250	23.4 dB
Dir 1	106	48	24	20	42	20	42.625	
420-26								
Refl	0	48	24	20	42	20	65.625	8.6 dBi
DE	72	48	24	20	42	20	53.375	23.4 dB
Dir 1	60	48	24	20	42	20	51.750	
Dir 2	174	48	24	20	42	20	38.625	

Note: For all antennas, tube diameters (inches) are: Seg 1=1.000, Seg 2=0.875, Seg 3=0.750, Seg 4=0.625, Seg 5=0.500. Seg 6=0.375.

right the first time, you won't have to take the antenna down after a few years and replace the hardware. Also, when repairing or modifying an installation, nothing is more frustrating than fighting rusty hardware at the top of a tower. Stainless steel hardware can also develop surface defects called *galling* that can cause threads on nuts and bolts to seize. On hardware ¼-inch and larger, the use of an anti-seize compound is recommended.

Project: Family of Computer-Optimized HF Yagis

Yagi designers are now able to take advantage of powerful personal computers and software to optimize their designs for the parameters of gain, F/R and SWR across frequency bands. Dean Straw, N6BV, has designed a family of Yagis for HF bands. These can be found in **Tables 22.11**, **22.12**, **22.13**, **22.14** and **22.15**, for the 10, 12, 15, 17 and 20 meter amateur bands, respectively.

For 12 through 20 meters, each design has been optimized for better than 20 dB F/R, and an SWR of less than 2:1 across the entire amateur frequency band. For the 10 meter band, the designs were optimized for the lower 800 kHz of the band, from 28.0 to 28.8 MHz. Each Yagi element is made of telescoping 6061-T6 aluminum tubing, with 0.058 inch thick walls. This type of element can be telescoped easily, using techniques shown in Fig 21.73. Measuring each element to an accuracy of ⅛ inch results in performance remarkably consistent with the computations, without any need for tweaking or fine-tuning when the Yagi is on the tower.

The dimensions shown are designed for specific telescoping aluminum elements, but the elements may be scaled to different sizes by using the information about tapering and scaling in *The ARRL Antenna Book*, although with a likelihood of deterioration in performance over the whole frequency band.

Each element is mounted above the boom with a heavy rectangular aluminum boom-to-element plate, by means of galvanized U-bolts with saddles, as shown in Fig 21.75. This method of element mounting is rugged and stable, and because the element is mounted away from the boom, the amount of element detuning due to the presence of the boom is minimal. The element dimensions given in each table already take into account any element detuning due to the mounting plate. The element mounting plate for all the 10 meter Yagis is a 0.250-inch thick flat aluminum plate, 4 inches wide by 4 inches long. For the 12 and 15 meter Yagis, a 0.375-inch thick flat aluminum plate, 5 inches wide by 6 inches long is used, and for the 17 and 20 meter Yagis, a 0.375-inch thick flat aluminum plate, 6 inches wide by 8 inches long is used. Where the plate is rectangular, the long dimension is in line with the element.

Each design table shows the dimensions for *one-half* of each element, mounted on one side of the boom. The other half of each element is the same, mounted on the other side of the boom. Use a tubing sleeve inside the center portion of the element so that the element is not crushed by the mounting U-bolts. Each telescoping section is inserted 3 inches into the next size of tubing. For example, in the 310-08 design for 10 meters (3 elements on an 8-ft boom), the reflector tip, made out of ½-inch OD tubing, sticks out 66.75 inches from the ⅝-inch OD tubing. For each 10 meter element, the overall length of each ⅝-inch OD piece of tubing is 21 inches, before insertion into the ¾-inch piece. Since the ¾-inch OD tubing is 24 inches long on each side of the boom, the center portion of each element is actually 48 inches of uncut ¾-inch OD tubing.

The boom for all these antennas should be constructed with at least 2-inch-OD tubing, with 0.065-inch wall thickness. Because each boom has three inches of extra length at each end, the reflector is actually placed three inches from one end of the boom. For the 310-08 design, the driven element is placed 36 inches ahead of the reflector, and the director is placed 54 inches ahead of the driven element. The antenna is attached to the mast with the *boom-to-mast* mounting plate shown in Fig 21.74.

Each antenna is designed with a driven element length appropriate for a gamma or T matching network, as shown in **Fig 21.76**. The variable gamma or T capacitors can be housed in small plastic enclosures for weatherproofing; receiving-type variable capacitors with close plate spacing can be used at powers up to a few hundred watts. Maximum capacitance required is usually 140 pF at 14 MHz and proportionally less at the higher frequencies.

The driven-element's length may require slight readjustment for best match, particularly if a different matching network is used. *Do not change either the lengths or the telescoping tubing schedule of the parasitic elements* — they have been optimized for best performance and will not be affected by tuning of the driven element.

TUNING ADJUSTMENTS

To tune the gamma match, adjust the gamma capacitor for best SWR, then adjust the position of the shorting strap or bar that connects the gamma rod to the driven element. Repeat this alternating sequence of adjustments until a satisfactory SWR is reached.

To tune the T-match, the position of the shorting straps and C1 and C2 are adjusted alternately for a best SWR. To maintain balance of the antenna, the position of the straps and capacitor settings should be the same for each side and adjusted together. A coaxial 4:1 balun transformer is shown at 21.76C. A toroidal balun can be used in place of the coax model shown. The toroidal version has a broader frequency range than the coaxial one. The T match is adjusted for 200 Ω and the balun steps this balanced value down to 50 Ω, unbalanced. Or the T match can be set for 300 Ω, and the balun used to step this down to 75 Ω unbalanced.

Dimensions for the gamma and T match rods will depend on the tubing size used, and the spacing of the parasitic elements of the beam. Capacitors C1 and C2 can be 140 pF for 14-MHz beams. Somewhat less capacitance will be needed at 21 and 28 MHz.

Preliminary matching adjustments can be done on the ground. The beam should be aligned vertically so that the reflector element is closest to and a few feet off the ground, with the beam pointing upward. The matching system is then adjusted for best SWR. When the antenna is raised to its operating height, only slight touch-up of the matching network may be required.

A *choke balun* (see the **Transmission Lines** chapter) should be used to isolate the coaxial feed line shield from the antenna. Secure the feed line to the boom of the antenna between the feed point and the supporting mast.

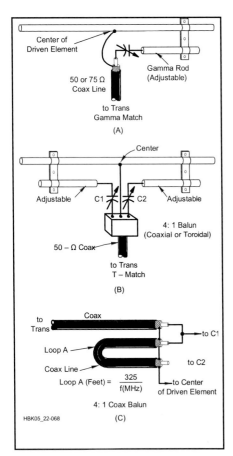

Fig 21.76 — Illustrations of gamma and T matching systems. At A, the gamma rod is adjusted along with the capacitor until the lowest SWR is obtained. A T match is shown at B. It is the same as two gamma-match rods.

21.7 Quad and Loop Antennas

One of the more effective DX antennas is the *quad*. It consists of two or more loops of wire, each supported by a bamboo or fiberglass cross-arm assembly. The loops are ¼-λ per side (one full wavelength overall). One loop is driven and the other serves as a parasitic element — usually a reflector. The design of the quad is similar to that of the Yagi, except that the elements are loops instead of dipoles. A two-element quad can achieve better F/R, gain and SWR across a band, at the expense of greater mechanical complexity compared to a two-element Yagi and very nearly the same performance as a three-element Yagi. A type of Yagi called the *quagi* has also been constructed with a quad element as the driven element. The larger quad driven element results in somewhat better SWR bandwidth, but gain and F/B are approximately the same as on regular Yagi antennas.

A variation of the quad is called the *delta loop*. The electrical properties of both antennas are the same. Both antennas are shown in **Fig 21.77**. They differ mainly in their physical properties, one being of plumber's delight construction, while the other uses insulating support members. One or more directors can be added to either antenna if additional gain and directivity are desired, though most operators use the two-element arrangement.

It is possible to interlace quads or deltas for two or more bands, but if this is done the lengths calculated using the formulas given in Fig 21.77 may have to be changed slightly to compensate for the proximity effect of the second antenna. Using a tuning capacitor as shown in the following project allows the antenna to be adjusted for peak performance without cumbersome adjustment of wire lengths.

If multiple arrays are used, each antenna should be tuned separately for maximum forward gain, or best front-to-rear ratio, as observed on a field-strength meter. The reflector stub on the quad should be adjusted for this condition. The resonance of the antenna can be found by checking the frequency at which the lowest SWR occurs. By lengthening or shortening it, the driven element length can be adjusted for resonance in the most-used portion of the band.

A gamma match can be used at the feed point of the driven element to match the impedance to that of coaxial cable. Because the loop's feed point impedance is *higher* than that of 50-Ω coaxial cable, a *synchronous transmission line transformer* or *Q-section* (see the **Transmission Lines** chapter) with an impedance intermediate to that of the loop and the coaxial cable can be used.

Project: Five-Band, Two-Element HF Quad

Two quad designs are described in this article, both nearly identical. One was constructed by KC6T from scratch, and the other was built by Al Doig, W6NBH, using modified commercial triband quad hardware. The principles of construction and adjustment are the same for both models, and the performance results are also essentially identical. One of the main advantages of this design is the ease of (relatively) independent performance adjustments for each of the five bands. These quads were described by William A. Stein, KC6T, in *QST* for April 1992. Both models use 8-ft-long, 2-inch diameter booms, and conventional X-shaped spreaders (with two sides of each quad loop parallel to the ground).

These designs can also be simplified to monoband quads by using the formulas in Fig 21.77 for loop dimensions and spacing. It is recommended to the antenna builder unfamiliar with quads that a monoband quad be attempted first in order to become acquainted with the techniques of building a quad. Once comfortable with constructing and erecting the quad, success with a multiband design is much easier to achieve.

THE FIVE-BAND QUAD AS A SYSTEM

Unless you are extraordinarily lucky, you should remember one general rule: Any quad must be adjusted for maximum performance after assembly. Simple quad designs can be tuned by pruning and restringing the elements to control front-to-rear ratio and SWR at the desired operating frequency. Since each element of this quad contains five concentric loops, this adjustment method could lead to a nervous breakdown!

Fig 21.78 shows that the reflectors and

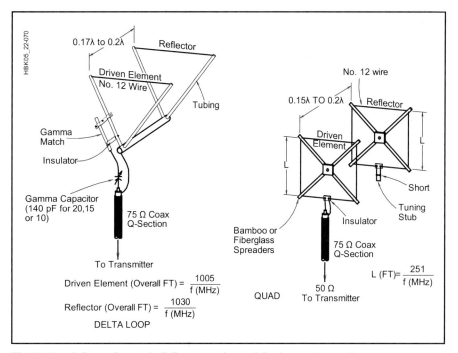

Fig 21.77 — Information on building a quad or a delta-loop antenna. The antennas are electrically similar, but the delta-loop uses plumber's delight construction. The λ/4 length of 75-Ω coax acts as a synchronous transmission-line transformer from approximate 100-Ω feed point impedance of quad to the 50-Ω feed line.

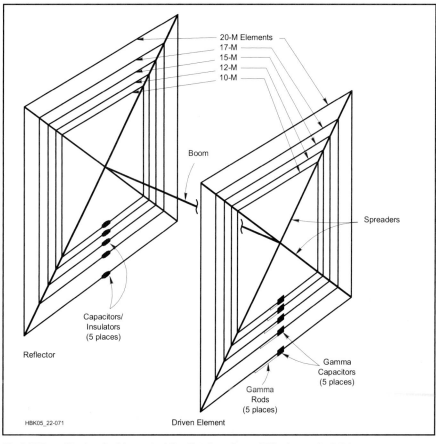

Fig 21.78 — Mechanical layout of the five-band quad. The boom is 8 ft long; see Table 21.16 for all other dimensions.

driven elements are each independently adjustable. After assembly, adjustment is simple, and although gamma-match components on the driven element and capacitors on the reflectors add to the antenna's parts count, physical construction is not difficult. The reflector elements are purposely cut slightly long (except for the 10 meter reflector), and electrically shortened by a tuning capacitor. The driven-element gamma matches set the lowest SWR at the desired operating frequency.

As with most multiband directive antennas, the designer can optimize any two of the following three attributes at the expense of the third: forward gain, front-to-rear ratio and bandwidth (where the SWR is less than 2:1). These three characteristics are related, and changing one changes the other two. The basic idea behind this quad design is to permit (without resorting to trimming loop lengths, spacing or other gross mechanical adjustments):

• The forward gain, bandwidth and front-to-rear ratio may be set by a simple adjustment after assembly. The adjustments can be made on a band-by-band basis, with little or no effect on previously made adjustments on the other bands.

• Setting the minimum SWR in any portion of each band, with no interaction with previously made front-to-back or SWR adjustments.

The first of the two antennas described, the KC6T model, uses aluminum spreaders with PVC insulators at the element attachment points. (The author elected not to use fiberglass spreaders because of their high cost.) The second antenna, the W6NBH model, provides dimensions and adjustment values for the same antenna, but using standard tri-band quad fiberglass spreaders and hardware. If you have a tri-band quad, you can easily adapt it to this design. When W6NBH built his antenna, he had to shorten the 20 meter reflector because the KC6T model uses a larger 20 meter reflector than W6NBH's fiberglass spreaders would allow. Performance is essentially identical for both models.

MECHANICAL CONSIDERATIONS

Even the best electrical design has no value if its mechanical construction is lacking. Here are some of the things that contribute to mechanical strength: The gamma-match capacitor KC6T used was a small, air-variable, chassis-mount capacitor mounted in a plastic box (see **Fig 21.79**). A male UHF connector was mounted to the box, along with a screw terminal for connection to the gamma rod. The terminal lug and wire are for later connection to the driven element. The box came from a local hobby shop, and the box lid was replaced with a piece of $1/32$-inch ABS plastic, glued in place after the capacitor, connector and wiring had been installed. The capacitor can be adjusted with a screwdriver through an access hole. Small vent (drain) holes were drilled near corresponding corners of each end.

Enclose the gamma-match capacitor in such a manner that you can tape unwanted openings closed so that moisture can't be directly blown in during wind and rainstorms. Also, smaller boxes and sturdy mounts to the driven element ensure that you won't pick up gamma capacitor assemblies along with the leaves after a wind storm.

Plastic gamma-rod insulators/standoffs were made from $1/32$-inch ABS, cut $1/2$-inch wide with a hole at each end. Use a knife to cut from the hole to the side of each insulator so that one end can be slipped over the driven element and the other over the gamma rod. Use about four such insulators for each gamma rod, and mount the first insulator as close to the capacitor box as possible. Apply five-minute epoxy to the element and gamma rod at the insulator hole to keep the insulators from sliding. If you intend to experiment with gamma-rod length, perform this gluing operation after you have made the final gamma-rod adjustments.

ELEMENT INSULATORS

As shown in Fig 21.78, the quad uses insulators in the reflectors for each band to break the loop electrically, and to allow reflector adjustments. Similar insulators were used to

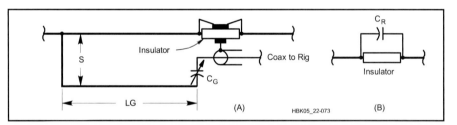

Fig 21.80 — Gamma-match construction details at A and reflector-tuning capacitor (C_R) attachment schematic at B. The gamma matches consist of matching wires (one per band) with series capacitors (C_g). See Table 21.16 for lengths and component specifications.

Fig 21.79 — Photo of one of the feed point gamma-match capacitors.

Table 21.16
Element Lengths and Gamma-Match Specifications of the KC6T and W6NBH Five-Band Quads

KC6T Model

Band (MHz)	Driven Element Length (in)	Gamma Match Length (in)	Gamma Match Spacing (in)	Gamma Match C_g (pF)	Reflector Length (in)	C_R (pF)
14	851.2	33	2	125	902.4	68
18	665.6	24	2	110	705.6	47
21	568	24	1.5	90	604.8	43
24.9	483.2	29.75	1	56	514.4	33
28	421.6	26.5	1	52	448.8	(jumper)

W6NBH Model

Band (MHz)	Driven Element Length (in)	Gamma Match Length (in)	Gamma Match Spacing (in)	Gamma Match C_g (pF)	Reflector Length (in)	C_R (pF)
14	851.2	31	2	117	890.4	120
18	665.6	21	2	114	705.6	56
21	568	26	1.5	69	604.8	58
24.9	483.2	15	1	75.5	514.4	54
28	421.6	18	1	41	448.8	(jumper)

break up each driven element so that element impedance measurements could be made with a noise bridge. After the impedance measurements, the driven-element loops are closed again. The insulators are made from ¼ × 2 × ¾-inch phenolic stock. The holes are ½-inch apart. Two terminal lugs (shorted together at the center hole) are used in each driven element. They offer a convenient way to open the loops by removing one screw. **Fig 21.80** shows these insulators and the gamma-match construction schematically. **Table 21.16** lists the component values, element lengths and gamma-match dimensions.

ELEMENT-TO-SPREADER ATTACHMENT

Probably the most common problem with quad antennas is wire breakage at the element-to-spreader attachment points. There are a number of functional attachment methods; **Fig 21.81** shows one of them. The attachment method with both KC6T and W6NBH spreaders is the same, even though the spreader constructions differ. The KC6T model uses #14 AWG, 7-strand copper wire; W6NBH used #18 AWG, 7-strand wire. At the point of element attachment (see **Fig 21.82**), drill a hole through both walls of the spreader using a #44 (0.086-inch) drill. Feed a 24-inch-long piece of antenna wire through the hole and center it for use as an attachment wire.

After fabricating the spider/spreader assembly, lay the completed assembly on a flat surface and cut the element to be installed to the correct length, starting with the 10 meter element. Attach the element ends to the insulators to form a closed loop before attaching the elements to the spreaders. Center the insulator between the spreaders on what will become the bottom side of the quad loop, then carefully measure and mark the element-mounting-points with fingernail polish (or a similar substance). Do *not* depend on the at-rest position of the spreaders to guarantee that the mounting points will all be correct.

Holding the mark at the centerline of the spreader, tightly loop the attachment wire around the element and then gradually space out the attachment-wire turns as shown. The attachment wire need not be soldered to the element. The graduated turn spacing minimizes the likelihood that the element wire will flex in the same place with each gust of wind, thus reducing fatigue-induced wire breakage.

FEEDING THE DRIVEN ELEMENTS

Each driven element is fed separately, but feeding five separate feed lines down the tower and into the shack would be costly and mechanically difficult. The ends of each of these coax lines also require support other than the tension (or lack of thereof) provided by the driven element at the feed point. It is best to use a remote coax switch on the boom approximately 1 ft from the driven-element spider-assembly attachment point.

If the gamma match is used, the cables connecting the gamma-match capacitors and the coax switch help support the driven elements and gamma capacitors. The support can be improved by taping the cables together in several places. A single coaxial feed line (and a control cable from the remote coax switch, if yours requires one) is the only required cabling from the antenna to the shack.

If the synchronous transmission line transformer is used, the ¼-wavelength of cable required is of convenient lengths that allow a remote antenna switch to be mounted on the antenna boom and the matching sections connected between the switch and driven elements. The drawback of the matching section technique is that it is not easily adjustable.

THE KC6T MODEL'S COMPOSITE SPREADERS

If you live in an area with little or no wind, spreaders made from wood or PVC are practical, but if you live where winds can reach 60 to 80 mi/h, strong, lightweight spreaders are a must. Spreaders constructed with electrical conductors (in this case, aluminum tubing) can cause a myriad of problems with unwanted resonances, and the problem gets worse as the number of bands increases.

To avoid these problems, this version uses composite spreaders made from machined PVC insulators at the element-attachment points. Aluminum tubing is inserted into (or over) the insulators 2 inches on each end. This spreader is designed to withstand 80 mi/h winds. The overall insulator length is

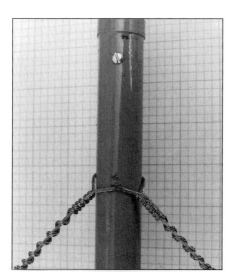

Fig 21.81 — Attaching quad wires to the spreaders must minimize stress on the wires for best reliability. This method (described in the text) cuts the chances of wind-induced wire breakage by distributing stress.

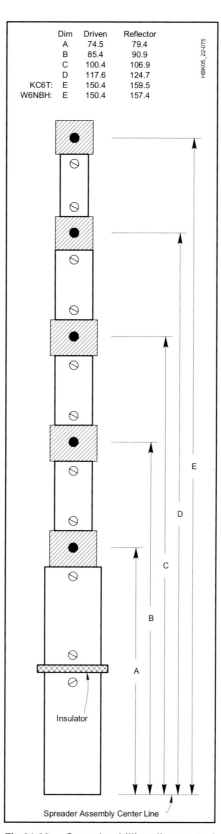

Dim	Driven	Reflector
A	74.5	79.4
B	85.4	90.9
C	100.4	106.9
D	117.6	124.7
KC6T: E	150.4	159.5
W6NBH: E	150.4	157.4

Fig 21.82 — Spreader-drilling diagram and dimensions (in.) for the five-band quad. These dimensions apply to both spreader designs described in the text, except that most commercial spreaders are only a bit over 13 ft (156 inches) long. This requires compensation for the W6NBH model's shorter 20 meter reflector as described in the text.

Antennas 21.45

designed to provide a 3-inch center insulator clear of the aluminum tubing. The aluminum tubing used for the 10 meter section (inside dimension "A" in Fig 21.82) is 1⅛-inch diameter × 0.058-inch wall. The next three sections are ¾-inch diameter × 0.035-inch wall, and the outer length is made from ½-inch diameter × 0.035-inch wall. The dimensions shown in Fig 21.82 are *attachment point* dimensions only.

Attach the insulators to the aluminum using #6 sheet metal screws. Mechanical strength is provided by Devcon no. S 220 Plastic Welder Glue (or equivalent) applied liberally as the aluminum and plastic parts are joined. Paint the PVC insulators before mounting the elements to them. Paint protects the PVC from the harmful effects of solar radiation. As you can see from Fig 21.82, an additional spreader insulator located about halfway up the 10 meter section (inside dimension "A") removes one of the structure's electrical resonances not eliminated by the attachment-point insulators. Because it mounts at a relatively high-stress point in the spreader, this insulator is fabricated from a length of heavy-wall fiberglass tubing.

Composite spreaders work as well as fiberglass spreaders, but require access to a well-equipped shop, including a lathe. The main objective of presenting the composite spreader is to show that fiberglass spreaders aren't a basic requirement — there are many other ways to construct usable spreaders. If you can lay your hands on a used multiband quad, even one that's damaged, you can probably obtain enough spreaders to reduce construction costs considerably.

GAMMA ROD

The gamma rod is made from a length of #12 AWG solid copper wire (W6NBH used #18 AWG, 7-strand wire). Dimensions and spacings are shown in Table 21.16. If you intend to experiment with gamma-rod lengths and capacitor settings, cut the gamma-rod lengths about 12 inches longer than the length listed in the table. Fabricate a sliding short by soldering two small alligator clips back-to-back such that they can be clipped to the rod and the antenna element and easily moved along the driven element. Note that gamma-rod spacing varies from one band to another. When you find a suitable shorting-clip position, mark the gamma rod, remove the clip, bend the gamma rod at the mark and solder the end to the element.

THE W6NBH MODEL

As previously mentioned, this model uses standard 13-ft fiberglass spreaders, which aren't quite long enough to support the larger 20 meter reflector specified for the KC6T model. The 20 meter W6NBH reflector loop is cut to the dimensions shown in Table 21.16, 12 inches shorter than that for the KC6T model. To tune the shorter reflector, a six-inch-long stub of antenna wire (conductors spaced two inches) hangs from the reflector insulator, and the reflector tuning capacitor mounts on another insulator at the end of this stub.

GAMMA-MATCH AND REFLECTOR-TUNING CAPACITOR

Use an air-variable capacitor of your choice for each gamma match. Approximately 300 V can appear across this capacitor (at 1500 W), so choose plate spacing appropriately. If you want to adjust the capacitor for best match and then replace it with a fixed capacitance, remember that several amperes of RF will flow through the capacitance. If you choose disc-ceramic capacitors, use a parallel combination of at least four 1-kV units of equal value. Any temperature coefficient is acceptable. NP0 units are not required. Use similar components to tune the reflector elements.

ADJUSTMENTS

Well, here you are with about 605 ft of wire. Your antenna will weigh about 45 pounds (the W6NBH version is slightly lighter) and have about nine square ft of wind area. If you chose to, you can use the dimensions and capacitance values given, and performance should be excellent. If you adjust the antenna for minimum SWR at the band centers, it should cover all of the lower four bands and 28 to 29 MHz with SWRs under 2:1; front-to-back ratios are given in **Table 21.17**.

Instead of building the quad to the dimensions listed and hoping for the best, you can adjust your antenna to account for most of the electrical environment variables of your installation. The adjustments are conceptually simple: First adjust the reflector's electrical length for maximum front-to-rear ratio (if you desire good gain, and are willing to settle for a narrower than maximum SWR bandwidth), or accept some compromise in front-to-rear ratio that results in the widest SWR bandwidth. You can make this adjustment by placing an air-variable capacitor (about 100-pF maximum) across the open reflector loop ends, one band at a time, and adjusting the capacitor for the desired front-to-rear ratio. The means of doing this will be discussed later.

During these reflector adjustments, the driven-element gamma-match capacitors may be set to any value and the gamma rods may be any convenient length (but the sliding-short alligator clips should be installed somewhere near the lengths specified in Table 21.16). After completing the front-to-rear adjustments, the gamma capacitors and rods are adjusted for minimum SWR at the desired frequency.

ADJUSTMENT SPECIFICS

Make a calibrated variable capacitor (with a hand-drawn scale and wire pointer). Calibrate the capacitor using your receiver, a known-value inductor and a grid-dip meter (plus a little calculation) or SWR analyzer.

Adjust each band by feeding it separately if the gamma match technique is used. If transmission line matching sections are used, they must all be connected to the remote switch when adjusting the antenna because each unused section acts as a short coaxial stub, adding reactance at the connection to the antenna.

To adjust front-to-back ratio, simply clip the (calibrated) air-variable capacitor across the open ends of the desired reflector loop. Connect the antenna to a portable receiver with an S meter. Point the back of the quad at a signal source, and slowly adjust the capacitor for a dip in the S-meter reading.

After completing the front-to-back adjustments, replace the variable capacitor with an appropriate fixed capacitor and seal the connections against the weather. Then move to the driven-element adjustments. Connect the coax through the SWR bridge to the 10 meter gamma-match capacitor box. Use an SWR bridge that requires only a watt or two (not more than 10 W) for full-scale deflection in the calibrate position on 10 meters. Using the minimum necessary power, measure the SWR. Go back to receive and adjust the capacitor until (after a number of transmit/receive cycles) you find the minimum SWR. If it is too high, lengthen or shorten the gamma rod by means of the sliding alligator-clip short and make the measurements again.

Stand away from the antenna when making transmitter-on measurements. The adjustments have minimal effect on the previously made front-to-rear settings, and may be made in any band order. After making all the adjustments and sealing the gamma capacitors, reconnect the coax harness to the remote coax switch.

Adjusting the SWR when using transmission-line matching sections requires first measuring the impedance of each loop at the feed point. To avoid standing next to the antenna while making the measurement, connect the test equipment to the antenna using a ½-λ piece of transmission line of any

Table 21.17
Measured Front-to-Back Ratios

Band	KC6T Model	W6NBH Model
14	25 dB	16 dB
18	15 dB	10 dB
21	25 dB	>20 dB
24.9	20 dB	>20 dB
28	20 dB	>20 dB

characteristic impedance. Cut a ¼-λ section of transmission line with an appropriate impedance a few percent longer than the exact value. Attach the matching section to the loop feed point. Measure the resulting impedance at the output of the matching section and trim its length to place the minimum SWR point at the desired frequency. Using this technique, it is not likely that an SWR of 1:1 can be obtained, but values below 1.5:1 should be attainable.

Project: Wire Quad for 40 Meters

Many amateurs yearn for a 40 meter antenna with more gain than a simple dipole. While two-element rotatable 40 meter beams are available commercially, they are costly and require fairly hefty rotators to turn them. This low-cost, single-direction quad is simple enough for a quick Field Day installation, but will also make a home station very competitive on the 40 meter band.

This quad uses a two-inch outside diameter, 18-ft boom, which should be mounted no less than 60 ft high, preferably higher. (Performance tradeoffs with height above ground will be discussed later.) The basic design is derived from the N6BV 75/80 meter quad described in *The ARRL Antenna Compendium, Vol 5*. However, since this simplified 40 meter version is unidirectional and since it covers only one portion of the band (CW or Phone, but not both), all the relay-switched components used in the larger design have been eliminated.

While this antenna is shown as tower-supported with a metal boom, the same antenna can be suspended between trees or towers. The boom is then replaced by a rope stretched between the supports with the tops of each loop attached to the rope with an insulator.

The layout of the simple 40 meter quad at a boom height of 70 ft is shown in **Fig 21.83**. The wires for each element are pulled out sideways from the boom with black ⅛-inch Dacron rope designed specifically to withstand both abrasion and UV radiation. The use of the proper type of rope is very important — using a cheap substitute is not a good idea. You will not enjoy trying to retrieve wires that have become, like Charlie Brown's kite, hopelessly entangled in nearby trees, all because a cheap rope broke during a windstorm! At a boom height of 70 ft, the quad requires a *wingspread* of 140 ft for the side ropes. This is the same wingspread needed by an inverted-V dipole at the same apex height with a 90° included angle between the two legs.

The shape of each loop is rather unusual, since the bottom ends of each element are brought back close to the supporting tower. (These element ends are insulated from the tower and from each other). Having the elements near the tower makes fine-tuning adjustments much easier — after all, the ends of the loop wires are not 9 feet out, on the ends of the boom! The feed point resistance with this loop configuration is close to 50 Ω, meaning that no matching network is necessary. By contrast, a more conventional diamond or square quad-loop configuration exhibits about a 100-Ω resistance.

Another bonus to this loop configuration is that the average height above ground is higher, leading to a slightly lower angle of

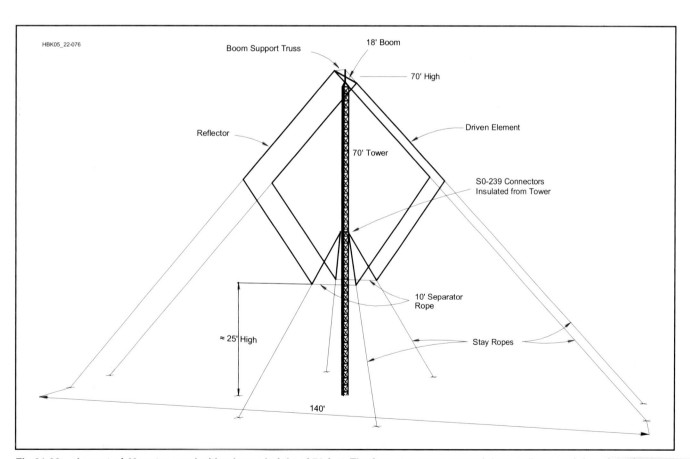

Fig 21.83 — Layout of 40 meter quad with a boom height of 70 feet. The four stay ropes on each loop pull out each loop into the desired shape. Note the 10-foot separator rope at the bottom of each loop, which helps it hold its shape. The feed line is attached to the driven element through a choke balun, consisting of 10 turns of coax in a 1-foot diameter loop. You could also use large ferrite beads over the feed line coax, as explained in the Transmission Lines chapter. Both the driven element and reflector loops are terminated in SO-239 connectors tied back to (but insulated from) the tower. The reflector SO-239 has a shorted PL-259 normally installed in it. This is removed during fine-tuning of the quad, as explained in the text.

radiation for the array and less loss because the bottom of each element is raised higher above lossy ground. The drawback to this unusual layout is that four more *tag-line* stay ropes are necessary to pull the elements out sideways at the bottom, pulling against the 10-foot separator ropes shown in Fig 21.83.

CONSTRUCTION

You must decide before construction whether you want coverage on CW (centered on 7050 kHz) or on phone (centered on 7225 kHz), with roughly 120 kHz of coverage between the 2:1 SWR points. If the quad is cut for the CW portion of the band, it will have less than about a 3.5:1 SWR at 7300 kHz, as shown in **Fig 21.84**. The pattern will deteriorate to about a 7 dB F/B at 7300 kHz, with a reduction in gain of almost 3 dB from its peak in the CW band. It is possible to use a quad tuned for CW in the phone band if you use an antenna tuner to reduce the SWR and if you can take the reduction in performance. To put things in perspective, a quad tuned for CW but operated in the phone band will still work about as well as a dipole.

Next, you must decide where you want to point the quad. A DXer or contester in the USA might want to point this single-direction design to cover Europe and North Africa. For Field Day, a group operating on the East Coast would simply point it west, while their counterparts on the West Coast would point theirs east.

The mechanical requirements for the boom are not severe, especially since a top truss support is used to relieve stress on the boom due to the wires pulling on it from below. The boom is 18 ft long, made of two-inch diameter aluminum tubing. You can probably find a suitable boom from a scrapped tri-band or monoband Yagi. You will need a suitable set of U-bolts and a mounting plate to secure the boom to the face of a tower. Or perhaps you might use lag screws to mount the boom temporarily to a suitable tree on Field Day! On a 70-ft high tower, the loop wires are brought back to the tower at the 37.5-ft level and tied there using insulators and rope. The lowest points of the loops are located about 25 ft above ground for a 70-ft tower. **Fig 21.85** gives dimensions for the driven element and reflector for both the CW and the Phone portions of the 40 meter band.

GUY WIRES

Anyone who has worked with quads knows they are definitely three-dimensional objects! You should plan your installation carefully, particularly if the supporting tower has guy wires, as most do. Depending on where the guys are located on the tower and the layout of the quad with reference to those guys, you will probably have to string the quad loops over certain guys (probably at the top of the tower) and under other guys lower down.

It is very useful to view the placement of guy wires using the View Antenna function in the *EZNEC* modeling program. This allows you to visualize the 3-D layout of an antenna. You can Rotate yourself around the tower to view various aspects of the layout. *EZNEC* will complain about grounding wires directly but will still allow you to use the View Antenna function. Note also that it is best to insulate guy wires to prevent interaction between them and the antennas on a tower, but this may not be necessary for all installations.

FINE TUNING, IF NEEDED

We specify stranded #14 AWG hard-drawn copper wire for the elements. During the course of installation, however, the loop wires could possibly be stretched a small amount as you pull and yank on them, trying to clear various obstacles. This may shift the frequency response and the performance slightly, so it is useful to have a tuning procedure for the quad when it is finally up in the air.

The easiest way to fine-tune the quad while on the tower is to use a portable, battery-operated antenna analyzer to adjust the reflector and the driven element lengths for specific resonant frequencies. You can eliminate the influence of mutual coupling to the other element by open-circuiting the other element.

For convenience, each quad loop should be connected to an SO-239 UHF female connector that is insulated from but tied close to the tower. You measure the driven element's resonant frequency by first removing the shorted PL-259 normally inserted into the reflector connector. Similarly, the reflector's resonant frequency can be determined by removing the feed line normally connected to the driven element's feed point.

Obviously, it's easiest if you start out with extra wire for each loop, perhaps six inches extra on each side of the SO-239. You can then cut off wire in ½-inch segments equally on each side of the connector. This procedure is easier than trying to splice extra wire while up on the tower. Alligator clips are useful during this procedure, but just don't lose your hold on the wires! You should tie safety strings from each wire back to the tower. Prune the wire lengths to yield the resonant frequencies (±5 kHz) shown in Fig 21.85 and then solder things securely. Don't forget to reinsert the shorted PL-259 into the reflector SO-239 connector to turn it back into a reflector.

HIGHER IS BETTER

This quad was designed to operate with the boom at least 60 ft high. However, it will work considerably better for DX work if you can put the boom up even higher.

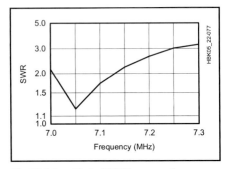

Fig 21.84 — Plot of SWR versus frequency for a quad tuned for CW operation.

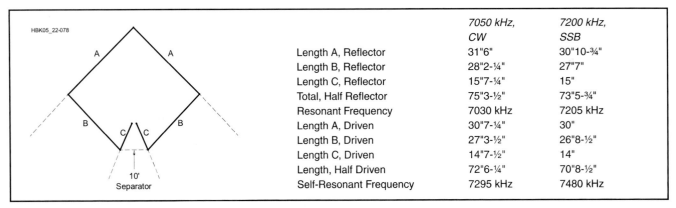

	7050 kHz, CW	7200 kHz, SSB
Length A, Reflector	31"6"	30"10-¾"
Length B, Reflector	28"2-¼"	27"7"
Length C, Reflector	15"7-¼"	15"
Total, Half Reflector	75"3-½"	73"5-¾"
Resonant Frequency	7030 kHz	7205 kHz
Length A, Driven	30"7-¼"	30"
Length B, Driven	27"3-½"	26"8-½"
Length C, Driven	14"7-½"	14"
Length, Half Driven	72"6-¼"	70"8-½"
Self-Resonant Frequency	7295 kHz	7480 kHz

Fig 21.85 — Dimensions of each loop, for CW or Phone operation.

Fig 21.86 shows the elevation patterns for four antennas: a reference inverted-V dipole at 70 ft (with a 90° included angle between the two legs), and three quads, with boom height of 70, 90 and 100 ft respectively. At an elevation angle of 20°, typical for DX work on 40 meters, the quad at 100 ft has about a 5 dB advantage over an inverted-V dipole at 70 ft, and about a 3 dB advantage over a quad with a boom height of 70 ft.

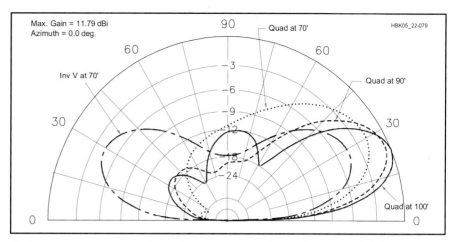

Fig 21.86 — Comparisons of the elevation patterns for quads at boom heights of 70, 90 and 100 ft, referenced to an inverted-V dipole at 70 ft.

21.7.1 Loop Antennas

The loop antennas described in this section are continuous loops at least one wavelength in circumference and formed into open shapes with sides that are approximately equal, such as triangles, diamonds, squares, or circles. Smaller loops used for receiving purposes are discussed in the *ARRL Antenna Book*. Loops with ratios of side lengths greater than 2 or 3:1 begin to have special characteristics beyond the scope of this chapter.

A 1-λ loop can be thought of as two ½-λ dipoles with their ends connected together and pulled apart into an open shape as described above. The feed point of one dipole is replaced with a short circuit so that there is only one feed point on the antenna. As such, the current and voltage distribution around the loop is an extension of Fig 21.1. Three typical loop shapes and the current distributions on them are shown in **Fig 21.87**. Note that the current flow reverses at points ¼-λ to either side of the feed point. That means the current direction opposite the feed point is the same as at the feed point.

The maximum radiation strength of a 1-λ loop is perpendicular to the plane of the loop and minimum in the plane of the loop. If the loop is horizontal, the antenna radiates best straight up and straight down and poorly to the sides. The gain of a 1-λ loop in the direction of maximum radiation is approximately 1 dBd.

If the plane of the three loops shown in Fig 21.87 is vertical, the radiation is horizontally polarized because the fields radiated by the vertical components of current are symmetrical and opposing, so they cancel, leaving only the horizontally polarized fields that reinforce each other perpendicular to the loop plane. If the feed point of the antenna is moved to a vertical side or the antenna is rotated 90°, it is the horizontally polarized fields that will cancel, leaving a vertically polarized field, still maximum perpendicular to the plane of the loop. Feeding the loop at some other location, rotating the loop by some intermediate value, or constructing the loop in an asymmetrical shape will result in polarization somewhere between vertical and horizontal, but the maximum radiation will still occur perpendicular to the plane of the loop.

In contrast to straight-wire antennas, the electrical length of the circumference of a 1-λ loop is shorter than the actual length. For a loop made of bare #18 AWG wire and operating at a frequency of 14 MHz, so that the length-to-diameter ratio is very large, the loop will be close to resonance when:

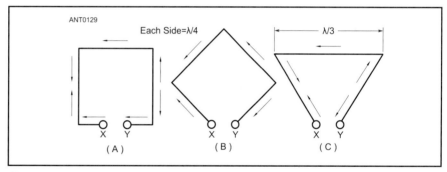

Fig 21.87 — At A and B, loops have sides ¼-λ long, and at C having sides 1/3-λ long for a total conductor length of 1 λ. The polarization depends on the orientation of the loop and on the position of the feed point (terminals X-Y) around the perimeter of the loop.

$$\text{Length (feet)} = \frac{1032}{f(\text{MHz})} \qquad (7)$$

The radiation resistance of a resonant 1-λ loop is approximately 120 Ω under these conditions. Since the loop dimensions are larger than those of a ½-Ω dipole, the radiation efficiency is high and the SWR bandwidth of the antenna significantly larger than for the dipole.

The loop antenna is resonant on all frequencies at which it is an integral number of wavelengths in circumference; f_0, $2f_0$, $3f_0$, etc. That means an 80-meter 1-λ loop will also have a relatively low feed point impedance on 40, 30, 20, 15, 12, and 10 meters. As each side of the loop becomes longer electrically, the radiation pattern of the loop begins to develop nulls perpendicular to the plane of the loop and lobes that are closer to the plane of the loop. A horizontal diamond-shaped loop with legs more than a wavelength long is a *rhombic* antenna and can develop significant gain along the long axis of the antenna. (The diamond-shaped rhombic is the origin of the symbol of the ARRL and many other radio organizations.)

Project: Vertical Loop Antenna for 28 MHz

This simple antenna provides gain over a dipole or inverted-V. Its size makes it an excellent antenna with which to experiment. The antenna can easily be scaled to lower (and higher) frequencies: Multiply each dimension by 28.4 / f (MHz), where f is the desired operating frequency in MHz.

The shape of the loop is such that it develops 2.1 dB gain over a dipole at low radiation angles with the top mounted one wavelength or more above ground. The antenna is simple to feed — no matching network is necessary. When fed with 50-Ω coax, the SWR is close to 1:1 at the design frequency, and is less than 2:1 from 28.0-28.8 MHz for an antenna resonant at 28.4 MHz. (If the loop is scaled to resonate at lower frequencies, the effects of ground will affect the antenna's resonant frequency and feed point impedance, but not drastically — be prepared to adjust the dimensions.)

The antenna is made from #12 AWG wire (see **Fig 21.88**) and is fed at the center of the bottom wire. Coil the coax into a few turns near the feed point to provide a simple choke balun. (see the **Transmission Lines** chapter for more information on choke baluns.) A coil diameter of about a foot will work fine. You can support the antenna on a mast with spreaders made of bamboo, fiberglass, wood, PVC or other non-conducting material. You can also use aluminum tubing both for support and conductors, but you may have to re-adjust the antenna dimensions for resonance.

This rectangular loop has two advantages over a resonant square loop. First, a square loop has just 1.1 dB gain over a dipole — a power increase of only 29%. Second, the input impedance of a square loop is about 125 Ω. You must use a matching network to feed a square loop with 50-Ω coax. The rectangular loop achieves gain by compressing its radiation pattern in the elevation plane. This happens because the top and bottom of the loop (where the current maxima are located) are farther apart than for a square loop. The antenna's beamwidth is slightly higher than that of a dipole (it's about the same as that of an inverted-V). A broad pattern is an advantage for a general-purpose, fixed antenna. The rectangular loop provides a bidirectional gain over a wide range of directions.

Mount the loop as high as possible. To provide 1.7 dB gain at low angles over an inverted-V, the top wire must be at least 30 ft high (about one wavelength). The loop will work at lower heights, but its gain advantage disappears. For example, at 20 ft the loop provides the same gain at low angles as an inverted-V.

Project: Multiband Horizontal Loop Antenna

Along with the multiband, non-resonant dipole, many amateurs operate on HF with great success using a horizontal loop antenna.

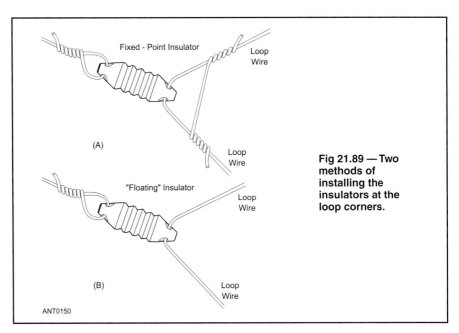

Fig 21.89 — Two methods of installing the insulators at the loop corners.

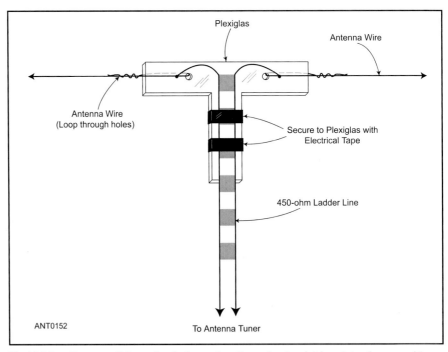

Fig 21.90 — One possible method of constructing a feed point insulator for use with open-wire line.

Fig 21.88 — Construction details of the 10 meter rectangular loop antenna.

All that is required are at least three supports able to hold the corners of the antenna 20 or more feet above the ground (and even that is negotiable) and enough room for a loop of wire one wavelength or more in circumference at the lowest frequency of operation. (Smaller loops can be used with an impedance-matching unit.)

Start by calculating the total length of wire you need using Equation 7. You'll need one insulator for each support and lengths of rope that are at least twice the height to which the insulator will be raised. You can feed the antenna at one of the corner insulators or anywhere along the wire with a separate insulator. Examples of corner insulators are shown in **Fig 21.89**. Using floating insulators allows the wire to move as the antenna flexes. One of the insulators should be of the fixed type, or the antenna can be fed at one corner with the loop wires attached to a pair of insulators sharing a common support rope. This holds the antenna feed point in place.

If the loop is only going to be used on the band for which it is resonant, coaxial cable can be used as the feed line, since SWR will be low. A choke balun at the feed point is recommended. For multiband use, open-wire feed line should be used, with an impedance-matching unit in the shack. A feed point insulator for open-wire line is shown in **Fig 21.90**.

On its fundamental frequency, the antenna's maximum radiation will be straight up, making it most useful for regional communications at high elevation angles with the occasional DX contact. At higher frequencies, the loop will radiate more strongly at lower angles for better signal strengths at long distances.

21.8 HF Mobile Antennas

HF mobile operation has been part of Amateur Radio since the 1930s. Once primarily an HF endeavor, the FM explosion of the 1970s seemed to displace HF mobiling in favor of local repeater and "mag-mount" VHF ground planes. HF mobiling is back, though, in a big way and with the excellent equipment and antennas available, here to stay. Material in this section was contributed and updated by Alan Applegate, KØBG.

High frequency (HF) mobile antennas come in every size and shape imaginable, from simple whips to elaborate, computer-controlled behemoths. Regardless of the type and construction, an HF mobile antenna should have a few important attributes.

1) Sturdiness: It should stay upright at highway speeds with a minimum of sway.
2) Mechanically stable: Sudden stops or sharp turns won't cause it to sway about, endangering others.
3) Flexibly mounted: Permits springing around branches and obstacles at low speeds.
4) Weatherproof: Withstands the effects of wind, rain, snow and ice at high speed.
5) Be tunable to different HF bands without stopping the vehicle if multiband operation is desired.
6) Can be mounted without altering the vehicle or its safety equipment.
7) Be easily removable when required.
8) Be as efficient as possible.

Of all the antenna choices available, the *whip* antenna has passed the test of time as providing all of these attributes in one way or another. The following sections discuss the different types of whips, how they are attached to and interact with the vehicle, and how they are connected to the transmitter.

21.8.1 Simple Whips

The simplest of antennas is a quarter-wave whip, but it's only practical on the upper HF bands due to the required length. For example, a 10 meter quarter-wavelength antenna is about 8 feet long. It doesn't require a loading coil, so its efficiency is close to 90%. The reason efficiency isn't 100% is because of resistive losses in the whip itself, stray capacitance losses in the mounting hardware and ground losses which we'll cover later. The end result is that the feed point impedance at the antenna's base is very close to 50 Ω and a good match to modern solid state transceivers.

WHIP RADIATION RESISTANCE

The power radiated by the antenna is equal to the radiation resistance times the square of the antenna current. The radiation resistance, R_r, of an electrically small antenna is given by:

$$R_r = 395 \times \left(\frac{h}{\lambda}\right)^2 \qquad (8)$$

where
 h = radiator height in meters
 λ = wavelength in meters = 300 / f in MHz

The efficiency of the antenna, η, equals the radiation resistance, R_r, divided by the resistive component of the feed point impedance, R_{fp}, which for actual antennas includes ground losses and losses in the antenna:

$$\eta = \frac{R_r}{R_{fp}} \times 100\% \qquad (9)$$

Since an electrically short antenna has a low radiation resistance, careful attention must be paid to minimizing losses in the antenna system that can greatly reduce the antenna's effectiveness.

21.8.2 Coil-Loaded Whips

As we move lower in frequency, the physical length has to increase, but there is a limit. In most localities, the maximum height at the tip of the antenna needs to be less than 13.5 feet. This generally limits an antenna to 10.5 feet (3.2 meters) for an average installation on a vehicle. In some areas even this is too long, while in others 16 feet isn't a problem. A 10.5-foot antenna is shorter than the resonant length for all HF bands except 10 and 12 meters and thus the feed point impedance will have some capacitive reactance. In order to keep within the 13.5 foot limit, we must add a loading coil to bring the antenna to resonance.

The capacitance in pF of an electrically small antenna is given approximately by:

$$C = \frac{55.78 \times h}{((\text{den1}) \times (\text{den2}))} \qquad (10)$$

where
 (den1) = (ln (h/r) − 1)
 (den2) = (1 − (f × h/75)^2)
 ln = natural logarithm
 r = conductor radius in meters
 f = frequency in MHz

Radiation resistance rises in a nonlinear fashion and the capacitance drops just as dramatically with increase in the ratio h/λ. **Fig 21.91** shows the relationship of capacitance to height for our 10.5 foot antenna. This can be used for estimating antenna capacitance for other heights. **Fig 21.92** shows

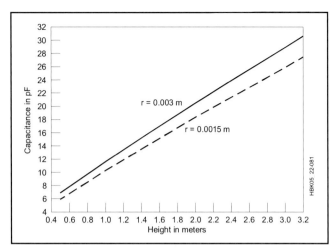

Fig 21.91 — Relationship at 3.5 MHz between vertical radiator length and capacitance. The two curves show that the capacitance is not very sensitive to radiator diameter.

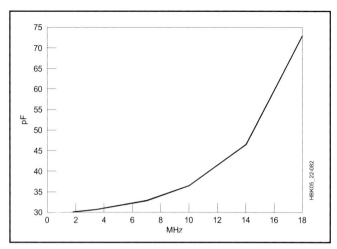

Fig 21.92 — Relationship between frequency and capacitance for a 3.2-meter vertical whip.

Table 21.18
Characteristics of a 10.5-ft Whip Antenna

Freq (MHz)	C (pF)	R_r (Ω)	Impedance (Ω)	Efficiency (%)	L (μH)
1.8	30.1	0.146	13.72 $-j$2716	1.064	240
3.5	30.6	0.55	7.43 $-j$1375	7.4	62.5
7	32.8	2.2	7.04 $-j$644	31.2	14.6
10	36.5	4.5	6.5 $-j$408	69.2	6.49
14	46.5	8.8	10 $-j$232	88	2.64

that capacitance is not very sensitive to frequency for h/λ less than 0.075 which occurs at 8 MHz in this case. However, the sensitivity increases rapidly as frequency is lowered.

The required loading coil can take many forms, and it may be placed almost anywhere along the length of the radiating element. It cancels out the capacitive reactance (–j) by introducing an equal but opposite inductive reactance (+j). Some coils are mounted at the base of the mast — a *base-loaded* antenna — and some are mounted near the center (*center-loaded*) or the top (*top-loaded*).

As the coil is moved higher, the radiation resistance increases (a good thing), but the necessary coil reactance also increases as do resistive losses in the coil. Generally speaking, a center-loaded antenna requires twice the reactance of a base loaded antenna, thus the coil losses also double. Therefore, it becomes a balancing act to choose the optimal location and proper coil location requires a thorough understanding of the parameters involved.

Table 21.18 lists the characteristics of the 10.5 foot, base-loaded antenna with a coil Q of 200, a stray capacitance of 2 pF, a whip diameter of 0.003 meter, and the inductive reactance required to bring the antenna into resonance on 1.8 through 14 MHz. If we move the loading coil to the center, the requisite reactance doubles as do the coil resistive losses. The radiation resistance also increases, but only on that part of the antenna above the loading coil. Factor in the ground losses and the optimal position changes, but it is typically close to the center of the antenna.

It should be noted that stray capacitance mentioned above is primarily due to the mounting hardware. The total amount of stray capacitance may be much higher depending on where and how the antenna is mounted. The higher the stray capacitance, the less efficient the antenna will be.

Loading coil Q is especially important on the lower HF bands where coil losses can exceed ground losses. The factors involved include wire size, wire spacing, length-to-diameter ratio and the materials used in constructing the coil. All of the factors interact with one another, making coil design a compromise — especially when wind loading and weight become major considerations.

Antenna system Q is limited by the Q of the coil. The bandwidth between 2:1 SWR points of the system = 0.36 × f/Q. On 80 meters, the bandwidth of the 10.5 foot whip = 0.36 × 3.5/200 = 6.3 kHz. If we could double the Q of the coil, the efficiency would double and the bandwidth would be halved. The converse is also true. In the interest of efficiency, the highest possible Q should be used!

Another significant factor arises from high Q. Let's assume that we deliver 100 W on 80 meters to the 7.43 Ω at the antenna terminals. The current is 3.67 A and flows through the 1375-Ω reactance of the coil giving rise to 1375 × 3.67= 5046 V_{RMS} (7137 V_{peak}) across the coil! This is a significant voltage and may cause arcing if the coil is wet or dirty.

With only 30.6 pF of antenna capacitance, the presence of significant stray capacitance at the antenna base shunts currents away from the antenna. RG-58 coax presents about 21 pF/foot. A 1.5-foot length of RG-58 would halve the radiation efficiency of our example antenna. For cases like the whip at 3.5 MHz, the matching network has to be right at the antenna!

In general, the larger the coil, the higher the Q. The more mass within the field of the coil (metal end caps for example) the lower the Q. Short, fat coils are better than long, skinny ones. Practical mechanical considerations for coils with reactances above 1000 Ω require the length-to-diameter ratio to increase, lowering Q.

21.8.3 Base vs Center vs Continuous Loading

There are a few important aspects to be kept in mind when selecting or building an HF mobile antenna. As the antenna becomes longer, less loading inductance is required and the coil Q can become higher, improving efficiency. Also, for longer antennas, the better the mounting location has to be in order to optimize efficiency. We'll cover mounting and efficiency later.

Placing the loading coil at the base results in the current distribution shown in **Fig 21.93A**. If we move the coil to the center, the current curve looks like the one in Fig 21.93B. The location of the optimal posi-

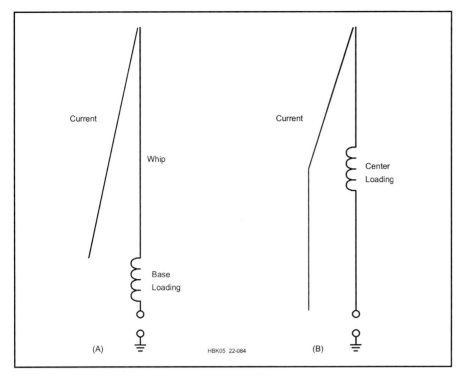

Fig 21.93 — Relative current distribution on a base-loaded antenna is shown at A and for a center-loaded antenna at B.

tion between the two extremes depends on the ground losses, and to a lesser degree on loading coil Q and overall length. For example, if the ground losses were zero, the best position would be at the bottom. As the ground losses increase, the optimal position gets closer to the center. If the ground losses are high enough, the optimal position is in the top one-third of the antenna's length, but efficiency is very poor.

Center-loading increases the current in the lower half of the whip as shown in Fig 21.93B. Capacitance for the section above the coil can be calculated just as for the base-loaded antenna. This permits calculation of the loading inductance. The center-loaded antenna is often operated without any base matching in which case the resistive component can be assumed to be 50 Ω for purposes of calculating the current rating and selecting wire size for the inductor. The reduced size of the top section results in reduced capacitance which requires a much larger loading inductor.

Because of the high value of inductance required for center-loading, high-Q coils are very large. The large wind resistance necessitates a very sturdy mount for operation at highway speed. One manufacturer of this type of coil does not recommend their use in rain or inclement weather. The higher Q of these large coils results in a lower feed point impedance, necessitating the use of a base matching element in the form of either a tapped inductor or a shunt capacitor to match to 50 Ω. Another manufacturer places the coil above the center and uses a small extendable whip or wand for tuning.

To minimize wind resistance, the coil lengths are several times their diameters. These antenna coils are usually close wound with enameled wire. The coils are covered with a heat-shrink tubing sleeve. If used in heavy rain or snow for extended periods water may get under the sleeve and seriously detune and lower the Q of the coils. The lower Q of these coils means that the antennas usually do not require a base matching element as the feed point impedance seems to be enough to 50 Ω.

In recent years, continuously-loaded antennas as shown in **Fig 21.94** have become popular. Proponents often believe the coil radiates thus avoiding the position conundrum; this is a myth as the coil does not radiate. Because the coil's length to diameter ratio is so large, its Q is relatively low. They do have a few attributes which add to their

Fig 21.94 — Continuously-loaded whip antennas are short and lightweight. The base section consists of a fiberglass tube wound with wire to form the loading inductor. At the top of the base section the length of a steel whip or stinger can be adjusted to bring the antenna to resonance. (W1ZR photo)

Table 21.19
HF Mobile Antenna Comparison

Antenna Type	Length	Frequency Coverage	Efficiency	Mounting Difficulty	Matching Required
Simple Whip	< 11 ft	15 m & up	Excellent	Easy	No
Base-Loaded	9 to 10.6 ft	160 - 6 m	Fair to good	Average	Yes
Center-Loaded	9 to 10.5 ft	160 - 6 m	Good to excellent	Average	Yes
Top-Loaded	<9 ft	160 - 6 m	Fair	Average	No
Continuous Loading	< 7 feet	80 - 6 m	Poor to fair	Easy	No
Remote-tuned Small	< 7 feet	80 - 6 m	Poor to fair	Easy	No
Remote-tuned Large	9 to 10.5 ft	160 - 6 m	Excellent	Difficult	Yes

Designing a Base Loading System

This design procedure was contributed by Jack Kuecken, KE2QJ. To begin, estimate the capacitance, capacitive reactance and radiation resistance as shown at the beginning of this section. Then calculate the expected loss resistance of the loading coil required to resonate the antenna. There is generally additional resistance amounting to about half of the coil loss which must be added in. As a practical matter, it is usually not possible to achieve a coil Q in excess of 200 for such applications.

Using the radiation resistance plus 1.5 times the coil loss and the power rating desired for the antenna, one may select the wire size. For high efficiency coils, a current density of 1000 A/inch^2 is a good compromise. For the 3.67 A of the example we need a wire 0.068-inch diameter, which roughly corresponds to #14 AWG. Higher current densities can lead to a melted coil.

Design the coil with a pitch equal to twice the wire diameter and the coil diameter approximately equal to the coil length. These proportions lead to the highest Q in air core coils.

The circuit of **Fig 21.A3** will match essentially all practical HF antennas on a car or truck. The circuit actually matches the antenna to 12.5 Ω *and the transformer boosts it up to 50 Ω*. Actual losses alter the required values of both the shunt inductor and the series capacitor. At a frequency of 3.5 MHz with an antenna impedance of 0.55 −j1375 Ω and a base capacitance of 2 pF results in the values shown in **Table 21.B**. Inductor and capacitor values are highly sensitive to coil Q. Furthermore, the inductor values are considerably below the 62.5 µH required to resonate the antenna.

This circuit has the advantage that the tuning elements are all at the base of the antenna. The whip radiator itself has minimal mass and wind resistance. In addition, the rig is protected by the fact that there is a dc ground on the radiator so any accidental discharge or electrical contact is kept out of the cable and rig. Variable tuning elements allow the antenna to be tuned to other frequencies.

Connect the antenna, L and C. Start with less inductor than required to resonate the antenna. Tune the capacitor to minimum SWR. Increase the inductance and tune for minimum SWR. When the values of L and C are right, the SWR will be 1:1.

Fig 21.A3 — The base-matched mobile whip antenna.

Table 21.B
Values of L and C for the Circuit of Fig 21.A3 on 3.5 MHz

Coil Q	L (µH)	C (pF)	System Efficiency (%)
300	44	11.9	8.3
200	29.14	35	3.72
100	22.2	58.1	1.4

Fig 21.95 — The screwdriver-style remotely-controlled whip antenna. A small motor in the base mast moves a coil past contacts at the top of the metal base section. The top whip section is attached to the top of the coil. As the coil moves out of the mast, more inductance is connected in series between the base section and top section. Screwdriver antennas are popular because they offer multiband coverage. (W1ZR photo)

popularity: Their input impedance is near 50 Ω in part because of their low-Q coils and they don't require matching once the length of the top whip or *stinger* is adjusted. They are light in weight (about a pound), short in length (typically 6 to 8 ft) and thus easy to mount. For temporary use, easy mounting may be more important than high efficiency. This type of antenna is most effective on 20 meters and higher frequency bands.

Table 21.19 compares different types of HF mobile antennas.

21.8.4 Remotely Controlled HF Mobile Antennas

Remotely controlled (motorized) HF mobile antennas are commonly referred to as *screwdriver antennas*. Don Johnson, W6AAQ (SK), is credited by many as the father of the screwdriver antenna. His design was not the first motorized antenna, but he certainly popularized it. There are now over 50 commercial versions available.

They're called screwdrivers because the first examples used a stripped-down rechargeable electric screwdriver assembly to adjust the resonant frequency of the antenna. The motor turns a threaded rod in and out of a nut attached to the bottom of the coil. This in turn moves the coil in and out of the lower mast section. Contacts at the top of the mast slide on the outside of the coil, thus adjusting the resonance point. Position sensors may be used to keep track of the location of the coil tap. Nowadays, calling them screwdrivers is a bit of a misnomer as the electric screwdriver motors have been replaced with much more reliable gear motors.

There are several remotely controlled HF mobile antennas that don't change length like true screwdrivers do. Both base and center loaded models are available (see **Fig 21.95**). Whether or not they're more efficient is dependent on the factors discussed in the previous section, rather than the method used to adjust the coil.

RF CHOKES

The motors and position sensors of all remotely-controlled antennas operate above RF ground potential. The amount of RF present on the leads depends on several factors, especially where and how the antenna is mounted and its overall (electrical) length. Thus the RF current coupled onto the leads must be minimized with an RF choke before entering the vehicle. An inadequate choke may result in erratic controller operation and possible interference with transceiver operation.

The choke should have an impedance of at least two orders of magnitude greater than the impedance of the circuit. In other words, at least 5 kΩ, and perhaps two or three times that in some cases (stubby antennas and poor mounting schemes are examples). Mix 31 ferrite split beads are ideal for this application, but it takes eight turns to obtain a 5 kΩ choking impedance. Depending on the wire size and insulation, you'll need to use the ½- or ¾-inch ID cores. Snap-on ferrite beads are available from most Amateur Radio dealers. (More information on this type of RF choke may be found in the **RF Techniques** chapter.)

The choke shown in **Fig 21.96** consists of 13 turns of #18 AWG wire, wound on a ¾-inch ID mix 31 split bead. It has an impedance of approximately 10 kΩ at 10 MHz. When winding the chokes, try not to overlap or twist the wires as this reduces the effectiveness.

21.8.5 Ground Losses

High frequency mobile ground loss data first appeared in a 1953 issue of *QST*, in an article written by Jack Belrose, VE3BLW (now VE2CV). In the article, Belrose said that the current flowing at the base of the antenna must be returned to the base of the antenna by currents induced in the ground beneath the radiator (antenna). These currents must be collected by the car body and through the capacitance of the car body to the ground. Since the maximum dimension of car body is considerably less than a quarter wavelength on most HF bands, only a portion of these currents will be collected by the car frame itself, and the rest will be collected by ground currents flowing through the capacitance of the car to the ground. Since the ground is not lossless, quite a large loss resistance (R_g) is found.

From that article, the accepted ground loss figure for HF mobile applications varies between 12 Ω (for 80 meters) and 2 Ω (for 10 meters). However, these figures do not include stray capacitance from the mounting location and method. Stray capacitance has the same effect as ground losses: reduced efficiency. As a result, in the real world, ground losses can be double the accepted values, reducing an otherwise efficient antenna to mediocrity.

21.8.6 Antenna Mounting

Here is an important caveat to keep in mind: While the roof of a vehicle is a very good place to mount an antenna, more and more new vehicles are equipped with side curtain air bags. They typically are mounted along the edges of the headliner, including the rear seat area if there is one. The wiring to these devices is routed through any one (or more) of the roof pillars. Extra care is required when installing antennas in vehicles so equipped. If you are the least bit apprehensive about your installation, seek professional help from your dealer or a qualified installer.

Mobile antenna mounting hardware runs the gamut from mundane to extravagant. Choosing the correct hardware is based on need, as well as on personal preference. There are enough variables with respect to mounting HF mobile antennas on modern vehicles that one could easily fill this *Handbook* and not cover them all. It is almost easier to explain *what not* to do, than explain *what* to do. In any case, there are a few very important aspects that need to be covered. With concerns similar to those of the antenna, the antenna mount should:

• be permanently mounted;
• be strong enough to support the antenna;
• have as much metal mass under it as possible;
• be well-grounded to the chassis or body;
• not interfere with doors, trunks, or access panels and
• be removable with minimal damage to the vehicle.

There are many reasons to permanently install any mobile antenna. Aside from maximizing performance by minimizing ground and stray capacitance losses, there is also a safety issue. If you need to use a temporary mount, use the multiple-magnet mounts for their superior holding strength. An HF mobile antenna attached to a vehicle traveling at highway speed by a single-magnet mount is a tenuous situation at best.

The type of mount is dictated by several conditions. These include a decision whether or not to drill holes and the size, weight and length of the antenna. If you're into daily HF mobile operation, you've already drilled the necessary holes. If you're not, a trunk lip, angle bracket, or license plate mount, and a lightweight, continuously-loaded antenna may meet your needs.

Ground plane losses directly affect a mobile antenna's efficiency. From this standpoint, mounts positioned high on the vehicle are preferred over a trailer hitch, bumper or other low-position mounting locations.

The ground plane of a mobile vertical antenna begins where the coax shield connects. When the feed line shield couples to the vehicle body capacitively as in a magmount, or when the mass of the vehicle is far below the feed point (long stalks attached to trailer hitch mounts), ground losses escalate dramatically. Running a ground strap to the nearest connecting point to the vehicle body does not eliminate ground losses. Remember that the ground connection is also part of the

Fig 21.96 — RF choke for screwdriver antenna control and power leads.

Fig 21.97 — A simple ball mount is sturdy but requires drilling holes in the vehicle body.

antenna and it is the efficiency of the whole system that is important.

While it is difficult to mount an HF antenna without at least part of the mast being close to the body, the coil must be kept free and clear. If it isn't, tuning problems and reduced efficiency will result. In some cases, vans and SUVs for example, front-mounting may become necessary to avoid coil-to-body interaction.

One drawback to trunk lip and similar clip mounts is the stress imposed on them as the lids and doors are open and closed. In most cases, angle brackets that attach to the inner surfaces with screws are a better choice.

Sometimes, the only solution is a custom-made bracket like the one shown later in this chapter. Here too, the need dictates the requirements, and there is no standard to follow except for those attributes mentioned above. Keep in mind that removing permanently installed antenna mounts such as the ball mount shown in **Fig 21.97** will always leave some body damage, but temporary ones do, too. It is only the severity that is in question, and that's in the eyes of the beholder.

21.8.7 Mobile HF Antenna Matching

Modern solid-state transceivers are designed for loads close to 50 Ω impedance. Depending on the design of the antenna (primarily depending on coil position and Q), overall length and the ground losses present, the input impedance is usually closer to 25 Ω but may vary from 18 Ω to more than 50 Ω. Note that a vehicle is an inadequate ground plane for any HF mobile antenna. Typical ground loss varies from 20 Ω (160 meters) to 2 Ω (10 meters). Stray capacitance losses may further increase the apparent ground losses.

It's important to remember two important facts. First, as the coil is moved past the center of the antenna toward the top, the coil's resistive losses begin to dominate, and the input impedance gets closer to 50 Ω. Second, short stubby antennas require more inductance than longer ones, which increases resistive losses in the coil (low Q). While no matching is required in either case, efficiency suffers, and may actually drop below 1% on the lower bands. Another way to look at the situation is that an antenna with no matching required implies a low efficiency.

Ground loss, coil position, coil Q, mast size, whip size and a few other factors determine the feed point impedance, which averages about 25 Ω for a typical quality antenna and mount. This represents an input SWR of 2:1, so some form of impedance matching is required for the transceiver. There are three ways to accomplish the impedance transformation: capacitive, transmission-line transformer and inductive

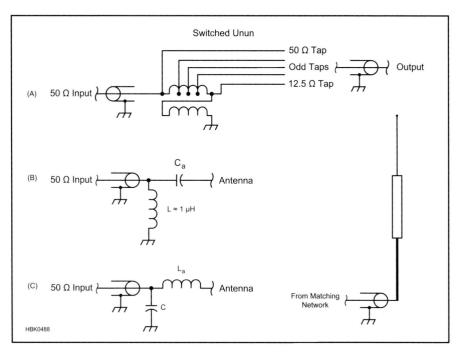

Fig 21.98 — A well-constructed and mounted HF mobile antenna will have an average input impedance around 25 Ω, requiring some matching to the feed line. The unun transmission line transformer (A) is a 4:1 configuration with taps added to match intermediate impedance values between 50 and 12.5 Ω. The high-pass L-network (B) uses some of the antenna's capacitive reactance as part of the network and the low-pass L-network (C) uses some of the antenna's inductive reactance as part of the network. Both (A) and (B) result in an antenna at dc ground, an important safety issue.

matching as shown in **Fig 21.98**. Each has its own unique attributes and drawbacks.

Transmission line transformers, in this case an unun in Fig 21.98A, do provide a dc ground for the antenna. They can be tapped or switched to match loads as low as a few ohms. Their broadband nature makes them ideal for HF mobile antenna matching. Since a remotely-controlled HF mobile antenna's input impedance varies over a wide range, transmission line transformers are best utilized for matching monoband antennas.

Inductive matching in Fig 21.98B borrows a little capacitance from the antenna (C_a) — the antenna is adjusted to a frequency slightly higher than the operating frequency, making the input impedance capacitive. This forms a high-pass L-network, which transforms the input impedance to the 50 Ω transmission line impedance. It is ideal for use with remotely-controlled antennas, as its reactance increases with frequency. By selecting the correct inductance, a compromise can be reached such that the impedance transformation will result in a low SWR from 160 through 10 meters. The approximate value is 1 μH, but may vary between 0.7 and 1.5 μH.

Adjusting the shunt coil can be done without transmitting by using an antenna analyzer and takes about 10 minutes. (Full instructions for properly adjusting a shunt coil may be found at **www.k0bg.com/coil.html**.) Because no further adjustment is necessary, shunt coil matching is ideal for remotely-controlled HF mobile antennas.

Capacitive matching in Fig 21.98C borrows a little inductance from the antenna (L_a) — the antenna is adjusted to a frequency slightly lower than the operating frequency, making the input impedance inductive. This forms a low-pass L-network, which transforms the input impedance to the 50 Ω transmission line impedance. While it works quite well, it has two drawbacks. First, capacitive matching presents a dc ground for the antenna, which tends to increase the static levels on receive. Second, the capacitance changes with frequency, so changing bands also requires a change in capacitance. This can be a nuisance with a remotely controlled antenna.

An important point should be made about dc grounding in addition to the static issue. If the antenna element should come in contact with a low-hanging high tension wire, or if lightning should strike it, dc grounding offers an additional level of protection for you and your transceiver.

21.8.8 Remotely-Tuned Antenna Controllers

There are three basic types of remote controllers: manual, position sensing and SWR sensing. Manual controllers consist of a DPDT center-off switch that changes the polarity of the current to the motor. Some commercial models include an interface with the radio that causes the radio to transmit a low-power carrier for tuning. Reading the SWR is left to the user. Some manual controllers incorporate a position readout to aid the operator in correctly positioning the antenna.

Position sensing controllers incorporate a magnet attached to the motor output shaft. The magnet opens and closes a reed switch. During set up, the antenna is set to one end of its range or the other. Then the resonant points are found (you have to do this yourself) and stored in multiple memory locations. As long as power remains applied to the controller, a simple button push will move the antenna to a specific preset point. Some controllers use band or frequency data from a port on the radio and reset the antenna to the nearest preset based on that information.

SWR-sensing controllers either read data from the radio or from a built-in SWR bridge. Depending on the make and model, a push of the radio's tuner button (or one on the controller) causes the radio to transmit at a reduced power setting. The controller then powers the antenna's tuning motor. When the preset SWR threshold is reached, the controller stops the transmission and shuts off the motor. **Fig 21.99** shows an example of a controller made to work with a specific transceiver.

Automatic controllers are far less distracting than manual ones. Most offer a parking function that collapses the coil of a screwdriver antenna into the mast (highest frequency position, lowest overall length). If you garage your vehicle, this is a welcome feature.

21.8.9 Efficiency

Length matters! All else being equal, a nine-foot antenna will be twice as efficient as a six-foot antenna, because radiation resistance relates directly to the square of the physical length. Further, longer antennas require less reactance to resonate, hence coil Q is higher, and resistive losses lower.

Mounting methodology matters! It is the mass under the antenna, not alongside, that counts. The higher the mounting, the less capacitive coupling there will be between the antenna and the surface of the vehicle and the lower ground losses will be.

Project: Mounts for Remotely-Tuned Antennas

Remotely tuned antennas have become very popular, but they all have one thing in common: they're difficult to mount. They require both a coaxial feed line and a dc power connection, and no one makes a universal mount for them. The short, stubby ones aren't any more difficult to mount than a small whip antenna, but the "full-sized" ones (8 ft and longer) require special consideration.

These antennas are heavy (up to 18 pounds), so the mounting medium must be extra strong, and well anchored. As a result, many hams opt for a bumper or trailer hitch mount, even though the low mounting position reduces efficiency.

For some, efficiency is paramount which dictates mounting the antenna as high as possible. Doing either low or high mount-

Fig 21.99 — A screwdriver antenna controller made to work with the IC-7000 transceiver.

Determining the Radiation Efficiency of a Center Loaded Mobile Whip

We can measure the radiation efficiency by measuring ground wave field strength E (dB referenced to µV/m). For the average radio amateur, a field strength meter is not a part of his ham shack gear. We can predict performance using readily available antenna modeling software (one of the many available versions of *NEC*) provided we have a measure of actual losses.

There are a number of loss parameters we do not know. We do not know the Q factor for the center loading coil (R_L), and we do not know the ground-induced loss resistance (R_g). In fact we do not know with certainty the radiation resistance (R_r), since the antenna sees an image of itself in the ground. *NEC* only gives us the sum of the various resistances.

$$R_{as} = R_r + R_C + R_g + R_L$$

R_C, the only parameter not discussed above, is the conductor loss resistance.

We need to know R_r if we are going to compute radiation efficiency, since radiation efficiency is given by:

$$\eta = \frac{R_r}{R_{as}}$$

So what do we do? We can measure R_{as} using the SWR analyzer, by adjusting the tuning so the reactance at the base of the antenna is equal to zero. We can then use *NEC* to predict the base impedance (resistive component), by changing the Q factor of the inductor so that R_{as} predicted equals R_{as} measured. We can then predict the ground field strength (dBµV/m) at say 100 m for a transmitter power of 1 kW. We then reference this predicted field strength to that for an electrically small lossless vertical antenna (129.54 dBµV/m at 100 m for 1 kW transmitter power — which corresponds to the commonly quoted value of 300 mVm at 1 km). This gives us a pretty good estimate of the radiation efficiency of our mobile whip.
— *Jack Belrose, VE2CV*

ing often requires custom fabrication. The accompanying photos illustrate the two different strategies that show Amateur Radio ingenuity at its finest.

Figs 21.100 and **21.101** depict the mobile installation of Fokko Vos, PA3VOS. Except for a Hi-Q heavy-duty quick-disconnect, the complete mount was custom engineered by Fokko. The mount bolts to a frame extension, which in turn is bolted to the undercarriage using existing bolts. Note that the rear hatch may be opened without the antenna being removed. Had the trailer hitch been used, this would not be the case.

Fig 21.102 depicts installation on a Ford F350-based motor home owned by Hal Wilson, KE5DKM. Hal designed the bracket, and had a local machine shop do the hard work. It is made of ¼-inch, high-strength aluminum, and the seams are welded. A powder-coat finish tops off the fabrication.

Shown here during installation, the bracket just fits into the right-side hood seam. The piece jutting out from the mount was to be used to further brace the mount. However, after all of the bolts were installed, additional bracing became unnecessary.

Project: Retuning a CB Whip Antenna

The most efficient HF mobile antenna is a full-size quarter-wavelength whip. Wouldn't it be nice if we could use one on every band? Alas, we cannot, but we can use one easily on 10 and 12 meters since the overall length will be less than 10 ft. If we start with a standard-length 102-inch whip and its base spring, we just need to shorten it a little for 10 meters, and lengthen it a little for 12 meters. Here's how to do it.

The formula for calculating the length of a ¼-λ antenna in feet is 234/f, where f is the frequency (MHz). Since the formula is for wire antennas, and the whip is larger in diameter, the resulting length will be slightly too long. This is a good thing because it is easier to remove a little length than it is to add some. This makes tuning easier.

Using the formula, we discover the needed length for 10 meters (28.5 MHz) is 98.5 inches. Thus, we need to remove 3.5 inches and account for the length of the base spring (about 6 inches), for a total of 9.5 inches to be removed. This is best accomplished by filing a notch on opposite sides of the tip of the whip, and snapping it in two. Protect your eyes when you do this, as shards and splinters can fly off the broken ends. If a fiberglass whip is being modified, clip the internal wire at the top of the remaining base section. Remove the plastic cap from the discarded top section and replace it over the new top of the antenna.

Depending on the mount used, the actual resonant frequency will be lower than 28.5 MHz as the mounts adds effective length. A standard CB antenna ball mount will easily support a whip. The finished antenna is shown in **Fig 21.103A**.

Once the antenna is mounted on the vehicle, simply trimming the overall length ½-inch at a time will eventually produce a low SWR at your desired frequency.

Fig 21.100 — The PA3VOS antenna mount easily supports a large screwdriver antenna and is offset to allow the hatch to open and close without removing the antenna.

Fig 21.101 — A close-up of the PA3VOS mount.

Fig 21.102 — The KE5DKM bracket mounts under the hood of a motor home.

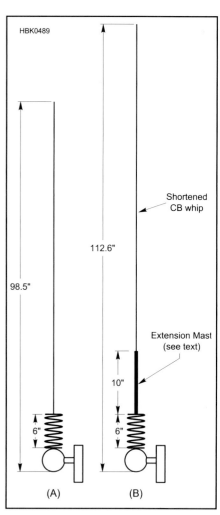

Fig 21.103 — CB whips are easily retuned for 10 meter (A) and 12 meter (B) mobile operation.

LENGTHENING FOR 12 METER OPERATION

Lengthening a CB whip for 12 meters requires a little more work. Thankfully, there's a easy solution if you have a CB radio or truck stop near you. Wilson and other vendors sell short masts designed for the CB market. They have the requisite ⅜ × 24 threads to accept a standard whip, and they come with a female-to-female coupler. The 10-inch model is ideal for our use, and costs under $10.

Using our formula 234/f, the overall length needs to be 112.6 inches for 24.93 MHz. Adding 6 inches for the spring, 10 inches. for the extension mast, and 102 inches. for the whip, gives us 118 inches. So we need to remove 5.4 inches from the whip. Then just trim off ½-inch at a time as above to resonate the antenna. The finished antenna is shown in Fig 21.103B.

FINISHING UP

There are two more things to consider. If a metal whip was modified, you'll need to replace the corona ball at the tip of the antenna. It helps reduce static from corona discharge. It's held on by a set screw and has little effect on tuning. Most CB shops sell them.

The other consideration is ground loss. Theoretically, a ¼-λ vertical will have an input impedance of 36 Ω. In a mobile installation, we have ground losses and stray capacitance losses in the mounting hardware. As a result, the real-world input impedance should be very close to 42 Ω, yielding a rather low SWR and good efficiency.

21.9 VHF/UHF Mobile Antennas

A common mistake is selecting a VHF/UHF mobile antenna based solely on its advertised gain with little consideration of the mounting method or its sturdiness.

Sturdiness is important given the unintended abuse to which mobile antennas are subjected. If you look closely at some antennas, you'll notice they have very small phasing coils, usually held together by small set screws. Smack one hard enough with a low-hanging limb, and your antenna will break. Do this to a simple quarter-wave whip and all you'll do is bend it. A little straightening and you're back on the air.

In urban areas, where higher angles of radiation are preferred, you're typically better off with a lower gain antenna such as a ¼-λ whip. If you're living in a suburban or rural area, gain antennas might have a slight edge. It all depends on the HAAT (height above average terrain) of the repeater being used with respect to the mobile station's HAAT. In mountainous areas you're better off with a unity gain antenna (¼-λ ground plane) as the repeaters are much higher in elevation.

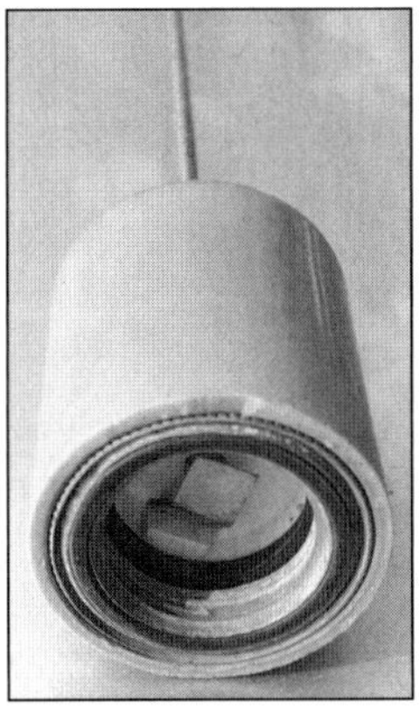

Fig 21.104 — The NMO (New Motorola) mount is widely used for VHF and UHF mobile antennas. It is available in mag-mount, trunk and lip mount and through-body mounting styles.

21.9.1 VHF/UHF Antenna Mounts

Without doubt, the best VHF/UHF mount ever devised is the NMO (New Motorola) shown in **Fig 21.104**. When properly installed in vehicle sheet metal, the mount will not leak even when the antenna is removed for car washing. SO-239 and ⅜-inch snap-in mounts often leak even with the antenna mounted.

Glass-mounted antennas are rather lossy, especially at lower VHF frequencies (2 meters). Many new vehicles use window glass with a metallic, anti-glare coating that interferes with capacitive coupling through the glass. These antennas also transfer mechanical abuse to the glass, risking breaking the glass the antenna is mounted on.

Setting the SWR of a VHF/UHF antenna is an important installation procedure, but not one to fret over afterwards. Unlike the HF bands, most VHF antennas will cover the whole band segment without the need to retune. That is to say, the SWR will be low across the FM portion of the respective bands. An in-line SWR meter is generally not required as the vertically-polarized antenna is not intended to be used in the weak-signal portion of the bands where horizontal polarization is the norm.

21.10 VHF/UHF Antennas

Improving an antenna system is one of the most productive moves open to the VHF enthusiast. It can increase transmitting range, improve reception, reduce interference problems and bring other practical benefits. The work itself is by no means the least attractive part of the job. Even with high-gain antennas, experimentation is greatly simplified at VHF and UHF because an array is a workable size, and much can be learned about the nature and adjustment of antennas. No large investment in test equipment is necessary.

Whether we buy or build our antennas, we soon find that there is no one *best* design for all purposes. Selecting the antenna best suited to our needs involves much more than scanning gain figures and prices in a manufacturer's catalog. The first step should be to establish priorities for the antenna system as a whole. Once the objectives have been sorted out in a general way, we face decisions on specific design features, such as polarization, length and type of transmission line, matching methods, and mechanical design.

21.10.1 Gain

As has been discussed previously, shaping the pattern of an antenna to concentrate radiated energy, or received signal pickup, in some directions at the expense of others is the only possible way to develop gain. Radiation patterns can be controlled in various ways. One is to use two or more driven elements, fed in phase. Such arrays provide gain without markedly sharpening the frequency response, compared to that of a single element. More gain per element, but with some sacrifice in frequency coverage, is obtained by placing parasitic elements into a Yagi array.

21.10.2 Radiation Pattern

Antenna radiation can be made omnidirectional, bidirectional, practically unidirectional, or anything between these conditions. A VHF net operator may find an omnidirectional system almost a necessity but it may be a poor choice otherwise. Noise pickup and other interference problems tend to be greater with omnidirectional antennas. Maximum gain and low radiation angle are usually prime interests of the weak-signal DX aspirant. A clean pattern, with lowest possible pickup and radiation off the sides and back, may be important in high-activity areas, where the noise level is high, or for challenging modes like EME (Earth-Moon-Earth).

21.10.3 Height Gain

In general, the higher a VHF antenna is installed, the better will be the results. If raising the antenna clears its view over nearby obstructions, it may make dramatic improvements in coverage. Within reason, greater height is almost always worth its cost, but height gain must be balanced against increased transmission line loss. Line losses can be considerable at VHF and above, and they increase with frequency. The best available line may be none too good, if the run is long in terms of wavelength. Consider line losses in any antenna planning.

21.10.4 Physical Size

A given antenna design for 432 MHz, say a 5-element Yagi on a 1-λ boom, will have the same gain as one for 144 MHz, but being only one-third the size it will intercept only one-ninth as much energy in receiving. Thus, to be equal in communication effectiveness, the 432-MHz array should be at least equal in physical size to the 144-MHz one, requiring roughly three times the number of elements. With all the extra difficulties involved in going higher in frequency, it is well to be on the big side in building an antenna for the UHF bands.

21.10.5 Polarization

Whether to position the antenna elements vertically or horizontally has been a question since early VHF operation. Originally, VHF communication was mostly vertically polarized, but horizontal gained favor when directional arrays became widely used. Tests of signal strength and range with different polarizations show little evidence on which to set up a uniform polarization policy. On long paths there is no consistent advantage, either way. Shorter paths tend to yield higher signal levels with horizontal in some kinds of terrain. Man-made noise, especially ignition interference, tends to be lower with horizontal polarization. Vertically polarized antennas, however, are markedly simpler to use in omnidirectional systems and in mobile work, resulting in a standardization on vertical polarization for mobile and repeater operation on FM and for digital communications. Horizontal polarization is the standard for weak signal VHF and UHF operation. (Circular polarization is preferred for satellite work as described below.) A loss in signal strength of 20 dB or more can be expected with cross-polarization so it is important to use antennas with the same polarization as the stations with which you expect to communicate.

21.10.6 Circular Polarization

Polarization is described as *horizontal* or *vertical*, but these terms have no meaning once the reference of the Earth's surface is lost. Many propagation factors can cause polarization change — reflection or refraction and passage through magnetic fields (Faraday rotation), for example. Polarization of VHF waves is often random, so an antenna capable of accepting any polarization is useful. Circular polarization, generated with helical antennas or with crossed elements fed 90° out of phase, will respond to any linear polarization.

The circularly polarized wave in effect threads its way through space, and it can be left- or right-hand polarized. These polarization senses are mutually exclusive, but either will respond to any plane (horizontal or vertical) polarization. A wave generated with right-hand polarization, when reflected from the moon, comes back with left-hand polarization, a fact to be borne in mind in setting up EME circuits. Stations communicating on direct paths should have the same polarization sense.

Both senses can be generated with crossed dipoles, with the aid of a switchable phasing harness. With helical arrays, both senses are provided with two antennas wound in opposite directions.

21.10.7 Transmission Lines

The most common type of transmission line at VHF through the low microwave bands is unbalanced coaxial cable. Small coax such as RG-58 or RG-59 should never be used in VHF work if the run is more than a few feet. Half-inch lines (RG-8 or RG-11) work fairly well at 50 MHz, and runs of 50 ft or less are acceptable at 144 MHz. Lines with foam rather than solid insulation have about 30% less loss. Low-loss cable is required for all but the shortest runs above 222 MHz and *waveguide* is used on microwave frequencies. (See the **Transmission Lines** chapter for a discussion of waveguides.)

Solid aluminum-jacketed *hardline* coaxial cable with large inner conductors and foam insulation are well worth the cost. Hardline can sometimes even be obtained for free from local Cable TV operators as *end runs* — pieces at the end of a roll. The most common CATV variety is ½-inch OD 75-Ω hardline. Hardline is considered *semi-rigid* in that it can be bent, but only with a large radius to avoid kinking and repeated bending should be avoided.

Waterproof commercial connectors for hardline are fairly expensive, but enterprising amateurs have *home-brewed* low-cost connectors. If they are properly waterproofed, connectors and hardline can last almost indefinitely. See *The ARRL Antenna Book* for details on connectors and techniques for working with hardline.

Properly-built open-wire line can operate with very low loss in VHF and even UHF installations. A line made of #12 AWG wire, spaced ¾-inch or less with Teflon spreaders, and running essentially straight from antenna to station, can be better than anything but the most expensive hardline at a fraction of the cost. Line loss under 2 dB per 100 feet at 432 MHz is readily obtained. This assumes the use of high-quality baluns to match into and out of the balanced line, with a short length of low-loss coax for the rotating section from the top of the tower to the antenna. Such an open-wire line could have a line loss under 1 dB at 144 MHz.

Effects of weather on transmission lines should not be ignored. A well-constructed open-wire line works well in nearly any weather, and it stands up well. TV-type twin-lead is almost useless in heavy rain, wet snow or icing conditions. The best grades of coax and hardline are impervious to weather. They can be run underground, fastened to metal towers without insulation, or bent into almost any convenient position, with no adverse effects on performance. However, beware of bargain coax. Lost transmitter power can be made up to some extent by increasing power, but once lost in the transmission line a weak signal can never be recovered in the receiver.

21.10.8 Impedance Matching

Theory and practice in impedance matching are discussed in detail in the **Transmission Lines** chapter, and in theory, at least, are the same for frequencies above 50 MHz. Practice may be similar, but physical size can be a major modifying factor in choice of methods.

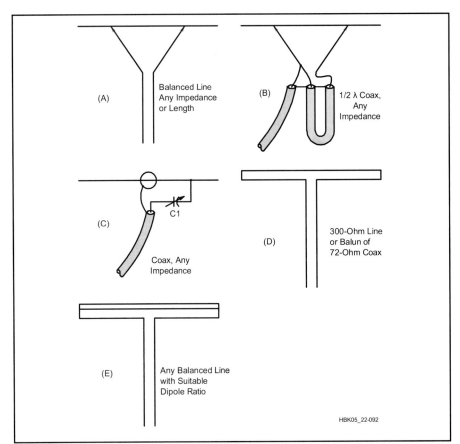

Fig 21.105 — Matching methods commonly used in VHF antennas. In the delta match, A and B, the line is fanned out to tap on the dipole at the points of best impedance match. The gamma match, C, is for direct connection of coax. C1 tunes out inductance in the arm. Folded dipole of uniform conductor size, D, steps up antenna impedance by a factor of four. Using a larger conductor in the unbroken portion of the folded dipole, E, gives higher orders of impedance transformation.

DELTA MATCH

Probably the first impedance match was made when the ends of an open line were fanned out and tapped onto a half-wave antenna at the points of most efficient power transfer, as in **Fig 21.105A**. Both the side length and the points of connection either side of the center of the element must be adjusted for minimum reflected power in the line, but the impedances need not be known. The delta makes no provision for tuning out reactance, so the length of the dipole is pruned for best SWR.

Once thought to be inferior for VHF applications because of its tendency to radiate if adjusted improperly, the delta has come back to favor now that we have good methods for measuring the effects of matching. It is very handy for phasing multiple-bay arrays with low-loss open lines, and its dimensions in this use are not particularly critical.

GAMMA MATCH

The gamma match is shown in Fig 21.105C and is covered in more detail in the preceding section on HF Yagi antennas and in the **Transmission Lines** chapter. The center of a half-wave dipole being electrically neutral, the outer conductor of the coax is connected to the element at this point, which may also be the junction with a metallic or non-conductive boom. The inner conductor is connected to the element at the matching point. Inductance of the connection to the element is canceled by means of C1. Both the point of contact with the element and the setting of the capacitor are adjusted for minimum SWR using an antenna analyzer or SWR bridge.

The capacitor C1 can be a variable unit during adjustment and then replaced with a suitable fixed unit when the required capacitance value is found. Maximum capacitance should be about 100 pF for 50 MHz and 35 to 50 pF for 144 MHz. The capacitor and arm can be combined with the arm connecting to the driven element by means of a sliding clamp, and the inner end of the arm sliding inside a sleeve connected to the inner conductor of the coax. It can be constructed from concentric pieces of tubing, insulated by plastic sleeving or shrink tubing. RF voltage across the capacitor is low, once the match is adjusted properly, so with a good dielectric, insulation presents no great problem. A clean, permanent, high-conductivity bond between arm and element is important, as the RF current is high at this point.

Because it is inherently somewhat unbalanced, the gamma match can sometimes introduce pattern distortion, particularly on long-boom, highly directive Yagi arrays. The T-match, essentially two gamma matches in series creating a balanced feed system, has become popular for this reason. (See the preceding discussion on T-matches in the HF Yagi section.) A coaxial balun like that shown in Fig 21.105B is used from the balanced T-match to the unbalanced coaxial line going to the transmitter. To maintain a symmetrical pattern, the feed line should be run along the antenna boom at the centerline of the elements to the mast. A choke balun is often used to minimize currents that might be induced on the outer surface of the feed line shield.

FOLDED DIPOLE

The impedance of a half-wave dipole feed point at its center is 72 Ω. If a single conductor of uniform size is folded to make a half-wave dipole, as shown in Fig 21.105D, the impedance is stepped up four times. Such a folded dipole can thus be fed directly with 300-Ω line with no appreciable mismatch. Coaxial feed line of 70 to 75 Ω impedance may then be used with a 4:1 impedance transformer. Higher impedance step-up can be obtained if the unbroken portion is made larger in cross-section than the fed portion, as in Fig 21.105E. The folded dipole is discussed further in the *ARRL Antenna Book*.

21.10.9 Baluns and Impedance Transformers

Conversion from balanced loads to unbalanced lines, or vice versa, can be performed with electrical circuits, or their equivalents made of coaxial line. A balun made from flexible coax is shown in **Fig 21.106A**. The looped portion is an electrical half-wave. This type of balun gives an impedance step-up of 4:1, 50 to 200 Ω, or 75 to 300 Ω typically. See the **RF Techniques** and **Transmission Lines** chapters for a detailed discussion of baluns and impedance transformers.

The physical length of the line section depends on the propagation factor of the line used, so it is best to check its resonant frequency, as shown at B. One end of the line is left open and an antenna analyzer used to find the lowest frequency at which the impedance at the other end of the line is a minimum, the frequency at which the section of line is ¼-λ long. Multiply the frequency by two to find the frequency at which the section is ½-λ long.

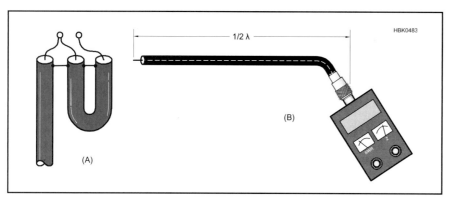

Fig 21.106 — Conversion from unbalanced coax to a balanced load can be done with a half-wave coaxial balun, A. The half-wave balun gives a 4:1 impedance step up. Electrical length of the looped section should be checked with an antenna analyzer with the far end of the line open, as in B. The lowest frequency at which the line impedance is a minimum is the frequency at which the line is ¼-λ long. Multiply that frequency by two to obtain the ½-λ frequency.

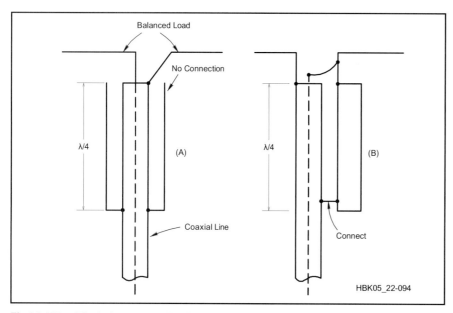

Fig 21.107 — The balun conversion function, with no impedance change, is accomplished with quarter-wave lines, open at the top and connected to the coax outer conductor at the bottom. The coaxial sleeve shown at A is preferred.

Coaxial baluns giving a 1:1 impedance transfer are shown in **Fig 21.107**. The coaxial sleeve, open at the top and connected to the outer conductor of the line at the lower end (Fig 21.107A) is the preferred type. A conductor of approximately the same size as the line is used with the outer conductor to form a quarter-wave stub, in Fig 21.107B. Another piece of coax, using only the outer conductor, will serve this purpose. Both baluns are intended to present a high impedance to any RF current that might otherwise tend to flow on the outer conductor of the coax. Choke baluns made of ferrite beads of the proper material type or mix may also be used. See the **RF Techniques** chapter for information about ferrite use at VHF and UHF.

Project: Simple, Portable Ground-Plane Antenna

This utility antenna is built on a coaxial connector. UHF connectors work well, but you may prefer to use type N or BNC connectors. With only two radials, it is essentially two dimensional, which makes it easier to store when not in use.

If the antenna is sheltered from weather, copper wire is sufficiently rigid for the radiating element and radials. Antennas exposed to the wind and weather can be made from brazing rod, which is available at welding supply stores. Alternatively, #12 or #14 AWG copper-clad steel wire could be used to construct this antenna.

The ground-plane antenna is shown in **Fig 21.108** and uses a female chassis-mount connector to support the element and two radials. To eliminate sharp ends, it's a good idea to bend the element and radial ends into a circle or to terminate them with a crimp terminal as in **Fig 21.109**. The crimp terminal approach is easier with stiff wire. Crimp and then solder the terminal to the wire. Make the overall length of the element and radials the same as shown in Fig 21.108, measuring to the outer tip of the loop or terminal.

Radials may be attached directly to the mounting holes of the coaxial connector. Bend a hook at one end of each radial for insertion through the connector. Solder the radials to the connector using a large soldering iron or propane torch.

Solder the element to the center pin of the connector. If the element does not fit inside the solder cup, use a short section of brass tubing as a coupler (a slotted ⅛-inch-ID tube will fit over an SO-239 or N-receptacle center pin).

If necessary, prune the antenna to raise the frequency of minimum SWR. Then adjust the radial droop angle for minimum SWR — this should not affect the frequency at which the minimum SWR occurs.

One mounting method for fixed-station antennas appears in Fig 21.108. The feed line and connector are inside the mast, and a hose clamp squeezes the slotted mast end to tightly grip the plug body. Once the antenna is mounted and tested, thoroughly seal the open side of the coaxial connector with silicone sealant, and weatherproof the connections with rust-preventative paint.

Project: Dual-Band Antenna for 146/446 MHz

This project by Wayde Bartholomew, K3MF (ex-WA3WMG), first appeared in *The ARRL Antenna Compendium, Volume 5*. This mobile whip antenna won't take long to build, works well and only requires one feed line for the two-band coverage.

Wayde used a commercial NMO-style base and magnetic mount. For the radiator and decoupling stub, he used brazing rod coated with a rust inhibitor after all the tuning was done. You can start with a 2 meter radiator that's 20.5 inches long. This is an inch longer than normal so that it may be pruned for best SWR.

Next, tack on the 6.5-inch long 70-cm decoupling stub. Trim the length of the

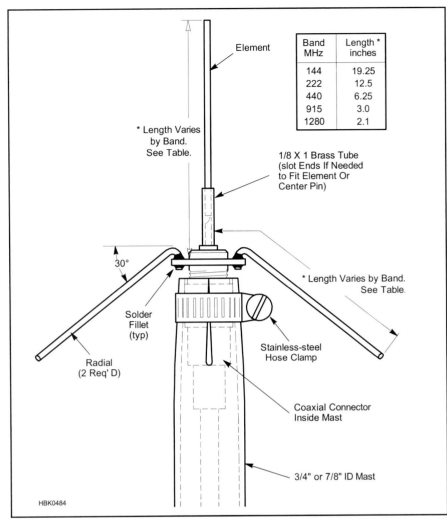

Band MHz	Length * inches
144	19.25
222	12.5
440	6.25
915	3.0
1280	2.1

Fig 21.108 — A simple ground-plane antenna for the 144, 222 or 440-MHz bands. The feed line and connector are inside the mast, and a hose clamp squeezes the slotted mast end to tightly grip the plug body. Element and radial dimensions given in the drawing are good for the entire band.

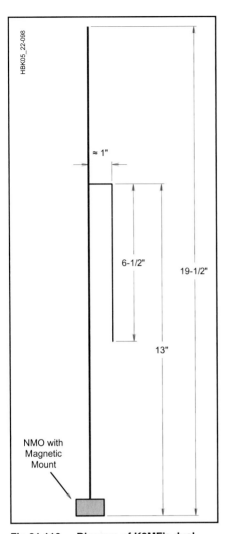

Fig 21.110 — Diagram of K3MF's dual-band 146/446-MHz mobile whip. Brazing rod is used for the 2 meter radiator and for the 70-cm decoupling stub.

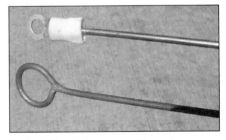

Fig 21.109 — Alternate methods for terminating element and radial tips on the simple ground-plane antenna. See text. (Photo by K8CH)

2 meter radiator for best SWR at 146 MHz and then tune the 70-cm stub on 446 MHz by moving it up and down along the antenna for best SWR. There should be no significant interaction between the adjustments for either frequency.

Final dimensions are shown in **Fig 21.110**. The SWR in the repeater portions of both bands is less than 2:1.

ADAPTING FOR FIXED-STATION USE

You can use the dual-band mobile whip as the radiating element for the ground-plane antenna in Fig 21.108. Don't change the 2 meter radials. Instead, add two 70-cm radials at right angles to the 2 meter set as in **Fig 21.111**. The antenna is no longer two-dimensional, but you do have two bands with one feed line *and* automatic band switching.

Fig 21.111 — K3MF's whip can be used to make a dual-band ground-plane antenna. Separate radials for 2 meters and 70-cm simplifies tuning. (Photo by K8CH)

Antennas   21.63

21.11 VHF/UHF Yagis

Without doubt, the Yagi is king of home-station antennas these days. Today's best designs are computer optimized. For years amateurs as well as professionals designed Yagi arrays experimentally. Now we have powerful (and inexpensive) personal computers and sophisticated software for antenna modeling. These have brought us antennas with improved performance, with little or no element pruning required. A more complete discussion of Yagi design can be found earlier in this chapter and in the *ARRL Antenna Book*.

21.11.1 Stacking Yagis

Where suitable provision can be made for supporting them, two Yagis mounted one above the other and fed in-phase may be preferable to one long Yagi having the same theoretical or measured gain. The pair will require a much smaller turning space for the same gain, and their lower radiation angle can provide interesting results. On long ionospheric paths a stacked pair occasionally may show an apparent gain much greater than the 2 to 3 dB that can be measured locally as the gain from stacking.

Optimum spacing for Yagis with booms longer than 1 λ is one wavelength, but this may be too much for many builders of 50-MHz antennas to handle. Worthwhile results are possible with separations of as little as ½ λ (10 ft), but ⅝ λ (12 ft) is markedly better. At 50 MHz, the difference between 12 and 20-ft spacing may not be worth the added structural problems.

The closer spacings give lowered measured gain, but the antenna patterns are cleaner (less power in the high-angle elevation lobes) than with 1-λ spacing. Extra gain with wider spacings is usually the objective on 144 MHz and higher bands, where the structural problems are not quite as severe as on 50 MHz.

One method for feeding two 50-Ω antennas, as might be used in a stacked Yagi array, is shown in **Fig 21.112**. The transmission lines from each antenna, with a balun feeding each antenna (not shown in the drawing for simplicity), to the common feed point must be equal in length and an odd multiple of ¼-λ. This line acts as a quarter-wave (Q-section) impedance transformer, raises the feed impedance of each antenna to 100 Ω, and forces current to be equal in each driven element. When the feed lines are connected in parallel at the coaxial tee connector, the resulting impedance is close to 50 Ω.

Project: Three and Five-Element Yagis for 6 Meters

Boom length often proves to be the deciding factor when one selects a Yagi design.

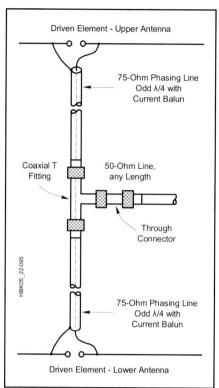

Fig 21.112 — A method for feeding a stacked Yagi array. Note that baluns at each antenna are not specifically shown. Good practice is to use choke baluns made up of ferrite beads slipped over the outside of the coax and taped to prevent movement. See the RF Techniques and Transmission Lines chapter for details.

Dean Straw, N6BV, created the designs shown in **Table 21.20**. Straw generated the designs in the table for convenient boom lengths (6 and 12 ft). The 3-element design has about 8 dBi gain, and the 5-element version has about 10 dBi gain. Both antennas exhibit better than 22 dB front-to-rear ratio, and both cover 50 to 51 MHz with better than 1.6:1 SWR.

Element lengths and spacings are given in the table. Elements can be mounted to the boom as shown in **Fig 21.113**. Two muffler clamps hold each aluminum plate to the boom, and two U bolts fasten each element to the plate, which is 0.25 inches thick and 4.4 inches square. Stainless steel is the best choice for hardware. However, galvanized hardware can be substituted. Automotive muffler clamps do not work well in this application, because they are not galvanized and quickly rust once exposed to the weather.

The driven element is mounted to the boom on a phenolic plate of similar dimension to the other mounting plates. A 12-inch piece of Plexiglas rod is inserted into the driven element halves. The Plexiglas allows the use of a single clamp on each side of the element and also seals the center of the elements against moisture. Self-tapping screws are used for electrical connection to the driven element.

Refer to **Fig 21.114** for driven element and hairpin match details. A bracket made from a piece of aluminum is used to mount

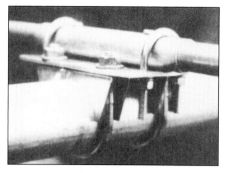

Fig 21.113 — The boom-to-element clamp. Galvanized U-bolts are used to hold the element to the plate, and 2-inch galvanized muffler clamps hold the plates to the boom.

Table 21.20
Optimized 6 Meter Yagi Designs

	Spacing From Reflector (in.)	Seg 1 Length (in.)	Seg 2 Length (in.)	Midband Gain F/R
306-06				
Refl	0	36	22.500	8.1 dBi
DE	24	36	16.000	28.3 dB
Dir 1	66	36	15.500	
506-12				
OD		0.750	0.625	
Refl	0	36	23.625	10.0 dBi
DE	24	36	17.125	26.8 dB
Dir 1	36	36	19.375	
Dir 2	80	36	18.250	
Dir 3	138	36	15.375	

Note: For all antennas, telescoping tube diameters (in inches) are: Seg1=0.750, Seg2=0.625.

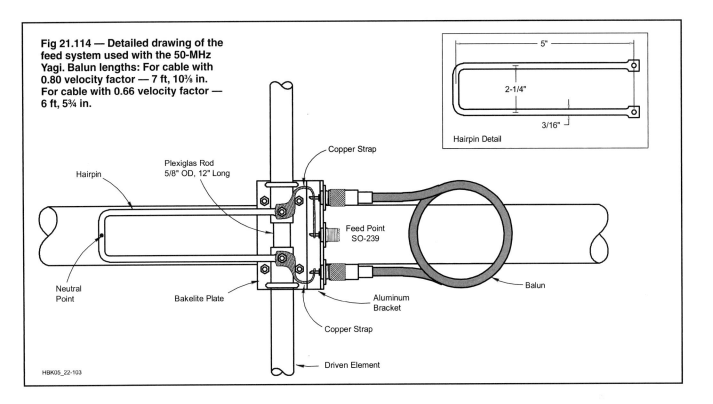

Fig 21.114 — Detailed drawing of the feed system used with the 50-MHz Yagi. Balun lengths: For cable with 0.80 velocity factor — 7 ft, 10⅜ in. For cable with 0.66 velocity factor — 6 ft, 5¾ in.

the three SO-239 connectors to the driven element plate. A 4:1 transmission-line balun connects the two element halves, transforming the 200-Ω resistance at the hairpin match to 50 Ω at the center connector. Note that the electrical length of the balun is $\lambda/2$, but the physical length will be shorter due to the velocity factor of the particular coaxial cable used. The hairpin is connected directly across the element halves. The exact center of the hairpin is electrically neutral and should be fastened to the boom. This has the advantage of placing the driven element at dc ground potential.

The hairpin match requires no adjustment as such. However, you may have to change the length of the driven element slightly to obtain the best match in your preferred portion of the band. Changing the driven-element length will not adversely affect antenna performance. *Do not adjust the lengths or spacings of the other elements — they are optimized already.* If you decide to use a gamma match, add three inches to each side of the driven element lengths given in the table for both antennas.

Project: Medium-Gain 2 Meter Yagi

This project was designed and built by L. B. Cebik, W4RNL (SK). Practical Yagis for 2 meters abound. What makes this one a bit different is the selection of materials. The elements, of course, are high-grade aluminum. However, the boom is PVC and there are only two #6 nut-bolt sets and two #8 sheet metal screws in the entire antenna. The remaining fasteners are all hitch-pin clips. The result is a very durable six-element Yagi that you can disassemble with fair ease for transport.

THE BASIC ANTENNA DESIGN

The 6-element Yagi presented here is a derivative of the *optimized wide-band antenna* (OWA) designs developed for HF use by NW3Z and WA3FET. **Fig 21.115** shows the general outline. The reflector and first director largely set the impedance. The next

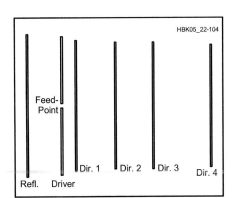

Fig 21.115 — The general outline of the 2 meter 6-element OWA Yagi. Dimensions are given in Table 21.21.

Table 21.21
2 Meter OWA Yagi Dimensions

Element	Element Length (in)	Spacing from Reflector (in)	Element Diameter (in)
Version described here:			
Refl.	40.52	—	0.1875
Driver	39.70	10.13	0.5
Alt. Driver	*39.96*	*10.13*	*0.1875*
Dir. 1	37.36	14.32	0.1875
Dir. 2	36.32	25.93	0.1875
Dir. 3	36.32	37.28	0.1875
Dir. 4	34.96	54.22	0.1875
Version using 1/8-inch diameter elements throughout:			
Refl.	40.80	—	0.125
Driver	40.10	10.20	0.125
Dir. 1	37.63	14.27	0.125
Dir. 2	36.56	25.95	0.125
Dir. 3	36.56	37.39	0.125
Dir. 4	35.20	54.44	0.125

2 directors contribute to setting the operating bandwidth. The final director (Dir. 4) sets the gain. This account is over-simplified, since every element plays a role in every facet of Yagi performance. However, the notes give some idea of which elements are most sensitive in adjusting the performance figures.

Designed using *NEC-4*, the antenna uses 6 elements on a 56-inch boom. **Table 21.21** gives the specific dimensions for the version described in these notes. The parasitic elements are all 3/16-inch aluminum rods. For ease of construction, the driver is ½-inch aluminum tubing. Do not alter the element diameters without referring to a source, such as RSGB's *The VHF/UHF DX Book*, edited by Ian White, G3SEK, (Chapter 7), for information on how to recalculate element lengths.

The driver is the simplest element to readjust. Table 21.21 shows an alternative driver using 3/16-inch diameter material. Of all the elements, the driver is perhaps the only one for which you may extrapolate reasonable lengths for other diameters from the two lengths and diameters shown. However, the parasitic elements may require more work than merely substituting one diameter and length for another. The lower portion of the table shows the design adjusted for ⅛-inch elements throughout. Not all element lengths change by the same amount using any single formula.

The OWA design provides about 10.2 dBi of free-space gain with better than 20 dB front-to-back (or front-to-rear) ratio across the entire 2 meter band. Azimuth (or E-plane) patterns show solid performance across the entire band. This applies not only to forward gain but rejection from the rear.

One significant feature of the OWA design is its direct 50-Ω feed point impedance that requires no matching network. Of course, a choke balun to suppress any currents on the feed line is desirable, and a simple ferrite bead balun (see the **Transmission Lines** and **Station Accessories** chapters) works well in this application. The SWR, shown in **Fig 21.116**, is very flat across the band and never reaches 1.3:1. The SWR and the pattern consistency together create a very useful utility antenna for 2 meters, whether installed vertically or horizontally. The only remaining question is how to effectively build the beam in the average home shop.

Table 21.22
Parts List for the 2 Meter OWA Yagi

Qty	*Item*
17'	0.1875" (3/16") 6061-T6 aluminum rod (Source: Texas Towers)
3.5'	0.5" (½") 6063-T832 aluminum tubing (Source: Texas Towers)
7'	Schedule 40, ½" PVC pipe (Source: local hardware store)
3	Schedule 40, ½" PVC Tee connectors (Source: local hardware store)
2	Schedule 40, ½" PVC L connectors (Source: local hardware store)
—	Miscellaneous male/female threaded pipe diameter transition fittings (Source: local hardware store)
1	Support mast
10	Stainless steel hitch-pin clips (hairpin cotter pins), 3/16" to ¼" shaft range, 0.04" "wire" diameter (McMasters-Carr part number 9239A024, or local hardware store)
2	Stainless steel #6 nut/bolt/lock-washer sets, bolt length 1" (Source: local hardware store)
2	Stainless steel #8 sheet metal screws (Source: local hardware store)
1	BNC connector (Source: local electronics outlet)
2"	1/16" thick aluminum L-stock, 1" per side (Source: local hardware store)
1	VHF bead-balun choke (Source: Wireman, Inc.)

THE BEAM MATERIALS

The boom is Schedule 40, ½-inch nominal PVC. Insulated booms are good for test antennas, since they do not require recalculating the element lengths due to the effects of a metal boom.

White PVC stands up for a decade of exposure in Tennessee, but apparently does not do as well in every part of the US. You may wish to use the gray electrical conduit version. If you use any other material for your boom, be sure that it is UV-protected. You'll find a parts list in **Table 21.22**. Sources for the parts are given in the table. However, you are encouraged to develop your own sources for antenna materials.

Fig 21.117 shows the element layout along the 56-inch boom. Centering the first element hole 1 inch from the rear end of the boom results in a succession of holes for the 3/16-inch pass-through parasitic elements. Only the driver requires special treatment.

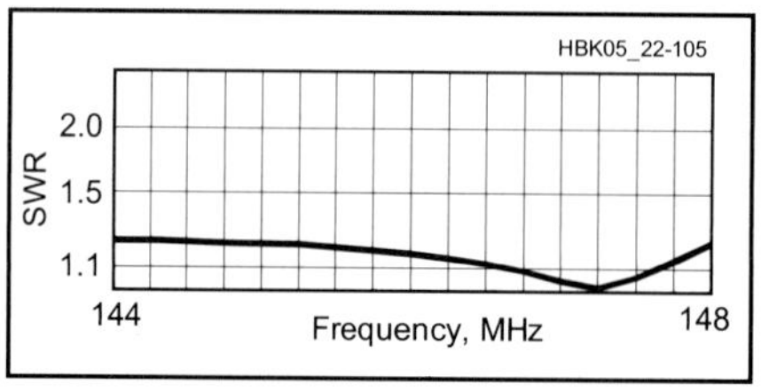

Fig 21.116 — SWR curve as modeled using *NEC-4* for the 2-m 6-element OWA Yagi.

We shall use a ⅜-inch hole to carry a short length of fiberglass rod that will support the two sides of the driver element. Note that the antenna uses a BNC connector, mounted on a small plate that we shall meet along the way.

The boom is actually a more complex structure than initially meets the eye. You need a support for the elements, and a means of connecting the boom to the mast. If you break the boom in the middle to install a Tee connector for the mast junction, you come very close to the 2nd director. **Fig 21.118** shows how to avoid the predicament.

Before drilling the boom, assemble it from common Schedule 40 ½-inch fittings and insert the lengths of PVC pipe. Fig 21.118 shows the dimensions for the center section of the boom assembly. However, PVC dimensions are always *nominal*, that is, meeting certain minimum size standards. So you may have to adjust the lengths of the linking pieces slightly to come up with a straight and true boom assembly.

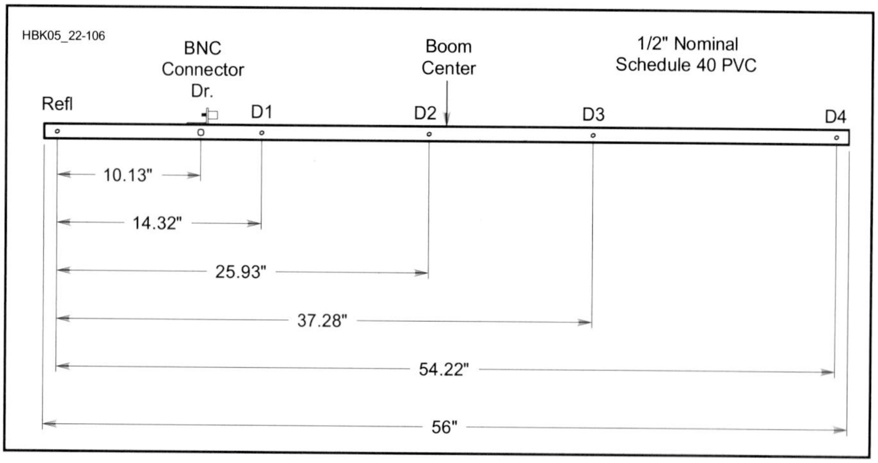

Fig 21.117 — Layout of elements along the PVC boom for the 2 meter 6-element OWA Yagi, showing placement of the BNC connector and the boom center.

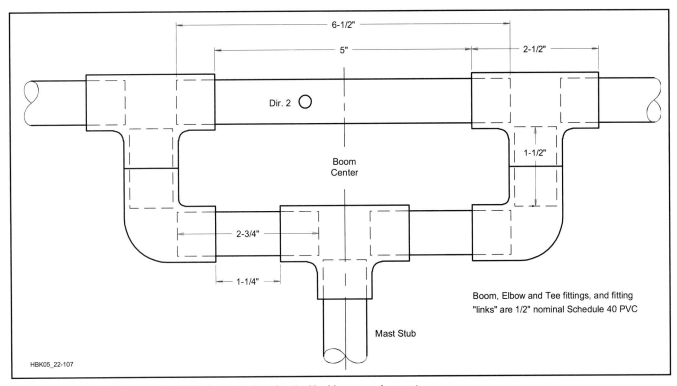

Fig 21.118 — Details of a parallel PVC pipe structure for the Yagi boom and mount.

Use scrap lumber to help keep everything aligned while cementing the pieces together. A 1×4 and a 1×6 nailed together along the edges produce a very good platform with a right-angle. Start with the two upper Tees and the Ls below each one. Dry-fit scrap PVC into the openings except for the short link that joins the fitting. Cement these in place and align them using the dry-fit pieces as guides to keep everything parallel. Next, cement the two short (2¾-inch) links into the third Tee. Then, cement one link into its L, using the dry-fit tube in the upper Tee as an alignment guide.

Before proceeding further, carefully measure the required length of PVC for the boom section between Tees. How well you measure here will determine whether the boom will be straight or whether it will bow up or down. Now, cement both the L and the Tee at the same time, pressing the cemented sections into the 2-board jig to assure alignment.

The final step in the process is to add the 23-inch boom end pieces to the open ends of the upper Tees. For the brief period in which the PVC cement is wet, it is possible to misalign the tubing. Dry-fit end caps on the boom ends and do the cement work using the 2-board jig. By pressing the assembly into the right angle of the boards, you can assure that you have a very true boom. When you've put the PVC cement back onto its shelf, your boom should be ready to drill.

Consider the boom-to-mast connection. The lower Tee in Fig 21.118 receives a short

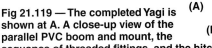

Fig 21.119 — The completed Yagi is shown at A. A close-up view of the parallel PVC boom and mount, the sequence of threaded fittings, and the hitch-pin clips used to secure parasitic elements is shown at B.

length of ½-inch nominal Schedule 40 PVC. This material has an outside diameter of about ⅞-inch, not a useful size for joining to a mast. However, PVC fittings have a handy series of threaded couplers that allow you to screw-fit a series of ever-larger sizes until you reach a more useful size. As **Fig 21.119B** shows, enough of these fittings will finish off with a 1¼-inch threaded female side and a 1¼-inch cement-coupling side. To this fitting, cement a length of 1¼-inch tubing that slides over a length of common TV mast. For a tight fit, wrap the TV mast with several layers of electrical tape in two places — one near the upper end of the PVC pipe section and the other close to where the PVC pipe ends. You

Antennas 21.67

may then use stainless steel through-bolts or set-screws to prevent the PVC assembly from turning.

BOOM AND ELEMENTS

Before installing the elements, you need to drill the holes in the boom. The two-board jig comes in handy once more. The key goals in the drilling process are to: A) precisely position the holes; B) create holes that are a fairly tight fit for the rod elements; and C) keep the elements aligned in a flat plane. For this purpose, a drill press is almost a necessity for all but those with the truest eyes.

Use the jig and a couple of clamps to hold the boom assembly in place. Because the assembly has two parallel sections, laying it flat will present the drill press with the correct angle for drilling through the PVC in one stroke. Drill the holes at pre-marked positions, remembering that the driver hole is ⅜ inch while all the others are 3/16 inch. Clean the holes, but do not enlarge them in the process.

By now you should have the rod and tube stock in hand. For antenna elements, don't rely on questionable materials that are designed for other applications. Rather, obtain 6063-T832 tubing and 6061-T6 rods from mail order sources, such as Texas Towers, McMasters-Carr, and others. These materials are often not available at local hardware depots.

Cut the parasitic elements to length and smooth their ends with a fine file or sandpaper. Find the center of each element and carefully mark a position about 1/16 in. outside where the element will emerge from each side of the boom. You'll drill small holes in these locations. You may wish to very lightly file a flatted area where the hole is to go to prevent the drill bit from slipping as you start the hole.

Drill 1/16-inch holes at each marked location all the way through the rod. De-burr the exit ends so that the rod will pass through the boom hole. These holes are the locations for hitch-pin clips. **Fig 21.120** shows the outline of a typical hitch-pin clip, which is also called a hairpin cotter pin in some catalogs and stores. Obtain stainless steel pins whose bodies just fit tightly over the rod when installed. Initially, install 1 pin per parasitic element. Slide the element through the correct boom hole and install the second pin. Although the upper part of the drawing shows a bit of room between the boom and pin, this space is for clarity. Install the pins as close to each side of the boom as you can.

Pins designed for a 3/16-inch rod are small enough that they add nothing significant to the element, and antenna tests showed that they did not move the performance curve of the antenna. Yet, they have held securely through a series of shock tests given to the prototype. These pins — in various sizes — offer the home builder a handy fastener that is applicable to many types of portable or field antennas. Although you may wish to use better fasteners when making permanent metal-to-metal connections, for joining sections of Field Day and similar antennas, the hitch-pin clips perform the mechanical function, while clean tubing sections themselves provide adequate electrical contact for a limited period of use.

THE DRIVER AND FEED LINE CONNECTOR

The final construction step is perhaps the one requiring the most attention to detail, as shown in **Fig 21.121**. The driver and feed point assembly consists of a 4- to 6-inch length of ⅜-inch fiberglass or other non-conductive rod, two sections of the driver element made from ½-inch aluminum tubing, a BNC connector, a home-made mounting plate, two sets of stainless steel #6 nuts, bolts, and lock-washers, and two stainless steel #8 sheet metal screws. Consult both the upper and lower portions of the figure, since some detail has been omitted from each one to show other detail more clearly.

First, trial fit the driver tubing and the fiberglass rod, marking where the rod exits the boom. Now pre-drill 5/64-inch holes through the tubing and the fiberglass rod. Do not use larger hardware, since the resulting hole will weaken the rod, possibly to the breaking

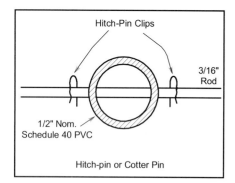

Fig 21.120 — The parasitic element mounting system, showing the placement of the hitch-pin clips and the shape of the clips.

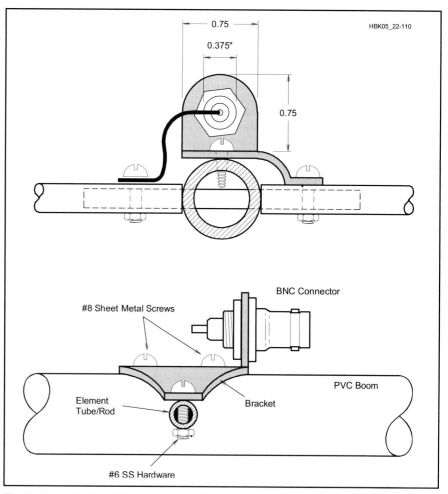

Fig 21.121 — Details of the feed point of the Yagi, showing the BNC connector, mounting plate, and connections to the ½-inch driver element halves placed over a central ⅜-inch fiberglass rod.

point. If you use an alternative plastic material, observe the same caution and be certain that the rod remains strong after drilling. Do not use wooden dowels for this application, since they do not have sufficient strength. Position the holes about ¼ to ⅜ inch from the tubing end where it presses against the boom. One hole will receive a solder lug and the other will connect to an extension of the BNC mounting plate.

Second, install the fiberglass rod through the boom. You can leave it loose, since the elements will press against the boom and hold it in place. Alternatively, you may glue it in place with a two-part epoxy. Slide the driver element tubes over the rod and test the holes for alignment by placing the #6 bolts in them.

Next, cut and shape the BNC mounting plate from 1/16-inch thick aluminum. The fitting is made from a scrap of L-stock 1 inch on a side. Before cutting the stock, drill the ⅜-inch hole needed for the BNC connector. Then cut the vertical portion. The horizontal portion requires a curved tab that reaches the bolt on one side of the boom. Use a bench vise to bend the tab in a curve and then flatten it for the bolt-hole. It takes several tries to get the shape and tab exact, so be patient. When the squared-edge piece finds its perfect shape, use a disk sander and round the vertical piece to follow the connector shape. Taper the tap edges to minimize excess material. The last step is to drill the mounting holes that receive the #8 sheet metal screws.

Mounting the assembly involves loosely attaching both the #6 and #8 hardware and alternately tightening up all pieces. Be certain that the side of the BNC connector that receives the coax points toward the mast. Next, mount the BNC connector. The shield side is already connected to one side of the driver. Mount the other side of the driver, placing a solder lug under the bolt head. Connect a short wire as directly as possible from the solder lug to the center pin of the BNC connector. After initial testing, you may coat all exposed connections with Plasti-Dip for weather protection.

TUNE-UP

Testing and tuning the antenna is a simple process if you build carefully. The only significant test that you can perform is to ensure that the SWR curve comes close to the one shown in Fig 21.116. If the SWR is high at 148 MHz but very low at 144 MHz, then you will need to shorten the driver ends by a small amount — no more than ⅛ inch per end at a time. Shaving the ends with a disk sander is most effective.

Using the antenna with vertical polarization will require good spacing from any support structure with metal vertical portions. One of the easiest ways to devise such a mounting is to create a PVC structure hat turns the entire boom by 90°. If you feel the need for added support, you can create can angular brace by placing 45° connectors in both the vertical and horizontal supports and running a length of PVC between them.

As an alternative, you can let the rear part of the boom be slightly long. To this end you can cement PVC fixtures — including the screw-thread series to enlarge the support pipe size. Create a smooth junction that you attach with a through-bolt instead of cement. By drilling one side of the connection with two sets of holes, 90° apart, you can change the antenna from horizontal polarization to vertical and back in short order.

The six-element OWA Yagi for 2 meters performs well. It serves as a good utility antenna with more gain and directivity than the usual three-element general-use Yagi. When vertically polarized, the added gain confirms the wisdom of using a longer boom and more elements. With a length under five feet, the antenna is still compact. The ability to disassemble the parts simplifies moving the antenna to various portable sites.

Project: Cheap Yagis by WA5VJB

If you're planning to build an EME array, don't use these antennas. But if you want to put together a VHF rover station with less than $500 in the antennas, read on as Kent Britain, WA5VJB, shows you how to put together a VHF/UHF Yagi with QRO performance at a QRP price. (This material is adapted from Kent's on-line paper "Controlled Impedance 'Cheap' Antennas" at www.wa5vjb.com/references.html.)

The simplified feed uses the structure of the antenna itself for impedance matching. So the design started with the feed and the elements were built around it. The antennas were designed with *YagiMax*, tweaked in *NEC*, and the driven elements experimentally determined on the antenna range.

Typically a high-gain antenna is designed in the computer, then you try to come up with a driven element matching arrangement for whatever feed point impedance the computer comes up with. In this design, compromises for the feed impedance, asymmetrical feed, simple measurements, wide bandwidth, the ability to grow with the same spacing, and trade-offs for a very clean pattern cost many dB of gain. But you can build these antennas for about $5!

Construction of the antennas is straightforward. The boom is ¾-inch square, or ½-inch by ¾-inch wood. To install an element, drill a hole through the boom and insert the element. A drop of cyanoacrylate "super glue," epoxy, or silicone adhesive is used to hold the elements in place. There is no boom-to-mast plate — drill holes in the boom and use a U-bolt to attach it to the mast!

The life of the antenna is determined by what you coat it with. The author had a 902-MHz version in the air varnished with polyurethane for two years with little deterioration.

The parasitic elements on prototypes have been made from silicon-bronze welding rod, aluminum rod, brass hobby tubing, and #10 or #12 AWG solid copper ground wire. So that you can solder to the driven element, use the welding rod, hobby tubing, or copper wire. The driven element is folded at one end with its ends inserted through the boom.

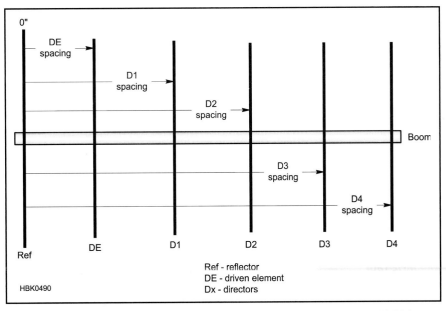

Fig 21.122 — Element spacing for the Cheap Yagis. Refer to Tables 21.23 to 21.30 for exact dimensions for the various bands.

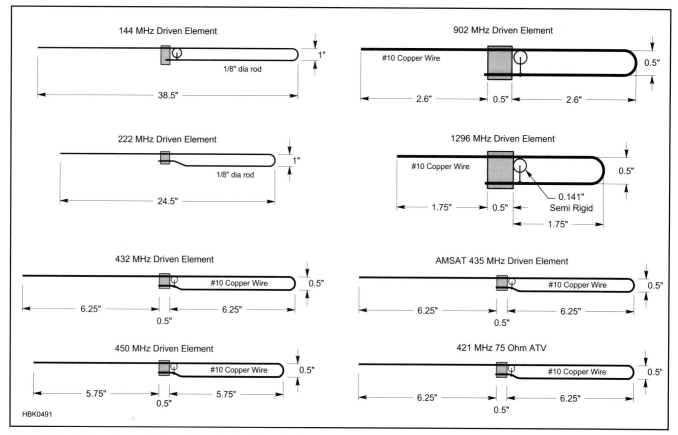

Fig 21.123 — Driven element dimensions for the Cheap Yagis. Attaching the coax shield to the center of the driven element is appropriate because that is the lowest impedance point of the element.

Fig 21.122 shows the basic plan for the antenna and labels the dimensions that are given in the table for each band. All table dimensions are given in inches.

Fig 21.123 shows how the driven element is constructed for each antenna. Trim the free end of the driven element to tune it for minimum SWR at the desired frequency. **Fig 21.124** shows how to attach coaxial cable to the feed point. Sliding a quarter-wave sleeve along the coax had little effect, so there's not much RF on the outside of the coax. You may use a ferrite bead choke balun if you like, but these antennas are designed for minimum expense!

Finally a bit of history on the design of these antennas. In 1993 at the Oklahoma City Central States VHF Society Conference, Arnie, CO2KK spoke on the difficulties building VHF antennas in non-industrialized nations. Just run down to the store and pick up some Delrin insulators and 0.141 inch Teflon coax? Arnie's tales were the motivation to use advanced technology to come up with something simple.

144 MHz Yagi

While others have reported good luck with 16-element long-boom wood antennas, six elements was about the maximum for most rovers. The design is peaked at 144.2 MHz, but performance is still good at 146.5 MHz. All parasitic elements are made from 3/16-inch aluminum rod and the driven element is made from 1/8-inch rod. Lengths and spacings are given in **Table 21.23**.

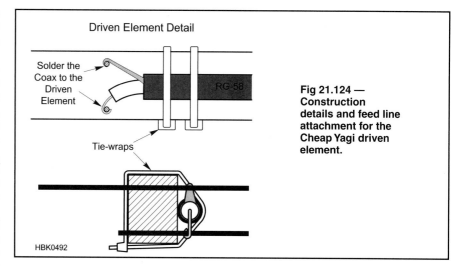

Fig 21.124 — Construction details and feed line attachment for the Cheap Yagi driven element.

222 MHz Yagi

This antenna is peaked at 222.1 MHz, but performance has barely changed at 223.5 MHz. You can drill the mounting holes to mount it with the elements horizontal or vertical. All parasitic elements are made from 3/16-inch aluminum rod and the driven element is made from 1/8-in. rod. Lengths and spacings are given in **Table 21.24**.

432 MHz Yagi

At this band the antenna is getting very practical and easy to build. All parasitic elements are made from 1/8-inch diameter rod and the driven element is made from #10 AWG solid copper wire. Lengths and spacings are given in **Table 21.25**.

435 MHz Yagi for AMSAT

KA9LNV provided help and motivation for these antennas. A high front-to-back ratio (F/B) was a major design consideration of all versions. The model predicts 30 dB F/B for the six-element and over 40 dB for the others. For gain, *NEC* predicts 11.2 dBi for the six-element, 12.6 dBi for the eight-element, and 13.5 dBi for the 10-element, and 13.8 dBi for the 11-element.

Using ¾-inch square wood for the boom makes it easy to build two antennas on the same boom for cross-polarization. Offset the two antennas 6½ inch along the boom and feed them in-phase for circular polarization, or just use one for portable operations. All parasitic elements are made from ⅛-inch diameter rod and the driven element is made from #10 AWG solid copper wire. Lengths and spacings are given in **Table 21.26**. The same element spacing is used for all four versions of the antenna.

450 MHz Yagi

For FM, this six-element Yagi is a good, cheap antenna to get a newcomer into a repeater or make a simplex-FM QSO during a contest. RadioShack ⅛ inch diameter aluminum ground wire (catalog #15-035) was used in the prototype for all the elements except the driven element, which is made from #10 AWG solid copper wire. Other ⅛-inch diameter material could be used. Lengths and spacings are given in **Table 21.27**.

902 MHz Yagi

This was the first antenna the author built using the antenna to control the driven element impedance. The 2.5-ft length has proven very practical. All parasitic elements are made from ⅛-inch-diameter rod and the driven element is made from #10 AWG solid copper wire. Lengths and spacings are given in **Table 21.28**.

1296 MHz Yagi

This antenna is the veteran of several "Grid-peditions" and has measured 13.5 dBi on the Central States VHF Society antenna range. Dimensions must be followed with great care. The driven element is small enough to allow 0.141-inch semi-rigid coax to be used. The prototype antennas use ⅛-inch silicon-bronze welding rod for the elements, but any ⅛-inch-diameter material can be used. The driven element is made from #10 AWG solid copper wire. Lengths and spacings are given in **Table 21.29**.

421.25 MHz 75-Ω Yagi for ATV

421 MHz vestigial sideband video is popular in North Texas for receiving the FM video input repeaters. These antennas are made for 421 MHz use and the driven element is designed for 75 Ω. RG-59 or an F adapter to RG-6 can be directly connected to a cable-TV converter or cable-ready TV on channel 57. All parasitic elements are made from ⅛-inch diameter rod and the driven element is made from #10 AWG solid copper wire. Lengths and spacings are given in **Table 21.30**. The same spacing is used for all versions.

Table 21.23
WA5VJB 144 MHz Yagi Dimensions

		Ref	DE	D1	D2	D3	D4
3-element	Length	41.0	—	37.0			
	Spacing	0	8.5	20.0			
4-element	Length	41.0	—	37.5	33.0		
	Spacing	0	8.5	19.25	40.5		
6-element	Length	40.5	—	37.5	36.5	36.5	32.75
	Spacing	0	7.5	16.5	34.0	52.0	70.0

Dimensions in inches.

Table 21.24
WA5VJB 222 MHz Yagi Dimensions

		Ref	DE	D1	D2	D3	D4
3-element	Length	26.0	—	23.75			
	Spacing	0	5.5	13.5			
4-element	Length	26.25	—	24.1	22.0		
	Spacing	0	5.0	11.75	23.5		
6-element	Length	26.25	—	24.1	23.5	23.5	21.0
	Spacing	0	5.0	10.75	22.0	33.75	45.5

Dimensions in inches.

Table 21.25
WA5VJB 432 MHz Yagi Dimensions

		Ref	DE	D1	D2	D3	D4	D5	D6	D7	D8	D9
6-element	Length	13.5	—	12.5	12.0	12.0	11.0					
	Spacing	0	2.5	5.5	11.25	17.5	24.0					
8-element	Length	13.5	—	12.5	12.0	12.0	12.0	12.0	11.25			
	Spacing	0	2.5	5.5	11.25	17.5	24.0	30.75	38.0			
11-element	Length	13.5	—	12.5	12.0	12.0	12.0	12.0	12.0	11.75	11.75	11.0
	Spacing	0	2.5	5.5	11.25	17.5	24.0	30.75	38.0	45.5	53.0	59.5

Dimensions in inches.

Table 21.26
WA5VJB 435 MHz Yagi Dimensions

		Ref	DE	D1	D2	D3	D4	D5	D6	D7	D8	D9
6-element	Length	13.4	—	12.4	12.0	12.0	11.0					
8-element	Length	13.4	—	12.4	12.0	12.0	12.0	12.0	11.1			
10-element	Length	13.4	—	12.4	12.0	12.0	12.0	12.0	11.75	11.75	11.1	
11-element	Length	13.4	—	12.4	12.0	12.0	12.0	12.0	11.75	11.75	11.75	11.1
	Spacing	0	2.5	5.5	11.25	17.5	24.0	30.5	37.75	45.0	52.0	59.5

Dimensions in inches.

Table 21.27
WA5VJB 450 MHz Yagi Dimensions

		Ref	DE	D1	D2	D3	D4
6-element	Length	13.0	—	12.1	11.75	11.75	10.75
	Spacing	0	2.5	5.5	11.0	18.0	28.5

Dimensions in inches.

Table 21.28
WA5VJB 902 MHz Yagi Dimensions

		Ref	DE	D1	D2	D3	D4	D5	D6	D7	D8
10-element	Length	6.2	—	5.6	5.5	5.5	5.4	5.3	5.2	5.1	5.1
	Spacing	0	2.4	3.9	5.8	9.0	12.4	17.4	22.4	27.6	33.0

Dimensions in inches.

Table 21.29
WA5VJB 1296 MHz Yagi Dimensions

		Ref	DE	D1	D2	D3	D4	D5	D6	D7	D8
10-element	Length	4.3	—	3.9	3.8	3.75	3.75	3.65	3.6	3.6	3.5
	Spacing	0	1.7	2.8	4.0	6.3	8.7	12.2	15.6	19.3	23.0

Dimensions in inches.

Table 21.30
WA5VJB 421.25 MHz 75-Ω Yagi Dimensions

		Ref	DE	D1	D2	D3	D4	D5	D6	D7	D8	D9
6-element	Length	14.0	—	12.5	12.25	12.25	11.0					
9-element	Length	14.0	—	12.5	12.25	12.25	12.0	12.0	11.25			
11-element	Length	14.0	—	12.5	12.25	12.25	12.0	12.0	12.0	11.75	11.75	11.5
	Spacing	0	3.0	6.5	12.25	17.75	24.5	30.5	36.0	43.0	50.25	57.25

Dimensions in inches.

21.12 Radio Direction Finding Antennas

Radio direction finding (RDF) is almost as old as radio communication. It gained prominence when the British Navy used it to track the movement of enemy ships in World War I. Since then, governments and the military have developed sophisticated and complex RDF systems. Fortunately, simple equipment, purchased or built at home, is quite effective in Amateur Radio RDF.

In European and Asian countries, direction-finding contests are foot races. The object is to be first to find four or five transmitters in a large wooded park. Young athletes have the best chance of capturing the prizes. This sport is known as *foxhunting* (after the British hill-and-dale horseback events) or *ARDF* (Amateur Radio direction finding).

In North America and England, most RDF contests involve mobiles — cars, trucks, and vans, even motorcycles. It may be possible to drive all the way to the transmitter, or there may be a short hike at the end, called a *sniff*.

These competitions are also called foxhunting by some, while others use *bunny hunting*, *T-hunting* or the classic term *hidden transmitter hunting*.

In the 1950s, 3.5 and 28 MHz were the most popular bands for hidden transmitter hunts. Today, most competitive hunts worldwide are for 144-MHz FM signals, though other VHF bands are also used. Some international foxhunts include 3.5-MHz events.

Even without participating in RDF contests, you will find knowledge of the techniques useful. They simplify the search for a neighborhood source of power-line interference or TV cable leakage. RDF must be used to track down emergency radio beacons, which signal the location of pilots and boaters in distress. Amateur Radio enthusiasts skilled in transmitter hunting are in demand by agencies such as the Civil Air Patrol and the US Coast Guard Auxiliary for search and rescue support. RDF is an important part of the evidence-gathering process in interference cases.

The most basic RDF system consists of a directional antenna and a method of detecting and measuring the level of the radio signal, such as a receiver with signal strength indicator. RDF antennas range from a simple tuned loop of wire to an acre of antenna elements with an electronic beam-forming network. Other sophisticated techniques for RDF use the Doppler effect or measure the time of arrival difference of the signal at multiple antennas.

All of these methods have been used from 2 to 500 MHz and above. However, RDF practices vary greatly between the HF and VHF/UHF portions of the spectrum. For practical reasons, high gain beams, Dopplers and switched dual antennas find favor on VHF/UHF, while loops and phased arrays are the most popular choices on 6 meters and below. Signal propagation differences between HF and VHF also affect RDF practices. But many basic transmitter-hunting techniques, discussed later in this chapter, apply to all bands and all types of portable RDF equipment.

21.12.1 RDF Antennas for HF Bands

Below 50 MHz, gain antennas such as Yagis and quads are of limited value for RDF. The typical tribander installation yields only a general direction of the incoming signal, due to ground effects and the antenna's broad forward lobe. Long monoband beams at greater heights work better, but still cannot achieve the bearing accuracy and repeatability of simpler antennas designed specifically for RDF.

RDF LOOPS

An effective directional HF antenna can be as uncomplicated as a small loop of wire or tubing, tuned to resonance with a capacitor. When immersed in an electromagnetic field, the loop acts much the same as the secondary winding of a transformer. The voltage at the output is proportional to the amount of flux passing through it and the number of turns. If the loop is oriented such that the greatest amount of area is presented to the magnetic field, the induced voltage will be the highest. If it is rotated so that little or no area is cut by the field lines, the voltage induced in the loop is zero and a null occurs.

To achieve this transformer effect, the loop must be small compared with the signal wavelength. In a single-turn loop, the conductor should be less than 0.08-λ long. For example, a 28-MHz loop should be less than 34 inches in circumference, giving a diameter of approximately 10 inches. The loop may be smaller, but that will reduce its voltage output. Maximum output from a small loop antenna is in directions corresponding to the plane of the loop; these lobes are very broad. Sharp nulls, obtained at right angles to that plane, are more useful for RDF.

For a perfect bidirectional pattern, the loop must be balanced electrostatically with respect to ground. Otherwise, it will exhibit two modes of operation, the mode of a perfect loop and that of a non-directional vertical antenna of small dimensions. This dual-mode condition results in mild to severe inaccuracy, depending on the degree of imbalance, because the outputs of the two modes are not in phase.

The theoretical true loop pattern is illustrated in **Fig 21.125A**. When properly balanced, there are two nulls exactly 180° apart. When the unwanted antenna effect is appreciable and the loop is tuned to resonance, the loop may exhibit little directivity, as shown in Fig 21.125B. By detuning the loop to shift the phasing, you may obtain a useful pattern similar to Fig 21.125C. While

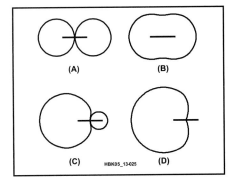

Fig 21.125 — Small loop field patterns with varying amounts of antenna effect — the undesired response of a loop acting merely as a mass of metal connected to the receiver antenna terminals. The horizontal lines show the plane of the loop turns.

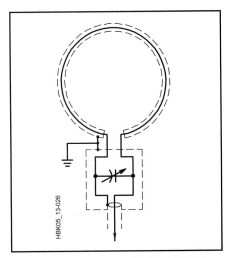

Fig 21.126 — Electrostatically-shielded loop for RDF. To prevent shielding of the loop from magnetic fields, leave the shield unconnected at one end.

not symmetrical, and not necessarily at right angles to the plane of the loop, this pattern does exhibit a pair of nulls.

By careful detuning and amplitude balancing, you can approach the unidirectional pattern of Fig 21.125D. Even though there may not be a complete null in the pattern, it resolves the 180° ambiguity of Fig 21.125A. Korean War-era military loop antennas, sometimes available on today's surplus market, use this controlled-antenna-effect principle.

An easy way to achieve good electrostatic balance is to shield the loop, as shown in **Fig 21.126**. The shield, represented by the dashed lines in the drawing, eliminates the antenna effect. The response of a well-constructed shielded loop is quite close to the ideal pattern of Fig 21.125A.

For 160 through 30 meters, single-turn loops that are small enough for portability are

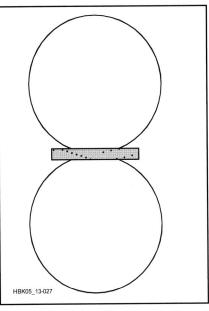

Fig 21.127 — Field pattern for a ferrite-rod antenna. The dark bar represents the rod on which the loop turns are wound.

usually unsatisfactory for RDF work. Multi-turn loops are generally used instead. They are easier to resonate with practical capacitor values and give higher output voltages. This type of loop may also be shielded. If the total conductor length remains below 0.08 λ, the directional pattern is that of Fig 21.125A.

FERRITE ROD ANTENNAS

Another way to get higher loop output is to increase the permeability of the medium in the vicinity of the loop. By winding a coil of wire around a form made of high-permeability material, such as ferrite rod, much greater flux is obtained in the coil without increasing the cross-sectional area.

Modern magnetic core materials make compact directional receiving antennas practical. Most portable AM broadcast receivers use this type of antenna, commonly called a *loopstick*. The loopstick is the most popular RDF antenna for portable/mobile work on 160 and 80 meters.

Like the shielded loop discussed earlier, the loopstick responds to the magnetic field of the incoming radio wave, and not to the electrical field. For a given size of loop, the output voltage increases with increasing flux density, which is obtained by choosing a ferrite core of high permeability and low loss at the frequency of interest. For increased output, the turns may be wound over two rods taped together. A practical loopstick antenna is described later in this chapter.

A loop on a ferrite core has maximum signal response in the plane of the turns, just as an air core loop. This means that maximum response of a loopstick is broadside to the axis of the rod, as shown in **Fig 21.127**. The loopstick may be shielded to eliminate the antenna effect; a U-shaped or C-shaped channel of aluminum or other form of "trough" is best. The shield must not be closed, and its length should equal or slightly exceed the length of the rod.

SENSE ANTENNAS

Because there are two nulls 180° apart in the directional pattern of a small loop or loopstick, there is ambiguity as to which null indicates the true direction of the target station. For example, if the line of bearing runs east and west from your position, you have no way of knowing from this single bearing whether the transmitter is east of you or west of you.

If bearings can be taken from two or more positions at suitable direction and distance from the transmitter, the ambiguity can be resolved and distance can be estimated by triangulation, as discussed later in this chapter. However, it is almost always desirable to be able to resolve the ambiguity immediately by having a unidirectional antenna pattern available.

You can modify a loop or loopstick antenna pattern to have a single null by adding a second antenna element. This element is called a *sense antenna*, because it senses the phase of the signal wavefront for comparison with the phase of the loop output signal. The sense element must be omnidirectional, such as a short vertical. When signals from the loop and the sense antenna are combined with 90° phase shift between the two, a heart-shaped (cardioid) pattern results, as shown in **Fig 21.128A**.

Fig 21.128B shows a circuit for adding a sense antenna to a loop or loopstick. For the best null in the composite pattern, signals from the loop and sense antennas must be of equal amplitude. R1 adjusts the level of the signal from the sense antenna.

In a practical system, the cardioid pattern null is not as sharp as the bidirectional null of the loop alone. The usual procedure when transmitter hunting is to use the loop alone to obtain a precise line of bearing, then switch in the sense antenna and take another reading to resolve the ambiguity.

PHASED ARRAYS AND ADCOCK ANTENNAS

Two-element phased arrays are popular for amateur HF RDF base station installations. Many directional patterns are possible, depending on the spacing and phasing of the elements. A useful example is two ½-λ elements spaced ¼-λ apart and fed 90° out of phase. The resultant pattern is a cardioid, with a null off one end of the axis of the two antennas and a broad peak in the opposite direction. The directional frequency range of

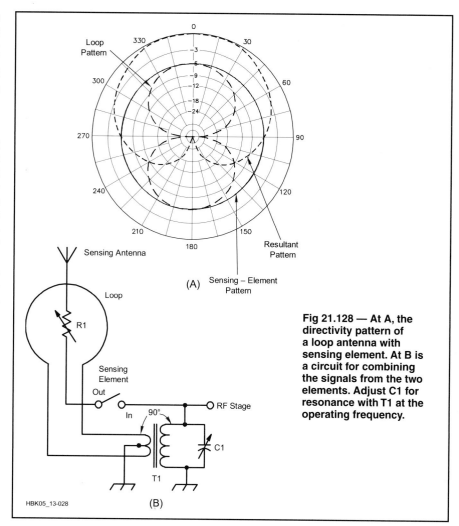

Fig 21.128 — At A, the directivity pattern of a loop antenna with sensing element. At B is a circuit for combining the signals from the two elements. Adjust C1 for resonance with T1 at the operating frequency.

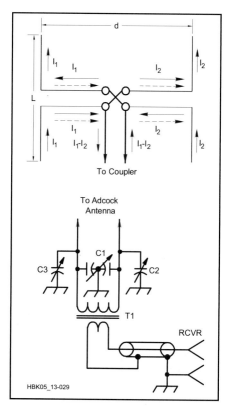

Fig 21.129 — A simple Adcock antenna and its coupler.

this antenna is limited to one band, because of the critical length of the phasing lines.

The best-known phased array for RDF is the Adcock, named after the man who invented it in 1919. It consists of two vertical elements fed 180° apart, mounted so the array may be rotated. Element spacing is not critical, and may be in the range from 0.1 to 0.75 λ. The two elements must be of identical lengths, but need not be self-resonant; shorter elements are commonly used. Because neither the element spacing nor length is critical in terms of wavelengths, an Adcock array may operate over more than one amateur band.

Fig 21.129 is a schematic of a typical Adcock configuration, called the H-Adcock because of its shape. Response to a vertically polarized wave is very similar to a conventional loop. The passing wave induces currents I1 and I2 into the vertical members. The output current in the transmission line is equal to their difference. Consequently, the directional pattern has two broad peaks and

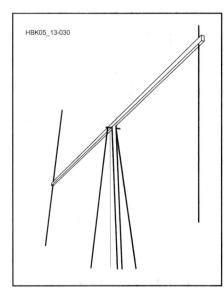

Fig 21.130 — An experimental Adcock antenna on a wooden frame.

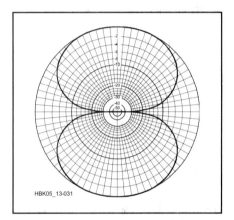

Fig 21.131 — The pattern of an Adcock array with element spacing of ½ wavelength. The elements are aligned with the vertical axis.

two sharp nulls, like the loop. The magnitude of the difference current is proportional to the spacing (d) and length (l) of the elements. You will get somewhat higher gain with larger dimensions. The Adcock of **Fig 21.130**, designed for 40 meters, has element lengths of 12 ft and spacing of 21 ft (approximately 0.15 λ).

Fig 21.131 shows the radiation pattern of the Adcock. The nulls are broadside to the axis of the array, becoming sharper with increased element spacing. When element spacing exceeds ¾ λ, however, the antenna begins to take on additional unwanted nulls off the ends of the array axis.

The Adcock is a vertically polarized antenna. The vertical elements do not respond to horizontally polarized waves, and the currents induced in the horizontal members by a horizontally polarized wave (dotted arrows in Fig 21.129) tend to balance out regardless of the orientation of the antenna.

Since the Adcock uses a balanced feed system, a coupler is required to match the unbalanced input of the receiver. T1 is an air-wound coil with a two-turn link wrapped around the middle. The combination is resonated with C1 to the operating frequency. C2 and C3 are null-clearing capacitors. Adjust them by placing a low-power signal source some distance from the antenna and exactly broadside to it. Adjust C2 and C3 until the deepest null is obtained.

While you can use a metal support for the mast and boom, wood is preferable because of its non-conducting properties. Similarly, a mast of thick-wall PVC pipe gives less distortion of the antenna pattern than a metallic mast. Place the coupler on the ground below the wiring harness junction on the boom and connect it with a short length of 300-Ω twin-lead-feed line.

LOOPS VS PHASED ARRAYS

Loops are much smaller than phased arrays for the same frequency, and are thus the obvious choice for portable/mobile HF RDF. For base stations in a triangulation network, where the 180° ambiguity is not a problem, Adcocks are preferred. In general, they give sharper nulls than loops, but this is in part a function of the care used in constructing and feeding the individual antennas, as well as of the spacing of the elements. The primary construction considerations are the shielding and balancing of the feed line against unwanted signal pickup and the balancing of the antenna for a symmetrical pattern. Users report that Adcocks are somewhat less sensitive to proximity effects, probably because their larger aperture offers some space diversity.

Skywave Considerations

Until now we have considered the directional characteristics of the RDF loop only in the two-dimensional azimuthal plane. In three-dimensional space, the response of a vertically oriented small loop is doughnut-shaped. The bidirectional null (analogous to a line through the doughnut hole) is in the line of bearing in the azimuthal plane and toward the horizon in the vertical plane. Therefore, maximum null depth is achieved only on signals arriving at 0° elevation angle.

Skywave signals usually arrive at nonzero wave angles. As the elevation angle increases, the null in a vertically oriented loop pattern becomes shallower. It is possible to tilt the loop to seek the null in elevation as well as azimuth. Some amateur RDF enthusiasts report success at estimating distance to the target by measurement of the elevation angle with a tilted loop and computations based on estimated height of the propagating ionospheric layer. This method seldom provides high accuracy with simple loops, however.

Most users prefer Adcocks to loops for skywave work, because the Adcock null is present at all elevation angles. Note, however, that an Adcock has a null in all directions from signals arriving from overhead. Thus for very high angles, such as under-250-mile skip on 80 and 40 meters, neither loops nor Adcocks will perform well.

ELECTRONIC ANTENNA ROTATION

State-of-the-art fixed RDF stations for government and military work use antenna arrays of stationary elements, rather than mechanically rotatable arrays. The best-known type is the *Wullenweber antenna*. It has a large number of elements arranged in a circle, usually outside of a circular reflecting screen. Depending on the installation, the circle may be anywhere from a few hundred feet to more than a quarter of a mile in diameter. Although the Wullenweber is not practical for most amateurs, some of the techniques it uses may be applied to amateur RDF.

The device, which permits rotating the antenna beam without moving the elements, has the classic name *radio goniometer*, or simply *goniometer*. Early goniometers were RF transformers with fixed coils connected to the array elements and a moving pickup coil connected to the receiver input. Both amplitude and phase of the signal coupled into the pickup winding are altered with coil rotation in a way that corresponded to actually rotating the array itself. With sufficient elements and a goniometer, accurate RDF measurements can be taken in all compass directions.

Beam Forming Networks

By properly sampling and combining signals from individual elements in a large array, an antenna beam is electronically rotated or steered. With an appropriate number and arrangement of elements in the system, it is possible to form almost any desired antenna pattern by summing the sampled signals in

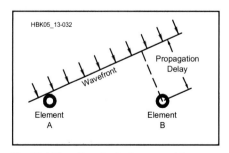

Fig 21.132 — One technique used in electronic beam forming. By delaying the signal from element A by an amount equal to the propagation delay, two signals are summed precisely in phase, even though the signal is not in the broadside direction.

appropriate amplitude and phase relationships. Delay networks and/or attenuation are added in line with selected elements before summation to create these relationships.

To understand electronic beam forming, first consider just two elements, shown as A and B in **Fig 21.132**. Also shown is the wavefront of a radio signal arriving from a distant transmitter. The wavefront strikes element A first, then travels somewhat farther before it strikes element B. Thus, there is an interval between the times that the wavefront reaches elements A and B.

We can measure the differences in arrival times by delaying the signal received at element A before summing it with that from element B. If two signals are combined directly, the amplitude of the sum will be maximum when the delay for element A exactly equals the propagation delay, giving an in-phase condition at the summation point. On the other hand, if one of the signals is inverted and the two are added, the signals will combine in a 180° out-of-phase relationship when the element A delay equals the propagation delay, creating a null. Either way, once the time delay is determined by the amount of delay required for a peak or null, we can convert it to distance. Then trigonometry calculations provide the direction from which the wave is arriving.

Altering the delay in small increments steers the peak (or null) of the antenna. The system is not frequency sensitive, other than the frequency range limitations of the array elements. Lumped-constant networks are suitable for delay elements if the system is used only for receiving. Delay lines at installations used for transmitting and receiving employ rolls of coaxial cable of various lengths, chosen for the time delay they provide at all frequencies, rather than as simple phasing lines designed for a single frequency.

Combining signals from additional elements narrows the broad beamwidth of the pattern from the two elements and suppress unwanted sidelobes. Electronically switching the delays and attenuations to the various elements causes the formed beam to rotate around the compass. The package of electronics that does this, including delay lines and electronically switched attenuators, is the beam-forming network.

21.12.2 Methods for VHF/UHF RDF

Three distinct methods of mobile RDF are commonly in use by amateurs on VHF/UHF bands: directional antennas, switched dual antennas and Dopplers. Each has advantages over the others in certain situations. Many RDF enthusiasts employ more than one method when transmitter hunting.

Fig 21.133 — The mobile RDF installation of WB6ADC features a thin wire quad for 144 MHz and a mechanical linkage that permits either the driver or front passenger to rotate the mast by hand.

Fig 21.134 — K0OV uses this mobile setup for RDF on several bands, with separate antennas for each band that mate with a common lower mast section, pointer and 360° indicator. Antenna shown is a heavy gauge wire quad for 2 meters.

DIRECTIONAL ANTENNAS

Ordinary mobile transceivers and handhelds work well for foxhunting on the popular VHF bands. If you have a lightweight beam and your receiver has an easy-to-read S-meter, you are nearly ready to start. All you need is an RF attenuator and some way to mount the setup in your vehicle.

Amateurs seldom use fractional wavelength loops for RDF above 60 MHz because they have bidirectional characteristics and low sensitivity, compared to other practical VHF antennas. Sense circuits for loops are difficult to implement at VHF, and signal reflections tend to fill in the nulls. Typically VHF loops are used only for close-in sniffing where their compactness and sharp nulls are assets, and low gain is of no consequence.

Phased Arrays

The small size and simplicity of two-element driven arrays make them a common choice of newcomers at VHF RDF. Antennas such as phased ground planes and ZL Specials have modest gain in one direction and a null in the opposite direction. The gain is helpful when the signal is weak, but the broad response peak makes it difficult to take a precise bearing.

As the signal gets stronger, it becomes possible to use the null for a sharper S-meter indication. However, combinations of direct and reflected signals (called *multipath*) will distort the null or perhaps obscure it completely. For best results with this type of antenna, always find clear locations from which to take bearings.

Parasitic Arrays

Parasitic arrays are the most common RDF antennas used by transmitter hunters in high competition areas such as Southern California. Antennas with significant gain are a necessity due to the weak signals often encountered on weekend-long T-hunts, where the transmitter may be over 200 miles distant. Typical 144-MHz installations feature Yagis or quads of three to six elements, sometimes more. Quads are typically home-built, using data from *The ARRL Antenna Book* and *Transmitter Hunting* (see Bibliography).

Two types of mechanical construction are popular for mobile VHF quads. The model of **Fig 21.133** uses thin gauge wire (solid or stranded), suspended on wood dowel or fiberglass rod spreaders. It is lightweight and easy to turn rapidly by hand while the vehicle moves. Many hunters prefer to use larger gauge solid wire (such as #10 AWG) on a PVC plastic pipe frame (**Fig 21.134**). This quad is more rugged and has somewhat wider frequency range, at the expense of increased weight and wind resistance. It can get mashed going under a willow, but it is easily reshaped and returned to service.

Yagis are a close second to quads in popularity. Commercial models work fine for VHF RDF, provided that the mast is attached at a good balance point. Lightweight and small-diameter elements are desirable for ease of turning at high speeds.

A well-designed mobile Yagi or quad installation includes a method of selecting wave polarization. Although vertical polarization is the norm for VHF-FM communications, horizontal polarization is allowed on many T-hunts. Results will be poor if a VHF RDF antenna is cross-polarized to the transmitting antenna, because multipath and scattered signals (which have indeterminate polarization) are enhanced, relative to the cross-polarized direct signal. The installation of Fig 21.133 features a slip joint at the boom-to-mast junction, with an actuating cord to rotate the boom, changing the polarization. Mechanical stops limit the boom rotation to 90°.

Parasitic Array Performance for RDF

The directional gain of a mobile beam (typically 8 dB or more) makes it unexcelled for both weak signal competitive hunts and for locating interference such as TV cable leakage. With an appropriate receiver, you can get bearings on any signal mode, including FM, SSB, CW, TV, pulses and noise. Because only the response peak is used, the null-fill problems and proximity effects of loops and phased arrays do not exist.

You can observe multiple directions of arrival while rotating the antenna, allowing you to make educated guesses as to which signal peaks are direct and which are from non-direct paths or scattering. Skilled operators can estimate distance to the transmitter from the rate of signal strength increase with distance traveled. The RDF beam is useful for transmitting, if necessary, but use care not to damage an attenuator in the coax line by transmitting through it.

The 3-dB beamwidth of typical mobile-mount VHF beams is on the order of 80°. This is a great improvement over 2-element driven arrays, but it is still not possible to get pinpoint bearing accuracy. You can achieve errors of less than 10° by carefully reading the S-meter. In practice, this is not a major hindrance to successful mobile RDF. Mobile users are not as concerned with precise bearings as fixed station operators, because mobile readings are used primarily to give the general direction of travel to "home in" on the signal. Mobile bearings are continuously updated from new, closer locations.

Amplitude-based RDF may be very difficult when signal level varies rapidly. The transmitter hider may be changing power, or the target antenna may be moving or near a well-traveled road or airport. The resultant rapid S-meter movement makes it hard to take accurate bearings with a quad. The process is slow because the antenna must be carefully rotated by hand to "eyeball average" the meter readings.

SWITCHED ANTENNA RDF UNITS

Three popular types of RDF systems are relatively insensitive to variations in signal level. Two of them use a pair of vertical di-

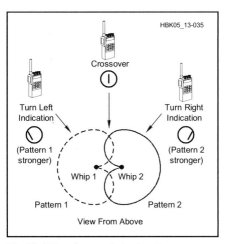

Fig 21.135 — In a switched pattern RDF set, the responses of two cardioid antenna patterns are summed to drive a zero center indicator.

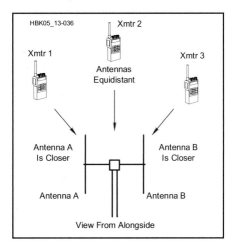

Fig 21.136 — A dual-antenna TDOA RDF system has a similar indicator to a switched pattern unit, but it obtains bearings by determining which of its antennas is closer to the transmitter.

pole antennas, spaced ½ λ or less apart, and alternately switched at a rapid rate to the input of the receiver. In use, the indications of the two systems are similar, but the principles are different.

Switched Pattern Systems

The switched pattern RDF set (**Fig 21.135**) alternately creates two cardioid antenna patterns with lobes to the left and the right. The patterns are generated in much the same way as in the phased arrays described above. PIN RF diodes select the alternating patterns. The combined antenna outputs go to a receiver with AM detection. Processing after the detector output determines the phase or amplitude difference between the patterns' responses to the signal.

Switched pattern RDF sets typically have a zero center meter as an indicator. The meter swings negative when the signal is coming from the user's left, and positive when the signal source is on the right. When the plane of the antenna is exactly perpendicular to the direction of the signal source, the meter reads zero.

The sharpness of the zero crossing indication makes possible more precise bearings than those obtainable with a quad or Yagi. Under ideal conditions with a well-built unit, null direction accuracy is within 1°. Meter deflection tells the user which way to turn to zero the meter. For example, a negative (left) reading requires turning the antenna left. This solves the 180° ambiguity caused by the two zero crossings in each complete rotation of the antenna system.

Because it requires AM detection of the switched pattern signal, this RDF system finds its greatest use in the 120-MHz aircraft band, where AM is the standard mode. Commercial manufacturers make portable RDF sets with switched pattern antennas and built-in receivers for field portable use. These sets can usually be adapted to the amateur 144-MHz band. Other designs are adaptable to any VHF receiver that covers the frequency of interest and has an AM detector built in or added.

Switched pattern units work well for RDF from small aircraft, for which the two vertical antennas are mounted in fixed positions on the outside of the fuselage or simply taped inside the windshield. The left-right indication tells the pilot which way to turn the aircraft to home in. Since street vehicles generally travel only on roads, fixed mounting of the antennas on them is undesirable. Mounting vehicular switched-pattern arrays on a rotatable mast is best.

Time-of-Arrival Systems

Another kind of switched antenna RDF set uses the difference in arrival times of the signal wavefront at the two antennas. This narrow-aperture Time-Difference-of-Arrival (TDOA) technology is used for many sophisticated military RDF systems. The rudimentary TDOA implementation of **Fig 21.136** is quite effective for amateur use. The signal from transmitter 1 reaches antenna A before antenna B. Conversely, the signal from transmitter 3 reaches antenna B before antenna A. When the plane of the antenna is perpendicular to the signal source (as transmitter 2 is in the figure), the signal arrives at both antennas simultaneously.

If the outputs of the antennas are alternately switched at an audio rate to the receiver input, the differences in the arrival times of a continuous signal produce phase changes that are detected by an FM discriminator. The resulting short pulses sound like a tone in the receiver output. The tone disappears when the antennas are equidistant from the signal source, giving an audible null.

The polarity of the pulses at the discriminator output is a function of which antenna is closer to the source. Therefore, the pulses can be processed and used to drive a left-right zero-center meter in a manner similar to the switched pattern units described above. Left-right LED indicators may replace the meter for economy and visibility at night.

RDF operations with a TDOA dual antenna RDF are done in the same manner as with a switched antenna RDF set. The main difference is the requirement for an FM receiver in the TDOA system and an AM receiver in the switched pattern case. No RF attenuator is needed for close-in work in the TDOA case.

Popular designs for practical do-it-yourself TDOA RDF sets include the Simple Seeker (described elsewhere in this chapter) and the W9DUU design (see article by Bohrer in the Bibliography). Articles with plans for the Handy Tracker, a simple TDOA set with a delay line to resolve the dual-null ambiguity instead of LEDs or a meter, are listed in the Bibliography.

Performance Comparison

Both types of dual antenna RDFs make good on-foot "sniffing" devices and are excellent performers when there are rapid amplitude variations in the incoming signal. They are the units of choice for airborne work. Compared to Yagis and quads, they give good directional performance over a much wider frequency range. Their indications are more precise than those of beams with broad forward lobes.

Dual-antenna RDF sets frequently give inaccurate bearings in multipath situations, because they cannot resolve signals of nearly equal levels from more than one direction. Because multipath signals are a combined pattern of peaks and nulls, they appear to change in amplitude and bearing as you move the RDF antenna along the bearing path or perpendicular to it, whereas a non-multipath signal will have constant strength and bearing.

The best way to overcome this problem is to take large numbers of bearings while moving toward the transmitter. Taking bearings while in motion averages out the effects of multipath, making the direct signal more readily discernible. Some TDOA RDF sets have a slow-response mode that aids the averaging process.

Switched antenna systems generally do not perform well when the incoming signal is horizontally polarized. In such cases, the bearings may be inaccurate or unreadable. TDOA units require a carrier type signal such as FM or CW; they usually cannot yield bearings on noise or pulse signals.

Unless an additional method is employed

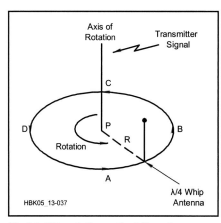

Fig 21.137 — A theoretical Doppler antenna circles around point P, continuously moving toward and away from the source at an audio rate.

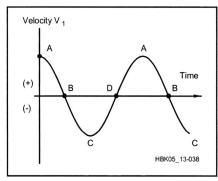

Fig 21.138 — Frequency shift versus time produced by the rotating antenna movement toward and away from the signal source.

to measure signal strength, it is easy to "overshoot" the hidden transmitter location with a TDOA set. It is not uncommon to see a TDOA foxhunter walk over the top of a concealed transmitter and walk away, following the opposite 180° null, because there is no display of signal amplitude.

DOPPLER RDF SETS

RDF sets using the Doppler principle are popular in many areas because of their ease of use. They have an indicator that instantaneously displays direction of the signal source relative to the vehicle heading, either on a circular ring of LEDs or a digital readout in degrees. A ring of four, eight or more antennas picks up the signal. Quarter-wavelength monopoles on a ground plane are popular for vehicle use, but half-wavelength vertical dipoles, where practical, perform better.

Radio signals received on a rapidly moving antenna experience a frequency shift due to the Doppler effect, a phenomenon well-known to anyone who has observed a moving car with its horn sounding. The horn's pitch appears higher than normal as the car approaches, and lower as the car recedes. Similarly, the received radio frequency increases as the antenna moves toward the transmitter and vice versa. An FM receiver will detect this frequency change.

Fig 21.137 shows a ¼-λ vertical antenna being moved on a circular track around point P, with constant angular velocity. As the antenna approaches the transmitter on its track, the received frequency is shifted higher. The highest instantaneous frequency occurs when the antenna is at point A, because tangential velocity toward the transmitter is maximum at that point. Conversely, the lowest frequency occurs when the antenna reaches point C, where velocity is maximum away from the transmitter.

Fig 21.138 shows a plot of the component of the tangential velocity that is in the direction of the transmitter as the antenna moves around the circle. Comparing Figs 21.137 and 21.138, notice that at B in Fig 21.138, the tangential velocity is crossing zero from the positive to the negative and the antenna is closest to the transmitter. The Doppler shift and resulting audio output from the receiver discriminator follow the same plot, so that a negative-slope zero-crossing detector, synchronized with the antenna rotation, senses the incoming direction of the signal.

The amount of frequency shift due to the Doppler effect is proportional to the RF frequency and the tangential antenna velocity. The velocity is a function of the radius of rotation and the angular velocity (rotation rate). The radius of rotation must be less than ¼ λ to avoid errors. To get a usable amount of FM deviation (comparable to typical voice modulation) with this radius, the antenna must rotate at approximately 30,000 RPM (500 Hz). This puts the Doppler tone in the audio range for easy processing.

Mechanically rotating a whip antenna at this rate is impractical, but a ring of whips, switched to the receiver in succession with RF PIN diodes, can simulate a rapidly rotating antenna. Doppler RDF sets must be used with receivers having FM detectors. The Dopple ScAnt and Roanoke Doppler (see Bibliography) are mobile Doppler RDF sets designed for inexpensive home construction.

Doppler Advantages and Disadvantages

Ring-antenna Doppler sets are the ultimate in simplicity of operation for mobile RDF. There are no moving parts and no manual antenna pointing. Rapid direction indications are displayed on very short signal bursts.

Many units lock in the displayed direction after the signal leaves the air. Power variations in the source signal cause no difficulties, as long as the signal remains above the RDF detection threshold. A Doppler antenna goes on top of any car quickly, with no holes to drill.

Many Local Interference Committee members choose Dopplers for tracking malicious interference, because they are inconspicuous (compared to beams) and effective at tracking the strong vertically polarized signals that repeater jammers usually emit.

A Doppler does not provide superior performance in all VHF RDF situations. If the signal is too weak for detection by the Doppler unit, the hunt advantage goes to teams with beams. Doppler installations are not suitable for on-foot sniffing. The limitations of other switched antenna RDFs also apply: (1) poor results with horizontally polarized signals, (2) no indication of distance, (3) carrier type signals only and (4) inadvisability of transmitting through the antenna.

Readout to the nearest degree is provided on some commercial Doppler units. This does not guarantee that level of accuracy, however. A well-designed four-monopole set is typically capable of ±5° accuracy on 2 meters, if the target signal is vertically polarized and there are no multipath effects.

The rapid antenna switching can introduce cross modulation products when the user is near strong off-channel RF sources. This self-generated interference can temporarily render the system unusable. While not a common problem with mobile Dopplers, it makes the Doppler a poor choice for use in remote RDF installations at fixed sites with high power VHF transmitters nearby.

MOBILE RDF SYSTEM INSTALLATION

Of these mobile VHF RDF systems, the Doppler type is clearly the simplest from a mechanical installation standpoint. A four-whip Doppler RDF array is easy to implement with magnetic mount antennas. Alternately, you can mount all the whips on a frame that attaches to the vehicle roof with suction cups. In either case, setup is rapid and requires no holes in the vehicle.

You can turn small VHF beams and dual-antenna arrays readily by extending the mast through a window. Installation on each model vehicle is different, but usually the mast can be held in place with some sort of cup in the arm rest and a plastic tie at the top of the window, as in **Fig 21.139**. This technique works best on cars with frames around the windows, which allow the door to be opened with the antenna in place. Check local vehicle codes, which limit how far your antenna may protrude beyond the line of the fenders. Larger antennas may have to be put on the passenger side of the vehicle, where greater overhang is generally permissible.

The window box (**Fig 21.140**) is an improvement over through-the-window mounts. It provides a solid, easy-turning mount for the mast. The plastic panel keeps out bad weather. You will need to custom-design the box for

Fig 21.139 — A set of TDOA RDF antennas is light weight and mounts readily through a sedan window without excessive overhang.

Fig 21.140 — A window box allows the navigator to turn a mast mounted antenna with ease while remaining dry and warm. No holes in the vehicle are needed with a properly designed window box.

your vehicle model. Vehicle codes may limit the use of a window box to the passenger side.

For the ultimate in convenience and versatility, cast your fears aside, drill a hole through the center of the roof and install a waterproof bushing. A roof-hole mount permits the use of large antennas without overhang violations. The driver, front passenger and even a rear passenger can turn the mast when required. The installation in Fig 21.134 uses a roof-hole bushing made from mating threaded PVC pipe adapters and reducers. When it is not in use for RDF, a PVC pipe cap provides a watertight cover. There is a pointer and 360° indicator at the bottom of the mast for precise bearings.

21.12.3 Direction-Finding Techniques

The ability to locate a transmitter quickly with RDF techniques is a skill you will acquire only with practice. It is very important to become familiar with your equipment and its limitations. You must also understand how radio signals behave in different types of terrain at the frequency of the hunt. Experience is the best teacher, but reading and hearing the stories of others who are active in RDF will help you get started.

Verify proper performance of your portable RDF system before you attempt to track signals in unknown locations. Of primary concern is the accuracy and symmetry of the antenna pattern. For instance, a lopsided figure-8 pattern with a loop, Adcock, or TDOA set leads to large bearing errors. Nulls should be exactly 180° apart and exactly at right angles to the loop plane or the array boom. Similarly, if feed-line pickup causes an off-axis main lobe in your VHF RDF beam, your route to the target will be a spiral instead of a straight line.

Perform initial checkout with a low-powered test transmitter at a distance of a few hundred feet. Compare the RDF bearing indication with the visual path to the transmitter. Try to "find" the transmitter with the RDF equipment as if its position were not known. Be sure to check all nulls on antennas that have more than one.

If imbalance or off-axis response is found in the antennas, there are two options available. One is to correct it, insofar as possible. A second option is to accept it and use some kind of indicator or correction procedure to show the true directions of signals. Sometimes the end result of the calibration procedure is a compromise between these two options, as a perfect pattern may be difficult or impossible to attain.

The same calibration suggestions apply for fixed RDF installations, such as a base station HF Adcock or VHF beam. Of course it does no good to move it to an open field. Instead, calibrate the array in its intended operating position, using a portable or mobile transmitter. Because of nearby obstructions or reflecting objects, your antenna may not indicate the precise direction of the transmitter. Check for imbalance and systemic error by taking readings with the test emitter at locations in several different directions.

The test signal should be at a distance of 2 or 3 miles for these measurements, and should be in as clear an area as possible during transmissions. Avoid locations where power lines and other overhead wiring can conduct signal from the transmitter to the RDF site. Once antenna adjustments are optimized, make a table of bearing errors noted in all compass directions. Apply these error values as corrections when actual measurements are made.

PREPARING TO HUNT

Successfully tracking down a hidden transmitter involves detective work — examining all the clues, weighing the evidence and using good judgment. Before setting out to locate the source of a signal, note its general characteristics. Is the frequency constant, or does it drift? Is the signal continuous, and if not, how long are transmissions? Do transmissions occur at regular intervals, or are they sporadic? Irregular, intermittent signals are the most difficult to locate, requiring patience and quick action to get bearings when the transmitter comes on.

Refraction, Reflections and the Night Effect

You will get best accuracy in tracking ground wave signals when the propagation path is over homogeneous terrain. If there is a land/water boundary in the path, the different conductivities of the two media can cause bending (refraction) of the wave front, as in **Fig 21.141A**. Even the most sophisticated RDF equipment will not indicate the correct bearing in this situation, as the equipment can only show the direction from which the signal is arriving. RDFers have observed this phenomenon on both HF and VHF bands.

Signal reflections also cause misleading bearings. This effect becomes more pronounced as frequency increases. T-hunt hiders regularly achieve strong signal bounces from distant mountain ranges on the 144-MHz band.

Tall buildings also reflect VHF/UHF signals, making mid-city RDF difficult. Hunting on the 440-MHz and higher amateur bands is even more arduous because of the plethora of reflecting objects.

In areas of signal reflection and multipath, some RDF gear may indicate that the signal is coming from an intermediate point, as in Fig 21.141B. High gain VHF/UHF RDF beams will show direct and reflected signals as separate S-meter peaks, leaving it to the operator to determine which is which. Null-based RDF antennas, such as phased arrays and loops, have the most difficulty with multi-path, because the multiple signals tend to make the nulls very shallow or fill them in entirely, resulting in no bearing indication at all.

If the direct path to the transmitter is masked by intervening terrain, a signal reflection from a higher mountain, building, water tower, or the like may be much stronger than the direct signal. In extreme cases, triangulation from several locations will appear to "confirm" that the transmitter is at the location of the reflecting object. The direct signal may not be detectable until you arrive at the reflecting point or another high location.

Objects near the observer such as concrete/steel buildings, power lines and chain-link fences will distort the incoming wavefront and give bearing errors. Even a dense grove of trees can sometimes have an adverse effect. It is always best to take readings in locations that are as open and clear as possible, and to take bearings from numerous positions for confirmation. Testing of RDF gear should also be done in clear locations.

Locating local signal sources on frequencies below 10 MHz is much easier during daylight hours, particularly with loop antennas. In the daytime, D-layer absorption minimizes skywave propagation on these frequencies. When the D layer disappears after sundown, you may hear the signal by a combination of ground wave and high-angle skywave, making it difficult or impossible to obtain a bearing. RDFers call this phenomenon the *night effect*.

While some mobile T-hunters prefer to go it alone, most have more success by teaming up and assigning tasks. The driver concentrates on handling the vehicle, while the assistant (called the "navigator" by some teams) turns the beam, reads the meters and calls out bearings. The assistant is also responsible for maps and plotting, unless there is a third team member for that task.

MAPS AND BEARING-MEASUREMENTS

Possessing accurate maps and knowing how to use them is very important for successful RDF. Even in difficult situations where precise bearings cannot be obtained, a town or city map will help in plotting points where signal levels are high and low. For example, power line noise tends to propagate along the power line and radiates as it does so. Instead of a single source, the noise appears to come from a multitude of sources. This renders many ordinary RDF techniques ineffective. Mapping locations where signal amplitudes are highest will help pinpoint the source.

Several types of area-wide maps are suitable for navigation and triangulation. Street and highway maps work well for mobile work. Large detailed maps are preferable to thick map books. Contour maps are ideal for open country. Aeronautical charts are also suitable. Good sources of maps include auto clubs, stores catering to camping/hunting enthusiasts and city/county engineering departments.

A *heading* is a reading in degrees relative to some external reference, such as your house

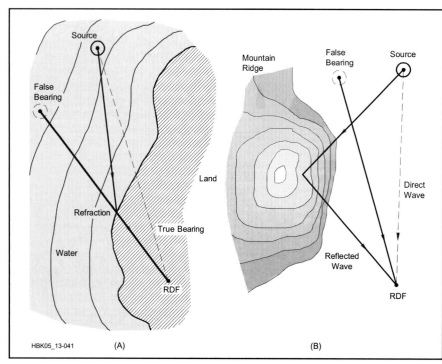

Fig 21.141 — RDF errors caused by refraction (A) and reflection (B). The reading at A is false because the signal actually arrives from a direction that is different from that to the source. At B, a direct signal from the source combines with a reflected signal from the mountain ridge. The RDF set may average the signals as shown, or indicate two lines of bearing.

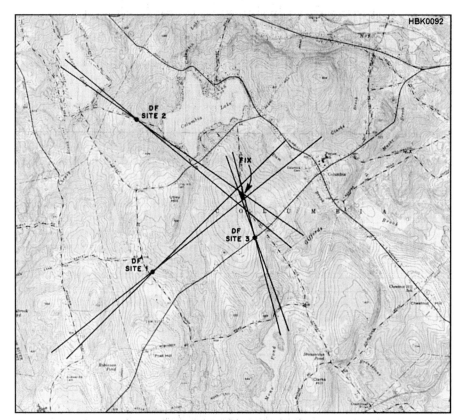

Fig 21.142 — Bearing sectors from three RDF positions drawn on a map for triangulation. In this case, bearings are from loop antennas, which have 180° ambiguity.

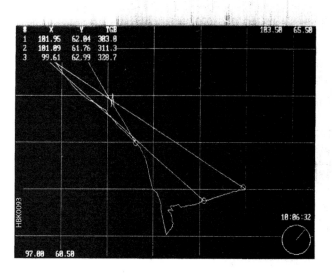

Fig 21.143 — Screen plot from a computerized RDF system showing three T-hunt bearings (straight lines radiating from small circles) and the vehicle path (jagged trace). The grid squares correspond to areas of standard topographic maps.

tion charts, respectively.)

Declination for your area is given on US Geological Survey (USGS) maps, though it undergoes long-term changes. Add the declination to your magnetic heading to get a true heading.

As an example, assume that the transmitted signal arrives at 30° with respect to the vehicle heading, that the compass indicates that the vehicle's heading is 15°, and the magnetic declination is +15°. Add these values to get a true transmitter bearing (that is, a bearing with respect to true north) of 60°.

Because of the large mass of surrounding metal, it is very difficult to calibrate an in-car compass for high accuracy at all vehicle headings. It is better to use a remotely mounted flux-gate compass sensor, properly corrected, to get vehicle headings, or to stop and use a hand compass to measure the vehicle heading from the outside. If you T-hunt with a mobile VHF beam or quad, you can use your manual compass to sight along the antenna boom for a magnetic bearing, then add the declination for true bearing to the fox.

Triangulation Techniques

If you can obtain accurate bearings from two locations separated by a suitable distance, the technique of *triangulation* will give the expected location of the transmitter. The intersection of the lines of bearing from each location provides a *fix*. Triangulation accuracy is greatest when stations are located such that their bearings intersect at right angles. Accuracy is poor when the angle between bearings approaches 0° or 180°.

There is always uncertainty in the fixes obtained by triangulation due to equipment limitations, propagation effects and measurement errors. Obtaining bearings from three or more locations reduces the uncertainty. A good way to show the probable area of the transmitter on the triangulation map is to draw bearings as a narrow sector instead of as a single line. Sector width represents the amount of bearing uncertainty. **Fig 21.142** shows a portion of a map marked in this manner. Note how the bearing from Site 3 has narrowed down the probable area of the transmitter position.

Computerized Transmitter Hunting

A portable computer is an excellent tool for streamlining the RDF process. Some T-hunters use one to optimize VHF beam bearings, generating a two-dimensional plot of signal strength versus azimuth. Others have automated the bearing-taking process by using a computer to capture signal headings from a Doppler RDF set, vehicle heading from a flux-gate compass, and vehicle location from a GPS receiver (**Fig 21.143**). The computer program can compute averaged headings from a Doppler set to reduce multipath effects.

or vehicle; a *bearing* is the target signal's direction relative to your position. Plotting a bearing on a hidden transmitter from your vehicle requires that you know the vehicle location, transmitter heading with respect to the vehicle and vehicle heading with respect to true north.

First, determine your location, using landmarks or a navigation device such as a GPS receiver. Next, using your RDF equipment, determine the bearing to the hidden transmitter (0 to 359.9°) with respect to the vehicle. Zero degrees heading corresponds to signals coming from directly in front of the vehicle, signals from the right indicate 90°, and so on.

Finally, determine your vehicle's true heading, that is, its heading relative to true north. Compass needles point to magnetic north and yield magnetic headings. Translating a magnetic heading into a true heading requires adding a correction factor, called *magnetic declination*, which is a positive or negative factor that depends on your location. (*Declination* is the term as denoted on land USGS topographic maps. *Deviation* and *Variation* are terms used on nautical and avia-

Provided with perfect position and bearing information, computer triangulation could determine the transmitter location within the limits of its computational accuracy. Two bearings would exactly locate a fox. Of course, there are always uncertainties and inaccuracies in bearing and position data. If these uncertainties can be determined, the program can compute the uncertainty of the triangulated bearings. A "smart" computer program can evaluate bearings, triangulate the bearings of multiple hunters, discard those that appear erroneous, determine which locations have particularly great or small multipath problems and even "grade" the performance of RDF stations.

By adding packet radio connections to a group of computerized base and mobile RDF stations, the processed bearing data from each can be shared. Each station in the network can display the triangulated bearings of all. This requires a common map coordinate set among all stations. The USGS Universal Transverse Mercator (UTM) grid, consisting of 1×1-km grid squares, is a good choice.

The computer is an excellent RDF tool, but it is no substitute for a skilled "navigator." You will probably discover that using a computer on a high-speed T-hunt requires a full-time operator in the vehicle to make full use of its capabilities.

SKYWAVE BEARINGS AND TRIANGULATION

Many factors make it difficult to obtain accuracy in skywave RDF work. Because of Faraday rotation during propagation, skywave signals are received with random polarization. Sometimes the vertical component is stronger, and at other times the horizontal. During periods when the vertical component is weak, the signal may appear to fade on an Adcock RDF system. At these times, determining an accurate signal null direction becomes very difficult.

For a variety of reasons, HF bearing accuracy to within 1 or 2° is the exception rather than the rule. Errors of 3 to 5° are common. An error of 3° at a thousand miles represents a distance of 52 miles. Even with every precaution taken in measurement, do not expect cross-country HF triangulation to pinpoint a signal beyond a county, a corner of a state or a large metropolitan area. The best you can expect is to be able to determine where a mobile RDF group should begin making a local search.

Triangulation mapping with skywave signals is more complex than with ground or direct waves because the expected paths are great-circle routes. Commonly available world maps are not suitable, because the triangulation lines on them must be curved, rather than straight. In general, for flat maps, the larger the area encompassed, and the greater the error that straight-line triangulation procedures will give.

A highway map is suitable for regional triangulation work if it uses some form of conical projection, such as the Lambert conformal conic system. This maintains the accuracy of angular representation, but the distance scale is not constant over the entire map.

One alternative for worldwide areas is the azimuthal-equidistant projection, better known as a great-circle map. True bearings for great-circle paths are shown as straight lines from the center to all points on the Earth. Maps centered on three or more different RDF sites may be compared to gain an idea of the general geographic area for an unknown source.

For worldwide triangulation, the best projection is the *gnomonic*, on which all great circle paths are represented by straight lines and angular measurements with respect to meridians are true. Gnomonic charts are custom maps prepared especially for government and military agencies.

Skywave signals do not always follow the great-circle path in traveling from a transmitter to a receiver. For example, if the signal is refracted in a tilted layer of the ionosphere, it could arrive from a direction that is several degrees away from the true great-circle bearing.

Another cause of signals arriving off the great-circle path is termed *sidescatter*. It is possible that, at a given time, the ionosphere does not support great-circle propagation of the signal from the transmitter to the receiver because the frequency is above the MUF for that path. However, at the same time, propagation may be supported from both ends of the path to some mutually accessible point off the great-circle path. The signal from the source may propagate to that point on the Earth's surface and hop in a sideways direction to continue to the receiver.

For example, signals from Central Europe have propagated to New England by hopping from an area in the Atlantic Ocean off the northwest coast of Africa, whereas the great-circle path puts the reflection point off the southern coast of Greenland. Readings in error by as much as 50° or more may result from sidescatter. The effect of propagation disturbances may be that the bearing seems to wander somewhat over a few minutes of time, or it may be weak and fluttery. At other times, however, there may be no telltale signs to indicate that the readings are erroneous.

CLOSING IN

On a mobile foxhunt, the objective is usually to proceed to the hidden T with minimum time and mileage. Therefore, do not go far out of your way to get off-course bearings just to triangulate. It is usually better to take the shortest route along your initial line of bearing and "home in" on the signal. With a little experience, you will be able to gauge your distance from the fox by noting the amount of attenuation needed to keep the S-meter on scale.

As you approach the transmitter, the signal will become very strong. To keep the S-meter on scale, you will need to add an RF attenuator in the transmission line from the antenna to the receiver. Simple resistive attenuators are discussed in another chapter.

In the final phases of the hunt, you will probably have to leave your mobile and continue the hunt on foot. Even with an attenuator in the line, in the presence of a strong RF field, some energy will be coupled directly into the receiver circuitry. When this happens, the S-meter reading changes only slightly or perhaps not at all as the RDF antenna rotates, no matter how much attenuation you add. The cure is to shield the receiving equipment. Something as simple as wrapping the receiver in foil or placing it in a bread pan or cake pan, covered with a piece of copper or aluminum screening securely fastened at several points, may reduce direct pickup enough for you to get bearings.

Alternatively, you can replace the receiver with a field-strength meter as you close in, or use a heterodyne-type active attenuator. Plans for these devices are at the end of this chapter.

The Body Fade

A crude way to find the direction of a VHF signal with just a hand-held transceiver is the body fade technique, so named because the blockage of your body causes the signal to fade. Hold your HT close to your chest and turn all the way around slowly. Your body is providing a shield that gives the hand-held a cardioid sensitivity pattern, with a sharp decrease in sensitivity to the rear. This null indicates that the source is behind you (**Fig 21.144**).

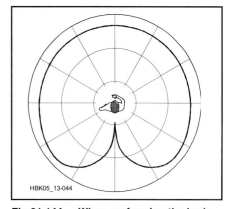

Fig 21.144 — When performing the body fade maneuver, a hand-held transceiver exhibits this directional pattern.

If the signal is so strong that you can't find the null, try tuning 5 or 10 kHz off frequency to put the signal into the skirts of the IF passband. If your hand-held is dual-band (144/440 MHz) and you are hunting on 144 MHz, try tuning to the much weaker third harmonic of the signal in the 440-MHz band.

The body fade null, which is rather shallow to begin with, can be obscured by reflections, multipath, nearby objects, etc. Step well away from your vehicle before trying to get a bearing. Avoid large buildings, chain-link fences, metal signs and the like. If you do not get a good null, move to a clearer location and try again.

Air Attenuators

In microwave parlance, a signal that is too low in frequency to be propagated in a waveguide (that is, below the *cutoff frequency*) is attenuated at a predictable logarithmic rate. In other words, the farther inside the waveguide, the weaker the signal gets. Devices that use this principle to reduce signal strength are commonly known as *air attenuators*. Plans for a practical model for insertion in a coax line are in *Transmitter Hunting* (see Bibliography).

With this principle, you can reduce the level of strong signals into your hand-held transceiver, making it possible to use the body fade technique at very close range. Glen Rick-

Fig 21.145 — The air attenuator for a VHF hand-held in use. Suspend the radio by the wrist strap or a string inside the tube.

erd, KC6TNF, documented this technique for *QST*. Start with a pasteboard mailing tube that has sufficient inside diameter to accommodate your hand-held. Cover the outside of the tube completely with aluminum foil. You can seal the bottom end with foil, too, but it probably will not matter if the tube is long enough. For durability and to prevent accidental shorts, wrap the foil in packing tape. You will also need a short, stout cord attached to the hand-held. The wrist strap may work for this, if long enough.

To use this air attenuation scheme for body fade bearings, hold the tube vertically against your chest and lower the hand-held into it until the signal begins to weaken (**Fig 21.145**). Holding the receiver in place, turn around slowly and listen for a sudden decrease in signal strength. If the null is poor, vary the depth of the receiver in the tube and try again. You do not need to watch the S-meter, which will likely be out of sight in the tube. Instead, use noise level to estimate signal strength.

For extremely strong signals, remove the "rubber duck" antenna or extend the wrist strap with a shoelace to get greater depth of suspension in the tube. The depth that works for one person may not work for another. Experiment with known signals to determine what works best for you.

Several RDF projects may be found on the *Handbook CD*.

21.13 Glossary

Antenna — An electrical conductor or array of conductors that radiates signal energy (transmitting) or collects signal energy (receiving).

Antenna tuner — A device containing variable reactances (and perhaps a balun) used to convert an antenna or feed line impedance to 50 Ω. (also called Transmatch, impedance-matching unit)

Apex angle — The included angle between the legs of an inverted-V antenna.

Azimuth (azimuthal) pattern — A radiation pattern in a plane oriented parallel to the Earth's surface or at a specified angle to the Earth's surface.

Balanced line — A symmetrical two-conductor feed line (also called open-wire line, ladder line, window line, twin-lead).

Balun — A device that transfers energy between a balanced and unbalanced system. A balun may or may not change the impedance ratio between the systems.

Base loading — Adding a coil to the base of a ground-plane antenna to increase its electrical length.

Beamwidth — The width in degrees of the major lobe of a directive antenna between the two angles at which the relative radiated power is equal to one-half its value (–3 dB) at the peak of the lobe.

Capacitance hat — A conducting structure with a large surface area that is added to an antenna to add capacitive reactance at that point on the antenna.

Center loading — Adding a coil near the center of a ground-plane antenna to increase its electrical length.

Coaxial cable (coax) — A coaxial transmission line with a center conductor surrounded by a layer of insulation and then a tubular shield conductor and covered by an insulating jacket. (see also — unbalanced line)

Delta loop — A full-wavelength loop shaped like a triangle or delta.

Delta match — Center-feed technique used with antenna elements that are not split at the center in which the transmission is spread apart and connected to the element symmetrically, forming a triangle or delta.

Dipole — An antenna, usually one-half wavelength long, divided into two parts at a feed point.

Directivity — The property of an antenna that concentrates the radiated energy to form one or more major lobes.

Director — An antenna element in a parasitic array that causes radiated energy from the driven element to be focused along the line from the driven element to the director.

Driven array — An array of antenna elements which are all driven or excited by means of a transmission line.

Driven element — An antenna element excited by means of a transmission line.

E-plane — The plane in which the electric field of an electromagnetic wave is maximum.

Efficiency — The ratio of useful output power to input power.

Elements — The conductive parts of an antenna system that determine the antenna's characteristics.

Elevation pattern — A radiation pattern in a plane perpendicular to the Earth's surface.

End effect — The effect of capacitance at the end of an antenna element that acts

to electrically lengthen the element.
Feed line — see transmission line
Front-to-back ratio — The ratio in dB of the radiation from an antenna in a favored direction to that in the opposite direction.
Front-to-rear ratio — The ratio in dB of the radiation from an antenna is a favored direction to an average of the radiation in the opposite direction across some specified angle.
Front-to-side ratio — The ratio in dB of the radiation from an antenna in a favored direction to that at right angles to the favored direction.
Gain — The increase in radiated power in the desired direction of the major lobe.
Gamma match — A matching system used with driven antenna elements in which a conductor is placed near the element and connected to the feed line with an adjustable capacitor at the end closest to the center and connected to the element at the other.
Ground plane — A system of conductors configured to act as a reflecting surface to an antenna element and connected to one side of the transmission line.
H-plane — The plane in which the magnetic field of an electromagnetic wave is maximum.
Hairpin match — A U-shaped conductor that is connected to the two inner ends of a split antenna element for the purpose of creating a match to a feed line.
Impedance — The ratio of voltage to current in a feed line or along an antenna.
Inverted-V — A dipole antenna supported at its mid-point with halves angled down toward the ground.
Isotropic — An imaginary antenna that radiates and receives equally well in all directions.
Ladder line — see balanced line.
Line loss — The power lost in a transmission line, specified in dB per unit of length.
Load — The electrical system or component to which power is delivered.
Lobe — A region in an antenna's radiation pattern between two nulls.
Matching — The process by which power at one impedance is transferred to a system at a different impedance.
Monopole — An antenna with a single-element that functions in concert with a ground-plane.
Null — A point of minimum radiation in an antenna's radiation pattern.
Open-wire line — see balanced line.
Parasitic array — A set of elements that form a radiation pattern through coupling and re-radiation of energy from one or more driven elements.
Polarization — The orientation of the electromagnetic field, usually referring to the orientation of the E field.
Q section — A quarter-wavelength section of transmission line used for impedance-matching purposes.
Quad — A directive antenna based on the Yagi with elements that consist of one-wavelength loops.
Radiation pattern — The characteristics of an antenna's distribution of energy in a single plane. (see also elevation pattern and azimuth pattern)
Radiation resistance — A resistance that represents the work done by the current in an antenna to radiate power.
Reflector — An antenna element in a parasitic array that causes radiated energy from the driven element to be focused along the line from the driven element away from the reflector.
Sense Antenna — An antenna added to a bidirectional array or loop that samples the incoming signal's phase for comparison to that of the main receiving antenna.
Stacking — Arranging two or more directive antennas such that their radiation pattern characteristics reinforce each other.
SWR — Standing-wave ratio. A measure of the match between a transmission line and a load such as an antenna.
T-match — A symmetrical version of the gamma match for a balanced antenna system.
Top loading — Addition of a reactance, usually capacitive, at the top of a ground-plane antenna so as to increase its electrical length.
Transmatch — see antenna tuner
Trap — A parallel LC-circuit used as an electrical switch to isolate sections of an antenna
Twin-lead — see balanced line.
Unipole — see monopole.
Yagi — A parasitic array consisting of a driven element and one or more director and reflectors.
Zepp — A half-wavelength antenna fed at one end by means of open-wire feed line.

21.14 References and Bibliography

J. S. Belrose, "Short Antennas for Mobile Operation," *QST*, Sep 1953, pp 30-35.

G. H. Brown, "The Phase and Magnitude of Earth Currents Near Radio Transmitting Antennas," *Proc IRE*, Feb 1935.

G. H. Brown, R. F. Lewis and J. Epstein, "Ground Systems as a Factor in Antenna Efficiency," *Proc IRE*, Jun 1937, pp 753-787.

G. H. Brown and O. M. Woodward, Jr, "Experimentally Determined Impedance Characteristics of Cylindrical Antennas," *Proc IRE*, April 1945.

C. L. Buchanen, W3DZZ, "The Multimatch Antenna System," *QST*, Mar 1955, p 22.

A. Christman, "Elevated Vertical Antenna Systems," *QST*, Aug 1988, pp 35-42.

R. B. Dome, "Increased Radiating Efficiency for Short Antennas," *QST*, Sep 1934, pp 9-12.

A. C. Doty, Jr, J. A. Frey and H. J. Mills, "Characteristics of the Counterpoise and Elevated Ground Screen," Professional Program, Session 9, Southcon '83 (IEEE), Atlanta, GA, Jan 1983.

A. C. Doty, Jr, J. A. Frey and H. J. Mills, "Efficient Ground Systems for Vertical Antennas," *QST*, Feb 1983, pp 20-25.

A. C. Doty, Jr, technical paper presentation, "Capacitive Bottom Loading and Other Aspects of Vertical Antennas," Technical Symposium, Radio Club of America, New York City, Nov 20, 1987.

A. C. Doty, Jr, J. A. Frey and H. J. Mills, "Vertical Antennas: New Design and Construction Data," *The ARRL Antenna Compendium, Volume 2* (Newington: ARRL, 1989), pp 2-9.

R. Fosberg, "Some Notes on Ground Systems for 160 Meters," *QST*, Apr 1965, pp 65-67.

G. Grammer, "More on the Directivity of Horizontal Antennas; Harmonic Operation — Effects of Tilting," *QST*, Mar 1937, pp 38-40, 92, 94, 98.

H. E. Green, "Design Data for Short and Medium Length Yagi-Uda Arrays," *Trans IE Australia*, Vol EE-2, No. 1, Mar 1966.

H. J. Mills, technical paper presentation, "Impedance Transformation Provided by Folded Monopole Antennas," Technical Symposium, Radio Club of America, New York City, Nov 20, 1987.

B. Myers, "The W2PV Four-Element Yagi," *QST*, Oct 1986, pp 15-19.

L. Richard, "Parallel Dipoles of 300-Ohm Ribbon," *QST*, Mar 1957.

J. H. Richmond, "Monopole Antenna on Circular Disc," *IEEE Trans on Antennas and Propagation*, Vol. AP-32, No. 12, Dec 1984.

W. Schulz, "Designing a Vertical Antenna," *QST*, Sep 1978, pp 19- 21.

J. Sevick, "The Ground-Image Vertical Antenna," *QST*, Jul 1971, pp 16-17, 22.

J. Sevick, "The W2FMI 20-Meter Vertical Beam," *QST*, Jun 1972, pp 14-18.

J. Sevick, "The W2FMI Ground-Mounted Short Vertical," *QST*, Mar 1973, pp 13-18, 41.

J. Sevick, "A High Performance 20-, 40- and 80-Meter Vertical System," *QST*, Dec 1973.

J. Sevick, "Short Ground-Radial Systems for Short Verticals," *QST*, Apr 1978, pp 30-33.

C. E. Smith and E. M. Johnson, "Performance of Short Antennas," *Proc IRE*, Oct 1947.

J. Stanley, "Optimum Ground Systems for Vertical Antennas," *QST*, Dec 1976, pp 13-15.

R. E. Stephens, "Admittance Matching the Ground-Plane Antenna to Coaxial Transmission Line," Technical Correspondence, *QST*, Apr 1973, pp 55-57.

D. Sumner, "Cushcraft 32-19 'Boomer' and 324-QK Stacking Kit," Product Review, *QST*, Nov 1980, pp 48-49.

W. van B. Roberts, "Input Impedance of a Folded Dipole," *RCA Review*, Jun 1947.

E. M. Williams, "Radiating Characteristics of Short-Wave Loop Aerials," *Proc IRE*, Oct 1940.

TEXTBOOKS ON ANTENNAS

C. A. Balanis, *Antenna Theory, Analysis and Design* (New York: Harper & Row, 1982).

D. S. Bond, *Radio Direction Finders*, 1st ed. (New York: McGraw-Hill Book Co).

W. N. Caron, *Antenna Impedance Matching* (Newington: ARRL, 1989).

L. B. Cebik, *ARRL Antenna Modeling Course* (Newington: ARRL, 2002).

K. Davies, *Ionospheric Radio Propagation* — National Bureau of Standards Monograph 80 2(Washington, DC: U.S. Government Printing Office, Apr 1, 1965).

R. S. Elliott, *Antenna Theory and Design* (Englewood Cliffs, NJ: Prentice Hall, 1981).

A. E. Harper, *Rhombic Antenna Design* (New York: D. Van Nostrand Co, Inc, 1941).

K. Henney, *Principles of Radio* (New York: John Wiley and Sons, 1938), p 462.

C. Hutchinson and R. D. Straw, *Simple and Fun Antennas for Hams* (Newington: ARRL, 2002).

H. Jasik, *Antenna Engineering Handbook*, 1st ed. (New York: McGraw-Hill, 1961).

W. C. Johnson, *Transmission Lines and Networks*, 1st ed. (New York: McGraw-Hill Book Co, 1950).

Johnson and Jasik, *Antenna Engineering Handbook*, 2nd ed. (New York: McGraw-Hill).

R. C. Johnson, *Antenna Engineering Handbook*, 3rd ed. (New York: McGraw-Hill, 1993).

E. C. Jordan and K. G. Balmain, *Electromagnetic Waves and Radiating Systems*, 2nd ed. (Englewood Cliffs, NJ: Prentice-Hall, Inc, 1968).

R. Keen, *Wireless Direction Finding*, 3rd ed. (London: Wireless World).

R. W. P. King, *Theory of Linear Antennas* (Cambridge, MA: Harvard Univ. Press, 1956).

R. W. P. King, H. R. Mimno and A. H. Wing, *Transmission Lines, Antennas and Waveguides* (New York: Dover Publications, Inc, 1965).

King, Mack and Sandler, *Arrays of Cylindrical Dipoles* (London: Cambridge Univ Press, 1968).

M. G. Knitter, ed., *Loop Antennas — Design and Theory* (Cambridge, WI: National Radio Club, 1983).

M. G. Knitter, ed., *Beverage and Long Wire Antennas* — Design and Theory (Cambridge, WI: National Radio Club, 1983).

J. D. Kraus, *Electromagnetics* (New York: McGraw-Hill Book Co).

J. D. Kraus, *Antennas*, 2nd ed. (New York: McGraw-Hill Book Co, 1988).

E. A. Laport, *Radio Antenna Engineering* (New York: McGraw-Hill Book Co, 1952).

J. L. Lawson, *Yagi-Antenna Design*, 1st ed. (Newington: ARRL, 1986).

P. H. Lee, *The Amateur Radio Vertical Antenna Handbook*, 2nd ed. (Port Washington, NY: Cowen Publishing Co., 1984).

D. B. Leeson, *Physical Design of Yagi Antennas* (Newington: ARRL, 1992).

A. W. Lowe, *Reflector Antennas* (New York: IEEE Press, 1978).

M. W. Maxwell, *Reflections — Transmission Lines and Antennas* (Newington: ARRL, 1990).

M. W. Maxwell, *Reflections II — Transmission Lines and Antennas* (Sacramento: Worldradio Books, 2001).

G. M. Miller, *Modern Electronic Communication* (Englewood Cliffs, NJ: Prentice Hall, 1983).

V. A. Misek, *The Beverage Antenna Handbook* (Hudson, NH: V. A. Misek, 1977).

T. Moreno, *Microwave Transmission Design Data* (New York: McGraw-Hill, 1948).

L. A. Moxon, *HF Antennas for All Locations* (Potters Bar, Herts: Radio Society of Great Britain, 1982).

Ramo and Whinnery, *Fields and Waves in Modern Radio* (New York: John Wiley & Sons).

V. H. Rumsey, *Frequency Independent Antennas* (New York: Academic Press, 1966).

P. N. Saveskie, *Radio Propagation Handbook* (Blue Ridge Summit, PA: Tab Books, Inc, 1980).

S. A. Schelkunoff, *Advanced Antenna Theory* (New York: John Wiley & Sons, Inc, 1952).

S. A. Schelkunoff and H. T. Friis, *Antennas Theory and Practice* (New York: John Wiley & Sons, Inc, 1952).

J. Sevick, *Transmission Line Transformers* (Atlanta: Noble Publishing, 1996).

H. H. Skilling, *Electric Transmission Lines* (New York: McGraw-Hill Book Co, Inc, 1951).

M. Slurzburg and W. Osterheld, *Electrical Essentials of Radio* (New York: McGraw-Hill Book Co, Inc, 1944).

G. Southworth, *Principles and Applications of Waveguide Transmission* (New York: D. Van Nostrand Co, 1950).

R. D. Straw, Ed., *The ARRL Antenna Book*, 21st ed. (Newington: ARRL, 2007).

F. E. Terman, *Radio Engineers' Handbook*, 1st ed. (New York, London: McGraw-Hill Book Co, 1943).

F. E. Terman, *Radio Engineering*, 3rd ed. (New York: McGraw-Hill, 1947).

S. Uda and Y. Mushiake, *Yagi-Uda Antenna* (Sendai, Japan: Sasaki Publishing Co, 1954). [Published in English — Ed.]

P. P. Viezbicke, "Yagi Antenna Design," NBS Technical Note 688 (US Dept of Commerce/National Bureau of Standards, Boulder, CO), Dec 1976.

G. B. Welch, *Wave Propagation and Antennas* (New York: D. Van Nostrand Co, 1958).

The GIANT Book of Amateur Radio Antennas (Blue Ridge Summit, PA: Tab Books, 1979).

IEEE Standard Dictionary of Electrical and Electronics Terms, 3rd ed. (New York: IEEE, 1984).

Radio Broadcast Ground Systems, available from Smith Electronics, Inc, 8200 Snowville Rd, Cleveland, OH 44141.

Radio Communication Handbook, 5th ed. (London: RSGB, 1976).

RDF BIBLIOGRAPHY

Bohrer, "Foxhunt Radio Direction Finder," *73 Amateur Radio*, Jul 1990, p 9.

Bonaguide, "HF DF — A Technique for Volunteer Monitoring," *QST*, Mar 1984, p 34.

DeMaw, "Maverick Trackdown," *QST*, Jul 1980, p 22.

Dorbuck, "Radio Direction Finding Techniques," *QST*, Aug 1975, p 30.

Eenhoorn, "An Active Attenuator for Transmitter Hunting," *QST*, Nov 1992, p 28.

Flanagan and Calabrese, "An Automated Mobile Radio Direction Finding System," *QST*, Dec 1993, p 51.

Geiser, "A Simple Seeker Direction Finder," *ARRL Antenna Compendium, Volume 3*, p 126.

Gilette, "A Fox-Hunting DF Twin'Tenna," *QST*, Oct 1998, pp 41-44.

Johnson and Jasik, *Antenna Engineering Handbook*, Second Edition, New York: McGraw-Hill.

Kossor, "A Doppler Radio-Direction Finder," *QST*, Part 1: May 1999, pp 35-40; Part 2: June 1999, pp 37-40.

McCoy, "A Linear Field-Strength Meter," *QST*, Jan 1973, p 18.

Moell and Curlee, *Transmitter Hunting: Radio Direction Finding Simplified*, Blue Ridge Summit, PA: TAB/McGraw-Hill. (This book, available from ARRL, includes plans for the Roanoke Doppler RDF unit and in-line air attenuator, plus VHF quads and other RDF antennas.)

Moell, "Transmitter Hunting — Tracking Down the Fun," *QST*, Apr 1993, p 48 and May 1993, p 58.

Moell, "Build the Handy Tracker," *73 Magazine*, Sep 1989, p 58 and Nov 1989, p 52.

O'Dell, "Simple Antenna and S-Meter Modification for 2-Meter FM Direction Finding," *QST*, Mar 1981, p 43.

O'Dell, "Knock-It-Down and Lock-It-Out Boxes for DF," *QST*, Apr 1981, p 41.

Ostapchuk, "Fox Hunting is Practical and Fun!" *QST*, Oct 1998, pp 68-69.

Rickerd, "A Cheap Way to Hunt Transmitters," *QST*, Jan 1994, p 65.

The "Searcher" (SDF-1) Direction Finder, Rainbow Kits.

RDF RESOURCES

Homing In
www.homingin.com — Web site by KØOV on direction finding techniques and activities

Amateur Radio Direction Finding (IARU Region II)
www.ardf-r2.org/en — ARDF activities and organizations in IARU Region II

Radio Direction Finding
en.wikipedia.org/wiki/Direction_finding — a general site on RDF with links to related subjects

DX Zone RDF Links
www.dxzone.com/catalog/Operating_Modes/Radio_Direction_Finding — a page of links to RDF articles and Web sites

Contents

22.1 Component Data
    22.1.1 EIA and Industry Standards
    22.1.2 Other Sources of Component Data
    22.1.3 The ARRL Technical Information Service (TIS)
    22.1.4 Definitions
    22.1.5 Surface-Mount Technology (SMT)
22.2 Resistors
    22.2.1 Resistor Types
    22.2.2 Resistor Identification
22.3 Capacitors
    22.3.1 Capacitor Types
    22.3.2 Electrolytic Capacitors
    22.3.3 Surface Mount Capacitors
    22.3.4 Capacitor Voltage Ratings
    22.3.5 Capacitor Identification
22.4 Inductors
22.5 Transformers
22.6 Semiconductors
    22.6.1 Diodes
    22.6.2 Transistors
    22.6.3 Voltage Regulators
    22.6.4 Analog and Digital Integrated Circuits
    22.6.5 MMIC Amplifiers
22.7 Tubes, Wire, Materials, Attenuators, Miscellaneous
22.8 Computer Connectors
22.9 RF Connectors and Transmission Lines
    22.8.1 UHF Connectors
    22.8.2 BNC, N, and F Connectors
22.10 Reference Tables

List of Figures
Fig 22.1 — Resistor wattages and sizes
Fig 22.2 — Surface mount resistors
Fig 22.3 — Power resistors
Fig 22.4 — Resistor value identification
Fig 22.5 — Common capacitor types and package styles
Fig 22.6 — Aluminum and tantalum electrolytic capacitors
Fig 22.7 — Aluminum electrolytic capacitor dimensions
Fig 22.8 — Surface-mount capacitor packages
Fig 22.9 — Surface-mount electrolytic packages
Fig 22.10 — Abbreviated EIA capacitor identification
Fig 22.11 — Obsolete capacitor color codes
Fig 22.12 — Complete EIA capacitor labeling scheme
Fig 22.13 — Obsolete JAN "postage stamp" capacitor labeling
Fig 22.14 — Color coding for cylindrical encapsulated RF chokes
Fig 22.15 — Color-coding for semiconductor diodes
Fig 22.16 — Axial-leaded diode packages and pad dimensions
Fig 22.17 — MMIC application
Fig 22.18 — MMIC package styles
Fig 22.19 — Installing PL-259 on RG-8 cable
Fig 22.20 — Installing PL-259 on RG-58 or RG-59 cable
Fig 22.21 — Installing crimp-on UHF connectors
Fig 22.22 — Installing BNC connectors
Fig 22.23 — Installing Type N connectors
Fig 22.24 — Installing Type F connectors

List of Tables
Resistors
Table 22.1 — Resistor Wattages and Sizes
Table 22.2 — SMT Resistor Wattages and Sizes
Table 22.3 — Power Resistors
Table 22.4 — Resistor Color Codes
Table 22.5 — EIA Standard Resistor Values
Table 22.6 — Mil–Spec Resistors

Capacitors
Table 22.7 — EIA Standard Capacitor Values
Table 22.8 — Ceramic Temperature Characteristics
Table 22.9 — Aluminum Electrolytic Capacitors Standard Sizes (Radial Leads)
Table 22.10 — Aluminum Electrolytic Capacitors EIA ±20%Standard Values
Table 22.11 — SMT Capacitor Two-Character Labeling
Table 22.12 — Surface Mount Capacitors EIA Standard Sizes
Table 22.13 — Capacitor Standard Working Voltages
Table 22.14 — European Marking Standards for Capacitors

Inductors
Table 22.15 — EIA Standard Inductor Values
Table 22.16 — Powdered-Iron Toroidal Cores: Magnetic Properties
Table 22.17 — Powdered-Iron Toroidal Cores: Dimensions
Table 22.18 — Ferrite Toroids: A_L Chart (mH per 1000 turns) Enameled Wire

Transformers
Table 22.19 — Power-Transformer Wiring Color Codes
Table 22.20 — IF Transformer Wiring Color Codes
Table 22.21 — IF Transformer Slug Color Codes
Table 22.22 — Audio Transformer Wiring Color Codes

Semiconductors
Table 22.23 — Semiconductor Diode Specifications
Table 22.24 — Package dimensions for small signal, rectifier and Zener diodes
Table 22.25 — Common Zener Diodes
Table 22.26 — Voltage-variable Capacitance Diodes
Table 22.27 — Three-terminal Voltage Regulators
Table 22.28 — Monolithic 50-Ω Amplifiers (MMIC gain blocks)
Table 22.29 — Small-Signal FETs
Table 22.30 — Low-Noise Bipolar Transistors
Table 22.31 — General-Purpose Bipolar Transistors
Table 22.32 — General-Purpose Silicon Power Bipolar Transistors
Table 22.33 — Power FETs or MOSFETs
Table 22.34 — RF Power Transistors
Table 22.35 — RF Power Transistors Recommended for New Designs
Table 22.36 — RF Power Amplifier Modules
Table 22.37 — Digital Logic Families
Table 22.38 — Operational Amplifiers (Op Amps)

Tubes, Wire, Materials, Attenuators, Miscellaneous
Table 22.39 — Triode Transmitting Tubes
Table 22.40 — Tetrode Transmitting Tubes
Table 22.41 — EIA Vacuum Tube Base Diagrams
Table 22.42 — Metal-oxide Varistor (MOV) Transient Suppressors
Table 22.43 — Crystal Holders
Table 22.44 — Copper Wire Specifications
Table 22.45 — Standard vs American Wire Gauge
Table 22.46 — Antenna Wire Strength
Table 22.47 — Guy Wire Lengths to Avoid
Table 22.48 — Aluminum Alloy Specifications
Table 22.49 — Impedance of Two-Conductor Twisted Pair Lines
Table 22.50 — Attenuation per foot of Two-Conductor Twisted Pair Lines
Table 22.51 — Large Machine-Wound Coil Specifications
Table 22.52 — Inductance Factor for Large Machine-Wound Coils
Table 22.53 — Small Machine-Wound Coil Specifications
Table 22.54 — Inductance Factor for Small Machine-Wound Coils
Table 22.55 — Measured Inductance for #12 Wire Windings
Table 22.56 — Relationship Between Noise Figure and Noise Temperature
Table 22.57 — Pi-Network Resistive Attenuators (50-Ω Impedance)
Table 22.58 — T-Network Resistive Attenuators (50-Ω Impedance)

Computer Connectors
Table 22.59 — Computer Connector Pin Outs

RF Connectors and Transmission Lines
Table 22.60 — Nominal Characteristics of Commonly Used Transmission Lines
Table 22.61 — Coaxial Cable Connectors

Reference Tables
Table 22.62 — US Customary Units and Conversion Factors
Table 22.63 — Metric System — International System of Units (SI)
Table 22.64 — Voltage-Power Conversion Table
Table 22.65 — Reflection Coefficient, Attenuation, SWR, and Return Loss
Table 22.66 — Abbreviations List

Chapter 22

Component Data and References

Radio amateurs have long been known for their electronic experimentation and homebrew building accomplishments. Using the wide variety of electronic components available, they design and build impressive radio equipment. With the industry growth of components for wireless communications, and surface mount technology (SMT), the choices available to the amateur today seem endless. For new and old homebrewers alike, selecting the proper component can seem a daunting task.

Fortunately, most amateurs tend to use a limited number of component types that have "passed the test of time," making component selection in many cases easy and safe. Others are learning to design and build using the new array of surface-mount technology components.

22.1 Component Data

This section, updated by Paul Harden, NA5N, provides reference information on the old and new components most often used by the Amateur Radio experimenter and homebrewer, and information for those wishing to learn more about component performance and selection.

22.1.1 EIA and Industry Standards

The American National Standards Institute (ANSI), the Electronic Industries Alliance (EIA), and the Electronic Components Association (ECA) establish the US standards for most electronic components, connectors, wire and cables. These standards establish component sizes, wattages, "standard values," tolerances and other performance characteristics. A branch of the EIA sets the standards for Mil-spec (standard military specification) and special electronic components used by defense and government agencies. The Joint Electron Devices Engineering Council (JEDEC), another branch of the EIA, develops the standards for the semiconductor industry. The EIA cooperates with other standards agencies such as the International Electrotechnical Commission (IEC), a worldwide standards agency. You can often find published EIA standards in the engineering library of a college or university.

And finally, the International Organization of Standardization (ISO), headquartered in Geneva, Switzerland, sets the global standards for nearly everything from paper sizes to photographic film speeds. ANSI is the US representative to the ISO.

These organizations, or their acronyms, are familiar to most of us. They are much more than a label on a component. EIA and other industry standards are what mark components for identification, establishes the "preferred standard values" and ensures their reliable performance from one unit to the next, regardless of their source. Standards require that a 1.2 kΩ 5% resistor from Ohmite Corp. has the same performance as a 1.2 kΩ 5% resistor from Vishay-Dale, or a 2N3904 to have the same performance characteristics and physical packaging whether from ON Semi or Gold Star.

Much of the component data in this chapter is devoted to presenting these component standards, physical dimensions and the various methods of component identification and marking. By selecting components manufactured under these industry standards, building a project from the *Handbook* or other source will ensure nearly identical performance to the original design.

22.1.2 Other Sources of Component Data

There are many sources you can consult for detailed component data but the best source of component information and data sheets is the Internet. Most manufacturers maintain extensive Web sites with information and data on their products. Often, the quickest route to detailed product information is to enter "data sheet" and the part number into an Internet search engine. Distributors such as Digi-Key and Mouser include links to useful information in their online catalogs as well. Some manufacturers still publish data books for the components they make, and parts catalogs themselves are often good sources of component data and application notes and bulletins.

Some of the tables printed in previous editions of this book have been moved to the accompanying CD-ROM to make room for new material. If a table or figure you need is missing, check the CD-ROM!

22.1.3 The ARRL Technical Information Service (TIS)

The ARRL Technical Information Service on the ARRL Web site (**www.arrl.org/technical-information-service**) provides technical assistance to members and nonmembers, including information about components and useful references. The TIS includes links to detailed, commonly needed information in many technical areas. Questions may also be submitted via email (**tis@arrl.org**); fax (860-594-0259); or mail (TIS, ARRL, 225 Main St, Newington, CT 06111).

22.1.4 Definitions

Electronic components such as resistors, capacitors, and inductors are manufactured with a *nominal* value — the value with which they are labeled. The component's *actual* value is what is measured with a suitable measuring instrument. If the nominal value is given as text characters, an "R" in the value (for example "4R7") stands for *radix* and is read as a decimal point, thus "4.7".

Tolerance refers to a range of acceptable values above and below the nominal component value. For example, a 4700-Ω resistor rated for ±20% tolerance can have an actual value anywhere between 3760 Ω and 5640 Ω. You may always substitute a closer-tolerance device for one with a wider tolerance. For most Amateur Radio projects, assume a 10% tolerance if none is specified.

The *temperature coefficient* or *tempco* of a component describes its change in value with temperature. Tempco may be expressed as a change in unit value per degree (ohms per degree Celsius) or as a relative change per degree (parts per million per degree). Except for temperature sensing components that may use Fahrenheit or Kelvin, Celsius is almost always used for the temperature scale. Temperature coefficients may not be linear, such as those for capacitors, thermistors, or quartz crystals. In such cases, tempco is specified by an identifier such as Z5U or C0G and an equation or graph of the change with temperature provided by the manufacturer.

22.1.5 Surface-Mount Technology (SMT)

"SMT" is used throughout this book to refer to components, printed-circuit boards or assembly techniques that involve surface-mount technology. SMT components are often referred to by the abbreviations "SMD" and "SMC," but all three abbreviations are considered to be effectively equivalent. *Through-hole* or *leaded* components are those with wire leads intended to be inserted into holes in printed-circuit boards or used in point-to-point wiring.

Many different types of electronic components, both active and passive, are now available in surface-mount packages. Each package is identified by a code, such as 1802 or SOT. Resistors in SMT packages are referred to by package code and not by power dissipation, as through-hole resistors are. The very small size of these components leaves little space for marking with conventional codes, so brief alphanumeric codes are used to convey the most information in the smallest possible space. You will need a magnifying glass to read the markings on the bodies of SMT components.

In many cases, vendors will deliver SMT components packaged in tape from master reels and the components will not be marked. This is often the case with SMT resistors and small capacitors. However, the tape will be marked or the components are delivered in a plastic bag with a label. Take care to keep the components separated and labeled or you'll have to measure their values one by one!

HAMCALC Calculators

The HAMCALC package of software calculators by George Murphy, VE3ERP, is very handy. Covering dozens of topics from antenna lengths to Z-matching, the package can be downloaded free of charge from **www.cq-amateur-radio.com/HamCalcem.html**.

22.2 Resistors

Most resistors are manufactured using EIA standards to establish common ratings for wattage, resistor values and tolerance regardless of the manufacturer. EIA marking methods for resistors utilize either an alphanumeric scheme or a color code to denote the value and tolerance.

In the earlier days of electronics, 10% and 20% tolerance resistors were the common and inexpensive varieties used by most amateurs. 1% tolerance resistors were considered the "precision resistors" and seldom used by the amateur due to their significantly higher cost.

Today, with improved manufacturing techniques, both 5% and 1% tolerance resistors are commonly available *and* inexpensive, with precision resistors to 0.1% not uncommon.

22.2.1 Resistor Types

The major resistor types are carbon composition, carbon film, metalized film and wire-wound, as described below. (For additional discussion of the characteristics of the different types of resistors, see the **Electrical Fundamentals** chapter.)

Carbon composition resistors are made from a slurry of carbon and binder material formulated to achieve the desired resistance when compressed into a cylinder and encapsulated. This yields a resistor with tolerances in the 5% to 20% range. "Carbon comp" resistors have a tendency to absorb moisture over time and to change value, but can withstand temporary "pulse" overloads that would damage or destroy a film-type resistor.

Carbon film resistors are made from a layer of carbon deposited on a dielectric film or substrate. The thickness of the carbon film is controlled to form the desired resistance with greater accuracy than for carbon composition. They are low cost alternatives to carbon composition resistors and are available with 1% to 5% tolerances.

Metalized film resistors replace carbon films with metal films deposited onto the dielectric using sputtering techniques to achieve very accurate resistances to 0.1% tolerances. Metal film resistors also generate less thermal noise than carbon resistors.

All three of these resistor types are normally available with power ratings from 1/10 W to 2 W. **Fig 22.1** and **Tables 22.1** and **22.2** provide the body sizes and lead or pad spacing for through-hole and SMT resistors.

For new designs, carbon film and metalized film resistors should be used for their improved characteristics and lower cost compared to the older carbon composition resistors. Metalized films have lower residual inductance and often preferred at VHF. Most surface mount resistors (shown in **Fig 22.2**) are metalized films.

Wire-wound resistors, as the name implies, are made from lengths of wire wound around an insulating form to achieve the desired resistance for power ratings above 2 W. Wire-wound resistors have high parasitic inductance, caused by the wire wrapped around a form similar to a coil, and thus should not be used at RF frequencies. **Fig 22.3** (A, B and D) show three types of wire-wound resistors with wattage ranges in **Table 22.3**.

An alternative to wire-wound resistors is the new generation of resistors known as *thick-film power resistors*. They are rated up to 100 W and packaged in a TO-220 or similar case which makes it easy to mount them on heat sinks and printed-circuit boards. Most varieties are non-inductive and suitable for RF use. Metal-oxide ("cement") resistors are also available in packages similar to that of Fig 22.3B. Similar to carbon composition resistors, metal-oxide resistors are non-inductive and useful at RF.

22.2.2 Resistor Identification

Resistors are identified by the EIA numerical or color code standard as shown in **Fig 22.4**. The EIA numerical code for resistor identification is widely used in industry. The nominal resistance, expressed in ohms, is identified by three digits for 2% (and greater) tolerance devices. The first two digits represent the significant figures; the last digit specifies the multiplier as the exponent of 10. (The multiplier is simply the number of zeros following the significant numerals.) For values less than 100 Ω, the letter R is substituted for one of the significant digits and represents a decimal point. An alphabetic character indicates the tolerance as shown in Table 22.2

For example, a resistor marked with "122J" would be a 1200 Ω, or a 1.2 kΩ 5% resistor. A resistor containing four digits, such as "1211," would be a 1210 Ω, or a 1.21 kΩ 1% precision resistor.

If the tolerance of the unit is narrower than ±2%, the code used is a four-digit code where the first three digits are the significant figures and the last is the multiplier. The

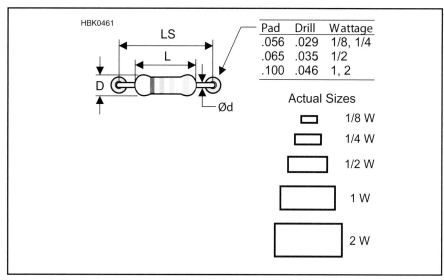

Fig 22.1 — Resistor wattages and sizes.

Table 22.1
Resistor Wattages and Sizes

Size	L	D	LS*	Ød	PCB Pad Size and Drill
1/8 W	0.165	0.079	0.25	0.020	0.056 round, 0.029 hole
1/4 W	0.268	0.098	0.35	0.024	0.056 round, 0.029 hole
1/2 W	0.394	0.138	0.60	0.029	0.065 round, 0.035 hole
1 W	0.472	0.197	0.70	0.032	0.100 round, 0.046 hole
2 W	0.687	0.300	0.90	0.032	0.100 round, 0.046 hole

Dimensions in inches.
*LS = Recommended PCB lead bend

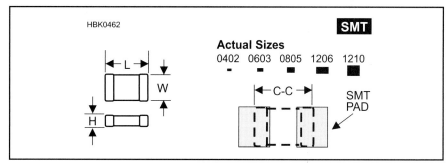

Fig 22.2 — Surface mount resistors.

Table 22.2
SMT Resistor Wattages and Sizes

Body Size	L	W	H	SMT Pad	C-C*
0402	0.039	0.020	0.014	0.025 × 0.035	0.050
0603	0.063	0.031	0.018	0.030 × 0.030	0.055
0805	0.079	0.049	0.020	0.040 × 0.050	0.075
1206	0.126	0.063	0.024	0.064 × 0.064	0.125
1210	0.126	0.102	0.024	0.070 × 0.100	0.150

Dimensions in inches.
*C-C is SMT pad center-to-center spacing

SMT Resistor Tolerance Codes

Letter	Tolerance
D	±0.5%
F	±1.0%
G	±2.0%
J	±5.0%

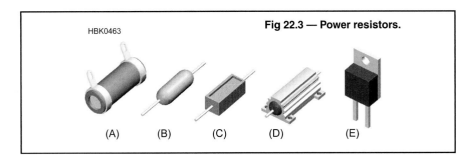

Fig 22.3 — Power resistors.

Table 22.3
Power Resistors

Fig 22.3	Power Resistor Type	Wattage Range
A	Wire-wound, ceramic core	10-300 W
B	Wire-wound, axial	3-10 W
C	Metal-oxide	5-25 W
D	Wire-wound, aluminum housing	3-50 W
E	Thick-film resistors*	15-100 W

*Wire-wound resistors are inductive, though seldom noted as such on the data sheets, and are not recommended for RF. Thick-film and metal-oxide power resistors are low inductance or noninductive.

letter R is used in the same way to represent a decimal point. For example, 1001 indicates a 1000-Ω unit, and 22R0 indicates a 22-Ω unit.

Here are some additional examples of resistor value markings:

Code	Value
101	10 and 1 zero = 100 Ω
224	22 and 4 zeros = 220,000 Ω
1R0	1.0 and no zeros = 1 Ω
22R	22.0 and no zeros = 22 Ω
R10	0.1 and no zeros = 0.1 Ω

The resistor color code, used only with through-hole components, assigns colors to the numerals one through nine and zero, as shown in **Table 22.4**, to represent the significant numerals, the multiplier and the tolerance. The color code is often memorized with a mnemonic such as "Big boys race our young girls, but Violet generally wins" to represent the colors black (0), brown (1), red (2), orange (3), yellow (4), green (5), blue (6), violet (7), gray (8) and white (9). You will no doubt discover other versions of this memory aid made popular over the years.

For example, a resistor with color bands black (1), red (2), red (2) and gold would be a 1200 Ω, or 1.2 kΩ 5% resistor, with the gold band signifying 5% tolerance.

The resistor color code should be memorized as it is also used for identifying capacitors, and inductors. It is also handy to use when connecting multi-conductor or ribbon cables.

Resistors are also identified by an "E" series classification, such as E12 or E48. The number following the letter E signifies the number of logarithmic steps per decade. The more steps per decade, the more choices of resistor values and tighter the tolerances can be. For example, in the E12 series, there are twelve resistor values between 1 kΩ and 10 kΩ with 10% tolerance; E48 provides 48 values between 1 kΩ and 10 kΩ at 1% tolerance. This system is often used with online circuit calculators to indicate the resistor accuracy and tolerance desired. The standard resistor values of the E12 (±10%), E24 (±5%), E48 (±2%) and E96 (±1%) series are listed in **Table 22.5**.

Resistors used in military electronics (Mil-spec) use the type identifiers listed in **Table 22.6**. In addition, Mil-spec resistors with paint-stripe value bands have an extra band indicating the reliability level to which they are certified.

Standard EIA Identification and Marking

5–20% Resistors

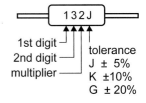

1st digit
2nd digit
multiplier

tolerance
J ± 5%
K ±10%
G ±20%

"Precision" Resistors

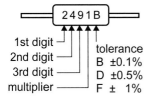

1st digit
2nd digit
3rd digit
multiplier

tolerance
B ±0.1%
D ±0.5%
F ± 1%

Examples:

"132J"=1300=1.3KΩ 5%
"510K"=51 =51Ω 10%
"2R2G"=2.2Ω 20%

"2491B"=2490=2.49K 0.1%
"5110D"=511 =511Ω 0.5%
"51R1F" =51.1=51.1Ω 1%

EIA Resistor Color Codes

Color Codes

Digit	Color
0	black (blk)
1	brown (brn)
2	red (red)
3	orange (org)
4	yellow (ylw)
5	green (grn)
6	blue (blu)
7	violet (vio)
8	gray (gry)
9	white (wht)

5–20% Resistors
(4-band color code)

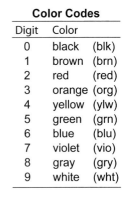

1st digit
2nd digit
multiplier

tolerance
gold ± 5%
silver ±10%
none ±20%

"Precision" Resistors
(5-band color code)

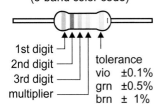

1st digit
2nd digit
3rd digit
multiplier

tolerance
vio ±0.1%
grn ±0.5%
brn ± 1%

Examples:

brn-org-org-gold = 13KΩ 5%
grn-brn-blk-silver = 51Ω 10%
brn-org-org-brn-brn = 1.33KΩ

Fig 22.4 — Resistor value identification.

Table 22.4
Resistor Color Codes

Color	Significant Figure	Decimal Multiplier	Tolerance (%)
Black	0	1	
Brown	1	10	1
Red	2	100	2
Orange	3	1,000	
Yellow	4	10,000	
Green	5	100,000	0.5
Blue	6	1,000,000	0.25
Violet	7	10,000,000	0.1
Gray	8	100,000,000	0.05
White	9	1,000,000,000	
Gold		0.1	5
Silver		0.01	10
No color			20

Table 22.5
EIA Standard Resistor Values

±10% (E12)	±5% (E24)	±2% (E48)		±1% (E96)			
100	100	100	316	100	178	316	562
120	110	105	332	102	182	323	576
150	120	110	348	105	187	332	590
180	130	115	365	107	191	340	604
220	150	121	383	110	196	348	619
270	160	127	402	113	200	357	634
330	180	133	422	115	205	365	649
390	200	140	442	118	210	374	665
470	220	147	464	121	215	383	681
560	240	154	487	124	221	392	698
680	270	162	511	127	226	402	715
820	300	169	536	130	232	412	732
	330	178	562	133	237	422	750
	360	187	590	137	243	432	768
	390	196	619	140	249	442	787
	430	205	649	143	255	453	806
	470	215	681	147	261	464	825
	510	226	715	150	267	475	845
	560	237	750	154	274	487	866
	620	249	787	158	280	499	887
	680	261	825	162	287	511	909
	750	274	866	165	294	523	931
	820	287	909	169	301	536	953
	910	301	953	174	309	549	976

Use Table 22.5 values for each decade.
Example: 133 = 13.3 Ω, 133 Ω, 1.33 kΩ, 13.3 Ω, 133 kΩ, 1.33MΩ

Table 22.6
Mil-Spec Resistors

Wattage	Metal Film Types	Fixed Film Types	Composition Types	
1/10 W	RN50			
1/8 W	RN55	RL05	RLR05	RCR05
1/4 W	RN60	RL07	RLR07	RCR07
1/2 W	RN65	RL20	RLR20	RCR20
1 W	RN75	RL32	RLR32	RCR32
2 W	RN80	RL42	RLR62	RCR42

Tolerance Codes
- B ±0.1%
- C ±0.25%
- D ±0.5%
- F ±1%
- G ±2%
- J ±5%
- K ±10%

Examples:
RN60D-2202F = 22 kΩ 1%
RL07S-471J = 470 Ω ±5%
RLR07C-471J = 470 Ω ±5%

Note: The RN Mil-Spec was discontinued in 1996 Still used by some manufacturers such as Vishay-Dale.

22.3 Capacitors

Capacitors exhibit the largest variety of electronic components. So many varieties and types are available that selecting the proper capacitor for a particular application can be overwhelming. Ceramic and film capacitors are the two most common types used by the amateur. (For additional information on the characteristics of the different types of capacitors, see the **Electrical Fundamentals** chapter.)

Though capacitors are classified by dozens of characteristics, the EIA has simplified the selection process by organizing ceramic capacitors into four categories called Class 1, 2, 3 and 4. Class 1 capacitors are the most stable and Class 4 the least preferred. Many catalogs now list ceramic capacitors by their class, greatly simplifying component selection.

For capacitors used in frequency-sensitive circuits, such as the frequency determining capacitors in oscillators or tuned circuits, select a Class 1 capacitor (C0G or NP0). For other applications, such as interstage coupling or bypass capacitors, components from Class 2 or Class 3 (X7R or Z5U) are usually sufficient. With modern manufacturing techniques, it is rare to find a Class 4 capacitor today.

Like resistors, capacitors are available in EIA standard series of values, E6 and E12, shown in **Table 22.7**. Most capacitors have a tolerance of 5% or greater. High-value capacitors used for filtering may have asymmetric tolerances, such as –5% and +10%, since the primary concern is for a guaranteed minimum value of capacitance.

22.3.1 Capacitor Types

Capacitor types can be grouped into ceramic dielectrics, film dielectrics and electrolytics. The type of capacitor can often be determined by their appearance, as shown in **Figs 22.5** and **22.6**. Capacitors with wire leads have two lead styles: *axial* and *radial*. Axial leads are aligned in opposite directions along a common axis, usually the largest dimension of the capacitor. Radial leads leave the capacitor body in the same direction and are usually arranged radially about the center of the capacitor body.

Disc (or *disk*) *ceramic* capacitors consist of two metal plates separated by a ceramic dielectric that establishes the desired capacitance. Due to their low cost, they are the most common capacitor type. The main disadvantage is their sensitivity to temperature changes (that is, a high temperature coefficient).

Monolithic ceramic capacitors are made by sandwiching layers of metal electrodes and ceramic layers to form the desired capacitance. "Monolithics" are physically smaller than disc ceramics for the same value of capacitance and cost, but exhibit the same high temperature coefficients.

Polyester film capacitors use layers of metal and polyester (Mylar) to make a wide

Table 22.7
EIA Standard Capacitor Values

±20% Capacitors (E6)

pF	pF	pF	µF	µF	µF	µF
1.0	10	100	0.001	0.01	0.1	1
1.5	15	150	0.0015	0.015	0.15	1.5
2.2	22	220	0.0022	0.022	0.22	2.2
3.3	33	330	0.0033	0.033	0.33	3.3
4.7	47	470	0.0047	0.047	0.47	4.7
6.8	68	680	0.0068	0.068	0.68	6.8

±10%, ±5% Capacitors (E12)

pF	pF	pF	µF	µF	µF	µF
1.0	10	100	0.001	0.01	0.1	1
1.2	12	120	0.0012	0.012	0.12	
1.5	15	150	0.0015	0.015	0.15	
1.8	18	180	0.0018	0.018	0.18	
2.2	22	220	0.0022	0.022	0.22	2.2
2.7	27	270	0.0027	0.027	0.27	
3.3	33	330	0.0033	0.033	0.33	3.3
3.9	39	390	0.0039	0.039	0.39	
4.7	47	470	0.0047	0.047	0.47	4.7
5.6	56	560	0.0056	0.056	0.56	
6.8	68	680	0.0068	0.068	0.68	
8.2	82	820	0.0082	0.082	0.82	

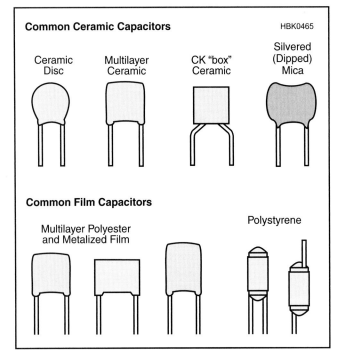

Fig 22.5 — Common capacitor types and package styles.

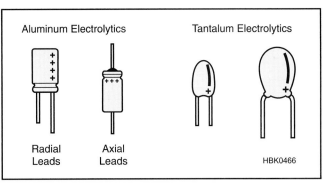

Fig 22.6 — Aluminum and tantalum electrolytic capacitors.

range of capacitances. They have poor temperature coefficients and are not recommended for high frequency use, but are suitable for low frequency and audio circuits.

To improve the performance of film capacitors, other dielectrics are used, such as polypropylene, polystyrene or polycarbonate film, or silvered-mica. These are very stable capacitors developed for RF use. Their main disadvantages are higher cost and lower working voltages than other varieties.

Capacitors are particularly sensitive to temperature changes because the physical dimensions of the capacitor determine its value. Standard temperature coefficient codes are shown in **Table 22.8**. Each code is made up of one character from each column in the table. For example, a capacitor marked Z5U is suitable for use between +10 and +85°C, with a maximum change in capacitance of −56% or +22%.

Capacitors with highly predictable temperature coefficients of capacitance are sometimes used in circuits whose performance must remain stable with temperature. If an application called for a temperature coefficient of −750 ppm/°C (N750), a capacitor marked U2J would be suitable. The older industry code for these ratings is being replaced with the EIA code shown in Table 22.8. NP0 (that is, N-P-zero) means "negative, positive, zero." It is a characteristic often specified for RF circuits requiring temperature stability, such as VFOs. A capacitor of the proper value marked C0G is a suitable replacement for an NP0 unit.

22.3.2 Electrolytic Capacitors

Aluminum electrolytic capacitors use aluminum foil "wetted" with a chemical agent and formed into layers to increase the effective area, and therefore the capacitance. Aluminum electrolytics provide high capacitance in small packages at low cost. Most varieties are polarized, that is, voltage should only be applied in one "direction." Polarized capacitors have a negative (−) and positive (+) lead. Standard dimensions of aluminum electrolytics are shown in **Fig 22.7** and **Table 22.9**. EIA standard values for aluminum electrolytics are given in **Table 22.10**.

Very old electrolytic capacitors should be used with care or, preferably, replaced. The wet dielectric agent can dry out during prolonged periods of non-use, causing the internal capacitor plates to form a short circuit when energized. Applying low voltage and gradually increasing it over a period of time may restore the capacitor to operation, but if the dielectric agent has dried out, the capacitor will have lost some or most of its value and will likely be lossy and prone to failure.

Tantalum electrolytic capacitors consist of a tantalum pentoxide powder mixed with a wet or dry electrolyte, then formed into a pellet or slug for a large effective area. Tantalums also provide high capacitance values in very small packages. Tantalums tend to be more expensive than aluminum electrolytic capacitors. Like the aluminum electrolytic capacitor, tantalum capacitors are also polarized for which care should be exercised. Some varieties of tantalums can literally explode or burst open if voltage is applied with reverse polarity or the voltage rating is exceeded. Tantalum electrolytics are used almost exclusively as high-value SMT components due to their small sizes. Capacitance values up to 1000 µF at 4 V are available with body sizes about a quarter-inch square.

Identifying the polarity markings of aluminum and tantalum electrolytics (shown in Fig 22.6) can be confusing. Most tantalum electrolytics are marked with a solid band indicating the positive lead. Aluminum electrolytics are available with bands or symbols

Table 22.8
Ceramic Temperature Characteristics

Common EIA Types:

EIA Class	EIA Code	Characteristics	Temp. Range*
1	C0G	0 ± 30 ppm/°C	−55 °C to +125 °C
2	Y5P	±10%	−30 °C to + 85 °C
2	X7R	±15%	−55 °C to +125 °C
2	Y5U	±20%	−10 °C to + 85 °C
2	Z5U	±20%	+10 °C to + 85 °C
2	Z5V	+80%, −20%	−30 °C to + 85 °C
3	Y5V	+80%, −20%	−10 °C to + 85 °C

Common Industry Types:

EIA Class	EIA Code	Characteristics	Temp. Range*
1	NP0	0 ± 30 ppm/°C	−55 °C to +125 °C
2	CK05	±10%	−55 °C to +125 °C

*Temp. range for which characteristics are specified and may vary slightly between different manufacturers

Temperature Coefficient Codes

Minimum Temperature	Maximum Temperature	Maximum capacitance change over temp range
X −55 °C	2 +45°C	A ±1.0%
Y −30 °C	4 +65°C	B ±1.5%
Z +10 °C	5 +85°C	C ±2.2%
6 +105 °C		D ±3.3%
7 +125 °C		E ±4.7%
		F ±7.5%
		P ±10%
		R ±15%
		S ±22%
		T −33%, +22%
		U −56%, +22%
		V −82%, +22%

Table 22.9
Aluminum Electrolytic Capacitors Standard Sizes (Radial Leads)

H	Dia	LS	Pad Size and Drill*
0.44	0.20	0.08	0.056 round, 0.029 hole
0.44	0.25	0.10	0.056 round, 0.029 hole
0.44	0.32	0.14	0.065 round, 0.029 hole
0.52	0.40	0.20	0.080 round, 0.035 hole
0.78	0.50	0.20	0.080 round, 0.035 hole
1.00	0.63	0.30	0.100 round, 0.035 hole
1.42	0.72	0.30	0.100 round, 0.035 hole
1.60	0.88	0.40	0.100 round, 0.035 hole

Dimensions in inches.
*Customary to make "+" lead square pad on PCB

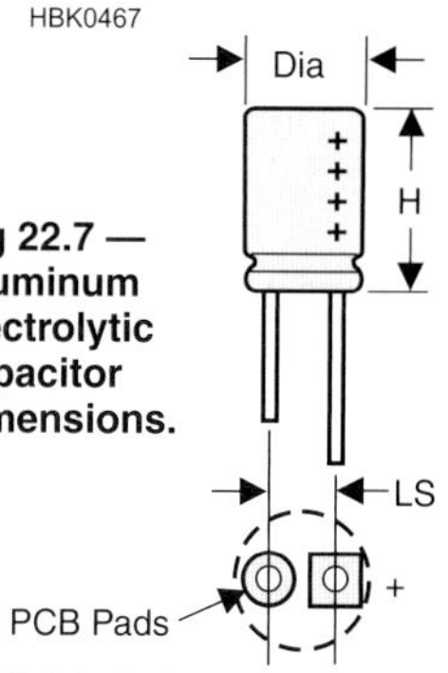

Fig 22.7 — Aluminum electrolytic capacitor dimensions.

Table 22.10
Aluminum Electrolytic Capacitors EIA ±20% Standard Values

µF	µF	µF	µF	µF
0.1	1.0	10	100	1000
0.22	2.2	22	220	2200
0.33	3.3	33	330	3300
0.47	4.7	47	470	4700
0.68	6.8	68	680	6800
0.82	8.2	82	820	8200

marking *either* the negative or positive lead. The positive lead of axial-lead capacitors is usually longer than the negative lead. Misidentifying the polarity of capacitors is a common error during assembly or repair.

22.3.3 Surface Mount Capacitors

SMT capacitors are generally film, ceramic or tantalum electrolytics. Body sizes are shown in **Fig 22.8** and **22.9**. Although the EIA scheme is the standard method of labeling capacitor value, you may encounter a two-character alphanumeric code (see **Table 22.11**) consisting of a letter indicating the significant digits and a number indicating the multiplier. The code represents the capacitance in picofarads. For example, a chip capacitor marked "A4" would have a capacitance of 10,000 pF, or 0.01 µF. A unit marked "N1" would be a 33-pF capacitor. If there is sufficient space on the device package, a tolerance code may be included. The standard SMT body sizes and pad spacing are provided in **Table 22.12**.

Table 22.11
SMT Capacitor Two-Character Labeling
Significant Figure Codes

Character	Significant Figures	Character	Significant Figures
A	1.0	T	5.1
B	1.1	U	5.6
C	1.2	V	6.2
D	1.3	W	6.8
E	1.5	X	7.5
F	1.6	Y	8.2
G	1.8	Z	9.1
H	2.0	a	2.5
J	2.2	b	3.5
K	2.4	d	4.0
L	2.7	e	4.5
M	3.0	f	5.0
N	3.3	m	6.0
P	3.6	n	7.0
Q	3.9	t	8.0
R	4.3	y	9.0
S	4.7		

Multiplier Codes

Numeric Character	Decimal Multiplier
0	1
1	10
2	100
3	1,000
4	10,000
5	100,000
6	1,000,000
7	10,000,000
8	100,000,000
9	0.1

Table 22.12
Surface Mount Capacitors — EIA Standard Sizes

Size	Length	Width	Height	C	SMT Pad	C-C*
0402	0.039	0.020	0.014	0.010	0.025 × 0.035	0.050
0603	0.063	0.031	0.018	0.014	0.030 × 0.030	0.055
0805	0.079	0.049	0.020	0.016	0.040 × 0.050	0.075
1206	0.126	0.063	0.024	0.020	0.064 × 0.064	0.125
1210	0.126	0.102	0.024	0.020	0.070 × 0.100	0.150

Surface Mount Electrolytic Capacitors — EIA Standard Sizes

Size	Length	Width	Height	C-C*	SMT Pad
A (1206)	0.126	0.063	0.063	0.110	0.055 × 0.060
B (1411)	0.138	0.110	0.075	0.136	0.075 × 0.090
C (2412)	0.236	0.126	0.098	0.265	0.090 × 0.120
D (2916)	0.287	0.169	0.110	0.250	0.100 × 0.100
E (2924)	0.287	0.236	0.142	0.250	0.100 × 0.100

Dimensions in inches.
*C-C is SMT pad center-to-center spacing

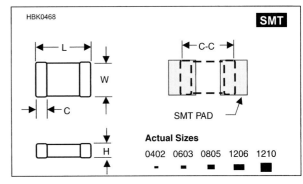

Fig 22.8 — Surface-mount capacitor packages.

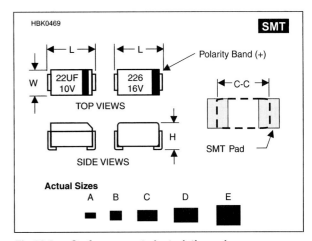

Fig 22.9 — Surface-mount electrolytic packages.

22.3.4 Capacitor Voltage Ratings

Capacitors are also rated by their maximum operating voltage. The importance of selecting a capacitor with the proper voltage rating is often overlooked. Exceeding the voltage rating, even momentarily, can cause excessive heating, a permanent shift of the capacitance value, a short circuit, or outright destruction. As a result, the voltage rating should be at least 25% higher than the working voltage across the capacitor; many designers use 50-100%.

Following the 25% guideline, filter capacitors for a 12-V system should have at least a 15-V rating (12 V × 1.25). However, 12-V systems such as 12-V power supplies and automotive 12-V electrical systems actually operate near 13.8 V and in the case of automotive systems, as high as 15 V. In such cases, capacitors rated for 15 V would be an insufficient margin of safety; 20 to 25-V capacitors should be used in such cases.

In large signal ac circuits, the maximum voltage rating of the capacitor should be based on the peak-to-peak voltages present. For example, the output of a 5-W QRP transmitter is 16 V_{RMS}, or about 45 V_{P-P}. Capacitors exposed to the 5 W RF power, such as in the output low-pass filter, should be rated well above 50 V for the 25% rule. A 100 W transmitter produces RF voltages of about 200 V_{P-P}.

Capacitors that are to be connected to primary ac circuits (directly to the ac line) for filtering or coupling *must* be rated for ac line use. These capacitors are listed as such in catalogs and are designed to minimize fire and other hazards in case of failure. Remember, too, that ac line voltage is given as RMS, with peak-to-peak voltage 2.83 times higher: 120 V_{RMS} = 339 V_{P-P}.

Applying peak-to-peak voltages approaching the maximum voltage rating will cause excessive heating of the capacitor. This, in turn, will cause a permanent shift in the capacitance value. This could be undesirable in the output low pass filter example cited above in trying to maintain the proper impedance match between transmitter and antenna.

Exceeding the maximum voltage rating can also cause a breakdown of the dielectric material in the capacitor. The voltage can jump between the plates causing momentary or permanent electrical shorts between the capacitor plates.

In electrolytic and tantalum capacitors, exceeding the voltage rating can produce extreme heating of the oil or wetting agent used as the dielectric material. The expanding gases can cause the capacitor to burst or explode.

These over-voltage problems are easily avoided by selecting a capacitor with a voltage rating 25-50% above the normal peak-to-peak

Table 22.13
Capacitor Standard Working Voltages

Ceramic	Polyester	Electrolytic	Tantalum
		6.3 V	6.3 V
		10 V	10 V
16 V		16 V	16 V
			20 V
25 V		25 V	25 V
		35 V	35 V
50 V	50 V	50 V	50 V
		63 V	63 V
100 V	100 V	100 V	
	150 V	150 V	
200 V	200 V		
	250 V	250 V	

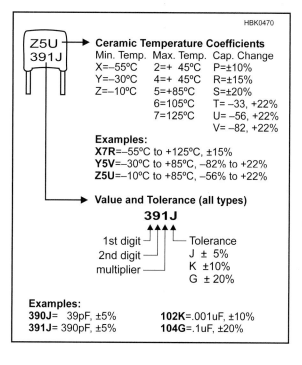

Fig 22.10 — Abbreviated EIA capacitor identification. This method is used on SMT capacitors. An "R" in the numeric field stands for "radix" and represents a decimal point, so that "4R7" indicates "4.7" for example.

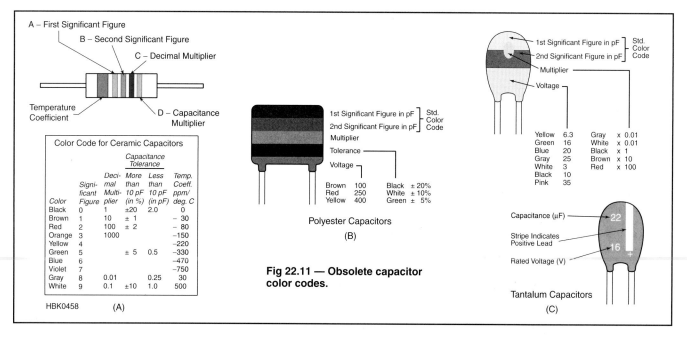

Fig 22.11 — Obsolete capacitor color codes.

operating voltage. **Table 22.13** lists standard working voltages for common capacitor types.

22.3.5 Capacitor Identification

Capacitors are identified by the EIA numerical or color code standard as shown in **Fig 22.10**. Since 2000, the EIA numerical code is the most dominant form of capacitor identification and is used on all capacitor types and body styles. Color coding schemes are becoming rare, used only by a few non-US manufacturers. Some thru-hole "gum drop" tantalum capacitors also still use the color codes of **Fig 22.11**. Electrolytic and tantalum capacitors are often labeled with capacitance and working voltage in µF and V as in Fig 22.11C.

Similar to the resistor EIA code, numerals are used to indicate the significant numerals and the multiplier, followed by an alphabetic character to indicate the tolerance. The multiplier is simply the number of zeros following the significant numerals. For example, a capacitor marked with "122K" would be a 1200 pF 10% capacitor. Additional digits and codes may be encountered as shown in **Fig 22.12**.

Military-surplus equipment using the obsolete "postage stamp" capacitors is still encountered in Amateur Radio. These capacitors used the colored dot method of value identification shown in **Fig 22.13**.

European manufacturers often use nanofarads or nF, such that 10 nF, or simply 10N, indicates 10 nanofarads. This is equivalent to 10,000 pF or 0.01 µF. This notational scheme, shown in **Table 22.14**, is more commonly found on schematic diagrams than actual part markings.

Table 22.14
European Marking Standards for Capacitors

Marking	Value
1p	1 pF
2p2	2.2 pF
10p	10 pF
100p	100 pF
1n	1 nF (= 0.001 µF)
2n2	2.2 nF (= 0.0022 µF)
10n	10 nF (= 0.01 µF)
100n	100 nF (= 0.1 µF)
1u	1 µF
5u6	5.6 µF
10u	10 µF
100u	100 µF

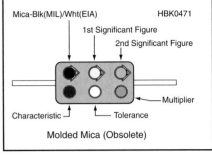

Fig 22.13 — Obsolete JAN "postage stamp" capacitor labeling.

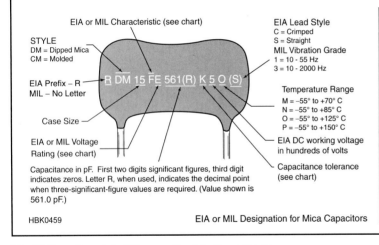

Fig 22.12 — Complete EIA capacitor labeling scheme.

22.4 Inductors

Inductors, both fixed and variable are available in a wide variety of types and packages, and many offer few clues as to their values. Some coils and chokes are marked with the EIA color code shown in Table 22.4. See **Fig 22.14** for another marking system for cylindrical encapsulated RF chokes. The body of these components is often green to identify them as inductors and not resistors. Measure the resistance of the component with an ohmmeter if there is any doubt as to the identity of the component. **Table 22.15** is a list of the EIA standard inductor values.

Table 22.16 lists the properties of common powdered-iron cores. Formulas are given for calculating the number of required turns based on a given inductance and for calculating the inductance given a specific number of turns. Most powdered-iron toroid cores that amateurs use are manufactured by Micrometals (**www.micrometals.com**). Paint is used to identify the material used in the core. The Micrometals color code is part of Table 22.16. **Table 22.17** gives the physical dimensions of powdered-iron toroids.

An excellent design resource for ferrite-based components is the Fair-Rite Materials Corp on-line catalog at **www.fair-rite.com**. The Fair-Rite Web site's Technical section also has free papers on the use of ferrites for EMI suppression and broadband transformers. The following list presents the general characteristics (material and composition and intended

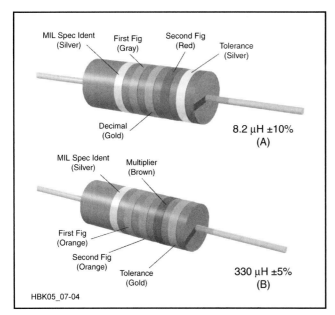

Fig 22.14 — Color coding for cylindrical encapsulated RF chokes. At A, an example of the coding for an 8.2-µH choke is given. At B, the color bands for a 330-µH inductor are illustrated. The color code is given in Table 22.4.

Table 22.15
EIA Standard Inductor Values

µH	µH	µH	mH	mH	mH
1.0	10	100	1.0	10	100
1.2	12	120	1.2	12	120
1.5	15	150	1.5	15	150
2.2	22	220	2.2	22	220
2.7	27	270	2.7	27	270
3.3	33	330	3.3	33	330
3.9	39	390	3.9	39	390
4.7	47	470	4.7	47	470
5.6	56	560	5.6	56	560
6.8	68	680	6.8	68	680
8.2	82	820	8.2	82	820

application) of Fair-Rite's ferrite materials:

• Type 31 (MnZn) — EMI suppression applications from 1 MHz up to 500 MHz.

• Type 43 (NiZn) — Suppression of conducted EMI, inductors and HF common-mode chokes from 20 MHz to 250 MHz.

• Type 44 (NiZn) — EMI suppression from 30 MHz to 500 MHz.

• Type 61 (NiZn) — Inductors up to 25 MHz and EMI suppression above 200 MHz.

• Type 67 (NiZn) — Broadband transformers, antennas and high-Q inductors up to 50 MHz.

• Type 73 (MnZn) — Suppression of conducted EMI below 50 MHz.

• Type 75 (MnZn) — Broadband and pulse transformers.

See **Table 22.18** for information about the magnetic properties of ferrite cores. Ferrite cores are not typically painted, so identification is often difficult. More information about the use of ferrites at RF is provided in the **RF Techniques** chapter.

Table 22.16
Powdered-Iron Toroidal Cores: Magnetic Properties

There are differing conventions for referring to the type of core material: #, mix and type are all used. For example, all of the following designate the same material: #12, Mix 12, 12-Mix, Type 12 and 12-Type.

Inductance and Turns Formula

The turns required for a given inductance or inductance for a given number of turns can be calculated from:

$$N = 100\sqrt{\frac{L}{A_L}} \qquad L = A_L\left(\frac{N^2}{10,000}\right)$$

where N = number of turns; L = desired inductance (µH); AL = inductance index (µH per 100 turns).*
Toroid diameter is indicated by the number following "T" — T-200 is 2.00 in. dia; T-68 is 0.68 in. diameter, etc.

AL Values

Size	26**	3	15	1	2	7	6	10	12	17	0
T-12	na	60	50	48	20	18	17	12	7.5	7.5	3.0
T-16	145	61	55	44	22	na	19	13	8.0	8.0	3.0
T-20	180	76	65	52	27	24	22	16	10.0	10.0	3.5
T-25	235	100	85	70	34	29	27	19	12.0	12.0	4.5
T-30	325	140	93	85	43	37	36	25	16.0	16.0	6.0
T-37	275	120	90	80	40	32	30	25	15.0	15.0	4.9
T-44	360	180	160	105	52	46	42	33	18.5	18.5	6.5
T-50	320	175	135	100	49	43	40	31	18.0	18.0	6.4
T-68	420	195	180	115	57	52	47	32	21.0	21.0	7.5
T-80	450	180	170	115	55	50	45	32	22.0	22.0	8.5
T-94	590	248	200	160	84	na	70	58	32.0	na	10.6
T-106	900	450	345	325	135	133	116	na	na	na	19.0
T-130	785	350	250	200	110	103	96	na	na	na	15.0
T-157	870	420	360	320	140	na	115	na	na	na	na
T-184	1640	720	na	500	240	na	195	na	na	na	na
T-200	895	425	na	250	120	105	100	na	na	na	na

*The units of AL (µH per 100 turns) are an industry standard; however, to get a correct result use A_L only in the formula above.
**Mix-26 is similar to the older Mix-41, but can provide an extended frequency range.

Magnetic Properties Iron Powder Cores

Mix	Color	Material	µ	Temp stability (ppm/°C)	f (MHz)	Notes
26	Yellow/white	Hydrogen reduced	75	825	dc - 1	Used for EMI filters and dc chokes
3	Gray	Carbonyl HP	35	370	0.05 - 0.50	Excellent stability, good Q for lower frequencies
15	Red/white	Carbonyl GS6	25	190	0.10 - 2	Excellent stability, good Q
1	Blue	Carbonyl C	20	280	0.50 - 5	Similar to Mix-3, but better stability
2	Red	Carbonyl E	10	95	2 - 30	High Q material
7	White	Carbonyl TH	9	30	3 - 35	Similar to Mix-2 and Mix-6, but better temperature stability
6	Yellow	Carbonyl SF	8	35	10 - 50	Very good Q and temperature stability for 20-50 MHz
10	Black	Powdered iron W	6	150	30 - 100	Good Q and stability for 40 - 100 MHz
12	Green/white	Synthetic oxide	4	170	50 - 200	Good Q, moderate temperature stability
17	Blue/yellow	Carbonyl	4	50	40 - 180	Similar to Mix-12, better temperature stability, Q drops about 10% above 50 MHz, 20% above 100 MHz
0	Tan	phenolic	1	0	100 - 300	Inductance may vary greatly with winding technique

Courtesy of Amidon Assoc and Micrometals
Note: Color codes hold only for cores manufactured by Micrometals, which makes the cores sold by most Amateur Radio distributors.

Table 22.17

Powdered-Iron Toroidal Cores: Dimensions

Toroid diameter is indicated by the number following "T" — T-200 is 2.00 in. dia; T-68 is 0.68 in. diameter, etc.
See Table 22.16 for a core sizing guide.

Red E Cores—500 kHz to 30 MHz ($\mu = 10$)

No.	OD (in)	ID (in)	H (in)
T-200-2	2.00	1.25	0.55
T-94-2	0.94	0.56	0.31
T-80-2	0.80	0.50	0.25
T-68-2	0.68	0.37	0.19
T-50-2	0.50	0.30	0.19
T-37-2	0.37	0.21	0.12
T-25-2	0.25	0.12	0.09
T-12-2	0.125	0.06	0.05

Black W Cores—30 MHz to 200 MHz ($\mu=6$)

No.	OD (In)	ID (In)	H (In)
T-50-10	0.50	0.30	0.19
T-37-10	0.37	0.21	0.12
T-25-10	0.25	0.12	0.09
T-12-10	0.125	0.06	0.05

Yellow SF Cores—10 MHz to 90 MHz ($\mu=8$)

No.	OD (In)	ID (In)	H (In)
T-94-6	0.94	0.56	0.31
T-80-6	0.80	0.50	0.25
T-68-6	0.68	0.37	0.19
T-50-6	0.50	0.30	0.19
T-26-6	0.25	0.12	0.09
T-12-6	0.125	0.06	0.05

Number of Turns vs Wire Size and Core Size

Approximate maximum number of turns—single layer wound—enameled wire.

Wire Size	T-200	T-130	T-106	T-94	T-80	T-68	T-50	T-37	T-25	T-12
10	33	20	12	12	10	6	4	1		
12	43	25	16	16	14	9	6	3		
14	54	32	21	21	18	13	8	5	1	
16	69	41	28	28	24	17	13	7	2	
18	88	53	37	37	32	23	18	10	4	1
20	111	67	47	47	41	29	23	14	6	1
22	140	86	60	60	53	38	30	19	9	2
24	177	109	77	77	67	49	39	25	13	4
26	223	137	97	97	85	63	50	33	17	7
28	281	173	123	123	108	80	64	42	23	9
30	355	217	154	154	136	101	81	54	29	13
32	439	272	194	194	171	127	103	68	38	17
34	557	346	247	247	218	162	132	88	49	23
36	683	424	304	304	268	199	162	108	62	30
38	875	544	389	389	344	256	209	140	80	39
40	1103	687	492	492	434	324	264	178	102	51

Actual number of turns may differ from above figures according to winding techniques, especially when using the larger size wires. Chart prepared by Michel J. Gordon, Jr, WB9FHC.
Courtesy of Amidon Assoc.

Table 22.18
Ferrite Toroids: A_L Chart (mH per 1000 turns) Enameled Wire

There are differing conventions for referring to the type of ferrite material: #, mix and type are all used. For example, all of the following designate the same ferrite material: #43, Mix 43, 43-Mix, Type 43, and 43-Type.

Fair-Rite Corporation (www.fair-rite.com) and Amidon (www.amidoncorp.com) ferrite toroids can be cross-referenced as follows:

For Amidon toroids, "FT-XXX-YY" indicates a ferrite toroid, with XXX as the OD in hundredths of an inch and YY the mix. For example, an FT-23-43 core has an OD of 0.23 inch and is made of type 43 material. Additional letters (usually "C") are added to indicate special coatings or different thicknesses.

For Fair-Rite toroids, digits 1 and 2 of the part number indicate product type (59 indicates a part for inductive uses), digits 3 and 4 indicate the material type, digits 5 through 9 indicate core size, and the final digit indicates coating (1 for Paralene and 2 for thermo-set). For example, Fair-Rite part number 5943000101 is equivalent to the Amidon FT-23-43 core.

Ferrite Toroids: A_L Chart (mH per 1000 turns)
Toroid diameter is specified as the outside diameter of the core. See Table 22.16 for a core sizing guide.

Core Size (in)	63/67-Mix $\mu = 40$	61-Mix $\mu = 125$	43-Mix $\mu = 850$	77 (72)-Mix $\mu = 2000$	J (75)-Mix $\mu = 5000$
0.23	7.9	24.8	188	396	980
0.37	19.7	55.3	420	884	2196
0.50	22.0	68.0	523	1100	2715
0.82	22.4	73.3	557	1170	NA
1.14	25.4	79.3	603	1270	3170
1.40	45	140	885	2400	5500
2.40	55	170	1075	2950	6850

31-Mix is an EMI suppression material and not recommended for inductive use.

Inductance and Turns Formula
The turns required for a given inductance or inductance for a given number of turns can be calculated from:

$$N = 1000 \sqrt{\frac{L}{A_L}} \quad L = A_L \left(\frac{N^2}{1,000,000} \right)$$

where N = number of turns; L = desired inductance (µH); A_L = inductance index (µH per 100 turns).

Ferrite Magnetic Properties

Property	Unit	63/67-Mix	61-Mix	43-Mix	77 (72)-Mix	J (75)-Mix	31-Mix
Initial perm.	(μ_i)	40	125	850	2000	5000	1500
Max. perm.		125	450	3000	6000	8000	Not spec.
Saturation flux density @ 10 oe	gauss	1850	2350	2750	4600	3900	3400
Residual flux density	gauss	750	1200	1200	1150	1250	2500
Curie temp.	°C	450	350	130	200	140	>130
Vol. resistivity	ohm/cm	1×10^8	1×10^8	1×10^5	1×10^2	5×10^2	3×10^3
Resonant circuit frequency	MHz	15-25	0.2-10	0.01-1	0.001-1	0.001-1	*
Specific gravity		4.7	4.7	4.5	4.8	4.8	4.7
Loss factor	$\frac{1}{\mu_i Q}$	110×10^{-6} @25 MHz	32×10^{-6} @2.5 MHz	120×10^{-6} @1 MHz	4.5×10^{-6} @0.1 MHz	15×10^{-6} @0.1 MHz	20×10^{-6} @0.1 MHz
Coercive force	Oe	2.40	1.60	0.30	0.22	0.16	0.35
Temp. Coef. of initial perm.	%/°C (20°-70°)	0.10	0.15	1.0	0.60	0.90	1.6

*31-Mix is an EMI suppression material and not recommended for inductive uses.

Ferrite Toroids—Physical Properties
All physical dimensions in inches.

OD (in)	ID (in)	Height (in)	A_e	ℓ_e	V_e
0.230	0.120	0.060	0.00330	0.529	0.00174
0.375	0.187	0.125	0.01175	0.846	0.00994
0.500	0.281	0.188	0.02060	1.190	0.02450
0.825	0.520	0.250	0.03810	2.070	0.07890
1.142	0.750	0.295	0.05810	2.920	0.16950
1.400	0.900	0.500	0.12245	3.504	0.42700
2.400	1.400	0.500	0.24490	5.709	1.39080

Different height cores may be available for each core size.
A_e — Effective magnetic cross-sectional area (in)^2
ℓ_e — Effective magnetic path length (inches)
V_e — Effective magnetic volume (in)^3
To convert from (in)^2 to (cm)^2, divide by 0.155
To convert from (in)^3 to (cm)^3, divide by 0.0610
Courtesy of Amidon Assoc. and Fair-Rite Corp.

22.5 Transformers

Many transformers, including power transformers, IF transformers, and audio transformers, are made to be installed on PC boards, and have terminals designed for that purpose. Some transformers are manufactured with wire leads that are color-coded to identify each connection. When colored wire leads are present, the color codes in **Tables 22.19**, **22.20** and **22.21** usually apply. In addition, many miniature IF transformers are tuned with slugs, color-coded to signify their application. **Table 22.22** lists application versus slug color.

Table 22.20
IF Transformer Wiring Color Codes

Plate lead:	Blue
B+ lead:	Red
Grid (or diode) lead:	Green
Grid (or diode) return:	Black

Note: If the secondary of the IF transformer is center-tapped, the second diode plate lead is green-and-black striped, and black is used for the center-tap lead.

Table 22.21
IF Transformer Slug Color Codes

Frequency	Application	Slug color
455 kHz	1st IF	Yellow
	2nd IF	White
	3rd IF	Black
	Osc tuning	Red
10.7 MHz	1st IF	Green
	2nd or 3rd IF	Orange, Brown or Black

Table 22.19
Power-Transformer Wiring Color Codes

Non-tapped primary leads:	Black
Tapped primary leads:	Common: Black
	Tap: Black/yellow striped
	Finish: Black/red striped
High-voltage plate winding:	Red
Center tap:	Red/yellow striped
Rectifier filament winding:	Yellow
Center tap:	Yellow/blue striped
Filament winding 1:	Green
Center tap:	Green/yellow striped
Filament winding 2:	Brown
Center tap:	Brown/yellow striped
Filament winding 3:	Slate
Center tap:	Slate/yellow striped

Table 22.22
Audio Transformer Wiring Color Codes

Plate lead of primary	Blue
B+ lead (plain or center-tapped)	Red
Plate (start) lead on center-tapped primaries	Brown (or blue if polarity is not important)
Grid (finish) lead to secondary	Green
Grid return (plain or center tapped)	Black
Grid (start) lead on center tapped secondaries	Yellow (or green if polarity not important)

Note: These markings also apply to line-to-grid and tube-to-line transformers.

22.6 Semiconductors

Most semiconductors are labeled with industry standard part numbers, such as 1N4148 or 2N3904, and possibly a date or batch code. You will also encounter numerous manufacturer-specific part numbers and the so-called "house numbers" (marked with codes used by an equipment manufacturer instead of the standard part numbers). In such cases, it is often possible to find the standard equivalent or a suitable replacement by using one of the semiconductor cross-reference directories available from various replacement-parts distributors. If you look up the house number and find the recommended replacement part, you can often find other standard parts that are replaced by that same part.

Information on the use of semiconductors, common design practices, and the necessary circuit design equations can be found in the chapters on **Analog Basics** and **Digital Basics**. Manufacturer Web sites are often a rich source of information on applying semiconductors, both in general and the specific devices they offer.

22.6.1 Diodes

The diode parameters of most importance are maximum forward current or power handling capacity, reverse leakage current, maximum peak inverse voltage (PIV), maximum reverse voltage and the forward voltage. (See **Table 22.23**) For switching or high-speed rectification applications, the time response parameters are also important.

Power dissipation in a diode is equal to the diode's forward voltage drop multiplied by the average forward current. Although fixed voltages are often used for diodes in small-signal applications (0.6 V for silicon PN-junction diodes, 0.3 V for germanium, for example), the actual forward voltage at higher currents can be significantly higher and must be taken into account for high-current applications, such as power supplies.

Most diodes are marked with a part number and some means of identifying the anode or cathode. A thick band or stripe is commonly used to identify the cathode lead or terminal. Stud-mount diodes are usually labeled with a small diode symbol to indicate anode and cathode. Diodes in axial lead packages are sometimes identified with a color scheme as shown in **Fig 22.15**. The common diode packaging standards are illustrated in **Fig 22.16** and the dimensions listed in **Table 22.24**. Many surface mount diodes are packaged in the same SMT packages as resistors.

Packages containing multiple diodes and rectifier bridge configurations are also

Table 22.23
Semiconductor Diode Specifications[†]
Listed numerically by device

Device	Type	Material	Peak Inverse Voltage, PIV (V)	Average Rectified Current Forward (Reverse) $I_O(A)(I_R(A))$	Peak Surge Current, I_{FSM} 1 s @ 25°C (A)	Average Forward Voltage, VF (V)
1N34	Signal	Ge	60	8.5 m (15.0 μ)		1.0
1N34A	Signal	Ge	60	5.0 m (30.0 μ)		1.0
1N67A	Signal	Ge	100	4.0 m (5.0 μ)		1.0
1N191	Signal	Ge	90	15.0 m		1.0
1N270	Signal	Ge	80	0.2 (100 μ)		1.0
1N914	Fast Switch	Si	75	75.0 m (25.0 n)	0.5	1.0
1N1183	RFR	Si	50	40 (5 m)	800	1.1
1N1184	RFR	Si	100	40 (5 m)	800	1.1
1N2071	RFR	Si	600	0.75 (10.0 μ)		0.6
1N3666	Signal	Ge	80	0.2 (25.0 μ)		1.0
1N4001	RFR	Si	50	1.0 (0.03 m)		1.1
1N4002	RFR	Si	100	1.0 (0.03 m)		1.1
1N4003	RFR	Si	200	1.0 (0.03 m)		1.1
1N4004	RFR	Si	400	1.0 (0.03 m)		1.1
1N4005	RFR	Si	600	1.0 (0.03 m)		1.1
1N4006	RFR	Si	800	1.0 (0.03 m)		1.1
1N4007	RFR	Si	1000	1.0 (0.03 m)		1.1
1N4148	Signal	Si	75	10.0 m (25.0 n)		1.0
1N4149	Signal	Si	75	10.0 m (25.0 n)		1.0
1N4152	Fast Switch	Si	40	20.0 m (0.05 μ)		0.8
1N4445	Signal	Si	100	0.1 (50.0 n)		1.0
1N5400	RFR	Si	50	3.0 (500 μ)	200	
1N5401	RFR	Si	100	3.0 (500 μ)	200	
1N5402	RFR	Si	200	3.0 (500 μ)	200	
1N5403	RFR	Si	300	3.0 (500 μ)	200	
1N5404	RFR	Si	400	3.0 (500 μ)	200	
1N5405	RFR	Si	500	3.0 (500 μ)	200	
1N5406	RFR	Si	600	3.0 (500 μ)	200	
1N5408	RFR	Si	1000	3.0 (500 μ)	200	
1N5711	Schottky	Si	70	1 m (200 n)	15 m	0.41 @ 1 mA
1N5767	Signal	Si		0.1 (1.0 μ)		1.0
1N5817	Schottky	Si	20	1.0 (1 m)	25	0.75
1N5819	Schottky	Si	40	1.0 (1 m)	25	0.9
1N5821	Schottky	Si	30	3.0		
ECG5863	RFR	Si	600	6	150	0.9
1N6263	Schottky	Si	70	15 m	50 m	0.41 @ 1 mA
5082-2835	Schottky	Si	8	1 m (100 n)	10 m	0.34 @ 1 mA

Si = Silicon; Ge = Germanium; RFR = rectifier, fast recovery.
[†]For package shape, size and pin-connection information see manufacturers' data sheets. Many retail suppliers offer data sheets to buyers free of charge on request. Data books are available from many manufacturers and retailers.

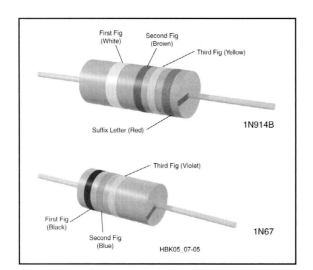

Fig 22.15 — Color-coding for semiconductor diodes. At A, the cathode is identified by the double-width first band. At B, the bands are grouped toward the cathode. Two-Fig designations are signified by a black first band. The color code is given in Table 22.4. The suffix-letter code is A-Brown, B-red, C-orange, D-yellow, E-green, F-blue. The 1N prefix is assumed.

Table 22.24
Package Dimensions for Small Signal, Rectifier and Zener diodes

Case	L	D	Ød	LS	PCB Pads*	Hole	Example
DO-35	0.166	0.080	0.020	0.30	0.056 dia	0.029	1N4148
DO-41	0.205	0.107	0.034	0.40	0.074 dia	0.040	1N4001
DO-201	0.283	0.189	0.048	0.65	0.150 dia	0.079	1N5401
DO-204	0.205	0.106	0.034	0.40	0.074 dia	0.040	1N4001

Dimensions in inches.
*Customary to make cathode lead square.

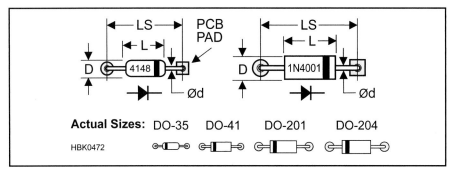

Fig 22.16 — Axial-leaded diode packages and pad dimensions.

commonly available. Full-wave bridge packages are labeled with tildes (~) for the ac inputs and + and – symbols for the rectifier outputs. High-power diodes are often packaged in TO-220 packages with two leads. The package may be labeled with a diode symbol, but if not you will have to obtain the manufacturer's data sheet to identify the anode and cathode leads.

The common 1N914, 1N4148, and 1N5767 switching diodes are suitable for most all small signal applications. The 1N4000 family is commonly used for ac voltage rectification up to 1000 volts PIV. Schottky diodes are used when low forward voltages are required, particularly at high-currents, and exhibit voltages of 0.1-0.2 V at low currents.

Zener diodes, used as voltage references, are manufactured in a wide range of voltages and power handling capacities. Power dissipation in a Zener diode is equal to the Zener voltage multiplied by the average reverse current. The Zener voltage has a significant temperature coefficient and also varies with reverse current. To avoid excessive variations in Zener voltage, limit the diode's power dissipation to no more than ½ of the rated value and for precision uses, ⅕ to ¹⁄₁₀ of the rated power dissipation is recommended. Common varieties of Zener diodes are listed in **Table 22.25**.

Voltage-variable capacitance diodes, also called Varicaps, varactors or tuning diodes, are used in oscillator and tuned circuits where a variable capacitor is needed. Operated with reverse bias, the depletion region forms a capacitor of variable width with a fairly linear voltage vs. capacitance function. Standard tuning diodes produce capacitances in the range of 5 to 40 pF. Hyper-abrupt tuning diodes produce variable capacitances to 100 pF or more for low frequency or wide tuning range applications. Some of the common voltage variable capacitance diodes are listed in **Table 22.26**.

Maximum capacitance occurs with minimum reverse voltage. As the reverse voltage is increased, the capacitance decreases. Tuning diodes are specified by the capacitance produced at two reverse voltages, usually 2 and 30 V. This is called the capacitance ratio and is specified in units of pF per volt. Beyond this range, capacitance change with voltage can become non-linear and may cause signal distortion.

All diodes exhibit some capacitance when reversed biased. Amateurs have learned to use reverse biased Zener and rectifier diodes to form tuning diodes with 20-30 pF maximum capacitance. These "poor man's" tuning diodes are widely used in homebrew projects. However, because the capacitance ratio varies widely from one diode to the next, requiring experimentation to find a suitable diode, they are seldom used in published construction articles.

Light emitting diodes (LED) are another common type of diode. The primary application is that of an illuminated visual indicator when forward biased. LEDs have virtually replaced miniature lamp bulbs for indicators and illumination. LED's are specified primarily by their color, size, shape and output light intensity.

The "standard" size LED is the T-1¾; 5 mm or 0.20 in. diameter. The "miniature" size is the T-1; 3 mm or 0.125 in. diameter. Today, the "standard" and "miniature" size is a bit of a misnomer due to the wide variety of LED sizes and shapes, including SMT varieties. However, the T-1 and T-1¾ remain the most common for homebrew projects due to their inexpensive availability and ease of mounting with a simple panel hole. Their long leads are ideal for prototyping.

22.6.2 Transistors

The information in the tables of transistor data includes the most important parameters for typical applications of transistors of that type. The meaning of the parameters and their relationship to circuit design is covered in the **Analog Basics** chapter or in the references listed at the end of that chapter.

The tables are organized by application; small-signal, general-purpose, RF power, and so on. Some obsolete parts are listed in these tables for reference in repair and maintenance of older equipment. Before using a device in a new design, it is recommended that you check the manufacturer's Web site to be sure that the device has not been replaced by a more capable part and that it is available for future orders.

22.6.3 Voltage Regulators

For establishing a well-regulated fixed voltage reference, the linear voltage regular ICs are often preferred over the Zener diode. Three-terminal voltage regulators require no external components and most have internal current limiting and thermal shutdown circuitry, making them virtually indestructible. (Three-terminal regulators are described in the **Analog Basics** chapter.) The specifications and packages for common voltage regulators are listed in **Table 22.27**.

For fixed-voltage positive regulators, the 7800 family in the TO-220 package is the most common and reasonably priced. They are available in a variety of voltages and supply up to 1 A of current or more, depending on the input voltage. The part number identifies the voltage. For example, a 7805 is a 5-V regulator and a 7812 is a 12-V regulator. The 78L00 low-power versions in the TO-98 package or SOT-89 surface-mount package provide load currents up to 100 mA. The 317 and 340 are the most common adjustable-voltage regulators. Integrated voltage regulators can be used with a pass transistor to extend their load current capability as described in the device data sheets.

Three-terminal regulators are selected primarily for output voltage and maximum load

Table 22.25
Common Zener Diodes
Power dissipation and (package style)

Voltage (V)	¼ W (DO-35)	0.35 W (SOT-23)	½ W (SOD-123)	½ W (DO-35)	1 W (DO41)
2.7	1N4618	MMBZ5223B	MMSZ5223B	1N5223B	—
3.3	1N4620	MMBZ5226B	MMSZ5226B	1N5226B	1N4728A
3.6	1N4621	MMBZ5227B	MMSZ5227B	1N5227B	1N4729A
3.9	1N4622	MMBZ5228B	MMSZ5228B	1N5228B	1N4730A
4.3	1N4623	MMBZ5229B	MMSZ5229B	1N5229B	1N4731A
4.7	1N4624	MMBZ5230B	MMSZ5230B	1N5230B	1N4732A
5.1	1N4625	MMBZ5231B	MMSZ5231B	1N5231B	1N4733A
5.6	1N4626	MMBZ5232B	MMSZ5232B	1N5232B	1N4734A
6.0	— —	MMBZ5233B	MMSZ5233B	1N5233B	—
6.2	1N4627	MMBZ5234B	MMSZ5234B	1N5234B	1N4735A
6.8	1N4099	MMBZ5235B	MMSZ5235B	1N5235B	1N4736A
7.5	1N4100	MMBZ5236B	MMSZ5236B	1N5236B	1N4737A
8.2	1N4101	MMBZ5237B	MMSZ5237B	1N5237B	1N4738A
9.1	1N4103	MMBZ5239B	MMSZ5239B	1N5239B	1N4739A
10	1N4104	MMBZ5240B	MMSZ5240B	1N5240B	1N4740A
11	1N4105	MMBZ5241B	MMSZ5241B	1N5241B	1N4741A
12	— —	MMBZ5242B	MMSZ5242B	1N5242B	1N4742A
13	1N4107	MMBZ5243B	MMSZ5243B	1N5243B	1N4743A
15	1N4109	MMBZ5245B	MMSZ5245B	1N5245B	1N4744A
18	1N4112	MMBZ5248B	MMSZ5248B	1N5248B	1N4746A
20	1N4114	MMBZ5250B	MMSZ5250B	1N5250B	1N4747A
22	1N4115	MMBZ5251B	MMSZ5251B	1N5251B	1N4748A
24	1N4116	MMBZ5252B	MMSZ5252B	1N5252B	1N4749A
27	1N4118	MMBZ5254B	MMSZ5254B	1N5254B	1N4750A
28	1N4119	MMBZ5255B	MMSZ5255B	1N5255B	—
30	1N4120	MMBZ5256B	MMSZ5256B	1N5256B	1N4751A
33	1N4121	MMBZ5257B	MMSZ5257B	1N5257B	1N4752A
36	1N4122	MMBZ5258B	MMSZ5258B	1N5258B	1N4753A
39	1N4123			1N5259B	1N4754A
43	—			1N5260B	1N4755A
47	1N4125			1N5261B	1N4756A
51	1N4126			1N5262B	1N4757A
56				1N5263B	1N4758A
60				1N5264B	—
62				1N5265B	1N4759A
68				1N5266B	1N4760A
75				1N5267B	1N4761A
82					1N4762A
91					1N4763A
100					1N4764A

current. Dropout voltage — the minimum voltage between input and output for which regulation can be maintained — is also very important. For example, the dropout voltage for the 5-V 78L05 is 1.7 V. Therefore, the input voltage must be at least 6.7 V (5 + 1.7 V) to ensure output voltage regulation. The maximum input voltage should also not be exceeded.

Make sure to check the pin assignments for all voltage regulators. While the fixed-voltage positive regulators generally share a common orientation of input, output, and ground, negative-voltage and adjustable regulators do not. Installing a regulator with the wrong connections will usually destroy it and may allow excessive voltage to be applied to the circuit it supplies.

22.6.4 Analog and Digital Integrated Circuits

Integrated circuits (ICs) come in a variety of packages, including transistor-like metal cans, dual and single in-line packages (DIPs and SIPs), flat-packs and surface-mount packages. Most are marked with a part number and a four-digit manufacturer's date code indicating the year (first two digits) and week (last two digits) that the component was made. As mentioned in the introduction to this chapter, ICs are frequently house-marked and cross-reference directories can be helpful in identification and replacement. Another very useful reference tool for working with ICs is IC Master (**www.icmaster.com**), a master selection guide that organizes ICs by type, function and certain key parameters.

A part number index is included, along with application notes and manufacturer's information for millions of devices.

IC part numbers provide a complete description of the device's function and ratings. For example, a 4066 IC contains four independent CMOS SPST switches. The 4066 is a CMOS device available from a number of different manufacturers in different package styles and ratings. The two- or three-letter prefix of the part number is generally associated with the part manufacturer. Next, the part type (4066 in this case) shows the function and pin assignments or "pin outs." Following the part type is an alphabetic suffix that describes the version of the part, package code, temperature range, reliability rating and possibly other information. For

Table 22.26
Voltage-Variable Capacitance Diodes[†]

Listed numerically by device

Device	Nominal Capacitance pF ±10% @ V_R = 4.0 V f = 1.0 MHz	Capacitance Ratio 2-30 V Min.	Q @ 4.0 V 50 MHz Min.	Case Style	Device	Nominal Capacitance pF ±10% @ V_R = 4.0 V f = 1.0 MHz	Capacitance Ratio 2-30 V Min.	Q @ 4.0 V 50 MHz Min.	Case Style
1N5441A	6.8	2.5	450		1N5471A	39	2.9	450	
1N5442A	8.2	2.5	450		1N5472A	47	2.9	400	
1N5443A	10	2.6	400	DO-7	1N5473A	56	2.9	300	DO-7
1N5444A	12	2.6	400		1N5474A	68	2.9	250	
1N5445A	15	2.6	450		1N5475A	82	2.9	225	
1N5446A	18	2.6	350		1N5476A	100	2.9	200	
1N5447A	20	2.6	350		MV2101	6.8	2.5	450	TO-92
1N5448A	22	2.6	350	DO-7	MV2102	8.2	2.5	450	
1N5449A	27	2.6	350		MV2103	10	2.0	400	
1N5450A	33	2.6	350		MV2104	12	2.5	400	
1N5451A	39	2.6	300		MV2105	15	2.5	400	
1N5452A	47	2.6	250		MV2106	18	2.5	350	TO-92
1N5453A	56	2.6	200	DO-7	MV2107	22	2.5	350	
1N5454A	68	2.7	175		MV2108	27	2.5	300	
1N5455A	82	2.7	175		MV2109	33	2.5	200	
1N5456A	100	2.7	175		MV2110	39	2.5	150	
1N5461A	6.8	2.7	600		MV2111	47	2.5	150	TO-92
1N5462A	8.2	2.8	600		MV2112	56	2.6	150	
1N5463A	10	2.8	550	DO-7	MV2113	68	2.6	150	
1N5464A	12	2.8	550		MV2114	82	2.6	100	
1N5465A	15	2.8	550		MV2115	100	2.6	100	
1N5466A	18	2.8	500						
1N5467A	20	2.9	500						
1N5468A	22	2.9	500	DO-7					
1N5469A	27	2.9	500						
1N5470A	33	2.9	500						

[†]For package shape, size and pin-connection information, see manufacturers' data sheets.

complete information on the part — any or all of which may be significant to circuit function — use the Web sites of the various manufacturers or enter "data sheet" and the part number into an Internet search engine.

When choosing ICs that are not exact replacements, be wary of substituting "similar" devices, particularly in demanding applications, such as high-speed logic, sensitive receivers, precision instrumentation and similar devices. In particular, substitution of one type of logic family for another — even if the device functions and pin outs are the same — can cause a circuit to not function or function erratically, particularly at temperature extremes. For example, substituting LS TTL devices for HCMOS devices will result in mismatches between logic level thresholds. Substituting a lower-power IC may result in problems supplying enough output current. Even using a faster or higher clock-speed part can cause problems if signals change faster or propagate more quickly than the circuit was designed for. Problems of this sort can be extremely difficult to troubleshoot unless you are skilled in circuit design. When necessary, you can add interface circuits or buffer amplifiers that improve the input and output capabilities of replacement ICs, but auxiliary circuits cannot improve basic device ratings, such as speed or bandwidth. Whenever possible, substitute ICs that are guaranteed or "direct" replacements and that are listed as such by the manufacturer.

ICs are available in different operating temperature ranges. Three standard ranges are common:
- Commercial: 0 °C to 70 °C
- Industrial: –25 °C to 85 °C
- Automotive: –40 °C to 85 °C
- Military: –55°C to 125°C

In some cases, part numbers reflect the temperature ratings. For example, an LM301A op amp is rated for the commercial temperature range; an LM201A op amp for the industrial range and an LM101A for the military range. It is usually acceptable, all other things being equal, to substitute ICs rated for a wider temperature range, but there are often other performance differences associated with the devices meeting wider temperature specifications that should be evaluated before making the substitution.

Table 22.27
Three-Terminal Voltage Regulators

Listed numerically by device

Device	Description	Package	Voltage	Current (A)
317	Adj Pos	TO-205	+1.2 to +37	0.5
317	Adj Pos	TO-204,TO-220	+1.2 to +37	1.5
317L	Low Current Adj Pos	TO-205,TO-92	+1.2 to +37	0.1
317M	Med Current Adj Pos	TO-220	+1.2 to +37	0.5
338	Adj Pos	TO-3	+1.2 to +32	5.0
350	High Current Adj Pos	TO-204,TO-220	+1.2 to +33	3.0
337	Adj Neg	TO-205	−1.2 to −37	0.5
337	Adj Neg	TO-204,TO-220	−1.2 to −37	1.5
337M	Med Current Adj Neg	TO-220	−1.2 to −37	0.5
309		TO-205	+5	0.2
309		TO-204	+5	1.0
323		TO-204,TO-220	+5	3.0
140-XX	Fixed Pos	TO-204,TO-220	Note 1	1.0
340-XX		TO-204,TO-220		1.0
78XX		TO-204,TO-220		1.0
78LXX		TO-205,TO-92		0.1
78MXX		TO-220		0.5
78TXX		TO-204		3.0
79XX	Fixed Neg	TO-204,TO-220	Note 1	1.0
79LXX		TO-205,TO-92		0.1
79MXX		TO-220		0.5

Note 1—XX indicates the regulated voltage; this value may be anywhere from 1.2 V to 35 V. A 7815 is a positive 15-V regulator, and a 7924 is a negative 24-V regulator.

The regulator package may be denoted by an additional suffix, according to the following:

Package	Suffix
TO-204 (TO-3)	K
TO-220	T
TO-205 (TO-39)	H, G
TO-92	P, Z

For example, a 7812K is a positive 12-V regulator in a TO-204 package. An LM340T-5 is a positive 5-V regulator in a TO-220 package. In addition, different manufacturers use different prefixes. An LM7805 is equivalent to a µA7805 or MC7805.

Common Voltage Regulators — Fixed Positive Voltage

Device	Output Voltage (V)	Output Current (A)	Load Regulation (mV)	Dropout Voltage (V)	Min. Input Voltage (V)	Max. Input Voltage (V)
Surface Mount SOT-89 Case						
78L05ACPK	5.0	0.1	60	1.7	7.0	20
78L06ACPK	6.2	0.1	80	1.7	8.5	20
78L08ACPK	8.0	0.1	80	1.7	10.5	23
78L09ACPK	9.0	0.1	90	1.7	11.5	24
78L12ACPK	12	0.1	100	1.7	14.5	27
78L15ACPK	15	0.1	150	1.7	17.5	30
TO-92 Case						
78L33ACZ	3.3	0.1	60	1.7	5.0	30
78L05ACZ	5.0	0.1	60	1.7	7.0	30
78L06ACZ	6.0	0.1	60	1.7	8.5	30
78L08ACZ	8.0	0.1	80	1.7	10.5	30
78L09ACZ	9.0	0.1	80	1.7	11.5	30
78L12ACZ	12	0.1	100	1.7	14.5	35
78L15ACZ	15	0.1	150	1.7	17.5	35
TO-220 Case						
78M05CV	5.0	0.5	100	2.0	7.0	35
7805ACV	5.0	1.0	100	2.0	7.0	35
7806ACV	6.0	1.0	100	2.0	8.0	35
7808ACV	8.0	1.0	100	2.0	10.0	35
7809ACV	9.0	1.0	100	2.0	11.0	35
7812ACV	12	1.0	100	2.0	14.0	35
7815CV	15	1.0	300	2.0	17.0	35

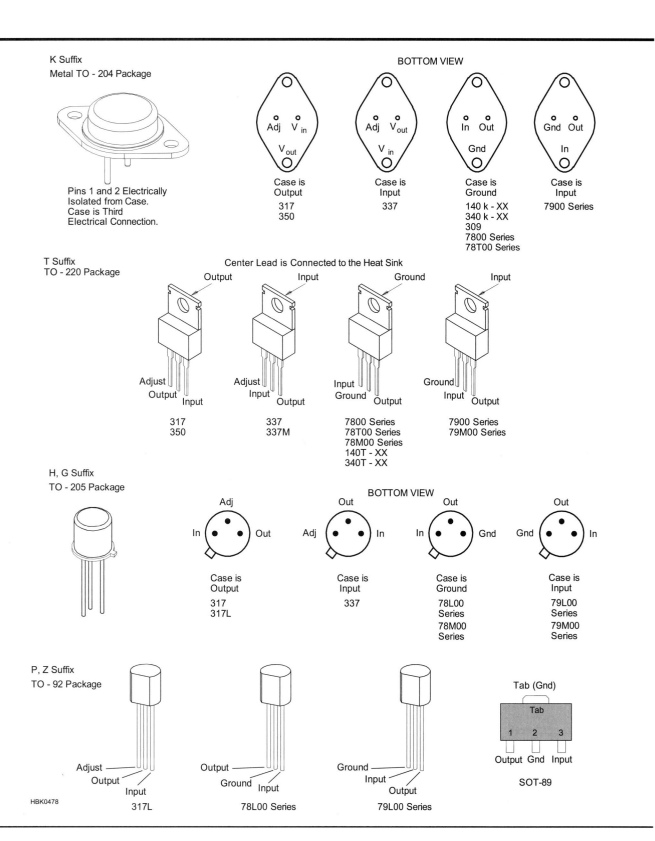

Table 22.28
Monolithic 50-Ω Amplifiers (MMIC Gain Blocks)

Device	Case Style	Freq. Range (MHz)	Gain at 100 MHz (dB)	Gain at 1000 MHz (dB)	Output Power P1dB (dBm)	IP3 (dB)	NF (dB)	DC Conditions Vb @ Ib
Avago Technologies								
MSA-0386	B	dc-2400	12.5	11.9	+10.0	+23.0	6.0	5.0 V @ 35 mA
MSA-0486	B	dc-3200	16.4	15.9	+12.5	+25.5	6.5	5.3 V @ 50 mA
MSA-0505	A	dc-2300	7.8	7.0	+18.0	+29.0	6.5	8.4 V @ 80 mA
MSA-0611	C	dc-700	19.5	12.0	+2.0	+14.0	3.0	3.3 V @ 16 mA
MSA-0686	B	dc-800	20.0	16.0	+2.0	+14.5	3.0	3.5 V @ 16 mA
MSA-0786	B	dc-2000	13.5	12.6	+2.0	+14.5	3.0	3.5 V @ 16 mA
MSA-0886	B	dc-1000	32.5	22.4	+5.5	+19.0	5.5	4.0 V @ 22 mA
Mini-Circuits "ERA" Series								
ERA-1+	A, B	dc-8000	12.2	12.1	+11.7	+26.0	5.3	3.6 V @ 40 mA
ERA-2+	A, B	dc-6000	16.2	16.0	+12.8	+26.0	4.7	3.6 V @ 40 mA
ERA-3+	A, B	dc-3000	22.9	22.2	+12.1	+23.0	3.8	3.5 V @ 35 mA
ERA-4+	A, B	dc-4000	13.8	13.7	+17.0	+32.5	5.5	5.0 V @ 65 mA
ERA-5+	A, B	dc-4000	20.2	19.8	+18.4	+33.0	4.5	4.9 V @ 65 mA
ERA-6+	A, B	dc-4000	11.1	11.1	+18.5	+36.5	8.4	5.2 V @ 70 mA
Mini-Circuits "MAR" Series								
MAR-1SM+	A, B	dc-1000	18.5	15.5	+1.5	+14.0	5.5	5.0 V @ 17 mA
MAR-2SM+	A, B	dc-2000	12.5	12.0	+4.5	+17.0	6.5	5.0 V @ 25 mA
MAR-3SM+	A, B	dc-2000	12.5	12.0	+10.0	+23.0	6.0	5.0 V @ 35 mA
MAR-4SM+	A, B	dc-1000	8.3	8.0	+12.5	+25.5	7.0	5.3 V @ 50 mA
MAR-6SM+	A, B	dc-2000	20.0	16.0	+2.0	+14.5	3.0	3.5 V @ 16 mA
MAR-7SM+	A, B	dc-2000	13.5	12.5	+5.5	+19.0	5.0	4.0 V @ 22 mA
MAR-8SM+	A, B	dc-1000	32.5	22.5	+12.5	+27.0	3.3	7.8 V @ 36 mA
Mini-Circuits "VAM" Series								
VAM-3+	C	dc-2000	11.5	11.0	+9.0	+22.0	6.0	4.7 V @ 35 mA
VAM-6+	C	dc-2000	19.5	15.0	+2.0	+14.0	3.0	3.3 V @ 16 mA
VAM-7+	C	dc-2000	13.0	12.0	+5.5	+18.0	5.0	3.8 V @ 22 mA
Mini–Circuits "GALI" Series								
GALI-1+	D	dc-8000	12.7	12.5	+10.5	+27.0	4.5	3.4 V @ 40 mA
GALI-2+	D	dc-8000	16.2	15.8	+12.9	+27.0	4.6	3.5 V @ 40 mA
GALI-3+	D	dc-3000	22.4	21.1	+12.5	+25.0	3.5	3.3 V @ 35 mA
GALI-39+	D	dc-7000	20.8	21.1	+10.5	+22.9	2.4	3.5 V @ 35 mA
GALI-4+	D	dc-4000	14.4	14.1	+17.5	+34.0	4.0	4.6 V @ 65 mA
GALI-5+	D	dc-4000	20.6	19.4	+18.0	+35.0	3.5	4.4 V @ 65 mA
GALI-6+	D	dc-4000	12.2	12.2	+18.2	+35.5	4.5	5.0 V @ 70 mA
GALI-S66+	D	dc-3000	22.0	20.3	+2.8	+18.0	2.7	3.5 V @ 16 mA
Mini–Circuits "RAM" Series								
RAM-1+	B	dc-1000	19.0	15.5	+1.5	+14.0	5.5	5.0 V @ 17 mA
RAM-2+	B	dc-2000	12.5	11.8	+4.5	+17.0	6.5	5.0 V @ 25 mA
RAM-3+	B	dc-2000	12.5	12.0	+10.0	+23.0	6.0	5.0 V @ 35 mA
RAM-4+	B	dc-1000	8.5	8.0	+12.5	+25.5	6.5	5.3 V @ 50 mA
RAM-6+	B	dc-2000	20.0	16.0	+2.0	+14.5	2.8	3.5 V @ 16 mA
RAM-7+	B	dc-2000	13.5	12.5	+5.5	+19.0	4.5	4.0 V @ 22 mA
RAM-8+	B	dc-1000	32.5	23.0	+12.5	+27.0	3.0	7.8 V @ 36 mA

Avago — www.avagotech.com
Mini-Circuits Labs — www.minicircuits.com

22.6.5 MMIC Amplifiers

Monolithic microwave integrated circuit (MMIC) amplifiers are single-supply 50-Ω wideband gain blocks offering high dynamic range for output powers to about +15 dBm. MMIC amplifiers are becoming increasingly popular in homebrew communications circuits. With bandwidths over 1 GHz, they are well suited for HF, VHF, UHF and lower microwave frequencies.

MMIC amplifiers produce power gains from 10 dB to 30 dB. They also have a high third-order intercept point (IP3), usually in the +20 to +30 dBm range, easing the concerns about amplifier compression for most applications. They are used for RF and IF amplifiers, local oscillator amplifiers, transmitter drivers, and other medium power applications in 50-Ω systems. MMICs are especially well suited for driving 50-Ω double-balanced mixers (DBM). **Fig 22.17** shows the typical circuit arrangement for most MMIC amplifiers.

MMICs are available in a variety of packages, mostly surface mount as shown in **Fig 22.18**, requiring very few external components. Vendor data sheets and application notes, found on the manufacturer's Web sites, should be used for the proper selection of the biasing resistor, coupling capacitors, and other design criteria. Some of the popular MMIC amplifiers are listed in **Table 22.28**.

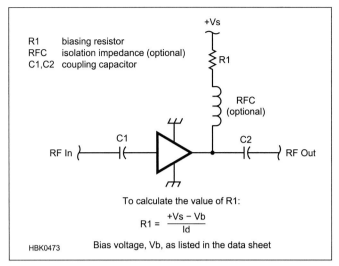

Fig 22.17 — MMIC application.

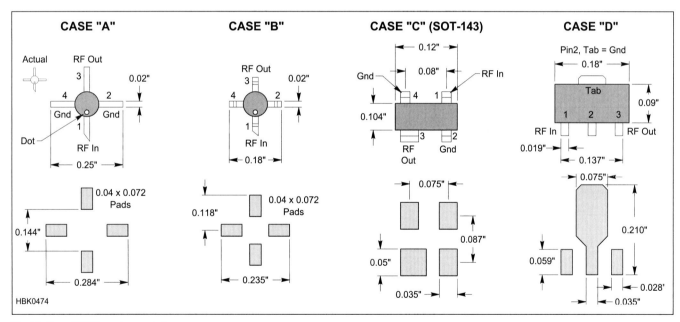

Fig 22.18 — MMIC package styles.

The main disadvantage of MMIC amplifiers is their relatively high current demands, usually in the 30 mA to 80 mA range per device, making them unsuitable for battery-powered portable equipment. On the other hand, the high current demand is what establishes their high gain and high IP3 characteristics with 50-Ω loads.

Another disadvantage is their wide gain-bandwidth. Their gain should be band-limited by input and output tuned circuits or filters to reduce the gain outside the desired ranges. For example, for an HF amplifier, 30 MHz low-pass filters can be used to reduce the gain outside the HF spectrum, or a band-pass filter used for the frequency band of interest.

Selecting the proper MMIC amplifier is fairly straightforward. First, select a device for the desired frequency bandwidth, gain, and output power. Ensure device current is compatible with the design application. Calculate the value for the bias resistor (R1 in Fig 22.17) based on the biasing voltage (V_b) listed in Table 22.28 and whatever value of supply voltage (V_s) is available.

With increasing availability and ease of use, there are many circuits where MMIC amplifiers can be used. There are many MMIC amplifiers that are relatively inexpensive for hobby use.

Table 22.29
Small-Signal FETs

Device	Type	Max Diss (mW)	Max V_{DS} (V)	$V_{GS(off)}$ (V)	Min gfs (μS)	Input C (pF)	Max ID (mA)[1]	f_{max} (MHz)	Noise Figure (typ)	Case	Base	Applications
2N4416	N-JFET	300	30	−6	4500	4	−15	450	4 dB @ 400 MHz	TO-72	1	VHF/UHF amp, mix, osc
2N5484	N-JFET	310	25	−3	2500	5	30	200	4 dB @ 200 MHz	TO-92	2	VHF/UHF amp, mix, osc
2N5485	N-JFET	310	25	−4	3500	5	30	400	4 dB @ 400 MHz	TO-92	2	VHF/UHF amp, mix, osc
2N5486	N-JFET	360	25	−2	5500	5	15	400	4 dB @ 400 MHz	TO-92	2	VHF/UHF amp. mix, osc
3N200 NTE222 SK3065	N-dual-gate MOSFET	330	20	−6	10,000	4-8.5	50	500	4.5 dB @ 400 MHz	TO-72	3	VHF/UHF amp, mix, osc
3N202 NTE454 SK3991	N-dual-gate MOSFET	360	25	−5	8000	6	50	200	4.5 dB @ 200 MHz	TO-72	3	VHF amp, mixer
MPF102 NTE451 SK9164	N-JFET	310	25	−8	2000	4.5	20	200	4 dB @ 400 MHz	TO-92	2	HF/VHF amp, mix, osc
MPF106 2N5484	N-JFET	310	25	−6	2500	5	30	400	4 dB @ 200 MHz	TO-92	2	HF/VHF/UHF amp, mix, osc
40673 NTE222 SK3050	N-dual-gate MOSFET	330	20	−4	12,000	6	50	400	6 dB @ 200 MHz	TO-72	3	HF/VHF/UHF amp, mix, osc
U304	P-JFET	350	−30	+10	27		−50	—	—	TO-18	4	analog switch chopper
U310	N-JFET	500 300	30 30	−6	10,000	2.5	60	450	3.2 dB @ 450 MHz	TO-52	5	common-gate VHF/UHF amp,
U350	N-JFET Quad	1W	25	−6	9000	5	60	100	7 dB @ 100 MHz	TO-99	6	matched JFET doubly bal mix
U431	N-JFET Dual	300	25	−6	10,000	5	30	100	—	TO-99	7	matched JFET cascode amp and bal mix
2N5670	N-JFET	350	25	8	3000	7	20	400	2.5 dB @ 100 MHz	TO-92	2	VHF/UHF osc, mix, front-end amp
2N5668	N-JFET	350	25	4	1500	7	5	400	2.5 dB @ 100 MHz	TO-92	2	VHF/UHF osc, mix, front-end amp
2N5669	N-JFET	350	25	6	2000	7	10	400	2.5 dB @ 100 MHz	TO-92	2	VHF/UHF osc, mix, front-end amp
J308	N-JFET	350	25	6.5	8000	7.5	60	1000	1.5 dB @ 100 MHz	TO-92	2	VHF/UHF osc, mix, front-end amp
J309	N-JFET	350	25	4	10,000	7.5	30	1000	1.5 dB @ 100 MHz	TO-92	2	VHF/UHF osc, mix, front-end amp
J310	N-JFET	350	25	6.5	8000	7.5	60	1000	1.5 dB @ 100 MHz	TO-92	2	VHF/UHF osc, mix, front-end amp
NE32684A	HJ-FET	165	2.0	−0.8	45,000	—	30	20 GHz	0.5 dB @ 12 GHz	84A		Low-noise amp

Notes:
[1] 25°C.
For package shape, size and pin-connection information, see manufacturers' data sheets.

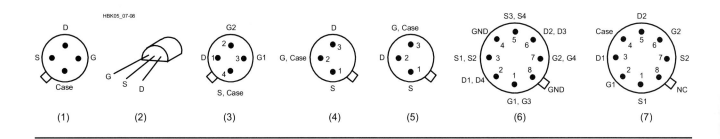

Table 22.30

Low-Noise Bipolar Transistors

Device	NF (dB)	F (MHz)	f_T (GHz)	I_C (mA)	Gain (dB)	F (MHz)	$V_{(BR)CEO}$ (V)	I_C (mA)	P_T (mW)	Case
MRF904	1.5	450	4	15	16	450	15	30	200	TO-206AF
MRF571	1.5	1000	8	50	12	1000	10	70	1000	Macro-X
MRF2369	1.5	1000	6	40	12	1000	15	70	750	Macro-X
MPS911	1.7	500	7	30	16.5	500	12	40	625	TO-226AA
MRF581A	1.8	500	5	75	15.5	500	15	200	2500	Macro-X
BFR91	1.9	500	5	30	16	500	12	35	180	Macro-T
BFR96	2	500	4.5	50	14.5	500	15	100	500	Macro-T
MPS571	2	500	6	50	14	500	10	80	625	TO-226AA
MRF581	2	500	5	75	15.5	500	18	200	2500	Macro-X
MRF901	2	1000	4.5	15	12	1000	15	30	375	Macro-X
MRF941	2.1	2000	8	15	12.5	2000	10	15	400	Macro-X
MRF951	2.1	2000	7.5	30	12.5	2000	10	100	1000	Macro-X
BFR90	2.4	500	5	14	18	500	15	30	180	Macro-T
MPS901	2.4	900	4.5	15	12	900	15	30	300	TO-226AA
MRF1001A	2.5	300	3	90	13.5	300	20	200	3000	TO-205AD
2N5031	2.5	450	1.6	5	14	450	10	20	200	TO-206AF
MRF4239A	2.5	500	5	90	14	500	12	400	3000	TO-205AD
BFW92A	2.7	500	4.5	10	16	500	15	35	180	Macro-T
MRF521*	2.8	1000	4.2	−50	11	1000	−10	−70	750	Macro-X
2N5109	3	200	1.5	50	11	216	20	400	2500	TO-205AD
2N4957*	3	450	1.6	−2	12	450	−30	−30	200	TO-206AF
MM4049*	3	500	5	−20	11.5	500	−10	−30	200	TO-206AF
2N5943	3.4	200	1.5	50	11.4	200	30	400	3500	TO-205AD
MRF586	4	500	1.5	90	9	500	17	200	2500	TO-205AD
2N5179	4.5	200	1.4	10	15	200	12	50	200	TO-206AF
2N2857	4.5	450	1.6	8	12.5	450	15	40	200	TO-206AF
2N6304	4.5	450	1.8	10	15	450	15	50	200	TO-206AF
MPS536*	4.5	500	5	−20	4.5	500	−10	−30	625	TO-226AA
MRF536*	4.5	1000	6	−20	10	1000	−10	−30	300	Macro-X

*denotes a PNP device

Complementary devices

NPN	PNP
2N2857	2N4957
MRF904	MM4049
MRF571	MRF521

For package shape, size and pin-connection information, see manufacturers' data sheets. Many retail suppliers and manufacturers offer data sheets on their Web sites.

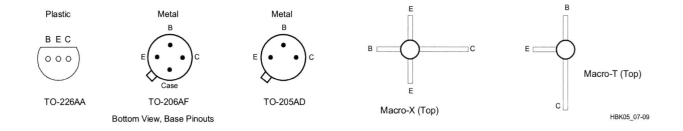

Table 22.31

General-Purpose Bipolar Transistors

Listed numerically by device

Device	Type	V_{CEO} Maximum Collector Emitter Voltage (V)	V_{CBO} Maximum Collector Base Voltage (V)	V_{EBO} Maximum Emitter Base Voltage (V)	I_C Maximum Collector Current (mA)	P_O Maximum Device Dissipation (W)	**Minimum DC Current Gain** $I_C = 0.1$ mA	**Minimum DC Current Gain** $I_C = 150$ mA	Current-Gain Bandwidth Product f_T (MHz)	Noise Figure NF Maximum (dB)	Base
2N918	NPN	15	30	3.0	50	0.2	20 (3 mA)	—	600	6.0	3
2N2102	NPN	65	120	7.0	1000	1.0	20	40	60	6.0	2
2N2218	NPN	30	60	5.0	800	0.8	20	40	250		2
2N2218A	NPN	40	75	6.0	800	0.8	20	40	250		2
2N2219	NPN	30	60	5.0	800	3.0	35	100	250		2
2N2219A	NPN	40	75	6.0	800	3.0	35	100	300	4.0	2
2N2222	NPN	30	60	5.0	800	1.2	35	100	250		2
2N2222A	NPN	40	75	6.0	800	1.2	35	100	200	4.0	2
2N2905	PNP	40	60	5.0	600	0.6	35	—	200		2
2N2905A	PNP	60	60	5.0	600	0.6	75	100	200		2
2N2907	PNP	40	60	5.0	600	0.4	35	—	200		2
2N2907A	PNP	60	60	5.0	600	0.4	75	100	200		2
2N3053	NPN	40	60	5.0	700	5.0	—	50	100		2
2N3053A	NPN	60	80	5.0	700	5.0	—	50	100		2
2N3563	NPN	15	30	2.0	50	0.6	20	—	800		1
2N3904	NPN	40	60	6.0	200	0.625	40	—	300	5.0	1
2N3906	PNP	40	40	5.0	200	0.625	60	—	250	4.0	1
2N4037	PNP	40	60	7.0	1000	5.0	—	50			2
2N4123	NPN	30	40	5.0	200	0.35	—	25 (50 mA)	250	6.0	1
2N4124	NPN	25	30	5.0	200	0.35	120 (2 mA)	60 (50 mA)	300	5.0	1
2N4125	PNP	30	30	4.0	200	0.625	50 (2 mA)	25 (50 mA)	200	5.0	1
2N4126	PNP	25	25	4.0	200	0.625	120 (2 mA)	60 (50 mA)	250	4.0	1
2N4401	NPN	40	60	6.0	600	0.625	20	100	250		1
2N4403	PNP	40	40	5.0	600	0.625	30	100	200		1
2N5320	NPN	75	100	7.0	2000	10.0	—	30 (1 A)			2
2N5415	PNP	200	200	4.0	1000	10.0	—	30 (50 mA)	15		2
MM4003	PNP	250	250	4.0	500	1.0	20 (10 mA)	—			2
MPSA55	PNP	60	60	4.0	500	0.625	—	50 (0.1 A)	50		1
MPS6531	NPN	40	60	5.0	600	0.625	60 (10 mA)	90 (0.1 A)			1
MPS6547	NPN	25	35	3.0	50	0.625	20 (2 mA)	—	600		1

Test conditions: $I_C = 20$ mA dc; $V_{CE} = 20$ V; $f = 100$ MHz

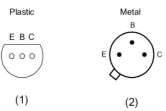

Bottom View, Base Pinouts

Table 22.32

General Purpose Silicon Power Transistors

TO-220 Case, Pin 1=Base, Pin 2, Case = Collector; Pin 3 = Emitter

TO-204 Case (TO-3), Pin 1=Base, Pin 2 = Emitter, Case = Collector;

NPN	PNP	I_C Max (A)	V_{CEO} Max (V)	h_{FE} Min	F_T (MHz)	Power Dissipation (W)
D44C8		4	60	100/220	50	30
	D45C8	4	60	40/120	50	30
TIP29		1	40	15/75	3	30
	TIP30	1	40	15/75	3	30
TIP29A		1	50	15/75	3	30
	TIP30A	1	60	15/75	3	30
TIP29B		1	80	15/75	3	30
TIP29C		1	100	15/75	3	30
	TIP30C	1	100	15/75	3	30
TIP47		1	250	30/150	10	40
TIP48		1	300	30/150	10	40
TIP49		1	350	30/150	10	40
TIP50		1	400	30/150	10	40
TIP110*		2	60	500	> 5	50
	TIP115*	2	60	500	> 5	50
TIP116		2	80	500	25	50
TIP31		3	40	25	3	40
	TIP32	3	40	25	3	40
TIP31A		3	60	25	3	40
	TIP32A	3	60	25	3	40
TIP31B		3	80	25	3	40
	TIP32B	3	80	25	3	40
TIP31C		3	100	25	3	40
	TIP32C	3	100	25	3	40
2N6124		4	45	25/100	2.5	40
2N6122		4	60	25/100	2.5	40
MJE1300		4	300	6/30	4	60
TIP120*		5	60	1000	> 5	65
	TIP125*	5	60	1000	> 10	65
	TIP42	6	40	15/75	3	65
TIP41A		6	60	15/75	3	65
TIP41B		6	80	15/75	3	65
2N6290		7	50	30/150	4	40
	2N6109	7	50	30/150	4	40
2N6292		7	70	30/150	4	40
	2N6107	7	70	30/150	4	40
MJE3055T		10	50	20/70	2	75
	MJE2955T	10	60	20/70	2	75
2N6486		15	40	20/150	5	75
2N6488		15	80	20/150	5	75
TIP140*		10	60	500	> 5	125
	TIP145*	10	60	600	> 10	125
2N3055A		15	60	20/70	0.8	115
2N3055		15	60	20/70	2.5	115
	MJ2955	15	60	20/70	2.5	115
2N6545		8	400	7/35	6	125
2N5039		20	75	20/100	—	140
2N3771		30	40	15	0.2	150
2N3789		10	60	15	4	150
2N3715		10	60	30	4	150
	2N3791	10	60	30	4	150
	2N5875	10	60	20/100	4	150
	2N3790	10	80	15	4	150
2N3716		10	80	30	4	150
	2N3792	10	80	30	4	150
2N3773		16	140	15/60	4	150
2N6284		20	100	750/18K	—	160
	2N6287	20	100	750/18K	—	160
2N5881		15	60	20/100	4	160
2N5880		15	80	20/100	4	160
2N6249		15	200	10/50	2.5	175
2N6250		15	275	8/50	2.5	175
2N6546		15	300	6/30	6-28	175
2N6251		15	350	6/50	2.5	175
2N5630		16	120	20/80	1	200
2N5301		30	40	15/60	2	200
2N5303		20	80	15/60	2	200
2N5885		25	60	20/100	4	200
2N5302		30	60	15/60	2	200
	2N4399	30	60	15/60	4	200
2N5886		25	80	20/100	4	200
	2N5884	25	80	20/100	4	200
MJ802		30	100	25/100	2	200
	MJ4502	30	100	25/100	2	200
MJ15003		20	140	25/150	2	250
	MJI5004	20	140	25/150	2	250
MJ15024		25	250	15/60	4	250

▨ = Complimentary pairs
* = Darlington transistor

Useful URLs for finding transistor/IC data sheets:

1. General-purpose substitution URL: www.nteinc.com
2. Philips semiconductors: www.semiconductors.philips.com
3. Mitsubishi: www.mitsubishichips.com
4. Motorola: www.freescale.com
5. STMicroelectronics: www.st.com
6. Toshiba: www.semicon.toshiba.co.jp/eng

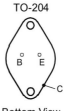

TO-204 Bottom View — TO-220 Front View

Table 22.33
Power FETs or MOSFETs

Device	Type	VDSS min (V)	RDS(on) max (Ω)	ID max (A)	PD max (W)	Case†	Mfr
BS250P	P-channel	45	14	0.23	0.7	E-line	Z
IRFZ30	N-channel	50	0.050	30	75	TO-220	IR
MTP50N05E	N-channel	50	0.028	25	150	TO-220AB	M
IRFZ42	N-channel	50	0.035	50	150	TO-220	IR
2N7000	N-channel	60	5	0.20	0.4	E-line	Z
VN10LP	N-channel	60	7.5	0.27	0.625	E-line	Z
VN10KM	N-channel	60	5	0.3	1	TO-237	S
ZVN2106B	N-channel	60	2	1.2	5	TO-39	Z
IRF511	N-channel	60	0.6	2.5	20	TO-220AB	M
MTP2955E	P-channel	60	0.3	6	25	TO-220AB	M
IRF531	N-channel	60	0.180	14	75	TO-220AB	M
MTP23P06	P-channel	60	0.12	11.5	125	TO-220AB	M
IRFZ44	N-channel	60	0.028	50	150	TO-220	IR
IRF531	N-channel	80	0.160	14	79	TO-220	IR
ZVP3310A	P-channel	100	20	0.14	0.625	E-line	Z
ZVN2110B	N-channel	100	4	0.85	5	TO-39	Z
ZVP3310B	P-channel	100	20	0.3	5	TO-39	Z
IRF510	N-channel	100	0.6	2	20	TO-220AB	M
IRF520	N-channel	100	0.27	5	40	TO-220AB	M
IRF150	N-channel	100	0.055	40	150	TO-204AE	M
IRFP150	N-channel	100	0.055	40	180	TO-247	IR
ZVP1320A	P-channel	200	80	0.02	0.625	E-line	Z
ZVN0120B	N-channel	200	16	0.42	5	TO-39	Z
ZVP1320B	P-channel	200	80	0.1	5	TO-39	Z
IRF620	N-channel	200	0.800	5	40	TO-220AB	M
MTP6P20E	P-channel	200	1	3	75	TO-220AB	M
IRF220	N-channel	200	0.400	8	75	TO-220AB	M
IRF640	N-channel	200	0.18	10	125	TO-220AB	M

Manufacturers: IR = International Rectifier; M = Motorola; S = Siliconix; Z = Zetex.

†For package shape, size and pin-connection information, see manufacturers' data sheets. Many retail suppliers offer data sheets to buyers free of charge on request. Data books are available from many manufacturers and retailers.

Table 22.34
RF Power Transistors

Device	Output Power (W)	Input Power (W)	Gain (dB)	Typ Supply Voltage (V)	Case	Mfr
1.5 to 30 MHz, HF SSB/CW						
2SC2086	0.3		13	1.2	TO-92	MI
BLV10	1		18	1.2	SOT123	PH
BLV11	2		18	12	SOT123	PH
MRF476	3	0.1	15	12.5-13.6	221A-04/1	MO
BLW87	6		18	1.2	SOT123	PH
2SC2166	6		13.8	1.2	TO-220	MI
BLW83	1.0		20	26	SOT123	PH
MRF475	1.2	1.2	10	12.5-13.6	221A-04/1	MO
MRF433	12.5	0.125	20	12.5-13.6	211-07/1	MO
2SC3133	1.3		14	1.2	TO-220	MI
MRF485	1.5	1.5	10	28	221A-04/1	MO
2SC1969	1.6		1.2	1.2	TO-220	MI
BLW50F	1.6		19.5	45	SOT123	PH
MRF406	20	1.25	12	12.5-13.6	221-07/1	MO
SD1285	20	0.65	15	12.5	M113	ST
MRF426	25	0.16	22	28	211-07/1	MO
MRF427	25	0.4	18	50	211-11/1	MO
MRF477	40	1.25	15	12.5-13.6	211-11/1	MO
MRF466	40	1.25	15	28	211-07/1	MO
BLW96	50		1.9	40	SOT121	PH
2SC3241	75		12.3	12.5	T-45E	MI
SDI405	75	3.8	13	12.5	M174	ST
2SC2097	75		12.3	13.5	T-40E	MI
MRF464	80	2.53	10	28	211-11/1	MO
MRF421	100	10	10	12.5-13.6	211-11/1	MO
SD1487	100	7.9	11	12.5	M174	ST
2SC2904	100		11.5	12.5	T-40E	MI
SD1729	130	8.2	12	28	M174	ST
MRF422	150	15	10	28	211-11/1	MO
MRF428	150	7.5	13	50	211-11/1	MO
SD1726	150	6	14	50	M174	ST
PT9790	150	4.8	15	50	211-11/1	MO
MRF448	250	15.7	12	50	211-11/1	MO
MRF430	600	60	10	50	368-02/1	MO
50 MHz						
MRF475	4	0.4	10	12.5-13.6	*221A-04/1*	MO
MRF497	40	4	10	12.5-13.6	221A-04/2	MO
SDI446	70	7	10	12.5	M113	ST
MRF492	70	5.6	11	12.5-13.6	211-11/1	MO
SD1405	100	20	7	12.5	M174	ST
VHF to 175 MHz						
2N4427	0.7		8	7.5	TO-39	PH
2N3866	1		10	28	TO-39	PH
BFQ42	1.5		8.4	7.5	TO-39	PH
2SC2056	1.6		9	7.2	T-41	MI
2N3553	2.5	0.25	10	28	79-04/1	MO
BF043	3		9.4	7.5	TO-39	PH
SD1012	4	0.25	12	12.5	M135	ST

Table 22.34 continued

Device	Output Power (W)	Input Power (W)	Gain (dB)	Typ Supply Voltage (V)	Case	Mfr
2SC2627	5		13	12.5	T-40	MI
2N5641	7	1	8.4	28	144B-05/1	MO
MRF340	8	0.4	13	28	221A-04/2	MO
BLW29	9		7.4	7.5	SOT120	PH
SD1143	10	1	10	12.5	M135	ST
2SC1729	1.4		10	13.5	T-31 E	MI
SD1014-02	15	3.5	6.3	12.5	M135	ST
BLVII	15		8	13.5	SOT123	PH
2N5642	20	3	8.2	28	145A-09/1	MO
MRF342	24	1.9	11	28	221A-04/2	MO
BLW87	25		6	13.5	SOT123	PH
2SC1946	28		6.7	13.5	T-31 E	MI
MRF314	30	3	10	28	211-07/1	MO
SD1018	40	14	4.5	12.5	M135	ST
2N5643	40	6.9	7.6	28	145A-09/1	MO
BLW40	40		10	12.5	SOT120	PH
MRF315	45	5.7	9	28	211-07/1	MO
PT9733	50	10	7	28	145A-09/1	MO
MRF344	60	15	6	28	221A-04/2	MO
2SC2694	70		6.7	12.5	T-40	MI
BLV75/12	75		6.5	12.5	SOT119	PH
MRF316	80	8	10	28	316-01/1	MO
SD1477	100	25	6	12.5	M111	ST
BLW78	100		6	28	SOT121	PH
MRF317	100	12.5	9	28	316-01/1	MO
TP9386	150	15	10	28	316-01/1	MO

220 MHz

Device	Output Power (W)	Input Power (W)	Gain (dB)	Typ Supply Voltage (V)	Case	Mfr
MRF207	1	0.15	8.2	12.5	79-04/1	MO
2N5109	2.5		11	12	TO-205AD	MO
MRF227	3	0.13	13.5	12.5	79-05/5	MO
MRF208	1.0	1	10	12.5	145A-09/1	MO
MRF226	1.3	1.6	9	12.5	145A-09/1	MO
2SC2133	30		8.2	28	T-40E	MI
2SC2134	60		7	28	T-40E	MI
2SC2609	100		6	28	T-40E	MI

UHF to 512 MHz

Device	Output Power (W)	Input Power (W)	Gain (dB)	Typ Supply Voltage (V)	Case	Mfr
2N4427	0.4		10	12.5	TO-39	PH
2SC3019	0.5		14	12.5	T-43	MI
MRF581	0.6	0.03	13	12.5	317-01/2	MO
2SC908	1		4	12.5	TO-39	MI
2N3866	1		10	28	TO-39	PH
2SC2131	1.4		6.7	13.5	TO-39	MI
BLX65E	2		9	12.5	TO-39	PH
BLW89	2		12	28	SOT122	PH
MRF586	2.5		16.5	1.5	79-04	MO
MRF630	3	0.33	9.5	12.5	79-05/5	MO
2SC3020	3	0.3	10	12.5	T-31 E	MI
BLW80	4		8	12.5	SOT122	PH
BLW90	4		11	12.5	SOT122	PH
MRF652	5	0.5	10	12.5	244-04/1	MO
MRF587	5		16.5	15	244A-01/1	MO
2SC3021	7	1.2	7.6	12.5	T-31 E	MI
BLW81	10		6	12.5	SOT122	PH
MRF653	10	2	7	12.5	244-04/1	MO
BLW91	10		9	28	SOT122	PH
MRF654	15	2.5	7.8	12.5	244-04/1	MO
2SC3022	18	6	4.7	12.5	T-31 E	MI
BLU20/12	20		6.5	12.5	SOT119	PH
BLX94A	25		6	28	SOT48/2	PH
2SC2695	28		4.9	13.5	T-31 E	MI
BLU30/12	30		6	12.5	SOT119	PH
BLU45/12	45		4.8	12.5	SOT119	PH
2SC2905	45		4.8	12.5	T-40E	MI
MRF650	50	15.8	5	12.5	316-01/1	MO
TP5051	50	6	9	24	333A-02/2	MO
BLU60/12	60		4.4	12.5	SOT119	PH
2SC3102	60	20	4.8	12.5	T-41 E	MI
BLU60/28	60		7	28	SOT119	PH
MRF658	65	25	4.15	12.5	316-01/1	MO
MRF338	80	15	7.3	28	333-04/1	MO
SD1464	100	28.2	5.5	28	M168	ST

UHF to 960 MHz

Device	Output Power (W)	Input Power (W)	Gain (dB)	Typ Supply Voltage (V)	Case	Mfr
MRF581	0.6	0.06	10	12.5	317-01/2	MO
MRF8372	0.75	0.11	8	12.5	751-04/1	MO
MRF557	1.5	0.23	8	12.5	317D-02/2	MO
BLV99	2		9	24	SOT172	PH
SD1420	2.1	0.27	9	24	M122	ST
MRF839	3	0.46	8	12.5	305A-01/1	MO
MRF896	3	0.3	10	24	305-01/1	MO
MRF891	5	0.63	9	24	319-06/2	MO
2SC2932	6		7.8	12.5	T-31 B	MI
SD1398	6	0.6	10	24	M142	ST
2SC2933	14	3	6.7	12.5	T-31 B	MI
SD1400-03	14	1.6	9.5	24	M118	ST
MRF873	15	3	7	12.5	319-06/2	MO
SD1495-03	30	6	7	24	M142	ST
SD1424	30	5.3	7.5	24	M156	ST
MRF897	30	3	10	24	395B-01/1	MO
MRF847	45	16	4.5	12.5	319-06/1	MO
BLV101A	50		8.5	26	SOT273	PH
SD1496-03	55	10	7.4	24	M142	ST
MRF898	60	12	7	24	333A-02/1	MO
MRF880	90	12.7	8.5	26	375A-01/1	MO
MRF899	150	24	8	26	375A-01/1	MO

Manufacturer codes:

MI = Mitsubishi; MO = Motorola; PH = Philips; ST = STMicroelectronics

There is a bewildering variety of package types, sizes and pin-out connections. (For example, for the 137 different transistors in this table there are 54 different packages.) See the data sheets on each manufacturer's Web pages for details.

Mitsubishi: **www.mitsubishichips.com**
Motorola: **www.freescale.com**
Philips semiconductors: **www.semiconductors.philips.com**
STMicroelectronics: **www.st.com**

Table 22.35
RF Power Transistors Recommended for New Designs

Device	Output Power (W)	Type	Gain (dB)	Typ Supply Voltage (V)	Case	Mfr
1.5 to 30 MHz, HF SSB/CW						
MRF171A	30	MOS	20	28	211-07/2	MO
BLF145	30	MOS	24	28	SOT123A	PH
MRF148A	30	MOS	18	50	211-07/2	MO
SD2918	30	MOS	18	50	M113	ST
SD1405	75	BJT	13	12.5	M174	ST
SD1733	75	BJT	14	50	M135	ST
SD1487	100	BJT	11	12.5	M174	ST
SD1407	125	BJT	15	28	M174	ST
SD1729	130	BJT	12	28	M174	ST
BLF147	150	MOS	17	28	SOT121B	PH
BLF177	150	MOS	20	50	SOT121B	PH
BLF175	150	MOS	24	50	SOT123A	PH
SD1726	150	BJT	14	50	M174	ST
SD1727	150	BJT	14	50	M164	ST
MRF150	150	MOS	17	50	211-07/2	MO
SD1411	200	BJT	16	40	M153	ST
SD1730	220	BJT	12	28	M174	ST
SD1731	220	BJT	13	50	M174	ST
SD1728	250	BJT	14.5	50	M177	ST
SD2923	300	MOS	16	50	M177	ST
SD2933	300	MOS	18	50	M177	ST
MRF154	600	MOS	17	50	368-03/2	MO
50 to 175 MHz						
BLF202	2	MOS	10	12.5	SOT409A	PH
BLF242	5	MOS	13	28	SOT123A	PH
SD1274	30	BJT	10	13.6	M135	ST
BLF245	30	MOS	13	28	SOT123	PH
SD1275	40	BJT	9	13.6	M135	ST
BLF246B	60	MOS	14	28	SOT161A	PH
SD1477	100	BJT	6	12.5	M111	ST
SD1480	100	BJT	9.2	28	M111	ST
SD2921	150	MOS	12.5	50	M174	ST
MRF141	150	MOS	13	28	211-11/2	MO
MRF151	150	MOS	13	50	211-11/2	MO
SD2931	150	MOS	14	50	M174	ST
BLF248	300	MOS	10	28	SOT262	PH
SD2932	300	MOS	15	50	M244	ST
VHF to 220 MHz						
MRF134	5	MOS	10.6	28	211-07/2	MO
MRF136	15	MOS	16	28	211-07/2	MO
MRF173	80	MOS	13	28	211-11/2	MO
MRF174	125	MOS	11.8	28	211-11/2	MO
BLF278	250	MOS	14	50	SOT261A1	PH
VHF to 470 MHz						
BLT50	1.2	BJT	10	7.5	SOT223	PH
SD2900	5	MOS	13.5	28	M113	ST
SD1433	10	BJT	7	12.5	M122	ST
SD2902	15	MOS	12.5	28	M113	ST
SD2904	30	MOS	10	28	M113	ST
SD2903	30	MOS	13	28	M229	ST
SD1488	38	BJT	5.8	12.5	M111	ST
SD1434	45	BJT	5	12.5	M111	ST
MRF392	125	BJT	8	28	744A-01/1	MO
SD2921	150	MOS	12.5	50	M174	ST
VHF to 512 MHz						
BLF521	2	MOS	10	12.5	SOT172D	PH
MRF158	2	MOS	17.5	28	305A-01/2	MO
MRF160	4	MOS	17	28	249-06/3	MO
BLF542	5	MOS	13	28	SOT171A	PH
VLF544	20	MOS	11	28	SOT171A	PH
MRF166C	20	MOS	16	28	319-07/3	MO
MRF166W	40	MOS	14	28	412-01/1	MO
BLF546	80	MOS	11	28	SOT268A	PH
MRF393	100	BJT	7.5	28	744A-01/1	MO
MRF275L	100	MOS	8.8	28	333-04/2	MO
BLF548	150	MOS	10	28	SOT262A	PH
MRF275G	150	MOS	10	28	375-04/2	MO
UHF to 960 MHz						
BLT70	0.6	BJT	6	4.8	SOT223	PH
BLT80	0.6	BJT	6	7.5	SOT223	PH
BLT71/8	1.2	BJT	6	4.8	SOT223	PH
BLT81	1.2	BJT	6	7.5	SOT223	PH
BLF1043	10	MOS	16	26	SOT538A	PH
BLF1046	45	MOS	14	26	SOT467C	PH
BLF1047	70	MOS	14	26	SOT541A	PH
BLF1048	90	MOS	14	26	SOT502A	PH

Notes:
Manufacturer codes: MI = Mitsubishi; MO = Motorola; PH = Philips; ST = STMicroelectronics

There is a bewildering variety of package types, sizes and pin-out connections. (For example, for the 71 different transistors in this table there are 35 different packages.) See the data sheets on each manufacturer's Web pages for details.

Mitsubishi: **www.mitsubishichips.com**
Motorola: **www.freescale.com**
Philips semiconductors: **www.semiconductors.philips.com**
STMicroelectronics: **www.st.com**

Table 22.36
RF Power Amplifier Modules
Listed by frequency

Device	Supply (V)	Frequency Range (MHz)	Output Power (W)	Power Gain (dB)	Package[†]	Mfr/ Notes
M57735	17	50-54	14	21	H3C	MI; SSB mobile
M57719N	17	142-163	14	18.4	H2	MI; FM mobile
S-AV17	16	144-148	60	21.7	5-53L	T, FM mobile
S-AV7	16	144-148	28	21.4	5-53H	T, FM mobile
MHW607-1	7.5	136-150	7	38.4	301K-02/3	MO; class C
BGY35	12.5	132-156	18	20.8	SOT132B	P
M67712	17	220-225	25	20	H3B	MI; SSB mobile
M57774	17	220-225	25	20	H2	MI; FM mobile
MHW720-1	12.5	400-440	20	21	700-04/1	MO; class C
MHW720-2	12.5	440-470	20	21	700-04/1	MO; class C
M57789	17	890-915	12	33.8	H3B	MI
MHW912	12.5	880-915	12	40.8	301R-01/1	MO; class AB
MHW820-3	12.5	870-950	18	17.1	301G-03/1	MO; class C

Manufacturer codes: MO = Motorola; MI = Mitsubishi; P = Philips; T = Toshiba.

[†]For package shape, size and pin-connection information, see manufacturers' data sheets. Many retail suppliers offer data sheets to buyers free of charge on request. Data books are available from many manufacturers and retailers.

Table 22.37
Digital Logic Families

Type	Propagation Delay for CL = 50 pF (ns) Typ	Propagation Delay for CL = 50 pF (ns) Max	Max Clock Frequency (MHz)	Power Dissipation (CL = 0) @ 1 MHz (mW/gate)	Output Current @ 0.5 V max (mA)	Input Current (Max mA)	Threshold Voltage (V)	Supply Voltage (V) Min	Supply Voltage (V) Typ	Supply Voltage (V) Max
CMOS										
74AC	3	5.1	125	0.5	24	0	V+/2	2	5 or 3.3	6
74ACT	3	5.1	125	0.5	24	0	1.4	4.5	5	5.5
74HC	9	18	30	0.5	8	0	V+/2	2	5	6
74HCT	9	18	30	0.5	8	0	1.4	4.5	5	5.5
4000B/74C (10 V)	30	60	5	1.2	1.3	0	V+/2	3	5 - 15	18
4000B/74C (5V)	50	90	2	3.3	0.5	0	V+/2	3	5 - 15	18
TTL										
74AS	2	4.5	105	8	20	0.5	1.5	4.5	5	5.5
74F	3.5	5	100	5.4	20	0.6	1.6	4.75	5	5.25
74ALS	4	11	34	1.3	8	0.1	1.4	4.5	5	5.5
74LS	10	15	25	2	8	0.4	1.1	4.75	5	5.25
ECL										
ECL III	1.0	1.5	500	60	—	—	-1.3	-5.19	-5.2	-5.21
ECL 100K	0.75	1.0	350	40	—	—	-1.32	-4.2	-4.5	-5.2
ECL100KH	1.0	1.5	250	25	—	—	-1.29	-4.9	-5.2	-5.5
ECL 10K	2.0	2.9	125	25	—	—	-1.3	-5.19	-5.2	-5.21
GaAs										
10G	0.3	0.32	2700	125	—	—	-1.3	-3.3	-3.4	-3.5
10G	0.3	0.32	2700	125	—	—	-1.3	-5.1	-5.2	-5.5

Source: Horowitz (W1HFA) and Hill, *The Art of Electronics—2nd edition*, page 570. © Cambridge University Press 1980, 1989. Reprinted with the permission of Cambridge University Press.

Table 22.38
Operational Amplifiers (Op Amps)
Listed by device number

Device	Type	Freq Comp	Max Supply* (V)	Min Input Resistance (MΩ)	Max Offset Voltage (mV)	Min dc Open-Loop Gain (dB)	Min Output Current (mA)	Min Small-Signal Bandwidth (MHz)	Min Slew Rate (V/µs)	Notes
101A	Bipolar	ext	44	1.5	3.0	79	15	1.0	0.5	General purpose
108	Bipolar	ext	40	30	2.0	100	5	1.0		
124	Bipolar	int	32		5.0	100	5	1.0		Quad op amp, low power
148	Bipolar	int	44	0.8	5.0	90	10	1.0	0.5	Quad 741
158	Bipolar	int	32		5.0	100	5	1.0		Dual op amp, low power
301	Bipolar	ext	36	0.5	7.5	88	5	1.0	10	Bandwidth extendable with external components
324	Bipolar	int	32		7.0	100	10	1.0		Quad op amp, single supply
347	BiFET	ext	36	106	5.0	100	30	4	13	Quad, high speed
351	BiFET	ext	36	106	5.0	100	20	4	13	
353	BiFET	ext	36	106	5.0	100	15	4	13	
355	BiFET	ext	44	106	10.0	100	25	2.5	5	
355B	BiFET	ext	44	106	5.0	100	25	2.5	5	
356A	BiFET	ext	36	106	2.0	100	25	4.5	12	
356B	BiFET	ext	44	106	5.0	100	25	5.0	12	
357	BiFET	ext	36	106	10.0	100	25	20.0	50	
357B	BiFET	ext	36	106	5.0	100	25	20.0	30	
358	Bipolar	int	32		7.0	100	10	1.0		Dual op amp, single supply
411	BiFET	ext	36	106	2.0	100	20	4.0	15	Low offset, low drift
709	Bipolar	ext	36	0.05	7.5	84	5	0.3	0.15	
741	Bipolar	int	36	0.3	6.0	88	5	0.4	0.2	
741S	Bipolar	int	36	0.3	6.0	86	5	1.0	3	Improved 741 for AF
1436	Bipolar	int	68	10	5.0	100	17	1.0	2.0	High-voltage
1437	Bipolar	ext	36	0.050	7.5	90		1.0	0.25	Matched, dual 1709
1439	Bipolar	ext	36	0.100	7.5	100		1.0	34	
1456	Bipolar	int	44	3.0	10.0	100	9.0	1.0	2.5	Dual 1741
1458	Bipolar	int	36	0.3	6.0	100	20.0	0.5	3.0	
1458S	Bipolar	int	36	0.3	6.0	86	5.0	0.5	3.0	Improved 1458 for AF
1709	Bipolar	ext	36	0.040	6.0	80	10.0	1.0		
1741	Bipolar	int	36	0.3	5.0	100	20.0	1.0	0.5	
1747	Bipolar	int	44	0.3	5.0	100	25.0	1.0	0.5	Dual 1741
1748	Bipolar	ext	44	0.3	6.0	100	25.0	1.0	0.8	Non-comp-ensated 1741
1776	Bipolar	int	36	50	5.0	110	5.0		0.35	Micro power, programmable
3140	BiFET	int	36	1.5 × 106	2.0	86	1	3.7	9	Strobable output
3403	Bipolar	int	36	0.3	10.0	80		1.0	0.6	Quad, low power
3405	Bipolar	ext	36		10.0	86	10	1.0	0.6	Dual op amp and dual comparator
3458	Bipolar	int	36	0.3	10.0	86	10	1.0	0.6	Dual, low power

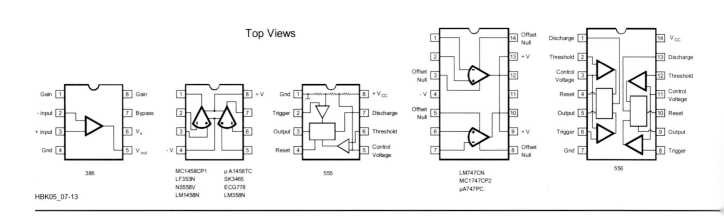

Top Views

386

MC1458CP1 / µA1458TC / LF353N / SK3465 / N5558V / ECG778 / LM1458N / LM358N

555

LM747CN / MC1747CP2 / µA747PC

556

Device	Type	Freq Comp	Max Supply* (V)	Min Input Resistance (MΩ)	Max Offset Voltage (mV)	Min dc Open-Loop Gain (dB)	Min Output Current (mA)	Min Small-Signal Bandwidth (MHz)	Min Slew Rate (V/μs)	Notes
3476	Bipolar	int	36	5.0	6.0	92	12		0.8	
3900	Bipolar	int	32	1.0		65	0.5	4.0	0.5	Quad, Norton single supply
4558	Bipolar	int	44	0.3	5.0	88	10	2.5	1.0	Dual, wideband
4741	Bipolar	int	44	0.3	5.0	94	20	1.0	0.5	Quad 1741
5534	Bipolar	int	44	0.030	5.0	100	38	10.0	13	Low noise, can swing 20V P-P across 600
5556	Bipolar	int	36	1.0	12.0	88	5.0	0.5	1	Equivalent to 1456
5558	Bipolar	int	36	0.15	10.0	84	4.0	0.5	0.3	Dual, equivalent to 1458
34001	BiFET	int	44	106	2.0	94		4.0	13	JFET input
AD745	BiFET	int	±18	104	0.5	63	20	20	12.5	Ultra-low noise, high speed
LT1001	Precision op amp, low offset voltage (15 μV max), low drift (0.6 μV/°C max), low noise (0.3 μV p-p)									
LT1007	Extremely low noise (0.06 μV p-p), very high gain (20 x 10^6 into 2 kΩ load)									
LT1360	High speed, very high slew rate (800 V/μs), 50 MHz gain bandwidth, ±2.5 V to ±15 V supply range									
NE5514	Bipolar	int	±16	100	1		10	3	0.6	
NE5532	Bipolar	int	±20	0.03	4	47	10	10	9	Low noise
OP-27A	Bipolar	ext	44	1.5	0.025	115		5.0	1.7	Ultra-low noise, high speed
OP-37A	Bipolar	ext	44	1.5	0.025	115		45.0	11.0	
TL-071	BiFET	int	36	10^6	6.0	91		4.0	13.0	Low noise
TL-081	BiFET	int	36	10^6	6.0	88		4.0	8.0	
TL-082	BiFET	int	36	10^6	15.0	99		4.0	8.0	Low noise
TL-084	BiFET	int	36	10^6	15.0	88		4.0	8.0	Quad, high-performance AF
TLC27M2	CMOS	int	18	10^6	10	44		0.6	0.6	Low noise
TLC27M4	CMOS	int	18	10^6	10	44		0.6	0.6	Low noise

*From −V to +V terminals

Top Views

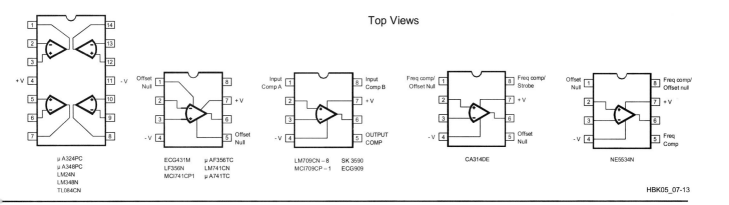

μA324PC / μA348PC / LM24N / LM348N / TL084CN

ECG431M / LF356N / MCI741CP1 — μAF356TC / LM741CN / μA741TC

LM709CN – 8 / MCI709CP – 1 — SK 3590 / ECG909

CA314DE

NE5534N

22.7 Tubes, Wire, Materials, Attenuators, Miscellaneous

Table 22.39
Triode Transmitting Tubes

The full 1988 *Handbook* table of power tube specifications and base diagrams can be viewed in pdf format on the *ARRLWeb* at www.arrl.org/hf-tube-amplifiers.

Type	Power Diss.(W)	Plate (V)	Plate (mA)	Grid dc (mA)	Freq (MHz)	Ampl Factor	Fil (V)	Fil (A)	C_{IN} (pF)	C_{GP} (pF)	C_{OUT} (pF)	Base Diagram	Service Class[1]	Plate (V)	Grid (V)	Plate (mA)	Grid dc (mA)	Input (W)	P-P (kΩ)	Output (W)
5675	5	165	30	8	3000	20	6.3	0.135	2.3	1.3	0.09	Fig 21	GG0	120	−8	25	4	¾	—	0.05
2C40	6.5	500	25	—	500	36	6.3	0.75	2.1	1.3	0.05	Fig 11	CTO	250	−5	20	0.3	¾	—	0.075
5893	8.0	400	40	13	1000	27	6.0	0.33	2.5	1.75	0.07	Fig 21	CT	350	−33	35	13	2.4	—	6.5
													CP	300	−45	30	12	2.0	—	6.5
2C43	12	500	40	—	1250	48	6.3	0.9	2.9	1.7	0.05	Fig 11	CTO	470	—	38[7]	—	¾	—	9[2]
811-A	65	1000	175	50	60	160	6.3	4.0	5.9	5.6	0.7	3G	CT	1500	−70	173	40	7.1	—	200
													CP	1250	−120	140	45	10.0	—	135
													B/CG	1250	0	21/175	28	12	—	165
													AB$_1$	1250	0	27/175	13	3.0	—	155
812-A	65	1500	175	35	60	29	6.3	4.0	5.4	5.5	0.77	3G	CT	1500	−120	173	30	6.5	—	190
													CP	1250	−115	140	35	7.6	—	130
													B[2]	1500	−48	28/310	270[4]	5.0	13.2	340
3CX100A5[6]	100	1000	125[5]	50	2500	100	6.0	1.05	7.0	2.15	0.035	—	AGG	800	−20	80	30	6	—	27
	70	600	100[5]										CP	600	−15	75	40	6	—	18
2C39	100	1000	60	40	500	100	6.3	1.1	6.5	1.95	0.03	—	G1C	600	−35	60	40	5.0	—	20
													CTO	900	−40	90	30	¾	—	40
													CP	600	−150	100[5]	50	¾	—	¾
AX9900, 5866	135	2500	200	40	150	25	6.3	5.4	5.8	5.5	0.1	Fig 3	CT	2500	−200	173	40	16	—	390
													CP	2000	−225	127	40	16	—	204
													B[2]	2500	−90	80/330	350[4]	14[3]	15.68	560
572B	160	2750	275	—	—	170	6.3	4.0	—	—	—	3G	CT	1650	−70	165	32	6	—	205
T160L	—	2400	—	—	—	—	—	—	—	—	—	—	B/GG[2]	2400	−2.0	90/500	—	100	—	600
8873	200	2200	250	—	500	160	6.3	3.2	19.5	7.0	0.03	Fig 87	AB$_2$	2000	—	22/500	98[3]	27[3]	—	505
8875	300	2200	250	—	500	160	6.3	3.2	19.5	7.0	0.03	—	AB$_2$	2000	—	22/500	98[3]	27[3]	—	505
833A	350	3300	500	100	30	35	10	10	12.3	6.3	8.5	Fig 41	CTO	2250	−125	445	85	23	—	780
													CTO	3000	−160	335	70	20	—	800
													CP	2500	−300	335	75	30	—	635
	450[6]	4000[6]	500	100	20[6]	35	10	10	12.3	6.3	8.5	Fig 41	CP	3000	−240	335	70	26	—	800
													B[2]	3000	−70	100/750	400[4]	20[4]	9.5	1650
8874	400	2200	350	—	500	160	6.3	3.2	19.5	7.0	0.03	—	AB$_2$	2000	—	22/500	98[3]	27[3]	—	505
3-400Z	400	3000	400	—	110	200	5	14.5	7.4	4.1	0.07	Fig 3	B/GG	3000	0	100/333	120	32	—	655
3-500Z	500	4000	400	—	110	160	5	14.5	7.4	4.1	0.07	Fig 3	B/GG	3000	—	370	115	30	5	750
3-600Z	600	4000	425	—	110	165	5	15.0	7.8	4.6	0.08	Fig 3	B/GG	3000	—	400	118	33	—	810
													B/GG	3500	—	400	110	35	—	950
3CX800A7	800	2250	600	60	350	200	13.5	1.5	26	—	6.1	Fig 87	AB$_2$GG[7]	2200	−8.2	500	36	16	—	750
3-1000Z	1000	3000	800	—	110	200	7.5	21.3	17	6.9	0.12	Fig 3	B/GG	3000	0	180/670	300	65	—	1360
3CX1200A7	1200	5000	800	—	110	200	7.5	21.0	20	12	0.2	Fig 3	AB$_2$GG	3600	−10	700	230	85	—	1500
8877	1500	4000	1000	—	250	200	5.0	10	42	10	0.1	—	AB$_2$	2500	−8.2	1000	—	57	—	1520

Table 22-40
Tetrode Transmitting Tubes
Also see www.arrl.org/hf-tube-amplifiers.

[1]Service Class Abbreviations:
AB$_2$GD=AB$_2$ linear with 50-Ω passive grid circuit.
B=Class-B push-pull
CP=Class-C plate-modulated phone
CT=Class-C telegraph

GG=Grounded-grid (grid and screen connected together)
[2]Maximum signal value
[3]Peak grid-grid volts
[4]Forced-air cooling required.

[5]Two tubes triode-connected, G2 to G1 through 20kΩ to G2.
[6]Typical operation at 175 MHz.
[7]±1.5 V.
[8]Values are for two tubes.
[9]Single tone.

[10]24-Ω cathode resistance.
[11]Base same as 4CX250B. Socket is Russian SK2A.
[12]Socket is Russian SK1A.
[13]Socket is Russian SK3A.

Type	Max. Plate Diss. (W)	Max. Plate Volts (V)	Max. Screen Diss. (W)	Max. Screen Volts (V)	Max. Freq. (MHz)	Filament Volts (V)	Amps (A)	C_{IN} (pF)	C_{GP} (pF)	C_{OUT} (pF)	Base	Serv. Class[1]	Plate (V)	Screen (V)	Grid (V)	Plate (mA)	Screen (mA)	Grid (mA)	P_{IN} (W)	P-P (kΩ)	P_{OUT} (W)
6146/ 6146A	25	750	3	250	60	6.3	1.25	13	0.24	8.5	7CK	CT	500	170	−66	135	9	2.5	0.2	—	48
6146A												CT	700	160	−62	120	11	3.1	0.2	—	70
8032	25	750	3	250	60	12.6	0.585	13	0.24	8.5	7CK	CT[6]	400	190	−54	150	10.4	2.2	3.0	—	35
6883												CP	400	150	−87	112	7.8	3.4	0.4	—	32
												CP	600	150	−87	112	7.8	3.4	0.4	—	52
6159B/	25	750	3	250	60	26.5	0.3	13	0.24	8.5	7CK	AB$_2$[8]	600	190	−48	28/270	1.2/20	22	0.3	5	113
												AB$_2$[8]	750	165	−46	22/240	0.3/20	2.6[2]	0.4	7.4	131
												AB$_1$[8]	750	195	−50	23/220	1/26	100[3]	0	8	120
807, 807W	30	750	3.5	300	60	6.3	0.9	12	0.2	7	5AW	CT	750	250	−45	100	6	3.5	0.22	—	50
5933												CP	600	275	−90	100	6.5	4	0.4	—	42.5
												AB$_1$	750	300	−35	15/70	3/8	753	0	—	72
1625	30	750	3.5	300	60	12.6	0.45	12	0.2	7	5AZ	B[5]	750	—	0	15/240	—	—	5.3[2]	6.65	120
6146B	35	750	3	250	60	6.3	1.125	13	0.22	8.5	7CK	CT	750	200	−77	160	10	2.7	0.3	—	85
8298A												CP	600	175	−92	140	9.5	3.4	0.5	—	62
												AB$_1$	750	200	−48	24/125	6.3	—	—	3.5	61
813	125	2500	20	800	30	10.0	5.0	16.3	0.25	14.0	5BA	CTO	1250	300	−75	180	35	12	1.7	—	170
												CTO	2250	400	−155	220	40	15	4	—	375
												AB1	2500	750	−95	25/145	27[2]	—	0	16	245
												AB$_2$[8]	2000	750	−90	40/315	1.5/58	230[3]	0.1[2]	—	455
												AB$_2$[8]	2500	750	−95	35/260	1.2/55	235[3]	0.35[2]	17	650
4CX250B	250	2000	12	400	175	6.0	2.9	18.5	0.04	4.7	—	CTO	2000	250	−90	250	25	27	2.8	—	410
												CP	1500	250	−100	200	25	17	2.1	—	250
												AB$_1$[8]	2000	350	−50	500	30	100	0	8.26	650
4-400A	400[4]	4000	35	600	110	5.0	14.5	12.5	0.12	4.7	5BK	CT/CP	4000	300	−170	270	22.5	10	10	—	720
												GG	2500	0	0	80/270[9]	55[9]	100[9]	39[9]	4.0	435
												AB$_1$	2500	750	−130	95/317	0/14	0	—	—	425
4CX400A	400	2500	8	400	500	6.3	3.2	24	0.08	7	See[11]	AB$_2$GD2200	325	−30	100/270	22	2	9	—	405	
												AB$_2$GD2500	400	−35	100/400	18	1	13	—	610	
4CX800A	800	2500	15	350	150	12.6	3.6	51	0.9	11	See[12]	AB$_2$GD2200	350	−56	160/550	24	1	32	—	750	
4-1000A	1000	6000	75	1000	—	7.5	21	27.2	0.24	7.6	—	CT	3000	500	−150	700	146	38	11	—	1430
8166		3000			110	6.0	9.0	81.5	0.01	11.8	—	CP	3000	500	−200	600	145	36	12	2.5	1390
		4000										AB$_2$	4000	500	−60	300/1200	0.95	2.6[2]	11	7	3000
		3000										GG	3000	0	0	100/700[9]	105[9]	170[9]	130[9]	3.1	1475
4CX1000A	1000	3000	12	400	110	6.0	10.0	81.5	0.02	11.8	—	AB$_1$[8]	2000	325	−55	500/2000	−4/60	—	—	3.85	2160
												AB$_1$[8]	2500	325	−55	500/2000	−4/60	—	—	1.9	2920
												AB$_1$	3000	325	−55	500/1800	−4/60	—	—	—	3360
4CX1500B	1500	3000	12	400	110	6.0	4.4	86	0.15	12	See[13]	AB$_1$	2750	225	−34	300/755	−14/60	0.95	1.5	—	1100
4CX1600B	1600	3300	20	350	250	12.6	20					AB$_2$GD2400	350	−53	500/1100	20	2	28	—	1600	
												AB$_2$GD2400	350	−70	200/870	48	2	83[10]	—	1500	
												AB$_2$GD3200	240	−57	200/740	21	1	33	—	1600	

Table 22.41

EIA Vacuum-Tube Base Diagrams

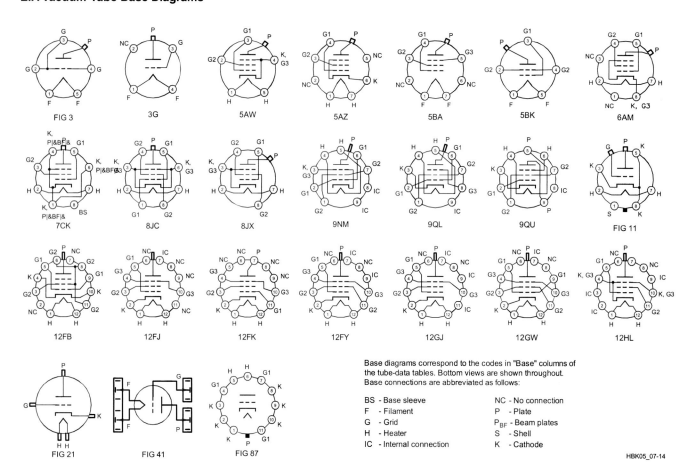

Base diagrams correspond to the codes in "Base" columns of the tube-data tables. Bottom views are shown throughout. Base connections are abbreviated as follows:

- BS - Base sleeve
- F - Filament
- G - Grid
- H - Heater
- IC - Internal connection
- NC - No connection
- P - Plate
- P_{BF} - Beam plates
- S - Shell
- K - Cathode

Alphabetical subscripts (D = diode, P = pentode, T = triode and HX = hexode) indicate structures in multistructure tubes. Subscript CT indicates filament or heater center tap.

Generally, when pin 1 of a metal-envelope tube (except all triodes) is shown connected to the envelope, pin 1 of a glass-envelope counterpart (suffix G or GT) is connected to an internal shield.

Table 22.42
Metal-Oxide Varistor (MOV) Transient Suppressors
Listed by voltage

Type No.	ECG/NTE†† no.	V acRMS	Maximum Applied Voltage V acPeak	Maximum Energy (Joules)	Maximum Peak Current (A)	Maximum Power (W)	Maximum Varistor Voltage (V)
V180ZA1	1V115	115	163	1.5	500	0.2	285
V180ZA10	2V115	115	163	10.0	2000	0.45	290
V130PA10A		130	184	10.0	4000	8.0	350
V130PA20A		130	184	20.0	4000	15.0	350
V130LA1	1V130	130	184	1.0	400	0.24	360
V130LA2	1V130	130	184	2.0	400	0.24	360
V130LA10A	2V130	130	184	10.0	2000	0.5	340
V130LA20A	524V13	130	184	20.0	4000	0.85	340
V150PA10A		150	212	10.0	4000	8.0	410
V150PA20A		150	212	20.0	4000	15.0	410
V150LA1	1V150	150	212	1.0	400	0.24	420
V150LA2	1V150	150	212	2.0	400	0.24	420
V150LA10A	524V15	150	212	10.0	2000	0.5	390
V150LA20A	524V15	150	212	20.0	4000	0.85	390
V250PA10A		250	354	10.0	4000	0.85	670
V250PA20A		250	354	20.0	4000	7.0	670
V250PA40A		250	354	40.0	4000	13.0	670
V250LA2	1V250	250	354	2.0	400	0.28	690
V250LA4	1V250	250	354	4.0	400	0.28	690
V250LA15A	2V250	250	354	15.0	2000	0.6	640
V250LA20A	2V250	250	354	20.0	2000	0.6	640
V250LA40A	524V25	250	354	40.0	4000	0.9	640

††ECG and NTE numbers for these parts are identical, except for the prefix. Add the "ECG" or "NTE" prefix to the numbers shown for the complete part number.

Table 22.43
Crystal Holders

Note: Solder Seal, Cold Weld, and Resistance Weld sealing methods are commonly available. All dimensions are in inches

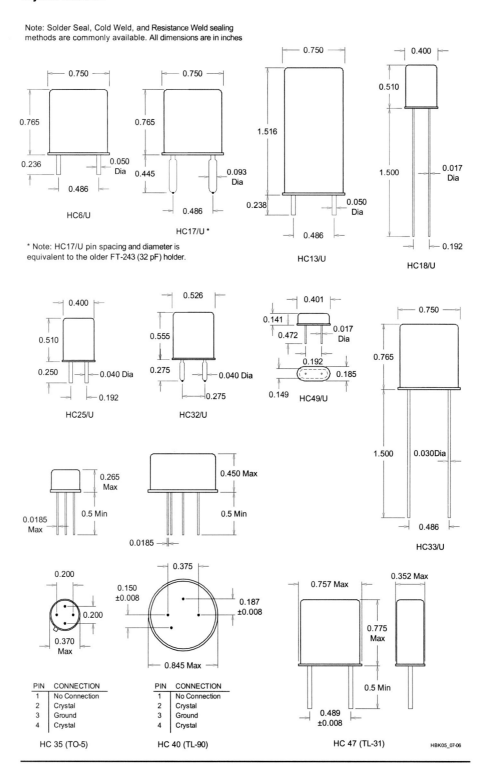

Table 22.44
Copper Wire Specifications

Bare and Enamel-Coated Wire

One mil = 0.001 inch

Wire Size (AWG)	Diam (Mils)	Area (CM[1])	Enamel Wire Coating Turns / Linear inch[2] Single	Heavy	Triple	Feet per Pound Bare	Ohms per 1000 ft 25° C	Current Carrying Capacity Continuous Duty[3] at 700 CM per Amp[4]	Open air	Conduit or bundles	Nearest British SWG No.
1	289.3	83694.49				3.948	0.1239	119.564			1
2	257.6	66357.76				4.978	0.1563	94.797			2
3	229.4	52624.36				6.277	0.1971	75.178			4
4	204.3	41738.49				7.918	0.2485	59.626			5
5	181.9	33087.61				9.98	0.3134	47.268			6
6	162.0	26244.00				12.59	0.3952	37.491			7
7	144.3	20822.49				15.87	0.4981	29.746			8
8	128.5	16512.25				20.01	0.6281	23.589			9
9	114.4	13087.36				25.24	0.7925	18.696			11
10	101.9	10383.61				31.82	0.9987	14.834			12
11	90.7	8226.49				40.16	1.2610	11.752			13
12	80.8	6528.64				50.61	1.5880	9.327			13
13	72.0	5184.00				63.73	2.0010	7.406			15
14	64.1	4108.81	15.2	14.8	14.5	80.39	2.5240	5.870	32	17	15
15	57.1	3260.41	17.0	16.6	16.2	101.32	3.1810	4.658			16
16	50.8	2580.64	19.1	18.6	18.1	128	4.0180	3.687	22	13	17
17	45.3	2052.09	21.4	20.7	20.2	161	5.0540	2.932			18
18	40.3	1624.09	23.9	23.2	22.5	203.5	6.3860	2.320	16	10	19
19	35.9	1288.81	26.8	25.9	25.1	256.4	8.0460	1.841			20
20	32.0	1024.00	29.9	28.9	27.9	322.7	10.1280	1.463	11	7.5	21
21	28.5	812.25	33.6	32.4	31.3	406.7	12.7700	1.160			22
22	25.3	640.09	37.6	36.2	34.7	516.3	16.2000	0.914		5	22
23	22.6	510.76	42.0	40.3	38.6	646.8	20.3000	0.730			24
24	20.1	404.01	46.9	45.0	42.9	817.7	25.6700	0.577			24
25	17.9	320.41	52.6	50.3	47.8	1031	32.3700	0.458			26
26	15.9	252.81	58.8	56.2	53.2	1307	41.0200	0.361			27
27	14.2	201.64	65.8	62.5	59.2	1639	51.4400	0.288			28
28	12.6	158.76	73.5	69.4	65.8	2081	65.3100	0.227			29
29	11.3	127.69	82.0	76.9	72.5	2587	81.2100	0.182			31
30	10.0	100.00	91.7	86.2	80.6	3306	103.7100	0.143			33
31	8.9	79.21	103.1	95.2		4170	130.9000	0.113			34
32	8.0	64.00	113.6	105.3		5163	162.0000	0.091			35
33	7.1	50.41	128.2	117.6		6553	205.7000	0.072			36
34	6.3	39.69	142.9	133.3		8326	261.3000	0.057			37
35	5.6	31.36	161.3	149.3		10537	330.7000	0.045			38
36	5.0	25.00	178.6	166.7		13212	414.8000	0.036			39
37	4.5	20.25	200.0	181.8		16319	512.1000	0.029			40
38	4.0	16.00	222.2	204.1		20644	648.2000	0.023			
39	3.5	12.25	256.4	232.6		26969	846.6000	0.018			
40	3.1	9.61	285.7	263.2		34364	1079.2000	0.014			
41	2.8	7.84	322.6	294.1		42123	1323.0000	0.011			
42	2.5	6.25	357.1	333.3		52854	1659.0000	0.009			
43	2.2	4.84	400.0	370.4		68259	2143.0000	0.007			
44	2.0	4.00	454.5	400.0		82645	2593.0000	0.006			
45	1.8	3.10	526.3	465.1		106600	3348.0000	0.004			
46	1.6	2.46	588.2	512.8		134000	4207.0000	0.004			

Teflon Coated, Stranded Wire

(As supplied by Belden Wire and Cable)

Size	Strands[5]	Turns per Linear inch[2] UL Style No. 1180	1213	1371
16	19×29	11.2		
18	19×30	12.7		
20	7×28	14.7	17.2	
20	19×32	14.7	17.2	
22	19×34	16.7	20.0	23.8
22	7×30	16.7	20.0	23.8
24	19×36	18.5	22.7	27.8
24	7×32		22.7	27.8
26	7×34		25.6	32.3
28	7×36		28.6	37.0
30	7×38		31.3	41.7
32	7×40			47.6

Notes

[1] A circular mil (CM) is a unit of area equal to that of a one-mil-diameter circle ($\pi/4$ square mils). The CM area of a wire is the square of the mil diameter.

[2] Figures given are approximate only; insulation thickness varies with manufacturer.

[3] Maximum wire temperature of 212°F (100°C) with a maximum ambient temperature of 135°F (57°C) as specified by the manufacturer. The *National Electrical Code* or local building codes may differ.

[4] 700 CM per ampere is a satisfactory design figure for small transformers, but values from 500 to 1000 CM are commonly used. The *National Electrical Code* or local building codes may differ.

[5] Stranded wire construction is given as "count" × "strand size" (AWG).

Table 22.45
Standard vs American Wire Gauge

SWG	Diam (in.)	Nearest AWG
12	0.104	10
14	0.08	12
16	0.064	14
18	0.048	16
20	0.036	19
22	0.028	21
24	0.022	23
26	0.018	25
28	0.0148	27
30	0.0124	28
32	0.0108	29
34	0.0092	31
36	0.0076	32
38	0.006	34
40	0.0048	36
42	0.004	38
44	0.0032	40
46	0.0024	—

Table 22.46
Antenna Wire Strength

American Wire Gauge	Recommended Tension[1] (pounds)		Weight (pounds per 1000 feet)	
	Copper-clad steel[2]	Hard-drawn copper	Copper-clad steel[2]	Hard-drawn copper
4	495	214	115.8	126
6	310	130	72.9	79.5
8	195	84	45.5	50
10	120	52	28.8	31.4
12	75	32	18.1	19.8
14	50	20	11.4	12.4
16	31	13	7.1	7.8
18	19	8	4.5	4.9
20	12	5	2.8	3.1

[1]Approximately one-tenth the breaking load. Might be increased 50% if end supports are firm and there is no danger of ice loading.
[2]"Copperweld," 40% copper.

Table 22.47
Guy Wire Lengths to Avoid

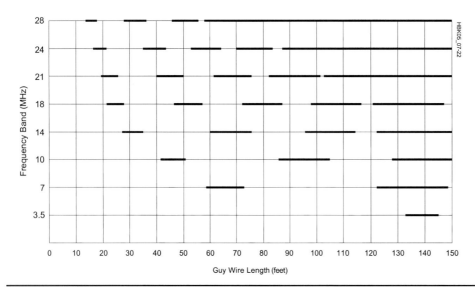

The black bars indicate ungrounded guy wire lengths to avoid for the eight HF amateur bands. This chart is based on resonance within 10% of any frequency in the band. Grounded wires will exhibit resonance at odd multiples of a quarter wavelength. *(Jerry Hall, K1TD)*

Table 22.48
Aluminum Alloy Specifications

Common Alloy Numbers

Type	Characteristic
2024	Good formability, high strength
5052	Excellent surface finish, excellent corrosion resistance, normally not heat treatable for high strength
6061	Good machinability, good weldability, can be brittle at high tempers
7075	Good formability, high strength

General Uses

Type	Uses
2024-T3	Chassis boxes, antennas, anything that will be bent or flexed repeatedly
7075-T3	
6061-T6	Mounting plates, welded assemblies or machined parts

Common Tempers

Type	Characteristics
T0	Special soft condition
T3	Hard
T6	Very hard, possibly brittle
TXXX	Three digit tempers—usually specialized high-strength heat treatments, similar to T6

Table 22.49
Impedance of Two-Conductor Twisted Pair Lines

	Twists per Inch				
Wire Size	2.5	5	7.5	10	12.5
#20	43	39	35		
#22	46	41	39	37	32
#24	60	45	44	43	41
#26	65	57	54	48	47
#28	74	53	51	49	47
#30			49	46	47

Measured in ohms at 14.0 MHz.

This illustrates the impedance of various two-conductor lines as a function of the wire size and number of twists per inch.

Table 22.50
Attenuation per Foot of Two-Conductor Twisted Pair Lines

	Twists per Inch				
Wire Size	2.5	5	7.5	10	12.5
#20	0.11	0.11	0.12		
#22	0.11	0.12	0.12	0.12	0.12
#24	0.11	0.12	0.12	0.13	0.13
#26	0.11	0.13	0.13	0.13	0.13
#28	0.11	0.13	0.13	0.16	0.16
#30			0.25	0.27	0.27

Measured in decibels at 14.0 MHz.

Attenuation in dB per foot for the same lines as shown above.

Table 22.51
Large Machine-Wound Coil Specifications

Coil Dia, Inches	Turns Per Inch	Inductance in μH Per Inch
1¼	4	2.75
	6	6.3
	8	11.2
	10	17.5
	16	42.5
1½	4	3.9
	6	8.8
	8	15.6
	10	24.5
	16	63
1¾	4	5.2
	6	11.8
	8	21
	10	33
	16	85
2	4	6.6
	6	15
	8	26.5
	10	42
	16	108
2½	4	10.2
	6	23
	8	41
	10	64
3	4	14
	6	31.5
	8	56
	10	89

Table 22.53
Small Machine-Wound Coil Specifications

Coil Dia, Inches	Turns Per Inch	Inductance in μH Per Inch
½ (A)	4	0.18
	6	0.40
	8	0.72
	10	1.12
	16	2.8
	32	12
⅝ (A)	4	0.28
	6	0.62
	8	1.1
	10	1.7
	16	4.4
	32	18
¾ (B)	4	0.6
	6	1.35
	8	2.4
	10	3.8
	16	9.9
	32	40
1 (B)	4	1.0
	6	2.3
	8	4.2
	10	6.6
	16	16.9
	32	68

Table 22.52
Inductance Factor for Large Machine-Wound Coils

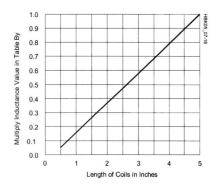

Factor to be applied to the inductance of large coils for coil lengths up to 5 inches.

Table 22.54
Inductance Factor for Small Machine-Wound Coils

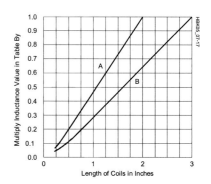

Factor to be applied to the inductance of small coils as a function of coil length. Use curve A for coils marked A, and curve B for coils marked B.

Table 22.55
Measured Inductance for #12 AWG Wire Windings

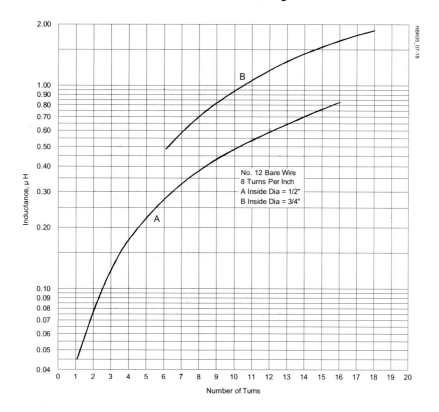

Values are for inductors with half-inch leads and wound with eight turns per inch.

Table 22.56
Relationship Between Noise Figure and Noise Temperature

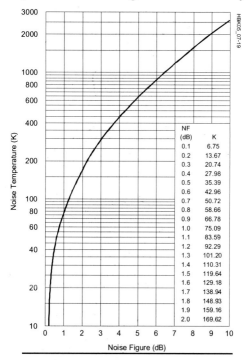

NF (dB)	K
0.1	6.75
0.2	13.67
0.3	20.74
0.4	27.98
0.5	35.39
0.6	42.96
0.7	50.72
0.8	58.66
0.9	66.78
1.0	75.09
1.1	83.59
1.2	92.29
1.3	101.20
1.4	110.31
1.5	119.64
1.6	129.18
1.7	138.94
1.8	148.93
1.9	159.16
2.0	169.62

Component Data and References

Table 22.57
Pi-Network Resistive Attenuators (50 Ω)

dB Atten.	R1 (Ohms)	R2 (Ohms)
1.0	870	5.77
2.0	436	11.6
3.0	292	17.6
4.0	221	23.8
5.0	178	30.4
6.0	150	37.4
7.0	131	44.8
8.0	116	52.8
9.0	105	61.6
10.0	96.2	71.2
11.0	89.2	81.7
12.0	83.5	93.2
13.0	78.8	106
14.0	74.9	120
15.0	71.6	136
16.0	68.8	154
17.0	66.4	173
18.0	64.4	195
19.0	62.6	220
20.0	61.1	248
21.0	59.8	278
22.0	58.6	313
23.0	57.6	352
24.0	56.7	395
25.0	56.0	443
30.0	53.2	790
35.0	51.8	1405
40.0	51.0	2500
45.0	50.5	4446
50.0	50.3	7906
55.0	50.2	14,058
60.0	50.1	25,000

Note: A PC board kit for the Low-Power Step Attenuator (Sep 1982 *QST*) is available from FAR Circuits. Project details are in the *Handbook* **template package STEP ATTENUATOR.**

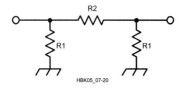

Table 22.58
T-Network Resistive Attenuators (50 Ω)

dB Atten.	R1 (Ohms)	R2 (Ohms)
1.0	2.88	433
2.0	5.73	215
3.0	8.55	142
4.0	11.3	105
5.0	14.0	82.2
6.0	16.6	66.9
7.0	19.1	55.8
8.0	21.5	47.3
9.0	23.8	40.6
10.0	26.0	35.1
11.0	28.0	30.6
12.0	30.0	26.8
13.0	31.7	23.5
14.0	33.3	20.8
15.0	35.0	18.4
16.0	36.3	16.2
17.0	37.6	14.4
18.0	38.8	12.8
19.0	40.0	11.4
20.0	41.0	10.0
21.0	41.8	9.0
22.0	42.6	8.0
23.0	43.4	7.1
24.0	44.0	6.3
25.0	44.7	5.6
30.0	47.0	3.2
35.0	48.2	1.8
40.0	49.0	1.0
45.0	49.4	0.56
50.0	49.7	0.32
55.0	49.8	0.18
60.0	49.9	0.10

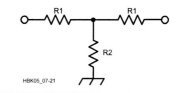

22.8 Computer Connectors

Most connections between computers and their peripherals are made with some form of multi-conductor cable. Examples include shielded, unshielded and ribbon cable. Common connectors used are the 9- and 25-pin D-Subminiature connector; the USB type-A, type-B and mini-USB connectors; the DIN and Miniature DIN connectors; and the 36-pin "Centronics" connector found on printers. **Table 22.59** shows a variety of computer connectors and pin outs, including some used for internal connections, such as power supplies and disk drives.

Table 22.59
ComputerConnector Pinouts

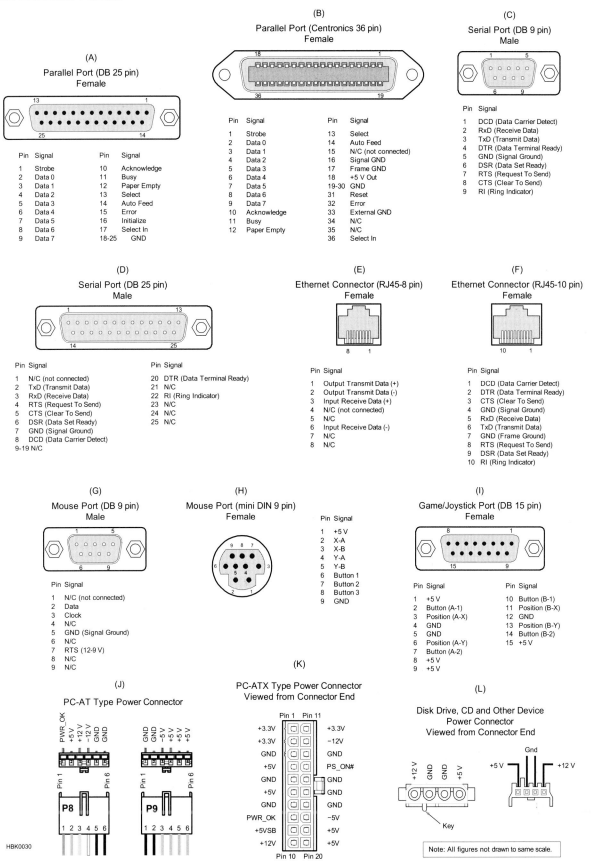

Note: All figures not drawn to same scale.

22.9 RF Connectors and Transmission Lines

There are many different types of transmission lines and RF connectors for coaxial cable, but the three most common for amateur use are the UHF, Type N and BNC families. The type of connector used for a specific job depends on the size of the cable, the frequency of operation and the power levels involved. **Table 22.60** shows the characteristics of many popular transmission lines, while **Table 22.61** details coax connectors.

22.9.1 UHF Connectors

The so-called UHF connector (the series name is not related to frequency) is found on most HF and some VHF equipment. It is the only connector many hams will ever see on coaxial cable. PL-259 is another name for the UHF male, and the female is also known as the SO-239. These connectors are rated for full legal amateur power at HF. They are poor for UHF work because they do not present a constant impedance, so the UHF label is a misnomer. PL-259 connectors are designed to fit RG-8 and RG-11 size cable (0.405-inch OD). Adapters are available for use with smaller RG-58, RG-59 and RG-8X size cable. UHF connectors are not weatherproof.

Fig 22.19 shows how to install the solder type of PL-259 on RG-8 cable. Proper preparation of the cable end is the key to success. Follow these simple steps. Measure back about ¾-inch from the cable end and slightly score the outer jacket around its circumference. With a sharp knife, cut through the outer jacket, through the braid and through the dielectric — almost to the center conductor. Be careful not to score the center conductor. Cutting through all outer layers at once keeps the braid from separating. (Using a coax stripping tool with preset blade depth makes this and subsequent trimming steps much easier.)

Pull the severed outer jacket, braid and dielectric off the end of the cable as one piece. Inspect the area around the cut, looking for any strands of braid hanging loose and snip them off. There won't be any if your knife was sharp enough. Next, score the outer jacket about ⁵⁄₁₆-inch back from the first cut. Cut through the jacket lightly; do not score the braid. This step takes practice. If you score the braid, start again. Remove the outer jacket.

Tin the exposed braid and center conductor, but apply the solder sparingly and avoid melting the dielectric. Slide the coupling ring onto the cable. Screw the connector body onto the cable. If you prepared the cable to the right dimensions, the center conductor will protrude through the center pin, the braid will show through the solder holes, and the body will actually thread onto the outer cable jacket. A very small amount of lubricant on the cable jacket will help the threading process.

Solder the braid through the solder holes. Solder through all four holes; poor connection

Fig 22.20 — Installing PL-259 plugs on RG-58 or RG-59 cable requires the use of UG-175 or UG-176 adapters, respectively. The adapter screws into the plug body using the threads of the connector that grip the jacket on larger cables. (*Courtesy Amphenol Electronic Components*)

83-1SP (PL-259) Plug with adapters (UG-176/U OR UG-175/U)

1. Cut end of cable even. Remove vinyl jacket 3/4" - don't nick braid. Slide coupling ring and adapter on cable.

2. Fan braid slightly and fold back over cable.

3. Position adapter to dimension shown. Press braid down over body of adapter and trim to 3/8". Bare 5/8" of conductor. Tin exposed center conductor.

4. Screw the plug assembly on adapter. Solder braid to shell through solder holes. Solder conductor to contact sleeve.

5. Screw coupling ring on plug assembly.

Fig 22.19 — The PL-259, or UHF, connector is almost universal for amateur HF work and is popular for equipment operating in the VHF range. Steps A through E are described in detail in the text.

to the braid is the most common form of PL-259 failure. A good connection between connector and braid is just as important as that between the center conductor and connector. Use a large soldering iron for this job. With practice, you'll learn how much heat to use. If you use too little heat, the solder will bead up, not really flowing onto the connector body. If you use too much heat, the dielectric will melt, letting the braid and center conductor touch. Most PL-259s are nickel plated, but silver-plated connectors are much easier to solder and only slightly more expensive.

Solder the center conductor to the center pin. The solder should flow on the inside, not the outside, of the center pin. If you wait until the connector body cools off from soldering the braid, you'll have less trouble with the dielectric melting. Trim the center conductor to be even with the end of the center pin. Use a small file to round the end, removing any solder that built up on the outer surface of the center pin. Use a sharp knife, very fine sandpaper or steel wool to remove any solder flux from the outer surface of the center pin. Screw the coupling ring onto the body, and you're finished.

Fig 22.20 shows how to install a PL-259 connector on RG-58 or RG-59 cable. An adapter is used for the smaller cable with standard RG-8 size PL-259s. (UG-175 for RG-58 and UG-176 for RG-59.) Prepare the cable as shown. Once the braid is prepared, screw the adapter into the PL-259 shell and finish the job as you would a PL-259 on RG-8 cable.

Fig 22.21 shows the instructions and dimensions for crimp-on UHF connectors that fit all common sizes of coaxial cable. While amateurs have been reluctant to adopt crimp-on connectors, the availability of good quality connectors and inexpensive crimping tools make crimp technology a good choice, even for connectors used outside. Soldering the center conductor to the connector tip is optional.

UHF connectors are not waterproof and must be waterproofed whether soldered or crimped as shown in the section of the **Safety** chapter on Antenna and Tower Safety.

22.9.2 BNC, N and F Connectors

The BNC connectors illustrated in **Fig 22.22** are popular for low power levels at VHF and UHF. They accept RG-58 and RG-59 cable, and are available for cable mounting in both male and female versions. Several different styles are available, so be sure to use the dimensions for the type you have. Follow the installation instructions carefully. If you prepare the cable to the wrong dimensions, the center pin will not seat properly with connectors of the opposite gender. Sharp scissors are a big help for trimming the braid evenly. Crimp-on BNC connectors are also available, with a large number of

(Text continues on page 22.51)

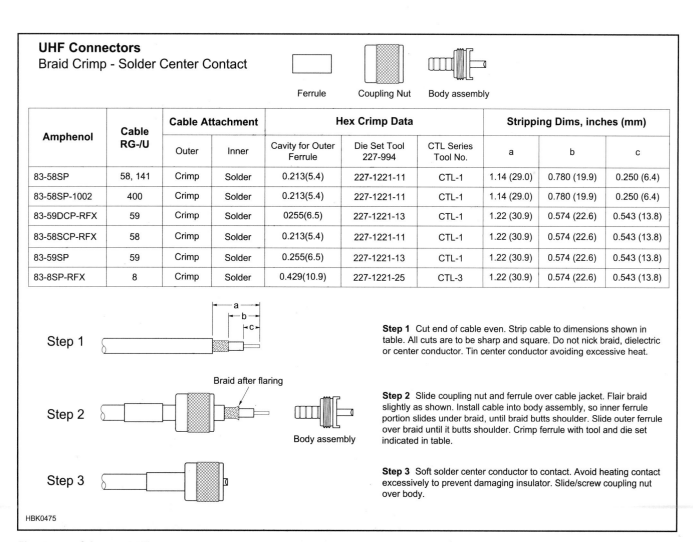

Fig 22.21 — Crimp-on UHF connectors are available for all sizes of popular coaxial cable and save considerable time over soldered connectors. The performance and reliability of these connectors is equivalent to soldered connectors, if crimped properly. (*Courtesy Amphenol Electronic Components*)

Table 22.60
Nominal Characteristics of Commonly Used Transmission Lines

RG or Type	*Part Number*	*Nom. Z_0 Ω*	*VF %*	*Cap. pF/ft*	*Cent. Cond. AWG*	*Diel. Type*	*Shield Type*	*Jacket Matl*	*OD inches*	*Max V (RMS)*	*Matched Loss (dB/100') 1 MHz*	*10*	*100*	*1000*
RG-6	Belden 1694A	75	82	16.2	#18 Solid BC	FPE	FC	P1	0.275	300	0.3	.7	1.8	5.9
RG-6	Belden 8215	75	66	20.5	#21 Solid CCS	PE	D	PE	0.332	2700	0.4	0.8	2.7	9.8
RG-8	Belden 7810A	50	86	23.0	#10 Solid BC	FPE	FC	PE	0.405	300	0.1	0.4	1.2	4.0
RG-8	TMS LMR400	50	85	23.9	#10 Solid CCA	FPE	FC	PE	0.405	600	0.1	0.4	1.3	4.1
RG-8	Belden 9913	50	84	24.6	#10 Solid BC	ASPE	FC	P1	0.405	300	0.1	0.4	1.3	4.5
RG-8	CXP1318FX	50	84	24.0	#10 Flex BC	FPE	FC	P2N	0.405	600	0.1	0.4	1.3	4.5
RG-8	Belden 9913F	50	83	24.6	#11 Flex BC	FPE	FC	P1	0.405	300	0.2	0.6	1.5	4.8
RG-8	Belden 9914	50	82	24.8	#10 Solid BC	FPE	FC	P1	0.405	300	0.2	0.5	1.5	4.8
RG-8	TMS LMR400UF	50	85	23.9	#10 Flex BC	FPE	FC	PE	0.405	600	0.1	0.4	1.4	4.9
RG-8	DRF-BF	50	84	24.5	#9.5 Flex BC	FPE	FC	PE	0.405	600	0.1	0.5	1.6	5.2
RG-8	WM CQ106	50	84	24.5	#9.5 Flex BC	FPE	FC	P2N	0.405	600	0.2	0.6	1.8	5.3
RG-8	CXP008	50	78	26.0	#13 Flex BC	FPE	S	P1	0.405	600	0.1	0.5	1.8	7.1
RG-8	Belden 8237	52	66	29.5	#13 Flex BC	PE	S	P1	0.405	3700	0.2	0.6	1.9	7.4
RG-8X	Belden 7808A	50	86	23.5	#15 Solid BC	FPE	FC	PE	0.240	300	0.2	0.7	2.3	7.4
RG-8X	TMS LMR240	50	84	24.2	#15 Solid BC	FPE	FC	PE	0.242	300	0.2	0.8	2.5	8.0
RG-8X	WM CQ118	50	82	25.0	#16 Flex BC	FPE	FC	P2N	0.242	300	0.3	0.9	2.8	8.4
RG-8X	TMS LMR240UF	50	84	24.2	#15 Flex BC	FPE	FC	PE	0.242	300	0.2	0.8	2.8	9.6
RG-8X	Belden 9258	50	82	24.8	#16 Flex BC	FPE	S	P1	0.242	300	0.3	0.9	3.2	11.2
RG-8X	CXP08XB	50	80	25.3	#16 Flex BC	FPE	S	P1	0.242	300	0.3	1.0	3.1	14.0
RG-9	Belden 8242	51	66	30.0	#13 Flex SPC	PE	SCBC	P2N	0.420	5000	0.2	0.6	2.1	8.2
RG-11	Belden 8213	75	84	16.1	#14 Solid BC	FPE	S	PE	0.405	300	0.1	0.4	1.3	5.2
RG-11	Belden 8238	75	66	20.5	#18 Flex TC	PE	S	P1	0.405	300	0.2	0.7	2.0	7.1
RG-58	Belden 7807A	50	85	23.7	#18 Solid BC	FPE	FC	PE	0.195	300	0.3	1.0	3.0	9.7
RG-58	TMS LMR200	50	83	24.5	#17 Solid BC	FPE	FC	PE	0.195	300	0.3	1.0	3.2	10.5
RG-58	WM CQ124	52	66	28.5	#20 Solid BC	PE	S	PE	0.195	1400	0.4	1.3	4.3	14.3
RG-58	Belden 8240	52	66	29.9	#20 Solid BC	PE	S	P1	0.193	1400	0.3	1.1	3.8	14.5
RG-58A	Belden 8219	53	73	26.5	#20 Flex TC	FPE	S	P1	0.195	300	0.4	1.3	4.5	18.1
RG-58C	Belden 8262	50	66	30.8	#20 Flex TC	PE	S	P2N	0.195	1400	0.4	1.4	4.9	21.5
RG-58A	Belden 8259	50	66	30.8	#20 Flex TC	PE	S	P1	0.192	1400	0.5	1.5	5.4	22.8
RG-59	Belden 1426A	75	83	16.3	#20 Solid BC	FPE	S	P1	0.242	300	0.3	0.9	2.6	8.5
RG-59	CXP 0815	75	82	16.2	#20 Solid BC	FPE	S	P1	0.232	300	0.5	0.9	2.2	9.1
RG-59	Belden 8212	75	78	17.3	#20 Solid CCS	FPE	S	P1	0.242	300	0.2	1.0	3.0	10.9
RG-59	Belden 8241	75	66	20.4	#23 Solid CCS	PE	S	P1	0.242	1700	0.6	1.1	3.4	12.0
RG-62A	Belden 9269	93	84	13.5	#22 Solid CCS	ASPE	S	P1	0.240	750	0.3	0.9	2.7	8.7
RG-62B	Belden 8255	93	84	13.5	#24 Flex CCS	ASPE	S	P2N	0.242	750	0.3	0.9	2.9	11.0
RG-63B	Belden 9857	125	84	9.7	#22 Solid CCS	ASPE	S	P2N	0.405	750	0.2	0.5	1.5	5.8
RG-142	CXP 183242	50	69.5	29.4	#19 Solid SCCS	TFE	D	FEP	0.195	1900	0.3	1.1	3.8	12.8
RG-142B	Belden 83242	50	69.5	29.0	#19 Solid SCCS	TFE	D	TFE	0.195	1400	0.3	1.1	3.9	13.5
RG-174	Belden 7805R	50	73.5	26.2	#25 Solid BC	FPE	FC	P1	0.110	300	0.6	2.0	6.5	21.3
RG-174	Belden 8216	50	66	30.8	#26 Flex CCS	PE	S	P1	0.110	1100	0.8	2.5	8.6	33.7
RG-213	Belden 8267	50	66	30.8	#13 Flex BC	PE	S	P2N	0.405	3700	0.2	0.6	2.1	8.0
RG-213	CXP213	50	66	30.8	#13 Flex BC	PE	S	P2N	0.405	600	0.2	0.6	2.0	8.2
RG-214	Belden 8268	50	66	30.8	#13 Flex SPC	PE	D	P2N	0.425	3700	0.2	0.7	2.2	8.0
RG-216	Belden 9850	75	66	20.5	#18 Flex TC	PE	D	P2N	0.425	3700	0.2	0.7	2.0	7.1
RG-217	WM CQ217F	50	66	30.8	#10 Flex BC	PE	D	PE	0.545	7000	0.1	0.4	1.4	5.2
RG-217	M17/78-RG217	50	66	30.8	#10 Solid BC	PE	D	P2N	0.545	7000	0.1	0.4	1.4	5.2
RG-218	M17/79-RG218	50	66	29.5	#4.5 Solid BC	PE	S	P2N	0.870	11000	0.1	0.2	0.8	3.4
RG-223	Belden 9273	50	66	30.8	#19 Solid SPC	PE	D	P2N	0.212	1400	0.4	1.2	4.1	14.5
RG-303	Belden 84303	50	69.5	29.0	#18 Solid SCCS	TFE	S	TFE	0.170	1400	0.3	1.1	3.9	13.5
RG-316	CXP TJ1316	50	69.5	29.4	#26 Flex BC	TFE	S	FEP	0.098	1200	1.2	2.7	8.0	26.1
RG-316	Belden 84316	50	69.5	29.0	#26 Flex SCCS	TFE	S	FEP	0.096	900	0.8	2.5	8.3	26.0
RG-393	M17/127-RG393	50	69.5	29.4	#12 Flex SPC	TFE	D	FEP	0.390	5000	0.2	0.5	1.7	6.1
RG-400	M17/128-RG400	50	69.5	29.4	#20 Flex SPC	TFE	D	FEP	0.195	1400	0.4	1.3	4.3	15.0
LMR500	TMS LMR500UF	50	85	23.9	#7 Flex BC	FPE	FC	PE	0.500	2500	0.1	0.4	1.2	4.0
LMR500	TMS LMR500	50	85	23.9	#7 Solid CCA	FPE	FC	PE	0.500	2500	0.1	0.3	0.9	3.3
LMR600	TMS LMR600	50	86	23.4	#5.5 Solid CCA	FPE	FC	PE	0.590	4000	0.1	0.2	0.8	2.7
LMR600	TMS LMR600UF	50	86	23.4	#5.5 Flex BC	FPE	FC	PE	0.590	4000	0.1	0.2	0.8	2.7
LMR1200	TMS LMR1200	50	88	23.1	#0 Copper Tube	FPE	FC	PE	1.200	4500	0.04	0.1	0.4	1.3
Hardline														
1/2"	CATV Hardline	50	81	25.0	#5.5 BC	FPE	SM	none	0.500	2500	0.05	0.2	0.8	3.2
1/2"	CATV Hardline	75	81	16.7	#11.5 BC	FPE	SM	none	0.500	2500	0.1	0.2	0.8	3.2
7/8"	CATV Hardline	50	81	25.0	#1 BC	FPE	SM	none	0.875	4000	0.03	0.1	0.6	2.9
7/8"	CATV Hardline	75	81	16.7	#5.5 BC	FPE	SM	none	0.875	4000	0.03	0.1	0.6	2.9
LDF4-50A	Heliax – ½"	50	88	25.9	#5 Solid BC	FPE	CC	PE	0.630	1400	0.02	0.2	0.6	2.4
LDF5-50A	Heliax – ⅞"	50	88	25.9	0.355" BC	FPE	CC	PE	1.090	2100	0.03	0.10	0.4	1.3
LDF6-50A	Heliax – 1¼"	50	88	25.9	0.516" BC	FPE	CC	PE	1.550	3200	0.02	0.08	0.3	1.1
Parallel Lines														
TV Twinlead (Belden 9085)		300	80	4.5	#22 Flex CCS	PE	none	P1	0.400	**	0.1	0.3	1.4	5.9
Twinlead (Belden 8225)		300	80	4.4	#20 Flex BC	PE	none	P1	0.400	8000	0.1	0.2	1.1	4.8
Generic Window Line		450	91	2.5	#18 Solid CCS	PE	none	P1	1.000	10000	0.02	0.08	0.3	1.1
WM CQ 554		440	91	2.7	#14 Flex CCS	PE	none	P1	1.000	10000	0.04	0.01	0.6	3.0
WM CQ 552		440	91	2.5	#16 Flex CCS	PE	none	P1	1.000	10000	0.05	0.2	0.6	2.6
WM CQ 553		450	91	2.5	#18 Flex CCS	PE	none	P1	1.000	10000	0.06	0.2	0.7	2.9
WM CQ 551		450	91	2.5	#18 Solid CCS	PE	none	P1	1.000	10000	0.05	0.02	0.6	2.8
Open-Wire Line		600	0.95-99***	1.7	#12 BC	none	none	none	**	12000	0.02	0.06	0.2	—

Approximate Power Handling Capability (1:1 SWR, 40°C Ambient):

	1.8 MHz	7	14	30	50	150	220	450	1 GHz
RG-58 Style	1350	700	500	350	250	150	120	100	50
RG-59 Style	2300	1100	800	550	400	250	200	130	90
RG-8X Style	1830	840	560	360	270	145	115	80	50
RG-8/213 Style	5900	3000	2000	1500	1000	600	500	350	250
RG-217 Style	20000	9200	6100	3900	2900	1500	1200	800	500
LDF4-50A	38000	18000	13000	8200	6200	3400	2800	1900	1200
LDF5-50A	67000	32000	22000	14000	11000	5900	4800	3200	2100
LMR500	18000	9200	6500	4400	3400	1900	1600	1100	700
LMR1200	52000	26000	19000	13000	10000	5500	4500	3000	2000

Legend:

**	Not Available or varies	N	Non-Contaminating
***	Varies with spacer material and spacing	P1	PVC, Class 1
ASPE	Air Spaced Polyethylene	P2	PVC, Class 2
BC	Bare Copper	PE	Polyethylene
CC	Corrugated Copper	S	Single Braided Shield
CCA	Copper Cover Aluminum	SC	Silver Coated Braid
CCS	Copper Covered Steel	SCCS	Silver Plated Copper Coated Steel
CXP	Cable X-Perts, Inc.	SM	Smooth Aluminum
D	Double Copper Braids	SPC	Silver Plated Copper
DRF	Davis RF	TC	Tinned Copper
FC	Foil + Tinned Copper Braid	TFE	Teflon®
FEP	Teflon® Type IX	TMS	Times Microwave Systems
Flex	Flexible Stranded Wire	UF	Ultra Flex
FPE	Foamed Polyethylene	WM	Wireman
Heliax	Andrew Corp Heliax		

Fig 22.22 — BNC connectors are common on VHF and UHF equipment at low power levels. (*Courtesy Amphenol Electronic Components*)

BNC CONNECTORS

Standard Clamp

1. Cut cable even. Strip jacket. Fray braid and strip dielectric. **Don't nick braid or center conductor.** Tin center conductor.

2. Taper braid. Slide nut, washer, gasket and clamp over braid. Clamp inner shoulder should fit squarely against end of jacket.

3. With clamp in place, comb out braid, fold back smooth as shown. Trim center conductor.

4. Solder contact on conductor through solder hole. Contact should butt against dielectric. Remove excess solder from outside of contact. Avoid excess heat to prevent swollen dielectric which would interfere with connector body.

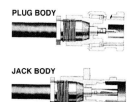

5. Push assembly into body. Screw nut into body with wrench until tight. **Don't rotate body on cable to tighten.**

Improved Clamp

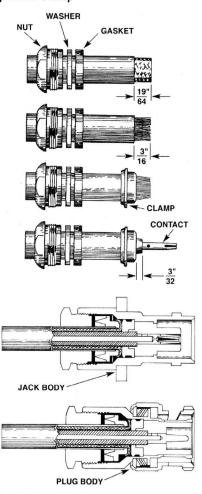

Follow 1, 2, 3 and 4 in BNC connectors (standard clamp) except as noted. Strip cable as shown. Slide gasket on cable *with groove facing clamp*. Slide clamp *with sharp edge facing gasket*. Clamp *should* cut gasket to seal properly.

C. C. Clamp

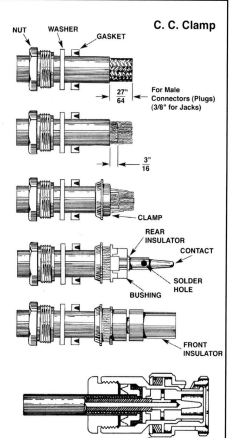

1. Follow steps 1, 2, and 3 as outlined for the standard-clamp BNC connector.

2. Slide on bushing, rear insulator and contact. The parts must butt securely against each other, as shown.

3. Solder the center conductor to the contact. Remove flux and excess solder.

4. Slide the front insulator over the contact, making sure it butts against the contact shoulder.

5. Insert the prepared cable end into the connector body and tighten the nut. Make sure the sharp edge of the clamp seats properly in the gasket.

Table 22.61
Coaxial Cable Connectors

UHF Connectors

Military No.	Style	Cable RG- or Description
PL-259	Str (m)	8, 9, 11, 13, 63, 87, 149, 213, 214, 216, 225
UG-111	Str (m)	59, 62, 71, 140, 210
SO-239	Pnl (f)	Std, mica/phenolic insulation
UG-266	Blkhd (f)	Rear mount, pressurized, copolymer of styrene ins.

Adapters

PL-258	Str (f/f)	Polystyrene ins.
UG-224,363	Blkhd (f/f)	Polystyrene ins.
UG-646	Ang (f/m)	Polystyrene ins.
M-359A	Ang (m/f)	Polystyrene ins.
M-358	T (f/m/f)	Polystyrene ins.

Reducers

UG-175	55, 58, 141, 142 (except 55A)
UG-176	59, 62, 71, 140, 210

Family Characteristics:

All are nonweatherproof and have a nonconstant impedance. Frequency range: 0-500 MHz. Maximum voltage rating: 500 V (peak).

N Connectors

Military No.	Style	Cable RG-	Notes
UG-21	Str (m)	8, 9, 213, 214	50 Ω
UG-94A	Str (m)	11, 13, 149, 216	70 Ω
UG-536	Str (m)	58, 141, 142	50 Ω
UG-603	Str (m)	59, 62, 71, 140, 210	50 Ω
UG-23, B-E	Str (f)	8, 9, 87, 213, 214, 225	50 Ω
UG-602	Str (f)	59, 62, 71, 140, 210	—
UG-228B, D, E	Pnl (f)	8, 9, 87, 213, 214, 225	—
UG-1052	Pnl (f)	58, 141, 142	50 Ω
UG-593	Pnl (f)	59, 62, 71, 140, 210	50 Ω
UG-160A, B, D	Blkhd (f)	8, 9, 87, 213, 214, 225	50 Ω
UG-556	Blkhd (f)	58, 141, 142	50 Ω
UG-58, A	Pnl (f)		50 Ω
UG-997A	Ang (f)		50 Ω

Panel mount (f) with clearance above panel

M39012/04-	Blkhd (f)	Front mount hermetically sealed
UG-680	Blkhd (f)	Front mount pressurized

N Adapters

Military No.	Style	Notes
UG-29,A,B	Str (f/f)	50 Ω, TFE ins.
UG-57A.B	Str (m/m)	50 Ω, TFE ins.
UG-27A,B	Ang (f/m)	Mitre body
UG-212A	Ang (f/m)	Mitre body
UG-107A	T (f/f/f)	—
UG-28A	T (f/f/f)	—
UG-107B	T (f/m/f)	—

Family Characteristics:

N connectors with gaskets are weatherproof. RF leakage: –90 dB min @ 3 GHz. Temperature limits: TFE: –67° to 390°F (–55° to 199°C). Insertion loss 0.15 dB max @ 10 GHz. Copolymer of styrene: –67° to 185°F (–55° to 85°C). Frequency range: 0-11 GHz. Maximum voltage rating: 1500 V P-P. Dielectric withstanding voltage 2500 V RMS. SWR (MIL-C-39012 cable connectors) 1.3 max 0-11 GHz.

BNC Connectors

Military No.	Style	Cable RG-	Notes
UG-88C	Str (m)	55, 58, 141, 142, 223, 400	

Military No.	Style	Cable RG-	Notes
UG-959	Str (m)	8, 9	
UG-260,A	Str (m)	59, 62, 71, 140, 210	Rexolite ins.
UG-262	Pnl (f)	59, 62, 71, 140, 210	Rexolite ins.
UG-262A	Pnl (f)	59, 62, 71, 140, 210	nwx, Rexolite ins.
UG-291	Pnl (f)	55, 58, 141, 142, 223, 400	
UG-291A	Pnl (f)	55, 58, 141, 142, 223, 400	nwx
UG-624	Blkhd (f)	59, 62, 71, 140, 210	Front mount Rexolite ins.
UG-1094A	Blkhd		Standard
UG-625B	Receptacle		
UG-625			

BNC Adapters

Military No.	Style	Notes
UG-491,A	Str (m/m)	
UG-491B	Str (m/m)	Berylium, outer contact
UG-914	Str (f/f)	
UG-306	Ang (f/m)	
UG-306A,B	Ang (f/m)	Berylium outer contact
UG-414,A	Pnl (f/f)	# 3-56 tapped flange holes
UG-306	Ang (f/m)	
UG-306A,B	Ang (f/m)	Berylium outer contact
UG-274	T (f/m/f)	
UG-274A,B	T (f/m/f)	Berylium outer contact

Family Characteristics:

Z = 50 Ω. Frequency range: 0-4 GHz w/low reflection; usable to 11 GHz. Voltage rating: 500 V P-P. Dielectric withstanding voltage 500 V RMS. SWR: 1.3 max 0-4 GHz. RF leakage –55 dB min @ 3 GHz. Insertion loss 0.2 dB max @ 3 GHz. Temperature limits: TFE: –67° to 390°F (–55° to 199°C); Rexolite insulators: –67° to 185°F (–55° to 85°C). "Nwx" = not weatherproof.

HN Connectors

Military No.	Style	Cable RG-	Notes
UG-59A	Str (m)	8, 9, 213, 214	
UG-1214	Str (f)	8, 9, 87, 213, 214, 225	Captivated contact
UG-60A	Str (f)	8, 9, 213, 214	Copolymer of styrene ins.
UG-1215	Pnl (f)	8, 9, 87, 213, 214, 225	Captivated contact
UG-560	Pnl (f)		
UG-496	Pnl (f)		
UG-212C	Ang (f/m)		Berylium outer contact

Family Characteristics:

Connector Styles: Str = straight; Pnl = panel; Ang = Angle; Blkhd = bulkhead. Z = 50 Ω. Frequency range = 0-4 GHz. Maximum voltage rating = 1500 V P-P. Dielectric withstanding voltage = 5000 V RMS SWR = 1.3. All HN series are weatherproof. Temperature limits: TFE: –67° to 390°F (–55° to 199°C); copolymer of styrene: –67° to 185°F (–55° to 85°C).

Cross-Family Adapters

Families	Description	Military No.
HN to BNC	HN-m/BNC-f	UG-309
N to BNC	N-m/BNC-f	UG-201,A
	N-f/BNC-m	UG-349,A
	N-m/BNC-m	UG-1034
N to UHF	N-m/UHF-f	UG-146
	N-f/UHF-m	UG-83,B
	N-m/UHF-m	UG-318
UHF to BNC	UHF-m/BNC-f	UG-273
	UHF-f/BNC-m	UG-255

Type N assembly instructions

HBK05_19-19

CLAMP TYPES

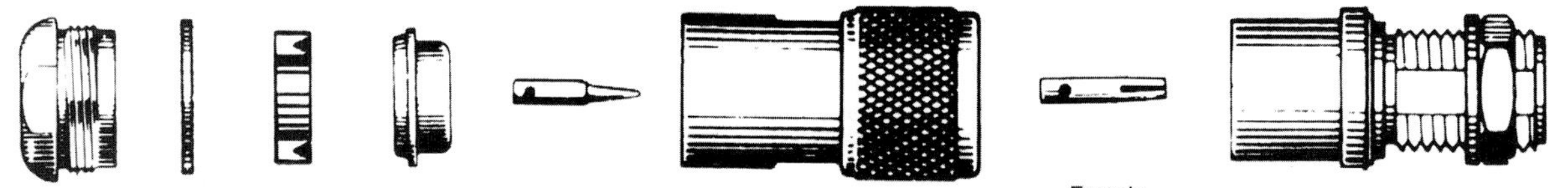

Nut Washer Gasket Clamp Male Contact Plug Body Female Contact Jack Body

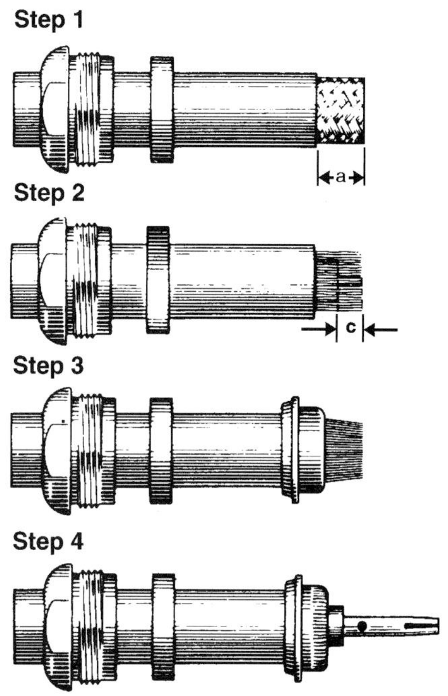

Step 1

Step 2

Step 3

Step 4

Step 5

Amphenol Number	Connector Type	Cable RG-/U	Strip Dims., inches (mm)	
			a	c
82-61	N Plug	8, 9, 144, 165, 213, 214, 216, 225	0.359(9.1)	0.234(6.0)
82-62	N Panel Jack		0.312(7.9)	0.187(4.7)
82-63	N Jack	8, 9, 87A, 144, 165, 213, 214, 216, 225	0.281(7.1)	0.156(4.0)
82-67	N Bulkhead Jack			
82-202	N Plug	8, 9, 144, 165, 213, 214, 216, 225	0.359(9.1)	0.234(6.0)
82-202-1006	N Plug	Belden 9913	0.359(9.1)	0.234(6.0)
82-835	N Angle Plug	8, 9, 87A, 144, 165, 213, 214, 216, 225	0.281(7.1)	0.156(4.0)
18750	N Angle Plug		0.484(12.3)	0.234(5.9)
34025	N Plug	58, 141, 142	0.390(9.9)	0.203(5.2)
34525	N Plug	59, 62, 71, 140, 210	0.410(10.4)	0.230(5.8)
35025	N Jack	58, 141, 142	0.375(9.5)	0.187(4.7)
36500	N Jack	59, 62, 71, 140, 210	0.484(12.3)	0.200(5.1)

Step 1 Place nut and gasket, with "V" groove toward clamp, over cable and cut off jacket to dim. a.

Step 2 Comb out braid and fold out. Cut off cable dielectric to dim. c as shown.

Step 3 Pull braid wires forward and taper toward center conductor. Place clamp over braid and push back against cable jacket.

Step 4 Fold back braid wires as shown, trim braid to proper length and form over clamp as shown. Solder contact to center conductor.

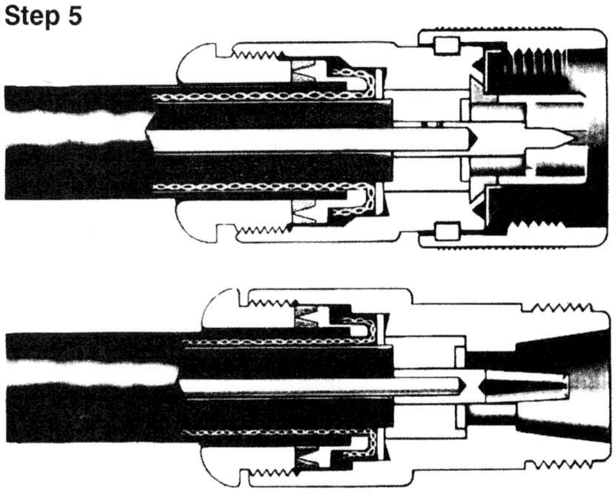

Step 5 Insert cable and parts into connector body. Make sure sharp edge of clamp seats properly in gasket. Tighten nut.

Fig 22.23 — Type N connectors are a must for high-power VHF and UHF operation. (*Courtesy Amphenol Electronic Components*)

(Continued from page 22.47)

variations, including a twist-on version. A guide to installing these connectors is available on the CD-ROM accompanying this book.

The Type N connector, illustrated in **Fig 22.23**, is a must for high-power VHF and UHF operation. N connectors are available in male and female versions for cable mounting and are designed for RG-8 size cable. Unlike UHF connectors, they are designed to maintain a constant impedance at cable joints. Like BNC connectors, it is important to prepare the cable to the right dimensions. The center pin must be positioned correctly to mate with the center pin of connectors of the opposite gender. Use the right dimensions for the connector style you have. Crimp-on N connectors are also available, again with a large number of variations. A guide to installing these connectors is available on the CD-ROM accompanying this book.

Type F connectors, used primarily on cable TV connections, are also popular for receive-

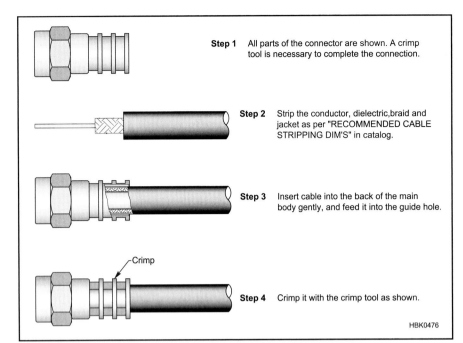

Fig 22.24 — Type F connectors, commonly used for cable TV connections, can be used for receive-only antennas with inexpensive RG-59 and RG-6 cable. (*Courtesy Amphenol Electronic Components*)

only antennas and can be used with RG-59 or the increasingly popular RG-6 cable available at low cost. Crimp-on connectors are the only option for these connectors and **Fig 22.24** shows a general guide for installing them. The exact dimensions vary between connector styles and manufacturers — information on crimping is generally provided with the connectors. There are two styles of crimp; ferrule and compression. The ferrule crimp method is similar to that for UHF, BNC, and N connectors in which a metal ring is compressed around the exposed coax shield. The compression crimp forces a bushing into the back of the connector, clamping the shield against the connector body. In all cases, the exposed center conductor of the cable — a solid wire — must end flush with the end of the connector. A center conductor that is too short may not make a good connection.

22.10 Reference Tables

Table 22.62
US Customary Units and Conversion Factors

Linear Units

12 inches (in) = 1 foot (ft)
36 inches = 3 feet = 1 yard (yd)
1 rod = 5½ yards = 16½ feet
1 statute mile = 1760 yards = 5280 feet
1 nautical mile = 6076.11549 feet

Area

1 ft² = 144 in²
1 yd² = 9 ft² = 1296 in²
1 rod² = 30¼ yd²
1 acre = 4840 yd² = 43,560 ft²
1 acre = 160 rod²
1 mile² = 640 acres

Volume

1 ft³ = 1728 in³
1 yd³ = 27 ft³

Liquid Volume Measure

1 fluid ounce (fl oz) = 8 fluid drams = 1.804 in
1 pint (pt) = 16 fl oz
1 quart (qt) = 2 pt = 32 fl oz = 57¾ in³
1 gallon (gal) = 4 qt = 231 in³
1 barrel = 31½ gal

Dry Volume Measure

1 quart (qt) = 2 pints (pt) = 67.2 in³
1 peck = 8 qt
1 bushel = 4 pecks = 2150.42 in³

Avoirdupois Weight

1 dram (dr) = 27.343 grains (gr) or (gr a)
1 ounce (oz) = 437.5 gr
1 pound (lb) = 16 oz = 7000 gr
1 short ton = 2000 lb, 1 long ton = 2240 lb

Troy Weight

1 grain troy (gr t) = 1 grain avoirdupois
1 pennyweight (dwt) or (pwt) = 24 gr t
1 ounce troy (oz t) = 480 grains
1 lb t = 12 oz t = 5760 grains

Apothecaries' Weight

1 grain apothecaries' (gr ap)
  = 1 gr t = 1 gr
1 dram ap (dr ap) = 60 gr
1 oz ap = 1 oz t = 8 dr ap = 480 gr
1 lb ap = 1 lb t = 12 oz ap = 5760 gr

Conversion
Metric Unit = Metric Unit × US Unit

(Length)

mm	25.4	inch
cm	2.54	inch
cm	30.48	foot
m	0.3048	foot
m	0.9144	yard
km	1.609	mile
km	1.852	nautical mile

(Area)

mm²	645.16	inch²
cm²	6.4516	in²
cm²	929.03	ft²
m²	0.0929	ft²
cm²	8361.3	yd²
m²	0.83613	yd²
m²	4047	acre
km²	2.59	mi²

(Mass) **(Avoirdupois Weight)**

grams	0.0648	grains
g	28.349	oz
g	453.59	lb
kg	0.45359	lb
tonne	0.907	short ton
tonne	1.016	long ton

(Volume)

mm³	16387.064	in³
cm³	16.387	in³
m³	0.028316	ft³
m³	0.764555	yd³
ml	16.387	in³
ml	29.57	fl oz
ml	473	pint
ml	946.333	quart
l	28.32	ft³
l	0.9463	quart
l	3.785	gallon
l	1.101	dry quart
l	8.809	peck
l	35.238	bushel

(Mass) **(Troy Weight)**

g	31.103	oz t
g	373.248	lb t

(Mass) **(Apothecaries' Weight)**

g	3.387	dr ap
g	31.103	oz ap
g	373.248	lb ap

Multiply →
Metric Unit = Conversion Factor × US Customary Unit

← **Divide**
Metric Unit ÷ Conversion Factor = US Customary Unit

Table 22.63

International System of Units (SI)—Metric Units

Prefix	Symbol			Multiplication Factor
exe	E	10^{18}	=	1,000,000 000,000,000,000
peta	P	10^{15}	=	1,000 000,000,000,000
tera	T	10^{12}	=	1,000,000,000,000
giga	G	10^{9}	=	1,000,000,000
mega	M	10^{6}	=	1,000,000
kilo	k	10^{3}	=	1,000
hecto	h	10^{2}	=	100
deca	da	10^{1}	=	10
		10^{0}	=	1
deci	d	10^{-1}	=	0.1
centi	c	10^{-2}	=	0.01
milli	m	10^{-3}	=	0.001
micro	μ	10^{-6}	=	0.000001
nano	n	10^{-9}	=	0.000000001
pico	p	10^{-12}	=	0.000000000001
femto	f	10^{-15}	=	0.000000000000001
atto	a	10^{-18}	=	0.000000000000000001

Linear

1 meter (m) = 100 centimeters (cm) = 1000 millimeters (mm)

Area

$1 \text{ m}^2 = 1 \times 10^4 \text{ cm}^2 = 1 \times 10^6 \text{ mm}^2$

Volume

$1 \text{ m}^3 = 1 \times 10^6 \text{ cm}^3 = 1 \times 10^9 \text{ mm}^3$
1 liter (l) = 1000 cm^3 = $1 \times 10^6 \text{ mm}^3$

Mass

1 kilogram (kg) = 1000 grams (g)
  (Approximately the mass of 1 liter of water)
1 metric ton (or tonne) = 1000 kg

Table 22.64
Voltage-Power Conversion Table
Based on a 50-ohm system

Voltage			Power	
RMS	Peak-to-Peak	dBmV	Watts	dBm
0.01 μV	0.0283 μV	−100	2×10^{-18}	−147.0
0.02 μV	0.0566 μV	−93.98	8×10^{-18}	−141.0
0.04 μV	0.113 μV	−87.96	32×10^{-18}	−134.9
0.08 μV	0.226 μV	−81.94	128×10^{-18}	−128.9
0.1 μV	0.283 μV	−80.0	200×10^{-18}	−127.0
0.2 μV	0.566 μV	−73.98	800×10^{-18}	−121.0
0.4 μV	1.131 μV	−67.96	3.2×10^{-15}	−114.9
0.8 μV	2.236 μV	−61.94	12.8×10^{-15}	−108.9
1.0 μV	2.828 μV	−60.0	20.0×10^{15}	−107.0
2.0 μV	5.657 μV	−53.98	80.0×10^{-15}	−101.0
4.0 μV	11.31 μV	−47.96	320.0×10^{-15}	−94.95
8.0 μV	22.63 μV	−41.94	1.28×10^{-12}	−88.93
10.0 μV	28.28 μV	−40.00	2.0×10^{-12}	−86.99
20.0 μV	56.57 μV	−33.98	8.0×10^{-12}	−80.97
40.0 μV	113.1 μV	−27.96	32.0×10^{-12}	−74.95
80.0 μV	226.3 μV	−21.94	128.0×10^{-12}	−68.93
100.0 μV	282.8 μV	−20.0	200.0×10^{-12}	−66.99
200.0 μV	565.7 μV	−13.98	800.0×10^{-12}	−60.97
400.0 μV	1.131 mV	−7.959	3.2×10^{-9}	−54.95
800.0 μV	2.263 mV	−1.938	12.8×10^{-9}	−48.93
1.0 mV	2.828 mV	0.0	20.0×10^{-9}	−46.99
2.0 mV	5.657 mV	6.02	80.0×10^{-9}	−40.97
4.0 mV	11.31 mV	12.04	320×10^{-9}	−34.95
8.0 mV	22.63 mV	18.06	1.28 μW	−28.93
10.0 mV	28.28 mV	20.00	1 2.0 μW	−26.99
20.0 mV	56.57 mV	26.02	8.0 μW	−20.97
40.0 mV	113.1 mV	32.04	32.0 μW	−14.95
80.0 mV	226.3 mV	38.06	128.0 μW	−8.93
100.0 mV	282.8 mV	40.0	200.0 μW	−6.99
200.0 mV	565.7 mV	46.02	800.0 μW	−0.97
223.6 mV	632.4 mV	46.99	1.0 mW	0
400.0 mV	1.131 V	52.04	3.2 mW	5.05
800.0 mV	2.263 V	58.06	12.80 mW	11.07
1.0 V	2.828 V	60.0	20.0 mW	13.01
2.0 V	5.657 V	66.02	80.0 mW	19.03
4.0 V	11.31 V	72.04	320.0 mW	25.05
8.0 V	22.63 V	78.06	1.28 W	31.07
10.0 V	28.28 V	80.0	2.0 W	33.01
20.0 V	56.57 V	86.02	8.0 W	39.03
40.0 V	113.1 V	92.04	32.0 W	45.05
80.0 V	226.3 V	98.06	128.0 W	51.07
100.0 V	282.8 V	100.0	200.0 W	53.01
200.0 V	565.7 V	106.0	800.0 W	59.03
223.6 V	632.4 V	107.0	1,000.0 W	60.0
400.0 V	1,131.0 V	112.0	3,200.0 W	65.05
800.0 V	2,263.0 V	118.1	12,800.0 W	71.07
1000.0 V	2,828.0 V	120.0	20,000 W	73.01
2000.0 V	5,657.0 V	126.0	80,000 W	79.03
4000.0 V	11,310.0 V	132.0	320,000 W	85.05
8000.0 V	22,630.0 V	138.1	1.28 MW	91.07
10,000.0 V	28,280.0 V	140.0	2.0 MW	93.01

$$1/d = \frac{\lambda/2}{d} = \frac{300}{2f \times d}$$

$$1/d = \frac{300}{2f \times d} = \frac{300}{2 \times 7.2 \times \frac{0.081 \text{ in}}{39.37 \text{ in/m}}} = 10,126$$

$$\text{Length (ft)} = \frac{492 \times 0.95}{f \text{ (MHz)}} = \frac{468}{f \text{ (MHz)}}$$

$$\text{Length (ft)} = \frac{492 \times K}{f \text{ (MHz)}}$$

Table 22.65
Reflection Coefficient, Attenuation, SWR and Return Loss

Reflection Coefficient (%)	Attenuation (dB)	Max SWR	Return Loss, dB	Reflection Coefficient (%)	Attenuation (dB)	Max SWR	Return Loss, dB
1.000	0.000434	1.020	40.00	45.351	1.0000	2.660	6.87
1.517	0.001000	1.031	36.38	48.000	1.1374	2.846	6.38
2.000	0.001738	1.041	33.98	50.000	1.2494	3.000	6.02
3.000	0.003910	1.062	30.46	52.000	1.3692	3.167	5.68
4.000	0.006954	1.083	27.96	54.042	1.5000	3.352	5.35
4.796	0.01000	1.101	26.38	56.234	1.6509	3.570	5.00
5.000	0.01087	1.105	26.02	58.000	1.7809	3.762	4.73
6.000	0.01566	1.128	24.44	60.000	1.9382	4.000	4.44
7.000	0.02133	1.151	23.10	60.749	2.0000	4.095	4.33
7.576	0.02500	1.164	22.41	63.000	2.1961	4.405	4.01
8.000	0.02788	1.174	21.94	66.156	2.5000	4.909	3.59
9.000	0.03532	1.198	20.92	66.667	2.5528	5.000	3.52
10.000	0.04365	1.222	20.00	70.627	3.0000	5.809	3.02
10.699	0.05000	1.240	19.41	70.711	3.0103	5.829	3.01
11.000	0.05287	1.247	19.17				
12.000	0.06299	1.273	18.42				
13.085	0.07500	1.301	17.66				
14.000	0.08597	1.326	17.08				
15.000	0.09883	1.353	16.48				
15.087	0.10000	1.355	16.43				
16.000	0.1126	1.381	15.92				
17.783	0.1396	1.433	15.00				
18.000	0.1430	1.439	14.89				
19.000	0.1597	1.469	14.42				
20.000	0.1773	1.500	13.98				
22.000	0.2155	1.564	13.15				
23.652	0.2500	1.620	12.52				
24.000	0.2577	1.632	12.40				
25.000	0.2803	1.667	12.04				
26.000	0.3040	1.703	11.70				
27.000	0.3287	1.740	11.37				
28.000	0.3546	1.778	11.06				
30.000	0.4096	1.857	10.46				
31.623	0.4576	1.925	10.00				
32.977	0.5000	1.984	9.64				
33.333	0.5115	2.000	9.54				
34.000	0.5335	2.030	9.37				
35.000	0.5675	2.077	9.12				
36.000	0.6028	2.125	8.87				
37.000	0.6394	2.175	8.64				
38.000	0.6773	2.226	8.40				
39.825	0.75000	2.324	8.00				
40.000	0.7572	2.333	7.96				
42.000	0.8428	2.448	7.54				
42.857	0.8814	2.500	7.36				
44.000	0.9345	2.571	7.13				

$$\rho = \frac{SWR - 1}{SWR + 1}$$

where $\rho = 0.01 \times$ (reflection coefficient in %)

$$\rho = 10^{-RL/20}$$

where RL = return loss (dB)

$$\rho = \sqrt{1 - (0.1^X)}$$

where $X = A/10$ and A = attenuation (dB)

$$SWR = \frac{1 + \rho}{1 - \rho}$$

Return loss (dB) = $-8.68589 \ln(\rho)$
where ln is the natural log (log to the base e)

Attenuation (dB) = $-4.34295 \ln(1 - \rho^2)$
where ln is the natural log (log to the base e)

Table 22.66
Abbreviations List

A
a—atto (prefix for 10^{-18})
A—ampere (unit of electrical current)
ac—alternating current
ACC—Affiliated Club Coordinator
ACSSB—amplitude-compandored single sideband
A/D—analog-to-digital
ADC—analog-to-digital converter
AF—audio frequency
AFC—automatic frequency control
AFSK—audio frequency-shift keying
AGC—automatic gain control
Ah—ampere hour
ALC—automatic level control
AM—amplitude modulation
AMRAD—Amateur Radio Research and Development Corporation
AMSAT—Radio Amateur Satellite Corporation
AMTOR—Amateur Teleprinting Over Radio
ANT—antenna
ARA—Amateur Radio Association
ARC—Amateur Radio Club
ARES—Amateur Radio Emergency Service
ARQ—Automatic repeat request
ARRL—American Radio Relay League
ARS—Amateur Radio Society (station)
ASCII—American National Standard Code for Information Interchange
ATV—amateur television
AVC—automatic volume control
AWG—American wire gauge
az-el—azimuth-elevation

B
B—bel; blower; susceptance; flux density, (inductors)
balun—balanced to unbalanced (transformer)
BC—broadcast
BCD—binary coded decimal
BCI—broadcast interference
Bd—baud (bids in single-channel binary data transmission)
BER—bit error rate
BFO—beat-frequency oscillator
bit—binary digit
bit/s—bits per second
BM—Bulletin Manager
BPF—band-pass filter
BPL—Brass Pounders League
BPL—Broadband over Power Line
BT—battery
BW—bandwidth
Bytes—Bytes

C
c—centi (prefix for 10^{-2})
C—coulomb (quantity of electric charge); capacitor
CAC—Contest Advisory Committee
CATVI—cable television interference
CB—Citizens Band (radio)
CBBS—computer bulletin-board service
CBMS—computer-based message system
CCITT—International Telegraph and Telephone Consultative Committee
CCTV—closed-circuit television
CCW—coherent CW
ccw—counterclockwise
CD—civil defense
cm—centimeter
CMOS—complementary-symmetry metal-oxide semiconductor
coax—coaxial cable
COR—carrier-operated relay
CP—code proficiency (award)
CPU—central processing unit
CRT—cathode ray tube
CT—center tap
CTCSS—continuous tone-coded squelch system
cw—clockwise
CW—continuous wave

D
d—deci (prefix for 10^{-1})
D—diode
da—deca (prefix for 10)
D/A—digital-to-analog
DAC—digital-to-analog converter
dB—decibel (0.1 bel)
dBi—decibels above (or below) isotropic antenna
dBm—decibels above (or below) 1 milliwatt
DBM—double balanced mixer
dBV—decibels above/below 1 V (in video, relative to 1 V P-P)
dBW—decibels above/below 1 W
dc—direct current
D-C—direct conversion
DDS—direct digital synthesis
DEC—District Emergency Coordinator
deg—degree
DET—detector
DF—direction finding; direction finder
DIP—dual in-line package
DMM—digital multimeter
DPDT—double-pole double-throw (switch)
DPSK—differential phase-shift keying
DPST—double-pole single-throw (switch)
DS—direct sequence (spread spectrum); display
DSB—double sideband
DSP—digital signal processing
DTMF—dual-tone multifrequency
DVM—digital voltmeter
DX—long distance; duplex
DXAC—DX Advisory Committee
DXCC—DX Century Club

E
e—base of natural logarithms (2.71828)
E—voltage
EA—ARRL Educational Advisor
EC—Emergency Coordinator
ECL—emitter-coupled logic
EHF—extremely high frequency (30-300 GHz)
EIA—Electronic Industries Alliance
EIRP—effective isotropic radiated power
ELF—extremely low frequency
ELT—emergency locator transmitter
EMC—electromagnetic compatibility
EME—earth-moon-earth (moonbounce)
EMF—electromotive force
EMI—electromagnetic interference
EMP—electromagnetic pulse
EOC—emergency operations center
EPROM—erasable programmable read only memory

F
f—femto (prefix for 10^{-15}); frequency
F—farad (capacitance unit); fuse
fax—facsimile
FCC—Federal Communications Commission
FD—Field Day
FEMA—Federal Emergency Management Agency
FET—field-effect transistor
FFT—fast Fourier transform
FL—filter
FM—frequency modulation
FMTV—frequency-modulated television
FSK—frequency-shift keying
FSTV—fast-scan (real-time) television
ft—foot (unit of length)

G
g—gram (unit of mass)
G—giga (prefix for 10^9); conductance
GaAs—gallium arsenide
GB—gigabytes
GDO—grid- or gate-dip oscillator
GHz—gigahertz (10^9 Hz)
GND—ground

H
h—hecto (prefix for 10^2)
H—henry (unit of inductance)
HF—high frequency (3-30 MHz)
HFO—high-frequency oscillator; heterodyne frequency oscillator
HPF—highest probable frequency; high-pass filter
Hz—hertz (unit of frequency, 1 cycle/s)

I
I—current, indicating lamp
IARU—International Amateur Radio Union
IC—integrated circuit
ID—identification; inside diameter
IEEE—Institute of Electrical and Electronics Engineers
IF—intermediate frequency

IMD—intermodulation distortion
in.—inch (unit of length)
in./s—inch per second (unit of velocity)
I/O—input/output
IRC—international reply coupon
ISB—independent sideband
ITF—Interference Task Force
ITU—International Telecommunication Union
ITU-T—ITU Telecommunication Standardization Bureau

J-K

j—operator for complex notation, as for reactive component of an impedance ($+j$ inductive; $-j$ capacitive)
J—joule (kg m^2/s^2) (energy or work unit); jack
JFET—junction field-effect transistor
k—kilo (prefix for 10^3); Boltzmann's constant (1.38x10^{-23} J/K)
K—kelvin (used without degree symbol) absolute temperature scale; relay
kB—kilobytes
kBd—1000 bauds
kbit—1024 bits
kbit/s—1024 bits per second
kbyte—1024 bytes
kg—kilogram
kHz—kilohertz
km—kilometer
kV—kilovolt
kW—kilowatt
kΩ—kilohm

L

l—liter (liquid volume)
L—lambert; inductor
lb—pound (force unit)
LC—inductance-capacitance
LCD—liquid crystal display
LED—light-emitting diode
LF—low frequency (30-300 kHz)
LHC—left-hand circular (polarization)
LO—local oscillator; Leadership Official
LP—log periodic
LS—loudspeaker
lsb—least significant bit
LSB—lower sideband
LSI—large-scale integration
LUF—lowest usable frequency

M

m—meter (length); milli (prefix for 10^{-3})
M—mega (prefix for 10^6); meter (instrument)
mA—milliampere
mAh—milliampere hour
MB—megabytes
MCP—multimode communications processor
MDS—Multipoint Distribution Service; minimum discernible (or detectable) signal
MF—medium frequency (300-3000 kHz)
mH—millihenry
MHz—megahertz

mi—mile, statute (unit of length)
mi/h (MPH)—mile per hour
mi/s—mile per second
mic—microphone
min—minute (time)
MIX—mixer
mm—millimeter
MOD—modulator
modem—modulator/demodulator
MOS—metal-oxide semiconductor
MOSFET—metal-oxide semiconductor field-effect transistor
MS—meteor scatter
ms—millisecond
m/s—meters per second
msb—most-significant bit
MSI—medium-scale integration
MSK—minimum-shift keying
MSO—message storage operation
MUF—maximum usable frequency
mV—millivolt
mW—milliwatt
MΩ—megohm

N

n—nano (prefix for 10^{-9}); number of turns (inductors)
NBFM—narrow-band frequency modulation
NC—no connection; normally closed
NCS—net-control station; National Communications System
nF—nanofarad
NF—noise figure
nH—nanohenry
NiCd—nickel cadmium
NM—Net Manager
NMOS—N-channel metal-oxide silicon
NO—normally open
NPN—negative-positive-negative (transistor)
NPRM—Notice of Proposed Rule Making (FCC)
ns—nanosecond
NTIA—National Telecommunications and Information Administration
NTS—National Traffic System

O

OBS—Official Bulletin Station
OD—outside diameter
OES—Official Emergency Station
OO—Official Observer
op amp—operational amplifier
ORS—Official Relay Station
OSC—oscillator
OSCAR—Orbiting Satellite Carrying Amateur Radio
OTC—Old Timer's Club
oz—ounce ($\frac{1}{16}$ pound)

P

p—pico (prefix for 10^{-12})
P—power; plug
PA—power amplifier
PACTOR—digital mode combining aspects of packet and AMTOR
PAM—pulse-amplitude modulation
PBS—packet bulletin-board system

PC—printed circuit
PD—power dissipation
PEP—peak envelope power
PEV—peak envelope voltage
pF—picofarad
pH—picohenry
PIC—Public Information Coordinator
PIN—positive-intrinsic-negative (semiconductor)
PIO—Public Information Officer
PIV—peak inverse voltage
PLC—Power Line Carrier
PLL—phase-locked loop
PM—phase modulation
PMOS—P-channel (metal-oxide semiconductor)
PNP—positive negative positive (transistor)
pot—potentiometer
P-P—peak to peak
ppd—postpaid
PROM—programmable read-only memory
PSAC—Public Service Advisory Committee
PSHR—Public Service Honor Roll
PTO—permeability-tuned oscillator
PTT—push to talk

Q-R

Q—figure of merit (tuned circuit); transistor
QRP—low power (less than 5-W output)
R—resistor
RACES—Radio Amateur Civil Emergency Service
RAM—random-access memory
RC—resistance-capacitance
R/C—radio control
RCC—Rag Chewer's Club
RDF—radio direction finding
RF—radio frequency
RFC—radio-frequency choke
RFI—radio-frequency interference
RHC—right-hand circular (polarization)
RIT—receiver incremental tuning
RLC—resistance-inductance-capacitance
RM—rule making (number assigned to petition)
r/min (RPM)—revolutions per minute
rms—root mean square
ROM—read-only memory
r/s—revolutions per second
RS—Radio Sputnik (Russian ham satellite)
RST—readability-strength-tone (CW signal report)
RTTY—radioteletype
RX—receiver, receiving

S

s—second (time)
S—siemens (unit of conductance); switch
SASE—self-addressed stamped envelope
SCF—switched capacitor filter
SCR—silicon controlled rectifier
SEC—Section Emergency Coordinator

SET—Simulated Emergency Test
SGL—State Government Liaison
SHF—super-high frequency (3-30 GHz)
SM—Section Manager; silver mica (capacitor)
S/N—signal-to-noise ratio
SPDT—single-pole double-throw (switch)
SPST—single-pole single-throw (switch)
SS—ARRL Sweepstakes; spread spectrum
SSB—single sideband
SSC—Special Service Club
SSI—small-scale integration
SSTV—slow-scan television
STM—Section Traffic Manager
SX—simplex
sync—synchronous, synchronizing
SWL—shortwave listener
SWR—standing-wave ratio

T

T—tera (prefix for 10^{12}); transformer
TA—ARRL Technical Advisor
TC—Technical Coordinator
TCC—Transcontinental Corps (NTS)
TCP/IP—Transmission Control Protocol/Internet Protocol
tfc—traffic
TNC—terminal node controller (packet radio)
TR—transmit/receive
TS—Technical Specialist
TTL—transistor-transistor logic
TTY—teletypewriter
TU—terminal unit
TV—television
TVI—television interference
TX—transmitter, transmitting

U

U—integrated circuit
UHF—ultra-high frequency (300 MHz to 3 GHz)
USB—upper sideband
UTC—Coordinated Universal Time (also abbreviated Z)
UV—ultraviolet

V

V—volt; vacuum tube
VCO—voltage-controlled oscillator
VCR—video cassette recorder
VDT—video-display terminal
VE—Volunteer Examiner
VEC—Volunteer Examiner Coordinator
VFO—variable-frequency oscillator
VHF—very-high frequency (30-300 MHz)
VLF—very-low frequency (3-30 kHz)
VLSI—very-large-scale integration
VMOS—V-topology metal-oxide-semiconductor
VOM—volt-ohmmeter
VOX—voice-operated switch
VR—voltage regulator
VSWR—voltage standing-wave ratio
VTVM—vacuum-tube voltmeter
VUCC—VHF/UHF Century Club
VXO—variable-frequency crystal oscillator

W

W—watt (kg $m^2 s^{-3}$), unit of power
WAC—Worked All Continents
WAS—Worked All States
WBFM—wide-band frequency modulation
WEFAX—weather facsimile
Wh—watthour
WPM—words per minute
WRC—World Radiocommunication Conference
WVDC—working voltage, direct current

X

X—reactance
XCVR—transceiver
XFMR—transformer
XIT—transmitter incremental tuning
XO—crystal oscillator
XTAL—crystal
XVTR—transverter

Y-Z

Y—crystal; admittance
YIG—yttrium iron garnet
Z—impedance; also see UTC

Numbers/Symbols

5BDXCC—Five-Band DXCC
5BWAC—Five-Band WAC
5BWAS—Five-Band WAS
6BWAC—Six-Band WAC
°—degree (plane angle)
°C—degree Celsius (temperature)
°F—degree Fahrenheit (temperature)
α—(alpha) angles; coefficients, attenuation constant, absorption factor, area, common-base forward current-transfer ratio of a bipolar transistor
β—(beta) angles; coefficients, phase constant, current gain of common-emitter transistor amplifiers
γ—(gamma) specific gravity, angles, electrical conductivity, propagation constant
Γ—(gamma) complex propagation constant
δ—(delta) increment or decrement; density; angles
Δ—(delta) increment or decrement determinant, permittivity
ϵ—(epsilon) dielectric constant; permittivity; electric intensity
ζ—(zeta) coordinates; coefficients
η—(eta) intrinsic impedance; efficiency; surface charge density; hysteresis; coordinate
θ—(theta) angular phase displacement; time constant; reluctance; angles
ι—(iota) unit vector
κ—(kappa) susceptibility; coupling coefficient
λ—(lambda) wavelength; attenuation constant
Λ—(lambda) permeance
μ—(mu) permeability; amplification factor; micro (prefix for 10^{-6})
μF—microfarad
μH—microhenry
μP—microprocessor
ξ—(xi) coordinates
π—(pi) ≈ 3.14159
ρ—(rho) resistivity; volume charge density; coordinates; reflection coefficient
σ—(sigma) surface charge density; complex propagation constant; electrical conductivity; leakage coefficient; deviation
Σ—(sigma) summation
τ—(tau) time constant; volume resistivity; time-phase displacement; transmission factor; density
ϕ—(phi) magnetic flux angles
Φ—(phi) summation
χ—(chi) electric susceptibility; angles
Ψ—(psi) dielectric flux; phase difference; coordinates; angles
ω—(omega) angular velocity $2\pi F$
Ω—(omega) resistance in ohms; solid angle

Contents

23.1 Electronic Shop Safety
    23.1.1 Soldering safety
    23.1.2 Chemicals
23.2 Tools And Their Use
    23.2.1 Sources of Tools
    23.2.2 Tool Descriptions and Uses
23.3 Soldering Tools and Techniques
    23.3.1 Soldering Irons
    23.3.2 Solder
    23.3.3 Soldering
    23.3.4 Desoldering
23.4 Surface Mount Technology (SMT)
    23.4.1 Equipment Needed
    23.4.2 Surface-Mount Parts
    23.4.3 SMT Soldering Basics
    23.4.4 Removing SMT Components

23.5 Electronic Circuits
    23.5.1 Electrostatic Discharge (ESD)
    23.5.2 Electronics Construction Techniques
    23.5.3 Printed-Circuit Boards
    23.5.4 From Schematic to Working Circuit
    23.5.5 Other Construction Techniques
23.6 Mechanical Fabrication
    23.6.1 Cutting and Bending Sheet Metal
    23.6.2 Finishing Aluminum
    23.6.3 Chassis Working
    23.6.4 Drilling Techniques
    23.6.5 Construction Notes

Chapter 23

Construction Techniques

Home construction of electronics projects can be a fun part of Amateur Radio. Some say that hams don't build things nowadays; this just isn't so! A recent ARRL survey shows that more than half of active hams build some electronic projects. When you go to any ham flea market, you see row after row of dealers selling electronic components; people are leaving those tables with bags of parts. They must be doing something with them!

Even experienced constructors will find valuable tips in this chapter. It discusses tools and their uses, electronic construction techniques, tells how to turn a schematic into a working circuit and then summarizes common mechanical construction practices. Originally written and compiled by Ed Hare, W1RFI, this chapter was updated by Jim Duffey, KK6MC, with contributions from Chuck Adams, K7QO.

23.1 Electronic Shop Safety

Building, repairing and modifying equipment in home workshops is a longstanding ham radio tradition. In fact, in the early days, building your own equipment was the only option available. While times and interests change, home construction of radio equipment and related accessories remains popular and enjoyable. Building your own gear need not be hazardous if you become familiar with the hazards, learn how to perform the necessary functions and follow some basic safe practices including the ones listed below. Let's start with our own abilities:

Consider your state of mind. Working on projects or troubleshooting (especially where high voltage is present) requires concentration. Don't work when you're tired or distracted. Be realistic about your ability to focus on the job at hand. Put another way, if we aren't able to be highly alert, we should put off doing hazardous work until we are able to focus on the hazards.

Think! Pay attention to what you are doing. No list of safety rules can cover all possibilities. Safety is always your responsibility. You must think about what you are doing, how it relates to the tools and the specific situation at hand. When working with tools, avoid creating situations in which you can be injured or the project damaged if things don't go "just right."

Take your time. If you hurry, not only will you make more mistakes and possibly spoil the appearance of your new equipment, you won't have time to think things through. Always plan ahead. Do not work with shop tools if you can't concentrate on what you are doing — it's not a race!

Protect yourself. Use of drills, saws, grinders and other wood- or metal-working equipment can release small fragments that could cause serious eye damage. Always wear safety glasses or goggles when doing work that might present a flying object hazard and that includes soldering, where small bits of molten solder can be flung a surprising distance. If you use hammers, wire-cutters, chisels and other hand tools, you will also need the protection that safety eyewear offers. Dress appropriately — loose clothing (or even hair) can be caught in exposed rotating equipment such as drill presses.

Don't work alone. Have someone nearby who can help if you get into trouble when working with dangerous equipment, chemicals or voltages.

Know what to do in an emergency. Despite your best efforts to be careful, accidents may still occur from time to time. Ensure that everyone in your household knows basic first aid procedures and understands how to summon help in an emergency. They should also know where to find and how to safely shut down electrical power in your shack and shop. Get medical help when necessary. Every workshop should contain a good first-aid kit. Keep an eye-wash kit near any dangerous chemicals or power tools that can create chips or splinters. If you become injured, apply first aid and then seek medical help if you are not sure that you are okay. Even a small burn or scratch on your eye can develop into a serious problem if not treated promptly.

What about the equipment and tools involved in shop work? Here are some basic safety considerations that apply to them, as well:

Read instructions and manuals carefully…and follow them. The manufacturers of tools are the most knowledgeable about how to use their products safely. Tap their knowledge by carefully reading all operating instructions and warnings. Avoiding injuries with power tools requires safe tool design as well as proper operation by the user. Keep the instructions in a place where you can refer to them in the future.

Respect safety features of the equipment you work on and use. Never disable any safety feature of any tool. If you do, sooner or later you or someone else will make the mistake the safety feature was designed to prevent.

Keep your shop or work area neat and organized. A messy shop is a dangerous shop. A knife left laying in a drawer can cut someone looking for another tool; a hammer left on top of a shelf can fall down at the worst possible moment; a sharp tool left on a chair can be a dangerous surprise for the weary constructor who sits down.

Keep your tools in good condition. Always take care of your investment. Store tools in a way to prevent damage or use by untrained persons (young children, for example). Keep the cutting edges of saws, chisels and drill bits sharp. Protect metal surfaces from corrosion. Frequently inspect the cords and plugs of electrical equipment and make any necessary repairs. If you find that your power cord is becoming frayed, repair it right away. One solution is to buy a replacement cord with a molded connector already attached.

Make sure your shop is well ventilated. Paint, solvents, cleaners or other chemicals can create dangerous fumes. If you feel dizzy, get into fresh air immediately, and seek medical help if you do not recover quickly.

Respect power tools. Power tools are not forgiving. A drill can go through your hand a lot easier than metal. A power saw can remove a finger with ease. Keep away from the business end of power tools. Tuck in your shirt, roll up your sleeves and remove your tie before using any power tool. If you have long hair, tie it back so it can't become entangled in power equipment.

23.1.1 Soldering Safety

Soldering requires a certain degree of practice and, of course, the right tools. What potential hazards are involved?

Since the solder used for virtually all electronic components is a lead-tin alloy, the first thing in most people's minds is lead, a well-known health hazard. There are two primary ways lead might enter our bodies when soldering: we could breathe lead fumes into our lungs or we could ingest (swallow) lead or lead-contaminated food. Inhalation of lead fumes is extremely unlikely because the temperatures ordinarily used in electronic soldering are far below those needed to vaporize lead. Nevertheless, since lead is soft and we may tend to handle it with our fingers, contaminating our food is a real possibility. For this reason, wash your hands carefully after any soldering (or touching of solder connections).

Handling lead or breathing soldering fumes is also hazardous. When smoking was common, it was not uncommon to see technicians who smoked light cigarettes with their soldering iron. This should not be done.

Soldering equipment gets *hot*! Be careful. Treat a soldering iron as you would any other hot object. A soldering iron stand is helpful, preferably one that has a cage that surrounds the hot tip of the iron. Here's a helpful tip — if the soldering iron gets knocked off the bench, train yourself not to grab for it because the chances are good that you'll grab the hot end!

Generally, solder used for electronic components contains a flux that assists the solder in making a thoroughly wetted joint. (Never use acid-flux, as described in the section on soldering.) When heated, the flux gives off a vapor in the form of a light gray smoke-like plume. This flux vapor, which often contains aldehydes, is a strong irritant and can cause potentially serious problems to persons who may have respiratory sensitivity conditions including those who suffer from asthma. In most cases it is relatively easy to use a small fan to move the flux vapor away from your eyes and face. Opening a window provides additional air exchange. In extreme cases use an organic vapor cartridge respirator.

Solvents are often used to remove excess flux after the parts have cooled to room temperature. Minimize skin contact with solvents by wearing molded gloves resistant to the solvent. If you use a solvent to remove flux, it is best to use the mildest one that does the job. Isopropyl alcohol, or rubbing alcohol, is often sufficient. Some water-soluble solder fluxes can be removed with water.

Observe these precautions to protect yourself and others:

• Properly ventilate the work area. If you can smell fumes, you are breathing them.

• Wash your hands after soldering, especially before handling food.

• Minimize direct contact with flux and flux solvents. Wear disposable surgical gloves when handling solvents.

For more information about soldering hazards and the ways to make soldering safer, see "Making Soldering Safer," by Brian P. Bergeron, MD, NU1N (Mar 1991 *QST*, pp 28-30) and "More on Safer Soldering," by Gary E. Meyers, K9CZB (Aug 1991 *QST*, p 42).

23.1.2 Chemicals

We can't seem to live without the use of chemicals, even in the electronics age and a number of chemicals are used every day by amateurs without causing ill effects. A sensible approach is to become knowledgeable of the hazards associated with the chemicals we use and treat them with respect. Here are a few key suggestions:

Read the information that accompanies the chemical and follow the manufacturer's recommended safety practices. If you would like more information than is printed on the label, ask for a material safety data sheet (MSDS). Material safety data sheets are available on the Internet from several sources, such as **www.meridianeng.com/datasheets.html**. Manufacturers of brand-name chemical products usually post an MSDS on their product Web sites.

Store chemicals properly, away from sunlight and sources of heat. Secure their containers to prevent spills and so that children and untrained persons will not gain access. Always keep containers labeled so there is no confusion about the contents. It is best to use the container in which the chemical was purchased. If you transfer solvents to other containers, such as wash bottles, label the new container with exactly what it contains.

Handle chemicals carefully to avoid spills. Clean up any spills or leaks promptly but don't overexpose yourself in the process. Never dispose of chemicals in household sinks or drains. Instead, contact your local waste plant operator, transfer station or fire department to determine the proper disposal procedures for your area. Many communities have household hazardous waste collection programs. Of course, the best solution is to only buy the amount of chemical that you will need, and use it all if possible. Always label any waste chemicals, especially if they are no longer in their original containers. Oil-filled capacitors and transformers were once commonly filled with oil containing PCBs. Never dispose of any such items that may contain PCBs in landfills.

Always use recommended personal protective equipment (such as gloves, face shield, splash goggles and aprons). If corrosives (acids or caustics) are splashed on you *immediately* rinse with cold water for a minimum of 15 minutes to flush the skin thoroughly. If splashed in the eyes, direct a gentle stream of cold water into the eyes for at least 15 minutes. Gently lift the eyelids so trapped liquids can be flushed completely. Start flushing before removing contaminated clothing. Seek professional medical assistance. If using hazardous chemicals, it is unwise to work alone since people splashed with chemicals need the calm influence of another person.

Food and chemicals don't mix. Keep food, drinks and cigarettes *away* from areas where chemicals are used and don't bring chemicals to places where you eat.

Table 23.1 summarizes the uses and hazards of chemicals used in the ham shack. It includes preventive measures that can minimize risk. Additional safe chemical handling information can be found on the Internet at **www2.umdnj.edu/eohssweb/aiha/technical/msds.htm**.

Table 23.1
Properties and Hazards of Chemicals often used in the Shack or Workshop

Generic Chemical Name	Purpose or Use	Hazards	Ways to Minimize Risks
Lead-tin solder	Bonding electrical components	Lead exposure (mostly from hand contact) Flux exposure (inhalation)	Always wash hands after soldering or touching solder. Use good ventilation.
Isopropyl alcohol	Flux remover	Dermatitis (skin rash)	Wear molded gloves suitable for solvents.
		Vapor inhalation	Use good ventilation and avoid aerosol generation.
		Fire hazard	Use good ventilation, limit use to small amounts, keep ignition sources away, dispose of rags only in tightly sealed metal cans.
Freons	Circuit cooling and general solvent	Vapor inhalation Dermatitis	Use adequate ventilation. Wear molded gloves suitable for solvents.
Phenols and methylene chloride	Enameled wire/ paint stripper	Strong skin corrosive	Avoid skin contact; wear suitable molded gloves; use adequate ventilation.
Beryllium oxide	Ceramic insulator found in some power transistors and vacuum tubes that conducts heat well	Toxic when in fine dust form and inhaled	Avoid grinding, sawing or reducing to dust form.
Beryllium metal	Lightweight metal, alloyed with copper	Same as beryllium oxide	Avoid grinding, sawing, welding or reducing to dust. Contact supplier for special procedures.
Various paints	Finishing	Exposures to solvents	Adequate ventilation; use respirator when spraying.
		Exposures to sensitizers (especially urethane paint)	Adequate ventilation and use respirator.
		Exposure to toxic metals (lead, cadmium, chrome, and so on) in pigments	Adequate ventilation and use respirator.
		Fire hazard (especially when spray painting)	Adequate ventilation; control of residues; eliminate ignition sources.
Ferric chloride	Printed circuit board etchant	Skin and eye contact	Use suitable containers; wear splash goggles and molded gloves suitable for acids.
Ammonium persulphate and mercuric chloride	Printed circuit board etchants	Skin and eye contact	Use suitable containers; wear splash goggles and molded gloves suitable for acids.
Epoxy resins	General purpose cement or paint	Dermatitis and possible sensitizer	Avoid skin contact. Mix only amount needed.
Sulfuric acid	Electrolyte in lead-acid batteries	Strong corrosive when on skin or eyes. Will release hydrogen when charging (fire, explosion hazard).	Always wear splash goggles and molded plastic gloves (PVC) when handling. Keep ignition sources away from battery when charging. Use adequate ventilation.

23.2 Tools and Their Use

All electronic construction makes use of tools, from mechanical tools for chassis fabrication to the soldering tools used for circuit assembly. A good understanding of tools and their uses will enable you to perform most construction tasks.

While sophisticated and expensive tools often work better or more quickly than simple hand tools, with proper use, simple hand tools can turn out a fine piece of equipment. **Table 23.2** lists tools indispensable for construction of electronic equipment. These tools can be used to perform nearly any construction task. Add tools to your collection from time to time, as finances permit.

23.2.1 Sources of Tools

Electronic-supply houses, mail-order/Web stores and most hardware stores carry the tools required to build or service Amateur Radio equipment. Bargains are available at ham flea markets or local neighborhood sales, but beware! Some flea-market bargains are really shoddy and won't work very well or last very long. Some used tools are offered for sale because the owner is not happy with their performance.

There is no substitute for quality! A high-quality tool, while a bit more expensive, will last a lifetime. Poor quality tools don't last long and often do a poor job even when brand new. You don't need to buy machinist-grade tools, but stay away from cheap tools; they are not the bargains they might appear to be.

CARE OF TOOLS

The proper care of tools is more than a matter of pride. Tools that have not been cared for properly will not last long or work well. Dull or broken tools can be safety hazards. Tools that are in good condition do the work for you; tools that are misused or dull are difficult to use.

Store tools in a dry place. Tools do not fit in with most living-room decors, so they are often relegated to the basement or garage. Unfortunately, many basements or garages are not good places to store tools; dampness and dust are not good for tools. If your tools are stored in a damp place, use a dehumidifier. Sometimes you can minimize rust by keeping your tools lightly oiled, but this is a second-best solution. If you oil your tools, they may not rust, but you will end up covered in oil every time you use them. Wax or silicone spray is a better alternative.

Store tools neatly. A messy toolbox, with tools strewn about haphazardly, can be more than an inconvenience. You may waste a lot

---

Table 23.2
Recommended Tools and Materials

Simple Hand Tools
Screwdrivers
Slotted, 3-in, ⅛-in blade
Slotted, 8-in, ⅛-in blade
Slotted, 3-in, 3/16-in blade
Slotted, stubby, ¼-in blade
Slotted, 4-in, ¼-in blade
Slotted, 6-in, 5/16-in blade
Phillips, 2½-in, #0 (pocket clip)
Phillips, 3-in, #1
Phillips, stubby, #2
Phillips, 4-in, #2
Phillips, 4-in, #2
Long-shank screwdriver with holding clip on blade
Jeweler's set
Right-angle, slotted and Phillips

Pliers, Sockets and Wrenches
Long-nose pliers, 6- and 4-in
Diagonal cutters, 6- and 4-in
Channel-lock pliers, 6-in
Slip-joint pliers
Locking pliers (Vise Grip or equivalent)
Socket nut-driver set, 1/16- to ½-in
Set of socket wrenches for hex nuts
Allen (hex) wrench set
Wrench set
Adjustable wrenches, 6- and 10-in
Tweezers, regular and reverse-action
Retrieval tool/parts holder, flexible claw
Retrieval tool, magnetic

Cutting and Grinding Tools
File set consisting of flat, round, half-round, and triangular. Large and miniature types recommended
Burnishing tool
Wire strippers
Wire crimper
Hemostat, straight
Scissors
Tin shears, 10-in
Hacksaw and blades

Hand nibbling tool (for chassis-hole cutting)
Scratch awl or scriber (for marking metal)
Heavy-duty jackknife
Knife blade set (X-ACTO or equivalent)
Machine-screw taps, #4-40 through #10-32 thread
Socket punches, ½ in, ⅝ in, ¾ in, 1⅛ in, 1¼ in, and 1½ in
Tapered reamer, T-handle, ½-in maximum width
Deburring tool

Miscellaneous Hand Tools
Combination square, 12-in, for layout work
Hammer, ball-peen, 12-oz head
Hammer, tack
Bench vise, 4-in jaws or larger
Center punch
Plastic alignment tools
Mirror, inspection
Flashlight, penlight and standard
Magnifying glass
Ruler or tape measure
Dental pick
Calipers
Brush, wire
Brush, soft
Small paintbrush
IC-puller tool

Hand-Powered Tools
Hand drill, ¼-in chuck or larger
High-speed drill bits, #60 through ⅜-in diameter

Power Tools
Motor-driven emery wheel for grinding
Electric drill, hand-held
Drill press
Miniature electric motor tool (Dremel or equivalent) and accessory drill press

Soldering Tools and Supplies
Soldering pencil, 30-W, ⅛-in tip

Soldering iron, 200-W, ⅝-in tip
Solder, 60/40, rosin core
Soldering gun, with assorted tips
Desoldering tool
Desoldering wick

Safety
Safety glasses
Hearing protector, earphones or earplugs
Fire extinguisher
First-aid kit

Useful Materials
Medium-weight machine oil
Contact cleaner, liquid or spray can
RTV sealant or equivalent
Electrical tape, vinyl plastic
Sandpaper, assorted
Emery cloth
Steel wool, assorted
Cleaning pad, Scotchbrite or equivalent
Cleaners and degreasers
Contact lubricant
Sheet aluminum, solid and perforated, 16- or 18-gauge, for brackets and shielding.
Aluminum angle stock, ½ × ½-in and ¼-in diameter round brass or aluminum rod (for shaft extensions)
Machine screws: Round-head and flat head, with nuts and lockwashers to fit. Most useful sizes: 4-40, 6-32 and 8-32, in lengths from ¼-in to 1½ in (Nickel-plated steel is satisfactory except in strong RF fields, where brass should be used.)
Bakelite, Lucite, polystyrene and copper-clad PC-board scraps.
Soldering lugs, panel bearings, rubber grommets, terminal-lug wiring strips, varnished-cambric insulating tubing, heat-shrinkable tubing
Shielded and unshielded wire
Tinned bare wire, #22, #14 and #12
Enameled wire, #20 through #30

of time looking for the right tool and sharp edges can be dulled or nicked by tools banging into each other in the bottom of the box. As the old adage says, every tool should have a place, and every tool should be in its place. If you must search the workbench, garage, attic and car to find the right screwdriver, you'll spend more time looking for tools than building projects.

SHARPENING

Many cutting tools can be sharpened. Send a tool that has been seriously dulled to a professional sharpening service. These services can sharpen saw blades, some files, drill bits and most cutting blades. Touch up the edge of cutting tools with a whetstone to extend the time between sharpening.

Sharpen drill bits frequently to minimize the amount of material that must be removed each time. Frequent sharpening also makes it easier to maintain the critical surface angles required for best cutting with least wear. Most inexpensive drill-bit sharpeners available for shop use do a poor job, either from the poor quality of the sharpening tool or inexperience of the operator. Also, drills should be sharpened at different angles for different applications. Commercial sharpening services do a much better job.

INTENDED PURPOSE

Don't use tools for anything other than their intended purpose! If you use a pair of wire cutters to cut sheet metal, pliers as a vise or a screwdriver as a pry bar, you ruin a good tool and sometimes the work piece, as well. Although an experienced constructor can improvise with tools, most take pride in not abusing them. Having a wide variety of good tools at your disposal minimizes the problem of using the wrong tool for the job.

23.2.2 Tool Descriptions and Uses

Specific applications for tools are discussed throughout this chapter. Hand tools are used for so many different applications that they are discussed first, followed by some tips for proper use of power tools.

SCREWDRIVERS AND NUTDRIVERS

For construction or repair, you need to have an assortment of screwdrivers. Each blade size is designed to fit a specific range of screw head sizes. Using the wrong size blade usually damages the blade, the screw head or both. You may also need stubby sizes to fit into tight spaces. Right-angle screwdrivers are inexpensive and can get into tight spaces that can't otherwise be reached.

Electric screwdrivers are relatively inexpensive and very useful, particularly for repetitive tasks. If you have a lot of screws to fasten, they can save a lot of time and effort. They come with a wide assortment of screwdriver and nutdriver bits. An electric drill can also function as an electric screwdriver, although it may be heavy and over-powered for many applications.

Keep screwdriver blades in good condition. If a blade becomes broken or worn out, replace the screwdriver. A screwdriver only costs a few dollars; do not use one that is not in perfect condition. Save old screwdrivers to use as pry bars and levers, but use only good ones on screws. Filing a worn blade seldom gives good results.

Nutdrivers, the complement to screwdrivers, are often much easier to use than a wrench, particularly for nuts smaller than $^3\!/_8$ inch. They are also less damaging to the nut than any type of pliers, with a better grip on the nut. A set of interchangeable nutdrivers with a shared handle is a very handy addition to the toolbox.

PLIERS AND LOCKING-GRIP PLIERS

Pliers and locking-grip pliers are used to hold or bend things. They are not wrenches! If pliers are used to remove a nut or bolt, the nut or the pliers is usually damaged. To remove a nut, use a wrench or nutdriver. There is one exception to this rule of thumb: To remove a nut that is stripped too badly for a wrench, use a pair of pliers, locking-grip pliers, or a diagonal cutter to bite into the nut and start it turning. Reserve an old tool or one dedicated to just this purpose as it is not good for the tool.

Pliers are not intended for heavy-duty applications. Use a metal brake to bend heavy metal; use a vise to hold a heavy component. If the pliers' jaws or teeth become worn, replace the tool.

There are many different kinds of fine pliers, usually called "needle-nose" pliers or something similar, that are particularly useful in electronics work. These are intended for light jobs, such as bending or holding wires or small work pieces. Two or three of these tools with different sizes of jaws will suffice for most jobs.

WIRE CUTTERS

Wire cutters are primarily used to cut wires or component leads. The choice of diagonal blades (sometimes called "dikes") or end-nip blades depends on the application. Diagonal blades are most often used to cut wires, while the end-nip blades are useful to cut off the leads of components that have been soldered into a printed-circuit board. Some delicate components can be damaged by cutting their leads with dikes. Scissors designed to cut wire can be used.

Wire strippers are handy, but you can usually strip wires using a diagonal cutter or a knife. This is not the only use for a knife, so keep an assortment handy.

Do not use wire cutters or strippers on anything other than wire! If you use a cutter to trim a protruding screw head, or cut a hardened-steel spring, you will usually damage the blades.

FILES

Files are used for a wide range of tasks. In addition to enlarging holes and slots, they are used to remove burrs, shape metal, wood or plastic and clean some surfaces in preparation for soldering. Files are especially prone to damage from rust and moisture. Keep them in a dry place. The cutting edge of the blades can also become clogged with the material you are removing. Use file brushes (also called file cards) to keep files clean. Most files cannot be sharpened easily, so when the teeth become worn, the file must be replaced.

DRILL BITS

Drill bits are made from carbon steel, high-speed steel or carbide. Carbon steel is more common and is usually supplied unless a specific request is made for high-speed bits. Carbon steel drill bits cost less than high-speed or carbide types; they are sufficient for most equipment construction work. Carbide drill bits last much longer under heavy use. One disadvantage of carbide bits is that they are brittle and break easily, especially if you are using a hand-held power drill. When drilling abrasive material, such as fiberglass, the carbide bits last much longer than the steel bits. Twist drills are available in a number of sizes listed in **Table 23.3**. Those listed in bold type are the most commonly used in construction of amateur equipment. You may not use all of the drills in a standard set, but it is nice to have a complete set on hand. You should also buy several spares of the more common sizes. Although Table 23.3 lists drills down to #54, the series extends to number #80. While the smaller sizes cannot usually be found in hardware stores or home improvement stores, they are commonly available through industrial tool suppliers and through various sources on the Internet.

SPECIALIZED TOOLS

Most constructors know how to use common tools, such as screwdrivers, wrenches and hammers. Although specialized tools usually do a job that can be done with other tools, once the specialty tool is used you will wonder how you ever did the job without it! Let's discuss other tools that are not so common.

A hand nibbling tool is shown in **Fig 23.1**. Use this tool to remove small "nibbles" of metal. It is easy to use; position the tool where you want to remove metal and squeeze the handle. The tool takes a small bite out of the metal. When you use a nibbler, be careful that

Table 23.3
Numbered Drill Sizes

No.	Diameter (Mils)	Will Clear Screw	Drilled for Tapping from Steel or Brass
1	228.0	12-24	—
2	221.0	—	—
3	213.0	—	14-24
4	209.0	12-20	—
5	205.0	—	—
6	204.0	—	—
7	201.0	—	—
8	199.0	—	—
9	196.0	—	—
10	193.5	—	—
11	191.0	10-24	—
		10-32	
12	189.0	—	—
13	185.0	—	—
14	182.0	—	—
15	180.0	—	—
16	177.0	—	12-24
17	173.0	—	—
18	169.5	—	—
19	166.0	8-32	12-20
20	161.0	—	—
21	159.0	—	10-32
22	157.0	—	—
23	154.0	—	—
24	152.0	—	—
25	149.5	—	10-24
26	147.0	—	—
27	144.0	—	—
28	140.0	6-32	—
29	136.0	—	8-32
30	128.5	—	—
31	120.0	—	—
32	116.0	—	—
33	113.0	4-40	—
34	111.0	—	—
35	110.0	—	—
36	106.5	—	6-32
37	104.0	—	—
38	101.5	—	—
39	099.5	3-48	—
40	098.0	—	—
41	096.0	—	—
42	093.5	—	—
43	089.0	—	4-40
44	086.0	2-56	—
45	082.0	—	—
46	081.0	—	—
47	078.5	—	3-48
48	076.0	—	—
49	073.0	—	—
50	070.0	—	2-56
51	067.0	—	—
52	063.5	—	—
53	059.5	—	—
54	055.0	—	—

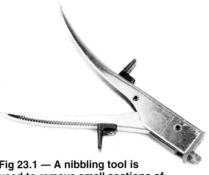

Fig 23.1 — A nibbling tool is used to remove small sections of sheet metal.

Fig 23.2 — A deburring tool is used to remove the burrs left after drilling a hole.

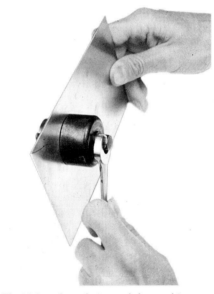

Fig 23.3 — A socket punch is used to easily punch a hole in sheet metal.

you don't remove too much metal, clip the edge of a component mounted to the sheet metal or grab a wire that is routed near the edge of a chassis. Fixing a broken wire is easy, but something to avoid if possible. It is easy to remove metal but nearly impossible to put it back. Do it right the first time!

Deburring Tool

A deburring tool is just the thing to remove the sharp edges left on a hole after drilling or punching operations. See **Fig 23.2**. Position the tool over the hole and rotate it around the hole edge to remove burrs or rough edges. As an alternative, select a drill bit that is somewhat larger than the hole, position it over the hole, and spin it lightly to remove the burr. Be sure to deburr both sides of the hole.

Socket or Chassis Punches

Greenlee is the most widely known of the socket-punch manufacturers. Most socket punches are round, but they do come in other shapes. To use one, drill a pilot hole large enough to clear the bolt that runs through the punch. Then, mount the punch as shown in **Fig 23.3**, with the cutter on one side of the sheet metal and the socket on the other. Tighten the nut with a wrench until the cutter cuts all the way through the sheet metal. These punches are often sold in sets at a significant discount to the same punches purchased separately.

Crimping Tools

The use of crimped connectors is common in the electronics industry. In many commercial and aerospace applications, soldered joints are no longer used. Hams have been reluctant to adapt crimped connections, largely due to mistrust of contacts that are not soldered, the use of cheap crimp connectors on consumer electronics, and the high cost of quality crimping tools or "crimpers." If high quality connectors and tools are used, the crimped connector will be as reliable a connection as a soldered one. The crimped connection is easier to make than a soldered one in most cases.

Crimped coaxial connectors are the most common crimped connector. MIL-spec or equivalent crimp connectors are available for the UHF, BNC, F and N-series MIL-spec connectors. Power connectors, such as the Anderson PowerPoles and Molex connectors are probably the second most commonly used crimped connections.

When purchasing a crimper, look for a ratcheting model with dies that are intended for the connectors you will be using. The common pliers-type crimper designed for household electrical terminals will have trouble crimping power connectors and is unsuitable for coaxial connectors. A good ratcheting crimper can be obtained for $50 to $100 with the necessary interchangeable dies. Large ratcheting crimpers suitable for the larger coaxial connectors can cost several hundred dollars. A good crimper and set of dies is an excellent investment for a club or group of like-minded hams.

Useful Shop Materials

Small stocks of various materials are used when constructing electronics equipment. Most of these are available from hardware or radio supply stores. A representative list is shown at the end of Table 23.2.

Small parts, such as machine screws, nuts, washers and soldering lugs can be economically purchased in large quantities (it doesn't pay to buy more than a lifetime supply). For items you don't use often, many radio supply stores or hardware stores sell small quantities and assortments. Stainless steel hardware can be kept on hand for outdoor use.

23.3 Soldering Tools and Techniques

Soldering is used in nearly every phase of electronic construction so you'll need soldering tools. This section discusses the tools and materials used in soldering.

23.3.1 Soldering Irons

A soldering tool must be hot enough to do the job and lightweight enough for agility and comfort. A heavy soldering gun useful for assembling wire antennas is too large for printed-circuit work, for example. A fine-tip soldering iron (sometimes called a "soldering pencil") works well for smaller jobs.

You may need an assortment of soldering irons to do a wide variety of soldering tasks. They range in size from a small 25-W iron for delicate printed-circuit work to larger 100 to 300-W soldering irons and guns. Small "pencil" butane-powered soldering irons and torches are also available, with a variety of soldering-iron tips. Battery powered irons are available too.

A 25-W pencil tip iron is adequate for printed circuit board work. A larger 40-W iron is necessary for larger jobs, such as soldering leads to panel connectors and making splices. These two irons should handle most electronic homebrew requirements. A larger 100-W iron is good for bigger jobs, such as soldering antenna connections or soldering to power cables.

You should get several different sizes and shapes of tips when you purchase an iron. While most people prefer a conical tip, the chisel tip is also useful. The lower wattage irons will likely have a good selection of tip sizes and geometries. Irons 100-W and larger usually have a non-interchangeable chisel tip. For printed circuit board work, a good rule of thumb is to use a tip whose point is the same size as the component leads you are soldering.

If you buy an iron for use on circuits that contain electrostatic sensitive components, get one that has a grounded tip. Otherwise, you risk electrostatic damage to the components. Such irons are usually specified as having a grounded tip, and will have a three-prong plug. It is usually not necessary to have a grounded tip on the 100-W iron as it is not used on sensitive components.

Soldering guns are used for larger jobs and are too large for most electronics work. Where the soldering iron tip has an internal heater, the soldering gun tip is heated directly by current flowing in the tip. The nuts connecting the tip to the iron can loosen with each heat cycle, so they need tightening periodically. Soldering guns are available that have high and low heat levels controlled by an extra trigger position. Soldering gun tips are usually copper and do not last as long as the iron-clad tips of the smaller irons.

If you do a lot of building, a *soldering station* with a temperature-controlled tip is a better investment than a simple iron. The iron tip temperature can be precisely controlled and can be varied to the type of work being done. Soldering stations generally reach operating temperature more quickly than conventional irons and can be turned down to idle when you are not soldering. Some even have a digital tip temperature display. Soldering stations are available from numerous manufacturers. It is best to buy from an established manufacturer to insure future availability of tips and other components. Reconditioned soldering stations are often available and are good value.

RF-heated irons and hot-air soldering tools are primarily in use by industry and particularly useful for soldering surface-mount devices. The cost of these stations is high, although they are available as used and reconditioned.

MAINTAINING SOLDERING IRONS

Keep soldering tools in good condition by keeping the tips well-tinned with solder. Tinning is performed by melting solder directly onto the tip and letting it form a coating. Do not keep the tips at full temperature for long periods when not in use.

After each use, remove the tip and clean off any scale that may have accumulated. Clean an oxidized tip by dipping the hot tip in sal ammoniac (ammonium chloride) and then wiping it clean with a rag. Sal ammoniac is somewhat corrosive, so if you don't wipe the tip thoroughly, it can contaminate electronic soldering. There are proprietary "tip tinner" products available that can also be used for this function.

If a tip becomes oxidized during use it can be restored to its shiny state by wiping the tip with a damp sponge or rag and then re-tinning. A gentle scraping is also useful for stubborn cases. A copper coil kitchen scrubber (do not use a scrubber that contains soap) works fine for this. Swipe the iron through the scrubber to clean it. A scrubber or sponge is most conveniently used by cramming it into a clean tuna can and placing it next to the solder station.

If a copper tip becomes pitted after repeated use, file it smooth and bright and then tin it immediately with solder. The solder prevents further oxidation of the tip. Modern soldering iron tips are nickel or iron clad and should not be filed, as the cladding protects the tip from pitting.

The secret of good soldering is to use the right amount of heat. Many people who have not soldered before use too little heat, dabbing at the joint to be soldered and making little solder blobs that cause unintended short circuits.

23.3.2 Solder

Solders have different melting points, depending on the ratio of tin to lead. Tin melts at 450 °F and lead at 621 °F. Solder made from 63% tin and 37% lead melts at 361 °F, the lowest melting point for a tin and lead mixture. Called 63-37 (or eutectic), this type of solder also provides the most rapid solid-to-liquid transition and the best stress resistance.

Solders made with different lead/tin ratios have a plastic state at some temperatures. If the solder is deformed while it is in the plastic state, the deformation remains when the solder freezes into the solid state. Any stress or motion applied to "plastic solder" causes a poor solder joint.

60-40 solder has the best wetting qualities. Wetting is the ability to spread rapidly, coat the surfaces to be joined, and bond materials uniformly. 60-40 solder also has a low melting point. These factors make it the most commonly used solder in electronics.

Some connections that carry high current can't be made with ordinary tin-lead solder because the heat generated by the current would melt the solder. Automotive starter brushes and transmitter tank circuits are two examples. Silver-bearing solders have higher melting points, and so prevent this problem. High-temperature silver alloys become liquid in the 1100 °F to 1200 °F range, and a silver-manganese (85-15) alloy requires almost 1800 °F.

Because silver dissolves easily in tin, tin-bearing solders can leach silver plating from components. This problem can be greatly reduced by partially saturating the tin in the solder with silver or by eliminating the tin. Commercial solders are available which incorporate these features. Tin-silver or tin-lead-silver alloys become liquid at temperatures from 430 °F for 96.5-3.5 (tin-silver), to 588 °F for 1.0-97.5-1.5 (tin-lead-silver). A 15.0-80.0-5.0 alloy of lead-indium-silver melts at 314 °F.

Rosin-core wire-type solder is formed into a tube with a flux compound inside. The resin (usually called "rosin" in solder) in a solder is a *flux*. Flux melts at a lower temperature than solder, so it flows out onto the joint before the solder melts to coat the joint surfaces. The solder used for surface-mount soldering (discussed later) is a cream or paste and flux, if used, must be added to the joint separately.

Flux removes oxide by suspending it in solution and floating it to the top. Flux is not a cleaning agent! Always clean the surfaces to be soldered before soldering. Flux is not a part of a soldered connection — it merely aids the soldering process.

After soldering, remove any remaining

flux. Rosin flux can be removed with isopropyl or denatured alcohol. A cotton swab is a good tool for applying the alcohol and scrubbing the excess flux away. Commercial flux-removal sprays are available at most electronic-part distributors. Water-soluble fluxes are also available.

Never use acid flux or acid-core solder for electrical work. It should be used only for plumbing or chassis work. If used on electronics, the flux will corrode and damage the equipment. For circuit construction, only use fluxes or solder-flux combinations that are labeled for electronic soldering.

LEAD-FREE SOLDER

In 2006, the European Union Restriction of Hazardous Substances Directive (RoHS) went into effect. This directive prohibits manufacture and import of consumer electronics which incorporate lead, including the common tin-lead solder used in electronic assembly. California recently enacted a similar RoHS law. As a result of these directives there has been a move to lead-free solders in commercial use. They can contain two or more elements that are not as hazardous as lead, including tin, copper, silver, bismuth, indium, zinc, antimony and traces of other metals. Two lead-free solders commonly used for electronic use are SnAgCu alloy SAC305 and tin-copper alloy Sn100. SAC305 contains 96.5% tin, 3% silver, and 0.5% copper and melts at 217 °C. Sn100 contains 99.3% tin, 0.6% copper, as well as traces of nickel and silver and melts at 228 °C. Both of these melting points are higher than the 176 °C melting point of 60-40 and 63-37 lead-bearing solder, but conventional soldering stations will be able to reach the melting points of the new solders easily. Tin-lead solders are still available, but the move away from them by commercial manufacturers will probably lead to the day when they will be unavailable to hams who build their own gear. Be prepared.

The new RoHS solders can be used in much the same manner as conventional solders. The resulting solder joint appears somewhat duller than a conventional solder joint, and the lead-free solders tend to wick higher than the lead-free solders. Due to the higher heat, it is important that the soldering iron tip be clean, shiny and freshly tinned so that heat is transferred to the joint to be soldered as quickly as possible to avoid excess heating of the parts being soldered. The soldering iron should be set to between 700 °F and 800 °F. Tips on soldering with the new lead-free solders can be found at **hackaday.com/2008/05/22/how-to-go-green-with-lead-free-solder**. Solder manufacturers also supply information on soldering with the new lead-free solders, such as at **documents.rs-components.com/EITC/UK/generalFiles/LF_hand_soldering.pdf** and **www.kester.com/en-us/leadfree**.

Most, if not all, RoHS-compliant components can be soldered with lead-tin solder. If the RoHS part has leads that are tinned or coated with an alloy to make soldering easier, it is necessary to use a hotter iron than would normally be required in lead-tin soldering. A soldering iron tip temperature of 315 °C (600 °F) or greater will be adequate for soldering RoHS parts with lead-tin solder. In contrast, a working tip temperature of 275 °C is generally adequate for working with conventional non-RoHS parts.

23.3.3 Soldering

The two key factors in quality soldering are time and temperature. Rapid heating is desired so that all parts of the joint are made hot enough for the solder to remain molten as it flows over the joint surfaces. Most unsuccessful solder jobs fail because insufficient heat has been applied. To achieve rapid heating, the soldering iron tip should be hotter than the melting point of solder and large enough that transferring heat to the cooler joint materials occurs quickly. A tip temperature about 100 °F (60 °C) above the solder melting point is right for mounting components on PC boards.

Use solder that is sized appropriately for the job. As the cross section of the solder decreases, so does the amount of heat required to melt it. Diameters from 0.025 to 0.040 inch are good for nearly all circuit wiring. Sensitive and smaller components can be damaged or surfaces re-oxidized if heat is applied for too long a period.

If you are a beginner, you may want to start with one of the numerous "Learn to Solder" kits available from many electronics parts and kit vendors. The kits come with a printed-circuit board, a basic soldering iron, solder, and the components to complete a simple electronics project.

Here's how to make a good solder joint. This description assumes that solder with a flux core is used to solder a typical PC board connection such as an IC pin.

• Prepare the joint. Clean all conductors thoroughly with fine steel wool or a plastic scrubbing pad. Clean the circuit board at the beginning of assembly and individual parts such as resistors and capacitors immediately before soldering. Some parts (such as ICs and surface-mount components) cannot be easily cleaned; don't worry unless they're exceptionally dirty.

• Prepare the tool. It should be hot enough to melt solder applied to its tip quickly (half a second when dry, instantly when wet with solder). Apply a little solder directly to the tip so that the surface is shiny. This process is called "tinning" the tool. The solder coating helps conduct heat from the tip to the joint and prevents the tip from oxidizing.

• Place the tip in contact with one side of the joint. If you can place the tip on the underside of the joint, do so. With the tool below the joint, convection helps transfer heat to the joint.

• Place the solder against the joint directly opposite the soldering tool. It should melt within a second for normal PC connections, within two seconds for most other connections. If it takes longer to melt, there is not enough heat for the job at hand.

• Keep the tool against the joint until the solder flows freely throughout the joint. When it flows freely, solder tends to form concave joints called "fillets" between the conductors. With insufficient heat solder does not flow freely; it forms convex shapes — blobs. Once solder shape changes from convex to concave, remove the tool from the joint. If a fillet won't form, the joint may need additional cleaning.

• Let the joint cool without movement at room temperature. It usually takes no more than a few seconds. If the joint is moved before it is cool, it may take on a dull, satin or grainy appearance that is characteristic of a "cold" solder joint. Reheat cold joints until the solder flows freely and hold them still until cool.

• When the iron is set aside, or if it loses its shiny appearance, wipe away any dirt with a damp cloth or sponge. If it remains dull after cleaning, tin it again.

Overheating a transistor or diode while soldering can cause permanent damage although as you get better at soldering, you'll be able to solder very quickly with little risk to the components. If the soldering iron will be applied for longer than a couple of seconds, use a small heat sink when you solder transistors, diodes or components with plastic parts that can melt. Grip the component lead with a pair of pliers up close to the unit so that the heat is conducted away (be careful — it is easy to damage delicate component leads). A small alligator clip also makes a good heat sink.

Mechanical stress can damage components, too. Mount components so there is no appreciable mechanical strain on the leads.

Soldering to hollow pins, such as found on connectors, can be difficult, particularly if the connector is used or has oxidized. Use a suitable small twist drill to clean the inside of the pin and then tin it. While the solder is still melted, clear the surplus solder from each pin with a whipping motion or by blowing through the pin from the inside of the connector. Watch out for flying hot solder — use safety goggles and protect the work surface! Do not perform this operation near open electronic equipment as the loose solder can easily form short circuits. If the pin surface is plated, file the plating from the pin tip. Then insert the wire and solder it. After soldering, remove excess solder with a file, if necessary.

When soldering to the pins of plastic connectors or other assemblies, heat-sink the pin with needle-nose pliers at the base where it comes in contact with the plastic housing. Do not allow the pin to overheat; it will loosen and become misaligned.

23.3.4 Desoldering

There are times when soldered components need to be removed. The parts may be bad, they may be installed incorrectly, or you may want to remove them for use in another project.

There are several techniques for desoldering. The easiest way is to use a desoldering braid. Desoldering braid is simply fine copper braid. It is available under a wide variety of trade names wherever soldering supplies are sold. The soldering braid is placed against the joint to be desoldered. A hot iron is pressed onto the braid. As the solder melts it is wicked into the braid and away from the joint. After all the leads have been treated in this manner the part can be removed. (A thin film of solder may remain, but is easily broken loose through the use of needle-nose pliers.) The part of the braid that wicked up the solder is clipped off. As copper is an excellent conductor of heat, the braid can get quite hot, so watch your fingers.

A desoldering vacuum pump can also be used. There are two types of desoldering vacuum tools, a simple rubber syringe bulb with a high temperature plastic tip and a desoldering pump. The desoldering pump is a simple manual vacuum pump consisting of a cylinder that contains a spring-loaded plunger attached to a metal rod inside a tip of high temperature plastic on the end of the pump. To desolder a joint, the plunger is pushed down, and locked in place. The tip is placed against the joint to be desoldered along with a soldering iron. When the solder melts, a button on the pump releases the plunger, which pulls the rod back, creating a vacuum that sucks the molten solder through the tip. The part being desoldered can then be removed.

The desoldering bulb employs a similar concept: heat the joint to be desoldered, squeeze the bulb, place the tip on the joint and release it to suck up the solder. Remove the part that was desoldered. If the first application of the desoldering pump doesn't suck up all the solder, reheat the joint and suck up the rest.

If the desoldering tool doesn't seem to be sucking up solder, the tip may be clogged with solder. The tip can be unclogged by pushing the solder through. You may have to clear the tip several times when doing a job that requires desoldering many joints. One can purchase small desoldering irons that contain a bulb on the handle that leads to a tip adjacent to the iron tip. This desoldering iron combination is somewhat easier to handle than the separate bulb and iron and does an effective job.

Desoldering stations are also available. One type contains a vacuum pump in a console much like a soldering station. A vacuum line is connected to the tip of the soldering iron. There is a valve trigger on the iron that is used to open the tip to the vacuum when the solder is melted. The solder is sucked up the line into a receptacle in the station. Another type of desoldering station that is commonly used in industry heats the joint with hot air. These are particularly convenient to use and have the advantage that the hot air can be used for soldering surface mount devices.

23.4 Surface Mount Technology (SMT)

Surface mount technology is not new — it was an established professional technique for years before its appearance in consumer and amateur electronics. This technique is particularly suitable for PC-board construction, although it can be applied to many other construction techniques. Surface-mounted components take up very little space on a PC board.

Today, nearly all consumer electronic devices are made with surface mount technology. Hams have lagged behind in adopting this technology largely due to the misconception that it is difficult, requires extensive practice and requires special equipment. In fact, surface mount devices can be soldered easily with commonly available equipment. There is no more practice required to become proficient enough to produce a circuit with surface-mount (SM) devices than there is with soldering through-hole (leaded) components.

There are several advantages to working with SMT:
• Projects are much more compact than if through-hole components are used
• SM parts are available that are not available in through-hole packages
• Fewer and fewer through-hole parts are being produced
• Equivalent SM components are often cheaper than through-hole parts, and
• SM parts have less self-inductance, less self-capacitance and better thermal properties.

There are several techniques that can be effectively used by hams to work with SM components; conventional soldering iron, hot air reflow and hot plate/hot air reflow. This section describes the soldering iron technique. On-line descriptions of reflow techniques are available for the advanced builder that wants to try them. As is the case for through-hole soldering, kits are available to teach the beginner how to solder SMT components.

The following material on working with surface-mount technology contains excerpts from a series of *QST* articles, "Surface Mount Technology — You Can Work with It! Part 1" by Sam Ulbing, N4UAU, published in the April 1999 issue. (The entire series of four articles is available on the CD-ROM included with this book. Additional information on SMT is available at **www.arrl.org/surface-mount-technology**.) Additional information and illustrations are presented in the sidebar on SMT construction.

23.4.1 Equipment Needed

You do not need lots of expensive equipment to work with SM devices.
• A fundamental piece of equipment for SM work is an illuminated magnifying glass as seen in the sidebar. You can use an inexpensive one with a 5-inch-diameter lens, and it's convenient to use the magnifier for all soldering work, not just for SM use. Such magnifiers are widely available from about $25. Most offer a 3× magnification and have a built-in circular light.
• A low-power, temperature-controlled soldering iron is necessary. Use a soldering iron with a grounded tip as most SM parts are CMOS devices and are subject to possible ESD (static) failure.
• Use of thin (0.020-inch diameter) rosin-core solder is preferred because the parts are so small that regular 0.031-inch diameter solder will flood a solder pad and cause bridging. Solder paste or cream can also be used.
• A flux pen comes in handy for applying just a little flux at a needed spot.
• Good desoldering braid is necessary to remove excess solder if you get too much on a pad. 0.100-inch wide braid works well.
• ESD protective devices such as wrist straps may be necessary if you live in a dry area and static is a problem.
• Tweezers help pick up parts and position them. Nonmagnetic, stainless-steel drafting dividers also work well. The nonmagnetic property of stainless steel means the chip doesn't get attracted to the dividers.

23.4.2 Surface-Mount Parts

Fig 23.4 shows some common SM parts. (The **Component Data and References** chapter has more information on component packages.) Resistors and ceramic capacitors come in many different sizes, and it is important to

Surface Mount Construction Techniques

"Oh no, this project uses SMT parts!" Some homebrewers recoil at the thought of assembling a kit that uses surface mount technology (SMT) components. They fear the parts are too small to see, handle, solder or debug when assembled. With experience, it isn't so difficult when using the right tools. Further, using SMT parts can make QRP projects smaller, lighter and more portable for optimized field use.

Two quite different projects will illustrate some successful SMT assembly techniques. One is a small DDS signal generator "daughtercard" kit that comes with an assortment of SMT capacitors, resistors and inductors, and an SOIC integrated circuit. The other example circuit is a small one-stage audio amplifier built "Manhattan-style"! Yes, you *can* homebrew using SMT parts — results can sometimes be even better than when using conventional leaded parts.

But first, here is some component history and what you need to do to get your work area ready for constructing an SMT project.

What is an SMT Component?

Resistors and capacitors with axial or radial leads have been most common over the years. Same too for integrated circuits arranged in dual inline package (DIP) format with rows of leads separated by a generous 0.3 inch or so. This open-leaded component and easily accessed IC pins made for easy circuit board assembly back in the Heathkit days. Although these types of components are still available today, parts miniaturization has brought about more compact and less expensive products. Discrete components packaging has shrunk to 0.12 × 0.06 inch, as shown in the "1206" capacitor in **Fig 23.A1** compared to a penny. Even smaller packages are common today, requiring much less pc board area for the same equivalent circuits. Integrated circuit packaging has also been miniaturized to create 10 × 5 mm SOIC packages with lead separations of 0.025 inch. You need some extra skills beyond what was necessary when assembling a Heath SB-104 transceiver back in 1974!

Preparing for the Job

The key to success with any construction project is selecting and using the proper tools. A magnifying lamp is essential for well-lighted, close-up work on the components. **Fig 23.A2** shows a convenient magnifying visor. Tweezers or fine-tipped pliers allow you to grab the small chip components with dexterity. Thinner solder (0.015 inch) than you might normally use is preferred because it melts more quickly and leaves a smaller amount of solder on the component lead. Use of a super fine-tipped soldering iron make soldering the leads of these small parts straightforward and easy. A clean work surface is of paramount importance because SMT components have a tendency to fly away even when held with the utmost care by tweezers — you'll have the best chance of recovering your wayward part if your table is clear. When the inevitable happens, you'll have lots of trouble finding it if the part falls onto a rug. It's best to have your work area in a room without carpeting, for this reason as well as to protect static-sensitive parts.

Assembling SMT Parts on a PC board

The first project example is the DDS Daughtercard — a small module that generates precision RF signals for a variety of projects. This kit has become immensely popular in homebrew circles and is supplied with the chip components contained in color-coded packaging that makes and easy job of identifying the little parts, a nice touch by a kit supplier.

Fig 23.A3 shows the DDS PC board, a typical layout for SMT components. All traces are on one side, since the component leads are not "through-hole." The little square pads are the places where the 1206 package-style chips will eventually be soldered.

The trick to soldering surface-mount devices to PC boards is to (a) pre-solder ("tin") one of the pads on the board where the component will ultimately go by placing a small blob of solder there;

Fig 23.A1 — SMT components are small. Clockwise from left: MMIC RF amp, 1206 resistor, SOIC integrated circuit, 1206 capacitor and ferrite inductor.

Fig 23.A2 — A magnifying visor is great for close-up work on a circuit board. These headsets are often available for less than $10 at hamfests and some even come with superbright LEDs mounted on the side to illuminate the components being soldered.

Fig 23.A3 — This DDS Daughtercard has all interconnections on the top side. Connections to the ground plane on the backside of the board are made by the use of "vias," wires through the PC board. Pin 28 of the SMT IC is shown being tack-soldered to hold it to the board, keeping all other pins carefully aligned on their pads. Then the other pins are carefully soldered, starting with pin 14 (opposite pin 28). Finally, pin 28 is reheated to ensure a good connection there. If you bridge solder across adjacent pads or pins, use solder wick or a vacuum solder sucker to draw off the excess solder.

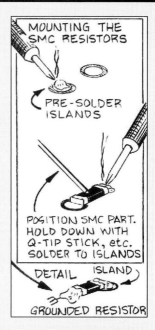

Fig 23.A4 — Attaching an SMT part. Things are a lot easier attaching capacitors, resistors and other discrete components compared to multi-pin ICs. Carefully hold the component in place and properly aligned using needle-nose pliers or tweezers and then solder one end of the component. Then reheat the joint while gently pushing down on the component with the pliers or a Q-tip stick to ensure it is lying flat on the board. Finally, solder the other side of the component.

Fig 23.A5 — The fully-populated DDS Daughtercard PC board contains a mix of SMT and through-hole parts, showing how both packaging technologies can be used together.

Fig 23.A6 — SMT resistors soldered to base board of the Audio Amp in the beginning stages of assembly.

Fig 23.A8 — Surface mount ICs can be mounted to general-purpose carrier boards, then attached as a submodule with wires to the base board of the homebrew project.

(b) carefully hold the component in place with small needle-nose pliers or sharp tweezers on the tinned pad; (c) reheat the tinned pad and component to reflow the solder onto the component lead, thus temporarily holding the component in place; and (d) solder the other end of the component to its pad. Finally, check all connections to make sure there are no bridges or shorts. **Fig 23.A4** illustrates how to use this technique. **Fig 23.A5** shows the completed DDS board.

Homebrewing with SMT Parts

The second project example is the K8IQY Audio Amp — a discrete component audio amplifier that is constructed "Manhattan-style." You glue little pads to the board wherever you need to attach component leads or wires. See **Fig 23.A6**.

Instead of using little squares or dots of PC board material for pads, you might decide to create isolated connection points by cutting an "island" in the copper using an end mill. No matter how the pads are created,

Fig 23.A7 — The completed homebrew Audio Amp assembly shows simple, effective use of SMT components used together with conventional leaded components when constructed "Manhattan-style."

SMT components may be easily soldered from pad-to-pad, or from pad-to-ground plane to build up the circuit. **Fig 23.A7** shows the completed board, combining SMT and leaded components.

Homebrewing with SOIC-packaged integrated circuits is a little trickier and typically requires the use of an "SOIC carrier board" such as the one shown in **Fig 23.A8**, onto which you solder your surface mount integrated circuit. You can then wire the carrier board onto your homebrew project, copper-clad base board or whatever you're using to hold your other circuit components.

Full details on the DDS Daughter-card, the K8IQY Islander Audio Amp, and the Islander Pad Cutter may be found online at **www.njqrp.org**.

So Start Melting Solder!

The techniques are easy, the SMT components are actually cheaper than conventional through-hole leaded components and you'll have a smaller, more portable project when you're done. Go for it! — *George Heron, N2APB*

know the part size for two reasons: Working with SM devices by hand is easier if you use the larger parts; and it is important that the PC-board pad size is larger than the part.

Tantalum capacitors are one of the larger SM parts. Their case code, which is usually a letter, often varies from manufacturer to manufacturer because of different thicknesses. The EIA code for ceramic capacitors and resistors is a measurement of the length and width in inches, but for tantalum parts, those measurements are in millimeters times 10! Keep in mind that tantalum capacitors are polarized; the case usually has a mark or stripe to indicate the positive end. Nearly any part that is used in through-hole technology is available in an SM package.

If you are a beginner, it is probably best to start with the larger sizes; 1206 for resistors and capacitors, SOT-23 for transistors, and SO-8 for ICs. When you get proficient with these parts you can move to smaller ones.

23.4.3 SMT Soldering Basics

Use a little solder to pre-tin the PC board pads. The trick is to add just enough solder so that when you reheat it, it flows to the IC, but not so much that you wind up with a solder bridge. Putting a (very) little flux on the board and the IC legs makes for better solder flow, providing a smooth layer. You can tell if you have the proper soldering-iron tip temperature if the solder melts within 1.5 to 3.5 seconds.

Place the part on the board and then use dividers (or fingers) to push and prod the chip into position. Because the IC is so small and light, it tends to stick to the soldering iron and pull away from the PC board. To prevent this, use the dividers to hold the chip down while tack-soldering two IC legs at diagonally opposite corners. After each tack, check that the part is still aligned. With a dry and clean soldering iron, heat the PC board near the leg. If you do it right, when the solder melts, it will flow to the IC.

The legs of the IC must lie flat on the board. The legs bend easily, so don't press down too hard. Check each connection with a continuity checker placing one tip on the board the other on the IC leg. Check all adjacent pins to ensure there's no bridging. It is easier to correct errors early on, so perform this check often.

If you find that you did not have enough solder on the board for it to flow to the part, add a little solder. It's best to put a drop on the trace near the part, then heat the trace and slide the iron and melted solder toward the part. This reduces the chance of creating a bridge. Soldering resistors and capacitors is similar to soldering an IC's leads, except the resistors and capacitors don't have exposed leads. The reflow method works well for these parts, too.

Attaching wires that connect to points off the board can be a bit of a challenge because even #24 AWG stranded wire is large in comparison to the SM parts. First, make sure all the wire strands are close together, then pre-tin the wire. Pre-tin the pad, carefully place the wire on the pad, then heat it with the soldering iron until the solder melts.

While SMT projects can be built with conventional solder and a fine-point soldering iron, if you move on to reflow techniques you will need to use solder paste. A small dot of solder paste is put on the board at each location a component will need to be soldered. The whole board is heated in an oven or with a heat gun, the solder melts and flows onto the pad and component contacts, then the board is cooled leaving all of the components soldered in place.

Only a small amount of paste is required. Kester Easy Profile 256 is a good solder paste to start with. Kester sells this in large quantities, at least for hams, and it is rather expensive in the smallest quantity they sell. Fortunately, it can be obtained in a small syringe with a fine needle-point applicator. As only a small amount is needed for each solder joint, this small amount will last through several medium sized projects. Solder paste must be kept cold or it deteriorates. Kept in a household refrigerator or freezer it has a shelf life of at least a year.

23.4.4 Removing SMT Components

The surface-mount ICs used in commercial equipment are not easy for experimenters to replace. They have tiny pins designed for precision PC boards. Sooner or later, you may need to replace one, though. If you do, don't try to get the old IC out in one piece! This will damage the IC beyond use anyway, and will probably damage the PC board in the process.

Although it requires a delicate touch and small tools, it's possible to change a surface mount IC at home. To remove the old one, use small, sharp wire cutters to cut the IC pins flush with the IC. This usually leaves just enough of the old pin to grab with tweezers. Heat the soldered connection with a small iron and use the pliers to gently pull the pin from the PC board. Use desoldering braid to remove excess solder from the pads and remove and flux with rubbing alcohol. Solder in the new component using the techniques discussed above.

To remove individual components, first remove excess solder from the pads by using desoldering braid. Then the component will generally come loose from the pads if gently lifted with a hobby knife or dental tool. If the component remains attached to the pad, touch the pad with the soldering iron and lift the component off the pad. It may take one or two attempts to free the component from all pads. In extreme cases, it may be necessary to flood the pad with solder to completely loosen the component.

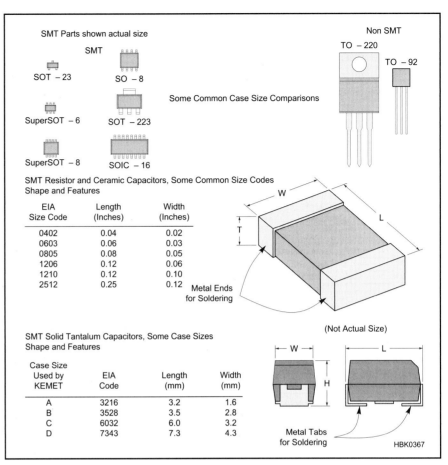

Fig 23.4 — Size comparisons of some surface-mount devices and their dimensions.

23.5 Electronic Circuits

Most of the construction projects undertaken by the average amateur involve electronic circuitry. The circuit is the "heart" of most amateur equipment. It might seem obvious, but in order for you to build it, the circuit must work! Don't always assume that a "cookbook" circuit that appears in an applications note or electronics magazine is flawless. These are sometimes design examples that have not always been thoroughly debugged. Many home-construction projects are one-time deals; the author has put one together and it worked. In some cases, component tolerances or minor layout changes might make it difficult to get a second unit working.

23.5.1 Electrostatic Discharge (ESD)

You need to take steps to protect the electronic and mechanical components you use in circuit construction. Some components can be damaged by rough handling. Dropping a ¼-W resistor causes no harm, but dropping a vacuum tube or other delicate subassemblies usually causes damage.

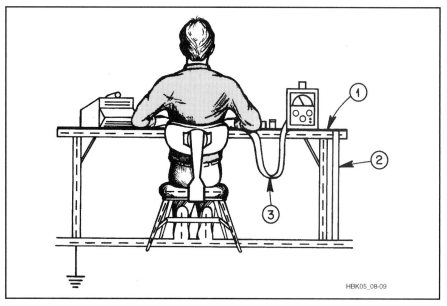

Fig 23.5 — A work station that has been set up to minimize ESD features (1) a grounded dissipative work mat and (2) a wrist strap that (3) grounds the worker through high resistance.

Some components are easily damaged by heat. Some of the chemicals used to clean electronic components (such as flux removers, degreasers or control-lubrication sprays) can damage plastic. Check them for safety before you use them.

Some components, especially high-impedance components such as FETs and CMOS gates, can be damaged by electrostatic discharge (ESD). Protect these parts from static charges. Most people are familiar with the static charge that builds up when one walks across a carpet then touches a metal object; the resultant spark can be quite lively. Walking across a carpet on a dry day can generate 35 kV! A worker sitting at a bench can generate voltages up to 6 kV, depending on conditions, such as when relative humidity is less than 20%.

You don't need this much voltage to damage a sensitive electronic component; damage can occur with as little as 30 V. The damage is not always catastrophic. A MOSFET can become noisy, or lose gain; an IC can suffer damage that causes early failure. To prevent this kind of damage, you need to take some precautions.

The energy from a spark can travel inside a piece of equipment to affect internal components. Protection of sensitive electronic components involves the prevention of static build-up together with the removal of any existing charges by dissipating any energy that does build up.

MINIMIZING STATIC BUILD-UP

Several techniques can be used to minimize static build-up. Start by removing any carpet in your work areas. You can replace it with special antistatic carpet, but that is expensive. It's less expensive to treat the carpet with antistatic spray, which is available from electronics wholesalers.

Even the choice of clothing you wear can affect the amount of ESD. Polyester has a much greater ESD potential than cotton.

Many builders who have their workbench on a concrete floor use a rubber mat to minimize the risk of electric shocks from the ac line. Unfortunately, the rubber mat increases the risk of ESD. An antistatic rubber mat can serve both purposes.

Many components are shipped in antistatic packaging. Leave components in their conductive packaging. Other components, notably MOSFETs, are shipped with a small metal ring that temporarily shorts all of the leads together. Leave this ring in place until the device is fully installed in the circuit.

Use antistatic bags to transport susceptible components or equipment. Keep your workbench free of objects such as paper, plastic and other static-generating items. Use conductive containers with a dissipative surface coating for equipment storage.

These precautions help reduce the build-up of electrostatic charges. Other techniques offer a slow discharge path for the charges or keep the components and the operator handling them at the same ground potential.

DISSIPATING STATIC

One of the best techniques is to connect the operator and the devices being handled to earth ground, or a common reference point. It is not a good idea to directly ground an operator working on electronic equipment, though; the risk of shock is too great. If the operator is grounded through a high-value resistor, ESD protection is still offered but there is no risk of shock.

The operator is usually grounded through a conductive wrist strap. 3M makes a grounding wrist band. This wrist band is equipped with a snap-on ground lead. A 1-MΩ resistor is built into the snap of the strap to protect the user should a live circuit be contacted. Build a similar resistor into any homemade ground strap.

The devices and equipment being handled are also grounded, by working on a charge-dissipating mat that is connected to ground. The mat should be an insulator that has been impregnated with a resistance material. Suitable mats and wrist straps are available from most electronics supply houses. **Fig 23.5** shows a typical ESD-safe work station.

The work area should also be grounded, directly or through a conductive mat. Use a soldering iron with a grounded tip to solder sensitive components. Most irons that have three-wire power cords are properly grounded. When soldering static-sensitive devices, use two or three jumpers: one to ground you, one to ground the work, and one to ground the iron. If the iron does not have a ground wire in the power cord, clip a jumper from the metal part of the iron near the handle to the metal box that houses the temperature con-

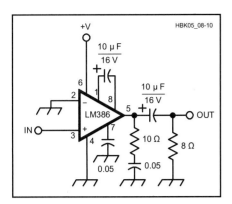

Fig 23.6 — Schematic diagram of the audio amplifier used as a design example of various construction techniques.

Fig 23.7 — The example audio amplifier of Fig 23.6 built using ground-plane construction.

trol. Another jumper connects the box to the work. Finally, a jumper goes from the box to an elastic wrist band for static grounding.

23.5.2 Electronics Construction Techniques

Kits are fun, but another facet of electronics construction is building and developing your own circuits, starting from circuit diagrams. Several different point-to-point wiring techniques or printed-circuit boards (PC boards) can be used to construct electronic circuits. Most circuit projects use a combination of techniques. The selection of techniques depends on many different factors and builder preferences.

For one-time construction, PC boards are really not necessary. It takes time to lay out, drill and etch a PC board. Alterations are difficult to make if you change your ideas or make a mistake. Most importantly, PC boards aren't always the best technique for building RF circuits.

The simple audio amplifier shown in **Fig 23.6** will be built using various point-to-point or PC-board techniques. This shows how the different construction methods are applied to a typical circuit. (Surface-mount techniques are discussed in the previous section.)

POINT-TO-POINT TECHNIQUES

Point-to-point techniques include all circuit construction techniques that rely on tie points and wiring, or component leads, to build a circuit. This is the technique used in most home-brew construction projects. It is sometimes used in commercial construction, such as old vacuum-tube receivers and modern tube amplifiers.

Point-to-point wiring is also used to connect the "off-board" components used in a printed-circuit project. It can be used to interconnect the various modules and printed-circuit boards used in more complex electronic systems. Most pieces of electronic equipment have at least some point-to-point wiring.

GROUND-PLANE CONSTRUCTION

A point-to-point construction technique that uses the leads of the components as tie points for electrical connections is known as "ground-plane construction," "dead-bug" or "ugly construction." (The term "ugly construction" was coined by Wes Hayward, W7ZOI.) "Dead-bug construction" gets its name from the appearance of an IC with its leads sticking up in the air. In most cases, this technique uses copper-clad circuit-board material as a foundation and ground plane on which to build a circuit using point-to-point wiring, so in this chapter it is called "ground-plane construction." An example is shown in **Fig 23.7**.

Ground-plane construction is quick and simple: You build the circuit on an unetched piece of copper-clad circuit board. Wherever a component connects to ground, you solder it to the copper board. Ungrounded connections between components are made point-to-point. Once you learn how to build with a ground-plane board, you can grab a piece of circuit board and start building any time you see an interesting circuit.

A PC board has strict size limits; the components must fit in the space allotted. Ground-plane construction is more flexible; it allows you to use the parts on hand. The circuit can be changed easily — a big help when you are experimenting. The greatest virtue of ground-plane construction is that it is fast.

Ground-plane construction is something like model building, connecting parts using solder almost — but not exactly — like glue. In ground-plane construction you build the circuit directly from the schematic, so it can help you get familiar with a circuit and how it works. You can build subsections of a large circuit on small ground-plane modules and string them together into a larger design.

Circuit connections are made directly, minimizing component lead length. Short lead lengths and a low-impedance ground conductor help prevent circuit instability. There is usually less inter-component capacitive coupling than would be found between PC-board traces, so it is often better than PC-board construction for RF, high-gain or sensitive circuits.

Use circuit components to support other circuit components. Start by mounting one component onto the ground plane, building from there. There is really only one two-handed technique to mount a component to the ground plane. Bend one of the component leads at a 90° angle, and then trim off the excess. Solder a blob of solder to the board surface, perhaps about 0.1 inch in diameter, leaving a small dome of solder. Using one hand, hold the component in place on top of the soldered spot and reheat the component and the solder. It should flow nicely, soldering the component securely. Remove the iron tip and hold the component perfectly still until the solder cools. You can then make connections to the first part.

Connections should be mechanically secure before soldering. Bend a small hook in the lead of a component, then "crimp" it to the next component(s). Do not rely only on the solder connections to provide mechanical strength; sooner or later one of these connections will fail, resulting in a dead circuit.

In most cases, each circuit has enough grounded components to support all of the components in the circuit. This is not always possible, however. In some circuits, high-value resistors can be used as standoff insulators. One resistor lead is soldered to the copper ground plane; the other lead is used as a circuit connection point. You can use ¼- or ½-W resistors in values from 1 to 10 MΩ. Such high-value resistors permit almost no current to flow, and in low-impedance circuits they act more like insulators than resistors. As a rule of thumb, resistors used as standoff insulators should have a value that is at least 10 times the circuit impedance at that circuit point.

Fig 23.8A shows how to use the standoff technique to wire the circuit shown at Fig 23.8C. Fig 23.8B shows how the resistor leads are bent before the standoff component is soldered to the ground plane. Components E1 through E5 are resistors that are used as standoff insulators. They do not appear in the schematic diagram. The base circuitry at Q1 of Fig 23.8A has been stretched out to reduce clutter in the drawing. In a practical circuit, all of the signal leads should be kept as short as possible. E4 would, therefore, be placed much closer to Q1 than the drawing indicates.

No standoff posts are required near R1 and R2 of Fig 23.8. These two resistors serve two purposes: They are not only the normal circuit resistances, but function as standoff posts as well. Follow this practice wherever a capacitor or resistor can be employed in the dual role.

WIRED TRACES — THE LAZY PC BOARD

If you already have a PC-board design, but don't want to copy the entire circuit — or you don't want to make a double-sided PC board — then the easiest construction technique is to use a bare board or perfboard and hardwire the traces.

Drill the necessary holes in a piece of single-sided board, remove the copper ground plane from around the holes, and then wire up the back using component leads and bits of wire instead of etched traces (**Fig 23.9**).

To transfer an existing board layout, make a 1:1 photocopy and tape it to your piece of PC board. Prick through the holes with an automatic (one-handed) center punch or by firm pressure with a sharp scriber, remove the photocopy and drill all the holes. Holes for ground leads are optional — you generally get a better RF ground by bending the component lead flat to the board and soldering it down. Remove the copper around the rest of the holes by pressing a drill bit lightly against the hole and twisting it between your fingers. A drill press can also be used, but either way, don't remove too much board material. Then wire up the circuit beneath the board. The results look very neat and tidy — from the top, at least!

Circuits that contain components originally designed for PC-board mounting are good candidates for this technique. Wired traces would also be suitable for circuits involving multi-pin RF ICs, double-balanced mixers and similar components. To bypass the pins of these components to ground, connect a miniature ceramic capacitor on the bottom of the board directly from the bypassed pin to the ground plane.

A wired-trace board is fairly sturdy, even though many of the components are only held in by their bent leads and blobs of solder. A drop of cyanoacrylate "super glue" can hold down any larger components, components with fragile leads or any long leads or wires that might move.

PERFORATED CONSTRUCTION BOARD

A simple approach to circuit building uses a perforated board (perfboard). Perfboard is available with many different hole patterns. Choose the one that suits your needs. Perfboard is usually unclad, although it is made with pads that facilitate soldering.

Circuit construction on perforated board is easy. Start by placing the components loosely on the board and moving them around until a satisfactory layout is obtained. Most of the construction techniques described in this chapter can be applied to perfboard. The audio amplifier of Fig 23.6 is shown constructed with this technique in **Fig 23.10**.

Perfboard and accessories are widely available. Accessories include mounting hardware and a variety of connection terminals for solder and solderless construction.

TERMINAL AND WIRE

A perfboard is usually used for this technique (**Fig 23.11**). Push terminals are inserted into the hole in a perfboard. Components can then be easily soldered to the terminals. As an alternative, drill holes into a bare or copper-clad board wherever they are needed. The components are usually mounted on one side of the board and wires are soldered to the bottom of the board, acting as wired PC-board "traces." If a component has a reasonably rigid lead to which you can attach other components, use that instead of a push terminal, a modification of the ground-plane

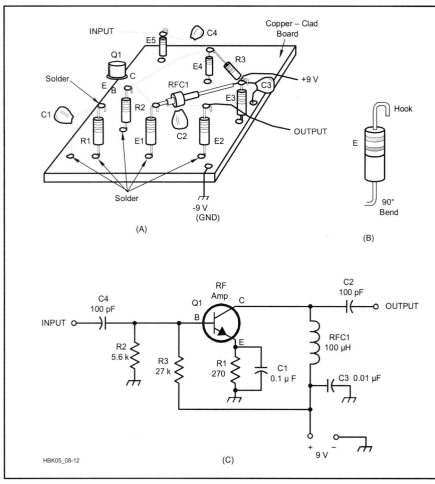

Fig 23.8 — Pictorial view of a circuit board that uses ground-plane construction is shown at A. A close-up view of one of the standoff resistors is shown at B. Note how the leads are bent. The schematic diagram at C shows the circuit displayed at A.

Fig 23.9 — The audio amplifier built using wired-traces construction.

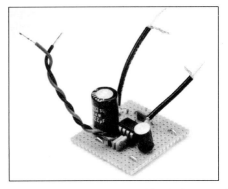
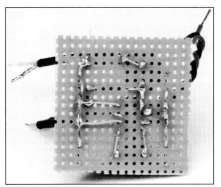

Fig 23.10 — The audio amplifier built on perforated board. Top view at left; bottom view at right.

Fig 23.12 — The audio amplifier built on a solderless prototyping board.

Fig 23.11 — The audio amplifier built using terminal-and-wire construction.

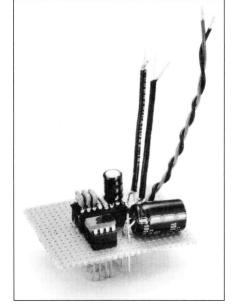
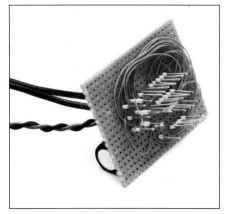

Fig 23.13 — The audio amplifier built using wire-wrap techniques.

construction technique.

If you are using a bare board to provide a ground plane, drill holes for your terminals with a high-speed PC-board drill and drill press. Mark the position of the hole with a center punch to prevent the drill from skidding. The hole should provide a snug fit for the push terminal.

Mount RF components on top of the board, keeping the dc components and much of the interconnecting wiring underneath. Make dc feed-through connections with terminals having bypass capacitors on top of the board. Use small solder-in feedthrough capacitors for more critical applications.

SOLDERLESS PROTOTYPE BOARD

One construction alternative that works well for audio and digital circuits is the solderless prototype board (protoboard), shown in **Fig 23.12**. It is usually not suitable for RF circuits, although audio circuits up to 100 kHz can usually be made with solderless prototype board.

A protoboard has rows of holes with spring-loaded metal strips inside the board. Circuit components and hookup wire are inserted into the holes, making contact with the metal strips. Components that are inserted into the same row are connected together.

Component and interconnection changes are easy to make.

Protoboards have some minor disadvantages. The boards are not good for building RF circuits; the metal strips add too much stray capacitance to the circuit. Large component leads can deform the metal strips.

WIRE-WRAP CONSTRUCTION

Wire-wrap techniques can be used to quickly construct a circuit without solder. Low- and medium-speed digital circuits are often assembled on a wire-wrap board. The technique is not limited to digital circuits, however. **Fig 23.13** shows the audio amplifier built using wire wrap. Circuit changes are easy to make, yet the method is suitable for permanent assemblies.

Wire wrap is done by wrapping a wire around a small square post to make each connection. A wrapping tool resembles a thick pencil. Electric wire-wrap guns are convenient when many connections must be made. The wire is almost always #30 wire with thin insulation. Two wire-wrap methods are used: the standard and the modified wrap (**Fig 23.14**). The modified wrap is more secure. The wrap-post terminals are square (wire wrap works only on posts with sharp corners). They should be long enough for at least two connections.

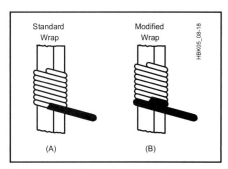

Fig 23.14 — Wire-wrap connections. Standard wrap is shown at A; modified wrap at B.

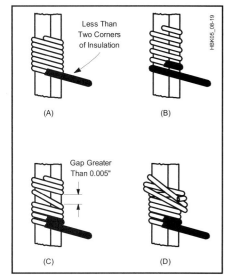

Fig 23.15 — Improper wire-wrap connections. Insufficient insulation for modified wrap is shown at A; a spiral wrap at B, where there is too much space between turns; an open wrap at C, where one or more turns are improperly spaced and an overwrap at D, where the turns overlap on one or more turns.

Fig 23.14 and **Fig 23.15** show proper and improper wire-wrap techniques. Mount small components on an IC header plug. Insert the header into a wire-wrap IC socket as shown in Fig 23.13. The large capacitor in that figure has its leads soldered directly to wire-wrap posts.

"READY-MADE" UTILITY PC BOARDS

"Utility" PC boards are an alternative to custom-designed etched PC boards. They offer the flexibility of perforated board construction and the mechanical and electrical advantages of etched circuit connection pads. Utility PC boards can be used to build anything from simple passive filter circuits to computers.

Circuits can be built on boards on which the copper cladding has been divided into connection pads. Power supply voltages can be distributed on bus strips. Boards like those shown in **Fig 23.16** are commercially available.

An audio amplifier constructed on a utility PC board is shown in **Fig 23.17**. Component leads are inserted into the board and soldered to the etched pads. Wire jumpers connect the pads together to complete the circuit.

Utility boards with one or more etched plugs for use in computer-bus, interface and general purpose applications are widely available. Connectors, mounting hardware and other accessories are also available. Check with your parts supplier for details.

23.5.3 Printed-Circuit Boards

PC boards are everywhere — in all kinds of consumer electronics, in most of your Amateur Radio equipment. They are also used in most kits and construction projects. A newcomer to electronics might think that there is some unwritten law against building equipment in any other way!

The misconception that everything needs to be built on a printed-circuit board is often a stumbling block to easy project construction. In fact, a PC board is probably the worst choice for a one-time project. In actuality, a moderately complex project (like a QRP transmitter) can be built in much less time using other techniques such as those described in the preceding section. The additional design, layout and manufacturing is usually much more work than it would take to build the project by hand.

So why does everyone use PC boards? The most important reason is that they are reproducible. They allow many units to be mass-produced with exactly the same layout, reducing the time and work of conventional wiring and minimizing the possibilities of wiring errors. If you can buy a ready-made PC board or kit for your project, it can save a lot of construction time. This is true because someone else has done most of the real work involved — designing the PC board layout and fixing any "bugs" caused by inter-trace capacitive coupling, ground loops and similar problems. In most cases, if a ready-made board is not available, ground-plane construction is a lot less work than designing, debugging and then making a PC board.

Using a PC board usually makes project construction easier by minimizing the risk of wiring errors or other construction blunders. Inexperienced constructors usually feel more confident when construction has been simplified to the assembly of components onto a PC board. One of the best ways to get started with home construction (to some the best part of Amateur Radio) is to start by assembling a few kits using PC boards. A list of kit manufacturers can be found on the QRP ARCI Web site, **www.qrparci.org,** under "QRP Links."

To make a PC board, resist material is applied to a copper-clad bare PC board, which is then immersed into an acid etching bath to remove selected areas of copper. In a finished board, the conductive copper is formed into a pattern of conductors or "traces" that form the actual wiring of the circuit. The following sections describe how to make your own PC boards.

Fig 23.16 — Utility PC boards like these are available from many suppliers.

Fig 23.17 — The audio amplifier built on a multipurpose PC breadboard. Top view at A; bottom view at B.

Construction Techniques   23.17

How to Buy Parts for Electronics Projects

The number one question received by the ARRL Technical Information Service starts out "Where can I buy…" It seems that one of the most perplexing problems faced by the would-be constructor is where to get parts. Sometimes you are lucky — the project author has made a kit or a PC board or key parts available. But not every project has a ready-made kit. If you would like to expand your construction horizons, you can learn to be your own "purchasing manager." That means searching out parts sources and dealing with them yourself.

In reality, it is not all that difficult to find most parts. If you are lucky, you may have a local source of electronics parts. Look in the Yellow Pages under "Electronic Equipment Suppliers" to find the local outlets. RadioShack is one local source found nearly everywhere. They carry an assortment of the more common electronic parts.

Unfortunately, though, in many areas of the country it's just not possible to find a local electronic parts supplier that can provide everything for your next project. This is not surprising. Years ago an electronics supplier had to stock a relatively small number of electronic components — resistors, capacitors, tube sockets, a few relays and variable resistors. Technology has increased the number of components by a few orders of magnitude.

Nowadays, the number of integrated circuits alone is enough to fill a multi-volume book. No single electronics supplier could possibly stock them all. It has become a mail-order world; the electronics world is no exception. If you can't find what you need locally to fill your electronic needs, the good news is that an astonishing variety of components is literally a mouse click or phone call away.

Almost all of these suppliers have detailed catalogs on their Web sites, and some offer printed catalogs. Bookmark the Web sites and collect their catalogs, and you'll have a ready source of parts for your next project. It's good to have a variety of sources, as you'll probably need to order from more than one company. (It's almost a corollary to Murphy's Law: No matter how wide a selection you find at one supplier, there's always at least one part you must buy somewhere else!)

When planning for a project you want to build, you need to convert the parts list and/or schematic into a parts-order list that shows the catalog number and quantity required of each component. This may require a similar list for each vendor when one supplier does not have all the parts.

Check the type, tolerance, power rating and other key characteristics of the parts. Group the parts by those parameters before grouping them by value. If all of the circuit's components are already grouped by value on the parts list, you can just count the number of each value. Each time you add parts to the order list, check them off the published parts list. Sometimes the parts list does not include common components like resistors and capacitors. If this is the case, make a copy of the schematic and check off the parts as you build your shopping list.

It's generally a good idea to locate sources for all needed parts before ordering anything. Sometimes a key component can be hard to find in small quantities, prohibitively expensive, or obsolete. Be prepared to substitute similar parts if the exact ones needed are not available.

Although you'll probably be able to order exactly the right number of each part for a project, buy a few extras of some parts for your junk box. It's always good to have a few extra parts on hand; you may break a component lead during assembly, or damage a solid-state component with too much heat or by wiring it in backwards. If you don't have extras, you'll need to order another part. Even if you don't need the extras for this project, they may come in handy, and you'll be encouraged to build another project! Pick up an extra toroid or two as well.

Now's the time to decide whether you're going to build your project with ground-plane construction or PC board. If you need a PC board, FAR Circuits and others have them for many ARRL book and *QST* projects. If one is available, that information is usually included in the article. If you're going to use ground-plane construction, buy a good-sized piece of single-sided copper-clad board — glass-epoxy board if you can. Phenolic board is inferior because it is brittle and deteriorates rapidly with soldering heat.

Don't forget an enclosure for your project. This is often overlooked in parts lists for most projects, because different builders like different enclosures. Make sure there's room in the box for all of the components used in your project. Some people like to cram projects in the smallest possible box, but miniaturization can be extremely frustrating if you're not good at it.

There are almost always a few items you can't get from one company and most have minimum orders. You may need to distribute your order between two or more companies to meet minimum-order requirements.

If you order enough parts, you'll soon find out which companies you like to deal with and which have slow service. It is frustrating to receive most of an order, then wait months for the parts that are on backorder. Fortunately, most of the larger suppliers indicate stock status on their Web sites. And most companies that cater to hams can tell you over the phone if a part is out of stock.

If you're ordering by mail and don't want the company to backorder your parts, write clearly on the order form, "Do not backorder parts." They will then ship the parts they have and leave you to order the rest from somewhere else.

If you are familiar with the Web sites, catalogs and policies of electronic-component suppliers, you will find that getting parts is not difficult. Concentrate on the fun part — building the circuit and getting it working. — *Bruce Hale, KB1MW*

PC BOARD MATERIALS AND SUPPLIES

PC Board Stock

PC board stock consists of a sheet made from insulating material, usually glass epoxy or phenolic, coated with conductive copper. Copper-clad stock is manufactured with phenolic, FR-4 fiberglass and Teflon base materials in thicknesses up to $\frac{1}{8}$ inch. The copper thickness varies. It is usually plated from 1 to 2 oz per square foot of bare stock.

Resists

Resist is a material that is applied to a PC board to prevent the acid etchant from eating away the copper on those areas of the board that are to be used as conductors. There are several different types of resist materials, both commercial and home brew. When resist is applied to those areas of the board that are to remain as copper traces, it "resists" the acid action of the etchant.

The PC board stock must be clean before any resist is applied. This is discussed later in the chapter. After you have applied resist, by whatever means, protect the board by handling it only at its edges. Do not let it get scraped. Etch the board as soon as possible, to minimize the likelihood of oxidation, moisture or oils contaminating the resist or bare board.

Resist Pens

Several electronics suppliers sell resist pens. Use a resist pen to draw PC-board artwork directly onto a bare board. Commercially available resist pens work well. Several types of permanent markers also function as resist, especially the "Sharpie" brand. They come in fine-point and regular sizes; keep two of each on hand.

Tape

To make a single PC board, Scotch, adhesive or masking tape, securely applied, makes a good resist. (Don't use drafting tape; its glue may be too weak to hold in the etching bath.) Apply the tape to the entire board, transfer the circuit pattern by means of carbon paper, then cut out and remove the sections of tape where the copper is to be etched away. An X-Acto hobby knife is excellent for this purpose.

Paint

Some paints are good resists. Exterior enamel works well. Nail polish is also good, although it tends to dry quickly so you must work fast. Paint the pattern onto the copper surface of the board to be etched. Use an artist's brush to duplicate the PC board pattern onto bare PC-board stock. Tape a piece of carbon paper to the PC-board stock. Tape the PC-board pattern to the carbon paper. Trace over the original layout with a ballpoint pen. The carbon paper transfers the outline of the pattern onto the bare board. Fill in the outline with the resist paint. After paint has been applied, allow it to dry thoroughly before etching.

Rub-On Transfer

Several companies produce rub-on transfer material that can also be used as resist. Patterns are made with various width traces and for most components, including ICs. As the name implies, the pad or trace is positioned on the bare board and rubbed to adhere to the board.

Etchant

Etchant is an acid solution that is designed to remove the unwanted copper areas on PC-board stock, leaving other areas to function as conductors. Almost any strong acid bath can serve as an etchant, but some acids are too strong to be safe for general use. Two different etchants are commonly used to fabricate prototype PC boards: ammonium persulphate and ferric chloride. The sidebar "Making PC Boards With Printed Artwork" describes a hydrogen peroxide-muriatic acid etchant. Ferric chloride is the most commonly used etchant and is usually sold ready-mixed. It is made from one part ferric chloride crystals and two parts water, by volume. No catalyst is required and it does not lose strength when stored.

Etchant solutions become exhausted as they are used. Keep a supply on hand. Dispose of the used solution safely; follow the instructions of your local environmental protection authority.

Most etchants work better if they are hot. A board that takes 45 minutes to etch at room temperature will take only a few minutes if the etchant is hot. Use a heat lamp to warm the etchant to the desired temperature. A darkroom thermometer is handy for monitoring the temperature of the bath.

Be careful! Do not heat your etchant above the recommended temperature, typically 160°F. If it gets too hot, it will probably damage the resist. Hot or boiling etchant is also a safety hazard.

Insert the board to be etched into the solution and agitate it continuously to keep fresh chemicals near the board surface. This speeds up the etching process. Normally, the circuit board should be placed in the bath with the copper side facing up.

After the etching process is completed, remove the board from the tray and wash it thoroughly with water. Use a household scrubbing pad, such as Scotchbrite, to rub off the resist. Using fine steel wool will also clean the board, but may leave fine steel particles behind.

WARNING: Use a glass or other non-reactive container to hold etching chemicals. Most etchants will react with a metal container. Etchant is caustic and can burn eyes or skin easily. Use rubber gloves and wear old clothing, or a lab smock, when working with any chemicals. If you get some on your skin, wash it with soap and cold water. Wear safety goggles (the kind that fit snugly on your face) when working with any dangerous chemicals. Read the safety labels and follow them carefully. If you get etchant in your eyes, wash immediately with large amounts of cool water and seek immediate medical help. Even a small chemical burn on your eye can develop into a serious problem.

PLANNING A LAYOUT BY HAND

A later section of this chapter explains how to turn a schematic into a working circuit. It is not as simple as laying out the PC board just like the circuit is drawn on the schematic. Read that section before you design a PC board.

Traditionally, amateurs have laid out artwork for the PC board by hand. While this is still quite viable, it can become tedious and error-prone for more complex circuits. There are many PC board layout software packages available for free or at very low cost. For example, the author of the sidebar on making PC boards from printed artwork, Chuck Adams, K7QO, uses *Eagle-5.3.0* layout software running under *Linux*. The free version can be used to lay out board sizes up to 3.25 by 4.0 inches with up to two layers. It has schematic capture, auto-router, and additional output formats to allow you to send files to a commercial PC board manufacturer if you want to do a club project requiring

---

On-Line PC Board Fabrication Services

In the past few years, on-line PC board fabrication services have become popular among hobbyists and professional designers. See the Computer-Aided Design chapter for a discussion of PCB design software and services. These services specialize in fast turnaround (two or three days, typically) of small boards in low quantities. Some accept artwork files in standard interchange formats and others have proprietary software packages. The cost per board is quite reasonable, considering the expense of maintaining the tools and techniques needed to construct boards in a home shop. The results are professional and high-quality.

One such service that was used by the developers of several projects in this *Handbook* is ExpressPCB (**www.expresspcb.com**). They offer double-sided and multilayer PCBs at rates based on the number square inches each board occupies. For example, if your PC occupies 3 × 5 inches, you can get two boards for about $75 — no solder mask or silk-screen of component placement is included — or four boards would cost about $100. Higher volumes are progressively less expensive per board. This makes club project boards quite attractive.

The software used to create the necessary artwork files is called *ExpressPCB* (for PC board layout) and *ExpressSCH* (for schematic capture). Both programs are available for download from **www.expresspcb.com** and work well together during the design process. While you can't send the layout artwork to other PC board fabricators, you can use schematic editor for all sorts of schematic jobs and save the files on your PC. You can also use the PC board artwork to etch boards at home, using the techniques described in this chapter.

ExpressPCB is typical of the on-line PC fabricator companies. There are quite a few such companies in business and you can find them by searching the Internet for "pcb fabrication". There are quite a range of capabilities and services, including technical design consulting and custom assembly services for small production runs.

---

many boards to be made. There are a number of such packages listed in the PCB section of the **Computer-Aided Design** chapter.

Rough Layout

Start by drawing a rough scale pictorial diagram of the layout. Draw the interconnecting leads to represent the traces that are needed on the board. Rearrange the layout as necessary to find an arrangement that com-

Making PC Boards With Printed Artwork

Chuck, K7QO, is a long time homebrewer. He operates what he builds and builds what he operates. Since he was first licensed 50 years ago, he has built rigs using nearly every conceivable construction technique available. He has recently developed this technique for fabricating printed circuit boards at home with easily available materials and a laser printer.

Materials

- ¼-cup and ½-cup plastic measuring cups (do NOT use these cups to prepare food)
- An ordinary clothes iron with a plain metallic ironing surface rather than one with a nonstick coating. The nonstick coating will wear off quickly with use in making printed circuit boards.
- Pyrex dish with 3-cup or 750-ml capacity with plastic/rubber snap-on cover.
- Rubber gloves, heavy duty and acid proof.
- Safety goggles
- Surgical gloves, disposable, for handling boards and chemicals.
- Small plastic tongs for lifting board from harsh chemicals.
- Disposable plastic food forks for moving board around in harsh chemicals.
- A larger glass container with screw-on cover for storage of used etchant.
- Bottle of 2% hydrogen peroxide.
- Clear enamel spray.
- Plastic scrubbing sponge such as ScotchBrite used on kitchen items. (Using steel wool leaves small iron filings behind.)
- Bottle of Tarn-X for cleaning copper.
- Large bag of large cotton balls. This for using both the Tarn-X to remove corrosion from the PC board material and for using acetone to remove the toner after the board is etched.
- Acetone, which can usually be found in the paint department of hardware stores or home improvement stores. Acetone is a hazardous chemical and should be used in a ventilated area and not breathed. Handle it with rubber gloves as it can cause skin irritation. Keep it tightly capped when not use.
- Muriatic acid. This can be found in the swimming pool section of home improvement stores or in the outdoor section as it is used for cleaning bricks, concrete, and pools.
- Glossy photo-printing paper. Commercial papers that work include Domtar Microprint (32 lb, 112 brightness, 8.5 × 11 in, 20 g/m^2) or Staples color laser photo supreme (SKU # 651611). Or you can use a commercial resist paper as discussed earlier in the section.

PC Board Material

One ounce or heavier copper-clad board on FR-4 substrate is best for homebrew PCBs. This is available surplus from a number of channels. The 0.060-inch thickness is best for all-around use.

While Chuck uses a band saw to cut the material into the smaller sizes, a shear, a paper cutter or a hacksaw also work well. The FR-4 substrate material is quite abrasive and will dull blades quickly. Use all safety precautions in using any cutting tool. Use a sanding block immediately after cutting to remove sharp edges along both top and bottom surfaces and the substrate.

Inexpensive and Effective Etchant

The secret to etching PC boards is the etchant. Chuck uses a mixture of hydrogen peroxide and muriatic acid. Contrary to what you might think, the acid is not doing the etching. It is the hydrogen peroxide. The oxygen in the peroxide oxidizes the copper and then dissolves into the acidic solution. While ferric chloride is often recommended for etching printed circuit boards, this mixture works as well or better and it is less expensive to make.

The proper ratio of the etching solution is 2:1 for 2 parts hydrogen peroxide to 1 part muriatic acid. To make the solution, while wearing the heavy rubber gloves and safety goggles, first pour ½ cup of hydrogen peroxide into the Pyrex container. The order is important. The peroxide goes into the Pyrex container first, then slowly add in ¼ cup of muriatic acid. This mixture will etch about two double-sided 3.25 × 3.0-inch boards.

Only make a single batch of the hydrogen peroxide/muriatic acid mixture. The mixture will not hold its etching strength for very long, so be frugal and don't make it unless you are going to use it. The bottle of hydrogen peroxide should always be tightly capped and stored in a cool dark place as it readily oxidizes and loses its strength.

pletes all of the circuit traces with a minimum number of jumper-wire connections. In some cases, however, it is not possible to complete a design without at least a few jumpers.

Layout

After you have completed a rough layout, redraw the physical layout on a grid. Graph paper works well for this. Use graph paper that has 10 lines per inch to draw artwork at 1:1 and estimate the distance halfway between lines for 0.05-inch spacing. Drafting templates are helpful in the layout stage. Local drafting-supply stores should be able to supply them. The templates usually come in either full-scale or twice normal size.

To lay out a double-sided board, ensure that the lines on both sides of the paper line up (hold the paper up to the light). You can then use each side of the paper for each side of the board. Remember that the side opposite the components (the "circuit side" or "foil side") will have pin arrangements that are reversed from the component top view! When using graph paper for a PC-board layout, include bolt holes, notches for wires and other mechanical considerations. Fit the circuit into and around them, maintaining clearance between parts.

Most through-hole IC pins are on 0.1-inch centers. Most modern components have leads on 0.1-inch centers. The rows of dual-inline-package (DIP) IC pins are spaced 0.3 or 0.4 inch. Measure the spacing for other components. Transfer the dimensions to the graph paper. It is useful to draw a schematic symbol of the component onto the layout.

The layout of a PCB for surface-mount devices is similar to that for through-hole components except that holes and pads are largely omitted. Traces end where parts are installed. SMT components have much smaller pin spacing than through-hole parts, so more care is required in the layout. If you lay out a board with computer software, the differences in laying out a board are largely transparent. One error that is frequently made by newcomers to SMT board layout is to make the board too compact and not leave room between components for probing.

Draw the traces and pads the way they will look. Using dots and lines is confusing. It's okay to connect more than one lead per pad, or run a lead through a pad, although using more than two creates a complicated layout. In that case, there may be problems with solder bridges that form short circuits. Traces can run under some components; it is possible to put two or three traces between 0.4-inch centers for a ¼-W resistor, for example.

Leave power-supply and other dc paths for last. These can usually run just about anywhere, and jumper wires are fine for these

Transferring the Layout to the Board

Design your layout. (Don't forget to reserve a small blank area to put your name, call sign and date!) Print the layout on the glossy paper. The printed layout should be a mirror of the final board. The software will do this automatically.

Clean the board with the plastic abrasive sponge. It should be shiny. Put some blank paper under the board to protect the underlying surface.

Carefully place the glossy image with the printed circuit board image on the board where you want it, face down with the toner in contact with the board surface. With the iron on its hottest setting and fully warmed up, carefully iron the image onto the board. Do not use the steam setting. Use heavy pressure; 20 lbs or more is adequate.

Use the pointy end of the iron to apply pressure everywhere on the paper. Don't scrimp here. Try to think about where all the traces are underneath the paper and aim for them all and especially the smallest ones. Patience is key. Be thorough and detailed. This makes a difference in how well your final board will turn out. Don't touch the paper when you are done as it will be hot. Take a break while the paper cools.

After it has cooled, put very hot water in a container and let the board and paper soak for about 15 to 30 minutes. Then come back and carefully peel the paper away from the board. Don't force it. It is easiest to do it from all four edges carefully. Don't touch the toner area. Handle the board by the edges.

Don't try to get all the paper the first time. Just avoid pulling the toner traces off the board. You should not see any toner on the paper being pulled from the board if you are doing this right.

With the heat used to apply the toner to the board, the board will darken and become oxidized. Don't worry about it—that part of the board will be etched away.

After you do the first paper removal, put the board back into some fresh *hot* water and let it soak some 15 minutes more. Then, using only the fingers and your skin, rub the rest of the paper off as much as possible. Don't use your fingernails or anything to help. Doing this under running lukewarm water from the faucet helps. It is not important that you get every bit of paper off the toner. It will look gray or white in some sections because the paper fibers are absorbed or stuck within the toner itself. Some people use an old toothbrush for this.

Etching the Board

Take the board and place it in the Pyrex dish with the peroxide/muriatic acid mix. You'll know you are doing this correctly and the mix is right when the board starts to darken almost immediately. This is the copper oxidizing. The clear liquid will start to turn a light green with dissolved copper in short order. Stay away from the fumes and stand upwind from any draft and do this in a well ventilated area.

Use the two plastic forks to gently push the board back and forth in the mixture, being extra careful not to touch the toner. If it comes lose, then that area of the copper will be etched away.

Move the board around until all the copper is etched away from the board except the areas covered by the toner. Don't go off and leave and forget about the etching process as it will ruin the board. Take the board out using the tongs and wash it under running water to stop the etching process.

Finishing the Board

To remove the toner after the etching process, use acetone and cotton balls, while wearing surgical gloves. A well-ventilated area is a must for this. Do it outdoors if possible. This procedure takes some practice, as sometimes the toner smears and sticks to the board a little if not enough acetone is used and kept on the surface. Don't overdo it, but practice on a few boards to get the hang of it.

After you get the toner off and rinse the board, use cotton balls and the Tarn-X to clean the copper surface.

Now the surface is exposed to the nasty atmosphere of the Earth. Protect it by very lightly coating the surface with the clear enamel in a clean, dust-free, well ventilated area also. This coating protects the surfaces and it also acts as a solder mask.

Chuck uses a 0.7 millimeter silicon carbide drill bit in a Dremel Moto-Tool mounted in a drill press to drill holes for through hole parts. A drill press is necessary as the drill bit is very small and easily damaged if you do this by hand. The 0.7 mm drill will work for most of the holes. One can go to a larger sized bit for parts that have larger leads.

You are now ready to stuff the board with your parts. Congratulations — you have just designed, fabricated and built a printed circuit board!

---

noncritical paths. (This is not the case on high-speed digital boards and if EMI from the board is to be minimized.)

Do not use traces less than 0.010-inch (10 mil) wide. If 1-oz PC board stock is used, a 10-mil trace can safely carry up to 500 mA. To carry higher current, increase the width of the traces in proportion. (A trace should be 0.200 inch wide to carry 10 A, for example.) Allow 0.1 inch between traces for each kilovolt in the circuit.

When doing a double-sided board, use pads on both sides of the board to connect traces through the board. Home-brew PC boards do not use plated-through holes (a manufacturing technique that has copper and tin plating inside all of the holes to form electrical connections). Use a through hole and solder the associated component to both sides of the board. Make other through-hole connections with a small piece of bus wire providing the connection through the board; solder it on both sides. This serves the same purpose as the plated-through holes found in commercially manufactured boards.

After you have planned the physical design of the board, decide the best way to complete the design. For one or two simple boards, draw the design directly onto the board, using a resist pen, paint or rub-on resist materials. To transfer the design to the PC board, draw light, accurate pencil lines at 0.1- or 0.05-inch centers on the PC board. Draw both horizontal and vertical lines, forming a grid. You only need lines on one side. For single-sided boards, use this grid to transfer the layout directly onto the board surface. To make drilling easier, use a center punch to punch the centers of holes accurately. Do this before applying the resist so the grid is visible.

When drawing a pad with plenty of room around it, use a pad about 0.05 to 0.1 inch in diameter. For ICs, or other close quarters, make the pad as small as 0.03 inch or so. A "ring" that is too narrow invites soldering problems; the copper may delaminate from the heat. Pads need not be round. It's okay to shave one or more edges if necessary, to allow a trace to pass nearby.

Draw the traces next. A drafting triangle can help. It should be spaced about 0.1 inch above the table, to avoid smudging the artwork. Use a 9-inch or larger triangle, with a rubber grommet taped to each corner (to hold it off the table). Select a sturdy triangle that doesn't bend easily.

Align the triangle with the grid lines by eye and make straight, even traces similar to the layout drawing. The triangle can help with angled lines, too. Practice on a few pieces of scrap board.

Make sure that the resist adheres well to the PC board. Most problems can be seen by eye;

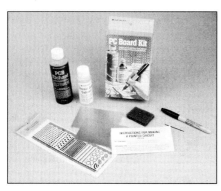

Fig 23.18 — PC-board materials are available from several sources. This kit includes layout materials, etchant, and PC board stock.

there can be weak areas or bare spots. If necessary, touch up problems with additional resist. If the board is not clean the resist will not adhere properly. If necessary, remove the resist, clean the board and start from the beginning.

Discard troublesome pens. Resist pens dry out quickly. Keep a few on hand, switch back and forth and put the cap back on each for a bit to give the pen a chance to recover.

Once all of the artwork on the board is drawn, check it against the original artwork. It is easy to leave out a trace. It is not easy to put copper back after a board is etched. In a pinch, replace the missing trace with a small wire.

Applied resist takes about an hour to dry at room temperature. Fifteen minutes in a 200°F oven is also adequate.

Special techniques are used to make double-sided PC boards. See the section on double-sided boards for a description.

MAKING A PC BOARD

Several techniques can be used to make PC boards. They usually start with a PC-board "pattern" or artwork. All of the techniques have one thing in common: this pattern needs to be transferred to the copper surface of the PC board. Unwanted copper is then removed by chemical or mechanical means. Most variations in PC-board manufacturing technique involve differences in resist or etchant materials or techniques.

No matter what technique you use, you should determine the required size of the PC board, and then cut the board to size. Trimming off excess PC-board material can be difficult after the components are installed.

The bare (unetched) PC-board stock should be clean and dry before any resist is applied. (This is not necessary if you are using stock that has been treated with pre-sensitized photoresist.) Wear rubber gloves when working with the stock to avoid getting fingerprints on the copper surface. Clean the board with soap and water, and then scrub the board with #000 steel wool. Rinse the board thoroughly then dry it with a clean, lint-free cloth. Keep the board clean and free of fingerprints or foreign substances throughout the entire manufacturing process.

No-Etch PC Boards

The most straightforward way to make very simple PC boards is to mechanically remove the unwanted copper. Use a grinding tool, such as the Moto-Tool manufactured by the Dremel Company (available at most hardware or hobby stores). Another technique is to score the copper with a strong, sharp knife, then remove unwanted copper by heating it with a soldering iron and lifting it off with a knife while it is still hot. This technique requires some practice and is not very accurate. It often fails with thin traces, so use it only for simple designs.

Photographic Process

Many magazine articles feature printed-circuit layouts. Some of these patterns are difficult to duplicate accurately by hand. A photographic process is the most efficient way to transfer a layout from a magazine page to a circuit board.

The resist ink, tape or dry-transfer processes can be time consuming and tedious for very complex circuit boards. As an alternative, consider the photo process. Not only does the accuracy improve, you need not trace the circuit pattern yourself!

A copper board coated with a light-sensitive chemical is at the heart of the photographic process. In a sense, this board becomes your photographic film.

Make a contact print of the desired pattern by transferring the printed-circuit artwork to special copy film. This film is attached to the copper side of the board and both are exposed to intense light. The areas of the board that are exposed to the light — those areas not shielded by the black portions of the artwork — undergo a chemical change. This creates a transparent image of the artwork on the copper surface.

Develop the PC board, using techniques and chemicals specified by the manufacturer. After the board is developed, etch it to remove the copper from all areas of the board that were exposed to the light. The result is a PC board that looks like it was made in a factory.

Materials and supplies for all types of PC-board manufacturing are available through a variety of electronic distributors. If you're looking for printed-circuit board kits (see **Fig 23.18**), chemicals, tools and other materials, review the articles on PC board construction in the ARRL Technical Information Service at **www.arrl.org/tis** in the Radio Technology area under Circuit Construction.

Iron-On Resist

An artwork positive is made using a standard photocopier or laser printer. A clothes iron transfers the printed resist pattern to the bare PC board. The technique for transferring the pattern to the blank PC board is described in the sidebar, "Making PC Boards With Printed Artwork".

The key to making high quality boards with the photocopy techniques is to be good at retouching the transferred resist. Fortunately, the problems are usually easy to retouch, if you have a bit of patience. A resist pen does a good job of reinforcing any spotty areas in large areas of copper.

Double-Sided PC Boards

All of the examples used to describe the above techniques were single-sided PC boards, with traces on one side of the board and either a bare board or a ground plane on the other side. PC boards can also have patterns etched onto both sides, or even have multiple layers. Most home-construction projects use single-sided boards, although some kit builders supply double-sided boards. Multilayer boards are rare in ham construction. One method for making double-sided boards is described in the sidebar, "Double-Sided PC Boards — by Hand!"

Tin Plating

Most commercial PC boards are tin-plated, to make them easier to solder. Commercial tin-plating techniques require electroplating equipment not readily available to the home constructor. Immersion tin plating solutions can deposit a thin layer of tin onto a copper PC board. Using them is easy; put some of the solution into a plastic container and immerse the board in the solution for a few minutes. The chemical action of the tin-plating solution replaces some of the copper on the board with tin. The result looks nearly as good as a commercially made board. Agitate the board or solution from time to time. When the tinning is complete, take the board out of the solution and rinse it for five minutes under running water. If you don't remove all of the residue, solder may not adhere well to the surface. Immersion tin plating solution is available from electronics vendors.

Drilling a PC Board

After you make a PC board using one of the above techniques, you need to drill holes in the board for the components. Use a drill press, or at least improvise one. Boards can be drilled entirely "free hand" with a hand-held drill but the potential for error is great. A drill press or a small Moto-Tool in an accessory drill press makes the job a lot easier. A single-sided board should be drilled after it is etched; the easiest way to do a double-sided board is to do it before the resist is applied.

To drill in straight lines, build a small movable guide for the drill press so you can slide

Double-Sided PC Boards — by Hand!

Forget those nightmares about expensive photoresists that didn't work; forget that business of fifty bucks a board! You don't need computer-aided design to make a double-sided PC board; just improve on the basics, and keep it simple. Anyone can make low-cost double-sided boards with traces down to 0.020 inch, with perfect front-to-back hole registration.

To make a double-sided board, drill the holes before applying the resist artwork; that is the only way to assure good front-to-back registration. The artwork on both sides can then be properly positioned to the holes. PC-board drilling was discussed earlier in the text.

After you have drilled the board, clean its surface thoroughly. After that, wear clean rubber or cotton gloves to keep it clean. One fingerprint can really mess up the application of resist or the etchant.

Tape the board to your work surface, making sure it can't move around. Transfer the artwork from your layout grid to the PC board, drawing by hand with a resist pen.

Allot enough time to finish at least one side of the artwork in one sitting. Start with the pads. To make a handy pad-drawing tool, press the tip of a regular-size Sharpie indelible ink felt-tip marker into one of the drilled PC-board holes. This "smooshes" the tip into the shape of the hole, leaving a flat shoulder to draw the pad. See **Fig 23.A9**. The diameter of the pad is determined by how hard the pen is pressed; pressing too hard forms a pad that is way too large for most applications. Practice on scrap board first. Use this modified pen to fill in all the holes and draw the pads at the same time. Use an unmodified resist pen to draw all of the traces and to touch up any voids or weak areas in the pads. For the rest of the drawing, the procedure described for single-sided boards applies to double-sided boards, too.

Fig 23.A9 — Make a permanent marker into a specialized PC-board drawing tool. Simply press the marker point into a drilled hole to form a modified point as shown. More pressure produces a wider shoulder that makes larger pads on the PC board.

After the resist is applied to the first side, carefully draw the second side. Inspect the board thoroughly; you may have scratched or smudged the first side while you were drawing the second.

Etching a double-sided board is not much different than etching a single-sided board, except that you must ensure that the etchant is able to reach both sides of the board. If you dunk the board in and out of the etchant solution, both sides are exposed to the etchant. If you use a tray, put some spacers on the bottom and rest the board on the spacers. (The spacers must be put on the board edges, not where you want to actually etch.) This ensures that etchant gets to both sides. If you use this method, turn the board over once or twice during the process. — Dave Reynolds, KE7QF

Or Photo-Etched

You can also make double-sided boards at home without drawing the layout by hand. This procedure can't produce results to match the finest professionally made double-sided boards, but it can make boards that are good enough for many moderately complex projects.

Start with the same sort of artwork used for single-sided boards, but leave a margin for taping at one edge. It is critical that the patterns for the two sides are accurately sized. The chief limiting factor in this technique is the requirement that matching pads on the two sides are positioned correctly. Not only must the two sides match each other, but they must also be the correct size for the parts in the project. Slight reproduction errors can accumulate to major problems in the length of a 40-pin DIP IC. One good tool to achieve this requirement is a photocopy machine that can make reductions and enlargements in 1% steps. Perform a few experiments to arrive at settings that yield accurately sized patterns.

Choose two holes at opposite corners of the etching patterns. Tape one of the two patterns to one side of the PC board. Choose some small wire and a drill bit that closely matches the wire diameter. For example, #20 AWG enameled wire is a close match for a #62 or a #65 drill, depending on the thickness of the wire's enamel coating. Drill through the pattern and the board at the two chosen holes. Drill the chosen holes through the second pattern. Place two pieces of the wire through the PC board and slide the second pattern down these wire "pins" to locate the pattern on the board. Tape the second pattern in position and remove the pins. From this point on, expose and process each side of the board as if it were a single-sided board, but take care when exposing each side to keep the reverse side protected from light.

one edge of the board against it and line up all of the holes on one grid line at a time. See **Fig 23.19**. This is similar to the "rip fence" set up by most woodworkers to make accurate and repeatable cuts with a table saw.

The drill-bit sizes available in hardware stores are too big for PC boards. You can use high-speed steel bits, but glass epoxy stock tends to dull these after a few hundred holes. (When your drill bit becomes worn, it makes a little "hill" around each drilled hole, as the worn bit pushes and pulls the copper rather than drilling it.) A PC-board drill bit, available from many electronic suppliers, will last for thousands of holes! If you are doing a lot of boards, it is clearly worth the investment. Small drill bits are usually ordered by number. Here are some useful numbers and their sizes:

Number	Diameter (in)
68	0.0310
65	0.0350
62	0.0380
60	0.0400

Use high RPM and light pressure to make good holes. Safety goggles and good lighting are a must! Count the holes on both the board and your layout drawing to ensure that none are missed. Use a larger-size drill bit, lightly spun between your fingers, to remove any burrs. Don't use too much pressure; remove only the burr.

PC-BOARD ASSEMBLY TECHNIQUES

Once you have etched and drilled a PC board you are ready to use it in a project. Several tools come in handy: needle-nose pliers, diagonal cutters, pocket knife, wire strippers, clip leads and soldering iron.

Cleanliness

Make sure your PC board and component leads are clean. Clean the entire PC board before assembly; clean each component before you install it. Corrosion looks dark instead of bright and shiny. Don't use sandpaper to clean your board. Use a piece of fine steel

Fig 23.19 — This home-built drill fence makes it easy to drill PC-board holes in straight rows.

Fig 23.20 — This is how to solder a component to a PC board. Make sure that the component is flush with the board on the other side.

wool or a Scotchbrite cleaning pad to clean component leads or PC board before you solder them together.

Installing Components

In a construction project that uses a PC board, most of the components are installed on the board. Installing components is easy — stick the components in the right board holes, solder the leads, and cut off the extra lead length. Most construction projects have a parts-placement diagram that shows you where each component is installed.

Getting the components in the right holes is called "stuffing" the circuit board. Inserting and soldering one component at a time takes too long. Some people like to put the components in all at once, and then turn the board over and solder all the leads. If you bend the leads a bit (about 20°) from the bottom side after you push them through the board, the components are not likely to fall out when you turn the board over.

Start with the shortest components — horizontally mounted diodes and resistors. Larger components sometimes cover smaller components, so these smaller parts must be installed first. Use adhesive tape to temporarily hold difficult components in place while you solder.

PC-Board Soldering

To solder components to a PC board, bend the leads at a slight angle; apply the soldering iron to one side of the lead, and flow the solder in from the other side of the lead. See **Fig 23.20**. Too little heat causes a bad or "cold" solder joint; too much heat can damage the PC board. Practice a bit on some spare copper stock before you tackle your first PC board project. After the connection is soldered properly, clip the lead flush with the solder.

Make sure you have the components in the right holes before you solder them. Components that have polarity, such as diodes, ICs and some capacitors must be oriented as shown on the parts-placement diagram.

23.5.4 From Schematic to Working Circuit

Turning a schematic into a working circuit is more than just copying the schematic with components. One thing is usually true — you can't build it the way it looks on the schematic. The schematic describes the electrical connections, but it does not describe the mechanical layout of the circuit. Many design and layout considerations that apply in the real world of practical electronics don't appear on the schematic.

HOW TO DESIGN A GOOD CIRCUIT LAYOUT

A circuit diagram is a poor guide toward a proper layout. Circuit diagrams are drawn to be readable and to describe the electrical connections. They follow drafting conventions that have very little to do with the way the circuit works. On a schematic, ground and supply voltage symbols are scattered all over the place. The first rule of RF layout is — *do not lay out RF circuits as their schematics are drawn!* How a circuit works in practice depends on the layout. Poor layout can ruin the performance of even a well-designed circuit.

The easiest way to explain good layout

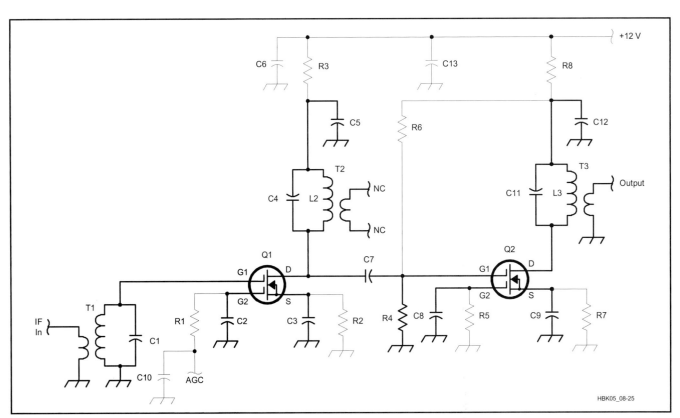

Fig 23.21 — The IF amplifier used in the design example. C1, C4 and C11 are not specified because they are internal to the IF transformers.

practices is to take you through an example. **Fig 23.21** is the circuit diagram of a two-stage receiver IF amplifier using dual-gate MOSFETs. It is only a design example, so the values are only typical. To analyze which things are important to the layout of this circuit, ask these questions:

• Which are the RF components, and which are only involved with LF or dc?

• Which components are in the main RF signal path?

• Which components are in the ground return paths?

Use the answers to these questions to plan the layout. The RF components that are in the main RF signal path are usually the most critical. The AF or dc components can usually be placed anywhere. The components in the ground return path should be positioned so they are easily connected to the circuit ground. Answer the questions, apply the answers to the layout and then follow these guidelines:

• Avoid laying out circuits so their inputs and outputs are close together. If a stage's output is too near a previous stage's input, the output signal can feedback into the input and cause problems.

• Keep component leads as short as practical. This doesn't necessarily mean as short as possible, just consider lead length as part of your design.

• Remember that metal transistor cases conduct, and that a transistor's metal case is usually connected to one of its leads. Prevent cases from touching ground or other components, unless called for in the design.

In our design example, the RF components are shown in heavy lines, though not all of these components are in the main RF signal path. The RF signal path consists of T1/C1, Q1, T2/C4, C7, Q2, T3/C11. These need to be positioned in almost a straight line, to avoid feedback from output to input. They form the backbone of the layout, as shown in **Fig 23.22A**.

The question about ground paths requires some further thought — what is really meant by "ground" and "ground-return paths"? Some points in the circuit need to be kept at RF ground potential. The best RF ground potential on a PC board is a copper ground plane covering one entire side. Points in the circuit that cannot be connected directly to ground for dc reasons must be bypassed ("decoupled") to ground by capacitors that provide ground-return paths for RF.

In Fig 23.22, the components in the ground-return paths are the RF bypass capacitors C2, C3, C5, C8, C9 and C12. R4 is primarily a dc biasing component, but it is also a ground return for RF so its location is important. The values of RF bypass capacitors are chosen to have a low reactance at the frequency in use; typical values would be 0.1 μF at LF, 0.01 μF

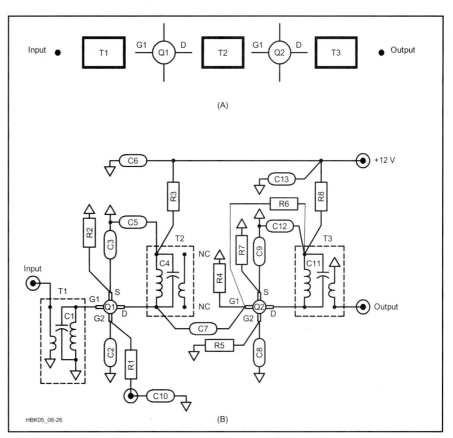

Fig 23.22 — Layout sketches. The preliminary line-up is shown in A; the final layout in B.

at HF, and 0.001 μF or less at VHF. Not all capacitors are suitable for RF decoupling; the most common are disc ceramic capacitors. RF decoupling capacitors should always have short leads. Surface mount capacitors with no leads are ideal for bypassing.

Almost every RF circuit has an input, an output and a common ground connection. Many circuits also have additional ground connections, both at the input side and at the output side. Maintain a low-impedance path between input and output ground connections. The input ground connections for Q1 are the grounded ends of C1 and the two windings of T1. The two ends of an IF transformer winding are generally not interchangeable; one is designated as the "hot" end, and the other must be connected or bypassed to RF ground.) The capacitor that resonates with the adjustable coil is often mounted inside the can of the IF transformer, leaving only two component leads to be grounded as shown in Fig 23.22B.

The RF ground for Q1 is its source connection via C3. Since Q1 is in a plastic package that can be mounted in any orientation, you can make the common ground either above or below the signal path in Fig 23.22B, although the circuit diagram shows the source at the bottom. The practical circuit works much better with the source at the top, because of the connections to T2.

It's a good idea to locate the hot end of the main winding close to the drain lead of the transistor package, so the other end is toward the top of Fig 23.22B. If the source of Q1 is also toward the top of the layout, there is a common ground point for C3 (the source bypass capacitor) and the output bypass capacitor C5. Gate 2 of Q1 can safely be bypassed toward the bottom of the layout.

C7 couples the signal from the output of Q1 to the input of Q2. The source of Q2 should be bypassed toward the top of the layout, in exactly the same way as the source of Q1. R4 is not critical, but it should be connected on the same side as the other components. Note how the pinout of T3 has placed the output connection as far as possible from the input. With this layout for the signal path and the critical RF components, the circuit has an excellent chance of working properly.

DC Components

The rest of the components carry dc, so their layout is much less critical. Even so, try to keep everything well separated from the main RF signal path. One good choice is to put the 12-V connections along the top of the layout, and the AGC connection at the bottom. The source bias resistors R2 and R7 can be placed alongside C3 and C9. The gate-2 bias resistors for Q2, R5 and R6 are not RF components so their locations aren't too critical. R7 has to cross the signal path in order to reach C12, however, and the best way to avoid signal pickup would be to mount R7 on the opposite side of the copper ground plane from the signal wiring. Generally speaking, ⅛-W or ¼-W metal-film or carbon-film resistors are best for low-level RF circuits.

Actually, it is not quite accurate to say that resistors such as R3 and R8 are not "RF" components. They provide a high impedance to RF in the positive supply lead. Because of R8, for example, the RF signal in T2 is conducted to ground through C5 rather than ending up on the 12-V line, possibly causing unwanted RF feedback. Just to be sure, C6 bypasses R3 and C13 serves the same function for R8. Note that the gate-1 bias resistor R6 is connected to C12 rather than directly to the 12-V supply, to take advantage of the extra decoupling provided by R8 and C13.

If you build something, you want it to work the first time, so don't cut corners! Some commercial PC boards take liberties with layout, bypassing and decoupling. Don't assume that you can do the same. Don't try to eliminate "extra" decoupling components such as R3, C6, R8 and C13, even though they might not all be absolutely necessary. If other people's designs have left them out, put them in again. In the long run it's far easier to take a little more time and use a few extra components, to build in some insurance that your circuit will work. For a one-time project, the few extra parts won't hurt your pocket too badly; they may save untold hours in debugging time.

A real capacitor does not work well over a large frequency range. A 10-μF electrolytic capacitor cannot be used to bypass or decouple RF signals. A 0.1-μF capacitor will not bypass UHF or microwave signals. Choose component values to fit the range. The upper frequency limit is limited by the series inductance, L_S. In fact, at frequencies higher than the frequency at which the capacitor and its series inductance form a resonant circuit, the capacitor actually functions as an inductor. This is why it is a common practice to use two capacitors in parallel for bypassing, as shown in **Fig 23.23**. At first glance, this might appear to be unnecessary. However, the self-resonant frequency of C1 is usually 1 MHz or less; it cannot supply any bypassing above that frequency. However, C2 is able to bypass signals up into the lower VHF range. (This technique should not be applied under all circumstances as discussed in the section on Bypassing in the **RF Techniques** chapter.)

Let's summarize how we got from Fig 23.21 to Fig 23.22B:
• Lay out the signal path in a straight line.
• By experimenting with the placement and orientation of the components in the RF signal path, group the RF ground connections for each stage close together, without mixing up the input and output grounds.
• Place the non-RF components well clear of the signal path, freely using decoupling components for extra measure.

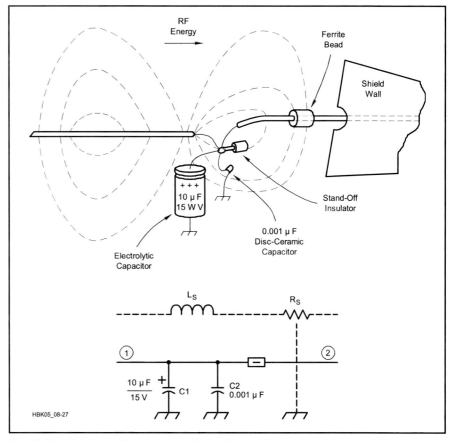

Fig 23.23 — Two capacitors in parallel afford better bypassing across a wide frequency range.

Practical Construction Hints

Now it's time to actually construct a project. The layout concepts discussed earlier can be applied to nearly any construction technique. Although you'll eventually learn from your own experience, the following guidelines give a good start:
• Divide the unit into modules built into separate shielded enclosures — RF, IF, VFO, for example. Modular construction improves RF stability, and makes the individual modules easier to build and test. It also means that you can make major changes without rebuilding the whole unit. RF signals between the modules can usually be connected using small coaxial cable.
• Use a full copper ground plane. This is your largest single assurance of RF stability and good performance.
• Keep inputs and outputs well separated for each stage, and for the whole unit. If possible, lay out all stages in a straight line. If an RF signal path doubles back or re-crosses itself it usually results in instability.
• Keep the stages at different frequencies well-separated to minimize interstage coupling and spurious signals.
• Use interstage shields where necessary, but don't rely on them to cure a bad layout.
• Make all connections to the ground plane short and direct. Locate the common ground for each stage between the input and the output ground. Single-point grounding may work for a single stage, but it is rarely effective in a complex RF system.
• Locate frequency-determining components away from heat sources and mount them so as to maximize mechanical strength.
• Avoid unwanted coupling between tuned circuits. Use shielded inductors or toroids rather than open coils. Keep the RF high-voltage points close to the ground plane. Orient air-wound coils at right angles to minimize mutual coupling.
• Use lots of extra RF bypassing, especially on dc supply lines.
• Try to keep RF and dc wiring on opposite sides of the board, so the dc wiring is well away from RF fields.
• Compact designs are convenient, but don't overdo it! If the guidelines cited above mean that a unit needs to be bigger, make it bigger.

COMBINING TECHNIQUES

You can use a mixture of construction techniques on the same board and in most cases you probably should. Even though you choose one style for most of the wiring, there will probably be places where other techniques would be better. If so, do whatever is best for that part of the circuit. The resulting hybrid may not be pretty (these techniques aren't called "ugly construction" for nothing), but it will work!

Mount dual-in-line package (DIP) ICs in an array of drilled holes, then connect them using wired traces as described earlier. It is okay to mount some of the components using a ground-plane method, push pins or even wire wrap. On any one board, you may use a combination of these techniques, drilling holes for some ICs, or gluing others upside down, then surface mounting some of the pins, and other techniques to connect the rest. These combination techniques are often found in a project that combines audio, RF and digital circuitry.

A Final Check

No matter what construction technique is chosen, do a final check before applying power to the circuit! Things do go wrong, and a careful inspection minimizes the risk of a project beginning and ending its life as a puff of smoke! Check wiring carefully. Make a photocopy of the schematic and mark each circuit on the schematic with a red X or use highlighter to mark the circuit when you've verified that it's connected to the right spot in the circuit.

Inspect solder connections. A bad solder joint is much easier to find before the PC board is mounted to a chassis. Look for any damage caused to the PC board by soldering. Look for solder "bridges" between adjacent circuit-board traces. Solder bridges (**Fig 23.24**) occur when solder accidentally connects two or more conductors that are supposed to be isolated. It is often difficult to distinguish a solder bridge from a conductive trace on a tin-plated board. If you find a bridge, re-melt it and the adjacent trace or

Fig 23.24 — A solder bridge has formed a short circuit between PC board traces.

traces to allow the solder's surface tension to absorb it. Double check that each component is installed in the proper holes on the board and that the orientation is correct. Make sure that no component leads or transistor tabs are touching other components or PC board connections. Check the circuit voltages before installing ICs in their sockets. Ensure that the ICs are oriented properly and installed in the correct sockets.

23.5.5 Other Construction Techniques

WIRING

Select the wire used in connecting amateur equipment by considering: the maximum current it must carry, the voltage its insulation must withstand and its use.

To minimize EMI, the power wiring of all transmitters should use shielded wire. Receiver and audio circuits may also require the use of shielded wire at some points for stability or the elimination of hum. Coaxial cable is recommended for all 50-Ω circuits. Use it for short runs of high-impedance audio wiring.

When choosing wire, consider how much current it will carry. Stranded wire is usually preferred over solid wire because stranded wire better withstands the inevitable bending that is part of building and troubleshooting a circuit. Solid wire is more rigid than stranded wire; use it where mechanical rigidity is needed or desired.

Wire with typical plastic insulation is good for voltages up to about 500 V. Use Teflon-insulated or other high-voltage wire for higher voltages. Teflon insulation does not melt when a soldering iron is applied. This makes it particularly helpful in tight places or large wiring harnesses. Although Teflon-insulated wire is more expensive, it is often available from industrial surplus houses. Inexpensive wire strippers make the removal of insulation from hookup wire an easy job. Solid wire is often used to wire HF circuits. Bare soft-drawn tinned wire, #22 to #12 AWG (depending on mechanical requirements) is suitable. Avoid kinks by stretching a piece 10 or 15 ft long and then cutting it into short, convenient lengths. Run RF wiring directly from point to point with a minimum of sharp bends and keep the wire well-spaced from the chassis or other grounded metal surfaces. Where the wiring must pass through the chassis or a partition, cut a clearance hole and line it with a rubber grommet. If insulation is necessary, slip spaghetti insulation or heat-shrink tubing over the wire. For power-supply leads, bring the wire through the chassis via a feedthrough capacitor.

In transmitters where the peak voltage does not exceed 500 V, shielded wire is satisfactory for power circuits. Shielded wire is not readily available for higher voltages — use point-to-point wiring instead. In the case of filament circuits carrying heavy current, it is necessary to use #10 or #12 AWG bare or enameled wire. Slip the bare wire through spaghetti then cover it with copper braid pulled tightly over the spaghetti. Slide the shielding back over the insulation and flow solder into the end of the braid; the braid will stay in place, making it unnecessary to cut it back or secure it in place. Clean the braid first so solder will take with a minimum of heat.

For receivers, RF wiring follows the methods described above. At RF, most of the current flows on the surface of the wire (a phenomenon called "skin effect"). Hollow tubing is just as good a conductor at RF as solid wire.

HIGH-VOLTAGE TECHNIQUES

High-voltage wiring requires special care. (Additional discussion of high-voltage construction can be found in the chapter on **Power Supplies**.) You need to use wire rated for the voltage it is carrying. Most standard hookup wire is inadequate. High-voltage wire is usually insulated with Teflon or special multilayer plastic. Some coaxial cable is rated at up to 3700 V.

Air is a great insulator, but high voltage can break down its resistance and form an arc. You need to leave ample room between any circuit carrying voltage and any nearby conductors. At dc, leave a gap of at least 0.1 inch per kilovolt. The actual breakdown voltage of air varies with the frequency of the signal, humidity and the shape of the conductors.

High voltage is also prone to corona discharge, a bleeding off of charge, primarily from sharp edges. For this reason, all connections need to be soldered, leaving only rounded surfaces on the soldered connection. It takes a little practice to get a "ball" of solder on each joint, but for voltages above 5 kV it is important.

Be careful working near high-voltage circuits! Most high-voltage power supplies can deliver a lethal shock.

CABLE ROUTING

Where power or control leads run together for more than a few inches, they present a better appearance when bound together in a single cable. Plastic cable ties or tubing cut into a spiral are used to restrain and group wiring. Check with your local electronic parts supplier for items that are in stock.

To give a commercial look to the wiring of any unit, route any dc leads and shielded signal leads along the edge of the chassis. If this isn't possible, the cabled leads should then run parallel to an edge of the chassis. Further, the generous use of the tie points mounted parallel to an edge of the chassis, for the support of one or both ends of a resistor or fixed capacitor, adds to the appearance of the finished unit. In a similar manner, arrange the small components so that they are parallel to the panel or sides of the chassis.

Tie Points

When power leads have several branches in the chassis, it is convenient to use fiber-insulated multiple tie points as anchors for junction points. Strips of this kind are also useful as insulated supports for resistors, RF chokes and capacitors. Hold exposed points of high-voltage wiring to a minimum; otherwise, make them inaccessible to accidental contact.

WINDING COILS

A detailed tutorial for winding coils by Robert Johns, W3JIP, from August 1997 *QST* can be found in the Radio Technology section of the ARRL TIS at **www.arrl.org/tis** under Circuit Construction. Understanding these techniques greatly simplifies coil construction.

Close-wound coils are readily wound on the specified form by anchoring one end of the length of wire (in a vise or to a doorknob) and the other end to the coil form. Straighten any kinks in the wire and then pull to keep the wire under slight tension. Wind the coil to the required number of turns while walking toward the anchor, always maintaining a slight tension on the wire.

To space-wind the coil, wind the coil simultaneously with a suitable spacing medium (heavy thread, string or wire) in the manner described above. When the winding is complete, secure the end of the coil to the coil-form terminal and then carefully unwind the spacing material. If the coil is wound under suitable tension, the spacing material can be easily removed without disturbing the winding. Finish space-wound coils by judicious applications of RTV sealant or hot-melt glue to hold the turns in place.

The "cold" end of a coil is the end at (or close to) chassis or ground potential. Wind coupling links on the cold end of a coil to minimize capacitive coupling.

Winding Toroidal Inductors

Toroidal inductors and transformers are specified for many projects in this *Handbook*. The advantages of these cores include compactness and a self-shielding property. **Figs 23.25** and **23.26** illustrate the proper way to wind and count turns on a toroidal core.

The task of winding a toroidal core, when more than just a few turns are required, can be greatly simplified by the use of a homemade bobbin upon which the wire is first wound. A simple yet effective bobbin can be fashioned from a wooden popsicle stick. Cut a "V" notch at each end and first wind the wire coil

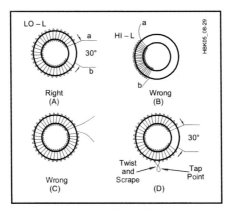

Fig 23.25 — The maximum-Q method for winding a single-layer toroid is shown at A. A 30° gap is best. Methods at B and C have greater distributed capacitance. D shows how to place a tap on a toroidal coil winding.

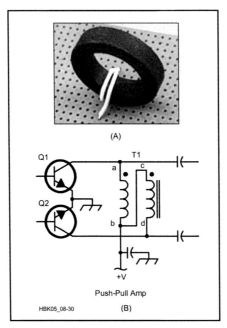

Fig 23.26 — A shows a toroidal core with two turns of wire (see text). Large black dots, like those at T1 in B, indicate winding polarity (see text).

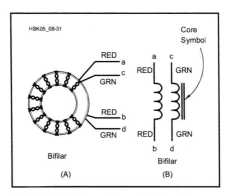

Fig 23.27 — Schematic and pictorial presentation of a bifilar-wound toroidal transformer.

on the popsicle stick lengthwise through the notches. Once this is done, the wound bobbin can be easily passed through the toroid's inside diameter. While firmly grasping one of the wire ends against the toroidal core, the bobbin can be moved up, around, and through the toroidal core repeatedly until the wire has been completely transferred from the bobbin. The choice of bobbin used is somewhat dependent on the inside diameter of the toroid, the wire size, and the number of turns required.

When you wind a toroid inductor, count each pass of the wire through the toroid center as a turn. You can count the number of turns by counting the number of times the wire passes through the center of the core. See Fig 23.26A.

Multiwire Windings

A bifilar winding is one that has two identical lengths of wire, which when placed on the core result in the same number of turns for each wire. The two wires are wound on the core side by side at the same time, just as if a single winding were being applied. An easier and more popular method is to twist the two wires (8 to 15 turns per inch is adequate), then wind the twisted pair on the core. The wires can be twisted handily by placing one end of each in a bench vise. Tighten the remaining ends in the chuck of a small hand drill and turn the drill to twist the pair.

A trifilar winding has three wires, and a quadrifilar winding has four. The procedure for preparation and winding is otherwise the same as for a bifilar winding. **Fig 23.27** shows a bifilar toroid in schematic and pictorial form. The wires have been twisted together prior to placing them on the core. It is helpful, though by no means essential, to use wires of different color for multifilar windings. It is more difficult to identify multiple windings on a core after it has been wound. Various colors of enamel insulation are available, but it is not easy for amateurs to find this wire locally or in small-quantity lots. This problem can be solved by taking lengths of wire (enameled magnet wire), cleaning the ends to remove dirt and grease, then spray painting them. Ordinary aerosol-can spray enamel works fine. Spray lacquer is not as satisfactory because it is brittle when dry and tends to flake off the wire.

The winding sense of a multifilar toroidal transformer is important in most circuits. Fig 23.26B illustrates this principle. The black dots (called phasing dots) at the top of the T1 windings indicate polarity. That is, points a and c are both start or finish ends of their respective windings. In this example, points a and d are of opposite polarity to provide push-pull voltage output from Q1 and Q2.

After you wind a coil, scrape the insulation off the wire before you solder it into the circuit.

Microwave Construction Techniques

Microwave construction is becoming more popular, but at these frequencies the size of physical component leads and PC-board traces cannot be neglected. Microwave construction techniques either minimize these stray values or make them part of the circuit design.

Microwave construction does not always require tight tolerances and precision construction. A fair amount of error can often be tolerated if you are willing to tune your circuits, as you do at MF/HF. This usually requires the use of variable components that can be expensive and tricky to adjust.

Proper design and construction techniques, using high precision, can result in a "no-tune" microwave design. To build one of these no-tune projects, all you need do is buy the parts and install them on the board. The circuit tuning has been precisely controlled by the board and component dimensions so the project should work.

One tuning technique you can use with a microwave design, if you have the suitable test equipment, is to use bits of copper foil or EMI shielding tape as "stubs" to tune circuits. Solder these small bits of conductor into place at various points in the circuit to make reactances that can actually tune a circuit. After their position has been determined as part of the design, tuning is accomplished by removing or adding small amounts of conductor, or slightly changing the placement of the tuning stub. The size of the foil needed depends on your ability to determine changes in circuit performance, as well as the frequency of operation and the circuit board parameters. A precision setup that lets you see tiny changes allows you to use very small pieces of foil to get the best tuning possible.

From a mechanical accuracy point of view, the most tolerant type of construction is waveguide construction. Tuning is usually accomplished via one or more screws threaded into the waveguide. It becomes unwieldy to use waveguide on the amateur bands below 10 GHz because the dimensions get too large.

At 24 GHz and above, even waveguide becomes small and difficult to work with. At these frequencies, most readily available coax connectors work unreliably, so these higher bands are really a challenge. Special SMA connectors are available for use at 24 GHz.

Modular construction is a useful technique for microwave circuits. Often, circuits are tested by hooking their inputs and output to known 50-Ω sources and loads. Modules are typically kept small to prevent the chassis and PC board from acting as a waveguide, providing a feedback path between the input and output of a circuit, resulting in instability.

At microwave frequencies, the mechanical aspects and physical size of circuits become very much a part of the design. A few millimeters of conductor has significant reactance at these frequencies. This even affects VHF and HF designs!

The traces and conductors used in an HF or VHF design resonate on microwave frequencies. If a high-performance FET has lots of gain in this region, a VHF preamplifier might also function as a 10-GHz oscillator if the circuit stray reactances were just right (or wrong!). You can prevent this by using shields between the input and output or by adding microwave absorptive material to the lid of the shielded module. (SHF Microwave sells absorptive materials.)

It is important to copy microwave circuits exactly, unless you really know what you are doing. "Improvements," such as better shielding or grounding can sometimes cause poor performance. It isn't usually attractive to substitute components, particularly with the active devices. It may look possible to substitute different grades of the same wafer, such as the ATF13135 and the ATF13335, but these are really the same transistor with different performance measurements. While two transistors may have exactly the same gain and noise figure at the desired operating frequency, often the impedances needed to maintain stability at other frequencies are be different. Thus, the "substitute" may oscillate, while the proper transistor would work just fine.

You can often substitute MMICs (monolithic microwave integrated circuits) for one another because they are designed to be stable and operate with the same input and output impedances (50 Ω).

The size of components used at microwaves can be critical — in some cases, a chip resistor 80 mils across is not a good substitute for one 60 mils across. Hopefully, the author of a construction project tells you which dimensions are critical, but you can't always count on this; the author may not know. It's not unusual for a person to spend years building just one prototype, so it's not surprising that the author might not have built a dozen different samples to try possible substitutions.

When using glass-epoxy PC board at microwave frequencies, the crucial board parameter is the thickness of the dielectric. It can vary quite a bit, in excess of 10%. This is not surprising; digital and lower-frequency analog circuits work just fine if the board is a little thinner or thicker than usual. Some of the board types used in microwave-circuit construction are a generic Teflon PC board, Duroid 5870 and 5880. These boards are available from Microwave Components of Michigan.

Proper connectors are a necessary expense at microwaves. At 10 GHz, the use of the proper connectors is essential for repeatable performance. Do not hook up microwave circuits with coax and pigtails. It might work but it probably can't be duplicated. SMA connectors are common because they are small and work well. SMA jacks are sometimes soldered in place, although 2-56 hardware is more common. — *Zack Lau, W1VT, ARRL Laboratory Engineer*

23.6 Mechanical Fabrication

Most projects end up in some sort of an enclosure, and most hams choose to purchase a ready-made chassis for small projects, but some projects require a custom enclosure. Even a ready-made chassis may require a fabricated sheet-metal shield or bracket, so it's good to learn something about sheet-metal and metal-fabrication techniques.

Most often, you can buy a suitable enclosure. These are sold by RadioShack and most electronics distributors. Select an enclosure that has plenty of room. A removable cover or front panel can make any future troubleshooting or modifications easy. A project enclosure should be strong enough to hold all of the components without bending or sagging; it should also be strong enough to stand up to expected use and abuse.

23.6.1 Cutting and Bending Sheet Metal

Enclosures, mounting brackets and shields are usually made of sheet metal. Most sheet metal is sold in large sheets, 4 to 8 ft or larger. It must be cut to the size needed.

Most sheet metal is thin enough to cut with metal shears or a hacksaw. A jigsaw or band saw makes the task easier. If you use any kind of saw, select a blade that has teeth fine enough so that at least two teeth are in contact with the metal at all times.

If a metal sheet is too large to cut conveniently with a hacksaw, it can be scored and broken. Make scratches as deep as possible along the line of the cut on both sides of the sheet. Then, clamp it in a vise and work it back and forth until the sheet breaks at the line. Do not bend it too far before the break begins to weaken, or the edge of the sheet might bend. A pair of flat bars, slightly longer than the sheet being bent, make it easier to hold a sheet firmly in a vise. Use "C" clamps to keep the bars from spreading at the ends.

Smooth rough edges with a file or by sanding with a large piece of emery cloth or sandpaper wrapped around a flat block.

23.6.2 Finishing Aluminum

Give aluminum chassis, panels and parts a sheen finish by treating them in a caustic bath. Use a plastic container to hold the solution. Ordinary household lye can be dissolved in water to make a bath solution. Follow the directions on the container. A strong solution will do the job more rapidly.

Stir the solution with a stick of wood until the lye crystals are completely dissolved. If the lye solution gets on your skin, wash with plenty of water. If you get any in your eyes, immediately rinse with plenty of clean, room-temperature water and seek medical help. It can also damage your clothing, so wear something old. Prepare sufficient solution to cover the piece completely. When the aluminum is immersed, a very pronounced bubbling takes place. Provide ventilation to disperse the escaping gas. A half hour to two hours in the bath is sufficient, depending on the strength of the solution and the desired surface characteristics.

23.6.3 Chassis Working

With a few essential tools and proper procedure, building radio gear on a metal chassis is a relatively simple matter. Aluminum is better than steel, not only because it is a superior shielding material, but also because it is much easier to work and provides good chassis contact when used with secure fasteners.

Spend sufficient time planning a project to save trouble and energy later. The actual construction is much simpler when all details are worked out beforehand. Here we discuss a large chassis-and-cabinet project, such as a high-power amplifier. The techniques are applicable to small projects as well.

Cover the top of the chassis with a piece of wrapping paper or graph paper. Fold the edges down over the sides of the chassis and fasten them with adhesive tape. Place the front panel against the chassis front and draw a line there to indicate the chassis top edge.

Assemble the parts to be mounted on the chassis top and move them about to find a satisfactory arrangement. Consider that some will be mounted underneath the chassis and ensure that the two groups of components won't interfere with each other.

Place controls with shafts that extend through the cabinet first, and arrange them so that the knobs will form the desired pattern on the panel. Position the shafts perpendicular to the front chassis edge. Locate any partition shields and panel brackets next, then sockets and any other parts. Mark the mounting-hole centers of each part accurately on the paper. Watch out for capacitors with off-center shafts that do not line up with the mounting holes. Do not forget to mark the centers of socket holes and holes for wiring leads. Make the large center hole for a socket *before* the small mounting holes. Then use the socket itself as a template to mark the centers of the mounting holes. With all chassis holes marked, center-punch and drill each hole.

Next, mount on the chassis the capacitors and any other parts with shafts extending to the panel. Fasten the front panel to the chassis temporarily. Use a machinist's square to extend the line (vertical axis) of any control shaft to the chassis front and mark the location on the front panel at the chassis line. If the layout is complex, label each mark with an identifier. Also mark the back of the front panel with the locations of any holes in the chassis front that must go through the front panel. Remove the front panel.

MAKING ENCLOSURES WITH PC BOARD MATERIAL

Much tedious sheet-metal work can be eliminated by fabricating chassis and enclosures from copper-clad printed-circuit board material. While it is manufactured in large sheets for industrial use, some hobby electronics stores and surplus outlets market usable scraps at reasonable prices. PC-board stock cuts easily with a small hacksaw. The nonmetallic base material isn't malleable, so it can't be bent. Corners are easily formed by holding two pieces at right angles and soldering the seam. This technique makes excellent RF-tight enclosures. If mechanical rigidity is required of a large copper-clad surface, solder stiffening ribs at right angles to the sheet.

Fig 23.28 shows the use of PC-board stock to make a project enclosure. This enclosure was made by cutting the pieces to size, then soldering them together. Start by laying the bottom piece on a workbench, then placing one of the sides in place at right angles. Tack-solder the second piece in two or three places, then start at one end and run a bead of solder down the entire seam. Use plenty of solder and plenty of heat. Continue with the rest of the pieces until all but the top cover is in place.

In most cases, it is better to drill all needed holes in advance. It can sometimes be difficult to drill holes after the enclosure is soldered together.

You can use this technique to build enclosures, subassemblies or shields. This technique is easy with practice; hone your skills on a few scrap pieces of PC-board stock.

23.6.4 Drilling Techniques

Before drilling holes in metal with a hand

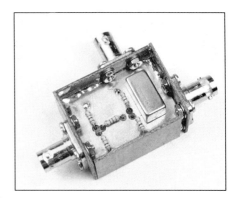

Fig 23.28 — A box made entirely from PC-board stock.

drill, indent the hole centers with a center punch. This prevents the drill bit from "walking" away from the center when starting the hole. Predrill holes greater than ½-inch in diameter with a smaller bit that is large enough to contain the flat spot at the large bit's tip. When the metal being drilled is thinner than the depth of the drill-bit tip, back up the metal with a wood block to smooth the drilling process.

The chuck on the common hand drill is limited to ⅜-inch bits. Some bits are much larger, with a ⅜-inch shank. If necessary, enlarge holes with a reamer or round file. For very large or odd-shaped holes, drill a series of closely spaced small holes just inside of the desired opening. Cut the metal remaining between the holes with a cold chisel and file or grind the hole to its finished shape. A nibbling tool also works well for such holes.

Use socket-hole punches to make socket holes and other large holes in an aluminum chassis. Drill a guide hole for the punch center bolt, assemble the punch with the bolt through the guide hole and tighten the bolt to cut the desired hole. Oil the threads of the bolt occasionally.

Cut large circular holes in steel panels or chassis with an adjustable circle cutter ("fly cutter") in a drill press at low speed. Occasionally apply machine oil to the cutting groove to speed the job. Test the cutter's diameter setting by cutting a block of wood or scrap material first.

Remove burrs or rough edges that result from drilling or cutting with a burr-remover, round or half-round file, a sharp knife or chisel. Keep an old chisel sharpened and available for this purpose.

RECTANGULAR HOLES

Square or rectangular holes can be cut with a nibbling tool or a row of small holes as previously described. Large openings can be cut easily using socket-hole punches.

23.6.5 Construction Notes

If a control shaft must be extended or insulated, a flexible shaft coupling with adequate insulation should be used. Satisfactory support for the shaft extension, as well as electrical contact for safety, can be provided by means of a metal panel bushing made for the purpose. These can be obtained singly for use with existing shafts, or they can be bought with a captive extension shaft included. In either case the panel bushing gives a solid feel to the control. The use of fiber washers between ceramic insulation and metal brackets, screws or nuts will prevent the ceramic parts from breaking.

PAINTING

Painting is an art, but, like most arts, successful techniques are based on skills that can be learned. The surfaces to be painted must be clean to ensure that the paint will adhere properly. In most cases, you can wash the item to be painted with soap, water and a mild scrub brush, then rinse thoroughly. When it is dry, it is ready for painting. Avoid touching it with your bare hands after it has been cleaned. Your skin oils will interfere with paint adhesion. Wear rubber or clean cotton gloves.

Sheet metal can be prepared for painting by abrading the surface with medium-grade sandpaper, making certain the strokes are applied in the same direction (not circular or random). This process will create tiny grooves on the otherwise smooth surface. As a result, paint or lacquer will adhere well. On aluminum, one or two coats of zinc chromate primer applied before the finish paint will ensure good adhesion.

Keep work areas clean and the air free of dust. Any loose dirt or dust particles will probably find their way onto a freshly painted project. Even water-based paints produce some fumes, so properly ventilate work areas.

Select paint suitable to the task. Some paints are best for metal, others for wood and so on. Some dry quickly, with no fumes; others dry slowly and need to be thoroughly ventilated. You may want to select rust-preventative paint for metal surfaces that might be subjected to high moisture or salts.

Most metal surfaces are painted with some sort of spray, either from a spray gun or from spray cans of paint. Either way, follow the manufacturer's instructions for a high-quality job.

PANEL LAYOUT AND LABELING

There are many ways to layout and label a panel. Some builders don't label any controls or jacks, relying on memory to figure what does what. Others use a marking pen to label controls and inputs. Decals and dry transfers have long been a staple of home brewing. Label makers that print on clear or colored tape are used by many.

With modern computers and available software, it is not hard to lay out professional looking panels. One can use a standard drawing program for the layout. The grids available on these drawing programs are sufficient to make sure that everything is lined up squarely. If the panel label is laid out before the panel is drilled for controls, a copy of the label can be used as a drill template.

Computer-aided design (CAD) programs can also be used to lay out and label panels, although they can have a steep learning curve and may be overkill for many applications.

WB8RCR has written two software programs, *Dial* and *Panel*, that are specifically designed for laying out panels and dials. There are *Windows* versions and platform independent versions available for download at **qslmaker.mi-nts.org**. These programs output a Postscript file.

These programs can be used in several ways. One can print out a mirror image of the layout on a transparency and then glue that to the front panel with the printing facing towards the panel. In this manner the transparency will protect the label. The panel layout can be printed out on card stock and affixed to the front panel with spray adhesive or self-adhesive contact paper can be used. If the printing is facing outward it can be sprayed with clear acrylic spray to protect it.

Surplus meters often find their way into projects. Unfortunately the meter faces usually do not have an appropriate scale for the project at hand. Relabeling meters has long been a mainstay to make home brew gear look professional. With the advent of computers this job has been made very easy. A software package, *MeterBasic*, included on the *Handbook* CD-ROM, is very easy to use and results in professional looking meters that indicate exactly what you want them to indicate.

SUMMARY

If you're like most amateurs, once you've got the building bug, you won't let your soldering iron stay cold for long. Starting is the hardest part. Now, the next time you think about adding another project to your station, you'll know where to start.

Contents

24.1 A 100-W Compact Z-Match Antenna Tuner

24.2 A Microprocessor-Controlled SWR Monitor

24.3 A 160- and 80-M Matching Network for Your 43-Foot Vertical

24.4 Switching the Matching Network for Your 43-Foot Vertical

24.5 An External Automatic Antenna Switch for Use With Yaesu or ICOM Radios

24.6 A Low-Cost Remote Antenna Switch

24.7 Audible Antenna Bridge

24.8 A Trio of Transceiver/Computer Interfaces

24.9 A Simple Serial Interface

24.10 USB Interfaces For Your Ham Gear

24.11 The Universal Keying Adapter

24.12 The TiCK-4 — A Tiny CMOS Keyer

24.13 The ID-O-Matic Station Identification Timer

24.14 An Audio Intelligibility Enhancer

24.15 An Audio Interface Unit for Field Day and Contesting

Chapter 24

Station Accessories

An Amateur Radio station is much more than the transceiver, amplifier and other major pieces of equipment found in a typical ham shack. In this chapter you will find a number of accessory items to make operation more convenient, or to tie other pieces of equipment together. Projects include antenna matching and switching solutions, transceiver/computer interfaces and other station accessories. Additional station accessory projects may be found on the *Handbook* CD-ROM.

24.1 A 100-W Compact Z-Match Antenna Tuner

A Z-Match tuner will match just about anything on all HF bands, and only uses two controls. They are well regarded, but acquiring the necessary air-wound inductors and variable capacitors can be difficult. In addition, air-wound inductors imply large tuners. The antenna tuner described here by Phil Salas, AD5X, is compact enough for portable HF operation, handles 100 W and uses readily available parts.

The idea for this project came from an article by Charles Lofgren, W6JJZ, in the *ARRL Antenna Compendium Vol 5*. In that article, the author suggested using a toroid core-based inductor. This would solve the air-core inductor/size problem. Suitable variable capacitors are available at a good price. The result is the compact unit shown in **Fig 24.1**.

Fig 24.1 — The Z-Match Tuner handles 100 W and can be used from 80 through 10 meters.

TUNER CONSTRUCTION

The final circuit shown in **Fig 24.2** is based on W6JJZ's article. The only real change was to go from two switch-selected output links (10 turns and 4 turns) to a single 8-turn output link, and to increase the variable capacitor size to 400-pF. The tuner provides a good match from 80-10 meters!

The tuner is built into a 3 × 5¼ × 5⅞ inch (HWD) aluminum box. Toroid T1 is supported by its leads and some hot glue between the inductor and the frame of C2. Also put a little hot glue between T1 and the side of the case.

The variable capacitors must be insulated from ground, therefore mount both capacitors on a piece of single-sided circuit- or perf-board that is cut just wide enough to fit into the aluminum case. Then mount this capacitor/perf-board assembly in the case with stand-off screws. Capacitor shaft couplings are made from a ⅛-NPT brass nipple, available from the plumbing section of most hardware stores. These nipples have a ¼-inch inside diameter. Cut a 1-inch nipple in half to make two couplers. Then drill and tap holes for two #6 screws in each piece. See **Fig 24.3** for capacitor mounting and shaft coupler details. For the insulated shafts, use ¼-inch diameter nylon rods available from many hardware stores.

OPERATION

Tuning the Z-Match tuner is very easy. First adjust C2 for maximum receiver noise. Then apply some RF power and adjust C1 and C2 for minimum SWR. If you need more capacitance for matching, use S1 to switch in the extra section of C1, or switch in a fixed mica capacitor across C1. Balanced feed lines terminated in banana plugs can plug right into the center pin of the output SO-239 and the adjacent banana jack. To feed coax, ground one end of the output link with switch S2.

OPTICAL HF SWR METER FOR THE Z-MATCH TUNER

While you can use any external SWR meter with the Z-Match tuner — including the SWR meter built into most rigs — the author built an optical SWR meter into the same case for convenience. Refer to **Fig 24.4A**. It works well with high intensity LEDs rather than conventional meters. The SWR meter is built on a small piece of perf-board and mounted to the solder lug on the tuner's input SO-239. A little hot glue between the perf-board and the back of the chassis adds stability (**Fig 24.5**).

This broadband circuit works well at the 100-W power level through at least 30 MHz. With careful lead control, it should work up through 6 meters. The transformer is an FT37-43 ferrite core wound with 10 bifilar turns of #26 enameled wire. The primary is just the single wire passing through the center of the toroid. To calibrate the SWR bridge, connect the output to a resistive 50-Ω load. Apply RF power on any HF band and adjust the 20-pF variable capacitor until the REFL LED goes out.

The FWD LED gives an indication of transmitter forward power. You may want to increase the value of the 4.7 kΩ resistor in the FWD circuit if the green LED is too bright. Or you could eliminate this LED and just use the REFL LED.

A BAR-GRAPH DISPLAY FOR THE OPTICAL SWR METER

If you can supply dc power to your Z-Match Tuner and SWR meter, you may want to add the bar graph display as shown in Fig 24.4B to display SWR reflected power. A version of the tuner with bar graph option is shown in **Fig 24.6** The nice thing about using the bar graph display is that it seems easier to null the reflected power because the display gives more of an analog meter "feel." Adjust the Z-Match tuner for minimum brightness of the REFL LED. When this occurs, the SWR should be less than 1.5:1.

This tuner addresses the issues of inductor size and finding reasonably priced multi-section variable capacitors. The result is a wide-band, easily adjustable tuner for either portable or base station operation.

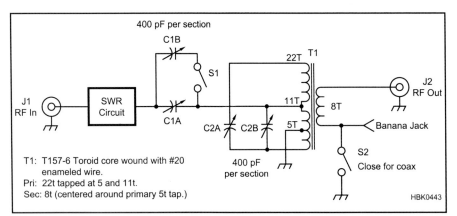

Fig 24.2 — Schematic of the 100-W Z-Match tuner.

C1, C2 — 400 pF per section, 2-section variable capacitor (Oren Elliott S2-399, www.orenelliottproducts.com). If you substitute another variable capacitor, use one with at least 350 pF per section, rated at 500 V RMS minimum.
J1, J2 — SO-239 connectors.
S1, S2 — SPDT mini-toggle switch.
T1 — Primary: 22 turns #20 AWG enam wire tapped at 5 and 11 turns. Secondary: 8 turns, with 4 turns either side of the primary 5 turn tap. Wound on T157-6 toroid (Amidon Associates, www.amidon-inductive.com).
Enclosure (Eagle, 40UB103 from www.mouser.com).

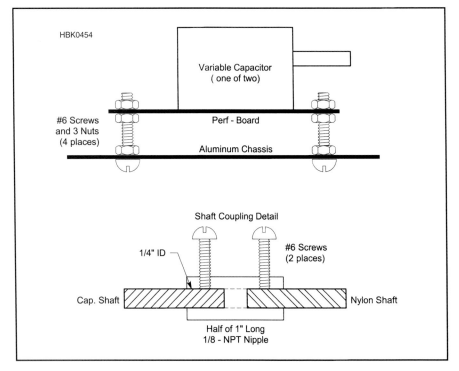

Fig 24.3 — The variable capacitors are insulated from the chassis by mounting them on perforated board. A 1/8-NPT nipple is made into a shaft coupling to couple the insulating shafts to the main capacitor shafts.

Fig 24.4 — Optical SWR meter schematic with the bar graph display modification (see text). The basic LED version is shown in A and the bar graph addition in B.

C3 — 2-20 pF variable capacitor (Mouser 24AA113 4-34 pF variable is suitable).
D1 — Red LED, high-intensity.
D2 — Green LED, high-intensity.
L1 — 10 bifilar turns of #26 enam wire on FT37-43 toroid core (Amidon Associates, www.amidon-inductive.com).
U1 — Display driver IC, LM3914 (All Electronics, www.allelectronics.com).
U2 — 10-LED bar graph display (Mouser 859-LTL-1000HR). Any 10-segment LED bar graph will work, but verify pinout.

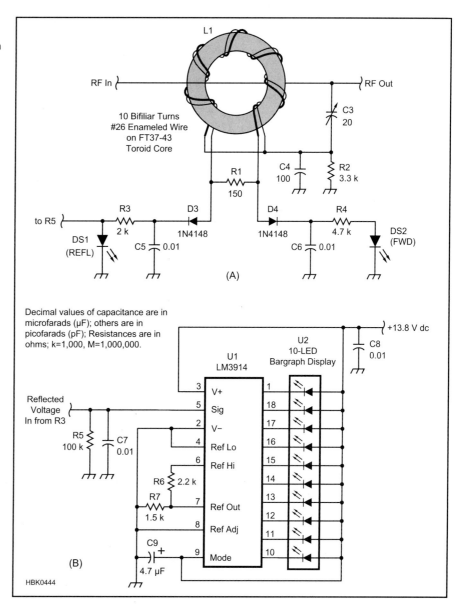

Fig 24.5 — The SWR meter circuit board is soldered directly to the RF input connector and attached to the back panel with hot glue.

Fig 24.6 — The bar graph display of reflected power makes tuning easier and more intuitive.

Station Accessories 24.3

24.2 A Microprocessor Controlled SWR Monitor

This update of the basic SWR meter, designed by Larry Coyle, K1QW, uses a microprocessor to drive an analog panel meter, exploiting the best features of both analog and digital technology. The instrument also displays forward/reflected/net power and return loss (see the **Transmission Lines** chapter) with peak-hold capability. All display selections are selected by front panel switches. [*More information is available on K1QW's Web site, www.lcbsystems.com, and a complete package of information on this project with supporting files and photos is available on the CD-ROM accompanying this Handbook.*]

OVERVIEW AND BLOCK DIAGRAM

After a bit of work, the author came up with the SWR/Power/Return Loss monitor shown in block diagram form in **Fig 24.7**. There are three interacting assemblies: SWR sense head, front panel assembly and signal processor board. (The latter two make up the Display and Signal Processor unit.) The SWR sense head is a conventional Stockton dual-transformer bridge with Schottky diode detectors. This component converts the forward and reflected RF amplitudes to dc and connects to the Display and Signal Processor by an ordinary stereo audio cable.

The user interface to the SWR monitor is familiar and intuitive. As you can see in **Fig 24.8A**, a front-panel rotary switch selects the quantity to be measured and an analog meter displays the result.

- Forward power on a linear scale
- Reflected power as a percentage of forward power
- Net power delivered to the load
- SWR on a linear scale
- Return loss on a linear scale

Some additional features are:
- Peak-hold front panel switch
- A front panel switch to multiply the meter sensitivity by 3X
- Data logging to a computer using the USB port.
- Transmitter Lockout (XLO) at high SWR

The rear panel, **Fig 24.8B**, shows the power switch and external connections

Transmitter Lockout (XLO) is a protective feature that lights a LED on the front panel and energizes a relay when the SWR exceeds a preset level. This opens a set of relay contacts which are connected to a pair of banana jacks on the rear panel. Once triggered, the relay remains energized for either three or nine seconds, depending on the configuration of the DIP switch located on the signal processor board. The same DIP switch also allows you to select the SWR level at which the XLO relay triggers — 1.5, 2, 3 or 4:1. By connecting the relay contacts in series with your rig's PTT or keying line, transmitting into high SWR can be prevented.

The two dc signals from the sense head make up the analog input to the signal processor board. The analog output for driving the meter is a pulse width modulated (PWM) signal generated by the microprocessor and smoothed by a simple resistor-capacitor (RC) filter.

THE MICROPROCESSOR

The Parallax "Propeller" microprocessor (**www.parallax.com**) is unique in that it's really eight processors ("cogs") in one package, each one able to run its own separate program during its own time slot, controlled by a central "hub." This may sound challenging, but this actually makes programming the chip

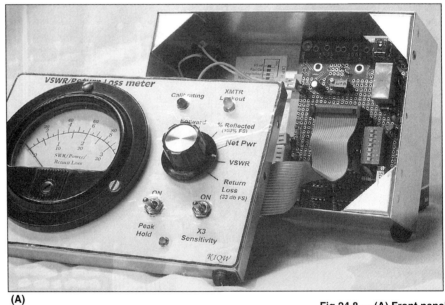

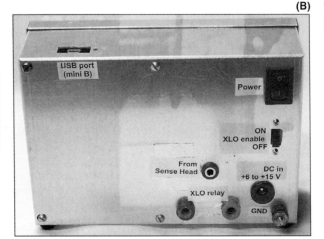

Fig 24.8 — (A) Front panel of the display and signal processor unit. The meter face and the front panel lettering were created using Microsoft *Visio*, printed on heavy paper stock and attached with contact cement. With the front cover moved aside, you can see the ribbon cable carrying signals between the front panel assembly and the processor board. (B) Rear panel of the display and signal processor. The mini-B USB connector is accessible through the opening on the top surface of the enclosure.

Fig 24.7 — This block diagram shows how the SWR sense head and the signal processor box are connected by an ordinary two-conductor shielded stereo audio cable. The signal processor box contains the processor board and the front panel assembly.

much easier. If you need to carry out several tasks at once — and most microprocessor applications do — just assign each job to its own individual cog and work on them separately. The central hub makes sure each cog gets its turn and keeps things synchronized.

The project uses the 3 × 4 inch Propeller USB Development Board from Parallax. It includes the Propeller chip, an Electrically-Erasable Programmable Read-Only Memory (EEPROM) for program storage, a 5 MHz crystal, two voltage regulators and, a USB mini-B connector with all the necessary USB interface circuitry.

The Propeller is programmed in a language called "Spin" which should make anyone who is familiar with the BASIC computer language feel at home. All the tools you need to write software for and to program the Propeller is available for free on the Parallax Web site, including an extensive library of pre-built code routines that you can just drop into your own programs.

Software is loaded onto the development board using the USB port from a host computer with a USB port. When not in use as a programming port, the same USB port is used as a serial data port for passing data between the SWR monitor and a computer running a terminal program such as HyperTerminal.

MICROPROCESSOR BOARD CIRCUITRY

The development board includes a prototyping area large enough to hold all the parts of the SWR monitor except for front and rear panel components. Since most of the parts come assembled on the development board, construction of the digital board is quite easy. As you can see in **Fig 24.9A**, the electrical schematic, only two integrated circuits and a few analog components are required.

The dc signals representing forward and reflected power from the SWR sense head are received by dual op-amp U2. Each section of U2 is configured as a unity-gain voltage follower with a compensating diode (D1 and D2) to balance out the dc offset introduced by the Schottky rectifier diodes in the sensing head. (The technique of using a diode in the feedback path of an op amp to compensate for the non-linearity of an RF detector diode is described in detail in "A Compensated, Modular RF Voltmeter" in the **Test Equipment and Measurements** chapter of this book.) Each section also has an RC low-pass filter at its input to filter out unwanted RF — always a possibility around a ham shack.

The next stage in the signal path is U3, an analog multiplexer (MUX). This chip acts as a three-position switch under control of the microprocessor. (A fourth channel is unused.) The forward and reflected signals (on pins 14 and 15) are sampled alternately, 20 times per second, as long as the program is running. The

Table 24.1A
Digital Inputs

Digital Input	Originates at	What it does
X3SEL	front panel - toggle switch	increases meter sensitivity by 3X
PEAKENBL	front panel - toggle switch	enables peak hold function on the meter
RLOSSEL	front panel – rotary switch	displays return loss on the meter
SWRSEL	front panel – rotary switch	displays SWR on meter
NETSEL	front panel – rotary switch	displays net power on meter
REFSEL	front panel – rotary switch	displays percent reflected power on meter
FWDSEL	front panel – rotary switch	displays forward power on meter
HIREFSEL	proc board - DIP switch	drives meter to full scale for calibration (must be OFF for normal operation)
FASTCAL	proc board - DIP switch	speeds up meter response for calibration (must be OFF for normal operation)
XLOLONG	proc board - DIP switch	enables long xmtr lockout (about 9 seconds)
XLTHRESH0	proc board - DIP switch	XLO threshold bit 0
XLTHRESH1	proc board - DIP switch	XLO threshold bit 1
XLOCKENBL	rear panel switch	enables xmtr lockout function
DIAG	proc board - jumper JP1	displays raw input levels on terminal
W100	proc board - jumper JP2	changes terminal display to 0-100 watt range (required when using 100-watt sense head)

Table 24.1B
Digital Signals for SWR Threshold Select

XLTHRESH1	XLTHRESH0	Selects SWR threshold level
OFF	OFF	1.5:1
OFF	ON	2:1
ON	OFF	3:1
ON	ON	4:1

third input to the MUX (pin 12) is grounded, and is also sampled by the microprocessor from time to time. This gives a measure of any dc offsets introduced by the analog-to-digital conversion and is used to correct the forward and reflected readings in software. This automatic zero adjustment takes place every five seconds or so and is indicated by a brief flash from the front-panel "Calibrating" LED that also serves as a "digital heartbeat" to show that the software is running.

The MUX output (pin 13) is passed to a potentiometer voltage divider to allow the scale factor to be set during calibration. From there it goes to the Propeller microprocessor itself where a delta-sigma analog-to-digital conversion takes place. (See the **Analog Basics** chapter for information on analog-to-digital conversion.)

DIGITAL INPUT/OUTPUT

The microprocessor responds to the user's inputs by monitoring the state of 13 digital input pins. Most of these pins are controlled by the rotary and toggle switches on the front panel. The panel connects to the processor board by a short ribbon cable as shown in Fig 24.8A. See Fig 24.9B for a schematic of the separate front panel assembly. The ribbon cable also carries power for the LEDs. On the processor board there are a multi-section dual in-line package (DIP) switch, S2, and two jumpers, JP1 and JP2, which select among the various operating modes. Lastly, the Xmtr Lockout (XLO) switch is mounted on the rear panel and visible in Fig 24.8B. These inputs to the microprocessor and the various modes of operation they control are summarized in **Table 24.1A**. The digital inputs are all active low; ie, the function is enabled when the signal line is at a low logic level. XLTHRESH1 and XLTHRESH0 form a two-bit code to select one of four SWR levels. When this SWR level is exceeded and if XLO is enabled, the transmitter lockout function is triggered.

There are only six digital *outputs*:
- Two address lines for the analog MUX to select among the three input channels
- Two lines to power the Calibrating and XLO LEDs located on the front panel
- One PWM signal to drive the front-panel analog meter
- One line to control the XLO relay.

Fig 24.8A shows the built-up processor board inside the 2 × 4 × 6 inch aluminum enclosure connected to the front panel with the ribbon cable. Here you can also see some of the wiring to the rear panel components. The USB connector is at the top edge of the circuit board and is accessible through a rectangular hole nibbled out of the top surface of the aluminum box.

THE RF SENSE HEAD

The SWR sense head is the business end of the SWR meter where the forward and reflected amplitude components of the RF present on the transmission line are separated and converted to dc. For this project I adopted a straightforward Stockton directional coupler circuit, with Schottky diode detectors and, unlike other designs, it requires no

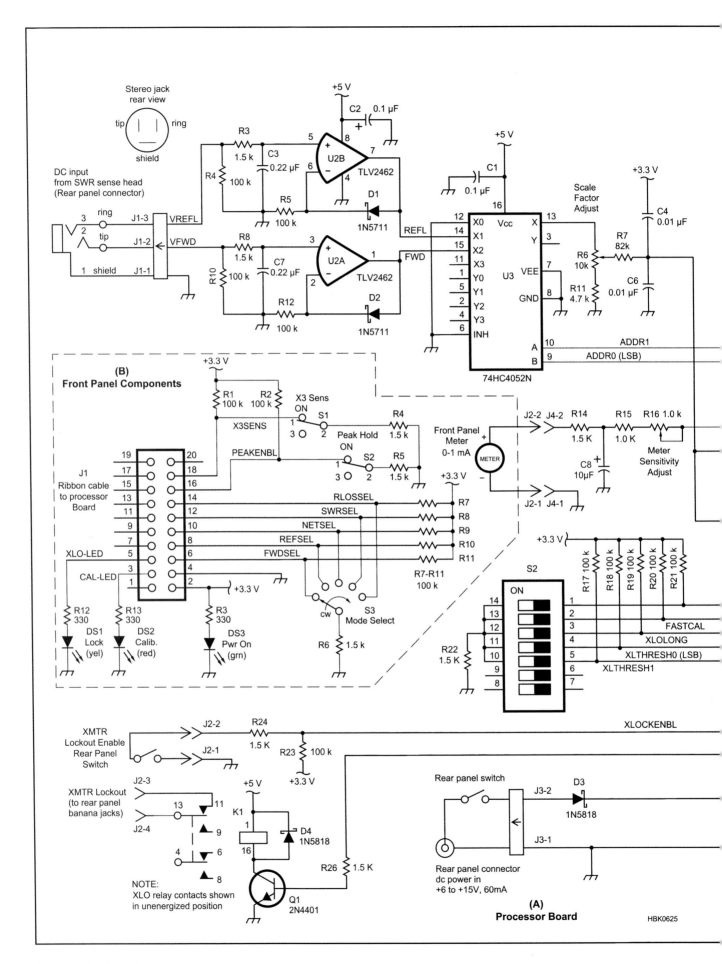

Fig 24.9 — (A) Processor board schematic. All of the parts shown here, except for the rear panel components, are mounted on the Parallax Propeller Development Board. The parts within the dotted line are already installed at the factory, and are shown here only for completeness. (B) Front panel schematic. Switches and LEDs connect to the processor board via a 20-pin dual-row header, J1, and a ribbon cable. For convenience in wiring, the analog meter connects to the processor board by a separate 2-pin connector. (See Parts List on page 24.8.)

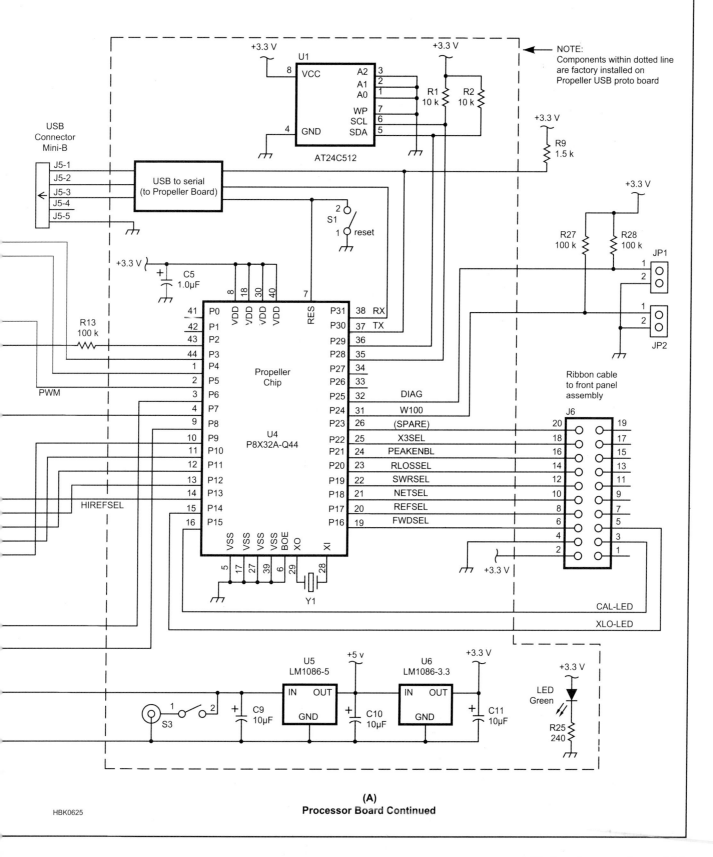

(A) Processor Board Continued

Station Accessories

Parts List for Figure 29 on previous 2 pages.

Parts List – Processor Board

C3, C7 — 0.22 µF ceramic capacitor
C1, C2 — 0.1 µF ceramic capacitor
C4, C6 — 0.01 µF ceramic capacitor
C8-C11 — 10 µF, 16 V electrolytic capacitor
D1, D2 — 1N5711 Schottky diode
D3, D4 — 1N5818 Schottky diode
J6 — 20-pin dual row header
JP1, JP2 — single jumper header
K1 — SPDT 5 V relay
Q1 — 2N4401 NPN transistor
R3, R8, R9, R14, R22, R24, R26 — 1.5 kΩ, ⅛ W, 5% resistor
R4, R5, R10, R12, R13, R17, R18, R19, R20, R21, R23, R27, R28 — 100 kΩ, ⅛ W, 5% resistor
R6 — 10 kΩ trim pot, 25 turns
R7 — 82 kΩ, ⅛ W, 5% resistor
R11 — 4.7 kΩ, ⅛ W, 5% resistor
R15 — 1 kΩ, ⅛ W, 5% resistor
R16 — 1.0 kΩ trim pot, 25 turns
S2 — 7-position DIP switch
U2 — TLV2462CP dual op amp
U3 — 74HC4052N dual 4-channel MUX
Parallax Proto USB Development Board, Part # 32812

Parts List – Front Panel

DS1 — yellow LED
DS2 — red LED
DS3 — green LED
J1 — 20-pin dual row header
R1, R2, R7, R8, R9, R10, R11 — 100 kΩ, ⅛ W, 5% resistor
R4, R5, R6 — 1.5 kΩ, ⅛ W, 5% resistor
R3, R12, R13 — 330 Ω, ⅛ W, 5% resistor
S1, S2 — SPST toggle switch
S3 — 5-position rotary switch
0-1 mA dc meter

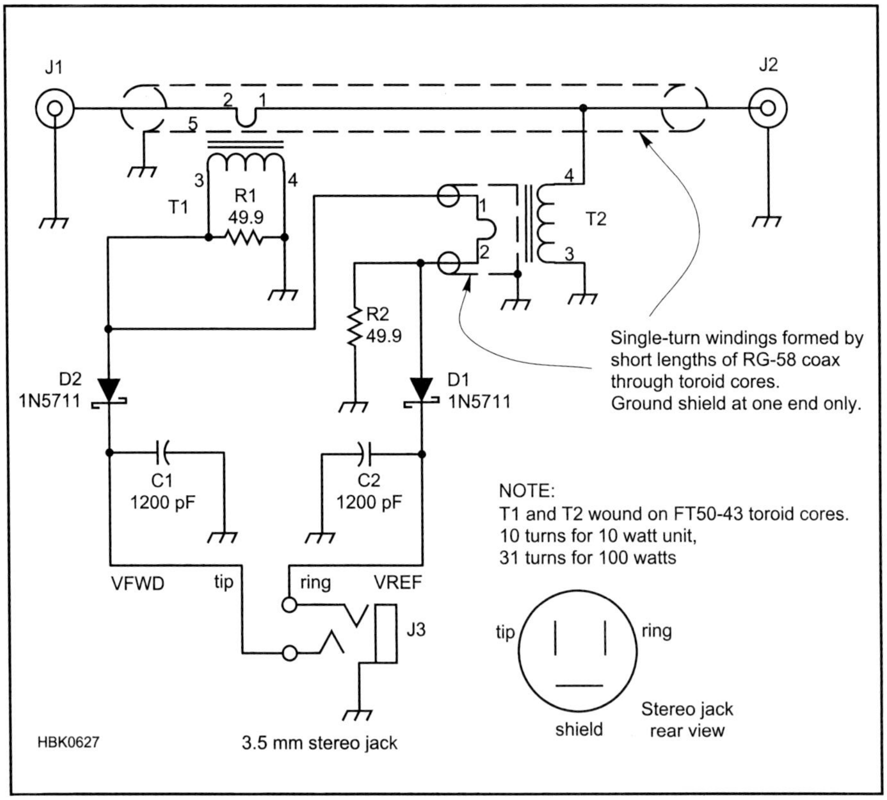

Fig 24.10 — SWR sense head schematic. All components except for the transformers are surface mount parts. R1 and R2 are size 1210, rated at ½ W. Transformers T1 and T2 are wound on FT50-43 ferrite toroid cores. For the 10-W sense head, use 10 turns of 24 AWG wire. For the 100-W head, use 31 turns.

balancing trimmers. The schematic is given in **Fig 24.10** and the circuit is described in the article by J. Grebenkemper, KA3BLO, "The Tandem Match — an Accurate Directional Wattmeter," *QST*, Jan 1987, pp 18-26 (also published in the 2009 and earlier editions of *The ARRL Handbook*).

The accuracy of this circuit can be greatly improved by taking care to match the diodes in the sensing head with the corresponding compensation diodes in the op amp input stages on the signal processor board (D1 and D2 in Fig 24.9A). The easiest way to do this is with the diode test function included on most utility digital multimeters which passes a small dc current through the diode being tested and displays the forward voltage drop. The high-frequency range can be extended by using surface-mount components in the sensing head.

There are two versions of the SWR sensing head. The QRP version has a maximum range of 10 W. The other can handle up to 100 W. The only difference between the two versions is the number of turns on the toroid transformers (see the parts list in Fig 24.10). The high frequency performance of the 100-W head at 50 MHz proved to be poorer than the 10-W unit. I attribute this to the added capacitance of the transformer windings (the higher power transformers have three times as many turns as the low-power ones).

The overall accuracy of the monitor compares fairly well to commercial devices. **Fig 24.11A** and **24.11B** show performance data I took using dummy loads of 50 and 96.7 Ω. At five watts input, using the 10-W sensing head, the SWR readings were within 3% over a range of 1.8 to 30 MHz, and within 5% at 50 MHz. Fig 24.11B shows good accuracy down to 500 mW input (great for QRP rigs!) and is even usable down to 250 mW.

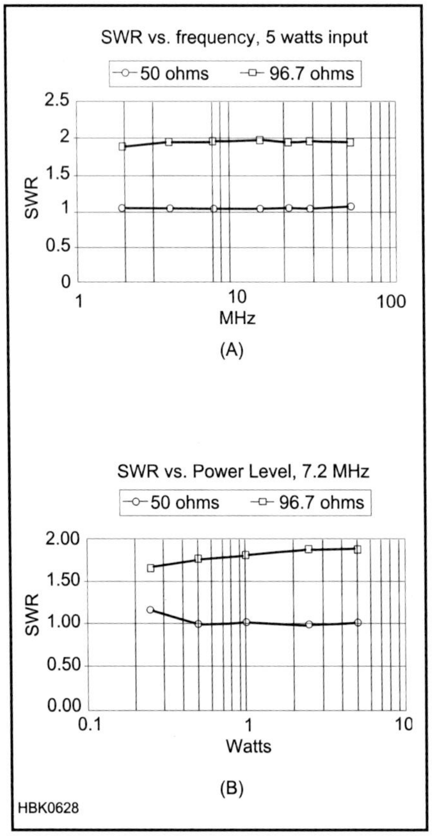

Fig 24.11 — (A) SWR plotted against frequency, up to 50 MHz. (B) SWR as a function of input power level at a fixed frequency of 7.2 MHz. The data shown here were taken using the 10-W sense head, with dummy loads measured at 50 and 96.7 Ω.

24.4.6 Software

For those who want to duplicate this project, the Spin language software listing, ready to program into the Parallax USB Development Board, is available in the package of

supporting files on the CD-ROM. There are also details of the subroutines handled by each of the Propeller cogs plus additional text and graphic files covering operation and features of the instrument.

As explained above, the organization of the Propeller into several peripheral processors (cogs) makes it easy to follow the program flow. The main program object (MAIN) runs in its own cog and is the first to be executed when power is applied (or whenever the chip is reset). Its first order of business is to set individual microprocessor pins to be inputs or outputs, as required.

Next, MAIN starts up three separate cogs for (1) analog-to-digital conversion (the ADC3CH cog), (2) generating the pulse width modulated waveform to drive the analog panel meter (the PWM cog) and (3) communicating with the terminal via the USB port (the MONITOR cog). They run continuously, swapping data with the MAIN program by means of variables stored in memory. After these three cogs are started, the MAIN program object enters a loop, endlessly repeating the following steps:

1) Read forward and reflected amplitudes and SWR data generated by the ADC3CH object.

2) Calculate forward power, percent reflected power, net power, SWR and return loss in decibels.

3) Display the quantity selected by the front panel switch on the panel meter using the PWM object.

4) Convert power levels, SWR and return loss information to ASCII strings and send them to the MONITOR object, which writes to the USB port.

This is a shining example of building on the accomplishments of others who have made their work freely available. Without such an extensive library of pre-tested routines on hand, the author probably would never have even started this project.

24.3 A 160- and 80-Meter Matching Network for Your 43-Foot Vertical

Many amateurs are using 43-ft vertical antennas for multiband operation. These antennas have higher radiation resistance than shorter verticals, which minimizes ground losses. This is especially important when you have an electrically short antenna — a characteristic of even a 43-ft antenna on 160 and 80 meters.

When fed with a 1:4 unun (unbalanced-to-unbalanced transformer), a 43-ft antenna has a reasonable compromise SWR on 60-10 meters, which means that cable and unun losses are pretty much negligible on these bands. Performance on 160 meters, and to a lesser extent on 80 meters, is not as good unless you provide matching right at the antenna. The high capacitive reactance and low radiation resistance of a 43-ft antenna on 160 and 80 meters make the mismatch so bad that it is almost impossible to match from your shack if you are using low-loss coax. If you can match the antenna system from your shack, you will lose a lot of power in your coax and unun because of the very bad mismatch at the antenna. Phil Salas, AD5X, describes two impedance matching devices for use with 43-ft verticals designed to significantly reduce SWR-related losses and to "help out" inside tuners on 160 and 80 meters.

THE MATCHING REQUIREMENT

Antenna analyzer measurements of the author's 43-ft vertical antenna indicate a capacitive reactance of about 580 Ω on 160 meters. This value will vary with different kinds of 43-ft verticals, proximity to other objects and other factors unique to each installation but is usually in the 550-650 Ω range. With this amount of capacitive reactance, approximately 50 µH of inductance is needed to resonate the antenna. On 80 meters, approximately 9 µH is needed. A 50 µH high-Q inductor is large, but a toroidal inductor will be more compact.

Fig 24.12 shows the circuit of a compact design that handles 1500 W SSB or CW on 80 and 160 meters. **Table 24.2** lists the necessary parts. The inductor consists of 35 turns of #14 AWG solid copper insulated house wiring wound on a T400A-2 toroid core, with the feed tapped 2 turns from the ground end for 80 meters, and 3 turns from the ground end for 160 meters. Start with 38 turns and then remove turns as necessary to get the network to resonate where you want it in the 160 meter band (more on this later).

The toroid assembly fits in a 6 × 6 × 4-inch NEMA enclosure. The assembly is secured with a 2.5-inch long #10 screw and associated hardware along with a 2 × 4 inch piece of unplated fiberglass PC board material (**Fig 24.13**). Before mounting the toroid, prepare it by scraping the insulation off the outside of the wire at turns 2, 3 and 11-13. Because of the high voltages possible with 1500 W (especially on 160 meters), wrap the toroid with two layers of 3M #27 glass-cloth electrical tape for added insulation between the wire and the toroid core.

To select between 160- and 80-meter operation, the appropriate connections are made with external jumpers across binding

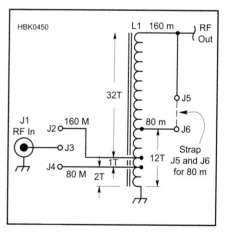

Fig 24.12 — 160/80-meter impedance matching network for use with a 43-ft vertical. See Table 24.2 for a parts list.

Table 24.2
Parts List: 160/80-Meter Impedance Matching Assembly

Qty	Description	Source/Part Number
1	6×6×4 inch NEMA enclosure	Hardware store/home center
1	T400A-2 powdered iron toroid (L1)	Amidon T400A-2
1	SO-239 connector (J1)	Mouser 601-25-7350
4	Black binding post (J2, J4-J6)	Mouser 164-R126B-EX
1	Red binding post (J3)	Mouser 164-R126R-EX
4	Banana plug	Mouser 174-R802-EX
1 roll	3M #27 glass tape (for L1)	Hardware store/home center

Misc: #14 AWG house wiring, stainless steel hardware.

posts as shown in **Fig 24.14**. Stainless steel #8 hardware (screws, washers, lockwashers, nuts) is used for the matching unit ground and RF output. Internal to the matching unit, a 2-inch-wide strip of aluminum duct repair tape makes a good low-impedance ground between the coax connector and the ground screw on the bottom of the case. Finally, use #14 AWG stranded insulated wire for all internal connections.

TUNING THE MATCHING NETWORK TO RESONANCE

Your particular installation will almost certainly require you to change the resonant frequency of the matching network. This is because there will be variations of the antenna impedance based on your particular installation and desired operating frequency range. The design is such that the overall inductance is too large for 160 meters, so the network should resonate at or below the lower band edge. Therefore, you will need to remove one or more of the upper inductor turns to resonate the network for the desired frequency on 160 meters. To do this, first solder wires from the turn 2 and 3 tap points on the coil to the two outer binding posts by the RF IN connector. The input tap points tend to be fairly non-critical and will probably always be the same for all installations.

Now solder a short wire from the RF IN center pin to the middle binding post. Next, externally jumper the middle binding post to the 160 meter binding post (turn 3). Connect the matching assembly to the base of your 43-ft vertical and use an antenna analyzer to find the minimum SWR point on 160 meters. If the resonant frequency is too low, remove a turn of wire and measure againch You should see a 50 kHz upward move in frequency per turn of wire removed.

When you have reached the desired resonant point on 160 meters, it is time to move to 80 meters. Externally jumper the input tap middle binding post to the 80-meter binding post (turn 2), and use a clip lead to short from the top of the coil to about turn 12. Again, find the frequency where minimum SWR occurs. Move the tap point up or down until you reach the desired resonance point. Solder a wire from this tap point to one of the binding posts. Solder another wire from the top of the coil to another binding post. Now you will be able to externally jumper these binding posts to go from 160 to 80 meters.

Fig 24.15 shows the author's final results for 160 and 80 meters as measured with an antenna analyzer connected directly to the matching network input at the base of the antenna. With resonance set for the CW end of these bands, the 2:1 SWR bandwidth on 160 meters is about 50 kHz and about 150 kHz on 80 meters. Even a 3:1 SWR on these bands results in negligible SWR-related cable losses and is easily matched with an in-shack tuner.

OPERATION

Using the matching unit is simple. Just disconnect your normal unun when you want to operate on 160 or 80 meters and connect this matching unit to the base of the antenna. Select either 160 or 80 meters with the external straps. You can connect both the unun and this matching unit to the antenna at the same time, and just leave off the ground wire from the unit that is not used. The matching unit connected to the base of a 43-ft vertical is shown in **Fig 24.16**.

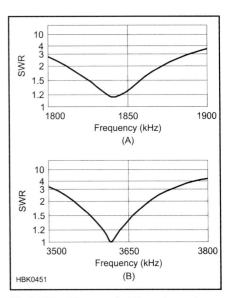

Fig 24.15 — Measured at the antenna base, the 2:1 SWR bandwidth is about 50 kHz for 160 meters and 150 kHz for 80 meters.

Fig 24.13 — Mounting details of the toroidal inductor for the 160/80-meters matching unit. The input side of the enclosure is on the left. Coax to the station connects to the RF IN SO-239, and a strap between the center binding post and one of the outer posts selects the input tap for operation on 160 or 80 meters. On the right side, a jumper between the two binding posts shorts a section of the coil for 80-meter operation. The RF OUT connection to the base of the antenna is made with via the #8 machine screw and hardware at the top, and the ground connection is on the bottom.

(A)

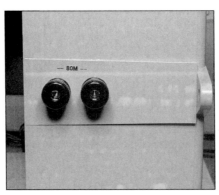

(B)

Fig 24.14 — Binding posts are used for jumpers to select 160 or 80 meter operation.

Fig 24.16 — The finished matching unit at the base of the author's 43-ft antenna (strapped for 80 meter operation).

24.4 Switching the Matching Network for Your 43-Foot Vertical

Another project in this chapter describes a simple 160 and 80 meter impedance matching network for installation at the base of a 43-ft vertical antenna. While that unit is very effective and inexpensive to build, it is also a little inconvenient in that you must connect it when it is needed, and you must also manually enable 160 or 80 meter operation using jumpers. Phil Salas, AD5X, enhances the original design with a more versatile matching assembly that can be controlled remotely for operation on all bands.

FIRST — A WORD ABOUT RF VOLTAGES

This unit uses relays for selecting the different bands, so a discussion of RF voltages is appropriate. RF voltages at the base of an untuned vertical can be quite high. As the antenna becomes shorter, the capacitive reactance becomes higher and so the resultant voltage drop across the combination of reactance and radiation resistance increases. With the 43-ft vertical, the worst-case situation will occur on 160 meters where the capacitive reactance is approximately 600 Ω. The radiation resistance is approximately 3 Ω.

With a perfect ground (no ground loss) and 1500 W of power properly matched to the antenna, peak RF voltage works out to about 19,000 V_{pk}:

$$I = \sqrt{1500 \text{ W} / 3 \text{ } \Omega} = 22.4 \text{ A}$$
$$|Z| = \sqrt{3^2 + 600^2} = 600 \text{ } \Omega$$
$$V_{RMS} = 22.4 \times 600 = 13,440 \text{ V}$$
$$V_{pk} = 13,440 \times 1.414 = 19,004 \text{ V}$$

Very few hams have a ground system that is lossless. With a ground loss of 10 Ω (better than most hams have), it works out to about 9100 V_{pk}. With the author's 600-W amplifier, assuming a ground loss of 10 Ω, it's about 5800 V_{pk}.

The author experimented with two different relays from Array Solutions. The RF-10 DPDT relay has 1.7 kV_{pk} contact-to-contact and 3.1 kV_{pk} contact-to-coil voltage breakdown ratings. This relay proved to be a good solution for up to about 500 W on 160 meters when the two sets of contacts are put in series. The second relay is the RF-3PDT-15 which has 3.1 kV_{pk} contact-to-contact and 5.3 kV_{pk} contact-to-coil voltage breakdown ratings. This relay has nearly twice the contact breakdown rating of the RF-10, and it's 3PDT so an additional set of contacts can be put in series to increase the breakdown voltage. This relay can be used in a full legal limit application if applied properly in the circuit (more on this later).

Depending on your power level and ground losses, the less expensive and smaller RF-10 may be all you need. Make the calculations to determine which relay is more appropriate for you. If you look for alternative relays, pay attention to the breakdown voltage rating; it's hard to find relays with suitable ratings for this application. The Array Solutions relays have 30 A contacts, but 15 A or more should be sufficient.

THE ALL-BAND MATCHING SOLUTION

As explained in the companion project describing a simple matching network, when fed with a 1:4 unun (unbalanced-to-unbalanced transformer), a 43-ft antenna has a reasonable compromise SWR on 60-10 meters. The antenna requires approximately 50 μH of inductance to resonate on 160 meters and 9 μH on 80 meters. For convenient all-band operation from the shack, the solution shown in **Fig 24.17** uses two relays to switch in the unun for 60-10 meter operation or the appropriate inductance for 160 and 80 meters. The unit will handle up to 1500 W from 160-10 meters.

Relay control requires inputs of 0 V, +12 V dc or –12 V dc. The matching unit operates as follows:

- With no control voltage applied, the 1:4 unun is connected to the antenna for 60-10 meter operation, and the inductor is disconnected, so the original antenna compromise SWR on these bands is preserved.

- With +12 V applied for 80-meter operation, K1 and K2 close. Contact K1C connects the unun secondary to L1 through K2C at the 200 Ω point (6 turns from the ground end). Contacts K2A and K2B short out the top part

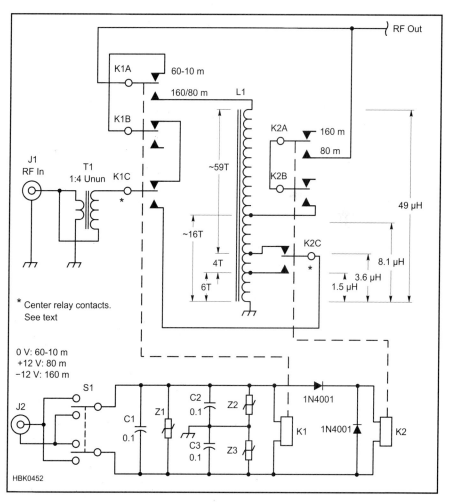

Fig 24.17 — This switched impedance matching network for use with a 43-ft vertical allows operation on 160-10 meters and is controlled from the station operating position. See Table 24.3 for a parts list.

of L1 (about 16 turns from the ground end), resonating the antenna on 80 meters.

• When the voltage is reversed (–12 V) for 160-meter operation, K1 stays closed but K2 opens. K2C now connects the unun secondary to L1 at the 200 Ω point for 160 meters (10 turns from the ground end). Contacts K2A and K2B open, removing the short and allowing the entire coil to be used to resonate the antenna on 160 meters.

Connecting the unun to L1 at the 200 Ω point keeps the unun secondary voltage reasonable, and the feed line and unun losses very low as well.

CONSTRUCTION

Fig 24.18 shows the layout. The matching unit is built in 8 × 8 × 4-inch NEMA weatherproof box available from most home centers. The author used an MFJ 404-0669 air-wound coil, and this box is too small to fit the entire inductor length needed. The inductor is split into two pieces; the long section consists of 61 turns, and the short section consists of 12 turns. As drawn in the schematic, the longer coil section is at the "bottom" (one end connects to ground), while the shorter coil is at the "top" (one end connects to RF OUT). If you have room for a larger box, a 12 × 12

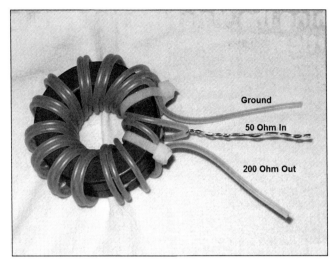

Fig 24.19 — Connections for a voltage unun (MFJ 10-10989D shown).

× 4-inch NEMA box fits the full inductor length without cutting.

Two Array Solutions RF-3PDT-15 relays are mounted to the case next to each other in the upper center. These are enclosed relays with all connections on one side, making for easy assembly. The 2.1 × 5.5 mm dc power jack and SO-239 RF IN connector are in the upper left corner. The MOVs and bypass capacitors for the control line are mounted on a terminal strip near the power connector.

The unun is mounted to the side of the enclosure using a machine screw and a piece of PC board material with the copper removed. The antenna is unbalanced, so a current balun or voltage unun (*not* a voltage balun) is typically used. A voltage unun should be wired as shown in **Fig 24.19**.

All internal wiring consists of insulated #14 AWG stranded wire. Attach the wires to the coil tap points using the MFJ coil clips called out in the parts list. (You can solder the wires directly to the coil, but this is difficult because of the size of the coil wire and the spacing of the turns.) Use #8 stainless steel screws, washers, lock-washers and nuts for coil mounting, and for the ground and RF OUT (antenna feed) terminals. **Table 24.3** lists the parts and part sources.

Relay Connections

As discussed earlier, the voltage across the full coil can be very high on 160 meters. Connecting the relay contacts in series as shown in the schematic increases the overall breakdown voltage. Also, the tap points on the inductor provide additional voltage-above-ground, which helps with the overall breakdown voltage.

The main concern is with the contact-to-

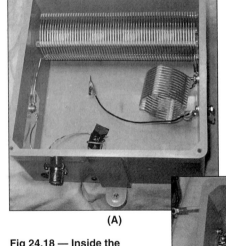

(A)

(B)

Fig 24.18 — Inside the enclosure. At A, the air-wound inductor has been cut into two pieces (see text) and mounted. The wire from the shorter piece will attach to the RF OUT terminal. At B, the other components have been added. The unun is mounted on the left side of the case. The relays are at the top center, and the MOVs and bypass capacitors are mounted on a terminal strip at the upper left, next to the RF IN and dc control connectors.

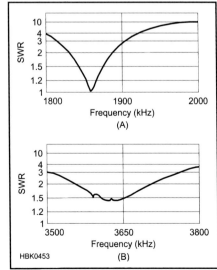

Fig 24.20 — Measured at the antenna base, the 2:1 SWR bandwidth is about 50 kHz for 160 meters and 150 kHz for 80 meters.

Table 24.3
Parts List: Switchable 160-10 Meter Impedance Matching Assembly

Qty	Description	Source/Part Number
1	8×8×4-inch NEMA box	Hardware store/home center
1	2-inch diam × 12-inch long coil, air wound with #12 AWG wire (79 µH, L1)	MFJ 404-0669
2	3PDT power relay (K1, K1)	Array Solutions RF-3PDT-15
4	Coil clips	MFJ 755-4001
1	SO-239 connector (J1)	MFJ-7721
1	1:4 unun, 1.5 kW rating (T1)	MFJ 10-10989D
1	2.1×5.5 mm dc jack (J2)	Mouser 163-1060-EX
3	18 V dc MOV (Z1-Z3)	Mouser 576-V22ZA2P
3	0.1 µf capacitor (C1-C3)	Mouser 581-SR215C104KAR
1	6-lug terminal strip	Mouser 158-1006
5	Micro-gator clips	Mouser 548-34
5	Test clips	Mouser 13AC130
1	2×1.38×0.8-inch plastic box (for control switch)	All Electronics 1551-GBK
1	DPDT center-off switch (S1)	All Electronics MTS-12
1	12-V dc, 1-A wall transformer	All Electronics DCTX-1218

Misc: #14 AWG house wiring, stainless steel hardware.

coil breakdown rating of 5.3 kV peak. The outer contacts are connected via insulated internal wires that are well separated from the coil by 0.1-0.2 inch. The problem is with the common wire for the center contacts, which is in contact with the coil. While the coil and common wire are both insulated, there is no air gap separation between them; this close proximity is what determines the relay breakdown voltage rating. Wired as shown in Fig 24.17, the center relay contacts are used for the lowest voltage points.

The switched control voltage input uses a 12-V, 1-A wall transformer supply. It's a good idea to keep the ±12 V control voltages separate and isolated from the regular station power supply to eliminate any possibility of shorting the main power supply when the control voltage polarity is flipped. The switching is easily done with a DPDT center-off switch (S1).

MATCHING NETWORK RESONANCE

Some initial adjustment is required to tune the resonant frequency of the matching network to account for variations in the antenna impedance based on your particular installation and desired operating frequency range. The overall starting inductance is too large for 160 meters, so the network will resonate at or below the lower band edge. Therefore, you can simply short one or more turns at the upper (RF OUT) end of the inductor to tune the network higher in frequency.

For the initial adjustments, don't connect the wire leads that attach between the relay contacts and the tap points on the coil. Instead, build short jumpers using the test clips and micro clips (perfect for the coil turn taps) called out in the parts list, and attach these between the relay contacts and coil using the suggested tap points shown on the schematic.

Next, connect the matching assembly to the base of your 43-ft vertical, enable 160 meter operation by applying −12 V dc and jumper turns from the RF OUT end of the inductor with a short clip lead until you find the desired 160 meter resonance point. Next, move the 160 meter relay tap point (from the unun secondary through K2C) until you get minimum SWR. (The starting point is 10 turns from the ground end.) Now permanently short the inductor turns by soldering a piece of #16 AWG buss wire across them. As shown in the photos, the author needed to short six turns (on the short coil).

Next apply +12 V to the relay assembly to enable 80-meter operation. Select the coil shorting point (from K2B — start 16 turns from the ground end) for your desired resonant frequency. Then adjust the tap point (from K2C — start 6 turns from ground) for best SWR. Finally remove the test clips leads, attach the coil clips, and solder wires between the coil clips and relay.

Fig 24.20 shows the results for 160 and 80 meters as measured with an antenna analyzer connected directly to the matching network input at the base of the author's antenna. With resonance set for the CW end of these bands, the 2:1 SWR bandwidth on 160 meters is about 50 kHz and about 150 kHz on 80 meters. Even a 3:1 SWR on these bands results in negligible SWR-related cable losses and is easily matched with an in-shack tuner.

OPERATION

Operation of this matching unit couldn't be simpler. When no control voltage is applied, the unun is connected and the antenna functions as it always has on 60-10 meters. For 80-meter operation, apply +12 V dc, and for 160-meter operation apply −12 V dc. The matching unit connected to the base of the author's 43-ft vertical is shown in **Fig 24.21**.

The matching network discussed in this article will permit very effective operation of your 43-ft vertical on all bands from 160-10 meters. Feel free to experiment a little. You might prefer to use the toroid from the basic matching network in the companion project instead of the air-core inductor. And if you are not running more than about 500 W, the less expensive RF-10 relays are all you need.

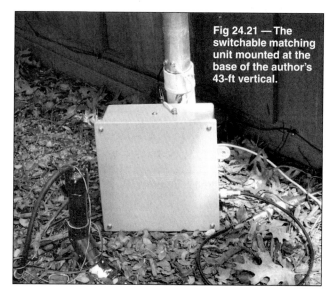

Fig 24.21 — The switchable matching unit mounted at the base of the author's 43-ft vertical.

24.5 An External Automatic Antenna Switch for Use With Yaesu or ICOM Radios

This antenna-switching-control project involves a combination of ideas from several earlier published articles.[1,2,3] This system was designed to mount the antenna relay box outside the shack, such as on a tower. With this arrangement, only a single antenna feed line needs to be brought into the shack. **Fig 24.22** shows the control unit and relay box, designed and built by Joe Carcia, NJ1Q. Either an ICOM or Yaesu HF radio will automatically select the proper antenna. In addition, a manual switch can override the ICOM automatic selection. That feature also provides a way to use the antenna with other radios. The antenna switch is not a two-radio switch, though. It will only work with one radio at a time.

Many builders may want to use only the ICOM or only the Yaesu portion of the interface circuitry, depending on the brand of radio they own. The project is a "hacker's dream." It can be built in a variety of forms, with the only limitations being the builder's imagination.

CIRCUIT DESCRIPTION

Fig 24.23 is a block diagram of the complete system. An ICOM or Yaesu HF radio connects to the appropriate decoder via the accessory connector on the back of the radio. Some other modern rigs have an accessory connector used for automatic bandswitching of amplifiers, tuners and other equipment. For example, Ten-Tec radios apply a 10 to 14-V dc signal to pins on the DB-25 interface connector for the various bands. Other radios use particular voltages on one of the accessory-connector pins to indicate the selected band. Check the owner's manual of your radio for specific information, or contact the manufacturer's service department for more details. You may be able to adapt the ideas presented in this project for use with other radios.

A single length of coax and a multiconductor control cable run from the rig and decoder/control box to the remotely located switch unit. The remote relay box is equipped with SO-239 connectors for the input as well as the output to each antenna. You can use any type of connectors, though.

ICOM radios use an 8-V reference and a voltage divider system to provide a stepped band-data output voltage. **Table 24.4** shows the output voltage at the accessory socket when the radio is switched to the various bands. Notice that seven voltage steps can be used to select different antennas. The ICOM accessory connector pin assignments needed for this project are:

Pin 1	+8 V reference
Pin 2	Ground
Pin 4	Band signal voltage
Pin 7	+12 (13.8) V supply

Yaesu radios provide the band information

Fig 24.22 — Automatic antenna switch relay box (left) and control box (right).

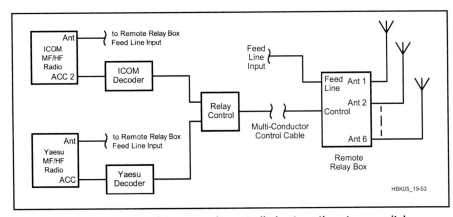

Fig 24.23 — Block diagram of the remotely controlled automatic antenna switch.

Table 24.4
ICOM Accessory Connector Output Voltages By Band

Band (MHz)	Output Voltage
1.8	7 – 8.0
3.5	6 – 6.5
7	5 – 5.5
14	4 – 4.5
18, 21	3 – 3.5
24, 28	2 – 2.5
10	0 – 1.2

Note: The voltage step between bands is not constant, but close to 1.0 V, and the 10-MHz band is not in sequence with the others.

Table 24.5
Yaesu Band Data Voltage Output (BCD)

Band	A (1)	B (2)	C (4)	D (8)	(BCD Equiv.)
1.8	5V	0V	0V	0V	1
3.5	0V	5V	0V	0V	2
7.0	5V	5V	0V	0V	3
10.1	0V	0V	5V	0V	4
14	5V	0V	5V	0V	5
18.068	0V	5V	5V	0V	6
21	5V	5V	5V	0V	7
24.89	0V	0V	0V	5V	8
28	5V	0V	0V	5V	9

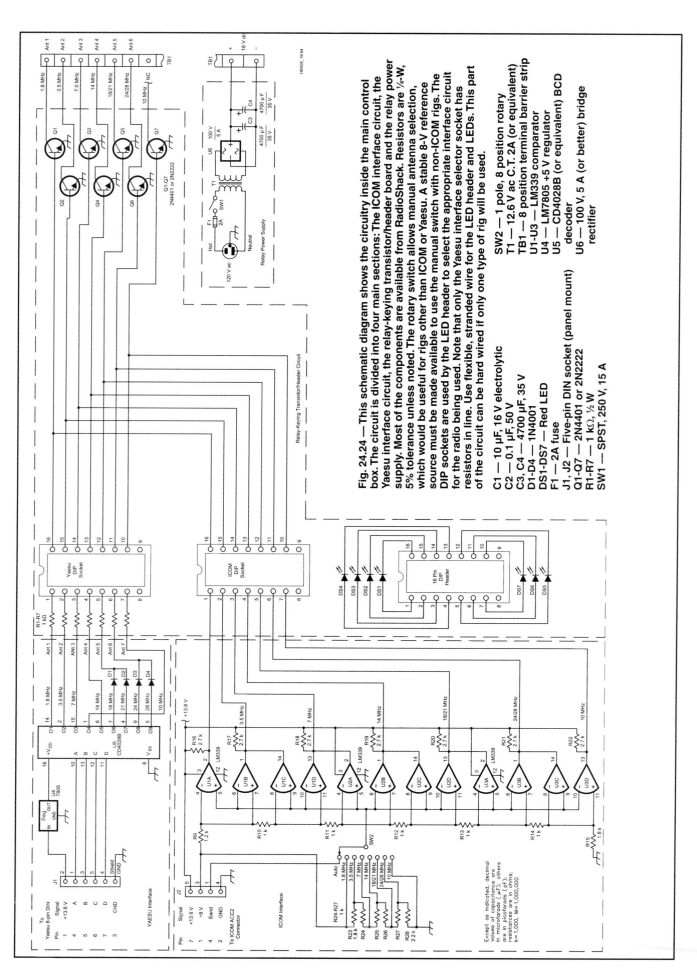

Fig. 24.24 — This schematic diagram shows the circuitry inside the main control box. The circuit is divided into four main sections: The ICOM interface circuit, the Yaesu interface circuit, the relay-keying transistor/header board and the relay power supply. Most of the components are available from RadioShack. Resistors are ½-W, 5% tolerance unless noted. The rotary switch allows manual antenna selection, which would be useful for rigs other than ICOM or Yaesu. A stable 8-V reference source must be made available to use the manual switch with non-ICOM rigs. The DIP sockets are used by the LED header to select the appropriate interface circuit for the radio being used. Note that only the Yaesu interface selector socket has resistors in line. Use flexible, stranded wire for the LED header and LEDs. This part of the circuit can be hard wired if only one type of rig will be used.

C1 — 10 µF, 16 V electrolytic
C2 — 0.1 µF, 50 V
C3, C4 — 4700 µF, 35 V
D1-D4 — 1N4001
DS1-DS7 — Red LED
F1 — 2A fuse
J1, J2 — Five-pin DIN socket (panel mount)
Q1-Q7 — 2N4401 or 2N2222
R1-R7 — 1 kΩ, ½ W
SW1 — SPST, 250 V, 15 A
SW2 — 1 pole, 8 position rotary
T1 — 12.6 V ac C.T. 2A (or equivalent)
TB1 — 8 position terminal barrier strip
U1-U3 — LM339 comparator
U4 — LM7805 +5 V regulator
U5 — CD4028B (or equivalent) BCD decoder
U6 — 100 V, 5 A (or better) bridge rectifier

as binary coded decimal (BCD) data on four lines. Nine different BCD values allow you to select a different antenna for each of the MF/HF bands. **Table 24.5** shows the BCD data from Yaesu radios for the various bands. The Yaesu 8-pin DIN accessory connector pin assignments needed for this project are:

Pin 1   +12 (13.8) V supply
Pin 3   Ground
Pin 4   Band Data A
Pin 5   Band Data B
Pin 6   Band Data C
Pin 7   Band Data D

Fig 24.24 is the schematic diagram for the control box. We will discuss each part of the control circuit later in this description. First, let's turn our attention to the external antenna box.

EXTERNAL ANTENNA BOX

Only the number of control lines going out to the relay box limits the number of antennas this relay box will switch. The unit shown in the photos has 10 SO-239 connectors, to switch the common feed line to any of nine antennas. Many hams will use an eight-conductor rotator cable (such as Belden 9405) to the relay box. Using eight wires, we can control seven relays (six for antennas and one to ground the feed line for lightning protection) plus the relay coil power supply and ground lead. The photos show eight relays, and the box can be expanded further if desired. There is also a connector for power and control lines. Use of a DB-15 allows for the addition of more relays and control lines later. A DB-9 connector would be suitable for use with the eight-conductor control cable, or you may wish to use a weatherproof connector. **Fig 24.25** shows the relay box schematic diagram.

Since the box will be located outside, use a weatherproof metal box — a Hammond Manufacturing, type 1590Z150, watertight aluminum box is shown. It's about 8.5×4.25×3.125 inches. This is a rather hefty box, meant to be exposed to years of various weather conditions. You can, however, use almost anything.

The coax connectors are mounted so each particular antenna connector is close to the relay, without too much crowding. For added weather protection (and conductivity), apply Penetrox to the connector flange mount, including the threads of the mounting screws. On the power/control line connector, use Coax Seal or other flexible cable sealant.

The aluminum angle stock on either side of the box is for mounting to a tower leg. The U-bolts should be of the proper size to fit the tower leg. They should also be galvanized or made of stainless steel.

ANTENNA RELAYS

One of the more difficult parts of this project was the modification of the relays (DPDT Omron LY2F-DC12). To improve isolation, the moveable contacts (armature) are wired in parallel and the connecting wire is routed through a hole in the relay case.

Remove the relay from its plastic case. Unsolder and remove the small wires from the armatures. Carefully solder a jumper across the armature lugs with #20 solid copper. Then solder a piece of very flexible wire (such as braid from RG-58 cable) to either armature lug. Obviously, the location of the wire depends on which side you wish to connect the SO-239. You will also need to make a hole in the plastic case that is large enough to accommodate the armature wire without placing any strain on the free movement of the armature. Slip a length of insulating tubing

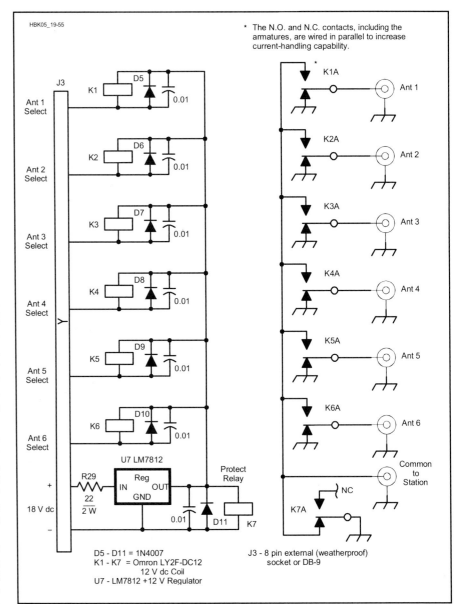

Fig 24.25—The schematic diagram for the external antenna relay box. All relays are DPDT, 250 V ac, 15 A contacts. R29 is used to limit the regulator current. Mount the regulator using TO-220 mounting hardware, with heatsink compound. With the exception of the normally closed and normally open contacts, all wiring is #22 solid copper wire.

D5-D11 — 1N4007.
J3 — 8 pin external weatherproof connector (or DB-9 with appropriate weather sealant).

K1-K7 — Omron LY2F-DC12 with 12 V coil.
U7 — LM7812 +12 V regulator.

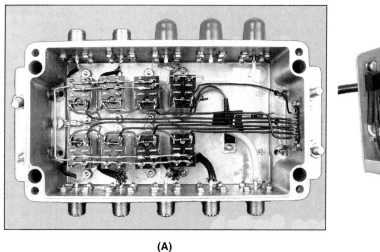

Fig 24.26 — At A, the external relay box. The LM7812 regulator is mounted to the bottom of the box. The relay normally open and normally closed contacts are wired in parallel using #12 solid copper. The two extra flange-mount SO-239 connectors at the upper right are for future expansion. At B, inside the control box. Components are mounted on RadioShack Universal Project Board. The bottom board in the enclosure as well as the right half of the middle section hold the ICOM circuit. The Yaesu interface is on the left side of the middle section. The top circuit board holds the DIP sockets and relay-selection transistors. All high voltage leads are insulated. The 7805 5-V regulator is mounted on the back panel using TO-220 mounting hardware, with heatsink compound.

over this wire to prevent it from shorting to the aluminum box.

The normally open and normally closed contacts are also wired in parallel. This can be done on the lugs themselves. For this, use #12 solid copper wire.

Mount the relays in the aluminum box, oriented so they can be wired together without difficulty. (See **Fig 24.26A**.) With the exception of the wire used for the relay coils (#22 solid wire), use #12 solid copper wire for the rest of the connections.

To eliminate the possibility of spikes or "back emf," a 1N4007 diode is soldered across the coil contacts of each relay. In addition, 0.01-µF capacitors across the diodes will reduce the possibility of stray RF causing problems with the relay operation.

Since the cable run from the shack to the tower can be quite long, consideration has to be given to the voltage drop that may occur. The relays require 12 V dc. The prototype used a 12 V dc regulator in the relay box, fed with 18 V dc (at 2 A) from the control box. If the cable run is not that long, however, you could just use a 12-V supply and skip the regulator.

One of the relays is used for lightning protection. When not in use, the relay grounds the line coming in from the shack. When the control box is activated, it applies power to this relay, thus removing the ground on the station feed line. All the antenna lines are grounded through the normally closed relay contacts. They remain grounded until the relay receives power from the control box.

CONTROL BOX

This is the heart of the system. The 18 V dc power supply for the relays is located in this box, in addition to the Yaesu and ICOM decoder circuits and the relay-control circuitry. All connections to the relay box are made via an 8-position terminal barrier strip mounted on the back of the control box.

The front of the box has LEDs that indicate the selected antenna. A rotary switch can be used for manual antenna selection. The power switch and fuse are also located on the front panel.

The wiring schemes on the Yaesu and ICOM ACC sockets are so different that the unit shown in the photos has a 5-pin DIN connector for each rig on the control box. Since there is only one set of LEDs, use an 8-pin DIP header to select the appropriate control circuit for each radio. See Fig 24.26B. The unit is built with point-to-point wiring on Radio Shack Universal Project Boards.

ICOM Circuitry

This circuit originally appeared in April 1993 *QST* and is modified slightly for this application. The original circuit allowed for switching between seven antennas (from 160 to 10 meters). The Band Data signal from the ICOM radios goes to a string of LM339 comparators. Resistors R9 through R15 divide the 8-V reference signal from the rig to provide midpoint references between the band signal levels. The LM339 comparators decide which band the radio is on. A single comparator selects the 1.8 or 10 MHz band because those bands are at opposite ends of the range. The other bands each use two comparators. One determines if the band signal is above the band level and the other determines if it is below the band level. If the signal is between those two levels, the appropriate LED and relay-selection transistor switch is turned on.

The ICOM circuit allows for manual antenna selection. The 8-V reference is normally taken directly from the ICOM ACC socket. If this circuit is to be used with other equipment, then a regulated 8-V source should be provided.

Yaesu Circuitry

The neat thing about Yaesu band data is that it's in a binary format. This means you can use a simple BCD decoder for band switching. The BCD output ranges from 1 to 9. In essence, you can switch between 9 antennas (or bands). Since the relay box switches just six antennas, steering diodes (D1 through D4 in Fig 24.24) allow the use of one antenna connection for multiple bands. One antenna connection is used for 17 and 15 meters, and another connection for 12 and 10 meters because the ICOM band data combines those bands. There is no control line or relay for a 30-meter antenna with this version of the project.

DIP Sockets and Header

A RadioShack Universal Project Board holds the DIP sockets along with the relay keying transistors. This board is shown as the top layer in Fig 24.26B. The Yaesu socket has 1-kΩ resistors wired in series with each input pinch The other header connects directly to the ICOM circuitry.

The DIP header is used to switch the keying transistors between the ICOM and Yaesu circuitry. The LEDs are used to indicate antenna number. Use stranded wire (for its flexibility) when connecting to the LEDs.

Relay Keying Transistors

Both circuits use the same transistor-keying scheme, so you need only one set of transistors. Each transistor collector connects to the terminal barrier strip. The emitters are grounded, and the bases are wired in parallel to the two 16-pin DIP sockets. The band data turns on one of the transistors, effectively grounding that relay-control lead. Current flows through the selected relay coil, switching that relay to the normally open position and connecting the station feed line to the proper antenna.

Power Supply

The power supply is used strictly for the relays. Other power requirements are taken from the rig used. There is room here for variations on the power supply theme; the parts used in this version were readily available.

Notes

[1]"An Antenna Switching System for Multi-Two and Single-Multi Contesting," by Tony Brock-Fisher, K1KP, January 1995 *NCJ*.
[2]"A Remotely Controlled Antenna Switch," by Nigel Thompson, April 1993 *QST*.
[3]*NA* Logging Program Section 11.

24.6 A Low-Cost Remote Antenna Switch

Getting multiple antenna feed lines into the shack is a common problem that can be easily solved with a suitable remote antenna switch. The project described here by Bill Smith, KO4NR, provides an easy and inexpensive solution. It originally appeared in April 2005 *QST*.

RELAY SELECTION

Interesting innovations in printed circuit board (PCB) power relay design have produced a number of compact units that exhibit high dielectric strength and can carry impressive amounts of current.

Although not factory tested for RF use, the American Zettler AZ755 series PCB relays work very well in this application. (See **www.americanzettler.com** for datasheets and information.) The AZ755 is rated for 480 W switched power with a resistive load and a maximum switched current of 20 A. Despite the relay's small physical size, the dielectric strength between the contacts and coil is 5 kV RMS, with an impressive 1 kV RMS between the open contacts. This means that

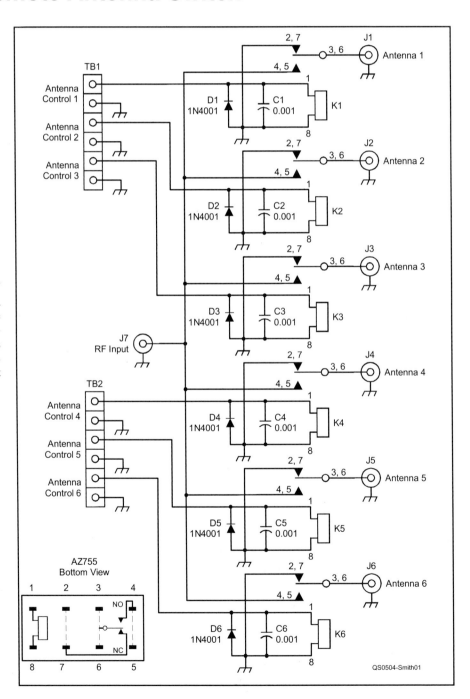

Fig 24.27 — Schematic and parts list for the remote antenna switch. Part numbers indicated with an M are available from Mouser Electronics (www.mouser.com). PC boards are available from FAR Circuits (www.farcircuits.net).

C1-C6 — 0.001 µF, 50 V disc ceramic (M 140-50P5-102K-TB).
D1-D6 — 1N4001 (M 512-1N4001).
J1-J7 — SO-239A chassis mount coaxial connector with silver-plated 4-hole square flange and Teflon insulation (Amphenol 83-798 available as Mouser 523-83-798).
K1-K6 — SPDT PC board power relay with 12 V dc coil (American Zettler AZ755-1C-12DE). Available from www.relaycenter.com. See text.
TB1, TB2 — PC board terminal block with 6 contacts (M 651-1729050).

Fig 24.28 — A completed remote antenna switch. The RF INPUT connector is in the center, with three ANTENNA connectors on each side. The control cable connects to the two terminal blocks.

Fig 24.29 — Connector flanges are soldered to the board's ground plane top and bottom. The ground ends of D1-D6 and C1-C6 are also soldered top and bottom. The relays have two pins each for the contact connections. Solder all pins and install eyelets supplied with the FAR Circuits board to ensure good connections for the relay common and normally closed pins.

the relay is resistant to a flashover that could damage the coil or pit the contacts.

The AZ755 series relays are offered in a wide variety of configurations. This project uses part number AZ755-1C-12DE. This model has a 12 V dc coil, but you can use any of the available coil voltages in the series. The contact style is Form C, which is SPDT. The E suffix indicates that the relay epoxy is sealed for better protection from dirt and moisture. For relays that are not epoxy sealed, drop the E from the part number.

CIRCUIT DESIGN

Fig 24.27 shows the circuit, which handles up to six antennas. The common contacts of the relays (K1-K6) are connected to SO-239 connectors (J1-J6) for the antenna feed lines. The normally open (NO) contacts are all connected to the RF INPUT connector, J7. The normally closed (NC) contacts are all connected to ground so that the antennas are grounded when not in use. To select an antenna, apply 12 V dc to the appropriate ANTENNA CONTROL terminal to energize the relay and connect the ANTENNA to the RF INPUT.

C1-C6 help to keep stray RF out. In addition, D1-D6 are installed across the coils to prevent voltage spikes when the power is removed from the coil. The finished board (available from FAR Circuits) with all components mounted is shown in **Fig 24.28**.

ASSEMBLY NOTES

The design is simple and assembly doesn't require special tools. In addition to the PC board and parts, you'll need a suitable enclosure to keep the board dry and pests away. The author used a plastic children's lunch box that hangs under his deck.

The most difficult part of the project is drilling the holes in the enclosure for the SO-239 connectors and getting everything to line up. You may find it easier to install the connectors in the enclosure first, and then solder them to the board. (Use the board to mark the center line and location of the connectors on the enclosure.) After the connectors are tacked in place, remove the screws holding the SO-239s to the enclosure and remove the total assembly. This ensures a perfect fit when reassembling.

Make sure the SO-239s are all the same type and brand to ensure a uniform fit. The board was designed around the Amphenol connectors recommended in the parts list. They have silver-plated center pins and flanges, and Teflon insulation. The silver plating makes the connectors easier to solder than nickel-plated connectors, and the Teflon insulation is much less prone to melting than the plastic often found on inexpensive connectors. Other SO-239 connectors may fit with possible modification for a good fit.

Use a *hot* soldering iron when soldering the flange of the SO-239 to the board's ground plane as shown in **Fig 24.29**. Be sure to solder top and bottom.

The PC board is double-sided, and FAR Circuits supplies eyelets with the board to use in the larger holes for the common and normally closed relay pins. They provide

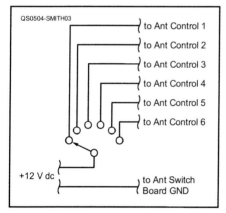

Fig 24.30 — Antennas are selected with a simple 12-V supply and rotary switch.

for a better connection to the component side of the board. In addition, soldering a short length of #14 bare copper wire into the center pin of each SO-239 will give you a better connection to the circuit trace on the board. Be sure to solder the ground ends of C1-C6 and D1-D6 on both the top and bottom sides of the board.

POWERING IT UP

The switch is controlled with a 12-V dc power supply and small ceramic rotary switch as shown in **Fig 24.30**. The switch simply sends 12 V to energize one of the relay coils and select the desired antenna. When 12 V is removed, the normally closed relay contacts connect all feed lines to ground. Connect the

PC board's ground plane to a ground rod for lightning protection and as a means of eliminating static.

The control cable can be the 8-conductor variety usually used for rotators. Long runs of wire may require large conductors to prevent unacceptable voltage drops at the relay coils, but the relays will work over a fairly wide range of coil voltages so it's not critical.

TESTING

Initial testing should be conducted on the bench with an antenna analyzer or a transceiver (internal antenna tuner off) and power/SWR meter connected to the RF INPUT jack. Connect a 50 Ω load to one of the antenna jacks on the switch board. Apply a small of RF power. The SWR should be close to 1:1 (similar to readings without the switch in the line).

If everything looks okay, increase the power and check the SWR through the switch on all bands that you plan to use it on. Next, move the wattmeter to the switch output and verify that the power on that side is about the same as the power at the input. There should be little or no measurable loss through the switch. Repeat these tests for all of the switch positions.

If you run an amplifier, test the switch at high power. The author used the switch at 1-1.3 kW during normal intermittent operation (SSB and CW) with no problems.

Although they are not designed for RF, the relays perform well. The board exhibits low SWR, low insertion loss and good isolation over a wide frequency range. The ARRL Lab tested the completed antenna switch board. Insertion loss measured <0.1 dB for 2-50 MHz (for all ports to common). SWR measured 1.1:1 or less from 2-28 MHz, 1.2:1 or less on 50 MHz. Isolation was >60 dB for 2-28 MHz, except for the two innermost ports, which were 50 dB at 28 MHz. Worst-case isolation on 50 MHz was 45 dB.

24.7 Audible Antenna Bridge

The audible antenna bridge (AAB) shown in **Fig 24.31** was designed by Rod Kreuter, WA3ENK. This project would help a visually impaired ham, but sighted hams will also find it very useful. (For example, tuning a mobile screwdriver antenna while you're driving.) The AAB is rugged and easy to use, both in the harsh light of day and in the dark of night. It's also inexpensive to build, and a kit is available.[1]

An absorptive type of SWR meter, the AAB is used to indicate how close the match is between your transmitter and your antenna. Its output is a tone whose pitch is proportional to SWR. It also sends the SWR value in Morse code.

Unlike most SWR meters on the market, this one actually absorbs power from your transmitter while you're tuning, helping to protect your transmitter during tune-up. Many transmitters have SWR foldback capability and will decrease the output power if the antenna match is poor, but some do not.

By absorbing power in the AAB, no matter what you do at the antenna port, the transmitter never sees an SWR greater than 2:1. The trade-off is that 75% of your transmitter power ends up as heat in the bridge, so you need to switch out the AAB after you have tuned your antenna.

THEORY OF OPERATION

The AAB works on a principle taken right from the *ARRL Antenna Book*. The bridge consists of four impedances, or arms. See **Fig 24.32**. If three of the arms are 50 Ω resistors and the fourth is an antenna (or an antenna and tuner combination), then the bridge will be balanced when the antenna port is also 50 Ω. Diodes D1 and D2 sample the forward and reverse voltages, respectively. A peripheral interface controller (PIC) with a 10 bit analog to digital converter (A/D) digitizes these voltages and calculates the SWR.

Fig 24.31 — Front view of the audible antenna bridge.

The SWR is used to control a hardware pulse width modulator that produces the output tone. It's also used by the Morse code routine to send the actual value of SWR, at 15 WPM with 5 WPM spacing.

The power that the AAB can handle during tune-up is a function of the resistors used in the three reference arms of the bridge. The 30 W noninductive resistors, R1-R3, are made by a number of companies but are not cheap. The bridge resistors must be noninductive. Don't use wire wound resistors!

Antenna tuning should be done with minimal power, and the AAB will work down to about 1 W. The 30 W resistors will handle tune-ups with about 60 W or perhaps 100 W for a short time. If you always remember to turn down your transmitter power, and you tune up quickly, the bridge resistors could be perhaps 5 W for normal rigs and 1 W for QRP rigs.

BUILDING THE AAB

A PC board is a nice way to build the AAB, but you could easily construct such a simple circuit using ugly construction. Keep the wires short in the RF section. See **Fig 24.33**.

Use a heat sink for the noninductive resistors. Like many power devices, the power rating depends on getting the heat out! These resistors get really hot running at 10 or 20 W. See **Fig 24.34**. Use a big heat sink with heat sink compound, give yourself plenty of room in the cabinet and drill some holes to let the heat escape.

Switch S1 chooses TUNE/OPERATE and S2, the INQUIRE button, is used to report on various conditions, such as the SWR value in Morse code. You can eliminate the TUNE/OPERATE switch, but remember to remove the AAB from your feed line after tuning up.

To use the AAB in a QRP station, you may want to reduce resistors R4 and R5. They are sized so that a 25 W transmitter will not saturate the analog to digital converter (ADC) on the PIC when using fresh batteries (more on this later). A good value for a 5 W transmitter would be about 20 kΩ. So long as the voltages at pin 12 and pin 13 of U1 are below the battery voltage, the system will work fine.

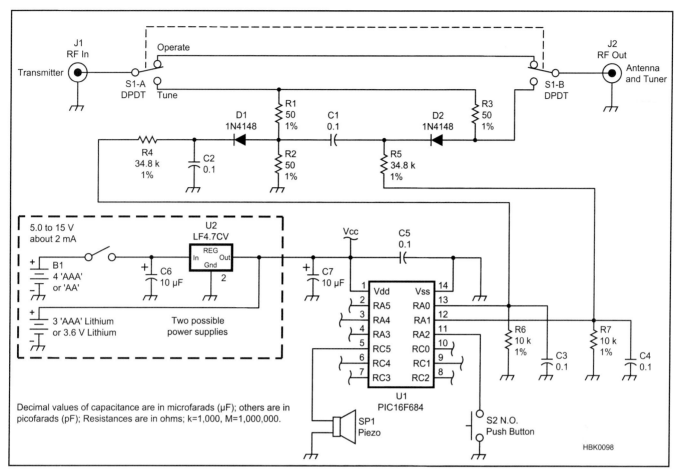

Fig 24.32 — Schematic and parts list of the AAB.

C1-C5 — 0.1 µF ceramic.
C6, C7 — 10 µF, 25 V electrolytic.
D1, D2 — 1N4148.
J1, J2 — SO-239 UHF coaxial jack.
R1 -R3 — Resistor, 50 Ω, 1%, 30 W Caddock noninductive (Mouser 684-MP930-50).

R4, R5 — Resistor, 34.8 kΩ, 1%.
R6, R7 — Resistor, 10 kΩ, 1%.
S1 — DPDT (Digi-Key SW333-ND).
S2 — Momentary contact (Mouser 612-TL1105LF250Q).
SP1 — Piezoelectric buzzer EFB-RD24C411 (Digi-Key P9924-ND).

U1 — PIC16F684 (Digi-Key PIC16F684-1/P-ND). Must be programmed before use. See Note 1.
U2 — LF4.7CV regulator (Mouser 511-LF47CV).

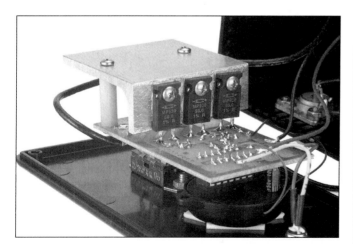

Fig 24.33 — Close-up view of three Caddock TO-220 style bridge resistors mounted on the heat sink.

Fig 24.34 — Interior view of the AAB, showing the battery pack at the bottom of the enclosure.

Station Accessories 24.21

POWERING THE AAB

To save power the AAB uses a sleep mode. If you haven't pushed the INQUIRE button for about 5 minutes, the AAB will go to sleep after sending the letter N for "night night." During sleep the current draw is about 2 µA. Pushing the switch while it's sleeping will wake up the AAB and it will send the letter U ("up").

The first four prototypes used four AAA batteries and a low-dropout 4.7 V regulator. This made for a nice stiff supply and very good battery life, with only one problem. Even though the PIC uses less than 2 µA during sleep, the regulator uses 500 µA. Battery data says that alkaline AA batteries will last 6 months at this level and AAA batteries about half that long. Consider adding an ON/OFF switch if you go this route.

The other option is to use batteries without a regulator. Normally you couldn't get away with this without providing a separate voltage reference for the ADC in the PIC. In this case it doesn't matter as much because SWR is a ratio of two voltages.

What does matter is the battery voltage with respect to the voltage produced by the RF signal. Diodes D1 and D2 rectify the RF and produce a dc voltage proportional to RF power. This voltage is divided by resistors R4, R6 and R5, R7. The voltage at the ADC inputs of the PIC, pins 12 and 13, must be less than the power supply voltage. With a regulated supply this isn't much of a problem and if you're careful it can also work with unregulated batteries.

Microchip recommends that these parts be run on 2.5 to 5.5 V if you need the 10 bit ADC to be accurate. Many battery combinations will provide this, but consider lithium batteries instead of alkaline. The voltage from an alkaline battery begins to fall very early in its life and continues at a downward slope until it is exhausted. A lithium battery, on the other hand, provides a nearly constant voltage throughout its useful life. A 3.6 V AA size lithium should last a few years, but with the lower output voltage you may want to reduce R4 and R5. The final prototype used three AAA lithium batteries (each about 1.7 V when new), with no regulator or ON/OFF switch, which should last at least two years.

U1's ADC converter will saturate at some level. With new batteries, let's say this level is 25 W when the power supply is 4.5 V. When the power supply has fallen to 3.5 V, the RF level that will saturate the converter will be 15 W. So if you find that the output of your rig, which used to be fine, is now saturating the ADC, it's time to change batteries.

USING THE AAB

Connect your transmitter to the INPUT and your antenna/tuner to the OUTPUT. Turn your transmit power down to 5 or 10 W and put S1 in the TUNE position. Wake the AAB up by pushing the INQUIRE button. Key your transmitter in a mode with some carrier power. The AAB should start to produce a steady tone. Adjust your tuner or antenna for the lowest tone pitch. Press the INQUIRE button for the numeric SWR value.

When the AAB is operating normally, it produces a tone proportional to SWR. The tone ranges from 250 Hz at an SWR of 1.0:1, rising to 4450 Hz at an SWR of about 15:1. Errors can occur and at times the AAB can't make a measurement. If an error occurs, either a high pitched tone or no tone is produced. Pushing the INQUIRE button will give you more information. Five different conditions can occur:

Normal operation. A tone proportional to SWR is produced. Press INQUIRE and a two digit SWR value is sent separated by the letter R (Morse for the decimal point character).

Error 1. The forward power is too low for a reliable measurement. No tone is produced. INQUIRE button sends letter L (low).

Error 2. Forward power is too high (almost saturating the ADC). A pulsing tone proportional to SWR is produced. However, if the ADC is saturated this value might be meaningless. INQUIRE button sends letter S (saturated).

Error 3. SWR is greater than 9.9:1. A tone proportional to SWR is produced. INQUIRE button sends letter H (high).

Error 4. The reverse voltage is greater than the forward voltage. Two things can cause this error. You might be near a high power transmitter, or you connected the transmitter to the output port. No tone is produced. INQUIRE button sends letter F (fault).

The PIC used in this design could be used in a more standard SWR meter, such as those with a transformer type sensor. As long as the meter produces a forward and reverse voltage that you can scale to be from 0 to 3 up to 5 V (depending on your power supply choice), it will work just fine. However, the driving impedance of the voltage must be 10 kΩ or less.

The author thanks the Ham Radio Committee of the National Federation of the Blind for inspiration, and Tom Fowle, WA6IVG, and Bill Gerrey, WA6NPC, of the Smith-Kettlewell Research Institute for testing prototypes.

[1]A kit of parts including a printed circuit board is available from Q-Sat, 319 McBath St, State College, PA 16801. The price is currently $30, postpaid. PC boards are available for $7 postpaid and a programmed PIC is available for $6 postpaid. Source code for an unprogrammed PIC is available on the ARRL Web site. Go to **www.arrl.org/qst-in-depth** and look for Kreuter0607.zip in the 2007 section.

24.8 A Trio of Transceiver/Computer Interfaces

Virtually all modern Amateur Radio transceivers (and many general-coverage receivers) have provisions for external computer control. Most hams take advantage of this feature using software specifically developed for control, or primarily intended for some other purpose (such as contest logging), with rig control as a secondary function.

Unfortunately, the serial port on most radios cannot be directly connected to the serial port on most computers. The problem is that most radios use TTL signal levels while most computers use RS-232-D.

The interfaces described here simply convert the TTL levels used by the radio to the RS-232-D levels used by the computer, and vice versa. Interfaces of this type are often referred to as level shifters. Two basic designs, one having a couple of variations, cover the popular brands of radios. This article, by Wally Blackburn, AA8DX, first appeared in February 1993 *QST*.

TYPE ONE: ICOM CI-V

The simplest interface is the one used for the ICOM CI-V system. This interface works with newer ICOM and Ten-Tec rigs. **Fig 24.35** shows the two-wire bus system used in these radios.

This arrangement uses a CSMA/CD (carrier-sense multiple access/collision detect) bus. This refers to a bus that a number of stations share to transmit and receive data. In effect, the bus is a single wire and common ground that interconnect a number of radios and computers.

The single wire is used for transmitting and receiving data. Each device has its own unique digital address. Information is transferred on the bus in the form of packets that include the data and the address of the intended receiving device.

The schematic for the ICOM/Ten-Tec interface is shown in **Fig 24.36**. It is also the Yaesu interface. The only difference is that the transmit data (TxD) and receive data (RxD) are jumpered together for the ICOM/Ten-Tec version.

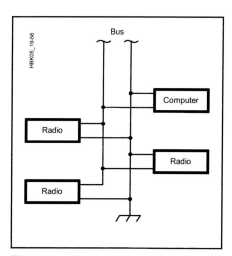

Fig 24.35 — The basic two-wire bus system that ICOM and newer Ten-Tec radios share among several radios and computers. In its simplest form, the bus would include only one radio and one computer.

The signal lines are active-high TTL. This means that a logical one is represented by a binary one (+5 V). To shift this to RS-232-D it must converted to –12 V while a binary zero (0 V) must be converted to +12 V. In the other direction, the opposites are needed: –12 V to +5 V and +12 V to 0 V.

U1 is used as a buffer to meet the interface specifications of the radio's circuitry and provide some isolation. U2 is a 5-V-powered RS-232-D transceiver chip that translates between TTL and RS-232-D levels. This chip uses charge pumps to obtain ±10 V from a single +5-V supply. This device is used in all three interfaces.

A DB25 female (DB25F) is typically used at the computer end. Refer to the discussion of RS-232-D earlier in the chapter for 9-pin connector information. The interface connects to the radio via a ⅛-inch phone plug. The sleeve is ground and the tip is the bus connection.

It is worth noting that the ICOM and Ten-Tec radios use identical basic command sets (although the Ten-Tec includes additional commands). Thus, driver software is compatible. The manufacturers are to be commended for working toward standardizing these interfaces somewhat. This allows Ten-Tec radios to be used with all popular software that supports the ICOM CI-V interface. When configuring the software, simply indicate that an ICOM radio (such as the IC-735) is connected.

TYPE TWO: YAESU INTERFACE

The interface used for Yaesu rigs is identical to the one described for the ICOM/Ten-Tec, except that RxD and TxD are not jumpered together. Refer to Fig 24.36. This arrangement uses only the RxD and TxD lines; no flow control is used.

The FT-990 and FT-1000/FT-1000D use an emitter follower as the TTL output. If the input

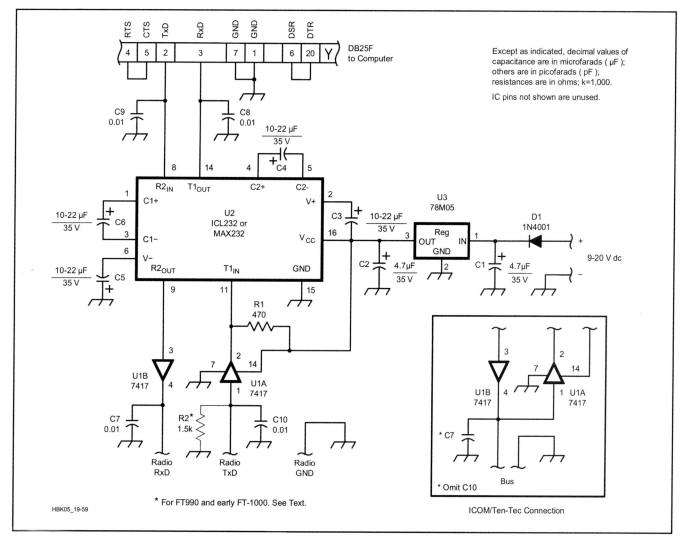

Fig 24.36 — ICOM/Ten-Tec/Yaesu interface schematic. The insert shows the ICOM/Ten-Tec bus connection, which simply involves tying two pins together and eliminating a bypass capacitor.

C7-C10 — 0.01-µF ceramic disc. U1 — 7417 hex buffer/driver. U2 — Harris ICL232 or Maxim MAX232.

impedance of the interface is not low enough to pull the input below threshold with nothing connected, the FT-990/FT-1000 (and perhaps other radios) will not work reliably. This can be corrected by installing a 1.5 kΩ resistor from the serial out line to ground (R2 on pin 1 of U1A in Fig 24.36.) Later FT-1000D units have a 1.5 kΩ pulldown resistor incorporated in the radio from the TxD line to ground and may work reliably without R2.

The same computer connector is used, but the radio connector varies with model. Refer to the manual for your particular rig to determine the connector type and pin arrangement.

TYPE THREE: KENWOOD

The interface setup used with Kenwood radios is different in two ways from the previous two: Request-to-Send (RTS) and Clear-to-Send (CTS) handshaking is implemented and the polarity is reversed on the data lines. The signals used on the Kenwood system are active-low. This means that 0 V represents a logic one and +5 V represents a logic zero. This characteristic makes it easy to fully isolate the radio and the computer since a signal line only has to be grounded to assert it. Optoisolators can be used to simply switch the line to ground.

The schematic in **Fig 24.37** shows the Kenwood interface circuit. Note the different grounds for the computer and the radio. This, in conjunction with a separate power supply for the interface, provides excellent isolation.

The radio connector is a 6-pin DIN plug. The manual for the rig details this connector and the pin assignments.

Some of the earlier Kenwood radios require additional parts before their serial connection can be used. The TS-440S and R-5000 require installation of a chipset and some others, such as the TS-940S require an internal circuit board.

CONSTRUCTION AND TESTING

The interfaces can be built using a PC board, breadboarding, or point-to-point wiring. PC boards and MAX232 ICs are available from FAR Circuits. The PC board template is available on the CD-ROM.

It is a good idea to enclose the interface in a metal case and ground it well. Use of a separate power supply is also a good idea. You may be tempted to take 13.8 V from your radio — and it works well in many cases: but you sacrifice some isolation and may have noise problems. Since these interfaces draw only 10 to 20 mA, a wall transformer is an easy option.

The interface can be tested using the data

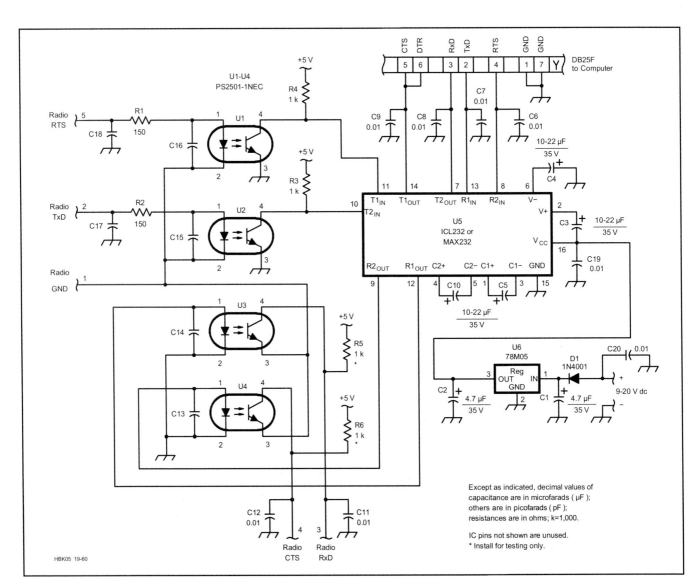

Fig 24.37 — Kenwood interface schematic.
C6-C9, C11, C12, C17, C18 — 0.01-µF ceramic disc.
C13-C16, C19-C21 — 0.01 µF ceramic disc.
U1-U4 — PS2501-1NEC (available from Digi-Key).
U5 — Harris ICL232 or Maxim MAX232.

Table 24.6
Kenwood Interface Testing

Apply	Result
GND to Radio-5	−8 to −12 V at PC-5
+5 V to Radio-5	+8 to +12 V at PC-5
+9 V to PC-4	+5 V at Radio-4
−9 V to PC-4	0 V at Radio-4
GND to Radio-2	−8 to −12 at PC-3
+5 V to Radio-2	+8 to +12 V at PC-3
+9 V to PC-2	+5 V to Radio-3
−9 V to PC-2	0 V at Radio-3

Table 24.7
ICOM/Ten-Tec Interface Testing

Apply	Result
GND to Bus	+8 to +12 V at PC-3
+5 V to Bus	−8 to −12 V at PC-3
−9 V to PC-2	+5 V on Bus
+9 V to PC-2	0 V on Bus

Table 24.8
Yaesu Interface Testing

Apply	Result
GND to Radio TxD	+8 to +12 V at PC-3
+5 V to Radio TxD	−8 to −12 V at PC-3
+9 V to PC-2	0 V at Radio RxD
−9 V to PC-2	+5 V at Radio RxD

in **Tables 24.6, 24.7** and **24.8**. Remember, all you are doing is shifting voltage levels. You will need a 5-V supply, a 9-V battery and a voltmeter. Simply supply the voltages as described in the corresponding table for your interface and check for the correct voltage on the other side. When an input of −9 V is called for, simply connect the positive terminal of the battery to ground.

During normal operation, the input signals to the radio float to 5 V because of pullup resistors inside the radio. These include RxD on the Yaesu interface, the bus on the ICOM/Ten-Tec version, and RxD and CTS on the Kenwood interface. To simulate this during testing, these lines must be tied to a 5-V supply through 1-kΩ resistors. Connecting these to the supply without current-limiting resistors will damage the interface circuitry. R5 and R6 in the Kenwood schematic illustrate this. They are not shown (but are still needed) in the ICOM/Ten-Tec/Yaesu schematic. Also, be sure to note the separate grounds on the Kenwood interface during testing.

Another subject worth discussing is the radio's communication configuration. The serial ports of both the radio and the computer must be set to the same baud rate, parity, and number of start and stop bits. Check your radio's documentation and configure your software or use the PC-DOS/MS-DOS MODE command as described in the computer manual.

24.9 A Simple Serial Interface

A number of possibilities exist for building an RS-232-to-TTL interface level converter to connect a transceiver to a computer serial port. The interface described here by Dale Botkin, NØXAS, is extremely simple, requiring only a few common parts.

Most published serial interface circuits fall into one of the following categories:

• Single chip, MAX232 type. These are relatively expensive, require several capacitors, take up lots of board space and usually require ordering parts.

• Single chip, MAX233 type. This version doesn't need capacitors. Otherwise it shares most of the characteristics of the MAX232 type device, but is even more expensive.

• Transistors. Usually transistor designs use a mix of NPN and PNP devices, four to eight or more resistors, and usually capacitors and diodes. This is just too many parts!

• Other methods. Some designs use series resistors, or TTL logic inverters with resistors. The idea is that the voltages may be out of specification for the chip, but if the current is low enough it won't damage the device. This is not always safe for the TTL/CMOS device. With just a series resistor the serial data doesn't get inverted, so it requires the use of a software USART. It can work, but it's far from ideal.

A SIMPLER APPROACH

The EIA-232 specification specifies a

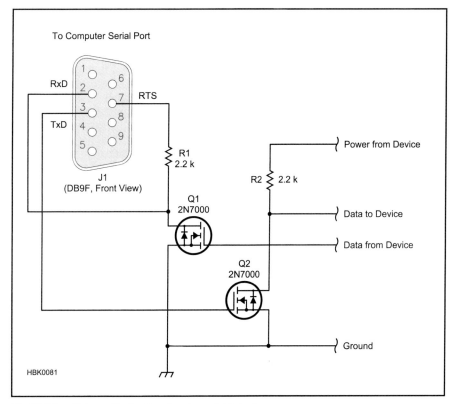

Fig 24.38 — Schematic for the simple serial interface.
J1 — DB9F or connector to fit device serial port. Note that pin assignments for RxD, TxD and RTS are different on DB25 connectors.
Q1, Q2 — Small signal MOSFET, 2N7000 or equiv.

signal level between ±3 V and ±12 V for data. The author looked at data sheets for devices ranging from the decades-old 1489 line receiver to the latest single-chip solutions. Not a single device found actually required a negative input voltage! In fact, the equivalent schematic shown for the old parts clearly shows a diode that holds the input to the first inverter stage at or near ground when a negative voltage is applied. The data sheets show that a reasonable TTL-level input will drive these devices perfectly well.

Relieved of the burden of exactly replicating the PC-side serial port voltages, the interface shown in **Fig 24.38** is designed for minimum parts count and low cost — under $1.

Why use MOSFETs for Q1 and Q2? In switching application such as this, there are several advantages to using small signal MOSFETs over the usual bipolar choices like the 2N2222 or 2N3904. First, they need no base current limiting resistors — the MOSFET is a voltage operated device, and the gate is isolated from the source and drainch The gate on a 2N7000 is rated for ±20 V, so you don't need a clamping diode to prevent negative voltages from reaching the gate. They can also be had just as cheaply as their common bipolar cousins. (If sourcing 2N7000s is a challenge in Europe, the BS170 is basically identical except for the pins being reversed.) Now all that is needed is a drain limiting resistor, since the gate current is as close to zero as we are likely to care about.

Q1 gets its input from the TTL-level device. Q1 and R1 form a simple inverter circuit, with the inverted output going to pin 2 (RxD) of the serial connector. The pull-up voltage for RxD comes from pin 7 (RTS) of the serial connector, so the voltage will be correct for whatever device is connected — be it a PC or whatever else you have. Q2 and R2 are another inverter, driven by TxD from pin 3 of the serial connector. The pull-up voltage for R2 should be supplied by the low-voltage device — your radio or microcontroller project.

There you go. Two cheap little MOSFETs, two resistors. The 2.2 kΩ, ¼ W resistors, seem to work with every serial port tried so far. The author has sold several hundred kits using this exact circuit for the serial interface without issues from any users. It's been tested with laptops, desktops, USB-to serial-converters, a serial LCD display and even a Palm-III with an RS-232 serial cable. It's also been used with very slight modifications in computer interfaces for VX-7R and FT-817 transceivers. It's simple, cheap and works great — what more could you ask?

24.10 USB Interfaces For Your Ham Gear

Over the past several years, serial ports have been slowly disappearing from computers. The once-ubiquitous serial port, along with other interfaces that were once a basic part of every computer system, has largely been replaced by Universal Serial Bus (USB) ports. This presents problems to the amateur community because serial ports have been and continue to be the means by which we connect radios to computers for control, programming, backing up stored memory settings, contest logging and other applications. Radios equipped with USB ports are rare indeed, and many of the optional computer interface products sold by ham manufacturers are serial-only. The project described here by Dale Botkin, N0XAS, offers a solution.

THE UNIVERSAL SERIAL BUS

So what is USB, and how can it be used in the ham shack? The USB standard came about during the early 1990s as a replacement for the often inconvenient and frustrating array of serial, parallel, keyboard, mouse and various other interfaces commonly used in personal computer systems. The idea is to use a single interface that can handle multiple devices sending and receiving different types of data, at different rates, over a common bus. In practice, it works quite well because of the very well defined industry specification for signal levels, data formats, software drivers and so forth.

At least two and sometimes as many as six or more USB ports are now found as standard equipment on nearly all new computer systems. Support for USB ports and generic USB devices is built into most operating systems including *Windows*, *Linux*, *Solaris*, *Mac OS* and *BSD* to name a few. USB use has spread beyond computers, and the standard USB "B" or the tiny 4-conductor "Mini B" connectors are now found on all manner of electronic devices including cameras, cell phones, PDAs, TVs, CD and DVD recorders, external data storage drives and even digital picture frames. USB is indeed becoming *universal*, as its name implies — everywhere, it seems, except for Amateur Radio gear!

The EIA-232 serial interface specification we have used for decades is just that — an electrical interface specification that defines voltage levels used for signaling. The asynchronous serial format most commonly used with serial ports defines how data is sent over whatever interface is being used. Any two devices using the same protocol and format can talk to each other, whether they are both computers, both peripheral devices (like printers, terminals or TNCs) or any combination. All that is required is the proper cable and the selection of matching parameters (speed, data bits, handshaking) on each end of the connection.

This is not the case with USB devices. USB uses a tiered-star architecture, with well defined roles for the host controller (your computer) and devices. Devices cannot communicate directly with each other, only with a host.

See **Fig 24.39**. At the top of the stack is the *host controller device* (HCD), usually your computer, which controls all communication within the USB network. The computer may have several host controllers, each with one or more ports. The host controller discovers, identifies and talks to up to 127 devices using a complex protocol. No device can communicate directly with another device; all communication is initiated and controlled by the host controller. As a result, you cannot simply connect two USB devices together and expect them to communicate. (Some USB printers will talk directly to digital cameras because the printer also has a host controller built in.)

As you have probably already figured out, moving from serial to USB is not a simple

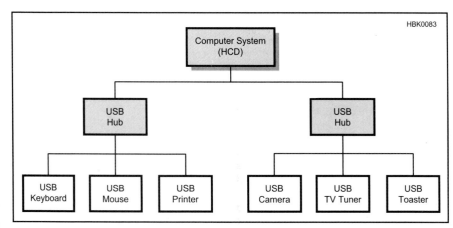

Fig 24.39 — Typical USB architecture.

matter of converting voltage levels or even data formats. USB devices all need some embedded "smarts," usually in the form of either a dedicated USB interface chip or a microcontroller using firmware to talk to the USB host. Where we could once use simple transistor circuits to adapt RS-232 serial to the TTL voltage levels used by many radios, we now must add the complete USB interface into the picture.

The USB specification is freely available and well documented, and there are books dealing with designing the software to use the interface. If you're starting from scratch it is a very daunting task to even understand exactly how USB works, let alone design and build your own USB-equipped device. If you're willing and dedicated enough, you can implement your own USB device using a fairly broad range of available microcontrollers. There are also dedicated function chips that implement a complete USB interface — or parts of it — in one or two devices.

Fortunately, several manufacturers have stepped up to make using USB relatively easy for the experimenter. One of the most commonly used solutions is from Future Technology Devices International (FTDI). FTDI manufactures a line of USB interface chips that plug into the Universal Serial Bus on one side and present a serial or parallel interface on the other side. The chips are quite versatile and are commonly used to add USB functionality to a wide range of devices ranging from cell phones to USB-to-serial cables. They are also available in several forms such as plug-in modules with a USB connector and the interface logic, all on a small PC board with pins that can be plugged into a standard IC socket.

CONSTRUCTION

This project uses an FT232R series chip from FTDI to bridge the USB interface from your computer to a TTL-level serial interface that can be adapted to use with many different ham rigs. The chip has built-in logic to handle the USB interface, respond to queries and commands from the computer, and send and receive serial data. By using special drivers available from FTDI (and built into *Windows* and many other systems), it can appear to your programs as a standard serial COM port.

Once the conversion to serial data is done, all that remains is to adapt the signals to the particular interface used by your equipment. In many cases, this serial data can simply be connected directly to the rig using the appropriate connector. In others it may require some extra circuitry to accommodate various interface schemes used by manufacturers. The most common is an open-collector bus type interface, which is easily accommodated using a few inexpensive parts.

Construction of this project can be handled in a couple of different ways. The FT232R series chips are available from various suppliers at low cost. Unfortunately they are only available in SSOP and QFLP surface mount packages that can be difficult to work with unless you use a custom printed circuit board. (An additional letter is tacked onto the part number to identify the package. For example, the FT232RL uses a 28-pin SSOP package.) There are SSOP-to-through-hole adapter boards available, but they are relatively expensive and still require some fine soldering.

Fortunately FTDI also sells a solution in the form of the TTL-232R (**Fig 24.40**). It incorporates a USB "A" connector, an FT232RQ chip and the required power conditioning on a tiny PCB encapsulated in plastic, all about the size of a normal USB connector. A 6-ft long cable terminates in a 6-position inline header that provides ground, +5 V, and TTL level TxD, RxD, RTS and CTS signals.

You can connect these signals directly to your own microprocessor projects and have an "instant" USB connection. You can also build a variety of interface cables or boards for use with different radios. This arrangement is ideal for the experimenter and the person who might need several different interfaces for use with several different rigs — like many of us! The cost of the TTL-232R cable is quite reasonable, in fact lower than the cost of having a prototype PCB made to build your own equivalent device. Of course that option is still available to those who need more flexibility.

USING USB IN THE SHACK

With the USB-to-serial conversion handled so easily, the only remaining task is to adapt the TTL-level serial interface of the FT232R to the interface used by your particular radio. In the case of some rigs, such as the Yaesu FT-817, Kenwood TM-G707A and many others, it's simply a matter of wiring up the right connector, as shown in **Fig 24.41**.

For communicating with individual ICOM and Ten-Tec rigs, one simple solution is to simply connect the transmit and receive data signals together and wire up a ⅛-inch stereo plug as shown in **Fig 24.42**. This will work with a single device connected, but lacks the open-collector drivers needed to be compatible with attaching multiple devices to the CI-V bus.

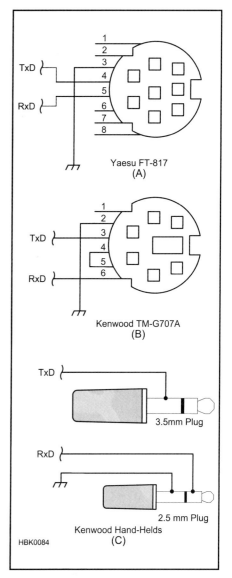

Fig 24.41 — Using the FTDI USB-to-serial interface with some radios is as simple as wiring up a connector. Connections for the Yaesu FT-817 are shown at A, the Kenwood TM-G707A (PG-4 type) at B, and Kenwood hand-helds (PG-5 type) at C. Ground, RxD and TxD connections are made to the 6-pin connector on the TTL-232R cable.

Fig 24.40 — The FTDI TTL-232R cable simplifies construction.

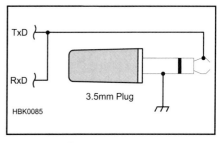

Fig 24.42 — The transmit and receive data signals can be connected together for use with a single Ten-Tec or ICOM CI-V device.

Station Accessories   24.27

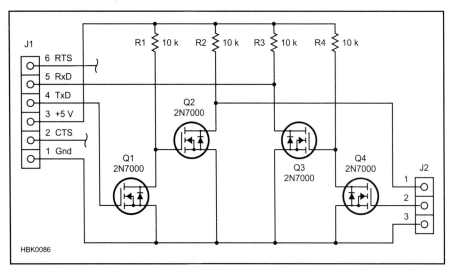

Fig 24.43 — Schematic of the basic interface buffer board.
J1 — 6-pin, 0.1-in pitch header to match TTL-232R cable.
J2 — 3-pin header to match radio cable.
Q1-Q4 — Small-signal MOSFET, 2N7000, BS-170 or equiv.
R1-R4 — 10 kΩ, ¼ W.

Fig 24.44 — The completed buffer board. Adapter cables for various radios plug into the header on the right edge of the board.

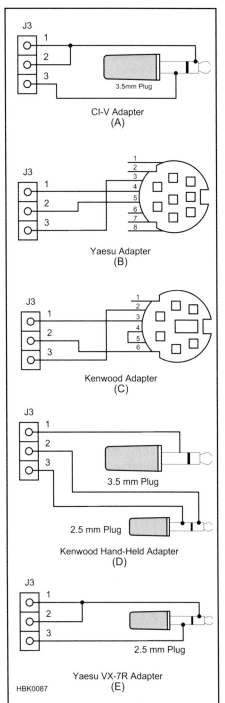

Fig 24.45 — Adapters for using the buffer board with various radios. J3 matches the adapter used on the buffer board.

A LITTLE MORE VERSATILITY

Many hand-held transceivers use a single data line for both transmit and receive. This method is very similar to CI-V, but requires an open-collector or open-drain configuration for a single device. Some additional buffering is required to provide an interface to the rig.

Fig 24.43 shows the design of the general-purpose buffer board for use with various radios. This circuit simply provides a noninverting, open-drain buffer for data in each direction. Adapters for various rigs can be plugged into J2 with no further changes needed. Use of MOSFETs instead of their bipolar counterparts simply reduces the parts count. No additional resistors are needed to limit base current, as you would need when using 2N2222 or similar transistors. Parts can be mounted on a small project board as shown in **Fig 24.44**.

This arrangement turns out to be extremely versatile and will work with CI-V, FT-817, Yaesu VX-series handhelds and the other rigs the author was able to test as well. All that is required is the proper adapter cable from the base interface board to the rig. Some examples are shown in **Fig 24.45**.

ADDING PTT

Handling PTT is quite simple, and it can be done either with or without isolation as shown in **Fig 24.46**. For a non-isolated PTT line, we can simply drive the gate of a 2N7000 or similar MOSFET from the FT232RL's RTS output. If you prefer an optically isolated PTT output, all you need to add is one resistor and an optocoupler such as a 4N27 or PS-7142. Note that the TTL-232R provides a 100 Ω resistor in series with the RTS signal (the same is true for TxD), so a smaller value resistor (R1) can be used for the optocoupler's input than you would normally use.

While connecting the data lines directly to the rig is a suitable solution for many applications, some Kenwood HF rigs require a different approach. They require inverted TTL-level data, and may experience "hash" noise from the computer when not using an isolated interface. **Fig 24.47** shows an optically isolated method of connecting to a Kenwood rig such as the TS-850S. Note that this has not been tested but the design is similar to numerous examples of serial interfaces for the TS-850S and similar radios.

It's probably only a matter of time before

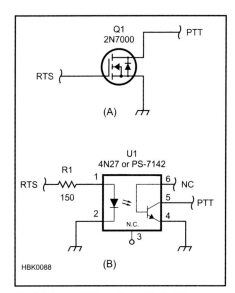

Fig 24.46 — PTT can be added either without isolation (A) or with an optoisolator (B).

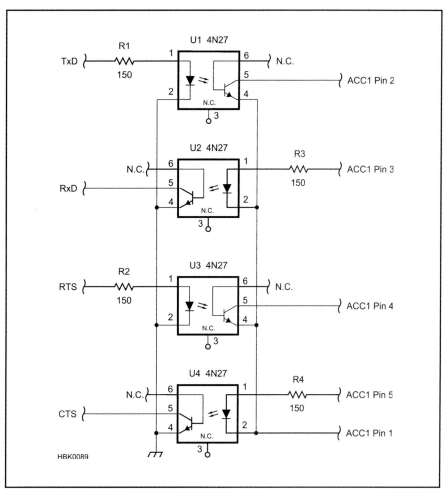

Fig 24.47 — Isolated Kenwood interface for the TS-850S and similar HF transceivers requiring inverted TTL-level data and isolated lines.

we start seeing the tiny USB "Mini B" connector appear on new ham gear. It has become inexpensive and easy to add USB to pretty much any product, and as serial ports become more and more rare the manufacturers will eventually face overwhelming consumer demand to switch. In the meantime, you can do a little homebrewing and use your new, serial port free computer to talk to all of your ham gear for very little cost.

24.11 The Universal Keying Adapter

When Dale Botkin, N0XAS, was about to start restoring his old Heathkit HW-16, he decided it was time to explore what would be needed to key this tube rig using an electronic keyer. He also remembered — not fondly — having been "bitten" a few times when he touched metal parts of the key. There was some substantial voltage on the key, and he wanted to avoid touching it againch The key jack of this rig has a negative voltage, something like −85 V or more. Even the 2N7000 output of a solid-state keyer, rated at 60 V, would be no match for that. The Universal Keying Adapter (UKA) shown in **Fig 24.48** bridges the gap between a modern keyer and a classic tube-type transmitter, and it can also be used to key older tube-type amplifier TR switching lines with solid-state transceivers that can't handle the voltage or current requirements.

KEYING SCHEMES

Older gear generally uses one of two keying

Fig 24.48 — This version of the Universal Keying Adapter 2 is available as a kit from www.hamgadgets.com. Parts count is low, and it can easily be built on a small project board.

schemes, grid-block or cathode keying. Grid-block keying requires that your key handle fairly high negative bias voltages, often in the 150 V range. Cathode keying is more demanding, with voltages up to +350 V or so. Most modern solid state keyers use a simple transistor output circuit suitable only for low-voltage, positive keying. Obviously, neither grid-block nor cathode keyed transmitters should be connected to such an output! At the very least it's going to damage the keyer.

Most grid-block keying adapters use one or two bipolar transistors and are designed specifically to key grid-block rigs. This is fine as far as it goes, but also requires that you have one setup for solid state rigs, another for grid-blocked rigs, and still another if you have a cathode keyed boat anchor as well. It's not an optimum solution. Of course one could use a relay to key just about anything, but relay contact noise would quickly get really irritating. Relays also eventually wear out, contacts get dirty and the coils can use quite a bit of current, meaning battery powered keyers will suffer from very limited battery life.

Solid-State Relays

Solid state relays exist, but they can be expensive. The author finally hit upon what seems to be an ideal solution. It's cheap, small, relatively low current, uses low voltage control and is capable of switching fairly high ac or dc voltages. Perfect! Crydom, NEC, Fairchild, Omron and other manufacturers make very similar small, inexpensive MOSFET solid state relays. Suitable model numbers include the NEC PS7142 and PS7342, Fairchild HSR412, Omron G3VM-401B and others. Depending on the part selected they can handle up to a couple hundred milliamps and will switch up to 400 V ac or dc. All of this in a 6-pin DIP form factor, and for just a few bucks!

The MOSFET solid-state relay (SSR) is very similar to an optoisolator, but somewhat more versatile. Ac or dc loads can be switched, and the allowable load voltages are much higher than regular optoisolators. Driving the relay input involves supplying a few mA of current to turn on an LED inside the device. This requires about 1.4 V at a recommended forward current of 10 to 30 mA. The output is a pair of MOSFETs with common sources and a photo gate rather than a hard-wired gate. Turn on the LED and the MOSFETs conduct, turn off the LED and the MOSFETs shut off. Just like a relay, it's simple and elegant.

KEYING ADAPTER CIRCUITS

For a simple keying adapter arrangement, all that is needed is the MOSFET solid-state relay, one resistor, and a dc power source such as a battery. See **Fig 24.49**. The keying input from your hand key or electronic keyer completes the input circuit through the current limiting resistor and the input side of the SSR, turning the output on. In this example, the value of R1 is determined using Ohm's Law to give between 10 and 30 mA of current at the desired input voltage. For example, for use with two AA alkaline batteries, a range of 100 to 300 Ω is okay; 150 or 220 Ω will work reliably with NiCd or NiMH cells as well. If you plan to use a 13.8 V dc power supply, a 1 kΩ resistor should be fine for R1.

It soon became apparent that other uses existed in addition to just isolating the key or keyer from a transmitter. The addition of a simple transistor inverter allows the use of computer keying in parallel with input from a straight key, bug or keyer. This is handy, for example, to contesters using computers for keying who may need to send some information by hand from time to time. See **Fig 24.50**. The use of a 2N7000 MOSFET instead of the common NPN transistor makes it easier to accommodate both serial and parallel port use, since the gate will withstand up to 20 V positive or negative and does not need to be current limited. In this example, the value of R1 should be determined as mentioned earlier. R2's purpose is to keep the MOSFET gate from floating, so its value is not critical but

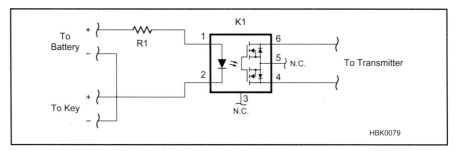

Fig 24.49 — The original Universal Keying Adapter circuit used a solid-state relay and a battery for power.

K1 — Optically coupled solid-state relay, Fairchild HSR412 or equivalent.
R1 — ¼ W resistor, varies with supply voltage (see text). Use 150 or 220 Ω for two AA cells; 1 kΩ for 13.8 V dc power source.

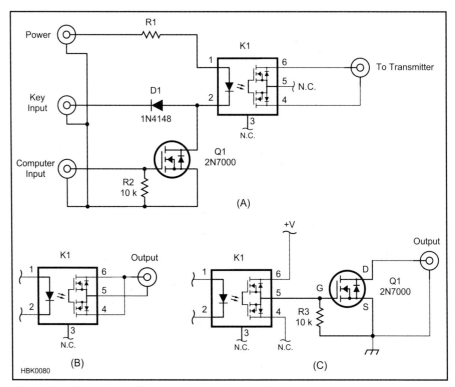

Fig 24.50 — The Universal Keying Adapter circuit with provisions for using computer keying in parallel with a hand key or keyer. Two methods of increasing the load current capacity are shown at B and C (see text for discussion).

K1 — Optically coupled solid-state relay, Fairchild HSR412 or equivalent.
R1 — ¼ W resistor, varies with supply voltage (see text). Use 150 or 220 Ω for two AA cells; 1 kΩ for 13.8 V dc power source.

10 kΩ or 100 kΩ are good values.

The design eventually evolved to include a full-wave bridge rectifier allowing either ac or dc input, a Zener diode for voltage regulation and an LED keying indicator. The result was the Universal Keying Adapter 2, which is also available in kit form (see **www.hamgadgets.com**). The UKA-2 can accept any dc or ac power source up to about 30 V, and is adaptable to lower power battery operation by substituting or eliminating some of the power supply parts.

OTHER CONFIGURATIONS

It is worth noting that if only positive or only negative voltages will be used, the output current capacity can be greatly increased and the on-state resistance greatly decreased by connecting the two MOSFET drain outputs in parallel rather than serial. An example of this is shown in Fig 24.50B. For the HSR412, maximum load current increases from 140 mA for series output to 210 mA for parallel. An added feature of this arrangement is the presence of the intrinsic "body diodes" of the MOSFETS. If you are keying an amp that has an internal keying relay, this can serve to absorb some of the back EMF when the relay releases. Note that this arrangement requires that pin 5 of the IC be at a lower voltage than the other two pins — in other words, ground for positive keying, or –V for negative keying. Since the output is completely isolated from the input, the polarity of the output connector can be changed without the risk of exposing your other gear to dangerous voltages.

Although the IC solid-state relays are good for a wide range of uses, there are applications that present more of a load than the device is able to safely handle even with the outputs in parallel. For loads of more than the rated device current, an external MOSFET switch can be used. Again, the resulting output is optically isolated from the input. Resistor R3 shown in Fig 24.50C keeps the gate of the 2N7000 MOSFET low until the input is activated, so its value is not critical. The gate voltage +V needs only to be within the safe range for the device selected; in most cases 12 V will do fine.

The UKA design works well with both grid-block and cathode keyed rigs, as well as solid state. It's been tested with a TS-930SAT, Heath HW-16, FT-817, Rock-Mites, FT-480R and more. It has also been used with cathode keyed rigs such as the Heath DX-40, and many more examples of this circuit are in use keying various "boat anchor" rigs and amps. It works quite well to key older tube power amplifiers with solid-state transceivers that are not equipped with suitable keying relays or circuits. The input works well with any rig, key, electronic keyer, serial or parallel port tried so far.

24.12 The TiCK-4 — A Tiny CMOS Keyer

TiCK stands for "Tiny CMOS Keyer" and this is the fourth version of the chip. It is based on an 8-pin DIP microcontroller from Microchip Corporation, the PIC12CE674. This IC is a perfect candidate for all sorts of Amateur Radio applications because of its small size and high performance capabilities. This project originally used the TiCK-2 chip (based on the PIC12C509) and was described fully in Oct 1997 *QST* by Gary M. Diana, Sr, N2JGU, and Bradley S. Mitchell, WB8YGG. The keyer has the following features:

- Mode A and B iambic keying.
- Low current requirement to support portable use.
- Low parts count consistent with a goal for small physical size.
- Simple rig and user interfaces. The operator shouldn't need a manual and the rig interface should be paddles in, key line out.
- An audible sidetone for user-feedback functions and to support transceivers without a built-in sidetone.
- Paddle select, allowing the operator to swap the dot and dash paddles without having to rewire the keyer (or flip the paddles upside down!).
- Manual keying to permit interfacing a straight key (or external keyer) to the TiCK.

New for the TiCK-4 version are:
- Two 50-character message memories (up from one).
- Remembers operating parameters (but not message memories) when power is removed.

DESIGN

Fig 24.51 shows the schematic for the keyer. The PIC12CE674 has 3.6 kbytes of program read-only memory (ROM) and 128 bytes of random-access memory (RAM). This means that all the keyer *functions* have to fit within the ROM. The keyer *settings* such as speed, paddle selection, iambic mode and sidetone enable are stored in RAM. Unlike the previous TiCK keyers, this microcontroller also includes EEPROM memory that's used to store these operating parameters. So you can power the chip off, and upon power-up it will remember the last stored values for the settings.

A PIC12CE674 has eight pins. Two pins are needed for the dc input and ground connections. The IC requires a clock signal. Several clock-source options are available. You can use a crystal, RC (resistor and capacitor) circuit, resonator, or the IC's internal oscillator. The authors chose the internal 4-MHz oscillator to reduce the external parts count. Two I/O lines are used for the paddle input. One output feeds the key line, another output is required for the audio feedback (sidetone) and a third I/O line is assigned to a pushbutton.

USER INTERFACE

Using the two paddles and a pushbutton, you can access all of the TiCK's functions. Certain user-interface functions need to be more easily accessible than others; a prioritized list of functions (from most to least accessible) is presented in **Table 24.9**.

The TiCK employs a *single button interface* (SBI). This simplifies the TiCK PC board, minimizes the part count and makes for ease of use. Most other electronic keyers have multibutton user interfaces, which, if used infrequently, make it difficult to remember the commands. Here, a single button push takes you through the functions, one at a time, at a comfortable pace (based on the current speed of the keyer). Once the code for the desired function is heard, you simply let up on the button. The TiCK then executes the appropriate function, and/or waits for the appropriate input, either from the paddles or the pushbutton itself, depending on the function in question. Once the function is complete, the TiCK goes back into keyer mode, ready to send code through the key line.

The TiCK-4 IC generates a sidetone signal that can be connected to a piezoelectric element or fed to the audio chain of a transceiver. The latter option is rig-specific, but can be handled by more experienced builders.

THE TICK LIKES TO SLEEP

To meet the low-current requirement, the authors took advantage of the PIC12CE674's ability to *sleep*. In sleep mode, the processor shuts down and waits for input from either of

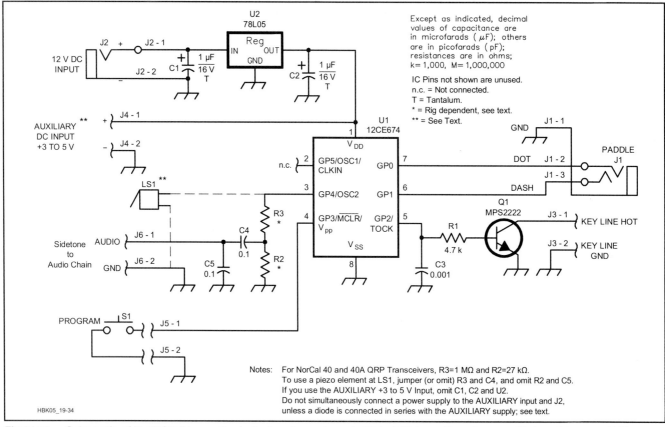

Fig 24.51 — Schematic of the TiCK-4 keyer. Equivalent parts can be substituted. Unless otherwise specified, resistors are ¼ W, 5%-tolerance carbon film units. The PIC12CE674 IC must be programmed before use; see the parts list for U1. RS part numbers in parentheses are RadioShack; M = Mouser (Mouser Electronics). Kits are available from Kanga US (www.kangaus.com).

C1, C2 — 1 µF, 16 V (or higher) tantalum (RS 272-1434; M 581-TAP105K035SCS).
J1 — 3-circuit jack (RS 274-249; M 161-3402).
J2 — 2-circuit jack (RS 274-251) or coaxial (RS 274-1563 or 274-1576).
J3 — 2-circuit jack (RS 274-251; M 16PJ135).

LS1 — Optional piezo element (RS 273-073).
Q1 — MPS2222A, 2N2222, PN2222, NPN (RS 276-2009; M 610-PN2222A).
S1 — Normally open momentary contact pushbutton (RS 275-1571).
U1 — Programmed PIC12CE674, available from Kanga US. Programmed ICs and data sheets are available, as are kits with the PC board and other parts. Source code is not available.
U2 — 78L05 5 V, 100 mA regulator (M 512-LM78L05ACZ) or LP2950CZ-5.0 low quiescent current regulator (Digi-Key LP2950CZ-5.0NS-ND).
Misc: PC board, 8-pin DIP socket, hardware, wire (use stranded #22 to #28).

Table 24.9
TiCK-4 User Interface Description

Action	TiCK-4 Response	Function
Press pushbutton briefly	none	Memory #1 playback.
Hold pushbutton down	E (dit)	Memory #2 playback.
Hold pushbutton down	S (dit-dit-dit)	Speed adjust: Press dit to decrease, dah to increase speed.
Hold pushbutton down	T (dah)	Tune: To unkey rig, press either paddle or pushbutton.
Hold pushbutton down	A (dit-dah)	ADMIN mode: Allows access to various TiCK IC setup parameters.
Press pushbutton briefly	I (dit-dit)	Input mode: Allows the user to enter message input mode.
Hold pushbutton down	1 (dit-dah-dah-dah-dah)	Message #1 input. Send message with paddle and push button momentarily when complete.
Hold pushbutton down	2 (dit-dit-dah-dah-dah)	Message #2 input. Send message with paddle and push button momentarily when complete.
Hold pushbutton down	P (dit-dah-dah-dit)	Paddle select: Press paddle desired to designate as dit paddle.
Hold pushbutton down	A (dit-dah)	Audio select: Press dit to enable sidetone, dah to disable. Default: enabled.
Hold pushbutton down	SK (dit-dit-dit dah-dit-dah)	Straight key select: Pressing either paddle toggles the TiCK to/from straight key/keyer mode. Default: keyer mode.
Hold pushbutton down	M (dah-dah)	Mode select: Pressing the dit paddle puts the TiCK into iambic mode A; dah selects iambic mode B (the default).
Hold pushbutton down	B (dah-dit-dit-dit)	Beacon mode: Pressing either paddle toggles the TiCK to/from Beacon/No-Beacon mode. Default: No-Beacon mode.
Hold pushbutton down	K (dah-dit-dah)	Keyer mode: If pushbutton is released, the keyer returns to normal operation.
Hold pushbutton down	S (dit-dit-dit)	Cycle repeats starting with Memory playback, Speed adjust, etc.

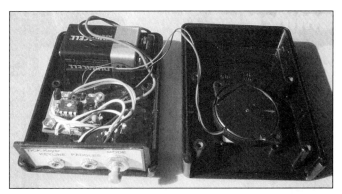

Fig 24.52 — Here's the completed TiCK-4 keyer complete with battery, mounted in a small case. (*Photos courtesy Bill Kelsey, N8ET*)

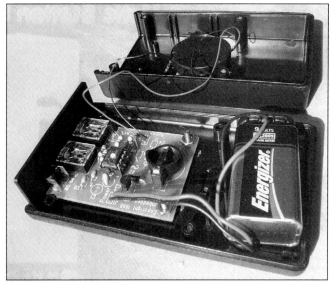

Fig 24.53 — The enhanced EMB version of the TiCK-4 includes a low-current regulator, battery backup and board-mounted jacks and switches.

the two paddles. While sleeping, the TiCK consumes about 1 µA. The TiCK doesn't wait long to go to sleep either: As soon as there is no input from the paddles, it's snoozing! This feature should be especially attractive to amateurs who want to use the TiCK in a portable station.

ASSEMBLING THE TICK

The TiCK-4's PC board size (1×1.2 inches) supports its use as an embedded and stand-alone keyer. See **Fig 24.52**. The PC board has two dc input ports, one at J2 for 7 to 25 V and another (J4, AUXILIARY) for 2.5 to 5.5 V. The input at J2 is routed to an on-board 5-V regulator (U2), while the AUXILIARY input feeds the TiCK directly. When making the dc connections, observe proper polarity: There is no built-in reverse-voltage protection at either dc input port.

The voltage regulator's bias current is quite high and will drain a 9 V battery quickly, even though the TiCK itself draws very little current. For this reason, the "most QRP way" to go may be to power the chip via the AUXILIARY power input and omit U2, C1 and C2. Another alternative is use of a low quiescent current 5 V regulator, such as the LP2950CZ-5.0, for U2.

When using the AUXILIARY dc input port, or if *both* dc inputs are used, connect a diode between the power source and the AUXILIARY power input pin, attaching the diode anode to the power source and the cathode to the AUXILIARY power input pinch This provides IC and battery protection and can also be used to deliver battery backup for your keyer settings.

SIDETONE

A piezo audio transducer can be wired directly to the TiCK-4's audio output. Pads and board space are available for voltage-divider components. This eliminates the need to interface the TiCK with a transceiver's audio chainch Use a piezo *element*, not a piezo *buzzer*. A piezo buzzer contains an internal oscillator and requires only a dc voltage to generate the sound, whereas a piezo element requires an external oscillator signal (available at pin 3 of U1).

If you choose to embed the keyer in a rig and want to hear the keyer's sidetone instead of the rig's sidetone, you may choose to add R2, R3, C4 and C5. Typically R3 should be 1 MΩ to limit current. R2's value is dictated by the amount of drive required. A value of 27 kΩ is a good start. C4 and C5 values of 0.1 µF work quite well. C4 and C5 soften the square wave and capacitively couple J6-1 to the square-wave output of pin 3. Decreasing the value of R2 decreases the amount of drive voltage, especially below 5 kΩ. Use a 20 kΩ to 30 kΩ trimmer potentiometer at R2 when experimenting.

IN USE

To avoid RF pickup, keep all leads to and from the TiCK as short as possible. The authors tested the TiCK in a variety of RF environments and found it to be relatively immune to RF. Make sure your radio gear is well grounded and avoid situations that cause an RF-hot shack.

The TiCK keys low-voltage positive lines, common in today's solid-state rigs. Don't try to directly key a tube rig because you will likely — at a minimum — ruin output transistor Q1.

To use the TiCK as a code-practice oscillator, connect a piezo element to the audio output at pin 3. If more volume is needed, use an audio amplifier, such as RadioShack's 277-1008.

The higher the power-supply voltage (within the specified limits), the greater the piezo element's volume. Use 5 V (as opposed to 3 V) if more volume is desired. Also, try experimenting with the location of the piezo element to determine the proper mounting for maximum volume.

SEVERAL VERSIONS

The TiCK keyer has been around for more than 10 years. The TiCK-4 is a direct plug-in replacement for the TiCK-1 and TiCK-2, so if you have built a keyer with an earlier version, you can upgrade to the TiCK-4 and gain the two memories and nonvolatile parameter memory features. The TiCK-3 is a surface-mount version with limited availability. An enhanced version, the TiCK-EMB-4, includes RF/static protection, low quiescent current regulator, battery backup and board-mounted jacks and switches. See **Fig 24.53**.

24.13 The ID-O-Matic Station Identification Timer

Would you rather not watch the clock when engaged in a QSO? Repeaters usually have an automatic 10-minute ID to remind you when it's time to identify your station, but on HF and simplex it's more of a problem. Dale Botkin, NØXAS, set out to design a simple 10-minute ID timer with a reminder beep and some sort of visual indicator, something along the lines of the ID timer portion of the old Heathkit SB-630. Along the way he added a few features to make the timer more useful, and as sometimes happens, things just kind of snowballed from there. The ID-O-Matic described here and shown in **Fig 24.54** is an automatic ID timer/annunciator with a rich set of features that make it useful for many applications.

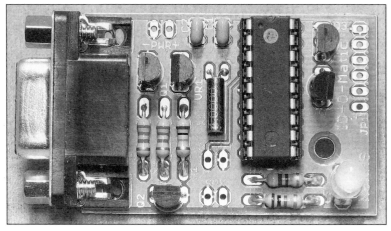

Fig 24.54 — This version of the ID-O-Matic is built from a kit offered by www.hamgadgets.com. The parts count is low, and it can easily be built on a prototype board.

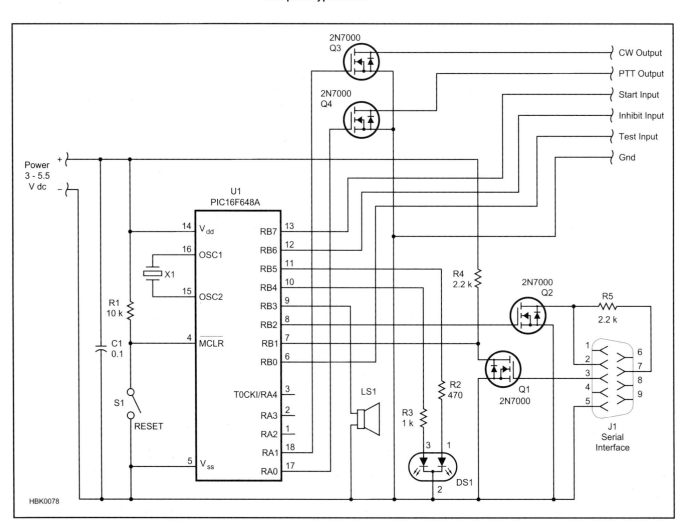

Fig 24.55 — Schematic of the ID-O-Matic. Resistors are ¼ W.

DS1 — Dual-color LED, MV5437.
J1 — DB9 female connector (Mouser 152-3409).
LS1 — Soberton GT-111P speaker or equiv (Mouser 665-AT-10Z or Digi-Key 433-1020-ND).
Q1-Q4 — 2N7000 MOSFET.
U1 — Microchip PIC16F648A. Must be programmed before use (see text). Programmed chips and parts kits are available from www.hamgadgets.com.
(ARRL now publishes a book about PIC Programming for beginners.)
X1 — 4 MHz cylindrical crystal (Mouser 520-ECS-400-18-10).
Enclosure: Pac-Tec HMW-ET (Mouser 76688-510).

FEATURE HIGHLIGHTS

There's an old saying that goes, "When all you have is a hammer, everything looks like a nail." While the author has a pretty well stocked junk box, he never liked building 555 timers and the like. He tends to use whatever is handy and can cobble together with an absolute minimum of parts, and likes to be able to make changes or add features later on. That often means writing firmware for a PIC microcontroller, and this time was no exception.

The basic device uses a single PIC 16F648A processor selected for its memory capacity, internal oscillator, hardware USART and other features. All features are implemented in program code, with just enough external hardware to provide a useful interface to the user. Audio and visual output is used to indicate the end of the time period, and one user control is all that is required other than a power switch.

Let's take a quick look first at the basic features. In its most basic timer mode, a single pushbutton is used to start the timing cycle by resetting the PIC. This starts a 10-minute timer running and lights the green section of a red-green dual color LED. The LED remains green until 60 seconds before the set period expires. With less than 60 seconds remaining, the LED turns yellow; this is accomplished by illuminating both the green and red halves. At 30 seconds the LED begins to blink, alternating between yellow and red. When the time period expires, the LED turns red and the timer starts beeping until you reset it with a pushbutton switch. Now the cycle starts again, ready to remind you in another 10 minutes.

So far so good, but using a PIC like this is a little bit of overkill. Of course no good project is complete until "feature creep" sets in! With the timing and audio portions of the program done, some other features were pretty simple to add. What if you want to ID at some other interval instead of 10 minutes, or have the device automatically ID in Morse code? What if you want to use it for a beacon or a repeater ID device? How about CW and PTT keying outputs for a fox hunt beacon?

Eventually the firmware evolved to have a fairly robust set of features. There is a serial interface for setup using a computer or terminal. You can set the timeout period anywhere from 1 to 32,767 seconds. You can set select a simple beeping alert, or set a Morse ID message with up to 60 characters. For repeater and fox hunt use, there are outputs for PTT and CW keying just in case they might be needed, as well as a couple of inputs intended for COR or squelch inputs if the chip is used in repeater mode. We'll cover all of that in a bit; first, let's look at the basic functions and the hardware design.

CIRCUIT DESIGN

As shown in **Fig 24.55**, the hardware is quite simple. In its basic form all you need is the PIC, a red and green dual-color LED, and a few other parts as seen in the schematic. Supply voltage can be 3 to 5.5 V, so you can use three AA alkaline cells. You can also add a 5 V regulator, allowing the use of a 12 V power supply. Despite their size, 9 V batteries have surprisingly low capacity and will only last a few days of continuous use in a project like this.

R1 is a pull-up resistor whose value is not critical; 10 kΩ is a good value. Its function is to hold the !RESET line high until S1 is pressed. R2 and R3 set the current for the red and green sections of the dual LED. They are different values because the red section of the LED is usually substantially brighter than the green. R4 pulls the serial RxD line high when no serial connection is present. For use in the shack to remind you to ID every 10 minutes, that's all you need.

The original prototype fits in a Pac-Tec enclosure model HMW-ET. For audio output, try a tiny Soberton GT-111P speaker. About the size of a small piezo element at 12 mm diameter and about 8 mm tall, this is actually a low current magnetic speaker that can be driven directly from the PIC's output pinch Switches can be whatever style you like. Many of the parts can be obtained at a local RadioShack, where you may also find a suitable enclosure and battery holder.

A level converter circuit allows you to connect a PC or terminal to set up some of the more advanced features. A single chip solution such as a MAX232 or similar chip could be used, but the circuit shown is effective, very low cost and easy to build. Most modern serial ports do not require full EIA-232 compliance, but will work fine without a negative voltage for the interface lines. Accordingly, the interface presented in the schematic simply inverts the polarity of the signals and also converts the ±12 V logic levels that may be present on the PC interface to levels appropriate for the PIC I/O pins. By using MOSFETs instead of the more traditional NPN and PNP transistors we can eliminate some of the parts usually seen in level shifting circuits like this. The only additional components are Q1, Q2 and R4, a current limiting resistor for the TxD line.

With a terminal or terminal emulator program you can set your own delay time, CW speed, ID message and select a simple beep or a CW ID. Inputs used for repeater operation (!INHIBIT and !START) are an exercise left to the builder. The thing to remember is that these inputs are active low, and the input voltage must be limited to no more than the PIC's V_{dd} supply voltage.

PUTTING THE ID-O-MATIC TO WORK

When power is applied, the PIC is set to beep at the end of a 10-minute time-out period. After a series of reminder beeps, it will automatically reset and begin a new timing cycle. To change any of the settings, simply connect a serial cable to your PC or a dumb terminal and the ID-O-Matic. Any serial communication program such as *PuTTY*, *HyperTerminal* or *Minicom* can be used. The communication settings are 9600 baud, 8 bits, no parity and no handshaking. Hit the ENTER key twice to view and edit the configuration. The program will step through a series of prompts as shown in **Fig 24.56**. At each prompt you may either enter a new setting or simply hit ENTER to keep the existing setting, shown in parenthesis.

There are a few optional inputs and outputs that could be used to make the timer suitable for use in a "fox" transmitter or repeater.

• Pin 6 (RB0) is the !TEST input. You can momentarily ground this input to hear your ID and/or beacon messages. If no messages have been stored, the ID-O-Matic will announce its firmware version. During normal operation, this pin will select the alternate ID message. This may be useful to indicate, for example, if a repeater site has switched to battery power, or if a "fox" transmitter has been located.

• Pin 12 (RB6) is the !INHIBIT input. This pin is normally held high by the PIC's internal weak pull-up resistors, and can be driven low by some external signal (squelch, for example) to delay the CW ID until the input goes high.

• Pin 13 (RB7) is the !START input, used only in repeater mode. Also pulled high internally, a LOW logic signal on this pin will start the timer. This can be handy to have your rig or repeater ID 10 minutes after a transmission, but not every 10 minutes. In repeater mode, the chip will ID a few seconds after the first

```
NOXAS ID-O-Matic V2.10

ID time (600):
Yellow time (60):
Blink time (30):
SPACE will delete ID string.
ID Msg ():
Beacon Msg ():
Alternate Msg ():
Auto CW ID (N):
CW Speed (15):
ID audio tone (753):
Repeater mode (N): Y
Courtesy beep tone (753):
Courtesy beep delay time (5):
Courtesy beep mult (1):
Beacon time (0):
PTT hang time (0):
PTT max time (0):
```

Fig 24.56 — Using *HyperTerminal* to set up the ID-O-Matic

low signal on the !START input. If repeated !START inputs are seen, subsequent IDs will occur at the programmed intervals.

• Pin 17 (RA0) is a PTT output. PTT goes high about 100 milliseconds before the audio starts and stays high until 100 ms after the end of the message. With the 2N7000, the PTT and CW outputs can handle up to 60 V at 200 mA.

• Pin 18 (RA1) is a CW output pin; it will output the same message as the audio output.

The original program code took several days to write and tweak, and has undergone numerous revisions since then. As with any project like this, the testing was the time consuming part. The author used the PIC C compiler from CCS, Inc. to write the program, but the logic is simple enough that porting to assembly, BASIC or a different C compiler should be pretty straightforward.

The code is relatively easy to adapt to your own equipment or requirements. For example, you may wish to move certain functions to different pins, change the polarity of various signals or change some of the timing defaults. The .LST file contains the C source with the ASM equivalents, and is a good starting point for someone wishing to understand the program logic in assembly rather than C. See the *ARRL Handbook CD* at the back of this book for these files.

The really interesting thing about this project is its versatility. Think of the ID-O-Matic as a general purpose PIC based project platform. With a serial interface, a few available inputs and outputs and a dual-color LED, the hardware can be adapted to many different uses simply by changing the PIC's program code. For example, the author built a 40 yard dash timer for his son's high school football team by substituting an infrared LED for the speaker, connecting an IR detector to one of the inputs and using another for a pushbutton switch. The serial interface is connected to a serial LCD display module. Within a couple of days he had a working device, customized with his son's team name, that can be used to time various events.

It's a good little project for beginning PIC users, and a useful station accessory that can be built in an evening and carried around in a shirt pocket.

24.14 An Audio Intelligibility Enhancer

Here's a simple audio processor (**Fig 24.57**) described by Hal Kennedy, N4GG, that can improve intelligibility, particularly for those with degraded high-frequency hearing. It might just give you an edge in that next DX pileup.

What we understand (intelligibility) of what we hear or detect is strongly affected by the end-to-end frequency response of a given communications circuit. Our equipment typically has a fixed audio bandwidth spanning about 300 Hz to 2700 Hz for SSB operation, but our hearing varies widely from operator to operator and it sometimes varies widely for an individual over the course of his or her life.

Speech intelligibility has been studied extensively since the 1940s. In brief, the "human speech signal" can be thought of as being made up of vowel and consonant sounds. The vowel sounds are low in frequency, with fundamental frequencies typically between 100 Hz and 400 Hz, and harmonics as high as 2 kHz. Consonant sounds occupy higher frequencies — typically from 2 kHz to as high as 9 kHz. (See "Speech Intelligibility Papers" at **www.meyersound.com/support/ papers/ speech/**). In spoken English, vowel sounds contain most of the energy in the speech signal, while consonant sounds are short, noise-like, and convey the majority of intelligibility. When consonants are not heard well, it sounds like the speaker is mumbling. It becomes difficult to distinguish *F* from *S* and *D* from *T*.

Frequently, hearing degradation manifests itself both as a loss of sensitivity over all frequencies and a frequency dependent loss where the sensitivity of our hearing degrades as the frequency increases. Sensitivity loss that is flat over frequency does not affect intelligibility, assuming that levels are high enough to be heard. A simple loss of sensitivity can be addressed by turning up the AF gain!

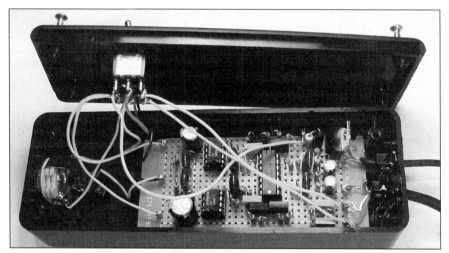

Fig 24.57 — The audio intelligibility enhancer is a compact audio processor that can help to compensate for degraded high-frequency hearing.

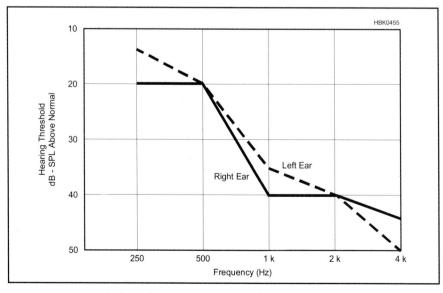

Fig 24.58 — The author's audiology exam results. Threshold sensitivity loss as a function of frequency is commonly associated with aging.

Unfortunately, turning up the gain does not compensate for frequency-dependent loss, and frequency-dependent loss does degrade intelligibility by reducing the ability to hear the consonants in human speech. A quick Web search for "human hearing" will yield a multitude of sites describing the basic concepts.

The author's audiology test results are shown in **Fig 24.58**. Audiologists refer to this pattern as *age induced hearing loss* — a roll-off of hearing sensitivity as frequency increases. Fortunately, intelligibility can be improved significantly for those with this common condition, by simply adding amplification that increases as a function of frequency.

Hearing deficiencies can be much more complex than the typical high frequency loss associated with aging. One's hearing can include notches and/or peaking as a function of frequency, as well as significant mismatch between the left and right ears. Most of these can be addressed through use of audio circuits whose gain approximates the inverse function of the hearing problem(s). Analog circuit designs become complex when more than a simple high-pass or single-notch filter is required. Modern high-end hearing aids use DSP circuits that break the audio spectrum into narrow frequency bands, each with programmable gain.

A professional audiology exam and threshold sensitivity test results are recommended as the starting point for determining how much, if any, audio shaping might improve your ham radio activities.

A VARIETY OF PRACTICAL OPTIONS

The author's operating interests require stereo operation to make significant use of a transceiver's dual receiver function when chasing DX, and to operate two radios simultaneously in contests. In each case operation is in stereo, with each ear dedicated to one of the two receivers. Monaural designs also have a potential drawback, in that the frequency response adjustments apply equally to both ears. In some cases the frequency response of each ear is different.

Stereo equalizers have been around for decades. The most basic are four-band graphic equalizers; they have adjustments to provide up to about 12 dB of gain or attenuation across frequency bands that are centered at four frequencies, each frequency typically one octave above the next (an octave is a doubling of frequency). Seven, 10, 12 and 15 band equalizers are readily available, allowing for finer tailoring of amplification characteristics as the number of bands go up and the frequency range for each band gets smaller.

Parametric equalizers are also available — these differ from graphic equalizers in that the center frequency and bandwidth of each filter element is adjustable. Prices are very reasonable.

An equalizer is a near-perfect solution to compensating for hearing deficiencies, and the ability to readily adjust gain at multiple frequencies allows optimizing via experimentation. Many equalizers are strictly low-level devices and lack sufficient output power to directly drive headphones, requiring an amplifier if the equalizer is to be inserted between rig and headphones.

Some transceivers include adjustable audio DSP circuits for receive as well as transmit audio. It may be possible to adjust the audio response of your transceiver to improve intelligibility without the use of an external device.

CIRCUIT DESCRIPTION

The heart of this audio intelligibility enhancer (AIE) is what audiophiles refer to as a *high-pass shelving filter*. Shelving filters are sometimes used in high-end audio systems to compensate for deficiencies in crossover networks or loudspeakers. The basic circuit and its frequency response are shown in **Fig 24.59**. Note that the shape of this response curve is the inverse of the hearing response of Fig 24.58. Referring to Fig 24.59, the circuit provides a gain, A_{V1}, equal to 1.0 for frequencies somewhat below f_1 and a higher gain, A_{V2}, that is given by $A_{V2} = 1 + R2/R1$ for frequencies somewhat above f_2. Frequency f_1 is given by $f_1 = 1/[2\pi C(R1+R2)]$, and frequency $f2 = 1/(2\pi C R1)$. Conveniently available component values of R1 = 1.33 kΩ, R2 = 6.65 kΩ, and C = 0.047 µF yield a high frequency gain, $A_{V2} = 6.0$, and cut-on and cut-off frequencies of $f_1 = 425$ Hz, and $f_2 = 2550$ Hz. With these values, the circuit gain rises at an average of 3 dB per octave from 300 Hz to 2700 Hz — the frequency range for SSB transmission. For those interested in modifying the design, or using a shelving filter for other purposes, it's important to note that rearranging the equations yields: $f_2/f_1 = A_{V2}$. The ratio of the two break frequencies always equals the maximum gain.

Fig 24.59 also shows a feedback capacitor, C_f, which, while not contributing to the shelving filter function, is included as good design practice. The LM324 op amps used to implement the filters have gain-bandwidth products in excess of 1 MHz. Operating them with gain at frequencies above the useful limits of our hearing (and all the way out to beyond 1 MHz) is an open invitation for RFI problems and unwanted high frequency noise. So, feedback capacitor C_f provides high frequency gain roll-off above the frequency $f = 1/(2\pi C_f R2)$ which for the value of C_f in this design (0.0015 µF) is 16.8 kHz.

Fig 24.60 is the complete schematic for the AIE. Two shelving filters, as described above, are used in series in each audio channel. Two filters in series provide sufficient gain change versus frequency to approximately match the author's high frequency hearing loss, which has a slope of about –7.5 dB/octave across the audiologist's measuring points from 250 Hz to 4 kHz. Two filters in series provide

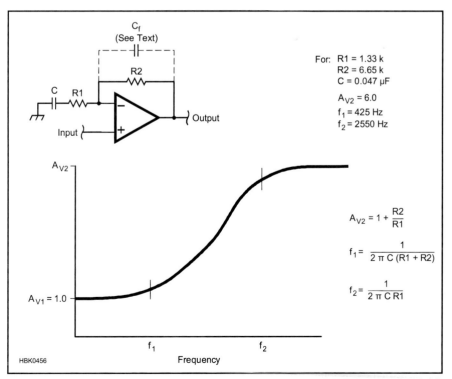

Fig 24.59 — Schematic and frequency response for a high-pass audio shelving filter.

+6 dB/octave gain change from 300 Hz to 2700 Hz and 31 dB of total gain ($A_{V\text{total}} = A_{V2} \times A_{V2} = 6 \times 6 = 36$ or about 31 dB). Some tips for tailoring the circuit to your hearing are included in the next section.

The shelving filters are followed by LM386 audio amplifiers (U2, U3). These provide sufficient output power to drive headphones or small speakers. LM386s have a built-in voltage gain of 20, so they are preceded by 20:1 voltage dividers (R7+R8, R17+R18) to maintain an end-to-end gain of 1 at low frequencies.

Voltage regulator U4 provides regulated and filtered dc voltage to the shelving filters and the output stages. Current draw for the AIE is approximately 35 mA from a +13.6 V dc source, if the LED pilot light (D1) is included. D1 draws 10 mA by itself, however, and R21 and D1 can be deleted if current consumption is a concern. Supply voltage is not critical — anything from 12 to 18 V dc will work.

Capacitors C1 and C10 are used to prevent

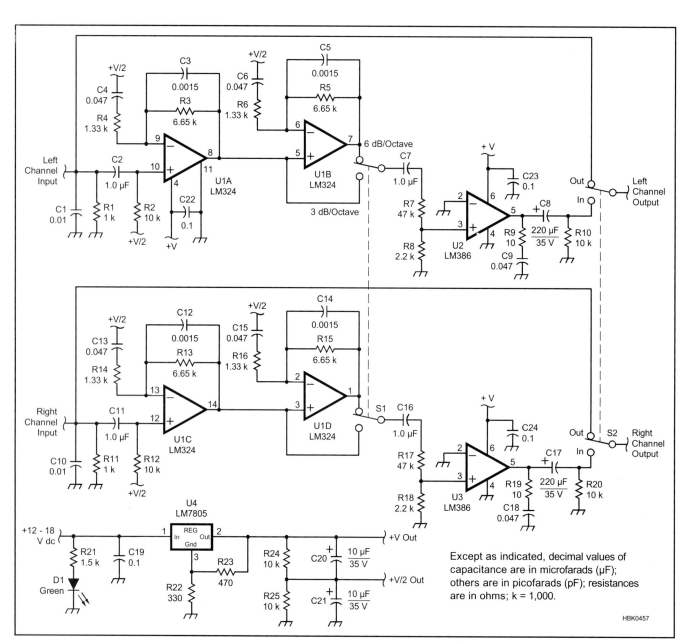

Fig 24.60 — The audio intelligibility enhancer schematic. Resistors are ¼ W. Parts are available from RadioShack, (www.radioshack.com) and Mouser Electronics (www.mouser.com) as well as many other suppliers. The part numbers shown in parentheses are RadioShack unless otherwise noted.

C1, C10 — 0.01 µF, 500 V ceramic (272-131).
C2, C7, C11, C16 — 1.0 µF, 250 V metal film (272-1055).
C3, C5, C12, C14 — 0.0015 µF, 100 V ceramic (Mouser 80-CK12BX152K).
C4, C6, C9, C13, C15, C18 — 0.047 µF, 50 V, metal film (272-1068).
C8, C17 — 220 µF, 35 V, electrolytic (272-1029).
C19, C22, C23, C24 — 0.1 µF, 50 V ceramic (272-109).
C20, C21 — 10 µF, 35 V, electrolytic (272-1025).
D1 — LED with holder, green (276-069).
S1, S2 — Miniature DPDT (275-626).
U1 — LM324 quad operational amplifier (276-1711).
U2, U3 — LM386 audio amplifier (276-1731).
U4 — LM7805, +5 V regulator (276-1770).
Project box, 6×2×1 in (270-1804) (see text).
Grid-style PC board, 0.1-in centers (276-158) (see text).

RFI, by bypassing to ground RF that may enter at the inputs. The original AIE without these capacitors suffered interference from loud distorted sounds (unrectified SSB RF) during transmit. C1 and C10 eliminated RFI problems, but they could be insufficient in the presence of large RF fields. If necessary, additional RF filtering can be achieved by adding RF chokes in series with both audio input and output lines.

TAILORING THE DESIGN TO YOUR HEARING

The AIE design presented here is intended to address high frequency hearing loss, and not other hearing deficencies. This AIE design does, however, have a reasonable amount of flexibility for tailoring to an individual's hearing.

The design as it is presented in Fig 24.60 includes two shelving filters per channel and switch S1. S1 allows selection of one (3 dB/octave amplification) or two (6 dB/octave amplification) filter stages. If you are the only operator of your station and you know your hearing deficiency is less than about 5 dB/octave, the switch and the second filter in each channel can simply be eliminated. If you are unsure of how much high frequency amplification you might need, or in the case where more than one operator may be sharing a station, the switch and the second stages will prove useful. S1 could also be replaced with two switches — one for each channel — if your high frequency hearing is not similar in both ears.

For gain slopes other than 3 dB/octave or 6 dB/octave, one or both filters in either channel can be modified using the equations given earlier. Taking this further, one or both of the op amps in either channel do not have to be shelving filters — they can be used to implement a variety of other filter functions including notching, peaking, or multi-pole functions. Op-amp filter design is covered extensively in circuit design handbooks and on the Internet.

In addition to tailoring the gain change versus frequency, the overall gain of each channel can be adjusted to compensate for the sensitivity differences between ears. Overall gain is adjusted by changing the voltage dividers R7+R8 and R17+R18. R8 and R18 can also be replaced with 5-kΩ potentiometers if adjustable gain is desired.

When considering changing the design, it's important to know that human speech recognition is remarkably tolerant of deviations from an ideal amplitude versus frequency response curve. Also, personal preference comes into play — you may hear more comfortably with less than complete compensation for high frequency hearing loss. Although 6 dB/octave compensation closely matches the author's hearing loss, he usually has the AIE switched to 3 dB/octave compensation. At this setting, intelligibility is improved considerably.

CONSTRUCTION

With a few exceptions, the layout of the component parts is not critical. The leads on the IC power decoupling capacitors (C22, C23, C24) and all leads to the op amp inputs should be kept as short as possible. Be sure to make the U1 power connections at pins 4 and 11.

The completed AIE is shown in **Figs 24.61** and **24.62**. The prototype is built on a grid style PC board with 0.1 inch centers — these are available from RadioShack — cut to fit snugly in a 6×2×1 inch plastic project box. Parts numbers for the PC board and the project box are included in the parts list. A PC board is available from FAR Circuits (**www.farcircuits.net**).

Two cautions are in order. First, building an AIE as small as the prototype requires small tools, a large magnifying glass and patience. Since the circuit design is not particularly sensitive to layout, consider using a larger layout and a larger enclosure. The second caution is with regard to using a plastic enclosure. Susceptibility to RFI and 60 Hz hum pickup will be greatly reduced by using a metal enclosure. If you have lots of RF in your shack you may have no choice but to use a metal enclosure.

Several of the components are not available from RadioShack, but they are readily available from mail order suppliers and can be found in a well-stocked junk box. Several substitutions can be made to yield an "all Radio Shack" design. The 1.33-kΩ resistors can be made from a 1-kΩ and a 330 Ω resistor in series; the 6.65-kΩ resistors can be made from two 3.3-kΩ resistors in series and the 1.5-kΩ resistor can be made from a 1-kΩ and a 470 Ω resistor in series. The value of the 0.0015 µF feedback capacitors is not critical (C3, C5,

Fig 24.61 — A close-up of the AIE before mounting in the project box. It all fits neatly within a 6×2×1 inch enclosure if you have the patience to build within tight spaces.

Fig 24.62 — The AIE takes up little desk space. It plugs inline between the transceiver and headphones and requires 13.6 V dc. In the author's shack, the unit is powered from a transceiver accessory jack and is always switched on.

C12, C14) — 0.001 µF disk ceramics can be substituted (RadioShack 272-126).

HOW WELL DOES IT WORK?

Intelligibility is absolutely crucial to maintaining high QSO rates and correct logging in contests, and in pulling the weak ones out of the noise when DXing or operating in less-than-ideal conditions. The AIE has provided the author with an immense improvement in both activities. It's just as helpful for reducing listening fatigue during ragchewing or general listening as well.

A real surprise was how much intelligibility improved on CW. This may seem counterintuitive at first, if you view "continuous wave" as comprising a single frequency. From a communications standpoint, however, the CW waveform is not continuous — it is comprised of elements — dots and dashes. The leading and trailing edges of each element are made up of frequencies higher than the carrier frequency, and hearing these helps the listener separate one element from the next. When the rise and/or fall times of the elements are very fast compared to the carrier frequency, the high frequency content of the waveform becomes sufficiently large to cause key clicks. Very slow rise and/or fall times produce the opposite of clicks — a sound that experienced CW operators describe as "mushy." Mushy code is very difficult to copy, particularly at high speeds, because the ability to decipher when an element has started or stopped has been reduced. Degraded high frequency hearing has the same detrimental effect on CW intelligibility as slowing down the waveform edges — it makes a properly formed CW waveform sound mushy. The AIE restores perception of the waveform edges in CW, just as it does for speech.

How much an AIE or any other approach to compensating for your hearing deficiencies will help you depends, to a large degree, on the state of your hearing. The more you have lost, the more you have to gain back. Audio shaping may also offer some intelligibility improvement if your hearing is within normal limits. Typical receiver bandwidths are narrower than the audio range needed to fully support intelligibility — some of this can be gained back with high frequency amplification.

One cautionary note from Jim Collins, N9FAY, a designer in the hearing aid industry. At low volume levels, a device such as the AIE is helpful in improving intelligibility. At high volume levels, the audio will start to sound "tinny." This happens because as sound levels increase, the corrective frequency curve must flatten. At upper sound levels of comfortable hearing, it becomes almost a flat line with unity gain, In other words, the hearing response is close to normal at high sound levels. If a device like the AIE is used for prolonged periods at high volume levels, it may accelerate hearing loss at upper frequencies. For more discussion, see Technical Correspondence in May 2005 *QST*, p 70.

24.15 An Audio Interface Unit for Field Day and Contesting

For Field Day and multioperator contest operation, it's sometimes desirable for a logger or observer to listen to a transceiver in addition to the main operator. The project shown in **Fig 24.63**, by John Raydo, KØIZ, makes two-person operation more efficient. Dual stereo headphone jacks, volume controls and an external speaker jack are provided, as well as jacks for both dynamic and electret microphones and an intercom feature for operator and logger to more easily communicate via headset mics.

There are times when the operator and logger need to talk to each other, for example to resolve a call-sign ambiguity. With this unit the operator can initiate two-way intercom communication by closing an intercom switch. The logger, however, can only initiate one-way intercom to the operator. This is so that the logger can't inadvertently cut off the operator during a contact. To respond, the operator closes his intercom switch. In both instances, radio volume is reduced for "talk-over" during intercom use.

Both the operator and logger need headsets with microphones. The unit has two intercom switches plus jacks for optional foot switches for hands-free operation. Here's how it works:

Operator communicating with the logger: The operator presses and holds the OPERATOR TALK switch (S3 on the unit or by using a foot switch connected to J4). This disconnects the operator mic from radio. Both mics are con-

Fig 24.63 — Front view of the audio interface box. Operator controls are on the left and logger controls are on the right.

nected via an internal amplifier to both headphones. Radio volume is reduced (adjustable using the RADIO VOL control). Operator and logger can talk to each other.

Logger communicating with the operator: The logger presses and holds the LOGGER TALK switch (S4 on the unit or by using a foot switch connected to J5). Logger mic is connected via an internal amplifier to both headphones. Operator mic remains connected to radio and can still transmit on the radio. Radio volume is reduced. The operator can respond to the logger by pressing the OPERATOR TALK switch.

CIRCUIT DESCRIPTION

The circuit diagram shown in **Fig 24.64** is designed for use with dynamic microphones such as the HC4 and HC5 elements used in various Heil headsets, or for Heil iC type headsets that use an electret element. The electret mic jacks supply the required dc for that type microphone. Mic bias current is limited to less than 1 mA to avoid damage if a dynamic mic is accidentally plugged into the wrong jack. The higher electret output is reduced by resistors R11 and R28.

Both mics have preamplifiers with 34 dB gain set by the ratio of the resistors R4/R3 and R9/R8. The INTERCOM volume controls, R14 and R15, allow operator and logger to individually adjust their intercom volumes. The preamps feed a single 2-W intercom amplifier.

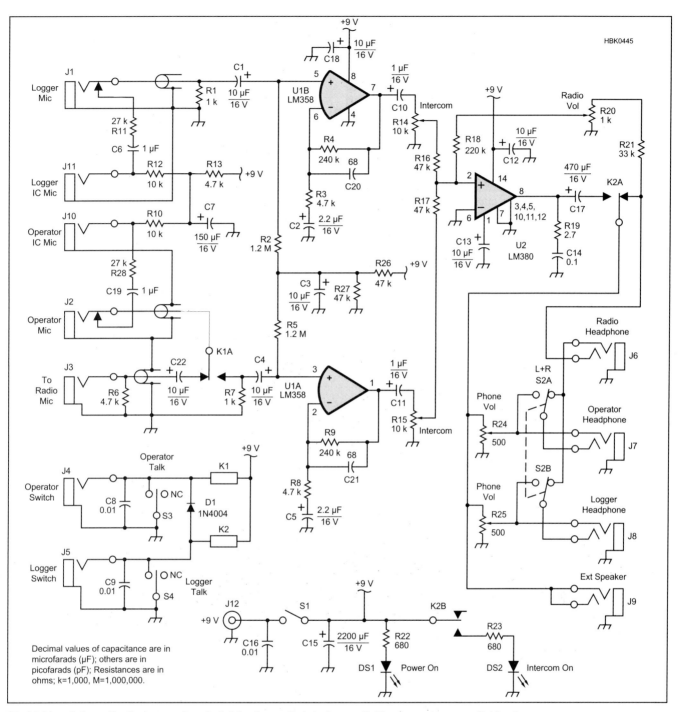

Fig 24.64 — Schematic diagram and parts list for the audio interface unit. Mouser parts are available at www.mouser.com, RadioShack parts at www.radioshack.com. Resistors are ⅛ W unless notes (¼ W are fine too.) A PC board is available from FAR Circuits (www.farcircuits.net).

C1, C3, C4, C12, C13, C18, C22 — 10 μF, 16 V electrolytic.
C2, C5 — 2.2 μF, 16 V electrolytic.
C6, C10, C11, C19 — 1 μF, 16 V electrolytic.
C7 — 150 μF, 16 V electrolytic.
C8, C9, C16 — 0.01 μF, 50 V ceramic.
C14 — 0.1 μF, 50 V ceramic.
C15 — 2200 pF, 16 V electrolytic.
C17 — 470 μF, 18 V electrolytic.
C20, C21—68 pF ceramic.
D1 — 1N4004.
DS1 — Red LED.
DS2 — Yellow LED.
J1-J5, J10, J11 — 3.5 mm mono jack, Mouser 16PJ135. Note: This is a switched jack. The switch contact is not connected on J3-J5, J10 and J11.
J6, J9 — 3.5 mm stereo jack, Mouser 161-3402E.
J7, J8 — Three conductor ¼ inch jack, Mouser 502-12B.
J12 — 2.5 mm jack for dc input, Mouser 502-712A.
K1, K2 — DPDT relay, 9 V coil, Mouser 653-G6A-274P40-DC9.
R14, R15 — 10-kΩ potentiometer, Mouser 31JA401-F.
R20 — 1-kΩ potentiometer, Mouser 31JN301-F.
R24, S1 — 500-Ω potentiometer with SPST switch, Mouser 31VM205-F.
R25 — 500-Ω, potentiometer, Mouser 31VA205-F.
S2 — DPDT toggle switch, Mouser 10TC260.
S3, S4 — SPDT momentary contact switch, center off, with long handle, Mouser 10TA445.
U1 — Dual op amp, LM358, Mouser 595-LM358P.
U2 — Audio amplifier, LM380 or NTE740A, Mouser 526-NTE740A.
Project box, 8 × 4.25 × 2.25 in, Mouser 616-72004-510-000.
Circuit board, RadioShack 276-158.

Station Accessories 24.41

Fig 24.65 — Rear view of the audio interface box. Operator jacks are on the right, logger on the left. Connections to the radio are in the center.

Fig 24.66 — Inside view of the audio interface box. A RadioShack project board is used for most components.

Closing the OPERATOR TALK switch transfers the operator mic from radio to intercom preamp and amplifier via K1 and then to the headphones via K2. Pressing the LOGGER TALK switch will also activate K2 (but not K1). During intercom use, audio from the radio passes through control R20 (RADIO VOL) at reduced volume. S2 (L+R) allows selection of both left and right sides of the headphones or only the right side, for the intercom. The operator and logger each have individual headphone PHONE VOL controls R24 and R25 (the operator control also incorporates the POWER switch, S1).

CONSTRUCTION

The unit shown is built in a PacTec plastic box. The front panel has the control switches and volume controls described in the previous section. The back panel (**Fig 24.65**) has ¼-inch stereo jacks for two pair of headphones, two 3.5-mm mono jacks for headset microphones, 3.5-mm stereo and mono jacks for connection to the radio's headphone and mic jacks, one stereo 3.5-mm jack for a speaker, two 3.5-mm mono jacks for foot switches and a jack for a 9-V power supply (a wall wart is okay). A 12-V dc supply can be substituted if K1 and K2 are changed to 12-V units and the LED series resistors R22 and R23 are changed to 1 kΩ.

Other than controls and jacks, most components are mounted on a RadioShack prototype circuit board supported by four ½ inch spacers and shown in the box in **Fig 24.66**. A suggested parts layout is shown in **Fig 24.67**. Point-to-point wiring is used between circuit board pads and is not particularly critical. A PC board for this project has been developed by FAR Circuits (**www.farcircuits.net**).

Connections to both mic jacks and the radio mic jack use shielded cable to reduce hum and risk of interference pickup. Connect the shields on the operator and radio mic cables to a common ground point on the circuit board near U1. Bypass capacitors C8, C9 and C16 and resistor R6 are mounted at their respective jacks. Rub-on lettering, followed by a couple light coats of clear plastic spray, helps give a professional appearance.

Connecting an external speaker in addition to two headphones might be too much of a load for some radios and cause reduced volume and/or distortion. Try an inexpensive amplified computer speaker plugged into the audio interface external speaker jack to reduce the load.

You will need two cables to connect the unit to your radio. A 3.5 mm mono plug-to-plug shielded cable for the mic circuit is compatible with a Heil headset mic. A 3.5 mm stereo plug-to-plug cable (RadioShack 42-2387 plus 274-367 adapter) works for the earphones.

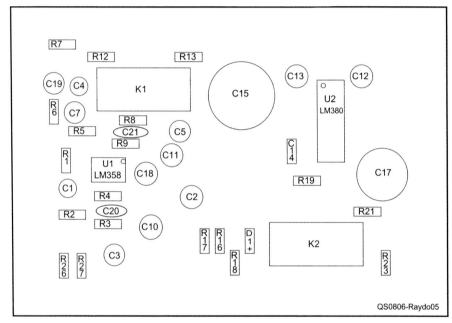

Fig 24.67 — Suggested parts layout for audio interface box.

Contents

25.1 Test and Measurement Basics
    25.1.1 A Short History of Standards and Traceability
    25.1.2 Basic Units: Frequency and Time
    25.1.3 Time and Frequency Calibration

25.2 DC Instruments and Circuits
    25.2.1 Basic Meters
    25.2.2 Voltmeters
    25.2.3 DC Measurement Circuits
    24.2.4 Electronic Voltmeters

25.3 AC Instruments and Circuits
    25.3.1 Calorimetric Meters
    25.3.2 Thermocouple Meters and RF Ammeters
    25.3.3 Rectifier Instruments
    25.3.4 RF Voltage
    25.3.5 RF Power
    25.3.6 *Project*: The Microwatter
    25.3.7 Directional Wattmeters
    25.3.8 AC Bridges

25.4 Frequency Measurement
    25.4.1 Frequency Marker Generators
    25.4.2 *Project*: A Marker Generator with Selectable Output
    25.4.3 Dip Meters
    25.4.4 *Project*: A Dip Meter with Digital Display
    25.4.5 Frequency Counters

25.5 Other Instruments and Measurements
    25.5.1 RF Oscillators for Circuit Alignment
    25.5.2 Audio-Frequency Oscillators
    25.5.3 *Project*: A Wide-Range Audio Oscillator
    25.5.4 *Project*: Measure Inductance and Capacitance with a DVM
    25.5.5 Oscilloscopes
    25.5.6 *Project*: An HF Adapter for Narrow-Bandwidth Oscilloscopes
    25.5.7 *Project*: A Calibrated Noise Source
    25.5.8 *Project*: A Signal Generator for Receiver Testing
    25.5.9 *Project*: Hybrid Combiners for Signal Generators
    25.5.10 *Project*: A Compensated, Modular RF Voltmeter

25.6 Receiver Performance Tests
    25.6.1 Receiver Sensitivity
    25.6.2 Dynamic Range
    25.6.3 Two-Tone IMD Test
    25.6.4 Third-Order Intercept

25.7 Spectrum Analyzers
    25.7.1 Time and Frequency Domain
    25.7.2 Spectrum Analyzer Basics
    25.7.3 Spectrum Analyzer Performance Specifications
    25.7.4 Spectrum Analyzer Applications
    25.7.5 Extensions of Spectral Analysis

25.8 Transmitter Performance Tests
    25.8.1 Spurious Emissions
    25.8.2 Two-Tone IMD
    25.8.3 Carrier and Unwanted Sideband Suppression
    25.8.4 Phase Noise
    25.8.5 Tests in the Time Domain

25.9 Test Equipment and Measurements Glossary

Chapter 25

Test Equipment and Measurements

This chapter, written by ARRL Technical Advisor Doug Millar, K6JEY, covers the test equipment and measurement techniques common to Amateur Radio. With the increasing complexity of amateur equipment and the availability of sophisticated test equipment, measurement and test procedures have also become more complex. There was a time when a simple bakelite cased volt-ohm meter (VOM) could solve most problems. With the advent of modern circuits that use advanced digital techniques, precise readouts and higher frequencies, test requirements and equipment have changed.

25.1 Test and Measurement Basics

The process of testing requires a knowledge of what must be measured and what accuracy is required. If battery voltage is measured and the meter reads 1.52 V, what does this number mean? Does the meter always read accurately or do its readings change over time? What influences a meter reading? What accuracy do we need for a meaningful test of the battery voltage?

25.1.1 A Short History of Standards and Traceability

Since early times, people who measured things have worked to establish a system of consistency between measurements and measurers. Such consistency ensures that a measurement taken by one person could be duplicated by others — that measurements are reproducible. This allows discussion where everyone can be assured that their measurements of the same quantity would have the same result. In most cases, and until recently, consistent measurements involved an artifact: a physical object. If a merchant or scientist wanted to know what his pound weighed, he sent it to a laboratory where it was compared to the official pound. This system worked well for a long time, until the handling of the standard pound removed enough molecules so that its weight changed and measurements that compared in the past no longer did so.

Of course, many such measurements depended on an accurate value for the force of gravity. This grew more difficult with time because the outside environment— such things as a truck going by in the street — could throw the whole procedure off. As a result, scientists switched to physical constants for the determination of values. As an example, a meter was defined as a stated fraction of the circumference of the Earth over the poles.

Generally, each country has an office that is in charge of maintaining the integrity of the standards of measurement and is responsible for helping to get those standards into the field. In the United States that office is the National Institute of Standards and Technology (NIST), formerly the National Bureau of Standards. The NIST decides what the volt and other basic units should be and coordinates those units with other countries. For a modest fee, NIST will compare its volt against a submitted sample and report the accuracy of the sample. In fact, special batteries arrive there each day to be certified and returned so laboratories and industry can verify that their test equipment really does mean 1.527 V when it says so.

25.1.2 Basic Units: Frequency and Time

Frequency and time are the most basic units for many purposes and the ones known to the best accuracy. The formula for converting one to the other is to divide the known value into 1. Thus the time to complete a single cycle at 1 MHz = 0.000001 s.

The history of the accuracy of time keeping, of course, begins with the clock. Wooden clocks, water clocks and mechanical clocks were ancestors to our current standard: the electronic clock based on frequency. In the 1920s, quartz crystal controlled clocks were developed in the laboratory and used as a standard. With the advent of radio communication time intervals could be transmitted by radio, and a very fundamental standard of time and frequency could be used locally with little effort. Today transmitters in several countries broadcast time signals on standard calibrated frequencies. **Table 25.1** contains the locations and frequencies of some of these stations.

In the 1960s, Hewlett-Packard began selling self-contained time and frequency standards called cesium clocks. In a cesium clock a crystal frequency is generated and multiplied to microwave frequencies. That energy is passed through a chamber filled with cesium gas. The gas acts as a very narrow band-pass filter. The output signal is detected and the crystal oscillator frequency is adjusted automatically so that a maximum of energy is detected. The output of the crystal is thus linked to the stability of the cesium gas and is usually accurate to several parts in 10^{-12}. This is much superior to a crystal oscillator alone; but at close to $40,000 each, cesium frequency standards are a bit extravagant for amateur use.

A rubidium frequency standard is an alternative to the cesium clock. They are not quite as accurate as the cesium, but they are much less expensive, relatively quick to warm up and can be quite small. Older models occasionally appear surplus. As with any precision instrument, it should be checked over and calibrated before use.

Most hams do not have access to cesium or rubidium standards—or need them. Instead we use crystal oscillators. Crystal oscillators provide three levels of stability. The least accurate is a single crystal mounted on a circuit board. The crystal frequency is affected by the temperature environment of the equipment, to the extent of a few parts per million (ppm) per degree Celsius. For example, the frequency of a 10-MHz crystal with temperature stability rated at 3 ppm might vary 60 Hz when temperature of the crystal changes by 2°C. If the crystal oscillator is followed by a frequency multiplier, any variation in the crystal frequency is also multiplied. Even so, the accuracy of a simple crystal oscillator is sufficient for most of our needs and most amateur equipment relies on this technique. For a discussion of crystal oscillators and temperature compensation, look in the **Oscillators and Synthesizers** chapter of this *Handbook*.

The second level of accuracy is achieved when the temperature around the crystal is stabilized, either by an "oven" or other nearby components. Crystals are usually designed to stabilize at temperatures far above any reached in normal operating environments. These oscillators are commonly good to 0.1 ppm per day and are widely used in the commercial two-way radio industry.

The third accuracy level uses a double oven with proportional heating. The two ovens compensate for each other automatically and provide excellent temperature stability. The

Table 25.1
Standard Frequency Stations
(Note: In recent years, frequent changes in these schedules have been common.)

Call Sign	Location	Frequency (MHz)
BSF	Taiwan	5, 15
CHU	Ottawa, Canada	3.330, 7.850, 14.670
FFH	France	2.500
IAM/IBF	Italy	5.000
JJY	Japan	2.5, 5, 8, 10, 15
LOL	Argentina	5, 10
RID	Irkutsk	5.004, 10.004, 15.004
RWM	Moscow	5, 4.996, 9.996, 14.996
WWV/WWVH	USA	2.5, 5, 10, 15, 20
VNG	Australia	2.5, 5
ZSC	South Africa	4.291, 8.461, 12.724 (part time)

ovens must be left on continuously, however, and warm-up requires several days to two weeks.

Crystal *aging* also affects frequency stability. Some crystals change frequency over time (age) so the circuit containing the crystal must contain components to compensate for this change. Other crystals become more stable over time and become excellent frequency standards. Many commercial laboratories go to the expense of buying and testing several examples of the same oscillator and select the best one for use. As a result, many surplus oscillators are surplus for a reason. Nevertheless, a good stable crystal oscillator can be accurate to 1×10^{-9} per day and very appropriate for amateur applications.

25.1.3 Time and Frequency Calibration

Many hams have digital frequency counters, which range from surplus lab equipment to new highly integrated instruments with nearly everything on one chip. Almost all of these are very precise and display nine or more digits. Many are even quite stable. Nonetheless, a 10-MHz oscillator accurate to 1 ppm per month can vary ±10 Hz in one month. This drift rate may be acceptable for many applications, but the question remains: How accurate is it?

This question can be answered by calibrating the oscillator. There are several ways to perform this calibration. The most accurate method compares the unit in question by leaving the oscillator operating, transporting it to an oscillator of known frequency and then making a comparison. A commonly used comparison method connects the output of the calibrated oscillator into the horizontal input of a high frequency oscilloscope, and the oscillator to be measured to the vertical input. Assuming the frequencies of both oscillators are very close, the resulting Lissajous pattern will rotate at the frequency difference between them. By watching the pattern pass a fixed point on the display, one can calculate the resulting drift in parts per million per minute (ppm/min).

Another technique of oscillator calibration uses a VLF phase comparator. This is a special direct-conversion receiver that picks up the signal from WWVB on 60 kHz. Phase comparison is used to compare WWVB with the divided frequency of the oscillator being tested. Many commercial units have a small strip chart printer attached and switches to determine the receiver frequency. Since these 60-kHz VLF Comparator receivers have been largely replaced by units that use Loran signals or rubidium standards, they can be found at very reasonable prices. A very effective 60-kHz antenna can be made by attaching an audio transformer with the low-impedance winding connected to the receiver antenna terminals by way of a series dc blocking capacitor. The high-impedance winding is then connected between ground and a random length of wire. A typical VLF Comparator can track an oscillator well into a few parts in 10^{-10}. This technique directly compares the oscillator with an NIST standard and can even characterize oscillator drift characteristics in ppm per day or week.

Another fairly direct method compares an oscillator with one of the WWV HF signals. The received signal is not immensely accurate, but if the oscillator of a modern HF transceiver is carefully compared, it will be accurate enough for all but the most demanding work.

The last and least accurate way to calibrate an oscillator is to compare it with another oscillator or counter owned by you or another local ham. Unless the calibration of the other oscillator or counter is known, this comparison could be very misleading. True accuracy is not determined by the label of a famous company or impressive looks. Metrologists (people who calibrate and measure equipment) spend more time calibrating oscillators than any other piece of equipment.

25.2 DC Instruments and Circuits

This section discusses the basics of analog and digital dc meters. It covers the design of range extenders for current, voltage and resistance; construction of a simple meter; functions of a digital volt-meter (DVM) and procedures for accurate measurements.

25.2.1 Basic Meters

In measuring instruments and test equipment suitable for amateur purposes, the ultimate readout is generally based on a measurement of direct current. There are two basic styles of meters: analog meters that use a moving needle display, and digital meters that display the measured values in digital form. The analog meter for measuring dc current and voltage uses a magnet and a coil to move a pointer over a calibrated scale in proportion to the current flowing through the meter.

The most common dc analog meter is the D'Arsonval type, consisting of a coil of wire to which the pointer is attached so that the coil moves (rotates) between the poles of a permanent magnet. When current flows through the coil, it sets up a magnetic field that interacts with the field of the magnet to cause the coil to turn. The design of the instrument normally makes the pointer move in direct proportion to the current.

DIGITAL MULTIMETERS

In recent years there has been a flood of inexpensive digital multimeters (DMMs) ranging from those built into probes to others housed in large enclosures. They are more commonly referred to as digital voltmeters (DVMs) even though they are multimeters; they usually measure voltage, current and resistance. After some years of refining circuits such as the "successive approximation" and "dual slope" methods, most meters now use the dual-slope method to convert analog voltages to a digital reading. DVMs have basically three main sections as shown in **Fig 25.1**.

The first section scales the voltage or current to be measured. It has four main circuits:

- a chain of multiplier resistors that reduce the input voltage to 0-1 V,
- a converter that changes 0-1 V ac to dc,
- an amplifier that raises signals in the 0-100 mV range to 0-1 V and
- a current driver that provides a constant current to the multiplier chain for resistance measurements.

The second section is an integrator. It is usually based on an operational amplifier that is switched by a timing signal. The timing signal initially shorts the input of the integrator to provide a zero reference. Next a reference voltage is connected to charge the capacitor for a determined amount of time. Finally the last part of the timing cycle allows the capacitor to discharge. The time it takes the capacitor to discharge is proportional to either the input voltage (V_{in}, after it was scaled into the range of 0 to 1 V) or 1 minus V_{in}, depending on the meter design. This discharge time is measured by the next section of the DVM, which is actually a frequency counter. Finally, the output of the frequency counter is scaled to the selected range of voltage or current and sent to the final section of the DVM — the digital display.

Since the timing is quite fast and the capacitor is not used long enough to drift much in value, the components that most determine accuracy are the reference voltage source and the range multiplier resistors. With the availability of integrated resistor networks that are deposited or diffused onto the same substrate, drift is automatically compensated because all branches of a divider drift in the same direction simultaneously. The voltage sources

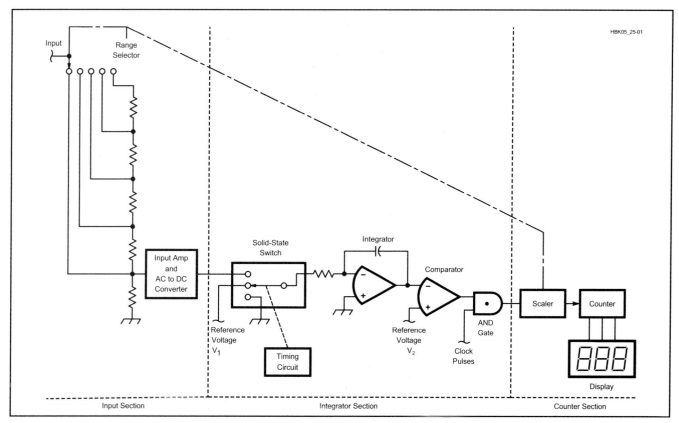

Fig 25.1 — A typical digital voltmeter consists of three parts: the input section for scaling, an integrator to convert voltage to pulse count, and a counter to display the pulse count representing the measured quantity.

are generally Zener diodes on substrates with accompanying series resistors. Often the resistor and Zener have opposite temperature characteristics that cancel each other. In more complex DVMs, extensive digital circuitry can insert values to compensate for changes in the circuit and can even be automatically calibrated remotely in a few moments.

Liquid crystal displays (LCD) are commonly used for commercial DVMs. As a practical matter they draw little current and are best for portable and battery-operated use. The usual alternative, light emitting diode (LED) displays, draw much more current but are better in low-light environments. Some older surplus units use gas plasma displays (orange-colored digits). You may have seen plasma displays on gas-station pumps. They are not as bright as LEDs, but are easier to read. On the down side, plasma displays require high-voltage power supplies, draw considerable current and often fail after 10 years or so.

The advantages of DVMs are high input resistance (10 MΩ on most ranges), accurate and precise readings, portability, a wide variety of ranges and low price. There is one disadvantage, however: Digital displays update rather slowly, often only one to two times per second. This makes it very difficult to adjust a circuit for a peak (maximum) or null (minimum) response using only a digital display. The changing digits do not give any clue of the measurement trend and it is easy to tune through the peak or null between display updates. In answer, many new DVMs are built with an auxiliary bar-graph display that is updated constantly, thus providing instantaneous readings of relative value and direction of changes.

CURRENT RANGES

The sensitivity of an analog meter is usually expressed in terms of the current required for full-scale deflection of the pointer. Although a very wide variety of ranges is available, the meters of interest in amateur work give maximum deflection with currents measured in microamperes or milliamperes. They are called microammeters and milliammeters, respectively.

Thanks to the relationships between current, voltage and resistance expressed by Ohm's Law, it is possible to use a single low-range instrument (for example, 1 mA or less for full-scale pointer deflection) for a variety of direct-current measurements. Through its ability to measure current, the instrument can also be used indirectly to measure voltage. In the same way, a measurement of both current and voltage will obviously yield a value of resistance. These measurement functions are often combined in a single instrument: the volt-ohm-milliammeter or VOM, a multirange meter that is one of the most useful pieces of test equipment an amateur can possess.

ACCURACY

The accuracy of a D'Arsonval-movement dc meter is specified by the manufacturer. A common specification is ±2% of full scale, meaning that a 0-100 µA meter, for example, will be correct to within 2 µA at any part of the scale. There are very few cases in amateur work where accuracy greater than this is needed. When the instrument is part of a more complex measuring circuit, however, the design and components can each cause error that accumulates to reduce the overall accuracy.

EXTENDING CURRENT RANGE

Because of the way current divides between two resistances in parallel, it is possible to increase the range (more specifically to decrease the sensitivity) of a dc current meter. The meter itself has an inherent resistance (its internal resistance) which determines the full-scale current passing through it when its rated voltage is applied. (This rated voltage is on the order of a few millivolts.) When an external resistance is connected in parallel with the meter, the current will divide between the two and the meter will respond only to that part of the current that flows through its movement. Thus, it reads only part of the total current; the effect makes more total current necessary for a full-scale meter reading. The added resistance is called a "shunt."

We must know the meter's internal resistance before we can calculate the value for a shunt resistor. Internal resistance may vary from a fraction of an ohm to a few thousand ohms, with greater resistance values associated with greater sensitivity. When this resistance is known, it can be used in the formula below to determine the required shunt for a given multiplication:

$$R = \frac{R_m}{n-1} \qquad (1)$$

where
R = shunt resistance, ohms
R_m = meter internal resistance, ohms
n = the factor by which the original meter scale is to be multiplied.

Often the internal resistance of a particular meter is unknown (when the meter is purchased at a flea market or is taken from a commercial piece of equipment, for example). Unfortunately, the internal resistance of a meter cannot be measured directly with an ohmmeter without risk of damage to the meter movement.

Fig 25.2 shows a method to safely measure the internal resistance of a linearly calibrated meter. It requires a calibrated meter that can measure the same current as the unknown meter. The system works as follows: S1 is switched off and R2 is set for maximum resistance. A supply of constant voltage is connected to the supply terminals (a battery

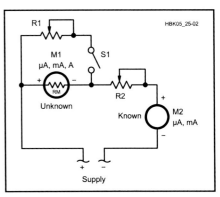

Fig 25.2 — This test setup allows safe measurement of a meter's internal resistance. See text for the procedure and part values.

will work fine) and R2 is adjusted so that the unknown meter reads exactly full scale. Note the current shown on M2. Close S1 and alternately adjust R1 and R2 so that the unknown meter (M1) reads exactly half scale and the known meter (M2) reads the same value as in the step above. At this point, half of the current in the circuit flows through M1 and half through R1. To determine the internal resistance of the meter, simply open S1 and read the resistance of R1 with an ohm-meter.

The values of R1 and R2 will depend on the meter sensitivity and the supply voltage. The maximum resistance value for R1 should be approximately twice the expected internal resistance of the meter. For highly sensitive meters (10 µA and less), 1 kΩ should be adequate. For less sensitive meters, 100 Ω should suffice. Use no more supply voltage than necessary.

The value for minimum resistance at R2 can be calculated using Ohm's Law. For example, if the meter reads 0 to 1 mA and the supply is a 1.5-V battery, the minimum resistance required at R2 will be:

$$R2 = \frac{1.5}{0.001}$$

R2 (min) = 1500 Ω

In practice a 2- or 2.5-kΩ potentiometer would be used.

MAKING SHUNTS

Homemade shunts can be constructed from several kinds of resistance wire or from ordinary copper wire if no resistance wire is available. The copper wire table in the **Component Data and References** chapter of this *Handbook* gives the resistance per 1000 ft for various sizes of copper wire. After computing the resistance required, determine the smallest wire size that will carry the full-scale current, again from the wire table. Measure off enough wire to give the required resistance. A high-resistance 1- or 2-W carbon-composition resistor makes an excellent form on which to wind the wire, as the high resistance does not affect the value of the shunt. If the shunt gets too hot, go to a larger diameter wire of a greater length.

25.2.2 Voltmeters

If a large resistance is connected in series with a meter that measures current, as shown in **Fig 25.3**, the current multiplied by the resistance will be the voltage drop across the resistance. This is known as a multiplier. An instrument used in this way is calibrated in terms of the voltage drop across the multiplier resistor and is called a voltmeter.

SENSITIVITY

Voltmeter sensitivity is usually expressed in ohms per volt (Ω/V), meaning that the meter full-scale reading multiplied by the sensitivity will give the total resistance of the voltmeter. For example, the resistance of a 1 kΩ/V voltmeter is 1000 times the full-scale calibration voltage. Then by Ohm's Law the current required for full-scale deflection is 1 milliampere. A sensitivity of 20 kΩ/V, a commonly used value, means that the instrument is a 50-μA meter.

As voltmeter sensitivity (resistance) increases, so does accuracy. Greater meter resistance means that less current is drawn from the circuit and thus the circuit under test is less affected by connection of the meter. Although a 1000-Ω/V meter can be used for some applications, most good meters are 20 kΩ/V or more. Vacuum-tube voltmeters (VTVMs) and their modern equivalent FET voltmeters (FETVOMs) are usually 10-100 MΩ/V and DVMs can go even higher.

MULTIPLIERS

The required multiplier resistance is found by dividing the desired full-scale voltage by the current, in amperes, required for full-scale deflection of the meter alone. To be mathematically correct, the internal resistance of the meter should be subtracted from the calculated value. This is seldom necessary (except perhaps for very low ranges) because the meter resistance is usually very low compared with the multiplier resistance. When the instrument is already a voltmeter with an internal multiplier, however, the meter resistance is significant. The resistance required to extend the range is then:

$$R = R_m (n - 1) \qquad (2)$$

where
 R_m = total resistance of the instrument
 n = factor by which the scale is to be multiplied

For example, if a 1-kΩ/V voltmeter having a calibrated range of 0 to 10 V is to be extended to 1000 V, R_m is 1000 × 10 = 10 kΩ, n is 1000/10 = 100 and R = 10,000 × (100 – 1) = 990 kΩ.

When extending the range of a volt-meter or converting a low-range meter into a voltmeter, the rated accuracy of the instrument is retained only when the multiplier resistance is precise. High-precision, hand-made and aged wire-wound resistors are used as multipliers of high-quality instruments. These are relatively expensive, but the home constructor can do well with 1% tolerance metal-film resistors. They should be derated when used for this purpose. That is, the actual power dissipated in the resistor should not be more than $\frac{1}{10}$ to $\frac{1}{4}$ the rated dissipation. Also, use care to avoid overheating the resistor body when soldering. These precautions will help prevent permanent change in the resistance of the unit.

Many DVMs use special resistor groups that have been etched on quartz or sapphire and laser trimmed to value. These resistors are very stable and often quite accurate. They can be bought new from various suppliers. It is also possible to "rescue" the divider/multiplier resistors from an older DVM that no longer functions and use them as multipliers. Look for a series of four or five resistors that add up to 10 MΩ: 0.9, 9, 90, 900, 9,000, 90,000 and 900,000 Ω. There is usually another 1-MΩ resistor in series to isolate the meter from the circuit under test. A few of these high-accuracy resistors in "odd" values can help calibrate less-expensive instruments.

DC VOLTAGE STANDARDS

For a long time NIST has statistically compared a bank of special Weston Cell or cadmium sulfate batteries to arrive at the standard volt. By using a special tapped resistor, a 1.08-V battery can be compared to other voltages and instruments compared. However, these are very high-impedance batteries that deliver almost no current and are relatively temperature sensitive. They are made up of a solution of cadmium and mercury in opposite legs of an "H" shaped glass container. You can read much more about them in *Calibration—Philosophy and Practice*, published by the John Fluke Co of Mount Lake Terrace, Washington.

Hams often use an ordinary flashlight battery as a convenient voltage reference. A fresh *D cell* usually provides 1.56 V under no load, as would be measured by a DVM. The Heath Company, which supplied thousands of kits to the ham community for many years, used such batteries as the calibration references for many of their kits.

Recently, NIST has been able to use a microwave to voltage converter called a "Josephson Junction" to determine the value of the volt. The converter transfers the accuracy of a frequency standard to the accuracy of the voltage that comes out of it. The converter generates only 5 mV, however, which then must be scaled to the standard 1-V level. One problem with high-accuracy measurements is stray noise (low-level voltages) that creates a floor below which measurements are meaningless. For that reason, meters with five or more digits must be very quiet and any comparisons must be made at a voltage high enough to be above the noise.

25.2.3 DC Measurement Circuits

CURRENT MEASUREMENT WITH A VOLTMETER

A current-measuring instrument should have very low resistance compared with the resistance of the circuit being measured; otherwise, inserting the instrument will alter the current from its value when the instrument is removed. The resistance of many circuits in radio equipment is high and the circuit operation is affected little, if at all, by adding as much as a few hundred ohms in series. [Even better, use a resistor that is part of the working circuit if one exists. Unsolder one end of the resistor, measure its resistance, reinstall it and then make the measurement. — *Ed.*] In such cases the voltmeter method of measuring current in place of an ammeter, shown in **Fig 25.4**, is frequently convenient. A voltmeter (or low-range milliammeter provided with a multiplier and operating as a voltmeter) having a full-scale voltage range of a few volts is used to measure the voltage drop across a suitable value of resistance acting as a shunt.

The value of shunt resistance must be calculated from the known or estimated maximum current expected in the circuit (allowing a safe margin) and the voltage required for full-scale movement of the meter with its multiplier. For example, to measure a current estimated at 15 A on the 2-V range of a DVM, we need to solve Ohm's Law for the value of R:

$$R_{SHUNT} = \frac{2\text{ V}}{15\text{ A}} = 0.133\ \Omega$$

This resistor would dissipate $15^2 \times 0.133 = 29.92$ W. For a short-duration measurement, 30 1.0-Ω, 1-W resistors could

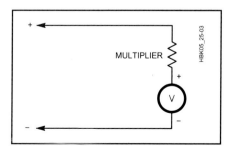

Fig 25.3 — A voltmeter is constructed by placing a current-indicating instrument in series with a high resistance, the "multiplier."

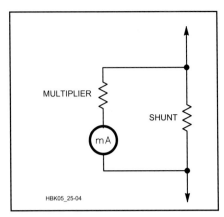

Fig 25.4 — A voltmeter can be used to measure current as shown. For reasonable accuracy, the shunt should be 5% of the circuit impedance or less, and the meter resistance should be 20 times the circuit impedance or more.

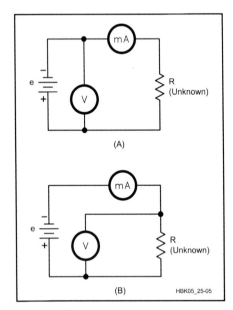

Fig 25.5 — Power or resistance can be calculated from voltage and current measurements. At A, error introduced by the ammeter is dominant. At B, error introduced by the voltmeter is dominant. The text gives an example.

The setup for measuring power is shown in **Fig 25.5A**, where R is any dc load, not necessarily an actual resistor. In this measurement it is always best to use the lowest voltmeter or ammeter scale that allows reading the measured quantity. This results in the percentage error being less than if the meter was reading in the very lowest part of the selected scale.

RESISTANCE

If both voltage and current are measured in a circuit such as that in Fig 25.5, the value of resistance R (in case it is unknown) can be calculated from Ohm's Law. For reasonable results, two conditions should be met:

1. The internal resistance of the current meter should be less than 5% of the circuit resistance.
2. The input impedance of the voltmeter should be greater than 20 times the circuit resistance.

These conditions are important because both meters tend to load the circuit under test. The current meter resistance adds to the unknown resistance, while the voltmeter resistance decreases the unknown resistance as a result of their parallel connection.

OHMMETERS

Although Fig 25.5B suffices for occasional resistance measurements, it is inconvenient when we need to make frequent measurements over a wide range of resistance. The device generally used for this purpose is the ohmmeter. Its simplest form is a voltmeter (or milliammeter, depending on the circuit used) and a small battery. The meter is calibrated so that the value of an unknown resistance can be read directly from the scale. **Fig 25.6** shows some typical ohmmeter circuits. In the simplest circuit, Fig 25.6A, the meter and battery are connected in series with the unknown resistance. If a given movement of the meter's needle is obtained with terminals A-B shorted, inserting the resistance to be measured will cause the meter reading to decrease. When the resistance of the voltmeter is known, the following formula can be applied.

$$R = \frac{eR_m}{E} - R_m \quad (3)$$

where
R = unknown resistance, ohms
e = voltage applied (A-B shorted)
E = voltmeter reading with R connected
R_m = resistance of the voltmeter.

The circuit of Fig 25.6A is not suited to measuring low values of resistance (less than 100 Ω or so) with a high-resistance voltmeter. For such measurements the circuit of Fig 25.6B is better. The unknown resistance is

$$R = \frac{I_2 R_m}{I_1 - I_2} \quad (4)$$

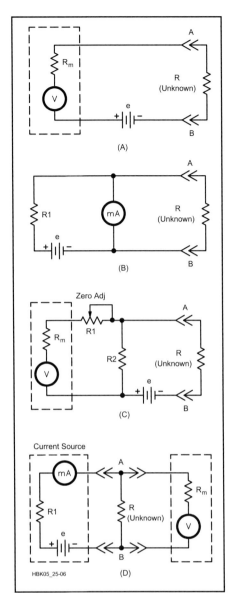

Fig 25.6 — Here are several kinds of ohmmeters. Each is explained in the text.

where
R = unknown resistance, ohms
R_m = the internal resistance of the milliammeter, ohms
I_1 = current with R disconnected from terminals A-B, amps
I_2 = current with R connected, amps.

This formula is based on the assumption that the current in the complete circuit will be essentially constant whether or not the unknown terminals are short circuited. This requires that R1 be much greater than R_m. For example, 3000 Ω for a 1-mA meter with an internal resistance of perhaps 50 Ω. In this case, a 3-V battery would be necessary in order to obtain a full-scale deflection with the unknown terminals open. R1 can be an adjustable resistor, to permit setting the open-terminal current to exact full scale.

be parallel connected in two groups of 15 (0.067 Ω per group) that are series connected to yield 0.133 Ω. For long-duration measurements, 2- to 5-W resistors would be better.

POWER

Power in direct-current circuits is usually determined by measuring the current and voltage. When these are known, the power can be calculated by multiplying voltage, in volts, by the current, in amperes. If the current is measured with a milliammeter, the reading of the instrument must be divided by 1000 to convert it to amperes.

A third circuit for measuring resistance is shown in Fig 25.6C. In this case a high-resistance voltmeter is used to measure the voltage drop across a reference resistor, R2, when the unknown resistor is connected so that current flows through it, R2 and the battery in series. With suitable R2s (low values for low-resistance, high values for high-resistance unknowns), this circuit gives equally good results for resistance values in the range from one ohm to several megohms. The voltmeter resistance, R_m, must be much greater (50 times or more) than that of R2. A 20-kΩ/V instrument (50-μA movement) is generally used. If the current through the voltmeter is negligible compared with the current through R2, the formula for the unknown is

$$R = \frac{eR2}{E} - R2 \quad (5)$$

where

R and R2 are in ohms
e = voltmeter reading with R removed and A shorted to B.
E = voltmeter reading with R connected.

R1 sets the voltmeter reading exactly to full scale when the meter is calibrated in ohms. A 10-kΩ pot is suitable with a 20-kΩ/V meter. The battery voltage is usually 3 V for ranges to 100 kΩ and 6 V for higher ranges.

FOUR-WIRE RESISTANCE MEASUREMENTS

In situations where a very low resistance, like a 50-Ω dummy load, is to be measured, the resistance of the test leads can be significant. The lead resistance can be as much as several tenths of an ohm which will cause measurements to read high by that amount.

To compensate for lead resistance, some meters allow for four-wire measurements. Briefly, two wires from the current source and two wires from the measuring circuit exit the meter case separately and connect directly to the unknown resistance (see Fig 25.6D). This eliminates the voltage drop in the current-source leads from the measurement. In practice, four-wire systems use special test clips that are similar to alligator clips, except that the jaws are insulated from each other and a meter lead is attached to each jaw. In some meters, an additional control allows the operator to short the test leads together and adjust the meter for a zero reading before making low-resistance measurements.

BRIDGE CIRCUITS

Bridges are an important class of measurement circuits. They perform measurement by comparison with some known component or quantity, rather than by direct reading. VOMs, DVMs and other meters are convenient, but their accuracy is limited. The accuracy of manufactured analog meters is determined at the factory, while digital meters are accurate only to some percentage ±1 in the least-significant digit. The accuracy of comparison measurements, however, is determined only by the comparison standard and bridge sensitivity.

Bridge circuits are useful across most of the frequency spectrum. Most amateur applications are at RF, as shown later in this chapter. The principles of bridge operation are easier to understand at dc, however, where bridge operation is simple.

THE WHEATSTONE BRIDGE

A simple resistance bridge, known as the Wheatstone bridge, is shown in **Fig 25.7**. All other bridge circuits are based on this design. The four resistors, R1, R2, R3 and R4 in Fig 25.7A, are known as the bridge arms. For the voltmeter reading to be zero (null) the voltage dividers consisting of the pairs R1-R3 and R2-R4 must be equal. This means

$$\frac{R1}{R3} = \frac{R2}{R4} \quad (6)$$

When this occurs the bridge is said to be *balanced*.

The circuit is usually drawn as shown at Fig 25.7B when used for resistance measurement. Equation 6 can be rewritten

$$RX = RS\left(\frac{R2}{R1}\right) \quad (7)$$

RX is the unknown resistor. R1 and R2 are usually made equal; then the calibrated adjustable resistance (the standard), RS, will have the same value as RX when RS is set to show a null on the voltmeter.

Note that the resistance ratios, rather than the actual resistance values, determine the voltage balance. The values do have important practical effects on the sensitivity and power consumption, however. Bridge sensitivity is the ability of the meter to respond to slight unbalance near the null point; a sharper null means a more accurate setting of RS at balance. Bridge circuits with narrower bandwidths tend to exhibit a sharper, more precise null, and therefore potentially greater measurement accuracy in contrast to a bridge with a wider bandwidth and a poorly defined or broader null.

The Wheatstone bridge is rarely used by amateurs for resistance measurement, since it is easier to measure resistances with VOMs and DVMs. Nonetheless, it is worthwhile to understand its operation as the basis of more complex bridges.

25.2.4 Electronic Voltmeters

We have seen that the resistance of a simple voltmeter (as in Fig 25.3) must be extremely

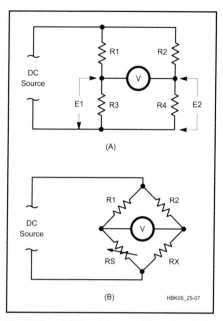

Fig 25.7 — A Wheatstone bridge circuit. A bridge circuit is actually a pair of voltage dividers (A). B shows how bridges are normally drawn.

high in order to avoid "loading" errors caused by the current that necessarily flows through the meter. The use of high-resistance meters tends to cause difficulty in measuring relatively low voltages because multiplier resistance progressively lessens as the voltage range is lowered.

Voltmeter resistance can be made independent of voltage range by using vacuum tubes, FETs or op amps as dc amplifiers between the circuit under test (CUT) and the indicator, which may be a conventional meter movement or a digital display. Because the input resistance of the electronic devices mentioned is extremely high (hundreds of megohms) they have negligible loading effect on the CUT. They do, however, require a closed dc path in their input circuits (although this path can have very high resistance). They are also limited in the voltage level that their input circuits can handle. Because of this, the device actually measures a small voltage across a portion of a high-resistance voltage divider connected to the CUT. Various voltage ranges are obtained from appropriate taps on the voltage divider.

In the design of electronic voltmeters it has become standard practice to use a voltage divider with a total resistance of 10 MΩ, tapped as required, in series with a 1-MΩ resistor incorporated in the meter. The total voltmeter resistance, including probe, is therefore 11 MΩ. The 1-MΩ resistor serves to isolate the voltmeter circuit from the CUT.

25.3 AC Instruments and Circuits

Most ac measurements differ from dc measurements in that the accuracy of the measurement depends on the purity of the sine wave. It is fairly easy to measure an ac voltage to between 1% and 5%, but getting down to 0.01% is difficult. Measurements to less than 0.01% must be left to precision laboratories. In general, amateurs measure ac voltages in household circuits, audio stages and RF power measurements, and 1% to 5% accuracy is usually close enough.

This section covers basic measurements, the nature of sine waves and meters. There are four common ways to measure ac voltage:
- Use a rectifier to change the ac to dc and then measure the dc.
- Heat a resistor in a Wheatstone bridge with the ac and measure the bridge unbalance.
- Heat a resistor surrounded by oil and measure the temperature rise.
- Use electronic circuits (such as multipliers and logarithmic amplifiers) with mathematical ac-to-dc conversion formulas. This method is not common, but it's interesting.

25.3.1 Calorimetric Meters

In a calorimetric meter, power is applied to a resistor that is immersed in the flow path of a special oil. This oil transfers the heat to another resistor that is part of a bridge. As the resistor heats, its resistance changes and the bridge becomes unbalanced. An attached meter registers the unbalance of the bridge as ac power. This type of meter is accurate for both dc and ac. They frequently operate from dc well into the GHz range. For calibration, an accurate dc voltage is applied and the reading is noted; a similar ac voltage is applied and the readings are compared. Some calorimetric meters are complicated, but others are simple.

25.3.2 Thermocouple Meters and RF Ammeters

In a thermocouple meter, alternating current flows through a low-resistance heating element. The power lost in the resistance generates heat that warms a thermocouple, which consists of a pair of junctions of two different metals. When one junction is heated a small dc voltage is generated in response to the difference in temperature of the two junctions. This voltage is applied to a dc milliammeter that is calibrated in suitable ac units. The heater-thermocouple/dc-meter combination is usually housed in a regular meter case.

Thermocouple meters are available in ranges from about 100 mA to many amperes. Their useful upper frequency limit is in the neighborhood of 100 MHz. Amateurs use these meters mostly to measure current through a known load resistance and calculate the RF power delivered to the load.

25.3.3 Rectifier Instruments

The response of a rectifier RF ammeter is proportional (depending on the design) to either the peak or average value of the rectified ac wave, but never directly to the RMS value. These meters cannot be calibrated in RMS without knowing the relationship that exists between the real reading and the RMS value. This relationship may not be known for the circuit under test.

Average-reading ac meters work best with pure sine waves and RMS meters work best with complicated wave forms. Since many practical measurements involve nonsinusoidal forms, it is necessary to know what your instrument is actually reading, in order to make measurements intelligently. Most VOMs and VTVMs use averaging techniques, while DVMs may use either one. In all cases, check the meter instruction manual to be sure what it reads.

PEAK AND AVERAGE WITH SINE-WAVE RECTIFICATION

Peak, average and RMS values of ac waveforms are discussed in the **Electrical Fundamentals** chapter. Because the positive and negative half cycles of the sine wave have the same shape, half-wave rectification of either the positive half or the negative half gives exactly the same result. With full-wave rectification, the peak reading is the same, but the average reading is doubled, because there are twice as many half cycles per unit of time.

ASYMMETRICAL WAVE FORMS

A nonsinusoidal waveform is shown in **Fig 25.8A**. When the positive half cycles of this wave are rectified, the peak and average values are shown at B. If the polarity is reversed and the negative half cycles are rectified, the result is shown in Fig 25.8C. Full-wave rectification of such a lopsided wave changes the average value, but the peak reading is always the same as that of the half cycle that produces the highest peak in half-wave rectification.

EFFECTIVE-VALUE CALIBRATION

The actual scale calibration of commercially made rectifier voltmeters is very often (almost always, in fact) in terms of RMS values. For sine waves, this is satisfactory and useful because RMS is the standard measurement at power-line frequencies. It is also useful for many RF applications when the waveform is close to sinusoidal. In other cases, particularly in the AF range, the error may be considerable when the waveform is not pure.

TURN-OVER

From Fig 25.8 it is apparent that the calibration of an average-reading meter will be the same whether the positive or negative sides are rectified. A half-wave peak-reading instrument, however, will indicate different values when its connections to the circuit are reversed (turn-over effect). Very often readings are taken both ways, in which case the sum of the two is the peak-to-peak (P-P) value, a useful figure in much audio and video work.

AVERAGE- VS PEAK-READING CIRCUITS

For traditional analog displays, the basic difference between average- and peak-reading rectifier circuits is that the output is not filtered for averaged readings, while a filter capacitor is charged to the peak value of the output voltage in order to measure peaks. **Fig 25.9A** and **B** show typical average-reading circuits, one half-wave and the other full-wave. In the absence of dc filtering, the meter responds to wave forms such as those shown at B, C and D in Fig 25.8; and since the inertia of the pointer system makes it unable to follow the rapid variations in current, it averages them out mechanically.

In Fig 25.9A, D1 actuates the meter; D2 provides a low-resistance dc return in the meter circuit on the negative half cycles. R1 is the voltmeter multiplier resistance. R2 forms a voltage divider with R1 (through D1) that prevents more than a few ac volts from appear-

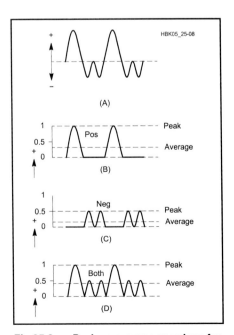

Fig 25.8 — Peak vs average ac values for an asymmetrical wave form. Note that the peak values are different with positive or negative half-cycle rectification.

ing across the rectifier-meter combination. A corresponding resistor can be used across the full-wave bridge circuit.

In these two circuits there is no provision to isolate the meter from any dc voltage in the circuit under test. The resulting errors can be avoided by connecting a large nonpolarized capacitor in series with the hot lead. The reactance must be low compared with the meter impedance (at the lowest frequency of interest, more on this later) in order for the full ac voltage to be applied to the meter circuit. Some meters may require as much as 1 µF at line (60 Hz) frequencies. Such capacitors are usually not included in VOMs.

Voltage doubler and shunt peak-reading circuits are shown in Fig 25.9C and D. In both circuits, C1 isolates the rectifier from dc voltage in the circuit under test. In the voltage-doubler circuit, the time constant of the C2-R1-R2 combination must be very large compared with the period of the lowest ac frequency to be measured; similarly with C1-R1-R2 in the shunt circuit. This is so because the capacitor is charged to the peak value (V_{P-P} in C, V_P in D) when the ac wave reaches its maximum and then must hold the charge (so it can register on a dc meter) until the next maximum of the same polarity. If the time constant is 20 times the ac period, the charge will have decreased by about 5% when the next charge occurs. The average voltage drop will be smaller, so the error is appreciably less. The error will decrease rapidly with increasing frequency (if there is no change in the circuit values), but it will increase at lower frequencies.

In Fig 25.9C and D, R1 and R2 form a voltage divider that reduces the voltage to some desired value. For example, if R1 is 0 Ω in the voltage doubler, the voltage across R2 is approximately V_{P-P}; if R1 = R2, the output is approximately V_P (as long as the waveform is symmetrical).

The most common application of the shunt circuit is an RF probe to read V_{RMS}. In that case, R2 is the input impedance of a VTVM or DVM: 11 MΩ. R1 is chosen so that 71% of the peak value appears across R2. This converts the peak reading to RMS for sine-wave ac. R1 is therefore approximately 4.7 MΩ, making the total resistance nearly 16 MΩ. A capacitance of 0.05 µF is sufficient for low audio frequencies under these conditions. Much smaller values of capacitance may be used at RF.

VOLTMETER IMPEDANCE

The impedance of a voltmeter at the frequency being measured may have an effect on the accuracy similar to that caused by the resistance of a dc voltmeter, as discussed earlier. The ac meter is a resistance in parallel with a capacitance. Since the capacitive re-

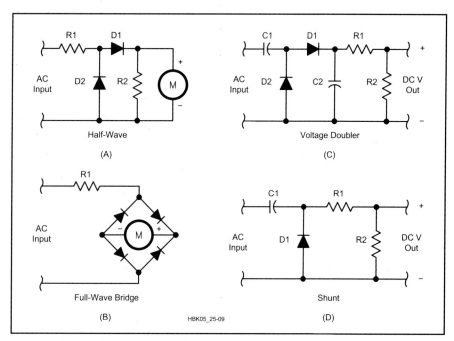

Fig 25.9 — Half (A) and full-wave (B) bridge rectifiers for average-reading, analog-display meters. Peak-reading circuits: a voltage doubler (C) and a shunt circuit (D). All circuits are discussed in the text.

actance decreases with increasing frequency, the impedance also decreases with frequency. The dc current drawn by the rectifier increases with voltage level in a non-linear manner, particularly at very low voltages (10 V or less) depending on the sensitivity of the meter and the kind of rectifier used.

In Fig 25.9A the dc load is essentially the meter resistance, which is generally quite low compared with the multiplier resistance R1. Hence, the total resistance will be about the same as the multiplier resistance. The capacitance depends on the components and construction, test-lead length and location, and other such factors. In general, the capacitance has little or no effect at lower line and audio frequencies, but ordinary VOMs lose accuracy at high audio frequencies and are of little use at RF. Rectifiers with very low inherent capacitance are used at RF and they are usually located at the probe tip to reduce losses.

Similar limitations apply to peak-reading circuits. In the shunt circuit, the resistive part of the impedance is smaller than in the voltage-doubler circuit because the dc load resistance, R1/R2, is directly across the circuit under test and in parallel with the diode ac load resistance. In both peak-reading circuits the effective capacitance may range from 1 or 2 to a few hundred pF, with 100 pF typical in most instruments.

SCALE LINEARITY

Fig 25.10 shows a typical current/voltage chart for a small semiconductor diode, which shows that the forward dynamic resistance of the diode is not constant, but rapidly decreases as the forward voltage increases from zero. The change from high to low resistance happens at much less than 1 V, but is in the range of voltage needed for a dc meter. With an average-reading circuit the current tends to be proportional to the square of the applied voltage. This makes the readings at the low end of the meter scale very crowded. For most measurement purposes, however,

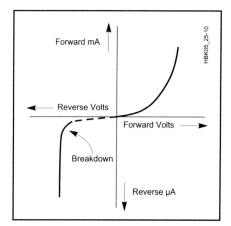

Fig 25.10 — Voltage vs current characteristics for a typical semiconductor diode. Actual values vary with different part numbers, but the forward current will always be increasing steeply with 1 V applied. Note that the forward current scale (mA) is 1000 times larger than that for reverse leakage current (µA). Breakdown voltage varies from 15 V to several hundred volts.

Sources for RF Ammeters

When it comes to getting your own RF ammeter, there's good news and bad news. First, the bad news. New RF ammeters are expensive, and even surplus pricing can vary widely between $10 and $100 in today's market. AM radio stations are the main users of new units. The FCC defines the output power of AM stations based on the RF current in the antenna, so new RF ammeters are made mainly for that market. They are quite accurate, and their prices reflect that!

The good news is that used RF ammeters are often available. For example, Fair Radio Sales in Lima, Ohio has been a consistent RF-ammeter source. Ham flea markets are also worth trying. Some grubbing around in your nearest surplus store or some older ham's junk box may provide just the RF ammeter you need.

RF Ammeter Substitutes

Don't despair if you can't find a used RF ammeter. It's possible to construct your own. Both hot-wire and thermocouple units can be homemade.

Pilot lamps in series with antenna wires, or coupled to them in various ways, can indicate antenna current* or even forward and reflected power.[†]

Another approach is to use a small low-voltage lamp as the heat/light element and use a photo detector driving a meter as an indicator. (Your eyes and judgment can serve as the indicating part of the instrument.) A feed-line balance checker could be as simple as a couple of lamps with the right current rating and the lowest voltage rating available. You should be able to tell fairly well by eye which bulb is brighter or if they are about equal. You can calibrate a lamp-based RF ammeter with 60-Hz or dc power.

As another alternative, you can build an RF ammeter that uses a dc meter to indicate rectified RF from a current transformer that you clamp over a transmission line wire.[††]

Copper-Top Battery Testers as RF Ammeters

Finally, there are the *free* RF ammeters that come as the testers with Duracell batteries! For 1.5-V cells, these are actually 3 to 5-Ω resistors with built-in liquid-crystal displays. The resistor heats the liquid-crystal strip; the length of the "lighted" portion (heat turns the strip clear, exposing the fluorescent ink beneath) indicates the magnitude of the current.

Despite their "+" and "–" markings, these indicators are not polarized. Their resistance is low enough to have relatively little effect on a 50-Ω system. (For example, putting one in series with a 50-Ω dummy load would increase the system SWR from 1 to 1.1:1. These testers can measure about 200 to 400 mA. (You can achieve higher ranges by means of a shunt.) Best of all, if you burn out one of these "meters" during your tests, you can replace it at any drugstore, hardware store or supermarket for a few dollars, with some batteries thrown in free. — *John Stanley, K4ERO*

*F. Sutter, "What, No Meters?," *QST*, Oct 1938, p 49.
[†]C. Wright, "The Twin-Lamp," *QST*, Oct 1947, pp 22-23, 110 and 112.
[††]Z. Lau, "A Relative RF Ammeter for Open-Wire Lines," *QST*, Oct 1988, pp 15-17.

---

it is far more desirable for the output to be linear (that is, for the reading to be directly proportional to the applied voltage), which means that the markings on the meter are more evenly spaced.

To obtain that kind of linearity it is necessary to use a relatively large load resistance for the diode: Large enough that this resistance, rather than the diode resistance, will determine how much current flows. With this technique you can have a linear reading meter, but at the expense of sensitivity. The resistance needed depends on the type of diode; 5 kΩ to 50 kΩ is usually enough for a germanium rectifier, depending on the dc meter sensitivity, but several times as much may be needed for silicon diodes. Higher resistances require greater meter sensitivity; that is, the basic meter must be a microammeter rather than a low-range milliammeter.

REVERSE CURRENT

When semiconductor diodes are reverse biased, a small leakage current flows. This reverse current flows during the half cycle when the diode should appear open, and the current causes an error in the dc meter reading. The quantity of reverse current is indicated by a diode's *back resistance* specification. This back resistance is so high that reverse current is negligible with silicon diodes, but back resistance may be less than 100 kΩ for germanium diodes.

The practical effect of semiconductor back resistance is to limit the amount of resistance that can be used in the dc load. This in turn affects the linearity of the meter scale. For practical purposes, the back resistance of vacuum-tube diodes is infinite.

25.3.4 RF Voltage

Special precautions must be taken to minimize the capacitive component of the voltmeter impedance at RF. If possible, the rectifier circuit should be installed permanently at the point where the RF voltage is to be measured, using the shortest possible RF connections. The dc meter can be remotely located, however.

For general RF measurements an RF probe is used in conjunction with a 10 MΩ electronic voltmeter. The circuit of **Fig 25.11**, which is basically the shunt peak-reading circuit of Fig 25.9D, is generally used. The series resistor, which is installed in the probe close to the rectifier, prevents RF from being fed through the probe cable to the electronic voltmeter. (In addition, the capacitance of the coaxial cable serves as a bypass for any RF on the lead.) This resistor, in conjunction with the 10-MΩ divider resistance of the electronic voltmeter, reduces the peak rectified voltage to a dc value equivalent to the RMS of the RF signal. Therefore, the RF readings are consistent with the regular dc calibration.

Of the diodes readily available to amateurs, the germanium point-contact or Schottky diode is preferred for RF-probe applications. It has low capacitance (on the order of 1 pF) and

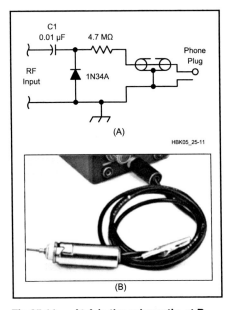

Fig 25.11 — At A is the schematic, at B a photo, of an RF probe for electronic voltmeters. The case of this probe is a seven-pin ceramic tube socket and a 2¼-inch tube shield. A grommet protects the cable where it leaves the tube shield, and an alligator clip on the cable braid connects the probe to the ground of the circuit under test.

in high-back-resistance Schottky diodes, the reverse current is not serious. The principal limitation is that its safe reverse voltage is only about 50 to 75 V, which limits the applied voltage to 15 or 20 V RMS. Diodes can be series connected to raise the overall rating. At RF, however, it is more common to use capacitors or resistors as voltage dividers and apply the divider output to a single diode.

25.3.5 RF Power

RF power can be measured by means of an accurately calibrated RF voltmeter connected across a dummy load in which the power is dissipated. If the load is a known pure resistance, the power, by Ohm's Law, is equal to E^2/R, where E is the RMS voltage.

THE HEWLETT-PACKARD 410B/C VTVM

The Hewlett-Packard 410B and 410C VTVMs have been standards of bench measurement for industry, and they are now available as industrial surplus. These units are not only excellent VTVMs, but they are also good wide-range RF power meters. Both models use a vacuum-tube detector mounted in a low-loss probe for ac measurements. With an adapter that allows the probe to contact the center conductor of a transmission line, it will give very good RF voltage measurements from 50 mV to 300 V and from 20 Hz to beyond 500 MHz. Very few other measuring instruments provide this range in a single sensor/meter. In addition to the 410B or C, you will also need a probe adapter HP model #11042A.

Do not take the probe apart for inspection because that can change the calibration. You can quickly check the probe by feeling it after it has been warmed up for about 15 minutes. If the body of it feels warm it is probably working. Inside, the 410B is quite different from the C model, with the B being simpler. The meter scales are also different; the 410C offers better resolution and perhaps better accuracy.

To make RF power measurements, remove the ac probe tip and twist lock the probe into the 11042A probe adapter. Attach the output of the adapter to a dummy load and use the formula

$$P = \frac{E^2}{R} \quad (8)$$

where
  P = power, in watts
  E = value given by the meter, in volts
  R = resistance of the dummy load, in ohms.

The resistance of the dummy load should be accurately known to at least ±0.1 Ω, preferably measured with a four-wire arrangement as described in the section on ohmmeters. For frequent measurements, make a chart of voltage vs power at your most often used wattages.

25.3.6 *Project:* The Microwatter

This simple, easy to build terminating microwattmeter by Denton Bramwell, W7DB appeared in June 1997 *QST*.[1] It can measure power levels from below –50 dBm (10 nW) to -20 dBm (10 μW) with an accuracy of within 1 dB at frequencies from below the broadcast band up to 2 meters. Beyond that, it's still capable of making *relative* power measurements. The range of the meter can be extended by using an attenuator or by adding a broadband RF amplifier.

There are many uses for this meter. Combined with a signal generator, you can use it to characterize crystal and LC filters, for direct, on-air checks of field strength and measuring antenna patterns. If you are fortunate enough to have a 50-Ω oscilloscope probe (or a high-impedance probe), you can also use the Microwatter as an RF voltmeter for circuit testing.

DESIGN

The circuit concept is quite simple (see **Fig 25.12**). Two nearly identical diodes (D1 and D2) are biased on by a small dc current. RF energy is coupled to one of the diodes, which then acts as a square-law detector. The difference between the voltages present on the two diodes is amplified by a differential amplifier (U1) and applied to an analog voltmeter equipped with hand-calibrated ranges. Getting adequate stability requires careful selection of parts.

D1 and D2 must have a low junction capacitance and a fast response in order to provide the desired bandwidth. Their temperature coefficients must also be extremely well matched. The solution is to use a very old part: the CA3039 diode array.[2] The diodes in this device are very fast silicon types, with exceptional thermal matching.[3] Any two diodes of the array—except the one tied to the substrate—can be used for D1 and D2. Unused IC pins can be cut off or bent out of the way. Similarly rated arrays of hot-carrier diodes may provide a better frequency response.

The Burr-Brown INA2128 provides two differential amplifiers with exceptional specifications: low thermal drift, superb common-mode rejection and low noise. In this circuit, one half of the INA2128 provides dc amplification of the detected voltage; the other half serves as a low-impedance offset-voltage source used for error nulling.

The RF-input terminating resistor is composed of two 100-Ω resistors (R3 and R4), providing a cleaner 50-Ω termination than a single resistor does. If you have them, use chip resistors. If not, clip the leads off two standard 100-Ω resistors and carefully scrape away any paint at the ends. After installing the resistors, verify the input resistance with an ohmmeter.

A stable power source is required. Operating the circuit from an unregulated battery supply allows slight relative changes in the local ground as the terminal voltage drops, creating drift in the reading. The Zener diodes (D3 and D4) following regulator U2 provide a simple, dynamic means of splitting the 12-V supply into the three voltages required to power the circuit.

C1 and C11 should be as close as physically possible to U1 pins 1 and 2. In combination with R1 and R5, these capacitors provide good immunity to unwanted RF. R14, the OFFSET NULL potentiometer, allows nulling the dc amplifier offset and any offset from slight mismatches in the CA3039 diodes. Use a multiturn potentiometer for R14—10 turns or more. Whether you use a panel-mount pot, or a PC-board-mounted trimmer as I did, R14 must be accessible. On the most-sensitive range, the instrument must be zeroed before each use.

The forward voltage drop difference between any two diodes in the CA3039 is specified as "typically less than 0.5 mV." If your device is typical, the nominal values of 5.1 kΩ for R10 and 3.9 kΩ for R9 will do just fine. The manufacturer does not reject parts until the forward voltage drop difference is *5 mV*—so the 0.5 mV figure isn't totally dependable. You can always rotate the diode array so that pins 1, 2, 3 and 4 occupy the spots designated for pins 5, 6, 7 and 8. That gives you a second set of diodes from which to choose. The simplest general solution for a homebrew project is to use nonidentical values for R9 and R10.

If your diode array isn't typical, your Microwatter won't zero properly. The solution is simple. Remove R14 from the circuit. Temporarily connect one end of a 10-kΩ potentiometer to the +8.1 V supply and attach the other end to the –3.9 V supply. Connect the pot arm to the PC-board point of R14's arm. With this arrangement, you'll be able to zero your Microwatter, although the setting will be sensitive and might not produce an exact zero. Once you have the zero point, turn off the Microwatter and remove the 10-kΩ pot carefully without disturbing its setting. Measure the resistance between the pot arm and end that was connected to the +8.1 V point—this is the value of R10. R9's value is that measured between the pot arm and the end previously connected to the –3.9 V point. The combined value of these two resistors and that of R14 provides a total resistance of about 10.5 kΩ, and the resistance ratios will allow easy meter zeroing.

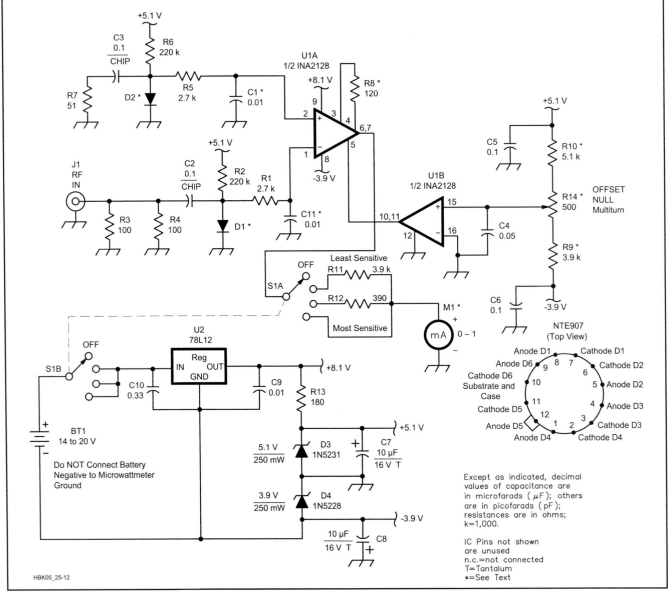

Fig 25.12 — Schematic of the Microwatter circuit. Equivalent parts can be substituted. Unless otherwise specified, resistors are ¼-W, 5%-tolerance carbon-composition or film units.

D1, D2 — Part of CA3039 or NTE907 diode array; see text and Notes 1 and 2. The NTE907 is available from Mouser Electronics.

M1 — 1-mA meter movement, 50-Ω internal resistance; see text.

R14 — 500-Ω, 10-turn (or more) potentiometer; see text.

S1 — 2-pole, 4-position rotary switch.

U1 — Burr Brown amplifier INA2128P; Digi-Key INA2128P; available from Digi-Key Corp.

U2 — 78L12, positive 12-V, 100-mA voltage regulator; Digi-Key LM78L12ACZ.

THE METER MOVEMENT

The specified meter movement has a 50-Ω internal resistance. Hence, on the most sensitive scale, a 50 mV output from U1 provides full-scale deflection. The middle and upper scales require 440 mV and 3.95 V, respectively, for full-scale deflection. Here's how you can use a meter that has a different internal resistance or current sensitivity.

Take, for example, a 25-μA meter movement with an internal resistance of 1910 Ω. A signal level of 47.8 mV (25 μA × 1910 Ω) provides full-scale deflection, so this meter can be substituted directly for the 1-mA, 50-Ω meter on the most sensitive scale. On the middle and top scales, the values of R12 and R11 should be 15.69 kΩ and 156.1 kΩ, or values close to those.

If the most sensitive meter you can find requires 80 to 100 mV to drive it to full scale deflection, don't worry. The Microwatter's most-sensitive scale will be compressed, but quite usable. The middle and top scales will still run full scale, probably not even requiring a change in the values of R11 and R12, if you use a 1-mA movement.

CONSTRUCTION AND CALIBRATION

For the prototype, components are mounted on double-sided PC board (see Note 1) with one side acting as a ground-plane. The cabinet is a 5 × 7 × 3-inch aluminum box, primed and painted gray (see **Fig 25.13**). Power is supplied by a set of NiCd batteries glued to the inside rear of the cabinet. The

Fig 25.13 — Photo of the Microwatter.

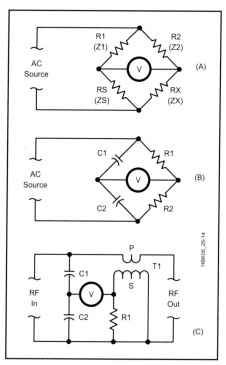

Fig 25.14 — A shows a bridge circuit generalized for ac or dc use. B is a form of ac bridge for RF applications. C is an SWR bridge for use in transmission lines.

PC board is small enough to be mounted on the back of the input BNC connector without other support. Once you have completed PC-board assembly, remove the solder flux from the board.

To calibrate the Microwatter, you'll need a 50-Ω RF source with a known output level. A suitable signal source can be made from a crystal oscillator and an attenuator, using at least 6 dB of attenuation between the oscillator and the Microwatter at all times. The output of such a system can be calibrated with an oscilloscope, or with a simple RF meter. Any frequency in the mid-HF range is suitable for calibration— W7DB used 10 MHz.

Remove the meter-face cover and prepare the face for new markings. You can paint it white and use India ink to make the meter scale arcs and incremental marks. Before calibrating the Microwatter, let it stabilize at room temperature for an hour, and turn it on at least 15 minutes before use. (At these low power levels, even the heat generated by the small dc bias on the diodes needs time to stabilize.) Set the Microwatter to its most sensitive scale. Use R14 to set the meter needle to your chosen meter zero point and mark that point with a pencil. Then apply a –40 dBm signal and mark the top end of the scale. Decrease the input signal level in 1-dB steps, marking each step. Recheck your meter zeroing, switch the Microwatter to the middle scale, apply a –30 dBm signal and mark this point. Again, decrease drive in 1-dB steps, marking each. Repeat with the least-sensitive scale, starting with –20 dBm. Once this is done, replace the meter-face cover and your Microwatter is ready to use.

25.3.7 Directional Wattmeters

Directional wattmeters of varying quality are commonly used by the amateur community. The high quality standard is made by the Bird Electronic Corporation, who call their proprietary line THRULINE. The units are based on a sampling system built into a short piece of 50-Ω transmission line with plug-in elements for various power and frequency ranges.

25.3.8 AC Bridges

In its simplest form, the ac bridge is exactly the same as the Wheatstone bridge discussed earlier in the dc measurement section of this chapter. However, complex impedances can be substituted for resistances, as suggested by **Fig 25.14A**. The same bridge equation holds if Z (complex impedance) is substituted for R in each arm. For the equation to be true, however, both phase angles and magnitudes of the impedances must balance; otherwise, a true null voltage is impossible to obtain. This means that a bridge with all "pure" arms (pure resistance or reactance) cannot measure complex impedances; a combination of R and X must be present in at least one arm aside from the unknown.

The actual circuits of ac bridges take many forms, depending on the intended measurement and the frequency range to be covered. As the frequency increases, stray effects (unwanted capacitances and inductances) become more pronounced. At RF, it takes special attention to minimize them.

Most amateur built bridges are used for RF measurements, especially SWR measurements on transmission lines. The circuits at Fig 25.14B and C are favorites for this purpose. The theory behind SWR measurement is covered in the **Transmission Lines** chapter and more fully in *The ARRL Antenna Book*.

Fig 25.14B is useful for measuring both transmission lines and lumped constant components. Combinations of resistance and capacitance are often used in one or more arms; this may be required for eliminating the effects of stray capacitance. The bridge shown in Fig 25.14C is used only on transmission lines and only on those lines having the characteristic impedance for which the bridge is designed.

Using an ac bridge to measure impedance at RF requires an awareness of the limitations of instrumentation and of the effects of stray and parasitic impedances on the measurement, particularly for impedances much higher than 50 Ω. The layout of the test fixture, the arrangement of cables and enclosures, and nearby conductors can all affect the measurement. Extreme care is required to obtain repeatable, reliable measurements. See the section Transmitting Ferrite Choke Baluns in the **Transmission Lines** chapter for examples of the issues associated with this type of measurement.

25.3.9 Component Measurements

Many multimeters now offer the ability to measure inductance and capacitance directly. The measurement is made by exciting the unknown component with an audio-frequency ac signal and calculating component value from the resulting phase difference between the voltage and current through the component. Accuracy of the meters is generally 10% or better. The measurements are made at a very low frequency (usually 10 kHz or lower) so RF parameters such as self-resonant frequency or Q cannot be measured.

A similar type of meter is used to measure the *equivalent series resistance* (ESR) and *equivalent series inductance* (ESL) of capacitors. When used in high-current, high-frequency applications, such as switchmode power supplies where the switched currents have components into the MHz range, losses can occur in the capacitors due to ESR or reactive effects from ESL can upset the operation of the power conversion circuits.

Notes

[1] A PC board and some components for this project are available from FAR Circuits. A PC-board template package is not available.
[2] The NTE907 replacement is available from Mouser Electronics.
[3] The absolute value of the difference in forward drop between any two diodes is 1 mV/°C.

25.4 Frequency Measurement

The FCC Rules for Amateur Radio require that transmitted signals stay inside the frequency limits of bands consistent with the operator's license privileges. The exact frequency within this band need not be known, as long as it is within the limits. On these limits there are no tolerances: Individual amateurs must be sure that their signal stays safely inside. See *FCC Rules and Regulations for the Amateur Service*, published by the ARRL.

Staying within these limits is not difficult; many modern transceivers do so automatically, within limits. If your radio uses a PLL synthesized frequency source, just tune in WWV or another frequency standard occasionally.

Checks on older equipment require some simple equipment and careful adjusting. The equipment commonly used is the frequency marker generator and the method involves use of the station receiver, as shown in **Fig 25.15**.

25.4.1 Frequency Marker Generators

A marker generator, in its simplest form, is a high-stability oscillator that generates a series of harmonic signals. When an appropriate fundamental is chosen, harmonics fall near the edges of the amateur frequency allocations.

Most US amateur band and subband limits are exact multiples of 25 kHz. A 25-kHz fundamental frequency will therefore produce the right marker signals if its harmonics are strong enough. But since harmonics appear at 25-kHz intervals throughout the spectrum, there is still a problem of identifying particular markers. This is easily solved if the receiver has reasonably good calibration. If not, most marker circuits provide a choice of fundamental outputs, say 100 and 50 kHz as well as 25 kHz. Then the receiver can be first set to a 100-kHz interval. From there, the desired 25-kHz (or 50-kHz) points can be counted. Greater frequency intervals are rarely required. Instead, tune in a signal from a station of known frequency and count off the 100-kHz points from there.

TRANSMITTER CHECKING

To check transmitter frequency, tune in the transmitter signal on a calibrated receiver and note the dial setting at which it is heard. To start, reduce the transmitter to its lowest possible level to avoid receiver overload. Also, place a direct short across the receiver antenna terminals, reduce the RF gain to minimum and switch in any available receiver attenuators to help prevent receiver IMD, overload and possible false readings.

Place the transmitter on standby (not trans-

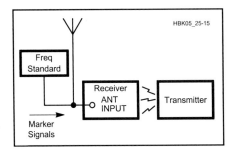

Fig 25.15 — A setup for checking transmitter frequency. Use care to ensure that the transmitter does not overload the receiver. False signals would result; see text.

mitting) and use the marker generator as a signal source. Tune in and identify the nearest marker frequencies above and below the transmitter signal. The transmitter frequency is between these two known frequencies. If the marker frequencies are accurate, this is all you need to know, except that the transmitter frequency must not be so close to a band (or subband) edge that sidebands extend past the edge.

If the transmitter signal is inside a marker at the edge of an assignment, to the extent that there is an audible beat note with the receiver BFO turned off, normal CW sidebands are safely inside the edge. (So long as there are no abnormal sidebands such as those caused by clicks and chirps.) For phone the safety allowance is usually taken to be about 3 kHz, the usual width of one sideband. A frequency difference of this much can be estimated by noting the receiver dial settings for the two 25-kHz markers that are either side of the signal and dividing 25 by the number of dial divisions between them. This will give the number of kilohertz per dial division. It is a prudent practice to allow an extra kHz margin when setting the transmitter close to a band or subband edge (5-kHz is a safe HF margin for most modes on modern transmitters).

TRANSCEIVERS

The method described above is good when the receiver and transmitter are separate pieces of equipment. When a transceiver is used and the transmitting frequency is automatically the same as that to which the receiver is tuned, setting the tuning dial to a spot between two known marker frequencies is all that is required. The receiver incremental tuning control (RIT) must be turned off.

The proper dial settings for the markers are those at which, with the BFO on, the signal is tuned to zero beat (the spot where the beat note disappears as tuning makes its pitch progressively lower). Exact zero beat can be determined by a very slow rise and fall of background noise, caused by a beat of a cycle or less per second. In receivers with high selectivity it may not be possible to detect an exact zero beat, because low audio frequencies from beat notes may be prevented from reaching the speaker or headphones.

Most commercial equipment has some way to match either the equipment's internal oscillator or marker generator with the signal received from WWV on one of its short-wave frequencies. It is a good idea to do this check on a new piece of gear. A recheck about a month later will show if anything has changed. Normal commercial equipment drifts less than 1 kHz after warm up.

Also check the dial linearity of equipment that has an analog dial or subdial. Often analog dials do not track frequency accurately across an entire band. Such radios usually provide for pointer adjustment so that dial error can be minimized at the most often used part of a band.

FREQUENCY-MARKER CIRCUITS

The frequency in most amateur frequency markers is determined by a 100-kHz or 1-MHz crystal. Although the marker generator should produce harmonics every 25 kHz and 50 kHz, crystals (or other high-stability resonators) for frequencies lower than 100 kHz are expensive and rare. There is really no need for them, however, since it is easy to divide the basic frequency down to the desired frequency; 50- and 25-kHz steps require only two successive divisions by two (from 100 kHz). In the division process, the harmonics of the generator are strengthened so they are useful up to the VHF range. Even so, as frequency increases the harmonics weaken.

Current marker generators are based on readily available crystals. A 1-MHz basic oscillator would first be divided by 10 to produce 100 kHz and then followed by two successive divide-by-two stages to produce 50 kHz and 25 kHz.

25.4.2 *Project:* A Marker Generator with Selectable Output

Fig 25.16 shows a marker generator with selectable output for 100, 50 or 25-kHz intervals. It provides marker signals well up into the 2-m band. The project was first built by Bruce Hale, KB1MW, in the ARRL Lab. A more detailed presentation appeared in the **Station Accessories** chapter of *Handbooks* from 1987 through 1994. An etching pattern and parts-placement diagram are available for this project. See the **Project Templates**

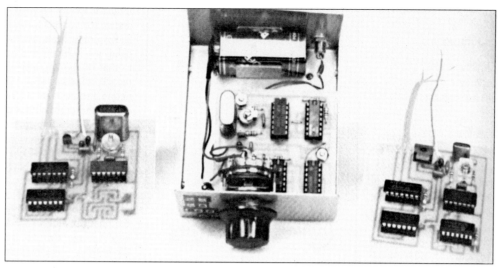

Fig 25.16A—The generator enclosure and views of the circuit board.

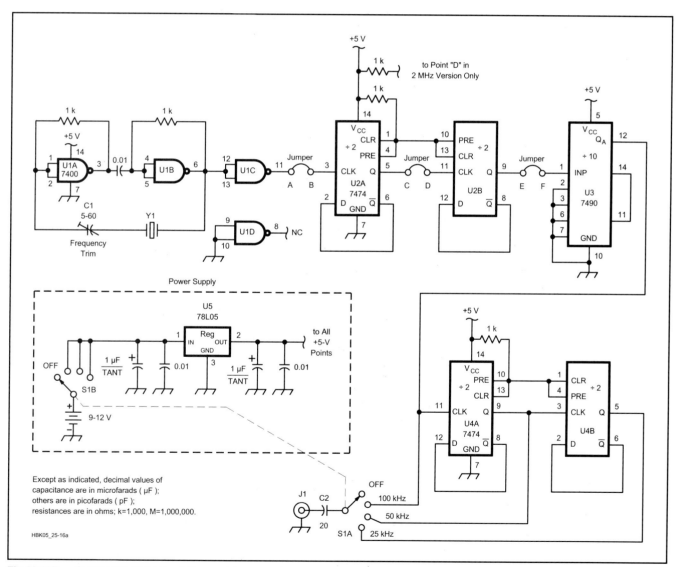

Fig 25.16B — Schematic diagram of the marker generator.

C1 — 5-60 pF miniature trimmer capacitor.
C2 — 20-pF disc ceramic.
U1 — 7400 or 74LS00 quad NAND gate.

U2, U4 — 7474 or 74LS74 dual D flip-flop.
U3 — 7490 or 74LS90 decade counter.
U5 — 78L05, 7805 or LM340-T5 5-V voltage regulator.

Y1 — 1, 2 or 4-MHz crystal in HC-25, HC-33 or HC-6 holder.

Test Equipment and Measurements 25.15

Table 25.2
Marker Generator Jumper Placement

Crystal Frequency	Jumper Placement
1 MHz	A to F (U2 not used)
2 MHz	A to B
	C to F
	D to +5 V, via 1-kΩ resistor
4 MHz	A to B
	C to D
	E to F

section on the *Handbook* CD-ROM.

A 1, 2 or 4-MHz computer-surplus crystal is suitable for Y1. Several prototypes were built with such crystals and all could be tuned within 50 Hz at 100 kHz. The marker division ratio must be chosen once the crystal frequency is selected. This is accomplished by means of several jumpers that disable part or all of U2A, the dual flip-flop IC. When Y1 is a 1-MHz crystal, U2 may be omitted entirely. **Table 25.2** gives the jumper placement for each crystal frequency.

The prototypes were built with TTL logic ICs. Unused TTL gate inputs should always be connected either to ground or to V_{CC} through a pull-up resistor. If low-power Schottky (they have "LS" as part of their numerical designator) parts are available for U1 through U4, use them. They draw much less current than plain TTL, and will greatly extend battery life.

25.4.3 Dip Meters

This device is often called a grid-dip meter or oscillator (from vacuum-tube days) or a transistor dip meter. Most dip meters can also serve as absorption frequency meters (in this mode measurements are read at the current peak, rather than the dip). Further, some dip meters have a connection for headphones. The operator can usually hear signals that do not register on the meter. Because the dip meter is an oscillator, it can be used as a signal generator in certain cases where high accuracy or stability are not required.

A dip meter may be coupled to a circuit either inductively or capacitively. Inductive coupling results from the magnetic field generated by current flow. Therefore, inductive coupling should be used when a conductor with relatively high current is convenient. Maximum inductive coupling results when the axis of the pick-up coil is placed perpendicular to a nearby current path (see **Fig 25.17**).

Capacitive coupling is required when current paths are magnetically confined or shielded. (Toroidal inductors and coaxial cables are common examples of magnetic self shielding.) Capacitive coupling depends

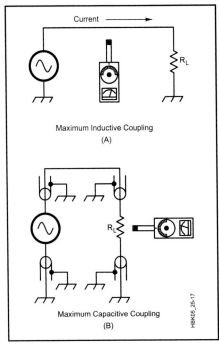

Fig 25.17 — Dip meter coupling. (A) uses inductive coupling and (B) uses capacitive coupling.

on the electric field produced by voltage. Use capacitive coupling when a point of relatively high voltage is convenient. (An example might be the output of a 12-V powered RF amplifier. *Do not* attempt dip-meter measurements on true high-voltage equipment such as vacuum-tube amplifiers or switching power supplies while they are energized.) Capacitive coupling is maximum when the end of the pick-up coil is near a point of high voltage (see Fig 25.17). In either case, the circuit under test is affected by the presence of the dip meter. Always use the minimum coupling that yields a noticeable indication.

Use the following procedure to make reliable measurements. First, bring the dip meter gradually closer to the circuit while slowly varying the dip-meter frequency. When a current dip occurs, hold the meter steady and tune for minimum current. Once the dip is found, move the meter away from the circuit and confirm that the dip comes from the circuit under test (the current reading should increase with distance from the circuit until the dip is gone). Finally, move the meter back toward the circuit until the dip is just noticed. Retune the meter for minimum current and read the dip-meter frequency with a calibrated receiver or frequency meter.

The current dip of a good measurement is smooth and symmetrical. An asymmetrical dip indicates that the dip-meter oscillator frequency is being significantly influenced by the test circuit. Such conditions do not yield usable readings.

A measurement of effective unloaded inductor Q can be made with a dip meter and an RF voltmeter (or a dc voltmeter with an RF probe). Make a parallel resonant circuit using the inductor and a capacitance equal to that of the application circuit. Connect the RF voltmeter across this parallel combination and measure the resonant frequency. Adjust the dip-meter/circuit coupling for a convenient reading on the voltmeter, then maintain this dip-meter/circuit relationship for the remainder of the test. Vary the dip meter frequency until the voltmeter reading drops to 0.707 times that at resonance. Note the frequency of the dip meter and repeat the process, this time varying the frequency on the opposite side of resonance. The difference between the two dip meter readings is the test-circuit bandwidth. This can be used to calculate the circuit Q:

$$Q = \frac{f_0}{BW} \qquad (9)$$

where
 f_0 = operating frequency,
 BW = measured bandwidth in the same units as the operating frequency.

When purchasing a dip meter, look for one that is mechanically and electrically stable. The coils should be in good condition. A headphone connection is helpful. Battery-operated models are convenient for antenna measurements.

25.4.4 *Project:* A Dip Meter with Digital Display

An up-to-date dip meter was described by Larry Cicchinelli in the October 1993 issue of *QEX*. It consists of a dip meter with a three-digit frequency display. The analog portion of the circuit consists of an FET oscillator, voltage-doubler detector, dc-offset circuit and amplifier. The digital portion of the circuit consists of a high-impedance buffer, prescaler, counter, display driver, LED display and control circuit.

CIRCUIT DESCRIPTION

The dip meter shown in **Fig 25.18** has four distinct functional blocks. The RF oscillator

Table 25.3
Calculated Coil Data for the Dip Meter

Frequency (MHz)	L (µH)	Turns
1.7 to 3.1	48.6	52
2.8 to 5.9	16.3	23
5.6 to 11.9	4.0	9
9.7 to 20.7	1.3	5
19.0 to 45.0	0.3	2

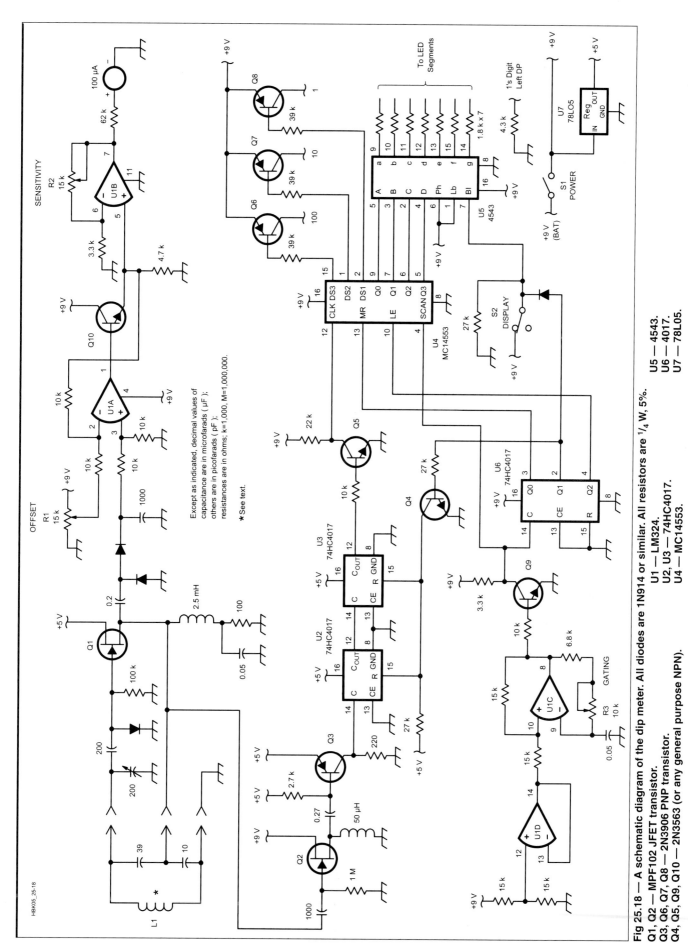

Fig 25.18 — A schematic diagram of the dip meter. All diodes are 1N914 or similar. All resistors are 1/4 W, 5%.

Q1, Q2 — MPF102 JFET transistor.
Q3, Q6, Q7, Q8 — 2N3906 PNP transistor.
Q4, Q5, Q9, Q10 — 2N3563 (or any general purpose NPN).

U1 — LM324.
U2, U3 — 74HC4017.
U4 — MC14553.

U5 — 4543.
U6 — 4017.
U7 — 78L05.

Test Equipment and Measurements 25.17

is a standard Colpitts using a common junction FET, Q1, as the active element. Its range is about 1.7 to 45 MHz. The 200-pF tuning capacitor gives a 2:1 tuning range. A 2:1 frequency range requires a capacitor with a 4:1 range. The sum of the minimum capacitance of the variable, the capacitors across the inductor and the strays must therefore be in the order of 70 pF. The values of L1 were determined experimentally by winding the coils and observing the lower and upper frequency values. [**Table 25.3** shows winding data calculated from the author's schematic. — *Ed.*] The coil sizes were experimentally determined, and the coils are constructed on 1¼-inch-diameter plug-in coil forms with #20 enameled wire. They are close wound. The number of turns shown are only a starting point; you may need to change them slightly in order to cover the desired frequency range.

The tapped capacitors are mounted inside the coil forms so that their values could be different for each band if required. The frequency spread of the lowest band is less than 2:1 because the tapped capacitor values are larger than those for the other bands. The frequency spread of the highest band is greater than 2:1 because its capacitors are smaller.

The analog display circuit begins with a voltage-doubler detector in order to get higher sensitivity. It drives a dc-offset circuit, U1A. R1 inserts a variable offset that is subtracted from the detector voltage. This allows the variable gain stage, U1B, to be more sensitive to variations in the detector output voltage. Q9 follows U1C to get an output gating voltage closer to ground. The resistor in series with the meter is chosen to limit the meter current to a safe value. For example, if a 1-mA meter is used, the resistor should be 8.2 kΩ.

The prescaler begins with a high-impedance buffer and amplifier, Q2 and Q3. If you are going to use the meter for the entire frequency range described, take care in the layout of both the oscillator and buffer/amplifier circuits. The digital portion of the prescaler is a divide-by-100 circuit consisting of two divide-by-10 devices, U2 and U3. The devices used were selected because they were available. Any similar devices may be used as long as the reset circuit is compatible. Q5 is a level translator that shifts the 5-V signal to 9 V.

The first part of the digital display block is the oscillator circuit of U1C, which creates the gate time for the frequency counter. R3 adjusts the oscillator to a frequency of 500 Hz, yielding a 1-ms gate. The best way to set this frequency is to listen for the dip-meter output on a communications receiver and adjust R3 until the display agrees with the receiver. Once this calibration has been made for one of the bands, all the bands are calibrated.

U1D gives a low-impedance voltage reference for U1C. Q9 was added to the output of the oscillator to remove a small glitch, which can cause the counter to trigger incorrectly. This type of oscillator has the advantage of simplicity. This circuit is fairly stable, easy to adjust and has a low parts count.

The digital system controller, U6, is a divide-by-10 counter. It has 10 decoded outputs, each of which goes high for one period of the input clock. The Q0 output is used to reset the frequency counter, U4. The Q1 output is used to enable the prescaler and disable the display and the Q2 output latches the count value into the frequency counter. Since the prescaler can only count while Q1 is high, it will be enabled for only 1 ms. Normally a 1-ms gate will yield 1-kHz resolution. Since the circuit uses a divide-by-100 prescaler, the resolution becomes 100 kHz.

The frequency counter, U4, is a three-digit counter with multiplexed BCD outputs. The clock input is driven from the prescaler, hence it is the RF oscillator frequency divided by 100. This signal is present for only 1 ms out of every 10 ms. The digit scanning is controlled by the 500-Hz oscillator of U1C.

U5 is a BCD-to-seven-segment decoder/driver. Its outputs are connected to each of the three common-anode, seven-segment displays in parallel. Only the currently active digit will be turned on by the digit strobe outputs of U5, via Q6, Q7 and Q8. The diode connected to the blanking input of U5 disables the display while U4 is counting. U7 is a 5-V regulator that allows the use of a single 9-V battery for both the circuit and the LEDs. S2 turns on the displays once the unit has been adjusted for a dip.

The circuit draws about 20 mA with the LEDs off and up to 35 mA with the LEDs on.

Many of the resistor values are not critical, and those used were chosen based upon availability; the op-amp circuits depend primarily on resistance ratios. The resistor at the collector of Q3 is critical and should not be varied. Use 0.27-μF monolithic capacitors. They have the required good high-frequency characteristics over the range of the meter.

Most of the parts can be purchased from Digi-Key. They did not have the '4543 IC, which was purchased from Hosfelt Electronics in Steubenville, OH. The 74HC4017 may be substituted with a 74HCT4017. The circuit is built on a 4-inch-square perf board (with places for up to 12 ICs) and is housed in a 7 × 5 × 3-inch minibox.

OPERATION

To use the unit, set the gain, R2, fully clockwise for maximum sensitivity. With this setting, the output of the offset circuit (Q10 emitter) is at ground. As R1 is rotated, the voltage on the arm approaches and then becomes less than, the detector output. At this point the meter will start to deflect upward. Adjust R1 so that the meter reads about center scale. (Manual adjustment allows for variations in the output level of the RF oscillator.) As L1 is brought closer to the circuit under test, the meter will deflect downward as energy is absorbed by the circuit. For best results use the minimum possible coupling to the circuit being tested. If the dip meter is overcoupled to the test circuit, the oscillator frequency will be pulled.

25.4.5 Frequency Counters

One of the most accurate means of measuring frequency is a frequency counter. This instrument is capable of numerically displaying the frequency of the signal supplied to its input. For example, if an oscillator operating at 8.244 MHz is connected to a counter input, 8.244 would be displayed. At present, counters are usable well up into the gigahertz range. Most counters that are used at high frequencies make use of a prescaler ahead of a basic low-frequency counter. A prescaler divides the high-frequency signal by 10, 100, 1000 or some other fixed amount so that a low-frequency counter can display the operating frequency.

The accuracy of the counter depends on its internal crystal reference. A more accurate crystal reference yields more accurate readings. Crystals for frequency counters are manufactured to close tolerances. Most counters have a trimmer capacitor so that the crystal can be set exactly on frequency. Crystal frequencies of 1 MHz, 5 MHz or 10 MHz have become more or less standard. For calibration, harmonics of the crystal can be compared to a known reference station, such as those shown in Table 25.1, or other frequency standard and adjusted for zero beat.

Many frequency counters offer options to increase the accuracy of the counter timebase; this directly increases the counter accuracy. These options usually employ temperature-compensated crystal oscillators (TCXOs) or crystals mounted in constant temperature ovens that keep the crystal from being affected by changes in ambient (room) temperature. Counters with these options may be accurate to 0.1 ppm (part per million) or better. For example, a counter with a timebase accuracy of 5.0 ppm and a second counter with a TCXO accurate to 0.1 ppm are available to check a 436-MHz CW transmitter for satellite use. The counter with the 5-ppm timebase could have a frequency error of as much as 2.18 kHz, while the possible error of the counter with the 0.1 ppm timebase is only 0.0436 kHz.

25.5 Other Instruments and Measurements

This section covers a variety of test equipment that is useful in receiver and transmitter testing. It includes RF and audio generators, an inductance meter, a capacitance meter, oscilloscopes, spectrum analyzers, a calibrated noise source, and hybrid combiners. A number of applications of this equipment for basic transmitter and receiver testing is also included.

25.5.1 RF Oscillators for Circuit Alignment

Receiver testing and alignment uses equipment common to ordinary radio service work. Inexpensive RF signal generators are available, both complete and in kit form. However, any source of signal that is weak enough to avoid overloading the receiver usually will serve for alignment work. The frequency marker generator is a satisfactory signal source. In addition, its frequencies, although not continuously adjustable, are far more precise, since the usual signal-generator calibration is not highly accurate. When buying a used or inexpensive signal generator, look for these attributes: output level is calibrated, the output doesn't "ring" too badly when tapped, and doesn't drift too badly when warmed up. Many military surplus units are available that can work quite well. Commercial units such as the HP608 are big and stable, and they may be inexpensive.

25.5.2 Audio-Frequency Oscillators

An audio signal generator should provide a reasonably pure sine wave. The best oscillator circuits for this use are RC coupled, operating as close to a class-A amplifier as possible. Variable frequencies covering the entire audio range are needed for determining frequency response of audio amplifiers.

An oscillator generating one or two frequencies with good waveform is sufficient for most phone-transmitter testing and simple troubleshooting in AF amplifiers. A two-tone (dual) oscillator is very useful for testing and adjusting sideband transmitters.

A circuit of a simple RC oscillator that is useful for general testing is given in **Fig 25.19**. This Twin-T arrangement gives a waveform that is satisfactory for most purposes. The oscillator can be operated at any frequency in the audio range by varying the component values. R1, R2 and C1 form a low-pass network, while C2, C3 and R3 form a high-pass network. As the phase shifts are opposite, there is only one frequency at which the total phase shift from collector to base is 180°: Oscillation will occur at this frequency. When C1 is about twice the capacitance of C2 or C3 the best operation results. R3 should have a resistance about 0.1 that of R1 or R2 (C2 = C3 and R1 = R2). Output is taken across C1, where the harmonic distortion is least. Use a relatively high impedance load — 100 kΩ or more.

Most small-signal AF transistors can be used for Q1. Either NPN or PNP types are satisfactory if the supply polarity is set correctly. R4, the collector load resistor may be changed a little to adjust the oscillator for best output waveform.

25.5.3 *Project:* A Wide-Range Audio Oscillator

A wide-range audio oscillator that will provide a moderate output level can be built from a single 741 operational amplifier (**Fig 25.20**). Power is supplied by two 9-V batteries from which the circuit draws 4 mA. The frequency range is selectable from 8 Hz to 150 kHz. Distortion is approximately 1%. The output level under a light load (10 kΩ) is 4 to 5 V. This can be increased by using higher battery voltages, up to a maximum of plus and minus 18 V, with a corresponding adjustment of R_F.

Pin connections shown are for the TO-5 case and the eight-pin DIP package. Variable resistor R_F is trimmed for an output level of about 5% below clipping as seen on an oscilloscope. This should be done for the temperature at which the oscillator will

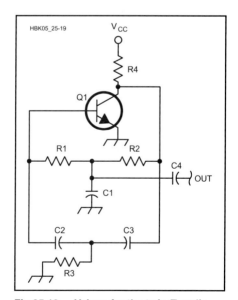

Fig 25.19 — Values for the twin-T audio oscillator circuit range from 18 kΩ for R1-R2 and 0.05 µF for C1 (750 Hz) to 15 kΩ and 0.02 µF for 1800 Hz. For the same frequency range, R3 and C2-C3 vary from 1800 Ω and 0.02 µF to 1500 Ω and 0.01 µF. R4 is 3300 Ω and C4, the output coupling capacitor, can be 0.05 µF for high-impedance loads.

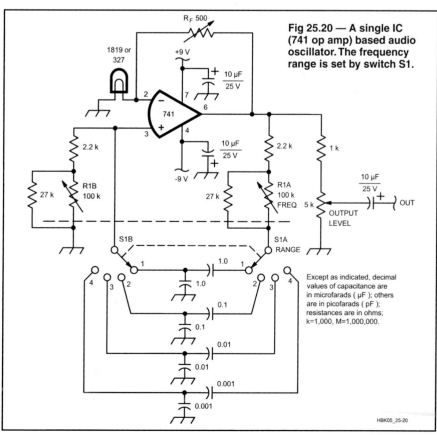

Fig 25.20 — A single IC (741 op amp) based audio oscillator. The frequency range is set by switch S1.

normally operate, as the lamp is sensitive to ambient temperature. This unit was originally described by Shultz in November 1974 *QST*; it was later modified by Neben as reported in June 1983 *QST*.

25.5.4 *Project:* Measure Inductance and Capacitance With A DVM

Many of us have a DVM (Digital voltmeter) or VOM (volt-ohm meter) in the shack, but few of us own an inductance or capacitance meter. If you have ever looked into your junk box and wanted to know the value of the unmarked parts, these simple circuits will give you the answer. They may be built in one evening (**Fig 25.21** and **Fig 25.22**), and will adapt your DVM or VOM to measure inductance or capacitance. The units are calibrated against a known part. Therefore, the overall accuracy depends only on the calibration values and not on the components used to build the circuits. If it is carefully calibrated, an overall accuracy of 10% may be expected if used with a DVM and slightly less with a VOM.

CONSTRUCTION

The circuits may be constructed on a small perf board (RadioShack dual mini board, #276-168), or if you prefer, on a PC board. A template and parts layout may be obtained from the ARRL.[1] Layout is non-critical—almost any construction technique will suffice. Wire-wrapping or point-to-point soldering may be used.

The circuits are available in kit form from Electronic Rainbow Inc., Indianapolis, IN. The IA inductance adapter kit is sold separately, but a cabinet is also available. The PC board is 1.75 × 2.5 inches. The CA-1 capacitance adapter

[1]See the **Project Templates** section of the *Handbook* CD-ROM.

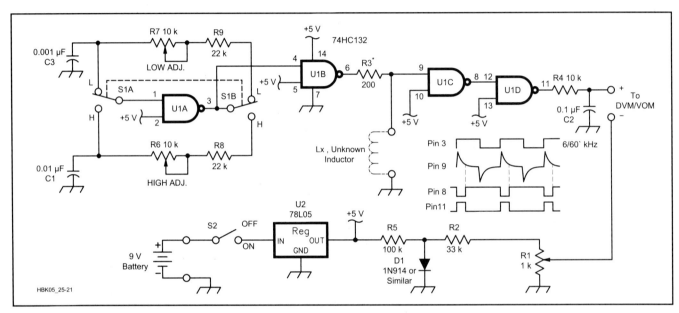

Fig 25.21 — All components are 10% tolerance. 1N4148 or equivalent may be substituted for D1. A LM7805 may be substituted for the 78L05. All fixed resistors are ¼-W carbon composition. Capacitors are in µF. R3 value may need to be increased or decreased slightly if calibration cannot be achieved as described in text.

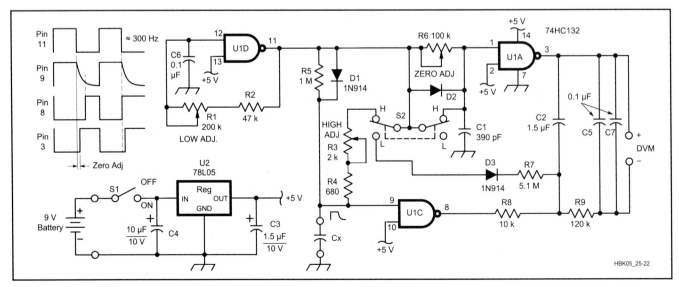

Fig 25.22 — All components are 10% tolerance. An LM7805 may be substituted for the 78L05. All fixed resistors are ¼-W carbon composition. Capacitors are in µF unless otherwise indicated.

kit comes with a 1.80 × 2.0-inch PC board.

INDUCTANCE ADAPTER FOR DVM/VOM

Description

The circuit shown in Fig 25.21 converts an unknown inductance into a voltage that can be displayed on a DVM or VOM. Values between 3 and 500 µH are measured on the low range and from 100 µH to 7 mH on the high range. NAND gate U1A is a two frequency RC square-wave oscillator. The output frequency (pin 3) is approximately 60 kHz in the low range and 6 kHz in the high range. The square-wave output is buffered by U1B and applied to a differentiator formed by R3 and the unknown inductor, LX. The stream of spikes produced at pin 9 decay at a rate proportional to the time constant of R3-LX. Because R3 is a constant, the decay time is directly proportional to the value of LX. U1C squares up the positive going spikes, producing a stream of negative going pulses at pin 8 whose width is proportional to the value of LX.

They are inverted by U1D (pin 11) and integrated by R4-C2 to produce a steady dc voltage at the + output terminal. The resulting dc voltage is proportional to LX and the repetition rate of the oscillator. R6 and R7 are used to calibrate the unit by setting a repetition rate that produces a dc voltage corresponding to the unknown inductance. D1 provides a 0.7 volt constant voltage source that is scaled by R1 to produce a small offset reference voltage for zeroing the meter on the low inductance range.

When S1 is low, *mV* corresponds to *µH* and when high, *mV* corresponds to 10 *µH*. A sensitive VOM may be substituted for the DVM with a sacrifice in resolution.

Test and Calibration

Short the LX terminals with a piece of wire and connect a DVM set to the 200-mV range to the output. Adjust R1 for a zero reading. Remove the short and substitute a known inductor of approximately 400 µH. Set S1 to the low (in) position and adjust R7 for a reading equal to the known inductance. Switch S1 to the high position and connect a known inductor of about 5 mH. Adjust R6 for the corresponding value. For instance, if the actual value of the calibration inductor is 4.76 mH, adjust R7 so the DVM reads 476 mV.

CAPACITANCE ADAPTER FOR DVM/VOM

Description

The circuit shown in Fig 25.22 measures capacitance from 2.2 to 1000 pF in the low range, from 1000 pF to 2.2 µF in the high range. U1D of the 74HC132 (pin 11) produces a 300 Hz square-wave clock. On the rising edge CX rapidly charges through D1. On the falling edge CX slowly discharges through R5 on the low range and through R3-R4 on the high range. This produces an asymmetrical waveform at pin 8 of U1C with a duty cycle proportional to the unknown capacitance, CX. This signal is integrated by R8-R9-C2 producing a dc voltage at the negative meter terminal proportional to the unknown capacitance. A constant reference voltage is produced at the positive meter terminal by integrating the square-wave at U1A, pin 3. R6 alters the symmetry of this square-wave producing a small change in the reference voltage at the positive meter terminal. This feature provides a zero adjustment on the low range. The DVM measures the difference between the positive and negative meter terminals. This difference is proportional to the unknown capacitance.

Test and Calibration

Without a capacitor connected to the input terminals, set SW2 to the low range (out) and attach a DVM to the output terminals. Set the DVM to the 2-volt range and adjust R6 for a zero meter reading. Now connect a 1000 pF calibration capacitor to the input and adjust R1 for a reading of 1.00 volt. Switch to the high range and connect a 1.00 µF calibration capacitor to the input. Adjust R3 for a meter reading of 1.00 volts. The calibration capacitors do not have to be exactly 1000 pF or 1.00 µF, as long as you know their exact value. For instance, if the calibration capacitor is known to be 0.940 µF, adjust the output for a reading of 940 mV.

25.5.5 Oscilloscopes

Most engineers and technicians will tell you that the most useful single piece of test and design equipment is the triggered-sweep oscilloscope (commonly called just a "scope"). This section was written by Dom Mallozzi, N1DM.

Oscilloscopes can measure and display voltage relative to time, showing the waveforms seen in electronics textbooks. Scopes are broken down into two major classifications: analog and digital. This does not refer to the signals they measure, but rather to the methods used inside the scope to process signals for display.

ANALOG OSCILLOSCOPES

Fig 25.23 shows a simplified diagram of a triggered-sweep oscilloscope. At the heart of nearly all scopes is a cathode-ray tube (CRT) display. The CRT allows the visual display of an electronic signal by taking two electric signals and using them to move (deflect) a beam of electrons that strikes the screen. Unlike a television CRT, an oscilloscope uses electrostatic deflection rather than magnetic deflection. Wherever the beam strikes the phosphorescent screen of the CRT it causes a small spot to glow. The exact location of the spot is a result of the voltage applied to the vertical and horizontal inputs.

All of the other circuits in the scope are used to take the real-world signal and convert it to a form usable by the CRT. To trace how a signal travels through the oscilloscope circuitry start by assuming that the trigger select switch is in the INTERNAL position.

The input signal is connected to the input COUPLING switch. The switch allows selection of either the ac part of an ac/dc signal or the total signal. If you wanted to measure, for example, the RF swing at the collector of an output stage, including the dc voltage level for reference, you would use the dc-coupling mode. In the ac position, dc is blocked from reaching the vertical amplifier chain so that you can measure a small ac signal superimposed on a much larger dc level. For example, you might want to measure a 25 mV 120-Hz ripple on a 13 Vdc power supply. You should not use ac-coupling at frequencies below the low-frequency cut-off of the instrument, typically around 30 Hz, because the value of the blocking capacitor represents a considerable series impedance to very low-frequency signals.

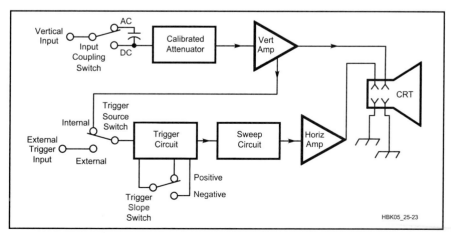

Fig 25.23 — Typical block diagram of a simple triggered-sweep oscilloscope.

After the coupling switch, the signal is connected to a calibrated attenuator. This is used to reduce the signal to a level that can be tolerated by the scope's vertical amplifier. The vertical amplifier boosts the signal to a level that can drive the CRT and also adds a bias component to locate the waveform on the screen.

A small sample of the signal from the vertical amplifier is sent to the trigger circuitry. The trigger circuit feeds a start pulse to the sweep generator when the input signal reaches a certain level. The sweep generator gives a precisely timed signal that looks like a triangle (see **Fig 25.24**). This triangular signal causes the scope trace to sweep from left to right, with the zero-voltage point representing the left side of the screen and the maximum voltage representing the right side of the screen.

The sweep circuit feeds the horizontal amplifier that, in turn, drives the CRT. It is also possible to trigger the sweep system from an external source (such as the system clock in a digital system). This is done by using an external input jack with the trigger select switch in the EXTERNAL position.

The trigger system controls the horizontal sweep. It looks at the trigger source (internal or external) to find out if it is positive- or negative-going and to see if the signal has passed a particular level. **Fig 25.25A** shows a typical signal and the dotted line on the figure represents the trigger level. It is important to note that once a trigger circuit is "fired" it cannot fire again until the sweep has moved all the way across the screen from left to right. In normal operation, the TRIGGER LEVEL control is manually adjusted until a stable display is seen. Some scopes have an AUTOMATIC position that chooses a level to lock the display in place without manual adjustment.

Fig 25.25B shows what happens when the level has not been properly selected. Because there are two points during a single cycle of the waveform that meet the triggering requirements, the trigger circuit will have a tendency to jump from one trigger point to another. This will make the waveform jitter from left to right. Adjustment of the TRIGGER control will fix this problem.

The horizontal travel of the trace is calibrated in units of time. If the time of one cycle is known, we can calculate the frequency of the waveform. In **Fig 25.26**, for example, if the SWEEP speed selector is set at 10 μs/division and we count the number of divisions (vertical bars) between peaks of the waveform (or any similar well defined points that occur once per cycle) we can find the period of one cycle. In this case it is 80 μs. This means that the frequency of the waveform is 12,500 Hz (1/80 μs). The accuracy of the measured frequency depends on the accuracy of the scope's sweep oscillator (usually approximately 5%) and the linearity of the ramp generator. This accuracy cannot compete with even the least-expensive frequency counter, but the scope can still be used to determine whether a circuit is functioning properly.

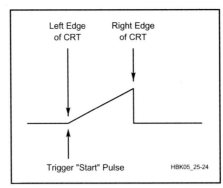

Fig 25.24 — The sweep trigger starts the ramp waveform that sweeps the CRT electron beam from side to side.

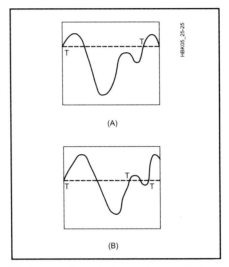

Fig 25.25 — In order to produce a stable display the selection of the trigger point is very important. Selecting the trigger point in A produces a stable display, but the trigger shown at B will produce a display that "jitters" from side to side.

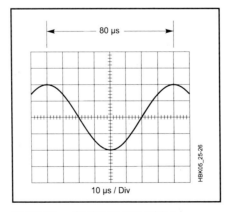

Fig 25.26 — Oscilloscopes with a calibrated sweep rate can be used to measure frequency. Here the waveform shown has a period of 80 microseconds (8 divisions × 10 μs per division) and therefore a period of 1/80 μs or 12.5 kHz.

DUAL-TRACE OSCILLOSCOPES

Dual-trace oscilloscopes can display two waveforms at once. This type of scope has two vertical input channels that can be displayed either alone, together or one after the other. **Fig 25.27** shows a simplified block diagram of a dual-trace oscilloscope. The only differences between this scope and the previous example are the additional vertical amplifier and the "channel switching circuit." This block determines whether we display channel A, channel B or both (simultaneously). The dual display is not a true dual display (there is only one electron gun in the CRT) but the dual traces are synthesized in the scope.

There are two methods of synthesizing a dual-trace display from a single-beam scope. These two methods are referred to as *chopped mode* and *alternate mode*. In the chopped mode a small portion of the channel A waveform is written to the CRT, then a corresponding portion of the channel B waveform is written to the CRT. This procedure is continued until both waveforms are completely written on the CRT. The chopped mode is especially useful where an actual measure of the phase difference between the two waveforms is required. The chopped mode is usually most useful on slow sweep speeds (times greater than a few microseconds per division).

In the alternate mode, the complete channel A waveform is written to the CRT followed immediately by the complete channel B waveform. This happens so quickly that it appears that the waveforms are displayed at the same time. This mode of operation is not useful at very slow sweep speeds, but is good at most other sweep speeds.

Most dual-trace oscilloscopes also have a feature called "X-Y" operation. This feature allows one channel to drive the horizontal amplifier of the scope (called the X channel) while the other channel (called Y in this mode of operation) drives the vertical amplifier. Some oscilloscopes also have an external Y input. X-Y operation allows the scope to display Lissajous patterns for frequency and phase comparison and to use specialized test adapters such as curve tracers or spectrum analyzer front ends. Because of frequency limitations of most scope horizontal amplifiers the X channel is usually limited to a 5 or 10-MHz bandwidth.

DIGITAL OSCILLOSCOPES

The classic analog oscilloscope just discussed has existed for over 50 years. In the

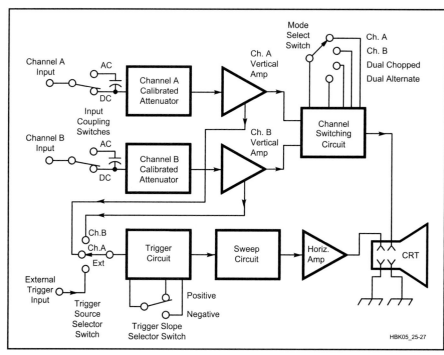

Fig 25.27 — Simplified Dual-trace oscilloscope block diagram. Note the two identical input channels and amplifiers.

last 15 years, the digital oscilloscope has advanced from a specialized laboratory device to a very useful general-purpose tool, with a price attractive to an active experimenter. It uses digital circuitry and microprocessors to enhance the processing and display of signals. These result in dramatically improved accuracy for both amplitude and time measurements. When configured as a digital storage oscilloscope (DSO) it can read a stored waveform for as long as you wish without time limitations incurred by an analog type of storage scope.

Examine the simplified block diagram shown in **Fig 25.28**. After the signal goes through the vertical input attenuators and amplifiers, it arrives at the analog-to-digital converter (ADC). The ADC assigns a digital value to the level of the analog input signal and puts this in a memory similar to computer RAM. This value is stored with an assigned time, determined by the trigger circuits and the crystal timebase. The digital oscilloscope takes discrete amplitude samples at regular time intervals. If you were to take this data directly from memory, it would put a series of dots on the screen. You would then have to connect the dots to reconstruct the original waveform. The digital scope's microprocessor does this for you by mathematically processing the signal while reading it back from the memory and driving a digital-to-analog converter (DAC), which then drives the vertical deflection amplifier. A DAC also takes the digital stored time data and uses it to drive the horizontal deflection amplifier.

For the vertical signals you will see manufacturers refer to "8-bit digitizing," or perhaps "10-bit resolution." This is a measure of how many digital levels are shown along the vertical (voltage) axis. More bits give you better resolution and accuracy of measurement. An 8-bit vertical resolution means each vertical screen has 2^8 (or 256) discrete values; similarly, 10 bits resolution yields 2^{10} (or 1024) discrete values.

It is important to understand some of the limitations resulting from sampling the signal rather than taking a continuous, analog measurement. When you try to reconstruct a signal from individual discrete samples, you must take samples at least twice as fast as the highest frequency signal being measured. If you digitize a 100-MHz sine wave, you should take samples at a rate of 200 million samples a second (referred to as 200 Megasamples/second). Actually, you really would like to take samples even more often, usually at a rate at least five times higher than the input signal.

If the sample rate is not high enough, very fast signal changes between sampling points will not appear on the display. For example, **Fig 25.29** shows one signal measured using both analog and digital scopes. The large spikes seen in the analog-scope display are not visible on the digital scope. The sampling frequency of the digital scope is not fast enough to store the higher frequency components of the waveform. If you take samples at a rate less than twice the input frequency, the reconstructed signal has a wrong apparent frequency; this is referred to as *aliasing*. In Fig 25.29 you can see that there is about one sample taken per cycle of the input waveform. This does not meet the

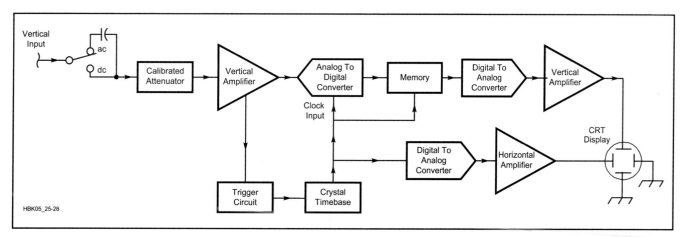

Fig 25.28 — Simplified block diagram of a digital oscilloscope. Note: The microprocessors are not shown for clarity. From "ABC's of Oscilloscopes," copyright Fluke Corporation (reproduced with permission).

2:1 criteria established above. The result is that the scope reconstructs a waveform with a different apparent frequency.

Many older digital scopes had potential problems with aliasing. Newer scopes use advanced techniques to check themselves. A simple manual check for aliasing is to use the highest practical sweep speed (shortest time per division) and then to change to other sweep speeds to verify that the apparent frequency doesn't change.

LIMITATIONS

Oscilloscopes have fundamental limits, primarily in frequency of operation and range of input voltages. For most purposes the voltage range of a scope can be expanded by the use of appropriate probes. The frequency response (also called the bandwidth) of a scope is usually the most important limiting factor. At the specified maximum response frequency, the response will be down 3 dB (0.707 voltage). For example, a 100-MHz 1-V sine wave fed into a 100-MHz bandwidth scope will read approximately 0.707 V on the scope display. The same scope at frequencies below 30 MHz (down to dc) should be accurate to about 5%.

A parameter called *rise time* is directly related to bandwidth. This term describes a scope's ability to accurately display voltages that rise very quickly. For example, a very sharp and square waveform may appear to take some time in order to reach a specified fraction of the input voltage level. The rise time is usually defined as the time required for the display to show a change from the 10% to 90% points of the input waveform, as shown in **Fig 25.30**. The mathematical definition of rise time, assuming a single-pole response in the oscilloscope circuitry, is given by:

$$t_r = \frac{0.35}{BW} \qquad (10)$$

where
t_r = rise time, μs
BW = bandwidth, MHz

It is also important to note that all but the most modern (and expensive) scopes are not designed for precise measurement of either time or frequency. At best, they will not have better than 5% accuracy in these applications. This does not change the usefulness of even a moderately priced oscilloscope, however. The most important value of an oscilloscope is that it presents an image of what is going on in a circuit and quickly shows which component or stage is at fault. It can show modulation levels, relative gain between stages and oscillator output.

OSCILLOSCOPE PROBES

Oscilloscopes are usually connected to a circuit under test with a short length of shielded cable and a probe. At low frequencies, a piece of small-diameter coax cable and some sort of insulated test probe might do. Unfortunately, at higher frequencies the capacitance of the cable would produce a capacitive reactance much less than the one-megohm input impedance of the oscilloscope. In addition each scope has a certain built-in capacitance at its input terminals (usually between 5 and 35 pF). These two capacitances cause problems when probing an RF circuit with a relatively high impedance.

The simplest method of connecting a signal to a scope is to use a specially designed probe. The most common scope probe is a ×*10 probe* (called a times ten probe). This probe forms a 10:1 voltage divider using the built-in resistance of the probe and the input resistance of the scope. When using a ×10 probe, all voltage readings must be multiplied by 10. For example, if the scope is on the 1 V/division range and a ×10 probe were in use, the signals would be displayed on the scope face at 10 V/division.

Unfortunately a resistor alone in series with the scope input seriously degrades the scope's rise-time performance and therefore its bandwidth. Since the scope input looks like a parallel RC circuit, the series resistor feeding it causes a significant reduction in available charging current from the source. This may be corrected by using a compensating capacitor in parallel with the series resistor. Thus two dividers are formed: one resistive voltage divider and one capacitive voltage divider. With these two dividers connected in parallel and the RC relationships shown in **Fig 25.31**, the probe and scope should have a flat response curve through the whole bandwidth of the scope.

To account for manufacturing tolerances in the scope and probe the compensating capacitor is made variable. Most scopes provide a "calibrator" output that produces a known-frequency square wave for the purpose of adjusting the compensating capacitor in a probe. **Fig 25.32** shows possible responses when the probe is connected to the oscilloscope's calibrator jack.

If a probe cable is too short, do not attempt to extend the length of the cable by adding a piece of common coaxial cable. The cable usually used for probes is much different from common 50 or 75-Ω coax. In addition the compensating capacitor in the probe is chosen to compensate for the provided length of cable. It usually will not have enough range to compensate for extra lengths.

The shortest ground lead possible should be used from the probe to the circuit ground. Long ground leads are inductors at high frequencies. In these circuits they cause ringing and other undesirable effects.

THE MODERN SCOPE

For many years a scope (even a so-called portable) was big and heavy. Computers and modern ICs have reduced the size and weight. Modern scopes can take other forms than the traditional large cabinet with built-in CRT. Some modern scopes use an LCD display for true portability. Some scopes take the form of a card plugged into a PC, where they use the PC and monitor for display. Even if they don't use a PC for a display, many scopes can attach

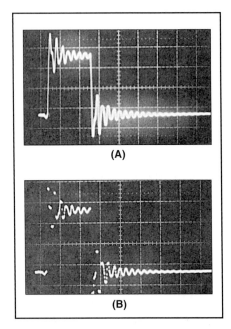

Fig 25.29 — Comparison of an analog scope waveform (A) and that produced by a digital oscilloscope (B). Notice that the digital samples in B are not continuous, which may leave the actual shape of the waveform in doubt for the fastest rise time displays the scope is capable of producing.

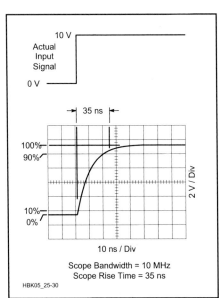

Fig 25.30 — The bandwidth of the oscilloscope vertical channel limits the rise time of the signals displayed on the scope.

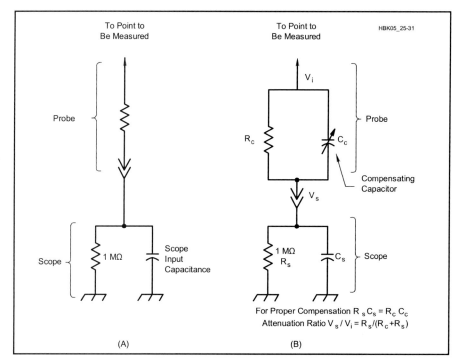

Fig 25.31 — Uncompensated probes such as the one at A are sufficient for low-frequency and slow-rise-time measurements. However, for accurate display of fast rise times with high-frequency components the compensated probe at B must be used. The variable capacitor is adjusted for proper compensation (see text for details).

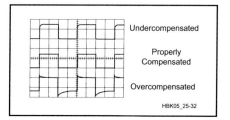

Fig 25.32 — Displays of a square-wave input illustrating undercompensated, properly compensated and overcompensated probes.

This circuit, with a 25-MHz LO frequency, is useful on frequencies in the 20 to 30-MHz range with transmitters of up to 50-W power output. By changing the frequency of the LO, any frequency in the range of the coupler can be displayed on a 5-MHz-bandwidth oscilloscope. The frequency displayed will be the difference between the LO and the input signal. As an example, a 28.1-MHz input and a 25-MHz LO will be seen as a 3.1-MHz signal on the oscilloscope.

More attenuation will be required for higher-power transmitters. This circuit was described by Kenneth Stringham Jr, AE1X, in the "Hints and Kinks" column of February 1982 *QST*.

See also the Oscilloscope Bibliography on page 25.48.

to a PC and download their data for storage and analysis using advanced mathematical techniques. Many high-end scopes now incorporate nontraditional functions, such as Fast Fourier Transforms (FFT). This allows limited spectrum analysis or other advanced mathematical techniques to be applied to the displayed waveform.

BUYING A USED SCOPE

Many hams will end up buying a used scope due to price. If you buy a scope and intend to service it yourself, be aware all scopes that use tubes or a CRT contain lethal voltages. Treat an oscilloscope with the same care you would use with a tube-type high-power amplifier. The CRT should be handled carefully because if dropped it will crack and implode, resulting in pieces of glass and other materials being sprayed everywhere in the immediate vicinity. You should wear a full-face safety shield and other appropriate safety equipment to protect yourself.

Another concern when servicing an older scope is the availability of parts. Many scopes made since about 1985 have used special ICs, LCDs and microprocessors. Some of these may not be available or may be prohibitive in cost. You should buy a used scope from a reputable vendor— even better yet, try it out before you buy it. Make sure you get the operators manual also.

25.5.6 *Project:* An HF Adapter for Narrow-Bandwidth Oscilloscopes

Fig 25.33 shows the circuit of a simple piece of test equipment that will allow you to display signals that are beyond the normal bandwidth of an inexpensive oscilloscope. This circuit was built to monitor modulation of a 10-m signal on a scope that has a 5-MHz upper-frequency limit. This design features a Mini-Circuits Laboratory SRA-1 mixer. Any stable oscillator or VFO with an output of 10 dBm can be used for the local oscillator (LO), which mixes with the HF signal to produce an IF in the bandwidth of the oscilloscope.

The mixer can handle RF signal levels up to –3 dBm without clipping, so this was set as an upper limit for the RF input. A toroidal transformer coupler is constructed by winding a 31-turn secondary of #28 AWG wire on an FT-37-75 core for RG-174 or an FT-50-75 core for use with slightly larger coax such as RG-58. The primary is the piece of coaxial cable passed through the core center. The coupler gives 30 dB of attenuation and has a flat response from 0.5 to 100 MHz. An additional 20-dB of attenuation was added for a total of 50 dB before the mixer. One-watt resistors will do fine for the attenuator. The completed adapter should be built into a shielded box.

25.5.7 *Project:* A Calibrated Noise Source

NOISE FIGURE MEASUREMENT

One of the most important measurements in communications is the noise figure of a receiving setup. Relative measurements are often easy, while accurate ones are more difficult and expensive. One EME (moon bounce) station checks noise and system performance by measuring the noise of the sun reflected off the moon. While the measurement source (use of the sun and moon) is not expensive, the measuring equipment on 2 m consists of 48 antennas (each over 30 ft long). This measurement equipment is not for everyone!

The rest of us use more conventional noise sources and measuring techniques. Coverage of noise and noise figure appear in the **RF Techniques** chapter of this *Handbook*.

Most calibrated and stable noise sources are expensive, but not this unit developed by Bill Sabin, WØIYH. It first appeared in May 1994 *QST*. When hams use a noise source, it is usually included in an RF bridge used to measure impedances and adjust antenna tuners. A somewhat different device (an *accurately calibrated and stable* noise source) is also useful. Combining a broadband RF noise source of known power output and a known output impedance with a true-RMS voltmeter, results in an excellent instrument for making interesting and revealing measurements on a variety of circuits hams commonly use. (Later on, some examples will be described.) The true-RMS voltmeter can be an RF voltmeter, a spectrum analyzer or an AF voltmeter at the output of a linear receiver.[1]

Calibrated noise generators and noise-figure meters are available at medium to astronomical prices. Here is a low-cost approach which can be used with reasonable confidence for many amateur applications where accuracy to tenths of a decibel is not needed, but where precision (repeatability) and comparative measurements are much more important. PC boards are available for this project.[2]

Semiconductor Noise Diodes

Any Zener diode can be used as a source of noise. If, however, the source is to be calibrated and used for reliable measurements, avalanche diodes specially designed for this purpose are preferable by far.[3] A good noise diode generates its noise through a carefully controlled *bulk avalanche* mechanism which exists *throughout* the PN junction, not merely at the junction surfaces where unstable and unreliable surface effects predominate due to local breakdown and impurity.[4] A true noise diode has a very low *flicker noise* (1/f) effect and tends to create a uniform level of truly *Gaussian noise* over a wide band of frequencies.[5] In order to maximize its bandwidth, the diode also has very low junction capacitance and lead inductance.

This project uses the NOISE/COM NC302L diode. It consists of a glass, axial-lead DO-35 package and is rated for use from 10 Hz to 3 GHz, if appropriate construction methods are followed. Prior to sale, the diodes are factory aged for 168 hours and are well stabilized. NOISE/COM has kindly agreed to make these diodes available to amateur experimenters for the special price of $10 each; the usual low-quantity price is about $25.[6]

NOISE SOURCE DESIGN

The noise source presents two kinds of available output power. One is the thermal noise (–174 dBm/Hz at room temperature) when the diode is turned off. This is called N_{OFF}. The other is the sum of this same thermal noise and an "excess" noise, N_E, which is created by the diode when turned on, called N_{ON} (equivalent to $N_{OFF} + N_E$). For accurate measurements, the output impedance of the test apparatus must be the same (on or off) so that the device under test (DUT) always sees the same generator impedance. In Amateur Radio work, this impedance is usually 50 Ω, resistive. The circuit design must guarantee this condition.

For maximum frequency coverage, a PC-

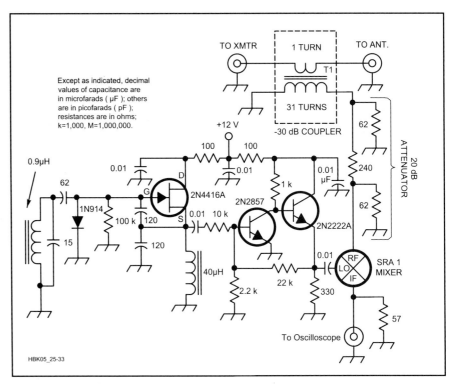

Fig 25.33 — This adapter displays HF signals on a narrow-bandwidth oscilloscope. It uses a 10-dBm 25-MHz LO, –30-dB coupler, 20-dB attenuator and diode-ring mixer. See text for further information.

board layout and coax connector suitable for use at microwaves are needed. For lower frequency usage, a less stringent approach can be employed. Two noise sources are presented here. One is for the 0.5 to 500-MHz region and uses conventional components that many amateurs already have. The other is for the 1-MHz to 2.5-GHz range; it uses chip components and an SMA connector.

CIRCUIT DIAGRAM AND CONSTRUCTION

Figs 25.34 and **25.35** show the simple schematics of the two noise sources. In series with the diode is a 46.4-Ω resistor that combines with the dynamic resistance of the diode in the avalanche, noise generator mode (about 4 Ω) to total about 50 Ω. When the applied voltage polarity is reversed, the diode is forward conducting and its dynamic resistance is still about 4 Ω, but the avalanche noise is now turned off. As a result, the noise source output impedance is always about 50 Ω. The 5-dB pad reduces the effect of any small impedance differences, so that the output impedance is nearly constant from the *on* to the *off* condition, and the SWR is less than 2:1.

Consider the noise situation of the noise diode when it is forward conducting. The resistance of the forward biased PN junction is a *dynamic* resistance. This dynamic resistance is *not* a source of thermal noise, since it is not an actual physical resistance such as in a resistor or lossy network. However, the 0.6-V forward drop across the PN junction does produce a shot noise effect. The mathematics of this shot noise shows that the noise power associated with this effect is only about 50% of the thermal noise power that would be available from a physical resistor having the same value as the dynamic resistance. Therefore, the forward biased junction does *not* add excess noise to the system.[7] There is an 1/f noise effect associated with this shot noise in the diode, but its corner frequency is at about 100 kHz and of no importance at higher frequencies. Also, the small amount of bulk resistance contributes a little thermal noise.

In order to maximize the unit's flatness and frequency response bandwidth, noise-source construction methods should aim for RF circuit lead lengths as close to zero as possible as well as minimum inductance in the ground path and the coupling capacitors. The power-supply voltage must be clean, well bypassed and set accurately. **Fig 25.36** shows a 0.5 to 500-MHz unit. This construction method satisfies quite well the electrical requirements wanted for this model. At 500 MHz, the return loss with respect to 50 Ω at the output jack decreased to 10 dB. A calibration chart

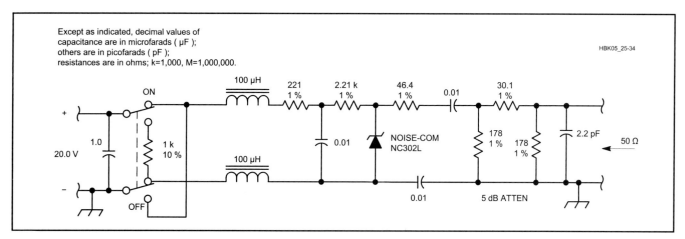

Fig 25.34 — Schematic of the 0.5 to 500-MHz calibrated noise source. Resistors are ⅛-W, 1%-tolerance metal-film units. 1% resistors are available from Digi-Key.

(**Fig 25.37**) is attached to the unit's top for easy reference. **Fig 25.38** shows the inside of the 1-MHz to 2.5-GHz noise source.

CALIBRATING THE NOISE SOURCE

If the construction is solid, the calibration should last for a long time. There are two ways to calibrate the noise source. If the unit *has been carefully constructed and its correct operation verified*, NOISE/COM will calibrate home-built units over the desired frequency range for $25 plus return shipping charges. Note that one factory calibrated unit can be used as a reference for many home calibrated units. **Fig 25.39** shows the NOISE/COM calibration data for both models of prototype noise sources, including SWR data. The noise data is strictly valid only at room temperature, so it's necessary to avoid extreme temperature environments.

The second calibration method requires a signal generator with known output levels at the various desired calibration frequencies. One approach is to build a tunable weak-signal oscillator that can be compared to some accessible high-quality signal generator, using a sensitive receiver as a detector.[8] The level of the signal source in dBm is needed.

Access to a multistage attenuator is also desirable. Build the attenuator using the nearest 1% values of metal-film resistors, so that systematic errors are minimized. A total attenuation of 25 dB in 0.1-dB steps is desirable. Attenuator construction must be appropriate for use at the intended frequency range. In some cases, a high-frequency correction chart may be needed.

With the calibrated signal source and the attenuator feeding the receiver in an SSB or CW mode the techniques discussed in the reference of Note 1 should be used to determine the excess noise (N_E) of the noise source and the noise bandwidth (B_N) of the receiver.

EXCESS NOISE RATIO

A few words about excess noise ratio (ENR) are needed. It is defined as the ratio of excess noise to thermal noise. That is,

$$\text{ENR} = \frac{N_{ON} - N_{OFF}}{N_{OFF}} = \frac{N_E}{N_{OFF}} \quad (12)$$

When the noise source is turned on, its output is $N_{OFF} + N_E$. The ratio of N_{ON} to N_{OFF} is then

$$\frac{N_{ON}}{N_{OFF}} = \frac{N_{OFF} + N_E}{N_{OFF}} = 1 + \frac{N_E}{N_{OFF}} = 1 + \text{ENR} \quad (13)$$

Therefore, ENR is a measure of how much the noise increases and the noise generator can be calibrated in terms of its ENR.

Normalizing ENR to a 1-Hz bandwidth and converting to decibels, this is

$$\text{ENR (dB)} = 174 \text{ (dBm/Hz)} + \frac{N_E \text{ (dBm)}}{B_N \text{ (Hz)}} \quad (14)$$

Prepare a calibration chart and attach it to the top of the unit (see Fig 25.37). If the unit is to be factory calibrated, first perform the calibration procedure to ensure everything

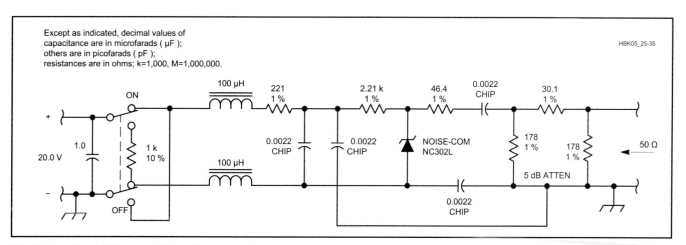

Fig 25.35 — The 1-MHz to 2.5-GHz calibrated noise source uses 1%-tolerance, 0.1-W chip resistors and chip capacitors. 1% resistors are available from Digi-Key.

is working properly. Remember, a factory calibrated unit can be used as a reference for other home calibrated units, once the calibration-transfer procedures have been worked out. This requires some careful thinking and proper techniques. Generally speaking, a NOISE/COM calibration is the best choice.

NOISE-FIGURE MEASUREMENT

The thermal noise power available from the attenuator remains constant for any value of attenuator setting. But the excess noise and therefore the ENR (in dB) due to the noise diode is equal to the calibration point of the source minus the setting (in dB) of the attenuator.

The noise-figure measurement of a device under test (DUT) uses the Y method and the setup in **Fig 25.40**. If the DUT has a noise-generator input and a true-RMS noise-measuring instrument at the output, then the total output noise (including the contribution of the measuring instrument) with the noise generator turned off is

$$N_{OFF\,(TOT)} = kTB_N\, F_{TOT}\, G_{DUT}\, G_{NMI} \tag{15}$$

where

kTB_N = thermal noise,
G_{DUT} = gain of the DUT,
G_{NMI} = gain of the noise-measuring instrument, and
F_{TOT} = noise factor of the combination of the DUT and the noise-measuring istrument.

When the noise generator is turned on, the output noise is

$$N_{ON\,(TOT)} = kTB_N\, F_{TOT}\, G_{DUT}\, G_{NMI} + (ENR)\, kTB_N\, G_{DUT}\, G_{NMI} \tag{16}$$

Where the last term is the contribution of excess noise by the noise generator. Note that none of these values is in dB or dBm.

If we divide equation 16 by equation 15 and say that the ratio

$$\frac{N_{ON\,(TOT)}}{N_{OFF\,(TOT)}} = Y \tag{17}$$

then,

$$Y = \frac{F_{TOT} + ENR}{F_{TOT}} = 1 + \frac{ENR}{F_{TOT}}$$

Note that kTB_N, G_{DUT} and G_{NMI} disappear, so that these quantities need not be known to measure noise factor. If we solve equation 17 for F_{TOT}, we get the noise factor

$$F_{TOT} = \frac{ENR}{Y - 1} \tag{18}$$

If the noise output doubles (increases by 3 dB) when we turn on the noise source, then Y = 2 and the noise factor is numerically equal to the excess noise ratio (ENR). If the attenuator steps are not fine enough or if the attenuator is not reliable over the entire frequency range, use equation 18 to get a better answer. (It's much simpler to use a good fine-step attenuator.) The value of F_{TOT} is that of the DUT in cascade with the noise-measuring instrument. To find F_{DUT}, we must know the noise factor F_{NMI} of the noise-measuring instrument and G_{DUT} and then use the Friis formula, unless G_{DUT} is very large (as it would be if the DUT were a high-gain receiver (see Note 1).

$$F_{DUT} = F_{TOT} - \frac{F_{NMI} - 1}{G_{DUT}} \tag{19}$$

The validity of equation 19 (if we need to use it) requires that the noise bandwidth of the noise-measuring instrument be less than the noise bandwidth of the DUT (see the reference of Note 1). Verify this before proceeding.

There's another advantage to using the power-doubling method. If the 3-dB attenuator of Fig 25.40 is used to maintain a constant noise level into the following stages and the

Fig 25.36 — An inside view of the 0.5 to 500-MHz noise source.

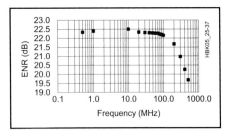

Fig 25.37 — Sample calibration chart for the 0.5 to 500-MHz noise source.

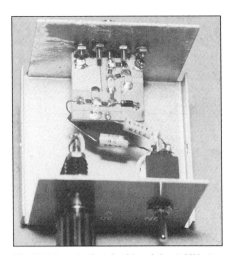

Fig 25.38 — A view inside of the 1-MHz to 2.5-GHz noise source.

Frequency (MHz)	0.5 to 500 MHz Unit		1.0 to 2500 MHz Unit	
	ENR (dB)	SWR	NR (dB)	SWR
0.5	22.33	1.03		
1	22.38	1.03	21.38	1.03
10	22.45	1.04	21.46	1.03
20	22.35	1.06		
30	22.32	1.06		
40	22.32	1.09		
50	22.30	1.11		
60	22.29	1.12		
70	22.25	1.15		
80	22.22	1.17		
90	22.20	1.20		
100	22.15	1.23	21.80	1.07
200	21.65	1.42		
300	20.96	1.62		
400	20.25	1.70		
500	19.60	1.90	20.71	1.44
1000			20.12	1.86
1500			20.00	2.06
2000			20.70	2.14
2500			21.51	1.88

Fig 25.39 — NOISE/COM calibration data for both prototype noise sources. The data is not universal; it varies from unit to unit.

RMS meter, this means that the noise factor, using the calibration scale and the input attenuator (without using equation 18), is

$$F_{DUT} = ENR + \frac{1}{G_{DUT}} \quad (20)$$

If G_{DUT} is large, then the last term can be neglected. If G_{DUT} is small, we need to know its value. However, we do not need to know the noise factor F_{NMI} of the circuitry after the DUT, as we did in the previous discussion.

The 3-dB attenuator method also removes all restrictions regarding the type of noise measuring instrument, since the meter reading is now used only as a reference point. This last statement applies only when two noise (or two signal generator) inputs are being compared.

FREQUENCY RESPONSE MEASUREMENTS

The noise generator, in conjunction with a spectrum analyzer, is an excellent tool for measuring the frequency response of a DUT, if the noise source is much stronger than the internal noise of the DUT and that of the spectrum analyzer. Many spectrum analyzers are not equipped with tracking generators, which can be quite expensive for an amateur's budget.

The spectrum analyzer needs to be calibrated for a noise input, if accurate amplitude measurements are needed, because it responds differently to noise signals than to sine-wave signals. The envelope detection of noise, combined with the logarithmic amplification of the spectrum analyzer, creates an error of about 2.5 dB for a noise signal (the noise is that much greater than the instrument indicates). Also, the noise bandwidth of the IF filter is different from its resolution bandwidth. Some modern spectrum analyzers have internal DSP algorithms that make the corrections so that external noise sources and also carrier-to-noise ratios, normalized to some noise bandwidth like 1.0 Hz, can be measured with fair accuracy if the input noise is a few decibels above thermal. One example is the Tektronix Model 2712. If only relative response readings are needed, then these corrections are not needed.

Also, the noise source itself can be used to establish an accurate reference level (in dBm) on the screen. An accurate, absolute measurement with the DUT in place will then be this reference level (in dBm), plus the increment in decibels produced by the DUT.

The noise-generator output can be viewed as a collection of sine waves separated by, say, 1 Hz. Each separated frequency "bin" has its own Gaussian amplitude and random phase with respect to all the others. So the DUT is simultaneously looking at a collection or "ensemble," of input signals. As the spectrum analyzer frequency sweeps, it looks simultaneously at all of the DUT frequencies that fall with-in the spectrum analyzer's IF noise bandwidth. The spectrum display is thus the "convolution" of the IF filter frequency response and the DUT frequency response. If the DUT is a narrow filter, a very narrow resolution and a slow sweep are needed in the spectrum analyzer. In addition, the analyzer's video, or post-detection, filter has a narrow bandwidth and also requires some settling time to get an accurate reading. So, some experience and judgment are required to use a spectrum analyzer this way.

USING YOUR STATION RECEIVER

Your station receiver can also be used as a spectrum analyzer. Place a variable attenuator between the DUT and the receiver. As you tune your receiver, in a narrow CW mode, adjust the attenuator for a constant reference level receiver output. The attenuator values are inversely related to the frequency response.

A calibrated noise source with an adjustable attenuator that can be easily switched into a receiver antenna jack is an excellent tool for measuring antenna noise level or incoming weak signal level (in dBm) or for establishing correct receiver operation.

The noise source can also be combined with a locally generated data-mode waveform of a known dBm value to get an approximate check on modem performance or to make adjustments that might assure correct operation of the system. The rigorous evaluation of system performance requires special equipment and techniques that may be unavailable at most amateur stations. Or, you could evaluate the intelligibility improvement of your SSB transmitter's speech processor in a noise background.

SUMMARY

The calibrated, flat-spectrum noise generator described in this article is quite a useful instrument for amateur experimenters. Its simplicity and low cost make it especially attractive. Getting a good calibration is the main challenge, but once it is achieved, the calibration lasts a long time, if the right diode is used. The ENR of the units described here is in the range of 20 dB. Use of a high-quality, external, 10-dB attenuator barrel will get into the range of 10-dB ENR. If the unit is sent to NOISE/COM the attenuator should also be sent, with the request that it be included in the calibration. That attenuator then "belongs" to the noise source and should be so tagged. If the attenuator is of high quality, the output SWR will also be improved. NOISE/COM suggests periodic recalibration, at your discretion.

Notes

[1] W. Sabin, "Measuring SSB/CW Receiver Sensitivity," QST, Oct 1992, pp 30-34. See also Technical Correspondence, QST, Apr 1993, pp 73-75.
[2] PC boards are available from FAR Circuits, Dundee, IL. A PC board template for the Sabin noise source is available on the CD that comes with this book.
[3] The term Zener diode is commonly used to denote a diode that takes advantage of avalanche effect, even though the Zener effect and the avalanche effect are not exactly the same thing at the device-physics level.
[4] The term bulk avalanche refers to the avalanche multiplication effect in a PN junction. A carrier (electron or hole) with sufficient energy collides with atoms and causes more carriers to be knocked loose. This effect "avalanches" and it occurs throughout the volume of the PN junction. This mechanism is responsible for the high-quality noise generation in a true noise diode.
[5] Gaussian noise refers to the instantaneous values of a noise voltage. These values conform to the Gaussian probability density function of statistics.
[6] NOISE/COM Co, Paramus, NJ.
[7] Motchenbacher and Fitchen, Low Noise Electronic Design (New York: Wiley & Sons, 1973), p 22.
[8] W. Hayward and D. DeMaw, Solid State Design for the Radio Amateur (Newington: ARRL, 1986).

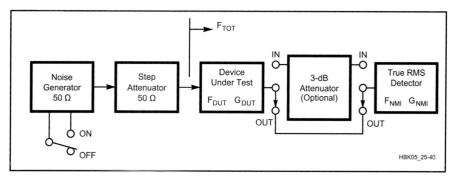

Fig 25.40 — Setup for measuring noise figure of a device under test (DUT).

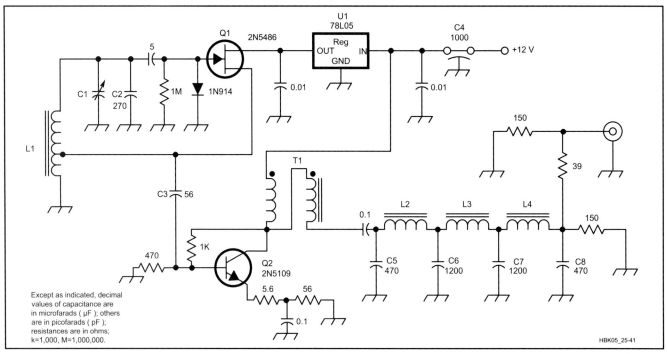

Fig 25.41 — Schematic diagram of the LC oscillator operating at 3.7 MHz. All resistors are ¼-W, 5% units.

C1 — 1.4 to 9.2-pF air trimmer (value and type not critical).
C2 — 270-pF silver-mica or NP0 capacitor. Value may be changed slightly to compensate for variations in L1.
C3 — 56-pF silver mica or NP0 capacitor. Value may be changed to adjust output power.
C4 — 1000 pF solder-in feedthrough capacitor. Available from Microwave Components of Michigan.
C5-C8 — Silver-mica, NP0 disc or polystyrene capacitor.
D1 — 1N914, 1N4148.
L1 — 31t #18 enameled wire on T-94-6 core. Tap 8 turns from ground end (7.5 µH).
L2, L4 — 21t #22 enameled wire on a T-50-2 core (2.5 µH, 2.43 µH ideal).
L3 — 23t #22 enameled wire on a T-50-2 core. (2.9 µH, 3.01 µH ideal).
T1 — 7t #22 enameled wire bifilar wound on an FT-37-43 core.
Q1 — 2N5486 JFET. MPF102 may give reduced output.
Q2 — 2N5109.
U1 — 78L05 low-current 5-V regulator.

Return Loss Bridges

Return loss is a measure of how closely one impedance matches a reference impedance in phase angle and magnitude. If the reference impedance equals the measured impedance level with a 0° phase difference it has a return loss of infinity. **Fig A** shows basic return-loss measurement setups. Return-loss bridges are good for measuring filter response because return loss measurements are a more sensitive measure of passband response than insertion-loss measurements.

A 100 Hz to 100-kHz Return-Loss Bridge

Ed Wetherhold, W3NQN, has developed a low-frequency return-loss bridge (RLB) that can be adapted to different impedance levels. (See **Fig B**.) This bridge is used primarily for testing passive-LC filters that have been designed to work at a certain impedance level. Return-loss measurements require that the signal generator and RLB match the specific filter impedance level.

The characteristic impedance of this RLB is set by the values of four resistors (R1 = R2 = R3 = R4 = characteristic impedance). Ed mounted the four resistors on a plug-in module and placed

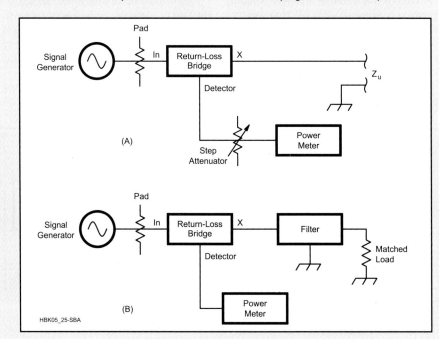

Fig A — A shows a setup to measure an unknown impedance with a return loss bridge. B is for measuring filter response.

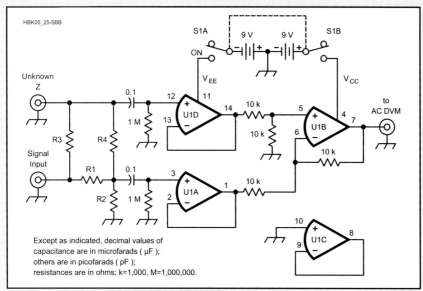

Fig B — An active low-frequency RLB. U1 is an LM324 quad op amp. R1, R2, R3 and R4 are ¼-W, 1% resistors with the same value as the output impedance of the signal generator and the input impedance of the test circuit.

Table A

Performance and Test Data for the 100-Hz to 100-kHz RLB

Bridge and signal-generator impedance = 500 Ω

Bridge Directivity

Frequency	Return Loss (dB, Unknown = 500 Ω)
100 Hz to 10 kHz	> 45
10 kHz to 80 kHz	40
80 kHz to 100 kHz	30

Return Loss of Known Resistive Loads

$R_{LOAD} = LF \times Z_{BRIDGE}$

where
R_{LOAD} = Load resistance, ohms
LF = Load factor
Z_{BRIDGE} = characteristic bridge impedance, ohms.

LF	Return Loss (dB)
5.848	3
3.009	6
1.925	10
1.222	20
1.065	30

their interconnections on the socket. Additional modules may be built for any impedance.

Table A provides computed values of return loss for several known loads and a 500-Ω RLB. If you build this RLB, use the values in the table to check its operation. For this frequency range, use an ac voltmeter in place of the power meter shown in Fig A. Choose one with good ac response well above 100 kHz.

Ed originally described this bridge circuit in an article published in 1993. That article provides complete construction details. Reprints are available for $3 from Ed Wetherhold (W3NQN), 1426 Catlyn Place, Annapolis, MD 21401-4208. The RLB shown is useful from 100 Hz to 100 kHz.

An RF Return-Loss Bridge

At HF and higher frequencies, return-loss bridges are used as shown in Fig A for making measurements in RF circuits. The schematic of a simple bridge is shown in **Fig C**. (Notice that the circuit is identical to that of a hybrid combiner.) It is built in a small box with short leads to the coax connectors. Either 49.9-Ω 1% metal-film or 51-Ω ¼-W carbon resistors may be used. The transformer is wound with 10 bifilar turns of #30 enameled wire on a high permeability ferrite core such as an FT-23-43 or similar.

Apply the output of the signal generator to the RF INPUT port of the RLB. It may be necessary to attenuate the generator output to avoid overloading the amplifier under test. Connect the bridge DETECTOR port to a power meter through a step attenuator and leave the UNKNOWN port of the bridge open circuited. Set the step attenuator for a relatively high level of attenuation and note the power meter indication.

Now connect the unknown impedance, Z_u, to the bridge. The power meter reading will decrease. Adjust the step attenuator to produce the same reading obtained when the UNKNOWN port was open circuited. The difference between the two settings of the attenuator is the return loss, measured in dB.

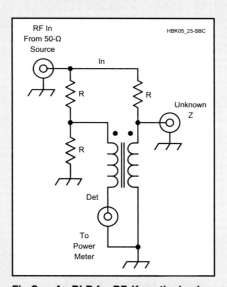

Fig C — An RLB for RF. Keep the lead lengths short. Wind the transformer on a high-permeability ferrite core. Use either 51-Ω carbon or 49.9-Ω 1% metal-film resistors.

The unknown impedance measured by this technique is not limited to amplifier inputs. Coax cable attached to an antenna, a filter, or any other fixed impedance device can be characterized by return loss. Return loss is measured in dB, and it is related to a quantity known as the voltage reflection coefficient, ρ:

$$RL = -20 \log |\rho|$$

$$|\rho| = 10^{\frac{-RL}{20}}$$

where
RL = return loss, dB
ρ = voltage reflection coefficient.

The relationship of return loss to SWR is:

$$SWR = \frac{1 + 10^x}{1 - 10^x}$$

$$x = \frac{-RL}{20}$$

where
For example, if RL = 20 dB
then

$$SWR = \frac{(1 + .1)}{(1 - .1)} = \frac{1.1}{0.9} = 1.22$$

25.5.8 Project: A Signal Generator for Receiver Testing

The oscillator shown in **Fig 25.41** and **Fig 25.42** was designed for testing high-performance receivers. Parts cost for the oscillator has been kept to a minimum by careful design. While the stability is slightly less than that of a well-designed crystal oscillator, the stability of the unit should be good enough to measure most amateur receivers. In addition, the ability to shift frequency is important when dealing with receivers that have spurious responses. More importantly, LC oscillators with high-Q components often have much better phase noise performance than crystal oscillators, because of power limitations in the crystal oscillators (crystals are easily damaged by excessive power).

The circuit is a Hartley oscillator followed by a class-A buffer amplifier. A 5-V regulator is used to keep the power supply output stable. The amplifier is cleaned up by a seven-element Chebyshev low-pass filter, which is terminated by a 6-dB attenuator. The attenuator keeps the filter working properly, even with a receiver that has an input impedance other than 50 Ω. A receiver designed to work with a 50-Ω system may not have a 50-Ω input impedance. The +4 dBm output is strong enough for most receiver measurements. It may even be too strong for some receivers. Note that sensitive components like crystal filters may require a step attenuator to lower the output level.

CONSTRUCTION

This unit is built in a box made of double-sided circuit board. Its inside dimensions are 1 × 2.2 × 5 inches (HWD). The copper foil of the circuit board makes an excellent shield, while the fiberglass helps temperature stability. Capacitor C2 should be soldered directly across L1 to ensure high Q. Since this is an RF circuit, leads should be kept short. While silver mica capacitors have slightly better Q, NP0 capacitors may offer better stability. Mounting the three inductors orthogonal (axis of the inductors 90° from each other) reduced the second-order harmonic by 2 dB when it was compared to the first unit that was made.

ALIGNMENT AND TESTING

The output of the regulator should be +5 V. The output of the oscillator should be +4 dBm (2.5 mW) into a 50-Ω load. Increasing the value of C3 will increase the power output to a maximum of about 10 mW. The frequency should be around 3.7 MHz. Additional capacitance across L1 (in parallel with C2) will lower the frequency if desired, while the trimmer capacitor (C1) specified will allow adjustment to a specific frequency. The drift of one of the first units made was 5 Hz over 25 minutes after a few minutes of warm up. If the warm-up drift is large, changing C2 may improve the situation somewhat. For most receivers, a drift of 100 Hz while you are doing measurements is not bad.

25.5.9 Project: Hybrid Combiners for Signal Generators

Many receiver performance measurements require two signal generators to be attached to a receiver simultaneously. This, in turn, requires a combiner that isolates the two signal generators (to keep one generator from being frequency or phase modulated by the other). Commercially made hybrid combiners are available from Mini-Circuits Labs, Brooklyn, NY.

Alternatively, a hybrid combiner is not difficult to construct. The combiners described here (see **Fig 25.43**) provide 40 to 50 dB of isolation between ports (connections) while attenuating the desired signal paths (each input to output) by 6 dB. The 50-Ω impedance of the system is kept constant (very important if accurate measurements are to be made).

The combiners are constructed in small boxes made from double-sided circuit-board material. Each piece is soldered to the next one along the entire length of the seam. This makes a good RF-tight enclosure. BNC coaxial fittings are used on the units shown. However, any type of coaxial connector can be used. Leads must be kept as short as possible and precision resistors (or matched units from the junk box) should be used. The circuit diagram for the combiners is shown in **Fig 25.44**.

Fig 25.43 — The hybrid combiner on the left is designed to cover the 1 to 50-MHz range; the one on the right 50 to 500 MHz.

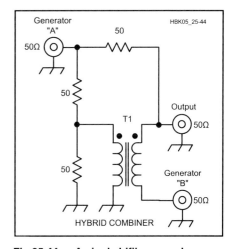

Fig 25.44 — A single bifilar wound transformer is used to make a hybrid combiner. For the 1 to 50-MHz model, T1 is 10 turns of #30 enameled wire bifilar wound on an FT-23-77 ferrite core. For the 50 to 500-MHz model, T1 consists of 10 turns of #30 enameled wire bifilar wound on an FT-23-67 ferrite core. Keep all leads as short as possible when constructing these units.

25.5.10 Project: A Compensated, Modular RF Voltmeter

This versatile and yet simple voltmeter measures levels from 100 mV to 300 V, from 30 MHz to audio frequencies, and is accurate to ±0.5 dB. The design allows for either a built-in analog meter or an externally connected

Fig 25.42 — A low-cost LC oscillator for receiver measurements. Toroidal cores are used for all of the inductances.

DVM—or both! This portable, simple, and inexpensive project was originally featured in the March/April 2001 issue of *QEX*. Like the original article, the project described below includes several refinements implemented by its author, Sid Cooper, K2QHE that evolved from a suggestion by then *QEX* Managing Editor, Robert Schetgen, KU7G, to increase the voltage range of the instrument.

The requirements for this RF voltmeter (RFVM, **Fig 25.46**) are many:
- Provide accurate, stable measurements
- Measure voltage at frequencies from audio through HF
- Measure voltage levels from QRP to QRO
- Measure voltages inside equipment or in coax
- Operate portably or from ac lines
- Be flexible: work with digital voltmeters already in the shack, or independently
- Be inexpensive

DESIGN APPROACH

To achieve stability, the RFVM uses op amps and any drift in the probe's diode detector is compensated by a matched diode in the RF op amp. This stretches the sensitivity to QRP levels. The dynamic range extends linearly from 0.1 V to 300 V (RMS) by using a series of compensated voltage multipliers.

The frequency response was flattened by using Schottky diodes, which easily reach from 60 Hz to 30 MHz. Both the basic probe and the multipliers very simply adapt to function as a probe, clip to various measurement points on the chassis or to screw onto UHF connectors. The active devices are only two op amps. They draw 800 µA of resting current and a maximum of 4.0 mA during operation from a 9-V battery, which is disconnected when a 9-V wall unit is plugged into a jack on the back panel. As designed, the RFVM has a small sloping panel cabinet with a microammeter display. A pair of front-panel banana OUTPUT jacks can be used with a DVM; this eliminates one op amp, the meter and a multiposition switch. Since the meter and switch are the most expensive parts (when they are not bought at a hamfest), how low-cost can this RF voltmeter get?

THE PROBE

The low-voltage probe is a detector circuit that uses a Schottky diode and a high-impedance filter circuit (**Fig 25.45**). The diode is matched with the one in the feedback circuit of the CA3160 shown in **Fig 25.64**. This match reduces the diode's threshold voltage from about 0.34 V to less than 0.1 V, making the voltage drop comparable to that of a germanium diode. Since I think 0.1 V adequately covers QRP requirements, I made no further tests with germanium diodes to determine how much lower in voltage we could go.[1]

Use the low voltage probe and the output of the CA3160 in the meter unit (Fig 25.64) to find a matched diode pair. Build the CA3160 circuit first for this purpose. Select a diode and place it in the feedback loop of the CA3160, then test each of the remaining diodes in the probe with the three pots set about midrange. Test each diode, first at probe inputs of 100 mV, then at 3.00 V, at 400 Hz. Record the dc output from the CA3160, using the OUTPUT terminals. A bag of 20 diodes from Mouser Electronics[2] contained seven matched pairs, with identical readings at both low and high voltages. My tests used the ac scale (good to 500 Hz) of a Heath 2372 DVM to read the input voltage, and its dc scale to read the output. The number of matched pairs is surprising, but the diodes in the bag may all be from the same production run.

RF PROBE ASSEMBLY

After selecting matched diodes, all components of the RF probe are mounted on a piece of perf board that is inserted in an aluminum tube (**Fig 25.47**). Since the circuit board is very light, it is supported on each end only by its input and output wires. The input RF wire is soldered to the center pin of the SO-239 connector. The two dc-output wires are held to the board by a small tie wrap, consisting of a piece of the #26 insulated wire, after having been passed through the plug cap and grommet. Cut a slit in the aluminum tube, perpendicular to its end. Make it at least 1/8-inch long and 1/32-inch wide to allow a bare #26 ground wire to pass through it.

Let's look at how the probe is assembled. The aluminum tube just fits around the back end of the SO-239 connector. When the four #6-32 screws, nuts and washers are installed in the four holes of the SO-239 connector, the flat face of the nut bears down strongly on the aluminum tube and rigidly holds it in place. Most standard SO-239 connector

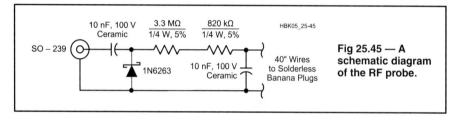

Fig 25.45 — A schematic diagram of the RF probe.

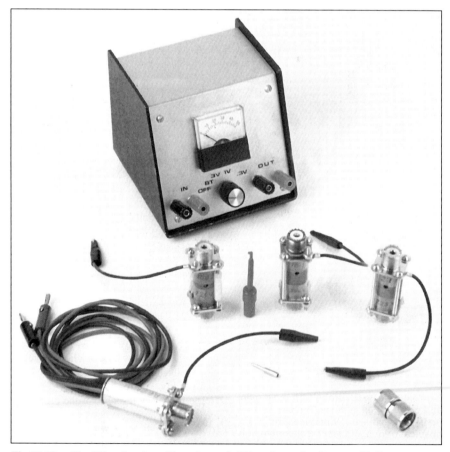

Fig 25.46 — The RF voltmeter with meter unit, RF probe and voltage multipliers.

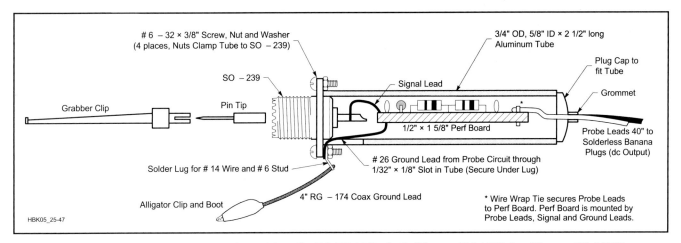

Fig 25.47 — RF probe construction showing the grabber clip (RS 270-334), pin tip (Mouser 534-1600), lug (Mouser 571-34120), for #14-#16 wire and a #6 stud), alligator clip (RS 270-1545), plug cap (Mouser 534-7604), grommet (Mouser 5167-208) and 4-in ground lead made from shield of RG-174.

holes easily accept #6-32 screws; if yours do not, enlarge them with a #27 bit. The slit cut in the aluminum tube allows the bare ground wire from the perf board to pass from inside the tube to the outside and around the screw that secures the ground lead of the probe. As can be seen in Fig 25.47, the threaded part of the SO-239 connector faces away from the aluminum tube.

This probe allows easy voltage measurements in a coaxial cable when the probe SO-239 is secured to a mating T connector. To make a probe reading at any point on a chassis or PC board with this connector, simply insert the larger end of a nicely mating pin tip into the SO-239. The pin tip's diameter is 0.08 inches at the small end and — fittingly — 0.14 inches at the mating end. I made this purely fortuitous discovery while rummaging through a very ancient junk box. Further, to make measurements with the probe clipped to a part or test point on a PC board or chassis, a clip or grabber can be mated to the pin tip that protrudes from the SO-239. This is shown in Figs 25.47 and **25.48**. All this may have been known to people in the connector industry, but it appears not to have been known or used elsewhere. **Fig 25.49** shows the grabber connected to the RF probe using the pin tip to join the two.

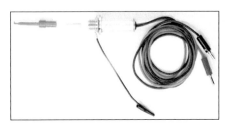

Fig 25.48 — A photo of the grabber clip, pin tip and RF probe.

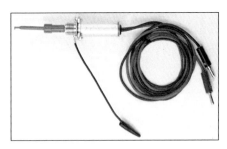

Fig 25.49 — The RF probe with the pin tip and grabber clip in place. The pin tip is not visible, but it provides the physical bridge and electrical connection between the clip and SO-239 of the RF probe.

PROBE MEASUREMENTS

The probe can accept RF signals up to 20 V (RMS) when there is no dc voltage or a combination of dc voltage and peak ac of about 28 V without exceeding the reverse-voltage rating of the diode. I made linearity measurements of the RF probe from 100 mV to 8.0 V at 400 Hz. (The range was limited by my available signal generator.) **Fig 25.50** is a plot of the error at the probe output versus the input amplitude. This was measured using a digital voltmeter with a 10-MΩ input resistance. This input resistance and the 4.1-MΩ resistor in the probe converts the peak voltage of a sine wave signal to the RMS reading of the DVM. The error is –44% at 100 mV, then 10% (0.9 dB) at 1.0 V and finally 2.9% at 8.0 V. The RF probe and DVM are obviously intended for higher voltage readings where the diode voltage drop doesn't affect the accuracy significantly.

The frequency response of the probe was measured at the midpoint of each ham band from 1.9 to 30 MHz with 650 mV input (see **Fig 25.51**). This time, because the input voltage was low and the absolute error would have been high, the probe was connected to the meter unit, which will be described later. The compensating diode loop of the meter unit reduces the error that would otherwise be read by the DVM from 90 mV (14%) to 4 mV (0.6%). The frequency response, however, is controlled by the probe. Use of the meter unit—which operates entirely on the dc signal—does not affect the frequency response, but it does improve the sensitivity of the readings. The measured response shown goes from –3.6% (–0.32 dB) to +1.2% (+0.1 dB) [A larger input capacitor in the probe will improve accuracy at lower frequencies. Pat McGuire, NO5O, reports good results from 50 kHz to 30 MHz by changing the input capacitor from 10 nF to 0.1 µF and performing the calibration described later at 50 kHz instead of 400 Hz. — Ed].

MULTIPLIERS

To extend the voltage range of the RF probe, use a 10× multiplier, which is a compensated divider, shown in **Fig 25.52**. The electrical design is straightforward and includes a small trimmer capacitor to adjust for a flat frequency response. These components are very lightweight, so they require no perf board and are easily supported by their leads, which are anchored at the two SO-239 connectors. To make measurements, the input connector can be mated to connectors in a coaxial cable, a pin tip can be inserted for probing PC boards or a grabber can be added for connections to components on a chassis. The multiplier can connect to the RF probe either through a male-to-male UHF fitting or by a piece of coax with a UHF male connector at each end.

The typical way to adjust the trimmer requires a square-wave input and an oscilloscope on the output. The trimmer is then set to produce a flat square-wave output with no overshoots on the leading edges and no droop across the top. An RF signal genera-

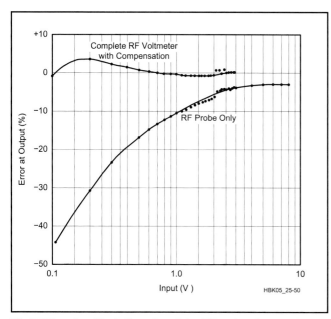

Fig 25.50 — Amplitude linearity of the RF probe alone and in combination with the RF voltmeter.

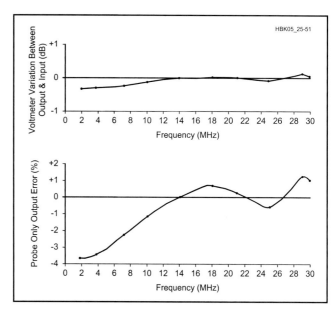

Fig 25.51 — A frequency-response plot for the RF voltmeter. The response of the probe determines the overall system response.

tor that covers 1.9 MHz to 30 MHz can be used instead, if a square wave generator is not available. Since this is a 10× multiplier, it is reasonable to expect the input to handle up to 200 V (RMS), but the limit is 150 V because the trimmer capacitor is rated at 225 V dc. You may be able to find small trimmers with higher voltage ratings that will take the multiplier to 200 V and still fit inside the tube housing.

Multiplier construction follows the same methods used for the RF probe but with a few differences. My first model used an aluminum tube like that in the RF probe with #6-32 screws to hold it all together, but the frequency response began to drop off at 21 MHz. By replacing the aluminum tube with a phenolic tube[3] having the same dimensions, the response is adequate to 30 MHz.

Fig 25.53 shows construction methods. Two slits at the ends of the tube allow the ground wires at the input and output ends to exit and wrap around the screws of each connector. A hole in the phenolic tube allows adjustment of the trimmer capacitor.

Fig 25.54 shows the exploded view of the pin tip and grabbers at the input end of the

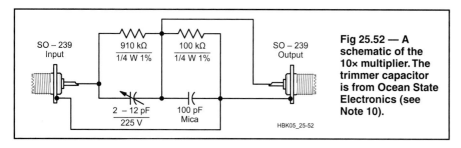

Fig 25.52 — A schematic of the 10× multiplier. The trimmer capacitor is from Ocean State Electronics (see Note 10).

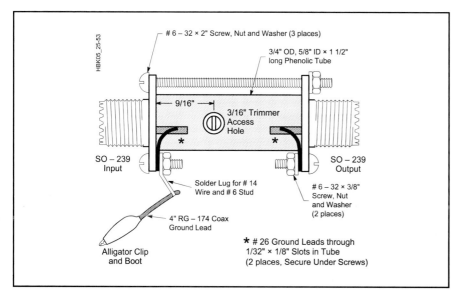

Fig 25.53 — Construction of the 10× multiplier. The phenolic tube is US Plastics #47081, see Note 3. The lug is Mouser 571-34120.

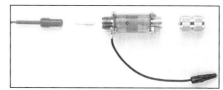

Fig 25.54 — A view showing the 10× multiplier with grabber, pin tip and a male UHF coupler that connects the multiplier to the RF probe.

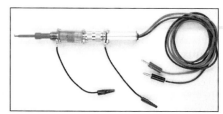

Fig 25.55 — The complete RF probe and 10× multiplier assembly.

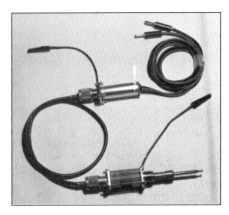

Fig 25.56 — The RF probe and multiplier can be separated by a piece of coax less than 12 inches long if it makes measurements more convenient.

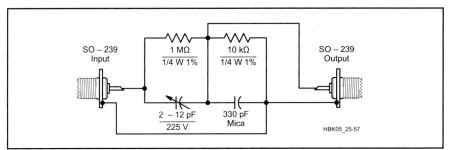

Fig 25.57 — Schematic of the 100× multiplier (maximum allowable input is 150 V RMS). The trimmer capacitor is from Ocean State Electronics (see Note 10).

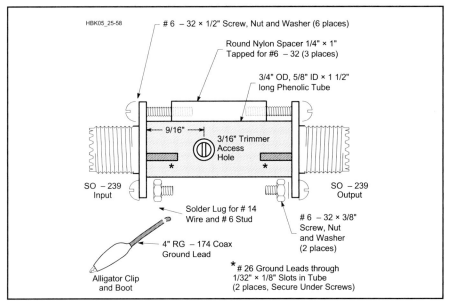

Fig 25.58 — 100× multiplier assembly showing the round Nylon spacer (Mouser 561-TSP10), phenolic tube (US Plastics #47081, see Note 3). The lug is Mouser 571-34120.

multiplier and a UHF male-to-male connector on the output side to mate with the RF probe. The assembled multiplier and probe are shown in **Fig 25.55**. The whole assembly is only 3½ inches long and is comfortable in the hand. If more flexibility is desired, the two pieces may be separated by as much as one foot of coax (**Fig 25.56**) without doing too much damage to the accuracy of the measurement.

The 100× multiplier follows the same design concepts as the 10× multiplier including the obligatory trimmer capacitor, as in **Fig 25.57**. Here too, the 225 V dc rating of the trimmer limits the maximum RF to 150 V RMS. So, this is not useful should it be used only with the probe and your DVM because the 10× multiplier already covers this range. Later, we will see how it is useful with the RF probe and the meter unit.

The 100× unit also uses a phenolic-tube housing with a hole to adjust the trimmer and a slit at each end to bring out the ground wires, which then wrap around screws in the connectors. In order to reach at least 30 MHz, however, the two-inch steel screws between the two SO-239 connectors must be insulated. Using plastic washers at the connector holes could do this job but would have required #4-40× 2-inch-long screws. After a long search at the biggest hardware stores and catalogs, I found nothing longer than 1½ inches. That's the best way to assemble this multiplier, if you can find the screws. Otherwise, use nylon spacers and shorter screws as shown in **Figs 25.58** and **25.59**. The pin tip and grabber at the input and UHF male-to-male connector or coaxial cable at the output are used as with the 10× multiplier.

The only purpose of the 2× multiplier is to create a 200× multiplier when used with the 100× multiplier in series, so as to extend the voltage range from 150 V to 300 V (1800 W at 50 Ω only under matched conditions).[4] The trimmer in each multiplier now divides

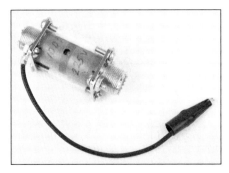

Fig 25.59 — The assembled 100× multiplier.

in half the input voltage. **Fig 25.60** shows the 2× multiplier with only one RC section since it relies on the RC sections in the 100× multiplier to complete the division (multiplication).

The construction, **Fig 25.61**, returns to the tubular format as before, but this time there are no slots to pass the ground wires; the four screws are able to do the job. The two multipliers are shown in series in **Fig 25.62**. The input capacitance for each multiplier is

Table 25.4
Multiplier Range versus Capacitance

Multiplier	10×	100×	200×
Input Cap (pF)	13.3	10.9	13.5

shown in **Table 25.4**.

When measurements are made on a load resistance of 50 Ω, the input capacity of the multipliers has no affect on readings below 30 MHz. When the impedance of the load resistance is larger, consider any error it introduces. To get a sense of the affects of the load resistance, use the equation below. It shows the relationship between the resistance, capacitance and frequency when the measured voltage is 3 dB down from what it would be if it were read by a meter with no input capacity:

$$f = \frac{1}{2\pi R_L C_P} \tag{21}$$

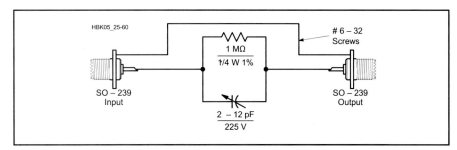

Fig 25.60 — A schematic diagram of the 2× multiplier. Its output connects to the input of the 100× multiplier for 200× measurements up to 300 V RMS. This multiplier is used *only* with the 100× multiplier.

Fig 25.62 — The 2× and 100× multipliers are joined by a male UHF coupler for readings up to 300 V (RMS, 1800 W, see Note 4).

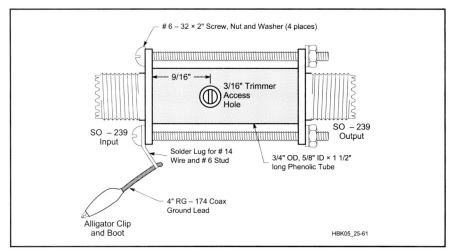

Fig 25.61 — 2× multiplier assembly showing the phenolic tube (US Plastics #47081, see Note 3). The lug is 16-14 #6 stud (Mouser 571-34120).

where

f = 3-dB loss frequency
R_L = load resistance
C_P = probe capacitance

For example, at a load of 1 kΩ and a probe capacity of 13.3 pF, the 3-dB down frequency is about 12 MHz, where the error is 30%. If an attenuator were used instead of the multiplier, it would not have this problem because it works with either a fixed load of 50 Ω at high frequencies or 600 Ω at low frequencies. Probes cannot select their frequency or load impedance and are thereby more flexible in use, so measurements must be made with consideration, but these multipliers would not have a problem at 50 Ω or 600 Ω either.

Table 25.5 summarizes the voltage ranges using only the multiplier with the RF probe and a DVM that has a 10-MΩ input resistance that's available in the shack.

The multipliers are intended for use at high voltages where safety precautions are a primary consideration to avoid personal injury. See the **Safety** chapter of this *Handbook*. In Tektronics' *ABC's of Probes*,[5] an entire section thoroughly covers the hazards and necessary precautions when making measurements with probes. It is worth the little effort to get a copy of it and also the Pomona Catalog,[6] which has some good information on probe use.

The multipliers have panel connectors at both ends and the RF probe has one at one end. Their grounds are connected and brought to the meter unit. This unit is grounded only when the ac-powered 9-V power supply that may be used with it, is grounded. When it is battery operated, the meter unit relies on the ground-clip connection to the ground of the equipment under test. This is satisfactory if there is no unknown break in the chain of ground connections that would make the panel connector hot. Furthermore, as insurance, it is good to wrap these connectors with vinyl tape or to cover the multiplier with a plastic boot to prevent contact with either hands or equipment under test. This is not shown in any photographs because it would have obscured the appearance and construction of the probe and multipliers.

The frequency responses of the multipliers were determined using an RF signal generator set at the midpoint of each ham band from 160 to 10 meters, including 30 MHz. An error of 5.42% (0.46 dB) was measured from 160 to 17 meters, which then decreased to 4.13% (0.35 dB) at 15 meters and then to zero through 30 MHz. The constant error from 160 to 17 meters is probably due to the inherent errors in the test equipment. The oscilloscope has a 60 MHz bandwidth, an input impedance of 1 MΩ and 30 pF, which introduces a load effect on the multiplier outputs. It also has a reading accuracy of ±3%. When the multiplier-components accuracy of 1% is included, it is not surprising to find the overall inaccuracy to be at most 5.42%.[7] The error of the multiplier itself could be inherently less than this. Before the overall frequency response was measured, the trimmer capacitor was adjusted with the input frequency set mid-frequency at 15 MHz. If your interest in RF probes and multipliers has been raised and you have more questions, see the references in Notes 5 and 6.

THE METER UNIT

The meter unit serves several purposes when used with the RF probe. It increases

---

Table 25.5

Multiplier versus Voltage Range

Probe Configuration	Range (V)	Maximum Voltage
Probe Alone*	1.0-20 V ac	20 V ac
With 10× Multiplier*	10-150 V ac	225 V (ac + dc)
With 100× and 2× Multiplier†	150-200 V ac	450 V (ac + dc)
With 100× and 2× Multiplier††	200-300 V ac	450 V (ac + dc)

*error less than −10% or −0.83 dB
†with error of −45% to −10% or less than 3.2 dB. A trimmer capacitor with a rating of 300 V dc increases the 10× range to 200 V ac and reduces the error from 45% to 10%. The RF probe and meter unit reduces the error to less than −3.5% or less than −0.3 dB.
††with error less than −6% or −0.54 dB

$$e_s - e_{d1} = e_p$$
$$e_o - e_{d2} = e_n$$
$$e_i = e_p - e_n = e_s - e_{d1} - e_o + e_{d2}$$
$$e_i = \frac{e_o}{G} = 0$$
$$e_s - e_o = e_{d1} - e_{d2}$$

e_s represents the input of the RF probe as a dc signal.
e_o represents the output from the high-gain CA3160 op amp.
e_{d1} is the voltage drop across the diode in the RF probe.
e_{d2} is the voltage drop across the diode in the feedback loop.
G = 320,000 for the CA3160.
The last equation shows that any difference between the signal and output results from unequal diode drops when the diodes are not matched.

Fig 25.63 — Basics of the diode-compensation method to improve measurement accuracy at low input voltages. It also provides a measure of temperature compensation and drift reduction. See Fig 25.50.

the accuracy of the probe and DVM from –45% to –3.5% at a voltage of 100 mV RF. At higher voltages, it maintains a minimum advantage of 5:1 in reducing the error, when the probe is used with a DVM. This is an increase in both sensitivity and accuracy. The usual non-linearity caused by the diode in the probe is reduced when used with the meter unit. The technique also incidentally provides temperature compensation for diode drift. The meter unit has a ±5% panel meter to display measurements, but it also contains a pair of OUTPUT terminals for a DVM. Add a DVM when more accuracy is desired at the low end of the range or when you want a bit more resolution. Finally, since the meter unit is fully portable, low-level field-strength measurements are possible with a whip antenna at the probe input connector. Due to its sensitivity and accuracy, the probe can be adapted for use in many places around the shack.

The non-linear response of the probe diode is compensated (improved) by a circuit in the meter unit. The feedback loop of a CA3160 op amp contains a diode matched to the one in the probe (see Note 1). **Fig 25.63** and its sequence of equations present a very simple sketch of how matched diodes do this when dc is applied through a diode. The final equation shows that any difference between the op amp input and output is due to a difference in voltage drops across the two diodes. When the diodes are matched, the error disappears. When RF is applied, the average currents through the diodes must be equal to keep the voltage drops equal. Articles by Kuzdrall (Note 1) Grebenkemper[8] and Lewallen[9] are first-class descriptions of the principles used in this RF voltmeter.

Fig 25.64 shows three pots for calibrating the meter unit. This should be performed at 400 Hz to avoid any effects due to RF. The 100-kΩ pot is typically used to null the offset of the CA3160, but is used here to initially set the offset to about 0.5 mV to 1.0 mV with 100 mV input. Then the 1-MΩ pot sets the output to 100 mV with 100 mV input, and the 10 kΩ pot sets the output to 3.0 V when the input is 3.0 V. Finally, the three pots are alternately adjusted until the 100 mV and 3.0 V set points occur together. The 100-kΩ pot is helpful in fine tuning the 100-mV point. A DVM was used at the op-

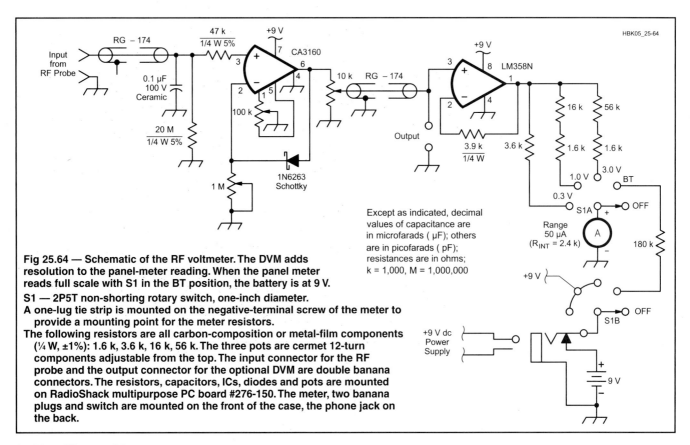

Fig 25.64 — Schematic of the RF voltmeter. The DVM adds resolution to the panel-meter reading. When the panel meter reads full scale with S1 in the BT position, the battery is at 9 V.
S1 — 2P5T non-shorting rotary switch, one-inch diameter.
A one-lug tie strip is mounted on the negative-terminal screw of the meter to provide a mounting point for the meter resistors.
The following resistors are all carbon-composition or metal-film components (¼ W, ±1%): 1.6 k, 3.6 k, 16 k, 56 k. The three pots are cermet 12-turn components adjustable from the top. The input connector for the RF probe and the output connector for the optional DVM are double banana connectors. The resistors, capacitors, ICs, diodes and pots are mounted on RadioShack multipurpose PC board #276-150. The meter, two banana plugs and switch are mounted on the front of the case, the phone jack on the back.

tional OUTPUT sockets during calibration.

A CA3160 was selected for the input op amp because of its high input impedance of 1.5 TΩ and high gain of 320,000. Although it has diodes that provide protection, I think it could be sensitive to electrostatic discharge, so handle it carefully. Use IC sockets to permit easy replacement of the ICs in case of damage. An LM358N IC follows the CA3160 (see Fig 25.64), primarily to drive the panel meter. If you use a DVM as the display, you can omit the LM358N circuit, meter and multi-position switch. The only functions lost are the battery-voltage check and power on/off switch. Add an on/off switch to the circuit when the multi-position switch is omitted. The CA3160 op amp circuit easily spans the range from 100 mV to 3.0 V without the need for a range switch.

CONSTRUCTION

A 5×5×4½-inch sloping-front instrument case was used to house the components of the meter unit. The meter on the front panel has a 2×2-inch face and requires a 1½-inch hole in the panel. Although a 50-µA meter from the junk box is used here, a 1-mA movement will work as well, provided the series resistors are changed accordingly.

The rotary switch has two poles and five positions for changing the meter range, testing the battery condition and switching the power off. Two sets of double banana binding posts are used, the INPUT pair accepts the dc signal from the RF probe. The OUTPUT pair provides a voltage for a DVM display, whether the panel meter is used or not. On the rear of the case, a miniature phone jack accepts 9-V power from either a battery or a 9-V dc supply. The ICs and other parts are mounted on a RadioShack multipurpose PC board that has very convenient holes and traces. The board is bolted to the back of the case via standoff insulators and wired to the front panel components (see **Fig 25.65**).

Notes

[1] J. A. Kuzdrall, "Linearized RF Detector Spans 50-to-1 Range," Analog Applications Issue, *Electronic Design*, June 27, 1994.
[2] Mouser Electronics, 2401 Hwy 287 N, Mansfield, TX 76063; tel 800-346-6873, fax 817-483-0931; E-mail **sales@mouser.com**; www.mouser.com.
[3] United States Plastics Corp, 1390 Neubrecht Road, Lima, OH 45801-3196, tel 419-228-2242, fax 419-228-5034.
[4] Load mismatch changes the measured voltage depending on the phase angle of the load impedance. In the worst-case, the mismatch multiplies the voltage by the square root of the SWR. With 1800 W, a 2:1 SWR (100-ohm load) increases the voltage from 300 to 424 V_{RMS} and a 3:1 SWR (150-ohm load) would give 520 V_{RMS}.
[5] *ABCs of Probes*, Tektronix Inc, Literature number 60W-6053-7, July 1998. Tektronix, Inc. Export Sales, PO Box 500 M/S 50-255 Beaverton, OR 97707-0001; 503-627-6877. Johnny Parham, "How to Select the Proper Probe," Electronic Products, July 1997.
[6] Pomona Test and Measurement Accessories catalog. ITT Pomona Electronics, 1500 E Ninth St, Pomona, CA 91766-3835.
[7] A. Frost, "Are You Measuring Your Circuit or Your Scope Probe?" *EDN*, July 22, 1999. E. Feign, "High-Frequency Probes Drive 50-Ω Measurements," *RF Design*, Oct 1998.
[8] J. Grebenkemper, KI6WX, "The Tandem Match — An Accurate Directional Wattmeter," *QST*, Jan 1987, pp 18-26; and "Tandem Match Corrections," *QST*, Jan 1988, p 49.
[9] R. Lewallen, W7EL, "A Simplified and Accurate QRP Directional Wattmeter," *QST*, Feb 1990, pp 19-23, 36.
[10] Ocean State Electronics, 6 Industrial Dr, PO Box 1458, Westerly, RI 02891; tel 800-866-6626, fax 401-596-3590.

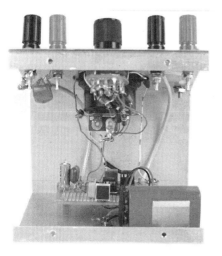

Fig 25.65 — Inside view of the meter unit. RG-174 connects the INPUT and OUTPUT banana binding posts to the circuit board. The rotary switch, one-lug tie strip and battery are also visible.

25.6 Receiver Performance Tests

Comparing the performance of one receiver to another is difficult at best. The features of one receiver may outweigh a second, even though its performance under some conditions is not as good as it could be. Although the final decision on which receiver to purchase will more than likely be based on personal preference and cost, there are ways to compare receiver performance characteristics. Some of the more important parameters are sensitivity, blocking dynamic range and two-tone IMD dynamic range.

Instruments for measuring receiver performance should be of suitable quality and calibration. Always remember that accuracy can never be better than the tools used to make the measurements. Common instruments used for receiver testing include:
- Signal generators
- Hybrid combiner
- Audio ac voltmeter
- Distortion meter (FM measurements only)
- Noise figure meter (only required for noise figure measurements)
- Step attenuators (10 dB and 1 dB steps are useful)

Signal generators must be calibrated accurately in dBm or mV. The generators should have extremely low leakage. That is, when the output of the generator is switched off, no signal should be detected at the operating frequency with a sensitive receiver. Ideally, at least one of the signal generators should be capable of amplitude modulation. A suitable lab-quality piece would be the HP-8640B, no longer manufactured, but a good item to scout on the surplus market.

While most signal generators are calibrated in terms of microvolts, the real concern is not with the voltage from the generator but with the power available. The unit that is used for most low-level RF work is the milliwatt, and power is often specified in decibels with respect to 1 mW (dBm). Hence, 0 dBm would be 1 mW. The dBm level, in a 50-Ω load, can be calculated with the aid of the following equation:

$$dBm = 10 \log_{10}\left[20(V)^2 \right] \quad (22)$$

where

dBm = power with respect to 1 mW
V = RMS voltage available at the output of the signal generator

The convenience of a logarithmic power unit such as the dBm becomes apparent when signals are amplified or attenuated. For example, a −107 dBm signal that is applied to an amplifier with a gain of 20 dB will result in an output increased by 20 dB. Therefore in this example (−107 dBm + 20 dB) = −87 dBm. Similarly, a −107 dBm signal applied to an attenuator with a loss of 10 dB will result in an output of (−107 dBm − 10 dB) or −117 dBm.

A hybrid combiner is a three-port device used to combine the signals from a pair of generators for all dynamic range measurements. It has the characteristic that signals applied at ports 1 or 2 appear at port 3 and

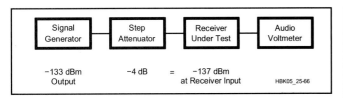

Fig 25.66 — A general test setup for measuring receiver MDS, or noise floor. Signal levels shown are for an example discussed in the text.

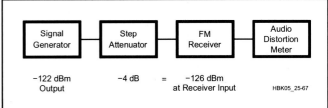

Fig 25.67 — FM SINAD test setup.

are attenuated by 3 dB. However, a signal from port 1 is attenuated 30 or 40 dB when sampled at port 2. Similarly, signals applied at port 2 are isolated from port 1 some 30 to 40 dB. The isolating properties of the box prevent one signal generator from being frequency or phase modulated by the other. A second feature of a hybrid combiner is that a 50-Ω impedance level is maintained throughout the system.

Audio voltmeters should be calibrated in dB as well as volts. This facilitates easy measurements and eliminates the need for cumbersome calculations. Be sure that the step attenuators are in good working order and suitable for the frequencies involved. A distortion meter, such as the Hewlett-Packard 339A, is required for FM sensitivity measurements and a noise figure meter, such as the Hewlett-Packard 8970A, is excellent for certain kinds of sensitivity measurements.

The ARRL Lab performs tests on commercial receivers as part of ARRL Product Review in *QST*. The *ARRL Lab Test Procedures Manual* is available as a PDF file for downloading from **www.arrl.org/product-review**.

25.6.1 Receiver Sensitivity

Several methods are used to determine receiver sensitivity. The mode under consideration often determines the best choice. One of the most common sensitivity measurements is minimum discernible signal (MDS) or noise floor. It is suitable for CW and SSB receivers.

This measurement indicates the minimum discernible signal that can be detected with the receiver. This level is defined as that which will produce the same audio-output power as the internally generated receiver noise. Hence, the term "noise floor."

To measure MDS, use a true-RMS voltmeter and a signal generator tuned to the same frequency as the receiver (see **Fig 25.66**). With the generator output at 0 or with maximum attenuation of its output note the voltmeter reading. Next increase the generator output level until the ac voltmeter at the receiver audio-output jack shows a 3-dB increase. The signal input at this point is the MDS. Be certain that the receiver is peaked on the generator signal. The filter bandwidth can affect the MDS. Always compare MDS readings taken with identical filter bandwidths. (A narrow bandwidth tends to improve MDS performance.) MDS can be expressed in μV or dBm.

In the hypothetical example of Fig 25.66, the output of the signal generator is –133 dBm and the step attenuator is set to 4 dB. Here is the calculation:

$$\text{Noise floor} = -133 \text{ dBm} - 4 \text{ dB} = -137 \text{ dBm} \quad (23)$$

where the noise floor is the power available at the receiver antenna terminal and 4 dB is the loss through the attenuator.

Receiver sensitivity is also often expressed as 10 dB S+N/N (a 10-dB ratio of signal + noise to noise) or 10 dB S/N (signal to noise). The procedure and measurement are identical to MDS, except that the input signal is increased until the receiver output increases by 10 dB for 10 dB S+N/N and 9.5 dB for 10 dB S/N (often called "10 dB signal to noise ratio"). AM receiver sensitivity is usually expressed in this manner with a 30% modulated, 1-kHz test signal. (The modulation in this case is keyed on and off and the signal level is adjusted for the desired increase in the audio output.)

SINAD is a common sensitivity measurement normally associated with FM receivers. It is an acronym for "*si*gnal plus *n*oise *a*nd *d*istortion." SINAD is a measure of signal quality:

$$\text{SINAD} = \frac{\text{signal} + \text{noise} + \text{distortion}}{\text{noise} + \text{distortion}} \quad (24)$$

where SINAD is expressed in dB. In this example, all quantities to the right of the equal sign are expressed in volts, and the ratio is converted to dB by multiplying the log of the fraction by 20.

$$\text{SINAD(dB)} = 10 \log \left(\frac{\text{Signal(V)}^2 + \text{Noise(V)}^2 + \text{Distortion(V)}^2}{\text{Noise(V)}^2 + \text{Distortion(V)}^2} \right)$$

Let's look at this more closely. We can consider distortion to be a part of the receiver noise because distortion, like noise, is an unwanted signal added to the desired signal by the receiving system. Then, if we assume that the desired signal is much stronger than the noise, SINAD closely approximates the signal to noise ratio. The common 12-dB SINAD specification therefore corresponds to a 4:1 S/N ratio (noise + distortion = 0.25 × signal).

The basic test setup for measuring SINAD is shown in **Fig 25.67**. The level of input signal is adjusted to provide 25% distortion (12 dB SINAD). Narrow-band FM signals, typical for amateur communications, usually have 3-kHz peak deviation when modulated at 1000 Hz.

Noise figure is another measure of receiver sensitivity. It provides a sensitivity evaluation that is independent of the system bandwidth. Noise figure is discussed further in the **Receivers** chapter.

25.6.2 Dynamic Range

Dynamic range is the ability of the receiver to tolerate strong signals outside of its bandpass range. Two kinds will be considered:

Blocking dynamic range (blocking DR) is the difference, in dB, between the noise floor and a signal that causes 1 dB of gain compression in the receiver. It indicates the signal level, above the noise floor, that begins to cause desensitization.

IMD dynamic range (IMD DR) measures the impact of two-tone IMD on a receiver. IMD is the production of spurious responses that results when two or more signals mix. IMD occurs in any receiver when signals of sufficient magnitude are present. IMD DR is the difference, in dB, between the noise floor and the strength of two equal incoming signals that produce a third-order product equal to the noise floor.

What do these measurements mean? When the IMD DR is exceeded, false signals begin to appear along with the desired signal. When the blocking DR is exceeded, the receiver begins losing its ability to amplify weak signals. Typically, the IMD DR is 20 dB or more below the blocking DR, so false signals appear well before sensitivity is significantly decreased. IMD DR is one of the most significant parameters that can be specified for a receiver. It is generally a conservative evaluation for other effects, such as blocking, which will occur only for signals well outside the IMD dynamic range of the receiver.

Both dynamic range tests require two signal generators and a hybrid combiner. When testing blocking DR (see **Fig 25.68**),

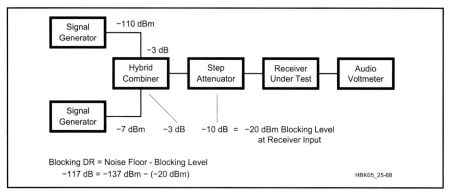

Fig 25.68 — Receiver Blocking DR is measured with this equipment and arrangement. Measurements shown are for the example discussed in the text. The audio voltmeter must be a true-RMS meter.

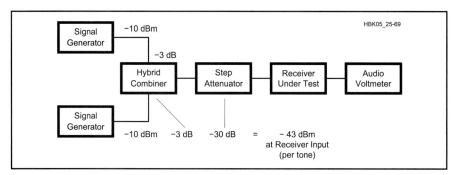

Fig 25.69 — Receiver IMD DR test setup. Signal levels shown are for the example discussed in the text. The audio voltmeter must be a true-RMS meter.

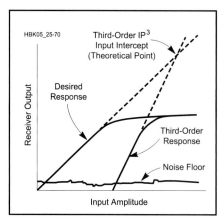

Fig 25.70 — A plot of the receiver characteristics that determine third-order input intercept, a measure of receiver performance.

one generator is set for a weak signal of roughly –110 dBm. The receiver is tuned to this frequency and peaked for maximum response. (ARRL Lab procedures require this level to be about 10 dB below the 1-dB compression point, if the AGC can be disabled. Otherwise, the level is set to 20 dB above the MDS.)

The second generator is set to a frequency 20 kHz away from the first and its level is increased until the receiver output drops by 1 dB, as measured with the ac voltmeter.

In the example shown, the output of the generator is –7 dBm, the loss through the combiner is fixed at 3 dB and the step attenuator is set to 10 dB. The 1-dB compression level is calculated as follows:

Blocking level =
$-7\text{ dBm} - 3\text{ dBm} - 10\text{ dB} = -20\text{ dBm}$
(25)

To express this as a dynamic range, the blocking level is referenced to the receiver noise floor (calculated earlier). Calculate it as follows:

Blocking DR =
blocking level – noise floor =
$-20\text{ dBm} - (-137\text{ dBm}) = 117\text{ dB})$ (26)

This value is usually expressed as an absolute value: 117 dB.

25.6.3 Two-Tone IMD Test

The setup for measuring IMD DR is shown in **Fig 25.69**. Two signals of equal level, spaced 20-kHz apart are injected into the receiver input. When we call these requencies f1 and f2, the so-called third-order IMD products will appear at frequencies of (2f1 – f2) and (2f2 – f1). If the two input frequencies are 14.040 and 14.060 MHz, the third-order products will be at 14.020 and 14.080 MHz. Let's talk through a measurement with these frequencies.

First, set the generators for f1 and f2. Adjust each of them for an output of –10 dBm. Tune the receiver to either of the third-order IMD products. Adjust the step attenuator until the IMD product plus noise produces an output 3 dB above the noise level as read on the ac voltmeter.

For an example, say the output of the generator is –10 dBm, the loss through the combiner is 3 dB and the amount of attenuation used is 30 dB. The signal level at the receiver antenna terminal that just begins to cause IMD problems is calculated as:

IMD level = $-10\text{ dBm} - 3\text{ dB} - 30\text{ dB}$ (27)
= –43 dBm

To express this as a dynamic range the IMD level is referenced to the noise floor as follows:

IMD DR = – noise floor – IMD level
= –137 dBm – (–43 dBm)
= – 94 dB   (28)

Therefore, the IMD dynamic range of this receiver would be 94 dB.

25.6.4 Third-Order Intercept

Another parameter used to quantify receiver performance is the third-order input intercept (IP³). This is the point at which the desired response and the third-order IMD response intersect, if extended beyond their linear regions (see **Fig 25.70**). Greater IP³ indicates better receiver performance. Calculate IP³ like this:

IP³ = 1.5 (IMD dynamic range in dB)
+ (MDS in dBm)   (29)

For our example receiver:

IP³ = 1.5 (94 dB) + (–137 dBm)
= +4 dBm

The preferred method for the third-order input intercept, however, is to use S5 signal levels instead of MDS levels. This is due to the tendency for a receiver to exhibit non-linearity at the noise floor level (i.e., near the level where MDS is typically defined).

Thus, Third Order Intercept =
(3 × (S5 IMD Level) – (S5 Reference))/2

The results of this method should show close correlation with the intercept point determined by the MDS test. If not, the test engineer determines (from further investigation) which method provides a more accurate result.

In the same way (using the S5 method),

Second Order Interrupt = 2 × (S5 IMD Level) – (S5 Reference)

The example receiver we have discussed here is purely imaginary. Nonetheless, its performance is typical of contemporary communications receivers.

EVALUATING THE DATA

Thus far, a fair amount of data has been gathered with no mention of what the numbers really mean. It is somewhat easier to understand exactly what is happening by arranging the data as shown in **Fig 25.71**. The base line represents power levels with a very small level at the left and a higher level (0 dBm) at the right.

The noise floor of our hypothetical receiver is at −137 dBm, the IMD level (the level at which signals will begin to create spurious responses) at −43 dBm and the blocking level (the level at which signals will begin to desensitize the receiver) at −20 dBm. The IMD dynamic range is some 23 dB smaller than the blocking dynamic range. This means IMD products will be heard long before the receiver begins to desensitize, some 23 dB sooner.

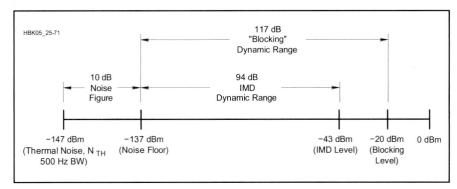

Fig 25.71 — Performance plot of the receiver discussed in the text. This is a good way to visualize the interaction of receiver-performance measurements. Note that −147 dBm is considered the lowest possible noise level in a 500 Hz receiver bandwidth.

25.7 Spectrum Analyzer

A spectrum analyzer is similar to an oscilloscope. Both visually present an electrical signal through graphic representation. The oscilloscope is used to observe electrical signals in the time domain (amplitude as a function of time). The time domain, however, gives little information about the frequencies that make up complex signals. Amplifiers, mixers, oscillators, detectors, modulators and filters are best characterized in terms of their frequency response. This information is obtained by viewing electrical signals in the *frequency domain* (amplitude as a function of frequency). One instrument that can display the frequency domain is the spectrum analyzer.

25.7.1 Time and Frequency Domain

To better understand the concepts of time and frequency domain, see **Fig 25.72**. The three-dimensional coordinates show time (as the line sloping toward the bottom right), frequency (as the line rising toward the top right) and amplitude (as the vertical axis). The two discrete frequencies shown are harmonically related, so we'll refer to them as f1 and 2f1.

In the representation of time domain at B, all frequency components of a signal are summed together. In fact, if the two discrete frequencies shown were applied to the input of an oscilloscope, we would see the solid line (which corresponds to f1 + 2f1) on the display.

In the frequency domain, complex signals (signals composed of more than one frequency) are separated into their individual

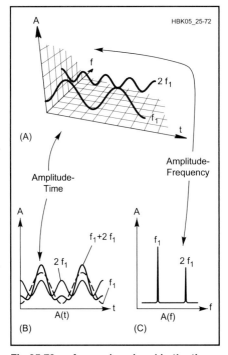

Fig 25.72 — A complex signal in the time and frequency domains. A is a three-dimensional display of amplitude, time and frequency. B is an oscilloscope display of time vs amplitude. C is spectrum analyzer display of the frequency domain and shows frequency vs amplitude.

frequency components. A spectrum analyzer measures and displays the power level at each discrete frequency; this display is shown at C.

The frequency domain contains information not apparent in the time domain and therefore the spectrum analyzer offers advantages over the oscilloscope for certain measurements. As might be expected, some measurements are best made in the time domain. In these cases, the oscilloscope is a valuable instrument.

25.7.2 Spectrum Analyzer Basics

There are several different types of spectrum analyzers, but by far the most common is nothing more than an electronically tuned superheterodyne receiver. The receiver is tuned by means of a ramp voltage. This ramp voltage performs two functions: First, it sweeps the frequency of the analyzer local oscillator; second, it deflects a beam across the horizontal axis of a CRT display, as shown in **Fig 25.73**. The vertical axis deflection of the CRT beam is determined by the strength of the received signal. In this way, the CRT displays frequency on the horizontal axis and signal strength on the vertical axis.

Most spectrum analyzers use an up-converting technique so that a fixed tuned input filter can remove the image. Only the first local oscillator need be tuned to tune the receiver. In the up-conversion design, a wideband input is converted to an IF higher than the highest input frequency. As with most up-converting communications receivers, it is not easy to achieve the desired ultimate selectivity at the first IF, because of the high frequency. For this reason, multiple conversions are used to generate an IF low enough so that the desired selectivity is practical. In the example shown, dual conversion is

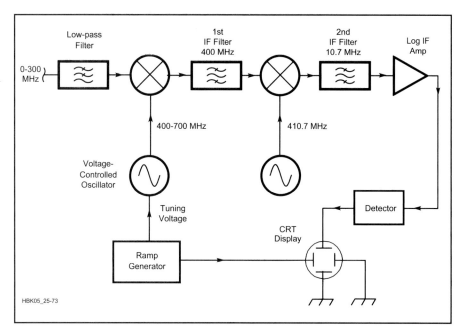

Fig 25.73 — A block diagram of a superheterodyne spectrum analyzer. Input frequencies of up to 300 MHz are up converted by the local oscillator and mixer to a fixed frequency of 400 MHz.

used: The first IF is at 400 MHz; the second at 10.7 MHz.

In the example spectrum analyzer, the first local oscillator is swept from 400 MHz to 700 MHz; this converts the input (from nearly 0 MHz to 300 MHz) to the first IF of 400 MHz. The usual rule of thumb for varactor tuned oscillators is that the maximum practical tuning ratio (the ratio of the highest frequency to the lowest frequency) is an octave, a 2:1 ratio. In our example spectrum analyzer, the tuning ratio of the first local oscillator is 1.75:1, which meets this specification.

The image frequency spans 800 MHz to 1100 MHz and is easily eliminated using a low-pass filter with a cut-off frequency around 300 MHz. The 400-MHz first IF is converted to 10.7 MHz where the ultimate selectivity of the analyzer is obtained. The image of the second conversion, (421.4 MHz), is eliminated by the first IF filter. The attenuation of the image should be great, on the order of 60 to 80 dB. This requires a first IF filter with a high Q; this is achieved by using helical resonators, SAW resonators or cavity filters. Another method of eliminating the image problem is to use triple conversion; converting first to an intermediate IF such as 50 MHz and then to 10.7 MHz. As with any receiver, an additional frequency conversion requires added circuitry and adds potential spurious responses.

Most of the signal amplification takes place at the lowest IF; in the case of the example analyzer this is 10.7 MHz. Here the communications receiver and the spectrum analyzer differ. A communications receiver demodulates the incoming signal so that the modulation can be heard or further demodulated for RTTY or packet or other mode of operation. In the spectrum analyzer, only the signal strength is needed.

In order for the spectrum analyzer to be most useful, it should display signals of widely different levels. As an example, signals differing by 60 dB, which is a thousand to one difference in voltage or a million to one in power, would be difficult to display. This would mean that if power were displayed, one signal would be one million times larger than the other (in the case of voltage one signal would be a thousand times larger). In either case it would be difficult to display both signals on a CRT. The solution to this problem is to use a logarithmic display that shows the relative signal levels in decibels. Using this technique, a 1000:1 ratio of voltage reduces to a 60-dB difference.

The conversion of the signal to a logarithm is usually performed in the IF amplifier or detector, resulting in an output voltage proportional to the logarithm of the input RF level. This output voltage is then used to drive the CRT display.

25.7.3 Spectrum Analyzer Performance Specifications

The performance parameters of a spectrum analyzer are specified in terms similar to those used for radio receivers, in spite of the fact that there are many differences between a receiver and a spectrum analyzer.

The sensitivity of a receiver is often specified as the minimum discernible signal, which means the smallest signal that can be heard. In the case of the spectrum analyzer, it is not the smallest signal that can be heard, but the smallest signal that can be seen. The dynamic range of the spectrum analyzer determines the largest and smallest signals that can be simultaneously viewed on the analyzer. As with a receiver, there are several factors that can affect dynamic range, such as IMD, second- and third-order distortion and blocking. IMD dynamic range is the maximum difference in signal level between the minimum detectable signal and the level of two signals of equal strength that generate an IMD product equal to the minimum detectable signal.

Although the communications receiver is an excellent example to introduce the spectrum analyzer, there are several differences such as the previously explained lack of a demodulator. Unlike the communications receiver, the spectrum analyzer is not a sensitive radio receiver. To preserve a wide dynamic range, the spectrum analyzer often uses passive mixers for the first and second mixers. Therefore, referring to Fig 25.73, the noise figure of the analyzer is no better than the losses of the input low-pass filter plus the first mixer, the first IF filter, the second mixer and the loss of the second IF filter. This often results in a combined noise figure of more than 20 dB. With that kind of noise figure the spectrum analyzer is obviously not a communications receiver for extracting very weak signals from the noise but a measuring instrument for the analysis of frequency spectrum.

The selectivity of the analyzer is called the resolution bandwidth. This term refers to the minimum frequency separation of two signals of equal level that can be resolved so there is a 3-dB dip between the two. The IF filters used in a spectrum analyzer differ from a communications receiver in that the filters in a spectrum analyzer have very gentle skirts and rounded passbands, rather than the flat passband and very steep skirts used on an IF filter in a high-quality communications receiver. This rounded passband is necessary because the signals pass into the filter passband as the spectrum analyzer scans the desired frequency range. If the signals suddenly pop into the passband (as they would if the filter had steep skirts), the filter tends to ring; a filter with gentle skirts is less likely to ring. This ringing, called scan loss, distorts the display and requires that the analyzer not sweep frequency too quickly. All this means that the scan rate must be checked periodically to be certain the signal amplitude is not affected by fast tuning.

25.7.2 Spectrum Analyzer Applications

Spectrum analyzers are used in situations where the signals to be analyzed are very

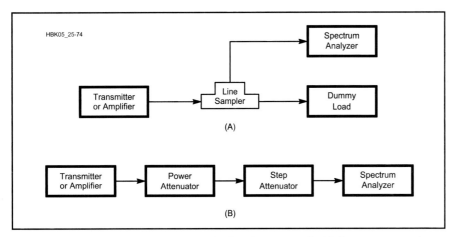

Fig 25.74 — Alternate bench setups for viewing the output of a high power transmitter or oscillator on a spectrum analyzer. A uses a line sampler to pick off a small amount of the transmitter or amplifier power. In B, most of the transmitter power is dissipated in the power attenuator.

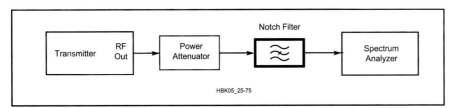

Fig 25.75 — A notch filter is another way to reduce the level of a transmitter's fundamental signal so that the fundamental does not generate harmonics within the analyzer. However, in order to know the amplitude relationship between the fundamental and the transmitter's actual harmonics and spurs, the attenuation of the fundamental in the notch filter must be known.

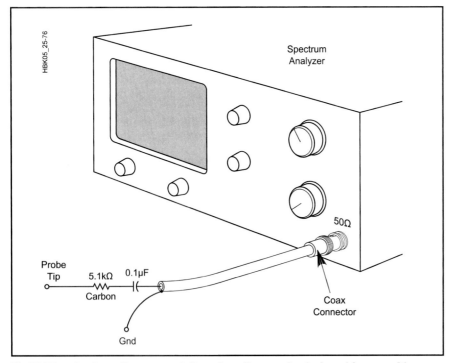

Fig 25.76 — A schematic representation of a voltage probe designed for use with a spectrum analyzer. Keep the probe tip (resistor and capacitor) and ground leads as short as possible.

complex and an oscilloscope display would be an indecipherable jumble. The spectrum analyzer is also used when the frequency of the signals to be analyzed is very high. Although high-performance oscilloscopes are capable of operation into the UHF region, moderately priced spectrum analyzers can be used well into the gigahertz region.

A spectrum analyzer can also be used to view very low-level signals. For an oscilloscope to display a VHF waveform, the bandwidth of the oscilloscope must extend from zero to the frequency of the waveform. If harmonic distortion and other higher-frequency distortions are to be seen the bandwidth of the oscilloscope must exceed the fundamental frequency of the waveform. This broad bandwidth can also admit a lot of noise power. The spectrum analyzer, on the other hand, analyzes the waveform using a narrow bandwidth; thus it is capable of reducing the noise power admitted.

Probably the most common application of the spectrum analyzer is the measurement of the harmonic content and other spurious signals in the output of a radio transmitter. **Fig 25.74** shows two ways to connect the transmitter and spectrum analyzer. The method shown at A should not be used for wide-band measurements since most line-sampling devices do not exhibit a constant-amplitude output over a broad frequency range. Using a line sampler is fine for narrow-band measurements, however. The method shown at B is used in the ARRL Lab. The attenuator must be capable of dissipating the transmitter power. It must also have sufficient attenuation to protect the spectrum analyzer input. Many spectrum analyzer mixers can be damaged by only a few milliwatts, so most analyzers have an adjustable input attenuator that will provide a reasonable amount of attenuation to protect the sensitive input mixer from damage. The power limitation of the attenuator itself is usually on the order of a watt or so, however. This means that 20 dB of additional attenuation is required for a 100-W transmitter, 30 dB for a 1000-W transmitter and so on, to limit the input to the spectrum analyzer to 1 W. There are specialized attenuators that are made for transmitter testing; these attenuators provide the necessary power dissipation and attenuation in the 20 to 30-dB range.

When using a spectrum analyzer it is very important that the maximum amount of attenuation be applied before a measurement is made. In addition, it is a good practice to start with maximum attenuation and view the entire spectrum of a signal before the attenuator is adjusted. The signal being viewed could appear to be at a safe level, but another spectral component, which is not visible, could be above the damage limit. It is also very important to limit the input power to the analyzer when pulse power is being measured.

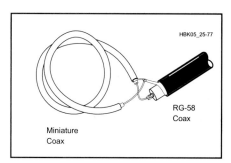

Fig 25.77 — A "sniffer" probe consisting of an inductive pick-up. It has an advantage of not loading the circuit under test. See text for details.

The average power may be small enough so the input attenuator is not damaged, but the peak pulse power, which may not be readily visible on the analyzer display, can destroy a mixer, literally in microseconds.

When using a spectrum analyzer it is necessary to ensure that the analyzer does not generate additional spurious signals that are then attributed to the system under test. Some of the spurious signals that can be generated by a spectrum analyzer are harmonics and IMD. If it is desired to measure the harmonic levels of a transmitter at a level below the spurious level of the analyzer itself, a notch filter can be inserted between the attenuator and the spectrum analyzer as shown in **Fig 25.75**. This reduces the level of the fundamental signal and prevents that signal from generating harmonics within the analyzer, while still allowing the harmonics from the transmitter to pass through to the analyzer without attenuation. Use caution with this technique; detuning the notch filter or inadvertently changing the transmitter frequency will allow potentially high levels of power to enter the analyzer. In addition, use care when choosing filters; some filters (such as cavity filters) respond not only to the fundamental but notch out odd harmonics as well.

It is good practice to check for the generation of spurious signals within the spectrum analyzer. When a spurious signal is generated by a spectrum analyzer, adding attenuation at the analyzer input will cause the internally generated spurious signals to decrease by an amount greater than the added attenuation. If attenuation added ahead of the analyzer causes all of the visible signals to decrease by the same amount, this indicates a spurious-free display.

The input impedance for most RF spectrum analyzers is 50 Ω; not all circuits have convenient 50-Ω connections that can be accessed for testing purposes, however. Using a probe such as the one shown in **Fig 25.76** allows the analyzer to be used as a troubleshooting tool. The probe can be used to track down signals within a transmitter or receiver, much like an oscilloscope is used. The probe shown offers a 100:1 voltage reduction and loads the circuit with 5000 Ω. A different type of probe is shown in **Fig 25.77**. This inductive pickup coil (sometimes called a "sniffer") is very handy for troubleshooting. The coil is used to couple signals from the radiated magnetic field of a circuit into the analyzer. A short length of miniature coax is wound into a pick-up loop and soldered to a larger piece of coax. The use of the coax shields the loop from coupling energy from the electric field component. The dimensions of the loop are not critical, but smaller loop dimensions make the loop more accurate in locating the source of radiated RF. The shield of the coax provides a complete electrostatic shield without introducing a shorted turn.

The sniffer allows the spectrum analyzer to sense RF energy without contacting the circuit being analyzed. If the loop is brought near an oscillator coil, the oscillator can be tuned without directly contacting (and thus disturbing) the circuit. The oscillator can then be checked for reliable starting and the generation of spurious sidebands. With the coil brought near the tuned circuits of amplifiers or frequency multipliers, those stages can be tuned using a similar technique.

Even though the sniffer does not contact the circuit being evaluated, it does extract some energy from the circuit. For this reason, the loop should be placed as far from the tuned circuit as is practical. If the loop is placed too far from the circuit, the signal will be too weak or the pick-up loop will pick up energy from other parts of the circuit and not give an accurate indication of the circuit under test.

The sniffer is very handy to locate sources of RF leakage. By probing the shields and cabinets of RF generating equipment (such as transmitters) egress and ingress points of RF energy can be identified by increased indications on the analyzer display.

One very powerful characteristic of the spectrum analyzer is the instrument's capability to measure very low level signals. This characteristic is very advantageous when very high levels of attenuation are measured. **Fig 25.78** shows the setup for tuning the notch and passband of a VHF duplexer. The spectrum analyzer, being capable of viewing signals well into the low microvolt region, is capable of measuring the insertion loss of the notch cavity more than 100 dB below the signal generator output. Making a measurement of this sort requires care in the interconnection of the equipment and a well designed spectrum analyzer and signal generator. RF energy leaking from the signal generator cabinet, line cord or even the coax itself, can get into the spectrum analyzer through similar paths and corrupt the measurement. This leakage can make the measurement look either better or worse than the actual attenuation, depending on the phase relationship of the leaked signal.

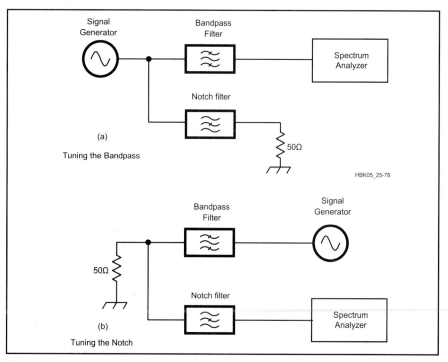

Fig 25.78 — Block diagram of a spectrum analyzer and signal generator being used to tune the band-pass and notch filters of a duplexer. All ports of the duplexer must be properly terminated and good quality coax with intact shielding used to reduce leakage.

25.7.5 Extensions of Spectral Analysis

What if a signal generator is connected to a spectrum analyzer so that the signal generator output frequency is exactly the same as the receiving frequency of the spectrum analyzer? It would certainly appear to be a real convenience not to have to continually reset the signal generator to the desired frequency. It is, however, more than a convenience. A signal generator connected in this way is called a tracking generator because the output frequency tracks the spectrum analyzer input frequency. The tracking generator makes it possible to make swept frequency measurements of the attenuation characteristics of circuits, even when the attenuation involved is large.

Fig 25.79 shows the connection of a tracking generator to a circuit under test. In order for the tracking generator to create an output frequency exactly equal to the input frequency of the spectrum analyzer, the internal local oscillator frequencies of the spectrum analyzer must be known. This is the reason for the interconnections between the tracking generator and the spectrum analyzer. The test setup shown will measure the gain or loss of the circuit under test. Only the magnitude of the gain or loss is available; in some cases, the phase angle between the input and output would also be an important and necessary parameter.

The spectrum analyzer is not sensitive to the phase angle of the tracking generator output. In the process of generating the tracking generator output, there are no guarantees that the phase of the tracking generator will be either known or constant. This is especially true of VHF spectrum analyzers/tracking generators where a few inches of coaxial cable represents a significant phase shift.

One effective way of measuring the phase angle between the input and output of a device under test is to sample the phase of the input and output of device under test and apply the samples to a phase detector. **Fig 25.80** shows a block diagram of this technique. An instrument that can measure both the magnitude and phase of a signal is called a vector network analyzer or simply a network analyzer. The magnitude and phase can be displayed either separately or together. When the magnitude and phase are displayed together the two can be presented as two separate traces, similar to the two traces on a dual-trace oscilloscope. A much more useful method of display is to present the magnitude and phase as a polar plot where the locus of the points of a vector having the length of the magnitude and the angle of the phase are displayed. Very sophisticated network analyzers can display all of the S parameters of a circuit in either a polar format or a Smith Chart format.

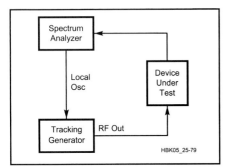

Fig 25.79 — A signal generator (shown in the figure as the "Tracking Generator") locked to the local oscillator of a spectrum analyzer can be used to determine filter response over a range of frequencies.

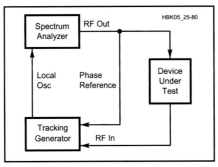

Fig 25.80 — A network analyzer is usually found in commercial communications development labs. It can measure both the phase and magnitude of the filter input and output signals. See text for details.

25.8 Transmitter Performance Tests

The test setup used in the ARRL Laboratory for measuring an HF transmitter or amplifier is shown in **Fig 25.81**. As can be seen, different power levels dictate different amounts of attenuation between the transmitter or amplifier and the spectrum analyzer.

The *ARRL Lab Test Procedures Manual* is available as a PDF file for downloading from **www.arrl.org/product-review**.

25.8.1 Spurious Emissions

Fig 25.82 shows the broadband spectrum of a transmitter, showing the harmonics in the output. The horizontal (frequency) scale is 5 MHz per division; the main output of the transmitter at 7 MHz can be seen about 1.5 major divisions from the left of the trace. Although not shown, a very large apparent signal is often seen at the extreme left of the trace. This occurs at what would be zero frequency and it is caused by the first local oscillator frequency being exactly the first IF. All up-converting superheterodyne spectrum analyzers have this IF feed-through; in addition, this signal is occasionally accompanied by a smaller spurious signal, generated within the analyzer. To determine what part of the displayed signal is a spurious response caused by IF feed-through and what is an actual input signal, simply remove the input signal and observe the trace. It is not necessary or desirable that the transmitter be modulated

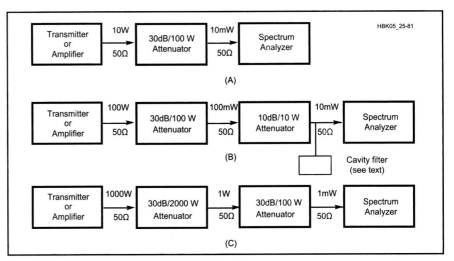

Fig 25.81 — These setups are used in the ARRL Laboratory for testing transmitters or amplifiers with several different power levels.

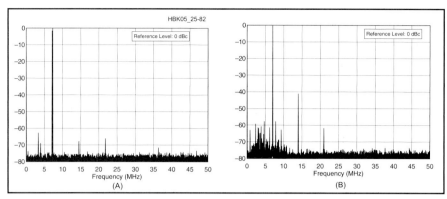
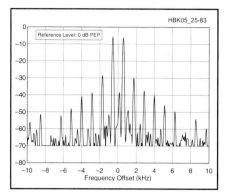

Fig 25.82 — Comparison of two different transmitters on the 40-m band as typically seen on a spectrum analyzer display. The display at the left shows a relatively clean transmitted signal but the transmitter at the right shows more spurious signal content. Horizontal scale is 5 MHz per division; vertical is 10 dB per division. According to current FCC spectral purity requirements both transmitters are acceptable.

Fig 25.83 — An SSB transmitter two-tone test as seen on a spectrum analy-zer. Each horizontal division represents 2 kHz and each vertical division is 10 dB. The third-order products are 30 dB below the PEP (top line), the fifth-order products are down 37 dB and seventh-order products are down 40 dB. This represents acceptable (but not ideal) performance.

for this broadband test.

Other transmitter tests that can be performed with a spectrum analyzer include measurement of two-tone IMD and SSB carrier and unwanted sideband suppression.

25.8.2 Two-Tone IMD

Investigating the sidebands from a modulated transmitter requires a narrow-band spectrum analysis and produces displays similar to that shown in **Fig 25.83**. In this example, a two-tone test signal is used to modulate the transmitter. The display shows the two test tones plus some of the IMD produced by the SSB transmitter. The test setup used to produce this display is shown in **Fig 25.84**.

In this example, a two-tone test signal with frequencies of 700 and 1900 Hz is used to modulate the transmitter. Set the transmitter output and audio input to the manufacturer's specifications. Each desired tone is adjusted to be equal in amplitude and centered on the display. The step attenuators and analyzer controls are then adjusted to set the two desired signals 6 dB below the 0-dB reference (top) line. The IMD products can then be read directly from the display in terms of "dB below Peak Envelope Power (PEP)." (In the example shown, the third-order products are 30 dB below PEP, the fifth-order products are 37 dB down, the seventh- order products are down 40 dB.)

25.8.3 Carrier and Unwanted Sideband Suppression

Single-tone audio input signals can be used with the same setup to measure unwanted sideband and carrier suppression of SSB signals. In this case, set the single tone to the 0-dB reference line. (Once the level is set, the audio can be disabled for carrier suppression measurements in order to eliminate IMD and other effects.)

25.8.4 Phase Noise

Phase/composite noise is also measured with spectrum analyzers in the ARRL Lab. This test requires specialized equipment and is included here for information purposes only. More information may be found in the **Oscillators and Synthesizers** chapter.

The purpose of the Composite-Noise test is to observe and measure the phase and amplitude noise, as well as any close-in spurious signals generated by a transmitter. Since phase noise is the primary noise component

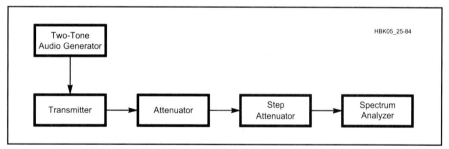

Fig 25.84 — The test setup used in the ARRL Laboratory to measure the IMD performance of transmitters and amplifiers.

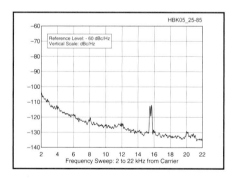

Fig 25.85 — Spectral display of an amateur transmitter output during composite-noise testing in the ARRL Lab. Power output is 100 W on the 14 MHz band. The carrier, off the left edge of the plot, is not shown. This plot shows composite transmitted noise 2 kHz to 22 kHz from the carrier. The reference level is 0 dBc, and the vertical scale is in dB.

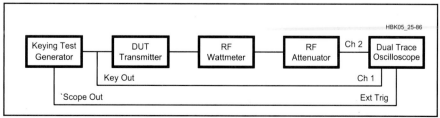

Fig 25.86 — CW keying waveform test setup.

in any well-designed transmitter, almost all of the noise observed during this test is phase noise.

This measurement is accomplished in the lab using a Hewlett-Packard phase noise test set under computer control. A spectrum analyzer (see **Fig 25.85**) displays the remaining noise and spurious signals from 100 Hz to 1 MHz from the carrier frequency.

25.8.5 Tests in the Time Domain

Oscilloscopes are used for transmitter testing in the time domain. Dual-trace instruments are best in most cases, providing easy to read time-delay measurements between keying input and RF- or audio-output signals. Common transmitter measurements performed with 'scopes include CW keying wave shape and time delay and SSB/FM transmit-to-audio turnaround tests (important for many digital modes).

A typical setup for measuring CW keying waveform and time delay is shown in **Fig 25.86**. A keying test generator is used to repeatedly key the transmitter at a controlled rate. The generator can be set to any reasonable speed, but ARRL tests are usually conducted at 20-ms on and 20-ms off (25 Hz, 50% duty cycle). **Fig 25.87** shows a typical display. The rise and fall times of the RF output pulse are measured between the 10% and 90% points on the leading and trailing edges, respectively. The delay times are measured between the 50% points of the keying and RF output waveforms. Look at the **Transmitters** chapter for further discussion of CW keying issues.

For voice modes (SSB/FM), a PTT-to-RF output test is similar to CW keying tests. It measures rise and fall times, as well as the on- and off-delay times just as in the CW test. See **Fig 25.88** for the test setup.

"Turnaround time" is the time it takes for a transceiver to switch from the 50% fall time of a keying pulse to 50% rise of audio output. The test setup is shown in **Fig 25.89**. Turn-around time measurements require extreme care with respect to transmitter output power, attenuation, signal-generator output and the maximum input signal that can be tolerated by the generator. The generator's specifications must not be exceeded and the input to the receiver must be at the required level, usually S9. Receiver AGC is usually off for this test, but experimentation with AGC and signal input level can reveal surprising variations. The keying rate must be considerably slower than the turn-around time; rates of 200-ms on/200-ms off or faster, have been used with success in Product Review tests at the ARRL Lab.

Turn-around time is an important consideration with some digital modes. AMTOR, for example, requires a turn-around time of 35 ms or less.

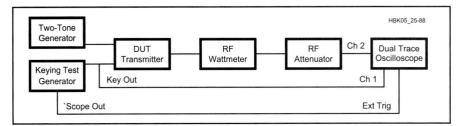

Fig 25.88 — PTT-to-RF-output test setup for voice-mode transmitters.

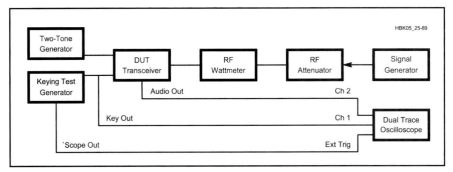

Fig 25.89 — Transmit-receive turn-around time test setup.

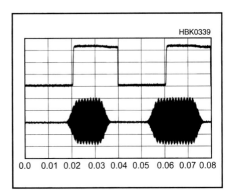

Fig 25.87 — Typical CW keying waveform for a modern Amateur Radio transceiver during testing in the ARRL Lab. This plot shows the first two dits in full-break-in (QSK) mode using external keying. Equivalent keying speed is 60 WPM. The upper trace is the actual key closure; the lower trace is the RF envelope. (Note that the first key closure starts at the left edge of the figure.) Horizontal divisions are 10 ms. The transceiver was being operated at 100 W output on the 14 MHz band.

OSCILLOSCOPE BIBLIOGRAPHY

R. vanErk, *Oscilloscopes*, Functional Operation and Measuring Examples, McGraw-Hill Book Co, New York, 1978.

V. Bunze, *Probing in Perspective* — Application Note 152, Hewlett-Packard Co, Colorado Springs, CO, 1972 (Pub No. 5952-1892).

The XYZs of Using a Scope, Tektronix, Inc, Portland, OR, 1981 (Pub No. 41AX-4758).

Basic Techniques of Waveform Measurement (Parts 1 and 2), Hewlett-Packard Co, Colorado Springs, CO, 1980 (Pub No. 5953-3873).

J. Millman, and H. Taub, *Pulse Digital and Switching Waveforms*, McGraw-Hill Book Co, New York, 1965, pp 50-54.

V. Martin, *ABCs of DMMs*, Fluke Corp, PO Box 9090, Everett, WA 98206.

25.9 Test Equipment and Measurements Glossary

Alternating current (ac) — The polarity constantly reverses, as contrasted to dc (direct current) where polarity is fixed.

Analog — Signals which have a full set of values. If the signal varies between 0 and 10 V all values in this range can be found. Compare this to a *digital* system.

Attenuator — A device which reduces the amplitude of a signal.

Average value — Obtained by recording or measuring N samples of a signal, adding up all of these values, and dividing this sum by N.

Bandwidth — A measure of how wide a signal is in frequency. If a signal covers 14,200 to 14,205 kHz its bandwidth is said to be 5 kHz.

BNC — A small bayonet-type connector used with coax cable.

Bridge circuit — Four passive elements, such as resistors, inductors, connected as a pair of voltage dividers with a meter or other measuring device across two opposite junctions. Used to indicate the relative values of the four passive elements. See the chapter discussion of Wheatstone bridges.

CMOS — A family of digital logic elements usually selected for their low power drain. See the **Digital Basics** chapter of this *Handbook*.

Coaxial cable (coax) — A cable formed of two conductors that share the same axis. The center conductor may be a single wire or a stranded cable. The outer conductor is called the shield. The shield may be flexible braid, foil, semirigid or rigid metal. For more information, look in the **Transmission Lines** chapter.

Combiners — See *Hybrid*.

D'Arsonval meter — A common mechanical meter consisting of a permanent magnet and a moving coil (with pointer attached).

Direct Current (dc) — The polarity is fixed for all time, as contrasted to ac (alternating current) where polarity constantly reverses.

Digital — A system that allows signals to assume a finite range of states. Binary logic is the most common example. Only two values are permitted in a binary system: one value is defined as a logical *1* and the other value as a logical *0*. See the *Handbook* chapter on **Digital Basics**.

Divider — A network of components that produce an output signal that is a fraction of the input signal. The ratio of the output to the input is the division factor. An analog divider divides voltage (a string of series connected resistors) or current (parallel connected resistors). Digital dividers divide pulse trains or frequency.

DMM (digital multimeter) — A test instrument that usually measures at least: voltage, current and resistance, and displays the result on a numeric digit display, rather than an analog meter.

Dummy antenna or dummy load — A resistor or set of resistors used in place of an antenna to test a transmitter without radiating any electromagnetic energy into the air.

DVM (digital volt meter) — See *DMM*.

FET voltmeter — See also *VTVM*. An updated version of a VTVM using field effect transistors (FETs) in place of vacuum tubes.

Flip-flop — A digital circuit that has two stable states. See the chapter on **Digital Basics**.

Frequency marker — Test signals generated at selected intervals (such as 25 kHz, 50 kHz, 100 kHz) for calibrating the dials of receivers and transmitters.

Fundamental — The first signal or frequency in a series of harmonically related signals. This term is often used to describe an oscillator, or transmitter's desired signal.

Harmonic — A signal occurring at some integral multiple (such as two, three, four) of a *fundamental* frequency.

Hybrid (hybrid combiners) — A device used to connect two signal generators to one receiver for test purposes, without the two generators affecting each other.

IC (integrated circuit) — A complete circuit built into a single electronic component.

LCD (liquid crystal display) — A low-power display device utilizing the physics of liquid crystals. They usually need either ambient light or backlighting to be seen.

LED (light emitting diode) — A diode that emits light when an appropriate voltage (usually 1.5 V at about 20 mA) is connected. They are used either as tiny pilot lights or in bar shapes to display letters and numbers.

Loran — A navigation system using very-low-frequency transmitters.

Marker — See *Frequency marker*.

Multiplier — A circuit that purposely creates some desired *harmonic* of its input signal. For example, a frequency multiplier that takes energy from a 3.5-MHz exciter and puts out RF at 7 MHz is a two times multiplier, usually called a frequency doubler.

N — A type of coaxial cable connector common at UHF and higher frequencies.

NAND — A digital element that performs the *not-and* function. See the **Digital Basics** chapter.

Noise (noise figure) — Noise is generated in all electrical circuits. It is particularly critical in those stages of a receiver that are closest to the antenna (RF amplifier and mixer), because noise generated in these stages can mask a weak signal. The noise figure is a measure of this noise generation. Lower noise figures mean that less noise is generated and weaker signals can be heard.

NOR — A digital element that performs the *not-or* function. See the **Digital Basics** chapter.

Null (nulling) — The process of adjusting a circuit for a minimum reading on a test meter or instrument. At a perfect null there is null, or no, energy to be seen.

Ohmmeter — A meter that measures the value of resistors. Usually part of a multi meter. See *VOM* and *DMM*.

Peak value — The highest value of a signal during the measuring time. If a measured voltage varies in value from 1 to 10 V over a measuring period, the peak value would be the highest measured, 10 V.

PL-259 — A connector used for coaxial cable, usually at HF. It is also known as a male UHF connector. It is an inexpensive and common connector, but it is not weatherproof, nor is its impedance constant over frequency.

Prescaler — A circuit used ahead of a counter to extend the counter range to higher frequencies. A counter capable of operating up to 50 MHz can count up to 500 MHz when used with a divide-by-10 prescaler.

Q — The ratio of the reactance to the resistance of a component or circuit. It provides a measure of bandwidth. Lower resistive losses make for a higher Q, and a narrower bandwidth.

RMS (root mean square) — A measure of the value of a voltage or current obtained by taking values from successive small time slices over a complete cycle of the waveform, squaring those values, taking the mean of the squares, and then the square root of the mean. Very significant when working with good ac sine waves, where the RMS of the sine wave is 0.707 of the peak value.

Scope — Slang for oscilloscope. See the Oscilloscopes section of this chapter.

Shunt — Elements connected in parallel.

Sinusoidal (sine wave) — The nominal waveform for unmodulated RF energy and many other ac voltages.

Spectrum — Used to describe a range of frequencies or wavelengths. The RF spectrum starts at perhaps 10 kHz and extends up to several hundred gigahertz. The light spectrum goes from infrared to ultraviolet.

Spurious emissions, or spurs — Unwanted energy generated by a transmitter or other circuit. These emissions include, but are not limited to, *harmonics*.

Thermocouple — A device made up of two different metals joined at two places. If one joint is hot and the other cold a voltage may be developed, which is a measure of the temperature difference.

Time domain — A measurement technique where the results are plotted or shown against a scale of time. In contrast to the frequency domain, where the results are plotted against a scale of frequency.

TTL (Transistor-transistor-logic) — A logic IC family commonly used with 5 V supplies. See the chapter on **Digital Basics**.

Vernier dial or vernier drive — A mechanical system of tuning dials, frequently used in older equipment, where the knob might turn 10 times for each single rotation of the control shaft.

VOM (volt-ohm-meter) — A multimeter whose design predates digital multi-meters (see **DMM**).

VTVM (vacuum tube voltmeter) — A meter that was developed to provide a high input resistance and therefore low current drain (loading) from the circuit being tested. Now replaced by the FET meter.

Wheatstone bridge — See Bridge circuit.

Contents

26.1 Test Equipment
    26.1.1 Senses
    26.1.2 Internal Equipment
    26.1.3 Bench Equipment
26.2 Where to Begin
    26.2.1 New Construction
    26.2.2 All Equipment
    26.2.3 Various Approaches
26.3 Testing Within a Stage
    26.3.1 Voltage Levels
    26.3.2 Noise
    26.3.3 Oscillations
    26.3.4 Amplitude Distortion
    26.3.5 Frequency Distortion
    26.3.6 Distortion Measurement
    26.3.7 Alignment
    26.3.8 Contamination
    26.3.9 Solder Bridges
    26.3.10 Arcing
    26.3.11 Replacing Parts
26.4 Typical Symptoms and Faults
    26.4.1 Power Supplies
    26.4.2 Amplifiers
    26.4.3 Oscillators
    26.4.4 Control Circuitry
    26.4.5 Digital Circuitry
26.5 Troubleshooting Hints and Maintenance
    26.5.1 Receivers
    26.5.2 Transmitters
    26.5.3 Transceivers
    26.5.4 Amplifiers

26.6 Components
    26.6.1 Check the Circuit
    26.6.2 Fuses
    26.6.3 Wires
    26.6.4 Connectors
    26.6.5 Resistors
    26.6.6 Capacitors
    26.6.7 Inductors and Transformers
    26.6.8 Relays
    26.6.9 Semiconductors
    26.6.10 Tubes
26.7 After the Repairs
    26.7.1 All Units
    26.7.2 For Transmitters Only
    26.7.3 Other Repaired Circuits
    26.7.4 Button It Up
26.8 Professional Repairs
    26.8.1 Packing It Up
26.9 Repair and Restoration of Vintage Equipment
    26.9.1 Component Replacement
    26.9.2 Powering Up the Equipment
    26.9.3 Alignment
    26.9.4 Using Vintage Receivers
26.10 References and Bibliography

Chapter 26

Troubleshooting and Maintenance

Traditionally, the radio amateur has maintained a working knowledge of electronic equipment. This knowledge, and the ability to make repairs with whatever resources are available, keeps amateur stations operating when all other communications fail. This troubleshooting ability is not only a tradition; it is fundamental to the existence of the service.

The sections on troubleshooting and repair, written by Ed Hare, W1RFI, tell you what to do when you are faced with equipment failure or a circuit that doesn't work. It will help you ask and answer the right questions: "Should I fix it or send it back to the dealer for repair? What do I need to know to be able to fix it myself? Where do I start? What kind of test equipment do I need?" The best answers to these questions will depend on the type of test equipment you have available, the availability of a schematic or service manual and the depth of your own electronic and troubleshooting experience.

The section on amplifier maintenance is based on the September 2003 *QST* article "Amplifier Care and Maintenance" by Ward Silver, NØAX. John Fitzsimmons, W3JN, contributed the section on Repair and Restoration of Vintage Equipment.

Not everyone is an electronics wizard; your gear may end up at the repair shop in spite of your best efforts. The theory you learned for the FCC examinations and the information in this *Handbook* can help you decide if you can fix it yourself. If the problem is something simple (and most are), why not avoid the effort of shipping the radio to the manufacturer? It is gratifying to save time and money, but, even better, the experience and confidence you gain by fixing it yourself may prove even more valuable.

Although some say troubleshooting is as much art as it is science, the repair of electronic gear is not magic. It is more like detective work. A knowledge of complex math is not required. However, you must have, or develop, the ability to read a schematic diagram and to visualize signal flow through the circuit.

SAFETY FIRST

Always! Death is permanent. A review of safety must be the first thing discussed in a troubleshooting chapter. Some of the voltages found in amateur equipment can be fatal! Only 50 mA flowing through the body is painful; 100 to 500 mA is usually fatal. Under certain conditions, as little as 24 V can kill.

Make sure you are 100% familiar with all safety rules and the dangerous conditions that might exist in the equipment you are servicing. Remember, if the equipment is not working properly, dangerous conditions may exist where you don't expect them. Treat every component as potentially "live."

Some older equipment uses "ac/dc" circuitry. In this circuit, one side of the chassis is connected directly to the ac line. This is an electric shock waiting to happen.

A list of safety rules can be found in **Table 26.1**. You should also read the **Safety** chapter of this *Handbook* before you proceed.

GETTING HELP

Other hams may be able to help you with your troubleshooting and repair problems, either with a manual or technical help. Check with your local club or repeater group. You may get lucky and find a troubleshooting "wizard." (On the other hand, you may get some advice that is downright dangerous, so be selective.) You can also place a classified ad in one of the ham magazines, perhaps when you are looking for a rare manual.

Your fellow hams in the ARRL Field organization may also help. Technical Coordinators (TCs) and Technical Specialists (TSs) are volunteers who are willing to help hams with technical questions. For the name and address of a local TC or TS, contact your Section Manager (listed in the front of any recent issue of *QST*).

THEORY

To fix electronic equipment, you need to understand the system and circuits you are troubleshooting. A working knowledge of electronic theory, circuitry and components is an important part of the process. If necessary, review the electronic and circuit theory explained in the other chapters of this book. When you are troubleshooting, you are looking for the unexpected. Knowing how circuits are supposed to work will help you to look for things that are out of place.

Table 26.1

Safety Rules

1. Keep one hand in your pocket when working on live circuits or checking to see that capacitors are discharged.
2. Include a conveniently located ground-fault current interrupter (GFCI) circuit breaker in the workbench wiring.
3. Use only grounded plugs and receptacles.
4. Use a GFCI protected circuit when working outdoors, on a concrete or dirt floor, in wet areas, or near fixtures or appliances connected to water lines, or within six feet of any exposed grounded building feature.
5. Use a fused, power limiting isolation transformer when working on ac/dc devices.
6. Switch off the power, *disconnect equipment from the power source, ground the output of the internal dc power supply*, and discharge capacitors when making circuit changes.
7. Do not subject electrolytic capacitors to excessive voltage, ac voltage or reverse voltage.
8. Test leads should be well insulated.
9. Do not work alone!
10. Wear safety glasses for protection against sparks and metal fragments.
11. Always use a safety harness when working above ground level.
12. Wear shoes with nonslip soles that will support your feet when climbing.
13. Wear rubber-sole shoes or use a rubber mat when standing on the ground or on a concrete floor.
14. Wear a hard hat when someone is working above you.
15. Be careful with tools that may cause short circuits.
16. Replace fuses only with those having proper ratings.

26.1 Test Equipment

Many of the steps involved in troubleshooting efficiently require the use of test equipment. We cannot see electrons flow. However, electrons do affect various devices in our equipment, with results we can measure.

Some people think they need expensive test instruments to repair their own equipment. This is not so! In fact, you probably already own the most important instruments. Some others may be purchased inexpensively, rented, borrowed or built at home. The test equipment available to you may limit the kind of repairs you can do, but you will be surprised at the kinds of repair work you can do with simple test equipment.

26.1.1 Senses

Although they are not "test equipment" in the classic sense, your own senses will tell you as much about the equipment you are trying to fix as the most-expensive spectrum analyzer. We each have some of these natural "test instruments."

Eyes — Use them constantly. Look for evidence of heat and arcing, burned components, broken connections or wires, poor solder joints or other obvious visual problems.

Ears — Severe audio distortion can be detected by ear. The "snaps" and "pops" of arcing or the sizzling of a burning component may help you track down circuit faults. An experienced troubleshooter can diagnose some circuit problems by the sound they make. For example, a bad audio-output IC sounds slightly different than a defective speaker.

Nose — Your nose can tell you a lot. With experience, the smells of ozone, an overheating transformer and a burned carbon-composition resistor each become unique and distinctive.

Finger — Carefully use your fingers to measure low heat levels in components. Small-signal transistors can be fairly warm to the touch; anything hotter can indicate a circuit problem. (Be careful; some high-power devices or resistors can get downright hot during normal operation.)

Brain — More troubleshooting problems have been solved with a VOM and a brain than with the most expensive spectrum analyzer. You must use your brain to analyze data collected by other instruments.

26.1.2 "Internal" Equipment

Some "test equipment" is included in the equipment you repair. Nearly all receivers include a speaker. An S meter is usually connected ahead of the audio chain. If the S meter shows signals, it indicates that the RF and IF circuitry is probably functioning. Analyze what the unit is doing and see if it gives you a clue.

Some older receivers include a crystal frequency calibrator. The calibrator signal, which is rich in harmonics, is injected in the RF chain close to the antenna jack and may be used for signal tracing and alignment.

26.1.3 Bench Equipment

Here is a summary of test instruments and their applications. Some items serve several purposes and may substitute for others on the list. The list does not cover all equipment available, only the most common and useful instruments. The theory and operation of much of this test equipment is discussed in more detail in the **Test Equipment and Measurements** chapter.

Multimeters — The multimeter is the most often used piece of test equipment. This group includes vacuum-tube voltmeters (VTVMs), volt-ohm-milliammeters (VOMs), field-effect transistor VOMs (FETVOMs) and digital multimeters (DMMs). Multimeters are used to read bias voltages, circuit resistance and signal level (with an appropriate probe). They can test resistors, capacitors (within certain limitations), diodes and transistors.

DMMs have become quite inexpensive. Their high input impedance, accuracy and

The Shack Notebook

If you don't keep a shack notebook, start one. A simple spiral notebook with notes about maintenance, wiring, color coding, antenna behavior and so forth can be a big time-saver. Be sure to note the date of each entry.

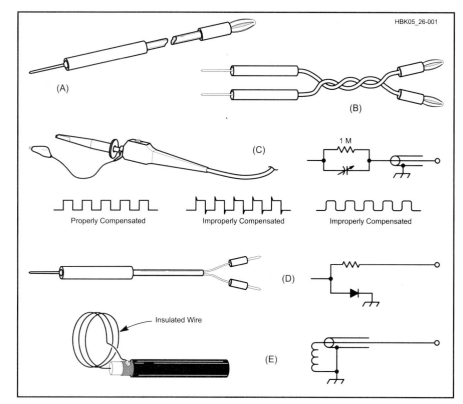

Fig 26.1 — An array of test probes for use with various test instruments.

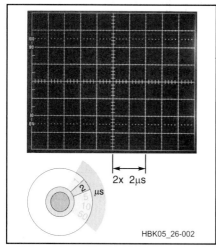

Fig 26.2 — An oscilloscope display showing the relationship between time-base setting and graticule lines.

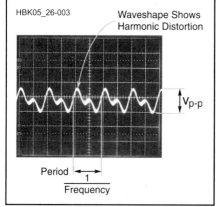

Fig 26.3 — Information available from a typical oscilloscope display of a waveform.

flexibility are well worth the cost. Many of them contain other test equipment as well, such as capacitance meters, frequency counters, transistor testers and even digital thermometers. Some DMMs are affected by RF, so most technicians keep an analog-display VOM on hand for use near RF equipment.

New analog meters are becoming less common, but many are available used. Look for meters with an internal resistance of 20 kΩ/V or higher. The 10 MΩ or better input impedance of DMMs, FETVOMs, VTVMs and other electronic voltmeters makes them the preferred instruments for voltage measurements.

Test leads — Keep an assortment of wires with insulated, soldered alligator clips. Commercially made leads have a high failure rate because they use small wire that is not soldered to the clips; it is best to make your own.

Open wire leads (**Fig 26.1A**) are good for dc measurements, but they can pick up unwanted RF energy. This problem is reduced somewhat if the leads are twisted together (Fig 26.1B). A coaxial cable lead is much better, but its inherent capacitance can affect RF measurements.

The most common probe is the low-capacitance (×10) probe shown in Fig 26.1C. This probe isolates the oscilloscope from the circuit under test, preventing the 'scope's input and test-probe capacitance from affecting the circuit and changing the reading. A network in the probe serves as a 10:1 divider and compensates for frequency distortion in the cable and test instrument.

Demodulator probes (see the **Test Equipment and Measurements** chapter and the schematic shown in Fig 26.1D) are used to demodulate or detect RF signals, converting modulated RF signals to audio that can be heard in a signal tracer or seen on a low-bandwidth 'scope.

You can make a probe for inductive coupling as shown in Fig 26.1E. Connect a two- or three-turn loop across the center conductor and shield before sealing the end. The inductive pick up is useful for coupling to high-current points.

RF power and SWR meters — Every shack should have one. It is used to measure forward and reflected RF power. A standing-wave ratio (SWR) meter can be the first indicator of antenna trouble. It can also be used between an exciter and power amplifier to spot an impedance mismatch.

Simple meters indicate relative power SWR and are fine for Transmatch adjustment and line monitoring. However, if you want to make accurate measurements, a calibrated wattmeter with a directional coupler is required.

Dummy load — A "dummy" or "phantom" load is a necessity in any shack. Do not put a signal on the air while repairing equipment. Defective equipment can generate signals that interfere with other hams or other radio services. A dummy load also provides a known, matched load (usually 50 Ω) for use during adjustments.

When buying a dummy load, avoid used, oil-cooled dummy loads unless you can be sure that the oil does not contain PCBs. This biologically hazardous compound was common in transformer oil until a few years ago. Mineral oil is fine.

Dummy loads in the shack are often required to dissipate a transmitter's or linear amplifier's full power output, i.e., up to 1500 W, at least for a short time. They are also used in lower power applications — typically smaller sized and rated, and can also be used as impedance standards or as terminations for low-power computer or communications lines.

Dip meter — This device is often called a transistor dip meter or a grid-dip oscillator from vacuum-tube days.

Most dip meters can also serve as an absorption frequency meter. In this mode, measurements are read at the current peak, rather than the dip. Some meters have a connection for headphones. The operator can usually hear signals that do not register on

Troubleshooting and Maintenance 26.3

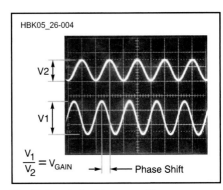

Fig 26.4 — A dual-trace oscilloscope display of amplifier input and output waveforms.

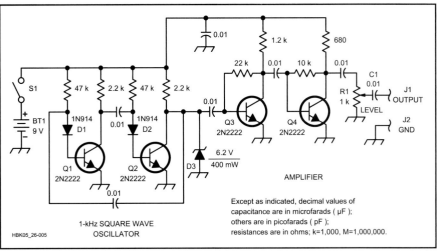

Fig 26.5 — Schematic of the AF/RF signal injector. All resistors are ¼-W, 5% carbon units, and all capacitors are disc ceramic.

BT1 — 9-V battery.
D1, D2 — Silicon switching diode, 1N914 or equiv.
D3 — 6.2-V, 400-mW Zener diode.
J1, J2 — Banana jack.

Q1-Q4 — General-purpose silicon NPN transistors, 2N2222 or similar.
R1 — 1-kΩ panel-mount control.
S1 — SPST toggle switch.

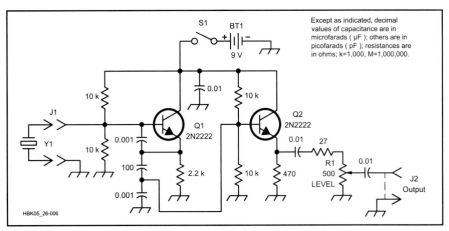

Fig 26.6 — Schematic of the crystal-controlled signal source. All resistors are ¼-W, 5% carbon units, and all capacitors are disc ceramic. A full-size etching pattern and parts-placement diagram can be found on the **Templates** Section of the *Handbook* CD-ROM.

BT1 — 9-V transistor radio battery.
J1 — Crystal socket to match the crystal type used.
J2 — RCA phono jack or equivalent.

Q1, Q2 — General-purpose silicon NPN transistors, 2N2222 or similar.
R1 — 500-Ω panel-mount control.
S1 — SPST toggle switch.
Y1 — 1 to 15-MHz crystal.

the meter. Because the dip meter is an oscillator, it can be used as a signal generator in certain cases where high accuracy or stability are not required.

When purchasing a dip meter, look for one that is mechanically and electrically stable. The coils should be in good condition. A headphone connection is helpful. Battery operated models are easier to use for antenna measurements. Dip meters are not nearly as common as they once were.[1]

Oscilloscope — The oscilloscope, or 'scope, is the second most often used piece of test equipment, although a lot of repairs can be accomplished without one. The trace of a 'scope can give us a lot of information about a signal at a glance.

The simplest way to display a waveform is to connect the vertical amplifier of the 'scope to a point in the circuit through a simple test lead. When viewing RF, use a low-capacitance probe that has been adjusted to match the 'scope. Select the vertical gain and timebase (horizontal scale, **Fig 26.2**) for the most useful displayed waveform.

A 'scope waveform shows voltage (if calibrated), approximate period (frequency is the reciprocal of the period) and a rough idea of signal purity (see **Fig 26.3**). If the 'scope has dual-trace capability (meaning it can display two signals at once), a second waveform may be displayed and compared to the first. When the two signals are taken from the input and output of a stage, stage linearity and phase shift can be checked (see **Fig 26.4**).

An important specification of an oscilloscope is its amplifier bandwidth. This tells us the frequency at which amplifier response has dropped 3 dB. The instrument will display higher frequencies, but its accuracy at higher frequencies is not known. Even well below its rated bandwidth a 'scope is not capable of much more than about 5% accuracy. This is adequate for most amateur applications.

An oscilloscope will show gross distortions of audio and RF waveforms, but it cannot be used to verify that a transmitter meets FCC regulations for harmonics and spurious emissions. Harmonics that are down only 20 dB from the fundamental would be illegal in most cases, but they would not change the oscilloscope waveform enough to be seen.

When buying a 'scope, get the greatest bandwidth you can afford. Old Hewlett-Packard or Tektronix 'scopes are usually quite good for amateur use.

Signal generator — Although signal generators have many uses, in troubleshooting they are most often used for signal injection (more about this later) and alignment.

An AF/RF signal-injector schematic is shown in **Fig 26.5**. If frequency accuracy is needed, the crystal-controlled signal source of **Fig 26.6** can be used. The AF/RF circuit provides usable harmonics up to 30 MHz, while the crystal controlled oscillator will function with crystals from 1 to 15 MHz. These two projects are not meant to compete with standard signal generators, but they are adequate for signal injection.[2] A better generator is required for receiver alignment or for receiver quality testing.

When buying a generator, look for one that can generate a sine wave signal. A good signal generator is double or triple shielded against leakage. Fixed-frequency audio should be available for modulation of the RF signal and for injection into audio stages. The most versatile generators can generate amplitude and frequency modulated signals.

Good generators have stable frequency controls with no backlash. They also have multiposition switches to control signal level. A switch marked in dBm is a good indication that you have located a high-quality test instrument. The output jack should be a coaxial connector (usually a BNC or N), not the kind used for microphone connections.

Some older, high-quality units are common. Look for World War II surplus units of the URM series, Boonton, GenRad, Hewlett-Packard, Tektronix, Measurements Inc or other well-known brand names. Some home-built signal generators may be quite good, but make sure to check construction techniques, level control and shielding quality.

Signal tracer — Signals can be traced with a voltmeter and an RF probe, a dip meter with headphones or an oscilloscope, but there are some devices made especially for signal tracing. A signal tracer is primarily a high-gain audio amplifier. It may have a built-in RF detector, or rely on an external RF probe. Most convert the traced signal to audio through a speaker.

The tracer must function as a receiver and detector for each frequency range in the test circuit. A high-impedance tracer input is necessary to prevent circuit loading.

A general-coverage receiver can be used to trace RF or IF signals, if the receiver covers the necessary frequency range. Most receivers, however, have a low-impedance input that severely loads the test circuit. To minimize loading, use a capacitive probe or loop pickup. When the probe is held near the circuit, signals will be picked up and carried to the receiver. It may also pick up stray RF, so make sure you are listening to the correct signal by switching the circuit under test on and off while listening.

Tube tester — Vacuum-tube testers used to be found in nearly every drug or department store. They are scarce now because tubes are no longer used in modern consumer or (most) amateur equipment. Older tube gear is found in many ham shacks or flea markets, though. There are many aficionados of vintage gear who enjoy working with old vacuum-tube equipment.

Most simple tube testers measure the cathode emission of a vacuum tube. Each grid is shorted to the plate through a switch and the current is observed while the tube operates as a diode. By opening the switches from each grid to the plate (one at a time), we can check for opens and shorts. If the plate current does not drop slightly as a switch is opened, the element connected to that switch is either open or shorted to another element. (We cannot tell an open from a short with this test.) The emission tester does not necessarily indicate the ability of a tube to amplify.

Other tube testers measure tube gain (transconductance). Some transconductance testers read plate current with a fixed bias network. Others use an ac signal to drive the tube while measuring plate current.

Most tube testers also check inter-element leakage. Contamination inside the tube envelope may result in current leakage between elements. The paths can have high resistance, and may be caused by gas or deposits inside the tube. Tube testers use a moderate voltage to check for leakage. Leakage can also be checked with an ohmmeter using the ×1M range, depending on the actual spacing of tube elements.

Transistor tester — Transistor testers are similar to transconductance tube testers. Device current is measured while the device is conducting or while an ac signal is applied at the control terminal. Commercial surplus units are often seen at ham flea markets. Some DMMs being sold today also include a built-in, simple transistor tester.

Most transistor failures appear as either an open or shorted junction. Opens and shorts can be found easily with an ohmmeter; a special tester is not required.

Transistor gain characteristics vary widely, however, even between units with the same device number. Testers can be used to measure the gain of a transistor. A tester that uses dc signals measures only transistor dc alpha and beta. Testers that apply an ac signal show the ac alpha or beta. Better testers also test for leakage.

In addition to telling you whether a transistor is good or bad, a transistor tester can help you decide if a particular transistor has sufficient gain for use as a replacement. It may also help when matched transistors are required. The final test is the repaired circuit.

Frequency meter — Most frequency counters are digital units, often able to show frequency to a 1-Hz resolution. Some older "analog" counters are sometimes found surplus, but a low-cost digital counter will outperform even the best of these old "classics."

Power supplies — A well-equipped test bench should include a means of varying the ac-line voltage, a variable-voltage regulated dc supply and an isolation transformer.

AC-line voltage varies slightly with load. An autotransformer with a movable tap lets you boost or reduce the line voltage slightly. This is helpful to test circuit functions with supply-voltage variations.

As mentioned earlier, ac/dc radios must be isolated from the ac line during testing and repair. Keep an isolation transformer handy if you want to work on table-model broadcast radios or television sets (check for other ac/dc equipment, too. Even some old phonographs or Amateur Radio transceivers used this dangerous circuit design).

A good multivoltage supply will help with nearly any analog or digital troubleshooting project. Many electronics distributors stock bench power supplies. A variable-voltage dc supply may be used to power various small items under repair or provide a variable bias supply for testing active devices. Construction details for a laboratory power supply appear in the **Power Supplies** chapter.

If you want to work on vacuum-tube gear, the maximum voltage available from the dc supply should be high enough to serve as a plate or a bias supply for common tubes (about 300 to 400 V ought to do it).

Accessories — There are a few small items that may be used in troubleshooting. You may want to keep them handy.

Many circuit problems are sensitive to temperature. A piece of equipment may work well when first turned on (cold) but fail as it warms up. In this case, a cold source will help you find the intermittent connection. When you cool the bad component, the circuit will suddenly start working again (or stop working). Cooling sprays are available from most parts suppliers.

A heat source helps locate components that fail only when hot. A small incandescent lamp can be mounted in a large piece of sleeve insulation to produce localized heat for test purposes.

A heat source is usually used in conjunction with a cold source. If you have a circuit that stops working when it warms up, heat the circuit until it fails, then cool the components one by one. When the circuit starts working again, the last component sprayed was the bad one.

A stethoscope (with the pickup removed — see **Fig 26.7**) or a long piece of sleeve insulation can be used to listen for arcing or sizzling in a circuit.

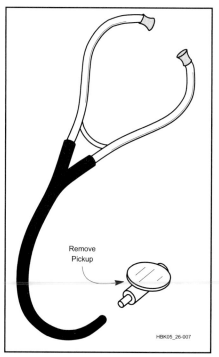

Fig 26.7 — **A stethoscope, with the pickup removed, is used to listen for arcing in crowded circuits.**

26.2 Where to Begin

26.2.1 New Construction

In most repair work, the technician is aided by the knowledge that the circuit once worked. It is only necessary to find the faulty part(s) and replace it. This is not so with newly constructed equipment. Repair of equipment with no working history is a special, and difficult, case. You may be dealing with a defective component, construction error or even a faulty design. Carefully checking for these defects can save you hours.

26.2.2 All Equipment

CHECK THE OBVIOUS

Try the easy things first. If you are able to solve the problem by replacing a fuse or reconnecting a loose cable, you might be able to avoid a lot of effort. Many experienced technicians have spent hours troubleshooting a piece of equipment only to learn the hard way that the on/off switch was "off" or the squelch control was set too high, or that they were not using the equipment properly.

Read the manual! Your equipment may be working as designed. Many electronic "problems" are caused by a switch that is set in the wrong position, or a unit that is being asked to do something it was not designed to do. Before you open up your equipment for major surgery, make sure you are using it correctly.

Next, make sure the equipment is plugged in, that the ac outlet does indeed have power, that the equipment is switched "on" and that all of the fuses are good. If the equipment uses batteries or an external power supply, make sure these are working.

Check that all wires, cables and accessories are working and plugged in to the right connectors or jacks. In a "system," it is often difficult to be sure which component or subsystem is bad. Your transmitter may not work on SSB because the transmitter is bad, but it could also be a bad microphone.

Connector faults are more common than component troubles. Consider poor connections as prime suspects in your troubleshooting detective work. Do a thorough inspection of the connections. Is the antenna connected? How about the speaker, fuses and TR switch? Are transistors and ICs firmly seated in their sockets? Are all interconnection cables sound and securely connected? Many of these problems are obvious to the eye, so look around carefully.

SIMPLIFY THE PROBLEM

If the broken equipment is part of a system, you need to find out exactly which part of the system is bad. For example, if your amateur station is not putting out any RF, you need to determine if it is a microphone problem, a transmitter problem, an amplifier problem or a problem somewhere in your station wiring. If you are trying to diagnose a bad channel on your home modular stereo system, it could be anything from a bad cable to a bad amplifier to a bad speaker.

Simplify the system as much as possible. To troubleshoot the "no-RF" problem, temporarily eliminate the amplifier from the station configuration. To diagnose the stereo system, start troubleshooting by checking just the amplifier with a set of known good headphones. Simplifying the problem will often isolate the bad component quickly.

DOCUMENTATION

Once you have determined that a piece of equipment is indeed broken, you need to do some preparation before you diagnose and fix it. First, locate a schematic diagram and service manual. It is possible to troubleshoot without a service manual, but a schematic is almost indispensable.

The original equipment manufacturer is the best source of a manual or schematic. However, many old manufacturers have gone out of business. Several sources of equipment manuals can be located by a Web search.

If all else fails, you can sometimes reverse engineer a simple circuit by tracing wiring paths and identifying components to draw your own schematic. If you have access to the databooks for the active devices used in the circuit, the pin-out diagrams and applications notes will sometimes be enough to help you understand and troubleshoot the circuit.

DEFINE PROBLEMS

To begin troubleshooting, define the problem accurately. Ask yourself these questions:

1. What functions of the equipment do not work as they should; what does not work at all?
2. What kind of performance can you realistically expect?
3. Has the trouble occurred in the past? (Keep a record of troubles and maintenance in the owner's manual, shack notebook or log book.)

Write the answers to the questions. The information will help with your work, and may help service personnel if their advice or professional service is required.

TAKE IT APART

All of the preparation work has been done. It is time to really dig in. You usually will have to start by taking the equipment apart. This is the part that can trap the unwary technician. Most experienced service technicians can tell you the tale of the equipment they took apart and were unable to easily put back together. Don't let it happen to you.

Take photos and lots of notes about the way you take it apart. Take notes about each component you remove. Write down the order in which you do things, color codes, part placements, cable routings, hardware notes and anything else you think you might need to be able to reassemble the equipment weeks from now when the back-ordered part comes in.

Put all of the screws in one place. A plastic jar with a lid works well; if you drop it the plastic is not apt to break and the lid will keep all the parts from flying around the work area (you will never find them all). It may pay to have a separate labeled container for each subsystem.

LOOK AROUND

Many service problems are visible, if you look for them carefully. Many a technician has spent hours tracking down a failure, only to find a bad solder joint or burned component that would have been spotted in careful inspection of the printed-circuit board. Start troubleshooting by carefully inspecting the equipment.

It is time consuming, but you really need to look at every connector, every wire, every solder joint and every component. A connector may have loosened, resulting in an open circuit. You may spot broken wires or see a bad solder joint. Flexing the printed-circuit board or tugging on components a bit while looking at their solder joints will often locate a defective solder job. Look for scorched components.

Make sure all of the screws securing the printed-circuit board are tight and making good electrical contact. (Do not tighten the adjusting screws, however! You will ruin the alignment.) See if you can find evidence of previous repair jobs; these may not have been done properly. Make sure that each IC is firmly seated in its socket. Look for pins folded underneath the IC rather than making contact with the socket. If you are troubleshooting a newly constructed circuit, make sure each part is of the correct value or type number and is installed correctly.

If your careful inspection doesn't reveal anything, it is time to apply power to the unit under test and continue the process. Observe all safety precautions while troubleshooting equipment. There are voltages inside some equipment that can kill you. If you are not qualified to work safely with the voltages and conditions inside of the equipment, do not proceed. See Table 26.1 and the **Safety** chapter.

OTHER SENSES

With power applied to the unit, listen for arcs and look and smell for smoke. If no problems are apparent, you will have to start testing the various parts of the circuit.

26.2.3 Various Approaches

There are two fundamental approaches to troubleshooting: the systematic approach and the instinctive approach. The systematic approach uses a defined process to analyze and isolate the problem. An instinctive approach relies on troubleshooting experience to guide you in selecting which circuits to test and which tests to perform. The systematic approach is usually chosen by beginning troubleshooters.

AT THE BLOCK LEVEL

The block diagram is a road map. It shows the signal paths for each circuit function. These paths may run together, cross occasionally or not at all. Those blocks that are not in the paths of faulty functions can be eliminated as suspects. Sometimes the symptoms point to a single block, and no further search is necessary.

In cases where more than one block is suspect, several approaches may be used. Each requires testing a block or stage. Signal injection, signal tracing, instinct or combination of all techniques may be used to diagnose and test electronic equipment.

SYSTEMATIC APPROACHES

The instinctive approach works well for those with years of troubleshooting experience. Those of us who are new to this game need some guidance. A systematic approach is a disciplined procedure that allows us to tackle problems in unfamiliar equipment with a reasonable hope of success.

There are two common systematic approaches to troubleshooting at the block level. The first is signal tracing; the second is signal injection. The two techniques are very similar. Differences in test equipment and the circuit under test determine which method is best in a given situation. They can often be combined.

POWER SUPPLIES

You may be able to save quite a bit of time if you test the power supply first. All of the other circuits may be dead if the power supply is not working. Power supply diagnosis is discussed in detail later in this chapter.

SIGNAL TRACING

In signal tracing, start at the beginning of a circuit or system and follow the signal through to the end. When you find the signal at the input to a specific stage, but not at the output, you have located the defective stage. You can then measure voltages and perform other tests on that stage to locate the specific failure. This is much faster than testing every component in the unit to determine which is bad.

It is sometimes possible to use over-the-air signals in signal tracing, in a receiver for example. However, if a good signal generator is available, it is best to use it as the signal source. A modulated signal source is best.

Signal tracing is suitable for most types of troubleshooting of receivers and analog amplifiers. Signal tracing is the best way to check transmitters because all of the necessary signals are present in the transmitter by design. Most signal generators cannot supply the wide range of signal levels required to test a transmitter.

Equipment

A voltmeter, with an RF probe, is the most common instrument used for signal tracing. Low-level signals cannot be measured accurately with this instrument. Signals that do not exceed the junction drop of the diode in the probe will not register at all, but the presence, or absence, of larger signals can be observed.

A dedicated signal tracer can also be used. It is essentially an audio amplifier. An experienced technician can usually judge the level and distortion of the signal by ear. You cannot use a dedicated signal tracer to follow a signal that is not amplitude modulated (single sideband is a form of AM). Signal tracing is not suitable for tracing CW signals, FM signals or oscillators. To trace these, you will have to use a voltmeter and RF probe or an oscilloscope.

An oscilloscope is the most versatile signal tracer. It offers high input impedance, variable sensitivity, and a constant display of the traced waveform. If the oscilloscope has sufficient bandwidth, RF signals can be observed directly. Alternatively, a demodulator probe can be used to show demodulated RF signals on a low-bandwidth 'scope. Dual-trace scopes can simultaneously display the waveforms, including their phase relationship, present at the input and output of a circuit.

Procedure

First, make sure that the circuit under test and test instruments are isolated from the ac line by transformers. Set the signal source to an appropriate level and frequency for the unit you are testing. For a receiver, a signal of about 100 µV should be plenty. For other circuits, use the schematic, an analysis of circuit function and your own good judgment to set the signal level.

In signal tracing, start at the beginning and work toward the end of the signal path. Switch on power to the test circuit and connect the signal-source output to the test-circuit input. Place the tracer probe at the circuit input and ensure that you can hear the test signal. Observe the characteristics of the signal if you

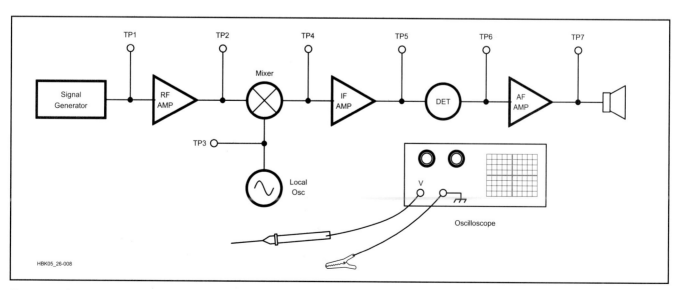

Fig 26.8 — Signal tracing in a simple receiver.

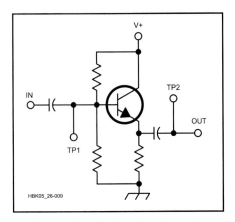

Fig 26.9 — The effect of circuit impedance on an oscilloscope display. Although the circuit functions as a current amplifier, the change in impedance from TP1 to TP2 results in the traces described. This is a common-collector amplifier.

are using a 'scope (see **Fig 26.8**). Compare the detected signal to the source signal during tracing.

Move the tracer probe to the output of the next stage and observe the signal. Signal level should increase in amplifier stages and may decrease slightly in other stages. The signal will not be present at the output of a "dead" stage.

Low-impedance test points may not provide sufficient signal to drive a high-impedance signal tracer, so tracer sensitivity is important. Also, in some circuits the output level appears low where there is an impedance change from input to output of a stage (see **Fig 26.9**). For example, the circuit in Fig 26.9 is a common-collector current amplifier with a high input impedance and low output impedance. The voltages at TP1 and TP2 are approximately equal and in phase.

There are two signals — the test signal and the local oscillator signal — present in a mixer stage. Loss of either one will result in no output from the mixer stage. Switch the signal source on and off repeatedly to make sure that the tracer reading varies (it need not disappear) with source switching.

SIGNAL INJECTION

Like signal tracing, signal injection is particularly suited to some situations. Signal injection is a good choice for receiver troubleshooting because the receiver already has a detector as part of the design. It is suitable for either high- or low-impedance circuits and can be used with vacuum tubes, transistors or ICs.

Equipment

If you are testing equipment that does not include a suitable detector as part of the circuit, some form of signal detector is required. Any of the instruments used for signal tracing are adequate.

Most of the time, your signal injector will be a signal generator. There are other injectors available, some of which are square-wave audio oscillators rich in RF harmonics (see Fig 26.5). These are usually built into a pen-sized case with a test probe at the end. These "pocket" injectors do have their limits because you can't vary their output level or determine their frequency. They are still useful, though, because most circuit failures are caused by a stage that is completely dead.

Consider the signal level at the test point when choosing an instrument. The signal source used for injection must be able to supply appropriate frequencies and levels for each stage to be tested. For example, a typical superheterodyne receiver requires AF, IF and RF signals that vary from 6 V at AF, to 0.2 mV at RF. Each conversion stage used in a receiver requires another IF from the signal source.

Procedure

If an external detector is required, set it to the proper level and connect it to the test circuit. Set the signal source for AF, and inject a signal directly into the signal detector to test operation of the injector and detector. Move the signal source to the input of the preceding stage, and observe the signal. Continue moving the signal source to the inputs of successive stages.

When you inject the signal source to the input of the defective stage, there will be no output. Prevent stage overload by reducing the level of the injected signal as testing progresses through the circuit. Use suitable frequencies for each tested stage.

Make a rough check of stage gain by injecting a signal at the input and output of an amplifier stage. You can then compare how much louder the signal is when injected at the input. This test may mislead you if there is a radical difference in impedance from stage input to output. Understand the circuit operation before testing.

Mixer stages present a special problem because they have two inputs, rather than one. A lack of output signal from a mixer can be caused by either a faulty mixer or a faulty local oscillator (LO). Check oscillator operation with a 'scope or absorption wavemeter, or by listening on another receiver. If none of these instruments are available, inject the frequency of the LO at the LO output. If a dead oscillator is the only problem, this should restore operation.

If the oscillator is operating, but off frequency, a multitude of spurious responses will appear. A simple signal injector that

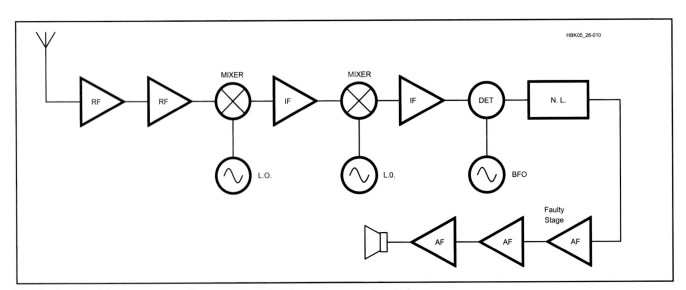

Fig 26.10 — The 14-stage receiver diagnosed by the "divide and conquer" technique.

produces many frequencies simultaneously is not suitable for this test. Use a well-shielded signal generator set to an appropriate level at the LO frequency.

DIVIDE AND CONQUER

Under certain conditions, the block search may be speeded by testing at the middle of successively smaller circuit sections. Each test limits the fault to one half of the remaining circuit (see **Fig 26.10**). Let's say the receiver has 14 stages and the fault is in stage 12. This approach requires only four tests to locate the faulty stage, a substantial saving of time.

This "divide and conquer" tactic cannot be used in equipment that splits the signal path between the input and the output. Test readings taken inside feedback loops are misleading unless you understand the circuit and the waveform to be expected at each point in the test circuit. It is best to consider all stages within a feedback loop as a single block during the block search.

Both signal tracing and signal injection procedures may be speeded by taking some diagnostic short cuts. Rather than check each stage sequentially, check a point halfway through the system. As an example:

An HF receiver is not working. There is absolutely no response from the speaker. First, substitute a suitable speaker — still no sound. Next, check the power supply — no problem there. No clues indicate any particular stage. Signal tracing or injection must provide the answer.

Get out the signal generator and switch it on. Set the generator for a low-level RF signal, switch the signal off and connect the output to the receiver. Switch the signal on again and place a high-impedance signal-tracer probe at the antenna connection. Instantly, the tracer emits a strong audio note. Good; the test equipment is functioning.

Move the probe to the input of the receiver detector. As the tracer probe touches the circuit the familiar note sounds. Next, set the tracer for audio and place the probe halfway through the audio chain. It is silent! Move the probe halfway back to the detector, and the note appears once again. Yet, no signal is present at the output of the stage. You now know that the defect is somewhere between the two points tested. In this case, the third audio stage is faulty.

THE INSTINCTIVE APPROACH

In an "instinctive" approach to troubleshooting, you rely on your judgment and experience to decide where to start testing, what and how to test. When you immediately check power supply voltages, or the ac fuse on a unit that is completely nonfunctional, that is an example of an instinctive approach. If you are faced with a receiver that has distorted audio and immediately start testing the speaker and audio output stage, or if you immediately start checking the filter and bypass capacitors in an audio stage that is oscillating or "motorboating" you are troubleshooting on instinct.

Most of our discussion on the instinctive approach is really a collection of tips and guidelines. Read them to build your troubleshooting skills.

The check for connector problems mentioned at the beginning of this section is a good idea. Experience has shown connector faults to be so common that they should be checked even before a systematic approach begins.

When instinct is based on experience, searching by instinct may be the fastest procedure. If your instinct is correct, repair time and effort may be reduced substantially. As experience and confidence grow, the merits of the instinctive approach grow with them. However, inexperienced technicians who choose this approach are at the mercy of chance.

26.3 Testing Within a Stage

Once you have followed all of the troubleshooting procedures and have isolated your problem to a single defective stage or circuit, a few simple measurements and tests will usually pinpoint one or more specific components that need adjustment or replacement.

First, check the parts in the circuit against the schematic diagram to be sure that they are reasonably close to the design values, especially in a newly built circuit. Even in a commercial piece of equipment, someone may have incorrectly changed them during attempted repairs. A wrong-value part is quite likely in new construction, such as a homebrew project.

26.3.1 Voltage Levels

Check the circuit voltages. If the voltage levels are printed on the schematic, this is easy. If not, analyze the circuit and make some calculations to see what the circuit voltages should be. Remember, however, that the printed or calculated voltages are nominal; measured voltages may vary from the calculations.

When making measurements, remember the following points:

- Make measurements at device leads, not at circuit-board traces or socket lugs.
- Use small test probes to prevent accidental shorts.
- Never connect or disconnect power to solid-state circuits with the switch on. Consider the effect of the meter on measured voltages. A 20-kΩ/V meter may load down a high-impedance circuit and change the voltage.

Voltages may give you a clue to what is wrong with the circuit. If not, check the active device. If you can check the active device in the circuit, do so. If not, remove it and test it, or substitute a known good device. After connections, most circuit failures are caused directly or indirectly by a bad active device. The experienced troubleshooter usually tests or substitutes these first.

Analyze the other components and determine the best way to test each. There is additional information about electronic components in the electronic-theory chapters and in the **Component Data and References** chapter.

There are two voltage levels in most circuits (V+ and ground, for example). Most component failures (opens and shorts) will shift dc voltages near one of these levels.

Typical failures that show up as incorrect dc voltages include: open coupling transformers; shorted capacitors; open, shorted or overheated resistors and open or shorted semiconductors.

26.3.2 Noise

A slight hiss is normal in all electronic circuits. This noise is produced whenever current flows through a conductor that is warmer than absolute zero. Noise is compounded and amplified by succeeding stages. Repair is necessary only when noise threatens to obscure normally clear signals.

Semiconductors can produce hiss in two ways. The first is normal — an even white noise that is much quieter than the desired signal. Faulty devices frequently produce excessive noise. The noise from a faulty device is usually erratic, with pops and crashes that are sometimes louder than the desired signal. In an analog circuit, the end result of noise is usually sound. In a control or digital circuit, noise causes erratic operation: unexpected switching and so on.

Noise problems usually increase with temperature, so localized heat may help you find the source. Noise from any component may be sensitive to mechanical vibration. Tapping various components with an insulated screwdriver may quickly isolate a bad part.

Noise can also be traced with an oscilloscope or signal tracer.

Nearly any component or connection can be a source of noise. Defective components are the most common cause of crackling noises. Defective connections are a common cause of loud, popping noises.

Check connections at cables, sockets and switches. Look for dirty variable-capacitor wipers and potentiometers. Mica trimmer capacitors often sound like lightning when arcing occurs. Test them by installing a series 0.01-µF capacitor. If the noise disappears, replace the trimmer.

Potentiometers are particularly prone to noise problems when used in dc circuits. Clean them with spray cleaner and rotate the shaft several times.

Rotary switches may be tested by jumpering the contacts with a clip lead. Loose contacts may sometimes be repaired, either by cleaning, carefully rebending the switch contacts or gluing loose switch parts to the switch deck. Operate variable components through their range while observing the noise level at the circuit output.

26.3.3 Oscillations

Oscillations occur whenever there is sufficient positive feedback in a circuit that has gain. (This can even include digital devices.) Oscillation may occur at any frequency from a low-frequency audio buzz (often called "motorboating") well up into the RF region.

Unwanted oscillations are usually the result of changes in the active device (increased junction or interelectrode capacitance), failure of an oscillation suppressing component (open decoupling or bypass capacitors or neutralizing components) or new feedback paths (improper lead dress or dirt on the chassis or components). It can also be caused by improper design, especially in home-brew circuits. A shift in bias or drive levels may aggravate oscillation problems.

Oscillations that occur in audio stages do not change as the radio is tuned because the operating frequency, and therefore the component impedances, do not change. However, RF and IF oscillations usually vary in amplitude as operating frequency is changed.

Oscillation stops when the positive feedback is removed. Locating and replacing the defective (or missing) bypass capacitor may effect an improvement. The defective oscillating stage can be found more reliably with a signal tracer or oscilloscope.

26.3.4 Amplitude Distortion

Amplitude distortion is the product of nonlinear operation. The resultant waveform contains not only the input signal, but new signals at other frequencies as well. All of the frequencies combine to produce the distorted waveform. Distortion in a transmitter gives rise to splatter, harmonics and interference.

Fig 26.11 shows some typical cases of distortion. Clipping (also called flat-topping) is the consequence of excessive drive. The corners on the waveform show that harmonics are present. (A square wave contains the fundamental and all odd harmonics.) These odd harmonics would be heard well away from the operating frequency, possibly outside of amateur bands. Key clicks are similar to clipping.

Harmonic distortion produces radiation at frequencies far removed from the fundamental; it is a major cause of electromagnetic interference (EMI). Harmonics are generated in nearly every amplifier. When they occur in a transmitter, they are usually caused by insufficient transmitter filtering (either by design, or because of filter component failure).

Incorrect bias brings about unequal amplification of the positive and negative wave sections. The resultant waveform is rich in harmonics.

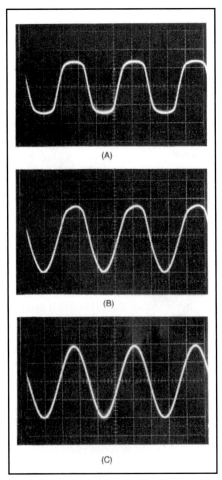

Fig 26.11 — Examples of distorted waveforms. The result of clipping is shown in A. Nonlinear amplification is shown in B. A pure sine wave is shown in C for comparison.

26.3.5 Frequency Distortion

If a "broadband" amplifier, such as an audio amplifier, doesn't amplify all frequencies equally, there is frequency distortion. In many cases, this "frequency distortion" is deliberate, as in a transmitter microphone amplifier that has been designed to pass only frequencies from 200 to 2000 Hz. In most cases, the amateur's ability to detect and measure distortion is limited by available test equipment.

26.3.6 Distortion Measurement

A distortion meter is used to measure distortion of AF signals. A spectrum analyzer is the best piece of test gear to measure distortion of RF signals. If a distortion meter is not available, an estimation of AF distortion can sometimes be made with a function generator (sine and square waves) and an oscilloscope.

To estimate the amount of frequency distortion in an audio amplifier, set the generator for a square wave and look at it on the 'scope. (Use a low-capacitance probe.) The wave should show square corners and a flat top. Next, inject a square wave at the amplifier input and again look at the input wave on the 'scope. Any new distortion is a result of the test circuit loading the generator output. (If the wave shape is severely distorted, the test is not valid.) Now, move the test probe to the test circuit output and look at the waveform. Refer to **Fig 26.12** to evaluate square-wave distortion and its cause.

The above applies only to audio amplifiers without frequency tailoring. In RF gear, the transmitter may have a very narrow audio passband, so inserting a square wave into the microphone input may result in an output that is difficult to interpret. The frequency of the square wave will have a significant effect.

Anything that changes the proper bias of an amplifier can cause distortion. This includes failures in the bias components, leaky transistors or vacuum tubes with interelectrode shorts. These conditions may mimic AGC trouble. Improper bias often results from an overheated or open resistor. Heat can cause resistor values to permanently increase. Leaky, or shorted capacitors and RF feedback can also produce distortion by disturbing bias levels. Distortion is also caused by circuit imbalance in Class AB or B amplifiers.

Oscillations in an IF amplifier may produce distortion. They cause constant, full AGC action, or generate spurious signals that mix with the desired signal. IF oscillations are usually evident on the S meter, which will show a strong signal even with the antenna disconnected.

26.3.7 Alignment

Alignment is rarely the cause of an electronics problem. As an example, suppose an AM receiver suddenly begins producing weak and distorted audio. An inexperienced person frequently suspects poor alignment as a common problem. Even though the manufacturer's instructions and the proper equipment are not available, our "friend" (this would never be one of US!) begins "adjusting" the transformer cores. Before long, the set is hopelessly misaligned. Now our misguided ham must send the radio to a shop for an alignment that was not needed before repairs were attempted.

Alignment does not shift suddenly. A normal signal tracing procedure would have shown that the signal was good up to the audio-output IC, but badly distorted after that. The defective IC that caused the problem would have been easily found and quickly replaced.

Look for the Obvious

The best example of how looking for the obvious can save a lot of repair time comes from my days as the manager of an electronics service shop. We had hired a young engineering graduate to work for us part time. He was the proud holder of a First-Class FCC Radiotelephone license (the predecessor to today's General Radiotelephone license). He was a likable sort, but, well . . . the chip on his young shoulder was a bit hard to take sometimes.

One day, I had asked him to repair a "tube-type" FM tuner. He had been poking around without success: hooking up a voltmeter, oscilloscope and signal generator, pretty much in that order. Finally, in total exasperation, he pronounced that the unit was beyond economical repair and suggested that I return it to the customer unfixed. The particular customer was a "regular," so I wanted to be sure of the diagnosis before I sent the tuner back. I told the tech I wanted to take a look at it before we wrote it off.

He started to expound loudly that there was no way that I, a lowly technician (even though I was also his boss) could find a problem that he, an engineering graduate and holder of a First Class . . . you get the idea. I did remind him gently that I was the boss, and he, realizing that I had him there, stepped aside, mumbling something about my suiting myself. He stepped back to gloat when I couldn't find it either.

I began by giving the tuner a thorough visual inspection. I looked it over carefully from stem to stern, while listening to our young apprentice proclaiming with certainty that one cannot fix electronic equipment by merely looking at it. I didn't see anything obviously wrong, so I decided to move wires and components around, looking for a bad solder joint or broken component. Of course, I had to listen to him telling me that one cannot possibly find bad components by touch. Unfortunately for our loud friend, he couldn't have been more wrong.

I grabbed hold of a ceramic bypass capacitor to give it a little wiggle, and much to my surprise it was hot enough to cause some real pain. I kept my composure; it was an opportunity for a good learning experience. Ceramic capacitors don't get very hot unless they are either shorted or very leaky. I kept silent and never let on that my finger "probe" had indeed located the bad part. I set the tuner down, sighed a bit, and then looked him right in the eye when I pointed to the capacitor and said "Change that part!"

They probably heard his bellowing in the next county! He went on and on about how there was just no way in the world that I could tell a good part from a bad part by just looking and touching things. He alternated between accusing me of pulling his leg and guessing, then back to just plain bellowing again. After letting this "source of great noise" run his course, I offered the ultimate shop challenge — I bet him a can of soda pop.

The traditional shop challenge did the trick. He smugly grabbed a replacement part from the bin and got out his soldering iron. In a matter of seconds (a new shop record, I believe) the capacitor was installed. He hooked the tuner up to a test amplifier and turned them on. After a couple of seconds, he smugly turned to me and started an "I told ya' so!" Just then, the last tube warmed up and the sounds of our local rock station blasted out of the speaker. He stopped in mid "told-ya" and stared at the tuner in disbelief. The tempo and pitch of his voice jumped by an order of magnitude as he asked me how I managed to fix the tuner without using even an ohmmeter to test a fuse. It was weeks before I told him — the soda pop tasted especially good.

The moral of the story is clear; sophisticated test equipment and procedures are useful in troubleshooting, but they are no substitute for the experience of a veteran troubleshooter.
— Ed Hare, W1RFI, ARRL Laboratory Supervisor

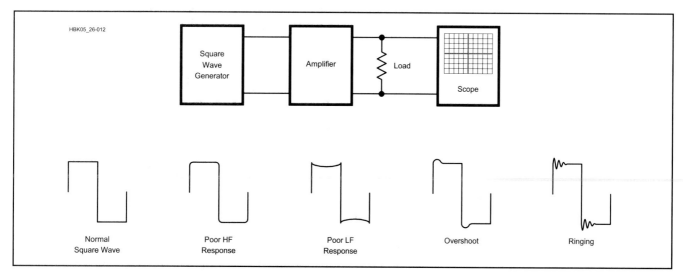

Fig 26.12 — Square-wave distortion and probable causes.

26.3.8 Contamination

Contamination is another common service problem. Cold soda pop spilled into a hot piece of electronics is an extreme example (but one that does actually happen).

Conductive contaminants range from water to metal filings. Most can be removed by a thorough cleaning. Any of the residue-free cleaners can be used, but remember that the cleaner may also be conductive. Do not apply power to the circuit until the area is completely dry.

Keep cleaners away from variable-capacitor plates, transformers and parts that may be harmed by the chemical. The most common conductive contaminant is solder, either from a printed-circuit board "solder bridge" or a loose piece of solder deciding to surface at the most inconvenient time.

26.3.9 Solder "Bridges"

In a typical PC-board solder bridge, the solder that is used to solder one component has formed a short circuit to another PC-board trace or component. Unfortunately, they are common in both new construction and repair work. Look carefully for them after you have completed any soldering, especially on a PC-board. It is even possible that a solder bridge may exist in equipment you have owned for a long time, unnoticed until it suddenly decided to become a short circuit.

Related items are loose solder blobs, loose hardware or small pieces of component leads that can show up in the most awkward and troublesome places.

26.3.10 Arcing

Arcing is a serious sign of trouble. It may also be a real fire hazard. Arc sites are usually easy to find because an arc that generates visible light or noticeable sound also pits and discolors conductors.

Arcing is caused by component failure, dampness, dirt or lead dress. If the dampness is temporary, dry the area thoroughly and resume operation. Dirt may be cleaned from the chassis with a residue-free cleaner. Arrange leads so high-voltage conductors are isolated. Keep them away from sharp corners and screw points.

Arcing occurs in capacitors when the working voltage is exceeded. Air-dielectric variable capacitors can sustain occasional arcs without damage, but arcing indicates operation beyond circuit limits. Transmatches working beyond their ability may suffer from arcing. A failure or high SWR in an antenna circuit may also cause transmitter arcing.

26.3.11 Replacing Parts

If you have located a defective component within a stage, you need to replace it. When replacing socket mounted components, be sure to align the replacement part correctly. Make sure that the pins of the device are properly inserted into the socket.

Some special tools can make it easier to remove soldered parts. A chisel-shaped soldering tip helps pry leads from printed-circuit boards or terminals. A desoldering iron or bulb forms a suction to remove excess solder, making it easier to remove the component. Spring-loaded desolder-ing pumps are more convenient than bulbs. Desoldering wick draws solder away from a joint when pressed against the joint with a hot soldering iron.

In all cases, remember that soldering tools and melted solder can be hot and dangerous! Wear protective goggles and clothing when soldering. A full course in first aid is beyond the scope of this chapter, but if you burn your fingers, run the burn immediately under cold water and seek first aid or medical attention. Always seek medical attention if you burn your eyes; even a small burn can develop into serious trouble.

26.4 Typical Symptoms and Faults

26.4.1 Power Supplies

Many equipment failures are caused by power-supply trouble. Fortunately, most power-supply problems are easy to find and repair (see **Fig 26.13**). First, use a voltmeter to measure output. Loss of output voltage is usually caused by an open circuit. (A short circuit draws excessive current that opens the fuse, thus becoming an open circuit.)

Most fuse failures are caused by a shorted diode in the power supply or a shorted power device (RF or AF) in the failed equipment. More rarely, one of the filter capacitors can short. If the fuse has opened, turn off the power, replace the fuse and measure the load-circuit dc resistance. The measured resistance should be consistent with the power-supply ratings. A short or open load circuit indicates a problem.

If the measured resistance is too low, check the load circuit with an ohmmeter to locate the trouble. (Nominal circuit resistances are included in some equipment manuals.) If the load circuit resistance is normal, suspect a defective regulator IC or problem in the rest of the unit. Electrolytic capacitors fail with long (two years) disuse; the electrolytic layer may be reformed as explained later in this chapter.

IC regulators can oscillate, sometimes causing failure. The small-value capacitors on the input, output or adjustment pins of the regulator prevent oscillations. Check or replace these capacitors whenever a regulator has failed.

AC ripple (hum) is usually caused by low-value filter capacitors in the power supply. Less likely, hum can also be caused by excessive load, a regulation problem or RF feedback in the power supply. Look for a defective filter capacitor (usually open or low-value), defective regulator or shorted filter choke. In older equipment, the defective filter capacitor will often have visible leaking electrolyte: Look for corrosion residue at the capacitor leads. In new construction projects make sure RF energy is not getting into the power supply.

Here's an easy filter-capacitor test: Temporarily connect a replacement capacitor (about the same value and working voltage) across the suspect capacitor. If the hum goes away, replace the bad component permanently.

Once the faulty component is found, inspect the surrounding circuit and consider what may have caused the problem. Sometimes one bad component can cause another to fail. For example, a shorted filter capacitor increases current flow and burns out a rectifier diode. While the defective diode is easy to find, the capacitor may show no visible damage.

SWITCHING POWER SUPPLIES

Switching power supplies are quite different from conventional supplies. In a "switcher," a switching transistor is used to change dc voltage levels. They usually have AF oscillators and complex feedback paths. Any component failure in the rectifiers, switch, feedback path or load usually results in a completely dead supply. Every part is suspect. While active device failure is still the number one suspect, it pays to carefully test all components if a diagnosis cannot be made with traditional techniques.

Some equipment, notably TVs and monitors, derive some of the power-supply voltages from the proper operation of other parts of the circuit. In the case of a TV or monitor, voltages are often derived by adding secondary low-voltage windings to the flyback transformer and rectifying the resultant ac voltage (usually about 15 kHz). These voltages will be missing if there is any problem with the circuit from which they are derived.

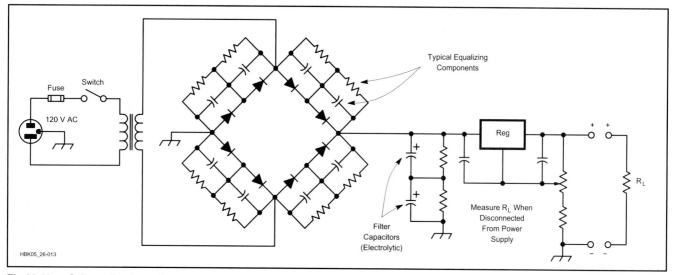

Fig 26.13 — Schematic of a typical power supply showing the components mentioned in the text.

26.4.2 Amplifiers

Amplifiers are the most common circuits in electronics. The output of an ideal amplifier would match the input signal in every respect except magnitude: No distortion or noise would be added. Real amplifiers always add noise and distortion.

GAIN

Gain is the measure of amplification. Gain is usually expressed in decibels (dB) over a specified frequency range, known as the bandwidth or passband of the amplifier. When an amplifier is used to provide a stable load for the preceding stage, or as an impedance transformer, there may be little or no voltage gain.

Amplifier failure usually results in a loss of gain or excessive distortion at the amplifier output. In either case, check external connections first. Is there power to the stage? Has the fuse opened? Check the speaker and leads in audio output stages, the microphone and push-to-talk (PTT) line in transmitter audio sections. Excess voltage, excess current or thermal runaway can cause sudden failure of semiconductors. The failure may appear as either a short, or open, circuit of one or more PN junctions.

Thermal runaway occurs most often in bipolar transistor circuits. If degenerative feedback (the emitter resistor reduces base-emitter voltage as conduction increases) is insufficient, thermal runaway will allow excessive current flow and device failure. Check transistors by substitution, if possible.

Faulty coupling components can reduce amplifier output. Look for component failures that would increase series, or decrease shunt impedance in the coupling network. Coupling faults can be located by signal tracing or parts substitution. Other passive component defects reduce amplifier output by shifting bias or causing active-device failure. These failures are evident when the dc operating voltages are measured.

In a receiver, a fault in the AGC loop may force a transistor into cutoff or saturation. Open the AGC line to the device and substitute a variable voltage for the AGC signal. If amplifier action varies with voltage, suspect the AGC-circuit components; otherwise, suspect the amplifier.

In an operating amplifier, check carefully for oscillations or noise. Oscillations are most likely to start with maximum gain and the amplifier input shorted. Any noise that is induced by 60-Hz sources can be heard, or seen with a 'scope synchronized to the ac line.

Unwanted amplifier RF oscillations should be cured with changes of lead dress or circuit components. Separate input leads from output leads; use coaxial cable to carry RF between stages; neutralize inter-element or junction capacitance. Ferrite beads on the control element of the active device often stop unwanted oscillations.

Low-frequency oscillations ("motorboating") indicate poor stage isolation or inadequate power-supply filtering. Try a better lead-dress arrangement and/or check the capacitance of the decoupling network (see **Fig 26.14**). Use larger capacitors at the power-supply leads; increase the number of capacitors or use separate decoupling capacitors at each stage. Coupling capacitors that are too low in value can also cause poor low-frequency response. Poor response to high frequencies is usually caused by circuit design.

AMPLIFIERS VS SWITCHES

To help you hone your skills, let's analyze a few simple circuits. There is often a big dif-

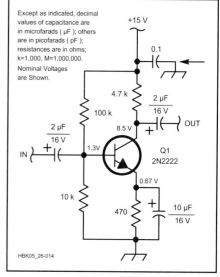

Fig 26.14 — The decoupling capacitor in this circuit is designated with an arrow.

ference in the performance of similar-looking circuits. Consider the differences between a common-emitter amplifier and a common-emitter switch circuit.

COMMON-EMITTER AMPLIFIER

Fig 26.15 is a schematic of a common-emitter transistor amplifier. The emitter, base and collector leads are labeled e, b and c, respectively. Important dc voltages are measured at these points and designated V_e, V_b and V_c. Similarly, the important currents are I_e, I_b and I_c. V+ indicates the supply voltage.

First, analyze the voltages and signal levels in this circuit. The "junction drop," is the

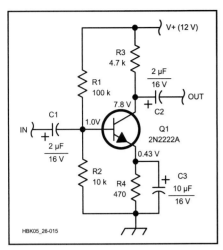

Fig 26.15 — A typical common-emitter audio amplifier.

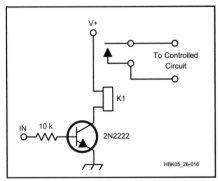

Fig 26.16 — A typical common-emitter switch or driver.

potential measured across a semiconductor junction that is conducting. It is typically 0.6 V for silicon and 0.2 V for germanium transistors.

This is a Class-A linear circuit. In Class-A circuits, the transistor is always conducting some current. R1 and R2 form a voltage divider that supplies dc bias (V_b) for the transistor. Normally, V_e is equal to V_b less the emitter-base junction drop. R4 provides degenerative dc bias, while C3 provides a low-impedance path for the signal. From this information, normal operating voltages can be estimated.

The bias and voltages will be set up so that the transistor collector voltage, V_c, is somewhere between V+ and ground potential. A good rule of thumb is that V_c should be about one-half of V+, although this can vary quite a bit, depending on component tolerances. The emitter voltage is usually a small percentage of V_c, say about 10%.

Any circuit failure that changes I_c (ranging from a shorted transistor or a failure in the bias circuit) changes V_c and V_e as well. An increase of Ic lowers V_c and raises V_e. If the transistor shorts from collector to emitter, V_c drops to about 1.2 V, as determined by the voltage divider formed by R3 and R4.

You would see nearly the same effect if the transistor were biased into saturation by collector-to-base leakage, a reduction in R1's value or an increase in R2's value. All of these circuit failures have the same effect. In some cases, a short in C1 or C2 could cause the same symptoms.

To properly diagnose the specific cause of low V_c, consider and test all of these parts. It is even more complex; an increase in R3's value would also decrease V_c. There would be one valuable clue, however; if R3 increased in value, I_c would not increase; V_e would also be low.

Anything that decreases I_c increases V_c. If the transistor failed "open," R1 increased in value, R2 were shorted to ground or R4 opened, then V_c would be high.

COMMON-EMITTER SWITCH

A common-emitter transistor switching circuit is shown in **Fig 26.16**. This circuit functions differently from the circuit shown in Fig 26.15. A linear amplifier is designed so that the output signal is a faithful reproduction of the input signal. Its input and output may have any value from V+ to ground.

The switching circuit of Fig 26.16, however, is similar to a "digital" circuit. The active device is either on or off, 1 or 0, just like digital logic. Its input signal level should either be 0 V or positive enough to switch the transistor on fully (saturate). Its output state should be either full off (with no current flowing through the relay), or full on (with the relay energized). A voltmeter placed on the collector will show either approximately +12 V or 0 V, depending on the input.

Understanding this difference in operation is crucial to troubleshooting the two circuits. If V_c were +12 V in the circuit in Fig 26.15, it would indicate a circuit failure. A V_c of +12 V in the switching circuit, is normal when V_b is 0 V. (If V_b measured 0.8 V or higher, V_c should be low and the relay energized.)

DC COUPLED AMPLIFIERS

In dc coupled amplifiers, the transistors are directly connected together without coupling capacitors. They comprise a unique troubleshooting case. Most often, when one device fails, it destroys one or more other semiconductors in the circuit. If you don't find all of the bad parts, the remaining defective parts can cause the installed replacements to fail immediately. To reliably troubleshoot a dc coupled circuit, you must test every semiconductor in the circuit and replace them all at once.

26.4.3 Oscillators

In many circuits, a failure of the oscillator will result in complete circuit failure. A transmitter will not transmit, and a superheterodyne receiver will not receive if you have an internal oscillator failure. (These symptoms do not always mean oscillator failure, however.)

Whenever there is weakening or complete loss of signal from a radio, check oscillator operation and frequency. There are several methods:

- Use a receiver with a coaxial probe to listen for the oscillator signal.
- A dip meter can be used to check oscillators. In the absorptive mode, tune the dip meter to within ±15 kHz of the oscillator, couple it to the circuit, and listen for a beat note in the dip-meter headphones.
- Look at the oscillator waveform on a 'scope. The operating frequency can't be determined with great accuracy, but you can see if the oscillator is working at all. Use a low capacitance (10×) probe for oscillator observations.
- Tube oscillators usually have negative grid bias when oscillating. Use a high-impedance voltmeter to measure grid bias. The bias also changes slightly with frequency.
- Emitter current varies slightly with frequency in transistor oscillators. Use a sensitive, high-impedance voltmeter across the emitter resistor to observe the current level. (You can use Ohm's Law to calculate the current value.)

Many modern oscillators are phase-locked loops (PLLs). A PLL is a marriage of an analog oscillator and digital control circuitry. Read the Digital Circuitry section in this chapter and the **Oscillators and Synthesizers** chapter of this book in order to learn PLL repair techniques.

To test for a failed oscillator tuned with inductors and capacitors, use a dip meter in the active mode. Set the dip meter to the oscillator frequency and couple it to the oscillator output circuit. If the oscillator is dead, the dip-meter signal will take its place and temporarily restore some semblance of normal operation. Tune the dip meter very slowly, or you may pass stations so quickly that they sound like "birdies."

STABILITY

We are spoiled; modern amateur equipment is very stable. Drift of several kilohertz per hour was once normal. You may want to modify old equipment for more stability, but drift that is consistent with the equipment design is not a defect. (This applies to new equipment as well as old.) It is normal for some digital displays to flash back and forth between two values for the least-significant digit.

Drift is caused by variations in the oscillator. Poor voltage regulation and heat are the most common culprits. Check regulation with a voltmeter (use one that is not affected by RF). Voltage regulators are usually part of the oscillator circuit. Check them by substitution.

Chirp is a form of rapid drift that is usually caused by excessive oscillator loading or poor power-supply regulation. The most common cause of chirp is poor design. If chirp appears suddenly in a working circuit, look for component or design defects in the oscillator or its buffer amplifiers. (For example, a shorted coupling capacitor increases loading drastically.) Also check lead dress, tubes and switches for new feedback paths (feedback defeats buffer action).

Frequency instability may also result from defects in feedback components. Too much feedback may produce spurious signals, while too little makes oscillator start-up unreliable.

Sudden frequency changes are frequently the result of physical variations. Loose components or connections are probable causes. Check for arcing or dirt on printed-circuit boards, trimmers and variable capacitors, loose switch contacts, bad solder joints or loose connectors.

FREQUENCY ACCURACY

Dial tracking errors may be associated with oscillator operation. Misadjustments in the frequency-determining components make dial accuracy worse at the ends of the dial. Tracking errors that are constant everywhere in the passband can be caused by misalignment or by slippage in the dial drive mechanism or indicator. This is usually cured by calibration of a simple mechanical adjustment.

In LC oscillators, tracking at the high-frequency end of the dial is controlled by trimmer capacitors. A trimmer is a variable capacitor connected in parallel with the main tuning capacitor (see **Fig 26.17**). The trimmer represents a higher percentage of the total capacitance at the high end of the tuning range. It has relatively little effect on tuning characteristics at the low-frequency end of the dial.

Low-end tracking is adjusted by a padder capacitor. A padder is a variable capacitor that is connected in series with the main tuning capacitor. Padder capacitance has a greater effect at the low-frequency end of the dial. The padder capacitor is often eliminated to save money. In that case, the low-frequency tracking is adjusted by the main tuning coil.

26.4.4 Control Circuitry

Semiconductors have made it practical to use diodes for switching, running only a dc lead to the switching point. This eliminates problems caused by long analog leads in the circuit. Semiconductor switching usually reduces the cost and complexity of switching components. Switching speed is increased; contact corrosion and breakage are eliminated. In exchange, troubleshooting is complicated by additional components such as voltage regulators and decoupling capacitors (see **Fig 26.18**). The technician must consider many more components and symptoms when working with diode and transistor switched circuits.

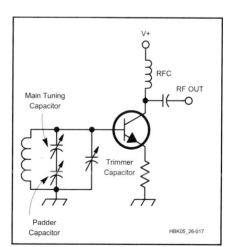

Fig 26.17 — A partial schematic of a simple oscillator showing the locations of the trimmer and padder capacitors.

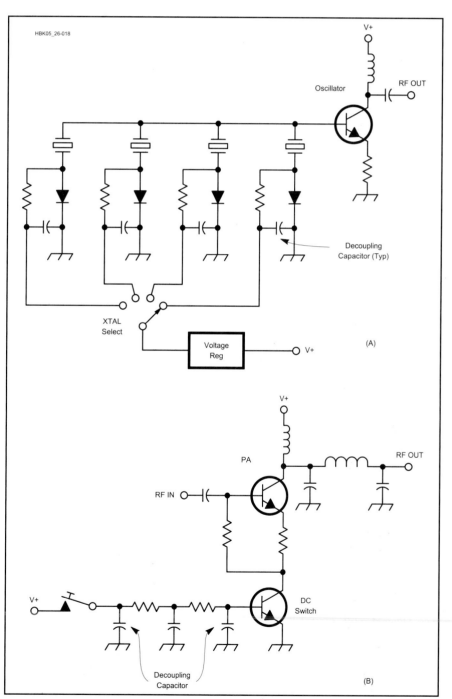

Fig 26.18 — Diode switching selects oscillator crystals at A. A transistor switch is used to key a power amplifier at B.

Mechanical switches are relatively rugged. They can withstand substantial voltage and current surges. The environment does not drastically affect them, and there is usually visible damage when they fail. Semiconductor switching offers inexpensive, high-speed operation. When subjected to excess voltage or current, however, most transistors and diodes silently expire. Occasionally, if the troubleshooter is lucky, one sends up a smoke signal to mark its passing.

Temperature changes semiconductor characteristics. A normally adequate control signal may not be effective when transistor beta is lowered by a cold environment. Heat may cause a control voltage regulator to produce an improper control signal.

A control signal is actually a bias for the semiconductor switch. Forward biased diodes and transistors act as closed switches; reverse biased components simulate open switches. If the control (bias) signal is not strong enough to completely saturate the semiconductor, conduction may not continue through a full ac cycle. Severe distortion can be the result.

When dc control leads provide unwanted feedback paths, switching transistors may become modulators or mixers. Additionally, any reverse biased semiconductor junction is a potential source of white noise.

MICROPROCESSOR CONTROL

Nearly every new transceiver is controlled by a miniature computer. Entire books have been written about microprocessor (µP) control. Many of the techniques are discussed in the Digital Circuitry section. Many microprocessor related problems end up back at the factory for service; however, the surface mounted components are just too difficult for most hams to replace. For successful repair of microprocessor controlled circuits, you should have the knowledge and test equipment necessary for computer repair. Familiarity with digital logic circuits and machine-language programming may also be desirable.

26.4.5 Digital Circuitry

The digital revolution has hit most ham shacks and amateur equipment. Microprocessors have brought automation to everything from desk clocks to ham transceivers and computer controlled EME antenna arrays. Although every aspect of their operation may be resolved to a simple 1 or 0, or tristate (an infinite impedance or open circuit), the symptoms of their failure are far more complicated. As with other equipment:
- Observe the operating characteristics.
- Study the block diagram and the schematic.
- Test.

- Replace defective parts.

Problems in digital circuits have two elementary causes. First, the circuit may give false counts because of electrical noise at the input. Second, the gates may lock in one state.

False counts from noise are especially likely in a ham shack. (A 15- to 20-µs voltage spike can trigger a TTL flip-flop.) Amateur Radio equipment often switches heavy loads; the attendant transients can follow the ac line or radiate directly to nearby digital equipment. Oscillation in the digital circuit can also produce false counts.

How these false counts affect a circuit is dependent on the design. A station clock may run fast, but a microprocessor controlled transceiver may "decide" that it is only a receiver. It might even be difficult to determine that there is a problem without a logic analyzer or a multitrace oscilloscope and a thorough understanding of circuit operation.

Begin by removing the suspect equipment from RF fields. If the symptoms stop when there is no RF energy around, you need to shield the equipment from RF.

In the mid '90s, microprocessors in general use ran clock speeds up to a few hundred megahertz. (They are increasing all the time.) It may be impossible to filter RF signals from the lines when the RF is near the clock frequency. In these cases, the best approach is to shield the digital circuit and all lines running to it.

If digital circuitry interferes with other nearby equipment, it may be radiating spurious signals. These signals can interfere with your Amateur Radio operation or other services. Digital circuitry can also be subject to interference from strong RF fields. Erratic operation or a complete "lock up" is often the result. *The ARRL RFI Book* has a chapter on computer and digital interference. That chapter discusses interference to and from digital devices and circuits.

LOGIC LEVELS

To troubleshoot a digital circuit, check for the correct voltages at the pins of each chip. The correct voltages may not always be known, but you should be able to identify the power pins (V_{cc} and ground).

The voltages on the other pins should be either a logic high, a logic low, or tristate (more on this later). In most working digital circuitry the logic levels are constantly changing, often at RF rates. A dc voltmeter may not give reliable readings. An oscilloscope or logic analyzer is usually needed to troubleshoot digital circuitry.

Most digital circuit failures are caused by a failed logic IC. In clocked circuits, listen for the clock signal with a coax probe and a suitable receiver. If the signal is found at the clock chip, trace it to each of the other ICs to be sure that the clock system is intact.

Some digital circuits use VHF clock speeds; an oscilloscope must have a bandwidth of at least twice the clock speed to be useful. If you have a suitable scope, check the pulse timing and duration against circuit specifications.

As in most circuits, failures are catastrophic. It is unlikely that an AND gate will suddenly start functioning like an OR gate. It is more likely that the gate will have a signal at its input, and no signal at the output. In a failed device, the output pin will have a steady voltage. In some cases, the voltage is steady because one of the input signals is missing. Look carefully at what is going into a digital IC to determine what should be coming out. Keep manufacturers' data books handy. These data books describe the proper functioning of most digital devices.

TRISTATE DEVICES

Many digital devices are designed with a third logic state, commonly called tristate. In this state, the output of the device acts as if it weren't there at all. Many such devices can be connected to a common "bus," with the devices that are active at any given time selected by software or hardware control signals. A computer's data and address busses are good examples of this. If any one device on the bus fails by locking itself on in a 0 or 1 logic state, the entire bus becomes nonfunctional. These tristate devices can be locked "on" by inherent failure or a failure of the signal that controls them.

SIMPLE GATE TESTS

Logic gates, flip-flops and counters can be tested (see **Fig 26.19**) by triggering them manually, with a power supply (4 to 5 V is a safe level). Diodes may be checked with an ohmmeter. Testing of more complicated ICs requires the use of a logic analyzer, multitrace scope or a dedicated IC tester.

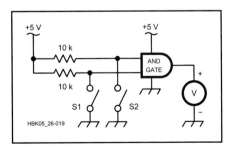

Fig 26.19 — This simple digital circuit can be tested with a few components. In this case, an AND gate is tested. Open and close S1 and S2 while comparing the voltmeter reading with a truth table for the device.

26.5 Troubleshooting Hints

Tables 26.2, 26.3, 26.4 and 26.5 list some common problems and possible cures. These tables are not all-inclusive. They are a collection of hints and shortcuts that may save you some troubleshooting time. If you don't find your problem listed, continue with systematic troubleshooting.

26.5.1 Receivers

A receiver can be diagnosed using any of the methods described earlier, but if there is not even a faint sound from the speaker, signal injection is not a good technique. If you lack troubleshooting experience, avoid following instinctive hunches. That leaves signal tracing as the best method.

The important characteristics of a receiver are selectivity, sensitivity, stability and fidelity. Receiver malfunctions ordinarily affect one or more of these areas.

SELECTIVITY

Tuned transformers or the components used in filter circuits may develop a shorted turn, capacitors can fail and alignment is required occasionally. Such defects are accompanied by a loss of sensitivity. Except in cases of catastrophic failure (where either the filter passes all signals, or none), it is difficult to spot a loss of selectivity. Bandwidth and insertion-loss measurements are necessary to judge filter performance.

SENSITIVITY

A gradual loss of sensitivity results from gradual degradation of an active device or long-term changes in component values. Sudden partial sensitivity changes are usually the result of a component failure, usually in the RF or IF stages. Complete and sudden loss of sensitivity is caused by an open circuit anywhere in the signal path or by a "dead" oscillator.

RECEIVER STABILITY

The stability of a receiver depends on its oscillators. See the Oscillators section elsewhere in this chapter.

DISTORTION

Receiver distortion may be the effect of poor connections or faulty components in the signal path. AGC circuits produce many receiver defects that appear as distortion or insensitivity.

AGC

AGC failure usually causes distortion that affects only strong signals. All stages operate at maximum gain when the AGC influence is removed. An S meter can help diagnose AGC failure because it is operated by the AGC loop.

An open AGC bypass capacitor causes feedback through the loop. This often results in a receiver "squeal" (oscillation). Changes in the loop time constant affect tuning. If stations consistently blast, or are too weak for a brief time when first tuned in, the time constant is too fast. An excessively slow time constant makes tuning difficult, and stations fade after tuning. If the AGC is functioning, but the "timing" seems wrong, check the large-value capacitors found in the AGC circuit — they usually set the AGC time constants. If the AGC is not functioning, check the AGC-detector circuit. There is often an AGC voltage that is used to control several stages. A failure in any one stage could affect the entire loop.

DETECTOR PROBLEMS

Detector trouble usually appears as complete loss or distortion of the received signal.

Table 26.2
Symptoms and Their Causes for All Electronic Equipment

Symptom *Cause*

Power Supplies
No output voltage — Open circuit (usually a fuse or transformer winding)
Hum or ripple — Faulty regulator, capacitor or rectifier, low-frequency oscillation

Amplifiers
Low gain — Transistor, coupling capacitors, emitter-bypass capacitor, AGC component, alignment
Noise — Transistors, coupling capacitors, resistors
Oscillations — Dirt on variable capacitor or chassis, shorted op-amp input
Untuned (oscillations do not change with frequency) — Audio stages
Tuned — RF, IF and mixer stages
Squeal — Open AGC-bypass capacitor
Static-like crashes — Arcing trimmer capacitors, poor connections
Static in FM receiver — Faulty limiter stage, open capacitor in ratio detector, weak RF stage, weak incoming signal
Intermittent noise — All components and connections, band-switch contacts, potentiometers (especially in dc circuits), trimmer capacitors, poor antenna connections
Distortion (constant) — Oscillation, overload, faulty AGC, leaky transistor, open lead in tab-mount transistor, dirty potentiometer, leaky coupling capacitor, open bypass capacitors, imbalance in tuned FM detector, IF oscillations, RF feedback (cables)Distortion (strong signals only)
Open AGC line, open AGC diode
Frequency change — Physical or electrical variations, dirty or faulty variable capacitor, broken switch, loose compartment parts, poor voltage regulation, oscillator tuning (trouble when switching bands)

No Signals
All bands — Dead VFO or heterodyne oscillator, PLL won't lock
One band only — Defective crystal, oscillator out of tune, band switch
No function control — Faulty switch, poor connection, defective switching diode or circuit

Improper Dial Tracking
Constant error across dial — Dial drive
Error grows worse along dial — Circuit adjustment

AM, SSB and CW signals may be weak and unintelligible. FM signals will sound distorted. Look for an open circuit in the detector near the detector diodes. If tests of the detector parts indicate no trouble, look for a poor connection in the power-supply or ground lead. A BFO that is "dead" or off frequency prevents SSB and CW reception. In modern rigs, the BFO frequency is either crystal controlled, or derived from the PLL.

RECEIVER ALIGNMENT

Unfortunately, IF transformers are as enticing to the neophyte technician as a carburetor is to a shade-tree mechanic. In truth, radio alignment (and for that matter, carburetor repair) is seldom required. Circuit alignment may be justified under the following conditions:
- The set is very old and has not been adjusted in many years.
- The circuit has been subject to abusive treatment or environment.
- There is obvious misalignment from a previous repair.
- Tuned-circuit components or crystals have been replaced.
- An inexperienced technician attempted alignment without proper equipment. ("But all the screws in those little metal cans were loose!")
- There is a malfunction, but all other circuit conditions are normal. (Faulty transformers can be located because they will not tune.)

Even if one of the above conditions is met, do not attempt alignment unless you have the proper equipment. Receiver alignment should progress from the detector to the antenna terminals. When working on an FM receiver, align the detector first, then the IF and limiter stages and finally the RF amplifier and local oscillator stages. For an AM receiver, align the IF stages first, then the RF amplifier and oscillator stages.

Both AM and FM receivers can be aligned in much the same manner. Always follow the manufacturer's recommended alignment procedure. If one is not available, follow these guidelines:

1. Set the receiver RF gain to maximum, BFO control to zero or center (if applicable to your receiver) and tune to the high end of the receiver passband.
2. Disable the AGC.
3. Set the signal source to the center of the IF passband, with no modulation and minimum signal level.
4. Connect the signal source to the input of the IF section.
5. Connect a voltmeter to the IF output as shown in **Fig 26.20**.
6. Adjust the signal-source level for a slight indication on the voltmeter.
7. Peak each IF transformer in order, from the meter to the signal source. The adjustments interact; repeat steps 6 and 7 until adjustment brings no noticeable improvement.
8. Remove the signal source from the IF-section input, reduce the level to minimum, set the frequency to that shown on the receiver dial and connect the source to the antenna terminals. If necessary, tune around for the signal — if the local oscillator is not tracking, it may be off.
9. Adjust the signal level to give a slight reading on the voltmeter.
10. Adjust the trimmer capacitor of the RF amplifier for a peak reading of the test signal. (Verify that you are reading the correct signal by switching the source on and off.)
11. Reset the signal source and the receiver tuning for the low end of the passband.
12. Adjust the local-oscillator padder for peak reading.
13. Steps 8 through 11 interact, so repeat them until the results are as good as you can get them.

26.5.2 Transmitters

Many potential transmitter faults are discussed in several different places in this chapter. There are, however, a few techniques used to ensure stable operation of RF amplifiers in transmitters that are not covered elsewhere.

High-power RF amplifiers often use parasitic chokes to prevent instability. Older parasitic chokes usually consist of a 51- to 100-Ω noninductive resistor with a coil wound around the body and connected to the leads. It is used to prevent VHF and UHF oscillations in a vacuum-tube amplifier. The suppressor is placed in the plate lead, close to the plate connection.

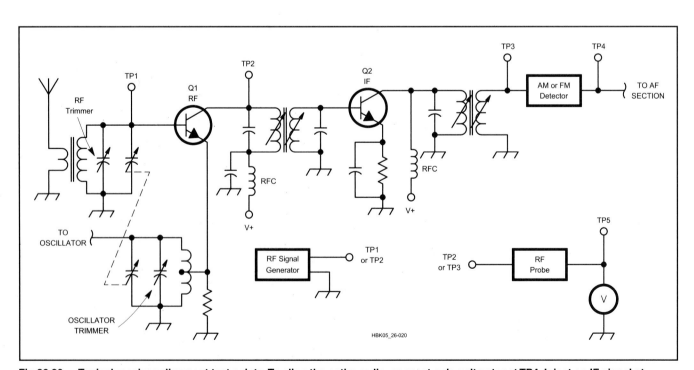

Fig 26.20 — Typical receiver alignment test points. To align the entire radio, connect a dc voltmeter at TP4. Inject an IF signal at TP2 and adjust the IF transformers. Move the signal generator to TP1 and inject an RF signal for alignment of the RF amplifier and oscillator stages. To align a single stage, place the generator at the input and an RF voltmeter (or demodulator probe and dc voltmeter) at the output: TP1/TP2 for RF, TP2/TP3 for IF.

Table 26.3
Receiver Problems

Symptom	Cause
Low sensitivity	Semiconductor degradation, circuit contamination, weak tube, alignment
Signals and calibrator heard weakly	
(low S-meter readings)	RF chain
(strong S-meter readings)	AF chain, detector
No signals or calibrator heard, only hissing	RF oscillators
Distortion	
On strong signals only	AGC fault
AGC fault	Active device cut off or saturated
Difficult tuning	AGC fault
Inability to receive	Detector fault
AM weak and distorted	Poor detector, power or ground connection
CW/SSB unintelligible	BFO off frequency or dead
FM distorted	Open detector diode

In recent years, problems with this style of suppressor have been discovered. Look at the **RF Power Amplifiers** chapter for information about suppressing parasitics.

Parasitic chokes often fail from excessive current flow. In these cases, the resistor is charred. Occasionally, physical shock or corrosion produces an open circuit in the coil. Test for continuity with an ohmmeter.

Transistor amplifiers are protected against parasitic oscillations by low-value resistors or ferrite beads in the base or collector leads. Resistors are used only at low power levels (about 0.5 W), and both methods work best when applied to the base lead. Negative feedback is used to prevent oscillations at lower frequencies. An open component in the feedback loop may cause low-frequency oscillation, especially in broadband amplifiers.

KEYING

The simplest form of modulation is on/off keying. Although it may seem that there cannot be much trouble with such an elementary form of modulation, two very important transmitter faults are the result of keying problems.

Key clicks are produced by fast rise and times of the keying waveform. Most transmitters include components in the keying circuitry to prevent clicks. When clicks are experienced, check the keying filter components first, then the succeeding stages. An improperly biased power amplifier, or a Class C amplifier that is not keyed, may produce key clicks even though the keying waveform earlier in the circuit is correct. Clicks caused by a linear amplifier may be a sign of low-frequency parasitic oscillations. If they occur in an amplifier, suspect insufficient power-supply decoupling. Check the power-supply filter capacitors and all bypass capacitors.

The other modulation problem associated with on/off keying is called back wave. Backwave is a condition in which the signal is heard, at a reduced level, even when the key is up. This occurs when the oscillator signal feeds through a keyed amplifier. This usually indicates a design flaw, although in some cases a component failure or improper keyed-stage neutralization may be to blame.

LOW OUTPUT POWER

Some transmitters automatically reduce power in the TUNE mode. Check the owner's manual to see if the condition is normal. Check the control settings. Transmitters that use broadband amplifiers require so little effort from the operator that control settings are seldom noticed. The CARRIER (or DRIVE) control may have been bumped. Remember to adjust tuned amplifiers after a significant change in operating frequency (usually 50 to 100 kHz). Most modern transmitters are designed to reduce power if there is high (say 2:1) SWR. Check these obvious external problems before you tear apart your rig.

Power transistors may fail if the SWR protection circuit malfunctions. Such failures occur at the "weak link" in the amplifier chain: It is possible for the drivers to fail without damaging the finals. An open circuit in the "reflected" side of the sensing circuit leaves the transistors unprotected, a short "shuts them down."

Low power output in a transmitter may also spring from a misadjusted carrier oscillator or a defective SWR protection circuit. If the carrier oscillator is set to a frequency well outside the transmitter passband, there may be no measurable output. Output power will increase steadily as the frequency is moved into the passband.

26.5.3 Transceivers

SWITCHING

Elaborate switching schemes are used in transceivers for signal control. Many transceiver malfunctions can be attributed to relay

Table 26.4
Transmitter Problems

Symptom	Cause
Key clicks	Keying filter, distortion in stages after keying
Modulation Problems	
Loss of modulation	Broken cable (microphone, PTT, power), open circuit in audio chain, defective modulator
Distortion on transmit	Defective microphone, RF feedback from lead dress, modulator imbalance, bypass capacitor, improper bias, excessive drive
Arcing	Dampness, dirt, improper lead dress
Low output	Incorrect control settings, improper carrier shift (CW signal outside of passband) audio oscillator failure, transistor or tube failure, SWR protection circuit
Antenna Problems	
Poor SWR	Damaged antenna element, matching network, feed line, balun failure (see below), resonant conductor near antenna, poor connection at antenna
Balun failure	Excessive SWR, weather or cold-flow damage in coil choke, broken wire
RFI	Arcing or poor connections anywhere in antenna system or nearby conductors

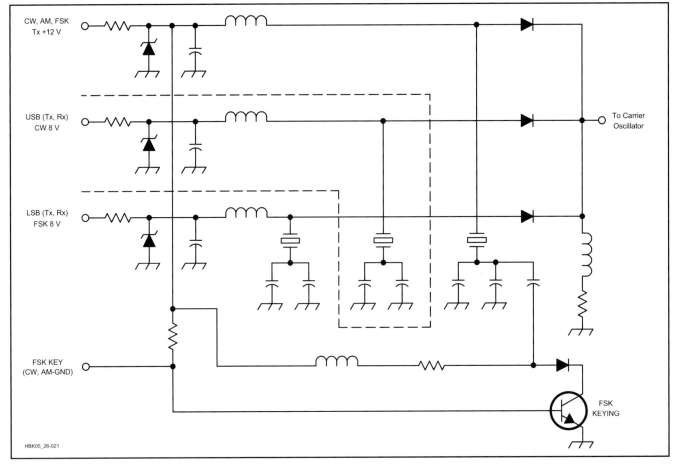

Fig 26.21 — Partial schematic of a transceiver oscillator. The symptoms described in the text are caused by one or more components inside the dashed lines or a faulty USB/CW control signal.

or switching problems. Suspect the switching controls when:
- The S meter is inoperative, but the unit otherwise functions. (This could also be a bad S meter.)
- There is arcing in the tank circuit. (This could also be caused by a bad antenna system.)
- Plate current is high during reception.
- There is excessive broadband PA noise in the receiver.

Since transceiver circuits are shared, stage defects frequently affect both the transmit and receive modes, although the symptoms may change with mode. Oscillator problems usually affect both transmit and receive modes, but different oscillators, or frequencies, may be used for different emissions. Check the block diagram.

For example, one particular transceiver uses a single carrier oscillator with three different crystals (see **Fig 26.21**). One crystal sets the carrier frequency for CW, AM and FSK transmit. Another sets USB transmit and USB/CW receive, and a third sets LSB transmit and LSB/FSK receive. This radio showed a strange symptom. After several hours of CW operation, the receiver produced only a light hiss on USB and CW. Reception was good in other modes, and the power meter showed full output during CW transmission. An examination of the block diagram and schematic showed that only one of the crystals (and seven support components) was capable of causing the problem.

VOX

Voice operated transmit (VOX) controls are another potential trouble area. If there is difficulty in switching to transmit in the VOX mode, check the VOX-SENSITIVITY and ANTI-VOX control settings. Next, see if the PTT and manual (MOX) transmitter controls work. If the PTT and MOX controls function, examine the VOX control diodes and amplifiers. Test the switches, control lines and control voltage if the transmitter does not respond to other TR controls.

VOX SENSITIVITY and ANTI-VOX settings should also be checked if the transmitter switches on in response to received audio.

Table 26.5
Transceiver Problems

Symptom	Cause
Inoperative S meter	Faulty relay
PA noise in receiver	
Excessive current on receive	
Arcing in PA	
Reduced signal strength on transmit and receive	IF failure
Poor VOX operation	VOX amplifiers and diodes
Poor VOX timing	Adjustment, component failure in VOX timing circuits or amplifiers
VOX consistently tripped by receiver audio	AntiVOX circuits or adjustment

Suspect the ANTI-VOX circuitry next. Unacceptable VOX timing results from a poor VOX-delay adjustment, or a bad resistor or capacitor in the timing circuit or VOX amplifiers.

ALIGNMENT

The mixing scheme of the modern SSB transceiver is complicated. The signal passes through many mixers, oscillators and filters. Satisfactory SSB communication requires accurate adjustment of each stage. Do not attempt any alignment without a copy of the manufacturer's instructions and the necessary test equipment.

26.5.4 Amplifiers

While this section focuses on vacuum-tube amplifiers using high-voltage (HV) supplies, it also applies to solid-state amplifiers that operate at lower voltages and generally have fewer points of failure. Amplifiers are simple, reliable pieces of equipment that respond well to basic care, regular maintenance and common sense. A well-maintained amplifier will provide reliable service and maximum tube lifetime.

SAFETY FIRST

It is important to review good safety practices. (See the **Safety** and **Power Supplies** chapters for additional safety information.) Tube amplifiers use power supply voltages well in excess of 1 kV and the RF output can be hundreds of volts, as well. Almost every voltage in an amplifier can be lethal! Take care of yourself and use caution!

Power Control — Know and control the state of both ac line voltage and dc power supplies. Physically disconnect line cords and other power cables when you are not working on live equipment. Use a lockout on circuit breakers. Double-check visually and with a meter to be absolutely sure power has been removed.

Interlocks — Unless specifically instructed by the manufacturer's procedures to do so, never bypass an interlock. This is rarely required except in troubleshooting and should only be done when absolutely necessary. Interlocks are there to protect you.

The One-Hand Rule — Keep one hand in your pocket while making any measurements on live equipment. The hand in your pocket removes a path for current to flow through you. It's also a good idea to wear shoes with insulating soles and work on dry surfaces. Current can be lethal even at levels of a few mA — don't tempt the laws of physics.

Patience — Repairing an amplifier isn't a race. Take your time. Don't work on equipment when you're tired or frustrated. Wait several minutes after turning the amplifier off to open the cabinet — capacitors can take several minutes to discharge through their bleeder resistors.

A Grounding Stick — Make the simple safety accessory shown in Fig 7.47 of the **Power Supplies** chapter and use it whenever you work on equipment in which hazardous voltages have been present. The ground wire should be heavy duty (#12 AWG or larger) due to the high peak currents (hundreds of amperes) present when discharging a capacitor or tripping a circuit breaker. When equipment is opened, touch the tip of the stick to every exposed component and connection that you might come in contact with. Assume nothing — accidental shorts and component failures can put voltage in places it shouldn't be.

The Buddy System and CPR — Use the buddy system when working around any equipment that has the potential for causing serious injury. The buddy needn't be a ham, just anyone who will be nearby in case of trouble. Your buddy should know how to remove power and administer basic first aid or CPR.

CLEANLINESS

The first rule of taking good care of an amplifier is cleanliness. Amplifiers need not be kept sparkling new, but their worst enemy is heat. Excess heat accelerates component aging and increases stresses during operation.

Outside the amplifier, prevent dust and obstructions from blocking the paths by which heat is removed. This means keeping all ventilation holes free of dust, pet hair and insects. Fan intakes are particularly susceptible to inhaling all sorts of debris. Use a vacuum cleaner to clean the amplifier and surrounding areas. Keep liquids well away from the amplifier.

Keep papers or magazines off the amplifier — even if the cover is solid metal. Paper acts as an insulator and keeps heat from being radiated through the cover. Amplifier heat sinks must have free air circulation to be effective. There should be at least a couple of inches of free space surrounding an amplifier on its sides and top. If the manufacturer recommends a certain clearance, mounting orientation or air flow, follow those recommendations.

Inside the amplifier, HV circuits attract dust that slows heat dissipation and will eventually build up to where it arcs or carbonizes. Use the vacuum cleaner to remove any dust or dirt. If you find insects (or worse) inside the amp, try to determine how they got in and plug that hole. Window screening works fine to allow airflow while keeping out insects. While you're cleaning the inside, perform a visual inspection as described in the next section.

Vacuuming works best with an attachment commonly known as a "crevice cleaner." **Fig 26.22** shows a crevice cleaning attachment being used with a small paintbrush to dislodge and remove dust. The brush will root dust out of tight places and off components without damaging them or pulling on connecting wires. Don't use the vacuum

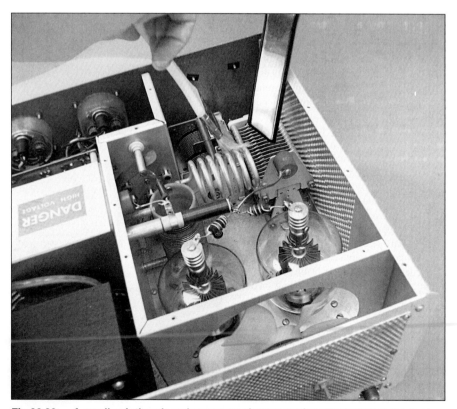

Fig 26.22 — A small paintbrush and a vacuum cleaner crevice attachment make dust removal easy.

cleaner's brush attachment; they're designed for floors, not electronics. Some vacuums also have a blower mechanism, but these rarely have enough punch to clean as thoroughly as a brush. Blowing dust just pushes the dust around and into other equipment.

If you can't get a brush or attachment close enough, a spray can of compressed air will usually dislodge dust and dirt so you can vacuum it up. If you use a rag or towel to wipe down panels or large components, be sure not to leave threads or lint behind. Never use a solvent or spray cleaner to wash down components or flush out crevices unless the manufacturer advises doing so — it might leave behind a residue or damage the component.

VISUAL INSPECTION

Remove any internal covers or access panels and...stop! Get out the chicken stick, clip its ground lead securely to the chassis and touch every exposed connection. Now, using a strong light and possibly a magnifier, look over the components and connections.

Amplifiers have far fewer components than transceivers, so look at every component and insulator. Look for cracks, signs of arcing, carbon traces (thin black lines), discoloration, loose connections, melting of plastic, and anything else that doesn't "look right." This is a great time to be sure that mounting and grounding screws are tight. Does anything smell burnt? Learn the smells of overheated components. Make a note of what you find, repair or replace — even if no action is required.

ELECTRICAL COMPONENTS

Let's start with the power supply. There are three basic parts to amplifier power supplies — the ac transformer and line devices, the rectifier/filter, and the metering/regulation circuitry. (See the **Power Supplies** chapter for more information.) Transformers need little maintenance except to be kept cool and be mounted securely. Line components such as switches, circuit breakers and fuses, if mechanically sound and adequately rated, are usually electrically okay, as well.

Rectifiers and HV filter capacitors require occasional cleaning. Look for discoloration around components mounted on a printed circuit board (PCB) and make sure that all wire connections are secure. HV capacitors are generally electrolytic or oil and should show no signs of leakage, swelling or outgassing around terminals.

Components that perform metering and regulation of voltage and current can be affected by heat or heavy dust. If there has been a failure of some other component in the amplifier — such as a tube — these circuits can be stressed severely. Resistors may survive substantial temporary overloads, but may show signs of overload such as discoloration or swelling.

Amplifiers contain two types of relays — control and RF. Control relays switch ac and dc voltages and do not handle input or output RF energy. The usual problem encountered with control relays is oxidation or pitting of their contacts. A burnishing tool can be used to clean relay contacts. In a pinch a strip of ordinary paper can be pulled between contacts gently held closed. Avoid the temptation to over-clean silver-plated relay and switch contacts. It is easy to remove contact plating with excessive polishing and while silver-plated relay and switch contacts may appear to be dark in color, oxidized silver (black) is still a good conductor. Once the silver's gone, it's gone; contact erosion will then be pervasive. If visual inspection shows heavy pitting or discoloration or resistance measurements show the relay to have intermittent contact quality, it should be replaced.

RF relays are used to perform transmit-receive (TR) switching and routing of RF signals through or around the amplifier circuitry. Amplifiers designed for full break-in operation will usually use a high-speed vacuum TR relay. Vacuum relays are sealed and cannot be cleaned or maintained. When you replace RF relays, use a direct replacement part or one rated for RF service with the same characteristics as the original.

Cables and connectors are subjected to heavy heat and electrical loads in amplifiers. Plastics may become brittle and connections may oxidize. Cables should remain flexible and not be crimped or pinched if clamped or tied down. Gently wiggle cables while watching the connections at each end for looseness or bending. Connectors can be unplugged and reseated once or twice to clear oxide on contact surfaces.

Carefully inspect any connector that seems loose. Be especially careful with connectors and cables in amplifiers with power supplies in separate enclosures from the RF deck. Those interconnects are susceptible to both mechanical and electrical stress and you don't want an energized HV cable loose on the operating desk. Check the electrical integrity of those cables and make sure they are tightly fastened.

As with relays, switches found in amplifiers either perform control functions or route RF signals. Adequately rated control switches, if mechanically sound, are usually okay. Band switches are the most common RF switch — usually a rotary phenolic or ceramic type. A close visual inspection should show no pitting or oxidation on the wiper (the part of the switch that rotates between contacts) or the individual contacts. Arcing or overheating will quickly destroy rotary switches. **Fig 26.23** is a photo of a heavy-duty band switch that has suffered severe damage from arcing. Slight oxidation is acceptable on silver-plated switches. Phosphor-bronze contacts can sometimes be cleaned with a light scrub from a pencil eraser, but plating can be easily removed, so use caution with this method and be sure to remove any eraser crumbs. Rotary switch contacts cannot be replaced easily although individual wafer sections may be replaced if an exact matching part can be obtained.

When replacing capacitors and resistors, be

Fig 26.23 — The band switch section on the left clearly shows the signs of destructive arcing. (*NØAX photo*)

sure to use an adequately rated part. Voltage and power-handling ratings are particularly important, especially for components handling high RF currents. An RF tank capacitor replacement should be checked carefully for adequate RF voltage and current ratings, not just dc. HV resistors are generally long and thin to prevent arcing across their surfaces. Even if a smaller (and cheaper) resistor has an equivalent power rating, resist the temptation to substitute it. In a pinch, a series string of resistors of the appropriate combined value can be used to replace one HV unit. Don't use carbon resistors for metering circuits, use metal or carbon film types. The carbon composition types are too unstable.

If you are repairing or maintaining an old amplifier and manufacturer-specific parts are no longer available, the ham community has many sources for RF and HV components. Fair Radio Sales (**www.fairradio.com**) and Surplus Sales of Nebraska (**www.surplus-sales.com**) are familiar names. Hamfests and Web sites often have amplifier components for sale. (See the **RF Power Amplifier** chapter's sidebar on using surplus or used parts for amplifiers.) You might consider buying a non-working amplifier of the same model for parts.

TUBES

Good maintenance of tubes starts with proper operation of the amplifier. Follow the manufacturer's instructions for input drive levels, duty cycles, tuning and output power level. Frequently check all metered voltages and current to be sure that the tubes are being operated properly and giving you maximum lifetime. Penta Labs' "Tube Maintenance & Education" (**www.pentalaboratories.com/maintenance.asp**) is an excellent Web page on maintaining power tubes.

The internal mechanical structures of tubes generally do not deal well with mechanical shock and vibration, so treat them gently. The manufacturer may also specify how the amplifier is to be mounted, so read the operating manual. Tubes generate a lot of heat, so it's important that whatever cooling mechanism employed is kept at peak efficiency. Airways should be clean, including between the fins on metal tubes. All seals and chimneys should fit securely and be kept clean. Wipe the envelope of glass tubes clean after handling them — fingerprints should be removed to prevent baking them into the surface. On metal tubes that use finger-stock contacts, be sure the contacts are clean and make good contact all the way around the tube. Partial contact or dirty finger stock can cause asymmetric current and heating inside the tube, resulting in warping of internal grids and possibly cause harmonics or parasitics.

Plate cap connections and VHF parasitic suppressors should be secure and show no signs of heating. Overheated parasitic suppressors may indicate that the neutralization circuit is not adjusted properly. Inspect socket contacts and the tube pins to be sure all connections are secure, particularly high-current filament connections. Removing and inserting the tubes once or twice will clean the socket contacts.

Adjustments to the neutralizing network, which suppresses VHF oscillations by negative feedback from the plate to grid circuit, are rarely required except when you are replacing a tube or after you do major rewiring or repair of the RF components. The manufacturer will provide instructions on making these adjustments. If symptoms of VHF oscillations occur without changing a tube, then perhaps the tube characteristics or associated components have changed. Parasitic oscillations in high-power amplifiers can be strong enough to cause arcing damage. Perform a visual inspection prior to readjusting the neutralizing circuit.

Metering circuits rarely fail, but they play a key part in maintenance. By keeping a record of "normal" voltages and currents, you will have a valuable set of clues when things go wrong. Record tuning settings, drive levels, and tube voltages and currents on each band and with every antenna. When settings change, you can refer back to the notebook instead of relying on memory.

MECHANICAL

Thermal cycling and heat-related stresses can result in mechanical connections loosening over time or material failures. Switch shafts, shaft couplings and panel bearings all need to be checked for tightness and proper alignment. All mounting hardware needs to be tight, particularly if it supplies a grounding path. Examine all panel-mounted components, particularly RF connectors, and be sure they're attached securely. BNC and UHF connectors mounted with a single nut in a round panel hole are notorious for loosening with repeated connect/disconnect cycles.

Rubber and plastic parts are particularly stressed by heat. If there are any belts, gears or pulleys, make sure they're clean and that dust and lint are kept out of their lubricant. Loose or slipping belts should be replaced. Check O-rings, grommets and sleeves to be sure they are not brittle or cracked. If insulation sleeves or sheets are used, check to be sure they are covering what they're supposed to. Never discard them or replace them with improperly sized or rated materials.

Enclosures and internal shields should all be fastened securely with every required screw in place. Watch out for loosely overlapping metal covers. If a sheet metal screw has stripped out, either drill a new hole or replace the screw with a larger size, taking care to maintain adequate clearance around and behind the new screw. Tip the amplifier from side to side while listening for loose hardware or metal fragments, all of which should be removed.

Clean the front and back panels to protect the finish. If the amplifier cabinet is missing a foot or an internal shock mount, replace it. A clean unit with a complete cabinet will have a significantly higher resale value so it's in your interest to keep the equipment looking good.

SHIPPING

When you are traveling with an amplifier or shipping it, some care in packing will prevent damage. Improper packing can also result in difficulty in collecting on an insurance claim, should damage occur. The original shipping cartons are a good method of protecting the amplifier for storage and sale, but they were not made to hold up to frequent shipping. If you travel frequently, it is best to get a sturdy shipping case made for electronic equipment. Pelican (**www.pelican-shipping-cases.com**) and Anvil (**www.anvilsite.com**) make excellent shipping cases suitable for carrying amplifiers and radio equipment.

Some amplifiers require the power transformer to be removed before shipping. Check your owner's manual or contact the manufacturer to find out. Failure to remove it before shipping can cause major structural damage to the amplifier's chassis and case.

Tubes should also be removed from their sockets for shipment. It may not be necessary to ship them separately if they can be packed in the amplifier's enclosure with adequate plastic foam packing material. If the manufacturer of the tube or amplifier recommends separate shipment, however, do it!

CLEANING AND MAINTENANCE PLAN

For amateur use, there is little need for maintenance more frequently than once per year. Consider the maintenance requirements of the amplifier and what its manufacturer recommends. Review the amplifier's manuals and make up a checklist of what major steps and tools are required.

TROUBLESHOOTING

A benefit of regular maintenance will be familiarity with your amplifier should you ever need to repair it. Knowing what it looks (and smells) like inside will give you a head start on effecting a quick repair.

The following discussion is intended to illustrate the general flow of a troubleshooting effort, not be a step-by-step guide. Before starting on your own amplifier, review the amplifier manual's "Theory of Operation" section and familiarize yourself with the schematic. If there is a troubleshooting procedure in the manual, follow it. **Fig 26.24** shows a general-purpose troubleshooting flow chart. Do not swap in a known-good tube or tubes until you are sure that a tube is actually defective. Installing a good tube in an amplifier with circuit problems can damage a good tube.

Many "amplifier is dead" problems turn out to be simply a lack of ac power. Before even opening the cabinet of an unresponsive amplifier, be sure that ac is really present at the wall socket and that the fuse or circuit breaker is really closed. If ac power is present at the wall socket, trace through any internal fuses, interlocks and relays all the way through to the transformer's primary terminals.

Hard failures in an HV power supply are rarely subtle, so it's usually clear if there is a problem. When you repair a power supply, take the opportunity to check all related components. If all defective components are not replaced, the failures may be repeated when the circuit is re-energized.

Rectifiers may fail open or shorted — test them using a DVM diode checker. An open rectifier will result in a drop in the HV output of 50 percent or more but will probably not overheat or destroy itself. A shorted rectifier failure is usually more dramatic and may cause additional rectifiers or filter capacitors to fail. If one rectifier in a string has failed, it may be a good idea to replace the entire string as the remaining rectifiers have been subjected to a higher-than-normal voltage.

HV filter capacitors usually fail shorted, although they will occasionally lose capacitance and show a rise in ESR (equivalent series resistance). Check the rectifiers and any metering components — they may have been damaged by the current surge caused from a short circuit.

Power transformer failures are usually due to arcing in the windings, insulation failures, or overheating. HV transformers can be disassembled and rewound by a custom transformer manufacturer.

Along with the HV plate supply, tetrode screen supplies occasionally fail, too. The usual cause is the regulation circuit that drops the voltage from the plate level. Operating without a screen supply can be damaging to a tube, so be sure to check the tube carefully after repairs.

If the power supply checks out okay and the tube's filaments are lit, check the resting or bias current. If it is excessive or very low, check all bias voltages and dc current paths to the tube, such as the plate choke, screen supply (for tetrodes) and grid or cathode circuits.

If you do not find power supply and dc problems, check the RF components or "RF deck." Check the input SWR to the amplifier. If it has changed then you likely have a problem in the input circuitry or one or more

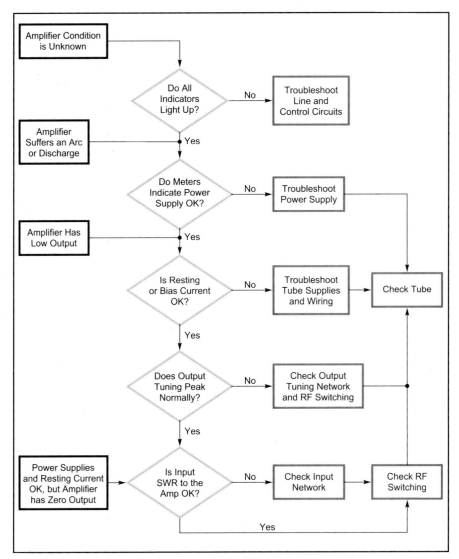

Fig 26.24 — This general-purpose flow chart will help identify amplifier problems. For solid-state units, substitute "Check Output Transistors" for "Check Tube."

tubes have failed. Perform a visual check of the input circuitry and the band switch, followed by an ohmmeter check of all input components.

If input SWR is normal and applying drive does not result in any change in plate current, you may have a defective tube, tube socket or connection between the input circuits and the tube. Check the TR control circuits and relay. If plate current changes, but not as much as normal, try adjusting the output tuning circuitry. If this has little or no effect, the tube may be defective or a connection between the tube and output circuitry may have opened. If retuning has an effect, but at different settings than usual, the tube may be defective or there may be a problem in the tuning circuitry. A visual inspection and an ohmmeter check are in order.

The key to finding the trouble with your amplifier is to be careful and methodical, and to avoid jumping to false conclusions or making random tests. The manufacturer's customer service department will likely be helpful if you are considerate and have taken careful notes detailing the trouble symptoms and any differences from normal operation. There may be helpful guidelines on the manufacturer's Web pages or from other Internet resources. Sometimes there is more than one problem — they work together to act like one very strange puzzle. Just remember that most problems can be isolated by careful, step-by-step tests.

26.6 Components

Once you locate a defective part, it is time to select a replacement. This is not always an easy task. Each electronic component has a function. This section acquaints you with the functions, failure modes and test procedures of resistors, capacitors, inductors and other components. Test the components implicated by symptoms and stage-level testing. In most cases, a particular faulty component will be located by these tests. If a faulty component is not indicated, check the circuit adjustments. As a last resort, use a shotgun approach — replace all parts in the problem area with components that are known to be good.

26.6.1 Check the Circuit

Before you install a replacement component of any type, you should be sure that another circuit defect didn't cause the failure. Check the circuit voltages carefully before installing any new component. Check the potential on each trace to the bad component. The old part may have "died" as a result of a lethal voltage. Measure twice — repair once! (With apologies to the old carpenter.) Of course, circuit performance is the final test of any substitution.

26.6.2 Fuses

Most of the time, when a fuse fails, it is for a reason — usually a short circuit in the load. A fuse that has failed because of a short circuit usually shows the evidence of high current: a blackened interior with little blobs of fuse element everywhere. Fuses can also fail by fracturing the element at either end. This kind of failure is not visible by looking at the fuse. Check even "good" fuses with an ohmmeter. You may save hours of troubleshooting.

For safety reasons, always use *exact* replacement fuses. Check the current and voltage ratings. The fuse timing (fast, normal or slow blow) must be the same as the original. Never attempt to force a fuse that is not the right size into a fuse holder. The substitution of a bar, wire or penny for a fuse invites a "smoke party."

26.6.3 Wires

Wires seldom fail unless abused. Short circuits can be caused by physical damage to insulation or by conductive contamination. Damaged insulation is usually apparent during a close visual inspection of the conductor or connector. Look carefully where conductors come close to corners or sharp objects. Repair worn insulation by replacing the wire or securing an insulating sleeve (spaghetti) or heat-shrink tubing over the worn area.

When wires fail, the failure is usually caused by stress and flexing. Nearly everyone has broken a wire by bending it back and forth, and broken wires are usually easy to detect. Look for sharp bends or bulges in the insulation.

When replacing conductors, use the same material and size, if possible. Substitute only wire of greater cross-sectional area (smaller gauge number) or material of greater conductivity. Insulated wire should be rated at the same, or higher, temperature and voltage as the wire it replaces.

26.6.4 Connectors

Connection faults are one of the most common failures in electronic equipment. This can range from something as simple as the ac-line cord coming out of the wall, to a connector having been put on the wrong socket, to a defective IC socket. Connectors that are plugged and unplugged frequently can wear out, becoming inter-mittent or noisy. Check connectors carefully when troubleshooting.

Connector failure can be hard to detect. Most connectors maintain contact as a result of spring tension that forces two conductors together. As the parts age, they become brittle and lose tension. Any connection may deteriorate because of nonconductive corrosion at the contacts. Solder helps prevent this problem but even soldered joints suffer from corrosion when exposed to weather.

The dissipated power in a defective connector usually increases. Signs of excess heat are sometimes seen near poor connections in circuits that carry moderate current. Check for short and open circuits with an ohmmeter or continuity tester. Clean those connections that fail as a result of contamination.

Occasionally, corroded connectors may be repaired by cleaning, but replacement of the conductor/connector is usually required. Solder all connections that may be subject to harsh environments and protect them with acrylic enamel, RTV compound or a similar coating.

Choose replacement connectors with consideration of voltage and current ratings. Use connectors with symmetrical pin arrangements only where correct insertion will not result in a safety hazard or circuit damage.

26.6.5 Resistors

Resistors usually fail by becoming an open circuit. More rarely they change value. This is usually caused by excess heat. Such heat may come from external sources or from power dissipated within the resistor. Sufficient heat burns the resistor until it becomes an open circuit.

Resistors can also fracture and become an open circuit as a result of physical shock. Contamination of a high-value resistor (100 kΩ or more) can cause a change in value through leakage. This contamination can occur on the resistor body, mounts or printed-circuit board. Resistors that have changed value should be replaced. Leakage is cured by cleaning the resistor body and surrounding area.

In addition to the problems of fixed-value resistors, potentiometers and rheostats can develop noise problems, especially in dc circuits. Dirt often causes intermittent contact between the wiper and resistive element. To cure the problem, spray electronic contact cleaner into the control, through holes in the case, and rotate the shaft a few times.

The resistive element in wire-wound potentiometers eventually wears and breaks from the sliding action of the wiper. In this case, the control needs to be replaced.

Replacement resistors should be of the same value, tolerance, type and power rating as the original. The value should stay within tolerance. Replacement resistors may be of a different type than the original, if the characteristics of the replacement are consistent with circuit requirements.

Substitute resistors can usually have a greater power rating than the original, except in high-power emitter circuits where the resistor also acts as a fuse or in cases where the larger size presents a problem.

Variable resistors should be replaced with the same kind (carbon or wire wound) and taper (linear, log, reverse log and so on) as the original. Keep the same, or better tolerance and pay attention to the power rating.

In all cases, mount high-temperature resistors away from heat-sensitive components. Keep carbon resistors away from heat sources. This will extend their life and ensure minimum resistance variations.

26.6.6 Capacitors

Capacitors usually fail by shorting, opening or becoming electrically (or physically) leaky. They rarely change value. Capacitor failure is usually caused by excess current, voltage, temperature or age. Leakage can be external to the capacitor (contamination on the capacitor body or circuit) or internal to the capacitor.

TESTS

The easiest way to test capacitors is out of circuit with an ohmmeter. In this test, the resistance of the meter forms a timing circuit

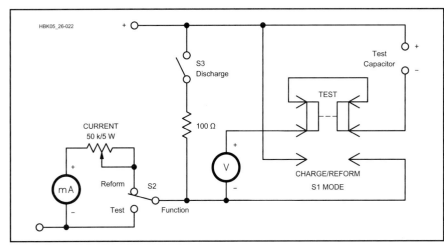

Fig 26.25 — A fixture for testing capacitors and reforming the dielectric of electrolytic capacitors. Use 12 V for testing the capacitor. Use the capacitor working voltage for dielectric reformation.

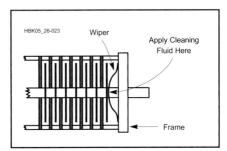

Fig 26.26 — Partial view of an air-dielectric variable capacitor. If the capacitor is noisy or erratic in operation, apply electronic cleaning fluid where the wiper contacts the rotor plates.

with the capacitor to be checked. Capacitors from 0.01 µF to a few hundred µF can be tested with common ohmmeters. Set the meter to its highest range and connect the test leads across the discharged capacitor. When the leads are connected, current begins to flow. The capacitor passes current easily when discharged, but less easily as the charge builds. This shows on the meter as a low resistance that builds, over time, toward infinity.

The speed of the resistance build-up corresponds to capacitance. Small capacitance values approach infinite resistance almost instantly. A 0.01-µF capacitor checked with an 11-MΩ FETVOM would increase from zero to a two-thirds scale reading in 0.11 s, while a 1-µF unit would require 11 s to reach the same reading. If the tested capacitor does not reach infinity within five times the period taken to reach the two-thirds point, it has excess leakage. If the meter reads infinite resistance immediately, the capacitor is open. (Aluminum electrolytics normally exhibit high-leakage readings.)

Fig 26.25 shows a circuit that may be used to test capacitors. To use this circuit, make sure that the power supply is off, set S1 to CHARGE and S2 to TEST, then connect the capacitor to the circuit. Switch on the power supply and allow the capacitor to charge until the voltmeter reading stabilizes. Next, switch S1 to TEST and watch the meter for a few seconds. If the capacitor is good, the meter will show no potential. Any appreciable voltage indicates excess leakage. After testing, set S1 to CHARGE, switch off the power supply, and press the DISCHARGE button until the meter shows 0 V, then remove the capacitor from the test circuit.

Capacitance can also be measured with a capacitance meter, an RX bridge or a dip meter. Some DMMs (digital multimeters) measure capacitance. Capacitance measurements made with DMMs and dedicated capacitance meters are much more accurate than those made with RX bridges or dip meters. To determine capacitance with a dip meter, a parallel-resonant circuit should be constructed using the capacitor of unknown value and an inductor of known value. The formula for resonance is discussed in the **Electrical Fundamentals** chapter of this book.

It is best to keep a collection of known components that have been measured on accurate L or C meters. Alternatively, a "standard" value can be obtained by ordering 1 or 2% components from an electronics supplier. A 10%-tolerance component can be used as a standard; however, the results will only be known to within 10%. The accuracy of tests made with any of these alternatives depends on the accuracy of the "standard" value component. Further information on this technique appears in Bartlett's article, "Calculating Component Values," in Nov 1978 *QST*.

CLEANING

The only variety of common capacitor that can be repaired is the air-dielectric variable capacitor. Electrical connection to the moving plates is made through a spring-wiper arrangement (see **Fig 26.26**). Dirt normally builds on the contact area, and they need occasional cleaning. Before cleaning the wiper/contact, use gentle air pressure and a soft brush to remove all dust and dirt from the capacitor plates. Apply some electronic contact cleaning fluid. Rotate the shaft quickly several times to work in the fluid and establish contact. Use the cleaning fluid sparingly, and keep it off the plates except at the contact point.

REPLACEMENTS

Replacement capacitors should match the original in value, tolerance, dielectric, working voltage and temperature coefficient. Use only ac-rated capacitors for line service. If exact replacements are not available, substitutes may vary from the original part in the following respects: Bypass capacitors may vary from one to three times the capacitance of the original. Coupling capacitors may vary from one half to twice the value of the original. Capacitance values in tuned circuits (especially filters) must be exact. (Even then, any replacement will probably require circuit realignment.)

If the same kind of capacitor is not available, use one with better dielectric characteristics. Do not substitute polarized capacitors for nonpolarized parts. Capacitors with a higher working voltage may be used, although the capacitance of an electrolytic capacitor used significantly below its working voltage will usually increase with time.

The characteristics of each type of capacitor are discussed in the **Analog Basics** and **RF Techniques** chapters. Consider these characteristics if you're not using an exact replacement capacitor.

26.6.7 Inductors and Transformers

The most common inductor or transformer failure is a broken conductor. More rarely, a short circuit can occur across one or more turns of a coil. In an inductor, this changes the value. In a transformer, the turns ratio and resultant output voltage changes. In high-power circuits, excessive inductor current can generate enough heat to melt plastics used as coil forms.

Inductors may be checked for open circuit failure with an ohmmeter. In a good inductor, dc resistance rarely exceeds a few ohms. Shorted turn and other changes in inductance show only during alignment or inductance measurement.

The procedure for measurement of inductance with a dip meter is the same as that given for capacitance measurement, except that a capacitor of known value is used in the resonant circuit.

Replacement inductors must have the same inductance as the original, but that is only the first requirement. They must also carry the same current, withstand the same voltage and present nearly the same Q as the original part. Given the original as a pattern, the amateur can duplicate these qualities for many inductors. Note that inductors with ferrite or iron-powder cores are frequency sensitive, so the replacement must have the same core material.

If the coil is of simple construction, with the form and core undamaged, carefully count and write down the number of turns and their placement on the form. Also note how the coil leads are arranged and connected to the circuit. Then determine the wire size and insulation used. Wire diameter, insulation and turn spacing are critical to the current and voltage ratings of an inductor. (There is little hope of matching coil characteristics unless the wire is duplicated exactly in the new part.) Next, remove the old winding — be careful not to damage the form — and apply a new winding in its place. Be sure to dress all coil leads and connections in exactly the same manner as the original. Apply Q dope to hold the finished winding in place.

Follow the same procedure in cases where the form or core is damaged, except that a suitable replacement form or core (same dimensions and permeability) must be found.

Ready-made inductors may be used as replacements if the characteristics of the original and the replacement are known and compatible. Unfortunately, many inductors are poorly marked. If so, some comparisons, measurements and circuit analysis are usually necessary.

When selecting a replacement inductor, you can usually eliminate parts that bear no physical resemblance to the original part. This may seem odd, but the Q of an inductor depends on its physical dimensions and the permeability of the core material. Inductors of the same value, but of vastly different size or shape, will likely have a great difference in Q. The Q of the new inductor can be checked by installing it in the circuit, aligning the stage and performing the manufacturer's passband tests. Although this practice is all right in a pinch, it does not yield an accurate Q measurement. Methods to measure Q appear in the **Test Equipment and Measurements** chapter.

Once the replacement inductor is found, install it in the circuit. Duplicate the placement, orientation and wiring of the original. Ground-lead length and arrangement should not be changed. Isolation and magnetic shielding can be improved by replacing solenoid inductors with toroids. If you do, however, it is likely that many circuit adjustments will be needed to compensate for reduced coupling and mutual inductance. Alignment is usually required whenever a tuned-circuit component is replaced.

A transformer consists of two inductors that are magnetically coupled. Transformers are used to change voltage and current levels (this changes impedance also). Failure usually occurs as an open circuit or short circuit of one or more windings.

Amateur testing of power transformers is limited to ohmmeter tests for open circuits and voltmeter checks of secondary voltage. Make sure that the power-line voltage is correct, then check the secondary voltage against that specified. There should be less than 10% difference between open-circuit and full-load secondary voltage.

Replacement transformers must match the original in voltage, volt-ampere (VA), duty cycle and operating-frequency ratings. They must also be compatible in size. (All transformer windings should be insulated for the full power-supply voltage.)

26.6.8 Relays

Although relays have been replaced by semiconductor switching in low-power circuits, they are still used extensively in high-power Amateur Radio equipment. Relay action may become sluggish. AC relays can buzz (with adjustment becoming impossible). A binding armature or weak springs can cause intermittent switching. Excessive use or hot switching ruins contacts and shortens relay life.

You can test relays with a voltmeter by jumpering across contacts with a test lead (power on, in circuit) or with an ohmmeter (out of circuit). Look for erratic readings across the contacts, open or short circuits at contacts or an open circuit at the coil.

Most failures of simple relays can be repaired by a thorough cleaning. Clean the contacts and mechanical parts with a residue-free cleaner. Keep it away from the coil and plastic parts that may be damaged. Dry the contacts with lint-free paper, such as a business card; then burnish them with a smooth steel blade. Do not use a file to clean contacts.

Replacement relays should match or exceed the original specifications for voltage, current, switching time and stray impedance (impedance is significant in RF circuits only). Many relays used in transceivers are specially made for the manufacturer. Substitutes may not be available from any other source.

Before replacing a multicontact relay, make a drawing of the relay, its position, the leads and their routings through the surrounding parts. This drawing allows you to complete the installation properly, even if you are distracted in the middle of the operation.

26.6.9 Semiconductors

DIODES

The primary function of a diode is to pass current in one direction only. They can be easily tested with an ohmmeter.

Signal or switching diodes — The most common diode in electronics equipment, they are used to convert ac to dc, to detect RF signals or to take the place of relays to switch ac or dc signals within a circuit. Signal diodes usually fail open, although shorted diodes are not rare. They can easily be tested with an ohmmeter.

Power-rectifier diodes — Most equipment contains a power supply, so power-rectifier diodes are the second-most common diodes in electronic circuitry. They usually fail shorted, blowing the power-supply fuse.

Other diodes — Zener diodes are made with a predictable reverse-breakdown voltage and are used as voltage regulators. Varactor diodes are specially made for use as voltage controlled variable capacitors. (Any semiconductor diode may be used as a voltage-variable capacitance, but the value will not be as predictable as that of a varactor.) A Diac is a special-purpose diode that passes only pulses of current in each direction.

Diode tests — There are several basic tests for most diodes. First, is it a diode? Does it conduct in one direction and block current flow in the other? An ohmmeter is suitable for this test in most cases. An ohmmeter will read high resistance in one direction, low resistance in the other. Make sure the meter uses a voltage of more than 0.7 V and less than 1.5 V to measure resistance. Use a good diode to determine the meter polarity.

Diodes should be tested out of circuit. Disconnect one lead of the diode from the circuit, then measure the forward and reverse resistance. Diode quality is shown by the ratio of reverse to forward resistance. A ratio of 100:1 or greater is common for signal diodes. The ratio may go as low as 10:1 for old power diodes.

The first test is a forward-resistance test. Set the meter to read ×100 and connect the test probes across the diode. When the negative terminal of the ohmmeter battery is connected to the cathode, the meter will typically show about 200 to 300 Ω (forward resistance) for a good silicon diode, 200 to 400 Ω for a good germanium diode. The exact value varies quite a bit from one meter to the next.

Next, test the reverse resistance. Reverse the lead polarity and set the meter to ×1M (times one million, or the highest scale available on the meter) to measure diode reverse resistance. Good diodes should show 100 to 1000 MΩ for silicon and 100 kΩ to 1 MΩ for germanium. When you are done, mark the meter lead polarity for future reference.

This procedure measures the junction resistances at low voltage. It is not useful to test Zener diodes. A good Zener diode will not conduct in the reverse direction at voltages below its rating.

We can also test diodes by measuring the voltage drop across the diode junction while the diode is conducting. (A test circuit is shown in **Fig 26.27**.) To test, connect the diode, adjust the supply voltage until the current through the diode matches the manufacturer's specification and compare the junction drop to that specified. Silicon junctions usually show about 0.6 V, while germanium is typically 0.2 V. Junction voltage-drop increases with current flow. This test can be used to match diodes with respect to forward resistance at a given current level.

A final simple diode test measures leakage current. Place the diode in the circuit described above, but with reverse polarity. Set the specified reverse voltage and read the leakage current on a milliammeter. (The currents and voltages measured in the junction voltage-drop and leakage tests vary by several orders of magnitude.)

The most important specification of a Zener diode is the Zener (or avalanche) voltage. The Zener-voltage test also uses the circuit of Fig 26.27. Connect the diode in reverse. Set the voltage to minimum, then gradually increase it. You should read low current in the reverse mode, until the Zener point is reached. Once the device begins to conduct in the reverse direction, the current should increase dramatically. The voltage shown on the voltmeter is the Zener point of the diode. If a Zener diode has become leaky, it might show in the leakage-current measurement, but substitution is the only dependable test.

Replacement diodes — When a diode fails, check associated components as well. Replacement rectifier diodes should have the same current and peak inverse voltage (PIV) as the original. Series diode combinations are often used in high-voltage rectifiers, with resistor and capacitor networks to distribute the voltage equally among the diodes.

Switching diodes may be replaced with diodes that have equal or greater current ratings and a PIV greater than twice the peak-to-peak voltage encountered in the circuit. Switching time requirements are not critical except in RF, logic and some keying circuits. Logic circuits may require exact replacements to assure compatible switching speeds and load characteristics. RF switching diodes used near resonant circuits must have exact replacements as the diode resistance and capacitance will affect the tuned circuit.

Voltage, current and capacitance characteristics must be considered when replacing varactor diodes. Once again, exact replacements are best. Zener diodes should be replaced with parts having the same Zener voltage and equal or better current, power, impedance and tolerance specifications. Check the associated current-limiting resistor when replacing a Zener diode.

BIPOLAR TRANSISTORS

Transistors are primarily used to switch or amplify signals. Transistor failures occur as an open junction, a shorted junction, excess leakage or a change in amplification performance.

Most transistor failure is catastrophic. A transistor that has no leakage and amplifies at dc or audio frequencies will usually perform well over its design range. For this reason, transistor tests need not be performed at the planned operating frequency. Tests are made at dc or a low frequency (usually 1000 Hz). The circuit under repair is the best test of a potential replacement part. Swapping in a replacement transistor in a failed circuit will often result in a cure.

A simple and reliable bipolar-transistor test can be performed with the transistor in a circuit and the power on. It requires a test lead, a 10-kΩ resistor and a voltmeter. Connect the voltmeter across the emitter/collector leads and read the voltage. Then use the test lead to connect the base and emitter (**Fig 26.28A**). Under these conditions, conduction of a good transistor will be cut off and the meter should show nearly the entire supply voltage across the emitter/collector leads. Next, remove the clip lead and connect the 10-kΩ resistor from the base to the collector. This should bias the transistor into conduction and the emitter/collector voltage should drop (Fig 26.28B). (This test indicates transistor response to changes in bias voltage.)

Transistors can be tested (out of circuit) with an ohmmeter in the same manner as diodes. Look up the device characteristics before testing and consider the consequences of the ohmmeter-transistor circuit. Limit junction current to 1 to 5 mA for small-signal transistors. Transistor destruction or inaccurate measurements may result from careless testing.

Use the ×100 Ω and ×1000-Ω ranges for small-signal transistors. For high-power transistors use the ×1 Ω and ×10-Ω ranges. The reverse-to-forward resistance ratio for good transistors may vary from 30:1 to better than 1000:1.

Germanium transistors sometimes show high leakage when tested with an ohmmeter. Bipolar transistor leakage may be specified from the collector to the base, emitter to base or emitter to collector (with the junction reverse biased in all cases). The specification may be identified as I_{cbo}, I_{bo}, collector cutoff current or collector leakage for the base-collector junction, I_{ebo}, and so on for other junctions.[3] Leakage current increases with junction temperature.

A suitable test fixture for base-collector leakage measurements is shown in **Fig 26.29**. Make the required connections and set the voltage as stated in the transistor specifications and compare the measured leakage current with that specified. Small-signal germanium transistors exhibit I_{cbo} and I_{ebo} leakage currents of about 15 μA. Leakage increases to 90 μA or more in high-power components. Leakage currents for silicon transistors are seldom more than 1 μA. Leakage current tends to double for every 10°C increase above 25°C.

Breakdown-voltage tests actually measure leakage at a specified voltage, rather than true breakdown voltage. Breakdown voltage is known as BV_{cbo}, BV_{ces} or BV_{ceo}. Use the same test fixture shown for leakage tests, adjust the power supply until the specified

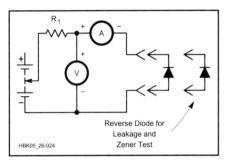

Fig 26.27 — A diode conduction, leakage and Zener-point test fixture. The ammeter should read mA for conduction and Zener point, μA for leakage tests.

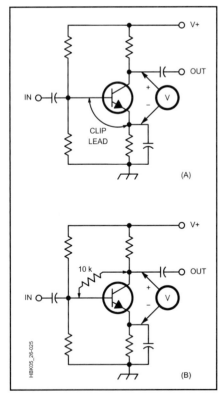

Fig 26.28 — An in-circuit semiconductor test with a clip lead, resistor and voltmeter. The meter should read V+ at (A). During test (B) the meter should show a decrease in voltage, ranging from a slight variation down to a few millivolts. It will typically cut the voltage to about half of its initial value.

leakage current flows, and compare the junction voltage against that specified.

A circuit to measure dc current gain is shown in **Fig 26.30**. Transistor gain can range from 10 to over 1000 because it is not usually well controlled during manufacture. Gain of the active device is not critical in a well-designed transistor circuit.

The test conditions for transistor testing are specified by the manufacturer. When testing, do not exceed the voltage, current (especially in the base circuit) or dissipated-power rating of the transistor. Make sure that the load resistor is capable of dissipating the power generated in the test.

While these simple test circuits will identify most transistor problems, RF devices should be tested at RF. Most component manufacturers include a test-circuit schematic on the data sheet. The test circuit is usually an RF amplifier that operates near the high end of the device frequency range.

Semiconductor failure is sometimes the result of environmental conditions. Open junctions, excess leakage (except with germanium transistors) and changes in amplification performance result from overload or excessive current. Electrostatic discharge can destroy a semiconductor in microseconds. Shorted junctions are caused by voltage spikes. Check surrounding parts for the cause of the transistor's demise, and correct the problem before installing a replacement.

JFETS

Junction FETs can be tested with an ohmmeter in much the same way as bipolar transistors (see text and **Fig 26.31**). Reverse leakage should be several megohms or more. Forward resistance should be 500 to 1000 Ω.

MOSFETS

MOS (metal-oxide semiconductor) layers are extremely fragile. Normal body static is enough to damage them. Even "gate protected" (a diode is placed across the MOS layer to clamp voltage) MOSFETs may be destroyed by a few volts of static electricity.

Make sure the power is off, capacitors discharged and the leads of a MOSFET are shorted together before installing or removing it from a circuit. Use a voltmeter to be sure the chassis is near ground potential, then touch the chassis before and during MOSFET installation and removal. This assures that there is no difference of potential between your body, the chassis and the MOSFET leads. Ground the soldering-iron tip with a clip lead when soldering MOS devices. The FET source should be the first lead connected to and the last disconnected from a circuit. The insulating layers in MOSFETs prevent testing with an ohmmeter. Substitution is the only practical means for amateur testing of MOSFETs.

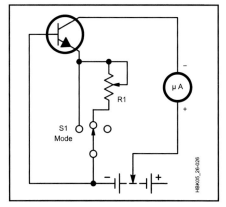

Fig 26.29 — A test circuit for measuring collector-base leakage with the emitter shorted to ground, open or connected to ground through a variable resistance, depending on the setting of S1. See the transistor manufacturer's instructions for test conditions and the setting of R1 (if used). Reverse battery polarity for PNP transistors.

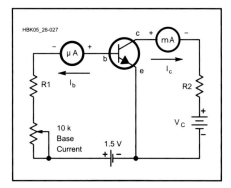

Fig 26.30 — A test circuit for measuring transistor beta. Values for R1 and R2 are dependent on the current range of the transistor tested. Reverse the battery polarity for PNP transistors.

FET CONSIDERATIONS

Replacement FETs should be of the same kind as the original part: JFET or MOSFET, P-channel or N-channel, enhancement or depletion. Consider the breakdown voltage required by the circuit. The breakdown voltage should be at least two to four times the power-supply and signal voltages in amplifiers. Allow for transients of ten times the line voltage in power supplies. Breakdown voltages are usually specified as $V_{(BR)GSS}$ or $V_{(BR)GDO}$.

The gate-voltage specification gives the gate voltage required to cut off or initiate channel current (depending on the mode of operation). Gate voltages are usually listed as $V_{GS(OFF)}$, V_p(pinch off), V_{TH} (threshold) or $I_{D(ON)}$ or I_{TH}.

Dual-gate MOSFET characteristics are more complicated because of the interaction of the two gates. Cutoff voltage, breakdown

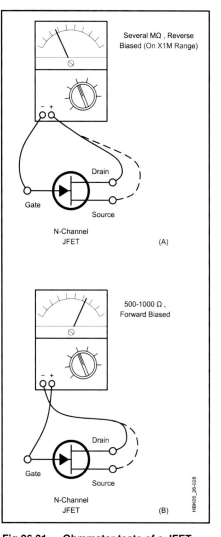

Fig 26.31 — Ohmmeter tests of a JFET. The junction is reverse biased at A and forward biased at B.

voltage and gate leakage are the important traits of each gate.

INTEGRATED CIRCUITS

The basics of integrated circuits are covered in earlier chapters of this book. Amateurs seldom have the sophisticated equipment required to test ICs. Even a multitrace 'scope can view only their simplest functions. We must be content to check every other possible cause, and only then assume that the problem lies with an IC. Experienced troubleshooters will tell you that — most of the time anyway — if a defective circuit uses an IC, it is the IC that is bad.

Linear ICs — There are two major classes of ICs: linear and digital. Linear ICs are best replaced with identical units. Original equipment manufacturers are the best source of a replacement; they are the only source with a reason to stockpile obsolete or custom-made items. If substitution of an IC is unavoidable, first try the cross-reference guides published by several distributors. You can also look in manufacturers' databooks and compare pinouts and other specifications.

Digital ICs — It is usually not a good idea to substitute digital devices. While it may be okay to substitute an AB74LS00YZ from manufacturer "A" with a CD74LS00WX from a different manufacturer, you will usually not be able to replace an LS (low-power Schottky) device with an S (Schottky), C (CMOS) or any of a number of other families. The different families all have different speed, current-consumption, input and output characteristics. You would have to analyze the circuit to determine if you could substitute one type for another.

SEMICONDUCTOR SUBSTITUTION

In all cases try to obtain exact replacement semiconductors. Specifications vary slightly from one manufacturer to the next. Cross-reference equivalents are useful, but not infallible. Before using an equivalent, check the specifications against those for the original part. When choosing a replacement, consider:
- Is it silicon or germanium?
- Is it a PNP or an NPN?
- What are the operating frequency and input/output capacitance?
- How much power does it dissipate (often less than $V_{max} \times I_{max}$)?
- Will it fit the original mount?
- Are there unusual circuit demands (low noise and so on)?
- What is the frequency of operation?

Remember that cross-reference equivalents are not guaranteed to work in every application. There may be cases where two dissimilar devices have the same part number, so it pays to compare the listed replacement specifications with the intended use. If "the book" says to use a diode in place of an RF transistor, it isn't going to work! Derate power specifications, as recommended by the manufacturer, for high-temperature operation.

26.6.10 Tubes

The most common tube failures in amateur service are caused by cathode depletion and gas contamination. Whenever a tube is operated, the coating on the cathode loses some of its ability to produce electrons. It is time to replace the tube when electron production (cathode current, I_c) falls to 50 - 60% of that exhibited by a new tube.

Gas contamination in a tube can often be identified easily because there may be a greenish or whitish-purple glow between the elements during operation. (A faint deep-purple glow is normal in most tubes.) The gas reduces tube resistance and leads to runaway plate current evidenced by a red glow from the anode, interelectrode arcing or a blown power-supply fuse. Less common tube failures include an open filament, broken envelope and interelectrode shorts.

The best test of a tube is to substitute a new one. Another alternative is a tube tester; these are now rare. You can also do some limited tests with an ohmmeter. Tube tests should be made out of circuit so circuit resistance does not confuse the results.

Use an ohmmeter to check for an open filament (remove the tube from the circuit first). A broken envelope is visually obvious, although a cracked envelope may appear as a gassy tube. Interelectrode shorts are evident during voltage checks on the operating stage. Any two elements that show the same voltage are probably shorted. (Remember that some inter-electrode shorts, such as the cathode-suppressor grid, are normal.)

Generally, a tube may be replaced with another that has the same type number. Compare the data sheets of similar tubes to assess their compatibility. Consider the base configuration and pinout, inter-electrode capacitances (a small variation is okay except for tubes in oscillator service), dissipated power ratings of the plate and screen grid and current limitations (both peak and average). For example, the 6146A may be replaced with a 6146B (heavy duty), but not vice versa.

In some cases, minor type-number differences signify differences in filament voltages, or even base styles, so check all specifications before making a replacement. (Even tubes of the same model number, prefix and suffix vary slightly, in some respects, from one supplier to the next.)

26.7 After the Repairs

Once you have completed your troubleshooting and repairs, it is time to put the equipment back together. Take a little extra time to make sure you have done everything correctly.

26.7.1 All Units

Give the entire unit a complete visual inspection. Look for any loose ends left over from your troubleshooting procedures — you may have left a few components temporarily soldered in place or overlooked some other repair error. Look for cold solder joints and signs of damage incurred during the repair. Double check the position, leads and polarity of components that were removed or replaced.

Make sure that all ICs are properly oriented in their sockets and all of the pins are properly inserted in the IC socket or printed-circuit board holes. Test fuse continuity with an ohmmeter and verify that the current rating matches the circuit specification.

Look at the position of all of the wires and components. Make sure that wires and cables will be clear of hot components, screw points and other sharp edges. Make certain that the wires and components will not be in the way when covers are installed and the unit is put back together.

Separate the leads that carry dc, RF, input and output as much as possible. Plug-in circuit boards should be firmly seated with screws tightened and lock washers installed if so specified. Shields and ground straps should be installed just as they were on the original.

26.7.2 For Transmitters Only

Since the signal produced by an HF transmitter can be heard the world over, a thorough check is necessary after any service has been performed. Do not exceed the transmitter duty cycle while testing. Limit transmissions to 10 to 20 seconds unless otherwise specified by the owner's manual.

1. Set all controls as specified in the operation manual, or at midscale.
2. Connect a dummy load and a power meter to the transmitter output.
3. Set the drive or carrier control for low output.
4. Switch the power on.
5. Transmit and quickly set the final-amplifier bias to specifications.
6. In narrowband equipment, slowly tune the output network through resonance. The current dip should be smooth and repeatable. It should occur simultaneously with the maximum power output. Any sudden jumps or wiggles of the current meter indicate that the amplifier is unstable. Adjust the neutralization circuit (according to the manufacturer's instructions) if one is present or check for oscillation. An amplifier usually requires neutralization whenever active devices, components or lead dress (that affect the output/input capacitance) are changed.
7. Check to see that the output power is consistent with the amplifier class used in the PA (efficiency should be about 25% for

Class A, 50 to 60% for Class AB or B, and 70 to 75% for Class C).

8. Repeat steps 4 through 6 for each band of operation from lowest to highest frequency.

9. Check the carrier balance (in SSB transmitters only) and adjust for minimum power output with maximum RF drive and no microphone gain.

10. Adjust the VOX controls.

11. Measure the passband and distortion levels if equipment (wideband 'scope or spectrum analyzer) is available.

26.7.3 Other Repaired Circuits

After the preliminary checks, set the circuit controls per the manufacturer's specifications (or to midrange if specifications are not available) and switch the power on. Watch and smell for smoke, and listen for odd sounds such as arcing or hum. Operate the circuit for a few minutes, consistent with allowable duty cycle. Verify that all operating controls function properly.

Check for intermittent connections by subjecting the circuit to heat, cold and slight flexure. Also, tap or jiggle the chassis lightly with an alignment tool or other insulator.

If the equipment is meant for mobile or portable service, operate it through an appropriate temperature range. Many mobile radios do not work on cold mornings, or on hot afternoons, because a temperature-dependent intermittent was not found during repairs.

26.7.4 Button It Up

After you are convinced that you have repaired the circuit properly, put it all back together. If you followed the advice in this book, you have all the screws and assorted doodads in a secure container. Look at the notes you took while taking it apart; put it back together in the reverse order. Don't forget to reconnect all internal connections, such as ac-power, speaker or antenna leads.

Once the case is closed, and all appears well, don't neglect the final, important step — make sure it still works. Many an experienced technician has forgotten this important step, only to discover that some minor error, such as a forgotten antenna connector, has left the equipment nonfunctional.

26.8 Professional Repairs

This chapter does not tell you how to perform all repairs. Repairs that deal with very complex and temperamental circuits, or that require sophisticated test equipment, should be passed on to a professional.

The factory authorized service personnel have a lot of experience. What seems like a servicing nightmare to you is old hat to them. There is no one better qualified to service your equipment than the factory.

If the manufacturer is no longer in business, check with your local dealer or look in the classified ads in electronics and Amateur Radio magazines. You can usually find one or more companies that service "all makes and models." Your local TV shop might be willing to tackle a repair, especially if you have located a schematic.

If you are going to ship your equipment somewhere for repair, notify the repair center first. Get authorization for shipping and an identification name or number for the package.

26.8.1 Packing It Up

You can always blame shipping damage on the shipper, but it is a lot easier for all concerned if you package your equipment properly for shipping in the first place. Firmly secure all heavy components, either by tying them down or blocking them off with shipping foam. Large vacuum tubes should be wrapped in packing material or shipped separately. Make sure that all circuit boards and parts are firmly attached.

Use a box within a box for shipping. (See **Fig 26.32**.) Place the equipment and some packing material inside a box and seal it with tape. Place that box inside another that is at least six inches larger in each dimension. Fill the gap with packing material, seal, address and mark the outer box. Choose a good freight carrier and insure the package.

Don't forget to enclose a statement of the trouble, a short history of operation and any test results that may help the service technician. Include a good description of the things you have tried. Be honest! At current repair rates you want to tell the technician everything to help ensure an efficient repair.

Even if you ended up sending it back to the factory, you can feel good about your experience. You learned a lot by trying, and

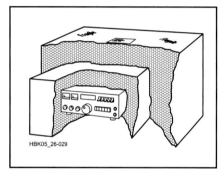

Fig 26.32 — Ship equipment packed securely in a box within a box.

you have sent it back knowing that it really did require the services of a "pro." Each time you troubleshoot and repair a piece of electronic circuitry, you learn something new. The down side is that you may develop a reputation as a real electronics whiz. You may find yourself spending a lot of time at club meetings offering advice, or getting invited over to a lot of shacks for a late-evening pizza snack. There are worse fates.

26.9 Repair and Restoration of Vintage Equipment

When purchasing a classic receiver or transmitter, unless you absolutely know otherwise, assume the radio will need work. Often you can get a top-of-the-line radio needing a bit of repair or clean-up inexpensively. Don't worry — these radios were designed to be repaired by their owners — and curiously, except for cosmetic parts such as cabinets and knobs, parts are much easier to find for 60-year old radios than a 20-year old imported transceiver!

Chances are the radio has gone for years without use. Even if it has been recently used, don't completely trust components that might be 60 or more years old. Don't start by plugging in your new acquisition! To do so might damage a hard-to-replace power transformer, or cause a fire.

Instead, if the radio didn't come with its owner's manual, get one. K4XL's Boat Anchor Manual Archive (**www2.faculty.sbc.edu/kgrimm/boatanchor**) is probably *the* best free resource for these manuals. Many other vendors offer high-quality reprints. Armed with the manual, remove the radio from its cabinet. You very likely will find evidence of unsightly repairs, modifications, or even dangling wires. While modifications aren't necessarily bad, they can certainly add some drama to any necessary subsequent troubleshooting. It's up to you to reverse or remove them.

26.9.1 Component Replacement

Correct any obvious problems such as dangling components. Replace the line cord with a 3-wire, grounded plug for safety. If the radio is one with a "live" chassis, you should operate it from an isolation transformer for safety. It's also a good idea to add a fuse, if the radio doesn't originally have one. Are we ready to give it the "smoke test"? Not so fast!

"RECAPPING"

Obviously, aged components deteriorate, and capacitors are particularly prone to developing leakage or short-circuits with age. There are as many opinions on capacitor replacement as there are radio collectors, but *at the very least* you should replace the electrolytic filter capacitors. Here's why: they *will* short circuit sometime, and when they do, they'll probably take the rectifier tube and the power transformer with them. Modern high voltage electrolytic capacitors are reliable and much smaller than their classic counterparts. You can mount the new capacitors under the chassis by mounting a new terminal strip (do *not* just wire them to the old capacitor terminals), you can re-stuff the old capacitor's can with new capacitors, or you can buy a new can from places such as **www.hayseedhamfest.com** or Antique Electronics Supply (**www.tubesandmore.com**). In any event, follow the manufacturer's schematic — don't assume that the "−" (minus) end of the capacitor goes to ground, as in some radios the ground path is through a resistor so as to develop bias for the audio output stage or RF gain circuit. Observe the polarity or you'll soon be cleaning up a stinky mess!

Old paper-wax and black plastic tubular capacitors should also be replaced. Again, a short circuit in one of them could take out other components, too. Modern film capacitors of the appropriate voltage are great replacements. Opinions vary as to whether all should be replaced, but replacements are cheap and you have the radio apart now, so why not?

REMOVING AND REPLACING COMPONENTS

Replacing capacitors and/or other components isn't difficult, unless they are buried under other components. The Hallicrafters SX-28 and SX-42 receivers are examples of receivers that have extremely difficult to reach components. Like the "re-capping" question, there are different schools of thought on the "proper" component replacement method. You can use solder wick and/or a desoldering tool to remove the solder from a terminal, unwrap the wires, and install the new component by wrapping the lead around the terminal and soldering it securely. The proponents of this method point out that this is the preferred military and commercial method. I find it often will needlessly damage other components such as tube sockets and create solder droplets inside of the radio.

Back in the day, radio repairmen clipped out a component leaving a short stub of wire, made little coils in the new lead, then soldered the coiled lead to the old stub. This is a much faster, easier and neater method.

26.9.2 Powering Up the Equipment

Get out your volt-ohm meter and measure the resistance from the B+ line to ground. Filter capacitors will cause a initial low-resistance reading that increases as the capacitors charge. If the resistance stays low or does not increase beyond tens of kΩ, find the short circuit before you proceed. Now it's time to plug in the radio. It's best to use a variable transformer such as a Variac and ramp up the voltage slowly, or use a "dim bulb tester" (a 100-W light-bulb wired in series with one leg of the ac power). Turn on the radio, and watch for any sparking, flashing or a red glow from the plates of the rectifier tube, or smoke. If any of these occur, immediately remove power and correct the problem. Observe that the tube filaments should light (although you won't see the glow from metal tubes, you should be able to feel them warm up). Again, any tubes that fail to light should be replaced before you continue.

Now hook up a speaker and antenna, and test the radio. With any luck you'll be greeted by a perfectly-performing radio. Seldom, however, is that the case. You may encounter any number of problems at this point. Dirty bandswitches and other controls manifest themselves by intermittently cutting out; they can be cleaned by DeOxit contact cleaner applied with a cotton swab (don't spray the switch directly!). Scratchy volume or RF gain controls can be cleaned with some DeOxit; in some cases you might need to remove the control and uncrimp the cover to reveal the carbon element inside.

If a receiver is totally dead at this point but the filaments and dial lights are lit, double-check to see that the "Receive-Standby" switch is in the receive position, and any battery plug or standby switch jumpers (as described in the manual) are in their correct place.

Although comprehensive troubleshooting is covered elsewhere in this chapter, the next step is comparing voltages with those stated in the user manual. If the manual doesn't have a voltage table denoting the expected voltage at each tube pin, expect between 200-350 V at the tube plate terminals, a few volts at the cathode (unless it's directly grounded), 70-200 V at the screen, and slightly negative voltage at the grid. If you're faced with this situation and a newcomer to troubleshooting vintage gear, help can be found at **AMFone.net**, **antiqueradios.org**, or other forums that cater to boat-anchors and/or vintage radio repair and restoration.

26.9.3 Alignment

Over the years hams have been cautioned that alignment is usually the *last* thing that should be attempted to repair a radio. In general this is true — but it's also a certainty that a 50 year old radio *will* need alignment in order for it to perform at its best. In any case, replace the capacitors and any other faulty components before you attempt alignment — it'll never be right if it still has bad parts! You'll need a good signal generator and a volt-ohm-meter or oscilloscope. Follow the manufacturer's instructions, and with care you'll be rewarded with a radio that performs as good as it did when it was new. Additional information on alignment can be found in the section "Testing Within A Stage."

26.9.4 Using Vintage Receivers

Connect a speaker, preferably the same impedance as the output impedance of the receiver. Some receivers have a 600-Ω and 3.2- or 4-Ω output. An 8-Ω speaker is fine — connect it to the low impedance output. Do not operate the receiver without a speaker, however, as the audio output transformer could be damaged by high voltage transients with no load. Alternatively, plug in a pair of headphones, keeping in mind that old receivers usually have high impedance headphone outputs and new headphones are usually low impedance: they'll work fine, but the volume may be considerably lower with the newer headphones.

If you're going to use the receiver in conjunction with a transmitter, you need to be able to mute the receiver while you're transmitting — otherwise, you'll end up with copious feedback from the receiver. Most receivers have mute terminals — some mute with a closed switch, others mute on open. Figure out which method your receiver and transmitter use. You'll need a relay if the receiver mute arrangement doesn't match that of the transmitter.

Some receivers — such as the older Hammarlund Super Pros, and pre-WWII Hallicrafters models — use the mute terminals to open the B+ when putting the receiver in standby mode. This is extremely dangerous with 300 V or so on exposed terminals! An easy modification will save you from an almost certain shock. Open up the receiver and remove the wires from the standby terminals. Solder them together and insulate the connection with electrical tape or a wire nut. Better, solder them to an unused, ungrounded terminal if there's one handy. Next, examine the RF gain control and notice that one terminal is probably grounded. Cut this wire and solder a 47-kΩ resistor between ground and the RF gain control terminal. Connect a pair of wires from the terminals of the mute connection across the 47-kΩ resistor just installed. Now, with the mute terminals open the RF gain is all the way down so the receiver is essentially muted. Short the terminals to receive. The voltage here is low and not dangerous.

Next, connect an antenna and antenna relay in the same manner and let's tune the bands. You'll find that the best fidelity from the receiver occurs at its maximum bandwidth. The crystal filter, if fitted, can help notch out heterodynes as can tuning the receiver slightly higher or lower. The bandspread control can be used to "fine tune." Now, just enjoy using your classic, vintage equipment!

26.10 References and Bibliography

J. Bartlett, "Calculating Component Values," *QST*, Nov 1978.

J. Carr, *How to Troubleshoot and Repair Amateur Radio Equipment*, Blue Ridge Summit, PA: TAB Books Inc, 1980.

D. DeMaw, "Understanding Coils and Measuring their Inductance," *QST*, Oct 1983.

H. Gibson, *Test Equipment for the Radio Amateur*, London, England: Radio Society of Great Britain, 1974.

C. Gilmore, *Understanding and Using Modern Electronic Servicing Test Equipment*, TAB Books, Inc, 1976.

F. Glass, *Owner Repair of Amateur Radio Equipment*, Los Gatos, CA: RQ Service Center, 1978.

R. Goodman, *Practical Troubleshooting with the Modern Oscilloscope*, TAB Books, Inc, 1979.

A. Haas, *Oscilloscope Techniques*, New York: Gernsback Library, Inc, 1958.

C. Hallmark, *Understanding and Using the Oscilloscope*, TAB Books, Inc, 1973.

A. Helfrick, *Amateur Radio Equipment Fundamentals*, Englewood Cliffs, NJ: Prentice-Hall Inc, 1982.

K. Henney, and C. Walsh, *Electronic Components Handbook*, New York: McGraw-Hill Book Company, 1957.

L. Klein, and K. Gilmore, *It's Easy to Use Electronic Test Equipment*, New York: John R. Rider Publisher, Inc (A division of Hayden Publishing Company), 1962.

J. Lenk, *Handbook of Electronic Test Procedures*, Prentice-Hall Inc, 1982.

G. Loveday, and A. Seidman, *Troubleshooting Solid-State Circuits*, New York: John Wiley and Sons, 1981.

A. Margolis, *Modern Radio Repair Techniques*, TAB Books Inc, 1971.

H. Neben, "An Ohmmeter with a Linear Scale," *QST*, Nov 1982.

H. Neben, "A Simple Capacitance Meter You Can Build," *QST*, Jan 1983.

F. Noble, "A Simple LC Meter," *QST*, Feb 1983.

J. Priedigkeit, "Measuring Inductance and Capacitance with a Reflection-Coefficient Bridge," *QST*, May 1982.

H. Sartori, "Solid Tubes — A New Life for Old Designs," *QST*, Apr 1977; "Questions on Solid Tubes Answered," Technical Correspondence, *QST*, Sep 1977.

B. Wedlock, and J. Roberge, *Electronic Components and Measurements*, Prentice-Hall Inc, 1969.

"Some Basics of Equipment Servicing," series, *QST*, Dec 1981-Mar 1982; Feedback May 1982.

Notes

[1]The ARRL has prepared a list of dip-meter sources. These are available on the **CD-ROM** included with this *Handbook*).

[2]More information about the signal injector and signal sources appears in "Some Basics of Equipment Servicing," February 1982 *QST* (Feedback, May 1982).

[3]The term "I_{cbo}" means "Current from collector to base with emitter open." The subscript notation indicates the status of the three device terminals. The terminals measured are listed first, with the remaining terminal listed as "s" (shorted) or "o" (open).

Contents

27.1 Managing Radio Frequency Interference
    27.1.1 Responsibility for Radio Frequency Interference
    27.1.2 Proper Station Operation
    27.1.3 Personal Diplomacy
    27.1.4 Interference To a Neighbor's Equipment
    27.1.5 Interference From a Neighbor's Equipment
    27.1.6 Being Prepared
    27.1.7 Contacting Your Neighbor

27.2 FCC Rules and Regulations
    27.2.1 FCC Part 97 Rules
    27.2.2 FCC Part 15 Rules

27.3 Elements of RFI
    27.3.1 Source-Path-Victim
    27.3.2 Differential-Mode vs Common-Mode Signals
    27.3.3 Types of RFI
    27.3.4 Spurious Emissions
    27.3.5 Fundamental Overload
    27.3.6 External Noise Sources
    27.3.7 Intermodulation Distortion
    27.3.8 Ground Connections

27.4 Identifying the Type of RFI Source
    27.4.1 Identifying Noise from Part 15 Devices
    27.4.2 Identifying Power-line and Electrical Noise
    27.4.3 Identifying Intermodulation Distortion

27.5 Locating Sources of RFI
    27.5.1 Noise Sources Inside Your Home
    27.5.2 Noise Sources Outside Your Home
    27.5.3 Approaching Your Neighbor
    27.5.4 Radio Direction Finding

27.6 Power-line Noise
    27.6.1 Before Filing a Complaint
    27.6.2 Filing a Complaint
    27.6.3 Techniques for Locating Power-line Noise Sources
    27.6.4 Amateur Power-line Noise Locating Equipment
    27.6.5 Signature or Fingerprint Method
    27.6.6 Locating the Source's Power Pole or Structure
    27.6.7 Pinpointing the Source on a Pole or Structure
    27.6.8 Common Causes of Power-line Noise
    27.6.9 The Cooperative Agreement

27.7 Elements of RFI Control
    27.7.1 Differential- and Common-Mode Signal Control
    27.7.2 Shields and Filters
    27.7.3 Common-Mode Chokes

27.8 Troubleshooting RFI
    27.8.1 General RFI Troubleshooting Guidelines
    27.8.2 Transmitters
    27.8.3 Television Interference (TVI)
    27.8.4 Cable TV
    27.8.5 DVD and Video Players
    27.8.6 Non-radio Devices

27.9 Automotive RFI
    27.9.1 Before Purchasing a Vehicle
    27.9.2 Transceiver Installation Guidelines
    27.9.3 Diagnosing Automotive RFI
    27.9.4 Eliminating Automotive RFI
    27.9.5 Electric and Hybrid-Electric Vehicles
    27.9.6 Automotive RFI Summary

27.10 RFI Projects
    27.10.1 Project: RF Sniffer

27.11 RFI Glossary

27.12 References and Bibliography

Chapter 27

RF Interference

This chapter is a complete revision of the "Interference" chapter in editions prior to 2011. A team of knowledgeable experts supported the greatly expanded scope. The topics listed for each of these individuals represent only their primary area of contribution.

ARRL Lab Staff:
Ed Hare, W1RFI, *Lab Manager*
Mike Gruber, W1MG
General RFI Management and Troubleshooting, FCC Regulations and Responsibilities

Ron Hranac, NØIVN –
Cable and Digital TV

Mark Steffka, WW8MS and
Jeremy Campbell, KC8FEI –
Automotive RFI

Mike Martin, K3RFI –
Power-line Noise

Amateurs live in an increasingly crowded technological environment. As our lives become filled with technology, every lamp dimmer, garage-door opener or other new technical "toy" contributes to the electrical noise around us. Many of these devices also "listen" to that growing noise and may react to the presence of their electronic siblings. The more such devices there are, the higher the likelihood that the interactions will be undesirable.

What was once primarily a conversation about "interference" has expanded to include power systems, shielding, intentional and unintentional radiators, bonding and grounding, and many other related topics and phenomena. These are all grouped together under the general label of *electromagnetic compatibility (EMC)*. The scope of EMC includes all the ways in which electronic devices interact with each other and their environment.

The general term for interference caused by signals or fields is *electromagnetic interference* or *EMI*. This is the term you'll encounter in the professional literature and standards. The most common term for EMI involving amateur signals is *radio frequency interference (RFI)* and when a television or video display is involved, *television interference (TVI)*. RFI is the term used most commonly by amateurs. Whether it's called EMI, RFI or TVI, unwanted interaction between receivers and transmitters has stimulated vigorous growth in the field of electromagnetic compatibility! (This chapter will use the term RFI to refer to all types of interference to or from amateur signals, except where noted.)

This chapter begins with an overview of dealing with interference and includes relevant FCC regulations. This section is an excellent resource when you are confronted with an interference problem and are wondering "What do I do now?" The information here is based on the experiences of ARRL Lab staff in assisting amateurs with RFI problems.

The second part of this chapter is a discussion on identifying and locating RFI-related noise and signal sources then presents some effective ways of resolving the problem. A glossary of RFI terminology concludes the chapter.

The material in this chapter may provide enough information for you to solve your problem, but if not, the ARRL Web site offers extensive resources on RF interference at **www.arrl.org/radio-frequency-interference-rfi**. Many topics covered in this chapter are covered in more detail in the *ARRL RFI Book* from a practical amateur perspective.

Throughout this chapter you'll also find references to "Ott," meaning the book *Electromagnetic Compatibility Engineering* by EMC consultant Henry Ott, WA2IRQ. EMC topics are treated in far greater depth in Ott's book than is possible in this *Handbook*. Readers interested in the theory of EMC, analysis of EMC mechanisms, test methodology and EMC standards can purchase a copy through the ARRL Publication Sales department or the ARRL Web site.

27.1 Managing Radio Frequency Interference

Sooner or later, nearly every Amateur Radio operator will have a problem with RFI, but temper your dismay. Most cases of interference can be cured! Before diving into the technical aspects of interference resolution, consider the social aspects of the problem. A combination of "diplomacy" skills and standard technical solutions are the most effective way to manage the problem so that a solution can be found and applied. This section discusses the overall approach to solving RFI problems. Specific technical causes and solutions are described in subsequent sections.

27.1.1 Responsibility for Radio Frequency Interference

When an interference problem occurs, we may ask "Who is to blame?" The ham and the other party often have different opinions. It is natural (but unproductive) to assign blame instead of fixing the problem.

No amount of wishful thinking (or demands for the "other guy" to solve the problem) will result in a cure for interference. Each party has a unique perspective on the situation and a different degree of understanding of the personal and technical issues involved. On the other hand, each party has certain responsibilities and should be prepared to address them fairly. (Given the realities of amateur operation, one of the parties is likely to be a neighbor to the amateur and so the term "neighbor" is used in this chapter.)

Always remember that every interference problem has two components — the equipment that is involved and the people who use it. A solution requires that we deal effectively with both the equipment and the people. The ARRL recommends that the hams and their neighbors cooperate to find solutions. The FCC also shares this view. It is important therefore to define the term "interference" without emotion.

27.1.2 Proper Station Operation

A radio operator is responsible for the proper operation of the radio station. This responsibility is spelled out clearly in Part 97 of the FCC regulations. If interference is caused by a spurious emission from your station, you must correct the problem there.

Fortunately, most cases of interference are not the fault of the transmitting station. If an amateur signal is the source of interference, the problem is usually caused by fundamental overload — a general term referring to interference caused by the intended, fundamental signal from a transmitter. If the amateur station is affected by interference, electrical noise is most often the culprit. Typical sources include power lines and consumer devices.

27.1.3 Personal Diplomacy

Whether the interference is to your station or from your station, what happens when you first talk to your neighbor sets the tone for all that follows. Any technical solutions cannot help if you are not allowed in your neighbor's house to explain them! If the interference is not caused by spurious emissions from your station, however, you should be a locator of solutions, not a provider of solutions.

Your neighbor will probably not understand all of the technical issues — at least not at first. Understand that, regardless of fault, interference is annoying whether your signals are causing interference to the neighbor or a device owned by the neighbor is causing interference to you.

Let your neighbor know that you want to help find a solution and that you want to begin by talking things over. Talk about some of the more important technical issues, in non-technical terms. Explain that you must also follow technical rules for your signal, such as for spurious emissions, and that you will check your station, as well.

27.1.4 Interference To a Neighbor's Equipment

Your transmitted signals can be the source of interference to a neighbor's equipment. Assure your neighbor that you will check your station thoroughly and correct any problems. You should also discuss the possible susceptibility of consumer equipment. You may want to print a copy of the RFI information found on the ARRL Web site at **www.arrl.org/information-for-the-neighbors-of-hams**. (This document is also included on the CD-ROM accompanying this book.)

Your neighbor will probably feel much better if you explain that you will help find a solution, even if the interference is not your fault. This offer can change your image from neighborhood villain to hero, especially if the interference is not caused by your station. (This is often the case.)

Here is a good analogy: If you tune your TV to channel 3, and see channel 8 instead, you would likely decide that your TV set is broken. Now, if you tune your TV to channel 3, and see your local shortwave radio station (quite possibly Amateur Radio), you shouldn't blame the shortwave station without some investigation. In fact, many televisions respond to strong signals outside the television bands. They may be working as designed, but require added filters and/or shields to work properly near a strong, local RF signal.

27.1.5 Interference From a Neighbor's Equipment

Your neighbor is probably completely unaware that his or her equipment can interfere with your operation. You will have to explain that some home electronics equipment can generate radio signals many times stronger than the weak signals from a distant transmitter. Also explain that there are a number of ways to prevent those signals from being radiated and causing interference. If the equipment causing the problem can be identified, the owner's manual or manufacturer of the device may provide information on the potential for RFI and for its elimination.

As with interference appearing to be caused by your station, explain that your intent is to help find a solution. Without further investigation it is premature to assume that the neighbor's equipment is at fault or that FCC regulations require the neighbor to perform any corrective action. Working together to find a mutually acceptable solution is the best strategy.

27.1.6 Being Prepared

In order to troubleshoot and cure RFI, you need to learn more than just the basics. This is especially important when dealing with your neighbor. If you visit your neighbor's house and try a few dozen things that don't work (or make things worse), your neighbor may lose confidence in your ability to help cure the problem. If that happens, you may be asked to leave.

Start by carefully studying the technical sources and cures for RFI in this book and in other references, such as the *ARRL RFI Book*. Review some of the ARRL Technical Information Service and *QST* articles about interference. If you are unfamiliar with any of the terms in this chapter, refer to the glossary.

LOCAL HELP

If you are not an expert (and even experts can use moral support), you should find some local help. Fortunately, such help is often available from your ARRL Section's Technical Coordinator (TC). The TC knows of any local RFI committees and may have valuable contacts in the local utility companies. Even an expert can benefit from a TC's help. The easiest way to find your TC is through your ARRL Section Manager (SM). There is a list of SMs on the ARRL Web site or in any

> **Warning: Performing Repairs**
>
> You are the best judge of a local situation, but the ARRL strongly recommends that you do not work on your neighbor's equipment. The minute you open up a piece of equipment, you may become liable for problems. Internal modifications to your neighbor's equipment may cure the interference problem, but if the equipment later develops some other problem, you may be blamed, rightly or wrongly. In some states, it is even *illegal* for you to do *any* work on electronic equipment other than your own.

recent *QST*. He or she can quickly put you in contact with the best source of local help.

Even if you can't secure the help of a local expert, a second ham can be a valuable asset. Often a second party can help defuse any hostility. When evaluating and solving RFI problems involving your station, it is very important for two hams to be part of the process. One can operate your station and the other can observe symptoms, and, when appropriate, try solutions.

PREPARE YOUR HOME AND STATION

The first step toward curing the problem is to make sure your own signal is clean and that devices in your home are not causing any problems. Eliminate all interference issues in your own home to be sure your station is operating properly and that your own electronic equipment is not being interfered with or causing interference to your station!

This is also a valuable troubleshooting tool for situations in which your station is suspected of being the source of interference: If you know your signals are "clean," you have cut the size of the problem in half! If the FCC ever gets involved, you can demonstrate that you are not interfering with your own electronics.

Apply RFI cures to your own consumer electronics and computer equipment. What you learn by identifying and eliminating interference in your own home will make you better prepared to do so in your neighbor's home. When your neighbor sees your equipment working well, it also demonstrates that filters work and cause no harm.

To help build a better relationship, you may want to show your station to your neighbor. A well-organized and neatly-wired station inspires confidence in your ability to solve the RFI problem. Clean up your station and clean up the mess! A rat's nest of cables, unsoldered connections and so on can contribute to RFI. Grounding is typically not a cure for RFI, but proper grounding will improve lightning safety and can greatly reduce hum and buzz from power systems. Make sure cable shields are connected properly and that RF current picked up from your transmitted signal by audio and power wiring is minimized.

Install a low-pass or band-pass transmit filter. (In the unlikely event that the FCC becomes involved, they will ask you about filtering.) Show your neighbor that you have installed the necessary filter on your transmitter. Explain that if there is still interference, it is necessary to try filters on the neighbor's equipment, too.

Operating practices and station-design considerations can cause interference to TV and FM receivers. Don't overdrive a transmitter or amplifier; that can increase its harmonic output.

Along with applying some of the interference-reducing solutions in this chapter, you can also consider steps to reduce the strength of your signal at the victim equipment. This includes raising, moving, or re-orienting the antenna, or reducing transmit power. The use of a balun and properly balanced feed line will minimize radiation from your station feed line. (See the **Transmission Lines** chapter.) Changing antenna polarization may help, such as if a horizontal dipole is coupling to a cable TV service drop. Using different modes, such as CW or FM, may also change the effects of the interference. Although the goal should be for you to operate as you wish with no interference, be flexible in applying possible solutions.

27.1.7 Contacting Your Neighbor

Now that you have learned more about RFI, located some local help (we'll assume it's the TC) and done all of your homework, make contact with your neighbor. First, arrange an appointment convenient for you, the TC and your neighbor. After you introduce the TC, allow him or her to explain the issues to your neighbor. Your TC will be able to answer most questions, but be prepared to assist with support and additional information as required.

Invite the neighbor to visit your station. Show your neighbor some of the things you do with your radio equipment. Point out any test equipment you use to keep your station in good working order. Of course, you want to show the filter you have installed on your transmitter's output.

Next, have the TC operate your station on several different bands while you show your neighbor that your home electronics equipment is working properly when your station is transmitting. Point out the filters or chokes you have installed to correct any RF susceptibility problems. If the interference is coming from the neighbor's home, show it to the neighbor and explain why it is a problem for you.

At this point, tell your neighbor that the next step is to try some of the cures seen in your home and station on his or her equipment. This is a good time to emphasize that the problem is probably not your fault, but that you and the TC will try to help find a solution anyway.

Study the section on Troubleshooting RFI before deciding what materials and techniques are likely to be required. You and the TC should now visit the neighbor's home and inspect the equipment installation.

AT YOUR NEIGHBOR'S HOME

Begin by asking when the interference occurs, what equipment is involved, and what frequencies or channels are affected, if appropriate. The answers are valuable clues. Next, the TC should operate your station while you observe the effects. Try all bands and modes that you use. Ask the neighbor to demonstrate the problem. Seeing your neighbor's interference firsthand will help all parties feel more comfortable with the outcome of the investigation.

If it appears that your station is involved, note all conditions that produce interference. If no transmissions produce the problem, your station *may* not be the source. (It's possible that some contributing factor may have been missing in the test.) Have your neighbor keep notes of when and how the interference appears: what time of day, what channels or device was being interfered with, what other equipment was in use, what was the weather? You should do the same whenever you operate. If you can readily reproduce the problem, you can start to troubleshoot the problem. This process can yield important clues about the nature of the problem.

The tests may show that your station isn't involved at all. A variety of equipment malfunctions or external noise can look like interference. Some other nearby transmitter or noise source may be causing the problem. You may recognize electrical noise or some kind of equipment malfunction. If so, explain your findings to the neighbor and suggest that he or she contact appropriate service personnel.

If the interference is to your station from equipment that may be in the neighbor's home, begin by attempting to identify which piece of equipment is causing the problem. Describe when you are receiving the interference and what pattern it seems to have (continuous, pulsed, intermittent, certain times of day, and so on).

Confirm that a specific piece of equipment is causing the problem. The easiest way to do this is to physically unplug the power source or by removing the batteries while you or the TC observe at your station. (Remember that turning a piece of equipment OFF with its ON/OFF switch may not cause the equipment to completely power down.) Removing cables from a powered-up piece of equipment may serve to further isolate the problem.

At this point, the action you take will depend on the nature of the problem and its source. Techniques for dealing with specific interference issues are discussed in the following sections of the chapter. If you are unable to determine the exact nature of the problem, develop a plan for continuing to work with the neighbor and continue to collect information about the behavior of the interference.

Product Review Testing and RFI

The ARRL Laboratory mostly considers RFI as an outside source interfering with the desired operation of radio reception. Power line noise, power supplies, motor controls and light dimmers are all *outside* a radio receiver and antenna system and can be a major annoyance that distracts from the pleasure of operating. Have you ever considered RFI generated *inside* your own receiver or by another legally operating Amateur Radio station?

Harmonics

RFI is just that: radio frequency interference. For instance, a harmonic from another radio amateur's transmitter could be interfering with the desired signal you're tuned to. One might think in this day and age, harmonics are minimal and do not cause interference. I ask you to reconsider.

The maximum spurious output of a modern amateur transmitter must be 43 dB lower than the output on the carrier frequency at frequencies below 30 MHz. While that figure may be "good enough" for an FCC standard, it's not good enough to eliminate the possibility of causing interference to other radio amateurs or possibly other services. Here at the ARRL Laboratory, the measurement of harmonic emission level is a measurement I make of RFI generated *outside* your receiver.

A radio amateur contacted me, concerned about a report that he was causing interference to operators on the 80 meter CW band while operating at full legal limit power during a 160 meter CW contest. Knowing FCC rule part 97.307, he made the effort to measure his 80 meter harmonic. Easily meeting FCC standards at 50 dB below carrier output on 160, he wondered how his transmitter could cause interference.

Here's a breakdown of power output and signal reduction:
0 dB down from 1500 W = 1500 W
10 dB down from 1500 W = 150 W
20 dB down from 1500 W = 15 W
30 dB down from 1500 W = 1.5 W
40 dB down from 1500 W = 150 mW
43 dB down from 1500 W = 75.2 mW (this is the FCC legal limit)
50 dB down from 1500 W = 15 mW

While 15 mW may seem too low of a power to cause interference, it wasn't in this case; the interfering signal was reported to be S9. QRP enthusiasts know that at 15 mW, signals can propagate well with the right conditions. Using a single band, resonant antenna will reduce interference caused by harmonics located on other bands, but today many stations employ antennas resonant on more than one ham band.[1]

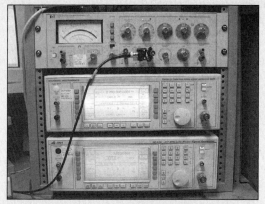

The Lab uses these signal generators to test receivers for internally generated intermodulation distortion, as well as other key performance parameters.

Signals Generated in the Receiver

What about RFI generated *inside* the receiver you're operating? You're tuning across the 15 meter band when, all of a sudden, you hear what seems to be an AM broadcast station. Is it a jammer? It is definitely interference — radio frequency interference — caused by two strong shortwave stations! In this particular case, a Midwest radio amateur experienced interference on 15 meters and figured out what was happening. One station was transmitting near 6 MHz and another transmitting above 15 MHz. These two strong stations added up to created a second order IMD (intermodulation distortion) product at the 1st IF stage, and this unwanted signal was passed along to subsequent stages and to the speaker. The RFI in this case was caused by insufficient receiver performance (second order IMD dynamic range) where the frequencies of the two stations added up to exactly the frequency that the operator was tuned to. Third-order IMD products from strong in-band signals are another form of RFI created within a radio receiver.

Nearby stations transmitting at or near the IF frequency will cause interference not because the transmitter is at fault, but because of a receiver's insufficient IF rejection. The same interference will be heard if a nearby transmitter is operating at an image frequency.

Power Supplies

RFI can also be created from another part of a radio system, such as an external power supply. In addition to transmitter and receiver testing, the Lab also measures the conducted emission levels of power supplies. This is an indication of the amount of RF at given frequencies conducted onto power lines from a power supply as described in this chapter.

Through our published Product Review test results in *QST* magazine, readers can compare the above figures of modern HF transceivers when considering the purchase of a new or used transceiver. Our published data tables spawn friendly competition between radio manufacturers who in turn, strive to perfect their circuit designs. The result is a better product for the manufacturer and a better product for you, the radio amateur. — *Bob Allison, WB1GCM, ARRL Test Engineer*

[1]In this case, the use of a bandpass filter designed for 160 meters would significantly attenuate the harmonic on 80 meters, eliminating the interference.

The ARRL Lab maintains an RF-tight screen room with a full suite of equipment for Product Review testing.

27.2 FCC Rules and Regulations

In the United States most unlicensed electrical and electronic devices are regulated by Part 15 of the FCC's rules. These are referred to as "Part 15 devices." Most RFI issues reported to the ARRL involve a Part 15 device. Some consumer equipment, such as certain wireless and lighting devices, is covered under FCC Part 18 which pertains to ISM (Industrial, Scientific and Medical) devices.

The Amateur service is regulated by FCC Part 97. (Part 97 rules are available online at **www.arrl.org/part-97-amateur-radio.** See also the sidebar "RFI-related FCC Rules and Definitions.") To be legal, the amateur station's signal must meet all Part 97 technical requirements, such as for spectral purity and power output.

As a result, it isn't surprising that most interference complaints involve multiple parts of the FCC rules. (The FCC's jurisdiction does have limits, though — ending below 9 kHz.) It is also important to note that each of the three parts (15, 18 and 97) specifies different requirements with respect to interference, including absolute emissions limits and spectral purity requirements. The FCC does not specify *any* RFI immunity requirements. Most consumer devices therefore receive no FCC protection from a legally licensed transmitter, including an amateur transmitter operating legally according to Part 97.

Licensed services are protected from interference to their signals, even if the interference is generated by another licensed service transmitter. For example, consider TVI from an amateur transmitter's spurious emissions, such as harmonics, that meet the requirements of Part 97 but are still strong enough to be received by nearby TV receivers. The TV receiver itself is not protected from interference under the FCC rules. However, within its service area the licensed TV broadcast signal is protected from harmful interference caused by spurious emissions from other licensed transmitters. In this case, the amateur transmitter's interfering spurious emission would have to be eliminated or reduced to a level at which harmful interference has been eliminated.

27.2.1 FCC Part 97 Rules

While most interference to consumer devices may be caused by a problem associated with the consumer device as opposed to the signal source, all amateurs must still comply with Part 97 rules. Regardless of who is at fault, strict conformance to FCC requirements, coupled with a neat and orderly station appearance, will go far toward creating a good and positive impression in the event of an FCC field investigation. Make sure your station and signal exhibit good engineering and operating practices.

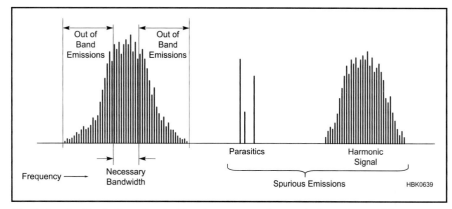

Fig 27.1 — An illustration of out-of-band versus spurious emissions. Some of the modulation sidebands are outside the necessary bandwidth. These are considered out-of-band emissions, not spurious emissions. The harmonic and parasitic emissions shown here are considered spurious emissions; these must be reduced to comply with §97.307.

What If the Police Are Called?

Many amateurs have had a similar experience. You are enjoying some time in front of the radio when the doorbell rings. When you answer the door, you find an irate neighbor has called the police about your transmissions interfering with their stereo (or cordless telephone or other home electronics). The officer tells you that you are interfering with your neighbor and orders you to stop transmissions immediately.

The bad news is you are in the middle of a bad situation. The good news is that most cases of interference can be cured! The proper use of "diplomacy" skills to communicate with a neighbor and standard technical cures will usually solve the problem. Even more good news is that if you are operating in accordance with your license and employing good engineering practices, the law and FCC rules are on your side.

Most RFI is caused by the unfortunate fact that most consumer equipment lacks the necessary filtering and shielding to allow it to work well near a radio transmitter. The FCC does not regulate the immunity of equipment, however, so when interference is caused by consumer-equipment fundamental overload, there is no FCC rules violation, and licensed stations have no regulatory responsibility to correct interference that may result.

Further, in 1982, Congress passed Public Law 97-259. This law is specific and reserves exclusive jurisdiction over RFI matters to the Federal Communications Commission. This national law preempts any state or local regulations or ordinances that attempt to regulate or resolve RFI matter. This is a victory for amateurs (and other services operating with the legal and technical provisions of their licenses).

Simply put, 97-259 says that cities and towns may not pass ordinances or regulations that would prohibit someone from making legal radio transmissions. But what do you do when your neighbor (or the police) confront you about RFI to their consumer electronics? First and foremost, remain calm. In all likelihood the officer or your neighbor has probably never heard of 97-259. Don't get defensive and get drawn into an argument. Don't make comments that the problem is with the neighbor's "cheap" equipment. While inexpensive radios are usually big culprits, any home electronics are potential problems due to inadequate technical designs.

Begin by listening to the complaint. Explain that while you understand, you are operating your equipment within its technical specifications. If your equipment doesn't interfere with your own home electronics, offer to demonstrate that to the officer. Also explain the basics of PL 97-259. If the officer (or neighbor) continues to insist that regardless of the law that you cease, consider temporarily complying with his or her request, with the understanding that you are doing so until the matter is resolved.

Work with your neighbors to understand that steps can be taken that should help resolve the problems (for example, placing toroids and filters on the consumer electronics). The ARRL Web has lots of helpful information as you work to resolve the problems. If your club has a local RFI committee or ARRL Technical Specialist, get them involved — their expertise can really be helpful. But above all, remember that when you practice easy, level-headed "diplomacy" you can usually keep the situation from escalating. — *Dan Henderson, N1ND, ARRL Regulatory Information Manager*

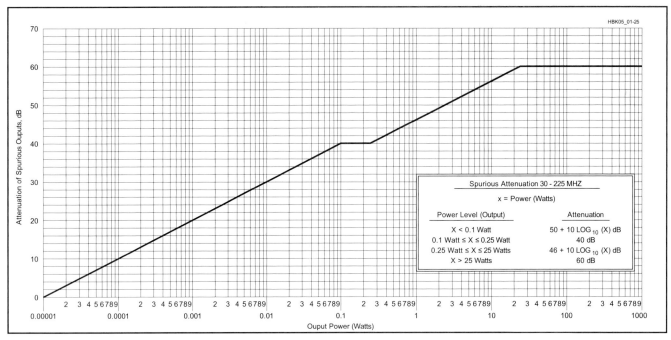

Fig 27.2 — Required attenuation of spurious outputs, 30-225 MHz.

The bandwidth of a signal is defined by §97.3(a)(8) while the paragraphs of §97.307 define the technical standards amateur transmissions must meet. Paragraph (c) defines the rules for interference caused by spurious emissions. As illustrated in **Fig 27.1**, modulation sidebands outside the necessary bandwidth are considered *out-of-band* emissions, while harmonics and parasitic signals are considered *spurious emissions*. Paragraphs (d) and (e) specify absolute limits on spurious emissions, illustrated in **Fig 27.2**. Spurious emissions must not exceed these levels, whether or not the emissions are causing interference. If spurious emissions from your transmitter are causing interference, it's your responsibility to clean them up.

Strict observance of these rules can not only help minimize interference to the amateur service, but other radio services and consumer devices as well.

27.2.2 FCC Part 15 Rules

In the United States, most unlicensed devices are regulated by Part 15 of the FCC's rules. While understanding these rules doesn't necessarily solve an RFI problem, they do provide some important insight and background on interference to and from a Part 15 device. (Part 18 devices and rules are similar in some respects) and will not be discussed separately — see the sidebar.)

There are literally thousands of Part 15 devices with the potential to be at the heart of an RFI problem. A Part 15 device can be almost anything not already covered in another Part of the FCC rules. In fact, many Part 15 devices may not normally even be associated with electronics, RF or in some cases, electricity. While televisions, radios, telephones and even computers obviously constitute a Part 15 device, the rules extend to anything that is capable of generating RF, including electric motors and consumer devices such as baby monitors, wireless microphones and intercoms, RF remote controls, garage door openers, etc. With so many Part 15 consumer devices capable of generating and responding to RF, it isn't surprising therefore that most reported RFI problems involving Amateur Radio also involve a Part 15 device.

TYPES OF PART 15 DEVICES

Part 15 describes three different types of devices that typically might be associated with an RFI problem. A fourth type of device, called a *carrier current device*, uses power lines and wiring for communications purposes. As we'll see, the rules are different for each type.

Intentional Emitters — Intentionally generate RF energy and radiate it. Examples include garage door openers, cordless phones and baby monitors.

Unintentional Emitters — Intentionally generate RF energy internally, but do not intentionally radiate it. Examples include computers and network equipment, superheterodyne receivers, switchmode power supplies and TV receivers.

Incidental Emitters — Generate RF energy only as an incidental part of their normal operation. Examples include power lines, arcing electric fence, arcing switch contacts, dc motors and mechanical light switches.

Carrier Current Devices — Intentionally generate RF and conduct it on power lines and/or house wiring for communications purposes. Examples include Powerline and X.10 networks, Access or In-House Broadband-Over-Power-Line (BPL), campus

---

FCC Part 18, Consumer Devices

Some consumer devices are regulated by Part 18 of the FCC Rules which pertains to the Industrial, Scientific and Medical (ISM) bands. Some lighting devices, such as the now ubiquitous compact fluorescent lamp (CFL), and home microwave ovens operate under Part 18.

Consumer Part 18 devices are generators of RF but not for communications purposes and can cause interference in some cases. However, there are no rules that protect them from interference. The purpose of Part 18 is to permit those devices to operate and to set up rules prohibiting interference.

From the standpoint of an RFI problem, Part 18 rules aren't much different than Part 15. Like a Part 15 device, a Part 18 device is also required to meet specified emissions limits. Furthermore, it must not cause harmful interference to a licensed radio service.

RFI-related FCC Rules and Definitions

Here are some of the most important rules and definitions pertaining to RFI and the Amateur Radio Service. Definitions in Part 2 are used in regulations that apply to all radio services.

§2.1 Definitions

Harmful Interference. Interference which endangers the functioning of a radionavigation service or of other safety services or seriously degrades, obstructs, or repeatedly interrupts a radiocommunication service operating in accordance with [the ITU] Radio Regulations.

Interference. The effect of unwanted energy due to one or a combination of emissions, radiations, or inductions upon reception in a radiocommunication system, manifested by any performance degradation, misinterpretation, or loss of information which could be extracted in the absence of such unwanted energy.

Out-of-band Emission. Emission on a frequency or frequencies immediately outside the necessary bandwidth which results from the modulation process, but excluding spurious emissions.

§97.3 Definitions

(a) The definitions of terms used in Part 97 are:

(8) Bandwidth. The width of a frequency band outside of which the mean power of the transmitted signal is attenuated at least 26 dB below the mean power of the transmitted signal within the band.

(23) Harmful interference. (see the previous Part 2 definition)

(42) Spurious emission. An emission, on frequencies outside the necessary bandwidth of a transmission, the level of which may be reduced without affecting the information being transmitted.

§97.307 Emission standards

(a) No amateur station transmission shall occupy more bandwidth than necessary for the information rate and emission type being transmitted, in accordance with good amateur practice.

(b) Emissions resulting from modulation must be confined to the band or segment available to the control operator. Emissions outside the necessary bandwidth must not cause splatter or key-click interference to operations on adjacent frequencies.

(c) All spurious emissions from a station transmitter must be reduced to the greatest extent practicable. If any spurious emission, including chassis or power line radiation, causes harmful interference to the reception of another radio station, the licensee of the interfering amateur station is required to take steps to eliminate the interference, in accordance with good engineering practice.

(d) For transmitters installed after January 1, 2003, the mean power of any spurious emission from a station transmitter or external RF amplifier transmitting on a frequency below 30 MHz must be at least 43 dB below the mean power of the fundamental emission. For transmitters installed on or before January 1, 2003, the mean power of any spurious emission from a station transmitter or external RF power amplifier transmitting on a frequency below 30 MHz must not exceed 50 mW and must be at least 40 dB below the mean power of the fundamental emission. For a transmitter of mean power less than 5 W installed on or before January 1, 2003, the attenuation must be at least 30 dB. A transmitter built before April 15, 1977, or first marketed before January 1, 1978, is exempt from this requirement.

(e) The mean power of any spurious emission from a station transmitter or external RF power amplifier transmitting on a frequency between 30-225 MHz must be at least 60 dB below the mean power of the fundamental. For a transmitter having a mean power of 25 W or less, the mean power of any spurious emission supplied to the antenna transmission line must not exceed 25 µW and must be at least 40 dB below the mean power of the fundamental emission, but need not be reduced below the power of 10 µW. A transmitter built before April 15, 1977, or first marketed before January 1, 1978, is exempt from this requirement.

---

radio-broadcast systems, and other power-line communications devices.

PART 15 SUMMARY

FCC's Part 15 rules pertain to unlicensed devices and cover a lot of territory. Although reading and understanding Part 15 can appear rather formidable — especially at first glance — the rules pertaining to RFI can be roughly summarized as follows:

• Part 15 devices operate under an unconditional requirement to not cause harmful interference to a licensed radio service, such as Amateur Radio. If such interference occurs, the operator of the Part 15 device is responsible for eliminating the interference.

• Part 15 devices receive no protection from interference from a licensed radio service. There are no FCC rules or limits with regard to Part 15 device RFI immunity.

When is the operator of a licensed transmitter responsible for interference to a Part 15 device?

• The rules hold the transmitter operator responsible if interference is caused by spurious emissions such as a harmonic. An example would be a harmonic from an amateur's transmitter interfering with a cordless telephone. In this case, the transmitter is generating harmful RF energy beyond its permitted bandwidth. A cure must be installed at the transmitter.

• The transmitter operator is not responsible when a Part 15 device is improperly responding to a legal and intentional output of the transmitter. An example of this case would be interference to a cordless telephone operation by the strong-but-legal signal from a nearby amateur transmitter. In this case, the Part 15 device is at fault and the cure must be installed there. It is important to note that this situation is typical of most interference to Part 15 devices.

Even though the causes and cures for these situations are different, the common element for all three situations is the need for personal diplomacy in resolving the problem.

PART 15 MANUFACTURER REQUIREMENTS

Under FCC rules, both device manufactures and operators of those devices share responsibility for addressing an RFI problem. The rules for manufacturers are primarily designed to reduce the possibility of harmful interference. They do not however completely eliminate the possibility of an RFI problem. If and when interference does occur, the rules are designed to minimize and confine the scope of problems such that they can be addressed on a case by case basis. Responsibility then falls on the device operator to correct the problem or cease using the device.

Manufacturers are subject to requirements that they use good engineering practice to help minimize the potential for interference. In addition, they must meet certain absolute conducted and radiated emissions limits for intentional and unintentional emitters. (See the sidebar for limits on conducted and radiated emissions.) These limits are high enough that S9+ interference levels can occur nearby, depending on frequency, distance and other factors. In fact, most reported Part 15 consumer products causing harmful interference to Amateur Radio are legal and meet these required limits. Therefore, the fact that a particular device is causing harmful interference is not in itself evidence or proof of a rules violation with regard to emissions limits.

With the exception of intentional emitters

and carrier-current devices, there are no absolute radiated emissions limits below 30 MHz. The size of a Part 15 device is usually small relative to the wavelength at these frequencies. It is typically too small to be an effective antenna at these longer wavelengths. Therefore, under the FCC rules, only conducted emissions are specified below 30 MHz. (Note that cables and wiring connected to the devices are often effective at radiating signals and are frequent sources of radiated RFI.)

In general, radiated emissions limits are specified only at frequencies above 30 MHz. At the shorter wavelengths above 30 MHz, the device itself is large enough to be a radiator. Wiring connected to it can also be an effective antenna for radiating noise.

Although incidental radiators do not have any absolute emissions limits, as for all Part 15 devices, manufacturers must still employ good engineering practice to minimize the potential for interference.

The FCC also requires manufacturers to add information as a label to most Part 15

Part 15 Absolute Emissions Limits for Unintentional Emitters

§15.107 Conducted limits

(a) Except for Class A digital devices, for equipment that is designed to be connected to the public utility (ac) power line, the radio frequency voltage that is conducted back onto the ac power line on any frequency or frequencies within the band 150 kHz to 30 MHz shall not exceed the limits in the following table, as measured using a 50 µH/50 ohms line impedance stabilization network (LISN). Compliance with the provisions of this paragraph shall be based on the measurement of the radio frequency voltage between each power line and ground at the power terminal. The lower limit applies at the band edges.

Conducted Limits — Non Class-A Digital Devices

Frequency of emission (MHz)	Conducted limit (dBµV) Quasi-peak	Average
0.15–0.5	66 to 56*	56 to 46*
0.5–5	56	46
5–30	60	50

*Decreases with the logarithm of the frequency.

(b) For a Class A digital device that is designed to be connected to the public utility (ac) power line, the radio frequency voltage that is conducted back onto the ac power line on any frequency or frequencies within the band 150 kHz to 30 MHz shall not exceed the limits in the following table, as measured using a 50 µH/50 ohms LISN. Compliance with the provisions of this paragraph shall be based on the measurement of the radio frequency voltage between each power line and ground at the power terminal. The lower limit applies at the boundary between the frequency ranges.

Conducted Limits — Class-A Digital Devices

Frequency of emission (MHz)	Conducted limit (dBµV) Quasi-peak	Average
0.15–0.5	79	66
0.5–30	73	60

(c) The limits shown in paragraphs (a) and (b) of this section shall not apply to carrier current systems operating as unintentional radiators on frequencies below 30 MHz. In lieu thereof, these carrier current systems shall be subject to the following standards:
 (1) For carrier current systems containing their fundamental emission within the frequency band 535-1705 kHz and intended to be received using a standard AM broadcast receiver: no limit on conducted emissions.
 (2) For all other carrier current systems: 1000 µV within the frequency band 535–1705 kHz, as measured using a 50 µH/50 ohms LISN.
 (3) Carrier current systems operating below 30 MHz are also subject to the radiated emission limits in §15.109(e).
(d) Measurements to demonstrate compliance with the conducted limits are not required for devices which only employ battery power for operation and which do not operate from the ac power lines or contain provisions for operation while connected to the ac power lines. Devices that include, or make provision for, the use of battery chargers which permit operating while charging, ac adaptors or battery eliminators or that connect to the ac power lines indirectly, obtaining their power through another device which is connected to the ac power lines, shall be tested to demonstrate compliance with the conducted limits.

§ 15.109 Radiated emission limits

(a) Except for Class A digital devices, the field strength of radiated emissions from unintentional radiators at a distance of 3 meters shall not exceed the following values:

Radiated Limits — Non Class-A Digital Devices

Frequency of emission (MHz)	Field Strength (µV/meter)
30–88	100
88–216	150
216–960	200
Above 960	500

(b) The field strength of radiated emissions from a Class A digital device, as determined at a distance of 10 meters, shall not exceed the following:

Radiated Limits — Class-A Digital Devices

Frequency of emission (MHz)	Field Strength (µV/meter)
30–88	90
88–216	150
216–960	210
Above 960	300

(c) In the emission tables above, the tighter limit applies at the band edges. Sections 15.33 and 15.35 which specify the frequency range over which radiated emissions are to be measured and the detector functions and other measurement standards apply.
(d) For CB receivers, the field strength of radiated emissions within the frequency range of 25–30 MHz shall not exceed 40 microvolts/meter at a distance of 3 meters. The field strength of radiated emissions above 30 MHz from such devices shall comply with the limits in paragraph (a) of this section.
(e) Carrier current systems used as unintentional radiators or other unintentional radiators that are designed to conduct their radio frequency emissions via connecting wires or cables and that operate in the frequency range of 9 kHz to 30 MHz, including devices that deliver the radio frequency energy to transducers, such as ultrasonic devices not covered under part 18 of this chapter, shall comply with the radiated emission limits for intentional radiators provided in §15.209 for the frequency range of 9 kHz to 30 MHz. As an alternative, carrier current systems used as unintentional radiators and operating in the frequency range of 525 kHz to 1705 kHz may comply with the radiated emission limits provided in §15.221(a). At frequencies above 30 MHz, the limits in paragraph (a), (b), or (g) of this section, as appropriate, apply.

RFI-related Part 15 FCC Rules and Definitions

The FCC's Part 15 rules are found in Title 47 section of the Code of Federal Regulations (CFR). They pertain to unlicensed devices. Here are some of the more important Part 15 rules and definitions pertaining to RFI.

§15.3 Definitions.

(m) *Harmful interference.* Any emission, radiation or induction that endangers the functioning of a radionavigation service or of other safety services or seriously degrades, obstructs or repeatedly interrupts a radio communications service operating in accordance with this chapter.

(n) *Incidental radiator.* A device that generates radio frequency energy during the course of its operation although the device is not intentionally designed to generate or emit radio frequency energy.

(o) *Intentional radiator.* A device that intentionally generates and emits radio frequency energy by radiation or induction.

(z) *Unintentional radiator.* A device that intentionally generates radio frequency energy for use within the device, or that sends radio frequency signals by conduction to associated equipment via connecting wiring, but which is not intended to emit RF energy by radiation or induction.

(t) *Power line carrier systems.* An unintentional radiator employed as a carrier current system used by an electric power utility entity on transmission lines for protective relaying, telemetry, etc. for general supervision of the power system. The system operates by the transmission of radio frequency energy by conduction over the electric power transmission lines of the system. The system does not include those electric lines which connect the distribution substation to the customer or house wiring.

(ff) *Access Broadband over Power Line (Access BPL).* A carrier current system installed and operated on an electric utility service as an unintentional radiator that sends radio frequency energy on frequencies between 1.705 MHz and 80 MHz over medium voltage lines or over low voltage lines to provide broadband communications and is located on the supply side of the utility service's points of interconnection with customer premises. Access BPL does not include power line carrier systems as defined in §15.3(t) or In-House BPL as defined in §15.3(gg).

(gg) *In-House Broadband over Power Line (In-House BPL).* A carrier current system, operating as an unintentional radiator, that sends radio frequency energy by conduction over electric power lines that are not owned, operated or controlled by an electric service provider. The electric power lines may be aerial (overhead), underground, or inside the walls, floors or ceilings of user premises. In-House BPL devices may establish closed networks within a user's premises or provide connections to Access BPL networks, or both.

Some of the most important Part 15 rules pertaining to radio and television interference from unintentional and incidental radiators include:

§15.5 General conditions of operation.

(b) Operation of an intentional, unintentional, or incidental radiator is subject to the conditions that no harmful interference is caused and that interference must be accepted that may be caused by the operation of an authorized radio station, by another intentional or unintentional radiator, by industrial, scientific and medical (ISM) equipment, or by an incidental radiator.

(c) The operator of the radio frequency device shall be required to cease operating the device upon notification by a Commission representative that the device is causing harmful interference. Operation shall not resume until the condition causing the harmful interference has been corrected.

§15.13 Incidental radiators.

Manufacturers of these devices shall employ good engineering practices to minimize the risk of harmful interference.

§15.15 General technical requirements.

(c) Parties responsible for equipment compliance should note that the limits specified in this part will not prevent harmful interference under all circumstances. Since the operators of Part 15 devices are required to cease operation should harmful interference occur to authorized users of the radio frequency spectrum, the parties responsible for equipment compliance are encouraged to employ the minimum field strength necessary for communications, to provide greater attenuation of unwanted emissions than required by these regulations, and to advise the user as to how to resolve harmful interference problems (for example, see Sec. 15.105(b)).

§15.19 Labeling requirements.

(a) In addition to the requirements in part 2 of this chapter, a device subject to certification, notification, or verification shall be labeled as follows:

(3) All other devices shall bear the following statement in a conspicuous location on the device:

This device complies with part 15 of the FCC Rules. Operation is subject to the following two conditions: (1) This device may not cause harmful interference, and (2) this device must accept any interference received, including interference that may cause undesired operation.

And the following requirements apply for consumer and residential Class B digital devices. Different requirements apply for Class A digital devices which can only be used in industrial and similar environments:

§15.105 Information to the user.

(b) For a Class B digital device or peripheral, the instructions furnished the user shall include the following or similar statement, placed in a prominent location in the text of the manual:

This equipment has been tested and found to comply with the limits for a Class B digital device, pursuant to part 15 of the FCC Rules. These limits are designed to provide reasonable protection against harmful interference in a residential installation. This equipment generates, uses and can radiate radio frequency energy and, if not installed and used in accordance with the instructions, may cause harmful interference to radio communications. However, there is no guarantee that interference will not occur in a particular installation. If this equipment does cause harmful interference to radio or television reception, which can be determined by turning the equipment off and on, the user is encouraged to try to correct the interference by one or more of the following measures:

Reorient or relocate the receiving antenna.

Increase the separation between the equipment and receiver.

Connect the equipment into an outlet on a circuit different from that to which the receiver is connected.

Consult the dealer or an experienced radio/TV technician for help.

devices or as text in the device's operating manual. This information attests to the potential for interference and to the responsibility of the device operator. It must be placed in a conspicuous location on the device or in the manual and contain the following statement:

This device complies with part 15 of the FCC Rules. Operation is subject to the following two conditions: (1) This device may not cause harmful interference, and (2) this device must accept any interference received, including interference that may cause undesired operation.

Owners of Part 15 devices are frequently unaware of this information and surprised to find it on the device or in its manual. Reading this label can be an important step in resolving RFI issues. Additional details regarding labeling requirements can be found in the sidebar on Part 15 Rules.

EQUIPMENT AUTHORIZATION

FCC regulations do not require Part 15 devices to be tested by the FCC. In fact, very few devices must actually undergo FCC testing. In most cases, the requirements are met by the manufacturer testing the device and the test results either kept on file or sent to the FCC, depending on the type of device involved. Here is some general information concerning various FCC approval processes for RF devices:

• *Certification:* Requires submittal of an application that includes a complete technical description of the product and a measurement report showing compliance with the FCC technical standards. Certification procedures have now largely replaced the once familiar Type Acceptance, which is no longer used by the FCC. Devices subject to certification include: low-power transmitters such as cordless telephones, security alarm systems, scanning receivers, super-regenerative receivers, Amateur Radio external HF amplifiers and amplifier kits, and TV interface devices such as DVD players.

• *Declaration of Conformity (DoC)*: Is a declaration that the equipment complies with FCC requirements. A DoC is an alternative to certification since no application to FCC is required, but the applicant must have the device tested at an accredited laboratory. A Declaration of Conformity is the usual approval procedure for Class B personal computers and personal computer peripherals.

• *Notification*: Requires submittal to the FCC of an abbreviated application for equipment authorization which does not include a measurement report. However, a measurement report showing compliance of the product with the FCC technical standards must be retained by the applicant and must be submitted upon request by the Commission. Devices subject to notification include: point-to-point microwave transmitters, AM, FM and TV broadcast transmitters and other receivers (except as noted elsewhere).

• *Verification*: Verification is a self-approval process where the applicant performs the necessary tests and verifies that they have been done on the device to be authorized and that the device is in compliance with the technical standards. Verified equipment requires that a compliance label be affixed to the device as well as information included in the operating manual regarding the interference potential of the device. Devices subject to verification include: business computer equipment (Class A); TV and FM receivers; and non-consumer Industrial, Scientific and Medical Equipment.

PART 15 OPERATOR REQUIREMENTS

All Part 15 devices are prohibited from causing harmful interference to a licensed radio service — including the Amateur Radio Service. This is an absolute requirement without regard to the emitter type or a manufacturer's conformance to emissions limits or other FCC technical standards. It is important to note that the manufacturer's requirements are not sufficient to prevent harmful interference from occurring under all circumstances. If and when a Part 15 device generates harmful interference, it becomes the responsibility of the device operator to correct the problem. Upon notice from the FCC, the device operator may also be required to cease using the device until such time as the interference has been corrected.

27.3 Elements of RFI

27.3.1 Source-Path-Victim

All cases of RFI involve a *source* of radio frequency energy, a device that responds to the electromagnetic energy (*victim*), and a transmission *path* that allows energy to flow from the source to the victim. Sources include radio transmitters, receiver local oscillators, computing devices, electrical noise, lightning and other natural sources. Note that receiving unwanted electromagnetic energy does not necessarily cause the victim to function improperly.

A device is said to be *immune* to a specific source if it functions properly in the presence of electromagnetic energy from the source. In fact, designing devices for various levels of immunity is one aspect of electromagnetic compatibility engineering. Only when the victim experiences a *disturbance* in its function as a consequence of the received electromagnetic energy does RFI exist. In this case, the victim device is *susceptible* to RFI from that source.

There are several ways that RFI can travel from the source to the victim: *radiation*, *conduction*, *inductive coupling* and *capacitive coupling*. *Radiated RFI* propagates by electromagnetic radiation from the source through space to the victim. *Conducted RFI* travels over a physical conducting path between the source and the victim, such as wires, enclosures, ground planes, and so forth. Inductive coupling occurs when two circuits are magnetically coupled. Capacitive coupling occurs when two circuits are coupled electrically through capacitance. Typical RFI problems you are likely to encounter often include multiple paths, such as conduction and radiation. (See the section Shields and Filters, also Ott, sections 2.1-2.3.)

27.3.2 Differential-Mode vs Common-Mode Signals

The path from source to victim almost always includes some conducting portion, such as wires or cables. RF energy can be conducted directly from source to victim, be conducted onto a wire or cable that acts as an antenna where it is radiated, or be picked up by a conductor connected to the victim that acts like an antenna. When the noise signal is traveling along the conducted portion of the path, it is important to understand the differences between *differential-mode* and *common-mode* conducted signals (see **Fig 27.3**).

Differential-mode currents usually have two easily identified conductors. In a two-wire transmission line, for example, the signal leaves the generator on one wire and returns on the other. When the two conductors are in close proximity, they form a transmission line and the two signals have opposite polarities as shown in Fig 27.3A. Most desired signals, such as the TV signal inside a coaxial cable or an Ethernet signal carried on CAT5 network cable, are conducted as differential-mode signals.

A common-mode circuit consists of several wires in a multi-wire cable acting as if they were a single current path as in Fig 27.3B. Common-mode circuits also exist when the outside surface of a cable's shield acts as a conductor as in Figs 27.3C and 27.3D. (See the chapter on **Transmission Lines** for a discussion about isolation between the shield's inner and outer surfaces for RF signals.) The return path for a common-mode signal often involves Earth ground.

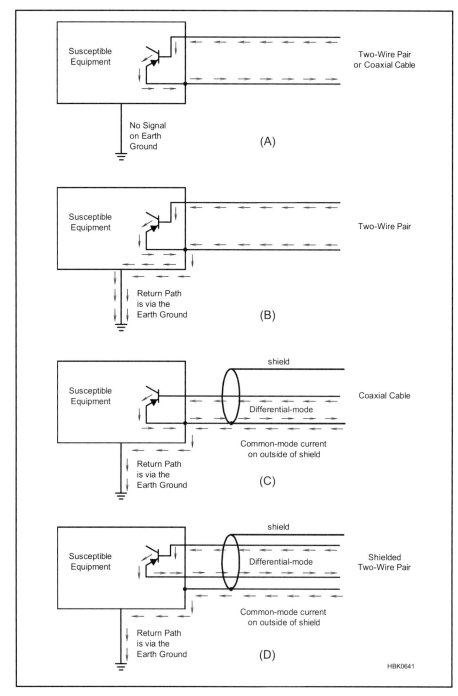

Fig 27.3 — Typical configurations of common-mode and differential-mode current. The drawing in A shows the currents of a differential-mode signal while B shows a common-mode signal with currents flowing equally on all of the source wires. In C, a common-mode signal flows on the outside of a coaxial cable shield with a differential-mode signal inside the cable. In D, the differential-mode signal flows on the internal wires while a common-mode signal flows on the outside of the cable shield.

27.3.3 Types of RFI

There are four basic types of RFI that apply to Amateur Radio. The first two occur in the following order of likelihood when the interfering source is an amateur transmitter intentionally generating a radio signal:

1) Fundamental Overload — Disruption or degradation of a device's function in the presence of a transmitter's fundamental signal (the intended signal from the transmitter).

2) Spurious Emissions — Reception of a radio signal interfered with by spurious emissions from the transmitter as defined in the previous section on Part 97 definitions and Fig 27.1.

The second two types of RFI occur, again in order of likelihood, when the reception of a desired signal is interfered with by RF energy received along with the desired signal.

3) External Noise Sources — Reception of a radio signal interfered with by RF energy transmitted incidentally or unintentionally by a device that is not a licensed transmitter

4) Intermodulation — Reception of a radio signal interfered with by intermodulation distortion (IMD) products generated inside or outside of the receiver

As an RFI troubleshooter, start by determining which of these is involved in your interference problem. Once you know the type of RFI, selecting the most appropriate cure for the problem becomes much easier.

27.3.4 Spurious Emissions

All transmitters generate some (hopefully few) RF signals that are outside their intended transmission bandwidth — out-of-band emissions and spurious emissions as illustrated in Fig 27.1. Out-of-band signals result from distortion in the modulation process or consist of broadband noise generated by the transmitter's oscillators that is added to the intended signal. Harmonics, the most common spurious emissions, are signals at integer multiples of the operating (or fundamental) frequency.

Transmitters may also produce broadband noise and/or parasitic oscillations as spurious emissions. (Parasitic oscillations are discussed in the **RF Power Amplifiers** chapter.) Overdriving an amplifier often creates spurious emissions. Amplifiers not meeting FCC certification standards but sold illegally are frequent sources of spurious emissions.

Regardless of how the unwanted signals are created, if they cause interference, FCC regulations require the operator of the transmitter to correct the problem. The usual cure is to adjust or repair the transmitter or use filters at the transmitter output to block the spurious emissions from being radiated by the antenna.

27.3.5 Fundamental Overload

Most cases of interference caused by an amateur transmission are due to *fundamental overload*. (See the sidebar "Fundamental Overload of Radio Receivers.") The world is filled with RF signals. Properly designed radio receivers of any sort should be able to select the desired signal, while rejecting all others. Unfortunately, because of design deficiencies such as inadequate shields or filters, some radio receivers are unable to reject strong out-of-band signals. Electronic equipment that is not a radio receiver can also suffer from fundamental overload from similar design shortcomings. Both types of fundamental overload are common in consumer electronics.

Fundamental Overload of Radio Receivers

When used in a discussion of radio (and TV) receiver performance, "fundamental overload" specifically refers to the effects of a signal at the receiver input that is too strong for the receiver circuits to process properly. In other words, the term means "overload of a radio receiver due to the strength of a transmitted signal's fundamental or intended component." For example, in a superheterodyne receiver, fundamental overload would be caused by a signal that drives an input amplifier stage into clipping or cutoff or both.

Reducing the level of the offending strong signal returns the receiver to normal operation and causes the undesirable effects to disappear. An attenuator is often a weapon of choice when encountering this type of problem. This type of fundamental overload is discussed in the **Receivers** chapter.

A strong signal can enter equipment in several different ways. Most commonly, it is conducted into the equipment by connecting wires and cables. Possible RFI conductors include antennas and feed lines, interconnecting cables, power cords, and ground wires. TV antennas and feed lines, telephone or speaker wiring and ac power cords are the most common points of entry.

If the problem is a case of fundamental overload, significant improvement can often be observed just by moving the victim equipment and the signal source farther away from each other. The effect of an interfering signal is directly related to its strength, diminishing with the square of the distance from the source. If the distance from the source doubles, the strength of the electromagnetic field decreases to one-fourth of its power density at the original distance from the source. This characteristic can often be used to help identify an RFI problem as fundamental overload. If reducing the strength of the signal source causes the same effect that is also a signature of fundamental overload.

27.3.6 External Noise Sources

Most cases of interference to the Amateur Service reported to the FCC are eventually determined to involve some sort of external noise source, rather than signals from a radio transmitter. Noise in this sense means an RF signal that is not essential to the generating device's operation. The most common external noise sources are electrical, primarily power lines. Motors and switching equipment can also generate electrical noise.

External noise can also come from unlicensed Part 15 RF sources such as computers and networking equipment, video games, appliances, and other types of consumer electronics. Regardless of the source, if you determine the problem to be caused by external noise, elimination of the noise must take place at the source. As an alternative, several manufacturers also make noise canceling devices that can help in some circumstances.

27.3.7 Intermodulation Distortion

As discussed in the chapter on **Receivers**, intermodulation distortion (IMD) is caused by two signals combining in such a way as to create *intermodulation products* — signals at various combinations of the two original frequencies. The two original signals may be perfectly legal, but the resulting *intermodulation distortion products* may occur on the frequencies used by other signals and cause interference in the same way as a spurious signal from a transmitter. Depending on the nature of the generating signals, "intermod" can be intermittent or continuous. IMD can be generated inside a receiver by large signals or externally by signals mixing together in non-linear junctions or connections.

27.3.8 Ground Connections

An electrical ground is not a huge sink that somehow swallows noise and unwanted signals. Ground is a *circuit* concept, whether the circuit is small, like a radio receiver, or large, like the propagation path between a transmitter and cable-TV installation. Ground forms a universal reference point between circuits.

While grounding is not a cure-all for RFI problems, ground is an important safety component of any electronics installation. It is part of the lightning protection system in your station and a critical safety component of your house wiring. Any changes made to a grounding system must not compromise these important safety considerations. Refer to the **Safety** chapter for important information about grounding.

Many amateur stations have several connections referred to as "grounds"; the required safety ground that is part of the ac wiring system, another required connection to the Earth for lightning protection, and perhaps another shared connection between equipment for RFI control. These connections can interact with each other in ways that are difficult to predict.

Rearranging the station ground connections may cure some RFI problems in the station by changing the RF current distribution so that the affected equipment is at a low-impedance point and away from RF "hot spots." Creating a low-impedance connection

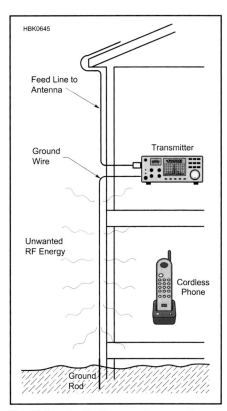

Fig 27.4 — An Earth ground connection can radiate signals that might cause RFI to nearby equipment. This can happen if the ground connection is part of an antenna system or if it is connected to a coaxial feed line carrying RF current on the outside of the shield.

between your station's equipment is easy to do and will help reduce voltage differences (and current flow) between pieces of equipment. However, a station ground is not always the cure-all that some literature has suggested.

LENGTH OF GROUND CONNECTIONS

The required ground connection for lightning protection between the station equipment and an outside ground rod is at least several feet long in most practical installations. (See the **Safety** chapter for safety and lightning protection ground requirements.)

In general, however, a long connection to Earth should be considered as part of an RFI problem only to the extent that it is part of the antenna system. For example, should a longwire HF antenna end in the station, a ground connection of *any* length is a necessary and useful part of that antenna and will radiate RF.

At VHF a ground wire can be several wavelengths long — a very effective antenna for any harmonics that could cause RFI! For example, in **Fig 27.4**, signals radiated from the required safety ground wire could very easily create an interference problem in the downstairs electronic equipment.

GROUND LOOPS

A *ground loop* is created by a continuous conductive path around a series of equipment enclosures. While this does create an opportunity for lightning and RFI susceptibility, the ground loop itself is rarely a cause of problems at RF. Ground loops are usually associated with problems of audio hum or buzz caused by coupling to power-frequency magnetic fields and currents. To avoid low-frequency ground-loop issues, use short, properly-shielded cables that are the minimum length required to connect the equipment and bundle them together to minimize the area of any enclosed loop. Ground-loop problems at RF are minimized by the use of a single-point or star ground system as shown in **Fig 27.5**.

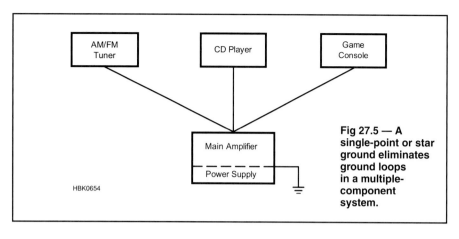

Fig 27.5 — A single-point or star ground eliminates ground loops in a multiple-component system.

27.4 Identifying the Type of RFI Source

It is useful to place an offending noise source into one of several broad categories at the early stages of any RFI investigation. Since locating and resolution techniques can vary somewhat for each type, the process of locating and resolving RFI problems should begin with identifying the general type of RFI source.

It is often impossible to identify the exact type of device generating the RFI from the sound of the interference. Because there are many potential sources of RFI, it is often more important to obtain and interpret clues from the general noise characteristics and the patterns in which it appears.

A source that exhibits a repeatable pattern during the course of a day or week, for example, suggests something associated with human activity. A sound that varies with or is affected by weather suggests an outdoor source. Noise that occurs in a regular and repeating pattern of peaks and nulls as you tune across the spectrum, every 50 kHz for example, is often associated with switchmode power supply or similar pulsed-current devices. A source that exhibits fading or other sky wave characteristics suggests something that is not local. A good ear and careful attention to detail will often turn up some important clues. A detailed RFI log can often help, especially if maintained over time.

Noise can be characterized as broadband or narrowband — another important clue. *Broadband noise* is defined as noise having a bandwidth much greater than the affected receiver's operating bandwidth and is reasonably uniform across a wide frequency range. Noise from arcs and sparks, such as power-line noise, tend to be broadband. *Narrowband noise* is defined as noise having a bandwidth less than the affected receiver's bandwidth. Narrowband noise is present on specific, discrete frequencies or groups of frequencies, with or without additional modulation. In other words, if you listened to the noise on an SSB receiver, tuning would cause its sound to vary, just like a regular signal. Narrowband noise often sounds like an unmodulated carrier with a frequency that may drift or suddenly change. Microprocessor clock harmonics, oscillators and transmitter harmonics are all examples of narrowband noise.

27.4.1 Identifying Noise from Part 15 Devices

The most common RFI problem reported to the ARRL comes from an unknown and unidentified source. Part 15 devices and other consumer equipment noise sources are ubiquitous. Although the absolute signal level from an individual noise source may be small, their increasing numbers makes this type of noise a serious problem in many suburban and urban areas. The following paragraphs describe several common types of electronic noise sources.

Electronic devices containing oscillators, microprocessors, or digital circuitry produce RF signals as a byproduct of their operation. The RF noise they produce may be radiated from internal wiring as a result of poor shielding. The noise may also be conducted to external, unshielded or improperly shielded wiring as a common-mode signal where it radiates noise. Noise from these devices is usually narrowband that changes characteristics (frequency, modulation, on-off pattern) as the device is used in different ways.

Another major class of noise source is equipment or systems that control or switch large currents. Among them are variable-speed motors in products as diverse as washing machines, elevators, and heating and cooling systems. Charging regulators and control circuitry for battery and solar power systems are a prolific source of RF noise. So are switchmode power supplies for computers and low-voltage lighting. This type of noise is only present when the equipment is operating.

Switchmode supplies, solar controllers and inverters often produce noise signals every N kHz, with N typically being from 5 to 50 or more kHz, the frequency at which current is switched. This is different from noise produced by spark or arc sources that is uniform across a wide bandwidth. This pattern is often an important clue in distinguishing switching noise from power-line or electrical noise.

Wired computer networks radiate noise directly from their unshielded circuitry and from network and power supply cables. The noise takes two forms — broadband noise and modulated carriers at multiple frequencies within the amateur bands. As an example, Ethernet network interfaces often radiate signals heard on a receiver in CW mode. 10.120, 14.030, 21.052 and 28.016 MHz have been reported as frequencies of RFI from Ethernet networks. Each network interface uses its own clock, so if you have neighbors with networks you'll hear a cluster of carriers around these frequencies, ±500 Hz or so.

In cable TV systems video signals are converted to RF across a wide spectrum and distributed by coaxial cable into the home. Some cable channels overlap with amateur bands, but the signals should be confined within the cable system. No system is perfect, and it is common for a defective coax connection to allow leakage to and from the cable. When this happens, a receiver outside the cable will hear RF from the cable and the TV receiver may experience interference from local transmissions. Interference to and from cable TV signals is discussed in detail later in this chapter.

27.4.2 Identifying Power-line and Electrical Noise

POWER-LINE NOISE

Next to external noise from an unknown source, the most frequent cause of an RFI problem reported to the ARRL involving a known source is power-line noise. (For more information on power-line noise, see the book *AC Power Interference Handbook*, by Marv Loftness KB7KK.) Virtually all power-line noise originating from utility equipment is caused by spark or arcing across some hardware connected to or near a power line. A breakdown and ionization of air occurs and current flows across a gap between two conductors, creating RF noise as shown in **Fig 27.6**. Such noise is often referred to as "gap noise" in the utility power industry. The gap may be caused by broken, improperly installed or loose hardware. Typical culprits include insufficient and inadequate hardware spacing such as a gap between a ground wire and a staple. Contrary to common misconception, corona discharge is rarely, if ever, a source of power-line noise.

While there may not be one single conclusive test for power-line noise, there are a number of important tell-tale signs. On an AM or SSB receiver, the characteristic raspy buzz or frying sound, sometimes changing in intensity as the arc or spark sputters a bit, is often the first and most obvious clue.

Power-line noise is typically a broadband type of interference, relatively constant across a wide spectrum. Since it is broadband noise, you simply can't change frequency to eliminate it. Power-line noise is usually, but not always, stronger on lower frequencies. It occurs continuously across each band and up through the spectrum. It can cause interference from the AM broadcast band through UHF, gradually tapering off as frequency increases. If the noise is not continuous across all of an amateur band, exhibits a pattern of peaks and nulls at different frequencies, or repeats at some frequency interval, you probably do not have power-line noise.

The frequency at which power-line noise diminishes can also provide an important clue as to its proximity. The closer the source, the higher the frequency at which it can be received. If it affects VHF and UHF, the source is relatively close by. If it drops off just above or within the AM broadcast band, it may be located some distance away — up to several miles.

Power-line noise is often affected by weather if the source is outdoors. It frequently changes during rain or humid conditions, for example, either increasing or decreasing in response to moisture. Wind may also create fluctuations or interruptions as a result of line and hardware movement. Temperature effects can also result from thermal expansion and contraction.

Another good test for power-line noise requires an oscilloscope. Remember that power-line noise occurs in bursts and usually at a rate of 120 bursts per second. Observe the suspect noise from your radio's audio output. (Note: The record output jack works best if available). Use the AM mode with wide filter settings and tune to a frequency without a station so the noise can be heard clearly. Use the LINE setting of the oscilloscope's trigger subsystem to synchronize the sweep to the line. Power-line noise bursts will remain stable on the display and should repeat every 8.33 ms (a 120-Hz repetition rate) or less commonly, 16.67 ms (60 Hz), if the gap is only arcing once per cycle. (This assumes the North American power-line frequency of 60 Hz.) See **Fig 27.7** for an explanation. If a noise does not exhibit either of these characteristics, it is probably

---

Keeping an RFI Log

The importance of maintaining a good and accurate RFI log cannot be overstated. Be sure to record time and weather conditions. Correlating the presence of the noise with periods of human activity and weather often provide very important clues when trying to identify power-line noise. It can also be helpful in identifying noise that is being propagated to your station via sky wave. A log showing the history of the noise can also be of great value should professional services or FCC involvement become necessary at some point.

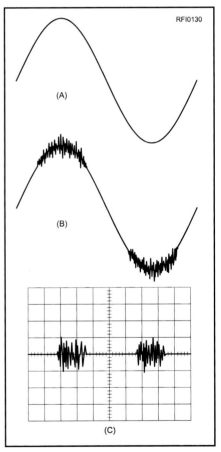

Fig 27.7 — The 60-Hz signal found on quiet power lines is almost a pure sine wave, as shown in A. If the line, or a device connected to it, is noisy, this will often add visible noise to the power-line signal, as shown in B. This noise is usually strongest at the positive and negative peaks of the sine wave where line voltage is highest. If the radiated noise is observed on an oscilloscope, the noise will be present during the peaks, as shown in C.

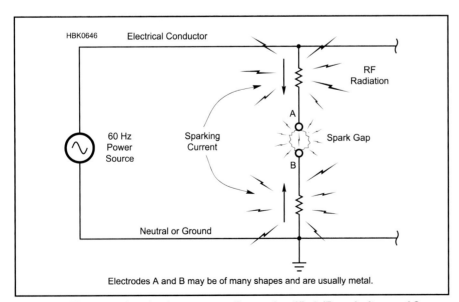

Fig 27.6 — The gap noise circuit on a power line — simplified. (From Loftness, *AC Power Interference Handbook*)

not power-line noise.

If a local TV station is transmitting analog TV signals on a lower VHF channel (very few remain as of mid-2010), additional clues may be obtained by viewing the noise pattern on an analog TV set using an antenna (not a cable TV connection). Power-line noise usually appears as two horizontal bars that drift slowly upward on the screen through the picture. (This is due to the difference between the NTSC signal's 59.94 Hz field rate and the 60 Hz power-line frequency.) As one bar rolls off the screen at the top of the display, a new one simultaneously forms at the bottom. In cases where the noise is occurring at 60 bursts per second, there will be only one bar on the display. In addition, the power-line noise bursts may have slightly different characteristics at the positive and negative peaks. This can cause each half of the cycle to have a slightly different pattern on the screen.

ELECTRICAL NOISE

Electrical noise sounds like power-line noise, but is generally only present in short bursts or during periods when the generating equipment or machinery is in use. Noise that varies with the time of day, such as daytime-only or weekends-only, usually indicates some electrical device or appliance being used on a regular basis and not power-line noise. Unless it is associated with climate control or HVAC system, an indoor RFI source of electrical noise less likely to be affected by weather than power-line noise.

ELECTRIC FENCES

A special type of electrical noise that is easy to identify is the "pop…pop…pop" of an electric fence. High voltage is applied to the fence about once a second by a charging unit. Arcs will occur at corroded connections in the fence, such as at a gate hook or splice. If brush or weeds touch the fence, the high-voltage will cause an arc at those points until the vegetation burns away (the arc will return when the vegetation re-grows). Each arc results in a short burst of broadband noise, received as a "tick" or "pop" in an HF receiver.

27.4.3 Identifying Intermodulation Distortion

Intermodulation distortion (IMD) often occurs within a receiver when two strong signals combine to produce intermodulation products that can interfere with a desired signal. Since the products are generated internally to the receiver, the strong signals must be filtered out or attenuated before they can enter the receiver circuits in which IMD products are generated.

Mixing of signals can also occur in any non-linear junction where the original signals are both strong enough to cause current to flow in the junction. This is a particular problem at multi-transmitter sites, such as broadcast facilities and industrial or commercial communications sites. Non-linear junctions can be formed by loose mechanical contacts in metal hardware, corroded metal junctions, and by semiconductor junctions connected to wires that act as antennas for the strong signals. Non-linear junctions also detect or demodulate RF signals to varying degrees, creating interfering audio or dc signals in some cases. IMD products generated externally cannot be filtered out at the receiver and must be eliminated where the original signals are being combined.

An intermodulation product generated externally to a receiver often appears as an intermittent transmission, similar to a spurious emission. (See the **Receivers** chapter for more information on intermodulation.) It is common for strong signals from commercial paging or dispatch transmitters sharing a common antenna installation to combine and generate short bursts of voice or data signals. AM broadcast transmissions can combine to produce AM signals with the modulation from both stations audible.

Like external intermodulation products, those generated by a receiver acting non-linearly appear as combinations of two or more strong external signals. For example, intermodulation from two SSB or FM voice signals produces somewhat distorted signals with the modulating signals of both stations. Since the two signals are not synchronized, the intermodulation products come and go unpredictably, present only when both of the external signals are present.

Intermodulation within a receiver can often be detected simply by activating the receiver's incoming signal attenuator. If attenuating the incoming signals causes the intermodulation product to by reduced in strength by a greater amount than that of the applied attenuation, intermodulation in the receiver is a strong possibility. Since receivers vary in their IMD performance, differences in the interfering signal strength between receivers is also an indication of intermodulation.

The following simple "attenuator test" can be used to identify an IMD product, even in cases where it appears similarly in multiple receivers:

• If your receiver does not have one, install a step attenuator at its RF input. If you use an attenuator internal to your receiver, it must attenuate the RF at the receiver's input.

• Tune the receiver to the suspected intermod product with the step attenuator set to 0 dB. Note the signal level.

• Add a known amount of attenuation to the signal. Typically 10 or 20 dB makes a good starting point for this test.

• If the suspect signal drops by more than the amount of added attenuation, the suspect signal is an IMD product. For example, if you add 10 dB of attenuation, and the signal drops by 30 dB, you have identified an IMD product.

• You can also compare the reduction in signal level between the suspect IMD product and a known genuine signal with and without the added attenuation. Use a known genuine signal that is about the same strength as the suspect signal with no added attenuation. If the suspect signal drops by the same as the added attenuation, it is not an IMD product.

Intermodulation distortion can be cured in a number of ways. The goal is to reduce the strength of the signals causing IMD so that the receiver circuits can process them linearly and without distortion.

If IMD is occurring in your receiver, filters that remove the strong unwanted signals causing the IMD while passing the desired signal are generally the best approach since they do not compromise receiver sensitivity. (The chapter on **Receivers** discusses how to add additional filtering to your receiver.) Turning off preamplifiers, adding or increasing the receiver's attenuation, and reducing its RF gain will reduce the signal strength in your receiver. Antenna tuners and external band-pass filters can also act as a filter to reduce IMD from out-of-band signals. Directional beams and antennas with a narrower bandwidth can also help, depending on the circumstances of your particular problem.

IMD can also be created in a broadcast FM or TV receiver or preamplifier. The solutions are the same — add suitable filters or reduce overall signal levels and gain so that the strong interfering signal can be processed linearly.

27.5 Locating Sources of RFI

Locating an offending device or noise source might sometimes seem like trying to find a needle in a haystack. With a little patience and know-how, it is often possible to find the source of a problem in relatively short order. RF detective work is often required and some cases require a little more perseverance than others. In any case, armed with some background and technique, it is often easier to find an offending source than the first-time RFI investigator might expect.

Whenever an unknown source of interference becomes an issue, begin the process of identifying the source by verifying that the problem is external to your radio. Start by removing the antenna connection. If the noise disappears, the source is external to your radio and you are ready to begin hunting for the noise source.

27.5.1 Noise Sources Inside Your Home

Professional RFI investigators and the experiences of the ARRL RFI desk confirm that most RFI sources are ultimately found to be in the complainant's home. Furthermore, locating an in-house source of RFI is so simple it that it makes sense to start an investigation by simply turning off your home's main circuit breaker while listening to the noise with a battery-powered portable radio. (Don't forget that battery-powered equipment may also be a noise source — remove the batteries from consumer devices, as well.) If the noise goes away, you know the source is in your residence. After resetting the breaker, you can further isolate the source by turning off individual breakers one at a time. Once you know the source circuit, you can then unplug devices on that circuit to find it.

CAUTION: *Do not attempt to remove cartridge fuses or operate exposed or open-type disconnects if it is possible to make physical contact with exposed electrical circuits.*

27.5.2 Noise Sources Outside Your Home

It is often possible to locate a noise source outside your home with a minimum of equipment and effort. Because of Part 15's absolute emissions limits, most Part 15 noise sources are within a few hundred feet of the complainant's antenna. They are also often on the same power transformer secondary system as the complainant. This typically reduces the number of possible residences to relatively few. If the noise source is not compliant with Part 15 limits, it may be blocks or even thousands of feet from your station.

Electrical noise sources in a home, such as an arcing thermostat or a noisy washing machine controller, can also be tracked down in the same way as noise from consumer electronic devices. Electrical noise from an incidental emitter, such as a power line, can propagate much farther than noise from an otherwise legal unintentional emitter. Some Part 15 devices, battery chargers for electric scooters and wheelchairs, for example, are notorious for exceeding Part 15 absolute emissions limits on conducted noise.

The following procedure can be used to trace a noise source to a private home, town house, apartment, or condominium. The number of homes that could be host to a source generating noise could make searching house by house impractical. In such cases, use noise tracking techniques discussed in the following sections to narrow the search to a more reasonable area.

1) Verify that the noise is active before attempting to locate it. Don't forget this all-important first step. You cannot find the source when it's not present.

2) If possible, use a beam to record bearings to the noise before leaving your residence. Walk or drive through the neighborhood with particular emphasis in the direction of the noise, if known. Try to determine the rough geographic area over which the noise can be heard. If the geographic area over which you can hear the noise is confined to a radius of several hundred feet or less, or it diminishes quickly as you leave your neighborhood, this confirms you are most likely dealing with a Part 15 consumer device.

3) Since the noise will be strongest at an electrical device connected to the residence containing the source, you want to measure the noise at a device common to the exterior of all the potential homes. Suitable devices include electric meters, main service breakers (whether outside or in a utility room), front porch lights, electric lamp posts, outside air conditioner units, or doorbell buttons. Whatever radiator you choose, it should be accessible at each home. The device you select to test as the noise radiator will be referred to in these instructions as the "radiator." Using the same type of device as a test point at each home helps obtain consistent results.

4) You are now ready to compare the relative signal strengths at the radiator on each of the potential source residences. Use a detector suitable to receive the noise, typically a battery-powered receiver. Preferably, the receiver should have a variable RF gain control. An external step attenuator will also work if the antenna is external to the radio. If the antenna can be removed, a probe can also be made from a small piece of wire or paper clip to reduce the receiver's sensitivity. Start by holding the detector about two inches from the radiator at the residence where the noise source may be located. Turn the detector's RF gain control down to a point where you can just barely hear the noise. Alternately, increase the attenuation if using an external step attenuator. Record the RF gain or attenuator setting for each test.

5) Proceed to the next residence. Again hold the detector approximately two inches from the radiator. (The detector should be placed at the same location at each residence, as much as is practical.) Since you had previously set the detector to just barely hear the noise at the residence having the interference problem, you can move on to the next residence if you do not hear the noise. Remember, in order for your detector to hear the noise at then next house, the noise level will have to be the same or higher than the previous location. If you need to increase the detector's sensitivity to be able to hear the noise, you are moving away from the noise source.

6) When you reach the next residence, if the level is lower or not heard, you're moving further from the source. Continue your search to residences in other directions or across the street. If the level is higher, then you're headed in the right direction. Be sure to turn the gain control down to the point of just barely hearing the noise as its strength increases.

7) Continue on to the next house, repeating the previous steps as necessary. The residence with the source will be the one with the strongest noise at the radiator.

Depending on the circumstances of a particular situation, it may be possible to first isolate the power pole to which the source residence is connected. Walk or drive along the power lines in the affected area while listening to the noise with a battery-powered radio. Continue to decrease the receiver's RF gain as the noise gets louder, thus reducing the area over which you can hear it. Finally, isolate the loudest pole by reducing the RF gain to a point at which you can hear it at only one pole. Once the pole has been isolated, look to see which houses are connected to its transformer. Typically this will reduce the number of potential residences to a very small number.

CAUTION: *Always observe good safety practices! Only qualified people familiar with the hazards of working around energized electrical equipment should inspect power-line or other energized circuitry.*

When attempting to isolate the pole, it is often best to use the highest frequency at which you can hear the noise. Noise can exhibit peaks and nulls along a power line that are a function of its wavelength. Longer wavelengths can therefore make it difficult to pinpoint a particular point along a line. Furthermore, longer wavelength signals typically propagate further along power lines. You can often reduce your search area by simply increasing the frequency at which you look for the noise.

In some cases, tuning upward in frequency can also be used to attenuate noise. This can be especially helpful in cases where your receiver does not have an RF gain control. As mentioned previously, switchmode power supplies typically generate noise that exhibits a regular and repeating pattern of peaks and nulls across the spectrum. While a typical interval might be every 50 kHz or so, the noise will often start to diminish at the highest frequencies. The peaks in some cases might drift over time, but tuning to the highest frequency at which you can hear the noise will often attenuate it enough to help locate it. If the peak drifts however, be sure to keep your receiver set on the peak as you attempt to locate the source.

Under FCC rules, the involved utility is responsible for finding and correcting harmful interference that is being generated by its own equipment. In cases where a utility customer is using an appliance or device that generates noise, the operator of the device is responsible for fixing it — even if the noise is conducted and radiated by the power company's power lines.

27.5.3 Approaching Your Neighbor

Once you identify the source residence and approach your neighbor, the importance of personal diplomacy simply cannot be overstated. The first contact regarding an RFI problem between a ham and a neighbor is often the most important; it is the start of all future relations between the parties. The way you react and behave when you first discuss the problem can set the tone for everything that follows. It is important, therefore, to use a diplomatic path from the very start. A successful outcome can depend upon it!

A self-help guide for the consumer published jointly by the ARRL and the Consumer Electronics Association (CEA) often proves helpful when discussing an interference problem with a neighbor. Entitled *What To Do if You Have an Electronic Interference Problem*, it may be printed and distributed freely. It is available on the ARRL Web site at **www.arrl.org/information-for-the-neighbors-of-hams** and also on the CD-ROM accompanying this book. Be sure to download and print a copy for your neighbor before you approach him or her.

With the noise active and with a copy of the pamphlet handy, approach your neighbor with a radio in hand, preferably an ordinary AM broadcast or short-wave receiver. Let them hear it but not so loud that it will be offensive. Tell them this is the problem you are experiencing and you believe the source may be in their home. Don't suggest what you think the cause is. If you're wrong, it often makes matters worse. Give them the pamphlet and tell them it will only take a minute to determine whether the source is in their home. Most neighbors will agree to help find the source, and if they agree to turn off circuit breakers, it can be found very quickly. Start with the main breaker to verify you have the correct residence, then the individual breakers to find the circuit. The procedure then becomes the same as described for your own residence.

27.5.4 Radio Direction Finding

Radio direction finding (RDF) can be a highly effective method to locate an RFI source although it requires more specialized equipment than other methods. Professional interference investigators almost always use radio direction finding techniques to locate power-line noise sources. See the **Antennas** chapter for more information on direction finding antennas.

A good place to start, whenever possible, is at the affected station. Use an AM receiver, preferably one with a wide IF bandwidth. An RF gain control is particularly helpful but an outboard step attenuator can be a good substitute. If there is a directional beam capable of receiving the noise, use it on highest frequency band at which the noise can be heard using the antenna. If you can hear the noise at VHF or UHF, you'll typically want to use those frequencies for RDF.

Select a frequency at which no other stations or signals are present and the antenna can discern a directional peak in the noise. Rotate the beam as required to get a bearing on the noise, keeping the RF gain at a minimum. Repeat with a complete 360° sweep using the minimum RF gain possible to hear the noise in its loudest direction. Try to decrease the RF gain to a point at which the noise clearly comes from one and only one direction. You can simultaneously increase the AF gain as desired to hear the noise.

Distant sources, including power-line noise, are generally easier to RDF at HF than nearby sources. Whenever possible, it's almost always better to use VHF or UHF when in close proximity to a source. Tracking a source to a specific residence by RDF at HF is sometimes possible. Such factors as balance and geometry of a home's internal wiring, open switch circuits and distance, may cause the residence to appear somewhat as a point source.

If the search is being conducted while mobile or portable, VHF and UHF are typically the easiest and most practical antennas. Small handheld Yagi antennas for 2 meters and 440 MHz are readily available and can serve double duty when operating portable. Many handheld receivers can be configured to receive AM on the VHF bands. Be sure to check your manual for this feature. VHF Aircraft band or "Air band" receivers are also a popular choice since they receive AM signals.

Using RDF to locate an HF noise source while in motion presents significant challenges. Conducted emissions are typical from a consumer device or appliance. In this case, the emissions can be conducted outside the residence and on to the power line. The noise can then propagate along neighborhood distribution lines, which in turn acts as an antenna. The noise can often exhibit confusing peaks and nulls along the line, and if in the vicinity of a power line radiating it, RDF can be extremely difficult, if not impossible. Depending on the circumstances, you could literally be surrounded by the near field of an antenna! You would generally want to stay away from power lines and other potential radiators when searching at HF.

Antennas for HF RDF while walking typically include small loops and ferrite rod antennas. In some cases, a portable AM broadcast radio with a ferrite rod antenna can be used for direction finding. An HF dipole made from a pair of whip antennas may be able to be used to get an approximate bearing toward the noise. Mount the dipole about 12 feet above ground (remembering to watch out for overhead conductors!) and rotate to null out the noise. For all three types of antenna, there will be two nulls in opposite directions. Note the direction of the null. Repeat this procedure from another location then triangulate to determine the bearing to the noise.

27.6 Power-line Noise

This chapter's section "Identifying RFI Source Types" describes power-line noise, its causes, and methods to identify it. Power-line noise is a unique problem in several respects. First and foremost, the offending source is never under your direct control. You can't just simply "turn it off" or unplug the offending device. Nor will the source be under the direct control of a neighbor or someone you are likely to know. In the case of power-line noise, the source is usually operated by a company, municipality, or in some cases, a cooperative. Furthermore, shutting down a power line is obviously not a practical option.

Another unique aspect of power-line noise is that it almost always involves a defect of some sort. The cure for power-line noise is to fix the defect. This is almost always a utility implemented repair and one over which you do not have any direct control.

FCC rules specify that that the operator of a device causing interference is responsible for fixing it. Whenever encountering a power-line noise problem, you will be dealing with a utility and won't have the option of applying a relatively simple technical solution to facilitate a cure, as you would if the device were located in your home. Utilities have a mixed record when it comes to dealing with power-line noise complaints. In some cases, a utility will have a budget, well-trained personnel, and equipment to quickly locate and address the problem. In other cases, however, the utility is simply unable to effectively deal with power-line noise complaints or even denies their equipment can cause RFI.

What does this mean for an amateur with a power-line noise complaint? Utilities can be of any size from large corporations to local cooperatives or city-owned systems. Regardless of the category in which your utility may fall, it must follow Part 15 of the FCC rules. Dealing with a company, coop or municipality, however, as opposed to a device in your home, or a nearby neighbor that you know personally, can present its own set of unique challenges. Multiple parties and individuals are often involved, including an RFI investigator, a line crew and associated management. In some cases, the utility may never have received an RFI complaint before yours.

27.6.1 Before Filing a Complaint

Obviously, before filing a complaint with your local utility, it is important to verify the problem as power-line noise as best as possible and verify that it is not caused by a problem with electrical equipment in your home. Other sources, such as lighting devices and motors, can mimic power-line noise, especially to an untrained ear. Don't overlook these important steps. Attempting to engage your utility in the resolution of an RFI problem can not only waste time but can be embarrassing if the source is right in your own home!

Utilities are not responsible for noise generated by customer-operated devices — *even if the noise is being radiated by the power lines.* They are responsible for fixing only that noise which is being generated by their equipment.

27.6.2 Filing a Complaint

Once you have verified the problem to be power-line noise (see this chapter's section on Identifying Power-line and Electrical Noise) and that it is not coming from a source in your home or a nearby residence, contact your utility's customer service department. In addition to your local phone book, customer service phone numbers are included on most power company Web sites.

It is important to maintain a log during this part of the process. Be sure to record any "help ticket" numbers that may be assigned to your complaint as well as names, dates and a brief description of each conversation you have with electric company personnel. If you identify specific equipment or power poles as a possible noise source, record the address and any identifying numbers on it.

Hopefully, your complaint will be addressed in a timely and professional manner. Once a noise source has been identified, it is up to the utility to repair it within a reasonable period. You and the utility may not agree on what constitutes a reasonable period, but attempt to be patient. If no action is taken after repeated requests, reporting the complaint to the ARRL and requesting assistance may be in order. (Before contacting the ARRL review The Cooperative Agreement, a section of this chapter.)

It is also important to cooperate with utility personnel and treat them with respect. Hostile and inappropriate behavior is almost always counter-productive in these situations. Remember, you want utility and other related personnel to help you — not avoid you. Even if the utility personnel working on your case seem unqualified, hostile behavior has historically never been a particularly good motivator in these situations. In fact, most protracted power-line noise cases reported to the ARRL began with an altercation in the early stages of the resolution process. In no case did it help or expedite correction of the problem.

27.6.3 Techniques for Locating Power-line Noise Sources

Radio direction finding (RDF) techniques typically offer the best and most efficient approach to locating most power-line noise sources. It is the primary method of choice used by professionals. While RDF is usually the most effective method, it also requires some specialized equipment, such as a hand-held beam antenna. Although specialized professional equipment is available for RDF, hams can also use readily available amateur and homebrew equipment successfully. The CD-ROM accompanying this *Handbook* includes some power-line noise locating equipment projects you can build.

Although it is the utility's responsibility to locate a source of noise emanating from its equipment, many companies simply do not possess the necessary expertise or equipment to do so. As a practical matter, many hams have assisted their utility in locating noise sources. In some cases, this can help expedite a speedy resolution.

There is a significant caveat to this approach however. Should you mislead the power company into making unnecessary repairs, they will become frustrated. This expense and time will be added to their repair list. Do not make a guess or suggestions if you don't know what is causing the noise. While some power companies might know less about the locating process than the affected ham, indiscriminate replacement of hardware almost always makes the problem worse. Nonetheless, depending on your level of expertise and the specifics of your situation, you may be able to facilitate a speedy resolution by locating the RFI source for the utility.

27.6.4 Amateur Power-line Noise Locating Equipment

Much of the equipment that an amateur would use to locate power-line noise has previously been described in the "Radio Direction Finding" section. Before discussing how to locate power-line noise sources, here are a few additional equipment guidelines:

Receiver — You'll need a battery-operated portable radio capable of receiving VHF or UHF in the AM mode. Ideally, it should also be capable of receiving HF frequencies, especially if the interference is a problem at HF and not VHF. Some amateurs also use the aircraft band from 108 to 137 MHz. The lower frequencies of this band can sometimes enable an RFI investigator to hear the noise at greater distances than on 2 meters or 70 cm. An RF gain control is essential but an outboard step attenuator can be used as a substitute. A good S-meter is also required.

Attenuator — Even if your receiver has an RF gain control, an additional outboard step attenuator can often be helpful. It can not only minimize the area of a noise search

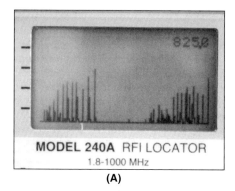
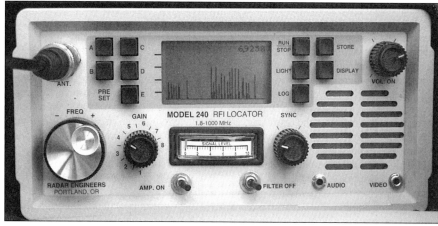

Fig 27.8 — An unknown noise source at the complainant's antenna is shown in A. During the RFI investigation, noise signatures not matching this pattern can be ignored, such as shown in B. Once the matching signature originally shown in A is found, the offending noise source has been located.

but also provide added range for the RF gain control. As with other RFI sources — you'll need to add more and more attenuation as you approach the source.

VHF/UHF Antennas — You'll need a hand-held directional beam antenna. A popular professional noise-locating antenna is an eight-element Yagi tuned for 400 MHz. Since power-line noise is a broadband phenomenon, the exact frequency is not important. Either a 2 meter or 70 cm Yagi are capable of locating a power-line noise source on a specific power pole.

Although professional grade antennas can cost several hundred dollars, some hams can build their own for a lot less. See the CD-ROM accompanying this book for the article, "Adapting a Three-Element Tape Measure Beam for Power-line Noise Hunting," by Jim Hanson, W1TRC. This low cost and easy to build antenna for locating power-line noise can be adapted for a variety of frequencies and receivers. Commercial 2 meter and 70 cm antennas for portable use are also suitable if a handle is added, such as a short length of PVC pipe.

Before using an antenna for power-line noise locating, determine its peak response frequency. Start by aiming the antenna at a known power-line noise source. Tune across its range and just beyond. Using minimum RF gain control, find its peak response. Label the antenna with this frequency using a piece of tape or marking pen. When using this antenna for noise locating, tune the receiver to this peak response frequency.

If you don't have a VHF or UHF receiver that can receive AM signals, see the CD-ROM accompanying this book for the article, "A Simple TRF Receiver for Tracking RFI," by Rick Littlefield, K1BQT. It describes the combination of a simple 136 MHz beam and receiver for portable RFI tracking.

HF Antennas — Depending on the circumstances of a particular case, a mobile HF whip such as a 7 or 14 MHz model can be helpful. Magnet-mount models are acceptable for temporary use. An RFI investigator can typically get within VHF range by observing the relative strength of the noise from different locations. Driving in a circle centered on the affected station will typically indicate the general direction in which the noise is strongest. As with beam antennas, determine the peak response frequency for best results.

Ultrasonic Pinpointer — Although an ultrasonic pinpointer is not necessary to locate the pole or structure containing the source, some hams prefer to go one more step by finding the offending noise source on that structure. Guidelines for the use of an ultrasonic device are described later in this section.

Professional-grade ultrasonic locators are often beyond the budget of the average ham. Home brewing options however, can make a practical ultrasonic locator affordable in most situations — and make a great weekend project too. See the CD-ROM accompanying this book for "A Home-made Ultrasonic Power Line Arc Detector" by Jim Hanson, W1TRC.

Oscilloscope — A battery-powered portable oscilloscope is only required for signature analysis. See the next section, Signature or Fingerprint Method, for details.

Thermal/Infrared Detectors and Corona Cameras — This equipment is not recommended for the sole purpose of locating power-line noise sources. It is rare that an RFI source is even detectable using infrared techniques. Although these are not useful tools for locating noise sources, many utilities still use them for such purposes with minimal or no results. Not surprisingly, ARRL experience has shown that these utilities are typically unable to resolve interference complaints in a timely fashion.

27.6.5 Signature or Fingerprint Method

Each sparking interference source exhibits a unique pattern. By comparing the characteristics between the patterns taken at the affected station with those observed in the field, it becomes possible to conclusively identify the offending source or sources from the many that one might encounter. It therefore isn't surprising that a pattern's unique characteristic is often called its "fingerprint" or "signature." See **Fig 27.8** for an example.

This is a very powerful technique and a real money saver for the utility. Even though there may be several different noise sources in the field, this method helps identify only those sources that are actually causing the interference problem. The utility need only correct the problem(s) matching the pattern of noise affecting your equipment.

You as a ham can use the signature method by observing the noise from your radio's audio output with an oscilloscope. Record the pattern by drawing it on a notepad or taking a photograph of the screen. Take the sketch or photograph with you as you hunt for the source and compare it to signatures you might observe in the field.

Professional interference-locating receivers, such as the Radar Engineers Model 240A shown in Fig 27.8B, have a built in oscilloscope display and waveform memory. This is the preferred method used by professional interference investigators. These receivers provide the ability to switch between the patterns saved at the affected station and those from sources located in the field.

Once armed with the noise fingerprint taken at the affected station, you are ready to begin the hunt. If you have a directional beam, use it to obtain a bearing to the noise. If multiple sources are involved, you'll need to record the bearings to each one. Knowing how high in frequency a particular noise can be heard also provides a clue to its proximity. If the noise can be heard at 440 MHz, for example, the source will typically be within walking distance. If it diminishes beginning 75 or 40 meters, it can be up to several miles away.

Since each noise source will exhibit unique characteristics, you can now match this noise "signature" with one from the many sources you may encounter in the investigation. Compare such characteristics as the duration of each noise burst, pulse shape, and number of pulses.

If you have a non-portable oscilloscope, you may still be able to perform signature matching by using an audio recorder. Make a high quality recording of the noise source at your station and at each suspected noise source in the field, using the same receiver if possible. Replay the sounds for signature analysis.

27.6.6 Locating the Source's Power Pole or Structure

Start your search in front of the affected station. If you can hear the noise at VHF or UHF, begin with a handheld beam suitable for these frequencies. As discussed previously, the longer wavelengths associated with the AM broadcast band and even HF, can create misleading "hotspots" along a line when searching for a noise source as shown in **Fig 27.9**.

As a general rule, only use lower frequencies when you are too far away from the source to hear it at VHF or UHF. Generally work with the highest frequency at which the noise can be heard. As you approach the source, keep increasing the frequency to VHF or UHF, depending on your available antennas. Typically, 2 meters and 70 cm are both suitable for isolating a source down to the pole level.

If you do not have an initial bearing to the noise and are unable to hear it with your portable or mobile equipment, start traveling in a circular pattern around the affected station, block-by-block, street-by-street, until you find the noise pattern matching the one recorded at the affected station.

Once in range of the noise at VHF or UHF, start using a handheld beam. You're well on your way to locating the structure containing the source. In many cases, you can now continue your search on foot. *Again, maintain minimum RF gain to just barely hear the noise over a minimum area. This important step is crucial for success.* If the RF gain is too high, it will be difficult to obtain accurate bearings with the beam.

Power-line noise will often be neither vertically nor horizontally polarized but somewhere in between. Be sure to rotate the beam's polarization for maximum noise response. Maintain this same polarization when comparing poles and other hardware.

27.6.7 Pinpointing the Source on a Pole or Structure

Once the source pole has been identified, the next step becomes pinpointing the offending hardware on that pole. A pair of binoculars on a dark night may reveal visible signs of arcing and in some cases you may be able to see other evidence of the problem from the ground. These cases are rare. More than likely, a better approach will be required. Professional and utility interference investigators typically have two types of specialized equipment for this purpose:

• An ultrasonic dish or pinpointer. The RFI investigator, even if not a lineman, can pinpoint sources on the structure down to a component level from the ground using this instrument. See **Fig 27.10**.

• An investigator can instruct the lineman on the use of a hot stick-mounted device used to find the source. This method is restricted to only qualified utility personnel, typically from a bucket truck. See **Fig 27.11**.

Both methods are similar but hams only have one option — the ultrasonic pinpointer.

Caution — Hot sticks and hot stick mounted devices are not for hams! Do not use them. Proper and safe use of a hot stick requires specialized training. In most localities, it is generally unlawful for anyone unqualified by a utility to come within 10 feet of an energized line or hardware. This includes hot sticks.

ULTRASONIC PINPOINTER TIPS

An ultrasonic dish is the tool of choice for pinpointing the source of an arc from the ground. While no hot stick is required, an unobstructed direct line-of-sight path is re-

Fig 27.10 — The clear plastic parabolic dish is an "ear" connected to an ultrasonic detector that lets utility personnel listen for the sound of arcs.

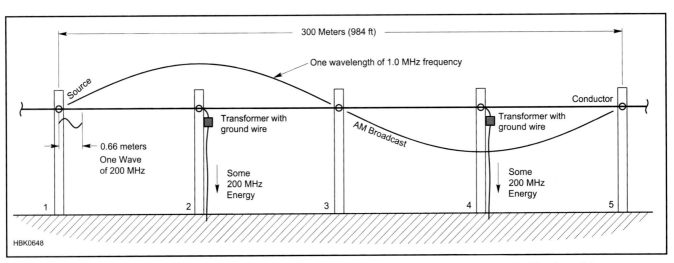

Fig 27.9 — Listening for noise signals on a power distribution line at 1 MHz vs 200 MHz can result in identification of the wrong power pole as the noise source. (From Loftness, *AC Power Interference Handbook*)

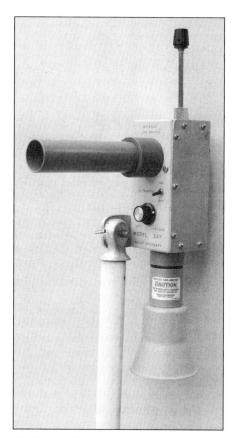

Fig 27.11 — The Radar Engineers Model 247 Hotstick Line Sniffer is an RF and ultrasonic locator. It is used by utility workers to pinpoint the exact piece of hardware causing a noise problem. Mounted on a hotstick, the sniffer is used from the pole or from a bucket.

quired between the arc and the dish. This is not a suitable tool for locating the structure containing the source. It is only useful for pinpointing a source once its pole or structure has been determined.

Caveat: Corona discharge, while typically not a source of RF power-line noise, can and often is a significant source of ultrasonic sound. It can often be difficult to distinguish between the sound created by an arc and corona discharge. This can lead to mistakes when trying to pinpoint the source of an RFI problem with an ultrasonic device.

The key to success, just as with locating the structure, is using gain control effectively. Use minimum gain after initially detecting the noise. If the source appears to be at more than one location on the structure, reduce the gain. In part, this will eliminate any weaker noise signals from hardware not causing the problem.

27.6.8 Common Causes of Power-line Noise

The following are some of the more common power-line noise sources. They're listed in order from most common to least common. Note that some of the most common sources are not connected to a primary conductor. This in part is due to the care most utilities take to ensure sufficient primary conductor clearance from surrounding hardware. Note, too, that power transformers do not appear on this list:

- Loose staples on ground conductor
- Loose pole-top pin
- Ground conductor touching nearby hardware
- Corroded slack span insulators
- Guy wire touching neutral
- Loose hardware
- Bare tie wire used with insulated conductor
- Insulated tie wire on bare conductor
- Loose cross-arm braces
- Lightning arrestors

27.6.9 The Cooperative Agreement

While some cases of power-line noise are resolved in a timely fashion, the reality is that many cases can linger for an extended period of time. Many utilities simply do not have the expertise, equipment or motivation to properly address a power-line noise complaint. There are often no quick solutions. Patience can often be at a premium in these situations. Fortunately, the ARRL has a Cooperative Agreement with the FCC that can help. While the program is not a quick or easy solution, it does offer an opportunity and step-by-step course of action for relief.

It emphasizes and provides for voluntary cooperation without FCC intervention.

Under the terms of the Cooperative Agreement, the ARRL provides technical help and information to utilities in order to help them resolve power-line noise complaints. It must be emphasized that the ARRL's role in this process is strictly a technical one — it is not in the enforcement business. In order to participate, complainants are required to treat utility personnel with respect, refrain from hostile behavior, and reasonably cooperate with any reasonable utility request. This includes making his or her station available for purposes of observing and recording noise signatures. The intent of the Cooperative Agreement is to solve as many cases as possible before they go to the FCC. In this way, the FCC's limited resources can be allocated where they are needed the most — enforcement.

As the first step in the process, the ARRL sends the utility a letter advising of pertinent Part 15 rules and offering assistance. The FCC then requires a 60-day waiting period before the next step. If by the end of 60 days the utility has failed to demonstrate a good faith effort to correct the problem, the FCC then issues an advisory letter. This letter allows the utility another 60-day window to correct the problem.

A second FCC advisory letter, if necessary, is the next step. Typically, this letter provides another 20- or 30-day window for the utility to respond. If the problem still persists, a field investigation would follow. At the discretion of the Field Investigator, he or she may issue an FCC Citation or Notice of Apparent Liability (NAL). In the case of an NAL, a forfeiture or fine can result.

It is important to emphasize that the ARRL Cooperative Agreement Program does not offer a quick fix. There are several built-in waiting periods and a number of requirements that a ham must follow precisely. It does however provide a step-by-step and systematic course of action under the auspices of the FCC in cases where a utility does not comply with Part 15. Look for complete details, including how to file a complaint, in the ARRL's Power-line Noise FAQ Web page at **www.arrl.org/power-line-noise-faq-page**.

27.7 Elements of RFI Control

27.7.1 Differential- and Common-Mode Signal Control

The path from source to victim almost always includes some conducting portion, such as wires or cables. RF energy can be conducted directly from source to victim, be conducted onto a wire or cable that acts as an antenna where it is radiated, or be picked up by a conductor connected to the victim that acts like an antenna. When the energy is traveling along the conducting portion of its path, it is important to understand whether it is as a differential- or common-mode signal.

Removing unwanted signals that cause RFI is different for each of these conduction modes. Differential-mode cures (a high-pass filter, low-pass filter, or a capacitor across the ac power line, for example) do not attenuate common-mode signals. Similarly, a common-mode choke will not affect interference resulting from a differential-mode signal.

It's relatively simple to build a differential-mode filter that passes desired signals and blocks unwanted signals with a high series impedance or presents a low-impedance to a signal return line or path. The return path for common-mode signals often involves Earth ground, or even the chassis of equipment if it is large enough to form part of an antenna at the frequency of the RFI. A differential-mode filter is not part of this current path, so it can have no effect on common-mode RFI.

In either case, an exposed shield surface is a potential antenna for RFI, either radiating or receiving unwanted energy, regardless of the shield's quality. In this way, a coaxial cable can act as an antenna for RFI if the victim device is unable to reject common-mode signals on the cable's shield. This is why it is important to connect cable shields in such a way that common-mode currents flowing on the shields are not allowed to enter the victim device.

27.7.2 Shields and Filters

Breaking the path between source and victim is often an attractive option, especially if either is a consumer electronics device. Remember, the path will involve one or more of three possibilities — radiation, conduction, and inductive or capacitive coupling. Breaking the path of an RFI problem can require analysis and experimentation in some cases. Obviously you must know what the path is before you can break it. While the path may be readily apparent in some cases, more complex situations may not be so clear. Multiple attempts at finding a solution may be required.

SHIELDS

Shielding can be used to control radiated emissions — that is, signals radiated by wiring inside the device — or to prevent radiated signals from being picked up by signal leads in cables or inside a piece of equipment. Shielding can also be used to reduce inductive or capacitive coupling, usually by acting as an intervening conductor between the source and victim.

Shields are used to set boundaries for radiated energy and to contain electric and magnetic fields. Thin conductive films, copper braid and sheet metal are the most common shield materials for the electric field (capacitive coupling), and for electromagnetic fields (radio waves). At RF, the small skin depth allows thin shields to be effective at these frequencies. Thicker shielding material is needed for magnetic field (inductive coupling) to minimize the voltage caused by induced current. At audio frequencies and below, the higher skin depth of common shield materials is large enough (at 60 Hz, the skin depth for aluminum is 0.43 inches) that high-permeability materials such as steel or mumetal are required.

Maximum shield effectiveness usually requires solid sheet metal that completely encloses the source or susceptible circuitry or equipment. Small discontinuities, such as holes or seams, decrease shield effectiveness. In addition, mating surfaces between different parts of a shield must be conductive. To ensure conductivity, file or sand off paint or other nonconductive coatings on mating surfaces.

The effectiveness of a shield is determined by its ability to reflect or absorb the undesired energy. Reflection occurs at a shield's surface. In this case, the shield's effectiveness is independent of its thickness. Reflection is typically the dominant means of shielding for radio waves and capacitive coupling, but is ineffective against magnetic coupling. Most RFI shielding works, therefore, by reflection. Any good conductor will serve in this case, even thin plating.

Magnetic material is required when attempting to break a low-frequency inductive coupling path by shielding. A thick layer of high permeability material is ideal in this case. Low frequency magnetic fields are typically a very short range phenomenon. Simply increasing the distance between the source and victim may help avoid the expense and difficulty of implementing a shield.

Adding shielding may not be practical in many situations, especially with many consumer products, such as a television. Adding a shield to a cable can minimize capacitive coupling and RF pickup, but it has no effect on magnetic coupling. Replacing parallel-conductor cables (such as zip cord) with twisted-pair is quite effective against magnetic coupling and also reduces electromagnetic coupling.

Additional material on shielding may be found in Chapter 2 of Ott.

FILTERS

Filters and chokes can be very effective in dealing with a conducted emissions problem. Fortunately, filters and chokes are simple, economical and easy to try. As we'll see, use of common-mode chokes alone can often solve many RFI problems, especially at HF when common-mode current is more likely to be the culprit.

A primary means of separating signals

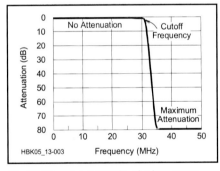

Fig 27.12 — An example of a low-pass filter's response curve.

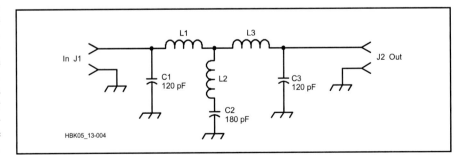

Fig 27.13 — A low-pass filter for amateur transmitting use. Complete construction information appears in the Transmitters chapter of *The ARRL RFI Book*. A high-performance 1.8-54 MHz filter project can be found in the RF and AF Filters chapter of this *Handbook*.

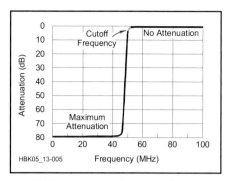

Fig 27.14 — An example of a high-pass filter's response curve.

Fig 27.15 — A differential-mode high-pass filter for 75-Ω coaxial cable. It rejects HF signals picked up by a TV antenna or that leak into a cable-TV system. All capacitors are high-stability, low-loss, NP0 ceramic disc components. Values are in pF. The inductors are all #24 AWG enameled wire on T-44-0 toroid cores. L4 and L6 are each 12 turns (0.157 µH) and L5 is 11 turns (0.135 µH).

relies on their frequency difference. Filters offer little opposition to signals with certain frequencies while blocking or shunting others. Filters vary in attenuation characteristics, frequency characteristics and power-handling capabilities. The names given to various filters are based on their uses. (More information on filters may be found in the **RF and AF Filters** chapter.)

Low-pass filters pass frequencies below some cutoff frequency, while attenuating frequencies above that cutoff frequency. A typical low-pass filter curve is shown in **Fig 27.12**. A schematic is shown in **Fig 27.13**. These filters are difficult to construct properly so you should buy one. Many retail Amateur Radio stores that advertise in *QST* stock low-pass filters.

High-pass filters pass frequencies above some cutoff frequency while attenuating frequencies below that cutoff frequency. A typical high-pass filter curve is shown in **Fig 27.14**. **Fig 27.15** shows a schematic of a typical high-pass filter. Again, it is best to buy one of the commercially available filters.

Bypass capacitors can be used to cure differential-mode RFI problems by providing a low-impedance path for RF signals away from the affected lead or cable. A bypass capacitor is usually placed between a signal or power lead and the equipment chassis. If the bypass capacitor is attached to a shielded cable, the shield should also be connected to the chassis. Bypass capacitors for HF signals are usually 0.01 µF, while VHF bypass capacitors are usually 0.001 µF. Leads of bypass capacitors should be kept short, particularly when dealing with VHF or UHF signals.

AC-line filters, sometimes called "brute-force" filters, are used to remove RF energy from ac power lines. Both common-mode and differential-mode noise can be attenuated by a commercially built line filter. A typical schematic is shown in **Fig 27.16**. Products from Corcom, (**www.corcom.com**) and Delta Electronics, (**www.delta.com.tw**) are widely available and well documented on their Web sites. Industrial Communications Engineers (**www.iceradioproducts.com**) sells stand-alone AC-line filters with ac plugs and sockets.

Warning: Bypassing Speaker Leads

Older amateur literature might suggest connecting a 0.01-µF capacitor across an amplifier's speaker output terminals to cure RFI from common-mode signals on speaker cables. Don't do this! Doing so can cause some modern solid-state amplifiers to break into a destructive, full-power, sometimes ultrasonic oscillation if they are connected to a highly capacitive load. Use common-mode chokes and twisted-pair speaker cables instead.

AC-line filters come in a wide variety of sizes, current ratings, and attenuation. In general, a filter must be physically larger to handle higher currents at lower frequencies. The Corcom 1VB1, a compact filter small enough to fit in the junction box for many low voltage lighting fixtures, provides good common mode attenuation at MF and HF and its 1 A at 250 V ac rating is sufficient for many LV lighting fixtures. In general, you will get more performance from a filter that is physically small if you choose the filter with the lowest current rating sufficient for your application. (Section 13.3 of Ott covers ac-line filters.)

Any wiring between a filter and the equipment being filtered acts as an antenna and forms an inductive loop that degrades the performance of the filter. All such wiring should be as short as possible, and should be twisted. Always bond the enclosure of the filter to the enclosure of the equipment by the shortest possible path. Some commercial filters are built with an integral IEC power socket, and can replace a standard IEC connector if there is sufficient space behind the panel. (IEC is the International Electrotechnical Commission, an international standards organization that has created specifications for power plugs and sockets.) Such a filter is bonded to the chassis and interconnecting leads are shielded by the chassis, optimizing its performance.

A capacitor between line and neutral or between line and ground at the noise source or at victim equipment can solve some RFI problems. ("Chassis ground" in this sense is not "Earth," it is the power system equipment ground — the green wire — at the equipment.) Power lines, cords, and cables are often subjected to short-duration spikes of very high voltage (4 kV). Ordinary capacitors are likely to fail when subjected to these voltages, and the failure could cause a fire. Only Type X1, X2, Y1 and Y2 capacitors should be used on power wiring. AC-rated capacitors can safely handle the current that flows through them when they are placed across an ac line along with the typical voltage surges that occur from time to time. Type X1 and X2 capacitors are rated for use between line and neutral, and are available in values between 0.1 µF and 1 µF. Type X2 capacitors are tested to withstand 2.5 kV, type X1 capacitors are tested to 4 kV. Type Y1 and Y2 capacitors are rated for use between line and ground; Y1 capacitors are impulse tested to 8 kV; Type Y2 to 5 kV. Note that 4700 pF is the largest value permitted to be used between line and ground — larger values can result in excessive leakage currents.

27.7.3 Common-Mode Chokes

Common-mode chokes on ferrite cores are the most effective answer to RFI from

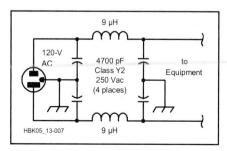

Fig 27.16 — A "brute-force" ac-line filter.

Fig 27.17 — A common-mode RF choke wound on a toroid core is shown at top left. Several styles of ferrite cores for common-mode chokes are also shown.

a common-mode signal. Differential-mode filters are *not* effective against common-mode signals. (AC-line filters often perform both common- and differential-mode filtering.) Common-mode chokes work differently, but equally well, with coaxial cable and paired conductors. (Additional material on common-mode chokes is found in sections 3.5 and 3.6 of Ott.)

The most common form of common-mode choke is multiple turns of cable wound on a magnetic toroid core, usually ferrite, as shown in **Fig 27.17**. The following explanation applies to chokes wound on rods as well as toroids, but avoid rod cores since they may couple to nearby circuits at HF.

Most of the time, a common-mode signal on a coaxial cable or a shielded, multi-wire cable is a current flowing on the outside of the cable's shield. By wrapping the cable around a magnetic core the current creates a flux in the core, creating a high impedance in series with the outside of the shield. (An impedance of a few hundred to several thousand ohms are required for an effective choke.) The impedance then blocks or reduces the common-mode current. Because equal-and-opposite fields are coupled to the core from each of the differential-mode currents, the common-mode choke has no effect on differential-mode signals inside the cable.

When the cable consists of two-wire, unshielded conductors such as zip cord or twisted-pair, the equal-and-opposite differential currents each create a magnetic flux in the core. The equal-and-opposite fluxes cancel each other and the differential-mode signal experiences zero net effect. To common-mode signals, however, the choke appears as a high impedance in series with the signal: the higher the impedance, the lower the common-mode current.

It is important to note that common-mode currents on a transmission line will result in radiation of a signal from the feed line. (See the sidebar for an explanation of balanced vs unbalanced transmission lines.) The radiated

Feed Line Radiation from Balanced vs Unbalanced Transmission Lines

Q. What is meant by the terms "balanced" and "unbalanced" when referring to transmission lines?

A. The physical differences between balanced and unbalanced feed lines are obvious. Balanced lines are parallel-type transmission lines, such as ladder line or twin lead. The two conductors that make up a balanced line run side-by-side, separated by an insulating material (plastic, air, whatever). Unbalanced lines, on the other hand, are coaxial-type feed lines. One of the conductors (the shield) completely surrounds the other (the center).

In an ideal world, both types of transmission lines would deliver RF power to the load (typically your antenna) without radiating any energy along the way. It is important to understand, however, that both types of transmission lines require a balanced condition in order to accomplish this feat. That is, the currents in each conductor must be equal in magnitude, but opposite in polarity.

The classic definition of a balanced transmission line tells us that both conductors must be symmetrical (same length and separation distance) relative to a common reference point, usually ground. It's fairly easy to imagine the equal and opposite currents flowing through this type of feeder. When such a condition occurs, the fields generated by the currents cancel each other-hence, no radiation. An imbalance occurs when one of the conductors carries more current than the other. This additional "imbalance current" causes the feed line to radiate.

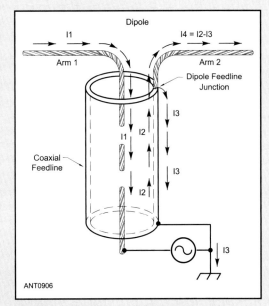

Fig 27.A — Various current paths are present at the feed point of a balanced dipole fed with unbalanced coaxial cable. The diameter of the coax is exaggerated to show the currents clearly.

Things are a bit different when we consider a coaxial cable. Instead of its being a symmetrical line, one of its conductors (usually the shield), is grounded. In addition, the currents flowing in the coax are confined to the outside portion of the center conductor and the inside portion of the shield.

When a balanced load, such as a resonant dipole antenna, is connected to unbalanced coax, the outside of the shield can act as an electrical third conductor (see Fig 27.A). This phantom third conductor can provide an alternate path for the imbalance current to flow. Whether the small amount of stray radiation that occurs is important or not is subject to debate. In fact, one of the purposes of a balun (a contraction of balanced to unbalanced) is to reduce or eliminate imbalance current flowing on the outside of the shield.

Warning: Surplus Ferrite Cores

Don't use a core to make a common-mode choke if you don't know what type of material it is made of. Such cores may not be effective in the frequency range you are working with. This may lead to the erroneous conclusion that a common-mode choke doesn't work when a core with the correct material would have done the job.

signal can then cause RFI in nearby circuits. This is most common when using coaxial cable as a feed line to a balanced load, such as the dipole in the sidebar. Using a common-mode choke to reduce common-mode feed line currents can reduce RFI caused by signals radiated from the feed line's shield.

The optimum core size and ferrite material is determined by the application and frequency. For example, an ac cord with a plug attached cannot be easily wrapped on a small ferrite core. The characteristics of ferrite materials vary with frequency, as shown by the graph in **Fig 27.18**. Type 31 material is a good all-purpose material for HF and low-VHF applications, especially at low HF frequencies. Type 43 is widely used for HF through VHF and UHF. (See the **Component Data and References** chapter for a table of ferrite materials and characteristics.)

Ferrite beads and clamp-on split cores are also used for EMI control at VHF and UHF, both as common-mode chokes and low-pass filters. (These are essentially single-turn chokes as the cable passes just once through the bead or core.) Multi-turn chokes are required for effective suppression at HF. It is usually more effective to form a common-mode choke by wrapping about multiple turns of wire or coaxial cable around a 1- to 2-inch OD core of the correct material.

Common-mode chokes can be used on single conductors unless the desired signal is in the RF spectrum. A common-mode choke creates a high resistive impedance at radio frequencies. At audio frequencies, however, it looks like a relatively small inductance, and is unlikely to have any effect on audio-frequency signals.

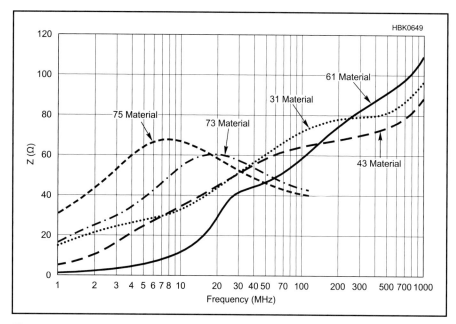

Fig 27.18 — Impedance vs. frequency plots for "101" size ferrite beads illustrate the effect of various ferrite materials across different frequency ranges. A 3.50 mm × 1.30 mm × 6.00 mm bead (Fair-Rite 301 size) was used for the above curve for material comparison, however all materials are not available in all shapes and sizes. Type 73 material is only available in smaller cores, type 31 is only available in larger cores, and type 75 is currently only available as a toroid core. (Graph provided courtesy of the Fair-Rite Corporation)

27.8 Troubleshooting RFI

Troubleshooting an RFI problem is a multi-step process, and all steps are equally important. First you must determine the type(s) of noise source(s) that are involved. Next, diagnose the problem by locating the noise source and the means by which it creates the noise. The final step is to identify the path by which the noise or signals reach the victim device. Only then can you cure the problem by breaking the path from source to victim.

Each step in troubleshooting an RFI problem involves asking and answering several questions: Is the problem caused by harmonics, fundamental overload, conducted emissions, radiated emissions or a combination of all of these factors? Should it be fixed with a low-pass filter, high-pass filter, common-mode chokes or ac-line filter? How about shielding, isolation transformers, a different ground or antenna configuration? By the time you finish with these questions, the possibilities could number in the millions. You probably will not see your exact problem and cure listed in this book or any other. You must not only diagnose the problem but find a cure as well!

Now that you have learned some RFI fundamentals, you can work on specific technical solutions. A systematic approach will identify the problem and suggest a cure. Most RFI problems can be solved in this way by the application of standard techniques. The following sections suggest specific approaches for different types of common interference problems. This advice is based on the experience of the ARRL RFI Desk, but is not guaranteed to provide a solution to your particular problem. Armed with your RFI knowledge, a kit of filters and tools, your local TC and a determination to solve the problem, it is time to begin. You should also get a copy of the *ARRL RFI Book*. It's comprehensive and picks up where this chapter leaves off.

27.8.1 General RFI Troubleshooting Guidelines

Before diving into the problem, take a step back and consider some of these "pre-troubleshooting steps."

Is It Really EMI? — Before trying to solve a suspected case of EMI, verify that the symptoms actually result from external causes. A variety of equipment malfunctions or external noise can look like interference.

Is It Your Station? — "Your" EMI problem might be caused by another ham or a radio transmitter of another radio service, such as a local CB or police transmitter. If it appears that your station is involved, operate your station on each band, mode and power level that you use. Note all conditions that produce interference. If no transmissions produce the problem, your station *may* not be the cause. (Although some contributing factor may have been missing in the test.) Have your neighbor keep notes of when and how the interference appears: what time of day, what station, what other appliances were in use, what was the weather? You should do the same whenever you operate. If you can readily reproduce the problem with your station, you can start to troubleshoot the problem.

Take One Away — Can you remove the source or victim entirely? The best cure for an RFI problem is often removing the source of the noise. If the source is something broken, for example, the usual solution is to repair it.

Power-line noise and an arcing electric fence usually fall into this category. If a switchmode power supply is radiating noise, replace it with a linear supply. Victim devices can sometimes be replaced with a more robust piece of equipment, as well.

Look Around — Aside from the brain, the eyes are a troubleshooter's best tool. Installation defects contribute to many RFI problems. Look for loose connections, shield breaks in a cable-TV installation or corroded contacts in a telephone installation. Have these fixed these first. Look for wiring connected to the victim equipment that might be long enough to be resonant on one or more amateur bands. If so, a common-mode choke may be an easy cure. Ideally you'll generally want place the choke as close to the victim device as practical. If this placement proves too difficult or additional suppression is required, chokes placed at the middle of the wiring may help break up resonances. These are just a few of the possible deficiencies in a home installation.

At Your Station — Make sure that your own station and consumer equipment are clean. This cuts the size of a possible interference problem from your station in half! Once this is done, you won't need to diagnose or troubleshoot your station later. Also, any cures successful at your house may work at your neighbor's as well. If you do have problems in your own home, continue through the troubleshooting steps and specific cures and take care of your own problem first.

Simplify the Problem — Don't tackle a complex system — such as a telephone system in which there are two lines running to 14 rooms — all at once. You could spend the rest of your life running in circles and never find the true cause of the problem.

There's a better way. In our hypothetical telephone system, first locate the telephone jack closest to the telephone service entrance. Disconnect the lines to more remote jacks and connect one RFI-resistant telephone at the remaining jack. (Old-style rotary-dial phones are often quite immune to RF.) If the interference remains, try cures until the problem is solved, then start adding lines and equipment back one at a time, fixing the problems as you go along. If you are lucky, you will solve all of the problems in one pass. If not, at least you can point to one piece of equipment as the source of the problem.

Multiple Causes — Many RFI problems have multiple causes. These are usually the ones that give new RFI troubleshooters the most trouble. For example, consider a TVI problem caused by the combination of harmonics from the transmitter due to an arc in the transmitting antenna, an overloaded TV preamplifier, fundamental overload generating harmonics in the TV tuner, induced and conducted RF on the ac-power connections, and a common-mode signal picked up on the shield of the TV's coaxial feed line. You would never find a cure for this multiple-cause problem by trying only one cure at a time.

In this case, the solution requires that all of the cures be present at the same time. When troubleshooting, if you try a cure, leave it in place even if it doesn't solve the problem. When you add a cure that finally solves the problem entirely, start removing the "temporary" attempts one at a time. If the interference returns, you know that there were multiple causes.

Take Notes — In the process of troubleshooting an RFI problem, it's easy to lose track of what remedies were applied, to what equipment, and in what order. Configurations of equipment can change rapidly when you're experimenting. To minimize the chances of going around in circles or getting confused,

Table 27.1
RFI Survival Kit

Quantity	Item
(2)	300-Ω high-pass filter
(2)	75-Ω high-pass filter
(2)	Commercially available clamp-on ferrite cores: #31 and #43 material, 0.3" ID
(12)	Assorted ferrite cores: #31 and #43 material, FT-140 and FT-240 size
(3)	Telephone RFI filters
(2)	Brute-force ac line filters
(6)	0.01-μF ceramic capacitors
(6)	0.001-μF ceramic capacitors

Miscellaneous:
- Hand tools, assorted screwdrivers, wire cutters, pliers
- Hookup wire
- Electrical tape
- Soldering iron and solder (use with caution!)
- Assorted lengths 75-Ω coaxial cable with connectors
- Spare F connectors, male, and crimping tool
- F-connector female-female "barrel"
- Clip leads
- Notebook and pencil
- Portable multimeter

"Keeping It Simple"

Filters and chokes are the number one weapons of choice for many RFI problems whether the device is the source or the victim. They are relatively inexpensive, easy to install, and do not require permanently modifying the device.

Common-mode choke — Making a common-mode choke is simple. Select the type of core and ferrite material for the frequency range of the interference. (Type 31 is a good HF/low-VHF material, type #43 from 5 MHz through VHF) Wrap several turns of the cable or wire pairs around the toroid. Six to 8 turns is a good start at 10-30 MHz and 10 to 15 turns from 1.8 to 7 MHz. (Ten to 15 turns is probably the practical limit for most cables.) Ferrite clamp-on split cores and beads that slide over the cable or wire are not as effective as toroid-core chokes at HF but are the right solution at VHF and higher frequencies. For a clamp-on core, the cable doesn't even need to be disconnected from its end. Use type 31 or type 43 material at VHF, type 61 at UHF. At 50 MHz, use two turns through type 31 or 43 cores.

"Brute-Force" ac-line filters — RF signals often enter and exit a device via an ac power connection. "Brute-force" ac-line filters are simple and easy to install. Most ac filters provide both common- and differential-mode suppression. It is essential to use a filter rated to handle the device's required current.

General installation guidelines for using chokes and filters

1. If you have a brute-force ac-line filter, put one on the device or power cord. If RFI persists, add a common-mode choke to the power cord at the device.
2. Simplify the problem by removing cables one at a time until you no longer detect RFI. Start with cables longer than 1/10th-wavelength at the highest frequency of concern. If the equipment can't operate without a particular cable, add common-mode chokes at the affected or source device.
3. Add a common-mode choke to the last cable removed and verify its effect on the RFI. Some cables may require several chokes in difficult cases.
4. Begin reconnecting cables one at a time. If RFI reappears, add a common-mode choke to that cable. Repeat for each cable.
5. Once the RFI goes away, remove the common-mode chokes you added one at a time. If the RFI does not return, you do not need to reinstall the choke. If the RFI returns after removing a choke, reinstall it. Keep only those chokes required to fix the problem.

take lots of notes as you proceed. Sketches and drawings can be very useful. When you do find the cause of a problem and a cure for it, be sure to write all that down so you can refer to it in the future.

RFI Survival Kit — **Table 27.1** is a list of the material needed to troubleshoot and solve most RFI problems. Having all of these materials in one container, such as a small tackle or craft box, makes the troubleshooting process go a lot smoother.

27.8.2 Transmitters

We start with transmitters not because most interference comes from transmitters, but because your station transmitter is under your direct control. Many of the troubleshooting steps in other parts of this chapter assume that your transmitter is "clean" (free of unwanted RF output).

Start by looking for patterns in the interference. Problems that occur only on harmonics of a fundamental signal usually indicate the transmitter is the source of the interference. Harmonics can also be generated in nearby semiconductors, such as an unpowered VHF receiver left connected to an antenna, rectifiers in a rotator control box, or a corroded connection in a tower guy wire. Harmonics can also be generated in the front-end components of the TV or radio experiencing interference.

If HF transmitter spurs at VHF are causing interference, a low-pass filter at the transmitter output will usually cure the problem. Install the filter after the amplifier (if used) and before the antenna tuner. (A second filter between the transmitter and amplifier may occasionally help as well.) Install a low-pass filter as your first step in any interference problem that involves another radio service.

Interference from non-harmonic spurious emissions is extremely rare in commercial radios. Any such problem indicates a malfunction that should be repaired.

27.8.3 Television Interference (TVI)

An analog TV signal must have about a 45 to 50 dB signal-to-noise ratio to be classified as good-quality viewing. If interference is present due to a single, discrete signal (for example, a CW signal) the signal-to-interference ratio must be 57 dB or greater, depending on the frequency of the interference within the affected channel. Digital TV has somewhat better immunity but for both formats, clear reception requires a strong signal at the TV antenna-input connector so the receiver must be in what is known as a *strong-signal area*.

TVI to a TV receiver (or a video monitor) normally has one of the following causes:

• Spurious signals within the TV channel coming from your transmitter or station.

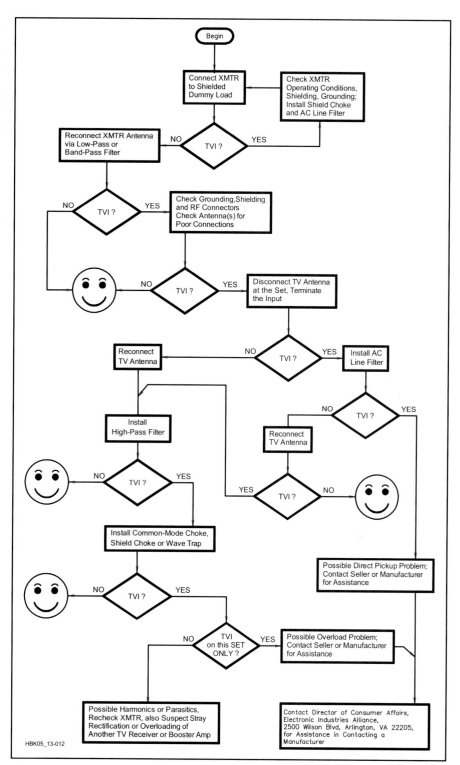

Fig 27.19 — TVI troubleshooting flowchart.

• The TV set may be overloaded by your transmitter's fundamental signal.

• Signals within the TV channel from some source other than your station, such as electrical noise, an overloaded mast-mounted TV preamplifier or a transmitter in another service.

• The TV set might be defective or misadjusted, making it look like there is an interference problem.

All of these problems are made potentially more severe because TV receiving equipment is hooked up to *two* antenna systems: (1) the incoming antenna or cable feed line and (2) the ac power line and interconnecting cables. These two antenna systems can couple significant levels of fundamental or harmonic energy into the TV set or video display! The

TVI Troubleshooting Flowchart in **Fig 27.19** is a good starting point.

The problem could also be caused by direct pickup of the transmitted signal by an unshielded TV or device connected to the TV.

Certain types of television receivers and video monitors are reported to cause broadband RF interference *to* amateur signals — large-screen plasma display models seem to be the most frequent offender — and this may be difficult to cure due to the nature of the display technology. Fortunately, less-expensive, more power-efficient, and RF-quieter LCD technology seems to be displacing plasma technology.

The manufacturer of the TV or video equipment can sometimes help with an interference problem. The Electronic Industries Alliance (EIA) can also help you contact equipment manufacturers. Contact them directly for assistance in locating help at **www.eia.org**.

COMMON SOURCES OF TVI

HF transmitters — A nearby HF transmitter is most likely to cause fundamental overload. This is usually indicated by interference to all channels, or at least all VHF channels. To cure fundamental overload from an HF transmitter to an antenna-connected TV, install a high-pass filter directly at the TV set's antenna input. (Do not use a high-pass filter on a cable-TV input because the HF range is used for data and other system signals.)

A strong HF signal can also result in a strong common-mode signal on the TV's feed line. A common-mode choke will block signals on the outside of the feed line shield, leaving the desired signals inside the feed line unaffected. **Fig 27.20** shows how a common-mode choke is constructed for a coaxial feed line.

These two filters can probably cure most cases of TVI! **Fig 27.21** shows a "bulletproof" installation for both over-the-air and cable TV receivers. If one of these methods doesn't cure the problem, the problem is likely direct pickup in which a signal is received by the TV set's circuitry without any conducting path being required. In that case, don't try to fix it yourself — it is a problem for the TV manufacturer.

High-pass filters *should not* be used in a cable TV feed line (Fig 27.21A) with two-way cable devices such as cable modems, set-top boxes, and newer two-way CableCARD-equipped TVs. The high-pass filter may prevent the device from communicating via the cable network's upstream signal path.

VHF Transmitters —Most TV tuners are not very selective and a strong VHF signal, including those from nearby FM and TV transmitters, can overload the tuner easily, particularly when receiving channels 2-13 over the air and not by cable TV. In this case, a VHF notch or stop-band filter at the TV can help by attenuating the VHF fundamental signal that gets to the TV tuner. Winegard (**www.winegard.com**) and Scannermaster (**www.scannermaster.com**) sell tunable notch filters. A common-mode choke may also be necessary if the TV is responding to the common-mode VHF signal present on the TV's feed line.

TV Preamplifiers — Preamplifiers are only needed in weak-signal areas and they often cause trouble, particularly when used unnecessarily in strong-signal areas. They are subject to the same overload problems as TVs and when located on the antenna mast it can be difficult to install the appropriate cures. You may need to install a high-pass or notch filter at the input of the preamplifier, as well as a common-mode choke on the input, output and power-supply wiring (if separate) to effect a complete cure. All filters, connections, and chokes must be weatherproofed. Secure the coax tightly to a metal mast to minimize common-mode current. (Do not secure twin-lead to a metal support.) Use two 1-inch long type 43 clamp-on ferrite cores if VHF signals are causing the interference and type 61 mate-

Fig 27.20 — To eliminate HF and VHF signals on the outside of a coaxial cable, use an 1- to 2-inch OD toroid core and wind as many turns of the cable on the core as practical.

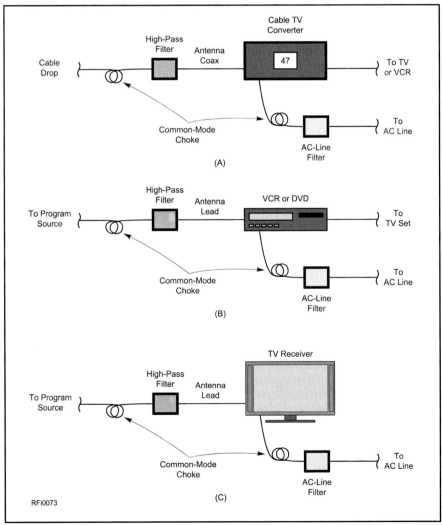

Fig 27.21 — Installing common-mode chokes and high-pass filters will cure most fundamental overload interference from HF sources. This technique does not address direct pickup or spurious emission problems.

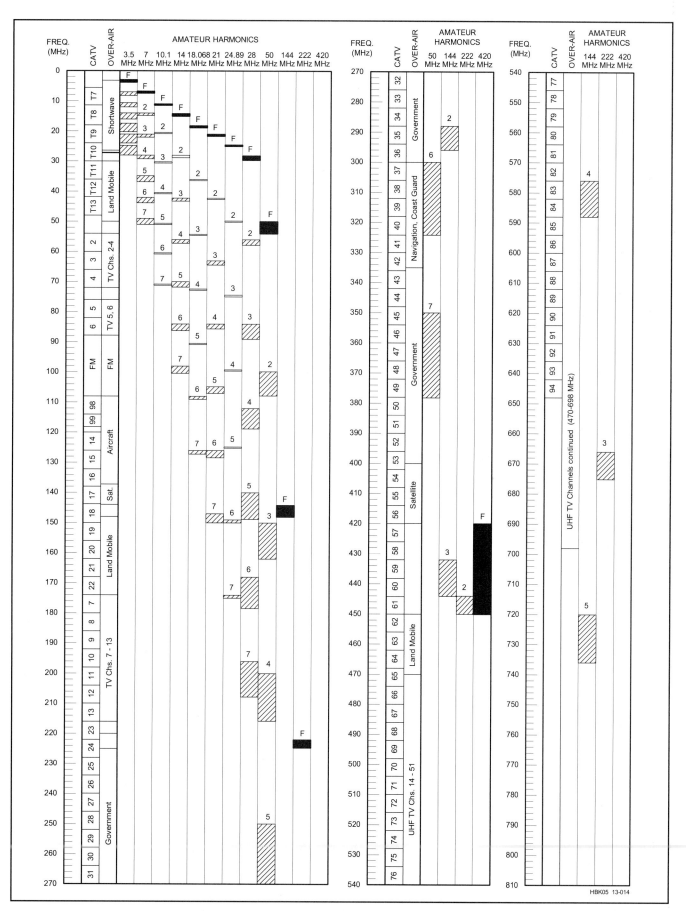

Fig 27.22 — This chart shows CATV and broadcast channels used in the United States and their relationship to the harmonics of MF, HF, VHF and UHF amateur bands. Over-the-air UHF TV channels 52-69 (698-890 MHz) have been reallocated to other services. (F denotes a fundamental frequency for amateur signals.)

RF Interference 27.29

rial for UHF. HF choke design is discussed in the section on Common-Mode Chokes.

Spurious Emissions — You are responsible for spurious emissions produced by your station. If your station is generating any interfering spurious signals, the problem must be cured there. Start by analyzing which TV channels are affected. The TV Channel Chart in **Fig 27.22** shows the relationship of the amateur allocations and their harmonics to over-the-air and cable channels. Each channel is 6 MHz wide. If the interference is only on channels that are multiples of your transmitting frequency, you probably have interference caused by harmonics of your transmitted signal.

It is not certain that these harmonics are coming from your station, however. Harmonics can be generated by overloaded preamplifiers or tuner input circuits. Harmonics can also be generated by non-linear junctions near your station transmitter or very near the TV receiver antenna. (See the section on Intermodulation Distortion.) If your transmitter and station check "clean" — check to see if you have interference on a TV set in your own home — then you must look elsewhere for the source of the harmonics.

Electrical Noise — Electrical noise on an analog TV screen generally appears as shown in **Fig 27.23**. Because the noise is nearly synchronized with the ac line frequency, noise artifacts move upward slowly on the screen. Digital TV signals are fairly resistant to electrical noise, but in extreme cases can cause the picture to freeze or fail to be displayed as discussed in the following section on Digital TV.

On an AM receiver (including SSB or CW receivers), electrical noise usually sounds like a buzz, sometimes changing in intensity as the arc or spark sputters a bit. If you have a problem with electrical noise, refer to the section on Electrical Noise.

ANALOG TV RECEIVERS

Even though over-the-air TV broadcasting largely switched to a digital format in 2009, millions of analog TV receivers are still in use for cable TV, satellite TV, with converter boxes for digital broadcast signals, and for displaying video from DVDs and other video sources. Older VCR and DVD players may also include an analog TV tuner to receive analog TV signals.

Interference to video displays and monitors that do not receive RF signals from an antenna or RF modulator should be assumed to be common-mode interference or direct pickup. The same applies to interference to a TV set displaying video signals (not through the antenna input). Interference that is present only on the audio is probably a case of common-mode RFI. (See the Stereos and Home Entertainment Systems section of this chapter.)

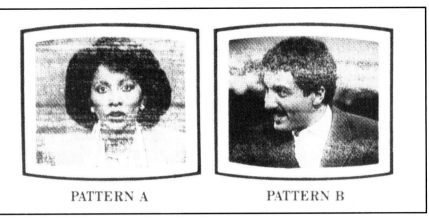

Fig 27.23 — Two examples of TVI from electrical noise to analog TV receivers.

Digital TV Basics

DTV operates on the same 6 MHz-wide channels used for analog TV. However, instead of each channel slot carrying an analog NTSC television signal (visual carrier, color sub-carrier, & aural carrier), the channel slot carries an 8-VSB digitally modulated signal. The 6 MHz-wide over-the-air channel slots themselves didn't change, just the signals carried in them! So, Ch. 2 is still 54-60 MHz, Ch. 3 is 60-66 MHz, and so forth, up to channel 51. Ch. 2-6 have largely been abandoned in the US as stations moved to broadcast digital TV on UHF channels. UHF channels 52-69 will be reallocated to other uses.

The designation "8-VSB" refers to 8-level vestigial sideband modulation. This is similar to 256-QAM, which means 256-state *quadrature amplitude modulation* — the 256 "states" are 256 combinations of signal phase and amplitude values that represent the 256 different transmitted symbols — a digital format used in cable TV networks (64-QAM is also used by cable companies). In the case of 8-VSB, the "8" refers to the eight-level baseband DTV signal that amplitude modulates an IF signal. For more info about 8-VSB modulation, see the online article, "What Exactly Is 8-VSB Anyway?" at **www.broadcast.net/~sbe1/8vsb/8vsb.htm**.

Digitally modulated signals used to transport video — whether 8-VSB for over-the-air or 64-QAM or 256-QAM in cable networks — are noise-like signals over their 6 MHz bandwidth. If a digital TV signal interferes with analog radio communications such as FM or SSB, the effect is similar to degraded signal-to-noise ratio. Indeed, a digital TV signal can be thought of as a 6 MHz wide "pile of noise." On a spectrum analyzer, it looks like a "haystack" as shown in **Fig 27.B** Of course, the IF bandwidth of the amateur receiver is quite narrow relative to the 6 MHz-wide digital TV channel, so the actual noise level seen by the receiver (assuming no overload or IMD problems) will in part be determined by the receiver's IF bandwidth. Still, the effect on the receiver is what amounts to an elevated noise floor, similar to the effects of wide-spectrum broadband noise.

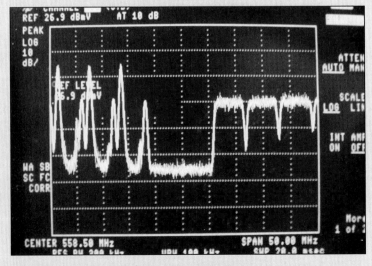

Fig 27.B — The spectrum of three analog NTSC TV channels (left of center) and three QAM signals (right of center). The digital channel power of the QAM "haystacks" is about 6 dB below the analog visual carrier PEP.

DIGITAL TV (DTV) RECEIVERS

In 2009, nearly all over-the-air TV broadcasters in the US, with the exception of low-power TV stations and translators, switched from the older NTSC analog format to a new digital format called DTV. The FCC's Digital TV Transition Web page (**www.dtv.gov**) has more information on this transition, including an FAQ page. Digital TV signals can operate with much lower signal-to-noise ratios, but are still susceptible to interference.

Interference to digital TV signals from amateur signals — narrowband interference, for instance, a CW carrier — to a 6 MHz-wide digital TV signal generally has two effects. If the interfering signal is strong enough, it will cause degraded *modulation error ratio* (MER) and degraded *bit error rate* (BER) in the digital video signal. If the amplitude of the interference is sufficient, the digital receiver's *forward error correction* (FEC) circuitry will be unable to fix the broken bits, and the digital video signal will "crash." (See the **Digital Modes** chapter for more information on coding and error correction in digital protocols.)

TV viewers watching any of the multiple video streams that may be contained within the digital video signal won't see any problems in the picture (or hear anything wrong in the sound), until the so-called "crash point" is reached. At that point, the picture will begin to show intermittent "tiling" (the picture breaking up into small squares) or blocking (freezing) in the image. As the amplitude of the interfering signal increases perhaps another 0.5 dB to 1 dB, the crash point or "digital cliff" is reached, and the picture and sound are gone! As you can see, there is a tiny window between receiving a perfect picture and receiving no picture. The same effect is produced by signal fading and may be difficult to distinguish from RFI.

Interference to the digital signal does not make its presence known through visual or audible artifacts such as streaks, lines, or tearing in the picture, or garbled audio. This means that a viewer experiencing interference may not be able to identify its source, but troubleshooting interference may also become more difficult. Nevertheless, the more robust digital modulation is often less susceptible to interference from narrowband amateur signals. (See the sidebar "Report from the ARRL Lab Regarding Amateur Radio Operation and Converter Boxes.") A clue to the source of the interference is that interference caused by an amateur signal will occur in sync with the amateur's transmissions while other types of interference will have no correlation.

The techniques for curing interference between amateur and digital TV signals are largely the same as for analog TV. Fundamental overload generally responds well to filters in the antenna or RF inputs. Interference caused by spurious emissions from the amateur station can be eliminated by filtering at the amateur transmitter. Common-mode problems in which RF signals are conducted into the television receiver's circuitry by external audio, video, and power cables are no more or less likely than for analog TV sets and can be addressed as described elsewhere in this chapter.

27.8.4 Cable TV

Cable TV has generally benefited Amateur Radio with respect to TVI. The cable system delivers a strong, consistent signal to the TV receiver, reducing susceptibility to interference from amateur signals. It is also a shielded system so an external signal shouldn't be able to cause interference. Most cable companies are responsible about keeping signal leakage (*egress*) and *ingress* — the opposite of leakage — under control, but problems do happen. Cable companies are not responsible for direct pickup or common-mode interference problems, but are responsible for leakage, ingress, and any noise radiated by common-mode currents from their equipment.

Cable companies are able to take advantage of something known as frequency reuse. That is, all radio frequencies higher than 5 MHz are used to transmit TV signals. The latter is possible because the cables and components used to transport signals to and from paying subscribers comprise what is known as a closed network. In other words, a cable company can use frequencies inside of its cables that may be used for entirely different purposes in the over-the-air environment. As long as the shielding integrity of the cable network is maintained, the cable company's signals won't interfere with over-the-air services, and vice-versa.

The reality is that the shielding integrity of a cable network *is* sometimes compromised, perhaps because of a loose or damaged connector, a cracked cable shield, rodent damage, poorly shielded customer premises equipment (CPE) such as cable-ready TVs and VCRs, and problems that may happen when someone tries to steal cable service! §76.605(a)(12) of the FCC Rules defines the maximum allowable signal leakage (*egress*) field strength at specified measurement distances, and §76.613 covers harmful interference. FCC Rules also mandate that cable

---

Report from the ARRL Lab Regarding Amateur Radio Operation and Converter Boxes

Upon receiving an email from a concerned member, I thought I would test out a digital TV to analog TV converter box that one of our Lab staffers had. It is a Zenith brand, identical to an Insignia brand I have at home.

To test the box by itself, I took the RF output from the converter box and fed it directly via coax to the analog TV RF input, using broadcast channel 3 on the TV.

Fed directly to the antenna input of the converter box, I combined an off-air digital signal with an analog RF signal representing a very strong nearby ham transmitter (CW and amplitude modulation). Watching and listening to the television, I varied my signal generator's frequency through the MF, HF, VHF and UHF spectrum at 0 dBm (equivalent to a signal stronger than 70 dB over S9 on some S meters) with no signal break-up or other problems. The TV signal strength meter was typically about 2/3 scale on all channels, just enough for the off-air TV signal to come in. The test was repeated watching both VHF and UHF DTV channels.

Problems occurred when interfering signals reached the +10 dBm range (extremely strong), where you would expect any receiver to have blocking issues. Another Lab staff member and I both concluded the box we tested was very good at rejecting strong signals.

For a real life situation, we observed digital TV signals when W1AW was fired up on 20 m, with 1500 W on SSB, right across the parking lot! This was a good, real-life overload test. No pixelation or loss of picture occurred. Please note our TV receiving antenna was a multiband HF vertical on the roof of ARRL HQ.

While this limited test may not be representative of all converter boxes, TVI from radio amateurs is still possible, but not likely, through the TV antenna. Many DTV problems occur with RF getting into interconnecting cables or power lines and can be solved on an as needed basis, by trying the same methods used with analog TVs. Solutions to such problems are described in this chapter and in more detail in The ARRL *RFI Book*.

Off-air DTV needs a good signal to be seen. Either you have it, or you don't, period. There is no in-between like analog. Your neighbor may not understand that DTV signal breakup may have nothing to do with your transmissions. As described at the beginning of this chapter, try to be a good neighbor and ambassador of ham radio by working to help with your neighbor to improve reception.

Note: I have not had any interference to Amateur Radio from my converter box at my home. — *Bob Allison, WB1GCM, ARRL Test Engineer*

Table 27.2

Amateur Radio Bands Relative to Cable TV Downstream Channels

Amateur Band	Over-The-Air Frequency Range	Cable Channel	Cable Frequency Range
6 meters	50-54 MHz	Below Ch. 2	50-54 MHz, sometimes used for narrowband telemetry carriers
2 meters	144-148 MHz	Ch. 18	144-150 MHz
1.25 meters	222-225 MHz	Ch. 24	222-228 MHz
70 cm	420-450 MHz	Ch. 57	420-426 MHz
		Ch. 58	426-432 MHz
		Ch. 59	432-438 MHz
		Ch. 60	438-444 MHz
		Ch. 61	444-450 MHz
33 cm	902-928 MHz	Ch. 142	900-906 MHz
		Ch. 143	906-912 MHz
		Ch. 144	912-918 MHz
		Ch. 145	918-924 MHz
		Ch. 146	924-930 MHz

operators "…shall provide for a program of regular monitoring for signal leakage by substantially covering the plant every three months," and leaks greater than 20 microvolts per meter (µV/m) at a 10 ft. measurement distance repaired in a reasonable period of time. As well, an annual "snapshot" of leakage performance must be characterized via a flyover measurement of the cable system, or a ground-based measurement of 75% of the network.

CABLE TV FREQUENCY USAGE

A typical modern North American cable network is designed to use frequencies in the 5 to 1002 MHz spectrum. Signals that travel from the cable company to the subscriber occupy frequencies from just above 50 MHz to as high as 1002 MHz range (this is the downstream or forward spectrum), and signals that travel from the subscriber to the cable company are carried in the 5 to as high as 42 MHz range, known as the upstream or return spectrum. The downstream is divided into 6 MHz-wide channel slots, which may carry analog NTSC television signals or 64- or 256-QAM digitally modulated signals used for digital video, high-speed data, and telephone services. Upstream signals from cable modems and two-way set-top boxes are generally carried on specific frequencies chosen by the cable company. **Table 27.2** summarizes cable downstream channel allocations that overlap Amateur Radio bands. The complete North American channel plan is shown in Fig 27.22.

Cable channels below about 550 MHz may carry analog NTSC signals or digital signals. Cable channels above 550 MHz generally carry digital signals, although there are exceptions. Cable channels above 870 MHz almost always carry only digital signals.

COMMON MECHANISMS FOR LEAKAGE AND INGRESS

As noted previously, cable TV leakage and ingress occur when the shielding integrity of the cable network is compromised. A large cable system that serves a major metropolitan area has literally millions of connectors, tens of thousands of miles of coaxial cable, thousands of amplifiers, hundreds of thousands of passives (splitters, directional couplers, and similar devices), and uncountable customer premises equipment connected to the cable network! Any of these may be a source of leakage and ingress.

ANALOG TV CHANNEL LEAKAGE SYMPTOMS

When signal leakage does happen, interference to the Amateur Radio service may occur. And where there is leakage, there is probably ingress. That compounds the problem, because not only are you experiencing interference, but when you transmit you may interfere with cable service in the area. One of the most common signs of possible leakage is interference to the 2 meter amateur band, especially in the vicinity of standard (STD) cable channel 18's visual carrier on 145.25 MHz. If you suspect cable leakage, listen for the telltale buzz from the video signal on or near 145.25 MHz (sometimes buzz may not be heard), and check other STD, incrementally related carrier (IRC), and harmonically related carrier (HRC) visual carrier frequencies on nearby channels listed in **Table 27.3** using a wide range receiver or scanner. (Leakage of a digital TV signal on cable channel 18 sounds like broadband noise over the 144-150 MHz range.) Also listen for TV channel sound on the FM aural carriers 4.5 MHz above the visual carriers.

DIGITAL SIGNAL LEAKAGE?

The digitally modulated signals carried in a cable TV network use 64-QAM or 256-QAM, the latter more common. If a QAM signal were to leak from a cable TV network, it is possible for interference to an over-the-air service to occur, but very unlikely to be identified as from a digital TV signal. The reason for this is that a QAM signal is noise-like, and sounds like normal background noise or hiss on a typical amateur receiver. The QAM signal's digital channel power — its average power over the entire occupied bandwidth — is typically 6 to 10 dB lower than what an analog TV signal's visual carrier peak envelope power (PEP) would be on the same channel. As well, a QAM signal occupies most of the 6 MHz channel slot, and there are no carriers per se within that channel bandwidth. Note that over-the-air 8-VSB digital TV broadcast signals transmit a pilot carrier near the lower end of the digital "haystack," but the QAM format used by cable operators has no comparable pilot carrier.

What makes the likelihood of interference occurring (or not occurring) has in large part to do with the behavior of a receiver in the presence of broadband noise. While each downstream cable TV QAM signal occupies close to 6 MHz of RF bandwidth, the IF

Table 27.3

VHF Midband Cable Channels

Cable Channel	Visual Carrier Frequency, MHz (STD)	Visual Carrier Frequency, MHz (IRC)	Visual Carrier Frequency, MHz (HRC)
98	109.2750	109.2750	108.0250*
99	115.2750	115.2750	114.0250*
14	121.2625	121.2625	120.0060
15	127.2625	127.2625	126.0063
16	133.2625	133.2625	132.0066
17	139.25	139.2625	138.0069
18	145.25	145.2625	144.0072
19	151.25	151.2625	150.0075
20	157.25	157.2625	156.0078
21	163.25	163.2625	162.0081
22	169.25	169.2625	168.0084

*Excluded from HRC channel set because of FCC frequency offset requirements

bandwidth of a typical amateur FM receiver might be approximately 20 kHz. Thus, the noise power in the receiver will be reduced by $10\log_{10}(6,000,000/20,000) = 24.77$ dB because of the receiver's much narrower IF bandwidth compared to the QAM signal's occupied bandwidth. In addition, there is the 6 to 10 dB reduction of the digital signal's average signal PEP.

Field tests during 2009 confirmed this behavior, finding that a leaking QAM signal would not budge the S-meter of a Yaesu FT-736R at low to moderate field strength leaks, even when the receiver's antenna — a resonant half-wave dipole — was located just 10 feet from a calibrated leak. In contrast, a CW carrier that produced a 20 µV/m leak resulted in an S-meter reading of S9+15 dB, definitely harmful interference! When the CW carrier was replaced by a QAM signal whose digital channel power was equal to the CW carrier's PEP and which produced the same leakage field strength (the latter integrated over the full 6 MHz channel bandwidth), the S-meter read <S1. When the leakage field strength was increased to 100 µV/m, the CW carrier pegged the S-meter at S9+60 dB, while the QAM signal was S3 in FM mode and between S1 and S2 in USB mode. It wasn't until the leaking QAM signal's field strength reached several hundred µV/m that the "noise" (and it literally sounded like typical white noise) could be construed to be harmful interference.

LOCATING LEAKAGE SOURCES

When a cable company technician troubleshoots signal leakage, the process is similar to Amateur Radio fox hunting. The technician uses radio direction finding techniques that may include equipment such as handheld dipole or Yagi antennas, Doppler antenna arrays on vehicles, near-field probes, and commercially manufactured signal leakage detectors. Many leakage detectors incorporate what is known as "tagging" technology to differentiate a leaking cable signal from an over-the-air signal or electrical noise that may exist on or near the same frequency. Most leakage detection is done on a dedicated cable channel in the 108-138 MHz frequency range.

ELIMINATING LEAKAGE

A large percentage of leakage and ingress problems are not the result of a single shielding defect, although this does happen. For example, a squirrel might chew a hardline feeder cable, or a radial crack might develop in the shield as a consequence of environmental or mechanical damage. Most often, leakage and ingress are caused by several small shielding defects in an area: loose or corroded hardline connectors and splices, old copper braid subscriber drop cabling, improperly installed F connectors, subscriber-installed substandard "do-it-yourself" components, and the previously mentioned poorly shielded cable-ready TVs and other *customer premises equipment* (*CPE*).

After the cable technician locates the source(s) of the leakage, it is necessary to repair or replace the culprit components or cabling. In the case of poorly shielded TVs or VCRs, the cable technician cannot repair those devices, only recommend that they be fixed by a qualified service shop. Often the installation of a set-top box will take care of a cable-ready CPE problem because the subscriber drop cabling is no longer connected directly to the offending device. The latter usually resolves direct pickup problems, too, because the cable-ready TV must be tuned to channel 3 or 4 to receive the set-top signals.

VERIFYING AN RFI SOURCE TO BE LEAKAGE

Spurious signals, birdies, harmonics, intermodulation, electrical noise, and even interference from Part 15 devices are sometimes mistaken for cable signal leakage. One of the most common is emissions from Part 15 devices that become coupled to the cable TV coax shield in some way. Non-leakage noise or spurious signals may radiate from the cable TV lines or an amplifier location, but only because the outer surface of the cable shield is carrying the coupled interference as a common-mode current.

When interference from what sounds like a discrete carrier in the downstream spectrum is received, note the frequency and compare it to known analog TV channel visual and aural carrier frequencies. Consider the following example in which an interfering signal is being heard in the 70 cm Amateur band and it falls within cable channel 60 (438-444 MHz). When a cable operator carries an analog TV channel that complies with the STD channel frequency plan (Table 27.3), the visual carrier is at 439.25 MHz. The aural carrier falls 4.5 MHz above the visual carrier, or 443.75 MHz. If the cable operator were using HRC channelization rather than STD channelization, channel 60's visual carrier would be at 438.0219 MHz and its aural carrier at 442.5219 MHz. Can you hear anything at either visual carrier frequency? What about the aural carrier frequencies? Since the visual carrier is the strongest part of an analog TV channel, that's the best place to listen for leakage.

The color subcarrier of STD channel 60 falls at 439.25 + 3.58 = 442.83 MHz, but this is a double sideband suppressed carrier. There are horizontal sidebands spaced every 15.734 kHz from the visual carrier (and also from the color subcarrier), but these are generally very low level. Analog TV channel visual carriers use AM, and the aural carriers use FM. If the interference is from the aural carrier of an analog TV channel, you may be able to hear the channel's audio. If the interfering carrier does not fall on expected cable frequencies, it may not be leakage.

A common non-leakage interference that may radiate from a cable network is broadband electrical interference or other noise in the MF and lower end of the HF spectrum. A common misconception is that since cable companies carry digital signals on frequencies that overlap portions of the over-the-air spectrum below 30 MHz, any "noise" that radiates from the cable plant must be leaking digital signals. This type of interference is almost always power-line electrical interference or other noise that is coupled to the cable network's shield as a common-mode signal.

Upstream digital signals from cable modems, which have channel bandwidths of 1.6, 3.2 or 6.4 MHz, are typically transmitted in the roughly 20 to 40 MHz range, and are bursty in nature rather than continuous like downstream digital signals. Set-top box upstream telemetry carriers are narrowband frequency shift keying (FSK) or quadrature phase shift keying (QPSK) carriers usually in the approximately 8 to 11 MHz range. This type of interference may respond to common-mode chokes. Radiated non-leakage noise-like interference from the outside of the hardline coaxial cables that happens outdoors, affecting MF and/or HF reception in the neighborhood, cannot be fixed using chokes. The source of the interference must be identified and repaired. A little RFI detective work may be necessary. Refer to the various RFI troubleshooting sections of this chapter.

If normal leakage troubleshooting techniques do not clearly identify the source of the interference, sometimes the cable company may temporarily shut off its network in the affected neighborhood. If the interference remains after the cable network is turned off, it is not leakage, and the cable company is not responsible for that type of interference. If the interference disappears when the cable network is turned off, then it most likely is leakage or something related to the cable network. Turning off even a small portion of the cable network is a last resort and may not be practical because of the service disruption to subscribers. It may be easier for the cable company to temporarily shut off a suspect cable channel briefly. Here, too, if the interference remains after the channel is turned off, the interference is not leakage.

HOW TO REPORT LEAKAGE

If you suspect cable signal leakage is causing interference to your Amateur station, *never attempt your own repairs to any part of the cable network, even the cabling in your own home*! Document what you have observed. For instance, note the frequency or frequencies involved, the nature of the interference, any changes to the interference with time of day, how long it has been occurring, and so forth. If you have fox-hunting skills and equipment, you might note the probable source(s) of the interference or at least the direction from which it appears to be originating.

Next, contact the cable company. You will most likely reach the cable company's customer service department, but ask to speak with the local cable system's Plant Manager (may also be called Chief Engineer, Director of Engineering, Chief Tech, VP of Engineering, or similar), and explain to him or her that you are experiencing what you believe to be signal leakage-related interference. If you cannot reach this individual, ask that a service ticket be created, and a technician familiar with leakage and ingress issues be dispatched. Share the information you have gathered about the interference. And as with all RFI issues, remember diplomacy!

In the vast majority of cases when cable leakage interference to Amateur Radio occurs, it is able to be taken care of by working with local cable system personnel. Every now and then for whatever reason, the affected ham is unable to get the interference resolved locally. Contact the ARRL for help in these cases.

27.8.5 DVD and Video Players

A DVD or similar video player usually contains a television tuner or has a TV channel output. It is also connected to an antenna or cable system and the ac line, so it is subject to all of the interference problems of a TV receiver. Start by proving that the video player is the susceptible device. Temporarily disconnect the device from the television or video monitor. If there is no interference to the TV, then the video player is the most likely culprit.

Next, find out how the interfering signal is getting into the video player. Temporarily disconnect the antenna or cable feed line from the video player. If the interference goes away, then the antenna line is involved. In this case, you can probably fix the problem with a common-mode choke or high-pass filter.

Fig 27.21 shows a bulletproof video player installation. If you have tried all of the cures shown and still have a problem, the player is probably subject to direct pickup. In this case, contact the manufacturer through the EIA.

Some analog-type VCRs are quite susceptible to RFI from HF signals. The video baseband signal extends from 30 Hz to 3.5 MHz, with color information centered around 3.5 MHz and the FM sound subcarrier at 4.5 MHz. The entire video baseband is frequency modulated onto the tape at frequencies up to 10 MHz. Direct pickup of strong signals by VCRs is a common problem and may not be easily solved, short of replacing the VCR with a better-shielded model.

27.8.6 Non-radio Devices

Interference to non-radio devices is not the fault of the transmitter. (A portion of the *FCC Interference Handbook,* 1990 Edition, is shown in **Fig 27.24**. Although the FCC no longer offers this *Handbook,* an electronic version is available from ARRL at **www.arrl.org/files/file/Technology/FCC RFI Information/tvibook.pdf**.) In essence, the FCC views non-radio devices that pick up nearby radio signals as improperly functioning; contact the manufacturer and return the equipment. The FCC does not require that non-radio devices include RFI protection and they don't offer legal protection to users of these devices that are susceptible to interference.

TELEPHONES

Telephones have probably become the number one, non-TVI interference problem of Amateur Radio. However, most cases of telephone interference can be cured by correcting any installation defects and installing telephone RFI filters where needed.

Telephones can improperly function as radio receivers. Semiconductor devices inside many telephones act like diodes. When such a telephone is connected to the telephone wiring (a large antenna) an AM radio receiver can be formed. When a nearby transmitter goes on the air, these telephones can be affected.

Troubleshooting techniques were discussed earlier in the chapter. The suggestion to simplify the problem applies especially to telephone interference. Disconnect all telephones except one, right at the service entrance if possible, and start troubleshooting the problem there.

If any single device or bad connection in the phone system detects RF and puts the detected signal back onto the phone line as audio, that audio cannot be removed with filters. Once the RF has been detected and turned into audio, it cannot be filtered out because the interference is at the same frequency as the desired audio signal. To effect a cure, you must locate the detection point and correct the problem there.

Defective telephone company lightning arrestors can act like diodes, rectifying any nearby RF energy. Telephone-line amplifiers or other electronic equipment may also be at fault. Do not attempt to diagnose or repair any telephone company wiring or devices on the "telco" side of your service box or that were installed by the phone company. Request a service call from your phone company.

Inspect the telephone system installation. Years of exposure in damp basements, walls or crawl spaces may have caused deterioration. Be suspicious of anything that is corroded or discolored. In many cases, homeowners have installed their own telephone

---

PART II

INTERFERENCE TO OTHER EQUIPMENT

CHAPTER 6

TELEPHONES, ELECTRONIC ORGANS, AM/FM RADIOS, STEREO AND HI-FI EQUIPMENT

Telephones, stereos, computers, electronic organs and home intercom devices can receive interference from nearby radio transmitters. When this happens, the device improperly functions as a radio receiver. Proper shielding or filtering can eliminate such interference. The device receiving interference should be modified in your home while it is being affected by interference. This will enable the service technician to determine where the interfering signal is entering your device.

The device's response will vary according to the interference source. If, for example, your equipment is picking up the signal of a nearby two-way radio transmitter, you likely will hear the radio operator's voice. Electrical interference can cause sizzling, popping or humming sounds.

Fig 27.24 — Part of page 18 from the FCC *Interference Handbook* (1990 edition) explains the facts and places responsibility for interference to non-radio equipment.

wiring, often using substandard wiring. If you find sections of telephone wiring made from nonstandard cable, replace it with standard twisted-pair telephone or CAT5 cable. If you do use telephone cable, be sure it is high-quality twisted-pair to minimize differential-mode pickup of RF signals.

Next, evaluate each of the telephone instruments. If you find a susceptible telephone, install a telephone RFI filter on that telephone, such as those sold by K-Y Filters. (**www.ky-filters.com**) If the home uses a DSL broadband data service, be sure that the filters do not affect DSL performance by testing online data rates with and without a filter installed at the telephone instrument.

If you determine that you have interference only when you operate on one particular ham band, the telephone wiring is probably resonant on that band. Install common-mode chokes on the wiring to add a high impedance in series with the "antenna." A telephone RFI filter may also be needed. (See the section on DSL Equipment for filtering suggestions.)

Telephone Accessories — Answering machines and fax machines are also prone to interference problems. All of the troubleshooting techniques and cures that apply to telephones also apply to these telephone devices. In addition, many of these devices connect to the ac mains. Try a common-mode choke and/or ac-line filter on the power cord (which may be an ac cord set, a small transformer or power supply).

Cordless Telephones — A cordless telephone is an unlicensed *radio* device that is manufactured and used under Part 15 of the FCC regulations. The FCC does not intend Part 15 devices to be protected from interference. These devices usually have receivers with very wide front-end filters, which make them very susceptible to interference.

A likely path for interference to cordless phones is as common-mode current on the base unit's connecting cables that will respond to common-mode chokes. In addition, a telephone filter on the base unit and an ac line filter may help. The best source of help is the manufacturer but they may point out that the Part 15 device is not protected from interference.

Cordless phone systems operating at 900 MHz and higher frequencies are often less susceptible to interference than older models that use 49 MHz. It may be easiest to simply replace the system with a model less susceptible to interference.

AUDIO EQUIPMENT

Consumer and commercial audio equipment such as stereos, home entertainment systems, intercoms and public-address systems can also pick up and detect strong nearby transmitters. The FCC considers these non-radio devices and does not protect them from

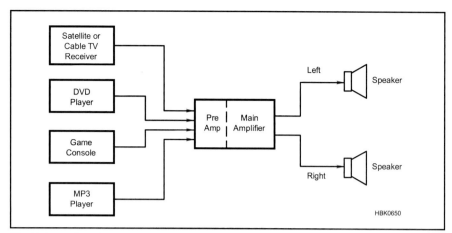

Fig 27.25 — A typical modern home-entertainment system.

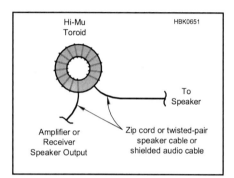

Fig 27.26 — Making a speaker-lead common-mode choke. Use ferrite material appropriate for the frequency of the RF interference.

licensed radio transmitters that may interfere with their operation. The RFI can be caused by one of several things: pickup on speaker leads or interconnecting cables, pickup by the ac mains wiring or direct pickup. If the interference involves wiring connected to the affected device, common-mode chokes are the most likely solution.

Use the standard troubleshooting techniques discussed earlier in this chapter to isolate problems. In a multi-component home entertainment system (as in **Fig 27.25**), for example, you must determine what combination of components is involved with the problem. First, disconnect all auxiliary components to determine if there is a problem with the main receiver/amplifier. Long speaker or interconnect cables are prime suspects.

Stereos and Home Entertainment Systems — If the problem remains with the main amplifier isolated, determine if the interference level is affected by the volume control. If so, the interference is getting into the circuit *before* the volume control, usually through accessory wiring. If the volume control has no effect on the level of the interfering sound, the interference is getting in *after* the control, usually through speaker wires.

Speaker wires are often effective antennas on HF and sometimes into VHF and above. The speaker terminals are often connected directly to the output amplifier transistors. Modern amplifier designs use a negative feedback loop to improve fidelity. This loop can conduct the detected RF signal back to the high-gain stages of the amplifier. The combination of all of these factors often makes the speaker cables the dominant receiving antenna for RFI.

There is a simple test that will help determine if the interfering signal is being coupled into the amplifier by the speaker leads. Temporarily disconnect the speaker leads from the amplifier, and plug in a test set of headphones with short leads. If there is no interference with the headphones, filtering the speaker leads will likely cure the problem.

Start by applying common-mode chokes. **Fig 27.26** shows how to wrap speaker wires around an large (2-inch O.D. or larger) ferrite core to cure speaker-lead RFI. Type 31 material is preferred at HF. (See the section on Common-Mode Chokes in this chapter.)

In some cases, the speaker wires may be picking up RF as a differential-mode signal. To reduce differential-mode pickup, replace the zip cord speaker wire with twisted-pair wire. (#16 AWG will work for most systems with higher-power systems requiring #12 AWG.)

Powered Speakers — A powered speaker is one that has its own built-in power amplifier. Powered subwoofers are common in home entertainment systems and small powered speakers are often used with computer and gaming systems. If a speaker runs on batteries and/or an external power supply, or is plugged into mains power, it is a powered loudspeaker. Powered loudspeakers are notoriously susceptibility to common-mode interference from internally misconnected cable shields and poor shielding. Apply suitable common-mode chokes to all wiring, including power wiring. If the RFI persists, try an RF filter at the input to the speaker, such as the LC low-

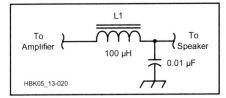

Fig 27.27 — A low-pass LC filter.

pass filter in **Fig 27.27**. Unshielded speakers may not be curable, however.

Intercoms and Public-Address Systems — RFI to these systems is nearly always caused by common-mode current on interconnect wiring. Common-mode chokes are the most likely cure, but you may also need to contact the manufacturer to see if they have any additional, specific information. Wiring can often be complex, so any work on these systems should be done by a qualified sound contractor.

COMPUTERS AND OTHER UNLICENSED RF SOURCES

Computers and microprocessor-based devices such as video games or audio players can be sources or victims of interference. These devices contain oscillators that can and do radiate RF energy. In addition, the internal functions of a computer generate different frequencies, based on the various data signals. All of these signals are digital — with fast rise and fall times that are rich in harmonics.

Computing devices are covered under Part 15 of the FCC regulations as unintentional emitters. As for any other unintentional emitter, the FCC has set absolute radiation limits for these devices. As previously discussed in this chapter, FCC regulations state that the operator or owner of Part 15 devices must take whatever steps are necessary to reduce or eliminate any interference they cause to a licensed radio service. This means that if your neighbor's video game interferes with your radio, the neighbor is responsible for correcting the problem. (Of course, your neighbor may appreciate your help in locating a solution!)

The FCC has set up two tiers of limits for computing devices. Class A is for computers used in a commercial environment. FCC Class B requirements are more stringent — for computers used in residential environments. If you buy a computer or peripheral, be sure that it is Class B certified or it will probably generate interference to your amateur station or home-electronics equipment.

If you find that your computer system is interfering with your radio (not uncommon in this digital-radio age), start by simplifying the problem. Temporarily remove power from as many peripheral devices as possible and disconnect their cables from the back of the computer. (It is necessary to physically

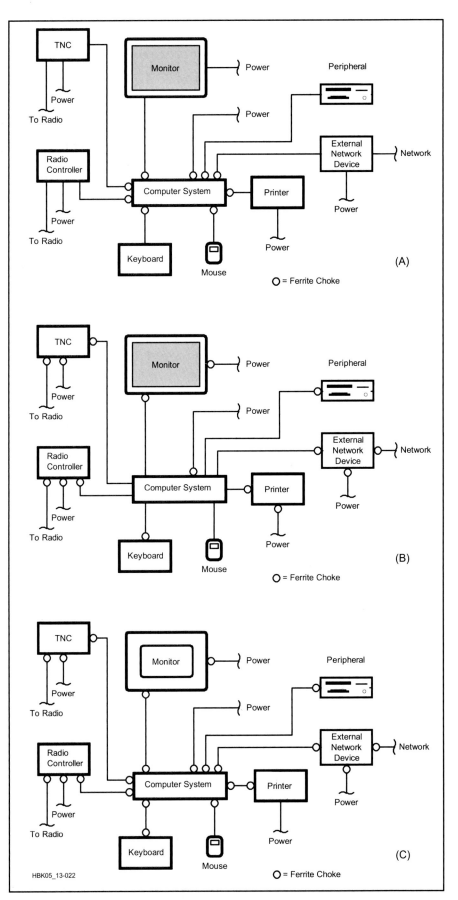

Fig 27.28 — Where to locate ferrites in a computer system. At A, the computer is noisy, but the peripherals are quiet. At B, the computer is quiet, but external devices are noisy. At C, both the computer and externals are noisy.

remove the power cable from the device, since many devices remain in a low-power state when turned off from the front panel or by a software command.) If possible, use just the computer, keyboard and monitor. This test may identify specific peripherals as the source of the interference.

Ensure that all peripheral connecting cables are shielded. Replace any unshielded cables with shielded ones; this often significantly reduces RF noise from computer systems. The second line of defense is the common-mode choke. The choke should be installed as close to the computer and/or peripheral device as practical. **Fig 27.28** shows the location of common-mode chokes in a complete computer system where both the computer and peripherals are noisy.

Switchmode power supplies in computers are often sources of interference. A common-mode choke and/or ac-line filter may cure this problem. In extreme cases of computer interference you may need to improve the shielding of the computer. (Refer to the *ARRL RFI Book* for more information about this.) Don't forget that some peripherals (such as modems) are connected to the phone line, so you may need to treat them like telephones.

27.9 Automotive RFI

Automobiles have evolved from a limited number of primitive electrical components to high technology, multi-computer systems on-wheels. Every new technology deployed can potentially interfere with amateur equipment.

Successful mobile operation depends on a multitude of factors such as choosing the right vehicle, following installation guidelines, troubleshooting and deploying the appropriate RFI fixes as needed.

A number of these factors will be covered in this section. Additionally, newly emerging electrical and hybrid-electric vehicles will be discussed, which pose unique challenges to amateur equipment installations and operation.

27.9.1 Before Purchasing a Vehicle

When shopping for a new vehicle intended for a mobile amateur installation, begin with research. A wealth of information is available on the Internet, and specifically at **www.arrl.org/automotive**, where the ARRL has compiled years of data from automotive manufacturers and other hams. Email reflectors and Web sites may provide information from hams willing to share their experiences concerning mobile communications in their own vehicles that may be the very make and model you were considering.

Armed with research, your next stop is your dealer. The manufacturer of each vehicle is the expert on how that vehicle will perform. The dealer should have good communication with the manufacturer and should be able to answer your questions. Ask about service bulletins and installation guidelines. You can also ask your dealer about fleet models of their vehicles. Some manufacturers offer special modifications for vehicles intended for sale to police, taxicabs and other users who will be installing radios (usually operating at VHF and UHF) in their cars.

When shopping for a vehicle, it is useful to take along some portable (preferably battery operated) receivers or scanners and have a friend tune through your intended operating frequencies while you drive the vehicle. This will help identify any radiated noise issues associated with that model vehicle, which can be more difficult to resolve than conducted noise. If you intend to make a permanent transceiver installation, give some consideration to how you will mount the transceiver and route the power and/or antenna cables. While looking for ways to route the wiring, keep in mind that some newer cars have the battery located in the trunk or under the rear seat, and that may make power wire routing easier.

Test the car before you buy it. A dealer expects you to take the car for a test ride; a cooperative dealer may let you test it for radio operation, too. A fair amount of checking can easily be performed without digging too deeply into the car. Check the vehicle for noise with a portable receiver on VHF, where your handheld transceiver will do the job nicely. On HF, you can usually locate noise with a portable short-wave receiver, or operate your HF transceiver with a portable antenna and cigarette-lighter plug. With the engine running, tune across the bands of interest. You may hear some noise, or a few birdies, but if the birdies don't fall on your favorite frequency, this is an encouraging sign! Check with the vehicle completely off, with the key in the ignition, and with the vehicle running — electronic subsystems operate in different ways with the vehicle running or not running.

To test the vehicle for susceptibility to your transmitted signal, you must transmit. It is important to note that without a full and complete installation, you will not be able to fully assess the effects of full-power transmissions on a vehicle. Any testing done with temporary equipment installations cannot be considered an absolute guarantee because an installed transmitter connected directly to the vehicle's power source may cause the vehicle to act differently.

To perform transmit tests, bring your radio and a separate battery (if permitted by dealer) so you can transmit at full power while in motion without having to run cables to the vehicle battery. Use a magnet-mount antenna (several *QST* advertisers sell mounts suitable for HF) for temporary testing. (Use the magnet-mount carefully; it is possible to scratch paint if any particles of dirt get on the bottom of the magnets.) Transmit on each band you will use to see if the RF has any effect on the vehicle. Lack of response to your transmissions is a good sign, but does not mean the vehicle is immune to RF as a permanent installation will result in different (likely stronger) field strengths and distributions in and around the vehicle and a permanent antenna more effectively coupled to the vehicle.

On both transmit and receive, you may want to experiment with the placement of the antenna. Antenna placement plays an important role in operation, and you may be able to find an optimal location for the antenna that predicts good performance with a permanent installation.

27.9.2 Transceiver Installation Guidelines

While most amateurs are familiar with the process of installing a transceiver, there are preferred practices that will help minimize potential problems. These include support from the automotive dealer, typical "best practices" installations, and consideration of special situations.

The first step is to ensure that your installation complies with both the vehicle manufacturer's and radio transceiver manufacturer's installation guidelines. Links to domestic automotive manufacturer installation guidelines are found at **www.arrl.org/automotive**. Automotive manufacturers that import vehicles for sale here do not publish installation guidelines because their vehicles are not typically used in police, fire and taxicab applications within the US.

The installation guidelines of different manufacturers vary as to how to install a radio transceiver's power leads. Most manufacturers recommend that the positive and negative leads from the radio be run directly to the battery. This minimizes the potential for the interaction between the radio's negative lead currents and vehicle electronics. If the manufacturer recommends that both wires be connected to the battery, they will also require that both wires be fused. This is necessary because, in the unlikely event that the connection between the battery and the engine block were to fail, excessive current could be drawn on the radio's negative lead when the vehicle starter is engaged.

Some vehicles provide a "ground block" near the battery for a negative cable to be connected. On these vehicles, run the negative power lead, un-fused, to the "ground block." When this technique is recommended by the manufacturer, the interaction between the power return currents and vehicle electronics has been evaluated by the manufacturer. In all cases, the most important rule to remember is this: If you want the manufacturer to support your installation, do it exactly the way the installation guidelines tell you to do it!

If no installation guidelines are available for your vehicle, the practices outlined below will improve compatibility between in-vehicle transceivers and vehicle electronics:

1) Transceivers
- Transceivers should be mounted in a location that does not interfere with vehicle operator controls and visibility, provides transceiver ventilation, and be securely mounted.
- Ensure all equipment and accessories are removed from the deployment path of the airbag and safety harness systems.

2) Power Leads
- The power leads should be twisted together from the back of the rig all the way to the battery. This minimizes the area formed by the power leads, reducing susceptibility to transients and RFI.
- Do not use the vehicle chassis as a power return.
- The power leads should be routed along the body structure, away from vehicle wiring harnesses and electronics.
- Any wires connected to the battery should be fused at the battery using fuses appropriate for the required current.
- Use pass-through grommets when routing wiring between passenger and engine compartments.
- Route and secure all under hood wiring away from mechanical hazards.

3) Coaxial Feed Lines
- The coaxial feed line should have at least 95% braid coverage. The cable shield should be connected to every coaxial connector for the entire circumference (no "pigtails").
- Keep antenna feed lines as short as practical and avoid routing the cables parallel to vehicle wiring.

4) Antennas
- Antenna(s) should be mounted as far from the engine and the vehicle electronics as practical. Typical locations would be the rear deck lid or roof. Metal tape can be used to provide an antenna ground plane on non-metallic body panels.
- Care should be used in mounting antennas with magnetic bases, since magnets may affect the accuracy or operation of the compass in vehicles, if equipped.
- Since the small magnet surface results in low coupling to the vehicle at HF, it is likely that the feed line shield will carry substantial RF currents. A large (2-inch OD or larger toroid) common-mode choke at the antenna will help reduce this current, but will also reduce any radiation produced by that current.
- Adjust the antenna for a low SWR.

27.9.3 Diagnosing Automotive RFI

Most VHF/UHF radio installations should result in no problems to either the vehicle systems or the transceiver, while HF installations are more likely to experience problems. In those situations where issues do occur, the vast majority are interference to the receiver from vehicle on-board sources of energy that are creating emissions within the frequency bands used by the receiver. Interference to one of the on-board electronic systems can be trivial or it can cause major problems with an engine control system.

The dealer should be the first point of contact when a problem surfaces, because the dealer should have access to information and factory help that may solve your problem. The manufacturer may have already found a fix for your problem and may be able to save your mechanic a lot of time (saving you money in the process). If the process works properly, the dealer/customer-service network can be helpful. In the event the dealer is unable to solve your problem, the next section includes general troubleshooting techniques you can perform independently.

GENERAL TROUBLESHOOTING TECHNIQUES

An important aspect is to use the source-path-victim model presented earlier in this chapter. The path from the source to the receiver may be via radiation or conduction. If the path is radiation, the electric field strength (in V/m) received is reduced as a function of the distance from the source to the receiver. In most cases, susceptible vehicle electronics is in the near-field region of the radiating source, where the electric and magnetic fields can behave in complex ways. In general, however, the strength of radiated signals falls off with distance.

The best part of all this is that with a general-coverage receiver or spectrum analyzer, a fuse puller and a shop manual, the vehicle component needing attention may be identified using a few basic techniques. The only equipment needed could be as simple as:
- A mobile rig, scanner or handheld transceiver, or
- Any other receiver with good stability and an accurate readout, and
- An oscilloscope for viewing interference waveforms

BROADBAND NOISE

Automotive broadband noise sources include:
- Electric motors such as those that operate fans, windows, sunroof, AM/FM antenna deployment, fuel pumps, etc.
- Ignition spark

If you suspect electric motor noise is the cause of the problem, obtain a portable AM or SSB receiver to check for this condition. Switch on the receiver and then activate the electric motors one at a time. When a noisy motor is switched on, the background noise increases. It may be necessary to rotate the radio, since portable AM radios use a directional ferrite rod antenna.

To check whether fuel pumps, cooling fans, and other vehicle-controlled motors are the source of noise, pull the appropriate fuse and see whether the noise disappears.

A note concerning fuel pumps: virtually every vehicle made since the 1980s has an electric fuel pump, powered by long wires. It may be located inside the fuel tank. Don't overlook this motor as a source of interference just because it may not be visible. Electric fuel-pump noise often exhibits a characteristic time pattern. When the vehicle ignition switch is first turned on, without engaging the starter, the fuel pump will run for a few seconds, and then shut off when the fuel system is pressurized. At idle, the noise will generally follow the pattern of being present for a few seconds before stopping, although in some vehicles the fuel pump will run almost continuously if the engine is running.

NARROWBAND NOISE

Automotive narrowband noise sources include:
- Microprocessor based engine control systems
- Instrument panel
- RADAR obstacle detection
- Remote keyless entry
- Key fob recognition systems
- Tire pressure monitoring systems
- Global positioning systems
- Pulse width modulating motor speed controls
- Fuel injectors
- Specialized electric traction systems found in newer hybrid/electric vehicles

Start by moving the antenna to different

locations. Antenna placement is often key to resolving narrowband RFI problems. However, if antenna location is not the solution, consider pulling fuses. Tune in and stabilize the noise, then find the vehicle fuse panels and pull one fuse at a time until the noise disappears. If more than one module is fed by one fuse, locate each module and unplug it separately. Some modules may have a "keep-alive" memory that is not disabled by pulling the fuse. These modules may need to be unplugged to determine whether they are the noise source. Consult the shop manual for fuse location, module location, and any information concerning special procedures for disconnecting power.

A listening test may verify alternator noise, but if an oscilloscope is available, monitor the power line feeding the affected radio. Alternator whine appears as full-wave rectified ac ripple and rectifier switching transients superimposed on the power system's dc power voltage (see **Fig 27.29**).

Alternators rely on the low impedance of the battery for filtering. Check the wiring from the alternator output to the battery for corroded contacts and loose connectors when alternator noise is a problem.

Receivers may allow conducted harness noise to enter the RF, IF or audio sections (usually through the power leads), and interfere with desired signals. Check whether the interference is still present with the receiver powered from a battery or power supply instead of from the vehicle. If the interference is no longer present when the receiver is operating from a battery or external supply, the interference is conducted via the radio power lead. Power line filters installed at the radio may resolve this problem.

27.9.4 Eliminating Automotive RFI

The next section includes various techniques to resolve the more common RFI problems. As a caveat, performing your own RFI work, in or out of warranty, you assume the same risks as you do when you perform any other type of automotive repair. Most state laws (and common sense) say that those who work on cars should be qualified to do so. In most cases, this means that work should be done either by a licensed dealer or automotive repair facility.

CONDUCTED INTERFERENCE

To reduce common-mode current, impedance can be inserted in series with the wiring in the form of common-mode chokes. (See this chapter's section on Common-Mode Chokes.) Wire bundles may also be wound around large toroids for the same effect.

Mechanical considerations are important in mobile installations. A motor vehicle is

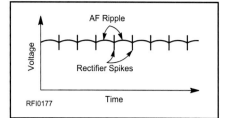

Fig 27.29 — Alternator whine consists of full-wave rectified ac, along with pulses from rectifier switching, superimposed on the vehicle's dc power voltage.

subject to a lot of vibration. If a choke is installed on a wire, this vibration may cause the choke to flex the wire, which may ultimately fail. It is critical that any additional shielding and/or chokes placed on wiring have been installed by qualified personnel who have considered these factors. These must be properly secured, and sometimes cable extenders are required to implement this fix.

RFI TO ON-BOARD CONTROL SYSTEMS

RFI to a vehicle's on-board control and electronic modules should be treated with common-mode chokes at the connection to the module. Some success has been reported by using braid or metal foil to cover a wire bundle as a shield and connecting the shield to the vehicle chassis near the affected module. Vehicle electronic units should not be modified except by trained service personnel according to the manufacturer's recommendations. The manufacturer may also have specific information available in the form of service bulletins.

FILTERS FOR DC MOTORS

If the motor is a conventional brush- or commutator-type dc motor, the following cures shown in **Fig 27.30** are those generally used. As always, the mechanic should consult with the vehicle manufacturer. To diagnose motor noise, obtain an AM or SSB receiver to check the frequency or band of interest. Switch on the receiver, and then activate the electric motors one at a time. When a noisy motor is switched on, the background noise increases as well.

The pulses of current drawn by a brush-commutator motor generate broadband RFI that is similar to ignition noise. However, the receiver audio sounds more like bacon frying rather than popping. With an oscilloscope displaying receiver audio, the noise appears as a series of pulses with random space between the pulses. Such broadband noise generally has a more pronounced effect on AM receivers than on FM. Unfortunately, the pulses may affect FM receivers by increasing the "background noise level" and will reduce perceived receiver sensitivity because of the degraded signal-to-noise ratio.

ALTERNATOR AND GENERATOR NOISE

As mentioned previously, brush-type motors employ sliding contacts which can generate noise. The resulting spark is primarily responsible for the "hash" noise associated with these devices. Hash noise appears as overlapping pulses on an oscilloscope connected to the receiver audio output. An alternator also has brushes, but they do not interrupt current. They ride on slip rings and supply a modest current, typically 4 A to the field winding. Hence, the hash noise produced by alternators is relatively minimal.

Generators use a relay regulator to control field current and thus output voltage. The voltage regulator's continuous sparking creates broadband noise pulses that do not overlap in time. They are rarely found in modern automobiles.

Alternator or generator noise may be conducted through the vehicle wiring to the power

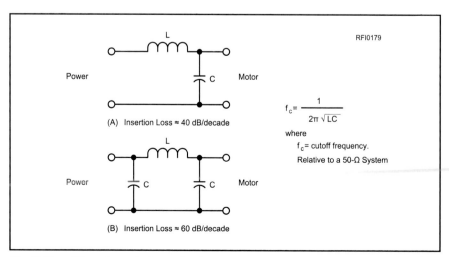

Fig 27.30 — Filters for reducing noise from dc motors

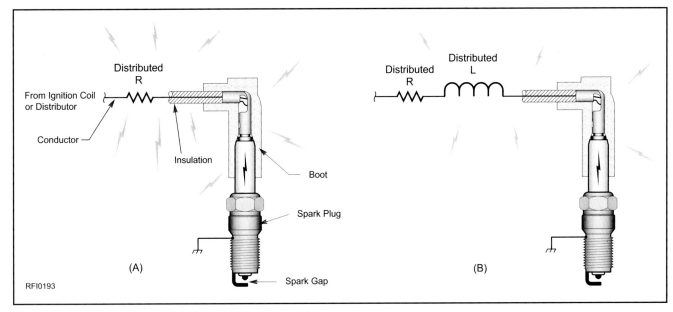

Fig 27.31 — Ignition noise suppression methods.

input of mobile receivers and transmitters and may then be heard in the audio output. If alternator or generator noise is suspected and an oscilloscope is not available, temporarily remove the alternator belt as a test. (This may not be possible in vehicles with a serpentine belt.)

IGNITION NOISE

Ignition noise is created by fast-rise-time pulses of coil current discharging across air gaps (distributor and spark plug). The theoretical models (zero rise time) of such pulses are called impulse functions in the time domain. When viewed in the frequency domain, the yield is a constant spectral energy level starting nearly at 0 Hz and theoretically extending up in frequency to infinity. In practice, real ignition pulses have a finite rise time, so the spectral-energy envelope decreases above some frequency.

It turns out that noise generated by ignition sparks and fuel injector activation manifest themselves as a regular, periodic "ticking" in the receiver audio output, which varies with engine RPM. If an oscilloscope were connected to the audio output, a series of distinct, separate pulses would appear. At higher speeds it sounds somewhat musical, like alternator whine, but with a harsher note (more harmonic content).

A distinguishing feature of ignition noise is that it increases in amplitude under acceleration. This results from the increase in the required firing voltage with higher cylinder pressure. (Noise at higher frequencies may also be reproduced better by the audio circuits.) Since ignition noise is usually radiated noise, it should disappear when the antenna element is disconnected from the antenna mount. The radiation may be from either the secondary parts of the system or it may couple from the secondary to the primary of the coil and be conducted for some distance along the primary wiring to the ignition system, then radiated from the primary wiring.

Two main methods are employed to suppress this noise — one involves adding an inductance, and the other involves adding a resistance — both in the secondary (high voltage) wiring. This is shown in **Fig 27.31**. The addition of these elements does not have a measurable effect on the engine operation, because the time constants involved in the combustion process are much longer than those associated with the suppression components. (Note that modifying your vehicle's ignition system may be considered as tampering with your vehicle's emission control system and may affect your warranty coverage — work with your dealer or limit your efforts to changing spark plug wires or possibly shielding them.)

The resistance method suppresses RFI by dissipating energy that would have been radiated and/or conducted. Even though the amount of energy dissipated is small, it is still enough to cause interference to sensitive amateur installations. The other method uses inductance and even though the energy is not dissipated, suppression occurs because the inductor will store the pulse energy for a short time. It then releases it into the ignition burn event, which is a low impedance path, reducing the RFI.

For traditional "Kettering" inductive discharge ignition systems, a value of about 5 kΩ impedance (either real and/or reactive) in the spark plug circuit provides effective suppression and, with this value, there is no detectable engine operation degradation. (Capacitor discharge systems, in comparison, are required to have very low impedance on the order of tens of ohms in order to not reduce spark energy, so they are not tolerant of series impedance). Most spark plug resistances are designed to operate with several kV across the plug gap, so a low-voltage ohmmeter may not give proper resistance measurement results.

The term "resistor wire" is somewhat misleading. High-voltage ignition wires usually contain both resistance and inductance. The resistance is usually built into suppressor spark plugs and wires, while there is some inductance and resistance in wires, rotors and connectors. The elements can be either distributed or lumped, depending on the brand, and each technique has its own merit. A side benefit of resistance in the spark plug is reduced electrode wear.

COIL-ON-PLUG IGNITION NOISE

Many newer spark-ignition systems incorporate a "coil on plug" (COP) or "coil near plug" (CNP) approach. There are advantages to this from an engine operation standpoint, and this approach may actually reduce some of the traditional sources of ignition system RFI. This is because of the very short secondary wires that are employed (or perhaps there are no wires — the coil is directly attached to the spark plug). This reduces the likelihood of coupling from the secondary circuit to other wires or vehicle/engine conductive structures.

There will always be some amount of energy from the spark event that will be conducted along the lowest impedance path. It may mean that the energy that would have been in the secondary circuit will be coupled back on the primary wiring harness attached to the coils.

This means that the problem may go from a radiated to a conducted phenomenon.

The fix for this in some cases may actually be easier or harder than one might think. Two approaches that have been used with success are ferrite cores and bypass capacitors.

Ferrite cores are recommended as the first choice, since they require no electrical modification to the vehicle. Ferrite clamp-on split cores are added to the 12-V primary harness attached to the coils. Depending on the frequency of the noise and selection of the ferrite material, there can be significant improvement (as much as 10 dB). Key to optimizing the amount of suppression is to determine where the noise "peaks" and selecting the correct ferrite material for that frequency range (see this chapter's section Using Ferrite for RFI Suppression).

The second method is to add a bypass capacitor between the primary wire of the 12-V coil and ground in the harness near the coil assemblies (there may be 2, 3 or 4 coils). This must be done carefully because it could affect the functionality of the ignition system and — perhaps most importantly — may void the vehicle warranty. This "bypass" capacitor performs the same function that bypass capacitors in any other application perform — separating the noise from the intended signal/power.

27.9.5 Electric and Hybrid-Electric Vehicles

Electric vehicles (EV) and hybrid-electric vehicles (HEV) are quickly becoming a practical means of transportation. EV/HEVs are advanced vehicles that pose unique challenges for amateur equipment. While EV/HEVs provide improved emissions and fuel economy, EV/HEVs utilize switched high voltage and high current to control propulsion. The switching techniques used generate RFI within much of our frequency bands — a cause for concern, particularly for HF operation.

This section is designed to enlighten vehicle owners to the challenges and to make suggestions when installing mobile amateur equipments in an EV/HEV.

EV AND HEV ARCHITECTURE

Most EVs and HEVs have similar electrical traction system (ETS) architectures consisting of a high voltage battery supplying energy to an inverter which controls an electric motor within a transmission connected to the drive wheels. The main difference between the two is that an HEV includes an internal combustion engine to aid in propulsion and a pure EV is strictly electrically powered.

The heart of the ETS is a device called an inverter. It simply converts dc voltage from the high voltage battery (typical voltage range from 42 to 350 V dc) to an ac waveform supplying the electric motor. This dc-to-ac conversion is performed by a matrix of six transistor switches. The switches chop the dc voltage into systematically varying pulses called pulse-width-modulation (PWM) to form an adjustable frequency and RMS voltage suitable to power electric motor.

In most cases, the ac voltage from the inverter is a three-phase waveform similar to industrial applications because three-phase motors can be smaller, more efficient, and provide greater torque than single-phase motors.

IMPORTANT — Bright orange cables connect the battery pack to the inverter and the inverter to the drive motor, transferring voltage and current to and from the inverter. Because of the non-sinusoidal waveforms being transferred, these cables are shielded and terminated at each end. Under no condition should these cables be disconnected or modified, because the high voltage system depends on a delicate balance of sensors and safety mechanisms. Possible malfunction and damage to the ETS may occur if modified.

EV AND HEV RFI CONCERNS

The inverter uses PWM to convert dc battery voltage to an ac waveform. The phase-to-phase terminal voltage appears in **Fig 27.32** as rectangular blocks with positive and negative amplitude equal to the battery voltage. For example, a 350 V dc battery pack will provide 700 V peak-to-peak at the motor terminals. In Fig 27.32, the same terminal voltage signal is sent through a low pass filter to show how PWM forms a sinusoidal waveform. Each pulse is essentially a square wave. Harmonics

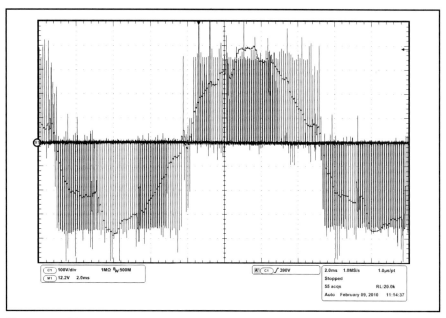

Fig 27.32 — Phase-to-phase motor terminal voltage.

from these pulses fall within most of our amateur HF bands, affecting radio performance. Because EV/HEV systems are evolving rapidly, check the ARRL's Automotive RFI Web page (**www.arrl.org/automotive**) for more information.

EV AND HEV RFI REMEDIES

Troubleshooting techniques described earlier apply in diagnosing RFI from EV/HEV systems. Limited RFI remedies are available associated to components within the ETS. Work with your dealer when you suspect the ETS as the RFI source. Do not attempt to modify or repair your ETS; the dealer's service center is most qualified to inspect and repair your EV/HEV electrical traction system.

During installation, mobile equipment power cables and antenna coaxial cables should be routed as far as possible from the bright orange cables. Common-mode chokes can decrease noise on 12-V dc power cables. Additionally, antenna placement plays a critical role in mobile equipment performance. Areas such as the top of a roof or trunk sometimes provide additional shielding.

27.9.6 Automotive RFI Summary

Most radio installations should result in no problems to either the vehicle systems or any issues with the transceiver. However, manufacturer, make and models differ, thus introducing challenges during amateur equipment installations.

Begin by researching your vehicle of interest and visiting the dealer. Insist on transmitting and receiving your favorite frequencies as you test drive. Request information pertaining to the manufacturer's transceiver installation guidelines. If manufacturer information is not available, follow the guidelines described earlier.

After installation, RFI problems may appear. Report you problem to the dealer, because they have access to manufacturer service bulletins which may describe a repair solution. Additional troubleshooting and remedies are also described previously to assist in successful communication.

Limited RFI remedies are available associated to components within the ETS. Work with your dealer when you suspect the ETS is the RFI source. Do not attempt to modify or repair your ETS; the dealer's service center is most qualified to inspect and repair your EV/HEV electrical traction system.

Lastly, the latest version of the *ARRL RFI Book* contains additional information on RFI in automobiles. More details are given about noise sources, troubleshooting techniques, a troubleshooting flow chart, additional filtering techniques, and information on EV/HEVs.

27.10 RFI Projects

Note: Additional RFI projects appear on the CD accompanying this book.

27.10.1 *Project*: RF Sniffer

Every home is full of electrical equipment capable of emitting electromagnetic radiation to interfere with radio amateurs trying to listen to signals on the bands. This project detects the radiation that causes problems to the amateur, and the noise can be heard. This device will allow you to demonstrate the "noise" with which we have to contend.

CONSTRUCTION

The circuit (**Fig 27.33**) uses a telephone pick-up coil as a detector, the output of which is fed into a LM741 IC preamplifier, followed by a LM386 IC power amplifier. See **Table 27.4** for the complete component list.

The project is built on a perforated board (**Fig 27.34**), with the component leads pushed through the holes and joined with hook-up wire underneath. There is a wire running around the perimeter of the board to form an earth bus.

Build from the loudspeaker backwards to

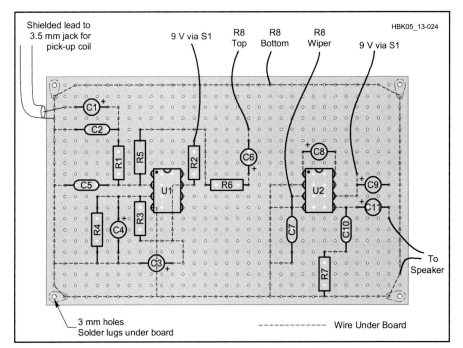

Fig 27.34 — The project is built on perforated board with point-to-point wiring underneath.

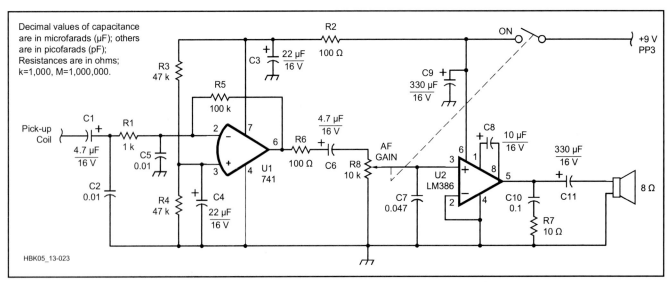

Fig 27.33 — The detector works by receiving stray radiation on a telephone pick-up coil and amplifying it to loudspeaker level.

Table 27.4
Components List

Resistors	Value
R1	1 kΩ
R2, R6	100 Ω
R3, R4	47 kΩ
R5	100 kΩ
R7	10 Ω
R8	10 kΩ, with switch

Capacitors	Value
C1, C6	4.7 µF, 16 V electrolytic
C2, C5	0.01 µF
C3, C4	22 µF, 16 V electrolytic
C7	0.047 µF
C8	10 µF, 16 V electrolytic
C9, C11	330 µF, 16 V electrolytic
C10	0.1 µF

Semiconductors	
U1	LM741
U2	LM386

Additional Items
LS1   Small 8-Ω loudspeaker
Perforated board
9 V battery and clip
3.5 mm mono-jack socket
Case
Telephone pick-up coil

Table 27.5
Readings (pick-up coil near household items)

29-MHz oscilloscope	0.56 V
Old computer monitor	0.86 V
Old computer with plastic case	1.53 V
New computer monitor	0.45 V
New tower PC with metal case	0.15 V
Old TV	1.2 V
New TV	0.4 V
Plastic-cased hairdryer	4.6 V
Vacuum cleaner	3.6 V
Drill	4.9 V

R8, apply power and touch the wiper of R8. If everything is OK you should hear a loud buzz from the speaker. Too much gain may cause a feedback howl, in which case you will need to adjust R8 to reduce the gain.

Complete the rest of the wiring and test with a finger on the input, which should produce a click and a buzz. The pick-up coil comes with a lead and 3.5 mm jack, so you will need a suitable socket.

RELATIVE NOISES

Place a high-impedance meter set to a low-ac-voltage range across the speaker leads to give a comparative readout between different items of equipment in the home. Sample readings are shown in **Table 27.5**.

27.11 RFI Glossary

Balanced circuit — A circuit whose two conductors have equal impedance to a common reference, such as a reference plane or circuit common.

Bond — (noun) A low-impedance, mechanically robust, electrical connection.

Common-mode — A voltage or current that is equal, in phase, and has the same polarity on all conductors of a cable. Common-mode current excites a cable as an antenna, and a cable acting as a receiving antenna produces common-mode current.

Conducted RFI — RFI received via a conducting path.

Coupled RFI — RFI received via inductive or capacitive coupling between conductors.

Differential mode — A signal that that is exists and is transmitted as a voltage *between* two conductors of a cable. At any instant, signal current on one conductor is equal to but of the opposite polarity to the current on the other conductor. Ordinary connections between equipment in systems are differential mode signals.

Disturbance — The improper operation of a device as a result of interference.

Electric field — The field present between two or more conductive objects as a result of potential difference (voltage) between those objects.

Electromagnetic field — The combination of a magnetic field and electric field in which the fields are directly related to each other, are at right angles to each other, and move through space as radio waves in a direction that is mutually perpendicular to both fields. An electrical conductor designed to produce electromagnetic fields when carrying an RF current is called an antenna.

Equipment ground — The connection of all exposed parts of electrical equipment to the Earth, or to a body that serves in place of the Earth.

Fundamental overload — 1. (Receiver Performance) Interference to a receiver caused by a signal at its input whose amplitude exceeds the maximum signal-handling capabilities of one or more receiver stages. 2. (RF interference) — Any disruption to the function of any RFI victim caused by the fundamental component of a transmitted signal or intended in-band output of a transmitter.

Ground — 1. A low impedance electrical connection to the Earth, or to a body that serves in place of the Earth. 2. A common signal connection in an electrical circuit.

Immunity — The ability of a device to function properly in the presence of unwanted electromagnetic energy. (After Ott, section 1.3)

Intentional radiator — A device that uses radio waves to transmit information by antenna action. A radio transmitter, with its associated antenna, is an intentional radiator.

Interference — 1. Disruption of a device's normal function as a result of an electromagnetic field, voltage, or current. 2. Disruption by a signal or noise of a receiver's ability to acquire and process a desired signal.

Magnetic field — The field produced by a permanent magnet or current flow through a conductor.

Path — The route by which electromagnetic energy is transferred from a transmitter to a receiver or from a source to a victim.

Radiated RFI — RFI received through radiation.

Shielding — A conductive barrier or enclosure interposed between two regions of space with the intent of preventing a field in one region from reaching the other region.

Source — A device that produces an electromagnetic, electric, or magnetic field, voltage, or current. If RFI is the result, the source is an *RFI source*.

Spurious emission — An emission outside the bandwidth needed for transmission of the mode being employed, the level of which may be reduced without reducing the quality of information being transmitted. Spurious emissions are most commonly the products of distortion (harmonics, intermodulation), of circuit instability (oscillation, including RF feedback), or of digital transmission with excessively fast rise times (including key clicks). Phase noise, such as that produced by a frequency synthesizer is also a spurious emission.

Susceptibility — The capability of a device to respond to unwanted electromagnetic energy. (After Ott, section 1.3)

System ground — A bond between one current-carrying conductor of the power system and the Earth.

Unintentional radiator — A device that produces RF as part of its normal operation but does not intentionally radiate it.

Victim — A device that receives interference from a *source*.

27.12 References and Bibliography

M. Gruber, ed, *The ARRL RFI Book*, Third Edition, ARRL, 2010

M. Loftness, *AC Power Interference*, Third Edition, Revised, distributed by the ARRL

H. Ott, *Electromagnetic Compatibility Engineering*, John Wiley and Sons, 2009

C. Paul, *Introduction to Electromagnetic Compatibility*, Wiley-Interscience, 1992

AES48-2005: AES standard on interconnections — Grounding and EMC practices — Shields of connectors in audio equipment containing active circuitry

Contents

28.1 Electrical Safety
    28.1.1 Station Concerns
    28.1.2. Do-It-Yourself Wiring
    28.1.3 National Electrical Code (NEC)
    28.1.4 Station Power
    28.1.5 Connecting and Disconnecting Power
    28.1.6 Ground-Fault Circuit Interrupters
    28.1.7 Low-Voltage Wiring
    28.1.8 Grounds
    28.1.9 Ground Conductors
    28.1.10 Antennas
    28.1.11 Lightning Transient Protection
    28.1.12 Other Hazards in the Shack
    28.1.13 Electrical Safety References
28.2 Antenna and Tower Safety
    28.2.1 Legal Considerations
    28.2.2 Antenna Mounting Structures
    28.2.3 Tower Construction and Erection
    28.2.4 Antenna Installation
    28.2.5 Weatherproofing Cable and Connectors
    28.2.6 Climbing Safety
    28.2.7 Antenna and Tower Safety References
28.3 RF Safety
    28.3.1 How EMF Affects Mammalian Tissue
    28.3.2 Researching Biological Effects of EMF Exposure
    28.3.3 Safe Exposure Levels
    28.3.4 Cardiac Pacemakers and RF Safety
    28.3.5 Low-Frequency Fields
    28.3.6 Determining RF Power Density
    28.3.7 Further RF Exposure Suggestions

Chapter 28

Safety

This chapter focuses on how to avoid potential hazards as we explore Amateur Radio and its many facets. The first section, updated by Jim Lux, W6RMK, details electrical safety, grounding and other issues in the ham shack. The following section on antenna and tower safety was written by Steve Morris, K7LXC, a professional tower climber and antenna installer with many years of experience. Finally, the ARRL RF Safety Committee explains good amateur practices, standards and FCC regulations as they apply to RF safety.

Safety First — Always

We need to learn as much as possible about what could go wrong so we can avoid factors that might result in accidents. Amateur Radio activities are not inherently hazardous, but like many things in modern life, it pays to be informed. Stated another way, while we long to be creative and innovative, there is still the need to act responsibly. Safety begins with our attitude. Make it a habit to plan work carefully. Don't be the one to say, "I didn't think it could happen to me."

Having a good attitude about safety is not enough, however. We must be knowledgeable about common safety guidelines and follow them faithfully. Safety guidelines cannot possibly cover all situations, but if we approach each task with a measure of common sense, we should be able to work safely.

Involve your family in Amateur Radio. Having other people close by is always beneficial in the event that you need immediate assistance. Take the valuable step of showing family members how to turn off the electrical power to your equipment safely. Additionally, cardiopulmonary resuscitation (CPR) training can save lives in the event of electrical shock. Classes are offered in most communities. Take the time to plan with your family members exactly what action should be taken in the event of an emergency, such as electrical shock, equipment fire or power outage. Practice your plan!

28.1 Electrical Safety

The standard power available from commercial mains in the United States for residential service is 120/240-V ac. The "primary" voltages that feed transformers in our neighborhoods may range from 1300 to more than 10,000 V. Generally, the responsibility for maintaining the power distribution system belongs to a utility company, electric cooperative or city. The "ownership" of conductors usually transfers from the electric utility supplier to the homeowner where the power connects to the meter or weather head. If you are unsure of where the division of responsibility falls in your community, a call to your electrical utility will provide the answer. **Fig 28.1** shows the typical division of responsibility between the utility company and the homeowner. This section is concerned more with wiring practices in the shack, as opposed to within the equipment in the shack.

There are two facets to success with electrical power: safety and performance. Since we are not professionals, we need to pursue safety first and consult professionals for alternative solutions if performance is unacceptable. The ARRL's Volunteer Consulting Engineers program involves professional engineers who may be able to provide advice or direction on difficult problems.

28.1.1 Station Concerns

There never seem to be enough power outlets in your shack. A good solution for small scale power distribution is a switched power strip with multiple outlets. The strip should be listed by a nationally recognized testing laboratory (NRTL) such as Underwriters Lab, UL, and should incorporate a circuit breaker. See the sidebars "What Does UL Listing Mean?" and "How Safe are Outlet Strips?" for warnings about poor quality products. It is poor practice to "daisy-chain" several power strips

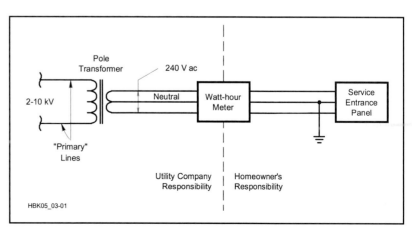

Fig 28.1 — Typical division of responsibility for maintenance of electrical power conductors and equipment. The meter is supplied by the utility company.

What Does UL Listing Mean?

UL is one of several nationally recognized testing laboratories (NRTLs), and probably the most well known. Listing *does not* mean what most consumers expect it to mean! More often than not the listing *does not* relate to the performance of the listed product. The listing simply indicates that a sample of the device meets certain manufacturers' construction criteria. Similar devices from the same or different manufacturers may differ significantly in overall construction and performance even though all are investigated and listed against the same UL product category. There is also a difference between a listed device and a listed component.

Many local laws and regulations, as well as the *NEC*, require that equipment and components used in electrical installations be listed by a NRTL. Some jurisdictions (Los Angeles County) require that any electrical equipment sold to consumers be listed.

The consumer must also be aware of the fine distinctions in advertising between a device or component that is advertised as "listed" or "designed to meet" or "meets." The latter two may not actually have been tested, or if tested, may have been tested by the manufacturer, and not an independent body.

It's also important to know that in some cases UL (and other standards organizations) only publish a standardized test procedure, but don't necessarily list or test the devices. Many standards also define varying levels of compliance, so knowing that your device meets some part of the standard may not be enough to know whether it meets *your* particular needs.

and may actually be a code violation. If you need more outlets than are available on a strip, have additional wall outlets installed.

Whether you add new outlets or use power strips, be sure not to overload the circuit. National and local codes set permissible branch capacities according to a rather complex process. Here's a safe rule of thumb: consider adding a new circuit if the total load is more than 80% of the circuit breaker or fuse rating. (This assumes that the fuse or breaker is correct. If you have any doubts, have an electrician check it.)

28.1.2. Do-It-Yourself Wiring

Amateurs sometimes "rewire" parts of their homes to accommodate their hobby. Most lo-cal codes *do* allow for modification of wiring (by building owners), so long as the electrical codes are met. Before making changes to your wiring, it would be wise to determine what rules apply and what agency has the authority to enforce them. This is called the *authority having jurisdiction* (AHJ) and it varies from location to location. Also see the following section on the National Electrical Code.

Generally, the building owner must obtain an electrical permit before beginning changes or additions to permanent wiring. Some jobs may require drawings of planned work. Often the permit fee pays for an inspector to review the work. Considering the risk of injury or fire if critical mistakes are left uncorrected, a permit and inspection are well worth the effort. *Don't take chances* — seek assistance from the building officials or an experienced electrician if you have *any* questions or doubts about proper wiring techniques.

Ordinary 120-V circuits are the most common source of fatal electrical accidents. Line voltage wiring must use an approved cable, be properly installed in conduit and junction boxes, within a chassis with a cover or lid, or other means described in the electrical code. Remember that high-current, low-voltage power sources, such as car batteries and high-current power supplies, can be just as dangerous as high-voltage sources, from melting metal, sparks and short circuits.

Never work on electrical wiring with the conductors energized! Switch off the circuit breaker or remove the fuse and take positive steps to ensure that others do not restore the power while you are working, such as using a circuit-breaker lockout. (**Fig 28.2** illustrates one way to ensure that power will be off until you want it turned on.) Check the circuit with an ac voltmeter to be sure that it is "dead" *each time you begin work.*

Before restoring power, check the wiring with an ohmmeter: From the installed outlet, there should be good continuity between the neutral conductor (white wire, "silver" screw) and the grounding conductor (green or bare wire, green screw). An ohmmeter should indicate a closed circuit between the conductors. (In the power line, high voltage world, line workers apply a shorting jumper before starting work so if the power does get reapplied, the safety jumper takes the hit.)

There should be no continuity between the hot conductor (black wire, "brass" screw) and the grounding conductor or the neutral conductor. With all other loads removed from the circuit (by turning off or unplugging them), an ohmmeter should indicate an *open* circuit between the hot wire and either of the other two conductors.

Commercially available plug-in testers are a convenient way to test regular three-wire receptacles, but can't distinguish between the neutral and ground being reversed

How Safe Are Outlet Strips?

The switch in outlet strips is generally *not* rated for repetitive *load break* duty. Early failure and fire hazard may result from using these devices to switch loads. Misapplications are common (another bit of bad technique that has evolved from the use of personal computers), and manufacturers are all too willing to accommodate the market with marginal products that are "cheap."

Nonindicating and poorly designed surge protection also add to the safety hazard of using power strips. MOVs with too low a threshold voltage often fail in a manner that could cause a fire hazard, especially in outlet strips that have nonmetallic enclosures, because day-to-day voltage surges that are otherwise unexceptional degrade the MOV each time, until it fails shorted.

A lockable disconnect switch or circuit breaker is a better and safer station master switch.

Fig 28.2 — If the switch box feeding power to your shack is equipped with a lock-out hole, use it. With a lock through the hole on the box, the power cannot be accidentally turned back on. *(Photo courtesy of American ED-CO)*

28.1.3 National Electrical Code (NEC)

Fortunately, much has been learned about how to harness electrical energy safely. This collective experience has been codified into the *National Electrical Code*, or *NEC*, simply known as "the code." The code details safety requirements for many kinds of electrical installations. Compliance with the *NEC* provides an installation that is *essentially* free from hazard, but not necessarily efficient, convenient or adequate for good service (paraphrased from NEC Article 90-1a and b). While the *NEC* is national in nature and sees wide application, it is not universal.

Local building authorities set the codes for their area of jurisdiction. They often incorporate the *NEC* in some form, while considering local issues. For example, Washington State specifically exempts telephone, telegraph, radio and television wires and equipment from conformance to electrical codes, rules and regulations. However, some local jurisdictions (city, county and so on) do impose a higher level of installation criteria, including some of the requirements exempted by the state.

Code interpretation is a complex subject, and untrained individuals should steer clear of the *NEC* itself. The *NEC* is not written to be understood by do-it-yourselfers, and one typically has to look in several places to find *all* the requirements. (For instance, Articles 810, 250, *and* 100 all contain things applicable to typical Amateur Radio installations.) Therefore, the best sources of information about code compliance and acceptable practices are local building officials, engineers and practicing electricians.

The Internet has a lot of information about electrical safety, the electrical code, and wiring practices, but you need to be careful to make sure the information you are using is current and not out of date. The ARRL Volunteer Consulting Engineer (VCE) program can help you find a professional who understands the amateur radio world, as well as the regulatory environment. There are also a variety of Web sites with useful information (such as www.mikeholt.com), but you need to be aware that advice may be specific to a particular installation or jurisdiction and not applicable for yours. With that said, let's look at a few *NEC* requirements for radio installations.

HOMEBREW AND "THE CODE"

In many cases, there are now legal requirements that electrical equipment have been listed by an NRTL, such as Underwriter Laboratories. This raises an issue for hams and homebrew gear, since it's unlikely you would take your latest project down to a test lab and pay them to evaluate it for safety.

For equipment that is not permanently installed, there's not much of an issue with homebrew, as far as the code goes, because the code doesn't deal with what's inside the equipment. For a lot of low voltage equipment, the code rules are fairly easy to meet, as well, as long as the equipment is supplied by a listed power source of the appropriate type.

The problem arises with permanent installations, where the scope of the code and local regulations is ever increasing. Such things as solar panel installations, standby generators,

About the National Electrical Code

Exactly how does the National Electrical Code become a requirement? How is it enforced?

Cities and other political subdivisions have the responsibility to act for the public safety and welfare. To address safety and fire hazards in buildings, regulations are adopted by local laws and ordinances usually including some form of permit and accompanying inspections. Because the technology for the development of general construction, mechanical and electrical codes is beyond most city building departments, model codes are incorporated by reference. There are several general building code models used in the US: Uniform, BOCA and Southern Building Codes are those most commonly adopted. For electrical issues, the *National Electrical Code* is in effect in virtually every community. City building officials will serve as "the authority having jurisdiction" (AHJ) and interpret the provisions of the *Code* as they apply it to specific cases.

Building codes differ from planning or zoning regulations: Building codes are directed only at safety, fire and health issues. Zoning regulations often are aimed at preservation of property values and aesthetics.

The *NEC* is part of a series of reference codes published by the National Fire Protection Association, a non-profit organization. Published codes are regularly kept up-to-date and are developed by a series of technical committees whose makeup represents a wide consensus of opinion. The *NEC* is updated every three years. It's important to know which version of the code your local jurisdiction uses, since it's not unusual to have the city require compliance to an older version of the code. Fortunately, the *NEC* is usually backward compatible: that is, if you're compliant to the 2008 code, you're probably also compliant to the 1999 code.

Do I have to update my electrical wiring as code requirements are updated or changed?

Generally, no. Codes are typically applied for new construction and for renovating existing structures. Room additions, for example, might not directly trigger upgrades in the existing service panel unless the panel was determined to be inadequate. However, the wiring of the new addition would be expected to meet current codes. Prudent homeowners, however, may want to add safety features for their own value. Many homeowners, for example, have added GFCI protection to bathroom and outdoor convenience outlets.

personal computers and home LANs all have received increased attention in local codes.

28.1.4 Station Power

Amateur Radio stations generally require a 120-V ac power source, which is then converted to the proper ac or dc levels required for the station equipment. In residential systems voltages from 110 V through 125 V are treated equivalently, as are those from 220 V through 250 V. Amateurs setting up a station in a light industrial or office environment may encounter 208 V line voltage. Most power supplies operate over these ranges, but it's a good idea to measure the voltage range at your station. (The measured voltage usually varies by hour, day, season and location.) Power supply theory is covered in the **Power Supplies** chapter.

Modern solid state rigs often operate from dc power, provided by a suitable dc power supply, perhaps including battery backups. Sometimes, the dc power supply is part of the rig (as in a 50-V power supply for a solid-state linear). Other times, your shack might have a 12-V (13.8 V) bus that supplies many devices. Just because it's low voltage doesn't mean that there aren't aspects of the system that raise safety concerns. A 15-A, 12-V power supply can start a fire as easily as a 15-A, 120-V branch circuit.

28.1.5 Connecting and Disconnecting Power

Something that is sometimes overlooked is that you need to have a way to safely disconnect all power to everything in the shack. This includes not only the ac power, but also battery banks, solar panels, and uninterruptible power supplies (UPS). Most hams won't have the luxury of a dedicated room with a dedicated power feed and the "big red switch" on the wall, so you'll have several switches and cords that would need to be disconnected.

The realities of today's shacks, with computers, multiple wall transformers ("wall-warts"), network interfaces and the radio equipment itself makes this tricky to do. One convenient means is a switched outlet strip, as used for computer equipment, if you have a limited number of devices. If you need more switched outlets, you can control multiple low-voltage controlled switched outlets from a common source. Or you can build or buy a portable power distribution box similar to those used on construction sites or stage sets; they are basically a portable subpanel with individual circuit breakers (or GFCIs, discussed later) for each receptacle, and fed by a suitable cord or extension cord. No matter what scheme you use, however, it's important that it be labeled so that someone else will know what to do to turn off the power.

International Power Standards

The power grid of the United States and Canada uses a frequency of 60 Hz and the voltage at ac power outlets is 120 V. This is also the case in other North American countries. If you travel, though, you'll encounter 220 V and 50 Hz with quite an array of plugs and sockets and color codes. If you are planning on taking amateur radio equipment with you on a vacation or DXpedition, you'll need to be prepared with the proper adapters and/or transformers to operate your equipment.

A table of international voltage and frequencies is provided on the CD-ROM accompanying this book, along with a figure showing the most common plug and socket configurations.

AC LINE POWER

If your station is located in a room with electrical outlets, you're in luck. If your station is located in the basement, an attic or other area without a convenient 120-V source, you may need to have a new line run to your operating position.

Stations with high-power amplifiers should have a 240-V ac power source in addition to the 120-V supply. Some amplifiers may be powered from 120 V, but they require current levels that may exceed the limits of standard house wiring. To avoid overloading the circuit and to reduce household light dimming or blinking when the amplifier is in use, and for the best possible voltage regulation in the equipment, it is advisable to install a separate 240 or 120-V line with an appropriate current rating if you use an amplifier.

The usual circuits feeding household outlets are rated at 15 or 20 A., This may or may not be enough current to power your station. To determine how much current your station requires, check the VA (volt-amp) ratings for each piece of gear. (See the **Electrical Fundamentals** chapter for a discussion of VA.) Usually, the manufacturer will specify the required current at 120 V; if the power consumption is rated in watts, divide that rating by 120 V to get amperes. Modern switching power supplies draw more current as the line voltage drops, so if your line voltage is markedly lower than 120 V, you need to take that into account.

Note that the code requires you to use the "nameplate" current, even if you've measured the actual current, and it's less. If the total current required is near 80% of the circuit's rating (12 A on a 15-A circuit or 16 A on a 20-A circuit), you need to install another circuit. Keep in mind that other rooms may be powered from the same branch of the electrical system, so the power consumption of any equipment connected to other outlets on the branch must be taken into account. If you would like to measure just how much power your equipment consumes, the inexpensive Kill-A-Watt meters by P3 International (**www.p3international.com**) measure volts, amps, VA and power factor.

If you decide to install a separate 120-V line or a 240-V line, consult the local requirements as discussed earlier. In some areas, a licensed electrician must perform this work. Others may require a special building permit. Even if you are allowed to do the work yourself, it might need inspection by a licensed electrician. Go through the system and get the necessary permits and inspections! Faulty wiring can destroy your possessions and take away your loved ones. Many fire insurance policies are void if there is unapproved wiring in the structure.

If you decide to do the job yourself, work closely with local building officials. Most home-improvement centers sell books to guide do-it-yourself wiring projects. If you have any doubts about doing the work yourself, get a licensed electrician to do the installation.

THREE-WIRE 120-V POWER CORDS

Most metal-cased electrical tools and appliances are equipped with three-conductor power cords. Two of the conductors carry power to the device, while the third conductor is connected to the case or frame. **Fig 28.3** shows two commonly used connectors.

When both plug and receptacle are properly wired, the three-contact polarized plug bonds the equipment to the system ground. If an internal short from line to case occurs, the "ground" pin carries the fault current and hopefully has a low enough impedance to trip the branch circuit breaker or blow the fuse in the device. A second reason for grounding the case is to reduce the possibility of shock for a user simultaneously connected to ground and the device. In modern practice, however, shock prevention is often done with GFCI circuit breakers as described below. These devices trip at a much lower level and are more reliable. Most commercially manufactured test equipment and ac-operated amateur equipment is supplied with three-wire cords.

It's a good idea to check for continuity from case to ground pin, particularly on used equipment, where the ground connection might have been broken or modified by the previous owner. If there is no continuity, have the equipment repaired before use.

Use such equipment only with properly installed three-wire outlets. If your house does not have such outlets, either consult a local electrician to learn about safe alternatives or have a professional review information you might obtain from online or other sources.

Equipment with plastic cases is considered "double insulated" and fed with a two-wire

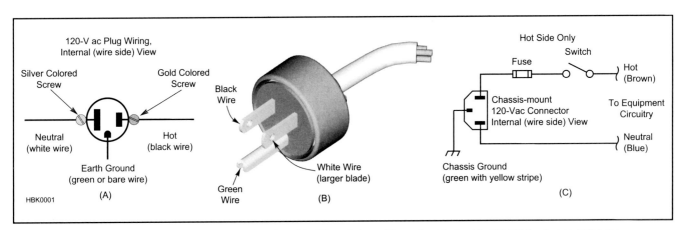

Fig 28.3 — 120 V ac plug wiring as viewed from the wire side (A) and viewed from the blade side (B). Wiring for an IEC type chassis connector is shown at C.

cord. Such equipment is safe because both conductors are insulated from the user by two layers. Nonetheless, there is still a hazard if, say, a double insulated drill were used to drill an improperly grounded case of a transmitter that was still plugged in. Remember, all insulation is prey to age, damage and wear that may erode its initial protection.

TRANSFER SWITCHES AND GENERATORS

More hams are adding standby generators and using alternate power sources such as solar panels or wind turbines, not as standalone systems like at Field Day, but interconnected with their home electrical system. These present some potential safety problems, such as preventing the local generator from "backfeeding" the utility's system during a power failure, and the fact that a solar panel puts out power whenever there is light falling on it.

For generators, the recommended approach is to use a *transfer switch*, which is a multipole switch that connects a selection of the house's circuits to the generator, rather than the utility power. The *NEC* and local regulations should be consulted for transfer switch selection and connection. The required wiring practices for permanently installed (stationary) generators are different from those for portable generators. Some issues that need to be considered are whether the neutral should be switched (many transfer switches do not switch the neutral, only the hot wire), and how the generator chassis is bonded to the building's grounding/bonding system. Most proper transfer switches are of the ON-OFF-ON configuration, with a mechanical interlock that prevents directly switching from one source to the other in a single operation.

The most dangerous thing to do with a generator is to use a so-called "suicide cord" with a male plug at each end: one end plugged into the generator's output receptacle and the other plugged into a convenient receptacle in the home. This is frequently illegal and at any rate should be avoided because of the inherent danger of having exposed, live contacts and the ease of overloading the circuit being fed.

Back-feeding your home's power panel should *never* be done unless the main breakers are in the OFF position or preferably removed. If your home's circuit-breaker panel does not have main breakers that can disconnect the external power line, *do not* use this technique to connect your generator to the home's wiring. Connect appliances to the generator directly with extension cords.

28.1.6 Ground-Fault Circuit Interrupters

GFCIs are devices that can be used with common 120-V circuits to reduce the chance of electrocution when the path of current flow leaves the branch circuit (say, through a person's body to another branch or ground). The *NEC* requires GFCI outlets in all wet or potentially wet locations, such as bathrooms, kitchens, and any outdoor outlet with ground-level access, garages and unfinished basements. Any area with bare concrete floors or concrete masonry walls should be GFCI equipped. GFCIs are available as portable units, duplex outlets and as individual circuit breakers. Some early units may have been sensitive to RF radiation but this problem appears to have been solved. Ham radio shacks in potentially wet areas (basements, out buildings) should be GFCI equipped. **Fig 28.4** is a simplified diagram of a GFCI.

Older equipment with capacitors in the 0.01 µF to 0.1 µF range connected between line inputs and chassis as an EMI filter (or that has been modified with bypass capacitors) will often cause a GFCI to trip, because of the leakage current through the capacitor. The must-trip current is 5 mA, but many GFCIs trip at lower levels. At 60 Hz, a 0.01 µF capacitor has an impedance of about 265 kΩ, so there could be a leakage current of about 0.5 mA from the 120-V line. If you had several pieces of equipment with such capacitors, the leakage current will trip the GFCI.

28.1.7 Low-Voltage Wiring

Many ham shacks use low-voltage control wiring for rotators or antenna relays. The electrical code isn't consistent in what it calls low voltage, but a guideline is "less than 50 V." Article 725 of the code contains most of the rules for low voltage/low power remote control and signaling, which is what hams are typically doing. These circuits are divided into three classes, with Class 1 being further subdivided, as shown in **Table 28.1**. There used to be code rules defining the classes in terms of power and voltage, but these days, the code is written so that the class of the circuit is defined by the power source, which has to be listed and labeled with the class.

Table 28.1
Traditional Divisions Among the Classes of Circuits

Class	Power	Notes
Class 1		
Power Limited	<30V, <1000VA	Transformer protected per Article 450. If not transformer, other overcurrent and fault protection requirements apply
Remote Control and Signaling	<600V	No limit on VA. Transformers protected as defined in Article 450
Class 2	Power supply <100VA Voltage <30V	
Class 3	Power supply <100VA Voltage <100V	

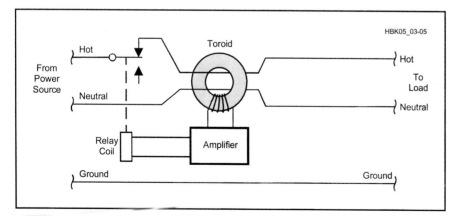

Fig 28.4 — Simplified diagram of a 120-V ac ground fault circuit interrupter (GFCI). When a stray current flows from the load (or outlet) side to ground, the toroidal current becomes unbalanced allowing detection, amplification and relay actuation to immediately cut off power to the load (and to the stray path!) GFCI units require a manual reset after tripping. GFCIs are required in wet locations (near kitchen sinks, in garages, in outdoor circuits and for construction work.) They are available as portable units or combined with over-current circuit breakers for installation in entrance panels.

That is, if you have something powered by a wall transformer that is listed and labeled as Class 2, the circuit is Class 2.

A typical example of a Class 1 Power Limited circuit that you might find in your home is 12-V low-voltage garden lighting or halogen lightning systems. A lot of amateur homebrew gear probably is also in this class, although because it's not made with "listed" components, it technically doesn't qualify. The other Class 1 would apply to a circuit using an isolation transformer of some sort.

Class 2 is very common: doorbells, network wiring, thermostats, and so on are almost all Class 2. To be Class 2, the circuit must be powered from a listed power supply that's marked as being Class 2 with a capacity less than 100 VA. For many applications that hams encounter, this will be the familiar "wall wart" power supply. If you have a bunch of equipment that runs from dc power, and you build a dc power distribution panel with regulators to supply them from a storage battery or a big dc power supply, you're most likely not Class 2 anymore, but logically Class 1. Since your homebrew panel isn't likely to be listed, you're really not even Class 1, but something that isn't covered by the code.

A common example of a Class 3 circuit that is greater than 30 V is the 70-V audio distribution systems used in paging systems and the like. Class 3 wiring must be done with appropriately rated cable

WIRING PRACTICES

Low voltage cables must be separated from power circuits. Class 2 and 3 cannot be run with Class 1 low voltage cables. They can't share a cable tray or the same conduit. A more subtle point is that the 2005 code added a restriction [Article 725.56(F)] that audio cables (speakers, microphone, etc.) cannot be run in the same conduit with other Class 2 and Class 3 circuits (like network wiring).

Low voltage and remote control wiring should not be neglected from your transient suppression system. This includes putting appropriate protective devices where wiring enters and leaves a building, and consideration of the current paths to minimize loops which can pick up the field from transient (or RF from your antenna).

28.1.8 Grounds

As hams we are concerned with at least four kinds of things called "ground," even if they really aren't ground in the sense of connection to the Earth. These are easily confused because we call each of them "ground."
1) Electrical safety ground (bonding)
2) RF return (antenna ground)
3) Common reference potential (chassis ground)
4) Lightning and transient dissipation ground

IEEE Std 1100-2005 (also known as the "Emerald Book," see the Reference listing, section 28.1.13) provides detailed information from a theoretical and practical standpoint for grounding and powering electrical equipment, including lightning protection and RF EMI/EMC concerns. It's expensive to buy, but is available through libraries.

ELECTRICAL SAFETY GROUND (BONDING)

Power-line ground is required by building codes to ensure the safety of life and property surrounding electrical systems. The *NEC* requires that all grounds be *bonded* together; this is a very important safety feature as well as an *NEC* requirement.

The usual term one sees for the "third prong" or "green-wire ground" is the "electrical safety ground." The purpose of the third, non-load current carrying wire is to provide a path to insure that the overcurrent protection will trip in the event of a line-to-case short circuit in a piece of equipment. This could either be the fuse or circuit breaker back at the main panel, or the fuse inside the equipment itself.

There is a secondary purpose for shock reduction: The conductive case of equipment is required to be connected to the bonding system, which is also connected to earth ground at the service entrance, so someone who is connected to "earth" (for example, standing in bare feet on a conductive floor) that touches the case won't get shocked.

An effective safety ground system is necessary for every amateur station, and the code requires that all the "grounds" be bonded together. If you have equipment at the base of the tower, generally, you need to provide a separate bonding conductor to connect the chassis and cases at the tower to the bonding system in the shack. The electrical safety ground provides a common reference potential for all parts of the ac system. Unfortunately, an effective bonding conductor at 60 Hz may present very high impedance at RF because of the inductance, or worse yet, wind up being an excellent antenna that picks up the signals radiated by your antenna.

RF GROUND

RF ground is the term usually used to refer to things like equipment enclosures. It stems from days gone by when the long-wire antenna was king. At low enough frequencies, a wire from the chassis or antenna tuner in the shack to a ground rod pounded in outside the window had low RF impedance. The RF voltage difference between the chassis

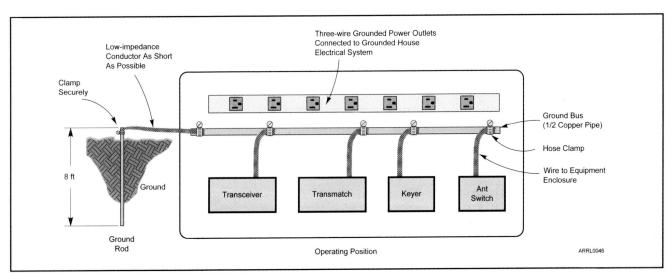

Fig 28.5 — An effective station ground bonds the chassis of all equipment together with low-impedance conductors and ties into a good Earth ground.

and "Earth ground" was small. And even if there were small potentials, the surrounding circuitry was relatively insensitive to them.

Today, though, we have a lot of circuits that are sensitive to interfering signals at millivolt levels, such as audio signals to and from sound cards. The summary is that we shouldn't be using the equipment enclosures or shielding conductors as part of the RF circuit.

Instead, we design our systems to create a common reference potential, called the "reference plane," and we endeavor to keep equipment connected to the reference plane at a common potential. This minimizes RF current that would flow between pieces of equipment. (See the **EMC** chapter for more information.)

Some think that RF grounds should be isolated from the safety ground system — *that is not true! All grounds, including safety, RF, lightning protection and commercial communications, must be bonded together in order to protect life and property*. The electrical code still requires that antenna grounds be interconnected (bonded) to the other "grounds" in the system, although that connection can have an RF choke. Remember that the focus of the electrical code bonding requirement is safety in the event of a short to a power distribution line or other transient.

COMMON REFERENCE POTENTIAL (CHASSIS GROUND)

For decades, amateurs have been advised to bond all equipment cabinets to an RF ground located near the station. That's a good idea, but it's not easily achieved. Even a few meters of wire can have an impedance of hundreds of ohms (1 µH/meter = 88 Ω/meter at 14 MHz). So a better approach is to connect the chassis together in a well-organized fashion to ensure that the chassis-to-chassis connections don't carry any RF current at all as in **Fig 28.5**. (See the **EMC** chapter for more information.)

LIGHTNING DISSIPATION GROUND

Lightning dissipation ground is concerned with conducting currents to the surrounding earth. There are distinct similarities between lightning dissipation ground systems and a good ground system for a vertical antenna. Since the lightning impulse has RF components around 1 MHz, it is an RF signal, and low inductance is needed, as well as low resistance.

The difference is that an antenna ground plane may handle perhaps a few tens of amps, while the lightning ground needs to handle a peak current of tens of kiloamperes.

A typical lightning stroke is a pulse with a rise time of a few microseconds, a peak current of 20-30 kA, and a fall time of 50 µs. The average current is not all that high (a few hundred amps), so the conductor size needed to carry the current without melting is surprisingly small.

However, large conductors are used in lightning grounds for other reasons: to reduce inductance, to handle the mechanical forces from the magnetic fields, and for ruggedness to prevent inadvertent breakage. A large diameter wire, or even better, a wide flat strap, has lower inductance. The voltage along a wire is proportional to the change in current and the inductance:

$$V = L \frac{di}{dt}$$

where
 di/dt = rate of change in current, about 20kA/2µs for lightning, or 10^9 A/s, and
 L = the inductance.

Consider a connection box on a tower that contains some circuitry terminating a control cable from the shack, appropriately protected internally with overvoltage protection. If the connection from the box to ground is high inductance, the lightning transient will raise the box potential (relative to the wiring coming from the shack), possibly beyond the point where the transient suppression in the box can handle it. Lowering the inductance of the connection to ground reduces the potential.

The other reason for large conductors on lightning grounds is to withstand the very high mechanical forces from the high currents. This is also the reason behind the recommendation that lightning conductors be run directly, with minimal bends, large radii for bends that are needed, and certainly no loops. A wire with 20,000 A has a powerful magnetic field surrounding it, and if current is flowing in multiple wires that are close to each other, the forces pushing the wires together or apart can actually break the conductors or deform them permanently.

The force between two conductors carrying 20,000 A, spaced a centimeter apart, is 8000 Newtons/meter of length (over 500 pounds/foot). Such forces can easily break cable strands or rip brackets and screws out. This problem is aggravated if there are loops in the wire, since the interaction of the current and its magnetic field tends to make the loop get larger, to the point where the wire will actually fail from the tension stresses.

GROUNDING METHODS

Earth ground usually takes one of several forms, all identified in the *NEC* and NFPA 780. The preferred earth ground, both as required in the NEC, and verified with years of testing in the field, is a concrete encased grounding electrode (CEGR), also known as a *Ufer ground*, after Herb Ufer, who invented it as a way to provide grounding for military installations in dry areas where ground rods are ineffective. The CEGR can take many forms, but the essential aspect is that a suitable conductor at least 20 feet long is encased in concrete which is buried in the ground. The conductor can be a copper wire (#8 AWG at least 20 ft long) or the reinforcing bars (rebar) in the concrete, often the foundation footing for the building. The connection to the rebar is either with a stub of the rebar protruding through the concrete's top surface or the copper wire extending through the concrete. There are other variations of the CEGR described in the *NEC* and in the electrical literature, but they're all functionally the same: a long conductor embedded in a big piece of concrete.

The electrode works because the concrete has a huge contact area with the surrounding soil, providing very low impedance and, what's also important, a low current density, so that localized heating doesn't occur. Concrete tends to absorb water, so it is also less susceptible to problems with the soil drying out around a traditional ground rod.

Ground rods are a traditional approach to making a suitable ground connection and are appropriate as supplemental grounds, say at the base of a tower, or as part of an overall grounding system. The best ground rods to use are those available from an electrical supply house. The code requires that at least 8 ft of the rod be in contact with the soil, so if the rod sticks out of the ground, it must be longer than 8 ft (10 ft is standard). The rod doesn't have to be vertical, and can be driven at an angle if there is a rock or hard layer, or even buried laying sideways in a suitable trench, although this is a compromise installation. Suitable rods are generally 10 ft long and made from steel with a heavy copper plating. Do not depend on shorter, thinly plated rods sold by some home electronics suppliers, as they can quickly rust and soon become worthless.

If multiple ground rods are installed, they should be spaced by at least half the length of the rod, or the effectiveness is compromised. IEEE Std 142 and IEEE Std 1100 (see the Reference listing) and other references have tables to give effective ground resistances for various configurations of multiple rods.

Once the ground rods are installed, they must be connected with either an exothermic weld (such as CadWeld) or with a listed pressure clamp. The exothermic weld is preferred, because it doesn't require annual inspection like a clamp does. Some installers use brazing to attach the wiring to the ground rods. Although this is not permitted for a primary ground, it is acceptable for secondary or redundant grounds. Soft solder (tin-lead, as used in plumbing or electrical work) should never be used for grounding conductors because it gets brittle with temperature cycling and can melt out if a current surge (as from a lightning strike) heats the conductor. Soft solder is specifically prohibited in the code.

Building cold water supply systems were used as station grounds in years past, but this is no longer recommended or even permitted in some jurisdictions, because of increased used of plastic plumbing both inside and outside houses and concerns about stray currents causing pipe corrosion.. If you do use the cold water line, perhaps because it is an existing grounding electrode, it must be bonded to the electrical system ground, typically at the service entrance panel.

28.1.9 Ground Conductors

The code is quite specific as to the types of conductors that can be used for bonding the various parts of the system together. Grounding conductors may be made from copper, aluminum, copper-clad steel, bronze or similar corrosion-resistant materials. Note that the sizes of the conductors required are based largely on mechanical strength considerations (to insure that the wire isn't broken accidentally) rather than electrical resistance. Insulation is not required. The "protective grounding conductor" (main conductor running to the ground rod) must be as large as the antenna lead-in, but not smaller than #10 AWG. The grounding conductor (used to bond equipment chassis together) must be at least #14 AWG. There is a "unified" grounding electrode requirement — it is necessary to bond *all* grounds to the electric service entrance ground. All utilities, antennas and any separate grounding rods used must be bonded together. **Fig 28.6** shows correct (A) and incorrect (B) ways to bond ground rods. **Fig 28.7** demonstrates the importance of correctly bonding ground rods. (Note: The *NEC* requirements do not address effective RF grounds. See the **EMC** chapter of this book for information about RF grounding practices, but keep in mind that RFI is not an acceptable reason to violate the *NEC*.) For additional information on good grounding practices, the IEEE "Emerald Book" (IEEE STD 1100-2005) is a good reference. It is available through libraries.

Additionally, the *NEC* covers safety inside the station. All conductors inside the building must be at least 4 inches away from conductors of any lighting or signaling circuit except when they are separated from other conductors by conduit or insulator. Other code requirements include enclosing transmitters in metal cabinets that are bonded to the grounding system. Of course, conductive handles and knobs must be grounded as well.

28.1.10 Antennas

Article 810 of the *NEC* includes several requirements for wire antennas and feed lines

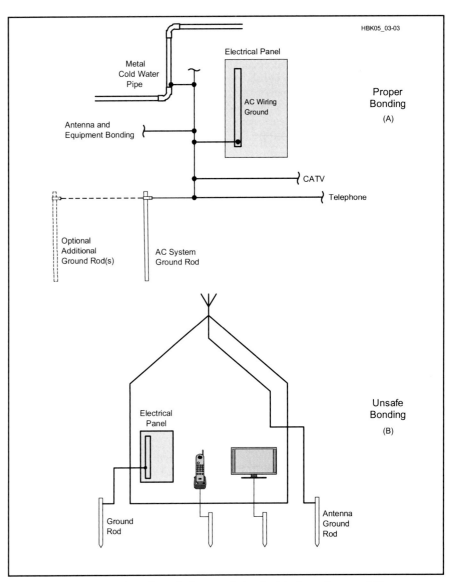

Fig 28.6 — At A, proper bonding of all grounds to electrical service panel. The installation shown at B is unsafe — the separate grounds are not bonded. This could result in a serious accident or electrical fire.

that you should keep in mind when designing your antenna system. The single most important thing to consider for safety is to address the potential for contact between the antenna system and power lines. As the code says, "One of the leading causes of electric shock and electrocution, according to statistical reports, is the accidental contact of radio, television, and amateur radio transmitting and receiving antennas and equipment with light or power conductors." (See Article 810.13, Fine Print Note.) The requirements in the code for wire sizes, bonding requirements, and installation practices are mostly aimed at preventing tragedy, by avoiding the contact in the first place, and by mitigating the effects of a contact if it occurs.

Article 820 of the *NEC* applies to Cable TV installations, which almost always use coaxial cable, and which require wiring practices different from Article 810 (for instance, the coax shield can serve as the grounding conductor). Your inspector may look to Article 820 for guidance on a safe installation of coax, since there are many more satellite TV and cable TV installations than Amateur Radio. Ultimately, it is the inspector's call as to whether your installation is safe.

Article 830 applies to Network Powered Communication Systems, and as amateurs do things like install 802.11 wireless LAN equipment at the top of their tower, they'll have to pay attention to the requirements in this Article. The *NEC* requirements discussed in these sections are not adequate for lightning protection and high-voltage transient events. See the section "Lightning/Transient Protection" later in this chapter for more information.

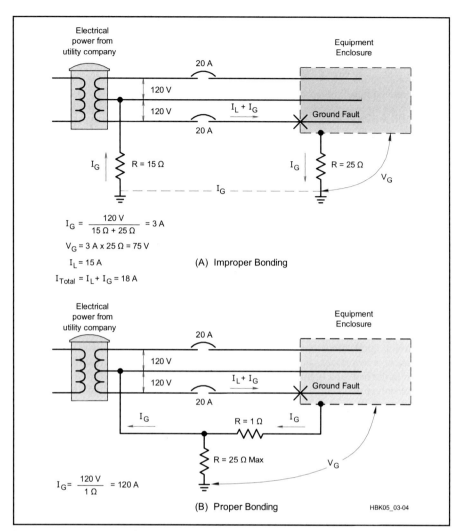

Fig 28.7 — These drawings show the importance of properly bonded ground rods. In the system shown in A, the 20-A breaker will not trip. In the system in B, the 20-A circuit breaker trips instantly. There is an equipment internal short to ground — the ground rod is properly bonded back to the power system ground. Of course, the main protection should be in a circuit ground wire in the equipment power cord itself!

ANTENNA CONDUCTORS

Transmitting antennas should use hard-drawn copper wire: #14 AWG for unsupported spans less than 150 feet, and #10 AWG for longer spans. Copper-clad steel, bronze or other high-strength conductors must be #14 AWG for spans less than 150 feet and #12 AWG for longer spans. Open-wire transmission line conductors must be at least as large as those specified for antennas. Stealth antennas made with light-gauge wire are not code-compliant.

LEAD-INS

There are several *NEC* requirements for antenna lead-in conductors (transmission lines are lead-in conductors). For transmitting stations, their size must be equal to or greater than that of the antenna. Lead-ins attached to buildings must be firmly mounted at least 3 inches clear of the surface of the building on nonabsorbent insulators. Lead-in conductors must enter through rigid, noncombustible, nonabsorbent insulating tubes or bushings, through an opening provided for the purpose that provides a clearance of at least two inches; or through a drilled windowpane. All lead-in conductors to transmitting equipment must be arranged so that accidental contact is difficult. As with stealth antennas, installations with feed lines smaller than RG-58 are likely not code compliant.

ANTENNA DISCHARGE UNITS (LIGHTNING ARRESTORS)

All antenna systems are required to have a means of draining static charges from the antenna system. A listed antenna discharge unit (lightning arrestor) must be installed on each lead-in conductor that is not protected by a permanently and effectively grounded metallic shield, unless the antenna itself is permanently and effectively grounded, such as for a shunt-fed vertical. Note that the usual transient protectors are *not* listed antenna discharge units.(The code exception for shielded lead-ins does *not* apply to coax, but to shields such as thin-wall conduit. Coaxial braid is neither "adequate" nor "effectively grounded" for lightning protection purposes.) An acceptable alternative to lightning arrestor installation is a switch (capable of withstanding many kilovolts) that connects the lead-in to ground when the transmitter is not in use.

ANTENNA BONDING (GROUNDING) CONDUCTORS

In general the code requires that the conductors used to bond the antenna system to ground be at least as big as the antenna conductors, but also at least #10 AWG in size. Note that the antenna grounding conductor rules are different from those for the regular electrical safety bonding, or lightning dissipation grounds, or even for CATV or telephone system grounds.

MOTORIZED CRANK-UP ANTENNA TOWERS

If you are using a motorized crank-up tower, the code has some requirements, particularly if there is a remote control. In general, there has to be a way to positively disconnect power to the motor that is within sight of the motorized device, so that someone working on it can be sure that it won't start moving unexpectedly. From a safety standpoint, as well, you should be able to see or monitor the antenna from the remote control point.

28.1.11 Lightning/Transient Protection

Nearly everyone recognizes the need to protect themselves from lightning. From miles away, the sight and sound of lightning boldly illustrates its destructive potential. Many people don't realize that destructive transients from lightning and other events can reach electronic equipment from many sources, such as outside antennas, power, telephone and cable TV installations. Many hams don't realize that the standard protection scheme of several decades, a ground rod and simple "lightning arrestor" is *not* adequate.

Lightning and transient high-voltage protection follows a familiar communications scenario: identify the unwanted signal, isolate it and dissipate it. The difference here is that the unwanted signal is many megavolts at possibly 200,000 A. What can we do?

Effective lightning protection system design is a complex topic. There are a variety of system tradeoffs which must be made and which determine the type and amount of protection needed. A amateur station in a home is a very different proposition from an air traffic control tower which must be available 24 hours a day, 7 days a week. Hams can easily follow some general guidelines that will protect their stations against high-voltage events that are induced by nearby lightning strikes or that arrive via utility lines. Let's talk about where to find professionals first, and then consider construction guidelines.

PROFESSIONAL HELP

Start with your local government. Find out what building codes apply in your area and have someone explain the regulations about antenna installation and safety. For more help, look in your telephone yellow pages for professional engineers, lightning protection suppliers and contractors.

Companies that sell lightning-protection products may offer considerable help to apply their products to specific installations. One such source is PolyPhaser Corporation. Look under "References" later in this chapter for a partial list of PolyPhaser's publications.

CONSTRUCTION GUIDELINES

Bonding Conductors

Copper strapping (or *flashing*) comes in a number of sizes. 1.5 inches wide and 0.051 inch thick or #6 AWG stranded wire, is the *minimum* recommended grounding conductor for lightning protection. Do not use braided strap because the individual strands oxidize over time, greatly reducing the effectiveness of braid as an ac conductor.

Use bare copper for buried ground wires. (There are some exceptions; seek an expert's advice if your soil is corrosive.) Exposed runs above ground that are subject to physical damage may require additional protection

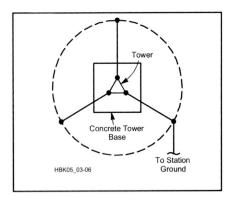

Fig 28.8 — Schematic of a properly grounded tower. A bonding conductor connects each tower leg to a ground rod and a buried (1 foot deep) bare, tinned copper ring (dashed line), which is also connected to the station ground and then to the ac safety ground. Locate ground rods on the ring, as close as possible to their respective tower legs. All connectors should be compatible with the tower and conductor materials to prevent corrosion. See text for conductor sizes and details of lightning and voltage transient protection.

(a conduit) to meet code requirements. Wire size depends on the application but never use anything smaller than #6 AWG for bonding conductors. Local lightning-protection experts or building inspectors can recommend sizes for each application.

Tower and Antennas

Because a tower is usually the highest metal object on the property, it is the most likely strike target. Proper tower grounding is essential to lightning protection. The goal is to establish short multiple paths to the Earth so that the strike energy is divided and dissipated.

Connect each tower leg and each fan of metal guy wires to a separate ground rod. Space the rods at least 6 ft apart. Bond the leg ground rods together with #6 AWG or larger copper bonding conductor (form a ring around the tower base, see **Fig 28.8**). Connect a continuous bonding conductor between the tower ring ground and the entrance panel. Make all connections with fittings approved for grounding applications. *Do not use solder for these connections.* Solder will be destroyed in the heat of a lightning strike.

Because galvanized steel (which has a zinc coating) reacts with copper when combined with moisture, use stainless steel hardware between the galvanized metal and the copper grounding materials.

To prevent strike energy from entering a shack via the feed line, ground the feed line *outside* the home. Ground the coax shield *to the tower* at the antenna and the base to keep the tower and line at the same potential. Several companies offer grounding blocks that make this job easy.

All grounding media at the home must be bonded together. This includes lightning-protection conductors, electrical service, telephone, antenna system grounds and underground metal pipes. Any ground rods used for lightning protection or entrance-panel grounding should be spaced at least 6 feet from each other and the electrical service or other utility grounds and then bonded to the ac system ground as required by the *NEC*.

A Cable Entrance Panel

The basic concept with transient protection is to make sure that all the radio and other equipment is tied together and "moves together" in the presence of a transient voltage. It's not so important that the shack be at "ground" potential, but, rather, that everything is at the *same* potential. For fast rise-time transients such as the individual strokes that make up a lightning strike, even a short wire has enough inductance that the voltage drop along the wire is significant, so whether you are on the ground floor, or the 10th floor of a building, your shack is "far" from Earth potential.

The easiest way to ensure that everything

Voltage Rise On Wires With Fast Transients

A rule of thumb is that a single wire has an inductance of about 1 µH per meter of length. The voltage across an inductor V= L di/dt. (di/dt is the change in current per unit of time.) A lightning stroke has a rise time of about 1-2 µs, so the current might go from zero to 10 kA in a microsecond or two, a di/dt of over 1 kA/µs (10^9 A/s). An inductance as low as 1 µH would create a voltage of 1000 volts from this current transient.

is at the same potential is to tie all the signals to a common reference. In large facilities, this reference would be provided by a grid of large diameter cables under the floor, or by wide copper bars, or even a solid metal floor. A more practical approach for smaller facilities like a ham shack is to have a single "tie point" for all the signals. This is often, but erroneously, called a "single-point ground", but what's really important is not only the shields (grounds) for the signals, but the signal wires as well are referenced to that common potential.

We want to control the flow of the energy in a strike and eliminate any possible paths for surges to enter the building. This involves routing the feed lines, rotator control cables, and so on at least six feet away from other nearby grounded metal objects.

A commonly used approach to ensuring that all the connections are tied together is to route all the signals through a single "entrance panel" which will serve as the "single point ground" although it may not actually be at ground potential. A convenient approach is to use a standard electrical box installed in the exterior wall

Both balanced line and coax arrestors should be mounted to a secure ground connection on the *outside* of the building. The easiest way to do this is to install a large metal enclosure or a metal panel as a bulkhead and grounding block. The panel should be connected to the lightning dissipation ground with a short wide conductor (for minimum impedance), and, like all grounds, bonded to the electrical system's ground. Mount all protective devices, switches and relay disconnects on the outside facing wall of the bulkhead. The enclosure or panel should be installed in a way that if lightning currents cause a component to fail, the molten metal and flaming debris do not start a fire.

Every conductor that enters the structure, including antenna system control lines, should have its own surge suppressor on an entrance panel. Suppressors are available from a number of manufacturers, including

> **Suppliers of Lightning Protection Equipment**
>
> For current vendor contact information, use your favorite Internet search tool.
> • Alpha Delta Communications: Coax lightning arrestors, coax switches with surge protectors.
> • The Wireman: copper wire up to #4 AWG, 2-inch flat copper strap, 8-ft copper clad ground rods and 1 × ¼-inch buss bar.
> • ERICO International Corporation: CadWeld bonding system and lightning protection equipment.
> • Harger Lightning & Grounding: lightning protection components.
> • Industrial Communication Engineers, Ltd (ICE): Coax lightning arrestors.
> • PolyPhaser Corporation: Many lightning protection products for feed lines, towers, equipment, and so on.
> • Zero Surge Inc: Power line surge protector.

Industrial Communication Engineers (ICE) and PolyPhaser, as well as the usual electrical equipment suppliers such as Square-D.

Lightning Arrestors

Feed line lightning arrestors are available for both coax cable and balanced line. Most of the balanced line arrestors use a simple spark gap arrangement, but a balanced line *impulse* suppressor is available from ICE.

DC blocking arrestors for coaxial cable have a fixed frequency range. They present a high-impedance to lightning (less than 1 MHz) while offering a low impedance to RF.

DC continuity arrestors (gas tubes and spark gaps) can be used over a wider frequency range than those that block dc. Where the coax carries supply voltages to remote devices (such as a mast-mounted preamp or remote coax switch), dc-continuous arrestors *must* be used.

28.1.12 Other Hazards in the Shack

UPS AND ALTERNATE ENERGY SOURCES

Many hams have alternate energy sources for their equipment, or an uninterruptible power supply (UPS), so that they can keep operating during a utility power outage. This brings some additional safety concerns, because it means that the "turning off the breaker" approach to make sure that power is disconnected might not work.

In commercial installations, fire regulations or electrical codes often require that the emergency power off (EPO) system (the big red button next to the door) also disconnect the batteries of the UPS system, or at least, disable the ac output. This is so that firefighters who may be chopping holes with conductive tools or spraying conductive water don't face the risk of electrocution. (According to NEC, Articles 645-10 and 645-11, UPSs above 750 VA installed within information technology rooms must be provided with a means to disconnect power to all UPS supply and output circuits. This disconnecting means shall also disconnect the battery from its load. The code further requires that the control for these disconnecting means shall be grouped and identified and shall be readily accessible at the principal exit doors.)

A similar problem exists with solar panel installations. Just because the breaker is turned off doesn't mean that dangerous voltages don't exist on the solar panel. As long as light is falling on them, there is voltage present. With no load, even a relatively dim light falling on part of the panels might present a shock or equipment damage hazard. Modern grid-tie solar systems with no batteries often have the panels wired in series, so several hundred volts is not unusual.

Recent revisions of the *NEC* have addressed many of the aspects of photovoltaic (PV) installations that present problems with disconnects, bonding, and grounding. Consulting your local authorities is always wise, and there are several organizations such as the Southwest Technology Development Institute at New Mexico State University that have prepared useful information (see the references at the end of this section). In general, PV systems at 12 or 24 V aren't covered by the *NEC*.

ENERGIZED CIRCUITS

Working with energized circuits can be very hazardous since, without measuring devices, we can't tell which circuits are live. The first thing we should ask ourselves when faced with troubleshooting, aligning or other "live" procedures is, "Is there a way to reduce the hazard of electrical shock?" Here are some ways of doing just that.

1) If at all possible, troubleshoot with an ohmmeter. With a reliable schematic diagram and careful consideration of how various circuit conditions may reflect resistance readings, it will often be unnecessary to do live testing.

2) Keep a fair distance from energized circuits. What is considered "good practice" in terms of distance? The *NEC* specifies minimum working space around electric equipment depending on the voltage level. The principle here is that a person doing live work needs adequate space so they are not forced to be dangerously close to energized equipment.

3) If you need to measure the voltage of a circuit, install the voltmeter with the power safely off, back up, and only then energize the circuit. Remove the power before disconnecting the meter.

4) If you are building equipment that has hinged or easily removable covers that could expose someone to an energized circuit, install interlock switches that safely remove power in the event that the enclosure is opened with the power still on. Interlock switches are generally not used if tools are required to open the enclosure.

5) Never assume that a circuit is at zero potential even if the power is switched off and the power cable disconnected. Capacitors can retain a charge for a considerable period of time and may even partially recharge after being discharged. Bleeder resistors should be installed, but don't assume they have discharged the voltage. Instead, after power is removed and disconnected use a "shorting stick" to ground all exposed conductors and terminals to ensure that voltage is not present. If you will be working with charged capacitors that store more than a few joules of energy, you should consider using a "discharging stick" with a high wattage, low value resistor in series to ground that limits the discharge current to around 5-10 A. A dead short across a large charged capacitor can damage the capacitor because of internal thermal and magnetic stress. Avoid using screwdrivers, as this brings the holder too close to the circuit and could ruin the screwdriver's blade. For maximum protection against accidentally energizing equipment, install a shorting lead between high-voltage circuits and ground while you are working on the equipment.

6) Shorting a series string of capacitors does not ensure that the capacitors are discharged. Consider two 400 μF capacitors in series, one charged to +300 V and the other to –300 V with the midpoint at ground. The net voltage across the series string is zero, yet each has significant (and lethal) energy stored in it. The proper practice is to discharge each capacitor in turn, putting a shorting jumper on it after discharge, and then moving to the next one.

7) If you must hold a probe to take a measurement, always keep one hand in your pocket. As mentioned in the sidebar on high-voltage hazards, the worst path current could take through your body is from hand to hand since the flow would pass through the chest cavity.

8) Make sure someone is in the room with you and that they know how to remove the power safely. If they grab you with the power still on they will be shocked as well.

9) Test equipment probes and their leads must be in very good condition and rated for the conditions they will encounter.

10) Be wary of the hazards of "floating" (ungrounded) test equipment. A number of options are available to avoid this hazard.

11) Ground-fault circuit interrupters can offer additional protection for stray currents

that flow through the ground on 120-V circuits. Know their limitations. They cannot offer protection for the plate supply voltages in linear amplifiers, for example.

12) Older radio equipment containing ac/dc power supplies have their own hazards. If you are working on these live, use an isolation transformer, as the chassis may be connected directly to the hot or neutral power conductor.

13) Be aware of electrolytic capacitors that might fail if used outside their intended applications.

14) Replace fuses only with those having proper ratings. The rating is not just the current, but also takes into account the speed with which it opens, and whether it is rated for dc or ac. DC fuses are typically rated at lower voltages than those for ac, because the current in ac circuits goes through zero once every half cycle, giving an arc time to quench. Switches and fuses rated for 120 V ac duty are typically not appropriate for high-current dc applications (such as a main battery or solar panel disconnect).

28.1.13 Electrical Safety References

ARRL Technical Information Service Web page on electrical safety in the Technology area of the ARRL Web site.

Block, R.W., "Lightning Protection for the Amateur Radio Station," Parts 1-3 (Jun, Jul and Aug 2002 *QST*).

Federal Information Processing Standards (FIPS) publication 94: *Guideline on Electrical Power for ADP Installations.* FIPS are available from the National Technical Information Service.

IAEI: *Soares' Book on Grounding*, available from International Association of Electrical Inspectors (IAEI).

"IEEE Std 1100 - 2005 IEEE Recommended Practice for Powering and Grounding Electronic Equipment," *IEEE Std 1100-2005 (Revision of IEEE Std 1100-1999)*, pp 0_1-589, 2006. "This document presents recommended design, installation, and maintenance practices for electrical power and grounding (including both safety and noise control) and protection of electronic loads such as industrial controllers, computers, and other information technology equipment (ITE) used in commercial and industrial applications."

National Electrical Code, NFPA 70, National Fire Protection Association, Quincy, MA (**www.nfpa.org**).

PolyPhaser: The Grounds for Lightning and EMP Protection. PolyPhaser's quarterly newsletter, *Striking News*, contains articles on Amateur Radio station lightning protection in the February and May 1994 issues. Complimentary copies of these issues are available from PolyPhaser.

Solar energy Web sites — **www.nmsu.edu/~tdi/PV=NEC_HTML/pv-nec/pv-nec.html** and **www.solarabcs.org**

Standard for the Installation of Lightning Protection Systems, NFPA 780, National Fire Protection Association, Quincy, MA (**www.nfpa.org**).

Electrical Shock Hazards and Effects

What happens when someone receives an electrical shock?

Electrocutions (fatal electric shocks) usually are caused by the heart ceasing to beat in its normal rhythm. This condition, called ventricular fibrillation, causes the heart muscles to quiver and stop working in a coordinated pattern, in turn preventing the heart from pumping blood.

The current flow that results in ventricular fibrillation varies between individuals but may be in the range of 100 mA to 500 mA. At higher current levels the heart may have less tendency to fibrillate but serious damage would be expected. Studies have shown 60-Hz alternating current to be more hazardous than dc currents. Emphasis is placed on application of cardiopulmonary resuscitation (CPR), as this technique can provide mechanical flow of some blood until paramedics can "restart" the heart's normal beating pattern. Defibrillators actually apply a carefully controlled waveform to "shock" the heart back into a normal heartbeat. It doesn't always work but it's the best procedure available.

What are the most important factors associated with severe shocks?

You may have heard that the current that flows through the body is the most important factor, and this is generally true. The path that current takes through the body affects the outcome to a large degree. While simple application of Ohm's Law tells us that the higher the voltage applied with a fixed resistance, the greater the current that will flow. Most electrical shocks involve skin contact. Skin, with its layer of dead cells and often fatty tissues, is a fair insulator. Nonetheless, as voltage increases the skin will reach a point where it breaks down. Then the lowered resistance of deeper tissues allows a greater current to flow. This is why electrical codes refer to the term "high voltage" as a voltage above 600 V.

How little voltage can be lethal?

This depends entirely on the resistance of the two contact points in the circuit, the internal resistance of the body, and the path the current travels through the body. Historically, reports of fatal shocks suggest that as little as 24 V *could* be fatal under extremely adverse conditions. To add some perspective, one standard used to prevent serious electrical shock in hospital operating rooms limits leakage flow from electronic instruments to only 50 µA due to the use of electrical devices and related conductors inside the patient's body.

28.2 Antenna and Tower Safety

By definition, all of the topics in this book are about radio telecommunications. For those communications, both receive and transmit antennas are required and those antennas need to be up in the air in order to work effectively. While antennas are covered elsewhere, this section will cover many of the topics associated with getting those antennas up there, along with related safety issues. A more complete treatment of techniques used to erect towers and antennas is available in the *ARRL Antenna Book* and in the *Up The Tower: The Complete Guide To Tower Construction* by Steve Morris, K7LXC.

28.2.1 Legal Considerations

Some antenna support structures fall under local building regulations as well as neighborhood restrictions. Many housing developments have Homeowner's Associations (HOAs) as well as Covenants, Conditions and Restrictions (CC&Rs) that may have a direct bearing on whether a tower or similar structure can be erected at all. This is a broad topic with many pitfalls. Detailed background on these topics is provided in *Antenna Zoning for the Radio Amateur* by Fred Hopengarten, K1VR, an attorney with extensive experience in towers and zoning. You may also want to contact one of the ARRL Field Organization's Volunteer Counsels.

Even without neighborhood issues, a build-

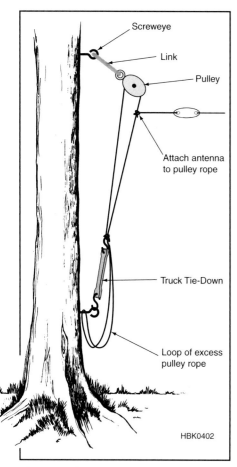

Fig 28.9 — Loop and halyard method of supporting wire antennas in trees. Should the antenna break, the continuous loop of rope allows antenna repair or replacement without climbing the tree.

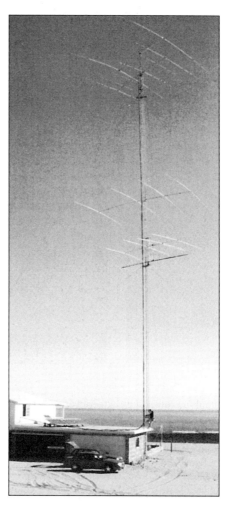

Fig 28.10 — A guyed tower with a good-sized load of antennas at XE2DV-W7ZR in Baja California, is shown at the left. At the right, the Trylon Titan self-supporting tower of W7WVF and N7YYG in Bandon, Oregon. (*K7LXC photos*)

ing permit is likely to be required. With the proliferation of cellular and other commercial wireless devices and their attendant RF sites, many local governments now require that the structures meet local building codes. Again, K1VR's book is extremely helpful in sorting all this out. Building permit applications may also require Professional Engineer (PE) calculations and stamp (certification). The ARRL Field Organization's Volunteer Consulting Engineer program may be useful with the engineering side of your project.

28.2.2 Antenna Mounting Structures

TREES AND POLES

The original antenna supports were trees: if you've got them, use them. They're free and unregulated, so it couldn't be easier. Single-trunked varieties such as fir and pine trees are easier to use than the multi-trunked varieties. Multi-trunked trees are not impossible to use — they just require a lot more work. For dipoles or other types of wire antennas, plan for the tree to support an end of the wire; trying to install an inverted V or similar configuration is almost impossible due to all of the intervening branches.

Install an eye-screw with a pulley at the desired height, trim away enough limbs to create a "window" for the antenna through the branches and then attach a rope halyard to the antenna insulator. Here's a useful tip: Make the *halyard* a continuous loop as shown in **Fig 28.9**. Since it's almost always the antenna wire that breaks, a continuous halyard makes it easy to reattach the wire and insulator. With just a single halyard, if the antenna breaks, the tree will have to be climbed to reach the pulley, then reinstall and attach the line. If you're unable to climb the support tree, contact a local tree service. Professional tree climbers are often willing to help out for a small fee.

Another way to get wires into trees is with some sort of launcher. Using a bow-and-arrow is a traditional method of shooting a fishing line over a tree to pull up a bigger line. There are now commercial products available that are easier to use and reach higher in the tree. For example, wrist-rocket slingshots and compressed-air launchers can reach heights of more than 200 feet!

Wooden utility poles offer a tree-related alternative but are not cheap, require special installation with a pole-setting truck, and there is no commercial antenna mounting hardware available for them. That makes them a poor choice for most installations.

TOWERS

The two most important parameters to consider when planning a tower installation are the maximum local *wind speed* and the proposed antenna *wind load*. Check with your local building department to find out what the maximum wind speed is for your area. Another source is a list of maximum wind speeds for all counties in the US from the TIA-222, *Structural Standard for Antenna Supporting Structures and Antennas*. This is an expensive professional publication so it's not for everyone, but the list is posted on the Champion Radio Products Web site under

Fig 28.11 — The bottom of N6TV's crank-up tower is shown at left. The motor drive mechanism is on the left and a fishing net on the right catches and coils the feed lines and control cables as the tower is lowered. On the right, K6KR's fully loaded crank-up extended to its maximum height of 90 feet. (*K7LXC photos*)

"Tech Notes." Tower capacities are generally specified in square feet of antenna load and antenna wind load specifications are provided by the antenna manufacturer.

Before beginning, learn and follow K7LXC's Prime Directive of tower construction — "DO what the manufacturer says." (And DON'T do what the manufacturer doesn't say to do.) Professional engineers have analyzed and calculated the proper specifications and conditions for tower structures and their environment. Taking any shortcuts or making different decisions will result in a less reliable installation.

Towers come in the two varieties shown in **Fig 28.10** — guyed and self-supporting. Guyed towers require a bigger footprint because the guys have to be anchored away from the tower — typically 80% of the tower height. Self-supporting towers need bigger bases to counteract the overturning moment and are more expensive than a guyed tower because there is more steel in them (the cost of a tower is largely determined by the cost of the steel).

The most popular guyed towers are the Rohn 25G and 45G. The 25G is a light-duty tower and the 45G is capable of carrying fairly big loads. The online Rohn catalog (see the References) has most of the information you'll need to plan an installation.

Self-supporting towers are made by several manufacturers and allow building a tower up to 100 feet or higher with a small footprint. Rohn, Trylon and AN Wireless are popular vendors. Another type of self-supporting tower is the *crank-up*, shown in **Fig 28.11**. Using a system of cables, pulleys and winches, crank-up towers can extend from 20 feet to over 100 feet high. These are moderately complex devices. The largest manufacturer of crank-up towers is US Towers.

Another simple and effective way to get an antenna up in the air is with a *roof-mounted tower*, seen in **Fig 28.12**. These are four-legged aluminum structures of heights from four to more than 20 feet. While they are designed to be lag-screwed directly into the roof trusses, it is often preferable to through-bolt them into a long 2×4 or 2×6 that acts a backing plate, straddling three to four roof trusses. In any case, if it is not clear how best to install the tower on the structure, have a roofing professional or engineer provide advice. Working on a roof-mounted tower also requires extra caution because of climbing on a roof while also working on a tower.

28.2.3 Tower Construction and Erection

THE BASE

Once all the necessary materials and the required approvals have been gathered, tower

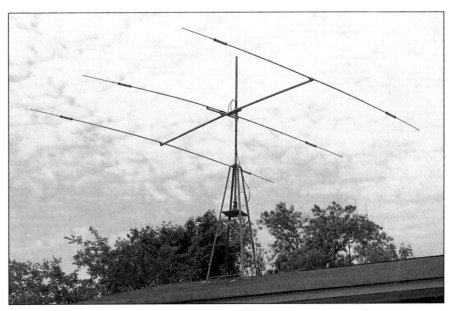

Fig 28.12 — The roof-mounted tower holding the AA2OW antenna system. (*AA2OW photo*)

installation can begin. Let's assume you and your friends are going to install it. The first job is to construct the base. A base for a guyed tower can be hand-dug as can the guy anchors. For a self-supporting tower, renting an excavator of some sort will make it much easier to move the several cubic yards of dirt.

Next, some sort of rebar cage will be needed for the concrete. Guyed towers only require rudimentary rebar while a self-supporting tower will need a bigger, heavier and more elaborate cage. Consult the manufacturer's specifications for the exact materials and dimensions.

Typical tower concrete specs call for 3000 psi (minimum) compressive strength concrete and 28 days for a full cure. A local pre-mix concrete company can deliver it. Pouring the concrete is easiest if the concrete truck can back up to the hole. If that's not possible, a truck- or trailer-mounted line pump can pump it up to 400 feet at minimal expense if using a wheelbarrow is not possible or practical. Packaged concrete from the hardware store mixed manually may also be used. Quikrete Mix #1101 is rated at 2500 psi after seven days and 4000 psi after 28 days.

TOOLS

Once the base and anchors are finished and the concrete has cured, the tower can be constructed. There are several tools that will make the job easier. If the tower is a guyed tower, it can be erected either with a crane or a *gin-pole*. The gin-pole, shown in **Fig 28.13**, is a pipe that attaches to the leg of the tower and has a pulley at the top for the haul rope. Use the gin-pole to pull up one section at a time (see below).

Another useful tool for rigging and hoisting is the *carabiner*. Shown in **Fig 28.14** (A and B), carabiners are oval steel or aluminum snap-links popularized by mountain climbers. They have spring-loaded gates and can be used for many tower tasks. For instance, there should be one at the end of the haul rope for easy and quick attachment to rotators, parts and tool buckets — virtually anything that needs to be raised or lowered. It can even act as a "third hand" on the tower.

Along with the carabineer, the *nylon loop sling* in Fig 28.14C can be wrapped around large or irregularly shaped objects such as antennas, masts or rotators and attached to ropes with carabiners. For a complex job, a professional will often climb with eight to ten slings and use every one!

A pulley or two will also make the job easier. At a minimum, one is needed for the haul rope at the top of the tower. A *snatch block* is also useful; this is a pulley whose top opens up to "snatch" (attach it to) the rope at any point. **Fig 28.15** shows two snatch-block pulleys used for tower work.

ROPES

Speaking of ropes, use a decent haul rope. Rope that is one-half inch diameter or larger affords a good grip for lifting and pulling. There are several choices of rope material. The best choice is a synthetic material such as nylon or Dacron. A typical twisted rope is fine for most applications. A synthetic rope

Fig 28.13 — A gin-pole consists of a leg clamp fixture, a section of aluminum mast and a pulley. It is used to lift the tower section high enough to be safely lowered into place and attached. (Based on Rohn EF2545.)

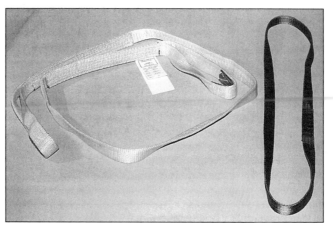

Fig 28.14 — (A) Oval mountain climbing type carabiners are ideal for tower workloads and attachments. The gates are spring loaded — the open gate is shown for illustration. (B) An open aluminum oval carabiner; a closed oval carabiner; an aluminum locking carabiner; a steel snaplink. (C) A heavy duty nylon sling on the left for big jobs and a lighter-duty loop sling on the right for everything else. (*K7LXC photos*)

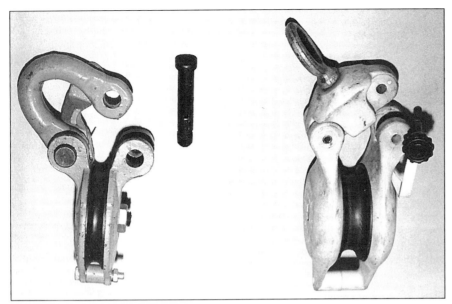

Fig 28.15 — Two snatch blocks; a steel version on the left and a lightweight high impact plastic one on the right in their open position. (*K7LXC photo*)

with a braid over the twisted core is known as *braid-on-braid* or *kernmantle*. While it's more expensive than twisted ropes, the outer braid provides better abrasion resistance. The least expensive type of rope is polypropylene. It's a stiff rope that doesn't take a knot as well as other types but is reasonably durable and cheap. **Table 28.2** shows the safe working load ratings for common types of rope.

When doing tower work, being able to tie knots is required. Of all the knots, the *bowline* is the one to know for tower work. The old "rabbit comes up through the hole, around the tree and back down the hole" is the most familiar method of tying a bowline. Most amateurs are knot-challenged so it's a great advantage to know at least this one.

INSTALLING TOWER SECTIONS

The easiest way to erect a tower is to use a crane. It's fast and safe but more expensive than doing it in sections by hand. To erect a tower by sections, a gin-pole is needed (see Fig 28.13). It consists of two pieces — a clamp or some device to attach it to the tower leg and a pole with a pulley at the top. The pole is typically longer than the work piece being hoisted, allowing it to be held above the tower top while being attached or manipulated.

With the gin-pole mounted on the tower, the haul rope runs up from the ground, through the gin-pole mast and pulley at the top of the gin-pole, and back down the tower. The haul rope has a knot (preferably a bowline) on the end for attaching things to be hauled up or down. A carabiner hooked into the bight of the knot can be attached to objects quickly so that you don't have to untie and re-tie the bowline with every use.

It's a good idea to pass the haul rope through a snatch-block at the bottom of the tower. This changes the direction of the rope from vertical to horizontal, allowing the ground crew to stand away from the tower (and the fall zone for things dropped off the tower) to manipulate the haul rope while also watching what's going on up on the tower.

GUYS

For guyed towers, an important construction parameter is guy wire material and *guy tension*. Do *not* use rope or any other material not rated for use as guy cable as permanent tower guys. Guyed towers for amateurs typically use either 3/16-inch or 1/4-inch *EHS* (extra high strength) steel guy cable. The only other acceptable guy material is Phillystran — a lightweight cable made of Kevlar fibers. Phillystran is available with the same breaking strength as EHS cable. The advantage of Phillystran is that it is non-conducting and does not create unwanted electrical interaction with antennas on the tower. It is an excellent choice for towers supporting multiple Yagi and wire antennas and does not have to be broken up into short lengths with insulators.

EHS wire is very stiff — to cut it, use a hand-grinder with thin cutting blades or a circular saw with a pipe-cutting aggregate blade. Be sure to wear safety glasses and gloves when cutting since there will be lots of sparks of burning steel being thrown off. Phillystran can be cut with a utility knife or a hot knife for cutting plastic.

If the guys are too loose, the result will be wind-induced shock loading. Guys that are too tight exert extra compressive load on the tower legs, reducing the overall capacity and reliability of the tower. The proper tension of EHS or Phillystran guys is 10% of the material's *ultimate breaking strength*. For 3/16-inch EHS the ultimate breaking strength is 4900 pounds and for 1/4-inch it's 6000 pounds so the respective guy tension should be 490 pounds and 600 pounds. The easiest to use, most accurate, and least expensive way to measure guy tension is by using a Loos tension gauge. It was developed for sailboat rigging so it's available at some marine supply stores or from Champion Radio Products.

Guy wires used to be terminated in a loop

Table 28.2
Rope types and safe working loads in pounds.
Ropes are three-strand twisted unless otherwise noted.

	Diameter (inches)			
	1/4	*3/8*	*1/2*	*5/8*
Manila	120	215	425	700
Nylon	180	400	700	1150
Polypropylene	210	450	710	1055
Nylon braid-on-braid	420	960	1630	2800
Dacron braid-on-braid	350	750	1400	2400

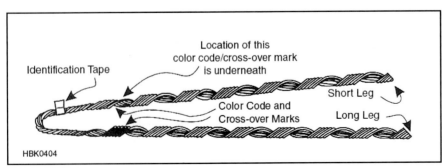

Fig 28.16 — A PreFormed Line Products Big Grip for guy wires.

with cable clamps but those have been largely replaced by pre-formed Big Grips, shown in **Fig 28.16**. These simply twist onto the guy wire and are very secure. They grip the guy cable by squeezing the cable as tension is applied. Be sure to use the right type of Big Grips for the thickness and material of the guy cable.

28.2.4 Antenna Installation

Now that the tower is up, install the antennas. VHF/UHF whips and wire antennas are pretty straightforward, but installing an HF Yagi is a more challenging proposition. With a self-supporting tower, there are no guy wires to contend with — generally, the antenna can just be hauled up the tower face. Sometimes it is that easy!

In most cases, short of hiring a crane, the easiest way to get a Yagi up and down a tower is to use the *tram* method. A single tramline is suspended from the tower to the ground and the load is suspended under the tramline. Another technique is the *trolley* method in which two lines are suspended from the tower to the ground and the antenna rides on top of the lines like a trolley car on tracks. Problems with the trolley technique include trying to get the lines to have the same tension, balancing the antenna so that it won't fall off of the lines, and the added friction of pulling the antenna up two lines. The tram method has none of these problems. **Fig 28.17** illustrates the tram method of raising antennas.

Tram and trolley lines are typically attached to the mast above the top of the tower. In the case of a big load, the lines may exert enough force to bend the mast. If in doubt, *back-guy* the mast with another line in the opposite direction for added support.

MASTS

A mast is a pipe that sticks out of the top of the tower and connects the rotator to the antenna. For small antenna loads and moderate wind speeds, any pipe will work. But as wind speed and wind load increase, more force will be exerted on the mast.

There are two materials used for masts — *pipe* and *tubing*. Pipe can be water pipe or conduit (EMT). Pipe is a heavy material with not much strength since its job is just to carry water or wires. Pipe is acceptable as mast material for small loads only. Another problem is that 1.5-in. pipe (pipe is measured by

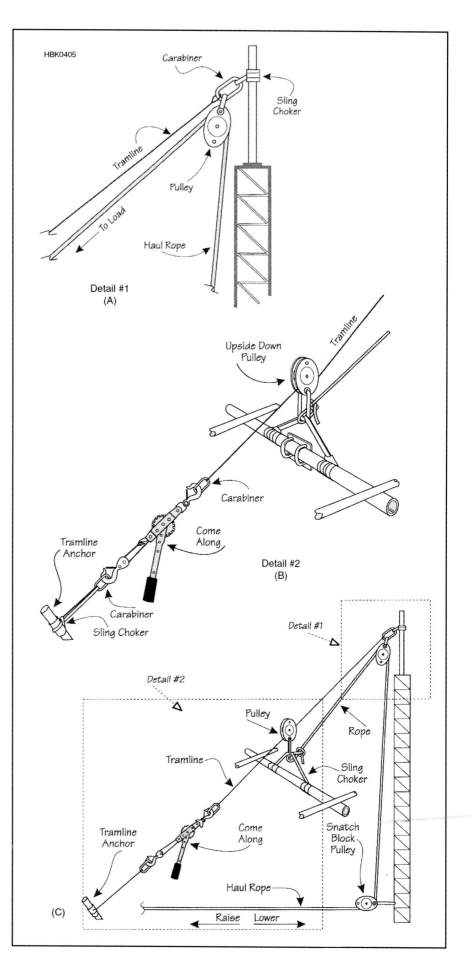

Fig 28.17 — At A, rigging the top of the tower for tramming antennas. Note the use of a sling and carabiner. (B) Rigging the anchor of the tramline. A come-along is used to tension the tramline. (C) The tram system for getting antennas up and down. Run the antenna part way up the tramline for testing before installation. It just takes a couple of minutes to run an antenna up or down once the tramline is rigged.

Table 28.3
Yield Strengths of Mast Materials

Material	Specification	Yield Strength (lb/in.^{2})
Drawn aluminum tube	6063-T5	15,000
	6063-T832	35,000
	6061-T6	35,000
	6063-T835	40,000
	2024-T3	42,000
Aluminum pipe	6063-T6	25,000
	6061-T6	35,000
Extruded alum. tube	7075-T6	70,000
Aluminum sheet and plate	3003-H14	17,000
	5052-H32	22,000
	6061-T6	35,000
Structural steel	A36	33,000
Carbon steel, cold drawn	1016	50,000
	1022	58,000
	1027	70,000
	1041	87,000
	1144	90,000
Alloy steel	2330 cold drawn	119,000
	4130 cold worked	75,000
	4340 1550 °F quench 1000 °F temper	162,000
Stainless steel	AISI 405 cold worked	70,000
	AISI 440C heat-treated	275,000

(From *Physical Design of Yagi Antennas* by David B. Leeson, W6NL)

its inside diameter or ID) is only 1.9-in. OD. Since most antenna boom-to-mast hardware is designed for a 2-in. mast, the less-than-perfect fit may lead to slippage.

For any larger load use carbon-alloy steel tubing rated for high strength. A moderate antenna installation in an 80 MPH wind might exert 40,000 to 50,000 pounds per square inch (psi) on the mast. Pipe has a yield strength of about 35,000 psi, so you can see that pipe is not adequately rated for this type of use. Chromoly steel tubing is available with yield strengths from 40,000 psi up to 115,000 psi but it is expensive. **Table 28.3** shows the ratings of several materials used as masts for amateur radio antennas.

Calculating the required mast strength can be done by using a software program such as the *Mast, Antenna and Rotator Calculator (MARC)* software. (See the References.) The software requires as inputs the local wind speed, antenna wind load, and placement on the mast. The software then calculates the mast bending moment and will recommend a suitable mast material.

28.2.5 Weatherproofing Cable and Connectors

The biggest mistake amateurs make with coaxial cable is improper weatherproofing. (Coax selection is covered in the chapter on **Transmission Lines**.) **Fig 28.18** shows how to do it properly. First, use high-quality electrical tape, such as 3M Scotch 33+ or Scotch 88. Avoid inexpensive utility tape. After tightening the connector (use pliers carefully to seat threaded connectors — hand-tight isn't good enough), apply two wraps of tape around the joint.

When you're done, sever the tape with a knife or tear it very carefully — *do not* stretch the tape until it breaks. This invariably leads to "flagging" in which the end of the tape loosens and blows around in the wind. Then let the tape relax before finishing the wrap.

Next put a layer of butyl rubber *vapor wrap* over the joint. (This tape is also available in the electrical section of the hardware store.) Finally, add two more layers of tape over the vapor wrap, creating a professional-quality joint that will never leak. Finally, if the coax is vertical, be sure to wrap the final layer so that the tape is going *up* the cable as shown in Fig 28.18. In that way, the layers will act like roofing shingles, shedding water off the connection. Wrapping it top to bottom will guide water between the layers of tape.

28.2.6 Climbing Safety

Tower climbing is a potentially dangerous activity, so you'll need to use the proper safety equipment and techniques. OSHA, the Federal Occupational Safety and Health Administration, publishes rules for workplace safety. Although amateurs are not bound by those rules, you'll be much better off by following them.

First, if you are still climbing with a waist-only safety belt and leather positioning lanyard, throw them away! They are illegal and dangerous. The most important piece of climbing equipment is a *fall arrest harness* (FAH). Along with the waist safety belt and D-rings, it also has adjustable suspenders, leg loops and a D-ring between the shoulder blades for a *fall arrest (FA) lanyard*. **Fig 28.19** shows how to wear the harness for tower climbing.

A tower climber must be attached to the tower 100% of the time. One method of attachment is the *waist-positioning lanyard* shown in **Fig 28.20**. **Fig 28.21** shows a climber properly positioned in the harness on the tower with the positioning lanyard holding the climber in working position. The lanyard can be fixed-length or an adjustable rope or webbing type. Fixed-length varieties are usually too short or too long for optimum placement, but they are the least expensive. All fall-arrest equipment complies with OSHA rules, so what is actually used depends mostly on personal choice and budget.

There are two choices for FA lanyards. One is fixed-length and the other is *shock-absorbing*. A falling person generates a lot of force in a short time so keep the FA lanyard attached above you with enough slack to allow movement around the tower. That minimizes the distance of any fall. The stitched loops of a shock-absorbing FA lanyard yield gradually, decelerating to avoid a sudden stop at the end of a fall. The shock-absorbing lanyard costs more than a fixed-length lanyard.

Suitably equipped, start climbing the tower. In order to be attached to the tower 100% of the time, attach the FA lanyard as high above you as you can and then climb up to it. Attach the positioning lanyard, detach the FA lanyard, and attach it above you again.

The positioning lanyard can also be moved up as you climb. As you reach the FA lanyard attachment point, unhook it, reposition, and repeat as many times as necessary. There are several varieties of FA lanyards including models with dual lanyards, thus allowing the lanyards to be alternated during climbing.

GROUND CREW SAFETY

The climber on the tower is the boss. Before tower work starts, have a safety meeting with the ground crew. Explain what is going to be done and how to do it as well as introducing them to any piece of hardware with which they may not be familiar (for example, carabiners, slings or come-along winches).

As part of the ground crew, there are a few rules to follow:

1) The climber on the tower is in charge.

2) Don't do anything unless directed by the climber in charge on the tower. This includes handling ropes, tidying up, moving hardware, and so on.

3) If not using radios to communicate, when talking to the climber on the tower, look up and talk directly to him or her in a

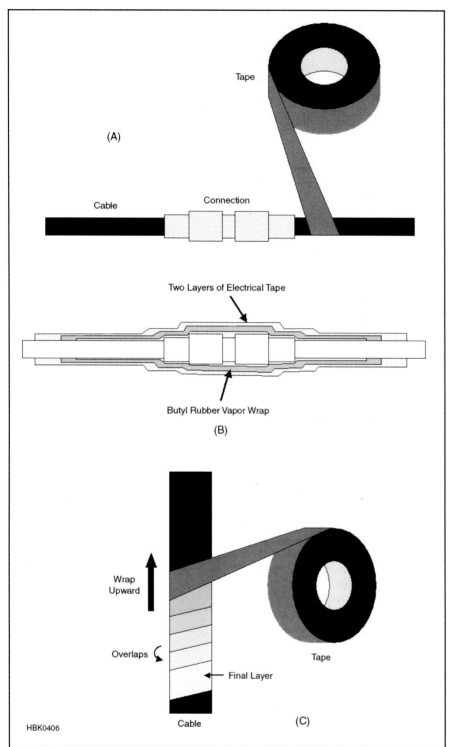

Fig 28.18 — Waterproofing a connector in three steps. At A, cover the connectors with a layer of good-quality electrical tape. B shows a layer of butyl rubber vapor wrap between the two layers of electrical tape. C shows how to wrap tape on a vertical cable so that the tape sheds water away from the connection. (Drawing (C) reprinted courtesy of *Circuitbuilding for Dummies*, Wiley Press)

Fig 28.19 — (A) The well-dressed tower climber. Note the waist D-rings for positioning lanyard attachment as well as the suspenders and leg loops. At (B) is an adjustable positioning lanyard. The climber also has working boots, gloves, safety glasses and hardhat. (*K7LXC photos*)

Safety 28.19

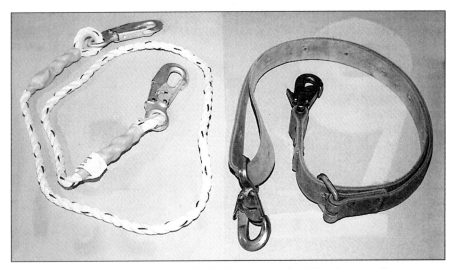

Fig 28.20 — A fixed-length rope positioning lanyard on the left and a versatile adjustable lanyard on the right. (*K7LXC photos*)

Fig 28.21 — The fall-arrest lanyard is above the climber so that the climber can climb up to it. The fall-arrest and positioning lanyards are then "leapfrogged" so that the climber remains attached to the tower 100 percent of the time. (*K7LXC photos*)

loud voice. The ambient noise level is higher up on the tower because of traffic, wind and nearby equipment.

4) Communicate with the climber on the tower. Let him or her know when you're ready or if you're standing by or if there is a delay. Advise the climber when lunch is ready!

28.2.7 Antenna and Tower Safety References

AN Wireless — **www.anwireless.com**
ARRL Volunteer Counsel program — **www.arrl.org/volunteer-counsel-program**
ARRL Volunteer Consulting Engineer program — **www.arrl.org/volunteer-consulting-engineer-program**
Brede, D., W3AS, "The Care and Feeding of an Amateur's Favorite Antenna Support — the Tree," *QST*, Sep 1989, pp 26-28, 40.
Champion Radio Products — **www.championradio.com**
Hopengarten, F., K1VR, *Antenna Zoning for the Radio Amateur* (ARRL, 2002)
Knot-tying Web site — **www.animatedknots.com**
Loos tension gauge — **www.championradio.com/rigging.html**
MARC software — **www.championradio.com/misc.html**
Morris, S., K7LXC, *Up The Tower: The Complete Guide To Tower Construction* (Champion Radio Products, 2009)
Rohn Tower — **www.rohnnet.com**
Straw, R.D., N6BV, ed., *The ARRL Antenna Book*, 21st ed. (ARRL, 2007)
Trylon — **www.trylon.com**
US Towers — **www.ustower.com**

28.3 RF Safety

Amateur Radio is basically a safe activity. In recent years, however, there has been considerable discussion and concern about the possible hazards of electromagnetic fields (EMF), including both RF energy and power frequency (50-60 Hz) EMF. FCC regulations set limits on the maximum permissible exposure (MPE) allowed from the operation of radio transmitters. Following these regulations, along with the use of good RF practices, will make your station as safe as possible. This section, written by the ARRL RF Safety Committee (see sidebar), deals with the topic of electromagnetic safety.

28.3.1 How EMF Affects Mammalian Tissue

All life on Earth has adapted to live in an environment of weak, natural, low frequency electromagnetic fields, in addition to the Earth's static geomagnetic field. Natural low-frequency EM fields come from two main sources: the sun and thunderstorm activity. During the past 100 years, man-made fields at much higher intensities and with different spectral distributions have altered our EM background. Researchers continue to look at the effects of RF exposure over a wide range of frequencies and levels.

Both RF and power frequency fields are classified as *nonionizing radiation* because the frequency is too low for there to be enough photon energy to ionize atoms. *Ionizing radiation*, such as X-rays, gamma rays and some ultraviolet radiation, has enough energy to knock electrons loose from atoms. When this happens, positive and negative *ions* are formed. Still, at sufficiently high power densities, nonionizing EMF poses certain health hazards.

It has been known since the early days of radio that RF energy can cause injuries by

The ARRL RF Safety Committee

The ARRL maintains an RF Safety Committee composed of scientific and medical experts in the many aspects of the study of RF safety. The RFSC serves as a resource to the ARRL Board of Directors and to the Amateur Radio community with regard to RF safety-related questions and problems. It monitors and analyzes relevant research and its members participate in standards coordinating committees and other expert committees related to this subject. The RFSC is responsible for writing about RF safety in ARRL publications and is consulted to confirm the accuracy of RF safety-related issues in articles submitted to *QST* and *QEX*. The RFSC participates in generating the RF safety questions for FCC amateur question pools and works with the FCC in developing its environmental regulations.

This section was written by the ARRL RFSC, chaired by Gregory Lapin, PhD, PE, N9GL with input from committee members Robert Gold, MD, WØKIZ, A. William Guy, PhD, W7PO, William Kaune, PhD, W7IEQ, William Raskoff, MD, K6SQL, James Ross, MD, MPH, W4GHL, Kai Siwiak, PE, PhD, KE4PT, and Bruce Small, MD, KM2L. The committee is aided in its tasks by its ARRL Staff Liaison, Ed Hare, W1RFI and its ARRL Board Liaison, Howard Huntington, K9KM. In the preparation of this section, the RFSC gratefully acknowledges the editorial assistance of C-K Chou, PhD and Mays Swicord, PhD, both of Motorola Labs and Robert Cleveland, PhD and Ed Mantiply, both of the FCC Office of Engineering and Technology.

heating body tissue. Anyone who has ever touched an improperly grounded radio chassis or energized antenna and received an *RF burn* will agree that this type of injury can be quite painful. Excessive RF heating of the male reproductive organs can cause sterility by damaging sperm. Other health problems also can result from RF heating. These heat related health hazards are called *thermal effects*. A microwave oven is an application that puts thermal effects to practical use.

There also have been observations of changes in physiological function in the presence of RF energy levels that are too low to cause heating. These functions generally return to normal when the field is removed. Although research is ongoing, no harmful health consequences have been linked to these changes.

In addition to the ongoing research, much else has been done to address this issue. For example, FCC regulations set limits on exposure from radio transmitters. The Institute of Electrical and Electronics Engineers, the American National Standards Institute and the National Council for Radiation Protection and Measurement, among others, have recommended voluntary guidelines to limit human exposure to RF energy. The ARRL maintains an RF Safety Committee, consisting of concerned scientists and medical doctors, who volunteer to serve the radio amateur community to monitor scientific research and to recommend safe practices.

THERMAL EFFECTS OF RF ENERGY

Body tissues that are subjected to *very high* levels of RF energy may suffer serious heat damage. These effects depend on the frequency of the energy, the power density of the RF field that strikes the body and factors such as the polarization of the wave and the grounding of the body.

At frequencies near the body's natural resonances RF energy is absorbed more efficiently. In adults, the primary resonance frequency is usually about 35 MHz if the person is grounded, and about 70 MHz if insulated from the ground. Various body parts are resonant at different frequencies. Body size thus determines the frequency at which most RF energy is absorbed. As the frequency is moved farther from resonance, RF energy absorption becomes less efficient. *Specific absorption rate (SAR)* is a measure that takes variables such as resonance into account to describe the rate at which RF energy is absorbed in tissue, typically measured in watts per kilogram of tissue (W/kg).

Maximum permissible exposure (MPE) limits define the maximum electric and magnetic field strengths, and the plane-wave equivalent power densities associated with these fields, that a person may be exposed to without harmful effect, and are based on whole-body SAR safety levels. The safe exposure limits vary with frequency as the efficiency of absorption changes. The MPE limits Safety factors are included to insure that the MPE field strength will never result in an unsafe SAR.

Thermal effects of RF energy are usually not a major concern for most radio amateurs because the power levels normally used tend to be low and the intermittent nature of most amateur transmissions decreases total exposure. Amateurs spend more time listening than transmitting and many amateur transmissions such as CW and SSB use low-duty-cycle modes. With FM or RTTY, though, the RF is present continuously at its maximum level during each transmission. It is rare for radio amateurs to be subjected to RF fields strong enough to produce thermal effects, unless they are close to an energized antenna or unshielded power amplifier. Specific suggestions for avoiding excessive exposure are offered later in this chapter.

ATHERMAL EFFECTS OF EMF

Biological effects resulting from exposure to power levels of RF energy that do not generate measurable heat are called *athermal effects*. A number of athermal effects of EMF exposure on biological tissue have been seen in the laboratory. However, to date all athermal effects that have been discovered have had the same features: They are transitory, or go away when the EMF exposure is removed, and they have not been associated with any negative health effects.

28.3.2 Researching Biological Effects of EMF Exposure

The statistical basis of scientific research that confuses many non-scientists is the inability of science to state unequivocally that EMF is safe. Effects are studied by scientists using statistical inference where the "null hypothesis" assumes there is no effect and then tries to disprove this assumption by proving an "alternative hypothesis" that there is an effect. The alternative hypothesis can never be entirely disproved because a scientist cannot examine every possible case, so scientists only end up with a *probability* that the alternative hypothesis is *not* true. Thus, to be entirely truthful, a scientist can never say that something was proven; with respect to low-level EMF exposure, no scientist can guarantee that it is absolutely safe. At best, science can only state that there is a very low probability that it is unsafe. While scientists accept this truism, many members of the general public who are suspicious of EMF and its effects on humans see this as a reason to continue to be afraid.

There are two types of scientific study that are used to learn about the effects of EMF exposure on mammalian biology: laboratory and epidemiological.

LABORATORY STUDY

Scientists conduct laboratory research using animals to learn about biological mechanisms by which EMF may affect mammals. The main advantage of laboratory studies on the biological effects of EMF is that the exposures can be controlled very accurately.

Some major disadvantages of laboratory study also exist. EMF exposure may not affect the species of animals used in the investigations the same way that humans may respond. A common example of this misdirection occurred with eye research. Rabbits had been used for many years to determine that exposure of the eyes to high levels of EMF could cause cataracts. The extrapolation of these results to humans led to the fear that use of radio would harm one's vision. However, the rabbit's eye is on the surface of its skull while the human eye is buried deep within the bony orbit in the skull. Thus, the human eye receives much less exposure from EMF and is less likely to be damaged by the same exposures that had been used in the laboratory experiments on rabbits.

Some biological processes that affect tissue can take many years to occur and laboratory experiments on animals tend to be of shorter duration, in part because the life spans of most animals are much shorter than that of humans. For instance, a typical laboratory rat can be studied at most for two years, during which it progresses from youth to old age with all of the attendant physiological changes that come from normal aging. A disease process that takes multiple exposures over many years to occur is unlikely to be seen in a laboratory study with small animals.

EPIDEMIOLOGICAL RESEARCH

Epidemiologists look at the health patterns of large groups of people using statistical methods. In contrast to laboratory research, epidemiological research has very poor control of its subjects' exposures to EMF but it has the advantages of being able to analyze the effects of a lifetime of exposure and of being able to average out variations among large populations of subjects. By their basic design, epidemiological studies do not demonstrate cause and effect, nor do they postulate mechanisms of disease. Instead, epidemiologists look for associations between an environmental factor and an observed pattern of illness. Apparent associations are often seen in small preliminary studies that later are shown to have been incorrect. At best, such results are used to motivate more detailed epidemiological studies and laboratory studies that narrow down the search for cause-and-effect.

Some preliminary studies have suggested a weak association between exposure to EMF at home or at work and various malignant conditions including leukemia and brain cancer. A larger number of equally well-designed and performed studies, however, have found no association. Risk ratios as high as 2 have been observed in some studies. This means that the number of observed cases of disease in the test group is up to 2 times the "expected" number in the population. Epidemiologists generally regard a risk ratio of 4 or greater to be indicative of a strong association between the cause and effect under study. For example, men who smoke one pack of cigarettes per day increase their risk for lung cancer tenfold compared to nonsmokers and two packs per day increases the risk to more than 25 times the nonsmokers' risk.

Epidemiological research by itself is rarely conclusive, however. Epidemiology only identifies health patterns in groups — it does not ordinarily determine their cause. There are often confounding factors. Most of us are exposed to many different environmental hazards that may affect our health in various ways. Moreover, not all studies of persons likely to be exposed to high levels of EMF have yielded the same results (see sidebar on preliminary epidemiological studies).

28.3.3 Safe Exposure Levels

How much EMF energy is safe? Scientists and regulators have devoted a great deal of effort to deciding upon safe RF-exposure limits. This is a very complex problem, involving difficult public health and economic considerations. The recommended safe levels have been revised downward several times over the years — and not all scientific bodies agree on this question even today. The latest Institute of Electrical and Electronics Engineers (IEEE) C95.1 standard for recommended radio frequency exposure limits was published in 2006, updating one that had previously been published in 1991 and adopted by the American National Standards Institute (ANSI) in 1992. In the new standard changes were made to better reflect the current research, especially

---

Preliminary Epidemiology

Just about every week you can pick up the newspaper and see a screaming banner headline such as: "Scientists Discover Link Between Radio Waves and Disease." So why are you still operating your ham radio? You've experienced the inconsistency in epidemiological study of diseases. This is something that every radio amateur should understand in order to know how to interpret the real meaning of the science behind the headlines and to help assuage the fears that these stories elicit in others.

Just knowing that someone who uses a radio gets a disease, such as cancer, doesn't tell us anything about the cause-and-effect of that disease. People came down with cancer, and most other diseases, long before radio existed. What epidemiologists try to identify is a group of people who all have a common exposure to something and all suffer from a particular disease in higher proportion than would be expected if they were not exposed. This technique has been highly effective in helping health officials notice excesses of disease due to things such as poisoning of water supplies by local industry and even massive exposures such as smoking. However, epidemiology rarely proves that an exposure causes a disease; rather it provides the evidence that leads to further study.

While the strength of epidemiology is that it helps scientists notice anomalies in entire populations, its weakness is that it is non-specific. An initial epidemiological study examines only two things: suspected exposures and rates of diseases. These studies are relatively simple and inexpensive to perform and may point to an apparent association that then bears further study. For instance, in one study of the causes of death of a selection of Amateur Radio operators, an excess of leukemia was suggested. The percentage of ham radio operators who died of leukemia in that study was higher than expected based on the percentage of the rest of the population that died of leukemia. By itself, this has little meaning and should not be a cause for concern, since the study did not consider anything else about the sample population except that they had ham licenses. Many other questions arise: Were the study subjects exposed to any unusual chemicals? Did any of the study subjects have a family history of leukemia? Did the licensed amateurs even operate radios, what kind and how often? To an epidemiologist, this result might provide enough impetus to raise the funds to gather more specific information about each subject and perform a more complete study that strengthens the apparent associations. However, a slight excess of disease in a preliminary study rarely leads to further study. Commonly, an epidemiologist does not consider a preliminary study to be worth pursuing unless the ratio of excess disease, also called the risk ratio, is 4:1 or greater. Unfortunately, most news reporters are not epidemiologists and do not understand this distinction. Rather, a slight excess of disease in a preliminary study can lead to banner headlines that raise fear in the society, causing unreasonable resistance to things like cell phones and ham radios.

Headlines that blow the results of preliminary epidemiological studies out of proportion are rarely followed by retractions that are as visible if the study is followed up by one that is more complete and shows no association with disease. In the case of the aforementioned epidemiological study of hams' licensing and death records, overblown publicity about the results has led to the urban legend that ham radio operators are likely to come down with leukemia. Not only is this an unfounded conclusion due to the preliminary nature of the original study, but a similar study was recently performed by the National Cancer Institute using a far larger number of subjects and no significant excess of any disease was found. Hams should be able to recognize when sensationalistic headlines are based on inconclusive science and should be prepared to explain to their families, friends and neighbors just how inconclusive such results are.

Where Do RF Safety Standards Come From?

So much of the way we deal with RF Safety is based on "Safety Standards." The FCC environmental exposure regulations that every ham must follow are largely restatements of the conclusions reached by some of the major safety standards. How are these standards developed and why should we trust them?

The preeminent RF safety standard in the world was developed by the Institute of Electrical and Electronics Engineers (IEEE). The most recent edition is entitled *C95.1-2005: IEEE Standard for Safety Levels with Respect to Human Exposure to Radio Frequency Electromagnetic Fields, 3 kHz to 300 GHz*. The IEEE C95.1 Standard has a long history. The first C95.1 RF safety standard was released in 1966, was less than 2 pages long and listed no references. It essentially said that for frequencies between 10 MHz and 100 GHz people should not be exposed to a power density greater than 10 mW/cm^2. The C95.1 standard was revised in 1974, 1982, 1991 and 2005. The latest (2005) edition of the standard was published in 2006, is 250 pages long and has 1143 references to the scientific literature. Most of the editions of the IEEE C95.1 standard were adopted by the American National Standards Institute (ANSI) a year or two after they were published by IEEE. The 2005 edition was adopted by ANSI in 2006.

The committee at IEEE that developed the latest revision to C95.1 is called International Committee on Electromagnetic Safety Technical Committee 95 Subcommittee 4 and had a large base of participants. The subcommittee was co-chaired by C-K Chou, Ph.D., of Motorola Laboratories, and John D'Andrea, PhD, of the U.S. Naval Health Research Center. The committee had 132 members, 42% of whom were from 23 countries outside the United States. The members of the committee represented academia (27%), government (34%), industry (17%), consultants (20%) and the general public (2%).

Early editions of C95.1 were based on the concept that heat generated in the body should be limited to prevent damage to tissue. Over time the standard evolved to protect against *all known adverse biological effects* regardless of the amount of heat generated. The 2005 revision was based on the principles that the standard should protect human health yet still be practical to implement, its conclusions should be based solely on scientific evidence and wherever scientifically defensible it should be harmonized with other international RF safety standards. It based its conclusions on 50 years of scientific study. From over 2500 studies on EMF performed during that time, 1300 were selected for their relevance to the health effects of RF exposure. The science in these studies was evaluated for its quality and methodology and 1143 studies were referenced in producing the latest standard.

Other major standards bodies have published similar standards. The National Council for Radiation Protection and Measurement (NCRP) published its safety standard entitled, *Report No. 86: Biological Effects and Exposure Criteria for Radiofrequency Electromagnetic Fields* in 1986. The International Commission on Non-Ionizing Radiation Protection (ICNIRP) published its safety standard entitled *Guidelines for Limiting Exposure to Time-Varying Electric, Magnetic, and Electromagnetic Fields (Up to 300 GHz)* in 1998.

---

related to the safety of cellular telephones. At some frequencies the new standard determined that higher levels of exposure than previously thought are safe (see sidebar, "Where Do RF Safety Standards Come From?").

The IEEE C95.1 standard recommends frequency-dependent and time-dependent maximum permissible exposure levels. Unlike earlier versions of the standard, the 1991 and 2006 standards set different RF exposure limits in *controlled environments* (where energy levels can be accurately determined and everyone on the premises is aware of the presence of EM fields) and in *uncontrolled environments* (where energy levels are not known or where people may not be aware of the presence of EM fields). FCC regulations adopted these concepts to include controlled/occupational and uncontrolled/general population exposure limits.

The graph in **Fig 28.22** depicts the 1991

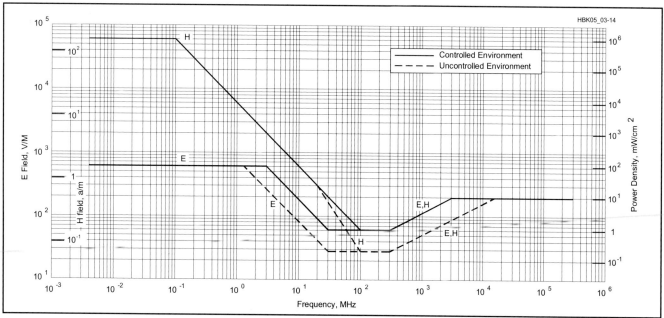

Fig 28.22 — 1991 RF protection guidelines for body exposure of humans. It is known officially as the "IEEE Standard for Safety Levels with Respect to Human Exposure to Radio Frequency Electromagnetic Fields, 3 kHz to 300 GHz."

FCC RF Exposure Regulations

FCC regulations control the amount of RF exposure that can result from your station's operation (§§97.13, 97.503, 1.1307 (b)(c)(d), 1.1310, 2.1091 and 2.1093). The regulations set limits on the maximum permissible exposure (MPE) allowed from operation of transmitters in all radio services. They also require that certain types of stations be evaluated to determine if they are in compliance with the MPEs specified in the rules. The FCC has also required that questions on RF environmental safety practices be added to Technician and General license examinations.

THE RULES
Maximum Permissible Exposure (MPE)

All radio stations regulated by the FCC must comply with the requirements for MPEs, even QRP stations running only a few watts or less. The MPEs vary with frequency, as shown in **Table A**. MPE limits are specified in maximum electric and magnetic fields for frequencies below 30 MHz, in power density for frequencies above 300 MHz and all three ways for frequencies from 30 to 300 MHz. For compliance purposes, all of these limits must be considered *separately*. If any one is exceeded, the station is not in compliance. In effect, this means that both electric and magnetic field must be determined below 300 MHz but at higher frequencies determining either the electric or magnetic field is normally sufficient.

The regulations control human exposure to RF fields, not the strength of RF fields in any space. There is no limit to how strong a field can be as long as no one is being exposed to it, although FCC regulations require that amateurs use the minimum necessary power at all times (§97.311 [a]).

Table A
(From §1.1310) Limits for Maximum Permissible Exposure (MPE)

(A) Limits for Occupational/Controlled Exposure

Frequency Range (MHz)	Electric Field Strength (V/m)	Magnetic Field Strength (A/m)	Power Density (mW/cm^2)	Averaging Time (minutes)
0.3-3.0	614	1.63	(100)*	6
3.0-30	1842/f	4.89/f	(900/f^2)*	6
30-300	61.4	0.163	1.0	6
300-1500	—	—	f/300	6
1500-100,000	—	—	5	6

f = frequency in MHz
* = Plane-wave equivalent power density (see Note 1).

(B) Limits for General Population/Uncontrolled Exposure

Frequency Range (MHz)	Electric Field Strength (V/m)	Magnetic Field Strength (A/m)	Power Density (mW/cm^2)	Averaging Time (minutes)
0.3-1.34	614	1.63	(100)*	30
1.34-30	824/f	2.19/f	(180/f^2)*	30
30-300	27.5	0.073	0.2	30
300-1500	—	—	f/1500	30
1500-100,000	—	—	1.0	30

f = frequency in MHz
* = Plane-wave equivalent power density (see Note 1).

Note 1: This means the equivalent far-field strength that would have the E or H-field component calculated or measured. It does not apply well in the near field of an antenna. The equivalent far-field power density can be found in the near or far field regions from the relationships: $P_d = |E_{total}|^2 / 3770$ mW/cm^2 or from $P_d = |H_{total}|^2 \times 37.7$ mW/cm^2.

Environments

The FCC has defined two tiers of exposure limits — *occupational/controlled limits* and *general population/uncontrolled limits*. Occupational/controlled limits apply when people are exposed as a condition of their employment and when they are aware of that exposure and can take steps to minimize it, if appropriate. General population/uncontrolled limits apply to exposure of the general public or people who are not normally aware of the exposure or cannot exercise control over it. The limits for general population/uncontrolled exposure are more stringent than the limits for occupational/controlled exposure. Specific definitions of the exposure categories can be found in Section 1.1310 of the FCC rules.

Although occupational/controlled limits are usually applicable in a workplace environment, the FCC has determined that they generally apply to amateur operators and members of their immediate households. In most cases, occupational/controlled limits can be applied to your home and property to which you can control

IEEE standard (which is still used as the basis of FCC regulation). It is necessarily a complex graph, because the standards differ not only for controlled and uncontrolled environments but also for electric (E) fields and magnetic (H) fields. Basically, the lowest E-field exposure limits occur at frequencies between 30 and 300 MHz. The lowest H-field exposure levels occur at 100-300 MHz. The ANSI standard sets the maximum E-field limits between 30 and 300 MHz at a power density of 1 mW/cm^2 (61.4 V/m) in controlled environments — but at one-fifth that level (0.2 mW/cm^2 or 27.5 V/m) in uncontrolled environments. The H-field limit drops to 1 mW/cm^2 (0.163 A/m) at 100-300 MHz in controlled environments and 0.2 mW/cm^2 (0.0728 A/m) in uncontrolled environments. Higher power densities are permitted at frequencies below 30 MHz (below 100 MHz for H fields) and above 300 MHz, based on the concept that the body will not be resonant at those frequencies and will therefore absorb less energy.

In general, the ANSI/IEEE standard requires averaging the power level over time periods ranging from 6 to 30 minutes for power-density calculations, depending on the frequency and other variables. The ANSI/IEEE exposure limits for uncontrolled environments are lower than those for controlled environments, but to compensate for that the standard allows exposure levels in those environments to be averaged over much longer time periods (generally 30 minutes). This long averaging time means that an intermittent RF source (such as an Amateur Radio transmitter) will result in a much lower exposure than a continuous-duty station, with all other parameter being equal. Time averaging is based on the concept that the human body can withstand a greater rate of body heating (and thus, a higher level of RF energy) for a short time.

Another national body in the United States, the National Council for Radiation Protection and Measurement (NCRP), also has adopted recommended exposure guidelines. NCRP urges a limit of 0.2 mW/cm^2 for nonoccupational exposure in the 30- 300 MHz range. The NCRP guideline differs from IEEE in

physical access. The general population/uncontrolled limits are intended for areas that are accessible by the general public, such as your neighbors' properties.

The MPE levels are based on average exposure. An averaging time of 6 minutes is used for occupational/controlled exposure; an averaging period of 30 minutes is used for general population/uncontrolled exposure.

Station Evaluations

The FCC requires that certain amateur stations be evaluated for compliance with the MPEs. Although an amateur can have someone else do the evaluation, it is not difficult for hams to evaluate their own stations. The ARRL book *RF Exposure and You* contains extensive information about the regulations and a large chapter of tables that show compliance distances for specific antennas and power levels. Generally, hams will use these tables to evaluate their stations. Some of these tables have been included in the FCC's information — *OET Bulletin 65* and its *Supplement B* (available for downloading at the FCC's RF Safety Web site). If hams choose, however, they can do more extensive calculations, use a computer to model their antenna and exposure, or make actual measurements.

Categorical Exemptions

Some types of amateur stations do not need to be evaluated, but these stations must still comply with the MPE limits. The station licensee remains responsible for ensuring that the station meets these requirements.

The FCC has exempted these stations from the evaluation requirement because their output power, operating mode and frequency are such that they are presumed to be in compliance with the rules.

Stations using power equal to or less than the levels in **Table B** do not have to be evaluated on a routine basis. For the 100-W HF ham station, for example, an evaluation would be required only on 12 and 10 meters.

Hand-held radios and vehicle-mounted mobile radios that operate using a push-to-talk (PTT) button are also categorically exempt from performing the routine evaluation.

Repeater stations that use less than 500 W ERP or those with antennas not mounted on buildings; if the antenna is at least 10 meters off the ground, also do not need to be evaluated.

Correcting Problems

Most hams are already in compliance with the MPE requirements. Some amateurs, especially those using indoor antennas or high-power, high-duty-cycle modes such as a RTTY bulletin station and specialized stations for moon bounce operations and the like may need to make adjustments to their station or operation to be in compliance.

The FCC permits amateurs considerable flexibility in complying with these regulations. As an example, hams can adjust their operating frequency, mode or power to comply with the MPE limits. They can also adjust their operating habits or control the direction their antenna is pointing.

More Information

This discussion offers only an overview of this topic; additional information can be found in *RF Exposure and You* and on the ARRL Web site at **www.arrl.org/rf-exposure**. The ARRL Web site has links to the FCC Web site, with OET Bulletin 65 and Supplement B and links to software that hams can use to evaluate their stations.

Table B
Power Thresholds for Routine Evaluation of Amateur Radio Stations

Wavelength Band	Evaluation Required if Power* (watts) Exceeds:
MF	
160 m	500
HF	
80 m	500
75 m	500
40 m	500
30 m	425
20 m	225
17 m	125
15 m	100
12 m	75
10 m	50
VHF (all bands)	50
UHF	
70 cm	70
33 cm	150
23 cm	200
13 cm	250
SHF (all bands)	250
EHF (all bands)	250
Repeater stations (all bands)	Non-building-mounted antennas: height above ground level to lowest point of antenna < 10 m *and* power > 500 W ERP Building-mounted antennas: power > 500 W ERP

*Transmitter power = Peak-envelope power input to antenna. For repeater stations only, power exclusion based on ERP (effective radiated power).

---

that it takes into account the effects of modulation on an RF carrier.

The FCC MPE regulations are based on a combination of the 1992 ANSI/IEEE standard and 1986 NCRP recommendations. The MPE limits under the regulations are slightly different than the ANSI/IEEE limits and do not reflect all the assumptions and exclusions of the ANSI/IEEE standard.

28.3.4 Cardiac Pacemakers and RF Safety

It is a widely held belief that cardiac pacemakers may be adversely affected in their function by exposure to electromagnetic fields. Amateurs with pacemakers may ask whether their operating might endanger themselves or visitors to their shacks who have a pacemaker. Because of this, and similar concerns regarding other sources of EM fields, pacemaker manufacturers apply design methods that for the most part shield the pacemaker circuitry from even relatively high EM field strengths.

It is recommended that any amateur who has a pacemaker, or is being considered for one, discuss this matter with his or her physician. The physician will probably put the amateur into contact with the technical representative of the pacemaker manufacturer. These representatives are generally excellent resources, and may have data from laboratory or "in the field" studies with specific model pacemakers.

One study examined the function of a modern (dual chamber) pacemaker in and around an Amateur Radio station. The pacemaker generator has circuits that receive and process electrical signals produced by the heart, and also generate electrical signals that stimulate (pace) the heart. In one series of experiments, the pacemaker was connected to a heart simulator. The system was placed on top of the cabinet of a 1-kW HF linear amplifier during SSB and CW operation. In another test, the system was placed in close proximity to several 1 to 5-W 2-meter hand-held transceivers. The test pacemaker was connected to the heart simulator in a third test, and then placed on the ground 9 meters below and 5 meters in front of a three-element Yagi HF antenna.

No interference with pacemaker function was observed in these experiments.

Although the possibility of interference cannot be entirely ruled out by these few observations, these tests represent more severe exposure to EM fields than would ordinarily be encountered by an amateur — with an average amount of common sense. Of course prudence dictates that amateurs with pacemakers, who use handheld VHF transceivers, keep the antenna as far as possible from the site of the implanted pacemaker generator. They also should use the lowest transmitter output required for adequate communication. For high power HF transmission, the antenna should be as far as possible from the operating position, and all equipment should be properly grounded.

28.3.5 Low-Frequency Fields

There has been considerable laboratory research about the biological effects of power line EMF. For example, some separate studies have indicated that even fairly low levels of EMF exposure might alter the human body's circadian rhythms, affect the manner in which T lymphocytes function in the immune system and alter the nature of the electrical and chemical signals communicated through the cell membrane and between cells, among other things. Although these studies are intriguing, they do not demonstrate any effect of these low-level fields on the overall organism.

Much of this research has focused on low-frequency magnetic fields, or on RF fields that are keyed, pulsed or modulated at a low audio frequency (often below 100 Hz). Several studies suggested that humans and animals could adapt to the presence of a steady RF carrier more readily than to an intermittent, keyed or modulated energy source.

The results of studies in this area, plus speculations concerning the effect of various types of modulation, were and have remained somewhat controversial. None of the research to date has demonstrated that low-level EMF causes adverse health effects.

Given the fact that there is a great deal of ongoing research to examine the health consequences of exposure to EMF, the American Physical Society (a national group of highly respected scientists) issued a statement in May 1995 based on its review of available data pertaining to the possible connections of cancer to 60-Hz EMF exposure. Their report is exhaustive and should be reviewed by anyone with a serious interest in the field. Among its general conclusions are the following:

1. The scientific literature and the reports of reviews by other panels show no consistent, significant link between cancer and power line fields.

2. No plausible biophysical mechanisms for the systematic initiation or promotion of cancer by these extremely weak 60-Hz fields have been identified.

3. While it is impossible to prove that no deleterious health effects occur from exposure to any environmental factor, it is necessary to demonstrate a consistent, significant, and causal relationship before one can conclude that such effects do occur.

In a report dated October 31, 1996, a committee of the National Research Council of the National Academy of Sciences has concluded that no clear, convincing evidence exists to show that residential exposures to electric and magnetic fields (EMF) are a threat to human health.

A National Cancer Institute epidemiological study of residential exposure to magnetic fields and acute lymphoblastic leukemia in children was published in the *New England Journal of Medicine* in July 1997. The exhaustive, seven-year study concludes that if there is any link at all, it is far too weak to be of concern.

In 1998, the US National Institute on Environmental Health Sciences organized a working group of experts to summarize the research on power-line EMF. The committee used the classification rules of the International Agency for Research on Cancer (IARC) and performed a meta-analysis to combine all past results as if they had been performed in a single study. The NIEHS working group concluded that the research did not show this type of exposure to be a carcinogen but could not rule out the possibility either. Therefore, they defined power-line EMF to be a Class 2b carcinogen under the IARC classification. The definition, as stated by the IARC is: "Group 2B: The agent is possibly carcinogenic to humans. There is limited pidemiological evidence plus limited or inadequate animal evidence." Other IARC Class 2b carcinogens include automobile ex-

Table 28.4
Typical 60-Hz Magnetic Fields Near Amateur Radio Equipment and AC-Powered Household Appliances
Values are in milligauss.

Item	Field	Distance
Electric blanket	30-90	Surface
Microwave oven	10-100	Surface
	1-10	12 in.
IBM personal computer	5-10	Atop monitor
	0-1	15 in. from screen
Electric drill	500-2000	At handle
Hair dryer	200-2000	At handle
HF transceiver	10-100	Atop cabinet
	1-5	15 in. from front
1-kW RF amplifier	80-1000	Atop cabinet
	1-25	15 in. from front

(Source: measurements made by members of the ARRL RF Safety Committee)

Table 28.5
Typical RF Field Strengths Near Amateur Radio Antennas
A sampling of values as measured by the Federal Communications Commission and Environmental Protection Agency, 1990

Antenna Type	Freq (MHz)	Power (W)	E Field (V/m)	Location
Dipole in attic	14.15	100	7-100	In home
Discone in attic	146.5	250	10-27	In home
Half sloper	21.5	1000	50	1 m from base
Dipole at 7-13 ft	7.14	120	8-150	1-2 m from earth
Vertical	3.8	800	180	0.5 m from base
5-element Yagi at 60 ft	21.2	1000	10-20	In shack
			14	12 m from base
3-element Yagi at 25 ft	28.5	425	8-12	12 m from base
Inverted V at 22-46 ft	7.23	1400	5-27	Below antenna
Vertical on roof	14.11	140	6-9	In house
			35-100	At antenna tuner
Whip on auto roof	146.5	100	22-75	2 m antenna
			15-30	In vehicle
			90	Rear seat
5-element Yagi at 20 ft	50.1	500	37-50	10 m antenna

Table 28.6
RF Awareness Guidelines

These guidelines were developed by the ARRL RF Safety Committee, based on the FCC/EPA measurements of Table 28.4 and other data.

- Although antennas on towers (well away from people) pose no exposure problem, make certain that the RF radiation is confined to the antennas' radiating elements themselves. Provide a single, good station ground (earth), and eliminate radiation from transmission lines. Use good coaxial cable or other feed line properly. Avoid serious imbalance in your antenna system and feed line. For high-powered installations, avoid end-fed antennas that come directly into the transmitter area near the operator.

- No person should ever be near any transmitting antenna while it is in use. This is especially true for mobile or ground-mounted vertical antennas. Avoid transmitting with more than 25 W in a VHF mobile installation unless it is possible to first measure the RF fields inside the vehicle. At the 1-kW level, both HF and VHF directional antennas should be at least 35 ft above inhabited areas. Avoid using indoor and attic-mounted antennas if at all possible. If open-wire feeders are used, ensure that it is not possible for people (or animals) to come into accidental contact with the feed line.

- Don't operate high-power amplifiers with the covers removed, especially at VHF/UHF.

- In the UHF/SHF region, never look into the open end of an activated length of waveguide or microwave feed-horn antenna or point it toward anyone. (If you do, you may be exposing your eyes to more than the maximum permissible exposure level of RF radiation.) Never point a high-gain, narrow-bandwidth antenna (a paraboloid, for instance) toward people. Use caution in aiming an EME (moonbounce) array toward the horizon; EME arrays may deliver an effective radiated power of 250,000 W or more.

- With hand-held transceivers, keep the antenna away from your head and use the lowest power possible to maintain communications. Use a separate microphone and hold the rig as far away from you as possible. This will reduce your exposure to the RF energy.

- Don't work on antennas that have RF power applied.

- Don't stand or sit close to a power supply or linear amplifier when the ac power is turned on. Stay at least 24 inches away from power transformers, electrical fans and other sources of high-level 60-Hz magnetic fields.

---

haust, chloroform, coffee, ceramic and glass fibers, gasoline and pickled vegetables.

Readers may want to follow this topic as further studies are reported. Amateurs should be aware that exposure to RF and ELF (60 Hz) electromagnetic fields at all power levels and frequencies has not been fully studied under all circumstances. "Prudent avoidance" of any avoidable EMF is always a good idea. Prudent avoidance doesn't mean that amateurs should be fearful of using their equipment. Most amateur operations are well within the MPE limits. If any risk does exist, it will almost surely fall well down on the list of causes that may be harmful to your health (on the other end of the list from your automobile). It does mean, however, that hams should be aware of the potential for exposure from their stations, and take whatever reasonable steps they can take to minimize their own exposure and the exposure of those around them.

Although the FCC doesn't regulate 60-Hz fields, some recent concern about EMF has focused on 60 Hz. Amateur Radio equipment can be a significant source of 60 Hz fields, although there are many other sources of this kind of energy in the typical home. Magnetic fields can be measured relatively accurately with inexpensive 60-Hz meters that are made by several manufacturers.

Table 28.4 shows typical magnetic field intensities of Amateur Radio equipment and various household items.

28.3.6 Determining RF Power Density

Unfortunately, determining the power density of the RF fields generated by an amateur station is not as simple as measuring low-frequency magnetic fields. Although sophisticated instruments can be used to measure RF power densities quite accurately, they are costly and require frequent recalibration. Most amateurs don't have access to such equipment, and the inexpensive field-strength meters that we do have are not suitable for measuring RF power density.

Table 28.5 shows a sampling of measurements made at Amateur Radio stations by the Federal Communications Commission and the Environmental Protection Agency in 1990. As this table indicates, a good antenna well removed from inhabited areas poses no hazard under any of the ANSI/IEEE guidelines. However, the FCC/EPA survey also indicates that amateurs must be careful about using indoor or attic-mounted antennas, mobile antennas, low directional arrays or any other antenna that is close to inhabited areas, especially when moderate to high power is used.

Ideally, before using any antenna that is in close proximity to an inhabited area, you should measure the RF power density. If that is not feasible, the next best option is make the installation as safe as possible by observing the safety suggestions listed in **Table 28.6**.

It also is possible, of course, to calculate the probable power density near an antenna using simple equations. Such calculations have many pitfalls. For one, most of the situations where the power density would be high enough to be of concern are in the near field. In the near field, ground interactions and other variables produce power densities that cannot be determined by simple arithmetic. In the far field, conditions become easier to predict with simple calculations.

The boundary between the near field and the far field depends on the wavelength of the transmitted signal and the physical size and configuration of the antenna. The boundary between the near field and the far field of an antenna can be as much as several wavelengths from the antenna.

Computer antenna-modeling programs are another approach you can use. *MININEC* or other codes derived from *NEC* (Numerical Electromagnetics Code) are suitable for estimating RF magnetic and electric fields around amateur antenna systems.

These models have limitations. Ground interactions must be considered in estimating near-field power densities, and the "correct ground" must be modeled. Computer modeling is generally not sophisticated enough to predict "hot spots" in the near field — places where the field intensity may be far higher than would be expected, due to reflections from nearby objects. In addition, "nearby objects" often change or vary with weather or the season, therefore the model so laboriously crafted may not be representative of the actual situation, by the time it is running on the computer.

Intensely elevated but localized fields often can be detected by professional measuring instruments. These "hot spots" are often found near wiring in the shack, and metal objects such as antenna masts or equipment cabinets. But even with the best instrumentation, these measurements also may be misleading in the near field. One need not make precise measurements or model the exact antenna system, however, to develop some idea of the relative fields around an antenna. Computer modeling using close approximations of the geometry and power input of the antenna will generally suffice. Those who are familiar with *MININEC* can estimate their power densities by computer modeling, and those who have access to professional power-density meters can make useful measurements.

While our primary concern is ordinarily the intensity of the signal radiated by an antenna, we also should remember that there are other potential energy sources to be considered. You also can be exposed to excessive RF fields directly from a power amplifier if it is operated without proper shielding. Transmission lines also may radiate a significant amount of energy under some conditions. Poor microwave waveguide joints or improperly assembled connectors are another source of incidental exposure.

28.3.7 Further RF Exposure Suggestions

Potential exposure situations should be taken seriously. Based on the FCC/EPA measurements and other data, the "RF awareness" guidelines of Table 28.6 were developed by the ARRL RF Safety Committee. A longer version of these guidelines, along with a complete list of references, appeared in a *QST* article by Ivan Shulman, MD, WC2S ("Is Amateur Radio Hazardous to Our Health?" *QST*, Oct 1989, pp 31-34).

In addition, the ARRL has published a book, *RF Exposure and You* that helps hams comply with the FCC's RF-exposure regulations. The ARRL also maintains an RF-exposure news page on its Web site. See **www.arrl.org/rf-exposure**. This site contains reprints of selected *QST* articles on RF exposure and links to the FCC and other useful sites.

SUMMARY

The ideas presented in this chapter are intended to reinforce the concept that ham radio, like many other activities in modern life, does have certain risks. But by understanding the hazards and how to deal effectively with them, the risk can be minimized. Common-sense measures can go a long way to help us prevent accidents. Traditionally, amateurs are inventors, and experimenting is a major part of our nature. But reckless chance-taking is never wise, especially when our health and well-being is involved. A healthy attitude toward doing things the right way will help us meet our goals and expectations.

Contents

29.1 Fixed Stations
- 29.1.1 Selecting a Location
- 29.1.2 Station Ground
- 29.1.3 Station Power
- 29.1.4 Station Layout
- 29.1.5 Interconnecting Your Equipment
- 29.1.6 Documenting Your Station
- 29.1.7 Interfacing High-Voltage Equipment to Solid-State Accessories

29.2 Mobile Installations
- 29.2.1 Installation
- 29.2.2 Coax
- 29.2.3 Wiring
- 29.2.4 Amplifiers
- 29.2.5 Interference Issues
- 29.2.6 Operating

29.3 Portable Installations
- 29.3.1 Portable AC Power Sources
- 29.3.2 Portable Antennas

29.4 Remote Stations
- 29.4.1 Introduction
- 29.4.2 Evaluation and Planning
- 29.4.3 Controlling Your Remote Site HF Station
- 29.4.4 A Basic Remote HF Station
- 29.4.5 Remote HF Station Resources
- 29.4.6 Remote Station Glossary

29.5 References and Bibliography

Chapter 29

Assembling a Station

Although many hams never try to build a major project, such as a transmitter, receiver or amplifier, they do have to assemble the various components into a working station. There are many benefits to be derived from assembling a safe, comfortable, easy-to-operate collection of radio gear, whether the shack is at home, in the car or in a field. This chapter will detail some of the "how tos" of setting up a station for fixed, mobile and portable operation. Such topics as station location, finding adequate power sources, station layout and cable routing are covered. It includes contributions from Wally Blackburn, AA8DX, a section on mobile installations from Alan Applegate, K0BG. and information on gasoline generators from Kirk Kleinschmidt, NT0Z. Rick Hilding, K6VVA, contributed the section on remote stations.

29.1 Fixed Stations

Regardless of the type of installation you are attempting, good planning greatly increases your chances of success. Take the time to think the project all the way through, consider alternatives, and make rough measurements and sketches during your planning and along the way. You will save headaches and time by avoiding "shortcuts." What might seem to save time now may come back to haunt you with extra work when you could be enjoying your shack.

One of the first considerations should be to determine what type of operating you intend to do. While you do not want to strictly limit your options later, you need to consider what you want to do, how much you have to spend and what room you have to work with. There is a big difference between a casual operating position and a "big gun" contest station, for example.

29.1.1 Selecting a Location

Selecting the right location for your station is the first and perhaps the most important step in assembling a safe, comfortable, convenient station. The exact location will depend on the type of home you have and how much space can be devoted to your station. Fortunate amateurs will have a spare room to devote to housing the station; some may even have a separate building for their exclusive use. Most must make do with a spot in the cellar or attic, or a corner of the living room is pressed into service.

Examine the possibilities from several angles. A station should be comfortable; odds are good that you'll be spending a lot of time there over the years. Some unfinished basements are damp and drafty — not an ideal environment for several hours of leisurely hamming. Attics have their drawbacks, too; they can be stifling during warmer months. If possible, locate your station away from the heavy traffic areas of your home. Operation of your station should not interfere with family life. A night of chasing DX on 80 m may be exciting to you, but the other members of your household may not share your enthusiasm.

Keep in mind that you must connect your station to the outside world. The location you choose should be convenient to a good power source and an adequate ground. If you use a computer and modem, you may need access to a telephone jack. There should be a fairly direct route to the outside for running antenna feed lines, rotator control cables and the like.

Although most homes will not have an "ideal" space meeting all requirements, the right location for you will be obvious after you scout around. The amateurs whose stations are depicted in **Figs 29.1**

Fig 29.1 — Randy, K5ZD, has arranged his equipment for efficient contest operation. The computer keyboard and monitor occupy center stage, flanked by two transceivers and amplifiers and various accessories and control switches. (*Photo courtesy Andrew Thompson*)

Fig 29.2 — Ward, N0AX, has had great success with this simple low power/QRP station tucked away in the corner of a room. All equipment fits on a computer cart and small rolling cabinet that make it easy to do station maintenance, as well as comfortable to use.

Fig 29.3 — Spreading out horizontally, John Sluymer, VE3EJ, has arranged his effective contest and DXing station to keep all of the controls at the same level for easy adjustment. (*VE3EJ photo*)

through **29.3** all found the right spot for them. Weigh the trade-offs and decide which features you can do without and which are necessary for your style of operation. If possible pick an area large enough for future expansion.

29.1.2 Station Ground

Grounding is an important factor in overall station safety, as detailed in the **Safety** chapter. An effective ground system is necessary for every amateur station. The mission of the ground system is twofold. First, it reduces the possibility of electrical shock if something in a piece of equipment should fail and the chassis or cabinet becomes "hot." If connected to a properly grounded outlet, a three-wire electrical system grounds the chassis. Much amateur equipment still uses the ungrounded two-wire system, however. A ground system to prevent shock hazards is generally referred to as *dc ground*.

The second job the ground system must perform is to provide a low-impedance path to ground for any stray RF current inside the station. Stray RF can cause equipment to malfunction and contributes to RFI problems. This low-impedance path is usually called *RF ground*. In most stations, dc ground and RF ground are provided by the same system.

GROUND NOISE

Noise in ground systems can affect our sensitive radio equipment. It is usually related to one of three problems:

1) Insufficient ground conductor size
2) Loose ground connections
3) Ground loops

These matters are treated in precise scientific research equipment and certain industrial instruments by attention to certain rules. The ground conductor should be at least as large as the largest conductor in the primary power circuit. Ground conductors should provide a solid connection to both ground and to the equipment being grounded. Liberal use of lock washers and star washers is highly recommended. A loose ground connection is a tremendous source of noise, particularly in a sensitive receiving system.

Ground loops should be avoided at all costs. A short discussion here of what a ground loop is and how to avoid them may lead you down the proper path. A ground loop is formed when more than one ground current is flowing in a single conductor. This commonly occurs when grounds are "daisy-chained" (series linked). The correct way to ground equipment is to bring all ground conductors out radially from a common point to either a good driven earth ground or a cold-water system. If one or more earth grounds are used, they should be bonded back to the service entrance panel. Details appear in the **Safety** chapter.

Ground noise can affect transmitted and received signals. With the low audio levels required to drive amateur transmitters, and the ever-increasing sensitivity of receivers, correct grounding is critical.

29.1.3 Station Power

Amateur Radio stations generally require a 120-V ac power source. The 120-V ac is then converted to the proper ac or dc levels required for the station equipment. RF power amplifiers typically require 240 V ac for best operation.

Power supply theory is covered in the **Power Supplies** chapter, and safety issues and station wiring are covered in the **Safety** chapter. If your station is located in a room with electrical outlets, you're in luck. If your station is located in the basement, an attic or another area without a convenient 120-V source, you will have to run a line to your operating position.

SURGE PROTECTION

Typically, the ac power lines provide an adequate, well-regulated source of electrical power for most uses. At the same time, these lines are fraught with frequent power surges that, while harmless to most household equipment, may cause damage to more sensitive devices such as computers or test equipment. A common method of protecting these devices is through the use of surge protectors. More information on these and lightning protection is in the **Safety** chapter.

29.1.4 Station Layout

Station layout is largely a matter of personal taste and needs. It will depend mostly on the amount of space available, the equipment involved and the types of operating to be done. With these factors in mind, some basic design considerations apply to all stations.

THE OPERATING TABLE

The operating table may be an office or computer desk, a kitchen table or a custom-made bench. What you use will depend on space, materials at hand and cost. The two most important considerations are height and size of the top. Most commercial desks are about 29 inches above the floor. This is a comfortable height for most adults. Heights much lower or higher than this may cause an awkward operating position.

The dimensions of the top are an important consideration. A deep (36 inches or more) top will allow plenty of room for equipment interconnections along the back, equipment about midway and room for writing toward the front. The length of the top will depend on the amount of equipment being used. An office or computer desk makes a good operating table. These are often about 36 inches deep and 60 inches wide. Drawers can be used for storage of logbooks, head-

Fig 29.4 — Top contest operator Pat Barkey, N9RV, operates this efficient station laid out vertically to make the most of available space. (*N9RV photo*)

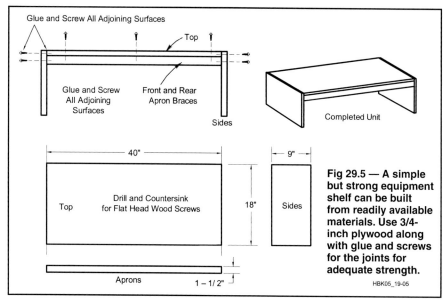

Fig 29.5 — A simple but strong equipment shelf can be built from readily available materials. Use 3/4-inch plywood along with glue and screws for the joints for adequate strength.

phones, writing materials, and so on. Desks specifically designed for computer use often have built-in shelves that can be used for equipment stacking. Desks of this type are available ready-to-assemble at most discount and home improvement stores. The low price and adaptable design of these desks make them an attractive option for an operating position. An example is shown in **Fig 29.4**.

ARRANGING THE EQUIPMENT

No matter how large your operating table, some vertical stacking of equipment may be necessary to allow you to reach everything from your chair. Stacking pieces of equipment directly on top of one another is not a good idea because most amateur equipment needs airflow around it for cooling. A shelf like that shown in **Fig 29.5** can improve

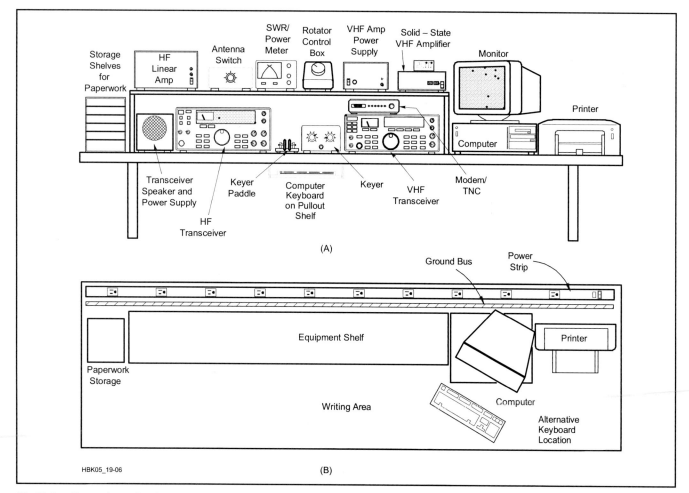

Fig 29.6 — Example station layout as seen from the front (A) and the top (B). The equipment is spaced far enough apart that air circulates on all sides of each cabinet.

Assembling a Station 29.3

equipment layout in many situations. Dimensions of the shelf can be adjusted to fit the size of your operating table.

When you have acquired the operating table and shelving for your station, the next task is arranging the equipment in a convenient, orderly manner. The first step is to provide power outlets and a good ground as described in a previous section. Be conservative in estimating the number of power outlets for your installation; radio equipment has a habit of multiplying with time, so plan for the future at the outset.

Fig 29.6 illustrates a sample station layout. The rear of the operating table is spaced about 1½ ft from the wall to allow easy access to the rear of the equipment. This installation incorporates two separate operating positions, one for HF and one for VHF. When the operator is seated at the HF operating position, the keyer and transceiver controls are within easy reach. The keyer, keyer paddle and transceiver are the most-often adjusted pieces of equipment in the station. The speaker is positioned right in front of the operator for the best possible reception. Accessory equipment not often adjusted, including the amplifier, antenna switch and rotator control box, is located on the shelf above the transceiver. The SWR/power meter and clock, often consulted but rarely touched, are located where the operator can view them without head movement. All HF-related equipment can be reached without moving the chair.

This layout assumes that the operator is right-handed. The keyer paddle is operated with the right hand, and the keyer speed and transceiver controls are operated with the left hand. This setup allows the operator to write or send with the right hand without having to cross hands to adjust the controls. If the operator is left-handed, some repositioning of equipment is necessary, but the idea is the same. For best results during CW operation, the paddle should be weighted to keep it from "walking" across the table. It should be oriented such that the operator's entire arm from wrist to elbow rests on the tabletop to prevent fatigue.

Some operators prefer to place the station transceiver on the shelf to leave the table top clear for writing. This arrangement leads to fatigue from having an unsupported arm in the air most of the time. If you rest your elbows on the tabletop, they will quickly become sore. If you rarely operate for prolonged periods, however, you may not be inconvenienced by having the transceiver on the shelf. The real secret to having a clear table top for logging, and so on, is to make the operating table deep enough that your entire arm from elbow to wrist rests on the table with the front panels of the equipment at your fingertips. This leaves plenty of room for paperwork, even with a microphone and

Fig 29.7 — **It was back to basics for Elias, K4IX, during a recent Field Day.**

Fig 29.8 — **Richard, WB5DGR, uses a homebrew 1.5-kW amplifier to seek EME contacts from this nicely laid out station.**

keyer paddle on the table.

The VHF operating position in this station is similar to the HF position. The amplifier and power supply are located on the shelf. The station triband beam and VHF beam are on the same tower, so the rotator control box is located where it can be seen and reached from both operating positions. This operator is active on packet radio on a local VHF repeater, so the computer, printer, terminal node controller and modem are all clustered within easy reach of the VHF transceiver.

This sample layout is intended to give you ideas for designing your own station. Study the photos of station layouts presented here, in other chapters of this *Handbook* and in *QST*. Visit the shacks of amateur friends to view their ideas. Station layout is always changing as you acquire new gear, dispose of old gear, change operating habits and interests or become active on different bands. Configure the station to suit your interests, and keep thinking of ways to refine the layout. **Figs 29.7** and **29.8** show station arrangements tailored for specific purposes.

Equipment that is adjusted frequently sits on the tabletop, while equipment requiring infrequent adjustment is perched on a shelf. All equipment is positioned so the operator does not have to move the chair to reach anything at the operating position.

ERGONOMICS

Ergonomics is a term that loosely means "fitting the work to the person." If tools and equipment are designed around what people can accommodate, the results will be much more satisfactory. For example, in the 1930s research was done in telephone equipment manufacturing plants because use of long-nosed pliers for wiring switchboards required considerable force at the end of the hand's range of motion. A simple tool redesign resolved this issue.

Considerable attention has been focused on ergonomics in recent years because we have come to realize that long periods of time spent in unnatural positions can lead to repetitive-motion injuries. Much of this attention has been focused on people whose job tasks have required them to operate computers and other office equipment. While most Amateur Radio operators do not devote as much time to their hobby as they might in a full-time job, it does make sense to consider comfort and flexibility when choosing furniture and arranging it in the shack or workshop. Adjustable height chairs are available with air cylinders to serve as a shock absorber. Footrests might come in handy if the chair is so high that your feet cannot support your lower leg weight. The height of tables and keyboards often is not adjustable.

Placement of computer screens should take into consideration the reflected light coming from windows. It is always wise to build into your sitting sessions time to walk around and stimulate blood circulation. Your muscles are less likely to stiffen, while the flexibility in your joints can be enhanced by moving around.

Selection of hand tools is another area where there are choices to make that may affect how comfortable you will be while working in your shack. Look for screwdrivers with pliable grips. Take into account how heavy things are before picking them up — your back will thank you.

FIRE EXTINGUISHERS

Fires in well-designed electronic equipment are not common but are known to occur. Proper use of a suitable fire extinguisher can make the difference between a small fire with limited damage and loss of an entire home. Make sure you know the limitations of your extinguisher and the importance of reporting the fire to your local fire department immediately.

Several types of extinguishers are suitable for electrical fires. The multipurpose dry chemical or "ABC" type units are relatively inexpensive and contain a solid powder that is nonconductive. Avoid buying the smallest size; a 5-pound capacity will meet most requirements in the home. ABC extinguishers are also the best choice for kitchen fires (the most common location of home fires). One disadvantage of this type is the residue left behind that might cause corrosion in electrical connectors. Another type of fire extinguisher suitable for energized electrical equipment is the carbon dioxide unit. CO_2 extinguishers require the user to be much closer to the fire, are heavy and difficult to handle, and are relatively expensive. For obvious reasons, water extinguishers are not suitable for fires in or near electronic equipment.

AIDS FOR HAMS WITH DISABILITIES

A station used by an amateur with physical disabilities or sensory impairments may require adapted equipment or particular layout considerations. The station may be highly customized to meet the operator's needs or just require a bit of "tweaking."

The myriad of individual needs makes describing all of the possible adaptive methods impractical. Each situation must be approached individually, with consideration to the operator's particular needs. However, many types of situations have already been encountered and worked through by others, eliminating the need to start from scratch in every case.

An excellent resource is the Courage Handi-Ham System. The Courage Handi-Ham System, a part of the Courage Center, provides a number of services to hams (and aspiring hams) with disabilities. These include study materials and a wealth of useful information on their comprehensive Web site. Visit **www.handiham.org** for more information.

29.1.5 Interconnecting Your Equipment

Once you have your equipment and get it arranged, you will have to interconnect it all. No matter how simple the station, you will at least have antenna, power and microphone or key connections. Equipment such as amplifiers, computers, TNCs and so on add complexity. By keeping your equipment interconnections well organized and of high quality, you will avoid problems later on.

Often, ready-made cables will be available. But in many cases you will have to make your own cables. A big advantage of making your own cables is that you can customize the length. This allows more flexibility in arranging your equipment and avoids unsightly extra cable all over the place. Many manufacturers supply connectors with their equipment along with pinout information in the manual. This allows you to make the necessary cables in the lengths you need for your particular installation.

Always use high quality wire, cables and connectors in your shack. Take your time and make good mechanical and electrical connections on your cable ends. Sloppy cables are often a source of trouble. Often the problems they cause are intermittent and difficult to track down. You can bet that they will crop up right in the middle of a contest or during a rare DX QSO! Even worse, a poor quality connection could cause RFI or even create a fire hazard. A cable with a poor mechanical connection could come loose and short a power supply to ground or apply a voltage where it should not be. Wire and cables should have good quality insulation that is rated high enough to prevent shock hazards.

Interconnections should be neatly bundled and labeled. Wire ties, masking tape or paper labels with string work well. See **Fig 29.9**. Whatever method you use, proper labeling makes disconnecting and reconnecting equipment much easier. **Fig 29.10** illustrates the number of potential interconnections in a modern, full-featured transceiver.

WIRE AND CABLE

The type of wire or cable to use depends on the job at hand. The wire must be of suf-

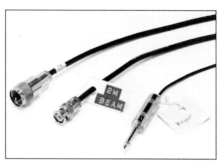

Fig 29.9 — Labels on the cables make it much easier to rearrange things in the station. Labeling ideas include masking tape, cardboard labels attached with string and labels attached to fasteners found on plastic bags (such as bread bags).

ficient size to carry the necessary current. Use the tables in the **Component Data and References** chapter to find this information. Never use underrated wire; it will be a fire hazard. Be sure to check the insulation too. For high-voltage applications, the insulation must be rated at least a bit higher than the intended voltage. A good rule of thumb is to use a rating at least twice what is needed.

Use good quality coaxial cable of sufficient size for connecting transmitters, transceivers, antenna switches, antenna tuners and so on. RG-58 might be fine for a short patch between your transceiver and SWR bridge, but is too small to use between your legal-limit amplifier and Transmatch. For more information, see the **Transmission Lines** chapter.

Hookup wire may be stranded or solid. Generally, stranded is a better choice since it is less prone to break under repeated flexing. Many applications require shielded wire to reduce the chances of RF getting into the equipment. RG-174 is a good choice for control, audio and some low-power applications. Shielded microphone or computer cable can be used where more conductors are necessary.

CONNECTORS

Connectors are a convenient way to make an electrical connection by using mating electrical contacts. There are quite a few connector styles, but common terms apply to all of them. Pins are contacts that extend out of the connector body, and connectors in which pins make the electrical contact are called "male" connectors. Sockets are hollow, recessed contacts, and connectors with sockets are called "female." Connectors designed to attach to each other are called "mating connectors." Connectors with specially shaped bodies or inserts that require a complementary shape on a mating connector are called "keyed connectors." Keyed connectors ensure that the connectors can only go together one way, reducing the possibility

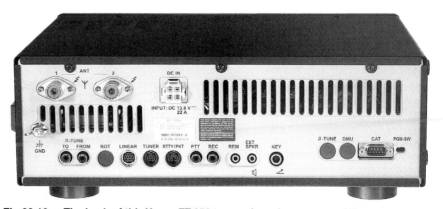

Fig 29.10 — The back of this Yaesu FT-950 transceiver shows some of the many types of connectors encountered in the amateur station. Note that this variety is found on a single piece of equipment.

Assembling a Station 29.5

of damage from incorrect mating.

Plugs are connectors installed on the end of cables and *jacks* are installed on equipment. *Adapters* make connections between two different styles of connector, such as between two different families of RF connectors. Other adapters join connectors of the same family, such as double-male, double-female and gender changers. *Splitters* divide a signal between two connectors.

While the number of different types of connectors is mind-boggling, many manufacturers of amateur equipment use a few standard types. If you are involved in any group activities such as public service or emergency-preparedness work, check to see what kinds of connectors others in the group use and standardize connectors wherever possible. Assume connectors are not waterproof, unless you specifically buy one clearly marked for outdoor use (and assemble it correctly).

Power Connectors

Amateur Radio equipment uses a variety of power connectors. Some examples are shown in **Fig 29.11**. Most low power amateur equipment uses coaxial power connectors. These are the same type found on consumer electronic equipment that is supplied by a wall transformer or "wall wart" style of power supply. Transceivers and other equipment that requires high current in excess of a few amperes often use Molex connectors (**www.molex.com** — enter "MLX" in the search window) with a white, nylon body housing pins and sockets crimped on to the end of wires.

An emerging standard, particularly among ARES and other emergency communications groups, is the use of Anderson PowerPole connectors (**www.andersonpower.com** — click "Product Brands"). These connectors are "sexless" meaning that any two connectors of the same series can be mated — there are no male or female connectors. By standardizing on a single connector style, equipment can be shared and replaced easily in the field.

Molex and PowerPole connectors use crimp terminals (both male and female) installed on the end of wires. A special crimping tool is used to attach the wire to the terminal and the terminal is then inserted into the body of the connector. Making a solid connection requires the use of an appropriate tool — do not use pliers or some other tool to make a crimp connection.

Some equipment uses terminal strips for direct connection to wires or crimp terminals, often with screws. Other equipment uses spring-loaded terminals or binding posts to connect to bare wire ends. **Fig 29.12** shows some common crimp terminals that are installed on the ends of wires using special tools.

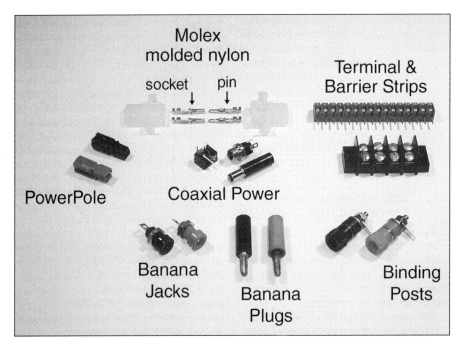

Fig 29.11 — These are the most common connectors used on amateur equipment to make power connections. (Courtesy of Wiley Publishing, *Ham Radio for Dummies*, or *Two-Way Radios and Scanners for Dummies*)

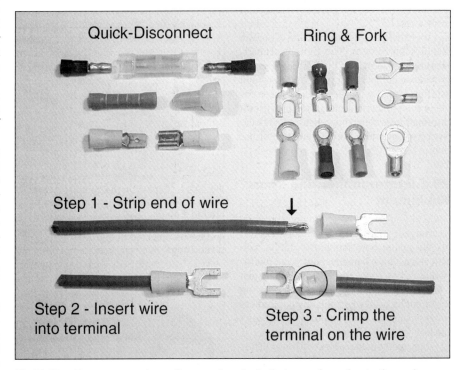

Fig 29.12 — Power connectors often use terminals that are crimped onto the end of wires with special crimping tools. (Courtesy of Wiley Publishing, *Ham Radio for Dummies*, or *Two-Way Radios and Scanners for Dummies*)

Audio and Control Connectors

Consumer audio equipment and Amateur Radio equipment share many of the same connectors for the same uses. Phone plugs and jacks are used for mono and stereo audio circuits. These connectors, shown in **Fig 29.13** come in ¼ inch, ⅛ inch (miniature) and subminiature varieties. The contact at the end of the plug is called the tip and the connector at the base of the plug is the sleeve. If there is a third contact between the tip and sleeve, it is the ring (these are "stereo" phone connectors).

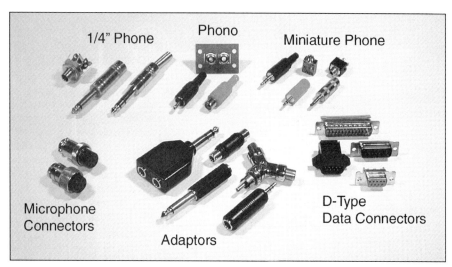

Fig 29.13 — Audio and data signals are carried by a variety of different connectors. Individual cable conductors are either crimped or soldered to the connector contacts. (*N0AX photo*)

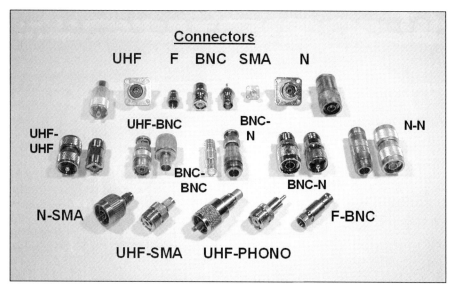

Fig 29.14 — Each type of RF connector is specially made to carry RF signals and preserve the shielding of coaxial cable. Adapters are available to connect one style of connector to another. (*N0AX photo*)

Phono plugs and jacks (sometimes called RCA connectors since they were first used on RCA brand equipment) are used for audio, video and other low-level RF signals. They are also widely used for control signals.

The most common microphone connector on mobile and base station equipment is an 8-pin round connector. On older transceivers you may see 4-pin round connectors used for microphones. RJ-45 modular connectors (see the section on telephone connectors below) are often used in mobile and smaller radios.

RF Connectors

Feed lines used for radio signals require special connectors for use at RF frequencies. The connectors must have approximately the same characteristic impedance as the feed line they are attached to or some of the RF signal will be reflected by the connector. Inexpensive audio and control connectors cannot meet that requirement, nor can they handle the high power levels often encountered in RF equipment. Occasionally, phono connectors are used for HF receiving and low-power transmitting equipment.

By far, the most common connector for RF in amateur equipment is the UHF family shown in **Fig 29.14**. (The UHF designator has nothing to do with frequency.) A PL-259 is the plug that goes on the end of feed lines, and the SO-239 is the jack mounted on equipment. A "barrel" (PL-258) is a double-female adapter that allows two feed lines to be connected together. UHF connectors are typically used up to 150 MHz and can handle legal-limit transmitter power at HF.

UHF connectors have several drawbacks including lack of weatherproofing, poor performance above the 2-meter band and limited power handling at higher frequencies. The Type-N series of RF connectors addresses all of those needs. Type-N connectors are somewhat more expensive than UHF connectors, but they require less soldering and perform better in outdoor use since they are moisture resistant. Type-N connectors can be used to 10 GHz.

For low-power uses, BNC connectors are often used. BNC connectors are the standard for laboratory equipment, as well, and they are often used for dc and audio connections. BNC connectors are common on handheld radios for antenna connections. The newest handheld transceivers often use small, screw-on SMA type connectors for their antennas, though.

The type of connector used for a specific job depends on the size of the cable, the frequency of operation and the power levels involved. More information on RF connectors may be found in the **Component Data and References** chapter.

Data Connectors

Digital data is exchanged between computers and pieces of radio equipment more than ever before in the amateur station. The connector styles follow those found on computer equipment.

D-type connectors are used for RS-232 (COM ports) and parallel (LPT port) interfaces. A typical D-type connector has a model number of "DB" followed the number of connections and a "P" or "S" depending on whether the connector uses pins or sockets. For example, the DB-9P and DB-9S are used for PC COM1 and COM2 serial ports.

USB connectors are becoming more popular in amateur equipment as the computer industry moves to eliminate the bulkier and slower RS-232 interface. A number of manufacturers make devices for interfacing transceivers and station equipment through computer USB ports.

Null modem adapters or cables are used to connect data circuits when direct connections between the data interfaces would connect outputs to outputs and inputs to inputs.

Pinouts for various computer connectors are shown in the **Component Data and References** chapter. More information about interfacing may be found in the **Digital Basics** chapter, and several practical interface projects are shown in the **Station Accessories** and **Digital Communications** chapters.

Telephone and Computer Network Connectors

Modular connectors are used for telephone

and computer network connections. Connector part numbers begin with "RJ." The connectors are crimped on to multiconductor cables with special tools. The RJ11 connector is used for single- and double-line telephone system connection with 4 or 6 contacts. The RJ10 is a 4-contact connector for telephone handset connections. Ethernet computer network connections are made using RJ45 connectors with 8 contacts.

29.1.6 Documenting Your Station

An often neglected but very important part of putting together your station is properly documenting your work. Ideally, you should diagram your entire station from the ac power lines to the antenna on paper and keep the information in a special notebook with sections for the various facets of your installation. Having the station well documented is an invaluable aid when tracking down a problem or planning a modification. Rather than having to search your memory for information on what you did a long time ago, you'll have the facts on hand.

Besides recording the interconnections and hardware around your station, you should also keep track of the performance of your equipment. Each time you install a new antenna, measure the SWR at different points in the band and make a table or plot a curve. Later, if you suspect a problem, you'll be able to look in your records and compare your SWR with the original performance.

In your shack, you can measure the power output from your transmitter(s) and amplifier(s) on each band. These measurements will be helpful if you later suspect you have a problem. If you have access to a signal generator, you can measure receiver performance for future reference.

29.1.7 Interfacing High-Voltage Equipment to Solid-State Accessories

Many amateurs use a variety of equipment manufactured or home brewed over a considerable time period. For example, a ham might be keying a '60s-era tube rig with a recently built microcontroller-based electronic keyer. Many hams have modern solid-state radios connected to high-power vacuum-tube amplifiers.

Often, there is more involved in connecting HV (high-voltage) vacuum-tube gear to solid-state accessories than a cable and the appropriate connectors. The solid-state switching devices used in some equipment will be destroyed if used to switch the HV load of vacuum tube gear. The polarity involved is important too. Even if the voltage is low enough, a key-line might bias a solid-state device in

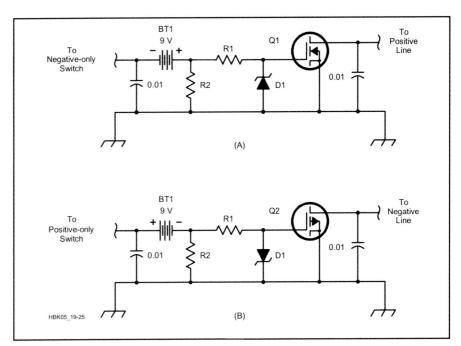

Fig 29.15 — Level-shifter circuits for opposite input and output polarities. At A, from a negative-only switch to a positive line; B, from a positive-only switch to a negative line.

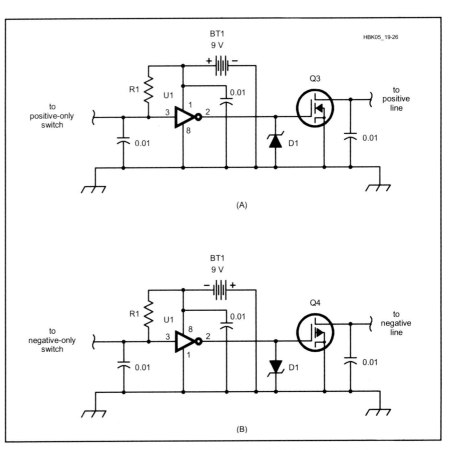

Fig 29.16 — Circuits for same-polarity level shifters. At A, for positive-only switches and lines; B, for negative-only switches and lines.

BT1 — 9-V transistor-radio battery.
D1 — 15-V, 1-W Zener diode (1N4744 or equiv).
Q3 — IRF620.
Q4 — IRF220 (see text).
R1 — 10 kΩ, 10%, ¼ W.
U1 — CD4049 CMOS inverting hex buffer, one section used (unused sections not shown; pins 5, 7, 9, 11 and 14 tied to ground).

such a way as to cause it to fail. What is needed is another form of level converter.

MOSFET LEVEL CONVERTERS

While relays can often be rigged to interface the equipment, their noise, slow speed and external power requirement make them an unattractive solution in some cases. An alternative is to use power MOSFETs. Capable of handling substantial voltages and currents, power MOSFETs have become common design items. This has made them inexpensive and readily available.

Nearly all control signals use a common ground as one side of the control line. This leads to one of four basic level-conversion scenarios when equipment is interconnected:

1) A positive line must be actuated by a negative-only control switch.
2) A negative line must be actuated by a positive-only control switch.
3) A positive line must be actuated by a positive-only control switch.
4) A negative line must be actuated by a negative-only control switch.

In cases 3 and 4 the polarity is not the problem. These situations become important when the control-switching device is incapable of handling the required open-circuit voltage or closed-circuit current.

Case 1 can be handled by the circuit in **Fig 29.15A**. This circuit is ideal for interfacing keyers designed for grid-block keying to positive CW key lines. A circuit suitable for case 2 is shown in Fig 29.15B. This circuit is simply the mirror image of that in Fig 29.15A with respect to circuit polarity. Here, a P-channel device is used to actuate the negative line from a positive-only control switch.

Cases 3 and 4 require the addition of an inverter, as shown in **Fig 29.16**. The inverter provides the logic reversal needed to drive the gate of the MOSFET high, activating the control line, when the control switch shorts the input to ground.

Almost any power MOSFET can be used in the level converters, provided the voltage and current ratings are sufficient to handle the signal levels to be switched. A wide variety of suitable devices is available from most large mail-order supply houses.

29.2 Mobile Installations

Solid-state electronics and miniaturization have allowed mobile operators to equip their vehicles with stations rivaling base station installations. Indeed, it is possible to operate from 160 meters through 70 cm with one compact transceiver. Adding versatility, most designed-for-mobile transceivers are set up so that the main body of the radio can be safely tucked under a seat, with the operating "head" conveniently placed for ease of use as shown in **Fig 29.17**.

It is not uncommon for power levels to exceed 100 W on HF, and 50-75 W on VHF. With proper antenna selection and placement (see the **Antennas** chapter), mobile stations can work the world, just like their base station counterparts. The only real difference between them is that you're trying to drive at the same time you are operating, and safe operating requires attention to the details.

For some of us living in antenna-restricted areas, mobile operating may offer the best solution for getting on the air. For others it is an enjoyable alternative to home-station operation. No matter which category you're in, you can enjoy success if you plan your installation with safety and convenience in mind.

29.2.1 Installation

Installing Amateur Radio equipment in modern vehicles can be quite challenging, yet rewarding, if a few basic rules are followed. For example, avoid temporary mounting schemes such as hook-and-loop fasteners, magnets, elastic cords and wedged-in blocks of wood. If you're not into building a specific mount for your vehicle, there are a plethora of no-holes-needed mounts available from Amateur Radio dealers. Some mounts are even designed for a specific transceiver make and model. **Table 29.1** lists some suppliers.

The main points to keep in mind when choosing a mounting location are safety (which should always be considered first), convenience and lack of distraction. The radio mounting location must avoid SRS (airbag) deployment zones — virtually eliminating the top of the dash in most modern vehicles — as well as vehicle controls. Microphone and power cabling should be placed out of the way and properly secured. The transceiver's controls should be convenient to use and to view. See **Fig 29.18** for examples.

Mounting radios inside unvented center consoles and overhead bins should also be avoided. Modern mobile transceivers designed for remote mounting allow the main body to be located under a seat, in the trunk, or in another out-of-the-way place (**Fig 29.19**) but be sure there is plenty of ventilation.

If you drive off-road or on rough roads, you may also wish to consider using shock mounts, also known as "Lord Mounts" for their original manufacturer. For more information, see the product information on "Plateform Mounts" from the Astrotex Company at **www.astrotex.com/lord.htm**.

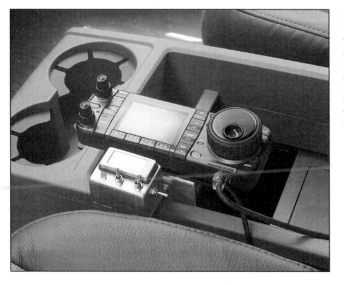

Fig 29.17 — In this mobile installation, the transceiver control head is mounted in the center console, next to a box with switches for adjusting the antenna.

Table 29.1
Mobile Mount Sources
Gamber Johnson — **www.gamberjohnson.com**
Havis-Shields — **www.havis.com**
Jotto Desk — **www.jottodesk.com**
PanaVise Products — **www.panavise.com**
RAM Mounting Systems — **www.ram-mount.com**

Fig 29.18 — At A, the transceiver control head is attached to one of many available mounts designed for this purpose. Mounts are typically highly adjustable, allowing the control head or radio to be positioned close to the operator. An antenna controller is mounted below the microphone. At B, HF and VHF transceiver control heads and the microphone are all mounted to the dashboard, within easy reach.

Fig 29.19 — The main body of the radio may be mounted in the vehicle's trunk or other out-of-the-way spot. Allow for plenty of ventilation.

29.2.2 Coax

Cable lengths in mobile installations seldom exceed 15 ft, so coax losses are not a major factor except for 70 cm and above. Good quality RG-58A, or RG-8X size coax is more than adequate for HF and VHF. While there is nothing wrong with using RG-8 size coax (0.405 inch), it is stiffer and has a larger bending radius, making it harder to work with in most mobile applications.

There are some caveats when selecting coax. Avoiding solid center conductors such as standard RG-58. It has a propensity to kink, is prone to vibration and can be difficult to solder properly. Both RG-58A and RG-8X use foam dielectric, and care is needed when soldering PL-259 connectors — especially when reducers are being used. The **Component Data and References** chapter illustrates the correct installation procedure. When applying heat, use a large-wattage iron with enough latent heat to flow the solder quickly. Soldering guns shouldn't be used for this reason.

29.2.3 Wiring

Proper wiring is an essential part of any mobile installation. Consider the following points when selecting materials and planning the cable routing.

• Wire needs to be correctly sized and fused, and of stranded construction.

• All cables need to be protected from abrasion, heat and chemicals.

• Wiring needs to be neat and tidy to avoid interaction with passengers and mechanical devices. This means excess wiring should be shortened and/or bundled with appropriate wire ties. **Fig 29.20** shows a typical vehicle wiring trough.

Power cables should be connected directly to the battery following manufacturers' recommendations, with the requisite positive and negative lead fuses located close by. **Fig 29.21** shows a fuse block. Accessory (cigarette lighter) sockets and power taps shouldn't be used except for very low current loads (<5 A), and then only with care. It pays to remember that a vehicle fire is both costly and dangerous! More information may be found at **www.fordemc.com/docs/download/Mobile_Radio_Guide.pdf** and **service.gm.com/techlineinfo/radio.html**.

Proper wiring also minimizes losses and helps prevent ground loops. Modern solid-state transceivers will operate effectively down to 12.0 V dc (engine off). If the voltage drops below 11.6 V under load, most transceivers will shut down or operate incorrectly. The vehicle chassis should not be used for ground returns; doing so can create a ground loop.

Firewall access can be easy in some vehicles, and nearly impossible in others. Using factory wiring grommets should be avoided, unless they're not being used. In some cases, the only alternative is to drill your own hole. If you have any questions or concerns, have your local mobile sound shop or two-way radio dealer install the wiring for you.

Power for ancillary equipment (wattmeters, remotely tuned antennas and so on) should follow the same wiring rules. The use of a multiple outlet power distribution panel such as a RigRunner (**www.westmountain-**

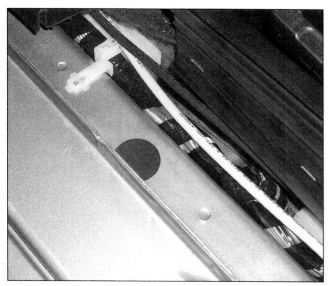

Fig 29.20 — Most vehicles have wiring troughs hidden behind interior body panels.

Fig 29.21 — Wiring attached to a fuse block.

radio.com) is also recommended. They're convenient, and offer a second level of protection.

WIRE SIZE

The **Component Data and References** chapter lists the current-handling capabilities of various gauges or wire and cable. The correct wiring size is one that provides a low voltage drop (less than 0.5 V under full load). Don't go by maximum current-current carrying capacity.

Here's the formula for calculating the cable assembly voltage drop (V_d):

$$V_d = [(R_W \times 2\,\ell \times 0.001) + 2\,k] \times I$$

where

R_W = resistance value (Ω per 1000 ft) from the **Component Data and References** chapter.

ℓ = overall length of the cable assembly including connectors, in feet.

k = nominal resistive value for one fuse and its holder. Note: Most power cables have two fuses. If yours doesn't, use 1 k in the formula. If you don't know the fuse and holder resistance, use a conservative value of 0.002 Ω.

I = peak current draw in amperes for a SSB transceiver, or steady state for an FM radio.

For example, the peak current draw for a 100 W transceiver is about 22 A, and a typical power cable length is 10 ft. Using the resistive values for 1000 feet of #10 AWG wire (0.9987 Ω), and a conservative value for the fuse resistances (0.002 Ω each), the calculated drop will be 0.527 V.

It's important to reiterate that the wire size should be selected for minimum voltage drop, not maximum power handling capability. The

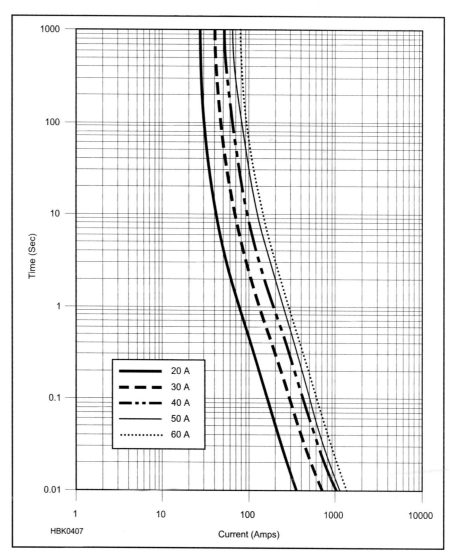

Fig 29.22 — Chart of opening delay versus current for five common sizes of Maxi fuses, the plastic-body high-current fuses common in vehicles. (*Based on a chart from Littelfuse Corp*)

voltage drop is often referred to as "I²R loss" — the current in amperes, squared, times the resistance — and should be held to a minimum whenever possible. In cabling, excessive I²R losses can cause the wire to overheat with predictable results.

The insulation material of wire used in mobile installations should have a temperature rating of at least 90 °C, and preferably 105 °C. It should be protected with split-loom covering whenever possible, especially under-hood wiring.

Selecting the correct size fuse is also important. The average current draw for any given fuse should not exceed 60% of its rating. Thus, the correct fuse rating for a 22-A load is 30 A. That same 30-A fuse will handle a 40-A load for about 120 seconds, and a 100-A load for about 2 seconds. Therefore, it pays to be conservative when selecting the carrying capacities of both wire and fuses. **Fig 29.22** shows the characteristics of several sizes of automotive fuses.

29.2.4 Amplifiers

Mobile HF amplifiers have been around for many years, and with the advent of high-power solid state devices they are common. However, running high power in a mobile environment requires careful planning. Considerations include, but are not limited to:
- alternator current ratings and battery capacity
- wiring (in addition to safe current ratings, excessive voltage drop will increase IMD)
- antenna and feed line power ratings
- placement and secure mounting in the vehicle
- wiring and placing of remote controls

See **www.k0bg.com/amplifiers.html** for more information on these topics.

Before purchasing an amplifier, take a close look at your antenna installation and make sure it is operating efficiently. Mating an amplifier to a poor antenna installation is counterproductive. Here's a rule of thumb applicable to any type of antenna: If the *unmatched* input SWR is less than 1.7:1 on 17 m or any lower frequency band, then it isn't mounted correctly, and/or you need a better antenna. Whatever antenna you use, it must be capable of handling the amplifier power level — 500 W or more. More information on HF mobile antennas and installation techniques may be found in the **Antennas** chapter.

Mobile amplifiers for VHF/UHF operation are not as popular as they once were because most mobile transceivers have adequate output power (about 50 W). Boosting this to 150-300 W or more should be done with caution. Mobile VHF/UHF antennas for high power (>100 W) are rare, so check antenna ratings carefully. Those that are available need to be permanently mounted, and preferably on the roof to avoid inadvertent contact.

With any high-power mobile installation, pay careful attention to RF safety. More information on RF exposure is in the **Safety** chapter.

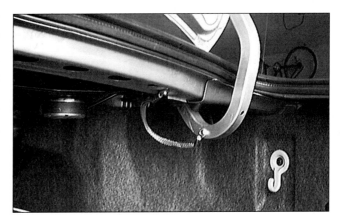

Fig 29.23 — Bonding vehicles parts — in this case, the trunk lid to the main body — can help reduce interference.

29.2.5 Interference Issues

In a mobile installation, radio frequency interference falls under two basic categories: egress (interference from the vehicle to your amateur station) and ingress (from your amateur gear to the vehicle). Most hams are familiar with ignition interference as it is the most common form of egress. Auto sound system feedback is a common form of ingress.

Both types of interference have unique solutions, but they have at least one in common, and that's bonding. Bonding refers to strapping the various bolted-on pieces to the frame or chassis of the vehicle. For example, the exhaust system is isolated from the structure of the vehicle, and acts like an antenna for the RFI generated by the ignition system. It should be strapped in at least three places. **Figs 22.23** shows an example of bonding. More on these techniques is available at **www.k0bg.com/bonding.html**.

Other RFI egress problems are related to fuel pumps, HVAC and engine cooling fans, ABS sensors, data distribution systems and control system CPUs. These are best cured at the source by liberal use of snap-on ferrite cores on the wiring harnesses of the offending devices. Snap-on cores come is a variety of sizes and formulations called mixes. The best all-around ferrite core material for mobile RFI issues is Mix 31. Suitable cores are available from most Amateur Radio dealers. Unknown surplus units typically offer little HF attenuation and should be avoided. See the **Electromagnetic Compatibility** and **Component Data and References** chapters for more information on ferrite cores.

Alternator whine is another form of RFI egress. It is typically caused by an incorrectly mounted antenna resulting in a ground loop, rather than a defective alternator diode. Brute-force filters only mask the problem, and increase I²R losses. Additional information on proper antenna mounting is in the **Antennas** chapter.

RFI ingress to the various on-board electronic devices is less common. The major causes are unchecked RF flowing on the control wires, and common mode currents flowing on the coax cable of remotely-tuned HF antennas. Again, this points out the need to properly mount mobile antennas.

For more information on RFI issues, see the **Electromagnetic Compatibility** chapter, *The ARRL RFI Handbook*, and the ARRL Technical Information Service (**www.arrl.org/tis**).

29.2.6 Operating

The most important consideration while operating mobile-in-motion is safety! Driver distraction is a familiar cause of vehicle crashes. While Amateur Radio use is far less distracting than cell phones, there are times when driving requires all of our attention. When bad weather, excessive traffic or a construction zone require extra care, play it safe and hang up the microphone and turn off the radio!

In addition to properly installing gear, a few operating hints can make your journey less distracting. One of those is familiarization with your transceiver's menu functions and its microphone keys (if so equipped). Even then, complicated programming or adjustments are not something to do while underway.

Logging mobile contacts has always been difficult. Compact digital voice recorders have made that function easy and inexpensive. Units with up to 24 hours of recording time are under $50.

For maximum intelligibility at the other end, avoid excessive speech processing and too much microphone gainch Don't shout into the mic. It's human nature to increase your oral volume level when excited or when the background level increases. In the closed cabin of a vehicle, your brain interprets the reflected sound from your own voice as an increase in background level. Add in a little traffic noise, and by the end of your transmission you're in full shout mode! One solution is to use a headset and the transceiver's built-in monitor function. Doing so gives you direct feedback (not a time-delayed echo), and your brain won't get confused.

Note that headset use is not legal in some jurisdictions, and never legal if both ears are covered. Even more convenient are cordless Bluetooth headsets. Amateur manufacturers are adopting the technology, and at least one aftermarket unit is available. The best part is, they leave your hands free to drive.

Overcoming vehicle ambient noise levels often requires the use of an external speaker, and all too often it is an afterthought. Selecting a speaker too small accentuates high frequency hash which makes reception tiresome. Using adapters to interface with vehicle stereo systems isn't productive for the same reason. Selecting one too large, and mounting becomes a safety issue.

For best results, use at least a 4-inch. speaker. Rather than mount it out in the open, mount it under the seat, out of the way. This also attenuates the high frequencies, enhances the lows, and increases intelligibility.

29.3 Portable Installations

Many amateurs experience the joys of portable operation once each year in the annual emergency exercise known as Field Day. Setting up an effective portable station requires organization, planning and some experience. For example, some knowledge of propagation is essential to picking the right band or bands for the intended communications link(s). Portable operation is difficult enough without dragging along excess equipment and antennas that will never be used.

Some problems encountered in portable operation that are not normally experienced in fixed-station operation include finding an appropriate power source and erecting an effective antenna. The equipment used should be as compact and lightweight as possible. A good portable setup is simple. Although you may bring gobs of gear to Field Day and set it up the day before, during a real emergency speed is of the essence. The less equipment to set up, the faster it will be operational.

29.3.1 Portable AC Power Sources

There are three popular sources of ac power for use in the field: batteries, dc-to-ac inverters, and generators. Batteries and inverters are covered in detail in the **Power Supplies** chapter. This section will focus on gasoline generators.

Essentially, a generator is a motor that's operating "backward." When you apply electricity to a motor, it turns the motor's shaft (allowing it to do useful work). If you need more rotational power, add more electricity or wind a bigger motor. Take the same motor and physically rotate its shaft and it generates electricity across the same terminals used to supply power when using the motor as a motor. Turn the shaft faster and the voltage and frequency increase. Turn it slower and they decrease. To some degree, all motors are generators and all generators are motors. The differences are in the details and in the optimization for specific functions.

A "motor" that is optimized for generating electricity is an alternator — just like the one in your car. The most basic generators use a small gas engine to power an ac alternator, the voltage and frequency of which depends on rotational speed. Because the generator is directly coupled to the engine, the generator's rotational speed is determined by the speed of the engine. If the engine is running too fast or too slow, the voltage and frequency of the output will be off. If everything is running at or near the correct speed, the voltage and frequency of the output will be a close approximation of the power supplied by the ac mains — a 120 V ac sine wave with a frequency of 60 Hz. Most standard consumer generators use two pole armatures that run at 3600 RPM to produce a 60 Hz sine wave.

VOLTAGE REGULATION

There are several electronic and mechanical methods used to "regulate" the ac output — to keep the voltage and frequency values as stable as possible as generator and engine speeds vary because of current loads or other factors. Remember, a standard generator *must* turn at a specific speed to maintain output regulation, so when more power is drawn from the generator, the engine must supply more torque to overcome the increased physical/magnetic resistance in the generator's core — the generator *can't* simply spin faster to supply the extra oomph.

Most generators have engines that use mechanical or vacuum "governors" to keep the generator shaft turning at the correct speed. If the shaft slows down because of increasing generator demand, the governor "hits the gas" and draws energy stored in a heavy rotating flywheel, for example to bring (or keep) the shaft speed up to par. The opposite happens if the generator is spinning too fast.

In addition to mechanical and vacuum speed regulating systems, generators that are a step up in sophistication additionally have electronic automatic voltage regulation (AVR) systems that use special windings in the generator core (and a microprocessor or circuit to monitor and control them) to help keep things steady near 120 V and 60 Hz. AVR systems can respond to short term load changes much more quickly than mechanical or vacuum governors alone. A decade ago AVR generators were the cream of the crop. Today, they're mostly used in medium to large units that can't practically employ inverters to maintain the best level of output regulation. You'll find them in higher quality 5 to 15 kW "home backup systems" and in many recreational vehicles.

Isolate Source and Load

Basic, inexpensive generators are intended to power lights, saws, drills, ac motors, electric frying pans and other devices that can reliably be run on "cruddy power." If you want the highest margin of safety when powering computers, transceivers and other sensitive electronics, a portable *inverter generator* is the best way to go. Some popular examples are shown in **Fig 29.24**, and their key specifications are shown in **Table 29.2**. Available in outputs ranging from 1 to 5 kW, these generators use one or more of the mechanical regulation systems mentioned previously, but their ultimate benefit comes from the use of a built-in ac-dc-ac inverter system that produces beautiful — if not perfect — 60 Hz sine waves at 120 V ac, with a 1% to 2% tolerance, even under varying load conditions.

Instead of using two windings in the generator core, an inverter generator uses 24 or more windings to produce a high frequency ac waveform of up to 20 kHz. A solid state inverter module converts the high frequency ac to smooth dc, which is in turn converted to clean, tightly regulated 120 V ac power. And that's not all. Most inverter generators are compact, lightweight and quiet.

GENERATOR CONSIDERATIONS

In addition to capacity and output regulation, other factors such as engine type, noise level, fuel options, fuel capacity, run time, size, weight, cost or connector type, may fac-

Fig 29.24 — Modern inverter generators from McCullough, Honda, Yamaha and Subaru/Robin.

See Table 29.2 for a partial list of specifications.

tor in your decision. Consider additional uses for your new generator beyond Field Day or other portable operation.

Capacity

Your generator must be able to safely power all of the devices that will be attached to it. Simply add up the power requirements of *all* the devices, add a reasonable safety margin (25 to 30%) and choose a suitably powerful generator that meets your other requirements.

Some devices — especially electric motors — take a lot more power to start up than they do to keep running. A motor that takes 1000 W to run may take 2000 to 3000 W to start. Many items don't require extra start-up power, but be sure to plan accordingly.

Always plan to have more capacity than you require or, conversely, plan to use less gear than you have capacity for. Running on the ragged edge is bad for your generator *and* your gear. Some generators are somewhat overrated, probably for marketing purposes. Give yourself a margin of safety and don't rely on built-in circuit breakers to save your gear during overloads. When operating at or beyond capacity, a generator's frequency and voltage can vary widely before the current breaker trips.

Size and Weight

Size and weight vary according to power output — low power units are lightweight and physically small, while beefier models are larger, weigh more and probably last longer. Watt for watt, however, most modern units are smaller and lighter than their predecessors. Models suitable for hamming typically weigh between 25 and 125 pounds.

Engines and Fuel

Low end generators are typically powered by low-tolerance, side valve engines of the type found in discount store lawnmowers. They're noisy, need frequent servicing and often die quickly. Better models have overhead valve (OHV) or overhead cam (OHC) engines, pressure lubrication, low oil shutdown, cast iron cylinder sleeves, oil filters, electronic ignition systems and even fuel injection. These features may be overkill for occasional use but desirable for more consistent power needs.

Run Time

Smaller generators usually have smaller gas tanks, but that doesn't necessarily mean they need more frequent refueling. Some small generators are significantly more efficient than their larger counterparts and may run for half a day while powering small loads. As with output power, run times for many units are somewhat exaggerated and are usually spec'd for 50% loads. If you're running closer to max capacity, your run times may be seriously degraded. The opposite is also true. Typical generators run from three to nine hours on a full tank of gas at a 50% load.

Noise

Except for ham-friendly inverter units — which are eerily quiet thanks to their high tech, sound dampening designs — standard generators are almost always too loud. Noise levels for many models are stated on the box, but try to test them yourself or talk to someone who owns the model you're interested in before buying. Environmental conditions, distance to the generator and the unit's physical orientation can affect perceived noise levels.

Generators housed in special sound dampened compartments in large boats and RVs can be much quieter than typical "outside" models. However, they are expensive and heavy, use more fuel than compact models, and most don't have regulation specs comparable to inverter models.

Regulation

For hams, voltage and frequency regulation are the biggies. AVR units with electronic output regulation (at a minimum) and inverter

Table 29.2
Specifications of the Inverter Generators Shown in Fig 29.24

Make and Model	Output (W) (Surge/Cont)	Run Time (h) Full / 25% Load	Noise Range (dBA @21 feet)	Engine Type	Weight (Pounds)	Notes*
McCulloch FDD210M0	N/A / 1800	4 / N/A	60-70	105.6 cc, 3 HP	65 (shipping)	a,b
Honda EU2000i	2000 / 1600	4 /15	53-59	100 cc, OHC	46	a,b,c
Yamaha EF2400iS	2400 / 2000	N/A / 8.6	53-58	171 cc, OHV	70	a,b,c
Subaru/Robin R1700i	N/A / 1650	N/A / 8.5	53-59	2.4 HP, OHV	46	a,b,c

*a — has 12 V dc output; b — has "smart throttle" for better fuel economy; c — has low oil alert/shutdown

generators are highly desirable and should be used exclusively, if only for peace of mind.

Unloaded standard generators can put out as much as 160 V ac at 64 Hz. As loads increase, frequency and voltage decrease. Under full load, output values may fall as low as 105 V at 56 Hz. Normal operating conditions are somewhere in between.

Some hams have tried inserting uninterruptible power supplies (UPSs) between the generator and their sensitive gear. These devices are often used to maintain steady, clean ac power for computers and telecommunication equipment. As the mains voltage moves up and down, the UPS's Automatic Voltage Regulation (AVR) system bucks or boosts accordingly. The unit's internal batteries provide power to the loads if the ac mains (or your generator) go down.

In practice, however, most UPSs can't handle the variation in frequency and voltage of a generator powered system. When fed by a standard generator, most UPSs constantly switch in and out of battery power mode — or don't *ever* switch back to ac power. When the UPS battery goes flat, the unit shuts off. Not *every* UPS and *every* generator lock horns like this, but an inverter generator is a better solution.

DC Output

Some generators have 12 V dc outputs for charging batteries. These range from 2 A trickle chargers to 100 A powerhouses. Typical outputs run about 10 to 15 A. As with the ac outputs, be sure to test the dc outputs for voltage stability (under load if possible) and ripple. Car batteries aren't too fussy about a little ripple in the charging circuit, but your radio might not like it at all!

Miscellaneous

Other considerations include outlets (120 V, 240 V and dc output), circuit breakers (standard or ground fault interrupter type), fuel level gauges, handles (one or two), favorite brands, warranties, starters (pull or electric), wheels, handles or whatever you require.

SETUP, SAFETY AND TESTING

Before starting the engine, read the user manual. Carefully follow the instructions regarding engine oil, throttle and choke settings (if any). Be sure you understand how the unit operates and how to use the receptacles, circuit breakers and connectors.

Make sure the area is clean, dry and unobstructed. Generators should *always* be set up *outdoors*. Do not operate gas powered engines in closed spaces, inside passenger vans, inside covered pickup beds, etc. If rain is a possibility, set up an appropriate canopy or other *outdoor protective structure*. Operating generators and electrical devices in the rain or snow can be dangerous. Keep the generator and any attached cords dry!

Exhaust systems can get hot enough to ignite certain materials. Keep the unit several feet away from buildings, and keep the gas can (and other flammable stuff) at a safe distance. Don't touch hot engines or mufflers!

When refueling, shut down the generator and let things cool off for a few minutes. Don't smoke, and don't spill gasoline onto hot engine parts. A flash fire or explosion may result. Keep a small fire extinguisher nearby. If you refuel at night, use a light source that isn't powered by the generator and can't ignite the gasoline.

Testing

Before starting (or restarting) the engine, *disconnect* all electrical loads. Starting the unit while loads are connected may not damage the generator, but your solid state devices may not be so lucky. After the engine has warmed up and stabilized, test the output voltage (and frequency), if possible, *before* connecting loads.

Because unloaded values may differ from loaded values, be sure to test your generator under load (using high wattage quartz lights or an electric heater as appropriate). Notice that when you turn on a hefty load, your generator will "hunt" a bit as the engine stabilizes. Measure ac voltage and frequency again to see what the power conditions will be like under load. See your unit's user manual or contact the manufacturer if adjustments are required.

Safety Grounds and Field Operation

Before we can connect *real* electrical loads in a Field Day situation we need to choose a grounding method — a real controversy among campers, RVers and home power enthusiasts.

To complicate matters, most generators have ac generator grounds that are connected to their metal frames, but some units do not bond the ac neutral wires to the ac ground wires (as in typical house wiring). Although they will probably safely power your ham station all day long, units with unbonded neutrals may appear defective if tested with a standard ac outlet polarity tester.

Some users religiously drive copper ground rods into the ground or connect the metal frames of their generators to suitable existing grounds, while others vigorously oppose this method and let their generators float with respect to earth ground (arguing that if the generator isn't connected to the earth, you can't complete the path to earth ground with your body should you encounter a bare wire powered by the generator; no path, no shock). Some user manuals insist on the ground connection, while others don't.

You can follow your unit's user manual, check your local electric code, choose a grounding method based on personal preference or expert advice, or do further research. Either method may offer better protection depending on exact circumstances.

Regardless of the grounding method you choose, a few electrical safety rules remain the same. Your extension cords *must* have intact, waterproof insulation, three "prongs" and three wires, and must be sized according to loads and cable runs. Use 14 to 16 gauge, three wire extension cords for low wattage runs of 100 feet or less. For high wattage loads, use heavier 12 gauge, three wire cords designed for air compressors, air conditioners or RV service feeds. If you use long extension cords to power heavy loads, you may damage your generator or your radio gear. When it comes to power cords, think *big*. Try to position extension cords so they won't be tripped over or run over by vehicles. And don't run electrical cords through standing water or over wet, sloppy terrainch

During portable operations, try to let all operators know when the generator will be shut down for refueling so radio and computer gear can be shut down in a civilized manner. Keep the loads disconnected at the generator until the generator has been refueled and restarted.

29.3.2 Portable Antennas

An effective antenna system is essential to all types of operation. Effective portable antennas, however, are more difficult to devise than their fixed-station counterparts. A portable antenna must be light, compact and easy to assemble. It is also important to remember that the portable antenna may be erected at a variety of sites, not all of which will offer ready-made supports. Strive for the best antenna system possible because operations in the field are often restricted to low power by power supply and equipment considerations. Some antennas suitable for portable operation are described in the **Antennas** chapter.

ANTENNA SUPPORTS

While some amateurs have access to a truck or trailer with a portable tower, most are limited to what nature supplies, along with simple push-up masts. Select a portable site that is as high and clear as possible. Elevation is especially important if your operation involves VHF. Trees, buildings, flagpoles, telephone poles and the like can be pressed into service to support wire antennas. Drooping dipoles are often chosen over horizontal dipoles because they require only one support.

An aluminum extension ladder makes an effective antenna support, as shown in **Fig 29.25**. In this installation, a mast, rotator and beam are attached to the top of the

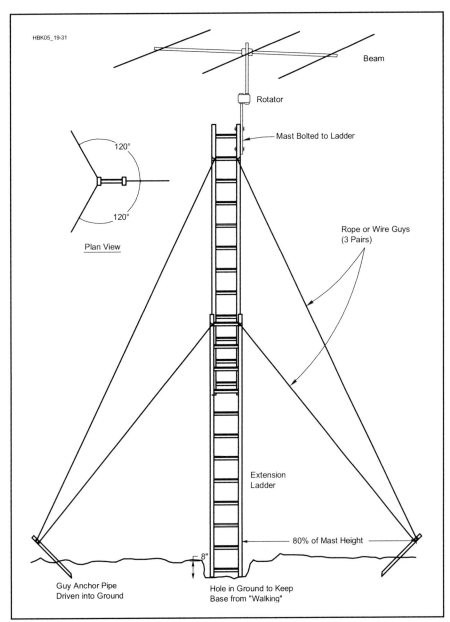

Fig 29.25 — An aluminum extension ladder makes a simple but sturdy portable antenna support. Attach the antenna and feed lines to the top ladder section while it is nested and lying on the ground. Push the ladder vertical, attach the bottom guys and extend the ladder. Attach the top guys. Do not attempt to climb this type of antenna support.

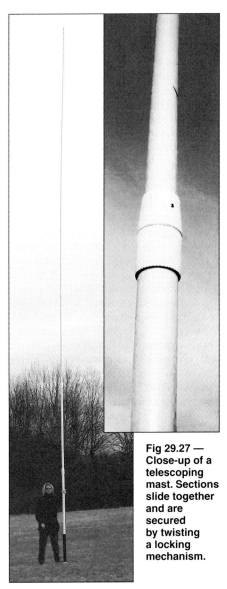

Fig 29.27 — Close-up of a telescoping mast. Sections slide together and are secured by twisting a locking mechanism.

Fig 29.26 — Telescoping fiberglass poles can be used to support a variety of wire antennas or small VHF/UHF Yagis. This one is 40 feet long, yet collapses to 8 feet for storage.

second ladder section with the ladder near the ground. The ladder is then pushed vertical and the lower set of guy wires attached to the guy anchors. When the first set of guy wires is secured, the ladder may be extended and the top guy wires attached to the anchors. Do not attempt to climb a guyed ladder.

Telescoping fiberglass poles (**Figs 29.26** and **29.27**) are popular for supporting wire verticals, inverted Vs and small VHF/UHF antennas. These poles can extend up to 40 ft in length, yet retract to 4 to 8 feet for easy transport. They typically weigh less than 20 pounds, and some are much lighter.

Figs 29.28 and **29.29** illustrate two methods for mounting portable antennas described by Terry Wilkinson, WA7LYI. Although the antennas shown are used for VHF work, the same principles can be applied to small HF beams as well.

In Fig 29.28A, a 3-ft section of Rohn 25 tower is welded to a pair of large hinges, which in turn are welded to a steel plate measuring approximately 18 × 30 inches. One of the rear wheels of a pickup truck is "parked" on the plate, ensuring that it will not move. In Fig 29.28B, quad array antennas for 144 and 222 MHz are mounted on a Rohn 25 top section, complete with rotator and feed lines. The tower is then pushed up into place using the hinges, and guy ropes, anchored to heavy-duty stakes driven into the ground, complete the installation. This method of portable tower installation offers an exceptionally easy-to-erect, yet sturdy, antenna support. Towers installed in this manner may be 30 or 40 ft high; the limiting factor is the number of "pushers" and "rope pullers" needed to get into the air. A portable station located in the bed of the pickup truck completes the installation.

The second method of mounting portable beams described by WA7LYI is shown in Fig 29.29. This support is intended for use with small or medium-sized VHF and UHF ar-

Fig 29.28 — The portable tower mounting system by WA7LYI. At A, a truck is "parked" on the homemade base plate to weigh it down. At B, the antennas, mast and rotator are mounted before the tower is pushed up. Do not attempt to climb a temporary tower installation.

hold the rotator in place, and the antennas are mounted on a 10 or 15-ft mast section. This system is easy to make and set up.

TIPS FOR PORTABLE ANTENNAS

Any of the antennas described in the **Antennas** chapter or available from commercial manufacturers may be used for portable operation. Generally, though, big or heavy antennas should be passed over in favor of smaller arrays. The couple of decibels of gain a 5-element, 20-m beam may have over a 3-element version is insignificant compared to the mechanical considerations. Stick with arrays of reasonable size that are easily assembled.

Wire antennas should be cut to size and tuned prior to their use in the field. Be careful when coiling these antennas for transport, or you may end up with a tangled mess when you need an antenna in a hurry. The coaxial cable should be attached to the center insulator with a connector for speed in assembly. Use RG-58 for the low bands and RG-8X for higher-band antennas. Although these cables exhibit higher loss than standard RG-8, they are far more compact and weigh much less for a given length.

Beam antennas should be assembled and tested before taking them afield. Break the beam into as few pieces as necessary for transportation and mark each joint for speed in reassembly. Hex nuts can be replaced with wing nuts to reduce the number of tools necessary.

rays. The tripod is available from any dealer selling television antennas; tripods of this type are usually mounted on the roof of a house. Open the tripod to its full size and drive a pipe into the ground at each leg. Use a hose clamp or small U-bolt to anchor each leg to its pipe.

The rotator mount is made from a 6-inch-long section of 1.5-inch-diameter pipe welded to the center of an "X" made from two 2-ft-long pieces of concrete reinforcing rod (rebar). The rotator clamps onto the pipe, and the whole assembly is placed in the center of the tripod. Large rocks placed on the rebar

  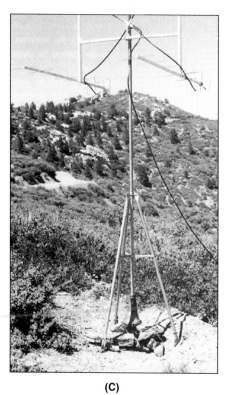

Fig 29.29 — The portable mast and tripod by WA7LYI. At A, the tripod is clamped to stakes driven into the ground. The rotator is attached to a homemade pipe mount. At B, rocks piled on the rotator must keep the rotator from twisting and add weight to stabilize the mast. At C, a 10-ft mast is inserted into the tripod/rotator base assembly. Four 432-MHz Quagis are mounted at the top.

29.4 Remote Stations

The following section was contributed by Rick Hilding, K6VVA. Rick has constructed his own competitive remote HF contest station and presents issues for the remote station builder to consider. He also wishes to acknowledge the assistance and input of N6RK, K6TU, W6OOL and SM2O. Although this section focuses on "remoting" HF equipment, the same considerations apply to most remote stations, except repeater installations.

29.4.1 Introduction

Remote stations (stations operated by remote control) have been a part of Amateur Radio for decades, but usually in the form of VHF/UHF repeaters. In the past, a few remote HF stations have been developed in impressive locations with significantly more land, equipment and expense than for a repeater installation. Prior to the expansion of Internet technologies, some of these remote pioneers utilized VHF/UHF links or commercial microwave equipment for connectivity between the remote and home stations. There has been a significant increase in the number of remote stations over the last five years, and the trend is predicted to accelerate.

Restrictions on antennas and interference issues have been placing increasing restrictions on HF operating. With today's technology, however, a remote HF station far from these problems may be as simple as a 100-W HF radio and all-band dipole or vertical with connectivity to the operator via the Internet. There is a flavor of remote station to meet almost every taste and budget.

This overview of remote HF station construction looks at three aspects of the project: planning the facilities, connectivity and remote station equipment. All three are equally important for a successful project and will be different for each installation. There is no cookbook approach to creating a remote station, so this discussion is intended to introduce topics that must be considered in the circumstances encountered by each station builder.

29.4.2 Evaluation and Planning

OPERATING REQUIREMENTS

Will your remote station be only for remote control from a home QTH or other single control point? Will it serve a dual purpose so you (and possibly others) can also operate at the remote site during favorable conditions? It is important to adequately plan physical space requirements for equipment and operator needs.

Is your remote HF station to be only for casual HF operating or will it be used for competitive contesting and DXing? If the latter, then the issue of *latency* (described below) will be much more significant.

Are you interested in operating on all HF bands including 60, 30, 17 and 12 meters? Will you operate on both CW and SSB? Digital modes? Remember that on the low bands, you may need to deal with the SWR variations of an antenna across an entire band.

Will your remote HF station be low power with a single transceiver or do you want to also use an amplifier? If you're planning a competitive SO2R (single operator, two radio) contest station, your needs are going to be much more complex involving band decoders, automatic dual-radio and antenna switching mechanisms, etc.

Do you plan to use a single multi-band antenna, or maybe a tribander and wire antennas for the low bands, or are you planning to install the "Mother of all Antenna Farms"? Make sure you have enough physical space available at the intended site.

PURCHASE VS RENT OR LEASE

If you don't already own or have the funds to purchase the property for your remote site, there is still hope! Although having direct control over every aspect of what you want to do is most desirable, you may want to consider a rental or lease with a cooperative land owner. In either case, having a detailed agreement or contract prepared by a competent attorney is a must. Be sure to include every contingency you can think of and especially what happens if the land owner decides to sell.

PAVED ROAD ACCESS

This is one of the most important things to consider in any remote site selection. Will you be able to get there safely, rain or shine? Pavement or a good all-weather, well-maintained gravel road should be available. Having more than one means of getting in or out should emergencies arise while you are at the site is also an important consideration. These concerns must be addressed before agreeing to purchase, rent or lease a remote site.

GEOLOGICAL AND OTHER HAZARDS

Evaluate the geological factors involved. In a hilltop or mountaintop site, is there evidence of soil movement that could affect access or damage your equipment shelter or tower installations? Does the property sit near an earthquake fault? Is there a significant risk of fire during fire season? Is the site exposed to high wind or storms? In flat-land sites, is there a risk of flooding or drainage problems? Remember, you won't be there to rescue your equipment! In consideration of both your initial construction activities and ongoing maintenance tasks, are there any wildlife hazards from animals or insects? What about allergies or reactions to plants such as poison ivy or poison oak? Diligently investigate all aspects of a potential remote site *before* you end up with bad surprises down the line. For a reasonable fee, there are companies from which you can secure a Geological and Property Hazards report to aid in your assessment.

RAW SITE PREPARATION

It you are fortunate enough to locate suitable property, be sure to evaluate the site preparation needs. How much time and money will be necessary to develop the remote site to meet your specific needs? Is there brush and vegetation to be dispensed with? Will trees need to be cut down? Will bulldozers, backhoes and trenchers need access to the site? Cement trucks? Service vehicles? Diligent evaluation of all costs involved will reduce future stress.

SITES WITH EXISTING FACILITIES

Finding a site with a self-supporting tower or two in place that can accommodate your needs, along with a building large enough for your equipment and operational needs can be like finding a pot of remote HF gold. However, unless you are experienced with towers and construction, hire professionals to evaluate everything. If you are considering co-locating at a remote site with existing antennas and equipment in use, consider hiring a professional to help you evaluate the likelihood of problems. Birdies and intermodulation could end up being a source of potential conflict and stress if problems can't be resolved.

TERRAIN CONSIDERATIONS

Evaluate the terrain on the bands you wish to operate. Dean Straw, N6BV, has developed a very useful program called *HFTA* (*High Frequency Terrain Analysis*) that is included with the *ARRL Antenna Book*, of which he is a former editor. Use *HFTA* to assist you in placement of your antennas. Don't expect excellent ground/soil conductivity if your remote site is on top of a rocky mountaintop. If you plan to use vertical antennas, make sure there is sufficient room on the property to place the radials, whether on a mountaintop or flat area.

NOISE LEVELS

Since one of the reasons for building a remote station is to lower noise levels, you'll want to carefully evaluate existing and potential noise sources. Start by visiting the site with a receiver in several different types of weather and listening on all the bands you intend to use with a full-size dipole antenna. If noise levels are low, this is no guarantee that problems won't crop up in different seasons or from hardware failures. You should avoid nearby substations or high-voltage and power distribution lines. If you will be sharing a site with other users, see the discussion below on noise and interference from and to their equipment.

TELEPHONE OR INTERNET ACCESS

If you are planning to operate your remote station via the Internet, unless wireless Internet service is available, having suitable telephone line access is required. Be sure that broadband service such as DSL or cable TV can reach your site by asking the local service provider to do a site evaluation. Satellite Internet downlink speed may be fast but uplink speed is often quite a bit lower. Be sure the minimum link speed is adequate in both directions.

If your control point is close enough to the remote site, another option is to implement your own private wireless "bridge" system. Even if multiple "hops" might be involved, with the right equipment you can potentially have better results than using even broadband Internet.

ELECTRICAL POWER

Having electrical power already available at your remote site can certainly save a lot of money. The most ideal situation is to have underground power to the remote site. Be sure to evaluate the full costs of installing ac power if it is not already present at the site or on the property. Truly remote sites may require wind, solar or generator power systems like

Fig 29.30 — Solar panels can be used to charge a bank of batteries to supply power. Effective off-grid power requires an accurate and conservative estimate of power needs and available energy sources. (*K6VVA photo*)

that shown in **Fig 29.30**.

If your remote site will require solar/wind/battery power, it is critical that you properly assess sunlight path and duration, wind patterns, and provide sufficient battery storage to meet your intended operating needs during lengthy periods of inclement weather. The article "Designing Solar Power System for FM Translators" is a valuable resource for planning of remote HF station off-grid power.[1] You must thoroughly account for the power requirements of each and every piece of equipment that will use power.

There are various estimates for the duty cycle of HF transmitters (or transceivers). A conservative duty cycle of 50% for CW/SSB

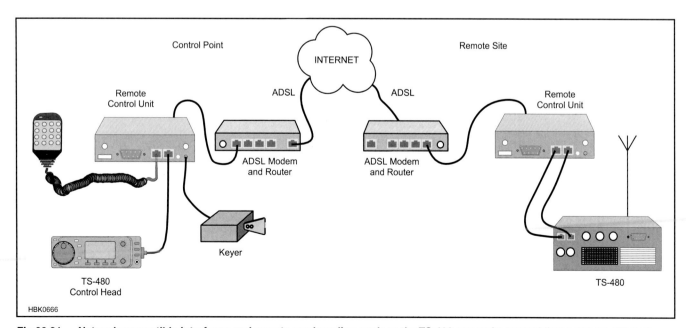

Fig 29.31 — Network-compatible interfaces and remote-ready radios such as the TS-480 can make assembling a remote HF station easier. The control head for the TS-480 can be used as the operator interface at the control point.

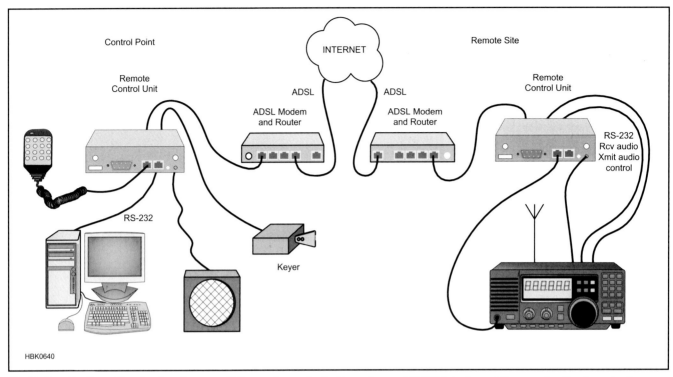

Fig 29.32 — Radios without network connectivity can be controlled by using an RS-232 interface at the remote site. A computer running control software at the control point forms the operator interface.

is recommended for power capacity evaluation. (RTTY/Digital/AM/FM duty cycles will likely be higher.) Internet switches, routers, rotators, computers, and anything electrical will also consume power. A supplementary diesel or propane remote-start generator to a basic off-grid system may be an appropriate option, particularly if you plan on using an amplifier. If your remote site is in a rugged terrain rural area, make sure a diesel or propane service provider vehicle can get to the location *before* you make final plans! It may be possible to use smaller propane tanks you can haul in a truck, but any manifold-type system necessitates insuring that the proper BTU requirements can be met based upon the propane generator manufacturer's ratings. Installing a solar and wind remote battery monitor system is advisable.

When you have made an accurate assessment of your complete system needs, add a generous additional contingency factor. It is also advisable to get quotes on your proposed system from at least two different "green power" equipment suppliers, and to make sure they fully understand your requirements and intended use *before* you purchase anything.

INSURANCE AND SECURITY

You've heard the expression "Stuff Happens." Unless you have blanket coverage in an existing homeowner's insurance policy that includes your remote site equipment, look into the ARRL "All-Risk" Ham Radio Equipment Insurance Plan. If you will be hiring others to do work at your remote site, be sure that liability coverage issues are addressed. The same applies to any guests or visitors such as fellow hams. Have a competent attorney draw up a "Liability Release" agreement with strong "Hold Harmless" language.

If your remote site is yours alone and not a co-habitation arrangement with other services, be sure and evaluate all security needs (a necessity even if co-located). You might need one or more security gates, Web-cam surveillance and auto-remote notification of security breaches.

PERSONAL SAFETY

Regardless of your age, it is prudent to have someone with you at the site. You could severely cut yourself, or fall and break a leg. If there are wildlife hazards in the area, take appropriate steps to protect yourself.

Carry a first aid kit and have one readily available at your remote site. Keep some emergency supplies such as various sealed food items in a varmint-proof container and a case of bottled water in the event you find yourself stranded.

29.4.3 Controlling Your Remote Site HF Station

Almost any modern HF transceiver can be controlled remotely by software. Several packages are listed in the Remote Station Resource spreadsheet on the CD-ROM included with this book. (The resource listing also includes vendors of remote control interface equipment.) The manufacturer of your radio may also offer remote control software.

Fig 29.31 illustrates an example of basic remote HF station control using the Remoterig control interfaces.[2] ADSL is used to connect the interfaces via the Internet. A suitable radio, microphone and CW paddle are connected to the interface on each end. The Kenwood TS-480 radio in the figure is designed for simple connectivity to a network-style interface. Other radios may require an RS-232 interface as shown in **Fig 29.32**. The radio in this system is located entirely at the remote site and a computer running control software creates the operator's interface at the control point.

A popular alternative for Contesters and DX'ers is to run the contest/logging software on a computer at the remote site. With a WKUSB (Winkey USB) unit for CW keying on the remote end, things can work very well. *Windows XP Pro* has self-contained server capabilities for *Windows* Remote Desktop, and the WinRDP client can be used for access from the control point computer. For *Windows XP Home* operating systems, VNC (Virtual Network Computing — **en.wikipedia.org/wiki/Virtual_Network_Computing**) utility software is available from several companies.

CONNECTING TO YOUR REMOTE SITE HF STATION

The most common way of connecting to

a remote site HF station is via the Internet. You will want to learn computer networking basics, find a ham friend with the necessary skills, or hire a computer networking consultant to help you. Terms like IP addresses, routers, port forwarding, DMZ, bandwidth, latency and others found in the Remote Station Glossary below will become very familiar. A convenient list of other computer network terms can be found online at **www.onlinecomputertips.com/networking/terminology.html**.

If your remote site has no telephone or Internet service and you have a line-of-sight path between your home and the site, one logical choice for connectivity is to use a pair of wireless network bridge units and a pair of solid state drive netbook computers with low power consumption. **Fig 29.33** shows an example of Airaya 5 GHz wireless bridges and ASUS Eee PC netbook computers.

If a direct line of site doesn't exist, multiple bridge units can be used to overcome various terrain challenges. Of course, overall latency will increase somewhat, and the additional locations would require a power source to operate the units. A word of caution about using unlicensed equipment and spectrum — you run the risk of potential interference from other users on the band.

UNDERSTANDING INTERNET LATENCY

High latency can seriously impair the usability of your remote station. For casual operating, you may be able to tolerate a latency of 250 ms or 300 ms. Serious contesters and DXers will most likely find this to be extremely frustrating, especially on CW. If you are a high-speed CW contest operator, a very low latency is required to be competitive.

ASSESSING LATENCY

You can easily evaluate basic Internet latency yourself. In a *Windows* (PC) environment, under Start | All Programs | Accessories you will find Command Prompt. From within this DOS-like window you can run the *ping*, *tracert* and or *pathping* commands to access a particular IP address and assess the latency. This is critical to know in advance if possible or to troubleshoot connectivity issues once you are operational).

For the purposes of conducting Internet system audio latency tests, N6RK's method illustrated as a block diagram in **Fig 29.34** is very effective. Using an HF receiver at the remote site and one at the control point, the recorded latency can easily be determined. **Fig 29.35** shows a latency of 40 ms for the return audio over K6VVA's private control link using a Remoterig audio quality setting that requires a network speed of about 180 kbps. The audio is crystal clear and at 40 ms, little difference is noticed from having the TS-480 control head connected directly to the rig body at home.

ESTIMATING INTERNET BANDWIDTH REQUIREMENTS

It is important to have a good idea of the actual bandwidth requirements for the effective operation of your remote HF station via the Internet. The higher the quality of the audio you need, the higher the bandwidth required. Unless you plan to use a private wireless bridge link, Internet speeds will determine your available bandwidth. Many remote HF station operators have found Skype (**www.skype.com**) a workable solution for remote audio routing as well as IP SOUND (**ip-sound.software.informer.com**). Another way to completely avoid Internet latency and bandwidth issues for audio is to use POTS (Plain Old Telephone System) connectivity with a dedicated analog phone line that might also be shared with a DSL data service.

Sharing an Internet connection can also cause reduction in bandwidth. If others in

Fig 29.33 — Solid-state disk drives in netbook-style computers and wireless point-to-point Ethernet bridges can be used to form a low-power, high-reliability computer-to-computer control link.

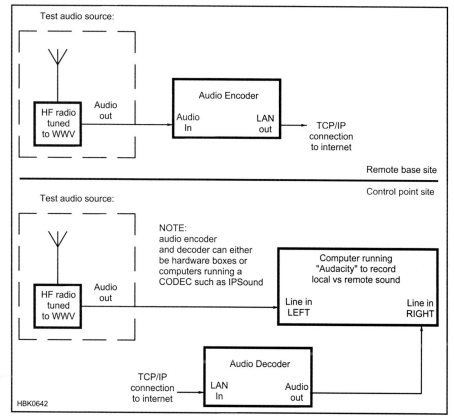

Fig 29.34 — By listening to a distant radio station at both the remote site and the control point, the delay in received audio from the remote site determines latency between the sites. (System design by N6RK.)

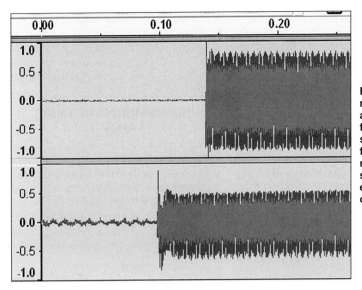

Fig 29.35 — The recordings of audio recorded from the remote site link and from the local receiver clearly show the estimated 40 ms of latency.

your home are online at the same time, their activity will reduce bandwidth available to you. A dedicated line or service upgrade to support remote HF station operation may be required unless you can account for all the concurrent household bandwidth requirements and determine that your service is capable of handling it all. A router with "QoS" (quality of service) prioritizing capabilities might solve or at least minimize some of the multi-user problems on a single line Internet connection.

CW and SSB do not pose the same the bandwidth problems as RTTY, unless pan-adapter/waterfall type displays are used. If you plan to use a Web-cam or two for monitoring and security purposes, you must account for that bandwidth in your overall connectivity planning, as well. Remember that screen changes of any type consume bandwidth.

29.4.4 A Basic Remote HF Station

There are several ways to approach implementing a remote HF station. One path is to start small with a single transceiver and a multi-band dipole, vertical and/or tribander. You can be up and running while working on your long-range plan.

K6VVA's SO2R remote station began with a single Kenwood TS-480 with the W4MQ IRT system cables and hardware, LDG AT-200PRO tuner, a refurbished TH3, Jr. tribander on a 27-foot mast and a Carolina Windom 160. The ASUS Eee PC with *Windows XP Pro* enabled use of the *Windows* Remote Desktop server and also contained the *TR4W* contest logging software which generated CW via a WKUSB unit. The initial equipment placement was in a small used camper which also housed the solar-power storage batteries in plastic battery cases, the Airaya wireless bridge with a small panel antenna mounted on the back of the camper. (You can see some of this equipment in **Fig 29.36**.)

Eventually, the remote site grew to include a larger Tuff Shed mini-barn with interior alterations to provide for a separate battery and power system room, an interior operating "shack" with view window of a nearby reservoir and custom built-in operating desk, as well as an overhead sleeping loft and front entry area.

MURPHY'S LAW FOR REMOTE HF STATIONS

Murphy seems to visit remote HF stations more than home QTH stations. Plan on it. Don't be scared, but rather be prepared! A must for your remote site station is at least one fire extinguisher and regular checks that it is fully charged and ready for use. Remember to include adequate station grounding and lightning protection in your remote HF station planning!

Troubleshooting remote HF station problems will present more challenges in general. It would be nice to have backup redundancy for everything at the remote site (including radios, amplifiers, and the like), but for most this will not be practical. If you start out with quality equipment, new coax and cables, and invest with long term thinking in mind, you'll avoid creating the opportunity for problems.

When strange glitches occur while you are on-air, I'd recommend first taking a look at your latency and bandwidth to see if there are any issues which have developed in the main connectivity pipeline.

29.4.5 Remote HF Station Resources

There is no current "Bank of Remote HF"

Fig 29.36 — The K6VVA remote station

from which to make immediate withdrawals, however you should find the RemoteOperating mailing list a helpful resource (**mailman.qth.net/mailman/listinfo/remoteoperating**). The ARRL has also published *Remote Operating for Amateur Radio*, a book on remote stations.[3]

Although certainly not a complete list, the CD-ROM included with this book includes a spreadsheet with listings for vendors of a software and equipment mentioned in this section. Don't forget to do an Internet search for "Remote HF Stations", "Remote Base" and derivations thereof. You never know what might show up! By the time this book appears in print, there will no doubt be a number of new items available for this rapidly expanding area of station design.

29.4.5 Remote Station Glossary

ADSL — Asymmetric Digital Subscriber Line. A common form of Internet connectivity which also allows for subscriber telephone use at the same time.

Band decoder — An interface device that reads frequency data from a modern transceiver and facilitates automatic control switching of other equipment such as an HF amplifier.

Bandwidth — Typically expressed in Mbps or kbps, this is used to represent both the capability of an Internet connection for data transfer as well as the amount of data to be transferred.

Charge controller — In solar and wind power generation, this device also serves as a regular to prevent overcharging of storage batteries.

Control point — Wherever the operator will be controlling the *remote site* station. This will normally be a home QTH, but could be anywhere connectivity to the remote site is available.

DHCP — Dynamic Host Configuration Protocol. This functionality is incorporated into most routers and provides automatic assignment of IP addresses to computer devices on a LAN where fixed or static IP addresses are not required.

Duty cycle — A device which is constantly "On" would normally be considered to have a 100% duty cycle. Morse code keying has spaces between the elements and the characters, therefore the duty cycle would be considerably less.

Firewall — Hardware or software capabilities in computers and routers which can be configured to minimize or completely block unauthorized users from access.

IP address — Every computer or device accessible via the Internet or a LAN has a unique numeric address, such as 192.168.1.1.

kbps — Kilobits per second (one thousand bits per second). Typically used as a reference to bandwidth data rate capability or actual usage.

LAN — Local Area Network. A home LAN enables multiple computers to connect to a single Internet service.

Latency — The delays involved in data routing from one point to another, but also a factor in A/D conversion processes.

Liability release — A written document to release one party or entity from getting sued in the event of injury.

Manifold — For propane generators, a distribution system which enables the use of multiple smaller tanks instead of one large tank.

Mbps — Megabits per second (one million bits per second). Typically used as a reference to bandwidth data rate capability or actual usage.

ms — Milliseconds. The unit of time typically used for measuring latency.

Netbook — A new generation of portable computers smaller than a traditional laptop or notebook variety. Many netbooks now use solid-state instead of rotating hard drives.

Off-grid — Generally refers to living off a main electrical power grid, but is also synonymous with extremely rugged, rural remote HF station locations at which solar/wind/battery and/or remote start gas/diesel/propane power must be used.

Packet — A formatted set of digital information. All information sent via the Internet is first converted into packets for transmission.

Pathping — A hybrid utility of the *path* and *tracert* functions which requires more time for analysis between two Internet connection points, but results in a more detailed analysis.

Ping — A utility used to ascertain the availability of another computer device over the Internet or LAN and also measure the round trip time required for the connection.

Port forwarding — Functionality most commonly used in routers to direct or re-direct incoming Internet traffic to specific destinations on a local network or LAN.

POTS — Plain Old Telephone Service. This generally refers to the use of a regular analog telephone system for purposes of remote audio and control signal routing in lieu of the Internet.

Remote site — The actual physical location of the transmitter, antennas and related equipment necessary to generate an RF signal on the HF bands. Some have used the term *Remote Base*.

Router — A device which allows traffic on a single Internet service line to be selectively distributed to multiple computer devices on a LAN. Many routers provide for assignment of local IP addresses as well as automatically via DHCP.

SmartPhone — A generic term for the new generation of portable phones with integrated computer capabilities.

Switch — A device which allows multiple computer devices to share the same Internet or LAN connection.

Tracert — A utility for tracing the path taken by packets across the Internet and latency assessing analysis along the route.

Waterfall — A graphical display of digital information.

WiFi — A form of network connectivity without a physical wired connection, although with limited range. Also known by its controlling standard, IEEE 802.11.

Wireless bridge — Low powered transmitter-receiver devices for providing point-to-point digital communications, usually Ethernet, such as for interfacing Internet services to areas without traditional means.

WOL — Wake-on-LAN functionality enables a user to remotely turn on a computer.

29.5 References and Bibliography

REFERENCES

[1]Sepmeier, RBR.com, **www.rbr.com/radio/ENGINEERING/95/13603.html**

[2]Remoterig by SM2O, **www.remoterig.com**.

[3]S. Ford, WB8IMY, *Remote Operating for Amateur Radio*, ARRL, 2010.

BIBLIOGRAPHY

R. Hilding, K6VVA, "How Not To Build A Remote Station — Part 1", Jan/Feb 2010, *National Contest Journal*, pp 8-10

R. Hilding, K6VVA, "How Not To Build A Remote Station — Part 2", Mar/Apr 2010, *National Contest Journal*, pp 19-23

"Link and Remote Control" — **www.arrl.org/link-remote-control**

Contents

30.1 Amateur Satellite History
    30.1.1 AMSAT
    30.1.2 Amateur Satellites Today
30.2 Satellite Transponders
    30.2.1 "Bent Pipe" Transponders
    30.2.2 Linear Transponders
    30.2.3 Digital Transponders
30.3 Satellite Tracking
    30.3.1 Satellite Orbits
    30.3.2 Satellite Tracking Software
    30.3.3 Getting Started With Software
    30.3.4 Orbital Elements
    30.3.5 Other Tracking Considerations
30.4 Satellite Ground Station Antennas
    30.4.1 Antenna Polarization
    30.4.2 Omnidirectional Antennas
    30.4.3 Directional Antennas
    30.4.4 Antenna Mounting
    30.4.5 Antenna Rotators
30.5 Satellite Ground Station Equipment
    30.5.1 Receive and Transmit Converters
    30.5.2 Transceivers
    30.5.3 VHF/UHF RF Power Amplifiers
    30.5.4 Typical Station Designs
30.6 Satellite Antenna Projects
    30.6.1 An EZ-Lindenblad Antenna for 2 Meters
    30.6.2 The W3KH Quadrifilar Helix
30.7 Satellite References and Bibliography
30.8 Earth-Moon-Earth (EME) Communication
    30.8.1 Background
30.9 EME Propagation
    30.9.1 Path Loss
    30.9.2 Echo Delay and Time Spread
    30.9.3 Doppler Shift and Frequency Spread
    30.9.4 Atmospheric and Ionospheric Effects
30.10 Fundamental Limits
    30.10.1 Background Noise
    30.10.2 Antenna and Power Requirements
    30.10.3 Coding and Modulation
30.11 Building an EME Station
    30.11.1 Antennas
    30.11.2 Feed lines, Preamplifiers and TR Switching
    30.11.3 Transmitters and Power Amplifiers
30.12 Getting Started
    30.12.1 Tracking the Moon
    30.12.2 System Evaluation
    30.12.3 Operating Procedures
    30.12.4 Finding QSO Partners
30.13 EME Notes and References
30.14 Glossary of Space Communications Terminology

Chapter 30

Space Communications

Radio amateurs point their antennas skyward to talk via communications satellites built by groups of interested and talented hams. In the first part of this chapter, Steve Ford, WB8IMY, provides some background and shows what you need to get involved with amateur satellites. The EME section, written by Joe Taylor, K1JT, discusses the requirements for successful contacts via the EME path. K1JT gratefully acknowledges input and advice for the EME section from Allen Katz, K2UYH, Marc Franco, N2UO, Peter Blair, G3LTF, Graham Daubney, F5VHX, Lionel Edwards, VE7BQH, Ian White, GM3SEK, Leif Asbrink, SM5BSZ and Rex Moncur, VK7MO.
A glossary of space communications terms may be found at the end of this chapter.

30.1 Amateur Satellite History

Many people are astonished to discover that Amateur Radio satellites are not new. In fact, the story of Amateur Radio satellites is as old as the Space Age itself.

The Space Age is said to have begun on October 4, 1957. That was the day when the Soviet Union shocked the world by launching Sputnik 1, the first artificial satellite. Hams throughout the world monitored Sputnik's telemetry beacons at 20.005 and 40.002 MHz as it orbited the Earth. During Sputnik's 22-day voyage, Amateur Radio was in the media spotlight since hams were among the few civilian sources of news about the revolutionary spacecraft.

Almost four months later, the United States responded with the launch of the Explorer 1 satellite on January 31, 1958. At about that same time, a group of Amateur Radio operators on the West Coast began considering the possibility of a ham satellite. This group later organized itself as Project OSCAR (OSCAR is an acronym meaning Orbiting Satellite Carrying Amateur Radio) with the expressed aim of building and launching amateur satellites. (See the sidebar "When Does a Satellite Become an OSCAR?")

After a series of high-level exchanges with the American Radio Relay League and the United States Air Force, Project OSCAR secured a launch opportunity. The first Amateur Radio satellite, known as OSCAR 1, would "piggyback" with the Discoverer 36 spacecraft being launched from Vandenberg Air Force Base in California. Both "birds" (as satellites are called among their builders and users) successfully reached low Earth orbit on the morning of December 12, 1961.

OSCAR 1 weighed only 10 pounds. It was built, quite literally, in the basements and garages of the Project OSCAR team. It carried a small beacon transmitter that allowed ground stations to measure radio propagation through the ionosphere. The beacon also transmitted telemetry indicating the internal temperature of the satellite.

OSCAR 1 was an overwhelming success. More than 570 amateurs in 28 countries forwarded observations to the Project OSCAR data collection center. OSCAR 1 lasted only 22 days in orbit before burning up as it reentered the atmosphere, but Amateur Radio's "low tech" entry into the high tech world of space travel had been firmly secured. When scientific groups asked the Air Force for advice on secondary payloads, the Air Force suggested they study the OSCAR design. What's more, OSCAR 1's bargain-basement procurement approach and management philosophy would become the hallmark of all the OSCAR satellite projects that followed, even to this day.

Since then, amateurs have successfully built and launched dozens of satellites, each one progressively more sophisticated than the last.

30.1.1 AMSAT

Much of the amateur satellite progress has been spearheaded by the Radio Amateur Satellite Corporation, better known as *AMSAT*. The original AMSAT was formed in Washington, DC in 1969 as an organization dedicated to fostering an Amateur Radio presence in space. The AMSAT model quickly became international with many countries creating their own AMSAT organizations, such as AMSAT-UK in England, AMSAT-DL in Germany, BRAMSAT in Brazil and AMSAT-LU in Argentina. All of these organizations operate independently but may cooperate on large satellite projects and other items of interest to the worldwide Amateur Radio satellite community. Because of the many AMSAT organizations now in existence, the US AMSAT organization is frequently designated AMSAT-NA.

Since the very first OSCAR satellites were launched in the early 1960s, AMSAT's inter-

When Does a Satellite Become an OSCAR?

While worldwide AMSAT organizations are largely responsible for the design and construction of the modern day Amateur Radio satellites, the original "OSCAR" designation is still being applied to many satellites carrying Amateur Radio. However, most Amateur Radio satellites are not usually assigned their sequential OSCAR numbers until *after* they successfully achieve orbit and become operational. Even then, an OSCAR number is only assigned after its sponsor formally requests one.

For example, let's make up a satellite and call it ROVER. The ROVER spacecraft won't receive an OSCAR designation until (1) it reaches orbit and (2) its sponsor submits a request. Now let's presume that ROVER makes it into orbit and the OSCAR request is made and granted. ROVER is now tagged as OSCAR 99 and its full name becomes ROVER-OSCAR 99. You'll find, however, that many hams will abbreviate the nomenclature. Some will simply call the satellite ROVER, or OSCAR 99. They may even abbreviate its full name to just RO-99.

If a satellite subsequently fails in orbit, or it re-enters the Earth's atmosphere, its OSCAR number is usually retired, never to be issued again.

Table 30.1
Currently Active Amateur Radio Satellites

Satellite	Uplink (MHz)	Downlink (MHz)	Mode
AMSAT-OSCAR 7		29.502	Beacon
		145.975	Beacon
		435.100	Beacon
	145.850 - 145.950	29.400 - 29.500	SSB/CW, non-inverting
	432.125 - 432.175	145.975 - 145.925	SSB/CW, inverting
AMSAT-OSCAR 27	145.850	436.795	FM repeater
AMSAT-OSCAR 51	145.920	435.300	FM repeater (67 Hz CTCSS required)
	145.880	435.150	USB/FM voice
	145.880	2401.200	FM voice
	145.860	435.150	9600 bps digital
	145.860	2401.200	9600 bps digital
	1268.700	2401.200	FM voice
	1268.700	435.150	FM voice
ARISS (ISS)	144.490	145.800	Crew contact, FM (Region 2/3)
	145.200	145.800	Crew contact, FM (Region 1)
	145.990	145.800	Packet BBS
	145.825	145.825	APRS digipeater
	—	144.490	SSTV downlink
	437.800	145.800	FM repeater
BEESAT	—	436.000	CW and GMSK beacon
CAPE-1	—	435.245	CW beacon
Castor	—	145.825	AX.25 beacon
Compass-One	—	437.275 (CW)	Telemetry
	—	437.405 (packet)	Telemetry (incl images)
CubeSat-OSCAR 57	—	436.8475 (CW)	Telemetry
	—	437.490 (packet)	Telemetry
CubeSat-OSCAR 58	—	437.465 (CW)	Telemetry
	—	437.345 (packet)	Telemetry
CubeSat-OSCAR 65	—	437.275	CW beacon
	—	437.475	AX.25 beacon
	1267.6000	437.475	GMSK 9600 bps
CubeSat-OSCAR 66	—	437.485 (packet)	Telemetry (incl images)
Cute-1–OSCAR 55	—	437.400	AFSK 1200 bps beacon
	—	436.8375	CW beacon
Dutch-OSCAR 64	—	145.870 (packet)	Telemetry
	435.530 - 435.570	145.880 - 145.920	SSB/CW, inverting
Fuji-OSCAR 29	145.900-146.000	435.800-435.900	SSB/CW
	—	435.7950	CW beacon
GeneSat-1	—	437.975	AFSK 1200 bps beacon and telemetry
Hope-OSCAR 68	145.825	435.675	FM repeater
	145.925 – 145.975	435.765 – 435.715	SSB/CW inverting
	145.825	435.675	1200 baud packet
	—	435.790	CW beacon
ITUpSAT1	—	437.325	FM CW and GFSK beacon
KKS-1	—	437.385	CW beacon
	—	437.445	AX.25 beacon
PRISM	—	437.250	CW and 1200 bps AFSK beacon
RS-22	—	435.352	CW beacon
RS-30	—	435.215, 435.315	CW beacon
Saudi-OSCAR 50	145.850	436.795	FM Repeater (67 Hz CTCSS required)
SwissCube beacon	—	437.505	CW and 1200 bps FSK
Sumbandila-OSCAR 67	145.875	435.345	FM repeater
STARS	—	437.275	Beacon
	—	437.305	Beacon
	—	437.465	Beacon
	—	437.850	Beacon
VUSat-OSCAR 52	—	145.860	CW Beacon
	—	145.936	Carrier Beacon
	435.220 - 435.280	145.870 - 145.930	SSB/CW, inverting

Information from AMSAT — 14 June 2010. For information on current satellite status and satellite development programs, see **www.amsat.org**.

national volunteers have pioneered a wide variety of new communications technologies. These breakthroughs have included some of the very first satellite voice transponders as well as highly advanced digital "store-and-forward" messaging transponder techniques. All of these accomplishments have been achieved through close cooperation with international space agencies that often have provided launch opportunities at significantly reduced costs in return for AMSAT's technical assistance in developing new ways to launch paying customers.

AMSAT's major source of operating revenue is obtained by offering memberships in the various international AMSAT organizations. Membership is open to radio amateurs and to others interested in the amateur exploration of space. Modest donations are also sought for tracking software and other satellite related publications. In addition, specific spacecraft development funds are established from time to time to help fund major AMSAT spacecraft projects through donations. In addition to money, AMSAT makes creative use of leftover materials donated from aerospace industries worldwide.

AMSAT-NA has just one paid employee. The rest of the organization, from the president on down to the workers designing and building space hardware, all donate their time and talents to the organization.

30.1.2 Amateur Satellites Today

When this edition of the *Handbook* went to press, there were 15 active Amateur Radio satellites in orbit (see **Table 30.1**), not counting the Amateur Radio presence aboard the International Space Station. All of these are low Earth orbiting (LEO) spacecraft that function either as communication relays or research platforms.

Hams have been longing for a high-orbiting "DX" satellite that would relay conversations across entire hemispheres. The last such satellite was OSCAR 40, which became silent in 2004. Unfortunately, it is becoming difficult for non-profit amateur satellite organizations to build such complex, powerful birds while relying solely on member dues and donations, not to mention thousands of hours of volunteer labor. To make matters worse, launch opportunities for large satellites have become increasingly scarce and prohibitively expensive.

The German AMSAT organization, AMSAT-DL, is building a high-orbit satellite that is presently known as Phase 3 Express. AMSAT-NA has been working on their version of a DX bird known as Eagle. Neither satellite is complete at the time of this writing.

For the foreseeable future, Amateur Radio satellite activity will likely be confined to spacecraft in low Earth orbits.

30.2 Satellite Transponders

In its most basic form, a *transponder* is the satellite communications system that receives signals from the Earth, alters their frequencies, amplifies them, and sends them back to Earth. The word originates from *transmitter* and *responder*. The transponder is at the heart of every satellite's ability to relay signals. There are three transponder types currently in use.

30.2.1 "Bent Pipe" Transponders

A bent-pipe transponder is the simplest transponder design in terms of function. It receives a signal at one frequency and simultaneously retransmits it at another. The metaphor is that of a U-shaped pipe that captures an object coming toward it and redirects the object back to its source (**Fig 30.1**).

A terrestrial FM repeater is a typical example of a bent-pipe transponder. Earthbound repeaters monitor one frequency and retransmit on another, usually (though not always) within the same band. Thanks to their elevated antenna systems, sensitive receivers and considerable output power, terrestrial repeaters can extend the coverage of mobile or handheld FM transceivers over hundreds or even thousands of square miles.

Bent-pipe satellite transponders also function as FM repeaters (although they could just as easily relay other modulation modes such as SSB). Satellite repeaters lack the output power of terrestrial repeaters because power generation in space is a difficult proposition. But what satellite repeaters lack in output power they more than make up for in antenna elevation. Even something as meager as a ¼ λ whip antenna can become a transmitting and receiving powerhouse when it is hundreds of miles above the planet.

The principle advantage of a bent-pipe transponder operating in the FM mode is that it is readily compatible with common Amateur Radio FM transceivers. Satellites such as AMSAT-OSCAR 51 and AMRAD-OSCAR 27 can be easily worked with the same transceivers you'd otherwise use to chat through a local FM repeater. It isn't uncommon to hear mobile and even handheld portable stations communicating through these satellites.

The principle *disadvantage* of a bent-pipe transponder is that it can relay only one signal at a time. Multiple signals on the satellite uplink frequency interfere with each other, usually resulting in an unintelligible cacophony on the downlink. Thanks to the "capture effect" common to FM receivers, only the strongest signal will come through clearly. With only 10 or 15 minutes available during typical low-orbit passes, an inconsiderate operator running high power can use the capture effect to monopolize the satellite, effectively shutting out all other stations.

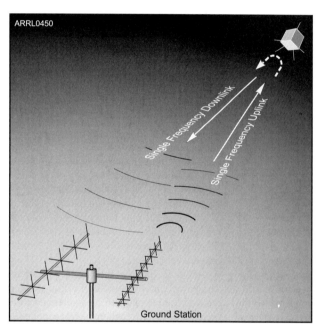

Fig 30.1 — A bent-pipe transponder gets its name from the way it functions. It takes the received uplink signal and instantly relays it on the downlink back to its source. Bent-pipe transponders are most often found on satellites that function primarily as FM repeaters.

The solution to the bent-pipe problem is to have a satellite that can relay more than one signal at a time. That requires a completely different sort of transponder.

30.2.2 Linear Transponders

Unlike a bent-pipe transponder that can

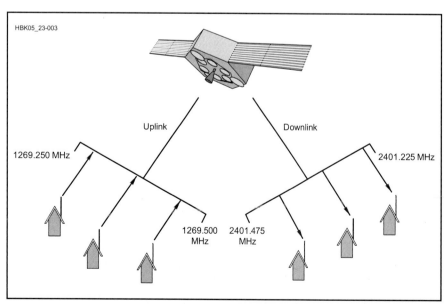

Fig 30.2 — Linear transponders repeat all signals received in an entire slice of spectrum. This allows the satellite to relay many conversations simultaneously.

relay only one signal at a time, a linear transponder receives signals in a narrow slice of the RF spectrum (the *passband*), shifts the frequency of the passband, amplifies all signals linearly, and then retransmits everything back to Earth. Rather than relaying only one signal, a satellite equipped with a linear transponder can relay many signals at once (**Fig 30.2**).

From an operator point of view, linear transponders have an enormous advantage. With single-channel bent-pipe transponders, the user is obligated to finish a conversation quickly and clear the frequency so that someone else can make a contact. Not so with linear transponders. With the ample spectrum available through a linear transponder, there is more than enough "room" for everyone to communicate for as long as they please, or at least until the satellite travels out of range.

A linear transponder can be used with any type of signal when real time communication is desired. From the standpoint of conserving valuable spacecraft resources such as power and bandwidth, however, the preferred user modes are SSB and CW. Transponders are specified by first giving the approximate input frequency followed by the output frequency. For example, a 146/435-MHz transponder has an input passband centered near 146 MHz and an output passband centered near 435 MHz. The same transponder could be specified in wavelengths, as a 2 meter/70 cm unit.

TRANSPONDER MODES

To keep linear transponder specifications as simple as possible, the satellite community often uses so-called *Mode* designators. In the early years of amateur satellite operation, these designations were rather arbitrarily assigned and bore little resemblance to the actual frequencies in use. Fortunately, the amateur satellite community has since settled on

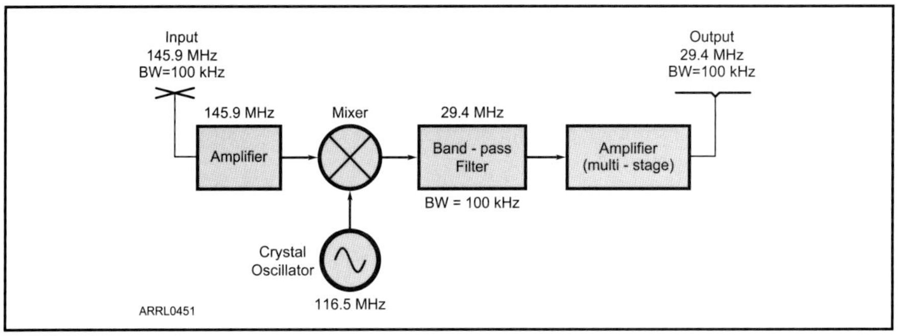

Fig 30.3 — A simplified block diagram of a linear transponder.

a newer series of transponder specifications that are far more intuitive. For instance, in our previous example, the 2 meter band is tagged with the label "V" (for VHF) while the 70 cm band is labeled "U" (for UHF). Therefore, a transponder that listens on 2 meters and repeats on 70 cm is today usually called a Mode V/U transponder. See **Table 30.2**.

Transponder design is, in many respects, similar to receiver design. Input signals are typically on the order of 10^{-13} W and the output level can be up to several watts. A major difference, of course, is that the transponder output is at radio frequency while the receiver output is at audio frequency. A block diagram of a simple transponder is shown in **Fig 30.3**. For several reasons, flight-model transponders are more complex than the one shown.

In spacecraft applications a key characteristic of a linear amplifier is its overall efficiency (RF output/dc input). Once we reach power levels above a few watts the use of class A, AB or B amplifiers cannot be tolerated. The power and bandwidth of a transponder must be compatible with each other and with the mission. That is, when the transponder is fully loaded with equal-strength signals, each signal should provide an adequate signal-to-noise ratio at the ground. Selecting appropriate values accurately on a theoretical basis using only link calculations is error prone. Experience with a number of satellites, however, has provided AMSAT with a great deal of empirical data from which it's possible to extrapolate accurately to different orbits, bandwidths, power levels, frequencies and antenna characteristics.

In general, low-altitude (300 to 1600 km) satellites that use passive magnetic stabilization and omnidirectional antennas can provide reasonable downlink performance with from 1 to 10 W PEP at frequencies between 29 and 435 MHz, using a 50- to 100-kHz-wide transponder. A high-altitude (35,000 km) spin-stabilized satellite that uses modest (7 to 10 dBi) gain antennas should be able to provide acceptable performance with 35 W PEP using a 300-kHz-wide transponder downlink at 146 or 435 MHz. Transponders are usually configured to be *inverting* in order to minimize Doppler shift. This means, for

example, that a signal with a frequency at the low end of the uplink passband shifts to a corresponding point on the high end of the downlink passband. In the case of an SSB signal, the sideband modulation inverts as well — lower sideband on the uplink becomes upper sideband on the downlink.

DYNAMIC RANGE

The dynamic range problem for transponders is quite different from that for HF receivers. At first glance it may seem that the situation faced by satellite transponders is simpler. After all, an HF receiver must be designed to handle input signals differing in strength by as much as 100 dB, while a low-altitude satellite will encounter signals in its passband differing by perhaps 40 dB. Good HF receivers solve the problem by filtering out all but the desired signal before introducing significant gain. A satellite, however, has to accommodate all users simultaneously. The maximum overall gain can, therefore, be limited by the strongest signal in the passband.

Considering the state-of-the-art in transponder design and available spacecraft power budgets, an effective dynamic range of about 25 dB is about the most that can be currently obtained. In earlier satellites, the receiver AGC was normally adjusted to accommodate the loudest user. As a result, stations 25 dB weaker were not able to put a usable signal through the spacecraft even though they might be capable of doing so when the AGC is not activated. In the ideal situation users would adjust uplink power so that spacecraft AGC is never activated.

The "power-hog" problem is a serious one. A single station running excessive power can effectively swamp a linear transponder. Far too often, an inexperienced operator may believe that cranking up transmit power will improve downlink signal strength. Instead, that increase in power only serves to depress the downlink signal levels of all other stations.

In the short-lived OSCAR 40 satellite, the designers tried an innovative approach to the power-hog problem with the addition of LEILA (*LEI*stungs *L*imit *A*nzeige [Power Limit Indicator]). LEILA's computer continuously monitored the 10.7 MHz trans-

---

Table 30.2

Satellite Uplink/Downlink Mode Designators

Satellite Band Designations
10 m (29 MHz): H
2 m (145 MHz): V
70 cm (435 MHz): U
23 cm (1260 MHz): L
13 cm (2.4 GHz): S
5 cm (5.6 GHz): C
3 cm (10 GHz): X

Common Operating Modes (Uplink/Downlink)
V/H (2 m/10 m)
H/V (10 m/2 m)
U/V (70 cm/2 m)
V/U (2 m/70 cm)
U/S (70 cm/13 cm)
U/L (70 cm/23 cm)
L/S (23 cm/13 cm)
L/X (23 cm/3 cm)
C/X (5 cm/3 cm)

ponder IF passband. When an uplink signal exceeded a predetermined level, the computer inserted a CW message over the offender's downlink. The message indicated to the transmitting station that the transponder was being overloaded, and served as an inducement for the offending station to reduce power until the CW signal disappeared. If the overloading signal continued, or exceeded an even higher preset level, LEILA activated a notch filter tuned to the offending station's frequency. LEILA turned out to be highly effective. If you suddenly heard CW being transmitted over your signal on the downlink, you knew you had to reduce power immediately or face the consequences.

Since the transponder is the primary mission subsystem, reliability is extremely important. One way to improve system reliability is to include at least two transponders on each spacecraft; if one fails, the other would be available full time. And there are significant advantages to *not* using identical units.

30.2.3 Digital Transponders

Digital transponders differ significantly from linear transponders. A digital transponder demodulates the incoming signal. The data can then be stored aboard the spacecraft (as in a packet mailbox) or used to immediately regenerate a digital downlink signal (as in a digipeater). The mailbox service is best suited to low altitude spacecraft. Digipeating is most effective on high altitude spacecraft. Like linear transponders, digital units are downlink limited. A key step in the design procedure is to select modulation techniques and data rates to maximize the downlink capacity. Using assumptions about the type of traffic expected, the designers select appropriate uplink parameters. An analysis suggests that the uplink data capacity should be about four or five times that of the downlink because of "collisions" among signals trying to access the transponder.

For mailbox operation designers often use a PACSAT (*packet radio satellite*) model with similar data rates for the uplink and downlink, and they couple a single downlink with four uplinks. Fuji-OSCARs 12 and 20, and the MicroSat series ran both links at 1200 bps. These PACSATs contained an FM receiver with a demodulator that accepted Manchester-encoded FSK on the uplink. To produce an appropriate uplink signal, ground stations needed FM transmitters and packet radio modems known as terminal node controllers (TNCs) that were capable of generating the Manchester-encoded signal for the uplink.

The PACSAT downlink used binary phase-shift keying (BPSK) at an output of either 1.5 or 4 W. This modulation method was selected because, at a given power level and bit rate, it provides a significantly better bit error rate than other methods that were considered. One way of receiving the downlink is to use an SSB receiver and pass the audio output to a PSK demodulator. The SSB receiver is just serving as a linear downconverter in this situation. Other methods of capturing the downlink are possible but the two proven systems now operating use this approach.

During the heydays of the PACSATs, there were several TNCs designed specifically for this application. Over time the PACSATs have gradually gone out of service and those specialized TNCs have all but disappeared from the marketplace. The remaining digital satellite transponders use 1200 bps AFSK or 9600 bps FSK data transmissions for both uplink and downlink. This means that ordinary packet radio TNCs — the kind currently used for various terrestrial applications — can be used with these digital satellites as well.

RUDAK

A discussion of digital transponders would not be complete without mentioning RUDAK. RUDAK is an acronym for Regenerativer Umsetzer fur Digitale Amateurfunk Kommunikation (Regenerative Transponder for Digital Amateur Communications). Early RUDAK systems took a different approach to achieving the desired ratio of uplink to downlink capacity. The one flown on OSCAR 13 used one uplink channel and one downlink channel with the data rate on the uplink (2400 bps) roughly six times that on the downlink (400 bps). The 400 bps rate on the downlink was chosen because this was the standard that had been used for downlinking high-orbiting satellite telemetry since the late 1970s. Users already capturing telemetry would be able to capture RUDAK transmissions from day one. Unfortunately, the RUDAK unit on OSCAR 13 failed during launch.

A system known as RUDAK II was flown on RS-14/AO-21 and actually consisted of two units. One unit was similar to the RUDAK flown on OSCAR 13. However, the other unit, known as the RUDAK Technology Experiment (or RTX), was a new experimental transponder using DSP technology. It was essentially a flying test bed for ideas being considered for future high-orbit missions.

The RUDAK-U system flown on OSCAR 40 contained two CPUs, one 153.6 kbaud modem, four hardwired 9600 baud modems and eight DSP modems capable of operating at speeds up to 56 kbaud. The great advantages of modern RUDAK systems are their extraordinary flexibility. Through the use of digital signal processing, the RUDAK is able to configure itself into any type of digital system desired. Had OSCAR 40 survived, the plan was to use its RUDAK system to create a highly capable digital communications platform in space.

30.3 Satellite Tracking

30.3.1 Satellite Orbits

Most active amateur satellites are in various types of low Earth orbits (LEOs) at altitudes that vary between approximately 350 km (the International Space Station) and 1300 km (OSCAR 29). There are satellites planned for future launch that will travel in the high Earth orbits (HEOs). At their closest approaches to Earth (their *perigees*), these satellites may come within 1000 km; at the highest points of their orbits (their *apogees*), they will travel as far as 50,000 km into space. Let's take a brief look at several of the most common orbits.

An *inclined* orbit is one that is inclined with respect to the Earth's equator. See **Fig 30.4A**. A satellite that is inclined 90° would be orbiting from pole to pole; smaller inclination angles mean that the satellite is spending more time at lower latitudes. The International Space Station, for example, travels in an orbit that is inclined about 50° to the equator. Satellites that move in these orbits frequently fall into the Earth's shadow (eclipse), so they must rely on battery systems to provide power when the solar panels are not illuminated. Depending on the inclination angle, some locations on the Earth will never have good access because the satellites will rarely rise above their local horizons. This was true, for example, in the days when the US Space Shuttles carried Amateur Radio operators. Shuttle orbits were usually inclined at low angles, which meant that Shuttles barely made it above the horizon for hams in the northern US and Canada.

A *sun-synchronous* orbit takes a satellite over the north and south poles. See Fig 30.4B. There are two advantages to a sun-synchronous orbit: (1) the satellite is available at approximately the same time of day, every day and (2) everyone, no matter where they are, will enjoy at least one high-altitude pass per day. OSCAR 51 is a good example of a satellite that travels in a sun-synchronous orbit.

A *dawn-to-dusk* orbit is a variation on the sun-synchronous model except that the satellite spends most of its time in sunlight and relatively little time in shadow.

OSCAR 27 travels in a dawn-to-dusk orbit. See Fig 30.4C.

The *Molniya* high Earth orbit (Fig 30.4D) was pioneered by the former Soviet Union. It is an elliptical orbit that carries the satellite far into space at its greatest distance from Earth. To observers on the ground, the satellite at apogee appears to hover for hours at a time before it plunges earthward and sweeps around the Earth at its closest approach. One great advantage of the Molniya orbit is that the satellite is capable of "seeing" an entire hemisphere of the planet while at apogee. Hams can use a Molniya satellite to enjoy long, leisurely conversations spanning thousands of kilometers here on Earth. When this book was written, however, there were no active ham satellites traveling in Molniya orbits.

30.3.2 Satellite Tracking Software

The key to communicating through a satellite is being able to track its movements and predict when it will appear above your local horizon. Fortunately, we have sophisticated software available that streamlines this task considerably.

You'll find satellite-tracking software written for *Windows*, *Mac* and *Linux* operating systems. Several popular applications are listed in **Table 30.3**. When computers were first employed to track amateur satellites, they provided only the most basic, essential information: when the satellite will be available (*AOS*, acquisition of signal), how high the satellite will rise in the sky and when the satellite is due to set below your horizon (*LOS*, loss of signal). Today we tend to ask

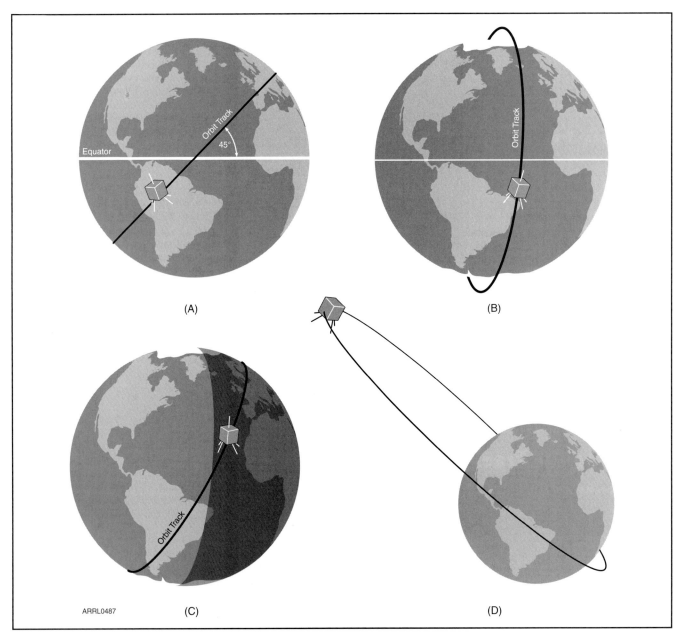

Fig 30.4 — At A, an *inclined* orbit is one that is inclined with respect to the Earth's equator (in this case, 45°). At B, a *sun-synchronous* orbit takes the satellite over the north and south poles, allowing every station in the world to enjoy at least one high-elevation pass per day. At C, a *dawn-to-dusk* orbit is a variation on the sun-synchronous model except that the satellite spends most of its time in sunlight and relatively little time in eclipse. At D, the *Molniya* orbit is an elliptical orbit that carries the satellite far into space at its greatest distance from Earth (apogee). To observers on the ground, the satellite at apogee appears to hover for hours at a time before it plunges earthward and (often) sweeps within 1000 km at its closest approach (perigee).

Table 30.3
A Sampling of Satellite Tracking Software

Name	Source	Operating System	Radio Control?	Antenna Control?
Nova	www.amsat.org (store)	Windows	No	Yes
SCRAP	www.amsat.org (store)	Windows	No	Yes
SatPC32	www.amsat.org (store)	Windows	Yes	Yes
SatScape	www.satscape.co.uk/classic.html	Windows	Yes	Yes
MacDoppler	www.dogparksoftware.com/MacDoppler.html	Mac OS	Yes	Yes
Ham Radio Deluxe	hrd.ham-radio.ch	Windows	Yes	No
Predict	www.qsl.net/kd2bd/predict.html	Linux	No	Yes
WinOrbit	www.sat-net.com/winorbit	Windows	No	No

a great deal more of our tracking programs. Modern applications still provide the basic information, but they usually offer many more features such as:

• The spacecraft's operating schedule, including which transponders and beacons are on.

• Predicted frequency offset (Doppler shift) on the link frequencies.

• The orientation of the spacecraft's antennas with respect to your ground station and the distance between your ground station and the satellite.

• Which regions of the Earth have access to the spacecraft; that is, who's in QSO range?

• Whether the satellite is in sunlight or being eclipsed by the Earth. Some spacecraft only operate when in sunlight.

• When the next opportunity to cover a selected terrestrial path (mutual window) will occur.

• Changing data can often be updated at various intervals such as once per minute — or even once per second.

A number of applications do even more. Some will control antenna rotators, automatically keeping directional antennas aimed at the target satellite. Other applications will also control the radio to automatically compensate for frequency changes caused by Doppler shifting.

Adding additional spacecraft to the scenario suggests more questions. Which satellites are currently in range? How long each will be accessible? Will any new spacecraft be coming into range in the near future? Obviously there is a great deal of information of potential interest. Programmers developing tracking software often find that the real challenge is not solving the underlying physics problems, but deciding what information to include and how to present it in a useful format. This is especially true since users have different interests, levels of expertise and needs. Some prefer to see the information in a graphical format, such as a map showing real-time positions for all satellites of interest. Others may prefer tabular data such as a listing of the times a particular spacecraft will be in range over the next several days.

There are also several Internet sites where you can do your tracking online. This eliminates all the hassles associated with acquiring and installing software. The currently available online tracking sites are not as powerful or flexible as the software you can install on your PC, however. One interesting site of this type is maintained by AMSAT-NA and you'll find it at **www.amsat.org/amsat-new/tools/predict/**.

30.3.3 Getting Started With Software

There are so many different types of satellite software, and they change so frequently, it would be foolhardy to attempt to give you detailed operational descriptions in any book. The book would be obsolete a month after it came off the press!

Even so, there are a number of aspects of satellite-tracking software that rarely change. For instance, most programs will ask you to enter your station location as part of the initial setup process. Some applications use the term "observer" to mean "station location," but the terms are synonymous for the sake of our discussion. Sophisticated programs will go as far as to provide you with a list of cities that you can select to quickly enter your location. Other programs will ask you to enter your latitude and longitude coordinates manually.

When entering latitude, longitude (and other angles), make sure you know whether the computer expects degree-minute or decimal-degree notation. Following the notation used by the on-screen prompt usually works. Also make sure you understand the units and sign conventions being used. For example, longitudes may be specified in negative number for locations west of Greenwich (0° longitude). Latitudes in the southern hemisphere may also require a minus sign. Fractional parts of a degree will have very little effect on tracking data so in most cases you can just ignore it.

Dates can also cause considerable trouble. Does the day or month appear first? Can November be abbreviated Nov or must you enter 11? The number is almost always required. Must you write 2010 or will 10 suffice? Should the parts be separated by colons, dashes or slashes? The list goes on and on. Once again, the prompt is your most important clue. For example, if the prompt reads "Enter date (DD:MM:YY)" and you want to enter February 9, 2010 follow the format of the prompt as precisely as possible and write 09:02:10.

When entering numbers, commas should never be used. For example, if a semi-major axis of 20,243.51 km must be entered, type 20243.51 with the comma and units omitted. It takes a little time to get used to the quirks of each software package, but you'll soon find yourself responding automatically.

Once you have your coordinates entered, you're still not quite done. The software now "knows" its location, but it doesn't know the locations of the satellites you wish to track. The only way the software can calculate the positions of satellites is if it has a recent set of *orbital elements*.

30.3.4 Orbital Elements

Orbital elements are sets of six numbers that completely describe the orbit of a satellite at a specific time. Although scientists may occasionally use different groups of six quantities, radio amateurs nearly always use the six known as Keplerian Orbital Elements, or simply *Keps*.

These orbital elements are derived from very precise observations of each satellite's orbital motion. Using precision radar and highly sensitive optical observation techniques, the North American Aerospace Defense Command (NORAD) keeps a very accurate catalog of almost everything in Earth orbit. Periodically, they issue the unclassified portions of this information to the National Aeronautics and Space Administration (NASA) for release to the general public. The information is listed by individual catalog number of each satellite and contains numeric data that describes, in a mathematical way, how NORAD observed the satellite moving around the Earth at a very precise location in space at a very precise moment in the past.

Without getting into the complex details of orbital mechanics, suffice it to say that your software simply uses the orbital element information NASA publishes that describe

Table 30.4
Examples of Elements Used for Satellite Tracking

(Download current elements for actual tracking use.)

NASA Two-Line Elements for OSCAR 27
AO-27
1 22825U 93061C 08024.00479406 -.00000064 00000-0 -86594-5 0 8811
2 22825 098.3635 349.6253 0008378 336.4256 023.6532 14.29228459747030

AMSAT Verbose Elements for OSCAR 51
Satellite: AO-51
Catalog number: 28375
Epoch time: 08024.16334624
Element set: 14
Inclination: 098.0868 deg
RA of node: 056.7785 deg
Eccentricity: 0.0083024
Arg of perigee: 232.8417 deg
Mean anomaly: 126.5166 deg
Mean motion: 14.40594707 rev/day
Decay rate: 7.0e-08 rev/day^2
Epoch rev: 18752
Checksum: 310

where a particular satellite was "then" to solve the orbital math and make a prediction (either graphically or in tabular format) of where that satellite ought to be "now." The "now" part of the prediction is based on the local time and station location information you've also been asked to load into your software.

Orbital elements are frequently distributed with additional numerical data (which may or may not be used by a software tracking program) and are commonly available in two forms, NASA format and AMSAT format.

UNDERSTANDING THE AMSAT FORMAT

Let's use the easier-to-understand AMSAT format example shown in **Table 30.4** to break down the meaning, line by line.

The first two entries identify the spacecraft. The first line is an informal *satellite name*. The second entry, *Catalog Number*, is a formal ID assigned by NASA.

The next entry, *Epoch Time*, specifies the time the orbital elements were computed. The number consists of two parts, the part to the left of the decimal point that describes the year and day, and the part to the right of the decimal point that describes the (very precise) time of day. For example, 96325.465598 refers to 1996, day 325, time of day .465598.

The next entry, *Element Set*, is a reference used to identify the source of the information. For example, 199 indicates element set number 199 issued by AMSAT. This information is optional.

The next six entries are the six key orbital elements.

Inclination describes the orientation of the satellite's orbital plane with respect to the equatorial plane of the Earth.

RAAN, Right Ascension of Ascending Node, specifies the orientation of the satellite's orbital plane with respect to fixed stars.

Eccentricity refers to the shape of the orbital ellipse. The closer this number is to 0, the more circular the orbit of the satellite tends to be. Conversely, an eccentricity value approaching 1 indicates the satellite is following a more elliptically shaped orbital path.

Argument of Perigee describes where the perigee of the satellite is located in the satellite orbital plane. Recall that a satellite's perigee is its closest approach to the Earth. When the argument of perigee is between 180 and 360° the perigee will be over the Southern Hemisphere. Apogee — a satellite's most distant point from the Earth — will therefore occur above the Northern Hemisphere.

Mean Anomaly locates the satellite in the orbital plane at the epoch. All programs use the astronomical convention for mean anomaly (MA) units. The mean anomaly is 0 at perigee and 180 at apogee. Values between 0 and 180 indicate that the satellite is headed up toward apogee. Values between 180 and 360 indicate that the satellite is headed down toward perigee.

Mean Motion specifies the number of revolutions the satellite makes each day. This element indirectly provides information about the size of the elliptical orbit.

Decay Rate is a parameter used in sophisticated tracking models to take into account how the frictional drag produced by the Earth's atmosphere affects a satellite's orbit. It may also be referred to as rate of change of mean motion, first derivative of mean motion, or drag factor. Although decay rate is an important parameter in scientific studies of the Earth's atmosphere and when observing satellites that are about to reenter, it has very little effect on day-to-day tracking of most Amateur Radio satellites. If your program asks for drag factor, enter the number provided. If the element set does not contain this information enter zero — you shouldn't discern any difference in predictions. You usually have a choice of entering this number using either decimal form or scientific notation. For example, the number –0.00000039 (decimal form) can be entered as –3.9e–7 (scientific notation). The e–7 stands for 10 to the minus seventh power (or 10^{-7}). In practical terms e–7 just means move the decimal in the preceding number 7 places to the left. If this is totally confusing, just remember that in most situations entering zero will work fine.

Epoch revolution is just another term for the expression "Orbit Number" that we discussed earlier. The number provided here does not affect tracking data, so don't worry if different element sets provide different numbers for the same day and time.

The *Checksum* is a number constructed by

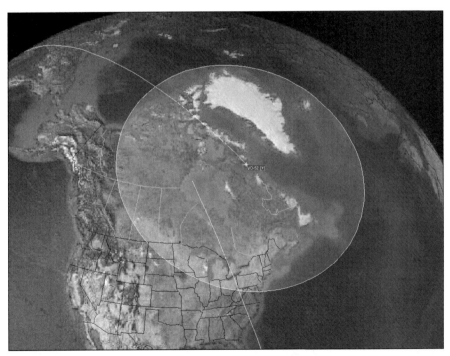

Fig 30.5 — This image shows the circular footprint of OSCAR 52 as depicted by *NOVA* satellite-tracking software. The footprint indicates the area of the Earth that is visible to the satellite at any given time.

the data transmitting station and used by the receiving station to check for certain types of transmission errors in data files. It does not bear any relationship to a satellite's orbit.

In the "old days" of satellite-tracking software you had to enter the orbital elements by hand. This was a tedious and risky process. If you entered an element number incorrectly, you would generate wildly inaccurate predictions.

Today, thankfully, most satellite-tracking programs have greatly streamlined the process. One method of entering orbital elements is to grab the latest set from the AMSAT-NA Web site at **www.amsat.org** (look under "Keps" in the main menu). You can download the element set as a text file and then tell your satellite-tracking program to read the file and create the database. Another excellent site is CelesTrak at **celestrak.com**. Your program will probably be able to read either the AMSAT or NASA formats.

If you're fortunate to own sophisticated tracking software such as *Nova*, and you have access to the Internet, the program will reach into cyberspace, download and process its Keps automatically. All it takes is a single click of your mouse button. Some programs can even be configured to download the latest Keps on a regular basis without any prompting from you.

30.3.5 Other Tracking Considerations

SATELLITE FOOTPRINTS

Many satellite tracking applications display not only the satellite track, but also the satellite *footprint*. A satellite footprint can be loosely defined as the area on the Earth's surface that is "illuminated" by the satellite's antenna systems at any given time. Another way to think of a footprint is to regard it as the zone within which stations can communicate with each other through the satellite.

Unless the satellite in question is *geostationary*, footprints are constantly moving. Their sizes can vary considerably, depending on the altitude of the satellite. The footprint of the low-orbiting International Space Station is about 600 km in diameter. In contrast, the higher orbiting OSCAR 52 has a footprint that is nearly 1500 km across. See the example of a satellite footprint in **Fig 30.5**. The amount of time you have available to communicate depends on how long your station remains within the footprint. This time can be measured in minutes, or in the case of a satellite in a highly elliptical orbit, hours.

It is worthwhile to note that the size and even the shape of a footprint can also vary according to the type of antenna the satellite is using. A highly directional antenna with a narrow beamwidth will create a small footprint even though the satellite is traveling in a high-altitude orbit. This usually isn't an issue for amateur satellites, however.

A SATELLITE'S "PHASE"

Some satellites use operating schedules to determine which transponders and antennas are active at any given time. It is based on the *phase* of the satellite's *mean anomaly*, or *MA*. Many software applications can use this information to provide detailed predictions that tell you not only when the satellite will appear, but also which transponders will be active at the time.

The expression "anomaly" is just a fancy term for *angle*. Astronomers have traditionally divided orbits into 360 mean-anomaly units, each containing an equal time segment. Because of the architecture of common microprocessors, it was much more efficient to design the computers controlling spacecraft to divide each orbit into 256 segments of equal time duration. Radio amateurs refer to these as mean anomaly or phase units. The duration of each segment is the satellites period divided by 256. For example, a mean anomaly unit for OSCAR 13 was roughly 2.68 minutes. At MA 0 (beginning of orbit) and MA 256 (end of orbit) the satellite is at perigee (its lowest point). At MA 128 (halfway through the orbit) the satellite is at apogee or high point. See **Fig 30.6**.

Because radio amateurs and astronomers use the term mean anomaly in a slightly different way, there's sometimes a question as to which system is being used. Any confusion is minor and usually easily resolved. Most OSCAR telemetry with real-time MA values and schedules use the 256 system. The term "phase," and the fact that no numbers larger than 256 ever appear, are significant hints. Computer tracking programs designed for non-radio amateur audiences generally use the traditional astronomical notation. It's easy to determine when this is the case because the mean anomaly column will contain entries between 257 and 360.

If a satellite is using an MA-based transponder schedule, the schedule will be posted on the AMSAT-NA Web site at **www.amsat.org**. (When this book was written there were no Amateur Radio satellites in orbit using MA scheduling, but this may change as currently planned satellites reach orbit.) Depending on its sophistication, your tracking software may allow you to enter this schedule data.

Schedules are generally modified every few months when satellite orientation is adjusted to compensate for changes in the sun

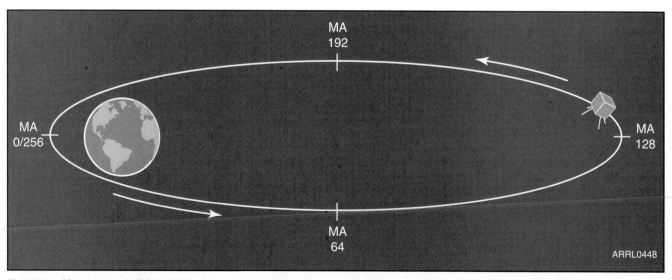

Fig 30.6 — Mean anomaly (MA) divides each orbit into 256 segments of equal time duration. Radio amateurs refer to these as "Mean Anomaly" or "Phase" units. The duration of each segment is the satellite's period divided by 256. For example, a mean anomaly unit for OSCAR 13 was roughly 2.68 minutes. At MA 0 (beginning of orbit) and MA 256 (end of orbit) the satellite is at perigee (its lowest point). At MA 128 (halfway through the orbit) the satellite is at apogee (or its highest point).

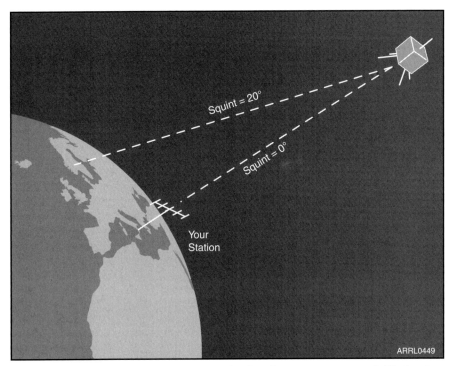

Fig 30.7 — The squint angle describes how the directive antennas on a satellite (such as Phase 3 satellites) are pointed with respect to your ground station. Squint angles can vary between 0° and 180°. A squint angle of 0° means the satellite antennas are pointed directly at you, which, in turn, means good link performance can usually be expected. When the squint angle is above 20° signal level begins to drop.

angle on the spacecraft. A typical schedule used for OSCAR 13, with its corresponding uplink (transmitting) and downlink (receiving) frequency band requirements, looked like this:

Off: from MA 0 until MA 49
Mode U/V (uplink on 70 cm/downlink on 2 meters): on from MA 50 until MA 128
Mode U/S (uplink on 70 cm/downlink on 2.4 GHz): on from MA 129 until MA 159
Mode U/V (uplink on 70 cm/downlink on 2 meters): on from MA 160 until MA 255

If you wanted to operate Mode U/S, you needed to be at the radio when the satellite was between MA129 and MA159 in its orbit.

SQUINT ANGLE

Another dilemma your software may help resolve is the *squint angle*. The squint angle describes how the directive antennas on a satellite are pointed with respect to your ground station. Squint angle can vary between 0° and 180°. A squint angle of 0° means the satellite antennas are pointed directly at you and that generally indicates that good link performance can be expected (**Fig 30.7**). When the squint angle is above 20°, signal level begins to drop and a disruptive amplitude flutter called spin modulation on uplinks and downlinks may become apparent.

Programs that include algorithms to calculate squint angle require information about the orientation or attitude of the satellite. This information is generally available from sources that provide the basic orbital elements and on telemetry sent directly from the satellite of interest. The parameters needed are labeled Bahn latitude and Bahn longitude. They are also known as BLAT and BLON or ALAT and ALON where the prefix "A" stands for attitude.

Programs that provide squint angle information may also contain a column labeled *Predicted Signal Level*. Values are usually computed using a simple prediction model that takes into account satellite antenna pattern, squint angle and spacecraft range. The model assumes a 0 dB reference point with the satellite overhead, at apogee and pointing directly at you. At any point on the orbit the predicted level may be several dB above (+) or below (−) this reference level.

You won't need to be concerned about squint angle for most LEO satellites. It only becomes a factor with the high-orbit birds since they generally use directive antennas. As with MA scheduling, you'll want to know that the satellite's antennas are pointing at your location before you fire up your equipment. With the right kind of software, you'll know well ahead of time.

30.4 Satellite Ground Station Antennas

Assembling a satellite ground station presents a different challenge compared to typical HF station setups. There are a number of variables that change depending on the types of satellites you hope to use. A ground station designed strictly for LEO satellites differs substantially from one intended for HEO birds. There are also issues surrounding computers and software, particularly if you plan to access digital satellites.

One aspect of your ground station with the greatest range of choice is your antenna system. To use most amateur satellites, you must transmit on one band (the uplink) and receive on another band (the downlink). Unless you are using a single, dual-band antenna, your satellite station will require at least two antennas. Some satellite operators have several antennas so that they can operate on various uplink/downlink band combinations as the need arises.

30.4.1 Antenna Polarization

The *polarization* of a satellite antenna is an important factor, regardless of whether it is used for transmitting or receiving. Polarization is determined by the position of the radiating element or wire with respect to the Earth. A radiator that is parallel to the Earth radiates horizontally, while a vertical radiator radiates a vertical wave. These are so-called *linearly polarized* antennas. If the radiating element is slanted above the Earth, it radiates waves that have *both* vertical and horizontal components.

For terrestrial VHF+ line-of-sight communication, polarization matching is important. If one station is using a horizontally polarized antenna and the other is using a vertically polarized antenna, the mismatch can result in a large signal loss under certain conditions. We don't worry about polarization mismatches on HF frequencies because whenever signals are refracted through the ionosphere, as HF signals usually are, polarization changes anyway.

The problem with applying polarization

concerns to spacecraft is that the orientation of a satellite's antennas relative to your ground station is constantly changing. This often results in fading when the polarization of its antennas conflict with yours. This problem doesn't plague satellites exclusively. Aircraft, automobiles and all other moving radio platforms can suffer the same effects.

Fortunately, there is a "cure" known as *circular polarization* (CP). With CP, wave fronts appear to rotate as they pass the receiving station, either clockwise (right-hand or RHCP) or counter clockwise (left-hand or LHCP). See **Fig 30.8**. The advantage of using circular polarization is that it can substantially reduce the effects of polarization conflict. Since the polarization of a CP antenna rotates through horizontal and vertical planes, the resulting pattern effectively "smoothes" the fading effects, generating consistent signals as a result.

That said, your satellite ground station antennas do *not* need to be circularly polarized to be effective. Linearly polarized antennas, either horizontal or vertical, are perfectly useful. Some ground station antenna designs use slanted or "crossed" elements to mix the horizontal and vertical polarization components as discussed earlier. The goal of fine-tuning your antenna polarization is to give your station an edge, something that is important when you are dealing with weak signals from deep space. But, while circular polarization of your antennas may give your station a definite advantage, it isn't an absolute requirement.

30.4.2 Omnidirectional Antennas

An ideal omnidirectional antenna radiates and receives signals in all directions equally. In the real world, most "omnidirectional" antennas have a certain amount of directivity. For example, a common ground plane antenna (**Fig 30.9**) is considered to be omnidirectional, but as you can see in **Fig 30.10** the radiation pattern is not uniform. Notice the deep null directly overhead. Signals will fade sharply as satellites move through this null, particularly during high-elevation passes.

Despite their low gain, omnidirectional antennas are attractive because they do not need to be aimed at their targets. This means that they don't require mechanical antenna rotators, which can add significant cost and complexity to a ground station. Omni antennas are also more compact than directional antennas for the same frequency. On the other hand, their low gain makes them practical only for LEO satellites that have sensitive receivers and relatively strong transmit signals.

Technically speaking, any type of omnidirectional antenna can be used for a satellite ground station. Hams have enjoyed satellite operations with simple ground planes, J-poles, "big wheels" and even automobile antennas. But for best results with an omni-based antenna system, you'll want a radiation pattern that minimizes the polarization conflicts and pattern nulls that can cause signals to fade. To this end, engineers have designed a number of omnidirectional antennas with these issues in mind.

THE EGGBEATER

The *eggbeater* antenna is a popular design named after the old-fashioned kitchen utensil

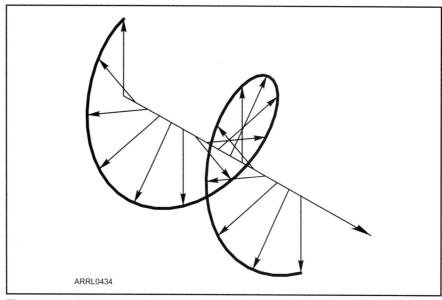

Fig 30.8 — A circularly polarized wave front describes a rotational path about its central axis, either clockwise (right-hand or RHCP) or counter clockwise (left-hand or LHCP).

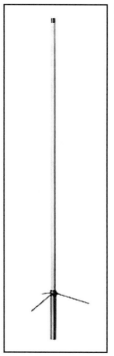

Fig 30.9 — An ordinary ground plane antenna. This particular model is designed for terrestrial communications on 2 m.

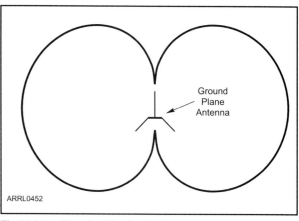

Fig 30.10 — The simplified radiation pattern of a ground plane antenna. Notice the deep null directly overhead.

it resembles (**Fig 30.11**). The antenna is composed of two full-wave loops of rigid wire or metal tubing. Each of the two loops has an impedance of 100 Ω, and when coupled in parallel they offer an ideal 50 Ω impedance for coaxial feed lines. The loops are fed 90° out of phase with each other and this creates a circularly polarized pattern.

An eggbeater may also use one or more parasitic reflector elements beneath the loops to focus more of the radiation pattern upward.

Space Communications 30.11

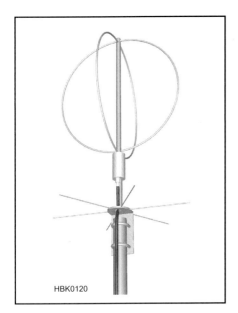

Fig 30.11 — The eggbeater antenna is a popular design named after the old-fashioned kitchen utensil it resembles. The antenna is composed of two full-wave loops of rigid wire or metal tubing.

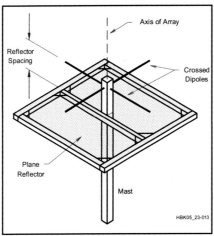

Fig 30.12 — The basic turnstile antenna consists of two horizontal half-wave dipoles mounted at right angles to each other (like the letter "X") in the same horizontal plane with a reflector screen beneath.

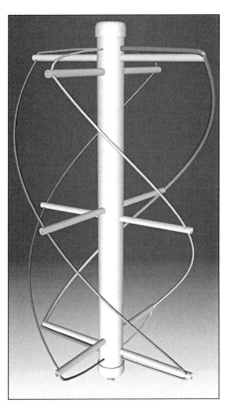

Fig 30.14 — The quadrifilar helicoidal antenna is comprised of four equal-length conductors (filars) wound in the form of a corkscrew (helix) and fed in quadrature.

This effect makes it a "gain" antenna, but that gain is at the expense of low-elevation reception. Toward the horizon an eggbeater is actually horizontally polarized. As the pattern rises in elevation, it becomes more and more right-hand circularly polarized. Experience has shown that eggbeaters seem to perform best when reflector elements are installed just below the loops.

Eggbeaters can be built relatively easily, but there are also a couple of commercial models available (see the advertising pages of *QST* magazine). The spherical shape of the eggbeater creates a fairly compact antenna when space is an issue, which is another reason why it is an attractive design.

THE TURNSTILE

The basic *turnstile* antenna consists of two horizontal half-wave dipoles mounted at right angles to each other (like the letter "X") in the same horizontal plane with a reflector screen beneath (**Fig 30.12**). When these two antennas are excited with equal currents 90° out of phase, their typical figure-eight patterns merge to produce a nearly circular pattern.

In order to get the radiation pattern in the upward direction for space communications, the turnstile antenna needs a reflector underneath. For a broad pattern it is best to maintain a distance of ⅜ λ at the operating frequency between the reflector and the turnstile. Homebrewed turnstile reflectors often use metal window-screen material that you can pick up at many hardware stores. (Make sure it is a metal, not plastic, screen material.)

Like their cousins the eggbeaters, turnstiles are relatively easy to build. In fact, building one may be your only choice since turnstiles

Fig 30.13 — The EZ Lindenblad antenna designed by Anthony Monteiro, AA2TX.

are rarely available off the shelf.

THE LINDENBLAD ANTENNA

In a *Lindenblad* antenna (**Fig 30.13**), each dipole element is attached to a section of shorted open-wire-line, also made from tubing, which serves as a balun transformer. A coaxial cable runs through one side of each open-wire line to feed each dipole. The four coaxial feed cables meet at a center hub section where they are connected in parallel to provide a four-way, in-phase power-splitting function. This cable junction is connected to another section of coaxial cable that serves as an impedance matching section to get a good match to 50 Ω. An innovative solution, the EZ-Lindenblad, is described later in this chapter. While certainly more elaborate than an eggbeater or turnstile, the Lindenblad creates a uniform circularly polarized pattern that is highly effective for satellite applications.

THE QUADRIFILAR HELICOIDAL ANTENNA

The quadrifilar helicoidal antenna (QHA) shown in **Fig 30.14** ranks among the best of the omnidirectional satellite antennas. It is comprised of four equal-length conductors (filars) wound in the form of a corkscrew (helix) and fed in quadrature. The result is a nearly perfect circularly polarized pattern.

QHAs can be challenging to build since the filar lengths and spacing have to be precise. Even so, homebrewing a QHA can save you a substantial amount of money. This antenna is available off the shelf (they are favorites for maritime satellite links), but they can be costly. A version that can be built from readily available materials is described later in this chapter.

30.4.3 Directional Antennas

Gain and directivity are important antenna factors. These factors are maximized in *directional* antennas. In fact, the design goal

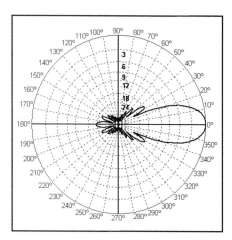

Fig 30.15 — The design goal of a directional antenna is to create a highly directional pattern along its axis. This example shows the radiation pattern of a high-gain Yagi antenna.

translates to an antenna that must be aimed at a satellite with a fair degree of accuracy. This can be a challenge when your target is a rapidly moving LEO bird.

The Yagi antennas you normally see are *linear* designs that are mounted in either horizontal or vertically polarized configurations. These same antennas can be successfully used for satellite work just as they are. However, there are ways you can optimize them to increase their effective performance.

The dipole elements of Yagi antennas radiate linearly polarized signals, but remember that the polarization direction really depends on the orientation of the antenna to the Earth. If two Yagi antennas are mounted on the same support boom, arranged for horizontal and vertical polarization, and combined with the correct phase difference (90°), a circularly polarized wave results. Because the electric fields of the antennas are identical in magnitude, the power from the transmitter will

of a directional antenna is to create a highly directional pattern along its axis (**Fig 30.15**). An ordinary flashlight is a reasonable analog for a directional antenna, although the pattern of a directional antenna isn't as well focused as a flashlight beam (antennas with parabolic reflectors come fairly close, though).

The chief advantage of a directional antenna is its considerable directivity and gain. When you are working with weak signals from spacecraft, you need all the gain you can get. Directional antennas are mandatory for high-altitude satellites when they are at apogee, nearly 50,000 km distant. They also are excellent for LEO birds, providing strong, consistent signals that omnidirectional antennas can rarely match.

But, a major disadvantage of a directional antenna is *also* its directivity! To achieve best results with a directional antenna, you must find a way to point it at the satellite you wish to work. This entails pointing it by hand, or by using an antenna rotator. If you were to simply leave a directional antenna fixed in one place, you would enjoy good signals only during the brief moments when satellites passed through the antenna's pattern. A rotator adds significant cost to a ground station and installing one isn't a trivial exercise. On the other hand, there is a way to reduce rotator cost, which we'll discuss later.

If you can afford a directional antenna and rotator system, you'll never regret the investment. When properly installed, it is vastly superior to any omnidirectional antenna you are likely to encounter.

YAGI ANTENNAS

The familiar Yagi antennas used for VHF+ terrestrial operation can be used for satellite applications as well. However, keep in mind that the greater the directivity, the more focused (narrow) the antenna pattern. This

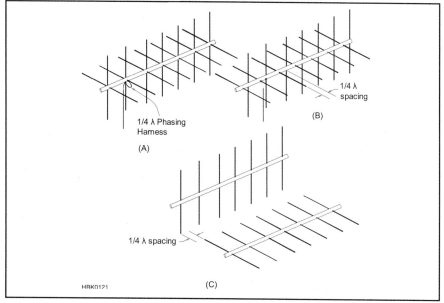

Fig 30.16 — Evolution of the circularly polarized Yagi. The simplest form of crossed Yagi, A, is made to radiate circularly by feeding the two driven elements 90° out of phase. Antenna B has the driven elements fed in phase, but has the elements of one bay mounted ¼ λ forward from those of the other. Antenna C offers elliptical (circular) polarization using separate booms. The elements in one set are mounted perpendicular to those of the other and are spaced ¼ λ forward from those of the other.

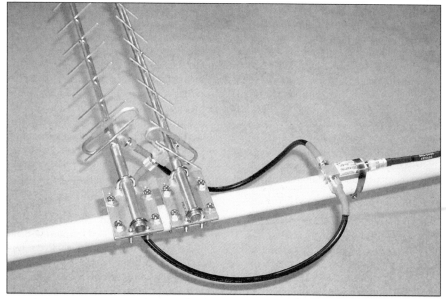

Fig 30.17 — An example of offset crossed-Yagi circularly polarized antennas with fixed polarization. This example shows a pair of Yagis for L band (1296 MHz) mounted on an elevation boom. (*KD1K photo*)

Space Communications 30.13

be equally divided between the two fields. Another way of looking at this is to consider the power as being divided between the two antennas; hence the gain of each is decreased by 3 dB when taken alone in the plane of its orientation. This design is known as a *crossed Yagi*.

A 90° phase shift must exist between the two antennas and the simplest way to obtain this shift is to use two feed lines. One feedline section is ¼ λ longer than the other, as shown in **Fig 30.16**. These separate feed lines are then paralleled to a common transmission line to the station. However, therein lies one of the headaches of this system. Assuming negligible coupling between the crossed antennas, the impedance presented to the common transmission line by the parallel combination is one half that of either section alone. (This is not true when there is mutual coupling between the antennas, as in phased arrays.) **Fig 30.17** shows a practical example of offset crossed Yagis for the 23 cm band.

So far we've been discussing single-band crossed Yagi designs to achieve circular polarization, or something close to it. If circular polarization isn't a top criterion, you may want to consider a *dual-band Yagi* that places linear Yagi antennas for two separate bands on the same support boom. While this design doesn't produce circular polarization, it partially makes up for that disadvantage in convenience and economics. The most common design combines antennas for 2 meters and 70 cm on the same boom (**Fig 30.18**). You can feed the antennas with two separate feed lines, or use single feed line and a *diplexer* (some manufacturers incorrectly call them *duplexers*) to separate the signals at the antenna, at the radio or at both locations (**Fig 30.19**). With a single 2 meter/70 cm Yagi antenna you can enjoy nearly all available amateur satellites. That single-purchase aspect makes them attractive for budget-conscious hams.

A type of Yagi that you may encounter for higher frequencies (typically 902 MHz and above) is the *loop Yagi* (**Fig 30.20**). In this design, the individual elements are bent into loops and mounted on a common boom. Despite their appearance, loop Yagis are linearly polarized antennas. Their advantage is that they create a substantial amount of gain in a relatively small physical space. Because of the sizes of the loops, however, loop Yagis are most practical and common at microwave frequencies.

HELICAL ANTENNAS

Another method to create a circularly polarized signal is by means of a helical antenna. The axial-mode helical antenna was introduced by Dr John Kraus, W8JK, in the 1940s. **Fig 30.21** shows an example of a microwave helical antenna. A larger helical for 70 cm is shown in **Fig 30.22**.

This antenna has two characteristics that

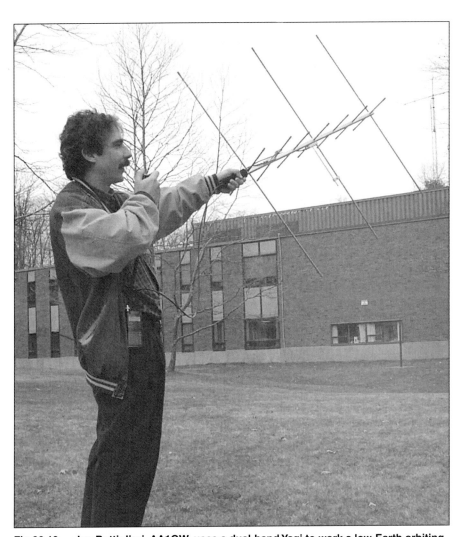

Fig 30.18 — Joe Bottiglieri, AA1GW, uses a dual-band Yagi to work a low-Earth orbiting FM repeater satellite with a handheld transceiver.

make it especially interesting and useful for satellite stations. First, the helix is circularly polarized. As discussed earlier, circular polarization is simply linear polarization that continually rotates as it travels through space. In the case of a helical antenna, this rotation is about the axis of the antenna. This can be pictured as the second hand of a watch moving at the same rate as the applied frequency, where the position of the second hand can be thought of as the instantaneous polarization of the signal. The second interesting property of the helical antenna is its predictable pattern, gain and impedance characteristics over a wide frequency range. This is one of the few antenna designs that have both broad bandwidth and high gain. The benefit of this property is that, when used for narrow-band applications, the helical antenna is very forgiving of mechanical inaccuracies.

Electrically, a helix looks like a large air-wound coil with one of its ends fed against a ground plane. The ground plane often consists of a screen of 0.8 to 1.1 λ diameter (or per side for a square ground plane). The circumference

Fig 30.19 — A diplexer made by the Comet Corporation. Diplexers can be used to separate or combine signals on different bands.

Fig 30.20 — This antenna stack includes a loop Yagi at the very top.

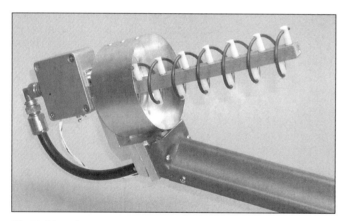

Fig 30.21 — A seven-turn LHCP helical antenna for a 2.4 GHz dish feed. This helical antenna uses a cupped reflector and has a preamplifier mounted directly to the antenna feed point. (*KD1K photo*)

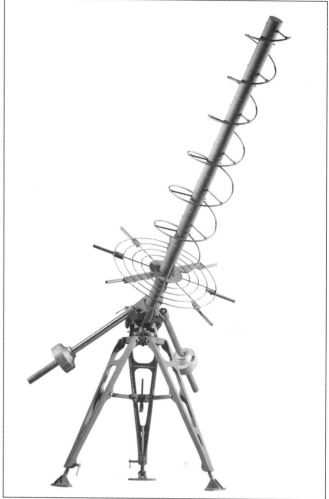

Fig 30.22 — A helical antenna for 70 cm.

of the coil form must be between 0.75 and 1.33 λ for the antenna to radiate in the axial mode. The coil should have at least three turns to radiate in this mode. The ratio of the spacing between turns should be in the range of 0.2126 to 0.2867 λ. The winding of the helix comes away from the cupped reflector with a counterclockwise winding direction for LHCP. A clockwise winding direction yields RHCP.

PARABOLIC DISH ANTENNAS

A number of modern ham satellites now include microwave transponders and this has created a great deal of interest in effective microwave antennas. From the ground station "point of view," antennas for the microwave bands are not only small, they pack a high amount of gain into their compact packages. Among the best high-performance antenna designs for microwave use are those that employ parabolic reflectors (so-called parabolic "dishes") to concentrate the transmitted and received energy.

Like a bulb in a flashlight, a parabolic antenna must have a *feed source* — the radiating and receiving part of the antenna—"looking" into the surface of the dish. Some dishes are designed so that the feed source is mounted directly in front of the dish. This is referred to as a *center-fed dish* (**Fig 30.23**). Other dishes are designed so that the feed source is off to one side, referred to as an *off-center-fed dish*, or just offset fed dish, as shown in **Fig 30.24**. The offset-fed dish may be considered a side section of a center-fed dish. The center-fed dish experiences some signal degradation due to blockage by the feed system, but this is usually an insignificantly small amount. The offset-fed dish is initially more difficult to aim, since the direction of reception is not the center axis, as it is for center-fed dishes.

The dish's parabola can be designed so the *focus point* — the point where the feed source must be — is closer to the surface of the dish, referred to as a *short-focal-length* dish, or further away from the dish's surface, referred to as a *long-focal-length* dish. To determine the exact focal length (F), measure the diameter of the dish (D) and the depth of the dish (d).

$$F = \frac{D^2}{16d}$$

The focal length divided by the diameter of the dish gives the *focal ratio*, commonly shown as f/D. Center-fed dishes usually have short-focal ratios in the range of f/D = 0.3 to 0.45. Offset-fed dishes usually have longer focal lengths, with f/D = 0.45 to 0.80. If you attach two small mirrors to the outer front

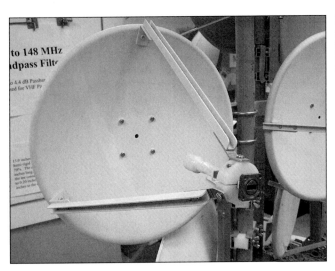

Fig 30.23 — As you can see, the feed source is directly in front, making this a center-fed dish.

Fig 30.24 — This is a good example of an offset-fed dish. Note the feed source at the bottom edge of the photo, pointing into the center of the parabola. Also note that this is a DirectTV satellite TV antenna that has been "repurposed" for Amateur Radio satellites!

Fig 30.25 — The 2.4 GHz satellite downlink antenna at ARRL Headquarters station W1AW is a so-called "barbeque grill" dish antenna originally designed for the Multichannel Multipoint Distribution Service (MMDS). These inexpensive antennas can be assembled and installed in minutes.

surface of a dish and then point the dish at the sun, you can easily find the focus point of the dish. This is where you want to place the feed source.

To invoke the flashlight analogy again, the feed source should evenly illuminate the entire dish, and none of the feed energy should fall outside the dish's reflecting surface. However, *no* feed system is perfect in illuminating a dish. These so-called "spillover losses" affect the gain by either under-illuminating or over-illuminating the dish. Typical dish efficiency is 50%. That's 3 dB of lost gain.

A great feed system for one dish can be a real lemon on another. For example, if you can get your hands on a surplus offset-fed satellite TV dish, you may find a helical radiator is the best feed source. Designs range anywhere from 2 to 6 turns. The two-turn helices are used for very short focal-length dishes in the $f/D = 0.3$ region, and the 6-turn helices are used with typically longer-focal-length (f/D ~ 0.6), offset-fed dishes. Generally speaking, helix feeds work poorly on the short-focal length dishes, but really perform well on the longer-focal length, offset-fed dishes.

Parabolic antennas can be made from commercial or military surplus, including satellite TV dishes that you can "repurpose" in the finest Amateur Radio tradition. You will also find ham dealers selling so-called *barbecue* dishes that were originally designed for the Multichannel Multipoint Distribution Service, or "wireless cable TV" (see **Fig 30.25**). MMDS antennas come with built-in feed sources at the focal points, so all you have to do is connect the feed line and go. These are among the most popular dish antennas for amateur satellite enthusiasts.

30.4.4 Antenna Mounting

One beneficial characteristic of satellite antennas is that they can be installed close to the ground as long as the antenna can "see" as much of the unobstructed sky as possible. Signals to and from the satellite will already be traversing hundreds or thousands of free space. So, adding a few more feet of elevation to your ground station antenna won't perceptibly improve the strength of those signals. What such an arrangement *will* do, however, is needlessly increase the length of coax (and corresponding line losses) between your transceiver and your antenna.

If your antennas already have an unobstructed view of the sky from the ground, you usually won't need a tower or roof installation unless trees or buildings surround the antennas and block their upward view. Or, to put it another way, your antenna support only needs to be high enough to make sure the back end of the antenna array is far enough off the ground to prevent people from walking into it while it's pointing straight up.

What's more, and as we have already discussed, satellite antennas (particularly those of the circularly polarized variety) along with their associated rotators (discussed below) tend to create a more complex antenna arrangement than that used for ordinary HF or VHF/UHF terrestrial operation. So, mounting them close to the ground makes performing any needed adjustments or repairs a whole lot simpler.

30.4.5 Antenna Rotators

If you plan to use directional antennas and don't wish to manipulate them by hand, you will need to install an *antenna rotator* (**Fig 30.26**). Making the correct decision as to how much capacity the rotator must have is very important to ensure trouble-free operation.

Rotator manufacturers generally provide antenna surface area ratings to help you choose a suitable model. The maximum antenna area is linked to the rotator's torque capability. Some rotator manufacturers provide additional information to help you select the right size of rotator for the antennas you plan to use. Hy-Gain provides an *Effective Moment* value. Yaesu calls theirs a *K-Factor*. Both of these ratings are torque values in foot-pounds. You can compute the Effective Moment of your antenna by multiplying the antenna turning radius by its weight. So long as the effective moment rating of the rotator is greater than or equal to the antenna value, the rotator can be expected to provide a useful service life.

There are several rotator grades available to amateurs. The lightest-duty rotator is the type typically used to turn TV antennas. These rotators will handle smaller satellite antennas such as crossed Yagis. The problem with TV rotators is that they lack braking or holding capability. High winds can turn the rotator motor via the gear train in a reverse fashion, requiring realignment. Broken gears sometimes result.

The next grade up from the TV class of rotator usually includes a braking arrangement, whereby the antenna is held in place when power is not applied to the rotator. Generally speaking, the brake prevents antenna misalignment or gear damage on windy days. If adequate precautions are taken, this type of rotator is capable of holding and turning a stack of satellite antennas, including a parabolic dish which, by its nature, presents considerable wind loading. Keep in mind that as rotators increase in power, they become more expensive.

AZIMUTH/ELEVATION ROTATOR

Perhaps the ultimate in operating convenience is the *azimuth/elevation (az/el) rotator*. This rotator is capable of moving your antennas horizontally (azimuth) and vertically (elevation) at the same time. There are well-designed models available from Yaesu (**Fig 30.27**) and Alfa Radio (**Fig 30.28**). You can operate these rotators manually, or connect them to your computer for automated tracking. The downside is that az/el rotators tend to be expensive, typically on the order of $600 or more at this writing.

If your budget can stand the strain, az/el rotators are clearly worth the investment. On the other hand, if cost is an issue, consider using a standard rotator instead. While a traditional rotator can only move your antennas in the azimuth plane (horizontally), you can strike a compromise by installing the antennas at a permanent 45° tilt (**Fig 30.29**). This configuration will allow you to work the vast majority of satellites with reasonable success. You won't be able to follow the satellite when it is overhead or near the horizon, but you'll enjoy the lion's share of every pass.

Regardless of which type you choose,

Fig 30.26 — A basic antenna rotator; the light-duty variety used to turn TV antennas. The wire exiting the bottom of the rotator housing is the multiconductor cable responsible for power and control.

Fig 30.28 — An az/el rotator made by Alfa Radio.

proper installation of the antenna rotator can provide many years of dependable service. Sloppy installation can cause problems such as a burned out motor, slippage, binding and even breakage of the rotator's internal gear and shaft castings or outer housing. Most rotators are capable of accepting mast sizes of different diameters, and suitable precautions must be taken to shim an undersized mast to ensure dead-center rotation.

If you decide to install your rotator on a tower, it is desirable to mount the rotator inside and below the top of the tower as far as possible. A long mast (10 feet or so) absorbs the torsion developed by the antenna during high winds, as well as during starting and

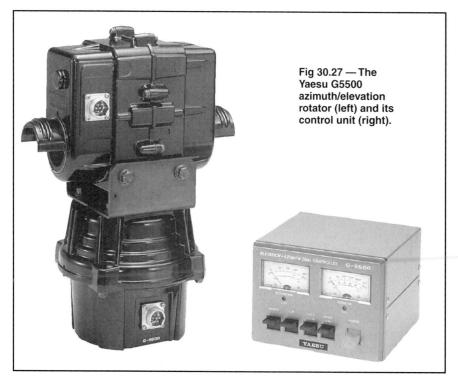

Fig 30.27 — The Yaesu G5500 azimuth/elevation rotator (left) and its control unit (right).

stopping. Another benefit is mitigating the effect of any misalignment among the rotator, mast and the top of the tower.

A tube at the top of the tower (a *sleeve bearing*) through which the mast protrudes almost completely eliminates any lateral forces on the rotator casing. All the rotator must do is support the downward weight of the antenna system and turn the array. Some installations use a *thrust bearing* mounted to the tower to support the weight of the antenna and mast.

Don't forget to provide a loop of coax to allow your antenna to rotate properly and allow water to drip off. Also, make sure you position the rotator loop so that it doesn't snag on anything.

COMPUTER CONTROL

As mentioned earlier, you can connect your rotator to your station computer and allow your satellite-tracking software to aim your antennas automatically (assuming your software supports rotator control). There used to be a number of commercial rotator/computer interface devices available for sale, but availability has dwindled over the years and those that remain tend to be expensive. An interface such as the Yaesu GS-232 (**Fig 30.30**) costs about $600 at the time of this writing. If you combine it with a Yaesu azimuth/elevation rotator, you will have invested $1300 total. Less expensive alternatives are now found as kits or homebrew devices. A good example is the G6LVB tracker interface at **www.g6lvb.com/Articles/LVBTracker/**. If you're willing to build it yourself, you can probably put together an LVB unit for less than $50. Do an Internet search and you'll no doubt uncover other homebrew interfaces.

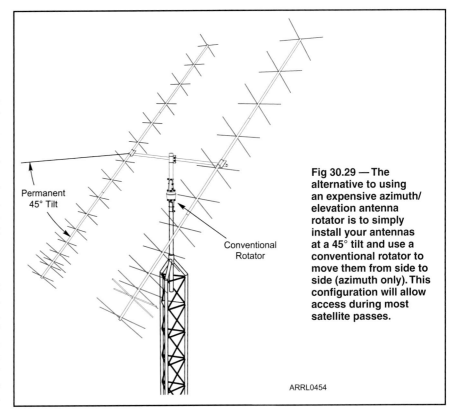

Fig 30.29 — The alternative to using an expensive azimuth/elevation antenna rotator is to simply install your antennas at a 45° tilt and use a conventional rotator to move them from side to side (azimuth only). This configuration will allow access during most satellite passes.

Fig 30.30 — The Yaesu GS-232 allows your computer and satellite tracking software to automatically control the movements of compatible azimuth/elevation antenna rotators.

30.5 Satellite Ground Station Equipment

30.5.1 Receive and Transmit Converters

Signals from satellites can be exquisitely weak, which means they need as much amplification as possible to be readable. Unfortunately, there are a number of factors that may conspire to weaken your radio's ability to render a decent received signal....

• *You're using omnidirectional antennas.* As discussed earlier, omni antennas lack much of the signal-capturing gain of directional antennas.

• *The feed line between the antennas and the radio is long and/or contains "lossy" coax.* Remember that even with the best coax, the longer the feed line the more signal you'll lose, especially at higher frequencies.

• *You're trying to communicate with a high-orbiting satellite at apogee.* When a signal travels up to 50,000 km to reach your station, even the gain of a directional antenna may not be sufficient.

PREAMPLIFIERS

The way to ensure that you have a useable received signal is to install a *receive preamplifier* (**Fig 30.31**) at the antenna. This is a high-gain, low-noise amplifier with a frequency response tailored for one band only.

When shopping for a receive preamplifier, you want the most amount of gain for the least amount of noise. Every preamplifier adds some noise to the system, but you want the least additional noise possible. A well-designed UHF preamplifier, for example, may have gain on the order of 15 to 25 dB and a *noise figure* (NF) of 0.5 to 2 dB (less is better).

If your antennas are outdoors, look for preamplifiers that are "mast mountable." These preamplifiers are housed in weatherproof enclosures.

You will need to devise a means to supply dc power to the remote preamplifier. This can be as simple as routing a two-conductor power cable to the device. Alternatively, preamplifiers can be powered by dc sent up the feed line itself. Some transceivers have the ability to insert 12 V dc on the feed line for this purpose. If not, you can use a *dc power inserter* to inject power at the station and/or recover it nearer the antenna (see **Fig 30.32**). Some preamplifier designs include feed line power capability, so all you need is an inserter at the "station end."

If your preamplifier is going to be installed in a feed line that will also be carrying RF power from the radio, you'll need a model that includes an internal relay to temporarily switch it out of the circuit to avoid damage to the preamplifier when you're transmitting. Some preamplifiers include this relay and nothing more; it is up to you to provide the means to energize the relay before you transmit. This is accomplished through a device known as a *TR sequencer*. A sequencer works with your transceiver to automatically switch the preamplifier out of the feed line before the radio can begin sending RF power (see **Fig 30.33**). A less complicated alternative is to purchase a preamplifier with *RF-sensed switching*. This design incorporates a sensor that detects the presence of RF from the radio and instantly switches the preamplifier out of harm's way. Note that RF-switched preamplifiers are rated according to the power they can safely handle. If you're transmitting 150 W, you'll need an RF-switched preamplifier rated for 150 W or more.

DOWNCONVERTERS

As the name implies, a *downconverter*, also known as a *receive converter*, converts

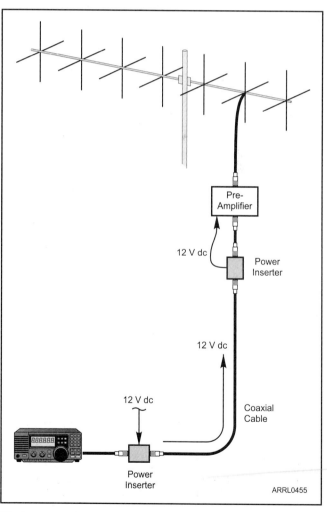

Fig 30.32 — A dc power inserter acts just as its name implies — it inserts a dc voltage to the coaxial cable at the station. The inserter is designed to block dc power from going "backward" to the radio. Instead, the power flows through the coax to the antenna where another inserter picks it off and supplies it to the device (a preamplifier, in this case). Both inserters pass RF with negligible loss.

Fig 30.31 — This receive preamplifier by Advanced Receiver Research gives signals a substantial boost before they travel down the coaxial cable to your radio.

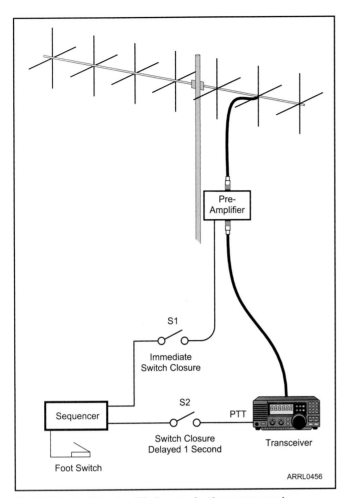

Fig 30.33 — In this simplified example, the sequencer is triggered when the operator presses the foot switch. It immediately closes switch S1, which activates a "bypass" relay in the receive preamplifier at the antenna, effectively removing it from the feed line circuit. One second later, the sequencer closes switch S2, which is connected to the transceiver PTT (Push To Talk) line, keying the radio and applying RF power.

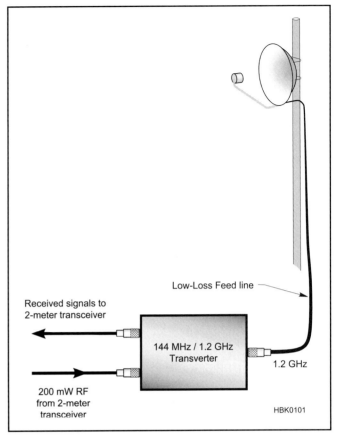

Fig 30.35 — In this example, the transverter is taking 1.2 GHz received signals and converting them to 2 m. When it's time to transmit, the transverter takes RF at 2 m and converts it to 1.2 GHz.

Fig 30.34 — This downconverter by Downeast Microwave converts signals at 2.4 GHz to 2 m. With its weatherproof enclosure, it is designed to be installed at the antenna.

one band of frequencies "down" to another. For example, a 2-to-10-meter downconverter would convert signals in a range from 144 to 146 MHz to 28 to 30 MHz.

In the days before HF/VHF/UHF transceivers, a downconverter was a popular means of receiving VHF and UHF signals by converting them to 10 meters (usually) for reception on an HF receiver. Today downconverters are used more often as a way to receive microwave signals, for example converting a range of frequencies at 2.4 GHz to 2 meters. A microwave downconverter works best when installed right at the antenna so that the microwave energy is immediately converted to a lower frequency before excessive feed line loss can occur. See **Fig 30.34**.

Like receive preamplifiers, downconverters are rated by their gain and NF (the lower the NF, the better). When it comes to installing a downconverter, all the same receive preamplifier issues apply. If the downconverter is installed outdoors it must be in a weatherproof enclosure. You must also supply dc power and be able to switch the downconverter out of the line if you are sending RF power to the same antenna.

TRANSVERTERS

Transverters are related to downconverters in that they not only convert received signals, they convert transmitted signals as well. You can use a transverter to generate, say, a 1.2 GHz uplink signal when powered by RF energy from a 2 meter transceiver. The same transverter can convert 1.2 GHz downlink signals to 2 meters as well (**Fig 30.35**).

Transmit/receive switching in most transverters is accomplished through the use of an internal switch that is keyed through a sequencer. Some models provide automatic RF-sensed switching.

When working with transverters, one issue is supplying a safe level of RF power to the input. If your transceiver pumps out 50 W of power at 2 meters, for example, this is way too much RF for most transverters to handle. Unless the transverter has a built-in RF power attenuator, it is designed to deal with RF power levels on the order of *milliwatts* (typically 200 to 300 mW). If you are lucky enough to own a transceiver that features a transmit transverter port, you can obtain your milliwatt power levels there. If not, you'll need to add an attenuator at the transverter input to reduce the RF output of your radio to safe levels.

30.5.2 Transceivers

There are many amateur transceivers that cover the VHF and UHF bands. Some popular radios offer all the HF bands as well. Some of these transceivers offer satellite-specific features, while others are oriented toward terrestrial operation.

WORKING WITH FM REPEATER AND DIGITAL SATELLITES

Almost any dual-band (2 meter/70 cm) FM transceiver will be adequate for operating FM repeater satellites such as OSCAR 51, and for digital operating with the International Space Station or LEO orbiting birds as well (**Fig 30.36**). Getting started with a dual-band FM radio and a repeater satellite is a worthwhile option.

Most modern dual-band rigs offer a "high power" output setting around 50 W. That's more than enough power to put a solid signal into a satellite with a directional antenna. It is also sufficient for omnidirectional antennas, including mobile antennas. (Yes, you can make contacts through FM repeater satellites while you drive!)

If you are considering digital operation, make sure to choose an FM transceiver that offers a data port. This will make it much easier to connect an external radio modem, such as a packet radio terminal node controller (TNC). There are even a few radios with TNCs already built in (**Fig 30.37**). Also, make sure the transceiver is rated to handle 1200 and 9600 baud data signals. Nearly all FM transceivers can work with 1200 baud, but 9600 baud is another matter. When in doubt, check the *QST* magazine Product Reviews. Look for the bit-error-rate (BER) test results.

You can use dual-band handheld transceivers to work the FM birds, but their RF output is so low (5 W or less) that you will definitely need to couple them to directional antennas to be heard consistently among the competing signals (**Fig 30.38**). What's more, at this writing, most of the newer handhelds will *not* allow so-called true (or "full") duplex operation. We will discuss what full-duplex means and why it is desirable for satellite work in a moment.

WORKING WITH LINEAR TRANSPONDER (SSB/CW) SATELLITES

FM signals tend to be wide and, by design, FM receivers are forgiving of frequency changes. That fortunate characteristic makes it easy to compensate for Doppler frequency shifting as an FM repeater satellite zips overhead.

SSB and CW signals are much narrower, however, and when you're working through a linear transponder satellite your signal is sharing the passband with several others. Not only do you need to adjust your receive (downlink) frequency almost continuously to keep the SSB voice or CW sounding "normal," you also have to stay on frequency to avoid drifting into someone else's conversation. The most

Fig 30.36 — The ICOM IC-2820 FM transceiver can receive and transmit on 2 m and 70 cm independently. With a dual-band FM rig like this, you can communicate through the FM repeater satellites, even while mobile.

Fig 30.37 — The Kenwood TM-D710 is a dual-band FM transceiver with a built-in packet radio Terminal Node Controller (TNC) for digital communication.

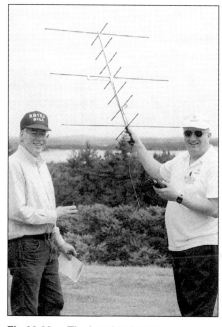

Fig 30.38 — The hand-held "Arrow" gain antenna is popular for FM repeater satellite operations. (*AMSAT photo*)

effective way to do this is to listen to your own signal coming through the satellite in real time while you are transmitting on the uplink. This type of operation is known as *full duplex*.

You will find many multimode (SSB, CW, FM) transceivers that boast a feature labeled "cross-band split" or even "cross-band duplex." Be careful, though. What you require is a radio that can transmit and receive on different bands *simultaneously*. Few amateur transceivers can manage such a trick!

As this book went to press, there were only two all-mode amateur transceivers on the market that were capable of full duplex operation: The VHF/UHF ICOM IC-9100 (**Fig 30.39**) and the all-band Kenwood TS-2000 (**Fig 30.40**). However, if you shop the used equipment market you'll find excellent satellite radios such as the Yaesu FT-736 (**Fig 30.41**), the ICOM IC-820 and IC-910H, the Yaesu FT-726 and FT-847 along with the Kenwood TS-790. All of these transceivers have full-duplex capability.

Another option that some hams have used successfully is to exploit computer control of their ordinary non-duplex multiband transceivers to compensate for Doppler shift. Some satellite-tracking programs can automatically change the frequency of your radio during a pass. They do this by mathematically calculating the frequency shift and tweaking your radio accordingly. This solution for frequency stability isn't as accurate as listening to your own downlink signal in full duplex, but it can work.

Another option is to use one transceiver for the uplink and a separate receiver or transceiver for the downlink. Some amateurs have even pressed old shortwave receivers into ser-

Fig 30.39 — The ICOM IC-9100 is an HF/VHF/UHF transceiver with 23 cm available as an option.

Fig 30.40 — The Kenwood TS-2000 offers full HF coverage, plus VHF and UHF with full-duplex satellite capability.

Fig 30.41 — The Yaesu FT-736 is a legendary satellite transceiver. Although no longer manufactured, the '736 is still available on the used market.

Fig 30.42 — A 150 W 2 meter RF power amplifier by Tokyo Hy-Power.

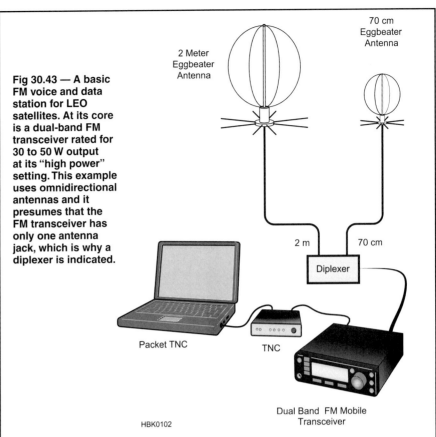

Fig 30.43 — A basic FM voice and data station for LEO satellites. At its core is a dual-band FM transceiver rated for 30 to 50 W output at its "high power" setting. This example uses omnidirectional antennas and it presumes that the FM transceiver has only one antenna jack, which is why a diplexer is indicated.

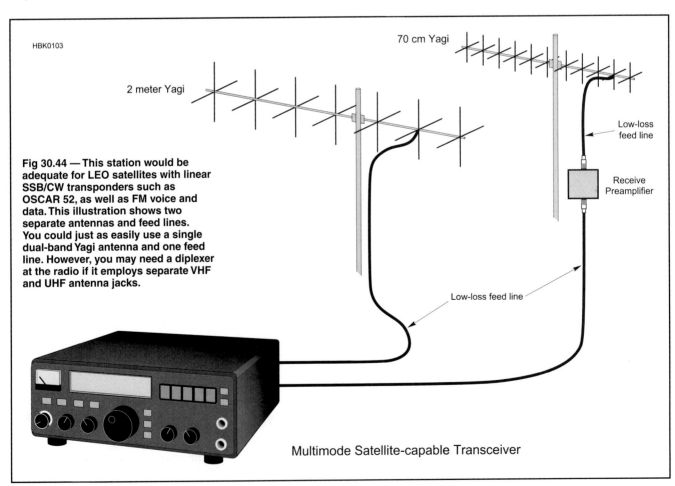

Fig 30.44 — This station would be adequate for LEO satellites with linear SSB/CW transponders such as OSCAR 52, as well as FM voice and data. This illustration shows two separate antennas and feed lines. You could just as easily use a single dual-band Yagi antenna and one feed line. However, you may need a diplexer at the radio if it employs separate VHF and UHF antenna jacks.

vice along with VHF or UHF downconverters. They transmit with a 2 m or 70 cm SSB radio on the uplink while listening to their shortwave receiver/downconverter combo on the downlink.

If you are working a linear transponder satellite with an uplink at 1.2 GHz or higher, you'll likely have to take the transverter approach to getting on the air, although there are some transceivers that offer optional 1.2 GHz modules. Fortunately, satellite builders are well aware that most hams own 2 meter and 70 cm transceivers and they design new birds accordingly. Even transponders with microwave downlinks are usually configured with uplinks on 2 meter or 70 cm. So, if you purchase a VHF/UHF transceiver all you need to add is a downconverter to receive the microwave downlink.

30.5.3 VHF/UHF RF Power Amplifiers

If your chosen transceiver offers at least 50 W output on the uplink band, you won't need an RF power amplifier to bring your signal to a level that can be "heard" by a LEO satellite, especially if you are using directional antennas.

On the other hand, if you are using omni antennas, 100 or 150 W output may help considerably. And if your target is an HEO satellite orbiting at 50,000 km, 100 W or more, along with a directional antenna, is *mandatory*. If your transceiver lacks the necessary punch for the application, the solution is an external RF power amplifier.

How much power should you buy? In most cases, a 100 or 150 W amplifier is a good choice (**Fig 30.42**). As you shop for amplifiers, take care to note the input and output specifications. How much RF at the input is necessary to produce, say, 150 W at the output? Can your radio supply that much power?

Another consideration is your dc power supply. While a 25 A 13.8 V dc supply is perfectly adequate to run a 100 W transceiver, if you also decide to add a 100 or 150 W amplifier to your satellite station, the current demands will increase considerably. A separate power supply may be required to provide an *additional* 20 A (or more) to safely power the amplifier when both the transceiver *and* the amplifier are transmitting at the same time.

30.5.4 Typical Station Designs

There are so many station equipment options available, it may be helpful to outline a few typical station configurations.

Fig 30.43 illustrates a basic FM voice and data station for LEO satellites. At its core is a dual-band FM transceiver rated for 30 to 50 W output. The example uses omnidirectional antennas and it presumes that the FM

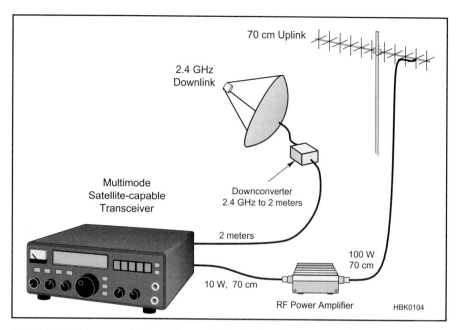

Fig 30.45 — This example illustrates a station designed to work a distant Phase III or IV satellite with a microwave downlink. It assumes that the uplink is at 70 cm and the downlink is at 2.4 GHz. Note the use of the downconverter to step the 2.4 GHz signal at the antenna down to 2 meters before it is fed to the radio.

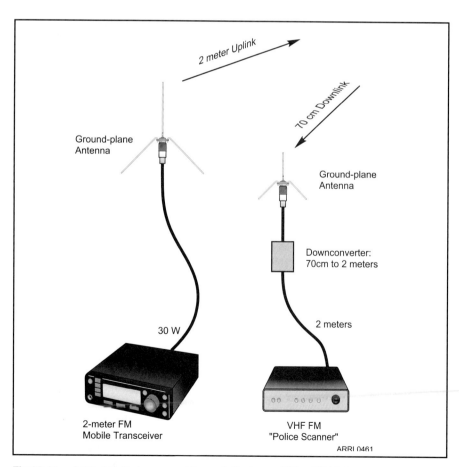

Fig 30.46 — Let's say that you want to work the low-orbiting FM birds, but you only have a 2 meter FM transceiver for the uplink and only a VHF FM "police scanner" receiver for the 70 cm downlink. You could use a downconverter for the 70 cm downlink, stepping the signal down to 2 meters for reception with the scanner.

transceiver has only one antenna jack, which is why a diplexer is indicated. Otherwise, you could run two separate coaxial feed lines back to the radio.

The performance of this station can be substantially enhanced by adding directional antennas or a dual-band Yagi but, of course, that will require an antenna rotator (human or mechanical!).

In **Fig 30.44** we've stepped up to a multiband, multimode transceiver with full duplex capability and directional antennas. This station would be excellent for LEO satellites with linear SSB/CW transponders such as OSCAR 52, as well as FM voice and data. By adding an RF power amplifier (assuming the output of the radio is less than 100 W), this station would be capable of working distant high-orbiting satellites as well.

Note that this illustration shows two separate antennas and feed lines. You could just as easily use a single dual-band Yagi antenna and one feed line. However, you may need a diplexer at the radio if it employs separate VHF and UHF antenna jacks.

Fig 30.45 addresses at least one method to work satellites with microwave downlinks. It assumes that the uplink is at 70 cm and the downlink is at 2.4 GHz. It also assumes that the satellite in question is a high-orbiting spacecraft (note the RF power amp on the uplink).

Note also the use of the downconverter to step the 2.4 GHz signal at the antenna to 2 meters before it is fed to the radio. You can actually use this same approach for any combination of bands and radios. For instance, let's say that you want to work the low-orbiting FM birds, but you have only a 2 meter FM transceiver for the uplink and a separate VHF FM police scanner receiver for the downlink. You could use a downconverter for the 70 cm downlink, stepping the signal down to 2 meters for reception with the scanner (**Fig 30.46**).

30.6 Satellite Antenna Projects

30.6.1 An EZ-Lindenblad Antenna for 2 Meters

This easy-to-build antenna project by Anthony Monteiro, AA2TX, works well for satellite or terrestrial communication. It is circularly polarized yet has an omnidirectional radiation pattern (no rotator needed to track satellites). With most of its gain at low elevation angles, it is ideal for accessing LEO amateur satellites. It is good for portable operation or general-purpose use at home stations because its circular polarization is compatible with the linearly polarized antennas used for FM/repeater and SSB or CW operation.

This type of antenna was devised by Nils Lindenblad of the Radio Corporation of America (RCA) around 1940. His idea was to employ four dipoles spaced equally around a λ/3 diameter circle with each dipole canted 30° from the horizontal. The dipoles are all fed in phase and are fed equal power. The spacing and tilt angles of the dipoles create the desired antenna pattern when the signals are all combined. After WWII, George Brown and Oakley Woodward, also of RCA, were tasked with finding ways to reduce fading on ground-to-air radio links at airports and built on Lindenblad's design.[1]

While the Brown and Woodward design is clever and worked well, it would be difficult for the home builder to duplicate because of the need for the four-way, in-phase, power splitting function. Since amateurs generally want to use 50 Ω coaxial cable feed line, we have to somehow provide an impedance match from the 50 Ω unbalanced coax to the four 75 Ω balanced dipole loads. Previous designs have used combinations of folded dipoles, open-wire lines, twin-lead feeds, balun transformers and special impedance matching cables to try to get a good match to 50 Ω. These in turn increase the complexity and difficulty of the construction.

THE EZ-LINDENBLAD

The key concept of the EZ-Lindenblad is to eliminate anything electrically or mechanically difficult. This leads to the idea of just feeding the four dipoles with coax cable

[1] G. Brown and O. Woodward Jr. *Circularly Polarized Omnidirectional Antenna*, RCA Review, Vol 8, no. 2, Jun 1947, pp 259-269.

Fig 30.47 — View of cross booms mounted to mast.

and soldering the cables to a connector with no impedance matching devices at all. This would certainly be *easy* but we also want the antenna to work! Without the extra impedance matching devices, how is it possible to get a good match to 50 Ω?

If we could get each of the four coax feed cables to look like 200 Ω at the connector, then the four in parallel would provide a perfect match to 50 Ω. We could do this if we used ¼ λ sections of 122 Ω coax to convert each 75 Ω dipole load to 200 Ω. Unfortunately, there is no such coax that is readily available.

But we can accomplish the same thing with ordinary 75 Ω, RG-59 coax if we run the cable with an intentional impedance mismatch. By forcing the standing wave ratio (SWR) on the cable to be equal to 200/75, or about 2.7:1, we can make each cable look exactly like 200 Ω at the connector as long as we make them the right length. It is easy to make the SWR equal 2.7:1 by just making the dipoles a little too short for resonance. An *EZNEC* (**www.eznec.com**) antenna model can be used to determine the exact dipole dimensions.

The conversion from the balanced dipole load to unbalanced coax cable can be accomplished by threading each cable through a ferrite sleeve making a choke balun. The only remaining issue is the required length of the feed cables. With a Smith Chart or software, we can easily determine the required length of 75 Ω coax to provide a 200 Ω load. An *EZNEC* antenna model was used to simulate cutting the dipole lengths until the SWR on the line reached 2.7:1. The model showed that the dipole load impedance would then be $49 - j55$ Ω. Plotting that value and the desired 200 Ω impedance at the connector on the chart and drawing a constant 2.7:1 SWR curve between the two impedance points, the length of the line needed is 0.374 λ. (For more information about this process, see **www.arrl.org/smith-chart** or the *TLW* software provided with *The ARRL Antenna Book*.)

The EZ-Lindenblad was designed for a center frequency of 145.9 MHz to optimize its performance in the satellite sub band. At 145.9 MHz, a wavelength is about 81 inches and since the coax used has a velocity factor of 0.78, we need to make the feed cables $81 \times 0.374 \times 0.78 = 23.6$ inches long.

It's imperative that stranded RG-59A foam PE dielectric coax be used. The solid-dielectric cable has a velocity factor of 66% and will not work. The author used part #RG59A-100 from **www.l-com.com**. This is a 100 ft roll; for one antenna you could buy L-Com part # CCF59-12, a 12 ft jumper, and cut it to the required length. Belden 9259 is another suitable cable.

CONSTRUCTION

This antenna was designed to be rugged and reliable yet easy to build using only hand tools with all of the parts readily available as well (see **Table 30.5**). Although not critical, the construction will be easier if the specified 17 gauge aluminum tubing is used since the inner wall of the tubes will be just slightly smaller than the outer wall of the PVC insert Ts used to connect them. If heavier gauge tubing is used, it will be necessary to file down the PVC insert Ts to make them fit inside the aluminum tubes.

Start by making a mounting bracket to mount the N-connector and the cross booms. Cut a ⅝ inch hole in one side of the short piece of angle stock and rivet or screw it to the bottom of the long piece of angle stock. The completed bracket with the connector and cables attached can be seen in **Fig 30.47**.

Next, cut the aluminum tubing to make the cross booms and dipole rods as shown in **Fig 30.48**. Drill holes for the sheet metal screws at each end of the cross booms but do not insert the screws yet. Attach the cross booms to the long section of angle stock with rivets or screws. One cross boom will mount just above the other as can be seen in Fig 30.47. The cross booms should be perpendicular to the mounting bracket so that they will be horizontal when the antenna is mounted to its mast. Make sure that the centers of the cross booms are aligned with each other so

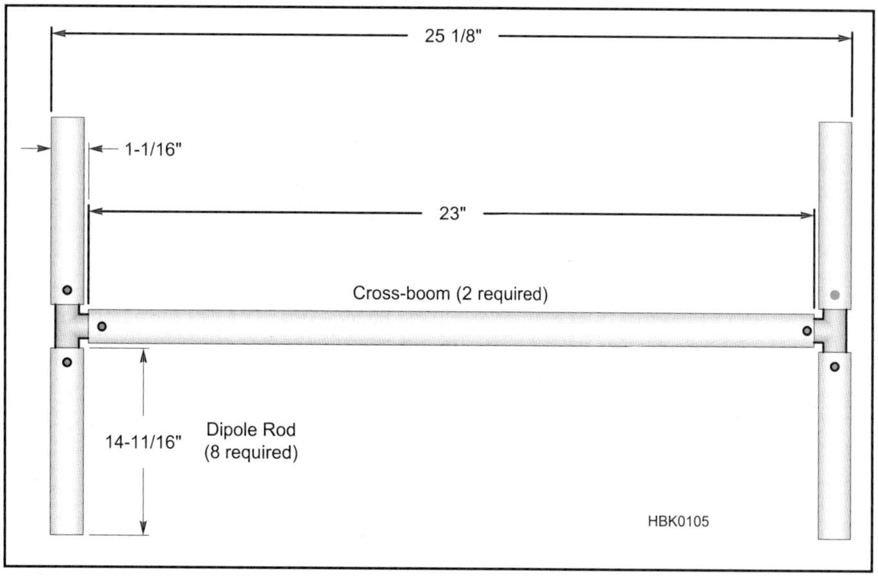

Fig 30.48 — Dimensions of the dipoles and cross booms. The antenna requires two cross booms, with a dipole at each end of each cross boom (four dipoles total).

Table 30.5
Required Materials

Quantity one, unless noted.

- Aluminum tubing, 17 gauge, 6 ft length of ¾ in OD, quantity 3. Available from Texas Towers, **www.texastowers.com**.
- Aluminum angle stock, 8 in length of 2 × 2 × 1⁄16 in
- Aluminum angle stock, 2 in length of 2 × 2 × 1⁄16 in for mounting connector.
- Screws, #8 × ½ in aluminum sheet metal, quantity 12.
- Screws, #8 × ½ in aluminum sheet metal or 3⁄16 in aluminum rivets, quantity 12.
- PVC insert T-connector, ½ × ½ × ½ in grey for irrigation polyethylene tubing. LASCO Fittings, Inc. Part# 1401-005 or equivalent. Available from most plumbing supply and major hardware stores, quantity 4.
- Plastic end caps (optional), black ¾ in, quantity 8.
- N-connector for RG-8 cable, single-hole, chassis-mount, female.
- Cable ferrite, Fair-Rite part #2643540002, quantity 4 (Mouser Electronics #623-2643540002).
- RG-59A polyethylene foam coax with stranded center conductor (Belden 9259 or equiv; *must* be foam PE dielectric with 78% velocity factor, not solid dielectric), 10 ft length.
- Copper braid, 4 in long piece.
- Ring terminal, uninsulated 22-18 gauge for 8-10 stud, quantity 4.
- Ring terminal, uninsulated 12-10 gauge for 8-10 stud, quantity 4.
- Heat shrink tubing for ¼ in cable, wire ties, electrical tape, as needed.
- Ox-Gard OX-100 grease for aluminum electrical connections.

that the ends of the cross booms are all 11.5 inches from the center cross.

Make the dipoles by inserting a PVC insert T into two dipole rods. It should be possible to gently tap in the rods with a hammer but it may be necessary to file down the insert T a little if the fit is too tight. Applying a little PVC cement to the insert T will soften the plastic and make it easier to insert into the aluminum tubing if the fit is too tight. The overall dipole length dimension is critical so take care to get this correct as shown in **Fig 30.49**.

Drill holes for screws in each dipole rod but do not insert the screws yet. The screws will be used to make the electrical connections to the dipoles at the center. The screw holes should be about ⅜ inches from the end of the tubing.

The dipole assemblies are attached by gently tapping the PVC insert-T into the end of each cross-boom with a hammer. The dimensions are shown in Fig 30.48.

Next, temporarily attach the mounting bracket to a support so that each of the cross booms is perfectly horizontal. Measure this with a protractor. Now, using the protractor, rotate the dipole assemblies to a 30° angle with the right-hand side of the nearest dipole tilting up when you are looking toward the center of the antenna. Drill a small hole through the existing cross-boom holes into the PVC insert-Ts and then use the sheet metal screws to fasten the dipole assemblies into place. For a nice finishing touch, the dipole ends can be fitted with ¾ inch black plastic end caps.

Next, make the four feed cables by cutting and stripping the RG-59A as shown in **Fig 30.50**. On the dipole connection side, unwrap the braid and form a wire lead. Apply the smaller ring terminal to the center conductor and use the larger ring terminal for the braid. At the other end of the cable, do not unwrap the braid but strip off the outer insulation. Slip a 1 inch piece of shrink wrap over the coax and apply to the dipole side. Next slip a cable ferrite over the cable and push all the way to the dipole end as far as it will go (ie,

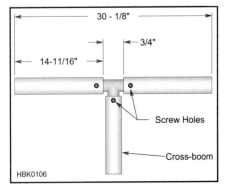

Fig 30.49 — Dipole assembly dimensions.

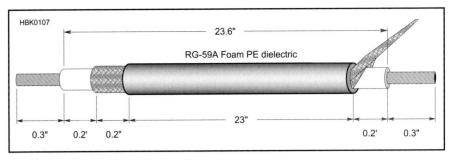

Fig 30.50 — Feed cables stripping dimensions.

Fig 30.51 — Close-up of dipole electrical connections.

up to the heat-shrink tubing.) The fit will be snug and you may need to put a little grease on the cable jacket to get it started.

Prepare each dipole for its feed cable by first cleaning the area around the screw holes with steel wool and then applying Oxy-Gard or other corrosion resistant electrical grease. The coax center conductor goes to the up side of the dipole and the braid goes to the down side. To make a connection, put a screw through the ring terminal and gently screw into the dipole tubing. Do not overtighten the screws or you will strip the tubing. **Fig 30.51** shows the completed connections.

Apply Oxy-Gard around the hole for the N connector. Take the 4 inch piece of braid and put the end of it through the hole for the N connector. This provides the ground connection. Secure the N connector in the mounting hole to clamp the braid. Use a wire tie or tape to hold the four feed cables together at the connector ends. Make sure to align the cables so that all the ground braids are together and the center conductors all extend out the same amount. Do not twist the center conductors together. Carefully push the four cable center conductors into the center terminal of the N connector and solder them in place. Wrap the exposed center conductors of the cables and the connector with electrical tape.

Take the piece of braid that is clamped to the N connector and wrap it around the four exposed ground braids of the coax cables.

Fig 30.52 — The EZ-Lindenblad as a portable or Field Day antenna. A 70 cm antenna is on the top.

Solder them all together. This will take a fair amount of heat but be careful not to melt the insulation. After this cools, apply electrical tape over all the exposed braid and fix with wire ties. Secure the cables to the cross booms with wire ties.

The mounting bracket provides a way to attach the antenna to a mast using whatever clamping mechanism is convenient (eg, U-bolts). The author's antenna was intended for portable operation and the bracket was drilled to accept two #8 machine screws. These screws pass through a portable mast and the antenna is secured with wing nuts for easy setup. The completed portable antenna

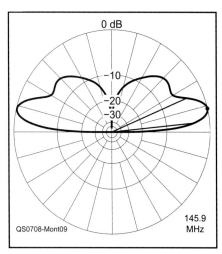

Fig 30.53 — *EZNEC* elevation radiation pattern of Lindenblad antenna.

is shown in **Fig 30.52**. The little antenna at the top is for 70 cm.

PERFORMANCE

The antenna impedance match to 50 Ω was tested using an MFJ-259B SWR meter, which was checked against an external frequency counter and precision 50 Ω load. The antenna provides an excellent match over the entire 2 meter band. This antenna was designed to safely handle any of the currently available VHF transceivers and tested by applying a 200 W signal key down for 9 minutes, then checking the ferrites and cables for temperature rise.

The antenna radiation pattern predicted by the *EZNEC* model is shown in **Fig 30.53**. This is the elevation plot with the antenna mounted at 6 ft above ground although it can be mounted higher if desired for better coverage to the horizon. As shown in the plot, the pattern favors the lower elevation angles. The −3 dB points are at 5° and 25° with the maximum gain of 4.8 dBic (dB with respect to an isotropic circularly polarized antenna) at around 13°. Most of the satellite pass elevations will be in this range and it is also the elevation at which the satellite provides the best chance for DX contacts. The antenna radiation is right-hand circularly polarized, which will work with virtually any LEO satellite that uses the 2 meter band.

The EZ-Lindenblad antenna has been used for SSB, FM and packet operation on a number of amateur satellites. A portable setup performed well on Field Day, an excellent test of any antenna as it is probably the busiest weekend of the year on the satellites.

30.6.2 The W3KH Quadrifilar Helix

If your existing VHF omnidirectional antenna coverage is "just okay," this twisted antenna project by Eugene F. Ruperto, W3KH, is probably just what you need! The ever-changing position of LEO satellites presents a problem for the Earth station equipped with a fixed receiving antenna: signal fading caused by the orientation of the propagated wavefront. This antenna provides a solution to the problem and can be used with weather satellites, or any of the polar-orbiting amateur satellites.

Several magazines have published articles on the construction of the quadrifilar helix antenna (QHA) originally developed by Dr Kilgus.[1] A particularly good reference is *Reflections* by Walt Maxwell, W2DU, who had considerable experience evaluating and testing this antenna while employed as an engineer for RCA.[2]

Part of the problem of replicating the antenna lies in its geometry. The QHA is difficult to describe and photograph. Some of the artist's renditions leave more questions than answers, and some connections between elements as shown conflicted with previously published data. However, those who have successfully constructed the antenna say it is *the* single-antenna answer to satellite reception for the low-Earth-orbiting satellites.

DESIGN CONSIDERATIONS

Experts imply that sophisticated equipment is necessary to adjust and test the antenna, but the author found it possible to construct successful QHAs by following a cookbook approach using scaled figures from a proven design. The data used as the design basis for the antenna described here were published in an article describing the design of a pair of circularly polarized S-band communication-satellite antennas for the Air Force and designed to be spacecraft mounted.[3] Using this antenna as a model, the author constructed QHAs for the weather-satellite frequencies and the polar-orbiting 2 meter and 70 cm amateur satellites with excellent results and without the need for adjustments and tuning. By following some prescribed universal calculations, a reproducible and satisfactory antenna can be built using simple tools.

UHF and microwave antennas require a high degree of constructional precision because of the antenna's small size. For instance, the antenna used for the Air Force at 2.2 GHz has a diameter of 0.92 inch and a length of 1.39 inches! On the other hand, a QHA for 137.5 MHz is 22.4 inches long and almost 15 inches in diameter; for 2 meters, the antenna is not much smaller. Antennas of this size are not difficult to duplicate.

ELECTRICAL CHARACTERISTICS

A half-turn ½ λ QHA has a theoretical gain of 5 dBi and a 3-dB beamwidth of about 115°, with a characteristic impedance of 40 Ω. The antenna consists basically of a four-element, half-turn helical antenna, with each pair of elements described as a *bifilar*, both of which are fed in phase quadrature. Several feed methods can be employed, all of which appear complicated except the infinite-balun design, which uses a length of coax as one of the four elements.

To produce the necessary 90° phase difference between the bifilar elements, either of two methods can be used. One is to use the same size bifilars, which essentially consist of two twisted loops with their vertical axes centered and aligned, and the loops rotated so that they're 90° to each other (like an egg-beater), and using a quadrature hybrid feed. Such an antenna requires *two* feed lines, one for each of the filar pairs.

The second and more practical method is the self-phasing system, which uses *different-size loops:* a larger loop designed to resonate *below* the design frequency (providing an inductive reactance component) and a smaller loop to resonate higher than the design frequency (introducing a capacitive-reactance component), causing the current to lead in the smaller loop and lag in the larger loop. The element lengths are 0.560 λ for the larger loop, and 0.508 λ for the smaller loop. According to the range tests performed by Maxwell, to achieve *optimum* circular polarization, the wire used in the construction of the bifilar elements should be 0.0088 λ in diameter.

Maxwell indicates that in the quadrifilar mode, the fields from the individual bifilar helices combine in optimum phase to obtain unidirectional end-fire gain. The currents in the two bifilars must be in quadrature phase. This 90° relationship is obtained by making their respective terminal impedances R +jX and R −jX where X = R, so that the currents in the respective helices are −45° and +45°. The critical parameter in this relationship is the terminal reactance, X, where the distributed inductance of the helical element is the primary determining factor. This assures the ±45° current relationship necessary to obtain true circular polarization in the combined fields and to obtain maximum forward radiation and minimum back lobe. Failure to achieve the optimum element diameter of 0.0088 λ results in a form of elliptical, rather than true circular polarization, and the performance may be *a few tenths of a decibel* below optimum, according to Maxwell's calculations. Using #10 wire translates roughly to an element diameter

[1]C. C. Kilgus, "Resonant Quadrafilar Helix;" *IEEE Transactions on Antennas and Propagation*, Vol AP-17, May 1969, pp 349-351.
[2]M.W. Maxwell, W2DU, *Reflections* (Newington: ARRL, 1990). [This book is out of print.]
[3]R. Brickner Jr and H. Rickert, "An S-Band Resonant Quadrifilar Antenna for Satellite Communication," RCA Corp, AstroElectronics Div.

Fig 30.54 — The quadrifilar helix antenna (QHA) pieces, ready for assembly.

of 0.0012 λ at 137.5 MHz — not ideal, but good enough.

To get a grasp of the QHA's topography, visualize the antenna as consisting of two concentric cylinders over which the helices are wound (see **Fig 30.54** through **Fig 30.58**). In two-dimensional space, the cylinders can be represented by two nested rectangles depicting the height and width of the cylinders. The width of the larger cylinder (or rectangle) can be represented by 0.173 λ and the width of the smaller cylinder represented by 0.156 λ. The length of the larger cylinder or rectangle can be represented by 0.260 λ, and the length of the smaller rectangle or cylinder can be represented by 0.238 λ. Using these figures, you should be able to scale the QHA to virtually any frequency. **Table 30.6** shows some representative antenna sizes for various frequencies, along with the universal parameters needed to arrive at these figures.

PHYSICAL CONSTRUCTION

Fig 30.59 shows the construction details. A 25-inch-long piece of schedule 40, 2-inch-diameter PVC pipe is used for the vertical member. The cross arms that support the helices are six pieces of ½-inch-diameter PVC tubing: three the width of the large rectangle or cylinder, and three the width of the smaller cylinder. Two cross arms are needed for the top and bottom of each cylinder. The cross arms are oriented perpendicularly to the vertical member and parallel to each other. A third cross arm is placed midway between the two at a 90° angle. This process is repeated for the smaller cylindrical dimensions using the three smaller cross arms with the top and bottom pieces oriented 90° to the large pieces.

Using ⅝ inch-diameter holes in the 2-inch pipe ensures a reasonably snug fit for the ½-inch-diameter cross pieces. Each cross arm is drilled (or notched) at its ends to accept the lengths of wire and coax used for the elements. Then the cross arms are centered and cemented in place with PVC cement. For the 137 and 146 MHz antennas, use #10 AWG copper clad antenna wire for three of the helices and a length of RG-8 for the balun, which is also the fourth helix. (Do not consider the velocity factor of the coax leg for length calculation.) For the UHF antennas, use #10 AWG soft-drawn copper wire and RG-58 coax. Copper

Fig 30.55 — The antenna with two of the four legs (filars) of one loop attached.

Fig 30.56 — This view shows the QHA with all four legs in place. The ends of the PVC cross arms that hold the coaxial leg are notched; the wire elements pass through holes drilled in the ends of their supporting cross arms.

Table 30.6
Quadrifilar Helix Antenna Dimensions

Freq (MHz)	Wavelength (λ) (inches)	Small Loop Leg Size (0.508 λ)	Small Loop Diameter (0.156 λ)	Small Loop Length (0.238 λ)	Big Loop Leg Size (0.560 λ)	Big Loop Diameter (0.173 λ)	Big Loop Length (0.261 λ)
137.5	85.9	43.64	13.4	20.44	48.10	14.86	22.33
146	80.9	41.09	12.6	19.25	45.30	14.0	21.03
436	27.09	13.76	4.22	6.44	15.17	4.68	7.04

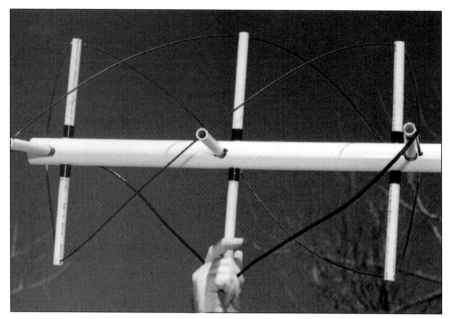

Fig 30.57 — Another view of the QHA.

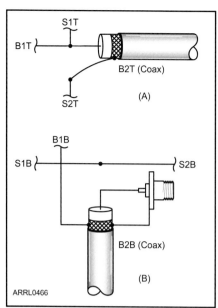

Fig 30.60 — At A, element connections at the top of the antenna. B shows the connections at the bottom of the antenna. The identifiers are those shown in Fig 30.59 and explained in the text.

Fig 30.58 — An end-on view of the top of the QHA prior to soldering the loops and installing the PVC cap.

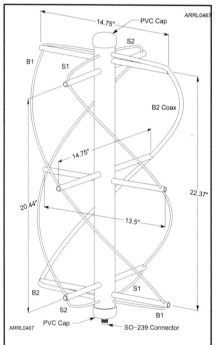

Fig 30.59 — Drawing of the QHA identifying the individual legs; see text for an explanation.

clad wire is difficult to work with, but holds its shape well. Smaller antennas can be built without the cross arms because the wire is sufficiently self-supporting.

To minimize confusion regarding the connections and to indicate the individual legs of the helices, label each loop or cylinder as B (for big) and S (for small); T and B indicate top and bottom. Each loop can be further split using leg designators as B1T and B1B, B2T and B2B, S1T and S1B and S2T and S2B, with B2 being the length of coax and the other three legs as wires. For righthand circular polarization (RHCP) wind the helices *counterclockwise* as viewed from the top. This is contrary to conventional axial mode helix construction. (For LHCP, the turns rotate *clockwise* as viewed from the top.) See **Fig 30.60** for the proper connections for the top view. When the antenna is completed, the view shows that there are two connections made to the center conductor of the coax (B2) top. These are B1T and S1T, for a total of three wires on one connection. S2T connects to B2T braid. The bottom of the antenna has S1B and S2B soldered together to complete the smaller loop. B1B and the braid of B2B are soldered together. Attach an SO-239 connector to the bottom by soldering the center conductor of B2B to the center of the connector and the braid of B2B to the connector's shell. The bottom now has two connections to the braid: one to leg B1B, the other to the shell of the connector. There's only one connection to the center conductor of B2B that goes to the SO-239 center pin.

Total price for all new materials-including the price of a suitable connector-should be in the neighborhood of $10 or less.

RESULTS

With a 70-foot section of RG-9 between the receiver and antenna, which is mounted about 12 feet above ground, and a preamp in the shack the author receives fade-free passes from the weather satellites. Although the design indicates a 3-dB beamwidth of 140°, an overhead pass provides useful data down to 10° above the horizon. The 70 cm antenna works fine for PACSATs, although Doppler effect makes manual tracking difficult. The weather-satellite antenna prototype worked better than expected and a number of copies built by others required no significant changes.

Thanks to Chris Van Lint, and Tom Loebl, WA1VTA, for supplying technical data to complete this project, and to Walt Maxwell, W2DU, for his review and technical evaluation and for sharing his technical expertise with the amateur satellite community.

30.7 Satellite References and Bibliography

Books

S. Ford, WB8IMY, *The ARRL Satellite Handbook* (Newington: ARRL, 2008)

Getting Started with Amateur Satellites, available from AMSAT-NA at **www.amsat-na.com/store/**.

Digital Satellite and Telemetry Guide, available from AMSAT-NA at **www.amsat-na.com/store/**.

Articles

P. Alvarez-Cienfuegos, EB4DKA and J. Montana, EA4CYQ, "A Better Way to Work Low Earth Orbit FM Satellites," *QST*, Feb 2009, pp 41-43

L.B. Cebik, W4RNL, "Circularly Polarized Aimed Satellite Antenna," Antenna Options, *QEX*, Sep/Oct 2007, pp 51-56.

P. Antoniazzi, IW2ACD, and M. Arecco, IK2WAQ, "Low-Profile Helix Feed for Phase 3E Satellites: System Simulation and Measurements," *QEX*, May/Jun 2006, pp 34-39.

M. Spencer, WA8SME, "A Satellite Tracker Interface," *QST*, Sep 2005, pp 64-67.

L. Smith, W5KQJ, "An Affordable Az-El Positioner for Small Antennas," *QST*, Jun 2003, pp 28-34.

30.8 Earth-Moon-Earth (EME) Communication

EME communication, also known as *moonbounce*, has become a popular form of amateur space communication. The EME concept is simple: the moon is used as a passive reflector for two-way communication between two locations on Earth (see **Fig 30.61**). With a total path length of about half a million miles, EME may be considered the ultimate DX. Very large path losses suggest big antennas, high power and the best low noise receivers; however, the adoption of modern coding and modulation techniques can significantly reduce these requirements from their levels of just 10 years ago. Even so, communication over the EME path presents unusual station design challenges and offers special satisfaction to those who can meet them.

30.8.1 Background

EME is a natural and passive propagation phenomenon, and EME QSOs count toward WAC, WAS, DXCC and VUCC awards. EME opens up the bands at VHF and above to a new frontier of worldwide DX.

Professional demonstrations of EME capability were accomplished shortly after WW II. Amateurs were not far behind, with successful reception of EME echoes in 1953 and pioneering two-way contacts made on the 1296, 144 and 432 MHz bands in the 1960s. Increased EME activity and advances to other bands came in the 1970s, aided by the availability of reliable low-noise semiconductor devices and significant improvements in the design of Yagi arrays and feed antennas for parabolic dishes. These trends accelerated further in the 1980s with the advent of low-noise GaAsFET and HEMT preamplifiers and computer-aided antenna designs, and again after 2000 with the introduction of digital techniques. See the sidebar, "Amateur EME Milestones."

EME QSOs have been made on all amateur bands from 28 MHz to 47 GHz. Many operators have made WAC, WAS and even DXCC on one or more of the VHF and UHF bands. EME is now within the grasp of most serious VHF and UHF operators.

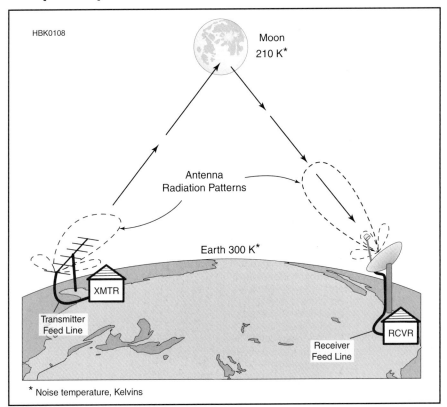

Fig 30.61 — Schematic representation of major system components for a (one-way) EME path. This illustration also shows some of the factors contributing to system noise temperature, which is discussed later in the chapter.

Amateur EME Milestones

Year	Event
1953	W3GKP and W4AO detect lunar echoes on 144 MHz
1960	First amateur 2-way EME contact: W6HB works W1FZJ, 1296 MHz
1964	W6DNG works OH1NL, 144 MHz
1964	KH6UK works W1BU, 432 MHz
1970	WB6NMT works W7CNK, 222 MHz
1970	W4HHK works W3GKP, 2.3 GHz
1972	W5WAX and K5WVX work WA5HNK and W5SXD, 50 MHz
1987	W7CNK and KA5JPD work WA5TNY and KD5RO, 3.4 GHz
1987	W7CNK and KA5JPD work WA5TNY and KD5RO, 5.7 GHz
1988	K5JL works WA5ETV, 902 MHz
1988	WA5VJB and KF5N work WA7CJO and KY7B, 10 GHz
2001	W5LUA works VE4MA, 24 GHz
2005	AD6FP, W5LUA and VE4MA work RW3BP, 47 GHz
2005	RU1AA works SM2CEW, 28 MHz
2009	GD0TEP works ZS6WAB, 70 MHz

30.9 EME Propagation

30.9.1 Path Loss

Path loss in free space is caused by nothing more than the spherical expansion of a radio wave as it propagates away from an antenna. An EME signal is attenuated as $1/d^2$ (inverse distance squared) over the quarter-million mile path to the moon, and again as $1/d^2$ on the return trip, for a net $1/d^4$ path loss. Radio waves incident on the surface of the moon are often said to be "reflected," although in fact they are partly absorbed and partly scattered by the irregular lunar surface. A full expression giving the EME path loss as a ratio of received power to transmitted power, assuming isotropic antennas at each end of the path, is

$$\ell = \frac{\eta r^2 \lambda^2}{64 \pi^2 d^4} \quad (1)$$

where
- r is the radius of the moon
- λ is the wavelength
- d is the distance to the moon
- η is the lunar reflection coefficient.

In this section we use the convention of lower-case letters to denote dimensionless ratios, and the corresponding upper-case letters to give equivalent values in dB. Thus, the EME path loss in dB is given for isotropic antennas by the expression

$$L = 10 \log \ell = 10 \log \left(\frac{\eta r^2 \lambda^2}{64 \pi^2 d^4} \right) \quad (2)$$

Inserting values $r = 1.738 \times 10^6$ m, $d = 3.844 \times 10^8$ m and $\eta = 0.065$ gives the average path losses quoted in **Table 30.7** for the principal amateur EME bands. The need to overcome these very large attenuations is of course the main reason why EME is so challenging. The moon's orbit is an ellipse, and its distance d varies by ±6.8% over each month. Because of the inverse-fourth-power law in Equations (1) and (2), this change results in path-loss variations of ±1.1 dB at the extremes of lunar distance, independent of frequency. The reflection of radio waves is of course not affected by the optical phases of the moon.

The dependence of path loss on λ^2 suggests that EME should be nearly 20 dB more difficult at 1296 MHz than at 144 MHz. This conclusion is misleading, however, because of the assumption of isotropic antennas. If one uses transmitting and receiving antennas of gain g_t and g_r, expressed as ratios, the expected power p_r received as a lunar echo may be written as the product

$$p_r = p_t g_t g_r \ell \quad (3)$$

Table 30.7
Two-Way EME Path Loss with Isotropic Antennas

Frequency (MHz)	Average Path Loss (dB)
50	−242.9
144	−252.1
222	−255.8
432	−261.6
902	−268.0
1296	−271.2
2304	−276.2
3456	−279.7
5760	−284.1
10368	−289.2
24048	−293.5

where p_t is the transmitted power. The standard expression for an antenna's power gain is

$$g = 4\pi A / \lambda^2$$

where A is the effective aperture or collecting area. Gain in dBi (dB over an isotropic antenna) may therefore be written as

$$G = 10 \log (4\pi A / \lambda^2)$$

With P_r and P_t expressed in dB relative to some reference power, for example 1 W, we have

$$P_r = P_t + G_t + L + G_r \quad (4)$$

Thus, assuming a fixed size of antenna, such as a parabolic dish or Yagi array of effective frontal area A, the frequency dependence is reversed: for a given transmitted power, lunar echoes would be 20 dB stronger for every decade increase in frequency, rather than 20 dB weaker. Most practical situations fall somewhere between these two extremes of frequency dependence.

For reasons explained in detail below, amateur EME communication is feasible with roughly comparable degrees of difficulty over nearly two decades of frequency, from 144 MHz to 10 GHz. Not surprisingly, some very different techniques must be mastered in order to do successful EME at the lower and upper extremes of this wide frequency range — so the final choice of band(s) for EME is often determined by the interests, skills and resources of an individual operator.

30.9.2 Echo Delay and Time Spread

Radio waves propagate at speed c, the speed of light, very nearly equal to 3×10^8 m/s. Propagation time to the moon and back is therefore 2d/c or about 2.4 s at perigee, 2.7 s at apogee and 2.56 s on average. The moon is nearly spherical, and its radius corresponds to r/c = 5.8 ms of wave travel time. The trailing parts of an echo, reflected from irregular surface features near the edge of the lunar disk, are delayed from the leading edge by as much as twice this value. In practice, most of the moon's surface appears relatively smooth at the radio wavelengths used for amateur EME. Lunar reflections are therefore quasi-specular, like those from a shiny ball bearing, and the power useful for communication is mostly reflected from a small region near the center of the disk. At VHF and UHF frequencies, the effective *time spread* of an echo amounts to no more than 0.1 ms.

Reflection from a smooth surface preserves linear polarization and reverses the sense of circular polarization. At shorter wavelengths the lunar surface appears increasingly rough, so reflections at 10 GHz and above contain a significant diffuse component as well as a quasi-specular component. The diffuse component is depolarized, and significant portions of it arise from regions farther out toward the lunar rim. The median time spread can then be as much as several milliseconds. In all practical cases, however, time spreading is small enough that it does not cause significant smearing of CW keying or intersymbol interference in the slowly keyed modulations commonly used for digital EME.

Time spreading does have one very significant effect. Signal components reflected from different parts of the lunar surface travel different distances and arrive at Earth with random phase relationships. As the relative geometry of the transmitting station, receiving station and reflecting lunar surface changes, signal components may sometimes add and sometimes cancel, creating large amplitude fluctuations. Often referred to as *libration fading*, these amplitude variations will be well correlated over a *coherence bandwidth* of a few kHz, the inverse of the time spread.

30.9.3 Doppler Shift and Frequency Spread

EME signals are also affected by Doppler shifts caused by the relative motions of Earth and moon. Received frequencies may be higher or lower than those transmitted; the shift is proportional to frequency and to the rate of change of total path length from transmitter to receiver. The velocities in question are usually dominated by the Earth's rotation, which at the equator amounts to about 460 m/s. For the self-echo or "radar" path, frequency shift will be maximum and positive at moonrise, falling through zero as the moon crosses the local meridian (north-south line) and a maximum negative value at moonset. The magnitude of shifts depends on station latitude, the declination of the moon and other geometrical factors. For two stations at different geographic locations the mutual Doppler shift is the sum of the individual (one-way) shifts. Maximum values are around 440 Hz at 144 MHz, 4 kHz at 1296 MHz and 30 kHz at 10 GHz.

Just as different reflection points on the lunar surface produce different time delays, they also produce different Doppler shifts. The moon's rotation and orbital motion are synchronized so that approximately the same face is always toward Earth. The orbit is eccentric, so the orbital speed varies; since the rotation rate does not vary, an observer on Earth sees an apparent slow "rocking" of the moon, back and forth.

Further aspect changes are caused by the 5.1° inclination between the orbital planes of Earth and moon. The resulting total line-of-sight velocity differences are around 0.2 m/s, causing a *frequency spread* of order 0.2 Hz at 144 MHz. Like all Doppler effects, these shifts scale with frequency. However, measured values of frequency spread increase slightly more rapidly than frequency to the first power because a larger portion of the lunar surface contributes significantly to echo power at higher frequencies. Linear scaling would suggest frequency spread around 15 Hz at 10 GHz, but measurements show it to be several times larger.

From a communication engineering point of view, libration fading is just another example of the so-called *Rayleigh fading* observed on any radio channel that involves multiple signal paths — such as ionospheric skywave, tropospheric scatter and terrain multipath channels with reflections from buildings, trees or mountains. Interference effects that cause signal fading depend on frequency spread as well as time spread. Signal amplitudes remain nearly constant over a *coherence time* given by the inverse of frequency spread. In general, fading rates are highest (shortest coherence times) when the moon is close to the local meridian and lowest near moonrise and moonset. They also depend on the moon's location in its elliptical orbit.

Typical coherence times are several seconds at 144 MHz, a few tenths of a second at 1296 MHz and 20 ms at 10 GHz. At 144 MHz, intensity peaks lasting a few seconds can aid copy of several successive CW characters, but at 432 MHz the timescale of peaks and dropouts is closer to that of single characters. At 1296 MHz the fading rates are often such that CW characters are severely chopped up, with dashes seemingly converted to several dits; while the extremely rapid fading at 10 GHz can give signals an almost "auroral" tone. Skilled operators must learn to deal with such effects as best they can. As described further below, modern digital techniques can use message synchronization as well as error-correcting codes and other diversity techniques to substantially improve the reliability of copy on marginal, rapidly fading EME signals.

30.9.4 Atmospheric and Ionospheric Effects

Propagation losses in the Earth's troposphere are negligible at VHF and UHF, although rain attenuation can be an important factor above 5 GHz. Tropospheric ducting of the sort that produces enhanced terrestrial propagation can bend signals so that the optimum beam heading for EME is directed away from the moon's center. Even under normal conditions, enough refraction occurs to allow radio echoes when the moon is slightly below the visible horizon. In practice, these problems are usually overshadowed by other complications of doing EME at very low elevations, such as blockage from nearby trees or buildings, increased noise from the warm Earth in the antenna's main beam and man-made interference.

The Earth's ionosphere causes several propagation effects that can be important to EME. These phenomena depend on slant distance through the ionospheric layer, which increases at low elevations. At elevation 10°, attenuation through the daytime ionosphere is generally less than 0.5 dB at 144 MHz, and nighttime values are at least 10 times lower. These numbers scale inversely as frequency squared, so ionospheric absorption is mostly negligible for EME purposes. Exceptions can occur at 50 MHz, and under disturbed ionospheric conditions at higher frequencies. Ionospheric refraction can also be important at 50 MHz, at very low elevations. Ionospheric scintillations (analogous to the "twinkling" of stars in the Earth's atmosphere) can exhibit significant effects at VHF and UHF, primarily on EME paths penetrating the nighttime geomagnetic equatorial zone or the auroral regions. Again, disturbed ionospheric conditions magnify the effects. The multipath time spread is very small, less than a microsecond, while frequency spread and fading rate can be in the fractional hertz to several hertz range. These scintillations can increase the fading rates produced by Earth rotation and lunar librations.

Much more important is the effect of Faraday rotation in the ionosphere. A linearly polarized wave will see its plane of polarization rotate in proportion to the local free-electron density, the line-of-sight component of the Earth's magnetic field and the square of wavelength. The effect is therefore greatest during the daytime, for stations well away from the equator, and at low frequencies. A mismatch $\Delta\theta$ between an incoming wave's polarization angle and that of the receiving antenna will attenuate received signal power by an amount $\cos^2\Delta\theta$. As shown in **Fig 30.62**, polarization losses increase rapidly when the misalignment exceeds 45°.

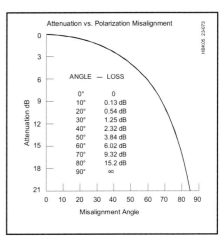

Fig 30.62 — Attenuation caused by misalignment of a linearly polarized signal with the polarization angle of a receiving antenna. The attenuation increases rapidly for alignment errors greater than 45°.

Because of the λ^2 dependence, Faraday rotation is generally important for EME operation only at 432 MHz and below. The effect is cumulative for an outgoing signal and its returning echo, so a station transmitting and receiving with the same linearly polarized antenna will see its own echoes disappear whenever the total Faraday rotation is close to an odd integral multiple of 90°. Faraday rotation in the daytime ionosphere can amount to as much as a full turn at 432 MHz and many turns at 144 MHz. At 432 MHz the rotation may be essentially constant over several hours; on lower bands significant changes can occur in 30 minutes or less. Variations are especially noticeable near sunrise or sunset at one end of the path, where ionization levels are changing rapidly.

The Earth's spherical shape determines the orientation in space of a wave emitted or received by an antenna with horizontal (or other locally referenced) polarization angle. As discussed in detail below, when combined with Faraday rotation this effect can cause users of fixed-linear-polarization antennas to experience apparent one-way propagation.

A polarized radio signal reflected from the moon's rough surface is partially scattered into other polarization states, and a disturbed ionosphere can sometimes generate a mixture of polarization angles. As a consequence, fading caused by 90° polarization misalignments will not always produce deep nulls. Measurements show that at UHF and below, the cross-polarized scattered signal is usually 15 dB or more below the principal polarization. On the other hand, at 10 GHz and higher, where the lunar surface is much rougher in terms of wavelength, cross-polarized diffuse echoes may be only a few dB below the principal reflected polarization. These comments apply to both linear and circular polarization.

30.10 Fundamental Limits

30.10.1 Background Noise

EME signals are always weak, so considerations of signal-to-noise ratio are paramount. (Noise is also discussed in the **RF Techniques** chapter.) A received signal necessarily competes with noise generated in the receiver as well as that picked up by the antenna, including contributions from the warm Earth, the atmosphere, the lunar surface, the diffuse galactic and cosmic background and possibly the sun and other sources. (Refer to Fig 30.61, and think of adding a warm atmosphere just above the Earth, the sun somewhere beyond the moon, and galactic and extragalactic noise sources at even greater distances, filling the whole sky.) If P_n is the total noise power collected from all such noise sources expressed in dBW, we can write the expected signal-to-noise ratio of the EME link as

$$SNR = P_r - P_n = P_t + G_t + L + G_r - P_n \quad (5)$$

Since isotropic path loss L is essentially fixed by choice of a frequency band (Table 30.7), optimizing the signal-to-noise ratio generally involves trade-offs designed to maximize P_r and minimize P_n — subject, of course, to such practical considerations as cost, size, maintainability and licensing constraints.

It is convenient to express P_n (in dBW) in terms of an equivalent system noise temperature T_s in kelvins (K), the receiver bandwidth B in Hz and Boltzmann's constant $k = 1.38 \times 10^{-23}$ J K^{-1}:

$$P_n = 10 \log(kT_sB) \quad (6)$$

The system noise temperature may in turn be written as

$$T_s = T_r + T_a \quad (7)$$

Here T_r is receiver noise temperature, related to the commonly quoted noise figure (NF) in dB by

$$T_r = 290 \left(10^{0.1NF} - 1\right) \quad (8)$$

Antenna temperature T_a includes contributions from all noise sources in the field of view, weighted by the antenna pattern. The lunar surface has a temperature around 210 K; since most antennas used for amateur EME have beamwidths greater than the moon's angular size, as well as sidelobes, the moon's effect will be diluted and noise from other sources will also be received. Sidelobes are important, even if many dB down from the main beam, because their total solid angle is large and therefore they are capable of collecting significant unwanted noise power.

At VHF the most important noise source is diffuse background radiation from our Galaxy, the Milky Way. An all-sky map of noise temperature at 144 MHz is presented in the top panel of **Fig 30.63**. This noise is strongest along the plane of the Galaxy and toward the galactic center. Galactic noise scales as frequency to the −2.6 power, so at 50 MHz the temperatures in Fig 30.63 should be multiplied by about 15, and at 432 divided by 17. At 1296 MHz and above galactic noise is negligible in most directions. During each month the moon follows a right-to-left path lying close to the ecliptic, the smooth solid curve plotted in Fig 30.63. Sky background temperature

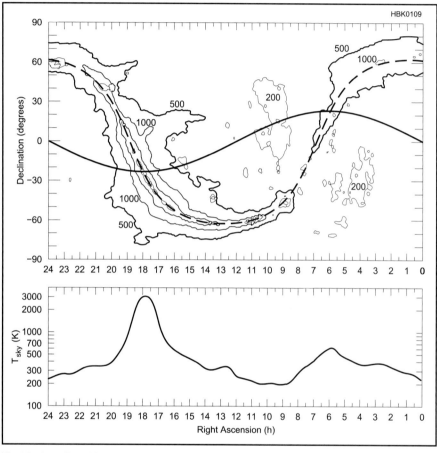

Fig 30.63 — *Top:* All-sky contour map of sky background temperature at 144 MHz. The dashed curve indicates the plane of our Galaxy, the Milky Way; the solid sinusoidal curve is the plane of the ecliptic. The Sun follows a path along the ecliptic in one year; the moon moves approximately along the ecliptic (± 5°) each month. Map contours are at noise temperatures 200, 500, 1000, 2000 and 5000 K. *Bottom:* One-dimensional plot of sky background temperature at 144 MHz along the ecliptic, smoothed to an effective beamwidth 15°.

Table 30.8
Typical Contributions to System Noise Temperature

Freq (MHz)	CMB (K)	Atm (K)	Moon (K)	Gal (K)	Side (K)	T_a (K)	T_r (K)	T_s (K)
50	3	0	0	2400	1100	3500	50	3500
144	3	0	0	160	100	260	50	310
222	3	0	0	50	50	100	50	150
432	3	0	0	9	33	45	40	85
902	3	0	1	1	30	35	35	70
1296	3	0	2	0	30	35	35	70
2304	3	0	4	0	30	37	40	77
3456	3	1	5	0	30	40	50	90
5760	3	3	13	0	30	50	60	110
10368	3	10	42	0	30	85	75	160
24048	3	70	170	0	36	260	100	360

behind the moon therefore varies approximately as shown in the lower panel of Fig 30.63, regardless of geographical location on Earth. For about five days each month, when the moon is near right ascension 18 hours and declination –28°, VHF sky background temperatures near the moon are as much as 10 times their average value, and conditions for EME on the VHF bands are poor.

By definition the sun also appears to an observer on Earth to move along the ecliptic, and during the day solar noise can add significantly to P_n if the moon is close to the sun or the antenna has pronounced sidelobes. At frequencies greater than about 5 GHz the Earth's atmosphere also contributes significantly. An ultimate noise floor of 3 K, independent of frequency, is set by cosmic background radiation that fills all space. A practical summary of significant contributions to system noise temperature for the amateur bands 50 MHz through 24 GHz is presented in **Table 30.8** and **Fig 30.64** and **Fig 30.65**, discussed in the next section.

30.10.2 Antenna and Power Requirements

The basic circumstances described so far ensure that frequencies from 100 MHz to 10 GHz are the optimum choices for EME and space communication. Over this region and a bit beyond, a wide variety of propagation effects and equipment requirements provide a fascinating array of challenges and opportunities for the EME enthusiast. The enormous path-loss variability encountered in terrestrial HF and VHF propagation does not occur in EME work, and some of the remaining, smaller variations — for example, those arising from changing lunar distance and different sky background temperatures — are predictable. We can therefore estimate with some confidence the minimum antenna sizes and transmitter powers required for EME communication on each amateur band.

The necessary information for this task is summarized in Table 30.8 and Figs 30.64 and 30.65. Columns 2 through 6 of the table give typical contributions to system noise temperature from the cosmic microwave background (CMB), the Earth's atmosphere, the warm surface of the moon, galactic noise entering through the main antenna beam and sky and ground noise from an antenna's side and rear lobes. Antenna temperature T_a is a combination of all these contributions, appropriately weighted by antenna pattern; the system noise temperature T_s is then the sum of T_a and receiver noise temperature T_r, referred to the antenna terminals. Numbers in Table 30.8 are based on the fundamentals described in the previous section and on hypothetical antennas and receivers that conform to good amateur practice in the year 2009;

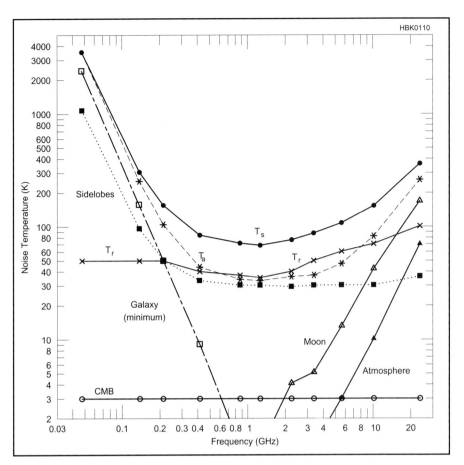

Fig 30.64 — Typical contributions to system noise temperature T_s as function of frequency. See text for definitions and descriptions of the various sources of noise.

they have been rounded to two significant figures. It is possible to do slightly better than these numbers — for example, by building antennas with lower sidelobe response or preamplifiers with still lower noise figure — but it's not easy!

The topmost curve in Fig 30.64 illustrates clearly why the frequency range from 100 MHz to 10 GHz is optimum for EME communication. Fig 30.65 shows that further reductions of T_s must come from lower T_r or better suppression of antenna sidelobes. There is nothing you can do about noise from the CMB, atmosphere, moon or Galaxy entering your main beam! For comparison, the very best professional receiving equipment achieves system noise temperatures around 20 K in the 1-2 GHz region — only a few dB better than current amateur practice. These systems generally use cryogenic receivers and very large dish antennas that can provide better suppression of sidelobes.

Having established reasonable target figures for system noise temperature, we can now proceed to estimate minimum antenna and power requirements for an EME-capable station on each amateur band. Rearrangement of Equations (5) and (6) yields the following relation for transmitter power P_t in dBW:

$$P_t = SNR - G_t - G_r - L + 10\log(kT_sB) \quad (9)$$

Values for L and T_s can be taken for each amateur band from Tables 30.7 and 30.8. For illustrative purposes let's assume SNR = 3 dB and B = 50 Hz, values appropriate for a good human operator copying a marginal CW signal. (The 50 Hz effective bandwidth may be established by an actual filter, or more commonly by the operator's "ear-and-brain" filter used together with a broader filter.) For antennas we shall assume bays of four long Yagis for the 50 through 432 MHz bands, parabolic dishes of diameter 3 m on 1296 and 2304 MHz and 2 m dishes on the higher bands. Representative gains and half-power beamwidths for such antennas are listed in columns 3 and 4 of **Table 30.9**. Column 5 then gives the necessary transmitter power in watts, rounded to two significant figures.

A station with these baseline capabilities should be self sufficient in terms of its ability to overcome EME path losses — and thus able to hear its own EME echoes and make CW contacts with other similarly equipped EME stations. Note that the quoted minimum values of transmitter power do not allow for feed line losses; moreover, a CW signal with SNR =

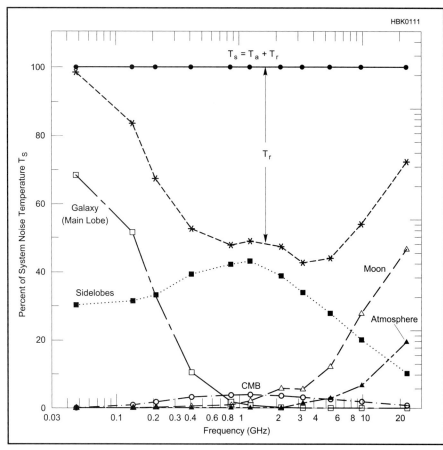

Fig 30.65 — Percentage contributions to system noise temperature as a function of frequency.

Fig 30.66 — TF/DL3OCH used a rural road sign to help him give contacts with Iceland to a number of EME operators on 1296 MHz. Bodo has activated many DXCC entities on 1296 MHz EME by using JT65, a single long Yagi and a 100 W solid state amplifier.

3 dB in 50 Hz bandwidth hardly represents "armchair copy." At the highest frequencies, issues of oscillator stability and Doppler spreading might make the assumed 50 Hz bandwidth unrealistically narrow, thus requiring somewhat more power or a larger antenna. On the other hand, lower power and smaller antennas can be sufficient for working stations with greater capabilities than those in the table.

Other factors can reduce the minimum power or antenna gains required for successful EME, at least some of the time. One possibility, especially effective at 50 and 144 MHz at low moon elevations, is to take advantage of reflections from reasonably smooth ground (or better still, water) in front of your antenna. Often referred to as *ground gain*, these reflections can add as much as 6 dB to an antenna's effective gain at elevations where the reflections are in phase with the direct signal. Another possibility is to use more efficient coding and modulation schemes than provided by Morse coded CW.

30.10.3 Coding and Modulation

International Morse code with on-off keying (OOK) is an excellent general purpose communication mode. It is easy to implement and performs well in weak-signal conditions. EME operating procedures for CW usually include multiple repetitions so that essential parts of a bare-minimum QSO can be assembled from fragments copied on signal peaks. However, modern communication theory points the way toward modulation schemes significantly more efficient than OOK, codes better than Morse, and error-control methods more effective than simple repetition. Amateur experiments with these ideas have led to the current popularity of digital EME on the VHF and lower UHF bands. In general, an efficient digital mode designed for basic communication with weak signals will compress user messages into a compact form and then add redundancy in the form of a mathematically defined error-correcting code (ECC). Such codes can ensure that full messages are recoverable with high confidence, even when significant portions of a transmission are lost or corrupted.

A number of distinct sources may contribute to the improved performance of such a mode over CW. Multitone FSK (MFSK) is a more efficient modulation than OOK, in part because each received symbol is roughly the equivalent of a full character, rather than a single dot or dash. For equivalent messages, MFSK can therefore be keyed much more slowly than CW and detected in a much smaller bandwidth. Morse code is self-synchronizing at the character level (if a signal is strong enough for letters to be recognized), but a Morse transmission contains no useful information for synchronizing a whole message. This fact makes it difficult to piece together copied fragments of a CW message being sent repeatedly.

In contrast, a synchronized digital transmission with ECC can encode the complete message into a new data format designed to enhance the probability that successful decoding will produce the message's full

Table 30.9
Typical Antenna and Power Requirements for CW EME

Freq (MHz)	Ant Type[1]	G (dBi)	HPBW (deg)	TxPwr (W)
50	4×12 m	19.7	18.8	1200
144	4×6 m	21.0	15.4	500
432	4×6 m	25.0	10.5	250
1296	3 m	29.5	5.5	160
2304	3 m	34.5	3.1	60
3456	2 m	34.8	3.0	120
5760	2 m	39.2	1.8	60
10368	2 m	44.3	1.0	25

[1]Example antennas for 50, 144 and 432 MHz are Yagi arrays with stated lengths; those for 1296 MHz and higher are parabolic dishes of specified diameter.

information content, with everything in its proper place. For the limited purpose of exchanging call signs, signal reports and modest amounts of additional information, digital EME contacts can be made at signal levels some 10 dB below those required for CW, while at the same time improving reliability and maintaining comparable or better rates of information throughput. Depending on your skill as a CW operator, the digital advantage may be even larger.

Thus, digital EME contacts are possible between similar stations with about 10 dB less power than specified in Table 30.9, or with 5 dB smaller antenna gains at both transmitter and receiver. An excellent example is the highly portable EME setup of DL3OCH, shown in **Fig 30.66**. With a single long Yagi (59 elements, 5 m boom, 21.8 dBi gain), a 100 W solid state amplifier and the JT65C digital mode,[1] this equipment has helped to provide dozens of new DXCC credits on 1296 MHz from countries with little or no regular EME activity.

30.11 Building an EME Station

30.11.1 Antennas

The antenna is arguably the most important element in determining an EME station's capability. It is not accidental that the baseline station requirements outlined in Table 30.9 use Yagi arrays on the VHF bands and parabolic dishes at 1296 MHz and above: one of these two antenna types is almost always the best choice for EME. The gain of a modern, well designed Yagi of length ℓ can be approximated by the equation

$$G = 8.1 \log (\ell / \lambda) + 11.4 \text{ dBi}, \qquad (10)$$

and stacks of Yagis can yield close to 3 dB (minus phasing line losses) for each doubling of the number of Yagis in the stack.

For comparison, the gain of a parabolic dish of diameter d with a typical feed arrangement yielding 55% efficiency is

$$G = 20 \log (d / \lambda) + 7.3 \text{ dBi} \qquad (11)$$

The gains of some nominal antennas of each type are illustrated graphically in **Fig 30.67**, which helps to show why Yagis are nearly always the best choice for EME on the VHF bands. They are light, easy to build and have relatively low wind resistance. Stacks of four Yagis are small enough that they can be mounted on towers for sky coverage free of nearby obstructions. Larger arrays of 8, 16 or even more Yagis are possible, although the complexity and losses in phasing lines and power dividers then become important considerations, especially at higher frequencies. Long Yagis are narrowband antennas, usable on just a single band.

We usually think of the linear polarization of a transmitted signal as being "horizontal" or "vertical." Of course, on the spherical Earth these concepts have meaning only locally. As seen from the moon, widely separated horizontal antennas may have very different orientations (see **Fig 30.68**). Therefore, in the absence of Faraday rotation an EME signal transmitted with horizontal polarization by station A will have its linear polarization misaligned at stations B and C by angles known as the *spatial polarization offset*. In Fig 30.68 the signal from A arrives with vertical polarization at B and at 45° to the horizon at C. Suppose C is trying to work A and $\theta_s = 45°$ is the spatial polarization offset from A to C. The return signal from C to A will be offset in the opposite direction, that is, by an amount $-\theta_s$

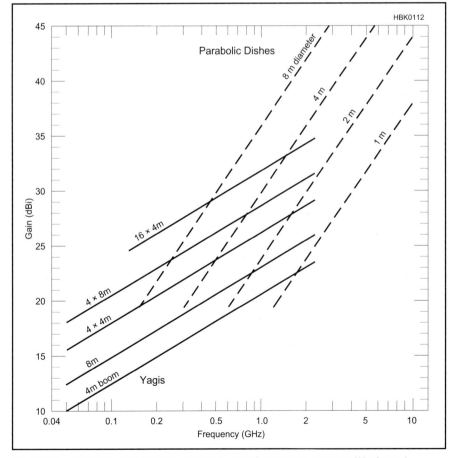

Fig 30.67 — Representative gains of practical Yagi antennas, arrays of Yagis and parabolic dishes as a function of frequency. Yagi arrays make the most cost-effective and convenient antennas for EME on the VHF bands, while parabolic dishes are generally the best choice above 1 GHz.

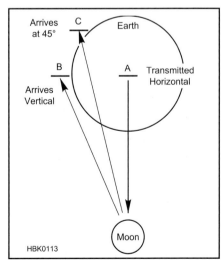

Fig 30.68 — The spherical Earth creates spatial polarization offsets for well-separated stations with horizon-oriented linear polarization. Here, a signal transmitted horizontally at A arrived with vertical polarization at B and midway between horizontal and vertical at C. When combined with Faraday rotation, offsets close to 45° can lead to apparent one-way propagation. See text for details.

Computer and Internet Resources for EME

Software for finding and tracking the moon:

MoonSked, by GM4JJJ, **www.gm4jjj.co.uk/MoonSked/moonsked.htm**
EME System, by F1EHN, **www.f1ehn.org**
EME2008, by VK3UM, **www.ve1alq.com/vk3um/index.html**
SkyMoon, by W5UN, **www.w5un.net**
GJTracker, by W7GJ, **www.bigskyspaces.com/w7gj/**
WSJT, by K1JT, **physics.princeton.edu/pulsar/K1JT/**
Web calculator: **www.satellite-calculations.com/Satellite/suncalc.htm**

Software for EME performance calculations:

EMECalc, by VK3UM, **www.ve1alq.com/vk3um/index.html**

Digital EME:

WSJT, MAP65: **physics.princeton.edu/pulsar/K1JT/**

Topical email reflectors:

Moon-Net: **www.nlsa.com/nets/moon-net-help.html**
Moon: **www.moonbounce.info/mailman/listinfo/moon**

Beginner information:

W5UN: **www.w5un.net**
EA6VQ: **www.vhfdx.net/jt65bintro.html**
W7GJ (EME on 50 MHz): **www.bigskyspaces.com/w7gj/**

Technical references:

SM5BSZ: **www.sm5bsz.com/linuxdsp/linrad.htm**
W1GHZ: **www.w1ghz.org/antbook/contents.htm**
GM3SEK: **www.ifwtech.co.uk/g3sek/eme/pol1.htm**
       and **www.ifwtech.co.uk/g3sek/stacking/stacking2.htm**
F5VHX: **www.rfham.com/g8mbi/g8mbi/pol1.htm**
Dubus: **www.marsport.org.uk/dubus/eme.htm**

Chatrooms and Loggers:

NØUK: **www.chris.org/cgi-bin/jt65emeA**
ON4KST: **www.on4kst.com/chat/**
HB9Q: **hb9q.ch/joomla/**

Monthly Newsletters:

144 MHz, by DF2ZC: **www.df2zc.de/newsletter/index.html**
432 and Above, by K2UYH: **www.nitehawk.com/rasmit/em70cm.html**

Solar Flux data:

Archival: **www.ngdc.noaa.gov/stp/SOLAR/ftpsolarradio.html#noonflux**
Current: **www.ips.gov.au/Solar/3/4/2**

Fig 30.70 — This 3 m TVRO dish with aluminum frame and mesh surface was outfitted for 1296 MHz EME as a joint effort by VA7MM and VE7CNF. The dual-circular polarization feed is a VE4MA/W2IMU design.

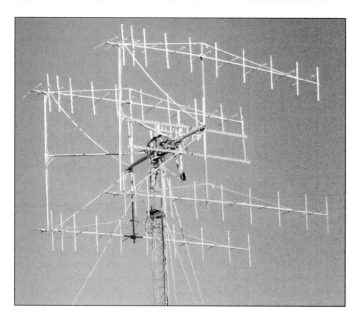

Fig 30.69 — Array of four 10-element, dual-polarization 144 MHz Yagis at KL7UW. Alaskan frost makes the horizontal and vertical elements stand out clearly. A pair of loop Yagis for 1296 MHz can be seen inside the 2 meter array.

$= -45°$. The Faraday rotation angle θ_F, on the other hand, has the same sign for transmission in both directions. Thus the net polarization shift from A to C is $\theta_F + \theta_s$, while that from C to A is $\theta_F - \theta_s$. If θ_F is close to any of the values $\pm 45°$, $\pm 135°$, $\pm 225°$, …, then one of the net polarization shifts is nearly 90° while the other is close to 0°. The result for stations with fixed linear polarization will be apparent one-way propagation: for example, A can copy C, but C cannot copy A.

Obviously no two-way contact can be made under these conditions, so the operators must wait for more favorable circumstances or else implement some form of polarization control or polarization diversity. One cost-effective solution is to mount two full sets of Yagi elements at right angles on the same boom. Arrays of such cross-polarized or "Xpol" Yagis make especially attractive EME antennas on the VHF and lower UHF bands because they offer a flexible solution to the linear-polarization misalignment problem. As an example, **Fig 30.69** shows the 4 × 10 element, dual-polarization EME array at KL7UW. This antenna and a 160 W solid-state amplifier have accounted for hundreds of EME contacts with the state of Alaska on 2 meters.

At 1296 MHz and above, gains of 30 dBi and more can be achieved with parabolic dishes of modest size. As a result, these antennas are almost always the best choice on these

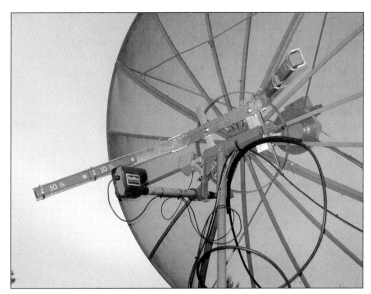

Fig 30.71 — Mounting arrangement, counterweights and az/el control system for the VA7MM 3 m dish.

bands. Their structure does not depend on any radio frequency resonances, so in many ways dishes are less critical to build than Yagis. Element lengths in high-gain Yagis must be accurate to better than 0.005 λ, while the reflecting surface of a dish need be accurate only to about 0.1 λ. Mesh surfaces are attractive at frequencies up to at least 5 GHz, because of their light weight and lower wind resistance. Openings in the mesh can be as large as 0.05 λ without allowing much ground noise to feed through the surface. A parabolic antenna has a single feed point, so there are no losses in phasing lines or power splitters. You can use a dish on several bands by swapping feeds, and with suitable feed designs you can produce either linear or circular polarization, including dual polarizations. A very attractive and convenient option is to transmit in one sense of circular polarization and receive in the opposite sense. Transmitting in right-hand circular (RHC) and receiving in LHC has become the standard for EME at 1296 and 2304 MHz, and will probably become the standard on higher bands as well.

As made clear in Fig 30.67, the 432 MHz band lies in a transition region where both Yagis and parabolic dishes have attractive features. Either four long Yagis or a 6 m dish can produce enough gain (about 25 dBi) to let you work many other EME stations on this band. Many linear-polarization systems are already in use — for good reason, since most amateur use of this band is for terrestrial communication — so converting everyone to circular polarization is impractical. Therefore, schemes have been devised to physically rotate dish feeds and even whole Yagi arrays to cope with the resulting polarization alignment problems. Another scheme is to use a dual-polarization dish feed or dual-polarization Yagis, as described above and increasingly used on 144 MHz. This approach has not yet gained wide popularity on 432 MHz, however.

ANTENNA PATTERN

A clean pattern with good suppression of side and rear lobes is important for all EME antennas — especially at 432 MHz and above, where excessive noise pickup through sidelobes can significantly increase T_s. For Yagi arrays you should use modern, computer-optimized designs that maximize G/T_s, the ratio of forward gain to system noise temperature. Be sure to pay attention to maintaining a clean pattern when stacking multiple antennas. First sidelobes within 10-15° of the main beam may not be a major problem, because

(A)

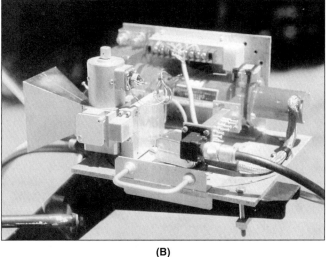

(B)

Fig 30.72 — N4MW outfitted this 2.6 m offset parabolic dish for 10 GHz EME. Equipment mounted at the focus (close-up at B) includes low noise preamplifier, transverter, traveling wave tube (TWT) power amplifier and feed horn for 10 GHz.

their solid angle is small and they will look mostly at cold sky when EME conditions are favorable. Side and rear lobes farther from the main beam should be suppressed as much as possible, however. Remember that even close-in sidelobes will degrade your receiving performance at low elevations.

For parabolic dishes, G/T_s is optimized by using a feed with somewhat larger taper in illumination at the edge of the dish than would yield the highest forward gain. Best forward gain is generally obtained with edge taper around -10 dB, while best G/T_s occurs around -15 dB. Edge taper of -12 dB is usually a good compromise. Some good reproducible designs for dish feeds are described in references at the end of this section.[2]

ANTENNA MOUNTS

EME antennas have high gain and narrow main beams that must be properly aimed at the moon in two coordinates. Although polar mounts (one axis parallel to the Earth's axis) have sometimes been used, by far the most popular mounting scheme today is the elevation-over-azimuth or *az/el* mount. Readily available computer software (see the sidebar, "Computer and Internet Resources for EME") can provide azimuth and elevation coordinates for the moon, and a small computer can also control antenna positioning motors to automate the whole pointing system.

For mechanical reasons it is desirable to place the antenna's center of gravity close to the intersection of the vertical (azimuth) and horizontal (elevation) axes. On the other hand, the mounting structure must not interfere with critical active regions of the antenna. Stacked Yagis are generally mounted so that metallic supporting members are perpendicular to the radiating elements or located at midpoints where the effective apertures of separate Yagis meet. Feed lines and conducting support members must not lie in the active planes containing Yagi elements, unless they run wholly along the boom. For dual-polarization Yagis, feed lines should be routed toward the rear of each Yagi and any mid-boom support members must be non-conducting. For EME there is nothing magical about using horizontal and vertical for the two orthogonal polarizations, and there are some advantages to mounting cross-Yagis with elements in the "×" rather than "+" orientation.[3]

Parabolic dishes are usually mounted from behind, with counterweights extending rearward to relieve torque imbalance on the elevation axis. Screw-jack actuators designed for positioning 1980s-style TVRO dishes can be readily adapted for elevation control.[4] Standard heavy-duty antenna rotators can be used for azimuth positioning of dishes up to about 3 m in size. Larger dishes may require heavier, one-of-a-kind designs for pointing control. **Figs 30.70** through **30.72** show examples of parabolic dishes in the 2-3 m range, neatly and successfully outfitted for EME at VA7MM and N4MW.

30.11.2 Feed lines, Preamplifiers and TR Switching

Any feed line between the antenna and receiver introduces attenuation and noise, so at UHF and above it is vital that the low-noise preamplifier (LNA) be mounted very close to the antenna terminals. At ambient temperature, every 0.1 dB of loss in front of the LNA adds at least 7 K to the effective T_r, and therefore to T_s. On bands where T_a is much lower than ambient, even 0.1 dB of attenuation can result in 0.5 dB loss of receiver sensitivity.

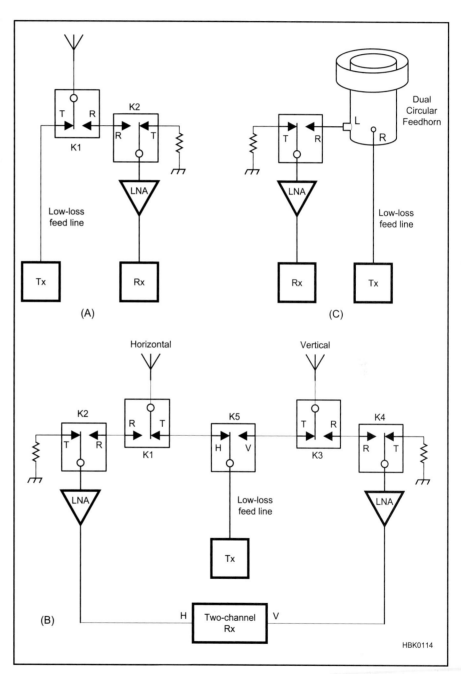

Fig 30.73 — Recommended "front ends" for EME systems include low-noise preamplifiers (LNAs) mounted at the antenna and use separate feed lines for transmitting and receiving. Relay K2 shown in (A) may be omitted if K1 provides adequate isolation of the LNA while transmitting. An arrangement like that in (B) is recommended for a dual-polarization antenna such as an array of cross-Yagis. Transmitter power is sent to one polarization or the other, but received signals in both polarizations are amplified and sent on to a dual-channel receiver. Part (C) shows a suitable arrangement for circular polarization implemented with a two-port, dual-circular-polarization feed horn. In this case, no high-power TR relay is required.

LNA gain should be sufficient to overcome feed line losses and dominate the noise contributed by subsequent stages by at least 15-20 dB. Current practices usually employ one or, especially at 432 MHz and above, two low-noise GaAsFET or HEMT transistors in the preamplifier, with only a simple noise impedance matching circuit between the first active device and the antenna. Only if severe out-of-band interference is present should a narrow filter be placed ahead of the first LNA. Bandpass filtering is often desirable between LNA stages, and may be used without significant impact on system noise temperature. The feedhorn of a dish antenna can have a valuable high-pass effect that attenuates signals at lower frequencies.

The same antenna is generally used for both transmitting and receiving, so the LNA must be out of the line and protected when transmitting. **Fig 30.73A** illustrates a preferred switching arrangement that uses two separate feed lines: a low-loss line to carry transmitter power and a relatively inexpensive feed line from LNA to receiver. A high-power relay K1 at the antenna handles TR switching; a second relay, K2, may be used to protect the LNA in case the isolation at K1's receive port is inadequate.

Additional relays are required in order to make best use of a dual-linear-polarization system. Fig 30.73B shows an arrangement that lets you select either horizontal or vertical polarization for transmitting (via relay K5) and use both polarizations simultaneously for receiving. A dual-channel receiver can form a linear combination of signals in the two channels to match the polarization of a desired signal exactly, whatever its angle. Such optimization is readily accomplished in a receiver whose last stages are defined and implemented in software, as in the highly effective *Linrad* system designed by SM5BSZ.[5] For digital EME using dual-polarization antennas, *Linrad* and a software program called *MAP65* make an especially powerful combination for digital EME.[6]

With circular polarization you may not need a high-power TR relay at all. Fig 30.73C shows a typical arrangement with transmitter connected to one port of a feed horn providing both senses of circular polarization, and the LNA to the other port. Since the isolation between ports may be only 20 or 30 dB, a low-power relay protects the LNA during transmit periods. In all of these schemes, suitable sequencing should be used to assure that the LNA is disconnected before transmission can begin. Many amateurs find it best to use the energized position of TR relays on receive. When the station is not in use, preamplifiers will then be disconnected from the antenna.

30.11.3 Transmitters and Power Amplifiers

FREQUENCY STABILITY

Weak signals are best detected in a bandwidth no wider than the signal itself. As a consequence, EME systems must use stable oscillators. For best results with CW your frequency drift over a minute or so should be no more than 10 Hz at the operating frequency; with digital modes the ideal target may be several times smaller. Most modern transceivers are stable enough at VHF, but some only marginally so at UHF.

The crystal oscillators used in transverters may need temperature compensation for adequate stability; better still, they can be phase-locked to an external high-stability reference oscillator. A number of digital EME QSOs have been made on the 23 cm and 13 cm bands using only 5-10 W transmitter power and parabolic dishes in the 2-3 m range — much smaller systems than the baseline examples listed in Table 30.9 — but such contacts generally depend on having stabilized local oscillators. Especially for digital EME, where detection bandwidths less than 10 Hz are used, unstable oscillators lead directly to loss of sensitivity.

POWER OUTPUT

After frequency stability, the most important specification for an EME transmitter is power output. The maximum power practically achievable by amateurs ranges from 1500 W on the VHF and lower UHF bands down to the 100 W range at 10 GHz. Fortunately, the required power levels for EME (Table 30.9) are compatible with these numbers. At 432 MHz and below, triode or tetrode vacuum tubes with external-anode construction can provide ample gain and power output. Some popular tubes include the 4CX250, 8930, 8874, 3CX800, 8877, GU-74B, GS-23B and GS-35B. Amplifiers using one or a pair of these (or other similar tubes) can provide output powers ranging up to 1000 or 1500 W on the 50 to 432 MHz bands. VHF and UHF power amplifiers based on solid state power devices have also become viable alternatives. Many amateurs have built amplifiers using these techniques, and commercial designs are available.

At frequencies above 1 GHz, transit-time limitations and physical structures prevent most high-power vacuum tubes from performing well. For many years planar triodes in the 2C39/7289/3CX100 family have been the mainstays of amateur 1296 MHz power amplifiers. Some higher power tubes for this band include the GI-7B, GS-15, TH347, TH308 and YL1050. The 2C39/7289/3CX100 tubes, as well as the GI-7B and GS-15, generally require water cooling at the power levels desirable for EME.

Solid-state power amplifiers are also available for the microwave bands. As prices come down and power levels increase, these units are becoming more popular with EME operators. Surplus solid-state amplifiers usable on the 902, 2304 and 3456 MHz bands have been attractive buys over the last few years, providing output power up to several hundred watts at reasonable cost. Two, four or even more units are sometimes used together to achieve higher output. At the highest EME frequencies surplus traveling wave tube (TWT) amplifiers can provide power levels up to several hundred watts.

RF SAFETY

One final point must be made in a discussion of high power amplifiers. Anyone who has thought about how a microwave oven works should know why RF safety is an important issue. Dangerous levels of radio frequency radiation exist inside and at short distances from power amplifiers and antennas. In normal operation, power density is highest in the immediate vicinity of small, low-gain antennas such as feeds for parabolic dishes. EME operators should be aware that ERP (effective radiated power) can be highly misleading as a guide to RF hazards. Somewhat counter-intuitively, a large, high-gain antenna helps to reduce local RF hazards by distributing RF power over a large physical area, thus reducing power density. Be sure to read the RF Safety section in the **Safety** chapter and pay attention to the RF protection guidelines there.

30.12 Getting Started with EME

Perhaps you already have a weak-signal VHF or UHF station with somewhat lesser capabilities than those recommended in Tables 30.8 and 30.9, and would like to make a few EME contacts before possibly undertaking the task of assembling a "real" EME station. On 144 and 432 MHz — probably the easiest EME bands on which to get started — you can visually point your single long Yagi at the rising or setting moon and work some of the larger EME stations. Your daily newspaper probably lists the approximate times of moon rise and moon set in your vicinity; many simple web-based calculators can give you that information as well as the moon's azimuth and elevation at any particular time (see the sidebar, "Computer and Internet Resources for EME").

These aids may be all you need to make your first EME contacts, especially if you take advantage of the weak-signal capabilities of an efficient digital mode. To optimize your chances of success, adapt your operating procedures to the prevailing standards that other EME operators will be using, as discussed below. For your first attempts you may want to make prearranged schedules with some established stations.

30.12.1 Tracking the Moon

For serious EME work you'll want a more general way of keeping your antenna pointed at the moon. The Earth and moon complete their mutual orbit every 27.3 days, one sidereal month. The lunar orbit is inclined to the plane of the ecliptic, the Earth-Sun orbital plane, by 5.1°. Since the Earth's equator is itself inclined at 23.5° to the ecliptic, the moon's path through the sky swings north and south of the equator by as much as 28° in a monthly cycle (see Fig 30.63).

Predicting the moon's exact position is a complicated problem in celestial mechanics, but is readily handled to sufficient accuracy by simple computer software. In general the problem can be reduced to 1) calculating the moon's position on the sky, as seen by a hypothetical observer at the center of the Earth; 2) applying a parallax correction to yield the lunar position at a specified location on the Earth's surface; and 3) converting the astronomical coordinates of right ascension and declination to azimuth and elevation at the specified terrestrial location.

Of course, the Earth's rotation and moon's orbital motion imply that the moon's direction is constantly changing. Suitable computer software can follow these changes and generate the necessary commands to keep your antenna pointed at the moon. Several free or inexpensive software packages with moon-tracking features are listed in the sidebar, "Computer and Internet Resources for EME," and such facilities are built into the programs *WSJT* and *MAP65* widely used for digital EME. For good results you should aim for pointing accuracies of about ¼ of your half-power beamwidth or better.

30.12.2 System Evaluation

Careful measurements of your EME system's performance can help to determine where station improvements can be made. Transmitter performance is essentially determined by power output, feed line losses and antenna gain. The first two can be measured in standard ways, but antenna gain is much more difficult. The most useful figure of merit for receiving performance is G/T_s, and while absolute measurements of either G or T_s separately are difficult, you can measure their ratio with useful accuracy and compare it with expectations. One technique particularly useful at 432 MHz and above is to use the sun as a broadband noise source. Point your antenna at cold sky and then at the sun, and measure y, the ratio of received noise power in the two directions. Operate your receiver at maximum bandwidth with the AGC off, and if possible make the observations with the sun at elevation 30° or higher.

To calculate G/T_s from y, you will need a contemporaneous estimate of solar flux density at your operating frequency. Daily measurements of solar flux are obtained at a number of standard frequencies and made available online (see the sidebar, "Computer and Internet Resources for EME"); you can interpolate an approximate flux value for the amateur band in question. Solar flux densities vary with sunspot activity, and day-to-day or even hour-to-hour variations are especially large at lower frequencies and near solar maximum. Representative monthly median values for six frequencies are presented in **Fig 30.74** over sunspot cycle 23. As a starting point, you can estimate a value of solar flux for your band and a similar point in the sunspot cycle directly from Fig 30.74. Then,

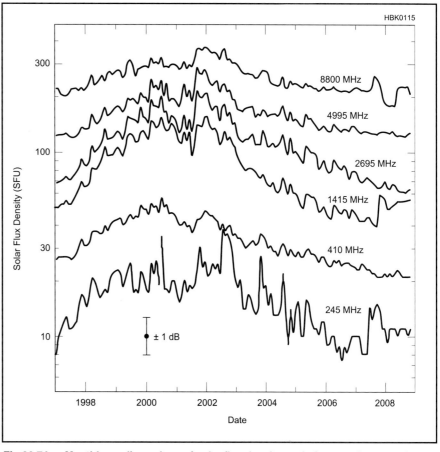

Fig 30.74 — Monthly median values of solar flux density at six frequencies, over the 11 years of solar cycle 23. Original data are from the Sagamore Hill Observatory in Massachusetts. At the lower frequencies, and especially near solar maximum, day-to-day and even hour-to-hour upward variations can be an order of magnitude larger than those in the monthly medians.

if S^* is the solar flux density in units of W m^{-2} Hz^{-1} and y is expressed as a dimensionless ratio, the corresponding value of G/T_s in dB is given by

$$\frac{G}{T_s} = 10 \log \left(\frac{8\pi k(y-1)}{S^* \lambda^2} \right) \quad (12)$$

Solar fluxes are usually quoted in Solar Flux Units (1 SFU = 10^{-22} W m^{-2} Hz^{-1}), and with S in those units Equation (12) can be reduced to the simpler relation

$$\frac{G}{T_s} = 10 \log \left(\frac{3.47 (y-1)}{S \lambda^2} \right) \quad (13)$$

Representative values of G/T_s and $Y_{sun} = 10 \log y$ for the example antennas of Table 30.9 are given in columns 4 and 5 of **Table 30.10**. The tabulated values of Y_{sun} are based on quiet-sun flux densities typical of years near sunspot minimum.

Of course, receiving your own echoes from the moon provides the best guarantee that your equipment is capable of EME communication with comparable stations. Transmitting a few dashes, then standing by to hear your lunar echo some 2.5 s later, brings a thrill that most hams never forget. With suitable signal-processing software, echoes can be detected and measured even with relatively low-power equipment. Version 4.9.8 of the *WSJT* program (see the sidebar, "Computer and Internet Resources for EME") includes an automated echo-testing facility that is useful for quantitative measurements even if your echoes are many dB below the audible threshold.

30.12.3 Operating Procedures

The more that's known about likely structure, content and timing of a transmitted message, the easier it is to copy. EME signals are often near the threshold of readability, so it is highly desirable to standardize operating procedures and message structure, and to provide transmissions with enough redundancy to capitalize on signal peaks and bridge likely gaps in reception.

You may wish to make your first EME QSOs with the aid of explicit schedules: that is, arrangements to attempt a contact with a particular station at a specified time and frequency. EME schedules usually state the duration of timed transmissions as well as starting time, transmitter frequency, and an indication of which station will transmit first. For a minimal QSO, message information is often reduced to the bare essentials of call signs, signal reports and acknowledgments. The signal report is sometimes reduced to a "yes or no" indication of whether both call signs have been successfully copied.

Remember to allow for Doppler shifts, especially at higher frequencies where the offset may exceed your received bandwidth. Most moon-tracking software used for EME can display the expected frequency shift for your own echoes as well as that for a distant station. In a scheduled QSO attempt, keep your transceiver set to the schedule frequency and use its RIT control to search for the other station around his expected Doppler shift. When looking for contacts at random, especially at 432 MHz and above, set your RIT to the expected Doppler shift of your own echoes. If a station you copy does likewise, you will find each other's signals on the same frequency as your own echoes. Many operators also benefit from using a software-driven panoramic spectral display of the "waterfall" variety, which can make signal acquisition much easier.

CW EME FORMAT

By convention, a minimal CW EME contact usually follows a format something like the sequence of messages in **Table 30.11**. If timed TR sequences are being used, the essential information is repeated for the full duration of a sequence.

The standard QSO procedures involve a number of different messages sent in sequence, and operators do not proceed to the next message until they have copied the essential information (call signs, signal report, acknowledgements) in previous messages. After call signs have been copied, a signal report is sent. Because CW dahs are easier to discern than dits, a default EME signal report (essentially meaning "I have copied both call signs") is the letter O. A station receiving call signs and O responds with RO, and a final acknowledgment of a valid contact is signified by sending RRR …. On 432 MHz and above, the letter M is sometimes used as an alternative signal report meaning "both call signs copied with difficulty." Of course, when signals are adequate for reasonably good copy normal RST signal reports can be used and other restrictions on message structure and timing relaxed.

In non-scheduled operation, it frequently happens (for example, in response to your CQ) that you can recognize and copy your own call more easily than the other station's call. The sequence YYY… (for "Your call") can be sent to ask a calling station to send his call only, repeating it many times. A contact is considered complete and valid when RRR has been received (after message number 5 in Table 30.11 has been received). However, at this point the other station does not know that his acknowledgment was copied, so it is normal to finish with something like message 6.

The conventional duration of transmit and receive periods is different on different bands and has evolved somewhat over time. On 50 and 144 MHz, stations usually transmit for one full minute and then receive for a full minute. On 432 MHz and above, schedules with 2.5-minute transmissions have been standard. The longer period gives stations with mechanically variable polarization adequate time to peak a received signal.

CW sending speed is generally around 12 to 15 WPM. Some operators find it helpful to use greater-than-normal spacing between complete letters. Keep in mind that characters sent too slowly may be chopped up by typical EME fading, while code sent too fast will be jumbled. When transmitting call sequences, send the other station's call once, followed by "DE" and your own call once or twice. Then pause and repeat the sequence. This cadence sets a rhythm so that the receiving operator can anticipate when the missing parts of a message can be expected to arrive. Send with proper spacing; the use of a programmable keyer is especially helpful and encouraged. A signal buried in the noise and accompanied by fading is hard enough to copy, without the added complication of irregular sending.

DIGITAL EME WITH JT65

Most digital EME is presently done with the JT65 protocol, as implemented in the computer programs *WSJT* and *MAP65* (see the sidebar, "Computer and Internet Resources for EME"). Like other popular digital modes, this one requires a personal computer with a sound card for audio input and output. JT65 uses structured message formats, a Reed Solomon error-correcting code with slightly more than five-fold redundancy and 1-minute TR

Table 30.10

Representative G/T_s and Y_{sun} for Example Antennas

Freq (MHz)	Ant Type[1]	G (dBi)	G/T_s (dB)	Y_{sun} (dB)
144	4×6 m	21.0	–3.8	4
432	4×6 m	25.0	5.8	12
1296	3 m	29.5	11.1	11
2304	3 m	34.5	15.6	11
3456	2 m	34.8	15.3	9
5760	2 m	39.2	18.8	9
10368	2 m	44.3	22.4	11

[1]Example antennas for 144 and 432 MHz are Yagi arrays with stated lengths; those for 1296 MHz and higher are parabolic dishes of specified diameter.

Table 30.11

Typical Messages in a Minimal EME CW Contact

Period	Message
1	CQ CQ CQ DE W6XYZ W6XYZ …
2	W6XYZ DE K1ABC K1ABC …
3	K1ABC DE W6XYZ OOO OOO …
4	W6XYZ DE K1ABC RO RO RO …
5	K1ABC DE W6XYZ RRR RRR …
6	W6XYZ DE K1ABC TNX 73 …

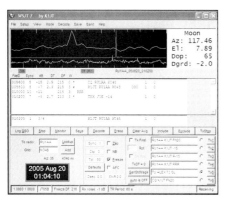

Fig 30.75 — Screen shot of computer program *WSJT*.

Table 30.12
Typical Messages in a JT65 EME Contact

Period	Using Shorthand Messages	Using Long-Form Messages
1	CQ W6XYZ CN87	CQ W6XYZ CN87
2	W6XYZ K1ABC FN42	W6XYZ K1ABC FN42
3	K1ABC W6XYZ CN87 OOO	K1ABC W6XYZ –21
4	RO	W6XYZ K1ABC R–19
5	RRR	K1ABC W6XYZ RRR
6	73	TNX RAY 73 GL

sequences. Signal peaks and dropouts due to multipath fading do not affect individual characters or words, but rather the probability of decoding the whole message. The modulation is 65-tone frequency shift keying (65-FSK) with computer-generated audio tones modulating a single-sideband transmitter in USB mode.

As expected from the theory described briefly above, results show that JT65 contacts can be made at signal levels about 10 dB less than those needed for CW. The detection bandwidth implemented in the receiving software for JT65 sub-modes A, B and C is 2.7, 5.4 and 10.8 Hz, respectively. These may be compared with minimum bandwidths of 25-50 Hz for 15 WPM CW. Requirements on oscillator stability are therefore somewhat more stringent for JT65 than for CW.

Basic procedures for minimal JT65 contacts are very similar to those for CW, and a typical message sequence is shown in **Table 30.12**. Standard messages include space reserved for Maidenhead grid locators, so this information is normally exchanged along with call signs. Following CW practice, you should proceed to the next message in the QSO sequence only when information from the previous step has been copied. A default signal report may be sent as OOO, but many operators prefer to send and receive numerical signal reports (giving the signal strength in dB, relative to noise power in a standard 2500 Hz bandwidth) as measured by the receiving software. Messages using such reports are shown in the final column of Table 30.12. Many additional details on the usage of JT65 can be found in documentation distributed with the *WSJT* program. **Fig 30.75** shows an example screen shot of K1JT finishing a contact with RU1AA via 144 MHz EME.

Keep in mind that digital EME techniques have been used for less than a decade and are likely to be still evolving. New protocols may be designed in the future using different types of ECC, different message structures, longer or shorter TR sequences, larger or smaller bandwidths, and so on. The latest information will likely be found online — for example, at some of the internet addresses listed in the sidebar, "Computer and Internet Resources for EME."

30.12.4 Finding QSO Partners

At the time of this writing (early 2009) levels of EME activity are highest on 144 MHz. Almost any time the moon is above your horizon (excluding a few days each month when it is near the plane of the Milky Way), you can find JT65 EME signals in the frequency range 144.100 to 144.160 MHz — often dozens of them, especially on weekends. Hundreds of operators, worldwide, are "on the moon" regularly using JT65 on 144 MHz, and some of the larger stations have made EME contacts with more than 2000 other stations. Operating frequencies of particular stations are often posted on real-time loggers (see the sidebar, "Computer and Internet Resources for EME"). Explicit schedules can also be made on the loggers, and this is a good way to initiate your first EME contacts. Don't compromise a true EME contact by resorting to the logger after a QSO attempt has been started, however.

CW activity on the 144 MHz band, as well as most EME activity on higher bands in any mode, tends to be concentrated in activity weekends scheduled once each month when the moon is in a favorable location. After 144 MHz, the bands with most EME activity are 1296 and 432 MHz. On those bands random CW (and occasionally SSB) activity is mostly found between 5 and 30 kHz above the nominal sub-band boundary (for example, between 1296.005 and 1296.030), and digital activity is between .060 and .090, concentrating around .065 or .070. A hundred or more stations are typically active on these bands during the annual ARRL International EME Competition and in other major operating events.

The higher microwave bands, especially 2.3 and 3.4 GHz, have at least several dozen regularly active stations. On these and higher bands activity tends to concentrate around .100. Keep in mind that different band segments may be assigned in different parts of the world.[7] For example, North American stations use 2304.100, but some Europeans work around 2320.100. EME activity on the 50, 222 and 902 MHz bands is mostly done with prearranged schedules. The sidebar, "Computer and Internet Resources for EME" lists some resources that can help you to find schedule partners and the dates of EME contests and activity weekends.

30.13 EME References

[1] J. Taylor, K1JT, "The JT65 Communications Protocol," *QEX*, Sep/Oct 2005, pp 3-12.

[2] M. Franco, N2UO, "Computer Optimized Dual Mode Circularly Polarized Feedhorn," in *Microwave Update 2008*, ARRL, 2008, and **ok1dfc.com/EME/technic/septum/N2UO%20opt.pdf**; P. Wade, W1GHZ, "VE4MA and Chaparral feeds with Septum Polarizers," **www.w1ghz.org/antbook/conf/VE4MA_Chaparral_septum_feeds.pdf**; G. Kollner, DL4MEA, "Designing a Super-VE4MA feed horn for 23 cm and 13 cm," *Dubus*, 1/2007, p 33.

[3] L. Åsbrink, SM5BSZ, "The '+' Configuration vs. the 'X' Configuration," **www.sm5bsz.com/polarity/simplesw.htm**.

[4] D. Halliday, KD5RO (now K2DH), "Microwave EME Using a Ten-Foot TVRO Antenna," *The ARRL UHF/Microwave Projects Manual*, Vol 1, 1996; M. Mattila, VE7CMK, and T. Haynes, VE7CNF, "Project Moonbounce: Earth-Moon-Earth Station VA7MM," **www3.telus.net/public/va7mm/eme/**.

[5] L. Åsbrink, SM5BSZ, "Linrad: New Possibilities for the Communications Experimenter," *QEX*, Part 1, Nov/Dec 2002 pp 37-41; Part 2, Jan/Feb 2003 pp 41-48; Part 3, May/Jun 2003 pp 36-43; Part 4, Sep/Oct 2003 pp 29-37; "Linrad with High Performance Hardware," *QEX*, Jan/Feb 2004, pp 20-31; "Linrad Home Page", **www.sm5bsz.com/linuxdsp/linrad.htm**.

[6] J. Taylor, K1JT, "MAP65: A Panoramic, Polarization-Matching Receiver for JT65," in *Microwave Update 2007*, ARRL, 2007, and **physics.princeton.edu/pulsar/K1JT/MAP65.pdf**.

[7] Y. Mataka, JA4BLC, "Full Spectrum Receiver for 13 cm EME," *Dubus* 3/2005, p 21.

30.14 Glossary of Space Communications Terminology

AMSAT — A registered trademark of the Radio Amateur Satellite Corporation, a nonprofit scientific/educational organization located in Washington, DC. It builds and operates Amateur Radio satellites and has sponsored the OSCAR program since the launch of OSCAR 5. (AMSAT, PO Box 27, Washington, DC 20044.)

Anomalistic period — The elapsed time between two successive perigees of a satellite.

Antenna temperature — Used in EME system evaluation, a measure of the noise picked up by the antenna from all noise sources in the field of view, weighted by the antenna pattern.

AO-# — The designator used for AMSAT OSCAR spacecraft in flight, by sequence number.

AOS — Acquisition of signal. The time at which radio signals are first heard from a satellite, usually just after it rises above the horizon.

Apogee — The point in a satellite's orbit where it is farthest from Earth.

Argument of perigee — The polar angle that locates the perigee point of a satellite in the orbital plane; drawn between the ascending node, geocenter, and perigee; and measured from the ascending node in the direction of satellite motion.

Ascending node — The point on the ground track of the satellite orbit where the sub-satellite point (SSP) crosses the equator from the Southern Hemisphere into the Northern Hemisphere.

Az-el mount — An antenna mount that allows antenna positioning in both the azimuth and elevation planes.

Azimuth — Direction (side-to-side in the horizontal plane) from a given point on Earth, usually expressed in degrees. North = 0° or 360°; East = 90°; South = 180°; West = 270°.

Circular polarization (CP) — Describes an electromagnetic wave in which the electric and magnetic fields are rotating. If the electric field vector is rotating in a clockwise sense, then it is called right-hand circular polarization and if the electric field is rotating in a counterclockwise sense, it is called left-hand circular polarization. Polarization sense is determined looking in the direction in which the wave is traveling.

Cross-polarized — Antennas that are aligned with their polarization at right angles or accept opposite senses of circular polarization.

Descending node — The point on the ground track of the satellite orbit where the sub-satellite point (SSP) crosses the equator from the Northern Hemisphere into the Southern Hemisphere.

Desense — A problem characteristic of many radio receivers in which a strong RF signal overloads the receiver, reducing sensitivity.

Doppler effect — An apparent shift in frequency caused by a change in distance between transmitter and receiver.

Downlink — The frequency on which radio signals originate from a satellite for reception by stations on Earth.

Earth station — A radio station, on or near the surface of the Earth, designed to transmit or receive to/from a spacecraft.

Eccentricity — The orbital parameter used to describe the geometric shape of an elliptical orbit; eccentricity values vary from e = 0 to e = 1, where e = 0 describes a circle and e approaches 1 for a very long, thin ellipse.

EIRP — Effective isotropic radiated power. Same as ERP except the antenna reference is an isotropic radiator.

Elevation — Angle above the local horizontal plane, usually specified in degrees. (0° = plane tangent to the Earth's surface at your location; 90° = straight up, perpendicular to the plane of the Earth).

EME — Earth-Moon-Earth (see also *Moonbounce*)

Epoch — The reference time at which a particular set of parameters describing satellite motion (*Keplerian elements*) are defined.

EQX — The reference equator crossing of the ascending node of a satellite orbit, usually specified in UTC and degrees of longitude of the crossing.

ERP — Effective radiated power. System power output after transmission-line losses and antenna gain (referenced to a dipole) are considered.

ESA — European Space Agency. A consortium of European governmental groups pooling resources for space exploration and development.

Faraday rotation — A rotation of the polarization of radio waves when the waves travel through the ionosphere, in the presence of the Earth's magnetic field.

FO-# — The designator used for Japanese amateur satellites, by sequence number. Fuji-OSCAR 12 and Fuji-OSCAR 20 were the first two such spacecraft.

Geocenter — The center of the Earth.

Geostationary orbit — A satellite orbit at such an altitude (approximately 22,300 miles) over the equator that the satellite appears to be fixed above a given point.

Ground gain — Reflections from the ground in front of an antenna can add as much as 6 dB to the effective gain at certain elevation angles.

Groundtrack — The imaginary line traced on the surface of the Earth by the subsatellite point (SSP).

Inclination — The angle between the orbital plane of a satellite and the equatorial plane of the Earth.

Increment — The change in longitude of ascending node between two successive passes of a specified satellite, measured in degrees West per orbit.

Iskra — Soviet low-orbit satellites launched manually by cosmonauts aboard Salyut missions. Iskra means "spark" in Russian.

JT65 — A digital mode used for EME communications.

Keplerian Elements — The classical set of six orbital element numbers used to define and compute satellite orbital motions. The set is comprised of inclination, Right Ascension of Ascending Node (RAAN), eccentricity, argument of perigee, mean anomaly and mean motion, all specified at a particular epoch or reference year, day and time. Additionally, a decay rate or drag factor is usually included to refine the computation.

LEO — Low Earth Orbit satellite such as the Phase 1 and Phase 2 OSCARs.

LHCP — Left-hand circular polarization.

Libration fading — Rapid fading of EME signals caused by changing relative geometry of reflecting features on the moon's surface and an observer on Earth.

LNA — Low noise amplifier — a preamplifier mounted as close to the antenna terminals as possible for satellite and EME communication.

LOS — Loss of signal — The time when a satellite passes out of range and signals from it can no longer be heard. This usually occurs just after the satellite goes below the horizon.

Mean anomaly (MA) — An angle that increases uniformly with time, starting at perigee, used to indicate where a satellite is located along its orbit. MA is usually specified at the reference epoch time where the Keplerian elements are defined. For AO-10 the orbital time is divided into 256 parts, rather than degrees of a circle, and MA (sometimes called phase) is specified from 0 to 255. Perigee is therefore at MA = 0 with apogee at MA = 128.

Mean motion — The Keplerian element to

indicate the complete number of orbits a satellite makes in a day.

Microsat — Collective name given to a series of small amateur satellites having store-and-forward capability (OSCARs 14-19, for example).

Molniya — Type of elliptical orbit, first used in the Russian Molniya series, that features a ground track that more or less repeats on a daily basis.

Moonbounce — A common name for EME communication in which signals are bounced off the moon before being received.

NASA — National Aeronautics and Space Administration, the US space agency.

Nodal period — The amount of time between two successive ascending nodes of satellite orbit.

Orbital elements — See *Keplerian Elements*.

Orbital plane — An imaginary plane, extending throughout space, that contains the satellite orbit.

OSCAR — Orbiting Satellite Carrying Amateur Radio.

PACSAT — Packet radio satellite (see *Microsat* and *UoSAT-OSCAR*).

Parabolic (dish) antenna — An antenna reflector that is a portion of a paraboloid of revolution. Used mainly at UHF and higher frequencies to obtain high gain and narrow beamwidth when excited by one of a variety of driven elements placed at the dish focus to illuminate the reflector.

Pass — An orbit of a satellite.

Passband — The range of frequencies handled by a satellite translator or transponder.

Path loss — The total signal loss between transmitting and receiving stations relative to the total radiated signal energy.

Perigee — The point in a satellite's orbit where it is closest to Earth.

Period — The time required for a satellite to make one complete revolution about the Earth. See *Anomalistic period* and *Nodal period*.

Phase 1 — The term given to the earliest, short-lived, low Earth orbit (LEO) OSCAR satellites that were not equipped with solar cells. When their batteries were depleted, they ceased operating.

Phase 2 — LEO OSCAR satellites. Equipped with solar panels that powered the spacecraft systems and recharged their batteries, these satellites have been shown to be capable of lasting up to five years (OSCARs 6, 7 and 8, for example).

Phase 3 — Extended-range, high-elliptical-orbit OSCAR satellites with very long-lived solar power systems (OSCARs 10 and 40, for example).

Phase 4 — Proposed OSCAR satellites in geostationary orbits.

Polar Orbit — A low, circular orbit inclined so that it passes over the Earth's poles.

Precession — An effect that is characteristic of AO-10 and AO-40 orbits. The satellite apogee SSP will gradually change over time.

Project OSCAR — The California-based group, among the first to recognize the potential of space for Amateur Radio; responsible for OSCARs I through IV.

QRP days — Special orbits set aside for very low power uplink operating through the satellites.

RAAN — Right Ascension of Ascending Node. The Keplerian element specifying the angular distance, measured eastward along the celestial equator, between the vernal equinox and the hour circle of the ascending node of a spacecraft. This can be simplified to mean roughly the longitude of the ascending node.

Radio Sputnik — Russian Amateur Radio satellites (see *RS #*).

Reference orbit — The orbit of Phase II satellites beginning with the first ascending node during that UTC day.

RHCP — Right-hand circular polarization.

RS # — The designator used for most Russian Amateur Radio satellites (RS-1 through RS-15, for example).

Satellite pass — Segment of orbit during which the satellite "passes" nearby and in range of a particular ground station.

Sidereal day — The amount of time required for the Earth to rotate exactly 360° about its axis with respect to the "fixed" stars. The sidereal day contains 1436.07 minutes (see *Solar day*).

Sidereal month — The Earth and moon complete their mutual orbit (with respect to fixed stars) every 27.3 days, one sidereal month.

Solar day — The solar day, by definition, contains exactly 24 hours (1440 minutes). During the solar day the Earth rotates slightly more than 360° about its axis with respect to "fixed" stars (see *Sidereal day*).

Spin modulation — Periodic amplitude fade-and-peak resulting from the rotation of a satellite's antennas about its spin axis, rotating the antenna peaks and nulls.

SSP — Subsatellite point. Point on the surface of the Earth directly between the satellite and the geocenter.

System noise temperature — Used in EME system evaluation, a measure of noise picked up by the antenna plus noise generated in the receiver.

Telemetry — Radio signals, originating at a satellite, that convey information on the performance or status of onboard subsystems. Also refers to the information itself.

Transponder — A device onboard a satellite that receives radio signals in one segment of the spectrum, amplifies them, translates (shifts) their frequency to another segment of the spectrum and retransmits them. Also called linear translator.

UoSAT-OSCAR (UO #) — Amateur Radio satellites built under the coordination of radio amateurs and educators at the University of Surrey, England.

Uplink — The frequency at which signals are transmitted from ground stations to a satellite.

Visibility Circle — The range of area on the Earth that are "seen" by a satellite. This is also called the "footprint" for that satellite.

Window — Overlap region between acquisition circles of two ground stations referenced to a specific satellite. Communication between two stations is possible when the subsatellite point is within the window.

WSJT — A suite of computer software that includes tools for EME communication.

Contents

31.1 Sound Card Modes
    31.1.1 Which Sound Card is Best?
    31.1.2 Sound Card Interfaces
    31.1.3 Transmit Audio Levels
    31.1.4 Sound Card Software

31.2 Packet Radio
    31.2.1 The Packet TNC
    31.2.2 Talking to a TNC
    31.2.3 TNCs and Radios
    31.2.4 TNC Timing
    31.2.5 Monitoring
    31.2.6 "Connected" vs "Unconnected"

31.3 The Automatic Packet/Position Reporting System (APRS)
    31.3.1 Setting Up an APRS Station
    31.3.2 Maps and APRS
    31.3.3 APRS Position Encoders
    31.3.4 APRS Networking Tips

31.4 PACTOR

31.5 High Speed Multimedia (HSMM)
    31.5.1 A Basic HSMM Radio Station
    31.5.2 Running High Power
    31.5.3 HSMM Antenna Options

31.6 Automatic Link Establishment (ALE)

31.7 D-STAR
    31.7.1 What is D-STAR?
    31.7.2 Digital Voice and Low-Speed Data (DV)
    31.7.3 High-Speed Data (DD)
    31.7.4 D-STAR Radios

31.8 APCO-25
    31.8.1 APCO-25 and Amateur Radio
    31.8.2 The APCO-25 Standard
    31.8.3 APCO-25 "Phases"

31.9 HF Digital Voice
    31.9.1 AOR and AMBE
    31.9.2 WinDRM
    31.9.3 FDMDV: Frequency Division Multiplex Digital Voice

31.10 EchoLink, IRLP and WIRES-II
    31.10.1 EchoLink
    31.10.2 IRLP
    31.10.3 WIRES-II

31.11 Digital Communications Glossary

31.12 Bibliography and References

Chapter 31

Digital Communications

In this chapter, Steve Ford, WB8IMY, discusses the techniques involved in assembling and configuring station components for operating on the various HF and VHF digital modes. Today's digital communication choices range from keyboard based modes like classic RTTY, packet radio and PSK31, to digital voice, local high speed multimedia networks and VHF/UHF networks linked by the Internet. Related information may be found in the **Modulation** and **Digital Modes** chapters.

Amateur digital communication has always been a niche activity, but it has grown substantially over the years. From the end of World War II until the early 1980s, *radioteletype*, better known as *RTTY*, was *the* Amateur Radio digital mode. If you had visited an amateur RTTY station prior to about 1977, you probably would have seen a mechanical teletype machine, complete with rolls of yellow paper. The teletype may have been connected to the transceiver through an interface known as a *TU*, or *terminal unit*. An oscilloscope would probably have graced the layout as well, used for proper tuning of the received signal.

When affordable microprocessor technology appeared in the late 1970s, terminal units evolved as well. Some included self-contained keyboards and video displays, making the mechanical teletype obsolete. As personal computers evolved, they became perfect companions for TUs. In this configuration, the PC functioned as a "dumb terminal," displaying the received data *from* the TU and sending data *to* the TU for transmission. TUs of this era offered ultra-sharp receive filters that allowed hams to copy weak signals in the midst of interference.

In the late 1980s, conventional terminal units began to yield to sophisticated microprocessor devices known as *multimode controllers*. As the name suggests, these compact units handle several different digital modes in one package, typically RTTY, packet, AMTOR and PACTOR. Like TUs, multimode controllers are stand-alone devices that communicate with a personal computer acting as a dumb terminal. All of the heavy lifting is being done by the controller and its self-contained software known as *firmware*.

In the early 1990s, sound cards appeared for personal computers. As sound cards became more powerful, hams began to realize their potential. With the right software, a sound card could take received audio directly from the radio and translate it into digital information. The same sound card could also create various forms of digital audio modulation for transmission. The first "sound card mode" was PSK31, developed by Peter Martinez, G3PLX. In the years that followed, sound cards became more powerful and versatile. Hams responded by developing more new digital modes to take advantage of the advances.

Hardware controllers are still with us, but they are primarily used for modes like packet and PACTOR that require more processing muscle and precision timing than a typical personal computer can provide on its own. Other amateur digital modes such as D-STAR depend on specially designed transceivers that combine the radio hardware with dedicated digital processing firmware.

31.1 Sound Card Modes

Sound card technology dominates the amateur HF digital communications world. Although the term sound *card* is commonly used, the techniques discussed in this section apply to motherboard-embedded sound chip sets, which are extremely common in computer systems today, and to external sound processing devices. We'll use the term "sound card" to refer to any of these hardware implementations. See **Fig 31.1**.

The sound card modes in use today include PSK, RTTY, MFSK16, Olivia, Hellschreiber and MT63. There are also sound-card applications for digital voice, discussed later in this chapter and for slow-scan TV, discussed in the **Image Communications** chapter.

The 31-baud version of PSK, known as PSK31, is by far the most popular mode for casual HF digital communication. The popularity of PSK31 notwithstanding, radioteletype (RTTY) still remains the chief HF mode for contesting and DXing. All other HF digital modes play minor

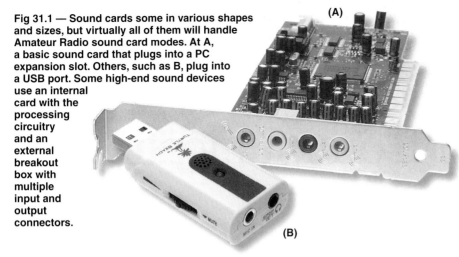

Fig 31.1 — Sound cards some in various shapes and sizes, but virtually all of them will handle Amateur Radio sound card modes. At A, a basic sound card that plugs into a PC expansion slot. Others, such as B, plug into a USB port. Some high-end sound devices use an internal card with the processing circuitry and an external breakout box with multiple input and output connectors.

roles, but each has characteristics that provide benefits depending on the use case. Olivia, MFSK16 and MT63, for example, provide more robust copy under poor conditions.

On VHF and above, the *WSJT* software suite is the sound card mode of choice for meteor scatter and moonbounce work. *WSJT* has also found experimental application on the HF bands. See the sidebar, "The WSJT Revolution."

Regardless of the mode in question, the sound card functions as the critical link. It is put to work as a digital-to-analog (D/A) and analog-to-digital (A/D) converter. In its A/D role, the sound card takes receiver audio and converts it to digital information. During transmission, the sound card is used as a D/A converter, taking digital information from the software application and creating a corresponding analog signal that is fed to the transceiver. (For more information on A/D and D/A converters, see the **Analog Basics** chapter.)

31.1.1 Which Sound Card is Best?

This is one of the most popular questions among HF digital operators. After all, the sound device is second only to the radio as the most critical link in the performance chain. Sound cards have traditional analog audio amplifiers, mixers and filters, all of which can introduce noise, distortion and crosstalk. A poor sound device will bury weak signals in noise of its own making and will potentially distort your transmit audio as well.

If you have a modest station and intend to enjoy casual chats and a bit of DXing, save your money. An inexpensive sound card, or the sound chipset that is probably on your computer's motherboard, is adequate for the task. There is little point in investing in a luxury sound card if you lack the radio or antennas to hear weak signals to begin with, or if they cannot hear you.

On the other hand, if you own the station hardware necessary to be competitive in digital DX hunting or contesting, a good sound card can give you a substantial edge. Other applications that require a lot of processing power, such as software-defined radio (SDR), require a high-performance sound card. Often vendors will offer a list of sound devices known to perform well with their equipment.

Sound cards convert analog audio signals to a set of digital samples, but this conversion from analog to digital isn't perfect, for several reasons. Here are some parameters to consider.

Sample Size

When a sound card takes a sample of the input voltage, it expresses it as a binary number with a certain number of bits. This is the *sample resolution,* or *sample size.* The sample resolution determines the number of steps between the smallest and the largest signal the card can measure. The greater the number of steps and the smaller they are, the more precise the samples will be. Larger steps introduce more *quantization noise,* so a sound card's signal-to-noise ratio is limited by the number of bits of resolution in each sample. For example, a card taking 8-bit samples measures only 256 voltage steps and cannot yield a signal-to-noise ratio better than about 49 dB. With 16-bit samples, there are 65,536 steps, and the ideal S/N rises to 98 dB.

Sample Rate

The clock that drives the A/D converter runs at a steady rate, known as the *sample rate.* As you might expect, a higher sample rate is required to accurately capture higher-frequency sounds. A waveform can be ac-

---

The *WSJT* Revolution

Hams routinely use meteors and the moon as radio reflectors for *meteor scatter* and *moonbounce* communication. You can read more about these activities in the chapters on **Propagation of RF Signals** and **Space Communications**. For many years, meteor scatter and moonbounce required large antennas and high power RF amplifiers. CW was the mode of choice since a concentrated, narrow-bandwidth signal had the best chance to survive the journey and still be intelligible at the receiving station. That changed in 2001 when Joe Taylor, K1JT, unveiled a suite of sound card applications known simply as *WSJT*.

By using the sound card and computer as powerful digital signal processors, *WSJT* greatly reduced the station hardware requirements, making it possible for amateurs with modest stations to make meteor scatter and moonbounce contacts. Hams have also experimented with using some *WSJT* modes on the HF bands to make contacts using extremely low power. *WSJT* does not support conversational contacts with lengthy exchanges of information. Instead, the software allows for basic information exchange sufficient to meet the requirements that a contact has taken place.

WSJT is available for both *Windows* and *Linux*. It is a software suite that supports five different operating modes: *FSK441* for meteor scatter; *JT65* for moonbounce, but also being used occasionally on HF; *JT6M*, optimized for meteor scatter on 6 m; *EME Echo* for measuring the echoes of your own signal from the moon; and *CW* for moonbounce communication using 15 WPM Morse code.

The software is available for free downloading from **physics.princeton.edu/pulsar/K1JT**.

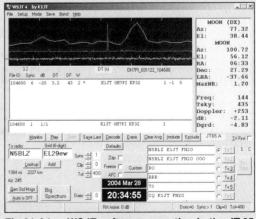

Fig 31.A1 — WSJT software operating in the JT-65 mode.

curately captured by sampling at twice the highest frequency of interest. Energy at higher frequencies produces *aliases*, so sound cards put a low-pass filter ahead of the A/D converter, running at a cutoff frequency equal to one-half the sample rate. These filters cannot be perfect, so there's bound to be either some high frequency roll-off, or some distortion due to aliases sneaking through.

Linearity

If the sample steps aren't all exactly the same size, or the clock drifts up and down a little bit in frequency, distortion is introduced. The ideal sound card would have a perfectly linear A/D converter and a perfectly stable clock, but of course these are impossible to achieve. Good-quality sound cards do, however, have crystal-controlled oscillators as their clock source.

Sample Rate Accuracy

Even if the clock runs at a stable frequency, we won't get the desired result if it's running at the *wrong* frequency. Sample rate accuracy can be important for analog modes that aren't continuously synchronized. For these modes, one sound card is generating a signal and the other is receiving it, and the two cards are expected to be running at exactly the same sample rate. Distortion, or even loss of data, can occur if the rates are slightly different.

Remember that sound cards work by taking the analog audio signal at the input and converting it to digital information by sampling the audio signal at a very high rate, typically between 8000 and 11025 Hz for most ham applications. Sound cards can have sampling rate errors that will seriously affect weak-signal copy. For example, the laptop on which this chapter is being written has an actual sampling rate of 11098.79 Hz instead of the nominal 11025 Hz. This error not only affects the apparent frequency of an incoming tone, but can keep a program that requires a high degree of frequency stability, such as MFSK16, from maintaining consistent framing on incoming data. The result will be poor or inconsistent copy.

VoIP applications such as EchoLink and IRLP send a continuous stream of data from one sound card to another over the Internet. If the sender and receiver aren't running at the same rate, it can cause audio drop-outs, as the buffer at the receiving end becomes empty or overflows.

WSJT requires a high degree of accuracy but works around this problem by measuring the actual sample rate for sound card input and output (on some computers they are not the same) and then doing appropriate manipulations of the data. Most software doesn't have active sample error correction, however. If your software and hardware support it, you might realize better performance at a higher sample rate such as 12 kHz.

Sample rate accuracy is usually less of an issue for modes such as RTTY, PSK31, Olivia and DominoEX that synchronize the receiver with the sender frequently. In these modes, the sound card's clock is still used as the timing reference, but exact sound card timing is far less critical.

SOUND CARD STATION SETUP

Fig 31.2 shows a typical sound card station setup. A simplified sound card is shown here, but even the simplest of sound cards can be complicated. They can have as few as two external connections but there may be as many as twelve or more. You may find ports labeled LINE IN, MIC IN, LINE OUT, SPEAKER OUT, PCM OUT, PCM IN, JOYSTICK, FIREWIRE, S/PDIF, REAR CHANNELS or SURROUND jacks, just to name a few.

Depending on your computer, you may be able to choose your receive audio connection from either MIC INPUT or LINE INPUT. Anything else you find is not an analog audio input. If you do have a choice, use the LINE INPUT for the receive audio from your radio. Although the MIC INPUT jack can be used, it will have much more gain than you need and you may find adjustment quite critical. Some sound cards have an "advanced" option to select a 20 dB attenuator that will reduce gain and make the MIC INPUT jack easier to use.

If your only goal is to decode received signals, you need nothing more than an audio cable between the transceiver and the sound card. You should not need ground isolation for receive audio as it is at a high level and is normally not susceptible to ground loops. You may also be able to choose from several outputs that appear on the radio. Your radio may have SPEAKER, HEADPHONE, LINE OUT, RECORD, PHONE PATCH and DATA OUTPUT jacks available. These may be fixed output or variable output (using the radio's volume control). Be careful with radio's DATA OUTPUT jack — it may not work on all modes.

For the sound card transmit connection, you will have a choice of the computer's HEADPHONE OUTPUT, LINE OUTPUT, SPEAKER OUTPUT or a combination of these. The SPEAKER OUTPUT is usually the best choice as it will drive almost anything you hook to it. The SPEAKER OUTPUT has a low source impedance making it less susceptible to load current and RF. Any one of these outputs will usually work fine, provided you do not load down a line or headphone output by using very low impedance speakers or headphones to monitor computer-transmitted audio. The transmit audio connection must have full ground isolation through your interface, especially if it drives the MIC INPUT of the radio. This usually means adding an isolation transformer, as discussed in the next section.

31.1.2 Sound Card Interfaces

In addition to providing audio connections between the sound card and your transceiver, you also need to provide a way for your computer to switch the radio between receive and transmit. This is where the *sound card interface* comes into play.

Commercial sound card interfaces such as the one shown in **Fig 31.3** match the audio levels, isolate the audio lines and provide transmit/receive switching, usually with your computer's serial (COM) or USB port. You can also make your own interface by simply isolating the audio lines and cobbling together a single-transistor switch to connect to your COM port. **Fig 31.4** shows some commonly used interface circuits.

In Fig 31.2 you'll notice that the transmit

Fig 31.2 — A typical interface connection between a computer and a transceiver. Note that the transmit audio connects to the radio through the interface, and TR keying is provided by the computer serial port. Newer sound card interfaces are often designed to work with computer USB ports.

Fig 31.3 — The MicroHam Digi Keyer is an example of a multifunction sound card interface. Some commercial devices include CW keyers, transceiver control interfaces and support for multiple transceivers.

audio connects to the radio through the interface. By doing so, the interface can provide isolation. Some interfaces also include a transmit audio adjustment, although this can also be accomplished at the computer, as described in the previous section.

It's important to mention that you may also be able to use the VOX function on your transceiver to automatically switch from receive to transmit when it senses the transmit audio from your sound card. This approach completely removes the need for a TR switching circuit, COM port and so on.

The weakness of this technique is that it will cause your radio to transmit when it senses *any* audio from your computer — including miscellaneous beeps, sounds or music. It's best to turn off these sounds from your computer so you don't transmit them accidentally. Another approach is to use a second sound card so that one can be used for regular computer audio applications and one dedicated to interfacing with the radio.

AFSK AND FSK

Most sound card modes rely on some form of frequency and/or phase-shift keying to create digitally modulated RF signals. This modulation takes place at audio frequencies with the sound card audio output applied directly to an SSB voice transceiver, either at the microphone jack or at a rear-panel accessory jack, and is called *audio frequency shift keying* (*AFSK*).

RTTY, PACTOR I and AMTOR signals can be sent using AFSK, and often are. It is also possible to transmit these modes by applying discrete binary data *directly* to the transceiver. This technique is known simply as direct *frequency-shift keying (FSK)*.

For example, each character in the Baudot RTTY code is composed of five bits. When modulated with AFSK, a "1" bit is usually represented by a 2125-Hz tone and is known as a *mark*. A "0" bit is represented by a 2295-Hz tone called a *space*. The difference between the mark and space is 170 Hz, called the *shift*. When applied to a single-sideband transceiver, this AFSK audio signal effectively generates an RF output signal that shifts back and forth between the mark and space frequencies.

A transceiver that supports FSK, however, can accept mark/space digital data directly from the computer and will use that information to automatically generate the frequency-shifting RF output. No audio signal is applied to the transceiver when operating FSK.

Is there an advantage to using AFSK or FSK? In years past, transceivers that did not support FSK operation often did not allow the use of narrow IF filtering. Those filters were reserved for CW, not the SSB voice mode used with AFSK. If you wanted to use RTTY with such a rig, you had to use AFSK and contend with the wider (2.4 kHz or so) SSB IF bandwidth, or else add an external audio filter. FSK-capable transceivers, on the other hand, allowed the RTTY operator to select narrow CW filters, reducing receive interference in crowded bands.

Many of today's transceivers offer adjustable-bandwidth digital signal processing (DSP) filters in the IF stages that can be used with any operating mode. This has effectively eliminated the FSK advantage, at least for receiving.

The appeal of FSK remains, however, when it comes to transmitting. A *properly modulated* AFSK signal is indistinguishable from FSK, but it is relatively easy to overdrive an SSB transmitter when applying an audio signal from a sound card (more on this in the next section). With FSK this is never an issue. You simply feed data from the computer to the radio; the radio does the rest.

This is why a number of RTTY operators still use FSK, and it's why transceiver manufacturers still offer FSK modes (sometimes labeled DATA or RTTY) in their products. Several sound card interfaces support FSK by providing a dedicated TTL circuit between the computer COM port, where the FSK data appears, and the transceiver FSK input. When used in this fashion, the sound card does not generate a transmit audio signal at its output. Instead, the RTTY software keys the various lines at the COM port to send the FSK data.

If the sound card interface you've chosen doesn't support FSK, you can build your own TTL interface using the circuit shown in Fig 31.4. This simple circuit uses a transistor that is keyed on and off by data pulses appearing on the COM port TxD pin (pin 3 on a 9-pin COM port).

It is important to note that FSK transceiver inputs can only be used for modes that are based on binary FSK, typically with 170 or 200-Hz shifts. These modes include RTTY,

Fig 31.4 — Some commonly used interface circuits. At A, isolating the sound card and transceiver audio lines. T1 and T2 are 1:1 audio isolation transformers such as the RadioShack 273-1374. At B, a simple circuit to use the computer COM port to key your transceiver PTT, and at C, a similar circuit for FSK keying. Q1 is a general purpose NPN transistor (MPS2222A, 2N3904 or equiv). At D, an optocoupler can be used to provide more isolation between radio and computer. On a DB9 serial port connector: RTS, pin 7; DTR, pin 4; TxD, pin 3; GND, pin 5.

PACTOR I and AMTOR. You cannot use the FSK input to transmit multi-frequency or phase-shift modulated signals such as MFSK16 or PSK31.

31.1.3 Transmit Audio Levels

When applying sound card audio to a transceiver, it is critical to maintain proper levels. By overdriving the transceiver's audio input you may create a wide, distorted RF signal that will be difficult, if not impossible, to decode at the receiving end. Such an overmodulated signal will also cause considerable interference on adjacent frequencies.

As you increase the transmit audio output from your sound card, pay careful attention to the ALC indicator on your transceiver. The ALC is the automatic level control that governs the audio drive level. When you see your ALC display indicating that audio limiting is taking place, or if the display indicates that you are exceeding the ALC "range," you are feeding too much audio to the transceiver.

Monitoring the ALC by itself is not always effective. Many radios can be driven to full output without budging the ALC meter out of its "nominal" range. Some radios become decidedly nonlinear when asked to provide SSB output beyond a certain level (sometimes this nonlinearity can begin at the 50% output level). We can ignore the linearity issue to a certain extent with an SSB voice signal, but not with digital modes because the immediate result, once again, is splatter.

The simplest method to tell if your signal is clean is to ask someone to give you an evaluation on the air. For example, PSK31 programs commonly use a waterfall audio that can easily detect "dirty" signals. The splatter appears as rows of lines extending to the right and left of your primary signal, as shown in **Fig 31.5**. (Overdriven PSK31 signals may also have a harsh, clicking sound.)

If you are told that you are splattering, ask the other station to observe your signal as you slowly decrease the audio level from the sound card or processor. When you reach the point where the splatter disappears, you're all set.

If PSK31 is your primary digital mode, an alternative solution for *Windows* users is the PSKMeter by Software Science. This clever device samples the RF from your transceiver and *automatically* adjusts the audio output of your sound card until your signal is clean. The RF sampling port of the PSKMeter connects to your antenna coax with a T connector. The data is fed to your computer through an available COM (serial) port, or a USB port if you have a USB-to-serial converter. The PSKMeter software then analyzes the sampled signal and adjusts the master output volume of your sound card accordingly. The PSKMeter is available in kit form and can be ordered on the Web at **www.ssiserver.com/info/pskmeter**.

Whether you choose automatic or manual adjustments, don't worry if you discover that you can only generate a clean signal at, say, 50 W output. With PSK31 and most other sound card modes, the performance differential between 50 W and 100 W is inconsequential.

CONSISTENT AUDIO LEVELS

One of the perennial problems with sound devices is the fact that optimum audio levels have a tendency to differ depending on the software you are using. The mixer levels you set correctly for one application may be wildly wrong for another. And what happens when another family member stops by the computer to listen to music or play a game? They're likely to change the audio levels to whatever suits them. When it comes time to use the computer again for HF digital operating, you may get an unpleasant surprise.

The commonsense answer is to simply check and reset the audio levels before you operate. There are some software applications that will "remember" your audio settings and reapply them for you. One example is *Quick-Mix* for *Windows*, which is available free on the Web at **www.ptpart.co.uk/quickmix**. *QuickMix* takes a "snapshot" of your sound card settings and saves them to a file that you name according to the application (for example, "PSK31"), which can be reloaded next time you operate using that mode.

An even more elegant solution is the free *Sound Card Manager* for *Windows* by Roger Macdonald, K7QV, which you can download at **www.romacsoftware.com/Sound-Management.htm**. Whenever you start an

Computer Power

You don't need an extremely powerful computer to enjoy digital operating. You can pick up an old 850 MHz Pentium laptop on eBay (**www.ebay.com**) for less than $300. You can probably find a similar desktop system for that price and even less. Either computer will be adequate for 90% of the software you are likely to use. Serious processing power only comes into play with software-defined radios or digital voice and image applications. For those endeavors, a 2 GHz system is best for smooth performance.

Every computer needs an operating system and that's where you may encounter some sticky issues. Until 2007, the most widely used operating system was *Windows XP*. Most Amateur Radio software available today was originally written for *XP*. Much of it will also run under *Windows Vista*, *XP's* replacement, but there are no guarantees. If you are upgrading to *Vista*, or if you've purchased a computer with *Vista* already installed, you'll need to test your ham applications for compatibility. One problem that crops up frequently involves the "Help" files that so many *Windows* programs use. *Vista* does not recognize the "old" *Windows* Help format, so when you click on Help (or a portion of the *Windows* application that uses the older Help format, such as the user manual), *Vista* may present you with an error message instead. No doubt ham programmers are already adapting to this change and re-writing their applications accordingly, but it may take some time for these revisions to make it into the marketplace.

Vista also requires more CPU power to run properly. Beware of buying an underpowered new or used computer with *Windows Vista* pre-installed (and beware of installing *Vista* on an anemic machine). To obtain decent performance from *Vista*, you need a minimum 2 GHz microprocessor and 1 GB of RAM.

The good news for *Windows* users is that *XP* is likely to be around for several years to come, although Microsoft will eventually stop supporting it. *Vista* is being replaced by *Windows 7*, which offers superior performance.

Of course, you don't have to use *Windows* to enjoy HF digital. The *Linux* operating system has been gaining in popularity and you can pick up *Linux* "distributions" on the Web free of charge. For a good example, check out *Ubuntu Linux* at **www.ubuntu.com**. Mac users will find *Mac OS* applications for digital modes such as APRS and many others. See Table 31.1.

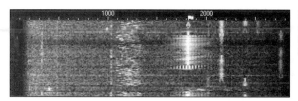

Fig 31.5 — The waterfall spectrum display built into popular sound card software can help detect overdriven PSK31 signals. Note the lines extending from the signal on the right, near the 2000 Hz marker, indicating splatter. The other signals are from properly adjusted transmitters.

Table 31.1
Sound Card Software

Windows
ALE
PCALE — www.hflink.com

Multimode
MixW — www.mixw.net
HamScope — www.qsl.net/hamscope
MultiPSK — f6cte.free.fr/index_anglais.htm
Ham Radio Deluxe — www.ham-radio-deluxe.com

Hellschreiber
IZ8BLY — xoomer.virgilio.it/aporcino/Hell/index.htm

MFSK16
IZ8BLY — xoomer.virgilio.it/aporcino/Stream/index.htm

PSK31
DigiPan — www.digipan.net
WinWarbler — www.dxlabsuite.com/winwarbler
W1SQLPSK — www.faria.net/w1sql
WinPSK — www.moetronix.com/ae4jy/winpsk.htm

Digital Voice
WinDRM — www.n1su.com/windrm

MacOS
Multimode
MultiMode — www.blackcatsystems.com/software/multimode.html
cocoaModem — homepage.mac.com/chen

Linux
Multimode
Fldigi — www.w1hkj.com/Fldigi.html

WINMOR
www.winlink.org/WINMOR

application (such as a piece of HF digital software), *Sound Card Manager* automatically reconfigures your sound card and then returns it to the default setting when you are done.

31.1.4 Sound Card Software

As you can see in **Table 31.1**, there is great deal of software available for sound-card-based digital modes. Some of the software applications are free, while others require registration and a fee.

There are applications dedicated to particular modes, such as *DigiPan* (PSK31) and *MMTTY* (RTTY). The trend in recent years has been to multimode programs that support many different digital modes in a single application.

Another trend has been toward *panoramic reception* where the software processes and displays all signals detected within the bandwidth of the received audio signal. The signal "signatures" are often shown within continuously scrolling *waterfall* displays like the one shown in **Fig 31.6**.

Panoramic reception is particularly popular among PSK31 operators because the narrow signals tend to cluster within a relatively small

Digital Communications and Public Service

In recent years, amateur digital communication has found frequent application in the realm of public service. There has been a proliferation of local and regional public-service networks, primarily using VHF packet as well as D-STAR, but three systems currently carry the lion's share of the long-haul traffic load.

Winlink 2000

Winlink 2000 is an Amateur Radio digital network with HF and VHF components. Both provide the ability to transfer e-mail to and from the Internet.

Winlink 2000 functions as a full-featured Internet-to-HF/VHF "star network" gateway system. The HF portion of the network is composed of Radio Message Server HF — *RMS HF* — stations. These stations are accessed through the use of PACTOR I, II or III and WINMOR using the multimode controller-based stations described in this chapter. On the VHF side, support is provided through *RMS Packet* stations. These packet radio stations can be accessed using common packet TNCs connected to FM voice transceivers.

Most VHF RMS Packet stations are on the air continuously. On the HF side, RMS HF stations throughout the world scan a variety of HF digital frequencies on a regular basis, listening on each frequency for about two seconds. By scanning through frequencies on several bands, these RMS HF stations can be accessed on whichever band is available at any given time.

All RMS Packet and HF stations link to a central group of redundant Common Message Servers. Since all e-mail messages coming to and from the Internet are stored on these servers and available to all RMS Packet and HF stations, Winlink 2000 users never have to designate a "home" BBS, as in the traditional packet radio world. You can send an e-mail from VHF to the Internet via a RMS Packet station in California and pick up the reply from an RMS HF station in South Africa.

In addition to text e-mail, Winlink 2000 is capable of handling file attachments such as images. However, the larger the attachment the longer it will take to send via radio. This is particularly true on VHF where most packet stations are still operating at only 1200 baud. See the Winlink 2000 Web site at **www.winlink.org**.

National Traffic System-Digital

The National Traffic System-Digital (NTSD) is based on the original Winlink structure now referred to as *Winlink Classic*. Unlike Winlink2000, which uses Internet linking between its Common Message Servers, NTSD is entirely RF based.

Most HF NTSD Mailbox Operations (MBOs) are accessed using PACTOR I or II. Designated NTSD operators in each region and local area relay messages, either between regions or to and from the local area stations. The hierarchical NTSD structure is discussed at **home.earthlink.net/~bscottmd/n_t_s_d.htm**.

Traditional packet radio is often used at the local level. In fact, NTS packet messages can be initiated and sent by any packet-capable operator. Messages for delivery are posted on cooperating NTS PBBSs (Packet Bulletin Board Systems). Messages come into these BBSs from the NTSD HF network, or from local packet networks in nearby sections or regions. In addition, any NTS voice messages that might not have been picked up on a voice net can be transcribed to text and posted to an NTS PBBS.

Global ALE High Frequency Network

Global ALE High Frequency Network (HFN) is an international Amateur Radio Service organization of ham operators dedicated to emergency/relief radio communications. The main purpose of the network is to provide efficient communications to remote areas of the world using amateur Automatic Link Establishment (ALE) as described in this chapter. The network supports direct station-to-station emergency text messaging as well as the ability to port text messages to and from the Internet. See **hflink.com/emcomm/**.

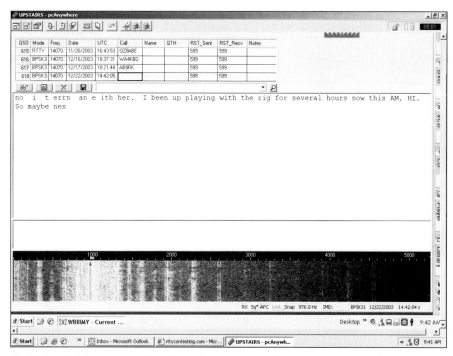

Fig 31.6 —Multimode sound-card applications typically support a variety of modes such as RTTY, PSK31, Olivia and MFSK16. Sent and received text appear in the white box. At the bottom of the screen is a waterfall tuning display. Tune the transceiver VFO to the calling frequency and then select stations to work by clicking on the signal trace in the display.

range of frequencies. By using panoramic reception, an operator can simply click the mouse cursor on one signal trace in the waterfall after another, decoding each one in turn.

The weakness of panoramic reception appears when wide IF filtering is used to display as many signals as possible. The automatic gain control (AGC) circuit in the receiver is acting on everything within that bandwidth, working hard to raise or lower the overall gain according to the overall signal strength. That's fine if all the signals are approximately the same strength, but if a very strong signal appears within the bandwidth, the AGC will *reduce* the gain to compensate. The result will be that many of the signals in the waterfall display will suddenly vanish, or become very weak, as the AGC drops the receiver gain. In cases where an extremely strong signal appears, *all* signals except the rock crusher may disappear completely.

The alternative is to use narrower IF or audio filter settings. These will greatly reduce the waterfall display width, but they will also remove or reduce strong nearby signals.

31.2 Packet Radio

Although the heyday of traditional packet radio has passed, it still survives in the form of networks devoted to specific applications such as emergency communication (see the sidebar "Digital Communications and Public Service") and DX spotting. A variant of packet radio also finds popular use in the Automatic Packet/Position Reporting System (APRS), addressed in a later section.

31.2.1 The Packet TNC

In the beginning, there was *X.25*, a protocol for wide-area digital networks that typically communicated over telephone lines. X.25 works by chopping data into strictly defined *packets* or *frames* of information. Each packet is sent to the destination device where it is checked for errors. If errors are discovered, the packet must be sent again to ensure that received data is 100% error free.

In the early 1980s, amateurs began adapting X.25 for over-the-air digital communications. The result was *AX.25*. The new AX.25 protocol worked in much the same way, although it identified each message by sender and destination station call signs and added a Secondary Station ID (SSID) in a range from 0 through 15.

To create and decode these AX.25 packets, hams invented the *terminal node controller*, or *TNC*. TNCs do much more than assemble and disassemble data; they are miniature computers unto themselves. A TNC is also programmed to work within a radio network where there may be other competing signals. For example, to maximize the throughput for everyone on the same frequency, a TNC is designed to detect the presence of other data signals. If it has a packet to send, but detects a signal on the frequency, it will wait until the frequency is clear. TNCs also have a variety of user adjustments and other features, such as a mailbox function to store messages when the operator is away.

Regardless of the changes in packet radio, the TNC is still a vital component. In essence, a TNC functions as a "radio modem." It acts the middleman between your radio and your computer. The TNC takes data from your computer, creates AX.25 packets and then transforms the AX.25 formatted data into audio signals for transmission by the radio. Working in reverse, the TNC demodulates the received audio, changes it back into data, disassembles the AX.25 packets and sends the result to the computer.

For 300 and 1200-baud applications, TNCs create signals for transmission using AFSK. The most common is 1200 baud packet, used primarily at VHF. When creating a 1200-baud signal, a *mark* or 1 bit is represented by a frequency of 1200 Hz. A *space* or 0 bit is represented by a frequency of 2200 Hz. The transition between each successive mark or space waveform happens at a rate of 1200 baud. The frequencies of 1200 and 2200 Hz fit within the standard narrowband FM audio passband used for voice, so that AFSK is accomplished by simply generating 1200 and 2200 Hz tones and feeding them into the microphone input of a standard FM voice transmitter.

FSK, not AFSK, is used for 9600 baud packet. The data signal is filtered and encoded and then applied directly to the FM transmitter, after the microphone amplifier stage, through a dedicated port.

A block diagram of a typical TNC is shown in **Fig 31.7**. You'll note that it has a serial interface connecting to a "terminal." Most commonly, the terminal is a full-fledged computer running terminal software. Data flows to the computer and vice versa via this interface. At the heart of the TNC is the microprocessor and the attendant *High-level Data Link Controller*, or *HDLC*. The microprocessor is the brain of the unit, but the HDLC is responsible for assembling and disassembling the packets. The modem is simply that — a modulator (changing data to audio tones) and demodulator (changing audio tones back to data).

You can still find TNCs for sale from several manufacturers. There are also several transceivers with packet TNCs built in.

Another solution is a software TNC known

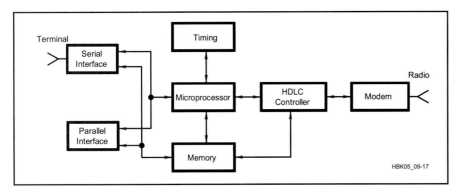

Fig 31.7 — A block diagram of a typical packet radio TNC.

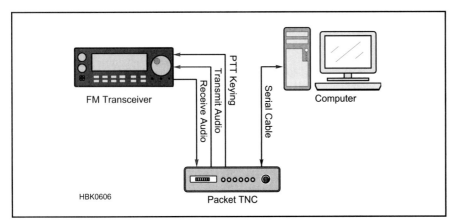

Fig 31.8 — A diagram of a packet radio station built around a hardware TNC.

input and the push-to-talk (PTT) line. (The TNC grounds the PTT line to key your transceiver.) Some transceivers also make receive audio available at the microphone jack for use with speaker/microphone combos. You can use this line to feed audio to the TNC. If it isn't available, you will have to make a separate connection to the transceiver's external speaker jack.

When you are setting up your TNC, be careful about the transmit audio level. Too high a level will create distorted signals that won't be decodable at the receiving station.

An easy way to check your transmitted signal is to use the TNC *calibrate* function. Get to the command mode (CONTROL-C) and enter **cmd: CALIBRATE**. Listen to your transmitted signal with another rig and raise the audio level from the TNC until the received volume seems to stop increasing. Now reduce the audio from the TNC until you can just hear a volume decrease in the receiver. Reduce it a bit more and you're done. Exit the calibrate function.

Some TNCs have an audio output adjustment on the board, some have an adjustment accessible through a hole in the side of the unit, and some have two fixed output levels selectable with a jumper. If one of these does not work, you may have to open up the transceiver and find the mic gain control. If this is necessary, be sure you adjust the mic gain control and *not* the deviation control (which can change the bandwidth of your signal).

as the *AGW Packet Engine*, or *AGWPE*, developed by George Rossopoylos, SV2AGW. The software is downloadable at **www.sv2agw.com/downloads**. With *AGWPE* and a sound card interface as described in this chapter, it is possible to operate packet radio without a hardware TNC — assuming that your primary packet software application includes the ability to interface with *AGWPE*.

31.2.2 Talking to a TNC

All hardware TNCs are functionally the same. When you buy a stand-alone TNC it usually includes a cable for connecting it to the radio, but you'll have to attach the appropriate mic and speaker jack connectors for your radio. You'll also have to furnish the cable that connects the TNC to your computer at the COM port. In most cases this is an RS-232 serial cable. See **Fig 31.8** for a typical setup. Most ham TNCs have yet to migrate to USB at the time of this writing. TNCs integrated into radios may need a cable between the radio's TNC port and a computer.

TNC manufacturers often supply the software necessary to communicate with the TNC, but any terminal program will work. You'll need to start the software and specify the COM port you'll be using, and set the baud and data parameters for that port. Refer to the manual for the specific program you've chosen. The baud rate of your computer must match the baud rate of your TNC. Some TNCs will automatically set their baud rate to match the computer; other TNCs have software commands or switches that determine the baud rate. Again, you'll need to refer to your manual for specific instructions. When setting the data parameters, 8-N-1 is normally used: 8 data bits, no parity, and 1 stop bit.

When you switch on the TNC, you should see some sort of "greeting" text on your screen. That's the first sign that all is well. If you see a bunch of gibberish, it means that the parameters of the TNC and computer don't agree and you'll have to make adjustments.

31.2.3 TNCs and Radios

Tones for 1200 baud can be fed directly into the microphone input of any VHF FM voice transceiver with an appropriate cable. Packet at 9600 baud requires an FM transceiver with a dedicated 9600-baud packet input, as noted earlier.

If you opt to craft your own cable, check your transceiver manual for the wiring diagram of the microphone jack. In most cases, there are separate connections for the audio

31.2.4 TNC Timing

The TNC's TXDELAY parameter specifies the delay interval between the time the TNC keys the radio and the time it starts sending data. Normally 300-400 ms are adequate, but some transceivers take longer to settle after the keying line is triggered. If you have a problem being heard and your audio seems normal, try increasing TXDELAY to 400-600 ms.

On a busy network, packets and packet acknowledgements fly back and forth at a furious rate. One way to keep interference to a minimum is to manipulate the RESP and DWAIT parameters in conjunction with PERSIST and SLOTTIME to allow staggered transmissions.

RESP is the time delay between reception of a packet and transmission of an acknowledgement. DWAIT sets the delay between the time when activity is last heard on the channel and the moment your radio transmits. You should set values of RESP and DWAIT to the values recommended in your area (check with the person managing the local network or PBBS). Your TNC probably accepts a value in "counts" rather than in milliseconds, so don't forget to convert by the proper value in order to arrive at the correct timing value in milliseconds. For example, if you have been asked to set DWAIT to 600 ms and the units of DWAIT for your TNC are 10 ms per count, then you would command DWAIT = 60.

Most TNCs contain commands called PERSIST and SLOTTIME, which help enormously in avoiding interference. PERSIST sets the probability that a packet will be transmitted during a given time interval called a SLOTTIME. The parameter SLOTTIME governs the interval between transmission timing "slots." Initially, PERSIST should be set to approximately 64 and SLOTTIME to a value of about 10, which is equivalent to 100 ms.

FRACK (frame acknowledgement) should be set to 6 and RETRY to 10. FRACK sets the number of seconds between retries and RETRY sets the number of times your TNC will try to send a packet and gain acknowledgement of it before it gives up and disconnects

31.2.5 Monitoring

Start by listening to an active packet frequency in your area. With the radio cable connected, turn on your radio and increase the receiver volume partway. Some TNCs include an LED indicator that shows that the TNC is receiving audio. Turn up the squelch control on the radio until the LED is extinguished. Tune the rig to any odd numbered frequency between 144.91 and 145.09, or between 145.61 and 145.79 MHz, and set the rig for simplex operation. Your best bet may be to search for a DX PacketCluster, or try monitoring APRS activity on 144.39 MHz. When you hear the buzzing packet signals and see text on your screen, you'll know you've hit the jackpot.

Depending on the type of activity you are monitoring, you may see what appears to be nonsense. If you are monitoring APRS, you'll see strings of numbers. These are latitude/longitude position reports. On PacketClusters, you'll see DX call signs and frequencies.

31.2.6 "Connected" vs "Unconnected"

When discussing TNCs and networks, it is important to understand the difference between connected and unconnected communication.

If you are simply monitoring local packet transmissions, your TNC is in an *unconnected* state. What you see is what you get. If a signal is garbled by noise or interference, you'll see nothing on your screen (unless you've enabled the PASSALL function, in which case you'll see gibberish). If you transmit an unconnected packet, the signal simply leaves your antenna destined for nowhere in particular. Some stations may decode it, some may not.

When your TNC is operating in a *connected* state, everything changes. When you are connected, your station is linked to another station in a "virtual" sense. In a connected state, every packet you send is intended specifically for the receiving station (even though others can see it).

When your TNC transmits a packet, it starts a countdown clock. If the clock reaches zero before your TNC receives an acknowledgement (known as an *ACK*) that the packet arrived without errors, it will send the same packet again. When the packet is finally acknowledged, the TNC will send the next packet. And so it goes, one packet after another. The operator at the other station may also be sending packets to you since this communication process can flow in both directions simultaneously.

The big advantage of the connected state is that data is delivered error-free. One packet station can connect to another directly, or through a series of relaying stations. Error-free can be a disadvantage, too. Specifically, a connected state works best when signals are strong and interference is minimal. Remember that if too many packets are lost — by either not arriving at all or arriving with errors — the link will fail. That's why AX.25 packet radio tends not to work well on the HF bands. With all the noise, fading and interference, packets are often obliterated enroute.

Unconnected packets are ideal for applications where you are transmitting essentially the same information over and over. Since unconnected packets can be decoded by any station, they are an excellent means of disseminating noncritical data (data that doesn't need guaranteed error-free delivery) throughout a given area. If a station fails to decode one packet, it merely waits for the next one. APRS uses exactly this approach.

31.3 The Automatic Packet/Position Reporting System (APRS)

The Automatic Packet/Position Reporting System, better known as *APRS* (**aprs.org**), is the brainchild of Bob Bruninga, WB4APR. In fact, APRS is a trademark registered by WB4APR. The original concept behind APRS involved tracking moving objects, and that's still its primary use today. However, APRS has been evolving to become an amateur network for other applications such as short text messaging, either from ham to ham, or between hams and non-hams through APRS Internet gateways.

Mobile APRS stations communicate their positions based on data provided by onboard Global Positioning System (GPS) receivers. The GPS receivers are attached to either packet radio TNCs or simplified packet devices known as *position encoders*, which in turn are connected to transceivers (see **Fig 31.9**). At receiving stations, various APRS software packages decode the position information and display the results as icons on computer-generated maps such as the one shown in **Fig 31.10**. When a station moves and transmits a

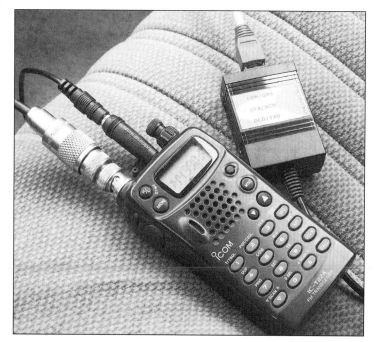

Fig 31.9 — A 2 meter FM handheld transceiver attached to an APRS position encoder.

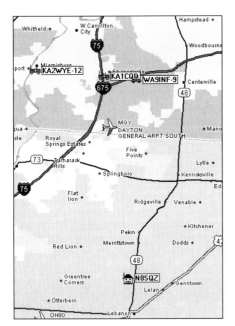

Fig 31.10 — A snapshot of APRS activity using *UI-View* software. Note the icons representing various mobile and fixed stations.

new position, the icon moves as well.

Virtually all APRS activity takes place today on 144.39 MHz using 1200-baud packet TNCs and ordinary FM voice transceivers. In areas where the APRS network is particularly active, you may hear traffic on 445.925 MHz as well. There is also some activity on HF.

31.3.1 Setting Up an APRS Station

If you own a 2 meter FM voice transceiver, you already have the primary component of your APRS station. Tune your radio to 144.39 MHz and listen for packet transmissions. If you hear them, it means you have APRS activity in your area.

To decode APRS packets, you'll need a TNC, and it doesn't necessarily have to be "APRS compatible." APRS compatibility is only a factor if you wish to connect the TNC to a GPS receiver to transmit position data.

If all you want to do is monitor APRS activity, you do *not* need a GPS receiver. If you want to participate in the local APRS network from a fixed (non-moving) station such as your home, you still do not need a GPS receiver. Just determine your home latitude/longitude coordinates and you can use them to establish the location of your home station on the network. There are numerous sites on the Internet that will convert your home address to a correct latitude and longitude.

The only APRS station that requires a GPS receiver is a *moving* station. The good news is that almost any GPS receiver will do the job. It does not have to be elaborate or expensive. The only requirement of an APRS-compatible GPS receiver is that it provide data output in *NMEA* (National Maritime Electronics Association) format. Note that many GPS receivers advertise the fact that they provide data output, but some do it in a proprietary format, not NMEA — check the manual. See **Fig 31.11** for a diagram of a typical APRS station with a GPS receiver.

APRS-compatible TNCs and tracking devices have standardized on the NMEA 0183 protocol. They expect data from the GPS receiver to be in that format so they can extract the necessary information and convert it into AX.25 packets for transmission. If the data from the GPS receiver is incompatible, the TNC or position encoder won't be able to make sense of it.

The final component of your APRS station is software. You'll need software to display the positions of APRS stations, along with other information contained in their transmissions. APRS software is also essential if you want to communicate over the APRS network from a fixed or portable station. Note, however, that APRS software is *not* necessary for mobile stations that wish to merely transmit APRS beacons for tracking purposes. That function is carried out automatically using the GPS receiver and APRS-compatible TNC or tracking device; it does not depend on software.

The most popular APRS *Windows* program is *UI-View*. *UI-View* was created by the late Roger Barker, G4IDE. You'll find it on the Web at **www.ui-view.org**. The 16-bit version is free for downloading. To use the 32-bit registered version, hams are asked to donate to their local cancer charities. Details are available on the *UI-View* Web site. For the Macintosh, there is *MacAPRS* available from **www.winaprs.com**. For *Linux* there is *Xastir*, available for download at **www.xastir.org**.

APRS software, regardless of the operating system, is designed to talk to the packet TNC, processing the incoming APRS data and creating icons on your computer screen. The application also uses the TNC to transmit APRS data. This means that the software and the TNC must communicate with each other at the same baud rate. Every APRS application has a setup menu to program the correct parameters for various TNCs.

Depending on the software, there may be other features such as logging or messaging. APRS software changes rapidly, so check the help file or manual for specific instructions.

31.3.2 Maps and APRS

No matter which APRS software you choose, one absolutely critical aspect is the mapping function itself. To get the most from APRS, your software maps must be as comprehensive as possible, preferably with the ability to show detail down to street level.

Downloadable APRS software applications generally do not come with detailed maps. Detailed map files are numerous and large, impractical to bundle with every APRS program. Instead, most applications are designed to import user-created custom maps, or to work with existing commercial mapping programs such as Microsoft *Streets*, Delorme *Street Atlas* and *Precision Mapping*.

UI-View, for example, has the ability to automatically load and display maps from *Precision Mapping*. You must purchase and install *Precision Mapping* on your PC, then download and install a small *Precision Mapping* "server" application into *UI-View*.

Each APRS transmission includes characters that define the type of map icon that will be displayed at the receiving end. If you are operating a fixed station, your APRS software will allow you to choose your icon. If you are a mobile station using a traditional TNC, you'll need to define your chosen icon in your beacon statement. APRS-compatible TNCs give you the ability to do this. APRS position encoders also allow you to choose your icon when you program the unit. Your mobile icon might be a car, boat or airplane.

31.3.3 APRS Position Encoders

You can create a mobile APRS station with a VHF FM transceiver, a TNC and a GPS

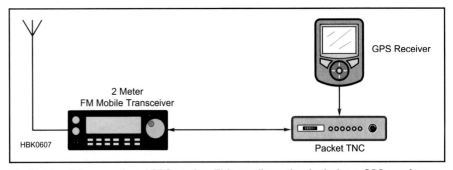

Fig 31.11 — Diagram of an APRS station. This configuration includes a GPS receiver, although the receiver is only required if the station is mobile. In mobile applications, a position encoder can be substituted for the packet TNC.

Fig 31.12 — Some APRS position encoders, such as this RPC Electronics RTrak, combine the GPS receiver, packet circuitry and 2 meter transmitter into a single device.

receiver. Wire everything together, connect an antenna and dc power and you're set. For hams on the go, however, it's common to replace the full-fledged TNC with an APRS *position encoder*, also referred to as an APRS *tracker*. A position encoder is a compact device designed for one purpose: to receive data from the GPS receiver, assemble APRS packets from the data and create modulated signals for use by the transmitter. Some position encoders include GPS receivers in their designs. You'll even find position encoders that are complete packages incorporating tiny GPS receivers and low power FM transmitters (**Fig 31.12**).

To use a position encoder you must program it the same way that you initially program a TNC. Like TNCs, position encoders connect to computer serial or USB ports for programming and most come with their own programming software. You must enter your call sign and other information such as your beacon interval (how often you want the position encoder to transmit your position). Most position encoders allow you to set the beacon interval to a certain amount of time (say, every two minutes). Some position encoders can be configured to transmit position beacons after a certain distance (every mile), or whenever the vehicle turns a corner.

31.3.4 APRS Networking Tips

One of the key features of APRS is that while it uses AX.25 to transport its messages, it essentially ignores all the AX.25 connection-oriented baggage. Unlike traditional packet radio, APRS stations do not establish "connections" with each other. Instead, APRS packets are sent to no one in particular, meaning to *everyone*.

Every APRS station has the ability to function as a digital repeater, or *digipeater*. So, if it receives a packet, it will retransmit the packet to others. As other digipeaters decode the same packet, they will also retransmit and spread it further. This is known as *flooding* and is illustrated in **Fig 31.13**.

As an APRS user, you can set up your station to address its packets through specific digipeaters according to their call signs. But when you're traveling, how do you determine which digipeaters you should use?

PATHS AND ALIASES

In the packet world, nodes and digipeaters can have *aliases*. A digipeater call sign may be W1AW-1, but it can also carry an alias, using the MYALIAS command in the digipeater TNC. Perhaps the digipeater alias would be NEWNG (meaning the ARRL HQ home town of Newington). You can route packets through the digipeater by addressing them to W1AW-1, or simply by addressing them to NEWNG. Any station that is set up to respond to an alias is capable of handling your packets automatically, even if you don't know its call sign.

Unlike typical packet use of aliases in which a given single station has a specific alias, APRS specifies standard digipeater aliases that nearly all stations use. This means that you can travel anywhere in the country and still participate in the APRS network without knowing digipeater call signs. (Otherwise, you'd have to reconfigure your TNC whenever you moved from one area to another.)

The most common APRS digipeater alias is *WIDEn-N*. The letters "n" and "N" represent numbers. The first (left-most) "n" designates how many WIDE digipeaters will relay your packets, assuming they can receive them.

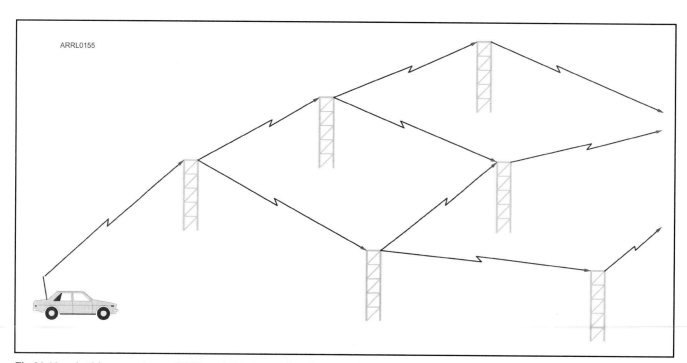

Fig 31.13 — In this example, an APRS packet is transmitted by a mobile station and is retransmitted by a nearby digipeater. Depending on how the mobile operator configured his TNC or tracker's path, the packet will be picked up and repeated by several other digipeaters. This is known as flooding.

WIDE2, for instance, is the same as saying that you want your packets relayed through *two* WIDE digipeaters. The second "N" is the *Secondary Station Identifier* (SSID). The SSID is used in APRS networks as a means of limiting how often (and how far) a packet can be repeated.

Here's how it works. Each time your packet traverses a WIDEn-N digipeater, the digipeater subtracts 1 from the SSID as it retransmits. The next digipeater deducts 1 and so on until the SSID reaches zero, at which time the packet will not be repeated again. This has the effect of limiting the flood radius. See **Fig 31.14**.

When you configure a TNC or position encoder for use with APRS, you can use these aliases to set up the paths for the beacon packets you'll be transmitting. In most devices this is accomplished with the UNPROTO parameter, sometimes simply referred to as the "Path." If you are a fixed station (a station at home, for instance), set your path as **WIDE2-2** (or with a traditional TNC UNPROTO statement, set it to APRS VIA WIDE2-2). This designates that your reports will be relayed by *two* WIDEn-N digipeaters and limits the spread beyond those repeaters to just two retransmissions. Set your TNC to beacon once every 30 minutes. That's sufficient for a fixed station.

If you are running APRS from a car, try **WIDE1-1,WIDE2-1** (or APRS VIA WIDE1-1,WIDE2-1). WIDE1-1 ensures that your packet will be picked up by at least one local digipeater or a home station acting as a fill-in digipeater and relayed at least once. WIDE2-1 gets your packet to another, presumably wider-coverage digipeater, but limits the retransmissions beyond this point. It's wasteful of the network to set up wide coverage for a station that is rapidly changing its position anyway. Mobile stations that are in motion should also limit their beacon rate to once every 60 seconds, or once per mile, whichever comes first.

Never invoke extremely wide coverage, such as a WIDE5-5 path, unless you are way out in the hinterlands and need every relaying station available to get your packets into the network.

Another popular alias is the "single state" — SSn-N — to limit the spread of your packets to a specific state or area within a state. To keep packets within the state of Connecticut, for example, you could use the SSn-N alias in a path statement, like this: CT1-1, CT2-2. This path assures that local Connecticut stations (CT1-1) will repeat the packets, and that broad-coverage stations (CT2-2) will relay them throughout a large portion of the state. APRS digipeaters outside Connecticut, however, will not respond to these packets because they won't recognize the CT2-2 alias (although they will recognize an alias for their state). The packets will still be heard across state borders but will not be digipeated or add to the packet activity in a neighboring state.

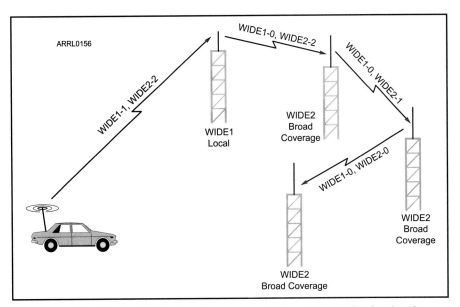

Fig 31.14 — By using the WIDEn-N system, we can limit packet flooding in a local network and greatly reduce congestion. The mobile station in this example has his path set as WIDE1-1,WIDE2-2. Notice how his packet propagates through the network and how the SSID number is reduced by one each time the packet is repeated through a digipeater with a corresponding alias. When it reaches the third WIDE2 digipeater, the counters all reach zero and digipeating stops.

31.4 PACTOR

PACTOR is a type of HF digital communication that, unlike many other HF modes, offers error-free text and file transfers. PACTOR communication resembles packet radio in that it establishes a "link" between two stations. Data is sent in discrete frames and each frame is acknowledged by the receiving station. This rapid back-and-forth exchange creates PACTOR's distinctive *chirp-chirp-chirp* signal. Through robust coding and sophisticated modulation techniques, PACTOR is often able to maintain the link even in the face of significant noise, fading and interference.

PACTOR I, introduced in the early 1990s, has been largely superseded by PACTOR II,

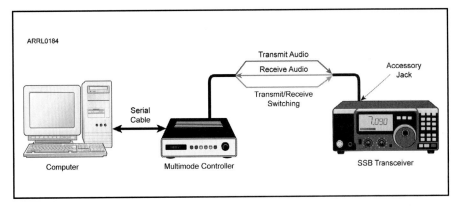

Fig 31.15 — A PACTOR station with a multimode controller. For PACTOR II and III operation, the controller must be a model manufactured by the SCS Corporation.

which provides faster throughput in difficult conditions. Both PACTOR I and II generate signals that occupy 500-Hz bandwidths. PACTOR III offers even more efficient communication, but its signal occupies a bandwidth in excess of 2000 Hz.

To set up a PACTOR station, you must purchase a stand-alone multimode controller that includes the PACTOR mode. As mentioned at the beginning of this chapter, multimode controllers are microprocessor-based devices that support several different digital modes in a single package. A typical multimode controller might include PACTOR, packet, RTTY and more. Controllers manufactured by Kantronics (**www.kantronics.com**) and Timewave (**www.timewave.com**) include PACTOR I. For PACTOR II or III, you must purchase a controller made by Special Communications Systems (**www.scs-ptc.com**), the original developers of PACTOR.

Fig 31.15 illustrates a typical PACTOR station built around a multimode controller. The computer is functioning only as a dumb terminal for the controller in this application, so it does not have to be particularly powerful. All that is required is basic terminal software as described in the section on packet TNCs. More sophisticated software applications can provide a smoother, easy-to-use interface, but these require more capable computers.

A PACTOR station requires an SSB transceiver capable of switching from transmit to receive within approximately 30 ms. Most modern HF transceivers can meet this requirement, but not all. *QST* magazine Product Reviews test this specification for all HF transceivers. The transmit audio supplied by the controller can be fed to the microphone or accessory jack to operate AFSK, which is the standard procedure, although PACTOR I can also operate using FSK.

The most popular use of PACTOR on amateur frequencies today is within the *Winlink 2000* network. PACTOR is also widely used as the HF digital arm of the National Traffic System-Digital (NTSD). See the sidebar "Digital Communications and Public Service."

31.5 High Speed Multimedia (HSMM)

Wireless networking using IEEE 802.11 standards has seen explosive growth during the last several years. Coffee shops, fast-food restaurants, hotels, airports and many other high-traffic locations now include wireless Internet (WiFi) hotspots. Some hotspots offer free Internet access while other charge an access fee, payable with your credit card.

Wireless networks are also popular at home. Establishing a home network is as simple as installing a *wireless router* to manage the data flow from the broadband Internet connection. The router allows one or more traditional desktop computers to tap the broadband connection through wired (usually Ethernet) access while simultaneously making the Internet available wirelessly to one or more laptops.

All these home networking devices — routers, wireless access cards, and so on — are unlicensed FCC Part 15-regulated transceivers with RF outputs measured in milliwatts. A number of their channel frequencies overlap two Amateur Radio bands: 2.4 and 5.8 GHz. This means that hams can put them to work as "transceivers" under our Part 97 rules. See **Table 31.2** for specific frequencies.

By using consumer-grade routers, low-loss coaxial cable and gain antennas, hams can quickly establish high-speed, long-distance wireless networks on these 2.4 or 5.8 GHz frequencies. This type of operating is referred to as *high speed multimedia*, or *HSMM*.

With HSMM amateurs have the opportunity to operate several different modes at the same time, and usually do. HSMM is generally IP-based, and given enough bandwidth, radio amateurs have the capability to do the same things with HSMM that are done on the Internet.

• *Audio*: This is technically digital voice, since it is two-way voice over an IP (VoIP) network similar to EchoLink and IRLP networks used to link many Amateur Radio repeaters over the Internet. (VoIP is covered in a later section.)

• *Video*: Motion and color video modes, are called amateur digital video (ADV). This is to distinguish it from digital amateur television (DATV). DATV uses hardware digital coder-decoders (CODEC) to achieve relatively high-definition video similar to *entertainment quality* TV. The usual practice in

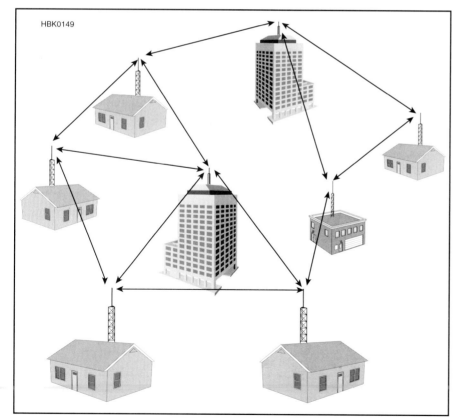

Fig 31.16 — An HSMM Mesh Network is highly decentralized. Individual mesh stations (nodes) act as repeaters to transmit data from nearby nodes. If one node drops out of the network, the other nodes will automatically re-route the traffic.

Table 31.2
Wireless Network Frequencies in Amateur Bands

Channel	Center Frequency	Channel	Center Frequency
1	2.412 GHz	132	5.660 GHz
2	2.417 GHz	136	5.680 GHz
3	2.422 GHz	140	5.700 GHz
4	2.427 GHz	149	5.745 GHz
5	2.432 GHz	153	5.765 GHz
6	2.437 GHz	157	5.785 GHz
		161	5.805 GHz
		165	5.825 GHz

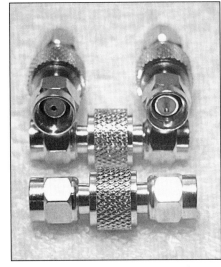

Fig 31.18 — RP connector adapters.

HSMM radio is to use a far less expensive (often free) PC-based software video CODEC to achieve *video communications quality* signals in much smaller bandwidths.

• *Text:* Text exchanges via a keyboard are often used in HSMM radio, but they are similarly called by their Internet or packet radio name: *Chat mode*, and if a server is available on the network, e-mail can also be exchanged.

• *Image*: Image file transfers using file transfer protocol (FTP) can also be done, just as on the Internet.

• *Motion video*: FTP of MPEG files can provide one-way video streaming of short video clips.

• *Remote control*: Individual devices or even complete stations can be remotely controlled.

• *Mesh networking*: A mesh is a wireless cooperative communication infrastructure in which each station functions as a relay, allowing the entire network to cover a large area. See **Fig 31.16**. This type of infrastructure can be decentralized (with no central server) or centrally managed (with a central server). Both are relatively inexpensive and very reliable. Individual mesh stations (nodes) act as repeaters to transmit data from nearby nodes. The reliability comes from the fact that each node is connected to several other nodes. If one node drops out of the network, due to hardware failure or any other reason, its neighbors simply find another route. Capacity can be increased by simply adding more nodes.

31.5.1 A Basic HSMM Radio Station

For the sake of simplicity, we'll discuss an HSMM setup with two stations operating in a host/client configuration. See **Fig 31.17**.

At the time of this writing, the most popular wireless router for HSMM applications is the Linksys model WRT54GL. It is a combination unit consisting of a wireless access point (AP) or hub coupled with a router. As with other routers, your host PC or laptop connects directly to it using a standard Ethernet cable. If the PC is also connected to the Internet, then it may also perform the function of a *gateway*.

The WRT54GL is a *Linux*-based model that supports firmware upgrades. HSMM-active amateurs have been creating their own WRT54GL firmware to support special applications (such as mesh networking). They are effectively "modifying" the WRT54GL in ways Linksys could not have imagined! You do not have the use a WRT54GL to explore HSMM, however. Any wireless router will do. The advantage of the WRT54GL is only that it is widely available and easily modifiable.

The first step in configuring your router for HSMM is to disconnect both flexible antennas that came with the unit and replace them with a high-gain directional antenna system. To connect a gain antenna, you are going to be become familiar with RP (reverse polarity) connectors. These connectors appear to be male connectors on the outside, but they have a socket rather than a pin for the center conductor. RP connectors are used by the manufacturers to discourage Part 15 owners from using the equipment in ways for which it was not WiFi certified.

As licensed radio amateurs, we can modify the system to accomplish our specific requirements within the amateur bands. To connect coaxial cable for an external antenna, you can use an adapter or short jumper cable assembly with appropriate connectors. The adapter shown in **Fig 31.18** has an RP connector on one end and a standard SMA connector on the other.

There are often two antenna ports on wireless devices. These are used for *receive space diversity*. The wireless device will normally automatically select whichever antenna is receiving the best signal at any specific moment. Which do you connect to?

The transmitted signal from the wireless router always goes out the same antenna port. It does not switch. In other words, most wireless devices use space diversity on the receive

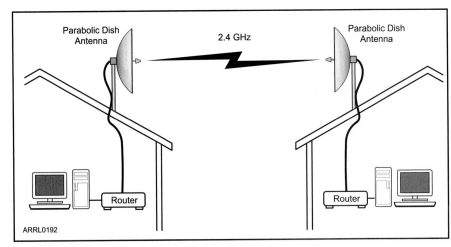

Fig 31.17 — This is a simplified diagram of a typical HSMM link between two stations. Note that high gain parabolic dish antennas are being used in this example.

side and not on transmit. Some access points/wireless routers will allow you to manually (via software) select the antenna port that is used for transmission. When there is no obvious choice, you will need to find some means of detecting which antenna is the transmit antenna port with RF output power present. That is the port to connect to your gain antenna.

Be sure to use low-loss coaxial cable, and keep the run as short as possible. Coaxial cable loss at these frequencies can be quite high, so you are often better off mounting the router directly at the antenna, connecting to it through a very short length of cable. From there you can run the Ethernet cable back to your computer.

SOFTWARE CONFIGURATION

You will need to use either the network configuration software from the router manufacturer or the configuration tool that is part of your operating system. Some routers offer user-friendly access through Internet browsers, which is particularly convenient. Regardless of the approach, the goal is to access and change your router's settings.

SSID: The AP/wireless router host software is provided with an *SSID* (Service Set Identifier) that many Part 15 stations turned off for somewhat higher security. But radio amateurs should leave it ON. Enter your call sign as the SSID and use it for the station identification. It constantly broadcasts your call thus, providing automatic and constant station identification. There are 32 characters available for use in this field so more information such as your group's name can be entered too, including spaces and punctuation. If the router asks if you want to enable the broadcast, click YES. (Note that SSID in the HSMM world has a somewhat different meaning than SSID in the packet radio community. Among packet users, SSID is defined as Secondary Station Identifier.)

ESSID: Some manufacturers use this term in place of SSID to put emphasis on the fact that the SSID is the name for your *network*, not for a specific wireless AP/router.

Access Point Name: When this field is made available (by default is it blank) it is for you to enter a description. This may be handy if you have deployed more than one AP in your network all with your call sign as the SSID. It would allow you to tell them apart. Otherwise, just leave it blank.

Channel: To avoid interfering with other services as much as possible, we need to also look at channel selection. Most HSMM activity is concentrated on 2.4 GHz, using channels 1 through 6, which are within the 2.4 GHz ham allocation.

It is probably best to avoid channel 6. It is the most common manufactures default channel setting and 80% or more of your neighbors will be using it for their household wireless local area network (WLAN). Channel 1 is used by most of the remaining manufacturers as their default channel, so avoid that too. The result is most radio amateurs use channels 3 or 4 depending whether there is a WISP (Wireless Internet Service Provider) operating in their area. Often a WISP will use one of these intermediate channels with a highly directive antenna for back-haul or other purposes. If so, you may wish to coordinate with the WISP and arrange to use some other channel rather than the one specifically used by the WISP. It is not a perfect solution because of all the overlap, but it is a good faith effort to keep most of your stronger signal out of anyone's home, business or governmental WLAN traffic.

WEP: This stands for Wired Equivalent Privacy. In spite of all the horror stories you may have read in the press, this encryption method provides more than adequate means to economically achieve authentication and thus keep the vast majority of free-loaders off your network. If you live in the country you may not need to enable this capability. In an urban environment, it is probably a good thing to do so that you need not constantly monitor every bit of traffic coming over the network to ensure that it originates from an Amateur Radio station. Mixing traffic with another service that shares the same frequency band is not a generally accepted practice except in times of emergency. Therefore, it is often necessary for HSMM radio stations to encrypt their transmissions. This is not to obscure the meaning of the transmission or hide the information. In amateur HSMM applications, the purpose of using WEP is only to block access by non-hams. Amateur HSMM networks openly publish their encryption keys for other hams to use.

Most wireless routers will allow for the use of multiple WEP keys, typically up to four. This will allow you to configure the device so that different client stations have different access authority. For example, club members may have one level of access, while a visiting radio amateur may be given a lesser access. Most HSMM radio groups have just one WEP key and everybody gets that one.

Remember that when it comes to the length of the WEP or other key used, our main purpose is to provide a simple and economical means of authentication already available on the wireless devices. In other words, it is to ensure that only Part 97 stations and not Part 15 stations auto-associate or auto-connect with our HSMM radio node. The shorter the WEP key, the better. This makes it easy to use and remember. During early HSMM radio experimenting around the year 2000, the shortest possible key (5 characters) was used: HSMM-

Authentication Type: Some routers will ask for the type of authentication you want to use such as *shared key*, *open system*, and *both*. Click on SHARED KEY because you will be sharing the WEP/WPA key with any and all radio amateurs who wish to access your HSMM radio node.

DHCP: Some routers will ask if you want Dynamic Host Configuration Protocol enabled. This is the function that assigns IP addresses. Unless you have another source of the DHCP function on your network, you will want to ENABLE this function.

Antenna Selection: A few wireless routers with dual antennas will ask you select an antenna. The default is normally receive space diversity. Because we are going to connect an outside gain antenna, you want to make a selection. Otherwise, you will need to identify which antenna is the actual transmit antenna and connect the feed line to that port, as discussed previously.

Mac Address Filter: Some wireless routers will allow for this security measure, but it is troublesome to administer it, so it is recommended that you not bother enabling this function. Use WEP or some other method of encryption using the guidelines discussed previously.

Output Power: Some wireless routers will allow you to set this power level, often up to 100 mW. As with all other radio amateur operations use only the minimum power needed to accomplish your mission.

THE FAR END OF THE LINK

The computer at the client end of the HSMM link need only have a wireless networking adapter, not a separate router. These transceivers/wireless adapter cards usually come in three forms:

1) One form is called a PC card. Earlier these were called PCMCIA cards, but more recent terminology is to simply call them laptop PC cards.

2) Another type of transceiver/adapter comes with a USB interface, such as the one shown in **Fig 31.19**. This is often considered a superior interface for most HSMM stations. The reason for this has nothing to do with the quality of the transceiver, but rather the fragile nature of the tiny connectors found on PC cards. They are not really designed for frequent plugging and unplugging. Without extreme care, they can be easily torn out.

Fig 31.19 — At the client end of the HSMM link, a simple wireless adapter will do. This USB model allows the user to remove the flexible antenna.

3) Linksys and other manufactures also produce similar cards for the expansion slot on the rear of your desktop PC too.

Regardless of the device you choose, it must have an antenna that is removable or has an external antenna port of some type.

To connect to the AP/wireless router in an HSMM radio network, the wireless computer user(s) at the far end must exit ad-hoc mode and enter what is called the *infrastructure mode*, in their operating software. Infrastructure mode requires that you specify the radio network your computer station is intended to connect to (the host station's call sign), so set your computer to recognize the SSID you assigned to the AP/wireless router to which you wish to connect.

As described previously, you may need to use an adapter to connect the card or interface antenna connector to the coaxial cable running to the antenna. At the other end, most 2.4 GHz antennas come with a standard female N-series connector.

Team up with a nearby radio amateur to test. Do your initial testing in the same room together making sure the link-up is working. Then as you increase distances going toward your separate station locations, you can coordinate using a suitable local FM simplex frequency. You will increasingly need this communication to assist with directional antenna orientation as you get farther apart.

31.5.2 Running High Power

It is tempting for some radio amateurs to think that if they run higher power they will get better range out of their HSMM radio station. This is not always the case. There are many factors involved in range determination when operating at 2.4 or 5.8 GHz. The first and most significant of these is the lay of the land (topology) and path obstructions. Running additional power is unlikely to correct for either of these impediments.

Second, running higher power to improve signal link margins often requires that this be done at both ends of the link to obtain meaningful results.

Third, most RF amplifiers for use with 802.11 are of the BDA (bidirectional amplifier) type, such as the one shown in **Fig 31.20**. They amplify both the incoming signal and the outgoing signal, and to get maximum effectiveness out of such devices they must be

Fig 31.20 — A 2.4 GHz bidirectional amplifier.

mounted as close as possible to the antenna.

Fourth, 802.11 signals from such inexpensive broadband devices often come with significant sidebands. This is not prime RF suitable for amplification. A tuned RF channel filter should be added to the system to reduce these sidebands and to avoid splatter.

Also, if your HSMM radio station is next door to an OSCAR satellite ground station or other licensed user of the band, you may need to take extra steps in order to avoid interfering with them, such as moving to channel 4 or even channel 5. In this case a tuned output filter may be necessary to avoid not only causing QRM, but also to prevent some of your now amplified sidebands from going outside the amateur band, which stops at 2450 MHz.

Do not use higher power as a substitute for higher antenna gain at both ends of the link. Add power only after all reasonable efforts have been taken to get the highest possible antenna system gain and directivity.

31.5.3 HSMM Antenna Options

There are a number of factors that determine the best antenna design for a specific HSMM radio application. Most commonly, HSMM stations use horizontal instead of vertical polarization because it seems to work better. In addition, most Part 15 stations are vertically polarized, so this is sometimes thought to provide another small barrier between the two different services sharing the band. With multipath propagation is it doubtful how much real isolation this polarization change actually provides.

More importantly, most HSMM radio stations use highly directional antennas (**Fig 31.21**) instead of omnidirectional antennas. Directional antennas provide significantly more gain and thus better signal-to-noise ratios, which in the case of 802.11 modulations means higher rate data throughput and greater range. Higher data throughput, in turn, translates into more multimedia radio capability. Highly directional antennas also help amateurs avoid interference from users in other directions.

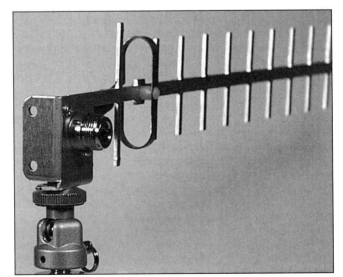

Fig 31.21 — Directional antennas, such as this MFJ 2.4 GHz Yagi, are best for HSMM links.

31.6 Automatic Link Establishment (ALE)

Automatic link establishment (ALE) is a "system" that incorporates digital techniques to establish a successful communications link. ALE is not a digital mode *per se*, but the primary software is sound card based.

To understand ALE, consider how HF propagation changes. A band can open over a particular path one hour, but close the next. A signal may be fully readable on one band, but inaudible on another. With that in mind, imagine two stations, one in New York City and one in Los Angeles, that wish to communicate on the HF bands (see **Fig 31.22**). Basic propagation guidelines give a general idea of which bands might be best, but operators at both ends of the path will have to experiment, calling on several bands until they find a frequency that supports a transcontinental path.

ALE automates this process, allowing the stations under computer control to automatically call each other on different bands until contact is established. In the example, the operator in New York City initiates the ALE call by entering the call sign of the receiving station into his ALE software, along with some likely frequencies. The call sign becomes the *Selective Call*, or SELCAL. The ALE software switches the transceiver to the first frequency in the queue and transmits the chosen SELCAL. If there is no response, the software will step the transceiver to another frequency and try again.

At the same time, the receiving station in Los Angeles has been automatically scanning specific ALE frequencies, listening for its call sign. When it finally decodes something that resembles its call sign, it stops scanning and listens since the ALE data burst contains several repetitions of the call. When it finally decodes its full call sign, the Los Angeles station transmits a "handshake" signal to the New York City station to confirm that a link has been established. All scanning stops and the ALE software sounds an alarm to call the operators to their stations.

ALE is a good tool for HF net operation in both routine and emergency applications where many stations may be standing by for calls. Think of ALE as being analogous to a VHF FM scanning transceiver with a digitally coded squelch that automatically scans several repeaters. Unless it hears the correct code, the radio remains silent. With amateur ALE you can call individual stations or entire groups of stations.

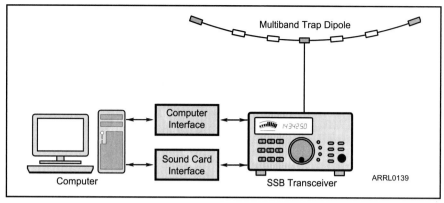

Fig 31.23 — An optimum ALE station has a computer-controlled transceiver and a multiband antenna system. In this example, the station is using a multiband trap dipole antenna. Note that a separate interface is needed between the computer and transceiver for frequency control.

In addition to its application as a selective calling system for HF voice, ALE can also be used as a pure digital mode for exchanging text and images.

AMATEUR ALE SOFTWARE

ALE has been used by the military and government for years, and hams have generally adopted the government ALE standards: FED-STD-1045 or MIL-STD 188-141. In 2001, ALE captured the attention of the Amateur Radio world with the release of *PCALE* software by Charles Brain, G4GOU. *PCALE* not only controls your transceiver, it uses your computer sound card to decode ALE signals and generate them for transmission. You'll find *PCALE* software on the Web at **www.hflink.com**. You must join the free HF Link group on Yahoo at **tech.groups.yahoo.com/group/hflink** before you can download the software.

From the standpoint of assembling a station for ALE, it works best when two conditions are satisfied: 1) you have a multiband HF antenna system; the more bands the better; and 2) your HF transceiver can be controlled by computer.

Although you can use ALE on just one frequency, the real power of ALE comes into play when your station is able to operate on multiple bands under computer control. A typical setup is shown in **Fig 31.23** and includes a radio control interface as well as a sound card interface. With *PCALE* controlling your radio, it will automatically change frequencies, monitoring as many bands and frequencies as you have programmed into the software. A list of ALE channels and information on ALE nets is available from **www.hflink.com**. Note that all ALE transmissions use USB.

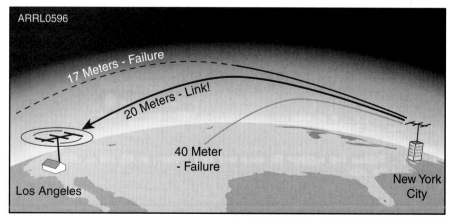

Fig 31.22 — The ALE calling station typically scans through a list of frequencies, calling on each one until a path to the receiving station is found.

31.7 D-STAR

The D-STAR digital protocol began in 2001 as a project funded by the Japanese Ministry of Posts and Telecommunications to investigate digital technologies for Amateur Radio. The research committee included representatives of the Japanese Amateur Radio manufacturers and the Japan Amateur Radio League (JARL). JARL is the publisher of the D-STAR protocol.

All radio manufacturers are free to develop D-STAR equipment, but at the time of this writing ICOM is the only company that has done so for the US amateur market. Although ICOM may develop D-STAR protocol enhancements peculiar to the function of its hardware, ICOM does not "own" the original D-STAR protocol.

31.7.1 What is D-STAR?

The primary application of D-STAR is digital voice, but the system is capable of handling any sort of data — including text, voice or images.

As shown in **Fig 31.24**, a D-STAR network can take several forms. D-STAR compatible transceivers can communicate directly (simplex), or through a D-STAR repeater for wide coverage. D-STAR signals cannot be repeated through traditional analog repeaters without modification.

The D-STAR system carries digitized voice and digital data, but does the job in two different ways. There is a combined voice-and-data mode (DV) and a high-speed data-only stream (DD). From the perspective of the D-STAR user, data and voice are carried at different rates and managed in different ways, but over the air they are transported as bit streams.

31.7.2 Digital Voice and Low-Speed Data (DV)

The D-STAR codec digitizes analog voice by using the AMBE 2020 codec. AMBE stands for Advanced Multiple Band Encoding and 2020 designates the particular variation used by D-STAR.

AMBE can digitize voice at several different rates. The D-STAR system uses a 2.4 kbit/s rate that offers a good compromise between intelligibility and the speed at which data must be transmitted via the radio link. In addition, AMBE adds information to the voice data that allows the receiving codec to correct errors introduced during transmission. The net result is that the digitized voice stream carries data at a rate of 3.6 kbit/s.

Interleaved with the digitized voice information, D-STAR's DV mode also carries 8-bit digital data at 1200 bit/s. This data is used for synchronization in the D-STAR protocol with the remaining bandwidth available for use by the radio manufacturer. ICOM uses the remaining bandwidth to carry repeats of the D-STAR RF header, the 20 character front-panel message, and serial data as described below.

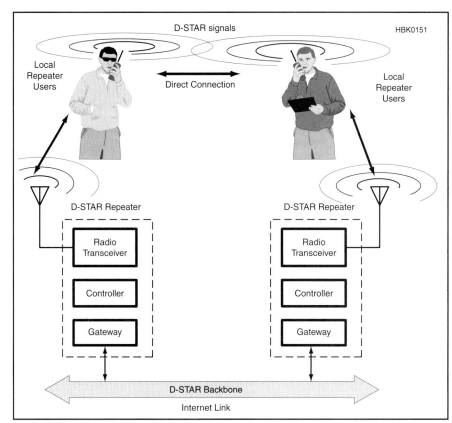

Fig 31.24 — A D-STAR network can take several forms. D-STAR compatible transceivers can communicate directly (simplex) or through a D-STAR repeater for wide coverage.

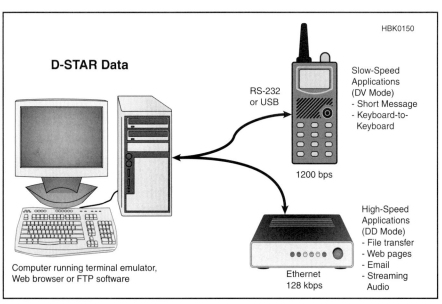

Fig 31.25 — Radios that support DV voice and data present an RS-232 or USB 2.0 interface to the user.

Radios that support DV voice and data provide an RS-232 or USB 2.0 interface to the user as shown in **Fig 31.25**. (The RS-232 interface is restricted to RxD, TxD and ground — "three-wire" connection.) Any computer terminal or program that can exchange data over those types of interfaces can use D-STAR's DV mode capabilities as a "radio cable."

Because D-STAR's DV mode handles the data stream in an unmodified "raw" format, it is up to the equipment or programs that are exchanging data to manage its flow. ICOM requires the sender and receiver to perform flow control by using special data characters. This is called *software flow control or xon/xoff flow control*.

Fig 31.26 — The ICOM IC-92AD is a 2 meter/70-cm handheld with D-STAR functionality.

31.7.3 High-Speed Data (DD)

D-STAR's high-speed data mode is called *D-STAR DD*. This mode does not digitize the analog voice signal. The data-only packets sent over the RF link at a raw data rate of 128 kbit/s half-duplex, but since that includes the packet header and the delay between packets, the *net data rate* is somewhat lower. As with the DV mode, data is transmitted without modification so flow control is left to the applications on each end. Radios supporting DD mode communications may also support DV mode.

Users connect to a radio supporting DD mode with an Ethernet interface via the usual RJ-45 modular jack found on computer networking equipment. The DD mode interface looks to computer equipment just like a customary IP network connection. Specifically, the DD mode interface is an Ethernet "bridge." This allows Web browsers and other Internet software to run normally, as if they were connected to standard computer network.

The net data rate of DD mode is comparable to or better than a high-speed dial-up Internet connection. Any streaming media mode that will run over dial-up Internet will likely perform well over D-STAR.

With its high signaling rate and 130-kHz bandwidth, FCC regulations restrict D-STAR DD operating to the 902 MHz and higher bands.

31.7.4 D-STAR Radios

When this book went to press, ICOM was the only manufacturer offering D-STAR compatible radios and repeaters. So, by default, the focus of any discussion of D-STAR hardware centers on ICOM gear.

ICOM D-STAR radios are available as mobile/base transceivers and also as handheld transceivers. With the exception of the ID-1 base/mobile transceiver, which operates at 1.2 GHz, ICOM D-STAR transceivers are designed to operate at 2 m, 70 cm or both. ICOM D-STAR transceivers also support both digital and analog operation, which means that they can also be used for traditional analog FM communication as needed. An example is shown in **Fig 31.26**.

To determine the most appropriate D-STAR radio for an application, the first step is to understand the requirements:

• Is DD mode operation required (high-speed data)? If so, the only radio supporting DD mode is the ID-1.

• What bands are required? That will depend on activity in your area. If data is to be transmitted while in motion, higher frequencies will result in fewer transmission errors, improving the net data exchange rate. (Note: DD operation is only permitted on 902 MHz and higher.)

• Is high-power required? Using higher power base/mobile radios will result in stronger signal strengths and fewer data transmission errors.

• If data is to be transmitted, what data interface does the computer have?

• Error correction for low-speed data (DV mode) is the responsibility of the data communications programs used to exchange data. Make sure to carefully evaluate the software you intend to use.

More information about D-STAR may be found in the **Digital Modes** and **Repeaters** chapters.

31.8 APCO-25

Unlike D-STAR, which is a digital standard devised by and for Amateur Radio, APCO-25 was developed specifically for local, state and federal public safety communications. "APCO" is the Associated Public-Safety Communications Officials, originally an association of police communication technicians, but now a private organization. APCO has a technical standards group responsible for planning the future needs of police (and more recently public safety) users. It was through this group that a standard for advanced narrow-band digital communications (voice or data) was developed. This standard is known as APCO Project 25, APCO-25, or simply P25.

The overall purpose of the APCO-25 standard is to make it possible for governments to shift from analog to digital communications with the least difficulty possible. This means placing a great deal of emphasis on *backward compatibility* (P25 radios include analog operation and newer P25 technology doesn't render older technology instantly obsolete) and *interoperability* (the ability for all P25 radios to communicate with each other). In the public safety world, interoperability is a key selling point so that various services and agencies can coordinate efforts.

31.8.1 APCO-25 and Amateur Radio

When this section was written, no one was making APCO-25 transceivers specifically for the Amateur Radio market, but that hasn't stopped some hams from exploring this mode. Because APCO-25 is an open, published standard, it *is* legal for Amateur Radio, but the trick is finding the means to adapt commercial APCO-25 gear to ham purposes.

The present-day Amateur Radio APCO-25 world looks a lot like the analog FM community in the early 1970s. Back then, none of the Amateur Radio manufacturers were making FM transceivers or repeaters. Hams were forced to modify existing commercial FM gear, which typically consisted of transceivers that had seen duty in police cruisers, taxi cabs and so on. Many of the FM "gurus" in those days were individuals who were employed by two-way radio service shops. These hams had easy access not only to test equipment, but also to the knowledge of how to modify commercial transceivers for ham applications. They built the first Amateur Radio FM repeaters by repurposing commercial two-way radio transmitters and receivers.

Amateur Radio APCO-25 enthusiasts today are treading the same path taken by analog FM pioneers more than 30 years ago. They are modifying commercial APCO-25 equipment for Amateur Radio and setting up APCO-25 repeater systems. Thanks to online sites such as eBay, it is relatively easy to track down surplus APCO-25 transceivers. Manufactured in both handheld and mobile configurations, these rigs are available for either VHF or UHF.

Modifying commercial APCO-25 radios has become a software hacking game. Since most functions of these radios are software defined, you can change their operating characteristics (including frequencies) with a computer and a compatible interface.

The first step is to obtain the programming software, which can be different for every brand. Some manufacturers provide programming software if you purchase the radio as new equipment. Others are highly restrictive and will not provide their software under any conditions. Motorola, a popular brand among amateur P25 users, uses different programs for every transceiver model. The programming software must be purchased from them at a cost of $250 to $300. To reprogram the Motorola transceivers (as well as many other brands), you may need a hardware device that is sometimes referred to as a Radio Interface Box, or RIB.

The downside of using surplus commercial equipment is that it can be expensive. At the time of this writing, used handheld APCO-25 transceivers were selling for as much as $700 at Internet auction sites. Modified APCO-25 repeaters can cost several thousand dollars.

The cost hurdle hasn't deterred hams from setting up APCO-25 networks. There are more than a dozen amateur APCO-25 repeater systems in operation throughout the United States. Many of these repeaters operate in mixed mode — analog and digital. Others are digital only. The ability to operate in mixed modes is one of the strengths of Amateur Radio APCO-25. An APCO-25 repeater, for instance, can support digital voice and data with APCO-25 transceivers, but can still relay analog FM traffic.

More ham-related programming information is available at the following Web sites, although it applies specifically to Motorola transceivers: **www.batlabs.com/newbie.html** and **www.batlabs.com/flash.html**.

31.8.2 The APCO-25 Standard

APCO-25 is comprised of a "Suite of Standards" that specifies eight open interfaces between the various components of a land mobile radio system.

- *Common Air Interface* (CAI) standard specifies the type and content of signals transmitted by compliant radios. One radio using CAI should be able to communicate with any other CAI radio, regardless of manufacturer
- *Subscriber Data Peripheral Interface* standard specifies the port through which mobiles and portables can connect to laptops or data networks
- *Fixed Station Interface* standard specifies a set of mandatory messages supporting digital voice, data, encryption and telephone interconnect necessary for communication between a Fixed Station and P25 RF Subsystem
- *Console Subsystem Interface* standard specifies the basic messaging to interface a console subsystem to a P25 RF Subsystem
- *Network Management Interface* standard specifies a single network management scheme which will allow all network elements of the RF subsystem to be managed
- *Data Network Interface* standard specifies the RF Subsystem's connections to computers, data networks, or external data sources
- *Telephone Interconnect Interface* standard specifies the interface to Public Switched Telephone Network (PSTN) supporting both analog and ISDN telephone interfaces.
- *Inter RF Subsystem Interface* (ISSI) standard specifies the interface between RF subsystems which will allow them to be connected into wide area networks.

You'll find more details about the APCO-25 standard on the Web at **www.apcointl.org/frequency/project25/index.html**.

31.8.3 APCO-25 "Phases"

The APCO-25 rollout was planned in "phases." Phase 1 radio systems operate in 12.5 kHz analog, digital or mixed mode. Phase 1 radios use continuous 4-level FM (C4FM) modulation for digital transmissions at 4800 baud and 2 bits per symbol, which yields 9600 bits per second total throughput. It is interesting to note that receivers designed for the C4FM standard can also demodulate the compatible quadrature phase shift keying (CQPSK) standard. The parameters of the CQPSK signal were chosen to yield the same signal deviation at symbol time as C4FM while using only 6.25 kHz of bandwidth. This is to pave the way for Phase 2, which is under development. Phase 1 is the current phase in use at the time this book was written and is likely to be in force for a number of years to come.

In a typical Phase 1 radio, the analog signal from the microphone is compressed and digitized by an Improved Multi-Band Excitation, or *IMBE*, vocoder. This is a proprietary device licensed by Digital Voice Systems Corporation. The IMBE vocoder converts the voice signal from the microphone into digital data at a rate of 4400 bit/s. An additional 2400 bit/s worth of signaling information is added, along with 2800 bit/s of forward error correction to protect the bits during transmission. The combined channel rate for IMBE in P25 radios is 9600 bit/s.

P25 radios are able to operate in analog mode with older analog radios and in digital mode with other P25 radios. If an agency wants to mix old analog radios with P25 radios, the system must use a control channel that both types of radios can understand. That means a trunked radio system.

31.9 HF Digital Voice

While D-STAR is the dominant digital voice/data system on the VHF+ bands, three very different systems can be found in operation on the HF bands. One uses dedicated hardware while the other two rely on sound cards and software. All require HF SSB transceivers, although they could just as easily be used on VHF SSB (or even FM) as well.

31.9.1 AOR and AMBE

The AOR Corporation was the first ham manufacturer to arrive on the scene with an HF

Fig 31.27 — The AOR ARD-9800 digital voice modem.

digital voice and data "modem" in 2004. The fundamental design is based on a Vocoder protocol created by Charles Brain, G4GUO. His protocol involves the use of Advanced Multi-Band Excitation, better known as *AMBE,* a proprietary speech coding standard developed by Digital Voice Systems. Brain's protocol operates at 2400 bit/s, with Forward Error Correction added to effectively produce a 3600 bit/s data stream. This data stream is then transmitted on 36 carriers, spaced 62.5 Hz apart, at 2 bits/symbol, 50 symbols/s using QPSK. This gives the protocol a total RF bandwidth of approximately 2250 Hz (compared to 2700-3000 Hz for an analog single sideband transmission).

The AOR units, such as the ARD9800 (**Fig 31.27**), are designed to be as "plug and play" as possible. You simply plug your microphone into the front-panel jack and then plug the modem into the microphone input of your transceiver. A front panel switch selects digital encoding or analog transmission, so the unit can be kept in the line when not used for digital conversations. On the receive side, the modem automatically detects the synch signal of an AMBE transmission and switches to digital mode automatically. In addition to voice, the AOR modems can send digital data (typically images).

AOR ON THE AIR

AOR digital voice is clear and quiet, an unusual thing to hear on an HF frequency. Tune around 14.236 MHz and you're likely to hear AOR AMBE signals. They sound like rough hissing noises.

During *QST* Product Review testing in 2004, the ARRL Lab discovered that decoding was solid down to about 10 dB S/N. Digital voice is an all-or-nothing proposition, however, and below that level the signals begin to break up. The result is absolute silence during those periods. Interference to the received signal produced a similar outcome.

The common approach on the air is to begin the conversation in analog SSB, then switch to digital. Each transmission starts with a one-second synch burst after the push-to-talk button is pressed. If the other station misses the sync signal, the audio doesn't decode and the transmission sounds like analog white noise. This means that for successful communication you must be on the correct frequency and ready to receive at the beginning of the transmission. However, if there is fading or interference during the synch transmission, you may be unable to decode the signal that follows.

A few additional steps will improve odds of success with AOR modems.

• Make sure that both stations are on exactly the same frequency, within about 100 Hz.

• Set IF receive filters to 3 kHz or wider.

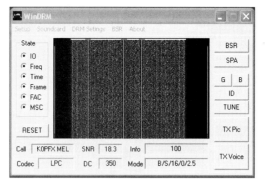

Fig 31.28 — *WinDRM* **digital voice/image software.**

• Don't overdrive the modem audio input.

• Turn off transceiver speech compression.

• Don't overdrive the radio. If the ALC meter shows any activity, turn down the output from the modem.

Transceiver duty cycle will be substantially higher when you're operating digital voice compared to normal SSB. To avoid damaging your radio, it is a good idea to reduce your output by 25 to 50%.

31.9.2 WinDRM

WinDRM began with *Dream,* a piece of open-source software developed by Volker Fisher and Alexander Kurpiers at the Darmstadt University of Technology in Germany. It was designed to decode Digital Radio Mondiale (DRM), a relatively new digital shortwave broadcast format. On the air, DRM presents as a wide, roaring signal. A commercial DRM signal is capable of carrying high-quality audio along with text and occasional images.

DRM uses Coded Orthogonal Frequency Division Multiplexing (COFDM) with Quadrature Amplitude Modulation (QAM). COFDM uses a number of parallel subcarriers to carry all the information, which makes it a reasonably robust mode for HF use.

Not long after *Dream* debuted, Francesco Lanza, HB9TLK, began adapting it for Amateur Radio. He redesigned *Dream* to support amateur DRM transmission and reception within a 2.5-kHz SSB transceiver bandwidth. That meant sacrificing some audio quality, along with the ability to simultaneously send images.

As the name implies, *WinDRM* is a *Windows* application (**Fig 31.28**). It is designed to use a computer sound card or on-board sound chipset to send and receive amateur DRM. Activity is primarily on HF. In addition to the software, you need a radio, computer and one of the sound-card interfaces discussed earlier in this chapter. Start by downloading and installing the *WinDRM* software at **n1su.com/windrm/download.html**. On this same Web page you will find documentation that explains the installation and operation.

To receive amateur DRM, all you need is a cable between the audio output of your radio and the LINE INPUT of your sound card. You may need to get into your sound card control software and boost the LINE INPUT gain. Using the *Windows* audio mixer, which you can access within *WinDRM,* bring up the **Recording Control** panel and make sure LINE is selected and that the "slider" control is up. If the *WinDRM* waterfall display is too bright, reduce the gain control.

Transmitting with *WinDRM* is somewhat more complicated in terms of your station setup, especially if you have only one sound card. The audio output from your sound card must be applied to your headset or PC speakers for receiving, but the *same* output must also feed your transceiver for transmitting (see **Fig 31.29**) so a switch for the audio output is needed. The easier, more elegant approach is to use two separate sound cards as shown in **Fig 31.30**. The second sound card doesn't need to be high-end; you can use an inexpensive USB sound card for this application.

As with the AOR modems, be careful not to overdrive your transceiver. If you see the ALC meter indicating excessive drive, reduce the sound card audio output.

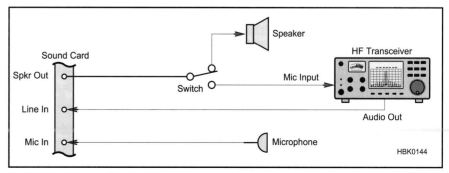

Fig 31.29 — The single sound card approach to using *WinDRM* **or** *FDMDV* **requires the means to switch the audio stream when switching from transmit to receive.**

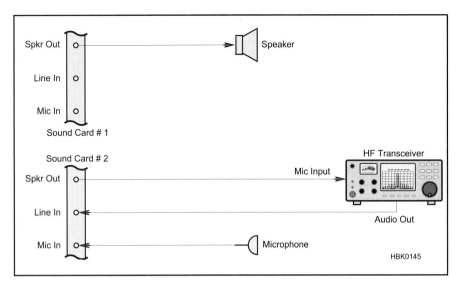

Fig 31.30 — The most elegant, easy-to-operate solution for *WinDRM* and *FDMDV* is to use two sound cards — one to process the received signal and the other to generate the transmit signal.

31.9.3 FDMDV: Frequency Division Multiplex Digital Voice

Like *WinDRM*, *FDMDV* is a software approach to digital voice that uses the sound card for processing the received signal and generating the transmit signal. The *FDMDV* software is available at **www.n1su.com/fdmdv**, along with the documentation. Station setup and hardware requirements are identical to *WinDRM*.

The advantages of FDMDV compared to WinDRM are twofold: 1) FDMDV can decode signals at substantially weaker levels and 2) FDMDV occupies an effective bandwidth of only 1100 Hz.

A popular frequency for HF digital voice is 14.236 MHz, USB. Information about digital voice nets and activity on other bands is available from **www.n1su.com** and other Web sites.

31.10 EchoLink, IRLP and WIRES-II

Worldwide communication on HF requires reliable propagation and more substantial radio and antenna requirements than a basic VHF FM setup. By using Internet links instead of HF radio links, a ham with a modest radio (even a handheld transceiver) can reliably communicate with stations hundreds or even thousands of miles away at any time of the day or night.

The most common form of amateur Internet linking involves the exchange of audio using *VoIP* — Voice over Internet Protocol — technology. This is the same technology used by Internet telephone services, and by online voice "chat" applications such as *Skype*, *TeamSpeak* and others. Three versions of Amateur Radio voice linking have become popular in the US: EchoLink, IRLP and WIRES-II.

31.10.1 EchoLink

EchoLink software, developed by Jonathan Taylor, K1RFD, is designed for *Windows* PCs, but there is a Mac version as well. With the software installed on a sound-card-equipped computer, any ham can create an EchoLink *node* that others can access by radio. Hams can also join the network directly without using radios. They simply plug microphone/headsets into their computers.

Each EchoLink node is assigned a number that can be up to six digits in length. RF users access the EchoLink network by first sending the DTMF code required to activate the link on a repeater or simplex node. Of course, this requires a transceiver equipped with a

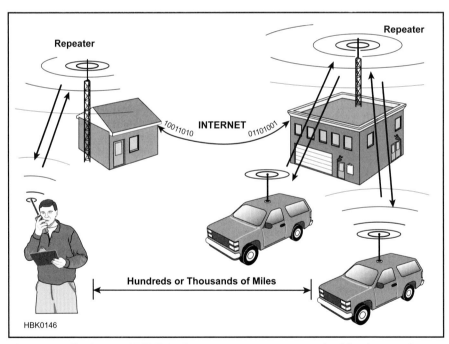

Fig 31.31 — The Internet VoIP link allows repeater users to speak with other amateurs who may be transmitting through other repeaters.

DTMF keypad. Once the link is up, they use the keypad again to transmit the number of the EchoLink repeater, link or user they wish to contact.

Hams who access the network directly (without a radio) select the person or system they wish to contact by simply clicking on its name in the software listing. The EchoLink network maintains continuously updated databases on several servers that indicate node and user locations and operational status. The network servers also support *conferencing* where many users can "meet" and speak together, either by direct access or via radio.

EchoLink supports communication among three different source groups:

• *Repeaters*: This is typically a VHF/UHF FM repeater with a computer at the site and a

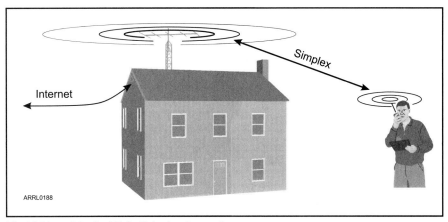

Fig 31.32 — A typical Echolink simplex node.

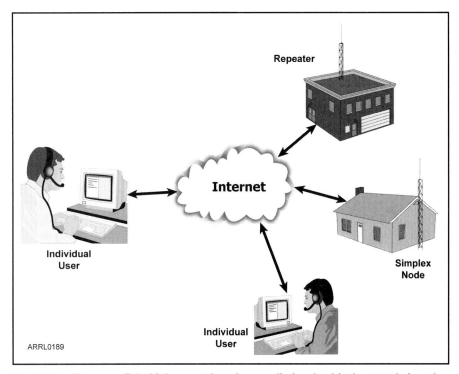

Fig 31.33 — For some EchoLink users, there is no radio involved (at least at their end of the conversation).

VoIP connection to the Internet via EchoLink. When the Echolink function is switched on, audio from the repeater is available on the EchoLink network, and vice versa. The Internet VoIP link allows repeater users to speak with other amateurs who may be transmitting through other repeaters, or who may be linking directly (see **Fig 31.31**).

• *Links*: Similar to repeater nodes, but operating on simplex frequencies without full repeater functionality (see **Fig 31.32**). These are typically FM transceivers connected to computers. Depending on the station setup, their coverage can be confined to a neighborhood, or it may encompass an entire county.

• *Users*: This is strictly a computer-to-computer VoIP connection (see **Fig 31.33**) with no RF involved. In other words, the operator is sitting in front of his computer wearing a microphone/headset.

All of the interfacing on the "RF side" of EchoLink is handled by connections to the computer sound card and serial port with a sound-card interface used to switch the radio between transmit and receive — exactly the same setup used by sound card based digital communications such as PSK31.

ECHOLINK SECURITY

Before being granted access, every ham who establishes an EchoLink node, or who connects to the network directly, must provide positive proof of identity and license during the EchoLink registration process. Details are available from **www.echolink.org/authentication**. After having been validated, each EchoLink node or user must provide a password, along with his or her call sign, to log in. Each time a connection is made for a QSO, the EchoLink servers verify both the sender and the receiver before communication can begin. Hams who use their radios to connect to the EchoLink network through a repeater or a simplex node do *not* have to be validated or provide a password. This is only necessary for the repeater or link stations, or for those who connect directly.

It is possible to configure EchoLink to accept connections only from certain types of stations: repeaters, links, users or all three. You can also set up a list of any number of "banned" call signs, which will not be allowed access. In addition, you can block or accept connections according to their international call sign prefix, in order to comply with reciprocal control-operator privileges or third-party traffic restrictions.

In Sysop mode, by default, EchoLink announces each station by call sign when the station connects. (The user can program a voice or CW ID that is generated automatically when needed.) The EchoLink software automatically generates detailed logs and (optionally) digital recordings of all activity on the link.

Unlike software such as e-mail programs, file-sharing programs and Web browsers, EchoLink does not have any way to pass files or "attachments" that might harm your computer. There are no known cases of EchoLink accepting or spreading a computer virus. Of course, any PC connected to the Internet should always be protected by some sort of Internet security hardware or software.

31.10.2 IRLP

IRLP stands for Internet Radio Linking Project. Like EchoLink, IRLP uses the Internet to establish VoIP links. The difference is that IRLP only permits access via RF nodes at repeaters or on simplex frequencies (all IRLP nodes are interlinked via a central Internet server). You must use a radio to access the IRLP network.

A typical IRLP node consists of a transceiver that is connected to the Internet through dedicated IRLP hardware and software. This requires a *Linux* computer running IRLP software and an IRLP interface board, or a turnkey "embedded" IRLP unit that combines the *Linux* computer, IRLP software and interface in a single package (**Fig 31.34**). The IRLP software controls the VoIP audio stream using carrier operated squelch (COS) or continuous tone coded subaudible squelch signals (CTCSS) from the transceiver. When COS is present, the computer detects it through the IRLP interface board.

The operator connects to the IRLP network (at a repeater, for instance) using DTMF sig-

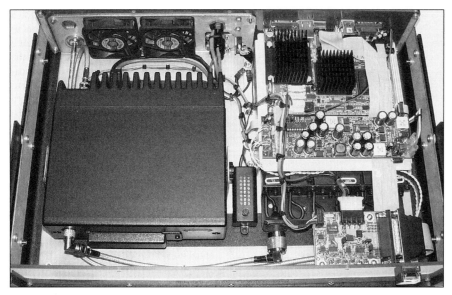

Fig 31.34 — The Micro-Node MN-3200 is an embedded IRLP node that combines the *Linux* computer, IRLP software and interface, along with a 70 cm repeater and duplexer in a single package.

nals sent over the radio. As with EchoLink, this means that the operator must have a radio with a DTMF keypad. The actual access code is determined by the owner of the node. The user sends the correct DTMF sequence to bring up the IRLP connection, then sends the sequence of the distant node he wishes to contact. That might be a repeater in another state or a distant simplex node. You can view a list of current IRLP nodes at **status.irlp.net/ statuspage.html**.

Here is an example of a typical IRLP connection:

• Press the transceiver PTT and identify your station. For example, "This is WB8IMY, Steve in Wallingford, Connecticut, USA connecting to node 8880."

• Next, press *#8880* (node 8880 in this example) on the DTMF keypad, release PTT and wait for the node connect announcement. The announcement indicates that your audio is now reaching the distant node. You can call a specific person, or just ask if anyone is around to chat.

• When finished, key PTT , make an announcement such as "WB8IMY releasing node 8880" and then press *#73* on the DTMF keypad to disconnect. The IRLP node will announce the disconnect and the node may now connect to a different node, reflector, or it will stand by awaiting another user.

It is interesting to note that there is a project underway (known as *EchoIRLP*) to bridge the IRLP and EchoLink networks.

31.10.3 WIRES-II

WIRES-II — Wide-Coverage Internet Repeater Enhancement System — is a VoIP network created by Vertex Standard (Yaesu). It is functionally similar to IRLP (including management via a central Internet server), but nodes and repeaters are connected via a Vertex Standard HRI-100 interface and a computer running under the *Windows* operating system. Vertex Standard transceivers are *not* required for WIRES-II communication, but as with IRLP, all access must be via RF.

There are two WIRES-II operational configurations using home stations (simplex nodes), repeaters or a combination of the two:

SRG — Sister Radio Group. You operate within WIRES-II in a small (10 node maximum) network that is ideal for closed-group operations. Within the network, all nodes operate using the same repeater node list, so you can link only to stations within this 10 node network. Because there are only ten nodes maximum, access to any of these nodes is possible using a single DTMF tone when calling. At the beginning of each transmission, this single DTMF tone locks communication between the calling node and the called node, but local (non-linked) transmissions are also possible, simply by omitting the DTMF tone at the beginning of the transmission.

FRG — Friends Radio Group. You may call any repeater registered with the WIRES-II FRG server. In the case of FRG operation, a six-digit DTMF code is required for access, and once the link is established this code need not be sent again (this is called the LOCK mode), unless the operator wants the ability to make non-linked transmissions (UNLOCK mode), in which case the six-digit code must be sent at the beginning of each transmission (using the DTMF Autodial feature of the transceiver, for example). Group calling to preset 10-repeater B, C, and D lists is also possible.

SRG is similar in philosophy to the local FM voice repeater where you tend to talk to the same group on a regular basis. FRG is similar in concept to IRLP, linking together a worldwide group of repeaters and base stations.

31.11 Glossary of Digital Communications Terms

AFSK — Audio frequency shift keying, a method of digital modulation in which audio tones of specific frequencies are used with an SSB voice transceiver.

ALE — Automatic link establishment. A process in which stations under computer control automatically call each other on different bands until contact is established.

AMBE — Advanced multi-band excitation, a proprietary speech coding standard developed by Digital Voice Systems.

AMTOR — Amateur teleprinting over radio, an amateur radioteletype transmission technique employing error correction as specified in several ITU-R Recommendations M.476-2 through M.476-4 and M.625.

APRS — Automatic Packet (or Position) Reporting System. A system of sending location and other data over packet radio to a common Web site for tracking and recording purposes.

ARQ — Automatic Repeat reQuest, an error-sending station, after transmitting a data block, awaits a reply (ACK or NAK) to determine whether to repeat the last block or proceed to the next.

ASCII — American National Standard Code for Information Interchange, a code consisting of seven information bits.

AX.25 — Amateur packet-radio link-layer protocol. Copies of protocol specification are available from ARRL HQ.

Baud — A unit of signaling speed equal to the number of discrete conditions or events per second. (If the duration of a pulse is 20 ms, the signaling rate is 50 bauds or the reciprocal of 0.02, abbreviated Bd).

Baudot code — A coded character set in which five bits represent one character. Used in the US to refer to ITA2.

BER — Bit error rate.

BERT — Bit-error-rate test.

Checksum — The output of an algorithm that allows the receiving system to detect errors in transmitted data.

CLOVER — Trade name of digital communications system developed by Hal Communications.

Collision — A condition that occurs when two or more transmissions occur at the same time and cause interference to the intended receivers.

CRC — Cyclic redundancy check, a mathematical operation. The result of the CRC is sent with a transmission block. The receiving station uses the received CRC to check data integrity.

Data modes — Computer-to-computer communication, such as by **packet radio** or **radioteletype** (**RTTY**), which can be used to transmit and receive computer characters, or digital information.

Digipeater — A station that relays digital data transmissions.

DRM — Digital Radio Mondiale, a digital modulation method used to transfer audio and data on HF bands.

EchoLink — A system of linking repeaters and computer-based users by using the Internet (also see **VoIP**).

Eye pattern — An oscilloscope display in the shape of one or more eyes for observing the shape of a serial digital stream and any impairments.

FEC — Forward error correction, an error-control technique in which the transmitted data is sufficiently redundant to permit the receiving station to correct some errors.

FSK — Frequency shift keying, a method of digital modulation in which individual bit values are represented by specific frequencies. If two frequencies are used, one is called *mark* and one *space*.

GPS — Global Positioning System

G-TOR — A digital communications system developed by Kantronics.

Hellschreiber — A facsimile system for transmitting text.

HSMM — High speed multimedia, a digital radio communication technique using spread spectrum modes primarily to simultaneously send and receive video, voice, text and data.

IEEE 802.11 — An IEEE standard for spread spectrum communication in the 2.4 GHz band at 1 Mbit/s and 2 Mbit/s data rates. 802.11 is also used as a general term for all spread spectrum devices operating under Part 15. (Also see **WiFi**.)

IEEE 802.11a — An IEEE standard for spread spectrum communication in the 5.8 GHz band at 6, 12, 16, 24, 36, 48, and 54 Mbit/s data rates.

IEEE 802.11b — An IEEE standard for spread spectrum communication in the 2.4 GHz band at 5.5 and 11 Mbit/s data rates in addition to being backward compatible with DSSS at 1 and 2 Mbit/s specified in 802.11.

IEEE 802.11g — An IEEE standard for spread spectrum communication in the 2.4 GHz band at 6, 12, 16, 24, 36, 48, and 54 Mbit/s data rates in addition to being backward compatible with DSSS at 1, 2, 5.5, and 11 Mbit/s specified in 802.11b.

IEEE 802.11n — An IEEE standard specifying data rates up to 250 Mbit/s and being backward compatible with 802.11a and 802.11g.

IEEE 802.16 — An IEEE standard specifying wireless last-mile broadband access. Also known as **WiMAX**.

IRLP — Internet Radio Linking Project, a system that uses the Internet to establish **VoIP** links among amateur stations who access the system via RF nodes at repeaters or on simplex frequencies.

Linux — A free Unix-type operating system originated by Linus Torvalds, et al. Developed under the GNU General Public License.

MFSK — Multi-frequency shift keying.

MFSK16 — A multi-frequency shift communications system

Modem — Modulator-demodulator, a device that connects between a data terminal and communication line (or radio).

MSK — Frequency-shift keying where the shift in Hz is equal to half the signaling rate in bits per second.

MT63 — A keyboard-to-keyboard mode similar to PSK31 and RTTY.

Multiple protocol controller (MPC) — A piece of equipment that can act as a **TNC** for several **protocols**.

Null modem — A device to interconnect two devices both wired as DCEs or DTEs; in EIA-232 interfacing, back-to-back DB25 connectors with pin-for-pin connections except that Received Data (pin 3) on one connector is wired to Transmitted Data (pin 3) on the other.

Packet radio — A digital communications technique involving radio transmission of short bursts (frames) of data containing addressing, control and error-checking information in each transmission.

PACTOR — Trade name of digital communications protocols offered by Special Communications Systems GmbH & Co KG (SCS).

Parity check — Addition of non-information bits to data, making the number of ones in a group of bits always either even or odd.

Position encoder — A device that receives data from a GPS receiver, assembles APRS packets from the data and creates modulated signals for use by the transmitter.

Project 25 — Digital voice system developed for APCO, also known as P25.

Protocol — A formal set of rules and procedures for the exchange of information within a network.

PSK — Phase-shift keying.

PSK31 — A narrow-band digital communications system.

Radioteletype (RTTY) — Radio signals sent from one teleprinter machine to another machine. Anything that one operator types on his teleprinter will be printed on the other machine. Also known as narrow-band direct-printing telegraphy.

Sound card — A computer sound processing device that may be included on the motherboard, as a plug-in card, or as a separate external device.

Sound card interface — A device used to connect a computer sound card to a transceiver to provide TR switching and transmit/receive audio connections.

Throb — A multi-frequency shift mode like MFSK16.

TNC — Terminal node controller, a device that assembles and disassembles packets (frames); sometimes called a PAD.

Terminal unit (TU) — An interface used between an old teletype unit and a transceiver.

Turnaround time — The time required to reverse the direction of a half-duplex circuit, required by propagation, modem reversal and transmit-receive switching time of transceiver.

VoIP — Voice over Internet Protocol, a technology used to exchange audio information over the Internet.

Waterfall — A continuously scrolling display where the software processes and displays signatures of all signals detected within the bandwidth of the received audio signal.

WEP — Wired Equivalent Privacy. An encryption algorithm used by the authentication process for authenticating users and for encrypting data payloads over a WLAN.

WEP Key — An alphanumeric character string used to identify an authenticating station and used as part of the data encryption algorithm.

WiFi — Wireless Fidelity. Refers to products certified as compatible by the WiFi Alliance. See **www.wi-fi.org**. This term is also applied in a generic sense to mean any 802.11 capability.

WiMAX — Familiar name for the IEEE 802.16 standard.

Wireless router — A low-power wireless device that manages data flow on a network that can include desktop or laptop computers, broadband Internet connections, printers and other compatible devices.

WIRES-II — Wide-Coverage Internet Repeater Enhancement System, a **VoIP** network created by Vertex Standard (Yaesu) that is functionally similar to **IRLP**.

WISP — Wireless Internet Service Provider

WLAN — Wireless Local Area Network.

WSJT — A suite of computer software for VHF/UHF digital communication including meteor scatter and moonbounce work.

31.12 Bibliography and References

BOOKS

R. Flickenger, *Building Wireless Community Networks* (O'Reilly & Associates, 2003).

S. Ford, WB8IMY, *ARRL VHF Digital Handbook* (ARRL, 2008).

S. Ford, WB8IMY, *ARRL HF Digital Handbook* (ARRL, 2007).

M. Gast; *802.11 Wireless Networks: The Definitive Guide* (O'Reilly & Associates, 2005).

B. Lafreniere, N6FN; *Nifty E-Z Guide to PSK31 Operation* (Nifty Ham Accessories, 2008).

D. Silage, *Digital Communication Systems Using* SystemVue (Charles River Media, 2006).

J. Taylor, K1RFD, *VoIP: Internet Linking for Radio Amateurs* (ARRL, 2009).

Radio Data Code Manual, (Klingenfuss; **www.klingenfuss.org**).

ARTICLES

S. Ford, WB8IMY, "Life Could Be a DReaM," *QST*, Apr 2007, pp 38-40.

M. Greenman, ZL1BPU, "MFSK for the New Millennium," *QST*, Jan 2001, pp 33-36.

G. Heron, N2APB "NUE-PSK Digital Modem," *QEX*, Mar/Apr 2008, pp 3-14.

S. Horzepa, WA1LOU, "Teaching an Old APRS New Tricks," *QST*, Feb 2006, pp 39-41.

M. Kionka, KI0GO, "Project 25 for Amateur Radio," *QST*, Sep 2008, pp 36-38.

P. Loveall, AE5PL "D-STAR Uncovered," *QEX*, Nov/Dec 2008, pp 50-56.

P. Martinez, G3PLX, "PSK31: A New Radio-Teletype Mode," *QEX*, Jul/Aug 1999, pp 3-9.

D. Schier, K1UHF, "The Ins and Outs of a Sound Card," *QST*, Oct 2003, pp 33-36.

J. Taylor, K1JT, "WSJT: New Software for VHF Meteor-Scatter Communication," *QST*, Dec 2001, pp 36-41.

J. Taylor K1RFD, "Computer Sound Cards for Amateur Radio," Product Review, *QST*, May 2007, pp 63-69.

Contents

32.1 Fast-Scan Amateur Television Overview
    32.1.1 ATV Activities
    32.1.2 Comparing Analog AM, FM and Digital ATV
    32.1.3 How Far Does ATV Go?
32.2 Amateur TV Systems
    32.2.1 Analog AM ATV
    32.2.2 Frequency Modulated ATV
    32.2.3 Digital ATV
    32.2.4 ATV Antennas
    32.2.5 ATV Identification
    32.2.6 ATV Repeaters
32.3 ATV Applications
    32.3.1 Public Service
    32.3.2 Radio Control Vehicles
    32.3.3 ATV from Near Space
32.4 Video Sources
32.5 ATV Glossary
32.6 Slow-Scan Television (SSTV) Overview
    32.6.1 SSTV History
32.7 SSTV Basics
    32.7.1 Computers and Sound Cards
    32.7.2 Transceiver Interface
    32.7.3 Transceiver Requirements
    32.7.4 SSTV Operating Practices
32.8 Analog SSTV
    32.8.1 Color SSTV Modes
    32.8.2 Analog SSTV Software
    32.8.3 Resources
32.9 Digital SSTV
    32.9.1 Digital SSTV Setup
    32.9.2 Operating Digital SSTV
    32.9.3 Future of SSTV
32.10 SSTV Glossary
32.11 Bibliography and References

# Chapter 32

Image Communications

This chapter covers two popular communication modes that allow amateurs to exchange still or moving images over the air. Advances in technology have made image communications easier and more affordable, resulting in a surge of interest.

The first part of this chapter, by Tom O'Hara, W6ORG, describes fast-scan television (FSTV), also called simply amateur television (ATV). ATV is full-motion video over the air, similar to what you see on your broadcast TV. Because of the wide bandwidth required for ATV signals, operation takes place on the UHF and microwave bands.

The second part of this chapter, prepared by Dave Jones, KB4YZ, describes slow-scan television (SSTV). Instead of full motion video, SSTV involves pictures that are transmitted at 8 seconds or longer per frame. SSTV is a narrow bandwidth image mode that is popular on the HF bands using SSB voice transceivers. SSTV operation can take place on FM, repeaters and satellites too.

32.1 Fast-Scan Amateur Television Overview

Fast-scan amateur television (FSTV or just ATV) is a wideband mode that is based on the analog *NTSC* (National Television System Committee) standards used for broadcast television in the US for many years, before most broadcasters switched to digital transmission in 2009. Analog AM and FM ATV use standard NTSC television scan rates. It is called "fast scan" only to differentiate it from slow-scan TV (SSTV) or digital TV (DTV). In fact, no scan conversions or encoders/decoders are necessary with analog ATV.

Any standard TV set capable of displaying analog NTSC broadcast or cable TV signals can display the AM Amateur Radio video and audio signals. New consumer digital television (DTV) sets sold in the US are designed to receive high definition television (HDTV) broadcasts using the 8-VSB (8-level Vestigial Side Band) standard from the Advanced Television Systems Committee (ATSC). DTV televisions will also include analog cable TV channel tuners until at least 2012. It is a good idea however, to not dispose of your old analog-only televisions, but keep them for ATV. Analog AM ATV on the 70 cm band will be the primary ATV system for some time.

To transmit ATV signals, standard RS-170 composite video (1-V peak-to-peak into 75 Ω) and line audio from home camcorders, cameras, DVD/VCRs or computers is fed directly into a transmitter designed for the ATV mode. The audio goes through a 4.5 MHz FM subcarrier generator in the ATV transmitter that is mixed with the video.

Picture quality is about equivalent to that of a VCR, depending on video signal level and any interfering carriers. All of the sync and signal-composition information is present in the composite-video output of modern cameras and camcorders. Most camcorders have an accessory cable or jacks that provide separate audio/video (A/V) outputs. Audio output may vary from one camera to the next, but usually it has been amplified from the built-in microphone to between 0.1 and 1 V P-P (into a 10-kΩ load).

32.1.1 ATV Activities

Amateurs regularly show themselves in the shack, zoom in on projects, show home video recordings, televise ham club meetings and share just about anything that can be shown live or by tape (see **Figs 32.1** and **32.2**, and application notes at **www.hamtv.com/info.html**). Whatever the camera "sees" and "hears" is faithfully transmitted, including full motion color and sound information. Computer graphics and video special effects are often transmitted to dazzle viewers. Several popular ATV applications are described in detail later in this chapter.

32.1.2 Comparing Analog AM, FM and Digital ATV

Receiving ATV using any of the three ATV modes is relatively easy using consumer televisions and receivers directly, or with the addition of an amateur band receive converter (downconverter). Much of the activity in an area depends on the first few hams who experiment with ATV, and on the cost and availability of equipment for others to see their first picture.

Analog AM TV on the 70 cm band is the easiest mode. The 70 cm ham band contains the same frequencies as cable channels 57 through 61, so a TV that has an analog cable

Fig 32.1 — Students enjoy using ATV to communicate school to school or between classrooms (top). The ATV view shows the aft end of the Space Shuttle cargo bay during a mission (bottom).

Fig 32.2 — The one-way ATV DX record is held by KC6CCC in San Clemente, California, for reception of 434-MHz video from KH6HME in Hawaii during a tropo opening in 1994. The distance is 2518 miles. See www.hamtv.com/atvdxrecord.html.

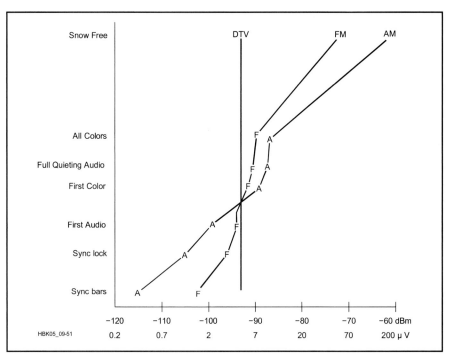

Fig 32.3 — Three approaches to ATV receiving. This chart compares AM, FM and digital ATV as seen on a TV receiver and monitor. Signal levels are into the same downconverter with sufficient gain to be at the noise floor. The FM receiver bandwidth is 17 MHz, using the US standard. The straight vertical line for DTV around −93 dBm illustrates the cliff effect described in the text.

Fig 32.4 — This graph shows the possible line-of-sight distances for P5 (snow free) video reception for various analog AM ATV transmitter levels in the 70 cm band. Power levels shown are in PEP. The Total Gain is calculated by adding the antenna gain (dBd) for both the receive and transmit antennas and then subtracting the feed line loss (in dB) at both ends. For other bands: 33 cm, subtract 6 dB; 23 cm, subtract 9 dB; and 13 cm, subtract 15 dB. For FM ATV (4 MHz deviation, 5.5 MHz sound), add 12 dB. For ATSC 8-VSB digital TV, the sudden loss of picture "cliff effect" distance is found by adding 26 dB. If the noise figure of the first stage in the downconverter is greater than 2 dB, subtract for each dB over 2. See the example in Table 32.1.

tuner can be used for receiving ATV signals as well. Simple 70 cm ATV transmitters are available.

FM ATV also uses simple transmitters, but most activity is on the 13 cm and 23 cm amateur bands. This is in part because of the wide occupied bandwidth needed for FM ATV signals, and in part because of the availability of receivers originally designed for commercial satellite or license free Part 15 use. The FM receivers have analog A/V outputs that must be connected to a video monitor.

Digital ATV signals using 8-VSB can also be received simply by using new televisions in the same manner as analog cable televisions, but the transmitters are complex. Transmission requires an MPEG-2 compression and digital transport encoder and a digital RF modulator/exciter. 8-VSB is an amplitude-modulated, 8-level baseband signal that is processed and filtered to occupy 5.38 MHz bandwidth. This signal fits in a standard 6-MHz channel with guard bands. RF amplifiers for DTV are more critical as to drive level, linearity and low intermodulation distortion than with analog AM or FM ATV modes. A good reference on the technical characteristics of an ATSC 8-VSB trans-

Table 32.1
ATV DX Graph Example

Transmit antenna	+10 dBd
Receive antenna	+12 dBd
Transmit feed line	−1 dB
Receive feed line	−2 dB
6 dB noise figure	−4 dB

Total Gain for 70 cm is 15 dB
With 20 W PEP, range is 35 miles (Fig 32.4)
For 23 cm (−9 dB) gain is 6 dB
With 1 W PEP, range is 3 miles (Fig 32.4)

mission can be found in an article by David Sparano entitled "What Exactly Is 8-VSB Anyway?" and available online from **www.broadcast.net/~sbe1/8vsb/8vsb.htm**.

PICTURE QUALITY

Experimentally, using the US standard, FM ATV gives increasingly better picture-to-noise (snow) ratios than AM analog ATV at receiver input signals greater than 5 µV. That's also about the signal level for the DTV "cliff effect" where the signal disappears. The DTV all-or-nothing cliff effect occurs because the digital signal detector and processing in the receiver requires a signal-to-noise ratio (SNR) of at least 15 dB for 8-VSB. Above 15 dB, you get an excellent picture, and at 14 dB SNR — nothing. Other DTV types — Digital Video Broadcast - Satellite (DVB-S) or Digital Video Broadcast - Terrestrial (DVB-T) — are a few dB better in the presence of noise.

Because of the wider noise bandwidth and FM threshold effect, AM analog video can be seen in the noise well before FM and DTV. For DX work, it has been shown that AM signals are recognizable in the snow at up to four times (12 dB) greater distance than FM or DTV signals, with all other factors equal. Above the FM threshold, however, FM rapidly overtakes AM. FM snow-free pictures occur above 50 µV, or four times farther away than with AM signals. The crossover point is near the signal level where sound and color begin to appear for all three systems. **Fig 32.3** compares analog AM, FM and digital ATV across a wide range of signal strengths.

32.1.3 How Far Does ATV Go?

The theoretical snow-free line-of-sight distance for 20-W PEP 70 cm analog AM ATV, given 15.8-dBd gain antennas and 2 dB of feed line loss at both ends, is 150 miles. (See **Fig. 32.4**.) In practice, direct line-of-sight ATV contacts seldom exceed 25 miles. Longer distances are possible with over-the-RF-horizon tropo openings, reflections, or through high hilltop repeaters. A 2518-mile reception record is shown in Fig 32.2. (See the **Propagation of Radio Signals** chapter.)

The antenna system is the most important part of an ATV system because it affects both receive and transmit signal strength. For best DX, use low-loss feed line and a broadband high-gain antenna, up as high as possible.

A snow-free, or "P5," picture rating (see **Fig 32.5**) requires at least 200 µV (−61 dBm) of signal at the input of the analog AM ATV receiver, depending on the system noise figure and bandwidth. The noise floor increases with bandwidth. Once the receiver system gain and noise figure reaches this floor, no additional gain will increase sensitivity. At 3-MHz bandwidth the noise floor is 0.8 µV (−109 dBm) at standard temperature in a perfect receiver. Analog TV luminance bandwidth rolls off around 3 MHz to separate it from the color subcarrier at 3.58 MHz. If you compare this 3 MHz bandwidth to an FM voice receiver with 15 kHz bandwidth, there is a 23 dB difference in the noise floor.

Much like the ear of an experienced SSB or CW operator, however, the eye can pick out sync bars in the noise below the noise floor. Sync lock and large, well contrasted objects or lettering can be seen between 1 and 2 µV. Color and subcarrier sound come out of the noise between 2 and 8 µV depending on their injection level at the transmitter and characteristics of your TV set. For the ATV DXer, using an older analog TV that does not go to blue screen or go blank (like a video squelch) with weak signals is a must, especially when rotating the antenna for best signal.

Operators must take turns transmitting on the few available channels. Two meter FM is used to coordinate ATV contacts, and the 2 meter link allows full-duplex audio communication between many receiving stations and the ATV transmitting station speaking on the sound subcarrier. This is great for interactive show and tell. It is also much easier to monitor a squelched 2 meter channel using an omnidirectional antenna rather than searching out each station by rotating a beam. Depending on the third-harmonic relationship to the video on 70 cm, 144.34 MHz and 146.43 MHz (simplex) are the most popular frequencies. They are often mixed with the subcarrier sound on ATV repeater outputs so all can hear the talkback.

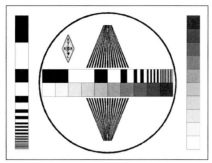
P5 — Excellent

P4 — Good

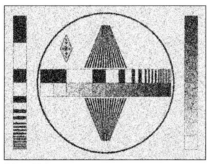
P3 — Fair

P2 — Poor

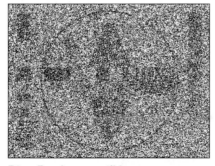
P1 — Barely perceptible
Fig 32.5 — An ATV quality reporting system.

32.2 Amateur TV Systems

Regardless of the type of ATV you're interested in — analog AM, FM or digital — you'll need to assemble the appropriate receiving and transmitting equipment and find other stations to work.

32.2.1 Analog AM ATV

Fig 32.6 shows the makeup of an analog AM TV channel (also called an NTSC channel) used for ATV, cable TV and for broadcast TV before the switch to digital. The channel is 6 MHz wide to accommodate the composite video, 3.58 MHz color subcarrier and 4.5 MHz sound subcarrier. Given the NTSC 525 horizontal line and 30 frames per second scan rates, the resulting horizontal resolution bandwidth is 80 lines per MHz. Therefore, with the typical TV set's 3-dB rolloff at 3 MHz (primarily in the IF filter), up to 240 vertical black lines can be seen. Color bandwidth in a TV set is less than this, resulting in up to 100 color lines. Lines of resolution are often confused with the number of horizontal scan lines per frame. The video quality should be every bit as good as on a home video recorder.

Most ATV is analog AM double sideband (DSB), with the widest component being the sound subcarrier out ±4.5 MHz. As can be seen in **Fig 32.7**, the video power density is down more than 30 dB at frequencies greater than 1 MHz from the carrier — more than 90% of the spectrum power is in the first 1 MHz on both sides of the carrier.

To fit within a 6 MHz wide channel, NTSC broadcast stations used an approach called vestigial sideband (VSB). VSB involves suppression, but not elimination, of the lower sideband so that it occupies less than 1 MHz. In addition, instead of combining a 4.5 MHz

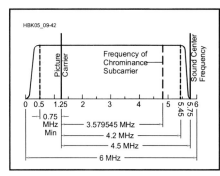

Fig 32.6 — An analog NTSC 6-MHz video channel with the video carrier 1.25 MHz up from the lower edge. The color subcarrier is at 3.58 MHz and the sound subcarrier at 4.5 MHz above the video carrier.

subcarrier to produce sound, broadcast stations used a second sound transmitter offset above the video carrier by 4.5 MHz so that the sound signal does not appear in the lower sideband.

DSB and VSB are both compatible with analog cable TV tuners, but the lower sound and color subcarriers are rejected in the TV set IF filter as unnecessary. In the case of VSB, less than 5% of the lower sideband energy is attenuated. The other significant energy frequencies are the sound (set in the ATV transmitter at 15 dB below the peak sync) and the color at 3.58 MHz (greater than 22 dB down).

AM ATV FREQUENCIES

The lowest frequency amateur band wide enough to support an analog AM ATV channel is 70 cm (420-450 MHz), and it is the most popular. With transmit power, antenna gains and feed line losses equal, decreasing frequency increases communication range. The 33 cm (902-928 MHz) band goes half the distance that 70 cm does, but this can be made up to some extent with high-gain antennas, which are physically smaller at the higher frequency. Depending on local band plan options, there is room for no more than two simultaneous AM ATV channels in the 33 cm and 70 cm bands without interference. If there is an in-band ATV repeater, simplex ATV operation shares space with the repeater input. Before transmitting, check with local ATV operators, repeater owners and frequency coordinators listed in the *ARRL Repeater Directory* for the coordinated frequencies used in your area.

The most popular in-band repeater output frequency is 421.25 MHz and is the same as cable channel 57. At least 12 MHz of separation is necessary for in-band repeaters because of TV-set adjacent-channel rejection and VSB filter characteristics. Cross-band ATV repeaters free up a channel on 70 cm for simplex and make it easier for repeater users to monitor their own repeated video with only proper antenna separation needed to prevent receiver desensitization.

Simplex, public service and R/C models use 426.25 MHz in areas with cross-band repeaters, or as an alternative to the main ATV activities on 434.0 or 439.25 MHz. The spectrum power density is so low at frequencies greater than 1 MHz from the video carrier of an AM analog ATV transmission that interference potential to other modes is low.

On the other hand, interference potential *to* ATV *from* other modes is high. Because a TV set receives a 6-MHz bandwidth, analog AM ATV is more susceptible to interference from many other sources than are narrower bandwidth modes. Interference 40 dB below the desired signal can be seen in video. Many of our UHF (and higher) amateur bands are shared with radar and other government radio positioning services. Signals from these services show up as horizontal bars in the picture. Interference from amateurs who are unaware of the presence of the ATV signal (or in the absence of a technically sound and publicized local band plan) can wipe out the sound or color, or can put diagonal lines in the picture.

Other stations operating on narrowband modes more than 1 MHz above or below the video carrier rarely experience interference from an AM ATV signal, or even know that the ATV transmitter is on the air, unless the narrowband station is operating on one of the subcarrier frequencies or the stations are too near one another.

If the band is full and the lower sideband color and sound subcarrier frequencies need

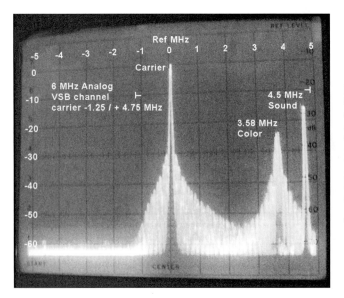

Fig 32.7 — Spectral display of a color analog AM VSB ATV signal. Spectrum power density varies with picture content, but typically 90% of the sideband power is within the first 1 MHz.

to be used by a dedicated link or repeater, a VSB filter in the antenna line can attenuate them another 20 to 30 dB, or the opposite antenna polarization can be used for more efficient packing of the spectrum.

Since most amateur linear amplifiers re-insert the lower sideband to within 10 dB of DSB, a VSB filter in the antenna line is the only cost-effective way to reduce the unnecessary lower sideband subcarrier energy if more than 1 W is used. In the more populated areas, 2 meter calling or coordination frequencies are often used to work out operating time shifts or other techniques to accommodate all users sharing or overlapping the same segment of the band.

RECEIVING AM ATV

Since the 70 cm band corresponds to cable TV channels 57 through 61, seeing your first ATV picture may be as simple as connecting a good outside 70 cm antenna (aligned for the customary local polarization) to a cable-ready TV set's antenna input jack. Cable channel 57 is 421.25 MHz, and each channel is progressively 6 MHz higher. (Note that analog cable channels and the old NTSC broadcast UHF channel frequencies are different above channel 13.) Check the *ARRL Repeater Directory* for a local ATV repeater output that falls on one of these cable channels. Cable-ready TVs may not be as sensitive as a low-noise downconverter designed just for ATV, but this low cost technique is well worth a try.

Most stations use a variable tuned downconverter specifically designed to convert the whole amateur band down to a VHF TV channel. Generally the 420 and 902 MHz bands are converted to TV channel 3 or 4, whichever is not used for over-the-air broadcast TV in the area. For 1240 MHz converters, channels 7 through 10 are used to get more image rejection.

The downconverter consists of a low-noise preamp, mixer and tunable or crystal-referenced local oscillator. Any RF at the input comes out at the lower frequencies. All signal processing of the AM video and FM sound is done in the TV set. A complete receiver with video and audio output would require all of the TV set's circuitry except the sweep and video display components. There is no picture quality gain by going directly from a receiver to a video monitor (as compared with a TV set) because IF and detector bandwidth are still the limiting factors.

A good low-noise amateur downconverter with 15 dB gain ahead of a TV set will give sensitivity close to the noise floor. A preamp located in the shack will not significantly increase sensitivity, but rather may reduce dynamic range and increase the probability of intermodulation interference. Sensitivity can best be increased by reducing feed line loss or increasing antenna gain. Or you can add an antenna-mounted preamp, which will eliminate the effects of loss in the feed line and loss through TR relays in the transmit linear amplifier. Each 6 dB total improvement — usually a combination of increased transmitter power, antenna gain or receiver sensitivity and reduced feed line loss — can double the line-of-sight distance or improve quality by 1 P unit (a measure of picture quality).

DRIVING AMPLIFIERS WITH AM ATV

Linear amplifiers for use with wideband AM video require some special design considerations compared to amplifiers used for FM and SSB voice operation. Many high-power amateur amplifiers would oscillate (and possibly self destruct) from high gain at low frequencies if feedback networks and power RF chokes did not protect them. These same stability techniques can affect operation over a 5-MHz video bandwidth. Sync, color and sound can be distorted unless the amplifier has been carefully designed for both stability and AM video modulation.

Several manufacturers offer special ATV amplifiers or standard models designed for all modes, including ATV. The collector and base bias supplies have a range of capacitors to keep the voltage constant under video modulation, while at the same time using the minimum-value low-resistance series inductors or chokes to prevent self-oscillation.

Almost all amateur linear power amplifiers exhibit some degree of gain compression from half-power to their full rated PEP output. For proper AM ATV operation, the amplifier can be driven to PEP output power of no more than its 1 dB gain compression level. If more power is needed, the analog AM ATV exciter/modulator can use a sync stretcher to maintain the proper transmitted video-to-sync ratio to compensate for higher outputs (see **Fig 32.8**).

To adjust a station with an ATV amplifier, disconnect both video and sound subcarrier and set the ATV transmitter's pedestal control for maximum power output. Then set the transmitter's RF power control to drive the amplifier to 90% of rated PEP output (this is the peak sync level). The 90% level is necessary to give some headroom for the 4.5 MHz sound subcarrier that is mixed and added with the video waveform. Once this peak sync level is set, use the blanking pedestal control to set amplifier output to 60% of this level. For example, a 100-W amplifier would first be set for 90 W with the RF power control and then 54 W with the pedestal control. Then the sound subcarrier can be turned back on and the video plugged in and video gain adjusted for best picture.

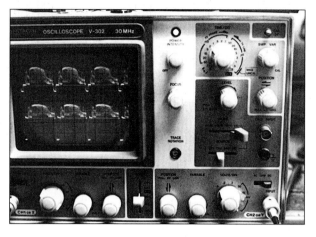

Fig 32.8 — An oscilloscope used to observe a video waveform. The lower trace is the video signal as it comes out of the sync stretcher. The upper trace is the detected RF signal from the amplifier.

If you could measure RF output with a peak-reading power meter made for video, the power would be between 90 and 100 W PEP. A dc oscilloscope connected to an RF diode detector in the antenna line shows that the sync and blanking pedestal power levels remain constant at their set levels regardless of video gain setting or average picture contrast. On an averaging power meter, however, it is normal with video to read about half the amplifier's rated power. The power reading will actually decrease with increasing video gain or with a change to a predominantly white picture. An averaging RF power meter cannot give an accurate measurement for an AM video signal given so many variables. If the amplifier drive level is properly set up according to the procedure outlined above, the actual sync tip PEP will remain constant.

32.2.2 Frequency Modulated ATV

In Europe, FM ATV on the 23 cm (1240 MHz) band is the standard because there is little room for ATV in the amateur 70 cm band. In the US, AM on 70 cm remains the most popular ATV mode because of equipment availability, lower cost, less occupied bandwidth and use of a standard analog TV set for receiving. However, FM ATV is gaining interest among experimenters and repeater owners for links and alternate inputs.

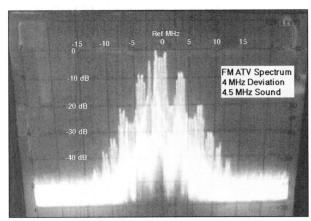

Fig 32.9 — Spectral display of an FM ATV transmission using 4 MHz deviation with a 4.5 MHz sound subcarrier. (Note that 5.5 MHz is the standard and would be 2 MHz wider.) Occupied bandwidth as defined in FCC Rule 97.3(a)(8) is −26 dB down from the mean power points.

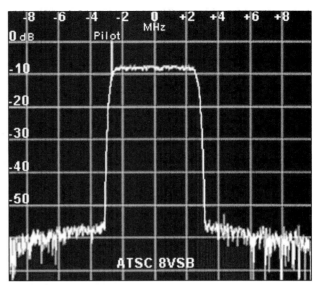

Fig 32.10 — Spectral display of an ATSC 8-VSB digital TV transmission. Note the pilot carrier unique to this mode of DTV.

As shown in **Fig 32.9**, FM ATV occupies 17 to 21 MHz depending on deviation and sound subcarrier frequency. While the 70 cm band is 30 MHz wide and thus theoretically could accommodate a 20 MHz wide FM ATV signal, the great interference potential to other users of the band precludes its use there. Most available FM ATV equipment is made for the 1.2, 2.4 and 10.25-GHz bands. The 3.3 and 5.6 GHz amateur bands are also being explored by using C-band satellite receivers and converted Part 15 equipment.

The US standard for FM ATV is 4 MHz deviation with the 5.5-MHz sound subcarrier set to 15 dB below the video level. Suggested frequencies are 1252 or 1255 MHz to stay away from FM voice repeaters and other users higher in the band, while keeping sidebands above the 1240 MHz band edge. Using the US standard, with Carson's rule for FM occupied bandwidth (see the **Modulation** chapter), it comes out to just under 20 MHz — so 1250 MHz would be the lowest possible frequency. Check with local frequency coordinators before transmitting because the band plan permits other modes in that segment.

C-band satellite TV receivers directly tune anywhere from 900 to 2150 MHz and may need a preamp added at the antenna for use on the 33 cm and 23 cm amateur bands. Early satellite TV receivers were made for antenna mounted LNBs (low noise block converters) with 40 dB or more gain. Satellite receivers are made for wider deviation (11 MHz) and need some video gain to give the standard 1-V P-P video output when receiving a signal with standard 4-MHz deviation. The additional video gain can often be had by adjusting an internal control or changing the gain with a resistor.

Some of the inexpensive Part 15 wireless video receivers in the 33 cm band use 4-MHz deviation FM video. Most of the 2.4-GHz Part 15 units are FM, so they can be used directly. On 2.4 GHz, some of the Part 15 frequencies are outside the ham band and care should be taken to use only those frequencies at least 8 MHz inside the 2390-2450 MHz ham band.

Part 15 receivers may or may not have the standard de-emphasis video network, however, so some circuitry may have to be added. Video pre-emphasis in the transmitter and de-emphasis in the receiver can double the communication distance or give 6 dB of SNR improvement by reducing the receiver noise bandwidth. Coordination among local stations is needed, though, because you cannot match stations that have the networks with those that don't without distorting the video greatly.

32.2.3 Digital ATV

The greatest advantage to running digital ATV is that you have a perfect P5 picture all the way down to 15 dB above the noise floor, and higher definition is possible in the same 6 MHz channel width used for analog AM ATV (see **Fig 32.10**). If you use the signal-to-noise level required to have snow-free ATV as a reference, 8-VSB DTV is 20 dB better than the 4 MHz deviation FM ATV and 32 dB better than analog AM ATV. This is a significant consideration for long-distance, point-to-point, line-of-sight paths and for paths requiring wide fade margins.

DTV can send much more usable picture and sound information in the same bandwidth as analog AM ATV under the theory that you don't have to send a continuous serial stream of a scanned picture at 30 interlaced frames per second. Rather, in DTV you send one whole frame and then send frames of only the changed pixels (or those pixels predicted to change, as explained in the next paragraph). Just the changes are sent a specified number of times (usually up to 15 times) before starting over again with a new complete frame.

There are complex algorithms based on how the human eye and mind perceive the picture. For instance, there is the probability that a pixel or group of pixels that are moving in a direction on the screen will occupy the adjacent pixel location at a certain time. According to Henry Ruh, AA9XW (a longtime ATVer and Chief Engineer at WYIN TV in Chicago), the coder/decoder (codec) software may expect a moving pool ball to continue in a straight line. If you watch closely, however, you may see the ball go deeper into the cushion than it really did until the next full frame. The bottom line is that the picture you see may be 10% actual data and 90% predictions.

DIGITAL ATV TRANSMITTER REQUIREMENTS

As can be seen in Fig 32.10, in an 8-VSB signal there is a pilot carrier 310 kHz up from the lower channel edge generated by a dc offset of the eight level modulation. The pilot carrier has a tight tolerance of 3 Hz and gives the RF PLL circuits in the 8-VSB receivers something to lock onto that is independent of the data being transmitted. Unlike 8-VSB, DVB-T or DVB-S signals do not have a pilot carrier (see **Fig 32.11**).

DTV transmissions are rated at average power. With 8-VSB modulation, the peak power is 7 dB greater than the average, or about 5 times. DTV transmission requires an amplifier with high linearity and good IMD performance to prevent distortion, noise and spectrum sideband growth. For proper opera-

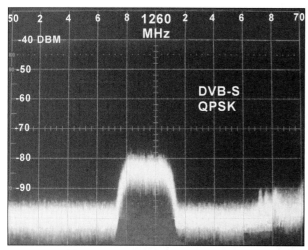

Fig 32.11 — A DVB-S signal from the WR8ATV repeater output as received on a spectrum analyzer. Note that the average power appears to be not too far above the noise floor, but gives a P5 picture. P units are measures of picture quality from P1 to P5 with P5 being the clearest and best picture.

Fig 32.12 — Art Towslee, WA8RMC, in his shack transmitting DTV through the local ATV repeater in Columbus, Ohio.

tion, amplifiers for 8-VSB operation are run at average powers that are 20% of their rated 1 dB gain compression levels. For example, an amplifier you might run at 100 W PEP output with analog AM ATV would run at 20 W average power or less for 8-VSB DTV. With DVB-T and DVB-S, power output can run slightly higher — up to 33% of an amplifier's rated 1 dB gain-compression level. Lower transmit power levels are not an issue, though. As described earlier, DTV requires less signal strength than analog AM ATV to produce a P5 picture.

With DTV, it's not a good idea to follow the amateur tradition of increasing amplifier drive to the point that a receiver starts to break up and then back the drive down a little. Although increased transmitter IMD may not affect the received signal until the combination of the received signal-to-noise and the added transmitter amplifier noise reaches the 15 dB cliff, the generated sideband levels can be quite large and interfere with other users.

An analog VSB filter in the antenna line can help keep the unwanted sideband growth down. As can be seen in Fig 32.10, DTV spectrum power density is high across the whole channel width and will show up as noise in a narrowband receiver tuned within the 6 MHz channel. For this reason, DTV should be used in the higher bands or 420-431 MHz segment of the 70 cm band so as not to interfere with FM voice repeaters, weak signal modes and satellite work.

DTV ISSUES

DTV using 8-VSB will stop working or freeze-frame with just a few miles per hour of antenna movement or multipath ghosting. Unlike analog ATV, DTV is impractical for mobile, portable, R/C vehicles and balloons. Developments in technology may allow mobile reception of over-the-air 8-VSB broadcast in the future. DVB-T does not have this problem, but is much more complicated and expensive to generate. DVB-S has some multipath susceptibility, but tests in Europe and the US show that it performs well under mobile conditions. Over time, DVB-S will likely become the most widely adopted amateur DTV system in the US, especially with the availability of inexpensive "free to air" satellite receivers.

Another unique factor with DTV communication is getting used to the digital processing delay when using a 2 meter talkback channel or even duplex audio and video. The delay can vary from a few hundred milliseconds to seconds.

AMATEUR DTV SYSTEMS

The first amateur DTV repeater in the US is in Columbus, Ohio (see **www.atco.tv**). This repeater uses DVB-S with QPSK modulation on 1260 MHz. Art Towslee, WA8RMC (shown in **Fig 32.12**) has a good simplified explanation of the various DVB modes on the ATCO Digital TV News Web page.

Nick Sayer, N6QQQ, has a blog in which he describes his experiments with 8-VSB on 33 cm — **nsayer.blogspot.com/search/label/ham**.

Early work by Uwe Kraus, DJ8DW and others is described at **www.von-info.ch/hb9afo/histoire/news043.htm**. Their experiments date to 1998, when they were the first to transmit moving color pictures with sound via a digital amateur television link over a distance of 62 miles. The early work used a 2 MHz bandwidth on 70 cm with 15 W output and 15 dB gain antennas.

Also see **www.von-info.ch/hb9afo/datv_e.htm** for sources and descriptions of DTV encoder and exciter boards. Most of these will take in a video signal and two audio signals and output a few milliwatts on an IF frequency to be mixed up to a ham band. The signal may also have a specific ham band output. Some have jumpers to select QPSK, GMSK, QAM or 8-VSB modulation as well as various forward error correction (FEC) and symbol rates.

With the shift to broadcast DTV in the US, it's possible that consumer RF modulators will start appearing for connecting video devices to TVs. This will bring the size and cost of chip sets down and make experimentation with 8-VSB and other DTV modes more economically practical for amateurs.

For more technical information on MPEG-2 compression see **en.wikipedia.org/wiki/MPEG-2**.

32.2.4 ATV Antennas

Foliage greatly attenuates signals at UHF, so place antennas above the treetops for the best results. Beams made for 432 MHz SSB/CW work or 440 MHz FM may not have enough SWR bandwidth to cover all the ATV frequencies for transmitting, but they are okay for reception. A number of manufacturers make beam antennas designed for ATV use that cover the whole band from 420 to 450 MHz. Use low-loss feed line and weatherproof all outside connectors with tape or coax sealer. Almost all ATV antennas use N connectors, which are more resistant to moisture contamination than other types. See the **Transmission Lines** and **Component Data and References** chapters for help with appropriate cable and connectors.

Antenna polarization varies from area

Fig 32.13 — At A, the simplest video identifier is to draw your call sign on a piece of paper and put it on the wall of your ham shack and in view of the camera. Doug Moon, K6KMN, also has his old auto call letter license plate as his ID. At B, the call sign, event name and location are overlaid on the video. During a public service event such as a long trail running race it is easy to forget to ID especially if operating full duplex and a lot is going on. See www.foothillflyers.org/hamtvac100.html for a description of ATV at the Angeles Crest 100 Mile Trail Race.

to area. It is common to find that the polarization was determined by the first local ATV operators based on antennas they had in place for other modes. Generally, UHF/microwave operators active on SSB, CW and digital modes have horizontally polarized antennas, while those into FM, public service or repeaters have vertical antennas. Check with local ATV operators before permanently locking down the antenna clamps. Circularly polarized antennas let you work all modes, including satellites.

32.2.5 ATV Identification

ATV identification can be on video or the sound subcarrier. A large high-contrast call-letter sign on the wall behind the operating table in view of the camera is the easiest way to fulfill the requirement (see **Fig 32.13A**). Transmitting stations fishing for DX during band openings often make up call sign IDs using fat black letters on a white background to show up best in the snow. Their city and 2 meter monitoring frequency (typically 144.34 or 146.43 MHz) are included at the bottom of the sign to make beam alignment and contact confirmation easier.

Quite often the transmission time exceeds 10 minutes, especially when transmitting demonstrations, public service events, space shuttle video, balloon flights or videotape. Intuitive Circuits makes a variety of boards that will overlay text on any video looped through them. Call sign characters and other information can be programmed into the board's non-volatile memory by on-board push buttons or an RS-232 line from a computer (depending on the version and model of the board). See Fig 32.13B.

One model will accept NMEA-0183 standard data from a GPS receiver and overlay latitude, longitude, altitude, direction and speed, as well as call letters, on the applied camera video. This is ideal for ATV rockets, balloons and R/C vehicles. The overlaid ID can be selected to be on, off or flashed on for a few seconds every 10 minutes to automatically satisfy the FCC ID requirement. The PC Electronics VOR-3 video operated relay board has an automatic nine-minute timer, and it also has an end-of-transmission hang timer that switches to another video source for ID.

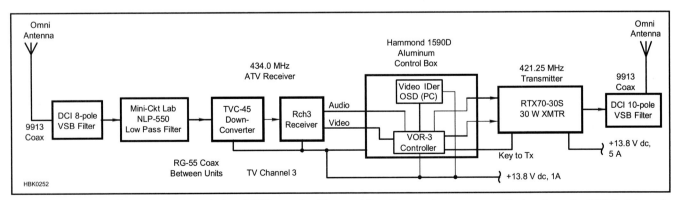

Fig 32.14 — A block diagram of a 70 cm in-band ATV repeater. The omnidirectional antennas are vertical and require 20 ft (minimum) of vertical separation to get >50 dB isolation to prevent receiver desensitization. Horizontal omnidirectional antennas require much more separation. A low pass filter on the receiver is also necessary because VSB or cavity type filters repeat a pass-band at odd harmonics and the third-harmonic energy from the transmitter may not be attenuated enough. P. C. Electronics makes the receiver, transmitter and VOR. Video ID can be done with a video overlay board like the Intuitive Circuits model OSD(PC) — by itself or even overlaid on a tower cam. Alternatively, an Intuitive Circuits ATVC-4+ ATV repeater controller board can do all the control box functions as well as remotely select from up to four video sources.

32.2.6 ATV Repeaters

There are two kinds of ATV repeaters: in-band and cross-band. In-band repeaters for 70 cm are more difficult to build and use, yet they are more popular because equipment for that band is more available and less expensive. Many ATV repeaters have added streaming video of their outputs to the Internet so users can see when they are active. Some repeaters even use the Internet to establish links. A good number of US repeaters can be found on the **www.batc.tv** Streaming Members Web site.

Why are 70 cm ATV repeaters more difficult to build than voice repeaters? The wide bandwidth of ATV makes for special filter requirements. Response across the 6-MHz passband must be as flat as possible with minimum insertion loss, but response must sharply roll off to reject other users as close as 12 MHz away. Special 6 to 10 pole interdigital or combline VSB filters are used to meet the requirement. An ATV duplexer can be used to feed one broadband omnidirectional antenna, but an additional VSB filter is needed in the transmitter line for sufficient attenuation to keep noise and IMD product energy from desensing the receiver.

A cross-band repeater requires less sophisticated filtering to isolate the transmitter and receiver because of the great frequency separation between the input and output. No duplexer is needed, only sufficient antenna spacing or low pass and/or high pass filters. In addition, a cross-band repeater makes it easier for users to see their own video. Repeater linking is easier too, if the repeater outputs alternate between the 23 cm and 33 cm bands.

Fig 32.15 — ATV repeater "tower cam" IDs have become popular so that users can see approaching weather. Some cameras have remote pan and tilt, so users can observe ice build-up on antennas or activity at the repeater site.

A 70 CM ATV REPEATER

Fig 32.14 shows a block diagram for a simple 70 cm in-band analog AM ATV repeater. No duplexer is shown because the antenna separation (>50 dB) and VSB filters provide adequate isolation.

To prevent unwanted key up from other signal sources, ATV repeaters use a video operated relay (VOR). The VOR senses the horizontal sync at 15,734 Hz in much the same manner that FM repeaters use CTCSS tones. Just as in voice repeaters, an ID timer monitors VOR activity and starts the repeater video ID generator every nine minutes, or a few seconds after a user stops transmitting.

A tower-mounted camera is often used in place of a video ID generator at repeaters (see **Fig 32.15**).

The repeater transmitter power supply should be separate from the supply for the rest of the equipment. With AM ATV the current varies greatly from maximum at the sync tip to minimum during white portions of the picture. Power supplies are not generally made to hold tight regulation with such great current changes at rates up to several megahertz. Even the power supply leads become significant inductors at video frequencies. They will develop a voltage across them that can be transferred to other modules on the same power supply line.

32.3 ATV Applications

32.3.1 Public Service

Video from an incident site back to an Emergency Operations Center (EOC) is a valuable addition to disaster communications. A real-time view of the incident scene can give commanders an immediate feel for the overall picture of what is going on or zoom in on a critical component rather than rely solely on descriptions relayed by voice or data. ATV has the added benefit of full duplex audio. The ATV transmitting station in the field talks on the sound subcarrier and the EOC can talk back at the same time on 2 meter FM voice. A camera and transmitter can be put on a robot and sent into a hazardous situation with no risk to people.

AN ATV "GO KIT"

Communications "Go-Kits" are popular with ARES and RACES groups. When an

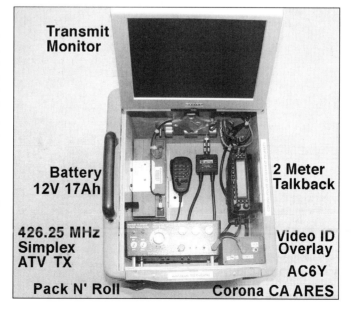

Fig 32.16—This portable ATV Go-Kit for emergency and public service work was built by Nick Klos, AC6Y. It includes a 70 cm ATV transmitter, 2 meter FM transceiver and battery in a Pack N' Roll container.

emergency occurs, hams can just grab their kit and go out in the field with everything they need. Nick Klos, AC6Y, put together three ATV Go-Kits for the police department ARES group in Corona, California. See **Fig 32.16**. Each kit uses an easy-to-transport Pack-N-Roll enclosure. (Pack-N-Rolls come in various sizes, have wheels, and have an extendable handle similar to airline carry-on luggage.) The ATV transmitter, 2 meter transceiver, battery and optional video monitor are fastened to plywood cut to fit the inside of the enclosure. A dc power supply can also be added if ac is available, or power can come from a dc extension cord with clips to connect to a car battery.

If the distance to be covered is short enough, a 2 meter/70 cm dual-band mobile vertical antenna connected through a duplexer and mounted on at least 10 ft of TV mast above the Go-Kit is sufficient for communication with a mobile command post. Longer or obstructed paths between the incident site and the EOC may require beams and more antenna placement care.

Multiple ATV Go-Kits allow for views of an incident site from different perspectives, including the police department helicopter and mobile or portable stations on the ground. The Go-Kits can also be used for parades, running and bike races, which are good operational practice for when the real emergency might happen.

PORTABLE REPEATERS

A portable ATV repeater for public service events can also be built in a Pack-N-Roll enclosure or a milk crate for easy transport and set up on the top of a building or hilltop by car or even helicopter. The path between an ATV station in the field and the EOC is rarely line-of-sight, so the portable ATV repeater can be placed at a high point that can be accessed from both locations.

The portable repeater shown in **Fig 32.17** is a cross-band unit with 70 cm input and 23 cm output. By using 70 cm input to the repeater, you get the best distance for the lowest power by those moving around an incident site with hard-hat cameras or portable units. Another advantage to using low-in and high-out at the repeater is that filtering is much easier because you don't have to contend with strong repeater transmitter harmonics in the receiver. Antenna separation of 5 ft is usually enough running a low-power, low-in/high-out portable repeater.

With weaker signals, a low-pass filter in the 70 cm receiver feed line and a high-pass filter in the 23 cm transmitter feed line may help to minimize desense. If a portable ATV repeater system is used at or near a communications site, band-pass filters in the antenna feed lines should be used to prevent intermodulation interference or receiver overload from nearby transmitters.

With compact 10-dBd gain corner reflectors or higher gain beams on 23 cm, you can get more than 5 miles line-of-sight from the repeater to the EOC with just 3 W of transmitter power. The portable repeater shown is self-contained with a 12-V, 17-Ah battery that will last about 12 hours. Alternatives include ac-operated 13.8 V, 3 A power supply or a dc cord clipped to a car battery. A local camera can be plugged in and used if needed. Details of the milk crate repeater can be found at **www.hamtv.com/info.html** under Portable Public Service Repeater.

32.3.2 Radio Control Vehicles

Some hams also enjoy operating radio control (R/C) vehicles, and ATV can add a new dimension to that hobby. The size, weight and cost of color cameras have come way down, and ATV transmitters are smaller as well, so ATV stations can more easily be integrated into R/C vehicles to give them eyes and in some cases ears.

AN EDUCATIONAL TOOL

Mark Spenser, WA8SME (ARRL Amateur Radio Education and Technology Coordinator) added ATV capability to a Parallax Board of Education Robot (BOE/BOT) as shown in **Fig 32.18**. With the add-on ATV board, the BOE/BOT simulates a Mars Lander. The robot and ATV board are also used to teach students how a TV remote control works, as well as demonstrate remote sens-

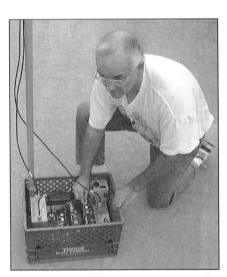

Fig 32.17 — At A, Tom O'Hara, W6ORG, adjusts a portable ATV repeater for public service work that can be built in a milk crate. The block diagram is similar to Fig 32.14 but without the VSB filters. The transmitter is a P. C. Electronics RTX23-3 on 1253.25 MHz, the receiver is on 426.25 MHz, and small beams or omni antennas are used. For more details on construction, visit www.hamtv.com. At B, the repeater is flown to a hilltop and quickly set up. Sharon Kelly, KF6OQO, is adjusting the antenna. At C, the video of a fire can be seen at an EOC on the other side of the hill while Gary Heston, W6KVC, talks back to the incident site ATV operator on 2 meters.

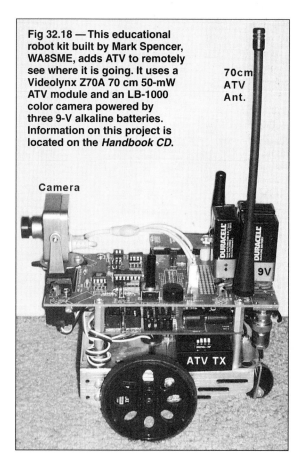

Fig 32.18 — This educational robot kit built by Mark Spencer, WA8SME, adds ATV to remotely see where it is going. It uses a Videolynx Z70A 70 cm 50-mW ATV module and an LB-1000 color camera powered by three 9-V alkaline batteries. Information on this project is located on the *Handbook CD*.

ing and data linking. This ATV application has encouraged students to use the kit as a science class project and become licensed hams in the process.

R/C VEHICLE TIPS

ATV and data transmitter modules have been put in R/C aircraft, blimps, cars and boats. A 50 mW transmitter on 70 cm is enough to give good video for ½ mile or so around an R/C flying field. An inverted ground-plane or vertical dipole is used typically on aircraft, but nulls in the radiation pattern can show up at steep bank angles. Ground-plane antennas are also used for receive because the pattern is broad. Mounting the receive antenna at least 10 ft above ground can help avoid multipath ghosting or signal blockage from nearby people and vehicles. Use of higher gain omnidirectional antennas is tempting for greater distance, but they have narrower radiation patterns and more nulls.

Operation at higher frequencies requires shorter antenna lengths that better might fit the physical size limitations of an R/C vehicle, but snow-free distance will be shorter given the same transmitter power and receive antenna gains.

If ac power is not available at the flying field for plugging in a TV, look for a portable television that can be powered from batteries. Some hams have used TV tuners designed for use with laptop computers. Position the screen to be north facing for least sun washout. It is best for even experienced R/C operators to not fly solely by video, but to have a copilot observe directly as the camera's field of view is limited.

Typical 72 MHz or 50 MHz R/C receivers are not designed to operate with a transmitter in the same vehicle. The R/C receiver front end could overload from ATV transmitter RF, so control distance testing is a must before the first flight. Testing can be done by having another ham (with a handheld radio for talkback) carry the R/C vehicle out to the normal limit of control distance. Then turn on the R/C transmitter and verify that the controls are working normally. Finally, turn on the ATV transmitter and again verify operation.

If the R/C receiver is captured by the ATV transmitter, then experiment with lowering power, changing antenna placement, or adding shielding or filtering. In the past, crystal oscillator leakage was often the culprit, rather than the 70 cm output. Shielding cured that. Interference may be less of a problem with a newer ATV transmitter that uses a PLL oscillator.

A PRACTICAL EXAMPLE

The added weight of the ATV system on the electric R/C helicopter shown in **Fig 32.19** is only 7 oz, but first you need to determine that the craft can lift the added load. The extra equipment must be mounted to maintain balance around the aircraft's center of gravity. Some testing with equivalent weights first can save equipment from possible damage. Note the stubby flexible antenna mounted upside down on one of the skid supports. This is counterbalanced by mounting the R/C receiving antenna on the opposite side.

A separate battery should be used for the ATV system rather than tapping into the motor or R/C receiver battery. The voltage will vary during the flight and could put motor noise in the video. A standard 9 V alkaline battery will last almost two hours continuous

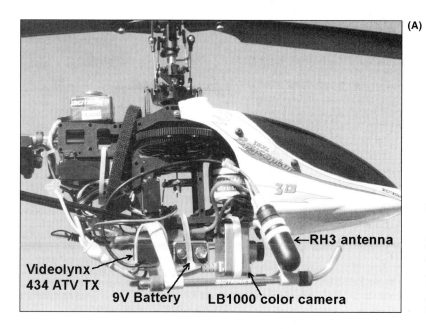

Fig 32.19 — At A, cable ties are used to fasten an LB-1000 color camera, Videolynx 434 ATV transmitter, 9-V battery and Diamond RH3 antenna to an R/C helicopter powered by an electric motor. At B, a computer with a TV tuner plugged into the USB port is used to view video coming from the ATV transmitter on the R/C helicopter — in this case, looking back at the R/C model operator.

duty with this system.

On larger R/C aircraft having sufficient payload, a GPS receiver and Intuitive Circuits video overlay board can put the speed, altitude and direction, along with call sign, over the camera video.

32.3.3 ATV from Near Space

Seeing the earth fall away via ATV as an amateur rocket or balloon rises can be quite exciting as shown in **Fig 32.20**. There are a number of groups sending up balloons and rockets with instrumentation, beacons and repeaters in addition to ATV. Balloon transmissions have been received as far as 500 miles away as the instrument package rises to 100,000 ft before the balloon pops from gas expansion and then parachutes back to earth. Radio direction finding clubs are often asked to help find where the balloon has come down and aid in recovery.

Rockets and balloons using ATV present a challenging optimum antenna radiation pattern problem similar to satellites that are predominantly flown overhead. For strongest picture and best distance (DX), the transmit and receive antennas should have their major power lobes pointed at each other and use the same polarization.

An omnidirectional vertical dipole antenna in the vehicle is fine for best DX on the horizon where the receiving station views the vehicle from a low angle. For example if a rocket is expected to go 1 mile vertically, receiving stations at 4 miles or more away will be within the maximum radiation lobe of the antenna and get the best video. Receiving stations at the launch site or closer than 4 miles away would be within pattern nulls.

If the rocket antenna is a ¼ wave vertical spike on the nose and the body of the rocket is the ground plane, the main lobe can have an up-tilt of 15 to 20° above the horizon and a significant amount of the signal will be wasted. Balloons can use an upside down ground plane hanging below the instrument package to give a down tilt to the earth below.

At the launch site, a horizontal dipole on the rocket (see **Fig 32.21**) and circularly polarized antenna on the ground works best. Balloons do well by hanging down an omni-directional wheel antenna that is horizontally polarized on the horizon, but circularly polarized directly below. Circular polarization at one end helps to minimize the signal strength variation as the vehicle spins.

A 1-W, 70 cm ATV transmitter board such as the P.C. Electronics TXA5-RC series is a good tradeoff between power and battery weight. A beam antenna for reception is necessary to make up for the low power if you want snow-free pictures from higher than 20,000 ft. Typically, ATV equipment is housed in an insulated enclosure suspended below the helium-filled balloon, along with a battery capable of lasting the expected length of flight time. Styrofoam is light and will retain some of the heat dissipated from the electronics so that the extreme cold at altitude does not cause the transmitter and battery to cease operation.

Balloon and rocket groups have a lot of how-to information and often announce when there will be a launch so that hams within a few hundred miles radius can try receiving the signals. Some groups are:

Bill Brown WB8ELK Balloon Flights — **fly.hiwaay.net/~bbrown**

Arizona Near Space Research ballooning — **www.ansr.org**

Amateur High Altitude Ballooning — **www.nearsys.com**

Edge of Space Science ballooning — **www.eoss.org**

N8UDK rocket ATV — **www.detroitatvrepeater.com/rockets.htm**

KC6CCC rocket ATV — **www.qsl.net/kc6ccc/rockets.htm**

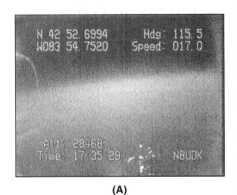

(A)

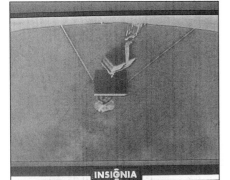

(B)

Fig 32.20 — At (A), an ATV camera view from a balloon at 93,399 ft shows the blackness of space, the curvature of the earth and hazy clouds below. Note that the standard NMEA altitude data output from GPS receivers is in meters. At (B), video received from a balloon on a 7-inch portable LCD TV at 20 miles slant distance. The camera is looking down at the instrument package, with clouds and the ocean below.

(A)

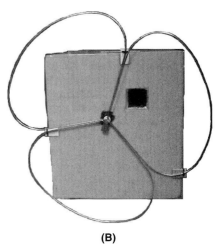

(B)

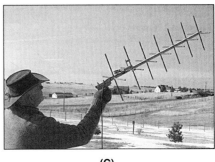

(C)

Fig 32.21 — At (A), a dipole antenna and a mirror are mounted on the outside of a 3.5-inch diameter amateur rocket. A video camera inside looks through the hole to the mirror so those receiving can see the view of the ground fall away as the rocket blasts off. At (B), N8UDK mounted an Old Antenna Lab 70 cm Little Wheel horizontal omni antenna to the bottom of his high-altitude balloon ATV payload. The camera is mounted inside the insulated foam package and looks out through the square cutout. At (C), a handheld antenna using circular polarization can be pointed at a balloon or rocket for maximum signal level as it travels. See www.hamtv.com/rocket.html.

32.4 Video Sources

Practically any video device that delivers a picture when plugged into the coaxial video jack of a video monitor can be plugged into an ATV transmitter and used to send video over the air. Suitable sources include the composite video from camcorders, video cameras, digital cameras, VCRs or computers. The standard A/V cable from a video device has a phono plug on the end and is often identified by yellow color coding (other colors are used for audio).

The cost of video cameras has come way down, thanks to the development of solid state imaging devices that vary in size, type and number of horizontal and vertical picture elements (pixels). Cameras widely available for home video work well for ATV.

VIDICON TUBES

The vidicon is a relatively simple, inexpensive vacuum tube imaging device that once was the standard for home video cameras, closed-circuit television and ATV applications. **Fig 32.22** shows a vidicon's physical construction. As with a cathode ray tube (CRT), an electron beam is created by a cathode and accelerated by grids. Horizontal and vertical deflection of the electron beam in a vidicon is accomplished with magnetic fields generated by coils on the outside of the tube. Varying the strength of the fields control the beam's position as it scans across the inside of the front of the vidicon, which is coated with a photoconductive screen layer. As the electron beam scans the screen, it charges each spot on the screen to the cathode voltage (about –20 V with respect to the signal electrode). Each spot acts like a leaky capacitor, discharging when illuminated by light hitting the front of the vidicon. The rate of discharge depends on the intensity of light illuminating the spot.

When it next scans across the spot, the electron beam will deposit enough electrons to recharge the spot to cathode potential. As it does so, the current flowing between the cathode and the screen increases, proportionally to how much the spot's "capacitor" discharged since last being scanned by the beam. Since this discharge depends on the amount of light hitting that spot on the screen, the changes in the beam current as it scans the screen correspond to the image being viewed by the screen. This creates a video signal that represents the scanned image. The variations in the beam current are very low (a fraction of a microamp), and the output impedance of the vidicon is very high, so the video circuitry must be designed with care to minimize hum and noise and the tube itself must be carefully shielded.

CHARGE-COUPLED DEVICES

A *charge-coupled device (CCD)*, in its simplest form, is made from a string of small capacitors with a MOSFET on its input and output. The first capacitor in the string stores a sample of the input voltage. When a control pulse biases the MOSFETs to conduct, the first capacitor passes its sampled voltage on to the second capacitor and takes another sample. With each successive control pulse, the input samples are passed to the next capacitor in the string. When the MOSFETs are biased off, each capacitor stores its charge. This process is sometimes described as a "bucket brigade," because the analog signal is sampled and then passed in stages through the CCD to the output.

Fig 32.23 shows a two-dimensional CCD that might be used in a digital camera. In a CCD used for imaging, an array of sensing elements called *pixels* are coated with light-sensitive material. The light-sensitive mate-

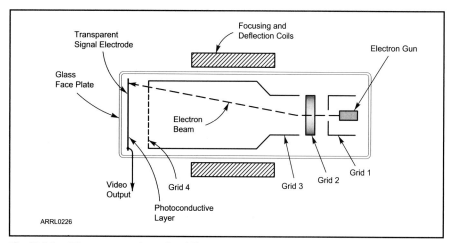

Fig 32.22 — The construction of a vidicon tube used for acquiring video images. Light striking the front of the tube causes the current in the electron beam from the cathode to photoconductive layer to change. These changes create the electronic video signal.

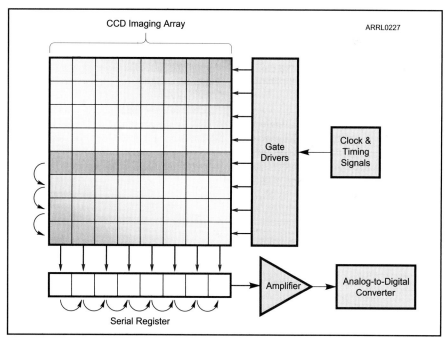

Fig 32.23 — The construction of a two dimensional CCD array that might be used in a digital camera. When a control pulse biases the array to conduct, the first row passes its sampled voltage on to the next row and takes another sample. With each successive control pulse, the input samples are passed to the next row until they reach the edge of the array. There they are transferred to a serial register and digitized.

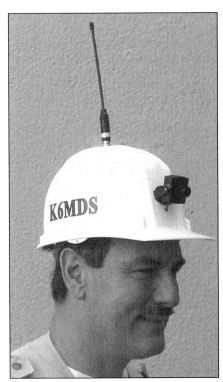

Fig 32.24 — Michael Saculla, K6MDS, built an ATV Hat Cam using a CMOS color camera along with a Videolynx 70 cm transmitter for public service applications.

Fig 32.25 — The tiny LB-1000 420-line CMOS color camera is typical of the many low-cost devices available. Intended primarily for security applications, they are popular with amateurs.

rial in each pixel *photocharges* a single CCD capacitor. The photogenerated charge on each capacitor is proportional to the amount of light striking the sensing element. The stored charges are then shifted out of the CCD array a line at a time as signals corresponding to each pixel. The resulting sequence of signal is similar to that of a vidicon scan. CCDs are the most common imaging device used in digital photographic cameras as well as video cameras, and they have very good low-light sensitivity.

Color CCDs use a screen, called a Bayer mask, over each square of four pixels. Each square has one for red, one for blue and two for green because the human eye is more sensitive to green. The luminance — gray scale — is sampled from every pixel. The possible lines of vertical resolution are determined by the number of pixels horizontally, but color resolution will be less because it comes from a square of four pixels.

When *lines of resolution* are given for a camera, this refers to the number of changes from vertical black lines to white that can be seen horizontally across the picture. While it is good to have a higher resolution camera — most are 380 lines or more — the picture will still be seen on a TV or monitor. In these display devices, response bandwidth limits luminance to about 240 lines and color to 100 lines in the NTSC system.

CMOS CAMERAS

CMOS active pixel sensors are a newer development and are replacing CCDs in imaging products. CMOS cameras have lower current requirements, lower price and less image lag, making them popular in cell phones. Amateurs use them for portable ATV applications such as the hat cam in **Fig 32.24**. A small CCD color camera may draw 120 mA while a comparable CMOS version draws 50 mA.

At first glance, board-level cameras are lighter for weight-sensitive applications like R/C and balloon vehicles, but those housed in a metal case are preferred to afford shielding from the effects of RF getting into the circuitry. The CMOS camera shown in **Fig 32.25** only weighs 3 oz in a 1.5-inch square metal case. It also includes a mic with line level audio output.

32.5 Glossary of ATV Terms

Aspect ratio — Image width divided by its height. Standard definition analog television uses an aspect ratio of 4:3 and broadcast HDTV uses 16:9.

ATSC — Advanced Television Systems Committee, the group that defined the set of standards and formats for the digital television formats implemented in the Broadcast Television Service. See **DTV**.

ATV — Amateur Television. Sending pictures by Amateur Radio. You'd expect this abbreviation to apply equally to fast-scan television (FSTV), slow-scan television (SSTV) and facsimile (fax), but it's generally applied only to FSTV.

Back porch — The blank part of a scan line immediately following the horizontal sync pulse. See **blanking**.

Black level — The signal level amplitude that corresponds to the black end of the video dynamic range.

Blanking — A "blacker-than-black" signal level that assures that the scanning trace cannot be seen as it is reset to scan another line or frame. In conventional television this is often referred to as the *blanking pedestal*, consisting of two segments — the **front porch** that precedes the vertical sync pulse and the **back porch** that follows the sync pulse.

Cathode ray tube (CRT) — A specialized electron tube, employing a phosphor-coated screen, used for image display. The classic TV "picture tube" used in older television sets and computer monitors, is an example of such a tube.

Charge-coupled device (CCD) — An integrated circuit that uses a combination of analog and digital circuitry to sample and store analog signal voltage levels, passing the voltages through a capacitor string to the circuit output.

Chrominance — The color component of a video signal. NTSC and PAL transmit color images as a black-and-white compatible luminance signal along with a color subcarrier. The subcarrier phase represents the hue and the subcarrier's amplitude is the saturation.

Color burst — Seven cycles of a 3.58 MHz subcarrier signal located on the back porch (see **blanking**) of an NTSC color TV signal waveform. This short burst is locked on to by a PLL in the TV receiver as the reference for the chroma in the video.

Color subcarrier — The modulated 3.58 MHz component of an NTSC color television signal that is used to convey the color or luminance image data.

Composite video — The standard 1 V peak to peak analog video into 75 Ω load signal consisting of color, video, blanking and sync found in consumer products and typically identified with a yellow color coded RCA connector. The sync tip is referenced at 0 V, blanking at 0.285 V, black level at 0.339 V and white at 1.0 V.

Compression — Various digital techniques to reduce the bandwidth, transmission rate or file size of an image.

Deflection — the circuits or other components controlling the vertical and horizontal sweep signals that move the scanning beam of a cathode ray tube image display.

DTV — Digital television, most commonly applied to a series of 16 digital formats (including HDTV or High Definition Television) that comprise the default commercial broadcast digital TV standards in the United States.

DVB-S — Digital Video Broadcast - Satellite.

DVB-T — Digital Video Broadcast - Terrestrial.

Field — Collection of top to bottom scan lines. When interlaced, a field does not contain adjacent scan lines and there is more than one field per frame.

Frame — One complete scanned image. NTSC has 525 lines per frame with about 483 usable after subtracting vertical sync and a few lines at the top containing various information.

Front porch — The blank part of a scan line just before the horizontal sync.

FSTV — Fast-Scan TV. Same as common, full-color, motion commercial broadcast TV.

Interlaced scanning — A scanning pattern, designed to reduce perceived flicker in broadcast television systems, in which the complete image frame is actually made up from two sequentially-scanned fields and thus improving the vertical resolution. The timing of the sequential scanning is such that the lines of the second field are interspersed between the lines of the first field. For example, NTSC sends a field with just the even lines in 1/60 second, then a field with just the odd lines in 1/60 second.

Luminance — The brightness component of a video signal and can refer to the white-to-black amplitude or the chroma weighted sum of the gamma-compressed R'G'B' components computed as Y (the luminance signal) = 0.59 G (green) + 0.30 R (red) + 0.11 B (blue).

MPEG — Motion Picture Experts Group, a set of digital image compression formats/standards for moving images.

NTSC — National Television System Committee. Analog broadcast television standard used in North America and Japan.

PAL — Phase alteration line. Analog television standard used in Germany and many other parts of Europe.

Progressive scanning — A scanning sequence in which all image lines are scanned sequentially to display the complete image frame. Used in some **DTV** broadcasts.

Pixel — Picture element, the dots that make up images on a monitor. Image resolution is directly related to the number of picture elements. If the size of a picture element is large relative to total image size, the image will be made up of a relatively small number of picture elements and thus have relatively poor resolution. If the size of a picture element is small relative to total image size, the image will consist of a large number of picture elements and thus demonstrate better resolution.

Raster — The pattern of scanning lines developed on the face of a cathode ray tube during the display of one image frame.

Resolution — The ability to see a number of vertical lines horizontally across the screen. The NTSC analog horizontal scan rate of 15734 Hz limits the luminance resolution to 80 lines per MHz of video response bandwidth. Digital TV's highest resolution is limited by the number of screen pixels. It takes 2 adjacent pixels, one black and one white for one line.

RGB — Red, Green, Blue. One of the models used to represent colors. Due to the characteristics of the human eye, most colors can be simulated by various blends of red, green, and blue light.

SECAM — Sequential color and memory. Analog television standard used in France and the Commonwealth of Independent States.

Sync — That part of a TV signal that indicates the beginning of a frame (vertical sync) or the beginning of a scan line (horizontal sync).

Vidicon tube — A type of photosensitive vacuum tube used in TV cameras.

White level — The white end of the video dynamic range in any image transmission format. The difference between white and black levels defines the video dynamic range of the mode.

8-VSB — 8-level Vestigial Side Band, the standard for broadcast **DTV** in the US.

32.6 Slow-Scan Television (SSTV) Overview

The previous sections discussed fast-scan amateur television, used to send wide-bandwidth full motion video in the 420 MHz and higher bands. In contrast, slow-scan television (SSTV) is a method of sending still images in a narrow bandwidth and is widely used on the HF bands, although SSTV is sent via FM repeaters and amateur satellites too. **Fig 32.26** shows a sample image.

Images are our most powerful communication tool. They can make us understand and remember better than any of our other senses. SSTV allows us to add images to our verbal communications via Amateur Radio. Working with the SSTV mode can provide much more than just swapping pictures. It

Fig 32.26 — Color SSTV image received from the International Space Station in Martin 1 mode 10/12/08 on 145.800 MHz FM.

provides a practical way to learn about radio propagation, computers and computer graphics. As with any activity, the more involved you become, the more knowledgeable you become about all the intricacies.

SSTV is also a great way to get others involved in Amateur Radio or enhance other activities. Consider adding SSTV capability to your emergency communications, public service, Field Day or Jamboree on the Air station.

32.6.1 SSTV History

SSTV originally involved the transmission of a visual image from a live video source. Images were black-and-white. Specific audio

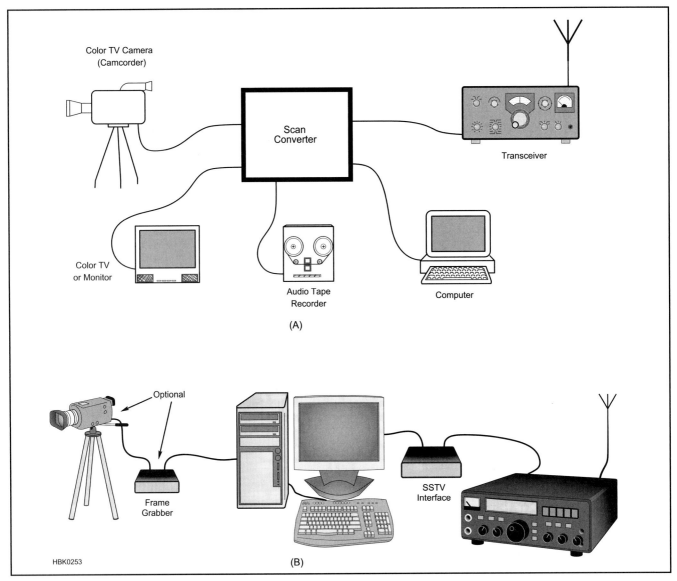

Fig 32.27 — Older SSTV stations required a dedicated hardware scan converter (A). Modern stations need only a sound card equipped PC and simple interface to send and receive SSTV (B).

tones represented black, white and shades of gray, and other audio tones were used for control signals. The receiving station converted the tones back into an image for display on a picture tube. It took 8 seconds to send a picture, and everything fit in the same bandwidth used for SSB transmissions.

Copthorne Macdonald, WA2BCW, now VY2CM, developed the first SSTV system. In 1958, MacDonald used a surplus long-persistence phosphor radar monitor to display SSTV images. The images were created using a vidicon camera or a flying-spot scanner. One frame of video had 120 lines and was sent at a rate of 15 lines per second (that's where the 8 seconds fits in). Three frames were sent back-to-back to maintain an image on the monitor. For many years, home-built systems were the only way to participate in SSTV. These were bulky, expensive and complex.

Robot Research produced SSTV equipment throughout the 1970s and 1980s. Their Robot 1200C scan converter, introduced in 1984, represented a giant leap forward for SSTV — all the SSTV functions were performed within a single box. (A *scan converter* is a device that converts signals from one TV standard to another. In this case, it converted SSTV signals to and from devices intended for fast scan television.) With updates and modifications, many 1200Cs are still in use today.

DOS-based PCs became a popular component for SSTV systems in the early 1990s. These hybrid systems were part hardware and part software. Most systems had an external box that processed the audio and performed the analog-to-digital conversion. In 1998, the Tasco Electronics TSC-70 Telereader color scan converter was used on the MIR space station to send pictures from space. Kenwood introduced Visual Communicator VC-H1, a portable SSTV unit that included a built-in camera and LCD monitor. Other scan converters were made, but none are produced any more.

The personal computer with sound card has become the hardware of choice to send and receive images, replacing specialized devices. Today's powerful PCs and SSTV software are the heart of a modern SSTV station. The PC processes the incoming and outgoing SSTV audio using the sound card, while the software manages the acquisition, storage, selection and editing of the SSTV pictures. If you have a PC in your ham station, adding software and an interface (to connect the computer sound card to the transceiver) is all that it takes to get started in this fascinating mode. **Fig 32.27** compares a current station SSTV configuration to one from the 1990s.

32.7 SSTV Basics

Traditional SSTV is an analog mode, but in recent years several digital SSTV systems have been developed. All make use of standard Amateur Radio transceivers — no special radio gear or antennas required. SSTV transmissions require only the bandwidth of SSB voice, and they are allowed on any frequencies in the HF and higher bands where SSB voice is permitted. SSTV does not produce full motion video or streaming video.

32.7.1 Computers and Sound Cards

A computer with sound card is required to use most of the software popular for SSTV. Software for the *Windows* operating system is the most popular, but other choices are available. Hardware and software setup is similar to that described for sound card modes in the chapter on **Digital Communications**; refer to that chapter for additional information.

The sound card characteristics discussed in the **Digital Communications** chapter apply for SSTV as well. Most sound cards will work, but sample rate accuracy is important for some SSTV modes (more on this later).

Within *Windows*, the Master Volume and Wave controls set the levels for the transmitted SSTV audio. All other mixer inputs should be muted. Equalizer and special effects should not be used. *Windows* sounds should be disabled to prevent them from going out along with the transmitted SSTV audio. The Recording control is used to select the input for receiving the SSTV audio. Adjust Line In or Microphone as needed for the proper level. Mic Boost should not be used.

32.7.2 Transceiver Interface

Interfacing the transceiver to the PC sound card can be as simple as connecting a couple of audio patch cables. Sound cards generally have stereo miniature phone jacks, so use matching plugs when making these connections. Only one audio channel is required or desired — the left channel. This is the tip connection on the plug. Use shielded cable with the shield connected to the sleeve (ground) on the plug. Ground loop problems may be avoided by using an audio isolation transformer.

The connection for receiving the SSTV audio from the transceiver may be made anywhere that received audio is available. The best choice is one that provides a fixed-level AF output. This will ensure that recording control levels will not have to be adjusted each time the volume on the transceiver is adjusted. You can use a headphone or speaker output if that's all that is available. If the received output level is enough for the LINE IN input on the sound card, use it, otherwise use the MICROPHONE input. If the level is too high for LINE IN, an attenuator may be required.

Use the LINE OUT connection on the sound card for the transmitted SSTV audio output. The cable for transmitted SSTV audio should go to the AFSK connection. If the transceiver does not have a jack for this, then the microphone jack must be used — requiring a more elaborate interface. TR keying can be provided using the transceiver's VOX. Other methods for activating transmit include manual switching, a serial port circuit, an external VOX circuit and or a computer-control command.

Commercially made and kit sound card interfaces are available. Most of the sound card interfaces for the digital modes are suitable for SSTV. The microphone will be used regularly between SSTV transmissions, so consider its use along with the ease of operation when setting up for SSTV.

For more information on interfaces and adjustments, see the **Digital Communications** chapter. Another useful resource is the sound card interface Web page by Ernie Mills, WM2U, at **www.qsl.net/wm2u/interface.html**.

32.7.3 Transceiver Requirements

An SSB voice transceiver with a stable VFO is necessary for proper SSTV operation. The VFO should be calibrated and adjusted to be within 35 Hz of the dial frequency (more on this later). The transceiver's audio bandwidth should not be constrained so as to infringe on the audio spectrum used by SSTV. Optional filters should be turned off unless they can be set to 3 kHz or wider. SSTV software has its own DSP signal processing tools, so they are not needed in the transceiver. Most of the other transceiver settings should be turned off, including transmit or receive audio equalization, noise blanker or noise reduction, compression or speech processing, and passband tuning or IF shift.

Adjust your transmitter output for proper operation with the microphone first, according to instructions in your manual. Then adjust the sound card output for desired drive level. For SSB operation, receiving stations should see about the same S-meter reading on the SSTV signal as for voice. Properly adjusted levels and clean audio quality will improve the reception of the transmitted signal as well as reduce interference on adjacent frequencies. The SSTV signal is 100% duty cycle. If your transceiver is not designed for extended full-power operation, reduce the power output using the *Windows* Volume Control.

SSTV may be transmitted using AM, FM or SSB. Use the same mode as you would use for voice operation on a given frequency. On HF, use the same sideband normally used for voice on that band. SSTV activity can be found on HF, VHF, UHF, repeaters, satellites, VoIP on the Internet and almost anywhere a voice signal can get through.

32.7.4 SSTV Operating Practices

Analog SSTV images are, in a sense, broadcast. They arrive as-is and do not require the recipient to establish a two-way connection. Most SSTV operation takes place on or near specific frequencies. Common analog SSTV frequencies include 3.845, 3.857 and 7.171 MHz in LSB and 14.227, 14.230, 21.340 and 28.680 MHz in USB. Establish a contact by voice first before sending SSTV. If no signals can be heard, on voice ask if there is anyone sending. It may be that someone is sending but you cannot hear them. If there is no response, then try sending a "CQ picture." Note that on popular frequencies, weak signals can often be heard. In this case, wait for traffic to finish before sending SSTV even if you got no response to your voice inquiry.

Receiving SSTV pictures is automatic as long as your software supports the mode used for transmission (more on SSTV modes in later sections). Once a picture is received, it will be displayed and may be saved.

When selecting images to send, consider appropriateness, picture quality and interest to the recipient. Choose an SSTV mode that is suitable for the image to be sent, band conditions, signal strengths and the recipient's receive capability. Announce the SSTV mode prior to sending. Avoid sending a CW ID unless required by regulations. Describe the picture only after it is confirmed that it was properly received.

To send SSTV, use your software to select and load a picture to send. It will be displayed in a transmit screen. Next, select the SSTV transmission mode. Click the transmit button, and your transceiver will go into transmit and send the SSTV audio. When the SSTV transmission is over, the transceiver returns to receive. Be sure to send the full frame or the next picture sent may not be received properly because it may not start scanning from the top.

SOURCE FOR IMAGES

The Internet is a popular source for images. With the unlimited number of images available, it is surprising how often the same internet pictures keep popping up on SSTV. Use a little imagination and come up with something original.

A digital camera is one of the best ways to create an original picture of your own. The subject matter could be almost anything that you might have available. Pictures of the shack, equipment and operator are always welcomed. A live camera or Web cam can provide an almost instant snapshot. Please keep your shirt on and comb your hair!

For those who like to discuss technical details, diagrams and schematics might be your ammunition. A flatbed scanner is ideal for importing diagrams, schematics and photographs. Screen shots of what is on the computer monitor can be saved simply by hitting the PRINT SCREEN key on the keyboard. Use the Paste function in *Windows* to transfer the image to your SSTV or image editing software.

Any image editing program can be used to make your own CQ picture, test pattern, video QSL card or 73 picture. Include your own personal drawings. Make your images colorful with lots of contrast to make them really stand out. You can also transmit pictures of your home, areas of local interest, other hobbies, projects, maps, cartoons and funny pictures.

It is common to have call sign, location and perhaps a short description as text on the pictures. If two calls are placed on an image, it is understood that the sender's call is placed last. Signal reports may also be included. Once the images are saved, they provide a convenient confirmation of contact.

32.8 Analog SSTV

With slow scan, the fastest frame rate is 8 seconds so full motion video is not possible. Analog SSTV uses only a single frequency at a time. A 1200-Hz sync pulse is sent at the beginning of each line. Pure black is represented by a 1500 Hz tone, and 2300 Hz is used for pure white. Tones between 1500 and 2300 Hz produce the grayscale. The frequency varies with the brightness level for each pixel. Since it is only the change in frequency that makes up the analog SSTV signal, it is considered frequency modulated. Amplitude is not important other than getting over the noise. Analog SSTV uses less bandwidth than voice.

32.8.1 Color SSTV Modes

Two popular modes for sending and receiving color SSTV pictures are called Martin and Scottie (named after their developers). Within each family are several different modes (Martin 1, Martin 2 and so forth). The various modes have different resolutions and scan rates.

Information about the size or resolution for each mode is generally available in the SSTV software. Better quality images will result when the source image is sized and cropped to the same dimensions used by the mode with which it is transmitted. Slower scan rates can provide better quality; those are the modes that take longer to send for the same resolution.

For example, Scottie 1 and Martin 1 are popular RGB (red, green, blue) color modes that take about two minutes to send a frame, with only one frame sent at a time. The image size sent is 320 × 256 — 320 pixels across with 256 lines. See **Figs 32.28** and **32.29**. True color is possible since each pixel is sent with 256 possible levels for each of the three color components (red, green and blue).

Each mode has a vertical interval signaling (VIS) code that identifies the mode being sent. The VIS codes use 1100 Hz for 1, 1300 Hz for 0, and 1200 Hz for start and stop. When these tones are received, it readies the system to receive in the proper SSTV mode.

For VIS detection to work, the receiver must be tuned within 70 Hz of the transmitted frequency. Two stations tuned exactly on the SSTV frequency but with VFO errors of +35 Hz and −35 Hz could successfully pass the VIS codes. As mentioned previously, each transceiver must have the VFO and display calibrated within 35 Hz to ensure VIS code detection with the transceiver set to the SSTV frequency. Using the sync pulse rate is another way to determine the mode when the VIS is not decoded. If a station sending is far off frequency, use the 1200 Hz sync signal in the spectrum or waterfall as a guide to adjust the VFO.

Received images are displayed in near real time as they are decoded. Almost any SSTV signal that is heard can produce an image, but it is rare to receive an image that is perfect. Changes in propagation or transceiver settings will become apparent as the image continues to scan down the screen. Noise will damage the lines received just as it occurs. Interference from other signals will distort the image or perhaps cause reception to stop. Signal fading may cause the image to appear grainy. Multipath will distort the vertical edges. Selective fading may cause patches of noise or loss of certain colors.

Images that are received with staggered edges are the result of an interruption of the sound card timing. Check with other operators to see if they also received the image with staggered edges. If not, then it may be your computer that has the problem and not the sending station. Some possible solutions are to close other programs, disable antivirus software and reboot the computer.

Under good conditions, images may come

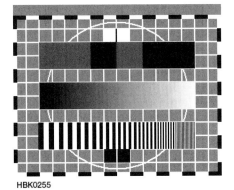

Fig 32.28 — This color test pattern is 320 × 256 resolution. The Robot modes from the 1980s only used 240 lines. CRTs were 4:3 aspect ratio, so the images were displayed in 320 × 240 resolution. When Scottie 1 and Martin 1 modes were added, they kept this resolution but added 16 header lines to the top to aid in synchronization. These 16 header lines, shown as the solid gray area at the top, are now used as part of the image and include the sender's call sign and other information.

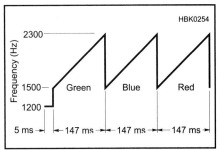

Fig 32.29 — This is a diagram of a single scan line of a Martin 1 transmission. The full frame includes 256 horizontal scan lines starting at the top. Before each line, a 1200 Hz tone for 5 ms is sent for synchronization. The pixels are scanned left to right and the colors are sent sequentially. First the green component is sent for each of the 320 pixels of one scan line — 256 values are possible from black to pure green. Next, is the blue component, and the red component is sent last (again, 256 possible values for each color). It takes 147 milliseconds to send each of the three color components. The result is 114 seconds to send the complete frame.

Table 32.2
SSTV Software

Windows
Analog SSTV
ChromaPIX — **barberdsp.com/cpix/chroma.htm**
MMSSTV — **mmhamsoft.amateur-radio.ca/mmsstv/index.htm**
Mscan— **mscan.com/?page_id=2**
ROY1 — **roy1.com/download_ita.htm**
SSTV32— **webpages.charter.net/jamie_5**
W95SSTV— **barberdsp.com/w95sstv/w95sstv.htm**

Multimode
JVComm32 — **jvcomm.de/index_e.html**
MixW — **www.mixw.net**
MultiPSK — **f6cte.free.fr/index_anglais.htm**

Digital SSTV
DIGTRX — **qslnet.de/member/py4zbz/hdsstv/HamDRM.htm**
EasyPal — **www.vk4aes.com**
WinDRM — **n1su.com/windrm**

MacOS
MultiMode — **www.blackcatsystems.com/software/multimode.html**
MultiScan — **web.me.com/kd6cji/MacSSTV**

Linux
MMSSTV — **mmhamsoft.amateur-radio.ca/mmsstv/index.htm**
QSSTV — **users.telenet.be/on4qz**

Amiga
AVT — **datapipe-blackbeltsystems.com/amiga.html**

For links to additional software and notes on setup and operation, see **www.tima.com/~djones**

---

in "closed circuit." This means that the quality appears nearly as good as a photograph. SSTV has a reporting system similar to the RST reporting system used on CW. For analog SSTV it is RSV — readability, signal strength and *video*. Video uses a scale of 1 to 5, so a report of 595 would be the same as closed circuit.

32.8.2 Analog SSTV Software

A variety of software programs are available for SSTV (see **Table 32.2**). Some multimode programs include SSTV and digital modes, while others are dedicated to SSTV. One very popular SSTV package is the *Windows* program *MMSSTV* by Mako Mori, JE3HHT. For more information or to download a copy, visit **mmhamsoft.amateur-radio.ca/mmsstv/index.htm**. *MMSSTV* will run on a 100 MHz PC, but some features may require a faster processor. Other SSTV software has similar features.

As explained in the **Digital Communications** chapter, sound card sample rate accuracy is important for some modes. Analog SSTV is one of them: pictures will appear *slanted* if the clock is off. See **Fig 32.30**. The *MMSSTV* Help file includes detailed information on several ways to do a quick and easy calibration (see the Slant Corrections section). The best method is the one that uses a time standard such as WWV. (Before performing this calibration procedure, you must have the sound card interface connected so *MMSSTV* can detect received audio.) After performing the clock calibration, chances are, the timing will also be correct for transmit. If not, *MMSSTV* provides a means for making a separate adjustment for transmit.

32.8.3 Resources

The International Visual Communications Association (IVCA) net meets each Saturday morning at 1500 UTC on 14.230 MHz USB. The net control will ask for check-ins and get each station to send a picture in Scottie 1 mode. Replays are sent in Scottie 2 mode. This will provide a way to see how well your pictures were received. Signal reports are welcome and brief questions may be asked. Anything more involved should be held until the net is over. Anyone may check-in to the net.

Information about the IVCA is available on Yahoo Groups: **groups.yahoo.com/group/ivca**. The *MMSSTV* Yahoo group is another valuable resource: **groups.yahoo.com/group/MM-SSTV**.

(A) (B)

Fig 32.30 — If the sound card clock is inaccurate, analog SSTV images may appear slanted (A). The same image is shown at B after calibrating the clock.

32.9 Digital SSTV

Several forms of digital SSTV have been developed, but the modulation method most widely used for digital SSTV as of 2009 is derived from the shortwave broadcast system Digital Radio Mondiale (DRM). (See the **Digital Communications** chapter for more information on the Amateur Radio voice applications of DRM.) *HamDRM* by Francesco Lanza, HB9TLK, is a variation of DRM that fits in a 2.5 kHz bandwidth and is used in various programs.

The DRM digital SSTV signal occupies the bandwidth between 350 and 2750 Hz (see **Fig 32.31**). As many as 57 subcarriers may be sent simultaneously, all at the same level. Three pilot carriers are sent at twice the level as the others. The subcarriers are modulated using *coded orthogonal frequency division multiplexing (COFDM)* and *quadrature amplitude modulation (QAM)*. Each *main service channel (MSC)* frame or segment has 400 ms duration, and several methods of error correction are used within the segments. (See the **Modulation** chapter for more information on ODFM and QAM.)

Digital SSTV using DRM is not a weak signal mode like the narrow bandwidth data modes such as PSK31. An S9 or better signal with little or no noise may be required before the software is able to achieve a sync lock. MSC sync lock is required before any data is received.

32.9.1 Digital SSTV Setup

A popular DRM digital SSTV program is *EasyPal* by Erik Sundstrup, VK4AES (**www.vk4aes.com**). Digital SSTV uses the same type of PC and sound card setup described in the analog SSTV section, but a more capable computer is required. For *EasyPal*, a 2 GHz or faster PC running *Windows XP* or newer operating system is required. As soon as the *EasyPal* software is installed, it is ready to receive pictures.

Unlike analog SSTV, the software detects and compensates automatically for clock timing differences so sound card calibration is not required. Software will also automatically adjust ±100 Hz for mistuned frequency.

With DRM SSTV the call sign is sent continuously. This may allow others to identify the transmitting station and perhaps turn an antenna in the right direction for better reception. Many submodes are available, with various transmission speeds and levels of robustness. The submode is automatically detected and receiving starts automatically. Decoding is done on the fly, so there is no waiting for the computer to finish processing before the image appears.

Power output may appear low as measured by a conventional wattmeter. The actual signal strength as seen by others should be about the same as the SSB voice signal. Avoid capacitors in the audio lines as they may interfere with the phase of the digital signal.

DRM audio levels are low, so there may be problems getting the signal to trigger a VOX circuit. Set the transceiver for full RF output. Then adjust the sound card Volume Control output until the transmitter shows little or no ALC indication. With an FM transmitter, keep the output level low to avoid overdeviation.

32.9.2 Operating Digital SSTV

Before jumping into digital SSTV, try analog SSTV first. Copy some pictures to see if the sound card setup works. The level adjustments for analog SSTV are not as critical as those for DRM.

Common DRM SSTV frequencies include 3.847, 7.173, 7.183 MHz in LSB and 14.233 MHz in USB. Tune your VFO to the whole number in kHz (for example, 14,233.0). If a sending station is far off frequency, use the pilot carriers as seen in the software spectrum display as a guide. Adjusting the VFO while receiving an image is not advised as it will delay synchronization.

The signal-to-noise ratio (SNR) as displayed in the software is a measure of the received signal quality. The higher the SNR the better — decoding will be more reliable. Under very good band conditions this number may exceed 18. In that case, a higher speed mode may work. Because of the way the software measures the SNR, the peak value displayed for SNR may require 20-30 seconds of reception. Adjustments made on either end may change the SNR. Submodes with less data per segment take longer to send, but they are more robust and allow for copy even if the SNR is low.

GETTING THE WHOLE PICTURE

Noise and fading may prevent 100% copy of all the segments. Any missing segments

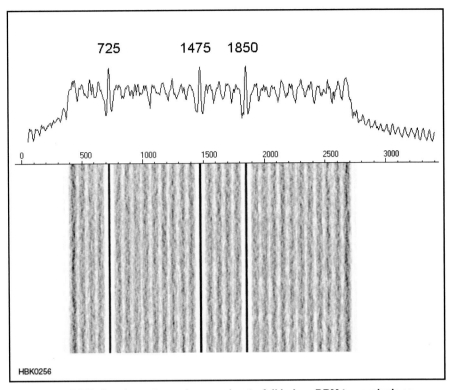

Fig 32.31 — DRM signal spectrum above and waterfall below. DRM transmissions may have slightly different characteristics due to the many possible submodes. All DRM transmissions used in Amateur Radio have three pilot carriers that have an amplitude twice that of the remaining subcarriers. These pilot carriers are found at 725, 1475 and 1850 Hz and aid in adjusting the frequency of the receiver to match the transmitted signal. The DRM TUNE signal also matches these three carriers. Software can compensate for mistune for as much as ±100 Hz, but only if the difference remains stable (little or no drift in frequency). The overall response should be flat, producing a level amplitude across the full spectrum provided that the signal is not affected by propagation. The waterfall should also be nearly the same intensity across the full width. The bandwidth can be as much as 2.4 kHz starting at 350 Hz and ending about 2750 Hz. Some roll-off can be tolerated.

Fig 32.32 — *EasyPal* DRM digital SSTV software is used to exchange high quality, color images. Several levels of error correction and various transmission speeds are available.

may be filled in later. Your software can send a *bad segment report (BSR)* that lists all the missing segments for a file that has only been partly received. In response, the other station can send a *FIX* which should complete the file transfer. If not, the BSR and FIX process may be repeated.

Digital SSTV is very interactive and may involve several stations on frequency sharing images. FIX transmissions intended for another station may provide some of your missing segments — or even all of them. A third party that copied the original transmission may also send a FIX or resend the image. Incremental file repair is possible even after several other transmissions are received. The more stations on frequency, the better the chance that one of them is in a position to help out with a FIX.

EasyPal has a feature to provide a higher level of error correction so that 100% copy of all the segments is not necessary to receive the complete file. The transmitted file is encoded with redundant data so that the original file may be recreated even though not all the segments were received. The receiving station must also be running *EasyPal*.

Encoded files have interleaved redundancy using Reed-Solomon (RS) error correction. Four different levels of RS encoding may be selected before the file is sent. Very Light Encode (RS1) is the lowest level. Transmission time for a file using RS1 will be increased by 13%. When receiving a file encoded with RS1, only 90% of the segments must be received before the file can be decoded and a picture displayed. This may happen even before the transmission is complete. Receiving RS encoded files is automatic; there is no need to select it for receive. Decoding of the file received with RS encoding will occur automatically. The use of RS encoding on the HF bands can reduce the need for BSRs and FIXs and has been found to make the file transfer process more efficient. Encoding may not be necessary on noise-free channels such as VHF FM.

Propagation only becomes a factor as it may take longer for the data to get through during poor conditions. The images received will be identical to the ones sent because the data in the files will also be identical. Replays will always be an exact copy. Multipath propagation does not disrupt DRM transmissions unless it is severe or results in selective fading.

Fig 32.32 shows an *EasyPal* screen following successful reception of an image.

IMAGE SIZE

Pictures of any size or resolution may be sent over digital SSTV. The sending station must pay careful attention to file size, though, or the transmission time may become excessively long. Compressing image files is necessary to get the transmit time down to a reasonable amount. Most images will be converted into JPEG 2000 (JP2), a lossy compression method that shows fewer artifacts. A slider varies the JP2 compression level, and a compromise must be made between image quality and file size. The smaller the file, the more visible the artifacts, but the faster it is sent.

Small image files may be sent without using compression. Some file types such as animated GIF files cannot be compressed, so they must be sent "as is."

A "busy picture" is one that shows lots of detail across most of the image area. This type of picture can be challenging to compress into a file size small enough to send that still maintains acceptable quality. Reducing the resolution by resizing and creating a much smaller image is the solution. Just about any busy picture can be resized down to 320 × 240 pixels, converted into JP2, and still look good when displayed on the receiving end.

About 2 minutes transmission time is the acceptable limit for the patience of most SSTV operators. A typical DRM digital SSTV transmission will take about 105 seconds for a file 23 kB in size, RS1 encoded and requiring 209 segments.

SENDING DIGITAL SSTV IMAGES

The ideal DRM signal will have a flat response across the 350 to 2750 Hz spectrum. The transceiver should be allowed to pass all frequencies within this bandwidth. In order to maintain the proper phase relationships with all the subcarriers, the signal must be kept linear. Avoid overdriving the transmitter and keep the ALC at the low end of the range. Eliminate hum and other stray signals in the audio.

The process of transmitting an image starts with selecting an image and resizing or compressing it if needed, as described in the previous section. Within *EasyPal*, when the transmit button is clicked, the image file will be RS encoded if that option is selected. Then the resulting file will be broken down into segments and sent using DRM.

In receiving DRM, the audio is decoded and segments that pass the error check will have their data stored in memory. When enough of the segments are successfully received, the RS file is decoded and the JPEG 2000 image file is created. The content of this file should be identical to the JPEG 2000 image file transmitted.

It can be quite gratifying to receive your first digital SSTV picture. A lot has to go just right, and there is little room for errors. Propagation and interference always play havoc. There is no substitute for a low noise location and good antenna when it comes to extracting the image from the ether. Be patient and when the right signal comes by you will see the all the lights turn green and the segment counter will keep climbing. You won't believe the quality of the pictures!

Help for all aspects of digital SSTV is available on Yahoo Groups DIGSSTV: **groups.yahoo.com/group/digsstv**.

32.9.3 Future of SSTV

Advancements in radio technology should make SSTV easier. Software defined radios (SDR) that make use of direct conversion of audio signals and use "virtual cables" internally may eliminate RF pickup and other forms of noise. Higher SNR should result. New transceivers that make use of optical cables may avoid problems with noise getting into or coming out of the sound card. All this could mean less distortion, lower noise levels and reduced hum and improved image quality. New digital SSTV systems will certainly be on the way. New features for DRM digital SSTV are being developed and new ways to use digital SSTV are discovered every day.

32.10 Glossary of SSTV Terms

AVT — Amiga Video Transceiver. 1) Interface and software for use with an Amiga computer; 2) a family of transmission modes first introduced with the AVT product.

Back porch — The blank part of a scan line immediately following the horizontal sync pulse.

Chrominance — The color component of a video signal. Robot color modes transmit pixel values as luminance (Y) and chrominance (R-Y [red minus luminance] and B-Y [blue minus luminance]) rather than RGB (red, green, blue).

Demodulator — For SSTV, a device that extracts image and sync information from an audio signal.

Field — Collection of top-to-bottom scan lines. When interlaced, a field does not contain adjacent scan lines and there is more than one field per frame.

Frame — One complete scanned image. The Robot 36-second color mode has 240 lines per frame

Frame Sequential — A method of color SSTV transmission that sent complete, sequential frames of red, then green and blue. Now obsolete.

Front porch — The blank part of a scan line just before the horizontal sync.

Interlace — Scan line ordering other than the usual sequential top to bottom. AVT "QRM" mode is the only SSTV mode that uses interlacing.

Line Sequential — A method of color SSTV transmission that sends red, green and blue information for each sequential scan line. This approach allows full-color images to be viewed during reception.

Luminance — The brightness component of a video signal. Usually computed as Y (the luminance signal) = 0.59 G (green) + 0.30 R (red) + 0.11 B (blue).

Martin — A family of amateur SSTV transmission modes developed by Martin Emmerson, G3OQD, in England.

Pixel — Picture element. The dots that make up images on a computer's monitor.

P7 monitor — SSTV display using a CRT having a very-long-persistence phosphor.

RGB — Red, Green, Blue. One of the models used to represent colors. Due to the characteristics of the human eye, most colors can be simulated by various blends of red, green, and blue light.

Robot — (1) Abbreviation for Robot 1200C scan converter; (2) a family of SSTV transmission modes introduced with the 1200C.

Scan converter — A device that converts one TV standard to another. For example, the Robot 1200C converts SSTV to and from FSTV.

Scottie — A family of amateur SSTV transmission modes developed by Eddie Murphy, GM3SBC, in Scotland.

SSTV — Slow Scan Television. Sending still images by means of audio tones on the MF/HF bands using transmission times of a few seconds to a few minutes.

Sync — That part of a TV signal that indicates the beginning of a frame (vertical sync) or the beginning of a scan line (horizontal sync).

VIS — Vertical Interval Signaling. Digital encoding of the transmission mode in the vertical sync portion of an SSTV image. This allows the receiver of a picture to automatically select the proper mode. This was introduced as part of the Robot modes and is now used by all SSTV software designers.

Wraase — A family of amateur SSTV transmission modes first introduced with the Wraase SC-1 scan converter developed by Volker Wraase, DL2RZ, of Wraase Elektronik, Germany.

DIGITAL SSTV TERMS

Bad segment report (BSR) — A DRM transmission that lists all the missing segments for a file that has only been partly received.

COFDM — Coded Orthogonal Frequency Division Multiplex, a method of using spaced subcarriers that are phased in such a way as to reduce the interference between them, plus coding to provide error correction and noise immunity.

Constellation — A set of points in the complex plane that represent the various combinations of phase and amplitude in a QAM or other complex modulation scheme.

Cyclic redundancy check (CRC) — A mathematical operation. The result of the CRC is sent with a transmission block. The receiving station uses the received CRC to check transmitted data integrity to determine if the data received is good or bad.

Digital Radio Mondiale (DRM) — A consortium of broadcasters, manufacturers, research and governmental organizations which developed a system for digital sound broadcasting in bands between 100 kHz and 30 MHz. Amateurs use a modified version for sending digital voice and images.

Error Protection — DRM submode selection. The HI level provides a greater amount of **FEC** used within the segment.

Fast access channel (FAC) — Auxiliary channel always modulated in 4QAM. Contains the submode and station information.

FIX — A DRM transmission that sends the data for all the segments for a received BSR.

FEC — Forward error correction, an error-control technique in which the transmitted data is sufficiently redundant to permit the receiving station to correct some errors.

LeadIn — The number of redundant segments sent at the beginning of the DRM file transmission to allow the receiving station to become synchronized.

Main service channel (MSC) — Contains the data. Can be modulated in 4QAM, 16QAM or 64QAM.

Mode — In digital SSTV, a particular submode of a DRM transmission. The amount of data in each segment is determined by the submode selected. Robustness varies for each submode.

QAM — Quadrature Amplitude Modulation. A method of simultaneous phase and amplitude modulation. The number that precedes it, for example 64QAM, indicates the number of discrete stages in each pulse.

Reed-Solomon error correction — A data encoding process that inserts redundant data so that errors in reading the data may be detected and corrected. *EasyPal* provides four levels ranging from Very Light Encode (RS1) to Heavy Encode (RS4).

Segment — One MSC frame of an DRM file transmission. Contains file name or transport ID and data.

TUNE — In DRM, a three-tone transmission used to set levels, check for IMD and adjust frequency.

32.11 Bibliography and References

ATV

Amateur Television Quarterly Magazine, **www.atvquarterly.com.**

Cerwin, S., WA5FRF, "Amateur Television from Model Planes and Rockets," *QST*, Sep 2000, pp 41-44.

CQ-TV, published by the British ATV Club, **www.batc.org.uk**.

Kramer, K., DL4KCK, "AGAF e.V. DATV-Boards-Instructions for Starting Up," *Amateur Television Quarterly Magazine*, Spring 2005.

PC Electronics Web site, **www.hamatv.com** (equipment, application notes).

Putman, P., KT2B, "Digital Television: Bigger, Wider, Sharper," *QST*, Aug 2000, pp 38-41.

Ruh, H., *ATV Secrets for the Aspiring ATVer*, Vol 1 and Vol 2. Available through *Amateur Television Quarterly Magazine*.

Seiler, T., HB9JNX/AE4WA, et al, "Digital Amateur TeleVision (D-ATV), proc. ARRL/TAPR Digital Communications Conference, **www.baycom.org/~tom/ham/dcc2001/datv.pdf**.

Sparano, D., "What Exactly Is 8-VSB Anyway?," **www.broadcast.net/~sbe1/8vsb/8vsb.htm**.

Taggart, R., "An Introduction to Amateur Television," *QST*, Part 1, Apr 1993, pp 19-23; Part 2, May 1993, pp 43-47; Part 3, Jun 1993, pp 35-41.

Taggart, R., WB8DQT, *Image Communications Handbook* (Newington: ARRL, 2002).

SSTV

Ely, R., WL7LP, MMSSTV.CHM, a Help file included with *MMSSTV* software.

Lanza, C., HB9TLK, "DRM Mode 'HAM' Specification," **www.qslnet.de/member/hb9tlk/drm_h.html**.

Macdonald, C., VY2CM, "The Early Days of Amateur Radio Slow-Scan TV," **www.wisdompage.com/EarlySSTV.html**.

Mori, M., JE3HHT, MMSSTV.TXT, a Help file included with *MMSSTV* software.

Taggart, R., WB8DQT, "Digital Slow-Scan Television," *QST*, Feb 2004, p 47-51.

Taggart, R., WB8DQT, *Image Communications Handbook* (ARRL, 2002).

Zurmely, R., PY4ZBZ, "HamDRM mode help," a Help file included with DIGTRX software.

Whitten, M., KØPFX, WinDRM Docs, **www.n1su.com/windrm/download.html**.

For links to more information on SSTV, visit the "CQ SSTV" Web site: **www.tima.com/~djones/**.

KENWOOD
Listen to the Future

AIRWAVE SUPERIORITY

Never before has a compact HT offered as many features, and such high powered performance as the TH-F6A. Arm yourself with one today and gain your own airwave superiority.

- Triband (144/220/440 MHz)
- Receives 2 frequencies simultaneously even on the same band
- 0.1~1300MHz high-frequency range RX (B band)[1]
- FM/FM-W/FM-N/AM plus SSB/CW receive
- Bar antenna for receiving AM broadcasts
- Special weather channel RX mode
- 435 memory channels, multiple scan functions
- 7.4V 1550mAh lithium-ion battery (std.) for high output[2] and extended operation
- 16-key pad plus multi-scroll key for easy operation
- Built-in charging circuitry for battery recharge while the unit operates from a DC supply
- Tough construction: meets MIL-STD 810 C/D/E standards for resistance to vibration, shock, humidity and light rain
- Large frequency display for single-band use
- Automatic simplex checker
- Wireless remote control function
- Battery indicator • Internal VOX • MCP software

[1] Note that certain frequencies are unavailable. [2] 5W output

TH-F6A
TRIBANDER

www.kenwoodusa.com

JQA-1205
ISO9001 Registered
Communications Equipment Division
Kenwood Corporation
ISO9001 certification

KENWOOD U.S.A. CORPORATION
Communications Sector Headquarters
3970 Johns Creek Court, Suite 100, Suwanee, GA 30024
Customer Support/Distribution
P.O. Box 22745, 2201 East Dominguez St., Long Beach, CA 90801-5745
Customer Support: (310) 639-4200 Fax: (310) 537-8235

ADS#28109

KENWOOD
Listen to the Future

All-Terrain Performance

On or off the road, Kenwood's new TM-271A delivers powerful mobile performance with 60W maximum output and other welcome features such as multiple scan functions and memory names. Yet this tough, MIL-STD compliant transceiver goes easy on you, providing high-quality audio, illuminated keys and a large LCD with adjustable green backlighting for simple operation, day or night.

144MHz FM TRANSCEIVER
TM-271A

- 200 memory channels (100 when used with memory names)
- Frequency stability better than ±2.5ppm (-20~+60°C)
- Wide/Narrow deviation with switchable receive filters
- DTMF microphone supplied
- NOAA Weather Band reception with warning alert tone
- CTCSS (42 subtone frequencies), DCS (104 codes)
- 1750Hz tone burst
- VFO scan, MHz scan, Program scan, Memory scan, Group scan, Call scan, Priority scan, Tone scan, CTCSS scan, DCS scan
- Memory channel lockout
- Scan resume (time-operated, carrier-operated, seek scan)
- Automatic repeater offset
- Automatic simplex checker
- Power-on message
- Key lock & key beep
- Automatic power off
- Compliant with MIL-STD 810 C/D/E/F standards for resistance to vibration and shock
- Memory Control Program (available free for downloading from the Kenwood Website: *www.kenwoodusa.com*)

KENWOOD U.S.A. CORPORATION
Communications Sector Headquarters
3970 Johns Creek Court, Suite 100, Suwanee, GA 30024
Customer Support/Distribution
P.O. Box 22745, 2201 East Dominguez St., Long Beach, CA 90801-5745
Customer Support: (310) 639-4200   Fax: (310) 537-8235

www.kenwoodusa.com
ADS#29708

JQA-1205 091-A
ISO9001 Registered
Communications Equipment Division
Kenwood Corporation
ISO9001 certification

HF/50MHz ALL MODE TRANSCEIVER TS-590

The All New TS-590S
High Performance RX, World Renowned Kenwood TX Audio

Kenwood has essentially redefined HF performance with the TS-590S compact HF transceiver. The TS-590S RX section sports IMD (intermodulation distortion) characteristics that are on par with those "top of the line" transceivers, not to mention having the best dynamic range in its class when handling unwanted, adjacent off-frequency signals.*

- HF-50MHz 100W
- Digital IF Filters
- Built-in Antenna Tuner
- Advanced DSP from the IF stage forward
- Heavy-duty TX section
- 500Hz and 2.7KHz roofing filters included
- 2 Color LCD

KENWOOD
Listen to the Future

www.kenwoodusa.com

KENWOOD U.S.A. CORPORATION
Communications Sector Headquarters
3970 Johns Creek Court, Suite 100, Suwanee, GA 30024
Customer Support/Distribution
P.O. Box 22745, 2201 East Dominguez St., Long Beach, CA 90801-5745
Customer Support: (310) 639-4200 Fax: (310) 537-8235 ADS#29210

JQA-1205 091-A
ISO9001 Registered
Communications Equipment Division
Kenwood Corporation
ISO9001 certification

* For 1.8/3.5/7/14/21 MHz Amateur bands, when receiving in CW/FSK/SSB modes, down conversion is automatically selected if the final passband is 2.7KHz or less.

Advertisers Index

Advertising Department Staff
Debra Jahnke, K1DAJ, *Sales Manager, Business Services*
Janet Rocco, W1JLR, *Account Executive*
Lisa Tardette, KB1MOI, *Account Executive*
Diane Szlachetka, KB1OKV, *Advertising Graphic Design*
Zoe Belliveau, W1ZOE, *Business Services Coordinator*

800-243-7768

Direct Line: 860-594-0207 ▪ Fax: 860-594-4285 ▪ e-mail: ads@arrl.org ▪ Web: www.arrl.org/ads

Advertiser	Page
Alpha Delta Communications, Inc. – www.alphadeltacom.com	A-19
Ameritron – www.ameritron.com	A-9
AOR – www.aorusa.com	A-7
ARRL – www.arrl.org	A-2, A-6, A-12, A-17, A-19, A-20
Array Solutions – www.arraysolutions.com	A-16
ASA Inc. – www.waterprooflogs.com	A-2
Associated Radio – www.associatedradio.com	A-22
Atria Technologies, Inc. – www.AtriaTechnologies.com	A-19
Austin Amateur Radio Supply – www.aaradio.com	A-22
bhi Ltd. – www.bhi-ltd.co.uk	A-18
Cable X-Perts, Inc. – www.CableXperts.com	A-17
Coaxial Dynamics – www.coaxial.com	A-2
Command Productions – 1-800-932-4268	A-13
Cushcraft Amateur Radio Antennas – www.cushcraftamateur.com	A-14
Diamond Antennas – www.diamondantenna.net	A-11
Elecraft – www.elecraft.com	A-21
Haggerty Radio Company – www.WA1FFL.com	A-11
hamcity.com – www.hamcity.com	Cover 2
HamPROs – see your local dealer	A-22
Ham Radio Outlet – www.hamradio.com	Cover 3
HamTestOnline – www.hamtestonline.com	A-11
Hot Press Ham Hats – www.hotpresstshirts.com	A-11
ICOM America – www.icomamerica.com	A-3, A-5
International Crystal Mfg. Co., Inc. – www.icmfg.com	A-13
Kenwood USA Corporation – www.kenwoodusa.com	A-i, A-ii, A-iii
Lentini Communications – www.lentinicomm.com	A-22
MFJ Enterprises – www.mfjenterprises.com	A-8
NCG Company – www.natcommgroup.com	A-4
Palstar – www.palstar.com	A-10
Pro.Sis.Tel – www. prosistel.net	A-6
Radio City, Inc. – www.radioinc.com	A-22
RadioWavz – www.radiowavz.com	A-6
RF Parts Company – www.rfparts.com	A-13, A-15, A-17
Second Hand Radio – www.secondhandradio.com	A-18
Tigertronics – www.tigertronics.com	A-18
Universal Radio, Inc – www.universal-radio.com	A-11, A-15, A-18 A-22
Yaesu USA – www.vertexstandard.com	A-23, A-24, A-25, A-26, A-27

If your company provides products or services of interest to our readers,
please contact the ARRL Advertising Department today for information on building your business.

Coaxial Dynamics
A CDI INDUSTRIES, INC. COMPANY
SPECIALISTS IN RF TEST EQUIPMENT & COMPONENTS

6800 Lake Abram Dr., Middleburg Heights, Ohio, USA 44130
1-800-COAXIAL (262-9425) • 440-243-1100 • Fax: 440-243-1101 • Web Site: www.coaxial.com

USB Wattmeter
Model 81041

The model 81041 is a portable, self-contained RF Wattmeter that features a studio-quality analog meter and USB interface. Numeric, analog meter, and bar graph data are simultaneously displayed on a PC's monitor. The functions indicated are Forward and Reflected Power, both in Watts and dBm, plus an automatic calculation of SWR and Return Loss.

The internal dual socket line section and forward / reflected switch gives the user the ability to display either forward or reflected on the analog meter, while both are displayed simultaneously on the PC.

Our use of a rugged shock mounted meter with a mirror-backed scale along with superior taut band technology, provides reliable and accurate readings of either forward or reflected power on the meter.

The 81041 uses standard elements to detect average RF power from 100 mW to 10 kW and from 2 MHz to 2.3 GHz. Software and a detachable six foot USB cable are included for a simple installation on any PC using Windows® Vista, 2000, XP or NT. No additional cables, AC or DC power adapters, batteries or custom remote sensors are required.

• Forward and Reflected Power in Watts and dBm •
• Automatically Calculates SWR and Return Loss • Internal Dual 7/8" Line Section •
• Quick Match Connectors • Uses Standard Plug-In Elements • Two Year Limited Warranty •

Dual Socket Wattmeter
Model 81021

The Model 81021 Average Reading Dual Socket Wattmeter allows you to measure both Forward and Reflected RF power with the flip of a switch. The Model 81021 uses standard Elements to accurately detect average RF power from 100mw to 10 kW over a frequency range of 0.45 MHz to 2.3 GHz.

Complete with an internal dual socket 7/8" Line Section and Quick Match RF connectors, Model 81021 offers the speed and reliability you expect from Coaxial Dynamics. A convenient front panel switch gives the user the ability to display Forward or Reflected power on the analog meter.

The Model 81021 is easy to use. No additional black boxes or delicate remote sensors are needed. Simply connect the Wattmeter in-line between the RF source and the Antenna or Load, insert the appropriate Elements and select either the Forward or Reflected switch position. The RF power is visually identified directly on the large 4 ½" mirrored scale.

Versatile and strong, the Model 81021 uses a heavy gauge metal case to protect the Wattmeter from impact shock and a leather strap makes for safe and comfortable handling. For added convenience, two sockets for storage of additional elements are located on the back of the unit.

Our use of a rugged shock mounted meter with a mirrored-backed scale along with superior taut band technology provides reliable and accurate readings, plus the integrity that satisfies both the US Navy and Canadian standards for bounce and vibration. This is your assurance of complete accuracy.

• Shock Mounted "Taut Band" Meter • Large 4 ½" Mirrored Scale •
• Internal Dual Socket 7/8" Line Section • Switch for Forward or Reflected Power •
• Quick Match Connectors • Uses Gold Plated Plug-In Elements • Two Year Limited Warranty •

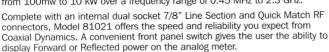
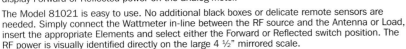

Rugged Waterproof All Weather Amateur Radio Log Books

ASA Inc. • Email: sales@waterprooflogbooks.com

WaterProofLogBooks.com

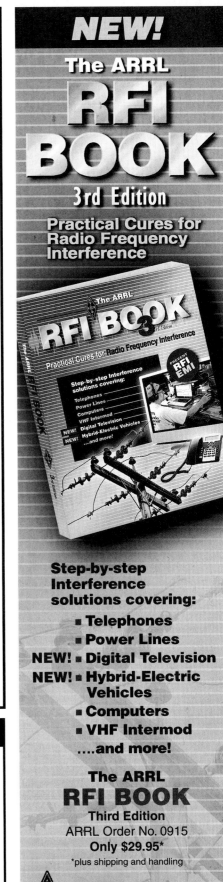

NEW!
The ARRL
RFI BOOK
3rd Edition

Practical Cures for Radio Frequency Interference

Step-by-step Interference solutions covering:
- Telephones
- Power Lines
- NEW! ■ Digital Television
- NEW! ■ Hybrid-Electric Vehicles
- Computers
- VHF Intermod
- ....and more!

The ARRL
RFI BOOK
Third Edition
ARRL Order No. 0915
Only $29.95*
*plus shipping and handling

ARRL The national association for AMATEUR RADIO™

SHOP DIRECT or call for a dealer near you.
ONLINE WWW.ARRL.ORG/SHOP
ORDER TOLL-FREE 888/277-5289 (US)

HB 08/2010

digital HFDSP

Icom's pioneering work in Digital Signal Processing lets today's ham isolate weak signals without needing to buy expensive optional filters

your **hobby.** your **life.** your love for **ham radio.**

IC-7800 HF + 6M

200 Watt Output (50W AM), Built-in Power Supply

RX: 0.030–60.000MHz*

Four 32 Bit IF-DSPs + 24 Bit AD/DA Converters

2 Completely Independent Receivers

+40dBm 3rd Order Intercept Point

3 Roofing Filters

Selectable, "Build Your Own" IF Filter Shapes

IC-7700 HF + 6M

200 Watt Output (50W AM), Built-in Power Supply

RX: 0.030–60.000MHz*

Two 32 Bit IF-DSPs + 24 Bit AD/DA Converters

Single Receive

+40dBm 3rd Order Intercept Point

3 Roofing Filters

Selectable, "Build Your Own" IF Filter Shapes

IC-7600 HF + 6M

100 Watt Output (30W AM)

RX: 0.030–60.000MHz*

USB Port for CI-V Format PC Control and Audio In/Out

Dual Receive

+30 dBm 3rd Order Intercept Point

3 Roofing Filters

Selectable, "Build Your Own" IF Filter Shapes

you can't work 'em if you can't hear 'em

*Frequency coverage may vary. Refer to owner's manual for exact specifications.
©2010-11 Icom America Inc. The Icom logo is a registered trademark of Icom Inc.
All specifications are subject to change without notice or obligation. 30493

Life is a JOURNEY. Enjoy the ride!

NEW! COMET CTC-50M
Window Gap Adapter!

Max Power: HF 100W PEP
VHF: 60W FM
UHF: 40W FM
900MHz - 1.3GHz: 10W
VSWR: <500MHz 1.3:1
         >500MHz 1.5:1
Impedance: 50Ohm
Length: 15.75"
Conn: 24k Gold Plated SO-239s

MALDOL HVU-8
Ultra-Compact 8 Band Antenna!

Unique ground radial system rotates 180 degrees around the base if building side mounting is required.

Max Power: HF 200W SSB/100W FM
6M - 70cm: 150W FM
TX: 80/40/20/15/10/6/2M/70cm
Impedance: 50 Ohm
Length: 8'6" approx
Weight: 5lbs 7oz
Conn: SO-239
Max Wind Speed: 92MPH

Each band tunes independently.
Approx 2:1 band-width:
80M 22kHz
40M 52kHz
20M 52kHz
15M 134kHz
10M 260kHz

COMET CHA-250B
Broadband HF Vertical!

3.5 - 57MHz with SWR of 1.6:1 or less!

- NO ANTENNA TUNER NEEDED
- NO RADIALS
- NO TRAPS
- NO COILS

If you suffer in an antenna restricted area, must manage with space restrictions or you simply want to operate incognito you will be forced to make significant antenna compromises. The CHA-250B makes the most of the situation, making operating HF easy!!

Max Power: 250W SSB/125W FM
TX: 3.5–57MHz
RX: 2.0–90MHz
Impedance: 50Ohm
Length: 23'5"
Weight: 7lbs 1 oz
Conn: SO-239
Max Wind Speed: 67MPH

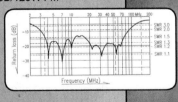

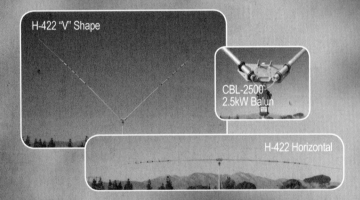

H-422 "V" Shape
CBL-2500 2.5kW Balun
H-422 Horizontal

NEW! COMET H-422
40/20/15/10M compact, broadband, rotatable dipole!

Assemble in either a "V or horizontal ("H") configuration. CBL-2500 2.5kW balun and heavy duty hardware included.

Max Power: 1000W SSB / 500W FM
SWR: Less than 1.5:1 at center frequency
Rotation Radius: "V" 12' 6"   "H" 17' 5"
Length: "V" 24' 5"   "H" 33' 10"
Weight: 11 lbs 14 ozs
Wind load: 3.01 sq feet
Max Wind Speed: 67 MPH

COMET and Maldol

For a complete catalog, call or visit your local dealer.
Or contact NCG Company. 15036 Sierra Bonita Lane, Chino, CA 91710
909-393-6133 • 800-962-2611 • FAX 909-393-6136 • www.natcommgroup.com

CONNECT ACROSS MILES INSTEAD OF METERS!

ANALOG & DIGITAL, ALL IN ONE

Every Icom portable and mobile transceiver that works D-STAR also offers world-class analog performance, all in the same rig. Keep in touch with friends via traditional FM while you advance in the hobby.

More info and downloads: www.icomamerica.com/amateur/DSTAR

IC-80AD NEXT GENERATION 2M/70CM DUAL BANDER

Join the future of amateur radio with the latest in D-STAR. This compact rig is perfect for those getting started on VHF/UHF bands, and who want to play with the new technology. In addition to the 5 watt output for 2m and 70cm, there are three additional output levels to extend your operational time between charges. To increase your fun, the '80AD has a wide band receiver that goes far beyond just ham bands.

Shown with Optional GPS Speaker Mic (HM-189GPS)

ID-880H VERSATILE 2M/70CM MOBILE

The mobile companion to the IC-80AD. Compact, dual band operation, remotable control head that fits just about anywhere, 50 full watts of output power on both VHF and UHF, easy-to-use menu system, wide band RX, and much more. It may be Icom's most affordable dual bander with D-STAR capability, but this workhorse is full of features you'd expect to find in expensive packages.

D-STAR Repeater System

ID-RP2C REPEATER CONTROLLER

The cornerstone of the D-STAR system. Handles up to four RF modules. Basic in-band or crossband operation. Linking capabilities through the Internet.

ID-RP2D 1.2GHZ DATA MODULE

Access point with a data rate of up to 128kbps. Perfect for email, web applications and support via Internet.

ID-RP2V 1.2GHZ DIGITAL VOICE MODULE

ID-RP2000V 2M DIGITAL VOICE MODULE

ID-RP4000V 70CM DIGITAL VOICE MODULE

©2010-11 Icom America Inc. The Icom logo is a registered trademark of Icom Inc. The D-PRS logo is a trademark of Icom Inc. All specifications are subject to change without notice or obligation. 30494

We build Antennas for you!!

There is a whole new world to explore, right from your own back yard!!!

(636)265-0448 Fax# (866)201-0593
sales@radiowavz.com

www.radiowavz.com

BIG BOY ROTATORS
The most powerfull antenna rotators available anywhere
Amateur, Commercial, Government and Military pourposes

Model D digital solid state controller, the ultimate in easy to set up controller

- Built in RS232 and Preset
- Programmable soft start-stop
- Programmable North or South stop
- 500° rotation (- 70° + 70° overlap)
- 1° degree accuracy and more…...

The **ProSisTel** rotators were designed to perform under tremendous stress with abnormally large antenna loads.

ProSisTel rotators give you: incredible starting and rotating torque, incredible braking resistence.
The rotators utilizing worm wheel technology far exceeds the specifications on any other amateur rotator made today.

Two years warranty

C.da Conghia 298
I-70043 Monopoli BA Italy
Phone ++39 0808876607
www.prosistel.net
Email: prosistel@prosistel.it

Are your an ARRL Member...Life Member?
www.arrl.org/membership

The ARRL Software Library for Hams v. 3.0

Quick access to utilities, applications and information:
- **Book excerpts** and videos
- **Contesting software**, including N1MM Logger
- **DX Bulletin Reader**
- **Weather Satellite** software
- **HF digital software** for PSK31, MFSK16, MT63, RTTY and more
- **WSJT** software for meteor scatter and moonbounce...**and more!**

The ARRL Software Library for Hams
CD-ROM, version 3.0
ARRL Order No. 1424
Only $19.95*
*plus shipping and handling

ARRL The national association for AMATEUR RADIO™
SHOP DIRECT or call for a dealer near you.
ONLINE WWW.ARRL.ORG/SHOP
ORDER TOLL-FREE 888/277-5289 (US)

HB 6/2009

AR5001D Professional Grade Wide Coverage Communications Receiver

The Legend Lives On!

The AR5001D delivers amazing performance in terms of accuracy, sensitivity and speed.

Available in both professional and consumer versions, the AR5001D features wide frequency coverage from 40 KHz to 3.15 GHz*, with no interruptions. Developed to meet the monitoring needs of security professionals and government agencies, the AR5001D can be controlled through a PC running Windows XP or higher. Up to three channels can be monitored simultaneously. Fast Fourier Transform algorithms provide a very fast and high level of signal processing, allowing the receiver to scan through large frequency segments quickly and accurately. AR5001D standard features include storage of up to 2000 frequencies, 45 MHz IF digital signal processing, direct digital sampling, a high performance analog RF front-end, a DDS local oscillator and advanced signal detection capabilities which can find hidden transmitters. With its popular analog signal meter and large easy-to-read digital spectrum display, the AR5001D is destined to become the choice of federal, state and local law enforcement agencies, the military, emergency managers, diplomatic service, news-gathering operations, and home monitoring enthusiasts.

Discover the next generation in AOR's legendary line of professional grade desktop communications receivers.

- Multimode receives AM, wide and narrow FM, upper and lower sideband and CW
- Up to 2000 alphanumeric memories (50 channels X 40 banks) can be stored
- Analog S-meter
- Fast Fourier Transform algorithms
- Operated by a Windows XP or higher computer through a USB interface using a provided software package that controls all of the receiver's functions
- An SD memory card port can be used to store recorded audio
- Analog composite video output connector
- CTCSS and DCS squelch operation
- Two selectable Type N antenna input ports
- Adjustable analog 45 MHz IF output with 15 MHz bandwidth
- Triple-conversion receiver exhibits excellent sensitivity
- Powered by 12 volts DC (AC Adapter included), it can be operated as a base or mobile unit
- Professional (government) version is equipped with a standard voice-inversion monitoring feature

Add to the capabilities of the AR5001D with options:

- Optional APCO-25 decoder
- Optional LAN interface unit enables control via the internet
- Optional I/Q output port allows capture of up to 1 MHz onto a computer hard drive or external storage device
- Optional AR-I/Q Windows software facilitates the easy storage and playback of transmissions captured within the selected spectrum in conventional modes, or, signals can be subjected to further analysis
- Optional GPS board can be used for an accurate time base and for time stamping digital I/Q data

The Serious Choice in Advanced Technology Receivers
AOR U.S.A., Inc.
20655 S. Western Ave., Suite 112
Torrance, CA 90501, USA
Tel: 310-787-8615 Fax: 310-787-8619
info@aorusa.com • www.aorusa.com

Cellular blocked for US consumer version. Unblocked version available to qualified purchasers with documentation. Specifications subject to change without notice or obligation

MFJ-259B *Antenna Analyzer*

MFJ-259B **$289⁹⁵**

The world's most popular antenna analyzer gives you a complete picture of your antenna performance 1.8 to 170 MHz. Super easy-to-use -- makes tuning your antenna quick and easy. Read antenna SWR, complex impedance, return loss, reflection coefficient. Determine velocity factor, coax cable loss in dB, length of coax and distance to a short or open in feet. Read inductance in uH and capacitance in pF at RF frequencies. Large easy-to-read two line LCD screen and side-by-side meters clearly display your information. Built-in frequency counter, Ni-Cad charger circuit, battery saver, low battery warning and smooth reduction drive tuning.

MFJ-99B, $93.80. Accessory Package for MFJ-259B. Includes MFJ-29C custom foam-filled carrying case, MFJ-66 dip meter adapters, MFJ-1312D 110VAC adapter and 10-pack high capacity Ni-MH MFJ *SuperCell*™ batteries. *Save $7!*

1.8-170 MHz *plus* 415-470 MHz *Antenna Analyzer*

All-in-one handheld antenna test lab lets you quickly check/tune HF, VHF, *UHF* antennas anywhere. Measures: SWR, Return Loss, Reflection Coefficient, R, X, Z, Phase Angle, Coax cable loss, Coax cable length, Distance to short/open in coax, Inductance, Capacitance, Resonant Frequency, Bandwidth, Q, Velocity Factor, Attenuation, more!

MFJ-269 **$389⁹⁵**

MFJ *Weather-Proof* Antenna *Feedthrough* Panel

Bring three coax-fed HF/VHF/UHF antennas, balanced line, random wire and ground into your hamshack without drilling through walls . . . **New!** MFJ-4602 **$69⁹⁵**

MFJ's Weather-proof *Antenna Feedthrough Panel* mounts in your window sill. Lets you feed three coax-fed antennas, balanced line, random wire and ground without drilling through walls.

Simply place in window sill and close window. One cut customizes it for any window up to 48 inches. Use horizontally or vertically. High-quality pressure-treated wood with excellent 3/4 inch thick insulating properties is painted with heavy coat of white outdoor enamel paint. Edges sealed by weather-stripping. Seals and insulates against all weather conditions. Gives years of trouble-free service. ³/₄Dx3¹/₂Hx48W in.

Inside/outside stainless steel plates bond all coax shields to ground. Stainless steel ground post brings outside ground connection inside. 3 *Teflon*® SO-239 coax connectors, *ceramic* balanced line/randomwire feedthru insulators.

Stainless Steel Bonding Plate | Indoor View | Outdoor View

300 Watt Automatic Tuner

MFJ-993B **$259⁹⁵**

The *world's best selling* automatic antenna tuner is *highly acclaimed the world over* for its ultra high-speed, wide matching range, reliability, ease-of-use! Matches virtually *any* antenna. Digital/Aural SWR/Wattmeter.

MFJ 300 Watt Tuner

MFJ-949E **$179⁹⁵**

More hams use MFJ-949s than any other antenna tuner in the world! Full 1.8-30 MHz coverage, custom inductor switch, 1000 Volt tuning capacitors, full-size lighted Cross-Needle SWR/Wattmeter, 8 pos. antenna switch, 3¹/₂Hx10⁵/₈Wx7D in.

World's *largest* Wattmeter

GIANT 6¹/₂ inch meter can be read across your shack! Extra-long scales for highly accurate measurements. *True* peak or average forward and reflected power readings. 20/200/2000 Watt ranges for accurate QRP or QRO operation. 1.8-30 MHz, SO-239 connectors. 7Wx5¹/₂Hx5D in.

MFJ-868 **$149⁹⁵**

Antenna Switches

MFJ-1702C 2 position antenna coax switch. Exclusive lightning arrestor and center pos. ground, lightning arrestor. 2.5 kW PEP, 1 kW CW. Loss below .2 dB. 50 dB isolation @ 450 MHz. 50 Ohms. 3x2x3".

MFJ-1702C **$39⁹⁵**

MFJ-1704. Like MFJ-1702C but has 4 positions. 6¹/₄x4¹/₄x1¹/₄ in. 50 load. SWR 1.1: 1 to 30 MHz; 1.5:1, 30-650 MHz. SO-239.

MFJ-1704 **$79⁹⁵**

All Band G5RV Antennas

MFJ-1778 **$44⁹⁵**

160-10 Meters with tuner. 102 foot.1.5 kW. Use as Marconi on 160 Meters with tuner/ground. Ladder line/SO-239. **MFJ-1778M, $39.95.** *Half-size*, 52 ft. G5RV-Jr. 60-10 Meters with tuner. 1.5 kW.

33 Foot *fiberglass* Mast

MFJ-1910 **$79⁹⁵**

33 foot super strong fiberglass *telescoping* mast collapses to 3.8 feet, weighs 3.3 lbs. Huge, strong 1³/₄" bottom section. Flexes to resist breaking. Resists UV. Put up *full size* inverted Vee dipole/vertical antenna in minutes and get *full size performance!*

MFJ 25Amp Switching Power Supply

Ham Radio's smallest and lightest 22 Amp continuous power supply is also its best selling! 22 Amps continuous/25 Amps max at 13.8VDC. 5-way binding posts on front, 5A quick connects on back. 85-135/170-260 VAC input. 2.9 lbs. 5³/₄Wx3Hx5³/₄D".

MFJ-4125 **$99⁹⁵**

MFJ Power Outlet Strip

MFJ-1118 **$84⁹⁵**

with versatile 5-way binding posts

Power two HF and/or VHF rigs and six accessories from your main 12 VDC supply. Built-in 0-25 VDC voltmeter. Two pairs 35 amp 5-way binding posts, fused and RF bypassed for transceivers. Six pairs RF bypassed binding posts provide 15 Amps for accessories. Master fuse, ON/OFF switch, "ON" LED. 12¹/₂x2³/₄x2¹/₂ in. Other power strip models available, some with *Powerpoles*™; see website.

Dry 1.5 kW *UHF/VHF/HF* Dummy Load

Air-cooled, non-inductive resistor in perforated metal housing; Full 1.5 kW for 10 seconds. 100 Watts continuous. Silk-screened derating curve. SWR below 1.1 to 30 MHz; below 1.5:1 30-650 MHz. SO-239 connector. 3x3x9 in.

MFJ-264 **$74⁹⁵**

MFJ-260C, $39.95. Like MFJ-264, but handles 300 Watts. 2¹/₄x2¹/₄x7 in.

MFJ-108B, $21.95. Dual clock, separate 24/12 hour displays. 4¹/₂Wx1D x2H".

Free MFJ Catalog
and Nearest Dealer . . . 800-647-1800

http://www.mfjenterprises.com
• 1 Year *No Matter What*™ warranty • 30 day money back guarantee (less s/h) on orders direct from MFJ

MFJ ENTERPRISES, INC.
300 Industrial Pk Rd, Starkville, MS 39759 **PH:** (662) 323-5869
Tech Help: (662) 323-0549
FAX:(662)323-6551 8-4:30 CST, Mon.-Fri. *Add shipping.*
Prices and specifications subject to change. (c) 2010 MFJ Enterprises, Inc.

MFJ . . . the world leader in amateur radio accessories! Visit wwww.mfjenterprises.com

AMERITRON... 800 Watts... $899!
More hams use Ameritron AL-811/H amplifiers than any other amplifier in the world!

AL-811H **$899** Suggested Retail *4-Tubes, 800 Watts*

AL-811 **$749** Suggested Retail *3-Tubes, 600 Watts*

Only the Ameritron AL-811H gives you four *fully neutralized* 811A transmitting tubes. You get absolute stability and superb performance on higher bands that can't be matched by un-neutralized tubes.

You get a quiet desktop linear that's so compact it'll slide right into your operating position -- you'll hardly know it's there... until QRM sets in. And you can conveniently plug it into your nearest 120 VAC outlet -- no special wiring needed.

You get all HF band coverage (with license) -- including WARC and most MARS bands at 100% rated output. Ameritron's *Adapt-A-Volt*™ hi-silicon core power transformer has a special buck-boost winding that lets you compensate for high/low power line voltages.

You also get efficient full size heavy duty tank coils, slug tuned input coils, operate/standby switch, transmit LED, ALC, dual illuminated meters, QSK with optional QSK-5, pressurized cooling that you can hardly hear, full height computer grade filter capacitors and more. 13 3/4 W x 8H x 16D inches.

AL-811, $749. Like AL-811H, but has three 811A tubes and 600 Watts output.

AMERITRON *no tune* Solid State Amplifiers

ALS-500M 500 Watt Mobile Amp

ALS-500M **$849** Suggested Retail

500 Watts PEP/400W CW output, 1.5-22 MHz, instant bandswitching, no tuning, no warm-up. SWR, load fault, thermal overload protected. On/Off/Bypass switch. Re-mote on/off control. DC current meter. Ex-tremely quiet fan. 13.8 VDC. 9Wx3 1/2 Hx15D in., 7 lbs. **ALS-500RC, $49,** Remote Head.

ALS-600 Station 600 Watt FET Amp
No tuning, no fuss, no worries -- just turn on and operate. 600 Watts PEP/500W CW, 1.5-22 MHz, instant bandswitching, SWR protected, extremely quiet, SWR/Wattmeter, ALC control. 120/220 VAC. Inrush protected. 9 1/2 Wx6Hx12D in. **ALS-600S, $1599,** ALS-600 with 10 lb. switching power supply.

ALS-600 **$1499** Suggested Retail

Desktop Kilowatt Amplifier

AL-80B **$1449** Suggested Retail

Whisper quiet *desktop* amp plugs into 120 VAC to give *full* kilowatt SSB PEP output. Ameritron's exclusive *DynamicALC*™ doubles average SSB power out and *Instantaneous RF Bias*™ gives cooler operation. All HF bands. 850 Watts CW out, 500 Watts RTTY out, extra heavy duty power supply, 3-500Z tube, 70% efficiency, tuned input, Pi/Pi-L output, inrush current protection, dual Cross-Needle meters, QSK compatible, 48 lbs. 14Wx8 1/2 Hx15 1/2 D in. **Two-year warranty.**

ALS-1300 Solid State 1200 Watt Amp

ALS-1300 **$2899** Suggested Retail

Ameritron's *highest power* solid state FET no-tune amplifier gives you instant bandswitching, no tuning, no warm-up, no tubes to baby and no fuss! Outstanding reliability is insured by using eight rugged MRF-150 power FET's mounted on the dual heavy duty heat sink. Run up to 1200 Watts of clean SSB output power (just 100 Watts drive gives full rated power) for continuous coverage between 1.5-22 MHz. Compact 10Wx6 1/2 Hx18D in.

Near Legal Limit™ Amplifier

AL-572 **$1695** Suggested Retail

New class of *Near Legal Limit*™ amplifier gives you 1300 Watt PEP SSB power output for *60%* of price of a full legal limit amp! 4 rugged 572B tubes. Instant 3-second warm-up, plugs into 120 VAC. Compact 14 1/2 Wx 8 1/2 Hx15 1/2 D inches fits on desktop. 160-15 Meters. 1000 Watt CW output. Tuned input, instantaneous RF Bias, dynamic ALC, parasitic killer, inrush protection, two lighted cross-needle meters, multi-voltage transformer.

Eimac 3CX800A7 Amplifiers

Suggested Retail
AL-800 **$2045** *1 Eimac®tube, 1250 W*
AL-800H **$3045** *2 Eimac®tubes, 1.5 kW Plus*
With Imported Tube
AL-800F, $1875
AL-800HF, $2745

These *compact* desktop amplifiers with 3CX800A7 tubes cover 160-15 Meters including WARC bands. Adjustable slug tuned input circuit, grid protection, front panel ALC control, vernier reduction drives, heavy duty 32 lb. silicone steel core transformer, high capacitance computer grade filter capacitors. Multi-voltage operation, dual lighted cross-needle meters. 14 1/4 Wx8 1/2 Hx16 1/2 D in.

AMERITRON *full legal limit* amplifiers
AMERITRON legal limit amps use a super heavy duty Peter Dahl Hypersil® power transformer capable of 2.5 kW!

Most powerful -- *3CX1500/8877*

AL-1500 **$3495** Eimac® Tube
AL-1500F **$3095** Suggested Retail Imported tube

Ameritron's *most powerful* amplifier uses the herculean 3CX1500/8877 ceramic tube. 65 watts drive gives you full legal output -- it's just loafing with a 2500 Watts power supply.

Toughest -- *3CX1200A7*

AL-1200 **$3459** Suggested Retail

Get ham radio's *toughest* tube with the Ameritron AL-1200 -- the *Eimac*® 3CX1200A7. It has a 50 Watt control grid dissipation. What makes the Ameritron AL-1200 stand out from other legal limit amplifiers? The answer: A super heavy duty power supply that loafs at full legal power -- it can deliver the power of more than 2500 Watts PEP two tone output for a half hour.

Classic -- *Dual 3-500Gs*

AL-82 **$2745** Suggested Retail

This linear gives you full legal output using a *pair* of *genuine* 3-500Gs. Competing linears using 3-500Gs *can't* give you 1500 Watts because their lightweight power supplies *can't* use these tubes to their full potential.

Call your dealer for your best price!
Free Catalog: 800-713-3550

AMERITRON
...the world's high power leader!
116 Willow Road, Starkville, MS 39759
TECH (662) 323-8211 • **FAX** (662) 323-6551
8 a.m. - 4:30 p.m. CST Monday - Friday
For power amplifier components call (662) 323-8211
http://www.ameritron.com
Prices and specifications subject to change without notice. ©2010 Ameritron.

Ameritron brings you the finest high power accessories!

ARB-704 amp-to-rig interface... $59.95

Protects rig from damage by keying line transients and makes hook-up to your rig easy!

RCS-4 Remote Coax Switch... $159.95
Use 1 coax for 4 antennas. No control cable needed. SWR <1.25, 1.5 - 60 MHz. Useable to 100 MHz.

RCS-8V Remote Coax Switch... $169.95
Replace 5 coax with 1! 1.2 SWR at 250 MHz. Useable to 450 MHz. < .1 dB loss, 1kW@ 150MHz.

RCS-10 Remote Coax Switch... $179.95
Replace 8 coax with 1! SWR<1.3 to 60 MHz. **RCS-10L, $219.95** with lightning arrestors.

AMERITRON... the world's high power leader! Visit www.ameritron.com

PALSTAR

DESIGNED AND MADE IN THE USA

COMING SOON

The HF-AUTOTUNE 1500Watt Discrete Antenna Tuner **TBA**

Palstar is hard at work reinventing the automatic tuner into the new *HF-AUTOTUNE*. It will feature motor-driven, continuously variable components for a perfect match every time. It will be capped off with a power rating of 1500 watts and compatible with all radios!

R30A General Coverage Shortwave Receiver $749.59

- 100kHZ-29.999 MHz, SSB (U/L)
- 20Hz/100Hz tuning steps synthesized
- Variable rate tuning with switchable tuning knob
- Collins torsional mode filters, AM and SSB
- 6-digit LCD Display
- 100 channel memory
- 9.13"W x 3.94"H x 9"D
- Weight: 3.9 lbs (1.8kg)

AT5K 3500 Watt Antenna Tuner $1195.00

- 3500 Watts single tone continuous
- Frequency coverage: 1.8-30MHz
- 160m to 10m
- Balanced output with high power 5kW TEFLON wire balun
- Peak and Peak Hold cross-needle metering
- Six-position switchable antenna input
- 16 1/4"W x 8 1/4"H x 19 3/4"D
- Weight: 30 lbs

LA-30 Rotating Loopstick Antenna $329.00

- MW: 410 to 2050 KHz (STD)
- LW: 110 to 550 KHz (OPT)
- Tropical: 1480 to 7500 KHz (OPT)
- Loopstick: Ferrite core, Litz wire
- Attenuator: -15 dB
- Gain: +15 dB
- 8.25"W x 4.75"H x 11.25"D
- Weight: 5 lbs

Commander VHF-144 2 Meter RF Amplifier $2995.00

- Freq. Range: 144-148 MHz
- Modes: USB, LSB, RTTY, FM, & CW
- RF Output: 1kW (SSB); 650W (FM or RTTY)
- High-speed Electronic bias switching
- True step-start control circuitry
- Power Req: 117/200/234VAC 50/60 Hz
- 14.5"W x 14.5"H x 6"D
- Weight: 43 lbs

VHF-2000 6 Meter RF Amplifier $3995.00

- Freq. Range: 50-54 MHz
- Modes: USB, LSB, RTTY, FM, & CW
- RF Output: 1500W, PEP SSB; 1500W FM or RTTY
- High-speed Electronic Bias Switching
- True step-start control circuitry
- Power Req: 200/234VAC 50/60Hz
- 18"W x 16"H x 7.75"D
- Weight: 75 lbs (with transformer)

HF-2500 1500 Watt RF Amplifier $3995.00

- 1500W Output (100% Duty Cycle)
- Two pulse rated ceramic metal triodes
- Bands: 160, 80, 40, 20, 17, 15, 12, & 10m
- High-speed Electronic bias switching
- True step-start control circuitry
- Power Req: 200/234VAC 50/60Hz
- 18.25"W x 8.5"H x 20.75"D
- Weight: 72 lbs (with transformer)

All Amplifiers use 3CPX800A7 Ceramic Triodes

Order Direct from Palstar at www.palstar.com, call us at 1-800-773-7931, or contact your local ham dealer for Palstar products. For Hams who demand THE BEST!
Palstar Inc. 9676 N. Looney Rd. Piqua, OH 45356

Available factory direct or at **HRO, AES**

www.palstar.com

& amateur radio dealers worldwide

IC-7600

The IC-7600 utilizes Icom's proven IF-DSP technology to provides professional performance in the 160 to 6 meter amateur bands. Three roofing filters, incredible WQVGA display, RTTY/PSK31, voice synthesizer, 101 memories, scan, PBT, 104dB dynamic range and +30dbm 3rd order IP and much more!

IC-7200

The **IC-7200** is an affordable, robustly built transceiver covering 160 to 6 meters with DSP filtering, PBT, notch, voice synth., 201 mems., atten., roofing filter, RIT, scanning, and 1 Hz tuning. Visit our website for details.

Universal Radio
6830 Americana Pkwy.
Reynoldsburg, OH 43068
◆ Orders: 800 431-3939
◆ Info: 614 866-4267
www.universal-radio.com

HamTestOnline™
Web-based training for the ham radio written exams

- Quick, easy way to learn.
- 100% guaranteed — you pass the exam or get your money back.
- Better than random practice tests.
- Presents concepts in logical order.
- Provides additional information.
- Better than books — question drill keeps you engaged.
- Uses the actual exam questions.
- Tracks your progress on each question.
- Focuses on your weak areas with "intelligent repetition".
- Encourages learning, not memorizing.

www.hamtestonline.com

AD9951-Based Direct-Digital VFO

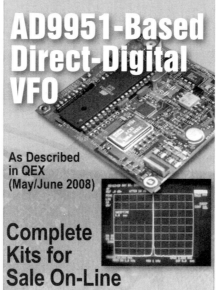

As Described in QEX (May/June 2008)

Complete Kits for Sale On-Line

See www.WA1FFL.com for Ordering and Technical Information.

- High SFDR. 0.5-30 MHz Coverage
- CAL, RIT, Transmit Offset
- 2K Flash EEPROM Memory Storage
- Surface-Mount Components Pre-Soldered
- Display and Shaft Encoder Included

As Demonstrated at Dayton and Boxboro 2010

HAGERTY
Radio Company
www.WA1FFL.com

HOTPRESS
HAM HATS
COMPUTERIZED EMBROIDERY
HAM HATS MADE BY HAMS
1-800-862-1221
www.hotpresstshirts.com

Advertising   A-11

See what's new @
http://www.icmfg.com

- Tight Tolerance
- Low Phase Noise
- Ovenized
- Custom Precision Crystals

ICM INTERNATIONAL CRYSTAL MANUFACTURING CO, INC
PO BOX 1768 • 10 NORTH LEE
OKLAHOMA CITY, OK 73101

800.725.1426 • www.icmfg.com

Join or Renew your ARRL Membership
www.arrl.org/join

EARN MORE MONEY
Get your dream job!

Be an FCC Licensed Wireless Technician!

Make up to $100,000 a year and more with NO college degree

Learn Wireless Communications and get your "FCC Commercial License" with our proven Home-Study Course!

- No need to quit your job or go to school.
- This course is easy, fast and low cost.
- No previous experience needed!
- Learn at home in your spare time!

Move to the front of the employment line in Radio-TV, Communications, Avionics, Radar, Maritime and more... even start your own business!

Call now for FREE info
800-932-4268 ext. 204

Or, email us:
fcc@CommandProductions.com

COMMAND PRODUCTIONS
Warren Weagant's FCC License Training
P.O. Box 3000, Dept. 204 • Sausalito, CA 94966
Please rush FREE info immediately!

NAME:
ADDRESS:
CITY/STATE/ZIP:

You may fax your request to: 415-332-1901

from MILLIWATTS to KILOWATTS
More Watts per Dollar

Taylor TUBES

Quality Transmitting & Audio Tubes
Eimac

- COMMUNICATIONS
- BROADCAST
- INDUSTRY
- AMATEUR

Svetlana

Immediate Shipment from Stock

3CPX800A7	3CX10000A7	4CX3000A	807
3CPX5000A7	3CX15000A7	4CX3500A	810
3CW20000A7	3CX20000A7	4CX5000A	811A
3CX100A5	4CX250B	4CX7500A	812A
3CX400A7	4CX250BC	4CX10000A	813
3CX400U7	4CX250BT	4CX10000D	833A
3CX800A7	4CX250FG	4CX15000A	833C
3CX1200A7	4CX250R	4CX20000A	845
3CX1200D7	4CX350A	4CX20000B	866-SS
3CX1200Z7	4CX350F	4CX20000C	5867A
3CX1500A7	4CX400A	4X150A	5868
3CX2500A3	4CX800A	YC-130	6146B
3CX2500F3	4CX1000A	YU-108	7092
3CX3000A7	4CX1500A	YU-148	3-500ZG
3CX6000A7	4CX1500B	572B	4-400A

– TOO MANY TO LIST ALL –

VISA MasterCard AmericanExpress

ORDERS ONLY:
800-RF-PARTS • 800-737-2787
Se Habla Español • We Export

TECH HELP / ORDER / INFO: 760-744-0700

FAX: 760-744-1943 or 888-744-1943

An Address to Remember:
www.rfparts.com

E-mail:
info@rfparts.com

RF PARTS COMPANY
Since 1967

Cushcraft R8 *8-Band Vertical*
Covers 6, 10, 12, 15, 17, 20, 30, and 40 Meters!

$539.95

The R-8 provides 360° (omni) coverage on the horizon and a low radiation angle in the vertical plane for a better DX.

The Cushcraft R8 is recognized as the industry gold standard for multi-band verticals, with thousands in use worldwide. Efficient, rugged, and built to withstand the test of time, the R8's unique ground-independent design has a well-earned reputation for delivering top DX results under tough conditions. Best of all, the R8 is easy to assemble, installs just about anywhere, and blends inconspicuously with urban and country settings alike.

Automatic Band Switching: The R8's famous "black box" matching network combines with traps and parallel resonators to cover 8 bands. You QSY instantly, without a tuner!

Rugged Construction: Thick fiberglass insulators, all-stainless hardware, and 6063 aircraft-aluminum tubing that is double or triple walled at key stress points handle anything Mother Nature can dish out.

Compact Footprint: Installs in an area about the size of a child's sandbox -- no ground radials to bury and all RF-energized surfaces safely out of reach.

Legal-Limit Power: Heavy-duty components are contest-proven to handle all the power your amplifier can legally deliver and radiating it as RF rather than heat.

The sunspot count is climbing and long-awaited band openings are finally becoming a reality. Now is the perfect time to discover why Cushcraft's R8 multi-band vertical is the premier choice of DX-wise hams everywhere!

R-8GK, $56.95. R-8 three-point guy kit for high winds.

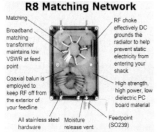
R8 Matching Network

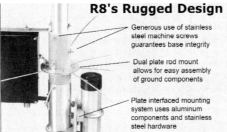
R8's Rugged Design

MA-5B *5-Band* Beam
Small Footprint -- Big Signal

MA-5B $499.95

The MA-5B is one of Cushcraft's most popular HF antennas, delivering solid *signal-boosting directivity* in a bantam-weight package. Mounts on roof using standard TV hardware. Perfect for exploring exciting DX without the high cost and heavy lifting of installing a large tower and full-sized array. Its 7 foot 3-inch boom has less than 9 feet of turning radius. Contest tough -- handles 1500 Watts.

The unique MA-5B gives you 5-bands, automatic band switching and easy installation in a compact 26-pound package. On 10, 15 and 20 Meters the end elements become a two-element Yagi that delivers solid power-multiplying gain over a dipole on all three bands. On 12 and 17 Meters, the middle element is a highly efficient trap dipole. When working DX, what really matters are the interfering signals and noise you *don't hear*. That's where the MA-5B's impressive side rejection and front-to-back ratio really shines. See *cushcraftamateur.com* for gain figures.

Cushcraft 10, 15 & 20 Meter Tribander Beams

Only the best tri-band antennas become DX classics, which is why the Cushcraft World-Ranger A4S, A3S, and A3WS go to the head of the class. For more than 30 years, these pace-setting performers have taken on the world's most demanding operating conditions and proven themselves every time. The key to success comes from attention to basics. For example, element length and spacing has been carefully refined over time, and high-power traps are still hand-made and individually tuned using laboratory-grade instruments. All this

A-4S $699.95

A-3S $599.95

attention to detail means low SWR, wide bandwidth, optimum directivity, and high efficiency -- important performance characteristics you rely on to maintain regular schedules, rack up impressive contest scores, and grow your collection of rare QSLs!

It goes without saying that the World-Ranger lineup is also famous for its rugged construction. In fact, the majority of these antennas sold years ago are still in service today! Conservative mechanical design, rugged over-sized components, stainless-steel hardware, and aircraft-grade 6063 make all the difference.

The 3-element A3S/A3WS and 4-element A4S are world-famous for powerhouse gain and super performance. **A-3WS, $499.95,** 12/17 M. **30/40 Meter** add-on kits available.

Cushcraft Dual Band Yagis
One Yagi for Dual-Band FM Radios

A270-10S $169.95

Dual-bander VHF rigs are the norm these days, so why not compliment your FM base station with a dual-band Yagi? Not only will you eliminate a costly feed line, you'll realize extra gain for digital modes like high-speed packet and D-Star! Cushcraft's A270-6S provides three elements per band and the A270-10S provides five for solid point-to-point performance. They're both pre-tuned and assembly is a snap using the fully illustrated manual.

A270-6S $129.95

Cushcraft Famous *Ringos* Compact FM Verticals

AR-2 $64.95 **AR-6 $99.95** **AR-10 $109.95**

W1BX's famous *Ringo* antenna has been around for a long time and remains unbeaten for solid reliability. The Ringo is broad-banded, lighting protected, extremely rugged, economical, electrically bullet-proof, low-angle, and more -- but mainly, it just plain works! To discover why hams and commercial two-way installers around the world still love this antenna, order yours now!

Free Cushcraft Catalog
and Nearest Dealer . . . 662-323-5803
Call your dealer for your best price!

Cushcraft
Amateur Radio Antennas
308 Industrial Park Road, Starkville, MS 39759 USA
Open: 8-4:30 CST, Mon.-Fri. **Add Shipping.**
• Sales/Tech: 662-323-5803 • FAX: 662-323-6551
http://www.cushcraftamateur.com
Prices/specifications subject to change without notice/obligation. © Cushcraft, 2010.

Cushcraft . . . Keeping you in touch around the globe!   *Visit www.cushcraftamateur.com*

Quality Radio Equipment Since 1942

AMATEUR TRANSCEIVERS

Universal Radio is your authorized dealer for all major amateur radio equipment lines including: Alinco, Icom, Kenwood and Yaesu.

SHORTWAVE RECEIVERS

Whether your budget is $50, $500 or $5000, Universal can recommend a shortwave receiver to meet your needs. Universal is the *only* North American authorized dealer for *all* major shortwave lines including: Icom, Yaesu, Microtelecom, Ten-Tec, Grundig, etón, Sony, Sangean, Palstar, RFspace and Kaito.

WIDEBAND RECEIVERS

Universal Radio carries a broad selection of wideband and scanner receivers. Alinco, AOR, Icom, Yaesu, GRE and Uniden Bearcat models are offered at competitive prices. Portable and tabletop models are available.

ANTENNAS AND ACCESSORIES
Your antenna is just as important as your radio. Universal carries antennas for every need, space requirement and budget. Please visit our website or request our catalog to learn more about antennas.

BOOKS
Whether your interest is amateur, shortwave or collecting; Universal has the very best selection of radio books and study materials.

■ USED EQUIPMENT
Universal carries an extensive selection of used amateur and shortwave equipment. Buy with confidence. All items have been tested by our service department and carry a 60 day warranty unless stated otherwise. Request our monthly printed used list or visit:
www.universal-radio.com

◆ VISIT OUR WEBSITE
Guaranteed lowest prices on the web? Not always. But we *do* guarantee that you will find the Universal website to be the **most** informative.
www.universal-radio.com

◆ VISIT OUR SHOWROOM
Showroom Hours
Mon.-Fri. 10:00 - 5:30
Saturday 10:00 - 3:00

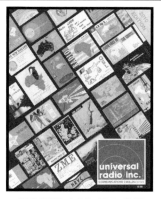

FREE 128 PAGE CATALOG
Our informative print catalog covers everything for the amateur, shortwave and scanner enthusiast. All items are also viewable at:
www.universal-radio.com

Universal Radio, Inc.
6830 Americana Pkwy.
Reynoldsburg, OH 43068

➤ 800 431-3939 Orders
➤ 614 866-4267 Information
➤ 614 866-2339 Fax
➤ dx@universal-radio.com

✓ Established in 1942.
✓ We ship worldwide.
✓ Visa, Master and Discover cards.
✓ Used equipment list available.
✓ Returns subject to 15% restocking fee.

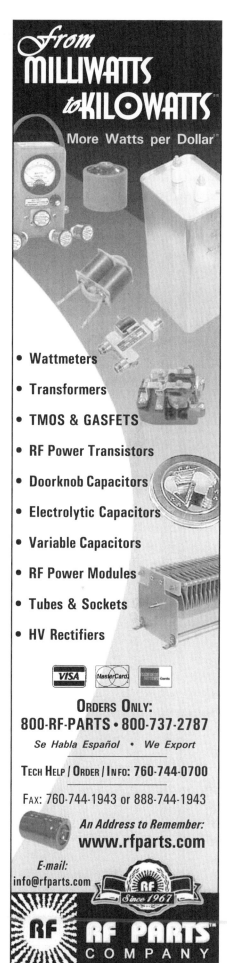

from MILLIWATTS to KILOWATTS
More Watts per Dollar

- Wattmeters
- Transformers
- TMOS & GASFETS
- RF Power Transistors
- Doorknob Capacitors
- Electrolytic Capacitors
- Variable Capacitors
- RF Power Modules
- Tubes & Sockets
- HV Rectifiers

ORDERS ONLY:
800-RF-PARTS • 800-737-2787

Se Habla Español • We Export

TECH HELP / ORDER / INFO: 760-744-0700

FAX: 760-744-1943 or 888-744-1943

An Address to Remember:
www.rfparts.com

E-mail: info@rfparts.com

RF PARTS COMPANY
Since 1967

Array Solutions — Your Source for Outstanding Radio Products

Top ranked test equipment from Array Solutions. For additional information visit: www.arraysolutions.com

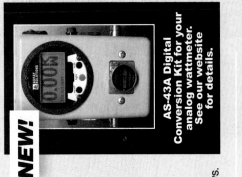

NEW! AS-43A Digital Conversion Kit for your analog wattmeter. See our website for details.

PowerMaster Wattmeter

A Wattmeter that is outstanding in functionality with NIST traceable accuracy.

- ALC-1 protection option, shuts down your transmitter quickly when necessary but allows for normal ALC operation.
- Fastest bar graph on the market, makes tuning an amp or a tuner a joy, faster than an analog meter.
- Displays SWR, peak power (numerically), and the visual bar graph simultaneously on the large, bright vacuum fluorescent display.
- PC application included – optional use. Native RS-232 connectivity – USB adapters available.
- Standard 3kW model covers 160 through 6m. Accuracy better than +/- 3%.
- Individual RF couplers available for 144, 220, and 432 MHz amateur bands.
- Optional 10kW and 20kW versions available (1 to 30 MHz).
- Calibrated couplers are available for VHF and UHF commercial applications.
- Optional 12VDC 1 Amp regulated power supply and rack mount options for one or two displays.

AIM 4170C Antenna Lab RF Analyzer

The AIM 4170C antenna analyzer measures the complex impedance (magnitude and phase) at each frequency of interest in the range of 5KHz to 180 MHz. A PC is used to calculate all RF parameters, including R +/-X, Magnitude and Phase, SWR, Return Loss, line loss, and more and plot the results in an easy to read graph and interactive Smith Chart.

Designed and priced for amateurs, but commercial lab-quality. Native RS-232; USB with adapter.

Vector Network Analyzer Model VNA 2180

Measures impedance magnitude, phase and transmission parameters for antennas, filters, and discrete components - using one or two ports.

- Frequency range is 5KHz to 180MHz.
- Data plots include: impedance, SWR, return loss, S11 and S21.
- Plots can be saved for before and after comparisons.
- Dual Smith charts with zoom and rotation.
- Analog/digital I/O port for accessories.
- Native USB connectivity.

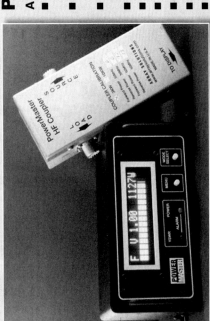

Other Quality Products from Array Solutions...

ACOM	Phillystran, Inc.	Tokyo Hy-Power	RigExpert	Prosistel Rotators	OptiBeam Antennas	Hofi®
Sales and Service for Amplifiers and Accessories	Official Worldwide Phillystran Distributor	Sales and Factory Authorized Service	Analyzers and Interfaces	Strongest Rotators on the Market	German Engineering means High Performance	Surge Arrestors & Antenna Switches

Array Solutions' products are in use at top DX and Contest stations worldwide as well as commercial and governmental installations. We provide RF solutions to the DoD, FEMA, Emcomm, UN, WFO, FAA and the State Dept. for products and installation of antennas systems, antenna selection, filtering, switching and grounding. We also offer RF engineering and PE consulting services.

www.arraysolutions.com
Phone 214-954-7140
sales@arraysolutions.com
Fax 214-954-7142

CABLE X-PERTS, INC.
Connecting You to the World...
1-800-828-3340

FOR PREMIUM ELECTRICAL PERFORMANCE FROM YOUR EQUIPMENT

We take great pride in our work!

Custom or Ready-Made Coaxial Assemblies

Visit us on-line for cable selection and great prices

www.CableXperts.com

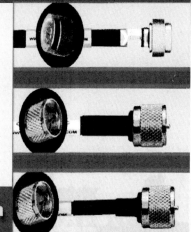

The ARRL Digital Technology for Emergency Communications Course *Version 1.1*

In this course, we will introduce you to all the ways Amateur Radio operators are using digital technology as a valuable emergency communications tool. The topics covered during the course include:

- Packet radio APRS
- Winlink 2000
- IRLP
- EchoLink and WIRES-II
- D-STAR
- APCO25
- HF sound card modes
- Automatic Link Establishment ("ALE")

We'll help you answer questions such as:

Can you transfer supply lists or personnel assignments between emergency operations sites?

Can you get critical e-mails to the Internet if a connection goes down?

Can you relay digital images of damage at specific locations?

Can you track the locations of emergency personnel and display them on computer maps?

...and more!

System Requirements (minimum) – Microsoft Windows® Vista/XP/2000/NT/98/95 or Apple OS X, 200 MHz processor, 32 MB RAM, sound card and speakers, 4-speed CD-ROM drive or higher. Requires Web browser; Microsoft Internet Explorer™ 6.0, Mozilla Firefox 2.0, Apple Safari 3.0—or later versions. Some documents require Adobe® Reader®.

Illustrations, screenshots, Internet links and audio are used to demonstrate transmission modes and equipment configurations. Bite-sized learning units make learning interesting and fun!

ARRL Order No. 1247
Only $49.95*
*plus shipping and handling

ARRL The national association for **AMATEUR RADIO**™

ARRL SHOP DIRECT or call for a dealer near you.
ONLINE WWW.ARRL.ORG/SHOP
ORDER TOLL-FREE 888/277-5289 (US)

HB 09/2009

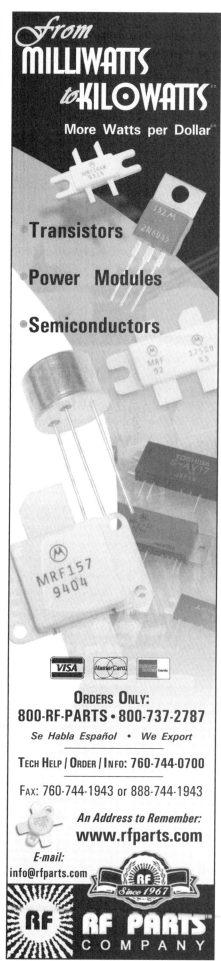

from **MILLIWATTS to KILOWATTS**
More Watts per Dollar

- Transistors
- Power Modules
- Semiconductors

ORDERS ONLY:
800-RF-PARTS • 800-737-2787
Se Habla Español • We Export

TECH HELP / ORDER / INFO: 760-744-0700

FAX: 760-744-1943 or 888-744-1943

An Address to Remember:
www.rfparts.com

E-mail: info@rfparts.com

RF PARTS COMPANY
Since 1967

The Yaesu FT-857D is the worlds's smallest HF/VHF/UHF multimode amateur transceiver covering 160 m to 70 cm with 100W on HF. With 60 meters and DSP2 built-in.

The FT-897D is a multi-mode high-power base/mobile transceiver covering 160 m to 70 cm including 60 meters. TCXO built-in. Visit www.universal-radio.com for details!

Universal Radio
6830 Americana Pkwy.
Reynoldsburg, OH 43068
◆ Orders: 800 431-3939
◆ Info: 614 866-4267
www.universal-radio.com

Tigertronics SignaLink™ USB

Only **$99.95** + S/H

Ask about our NEW Plug & Play Jumper Modules!

www.tigertronics.com

Nothing beats the **SignaLink USB's** combination of performance, value, and ease of use! Whether you're new to Digital operation, or an experienced user, the **SignaLink USB's** built-in sound card, front panel controls, and simplified installation will get the job done right the first time—and without breaking the bank! The **SignaLink USB** supports all sound card digital and voice modes, and works with all radios. It is fully assembled (made right here in the USA!) and comes complete with printed manual, software, and all cables. Visit our website today for all of the exciting details!

Order Toll Free!
800-822-9722
541-474-6700

Tigertronics   154 Hillview Drive   Grants Pass, Oregon 97527

used - surplus - obsolete - buy - sell - swap - all for free
no fees ~ no commissions ~ it's free !
www.SecondHandRadio.com
if it's electronic or electrical - find it here or sell it here

transmitters, receivers, vfo's, power supplies, antennas, tuners, amplifiers, keys
relays, video, audio, vintage, capacitors, inductors, i.c.'s, diodes, transistors, tubes
microphones, headphones, speakers, hardware, knobs, dials, connectors, switches, lamps, jacks, plugs
GPS, filters, heatsinks, chokes, transformers, resistors, solder, terminals, insulators, schematics, books, mixers, tools, sockets
potentiometers, varactors, signal generators, frequency counters, swr meters, analyzers, transistors, broken radios, subassemblies, chassis, dummy loads
enclosures, encoders, adapters, oscillators, optoelectronics, sensors, industrial control, solenoids, fans, batteries, wire, cable, coax, oscilloscopes, actuators, brushes, blowers, arc tubes, resonators, clamp meters, isolators
fork terminals, fold shielding tape, footswitches, fuse testers, fuse pullers, fiber optics, indicating relays, infrared sensors, incandescent, joysticks, ethernet, knob locks, keyboards, mice, liquid crystal displays, labeling systems, limit switches, lenses, lasers, modems, photovoltaics

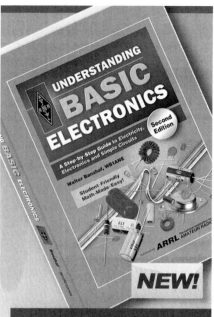

Understanding Basic Electronics
A Step-by-Step Guide to Electricity, Electronics, and Simple Circuits
Second Edition
By Walter Banzhaf, WB1ANE

Introducing **Understanding Basic Electronics**, second edition—your gateway into the exciting world of electricity and electronics. This book is written in a friendly, easy-to-understand style that beginners and nontechnical readers will enjoy. Real-world examples and clear illustrations make the study of electronics interesting and fun.

Contents:
- Electronics — What is it Good For? (Absolutely everything)
- Analog and Digital Electronic Circuits
- Electrical Terms: Voltage and Current
- Conductors, Insulators and Resistors
- Electricity and Magnetism
- Capacitors and Inductors
- Electrical Circuits — Series and Parallel
- Ohm's Law
- Energy and Power
- What is Alternating Current (AC)?
- Transformers
- Impedance
- Resonant Circuits
- Active Device Concepts: Semiconductors, Diodes, Transistors and ICs

...and much more!

Understanding Basic Electronics
ARRL Order No. 0823*
Special ARRL Member Price!
Only **$29.95*** (regular $32.95)

Also available:
Basic Antennas
ARRL Order No. 9994.....Only $29.95*
Basic Radio
ARRL Order No. 9558.....Only $29.95*

*plus shipping and handling

ARRL *The national association for* **AMATEUR RADIO**™
SHOP DIRECT *or call for a dealer near you.*
ONLINE WWW.ARRL.ORG/SHOP
ORDER TOLL-FREE 888/277-5289 (US)

HB 08/2010

ALPHA DELTA — The Leader of the "Pack"

Broadband (0-3 GHz) Coax Surge Protectors designed for peak efficiency and field reliability. The Model TT3G50 surge protector series offers a substantial improvement in performance and reliability compared to other designs, based on customer reports and military/government and commercial lab tests.

- **Broadband low loss design**, 0-3GHz (N types) in a single unit. No need for several bandpass models as in other designs.
- **Model TT3G50** precision cavity designs do **NOT** use, or need, internally soldered DC blocking components, or fragile neon type tubes. These can suffer field failures, requiring entire unit replacement.

- **DC blocked designs** require the entire unit to be removed from the sealed coax circuit if hit with a surge beyond its rating, and discarded. The Model TT3G50 stays in the coax circuit, solving a major field maintenance and cost issue. The Model **ARC-PLUG**™ hermetically sealed gas tube module in the Model TT3G50 is field replaceable with the twist of a knurled knob. No tools required!
- **Independent & MIL lab tests** show our design works as well or better than DC blocked designs. (Tested/approved by U.S. Navy, ARINC, Ft. Monmouth, U.S. Army Patriot Missile System, USAF, others)
- **Defense Logistics Agency (DLA)** has assigned NSNs (National Stock Numbers) to us after exhaustive MIL testing. See Cage Code 389A5.
- **Our design** permits control voltage thru-put instead of the "wire around" requirements of others.
- **ARC-PLUG**™ module and connectors **"O" ring sealed** for weather protection.
- **Various connector** combinations available. Manufactured in our **ISO-9001** certified **U.S. facility**, for highest quality, and **NOT** in some non-certified and often expensive overseas source.
- **Model ATT3G50** (200 watt rating, N connectors, 0 thru 3 GHz) .... $59.95 ea.
- **Model ATT3G50U** (200 watt rating, UHF connectors, 0 thru 500 MHz) ..... $49.95 ea.
- Add $10.00 s/h for U.S. orders. For OEM/bulk packed orders, use Model TT3G50 part numbers. For 2 kW rating, add suffix "HP" to part numbers. Same price. Call for OEM/export quotes.

Toll Free Order Line (888) 302-8777
(Add $10.00 ea. S/H in U.S., Exports quoted.)

ALPHA DELTA COMMUNICATIONS, INC.

www.alphadeltacom.com

 Call for addresses based on product category.
(606) 598-2029 • fax (606) 598-4413

Microcontroller Modules
8 & 32 bit, w/USB

Comm Modules
RS-232, Bluetooth USB

User I/O Modules
2x20 LCD, Keypad

And More

GREAT FOR:
- Home Projects
- Breadboards
- Experiments

BASIC On Board
Eliminates development tools

Schematics
Available on the web site

ATRIA Technologies Inc
www.AtriaTechnologies.com

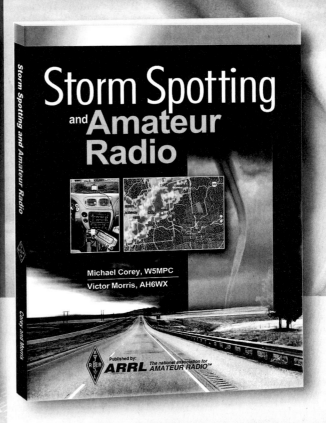

Top Performing Transceivers and Accessories from America's leading kit company.

The K3/P3
No other rig in this price class comes close to the K3's performance. It provides state-of-the-art performance as a primary home station, yet its size and weight make it ideal for DXpeditions, RV operation, and Field Day. Both Kit and Factory Assembled.

The K2
Our high-performance multiband 100W HF transceiver with features and specifications that exceed those of most high-end factory-assembled HF transceivers!

The KX1
An ultra-light, multi-band, transceiver with an internal battery and an automatic antenna tuner. This "CW-station-in-a-box" is a backpacker's dream!

The K1
A compact, high-performance 4 (K1-4) or 2 (K1-2) band CW transceiver kit that's easy to build. A great choice for first-time kit builders!

The T1 ATU
The pocket-size Elecraft stand-alone automatic antenna tuner. Can be used with any 0.5 to 20-watt HF transceiver! Available as a kit, or assembled!

Mini-module Kits
Great for first-time builders! This collection of miniature, practical station accessories includes antenna baluns, a dummy load, signal generators, and other test/alignment modules. See our web site for full list and details.

Visit our web page for details on our complete product line, on-line secure ordering, customer and magazine review reprints, downloadable manuals and more!

Hands-on Ham Radio™

www.elecraft.com • 831-662-8345
P.O. Box 69, Aptos, California 95001-0069

Elecraft is a registered trademark of Elecraft, Inc.

Be Prepared!

Visit HamPROS for some Great Deals on all the Latest Equipment

Photo courtesy of weatherphotographs.com

YAESU
Choice of the World's top DX'ers

It's Here! NEW!

FTDX-5000/D/MP Series
Shown above with SM5000 Station Monitor (optional on some versions).

FT-2000/FT-2000D
HF,6M, 100W/200W

NEW!

FT-950
100 Watts of power output on SSB, CW, and FM (25 Watts AM carrier).

FTM-350R
2M/440 50W
220MHz 1W
APRS & WIRES Ready
GPS (optional)

FT-450 & 450AT
state of the art
IF DSP technology,
world class performance,
easy to use

FT-897D
HF/6/2/440 all mode
Portable 100W HF/6,
50W 2M, 20W,
440MHz

FT-817ND
HF/6/2/440

FT-857D
HF/6/2/440 all mode

FTM-10R/10SR
144/440MHz 50/40W

FT-8800R
2M/440MHz

FT-8900R
10M/6M/2M/440MHz

FT-1900R
144-148MHz

FT-2900R
2 meter transceiver

FT-7900R
Dual bander

VX-8 Series

VX-8GR
Full 5 watts
144/440MHz
GPS unit and
Antenna
NEW!

VX-8DR
50/144/440MHz
GPS unit and
Antenna Optional,
Lithium Ion battery
VX-8DR is an APRS®
enhanced version
of the popular VX8R.

VX-3R
2M/440MHz

VX-7R/RB
Triple Band

FTDX-9000C/D/MP
400 watts, dual analog meter sets, LCD display,
memory card installed, main and sub receiver
VRF, full dual rx, external power supply.

ICOM

IC-7700
No compromise RX performance

IC-7600
USB, LSB, CW, RTTY,
PSK31, AM, FM

IC-7800
HF/6m @
200 Watts,
HF/50 MHz
All Band

IC-7200
HF/6 100W

IC-7000
HF/6M/2M/440
IF DSP

IC-718
HF 100W on HF,
VOX operation

D-STAR Compatible
ID-880H
2M/440 Dual Band

NEW!
IC-T70A
2M/440MHz
5W HT

IC-706MKIIG
HF/6M/2M/440

IC-R9500
144/430(440) MHz
and above 2GHz band,
+5dBm IP3 capability

D-STAR Compatible
IC-2820H
2M/440 dual bander

NEW!
IC-V80
2M 5.5W

*Restrictions apply, contact your HamPROS Dealer for details.

KENWOOD

TS-480HX/TS-480SAT
HF/6M
All-Mode Transceiver
100 Watts Output

TH-K2AT
2M HT, 5 Watts
w/NOAA warning

TM-271A
2M FM Transceiver,
136-174MHz
RX CTCSS

TM-V71A
2M/440MHz
50W Mobile

TH-F6A
144, 220
and 440

TS-2000/X
High performance

TM-D710A
2M/70cm
Mobile
50W/50W

Associated Radio
800-497-1457

Local (913) 381-5900
FAX (913) 648-3020
8012 Conser
Overland Park, KS 66204
www.associatedradio.com

Universal Radio, Inc.
800-431-3939

Local (614) 866-4267
FAX (614) 866-2339
6830 Americana Pkwy.,
Reynoldsburg, Ohio 43068
www.universal-radio.com

Lentini Communications, Inc.
800-666-0908

Local (860) 828-8005
FAX (860) 828-1170
221 Christian Lane – Unit A
Berlin, CT 06037
www.lenticomm.com

Austin Amateur Radio Supply
800-423-2604

Local (512) 454-2994
FAX (512) 454-3069
5325 North I-35
Austin, Texas 78723
www.aaradio.com

Radio City, Inc.
800-426-2891

Local (763) 786-4475
FAX (763) 786-6513
2663 County Road I
Mounds View, MN 55112
www.radioinc.com
Find us on Twitter:
twitter.com/mnradiocity

samlex america
High Quality
Photovoltaic
Modules in the
50W - 170W
Maximum
Output Power
Range

SEC-1223
23 Amp Switching Power Supply
This high efficiency UL-60950-1
approved AC-DC power converter

MasterCard, VISA, DISCOVER

We participate in all radio manufacturers' coupon, free accessory, and rebate programs, and stock a tremendous variety of radio and accessory products from essentially all of the major Amateur Radio manufacturers. If you don't see what you are looking for in this ad, give us a call!

Prices, products and policies may vary between dealer locations. Not all dealers have all product lines. All prices and policies subject to change. Not responsible for typographical errors.

We carry samlex america, ASTRON CORPORATION, MFJ, COMET, HEIL SOUND, LDG ELECTRONICS, ALINCO, W5YI and MORE!

Yaesu - offering the World's most complete line of home and field radios

We seek to combine the highest quality with the utmost in operating efficiency and technological advancement to enhance the final product that you will enjoy on the air.

HF/50 MHz Transceiver
FT DX 9000D 200 W Version

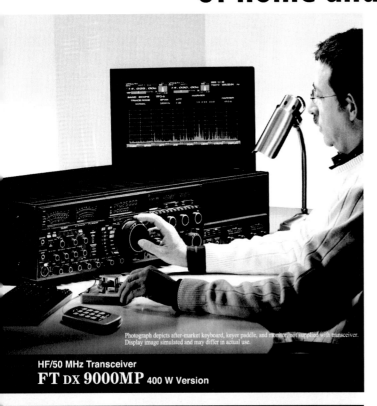

HF/50 MHz Transceiver
FT DX 9000MP 400 W Version

HF/50 MHz Transceiver
FT DX 9000 Contest Custom-Configurable Version

Loaded with Leading-edge Performance Capabilities...
The First Triumph in the 2nd Generation of the FT DX 9000 Lineage:
The Powerful FT-2000!

HF/50 MHz Tranceiver
FT 2000 100W Version (Internal Power Supply)

HF/50 MHz Transceiver
FT-2000D 200 W Version (External Power Supply)

Introducing the Yaesu FT-950 transceiver for DX enthusiasts Superb receiver performance
Direct lineage from the legendary FT DX 9000 and FT-2000

COMPACT HF TRANSCEIVER WITH IF DSP

Shown with after-market keyer paddle, keyboard, and monitor (not supplied). DMU-2000 Data Management Unit (option)

HF/50 MHz 100 W Transceiver
FT-950

Compact size: 9" X 3.3" x 8.5" and Light weight: 7.9 lb
HF/50 MHz 100 W All Mode Transceiver
FT-450 Automatic Antenna Tuner ATU-450 optional
■ **FT-450AT** With Built-in ATU-450 Automatic Antenna Tuner

YAESU
Choice of the World's top DX'ers

Vertex Standard
US Headquarters
10900 Walker Street
Cypress, CA 90630 (714)827-7600

For the latest Yaesu news, visit us on the Internet:
http://www.vertexstandard.com

Specifications subject to change without notice. Some accessories and/or options may be standard in certain areas. Frequency coverage may differ in some countries. Check with your local Yaesu Dealer for specific details.

The Hottest Field Gear Anywhere!

HF/VHF/UHF Portable Operation
Just Got a Lot More Powerful!
FT-897D TCXO DSP 60 m Band
HF/50/144/430 MHz
100 W All Mode Transceiver (144 MHz 50 W/430 MHz 20 W)

HF/VHF/UHF Multimode Mobile Transceiver,
now Including Built-in DSP
FT-857D DSP 60 m Band
HF/50/144/430 MHz
100 W All Mode Transceiver (144 MHz 50 W/430 MHz 20 W)

Automatic Matching for FT-897/857 Series Transceivers
FC-40
Automatic-Matching 200-Memory Antenna Tuner (160 m ~ 6 m Band)
WATERPROOF

Mobile Auto-Resonating 7~430 MHz for FT-897/857 Series Transceivers
ATAS-120A
Active Tuning Antenna System (no separate tuner required)

VHF/UHF Base RadialKit ATBK-100 for ATAS-120A.

ATAS-25 ATAS MICRO
Manually-Tuned Portable Antenna

REAL PERFORMANCE,
REALLY PORTABLE
FT-817ND
HF/50/144/430 MHz
5 W All Mode Transceiver (AM 1.5 W)
60 m Band

YAESU
Choice of the World's top DX'ers℠

Vertex Standard
US Headquarters
10900 Walker Street
Cypress, CA 90630 (714)827-7600

For the latest Yaesu news, visit us on the Internet:
http://www.vertexstandard.com

Specifications subject to change without notice. Some accessories and/or options may be standard in certain areas. Frequency coverage may differ in some countries. Check with your local Yaesu Dealer for specific details.

A TECHNOLOGY BREAKTHROUGH
New Advanced VX-8 Series GPS/APRS® Handheld Transceivers
Choose the Yaesu that meets your APRS® operating preferences in the field

50/144/(222)*/430 MHz
FM 5 W/AM 1W (50 MHz) Triple Band Handheld
VX-8DR *222 MHz/1.5 W (USA version)
(7.4V 1,100 mAh Lithium Ion battery/FNB-101LI and battery charger/NC-86A included)

144/430 MHz
FM 5 W Dual Band Handheld
VX-8GR
(7.4V 1,100 mAh Lithium Ion battery/FNB-101LI and battery charger/NC-86A included)

Actual Size

VX-8DR NEW
All-in-one Prestigious Tri-band Transceiver
Bluetooth® for hands-free Operation with optional accessories
Waterproof/Submersible IPX 7 rated - 3 ft for 30 minutes

VX-8GR NEW
144/430 MHz Dual Band Transceiver with GPS unit included
Built-in GPS Antenna - Waterproof
Wide Band Receive for 108-999 MHz (Cellular blocked – US Version)

Optional GPS and antenna unit for GPS/APRS® operation
Attached to the radio (microphone input) using the optional GPS Antenna Adapter CT-136

The optional GPS Antenna Unit FGPS-2 attached to the optional speaker Microphone MH-74A7A

 Bluetooth®

Supports APRS® communication by the Built-in Worldwide Standard AX.25 Data TNC

The VX-8 series radios are compatible with the world wide standard APRS® (Automatic Packet reporting System) using the GPS system to locate and exchange position information.

- SmartBeaconing™ Function
- Memories to list 50 stations
- Memories to store 30 APRS® messages
- DIGI-PATH routing indication function
- 8 DIGI-PATH routing settings
- GPS Compass Display - "Heading Up" or "North Up"
- APRS® Symbol Icon pre-set function
- Clearly displayed APRS® Beacon Messages
- Selective Message Received indicated by Flashing LED

APRS® is a registered trademark of Bob Bruninga WB4APR. SmartBeaconing™ from HamHUD Nichetronix

For the latest Yaesu news, visit us on the Internet:
http://www.yaesu.com

YAESU
Choice of the World's top DX'ers ℠

Vertex Standard US Headquarters
10900 Walker Street Cypress, CA 90630 (714) 827-7600

Specifications subject to change without notice. Some accessories and/or options may be standard in certain areas. Frequency coverage may differ in some countries. Check with your local Yaesu Dealer for specific details.

Heavy-Duty FM Dual Band Mobile with Exceptionally Wide Receiver Coverage*

*108 to 520 MHz/ 700 to 999.99 MHz (Cellular blocked)

• Separation Kit for Remote Mounting (optional separation kit YSK-7800 requires)

2 m/70 cm DUAL BAND

2 m/70 cm DUAL BAND FM TRANSCEIVER
FT-7900R new

The King of Mobile

75 WATTS

Commercial Grade Field Radio Submersible Construction

Compact Field Radio with Top Mounted LCD and Loud Audio

HEAVY-DUTY 75 W 2 m FM TRANSCEIVER
new **FT-2900R** 2 m MONO BAND

Best Selling, Reliable Mobile

55 WATTS

2 m MONO BAND

new

ULTRA-COMPACT
5 W 2 m FM HANDHELD TRANSCEIVER
FT-250R

ULTRA RUGGED 55 W 2 m FM TRANSCEIVER
new **FT-1900R** 2 m MONO BAND

VHF FM 5 W COMPACT HANDHELD TRANSCEIVER
FT-270R

2 m MONO BAND
new

YAESU
Choice of the World's top DX'ers

Vertex Standard
US Headquarters
10900 Walker Street

For the latest Yaesu news, visit us on the Internet:
http://www.yaesu.com

Specifications subject to change without notice. Some accessories and/or options may be standard in certain areas. Frequency coverage may differ in some countries. Check with your local Yaesu Dealer for specific details.

Index

Editor's Note: Except for commonly used phrases and abbreviations, topics are indexed by their noun names. Many topics are also cross-indexed.

The letters "ff" after a page number indicate coverage of the indexed topic on succeeding pages.

A separate Project index and Author index follow the main index.

555 timer IC: 3.56
8-level vestigial sideband (8-VSB): 8.18, 32.6
802.11 network: 4.29

A

A Index: 19.12
AADE filter design software: 11.33
Abbreviations list: 22.57
Absorption: 19.2
    Atmospheric: 19.23
    D-layer: 19.12
Absorption Glass Mat, battery (AGM): 7.37
AC circuit models: 5.18ff
AC coupling: 3.29
AC ground: 3.30
AC instrumaents: 25.8ff
    Average vs peak-reading: 25.8
    Bridges: 25.13
    Calorimetric meters: 25.8
    Dip meter: 25.16
    Directional wattmeters: 25.13
    Effective value, calibration: 25.8
    Measure inductance and capacitance with DVM: 25.20
    RF ammeter: 25.10
    RF power: 25.11
    RF voltage: 25.10
    Scale linearity: 25.9
AC power: 7.2ff
AC-DC power conversion: 7.4
    Comparison of rectifier circuits: 7.7
    Filter, capacitive and inductive: 7.5
    Full-wave bridge and center-tap rectifier: 7.6
    Half-wave rectifier: 7.5
    Ripple: 7.4
AC-line filter: 27.23
Active filter: 3.47
Adaptive filter (DSP): 15.17
Adaptive compression: 16.4
Adcock antenna: 21.74
Admittance (Y): 2.46
Advanced Circuits (PCB vendor): 6.32
AF (*see* Audio frequency)
AFSK (*see* Audio frequency shift keying)
AGC (*see* Automatic gain control)
ALC (*see* Automatic level control)
ALE (*see* Automatic link establishment)
Alias: 15.7
Alternating current (AC): 2.15
    Average values: 2.20
    Frequency: 2.16, 25.1ff
    Fundamental: 2.16
    Harmonic: 2.16
    Instantaneous values: 2.19
    Non-sinusoidal measurements: 2.21

Peak and peak-to-peak values: 2.19
Peak envelope power (PEP): 2.21
Peak power: 2.19
Period: 2.16
Periodic waveform: 2.16
Phase: 2.17
Root-mean-square (RMS): 2.20
Waveform measurements: 2.21
Waveforms: 2.15
Alternator: 2.16
Alternator transmitter: 13.2
Alternator whine: 27.39
Aluminum
    Alloy types and specifications: 22.41
    Tubing, standard sizes: 21.39
AM (*see* Amplitude modulation)
Amateur Radio
    American Radio Relay League (ARRL): 1.5
    Call sign: 1.6
    Contesting: 1.10
    Direction finding: 1.15
    DXing: 1.11
    Getting on the air: 1.8
    In the classroom: 1.17
    License: 1.3
    Log (on-air activity): 1.7
    Mobile operating: 1.14
    Modes: 1.9
    On-the-air activities: 1.9
    Operating awards: 1.12
    Public service: 1.15
    QRP: 1.13
    QSLs: 1.13
    Resources: 1.18
    Rules and regulations: 1.4
    Study guides: 1.4
    VHF, UHF, microwave: 1.14
Amateur television (ATV): 8.17, 32.1ff
    8-VSB: 8.18, 32.6
    Aircraft applications: 32.11
    Analog AM: 32.1, 32.4
    Antennas: 32.7
    Applications: 32.9ff
    Balloon flights: 32.12
    Communication distance: 32.2, 32.3
    Comparing ATV modes: 32.1
    Digital: 8.18, 32.2, 32.6, 32.7
    DVB-S: 8.18, 32.6
    DVB-T: 32.6
    Educational applications: 32.10
    Emergency communications: 32.9
    Equipment, AM: 32.5
    Equipment, FM: 32.6
    FM: 32.2, 32.5, 32.6
    Frequencies: 32.4

Glossary:	32.14
Identification:	32.8
NTSC channel:	8.17, 32.4
Part 15 receivers for UHF/microwaves:	32.6
Picture quality reporting system:	32.3
Radio control vehicles:	32.10ff
References:	32.23
Repeater:	32.9ff
Signal levels:	32.2
Vestigial sideband (VSB):	8.17, 32.4
Video sources:	32.13
American Radio Relay League (also see ARRL):	1.5
Ammeter, RF:	25.10
Ampere (A):	2.2
Ampere-hours (Ah):	2.11, 7.32
Amplification:	3.1
Linear:	17.2
Amplifier:	3.1
Bidirectional (WiFi):	31.16
Bode plot:	3.3
Buffer:	3.40, 9.20
Class:	3.30, 5.23, 17.2, 17.7ff, 17.29
Common-emitter/base/collector:	3.32, 3.33
Configuration:	3.30
Cutoff (corner) frequency:	3.5
Dynamic range:	3.2
Emitter-follower (EF):	3.32
Feedback:	5.23
Field-effect transistor:	3.37
Frequency response:	3.3, 3.29
Gain:	3.2
Half-power frequency:	3.5
High-frequency model:	3.30
Hybrid:	13.31
Input impedance:	3.33
Large-signal model:	3.30
Limiter:	3.5
Linear:	13.30, 17.2
Load line:	3.33
Log:	3.5
Low-frequency model:	3.30
Microphone:	13.13
Nonlinear:	13.31
Operating point:	3.33
Output impedance:	3.35
Quiescent (Q) point:	3.33
Slew rate:	3.2
Small-signal model:	3.30
Stability:	3.4
Summing:	3.4
TR switching:	14.18
Transconductance:	3.6
Transmitter:	13.30ff
Amplifier, RF power:	17.1ff
Amateur television (ATV), AM:	32.5
Automatic level control (ALC):	17.24
Broadband transformer:	17.29ff
Cold tuning:	17.11
Combiners and splitters:	17.31
Common cathode:	17.6
Component ratings:	17.13ff
Cooling methods:	17.18ff
DC blocking capacitor:	17.16
Earth-Moon-Earth (EME) communication:	30.40
Filament voltage:	17.16
Grid bias:	17.17
Grounded grid:	17.6
Harmonic rejection:	17.10ff
Impedance matching:	17.29ff
Low-pass filter design:	17.40ff
Maintenance:	26.21ff
Mobile operation:	29.12
Neutralization:	17.20ff
Parasitic oscillation:	17.22ff
Parasitic suppressor:	17.23
Pi network:	17.10
Pi-L network:	17.10
Plate voltage:	17.16
Protection and control circuits:	17.36ff
References:	17.56
RF choke:	17.15
Satellite ground station:	30.23
Screen voltage:	17.17
Solid state:	17.28ff
Solid state HF amplifier design:	17.32ff
Solid state power amplifier heat management:	17.43
Solid state, advantages:	17.28
Stabilization:	17.20ff, 17.31ff
Surplus parts:	17.44
Tank circuit:	17.9ff
Tank circuit design methods:	17.10ff
Transmission line transformer:	17.29
Troubleshooting:	26.21ff
Tube HF amplifier design:	17.24ff
Tuning:	17.26
Vacuum tube, advantages:	17.3
VHF/UHF tank circuit:	17.11
Amplitude modulation (AM):	3.5, 8.3, 10.5ff
Demodulation:	10.8
Modulator:	13.7
Receiver:	12.11
Transmitter:	13.7
AMSAT:	30.1
Analog (signal):	4.1
Analog switches and multiplexers:	3.57
Crosspoint switch:	3.57
Crosstalk:	3.57
Analog systems:	3.28
Analog signal:	3.1
Black box:	3.28
Buffering:	3.30
Cascading stages:	3.29
Interstage loading:	3.29
Network:	3.28
Port:	3.28
Signal processing:	3.1
Terminology:	3.6
Transfer function:	3.28
Analog-digital conversion:	3.49
Accuracy:	3.50
ADC, DAC:	3.49
Aliasing:	3.50, 15.7
Anti-aliasing filter:	15.7
Band-limiting:	3.50
Bandwidth:	3.50
Binary coded decimal (BCD):	3.49
Bipolar and Unipolar:	3.49
Choosing a converter:	3.55
Code:	3.49
Conversion rate:	3.50
Decimation:	15.8
Differential nonlinearity:	3.50
Digitization:	3.49
Effective number of bits (ENOB):	3.50
Full-scale (FS) value:	3.49
Full-scale error:	3.50
Hexadecimal:	3.49
Integral nonlinearity (INL):	3.50
Interpolation:	15.9
Jitter:	3.50
Least significant bit (LSB):	3.49
Linearity error:	3.50
Nyquist frequency:	3.50
Nyquist rate:	3.50
Nyquist Sampling Theorem:	3.50
Offset:	3.50
Over- and undersampling:	3.50
Percentage resolution:	3.49

Quantization:	15.9ff
Quantization code:	3.49
Quantization error:	3.50, 15.2, 15.10
Range:	3.49
Reconstruction filter:	15.8
Resolution:	3.49, 15.2
Sample:	3.50
Sample rate:	15.2, 15.6ff
Sampling:	15.6ff
Signal-to-noise and distortion ratio (SINAD):	3.50
Sinc function:	15.8
Spurious-free dynamic range (SFDR):	3.50, 15.2
Step size:	3.49
Total harmonic distortion+noise (THD+N):	3.50
Zero-stuffing:	15.9
Analog-to-digital converter (ADC):	15.2
Analog and digital ground:	3.53
Delta-encoded converters:	3.52
Dual-slope integrating converter:	3.52
Flash (direct-conversion) converter:	3.51
Input buffering and filtering:	3.53
Parallel I/O interfaces:	4.20
Sample-and-hold (S/H):	3.52
Serial I/O interfaces:	4.20
Sigma-delta converter:	3.52
Single-ended and differential inputs:	3.53
Successive-approximation converter (SAC):	3.51
AND gate:	4.5
Anderson PowerPole:	29.6
Angle, radians and degree-radian conversion:	2.17
Angle modulation:	10.9ff
Analog signals:	8.5
Demodulation:	10.12
Digital signals:	8.10
Equations:	10.9
Quadrature detection:	10.12
Transmitter:	13.10ff
Angular frequency (ω):	2.74
Ansoft Designer SV2:	6.2
Antenna	
Amateur television (ATV):	32.7
Angle of radiation:	21.5
ATV aboard aircraft:	32.12
Azimuth and elevation pattern:	21.4
Bandwidth:	21.3
Base loading:	21.29
Beam:	21.36
Beamwidth:	21.5
Cables and control wires on towers:	21.31
Capacitance hat:	21.29
Center loading:	21.28, 21.29
Construction, wire antennas:	21.8
Cross-polarization:	21.2
Delta loop:	21.42
Delta match:	21.61
Dipole:	21.6
Dipole, multiband:	20.6
Directivity:	21.1
E- and H-plane pattern:	21.4
Earth-Moon-Earth (EME) communication:	30.36ff
Effect of insulation:	21.8
Effects of ground:	21.3, 21.5, 21.26
End effect:	21.2, 21.6
Feed point:	21.7
Feed point impedance:	21.2
Folded dipole:	21.61
Front-to-back, front-to-side ratio:	21.4
Gain:	21.1
Gamma match:	21.27, 21.61
Ground plane:	21.27
Hairpin match:	21.38
Half-sloper:	21.33
High Speed Multimedia (HSMM):	31.16
Installation:	28.17
Inverted-L:	21.32
Inverted-V:	21.11
Isotropic radiator:	21.1
K factor:	21.3
Length-to-diameter ratio:	21.3
Lightning arrestor:	28.9
Loop:	21.49
Loop and halyard:	28.13
Losses, ground:	21.26
Magnet (mag) mount:	18.7
Marconi:	21.27
Mast material:	28.17
Mobile VHF/UHF:	21.59
Modeling:	21.14, 29.18
Monopole (unipole):	21.26
Multiband trap vertical:	21.31
National Electrical Code (NEC):	28.8
Near-vertical incidence skywave (NVIS):	21.5, 21.15
Omnidirectional:	21.5
Parasitic element:	21.37
Polarization:	21.1
Portable operation:	29.15ff
Quad:	21.42
Radials, ground:	21.26
Radiation pattern:	21.4
Radiation resistance:	21.2
Repeater operation (home station):	18.8
Safety:	28.13ff
Safety references:	28.20
Shunt-fed (series fed):	21.27
Slant-wire feed:	21.28
T:	21.32
Through-the-glass (mobile):	18.7
Top loading:	21.28, 21.29
Vertical:	21.26
VHF/UHF mobile:	18.7
Wire strength:	22.40
Zepp:	21.7
Zoning:	28.13
Antenna coupler:	20.13
Antenna switch	
Automatic:	24.14
Remote:	24.18
Antenna tuner (ATU):	20.13
Adjusting:	20.15
Location in the antenna system:	20.15
T configuration:	20.10, 20.14
Unbalanced-to-balanced:	20.13
Z-Match:	24.1
Antenna, mobile HF	
Base loading:	21.52
Center loading:	21.52
Efficiency:	21.58
Ground losses:	21.55
Matching:	21.56
Mounting:	21.55
Remotely controlled:	21.54, 21.57
Screwdriver:	21.54
Top loading:	21.52
Whip:	21.51
Whip, current distribution:	21.53
Antenna, satellite:	30.10ff
Crossed Yagi:	30.13
Eggbeater:	30.11
Helical:	30.14
Lindenblad:	30.14
Omnidirectional:	30.11
Parabolic dish:	30.15
Polarization:	30.10
Rotators:	30.27
Yagi:	30.13
Antenna, VHF/UHF	
Balun:	21.62
Circular polarization:	21.60

Effect of height:	21.60
Matching:	21.60
Radiation pattern:	21.60
Transmission lines:	21.60
Yagis:	21.64

Antenna, Yagi
Director:	21.37
Optimized designs:	21.40
Optimized wide-band antenna (OWA):	21.65
Plumbers delight construction:	21.38
Reflector:	21.37
Stacking:	21.64
VHF/UHF:	21.64
Yagi, construction:	21.38
Yagi, Yagi-Uda:	21.36ff

Anti-VOX:	14.18
Anticyclone:	19.24
APCO-25:	16.27, 31.19
Apparent power:	2.48
Application-specific integrated circuit (ASIC):	3.27, 15.3
APRS:	16.23, 31.9
Alias:	31.11
Automatic Identification System (AIS):	16.24
Beacon:	16.23
D-STAR and DPRS:	16.24
Digipeater:	31.12
Email:	16.24
GPS:	16.23
Igate:	16.23
Network:	31.11
Relay:	16.23
Secondary Station ID (SSID):	16.23
Setup:	31.10
Software:	16.23, 31.10
Tactical call:	16.23
Tracker:	16.23
Winlink:	16.24
Arcing:	5.10
Arithmetic logic unit (ALU):	15.4
ARRL:	1.5
RF Safety Committee:	28.21
Section Manager (SM):	27.2
Technical Coordinator (TC):	26.1, 27.2
Technical Information Service (TIS):	22.2
Technical Specialist (TS):	26.1
Volunteer Consulting Engineer (VCE):	28.1
ASAPS:	19.20, 19.21
Astrotex (shock mounts):	29.9
Asymmetrical waveforms:	25.8
Atmosphere:	19.6
Attenuation:	3.1
Conversion between reflection coefficient, SWR, and return loss:	22.56
Free space:	19.2
Of transmission lines:	20.4
Attenuation constant (a):	20.7
Attenuator:	3.3
Pi- and T-network values:	22.44
ATV (*see* Amateur television)	

Audio
Frequency response:	24.36
Radio interface:	24.40
Transmitted (sound card):	31.5
Audio amplifier, IC:	3.57
Audio equalizer, intelligibility enhancement:	24.36
Audio equipment RFI:	27.35
Audio frequency (AF):	2.18
Audio frequency (AF) oscillator:	25.19
Audio frequency shift keying (AFSK):	8.13, 31.4
Aurora:	19.14
Auroral E:	19.14
Authority having jurisdiction (AHJ):	28.2
Automatic gain control (AGC):	12.28ff, 14.7
Audio-derived:	12.30
DSP-based:	15.28
Pumping:	12.28
Response:	12.28
Time constant:	12.28
Automatic level control (ALC):	17.2, 17.24, 31.5
Solid state amplifier:	13.20
Transmitter:	13.19
Automatic link establishment (ALE):	16.15, 31.6
FSK:	16.15
Link quality analysis (LQA):	16.15
Scanning:	16.15
Automatic Packet/Position Reporting System (*see* APRS)	
Automatic repeat request (ARQ):	16.16
Automotive RFI:	27.37, 29.12
Alternator whine:	27.39
Electric vehicle (EV):	27.41
Hybrid-electric vehicle (HEV):	27.41
Ignition noise:	27.40
Autotransformer:	2.60, 2.65
Avalanche breakdown:	3.14
Average and peak reading:	25.8
Azimuth/elevation (az/el) antenna rotator:	30.17

B

Backlash:	9.18
Backscatter:	19.17
Balanced modulator:	8.5, 10.30ff, 13.8, 13.14
MC1496P:	10.33ff
Balloon, amateur television (ATV) applications:	32.11
Balun:	20.16, 21.8
Bazooka:	20.17
Bead:	20.24
Broadband:	20.23
Choke:	20.17ff
Coiled coaxial:	20.20
Current:	5.30, 20.18ff
Current (ferrite bead) balun:	20.19
Guanella:	5.29
Quarter-wave:	20.18
Transmission line:	5.29
Voltage:	20.18
Wound-coax ferrite choke:	20.24
Band stacking register:	14.7
Band-pass filter:	3.5
Bandset and bandspread:	9.17

Bandwidth
Antenna:	21.3
Digital signals:	8.16
Occupied:	8.20
Receiver:	12.13
Barkhausen criterion:	9.3
Battery:	7.32
Absorption Glass Mat (AGM):	7.37
Alkaline:	7.33
Battery pack:	7.33
Capacity:	7.32
Cell voltage:	7.32
Charging rate (C-rate):	7.35
Chemical hazards:	7.32
Cold-cranking amps (CCA):	7.37
Deep-cycle:	7.33, 7.37
Discharge curve:	7.34
Discharge planning:	7.34
Disposal and recycling:	7.33
Electrolyte:	7.33
Energy storage comparison:	7.34
Gel-cell:	7.33
Hydrometer:	7.33
Internal resistance:	7.34
Lead-acid:	7.33
Lithium:	7.33
Lithium-ion (Li-Ion):	7.34
Maintenance-free:	7.34

Marine: 7.37
  Mercury: 7.33
  Nickel-cadmium (NiCd): 7.33
  Nickel-metal hydride (NiMH): 7.34
  Primary: 7.33
  Reserve capacity (RC): 7.37
  Sealed lead-acid (SLA): 7.33
  Secondary: 7.33
  Self-discharge: 7.32
  Shelf-life: 7.32
  Silver oxide: 7.33
  Starter-lights-ignition (SLI): 7.37
  Vented-cell: 7.33
  Zinc-air: 7.33
 Battery charging
  Absorption mode: 7.53
  Bulk-mode: 7.53
  Constant-current charging: 7.36
  Fast-charge-rate charger: 7.36
  Float charge: 7.36, 7.53
  Overcharge mode: 7.36, 7.53
  Solar charging: 7.36
  Trickle charge: 7.35
 Battery Council Institute (BCI): 7.37
 Baud rate: 8.8
 Bazooka (balun): 20.17
 Beacon (radio propagation): 19.22
 Beat frequency: 2.17
 Beat frequency oscillator (BFO): 8.5, 12.4, 12.10
 Bermuda high: 19.25
 Beta (current gain): 3.18, 3.31
 Beta match: 20.10
 Bias (amplifier): 3.7, 17.2
  Bias point: 3.7
  Quiescent (Q) point: 3.7
 Bias (device), forward and reverse: 3.9
 Bilateral diode switch (*see* Diac)
 Bilateral networks: 5.26
 Bill of Materials (BOM): 6.33
 Binary
  Number: 4.2
  State: 4.3
 Binary coded decimal (BCD): 3.49, 4.3
 Binary phase-shift keying (BPSK): 8.10, 10.24
 Bipolar junction transistor (BJT): 3.18
  Active (linear) region: 3.18
  Breakdown region: 3.19
  Common-emitter/base/collector: 3.19, 3.30
  Current gain (β): 3.18
  Cutoff region: 3.18
  Dynamic emitter resistance, r_e: 3.32
  Early effect: 3.19
  Ebers-Moll model: 3.31
  Emitter, collector, and base: 3.18
  High-frequency model: 5.22
  Hybrid-pi model: 3.32, 5.22
  Model (CAD): 6.7
  Operating parameters: 3.19
  Phototransistor: 3.22
  Saturation region: 3.18
  Small-signal characteristics: 3.19
  Specifications: 22.25
 Bit rate: 8.14
 BJT (*see* Bipolar junction transistor)
 Bleeder resistor: 2.26, 7.13
  High-voltage supplies: 7.29, 7.30
 Blocking dynamic range (BDR): 12.25
 Blocking gain compression (BGC): 12.25
 BNC connector: 22.47, 29.7
 Bode plot: 3.3
 Boltzmann's constant (k): 5.31
 Boolean
  Algebra: 4.4
  Theorems: 4.6

 Boost converter: 7.23
 BPL (broadband over power lines): 27.5
 BPSK (*see* Binary phase-shift keying)
 Branch (circuit): 2.7
 Breakdown, dielectric: 5.10
 Break-in CW: 14.17
 Broadband noise: 27.13
 Buck converter: 7.21
  Critical inductance: 7.22
 Buck-boost converter: 7.23
 Buffer (digital), inverting and non-inverting: 4.6
 Buffer amplifier
  Cascode buffer: 3.40
  Darlington pair: 3.40
  Emitter-follower (EF): 3.40
  Source-follower: 3.40
 Bulk-mode charging: 7.53
  Bypass capacitor (RFI): 27.23
 Bypassing: 5.10
  Paralleling capacitors: 5.11

C

 Cable TV, frequencies and leakage: 27.32
 Calculators: 2.73
 Call sign: 1.6
 Calorimetric meters: 25.8
 Camera
  Charge-coupled device (CCD): 32.13
  CMOS: 32.14
  Slow-scan television (SSTV) images: 32.17
  Vidicon: 32.13
 Capacitance (C): 2.22
  Measurement with DVM: 25.20
  Motional: 9.25
 Capacitor: 2.22
  AC current and voltage: 2.26
  AC-rated: 27.23
  Axial and radial leads: 2.30, 5.5, 22.6
  Ceramic, disc and monolithic: 22.6
  Charging and discharging: 2.25
  Color codes, obsolete: 22.9
  DC blocking: 17.16
  Dielectric constant: 2.24
  Dissipation factor (DF): 2.28
  Equivalent (effective) series resistance (ESR): 2.28
  EIA identification and marking: 22.10
  Electrolytic, aluminum and tantalum: 22.7
  Equivalent (effective) series inductance (ESL): 5.5
  Film: 22.6
  High-voltage: 7.29
  Labeling: 22.8
  Loss angle or tangent (θ): 2.28
  Model (CAD): 6.5
  Parasitic inductance: 5.5
  Parasitic inductance model: 5.5
  Poly-chlorinated biphenyls (PCB): 7.30
  Ratings: 2.28
  Ratings, RF power amplifier: 17.14
  RC time constant (τ): 2.25
  Series and parallel capacitors: 2.24
  Standard values: 22.6
  Temperature characteristics: 22.7
  Temperature coefficient (tempco): 2.28
  Troubleshooting: 26.25
  Types: 2.29
  Voltage ratings: 22.9
 Capture effect: 8.7, 10.13
 Carabiner: 28.15
 Carrier squelch: 18.8
 Carrier-current device: 27.5
 Carson's rule: 8.6
 Cascode buffer: 3.40
 Cathode-ray tubes (CRT): 3.28

CDMA (*see* Code-division multiple access spread spectrum)
Ceramic microphone: 13.13
Cesium clock: 25.2
Characteristic (surge) impedance (Z_0): 20.3
Characteristic curve: 3.6, 17.4ff
Charge-coupled device (CCD): 3.28, 32.13
Chatter (comparator): 3.46
Chinook wind: 19.25
Chip rate: 8.14
Chip64 and Chip128: 16.10
    Spread spectrum: 16.10
Choke
    Common-mode: 27.23ff
Choke balun: 20.17ff
    Measuring impedance: 20.26
Chordal hop: 19.16
Circuit: 2.2ff
    Branches and nodes: 2.7
    Equivalent circuit: 2.9
    Open (short) circuit: 2.2
    Parallel (series) circuit: 2.2
Circuit breaker: 2.13, 7.2
    Ground fault circuit interrupter (GFCI or GFI): 7.2, 28.5
    Ratings: 2.13
Circuit simulation: 3.1, 6.1ff
    10-dB attenuator: 6.17
    7-MHz filter: 6.13ff, 6.17
    Active devices: 6.7
    Behavioral model: 6.24
    Capacitors: 6.5
    Defining gain: 6.15
    Device parameter sources: 6.9
    Inductors: 6.5
    Intermodulation modeling: 6.20
    Netlist: 6.10
    Phase-lead RC oscillator: 6.11ff
    Resistors: 6.3
    RF-fluent simulation: 6.17, 6.26
    S-parameters: 6.21
    Switchmode power conversion: 7.28
    Two-JFET VFO and buffer: 6.3ff
Circular mil: 22.39
Clamping (*see* Clipping)
Class (amplifier): 17.2, 17.7ff, 17.29
    A,B,AB,C,D,E,F: 3.30, 5.23
    Efficiency: 5.23
Clipper circuit: 3.17
Clipping: 3.5
Clock
    Frequency and period: 4.8
CLOVER (CLOVER-2000, -II): 16.19
    Coder efficiency: 16.20
    Modulation: 16.19
    Phase-shift modulation (PSM): 16.19
CMOS: 3.13
    Camera: 32.14
    Families: 4.17
    Integrated circuit (IC): 4.17
    Interface to TTL: 4.19
Coaxial cable (coax): 20.1
    Center insulator: 20.1
    Dielectric: 20.2
    Hardline: 20.1
    Jacket: 20.1, 20.6
    Shield: 20.1
    Types of: 20.6
Code-division multiple access spread spectrum (CDMA): 8.15
Codec: 16.4
Coded orthogonal frequency-division multiplexing
    (COFDM): 8.12, 32.20
Coded squelch (repeater): 18.8
COFDM (*see* Coded orthogonal frequency-division multiplexing)
Cohn ladder crystal filter: 11.32
Coil (*see* Inductor)

Collins system: 14.2
Color code
    AC wiring: 28.4
    Capacitor, obsolete: 22.9
    Diode: 22.16
    Inductor: 22.11
    Resistor: 22.3
    Transformer: 22.15
Combinational logic: 4.4ff
Combiner, power: 20.20
Common reference potential: 28.7
Common-base (CB) amplifier: 3.32, 3.36
    Current gain: 3.37
    Input impedance: 3.37
    Output impedance: 3.37
Common-collector (CC) amplifier: 3.32
    Input impedance: 3.36
    Output impedance: 3.36
    Power gain: 3.36
Common-drain (CD) amplifier: 3.39
    Output impedance: 3.39
Common-emitter (CE) amplifier: 3.32
    AC performance: 3.35, 3.36
    Frequency response: 3.36
    Input impedance: 3.33, 3.36
    Output impedance: 3.35
    Power gain: 3.36
    Voltage gain: 3.36
Common-gate (CG) amplifier: 3.39
    Input impedance: 3.40
    Output impedance: 3.40
    Voltage gain: 3.39
Common-mode: 27.9, 27.22
    Current: 20.16
    Common-mode rejection ratio (CMRR): 3.43
Common-source (CS) amplifier: 3.38
    AC performance: 3.39
    Input impedance: 3.38
    Output impedance: 3.38
    Self-bias: 3.39
    Source bypass: 3.39
    Source degeneration: 3.38
    Voltage gain: 3.38
Comparator: 3.45
    Hysteresis: 3.46
Compensation (frequency): 3.43
Compensation (network): 3.4
Complex frequency: 9.43
Complex loss coefficient (h): 20.7
Complex numbers: 2.74
Component models (CAD): 6.3ff
Compression (data): 16.4
Compression (signal): 3.5
Computer-aided circuit design (CAD) (*see* Circuit simulation)
Conductance (G): 2.3
    Leakage: 5.3
    Series and parallel: 2.8
    Siemens (S): 2.3
Conducted emissions: 27.6
Conduction current: 5.5
Conductor: 2.1
Connector: 29.5ff
    Audio: 29.6
    BNC: 22.47
    Coaxial, types: 22.50
    Computer, pinouts: 22.44
    Crimp: 29.6
    Data: 29.7
    F: 22.52
    N: 22.51
    Power, Anderson PowerPole: 29.6
    RF: 29.7
    RJ: 29.8
    UHF: 22.46

    Weatherproofing: 28.18
Constant-current charging: 7.36
Constellation diagram: 15.24
Construction techniques
    Circuit layout: 23.24
    Crimping tools: 23.6
    Drill sizes: 23.6
    Drilling: 23.30
    Electrostatic discharge (ESD): 23.13
    Enclosure fabrication: 23.30
    Ground-plane (dead-bug, ugly): 23.14
    High-voltage: 23.27
    Metalworking: 23.30
    Microwave: 23.29
    Nibbling tool: 23.5
    Painting: 23.31
    Panel layout: 23.31
    PCB fabrication services: 6.33ff, 23.19
    Perforated board: 23.15
    Point-to-point: 23.14
    Printed-circuit board (PCB): 23.17ff
    Purchasing parts: 23.18
    Socket or chassis punch: 23.6
    Soldering: 23.7
    Solderless breadboard or prototype board: 23.16
    Surface-mount technology (SMT): 23.9
    Terminal and wire: 23.15
    Tools: 23.4
    Winding coils: 23.28
    Wire-wrap: 23.16
    Wired traces (lazy PC board): 23.15
    Wiring: 23.27
Consumer Electronics Association (CEA): 27.17
Contesting (radiosport): 1.10
Continuous conduction mode (CCM): 7.22
Continuous tone-coded squelch system (CTCSS): 18.6, 18.8
    Tone frequencies: 18.8
Continuous wave (*see* CW)
Control interface
    Audio: 24.40
    Serial, RS-232: 24.25
    Transceiver: 24.22
    USB: 24.26
Control loop: 9.34
Conventional current: 2.2
Conversion
    Attenuation, SWR, return loss: 22.56
    Decibels and percentage: 2.73
    Degrees and radians: 2.17
    Power and voltage: 22.55
Conversion loss: 10.21
Convolution: 15.12
Cooperative agreement, utilities: 27.21
Coordination (repeater): 18.9
Corona discharge: 19.31, 27.21
Coronal hole: 19.11
Coronal mass ejection (CME): 19.11
Coulomb (C): 2.2
Counter
    Flip-flops: 4.11
    Frequency: 25.18
Coupling (inductive): 2.36
Courtesy tone: 18.4
Crimp connector: 29.6
Critical frequency: 19.8
Cross-reference, semiconductor: 3.8
Crosspoint switch: 3.57
Crosstalk: 3.57
Crowbar overvoltage protection (OVP): 7.20
Crystal detector: 12.2
Crystal, quartz: 9.23
    AT-cut: 9.23, 11.30
    Characterization: 11.33
    Control of transmitter: 13.2
    Equivalent circuit: 9.25, 11.30
    Equivalent (effective) series resistance (ESR): ... 9.26, 11.30
    Fundamental: 11.30
    Holders: 22.38
    Motional capacitance and inductance: 9.25
CTCSS (*see* Continuous tone-coded squelch system): 18.6
Current (I): 2.1
    Ampere (A): 2.2
    Conventional current: 2.2
    Current source: 2.9
    Electronic current: 2.2
Current balun: 5.30
Current divider: 2.7
Current gain (β): 3.18
Current inrush: 7.12
Current transfer ratio (CTR): 3.24
Cutoff (corner) frequency: 3.5
Cutoff (semiconductor): 3.11
CW: 8.7, 9.3
    Keying waveform: 13.6, 13.17ff
    Modulation: 10.6
    Receiver: 12.12
    Signal spectrum: 13.6
    Transmitter: 13.3, 13.5ff
    Waveshaping: 8.8

D
D'Arsonval: 25.3ff
D layer propagation: 19.12
D-STAR: 16.24, 16.26, 18.11ff, 31.18
    Backbone: 16.26
    Call sign routing: 16.26
    d*Chat: 16.27
    D-RATS: 16.27
    D-STAR TV: 16.27
    DD mode: 31.19
    Digital data (DD): 16.26
    Digital voice (DV): 16.26
    DPlus software: 16.27, 18.13
    DPRS: 16.27
    DV Dongle: 16.27
    DV mode: 31.18
    Ethernet bridge: 16.26
    Gateway: 16.26
    Hidden transmitter: 16.26
    ICOM: 16.27
    JARL: 16.26
    Network overview: 18.12
    References: 18.14
    Repeater hardware: 18.13
    Station ID: 18.11
    Station routing: 18.11
    Texas Interconnect Team: 16.26
    Trust server: 16.26
    Use with simplex FM radios: 16.27
D-type flip-flop: 4.10
Damping factor: 9.47
Darlington pair: 3.40
    Pass transistor: 7.16
Data compression: 16.4ff
Data converter: 15.2
Data sheet: 3.7
dBc: 9.8
dBd: 21.1
dBi: 21.1
DC channel resistance, r_{DS}: 3.19
DC coupling: 3.29
DC instruments: 25.3ff
    Accuracy: 25.4
    Basic meters: 25.3ff
    Current ranges: 25.4
    DC power: 25.6
    Extending current range: 25.4

Resistance measurement:	25.6
Shunts:	25.4
Voltmeters:	25.5ff

DCS (*see* Digital coded squelch)

Decibels:	2.72
Decimation:	15.8
Decoupling:	5.11
Deep-cycle battery:	7.37
Demodulator:	10.1ff
I/Q:	8.12, 15.22ff
DeMorgan's Theorem:	4.7
Depletion region:	3.9
Detection:	10.8
Detector:	12.2
Amplitude modulation (AM), DSP-based:	15.29
Quadrature:	10.12, 11.29
Tayloe:	11.29
Determinant:	5.32
Deviation:	10.9
Frequency:	8.6
Ratio:	8.6
Diac:	3.11
Diamagnetic:	2.31
Dielectric breakdown and arcing (strength):	5.10
Dielectric constant:	2.24
Transmission line:	20.2, 20.8
Difference amplifier:	3.45
Differential amplifier:	3.45
Differential phase-shift keying (DPSK):	8.10
Differential quadrature phase-shift keying (DQPSK):	8.12
Differential-mode (signal, current):	20.19, 27.9, 27.22
Digipeater:	16.23, 31.12
Digital (signal):	3.1, 4.1, 15.6

Digital audio

Audio formats:	16.3
Bit depth:	16.3
Bit rate:	16.4
Codec:	16.4
MIDI:	16.3
MP3:	16.4
Digital coded squelch (DCS):	18.9

Digital communications

AFSK	31.4
Amateur television (ATV):	32.2, 32.6, 32.7
APCO-25:	31.19
Automatic link establishment (ALE):	31.6, 31.17
Automatic Packet/Position Reporting System (APRS):	31.9
Channel capacity:	8.17
D-STAR:	31.18
Earth-Moon-Earth:	30.42
EchoLink:	31.22
FSK:	16.5, 31.4
Glossary:	31.25
High Speed Multimedia (HSMM):	31.13
Internet Radio Linking Project (IRLP):	31.23
Modulation:	8.7
National Traffic System (NTS):	31.6
P25:	31.19
Packet radio:	31.7ff
PACTOR:	31.12
Receiver:	12.12
References:	31.26
Slow-scan television (SSTV):	32.20ff
Sound card modes:	31.1
Voice (HF):	31.21
Voice over Internet Protocol (VoIP):	31.22
Winlink 2000:	31.6
WIRES-II:	31.24
Digital data (mode):	16.1
Automatic repeat request (ARQ):	16.2
Baud:	16.1
Baudot code:	16.2
Bit, bit rate (bps):	16.1
Bridge:	16.23
Checksum:	16.2
Code pages:	16.3
Control characters:	16.3
Cyclical redundancy check (CRC):	16.2
Data rate:	16.1
Dibit:	16.8
Emission designator:	16.1
Encryption:	16.5
Error detection and correction:	16.2
Error-correcting code (ECC):	16.2
Forward error correction (FEC):	16.2
Gray code:	16.9
Gross and net bit rate:	16.1
Hybrid ARQ:	16.2
Parity:	16.2
Symbol, symbol rate:	16.1
Throughput (bps):	16.2
Unicode:	16.3
Varicode:	16.2, 16.8
Viterbi algorithm, coding:	16.8
Digital downconverter (IC):	15.27

Digital image

Color depth:	16.3
High, true color:	16.3
JPEG:	16.4
Palette:	16.3
Pixel:	16.3
Raster:	16.3
Vector:	16.3

Digital logic

3.3 V – 5 V interface:	4.19
Bipolar families:	4.14
Combinational:	4.4ff
Comparing families of:	4.13
Families:	22.31
Gates:	4.5
Positive and negative:	4.7
Programmable devices:	4.22
Sequential:	4.8ff
Transistor-transistor logic (TTL):	4.14
Digital signal processing (DSP):	15.1ff
Adaptive filter:	15.17
Analytic signals:	15.21ff
Automatic gain control (AGC):	15.28
Embedded systems:	15.6
Filters:	15.11ff
Finite impulse response (FIR) filter:	15.11ff
Glossary:	15.32
Hilbert transformer:	15.23
I/Q demodulator:	15.22ff
I/Q modulator:	15.22ff
Infinite impulse response (IIR):	15.16ff
Microprocessors:	15.4
Modulation:	15.21ff
Modulator and demodulator:	10.14, 15.29
Receiver filter:	12.22
Receivers, overview:	12.6
References:	15.31
Signal generation:	13.29
Sine wave generation:	15.17ff
Speech processing:	15.30
SSB modulator and demodulator:	15.22ff
Tone decoder:	15.19
Use in transceivers:	14.4
Digital video:	16.3
Bit rate:	16.4
Codec:	16.5
Digital voice, HF:	16.15
AOR:	16.15
Dream:	16.15
FDMDV:	16.15
WinDRM:	16.15
Digital-to-analog converter (DAC):	15.2
Binary weighting:	3.54

Current output DAC: 3.54
Monotonicity: 3.54
Parallel I/O interfaces: 4.20
R-2R ladder DAC: 3.54
Serial I/O interfaces: 4.20
Settling time: 3.54
Summing: 3.54
Diode: 3.14, 22.15
AC circuit model: 5.19
Anode: 3.14
At high frequencies: 5.19
Cathode: 3.14
Color code: 22.16
Dynamic resistance: 5.19
Fast-recovery: 3.14, 7.11
Forward voltage: 3.15
Free-wheeling: 7.5, 7.22
Gunn: 9.32
Package dimensions: 22.17
Peak Inverse Voltage (PIV, PRV): 3.14
Photodiode: 3.22
Point-contact: 3.15
Ratings: 3.14
Schottky: 3.15, 7.11
Specifications: 22.16
Switching time: 5.19
Transient voltage suppressor (TVS): 3.16
Troubleshooting: 26.27
Vacuum tube: 17.4
Varactor: 3.15, 9.38
Zener: 3.16
Diode mixer: 10.18ff
Dip (grid dip) meter: 25.16, 26.3
Diplexer: 10.22ff
Diplexer Designer by W4ENE: 11.48
Dipole
Folded dipole: 21.12
Loaded: 21.11
Multiband: 21.13
Off-center-fed (OCF): 21.7
Sloping: 21.11
Trap: 21.13
Twin-folded, terminated (TFTD): 21.12
Vertical: 21.11, 21.33
Direct conversion receiver: 12.4ff
Direct current (DC): 2.15
Direct digital synthesis (DDS): 9.33, 9.50, 15.17, 15.28
Direct FM: 10.11
Direct sequence spread spectrum (DSSS): 8.15
Direction finding: 1.15
Directional wattmeter (*see also* Reflected power meter): 25.12
Discontinuous conduction mode (DCM): 7.22
Dish antenna, parabolic: 30.15
Displacement current: 5.5
Dissipation (power): 2.11
Distributed elements: 5.1
Dithering: 15.10
DominoEX: 16.10
Doppler shift
Earth-Moon-Earth (EME) communication: 30.31
Mobile communication: 19.27
Satellite communication: 30.4, 30.7, 30.21
Double conversion (receiver): 12.14
Double sideband (DSB): 8.4, 10.7, 13.7
Double-balanced mixer: 10.20ff
Downconverter: 30.19, 32.5
Dplus (*see* D-STAR)
DPRS (*see* D-STAR)
DPSK (*see* Differential phase-shift keying)
DQPSK (*see* Differential quadrature phase-shift keying)
Drift velocity: 3.12
Driver array IC: 3.55
DSP (*see* Digital signal processing)
DTMF (*see* Dual-tone multi-frequency)

Dual VFO: 14.6
Dual-tone multi-frequency (DTMF): 18.9
Tone frequencies: 18.9
Ducting: 19.24ff
Dummy load: 26.3
Duplexer: 18.4
DVB-S: 8.18, 32.6
DVB-T: 32.6
DVD (Digital Video Disk) player RFI: 27.34
DVM (digital voltmeter): 25.3ff, 25.4
Measuring inductance and capacitance: 25.20
DXing: 1.11
Dynamic microphone: 13.13
Dynamic range: 3.2, 12.1
Blocking: 12.25
Intermodulation distortion (IMD): 10.14, 10.16, 12.25
Receiver: 12.25ff
Test procedures, receiver: 10.16, 25.40
Dynamic resistance: 2.57, 5.19

E

E layer propagation (E skip): 19.12ff, 19.13
Eagle (CAD software): 6.2, 6.4, 6.31
Earth-Moon-Earth (EME) communication: 30.30ff
Antennas: 30.34, 30.36ff
Atmospheric and ionospheric effects: 30.32
Background noise: 30.33
Building a station: 30.36ff
Coding and modulation: 30.35
Doppler shift: 30.31
Echo delay: 30.31
Faraday rotation: 30.32
Feed lines: 30.39
Frequency spread: 30.31
Ground gain: 30.35
JT65: 30.42
Libration fading: 30.32
Operating procedures: 30.42
Path loss: 30.31
Power, transmitter: 30.34
Preamplifiers: 30.39
Propagation: 19.30, 30.31ff
References: 30.39
Resources (software, Internet): 30.37
Sky temperature: 30.33
System evaluation: 30.41
Time spread: 30.31
TR switching: 30.39
Tracking the moon: 30.41
Transmitter requirements: 30.40
EchoLink: 18.10, 31.22
Eddy current: 2.38, 2.64
Education
Amateur Radio in the classroom: 1.17
Amateur television (ATV) applications: 32.10
Effective series resistance (ESR): 2.28
Effective sunspot number (SSNe): 19.21
Efficiency (Eff, η): 2.11
Eggbeater antenna: 30.11
EHF (frequency range): 2.19
EIA-RS-232: 4.28
Electret microphone: 13.13
Electric charge: 2.1
Coulombs (C): 2.2
Electric fence RFI 27.15
Electric field: 21.1
Electromagnetic wave: 19.1
Electrostatic field: 2.22
RF exposure: 28.24
Electric vehicle (EV) RFI: 27.41
Electrical length (L_e, transmission line): 20.8
Electroluminescence: 3.23
Electromagnetic

Radiation: 19.1
    Spectrum: 19.1
    Wave: 19.1
Electromagnetic compatibility (EMC): 27.1
Electromagnetic interference (EMI): 27.1
Electromotive force (EMF): 2.1
    Voltage (V, E): 2.1
Electron: 2.1
    Free electron: 2.1, 3.8
    Mobility: 3.12
Electronic current: 2.2
Electronic design automation (EDA) (*see* Circuit simulation)
Electrostatic discharge (ESD): 3.21
    Prevention: 23.13
Elsie by W4ENE: 11.9
Embedded system (DSP): 15.6
EME (*see* Earth-Moon-Earth communication)
Emergency communication: 1.15
    Amateur television (ATV) applications: 32.9
    Using repeaters: 18.3
Emission designators: 8.1
    Table, ITU Radio Regulations: 8.2
Emitter-follower (*see* Common-collector amplifier)
Energy and work: 2.2, 2.11
    Ampere-hours (Ah): 2.11
    Energy density: 2.11
    Horsepower (hp): 2.11
    Joule (J): 2.2
    Potential energy: 2.22
    Watt (W): 2.11
    Watt-hour (Wh): 2.11
Envelope detection: 10.8
Equator, geomagnetic: 19.18
Equipment
    Arranging a station: 29.3
    Interconnection: 29.5
    Mobile mount: 29.9
Equivalent (effective) series resistance (ESR)
    Capacitor: 2.28
    Quartz crystal: 9.26, 11.30
Equivalent (effective) series inductance (ESL): 5.5
Ethernet: 4.29
    Bridge: 29.21
    RFI: 27.13
Express PCB (PCB vendor): 6.32
Eye diagram: 8.21

F

F connector: 22.52
F layer propagation: 19.16ff
Facsimile (Fax): 8.18, 16.12
    Formats: 16.12
    Index of cooperation: 16.12
Fading
    Digital signals: 19.28
    Ionospheric: 19.9
    Tropospheric: 19.26
FAI: 19.13
Fair-Rite Products Corp: 5.14
Fan-out: 4.14
Farad (F): 2.23
Faraday rotation: 19.19, 30.32
Fast Fourier transform (FFT): 8.12, 15.18, 16.10
Fast-scan television (FSTV) (*see* Amateur television)
FCC
    Citation: 27.21
    Field Inspector: 27.21
    Interference Handbook: 27.34
    Notice of Apparent Liability (NAL): 27.21
    RFI: 27.2ff
FCC Part 15: 27.4
    Absolute emission limits: 27.7
    Certification: 27.9
    Declaration of conformity: 27.9
    Notification: 27.9
    Operator requirements: 27.9
    RFI rules and definitions: 27.8
FCC Part 18: 27.4
FCC Part 2 definitions: 27.6
FCC Part 97: 27.2ff
    Spurious emission limits: 27.4
Feed line (*see* Transmission line)
Feed line radiation: 27.24
Feed point impedance: 20.7
Feedback: 3.4
Feld-Hell: 16.12
Ferri- and ferromagnetic: 2.31
Ferrite: 5.14
    Bead: 5.15
    Chokes: 5.18
    Equivalent circuits: 5.15, 5.17
    Permeability: 5.14
    Permeability vs frequency: 5.15
    Resonances of cores: 5.16
    Toroid cores: 22.14
    Type 31 material: 5.18
    Type or mix: 5.14, 22.11
    Use for EMI suppression: 5.18
FET (*see* Field-effect transistor)
FFT (*see* Fast Fourier transform)
Fiber optics: 3.24
Field
    Electric (electrostatic): 2.22, 19.1
    Magnetic: 2.31, 19.1
Field Day stub assembly: 20.9
Field programmable gate array (FPGA): 4.22, 15.3
Field-aligned irregularities (FAI): 19.13
Field-effect transistor (FET): 3.12
    Breakdown region: 3.20
    Channel: 3.12
    CMOS, NMOS, PMOS: 3.13
    Common-source/gate/drain: 3.20, 3.38
    Cutoff region: 3.20
    DC channel resistance, r_{DS}: 3.19
    Depletion mode: 3.13
    Enhancement mode: 3.13
    Forward transconductance, g_m: 3.19
    Gate leakage current, I_G: 3.19
    High-frequency circuit model: 5.22
    Junction breakpoint voltage: 3.19
    Junction FET (JFET): 3.12
    Mixer: 10.17ff
    Model (CAD): 6.8
    MOSFET: 3.13
    N-channel, P-channel: 3.12
    Ohmic region: 3.20
    On resistance, $r_{DS(ON)}$: 3.12
    Operating parameters: 3.20
    Phototransistor: 3.22
    Pinch-off: 3.13
    Saturation region: 3.20
    Small-signal model: 5.22
    Source, drain, and gate: 3.12
    Specifications: 22.24
    Troubleshooting: 26.29
Filter: 3.4
    Active: 3.47
    Active filter design tools: 11.21
    Active RC: 11.18, 11.20
    Adaptive (DSP): 15.17
    All-pass: 3.5
    All-pass active: 11.28
    Anti-aliasing: 15.7
    Attenuation: 11.2
    Band-pass: 3.5, 11.2
    Band-stop (band-reject): 11.2
    Bandwidth: 11.3, 11.6

Bessel (constant delay): 11.5
Brick wall response: 11.3
Butterworth (maximally flat): 11.3
Capacitor- and inductor-input: 11.6
Cauer: 11.4
Center frequency: 11.6
Chebyshev (equiripple): 11.3, 11.31
Cutoff frequency: 11.2
Denormalizing (scaling): 11.15
Digital: 15.11ff
Digital commutating: 10.28
Digital signals: 8.16
DSP: 11.19
DSP, use in transceivers: 14.4
Effect of component Q: 11.9
Effect of ripple: 11.11
Elliptic-function: 11.4
Family: 11.3
Family comparison: 11.5
Finite impulse response (FIR): 15.11ff
Group delay: 11.4
Helical resonator: 11.39
High-pass: 3.5, 11.2
IF: 13.15
Infinite impulse response (IIR): 15.16ff
Ladder: 11.6
Linear phase response: 11.32
Loop: 9.35
Low-pass: 3.5, 11.2
Lumped-element (LC): 11.5
Magnitude response: 11.2
Mesh capacitor-coupled: 11.9
Mid-band response: 3.5
Multiple-feedback: 11.20
Nodal-capacitor- and -inductor-coupled: 11.8
Normalized: 11.15
Normalized value tables: 11.12
Notch: 11.2
Order: 11.3
Overshoot: 11.5
Passband: 3.5, 11.2
Phase response: 11.2
Reconstruction: 15.8
Return loss: 11.11
RFI: 27.22
Ringing: 11.5
Ripple: 11.3
Ripple bandwidth: 11.3
Roll-off: 3.5
Rolloff: 11.3
Roofing: 12.15, 12.25, 12.27, 14.5
Sallen-Key: 11.20
Scaling, frequency and impedance: 11.15
Shape, response: 11.10
Stop band: 3.5, 11.2
Stop band depth and frequency: 11.4
Switched-capacitor (SCAF): 11.19
Topology: 11.7
Transformation, high-pass to band-stop: 11.7
Transformation, low-pass to band-pass: 11.6
Transition region: 11.2
Ultimate attenuation: 11.2
Use at VHF and UHF: 11.42
Vestigial sideband (VSB): 32.4
Filter, crystal: 11.30, 12.15ff
Cohn minimum-loss (min-loss): 11.32
Group delay: 11.32
Half-lattice: 11.31
Insertion loss: 11.31
Monolithic: 11.31
Motional parameters: 11.30
Overtone: 11.30
Parallel-resonator-end-section (PRES): 11.32
Pole-zero separation (PZ): 11.30

Shape factor: 11.31
Unloaded Q: 11.30
Filter, surface acoustic wave (SAW): 11.35
Filter, transmission line: 11.35
Band-pass: 11.37
Hairpin resonator: 11.37
Interdigital: 11.37
Microstrip: 11.35
Quarter-wave: 11.37
Stripline: 11.35
FilterPro by Texas Instruments: 11.21
Finite impulse response (FIR) filter: 15.11ff
Coefficient: 15.14
Tap: 15.12
Window function: 15.13
Window function equations: 15.15
Fire extinguisher: 29.4
FireWire (IEEE-1394): 4.29
Flat topping: 17.2
Flip-flop (digital): 4.9ff
Float charging: 7.36
Flyback converter: 7.23
Flywheel effect: 17.9
Foldback current limiting: 7.17
Footprint (satellite): 30.9
Forecast
    Propagation: 19.19ff
    Tropospheric ducting: 19.26
Forward bias: 3.9
Forward converter: 7.25
Forward resistance: 3.9
Fourier transform: 15.18
FR-4 (PC board): 6.33
Free space attenuation: 19.2
Frequencies
    Beacon: 19.22
    WiFi - shared Amateur Radio channels: 31.14
Frequency: 2.16
    Calibration: 25.2
    Counters: 25.18
    Deviation: 8.6, 10.9
    Discrimination: 10.11
    Marker generators: 25.14ff
    Measurement: 25.14ff
    Multiplier: 10.10
Frequency coordination
    Amateur television (ATV): 32.4
    Repeater: 18.9
Frequency hopping spread spectrum (FHSS): 8.15
Frequency modulation (FM): 8.6, 10.9
    Equipment, FM voice: 18.6
    Receiver: 12.12, 12.32ff
    Transmitter: 13.10
Frequency multiplier: 13.11
Frequency response: 3.3, 3.29
    Phase-locked loop (PLL): 9.37
Frequency shift keying (FSK): 8.10, 16.5, 31.4
FSK441: 16.12
    Bit sequences and messages: 16.13
    Meteor scatter: 16.12
    Ping Jockey: 16.13
    Synchronization to WWV: 16.13
FSTV (*see* Amateur television)
FTP: 16.23
Full-bridge converter: 7.25
    Quasi-square wave: 7.25
Full-wave bridge, center-tap rectifier: 7.6
    Output voltage: 7.6
    Voltage ratings: 7.6
Fundamental (signal): 2.16
Fundamental diode equation: 3.9
Fundamental overload: 27.11ff
Fuse: 2.13, 7.2
    Fuse holders: 2.14

High-voltage: 7.31
Ratings: 2.13

G

G-TOR: 16.17
   Golay FEC: 16.17
   Huffman A and B coding: 16.17
GaAs FET: 3.22
Gain
   Amplifier: 3.2, 3.6
   Antenna: 21.1
   Defining (circuit simulation): 6.15
   Noise: 5.31
   Voltage, current, and power: 3.2
Gain-bandwidth product (F_T, GBW): 3.29, 3.43, 5.22
Gallium Arsenide (GaAs): 3.12
Gap noise circuit: 27.14
Gate (digital): 4.5
Gate leakage current, I_G: 3.19
Gateway (D-STAR): 18.11
Gauss (G): 2.32
Gaussian filter: 8.10
Gaussian minimum-shift keying (GMSK): 8.10
GC Prevue (CAD software): 6.40
gEDA (CAD software): 6.2, 6.4
Generator, power: 28.5, 29.13ff
Generic array logic (GAL): 4.22, 15.3
Geomagnetic equator: 19.18
Geomagnetic storm: 19.11
Gerbv (CAD software): 6.40
Germanium (device): 3.9
Getter: 9.24
Gilbert (Gb): 2.32
Gilbert cell mixer: 10.5, 10.30ff, 10.33ff
Gin-pole: 28.15
Glossary
   AC theory and reactance: 2.21
   Amateur Radio, general terms: 1.18
   Amateur television (ATV): 32.14
   Analog electronic terms: 3.60
   Antenna: 21.83
   DC and basic electricity: 2.2
   Digital communications: 31.25
   Digital electronics: 4.30
   Digital modes: 16.28
   Digital signal processing (DSP): 15.32
   Filters: 11.53
   FM and Repeater: 18.11
   Modulation: 8.22
   Oscillators and synthesizers: 9.52
   Power supply: 7.37
   Propagation: 19.31
   Radio frequency interference (RFI): 27.43
   Remote stations: 29.22
   RF design techniques: 5.33
   Slow-scan television (SSTV): 32.22
   Software defined radio (SDR): 15.32
   Space Communications: 30.44
   Transmission lines: 20.30
GMSK (*see* Gaussian minimum-shift keying)
GNU PCB (CAD software): 6.31
Gray code: 16.9, 16.10
Gray (grey) line propagation: 19.17
Grid (vacuum tube): 17.5
   Bias: 17.17
   Control: 17.5
   Screen: 17.5
   Suppressor: 17.5
Ground: 29.2
   AC (signal) ground: 3.30
   Bonding: 28.6
   Cold water pipe: 28.7
   Common reference potential: 28.7
   Concrete encased grounding electrode (CEGR): 28.7
   Conductors: 28.7, 28.10
   Dissimilar metals: 28.10
   Earth: 28.7
   Galvanizing: 28.10
   Lightning dissipation: 28.7
   Methods of grounding: 28.7
   Radio frequency interference (RFI): 27.12
   RF ground: 28.6
   Safety ground: 27.12, 28.6
   Single-point: 28.10
   Soldering and brazing: 28.7
   Star: 27.13
Ground fault circuit interrupter (GFCI or GFI): .. 7.2, 28.5
Ground loop: 27.12
Ground rod: 28.7
Ground wave: 19.6
Grounded grid: 17.6
   Amplifier design example: 17.24ff
Guanella balun: 5.22
Guanella transformer: 20.20
Gunn diode: 9.32
Guy wire, lengths to avoid: 22.40

H

H (hybrid) parameters: 5.32
Half-bridge converter: 7.25
Half-power frequency: 3.5
Half-wave rectifier: 7.5
   Output voltage: 7.5
   Voltage ratings: 7.6
Ham Radio Deluxe digital communications software: 16.11
Hamfests: 1.8
Hardware description language (HDL): 15.3
Harmful interference: 27.4, 27.6
Harmonics: 2.16, 9.23, 13.33
   RFI: 27.4, 27.11
Harvard architecture: 15.4
Heat management: 2.65
   Forced-air and water cooling: 2.69
   Heat pipe cooling: 2.69
   Rectifiers: 2.68
   RF heating: 2.68
   RF power amplifier: 17.18ff
   Solid state power amplifier: 17.42
   Thermoelectric cooling: 2.69
   Transistor derating: 2.68
Heat sink: 2.66, 7.12
   Selection: 2.66
Helical antenna: 30.14
Helical by W4ENE: 11.41
Helical resonator: 11.39ff
   Coupling: 11.42
   Design nomograph: 11.40
   Insertion loss: 11.42
   Tuning: 11.41
Hellschreiber: 16.12
Henry (H): 2.34
Hertz (Hz): 2.16
Heterodyne
   Receiver: 12.3ff
   Transmitter: 13.4
Hexadecimal: 3.49, 4.3
HF (frequency range): 2.19
h_{FE}, h_{fe}: 3.18, 5.32
HFTA (HF Terrain Analysis) software: 29.18
High Speed Multimedia (HSMM): 31.13ff
   Amplifiers: 31.16
   Antenna: 31.16
   Router configuration: 31.15
   Service set identifier (SSID): 31.15
   Software configuration: 31.15
   Station setup: 31.14

WiFi frequencies: 31.14
Wired equivalent privacy (WEP): 31.15
High-level data link controller (HDLC): 31.7
High-pass filter: 3.5, 11.2
RFI: 27.3, 27.23
High-voltage power: 7.29
Bleeder resistor: 7.29, 7.30
Capacitors: 7.29
Construction techniques: 7.31, 23.27
Equalizing resistors: 7.29
Fuses: 7.31
Grounding stick (hook): 7.31, 28.11
Inductors: 7.30
Interlock: 7.32
Metering: 7.30
Transformers: 7.30
Hilbert transformer: 15.23
Hipot test: 7.30
Horizon (radio): 19.23
Horsepower (hp): 2.11
Hot, neutral, ground wiring: 7.2, 28.4
Hot stick: 27.20
Hot-carrier diode (*see* Schottky diode)
HSMM (*see* High Speed Multimedia)
Huffman coding: 16.4
Hybrid combiner: 25.32
Hybrid-electric vehicle (HEV)
Automotive RFI: 27.41
Hybrid-pi model: 3.32, 5.22
Hydrometer: 7.33
Hysteresis (comparator): 3.46
Hysteresis (magnetic): 2.34

I

I-V curves: 3.6
I/Q
Demodulator: 8.12, 15.22ff
Modulator: 8.11, 15.22ff
IC (*see* Integrated circuit)
Identification timer, ID-O-Matic: 24.34
IEC (International Electrotechnical Commission): 27.23
IEEE-1394 (FireWire): 4.29
IF (*see* Intermediate frequency)
Ignition noise: 27.40
Image
Frequency: 12.4
GIF: 16.3
Image-reject mixer: 12.15
Rejection: 12.25
Response: 12.13ff
Image communications: 32.1ff
Image modulation: 8.17
Imaginary numbers: 2.74
IMD (*see* Intermodulation distortion)
Immunity, RF: 27.4, 27.9
Impedance (Z): 2.43
Calculating from R and X: 2.45, 2.46
Equivalent series and parallel circuits: 2.47
Graphical representation: 2.44
Ohm's law for impedances: 2.47
Phase angle: 2.44
Polar form: 2.44
Polar-rectangular conversion: 2.45
Power factor: 2.48
Reactive power (VA): 2.48
Rectangular form: 2.44
Impedance inversion, networks: 5.26
Impedance matching: 5.25
Antenna to transmission line: 20.11
Conjugate matching: 20.9
L network: 5.25, 20.9
Network: 20.9
Pi network: 5.26, 17.10, 20.10
Pi-L network: 17.10
Quarter-wave (Q) section: 20.11
Resonating an antenna: 20.11
Series- and shunt-input circuits: 20.10
Solid state amplifier: 17.29
Switching: 24.11
T network: 5.26, 20.9, 20.14
Transformer: 20.12
Transmission line: 20.9, 20.13
Vertical antenna: 24.9
Impedance matching unit: 20.13
Adjusting: 20.15
Antenna tuner (ATU, Antenna coupler, Matchbox,
Transmatch): 20.13
Location in the antenna system: 20.15
T configuration: 20.14
Unbalanced-to-balanced: 20.13
Impedance transformer: 20.18ff
Incidental emitters: 27.5
Independent sideband (ISB): 10.7
Indirect FM: 10.11
Inductance (L): 2.34
Henry (H): 2.34
Induced voltage (back-voltage): 2.34
Measurement with DVM: 25.20
Motional: 9.25
Mutual inductance (M): 2.36
Straight wires: 2.51
Inductance index (A_L): 2.53
Inductor: 2.31
AC current and voltage: 2.38
Air core: 2.50
At radio frequencies: 5.6
Color code: 22.11
Coupling: 2.36
Eddy current: 2.38
Effects of coupling: 5.7
Effects of shielded enclosures: 5.7
Encapsulated: 2.49
Energizing and de-energizing: 2.38
Ferrite toroid cores: 22.14
Ferrite toroidal: 2.53
High-voltage: 7.30
Inductance index (A_L): 2.53
Iron core: 2.52
Kick-back: 2.37
Machine-wound coil specifications: 22.42
Model (CAD): 6.5
Parasitic capacitance: 5.6
Powdered-iron toroid cores: 22.12
Powdered-iron toroidal: 2.53
Ratings, RF power amplifier: 17.14
RL time constant (τ): 2.36
Saturation: 2.35
Series and parallel: 2.36
Series resistance: 5.6
Slug-tuned: 2.52
Standard values: 22.11
Swinging choke: 7.15
Troubleshooting: 26.25
Infinite impulse response (IIR) filter: 15.16ff, 15.17
Insertion loss (IL): 5.13
Instrumentation amplifier: 3.45
Insulated-gate FET (*see* MOSFET)
Insulator: 2.1
Integrated circuit (IC): 22.18
Application-specific IC (ASIC): 3.27
Bipolar families: 4.14
CMOS: 4.16
Comparing logic families: 4.13
Digital: 4.13ff
Latch-up: 3.27
Metal-oxide semiconductor (MOS): 4.16
Mixer: 10.30ff

NMOS: 4.16
Parasitic SCR: 3.27
PMOS: 4.16
Programmable-gate arrays (PGA): 3.27
Transistor-transistor logic (TTL): 4.14
Integrated circuits (Linear): 3.25
Hybrid circuits: 3.25
Monolithic: 3.26
Silicon-on-sapphire (SOS): 3.26
Substrate: 3.26
Integrated power amplifier module: 3.59
Integrator: 9.42
Intentional emitters: 27.5
Inter-IC Communication (I2C) bus: 4.20
Inter-symbol interference: 8.10, 8.16, 19.29
Intercept point (IP): 10.15, 10.16, 12.26
Interface
Between logic families: 4.18
Digital to analog I/O: 4.19
High voltage equipment: 29.8
Microcontroller I/O: 4.21
Personal computer: 4.27ff
Slow-scan television (SSTV): 32.17
Sound card: 31.3
Interference (see RF Interference)
Interlock: 7.32
Intermediate frequency (IF): 12.4, 12.10, 12.13ff
Filter: 13.15
Intermodulation distortion (IMD): 8.19, 10.4, 12.25
Circuit simulation example: 6.20
Dynamic range: 10.14, 10.16, 12.25
Dynamic range test procedures: 10.16, 25.40
Identifying: 27.15
RFI: 27.4, 27.11ff
Testing: 13.34
Transmitter: 13.34
Internal impedance
Voltage and current sources: 2.9
International power standards: 28.4
Internet Radio Linking Project (IRLP): 18.10, 31.23
Interpolation: 15.9
Inverse FFT: 8.12
Inversion: 19.24
Inverter (digital): 4.6
Inverters, DC-AC: 7.36
Sine-wave output: 7.37
Square-wave output: 7.36
Volt-ampere (VA) product, rating: 7.37
Inverting amplifier: 3.44
Ion: 2.1
IONCAP: 19.20, 19.21
Ionogram: 19.8
Ionosonde: 19.8
Ionosphere: 19.7ff
Ionospheric
Fading: 19.9
Forward scatter: 19.12
Refraction: 19.7
IR drop (see Voltage, Voltage drop)
IRLP (see Internet Radio Linking Project)
ISM (Industrial, Scientific, Medical): 27.4
Isolation
Summing amplifier: 3.4
ITU emission designators: 8.2

J

J-K flip-flop: 4.10
Josephson Junction: 25.5
Joules (J): 2.2
JPEG compression: 16.4
JT65: 16.13
Data compression: 16.13
Earth-moon-earth (EME): 16.13

Reed-Solomon code: 16.14
Synchronization: 16.14
JT6M: 16.13
FSK, 44-tone: 16.13
Ping Jockey: 16.13
WSJT: 16.13
Junction FET (JFET): 3.12

K

K index: 19.12
Key click: 8.7, 10.7, 13.33
Keyer, miniature, TiCK-4: 24.31
Keying, CW waveform: 10.7, 13.6, 13.17ff
Keying interface: 24.29
Kicad (CAD software): 6.2, 6.4, 6.31
Kirchoff's Current Law (KCL): 2.7
Kirchoff's Voltage Law (KVL): 2.7
Kit, transceiver: 14.19
Klystron: 9.32
Knife-edge diffraction: 19.2

L

L network: 5.25, 20.10
Large-signal model: 3.30
Laser diode: 3.23
Lasing current: 3.24
Monochromatic light: 3.24
Latch-up: 3.27
Leakage current: 7.29
Leakage flux: 2.63
Leakage path: 5.10
Leakage reactance: 2.64
Least significant bit (LSB): 3.49
Level converter: 29.9
LF (frequency range): 2.19
Libration fading: 30.32
Light-emitting diode (LED): 3.23
Circuit design: 3.24
Electroluminescence: 3.23
Lightning (source of noise): 19.31
Lightning protection
Cable entrance panel: 28.10
Lightning arrestor: 28.9
Suppliers: 28.11
Limiter (amplifier): 3.5, 8.7
Limiting: 3.5
Lindenblad antenna: 30.12
Line loss: 20.4
Dependence on frequency: 20.5
Line of sight propagation: 19.22
Linear (system): 3.1
Linear amplifier: 17.2
Liquid-crystal displays (LCD): 3.28
Litz wire: 2.50, 5.6
Log (on-air activity): 1.7
Log (logarithmic) amplifier: 3.48
Logbook of The World (LoTW): 1.12
Logical operations: 4.5ff
Long-path propagation: 19.17
Loss
Loss angle or tangent (θ): 2.28
Filter insertion loss: 5.13
Path (EME): 30.31
Losses (radiative): 5.10
Lossless and lossy compression: 16.4
Low-pass filter: 3.5, 11.2
DC motors: 27.39
RFI: 27.3, 27.23
Telephone RFI: 27.35
Lower sideband (LSB): 8.5, 13.8
Lowest usable frequency (LUF): 19.9
LTSpice by Linear Technologies: 3.1, 6.2, 9.21, 11.27, 17.11

LUF: 19.9
Lumped elements: 5.1
LZW compression: 16.4

M

M-factor: 19.9
Magnetic field: 2.31, 21.1
    Electromagnetic wave: 19.1
    Gauss (G): 2.32
    Hysteresis: 2.34
    Magnetic field strength (H): 2.32
    Magnetic flux: 2.32
    Magnetic flux density (φ): 2.32
    Magnetic flux linkage: 2.34
    Magnetomotive force (MMF, $\mathfrak{I}$): 2.32
    Mean magnetic path length: 2.32
    Oersted (Oe): 2.32
    Permeability (μ): 2.32
    RF exposure: 28.24
    Right-hand rule: 2.32
    Saturation: 2.33
    Tesla (T): 2.32
Magnetic flux
    Maxwell (Mx): 2.32
    Weber (Wb): 2.32
Magnetometer: 19.11
Magnetomotive force (MMF, $\mathfrak{I}$): 2.32
    Reluctance ($\mathfrak{R}$): 2.33
Magnetosphere: 19.6
Magnetron: 9.32
Maintenance and troubleshooting: 26.1ff
Marine battery: 7.37
Marine layer: 19.26
Martin (SSTV mode): 32.18
Master-slave flip-flop: 4.10
Matchbox: 20.13
Mathematics tutorials: 2.72
Maunder minimum: 19.10
Maximum usable frequency (MUF): 19.8
    Prediction: 19.19ff
Maxwell (Mx): 2, 32
Megger: 7.30
MESFET: 3.22
Metal-oxide semiconductor (MOS)
    Integrated circuit (IC): 4.16
Meteor
    Scatter: 19.15
    Shower: 19.15
Meter, internal resistance: 25.4
Metric units: 22.54
MF (frequency range): 2.19
MFSK (*see* Multi-level FSK)
Microcontroller: 4.23ff
    CPU size: 4.24
    Development tools: 4.25
    FLASH memory: 4.26
    I/O: 4.21
    I/O requirements: 4.24
    Memory: 4.24
    Peripherals: 4.24
    Programming: 4.25
    Resources: 4.26
    Selecting: 4.24
Microphone: 13.12ff
    Amplifier: 13.13
Microprocessor: 15.4
Microstrip: 11.35
Microwave
    Communication link: 12.40
    Construction techniques: 23.29
    Receiver: 12.37ff
Microwire: 4.20
Minimum discernable signal (MDS): 9.9, 25.40

Minimum-shift keying (MSK): 8.10
Mix (ferrite): 2.53, 5.14
Mixer: 10.1ff, 12.5
    Balanced, IC: 3.58
    Diode: 10.18ff
    Diplexer: 10.22
    Double-balanced: 10.20ff
    Dynamic range: 10.14ff
    Equations: 10.3
    FET: 10.17ff
    Gain controlled amplifier: 10.16ff
    Gilbert cell: 10.26ff
    Image-reject: 12.15
    MOSFET: 19.29
    NE602/SA602/SA612: 10.30ff
    Nonlinear signal combinations: 12.4
    References: 10.35
    Spurious signal: 13.19
    Switching: 10.18ff
    Tayloe: 10.28ff
    Transistor: 10.24ff
Mixing products: 10.4
MixW digital communications software: 16.11
Mobile operation: 29.9ff
    Amplifiers: 29.12
    Antenna: 21.51ff
    Automotive interference: 29.12
    Cables and wiring: 29.10ff
    Operating techniques: 29.12
    RFI: 27.37
    Shock mounts: 29.9
    VHF/UHF propagation: 19.27ff
Mode (digital): 16.1ff
Mode (satellite transponder): 30.4
Modem: 16.16
Modulation: 8.1ff
    Accuracy: 8.20
    Amateur television (ATV), AM: 32.4
    Amateur television (ATV), Digital: 32.6
    Amateur television (ATV), FM: 32.5, 32.6
    Amplitude (AM): 8.3, 10.5ff
    Amplitude modulation (AM): 3.5
    Analog: 8.3ff
    Angle: 8.5, 8.10, 10.9ff
    Continuous wave (CW): 8.7, 10.6
    Digital: 8.7
    Digital signal processing (DSP): 15.21ff
    Double sideband (DSB): 8.4
    Frequency (FM): 8.6, 13.10
    Glossary: 8.22
    Image: 8.17
    Impairments: 8.19ff
    In transmitters: 13.5
    Index (AM): 8.3
    Index (FM): 8.6
    Linearity: 10.6
    Multi-carrier: 8.12
    On-off keying (OOK): 8.7
    Phase (PM): 8.6, 13.10
    Products: 3.5
    Pulse: 8.8
    Quadrature: 8.10ff
    References: 8.24
    Single sideband (SSB): 8.5
    Slow-scan television (SSTV), Analog: 32.18
Modulator: 10.1ff, 10.35, 13.7
    Balanced: 8.5, 13.8, 13.14
    Balanced, IC: 3.58
    Balanced, MC1496: 10.33ff
    I/Q: 8.11, 15.22ff
Molex connector: 29.6
Monolithic microwave IC (MMIC): 3.58
    Specifications: 22.22
Monte Carlo analysis: 6.1

Morse code: 8.8
MOSFET: 3.13, 3.20
    Depletion mode: 3.21
    Electrostatic discharge (ESD): 3.21
    Enhancement mode: 3.21
    Model (CAD): 6.8
    Pass transistor: 7.17
    Power: 3.21
Motorboating: 9.14
MP3 compression: 16.4
MSFK16: 16.8
    Convolution code: 16.9
    Fast Fourier Transform (FFT): 16.10
    Gray code: 16.9
    Hamming distance: 16.10
    Interleaver: 16.9
    Modulation: 16.10
    Quadbit, nibble: 16.9
MSK (*see* Minimum-shift keying)
MT63: 16.11
    Error-correction: 16.11
    Latency: 16.11
    Navy MARS: 16.11
    Noise immunity: 16.11
    Tuning error: 16.11
    Walsh function: 16.11
MUF (*see* Maximum usable frequency)
Multi-level FSK (MFSK): 8.10
Multi-protocol controller (MPC): 16.16
Multimeter, input impedance: 26.2
Multimode communications processor (MCP): 16.16
Multipath propagation: 19.28
Multiplier, frequency: 13.11
MultiPSK digital communications software: 16.11
Multivibrator:
    Astable (free-running): 4.13
    Bistable: 4.12
    Monostable (one-shot): 4.12
Mutual inductance (M): 2.36, 2.61

N

N connector: 22.51, 29.7
NAND gate: 4.5
Narrowband noise: 27.13
National Electrical Code (NEC): 28.2
    Antennas: 28.8
National Institute of Standards and Testing (NIST): 25.1
National Recognized Testing Laboratories (NRTL): 28.2
National Traffic System (NTS): 1.16
    Digital: 31.6
Natural frequency: 9.47
NCDXF beacon: 19.22
NE602/SA602/SA612 mixer: 10.30ff
Near-vertical incidence skywave (NVIS): 19.9, 21.5
Neper: 20.7
Netlist: 6.10, 6.33
Network
    APRS: 31.11
    D-STAR overview: 18.12
    High Speed Multimedia (HSMM): 31.13
    Mesh: 31.13
    Wireless: 4.29
Neutralization: 17.20ff
NIST: 25.1, 25.5
Node (circuit): 2.7
Noise: 5.30
    Amplifier: 5.30
    Atmospheric: 12.22
    Bandwidth: 9.9, 12.22
    Broadband: 27.13
    Cosmic: 19.31
    Factor (F) or Figure (NF): 3.29, 5.30
    Galactic: 12.23
    Gap noise: 27.14
    Generator, use in transmission line measurements: 20.8
    Ignition: 27.40
    Immunity (digital IC): 4.14
    In transmitters: 13.19
    Johnson noise: 3.29
    Man-made: 12.23, 19.30
    Narrowband: 27.13
    Phase: 9.5, 12.23
    Power: 5.31
    Pulse: 12.31
    Signal-to-noise ratio (SNR): 3.29
    Sources (in receivers): 12.22
    Temperature: 30.33
    Thermal: 12.23
    Thermal noise: 3.29
Noise blanker: 12.31
Noise factor (F): 3.29, 5.30, 12.23
Noise figure (NF): 3.29, 5.30, 12.23
    Measurement: 25.28
    Relationship to noise temperature: 22.43
Noise gain: 5.31
Noise limiter: 12.31
Noise reduction: 12.31ff
    Digital signal processing (DSP): 12.32, 14.4
Noise source: 25.25
Noise temperature: 12.23, 12.40
    Relationship to noise figure: 22.43
Nominal value: 2.4, 22.2
Non-inverting amplifier: 3.44
Non-linear junction: 27.15
NOR gate: 4.5
Northern California DX Foundation (NCDXF): 19.22
Norton equivalent circuit: 2.10
    Norton's theorem: 2.10
Notch filter: 11.2
    DSP, use in transceivers: 14.4
NTS (*see* National Traffic System)
NTSC channel: 8.17, 32.4
Number systems: 4.2ff
    Binary: 4.2
    Binary coded decimal (BCD): 4.3
    Conversion techniques: 4.3
    Hexadecimal: 4.3
Numerically controlled oscillator (NCO): 15.17
NVIS (*see* Near-vertical incidence skywave)
Nyquist
    Criterion: 8.9, 15.6ff
    Filter: 8.16
    Frequency: 3.50, 8.9, 15.6ff
    Rate: 3.50, 15.6ff
Nyquist sampling theorem: 3.50
Nyquist stability criterion: 9.3, 9.42

O

O wave: 19.9
Occupied bandwidth: 8.20
Oersted (Oe): 2.32
OFDM (*see* Orthogonal frequency-division multiplexing)
Ohm (Ω): 2.3
Ohm's law: 2.3
    Power formulas: 2.12
Ohmmeter: 25.6
Olivia: 16.11
    MFSK: 16.11
    Orthogonal tones: 16.11
Open-collector output: 3.46
Operational amplifier (op-amp): 3.43
    Active filter: 3.47
    Common-mode rejection ratio (CMRR): 3.43
    Comparator: 3.45
    Compensation (frequency): 3.43
    Difference amplifier: 3.45

Differential amplifier: 3.45
Gain: 3.43
Gain-bandwidth product (F_T, GBW): 3.43
Input and output impedance: 3.43
Input bias current: 3.43
Input offset voltage: 3.43
Instrumentation amplifier: 3.45
Inverting and non-inverting amplifier: 3.44
Log amplifier: 3.48
Open-loop gain: 3.43
Peak detector: 3.48
Power-supply rejection ratio (PSRR): 3.43
Rail-to-rail: 3.43
Rectifier circuit: 3.47
Specifications: 22.32
Summing amplifier: 3.45
Summing junction: 3.44
Unity-gain buffer: 3.45
Use of feedback: 3.44
Virtual ground: 3.44
Voltage-current converter: 3.48
Voltage-follower circuit: 3.44, 3.45
Optoisolator: 3.24
   Circuit design: 3.24
   Current transfer ratio (CTR): 3.24
OR gate: 4.5
Orbit
   Elements: 30.7
   Satellite: 30.5
   Satellite footprint: 30.9
   Satellite phase: 30.9
   Squint angle: 30.10
OrCAD (CAD software): 6.2ff
Orthogonal frequency-division multiplexing (OFDM): 8.12, 32.20
OSCAR (satellite): 30.2
Oscillator
   Alignment, RF: 25.19
   AM-to-PM conversion: 9.6
   Audio frequency: 25.19
   Barkhausen criterion: 9.3
   Butler: 9.25
   Cavity: 9.31
   Colpitts: 9.12
   Components: 9.17
   Crystal test: 11.34
   Design and simulation: 9.21
   Dielectric resonator: 9.32
   Differential: 9.16
   Fundamental: 9.23
   Gunn diode: 9.32
   Harmonic: 9.23
   Hartley: 9.13, 9.17
   Klystron: 9.32
   Logic-gate crystal oscillator: 9.29
   Loop gain: 9.3
   Magnetron: 9.32
   Microwave: 9.31
   Motorboating: 9.14
   Numerically controlled (NCO): 15.17
   Oven-controlled crystal oscillator (OCXO): 9.24
   Overtone: 9.23
   Permeability tuning: 9.14
   Phase noise: 9.5
   Phase shift: 9.29
   Pierce: 9.29
   Pulling: 9.26
   Quartz crystal: 9.23
   RC: 9.15
   Reciprocal mixing: 9.7
   Reference: 9.34
   Relaxation: 9.14
   Shielding and isolation: 9.20
   Squegging: 9.14
   Start-up: 9.3
   Temperature compensation: 9.19
   Temperature-compensated crystal oscillator (TCXO): 9.27
   Transmission-line resonator: 9.31
   Traveling wave tube (TWT): 9.32
   Two-JFET VFO and buffer (CAD example): 6.3ff
   Vackar: 9.15
   Variable crystal oscillator (VXO): 9.28
   Variable frequency (VFO): 9.13
   VHF and UHF: 9.30
   Voltage-controlled (VCO): 9.35
   Yttrium-iron garnet (YIG): 9.32
Oscilloscope: 25.21ff
   Analog: 25.21
   Buying used: 25.25
   Digital: 25.22
   Dual-trace: 25.22
   Probes: 25.24
   References: 25.48
   Use for troubleshooting: 26.4
OSI networking model: 16.15
   Layers: 16.15
   Protocol stack: 16.15
Ott, Henry WA2IRQ: 27.1ff
Out-of-band emissions: 27.4
Overmodulation: 10.5
Overtone: 9.23
Ozonosphere: 19.6

P

P25: 16.27, 31.19
   APCO-25: 16.27
   Continuous 4-level FM (C4FM): 16.28
   Improved MultiBand Excitation (IMBE): 16.28
   Tetra: 16.27
Packet radio (AX.25): 16.22, 31.7ff
   APRS: 16.23, 31.9
   Cluster: 16.22
   Digipeater: 16.23
   FTP: 16.23
   High level data link controller (HDLC): 31.7
   KISS: 16.23
   Packet bulletin-board systems (PBBS): 16.22
   TCP/IP: 16.23
   Terminal node controller (TNC): 31.7
   TNC interface: 16.22
PACTOR: 16.16, 31.12
   Contact flow: 16.16
   Link establishment: 16.17
   PACTOR-I, -II, -III: 16.16
   Speed comparison: 16.17
Panadapter: 14.7
Parabolic dish antenna (satellite): 30.15
Parallel I/O interface: 4.20, 4.27ff
Parallel-bridge converter: 7.25
Paramagnetic: 2.31
Parasitic: 5.1
   Capacitance (stray): 5.3, 5.5
   Capacitance of inductors: 5.6
   Effect on filter performance: 5.13
   Effect on Q: 5.8
   Effects of parasitic characteristics: 5.3
   General component model: 5.3
   Inductance: 5.3
   Inductance of capacitors: 5.5
   Inductance of resistors: 5.4
   Inductance per inch of wire: 5.3
   Inter-electrode capacitance: 5.6
   Leakage conductance: 5.3
   Oscillation: 17.22ff, 17.32
   Package capacitance: 5.6
   Parasitic SCR: 3.27
   Radiative losses: 5.10
   Resistance: 5.3

Self-resonance: 5.9
Suppressor: 17.23, 17.32
Parasitics (oscillation): 27.4, 27.11
Part 15 devices: 27.4
Part 18 devices: 27.4
Path loss
    Earth-Moon-Earth (EME) communication: 30.31
PC board construction: 23.17ff
PC board layout: 6.31ff
    Annotation, forward and backward: 6.33
    Annular ring: 6.34
    Auto-placement: 6.36
    Auto-router (autorouting): 6.37
    Bill of Materials (BOM): 6.33
    Board house: 6.33
    Component and solder sides: 6.33
    Component footprint: 6.32
    Copper fill: 6.38
    Design Rule Check (DRC): 6.37
    Double- and multi-layer: 6.33
    Drill file: 6.39
    Excellon file: 6.39
    Gerber file (RS-274X): 6.39
    Net: 6.32
    Netlist: 6.33
    Panelizing: 6.34
    Photoplotter: 6.39
    Plated-through hole: 6.33
    RF interference: 6.37
    Routing (traces): 6.36
    Silkscreen: 6.34
    Software: 6.31ff
    Solder mask: 6.34
    Surface-mount component: 6.33
    Symbol library: 6.32
    Through-hole component: 6.33
    Tin plating: 6.34
    Trace current limits: 6.38
Peak and average reading meters: 25.8
Peak and peak-to-peak values: 2.19
Peak detection: 3.5
Peak detector: 3.48
Peak envelope power (PEP): 2.21
Peak inverse voltage (PIV, PRV): 3.14, 7.12
Peak power: 2.19
Peak surge current, I_{SURGE}: 7.12
Pedersen ray: 19.16
Pentode: 17.5
Performance requirements: 3.33, 11.11
Period: 2.16
Permeability (μ): 2.32
    Ferrite: 5.14
    Of magnetic materials: 2.33
Phase: 2.17
    Lag and lead: 2.17
    Phase shift: 3.3
Phase angle (impedance): 2.44
Phase constant (β): 20.7
Phase detector: 9.34, 9.39
Phase modulation (PM): 8.6, 10.9
    Transmitter: 13.10
Phase noise: 9.5, 12.23
    Measurement: 9.8
Phase-frequency detector: 9.40
Phase-locked loop (PLL): 9.34, 9.36
    Capture range: 9.35
    Control loop: 9.34
    Demodulator: 10.13
    Design example: 9.46
    Frequency divider: 9.35, 9.38
    Frequency response: 9.37
    Lock range: 9.35
    Loop bandwidth: 9.35
    Loop compensation amplifier: 9.41

Loop filter: 9.35
Loop stability: 9.43
Measurements: 9.48
Modulator: 10.11
Noise: 9.36, 9.45
Operation: 9.35
Phase detector: 9.34, 9.39
Prescaler: 9.39
Reference oscillator: 9.34
Settling time: 9.37
Voltage-controlled oscillator (VCO): 9.35, 9.38
Phasing line: 20.11
Phasing method (SSB generation): 8.5, 13.8
Phasor: 9.5
Phasor diagram: 15.22
Photoconductivity: 3.22
Photoconductor: 3.8, 3.22
Photodiode: 3.22
Photoelectricity: 3.22
Photoisolator (see Optoisolator)
Photoresistor: 3.22
Phototransistor: 3.22
    Optoisolator: 3.24
Photovoltaic cell: 3.22
    Conversion efficiency: 3.23
    Open-circuit, V_{OC}, or terminal, V_T, voltage: 3.22
    Photovoltaic potential: 3.22
    Short-circuit current, I_{SC}: 3.23
    Solar panel: 3.23
Pi network: 5.26
    Impedance inversion: 5.26
    RF power amplifier: 17.10
Pi-EL Design by W4ENE: 17.10
Pi-L network
    Component selection: 17.11
    RF power amplifier: 17.10
    Table of values: 17.12
Picket fencing: 19.28
Piezoelectric: 9.23
Pilot carrier: 13.12
PIN diode: 3.15
Pinch-off: 3.13
Pinouts: 3.7
PL-259 connector: 22.46, 29.7
Plasma frequency: 19.8
Plumbers delight antenna: 21.38
PN junction: 3.8
    Anode: 3.9
    Barrier (threshold) voltage: 3.9
    Cathode: 3.9
    Depletion region: 3.9
    Forward resistance: 3.9
    Forward voltage drop: 3.10
    Fundamental Diode Equation: 3.9
    Junction capacitance: 3.14
    Recombination: 3.8
    Response speed, recovery time: 3.14
    Reverse breakdown voltage: 3.14
    Reverse leakage current: 3.9, 3.15
    Reverse-bias saturation current, I_S: 3.10, 3.15
PNPN Diode: 3.11
Polar cap absorption (PCA): 19.11
Polar coordinates: 2.73
Polarity (phase): 2.17
Polarity (voltage): 2.1
Polarity protection circuit: 3.17
Polarization
    Antenna: 21.1
    Radio wave: 19.9
Pole (transfer function): 3.3, 9.42, 11.20, 11.30
Pole-zero diagram: 9.42
Portable operation: 29.13ff
    Antennas: 29.15ff
    Power sources: 29.13ff

Potentiometer: 2.6
    Taper (log, audio): 2.6
Powdered-iron core: 22.12
Power: 2.11
    Horsepower (hp): 2.11
    Joule (J): 2.2
    Measurement: 13.32
    Station: 29.2
    Watt (W): 2.11
    Watt-hour (Wh): 2.11
Power amplifier modules
    Specifications: 22.31
Power density
    Spectral: 5.31, 9.8
Power factor: 2.48
Power filtering
    Bleeder resistor: 7.13
    Capacitor-input: 7.14
    Choke inductor constant (A): 7.14
    Choke-input: 7.14
    Swinging choke: 7.15
Power strip, load rating: 28.2
Power supply
    Construction techniques: 7.39
    IEC connector: 7.39
    Primary circuit connection standard (IEC): 7.39
    Use for troubleshooting: 26.5
Power supply regulation: 7.13
    Crowbar OVP circuit: 7.20
    Foldback current limiting: 7.17
    Linear regulator: 7.15
    Load resistance: 7.13
    MOSFET: 7.17
    Over-voltage protection (OVP): 7.20
    Overcurrent protection: 7.17
    Pass transistor: 7.15
    Pass transistor power dissipation: 7.16
    Remote sensing: 7.16
    Series and shunt regulator: 7.15
    Three-terminal voltage regulator: 7.18
    Voltage regulation: 7.13
    Zener diode regulator: 7.15
Power-line noise: 27.14, 27.18ff
    Cooperative agreement: 27.21
    Corona discharge: 27.21
    Signature or fingerprint: 27.19
    Ultrasonic: 27.20
    Utility complaint: 27.18
PowerPole, Anderson: 29.6
Powerstat (transformer): 2.65
Preamplifier: 12.24
    10 GHz: 12.37
    430 MHz: 12.36
    Earth-Moon-Earth (EME) communication: 30.39
    Satellite: 30.19
Precipitation static: 19.31
Precision (value): 2.4
Precision rectifier circuit: 3.47
Precision resistor: 2.4, 22.2
Prescaler: 9.39
Preselector: 9.7, 11.1
Primary battery: 7.33
Processing gain: 8.14
Product detector: 8.5
Programmable gate array (PGA): 3.27, 4.22
Programmable logic device (PLD): 4.22, 15.3
Propagation: 19.1ff
    10 GHz communication link: 12.40
    Amateur television (ATV) distance: 32.2
    Chart: 19.19
    D layer: 19.12
    Delay (digital circuit): 4.4
    E layer: 19.12ff
    Earth-Moon-Earth (EME) communication: 30.31ff
    Emerging theories: 19.19
    F layer: 19.16ff
    Forecast: 19.19ff
    Glossary: 19.31
    References: 19.33
    Sky wave: 19.6ff
    Software: 19.20, 19.21
    Space communication: 19.29ff
    Summary by band: 19.4ff
    Tropospheric: 19.22ff
    VHF/UHF mobile: 19.27ff
Protocol, digital
    APRS 16.23
    Automatic repeat request (ARQ): 16.2, 16.16
    B2F: 16.24
    Broadcast: 16.16
    CLOVER: 16.19
    Connected and connectionless: 16.16
    D-STAR: 16.26
    G-TOR: 16.17
    Multicast: 16.16
    P25: 16.27
    Packet radio (AX.25): 16.22
    PACTOR: 16.16
    SMTP: 16.24
    Stack: 16.15
    TCP/IP: 16.16, 16.23
    TELNET: 16.24
    Terminal node controller (TNC): 16.16
    UDP: 16.16
    Unicast (point to point): 16.16
    Winlink 2000 (WL2K): 16.24
    WINMOR: 16.21
PSK31: 16.7, 31.1
    BPSK, QPSK: 16.8
    Convolution code: 16.8
    Dibit: 16.8
    Envelope modulation: 16.8
    Phase-shift modulation: 16.8
    PSK63, PSK125: 16.8
    Varicode: 16.8
    Viterbi algorithm: 16.8
Public service: 1.15
    Amateur television (ATV) applications: 32.9
    Digital communications: 31.6
Public-address equipment RFI: 27.36
Pull-up resistor: 3.46
Pulse modulation: 8.8
Pulse-amplitude modulation (PAM): 8.9
Pulse-position modulation (PPM): 8.9
Pulse-width modulation (PWM): 8.9
    Switchmode power conversion: 7.22, 7.26
Push-to-talk (PTT): 14.17

Q

Q section: 20.11
QRP: 1.13
QSL card: 1.13
Quadrature amplitude modulation (QAM): 8.11, 32.20
Quadrature detector: 10.12, 11.29
Quadrature modulation: 8.10ff
Quadrature phase shift keying (QPSK): 8.11, 10.24
Quadrifilar helicoidal antenna: 30.12, 30.27
Quality factor (Q): 2.48
    Amplifier tank circuit: 17.9
    Loaded and unloaded: 2.55, 5.13
    Of components: 2.48
Quantization: 15.6ff, 15.9ff
Quantization error: 3.50, 15.2, 15.10

R

Radians: 2.17, 2.74, 11.4

Radiated emissions: 27.6
Radiation
    Ionizing and non-ionizing: 28.20
Radiation inversion: 19.24
Radio direction finding (RDF): 1.15
    Adcock antenna: 21.74
    Antennas: 21.73
    Doppler shift: 21.78
    Electronic antenna rotation: 21.77
    Maps and bearings: 21.80
    RFI: 27.17
    Sense antenna: 21.74
    Switched antenna arrays: 21.77
Radio frequency (RF): 2.18
Radio frequency interference (RFI): 27.1ff
    AC line (brute-force) filter: 27.23
    Alternator whine: 27.39
    Audio equipment: 27.35
    Audio filter: 27.35
    Automotive: 27.37
    Bypass capacitors: 27.23
    Choke placement: 27.28, 27.35ff
    Common-mode: 27.9, 27.22
    Common-mode choke: 27.23ff
    Computers and accessories: 27.36
    Consumer Electronics Association (CEA): 27.17
    Cooperative agreement: 27.21
    Corona discharge: 27.21
    Differential mode: 27.9
    Differential-mode: 27.22
    Electric fence: 27.15
    Electric vehicle (EV): 27.41
    Electrical noise: 27.15
    Fundamental overload: 27.11ff
    Ground: 27.12
    Ground loop: 27.12
    Harmful interference: 27.4
    High-pass filter: 27.3, 27.23
    Hybrid-electric vehicle (HEV): 27.41
    Identifying: 27.13
    Ignition noise: 27.40
    Immunity: 27.4, 27.9
    Intermodulation distortion (IMD): 27.4, 27.11ff
    Locating source of: 27.16
    Low-pass filter: 27.3, 27.23
    Non-linear junction: 27.15
    Notch filter: 27.28
    Path: 27.9
    Power-line noise: 27.14, 27.18ff
    Product review testing: 27.4
    Public-address equipment: 27.36
    Source: 27.9
    Speakers: 27.35
    Survival kit: 27.26
    Susceptibility: 27.3ff
    Telephone: 27.34
    Transmitters: 27.27
    Troubleshooting: 27.25ff
    Ultrasonic location: 27.20
    Utility complaint: 27.18
    Victim: 27.9
Radio horizon: 19.23
Radio propagation: 19.1ff
Radio spectrum: 2.18
    Classifications: 2.19
Radio subsystem IC: 3.58
    Balanced mixers and modulators: 3.58
    Passive mixer: 3.58
Radiosport (contesting): 1.10
Radioteletype (RTTY): 16.5, 31.1, 31.4
    AFSK, FSK: 16.5
    Baudot code: 16.6
    Diddle character: 16.6
    F1B *versus* F2B: 16.6
    Fading: 16.7
    Inverted or reverse shift: 16.6
    Keyed AFSK: 16.6
    Mark: 16.6
    Minimal shift keyed (MSK): 16.7
    Shift, LTRS and FIGS: 16.6
    Shifts: 16.7
    Space: 16.6
    Spotting an RTTY signal: 16.6
    Start and stop bit: 16.6
    Terminal Unit (TU): 16.6
    Tone pair: 16.6
    Unshift-on-space (USOS): 16.7
    USB and LSB: 16.6
Rain scatter: 19.24
Ramp wave: 2.16
Rate of change (Δ): 2.34
Rayleigh fading: 19.27
RC time constant (τ): 2.25
Reactance (X): 2.27
    Capacitive (X_C): 2.27
    Chart *versus* frequency: 2.40
    Complex waveforms: 2.43
    Inductive (X_L): 2.38
    Like, in series and parallel: 2.39
    Ohm's law for reactance: 2.39
    Unlike, in series and parallel: 2.41
Reactance modulator: 10.10
Reactive power (VA): 2.48
Receive converter: 12.33
Receive incremental tuning (RIT): 14.3
Receiver: 12.1ff
    10 GHz: 12.40
    902 to 928 MHz (33 cm): 12.37
    Amateur television (ATV), AM: 32.5
    Amplitude modulation (AM): 12.11
    Automatic gain control (AGC): 12.28ff
    Bandwidth: 12.13
    Beat frequency oscillator (BFO): 12.4, 12.10
    CW: 12.12
    Data modes: 12.12
    Detector: 12.2
    Direct conversion: 12.4ff
    Double conversion: 12.14
    DSP filtering: 12.22
    Dynamic range: 12.1, 12.25ff, 25.40
    Dynamic range performance: 12.28
    Frequency modulation (FM): 12.12, 12.32ff
    General coverage (in transceivers): 14.6
    Heterodyne: 12.3ff
    Image frequency: 12.4
    Image rejection: 12.25
    Image response: 12.13ff
    Intercept point (IP): 10.16, 12.26
    Intermediate frequency (IF): 12.4, 12.10, 12.13ff
    Microwave: 12.37ff
    Noise: 12.22
    Noise reduction: 12.31ff
    Performance tests: 25.39ff
    References: 12.41
    Selectivity: 12.1
    Sensitivity: 12.1, 12.22ff, 25.40
    Single sideband (SSB): 12.11
    Software defined radio (SDR): 15.25ff
    Superheterodyne (superhet): 12.3, 12.10ff
    Third-order intercept point (IP3): 10.16, 12.26, 25.41
    Tuned radio frequency (TRF): 12.3
    Two-tone IMD test: 10.16, 25.41
    UHF techniques: 12.35ff
    Upconverting architecture: 14.4ff
    VHF/UHF: 12.32ff
Receiver subsystem IC: 3.59
Recombination: 3.8
Reconstruction filter: 15.8

Rectangular coordinates: 2.73
Rectification: 3.5, 7.4
Rectifier: 7.10
    Average dc current rating, I_0: 7.12
    Circuits: 3.17
    Copper oxide: 7.10
    Current inrush: 7.12
    Equalizing resistors: 7.11
    Fast-recovery: 7.11
    Heat sinking: 7.12
    Inrush current limiting: 7.12
    Instruments: 25.8
    Mercury vapor: 7.10
    Parallel diodes: 7.11
    Peak inverse voltage (PIV, PRV): 7.12
    Peak repetitive forward current, I_{REP}: 7.12
    Peak surge current, I_{SURGE}: 7.12
    Power dissipation: 7.12
    Ratings and protection: 7.11
    Reverse recovery time: 7.12
    Schottky diode: 7.11
    Selenium: 7.10
    Semiconductor: 7.11
    Series diode strings: 7.11
    Switching speed: 7.12
    Thermal resistance (θ): 7.12
    Vacuum tube: 7.10
Reed-Solomon code: 16.14
References
    Amateur television (ATV): 32.23
    Analog basics: 3.62
    Antenna safety: 28.20
    Antennas: 21.84
    Computer-Aided Circuit Design: 6.40
    D-STAR: 18.14
    Digital communications: 31.26
    Digital electronics: 4.31
    Digital signal processing (DSP): 15.31
    Earth-Moon-Earth (EME) communication: 30.43
    Electrical safety: 28.12
    Filters: 11.53
    Mixers, Modulators and Demodulators: 10.29ff
    Modulation: 8.24
    Oscillators and synthesizers: 9.53
    Oscilloscope: 25.48
    PC-board layout: 6.40
    Power supplies: 7.39
    Propagation: 19.33
    Radio frequency interference (RFI): 27.44
    Receiver: 12.41
    Remote stations: 29.23
    RF design techniques: 5.34
    RF power amplifier: 17.56
    Satellite: 30.30
    Slow-scan television (SSTV): 32.23
    Software defined radio (SDR): 15.31
    Transceivers: 14.27
    Transmission lines: 20.32
    Transmitters: 13.34
    Troubleshooting: 26.33
Reflection: 19.2
Reflection coefficient (G,ρ): 20.4
    Conversion between attenuation, SWR, return loss: 22.56
Refraction: 19.2
    Ionospheric: 19.7
    Tropospheric: 19.24
Refractive index: 19.2
Regeneration: 3.4
Regenerative detector: 12.2
Register
    Formed by flip-flops: 4.11
    Shift: 4.11
    Storage: 4.11
Relay: 2.14
    Configurations: 2.14
    Construction: 2.14
    Ratings: 2.14
    Troubleshooting: 26.27
    Types of relays: 2.14
Reluctance ($\mathcal{R}$): 2.33
Remote stations: 29.18ff
    Connecting and controlling: 29.20ff
    Electrical power: 29.19
    Ethernet bridge: 29.21
    Insurance and security: 29.20
    Internet access: 29.19
    Internet bandwidth: 29.21
    Latency: 29.18, 29.21
    Property and access: 29.18
    Requirements: 29.18
    Safety: 29.20
    Site preparation: 29.18
Repeater: 18.1ff
    AM and SSB: 18.2
    Amateur television (ATV): 18.2, 32.9ff
    Closed: 18.6
    Controller: 18.4
    Coordination: 18.2, 18.9
    D-STAR: 18.11ff
    D-STAR hardware: 18.13
    Digital: 18.2, 18.3
    Digital ATV: 32.7
    Duplexer: 18.4
    FM voice: 18.2, 18.4ff
    Glossary: 18.11
    History: 18.1
    Internet linking: 18.10
    Linking: 18.4, 18.9
    Offsets, standard: 18.9
    Operation (FM voice): 18.4
    Portable ATV: 32.10
    Rules and regulations: 18.3
    Tone access: 18.6, 18.8
Reserve capacity, battery (RC): 7.37
Resistance (R): 2.3
    Bridge circuits: 25.7
    Effect of temperature: 2.4
    Equivalent resistance: 2.7
    Four-wire resistance measurement: 25.7
    Ohmmeter: 25.6
    Radiation resistance: 21.2
Resistivity (ρ): 2.3
Resistor: 2.2, 2.5
    Bleeder: 7.13
    Carbon composition (carbon comp): 22.2
    Carbon film: 22.2
    Color code: 22.3
    EIA identification and marking: 22.4
    Equalizing: 7.29
    Metalized film: 22.2
    Model (CAD): 6.3
    Non-inductive: 5.4
    Package dimensions: 22.3
    Parallel resistors: 2.7
    Parasitic inductance: 5.4
    Parasitic inductance model: 5.4
    Power: 22.4
    Precision: 2.4, 22.2
    Pull-up: 3.46
    Resistor pack: 3.45
    Series resistors: 2.8
    Standard values: 22.5
    Temperature coefficient (tempco): 2.5
    Thick-film power: 22.3
    Thin-film resistor: 3.26
    Troubleshooting: 26.25
    Types of resistors: 2.5
    Wire-wound: 22.3

Resonance: 2.42, 9.23
   Antenna: 21.2
   Of ferrite cores: 5.16
Resonant circuits: 2.54
   Antiresonance: 2.57
   Circulating current: 2.59
   Parallel: 2.56
   Parallel, above and below resonance: 2.59
   Parallel, loaded Q: 2.59
   Parallel, use for impedance matching: 2.60
   Resonant frequency: 2.54
   Selectivity: 2.56
   Series: 2.55
   Series and parallel: 2.42
   Series, unloaded Q and bandwidth: 2.55
Resonant frequency: 2.54
Restoration of vintage equipment: 26.32
Return loss (RL): 5.33, 11.11, 20.4
   Bridge: 25.30
   Converting between attenuation, reflection coefficient,
     and SWR: 22.56
   Converting to SWR: 5.33
Reverse bias: 3.9
Reverse breakdown voltage: 3.14
Reverse leakage current: 3.9
RF ammeter: 25.10
RF amplifier
   Feedback: 5.23
   Power amplifiers: 17.1ff
RF choke: 2.53, 17.15, 22.10
RF connectors: 22.46ff, 29.7
RF exposure and safety: 28.20ff
   Cardiac pacemakers: 28.25
   Environments, controlled and uncontrolled: 28.23
   FCC exposure regulations: 28.24
   Low-frequency fields: 28.26
   Maximum permissible exposure (MPE): 28.21
   MPE limits table: 28.24
   Power density: 28.27
   Power threshold table: 28.25
   RF awareness guidelines: 28.27
   RF Exposure and You: 28.28
   RF safety standards: 28.23
   Specific absorption rate (SAR): 28.21
   Station evaluation: 28.25
   Thermal and athermal effects: 28.21
RF heating: 2.68, 5.8
RF Interference (RFI): 27.1ff
RF power amplifier: 17.1ff
RF sniffer: 27.42
RF transformers: 5.27
   Air-core resonant: 5.27
   Binocular core: 5.29
   Broadband ferrite: 5.28
   Transmission line: 5.29
Right-hand rule: 2.32
Ripple: 7.4
   Frequency: 7.13
   Voltage: 7.13
Rise (fall) time in transmission lines: 20.21
RJ connector: 29.8
RL time constant (τ): 2.36
Robot (remote control), ATV application: 32.11
Roll-off: 3.5
Roofing filter: 12.15, 12.25, 12.27, 14.5
Root-mean-square (RMS): 2.20
Rotator, antenna
   Azimuth/elevation: 30.17
   Satellite: 30.16
RS-232: 4.28
RS-274X: 6.39
RS-422: 4.29
RTTY (*see* Radioteletype)
Rubidium clock: 25.2

Run length encoding (RLE): 16.4
Ruthroff transformer: 20.21

S

S (scattering) parameters: 5.32, 6.21
Safe operating area (SOA): 3.14, 7.17
Safety
   Antenna: 28.13ff
   Battery chemical hazards: 7.32
   Battery disposal and recycling: 7.33
   Bleeder resistor: 7.30
   Chemical: 23.2
   Electrical: 28.1ff
   Generator: 29.15
   High-voltage: 7.32
   Interlock: 7.32
   Remote stations: 29.20
   RF exposure: 28.20ff
   Safety rules: 26.2
   Soldering: 23.2
   Tools: 23.1
Safety ground: 7.2
Safety, electrical: 28.1ff
   Back-feeding ac line: 28.5
   Class 1, 2, 3 wiring: 28.5
   Distribution box: 28.3
   Energized circuits: 28.11
   Lockout: 28.2
   National Electrical Code (NEC): 28.2
   References: 28.12
   Solar panel: 28.11
   Uninterruptible power supply (UPS): 28.11
Safety, RF exposure: 28.20ff
Sample rate: 15.2
Sampling, analog-digital conversion: 15.6ff
Satellite: 30.1ff
   Active satellite list: 30.2
   Amplifiers, RF power: 30.23
   AMSAT: 30.1
   Antenna: 30.10ff
   Antenna rotators: 30.16
   Antenna, eggbeater: 30.11
   Antenna, helical: 30.14
   Antenna, Lindenblad: 30.12
   Antenna, parabolic dish: 30.15
   Antenna, quadrifilar helicoidal: 30.12
   Antenna, turnstile: 30.12
   Antenna, Yagi: 30.13
   Downconverter: 30.19
   Orbit footprint: 30.9
   Orbit, phase: 30.9
   Orbit, squint angle: 30.10
   Orbital Elements: 30.7
   Orbits: 30.5
   OSCAR: 30.2
   Preamplifier: 30.19
   Propagation: 19.30
   References: 30.30
   Space weather: 19.22
   Station design: 30.23
   Station equipment: 30.19ff
   Tracking software: 30.6
   Transceiver: 30.21
   Transponder modes: 30.4
   Transponder, inverting: 30.4
   Transponders: 30.3ff
Saturation (magnetic): 2.33
Saturation (semiconductor): 3.11
Sawtooth wave: 2.16
Scaling: 3.1
Scan converter: 32.16
Scatter
   Forward: 19.12

Rain:	19.24
Tropospheric:	19.23
Scattering:	19.2
Schematic capture:	6.4
Schematic capture software	
Eagle:	6.4, 6.32
gEDA:	6.4
GNU PCB:	6.4
Kicad:	6.4, 6.32
Schematic diagram:	2.2
Schottky barrier:	3.15
Schottky diode:	3.15
Scintillation:	19.26, 19.29, 30.32
Scottie (SSTV mode):	32.18
SCR (*see* Silicon controlled rectifier)	
Screen grid:	17.5
SDR (*see* Software defined radio)	
Secondary battery:	7.33
Secondary emission:	17.5
Secondary station identifier (SSID):	31.12
Section Manager (SM):	27.2
SELCAL:	31.17
Selective fading:	8.4, 8.10
Selectivity:	12.1
Self-resonance:	5.9
Semiconductor	
Acceptor and donor impurities:	3.8
Compound:	3.8
Cutoff (semiconductor):	3.11
Depletion region:	3.9
Doping:	3.8
Extrinsic:	3.8
Free electron:	3.8
Hole:	3.8
Intrinsic:	3.8
Junction semiconductor:	3.9
Majority and minority carriers:	3.8
Monocrystalline:	3.8
N-type, P-type:	3.8
Photoconductor:	3.8
PN junction:	3.8
Polycrystalline:	3.8
Recombination:	3.8
Safe Operating Area (SOA):	3.13
Saturation:	3.11
Substrate:	3.12
Temperature effects:	3.13
Thermal runaway:	3.13
Troubleshooting:	26.27
Sense antenna:	21.74
Sensitivity (receiver):	12.1, 12.22ff
Sequential logic:	4.8ff
Serial I/O interface:	4.20ff, 4.27ff
IEEE-1394 (FireWire):	4.29
RS-232:	4.28
RS-422:	4.29
Universal serial bus (USB):	4.29
Serial Peripheral Interconnect (SPI):	4.20
Shack notebook:	26.2
SHF (frequency range):	2.19
Shield, RFI:	27.22
Shift (digital modes):	31.4
Shift register:	4.11
Shock hazard:	26.1
Shock mounts:	29.9
Short skip:	19.13
Shortwave broadcast bands:	19.22
Sidescatter:	19.17
Siemens (S):	2.3, 2.28, 2.39
Signal generator:	25.32
Crystal-controlled source:	26.6
Use for troubleshooting:	26.4
Signal injector:	26.4
Signal report, amateur television (ATV):	32.3
Signal tracing:	26.5
Signal-to-noise and distortion ratio (*see* SINAD)	
Signal-to-noise ratio (SNR):	3.29, 12.1
Significant figures:	2.75
Silicon (device):	3.9
Silicon controlled rectifier (SCR):	3.11
Anode gate:	3.11
Cathode gate:	3.11
Use in a crowbar circuit:	3.11
Use in ac power control:	3.11
Simplex:	18.2
SINAD:	3.50, 12.33, 25.40
Sinc function:	8.16, 15.8
Sine wave:	2.15, 2.16
Sine wave generation (DSP):	15.17ff
Single sideband (SSB):	8.5, 10.7
DSP modulator and demodulator:	15.22ff
Filter method:	8.5, 13.8
Lower sideband (LSB):	8.5, 13.8
Phasing method:	8.5, 13.8
Receiver:	12.11
Transmitter:	13.8ff
Upper sideband (USB):	8.5, 13.8
Weaver method:	8.5, 13.9, 15.24
Skew path:	19.18
Skin effect:	5.6, 5.8, 20.1, 21.8
Skin depth (δ):	5.8
Skip zone:	19.8
Sky temperature:	30.33
Sky wave propagation:	19.6ff
Slew rate:	3.2
Slow-scan television (SSTV):	8.18, 16.12, 32.15ff
Analog modes:	8.18, 32.18
Digital modes:	8.19, 32.20ff
Equipment:	32.17
Frame sequential:	8.18
Frequencies:	32.17
Glossary:	32.22
History:	32.15, 32.16
Image sources:	32.21
Line sequential:	8.18
Operating practices:	32.17, 32.20
Picture quality:	32.19
Redundant digital file transfer (RDFT):	8.19
References:	32.23
Resources:	32.19
Scan converter:	32.16
Software:	32.19
Station:	32.16
Vertical interval signaling (VIS):	8.19, 32.18
Small-signal model:	3.30
BJT:	3.30
FET:	3.37
Use in RF design:	5.22
Smith chart:	20.4
Snubber network:	2.37
Software	
AADE filter design software:	11.33
Ansoft Designer SV2:	6.2
APRS:	31.10
ASAPS:	19.20, 19.21
Automatic link establishment (ALE):	31.17
Circuit simulation:	6.1ff
Digital modes for sound card:	31.6
Diplexer Designer by W4ENE:	11.48
DPlus (D-STAR application):	18.13
Eagle:	6.2
Elsie by W4ENE:	11.9
FilterPro by Texas Instruments:	11.21
GC Prevue:	6.40
gEDA:	6.2
Gerbv:	6.40
Helical by W4ENE:	11.41
HFTA:	29.18

Entry	Page
IONCAP:	19.20, 19.21
JT65:	30.42
Kicad:	6.2
LTSpice:	6.2, 17.11
LTSpice by Linear Technologies:	11.27
Modulators and demodulators:	15.29
MUF prediction:	19.20, 19.21
OrCAD:	6.2
PC-board layout:	6.31ff
Pi-EL Design:	17.10
Satellite tracking:	30.6
Schematic capture:	6.4
Slow-scan television (SSTV):	32.19
SPICE:	6.1ff
SVC Filter Designer:	17.41
SVC Filter Designer by W4ENE:	11.11
TLW:	20.4
TubeCalculator:	17.7ff
VOACAP:	19.20, 19.21
W6ELProp:	19.20, 19.21
Webench by National Semiconductor:	11.21
WinCAP Wizard:	19.20, 19.21
WinDRM:	32.20
WSJT:	30.42
Software defined radio (SDR):	15.1, 15.25ff
Dynamic range:	15.26
Glossary:	15.32
Hardware:	15.25ff
Local oscillator (LO):	15.28
Modulators and demodulators:	15.29
Receivers, overview:	12.6
References:	15.31
Sampling at RF:	15.27
Software architecture:	15.28ff
Speech processing:	15.30
Software-defined radio (SDR):	14.14
Solar battery or cell (*see* Photovoltaic cell)	
Solar cycle:	19.10
Solar flare:	19.11
Solar flux:	19.10
Solar panel:	3.23
Solar power	
Battery charging:	7.36
Remote stations:	29.19
Solar wind:	19.6
Soldering:	23.7
Desoldering:	23.9
Lead-free:	23.8
Reduction of Hazardous Substances (RoHS):	23.8
Safety:	23.2
Solder types:	23.7
Surface-mount technology (SMT):	23.12
Solenoid:	2.14
Solenoidal coil:	2.31
Sound card:	31.2ff
Accuracy:	31.3
Aliasing:	31.3
Digital communications applications:	31.1ff
Interface:	31.3
Linearity:	31.3
Sample rate:	31.2
Sample size:	31.2
Setup:	31.3
Slow-scan television (SSTV) applications:	32.17
Software:	31.6
Source, RFI:	27.9
Source-follower (*see* Common-drain amplifier)	
Space weather satellite:	19.22
Speakers	
RFI:	27.35
Speaker driver:	3.58
Twisted-pair cable:	27.35
Spectral power density:	5.31
Spectrum, frequency range classifications:	2.19
Spectrum analyzer:	25.42ff
Performance specs:	25.43
Speech	
Clipping, AF:	13.13
Clipping, IF:	13.16
Compression:	13.14
Processing:	13.13
Processing, with ALC:	13.19
Processing, with digital signal processing (DSP):	15.30
SPICE (CAD software):	6.1ff
Spike (*see* Transient)	
Splatter:	8.20
Split operation:	18.2
Repeater offsets:	18.9
Splitter, power:	20.20
Sporadic E propagation:	19.13
Spread spectrum (SS):	8.13ff, 16.10
Chip rate:	8.14
Code-division multiple access (CDMA):	8.15
Direct sequence (DSSS):	8.15, 16.10
Frequency hopping (FHSS):	8.15
Processing gain:	8.14
Spreading:	16.11
Spurious emissions:	13.33, 27.4ff, 27.11
Spurious-free dynamic range (SFDR):	3.50, 12.27, 15.2
Square wave:	2.16
Square-law device:	3.37
Squegging:	9.14
Squelch:	18.8
Squint angle (satellite):	30.10
S-R flip-flop:	4.9
SSNe:	19.21
SSTV (*see* Slow-scan television)	
Standard frequency stations:	25.1
Standards (traceability history):	25.1ff
Standing-wave ratio (SWR, VSWR, ISWR):	20.4
Bridge:	20.24, 24.20
Conversion of attenuation, reflection coefficient,	
and return loss:	22.56
Converting to return loss:	5.33
Flat line:	20.4
Myths about:	20.15
Star ground:	27.13
Static (precipitation):	19.31
Station	
Accessory projects:	24.1ff
Assembly:	29.1ff
Documentation:	29.8
Layout:	29.2ff
Location:	29.1
Power:	29.2
Step recovery diode (SRD):	13.11
Storage register:	4.11
Stratosphere:	19.6
Stripline:	11.35
Stub (transmission line):	20.7
As filters:	20.7
Combinations:	20.9
Connecting:	20.8
Field Day:	20.9
Measuring:	20.8
Quarter- and half-wave:	20.7
Universal:	20.24
Sub-receiver:	14.7
Subaudible tone:	18.6, 18.8
Substitution guide (semiconductor):	3.8
Sudden ionospheric disturbance (SID):	19.11
Summing amplifier:	3.45
Sunspot:	19.10
Sunstone Circuits (PCB vendor):	6.32
Superheterodyne (superhet) receiver:	12.3, 12.10ff
Superposition:	3.1
Surface acoustic wave (SAW):	11.35
Surface-mount technology (SMT):	22.2, 23.9

Package types: 23.12
Soldering: 23.12
Susceptance (B): 2.28
Capacitive (B_C): 2.28
Inductive (B_L): 2.39
Susceptibility, RF: 27.3ff
SVC Filter Designer by W4ENE: 11.11, 17.41
Swinging choke: 7.15
Switch: 2.12
Diode: 3.17
Make and break: 2.13
Poles and positions: 2.13
Ratings: 2.12
Transmit-receive (TR): 14.18
Types of switches: 2.13
Switching circuit: 3.41
Circuit design: 3.41
High-side and low-side switching: 3.42
Power dissipation: 3.41
Reactive loads: 3.42
Transistor selection: 3.42
Switching mixer: 10.18ff
Switching time: 5.19
Switchmode power conversion: 7.21
Boost converter: 7.23
Bridge converter: 7.25
Buck converter: 7.21
Buck-boost converter: 7.23
Continuous conduction mode (CCM): 7.22
Design aids and tools: 7.28
Discontinuous conduction mode (DCM): 7.22
Flyback converter: 7.23
Forward converter: 7.25
Pulse-width modulation: 7.22, 7.26
RFI: 7.21, 27.13
Switching loss: 7.21
Switchmode power supply RFI: 27.13
SWR monitor: 24.4
Symbol library (PCB layout): 6.32
Symbol rate: 8.8
Synchronicity (logic): 4.8
Synchronous detector: 8.4
Synchronous transformer: 20.11
Synthesizer, frequency: 9.33ff
Direct and indirect: 9.33
Direct digital synthesis (DDS): 9.33, 15.17
Integrated circuit: 9.49
Multi-loop: 9.37
Phase-locked loop (PLL): 9.37
Rate multiplier synthesis: 9.33

T

T network: 5.26, 20.11
Designing: 20.11
Tank circuit: 17.9ff
Component ratings: 17.14ff
Design example: 17.27
Efficiency: 17.10
Flywheel effect: 17.9
Manual design methods: 17.11
Tayloe detector: 11.29
Tayloe mixer: 10.28ff
TCP/IP: 16.23
Technical Coordinator (TC): 27.2
Telephone
RFI: 27.34
RFI filter: 27.35
Television interference (TVI): 27.1, 27.27ff
Analog TV: 27.30
Cable TV: 27.31
Channel guide: 27.29
Digital TV: 27.30
DVD and VCRs: 27.34
Electrical noise: 27.30
Fundamental overload: 27.28
High-pass filter: 27.28
Preamplifiers: 27.28
Spurious emissions: 27.30
Temperature coefficient (tempco): 2.5, 22.2
Capacitor: 2.28
Temperature compensation: 2.69
Oscillator: 9.19
Temperature control: 2.70
Temperature inversion: 19.24
Temperature sensor: 3.58
Terminal node controller (TNC): 16.16, 31.7ff
APRS: 31.9
Modem: 16.16
Multi-protocol controller (MPC): 16.16
Multimode communications processor (MCP): 16.16
Setup: 31.8
Tesla (T): 2.32
Test equipment and measurement: 25.1ff
Test leads: 26.3
Test probes: 26.3
Demodulator: 26.3
Inductive pickup: 26.3
Tetrode: 17.5
Thermal conductivity (k): 2.66
Thermal resistance (θ): 2.66
Rectifier: 7.12
Thermal runaway: 3.13, 7.53
Thermionic emission: 17.3
Thermistors: 2.69
Thermocouple meters: 25.8
Thevenin equivalent circuit: 2.9
Thevenin's Theorem: 2.9
Thin-film resistor: 3.26
Third-order intercept point (IP3): 10.16, 12.26
Three-terminal voltage regulator: 7.18
Adjustable regulator: 7.19
Current source: 7.20
Increasing output current: 7.20
Low dropout regulator: 7.19
Throb: 16.11
Thyristor: 3.11
Time: 25.1ff
Calibration: 25.2
Time constant (τ)
RC: 2.25
RL: 2.36
Time domain: 2.17
Timer (multivibrator) IC: 3.56
Astable (free-running): 3.56
Monostable (one-shot): 3.56
TLW by N6BV: 20.4
TNC (*see* Terminal node controller)
Tolerance: 2.4, 22.2
Tone
CTCSS: 18.8
Decoder, DSP: 15.19
Repeater access: 18.8
Total harmonic distortion+noise (THD+N): 3.50
Tower
Base, concrete: 28.14
Carabiner: 28.15
Climbing harnesses and belts: 28.18
Climbing safety: 28.18
Crank-up: 28.9
Erection and maintenance: 28.15
Gin-pole: 28.15
Guy wires, lengths to avoid: 22.40
Guyed: 28.14
Guying and guy wires: 28.16
Roof-mounted: 28.14
Ropes: 28.15
Safety references: 28.20

Self-supporting: 28.14
TIA-222: 28.13
Wind load: 28.13
TR switching: 14.16
Amplifier compatibility: 14.18
Transceiver: 14.1ff
Architecture: 14.2
Control circuits: 14.16ff
DSP features: 14.4
Features in modern radios: 14.8ff
Handheld: 18.6
History: 14.1ff
Home station (FM voice): 18.8
Hybrid: 14.3
Mobile (FM voice): 18.6
Portable and mobile: 14.10
References: 14.27
Selection of: 14.8ff
Software-defined radio (SDR): 14.14, 15.25ff
TR switching: 14.16
Upconverting architecture: 14.4ff
VHF/UHF coverage: 14.15
Transconductance: 3.6
Forward transconductance, g_m: 3.19
Transducer: 2.16
Transequatorial propagation (TE): 19.18
Transfer characteristics: 3.3
Transfer function: 3.3, 3.28
Transfer switch: 28.5
Transformer: 2.61
50 Hz considerations: 7.3
Broadband (RF): 17.29
Color code: 22.15
Evaluating: 7.3
High-voltage: 7.30
Impedance ratio: 2.63
Interwinding capacitance: 2.64
Isolation: 7.3
Laminations: 2.62
Leakage flux: 2.63
Leakage reactance: 2.64
Losses: 2.63
Magnetizing inductance: 7.3
Power: 7.3
Power ratio: 2.63
Primary and secondary: 2.61
RF: 5.27
Shielding: 2.64
Step-up and step-down: 2.62
Troubleshooting: 26.25
Turns ratio: 2.62
Volt-ampere rating: 7.3
Voltage and current ratio: 2.62
Transient: 7.12
Transient protection
Gas tube: 7.12
Varistor: 7.12
Zener diode: 7.12
Transient-suppressor diode (TVS): 3.16
Transistor
Gain versus frequency: 5.22
Switching circuits: 3.41
Troubleshooting: 26.28
Transistor array IC: 3.55
Transistor-transistor logic (TTL): 4.14
Circuits: 4.15
Driving CMOS: 4.18
Three-state outputs: 4.16
Unused inputs: 4.16
Transition time (digital circuit): 4.4
Transmatch: 20.13
Transmission line: 20.1ff
Attenuation (loss): 20.4
Attenuation of two-conductor twisted pair: 22.41

Balanced and unbalanced load: 20.16
Balancing devices: 20.16
Balun: 5.29
Choosing: 20.6
Coaxial cable: 20.1, 21.7
Digital circuits: 20.27
Impedance of two-conductor twisted pair: 22.41
Impedance transformer: 20.7
Incident, forward, and reflected waves: 20.4
Ladder line: 20.6, 21.7
Load: 20.3, 20.16
Lumped-constant: 20.3
Matched and mismatched: 20.3
Matched- (ML) and Mismatched-line loss: 20.4
Microstrip: 11.36, 20.20
Open-wire: 20.1, 21.7
Parallel-conductor (twin-lead): 20.1
Phasing: 20.11
Quarter-wave (Q) section: 20.11
Radiation cancellation in: 20.1
Radiation from: 20.16
Specifications: 22.48
Stripline: 11.36
Stubs: 20.7
Synchronous transformer: 20.11
Termination: 20.3
Transformer: 5.29, 17.29, 20.18ff
Twin-lead: 20.1, 21.7
Unun: 5.29
Waveguide: 20.28
Weatherproofing: 28.18
Window line: 20.1, 21.7
Transmitter: 13.1ff
Alternator: 13.2
Amplitude modulation (AM): 13.7
Automatic level control (ALC): 13.19
Carrier and unwanted sideband suppression: 25.49
Frequency calibration: 13.32
Frequency modulation (FM): 13.10
Frequency stability: 13.32
History: 13.2ff
Modulation: 13.5
Multiband: 13.4
Oscillator: 13.2
Performance: 13.32ff
Performance tests: 25.46ff
Phase modulation (PM): 13.10
Phase noise: 25.47
Power amplifier: 13.30ff
References: 13.34
Spurious emission: 13.33, 25.46
Subsystem IC: 3.59
Superhet SSB/CW: 13.12
Two-tone IMD: 25.47
Transponder
Digital: 30.5
Dynamic range: 30.4
LEILA: 30.4
Linear: 30.3ff
Modes: 30.4
Satellite: 30.3ff
Transverse electromagnetic mode (TEM): 11.35
Transverter: 14.15, 30.20
Traveling wave tube (TWT): 9.32
Tri-state gate: 4.6
Triac: 3.11
Triangle wave: 2.16
Trickle charging: 7.36
Triggering, flip-flop: 4.9
Triode: 17.5
Troposphere: 19.6
Tropospheric
Bending: 19.26
Ducting: 19.24ff

Fading: 19.26
    Propagation: 19.22ff
    Refraction: 19.24
    Scatter: 19.23
Troubleshooting
    Accessories: 26.5
    Amplifier and switch circuits: 26.13
    Component replacement: 26.26
    Components: 26.25
    Control circuits: 26.15
    Defining problems: 26.6
    Digital circuits: 26.16
    Distortion: 26.10
    FET: 26.29
    Guide charts: 26.20
    Methods and practices: 26.6
    Oscillation: 26.10
    Oscillators: 26.14
    Professional repairs: 26.31
    Radio frequency interference (RFI): 27.25ff
    Receivers: 26.17
    References: 26.33
    Relays: 26.27
    RF Power Amplifiers: 16.21ff
    Semiconductors: 26.27
    Shack notebook: 26.2
    Signal injection and tracing: 26.7
    Symptoms and faults: 26.12
    Television interference (TVI): 27.27
    Transceivers: 26.18
    Transmitters: 26.18
    Vacuum tubes: 26.30
TTL (*see* transistor-transistor logic)
Tube (*see* Vacuum tube)
Tube tester: 26.5
TubeCalculator software: 17.7ff
Tuna Tin 2 transmitter: 13.6
Tuned circuit: 2.54ff
    Single-tuned circuit: 5.13
Tuned radio frequency (TRF) receiver: 12.3
Tuning, RF power amplifier: 17.26
Turn-over effect (instruments): 25.8
Turnstile antenna 30.12
Twisted-pair speaker cable: 27.35
Two-port network: 3.28, 5.31
    Parameters: 5.32
Two-tone IMD: 8.20, 10.16, 13.34, 25.41, 25.47

U

Ufer ground: 28.7
UHF (frequency range): 2.19
UHF connector: 22.46, 29.7
UHF receiver techniques: 12.35ff
Ultraviolet radiation: 19.6
Underwriter Labs (UL): 28.1
Unintentional emitters: 27.5
Uninterruptible power supply (UPS): 28.11
Unipolar transistor (*see* Field-effect transistor)
Units and conversion factors: 22.53
Unity-gain buffer: 3.45
Universal serial bus (USB): 4.29
Universal stub: 20.24
Unun (transmission line): 5.29
Upconverting receiver: 14.4ff
Upper sideband (USB): 8.5, 13.8

V

Vacuum tube: 17.3ff
    Cathode: 17.3
    Construction: 17.6
    Cooling methods: 17.6
    Grid (control, screen, suppressor): 17.5
    Grid dissipation: 17.13
    Nomenclature: 17.5
    Operating parameters: 17.7ff
    Pentode: 17.5
    Plate (anode): 17.3
    Plate dissipation: 17.13
    Ratings: 17.13
    Screen dissipation: 17.13
    Secondary emission: 17.5
    Specifications: 22.34
    Tetrode: 17.5
    Thermionic emission: 17.3
    Triode: 17.5
    Troubleshooting: 26.30
Varactor diode: 13.11
    Specifications: 22.19
Variable-frequency oscillator (VFO): 9.12
    Backlash: 9.18
Variac (transformer): 2.65
Varicode: 8.8, 16.3, 16.7ff
Varistor: 7.12
    Specifications: 22.37
VCR (Videotape Cassette Recorder) RFI: 27.34
Vector: 2.44
Velocity (radio wave): 19.1
Velocity factor (VF): 20.1
Vertical interval signaling (VIS): 8.19, 32.18
Vestigial sideband (VSB): 8.17, 32.4
VFO (*see* Oscillator)
VHF (frequency range): 2.19
VHF/UHF
    Amplifier tank circuit: 17.11
    FM voice equipment: 18.6
    Home station antenna (for repeaters): 18.8
    Mobile antennas: 18.7
    Preamplifier: 12.24
    Propagation (mobile): 19.27ff
    Receiver: 12.32ff
    Transverter: 14.15
Victim (RFI): 27.9
Vidicon: 32.13
Vintage equipment, repair and restoration: 26.32
Virtual height: 19.7
VLF (frequency range): 2.19
VLF phase comparator: 25.2
VOACAP: 19.20, 19.21
Voice over Internet Protocol (VoIP): 31.22
Voice-operated TR switching (VOX): 14.17ff
Volt-amp reactive (VAR): 2.48
Volt-ampere rating (transformer): 7.3
Voltage (V, E): 2.1
    Potential: 2.1
    Voltage drop: 2.9
    Voltage source: 2.9
    Volts (V): 2.1
Voltage divider: 2.8
Voltage multiplier: 7.7, 7.10
    Full-wave voltage doubler: 7.7, 7.8
    Half-wave voltage doubler: 7.7
    Voltage tripler and quadrupler: 7.8
Voltage reference IC: 3.55
Voltage regulation: 7.13
    Adjustable regulator: 7.19
    Dynamic and static regulation: 7.13
    Three-terminal voltage regulator: 7.18
Voltage regulator IC: 3.55, 22.17
    Specifications: 22.20
Voltage-controlled oscillator (VCO): 9.35
    Noise: 9.49
    Voltage-to-frequency gain: 9.44
Voltage-current converter: 3.48
Voltage-follower circuit: 3.45
Voltage-power conversion: 22.55
Voltage-to-frequency gain: 9.44

Voltmeter: 25.5
   Compensated, modular RF: 25.32ff
   Current measurement: 25.5
   DC voltage standards: 25.5
   Impedance: 25.9
   Multipliers: 25.5
   RF probe: 25.33
   RF probe multipliers: 25.34
   Sensitivity: 25.5
Volunteer Examiner (VE): 1.5
VOM (volt-ohm-meter): 25.3
Von Neumann architecture: 15.4
VOX: 14.17ff
VSB (*see* Vestigial sideband)

W

W3DZZ trap dipole: 21.22
W6AAQ screwdriver antenna: 21.54
W6ELProp: 19.20, 19.21
WA2IRQ, Henry Ott: 27.1ff
Waterfall display: 31.6
Watt (W): 2.11
Watt-hour (Wh): 2.11
Wattmeter
   Directional: 20.4
   Reflectometer: 20.4
Wave cyclone: 19.25
Waveguide: 20.28
   Coupling to: 20.29
   Cut-off frequency: 20.29
   Dimensions: 20.29
   Dominant mode: 20.29
   Modes (TM, TE): 20.29
Wavelength (λ): 2. 18, 19.1
Waveshaping, CW: 8.8, 10.7, 13.6, 13.17ff, 13.33
Weaver method (SSB generation): 8.5, 13.9, 15.24
Webench by National Semiconductor: 11.21
Weber (Wb): 2.32
Wheatstone bridge: 25.7
WiFi
   Amplifiers: 31.16
   Antenna: 31.16
   Router Configuration: 31.15
   Service set identifier (SSID): 31.15
   Shared Amateur Radio frequencies: 31.14
   Wired equivalent privacy (WEP): 31.15
WinCAP Wizard: 19.20, 19.21
WinDRM: 31.21
   Slow-scan television (SSTV): 8.19, 32.20ff
Winlink 2000 (WL2K): 16.24, 31.6
   Airmail: 16.25
   B2F: 16.24
   Common Message Servers (CMS): 16.24
   Paclink: 16.24
   PACTOR: 16.24
   PMBO: 16.24
   Radio Message Server (RMS): 16.24
   TELNET: 16.24
WINMOR: 16.21
   Comparison to PACTOR: 16.21
   SCAMP: 16.21
   Sound card: 16.21
   TNC interface: 16.21
Wire, specifications: 22.39
Wired equivalent privacy (WEP): 31.15
Wireless network: 4.29
WIRES-II: 31.24
WSJT: 16.13, 31.2
   JT65 (digital EME communication): 30.42
   Meteor scatter: 19.15
WSPR: 16.14
WWV, WWVH: 19.22, 25.2, 25.14
   Propagation forecast: 19.10, 19.12

X

X wave: 19.9
XOR gate: 4.5

Y

Y (admittance) parameters: 5.32
Yagi antenna: 21.36ff
   Satellite: 30.13
   Yagi-Uda antenna: 21.36

Z

Z (impedance) parameters: 5.32
Zener diode: 3.16
   Noise generation: 3.18
   Specifications: 22.18
   Temperature-compensated reference: 9.18
   Voltage regulator: 7.15
Zero (transfer function): 3.3, 9.42, 11.20, 11.30
Zero beat: 14.1
Zero bias: 17.2
Zero-stuffing: 15.9

Project Index

Amplifier, RF Power
   250-W Broadband Linear Amplifier: 17.32
   3CX1500D7 RF Linear Amplifier: 17.47
   6-Meter Kilowatt Amplifier: 17.54

Antenna tuner
   160 and 80 Meter Matching Network for Your
      43-Foot Vertical: 24.9
   Switching the Matching Network for Your 43-Foot Vertical: ...24.11
   Z-match: 24.1

Antenna, accessory
   Audible Antenna Bridge: 24.20
   External Automatic Antenna Switch for Use with Yaesu or ICOM
      Radios: 24.14
   Low-cost Remote Antenna Switch: 24.18
   Microprocessor-controlled SWR Monitor: 24.4
   Mounts for Remotely-Tuned Antennas: 21.57
   Transmitting Chokes: 20.20

Antenna, HF
   40-15 Meter Dual-Band Dipole: 21.16
   All-Wire 30 Meter CVD: 21.36
   Compact Vertical Dipole (CVD): 21.35
   Extended Double-Zepp For 17 Meters: 21.25
   Family of Computer-Optimized HF Yagis: 21.41
   Five-Band, Two-Element HF Quad: 21.43
   Half-Wave Vertical Dipole (HVD): 21.33
   Multiband Center-Fed Dipole: 21.15
   Multiband Horizontal Loop Antenna: 21.50
   Retuning a CB Whip Antenna: 21.58
   Top-Loaded Low-Band Antenna: 21.29
   Two W8NX Multiband, Coax-Trap Dipoles: 21.21
   Vertical Loop Antenna for 28 MHz: 21.49
   W4RNL Inverted-U Antenna: 21.17
   Wire Quad for 40 Meters: 21.47

Antenna, satellite
   EZ Lindenblad Antenna: 30.24
   W3KH Quadrifilar Helix Antenna: 30.27

Antenna, VHF/UHF
   A Medium-Gain 2 Meter Yagi: 21.65
   Cheap Yagis by WA5VJB: 21.69
   Dual-Band Antenna for 146/446 MHz: 21.62
   Simple, Portable Ground-Plane Antenna: 21.62
   Three and Five-Element Yagis for 6 Meters: 21.64

Computer interface
   Simple Serial Interface: 24.25
   Trio of Transceiver/Computer Interfaces: 24.22
   USB Interfaces for Your Ham Gear: 24.26

CW Keyer
   TiCK-4 —A Tiny CMOS Keyer: 24.31
   Universal Keying Adapter: 24.29

Filter, audio
   Audio Intelligibility Enhancer: 24.36

Filter, RF
   Broadcast-Band Reject Filter: 11.44
   Crystal Ladder Filter for SSB: 11.43
   Diplexer Filters: 11.46
   Field Day Stub Assembly: 20.9
   High-Performance, Low-Cost 1.8 to 54 MHz Low-Pass Filter:11.49
   Optimized Harmonic Transmitting Filters: 11.46
   Passive CW Filter: 11.18
   Wave Trap for Broadcast Stations: 11.45

Power supply
   12-V, 15-A Power Supply: 7.43
   13.8-V, 5-A Regulated Power Supply: 7.48
   Automatic Sealed-Lead-Acid Battery Charger: 7.52
   Four-Output Switching Bench Supply: 7.40
   High-Voltage Power Supply: 7.50
   Micro M+ PV Charge Controller: 7.51
   Overvoltage Crowbar Circuit: 7.58
   Overvoltage Protection for AC Generators: 7.57

Receiver
   10 GHz Preamplifier: 12.37
   430 MHz Preamplifier: 12.35
   MicroR2 SSB or CW Receiver: 12.16
   Rock Bending Receiver for 7 MHz: 12.5
   Rock-bending Receiver for 7 MHz: 12.5

RF Interference
   RF Sniffer: 27.41

Station accessory
   Audio Interface Unit for Field Day and Contesting: 24.40
   ID-O-Matic - Station Identification Timer: 24.34

Test equipment
   Calibrated Noise Source: 25.25
   Compensated, Modular RF Voltmeter: 25.32
   Dip Meter with Digital Display: 25.16
   HF Adaptor for Narrow-Bandwidth Oscilloscopes: 25.25
   Hybrid Combiners for Signal Generators: 25.32
   Marker Generator with Selectable Output: 25.14
   Microwatter: 25.11
   Return Loss Bridge: 25.30
   Signal Generator for Receiver Testing: 25.32
   Wide-Range Audio Oscillator: 25.19

Transceiver
   HBR-2000 High Performance Transceiver: 14.23
   TAK-40 SSB/CW Transceiver: 14.20

Transmitter
   MicroT2 Single-Band SSB Transmitter: 13.20
   MkII Universal QRP Transmitter: 13.25
   Tuna Tin 2 HF CW Transmitter: 13.5

Author Index

Author	Call	Topic	Section Ref	Page Ref
Adams, Chuck	K7QO	Making PC Boards With Printed Artwork	23.5.3	23.20
Allison, Bob	WB1GCM	RFI performance of digital TV converters	27.8.3	27.30
		Product Review Testing and RFI	27.1	27.4
Applegate, Alan	K0BG	Mobile stations	29.2	29.9ff
		Mobile battery selection	7.13.8	7.37
		Mobile antennas	21.8	21.51ff
Bartholomew, Wayde	K3MF	Dual-band antenna for 146/446 MHz	21.10	21.62
Belrose, John	VE2CV	Sloping antennas	21.5.1	21.33
		Mobile ground losses	21.8.5	21.55
		Mobile antenna efficiency	21.8.9	21.57
Blackburn, Wally	AA8DX	Station Construction	Chapter 29	29.1ff
Bloom, Alan	N1AL	Digital Signal Processing (DSP)	Chapter 15	15.1ff
		Modulation	Chapter 8	8.1ff
Botkin, Dale	N0XAS	Digital Basics	Chapter 4	4.1ff
Britain, Kent	WA5VJB	Cheap Yagi antennas	21.11	21.69
Brown, Jim	K9YC	Ferrite materials	5.4	5.14ff
		Ferrite transmitting chokes	20.5.4	20.21ff
		Measuring chokes	Chapter 25	CD-ROM
Buxton, Al	W8NX	Multiband, coax-trap dipoles	21.1	21.21
Campbell, Jeremy	KC8FEI	Automotive RF interference	27.9	27.36ff
Campbell, Rick	KK7B	Micro R2 Receiver	12.3.2	12.16
		RF semiconductor design	Chapter 5	5.1ff
		Binaural I-Q receiver	Chapter 12	CD-ROM
Cebik, LB (SK)	W4RNL	Inverted-U antenna	21.1	21.17
		Medium-gain 2 meter Yagi	21.11	21.65
Chen, Kok	W7AY	Unstructured Digital Modes	16.2	16.5ff
Cooknell, D.A.	G3DPM	Oscillator construction	9.3.2	9.16
Cutsogeorge, George	W2VJN	Transmission Lines	Chapter 20	20.1ff
Danzer, Paul	N1II	FM voice repeaters	Chapter 18	18.1ff
		Digital Basics	Chapter 4	4.1ff
		Parallel-port interface	Chapter 4	CD-ROM
Dehaven, Jerry	WA0ACF	Phase-locked loops	9.7	9.34ff
DeMaw, Doug (SK)	W1FB	Tuna Tin 2 transmitter	13.3.2	13.5
Doig, Al	W6NBH	Multiband quad antenna	21.7	21.43ff
Duffey, Jim	KK6MC	Circuit Construction	Chapter 23	23.1ff
Eenhoorn, Anjo	PA0ZR	Active attenuator	Chapter 27	CD-ROM
Fitzsimmons, John	W3JN	Restoring vintage equipment	26.9	26.32
Ford, Steve	WB8IMY	Space Communications	30.1-30.6	30.1ff
Ford, Steve	WB8IMY	Digital Communications	Chapter 31	31.1ff
Frey, Dick	K4XU	Solid-state amplifiers	17.10-17.11	17.29ff
Geiser, Dave	W5IXM	Simple seeker RFI receiver	Chapter 27	CD-ROM
Gordon-Smith, Dave	G3UUR	Crystal filters	11.6	11.30ff
Grebenkemper, John	KI6WX	Measuring phase noise	9.2.4	9.9
		Tandem match	Chapter 24	CD-ROM
Grover, Dale	KD8KYZ	PC-board CAD	6.3	6.31ff
Gruber, Mike	W1MG	RF Interference	Chapter 27	27.1ff
Grumm, Linley	K7HFD	Low-noise oscillator	9.3.2	9.15
Hallas, Joel	W1ZR	Receivers	Chapter 12	12.1ff
		Transmitters	Chapter 13	13.1ff
		Transceivers	Chapter 14	14.1ff
Halstead, Roger	K8RI	Surplus amplifier parts	17.11	17.44
		Amplifier tuning	17.9	17.26
Hansen, Markus	VE7CA	High-Performance HF Transceiver - HBR-2000	14.7.3	14.23
Harden, Paul	NA5N	Component Data and References	Chapter 22	22.1ff
Hare, Ed	W1RFI	Circuit Construction	Chapter 23	23.1ff
		Troubleshooting and Maintenance	Chapter 26	26.1ff
		RF Interference	Chapter 27	27.1ff
Hartnagel, Hans		The Dangers of Simple Usage of Microwave Software	Chapter 6	CD-ROM
Hayward, Roger	KA7EXM	JFET Hartley VFO	9.3.2	9.17

Author	Call	Topic	Section Ref	Page Ref
Hayward, Wes	W7ZOI	RF semiconductor design	Chapter 5	5.1ff
		JFET Hartley VFO	9.3.2	9.17
		Oscillator temperature compensation	9.3.5	9.19\
		Segment-tuned VCO	9.7.7	9.50
Henderson, Dan	N1ND	RFI Management	27.2	27.4
Henderson, Randy	WI5W	Rock-bending Receiver for 7 MHz	12.2.1	12.5
Hilding, Rick	K6VVA	Remote Stations	29.4	29.18ff
Honnaker, Scott	N7SS	Digital Modes	Chapter 16	16.1ff
Hood, Mike	KD8JB	All-copper, 2 meter J-pole antenna	Chapter 21	CD-ROM
Hranac, Ron	N0IVN	Cable and digital television	27.8.3-27.8.4	27.29ff
Hutchinson, Chuck	K8CH	Antennas	Chapter 21	21.1ff
Hutchinson, Sylvia	K8SYL	75 and 10 meter dipole	Chapter 21	CD-ROM
Jones, Dave	KB4YZ	Slow-scan television	32.6-32.10	32.15ff
Jones, Bill	K8CU	1.8 to 54 MHz Low-Pass Filter	11.11.6	11.49
Karlquist, Rick	N6RK	Audio latency measurement system	29.4.3	29.21
		Mixers, Modulators, and Demodulators	Chapter 10	10.1ff
Kay, Leonard	K1NU	Analog Basics	Chapter 3	3.1ff
		RF Techniques	Chapter 5	5.1ff
Kuecken, Jack	KE2QJ	Base-loading system for whip antennas	21.8.3	21.54
Kleinschmidt, Kirk	NT0Z	Generators	29.3	29.13
Lakhe, Rucha		Mathematical stability problems in modern non-linear simulations programs	Chapter 6	CD-ROM
Lapin, Greg	N9GL	Analog Basics	Chapter 3	3.1ff
Larkin, Bob	W7PUA	RF semiconductor design	Chapter 5	5.1ff
Lau, Zack	W1VT	ARRL phase noise measurement	9.2.5	9.10
Lewellan, Roy	W7EL	JFET Hartley VFO	9.3.2	9.17
Lindquist, Rick	WW3DE	What Is Amateur Radio	Chapter 1	1.1ff
Loveall, Pete	AE5PL	D-STAR repeaters	18.4	18.11
Luetzelschwab, Carl	K9LA	Propagation	Chapter 19	19.1ff
Lux, Jim	W6RMK	Electrical Safety	28.1	28.1ff
Martin, Mike	K3RFI	Power-line RF Noise	27.6	27.17ff
McClellan, Jim	N5MIJ	D-STAR repeaters	18.4	18.11
Millar, Doug	K6JEY	Test Equipment and Measurements	Chapter 25	25.1ff
Miller, Russ	N7ART	2-meter RF power amplifier	Chapter 17	CD-ROM
Moell, Joe	K0OV	Radio Direction-finding antennas and techniques	21.12	21.72ff
Morris, Steve	K7LXC	Tower and antenna safety	28.2	28.12ff
Mullett, Chuck	KR6R	CAD for power supplies	17.11	17.28
Newkirk, David	W9VES	Mixers, Modulators, and Demodulators	Chapter 10	10.1ff
Newkirk, David	W9VES	Computer-Aided Circuit Design	Chapter 6	6.1ff
O'Hara, Tom	W6ORG	Amateur television	32.1-32.5	32.1ff
Pearce, Gary	KN4AQ	FM voice repeaters	Chapter 18	18.1ff
Pittenger, Jerry	K8RA	3CX1500D7 RF Linear Amplifier	17.12	17.45ff
Pocock, Emil	W3EP	Propagation	Chapter 19	19.1ff
Rohde, Ulrich	N1UL	Measuring intermodulation performance	10.4	10.16
		Testing mixer performance	10.5	10.21\
		Mixer performance capability	Chapter 10	CD-ROM
		2 meter down-converter	12.7.4	12.34
		Simulation at RF	6.3	6.28ff
		Modified Vackar oscillator	9.3.1	9.15
		Butler oscillator	9.5.4	9.27
		Phase noise	Chapter 9	CD-ROM
		Oscillator design	Chapter 9	CD-ROM
		VHF/UHF Grounded-base Oscillator	Chapter 9	CD-ROM
		Using Simulation at RF	Chapter 6	CD-ROM
		The Dangers of Simple Usage of Microwave Software	Chapter 6	CD-ROM
		Mathematical stability problems in modern non-linear simulations programs	Chapter 6	CD-ROM

Author	Call	Topic	Section Ref	Page Ref
Sabin, Bill	WØIYH	Half-Lattice Single-Crystal Filter	11.6	11.31
		Conversion Loss in Mixers	10.5.2	10.19ff
		Diplexer Filter	11.11.5	11.46
		Superhet Receiver	12.3.2	12.15
		IF Speech Clipper	13.3.6	13.16
		CW Key Waveform	13.3.7	13.17
		Thermistor Uses	2.15.10	2.70
		Calibrated Noise Source	25.5.7	25.25
		Impedance matching networks	Chapter 20	CD-ROM
		Transmission-line transformers	20.5.2	20.19ff
Severns, Rudy	N6LF	Power Supplies	Chapter 7	7.1ff
Silver, Ward	NØAX	Filters	11.1-11.4	11.1ff
		Amplifier maintenance	26.5.4	26.21
		Electrical Fundamentals	Chapter 2	2.1ff
		Antennas	Chapter 21	21.1ff
		Analog Basics	Chapter 3	3.1ff
Stanley, John	K4ERO	Amplifiers, Vacuum-Tube	17.1-17.9	17.1ff
Steffka, Mike	WW8MS	Automotive RF Interference	27.9	27.36ff
Stein, William	KC6T	Multiband quad antenna	21.7	21.43ff
Stockton, David	GM4ZNX	Poles and zeros	3.1.2	3.3
		Oscillators	9.1-9.6	9.1ff
Stroud, Dick	W9SR	Top-loaded, low-band antenna	21.3.3	21.29
Straw, Dean	N6BV	Antenna modeling and design	Chapter 21	21.1ff
		HFTA antenna modeling software	29.4.2	29.18
		Transmission Lines	Chapter 20	20.1ff
Stuart, Ken	W3VVN	Power Supplies	Chapter 7	7.1ff
Tayloe, Dan	N7VE	Active Filters	11.5	11.18ff
		Tayloe mixer	10.5.5	10.28
Taylor, Joe	K1JT	Earth-Moon-Earth (EME) Communications	30.7-30.13	30.30ff
Taylor, Roger	K9ALD	DC circuits and resistance	Chapter 2	2.1ff
Telewski, Fred	WA7TZY	Synthesizers	9.7	9.33ff
		Poles and zeros	3.1.2	3.3
Tonne, Jim	W4ENE	Optimized Harmonic Transmitting Filters	11.11.4	11.46
		Filters	11.1-11.4	11.1ff
Tracy, Mike	KC1SX	ARRL phase noise measurement	9.2.5	9.10
Ulbing, Sam	N4UAU	Surface-mount technology	Chapter 23	CD-ROM
Veatch, Jim	WA2EUJ	TAK-40 SSB/CW Transceiver	14.7.2	14.19
Wade, Paul	W1GHZ	Noise theory	Chapter 5	CD-ROM
Wetherhold, Ed	W3NQN	Broadcast-band Reject Filter	11.11.2	11.45
		Passive CW Audio Filter	Chapter 11	CD-ROM
		Return Loss Bridge	Chapter 25	25.30
Yoshida, Wayne	KH6WZ	Surface-mount device desoldering	Chapter 23	CD-ROM
Youngblood, Gerald	K5SDR	Tayloe mixer	10.5.5	10.28

Notes

Notes

Notes

Notes

FEEDBACK

Please use this form to give us your comments on this book and what you'd like to see in future editions, or e-mail us at **pubsfdbk@arrl.org** (publications feedback). If you use e-mail, please include your name, call, e-mail address and the book title, edition and printing in the body of your message. Also indicate whether or not you are an ARRL member.

Where did you purchase this book? ☐ From ARRL directly   ☐ From an ARRL dealer

Is there a dealer who carries ARRL publications within:
☐ 5 miles   ☐ 15 miles   ☐ 30 miles   of your location?   ☐ Not sure.

License class:
☐ Novice   ☐ Technician   ☐ Technician with code   ☐ General   ☐ Advanced   ☐ Amateur Extra

Name _____   ARRL member? ☐ Yes ☐ No

_____   Call Sign _____

Address _____

City, State/Province, ZIP/Postal Code _____

Daytime Phone (   ) _____   Age _____

If licensed, how long? _____

Other hobbies _____

_____   E-mail _____

Occupation _____

For ARRL use only	2011 HBK
Edition	88 89 90 91 92
Printing	1 2 3 4 5 6 7 8 9 10 11 12

From _____
_____
_____

Please affix postage. Post Office will not deliver without postage.

EDITOR, THE ARRL HANDBOOK
ARRL—THE NATIONAL ASSOCIATION FOR AMATEUR RADIO
225 MAIN STREET
NEWINGTON CT 06111-1494

— please fold and tape —

About the Included CD-ROM

On the included CD-ROM you'll find this entire *Handbook*, including text, drawings, tables, illustrations and photographs — many in color. Using Adobe *Reader* (included separately on the CD-ROM), you can view and print the text of the book, zoom in and out on pages, and copy selected parts of pages to the clipboard. A powerful search engine — Reader Search — helps you find topics of interest. Also included is supplemental information and articles, PC board template packages for many projects, and companion software mentioned throughout. See the README file on the CD-ROM for more information.

INSTALLING YOUR ARRL HANDBOOK CD

Windows
1. Close any open applications and insert the CD-ROM into your CD-ROM drive. (Note: If your system supports "CD-ROM Insert Notification," the *ARRL Handbook CD* may automatically run the installation program when inserted.)
2. Select **Run** from the *Windows* **Start** menu.
3. Type **d:\setup** (where d: is the drive letter of your CD-ROM drive; if the CD-ROM is a different drive on your system, type the appropriate letter) and press **Enter**.
4. Follow the instructions to install the CD-ROM files on your system.

Macintosh
To install the CD files on the Mac, open the CD-ROM (ARRL 2011 Handbook) icon then click to go into the "Program Files" folder. Drag the "ARRL 2011 Handbook" folder icon to your hard drive.

To use the search capability in the *Handbook CD* you will need to have Adobe *Reader* installed. Mac Preview cannot access all of the features of the PDF files. If you need to install the Adobe *Reader* 9 program, it is located in the Adobe Reader folder of the CD. Select the appropriate folder and file for either Mac (Intel) or Mac (Power PC). After installation you will be then able to open the Handbook files in the Reader to view them.

USING YOUR ARRL HANDBOOK CD

Windows
Click on START, ALL PROGRAMS, then click ARRL SOFTWARE, and ARRL 2011 HANDBOOK CD to start the *Handbook CD* viewer. The ARRL 2011 HANDBOOK menu also has additional shortcuts for accessing the included Supplemental Files and Companion Software Files on the CD.

If you chose not to install the Handbook files on your hard drive, you will need to load the CD in the drive first, then go to MY COMPUTER, select your CD Drive, then click down into PROGRAM FILES, then ARRL 2011 HANDBOOK, and then click to start "intro.pdf".

Macintosh
Go to where you copied the ARRL 2011 Handbook folder onto your hard disk, then look for the "intro" file. Click to start this file in Adobe Reader.

If you chose not to install the Handbook files on your hard drive, you will need to load the CD in the drive first, then go to CD Drive icon (ARRL 2011 Handbook) on your desktop, then click down into PROGRAM FILES, then ARRL 2011 HANDBOOK, and then click to start "intro.pdf".

SUPPLEMENTAL FILES, TEMPLATES AND SOFTWARE

See the **Supplemental Files** and **Companion Software** sections in the *Handbook* CD-ROM Table of Contents for information about accessing the many supplemental files, template packages and other support features included on the CD.